HANDBOOK OF TELEMETRY

AND

REMOTE CONTROL

ELLIOT L. GRUENBERG, *Editor-in-Chief*

*Manager, Radiation Systems Research
Center for Exploratory Studies
International Business Machines Corporation
Yorktown Heights, New York*

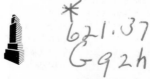

McGRAW-HILL BOOK COMPANY

New York   San Francisco   Toronto   London   Sydney

HANDBOOK OF TELEMETRY AND REMOTE CONTROL

Copyright © 1967 by McGraw-Hill, Inc. All Rights Reserved. Printed in the United States of America. This book, or parts thereof, may not be reproduced in any form without permission of the publishers. *Library of Congress Catalog Card Number* 65-24524

25075

1234567890MP72106987

CONTRIBUTORS

1402128

CHARLES D. ALBRIGHT, *Research Engineer, Data-Control Systems, Inc., Danbury, Conn. (FM-FM Telemetry Systems, except Transmitters, Chapter 6)*

WILLIAM B. BLESSER, *Professor, Mechanical Engineering, Polytechnic Institute of Brooklyn, Brooklyn, N.Y. (Components, Chapter 15)*

JOHN F. BRINSTER, *President, General Devices Corp., Princeton, N.J. (Commutation, Chapter 4; PDM Telemetry Systems, Chapter 7)*

LEO A. CHAMBERLIN, *Senior Engineer, Texas Instruments, Inc., Dallas, Tex. (Data-transmission-system Design, Chapter 8)*

JOHN W. CLARK, *Consultant in Ocean Engineering, San Diego, Calif. Formerly Hughes Aircraft Co., Fullerton, Calif. (Remote Handling, Chapter 15)*

GEORGE R. COOPER, *Professor, Electrical Engineering, Purdue University, Lafayette, Ind. (Sampling and Handling of Information, Chapter 2)*

CHARLES H. DOERSAM, JR., *Associate Professor, Systems Science, Polytechnic Institute of Brooklyn, Brooklyn, N.Y.; President, Doerco Consultants, Inc. (System Selection, Chapter 4; Automatic Checkout Equipment, Chapter 11)*

JOHN E. GAFFNEY, JR., *Development Engineer, International Business Machines Corporation, Rockville, Md. (Data-handling Equipment, except Automatic Checkout Equipment, Chapter 11; Industrial Telemetry and Remote Control, Chapter 13)*

GEORGE E. GOODE, *Member, Technical Staff, Texas Instruments, Inc., Dallas, Tex. (Data Recovery, Chapter 8)*

ELLIOT L. GRUENBERG, *Senior Engineer, International Business Machines Corporation, Yorktown Heights, N.Y. (Fundamentals, Chapter 1; High-efficiency Telemetry Systems, Phase-locked-loop Systems, Chapter 9; Space Telemetry, Chapter 14; Introduction to Remote Control, Chapter 15)*

DANIEL HOCHMAN, *Manager, Electronics Advanced Development, Lockheed Missiles and Space Company, Sunnyvale, Calif. (PAM-FM Telemetry, Chapter 9)*

CONRAD H. HOEPPNER, *President, Industrial Electronetics Corp., Melbourne, Fla. (Radio Telemetry, Chapter 4)*

JAMES B. KENDRICK, *TRW Space Systems, Redondo Beach, Calif. (Space Systems, Chapter 14)*

GRANINO A. KORN, *Professor, Electrical Engineering, The University of Arizona, Tucson, Ariz. (Appendix)*

CONTRIBUTORS

THERESA M. KORN, *Consultant, Tucson, Ariz. (Appendix)*

DAVID B. LEESON, *Applied Technology, Inc., Palo Alto, Calif. (Transmitters, Chapter 6)*

ALAN L. McBRIDE, *Member, Technical Staff, Texas Instruments, Inc., Dallas, Tex. (RF Link, System Error Analysis, Chapter 8)*

WILLIAM P. McGARRY, *Senior Research Engineer, Data-Control Systems, Inc., Danbury, Conn. (FM-FM Telemetry Systems, except Transmitters, Chapter 6)*

VERNON L. MILLER, *Chief, Range Instrumentations Systems Office, White Sands Missile Range, N. Mex. (Trajectory Instrumentation, Chapter 3)*

OWEN J. OTT, *Manager of Engineering, Data-Control Systems, Inc., Danbury, Conn. (FM-FM Telemetry Systems, except Transmitters, Chapter 6)*

GEORGE J. PASTOR, *Manager, Display Systems Department, Philco Houston Operations, Western Development Laboratories, Division of Philco Corp., Subsidiary of Ford Motor Company, Newport Beach, Calif. (Transducers, Chapter 3)*

EDWARD Y. POLITI, *President, Solid State Electronics Corp., Sepulveda, Calif. (Telemetry-system Component Design, except Power Sources for Remote Use, Chapter 10)*

HARRY H. ROSEN, *Special Subsystems Engineering, Reentry Systems Department, General Electric Company, Philadelphia, Pa. (Data Reduction, Chapter 12)*

RAYMOND A. RUNYAN, *Vice-President and General Manager, Data-Control Systems, Inc., Danbury, Conn. (FM-FM Telemetry Systems, except Transmitters, Chapter 6)*

GEORGE P. SARRAFIAN, *Manager, Apparatus Research and Development Laboratory, Space Missions Branch, Texas Instruments, Inc., Dallas, Tex. (Introduction, PCM Telemetry Systems, Chapter 8)*

HANS SCHARLA-NIELSEN, *Principal Engineer, Radiation, Inc., Melbourne, Fla. (Radio Links, Chapter 4)*

SIDNEY SHAPIRO, *Senior Engineer, International Business Machines Corporation, Bethesda, Md. (Satellite-relay Communications, Chapter 16)*

MARTIN L. SHOOMAN, *Associate Professor, Electrical Engineering, Polytechnic Institute of Brooklyn, Farmingdale, N.Y. (Feedback Systems, Chapter 15)*

JOE E. SMITH, *President, International Data Systems Corp., Dallas, Tex. (Signal Conditioning, Chapter 4)*

MORTON J. STOLLER, *Deceased, formerly NASA, Washington, D.C. (Satellite Telemetry, Chapter 14)*

JOHN G. TRUXAL, *Vice President for Educational Development, Polytechnic Institute of Brooklyn, Brooklyn, N.Y. (Principles of Vehicular Guidance and Control, Chapter 15)*

NORMAN D. WHEELER, *Engineer, Spacecraft Department, Missile and Space Division, General Electric Company, Philadelphia, Pa. (Power Sources for Remote Use, Chapter 10)*

LAVERGNE E. WILLIAMS, *Member, Technical Staff, Aerospace Corp., Eastern Test Range Office, Patrick Air Force Base, Melbourne, Fla. (Antennas, Chapter 4)*

PREFACE

For many years, there has been developing in industry an art of measurement and control from remote distances. Demand for this work has originated, and continues, in the gas and electric utilities and in manufacturing activities in which flow processes are predominant, such as the petroleum and chemical industries. More recently, the need has arisen to make scores of simultaneous measurements in the testing of aircraft engines, in wing stress analysis and in wind tunnel measurements. In a sense, these are also remote measurements, because the test points range over a wide area. The flight testing of aircraft has further stimulated this art. The latest and perhaps most important challenge has arisen in connection with the development of missiles and rockets. In this case, resort must be made to purely automatic methods of measurement, and these means must be perfected to withstand much more severe operating environments. They must provide large quantities of simultaneous measurements and at high speeds. Work is already under way to explore space with unmanned and manned vehicles. This work once again intensifies the need for better measuring tools, which of course must be provided by the arts of telemetry and telecontrol.

It is the object of this handbook to provide a comprehensive statement of the art, the first to our knowledge in this field. To be of greatest use, we have not restricted ourselves to narrow definitions of telemetry and remote control, but have included topics which are intimately related, such as electronic-position measuring systems, missile guidance techniques, flight-control systems. The topics bearing on range instrumentation, such as electronic position measuring systems are really only an extension of the meaning of telemetry to the measurement of distance.

The importance of data processing to the telemetry art has long been recognized. The advent of real-time computation increases the effectiveness of test-range operation tremendously and eliminates much data-reduction labor. Hence, considerable attention has been given to these topics.

The standard forms of telemetry which have been, and are being, employed for control of remote power station and power distribution, pipe-line metering and process control are also of interest. Techniques used in these cases are in many instances quite different from those used in radio telemetry. Both systems and techniques are described. Opportunity exists for cross fertilization of ideas in this field and those used in missile telemetry.

It is the intention of this handbook to provide a working tool for persons of the following professional interests: Telemetering-system design, missile-system engineering, missile-test engineering, range-instrumentation design, space research technicians, industrial-control design, flight test and wind tunnel technicians, process-control computer design, telemetry data processors and analysts, utility systems engineers, nuclear reactor test engineers and avionics engineers.

This handbook has been assembled by specialists active in a dynamic field. Emphasis has been placed upon principles of operation not likely to change in themselves over the years as rapidly as equipment details and facilities. Some allusions to these must be made but the fundamentals should remain useful even in this rapidly changing field for a reasonable time in the future.

Telemetry requires knowledge assembled from many disciplines. It is our hope that this handbook brings together the basic knowledge in such a way that it can be understood by the professional worker in this field, and that it is sufficiently self-contained so as to minimize cross referencing to other texts in other disciplines.

Application areas have been given secondary consideration to discussion of design principles. For this reason, some applications are slighted. Specifically, the fields of marine and medical or biological telemetry are not covered per se. It is expected that the design or system engineer will be able to find the design information needed when confronted with a new problem.

Remote control has not been treated in American texts as related to telemetry. Russian practice has been different, the two being integrated under the heading of "telemechanics." This handbook treats remote control extensively, adding the principles of feedback and control theory to those concepts of information and communications theory, which are more familiar to telemetry engineers. The two arts may grow closer together in the near future.

Telemetering has always been identified with communication from space vehicles. The use of such vehicles for communications relay posts in space has been a natural extension. Accordingly, a chapter on satellite communications has been included.

The dynamics and mission objectives of missile and space vehicles have an important bearing on the design and functioning of the telemetry and remote-control equipment. Such information has been included. The editor is particularly grateful for the use of such data supplied to him by TRW Systems, Redondo Beach, California.

A comprehensive mathematical appendix has been included, with the end in view of enhancing the self-contained feature of the book, and with the conviction that this field rests upon principles which require a good understanding of mathematics. Although this need not be called into play every day, resolution of important, though subtle, points may require resort to these first principles. The reader will find Chapter 2 a most useful starting point in many such instances.

Acknowledgment is gratefully extended to the contributing authors who provided material in the face of the demands of a busy profession which provides so many opportunities for other profitable uses of their time. Their help in this attempt to organize the body of knowledge in this relatively new field is a service to future progress in the field.

I am most grateful to individuals such as Andrew Viterbi, R. Rochelle, and H. Van Trees, who allowed me the use of their published work, which I found necessary to provide an extensive review of advanced telemetry and phase-locked-loop theory in Chapter 9. It was also most helpful that Rand Corp. and Siegfried Reiger (now of Communication Satellite Corp.) permitted use of Reiger's work on communications satellite systems.

Several organizations have been particularly helpful; among them, Texas Instruments, Inc., Dallas, Texas; Data-Control Systems, Inc., Danbury, Conn.; and General Devices Co., Princeton, N.J. I am particularly obliged to the International Business Machines Corp. for its cooperation which helped to bring this project to its conclusion by making available assistance on manuscript preparation. In particular, Barbara Bohachik was most helpful. I wish to thank Hunter McCartney for his assistance with some of the editorial problems. I am obliged most of all to my wife, Ruth, who helped so much in all phases of the work.

Elliot L. Gruenberg

CONTENTS

Contributors v
Preface vii

Chapter 1. **Fundamentals**
E. L. Gruenberg. 1-1

Chapter 2. **Sampling and Handling of Information**
G. R. Cooper 2-1

Chapter 3. **Sensing of Information**
G. J. Pastor and V. L. Miller 3-1

Chapter 4. **Radio-telemetry Systems**
C. H. Hoeppner, C. H. Doersam, Jr., J. H. Smith, J. F. Brinster, H. Scharla-Nielsen, and L. E. Williams 4-1

Chapter 5. **Standards for Telemetry** 5-1

Chapter 6. **FM-FM Telemetry Systems**
C. D. Albright, W. P. McGarry, O. J. Ott, R. A. Runyan, and D. B. Leeson 6-1

Chapter 7. **PDM Telemetry Systems**
J. F. Brinster. 7-1

Chapter 8. **PCM Telemetry Systems**
G. P. Sarrafian, L. A. Chamberlin, G. E. Goode, and A. L. McBride. 8-1

Chapter 9. **PAM and Advanced Telemetry Systems**
D. Hochman and E. L. Gruenberg 9-1

Chapter 10. **Telemetry-system Component Design**
E. Y. Politi and N. D. Wheeler 10-1

Chapter 11. **Data-handling Equipment**
J. E. Gaffney, Jr., and C. H. Doersam, Jr. 11-1

Chapter 12. **Data Reduction**
H. H. Rosen and the Staff of Reentry Systems Department, General Electric Company. 12-1

Chapter 13. **Industrial Telemetry and Remote Control**
J. E. Gaffney, Jr. 13-1

Chapter 14. **Space Systems and Telemetry**
Staff of TRW Space Systems, Inc., M. J. Stoller, and E. L. Gruenberg 14-1

Chapter 15.	**Remote Control**	
	J. G. Truxal, M. L. Shooman, W. B. Blesser, and J. W. Clark	**15-1**
Chapter 16.	**Satellite-relay Communications**	
	S. Shapiro	**16-1**
Appendix.	**Mathematical Formulas, Theorems, and Definitions**	
	G. A. Korn and T. M. Korn	**A-1**

Index follows the Appendix.

HANDBOOK OF TELEMETRY AND REMOTE CONTROL

Chapter 1

FUNDAMENTALS

ELLIOT L. GRUENBERG, *International Business Machines Corporation, Yorktown Heights, N.Y.*

INTRODUCTION

1	History of Telemetering	1-1
2	Definitions	1-2
3	Applications of Telemetering	1-6
4	Remote Control	1-6

SYSTEMS

5	Wired Telemetering Systems	1-6
6	Radio Telemetering Systems	1-8

UNITS AND CONVERSION FACTORS

7	Metric Systems of Absolute Units	1-9
8	Conversion Factors	1-9
9	Mathematics	1-24

INTRODUCTION

1 History of Telemetering

Telemetering literally means measuring at a distance, or remote measuring. The term could be applied to measurement of the position or location of an object at a distance. This latter art has become known as electronic trajectory measurement and is of particular use on the instrumentation ranges for observation of the position of pilotless vehicles. It will be treated more specifically in Chap. 3.

Telemetering has grown to mean more specifically the observation of variables and physical measurements at a remote location. Borden and Mayo-Wells[1] note that the first known use of telemetering was by Shilling in Russia in 1812 for the firing of mines. Since that time the art has developed toward industrial applications making use of wired techniques; radio means have become more largely used for aircraft and missile testing. Of late the techniques used for missile testing have been expanded to the obtaining of data from space vehicles. This latter development, although making use of the techniques developed for missiles, has required a return to the longer-lived components which were more typical of industrial applications. Far more reliable components than ever are now required.

A great impetus to radio telemetry came with the need to develop aircraft which required large amounts of data while in flight. Such data were at first recorded, but later direct broadcast via radio was found desirable. This became particularly important with the development of pilotless aircraft in the period right after World War II.[2]

At about the same time, missile development began to expand at an accelerated pace. In 1944, the V-2 rocket was perfected. From Sept. 6, 1944, to March, 1945, 1,027 of these rockets were fired from The Hague with 92.3 per cent launch success.[3] This rocket had a maximum range of 200 miles and a warhead of approximately 2,000 lb. It weighed 14 tons at takeoff and was 46 ft 11 in. tall and 5 ft 5 in. in diameter. Subsequently many of these rockets were fired in the United States for upper-atmosphere-research purposes. In 1954, the United States undertook actively the development of ICBM and IRBM missiles and at about the same time started the development of a satellite for the IGY program, which commenced in 1957. At the same time, such activity was undertaken in Russia. These events culminated in the successful launching in 1957 of Sputniks I and II, and in 1958 of Explorer I and Vanguard I. In 1958, the National Aeronautics and Space Administration was founded in the United States and space activities were accelerated. Since that time many space vehicles have been launched, including space radio transmitters. Many of these are enumerated in Ref. 3.

In the meantime, industrial telemetering was developing toward the use of digital computers. A typical modern system was the Stirlington Station of Louisiana Power and Light Company in 1958.

Milestones of significance to telemetering are listed in Table 1.

2 Definitions

A telemetering system is intended to provide measurements from a remote point. Therefore, telemetry encompasses disciplines of several fields. These include instrumentation, communications, information theory, and data processing.

There is no real distinction between the terms *telemetering* and *telemetry*. The term telemetry has been more closely associated with radio telemetering.

Originally, measurements were transmitted as indications of remote meter readings, and some of the original telemetering systems were in reality remote displays of these meter readings. More recently, the state of the art has developed such that data are transmitted from remote locations to recording or computing stations. Data are sensed by transducers or pickoffs and usually are transferred at the remote site to electrical signals. In modern practice these signals can be transmitted by telephone wires or by radio instead of by the special electrically wired systems once more generally used. In transmission, the telemetry signals may, in many cases, be indistinguishable from data-transmission signals. Data-transmission signals are those which can be transmitted from one computing site to another or from a punched-card reader to a computing center. Hence, except for the element of measuring, the field of telemetering is growing relatively indistinguishable from the field of data transmission.

In this handbook, telemetering will be used to cover not only the transmission system but the sensing and transducing systems on the transmitting end and the data-handling system on the receiving end. The data-handling system will be interpreted to include the receivers, recorders, and data-entry systems to digital processing systems.

It is desirable to distinguish *telemetering* from what is meant by *communications* systems. *Communications* systems are intended to convey any form of information from one point to another, as from a remote point to the receiver.

Table 1. Telemetering Milestones

1812	Shilling, a Russian, used telemetering in firing mines
1845	Takobi devised a military data-transmission system. Konstantinov and Pouli developed a telemeter for recording and analyzing the flight of cannonballs
1857	Warships had ship's telegraph between bridge and engine and gun locations
1874	Olland developed meteorological telemeter for measurements on Mont Blanc
1889	A patent covering inductive adjustment to an interrupted counter telemeter was issued to P. Moennith
1893	Tsiolkowski published his first works on rocketry
1901	C. J. A. Michalke patented the position motor, the forerunner of the selsyn
1901	First aircraft flight and the first demonstrations of practical transatlantic radio
1906	Golitsyn developed seismic telemetering at Pulkovo
1912	First telemeter system for load dispatching installed by the Commonwealth Edison System of Chicago
1913–1914	Telemetering used extensively in the Panama Canal
1913	First use of radio meteorological telemeter by Weather Bureau
1915	The NACA was founded
1922	Goddard tested the first liquid-fuel rocket
1930	Astin and Curtiss developed the radiosonde
1941–1945	FM-FM, PAM, and PDM telemetry was under development at the Palmer Laboratory at Princeton University and the Applied Physics Laboratory at Johns Hopkins University. These developments were for use in aircraft and missiles
1944	The V-2 rocket was perfected
1945	White Sands Proving Ground was established
1946	Experimental guided-missile group was established at Eglin Field, Florida, and the first missile launched at Naval Air Facility at Point Mugu, Calif.
1946	Project Rand said artificial satellite is feasible
1947	V-2 instrumented by NRL carried telemetry for upper-atmosphere research for first time
1948	PCM system was analyzed by Pierce, Shannon, and Oliver
1949	Navy developed a multichannel telemetry system
1950	First National Telemetering Conference
1951	Air Force Missile Test Center established at Cape Canaveral
1954	First PCM telemetry system
1957	The IGY (International Geophysical Year) Program started in July. Sputniks I and II were launched by Russia and Explorer I and Vanguard I by the United States
1958	SCORE was launched, the first satellite communications relay. It operated on frequencies of 132.435 and 132.905 Mc. It contained 35 lb of communications equipment
1958	An automatic power-distribution system was installed at Stirlington Station of Louisiana Power and Light Company
1958	The National Aeronautics and Space Agency (NASA) was established
1960	Pioneer V made first long-range radio transmission over 5 million miles
1962	First television communications relay by satellite
1962	Radio transmission was achieved from Mariner II over 51 million miles in the vicinity of the planet Venus. Commands to the space vehicle were transmitted over 41 million miles
1963	A communications relay operated from synchronous inclined orbit
1965	Digital radio transmission of pictures was achieved from Mariner IV in the vicinity of the planet Mars, a distance of over 150 million miles. The bit rate was $8\frac{1}{3}$ bits/sec. Each picture transmission consisted of 240,000 bits.

The sources of information for Table 1 are Refs. 1 and 3. No attempt has been made here to follow the proliferations of missile and astronautical events that have occurred. Only those events having an extraordinary impact upon telemetering have been listed.

As the term communications is usually applied only to the transmission system, the communications common carrier seldom is required to vouch for the accuracy of the information transmitted over its facilities nor does the common carrier assume responsibility for errors in transmission. The common carrier of course endeavors to prevent such errors, but it cannot be expected to have a system responsibility such as is inherent in a telemetry system, which must have in it means for error detection and recognition. Such means, for example, as calibration points and check bits represent methods that the telemetry system incorporates to overcome the error problem. Hence the telemetry system is distinguished from communications systems in the important aspect of maintaining overall system accuracy specifications.

Data is a term used mainly to distinguish voice and record information transmissions from transmissions of other types of information. Data includes both instructions and measurements which are transmitted. Telemetry is largely concerned with the transmission of *measurements* which can be encoded in either analog or digital fashion, the latter being the encoding technique for PCM.

Information refers to message content. Information can be applied as a description of either voice or data transmission. The importance of the term, however, is to put on a common basis the study of all methods of using a communications channel in order to establish how efficiently the channel is being used. In Chap. 2, this term will be defined more closely and its applications to different telemetry systems will be carefully explored. Information is a more fundamental concept than *modulation*, for example. *Modulation* techniques are limited to efficient methods of impressing the information on a particular communications media such as radio-frequency carriers. Information can be studied without regard to these physical limitations.

Telemetry systems are closely allied with instrumentation systems. The term *instrumentation* has been used in several senses. In one sense it includes the entire telemetry system. In a more restricted sense it refers to the system of measuring variables whether these variables are remotely located or not. The term is directly associated with the concept of measurement. For example, it includes the technique of measuring the position and velocity of a vehicle. This may be done by observations made by radar sets of the distance from the radar stations as well as the angle of observations. This measurement is made at a distance from the radar set and can indeed be considered a method of remote measurement. However, telemetry systems are not generally considered to include such measurements. A discussion of this type of measurement is included in Chap. 3, since it is closely associated with radio telemetry applications.

Remote control or *telecontrol* is an extension of the concept of remote measurement. In remote control, actuators are included to make a change at the remote location. This may be done by what is now termed a command link or through radar beams. In a sense inertially guided vehicles are remotely preprogrammed remote-controlled devices. These systems will be considered in Chap. 15.

A major distinction between instrumentation and telemetering is the element of *distance* from measured quantity to displayed quantity. Where special provision must be made to make the measurement available at a remote location, telemetering has been resorted to.

Another distinctive feature of telemetering is *multiplexing*. The older wired systems of telemetering did not include this feature to any great extent. However, a major factor in making remote tests in modern installations is the need to obtain up to hundreds of measurements. An example is the wind tunnel. Another example is the missile and rocket test stand. These measurements must be continued throughout burnout in many instances, hence the development of several schemes of multiplexing, foremost of which are the frequency- and time-multiplexing systems. This element of telemetering is so distinct that the major systems of telemetry are distinguished from one another in part by the mode of multiplexing. The pulsed systems, mainly PAM, PDM, and PCM, are all time-multiplex systems and the FM-FM system is a frequency-multiplex telemetry system.

A further detailed discussion of terminology, particularly in the radio telemetry field, is contained in the IRIG glossary section presented in Chap. 5.

Table 2. Applications of Telemetering

	Now	Future*
Business:		
Accounting	...	o
Branch tie-in	...	o
Burglar alarms	...	o
Fire alarms	...	o
Inventory control	...	o
Space reservation	x	
Stock ticker	x	
Telautograph	x	
Teleprinter	x	
Communications:		
Automatic teleprinter correction	x	
Broadcast remote control	x	
Telecasting remote control	...	o
Wire data circuits	...	o
Industrial:		
Data for automatic control	...	o
Fluid-depth measurement	x	
Measuring gas flow	x	
Pipeline control	x	
Rainfall reporting	x	
Reading gages by TV	x	
Military:		
Battleground television	...	o
Doppler devices	x	
ICBM test	...	o
Personnel data	x	
Radioactive monitoring	...	o
SAGE	x	
Search-rescue operations	...	o
Sonobuoy	x	
Target drone aircraft	x	
TV-guided missiles	x	
Ultrafax	...	o
Underwater-detection nets	...	o
Radio-guided missiles	x	
Research:		
Aircraft design	x	
Flight testing	x	
Landing on planets	...	o
Medical research	x	
Missiles	x	
Navigation-data send-back	...	o
Nuclear experimentation	x	
Projectiles	x	
Radiosondes	x	
Satellites	x	
Space probes	x	
Transportation:		
Automatic flight plans	...	o
Automatic railroading	...	o
Hyperbolic navigation	x	
Radar speed detection	x	
Radar traffic detection	...	o
Rho-theta navigation	x	
Train identification	x	

* Not presently in common use.

3 Applications of Telemetering

Telemetry has been typified by a wide variety of applications. This number can be expected to grow as the need for remote experimentation and space exploration grows. Industrial and business applications will grow because of the need for process control and business control. Table 2 presents a typical list of telemetering and remote-control applications.[4]

4 Remote Control

The art of remote control or telecontrol is not so well defined as that of telemetering. There are no official standards, for example, for remote control. Remote-control systems involve both a communication of information from a remote site and a return communication back to some actuating device. This return link is sometimes in radio telemetry called a *command link*. Such a system can be recognized to be a form of feedback system. However, in most cases, the speed of response of such a system has been so slow as not to warrant special attention to the feedback nature of this loop except for radio guidance developed for guided missiles. Events indicate the gradual evolution of *real-time* systems. These are systems in which actuation of controls can occur during the operation of the remote process or the flight of the remote vehicle.

Because of this gradual development, more understanding of the feedback nature of the problem will be required. These principles are dealt with in depth in Chap. 15, as well as the nature of remote-handling situations. It is interesting to note that a text has recently appeared attempting a unified treatment of radio control.[5]

SYSTEMS

5 Wired Telemetering Systems

Telemetering has been most consistently characterized as a system rather than a technique or a device. The concept includes the means of changing the variable measured, known also as the *measurand*, into an electrical signal, the method of transmitting this signal to the receiver, and the means of changing this signal into a useful form. Originally, this form was only a meter reading. More recently the output of the telemetering system may be a cathode-ray-tube display, a printer, or a magnetic-tape input to a data-processing system. Figure 1 after Lynch[1] shows the fundamental process, particularly as applied to process control. In such cases the need for remote control often exists. Here a return control signal to an actuator is required. The question whether to use proportional or on-off control is a serious engineering point of design depending on principles discussed in Chap. 15, relating to feedback control systems.

The older wired telemetering systems have been classified by the American Standards Association as:

Current
Voltage
Frequency
Position
Impulse or pulse

The basis of this classification has been the method of receiving the analog of the measurand at the remote point. The system problem has been to preserve the accuracy of the analog over the required distance at reasonable cost. The sources of error in most of these systems are signal drop along the transmission line, and power-supply fluctuations. To minimize these, the systems resort to servomechanisms.[1] An example, which might be classified as voltage, current, resistance, or position, depending on the way the measurand is detected at the remote station, is shown in Fig. 2. The tap on resistor R_B is caused to seek a position corresponding to the position of the movable arm of potentiometer R_A relatively independent of the supply voltage and line resistance. The servosystem is designed to move the arm of R_B to

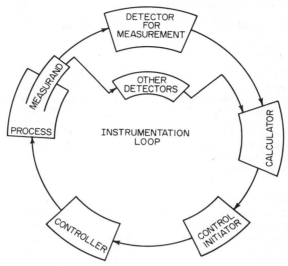

FIG. 1. Some typical nomenclature often used in reference to process-type instrumentation.

oppose the error signal at its input which is detected at the remote station, and which is derived from the summing network R_1, R_2, R_3. The measurand can be derived from the position of the potentiometer R_B shaft, or from the current or voltage from a standardized system as shown. R_4 provides compensation signal for receiver power

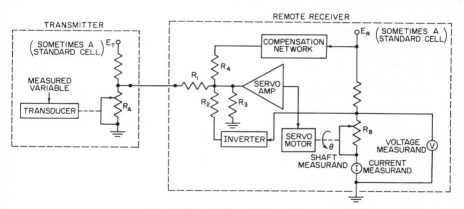

FIG. 2. Wired telemetering system.

supply. A calibration technique may be used to determine the effect of transmitter power-supply fluctuations, or a common supply may be used at the expense of an additional wire. The resistance of R_B to the tap could be measured as another method of obtaining an indication.

Another example of remote-position indication is the selsyn system, which requires an a-c power supply. In this method the rotor of a special a-c motor will follow the position of a remote motor provided the stators and rotors are properly interconnected.[1]

The limitations of the wired systems are:

1. A finite limit to the distance of transmission for a given accuracy
2. Low data rates
3. Lack of multiplexing capability
4. Unsuitability for direct entry into data processors
5. The need for wire or metallic connection

The last point of course rules out the wired system for applications such as aircraft and missiles and other mobile situations. The multiplexing limitation would require a wire metallic connection for each variable. Modern systems have need to measure hundreds of variables and enter measurements into high-speed computers. For these reasons many of the older telemetering systems have been supplanted by data-transmission systems over leased-wire or microwave links as discussed in Chap. 13. These systems can be conveniently treated in the same manner as radio-telemetry systems.

6 Radio Telemetering Systems

Radio telemetry systems are commonly classified in accordance with the methods of multiplexing and modulation employed. The older ASA classifications have little meaning for these systems since currents and voltages are not directly transmitted. An outstanding feature of such systems is the capability to multiplex the outputs of many transducers over the same transmission band.

The two basic methods for multiplexing are *frequency* division and *time* division. In frequency division the data signals from separate instruments are kept independent of each other by separate allocation of frequency channels, whereas in *time* division, separate periods of time are provided each channel. This latter implies data must be sampled on a time basis and referenced to a synchronizing signal. The latter is the method inherent in the pulse telemetry systems, whereas the former is used in the FM-FM (or FM/FM) system. This designation has the following meaning:

FM	FM
Frequency-subcarrier modulation	Carrier-frequency modulation

This is sometimes designated as FM/FM instead of FM-FM. There is no difference between the two.

Similarly the following designations have the given meanings:

PCM	FM
Pulse-code modulation	Frequency modulation
PAM	PM
Pulse-amplitude modulation	Phase modulation
PDM	FM
Pulse-duration modulation	Frequency modulation
or	
PWM	
Pulse-width modulation	

Thus the first term applies to the subcarrier or the initial modulation (or encoding) in the telemetry transmitter. The transducers in the pulse systems must be sampled in time and the signal sample then encoded in its special time slot before the final modulation.

The PCM system encodes the sample as a binary symbol; thus it is the only commonly used binary or digital telemetry system.

These four methods of subcarrier modulation represent the most standardized systems of radio telemetry. Standards have been developed for these systems and are presented in Chap. 5. The final carrier modulation may be AM, PM, or FM. The commonest is FM. Other subcarrier modulations are possible, and an extremely wide variety in this choice is possible. Comparisons of these modulations are made in Chap. 2. Spectrum occupancy and signal/noise performance at the receiver are principal factors in the choice of scheme. It is not necessary to use all possible modulation combinations since the four general ones can cover the great majority of applications.

Strictly speaking, frequency and time division are not the only multiplexing possibilities. All that is required is that there be a minimum of *crosstalk* between channels. This means that, for two channel signals $e_1(t)$ and $e_2(t)$,

$$\int_0^T e_1(t) e_2(t) \, dt = 0$$

over the time T required for a message sample transmission. Thus any *orthogonal* function will be satisfactory (Chap. 2). Ballard[6] has used Legendre polynomials as the basis of a system design. Exponential polynomials are also possibly useful for multiplexing.

Triple modulation is sometimes used, particularly
PAM-FM-FM
PDM-FM-FM

The permissible information capacity for the standard FM channels is given in Chap. 5. The data rate of slowly varying measurements permits sampling. Hence several transducers may share a common FM-FM channel after a time-division multiplex. PDM is commonly employed in such service.

More specialized systems include
SSB (single sideband)—FM
PFM (pulse-frequency modulation)

Single sideband-FM was devised to accommodate a larger range of vibration information than can FM-FM. Another method of achieving a more compact use of the spectrum than FM-FM is frequency translation—FM, which is discussed in Chap. 6.

Pulse-frequency modulation, used on the Vanguard satellite, is designed to provide a minimum-power telemetry transmitter by encoding the sampled information as one of a set of discrete frequencies. The receiver contains a corresponding set of discrete filters so that signal is detected as a choice of filter. This and other forms of orthogonal telemetry systems will be discussed in Chap. 9.

Miniaturization

An integral concern of radio telemetering has been the requirement to miniaturize the part of the system which senses, transduces, and transmits data at the remote and inaccessible or hazardous location. This requirement is by no means limited to missile and spacecraft telemetry (Chap. 14), but medical and industrial applications are emerging. Figure 3 shows a 24-channel telemetry transmitter for monitoring the temperature and vibration of the blades of high-speed gas or steam turbines *while in operation* to prevent fatigue failure. Each transmitter weighs 2.6 ounces and is mounted in a 7-in. O.D. ring around the turbine shaft and is 4 in. long.

This is also an example of the convenience of radio telemetry which avoids slip rings. Further discussion of miniaturization will be found in Chap. 10 and radio telemetry in Chap. 4.

Table 3. Metric Systems of Dimensions and Units*

Quantity and symbol	Mechanical cgs system		Absolute electrostatic (aesu) system		Absolute electromagnetic (aemu) system		The mks (Giorgi) system	
	Dimensions, MLT	Unit	Dimensions, MLTϵ	Unit	Dimensions, MLTμ	Unit	Dimensions, MLTI	Unit
Length, L	L	**Centimeter**	L	**Centimeter**	L	**Centimeter**	L	**Meter**
Time, T	T	**Second**	T	**Second**	T	**Second**	T	**Second**
Mass, M	M	**Gram**	M	**Gram**	M	**Gram**	M	**Kilogram**
Force, F	MLT^{-2}	Dyne	MLT^{-2}	Dyne	MLT^{-2}	Dyne	MLT^{-2}	Newton†
Work, energy, E	ML^2T^{-2}	Erg	ML^2T^{-2}	Erg	ML^2T^{-2}	Erg	ML^2T^{-2}	Joule
Viscosity, μ	$ML^{-1}T^{-1}$	Poise					$ML^{-1}T^{-1}$	No name
Permittivity, ϵ			aesuϵ	Statfarad/centimeter	$L^{-2}T^2\mu^{-1}$	No name	$M^{-1}L^{-3}T^4I^2$	Farad/meter
Permeability, μ			$L^{-2}T^2\epsilon^{-1}$	No name	aemuμ	Abhenry/centimeter	$MLI^{-2}T^{-2}$	Henry/meter
Charge, Q			$M^{1/2}L^{3/2}T^{-1}\epsilon^{1/2}$	Statcoulomb	$M^{1/2}L^{1/2}\mu^{-1/2}$	Abcoulomb	IT	Coulomb‡
Current, I			$M^{1/2}L^{3/2}T^{-2}\epsilon^{1/2}$	Statampere	$M^{1/2}L^{1/2}T^{-1}\mu^{-1/2}$	Abampere	I	**Ampere‡**
Potential, E			$M^{1/2}L^{1/2}T^{-1}\epsilon^{-1/2}$	Statvolt	$M^{1/2}L^{3/2}T^{-2}\mu^{1/2}$	Abvolt	$ML^2T^{-3}I^{-1}$	Volt
Resistance, R			$L^{-1}T\epsilon^{-1}$	Statohm	$L^{-1}T\mu$	Abohm	$ML^2T^{-3}I^{-2}$	Ohm

* The table gives selected units of the mechanical and electrical absolute systems. Additional derived units are given in the following conversion tables. Fundamental dimensions and units are shown in boldface. Consistent thermodynamic systems of units may be obtained by adding the degree centigrade as a fundamental unit to the cgs and mks systems.

† 1 newton = 1 joule/meter = 1 × 10⁵ dynes.

‡ The coulomb was generally used in the United States as the fundamental unit, but the Tenth General Conference on Weights and Measures, International Bureau of Weights and Measures, 1955, adopted the meter, kilogram, second, ampere, and degree Kelvin as systematic units. The coulomb is thus defined as one ampere-second instead of the ampere being defined as one coulomb per second.

FIG. 3. 24-channel telemetry transmitter for high-speed turbines. (*Courtesy, Industrial Electronics Corporation, Melbourne Florida.*)

UNITS AND CONVERSION FACTORS

The engineer engaged in telemetering activities may be called upon to instrument a wide variety of tests involving any conceivable physical unit measurable. Sensors for most of these measurements are discussed in Chap. 3. Here are presented relationships between systems of units and conversion factors for fundamental measurable physical quantities. These have been taken, with permission, from H. E. Etherington (ed.), "Nuclear Engineering Handbook," (McGraw-Hill Book Company, New York, 1958).

7 Metric Systems of Absolute Units

Table 3 provides the relationships of the metric system of units.

8 Conversion Factors

The conversion tables appear in the following sequence: mechanical, heat, electrical, and miscellaneous. In each category, the units generally regarded as fundamental are given first. Within each table, metric units are given at the left in order of ascending magnitude; units of conventional engineering follow in convenient groups also in order of increasing magnitude.

All factors are accurate to four significant figures. Quantities given to less than four significant figures are exact equivalents.

Use of Conversion Tables

To convert a quantity given in one unit to the equivalent expressed in another unit, locate 1 under the given unit and in the same horizontal row find the multiplier to convert to the desired unit.

Example: Suppose a volume of 25 ft^3 is to be expressed in cubic centimeters. The volume-conversion table (Table 6), fifth horizontal row, shows that 1 ft^3 is equal to 2.832×10^4 cm^3. A volume of 25 ft^3 therefore equals

$$25 \times 2.832 \times 10^4 = 7.080 \times 10^5 \text{ cm}^3$$

Example of Derivation of Conversion Factors

Suppose it is desired to derive the conversion factor f for converting kilowatts to foot-pounds per minute.

FUNDAMENTALS

$$1 \text{ kw} = f \text{ ft-lb}_F/\text{min}$$

$$f = \frac{1 \text{ kw} \times 1 \text{ min}}{1 \text{ ft} \times 1 \text{ lb}_F}$$

It is necessary to develop the units of this ratio to units for which conversion factors are known. It is assumed for the purpose of this example that equivalents of the kilowatt in foot-pound-minute units are not known.

$$1 \text{ kw} = 1{,}000 \text{ watts} = 1{,}000 \text{ joules/sec} = 10^{10} \text{ ergs/sec} = 10^{10} \text{ dyne-cm/sec}$$

$$f = \frac{10^{10} \times \text{dyne-cm/sec} \times 60 \text{ sec}}{30.48 \text{ cm} \times 1 \text{ lb}_F} = \frac{60}{30.48} \times 10^{10} \times \frac{1 \text{ dyne}}{1 \text{ lb}_F}$$

Known conversion factors can be substituted at any time, and from the force conversion table, $1 \text{ lb}_F = 4.448 \times 10^5$ dynes.

$$f = \frac{60}{30.48} \times 10^{10} \times \frac{1 \text{ dyne}}{4.448 \times 10^5 \text{ dynes}} = 4.425 \times 10^4$$

Table 4. Length

Centimeters	Meters	Inches	Feet
1	0.01	0.3937	0.03281
100	1	39.37	3.281
2.540	0.02540	1	0.08333
30.48	0.3048	12	1

U.S. Customary Measures
12 in. = 1 ft
3 ft = 1 yd
5,280 ft = 1,760 yards = 1 mile
1 mil = 0.001 in.

Nautical Measures
6 ft = 1 fathom
120 fathoms = 1 cable length
1 international nautical mile (adopted July 1, 1954, by Departments of Defense and Commerce) = 1,852 meters = 6,076.10333 ft (approx.)
1 U.S. nautical mile = 1 geographical mile = 1 min of longitude at equator = 6,080.20 ft
1 British nautical mile = 6,080 ft

Astronomical Measures
1 light year = 5.879×10^{12} miles = 9.461×10^{17} cm

Metric Measures
10 mm = 1 cm
10 cm = 1 dm
10 dm = 1 meter
1,000 meters = 1 km
$1 \mu = 1 \times 10^{-6}$ meter
 $= 1 \times 10^{-3}$ mm
$1 \text{ m}\mu = 1$ micromillimeter $= 1 \times 10^{-3} \mu = 1 \times 10^{-6}$ mm
$1 \mu\mu = 1 \times 10^{-6} \mu = 1 \times 10^{-9}$ mm
$1 \text{ A} = 1 \times 10^{-8}$ cm $= 1 \times 10^{-4} \mu = 0.1 \text{ m}\mu$
1 X-ray unit (XU) $= 1 \times 10^{-3} \text{ A} = 1 \times 10^{-11}$ cm

Table 5. Area

Square centimeters	Square meters	Square inches	Square feet
1	1×10^{-4}	0.1550	0.001076
1×10^4	1	1,550	10.76
6.452	6.452×10^{-4}	1	0.006944
929.0	0.09290	144	1

U.S. Customary Measures
144 in.² = 1 ft²
9 ft² = 1 yd²
4,840 yd² = 1 acre
640 acres = 1 mile²

Electrical Unit
1 cir mil (area of circle 0.001 in. in diameter) = 0.7854×10^{-6} in.²
 $= 5.067 \times 10^{-6}$ cm²

Metric Measures
1 ha = 10,000 meters² = 2.471 acres
1 acre = 0.4047 ha

UNITS AND CONVERSION FACTORS

Table 6. Volume

Cubic centimeters	Liters	Cubic meters	Cubic inches	Cubic feet	Cubic yards	U.S. gallons (liquid)	Imperial gallons
1	0.001000	1×10^{-6}	0.06102	3.531×10^{-5}	1.308×10^{-6}	2.642×10^{-4}	2.200×10^{-4}
1000	1	0.001000	61.03*	0.03532*	0.001308	0.2642	0.2200
1×10^6	1,000	1	6.102×10^4	35.31	1.308	264.2	220.0
16.39	0.01639	1.639×10^{-5}	1	5.787×10^{-4}	2.143×10^{-5}	0.004329	0.003605
2.832×10^4	28.32	0.02832	1,728	1	0.03704	7.481	6.229
7.646×10^5	764.5	0.7646	4.666×10^4	27	1	202.0	168.2
3785	3.785	0.003785	231	0.1337	0.004951	1	0.8327
4546	4.546	0.004546	277.4	0.1605	0.005946	1.201	1

Fluid Measures

Fundamental Standards of Fluid Capacity. The metric, United States, and British units are defined independently.

One *liter* is the volume of one kilogram of water at 3.98°C and 760 mm of mercury. This is equal to 1.000028 cubic decimeters,† the deviation from one cubic decimeter being the difference between the intended and actual mass of the standard kilogram.

    1 U.S. gal = 231 in.³ exactly (definition)

One Imperial gallon is the volume occupied by 10 pounds avoirdupois of distilled water at 62°F and 30 in. of mercury. This volume is 277.420 U.S. cubic inches.

Metric Fluid Measures

1,000 ml = 100 cl = 1 liter
1 ml = 1 cm³ (approx., see above)

U.S. and British Fluid Measures

4 gills (gi) = 1 liq pt
2 liq pt = 1 liq qt
4 liq qt = 1 gal

Note: The relationships among the gill, pint, quart, and gallon are the same in United States and British systems; the magnitudes of corresponding units are in the same ratio as the United States and the Imperial gallon given in the table.

One U.S. gallon of water at 62°F weighs 8.336 pounds avoirdupois (approximately).
One Imperial (British and Canadian) gallon of water at 62°F weighs, by definition, 10 pounds avoirdupois (exactly).

16 U.S. fl oz = 1 U.S. liq pt
(1 fl oz of water weighs 1.042 oz avdp at 62°F)
160 British fl oz = 1 Imperial gal
(1 fl oz of water weighs 1 oz avdp at 62°F)

Dry Measures

1 U.S. bu = 2,150.42 in.³
1 British bu = 2,219.34 in.³ (U.S.) = 8 Imperial gal‡

Shipping Measures

Internal capacity: 100 ft³ = 1 register ton
Cargo: 40 ft³ = 1 U.S. shipping ton
              42 ft³ = 1 British shipping ton

Lumber

One board foot is the volume of one square foot of board one inch thick (144 cubic inches), measured before dressing. The feet board measure (fbm), or board feet of flat lumber, is the area in square feet multiplied by the thickness in inches. The board foot measurement of dressed stock is calculated from the dimensions of the rough lumber required; thicknesses less than one inch are taken as one inch.

---

* The difference between the liter and 1,000 cm³ is insignificant but may lead to a fourth-figure difference in rounding off if the fifth figure is very close to 5.

† Units of Weight and Measure (United States Customary and Metric), Definitions and Tables of Equivalents, Natl. Bur. Standards Misc. Publ. 214, July 1, 1955.

‡ The Imperial gallon is both a dry and a liquid measure. The United States gallon is legally only a liquid measure—one-eighth of a United States bushel is sometimes called a "dry gallon" (equal to 1.16365 U.S. gal).

Table 7. Mass

Grams	Kilograms	Pounds (avdp)	Short tons	Long tons
1	0.001	0.002205	1.102×10^{-6}	9.842×10^{-7}
1000	1	2.205	0.001102	9.842×10^{-4}
453.6	0.4536	1	0.0005	4.464×10^{-4}
9.072×10^5	907.2	2000	1	0.8929
1.016×10^6	1016	2240	1.12	1

Metric Weights

$1{,}000{,}000$ μg (or γ) = $1{,}000$ mg = 1 g
$1{,}000$ g = 1 kg
$1{,}000$ kg = 1 metric ton or **tonne**

For precious stones: 1 carat = 1 metric carat = 200 mg

Avoirdupois Weights

16 oz (oz avdp) = 7,000 grains = 1 lb (**lb avdp**)
100 lb = 1 short cwt
2,000 lb = 1 short ton
112 lb = 1 long cwt
2,240 lb = 1 long ton

Engineer's unit: 1 slug (or geepound) = 32.174 lb

*Troy Weights**

12 oz (oz troy) = 5760 grains = 1 lb (**lb troy**)

* The grain is the same as the avoirdupois measure; 1 pound troy = 5,760/7,000 pound avoirdupois, 1 ounce troy = $(5{,}760 \times 16)/(7{,}000 \times 12)$ = 1.097 avoirdupois ounce.
"Pound" and "ounce" always mean the avoirdupois measures unless otherwise stated or implied by context. The *apothecary's pound* and *ounce* are the same as the corresponding **troy** masses.

Table 8. Time

1 mean solar (or tropical) year = 365.24 days = 8,766 hr
= 5.259×10^5 min = 3.156×10^7 sec
1 day = 24 hr = 1,440 min = 8.640×10^4 sec

Table 9. Force

Dynes	Newtons (joules/meter)	Grams	Poundals	Pounds	Torque
1	1×10^{-5}	0.001020	7.233×10^{-5}	2.248×10^{-6}	For equivalents of the dyne-centimeter, kilogram (force)-meter, newton-meter, and foot-pound, use Table 15 (Energy, Work, and Heat) conversion factors. For quantities measured in inch-pounds, first divide by 12 to convert to foot-pounds.
1×10^5	1	102.0	7.233	0.2248	
980.7	0.009807	1	0.07093	0.002205	
1.383×10^4	0.1383	14.10	1	0.03108	
4.448×10^5	4.448	453.6	32.17	1	

Table 10. Angular Measures

Plane Angle

Seconds	Minutes	Degrees	Right angles or quadrants	Revolutions or circumferences	Radians
1	0.01667	2.778×10^{-4}	3.086×10^{-6}	7.716×10^{-7}	4.848×10^{-6}
60	1	0.01667	1.852×10^{-4}	4.630×10^{-5}	2.909×10^{-4}
3600	60	1	0.01111	0.002778	0.01745
3.24×10^5	5400	90	1	0.25	1.571
1.296×10^6	2.16×10^4	360	4	1	6.283
2.063×10^5	3438	57.30	0.6366	0.1592	1

π radians = 180°
$1° = \dfrac{\pi}{180} = 0.0174533$ radians
100 centesimal minutes = 1 grade
100 grades = 1 right angle

Angular Velocity

Revolutions per second	Revolutions per minute	Radians per second
1	60	6.283
0.01667	1	0.1047
0.1592	9.549	1

Solid Angles

1 sphere (or steregon) = 4π (or 12.5664) steradians = 8 spherical right angles. A steradian is the solid angle subtended at the center of a sphere of radius r by an area r^2 of the spherical surface.

UNITS AND CONVERSION FACTORS

Table 11. Velocity

Centimeters per second	Meters per second	Feet per second	Feet per minute	Miles per hour
1	0.01	0.03281	1.969	0.02237
100	1	3.281	196.9	2.237
30.48	0.3048	1	60	0.6818
0.5080	0.005080	0.01667	1	0.01136
44.70	0.4470	1.467	88	1

Nautical Velocity
1 knot (U.S.) = 1 U.S. nautical mile/hr
= 1.152 statute miles/hr
= 1.689 fps
= 51.48 cm/sec
For other nautical miles see *Length conversion* factors

Table 12. Flow

Cubic centimeters per second	Cubic feet per minute	U.S. gallons per minute	Imperial gallons per minute
1	0.002119	0.01585	0.01320
472.0	1	7.481	6.229
63.09	0.1337	1	0.8327
75.77	0.1605	1.201	1

For other conversions involving no change of time unit use volume-conversion table (Table 6). 1 U.S. gpm = 8.02$1\rho$ lb/hr = 500.7ρ' lb/hr where ρ = density, lb/ft³, ρ' = density, g/cm³.

Mass Velocity

$$\text{Mass velocity} = \frac{\text{mass per unit of time}}{\text{cross-sectional area of stream}} = \text{velocity} \times \text{density}$$

The units are usually pound-mass, foot, hour (occasionally, pound-mass, foot, second).

Table 13. Density

Grams per cubic centimeter	Kilograms per cubic meter	Pounds per cubic inch	Pounds per cubic foot	Pounds per U.S. gallon	Pounds per Imperial gallon
1	1000	0.03613	62.43	8.345	10.02
0.001	1	3.613 × 10⁻⁵	0.06243	0.008345	0.01002
27.68	2.768 × 10⁴	1	1728	231	277.4
0.01602	16.02	5.787 × 10⁻⁴	1	0.1337	0.1605
0.1198	119.8	0.004329	7.481	1	1.201
0.09978	99.78	0.003605	6.229	0.8327	1

Density and Specific Gravity. Density is the mass per unit volume of a substance at a specified temperature. Specific gravity is the ratio of the weight of a substance at a specified temperature to the weight of an equal volume of a reference substance (usually water), also at a specified temperature. In notations such as 20°C/4°C or 20/4C, the first temperature refers to the material and the second to the reference substance. If the reference temperature is (as in the example) 4°C, specific gravity is numerically equal to density in grams per milliliter. The dimension of density is ML⁻³; specific gravity is a numeric.

Specific Gravity of Liquids by Hydrometer

Liquids lighter than water:

$$\text{Degrees API (for petroleum products)} = \frac{141.5}{\text{sp gr 60°F/60°F}} - 131.5$$

$$\text{Degrees Baumé (for other liquids)} = \frac{140}{\text{sp gr 60°F/60°F}} - 130$$

Liquids heavier than water:

$$\text{Degrees Baumé (all liquids)} = 145 - \frac{145}{\text{sp gr 60°F/60°F}}$$

Mass per Unit Length
1 g/cm = 0.005600 lb/in. = 0.06720 lb/ft
1 lb/in. = 178.6 g/cm
1 lb/ft = 14.88 g/cm

Mass per Unit Area
1 g/cm² = 2.048 psf
1 psf = 0.4882 g/cm²

Table 14. Pressure

Baryes or dynes per square centimeter	Kilograms per square centimeter	Newtons per square meter	Pounds per square inch	Pounds per square foot	Centimeters of mercury at 0°C	Inches of mercury at 0°C (32°F)	Inches of water at 60°F	Feet of water at 60°F	Atmospheres*
1	1.020 × 10^{-6}	0.1	1.450 × 10^{-5}	0.002089	7.501 × 10^{-5}	2.953 × 10^{-5}	4.018 × 10^{-4}	3.349 × 10^{-5}	9.869 × 10^{-7}
9.807 × 10^5	1	9.807 × 10^4	14.22	2048	73.56	28.96	394.1	32.84	0.9678
10	1.020 × 10^{-5}	1	1.450 × 10^{-4}	0.02089	7.501 × 10^{-4}	2.953 × 10^{-4}	4.018 × 10^{-3}	3.349 × 10^{-4}	9.869 × 10^{-6}
6.895 × 10^4	0.07031	6895	1	144	5.171	2.036	27.71	2.309	0.06805
478.8	4.882 × 10^{-4}	47.88	0.006944	1	0.03591	0.01414	0.1924	0.01603	4.725 × 10^{-4}
1.333 × 10^4	0.01360	1333	0.1934	27.85	1	0.3937	5.358	0.4465	0.01316
3.386 × 10^4	0.03453	3386	0.4912	70.73	2.540	1	13.61	1.134	0.03342
2488	0.002538	248.8	0.03609	5.197	0.1866	0.07348	1	0.08333	0.002456
2.986 × 10^4	0.03045	2986	0.4331	62.37	2.230	0.8818	12	1	0.02947
1.013 × 10^6	1.033	1.013 × 10^5	14.70	2116	76	29.92	407.2	33.93	1

Units of Atmospheric Pressure
1 barye = 1 dyne/cm²
1 bar = 10^6 dynes/cm²
(This is the internationally accepted meaning; but 1 bar is also sometimes used as a synonym of 1 barye or 1 dyne/cm².)

* The standard atmosphere is based on 76 cm (= 29.92 in.) of mercury at 0°C, at a location of standard gravity and is defined by the Tenth General Conference on Weights and Measures (1954) as 1,013,250 dynes/cm² or 101,325 newtons/meter². Hence 1 atm equals 1.01325 bars or 1.03323 kg force per square centimeter. (In power plant practice, steam condenser vacuum is referred to a 30-in. barometer instead of 29.92 in.)

Table 15. Energy, Work, Heat

Ergs (dyne-centimeters)	Kilogram (force)-meters	Joules (newton-meters, watt seconds)	Kilowatt-hours	Calories	Kilocalories	Foot-pounds	Horsepower-hours	British thermal units	Centigrade heat units
1	1.020 × 10^{-8}	1 × 10^{-7}	2.778 × 10^{-14}	2.388 × 10^{-8}	2.388 × 10^{-11}	7.376 × 10^{-8}	3.725 × 10^{-14}	9.478 × 10^{-11}	5.266 × 10^{-11}
9.807 × 10^7	1	9.807	2.724 × 10^{-6}	2.342	0.002342	7.233	3.653 × 10^{-6}	0.009295	0.005164
1 × 10^7	0.1020	1	2.778 × 10^{-7}	0.2388	2.388 × 10^{-4}	0.7376	3.725 × 10^{-7}	9.478 × 10^{-4}	5.266 × 10^{-4}
3.6 × 10^{13}	3.671 × 10^5	3.6 × 10^6	1	8.598 × 10^5	859.8	2.655 × 10^6	1.341	3412	1896
4.187 × 10^7	0.4269	4.187	1.163 × 10^{-6}	1	0.001	3.088	1.560 × 10^{-6}	0.003968	0.002205
4.187 × 10^{10}	426.9	4187	0.001163	1000	1	3088	0.001560	3.968	2.205
1.356 × 10^7	0.1383	1.356	3.766 × 10^{-7}	0.3238	3.238 × 10^{-4}	1	5.051 × 10^{-7}	0.001285	7.139 × 10^{-4}
2.685 × 10^{13}	2.737 × 10^5	2.685 × 10^6	0.7457	6.412 × 10^5	641.2	1.98 × 10^6	1	2544	1414
1.055 × 10^{10}	107.6	1055	2.931 × 10^{-4}	252.0	0.2520	778.2	3.930 × 10^{-4}	1	0.5556
1.899 × 10^{10}	193.7	1899	5.275 × 10^{-4}	453.6	0.4536	1401	7.074 × 10^{-4}	1.8	1

Energy, Work, Heat
1 liter-atm = 24.20 cal = 101.3 joules = 0.09604 Btu
1 (ft³) (atm) = 2.719 Btu
1 (ft³) (psi) = 0.1850 Btu

Table 16. Power

Ergs per second	Kilogram (force)-meters per second	Watts (joules per second)	Kilowatts	Calories per second	Kilocalories per second	Foot-pounds per second	Foot-pounds per minute	Horsepower	Btu (IT) per hour	Chu per hour
1	1.020×10^{-8}	1×10^{-7}	1×10^{-10}	2.388×10^{-8}	2.388×10^{-11}	7.376×10^{-8}	4.425×10^{-6}	1.341×10^{-10}	3.412×10^{-7}	1.896×10^{-7}
9.807×10^7	1	9.807	0.009807	2.342	2.342×10^{-3}	7.233	434.0	0.01315	33.46	18.59
1×10^7	0.1020	1	0.001	0.2388	2.388×10^{-4}	0.7376	44.25	0.001341	3.412	1.896
1×10^{10}	102.0	1000	1	238.8	0.2388	737.6	4.425×10^4	1.341	3412	1896
4.187×10^7	0.4269	4.187	0.004187	1	0.001	3.088	185.3	0.005615	14.29	7.937
4.187×10^{10}	426.9	4187	4.187	1000	1	3088	1.853×10^5	5.615	1.429×10^4	7937
1.356×10^7	0.1383	1.356	0.001356	0.3238	3.238×10^{-4}	1	60	0.001818	4.626	2.570
2.260×10^5	0.002304	0.02260	2.260×10^{-5}	0.005397	5.397×10^{-6}	0.01667	1	3.030×10^{-5}	0.07710	0.04284
7.457×10^9	76.04	745.7	0.7457	178.1	0.1781	550	3.3×10^4	1	2544	1414
2.931×10^6	0.02989	0.2931	2.931×10^{-4}	0.07000	7.0000×10^{-5}	0.2162	12.97	3.930×10^{-4}	1	0.5556
5.275×10^6	0.05379	0.5275	5.275×10^{-4}	0.1260	1.260×10^{-4}	0.3891	23.35	7.074×10^{-4}	1.8	1

Horsepower
1 hp = 550 ft-lb/sec (definition) = 33,000 ft-lb/min = 1,980,000 ft-lb/hr = 745.70 watts
1 electric hp = 746 watts (definition) = 1.0004 hp
1 metric hp = 75 kg (force)-meters/sec (definition)
 = 735.499 watts = 0.9863 hp
1 boiler hp = 34.5 lb water evaporated per hour from and at 212°F
 = 33,472 Btu/hr = 13.155 hp (heat equivalent)

Refrigeration
1 ton of refrigeration (U.S.) = 288,000 Btu/day = 12,000 Btu/hr = 200 Btu/min
 = 4.716 hp = 840.0 cal/sec = 0.9055 British ton of refrigeration
1 ton of refrigeration (British) = 321,200 Btu/day = 13,385 Btu/hr
 = 223.1 Btu/min = 5.260 hp = 936.9 kcal/sec
 = 1.115 U.S. tons of refrigeration
The units are approximately equal to the latent heat of fusion of one short ton and one long ton, respectively, of ice per 24 hr.

Table 17. Temperature

Centigrade to Fahrenheit
(Values are exact)

°C	0	10	20	30	40	50	60	70	80	90
0	32	50	68	86	104	122	140	158	176	194
100	212	230	248	266	284	302	320	338	356	374
200	392	410	428	446	464	482	500	518	536	554
300	572	590	608	626	644	662	680	698	716	734
400	752	770	788	806	824	842	860	878	896	914
500	932	950	968	986	1004	1022	1040	1058	1076	1094
600	1112	1130	1148	1166	1184	1202	1220	1238	1256	1274
700	1292	1310	1328	1346	1364	1382	1400	1418	1436	1454
800	1472	1490	1508	1526	1544	1562	1580	1598	1616	1634
900	1652	1670	1688	1706	1724	1742	1760	1778	1796	1814
1000	1832	1850	1868	1886	1904	1922	1940	1958	1976	1994
1100	2012	2030	2048	2066	2084	2102	2120	2138	2156	2174
1200	2191	2210	2228	2246	2264	2282	2300	2318	2336	2354
1300	2372	2390	2408	2426	2444	2462	2480	2498	2516	2534
1400	2552	2570	2588	2606	2624	2642	2660	2678	2696	2714

Interpolation Table

°C	1	2	3	4	5	6	7	8	9
°F	1.8	3.6	5.4	7.2	9.0	10.8	12.6	14.4	16.2

Example: 1054°C = 1922 + 7.2 = 1929.2°F.

Fahrenheit to Centigrade
(Centigrade temperatures to nearest 0.1°C)

°F	0	10	20	30	40	50	60	70	80	90
0	−17.8	−12.2	−6.7	−1.1	4.4	10.	15.6	21.1	26.7	32.2
100	37.8	43.3	48.9	54.4	60.	65.6	71.1	76.7	82.2	87.8
200	93.3	98.9	104.4	110.0	115.6	121.1	126.7	132.2	137.8	143.3
300	148.9	154.4	160.0	165.6	171.1	176.7	182.2	187.8	193.3	198.9
400	204.4	210.0	215.6	221.1	226.7	232.2	237.8	243.3	248.9	254.4
500	260.0	265.6	271.1	276.7	282.2	287.8	293.3	298.9	304.4	310.0
600	315.6	321.1	326.7	332.2	337.8	343.3	348.9	354.4	360.0	365.6
700	371.1	376.7	382.2	387.8	393.3	398.9	404.4	410.0	415.6	421.1
800	426.7	432.2	437.8	443.3	448.9	454.4	460.0	465.6	471.1	476.7
900	482.2	487.8	493.3	498.9	504.4	510.0	515.6	521.1	526.7	532.2
1000	537.8	543.3	548.9	554.4	560.0	565.6	571.1	576.7	582.2	587.8
1100	593.3	598.9	604.4	610.0	615.6	621.1	626.7	632.2	637.8	643.3
1200	648.9	654.4	660.0	665.6	671.1	676.7	682.2	687.8	693.3	698.9
1300	704.4	710.0	715.6	721.1	726.7	732.2	737.8	743.3	748.9	754.4
1400	760.0	765.6	771.1	776.7	782.2	787.8	793.3	798.9	804.4	810.0
1500	815.6	821.1	826.7	832.2	837.8	843.3	848.9	854.4	860.0	865.6
1600	871.1	876.7	882.2	887.8	893.3	898.9	904.4	910.0	915.6	921.1
1700	926.7	932.2	937.8	943.3	948.9	954.4	960.0	965.6	971.1	976.7
1800	982.2	987.8	993.3	998.9	1004.4	1010.0	1015.6	1021.1	1026.7	1032.2
1900	1037.8	1043.3	1048.9	1054.4	1060.0	1065.6	1071.1	1076.7	1082.2	1087.8
2000	1093.3	1098.9	1104.4	1110.0	1115.6	1121.1	1126.7	1132.2	1137.8	1143.3
2100	1148.9	1154.4	1160.0	1165.6	1171.1	1176.7	1182.2	1187.8	1193.3	1198.9
2200	1204.4	1210.0	1215.6	1221.1	1226.7	1232.2	1237.8	1243.3	1248.9	1254.4
2300	1260.0	1265.6	1271.1	1276.7	1282.2	1287.8	1293.3	1298.9	1304.4	1310.0
2400	1315.6	1321.1	1326.7	1332.2	1337.8	1343.3	1348.9	1354.4	1360.0	1365.6
2500	1371.1	1376.7	1382.2	1387.8	1393.3	1398.9	1404.4	1410.0	1415.6	1421.1

Interpolation Table

°F	1	2	3	4	5	6	7	8	9
°C	0.6	1.1	1.7	2.2	2.8	3.3	3.9	4.4	5.0

Example: 1054°F = 565.6 + 2.2 = 567.8°C.

Conversion Formulas

Centigrade to Fahrenheit: $t_F = \frac{9}{5} t_C + 32$
Fahrenheit to Centigrade: $t_C = \frac{5}{9}(t_F - 32)$
Degrees Kelvin: $t_K = t_C + 273.16$
Degrees Rankine: $t_R = t_F + 459.69$
  $t_R = \frac{9}{5} t_K$

UNITS AND CONVERSION FACTORS

Table 18. Specific Heat

1 cal/(g)(°C) = 1 kcal/(kg)(°C) = 1 Btu/(lb)(°F) = 1 chu/(lb)(°C)
= 4.187 joules/(g)(°C)

Specific heat was originally defined as the heat capacity of a substance relative to that of an equal mass of water at a specified temperature. The property was therefore a dimensionless ratio. The term is now generally used in the sense of heat capacity per unit mass per degree of temperature rise.

Table 19. Thermal Conductivity

Cal / (sec)(cm)(°C)	Watts / (cm)(°C)	Btu / (hr)(in.)(°F)	Btu / (hr)(ft)(°F)	Btu / (hr)(ft²)(°F/in.)
1	4.187	20.02	241.9	2903
0.2388	1	4.782	57.78	693.3
0.04961	0.2077	1	12	144
0.004134	0.01731	0.08333	1	12
3.445 × 10⁻⁴	0.001441	0.006944	0.08333	1

Table 20. Heat Capacity

Calorie per gram	Joule per gram	Btu per pound
1	4.187	1.8
0.2388	1	0.4299
0.5556	2.326	1

1 kcal/kg = 1 cal/g = 1 chu/lb

1 joule/(sec)(cm)(°C) = 1 watt/(cm)(°C)
1 joule/(sec)(m)(°C) = 1 watt/(m)(°C) = 0.01 watt/(cm)(°C)

Thermal conductivity is frequently stated as heat units per unit time, per unit area, per unit of temperature difference per unit thickness, written, for example, Btu/(hr)(ft²)(°F/ft). For consistent units the designation can be simplified as in the table. The inconsistent units shown in the last column are commonly used in practical problems involving heat conduction through plates, walls, insulation, etc. Conversion factors for the chu used in conjunction with the degree centigrade are the same as for the Btu used in conjunction with the degree Fahrenheit; e.g., the Btu/(hr)(ft)(°F) column serves also for chu/(hr)(ft)(°C).

Table 21. Heat Flux and Conductance

Heat Flux				Thermal Conductance		
Cal / (sec)(cm²)	Watts / cm²	Btu / (hr)(ft²)	Chu / (hr)(ft²)	Cal / (sec)(cm²)(°C)	Watts / (cm²)(°C)	Btu / (hr)(ft²)(°F)
1	4.187	1.327 × 10⁴	7373	1	4.187	7373
0.2388	1	3170	1761	0.2388	1	1761
7.535 × 10⁻⁵	3.155 × 10⁻⁴	1	0.5556	1.356 × 10⁻⁴	5.678 × 10⁻⁴	1
1.356 × 10⁻⁴	5.678 × 10⁻⁴	1.8	1			

1 joule/(sec)(cm²) = 1 watt/cm²
1 joule/(sec)(meter²) = 1 watt/meter²
= 1 × 10⁻⁴ watt/cm²

1 joule/(sec)(cm²)(°C) = 1 watt/(cm²)(°C)
1 joule/(sec)(meter²)(°C) = 1 watt/(meter²)(°C)
= 1 × 10⁻⁴ watt/(cm²)(°C)
1 chu/(hr)(ft²)(°C) = 1 Btu/(hr)(ft²)(°F)

Table 22. Internal Heat Generation and Power Density

Watts / cm³	Cal / (sec)(cm³)	Btu / (hr)(in.³)	Btu / (hr)(ft³)
1	0.2388	55.91	9.662 × 10⁴
4.187	1	234.1	4.045 × 10⁵
0.01788	0.004272	1	1728
1.035 × 10⁻⁵	2.472 × 10⁻⁶	5.787 × 10⁻⁴	1

1 watt/cm³ = 1 Mw/meter³ = 1.000 kw/liter
1 chu = 1.8 Btu

Table 23. Viscosity

Absolute Viscosity, μ

Centipoises	Poises	Kilogram-mass (sec)(meter)	Pound-mass (sec)(ft)	Pound-force (sec)(ft²)	Pound-mass (hr)(ft)
1	0.01	0.001	6.720×10^{-4}	2.089×10^{-5}	2.419
100	1	0.1	0.06720	0.002089	241.9
1000	10	1	0.6720	0.02089	2419
1488	14.88	1.488	1	0.03108	3600
4.788×10^4	478.8	47.88	32.17	1	1.158×10^5
0.4134	0.004134	4.134×10^{-4}	2.778×10^{-4}	8.634×10^{-6}	1

Alternative Dimensional Systems

The units of absolute viscosity are expressed alternatively in absolute units (dimensions M/TL) or in equivalent gravitational units (dimensions FT/L²).

$$1 \text{ poise} = 1 \text{ g/(sec)(cm)} = 1 \text{ dyne-sec/cm}^2$$
$$1 \text{ lb}_M/(\text{sec})(\text{ft}) = 1 \text{ poundal-sec/ft}^2$$
$$1 \text{ slug/(sec)(ft)} = 1 \text{ lb}_F\text{-sec/ft}^2$$
$$1 \text{ kg}_M/(\text{sec})(\text{meter}) = 1 \text{ newton-sec/meter}^2$$

All tabulated viscosity data in this handbook are expressed in *pound-mass per hour per foot* units (last column). Laboratory data are expressed formally in centipoises, a metric unit of convenient size for many liquids; water has a viscosity of 1 centipoise at a temperature of 20.20°C (68.4°F).

Kinematic Viscosity, ν

The kinematic viscosity is the absolute viscosity divided by the density of the fluid: $\nu = \mu/\rho$. The cgs unit of viscosity is the stoke.

$$1 \text{ stoke} = 1 \text{ cm}^2/\text{sec} = 0.1550 \text{ in.}^2/\text{sec} = 3.875 \text{ ft}^2/\text{hr}$$

Viscosity can be measured in the laboratory by the time in seconds for free discharge of a given volume of fluid through the capillary of an efflux viscometer (viscosimeter). The kinematic viscosity in stokes can be calculated from the efflux time t by the empirical formula

$$\nu = At - B/t \quad \text{cm}^2/\text{sec}$$

where A and B are 0.0022 and 1.8, respectively, on the Saybolt Universal scale (United States) and 0.0026 and 1.72, respectively, on the Redwood scale (British). The viscosity of reactor coolants is below the useful range of these and similar viscosity scales.

Table 24. Electrical Units

System	Quantity of electricity	Electric current	Potential difference and electromagnetic force	Resistance	Capacitance
Absolute (cgs) electrostatic, esu	$10^{-1}c$ statcoulombs	$10^{-1}c$ statamp	$10^8 c^{-1}$ statvolt	$10^9 c^{-2}$ statohm	$10^{-9} c^2$ statfarads
MKS or practical, prau	1 coulomb	1 amp	1 volt	1 ohm	1 farad
Absolute (cgs) electromagnetic, emu	10^{-1} abcoulomb	10^{-1} abamp	10^8 abvolts	10^9 abohms	10^{-9} abfarad

System	Magnetic field intensity	Magneto-motive force	Magnetic flux	Magnetic flux density	Inductance
Absolute (cgs) electrostatic, esu					$10^9 c^{-2}$ stathenry
MKS or practical, prau	1 praoersted	1 pragilbert	1 weber	1 weber/meter^2	1 henry
Absolute (cgs) electromagnetic, emu	10^{-3} oersted	10^{-1} gilbert	10^8 maxwells (or lines)	10^4 gauss	10^9 abhenrys

$c = 2.99793 \times 10^{10}$, a numeric equal in magnitude to the speed of light in centimeters per second. Approximately, $c = 2.998 \times 10^{10}$, $c^2 = 8.988 \times 10^{20}$, $c^{-1} = 0.3336 \times 10^{-10}$, $c^{-2} = 1.113 \times 10^{-21}$

Examples:

1. 500 amp = 500 × 10^{-1} abamp = 50 abamp.
2. 1 × 10^6 statcoulombs = (1 × 10^6)/(10^{-1} × 2.998 × 10^{10}) coulombs = **3.336 × 10^{-4} coulomb**.

$$1 \text{ gilbert} = \frac{1}{4\pi} \text{ ampere turns}$$

1 oersted = 1 gilbert/cm
1 gauss = 1 maxwell (or line)/cm^2

Atomic Units

The units of classical physics (the cgs units and the supplementary units of the absolute electrical systems) are generally used in theoretical physics. Since these classical units are inconveniently large for practical calculation on an atomic scale, convenient nonsystematic units are defined for limited use.

Atomic Mass Unit, amu. The amu, expressed in the physical scale, is defined as one-sixteenth of the mass of an O^{16} atom.

$$1 \text{ amu} = (1.659790 \pm 0.000044) \times 10^{-24} \text{ g}$$

Numerically, this is the reciprocal of Avogadro's number on the physical scale.

Electron Volt, ev. The electron volt is the kinetic energy acquired by a particle of unit charge (the charge of one electron) in falling through a potential difference of one (practical) volt.

$$1 \text{ Mev} = 1{,}000{,}000 \text{ ev}$$

Barn. This unit of area is a measure of nuclear cross section.

1 barn = 10^{-24} cm^2
1 millibarn = 0.001 barn

Table 25. Physical Constants*

Constant	Value
Boltzmann constant and related constants:	
Boltzmann constant, k	$(1.38044 \pm 0.00007) \times 10^{-16}$ erg/°C
	$(8.6167 \pm 0.0004) \times 10^{-5}$ ev/°C
	$(4.7871 \pm 0.0002) \times 10^{-5}$ ev/°F
Universal gas constant, $R = Nk$:	
Chemical scale	$(8.31470 \pm 0.00034) \times 10^{7}$ erg/(mole)(°C)
	2781.70 ± 0.11 ft-lb/(lb-mole)(°C)
	1545.39 ± 0.06 ft-lb/(lb-mole)(°F)
	$(8.20575 \pm 0.00034) \times 10^{-2}$ liter-atm/(mole)(°C)
	$1.98591 \pm >0.0008$ cal/(mole)(°C)
	$1.98591 \pm >0.0008$ chu/(lb-mole)(°C)
	$1.98591 \pm >0.0008$ Btu/(lb-mole)(°F)
Physical scale	$(8.31696 \pm 0.00034) \times 10^{7}$ erg/(mole)(°C)
Electronic data:	
Faraday constant, F:	
Chemical scale	96495.7 ± 1.1 coulombs/mole
Physical scale	9652.19 ± 0.11 emu/mole
	$(2.89366 \pm 0.00003) \times 10^{14}$ esu/mole
Electronic charge, $e = F/N$	$(1.60206 \pm 0.00003) \times 10^{-20}$ emu
	$(1.60206 \pm 0.00003) \times 10^{-19}$ coulomb
	$(4.80286 \pm 0.00009) \times 10^{-10}$ esu
Charge-to-mass ratio of the electron, e/m	$(5.27305 \pm 0.00007) \times 10^{17}$ esu/g
	$(1.75890 \pm 0.00002) \times 10^{7}$ emu/g
	$(1.75890 \pm 0.00002) \times 10^{8}$ coulombs/g
Gravitational constant, G	$(6.670 \pm 0.005) \times 10^{-8}$ dyne-cm²/gram²
	$(4.958 \pm 0.004) \times 10^{-15}$ lb-ft²/slug²
Planck's constant:	
Planck's constant, h	$(6.62517 \pm 0.00023) \times 10^{-27}$ erg sec
	$(4.13541 \pm 0.00007) \times 10^{-15}$ ev sec
Stefan-Boltzmann and related constants:	
Surface constant, $\sigma = (\pi^2/60)(k^4/\hbar^3 c^2)$	$(0.56687 \pm 0.00010) \times 10^{-4}$ erg/(cm²)(°K)⁴(sec)
	$(1.35393 \pm >0.00024) \times 10^{-12}$ cal/(cm²)(°K)⁴(sec)
	$(0.99831 \pm >0.00018) \times 10^{-8}$ chu/(ft²)(°K)⁴(hr)
	$(0.17118 \pm >0.00003) \times 10^{-8}$ Btu/(ft²)(°R)⁴(hr)
Density constant, $4\sigma/c$	$(7.5635 \pm 0.0013) \times 10^{-15}$ erg/(cm³)(°K)⁴
Wein's displacement law $\lambda_{max} T$	(0.289782 ± 0.000013) (cm)(°K)
Velocity of light:	
c	$(2.997930 \pm 0.000003) \times 10^{10}$ cm/sec

* Except where otherwise indicated, all data are taken directly from tables by E. R. Cohen et al., Analysis of Variance of the 1952 Data on the Atomic Constants and a New Adjustment, 1955, *Rev. Mod. Phys.*, **27**: 363–380 (1955).

Power

$$1 \text{ Mev/sec} = 1.60206 \times 10^{-13} \text{ watt} = 1.60206 \times 10^{-16} \text{ kw}$$
$$= 3.8264 \times 10^{-14} \text{ cal/sec} = 5.4678 \times 10^{-13} \text{ Btu/hr}$$
$$1 \text{ watt} = 6.24196 \times 10^{12} \text{ Mev/sec}$$

UNITS AND CONVERSION FACTORS

Table 26. Mass-Energy Equivalents*

Amu	Mev	Grams	Ergs
1	931.141 ± 0.010	$(1.659790 \pm 0.000044) \times 10^{-24}$	$(1.491750 \pm 0.000043) \times 10^{-3}$
$(1.073951 \pm 0.000011) \times 10^{-3}$	1	$(1.78252 \pm 0.00004) \times 10^{-27}$	$(1.60206 \pm 0.00003) \times 10^{-6}$
$(6.02486 \pm 0.00016) \times 10^{23}$	$(5.61000 \pm 0.00011) \times 10^{26}$	1	$(8.987584 \pm 0.000018) \times 10^{20}$
670.354 ± 0.019	$(0.624196 \pm 0.000012) \times 10^{6}$	$(1.112646 \pm 0.000002) \times 10^{-21}$	1

$\quad$ 1 ev $= (1.60206 \pm 0.00003) \times 10^{-12}$ erg
$\quad$ 1 Mev $= 1.60206 \times 10^{-13}$ joule $= 4.4508 \times 10^{-20}$ kwhr $= 1.5188 \times 10^{-16}$ Btu
$\qquad \qquad \qquad = 3.8264 \times 10^{-14}$ cal
1 g $= 8.988 \times 10^{13}$ joules $= 2.496 \times 10^{7}$ kwhr $= 2.147 \times 10^{13}$ cal $= 8.518 \times 10^{10}$ Btu
1 lb $= 4.077 \times 10^{16}$ joules $= 1.132 \times 10^{10}$ kwhr $= 9.737 \times 10^{15}$ cal $= 3.863 \times 10^{13}$ Btu
$\quad$ 1 ev/atom (or molecule) $= 23047$ cal/mole (chemical scale)
1 cal/mole (chemical scale) $= 4.3389 \times 10^{-5}$ ev/atom (or molecule)

* Equivalents are derived from the following values by Cohen et al.: $c = (2.997930 \pm 0.000003) \times 10^{10}$ cm/sec, 1 amu $= 931.141 \pm 0.010$, 1 g $= (6.02486 \pm 0.00016) \times 10^{23}$ amu, 1 ev $= (1.60206 \pm 0.00003) \times 10^{-12}$ erg.
The heat-conversion factor is taken as 4.18684 joules/cal, and the molar volume on the chemical scale as 22414.6 cm³.

Table 27. Rest Mass of Particles*

Particle	Amu	Mev	Grams
Electron	$(5.48763 \pm 0.00006) \times 10^{-4}$	0.510976 ± 0.000007	$(9.1083 \pm 0.0003) \times 10^{-28}$
Neutron	1.008982 ± 0.000003	939.505 ± 0.010	$(1.67470 \pm 0.00004) \times 10^{-24}$
Proton	1.007593 ± 0.000003	938.211 ± 0.010	$(1.67239 \pm 0.00004) \times 10^{-24}$
Hydrogen atom	1.008142 ± 0.000003	938.722 ± 0.010†	$(1.67330 \pm 0.00004) \times 10^{-24}$†

$\quad$ Hydrogen mass/proton mass $= 1.000544613 \pm 0.000000006$
$\quad$ Proton mass/electron mass $= 1836.12 \pm 0.02$
Reduced mass of electron in hydrogen atom $= (9.1034 \pm 0.0003) \times 10^{-28}$ g

* Except where otherwise indicated, all data are taken directly from tables by E. R. Cohen et al., Analysis of Variance of the 1952 Data on the Atomic Constants and a New Adjustment, 1955, *Revs. Mod. Phys.*, **27**: 363–380 (1955).
† Calculated from data by Cohen et al., *loc. cit.*

Table 28. Wavelength and Energy Relations*

Property	Value
Photons:	
$\quad$ Wavelength, λ, cm	$(12397.67 \pm 0.22) \times 10^{-8}/E$
$\quad$ Wave number, $\bar{\nu} = 1/\lambda$, cm^{-1}	$(8066.03 \pm 0.14)E$
Electrons:	
$\quad$ Compton wavelength, h/mc, cm	$(24.2626 \pm 0.0002) \times 10^{-11}$
$\quad$ de Broglie wavelength, h/mv, cm†	$(1.226378 \pm 0.000010) \times 10^{-7}/E^{1/2}$
Neutrons:	
$\quad$ Compton wavelength, h/M_nc, cm	$(13.1959 \pm 0.0002) \times 10^{-14}$
$\quad$ de Broglie wavelength, h/M_nv, cm†	$(3.95603 \pm 0.00005) \times 10^{-3}/v$
	$(2.86005 \pm 0.00004) \times 10^{-9}/E^{1/2}$
Velocity, v, cm/sec†	$(1.38320 \pm 0.00003)E^{1/2} \times 10^{6}$‡
Velocity of 0.025-ev neutron, meters/sec	2187.036 ± 0.012
Energy, E, ev†	$(5.22671 \pm 0.000006)v^2 \times 10^{-13}$‡
Energy of 2,200-meter/sec neutron, ev	0.0252973 ± 0.0000003
kT temperature for 2,200-m/sec neutrons, °C	20.426 ± 0.022‡

$\qquad \qquad v$ = velocity, cm/sec $\qquad E$ = energy, ev

* Except where otherwise indicated, all data are taken directly from tables by E. R. Cohen at al., Analysis of Variance of the 1952 Data on the Atomic Constants and a New Adjustment, 1955, *Revs. Mod. Phys.*, **27**: 363–380 (1955).
† Quantities involving velocity are for the nonrelativistic range.
‡ Calculated from data by Cohen et al., *loc. cit.*

REFERENCES ON UNITS

1. Units of Weight and Measure (United States Customary and Metric), Definitions and Tables of Equivalents, *Natl. Bur. Standards Publ.* 214, July 1, 1955.
2. Standard Time throughout the World, *Natl. Bur. Standards Circ.* 496, Aug. 1, 1950.
3a. Griffith, Ezer: "The Heat Unit," Institute of Mechanical Engineers, London, 1951.
 b. Stimson, H. F.: Heat Units and Temperature Scales for Calorimetry, *Am. J. Phys.*, **23**: 614–622 (1955).
4. Establishment and Maintenance of the Electrical Units, *Natl. Bur. Standards Circ.* 475, June 30, 1949.
5. Cohen, E. R., J. W. M. Du Mond, T. W. Layton, and J. S. Rollett: Analysis of Variance of the 1952 Data on the Atomic Constants and a New Adjustment, 1955, *Revs. Mod. Phys.*, **27**: 363–380 (1955).
6a. Birge, R. T.: A New Table of Values of the General Physical Constants, *Revs. Mod. Phys.*, **13**: 233 (1941).
 b. Birge, R. T.: The General Physical Constants as of August, 1941, *Phys. Soc. (London), Repts. Progr. in Phys.*, **8**: 90 (1942).
7. Nier, A. O.: A Redetermination of the Relative Abundances of the Isotopes of Carbon, Nitrogen, Oxygen, Argon, and Potassium, *Phys. Rev.*, **77**: 792 (1950).

BIBLIOGRAPHY ON UNITS

CF-51-8-10: "Manual of Pile Engineering," AEC Technical Information Service, Oak Ridge, Tenn., Dec. 28, 1951.
Cohen, E. R., K. M. Crowe, J. W. M. Du Mond: "The Fundamental Constants of Physics," Interscience Publishers, Inc., New York, 1957.
Eshbach, O. W. (ed.): "Handbook of Engineering Fundamentals," 2d ed., Sec. 1, pp. 148–166 and Sec. 3, pp. 01–35, John Wiley & Sons, Inc., New York, 1952.
International Bureau of Weights and Measures: Reports of *General Conferences*.
National Bureau of Standards: Various *Circulars* and *Miscellaneous Publications*.
Zimmerman, O. T., and I. Lavine: "Conversion Factors and Tables," 2d ed., Industrial Research Services, Inc., Dover, New Hampshire, 1955.

9 Mathematics

A condensation of important mathematical fundamentals and useful tables is presented in the Appendix. Background on random processes and Fourier analysis will be found in Chap. 2 and on control and feedback theory in Chap. 15.

REFERENCES

1. P. Borden, and W. J. Mayo-Wells, "Telemetering Systems," Reinhold Publishing Corporation, New York, 1960.
2. N. V. Kiebert, Jr., unpublished papers, 1947.
3. C. S. Sheldon, "A Chronology of Missile and Astronautic Events," Committee on Science and Astronautics, U.S. House of Representatives, Mar. 8, 1961.
4. A. A. McKensie and H. A. Manoogian, Telemetering, Electronic Data Trasmission, *Electronics*, April, 1956, pp. 155–180.
5. V. N. Tipugin and V. A. Vetsel, "Radio Control," Moscow, 1962. Translation available from Office of Technical Services, Department of Commerce.
6. A. H. Ballard, A New Multiplex Technique for Telemetry, *Proc. Natl. Telemetering Conf.* 1962.

Additional telemetry references are:

M. H. Nichols and L. L. Rauch, "Radio Telemetry," 2d ed., John Wiley & Sons, Inc., New York, 1956.
H. L. Stiltz (ed.), "Aerospace Telemetry," Prentice-Hall, Inc., Englewood, N.J., 1961.
Bendix-Pacific Division, "Telemetry Standards Handbook," Bendix-Pacific Division, North Hollywood, Calif., 1961.

Useful bibliographies covering the telemetry field may be found in the following:

W. V. Kiebert, Jr., A Bibliography of Telemetry, *IRE Trans. Telemetry Remote Control*, vol. TRC-4, pp. 10–19, June, 1958.

REFERENCES

F. E. Rock and K. M. Uglow, A survey of progress reported in 1956 and 1957 in *IRE Trans. Telemetry Remote Control*, vol. TRC-4, pp. 2–9, June, 1958.

Publications most specifically oriented to telemetry and remote control are:

1. National Telemetering Conference. This has been held annually since 1950. Proceedings are available at the meetings and from the sponsoring societies, which are American Institute of Astronautics and Aeronautics (AIAA), Institute of Electronic and Electrical Engineers (IEEE), and Instrument Society of America (ISA).
2. Transactions of the Professional Group on Space Electronics and Telemetry, published by the IEEE since 1954. This group has published proceedings of Annual Symposia on Telemetry and Space Electronics.

Chapter 2

SAMPLING AND HANDLING OF INFORMATION

GEORGE R. COOPER, *Purdue University, Lafayette, Ind.*

MATHEMATICAL REPRESENTATION OF SIGNAL AND ERROR

1	General Methods of Representation	2–4
2	Representation in Terms of Orthonormal Exponentials	2–5
3	Representation in Terms of Discrete Samples	2–5
4	Quantization	2–6
5	Representation in Terms of Sinusoids	2–7
6	Representation of Random Variables	2–10
7	Elementary Probability Theory	2–11
8	Correlation Functions	2–15
9	Spectral Density	2–16
10	Response of Linear Systems	2–19

CONCEPTS OF INFORMATION THEORY

11	Units of Information	2–21
12	The Discrete Information Source	2–22
13	The Discrete Noiseless Channel	2–24
14	The Discrete Channel with Noise	2–25
15	Fundamental Theorem for the Discrete Channel with Noise	2–28
16	The Continuous Information Source	2–28
17	The Continuous Channel	2–29
18	Information Rate for a Continuous Source	2–31

OPTIMUM CODING

19	Minimum-redundancy Codes	2–32
20	Error-correcting Codes	2–35

MODULATION AND MULTIPLEXING

21	Modulation of Sinusoidal Carriers	2–38
22	Pulse Modulation	2–39
23	Pulse-code Modulation (PCM)	2–41
24	Multiplexing	2–42
25	Frequency Spectra of Modulated Signals	2–43
26	Frequency Spectra of Multiplexed Signals	2–48
27	Bandwidth Considerations	2–49
28	Signal-to-Noise Ratio at Threshold	2–51
29	Information-handling Capabilities	2–54

LIST OF SYMBOLS

a	Effective pulse width
A_n	A weighting factor
b_i	Coefficient of exponential series
B	Signal bandwidth, cps
B_o	Bandwidth occupancy
B_R	Receiver bandwidth
C	Carrier amplitude (peak)
C	Channel capacity, bits/sec
C_f	Crest factor for random noise
D^2	Variance of correlation measurement
D_e^2	Mean-square deviation ratio
D_m	Maximum deviation ratio
ε	Mean-square error
E^2	Pulse energy
EBU	Efficiency of bandwidth utilization
F_n	Receiver noise figure
F_s	Subcarrier frequency spacing, cps
$F_u(\omega)$	Fourier transform of $u(t)$
h	Pulse height
$h(i)$	Information in state i
$h(x)$	Information in $x(t)$
H	Entropy
$H(x)$	Entropy of a source
$H(y)$	Entropy of received signal
$H_x(y), H_y(x)$	Conditional entropies
$H_{\max}$	Maximum entropy
$H_1(x)$	Entropy per sample point
$H_R(x)$	Entropy rate
I	Amount of information per symbol
IE	Information efficiency
$J_n(\ )$	Bessel function of first kind, order n
k	Boltzmann's constant
K_1, K_2	Constant factors in improvement ratio
L	Code length (average)
m	Number of message digits per code group
$m(t)$	Message function
M	Average message power
M_1, M_2	Modulation indexes in improvement ratio
$M(\omega)$	Characteristic function of a density function
n	Number of digits per code group
n	Number of symbols per second
n_c	Number of channels
N	Number of message states
N	Average noise power
N_1	Entropy power
$N(f)$	Noise spectral density (one-sided)
$N(T)$	Number of signals of duration T
p	Probability of correct transmission of a binary digit
$p(x), p_1(x_1)$	Probability density function
$p_{11}(x_1,t_1; y_1,t_2)$	Joint probability density function
$p_2(x_1,t_1; x_2,t_2), p(x_1,x_2)$	Second probability density function
$p(i)$	Probability of state i
$p(j|i), p(i|j)$	Conditional probability density functions
$P(x), P_1(x_1)$	Probability distribution functions
$P_{11}(x_1,t_1; y_1,t_2)$	Joint probability distribution function
$P_2(x_1,t_1; x_2,t_2), P(x_1,x_2)$	Second probability distribution function

P_c	Average carrier power
P_s	Average modulated signal power
P_t	Threshold power
q	Probability of incorrect transmission of a binary digit
R	Average information rate
R_{max}	Maximum value of average information rate
R_0	Improvement ratio (power)
$R_{xx}(\tau), R_{yy}(\tau)$	Autocorrelation functions
$R_{xy}(\tau), R_{yx}(\tau)$	Crosscorrelation functions
$S(t)$	Signal function
S_i/N_i	Signal-to-noise power ratio at receiver input
S_0/N_0	Signal-to-noise power ratio at output of each channel
T	Sampling interval
T	Absolute temperature
$u(t)$	Single-pulse time function
$U(t)$	Pulse-train time function
$w(t)$	Impulse response of a linear system
W	Message bandwidth, cps
W_E	Effective spectral bandwidth, cps
W_N	Equivalent noise bandwidth
$x(t)$	Any time function
$X(\omega)$	Fourier transform of $x(t)$
$X(s)$	Laplace transform of $x(t)$
X	Variable in channel-capacity equation
X_0	Root of channel-capacity equation
$\bar{x}$	Mean value of $x(t)$
$\overline{x^2}$	Mean-square value of $x(t)$
$y(t)$	Any time function
$\delta(\ )$	Delta function
δ_{ij}	Kronecker delta
η	Coding efficiency
$\rho_{xx}(\tau), \rho_{yy}(\tau)$	Autocorrelation functions
$\rho_{xy}(\tau), \rho_{yx}(\tau)$	Crosscorrelation functions
σ_x^2	Variance of $x(t)$
τ	Time translation
$\psi_n(t)$	An elementary time function
ϕ_c	Carrier phase
ϕ_m	Maximum phase deviation
$\Phi_{mm}(\omega)$	Message spectral density
$\Phi_{xx}(\omega), \Phi_{xx}(s)$	Spectral densities
$\Phi_{xy}(\omega), \Phi_{yx}(\omega)$	Cross-spectral densities
ω	Angular frequency, radians/sec
$\Delta\omega$	Angular-frequency deviation, radians/sec
ω_0	Fundamental frequency, radians/sec
ω_c	Carrier frequency, radians/sec
ω_m	Message frequency, radians/sec
ω_s	Pulse-repetition average frequency

MATHEMATICAL REPRESENTATION OF SIGNAL AND ERROR

The primary purpose of any telemetering system is to transfer information from one location to another. It is obviously desirable that this transfer be accomplished efficiently and with as little error as possible, and this, in turn, requires that it be possible to analyze the performance of the system and to optimize this performance in some respect. The use of analytical methods to carry out this analysis and optimization depends almost entirely upon the ability to represent the information and the error by appropriate mathematical models.

In carrying out a mathematical analysis it must always be recognized that the model being used is never exact and that the degree of approximation which can be tolerated depends both upon the desired accuracy of results and upon the mathematical competence of the analyst. Therefore, it is the dual purpose of this section to present, in compact form, certain mathematical tools which are useful in system analysis and to indicate the nature of the approximations involved in using these tools.

In a broad sense, the signals and errors encountered in telemetry systems can be classified as either deterministic or random. A deterministic time function is defined as one whose future value can be exactly predicted from a complete knowledge of its past history. Timing signals, standardizing and test signals, and frequency standards are typical examples. Such functions can be described by explicit mathematical expressions. A random time function cannot be exactly predicted from a knowledge of its past history and cannot be described by an explicit mathematical expression. The data being transmitted and the system errors generally fall into this category.

1 General Methods of Representation

The most general representation of a time-dependent signal is by means of an abstract symbol, and many general analytical results can be obtained without resorting to more specific representations. It is necessary, however, to attach some physical significance to this symbol, and the manner in which this is done forms the basis for classifying different methods of representation.

In the time-domain method of representation the abstract symbol $x(t)$ represents the magnitude of the signal at any arbitrary time t. A detailed analysis of any system usually requires the consideration of particular types of signals, and this makes it necessary to specify explicit functions of time for $x(t)$. Since there are a vast number of ways in which this might be done, even for the same physical signal, the eventual choice is usually motivated by considerations of mathematical convenience or easy visualization.

Mathematical convenience usually dictates that $x(t)$ be represented as a linear combination of elementary time functions. Thus, in general, the useful representations are in the form of a weighted sum of preselected waveforms $\psi_n(t)$.

$$x(t) = \sum_n A_n \psi_n(t) \qquad (1)$$

The problem of selecting the best $\psi_n(t)$ for a given application, and of determining the corresponding weighting factors A_n, has not been solved in general, although many specific results are available.

In a large measure, desirable forms for $\psi_n(t)$ depend upon whether the major interest is centered in the transmission properties of a system responding to $x(t)$ or in the information-bearing properties of $x(t)$. In the first case, the entire class of exponential functions is extremely useful because these functions are invariant in form under the operations of differentiation and integration by which the system response is described. In the second case, a sampled-data-type representation is more useful because the information content of a signal is directly related to the number of independent pieces of data required to describe it uniquely.

When the systems being considered are linear, with either constant or time-varying parameters, the use of sinusoidal functions—which form a subclass of the exponential functions—leads to operational methods in which the independent variable is frequency rather than time. This frequency-domain representation of signals is so widely used in investigating the transmission properties of linear systems that it has come to take precedence over the more fundamental time-domain representations for this application. A more complete discussion of the frequency-domain method is given in Sec. 4. Time-domain methods using nonsinusoidal exponentials are discussed in Sec. 2, and sampled-data representations, in the time domain, are considered in Sec. 3.

MATHEMATICAL REPRESENTATION OF SIGNAL AND ERROR

In addition to the representations already listed, many others have been investigated and found useful for special purposes. These include power series in t, Laguerre polynomials, Bessel functions, and hyperbolic secant functions. However, the use of such representations is not sufficiently widespread as yet to warrant further discussion of them here.

2 Representation in Terms of Orthonormal Exponentials

When the time function $x(t)$ satisfies the condition

$$\int_0^\infty x^2(t)\,dt < \infty \tag{2}$$

it can be shown that an infinite number of complete sets of exponential functions $\psi_n(t)$ will satisfy (1) in the interval $0 < t < \infty$.[1] From the standpoint of minimizing the mean-square error in the approximation (when only a finite number of terms are used), and of simplifying the determination of the weighting factors A_n, it is convenient to select a set which is orthogonal and normalized. Hence, let

$$\psi_n(t) = \sum_{i=1}^n b_i \epsilon^{-ict} \tag{3}$$

and require that

$$\int_0^\infty \psi_n^2(t)\,dt = 1 \qquad n = 1, 2, \ldots \tag{4}$$

$$\int_0^\infty \psi_m(t)\psi_n(t)\,dt = 0 \qquad m \neq n \tag{5}$$

The weighting factors are then given by

$$A_n = \int_0^\infty x(t)\psi_n(t)\,dt \tag{6}$$

The orthonormal functions $\psi_n(t)$ can be computed one by one by invoking relations (4) and (5) for each value of n in sequence, starting with $n = 1$. The first few members of this set (and usually all that are required for reasonable accuracy) are shown below. Numerical values of these functions are tabulated in the literature.[1] The value of c is chosen so that ϵ^{-ct} approaches zero at roughly the same rate as $x(t)$, but this value is not critical.

$$\psi_1(t) = \sqrt{2c}\,\epsilon^{-ct}$$
$$\psi_2(t) = \sqrt{c}\,(6\epsilon^{-2ct} - 4\epsilon^{-ct})$$
$$\psi_3(t) = \sqrt{6c}\,(10\epsilon^{-3ct} - 12\epsilon^{-2ct} + 3\epsilon^{-ct})$$
$$\psi_4(t) = \sqrt{2c}\,(70\epsilon^{-4ct} - 120\epsilon^{-3ct} + 60\epsilon^{-2ct} - 8\epsilon^{-ct})$$
$$\psi_5(t) = \sqrt{10c}\,(126\epsilon^{-5ct} - 280\epsilon^{-4ct} + 210\epsilon^{-3ct} - 60\epsilon^{-2ct} + 5\epsilon^{-ct})$$

As mentioned previously, the exponential representation is convenient for analyzing the transmission properties of a linear system to which the signal is being applied. In addition to this, however, the orthogonality of the above set of exponentials provides additional convenience in problems involving the synthesis of systems having a desired transmission characteristic or the optimization of systems with respect to some transmission criterion.

3 Representation in Terms of Discrete Samples

The representation of random time functions in terms of the value of the function at discrete instants of time is particularly convenient when the statistical aspects are important. The simplest, and most commonly quoted, statement of the basic sampling theorem assumes that the time function is band-limited; i.e., it contains no frequency components in excess of some specified value of W cps.[2] Under this condition, the theorem states that the time function is uniquely determined by its values at an infinite set of sample points spaced $1/2W$ sec apart.

This same concept can be expressed mathematically in the form of (1) by choosing $\psi_n(t)$ to be a time function which is band-limited and which is simply evaluated at the sampling points. Such a function is

$$\psi_n(t) = \frac{\sin \pi(2Wt - n)}{\pi(2Wt - n)} \tag{7}$$

which has a value of unity at the nth sample point and is zero at all other sample points. Hence the necessary weighting factor at the nth sample point is simply the sample value, and the time function can be expressed as

$$x(t) = \sum_{n=-\infty}^{\infty} x\left(\frac{n}{2W}\right) \frac{\sin \pi(2Wt - n)}{\pi(2Wt - n)} \tag{8}$$

Although the above statement of the sampling theorem assumed that the sample points were uniformly spaced, this is not a necessary requirement for a unique specification. In particular, it is possible to show that a unique specification is still obtained if a finite number of uniform sample points are allowed to migrate to new noncoincident positions.[3] Although such a modification complicates the mathematical expressions—(8) no longer applies—it has practical significance in justifying the uniqueness of so-called "natural sampling" which is typical of some pulse-modulation systems.

It is of interest to relate the mean-square value of the samples to the mean-square value of the time function from which they are derived. In particular, it can be shown that these two approach one another in the limit as the number of samples is indefinitely increased. Hence

$$\lim_{N \to \infty} \frac{1}{2N+1} \sum_{n=-N}^{N} x^2\left(\frac{n}{2W}\right) = \lim_{T \to \infty} \frac{1}{2T} \int_{-T}^{T} x^2(t)\, dt \tag{9}$$

The practical significance of this result is demonstrated in a later section.

Although (8) is exact for band-limited time functions of infinite duration, no practical time function satisfies either of these requirements. A time function of finite duration cannot be band-limited, and hence representation in the form of (8) will be an approximation. This approximation will have the correct value at all sample points (including those outside the interval of existence), but the value between sample points will not be correct in general. Although the error between sample points can become large in theory, as a practical matter it usually does not if the sampling rate is properly chosen. Generally, a sampling rate of $3W_E$ samples per second (as compared with $2W$ samples per second in the band-limited case) is adequate when W_E is the effective spectral bandwidth as defined in Sec. 9.

4 Quantization

Quantization is an operation by which a sample taken from a function having a continuous range of values is assigned a single discrete number.[4] Although this operation does not form the basis for a mathematical description of time functions, it is frequently associated with the sampling operation, and hence some of its consequences should be discussed.

Quantization is usually accomplished by dividing the range of possible x values into small intervals of width Δx. Any observed sample will fall within one of these small intervals and is assigned a discrete value corresponding to the *center* of this interval.

MATHEMATICAL REPRESENTATION OF SIGNAL AND ERROR

Hence the samples coming out of the quantizer will not have a continuous range of amplitudes but instead will have a finite set of discrete amplitudes.

Since the exact amplitude of a sample is lost by quantizing, there will always be some error introduced. The mean-square value of this error is related to the quantizing interval by

$$[\text{Mean-square error in } x(t)] = \frac{1}{12}(\Delta x)^2 \tag{10}$$

It is apparent, therefore, that Δx should be small and hence the number of quantizing intervals should be large. Usually, the number of intervals is chosen to be a power of 2, since quantization is normally a preliminary to binary coding. Thus, adding one digit to the code length doubles the number of intervals and reduces the mean-square error by a factor of 4.

5 Representation in Terms of Sinusoids

When $x(t)$ is periodic, it is convenient to let $\psi_n(t)$ be a set of harmonically related sinusoids. This choice leads immediately to the familiar Fourier-series representation.[5] An especially compact form of the Fourier series results if the sinusoids are expressed in exponential form. Thus, let

$$\psi_n(t) = \epsilon^{jn\omega_0 t} \tag{11}$$

where $\omega_0 = 2\pi/T$, radians/sec, T = period, sec.
Equation (1) then becomes

$$x(t) = \sum_{n=-\infty}^{\infty} A_n \epsilon^{jn\omega_0 t} \tag{12}$$

in which the weighting factors are given by

$$A_n = 1/T \int_{-\frac{T}{2}}^{\frac{T}{2}} x(t)\epsilon^{-jn\omega_0 t}\, dt \tag{13}$$

A necessary requirement on $x(t)$ is that

$$\int_{-\frac{T}{2}}^{\frac{T}{2}} |x(t)|\, dt < \infty$$

The usual sine and cosine forms of (12) and (3) can be obtained by using the relation

$$\epsilon^{\pm jn\omega_0 t} = \cos n\omega_0 t \pm j \sin n\omega_0 t \tag{14}$$

An important feature of the Fourier-series representation of periodic time functions arises when the summation of (12) is terminated at some finite value, say $n = \pm N$. The resulting approximation to $x(t)$ can be shown to possess the smallest mean-square error, for all values of N, of any possible sinusoidal representation. Furthermore, this mean-square error is given by

$$[\text{Mean-square error in } x(t)] = \sum_{n=-\infty}^{\infty} |A_n|^2 - \sum_{n=-N}^{N} |A_n|^2$$

$$= \frac{1}{T}\int_{-\frac{T}{2}}^{\frac{T}{2}} x^2(t)\, dt - \sum_{n=-N}^{N} |A_n|^2 \tag{15}$$

This result makes it possible to estimate the number of terms required in the Fourier series for a given accuracy.

The weighting factors given by (13) are usually complex and hence can be represented in terms of their magnitude and phase angle. Thus, in general

$$A_n = |A_n|\epsilon^{i\theta_n}$$
$$A_{-n} = |A_n|\epsilon^{-i\theta_n} \tag{16}$$

in which $|A_n|$ is one-half the peak amplitude of the nth harmonic and θ_n is the phase of the nth harmonic with respect to $\cos n\omega_0 t$. If this reference is understood, then a complete and unique specification of $x(t)$ can be obtained by specifying the *magnitude* and *phase* of all the harmonics. Since this specification does not involve t explicitly ($|A_n|$ and θ_n are constant with time) it is said to be a specification in the frequency domain.

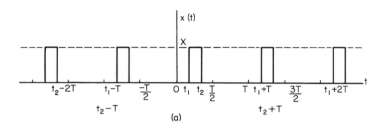

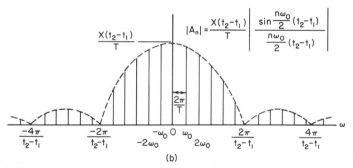

Fig. 1. Discrete frequency spectrum of periodic rectangular pulses. (a) Periodic rectangular pulses. (b) Discrete frequency spectrum showing harmonic amplitudes.

It is often of interest to know how the average power associated with a periodic $x(t)$ is distributed among the various harmonic components. Since the average power of each harmonic is related to its peak amplitude, a plot of $|A_n|$ as a function of frequency reveals the desired relationship. This plot will not be a continuous curve, however, since the only frequencies present are those which are multiples of the lowest-frequency component. Hence a convenient way of representing this is by means of the *discrete frequency spectrum* as shown in Fig. 1. It is customary to plot this spectrum for positive values of n only (since $|A_n| = |A_{-n}|$), but both positive and negative values are shown here in order to emphasize both the symmetry and the relationship to the continuous spectra which will be discussed. It is also possible to plot the phase angles θ_n in a similar manner, but this is not so common.

A sinusoidal representation of a nonperiodic time function can be constructed by considering a periodic function of similar form and allowing the period to increase without limit. When this is done, the frequency separation between harmonic components and the component amplitudes both become vanishingly small. It is

MATHEMATICAL REPRESENTATION OF SIGNAL AND ERROR   2-9

appropriate, therefore, not to consider discrete frequency components or a summation of these components but rather to consider a continuous distribution and an integration over this distribution. This concept leads to the Fourier-integral representation of a nonperiodic time function and can be justified rigorously by other means.[5]

The Fourier integral, or transform, of $x(t)$ is defined as

$$X(\omega) = \int_{-\infty}^{\infty} x(t)\epsilon^{-j\omega t}\, dt \qquad (17)$$

for time functions which satisfy the condition

$$\int_{-\infty}^{\infty} |x(t)|\, dt < \infty \qquad (18)$$

The corresponding summation (or inversion) integral is

$$x(t) = 1/2\pi \int_{-\infty}^{\infty} X(\omega)\epsilon^{j\omega t}\, d\omega \qquad (19)$$

and it is this expression which is analogous to (12). The factor $1/2\pi$ can appear in either (17) or (19) or can be divided between them in any fashion. The choice shown here, which is very convenient for much of the material which follows, yields an $X(\omega)$ in which the unit of bandwidth is the cycle per second.

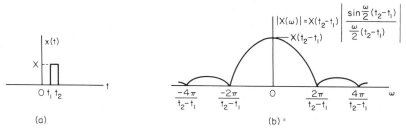

FIG. 2. Continuous frequency spectrum of a single rectangular pulse. (a) A single rectangular pulse. (b) Continuous frequency spectrum.

The function $X(\omega)$, which uniquely defines $x(t)$, is not a function of time and hence is said to be a frequency-domain representation of $x(t)$. It has the dimensions of an amplitude density [i.e., if $x(t)$ is measured in volts, $X(\omega)$ will have dimensions of volts/cps] and is usually a continuous function of ω. In general, $X(\omega)$ is complex and can be represented in terms of its magnitude and phase as

$$\begin{aligned} X(\omega) &= |X(\omega)|\epsilon^{j\theta(\omega)} \\ X(-\omega) &= |X(\omega)|\epsilon^{-j\theta(\omega)} \end{aligned} \qquad (20)$$

The magnitude function $|X(\omega)|$ yields the frequency spectrum and is usually a continuous function of ω. A plot of this function gives an indication as to how the energy in $x(t)$ is distributed with respect to frequency.

The frequency spectrum for a single rectangular pulse is shown in Fig. 2. It should be noted that this spectrum has the same form as the *envelope* of the discrete frequency spectrum for periodic rectangular pulses. A similar relationship between continuous and discrete spectra is generally obtained when a time function is periodically repeated.

There are many time functions of interest which do not satisfy the condition of (18). In such cases it is often possible to introduce a convergence factor of the form $\epsilon^{-\sigma|t|}$ so that $x(t)\epsilon^{-\sigma|t|}$ does satisfy condition (18). Additional convenience is afforded by incorporating the convergence factor in the definition of the transform. This step is easily accomplished by defining a *complex frequency* variable s of the form

$$s = \sigma + j\omega$$

The resulting transform now becomes

$$X(s) = \int_{-\infty}^{\infty} x(t)\epsilon^{-st}\, dt \tag{21}$$

which is the *bilateral Laplace transform*. The corresponding inversion integral is

$$x(t) = 1/2\pi j \int_{c-j\infty}^{c+j\infty} X(s)\epsilon^{st}\, ds \tag{22}$$

in which the value of c (which defines the path of integration in the complex s plane) may be different for positive and negative values of time.

The frequency spectrum can be obtained from the bilateral Laplace transform by letting $s = j\omega$ after $X(s)$ has been found. The resulting complex function has a magnitude $|X(j\omega)|$ which can be interpreted as a frequency spectrum in exactly the same way as the magnitude function $|X(\omega)|$ obtained from the Fourier transform. In fact, for time functions which are Fourier transformable, the two magnitude functions are identical.

6 Representation of Random Variables

The signals which are of greatest interest in any information-handling system are basically random in the sense that their future value cannot be predicted exactly from their past history. Noise which is introduced into the system is also random. Since it is not possible to write explicit mathematical equations for such random quantities, it is necessary to resort to other forms of mathematical description. These alternative forms are necessarily less precise, but in general, they convey all the information about the random quantity that is available.

There are two general methods of describing random functions mathematically. The first, and most basic, is a probabilistic description in which the random quantity is characterized by a *probability model*. Some of the elementary concepts of probability and a few of the more common probability models are discussed in Sec. 7. The second method, which is derived from the first, is a statistical description in which the random quantity is characterized in terms of its *statistical parameters*. These statistical parameters are usually averages of various functions of the random quantity and as such imply a less precise description than the probabilistic one. Nevertheless, the statistical description is generally more useful in the majority of engineering problems concerned with random signals and noise. The most important statistical parameters are discussed in Secs. 8 and 9.

The concept of probability is closely linked with the frequency of occurrence of some particular *event*. As applied to information-bearing signals and noise, the event may be the observation of a time function within a given range of values at a given time. The frequency of occurrence is obtained by considering all possible time functions that might have existed and determining what fraction of these will be observed in the given range of values at the given time. This fraction is the *probability* of the given event. The collection of all possible time functions constitutes the *random process* and, when the probability measure is specified, is referred to as an *ensemble*. The values of all members of the ensemble at some particular time constitute a *random variable*. Each member of the ensemble is referred to as a *sample function*.

On the basis of the above definitions it will be noted that the "randomness" is with respect to various members of the ensemble and not necessarily with respect to time. However, there is a special class of random processes, known as *ergodic* processes, in which each member of the ensemble exhibits the same characteristics with respect to time as each random variable does with respect to the ensemble. The ergodic property will be assumed to apply in all the discussion of random variables which follows.

It is generally difficult to justify the assumption of ergodicity on a rigorous basis.

From a practical standpoint the assumption is justified on the basis that it is the only possible assumption which will lead to reasonable convenience in analytical efforts. Since the complete ensemble of time functions can never be observed in a physical system, it is necessary to base the probability description upon the one or more time functions that are observed, or to assume the probability description in its entirety without any observations. In the first case ergodicity is implied; in the second case it would be assumed unless there are compelling reasons for not doing so.

7 Elementary Probability Theory

The discussion of probability theory in this section is aimed primarily at defining terms and establishing notation. More complete discussions are available many places in the literature.[1,6] The random process is designated as $\{x(t)\}$ and any sample function as $x(t)$. The random variable corresponding to observation of the random process at time t_1 is $x(t_1)$.

FIG. 3. Probability distribution and density functions. (a) A probability distribution function. (b) A probability density function.

The *first probability distribution function* of $\{x(t)\}$ is the most fundamental probability function and is defined by

$$P_1(x_1) \triangleq \text{Prob}\,[x(t_1) \leq x_1] \tag{23}$$

This function is always a monotone, nondecreasing function of x_1 such that

$$0 = P_1(-\infty) \leq P_1(x_1) \leq P_1(\infty) = 1 \tag{24}$$

If $P_1(x_1)$ is differentiable with respect to x_1, then the *first probability density function* may be defined as

$$p_1(x_1) \triangleq dP_1(x_1)/dx_1 \tag{25}$$

This function has the following properties:

$$p_1(x_1)\,dx_1 = \text{Prob}\,[x_1 < x(t_1) \leq x_1 + dx_1]$$

$$\int_a^b p_1(x_1)\,dx_1 = \text{Prob}\,[a < x(t_1) \leq b] \qquad a < b$$

$$\int_{-\infty}^{x_1} p_1(u)\,du = P_1(x_1)$$

$$\int_{-\infty}^{\infty} p_1(x_1)\,dx_1 = 1$$

For a general random process $P_1(x_1)$ and $p_1(x_1)$ would depend upon t_1 also. For the special class of *stationary* random processes, however, these functions are independent of t_1. Since ergodic processes are necessarily stationary (although the converse is not true), and since ergodicity has been assumed throughout, no dependence on t_1 has been implied in the above notation. The notation is frequently simplified still more by omitting all subscripts, even though this form is not mathematically complete. Thus common usage employs $P(x)$ and $p(x)$ to be equivalent to $P_1(x_1)$ and $p_1(x_1)$ as defined above. This usage, which is justified only by convenience, will be employed in the following sections. A typical probability distribution and density function are shown in Fig. 3.

The *second probability distribution function* may be defined for two random variables $x(t_1)$ and $x(t_2)$ as

$$P_2(x_1,t_1; x_2,t_2) \triangleq \text{Prob } [x(t_1) \leq x_1; x(t_2) \leq x_2] \tag{26}$$

The properties of this function are

$$0 \leq P_2(x_1,t_1; x_2,t_2) \leq 1$$
$$P_2(-\infty,t_1; x_2,t_2) = P_2(x_1,t_1; -\infty,t_2) = P_2(-\infty,t_1; -\infty,t_2) = 0$$
$$P_2(\infty,t_1; \infty,t_2) = 1$$
$$P_2(x_1,t_1; \infty,t_2) = P_1(x_1)$$
$$P_2(\infty,t_1; x_2,t_2) = P_1(x_2)$$

The *second probability density function* may be defined as

$$p_2(x_1,t_1; x_2,t_2) \triangleq \frac{\partial^2 P_2(x_1,t_1; x_2,t_2)}{\partial x_1 \, \partial x_2} \tag{27}$$

and has the following properties:

$$p_2(x_1,t_1; x_2,t_2) \, dx_1 \, dx_2 = \text{Prob } [x_1 < x(t_1) \leq x_1 + dx_1; x_2 < x(t_2) \leq x_2 + dx_2]$$

$$\int_{a_1}^{b_1} dx_1 \int_{a_2}^{b_2} p_2(x_1,t_1; x_2,t_2) \, dx_2 = \text{Prob } [a_1 < x(t_1) \leq b_1; a_2 < x(t_2) \leq b_2]$$

where $a_1 < b_1$ and $a_2 < b_2$

$$\int_{-\infty}^{\infty} dx_1 \int_{-\infty}^{\infty} p_2(x_1,t_1; x_2,t_2) \, dx_2 = 1$$
$$\int_{-\infty}^{\infty} p_2(x_1,t_1; x_2,t_2) \, dx_2 = p_1(x_1)$$
$$\int_{-\infty}^{\infty} p_2(x_1,t_1; x_2,t_2) \, dx_1 = p_1(x_2)$$

The second-order functions are frequently referred to as *joint* distribution and density functions. For stationary random processes they depend not upon the absolute values of t_1 and t_2 but only upon the time difference $t_2 - t_1$. In this case it is common to designate the joint functions as $P(x_1,x_2)$ and $p(x_1,x_2)$ and define the random variables x_1 and x_2 in such a way as to include the time difference.

Higher-order distribution and density functions can be defined in a manner analogous to the first two. The complete probability model of a random process consists of a specification of the distribution and density functions of all orders. For most engineering purposes, however, the first two are sufficient to obtain the most commonly used statistical parameters.

It is also possible to define joint distribution and density functions for random variables taken from two different random processes. These two random processes may be related in some way, as in the case of the input and output of a transmission system, or they may be completely unrelated, as in the case of signal and noise in a transmission system. Designating one random process as $\{x(t)\}$ and the other as $\{y(t)\}$, the joint distribution function may be defined as

$$P_{11}(x_1,t_1; y_1,t_2) \triangleq \text{Prob } [x(t_1) \leq x_1; y(t_2) \leq y_1] \tag{28}$$

The corresponding joint density function is

$$p_{11}(x_1,t_1; y_1,t_2) \triangleq \frac{\partial^2 P_{11}(x_1,t_1; y_1,t_2)}{\partial x_1 \, \partial y_1} \tag{29}$$

The properties of P_{11} and p_{11} are analogous to those of P_2 and p_2, respectively.

When the two random processes are completely unrelated they are said to be *statistically independent*. A necessary and sufficient condition for independence is the

MATHEMATICAL REPRESENTATION OF SIGNAL AND ERROR 2–13

factorability of the joint distribution and density functions. That is,

$$P_{11}(x_1,t_1;\ y_1,t_2) = P_1^x(x_1,t_1)P_1^y(y_1,t_2) \tag{30}$$
or
$$p_{11}(x_1,t_1;\ y_1,t_2) = p_1^x(x_1,t_1)p_1^y(y_1,t_2) \tag{31}$$

implies statistical independence. The superscripts signify that the first-order distribution and density functions need not be the same for the two random variables.

When the random processes are not statistically independent, the joint distribution functions may be expressed

$$\begin{aligned} P_{11}(x_1,t_1;\ y_1,t_2) &= P_1^x(x_1,t_1)P_1^y(y_1,t_2|x_1,t_1) \\ &= P_1^x(x_1,t_1|y_1,t_2)P_1^y(y_1,t_2) \end{aligned} \tag{32}$$

in which $P_1^y(y_1,t_2|x_1,t_1)$ and $P_1^x(x_1,t_1|y_1,t_2)$ are the *conditional probability* distribution functions defined by

$$P_1^y(y_1,t_2|x_1,t_1) = \text{Prob}\ [y(t_2) \leq y_1 \text{ given that } x(t_1) \leq x_1] \tag{33}$$
$$P_1^x(x_1,t_1|y_1,t_2) = \text{Prob}\ [x(t_1) \leq x_1 \text{ given that } y(t_2) \leq y_1] \tag{34}$$

In a similar manner the joint density function may be expressed as

$$\begin{aligned} p_{11}(x_1,t_1;\ y_1,t_2) &= p_1^x(x_1,t_1)p_1^y(y_1,t_2|x_1,t_1) \\ &= p_1^x(x_1,t_1|y_1,t_2)p_1^y(y_1,t_2) \end{aligned} \tag{35}$$

When the random processes are stationary it is common to suppress the time in the notation and omit the subscripts. For example, the two right-hand sides of (35) can be written as

$$p(x)p(y|x) = p(y)p(x|y) \tag{36}$$

This particular relationship is used extensively in the discussion of information theory which follows.

The probability density functions are useful in determining the *mathematical expectation* of any function of a random variable. Thus the mathematical expectation (or expected value) of $f(x)$, where x is the random variable, is

$$E[f(x)] = \int_{-\infty}^{\infty} f(x)p(x)\ dx \tag{37}$$

in the stationary case. In the case of two stationary random variables x_1 and x_2, the expected value of some function $g(x_1,x_2)$ is

$$E[g(x_1,x_2)] = \int_{-\infty}^{\infty} \int_{-\infty}^{\infty} g(x_1,x_2)p(x_1,x_2)\ dx_1\ dx_2 \tag{38}$$

Equation (37) above is particularly useful in determining such statistical parameters as the mean and mean-square values of a random variable. The mean value (or average value) is defined as

$$\bar{x} = E[x(t_1)] = \int_{-\infty}^{\infty} xp(x)\ dx \tag{39}$$

while the mean-square value is

$$\overline{x^2} = E[x^2(t_1)] = \int_{-\infty}^{\infty} x^2 p(x)\ dx \tag{40}$$

Note that these are the first and second moments of the probability density function. The *variance* of the random variable is defined as

$$\begin{aligned} \sigma_x^2 &= E\{[x(t_1) - \bar{x}]^2\} \\ &= \overline{x^2} - (\bar{x})^2 \end{aligned} \tag{41}$$

while σ_x is called the *standard deviation*.

The statistical parameters for an ergodic random process can also be obtained as time averages. Hence

$$\bar{x} = \lim_{T \to \infty} 1/2T \int_{-T}^{T} x(t)\, dt \tag{42}$$

and

$$\overline{x^2} = \lim_{T \to \infty} 1/2T \int_{-T}^{T} x^2(t)\, dt \tag{43}$$

will yield the same results as (39) and (40), and in some circumstances may be easier to evaluate. It should also be noted that a nonstationary random process cannot be ergodic and time averages like (42) and (43) may have no meaning.

Second-order statistical parameters such as the correlation function and spectral density may be obtained from the application of (38). These parameters are discussed in subsequent sections.

Although many types of probability density functions arise in communication systems, there are three which are particularly important. These three are tabulated below, along with a short discussion of the sources of such random variables, and are illustrated in Fig. 4.

Normal (Gaussian) Density Function

$$p(x) = (1/\sqrt{2\pi}\,\sigma_x)\epsilon^{-\frac{(x-\bar{x})^2}{2\sigma_x^2}} \qquad -\infty \leq x \leq \infty \tag{44}$$

where $\bar{x}$ = mean value
σ_x = standard deviation

The normal distribution almost always arises in physical systems when the random variable is the sum of a large number of independent disturbances.[1] Hence such sources of noise as thermal agitation in conductors, shot noise in vacuum tubes, current noise in semiconductors, and antenna noise due to thermal radiation and distant atmospheric disturbances all lead to normal density functions. Furthermore, almost any density function tends to approach the normal function when the random variable is passed through a linear system which reduces its bandwidth greatly. As a consequence the normal density function is the one most often encountered in the analysis of linear systems.

Rayleigh Density Function

$$p(y) = (y/\sigma_x^2)\epsilon^{-y^2/2\sigma_x^2} \qquad y \geq 0 \tag{45}$$

The Rayleigh density function is associated with the *envelope* of a random process having a normal distribution. In (45), σ_x^2 is the variance of the normal random variable $x(t)$, while its envelope $y(t)$ has a mean of

$$\bar{y} = \sqrt{\pi/2}\,\sigma_x$$

and a variance of

$$\sigma_y^2 = 0.429\sigma_x^2$$

The Rayleigh density function arises in connection with the output from linear envelope detectors.

Uniform Density Function

$$p(\theta) = 1/2\pi \qquad -\pi < \theta \leq \pi$$
$$ = 0 \qquad \text{elsewhere} \tag{46}$$

The uniform density function is commonly associated with the *phase* of a normal random process. As such, it arises in connection with frequency-modulation detectors. However, it may also be associated with other random variables and may be written more generally as

$$p(x) = 1/(x_2 - x_1) \quad x_1 < x \leq x_2$$
$$= 0 \quad \text{elsewhere} \tag{47}$$

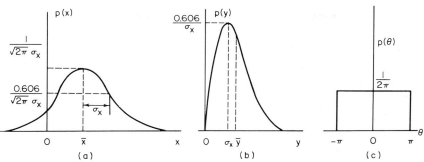

FIG. 4. Three of the most common probability density functions. (a) Normal density function. (b) Rayleigh density function. (c) Uniform density function.

8 Correlation Functions

The *autocorrelation function* of a stationary random process $\{x(t)\}$ is defined as[1]

$$R_{xx}(\tau) \triangleq \int_{-\infty}^{\infty} dx_1 \int_{-\infty}^{\infty} x_1 x_2 p(x_1, x_2) \, dx_2 \tag{48}$$

where $x_1 = x(t_1)$
$x_2 = x(t_1 + \tau)$

If the process is also ergodic, the autocorrelation function may be obtained from a time average. In this case

$$R_{xx}(\tau) = \lim_{T \to \infty} (1/2T) \int_{-T}^{T} x(t) x(t + \tau) \, dt \tag{49}$$

Note that if $x(t)$ has the units of volts, for example, $R_{xx}(\tau)$ will have units of (volts).[2] Hence it is related to the average power (in 1 ohm) associated with a time function and a translation of that time function. The important properties of autocorrelation functions may be listed as follows for random processes not containing periodic components:

(a) $R_{xx}(\tau) = R_{xx}(-\tau)$
(b) $R_{xx}(0) = \overline{x^2}$
(c) $|R_{xx}(\tau)| \leq R_{xx}(0)$
(d) $\lim_{\tau \to \infty} R_{xx}(\tau) = (\bar{x})^2$

A typical autocorrelation function exhibiting these properties is shown in Fig. 5. If the random process is periodic, the autocorrelation function will also be periodic with the same period.

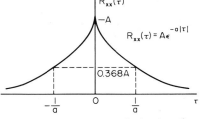

FIG. 5. A typical autocorrelation function.

The *normalized autocorrelation function* is frequently more convenient for analytical work. This quantity is defined as

$$\rho_{xx}(\tau) \triangleq (1/\sigma_x^2)\, [R_{xx}(\tau) - (\bar{x})^2] \tag{50}$$

and always has a value of unity at $\tau = 0$.

A *crosscorrelation function* can be defined for two stationary random processes $\{x(t)\}$ and $\{y(t)\}$. Thus

$$R_{xy}(\tau) \triangleq \int_{-\infty}^{\infty} dx_1 \int_{-\infty}^{\infty} x_1 y_2 p(x_1, y_2) \, dy_2 \tag{51}$$

where $x_1 = x(t_1)$
$y_2 = y(t_1 + \tau)$

Another crosscorrelation function for these processes can be defined as

$$R_{yx}(\tau) \triangleq \int_{-\infty}^{\infty} dx_2 \int_{-\infty}^{\infty} x_2 y_1 p(x_2, y_1) \, dy_1 \tag{52}$$

where $y_1 = y(t_1)$
$x_2 = x(t_1 + \tau)$

Note that the order in which the subscripts appear on the crosscorrelation function determines which process is shifted in time. These two crosscorrelation functions are related, however, by

$$R_{xy}(\tau) = R_{yx}(-\tau) \tag{53}$$

In general, crosscorrelation functions are not even functions of τ and do not necessarily have their largest value at $\tau = 0$. They can be related to the corresponding autocorrelation functions, however, by the inequality

$$|R_{xy}(\tau)| \leq [R_{xx}(0) R_{yy}(0)]^{1/2} \tag{54}$$

If two random processes with zero means are statistically independent the crosscorrelation functions will be zero for all values of τ. However, except for the special case of jointly Gaussian processes, zero crosscorrelation does *not* imply independence.

It is also possible to define a *normalized crosscorrelation function* as

$$\rho_{xy}(\tau) = (1/\sigma_x \sigma_y)[R_{xy}(\tau) - \bar{x}\bar{y}] \tag{55}$$

with an analogous definition for $\rho_{yx}(\tau)$. These functions cannot exceed unity and, in fact, may never be unity for any value of τ.

The crosscorrelation functions are particularly useful when dealing with the sum of two or more random processes. Thus, if

$$z(t) = x(t) + y(t)$$

it may be shown that

$$R_{zz}(\tau) = R_{xx}(\tau) + R_{yy}(\tau) + R_{xy}(\tau) + R_{yx}(\tau) \tag{56}$$

In general, the autocorrelation function of the sum of any number of random processes is the sum of all possible autocorrelation and crosscorrelation functions.

In many practical situations it is necessary to determine correlation functions experimentally. If the random process being observed is assumed ergodic, then (49) can be used to evaluate the autocorrelation function except that it is not practical to allow T to become infinite. Hence the measured autocorrelation function, for any finite T, will differ from the true value and the variance of this difference D^2 is bounded by

$$D^2 \leq 2/T \int_0^{\infty} R_{xx}^2(\tau) \, d\tau \tag{57}$$

Evaluation of (57) for some typical cases indicates that the experimental observation time $2T$ must be 50 to 100 times larger than the largest value of τ being considered if the rms error in the measurement is to be as small as 10 per cent.

9 Spectral Density

The usefulness of frequency-domain representations for deterministic time functions has been noted previously. This same concept can be extended to random variables by considering an *average power density* rather than an amplitude density. This

MATHEMATICAL REPRESENTATION OF SIGNAL AND ERROR

power-density spectrum is usually designated simply as the *spectral density* and specifies the average power (in 1 ohm) per unit bandwidth as a function of frequency.

Although it is possible, under suitable restrictions, to define the spectral density as an ensemble average of the square of the Fourier transform of the random variable,[1] it is usually easier to obtain it from the autocorrelation function. Thus, if a stationary random process $\{x(t)\}$ has an autocorrelation function of $R_{xx}(\tau)$, the spectral density is simply

$$\Phi_{xx}(\omega) \triangleq \int_{-\infty}^{\infty} R_{xx}(\tau) \epsilon^{-j\omega\tau} \, d\tau \tag{58}$$

It will be noted that this is just the direct Fourier transform of the autocorrelation function. Hence the corresponding inversion integral is

$$R_{xx}(\tau) = 1/2\pi \int_{-\infty}^{\infty} \Phi_{xx}(\omega) \epsilon^{j\omega\tau} \, d\omega \tag{59}$$

If $x(t)$ has the units of volts, $\Phi_{xx}(\omega)$ will have units of (volts)^2/cps. The spectral density corresponding to the autocorrelation function of Fig. 5 is shown in Fig. 6.

It can be shown that $\Phi_{xx}(\omega)$ is always a real, positive, and even function of ω. The mean-square value of the random variable can be obtained from the area under $\Phi_{xx}(\omega)$. That is,

$$\overline{x^2} = R_{xx}(0) = 1/2\pi \int_{-\infty}^{\infty} \Phi_{xx}(\omega) \, d\omega \tag{60}$$

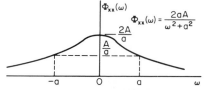

Fig. 6. Spectral density for the autocorrelation function of Fig. 5.

When the random process contains an average value or periodic components, the corresponding spectral density must have δ functions at the appropriate frequencies. The following example indicates the weight (or area) that must be associated with each δ function. Consider a random process having sample functions of the form

$$x(t) = X_0 + X_1 \cos(\omega_0 t + \theta) \tag{61}$$

where θ is the random variable and is uniformly distributed. The corresponding autocorrelation function is

$$R_{xx}(\tau) = X_0^2 + \tfrac{1}{2} X_1^2 \cos \omega_0 \tau \tag{62}$$

and the spectral density is

$$\Phi_{xx}(\omega) = 2\pi X_0^2 \delta(\omega) + (\pi/2) X_1^2 [\delta(\omega - \omega_0) + \delta(\omega + \omega_0)] \tag{63}$$

It will be noted that the weights on the δ functions have been selected such that (60) yields the proper mean-square value for *each* frequency component.

It is also possible to associate cross-spectral densities with two random processes. These may be defined in terms of the crosscorrelation functions as

$$\Phi_{xy}(\omega) \triangleq \int_{-\infty}^{\infty} R_{xy}(\tau) \epsilon^{-j\omega\tau} \, d\tau \tag{64}$$

$$\Phi_{yx}(\omega) \triangleq \int_{-\infty}^{\infty} R_{yx}(\tau) \epsilon^{-j\omega\tau} \, d\tau \tag{65}$$

The corresponding inversion integrals are

$$R_{xy}(\tau) = 1/2\pi \int_{-\infty}^{\infty} \Phi_{xy}(\omega) \epsilon^{j\omega\tau} \, d\omega \tag{66}$$

$$R_{yx}(\tau) = 1/2\pi \int_{-\infty}^{\infty} \Phi_{yx}(\omega) \epsilon^{j\omega\tau} \, d\omega \tag{67}$$

In general, the two cross-spectral densities are complex conjugates whose real parts are even functions of ω and whose imaginary parts are odd functions of ω. Thus the sum of the two cross-spectral densities for a given pair of random processes will always be a real, even function of ω. This sum arises when considering the sum of the processes. Thus if

$$z(t) = x(t) + y(t)$$

the spectral density of the sum is

$$\Phi_{zz}(\omega) = \Phi_{xx}(\omega) + \Phi_{xy}(\omega) + \Phi_{yx}(\omega) + \Phi_{yy}(\omega) \tag{68}$$

In general, the spectral density of the sum of any number of random processes will be the sum of all possible spectral and cross-spectral densities.

The experimental determination of spectral density is sometimes difficult to accomplish in a manner which ensures reliable results. Often it is more convenient to determine the autocorrelation function of the observed data and then employ (58) to obtain the spectral density.

The spectral densities may also be related to the correlation functions by means of the bilateral Laplace transform. In many cases this form may be more convenient

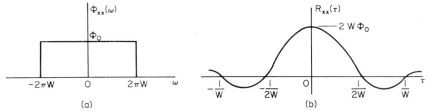

FIG. 7. Spectral density and autocorrelation function for band-limited white noise. (a) Spectral density. (b) Autocorrelation function.

because the inversion integrals can be immediately evaluated by application of the theory of residues. Since the modification to a bilateral Laplace transform is primarily a change in notation only, it is sufficient to illustrate it with a pair of equations corresponding to (58) and (59). Thus

$$\Phi_{xx}(s) = \int_{-\infty}^{\infty} R_{xx}(\tau) \epsilon^{-s\tau} \, d\tau \tag{69}$$

$$R_{xx}(\tau) = 1/2\pi j \int_{-j\infty}^{j\infty} \Phi_{xx}(s) \epsilon^{s\tau} \, ds \tag{70}$$

Similar expressions may be written for the cross-spectral densities.

A case of particular interest is that in which the spectral density is a constant independent of frequency. Such a random variable is usually called *white noise*, and although it does not exist in actuality, it is often convenient to assume a spectral density of this form whenever the random variable has a bandwidth which is large compared with that of the systems being considered. If the white-noise spectrum has a value of

$$\Phi_{xx}(\omega) = \Phi_0 \tag{71}$$

the corresponding autocorrelation function is a δ function of the form

$$R_{xx}(\tau) = \Phi_0 \delta(\tau) \tag{72}$$

A modified form of white noise which arises in the discussion of information theory is *band-limited white noise* in which the spectral density is constant over a finite range of frequencies and zero outside this range. For example, if

$$\Phi_{xx}(\omega) = \Phi_0 \qquad -2\pi W < \omega < 2\pi W$$
$$= 0 \qquad |\omega| > 2\pi W \tag{73}$$

the corresponding autocorrelation function is

$$R_{xx}(\tau) = 2W\Phi_0(\sin 2\pi W\tau / 2\pi W\tau) \tag{74}$$

These are shown in Fig. 7.

The discussion of sampling noted that a band-limited function could be completely described by samples taken $1/2W$ sec apart. From (74) above it is clear that such samples are uncorrelated since

$$R_{xx}(n/2W) = 2W\Phi_0(\sin n\pi / n\pi) = 0 \tag{75}$$

The concept of band-limited white noise also provides a means of specifying an *effective spectral bandwidth* for random processes whose spectral density approaches zero only at infinite frequency. Thus, if a general spectral density $\Phi_{xx}(\omega)$ has its maximum value at ω_0, the effective spectral bandwidth W_E is the width of a band-limited white-noise spectrum having the same maximum value and the same area (i.e., the same mean-square value). Hence the effective spectral bandwidth may be expressed as

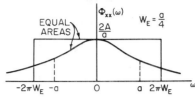

FIG. 8. Effective spectral bandwidth of the spectral density shown in Fig. 6.

$$W_E = 1/4\pi \int_{-\infty}^{\infty} \frac{\Phi_{xx}(\omega)}{\Phi_{xx}(\omega_0)} d\omega \quad \text{cps} \tag{76}$$

The effective spectral bandwidth of the spectral density shown in Fig. 6 is indicated in Fig. 8.

10 Response of Linear Systems

In general, the reason for making a mathematical description of signal and noise is to make it possible to determine analytically the response of physical systems to these quantities. In order to accomplish this it is also necessary to describe the system mathematically. In the case of linear systems, the mathematical description can be done in either the time domain or the frequency domain.

The time-domain description of a linear system is usually the unit-impulse response, or weighting function. This function $w(t)$ is defined as the output of the system when a unit impulse is applied to the input at $t = 0$. The output $y(t)$ resulting from any other input time function $x(t)$ may then be expressed in terms of the convolution integral. Thus, in general,[1]

$$y(t) = \int_{-\infty}^{t} w(t - \lambda)x(\lambda) d\lambda \tag{77}$$

or

$$y(t) = \int_{0}^{\infty} x(t - \lambda)w(\lambda) d\lambda \tag{78}$$

When the input is a stationary random process having specified statistical parameters, the output will also be a stationary random process for which the statistical parameters are to be determined. Equation (78) is usually more convenient for this purpose and leads to the following results for the mean and mean-square value of the output:

$$\bar{y} = \bar{x} \int_{0}^{\infty} w(\lambda) d\lambda \tag{79}$$

$$\overline{y^2} = \int_{0}^{\infty} d\lambda_1 \int_{0}^{\infty} R_{xx}(\lambda_1 - \lambda_2)w(\lambda_1)w(\lambda_2) d\lambda_2 \tag{80}$$

where $\bar{x}$ is the mean value of the input and $R_{xx}(\tau)$ is the autocorrelation function of the input. The autocorrelation function of the output may also be obtained from

$$R_{yy}(\tau) = \int_{0}^{\infty} d\lambda_1 \int_{0}^{\infty} R_{xx}(\tau + \lambda_1 - \lambda_2)w(\lambda_1)w(\lambda_2) d\lambda_2 \tag{81}$$

It may also be desired to obtain the crosscorrelation functions between the input and output. These are

$$R_{xy}(\tau) = \int_0^\infty R_{xx}(\tau - \lambda) w(\lambda)\, d\lambda \tag{82}$$

$$R_{yx}(\tau) = \int_0^\infty R_{xx}(\tau + \lambda) w(\lambda)\, d\lambda \tag{83}$$

When the input has a bandwidth which is wide compared with the bandwidth of the system it can be assumed with little error that the input is white noise for which the autocorrelation function is

$$R_{xx}(\tau) = \Phi_0 \delta(\tau) \tag{72}$$

For this simplified case the above results can be written as

$$\overline{y^2} = \Phi_0 \int_0^\infty w^2(\lambda)\, d\lambda \tag{84}$$

$$R_{yy}(\tau) = \Phi_0 \int_0^\infty w(\lambda) w(\lambda + \tau)\, d\lambda \tag{85}$$

$$R_{xy}(\tau) = \Phi_0 w(\tau) \qquad \tau \geq 0 \tag{86}$$

$$R_{yx}(\tau) = \Phi_0 w(-\tau) \qquad \tau \leq 0 \tag{87}$$

The frequency-domain description of a linear system is the frequency response, or transfer function. This function $W(s)$ is the Laplace transform of the impulse response $w(t)$. Thus

$$W(s) = \int_0^\infty w(t) \epsilon^{-st}\, dt \tag{88}$$

since $w(t)$ exists for positive time only. The transform of the output $Y(s)$ is simply

$$Y(s) = W(s) X(s) \tag{89}$$

where $X(s)$ is the transform of the input.

When the input is a stationary random process having a spectral density of $\Phi_{xx}(s)$, the output spectral density is

$$\Phi_{yy}(s) = \Phi_{xx}(s) W(s) W(-s) \tag{90}$$

The mean-square value of the output can be obtained from

$$\overline{y^2} = 1/2\pi j \int_{-j\infty}^{j\infty} \Phi_{xx}(s) W(s) W(-s)\, ds \tag{91}$$

Integrals of this type are tabulated in several places.[7]

The cross-spectral densities between the input and output may be expressed as

$$\Phi_{xy}(s) = \Phi_{xx}(s) W(s) \tag{92}$$
$$\Phi_{yx}(s) = \Phi_{xx}(s) W(-s) \tag{93}$$

If the input can be considered as white noise, the above results can be simplified to

$$\Phi_{yy}(s) = \Phi_0 W(s) W(-s) \tag{94}$$

$$\overline{y^2} = \Phi_0/2\pi j \int_{-j\infty}^{j\infty} W(s) W(-s)\, ds \tag{95}$$

$$\Phi_{xy}(s) = \Phi_0 W(s) \tag{96}$$
$$\Phi_{yx}(s) = \Phi_0 W(-s) \tag{97}$$

In such cases it is often convenient to define an *equivalent noise bandwidth* as the width of a transfer function of ideal rectangular shape and same *maximum* transmission as the actual system, which will pass the same average power as the actual system. Hence

$$W_N = \frac{1}{4\pi j} \int_{-\infty}^{\infty} \frac{W(s)W(-s)}{W(j\omega_0)W(-j\omega_0)} \, ds \quad \text{cps} \tag{98}$$

where $W(j\omega_0)$ is the maximum transmission of the system. The mean-square value of the output is

$$\overline{y^2} = 2\Phi_0 W_N W(j\omega_0) W(-j\omega_0) \tag{99}$$

For a low-pass system ω_0 is usually zero, while in a bandpass system ω_0 is usually the center of the passband.

CONCEPTS OF INFORMATION THEORY

Before different telemetering systems can be compared with respect to their information-handling capabilities, it is necessary to establish a quantitative measure for the information content of the signals. The purpose of this section is to establish such a measure on a conceptual basis, enumerate some of its properties, and discuss, in a general fashion, the limitations imposed by noise upon the information capacity of any system.

Information-handling systems are usually classified as discrete, continuous, or mixed. This classification is based upon the nature of both the information source and the signal which is applied to the transmission channel. Hence a discrete system is one in which both the data produced by the information source and the signal in the transmission channel can have only a countable number of specific values, while in the continuous system both these quantities may have a continuous range of values. In a mixed system, either quantity may be continuous while the other is discrete. All three possibilities have found use in present-day telemetering systems.

11 Units of Information

Before discussing the mathematical description of the information content associated with the original data or the transmitted signal, it is important to point out that this information content is not related to the usefulness or correctness of the data or signal since these attributes are usually not capable of quantitative evaluation. Thus the information capacity required to transmit useless or false data may be as large as, or even larger, than that required for useful and correct data.

One attribute of a signal which can often be evaluated quantitatively, and which also possesses an intuitive relation to information content, is the probability of that signal's occurring. Thus, if a signal is certain to occur, with probability 1, the receiver gains no information as a consequence of having received this signal. Such a signal should be assigned zero information content. On the other hand, if a signal is very improbable, then the receiver gains a great deal of information upon receiving it and, hence, the information content of this signal is large. Therefore, a good conceptual description of information content is to identify it with the uncertainty that exists about the signal *before* it is actually transmitted.

Another property which the unit of information should possess is that the information content associated with two or more independent signals should be the *sum* of the information contents of the individual signals. Since the joint probability associated with independent signals is given by the *product* of their individual probabilities, it follows that information should be a *logarithmic* function of probability. Thus, if a signal $x(t)$ has a probability $p(x)$, its information content is defined to be[8]

$$h(x) \triangleq -\log p(x) \tag{100}$$

Although any base might be used for the logarithm in (100), it is common to specify that one unit of information should be associated with a probability of one-half. On this basis, the logarithmic base becomes 2 and the unit of information is the binary digit, or *bit*. The base 10 is also used occasionally and the corresponding unit of information is the *Hartley*. Unless otherwise indicated, the base 2 will be used throughout this discussion.

12 The Discrete Information Source[9]

A discrete information source is one whose output can possess only a countable number of discrete values or states. In general, successive states are produced randomly according to certain probability relations which may depend upon the preceding states. The discrete source may occur when either the physical quantity being observed possesses discrete levels only or when the value of this quantity is rendered discrete by the action of some transducer.

For purposes of specifying the information content of such a source, let the ensemble of discrete values be designated as $\{x_i\}$, where i ranges from 1 to N, the total number of possible states. The discrete probability that the ith state will be produced by the source is designated as $p(i)$ and the information content associated with this state is, from (100),

$$h(i) = -\log p(i) \quad \text{bits} \tag{101}$$

Since the different states are not equally probable, the amount of information associated with each is not the same. Hence the *average* information per state may provide a more meaningful measure of the source information, and this quantity is determined from the mathematical expectation of $h(i)$. The computation of this mathematical expectation generally involves some specification of the dependence of any given state upon the preceding states.

The simplest situation arises when the various x_i are produced independently, that is, with no influence from previous states. In this case the mathematical expectation is simply

$$H = -\sum_{i=1}^{N} p(i) \log p(i) \quad \text{bits/state} \tag{102}$$

The similarity between this result and certain formulations in statistical mechanics has led to the adoption of the term *entropy* for H. It should be noted that, for the discrete source, H is always positive since $p(i)$ is positive and less than unity.

It can easily be shown that the maximum entropy that can be associated with N discrete states occurs when these states are all equally probable and that this maximum value is

$$H_{max} = \log N \quad \text{bits/state} \tag{103}$$

when $p(i) = 1/N$

The computation of entropy becomes more involved when there is statistical influence from one state to succeeding states. The simplest situation of this sort is the first-order Markov process in which the probability of any given state depends upon the first preceding state only. This is described by the conditional probability $p(j|i)$, that is, the probability of state j occurring given that state i has just occurred. The information content associated with this probability is

$$h(j|i) = -\log p(j|i) \quad \text{bits} \tag{104}$$

and the average information content per state (or entropy) is now given by

$$H = -\sum_{i=1}^{N}\sum_{j=1}^{N} p(j,i) \log p(j|i) \quad \text{bits/state} \tag{105}$$

where $p(j,i)$ is the joint probability of states i and j occurring and is given by

$$p(j,i) = p(i)p(j|i) \tag{106}$$

CONCEPTS OF INFORMATION THEORY 2-23

The entropy specified in (105) is often designated as the conditional entropy and for a given N is always less than (102) or (103).

It is possible, in principle, to obtain similar expressions for the entropy when the influence extends over a greater number of preceding states but the computations become very involved because of the large number of terms which arise.

The term *redundancy* is often applied to any measure of the entropy loss due to non-uniform probabilities or to influence between successive states. One common way of defining redundancy is

$$\text{Redundancy} \triangleq 1 - H/H_{max} \qquad (107)$$

where H is the actual entropy of the source and H_{max} is the maximum entropy which the source could have with the same number of states. The redundancy of a source is an important parameter in determining the type of coding which should be used in preparing the messages produced by the source for application to the transmission channel.

The rate at which a discrete source produces information can be determined from a knowledge of the entropy and the number of different states which it can assume per unit time. Thus, if a source of entropy H can produce n changes in state per second, the information rate R is simply

$$R = nH \quad \text{bits/sec} \qquad (108)$$

From (103) it is seen that for a source having N independent, equally probable states, the maximum rate at which information can be produced is given by

$$R_{max} = n \log N = \log N^n \quad \text{bits/sec} \qquad (109)$$

If the states are not independent or not equally probable, the information rate will be less than R_{max}.

A simple example will serve to illustrate the calculation of entropy for a discrete source. Consider a system in which the physical quantity being observed can exist at any one of 10 distinct levels and let the probability of observing any level be as follows:

Level	Probability
1	0.02
2	0.03
3	0.05
4	0.10
5	0.20
6	0.30
7	0.20
8	0.05
9	0.03
10	0.02
	1.00

If successive observations are assumed to be statistically independent, the entropy associated with this source is

$$H = - \sum_{i=1}^{10} p(i) \log p(i) = -[2(0.02 \log 0.02 + 0.03 \log 0.03$$
$$+ 0.05 \log 0.05 + 0.2 \log 0.2)$$
$$+ 0.1 \log 0.1 + 0.3 \log 0.3]$$
$$= 2.75 \text{ bits/state}$$

The maximum entropy which this source could have would occur if all levels were equally probable and would be

$$H_{max} = \log 10 = 3.32 \text{ bits/state}$$

Hence the redundancy of the source is $1 - 2.75/3.32 = 0.172$, or 17.2 per cent.

13 The Discrete Noiseless Channel[9]

A discrete noiseless channel is one in which only a countable number of distinct symbols can exist and no errors in transmission occur. The most common form of discrete channel is the binary channel, in which only two signal levels are possible, but use has also been made of ternary and quaternary channels and there are, of course, an infinite number of other possibilities. Under conditions of high signal-to-noise ratio, transmission errors become negligible and the noiseless condition is approached.

If the rate at which symbols can be transmitted in the discrete channel has an upper bound, then the rate at which information can be transmitted over this channel also has an upper bound even in the absence of noise. This situation is in contrast to that which exists in continuous channels and hence warrants special consideration.

The capacity of the discrete channel can be expressed in bits per unit time and is defined to be

$$C \triangleq \lim_{T \to \infty} [\log N(T)/T] \quad \text{bits/sec} \tag{110}$$

where $N(T)$ is the number of possible signals of duration T that could be transmitted. For example, assume that a telemetry system transmits n samples per second and each sample is coded into a group of 5 binary digits. Since there are 32 ways of arranging 5 binary digits, in a time T there will be 32^{nT} possible signals. Thus the channel capacity is

$$C = \lim_{T \to \infty} (\log 32^{nT}/T) = \lim_{T \to \infty} (nT \log 32/T) = 5n \text{ bits/sec}$$

a result which is intuitively correct.

When the channel symbols are not of equal duration, the evaluation of $N(T)$ is more involved but a general procedure is still available. Let the state of the channel be defined by the last symbol which was transmitted and let t_{ij} be the duration of the symbol which changes the state of the channel from i to j. Then it can be shown that the channel capacity C is given by

$$C = \log X_0 \quad \text{bits/sec} \tag{111}$$

where X_0 is the largest real root of the determinant equation

$$|X^{-t_{ij}} - \delta_{ij}| = 0 \tag{112}$$

in which δ_{ij} is the Kronecker delta defined by

$$\begin{aligned} \delta_{ij} &= 1 & i = j \\ &= 0 & i \neq j \end{aligned}$$

If there are restrictions on the possible sequences of channel symbols, then the time t_{ij} corresponding to unpermitted changes in state is taken as infinity.

As a simple illustration of the calculation of channel capacity, consider a noiseless channel in which there are only two symbols—a pulse having a duration of 0.01 sec and a space having a duration of 0.09 sec. Assume also that two spaces in sequence are not permitted. If the pulse is associated with the index number 1, while the space is associated with 2, then the transition times become

$$\begin{aligned} t_{11} &= 0.01 & t_{12} &= 0.09 \\ t_{21} &= 0.01 & t_{22} &= \infty \end{aligned}$$

and the determinant equation is

$$\begin{vmatrix} X^{-0.01} - 1 & X^{-0.09} \\ X^{-0.01} & X^{-\infty} - 1 \end{vmatrix} = 0 \tag{113}$$

or

$$X^{-0.01} - 1 + X^{-0.1} = 0 \tag{114}$$

The largest real root of (114) is found by trial and error to be

$$X_0 = (1.1978)^{100}$$

from which the channel capacity becomes

$$C = \log (1.1978)^{100} = 26.06 \text{ bits/sec}$$

CONCEPTS OF INFORMATION THEORY

The fundamental theorem for the discrete noiseless channel may now be stated: *If a discrete source has an entropy of H bits per state and a discrete channel has a capacity of C bits/sec, then it is possible to encode the output of the source into channel symbols in such a way as to transmit information at an average rate of $C/H - \epsilon$ states per second where ϵ is arbitrarily small. It is not possible to transmit at an average rate greater than C/H.* The major implications of this theorem are (1) some sort of coding operation is necessary if an arbitrary discrete source is to be transmitted over an arbitrary discrete channel with maximum information rate or, equivalently, with minimum channel capacity; and (2) there is a finite rate which cannot be exceeded regardless of the type of encoding employed. Unfortunately, the theorem does not give any explicit procedure for constructing codes which achieve the maximum information rate, and in general, such codes are not possible for finite-length sequences. There are intuitive coding procedures, however, which approach the ideal, and these are discussed below in the sections on optimum coding.

14 The Discrete Channel with Noise

In the case of the noiseless channel it was assumed that the received symbol was identical with the transmitted symbol. In a less ideal system, however, this may not be true because the presence of noise may result in the received symbol's being

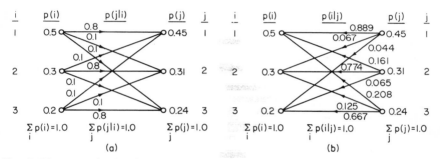

Fig. 9. Linear graphs showing the various probabilities associated with correct and erroneous transmission of three symbols. (a) The conditional probabilities $p(j|i)$. (b) The conditional probabilities $p(i|j)$.

interpreted as one different from that transmitted. Since the receiver does not know when such errors occur there remains some uncertainty about what was actually transmitted. This uncertainty can be interpreted as a reduction in the amount of information conveyed by the transmission channel, and it is the purpose of this section to show how this information loss can be expressed quantitatively.

Since transmission errors occur in a random fashion, they are best described on a probabilistic basis. Thus, if the index i is associated with the input symbols into the channel and the index j is associated with the output symbols from the channel, the conditional probability $p(j|i)$ is a measure of the probability of correct transmission when $i = j$ and a measure of the probability of errors when $i \neq j$. If $p(i)$ is the probability of the ith symbol being introduced into the channel, then the joint probability of both transmitting i and receiving j is

$$p(i,j) = p(i)p(j|i) = p(j)p(i|j) \qquad (115)$$

The conditional probability $p(i|j)$ is a measure of the uncertainty that exists at the receiver after having received j while $p(j)$ is the probability with which the various symbols are received and will often be different from $p(i)$.

It is often convenient to represent the various probabilities by means of a pair of directed linear graphs. Such a pair of graphs for a hypothetical situation involving three channel symbols is shown in Fig. 9. The advantage of the graphical representation is that the various probabilities needed for the computation of entropy are more readily evident.

In the calculation of entropy, it is convenient to use the symbol x to denote the input to the channel while the symbol y is used to denote the output from the channel. The four entropies, and their interpretation in terms of uncertainty, may now be listed (assuming channel symbols produced independently).

$H(x) \triangleq -\sum_i p(i) \log p(i)$ = uncertainty about the channel input *prior* to transmission

$H(y) \triangleq -\sum_j p(j) \log p(j)$ = uncertainty about the channel output *prior* to transmission

$H_x(y) \triangleq -\sum_i \sum_j p(i,j) \log p(j|i)$ = uncertainty at the input about the channel output after transmission

$H_y(x) \triangleq -\sum_i \sum_j p(i,j) \log p(i|j)$ = uncertainty at the output about the channel input after transmission

The last conditional entropy $H_y(x)$ is usually called the *equivocation*. As a result of the product law of probabilities, as expressed in (115), these four entropies are related by

$$H(x) + H_x(y) = H(y) + H_y(x) \tag{116}$$

The entropies corresponding to the probabilities shown in Fig. 9 may be calculated for purposes of illustration.

$H(x) = -(0.5 \log 0.5 + 0.3 \log 0.3 + 0.2 \log 0.2) = 1.483$ bits/symbol
$H(y) = -(0.45 \log 0.45 + 0.31 \log 0.31 + 0.24 \log 0.24) = 1.536$ bits/symbol
$H_x(y) = -[0.5(0.8 \log 0.8 + 0.1 \log 0.1 + 0.1 \log 0.1)$
$\qquad + 0.3(0.1 \log 0.1 + 0.8 \log 0.8 + 0.1 \log 0.1)$
$\qquad + 0.2(0.1 \log 0.1 + 0.1 \log 0.1 + 0.8 \log 0.8)]$
$\quad = 0.922$ bits/symbol
$H_y(x) = -[0.45(0.889 \log 0.889 + 0.067 \log 0.067 + 0.044 \log 0.044)$
$\qquad + 0.31(0.161 \log 0.161 + 0.774 \log 0.774 + 0.065 \log 0.065)$
$\qquad + 0.24(0.208 \log 0.208 + 0.125 \log 0.125 + 0.667 \log 0.667)]$
$\quad = 0.869$ bits/symbol

Note that the equality of (116) is satisfied.

The average amount of information conveyed by a channel symbol is defined to be the *difference* between the uncertainties about the channel input prior to transmission and after transmission. Thus

$$I = H(x) - H_y(x) \quad \text{bits/symbol} \tag{117}$$

or from (116)

$$I = H(y) - H_x(y) \quad \text{bits/symbol} \tag{118}$$

For the numerical example considered above this information is

$$I = 0.614 \text{ bits/symbol}$$

A particularly simple situation which has much practical application is that of the binary symmetric channel. In this case there are two input symbols, say 0 and 1, and they are introduced into the channel with equal probabilities. The probabilities of error are also equal and designated by q while the probabilities of correct transmission are designated by p. The linear graph for the binary symmetric channel is shown in Fig. 10. Because of the symmetry $p(i|j) = p(j|i)$ and $p(i) = p(j)$ and hence

$H(x) = H(y) = -(0.5 \log 0.5 + 0.5 \log 0.5) = 1.0$ bit/symbol
and $H_x(y) = H_y(x) = -(p \log p + q \log q)$ bits/symbol

The information per symbol conveyed by the binary symmetric channel is just

$$I = 1 + p \log p + q \log q \quad \text{bits/symbol} \tag{119}$$

The manner in which this information varies with q, the probability of error, is shown in Fig. 11. It is of interest to note that a channel that is always in error can convey as much information as one that is always correct.

When the channel symbols are of equal duration and can occur in any sequence, the information quantity I can be converted into a rate of transmission by multiplying it by the average number of symbols that are transmitted in the channel per unit time. Thus, if the channel can transmit n symbols per second, the rate of information transmission is simply

$$nI = n[H(x) - H_y(x)] \quad \text{bits/sec} \tag{120}$$

The maximum value which this rate can assume is defined to be the channel capacity C. Thus

$$C \triangleq \max \{n[H(x) - H_y(x)]\} \quad \text{bits/sec} \tag{121}$$

where the maximizing is accomplished by adjusting the probabilities $p(i)$ associated with the input symbols. When there are a large number of symbols, with different conditional probabilities associated with each pair, it becomes a very difficult task to determine the channel capacity. The problem becomes much simpler if there is a large amount of symmetry present, and numerous examples have been worked out in the literature.[9] The special case in which all input

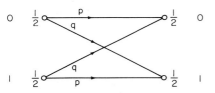
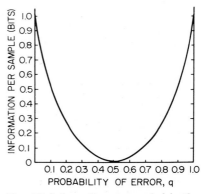

Fig. 10. Linear graph for the binary symmetric channel.

Fig. 11. Information per symbol in the binary symmetric channel as a function of the probability of error.

symbols are affected by noise in the same way is particularly simple, and the channel capacity may be expressed for the case of M symbols as

$$C = n\left[\log M + \sum_j p(j|i) \log p(j|i)\right] \quad \text{bits/sec} \tag{122}$$

The numerical example of Fig. 9 is a case for which (122) applies because the same set of $p(j|i)$ exists for all three input symbols. Thus the capacity of this hypothetical channel is

$$C = n(\log 3 + 0.8 \log 0.8 + 0.2 \log 0.1)$$
$$= 0.663n \text{ bits/sec}$$

This capacity would require the input symbols to be equally probable, i.e.,

$$p(i) = 0.333$$

for all i.

For the case of the general binary channel it can be shown that the rate of transmission is maximized when the two input symbols are equally probable and when the probabilities of error for the two symbols are the same. Hence the binary symmetric channel is the optimum form of binary channel and has a capacity of

$$C = n(1 + p \log p + q \log q) \quad \text{bits/sec} \tag{123}$$

Entropy tables[27] for calculating (119) and (123) may be found in the Appendix, pp. A-76 to A-94.

15 Fundamental Theorem for the Discrete Channel with Noise

One of the most important conclusions regarding the discrete channel with noise is summarized by the fundamental theorem which states: *If a discrete source produces information at a rate of R bits/sec and a discrete noisy channel has a capacity of $C \geq R$ bits/sec, then there exists a method of encoding the output of the source into channel symbols such that transmission over the channel can be accomplished with arbitrarily small equivocation. If $R > C$ it is possible to encode the source so that the equivocation is less than $R - C + \epsilon$, where ϵ is arbitrarily small. There is no method of encoding which gives an equivocation less than $R - C$.*

This theorem provides the rather startling result that it is possible to transmit information over a noisy channel with negligible error provided that the rate of the source is less than the capacity of the channel and that this is true regardless of the amount of noise present. Of course, if there is a large amount of noise present, the channel capacity becomes small.

In order to realize the condition of negligible error it is necessary to encode the output of the source into channel symbols in an optimum fashion, and the theorem does not specify how this can be accomplished. In fact, no general procedures are known but certain special cases which produce favorable results are discussed under Optimum Coding.

The theorem also gives assurance that no matter how elaborate the encoding operation might be, it is not possible to transmit information with negligible error when the rate of the source is greater than the channel capacity. Hence it becomes possible to compare the results obtained from any given coding scheme with the theoretically best results in order to judge the effectiveness of a practical system.

16 The Continuous Information Source

When an information source can assume a continuous range of values, the computation of information rate becomes much more involved. In fact, it is not possible to assign a unique information rate to any continuous source without first assigning an error criterion because it would require an infinite amount of information to specify exactly the output of the source at even one instant of time. Because of this difficulty, the discussion of information rate from a continuous source is deferred to Sec. 18.

In spite of the above difficulty, however, it is possible to calculate the entropy associated with a continuous source, although the quantity should perhaps be referred to as the "differential entropy" since it is not an absolute measure of information and, in fact, need not be positive. The computation of this entropy is quite involved in general but for the special case of a band-limited signal function with a flat spectral density out to some finite frequency of W cps it becomes straightforward. Only this case will be discussed initially.

It was pointed out in Sec. 3 that a band-limited signal function $x(t)$ could be uniquely represented in terms of its sample values at a set of points spaced $1/2W$ sec apart. The value of any given sample is a random variable having a probability density function $p(x)$. By analogy with the discrete case, the entropy associated with this sample point is defined as

$$H_1(x) \triangleq - \int_{-\infty}^{\infty} p(x) \log p(x)\, dx \quad \text{bits/sample} \tag{124}$$

in which subscript 1 is used to emphasize that only one sample point is being considered.

When the signal function is a sample of band-limited white noise, the random variables associated with different sample points are statistically independent. If, in addition, the signal function is a member of a stationary random process, the density functions associated with the various sample points are all the same. Hence the entropy per unit time can be obtained simply by multiplying (124) by $2W$, the number

CONCEPTS OF INFORMATION THEORY 2–29

of samples per unit time. Thus the entropy rate becomes

$$H_R(x) = 2WH_1(x) = -2W \int_{-\infty}^{\infty} p(x) \log p(x)\, dx \quad \text{bits/sec} \qquad (125)$$

It is of interest to determine the form of $p(x)$ which leads to a maximum value of $H_1(x)$. There are two cases which are of particular importance:
1. Signal limited to a finite interval; $x_1 \leq x(t) \leq x_2$. When the signal function can exist in a finite interval only, maximum entropy occurs when $p(x)$ is uniform in this interval. Hence

$$\begin{aligned} p(x) &= 1/(x_2 - x_1) & x_1 \leq x \leq x_2 \\ &= 0 & \text{elsewhere} \end{aligned} \qquad (126)$$

and the entropy is

$$H_1(x) = \log(x_2 - x_1) \quad \text{bits/sample} \qquad (127)$$

2. Signal limited to a finite average power; $\overline{x^2} = \sigma^2$. When the signal function can have any value but must possess a specified average power, maximum entropy occurs when $p(x)$ is Gaussian. Hence

$$p(x) = (1/\sqrt{2\pi}\,\sigma)\, \epsilon^{-x^2/2\sigma^2} \quad -\infty < x < \infty \qquad (128)$$

and the entropy is

$$H_1(x) = \log \sqrt{2\pi \epsilon \sigma^2} \quad \text{bits/sample} \qquad (129)$$

where ϵ is the base of the natural logarithms.

It is of interest to note in both the above examples that the entropy is a function of the *magnitude* of the signal function as well as its density function. Hence the entropy of a continuous function can be made any value simply by amplification or attenuation, that is, by a change in scale. The intuitive concept of information, however, indicates that a mere change in scale should not change the information content of a signal, and hence entropy cannot be an absolute measure of information in the continuous case. This is in distinct contrast to the discrete case, but as will be seen, the difficulty is easily resolved by considering only *differences* in entropy which, because of the logarithmic nature, are not a function of changes in scale.

When the spectral density of the signal function is not white and band-limited, the computation of entropy rate becomes more involved because there is usually correlation between adjacent samples. For many purposes it is sufficient to specify such signal functions in terms of their *entropy power*. The entropy power is defined to be the average power of a white, band-limited, Gaussian function which has the same entropy rate as the actual signal function. The entropy power is always less than or equal to the actual average power of the signal.

17 The Continuous Channel

The channel to be considered in this discussion is one in which there is noise present but no amplitude or frequency distortion. Thus the input to the channel is a time function $x(t)$ and the output from the channel $y(t)$ is simply the sum of the input and the noise. All these time functions will be assumed to be white and band-limited.

As in the case of the discrete channel, there are four entropies which can be defined. Thus

$$H_1(x) \triangleq -\int_{-\infty}^{\infty} p(x) \log p(x)\, dx \quad \text{bits/sec} \qquad (130)$$

$$H_1(y) \triangleq -\int_{-\infty}^{\infty} p(y) \log p(y)\, dy \quad \text{bits/sec} \qquad (131)$$

$$H_{1x}(y) \triangleq -\int_{-\infty}^{\infty}\int_{-\infty}^{\infty} p(x,y) \log p(y|x)\, dx\, dy \quad \text{bits/sec} \qquad (132)$$

$$H_{1y}(x) \triangleq -\int_{-\infty}^{\infty}\int_{-\infty}^{\infty} p(x,y) \log p(x|y)\, dx\, dy \quad \text{bits/sec} \qquad (133)$$

As before, $H_y(x)$ is the equivocation and the four entropies are related by

$$H_1(x) + H_{1x}(y) = H_1(y) + H_{1y}(x) \tag{134}$$

The information conveyed by the channel, per sample point, is again defined as

$$\begin{aligned} I_1 &\triangleq H_1(x) - H_{1y}(x) \\ &= H_1(y) - H_{1x}(y) \end{aligned} \quad \text{bits/sec} \tag{135}$$

It is important to note that the *difference* of entropies is an absolute measure of information even though the individual entropies are not. The rate at which information is conveyed by the channel is obtained by multiplying by $2W$.

The capacity of the continuous channel is defined as the maximum value of the information rate. Thus one form is

$$C \triangleq \max\ \{2W[H_1(y) - H_{1x}(y)]\} \quad \text{bits/sec} \tag{136}$$

where the maximizing is accomplished by adjusting $p(x)$. General methods for obtaining the channel capacity with arbitrary restrictions are not known, but certain special cases of importance have been evaluated.

The simplest case, and one of the most important, is that in which the noise in the channel is statistically independent of the signal and adds linearly to the signal. The conditional entropy $H_{1x}(y)$ then becomes the entropy of the noise only so that (136) can be written as

$$C = 2W\{\max\ [H_1(y)] - H_1(n)\} \quad \text{bits/sec} \tag{137}$$

In order to carry the evaluation of channel capacity farther it is necessary to impose some constraints on the received signal. The most common constraint is one on the average power, and for this case it is known that maximum entropy occurs when $p(y)$ is Gaussian. The received signal will be Gaussian if both the transmitted signal and the noise are Gaussian, and the entropies for these can be calculated easily.

In order to put the above situation on a quantitative basis, let the signal in the channel have an average power of S and the noise an average power of N. The average power of the received signal will therefore be $S + N$ and the entropies required in (137) are

$$H_1(y) = \log\sqrt{2\pi\epsilon(S+N)} \quad \text{bits/sec} \tag{138}$$

and

$$H_1(n) = \log\sqrt{2\pi\epsilon N} \quad \text{bits/sec} \tag{139}$$

as given by (129). Hence the channel capacity for an average power limitation is

$$\begin{aligned} C &= 2W[\log\sqrt{2\pi\epsilon(S+N)} - \log\sqrt{2\pi\epsilon N}] \\ &= W\log\,[(S+N)/N] \quad \text{bits/sec} \end{aligned} \tag{140}$$

This is the result which is usually quoted in the literature, often without adequately specifying the special conditions for which it applies. Because of the widespread misuse of (140) it is advisable to summarize once more the assumptions upon which it is based and the special conditions to which these assumptions lead. The assumptions are

1. The transmitted signal and noise are white and limited in bandwidth to W cps.

2. The transmitted signal and noise are from statistically independent, stationary random processes and add linearly in the channel.

3. The received signal is limited in average power.

The special conditions which are forced by these assumptions are

1. The transmitted signal samples must come from a Gaussian distribution.

2. The noise samples must come from a Gaussian distribution.

The channel capacity quoted above may be useful for comparison purposes even if the actual system does not satisfy the restrictions because it places an upper bound on the information-handling capabilities. Usually certain liberties must be taken in the interpretation of bandwidth, and this is often done on an intuitive basis.

Of greater importance than numerical comparisons, however, are the implications that can be obtained from the expression for channel capacity. In particular, (140) implies that it is possible to trade signal-to-noise ratio for bandwidth or vice versa. This is made more evident by expressing channel capacity in terms of bits per second per unit bandwidth. The result, which is plotted in Fig. 12, is

$$C/W = \log(1 + S/N) \quad \text{bits/cycle} \tag{141}$$

In order to accomplish this trade it is necessary to encode the message into a channel signal which comes as close to being white, band-limited, and Gaussian as possible. In other words, an efficient coding scheme tends to make the signal look like random noise. The degree to which this can be accomplished in practical systems will be discussed later.

When the special conditions assumed above are not satisfied, it is generally not possible to solve for the channel capacity explicitly. In some cases it is possible to obtain upper and lower bounds on the channel capacity, however, and these bounds are often sufficiently close to represent an adequate solution to the problem. In particular, if the noise in the channel is neither Gaussian nor white in the bandwidth W, then it will have an entropy power N_1 which is less than its actual average power N. The channel capacity is then bounded by

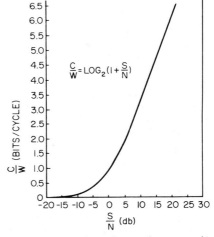

Fig. 12. The channel capacity per unit bandwidth as a function of signal-to-noise ratio.

$$W \log[(S + N_1)/N_1] \leq C \leq W \log[(S +)/N_1] \tag{142}$$

and these bounds approach one another for large signal-to-noise ratios.

If the noise in the channel is Gaussian but not white, its entropy power may be calculated from

$$N_1 = \exp\left[1/W \int_0^W \log WN(f)\, df\right] \tag{143}$$

where $N(f)$ is the noise spectral density within the bandwidth W. This value of N_1 may then be used in (142).

18 Information Rate for a Continuous Source

It has been mentioned previously that the entropy of a continuous source was not an absolute measure of information. This condition caused no difficulty in the computation of channel capacity because that quantity was defined as the difference between two entropies. The situation is not so simple, however, when it is desired to specify the information rate for a source because this is not a difference quantity and, in fact, is infinite if an exact specification is required.

As a practical matter, however, it is usually not necessary that the receiver reconstruct the output of the source exactly. Once an acceptable error has been specified, the information rate from the source becomes finite when measured with respect to this error. Hence the information rate from a continuous source is always defined relative to some error criterion.

Although there are many ways in which an error criterion might be defined, the easiest one to handle mathematically is the mean-square criterion. Thus the mean-square value of the difference between the system output and the source output is

$$\varepsilon = \lim_{T \to \infty} 1/T \int_0^T [x(t) - y(t)]^2\, dt \tag{144}$$

The information rate is then defined to be

$$R \triangleq \min \{2W[H(y) - H_x(y)]\} \quad \text{bits/sec} \tag{145}$$

where the minimizing is done with respect to $p(y|x)$ subject to the constraint that ε is held constant. This definition is a consequence of the fundamental theorem for continuous channels, which states: *If a continuous information source has a rate R for the specified error criterion, it is possible to encode the output of this source and transmit it over a channel of capacity C with an error arbitrarily close to the specified error if $R \leq C$. This is not possible if $R > C$.*

The evaluation of information rate for a continuous source is difficult to carry out except in special cases. One special case for which a result is available is that in which the source produces a message ensemble which is white, band-limited, Gaussian, and has an average power of M, and where there is a specified mean-square error ε. For this case

$$R = W \log (M/\varepsilon) \quad \text{bits/sec} \tag{146}$$

If the source is not white, but has an entropy power M_1, the information rate is bounded by

$$W \log (M_1/\varepsilon) \leq R \leq W \log (M/\varepsilon) \tag{147}$$

The entropy power can be determined from the spectral density of the source $M(f)$ by the relation

$$M_1 = \exp \left[1/W \int_0^W \log WM(f)\, df \right] \tag{148}$$

It should be noted that in all the above equations, W is the bandwidth of source in cps, and this is not necessarily the same as the bandwidth of the channel.

OPTIMUM CODING

The term "coding" is an ambiguous one which is used in many different ways. The usage to be employed in this section is that which is common to the field of information theory and does *not* correspond to the usage which is common in telemetry. The justification for this departure from telemetry terminology is that no other term appears to be equally suitable for the application being discussed, while the operation often referred to as coding in telemetry is equally well designated as modulation.

In the previous section, reference was made twice to the coding problem. In connection with the discrete noiseless channel, it was mentioned that coding was required to match the information source to the channel in order to make the information-transmission rate approach channel capacity. This type of coding will be referred to as minimum-redundancy coding because its objective is the *reduction* of redundancy that may exist in the source. In the case of the discrete channel with noise, coding may also be required to combat the effects of the noise. This type of coding will be designated as error-correction coding, and it accomplishes this objective by *increasing* the redundancy of the channel symbols. Surprisingly, it may be advantageous to employ both types of coding simultaneously.

19 Minimum-redundancy Codes

According to the previous definition, a discrete information source possesses N discrete states which occur at random and according to specified probability relations. In order to transmit the output of this source over a discrete channel it is necessary to convert the output into channel symbols. This conversion operation is the coding to which the present article pertains. There are obviously an infinite number of situations which might arise, but only those involving a binary channel, with the symbols

0 and 1, will be discussed. This limitation does not lose generality, because the binary symbols can be further encoded into any other set of symbols without any change in message entropy. In addition, the binary symbols are convenient to deal with analytically and are typical of practical discrete channels.

The most straightforward way of accomplishing the encoding is to assign an arabic integral value to each of the possible states and then represent each integer by its corresponding binary number. This is the practice that is most commonly carried out, although many modifications have been suggested to minimize the number of digits which change between adjacent states, or to simplify the conversion from binary to decimal or vice versa. The distinguishing characteristics of this type of encoding, from an information-theory standpoint, are that all code groups have the same number of binary digits and that no account has been taken of the statistical properties of the source. Hence such codes can be optimum only when N is a power of 2 when all states are equally probable. Since this condition rarely exists in an information source, a different approach to coding is necessary if transmission at or near channel capacity is to be achieved.

The basic philosophy of minimum-redundancy codes is to assign short code groups to message states with high probability of occurrence and long code groups to low-probability message states. This procedure tends to equalize the times that each state occupies the channel and to reduce the rate at which binary digits must be transmitted. Hence the redundancy of the coded message is smaller than the redundancy of the original message and the optimum code is the one providing the minimum value of redundancy.

So far as practical systems are concerned the major disadvantage of minimum-redundancy codes is that such coding necessarily involves the storage of part of the information. This is a consequence of observing state changes in the source at a constant rate but delivering code groups at a variable rate determined by the lengths of the groups (assuming no spaces between groups). There is also some delay involved in this procedure, but this is frequently of no consequence.

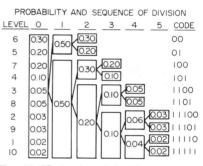

FIG. 13. Illustration of the Shannon-Fano method of coding.

There are several intuitive procedures by which nearly optimum codes may be constructed. One of the easiest to apply is the Shannon-Fano method,[9,10] which involves the following steps:

1. Tabulate the N message states in order of decreasing probability.
2. Divide this list into two groups of as nearly equal probability as possible.
3. To the top group assign a 0 (say) as the first digit and to the lower group assign a 1 as the first digit.
4. Divide each group into two subgroups of as nearly equal probability as possible.
5. To each of the top subgroups assign a 0 as the second digit and to each of the lower subgroups assign a 1 as the second digit.
6. Repeat the process until no more division is possible.

A simple example will serve to illustrate the procedure and indicate its effectiveness. The numerical values are taken from the example in Sec. 12, and the method of dividing into groups is indicated in Fig. 13. It is observed that the code sequences vary from 2 binary digits for the most probable messages to 5 binary digits for the least probable messages. The average length of code sequence is simply

$L = 2(0.3 + 0.2) + 3(0.2 + 0.1) + 4(0.05 + 0.05) + 5(0.03 + 0.03 + 0.02 + 0.02)$
$= 2.8$ binary digits

Note that, if code sequences of constant length had been used, each sequence would have had a length of 4 binary digits.

Variable-length codes must be constructed in such a way that there is no ambiguity regarding the places at which one sequence ends and another one starts. This requires that no sequence be identical to the first part of a sequence of greater length. The systematic procedure of the Shannon-Fano method ensures that this requirement is met. Unfortunately this method does not ensure that an optimum code has been obtained, although optimum behavior is approached as N approaches infinity and the codes are usually close to optimum for finite N.

A procedure which does yield an optimum code for N independent message states has been given by Huffman[11] and involves the following steps:

1. Tabulate the N message states in order of decreasing probability.
2. Combine the two least probable states into a single state and retabulate the $N - 1$ message states in order of decreasing probability.
3. Repeat the combining process until no further combination is possible.
4. Determine the number of times that each of the original message states enters into a combining operation. This number specifies the length of code required for that message state.
5. Form code groups of the proper length by any method which avoids ambiguity.

The previous example may be recoded by the Huffman method as shown in Fig. 14.

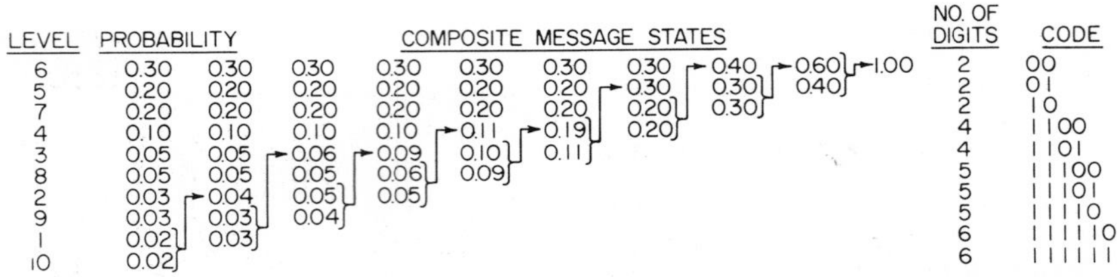

LEVEL	PROBABILITY			COMPOSITE MESSAGE STATES						NO. OF DIGITS	CODE
6	0.30	0.30	0.30	0.30	0.30	0.30	0.40	0.60	1.00	2	00
5	0.20	0.20	0.20	0.20	0.20	0.20	0.30	0.30	0.40	2	01
7	0.20	0.20	0.20	0.20	0.20	0.20	0.20	0.30		2	10
4	0.10	0.10	0.10	0.10	0.11	0.19	0.20			4	1100
3	0.05	0.05	0.06	0.09	0.10	0.11				4	1101
8	0.05	0.05	0.05	0.06	0.09					5	11100
2	0.03	0.04	0.05	0.05						5	11101
9	0.03	0.03	0.04							5	11110
1	0.02	0.03								6	111110
10	0.02									6	111111

FIG. 14. Illustration of the Huffman method of coding.

It will be noted that the number of digits in each sequence ranges from 2 to 6. The average length of code sequence is now

$$L = 2(0.3 + 0.2 + 0.2) + 4(0.1 + 0.05) + 5(0.05 + 0.03 + 0.03) + 6(0.02 + 0.02)$$
$$= 2.79 \text{ binary digits}$$

Although the reduction in average length is not impressive in this example, it may be significant in other cases.

It will be noted that even the optimum coding method of Huffman has not reduced the average code length to the entropy of the source, which was calculated in Sec. 12 as 2.75 bits/state. As a means of comparing various coding methods, it is common to define a *coding efficiency* η as

$$\eta = H/L \qquad (149)$$

where H is the actual entropy of the source and L is the average length of the code. The factor η also represents the fraction of channel capacity that is actually achieved by the coding scheme being considered in a noiseless channel. As a result of the fundamental theorem for noiseless channels, η has a maximum value of 1.00, but in general this value is approached only as N approaches infinity.

For the example being considered, the coding efficiency is

$$\eta = 2.75/2.79 = 0.985$$

for the Huffman coding scheme. This efficiency could be increased by coding all possible *pairs* of states rather than individual states. In this case there would be 100 such pairs but the coding procedure would be the same. Although only a very limited improvement in efficiency is possible in this example, had N been much smaller originally, the concept of encoding combinations of states could lead to substantial improvement.

20 Error-correcting Codes

In contrast to the previous article, where the objective of coding was to reduce redundancy, error-correcting codes increase redundancy. In this case the objective of coding is to introduce redundancy in a known fashion so that its presence can be used at the receiving point to correct errors that result from noise in the transmission channel. As in the previous case only binary codes will be considered and the channel will be assumed to be symmetric.

In spite of the extensive work which has been done on error-correcting codes, existing information on specific procedures for constructing such codes is still quite fragmentary. Most existing methods deal with *systematic codes*, i.e., codes in which each group contains exactly n binary digits of which exactly m digits are associated with the information being transmitted and the remaining digits represent the redundancy. Even in this special case, however, a general procedure for constructing optimum codes is not known.

The effectiveness of an error-correcting code may be measured in terms of the number of errors which can be corrected in each group of n binary digits. On this basis the simplest, and least effective, codes are the single-error-correcting codes due

Fig. 15. Chart showing the digits entering into each parity check for the Hamming code. This chart can be extended to any number of digits by following the same pattern. M = message digit. C = check digit.

to Hamming.[12] In such codes it is easily shown that the number of message digits m must be related to the total number of digits in each group n by the inequality

$$2^m \leq 2^n/(n+1) \qquad (150)$$

In general, the largest value of m satisfying (150) will be used. The remaining $m - n$ digits (usually referred to as check digits) are selected by performing a sequence of *parity checks* upon the message digits.

A parity check is performed by counting the number of 1's, say, in a selected group of digits and noting whether this number is even or odd. Such parity checks are used both to set the check digits in the transmitted code groups and to determine the location of errors in received code groups. For the sake of definiteness, only even parity checks will be used in this discussion.

In the Hamming code the check digits are usually, but not necessarily, located at position numbers that are powers of 2 (i.e., 1, 2, 4, 8, etc.). The state of each check digit is determined at the transmitter by making a parity check over the digit in question and certain succeeding message digits as indicated in Fig. 15. In each case the check digit is chosen to make the parity check even.

As an illustration of the use of Fig. 15 suppose it is desired to construct a Hamming code in which each code group contains exactly 4 message digits ($m = 4$). From (150) or from simply counting message digits along the bottom line of Fig. 15, it is determined that each code group must have a total of 7 digits and, hence, 3 check digits. Next assume that a particular piece of data calls for the message digits to be 1 0 1 1. The determination of the corresponding check digits may be described by the following tabulation, in which it may be noted that each parity check row involves only those positions blacked out in Fig. 15 and that check digit which produces an even number of 1's in that row.

Digit position	1	2	3	4	5	6	7
Message	...	...	1	...	0	1	1
First parity check	0	...	1	...	0	...	1
Second parity check	...	1	1	...	...	1	1
Third parity check	...	...	...	0	0	1	1
Transmitted sequence	0	1	1	0	0	1	1

At the receiving point exactly the same parity checks are performed, and the outcome of each parity check is used to form one digit of a *checking number*. Specifically, if the first parity check succeeds (i.e., if there are an even number of received 1's in the digits being checked), a 0 is entered in the right-hand position of the checking number; otherwise a 1 is entered here. The second parity check is used to establish the state of the next digit of the checking number, and so on until all parity checks have been made. The checking number thus formed indicates the *position* of any single error that might have occurred in transmission. Since the position of an error is known, it may be corrected. If no transmission errors occurred the checking number will have 0 in every position.

As an illustration of the correction procedure, assume that the specific sequence obtained above is received with an error in position 5. The tabulation below indicates the parity checks and the formation of the checking number. It will be noted that this checking number is the binary representation for 5, and hence the error is located.

Digit position	1	2	3	4	5	6	7	Checking number
Received sequence	0	1	1	0	1	1	1	
First parity check	0	...	1	...	1	...	1	1
Second parity check	...	1	1	...	...	1	1	0
Third parity check	...	...	...	0	1	1	1	1

Therefore, the checking number = 1 0 1 = 5.

While the Hamming code will correct any single error which occurs in a given code group it will not, in general, make a proper correction if more than one error occurs. It would be desirable to have similar explicit procedures for constructing multiple-error correcting codes, but such procedures are not known.

If error-correcting codes more effective than the Hamming code are desired it is possible to use the *group alphabets* devised by Slepian.[13] These codes are multiple-error-correcting in the sense that they maximize the probability of correct decoding. Thus a given group alphabet may correct all single errors and a large fraction of the double errors and triple errors but not all. Furthermore, this class of codes is sufficiently general to include many other methods, including the Hamming code and the Reed-Muller code, as special cases.

Procedures are available for constructing group alphabets of any size. However, there is no method of determining beforehand which of the many possible alphabets of a given size is best in the sense of providing the greatest probability of correct decoding. Hence the establishment of "best" group alphabets is usually carried out by searching among the most likely possibilities with the aid of a large-scale digital computer. By this means certain specific results have been obtained.

Figure 16 shows the construction of a "best" group alphabet containing 10 binary digits. The first five of the digits are determined by the information being transmitted while the last five are determined by the parity checks indicated in the figure. Similar parity checks are performed at the receiving end to determine what corrections must be made. In this particular case, all single errors will be corrected and 21 of the

Fig. 16. Chart showing the digits entering into each parity check for a 10-digit "best" group alphabet. M = message digit. C = check digit.

45 possible double errors will be corrected. The construction and characteristics of numerous other "best" group alphabets are tabulated in Ref. 13 and 14.

MODULATION AND MULTIPLEXING

The term "modulation," as used in this section, refers to a nonlinear operation in which message data are used to modify some characteristic of another signal known as the carrier. Although the carrier might be of almost any form, the only cases which will be considered are those in which it is either a sinusoid or a periodic sequence of pulses. In the latter case, the operation is frequently referred to as coding, but this terminology will not be employed here.

The term "multiplexing" refers to a sharing operation which makes it possible for a given telemetry system to transmit data from many different sensors simultaneously. The only multiplexing methods that are used to any extent are *frequency division* and *time division*, and the present discussion will be limited to these two.

There are two basic reasons for using modulation in a telemetry system. In the first place, it is an inherent part of multiplexing, regardless of the type of multiplexing being used, and would be required even in a direct-wire-transmission system. Secondly, many telemetry systems involve radio links in which modulation is required to place the transmission in that portion of the radio-frequency spectrum which is specified for either technical or legal reasons.

There are four items which are of major interest in connection with any specified combination of modulation and multiplexing methods. These are

1. The frequency spectrum occupied by the telemetry transmission
2. The signal-to-noise ratio achieved at the system output as a function of transmitted power
3. The improvement threshold above which the system must operate
4. The efficiency with which the system utilizes the information capacity of the channel

These four items will be discussed in the present section, and comparisons among the various systems will be tabulated.

21 Modulation of Sinusoidal Carriers

A sinusoid is completely described by its maximum amplitude, frequency, and phase with respect to a prescribed reference. Modulation is accomplished by modifying one or more of these characteristics in accordance with the message being transmitted. The most common situation is that in which only one characteristic is modified, and the present discussion will be limited to these three cases.

Amplitude Modulation (AM)

The defining equation for amplitude modulation is[4]

$$s(t) = C[1 + m(t)] \cos(\omega_c t + \phi_c) \qquad (151)$$

in which $s(t)$ = modulated signal
$m(t)$ = normalized message $[|m(t)| \leq 1]$
C = unmodulated carrier amplitude
ω_c = carrier angular frequency
ϕ_c = carrier phase

The carrier is fully modulated when the maximum value of $|m(t)|$ is equal to 1, and distortion occurs when this condition is exceeded.

The average unmodulated carrier power (on a 1-ohm basis) is

$$P_c = \tfrac{1}{2} C^2 \qquad (152)$$

while the average signal power is

$$P_s = (C^2/2)(1 + 2\bar{m} + \overline{m^2}) \qquad (153)$$

where $\bar{m}$ = mean value of message
$\overline{m^2}$ = mean-square value of message

It will be noted that the signal power increases with the degree of modulation and is bounded by

$$P_c \leq P_s \leq 4 P_c \qquad (154)$$

If the message $m(t)$ is sinusoidal with zero mean, the signal power is bounded by

$$P_c \leq P_s \leq \tfrac{3}{2} P_c \qquad (155)$$

and this upper bound is the one usually assumed.

Frequency Modulation (FM)

The defining equation for frequency modulation is[4]

$$s(t) = C \cos [\omega_c t + \Delta\omega \int m(t)\, dt + \phi_c] \qquad (156)$$

in which $s(t)$ = modulated signal
$m(t)$ = normalized message $[|m(t)| \leq 1]$
C = carrier amplitude
$\Delta\omega$ = maximum angular frequency deviation
ϕ_c = carrier phase

The instantaneous frequency deviation is $\Delta\omega m(t)$, and full modulation is said to occur when the maximum value of $|m(t)|$ is equal to 1.

In frequency modulation the average power is not changed when modulation occurs. Hence the average unmodulated carrier power and the average signal power are both given by

$$P_c = P_s = \tfrac{1}{2} C^2 \qquad (157)$$

Phase Modulation (PM)

The defining equation for phase modulation is[4]

$$s(t) = C \cos [\omega_c t + \phi_m m(t) + \phi_c] \tag{158}$$

in which $s(t)$ = modulated signal
$m(t)$ = normalized message $[|m(t)| \leq 1]$
C = carrier amplitude
ϕ_m = maximum phase deviation
ϕ_c = carrier phase

The instantaneous phase deviation is $\phi_m m(t)$, and full modulation occurs when the maximum value of $|m(t)|$ is equal to 1. Since a change in phase is equivalent to a change in frequency, it is possible to interpret phase modulation in terms of frequency modulation in which the message is $dm(t)/dt$ and the instantaneous frequency deviation is $\phi_m [dm(t)/dt]$.

Phase modulation is generally not suitable for situations in which $m(t)$ has a non-zero mean value which must be preserved in transmission. Such a component introduces a constant phase deviation which cannot be distinguished from ϕ_c or from phase shifts caused by the transmission system.

The average power is also constant in phase modulation and hence is

$$P_c = P_s = \tfrac{1}{2} C^2 \tag{159}$$

22 Pulse Modulation

The "carrier" in pulse modulation consists of a periodic sequence of identical pulses. These pulses are generally assumed to be rectangular, but they may have almost any shape and, in actuality, their shape is usually determined by the bandpass characteristics of the systems. For a single channel, the pulse rate corresponds to the sampling rate which is required to describe a message adequately. Each sample of the message is then used to modify some characteristic of one pulse.

Although there are many pulse characteristics which might be modified, the most common ones are the amplitude, the duration, and the position. In general, it is also necessary to provide for synchronization of a pulse-modulation channel, but this aspect will not be discussed here since it is not pertinent to the items being considered.

Pulse-amplitude Modulation (PAM)

The unmodulated pulse sequence $U(t)$ can be represented mathematically by a Fourier-series expansion.[4] Thus

$$U(t) = \sum_{n=-\infty}^{\infty} A_n \epsilon^{jn\omega_s t} \tag{160}$$

where

$$A_n = 1/T \int_{-\frac{T}{2}}^{\frac{T}{2}} u(t) \epsilon^{-jn\omega_s t} \, dt \tag{161}$$

$$\omega_s = 2\pi/T$$

in which $u(t)$ describes the shape of each pulse and T is the repetition period for the pulses. If the message to be transmitted $m(t)$ is band-limited to a range of 0 to W cps, then according to the sampling theorem (Sec. 3) it is necessary that $T \leq 1/2W$. The pulse-amplitude modulated signal may be expressed as

$$s(t) = m(t) U(t) = \sum_{n=-\infty}^{\infty} m(t) A_n \epsilon^{jn\omega_s t} \tag{162}$$

when double-polarity pulses are desired, or as

$$s(t) = [1 + m(t)]U(t) = \sum_{n=-\infty}^{\infty} [1 + m(t)]A_n \epsilon^{jn\omega_s t} \qquad (163)$$

for single-polarity pulses. In the latter case, full modulation occurs when the maximum value of $|m(t)|$ is equal to 1.

The average power of the unmodulated pulse sequence can be expressed as

$$P_c = (1/T)E^2 \qquad (164)$$

where E^2 is the energy contained in each pulse. Hence

$$E^2 = \int_{-\frac{T}{2}}^{\frac{T}{2}} u^2(t)\, dt \qquad (165)$$

The average signal power for double-polarity pulses is given by

$$P_s = (\overline{m^2}/T)E^2 \qquad (166)$$

where $\overline{m^2}$ is the mean-square value of the message, and by

$$P_s = (E^2/T)(1 + 2\bar{m} + \overline{m^2}) \qquad (167)$$

for single-polarity pulses, where $\bar{m}$ is the mean value of the message.

Pulse-duration Modulation (PDM)

An alternative representation for a periodic sequence of pulses is the direct summation of pulses. Thus

$$U(t) = \sum_{k=-\infty}^{\infty} u(t - kT) \qquad (168)$$

The modulation process alters $u(t)$ by shifting either the leading edge or the trailing edge of the pulse, or both. The amount of shift is proportional to the magnitude of the message sample observed at the time of the pulse. This may be indicated formally by writing

$$s(t) = \sum_{k=-\infty}^{\infty} u[(t - kT), m(kT)] \qquad (169)$$

A more precise representation requires a specification of the exact pulse shape.

As a typical illustration of the above representation, consider rectangular pulses with fixed leading edges and modulated trailing edges. An unmodulated pulse of width a may be represented as the difference of two step functions. Thus

$$u(t) = h[u_0(t) - u_0(t - a)] \qquad (170)$$

where
$$\begin{aligned} u_0(t) &= 0 \quad t < 0 \\ &= 1 \quad t \geq 0 \\ h &= \text{pulse height} \\ 0 &< a \leq \tfrac{1}{2}T \end{aligned}$$

The modulated signal may now be written as

$$s(t) = \sum_{k=-\infty}^{\infty} hu_0(t - kT) - hu_0\{t - [kT + a + am(kT)]\} \qquad (171)$$

Full modulation occurs when the maximum value of $|m(kT)|$ is equal to 1.

MODULATION AND MULTIPLEXING

The average power of the unmodulated pulse sequence is again

$$P_c = (1/T)E^2 \tag{172}$$

where
$$E^2 = \int_{-\frac{T}{2}}^{\frac{T}{2}} u^2(t)\, dt$$

For the rectangular pulses considered above, this becomes

$$P_c = ah^2/T \tag{173}$$

The average signal power is given by

$$P_s = (1 + \bar{m})(E^2/T) = (1 + \bar{m})(ah^2/T) \tag{174}$$

where $\bar{m}$ is the mean value of the message. It may be noted that the average signal power does not change with modulation when $\bar{m} = 0$.

Pulse-position Modulation (PPM)

Pulse-position modulation is a modified form of pulse-duration modulation in which the variable edge is replaced by a short pulse, thereby conserving signal power. The modulated signal may be represented by

$$s(t) = \sum_{k=-\infty}^{\infty} u\{t - [k + \tfrac{1}{2}m(kT)]T\} \tag{175}$$

Full modulation occurs when the maximum value of $|m(kT)|$ is equal to 1.

Since neither the pulse width nor the pulse height is changed as a result of modulation, the average power of the modulated signal is the same as the average power of the unmodulated pulse sequence. Hence

$$P_c = P_s = (1/T)E^2 = 1/T \int_{-\frac{T}{2}}^{\frac{T}{2}} u^2(t)\, dt \tag{176}$$

For rectangular pulses of height h and width a, the average power is

$$P_c = P_s = ah^2/T \tag{177}$$

23 Pulse-code Modulation (PCM)

The previously discussed pulse-modulation methods (PAM, PDM, PPM) were essentially analog in nature since modulation produced a continuous variation in some characteristic of the basic pulses. Pulse-code modulation, on the other hand, is essentially digital since only a finite number of message levels can be transmitted and these are represented by an appropriate binary code.

The normalized message function $m(t)$ is sampled every T sec and each sample is quantized into one of 2^n possible levels. The transmitted signal consists of a group of n binary digits (pulse or no pulse) which indicates the proper level. In general it is not possible to write a mathematical expression which uniquely relates the transmitted signal to the message function.

Likewise it is usually not possible to relate the average signal power to the mean-square value of the message function. However, it is often reasonable to assume that pulses are present one-half of the time so that a good approximation to the average signal power is given by

$$P_s = \tfrac{1}{2}E^2 = \tfrac{1}{2} \int_{-\frac{T}{2n}}^{\frac{T}{2n}} u^2(t)\, dt \tag{178}$$

SAMPLING AND HANDLING OF INFORMATION

where $u(t)$ is the shape of each pulse in the code group. If the pulses are rectangular with a height of h, and if there is no space between pulses, the average power is

$$P_s = \tfrac{1}{2} h^2 \tag{179}$$

The average power in the absence of modulation depends upon the particular code group assigned to the zero level. In most cases this is zero.

Although many types of binary codes are possible, the most common are the normal (or scalar) code and the reflected (or Gray) code. In the normal code each digit represents a power of 2. Thus, for example,

$$\begin{aligned} 1\ 0\ 1\ 1 &= 1 \times 2^3 + 0 \times 2^2 + 1 \times 2^1 + 1 \times 2^0 \\ &= 8 + 0 + 2 + 1 = 11 \end{aligned}$$

This code has the disadvantage that a change of "one" in the decimal equivalent may change several digits in the code group. In the reflected code, which does not have this disadvantage, each digit represents one less than a power of 2 and may be positive or negative. For example,

$$\begin{aligned} 1\ 1\ 1\ 0 &= 1(2^4 - 1) - 1(2^3 - 1) + 1(2^2 - 1) - 0(2^1 - 1) \\ &= 15 - 7 + 3 - 0 = 11 \end{aligned}$$

The pattern for constructing either code is clearly shown in the following tabulation:

Decimal	Normal binary	Reflected binary
0	0 0 0 0	0 0 0 0
1	0 0 0 1	0 0 0 1
2	0 0 1 0	0 0 1 1
3	0 0 1 1	0 0 1 0
4	0 1 0 0	0 1 1 0
5	0 1 0 1	0 1 1 1
6	0 1 1 0	0 1 0 1
7	0 1 1 1	0 1 0 0
8	1 0 0 0	1 1 0 0
9	1 0 0 1	1 1 0 1
10	1 0 1 0	1 1 1 1
11	1 0 1 1	1 1 1 0
12	1 1 0 0	1 0 1 0
13	1 1 0 1	1 0 1 1
14	1 1 1 0	1 0 0 1
15	1 1 1 1	1 0 0 0

24 Multiplexing

The primary purpose of multiplexing is to make it possible to transmit a number of different messages simultaneously with a given system. From a mathematical standpoint this purpose is accomplished if the messages can be converted by modulation into a set of orthogonal signals as defined in Sec. 2. The most obvious ways of making signals orthogonal are to separate them in frequency or to separate them in time, and these will be the only cases considered. It should be emphasized, however, that many other possibilities exist and there is no basic reason why orthogonal signals cannot occupy the same frequency region at the same time.[15] The use of frequency-division and time-division methods almost exclusively is motivated primarily by equipment considerations rather than theoretical considerations.

Frequency Division

Frequency-division multiplexing is accomplished by modulating sinusoidal carriers of different frequencies with the various messages. In a radio-telemetry system these subcarriers are in turn used to modulate another carrier at a higher frequency. Since each stage of the modulation can be done by any of the three methods previously mentioned, there are a total of nine ways of accomplishing the complete modulation. However, the use of PM with messages having a d-c component is usually undesirable, so that only six methods remain for consideration. These will be designated by stating first the modulation method for the subcarriers and then the method for the main carrier. Thus AM-FM implies amplitude modulation of the messages onto the subcarriers and frequency modulation of the subcarriers onto the main carrier.

An important consideration in any multiplexing system is that of "crosstalk" from one channel to another. In a frequency-division system the crosstalk results from modulation products from one message having frequency components falling within the frequency region assigned to other messages. The amount of crosstalk present in a given system depends upon nonlinearities in the system, bandwidths of the individual channels, and the choice of subcarrier frequencies. These items are discussed in the literature [16,17] and will not be considered further here because they are not pertinent to the inherent capabilities of frequency-division systems.

Time Division

Time-division multiplexing is accomplished by sampling the different messages at slightly different times and using these samples for one of the pulse-modulation methods previously discussed. In a radio-telemetry system the modulated pulses are then used to modulate a sinusoidal carrier. Thus there are 12 possible time-division multiplexing methods, and these will be designated by stating first the pulse-modulation method and then the sinusoidal-carrier-modulation method (i.e., PDM-FM, for example). Two of the possible methods (PPM-FM and PPM-PM) need not be considered, however, because the use of a continuous sinusoidal carrier defeats the fundamental advantage of PPM—that of the pulses being on only a small fraction of the time.

Crosstalk is also present in time-division systems and arises from pulse widening due to restricted bandwidth in the system or from inaccurate synchronizing. These problems are discussed in the literature[16] but are not pertinent here.

Composite and Hybrid Systems

Double multiplexing is frequently desirable in systems having a large number of channels, particularly if the data are slowly varying. Although there are a large number of possible combinations, only a few have found practical application. Usually time division is used for the initial message modulations and these are in turn applied to a frequency-division system (PAM-FM-FM, for example).

It is also possible to consider hybrid systems in which the subcarriers are modulated by different methods. For example, some subcarriers may be amplitude-modulated while others are frequency-modulated. In addition, some subcarriers may involve double multiplexing while others do not. There are obviously a tremendous number of variations which might be used.

Although it is out of the question to tabulate the characteristics of all possible composite, hybrid, and composite-hybrid systems, the general concepts to be outlined are applicable in any case and can be used as a general guide.

25 Frequency Spectra of Modulated Signals

A knowledge of the frequency spectrum produced by any method of modulation is essential for proper design of the system and for calculation of signal-to-noise ratios. Although there are many satisfactory ways of expressing information about the frequency spectrum, the spectral-density method will be used here because it is applicable to either periodic or random time functions.

When spectral density is defined as in Eq. (58) its value indicates the average power per unit bandwidth (in watts/cps) as a function of angular frequency ω. For purely random time functions with zero mean, the spectral density is bounded, continuous, real, positive, and even with respect to ω. In the case of periodic time functions the spectral density contains δ functions at the appropriate discrete frequencies.

The present discussion summarizes the spectral densities that result from single modulation methods with both sinusoidal and pulse carriers. Additional results which apply to multiplex systems are presented in the following section. Whenever possible the spectral density for the modulated signal is given for completely arbitrary message spectral densities. When this generality is not possible, typical specialized results are given.

In designating the spectral density for a message function $m(t)$, from any general information source, it is convenient to separate the discrete component which represents the average value from the remainder of the spectral density. Thus the total message spectral density $\Phi_{mm}(\omega)$ will be of the form

$$\Phi_{mm}(\omega) = 2\pi(\bar{m})^2\delta(\omega) + \Phi_m(\omega) \tag{180}$$

where $\bar{m}$ is the average value of the message. The mean-square value of the message is related to the spectral density by

$$\overline{m^2} = 1/2\pi \int_{-\infty}^{\infty} \Phi_{mm}(\omega)\, d\omega \tag{181}$$

The spectral density of the modulated signal is designated as $\Phi_{ss}(\omega)$ and is tabulated in Table 1 for the sinusoidal-carrier-modulation methods. In all cases the carrier has an unmodulated peak amplitude of C and an unmodulated angular frequency of ω_c. The following paragraphs contain special notation and supplementary information for the individual methods.

Table 1. Spectral Density of Modulated Sinusoids

Modulation	Message	Signal spectral density $\Phi_{ss}(\omega)$	
AM	General	$(\pi/2)C^2(1+\bar{m})^2[\delta(\omega-\omega_c)+\delta(\omega+\omega_c)] + (C^2/4)[\Phi_m(\omega-\omega_c)+\Phi_m(\omega+\omega_c)]$	
FM	Sinusoid	$\dfrac{\pi}{2}C^2 \sum_{n=-\infty}^{\infty} J_n^2(D_m)[\delta(\omega-\omega_c-n\omega_m)+\delta(\omega+\omega_c+n\omega_m)]$	
	Square wave	$\dfrac{\pi}{2}C^2 \sum_{n=-\infty}^{\infty} \left(\dfrac{D_m}{D_m+n}\right)^2 \left[\dfrac{\sin(D_m-n)(\pi/2)}{(D_m-n)(\pi/2)}\right]^2 [\delta(\omega-\omega_c-n\omega_m)$ $+ \delta(\omega+\omega_c+n\omega_m)]$	
	B, W, G $D_e \ll 1$	$\dfrac{C^2}{2}\left[\dfrac{\pi^2 D_e^2 W}{(\omega-\omega_c)^2+(\pi^2 D_e^2 W)^2} + \dfrac{\pi^2 D_e^2 W}{(\omega+\omega_c)^2+(\pi^2 D_e^2 W)^2}\right]$	(approx)
	B, W, G $D_e \gg 1$	$\dfrac{C^2}{4\sqrt{2\pi}\,(2\pi W D_e)}[\epsilon^{-\frac{(\omega-\omega_c)^2}{2(2\pi W D_e)^2}} + \epsilon^{-\frac{(\omega+\omega_c)^2}{2(2\pi W D_e)^2}}]$	(approx)
	B, S, G $D_e \ll 1$	$\begin{cases}\dfrac{\pi}{2}C^2\epsilon^{-3D_e^2}\left[\delta(\omega-\omega_c)+\delta(\omega+\omega_c)+\dfrac{3D_e^2}{4\pi W}\right],\text{ for }|\omega-\omega_c|<2\pi W \\ 0,\text{ for }|\omega-\omega_c|>2\pi W \end{cases}$	(approx)
	B, S, G $D_e \gg 1$	$\dfrac{C^2}{4\sqrt{2\pi}\,(2\pi W D_e)}[\epsilon^{-\frac{(\omega-\omega_c)^2}{2(2\pi W D_e)^2}} + \epsilon^{-\frac{(\omega+\omega_c)^2}{2(2\pi W D_e)^2}}]$	(approx)
PM	....	See discussion	

Amplitude Modulation (AM)

The principal effect of amplitude modulation is to translate the message spectral density from a region around zero frequency to a corresponding region around the carrier frequency without changing its shape.[4] This assumes that overmodulation does not occur.

Frequency Modulation (FM)

Simple general relationships connecting signal spectral density and message spectral density do not exist for frequency modulation. The six special cases listed in Table 1 were selected as being particularly pertinent to the telemetry field. The notation for each case is listed below:

Sinusoid[4]: $\quad \Phi_{mm}(\omega) = (\pi/2)[\delta(\omega - \omega_m) + \delta(\omega + \omega_m)]$

$D_m = \Delta\omega/\omega_m$ (maximum deviation ratio)
$\Delta\omega$ = maximum angular-frequency deviation
ω_m = angular frequency of message
$J_n(D_m)$ = Bessel function of first kind, order n

Square wave:[4]

$$\Phi_{mm}(\omega) = 2\pi \sum_{n=1}^{\infty} \left[\frac{\sin(n\pi/2)}{n\pi/2}\right]^2 [\delta(\omega - n\omega_m) + \delta(\omega + n\omega_m)]$$

$D_m = \Delta\omega/\omega_m$ (maximum-deviation ratio)
$\Delta\omega$ = maximum angular-frequency deviation
ω_m = fundamental square-wave frequency

Band-limited, white, Gaussian (B, W, G):[18]

$\Phi_{mm}(\omega) = 1/2\pi W \quad |\omega| \leq 2\pi W$
$\qquad\qquad\;\, = 0 \qquad\quad\; |\omega| > 2\pi W$

W = message bandwidth, cps
$D_e^2 = \overline{(\Delta\omega)^2}/(2\pi W)^2$ (mean-square deviation ratio)
$\overline{(\Delta\omega)^2}$ = mean-square angular-frequency deviation

Band-limited, square-law, Gaussian (B, S, G):[18]

$\Phi_{mm}(\omega) = (\omega/2\pi W)^2 \quad |\omega| \leq 2\pi W$
$\qquad\qquad\;\, = 0 \qquad\qquad\;\; |\omega| > 2\pi W$

W = message bandwidth, cps
$D_e^2 = \overline{(\Delta\omega)^2}/(2\pi W)^2$ (mean-square deviation ratio)
$\overline{(\Delta\omega)^2}$ = mean-square angular-frequency deviation

It will be shown later that the square-law spectrum is pertinent to multiplex systems.

The similarity of the signal spectral densities for the white message spectrum and for the square-law message spectrum (when $D_e \gg 1$) implies that the message amplitude probability density function is more significant than the message spectrum. This conclusion can be generalized in the *principle of adiabatic frequency sweeps*,[19] which states that, in frequency modulation of a carrier by a stationary random process, the resulting signal spectral density is proportional to the first-order probability density of the modulating process for sufficiently slow frequency deviations.

Phase Modulation (PM)

The spectral density for phase-modulated carriers can be obtained by considering the equivalent frequency modulation.[18] As noted in Sec. 21, the equivalent frequency-modulating message is the derivative of the phase-modulating message.

Hence the results tabulated in Table 1 for frequency modulation can be used for phase modulation with messages whose spectral densities are

$$\Phi_{mm}(\omega)_{PM} = (1/\omega^2)\Phi_{mm}(\omega)_{FM} \qquad (182)$$

In particular it may be noted that the band-limited square-law message spectrum in the FM case becomes the band-limited white message spectrum for PM.

The spectral densities for various pulse-modulation methods are shown in Table 2.[20] In all cases the unmodulated pulse-repetition period is T sec and each unmodulated pulse is described by the time function $u(t)$. In determining these spectral densities the assumption has been made that message samples are mutually independent, and this is equivalent to specifying that the message spectral density is band-limited and white. Under these conditions the resulting signal spectral densities depend only upon the pulse shape and the message amplitude probability density function. The pulse shape enters into the spectral density in terms of the single pulse Fourier transform by

$$F_u(\omega) = \int_{-\infty}^{\infty} u(t)\epsilon^{-j\omega t}\, dt \qquad (183)$$

The message amplitude probability density function $p(m)$ is used to define some expected values of the Fourier transform, as indicated below.

Table 2. Spectral Density of Modulated Pulses

Modulation	Signal spectral density $\Phi_{ss}(\omega)$
PAM	$\dfrac{1}{T}|F_u(\omega)|^2 \left[\overline{m^2} - (\overline{m})^2 + \dfrac{2\pi(\overline{m})^2}{T} \sum_{n=-\infty}^{\infty} \delta\left(\omega - \dfrac{2\pi n}{T}\right)\right]$
PPM	$\dfrac{1}{T}|F_u(\omega)|^2 \left[1 - |M(\omega)|^2 + \dfrac{2\pi|M(\omega)|^2}{T} \sum_{n=-\infty}^{\infty} \delta\left(\omega - \dfrac{2\pi n}{T}\right)\right]$
PDM (symmetrical)	$\dfrac{1}{T}\left[\overline{F_u^2(\omega)} - (\overline{F_u(\omega)})^2 + \dfrac{2\pi}{T}(\overline{F_u(\omega)})^2 \sum_{n=-\infty}^{\infty} \delta\left(\omega - \dfrac{2\pi n}{T}\right)\right]$
PDM (single edge)	$\dfrac{1}{T}\left[\overline{|F_u(\omega)|^2} - |\overline{F_u(\omega)}|^2 + \dfrac{2\pi}{T}|\overline{F_u(\omega)}|^2 \sum_{n=-\infty}^{\infty} \delta\left(\omega - \dfrac{2\pi n}{T}\right)\right]$
PCM	$\dfrac{1}{4T}|F_u(\omega)|^2 \left[1 + \dfrac{2\pi}{T} \sum_{n=-\infty}^{\infty} \delta\left(\omega - \dfrac{2\pi n}{T}\right)\right]$

Pulse-amplitude Modulation (PAM)

The continuous part of the spectral density is determined by the pulse shape and the message variance. Discrete components of the spectral density depend upon the mean value of the message $\overline{m}$ in the case of double-polarity pulses. In the case of single-polarity pulses, $\overline{m}$ must be interpreted as the sum of the message mean value and the unmodulated pulse amplitude.

The message variance $\overline{m^2} - (\overline{m})^2$ may be associated either with the continuous message function $m(t)$ or with samples taken from this function. This is a consequence of the result stated in Eq. (9) in the discussion of sampling.

Pulse-position Modulation (PPM)

In this case the spectral density depends upon both the pulse shape and the message amplitude density function. The latter enters into the function $M(\omega)$, which is defined by

$$M(\omega) \triangleq \int_{-\infty}^{\infty} p(m)\epsilon^{j\omega k m}\, dm \qquad (184)$$

where k is a constant which relates the message sample amplitude to the shift in pulse position.

Pulse-duration Modulation (PDM)

In symmetrical pulse-duration modulation it is assumed that the center of each pulse remains fixed and that the width of the pulse is varied in accordance with the

Table 3. Fourier Transforms for Several Types of Pulses

SHAPE	GRAPH	TIME FUNCTION, $\mu(t)$	FOURIER TRANSFORM $F_\mu(\omega) = \int_{-\infty}^{\infty} \mu(t)\epsilon^{-j\omega t}\, dt$
RECTANGULAR		$h,\quad \|t\| \le \frac{a}{2}$ $0,\quad \|t\| > \frac{a}{2}$	$ha\, \dfrac{\sin\left(\frac{\omega a}{2}\right)}{\left(\frac{\omega a}{2}\right)}$
TRIANGULAR		$-\frac{h}{a}(t-a),\quad 0\le t\le a$ $\frac{h}{a}(t+a),\quad -a\le t < 0$ $0,\quad \|t\| > a$	$ha\, \dfrac{\sin^2\left(\frac{\omega a}{2}\right)}{\left(\frac{\omega a}{2}\right)^2}$
TRAPEZOIDAL		$\frac{h}{ba}\left[t+\frac{a}{2}(1+b)\right],\ -\frac{a}{2}(1+b)\le t \le -\frac{a}{2}(1-b)$ $h,\quad \|t\| < \frac{a}{2}(1-b)$ $-\frac{h}{ba}\left[t-\frac{a}{2}(1+b)\right],\ \frac{a}{2}(1-b)\le t\le \frac{a}{2}(1+b)$ $0,\quad \|t\| > \frac{a}{2}(1+b)$	$ha\, \dfrac{\sin\left(\frac{\omega a}{2}\right)}{\left(\frac{\omega a}{2}\right)} \cdot \dfrac{\sin\left(\frac{\omega b a}{2}\right)}{\left(\frac{\omega b a}{2}\right)}$
COSINE SQUARED		$h\cos^2\left(\frac{\pi t}{2a}\right),\quad -a\le t\le a$ $0,\quad \|t\| > a$	$\dfrac{ha}{2}\, \dfrac{\sin(\omega a)}{(\omega a)}\left[1+\dfrac{2\pi^2 \omega a}{\pi^2 - \omega^2 a^2}\right]$
GAUSSIAN		$h\epsilon^{-.693\left(\frac{2t}{a}\right)^2},\quad -\infty < t < \infty$	$\dfrac{ha}{2}\sqrt{\dfrac{\pi}{.693}}\, \epsilon^{-\frac{\omega^2 a^2}{16(.693)}}$
ERROR FUNCTION		$h\left[\operatorname{erf}\dfrac{\sqrt{\pi}}{b}\left(\dfrac{t}{a}+\dfrac{1}{2}\right) - \operatorname{erf}\dfrac{\sqrt{\pi}}{b}\left(\dfrac{t}{a}-\dfrac{1}{2}\right)\right],$ $-\infty < t < \infty$	$ha\, \dfrac{\sin\left(\frac{\omega a}{2}\right)}{\left(\frac{\omega a}{2}\right)}\, \epsilon^{-\frac{\omega^2 a^2 b^2}{4\pi}}$

message sample values. The Fourier transform of each pulse is therefore a function of the message sample value and may be designated as $F_u(\omega,m)$. The mean and mean-square values of the Fourier transform enter into the signal spectral density and may be obtained from

$$\overline{F_u(\omega)} = \int_{-\infty}^{\infty} F_u(\omega,m) p(m)\, dm \qquad (185)$$

$$\overline{F_u{}^2(\omega)} = \int_{-\infty}^{\infty} F_u{}^2(\omega,m) p(m)\, dm \qquad (186)$$

In single-edge pulse-duration modulation it is assumed that one edge of each pulse remains fixed while the other edge is varied in accordance with the message sample values. Since the pulse is now unsymmetrical the corresponding Fourier transform is complex. Hence the mean-square value now becomes

$$\overline{|F_u(\omega)|^2} = \int_{-\infty}^{\infty} |F_u(\omega,m)|^2 p(m)\, dm \qquad (187)$$

Pulse-code Modulation (PCM)

The spectral density shown in Table 2 assumes single-polarity pulses which are either present or absent with equal probability. If double-polarity pulses are used (with equal probability for each polarity), then the continuous part of the spectral density remains unchanged but all the discrete components vanish.

Table 3 presents some of the more common pulse shapes and their corresponding Fourier transforms.

26 Frequency Spectra of Multiplexed Signals

A detailed specification of the frequency spectra of multiplexed signals is difficult to accomplish because it involves many parameters. However, the results presented in Sec. 25 can be used to make some rather general conclusions. The two situations of major interest are those in which the final modulation is either AM or FM. Since the PM case is related to the FM case in the manner discussed, it will not be considered separately.

Amplitude-modulation Systems

In frequency-division multiplex systems the modulating signal for the main carrier consists of a set of subcarriers which are modulated by AM or FM. In either case the spectral densities associated with the subcarriers do not overlap appreciably so that the spectral density of the modulating signal is simply the sum of the spectral densities of the subcarriers. The final modulation translates this spectral density to the frequency of the main carrier. This result is valid for both AM-AM and FM-AM systems.

In time-division multiplex systems the modulating signal for the main carrier is a set of modulated pulses. The spectral density for such a set of modulated pulses depends upon the shape and duration of the pulses and not upon the number of channels being sampled (except for the manner in which this number affects pulse duration). Hence, in PAM-AM, PDM-AM, PPM-AM, and PCM-AM systems the final spectral density consists of the pulse spectra translated to the frequency of the main carrier.

Frequency-modulation Systems

When the final modulation is FM the spectral density is not so easily related to that of channels. Fortunately, however, in a great many practical situations the modulating signal for the main carrier has a nearly Gaussian amplitude probability density, and in these cases the resulting final spectral density is Gaussian-shaped, for large deviation ratios, regardless of the channel spectral densities. This convenient result is a consequence of the *central-limit theorem*,[6] which states that the sum of a large number of independent random variables almost always has a probability density function which is nearly Gaussian. Empirical results indicate that even as few as five or six random variables are sufficient for a good approximation.

The above result can be applied directly to AM-FM, PAM-FM, and FM-FM systems. In the first two cases the modulating signal for the main carrier can be considered as being band-limited, white, and Gaussian. In the FM-FM case, because of the tapering usually employed on the subcarrier amplitudes, the sum of all channels is often more nearly band-limited, square-law, and Gaussian. In either case, however, the spectral densities shown in Table 1 apply.

In the PCM-FM case the final modulating signal can again be considered as band-limited, white, and Gaussian if the PCM pulses are passed through a band-limiting data filter before the final modulation. The Gaussian assumption does not follow from the central-limit theorem in this case but has been justified by computing the lower-order moments of the density function at the filter output.[21]

The PDM-FM case is considerably more complicated because the final modulating signal is neither Gaussian nor band-limited. A special case assuming rectangular pulses with single-edge modulation has been worked out for large deviation ratios.[21] The channel messages (and, hence, the pulse duration) were assumed Gaussian. The result is

$$\Phi_{ss}(\omega)_{PDM-FM} \cong \frac{C^2}{16W} \left\{ \left[\frac{\sin(\omega_1/4W)}{\omega_1/4W} \right]^2 + \left[\frac{\sin(\omega_2/4W)}{\omega_2/4W} \right]^2 + \left[\frac{\sin(\omega_3/4W)}{\omega_3/4W} \right]^2 + \left[\frac{\sin(\omega_4/4W)}{\omega_4/4W} \right]^2 \right\} \quad (188)$$

where W = message bandwidth (all channels), cps
$\omega_1 = \omega - (\omega_c + 2\pi DW)$
$\omega_2 = \omega - (\omega_c - 2\pi DW)$
$\omega_3 = \omega + (\omega_c + 2\pi DW)$
$\omega_4 = \omega + (\omega_c - 2\pi DW)$
$D = \Delta\omega/2\pi W \gg 1$ (deviation ratio)
$\Delta\omega$ = frequency deviation on either side of ω_c

It is of interest to note that this spectral density drops off approximately as $|\omega - \omega_c|^{-2}$ instead of exponentially as in the band-limited Gaussian cases.

PPM-FM is not useful because the presence of a continuous carrier destroys the basic power-saving advantage of using short pulses.

The concepts presented above can also be used in considering double-multiplex and hybrid systems. The number of possibilities in such systems is too great to make any systematic discussion of spectral density feasible.

27 Bandwidth Considerations

In comparing telemetry systems it is convenient to be able to assign a single number which describes the range of frequencies required by the system. When spectral densities extend to infinity, as is usually the case, this range of frequencies (or bandwidth) must be selected on an arbitrary basis, and there are many ways in which this might be done. Furthermore, it is usually necessary to specify several different bandwidths in a given system, and these may be defined in different ways. The discussion below considers those bandwidths which are pertinent to a single modulated carrier. The extension of these definitions to multiplex systems is straightforward.

Message Bandwidth (W cps)

The message bandwidth is that portion of the message spectrum which must be preserved to retain a given accuracy. In the case of sampled messages, the message bandwidth is just one-half the sampling rate. In continuous systems it will be assumed that the message is band-limited (by the use of a data filter, if necessary) so that W is well-defined.

Signal Bandwidth (B cps)

The signal bandwidth is that portion of the signal spectrum which must be preserved in order to recover the message with a given accuracy. The relationship between message bandwidth and signal bandwidth depends upon the type of modulation used and upon the criterion employed.

In the case of AM, a band-limited message implies a band-limited signal so that

$$B = 2W \qquad (189)$$

when double-sideband modulation is employed.

The situation is more complicated in the FM case because of the influence of the criterion employed. Many different criteria based on such things as sideband

Table 4. Signal Bandwidth for Various Modulation Methods

Modulation	Message	Signal bandwidth B, cps	Notes
AM	Band-limited, W	$2W$	
FM	Sinusoid, f_m Band-limited, W Gaussian	$2(D_m f_m + f_m)$ $2(3.3 D_e W + W)$	D_m = maximum deviation ratio D_e = rms deviation ratio
PAM, PDM, PPM, PCM	Sampled	$\dfrac{1}{a}$	a = effective pulse length (see Table 3)
PAM-AM, PDM-AM, PPM-AM, PCM-AM	Sampled	$\dfrac{2}{a}$	
PAM-FM, PCM-FM	Sampled	$2(3.3 D_e W + W)$	Sampling rate = $2W$ (n_c channels) $= \dfrac{2W}{n_c}$ (1 channel)
PDM-FM	Sampled	$2(\Delta f_p + W)$	Δf_p = peak frequency deviation

amplitude, signal energy, and peak frequency deviation have appeared in the literature. One which is particularly convenient, and appears to yield reasonable answers, is[22]

$$B = 2(\Delta f_p + W) \qquad (190)$$

where $2\Delta f_p$ is the peak-to-peak frequency deviation. When the peak amplitude of the message is well defined (as it is for sine waves, square waves, pulses, etc.), the peak frequency deviation is correspondingly well defined. When the message is random with a Gaussian amplitude distribution, the peak frequency deviation must be arbitrarily defined in terms of the rms frequency deviation. Thus

$$\Delta f_p = K D_e W \qquad (191)$$

where D_e is the rms deviation ratio used in Table 1. Values of K ranging from 2 to 4 have been used in the literature as a result of employing different criteria. The value $K = 3.3$, which will be used in this discussion, yields the approximate frequency at which signal spectral density drops to 0.001 of its maximum value.

In all the pulse-modulation methods, the bandwidth is determined primarily by the pulse length. Hence it is appropriate to define

$$B = 1/a \qquad (192)$$

where a is the effective pulse length as shown in Table 3. In all cases except the Gaussian pulse this corresponds to the frequency at which the spectral density goes through its first zero.

The above results are consolidated in Table 4.

Receiver Bandwidth (B_R cps)

The receiver bandwidth B_R must be greater than the signal bandwidth because of imperfect selectivity and nonlinear phase response of practical filters. In the absence of noise this bandwidth could always be made large enough to make signal distortion negligible. However, the presence of noise may place an upper limit on B_R, and for this reason it is convenient to consider B_R as the equivalent noise bandwidth of the receiver. [See Eq. (98) for a definition of equivalent noise bandwidth.]

Bandwidth Occupancy (B_o cps)

Bandwidth occupancy may be defined as that portion of the frequency spectrum which is unavailable for other use. This bandwidth is usually greater than receiver bandwidth and is determined by such factors as adjacent-channel signal power and spectral density, filter selectivity, permitted crosstalk level, modulation methods, and service range. Because of the many factors involved, it is difficult to include consideration of this bandwidth in a theoretical discussion of system capabilities. However, a fairly complete discussion of this problem is available in Refs. 22 and 23.

28 Signal-to-Noise Ratio at Threshold

The accuracy of any elemetry system is limited by the signal-to-noise ratio that can be attained at the final system output. However, a specification of this signal-to-noise ratio only does not give a complete picture of the many factors which enter into proper system design.

In the first place, it is necessary to know how the output signal-to-noise ratio is related to the signal-to-noise ratio at the input to the telemetry receiver. It is also necessary to know how the input signal-to-noise ratio depends upon transmitted power, but this item is influenced by so many factors which are not under the control of the system designer that it does not fall within the scope of the present discussion (refer to Chap. 4, Radio Telemetry). An item which is pertinent, however, is the input signal-to-noise ratio required to attain receiver threshold.

A convenient way of comparing the relative merits of various modulation methods is to tabulate the ratio of the output signal-to-noise ratio of the given system to the output signal-to-noise ratio of a single-channel AM system. In making this comparison, it is usually assumed that all channels are fully modulated with a sinusoidal message at the highest possible message frequency. The resulting ratio is known as the *improvement ratio* or the *wideband gain*. The latter term derives from the fact that any improvement in signal-to-noise ratio is a consequence of a modulation method which widens the signal spectrum.

A fully modulated, single-channel AM system has an output signal-to-noise power ratio S_0/N_0 given by[4]

$$S_0/N_0 = C^2/8W\Phi_n = S_i/N_i \tag{193}$$

where C = carrier amplitude
W = highest message frequency
Φ_n = noise spectral density at receiver input
S_i/N_i = receiver input signal-to-noise power ratio

A single-channel FM system, with a maximum deviation ratio of D_m, can be shown to have an output signal-to-noise power ratio of[4]

$$\frac{S_0}{N_0} = \frac{3C^2 D_m{}^2}{8W\Phi_n} \tag{194}$$

Hence the improvement ratio R_0 for FM is

$$R_0 = \frac{(S_0/N_0)_{FM}}{(S_0/N_0)_{AM}} = 3D_m^2 \qquad (195)$$

In general, the improvement ratio for any system (XM-YM) is obtained from[23]

$$R_0 = \frac{(S_0/N_0)_{XM-YM}}{(S_0/N_0)_{AM}} = (K_1 M_1 M_2 K_2)^2 \qquad (196)$$

where K_1 is a constant associated with the first modulation (XM) and is always unity except for FM when $K_1 = \sqrt{3}$. M_1 is the modulation index associated with each subcarrier and is unity for AM and D_m for FM. M_2 is the modulation index of the

Table 5. Improvement Ratio, Threshold Power, and Signal-to-Noise Ratio at Threshold for Various Multiplex Systems

Modulation	Improvement ratio R_0	Threshold power $P_t \times 10^{20}$	$\frac{S}{N}$ at threshold, $\left[\frac{S_0}{N_0}\right]_t$
AM-AM	$\frac{1}{16n_c}$	$3.2BF_n$	$\frac{1}{8}$
AM-FM	$\frac{3B^2}{64n_c^3 F_s^2}$	$3.2BF_n$	$\frac{3B^3}{128Wn_c^3 F_s^2}$
FM-AM	$\frac{3B^2}{256W^2 n_c^3}$	$3.2BF_n$	$\frac{3B^3}{512W^3 n_c^3}$
FM-FM	$\frac{9B^2}{256W^2 n_c^3}$	$3.2BF_n$	$\frac{9B^3}{512W^3 n_c^3}$
PAM-AM	$\frac{1}{n_c}$	$6.4BaWF_n$	$\frac{Ba}{n_c}$
PAM-FM	$\frac{\pi^2 B^2}{32W^2 n_c^3}$	$3.2BF_n$	$\frac{\pi^2 B^3}{64W^3 n_c^3}$
PDM-AM	$\frac{B}{6Wn_c^2}$	$25.6BaWF_n$	$\frac{4B^2 a}{6Wn_c^2}$
PDM-FM	$\frac{\pi^2 B}{3Wn_c^2}$	$3.2BF_n$	$\frac{\pi^2 B^2}{6W^2 n_c^2}$
PPM-AM	$\frac{4B^2}{81W^2 n_c^3}$	$25.6BaWF_n$	$\frac{16B^3 a}{81W^2 n_c^3}$
PCM-AM	....	$25.6BaWF_n$	$\frac{3}{2} 2^{2n}$
PCM-FM	....	$3.2BF_n$	$\frac{3}{2} 2^{2n}$

subcarrier onto the main carrier and is different for each case. K_2 is always $1/\sqrt{2}$ and results from the fact that subcarrier sidebands add linearly while noise adds in an rms fashion. A tabulation of such improvement ratios is given in the first column of Table 5. For convenience, the items in this table are expressed in terms of signal bandwidth B, which can, in turn, be related to message bandwidths and degree of modulation by the relations given in Table 4. The special assumptions and notation pertinent to the tabulation of R_0 are listed below:

1. Notation:

   B = signal bandwidth, cps = B_R, receiver bandwidth
   W = message bandwidth, cps
   F_s = subcarrier spacing, cps
   n_c = number of channels

2. It is assumed that all channels have been adjusted to have the same improvement ratio.

3. The amplitudes or frequency deviations of the subcarriers are assumed to add in a mean-square fashion.

4. The PCM cases are not included because their S_0/N_0 does not depend on S_i/N_i above threshold.

The *improvement threshold* of any system is the input signal power required at the receiver to realize the improvement ratio. It is typical of wideband systems to

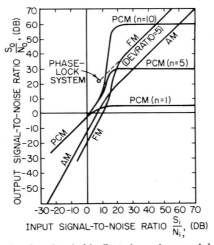

FIG. 17. Curves showing threshold effects in various modulation methods.

exhibit a pronounced threshold below which the output signal-to-noise ratio deteriorates more rapidly than the input signal-to-noise ratio. This effect is clearly indicated in the curves of Fig. 17.[24]

The threshold power for any receiver can be obtained from

$$P_t = kTBF_nC_f \quad \text{watts} \tag{197}$$

where k = Boltzmann's constant
   = 1.37×10^{-23} watt-sec/°K
   T = absolute temperature ≈ 292°K
   $B = B_R$ = receiver equivalent noise bandwidth, cps
   F_n = noise figure of receiver
   C_f = ratio of signal power to noise power at threshold

The determination of P_t depends upon what value is chosen for C_f, and this value is somewhat arbitrary. If the main carrier is present all the time, the usual assumption is that threshold occurs when the peak noise equals the peak carrier, and the peak noise is assumed to be four times the rms noise. This leads to a C_f of 8. In pulsed-carrier systems it is usually assumed that threshold occurs when the peak noise is one-half the peak signal. The factor C_f therefore becomes a function of duty factor and depends on pulse width and repetition time.

The second column of Table 5 tabulates the threshold powers for various systems. In the pulse-modulation systems the relation between bandwidth B and pulse width a is given in Table 4. For those systems in which the second modulation is FM, the tabulated values assume a C_f of 8. This value is unduly pessimistic if the FM receiver employs the *phase-lock loop* principle of operation, and in such cases the threshold power may be less than the value in the table by a factor of 30 to 50. The third column of Table 5 lists the output signal-to-noise ratio per channel when the input signal power is at its threshold value. The signal power required to achieve any desired value of output signal-to-noise ratio can be obtained from

$$P_s = P_t \frac{(S_0/N_0)}{(S_0/N_0)_t} \tag{198}$$

It is important to observe that in most cases, $(S_0/N_0)_t$ increases more rapidly with signal bandwidth B than does threshold power P_t. Hence the highest-quality system is obtained by increasing bandwidth until the available power reaches threshold. The other extreme occurs when the available power is reduced until the bandwidth required for threshold is the minimum allowable in terms of signal impairment. The first condition is suitable for high-accuracy short-range systems while the latter provides a system with maximum range.

A special situation arises in connection with PCM systems because the output signal-to-noise ratio in such a system depends entirely on the number of quantizing levels used whenever the system is above threshold. Thus, for any input signal-to-noise ratio above threshold, the output signal-to-noise power ratio is

$$S_0/N_0 = \frac{3}{2} 2^{2n} \tag{199}$$

where n is the number of digits per code group. A few values are tabulated below.

n	S_0/N_0
1	6
2	24
3	96
4	384
5	1,536
6	6,144
7	24,576

29 Information-handling Capabilities

Although the signal-to-noise ratios and bandwidths tabulated in the previous two articles are useful for comparing some aspects of the various modulation methods, they fail to convey a complete picture of the relative merits. In particular, these items do not immediately indicate the effectiveness with which the telemetry system utilizes its frequency-spectrum occupancy to achieve its primary purpose—the transmission of information.

One way of evaluating a given modulation method in a manner which comes closer to the above objective is in terms of its *efficiency of bandwidth utilization*. This efficiency compares the information capacity of the system output with the information capacity of the transmission channel and may be defined as

$$\text{EBU} \triangleq \frac{n_c W \log\,(1 + S_0/N_0)}{B \log\,(1 + S_i/N_i)} \tag{200}$$

in which n_c = number of channels
 W = message bandwidth of each channel
 S_0/N_0 = signal-to-noise power ratio in the output of each channel
 B = modulated signal bandwidth
 S_i/N_i = signal-to-noise ratio at the receiver input

It should be noted that the efficiency of bandwidth utilization does not consider the amount of information actually being conveyed by the system. Instead it compares the theoretical capabilities of two parts of the system and thus is a measure of the *potential* effectiveness of the wideband gain.

Numerical evaluation of (200) for the various modulation methods under realistic conditions of operation indicates that-the EBU ranges from about 0.25 for AM-AM and PCM-FM to about 0.03 for FM-FM.[16]

Another method of comparison which is conceptually different, but practically quite similar, is the *information efficiency*. This method compares the maximum information rate at the output of the system with the information capacity of the transmission channel. Hence the information efficiency may be defined as

$$\text{IE} \triangleq \frac{R}{C} = \frac{R}{B \log (1 + S_i/N_i)} \quad (201)$$

where R is the information rate and C is the channel capacity. The only way in which this quantity differs from the efficiency of bandwidth utilization is that the computation of R utilizes whatever a priori knowledge one has concerning the message data to establish a more realistic measure of the information actually conveyed. Since the information rate R can never exceed the information capacity of the system output, it follows that IE is always smaller than EBU.

In the case of PCM systems, the maximum rate at which information is conveyed can be determined from the rate at which digit pulses are transmitted and the probability of incorrectly receiving these pulses. This rate has been shown to be[25]

$$R \approx 2nn_c W(1 - \epsilon^{-0.46 S_p/N_p}) \quad (202)$$

where n is the number of digits per code group and S_p/N_p is the signal-to-noise ratio of the detected pulses (before decoding).

When the system output has a continuous range of amplitudes (that is, all systems. except PCM), the information rate may be treated by using the results of Sec. 18 In particular, when the output messages are white, band-limited, and Gaussian, the rate will be

$$R = n_c W \log (S_0/N_0) \quad (203)$$

It is clear that for large-output signal-to-noise ratios, IE approaches EBU and that for small signal-to-noise ratios ($S_0/N_0 \to 1$), IE approaches zero while EBU remains finite.

It is of interest to examine some of the reasons for the low information efficiencies of actual systems. Perhaps the most fundamental reason of all is that in an actual system the coding process is invariably of finite length. As discussed in Sec. 19, it is necessary for code lengths to approach infinity if all redundancy in the message is to be removed so that the information rate can be maximized. A second reason is that signals in the transmission channel are never Gaussian or white, and both these conditions are assumed in the development of the expression for channel capacity as done in Sec. 17. Finally, most practical systems fail to use all the information potentially available in the transmitted signal because only one characteristic of the signal, such as amplitude or phase, is modulated. Many methods for more efficient use of the signal have been suggested in the literature.[19,26]

REFERENCES

1. J. H. Laning, Jr., and R. H. Battin, "Random Processes in Automatic Control," McGraw-Hill Book Company, New York, 1956.
2. M. Schwartz, "Information Transmission, Modulation, and Noise," McGraw-Hill Book Company, New York, 1959.
3. J. L. Yen, On Nonuniform Sampling of Bandwidth-limited Signals, *IRE Trans. Circuit Theory*, vol. CT-3, pp. 251–257, December, 1956.

4. H. S. Black, "Modulation Theory," D. Van Nostrand Company, Inc., Princeton, N.J., 1953.
5. E. A. Guillemin, "The Mathematics of Circuit Analysis," John Wiley & Sons, Inc., New York, 1949.
6. W. B. Davenport, Jr., and W. L. Root, "Introduction to Random Signals and Noise," McGraw-Hill Book Company, New York, 1958.
7. G. C. Newton, Jr., L. A. Gould, and J. F. Kaiser, "Analytical Design of Linear Feedback Controls," John Wiley & Sons, Inc., New York, 1957.
8. P. M. Woodward, "Probability and Information Theory," McGraw-Hill Book Company, New York, 1953.
9. C. E. Shannon, A Mathematical Theory of Communication, *Bell System Tech. J.*, vol. 27, nos. 3, 4, pp. 379–423, 623–656, 1948.
10. R. M. Fano, The Transmission of Information, *MIT Res. Lab. Electron. Technical Report* 65, 1949.
11. D. A. Huffman, A Method for the Construction of Minimum-redundancy Codes, *Proc., IRE*, vol. 40, no. 9, pp. 1098–1101, September, 1952.
12. R. W. Hamming, Error Detecting and Error Correcting Codes, *Bell System Tech. J.*, vol. 29, no. 2, pp. 147–160, April, 1950.
13. D. Slepian, A Class of Binary Signaling Alphabets, *Bell System Tech. J.*, vol. 35, no.1, pp. 203–234, January, 1956.
14. A. B. Fontaine and W. W. Peterson, Group Code Equivalence and Optimum Codes, *IRE Trans. Circuit Theory*, vol. CT-6, pp. 60–70, May, 1959.
15. L. A. Zadeh, and K. S. Miller, Fundamental Aspects of Linear Multiplexing, *Proc. IRE*, vol. 40, no. 9, pp. 1091–1097, September, 1952.
16. M. H. Nichols, and L. L. Rauch, "Radio Telemetry," John Wiley & Sons, Inc., New York, 1956.
17. W. R. Bennett, Cross-modulation Requirements on Multi-channel Amplifiers below Overload, *Bell System Teech. J.*, vol. 19, pp. 587–610, 1940.
18. J. L. Stewart, The Power Spectrum of a Carrier Frequency Modulated by Gaussian Noise, *Proc. IRE*, vol. 42, no. 10, pp. 1539–1542, October, 1954.
19. D. Middleton, "An Introduction to Statistical Communication Theory," McGraw-Hill Book Company, New York, 1960.
20. H. Kaufman, and E. H. King, Spectral Power Density Functions in Pulse Time Modulation, *IRE Trans. Inform. Theory*, vol. IT-1, no. 1, pp. 40–46, March, 1955.
21. G. R. Cooper, unpublished intracompany memorandum, Aeronutronic Systems, Inc., Glendale, Calif., August, 1957.
22. C. B. Feldman, and W. R. Bennett, Band Width and Transmission Performance, *Bell System Tech. J.*, vol. 28, pp. 490–595, July, 1949.
23. V. D. Landon, Theoretical Analysis of Various Systems of Multiplex Transmission, *RCA Rev.*, vol. 9, nos. 2, 3, pp. 287–351, 433–482, 1948.
24. R. M. Page, Comparison of Modulation Methods, *Conv. Record IRE*, pt. 8, pp. 15–25, 1953.
25. Z. Jelonek, "A Comparison of Transmission Systems," London Symposium on Applications of Communication Theory, 1952, pp. 44–81.
26. N. M. Blackman, A Comparison of the Informational Capacities of Amplitude- and Phase-modulation Communication Systems, *Proc. IRE*, vol. 41, no. 6, pp. 748–759, June, 1953.
27. R. A. Fano, "Transmission of Information," Appendix B, pp. 351–383, The M.I.T. Press, Cambridge, Mass., 1961.

Chapter 3

SENSING OF INFORMATION

GEORGE J. PASTOR, *Aeronutronics, Ford Motor Company, Newport Beach, Calif.* (*Transducers*)

VERNON L. MILLER, *Integrated Range Mission, White Sands Missile Range, N. Mex.* (*Trajectory Instrumentation*)

TRANSDUCERS

1	Sensing of Physical Phenomena.................................	3–1
2	General Device and Performance Characteristics................	3–4
3	Description of Basic Transducers............................	3–8
4	Commercially Available Transducers..........................	3–42

TRAJECTORY INSTRUMENTATION

5	Sensing of Position...	3–42
6	Techniques for Electronic Trajectory Measurement.............	3–43
7	Electronic Trajectory-measurement Systems....................	3–46

TRANSDUCERS*

1 Sensing of Physical Phenomena

Many of the physical phenomena that modern man wants to know about are not convenient to measure directly. Empirically observed effects of the phenomena, however, provide a tool for measurement. A change in temperature, for example, will cause a change in the dimensions of a metal rod; consequently we have mercury thermometers. In this example the change in the energy content of the surrounding medium is manifested by a change in the length of the mercury column. The latter can be measured and compared with known and accepted standards. One can conclude that the thermometer column *senses* the temperature and converts it into a change in length.

As man reaches deeper into the universe it is becoming more difficult to make

---

* Parts of the material in this section were obtained by the author in connection with an Aeronutronic survey performed under U.S. Air Force Contract No. AF 33(616)-5928 administered by the Wright Air Development Center under the direction of C. W. Lucsinger and W. Michie.

measurements by such relatively direct methods, while the number of desired measurements keeps increasing.

To measure a remote variable an ingenious chain of devices have to be interconnected. Since information between remote points is normally transmitted by electromagnetic radiation, the radio link dominates the characteristics of this chain. On the front end of the chain one needs a device which converts the remote variable into a form easily applicable to the radio link, while on the receiving end, the information (measurement) has to be recovered and presented in an easily recognizable manner. The device that converts the remote variable into a form applicable to the radio link is the transducer. (The terms sensor, pickup, and pickoff are alternatives for "transducer" and will be used interchangeably here.) The complete chain is shown in Fig. 1. It must be recognized that the whole chain serves one purpose, that of delivering the remote measurement to the ground observer. The functions of the various "boxes" may be different and well defined, but compatibility of either box with at least the adjacent ones, if not with the whole system, is imperative. In this section transducers are discussed in view of the system, while the remaining components of the system will be described in later sections.

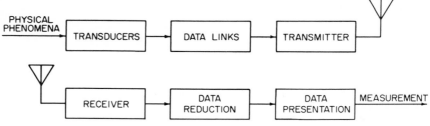

Fig. 1. Components of the data-handling chain.

A device which exhibits a change in some of the electrical parameters is the most suitable for applying modulating intelligence to the radio link. We shall consequently define the transducer as a device which converts a change in some form of energy (the variable) such as heat, radiation, sound, motion, etc., or a natural phenomenon or event, into a change in a measurable electrical parameter.

It is worth noting here that all transducers are affected by more than one of the above variables. It takes great engineering skill to make the response of the transducers sufficiently predominant to a single variable such that the other variables cause negligible errors. A simple example to illustrate this point is the resistance-wire strain gage. The resistance of the strain gage is a function of the stress applied to the wire; however, it is also a function of the wire temperature. To minimize the temperature error (the false "indicated strain") a wire with very low temperature coefficient of resistance must be used.

Because of the great increase in the number of unmanned vehicle flights in recent years a great number of new electromechanical transducers and new measurement concepts had to be invented. Typical telemetered measurements include:

Acceleration	Ice accumulation
Airspeed	Mach number
Air temperature	Rate of climb
Altitude	Pressure
Angle of attack	Rate of turn
Fuel consumption	Thrust
Heading	Vibration, etc.

Not only does the verification of the design techniques and assumptions require these measurements but a large class of vehicles depend on reliable and accurate measurements by transducers for guidance and control (celestial and inertial systems). While in this text transducers are considered from the telemetering point of view, it is recognized that the stringent guidance and control requirements gave added impetus to transducer development and refinement.

There are numerous ways to categorize transducers: (1) active or passive, (2) according to the quantity to be measured, (3) according to the changing output parameter.

An active transducer produces an output voltage or current that varies because of a changing quantity to be measured. Some of these require an external power source (phototransistor), while others do not (piezoelectric crystal).

A passive transducer exhibits a change in one of the passive electrical parameters such as resistance, capacitance, or inductance (self or mutual) because of a change in the physical quantity to be measured. These almost invariably require an external source of electrical energy to convert the change in parameter value into a change in voltage, current, or frequency.

Various transducers are tabulated in Table 1 according to their active or passive nature.

Table 1. Active and Passive Transducers

Active (voltage- or current-generating transducers)	Passive (variable-parameter transducers)
1. Piezoelectric	1. Variable resistance
2. Photoelectric	a. Nonmechanical resistance change
3. Thermoelectric	b. Change in internal structure
4. Magnetoelectric	c. Mechanically variable resistance
5. Electronic	2. Variable capacitance
6. Electrochemical	3. Variable inductance
7. Radioactive	4. Differential transformer
	5. Magnetostrictive

There are usually several types of transducers that may be used to measure the same physical quantity. Conversely each basic transducer can measure various physical phenomena with minor modifications.

Some typical measurements and the various basic transducers for each of the measurements are shown in Table 2.

Table 2. Typical Measurements and Basic Transducers[1]

Quantity to be measured	Type of transducer									
	Capacitance	Electronic	Inductance	Magnetoelectric	Magnetostrictive	Photoelectric	Piezoelectric	Radioactive	Resistive	Thermoelectric
Acceleration	x	x	x	x	x	...	x	...	x	
Displacement	x	x	x	x	...	x	x	x	x	
Flow	x	...	x	x	...	...	x	x	x	
Force	...	...	x	...	...	...	x	x	x	
Humidity	x	...	...	...	...	...	...	...	x	
Level	x	...	...	...	...	x	x	x	x	
Light	...	...	...	...	...	x	...	...	x	x
Mass	...	...	x	x	...	...	x	x	x	x
Pressure	x	x	x	x	x	...	x	x	x	x
Strain	...	...	x	...	...	...	...	...	x	
Temperature	...	...	...	...	...	x	...	x	x	x
Thickness	x	...	x	x	...	x	x	x		
Velocity	x	x	x	x	...	x	x	x	x	
Viscosity	x	...	...	...	x	...	x	...	x	

SENSING OF INFORMATION

Before each of these transducers is discussed in detail, it will be helpful to identify some of the general terms and concepts whereby the performance of a transducer is evaluated. These characteristics together with the requirements for accessory equipment form the basis for selecting a particular transducer for a given task.

2 General Device and Performance Characteristics

Input Range (Input Amplitude Range)

Any given transducer will perform properly only over a restricted range of the input variable. This range is usually limited either by loss of linearity or by the danger of permanent damage to the device. The input range is usually specified by the manufacturer in the units of the measured quantity.

Output Range

The output range is limited by the maximum and minimum in the electrical-parameter variations corresponding to the maximum and minimum values of the input phenomenon. When the transducer exhibits a change in one of the passive parameters for indication, the output range is frequently specified as a range of voltage or current change, corresponding to the input range, at a specified supply-source voltage. When the varying parameter is used to determine the frequency of oscillation of a (usually subcarrier) oscillator, the range of frequencies corresponding to the input-variable range may be given as the output range.

Output Impedance

Output impedance is the impedance seen between the output terminals of the transducer. It is very important because it determines the type of electric circuits that may be connected to the transducer. Also with the output voltage or current range known the output impedance determines the amount of power available from the transducer. The reactive part (if not negligible) of the output impedance determines partly the frequency characteristics of the transducer.

Transfer Characteristics

The transfer characteristics of the device relate the output (indication) value to the input (excitation) magnitude. The relationship may be directly proportional or described by some mathematical function.

For example, if a pressure gage has an input range of 0 to 10 psi, output range of 0 to 5 volts direct current, and the output is proportional to the input throughout the range, the transfer characteristic of the device is a straight line with slope of 5 volts d-c/10 psi = 0.5 volt/psi.

When the slope is a constant, the device is called linear. The slope of the transfer characteristic of a linear transducer is also called the constant of proportionality, or sensitivity. Linearity of the transfer characteristic is generally desirable since it facilitates simple data reduction.

When the transfer characteristic is nonlinear a plot of the variation of the output vs. variation of input must be given, unless a mathematical function can define this relationship accurately enough. This plot is the calibration curve of the transducer.

Linearity

The linearity of the device was defined above as the constant of proportionality of output and input variables. The term "linearity," however, is frequently used for *deviation from linearity*. Referring to Fig. 2, the solid curve represents the calibration curve of a transducer, and the dotted line represents the "best chord" one can draw such that the maximum horizontal distance between calibration curve and straight line is equal on both sides. Note that by the linear approximation the voltage V_i

corresponds to an indicated input pressure of P_i while according to the calibration curve the actual pressure is P_a. The difference

$$P_a - P_i = \Delta P$$

is the maximum deviation from linearity. Linearity is usually expressed as the ratio of ΔP and the maximum value of the input variable within the range in percentage form.

$$\text{Linearity} = \Delta P / P_{\max} \times 100 \text{ per cent}$$

Sometimes linearity is expressed as the per cent deviation based on the actual value of the input variable, which in the case of the above example becomes

$$\text{Linearity} = \Delta P / P_a \times 100 \text{ per cent}$$

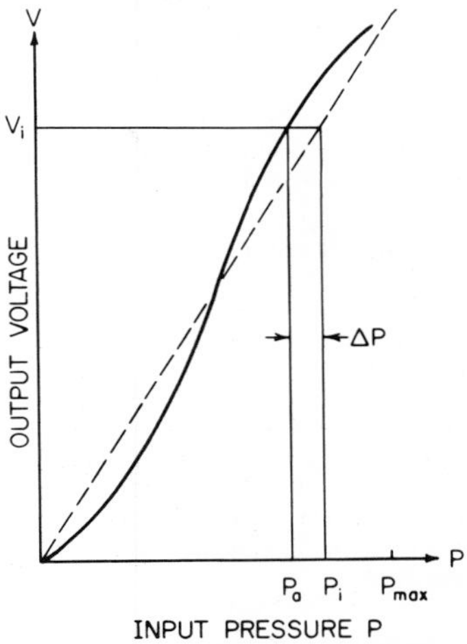

Sensitivity

"Sensitivity" is a frequently used alternate term for the linear transfer characteristic. It is usually expressed as the ratio of the change in the output corresponding to a given change in the input. However, sometimes it is defined as the full-scale output corresponding to the full-scale input. Note that if the device is not perfectly linear, sensitivity according to the first definition may vary, while the second definition gives the average sensitivity over the range.

FIG. 2. Deviation from linearity in a calibration curve.

Resolution

Resolution defines the smallest increment of the input variable to which the device has a detectable response. It is usually expressed as a percentage of full-scale value of the input variable; however, sometimes it is stated in the units of the input (or corresponding output) variable. In the case of wirewound potentiometers the resolution is expressed as one turn per total number of turns (the reciprocal of the total number of turns).

Drift

There are two types of changes in the transfer characteristics called drift. A change in the zero reading is called the zero drift, while a change in sensitivity is termed sensitivity drift. These are illustrated in Fig. 3a and b, respectively, while c

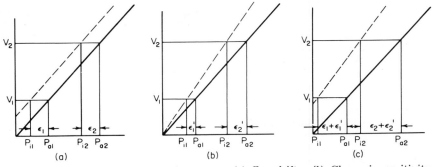

FIG. 3. Effects of drift on the calibration curve. (a) Zero drift. (b) Change in sensitivity. (c) Combined effect of a and b.

illustrates the combined result of a and b. It can be seen that the zero drift introduces a constant error throughout the operating range ($\epsilon_1 = \epsilon_2$). A change in sensitivity, on the other hand, makes the error a function of the magnitude of the variable, but the per cent error remains constant ($\epsilon_1/P_{a1} = \epsilon_2/P_{a2}$). In the case of the combination of the two, neither the absolute nor the per cent error is constant over the operating range.

The error due to zero drift can be easily canceled out by taking differential measurements whenever possible. To correct for sensitivity changes periodic two-point calibrations are required.

Error

The difference between the indicated value and the actual value of the input variable is an error (see Figs. 2 and 3). There are two classes of errors: random and systematic. Random errors unavoidably limit the accuracy of the transducer measurement, while systematic errors may be corrected for as described under drift and do not necessarily limit the accuracy.

Accuracy

The accuracy of a measurement is dependent on *all* errors present. A device may have a resolution of say 1 psi and sensitivity of 1 mv/psi. If the difference between indicated and actual values due to all sources of errors in the worst case is 5 mv, the accuracy is no better than ±5 psi; in other words whatever the indicated reading there is a ±5 psi uncertainty as to its accuracy. Accuracy also may be expressed as a per cent value based either on the maximum value of the range or on the value of the reading.

Frequency Range

The frequency range covers the frequencies of the input variable for which the tranducer response is uniform (flat) and the output is dependent only on the magnitude of the input. Frequency is inversely proportional to time. So far the time required by the transducer has not been considered, but obviously if the input variable changes faster than it takes the transducer output to reach the corresponding magnitude in the output, a decrease in sensitivity will result. This limits the high-frequency response. On the other hand, several transducers' output is dependent on a continuously changing input variable; in other words it lacks d-c response, and sensitivity to slowly varying inputs is lower than to fast ones.

Since any time-varying function may be resolved into a corresponding frequency spectrum, the frequency range is often specified as a rise time, time constant, response time, or two frequency limits in cycles per second. When transients are of interest the frequency response must be higher than the highest-frequency component of the input variable. (The rise time of the device must be negligible in comparison with the rise time of the input variable.) The frequency range of the input variable is not to be confused with the range of frequencies a subcarrier oscillator is deviated because of the amplitude change in the output of the transducer, or with the range of frequencies of an a-c supply voltage that is required by the transducer.

Resonant Frequency (Natural Frequency)

As was pointed out above, the frequency range within which the transducer may be considered linear is limited by the time it takes the transducer to respond to an excitation. The finite response time causes the output to decrease at high frequencies.

Resonance is another phenomenon that limits the useful frequency range of the transducer. Resonance is a phenomenon typical of all systems (electrical, mechanical, etc.) whose performance may be described by a second-order differential equation. In this case (assuming no damping) there is a particular frequency of excitation f_n at which the output is considerably greater than at frequencies far removed from f_n. The frequency at which the device *resonates* (or is *in resonance*) is dependent on the physical constants that are characteristic of the device, such as mass and spring constant. Obviously the designer of the transducer enjoys some degree of freedom to place the resonant frequency far from the intended frequency range of the transducer. Figure 4a illustrates this situation. Fortunately (for this application) in all lossy systems, consequently in all physical systems, there are factors opposing an unlimited growth in the output due to a finite input. This is represented by the amount of damping present in the system; the amount of damping can also be controlled within limits. The measure of the damping is the magnitude of the damping coefficient. In Fig. 4b frequency responses corresponding to various damping coefficients are shown. When one considers a device with finite damping the maximum response to a sinusoidal excitation does not necessarily occur at the resonant frequency but at somewhat lower frequency.[2] Critical damping is defined as that which results in the

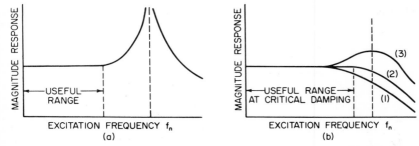

FIG. 4. Typical frequency response curve of a transducer. (a) Undamped resonance. (b) Same with damping. (1) Overdamped. (2) Critically damped. (3) Underdamped.

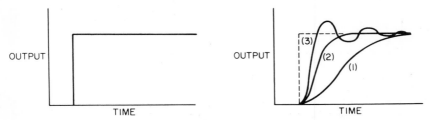

FIG. 5. Typical transient response of a transducer. (1) Overdamped. (2) Critically damped. (3) Underdamped.

widest band of frequencies within which the device is linear. (In this case the frequency response is similar to a simple low-pass filter.) The above definitions are not rigorous but they are practical and useful. In the case of critical damping (curve 2) it is shown how the useful frequency range of the device has been considerably extended. Considering the completely undamped case (which is only theoretically possible) the maximum input frequency is usually held to about one-tenth of the resonant frequency. (This is a widely used "rule of thumb.")

Turning our attention to the transient response of the device, critical damping represents the best rise time obtainable in the output without overshoot or oscillatory response. This case is shown as curve 2 in Fig. 5, with the other curves corresponding to under- and overdamped systems.

Rise Time

The rise time is usually defined as the time for the transducer output to reach from 10 to 90 per cent of the final (steady-state) value upon application of a step-excitation function to its input. Some other definitions are used occasionally, however.

Delay Time

Delay time is defined as the time between the application of a step in the excitation and the reaching of the 50 per cent value of the final output.

Response Time

The term "response time" is used rather loosely, sometimes in place of delay time, or sometimes redefined as the time for the output to reach 90 per cent of the final value from the time of the application of the step excitation.

Time Constant

The time constant τ of a device is the time it takes the output to reach $(1 - 1/\epsilon) \times 100$ per cent $= 63.2$ per cent of the final value.

The last four definitions are illustrated in Fig. 6.

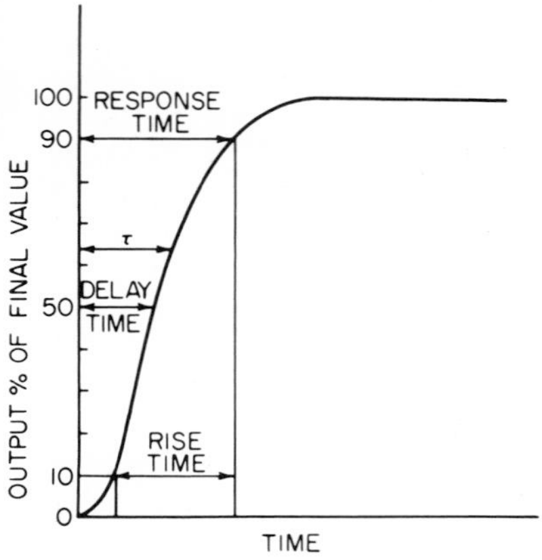

FIG. 6. Illustration of rise time, delay time, time constant, and response time.

Temperature Range

Since, as it was pointed out before, temperature does affect the performance of most transducers, the usable range of temperatures is usually specified by the manufacturer. This is the range within which all the other specifications are valid. Danger of destruction of the device (permanent damage) is very seldom the limiting factor while deterioration in performance certainly is.

Other Environmental Conditions

Many transducers contain moving parts and/or are inherently sensitive to vibration, shock, and acceleration. Because of this, and because of their frequent application under such severe environments, the limits for satisfactory performance are usually quoted by the manufacturer.

The remaining characteristics of transducers often influence the choice because of other practical rather than performance considerations. These are power consumption (if any), size and weight, and requirements for accessories. These will be kept in mind in the detailed description of the various transducers in the following paragraphs.

3 Description of Basic Transducers

Now that the general characteristics have been defined on the basis of which the performance of transducers can be evaluated and compared, the basic transducers will be described in this section.

It is believed that an organization of the description of transducers according to their principles of operation rather than by their end use is appropriate here since the aim is to have a ready reference to acquaint the reader with any given commercially available transducer for which the principle of operation is known. This principle of operation remains independent of the end use. If on the other hand the descriptions were to follow the end use, repetitions would become unavoidable and the reader would have to read through a great variety of devices until the one of interest could be found. The transducers are discussed in alphabetical order based on the key words describing their principle of operation.

Capacitance Transducers

The concept of capacitance is introduced in electrostatics, where it is defined as the ratio of charge to voltage between two conductors, or in equation form,

$$C = Q/V \qquad (1)$$

where C = capacitance, farads
Q = charge, coulombs
V = voltage, volts

When two parallel plates of effective area A are separated by distance d, with a dielectric medium between them with dielectric constant K, the capacitance of the device is given by

$$C = m(KA/d) \qquad (2)$$

where m is the constant of proportionality. The unit of capacitance is the farad, but the practical unit is the picofarad (pf), which is 10^{-12} farads.

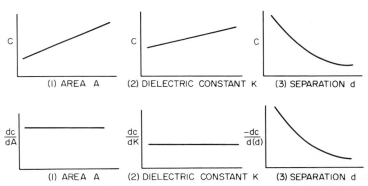

Fig. 7. Variable-capacitance performance curves. (a) Effect of varying (1) area, (2) dielectric constant, (3) separation between plates in a capacitor. (b) Change in sensitivity corresponding to the curves in a.

Equation (2) indicates that in making a variable-capacitance transducer one has three degrees of freedom:
1. Vary the effective area.
2. Vary the dielectric constant of the medium between the plates.
3. Vary the distance of separation.

The effects of these variations are shown in Fig. 7 for a mathematical model of a capacitor where the effective area is assumed to be the perpendicular projection of one plate on the other. (This assumes no electrostatic-field lines other than straight lines perpendicular to both plates.)

Figure 7b illustrates that varying the separation is the least desirable from the uniform-sensitivity point of view. Yet one of the oldest transducers, the condenser* microphone, uses exactly this principle.

In the microphone one plate of the capacitor is a thin diaphragm which vibrates because of the sound pressure. If the *change* in separation is small as compared with the separation, one is using a small segment of the calibration curve within which it may be assumed to be linear (constant sensitivity). The same principle is used to measure higher than audio-frequency pressures; the frequency limit is set by the resonant frequency of the diaphragm, which may be over 100 kc/sec.

There are several other means whereby the capacitor plate separation may be varied. If one of the plates is made to be movable, the capacitor may become a linear-displacement transducer or a liquid-level sensor.

* Condenser is the obsolete term for capacitor (still widely used).

The variation of the capacitance in the capacitor transducer may be made linear with respect to plate separation by using a three-plate capacitor with the center plate movable. In this case if the difference of the partial capacitances is used as the output variable the sensitivity remains constant.

The variation of the effective area of the capacitor plates is widely used in tuning radio receivers. The same principle is used to measure shaft rotation. When an a-c excitation is used the electrical phase shift of the output signal can be made equal (within very small tolerances) to the mechanical angular displacement of a shaft. By this method angular displacement, velocity, and acceleration may be measured with appropriate detecting methods.

The dielectric constant of the medium between the plates can be varied by passing materials between the plates or by allowing liquids to penetrate the dielectric. Flow, density, thickness, and length may be measured by this method.

From the description given so far it is apparent that the capacitor is a passive transducer (it does not generate a voltage or current); hence an energy-supply source is required. This supply may be either a d-c or an a-c voltage.

In case of d-c supply voltage the device will lack d-c response. (Static measurements cannot be made.) This is made evident by Fig. 8a, where the variation of capacitance is assumed to be sinusoidal at the input variable's angular frequency ω_i.

Fig. 8. Elementary capacitor transducers. (a) D-C application. (b) A-C application.

It can be seen that if the capacitance value remains constant at any value there is no current flow in the circuit; consequently the output voltage e_0 is zero. Upon varying the capacitance there will be current flow in the circuit since its effect is similar to applying a sinusoidal voltage to the circuit.

In the case of an a-c supply the capacitor is made part of a tuned circuit which is tuned to the supply frequency ω_c with $C = \Delta C_{max}$, where ΔC_{max} is larger than any expected value within the useful output range of the capacitor, or

$$\omega_c = \frac{1}{[L(C_o + \Delta C_{max})]^{1/2}} \tag{3}$$

In this case any lesser variation of ΔC will detune the circuit with a corresponding change in the output voltage along the nearly linear portion of the resonance curve. This arrangement is shown Fig. 8b.

The main advantage of capacitance transducers is their excellent sensitivity. In displacement transducers a sensitivity of 1 pf/0.0001 in. can be obtained by special construction,[3] and displacement measurements of one-tenth of an angstrom (1Å = 10^{-8} cm) were reported.[4] The force requirement to produce displacement is extremely small. The practical capacitance values are between a fraction of a picofarad and 1,000 pf, while the fractional variation of capacitance may be made to approach unity.

The main disadvantages of variable-capacitor transducers are as follows: The output impedance is, of course, mostly reactive and usually fairly high; it is a function of frequency. The dimensions for the high-sensitivity units are extremely critical and difficult to maintain; this also sacrifices ruggedness. The main shortcoming of the device is its sensitivity to the stray capacitive pickup due to long lead lengths and/or moving leads. Because of this either a sensitive preamplifier must be used very close to the transducer or some bridge-circuit technique must be applied to minimize the lead effects. In order to obtain high sensitivities, output signal amplitudes are usually compromised, requiring high-gain high-sensitivity electronic circuits with the transducer. The main sources of error are the variation of the dielectric constant of most materials due to temperature, humidity, and aging, and the changing of the insulation resistance holding the plates in place.

Frequently, when the variation of capacitance is an unsatisfactory output form, a capacitance-to-voltage transducer is utilized.[5]

Dielectric Transducer

The dielectric transducer is a variation of the capacitance transducer. It is discussed separately to illustrate and emphasize a principle. As was mentioned before, the performance of the previously discussed capacitor transducers is impaired by variation of the dielectric constant due to humidity. Any such detrimental effect can be used for measuring the cause of the effect, in this case humidity. The dielectric humidity transducer uses a special porous dielectric (aluminum oxide) to sense changes in humidity. Both the resistance and capacitance of the transducer are functions of humidity.[6]

Digital Transducers

Because of the recent advent of time-division multiplexed telemetering, such as PDM-FM and PCM-FM, great interest is expressed in digital transducers. Several kinds of devices are termed digital transducers; however, most frequently they are used only as operators on an existing auxiliary input which may be a train of pulses. Typical applications of this kind are where the varying output parameter of the transducer introduces a corresponding delay, or stretching in the reference pulse train. Frequently the transducer, which is basically an analog device, has a built-in digitizer. This approach tends to complicate the vehicle-borne system by duplication and lack of uniformity.

One notable exception to the above-described devices is the shaft-position encoder. This consists of a disk mounted on a rotating shaft. The disk has slits cut in concentric rings which are usually arranged to correspond to a straight binary representation of the shaft position. A number of light beams, corresponding to the number of concentric rings of slits, and consequently to the number of binary digits, in connection with a corresponding number of photosensitive detectors, accomplish the "reading out" of the binary representation of the shaft position. The shaft-position encoder has a very high capability of resolution; a 12-concentric-ring encoder can resolve an angle to $360°/2^{12} = 0.088°$. The size of the disk becomes the limiting factor in obtaining even higher resolutions. The shaft-position encoder has the obvious application of indicating angular displacement, velocity, or acceleration.

Eddy-current Transducer

Eddy-current transducers operate on the basis of two fundamental physical laws, Faraday's and Lenz's laws. According to the first, if a conductor moves in a magnetic field (or a changing magnetic field is applied perpendicularly to a stationary conductor) a potential difference (or current) will be induced perpendicular to both the direction of motion and the direction of the magnetic field. Lenz's law states that any induced physical phenomenon is such that it tends to reduce the phenomenon that causes its existence.

Consequently, if an alternating current flows in a coil and a conducting plate is brought in the vicinity of the coil perpendicular to both the axis of the coil and the resulting alternating magnetic field, eddy currents will be induced in the conducting plate. These currents generate a magnetic field which opposes the one responsible for their existence. This in effect reduces the original magnetic field. The reduced magnetic field represents a reduced inductive reactance for the coil which is measurable. The distance between the coil and conductive plate can be measured by this method.[3] An alternate method for sensing the output is to use pickup coils near the moving plate. The eddy currents will induce currents in the pickup coils which are proportional to the acceleration of the plate. In this latter case there is no need for an alternating magnetic field.

Generally, moving-coil magnetic devices can perform better than eddy-current transducers since much more mechanical power is required to actuate eddy-current devices than moving-coil pickups.[8]

The eddy-current displacement transducer is used for measuring small distances, from 10^{-5} to 10^{-2} in. The transfer characteristics are nonlinear.

Electrolytic Gages

The electrolytic gage operates on the principle that the resistance of the electrolyte between the electrodes is proportional directly to the distance between the electrodes and inversely to the cross-sectional area of the conducting path. Electrolytic gages may be used for displacement measurement by varying either the distance between the electrodes or the effective cross-sectional area of the electrolyte. Their main advantage is that the liquid does not offer appreciable resistance to the (slow) movement of electrodes; consequently very small force is required to produce a displacement. The size of the electrolytic gage may be made very small.

An obvious application of the electrolytic gage is for liquid-level indication and as a flowmeter. In the former the liquid level determines the cross-sectional area of the conducting path if the electrodes are held at a constant separation. The relationship between resistance and liquid level is an inverse, and consequently a nonlinear one.

Hygroscopic salts exhibit appreciable resistance change with humidity and are used as electrolytes for humidity transducers.

The major disadvantage of all electrolytic transducers is the high temperature dependence of the resistivity of the electrolyte. This is usually a decreasing function of increasing temperature; i.e., electrolytes have a negative temperature coefficient of resistance. This characteristic naturally may be used for temperature measurement. Other means, however, are far superior to it.

Another disadvantage of electrolytic gages is polarization, which may be overcome by the use of alternating current. In this case, however, the capacitance between the electrodes affects the gage performance adversely.

For telemetering purposes electrolytic gages are not very suitable, mainly because they are markedly affected by more than one physical variable and bulky accessories are required to overcome this.

Electronic Transducers

A great variety of transducers may be termed electronic. Two definitions will be used for electronic transducers here:

1. Devices which use an active electronic element such as the electron (vacuum or gas-filled) tube or semiconductor (photocell, crystal diode, transistor) to sense the input variable
2. Devices which convert energy usually carried by some form of radiation into electrical signals

The simplest form of the electronic transducer is the mechanical-displacement-input electron tube. It may be a diode or a triode; in either case the "plate resistance" of the tube is varied by the mechanical displacement of an electrode. This is the direct consequence of the "three-halves-power law" of the space-charge-limited current I, which is given by

$$I = 2.335 \times 10^{-6}(a/d^2)E^{3/2} \qquad (4)$$

where I = plate current, amp
  a = effective area of the cathode
  d = distance between electrodes
  E = applied potential difference between plate and cathode

Triodes generally give greater sensitivities. The linearity of these devices is usually about 2 per cent.[3]

The main advantage of these devices is the large output-voltage variation (of the order of ± 10 to 20 volts) produced by small displacements. The disadvantages are the need for a high-voltage d-c power supply, sensitivity change due to heater-current variation, and thermal drift.

The remaining electronic transducers are mainly used for radiation and light-intensity measurements. These are the photosensitive transducers, ionization chambers, Geiger-Mueller counters and their derivatives, and scintillation counters. These are discussed in separate paragraphs.

Geiger-Mueller Counters

The most frequently used radiation counter is the Geiger-Mueller tube. G-M tubes are built in many variations but in their basic form are constructed as shown in

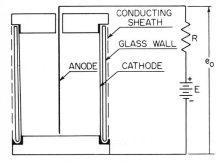

Fig. 9. The Geiger-Mueller counter.

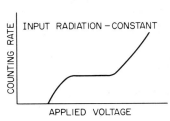

Fig. 10. Variation of counting rate with applied voltage in the G-M tube.

Fig. 9. The anode is a fine tungsten or platinum wire at the axis of the cylindrical cathode. Both anode and cathode are enclosed in a cylindrical glass envelope. Outside the glass envelope there is usually a thin conducting metal sheath electrically connected to the cathode to eliminate the electrostatic field across the glass envelope. The shield, glass wall, and cathode either are made thin enough, or a window is provided, for radiation to enter the tube. The tube is first evacuated and cleaned, then filled with a gas (hydrogen with argon, or organic vapors or argon with organic vapors) at a low pressure of about 10 cm of mercury. A high potential (1,000 to 1,500 volts) is applied to the electrodes corresponding to the plateau region of Fig. 10. This is important to make the number of counts independent of supply-voltage variations. The operation of the tube in the plateau region is briefly described as follows[10] (for detailed discussion the reader is referred to Refs. 3, 9, and 10): The incident ionizing particle upon each collision forms an electron-ion pair. The electron is attracted by the anode and the ion by the cathode. The drifting electron causes additional ionization by collision. The excited atoms again cause additional ionization in all directions because of photoemission. In the vicinity of the anode the field is strong enough to form new avalanches until the entire anode is surrounded by a cylinder of ions and the discharge is self-sustaining.

The positive ions move toward the cathode. They form a "virtual anode," and as they move radially outward the field strength in the counter decreases. The discharge ceases when the reduced field strength is below that needed to sustain the discharge. To avoid reignition of the discharge due to secondary emission caused by the positive ions hitting the cathode, "quenching" of the counter is employed by external means.

The output of the Geiger-Mueller counter is pulses with amplitudes ranging from 1 to 10 volts.

The highest counting rates are of the order of 10^3 to 10^4 counts/sec. The Geiger-Mueller counter can easily be made to discriminate against low-energy particles by making the window sufficiently thick or by the use of filters such that the penetration threshold is at a desired energy level.

The major disadvantage of the Geiger-Mueller counter is the relatively very high supply-voltage requirement. This, however, is somewhat balanced by the very small power consumption. The output impedance is high; consequently only very high input-impedance devices may be used following the G-M tube.

The G-M counter is basically a special application of ionization chambers which can measure radio-activity. Ionization chambers differentiate between various types of incident rays or particles by different pulse amplitudes. The frequency of pulses is a measure of the intensity of the radiation. The pulses produced are of the order of 10^{-5} to 10^{-3} volts, causing currents less than 10^{-6} amp.[11] Sensitive accessories make consideration of this type of counting prohibitive. To overcome this problem the pulses may be electronically integrated and thereby a time average of the incident radiation determined.

Proportional counters are essentially G-M counters operated below the plateau voltage. They are characterized by output pulses whose amplitude is proportional to the energy of the incoming particles. Their sensitivity is below that of the G-M counter but is better than that of the ionization chambers.

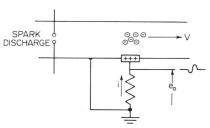

FIG. 11. Application of the electrostatic-induction principle to the gas-flow velocity transducer.

Induction (Electrostatic) Devices

Electrostatic induction generates an induced field within a conductor such that the sum of the applied and induced fields within the conductor is equal to zero. During the generation of the induced field the movement of charges represents a current flow which is detectable. This principle is applied to a gas (air) flow velocity transducer.[3] The operation of the device is described with reference to Fig. 11 as follows: A spark discharge or some other method produces ionization in the gas flowing through the duct. The passage of the ion cloud is detected by the induction electrode in which a charge buildup of opposite sign to that of the ion cloud will cause a bipolar current pulse in an associated resistance. Since current is defined as the rate of change of charge the current pulse will be negative in the sense shown during the positive-charge buildup and negative during discharge. All polarities are reversed if a positive ion cloud is considered. Since the ion cloud does not affect the gas-flow pattern, the velocity of the flow is inversely proportional to the time between the spark and the passage of the ion cloud opposite to the induction electrode.

Gas velocities between 20 mph and Mach 2 were successfully measured by this method.

The electrostatic-induction principle is also used for measuring space-charge concentration in a medium, surface charges, and surface potentials, but these measurements are usually outside the realm of telemetry.

Magnetic Transducers

There are a great variety of magnetic, or more precisely, electromagnetic transducers. Because of the large number of transducers using some electromagnetic principle for measurement, magnetic transducers will be further subdivided according to the variable parameter used. Some of the important definitions of electromagnetics will be very briefly presented before the description of transducers. These parameters are generally not so well known as the pure electrical ones; hence their description here is believed to be an aid to the understanding of the principles of operation of various magnetic transducers. For a more complete description the reader is referred to textbooks on magnetic circuits such as "Magnetic Circuits and Transformers" by MIT (John Wiley & Sons, Inc., New York, 1943) or John D. Kraus, "Electromagnetics," (McGraw-Hill Book Company, New York, 1953).

Definitions of Magnetic Terms

Ferromagnetism—A material is ferromagnetic if it is easily magnetized; i.e., it constrains most of the magnetic flux to within the physical dimensions of the material.

Paramagnetism—Materials which exhibit a strengthening of the magnetic field upon application of an external magnetic field are known as paramagnetic.

Diamagnetism—Materials which exhibit a weakening of the magnetic field upon application of an external magnetic field are known as diamagnetic.

Nonmagnetic—Materials that show negligibly weak magnetic effects are termed nonmagnetic. Vacuum is the only true nonmagnetic medium.

Magnetic Flux Density—The magnetic flux density (usually denoted by B) is the force produced by a unit current in an infinitesimal length of wire; the force is perpendicular to the direction of current. Another definition of magnetic flux density is the force on a unit magnetic pole (fictitious).

Magnetic Field Intensity—By analogy to electrostatics the magnetic flux density is often called magnetic field intensity.

Magnetic Flux—The fictitious force lines passing perpendicularly through a given area are called the magnetic flux (usually denoted by ψ). The magnetic flux is the dot product of the flux density and area vectors.

$$\psi = \mathbf{B} \cdot \mathbf{A} = |\mathbf{B}| \cdot |\mathbf{A}| \cos \theta \tag{5}$$

where B = magnitude of **B** (vector)
A = magnitude of **A** (vector)
θ = angle between the two vectors

Magnetizing Force (Magnetic Field)—The magnetizing force (denoted by H) is proportional to the magnetic flux density (B), and is given by

$$H = B/\mu \tag{6}$$

where μ = permeability of the medium.

Since by Ampere's law the line integral of H around a single closed path is equal to the current enclosed[12] the dimensions of H are current per length.

The frequent usage of "magnetic field intensity" for H is not appropriate in view of the analogy between B and E (magnetic and electric field intensities, respectively), rather than H and E.

Permeability—Permeability (denoted by μ) is the ratio of magnetic field intensity B to magnetizing force H. It is a characteristic of all materials. The permeability of vacuum is usually denoted by μ_0. Relative permeability is defined as $\mu_r = \mu/\mu_0$. The maximum relative permeability of ferromagnetic materials varies between 10^2 and 10^5. The permeability of ferromagnetic materials is not a constant but it is a function of the magnitude and time history of H.

Hysteresis Curve—The curve describing the variation of B with the variation of H in a ferromagnetic material is called the hysteresis curve. The slope of the curve is the permeability of the material.

Flux Linkage—I current passes through several turns of wire, the resulting flux lines pass through or *link* each turn (assuming that the turns are closely spaced). The total flux linkage λ then equals the product of the number of turns N and the total magnetic flux ψ, or

$$\lambda = N\psi \tag{7}$$

Inductance—Inductance L is defined as the ratio of the flux linkage λ to the current I producing the flux linkage, or

$$L = \lambda/I \tag{8}$$

This definition assumes constant permeability. If this is not the case inductance is given as a ratio of differentials, as

$$L = d\lambda/dI = N(d\psi/dI) \tag{9}$$

Inductance is a measure of the capability of a device to store magnetic energy. In this respect it is analogous to capacitance in electrostatics. The unit of inductance is the henry, but the more practical units are the milli- and microhenry (10^{-3} and 10^{-6} henrys, respectively).

Mutual Inductance—When two coils are linked by the flux lines produced by either one, the mutual inductance M is given by

$$M = N_1(d\psi_{21}/dI_1) = N_2(d\psi_{12}/dI_2) \tag{10}$$

where N_1 = number of turns in coil 1
N_2 = number of turns in coil 2
ψ_{21} = flux linking coils 2 and 1 due to current in coil 1
ψ_{12} = flux linking coils 1 and 2 due to current in coil 2
I_1 = current in coil 1
I_2 = current in coil 2

Magnetomotive Force—The product of the number of turns N of wire and the current I flowing in the wire is called the magnetomotive force (mmf,F)

$$F = NI \quad \text{amp-turns} \tag{11}$$

This quantity is analogous to the electromotive force (voltage source) in an electric circuit.

Magnetic Potential—The magnetic potential U is defined as the line integral of the perpendicular component of H along the magnetic path between two points. (This definition assumes that the path of integration avoids all currents.)

$$U = \int_a^b \mathbf{H} \cdot d\mathbf{l} = |H|\,|l|\cos\theta \quad \text{amp} \tag{12}$$

where $|\mathbf{H}|$ = magnitude of $\mathbf{H}$ (vector)
$d\mathbf{l}$ = infinitesimal line vector between points a and b
θ = angle between $\mathbf{H}$ and $d\mathbf{l}$

The magnetic potential is analogous to the electric potential or a voltage drop in a circuit due to the IR drop. By analogy to Kirchhoff's voltage law, the magnetomotive force equals the algebraic sum of the magnetic-potential drops in a closed magnetic path, i.e.,

$$F_{ab} = U_{ai} + U_{jk} \cdots U_{nb} \tag{13}$$

Reluctance and Permeance—Closely following the above-mentioned analogy of electric and magnetic circuits, the total reluctance $\mathcal{R}$ of a magnetic circuit is defined as the ratio of the magnetomotive force F and flux in the magnetic circuit (similar to the resistance in the electric case), or

$$\mathcal{R} = F/\psi \tag{14}$$

Similarly the reluctance $\mathcal{R}$ between two points (a,b) with magnetic-potential difference of $U_{a,b}$ is given by

$$\mathcal{R} = U_{ab}/\psi \tag{15}$$

For uniform cross section of area A, and uniform field H and separation l between points a and b, (14) reduces to

$$\mathcal{R} = l/\mu A \tag{16}$$

Comparing Eqs. (13), (14), and (15), it follows that

$$\Re = \sum_{j=1}^{j=n} \Re_j \qquad (17)$$

The reciprocal of reluctance is called permeance (denoted by $\mathcal{P}$) or

$$\mathcal{P} = 1/R = \mu A/l \qquad (18)$$

Differential Transformers. Differential transformers basically measure displacement, but in connection with an elastic member they can be made to measure force, pressure, or acceleration. Displacement measurements may be used to indicate strain, or by using electronic differentiating circuits the displacement measurement may be converted into velocity or acceleration indication. Either linear or angular displacements may be measured by appropriate differential transformers.

The linear differential transformer consists of three coils wound around a movable common core. Upon application of an a-c excitation voltage to the center coil the two outer coils develop voltages with opposite polarities. Upon series-connecting the two outer coils, these voltages cancel and no (or negligibly small) output exists for the balanced, or null, position. Displacing the core in either direction, the induced voltage is increased in the core toward which the displacement takes place while the induced voltage of the opposite core decreases. Figure 12a illustrates the variation of output voltage as a function of core displacement, while b shows the schematic circuit of the differential transformer.

Using high- and uniform-permeability core materials, such as nickel-iron alloys, results in excellent linearity within certain limits between output voltage and displacement of the core. The linear range of displacement may be from ±0.005 to ±1.0 in. The zero-reading error is of the order of 1 per cent while the linearity is of the order of ±0.5 per cent, which may be reduced to ±0.1 per cent by special design.

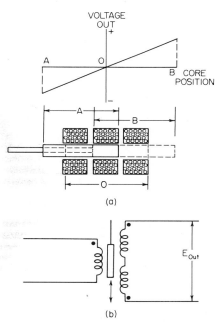

Fig. 12. Differential transformer. (a) Variation of output signal with core displacement. (b) Schematic diagram.

The linearity is almost independent of the load impedance provided the load impedance is at least equal to the output impedance of the transformer. The sensitivity of differential transformers varies between 0.2 and 6.0 mv per 0.001 in. displacement per volt excitation. The sensitivity of a given design is the function of

1. Magnitude of the excitation voltage
2. Frequency of the excitation voltage
3. Magnitude of the load impedance

The frequency dependence of the sensitivity is a slowly varying function; hence it is not critical. Excitation frequencies of 60 to 20,000 cps are being used for commercially available transformers.

The maximum frequency of the input (displacement) should be less than or equal to one-tenth of the excitation frequency.

The main advantages of the differential transformer are very good linearity; relatively low output impedance, which varies from a few ohms to several kilohms depending on the construction of the device and the excitation frequency; the ease with which the output range can be selected; and the magnitude of the output, which may be as high as 10 volts rms. The last two characteristics frequently eliminate the need for any additional connecting electronics between transducer and modulator or provide for simple connecting electronics such as a rectifier without need for amplification.

The differential-transformer performance is very sensitive to magnetic pickup. Consequently careful magnetic shielding is important; a decrease in insulation resistance due to humidity, corrosion, etc., will cause performance deterioration. Similarly increasing temperature will change the sensitivity of the transducer because of the increasing resistance of the coils, dimension changes, permeability change of the core, etc. The sensitivity change may be in either direction; consequently in severe environmental applications either careful electrical temperature compensation or previous calibration and temperature monitoring is required. It is claimed that commercially available units operate between -85 and $+450°F$.

Hall-effect Magnetic Transducer. The Hall effect was discovered in the late nineteenth century. It has, however, not played a significant role until the recent advent of semiconductors.[13]

The Hall effect occurs when a transverse magnetic field is applied to a current-carrying conductor. The Hall effect denotes the appearance of a transverse electric field which is normal to the original electric field causing the current flow. The resulting transverse electric field produces a measurable voltage drop across the conductor since the applied magnetic field forces the electrons to deflect within the bounds of the conductor. The directions of all vector quantities involved in the Hall effect are shown in Fig. 13. In steady state, the transverse electric field within the conductor resulting from the uneven distribution of electrons balances the deflecting Hall effect and no further charge accummulation occurs. The current then continues to flow longitudinally. The resultant electric field in this case is not parallel with the current flow but forms angle θ with the current vector.

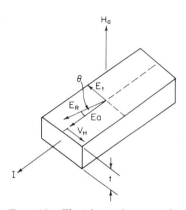

FIG. 13. Electric and magnetic fields due to Hall effect. H_a = applied magnetic field. E_a = applied electric field. E_t = transverse electric field due to Hall effect. E_R = resultant electric field. V_H = transverse Hall voltage. θ = angle between the resultant electric field and current.

The Hall coefficient gives the ratio of the transverse electric field and the product of longitudinal current density times the magnetic field. The transverse voltage then is given by

$$V_H = K(IB/t) \tag{19}$$

where V_H = transverse voltage
K = Hall coefficient
I = longitudinal current
t = thickness of the conductor

The value of the Hall coefficient varies widely with various materials. Semiconductors have Hall coefficients several orders of magnitude higher than most metals. Semiconductors, consequently, are preferred for this type of transducer. The Hall coefficient, however, changes with temperature and is also a function of the magnetic field intensity. The resistivity of semiconductors changes considerably with temperature, making individual calibration of these devices essential.

Hall-effect transducers may be used directly to measure magnetic fields of the order of 10,000 gauss or may be used to indicate induced magnetic fields in connection with other devices.

The output voltage is usually in the millivolt range. However, the output impedance is low, of the order of 100 ohms, which is characteristic of semiconductor devices.

Inductance Transducers. Most magnetic transducers operate basically through changes of self-inductance or mutual inductance. The dividing line between inductance gages and the other types of magnetic transducers is indeed very thin. A differentiation between inductance and induction transducers is made, however, because the two types operate on different physical principles. Induction transducers are discussed in the next paragraph.

Inductance is defined under Definitions of Magnetic Terms. It is one of the three fundamental passive electric-circuit parameters. The device whose inductance is under consideration is called the inductor. Most practical inductors are of the form of a single turn or several turns of wire. The inductance (denoted by L) of these practical inductors in general is given by

$$L = Kn^2\mu \tag{20}$$

where L = inductance, henrys
K = constant depending on the geometry of the coil
n = number of turns in the wire
μ = permeability of the medium inside and around the coil

When two inductors are serially connected the total inductance is given by the well-known relationship

$$L = L_1 + L_2 \pm 2M \tag{21}$$

where L_1 = inductance of coil 1
L_2 = inductance of coil 2
M = mutual inductance between coils 1 and 2

The plus sign is to be used if the windings are wound in the same sense, the minus sign if in the opposite sense.

The mutual inductance M of two inductors is given by

$$M = k\sqrt{L_1L_2} \tag{22}$$

where k = coefficient of coupling and depends on the geometry of the two coils

Equations (20), (21), and (22) show that there are three possible ways to vary the inductance (self or mutual) of inductors: (1) by varying the number of turns, (2) by varying the geometry, or (3) by varying the permeability of the surrounding medium. These three groups are discussed in greater detail as follows:

1. Varying the number of turns in an inductor is not very practical; however, varying the *effective* number of turns in an inductor can be easily done. If a wiper contact moves along the coil, the inductance between the wiper and fixed contact at one end of the inductor is a function of the distance between the wiper and the fixed contact. This device is the inductance potentiometer. Inductance potentiometers are used for indicating linear displacement if the inductor coil has a linear axis (solenoid), or for angular displacement, where the coil axis is circular (toroid). For further discussion of these devices refer to *potentiometers*, below.

2. Varying the geometry of a single inductor cannot be conveniently accomplished; consequently no such devices exist. But varying the geometry of two coils with respect to each other can be easily done, and the differential transformer (discussed under Differential Transformers) is a typical example of this method.

Another example is the variable-mutual-inductance transducer. In its simplest form this device consists of two coupled coils with air cores in close proximity. The mutual inductance of the coils is a function of the distance separating them. If an alternating potential is applied to one coil, the magnitude of the induced alternating current in the other is a measure of the mutual inductance and thereby indicates the distance between coils. This device is extremely inefficient since it depends on detecting small changes in relatively large quantities. The more practical evolution of this device is the differential transformer mentioned before.

3. Varying the permeability of the surrounding medium is the most convenient to accomplish; hence this is the most common way variable-inductance transducers are obtained. Since reluctance is the reciprocal of permeance a "variable-reluctance" gage is the same as a "variable-permeance" one. These are discussed in detail below.

The considerations above dispose of almost all inductance devices to be dealt with somewhere else. The only variable-inductance transducer that needs to be mentioned here is a liquid-level inductance gage. In this device[14] the liquid is permitted to rise in a nonmagnetic steel tube which forms the core of an inductor. If the liquid has different permeability from the air-filled steel tube, the height of the liquid will be indicated by a nonlinear variation of inductance of the inductor coil. It may be seen that this device could well be considered in group 3 as a variable-permeance transducer.

The detailed description of all remaining inductance transducers is included in other paragraphs. Only some general characteristics of the inductance devices will be mentioned here.

Since a steady-state d-c excitation is not affected by the inductance in the electric circuit and the transmission of rapid transients requires excessive bandwidth, inductance devices are normally used with a-c excitation (supply voltage).

In general the variation of both self- and mutual inductances is small compared with the total inductance. Consequently ratio or differential outputs are desirable. An additional advantage of this latter method is its lesser susceptibility to stray fields and variations due to changes in environmental conditions since their effects can be made to cancel in two coils.

An air-core inductor is desirable at high operating frequencies. In this case it can be made part of the r-f (radio-frequency) circuits whereby any connecting electronics is eliminated and the physical excitation (force, displacement, etc.) is directly transduced into a frequency-modulated r-f (or subcarrier) signal. The main disadvantage of the air-core inductor is that it generates a stray magnetic field well beyond the confines of the coil.

The disadvantage of stray fields may be overcome by using a core of high-permeability material such as iron (permeability of iron is 5,000 times that of air). The iron-core inductor is much less susceptible to stray fields than an air-core one but introduces other problems. Since permeability is not a constant, the inductance of the iron-core inductor will vary somewhat with current and it will be also frequency-sensitive. Iron-core inductors cannot be used at very high radio frequencies because of large hysteresis losses.

The main advantages of the variable-inductance devices are their low impedance levels (the d-c resistance of coils is usually of the order of a few ohms), low power losses, and high signal levels which frequently require no amplification at all.

The sensitivity of these devices can be greatly improved by the well-known bridge-circuit techniques.

Susceptibility to environmental changes can be controlled to a certain extent by differential or ratio connection, while additional compensation may be obtained by the serial insertion of small inductors with temperature coefficients opposite to that of the transducer, and temperature-sensitive resistors to compensate for the resistance change of the transducer.

Induction Transducers. Faraday's induction law states that

$$E = \oint \mathbf{E} \cdot d\mathbf{l} = -d/dt \int_s \mathbf{B} \cdot \mathbf{n}\, ds \qquad (23)$$

This equation can be simplified in most practical cases under consideration to

$$e_o = -n(d\phi/dt) \qquad (24)$$

where e_o = induced output voltage of a coil
n = number of turns of wire in the coil
$d\phi/dt$ = time derivative or change of the magnetic flux passing through the coil

This law forms, of course, the basis of the electric generator, most electric meters, and various displacement, velocity, and acceleration transducers.

The main advantage of the induction transducer is that it is a generator; i.e., it produces an electric signal directly from the mechanical input variable and thereby eliminates the need for an electrical-energy supply.

The induction transducer in its simplest form is a coil wound around a nonmagnetic hollow core in which a permanent magnet may be inserted (see Fig. 14). Any linear movement of the magnet will produce a changing flux within the coil, resulting in an output-voltage pulse. This method was successfully used to measure peak air pressures following detonations and obviously may be used for velocity measurements (since the output is proportional to $d\phi/dt$). If the permanent magnet is made to pass within several equally spaced serially wound coils, the acceleration is directly indicated by the change in spacing of the output pulses. The main difficulty of this method is reducing the friction between the permanent magnet and its housing.

The more common use of the induction transducer is in connection with continuous rotary motion. In this application a continuous a-c or d-c output signal is obtained. The device is then called a tachometer. Tachometers are essentially a-c or d-c generators. No appreciable power is derived from the device; rather the frequency of the a-c signal, or the magnitude of the (almost) open-circuited d-c voltage is a measure of the angular velocity. Care must be exercised to limit the output current to a small value such that it does not induce a counter emf (electromotive force) opposing the generated one which can introduce nonlinearities. A tachometer may use a permanent magnet or electromagnet for constant field generation within which a movable armature rotates. Another construction is that of fixed armature and rotating permanent magnet.

Tachometer sensitivities range between 1 and 20 mv/rpm; the maximum input range is of the order of 20,000 rpm.

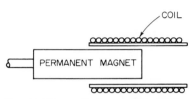

Fig. 14. Linear displacement induction transducer.

The main disadvantage of tachometers, especially in light and sensitive systems, is their inertia loading on the system. The output electrical energy plus the losses are all derived from the energy to be measured. The moving parts limit the life and the maximum input range of the device.

Other induction-type transducers are the various search coils for magnetic field measurements (magnetometer); these are basically the same as the elemetary induction transducer described before.

Induction transducers are being used for flow-velocity measurements with a turbine inserted in the liquid flow. To overcome the impeding of the flow by the turbine, a magnetic field may be applied across the conduit, and if the liquid has a measurable conductivity, a difference of potential will be developed across the conduit in a perpendicular direction to both flow and magnetic field.

Magnetoresistive Transducer. The operation of the magnetoresistive transducer is based on the phenomenon that metals, at low temperature, exhibit a change of resistivity due to an applied magnetic field.

Bismuth shows a marked change in resistivity even at room temperature; consequently it may be used for measuring the strength of the magnetic field.

Since the output variable, the change in resistivity, is highly temperature-sensitive magnetoresistive transducers are seldom used in telemetering.

Magnetostrictive or Magnetoelastic Transducers. Magnetostriction (alternately magnetorestriction), or the Joule effect, describes the phenomenon that the dimensions of a ferromagnetic sample change upon being subjected to a magnetic field. This is a reversible process; consequently, when a ferromagnetic sample is subjected to stress (resulting in strain) the permeability of the material changes. The latter phenomenon is called the Villari effect. Both effects are used for transducer purposes.

The extent of the Joule effect is illustrated in Fig. 15 for nickel.[15]

It is seen that the change is independent of the direction of the field, is a nonlinear function of the field, and exhibits a limiting value corresponding to the magnetic saturation of the sample. Among the large number of ferromagnetic materials that exhibit magnetostriction are the following:

Alfer (13 per cent aluminum, 87 per cent iron)
Monel metal
Permalloy 40 (40 per cent nickel, 60 per cent iron)
Cekas (11 per cent chromium, 59 per cent nickel, 28 per cent iron, 2 per cent manganese)
Other nickel-iron alloys (63 per cent nickel, 37 per cent iron typically)

Magnetostriction may be positive or negative, i.e., may result in expansion or contraction. It is a characteristic of the material. The extent of the magnetostrictive behavior, however, may be changed by heat-treating (elimination of internal stresses).

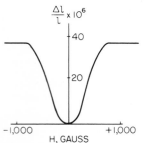

Fig. 15. Strain vs. magnetic field in nickel.

For practical transducers the Villari effect is used more frequently.[3] In its simplest form the magnetostrictive transducer consists of a highly magnetostrictive core with a coil wound around it. When an applied force causes stress in the core the resulting change in permeability causes a change in the inductance of the coil. The transfer characteristic is nonlinear, as shown in Fig. 16. The nonlinearity of the transfer characteristic is unavoidable since the typical magnetostrictive characteristics of most materials are highly nonlinear. In Fig. 17 the magnetostrictive characteristics of nickel and iron are shown qualitatively.

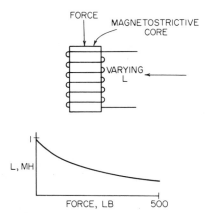

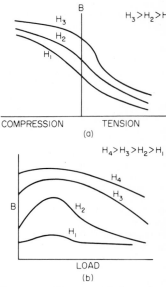

Fig. 16. Magnetostrictive transducer and its transfer characteristic.

Fig. 17. Magnetostrictive characteristics of nickel and iron. (a) Nickel. (b) Iron.

Magnetostrictive transducers are used mainly for measurement of forces (grams to tons). Care must be exercised to limit the extent of the deformity to the "reasonably linear" region of the transfer characteristics.

The high-frequency response of these transducers may be of the order of several 10^4 cps, which suggests their use for transient measurements.

The main disadvantages of the magnetostrictive transducers are (1) temperature sensitivity, which causes both zero drift and sensitivity changes, (2) hysteresis errors (of the order of 1 per cent), and (3) permanent changes due to excessive stress and aging.

Variable-reluctance (Permeance) Transducers. Referring to the discussion of Inductance Transducers above, a variable-reluctance gage operates on the principle that the variation of the permeance in the magnetic circuit will be manifested by a change in the effective inductance of an inductor. The change can be sensed by a change in the magnitude, frequency, or phase of an a-c signal derived from the inductor.

In most cases the bulk of the reluctance in the magnetic circuit is represented by a small air gap whose dimensions are being changed by the input variable. This method assures high sensitivity. For all variable-reluctance transducers displacement is the essential form of the input variable. Through displacement measurement, however, the variable-reluctance gage may be used to measure force, pressure, velocity, acceleration, etc.

The simplest form of the variable-reluctance gage which may be used for strain measurement is shown in Fig. 18a. The transfer characteristic is seen to be nonlinear, and the greatest sensitivity is obtained when the air gap approaches zero.

Obviously a differential or push-pull device will tend to compensate for these changes in sensitivity and therefore is much more desirable (see differential transformers and Fig. 19a and b).

The variable-reluctance gage shown in Fig. 18b is very similar to the induction transducer of Fig. 14 except that here the coil is excited by an a-c supply source. The moving iron core in this case causes a change in the effective inductance of the coil as shown in the transfer characteristic. Displacements up to several inches may be measured by this method.

A variable-reluctance thickness gage is shown in Fig. 18c. The reluctance of the test specimen is assumed to be much higher than that of the remaining magnetic path.

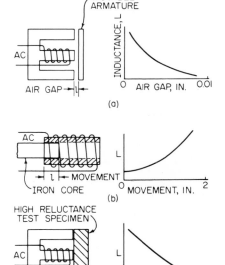

Fig. 18. Variable-reluctance transducers.

In this case the reluctance is proportional to the l/A (length over area) ratio of the test specimen. Since the area is proportional to t, the inductance of the coil L is inversely proportional to the thickness t, as shown in the transfer characteristic. Thicknesses from 0.001 to 0.1 in. with approximately 2 per cent accuracy may be measured by this method.

The push-pull versions of the devices of Fig. 18a and b are shown in Fig. 19a and b. These are inductance-ratio devices with two outputs. The linearity of the ratio output is greatly improved and temperature compensation is made much easier. An additional advantage of these devices is that they eliminate the force loading (due to Lenz's law) on the movable armature, which permits an additional increase in sensitivity.

The ratio-output devices point the way to the great variety of variable-reluctance transformers where the mutual inductance of two or more coils is also varied and whose most widely used prototype, the differential transformer, was discussed in detail above.

Saturable Reactors.* The saturable reactor operates by the principle that the permeability of a magnetic material (the core) varies with the amount of magnetization. Consequently, for a given magnetization any additional external field changes cause a variation in the differential permeability. In the near-saturation region of the permeability curve (the derivative of the hysteresis curve) small excitation-current changes will produce relatively large changes in permeability.

Saturable reactors are mainly used for control purposes. Their use for sensing is greatly limited by their notoriously poor high-frequency response.

Synchros.* Synchros convert a-c multiphase voltages into torque and angular position. The conversion is reversible, whereby the angular position of the synchro's shaft may be converted into voltages corresponding to the shaft angles. Synchros find extensive application in servomechanisms and in controls.

As transducers synchros may be used for sensing and measuring angular displacements up to 360°. They are usually extremely rugged and are not affected by temperature or humidity. The obtainable accuracies are usually within 0.25°.

Photoelectric Transducers

All devices whose principle of operation is based on the absorption of visible light energy are included here. A further subdivision based on the different physical reactions to the light absorption is necessary because of the large variety of photoelectric transducers.

In general the main advantage of photoelectric transducers is that light does not interfere or load in any way the measurand, or input variable. Almost all phenomena of interest can be measured by some application of a photoelectric transducer. Photoelectric transducers suffer generally from two major disadvantages, (1) the excessive cost of optical and supporting electronic instrumentation, and (2) their temperature sensitivity, which is a perennial consequence of the closeness of the visible and infrared bands of radiation.

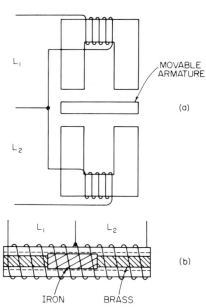

FIG. 19. Ratio output variable-reluctance gages.

Photoconductive Transducers. Photoconduction defines the phenomenon that the resistivity of most semiconductors decreases upon exposure to light. Consequently if a potential difference exists across a sample of semiconductor, the amount of current flow in the circuit is proportional to the intensity (and wavelength) of the incident light. A simple circuit to illustrate photoconductivity is shown in Fig. 20. The value of the external resistor R must be low because the photocurrent I_p (current induced by photoconduction) is also dependent on the voltage across the semiconductor E_S.

The most frequently used semiconductors for photoconductive transducers are

1. Selenium
2. Metal sulfides
3. Silicon
4. Germanium

* For an excellent detailed description and analysis, refer to J. G. Truxal (ed.), "Control Engineers' Handbook," McGraw-Hill Book Company, New York, 1958. See also Chap. 15.

Performance characteristics of these are tabulated in Table 3 reproduced from Ref. 3.*

Some of the column headings are somewhat further explained as follows: The transfer function may be either the variation of the resistance of a specific geometry or the resistivity of the material or the photocurrent variation of a specific sample with respect to variation in incident-light intensity. The transfer function is also dependent on the incident-light spectrum. Photoconductors can respond to wavelengths from thermal to x rays and even to electron bombardment.

The luminous sensitivity, the ratio of photocurrent to illumination in lumens, is a function of applied voltage; hence comparisons are of little value.

The condition in darkness is characterized by either the high "dark resistance" or the low leakage or "dark current." The condition when illuminated reflects the results characterized by the transfer function.

The quantum yield or amplification reflects the ratio of lifetime τ of the photon-liberated electron and the transit time T_r of carriers between the two ohmic contacts. If the lifetime (τ, time before recombination) is much larger than the time required for the generated free electrons to be swept to the electrodes by the applied field, considerable amplification is obtained.

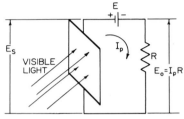

Fig. 20. Simple photoconductive transducer.

An obvious extension of the photoconductive semiconductors are the commercially available photodiodes and phototransistors. In the former if a hole-electron pair is generated in the vicinity of the p-n junction the photon-injected carrier will have a high probability of diffusing through the junction,[13] while in the latter the emitter is exposed to the incident light whereby normal transistor action takes place based on the photon-emitted carriers.

Photoemissive Tubes. There are three distinct types of photoemissive tubes: (1) high-vacuum diodes, (2) gas-filled diodes, and (3) photomultipliers.

These will be discussed in greater detail, but first some of the fundamentals of the photoemissive process are reviewed.[16]

Every metal is characterized by its "work function" E_W, which is the excess energy necessary to cause an electron to leave the surface of the metal. Einstein's equation

$$\tfrac{1}{2}mv^2 \leq hf - eE_W \qquad (25)$$

where m = electronic mass
v = velocity of the electron upon leaving the surface
h = Planck's constant
f = frequency of the incident light
e = electronic charge
E_W = work function of the metal

states that in order to liberate an electron with finite velocity, the photon must have an energy greater than eE_W. Since the photon has an associated energy content hf depending on the light frequency, there is a threshold frequency f_c for photoemission from a given metal, or

$$f_c = eE_W/h \qquad (26)$$

where the symbols are as defined above.

Consequently, photoemission at given light intensity is a function of the light frequency or wavelength. This selectivity alternately is termed photoelectric yield, relative response, quantum yield, spectral sensitivity, specific photosensitivity, etc. In either case it relates the photocurrent I_p to incident light in amperes per watt as a function of frequency.

* Reference 3 is an excellent book on transducers from which a large amount of information presented here was derived.

Table 3. Photoconductive-material Characteristics

Characteristic	Selenium	Cadmium sulfide	Cadmium selenide	Lead sulfide	Thallium sulfide	Silicon	Germanium
Transfer function R, ρ, or $I = f$ (incident light flux L)	Gray Se: I prop $L^{0.5}$; amorphous Se: I prop $L^{0.9}$	I proportional L^n; n, at low light level, 1–3; at high level, between 0.8 and 1	Current increases with light first steeply, later with gradually diminishing slope	Essentially linear	Current increases with light first steeply, later with gradually diminishing slope	Essentially linear	Essentially linear
Luminous sensitivity	Varies widely, 100–3,000 amp/lumen		Order of 10 ma/lumen	3 ma/lumen	1–1.5 ma/lumen	Order of 60 amp/lumen	30 ma/lumen at 2400°K color temperature
Condition in darkness	Gray Se, commercial cells, 10^4–10^7 ohms; amorphous Se layers, 10^8–10^{10} ohms	2×10^9 ohm-cm resistivity; powder can be made 10^6–10^{11} ohm-cm; commercial cells resistance 10^9 ohms	Dark resistance of commercial cells 10^{10} ohms	Technical cells 10^5–10^7 ohms	Varies widely, 10^4–10^8 ohms, frequently 10^6–10^7		Dark current 20 amp (ordinarily 1–5 amp)
Condition when illuminated	Gray Se: light flux of 0.1 lumen may reduce resistivity by factor of 3	Resistance at 1 ft-c reduced by factor 10^6	Resistance at 1 ft-c $5 \cdot 10^5$ ohms; resistance reduced from dark by factor of 10^6		Illumination by 0.25 ft-c decreases resistance by factor of 5	(I photo, I dark) $L = 12$–20	Photocurrent about 300 amp for 10 millilumens
Quantum yield or amplification	About 1	For X rays: 10^8 For α or β rays: 10^8–10^{10}			At max response, between 15 and 24		1
Voltage-current characteristics		For crystals essentially linear	Essentially linear			Essentially linear except at fields above 200 volts/cm	Current rises first with increased voltage, reaches saturation level which depends upon light level
Time response	Varies widely; gray Se 50 msec to minutes; amorphous 50 sec. Overswing at low voltages	Rise time for 10 ft-c, 30 msec; for 1 ft-c, 300 msec decay time for 10 ft-c, 10 msec; for 1 ft-c, 55 msec	1.5 msec at high light level, 15 msec at 1 ft-c sintered layer; 1 msec at 10 ft-c	0.1 msec	1.3 msec; varies widely with illumination and temperature	0.3 msec	Time constant, 10^{-6} sec
Effect of temperature	Dark current increases at higher temp; photocurrent increases steeply above 10°C	Sensitivity is minimum at 50°C; increases for decreased as well as increased temp. Temp hysteresis	Current increases 100% if temp varies from +25 to −25°C		Pronounced effect; dark resistance doubles if temperature reduced from 25 to 15°C		Temp increase from 20 to 55°C increases dark current by factor of 10, photocurrent by larger factor

3–26

Typical spectral-sensitivity curves are shown in Fig. 21.

The alkali metals show the highest photoelectric quantum yield. This is of the order of 5 to 30 per cent. The quantum yield depends upon both the composition and processing of the metal. Photoemission, however, is not restricted to metals only. Semiconductors also exhibit photoemission, but the photoconductive process dominates there and is more useful (because of the lower voltages necessary to collect the liberated carriers).

The performance of all photoemissive devices is limited by the dark current or noise. There are several sources for noise currents: thermionic emission, shot noise, Johnson's noise (the latter two are typical of all vacuum tubes), positive ion bombardment of the cathode, secondary emission, high field emission, etc.

The description of the three basic photoemissive tubes is as follows:

High-vacuum Diodes (Phototubes). The basic circuit of the vacuum photoemissive

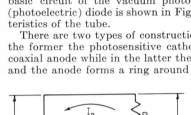

Fig. 21. Spectral sensitivity of the alkali metals.

(photoelectric) diode is shown in Fig. 22 with the typical voltage-current plate characteristics of the tube.

There are two types of constructions for phototubes: standard and end types. In the former the photosensitive cathode either completely or partially surrounds the coaxial anode while in the latter the cathode faces an opening on the end of the tube and the anode forms a ring around the opening. The walls of the tube are usually made of mica or quartz which do not inhibit the passage of ultraviolet or infrared light as conventional tube envelopes do.

There is a large variety of commercially available phototubes, typical plate-voltage ranges are from 10 to 18 to 10 to 200 volts for linear plate characteristics. The plate-current ranges are of the order of 0 to 20 μa corresponding to sensitivities of 100 μa/lumen or approximately 0.1 $\mu a/\mu w$ at the peak of sensitivity. As it may be seen from current values, vacuum phototubes are high-output-impedance devices and require voltage-sensitive connecting electronic circuitry.

The time response of these tubes is excellent (of the order of millimicroseconds), it is limited only by the typical tube capacitances (5 pf) and the transit time between cathode and plate.

Gas-filled Diodes (Phototubes). The sensitivity of the phototube is improved by filling the envelope at low pressure (0.5 mm Hg) with an inert gas such as argon or neon. The increased sensitivity occurs as the result of ionization by collision (Townsend discharge). Care must be exercised, however, never to permit a glow discharge which will permanently damage the tube because of cathode sputtering.

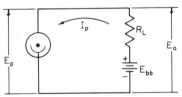

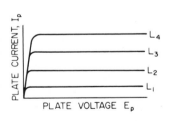

L_n = LIGHT FLUX $L_4 > L_3 > L_2 > L_1$

Fig. 22. Typical phototube circuit and characteristics.

Typical plate characteristics of a gas-filled phototube are shown in Fig. 23a, while a comparison of plate characteristics and sensitivity curves of two similarly constructed vacuum and gas-filled tubes is shown in Fig. 23b.

The photocurrent in the gas-filled tube is a function of the plate voltage because of the higher number of electrons from ionization by collision at higher accelerating potentials.

The frequency response of the gas-filled phototube is much poorer than that of the vacuum phototube because of the longer time required by the heavy positive ions to reach the cathode. This time is of the order of a millisecond corresponding to upper frequency cutoff of 1 to 5 kc.

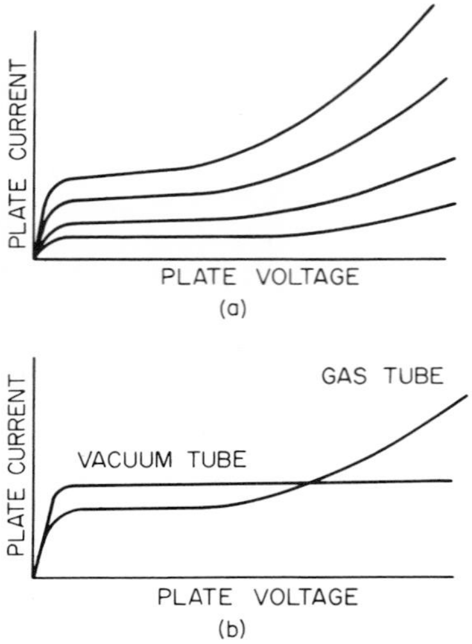

The linearity of the gas-filled phototube is usually poor (because of the voltage dependence of the plate current); consequently for transducer applications (except on-off switching) vacuum phototubes should be preferred.

Photomultipliers. A schematic diagram of the photomultiplier tube is shown in Fig. 24. Its series of electrodes are called the dynodes. The current-multiplication action of the tube is based on secondary emission caused by the electrons striking the successive dynodes. Nine to ten stages of dynodes are generally built in the photomultiplier. The dynodes are maintained at successively increasing potential, with 50 to 150 volts interelectrode potentials. This results in relatively very high voltage requirement (500 to 1,500 volts) but at low power levels. Pulse currents of the order of 1 amp may be obtained from a photomultiplier. Obviously the dark current is also multiplied, and to reduce it refrigeration is often used. This and its high-voltage requirement make the photomultiplier not very attractive for airborne applications, but it is widely used for measuring scintillations and in a large number of other industrial applications.

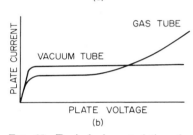

FIG. 23. Typical characteristics of a gas-filled phototube. (a) Gas-filled tube characteristics. (b) Comparison of similarly constructed vacuum and gas-filled tube characteristics at the same light intensity.

Photovoltaic Cells. Photovoltaic cells differ from the two previously discussed photoelectric devices mainly by being generator-type transducers; i.e., photovoltaic cells generate an emf (electromotive force) internally by the action of the incident light energy without the need of an external power source.

Various semiconductors such as selenium, germanium, silicon, and copper oxide are used for photovoltaic cells.

The cell normally consists of four distinct layers,[16] seen in Fig. 25a and b. These layers and their identifying letters in the figure are as follows:

M = metal electrode transparent to the incident light

BR = barrier layer, which is a thin insulating layer between the semiconductor and conducting base layer or metal electrode

S = semiconductor layer

BS = base layer, which is a good conductor and forms the second electrode

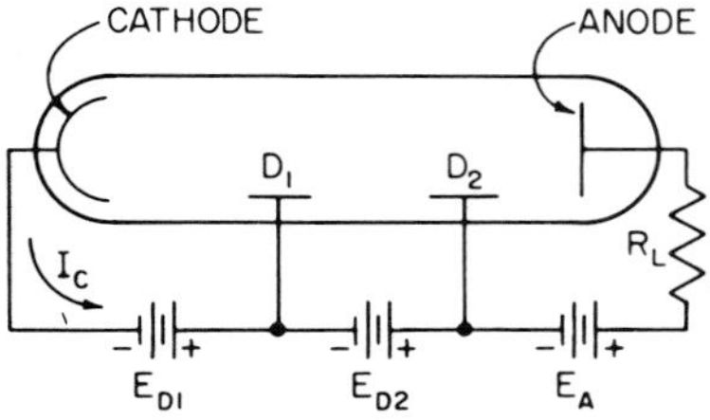

FIG. 24. Photomultiplier tube.

The copper–copper-oxide cell lends itself to both types of construction while selenium, germanium, and silicon cells are of the front-wall type.

The principle of operation of the photovoltaic cell is that the incident photon generates a hole-electron pair. The resulting field causes the electrons to flow through the barrier layer to one electrode and through the holes to the other. It is seen that this results in opposite conventional current flow for back-wall- and front-wall-type cells.

The transfer characteristic between generated emf and illumination is highly nonlinear, as shown in Fig. 26a. The generated emf is usually in the 100- to 600-mv range. The emf is independent of the illuminated area; however, the short-circuited current is a function of both the area and the intensity of illumination. Figure 26b illustrates that the short-circuited current of a cell with a given area is a linear function of the incident light flux but becomes increasingly nonlinear for increasing load resistance.

Selenium cells have sensitivities of the order of 1,000 μa/lumen while copper-oxide cells have only 100 μa/lumen.

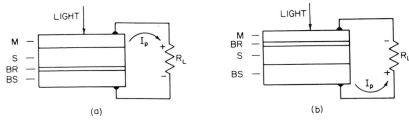

FIG. 25. Typical copper-oxide photovoltaic cell. (a) Back-wall type. (b) Front-wall type.

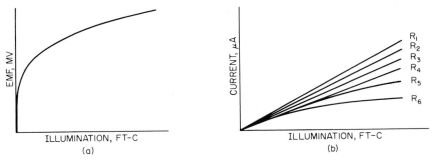

FIG. 26. Typical photovoltaic-cell performance curves. Increasing subscripts indicate logarithmically increasing load resistors. (a) Typical transfer characteristic. (b) Effect of the load resistance on the transfer characteristics.

The presence of a barrier layer accounts for an inherent capacitance across the cell which limits the high-frequency response of the cells. The approximate upper cutoff frequencies, in cps of the various cells are as follows:

Copper-oxide cells	100
Selenium	100
Germanium	10^6
Silicon	10^4

The main disadvantage of photovoltaic cells is their temperature sensitivity. This temperature dependence is also a function of the load resistance. The output current may vary from less than 0.1 to more than 1.0 per cent per degree centigrade, depending on load resistance values. Another disadvantage of photovoltaic cells is that they exhibit a fatigue effect, which requires frequent recalibration.

Piezoelectric Crystal

Several asymmetrical crystalline materials develop a potential difference between certain surfaces upon application of force or stress to the crystal. This phenomenon is known as piezoelectric effect. The materials that exhibit the piezoelectric effect are:
 Quartz
 Ammonium dihydrogen phosphate (ADP)
 Rochelle salt
 Tourmaline
 Barium titanates
Both natural and man-made piezoelectric crystals are available.

The piezoelectric effect is produced by the relative displacement of the asymmetrical charge distribution within the crystal. Since this distribution is dependent on the crystalline structure and orientation, natural piezoelectric crystals have a very high directional sensitivity. In synthetic crystals, such as rochelle salt and lately the barium titanates, the material can be formed into any desired shape, and any directional sensitivity may therefore be obtained or eliminated.

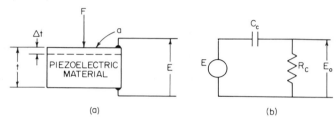

FIG. 27. Piezoelectric crystal. (a) Schematic. (b) Equivalent circuit.

The output voltage produced by the piezoelectric crystal is obtained[17] with reference to Fig. 27, in the following manner:

Young's modulus Y for the crystal is defined as

$$Y = \frac{F/a}{\Delta t/t} = \frac{Ft}{a\,\Delta t} \qquad (27)$$

where F = applied force
 a = area of the crystal
 t = thickness of the crystal
 Δt = change in thickness of the crystal

The induced charge across the surface is a characteristic function of the material and the applied force, i.e.,

$$Q = dF \qquad (28)$$

where Q = induced charge
 d = "piezo stress coefficient"
 F = applied force

Substituting (27) in (28) gives

$$Q = da\,Y(\Delta t/t) \qquad (29)$$

Since the two electrodes form a capacitor whose capacitance C_c is given by

$$C_c = k(a/t) \qquad (30)$$

where k = specific dielectric constant of the material, the developed voltage E is given by

$$E = Q/C_c \qquad (31)$$

Using (27), (29), and (30) in (31), one obtains

$$E = (dt/ka)F = VtP \qquad (32)$$

where V = voltage-sensitivity coefficient of the material, volts/cm/unit pressure
P = applied pressure, force/unit area

The numerical values of the various coefficients and Young's moduli applicable to the above derivation are tabulated in Table 4 for several piezoelectric crystals and materials (reproduced from Ref. 17).

Since the bulk material of the crystal has a finite conductivity the equivalent circuit of the piezoelectric crystal contains a loss or leakage resistance R_c, as shown in Fig. 27b. It is seen from the equivalent circuit that the piezoelectric crystal lacks d-c response; i.e., it cannot sense a nonvarying excitation. It has, on the other hand, an excellent high-frequency response up to about 20,000 cps. The high-frequency response is frequently limited by the wiring capacitances of the associated electronic circuits. Because of the high output impedance of the piezoelectric transducer, when it is used at a remote location, a preamplifier or impedance transformer in the vicinity of the crystal offers great advantages in preserving the high-frequency response of the crystal. Additional improvements in impedance matching may be obtained by parallel connection of several crystals, while increased sensitivity (and output impedance) is obtained by series connection.

The main advantages of piezoelectric transducers besides their good high-frequency characteristics are their simplicity, ruggedness, and marked insensitivity to aging.

The main disadvantage of piezoelectric transducers is the lack of d-c response which, however, may be circumvented by application of an external a-c carrier excitation. The thermal sensitivity of the man-made piezoelectric materials is generally higher than that of the natural crystals, and satisfactory performance is limited to temperatures below the Curie point, about 125°C for barium titanate. Most piezoelectric materials are sensitive to humidity and must be protected from a humid environment.

Resistance Devices

Resistance, which is the simplest electrical parameter, is a material property. Its magnitude is a function of the material composition, the geometry of the specimen, and environmental conditions which may change either or both of the previous two characteristics of a specimen.

The measurement of resistance follows the simplest of the basic laws of electricity (Ohm's law) and can be accomplished equally well by d-c or a-c methods.

The resistance of a given sample of material is given by two elementary equations of physics:

$$R = \rho(l/A) \tag{33}$$

and

$$R = R_0[1 + \alpha(T - T_0)] \tag{34}$$

where R = resistance of the sample at temperature T
ρ = resistivity of the material
l = length of the sample
A = cross-sectional area of the sample
R_0 = resistance of the sample at the reference temperature T_0
T = temperature corresponding to R
α = temperature coefficient of resistance for the material at T_0 (α is not necessarily a constant)

These two equations suggest a large number of transducer applications for the resistance device. These applications, which may be further subdivided on the basis of the type of resistance change that is used as an output indication, are as follows:

1. Change in resistance due to varying the mechanical contact with the device
2. Change in resistance due to the changing geometry of the device
3. Change in resistance due to the variation in resistivity of the device

Several transducers fall in each of the above categories; however, frequently the phenomena of 2 and 3 occur simultaneously, in which case nonlinearities and in general degradation of performance follow. These three categories and the most typical and

Table 4. Piezoelectric-crystal Characteristics

Characteristic	Bulk velocity V_s, cm/sec	Density	Piezo strain coefficient e, coulombs/sq cm	Young's modulus $E = V_s^2$, dynes/sq cm	Elastance (compliance) $S = 1/E$, sq cm/dyne	Piezo stress coefficient $d = cs$, coulombs/dyne	Specific dielectric constant K	Unit capacitance $C_0 = 8.85 \times 10^{-14} K$, farads/sq cm/cm	Unit open-circuit output $V = d/C_0$, volts/cm/dyne/sq cm
Quartz, X cut	5.7×10^5	2.65	0.176×10^{-4}	8.61×10^{11}	0.116×10^{-11}	0.0204×10^{-15}	4.5	0.396×10^{-12}	0.512×10^{-4}
ADP, 45° Z cut	4.92×10^5	1.80	0.493×10^{-4}	4.37×10^{11}	0.229×10^{-11}	0.113×10^{-15}	14	1.24×10^{-12}	0.91×10^{-4}
Rochelle, 45° X cut	2.4×10^4	1.77	5×10^{-4}	1.04×10^{11}	0.961×10^{-11}	4.81×10^{-15}	200	17.7×10^{-12}	2.72×10^{-4}
Rochelle, 45° Y cut	2.7×10^5	1.77	0.307×10^{-4}	1.29×10^{11}	0.775×10^{-11}	0.238×10^{-15}	10	0.885×10^{-12}	2.00×10^{-4}
Tourmaline, Z cut	7.54×10^5	3.0	0.333×10^{-4}	17.0×10^{11}	0.059×10^{-11}	0.0916×10^{-15}	5.5	0.487×10^{-12}	0.493×10^{-4}
Polycrystalline barium titanate	4.2×10^5	5.55	....	8.0×10^{11}	....	6×10^{-15}	1,200	106×10^{-12}	0.585×10^{-4}

commonly used transducers within them are discussed in greater detail in the following paragraphs.

Potentiometers. Devices which exhibit a change of resistance value between two contacts because of the mechanical motion of a contact are called potentiometers. A similar definition applies to inductance potentiometers, but since resistive potentiometers are used much more frequently, the latter will be considered here.

The simplest form of the potentiometer is a uniform-cross-section resistance wire or ribbon with a sliding contact as shown in Fig. 28a. From Eq. (33) it follows that the resistance measured between A and B is a linear function of the displacement of the sliding contact, assuming that the resistance of all connections and leads is negligible with respect to the resistance wire. Obviously this assumption becomes increasingly inaccurate as the sliding contact approaches point A; i.e., the resistance of the device approaches that of the external circuit. The main shortcoming of this device is the very low resistance (large currents) of the wire. More practical resistance-wire potentiometers consist of wire wound on an insulating core (the inductance of these coils must be neutralized by wiring half the coil in the opposite sense or by some other means). These potentiometers may be cylindrical or doughnut-shaped as shown in Fig. 28b, c, and d.

All potentiometers are basically displacement-measuring devices. It is seen in Fig. 28a and b that these devices indicate linear displacement while the potentiometers of Fig. 28c and d indicate angular position within a range of less than or more than 360°. The term electrical angle is frequently used to define the angle within which a unidirectional change in resistance value occurs. The electrical angle is less than 360° in c and is equal to 360° in d. In several instances the wiper will complete a revolution only if the shaft is rotated several turns; in this case the electrical angle is a multiple of 360°.

As it was stated before, potentiometers are basically displacement-sensitive devices; however, they are often used in connection with other sensors such as bellows, Bourdon tubes, or damped spring-mass systems which convert pressure, force, acceleration, or some other excitation into displacement variation.

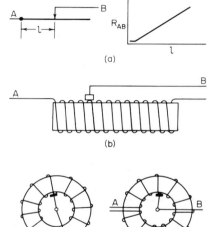

Fig. 28. Potentiometers.

Potentiometers can satisfy a great variety of desired transfer characteristics. By varying the length of the resistance wire per unit displacement of the sliding contact or by adding parallel resistors between different sections of the potentiometer, various nonlinear transfer characteristics are obtained. One of the most frequently used nonlinear potentiometers has a sine or cosine function as its transfer characteristics.

The main advantages of potentiometers are their relative simplicity, their flexibility in impedance levels and output signal ranges, and their suitability for d-c and a-c supply excitation.

The main limitations of potentiometers arise from their nonnegligible torque or force requirements and the noise that is generated within the device. The latter is due to

1. Nonuniform area of contact while the sliding contact is moving
2. Transients present during making and breaking contact with different loops of wire

3. Oxidation, foreign matter getting between coil and contact
4. Generation of emf between two dissimilar metals

Wirewound potentiometers were discussed above; however, others such as precision film and carbon-composition potentiometers are also commercially available. The main advantage of these is the improved resolution which becomes practically zero while with wirewound potentiometers it is the reciprocal of the effective number of turns of wire.

Strain Gages. Strain gages are possibly the most widely used transducer elements. Because of this, they are discussed in greater detail. Their main advantage lies in their small size, which makes them ideally suitable for remote sensing in missiles and space vehicles. There are various kinds of strain gages such as wire, film, foil, bonded, and unbonded. Their principle of operation, however, is the same; namely, the resistance of the element changes because of the stress-induced strain in the element.

To understand clearly the limitations of strain gages it is worthwhile to review the physics of a conductor under stress.

It is customary to define the "gage factor" G of the gage as

$$G = \frac{\Delta R/R}{\Delta L/L} \tag{35}$$

where $\Delta R/R$ is the fractional change of resistance and $\Delta L/L$ is the fractional change in length (which, of course, is the definition of strain). Within certain limits, G is nearly constant, and accuracies of measurement of the order of 0.1 per cent can be obtained. The following analysis is presented to demonstrate how the gage factor will vary outside these limits with temperature and state of stress.

The resistance of the element of wire is given by

$$R = \rho(L/A) \tag{36}$$

where ρ = resistivity of the material, ohm-in.
L = length of element, in.
A = cross-sectional area of element, sq in.

$$A = \pi D^2/4 \quad \text{for circular wire } (D = \text{diameter}) \tag{37}$$

If the primed and fractional terms in the following analysis refer respectively to the stressed and unstressed state of a circular element of wire, then the functional change in resistance is expressed as

$$\frac{\Delta R}{R} = -\left(1 - \frac{R'}{R}\right) = -\left(1 - \frac{\rho'}{\rho}\frac{L'}{L}\frac{A}{A'}\right) \tag{38}$$

where $L' = L + \Delta L$, $A' = A + \Delta A$, and $\rho' = \rho + \Delta \rho$, then

$$L'/L = 1 + \Delta L/L \tag{39}$$
$$A/A' = 1 - \Delta A/A = 1 - 2\Delta D/D \tag{40}$$
$$\rho'/\rho = 1 + \Delta \rho/\rho \tag{41}$$

Substituting these ratios into Eqs. (38) and (35) and neglecting all but first-order terms, it may be shown that

$$G = \frac{\Delta R/R}{\Delta L/L} = 1 - \frac{2\Delta D/D}{\Delta L/L} + \frac{\Delta \rho/\rho}{\Delta L/L} \tag{42}$$

For the circular wire, $(\Delta D/D)/(\Delta L/L) = -\mu$ (Poisson's ratio) is effectively a geometric constant. Hence, if only elastic straining is considered where the resistivity remains effectively constant, $\Delta \rho = 0$, then G remains invariant. Poisson's ratio for most alloy composition materials in the elastic range lies between 0.25 and 0.35, while for plastic materials or beyond the yield point in the plastic range $\mu \to 0.5$ or $G \to 2$. However, in general ρ does vary with state of stress and varies to an even greater extent with temperature; hence G must vary with variations in $\Delta\rho/\rho/\Delta L/L$. This ratio may be used as a criterion for the manufacture of gages with larger gage factors. Repeatability of the elastic gages requires that the applied stresses do not exceed the elastic limit of the wire. This limits the change in R to the order of a per cent. This small fractional change is one reason for using strain gages in bridge circuits. The other reason is temperature compensation. The resistance of any object depends on temperature, according to the equation

$$R = R_0[1 + \alpha(T - T_0)] \qquad (43)$$

The changes in resistance due to temperature are indistinguishable from those due to stresses. To circumvent this problem, an identical unloaded gage can be switched in for calibration purposes, or as it is customarily done, temperature effects may be canceled in a four-arm Wheatstone-bridge circuit. A deficiency in this method is that it depends on identical changes in gage characteristics due to temperature. In practice this is not necessarily the case.

In the preceding analysis, changes in geometry due to wire curvatures (where the strain-gage wire is bent back) were neglected. While this is justifiable for facilitating analysis, it may contribute to the nonlinearities of gage performance at high temperatures, with the exception of foil-type gages. The end loops in foil gages are purposely broadened for minimizing transverse and end effects.

For high-temperature applications, it is desirable to have a low temperature coefficient of resistance and desirable to have a high-resistivity wire. These characteristics for several currently used wires are shown in Table 5.

Table 5. Characteristics of Strain-gage Wires

Wire	Resistivity, ohms/cir mil-ft	Avg temp coefficient of resistance (in terms of apparent strain), μin./in.°F	G, gage factor
Advance..........	294	4	2
Nichrome V........	650	30	2
Pt Ir (180–120).....	200	80	6
Karma.............	800	2.5	2

Bonded gages are usually preferred for remote sensing. Unbonded gages are generally used for measuring very small forces. Bonding the strain-gage wire between two thin sheets of insulator facilitates attachment of the gage to structures and provides electrical and environmental insulation. Strict requirements are placed on the bonding material, since it partially determines the performance of the gage. The strength requirements of adhesives are sometimes referred to as "shear" and "peal." The successful transmittal of strain is almost exclusively dependent upon shear, except in the case of excessive bending where "peal" strength is required. The adhesion of the bonding material to both the surface and wire grid must be sufficient to transfer strain from the structure to each infinitesimal portion of the grid. It is to be noted that, when this happens, a parallel wire grid will indicate the average strain along the longitudinal axis of the wire. In order to measure localized strain, the gage length must be kept very short ($\frac{1}{32}$-in. gages are available). The bonding material must be free of creep at all values of expected stresses and temperatures, must prevent corrosion of the wire, must be nonhygroscopic, and must have a very high electrical resistivity which does not change significantly with temperature. Since strain gages are most frequently attached to metal, which is a conductor, this last requirement assumes the greatest importance.

Any analytical approach to assess the effect of both the bonding material and the method of curing and attachment on the high-temperature performance of strain gages is impossible because of the complexity of superimposed phenomena. The various empirical approaches used are well documented in the literature. Some illustrative examples are described here.

The synthetic-mica strain gage developed by G. A. Brewer[8] is one of the few gages which could fulfill the requirement for satisfactory performance up to 2000°F. At present, however, no available method has been found to attach it to steel for operation above 1500°F. The gage is fabricated by sandwiching the Nichrome V element between synthetic mica. At 2000°F and under 1,000 psi pressure, the sensing element reacts with the mica to form an integral gage. The best adherence was obtained with

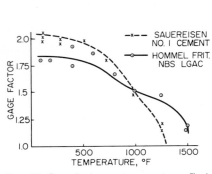

Fig. 29. Gage factor vs. temperature, G. A. Brewer synthetic-mica gage.

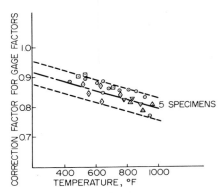

Fig. 30. Gage-correction factors vs. temperature, HT 100 gages.

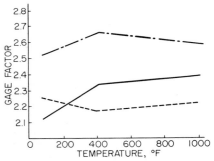

Fig. 31. Gage factors in tension.

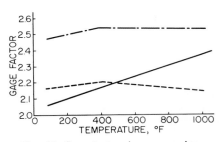

Fig. 32. Gage factors in compression.

fusing frits such as the National Bureau of Standards LGAC, while the best cements were the Sauereisen S-1 and S-31 (sodium silicate base). The variation of gage factor with temperature is shown in Fig. 29. While a gage alone with this kind of gage-factor variation is useless for in-flight measurements, the mere fact that it does perform at 1900°F holds great promise. If the deterioration of gage factor is consistently repeatable within reasonable tolerances, the simultaneous measurement of steady-state temperature at the gage location will enable one to apply a temperature-correction factor to the indicated strain and thereby obtain the actual strain with fair accuracy.

These considerations have led to the development of the combined "thermocouple strain gage."[19] Experimental tests show that curing methods affect the zero-drift characteristics of the gage. It is claimed that curing the HT-1200 type gage at 1250°F for 1 hr causes it to be drift-free up to 1000°F. An evaluation of several samples of the HT-100 (Karma wire) strain gage resulted in correction factors for the room-temperature gage factor as shown in Fig. 30. The effect of high temperatures on indicated strain is shown in Figs. 31 and 32. It is hoped that a new technique using a metallic-screen backing to facilitate high-temperature curing before installation of the gage will improve the uniformity of the gage performance.

An improved version of the bonded-wire strain gage is the recently developed and commercially available etched-foil strain gage. This type is produced by photo-etching techniques similar to those used in fabricating printed circuits. The lower temperature limit of these gages is $-300°F$, while for the upper temperature limits the manufacturer's specifications are reproduced in Table 6.

Table 6. Temperature Limits of the Baldwin-Lima-Hamilton SR-4 Foil Gages

Gage	Upper temp limit, °F		
	Static	Repeated dynamic	Random-peak (dynamic)
Foil:			
Constantan..................	600	600	600
Nichrome V*			
Base:			
Paper......................	160	180	200
Epoxy......................	160	180	200
Phenol resin (Bakelite)........	300	400	500
Cement:			
Nitrocellulose (DuPont)	160	180	200
Epoxy (CIBA-502)...........	160	180	200
Phenol resin (Bakelite)........	300	400	500
Ceramic (RX-1).............	600	600	600
Ceramic (Allen P-1)..........	1100	1300	1400

* The upper temperature limits of Nichrome V are set by the electrical conductivity of the ceramic cements.

Some test results[20] for a group of gages made from Alloy D developed by Armour Research Foundation and designed by the Enameled Metal Laboratory of the NBS, and an experimental group of gages similar to the HT-100 (mentioned earlier) are presented in Figs. 31 and 32. These curves are for NBS gages, and for High Temperature Instruments Corp. experimental gage. In all cases, Allen P-1 cement was used for attachment. Some of the conclusions were that the drift behavior of these gages between 400 and 900°F makes static measurements impractical, and the variation of gage factor with temperature depends on the thermal treatment of the wire prior to fabrication.

Another approach to high-temperature strain gages is the use of deposited metal or alloy film instead of wire. The film can be deposited on nonconducting test surfaces, thus eliminating cements or adhesives. This type of gage is under development at the Mechanical Instruments Laboratory of the NBS.

The performance-characteristic variations with temperature discussed previously were obtained under laboratory conditions, and the measurements were made at static temperature-equilibrium conditions. Even under these conditions, there is not a strain gage known today whose absolute reading at temperatures above 300 to 400°F can be accepted as reasonably accurate.

Attempting to measure stress transients with simultaneous heat-flux transients (which is frequently necessary) is indeed very difficult. One standard approach to overcome this problem is to obtain a sufficiently long thermal time constant to make the transducer at least temporarily insensitive to temperature in these. For the case of simultaneous temperature and strain monitoring, the uncertainties of the transient effects of two transducers make the correlation very difficult. It is believed that transient measurements at high temperatures present the most serious of the numerous problems strain-gage designers are faced with.

The high-temperature problems of strain gages overshadow the ones associated with cold temperatures. In many applications, however, where liquid-fuel propellants are employed parts of the vehicle structure will reach temperatures as low as $-300°F$; hence gage performance at this temperature must be satisfactory. The gage-factor variation of several commercially available Bakelite gages[21] is shown in Fig. 33. The change in gage factor between 0 and $-300°F$ is within 10 per cent. An additional conclusion of the above-referenced investigation was that the amount of zero shift after cycling between room temperature and $-320°F$ is small. The gages and the bonding cements do not exhibit adverse effects after subjection to temperatures as low as $-320°F$.

There has been a considerable increase in the use of semiconductors for strain gages.[25] Semiconductor strain gages have a much higher gage factor than conventional gages (100 and higher). Extremely small strains (microinches per inch) can be measured. Silicon is the most common material. Gage factor and sensitivity to temperature are both lowered by doping (increasing impurity concentration). Flexibility and linearity are also affected.

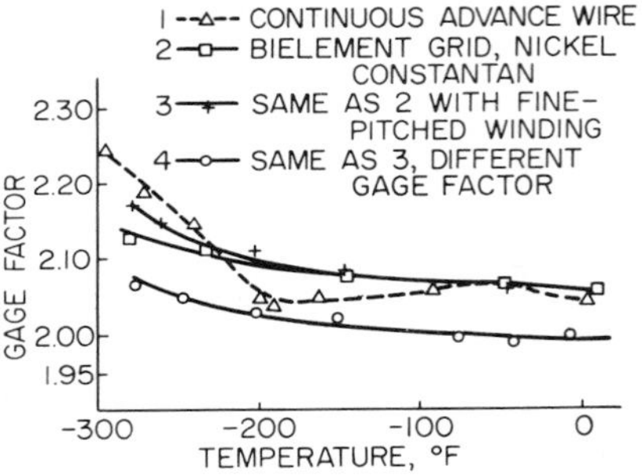

Variable-resistivity Gages. Resistivity of materials is affected by several environmental factors such as temperature, pressure, humidity, and incident radiation. Consequently various transducers operate on the principle of resistivity changes. Since conductivity is the reciprocal of resistivity the two principles of operation are identical. Accordingly some devices already described could be listed in this paragraph (photoconductive transducers, for example). The most significant application of the variable-resistivity gage is the resistance thermometer or resistive temperature gage.

FIG. 33. Gage factor vs. temperature, Bakelite gages.

Curves:
1 --△-- CONTINUOUS ADVANCE WIRE
2 --○-- BIELEMENT GRID, NICKEL CONSTANTAN
3 --+-- SAME AS 2 WITH FINE-PITCHED WINDING
4 --○-- SAME AS 3, DIFFERENT GAGE FACTOR

One type of variable-resistivity temperature gage utilizes the change in resistance of a conductor due to temperature. Since these changes are small the temperature-sensitive element is usually incorporated in a Wheatstone bridge which, through its balanced-type circuit, greatly increases the sensitivity of the device and the ease of detection. Performance of the bridge becomes very similar to the strain-gage bridges. The desired characteristics of the thermal-sensitive element are high resistivity to restrict the length of the elements to reasonable limits, and high temperature coefficient of resistance to increase sensitivity.

Platinum wire has been used traditionally as the thermal-sensitive element because of its repeatable resistance-temperature characteristics. The temperature coefficient of resistance of platinum, however, is rather low (0.003 at 20°C); similarly its resistivity is low (10 ohm-cm at 20°C). Other wires with higher resistivity and temperature coefficients of resistance, such as copper, exhibit poor repeatability and poor mechanical strength due to brittleness.

A recently developed device, the platinum-film temperature gage, is an improvement over the platinum-wire gage. This device is produced by depositing 2.5- by 10^{-7}-in.-thick films of platinum under controlled conditions. Its advantages are its high resistance in small volume, fast response (10^{-8} sec response for special units, and in general less than 1 msec response is claimed), and the wide choice of configurations which are attainable with resistance values up to 10,000 ohms. The platinum film is applied either by painting, by vacuum evaporation, or by sputtering. Good resistance to vibration and shock, and capability of experiencing 1000°F thermal shock without cracking are claimed. Operation between liquid-nitrogen temperature and 1000°F is normal, and higher-temperature operation is contemplated.

The principal limitation of these resistance temperature gages is range. When a resistance gage is available for a given range, however, it is preferable over thermocouples.

Thermistor semiconductors (such as various carbon compositions, silicon carbide, and silicon boron) exhibit a change in resistance with temperature according to the equation

$$R = Ae^{b/T} \tag{44}$$

where R = resistance, ohms
A, b = constants within a restricted temperature range which depend on the material properties
e = base of natural logarithms
T = temperature, °C

Differentiating Eq. (44), it can be shown that

$$dR/dT = -(b/T^2)R \tag{45}$$

From Eqs. (44) and (45), it can be seen that the resistance is a decreasing nonlinear function of temperature with the slope (sensitivity) negatively decreasing with increasing temperature. If one defines the sensitivity S as the ratio of the fractional change of resistance to the change in temperature

$$S = \frac{\Delta R/R}{\Delta T} \tag{46}$$

values of 2 to 5 per cent per degree centigrade are obtained in the usable range of temperatures. This is about ten times that of platinum. Until recently, the usable range was considered to be between -100 and $+600°F$, which compared unfavorably with other temperature devices. Other objectionable features of thermistors are (1) long time constants of the order of 0.5 sec (which is a function of the rate of heat flow around the thermistor); and (2) fragility of the unshielded small beads and the sluggishness of the more rugged bulb-mounted beads, rods, and disks. Opinions on reliability and stability vary, but some of the recently developed types are excellent.

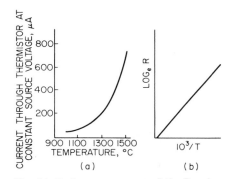

Fig. 34. Performance curves of the Russian 2700°F thermistor.

Very recently the lower temperature limit of practical thermistors[22] was extended to $-445°F$ (which is only 16°R above absolute zero!). This is a significant accomplishment in view of the fact that the semiconductor approaches the characteristics of an insulator in this range because of the exponential increase in resistance with decreasing temperature.

In Russia, thermistors usable up to 1500°C (2700°F) are being developed.[23] Since semiconductors approach the conductivity of metals at around 1000°C and become useless as thermistors, the Russian approach is to investigate the conductivity behavior of insulators at these temperatures and apply it to high-temperature thermistors. Performance of such a high-temperature thermistor made by sintering alumina tablets to silite (carborundum) electrodes is shown in Fig. 34. The maximum dispersion of the experimental points is within ±10 per cent of the curve. In detail b, Fig. 34, the natural logarithms of the computed resistance values corresponding to detail a are plotted vs. the reciprocal of the temperature term. This indicates that

$$\ln R = A(B/T) \tag{47}$$

or

$$R = Ae^{B/T} \tag{48}$$

which is the familiar expression for semiconductors.

Thermocouples

Temperature measurements made by thermocouples are probably the most widely used method at the present time. Thermocouples cover the widest range of temperatures and are the only electrical devices capable of measuring temperatures above 2000°F. Thermocouples operate on the basis of the well-known effect of thermoelectricity.[24] When two wires of different materials form a loop with two junctions of the dissimilar materials, a difference of junction temperatures will produce a voltage proportional to the temperature difference. The constant of proportionality depends on the type of materials used.

Table 7. Thermocouple Characteristics

Type		Composition,* %		Temp range, °F	Max temp, °F	Melting* temp, °F		Polarity*	Limit of error,† %
Platinum	Rhodium Platinum	100	10 90	0 to +2650	3100	3220	3320	− +	±0.5
Platinum	Rhodium Platinum	100	13 87	0 to +2650	3100			− +	±0.5
Chromel	Alumel	90 Ni 10 Cr	95 Ni 2 Al 2 Mn 1 Si	0 to +2200	2450	ca. 2600		+ −	±0.75
Iron	Constantan	99.9 Fe	55 Cu 45 Ni	−300 to +1400	1800	2790	2170	+ −	±0.4
Copper	Constantan	99.9 Cu	55 Cu	−300 to +700	1100	1980	2170	+ −	±0.4

* In order of listing under type.
† According to the Recommended Practice RP1.1 of the Instrument Society of America for new devices.

The most widely used metal and/or alloy thermocouples are tabulated in Table 7 with some of the pertinent characteristics of each. The corresponding open-circuit electromotive forces (emf) produced as a function of junction-temperature difference are shown in Fig. 35. The curves are not entirely linear; hence accurate calibration curves must be obtained prior to use.

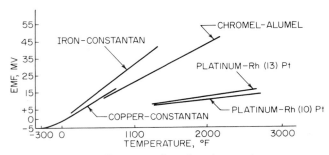

Fig. 35. Thermocouple emf vs. temperature.

Another advantage of thermocouples (besides the wide range) is the fast response times obtained. Exposed thermocouples with response time of 0.25 sec are available, and even faster responses can be obtained by special hypodermic construction.

The problem areas, and in general the disadvantages of thermocouples are discussed below:

Output signals, as shown in Fig. 35, are very small and require amplification.

Without amplification they are certainly not suitable for commutation, which is normally required because of the great number of measurements taken.

The reference junction must be maintained at a known and constant temperature since the indicated reading is, at most, as accurate as the reference temperature. Extremely good temperature control is required at the reference junction. This may involve complex thermal insulation and dynamic temperature-control circuitry. Even in this case, some means of temperature compensation, such as compensating resistors, is used to balance the cycling of the control unit. An alternate method is varying a constant small voltage in series with the thermocouples to compensate for reference-junction temperature variations.

Another source of inaccuracy in thermocouples is the susceptibility of the junction to corrosion and oxidization and to interdiffusion of the different alloys at elevated temperatures. Protection from corrosion and oxidization can be obtained by using gastight metallic or ceramic walls or tubes. These, however, cause a deterioration in the time response of the thermocouple. Since the thermocouple indicates its own tip (junction) temperature, a difference may exist between this temperature and the one it is to measure. Heat conduction of heavy wires, insulation, shields, etc., adversely affects the accuracy of readings. Thin wires, on the other hand, are much more susceptible to calibration changes due to mechanical working, twisting, or bending. Excessive vibration and repeated heating and cooling or short exposures beyond the operating temperatures will adversely affect thermocouple performance.

REFERENCES

1. Ronald K. Jurzen, Transducers in Measurement, *Electronics*, vol. 31, no. 27, July 4, 1958.
2. G. E. White, The Meaning of Natural Frequency, *Stratham Instrument Notes*, no. 12, November, 1949.
3. Kurt S. Lion, "Instrumentation in Scientific Research," McGraw-Hill Book Company, New York, 1959.
4. *Natl. Bur. Std. News Bull*, vol. 42, no. 1, 1958.
5. Paul S. Lederer, *Natl. Bur. Std. (U.S.) Rept.* 4541.
6. C. L. Cutting, A. C. Jason, and J. L. Wood, *J. Sci. Instr.*, vol. 32, p. 425, 1955.
7. Notes on Linear Variable Differential Transformers, *Schaevitz Engineering Bull.* AA-1A.
8. Howard C. Roberts, "Mechanical Measurements by Electrical Methods," The Instrument Publishing Co., Inc., 1951.
9. James M. Cork, "Radioactivity and Nuclear Physics," D. Van Nostrand Company, Inc., Princeton, N.J., 1950.
10. B. B. Rossi and H. H. Staub, "Ionization Chambers and Counters," McGraw-Hill Book Company, New York, 1949.
11. David B. Kret, "Transducers," Allen B. DuMont Laboratories, Inc., Passaic, N. J., 1953.
12. John D. Kraus, "Electromagnetics," McGraw-Hill Book Company, New York, 1953.
13. William Shockley, "Electrons and Holes in Semiconductors," D. Van Nostrand Company, Inc., Princeton, N.J.
14. E. W. Pulsford, *J. Sci. Instr.*, vol. 32, p. 362, 1955.
15. L. A. Petermann, Producing Motion with Magnetostrictive and Piezoelectric Transducers, *Elec. Mfg*, December, 1955.
16. J. Millman and S. Seely, "Electronics," 2d ed., McGraw-Hill Book Company, New York, 1951.
17. A. I. Dranetz, G. N. Howatt, and J. W. Crownover, Barium Titanates as Circuit Elements, *Tele-Tech*, April, 1949.
18. G. A. Brewer, "Synthetic Mica High Temperature Strain Gage Research," High Temperature Instruments Corp., ASTIA Document AD 155-560.
19. L. Herczeg and P. Beckman, "High Temperature Strain Gages," High Temperature Instruments Corp.
20. R. L. Bloss and C. H. Melton, Evaluation of Resistance Strain Gages at Elevated Temperatures, *Natl. Bur. Std. (U.S.) Rept.* 6117, ASTIA Document AD-202-419.
21. D. J. Madsen, Measuring Strain at Extreme Temperatures, *Elec. Equipment Engr.*, February, 1959.
22. H. B. Sachse and Z. W. Vollmer, Thermistor Sensing Elements for −445°F, *Electron. Ind.*, February, 1959.

23. N. T. Oreschhin (translated from Russian by R. C. Murray), "Thermistors for High Temperature," ASTIA Document AD-137-980.
24. W. G. Holzbock, "Instruments for Measurement and Control," Reinhold Publishing Corporation, New York, 1955.
25. Mills Dean and R. D. Douglas, "Semiconductor and Conventional Strain Gages," Academic Press, Inc., New York, 1962.

4 Commercially Available Transducers

The Instrument Society of America has undertaken the compilation of data on transducers. Such data is to be found in their publication "ISA Transducer Compendium," edited by Emil J. Munnar, published in 1963, and distributed by Plenum Press, New York. Other compilations have been made by the Allen B. DuMont Laboratories in 1953 and by Aeronutronic Systems, Inc. for the U.S. Air Force in 1958.

It is very difficult to keep such data complete and up to date. Improvements in materials change the sizes and weight of transducers, and occasionally an improved principle of measurement will greatly change transducer parameters. It is, therefore, most desirable that the ISA has undertaken this work; the reader is referred to that source or to the manufacturers for detailed information.

TRAJECTORY INSTRUMENTATION

5 Sensing of Position

The field of electronic trajectory-measurement systems is quite extensive. From its beginning with the first developments of pulse radar, Doppler, Raydist, and interferometer systems about twenty years ago, it has now become a family of many systems. This section deals with the uses of these systems, the basic principles they employ, and in brief form the techniques applied in the individual systems. Regarding the latter, a glossary-type description of each of the systems will be provided.

Consider first the role of the systems used for missile-trajectory measurement. Stated simply, these systems tell us where the missile is and what its motions are. The trajectory, of course, is the path of the missile and is ordinarily described by a series of space-position points measured with respect to some earth coordinate system. The velocity, acceleration, and sometimes higher-order derivatives are also needed and may be obtained either by direct measurement or by differentiation of position data. In addition to the foregoing, which are strictly parameters of the trajectory path, the trajectory-measurement systems often aid in other external performance measurements, such as attitude, angle of attack, roll, drag, and miss distance. The task facing the systems in obtaining these measurements cannot be simply stated. It may vary with type of missile, test purpose, and circumstances encountered. It seems that all too often the measurement requirement is just a little beyond our technological boundary.

Some examples will illustrate the severity of the measurements necessary to support our missile programs. First, consider the ballistic missiles, for these present some especially difficult problems. If a 5,000-mile missile has a velocity error at burnout of 9 ft/sec, which is about 0.02 per cent, or a heading error of $0.03°$, it will miss its target by 3 to 5 miles, depending on the trajectory. For the instrumentation systems to be of sufficient value in perfecting the missile system, it is usually necessary that they be five to ten times more accurate than the missile-system design tolerance. It follows then that the missile velocities have to be measured to better than 1 ft/sec, and it is not uncommon that this must be accomplished while the missile is 100 to 200 miles from the instrumentation sites. The shorter-range ballistic missiles are often just as difficult to accommodate because of their much smaller allowable target dispersions. Position accuracies to less than 1 ft and velocity accuracies to $\frac{1}{100}$ ft/sec are often necessary.

The requirements for the nonballistic missiles are usually not so stringent accuracy-wise, but many times the test circumstance complicates the task. Long-range missiles which hug the ground and missile target engagements which take place beyond the horizon are the most troublesome situations.

6 Techniques for Electronic Trajectory Measurement

There are now approximately thirty-five electronic trajectory-measurement systems in existence on the test ranges and the outlook is for quite a few more in the years to come. Each of these systems has distinctive characteristics and capabilities. However, even though the systems are outwardly dissimilar, they stem from common phenomena and are developed around a few basic measurement principles. Because of this basic similarity, it is possible to treat the systems categorically and thus simplify the individual system descriptions.

The applicable phenomena in basic physics are simply those concerning propagation of electromagnetic waves. The theories of their behavior are common and need not be presented here. Let it suffice that the most useful effects—those which permit the existence of the electronic trajectory-measurement systems—are that propagated waves have determinable frequencies, wavelengths, velocity, attenuation, and directivity. As a signal traverses some path between a transmitter and a receiver, these properties relate parameters of that path to the measurable quantities of propagation time, phase, delay frequency shift, and amplitude. This, then, is the foundation for the basic measurement techniques used in the electronic trajectory-measurement systems. Amplitude measurement, except as it applies to directivity in antenna point, is not a practical tool for accurate determination of propagation-path parameters. So, it will not be further discussed in this paper. The other measurable quantities—propagation time, phase delay, and frequency shift—are closely related and so are the techniques developed around them. For example, a measurement of propagation time is actually a direct time measurement of the phase delay of the signal. The two terms are essentially synonymous in that the measurements tell us the same thing. Frequency measurement, which is made to detect some change in frequency, is related to phase and time measurement in that it represents a phase derivative which can be integrated to produce the same information. It follows, then, that the basic techniques, of which there are three, all use some form of the phase-delay measurement. They are (1) phase-delay or direct propagation-time measurement, (2) phase-rate or Doppler-frequency measurement, and (3) relative phase-delay measurement. Each of these techniques, and the primary measures each provides, will now be briefly presented.

Phase-delay or Propagation-time Measurement

A measurement of phase delay or propagation time is directly relatable to the length of the transmission path. In missile instrumentation systems the measurement is made for a round-trip path, such as from a ground transmitter to the missile and return to the ground, as illustrated by Fig. 36, so that accurate elapsed-time and phase measurements may be obtained from a direct comparison of the times of occurrence of the transmitted and received signals. This comparison is ordinarily performed on the ground and the missile serves only as a signal relay by reflecting the signal or first amplifying and then retransmitting it. When the ground receiver is located with the transmitter, the round-trip distance is simply twice the distance from the transmitter to the missile and the direct range is thus easily obtained. In many applications, receivers are located remotely from the ground transmitter and comparison is made either at the remote receivers or back at the transmitter. These arrangements are illustrated by Fig. 37. For these cases, the measurement represents the distance from the transmitter to the missile to the remote receiver, and the total path traversed is commonly called "loop range."

If range measurements only are used for missile space-position determination, measurements from three or more separated locations are necessary. Position determination through three direct range measurements entails mathematical solution of the intersection of three spheroids. If loop range measurements rather than direct range measurements, are employed, the problem is finding mathematically the intersection of three ellipsoids.

Either pulsed or CW signals may be used for propagation-time measurement. The basic principle is essentially the same for both types; however, equipment and circuit techniques differ. When pulsed signals are employed, the elapsed time from transmission to reception is measured directly. Usually this is done by gating a fixed high-frequency clock. The elapsed transmission time thus obtained is directly convertible to the length of the transmission path. When CW signals are used to obtain a phase-delay measurement, the comparison is somewhat more complex. This is because the relatively short wavelengths of usable r-f signals result in many cycles of phase shift being incurred for normal distances measured. To avoid the ambiguity which would exist if the carrier alone were used, additional ranging frequencies are used to modulate the carrier. The phase delay of each of these is then measured. The modulating frequencies are selected so that the lowest will encounter less than one cycle of phase shift for the maximum distance to be measured and the highest will permit the desired range resolution. The number and spacing of other

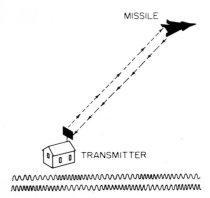

Fig. 36. Phase-delay range measurement.

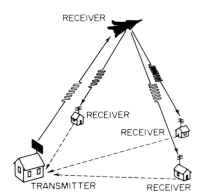

Fig. 37. Multistation phase-delay measurement.

frequencies needed is governed by the phase-resolution capability of the ambiguity-elimination circuitry of the particular system. Other types of modulation, such as linear frequency sweeping, may be used in lieu of fixed ranging frequencies. The phase measurement for such systems tends to become more complicated, but more usable information can be transmitted over a given bandwidth. This is increasingly important in the higher-frequency systems.

Phase-rate or Doppler Measurement

When there is relative motion between a signal source and a receiver, the received signal will have an apparent frequency shift. This, of course, is the Doppler principle. In trajectory-measurement systems it is conventional to extract the Doppler component of frequency so that this may be measured directly. This Doppler effect is then represented by a relatively low frequency signal which has a frequency directly proportional to the radial velocity of the source and which has an accumulated phase over any specified period proportional to the change of radial separation. To achieve accurate Doppler measurement, the measurement is usually obtained from a direct subtraction of transmitted and received frequencies for a round-trip path (as shown in Fig. 38) such as was described above for the phase-delay technique. The Doppler frequency thus detected is proportional to the rate of change of the path length from the transmitter to the missile to the receiver. When the transmitter and receiver are physically located together, the Doppler frequency is proportional to twice the radial velocity of the missile. When the ground receiver is remote from the transmitter, the Doppler frequency is proportional to the rate of change of the loop range from transmitter to missile to receiver.

Trajectory data are obtained from the Doppler signal by any of several methods. When velocity only is desired, it can be obtained directly from radial-velocity measurements either by combining three measurements as components of the velocity vector, or by using only one radial-velocity measurement with data from other sources which permit its adjustment to the total vector. More often the trajectory path is desired in the form of a series of space-position measurements. While these could be obtained from direct integration of velocities, it is more direct and more precise to compute direct or loop distances from accumulated Doppler cycles for each of a number of receiver stations. When this approach is employed, the product of the Doppler-type system is the same as that of the direct phase-measurement system. The significant difference of the two methods is only that the Doppler technique requires a continuous integration of phase shift while the phase-measurement technique uses direct measurement of phase. As we leave this technique, please note that phase-rate, or Doppler-type, systems depend on movement of the transmitting source as a basis of measurement. A Doppler-type system will not provide a sensible measurement in a static circumstance.

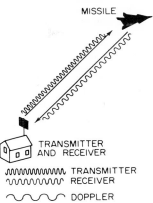

Fig. 38. Doppler measurement.

Relative Phase and Interferometer-type Measurement

The third basic technique makes use of relative, rather than absolute, phase measurement. Figure 39 illustrates this technique. In this case, the desired measurement is the difference in phase of a signal emanating from a missile as it is received at two or more ground receiver stations. The relative phase measurement is representative of the difference of the lengths of the transmission paths. One such phase-difference measurement tells only that the missile signal source lies on the surface of a hyperboloid which has as its transverse axis the line connecting the two receivers, and has a focus at one of the receiver stations as determined by the sign of the phase measurement. Location of the missile position involves the solution for the point of intersection of three hyperboloids obtained from phase measurements of three pairs of receivers.

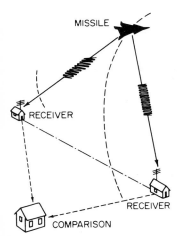

Fig. 39. Relative phase measurement.

In making relative phase or time measurements, either signals originating in the missile or signals originating on the ground and relayed by the missile may be used. While these can be either pulsed or CW signals, CW signals are almost always used because of the difficulty in obtaining adequate resolution in relative phase or time measurement with pulse techniques. A phase comparison may be made for the carrier itself or for other frequencies modulating the carrier. Nonambiguous measurements are possible if modulating frequencies are used and if the lowest of these has a wavelength longer than the distance between ground receivers. Where this is not the case, ambiguity will be present. When measurements are made using relatively short wavelength modulating frequencies or the carrier itself, it becomes necessary to begin at some reference point, such as a missile-launch position, and rely on continuous phase accumulation throughout the trajectory. The problem here is similar to the integration necessary when position is derived from the Doppler-type measurement.

When the separation of the receivers is very short relative to the distance to any measured point it can be assumed that the measured point is on the asymptotic cone of the hyperboloid. The system is then considered to have conical rather than hyperbolic geometry with each generated cone having its apex midway between the pair of receiving antennas. The relative phase measurement obtained is thus directly proportional to the direction cosine (see Fig. 40) of a line which lies on the surface of the cone and which extends outward from the midpoint of the baseline axis. The position of the missile can then be found by solving for the space intersection of three cones obtained from relative phase measurements from three antenna pairs. This short-baseline relative-phase-measurement arrangement approximates an interferometer, and this is the common designation for this type of system.

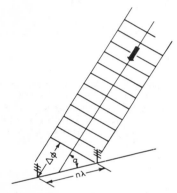

Fig. 40. Electronic interferometer.

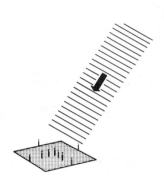

Fig. 41. Crossed-baseline nonambiguous interferometer.

Here again, the ambiguity problem arises. If the receiving antennas are separated by a number of signal wavelengths, then twice that number of possible cones will be generated. To obtain sufficient accuracy of the phase measurement, however, it is usually necessary to use antenna spacing of such a number of wavelengths. To solve the resulting ambiguity problems, it is common practice to use several antennas on one baseline, as shown by Fig. 41, with the longer spacing providing the desired direction-cosine resolution, and the shorter spacing being used to eliminate the ambiguity. The ambiguity could also be removed by use of multiple frequencies or modulated signals, but this is not done in practice because it would require either a number of widely separated frequencies or excessive bandwidths.

One other important aspect of the interferometer systems is that often two perpendicular intersecting baselines are established at one site location. These then yield two direction cosines which may be combined with a third obtained from some remote point or with a single range measurement for determination of the missile space position.

7 Electronic Trajectory-measurement Systems

Having now covered the basic measurement techniques in rather brief form, let us proceed to the systems individually. All these systems use the techniques, or combinations of the techniques, described above.

TRAJECTORY INSTRUMENTATION 3–47

DOVAP (Doppler Velocity and Position)

DOVAP is a multistation system employing the Doppler principle to obtain missile space-position measurements. Basically, the system consists of a ground reference transmitter operating at 36.9 Mc, a frequency-doubling missile transponder, and a number of remote ground receivers where the reference and return frequencies are compared and the Doppler frequency extracted. The Doppler signals are relayed to central recording stations. Cycles of each of the Doppler records are then incrementally accumulated from some reference point, which is usually the missile-launcher position. The accumulation at any particular time is representative of the change of loop range and so describes an ellipsoid of revolution, which has foci at the transmitter and a particular receiver. The missile space position for any instant of time is found by solving for the intersection of three such ellipsoids. Velocity and acceleration data are obtained from differentiation of the position data.

The DOVAP system is probably the most outstanding of the Doppler-type systems. Its techniques were used during World War II for measurement of bomb trajectories, and when missile programs began immediately following World War II, this was one of the first systems put into use. From that time until now, approximately 20 years, the system has operated reliably and produced highly precise data at White Sands Missile Range. It is today a major instrumentation system at White Sands, Atlantic Missile Range, and Fort Churchill.

TRIDOP

The TRIDOP system is similar in principle to DOVAP. It uses a reference transmitter at 132.48 Mc and a transponder frequency of 264.96 Mc. In TRIDOP, a timing signal is transmitted to the receiving stations where it is used as a time base for magnetic-tape recording and provides a 5-kc bias frequency for the Doppler data. This bias frequency prevents the recorded Doppler frequency from passing through zero and thus facilitates automatic cycle counting. TRIDOP was developed and installed at Pacific Missile Range in the early 1950s.

EXTRADOP (Extended-range DOVAP)

EXTRADOP is the name given to the Atlantic Missile Range extended-baseline DOVAP system. In this arrangement of DOVAP, a coherent reference frequency is transmitted by cable to receiver stations which are located beyond line of sight. AMR has used this method to tie together the mainland and island DOVAP networks.

UDOP (UHF Doppler)

UDOP is another outgrowth of the DOVAP system. It differs principally in the frequencies used, these being a reference frequency of 450 Mc and a transponder-return frequency of 900 Mc. The reference frequency is distributed to the ground receivers at 50 Mc and is multiplied to 900 Mc at each receiver station. UDOP was first developed at the Army Ballistic Missile Agency and then installed at Atlantic Missile Range, where it is expected ultimately to replace the vhf DOVAP system.

DOPLOC (Doppler Phase Lock)

DOPLOC is a passive Doppler-type system developed by the Ballistic Research Laboratories, Aberdeen Proving Ground, specifically for tracking orbiting satellites or missiles. It is a very long baseline, ultrasensitive system designed to track from horizon to horizon. The system can operate from signals either transmitted by, or reflected from, the object being tracked. To date, DOPLOC has been used on frequencies of 20, 108, 132, and 960 Mc. At each DOPLOC station, the signal received from the missile is compared with a locally generated signal to obtain a Doppler record. Analysis of one such Doppler record, in a time-history sense, enables determination of many of the parameters of a satellite orbit. Time of closest approach is immediately obtainable, and if the satellite velocity is known, its range at any time can be computed. A combination of records from several receiving stations allows accurate calculation of the trajectory or orbit.

MICROLOCK

MICROLOCK, a development of the Jet Propulsion Laboratory, is similar in principle to DOPLOC. The differences in the two systems are chiefly in the phase-lock techniques used. MICROLOCK is designed to track the 108-Mc satellite telemetry carrier and it records the telemetry data as well as the Doppler signal for measurement of the orbit. The MICROLOCK receivers are capable of tracking satellites carrying 10-mw transmitters to ranges of 2,000 nautical miles.

TRAC(E) (Tracking and Communication, Extraterrestrial)

TRAC(E) is the successor to MICROLOCK, built by the Jet Propulsion Laboratory, for space-probe tracking. The first TRAC(E) installations are designed for 960-Mc operation and obtain two angle measurements, Doppler and telemetry data. Angular measurements are made with 0.4 angular mil (0.022°) rms jitter at receiver threshold. The antenna system operates in any of three modes—manual, automatic, and acquisition. On Pioneer IV, the system was capable of operation to a range of 1,150,000 miles. Plans for future improvement include conversion of TRAC(E) to a two-way system that will measure two angles, range and range rate.

REFLECTION DOPPLER (Velocimeter, AN/TPS-5)

Reflection Doppler, sometimes referred to as Doppler radar, is a single-station instrument used for radial velocity measurement. The velocimeters and AN/TPS-5 are S-band directional instruments which beam a signal to the missile and obtain the Doppler effect from the reflected return signal. Reflection Doppler is a relatively short range system used primarily for obtaining velocity and acceleration measurements during the early phases of missile flight. The systems have been used at a number of missile ranges and several sled tracks.

PARDOP

PARDOP is not an operational system. It reached the experimental stage about 5 years ago, and because of some difficulties encountered in field tests and a dwindling requirement for the system, it was relegated to inactive status. It is included here because at this time the PARDOP principle is again being investigated. With the improved components that are now available, a system of the PARDOP type appears promising. PARDOP, in essence, is a passive DOVAP system operating in the 132-Mc region.

SPHEREDOP (Spherical Doppler)

SPHEREDOP is another system which has been used only experimentally. However, it was the forerunner of such systems as MICROLOCK and DOPLOC. SPHEREDOP requires carrying a stable oscillator aboard the missile and detecting the Doppler effect at a number of receiver stations by comparing the received signal with a reference frequency. Using a minimum of four receiver stations, the bias between the missile-borne oscillator and the ground reference frequency can be computed. Accumulations of the Doppler records from each receiver station then describe spheroids which intersect at the missile space position.

HYPERDOP (Hyperbolic Doppler)

This is a rearrangement of the SPHEREDOP system. In HYPERDOP, the Doppler signals detected at the various receiver stations are compared with each other so that accumulations for each pair of receivers represent a relative range measurement. Three pairs of receiver stations then yield three hyperboloids which intersect at the missile space position.

DME (Distance-measuring Equipment)

DME is the name generally applied to a single-station instrument which uses a measurement of phase delay to obtain range to the missile. A number of systems with different carrier and modulating frequencies come under this DME heading. One system in use, which is considered typical, has a ground transmitter operating at 461 Mc and a transponder-return frequency of 399 Mc. The transmitted carrier is modulated with frequencies of 491.76, 61.47, 7.684, and 0.9605 kc. The highest of these modulating frequencies has a wavelength of 2,000 ft and the lowest has a wavelength of 1,024,000 ft. Phase comparison of the transmitted and received signals is accomplished in a servo-type phase meter, which has a resolution of approximately 2°. Since the measured phase delay is proportional to twice the range, the range resolution is then about 5.5 ft with nonambiguous measurements to ranges of 512,000 ft.

SECOR (Sequential Ranging)

SECOR is a multistation system comprised of a minimum of three DME's operating in a sequential mode. One of the DME stations serves as a master to control the sequencing, and when three stations are used, each interrogates the missile transponder forty times per second. Both vhf and uhf SECOR systems have been developed by Cubic Corporation for Atlantic Missile Range. The vhf system was the earlier development and has now been deactivated. This system had a ground-transmitter frequency of 248.5 Mc (the same for all DME's) and a missile return frequency of 218.5 Mc. UHF SECOR is now in operation at the Eglin Gulf Test Range. The uhf system uses a ground-transmitter frequency of 512 Mc and a transponder return of 482 Mc. The modulating frequencies used in SECOR are 491.76, 61.47, 7.68375, 1.92094, and 0.192 to 094 kc. UHF SECOR is designed to have a phase resolution in the fine measurement of $\frac{1}{2}°$ with the modulating frequencies used. The distance measurements are nonambiguous to a range of 2,560,000 ft and have a resolution of 1.4 ft.

ROMOTAR

ROMOTAR is a multistation phase-comparison system comprised of one master transmitter-receiver station and three remote receiver stations. These yield one direct and three loop-range measurements. ROMOTAR carrier frequencies are in the 450- to 500-Mc region. The modulation frequencies used are 491.17, 49.117, and 4.9117 kc. The resolution of range measurement is 1.4 ft and measurements are nonambiguous to a range of 100,000 ft. The ROMOTAR system is used at the Salton Sea Test Range by Sandia Corporation.

DORAN (Doppler Ranging)

DORAN is a multistation system geometrically similar to DOVAP, but which uses phase-comparison techniques to provide loop-range measurements. The reference transmitter has a frequency of 132.48 Mc and the transponder-return frequency is 264.96 Mc. The modulating frequencies used are 749.44, 74.944, 7.4944, and 0.74944 kc. During evaluation, many troubles were encountered with DORAN, and at this time it is not an operational system. Many of the DORAN equipments are being used for passive tracking experiments at White Sands Missile Range.

MIRAN (Missile Ranging)

MIRAN is a multistation pulse-type ranging system. The system was in use at White Sands for a number of years but has recently been phased out of operation. It is being replaced by more advanced systems. MIRAN had a master transmitter-receiver station and a number of slave receivers. The master station transmitted at 600 Mc, with the signal being reradiated by the missile transponder at 580 Mc. The signal then received at each of the slave receiver stations was relayed back to the master transmitter, where elapsed time was measured. The elapsed-time measurement was proportional to the loop range, and three such loop ranges were used to obtain the missile space position.

RAYDIST

RAYDIST, a development of Hastings-RAYDIST, is one of the earliest of the relative phase-measurement systems. It has been used in both two- and three-dimensional forms, and while RAYDIST is principally a navigation system, it has been used for missile tracking. Components of the system used for missile instrumentation are a ground transmitter, a missile transmitter, and three receivers for a two-dimensional system, or four receivers for a three-dimensional system. The ground transmitter operates at a frequency of 4.135 Mc (permitting beyond-horizon measurement) and the missile transmitter is offset from this by 400 cycles. At each of the receiver stations, the 400-cycle beat is detected, and with one receiver serving as a master, a phase relationship is determined. In a three-dimensional system, the relative phase measurement is representative of a hyperboloid, and three hyperboloids thus determined permit location of the missile position. The RAYDIST system is dependent on receiving continuous phase data to avoid ambiguity.

AME (Angle-measuring Equipment)

AME generally applies to the interferometer type of angle-measuring equipment. As such, it is not a system but a system component. When two or more AME's are combined, or when AME's are combined with other components to form a complete measuring system, some other terminology is used. These various applications will be described later. The most common AME is that developed and in production at Cubic Corporation. It consists of two perpendicular-bisecting baselines, each having antennas with an effective spacing of 50, 5, and $\frac{1}{2}$ wavelengths. Direction cosine measurements to about 25 ppm are obtainable from the 50-wavelength baseline, with the shorter antenna spacings being used to resolve ambiguities. A crossed-baseline AME provides two direction cosines from which azimuth and elevation angles may be computed. Most AME's to date have been designed to operate in the frequency region of VHF telemetry, allowing the systems to obtain measurements on any missile carrying a crystal-controlled telemetry transmitter. This eliminates the need for any special airborne transmitter to serve the AME systems. Both electromechanical-servo phase meters and wholly electronic phase meters have been used with the AME systems. Real-time digital data are obtainable from either type. The wholly electronic phase meters are potentially more accurate and can be expected to be most widely used in the future.

COTAR (Correlation Tracking and Ranging)

COTAR is an AME-DME type of system now coming into rather wide use. The first system was developed by Cubic Corporation for the Atlantic Missile Range, and Cubic is now producing the system for other test ranges. The AME component of COTAR is the 50-wavelength crossed baselines, as described above. The DME is identical to the DME used in the vhf SECOR system. The measurements obtained at one COTAR station are two nonambiguous direction cosines and a slant range which are sufficient for computation of the missile trajectory. In the vhf COTAR system, the DME carrier serves as a signal source for the AME. However, the AME's may be used separately to track a telemetry carrier or other signal source aboard the missile and thus obtain partial trajectory measurements. While the COTAR systems in the field to date have been designed in the vhf region, uhf components have been developed and other systems using essentially identical techniques are in existence.

MOPTARS (Multiobject Phase-tracking and Ranging System)

MOPTARS is basically a COTAR system with multitarget-tracking capability. Multitarget operation is obtained by sequentially interrogating the various airborne transponders and simultaneously selecting the corresponding servo phase meters and associated readout circuits. MOPTARS, with its sequentially operating AME and DME, is designed to have a five-target tracking capability. However, the original system, developed by Cubic Corporation for the Eglin Gulf Test Range, was installed with only three-target capability. It is intended that it be expanded to five-target operation at a later date. Operating frequencies of MOPTARS are ground transmitter, 396.5 Mc; transponder, 366.5 Mc; and modulating frequencies the same as those in the SECOR system.

MIDAS (Missile-intercept Data-acquisition System)

MIDAS is another of the Cubic-developed systems using the interferometer AME. The system consists of two widely separated dual-channel AME stations. Each AME is capable of tracking two airborne objects which emit different frequencies in the 215- to 260-Mc region. Four direction cosines are then obtained for each object (two from each AME) from which trajectory, intercept and acquisition data are obtainable. At least two MIDAS systems are in the field, and one of these is presently being evaluated at Naval Ordnance Test Station.

MATTS (Multiple-airborne-target Trajectory System)

MATTS is a MIDAS system with three-target capability. The system requires a vhf source of different frequency in each of the airborne objects. MATTS is also a development of Cubic Corporation for Eglin Gulf Test Range.

DOETS (Dual-object Electronic Tracking System)

DOETS is identical to MIDAS and is located at the Eglin Gulf Test Range.

MIDOT (Multiple Interferometer Determination of Trajectory)

MIDOT consists of two or more interferometer-type crossed-baseline AME stations. Each AME has 100-wavelength perpendicular-bisecting baselines. Shorter baseline segments for ambiguity resolution are not used. The phase data from each of the AME stations are recorded with a synchronized time base and ambiguity is resolved in data processing. MIDOT operates in the vhf telemetry band and can use any airborne telemetry transmitter as a signal source. Recently, a DME has been added to the MIDOT system. The DME has a carrier frequency of about 200 Mc and uses ranging frequencies of 22.129 and 24.588 kc. MIDOT is a development of Sandia Corporation and has been used for some time at the Tonapah Range. Sandia has also installed MIDOT equipment at Point Arguelle for use as a backup range safety system.

MINITRACK

MINITRACK is the Naval Research Laboratory system designed for long-range tracking of satellites which carry low-power signal sources. A number of MINITRACK stations are in operation around the world. A MINITRACK station consists of an interferometer-type AME with directional antennas spaced at 50 wavelengths (at 108 Mc). Four additional antennas in combination with the directionality of the field are used for ambiguity resolution. The MINITRACK system was designed to track 10- to 15-mw signals to ranges of 1,500 miles and to provide data to an accuracy of 0.1 milliradian.

ITS (Integrated Trajectory System)

ITS is a combination AME-DME-Doppler system being developed by Cubic Corporation for White Sands Missile Range. The initial installation will consist of two AME's, each with two-target capability, and three DME's, each with three-target capability. The AME's are 128-wavelength crossed baselines with effective spacings of 16, 4, and ½ wavelengths for ambiguity resolution. The AME's will operate over a frequency range of 215 to 277 Mc. DME's have ground transmitters operating at 300 Mc and transponder-return frequencies of 266.66, 270.0, and 276.9 Mc, for the three targets, respectively. DME's can be operated either sequentially, as in the SECOR system, or in a continuous mode to provide loop-range rather than direct range measurements. DME modulation frequencies are 482, 30, 1.8, and 0.2 kc. Design accuracies of the system are 15 ppm for the AME, and 1.5 ft for the DME, with nonambiguous range measurements to 400 miles. A Doppler pickoff at the DME stations will increase the accuracy of velocity measurements. Ultimately, ITS will include a DOVAP system in the same frequency region as the DME's. The system will then be quite versatile. It will be able to track five objects simultaneously or provide higher accuracy for fewer objects. From an airborne DME transponder, distance measurements, angle measurements, and Doppler may be obtained. From an airborne Doppler transponder, Doppler and angle measurements may be obtained, or from a telemetry carrier, angle measurements may still be made and trajectory and miss distance obtained for two airborne objects.

AZUSA

AZUSA is a single-station system using crossed-baseline interferometer and phase-comparison DME techniques. It is a directional system with all antennas being parabolic dish type. The entire system is directed by a d-f antenna located at the intersection of the crossed baselines. The AME baselines are 800 wavelengths long with additional antennas at the 80-wavelength points, which, in combination with the d-f antenna, are used for ambiguity resolution. The ground transmitter operates at a nominal 5,060 Mc and for range measurement has modulation frequencies of 98.36 kc, 3.93 kc, and 157 cps. The transponder-return frequency is approximately 5,000 Mc. To eliminate errors in direction-cosine measurements, which may arise from transmitter variations or from Doppler effect, the transmitter signal is continuously regulated to produce a constant received frequency. The data obtained from the AZUSA system are direction cosines, range, and range rate. The latter is a radial-velocity measurement derived from the coherent carrier. During the past several years, AZUSA has become a highly reliable and highly accurate instrumentation system at the Atlantic Missile Range.

Convair has developed and operated the AZUSA at Atlantic Missile Range and recently has delivered a second model of the system, which is presently being evaluated. The Mark II AZUSA has antennas spaced at 50 meters and 5 meters and, in addition will have 500 meter-baselines for a measurement of cosine rate. The Mark II AZUSA is designed to have an absolute accuracy in direction cosine to 10 ppm and incremental accuracy to $\frac{1}{2}$ ppm.

MISTRAM (High-precision Trajectory-measuring System)

MISTRAM consists of two L-shaped baselines with antennas spaced at 10,000 and 100,000 ft. Using the 10,000-ft baseline, measurements of range, range rate, range difference, and range-rate difference are obtained. The 100,000-ft baseline is used for range-rate-difference measurements. The system uses an X-band ground transmitter which transmits two frequencies spaced approximately 256 Mc apart. One of these is swept in frequency by 8 Mc. The airborne transponder offsets the retransmitted frequency by 68 Mc. The 256-Mc carrier separation is used for fine-range measurement and the 8-Mc sweep is used to resolve ambiguity. Range-rate measurements are made on the carrier. MISTRAM is expected to produce velocity measurements accurate to 0.1 ft/sec for X and Y, and 0.4 ft/sec for Z. Later improvements planned for the system will increase the velocity accuracy to 0.05 ft/sec in all coordinates. MISTRAM is a development of the General Electric Defense Systems Division for the Atlantic Missile Range.

ELSSE (Electronic Sky Screen Equipment)

ELSSE systems are used for flight safety only and are the simplest of the interferometer-type systems. The systems use single baselines without ambiguity resolution and depend on receiving continuous data from the time of missile launch. The usual arrangement is to have one baseline directly behind the missile launcher and one baseline off to one side. These are used to measure target-line deviation and programming deviation, respectively. Both telemetry ELSSE and DOVAP ELSSE systems have been established. The telemetry ELSSE uses a 50-wavelength baseline and essentially is that portion of a COTAR system. The DOVAP ELSSE, which is also known as the Beat-Beat system, uses a 200-wavelength baseline. For both types of systems, a measured phase is converted directly to deviation angle, with an accuracy of about 1 per cent.

EMA (Electronic Missile Acquisition)

EMA is a crossed-baseline interferometer system used for pointing some other instrument requiring missile acquisition. The system consists of two crossed baselines which intersect at the $3\frac{3}{4}$-wavelength point at which a fifth antenna is located. Comparison of the $3\frac{3}{4}$-wavelength and $4\frac{1}{4}$-wavelength baseline segments yields an effective half wavelength which is used for ambiguity resolution. The EMA system is designed to track the return DOVAP frequency, which is 73.8. An analog computer is used to compare the direction-cosine data obtained directly with the pointing angles of the instrument being directed. Either manual correction of the pointed instrument or automatic tracking may be used.

SARA (Ship Angle and Range)

SARA is a system which thus far has existed only in concept. If activated, it will be a variation of the SECOR and COTAR systems operating in the 15- to 30-Mc region. The system would be used for precise navigation of three or more ships when it is necessary to have shipboard instrumentation sites.

AGAVE (Automatic Gimbled-antenna Vectoring Equipment)

The AGAVE is an automatic pointing instrument which uses phase nulling for driving a quad-helix array. It has an acquisition beam width of about 20° and a pointing accuracy of $\frac{1}{2}$°. These instruments have been used to point the radar systems which have provided tracking for Projects Mercury and Gemini. AGAVE is a Cubic Corporation development.

This completes the systems coverage. A few systems have been omitted intentionally because some are classified, and some are simply obscure and insignificant. Other important developments are no doubt in process.

Radar

In addition to the above systems, a very considerable amount of trajectory measurement work is accomplished by systems of pulsed instrumentation radars. An excellent source descriptive of radars and their parameters is Skolnik.[4] Such radars provide elevation and azimuth angle and range measurements with respect to the missile or space vehicle. The range measurement is usually the most accurate, depending as it does only upon the pulse-width equivalent in distance. Measurements of a few feet, relatively independent of range to target, are common.

Angular observations are not so accurate at long ranges, in terms of feet. Measurements can be made to a precision of 0.1 mil of arc; at 100 miles this represents a distance error of 50 ft. More precision is obtained by using systems of radar which view the target at different angles and make use of the more exact range-only information and accurate knowledge of the locations of the radar sites. Radars which have been used for this purpose are:

	Frequency band
AN/FPS-16	C
AN/MPQ-12	S
RAMPART	S

One of the most commonly used such radars is the AN/FPS-16. It is often used in conjunction with a vehicle-borne beacon to extend its range. The following is a description of its most important parameters:[5]

Frequency	5,400–5,900 Mc
Pulse width	0.25, 0.5, or 1 μsec
PRF	Any submultiple of 81.95 kc between 177 and 1,707 pulse groups per second
Power	250 kw tunable, 1 Mw and 3 Mw fixed frequency
Beam width	1.2°

The FPS-16 will skin-track a 1-sq-m target to a range of 200 miles with the 1-Mw transmitter and beacon-track to a range of several thousand miles when a suitable beacon with "nth-time-around-tracking" techniques or range-extension modifications are used. Two hundred nautical miles is with zero db signal-to-noise ratio.

The least significant octal bit is 0.048828125 mil in azimuth and elevation and either ½ or 1 yard in range.

The AN-FPS-16 antenna pedestal is mounted on a 12- by 12-ft concrete tower which extends about 27 ft above grade level. The center of the emplaced antenna is approximately 36 ft above grade level. The electronic equipment, auxiliary system, maintenance section, etc., are housed in a 66- by 30- by 24-ft-high two-story concrete block building. The building surrounds, but is not attached to, the pedestal tower. This method of construction places the tower within the air-conditioned environment of the equipment building and provides protection from solar radiation and other weather effects which would dilute the inherent accuracy of the system. Power requirements for each station are 120/208 volts ± 10 volts, 4 wires, 60 cycles, and 175 kva.

Tracking coverage:	
Azimuth	Continuous
Elevation	−10 to +85°
Range	500 yards to 32,000 nautical miles
Elevation motion	−10 to +190
Tracking rates:	
Azimuth	750 mils/sec (42°/sec) maximum
Elevation	400 mils/sec (22.5°/sec) maximum
Range	8,000 yards/sec maximum
Acceleration:	
Azimuth	550 mils/sec² maximum
Elevation	550 mils/sec² maximum
Range	4,000 yards/sec² maximum
Slewing rates:	
Azimuth	45°/sec maximum
Elevation	24°/sec maximum
Range	30,000 yards/sec maximum
Automatic search angle	A. Circle scan: 0.9, 2.3, or 3.1° at 1, ½, or ¼ cps
	B. Roster scan: 10 by 6° or 6 by 10° at ½ cps on 6° and 1 cpm on 10°
	C. Sector scan: azimuth or elevation: 10°, 20°, or 30° at 1, 2, or 4 cpm
Range search	±1,000 yards (adjustable up to ±5,000 yards)
Servo bandwidths	Continuously variable at console
	Azimuth 0.25 to 5 cps
	Elevation 0.25 to 5 cps
	Range 1 to 10 cps
Angle velocity lag (until out of beam)	At maximum servo bandwidth: 0.004 mil/mil/sec ($KV = 250$/sec)

TRAJECTORY INSTRUMENTATION 3-55

Range velocity lag (until out of range gate)....	A. At maximum bandwidth: 1 yard for 5,000 yards/sec (KV constant at 5,000/sec)
	B. At minimum bandwidth: 1 yard for 5,000 yards/sec (KV constant at 5,000/sec)
Angle acceleration lag (until out of beam).....	At maximum bandwidth: 0.1 mil/mil/sec/sec
Range acceleration lag (until out of gate)......	At maximum bandwidth: 1 yard for 250 yards/sec/sec
Mount, normal leveling......................	Elevation axis leveled (± 0.025 mil)
Accuracy (determined independently by WSMR and other test ranges)....................	Azimuth ± 0.1 mil
	Elevation ± 0.3 mil
	Range $\pm$ to 15 yards
Range for balance between range and angle errors	About 25 nautical miles
Antenna side lobes..........................	-25 db

A typical breakdown of the sources of error in tracking radars is shown in Table 8. These are fixed errors, as opposed to dynamic errors arising from operation of the tracking servos.

It is typical practice to extend the performance range of instrumentation radars by using cooperative repeating beacons in the tracked vehicle. A typical set of beacon parameters is shown below:

Motorola SST-131 Beacon

Transmitted power..............	400 watts peak
Pulse width...................	0.6 ± 0.1 μsec
Receiver sensitivity.............	-65 dbm min.
Antenna gain..................	$+6$ db max., $+21$ db min.

Figure 42 derived from the "Final Report, Instrumentation Radar AN/FPS-16 (XN-2)" AD 250 500, shows the signal-to-noise ratio at the FPS-16 and the gain margin at the transponder receiver for an FPS-16/beacon combination. The beacon

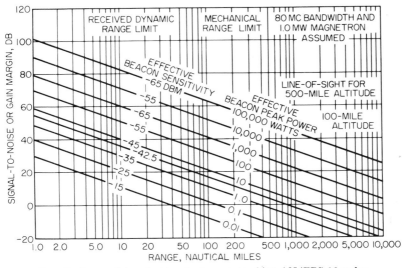

Fig. 42. Range determination for beacon tracking AN/FPS-16 radar.

Table 8. Breakdown of Fixed Radar Error, FPS-16

Error source	Angle, mils rms		Range, ft rms	
	Bias	Noise	Bias	Noise
Boresight and range zero setting	0.025	.....	1.0	
Boresight and range zero shift	0.04	.....	2.0	
Wind forces (50 mph)	0.02	0.002		
Servo noise, unbalance	0.01	0.02	0.5	0.5
Receiver delay	.....	.....	2.0	1.0
Subtotal: Radar-dependent tracking errors	0.052	0.02	3.0	1.3
Leveling and north alignment	0.01			
Mechanical deflections	0.01			
Orthogonality of axes	0.02			
Thermal distortion	0.01			
Range oscillator	.....	.....	3.0	0.6
Data-system zero setting	0.01	.....	1.0	
Bearing wobble	.....	0.005		
Data gear error	.....	0.03	...	1.0
Data takeoff	.....	0.025	...	1.2
Range resolver	.....	.....	...	4.5
Internal jitter	.....	.....	...	2.0
Subtotal: Radar-dependent translation errors	0.028	0.04	3.2	5.2
Tropospheric refraction	0.05	*	2.0	0.5
Ionospheric refraction	0.01	0.00	1.0	0.0
Subtotal: Propagation errors	0.05	*	2.2	0.5
Total fixed error	0.078	0.045	4.9	5.4

* Dependent upon local weather conditions.

antenna gain used in preparing Fig. 42 was 0 db, the FPS-16 noise figure was 11 db, and the FPS-16 antenna gain was 44.5 db.

To determine the maximum tracking range for accurate tracking, use the criterion of 12 db signal-to-noise ratio at the radar, and then use the beacon-range equation shown graphically in Fig. 42 to establish the range.

Trends

Two important trends in electronic-trajectory-system development warrant mention. The first, and obvious one, is the exploitation of techniques which enable the systems to track at longer ranges. To accommodate our outer-space missile programs, systems must be developed which will provide increases in accuracy and tracking ranges ten to a hundred times better than the systems we now have. Some remarkable advances have already been realized.

The second, and not so obvious, trend is the tendency toward the integrated-system concept. This trend is brought about by the need for more order and efficiency at the test ranges as they face the sizable workloads such as those which have existed during the past couple of years. Integration of systems means the combining of a number of systems or techniques into a single system, so that the techniques complement and aid each other. The resultant systems are more versatile and have greater yield than they would if summed separately. ITS and MISTRAM are examples of this trend.

At White Sands Missile Range, the ARTRAC Program (Advanced Range Testing, Reporting, and Control Plan) is the overall plan for integrating the entirety of instrumentation into a single unified system. Under ARTRAC, all the data-collection subsystems—electronic trajectory, telemetry, and optical—are joined as system components and will contribute to the measurement task. The combinations of components used, and the part played by each, vary as data requirements and other influences govern. Future tracking systems will be designed to be optimum for this integrated instrumentation complex.

Other test ranges have development plans which have goals similar to ARTRAC. It is probably safe to say that these programs will effect more order and efficiency at the test ranges and will demand and obtain much more from the contributing electronic tracking systems.

REFERENCES

The following bibliography, kept up to date by IRIG, will guide the reader to basic reports on the design, operation, and error analysis of the electronic trajectory-measuring systems discussed above:

1. Inter-Range Instrumentation Group (IRIG), "Bibliography of Reports on Data Acquisition Instrumentation"; Document No. 101-62, Secretariat, Range Commanders Conference, (Attn: ORDBS-IRM-RCC), White Sands Missile Range, New Mexico, 1962.

Additional documents on range instrumentation are:

2. ———, AFMTC Instrumentation Handbook AFMTC-TR-60-15, ASTIA 242 921, 2d ed., 1960.
3. IRIG, "Electronic Trajectory Systems Catalog," vols. I, II, IIA, and III, covering electronic trajectory systems, radar beacons, and miss distance indicators—issued by Office of the Assistant Secretary of Defense (Research and Engineering), Washington, D.C., 1958 (compiled by IRIG Electronic Trajectory Measurements Working Group).
4. Merrill I. Skolnik, "Introduction to Radar Systems," McGraw-Hill Book Company, New York, 1962.
5. C. R. Simmons and W. C. Young, "Technical Reference Publication No. 1 on Radar Instrumentation Systems," Analysis Branch, Data Reduction Division, Range Operations Directorate, White Sands Missile Range, New Mexico (undated).

Chapter 4

RADIO-TELEMETRY SYSTEMS

CONRAD H. HOEPPNER, *Industrial Electronetics Corp., Melbourne, Fla.* (*Radio Telemetry*)

C. H. DOERSAM, JR., *Polytechnic Institute of Brooklyn, Brooklyn, N.Y.* (*System Selection*)

JOE H. SMITH, *International Data Systems Corp., Dallas, Tex.* (*Signal Conditioning*)

JOHN F. BRINSTER, *General Devices Corp., Princeton, N.J.* (*Commutation*)

HANS SCHARLA-NIELSEN, *Radiation, Inc., Melbourne, Fla.* (*Radio Links*)

LAVERGNE E. WILLIAMS, *Aerospace Corp., Melbourne, Fla.* (*Antennas*)

RADIO TELEMETRY

1	Radio-telemetry Systems Applications	4–2
2	Radio-telemetry Systems Design Problems	4–5
3	Remote Measurement from Transducer to Analyst	4–7
4	Systems of Radio Telemetry	4–11
5	Trends in Telemetry	4–24

SYSTEM SELECTION

6	Basis for System Selection	4–26
7	Determining Telemetering-equipment Information-rate Capacity	4–30
8	Other Considerations in the Selection of Equipment	4–32
9	Adjusting Information and Equipment Capacity	4–33
10	Utilization of Government Test Facilities	4–35

SIGNAL CONDITIONING

11	System Considerations	4–35
12	Signal-conditioner Circuits	4–41

COMMUTATION

13	Principles of Commutation and Decommutation	4–53
14	Mechanical Commutation Devices	4–60

15	Electronic Commutation Devices	4–78
16	Comparison of Basic Commutator Types	4–85
17	Decommutation Devices	4–86
18	Commutation Terminology	4–88

RADIO LINKS

19	R-F Link System Design	4–92
20	Link Parameters	4–95

VEHICULAR ANTENNAS

21	Systems Considerations for Antenna Design	4–111
22	Discussion of Antenna Characteristics and Terms	4–112
23	Effect of Vehicle on Antenna Patterns	4–114
24	Specific Antenna Types	4–118
25	Calculation and Display of Vehicle R-F Link Coverage	4–131

It is the intent in this chapter to present a survey of radio-telemetry equipment used generally in aircraft, missiles, rockets, and satellites, and not point-to-point telemetry equipment tied together by wire lines. Only telemetry applications which are difficult or impossible to perform by means of wire lines are considered here, although some of the design concepts may be applied to such systems. In addition to the telemetry systems themselves, data-reduction and -processing problems are discussed briefly. All the systems treated are generally characterized by a large number of telemetry measurements transmitted over a single radio carrier. Specific examples from the simplest radio telemeter to a large versatile telemetry system will be considered as well as the characteristics and parameters of the principal telemetry systems.

The applicable standards of the IRIG are given as a unit in Chap. 5.

A simplified approach to telemetry-system selection is presented. This is broad enough to be used as a basis for any telemetry-system design, radio or wire, as it makes use of fundamental information concepts. (See also Chap. 2 for theory.)

Design considerations peculiar to radio-telemetry systems are discussed in subsequent sections of this chapter. These are signal conditioning and commutation in missiles and spacecraft, radio-links design, and antenna design. Subsequent chapters are devoted to more detailed presentation of the common forms of telemetry systems (Chaps. 6 through 9). More advanced systems are also discussed in Chap. 9. Methods more appropriate to industrial telemetry which usually employ wire, telephone, or microwave links are discussed in Chap. 13.

Radio transmitters and receivers are discussed in Chap. 6, where the discussion is limited to design considerations which are significantly affected by their use in radio telemetry and command links.

RADIO TELEMETRY

1 Radio-telemetry Systems Applications

The development of radio telemetry has been principally centered around the drone and missile programs of the armed forces. It began during World War II and has been growing and expanding to embrace aircraft testing as well as the testing of unmanned vehicles and spacecraft. Early drones were piloted aircraft with the pilots removed and autopilots with remote radio control substituted. During test phases a pilot was usually carried to perform takeoffs and landings and to observe the results and deficiencies of the control equipment. When the drone was used as a weapon the pilot was removed and the equipment functioned both automatically and by remote control. It soon became apparent that these missiles could be made smaller, of higher performance, and more economically if in their initial design no provision was made for a pilot.

One of the early missiles of pilotless design, designated the JB-2, was a copy of the German V-1 buzz bomb. Remote-control equipment from the drone programs was adapted to the JB-2. Telemetry was developed to measure the performance of the control equipment and the missile. With this evolution it is apparent that remote-control equipment preceded telemetry by some years, but it was an "on-off" system rather than one permitting proportional control. Furthermore, remote control was an intermittent function inasmuch as the vehicles were stabilized by internal automatic equipment. Telemetry, on the other hand, required proportional and linear transmission of measurements on a continuous basis. Consequently, remote-control systems were "adapted" in concept only and telemetering-equipment development proceeded independently.

Many functions needed to be measured in the early missile programs. Measurements were made of the servo control systems, the remote-control systems, the guidance systems, the missile structure, aerodynamics, propulsion, etc. In addition to the measurement of voltages, these measurements were usually linear motions or shaft positions, or rotations, thrust, acceleration, temperature, and pressure. The development of transducers for converting these quantities into electrical analogs followed.

Meanwhile, piloted aircraft became more and more complex and pilots' notes and on-board recording were supplemented by radio telemetry which furnished data that could be analyzed in the event an airplane crashed. Telemetry served also as an aid to the test pilot in the checking out of prototype aircraft. However, this kind of data taking for piloted aircraft has not yet been widely accepted. It was not until 1958 that the Federal Communications Commission first allocated radio frequencies for aircraft flight testing.

Another forerunner of radio telemetry was telemetry and supervisory control in electric and gas utility transmission and distribution systems. Very few of these developments, however, could be borrowed for telemetry in the missile and aircraft programs. The public-utility measurements were made slowly, requiring only a very narrow band of frequencies for intelligence. With wire connections, there were no problems of radio fading, and many of the systems could be used only with continuous links between transmitter and receiver. Fades such as were normally occurring in radio systems would render the data valueless. Transducers were large and weighty, made for durability and easy servicing. They were not considered expendable and were chosen largely with a view toward long life and reliability; their response was slow. Instead of techniques' being borrowed from the utility field, the reverse trend has now appeared and utility telemetering has borrowed from the techniques of radio telemetering developed for missile testing.

Real-time Data

In the early missile programs it was desired to obtain a visual indication of such things as missile speed and response to controls—a telemetry link was employed. Real-time data during flight test have largely disappeared during the last decade because of the rapidity of missile responses and the inability of an operator to analyze the data and take proper corrective measures. It is reappearing, however, with the advent of digital computers. Plans have been laid for using real-time telemetry to program tests of missiles through computers, thereby enabling greater exploitation of the test vehicle.

In the tactical use of missiles, the proper functioning of the guidance and the determination of hitting a target are important to determine whether or not another missile should be fired at the same target. Telemetry can provide this information. Telemetering will also be useful in threat evaluation and in various reconnaissance functions.

Another real-time telemetry function has been to transmit the acceleration of a ballistic missile to the ground where it may be integrated to determine the velocity and thus determine the propulsion cutoff time. Telemetry is also used extensively in preflight checkout of the missile, inasmuch as the measuring system has already been installed and fewer connections are required.

Postflight Data Analysis

External instrumentation such as radar and theodolites are used to determine the position of a test vehicle as a function of time. These position data can be correlated with the internal acceleration data from the telemetry by differentiating the position data and integrating the acceleration data with respect to time. This is a basic check of both the instrumentation and the missile performance. By far the largest use of telemetry data is the postflight analysis of internal instrumentation. Data taken by telemetry are used for (1) determining the performance of the vehicle, (2) gathering data for improving the design of the vehicle, (3) studying environmental conditions in the vehicle, (4) determining causes of failure, (5) determining the time occurrence of events, and (6) determining basic parameters of the upper atmosphere and the space beyond the atmosphere. Postflight data analysis takes many forms, as will be discussed in detail later in this chapter. In general, the measurements made in the vehicle are recorded as a function of time; external measurements such as position and velocity are also recorded as a function of time. In postflight computations the parameter time is eliminated and the internal measurements are plotted against altitude, velocity, or acceleration. Graphic plots are usually acceptable, but on occasions numeric prints of the data as well as preparation of data for entrance into general-purpose digital computers is required. The correlation of functions, or plotting of one function against another, is also important.

Preflight Missile and Equipment Checkout

Telemetry is used extensively for the preflight checkout of a missile and its accompanying equipment. This function is done by wires in many cases, but the telemetry transmitter is already connected to many of the points which are to be checked. Consequently, the use of telemetry for preflight checks is a simple and economical expedient. Checkout of missile functions by means of telemetry, of course, also checks out the telemetry equipment. Further, the telemetry may be received at a distance from the missile, and many of these preflight tests may be used as calibrations for the operational telemetry receiving stations. Calibration of the entire measuring system may be accomplished. For example, if a gas pressure is to be measured by telemetry a series of actual gas pressures may be applied to the transducer and simultaneously indicated by a precise standard gage. If this is done at many levels, both linearity and scale of the transducer and telemetry equipment are calibrated. This may be done for many channels of telemetry simultaneously and the calibrations recorded for subsequent correction of the data. In many cases, preflight checkout by means of telemetry is very convenient because telemetry equipment is usually operated from its own separate power supply and may be used in tests without having the complete missile system energized and operating.

Down-range Data Gathering

In the test of a moving vehicle such as a missile or an aircraft, a single receiving station is usually not adequate to receive data during the entire period of the flight. This is particularly true in the case of long-range low-altitude missiles which travel nearly in a straight line. Consequently, additional receiving stations are required along the flight path. These stations—"down-range stations"—must be simple, reliable, and easily maintained because they are usually located in remote and isolated places at which it is difficult to maintain qualified technical personnel. The data received at these stations must be electrically reproducible so that they may be transported back to the main data-processing center and used with the same equipment used by the receiving station located at the launching site. Magnetic-tape recorders are generally employed for this purpose.

Many problems are associated with down-range receiving stations. One problem is the difficulty of acquiring the telemetering signal as the vehicle comes over the horizon. In order to obtain good receiving-station sensitivity, large antennas are used. These antennas automatically track the transmitted signal once they are "locked on" but they are rather highly directive and it is difficult to make the initial signal acquisition.

Another problem is the transportation of the magnetic tape back to the main processing center. This may often cause considerable delay since wire-line or radio-transmission facilities are not used and transportation by boat and aircraft is usually provided. In the near future, it is probable that the data will be transmitted by means of wire or radio facilities.

The problem of calibrating the station is also a serious one. Because the radio signal is not received while the missile is on the ground, calibration cannot be made in conjunction with the transmitter as is done with the station near the launching site. Special calibrating equipment must be employed at the site and its corrections carefully correlated with those made at the launching site.

Another problem is the overlap of reception at two or more stations. When the same data are received at two or more stations, choice must be made of which data to use or whether to run a comparison. In the case of a fading signal, this may be a difficult choice to make, and very often the calibrations are not similar enough for the indiscriminate use of either station's data.

2 Radio-telemetry Systems Design Problems

One of the first considerations in the design of a telemetry system is the wide variation of input transducers that are encountered. These provide both high-level and low-level voltage outputs, both a-c and d-c, and have both high and low internal impedances. Some transducers such as thermocouples generate their own voltages while other transducers such as strain-gage bridges need external excitation in order to provide output. Some transducers vary in resistance proportional to the quantity being measured while others vary in reluctance.

Encountered more and more frequently are outputs from both digital and analog computers which must be transmitted accurately and at the proper instant. Many scientific devices for exploring the upper atmosphere and the space near the earth provide inputs to the telemeter. These include devices for measuring high-energy particle density, ion density, mass spectrometers, and the impact of micrometeorites; ionization manometers; and instruments for measuring the intensity and direction of the earth's magnetic field. Many other quantities such as temperature, pressure, shaft position, vibration, and acceleration are converted to voltage in different ways for transmission via the telemeter. Typical telemetry systems may range from a few dozen of these measurements to several hundreds. All must be properly assembled and "conditioned" for introduction into the telemetry link (see Secs. 11 and 12).

It is a common procedure to group inputs in accordance with their voltage level and impedance, and to multiplex these to a common channel. This is done either slowly or rapidly, as the expected data require. In the case of temperatures which vary rather slowly, only a few samples per second are required. In the case of vibration, several thousand samples per second may be required in order to preserve amplitude, frequency, and phase. Also, many measurements are made in a common location on a missile or test vehicle where it is convenient to group them for transmission by a single telemetry channel. In multistage rockets, telemetry channels allocated to one stage may be transferred to the second stage after the first is dropped. The levels of voltages, sampling rates or frequencies, accuracies, and numbers of channels are all tabulated and grouped, and then the telemetry system is assembled to transmit the composite measurements.

Measurements collected from all points of a missile are brought to a single transmission center. This is done by means of connected wiring within the missile. Many other equipments are also operating within the missile which generate extraneous signals and/or noise which may be picked up by these connecting wires. Consequently, extreme care must be exercised in installing a telemetry wiring harness for location, shielding, proper grounding, etc. In addition, the resistance and capacitance of the wiring may have definite effect on the output of the transducer and it must be balanced or otherwise compensated and calibrated. Also, care must be exercised that the telemetry wiring does not radiate spurious signals to surrounding sensitive equipment.

Environmental conditions within the test vehicle during flight are not only severe but are rapidly changing. Temperatures may increase and decrease several hundred degrees. Accelerations may change by an order of magnitude and vibrations may appear in many and varied combinations. The effect of these conditions upon the telemetry equipment must be foreseen in the design and layout of the equipment in order to minimize error. Often calibration is made during the flight of the missile to determine whether or not changes have occurred in the telemetry equipment as a result of the severe and changing environment.

Calibration is perhaps the most valuable tool of the telemetry engineer. It is of course desirable to make preflight, in-flight, and postflight calibrations of each channel from transducer input to the recorded record. This is seldom possible, however, and the engineer must resort to substitutes in order to approach this condition. Certainly, in most cases postflight calibration is impossible because the missile and its equipment have been destroyed. Preflight calibration is often practicable on only a portion of the telemetry system such as in the case of strain-gage calibration. Here the missile element being telemetered is usually not subjected to the ranges of strains expected to be encountered; instead, the strain gage is replaced by one or several dummy resistance bridges to calibrate only the remaining portion of the system. It is the rule, rather than the exception, that transducers are calibrated separately and their calibration added to that of the connected system. In-flight calibration, of course, has the difficulty that the measurement may be interrupted at a critical period for calibration. Calibration methods have been devised which do not interrupt measurements, but they are limited in range by the amplitude limits of each individual channel.

The capacity of a telemeter for handling information usually has a fixed limit determined by the bandwith of the radio link. This maximum may not be entirely utilized in a particular test, or, on the other hand, it may be insufficient for another test. When it is insufficient, two or more telemetry links are required. In most systems, there is provision for the flexible interchange between the number of channels and the number of samples per channel. In more recent telemetry systems such as the pulse-code-modulation system, there is a provision for the interchange of the number of channels, the number of samples per second per channel, and the accuracy of each channel. The requirements for number of channels, frequency response, and accuracy vary widely from program to program. Also, they vary as a single program progresses from research through development and into final use. In the early stages of a missile research program, a large number of accurate measurements of high-frequency response are usually required. Then, as the data become better and better determined, these measurements are replaced by monitoring-type measurements which determine causes of failures and the time of occurrence of events. However, when the end product is scientific exploration, the measurements may increase in complexity and accuracy as the program progresses. It is well to provide from the very inception of an experimental program a flexible telemetry system which may expand and contract with the changing need.

After the data have been received and recorded there are many different uses to which they are put. Surprisingly enough, much the greater portion of the data taken in a guided-missile program is discarded after an initial look without having further usefulness. Methods and procedures for discarding data are extremely important to the efficient use of a telemetry system. Data are discarded because (1) the measurements are as they were expected, (2) they were taken at a much higher rate than required, (3) the transducer or channel became inoperative, (4) the vehicle did not perform properly, or (5) the data expected were not detected. After the valueless data have been discarded, graphic plots, numeric prints, and computations are made of the remaining data and the results are usually presented in form permitting multiple topics for reports. The data are analyzed at the test center, at the vehicle contractor's plant, and at scientific centers throughout the nation. The discoveries are published and submitted to the cognizant agencies.

Other miscellaneous telemetry problems are associated with the radio link from the moving vehicle to the fixed ground receiving station (Secs. 19 and 20). The strength of the radio signal, of course, varies as an inverse function of distance between transmitter and receiver. Also, there may be multiple propagation paths which produce the familiar "ghosts" of television. Antennas at both the transmitter and receiver are to some degree directive and both problems of tracking the missile with the ground receiving antenna and eliminating nulls in the missile transmitting antenna are always present. Further, the polarization of signals may rotate because the missile tumbles and rolls so that a single-linearity polarized ground receiving antenna may not pick up sufficient signal even though the total signal strength at the location is adequate. In the past, telemetry radio links have been characterized by frequent fades giving data omissions often at times when the data are most desired. Much progress has been made in recent years to eliminate this unsatisfactory condition both by the employment of circularly polarized receiving antennas and by utilizing polarization diversity reception.

It must also be remembered that weight carried aloft by the missile seriously limits its range; therefore, low weight of the telemeter is vital. Even more important is the power requirement of the telemeter. Up until the time of solar cells it was necessary to take aboard the missile an entire primary supply of batteries for operating the telemeter. Constant research has greatly increased the watthours per pound of battery weight but it is still a serious limitation, particularly in long-range missiles. Every effort must be made to minimize power requirements.

3 Remote Measurement from Transducer to Analyst

The preceding paragraphs described a number of the requirements placed upon the telemetry system by the transducers and quantities being measured. Unfortunately, the development of telemetry has not been such as to satisfy all the requirements, and in many cases the telemetry system seriously limits the measurement. Consequently, a compromise is required between telemetry capabilities and the requirements of measurement. The shortcomings and limitations of the telemetry system place restrictions upon measurements above and beyond those encountered in the laboratory when the telemeter is not employed. In the first place, an electrical output from the measuring device is required in order that the intelligence may be placed on a radio link. Consequently, transducers are necessary which produce an electrical output in one form or another. Also, the telemetry system may not be perfectly stable down to zero frequency and transducers and methods of measurement must be chosen to minimize the effects of drift. Overmodulating the subcarrier or the time-division multiplexer may also affect adjacent channels, as well as produce erroneous data in the offending channel. If various measuring devices are switched, the switching transients must be minimized or the accuracy of the telemetry system may be impaired. When mechanical commutators or time multiplexers are employed, the measurement of the time occurrence of the event such as the impact of a cosmic particle or the receipt of a guidance pulse is made more difficult and the time ambiguity of the multiplexed system is a serious limitation.

The measurement of a large number of parameters requires extensive and bulky equipment unless the parameters can be combined in groups of similar inputs to minimize the signal conditioning required. This fact generally dictates a relatively standard transducer rather than an optimum one for each particular measurement.

The bandwidth of the measurement, or the frequency with which the measured quantity changes, is also seriously limited by the telemeter. In the FM-FM telemeter, the permissible bandwidth varies from a relatively low value on the lower subcarriers to a reasonably high value (2,300 cycles) per second on the upper subcarriers. The bandwidth of the measurement must not exceed these limitations or sidebands will be generated in adjacent channels, thereby reducing the accuracy of the other multiplexed measurements. In a time-multiplexed system, the problems of folded data are present whenever the rate of change is faster than one-half the sampling rate. When this occurs, it is not known whether the quantity has reversed itself several times between samples or if there has been no reversal at all. It is generally considered desirable to limit the bandwidth of the data so that this ambiguity is not present. However, with refined techniques of analysis, this is not a rigid requirement.

The form which the data end product takes is also a limitation on measurement. In general, time-history plots of the measured quantity are desired. In this case, the speed at which the recording medium moves is often a severe limitation. If sampling is not regular, demultiplexing difficulties are magnified.

The choice of measuring equipment is often limited by the ability to calibrate it. A calibration is usually made from a graphic plot and is applied to the data, which are then replotted in calibrated form. Calibration corrections selected by other means may be applied before plotting or printing data in many cases. Acquiring the calibration curves in the first place, however, is often a difficult procedure. The transducers are calibrated before they are installed, but the remainder of the data system must be calibrated by substitution. This requires accessibility for substitute transducers or signals which may be applied. Also for this purpose transducers which have simple simulators, such as resistors for a strain gage, are desirable.

Choice of Transducers

Within the limitations outlined above, the transducers must be chosen to match the particular telemeter employed. A variable-reluctance transducer may be quite satisfactory in an FM-FM system but may be very difficult to use with a time-multiplexed system. Analog transducers are chosen to match levels and impedances and must often be interleaved with digital transducers, such as shaft encoders and outputs from digital computers. These choices must be made to maximize the utility of the telemeter.

The accuracy of measurement may, in many cases, be limited by the telemeter rather than by the transducer. When this is the case, it is sometimes possible to use several transducers to spread the range of measurement over several telemetry channels. This is done in a manner similar to the display of a watthour meter in which the reading of each dial is transmitted over a separate channel. An example of this technique is the measurement of atmospheric pressure by means of a bellows gage at high pressure, a Pirani gage at medium pressure, and an ionization gage at low pressure. To detect system errors separate measurements of the same quantity are made by separate transducers and telemetry channels. This form of redundant measurement has unfortunately been used too little in radio telemetry.

Transducers must also be chosen to measure the desired quantity without measuring other effects to which they are subjected. In other words, a pressure transducer should measure pressure and not be affected by temperature changes to which it is subjected. The two principal offenders in this regard are temperature and acceleration. It is a major telemetry problem to select transducers which are free of temperature and acceleration effects unless they are used to measure those quantities. Other parameters which affect transducers to a lesser degree are pressure, humidity, aging, and vibration. Tests of all these quantities can be made before the transducer is mounted in the vehicle, and from the results of these tests the proper transducer can be chosen. Also, even though transducers are installed in groups with the same kind of measurements handled in the same manner, each transducer of a group may be subjected to separate environmental conditions; therefore, it is not usually possible to have a single "dummy" transducer calibrate out the environmental effects on the other live transducers.

Requirements for Radio-telemetry Measurements

The principal requirements for radio-telemetry measurements are generated by

1. The scientific exploration of the upper atmosphere and the space surrounding the earth
2. The gathering of test data on experimental vehicles
3. The gathering of data from operational weapons
4. Miscellaneous industrial tests and operations

These requirements, when interpreted in terms of the telemetry engineer, are described under three major headings, viz.

1. The number of channels of measurement required
2. The rate of change (frequency response) of the measured parameter in each channel
3. The accuracy of the data to be measured in each channel

Typical requirements, as shown in Table 1, may be met by one or more radio-telemetry systems and associated equipment. These data may be transmitted over one or more radio-frequency carriers.

Table 1. Typical Measurement Requirements

Quantity to be measured	Number of separate quantities	Frequency response,* cps	Accuracy ±
Temp, 0–100°C...	4	1	5°
Temp, 1500–2000°C...	1	5	10°
Temp, −100–40°C...	3	1	5°
Pressure, 1–15 psi...	3	20	½ lb
Pressure, 10^{-3}–10^{-8} mm Hg...	1	0.1	10^{-8} mm Hg
Pressure, 2,000–3,000 psi...	2	100	10 psi
Angular position, −10–10°...	6	10	0.1°
Angular position, 0–90°...	3	5	1.0°
Voltage, 0–10 mv...	20	250	100 μv
Voltage, 0–50 mv...	10	50	100 μv
Voltage, 0–5 volts...	30	10	0.1 volt
Voltage, −5–5 volts...	40	10	0.1 volt
Voltage, 0–300 volts...	2	500	10 volts
Alternating voltage, 400 cps phase preserved to ½°...	3	400	10 volts
Strain, 0–250 μin./in...	12	100	5 μin./in.
Vibration, 0–2,000 cps...	3	2,000	10% of full scale
Vibration, 0–100 cps...	9	100	5% of full scale
Turbine rpm, 0–5,000 rpm...	1	10	10 rpm
On-off functions...	23	Time resolution 0.01 sec	No error
Input counts, 0–1,024 counts read once per sec	2	10 pulses once per sec	No error

* Determined from rate of change to be expected.

Telemetry Data Records

In a telemetry system, it is nearly always necessary to change the form of the quantity being measured. That is, temperature is not reproduced as a temperature but rather as a printed number or a print on a graphic plot. Principal forms of presentation for telemetry data are

1. Visual indicators or dials
2. Oscilloscope presentations
3. Time-history plots (these may be records made by pen recorders, hot-wire recorders, oscillographs, or multistylus plotters)
4. X-Y plots in which one function is plotted against another on a permanent recording material
5. Numeric printing in which a number is printed to correspond to the units of the quantity being measured

Data Reduction

Useless data must be removed in order to make useful telemetry data more readily accessible. In order to eliminate the useless data, it is first necessary to inspect all the data to determine the channels and time periods which contain useless data. This is generally done by a method commonly referred to as "quick-look." For a "quick-look" a large number of channels are plotted as time histories to relatively small scales and a visual inspection of the large number of channels thus compressed is simple. Actually, the procedure is to choose the useful data since the *useless* data constitute the greater bulk of those taken. The useful data are chosen by noting each channel and the time of its occurrence. It is then reproduced at greater accuracy or expanded time scale from the primary electrical recording system.

Another problem in data reduction is that of eliminating redundant data. This is done by choosing the best data or averaging data from among several records. Redundant data are produced by using more than one channel or transducer to measure a single quantity, or using more than one ground station to receive data simultaneously, or by using several recorders at a single ground station, each recording the same data. It is necessary to edit these redundant data and decide which records are to be used and, if many records are to be used, how they will be averaged or displayed.

An excessive sampling rate may also generate more data than are required for analysis or display. This can be compressed by a reduced-speed plotter or a sampling technique which chooses only occasional data points among the large number recorded.

Another example of data reduction occurs in the event that many channels are monitored for malfunctions and the data received are exactly as expected. In this case, a simple inspection and analysis will reveal this fact and the data need not be displayed or recorded; the analysis simply states that no malfunctions occurred.

Smoothing may also be used to reduce the amount of data for analysis and display. Vibration occurring on an acceleration channel may be smoothed by a low-pass filter or, subsequently, by computation. In this process, the statistical averaging of a large number of samples may not only reduce the amount of data which need to be analyzed but also increase the accuracy of the data which have been smoothed. The computer, using the least-squares method of curve fitting, is often used for this function.

Data Analysis

Once a minimum amount of data has been selected and recorded, it must be read and analyzed. Either the reduced raw data or smoothed or averaged data may be employed for further analysis. Analysis usually involves computation or correlation and interpretation. As an example, it is desired to study vector acceleration. Three axes of acceleration are telemetered, the vector magnitude and angle are computed, and the results are plotted against time from launch. Also a simple method of correlation is to plot one function against another. For this, two or more functions are introduced to the computer at the same time a time interpolation is made and an X-Y plot output function is generated. The process of data computation by large-scale automatic digital computers, a subject beyond the scope of this chapter, has been well and adequately treated in several computer texts (refer to Chap. 12).

Once the data have been reduced and/or computed, it is necessary that they be printed and duplicated. Printers will be required to operate at very high speeds if it is necessary to print all the data as they are received. In fact, one telemetry system produces data in excess of 25,000 points per second. If the data are recorded electrically, they may be reproduced at reduced speed to match conventional printers and plotters. Even so, real-time data are often required, and even for the handling of postflight data, rapid printing and plotting are required to keep pace with the data production in order to shorten the testing cycle. For these reasons, high-speed plotters and printers such as the radiation multistylus recorder have been developed.

It is also necessary that a plotter be able to plot data with lasting accuracy. This may be done by printing the coordinate lines on the paper at the same time the function is laid down. Consequently, if the paper shrinks or expands with humidity and temperature, the scale changes with it, and the data may be read to their original plotted accuracy.

It is desirable that multiple copies of data be made. Copying can be done by an ozalid process if data are plotted on a reasonably transparent medium or by the xerographic process if the plot is made on an opaque medium. The photo-offset process, of course, is applicable to any kind of original medium and is utilized when very wide distribution is required. Since data distribution is usually somewhat limited, the cheaper ozalid process is predominant.

Once the data have been reduced, computed, correlated, and duplicated, they are available for detailed inspection and analysis. Two major agencies are interested in that analysis. One is the contractor, who is responsible for the development of the missile or for performing the scientific exploration; the other is the customer, usually a government agency, who desires to monitor the progress of the development which he is sponsoring. The contractor desires to determine the parameters of his vehicle to permit development of a vehicle to fulfill its mission and his contract. He is interested in finding defects so they may be corrected and in making improvements based upon the measured data. The customer desires to inspect data in order that he may monitor the progress through an independent observation rather than take the contractor's analysis. The customer also supplies the test site and facilities for the test, and he must make the determination as to whether or not these facilities are adequate and useful.

4 Systems of Radio Telemetry

In this Section five systems will be described. Of these, two systems represent the extremes, simplest and most complex, of telemetry development and three systems are in general use at missile test ranges. Of the first two systems, one is a simple, single-channel system, and the other is an advanced, comprehensive, and versatile multichannel system. The three other systems described fall between these extremes.

The Portel Telemetry System

A representative of the simple system is the Portel (portable telemetry) equipment which was designed to telemeter one channel of information over relatively short distances. In this system, the output from a resistance-bridge transducer is transmitted over a radio carrier to a receiving station where a permanent record is made by a moving-pen recorder. Figures 1 and 2 are block diagrams of the transmitting and receiving portions, respectively, of the Portel telemeter. (Figure 3 is a view of the transmitter.) The Portel system is an FM-FM telemeter which generates a frequency-modulated subcarrier which, in turn, frequency-modulates a radio carrier. Any unbalance of the resistance bridge changes the frequency of the subcarrier in proportion. Thus the frequency of the subcarrier represents the quantity being measured, viz., strain, temperature, etc. The subcarrier is converted to modulation of a radio-frequency oscillator and the modulated signal is fed to an antenna. Batteries supply power for the various conversions. A receiving antenna receives the signal and feeds it to the radio receiver which amplifies, detects, and demodulates it to reproduce the subcarrier. The subcarrier is fed to a signal converter which provides an electrical output proportional to the unbalance of the resistance bridge. An appropriate power supply here also supplies energy for these functions.

The transmitting unit is designed to be small, light, and efficient while the receiving station is of conventional rack-and-panel construction for mounting in mobile vehicles.

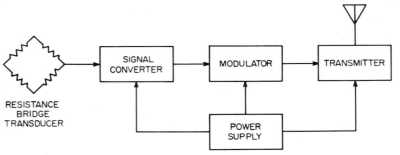

FIG. 1. Portel transmitter block diagram.

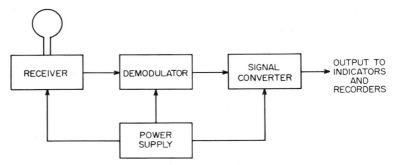

FIG. 2. Portel receiver block diagram.

It may be noted that one end cap of the transmitter is insulated from the remainder of the case in order that it may act as an antenna. The entire transmitter is constructed of solid-state components for a maximum of reliability and a minimum of battery drain. The battery supply shown provides 200 hr of operating time. The subcarrier frequency of this particular unit is 2,500 cycles, ±15 per cent. The sensitivity is such that a change in resistance of one of the arms of 1 part in 10^4 gives a change of 10 cycles in the subcarrier frequency. This corresponds to the measurement of a strain of 25 μin./in. in a 120-ohm gage with a gage factor of 2. The zero and full scale are set at the receiving station in an initial installation, and it is not necessary that the bridge be balanced.

FIG. 3. Portel transmitter. (*Courtesy of Radiation, Inc.*)

Figure 4 is a block diagram of the transmitter showing the operation of the subcarrier oscillator portion. Excitation is supplied to the bridge from an amplifier, and the output of the bridge (which is zero when the bridge is balanced) is mixed through a resistance-adding circuit with a signal from the amplifier shifted 90° from the bridge excitation. This combined voltage is further shifted 90° in a phase shifter so that a total of 180° phase shift is supplied to the input of the amplifier. The amplifier contains an odd number of stages giving another 180° net phase reversal; hence the total shift of the loop is 360° and the gain is sufficient to cause oscillation. This type of oscillator is conventionally referred to as a phase-shift oscillator. The gain is carefully controlled and stabilized to prevent oscillations at frequencies other than the design range. When the resistance bridge is unbalanced, both the amplitude and phase of the output of the resistance adder change. The change in amplitude is unimportant, but the change in phase causes a change of frequency such that at the new frequency the total phase shift of the loop remains at 360°. It is readily noted that a change of resistance producing an unbalance in one direction causes a leading phase shift while a change producing an unbalance in the other direction causes a lagging phase shift. Consequently, the frequency is increased or decreased depending upon the direction of the resistance unbalance. This shift is a linear function of the resistance unbalance.

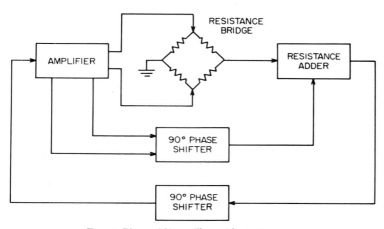

FIG. 4. Phase-shift oscillator block diagram.

An output from an amplitude-stabilized portion of the amplifier is fed to the modulator of the radio transmitter. This voltage is applied to a varicap, which is a portion of the transmitter's tank circuit, changing its capacitance which, in turn, changes the frequency of the oscillator. Output is coupled from the tank coil through a very small capacitor to the insulated end cap which is the transmitter antenna. Linearity of frequency modulation here is also good at small variations of capacitance. The oscillator is tunable from 88 to 108 Mc. By simple tank-circuit substitution, this frequency may be placed anywhere in the range from 1 to 200 Mc with approximately 20 per cent tuning range. Figure 5 is a circuit diagram of the transmitter.

The first version of the receiver was designed to be operated in very high a-c electric fields. Consequently, a shielded-loop antenna is used for input and the entire system is shielded to prevent noise from reaching the intermediate-frequency amplifier. A conventional FM tuner and ratio detector are used to reproduce the subcarrier. Satisfactory operation has been obtained with a common-mode a-c potential at the transducer of 10^8 greater than the transducer output. This factor can undoubtedly be much greater and is limited only by the shielding and isolation of the radio receiver.

The subcarrier output from the receiver is carefully clipped and limited and fed to a phase-locked loop detector. The phase locked loop was chosen because of its simplicity and its versatility. Also, by adjustment of the feedback filter in the loop (transfer function), the change in output with frequency may be varied. Thus, in a very simple manner, a rising or falling amplitude vs. frequency characteristic may be obtained to compensate for subsequent nonflat equipment. Furthermore, full-scale voltage may be easily varied and a zero voltage determined quite independently of the bridge balance. In other words, the bridge zero need not be the center frequency of the subcarrier, but the arbitrarily chosen zero remains stable and fixed at any chosen frequency.

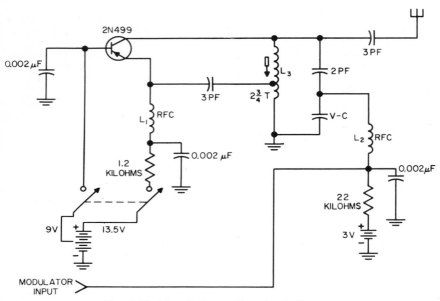

FIG. 5. Portel radio transmitter circuit diagram.

The PCM Telemetry System

At the other end of the telemetry-equipment scale lies at AKT-14/UKR-7 pulse-code-modulation telemetry system. This system was developed by Radiation, Inc., under contract for the U.S. Air Force. The salient features of the system are its inherent high accuracy and stability, its large information-handling capacity, its freedom from adjustment and calibration, and its ability to provide data in a form which may be directly assimilated by large general-purpose digital computers. Figures 6 and 7 are block diagrams of these equipments. The AKT-14 units are the transmitting portion of the system and the UKR-7 units are the receiving portion. The AKT-14 equipment is capable of working with a wide variety of different transducers. It supplies excitation for resistance bridges and the bridge outputs are fed into sequential multiplexer. It can handle both high-level and low-level voltage outputs from transducers. Digital transducers may be multiplexed into the system following the coder.

An a-c supply is used for bridge excitation. This voltage is amplified and sampled by the multiplexer. The samples are extremely short and correspond to the peak of the sine-wave output from the transducer. Each transducer is sampled sequentially with a sample length of approximately 4 μsec; 24,000 samples per second are taken by the multiplexer. All the input functions are presented sequentially to the coder which transforms the amplitude of the voltage to 8-bit binary code. This process provides a voltage resolution of 1 part in 256. The serial code is then presented to the modulator which, in turn, feeds a radio transmitter.

The modulator incorporates a special feature which permits the modulation bandwidth to be halved. Output code from the coder is fed to a bistable electronic switch whose state is changed when a yes bit occurs, but whose state remains unchanged for a no bit. The highest-bandwidth condition is that when all yes bits occur. This condition produces an output from the bistable electronic switch which is half the frequency of the original signal, and each change of state represents a yes bit. In other than the highest-bandwidth condition, a change of state continues to represent a yes bit and the absence of a change at the proper time represents a no bit.

The UKR-7 receiving system receives the transmitted binary-coded signal and feeds it first to a sync-restoring unit. This unit acts as a flywheel during fades in the

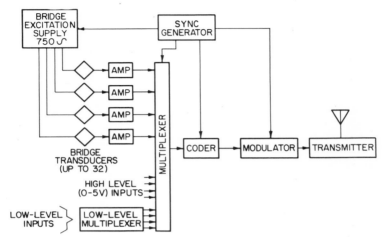

Fig. 6. AKT-14 block diagram.

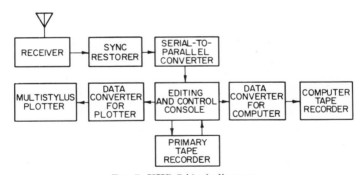

Fig. 7. UKR-7 block diagram.

radio signal to restore data immediately at the conclusion of the fade. It also provides the signals necessary to transform the serial data word to parallel words. This conversion permits recording on multitrack tape recorders and a simple means of plotting and providing digital-computer inputs. If a wideband (video) type of recorder were employed, it could record the serial output or even an intermediate carrier frequency of the receiver. However, these recorders are expensive and complex. Consequently, this system does not utilize them.

The editing and control console controls the distribution and processing of data. In general practice, all the data are recorded on the primary tape recorder. Also, all the data are plotted for quick-look purposes and then only particular chosen portions of data are reproduced from the primary tape recorder for further accurate plotting

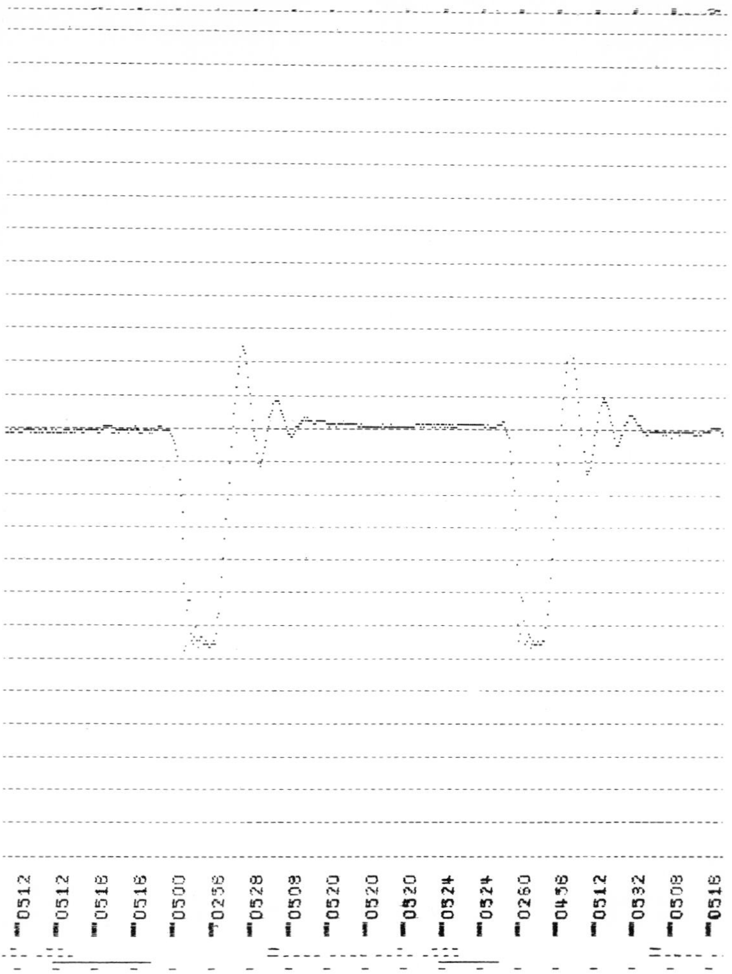

Fig. 8. A single data channel plotted and printed. (*Courtesy of Radiation, Inc.*)

Table 2

Parameter	AKT-14/UKR-7	PCM-FM
Channels..........................	32	32
Samples per sec per channel..........	750	750
Binary bits per channel..............	8	10
Bits between channels...............	1	1
Total sampling rate, samples/sec.....	24,000	24,000
Total bit rate, bits/sec..............	216,000	264,000

or conversion to computer inputs. The choice of data to be further processed once made is simply transmitted as commands by the editing and control console to provide the data in plotted form or grouped as necessary for further commutation in a digital computer. Typical data plots are shown in Figs. 8 to 10. The parameters of this equipment are shown in Table 2.

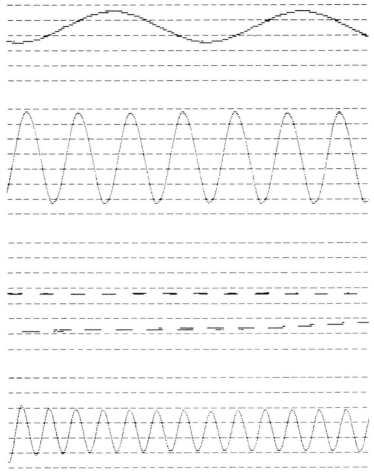

FIG. 9. Four data channels plotted. (*Courtesy of Radiation, Inc.*)

A subsequent PCM-FM telemetry system has also been developed with the second set of parameters in Table 2. This telemeter was developed for the Holloman Air Force Base by Radiation, Inc., and is used on the high-speed rocket sled for the acceleration testing of missile components.

The foregoing descriptions are brief accounts of the lowest to the highest capacity of the telemeters. The first system utilized a subcarrier for the transmission of intelligence and the second employed time-sequence sampling. The subcarrier method permits the use of other subcarriers at different frequencies and has been given the name of "frequency multiplexing." The time-sequence sampling has been named "time multiplexing." Both have their advantages and disadvantages. Of the three systems in general use at missile test ranges, one is a frequency-multiplexed system and two are time-multiplexed systems. To the frequency-multiplexed system, a time-multiplexed input is usually added to one or more of the subcarrier

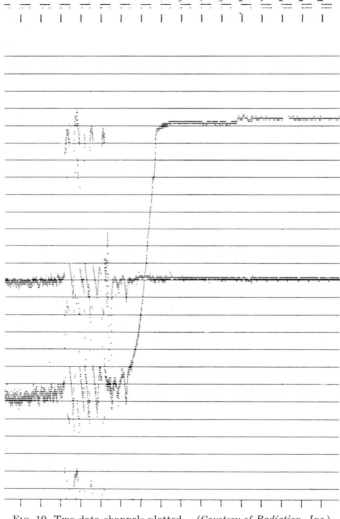

Fig. 10. Two data channels plotted. (*Courtesy of Radiation, Inc.*)

channels. This time-multiplexed input is from a mechanical switch or commutator which provides the sequential inputs to a single channel.

The Standard FM-FM Telemetry System

Figures 11 and 12 show block diagrams of the transmitter and receiver sections of an FM-FM telemetry system. Since there are many varieties of this equipment available, no attempt has been made to show a single complete system. As a matter of fact, a complete system made by a single manufacturer is seldom found in practice, but rather a user chooses components from a number of manufacturers and assembles his own system.

FM-FM telemetry has been in common use for approximately 10 years, and various parameters were "standardized" by the Research and Development Board of the U.S. Department of Defense and subsequently by the Inter-Range Instrumentation Group. These standards are presented in Chap. 5.

A wide variety of transducers may be utilized with the FM-FM system. Voltage- and current-generating transducers, resistance bridges, and even variable-reluctance

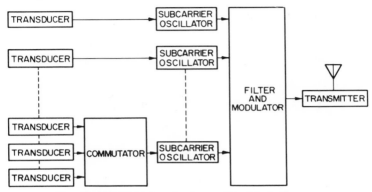

Fig. 11. FM-FM telemetry transmitter block diagram.

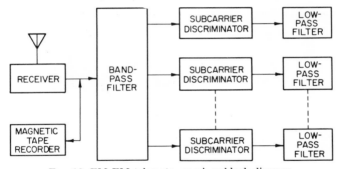

Fig. 12. FM-FM telemetry receiver block diagram.

transducers may be employed. The latter are generally used as a part of the resonant circuit of the subcarrier oscillator. Many forms of subcarrier oscillators are available, viz., voltage-controlled, current-controlled, resistance-bridge-controlled, and variable-reluctance-controlled oscillators. Vacuum-tube and transistor models of most types are available. They are usually made with output-level adjustments such that a properly emphasized composite signal may be presented to the radio modulator. A subcarrier oscillator, however, must be chosen for the particular frequency range desired. Many oscillators are available for only a portion of the standard range.

It will be noted from the block diagram of Fig. 11 that the outputs from the subcarrier oscillators are fed through filters to the radio modulator. These are not absolutely necessary but are highly desirable in order to perform two functions: (1) to isolate each subcarrier oscillator from the others at a different frequency so that its output is transmitted to the modulator instead of being dissipated in the other oscillators and (2) to prevent overmodulation or overly rapid modulation from generating signal frequencies in adjacent channels. These filters are bandpass filters with separate inputs and a common output.

It is necessary that the modulator preserve the linearity of the complex modulating wave. Intermodulation distortion results in spurious signals and data errors and can be prevented only by maintaining linearity in the modulation, the radio-frequency transmission, and the demodulation processes. (This is a restriction of minor importance in time-multiplexing systems.) It is further necessary for efficient FM operation to eliminate amplitude modulation. Also, many modulators are an inherent part of the frequency-stability system of the radio transmitter and they must be usable with crystal frequency-stabilized transmitters.

Various radio transmitters are in common use. Many have self-excited output stages. Others are crystal-controlled or stabilized through multiplier chains. Output power may vary from a few milliwatts to 100 watts and frequency deviation may vary from a few cycles per second to 500 kc/sec. Frequency-deviation ratios up to

Fig. 13. Portable telemetry receiving antenna. (*Courtesy of Radiation, Inc.*)

five are usually employed. Telemetering capabilities available at the various test ranges are given in the standards.

In many cases, more than one FM-FM system is used in a single test vehicle and more than one radio carrier is fed to a single antenna. When this is done, an r-f multiplexer is used to isolate one transmitter from the others. It also matches impedances between the transmitters and the antenna. Since projections from or cuts in a missile skin are undesirable aerodynamically and structurally, a single antenna system must usually serve all transmitters. Consequently, a best location is found for an antenna system and several transmitters are multiplexed into it. In the case of Sputnik III, however, four antennas each projecting over a foot from the nose surface, four antennas folding out from the base, and one nose antenna were utilized, greatly reducing the problems of fading, directivity, and radio multiplexing. Most significant were the four nonfolding antennas projecting from the nose surface. Subsequent United States satellites have been built according to this practice.

At the ground station, a large high-gain directive-receiving antenna tracks the moving missile. Circular polarization is usually employed to receive the signal regardless of the attitude of the missile. Figure 13 shows a typical antenna. A preamplifier is located near the antenna feed to minimize the thermal noise of the system.

Enough amplification is included to more than compensate for transmission-line losses. In typical stations, a radio-frequency demultiplexer, or channel splitter, is included to permit the simultaneous reception of up to eight telemetry carriers. As many receivers as are required for the transmissions are employed, and each feeds a separate head on a magnetic-tape recorder and/or separate sets of subcarrier filters. Either the direct signal or the playback of the tape is broken into separate subcarriers by the subcarrier filters, and each is fed to the appropriate subcarrier discriminator. The discriminators are of many types, including counting discriminators and phase-locked loops. Discriminators employing inductors are seldom used because of the large sizes required at the low subcarrier frequencies. The intelligence fidelity is maximized by the choice of a proper filter at the output of the subcarrier discriminator. Furthermore, the output filter removes the subcarrier frequency before the signal is fed to oscillographic or pen recorders. As previously stated, the composite subcarrier signal is also recorded on magnetic-tape recorders, and in general missile practice, the tape recorder may replace the entire ground station at outlying recording bases. When this is done, the recorded data are played back through the same filters and discriminators as above. However, the tape recorder adds an error component because of its inherent flutter and wow. This error is minimized by the design of recorders but is also compensated by the use of a compensating tone, other than a sub-carrier tone, which is recorded when the composite subcarriers are recorded. This compensating tone is then detected and processed in the same manner as the other subcarriers. The output, after demodulation, is fed in phase opposition to other sub-carriers. Because the phase delay through the filters and discriminators of each sub-carrier is different, the flutter output must be delayed individually and separately to match the delay of each subcarrier. Furthermore, the delay which must be matched varies over the subcarrier frequency range; consequently, precise compensation over the complete subcarrier deviation range cannot be achieved by this method.

More details of FM-FM systems are included in Chap. 6.

PDM-FM Telemetry System

The operation of pulse-duration-modulation systems is shown in the block diagrams of Figs. 14 and 15. This system has been employed at missile ranges, and test-range

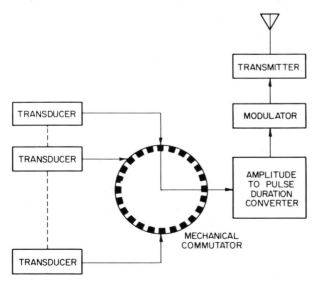

Fig. 14. PDM-FM telemetry transmitter block diagram.

standards similar to those for the FM-FM system have been prepared. These are included in Chap. 5.

Many of the parts of the PDM-FM system are common with those of the FM-FM system, and the preceding description of antennas, radio transmitters, tape recorders, etc., is applicable and will not be repeated here. The system is light, simple, and compact but not so versatile as the FM-FM system. Transducers are limited to those which produce 0- to 5-volt outputs; otherwise, amplifiers must be employed which provide a common level to the commutator. In general, the PDM-FM system may be compared with a single commutated channel of the FM-FM system, and it

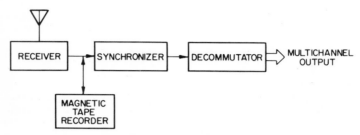

Fig. 15. PDM-FM telemetry receiver block diagram.

has often been employed as input to the 70-kc subcarrier of the FM-FM system. Variable-reluctance transducers cannot be used unless their output is first converted to a voltage (see Chap. 7).

PPM-AM Telemetry System

The pulse-position-modulation telemetry system has been employed in the United States principally by the Navy. It has not been chosen as a test-range standard, but instead, receiving equipment has been moved into the test ranges as required. There have been further developments of high-capacity PPM-AM telemetry systems which have not been put into use in the United States. The PPM system is very similar to the PDM system, except that, instead of using the entire duration of the pulse, only the beginning and the end of the pulse are transmitted as intelligence markers, and a conversion from voltage amplitude to pulse position is employed. It is also a time-multiplexed system, and equipments of up to 36 primary channels have been developed with a total sampling rate of 50,000 samples per second. The equipment presently employed by the Naval Research Laboratory is summarized in Table 3. Figures 16 and 17 are block diagrams of the transmitter and receiver. Other PPM-AM systems* developed prior to the AN/DKT-7(XN-2) are not described here.

Table 3. NRL Telemeter Transmitting Set AN/DKT-7(XN-2)

Type	PPM-AM
Channels	15
Sampling rate	312.5 samples/sec
Supercommutation	4 channels may be cross-strapped for 1,250 samples/sec
R-F frequency link	220–239 Mc
Transmitter power out	40 watts (peak pulse power)
Accuracy	$\pm 2\%$ ($\pm 1\%$ with external calibration)
Power in	28 volts d-c at 3.5 amp, 6.7 volts d-c at 11.3 amp
Commutation	Electronic
Signal input	0–5 volts
Pulse interval	200 μsec
Bandwidth	100 kc approx

*Naval Res. Lab. Repts. 2955, 3013, and 3030.

RADIO TELEMETRY   4-23

The AN/DKT-7(XN-2) telemetry transmitter utilizes cascaded electronic multiplexers to sample 15 voltages sequentially. The first multiplexer operates at a rate of 1,250 samples per second and the second multiplexer at 5,000 samples per second. Each of the 15 channels is sampled at a rate of 1,250 ÷ 4 or 312.5 samples per second and the sampled voltages are fed by the multiplexers to the voltage comparator where the conversion to pulse-position modulation is made.

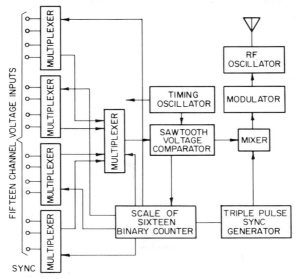

FIG. 16. PPM-AM telemetry transmitter block diagram.

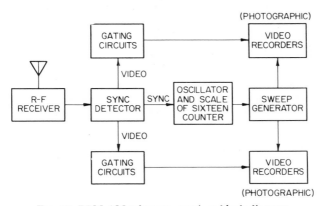

FIG. 17. PPM-AM telemetry receiver block diagram.

To convert the voltages to pulse-position modulation, a precision sawtooth voltage is generated which begins at zero volts and increases until it matches the sampled voltage. This point is detected by the voltage comparator and a video pulse is generated at that time. The generation of the video pulse occurs at a position in time after the beginning of the sawtooth voltage which is proportional to the sampled voltage, hence the name pulse-position modulation. This process is repeated for each of 15 channels and for the sixteenth channel three pulses of fixed widths and spacings are transmitted instead of data pulses. This "sync" signal or "frame pulse" permits a receiving-station multiplexer to be synchronized with the transmitting multiplexer and the channels to be separated at the video recorders.

At the receiving station, synchronization is accomplished by first separating the sync pulse from the data pulse. This is done by means of a preset discriminator which delays the first two pulses appropriately so that they may be brought into AND circuits to provide a single output signal. The time spacing is less than the spacing of the data pulses, thereby preventing any combination of data from producing a sync signal.

This sync signal synchronizes a local oscillator which feeds a scale of 16 counters to produce gating signals and sawtooth sweep voltages for the video recorders. Data pulses are applied to the intensity grid of the cathode-ray tubes and sweep signals are applied to the horizontal deflecting plates. Photographic film moves at uniform speed at a right angle to the sweep deflection to produce a time-history trace of data.

Either a single channel or many channels may be recorded on the same film.

5 Trends in Telemetry

As a result of the development of new techniques and components, many advances in the state of telemetry are now possible and new improved telemetry systems are gradually replacing older models. The significant advances will be enumerated, and their effects on telemetry-system performance will be described.

Transistorized Circuits

The development of the transistor, and particularly the silicon transistor, has been more significant to missile telemetry than to almost any other branch of electronics. It has permitted the reduction of size, weight, and power requirements—three factors which are of vital importance to missile operations. The replacement of the vacuum tube is a gradual process, however, since stable operation over a wide range of temperature is more difficult with transistors. High-frequency operation is just being achieved. The use of transistors, however, does make practical high-capacity systems of pulse-code-modulation telemetry which would otherwise be prohibitively large.

Large Automatic Tracking Antennas

The development of large immovable parabolic reflecting antennas was largely accomplished under studies of forward-scatter propagation and the antennas were later adapted to telemetry use. The high gain of the large reflector dictates that the beam width be relatively narrow and, therefore, tracking difficulties are presented in missile and satellite operations. It was not until the automatic tracking feature was added to the forward-scatter propagation antenna that it became practical for telemetering from guided missiles. Several automatic tracking telemetry antennas employing 85-ft reflectors as shown in Fig. 18 are in service. Their effect upon radio-telemetry reception is phenomenal. Basically, there is a 10-db improvement in reception over previous techniques. This has made continuous data reception possible where otherwise there were losses due to fading. On the other hand, for the same performance characteristics, the missile transmitter power may be reduced by a factor of 10.

Phase-locked Frequency-modulation Discriminators

Another recent improvement in telemetry receiving techniques has been the phase-locked FM discriminator, or detector. The phase-locked principle is one in which the frequency of a local oscillator is varied to correspond with the incoming frequency. This makes it possible to transmit the resulting beat frequency through a filter of narrow bandwidth. The local beat-frequency oscillator is voltage-controlled, the control voltage being the demodulated signal. The detector is a phase detector or multiplier instead of the conventional heterodyne detector. This technique has added another 6 db of improvement to telemetry receiving stations, and this improvement can be increased to 15 db if the phase-locked principle is also applied to the higher subcarriers. Of course, still greater improvement may be obtained if the bandwidth of the telemeter is curtailed. Tests have shown that with phase-locked receivers, a 10-mw telemetry transmitter transmitting only the lowest six standard

Fig. 18. 85-ft automatic tracking receiving antenna. (*Courtesy of Radiation, Inc.*)

subcarriers has been received well over several thousand miles. Refer to Chap. 9, Secs. 16 to 19, for phase-locked-loop design principles.

Solar Power Sources

Also recently developed are "power sources" which transform solar energy directly to electricity. These devices are particularly useful in satellite telemetry where an indefinite power-source life is required. Erosion from micrometeorites is not definitely determined, but a life of many years may be expected. The cells maintain full efficiency in outer space for they do not collect dirt to impair their operation. This power source permits an operating life of telemetry equipment that was impossible a short time ago.

Predetection Recording

In telemetry practice, it is usually very desirable to make a primary recording on magnetic tape. The data recorded on magnetic tape are then later reproduced for various processes. In the detection process, there is an irrevocable loss of signal-to-noise ratio which may be avoided by predetection recording and use of an optimum correlation detector. The kind of detector may be chosen after the signal has been studied to optimize signal in (1) noise, (2) multipath reflections, or (3) particular interferences. Another advantage of predetection recording is that systems of telemetry which now require different recorders for proper recording (FM-FM, PDM, and PCM) may be recorded on the same recorder. In fact, the typical video recorder used for television may be adapted to record simultaneously four or more of the typical telemetry signals without regard to the kind of modulation.

Automatic Data-reduction Equipment and Processing

With the advent of the general-purpose digital computer, telemetry data may be processed (multiplied, averaged, smoothed, calibrated, etc.) at high speed and with great accuracy. The principal developments in this field in recent years have been automatic equipment to bring telemetry data into and out of these digital computers. Also, many smaller special-purpose digital computers have evolved which handle only a specific process such as calibration. These equipments are available in many forms and offer a wide range of adaptability for information handling. The use of these machines for telemetry is discussed in Chap. 12.

SYSTEM SELECTION

6 Basis for System Selection

The need to transfer information is the basis for defining any telemetering problem. Information as used here is the uncertainty in the measurement of a value of the parameter. This implies that it is made at a certain time and to a certain accuracy. The accuracy of the measurement contributes to the amount of information. If a measurement is made to a higher degree of accuracy it contains a greater amount of information. The frequency of measurement contributes to the amount of information. If a parameter must be measured more often because of uncertainty about change in its value, it will be necessary to telemeter more measurements and consequently more information. Similarly the number of different parameters to be measured also contributes to the total amount of information to be telemetered.

Thus the information to be telemetered is a function of several measurable quantities, and we may write an expression for the information rate of the problem in terms of these. See Eq. (2) below.

In the situation where it is desired to measure the value of a variable parameter "continuously" there is always a finite rate at which the parameter under consideration can change. This rate may be described by the frequency of samples of measurements which is necessary to describe this most rapid change, within the accuracy desired. Discrete measurements (or samples of data) may theoretically be made with sufficient frequency to reproduce any physical phenomenon it is desired to measure in order to represent this phenomenon fully within the desired limits of accuracy.

It is also true that the continuous channel will not convey information changes which occur at such a rapid rate as to be beyond the information bandwidth of the channel. Experimentally this can be shown by increasing the frequency of information change beyond the limit of a channel to a point where nothing comes through. Prior to reaching this point the accuracy falls off rapidly in the usual information channels. Thus the information-transfer capacity R of a discrete or of a continuous channel may be expressed in terms of the accuracy and frequency of samples. This is particularly useful for the purpose of obtaining a common denominator in telemetering applications. For a more complete discussion of the points summarized here, together with proofs and exact numerical information, the reader is referred to Chap. 2, Secs. 11 through 18.

SYSTEM SELECTION

The relation between the sampling rate and the bandwidth may be given as

$$n = qB_i \tag{1}$$

where n = sampling rate, samples/sec.
B_i = message bandwidth (or maximum rate of change of information while preserving the specified accuracy in the information transfer)
q = arbitrary constant relating B_i and n

In practice a workable value for the constant q is 5. This means that for the maximum rate of change of the parameter being telemetered there will be five observations of the parameter in each full cycle of change at the maximum rate. It can be seen that this is an arbitrary value. Two is the theoretical minimum. Twelve is a high number which is sometimes used when greater accuracy is thought to be necessary. It is useful to observe at this point that in real life the maximum rate of change of a variable seldom persists for long without the parameter's exceeding the limits of its scale. Consequently 5 works out to be a good practical value of q.

Step 1 in the information-transfer-rate determination for any problem is to decide the desired maximum rate of change (information bandwidth) which it is desired to transmit for each variable parameter to be telemetered. This is now converted into the sample rate n for each parameter by the use of Eq. (1) with appropriate constant. For example, if a parameter such as roll rate of a missile is reasonably expected never to exceed 20° per second it may be decided that 30° per second capability should be provided as the maximum rate, sufficient to indicate a malfunction (for total destruction may be expected if 28° per second is reached). The minimum change in roll rate to be observed is to be 1° per second. The needed sample rate is calculated at 5 × 30 or 150 samples per second.

Once the equivalent information rate is determined for each parameter it is desired to transmit, in terms of the desired information about that parameter, *step* 2 is to decide to what accuracy it will be desirable to know the value of each variable. It should be noted here that for the purpose of this estimate the relation between sampling rate and accuracy from a dynamic-response viewpoint is ignored. It is intended only that this procedure be used as an approximate empirical aid to determine an "information-transfer" rate consistent with these definitions. This information-transfer rate will then be useful in the selection of a telemetering system which is defined on a comparable empirical basis to fulfill the needs of a specific application.

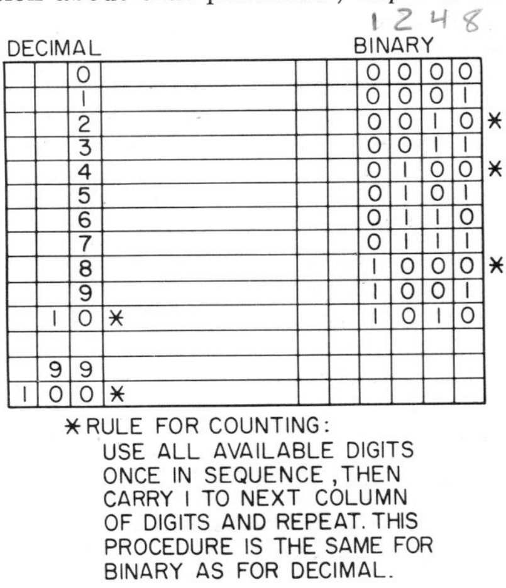

For any single sample the quantity of information involved in the sample is determined by the accuracy of the measurement being transmitted. The accuracy desired can most easily be expressed in binary terms since the binary bit is a convenient measure of information quantity.

DECIMAL			BINARY 1 2 4 8			
0			0	0	0	0
1			0	0	0	1
2			0	0	1	0 ✱
3			0	0	1	1
4			0	1	0	0 ✱
5			0	1	0	1
6			0	1	1	0
7			0	1	1	1
8			1	0	0	0 ✱
9			1	0	0	1
1	0	✱	1	0	1	0
9	9					
1	0	0	✱			

✱RULE FOR COUNTING:
USE ALL AVAILABLE DIGITS ONCE IN SEQUENCE, THEN CARRY 1 TO NEXT COLUMN OF DIGITS AND REPEAT. THIS PROCEDURE IS THE SAME FOR BINARY AS FOR DECIMAL.

FIG. 19. Binary-number representation.

The binary system is merely another way of expressing numerical information. The relation between binary numbers and decimal numbers is reviewed in Fig. 19 and in Fig. 20, which shows binary equivalents to the more familiar decimal notation. Thus, for example, 7 binary bits describe 128 possible discrete points which may be spread across the range of the variable. Figure 21 shows a variable which is to be measured and which appears somewhere in its range between zero and full scale. This range is divided into equal increments for the linear case. Since there are 128 increments the maximum error is one-half of an increment or 1 part in 256 (see Fig. 22a). Similarly some measurements may fall exactly on a quantized value, in which case the error is a minimum equal to zero (Fig. 22b). The average error then will fall halfway between for a resultant average error of 1 part in 512 for a 7-bit system (Fig. 22c). This is plotted linearly in Fig. 23 and on log scale in Fig. 24. Thus if two-tenths of 1 per cent is an acceptable average error then 7 bits per measured value is sufficient, or conversely 1 per cent maximum error can be obtained with 6 bits and

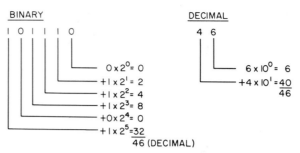

FIG. 20. Binary-decimal conversion.

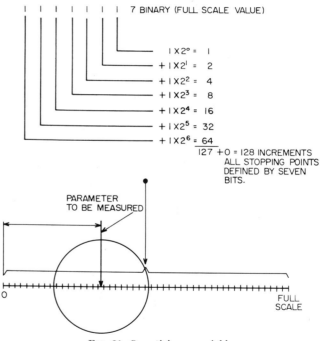

FIG. 21. Quantizing a variable.

1 per cent average error can be obtained with only 5 bits in 32 levels, a fact not usually appreciated.

Having determined the desired accuracy needed for each variable to be telemetered in this application we are now ready to proceed to *step* 3 to determine the total information rate. This in turn will define the information-channel capacity of the telemetering system needed. The total information rate R is simply the summation of the

SYSTEM SELECTION

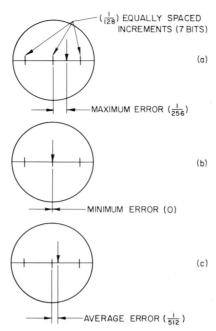

FIG. 22. Maximum, minimum, and average error.

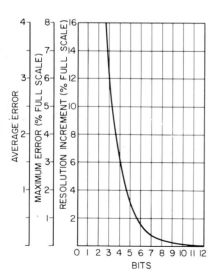

FIG. 23. Error vs. accuracy in bits (linear plot).

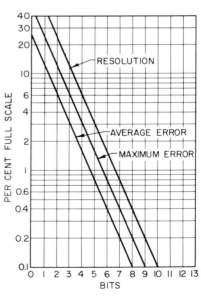

FIG. 24. Error vs. accuracy in bits (logarithmic plot).

separate products of bits and samples per second required for each variable:

$$R = \sum_{i=1}^{i=v} A_i n_i \qquad (2)$$

where v = number of different variables to be telemetered
   A_i = accuracy of each of the particular ith variable, expressed in the number of binary digits necessary
   n_i = effective sampling rate of the ith parameter, samples/sec
   R = total information rate, equivalent bits/sec

This expression provides a means for defining the problem in terms common to all telemetering problems. It is useful for all information-transfer systems regardless of whether or not the information is continuous or sampled. It is a measure of the amount and the rate at which information is to be transferred.

Step 4 is to define the capabilities of various system approaches on a common basis in order to provide a means for selection of a system which will provide best information-transfer-rate utilization. A perfect match is not usually attained since in practice the majority of the systems applications encountered contain a variety of variables each with different bandwidths (or required equivalent sampling rates) and each with different accuracy requirements. Similarly the existing telemetering methods do not always just happen to have exactly the correct and desired information accuracy or sampling rate (bandwidth) capacity for each variable. The net result is that extra bandwidth or information rate must be provided and is wasted. The difference between a successful telemetering application and a poor one is the efficiency of the utilization of the information rate provided by the system in meeting the information-rate requirements of the problem with the smallest possible waste of capacity. An expression for this efficiency may be set forth as

$$\varepsilon_{ta} = R/R_{ts} \qquad (3)$$

where ε_{ta} = efficiency of the telemetering application, per cent
   R_{ts} = information rate of the telemetering system, bits/sec

This section has described the method of obtaining an equivalent required information rate from an analysis of the data sources. The next section will indicate how to determine the equivalent information rate of a system and what effect this has on selection.

7 Determining Telemetering-equipment Information-rate Capacity

Many factors affect the selection of equipment. Paramount among these, however, is the necessity for the system at least theoretically to be capable of telemetering the desired information rate R of the source variables. Therefore, in evaluating the usefulness of various alternate schemes the first task is to determine their equivalent information rate R_{ts} (the information rate of a telemetering system). This must be slightly above or in the limit equal to R. Any large difference indicates a poor match between source and system and is therefore an inefficient design.

The R_{ts} may be calculated by a method which is nearly the converse of the one just described for the calculation of R. For telemetering systems with a single channel this is simply the *equivalent* bandwidth of the information channel, expressed in bits per second. Thus a single channel capable of "continuous" telemetering of a variable with a response up to 100 cps (within limits of acceptable accuracy) has an equivalent bits per second information rate of

$$R_{ts_i} = B_i A_i q_i \qquad (4)$$

where B_i = information bandwidth, cps (or maximum rate of change of information while preserving the specified accuracy)
   q_i = an arbitrary factor relating B_i to sampling rate and again is taken as 5 for the same reasons as before
   A_i = accuracy of the ith channel
   R_{ts_i} = information rate of the ith channel

The total information rate then is

$$R_{ts} = \sum_{i=1}^{i=n_c} R_{tsi} \tag{5}$$

where n_c = total number of channels

The typical telemetering-system equivalent capacities are shown in Fig. 25. The frequency-multiplexed systems (e.g., FM-FM) are treated as a number of parallel channels. Each channel has a different bandwidth which has an equivalent information rate in bits per second. These may be calculated and summed to determine the system value. In time-multiplexed systems such as PCM, PPM, subcommutated FM-FM, or PDM-FM the sample is either quantized in the case of PCM, PPM, or PDM-FM or is so short compared with the usual sampling interval (as in the case with the subcommutated FM) that it is effectively quantized. In this case the equivalent information rate is merely the product

$$R_{ts} = A_i n n_c \tag{6}$$

where A_i = accuracy expressed in bits for each sample
 n = sampling rate, samples/sec, in each channel
 n_c = number of channels in a frame or complete cycle

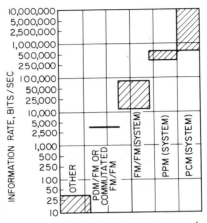

Fig. 25. Typical telemetry systems equivalent capacities.

In the case where the channels are not all sampled at the same rate the information rate may be calculated by summing the individual rates thus:

$$R_{ts} = \sum_{i=1}^{i=n_c} A_i n_i \tag{7}$$

where n_i = sampling rate of the ith channel, samples/sec

The data for Fig. 25 were derived on the basis of performance achievable in the field. Conservative values have been used, but since this is a comparison no particular system has been unduly penalized. The standard PDM-FM system provides a maximum of 900 samples per second and the accuracy is assumed to be approximately 1 to 2 per cent so that, referring to Fig. 23 or 24, 4 bits was the value used for A_i or 3,600 bits/sec.

The FM-FM standard system is calculated by a summation of the individual channels. These are given in Table 4. The R_{ts} for each channel is again based on an A_i of 1 to 2 per cent, which is the equivalent of 4 bits. As noted, the frequency response or B_i of each channel is based on a maximum deviation and a deviation ratio of 5 (Chap. 5). The range of approximately 14,000 to 80,000 for this system will depend on the selection of channels and their utilization.

The PPM information rates are based on the AN/AKT-11, which has an information rate of 50,000 samples per second at an accuracy of better than 1 per cent. Six bits per sample are assumed, which gives a bit rate of 300,000 per second. The higher limit of 500,000 is believed to be practical with techniques presently known.

The use of PCM is becoming more widespread, and Fig. 25 indicates one of the reasons. Presently available equipment will furnish 500,000 bits/sec and rates up to 10,000,000 bits/sec appear feasible.

It is interesting to note that the information rates of the other examples given all fall below the PDM system and, in the case of the single-bit system or alarm, fall

Table 4. FM/FM Equivalent Information Rates

Band	Center frequency, cps	Frequency response B_i	R_{ts}		
1	400	6	120		
2	560	8.4	168		
3	730	11	220		
4	960	14	280		
5	1,300	20	400		
6	1,700	25	500		
7	2,300	35	700		
8	3,000	45	900		
9	3,900	59	1,800		
10	5,400	81	1,620		
11	7,350	110	2,200		
12	10,500	160	3,200		
13	14,500	220	4,400		
14	22,000	330	6,600	13,288	subtotal
15	30,000	450	9,000		
16	40,000	600	12,000		
17	52,500	790	15,800		
18	70,000	1,050	21,000	64,400	subtotal
			80,288	77,688	total

very close to zero. This is a large class of problems in which a variety of techniques have been used over the past fifty years. In general the information rate of the means used to telemeter is quite high, merely because there is nothing available as an alternate. The burglar alarm or fire alarm is a very good example. In many such systems the channel which is kept open continuously is capable of several hundred cycles response, which is a bit rate of say 1,000. The typical alarm may be used as much as once a week, which is a bit rate of one per $60 \times 60 \times 24 \times 7$ or 605,000 sec for an efficiency of use of 0.6×10^{-9}. Other industrial applications tend to have equally poor system utilization efficiencies and efforts to improve them should be centered around this. In the case of the alarm, for example, improvement can be achieved by initiating the circuit only when an alarm is present and at other times using the circuit for other purposes. Telephone lines are being used more for data transmission in connection with computers and industrial process control. Refer to Chap. 13.

The variation in information capacity among the several telemetry systems comes about because of differences of spectrum utilization by the different modulation techniques. These factors are discussed more thoroughly in Chap. 2.

8 Other Considerations in the Selection of Equipment

The discussion to this point has been confined to the matching of information rate between problem and system. The reason why this is important is that there are many possible methods of telemetering and the information-rate analysis will permit a first approximation in selection. Having determined the information rate needed the number of possible choices for an efficient design is considerably reduced. We are now ready to consider the other factors which will influence a successful application. These are:

Physical size, weight, and volume. Will it fit? Will it operate in the available space, weight, and volume?

The environment. Will the equipment selected operate in the existing environment for the necessary time?

Power requirements. Is adequate power available? The different multiplex/modulation techniques having differing signal/noise performance (and conversely, error rates) and hence make different demands on power for a given transmission range. Comparisons of this sort are made in Chap. 2, Sec. 28, taking into account threshold effects which demand certain minimum received power levels before proper performance is obtained. A more refined level of telemetry-system selection considers this point (Chap. 9, Secs. 7 through 9).

The transmission path. Is the path and is the configuration of the transmitting and receiving points such as to permit adequate transmission of the information without fading or excessive noise? (Sec. 19.) Each of these limitations will manifest itself in reduced information-rate transfer.

Transducers and pickup. Is the system selected suitable for the transducers available for the parameters to be measured?

Alternates. How does the final installed system including power, transducers, and antennas compare with alternative systems of equivalent information-rate capability?

Efficiency. How well can the problem utilize the available information-rate capability? For example, if a number of continuous 100-cps information bandwidth signals are to be telemetered and the system under consideration has a number of fixed information bandwidth channels at between 10 and 1,000 cps this represents a poor match.

Standard. For many military applications standards have been set up. Test facilities are equipped to receive and perform data reduction on these standard systems and frequencies have been allocated for operation of these standards. (Refer to Chap. 5.) In this case the method and system selected will also be influenced by the availability of coverage from existing range instrumentation during the period in which it is desired to schedule the operation.

9 Adjusting Information and Equipment Capacity

If the problem requires a high information rate, as is usual in a military situation, then the telemetering applications engineer is probably stalemated. It creates a paradox of not being able to furnish sufficient telemetering within the available size and space utilizing available systems. It will therefore be necessary to make some compromises. These may be in two forms, the obvious ones which involve a reduction of information and a second set of possibilities which involve consideration of the relative priority of the information and utilization of methods of obtaining a greater quantity of useful information. When it is necessary to reduce the amount of information to fit the capacity of the available system the foregoing discussion is directly usable in establishing a reduced information rate. This may then be done by reducing either the accuracy of the data transmitted, the frequency response, bandwidth, or samples per second of the data transmitted and finally by reducing the number of different parameters to be telemetered.

The requirements for telemetered data are frequently specified by a group separate from those responsible for providing the telemetering system. Thus it is recommended that they not be accepted at their original specification. The group writing these specifications frequently lacks knowledge of the telemetering problems and may request more information than they actually need as a substitute for the more time-consuming and difficult process of analyzing their problem in more detail. Thus some time spent to analyze the requirements in greater detail at this point may reduce the requirements to a more realistic specification which will be equally useful to those requiring the data and which can more easily be met by the telemetering system.

The second method for reducing the telemetering requirements to fit the capacity of a system is to find a way of testing so as to obtain a higher percentage of useful information. Again this requires understanding of the problem but in this case from a different viewpoint. In this case the purpose of the test must be understood. While all the parameters may in fact be of equal interest and none can be eliminated, each requiring its channel, it may in fact turn out that it is not necessary to obtain information from all at the same time. Thus in some experiments certain parameters may be of interest only during an early phase of the test, say the first 30 sec. Thereafter the parameter may not be of significance and in fact may have ceased to exist. In this case it is no longer necessary to reserve the capacity for this parameter and it can be programmed or switched in time during various phases of the test so that in the early phase it telemeters one parameter and thereafter it is connected to telemeter other parameters which are of interest in a later phase.

Another way of reducing the required telemetering capacity is to study the *interdependency of the parameters* which are being telemetered. It may develop that one variable is a fixed function of one or more others, a function which if known and if it can be depended on may eliminate one or more channels of telemetering.

Still another way of reducing the amount of data, which is particularly applicable to wide information-rate bands, again depends on an understanding of the physical

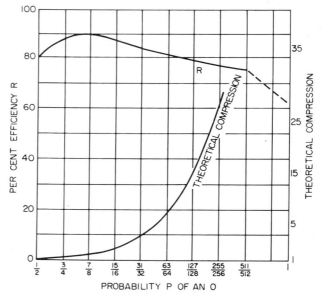

Fig. 26. Data compression by predictive coding. (*Derived from P. Elias, Predictive Coding, IRE Trans. Inform. Theory, March, 1955.*)

problem and a critical evaluation of what is actually intended or needed from the experiment. Vibration is such an example. In many cases the telemetering of a full vibration spectrum will require a rather wide information bandwidth. Frequently it is sufficient for the purpose of the test to provide two narrow channels, one telemetering some function of amplitude, for example, the maximum amplitude in a plane, and the other some function of frequency.

Frequently a careful analysis of the physical phenomena being observed will suggest methods whereby the bandwidth of the data may be reduced, thereby simplifying the telemetering problem.

Methods, sometimes called data compression or compaction, are presently being developed which operate on information-theory principles. P. Elias, for example, has examined predictive coding. In such a method, for the transmission of binary information, use is made of foreknowledge of the probability of a 1 or 0 being sent. If there is a high number of 0's compared with 1's, then it is possible to send a binary code representing the number of 0's with a saving of bandwidth. The degree of such savings from "run-length encoding" can be determined from Fig. 26, derived from Elias's work. The actual saving is the product of the value for the compression and the coding efficiency.

10 Utilization of Government Test Facilities

The United States government controls the testing of military missiles and space vehicles. This is accomplished at a number of different test sites. A number of examples may be cited. On the West Coast there is the Pacific Range which includes the Naval Air Missile Test Center at Point Mugu; Vandenburg Air Force Base at the site of Camp Cooke south of San Francisco; and in addition the Naval Ordnance Test Center at China Lake and Edwards Air Force Base at Muroc. In the Florida area the Air Force Missile Test Center operates a long range at Cape Kennedy from Patrick Air Force Base and controls the operation of a shorter range from Eglin Field. The Army operates a short range at White Sands Proving Ground which includes the facilities of the Holloman Air Force Base. NASA tests short-range vehicles on the Virginia coast at Wallops Island.

Each of these facilities has provided centralized telemetering receiving stations and provision for reducing the received data. It is mandatory to utilize these facilities, and in order to do so a considerable amount of coordination must be accomplished starting as far in advance of the actual test schedule as possible. A system must be selected which is compatible with the ground equipment installed at the test range to be used (refer to Chap. 5). This is subject to frequent change, and the reader should obtain the latest information from the range which it is planned to use.

The selection of the system will then include allocation of the necessary r-f frequencies and limitations of r-f power with times assigned for operation on these frequencies. In the case of FM-FM the utilization of channels will also be influenced by the availability of discriminators for the various channels and the coverage planned from the various receiving points. Each test area has technical assistance available for the purpose of furnishing this information and for obtaining an optimum utilization of the acility for each test operation.

SIGNAL CONDITIONING*

The airborne or spacecraft instrumentation system accumulates, conditions, and transmits measurement data from transducers and other electrical information sources located in the vehicle. The following pages will discuss exclusively the signal-conditioning system, also known as the signal processor or signal converter, and closely associated parameters.

The signal conditioner accepts signals from transducers such as thermocouples, strain gages, flowmeters, accelerometers, synchros, and potentiometers, or other electrical information sources. These signals are converted in form or level to a normalized output voltage in the nominal range of 0 to 5 volts. This normalized voltage varies in direct proportion to the intelligence received and is used for standardized modulation of r-f circuitry for remote transmission or may be fed directly to on-board recorders.

This section will discuss the application of solid-state devices to the design of basic circuits commonly used in signal conditioners. The vehicle-system characteristics which affect these transistor circuit designs are also briefly discussed.

11 System Considerations

The complete instrumentation system, in many cases, is used only to measure parameters of interest during the research and development phase of the vehicle.

---

* The author wishes to acknowledge the assistance of the following people who contributed significantly to the writing of this material: John Jones and Richard Routsong, International Data Systems, Inc.; Dudley Garner and Gerald Goodwin, Radiation, Inc.; Claxton Rae, Martin Company. He also wishes to express appreciation to International Data Systems, Inc., for the use of simplified schematics and other material used herein.

This equipment must eventually give way to the warhead, additional fuel capacity, or other payloads. This often means that primary consideration is given to permanent-type equipment such as the guidance and control system. Thus the airborne instrumentation system must occupy the remaining space and use the available power sources.

Since transducers and other information sources are widely dispersed, the signals to be monitored must be routed through many feet of wire. Consequently, these signals are vulnerable to the introduction of errors that are caused by the electrostatic and magnetic fields generated by the electrical power equipment on the vehicle. Further, this power equipment will ordinarily use the vehicle frame as a common return line. The current flowing through the frame results in a wide variation in "ground potential" throughout the vehicle. Unless extreme care is used in the design of the instrumentation system errors may be obtained which are greater than the signals that are to be measured.

With the foregoing considerations in mind, the discussion on signal conditioning will be covered from the viewpoints of the system engineer who writes the specification for the equipment and the equipment designer who designs equipment to meet the specifications. A typical signal-conditioner package is shown in Fig. 27.

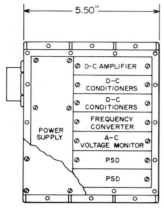

FIG. 27. Typical signal-conditioner package.

The signal conditioner is frequently required to recover millivolts of signals from volts of noise. The following paragraphs will cover some of the areas which must be carefully considered and defined by specifications, if a reasonable degree of accuracy is to be obtained. It will also be shown that overspecifying can be detrimental, since it may impose impossible or impractical limitations on the equipment itself.

Grounding

The missile electrical system in its simplest form has several power sources including a-c and d-c along with rotating equipment and r-f generating equipment. The error due to these power sources can be minimized only by proper system-grounding techniques in the following three discrete areas:

Power Ground. The power ground is the return path for all power used within the signal-conditioner package and carries all currents associated with the primary power supply. In general, the ground connection will be brought out separately for grounding at the vehicle central ground point. If this return is tied to the signal-conditioner case or vehicle frame, then it must be recognized that this point may vary considerably with reference to true ground. Any unregulated power supplies will contain that same variation. Consequently, errors will be caused in circuits which are sensitive to power-supply variations.

Thus, for optimum system performance, the following procedures for connecting the power lines should be used.

Provide a single, central ground point for the vehicle. All grounds—115 volts a-c, 400 cps; +28 volts; signal ground; etc.—should be connected at this point. They should not be connected common at another point in the system.

Power leads should be connected separately from the power source to each package of the system. This prevents power currents of one package from modulating the power to another package. Although this requirement is quite costly in additional vehicle wiring of large power lines, the maximum practical number of independent sets of power leads from a particular power source should be used.

The power ground and signal ground should be electrically isolated in each system component signal conditioner, telemetery package, etc. This eliminates large errors that are caused by power and signal currents flowing through a common conductor. Since the power currents (1 ma to 2 amp) for each package are normally quite large compared with the output or input signal currents (0.01 to 10 ma), a resistance of only 1 ohm in a common line will cause an error of 2 to 20 per cent of a 5-volt full-scale signal output. This isolation is readily achieved with an isolation transformer for a-c power sources and a d-c–d-c converter for d-c power sources.

Input Signal Grounds. If improper grounding and cabling is used from the transducer to the signal conditioner, it is impossible to recover accurately low-level signals (0- to 20-mv range) even if a signal-conditioner circuit with high accuracy and common-mode rejection capabilities is used.

Two primary sources of error exist before the signals are processed by the signal conditioner. The first of these, improper grounding, may be seen from Fig. 28. The thermocouple will normally be grounded to the vehicle frame. It can be seen that if any other point in the input circuit is grounded, then the apparent signal E_S will be equal to $E_t + E_g$, where E_g is the difference in ground potentials. Also, two input circuits cannot be connected together, since this would produce a second ground for a particular channel. Thus any low-level channel which is grounded externally to the signal-conditioner package cannot be grounded internally.

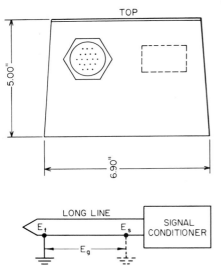

Fig. 28. Double ground error.

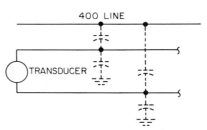

Fig. 29. Common-mode error.

The second error source is a-c "common-mode voltage," which is the voltage common to both lines of the input circuit. This voltage is usually caused by the fields generated by high-energy power conductors such as the 115 volts a-c, 400-cycle primary source. Figure 29 illustrates the nature of the common-mode error. In this case, the 400-cycle voltage will be capacitance-coupled to each of the input signal lines and from these lines to ground. The input circuit thus forms a bridge which becomes unbalanced by any unbalance or dissymmetry in the ground paths of the two input lines. The effect of a-c common-mode voltage can be reduced by the following methods:

1. Use minimum source impedance for the transducer. Essentially, this lowers the amount of common-mode voltage on either line.
2. Use ungrounded transducers.
3. Use either an isolated input circuit or an isolated amplifier.
4. Use shielded wiring, with proper grounding of the shield.

Shield Ground. As described previously, a-c common-mode voltages and noise "pickup" voltages on low-level signal lines are caused by large ground currents flowing in the vehicle frame, conductors carrying currents from power sources, and electrical power equipment. To achieve a reasonable degree of accuracy on these low-level signal channels, the signal lines must be protected. Some general rules for minimizing the effect of common-mode and noise voltages are:

1. Use well-shielded, tightly twisted pairs for input cabling on all critical low-level circuits.
2. Ground the cable shield at only one point and to the source of common-mode voltage if possible; otherwise, connect to the input shield lines as close to the transducer as possible.
3. Use a guard shield around the low-level section of the particular signal-conditioner circuit and tie this shield to the shield of the input cable.
4. Never leave the shield floating, and never ground the shield to the signal conditioner.

In summary, the following five grounds should be provided within the signal-conditioner package: a-c power ground, d-c power ground, input signal ground, shield ground, and telemetry signal ground or output ground.

Chassis isolation can be accomplished by isolating all signal and power circuitry from the package structure; and each internal circuit must be tied to the appropriate ground.

Circuit Protection

In practice, one of the commonest causes of equipment failure is the lack of adequate attention to compatibility between the signal conditioner and the vehicle in which it is to be used. It is usually simpler to correct compatibility problems by changing the signal conditioner, which is essentially the interface between the vehicle and the instrumentation package, than to change the vehicle itself. Therefore, from the signal-conditioner designer's point of view, it is well to anticipate the problems and incorporate proper corrective measures. The following paragraphs will discuss some of the common problems and recommended corrective measures.

Power-source Polarity Reversal. Disaster can be avoided here by incorporation of a power diode in series with the input power line. Thus, although the equipment will not function during power reversal, no harm will be done and the polarity error can be corrected.

Overvoltage on Input Circuits. Although most normal signals are in the low-voltage or millivolts range, it is not uncommon to have potentials such as shorts to the 28-volt supply or an open circuit in a bridge, either of which can apply full bridge voltage to the input circuit. Damage can be prevented by allowing a sufficient rating on the input components to accept these fault voltages. The consequent high current flow can be alleviated by providing appropriate input impedance.

Short Circuits on Output of Signal Conditioner. One of the commonest system failures is due to shorting the output-circuit test points during system checkout. This can be avoided by providing current limiting for test points. Care should be used in the physical and electrical location of test points, and isolation provided whenever possible.

Out-of-range Signals. Protection should be provided at the signal-conditioner output to limit the excursions of the output voltage as a result of abnormal variations in the input signal. A typical result of excessive output voltage would be the overmodulation of subcarrier oscillators in an FM system, in which case one oscillator channel could be deviated outside its band, resulting in loss of information on several channels.

Power-voltage Variation. In most cases the solid-state signal-conditioner modules will operate directly from the 28-volt power source, with possibly some internal regulation. It is not uncommon to encounter power-supply variations of several volts, with short-term transients that are even more severe. All circuits should be designed so that they will not be damaged by low or high voltage, and decoupling and component derating should be taken into consideration.

Specification Parameters

Since rigid requirements are normally imposed on the size, weight, reliability, and cost of all components of the vehicle, it is necessary that great care be given to provide optimum matching of the signal conditioner to the signal source and load. The specification must be precise, complete, and unambiguous. The following paragraphs will discuss some of the more critical items and their qualitative effect on system design and complexity.

Accuracy. The accuracy of each channel component as well as the overall system accuracy should be kept in mind when writing the specification for a signal conditioner. And, as can be seen from Fig. 30, the signal conditioner is but one of many error sources. It can be shown that the overall system error is the root mean square of the channel and system-component errors. The overall accuracy of the PDM system shown will generally be 1 to 2 per cent, and system-component accuracies are normally specified as approximately 1 per cent. However, if the transducer accuracy is only 3 per cent, it can be shown that the signal-conditioner accuracy can be relaxed to 2 per cent, and the resulting change in overall accuracy will be only 0.2 per cent.

This point is most evident in PCM telemetry where it is theoretically possible to achieve accuracies of $\frac{1}{4}$ per cent or better. It must be remembered that the worst component accuracy largely determines the overall accuracy, and it is very costly to overspecify.

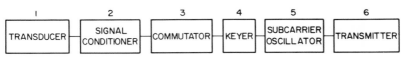

Fig. 30. Error diagram.

$$E_{cum} = \sqrt{E_1^2 + E_2^2 + E_3^2 + E_4^2 + E_5^2 + E_6^2}$$

Input Impedance. Transistor circuits are inherently limited as a high-impedance device; and again, the overall design may suffer if this item is overspecified. However, the input circuit affects accuracy directly if improper matching is provided; so an equitable compromise must be made in the case of high source impedances.

The loading effect of the signal conditioner is approximately $(Z_{in} \times 100 \text{ per cent})/(Z_s + Z_{in})$, where Z_s is the source impedance (usually resistive) and Z_{in} is the signal-conditioner input resistance.

In general, input impedances of 100 kilohms or lower can be readily achieved. One to five megohms is difficult to achieve, and higher impedances can be achieved only by resorting to complex circuits or restricted temperature ranges.

If the source impedance is high but relatively constant, it is frequently advisable to use a lower input impedance which is stable. This impedance will attenuate the signal source but by a fixed amount which may be calibrated.

Gain, Linearity, and Offset. Just as with the system error which was described earlier, the signal conditioner involves several components, all of which are potential sources of error. Each of the channel components must be clearly defined and specified with a tolerance that is commensurate with the system error.

A feedback loop is always used in a stable amplifier. Thus gain and gain stability are interrelated, and for a given amplifier, the gain is usually changed by varying the amount of feedback which in turn affects stability. Ideally, the gain range required of the amplifier or other signal conditioner should be as narrow as possible, in order that feedback may be optimized for maximum gain stability.

Linearity is a straightforward characteristic but it is usually determined by taking a set of input-output characteristics and comparing them with a straight line. The line must be drawn in such a manner as to allow for the plotting error as well as the residual or *offset* error as the input voltage approaches zero. A common way to specify the linearity error is that it is within 1 per cent scale of the best straight line between the zero signal and full-scale output.

Frequency Response. The frequency response of each channel must be specified in terms of per cent or decibel variation from the center-frequency amplitude. It must also specify the cutoff point and the rate of attenuation.

Table 5. Typical Telemetry Signal-conditioner Input and Output Data

Channel No.	Input reference	Input type - Differential	Input type - Single ended	Source impedance	Input impedance, kilohms min	Input measurement range	Output range, volts d-c	Frequency response	Required converter accuracy, %	Input max range	Type circuit required
1-8	0-3 volts 0-400 ~ CM	x	...	5 ohm	10	0 mv d-c +10 mv d-c	0 +5	0-5 ~	2	0-20 mv	D-C amp
9-12	Signal ground	...	x	10 kilohm	200	2 volts a-c < 0° rms 2 volts a-c < 180° rms (400 cps)	0 +5	0-5 ~	2	5 volts a-c < 0° rms 5 volts a-c < 180° rms	Phase-sensitive demodulator
13	Power ground	...	x	1 ohm	10	+22 volts d-c	0 +5	0-5 ~	3	0 volts d-c +50 volts d-c	D-C converter
14	Power ground	...	x	1 ohm	10	+32 volts d-c +42 volts d-c	0 +5	0-5 ~	3	0 volts d-c +50 volts d-c	D-C converter
15	Power ground	...	x	10 kilohm	200	0 volts 12 volts rms a-c (400 cps)	0 +5	0-5 ~	2	0 volts a-c 20 volts rms a-c	A-C converter
16	Power ground	...	x	1 ohm	5	85 volts rms 125 volts rms (360-440 cps)	0 +5	0-5 ~	2	0 volts 150 volts rms (340-480 ~)	A-C converter
17	Power ground	...	x	1 ohm	5	360 cps 440 cps (102-125 volts rms)	0 +5	0-5 ~	2	102-125 volts rms 340-480 cps	Frequency converter
18	0-5 volts 0-400 ~ CM	x	...	200 ohm	10	0 volts d-c +50 mv	0 +5	0-5 ~	2	0 volts d-c +100 m volts d-c	D-C amp
19	Signal ground	...	x	10 kilohm	200	−3.2 volts d-c +3.2 volts d-c	0 +5	0-5 ~	2	±10 volts d-c	D-C converter
20	Signal ground	...	x	10 kilohm	200	−7.2 volts d-c +9.6 volts d-c	0 +5	0-5 ~	2	±20 volts d-c	D-C converter

SIGNAL CONDITIONING   4-41

Since most of the circuits to be used employ some degree of feedback, it is desirable to limit the frequency band to the minimum required to pass the signal and to make the attenuation rate 6 db/octave if possible, so that simple filters may be used.

Common Mode. The specification for common mode should define the nature of the common-mode voltage source. This should include the level of the common-mode voltage, whether d-c or a-c, and frequency, as well as the source impedance and the unbalance to ground of the source.

12   Signal-conditioner Circuits

Table 5 illustrates the input-output characteristics of a typical requirement for instrumentation. The number of channels may vary from ten to several hundred. This table summarizes the nature of the intelligence from each channel source and provides the basic information necessary to design the signal-conditioner system.

The table also provides information with regard to ground requirements, as may be seen in the columns "input reference" and "input measurement range."

The output characteristics are usually standardized and are determined by the telemetry system or recorder to which the output signal is supplied.

Examination of Table 5 shows that the circuits, shown below, will fulfill all the requirements for conditioning the signals listed in the table. Special conditioners such as pulse-height analyzers, pulse measuring, and spectrum analyzers are sometimes required; however, the discussion of these special types is beyond the scope of this section.

These basic circuit types are:

D-C amplifier
Frequency converter
Phase-sensitive demodulator
D-C voltage monitor or converter
A-C voltage converter or discriminator

These circuits will be discussed in this section.

D-C Amplifiers

A large portion of the measurements to be made on a vehicle will be low-level signals and in general will require direct-coupled converters. Thus d-c amplifiers are used to normalize these inputs to a 0- to 5-volt output level. Also, these amplifiers must handle differential signals and present a very high degree of rejection to common-mode voltages.

Many different applications exist for amplifiers of d-c signals; consequently, many different circuit techniques are used. The specification should cover every interface consideration of the amplifier but should not restrict the equipment supplier by unnecessary requirements internal to the amplifier. A typical specification for the electrical requirements of a d-c amplifier may be as in Table 6.

Mechanical Chopper Amplifier. The specification in Table 6 for a d-c amplifier may be met by using a mechanical chopper amplifier, with the exception that the mechanical chopper's frequency response is limited to 100 cps. Normally, a 4:1 chopper-to-signal frequency ratio will provide a ±1 per cent frequency response. Mechanical choppers normally operate at 400 cps, and drive power may be provided from either an external or internal source. These choppers must have low noise characteristics and minimum contact bounce. The input transformer must provide the proper inductance to achieve the required input impedance and minimum capacitance that are required to prevent common-mode signals from entering the secondary. Figure 31 is a simplified diagram of a mechanical chopper amplifier. The input chopper is followed by an a-c amplifier. And a demodulator converts the alternating current back to direct current at the 0- to 5- volt level.

Table 6

Parameter	Specification requirement
Input signal	Low level, differential
Input signal range	±5 mv min
	±50 mv max
Input impedance	10 kilohms min
Source impedance	500 ohms max
Source impedance unbalance	100 ohms max
Output level	0 to +5 volts, phase-sensitive
Output impedance	1,000 ohms max
Gain range	50 to 1,000
Gain-adjustment resolution	±0.2% (at output)
Gain stability	±1%
Zero suppression	0 to +2.5 volts
Zero-suppression stability	±25 mv
Linearity	±1%, end points
Output ripple	25 mv peak-to-peak
Frequency response	±1%, 0–1,000 cps
Common-mode signal	±10 volts d-c
	2 volts peak-to-peak, a-c
Common-mode rejection	50,000 to 1, 0–500 cps with 100-ohm max unbalance source impedance
Operating temperature	−20 to +85°C
Input power	28 ± 4 volts d-c
	6.3 ± 0.5 volt a-c, 400 ± 20 cps

The diode demodulator output becomes nonlinear as it approaches zero, and it is not polarity-sensitive. Both conditions can be corrected by providing a synchronous double-pole chopper to demodulate the output in phase with the input. This is shown in Figure 32. In this type of amplifier the stability is dependent upon the a-c gain stability, the output filter design, and the chopper characteristics. Where high d-c

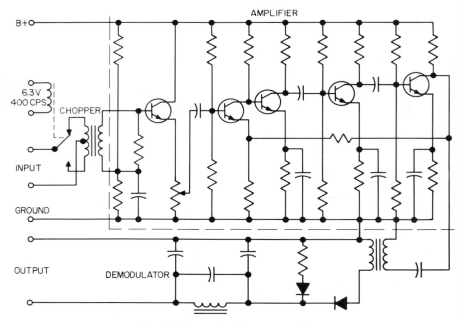

Fig. 31. Mechanical chopper amplifier.

gains (1,000) are required, the offset or zero-signal-level output depends largely on the chopper and amplifier noise.

Transistorized Chopper Amplifier. The transistorized chopper amplifier offers several advantages that a mechanical chopper cannot meet. These are (1) long life, (2) a lower power requirement, (3) higher frequency response, (4) smaller size, and

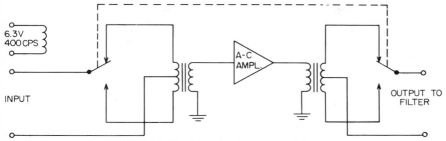

FIG. 32. Double-pole chopper amplifier.

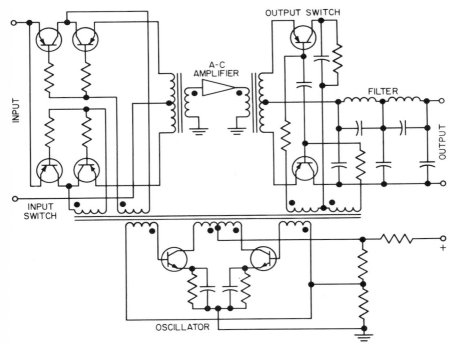

FIG. 33. Transistor chopper amplifier. Note: Polarity of transformers is indicated by solid dot.

sometimes (5) lower cost. Figure 33 shows a transistor chopper using the same a-c amplifier as the mechanical unit. The chopper and demodulator are driven by a Royer oscillator. This system can achieve much higher frequency response (3 to 5 kc) by operating at a 15- to 20- kc chopping rate. These higher-frequency amplifiers may be used for measurement of high-speed transients, vibration measurement, and other a-c measurements. Some design areas of the transistor chopper amplifier are as follows: (1) drift, (2) common mode, and (3) signal levels.

1. *Drift.* With proper selection of the transistors that are used in the inverted configuration, the onset of a four-transistor chopper can be matched to within 20 μv over the specified temperature range. The initial V_{ce} of each transistor is in the order of 0.8 to 1.0 mv. The V_{ce} temperature characteristics, of the transistor as well as the initial V_{ce} must be matched. These transistors must be at equal environmental temperatures and have a constant base drive that is in the order of 500 μa.

2. *Common Mode.* High common-mode signal rejection is obtained by reducing the primary to secondary capacitance of the input transformer and the chopper drive transformer. To achieve this low capacitance, the chopper drive transformer (saturable core) must be wound with maximum spacing between windings.

3. *Signal Levels.* Input signal levels in the order of 5 mv full scale can be amplified with the transistor chopper amplifier. But care should be taken to provide protection against transistor breakdown voltage when high-level or fault-condition inputs are encountered. The same is true for the demodulator transistors (they are usually the unmatched) which provide the high-level output signals.

Since the chopper drive oscillator is required, it is very simple to add another winding and rectifier circuit to provide an isolated bridge power supply. This individual bridge power supply will eliminate ground loops and improve the overall accuracy of a system.

Frequency Converter

Frequency measurements may be required in the determination of fuel flow, hydraulic flow, tachometer rate, signal frequencies, power frequencies, etc. These measurements may be broken down into two types, normal-range (wideband) and expanded-scale (narrowband). The normal range of conversion may be accomplished by a

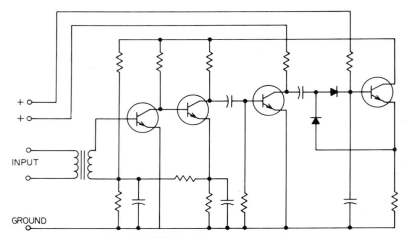

Fig. 34. Wideband frequency converter.

rate-counting circuit, and the expanded scale may be accomplished by a series-resonant circuit or by a gated-switch circuit. In addition to a variable-frequency input, the frequency converter may have input signals that vary in amplitude and shape.

Wideband Frequency Conversion. The wideband frequency converter must not be sensitive to input signal amplitude or shape. Therefore, the input stage following a transformer must be either a squaring amplifier or a one-shot device. Figure 34 shows a typical circuit. The input circuit is a squaring amplifier and is driven to both saturation and cutoff, producing an output that is essentially a rectangular wave of constant amplitude. This signal is applied to the rate-counting circuit (Fig 35). The voltage applied to the capacitor will vary from 0 to B+. During saturation, the capacitor is discharged through CR 1. During the opposite half-cycle, the capacitor is charged through the collector resistor CR2 and the output circuit load. The output voltage is then as follows:

$$E_{out} = KECF \qquad (8)$$

where K = const
E = peak-to-peak voltage at collector
C = capacitance (const)
F = input frequency

Therefore, E_{out} is a direct function of frequency. The overall circuit stability and linearity are dependent upon saturation and cutoff of the transistor. Saturation and cutoff are dependent upon enough gain to saturate even at minimum input signals, component stability, minimum amount of output voltage swing compared with the collector voltage swing, and a constant B+.

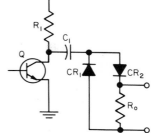

FIG. 35. Rate-counting circuit.

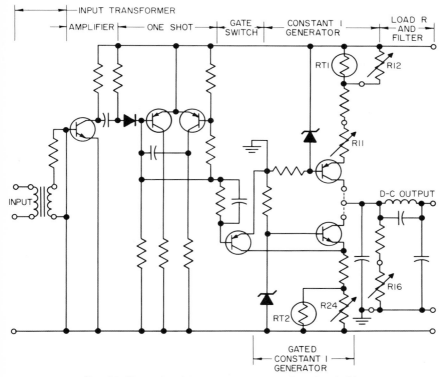

FIG. 36. Narrowband frequency converter (gated switch).

Narrowband Frequency Converter. The narrowband frequency converter is normally used to measure narrow changes about some center frequency, such as the 400-cycle power source. As an example, a variation from 380 to 420 cycles will produce an output voltage of 0 to 5 volts. Two examples of this type of circuit, the gated switch and resonant circuit, are described below.

1. *Gated-switch Narrowband Frequency Converter.* The schematic diagram of one type of narrowband frequency converter is shown in Fig. 36.

The input transformer may have either a step-up or a step-down turns ratio, depending on the amplitude of the input voltage whose frequency is to be measured. The amplifier transistor is quickly driven into saturation as the transformer output drives the base positive and is quickly cut off as the transformer output drives the base negative. Thus, with a sine-wave input, the amplifier output is virtually a square wave. The one-shot is triggered by a differentiated positive-going pulse which is coupled from the amplifier collector while the amplifier transistor is being cut off.

From the schematic it can be seen that the gating switch is driven by the output of the one-shot. With the one-shot in the quiescent state, the gating switch is in the OFF state. As the one-shot is triggered the gating switch is turned ON, clamping the emitter of the gated constant-current generator to ground potential and thus reducing the current to zero. The length of time that the gated constant-current generator remains OFF is determined by the one-shot's period, which is constant and independent of frequency.

Now, refer to Fig. 37a. For convenience, assume that the one-shot is designed so that the gated constant-current generator off time t_1 is equal to the on time t_2 at the lowest frequency to be measured. The period of one cycle of the input signal is $t_1 + t_2$ at the lowest frequency to be measured and $2t_1$ at the highest frequency to be

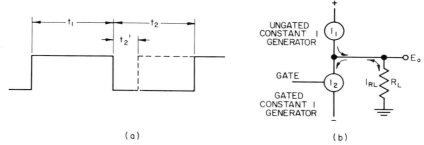

(a) (b)

FIG. 37. Generation of narrowband frequency-converter output.

t_1 = one-shot period = constant I generator OFF time
$t_1 + t_2$ = period of one cycle of input signal

measured. As the input frequency is increased t_1 remains constant while t_2 decreases to equal t_1 at the highest frequency to be measured. Thus the ratio of off time to on time increases, which is to say the gated constant-current generator is OFF for a greater percentage of time as the input signal frequency is increased.

Figure 37b is a diagram showing the functions of the constant-current generators and the load resistor R_L in generating the desired output voltage. The ungated constant-current generator supplies a current I_2 through R_L in the reverse direction, but only while the gating switch is turned ON. The resultant average current through the load resistor may be expressed as follows:

$$I_L = \frac{t_1 I_1 + t_2(I_1 - I_2)}{t_1 + t_2} \tag{9}$$

The output voltage $E_0 = I_L R_L$. At the minimum frequency to be measured, the gated constant-current generator is gated OFF 50 per cent of the time ($t_1 = t_2$), and if the desired output is 0 volts d-c, substituting in Eq. (9), $I_2 = 2I_1$. If the value of R_L is determined by considering the output impedance, and the desired value of E_o (at a given input frequency) is known, the value of I_1 and I_2 may be computed from the above equations.

The operation of the constant-current generators may be understood by again referring to Fig. 36. The base of the ungated constant-current generator transistor is held at a fixed potential by using a zener reference diode. Because the base is held constant, the emitter is also held constant and the emitter current is dependent upon the power-supply voltage and the emitter resistance. Therefore, the current may be varied by changing the emitter resistance. The gated constant-current generator functions in the same manner, except that the emitter is connected to a gating switch which is capable of switching the constant-current generator ON and OFF.

Potentiometer R11 of Fig. 36 is used to vary the current of the constant-current generator so that the output voltage is zero with the input frequency set at the lowest frequency to be measured. Potentiometer R16 is used to vary the value of the load resistance so that the output voltage will be equal to the desired value when the input frequency is set at the highest frequency to be measured. Potentiometers R12 and R24 are used in conjunction with sensistors RT1 and RT2 to prevent changes in the constant-current sources due to temperature variations.

The stability of the system depends on the period of the one-shot multivibrator, the constant-current generators, and the load resistor. Using sensistors and potentiometers to compensate the constant-current generator for changes in temperature, and by making the one-shot as stable as possible, the circuit described will maintain ± 1 per cent output stability over a temperature range of -55 to $+85°C$.

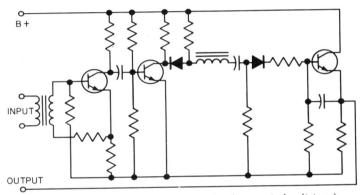

FIG. 38. Narrowband frequency converter (resonant-circuit type).

The linearity of the system is very good, as is shown by the equation for output voltage:

$$E_o = R_L \frac{t_1 I_1 + t_2(I_1 - I_2)}{t_1 + t_2} \tag{10}$$

or

$$-E_o = R_L \left[\frac{t_2 I_2}{t_1 + t_2} - I_1 \right] \tag{11}$$

This is the equation of a straight line, since t_1, I_1, R_L, and I_2 are constants which are independent of input frequency. $t_2/(t_1 + t_2)$, the duty cycle, is the independent variable.

Since the current through the load resistor is of a pulsating nature, an output filter is required to give the most optimum output ripple and rise-time characteristics.

2. *Resonant-circuit Narrowband Frequency Converter.* Another example of a narrowband frequency-converter circuit is shown in Fig. 38. The signal input level determines the gain required to feed a square wave into the resonant under minimum-signal conditions. A diode is inserted before the tuned circuit to remove the negative portion of the signal. The resonant circuit is tuned slightly above the maximum signal frequency, and as the signal frequency increases, the output amplitude increases proportionately. The output diode rectifies the voltage and a d-c output results. This d-c output is resistor-coupled to an emitter follower where the zero and gain may be adjusted by means of the base resistors.

This circuit requires extreme care in the design of the choke and selection of the series capacitor. It also requires a stable power supply because the output level is determined by the saturation and cutoff point of the last amplifier stage.

Phase-sensitive Demodulator

The purpose of the phase-sensitive demodulator (PSD) is to furnish a d-c output signal that is proportional to the amplitude and polarity of an a-c input signal. A reference signal of the same frequency as the input signal is necessary to determine the polarity of input signal. It should be emphasized that the output signal that is

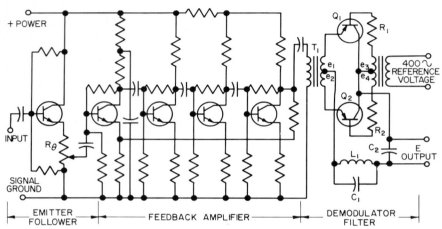

FIG. 39. Phase-sensitive demodulator.

normally required is proportional only to that component of the input signal which is in phase with the reference. That is,

$$E_{out} = KE_{in} \cos \theta \qquad (12)$$

where K = desired gain

θ = phase angle between input signal and reference

If there is a phase shift that is other than 180° between the input and reference, means should be provided to maintain the phase angle θ at 0 or 180°. Even though excellent quadrature ($KE_{in} \sin \theta$) rejection can be achieved, amplitude changes that are due to slight phase changes ($d \cos \theta)/d\theta$ become greater as θ approaches 90°. If a nominal phase shift is known and a variation in phase shift is expected, a simple RC network should be used to return θ to a nominal value of 0 or 180°.

Since it is a standard practice for signal-conditioning equipment to furnish only positive outputs, the output of a PSD is normally biased to yield a half-scale signal for zero input. With this arrangement, the output signal is described by

$$E_{out} = E_{bias} + KE_{in} \cos \theta \qquad (13)$$

where $E_{bias} = KE_{in}$ (full scale) $\cos \theta$

Figure 39 shows a typical transistorized PSD that satisfies these requirements. It includes a stabilized a-c amplifier and a synchronous detector or demodulator. The schematic diagram of Fig. 39 is a transistorized unit with design specification as in Table 7.

High input impedance is achieved by coupling the input signal directly into a high-gain low-leakage emitter-follower circuit. Present-day low-leakage silicon transistors

SIGNAL CONDITIONING

Table 7. Electrical Characteristics, PSD

Input signal	±25 mv rms (max gain) 400 cps
Input impedance	Greater than 100 kilohms
Output voltage	0 to 5 volts d-c or ±2.5 volts d-c
Reference frequency	400 cps
Reference voltage	15 volts rms (for extreme linearity and quadrature rejection a square-wave drive is recommended)
Gain stability	±1% of full scale
Linearity	±1% (best straight line with sine-wave reference)
Output impedance	2.5 kilohms
Output ripple	25 mv max
Response time	8 msec max

allow this circuit to operate reliably without employing complicated biasing schemes. Gain adjustment is provided by a 40-turn precision potentiometer R_o which is deliberately placed outside the amplifier feedback loop to allow a wide-gain adjustment range without sacrificing amplifier stability.

The amplifier employs emitter feedback to reduce the load on the input emitter follower. Approximately 40 db of feedback is present at 400 cps. The last stage is a-c coupled into the miniature output transformer to avoid d-c saturation. Until recently, a diode bridge was the most frequent solution for the demodulator circuit requirement in solid-state PSD's. The unit described here, however, uses a transistorized demodulator. Silicon signal-switching units are operated in an inverted configuration in order to offer much lower offset voltage and lower ON resistance or saturation resistance than typical diode switches. Unlike diode switches, no matching is required since the offset of a single transistor is less than 2 mv and remains relatively constant over a wide temperature range. If the output filter is properly designed the low ON resistance will guarantee a fast response time.

Referring to Fig. 39, the transistorized demodulator operates as follows: When reference signal e_3 is negative with respect to the collectors of Q_1 and Q_2, the base-collector junction of Q_1 is forward-biased and signal voltage e_2 appears across the output. During this condition, Q_2 is off, but when e_2 becomes negative, Q_2 conducts and e_2 appears across the output. The operation of the circuit is the same regardless of the polarity of e_1 and e_2. The tuned circuit L_1 and C_1 is resonant at the second harmonic of the signal frequency. Capacitor C_2 removes spikes and other harmonics.

D-C Voltage Monitor or Converter

This classification of circuits includes those which monitor the level or ON-OFF state of d-c power supplies and control voltages. Typical examples of these circuits are discussed below.

Battery Voltage Conditioner. This is one of the simpler d-c conditioner circuits and simply provides a 0- to 5-volt linear output which corresponds to the normal full-range variation in the 28-volt supply.

As shown in Fig. 40, this circuit employs a voltage-divider combination R_1 and R_2 such that

$$\Delta E_{in} \frac{R_2}{R_1 + R_2} = 5 \text{ volts} \qquad (14)$$

where ΔE_{in} = variation in input voltage

The zener diodes provide a constant voltage drop which cancels the steady-state voltage. Thus, if the input voltage is specified as 28 volts ± 5 volts, then $\Delta E_{in} = 10$ volts, and $R_2/(R_1 + R_2) = \frac{1}{2}$ or $R_2 = R_1$. The voltage at P_1 will be 14 volts ± 2½ volts and if an 11½-volt zener diode is used, E_o will be zero for the −2½-volt swing and 5 volts for the positive swing.

In some cases, such as low-current well-regulated power supplies, extremely high input-impedance requirements will be imposed.

D-C Step-function Converter. One other common measurement is one which furnishes an output voltage which indicates the various ON-OFF states of two or more input voltages.

A simple example of this is the circuit of Fig. 41, which provides four discrete levels of output in the range of 0 to 5 volts, one for each possible ON-OFF combination of two d-c input signals.

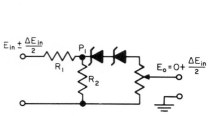

FIG. 40. Battery voltage conditioner.   FIG. 41. Step-function converter.

This technique employs a resistive ladder network with values of R such that when E_1 and E_2 are both 28 volts, the output is 5 volts. It can then be shown that the following conditions apply:

Input condition		Output voltage
E_1	E_2	
ON	ON	5 volts
ON	OFF	3.33 volts
OFF	ON	1.67 volts
OFF	OFF	0 volts

This technique may be expanded to monitor several inputs, but it becomes ambiguous if the per cent change in power-supply voltage approaches the system accuracy multiplied by the number of output steps.

A-C Voltage Converter or Discriminator

This classification of circuits monitors those signals which generally do not require amplification but must provide a d-c output over the normal range of variation of the a-c input. Two examples of this type of circuit are described below.

400-cycle A-C Voltage Converter. Figure 42 is a schematic of an a-c voltage converter which operates over a range of 185 to 125 volts at 400 cycles. This circuit employs a transformer to provide the appropriate secondary voltage that will produce 5 volts d-c at the rectifier output. The full-wave bridge rectifies this voltage and the bias supply R_1 adjusts the output to zero volts at the low-voltage end of the input signal. R_2 adjusts the output to 5 volts at the high end. This circuit is independent of frequency variation over the range of 380 to 420 cycles.

True RMS Voltmeter. The circuit described above provides an output which is proportional to the average of the a-c input signal. In some cases, it is desirable to determine accurately the true rms value of the input voltage. The block diagram of a unique circuit for meeting this requirement is shown in Fig. 43. This discriminator is completely transistorized and is designed to produce an output of 0 to 5 volts d-c for an input of 105 to 125 volts. The accuracy is better than 1.0 per cent for any waveform which contains frequencies between 25 cps and 20 kc.

The circuitry consists of a duty-cycle modulator, an amplitude modulator, an integrating network, a detector, and a differential amplifier from a constant-current source. With no input signal, the resulting output is a 70-kc symmetrical square wave. As a signal is applied, the duty cycle (the ratio of the ON time to the period) is proportioned to the applied voltage and maintains a constant amplitude.

The same signal is applied to the amplitude of the duty-cycle modulator. This combined operation produces a current proportional to the square of the input voltage.

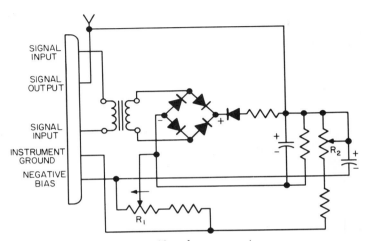

Fig. 42. 400-cycle a-c converter.

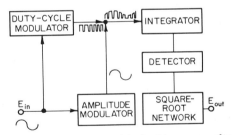

Fig. 43. Block diagram transistorized true rms voltmeter.

The current is then integrated and applied to a detector, causing the output of the detector to be proportional to the mean square of the input voltage. The signal is then put through a square-root network which produces an output proportional to the rms of the input.

Power Supply

The power-supply requirement for a signal conditioner is usually decided upon after a review of each type of circuit that will be used. At this time the voltage level, regulation, and stability requirements are established. The power supply will convert the unregulated vehicle a-c or d-c power to a regulated voltage output. This will be done by a transformer or a d-c to d-c converter; the rectified output will then be regulated to produce the required d-c voltage. Figure 44 shows a typical voltage regulator. The gain required in the reference amplifier for line and load regulation is expressed by factor $1/(1 - UB)$ where U is the loop gain and B is the divider ratio.

Modern power supplies have some features that make the overall system integration much less of a problem if it is given consideration during the design period.

Some of the features of a modern power supply are as follows:

High Efficiency. Proper transformer and d-c converter design as well as lower current-regulation requirements improve efficiency. Normally, an output-voltage-to-reference-voltage ratio that is high can achieve better efficiency as input variation that is too large can reduce overall efficiency.

Output Impedance. If a positive-feedback loop is added to the reference amplifier within the negative-feedback loop, an output impedance of zero can be achieved. This can be done by adding some resistance between the output and the input of the reference amplifier. This is shown in Fig. 45.

Line and Load Regulation. With very high gain within the reference amplifier, and with allowances for minimum input voltage, any desired regulation can be achieved.

Small Size. By using transistors, silicon controlled rectifiers, accurate reference diodes, and better transformer material, the size of the power supply will be smaller than conventional units. The effort to achieve small-sized area may lead to poor

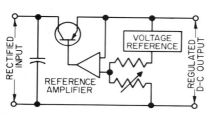

FIG. 44. Power-supply regulator, constant voltage–constant current.

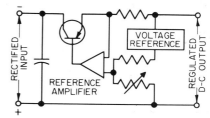

FIG. 45. Power-supply regulator, constant voltage.

thermal regulation if the proper heat sink is not available to dissipate and transfer the generated heat.

Short-circuit Protection. The short-circuit protection is achieved by adding a constant-current circuit such as in Fig. 18, which is the same circuit that makes possible a zero output impedance. The current can be limited at any preselected value.

Temperature Stability. A reference diode is used to achieve temperature stability that is in the order of 0.0005 per cent per degree centigrade or better. Secondly, the amplifier stability can be made to approach zero. Other factors generally do not affect temperature characteristics.

R-F Interference. The effect of this interference will be minimized if this problem is recognized during design and proper filtering is provided and high-frequency components are used. Filtering should also be added to prevent signals that are generated within the power supply from feeding back on the input lines to other systems in the vehicle. When currents caused by magnetic fields are encountered, shielding should be considered.

Heat Dissipation. After the total power-supply dissipation has been determined, careful consideration should be given during packaging to such factors as location of high-power components with respect to temperature-sensitive components, derating of components with the maximum operating temperature, plus the temperature rise expected; and for equipment applications where there is inadequate air flow, provision should be made for removing heat from the package.

Physical Location of Reference Point. When printed-circuit boards or wiring are used in a regulator, the reference point should be physically nearest the load. This eliminates the IR drop between the load and the reference.

COMMUTATION

As late as 1945, no practical mechanical devices existed to provide high-speed sequential sampling of transducer voltages for multiplex instrumentation purposes. Interest in multichannel data acquisition for military and scientific vehicles subsequently encouraged the rapid development of motor-driven multicontact switches which came to be known as commutators. Commutator technology including the required drive systems as well as the contact mechanism itself made rapid strides each year despite its highly specialized application.

The reliable sampling of up to 1,000 data points or more per second was first made possible for an extremely limited lifetime of from 1 to 5 hr. Over the past 10 years, however, this lifetime has been gradually extended to the present state of the art which makes possible several thousand hours of operation without servicing.

Meanwhile electronic methods of commutation, employing electronic gating techniques, were also developed for telemetry. Early commutators using electronic tubes were cumbersome, required considerable power, and were not suitable for sampling low-level signals without individual channel amplifiers. With the advent of small dependable solid-state elements, it has become possible to design switching circuitry to provide reliable and accurate multichannel sampling in very small physical packages requiring relatively low power.

Some of the basic principles involved in commutating and decommutating data signals and certain practical aspects of commutated systems are discussed here. The following sections also describe some of the principal circuit logic employed for types of commutators used today in instrumentation and telemetry. A glossary of commutation terms is also included and an applicable bibliography which contains references to more detailed and diverse types of commutation devices which cannot be discussed within the scope of this handbook. Detailed considerations regarding their application to particular requirements may often be found in the articles listed. A number of patents have been listed to provide the reader with supplementary technical details as well as a history of developments in the field.

13 Principles of Commutation and Decommutation

Basic Sampling Methods

Two basic types of sampling are used in telemetry. These are called *instantaneous sampling* and *interval sampling*. Instantaneous sampling is defined as a process in which relatively short samples of the intelligence signal are taken with the intention of using them to reconstruct the original form of the signal at a later time. The wave train thus produced is called a pulse-amplitude-modulated signal, commonly abbreviated PAM. Interval sampling, on the other hand, is a process of sampling an intelligence signal for an appreciable period of time in order to observe the fine character of the signal over that interval. Instantaneous samples are very small in duration compared with the period corresponding to the highest-frequency component of the intelligence to be sampled, whereas interval samples are generally much longer in duration than the longest period of interest in the intelligence signal. The former is most often used for time-division multiplex of low-frequency data involving automatic decommutation. The latter is used for periodic observation of a group of high-frequency intelligence channels of interest. Vibration data, for example, are sometimes observed with interval-sampling techniques.

The Sampling Principle

The sampling principle simply states that, for instantaneous sampling processes, more than two samples per cycle of intelligence are required in order to reconstruct the intelligence with fidelity. Another way of simply expressing the sampling theorem is to say that a signal which has no frequency components above a given value is uniquely determined by a series of points whose spacing in time is equivalent to half the period corresponding to that value. The basis for this relationship has been shown by many sources indicated in the attached bibliography and will not be repeated here (see also Chap. 9, Sec. 7). If one plots the Fourier spectrum for a set of instantaneous samples of a given intelligence signal, with modulation sidebands around each harmonic of the sampling frequency, it can be readily seen that the sidebands begin to

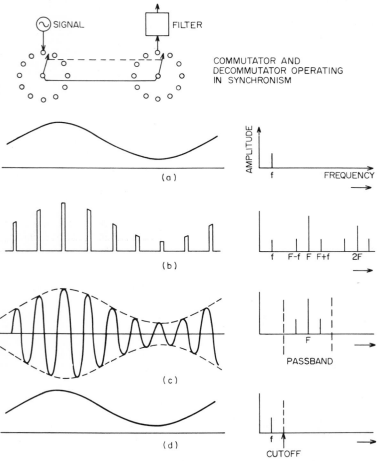

Fig. 46. Waveforms and spectra in the commutation process. (a) Signal (intelligence) of frequency f. (b) Single-channel PAM signal resulting from instantaneous sampling at F samples per second. (c) Output of bandpass filter. (d) Output of low-pass filter.

interfere or overlap as the intelligence signal frequency approaches one-half of the sampling frequency.

Figure 46 represents the basic commutation, decommutation, and recovery process. a shows the original intelligence signal of frequency f. b represents the same signal sampled at F times per second where F is greater than $2f$. As indicated in the schematic shown, only one channel is assumed to be involved in the commutation and decommutation process. c represents the output of a bandpass filter while d shows the original signal recovered by use of a low-pass filter. The intelligence may be recovered from c by further rectification and filtering, a technique successfully employed in early telemetry systems. The principal Fourier components are shown for clarification of the concept.

Multichannel Commutation

Time-division multiplex, in practice, involves a multiplicity of data signals such as shown in Fig. 47 for the case of three working channels. The transducers are sampled in sequence and the resulting waveform applied to the decommutator. The decommutator is assumed to be in speed and phase synchronization, producing the samples of the individual channels at corresponding output terminals. As in the

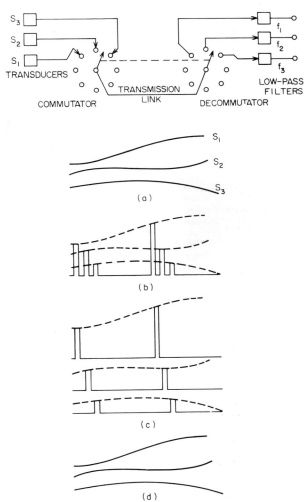

Fig. 47. The multichannel commutation and decommutation process. (a) Signals produced by transducers. (b) Transducer signals sampled in sequence by commutator. (c) Samples of the transducer signals separated by process of decommutation. (d) Original transducer signals completely reconstructed from samples by interpolation process.

single-channel case given previously, interpolation techniques are then used to reconstruct the original signals in each of the channels.

Multichannel systems using this basic process may become very complex. Transmission may take place with a format entirely different from that shown. Provisions must be made for synchronization. Commutation often involves multipole operation, subcommutation, and calibration. These subjects are considered in other sections of this handbook, particularly Chaps. 8 and 9.

Ambiguity in Commutation

It is clear that care must be exercised in limiting the frequency components contained in signals to be sampled. If this is not done, errors can be relatively large and the results meaningless. If, for example, in extreme cases, the intelligence frequency becomes equal to, or a multiple of, the sampling frequency the system output is the same as for a d-c intelligence signal as shown in Fig. 48a. Furthermore, as the frequency differs from such a multiple the resultant PAM waveform will appear as though a much lower frequency signal is being sampled. The apparent signal will be the difference between the intelligence frequency and the sampling frequency. This phenomenon of an apparent downward shift in spectrum is called the *aliasing* effect.

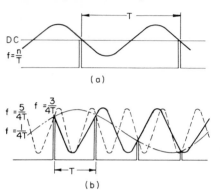

FIG. 48. Ambiguous effects in the sampling process. (a) Instantaneous sampling of d-c signals and signals which are multiples of the sampling rate. (b) One set of instantaneous samples can represent more than one signal waveform.

A vivid example of how a single set of instantaneous samples can represent two or more signal waveforms is given in Fig. 48b. Three sinusoidal signals are shown. Two are odd harmonics (third and fifth harmonics) of the first. The sampling rate shown is four times the frequency of the lowest-frequency signal. Other odd multiples of the first frequency would similarly result in the same sample values.

Remote Transmission of PAM Signals

Having produced a composite sequence of channel samples by the commutation process it is usually necessary to transmit them to remote locations. Let us briefly consider a few of the factors influencing transmission of a PAM signal of this character.

Three data-transmission media are generally used in telemetry, namely, the radio link, the wire link, and the tape recorder/playback systems. The radio link differs from the other two in that noise is often introduced together with the desired signal, and provision must therefore be made to obtain adequate signal-to-noise performance. This subject is discussed fully in other chapters and will therefore not be considered here. One important observation, however, might be noted. Wire links are characterized by a high degree of linearity and limited bandwidths whereas radio links are generally somewhat nonlinear but allow relatively large transmission bandwidths.

Although transmission performance is dependent on the exact modulation technique employed, there are certain system relationships which generally must be observed. Bandwidth restriction is one of the most important considerations since this factor always comes into play in achieving optimum signal-to-noise performance.

Excessive restriction of the transmission bandwidth for a PAM video signal produces crosstalk since the modulation of any channel then carries over into subsequent channels. If an ideal low-pass filter is employed, adjacent-channel crosstalk will vary greatly, with its cutoff value depending on the delayed contribution of its ringing effects as shown in Fig. 49b for the 50 per cent duty-cycle signal of Fig. 49a. As the cutoff frequency is varied, the crosstalk will be in and out of phase with the modulation of the earlier channel which is producing it. Crosstalk magnitude is dependent not only on cutoff frequency but also on the precise values of duty cycle of the PAM video signal. Full representation of this effect is therefore difficult to show easily, and the reader is referred to various references of the attached bibliography for quantitative details.

Figure 50 provides an approximation of crosstalk magnitude as a function of video bandwidth for 50 per cent duty cycle such as shown in Fig. 49a. Video frequency-cutoff values are given in terms of multiples of the commutation rate, which is the product of the number of channels and the frame rate. For simplification only a range of values is given corresponding to a broad region about each point of the abscissa plotted. Magnitude is considered without regard to phase. In general, one can assume that duty cycles appreciably larger or smaller will produce a greater amount of crosstalk. It is not only possible to arrange the filter cutoff and duty to give optimum results, but the filtered waveform may also be gated at the decommutator for optimum crosstalk reduction.

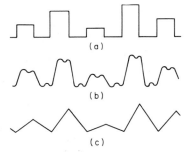

FIG. 49. Effects of bandwidth restriction on a PAM signal. (a) PAM signal with 50 per cent duty cycle. (b) Ringing effect of low-pass filter. (c) Integrating effect of sin ω/ω type filter.

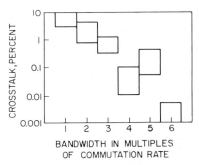

FIG. 50. Adjacent-channel crosstalk produced by video bandwidth restriction.

It is also possible to restrict the transmission bandwidth with other types of filters. One type considered to be advantageous in communication is the sin ω/ω type, sometimes called a finite-memory filter, which has an integrating effect on the signal waveform. If such a filter is designed to have its first node at twice the commutation rate it will produce an area-equivalent triangular waveform without crosstalk as indicated in Fig. 49c. This type of filter passes all the information content but its random-noise transfer characteristic is equivalent to a relatively narrow low-pass filter whose cutoff value is just equal to the commutation rate. Since the rms noise voltage is proportional to the square root of video bandwidth the signal-to-noise ratio for the low-pass filter case is not nearly so good. It is actually lower by the square root of the multiple of the commutation rate chosen for the low-pass-filter cutoff frequency.

Other characteristics of the transmission link are equally significant in preserving the characteristics of the original information to be transmitted. In unipolar instantaneous sampling of d-c and slowly varying signals, as commonly employed in telemetry, the composite signal results in components of similar frequency. Transmission of d-c samples must therefore be made through a link having d-c response, or effective d-c restoration. It is clear that lack of d-c characteristics causes severe crosstalk since the entire waveform shifts as symmetry varies according to the channel modulation. This type of crosstalk generally affects all channels as compared with the adjacent-channel effects discussed above.

Figure 51 is a simplified representation of how the average value of waveform shifts with modulation amplitude on its channels. When the first two channels shown are varied from full scale to 25 per cent value the symmetry of the waveform portion shown will shift with respect to the original average value. Therefore, unless the waveform is clamped as shown in b, a corresponding shift in level will take place in the unmodulated channels. An a-c link would be unusable without some modification. D-C restoration and factors affecting accuracy in restoration circuitry will not be considered further.

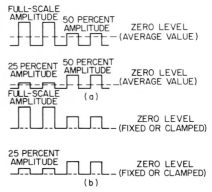

FIG. 51. Crosstalk due to lack of d-c response. (a) A-C response. (b) D-C response.

Decommutation and Interpolation

The PAM signal received at the recording station is separated into its individual channels, and each set of channel samples is then interpolated to produce the original channel intelligence. In order to decommutate the received signal properly it is necessary to provide a decommutator operating in exact speed and phase synchronism with the transmitting commutator. Synchronism is accomplished by extracting certain information contained in the transmitted signal marking the occurrence of each channel and frame. The synchronizing markers are generally distinguishable from the channel intelligence modulation by some unique characteristic such as amplitude or duration or both simultaneously.

In practice, a typical PAM multiplex signal has the format shown in Fig. 52. Full-scale signal modulation is restricted to only a portion of the available waveform amplitude range. A pedestal of approximately 25 per cent of the amplitude is thereby provided for indicating channel positions. The frame, only partially shown, is marked

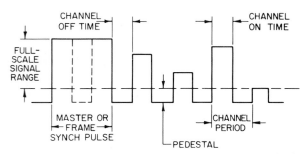

FIG. 52. Typical PAM format showing the inital portion of a frame.

by a two-channel master pulse of full-scale amplitude. Both sets of synchronizing pulses are easily extracted using simple solid-state circuitry. These pulse sets are used to drive electronic or electromechanical stepping commutators as described later to effect decommutation.

In the presence of noise the selection of synchronizing signals must be effected with optimum bandwidth restriction to achieve the greatest overall signal-to-noise performance. Techniques such as flywheel circuitry, missing pulse restorers, and phase-locked generators have all been used in practice for this purpose (see Chaps. 7, 8, and 9).

Interpolation

Much has been published on the mathematical analysis of interpolation, the process of reproducing the original intelligence from its periodic instantaneous samples. It is the intention here to review only a few of the principles involved in this process to orient the reader.

There are several ways by which a set of samples can be treated to create a useful form of the original signal they represent. The most common is that called minimum-bandwidth interpolation. This is achieved by means of an ideal low-pass filter as previously discussed. The output of such a filter is a smooth interpolation function representing the original intelligence in accordance with the sampling theorem. Realizable low-pass filters, as opposed to deal filters, require a finite frequency interval to go from the passband region to full attenuation. Perfect minimum-bandwidth recovery is therefore not possible in practice.

Two other recovery processes are worthy of mention. Linear, or polygonal, interpolation can be achieved by use of two cascaded sin ω/ω type filters whereas stair-step or step-function interpolation can be achieved with a single sin ω/ω filter. Actually these are but two of an infinite series approaching the ideal low-pass filter. Figure 53 is a comparative representation of the three interpolation methods mentioned above.

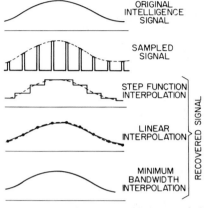

FIG. 53. Interpolation methods for sampled data.

The intelligence signal is superimposed as a discontinuous line without consideration of time delay to clarify their basic differences.

In actual systems, it is necessary to sample at a rate somewhat greater than that required by the sampling principle discussed above. Because of practical considerations, as many as 5 to 10 samples per cycle of highest-frequency intelligence are used in common systems employing a low-pass filter for interpolation (refer to Chap. 9, Sec. 7). In early systems, where the instantaneous values of the samples were merely recorded as a function of time after channel separation, the original intelligence had to be reconstructed by drawing the best curve through those points. This method of recovery is called optical interpolation. It is interesting to note that a study of errors in optical interpolation of a series of instantaneous samples indicates that from 8 to 10 samples per cycle are required to obtain 3 per cent accuracy.

The Aperture Effect

Filtering a single-channel PAM signal results in an output level proportional to the duty cycle of the frame. In the design of equipment for decommutation and interpolation it is often considered to be advantageous to increase the effective channel gain by lengthening the samples before filtering. This means that instead of an instantaneous value averaged over the period one obtains a value comparable with the frame period itself. Such a technique produces considerable distortion which is a function of frequency. This effect is described in Chap. 7 in more detail. It is clear that pulse lengthening can provide a true response only at d-c and very low frequencies beyond which it falls off quite rapidly. The distortion can, however, be compensated by a suitable modification or addition to the filter. A compensating network can be made to provide a reasonably flat response over the passband of most filters used for interpolation.

14 Mechanical Commutation Devices

Mechanical Commutators

The simplest form of mechanical commutator or sampling switch is one in which a rotating brush makes contact with a relatively large number of contact segments in sequence, connecting the channel information signals, in turn, to a common output. Similarly, the process may be reversed, such as in a second switch rotating in synchronism with the first, which distributes the composite information from the rotating brush to a multiplicity of independent outputs corresponding to the contact segments. The latter type of mechanism is called a decommutator or distributor. The principles involved have already been considered. It is the intention of this section to provide the reader with sufficient information on electrical and mechanical design features to gain an understanding of the capabilities and limitations of this type of device. Figure 54 shows the functional elements of such commutators.

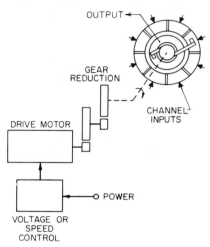

FIG. 54. Functional diagram of basic electromechanical commutator.

In practice, multichannel commutators for telemetry are generally much more complicated, having several synchronized poles, one or two usually providing timing or gating signals for the system. They are generally motor-driven or are mechanically synchronized with other equipment by connection to an available shaft. Their motion is generally continuous but intermittent or stepping commutators have recently become available. Mechanical commutators are principally characterized by their ability to provide direct physical contact in closing and opening circuits. Decommutators, as we shall discuss later, are usually electronic in design rather than mechanical in order to achieve the high response required for channel and frame synchronization.

Although there are many different ways in which electrical sampling may be accomplished mechanically, there are a few methods which have been commonly used. The types which have proved most acceptable in telemetry are distinguished principally by the basic brush technique employed. There are two broad categories which may be described as high-contact-resistance switches and low-contact-resistance switches.

Telemetering switches in the first category generally employ self-lubricating brushes in which graphite flakes provide the major lubrication. The second category involves a metal-to-metal contact arrangement which depends upon a more complex form of lubrication associated with metallurgical properties and brush configuration. Commutators employing self-lubricating brushes have been used most extensively in the past and relatively high contact resistance was generally acceptable (or was designed around) in most telemetry applications. Present-day commutators are largely of the metal-to-metal type, providing extremely low contact resistance compared with the resistance values associated with the source and load.

Other forms of electromechanical switching involving electrostatic, capacitative, magnetic, and liquid-conductor principles have not been made a significant part of this discussion since they represent the special cases in telemetry, often with questionable field history. Information on some of these types may be found in the publications listed in the attached bibliography.

Characteristics of Mechanical Switches

Most mechanical commutators in regular use involve raised contacts as opposed to contacts between which the insulation surface is flush. Early work showed this to be necessary for achieving long life and high reliability. This contact arrangement results in a shorting-type (or make-before-break) wiper action in which the wiper makes contact with a given segment before leaving the previous segment. It is clear that for nonshorting wiping action, it is necessary to have additional unused segments between adjacent used channel segments. In practice, therefore, the wiper must have appreciable cross-sectional area, and timing, among other properties, becomes a function of its flatness.

Commutator Configurations and Wiring Methods

Practical switches have been constructed with up to 500 contacts per pole and with several synchronized poles. The most general requirement, however, is for 100 channels or less with between two and four poles. It should be pointed out, by way of example, that a circle of 60 contact segments involves a diameter of from $\frac{1}{2}$ to 1 in. in modern designs depending on the duty cycle required. Where several larger circles of contacts are used on the same contact plate to provide additional poles having the same number of channels, it is necessary to use larger contacts to provide the same characteristics. In stacked multi-pole designs, the poles are sometimes arranged on separate plates in parallel fashion. In some instances, both concentric and stacked arrangements are used for a given switch package. The wiping surface used depends on the design concepts of the manufacturer.

Switch designs are in use employing the inside or outside cylindrical surfaces of the segmented rings or one of the plane faces. Flat face (plane) construction has been found to be most satisfactory in achieving desired switching characteristics and is more amenable to both versatility and miniaturization. Figure 55 clarifies these common wiping arrangements.

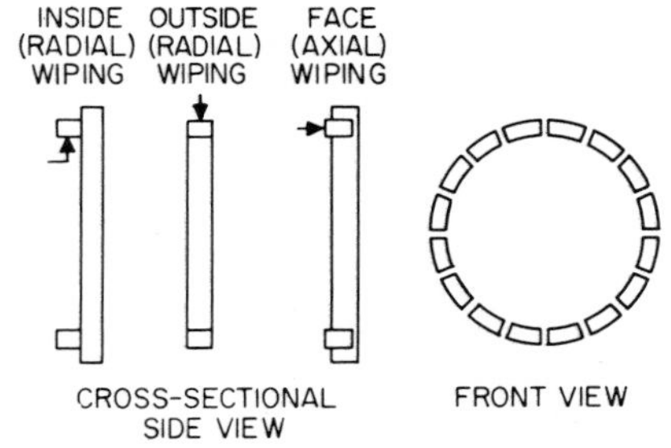

Based on a shorting-type wiper the contacts may be wired to achieve a variety of effects as indicated by the five examples shown in Fig. 56. Timing tolerances and measuring methods are considered below.

FIG. 55. Multicontact assemblies showing three common brush-wiping arrangements.

Timing and Timing Measurement

The timing of a multicontact switch is a function of a number of independent factors. It is primarily a function of the physical dimensions of the wiper and contact segments. It is directly a function of the accuracy of angular location of contacts and of the make-and-break character in the wiping process. In general, mechanical switches are designed to have a minimum of make-and-break uncertainty, or what is more commonly called *edge noise*. Since the latter, although it may be small, is present to some extent in every mechanical-switch design, it usually must be taken into account in computing the timing and timing tolerances for a given application. (In some instances, electronic gating of the sampled signal is provided, timed by an auxiliary pole, to achieve a much more regular and sharper rise time where it becomes essential to do so.) A wide range of duty cycles ("on" time relative to channel period) is achieved by appropriate selections of contact sizes. The easiest timing arrangement to achieve is generally one in which the "on" time of the nonshorting channels is some-

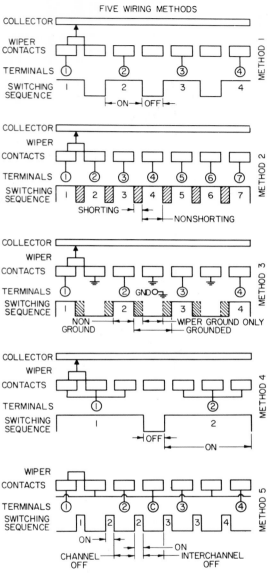

Fig. 56. Typical waveforms for common methods of wiring commutators. *Method 1.* Nonshorting operation. Alternate contacts connected, intermediate contacts not used. Two contacts required per channel. *Method 2.* Shorting operation. All contacts connected. One contact required per channel. *Method 3.* Grounding operating. Alternate contacts connected, intermediate contacts tied to ground. Provides grounded channels with two short contact ground time pulses per each channel. Two contacts required per channel. *Method 4.* Nonshorting operating. N number of adjacent contacts required per channel. *Method 5.* Shorting-type operation. Alternate contacts connected, intermediate contacts tied together and used as collector. No collector brush or ring required. Provides two short pulses per each channel. Two contacts required per channel.

what in excess of 50 per cent of the total channel period. This corresponds to the case of equal on an off segment sizes.

The phasing of corresponding contacts for two or more poles must take into account all the mechanical factors involved including fixed phasing errors, errors due to brush-location tolerances, errors in contact locations, uncertainty of make and break, etc. Timing analysis must also take into account the driving mechanism for the switch, which generally has appreciable inertia. If there is a gear train involved, its mechanical-transmission characteristic must also be considered in the timing analysis.

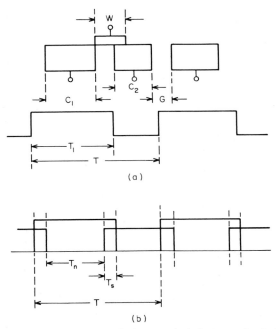

FIG. 57. Method of determining duty cycle for a typical electromechanical commutator.

$$\text{Duty cycle (nonshorting time)} = \frac{T_n}{T} = \frac{C_1 + 2G - W}{C_1 + C_2 + 2G}$$

Note: In shorting operation all contacts are most often used having equal lengths. The channel period then becomes just half of that indicated. The value $C + G$ is then used as the denominator in the above formula where $C = C_1 = C_2$. (a) Resulting waveform for nonshorting or break-before-make operation. (b) Resulting waveform for shorting or make-before-break operation.

The basic method of calculating commutator timing is given in Fig. 57. The wiper length W is the effective electrical length, which differs slightly from the apparent length as measured. Quantitative relationships are shown for both shorting and nonshorting operation.

Because of the variations in timing from channel to channel a composite or average-value measurement is often made together with deviations from nominal to define the character of a given commutator. An example of duty-cycle measurement using an oscilloscope is given in Fig. 58. This figure shows exaggerated timing variations in adjacent channels relating them to a common form of scope presentation. The scope is usually triggered by the channel leading edge and the on time measured directly in percentage of nominal channel period. This is simplified by adjusting the sweep to make 10 major divisions correspond with the nominal design value of channel period. The design value, also called the geometric channel period, is defined as the frame period divided by the number of channels.

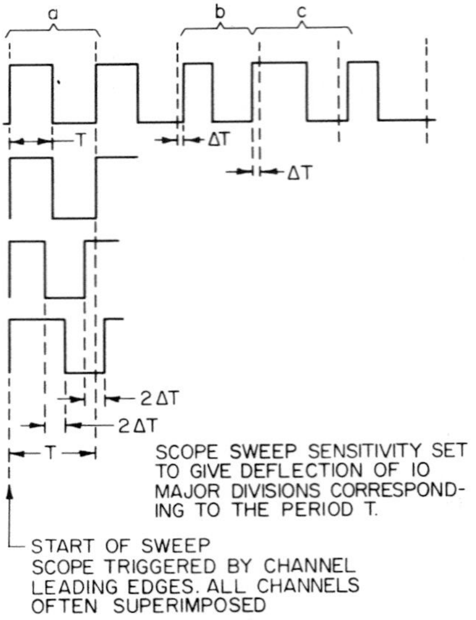

The extreme variation measured should include all contributory factors, particularly contact make-and-break uncertainty, errors in angular position of contact segments, and angular variations of the rotor movement as may be produced by non-uniform gearing or other problems in the drive system. The allowed tolerance in duty cycle is always expressed as a percentage of the nominal channel period. A common specified tolerance is plus or minus 5 per cent.

The duty cycle and its variation is, of course, only one of several measurements of concern. The variations in channel spacing are also of interest for many applications. Deviations in channel spacing are again referred to the ideal geometric design value of channel period. This tolerance is also expressed in percentage of channel period, indicating either the variation of channel center or the channel leading edge from the design-value reference. Because of the complexity of measurement, however, this expression of timing tolerance is not often used in practice. Figure 59 shows a means for measuring channel timing referred to a set of uniformly spaced pulses representing ideal design intervals. A digital instrument is used to adjust the pulse set to represent the number of channels in the frame. A variable time delay is provided to shift the relative phases in such a way as to effect the most convenient measurement according to the timing specifications provided.

Fig. 58. Common measurement of nominal channel duty cycle under conditions of extreme leading- and trailing-edge variation. (a) Normal design value (geometric channel period). (b) On time decreased at both leading and trailing edges by ΔT. (c) On time increased at both leading and trailing edges by ΔT.

Phasing of Poles

Up to now we have confined our timing discussion to a single pole of the commutator. In practice, many applications are concerned with the relative timing of from two to four poles operating simultaneously. Tolerance buildup must therefore be considered with great care in specifying, fabricating, and testing such commutators.

The phasing between two poles used for differential measurement (where corresponding channels of the poles are intended to be identical and in phase) is usually adjusted for best channel correspondence over the frame. In practice, this is done by applying voltages of easily distinguished levels to the poles such that three different levels appear in the resulting waveform. The high and low values correspond to intervals when one pole or the other is on alone. An intermediate level corresponds to the interval of simultaneous occurrence. The latter interval is called the coincident on time for two corresponding channels of the poles being compared.

Figure 60 shows the simplest phase-measuring technique applied to a case involving two poles out of phase by the interval ΔP. As in the example previously given, the leading- and trailing-edge variations ΔT for both poles are exaggerated.

This same measuring technique is used to measure the shorting overlap of adjacent

COMMUTATION

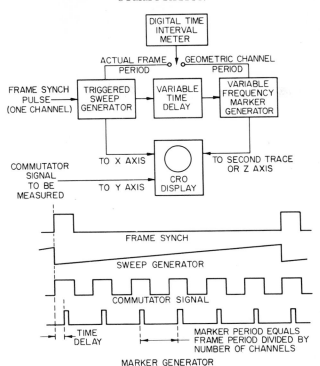

Fig. 59. Measurement of commutator timing with fixed channel references.

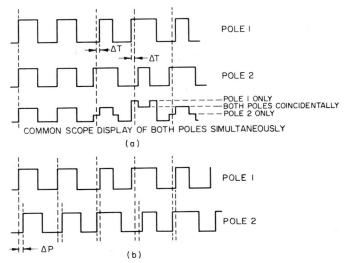

Fig. 60. The measurement of phase relationships between two poles having extreme leading- and trailing-edge variation. (a) Both poles nominally in phase. (b) Pole 2 nominally lags pole 1 by the amount ΔP.

contacts. In this case, different potentials are applied to alternate contacts or contact sets under examination obtaining a similar multilevel waveform showing overlap and nonoverlap intervals.

An auxiliary pole is frequently employed as a trigger or timing pole. In such cases the phase relationship between the channels of the auxiliary pole and one or more data poles becomes all important. Common requirements involve two principal relationships, namely, the phasing between the leading edges of corresponding channels of the two poles in question and the uniformity of the channel periods of the auxiliary pole used for timing. A typical case of complex phasing relationship occurs where there

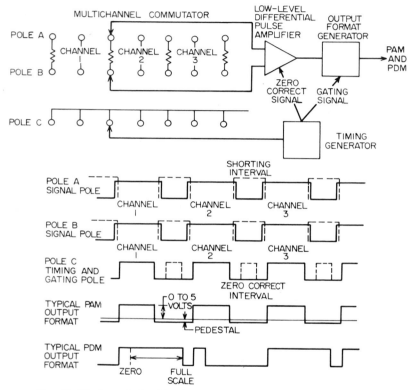

Fig. 61. Timing relationships for typical PAM and PDM multicoders.

are two differential data poles and a timing pole to be phased in accordance with critical system requirements. Commutator phasing requirements for low-level multicoders, discussed in Chap. 7, fall into such a category. Reference is made to Fig. 61 showing a typical timing requirement. In this case pole C must have a relatively uniform channel period. Its on time must also be uniform and must fall within the coincident on time of poles A and B. Similarly the zero correct interval sometimes developed from still another pole must fall properly within the coincident on time of the shorting intervals of A and B.

Sampling Rate

The sampling rate for mechanical switches is often specified in revolutions per second since one revolution usually corresponds to one frame. For analysis purposes, it is

COMMUTATION 4–67

best to specify operation in terms of the number of contacts per second since mechanical designs vary in size and hence the actual wiper velocities differ widely. Although much higher commutation rates have been achieved, standard switch types provide reliable sampling up to approximately 5,000 contacts per second. The majority of long-life telemetry applications involve commutation rates of 2,000 contacts per second or less.

Pedestal Insertion

A pedestal is normally provided in the transmitted waveform for purposes of decommutator synchronization. In simple high-level mechanically commutated systems this pedestal is inserted by connecting alternate (or interchannel) segments of

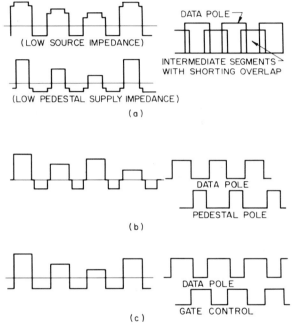

Fig. 62. Pedestal insertion methods. (a) Pedestal insertion with shorting pole (or intermediate segments). (b) Pedestal insertion with nonshorting poles. (c) Pedestal insertion with electronic gating.

the commutator to a negative-voltage source. This method produces the type of waveform shown in Fig. 62a. Two waveforms are shown here representing widely different values of source and negative supply impedances. The upper waveform shown assumes a relatively low data source impedance whereas the lower one assumes a relatively low pedestal supply impedance. The level of the shorting notch is shown to vary accordingly. A very low supply impedance virtually eliminates this notch for most sources commonly used in telemetry, but in some instances undesirable back effects on the transducer may result. A relatively high pedestal supply impedance on the other hand places the notch in the signal range, at all times resulting in uniform pedestals. The requirements for automatic decommutation may be met by either method.

It is often desired to insert the pedestal in a manner requiring no shorting action. This is done by supplying a second pole on the commutator phased to place the pedestal between the data samples without shorting as shown in *b*. Care must be taken to avoid unwanted pickup during the short interval (notch interval) when the commutator is open-circuited.

A third method used more commonly is that employing a simple buffer and format generator within the commutator case. (The commutator is then called a PAM multicoder.) In this case a second pole of the commutator specially designed to provide a very clean gate-control pulse is used to gate out the data as shown in *c*. Where a more precise wave shape is required the second pole actuates a one-shot which in turn controls the gate. Since the one-shot is usually set for a fixed time interval the duty cycle of the waveform will vary with the frame rate. This variation does not take place when the gating is accomplished by use of a second commutator pole directly.

The use of the buffer, as opposed to the simple commutator, simplifies application of the equipment. Not only is the data-shorting problem completely avoided but the buffer also provides for relatively high input impedance and low output impedance. In addition, the level of the pedestal may be fixed over a wide range, including positive or negative values, and the gain may be conveniently adjusted to suit the system.

Commutator Drive Systems and Speed Control

As explained above, mechanical commutators consist of three principal mechanical elements, the basic drive mechanism, a gear reduction, and the commutator assembly. The following is a review of drive systems and associated circuitry.

The common drive mechanisms of greatest interest are

1. D-C motor drive with unregulated voltage source
2. D-C motor drive with temperature-compensated voltage regulation
3. D-C motor drive with speed-sensing regulation
4. D-C motor drive with speed-sensing regulator and provision for external synchronization (speed and phase synchronization)
5. A-C motor drive
6. A-C motor drive with d-c to a-c inverter
7. D-C stepping drive

Simple permanent-magnet-type d-c motor drives have been most commonly used in the past. Voltage regulation has been provided when closer sampling-rate control is

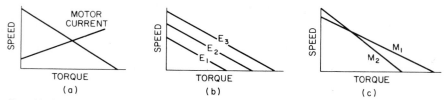

FIG. 63. Speed-torque characteristics of a miniature permanent magnet, d-c drive motor. (*a*) Speed-torque curve showing motor current characteristics. (*b*) Speed-torque curve for three different motor voltages. (*c*) Speed-torque curve for two values of magnetic field.

required. Solid-state regulators have, in fact, become so simple, reliable, and inexpensive that the best commutators made today contain them as standard practice. However, voltage regulation, even when completely compensated, is not always a satisfactory solution to the problem of speed control. Reference should be made to Fig. 63 showing a hypothetical speed-torque characteristic of a miniature permanent-magnet-type d-c motor.

Representative speed-torque curves are shown for different operating conditions. *a* shows the typical straight-line characteristics accompanied by a linear increase in current with load torque. *b* shows a simple family of curves for discrete voltage values. *c* shows how variations in the magnetic field strength can cause corresponding changes in slope of the characteristic. Certain pertinent observations may be made.

It is clear that regardless of how well the voltage is regulated the speed remains directly dependent on torque. Variations in torque due to factors such as wear, vibration, acceleration, and temperature will therefore cause corresponding speed changes. In a well-designed commutator, however, an acceptable tolerance can be maintained for most applications. This is particularly true where the regulator temperature characteristics are arranged to compensate for the inherent speed-temperature variations of the commutator. It can furthermore be observed that demagnetization of the permanent-magnet motor increases the slope of the curve and makes the speed-regulation problem more difficult. Although a magnetizer may be used for small motor-speed adjustments attempts at large variations after the permanent magnet has been fixed near saturation may lead to subsequent speed instability.

The most satisfactory form of speed regulation for this type of motor is one in which the speed is sensed and used to control the applied voltage in such a way as to maintain constant speed. Simple and reliable schemes have been developed to maintain speed to within 1 per cent under widely varying operating conditions. The speed is sensed by means of an auxiliary contact on the commutator or by means of a suitable magnetic pickup. The speed-sensing regulator may be expanded to provide for synchronizing the motor to a set of externally supplied pulses. Frame phasing may also be achieved if the application demands it.

A-C motors are used more extensively in larger ground-based commutators than in miniature airborne types. This results chiefly from the fact that they are considerably less efficient and consequently experience an appreciable rise in temperature. The d-c counterpart is often from five to ten times more efficient and considerably smaller for a given torque. One recognized advantage of the a-c motor is the elimination of motor brushes in the commutator drive system. Where efficiency is of no concern, a-c motors are sometimes driven from d-c to a-c inverters to avoid d-c motor problems. This matter is discussed further in a subsequent section on reliability.

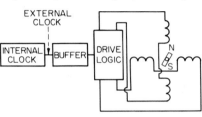

FIG. 64. Functional elements of the stepping commutator drive.

The electromagnetic stepper drive is unique in that it fulfills many of the needs not met with the other drive systems. It is digital in nature, stepping one channel for each clock pulse applied up to several hundred steps per second. It has no detents, brushes, or ratchets, allowing smoothly controlled high-speed sampling. Figure 64 shows the functional elements of the stepping drive system. Figure 65 shows the stepping commutator schematic and how it may be provided with automatic position-seeking or homing characteristics. Figure 66 shows several examples of how the digital character of the stepping commutator is advantageously used.

Life

Life must usually be defined in terms appropriate to each application. In general, the end of life occurs when the device has failed mechanically or electrically to meet the requirements specified. Most often failure is caused by failure to meet timing or contact resistance requirements. Mechanical failure occurs in the form of the mechanism wearing out after prolonged usage or by brush-wear particles causing conductive paths between insulated elements of the switch. It should be noted that motor-driven commutators of this type, properly designed, now have relatively long service-free life spans compared with common industrial multicontact devices. They are capable of making many millions of contacts before service or replacement is required. Depending upon the design and application specified, life varies from several hundred hours to well over a year of continuous operation.

An estimate of commutator life has been made, based on historical information

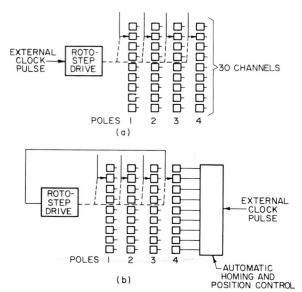

Fig. 65. Functional schematic of the stepping commutator. (a) Normal commutator operation. (b) Controlled stepping operation.

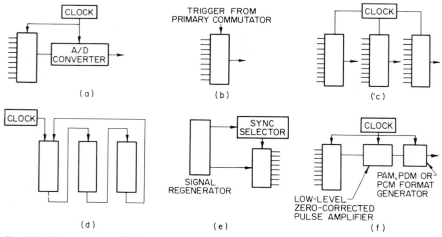

Fig. 66. Common uses of the stepping commutator in telemetry. (a) Digital instrumentation. (b) Subcommutator or calibrator. (c) Synchronized commutators (parallel mode). (d) Synchronized commutators (series mode). (e) Decommutator. (f) Low-level stepping multicoder.

available, to show the rapid advance in this technology over a relatively short period of time. Figure 67 shows such a curve of progress for a particular type of commutator operating at approximately 1,000 nonshorting channels per second (2,000 contacts per second). These data do not take into account detailed differences such as noise specifications, timing, contact resistance, environments, and package size. They suggest a high rate of improvement in all respects, from which we may anticipate new designs for more difficult requirements with increasing reliability.

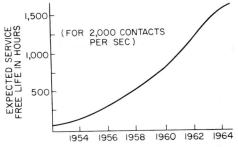

FIG. 67. Development progress for one class of commutator operating at 1,000 channels per second.

Size and Weight

The physical size and weight of sampling switches for telemetry depend strongly on the mechanical requirements involved. Apart from the number of contacts and poles which must be accommodated the construction must be such as to withstand the conditions of environment specified. Large commutators constructed for missile applications, where relatively high vibrational accelerations are involved, are necessarily made of strong castings or machined parts which appear at first to be unnecessarily bulky and heavy. On the other hand, where units are intended for aircraft telemetry or for ground use, the structures can be simpler and lighter in weight. The latest designs tend toward compact miniature construction having a high degree of ruggedness with much reduced weight.

Contact Resistance

Contact resistance is an intimate function of the wiping mechanism, of the nature of the materials used, of the normal brush force involved, an under some conditions, of the speed of operation. Any lubrication that may be used also influences the extent and character of contact resistance. In a general sense, contact resistance may be divided into two principal components, a d-c component which remains relatively constant over the channel period and an a-c component which consists of rapid resistance fluctuations, generally increasing with a changing character over a long period of operating time. For most applications the d-c component is negligible and the a-c component is the only portion of concern. Resistance variations in the measuring circuit result in corresponding variations in the load current and hence "noise" voltage.

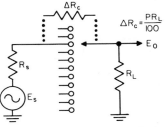

FIG. 68. Noise due to contact-resistance variation. E_s is a typical signal source with a resistance of R_s. E_o is the output signal across the load resistance R_L. P is the per cent noise in the output signal E_o. Noise due to variations in contact resistance ΔR_c on a peak-to-peak signal-to-noise basis is considered in the above relation. The formula gives the allowable value of ΔR_c for P per cent noise in the output signal when the load resistance is relatively high compared with R_s and R_c.

If the total circuit resistance is several orders of magnitude larger than the resistance variation of the switch, the resulting electrical noise will be relatively insignificant. This effect is clarified in Fig. 68 showing the quantitative relationship between effective contact resistance noise and circuit parameters.

Most high-level signal applications involve from 10,000 ohms to 1 megohm for the switch loads, allowing a measuring accuracy of a small fraction of a per cent with respect to resistance noise. For low-level transducer signals, the loads are often smaller and the commutator resistance variations become more significant. In the case of earlier switches, permitting relatively high contact resistance, values were generally in the region of from 1 to 100 ohms. In the case of newer low-contact-resistance devices, maximum values are generally in the region of $\frac{1}{10}$ to 1 ohm. It was stated above that the specific value of contact resistance is controlled largely by the wiper material and brush design employed as well as by the type of finish maintained on the contacting surfaces. Purely metallurgical effects, changing with time, are understood and controlled only after extensive experience with particular materials

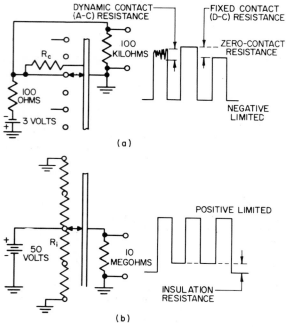

FIG. 69. Dynamic-resistance measurements in mechanical commutation. (a) Measurement of dynamic contact resistance R_c. (b) Measurement of intercontact insulation resistance R_i.

and mechanical designs. The achievement of uniform and reliable characteristics over long periods of time has therefore become an art with only a few organizations which have developed specialized capabilities in this industry. Another significant contact-resistance effect requires specific mention. Contact bounce is considered to be a short interval of high contact resistance or open circuit resulting from much reduced brush force at some point within the normal channel on time. This effect, due to faulty brush dynamics, is substantially eliminated in modern commutator design. Contact-resistance effects may be measured and observed with a circuit such as shown in Fig. 69a. R_c represents the total contact resistance of both the contact and collection brush interfaces in series.

Insulation Resistance

The basic insulation resistance of a mechanical commutator is a function of the structural materials employed for the contact plates and is generally adequate under all intended conditions of use. Insulation resistance between active contacts is, however, reduced with operation because of the accumulation of conducting wear particles. Despite the fact that much of the wear material produced is actually nonconductive some metallic particles also exist. The choice of wiper type is most important with respect to insulation resistance. Graphite brushes cause a slow reduction in value while metallic brushes eventually produce an abrupt short circuit. Taking all requirements into consideration, a good compromise is easily reached in the process of design for a given application. Figure 69b shows a simplified method of dynamic measurement of interchannel insulation resistance. The measuring accuracy of the switch must be considered in terms of errors due to this effect. Limits in source and load impedances may be imposed accordingly. Fifty and 100 megohms are typical minimum specified values of intercontact resistance.

Switching Noise

Except for operation at the very low signal levels and very high sampling rates no significant generated noise exists in a properly designed mechanical sampling switch

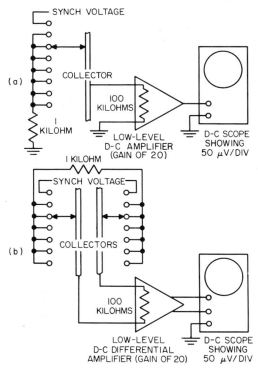

FIG. 70. Typical low-level noise-test methods. (a) Commutator pole used in single-ended measuring circuit. (b) Commutator poles used in double-ended (differential) measuring circuit.

for telemetry applications. Here again two components may be measured, a d-c component which may vary from channel to channel (sometimes called offset voltage) and an a-c or high-frequency component with a random character. Measurements indicate that a-c generated noise may be in the region of 10 to 50 μv for typical switches used in instrumentation. D-C or equivalent-channel offset potential may be of the same order of magnitude. Both appear to increase and change in character with commutation rate. Contrary to common belief, switch noise observed in practice is often due to the effects of the above-mentioned variation in contact resistance or to external electrical noise pickup. The latter may be induced by varying magnetic fields in the region of signal conductors or by capacitative coupling of high-frequency components in the vicinity. In new designs of very miniature construction, it is particularly difficult to eliminate magnetic effects on the measuring circuitry because of close physical proximity. These and other special factors must be considered in effective design and use of switches for low signal levels. Since most low-level systems involve differential measurement, some noise components may be canceled by using twisted leads for channel pairs. Magnetic shielding may also be used effectively in some cases for improved results. Figure 70 shows typical noise-measuring methods for commutators; *a* is used for tests of single poles whereas *b* is used for pairs of poles intended for differential measurement. Typical measurements for peak-to-peak noise are less than 50 μv for the differential case and less than 2 mv for the single-ended case.

Other Types of Sampling Switches

The type of multicontact switch discussed above involves a distinct wiping action. This process provides a very advantageous continuous cleaning mechanism but also a wearing process. The success of the switch design depends upon the establishment of a good wear equilibrium condition capable of achieving the service-free life desired. Most present-day telemetry requirements can be met with some specific design of this type.

Other switch mechanisms have been tried at these rates with varying degrees of success. The most practical of these include magnetically or cam-operated contact reeds, mercury jets, and the depression of a flexible conductor to make contact with a second conductor separated from it by a thin spacer. Each form has apparent advantages balanced by certain obvious disadvantages. In the final analysis, the successful commutator depends not only on the basic mechanism of design but very strongly on the techniques of fabrication. In the telemetry field, it has often been shown that a few units can be hand-tailored to meet a given set of requirements. However, commutator production at a given quality level requires a very special-purpose facility with a high degree of experience, highly organized inspection, and careful quality control. Constant testing, periodic design improvement, and the maximum possible standardization have also been proved requisites to achieve consistent predictable performance to the latest state of the art.

Reliability Considerations

Much has been debated with regard to the reliability of mechanical commutators. These devices differ from most other electromechanical contacting devices in that they involve relatively high contact switching rates combined with a high degree of timing accuracy and other difficult electrical-performance requirements. Failure to meet any one of the required specifications means a failure of the component.

It has been a widespread opinion in the past that mechanical commutation is somewhat unreliable. This reputation stemmed from the fact that many users attempted to apply these devices before adequate technology was developed in the industry. Furthermore, manufacturers in the past have not themselves always recognized some severe limitations in their own technology and have offered devices to the market which were seriously lacking in performance and reliability. The art is sufficiently unique so that it cannot be assumed that experience in a related technology can merely be extended in order to encompass telemetry commutation. Realization of this fact has often come too late, and there has been a large turnover of manufacturers.

The present-day commutator, however, is a well-developed precision instrument. Its performance capabilities far exceed that of even the recent past. Its complex characteristics are reproducible and operation is readily subject to statistical analysis from a reliability standpoint.

Factors relating to reliability are many but the principal considerations are strongly dependent on the manufacture. The commutator first must be designed with an intimate understanding of factors affecting mechanical life and performance. Control procedures must then be instituted to assure fabrication to the required designs. Testing, peculiar to this type of instrument, must finally be made in order to be assured that it falls into an acceptable category.

The application of reliability theory to component testing is well known. Only a few significant aspects are reviewed here to orient the reader. It is first important to realize that the basic reliability of components of this type can be determined only by lengthy empirical testing programs. Where a sufficient number of samples are involved a reasonable measure of reliability can be obtained.

MTBF, the inverse of failure rate, is now a common measure of performance. The concept of confidence level based on a chi-squared failure distribution is, however, necessary to present a complete picture. These parameters are readily related to allowed failures and required test hours, as indicated by the examples given in Fig. 71. Figure 71a shows the relationship between the allowed number of failures and the total test hours for a typical MTBF value of 2,000 hr. Figure 71b similarly shows the relationship between the allowed number of failures and MTBF values for a common test program of 30,000 unit hours. These relationships are shown for two widely different confidence levels of 60 and 99 per cent. Figure 72 shows the test hours required for a single failure in order to achieve a given MTBF for various confidence levels. Test hours are shown here in terms of multiples of MTBF for greater generality. These values may also be thought of as the number of units which must be tested over the specified MTBF interval with a single allowed failure.

Detailed and repetitive failure analysis is essential in advancing the state of the art

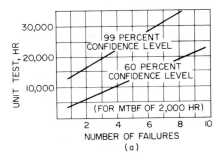

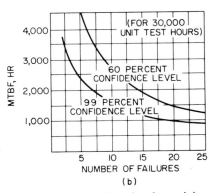

Fig. 71. Relationships in determining commutator reliability.

and can substantially improve the confidence level for a particular design. It has also often resulted in improved fabrication and test procedures and hence a more acceptable failure-rate level on production components. Failures encountered in new designs are especially subject to reduction by simple modifications. Figure 73 shows an example of a major improvement made in exactly this manner.

The high degree of overall commutator development has necessarily required substantial improvement of each of its component parts. The drive mechanism, for example, has been improved to maintain acceptable quality level of the component over long specified lifetimes. Three principal drive types were discussed above, namely, the d-c motor, the a-c synchronous motor, and the stepping motor combined with associated gear reductions. Concern has been expressed by many users, wishing to extend useful commutator life further, that d-c motors may be a restrictive factor because of motor brush wear. There is little reliable information of a specific nature on this subject. However, composite failure rates used in the motor industry are shown in Fig. 74 comparing brush and brushless types. It is of interest to note that failure rates differ by approximately a factor of 2 at any speed and that they are proportional to speed for both classes. Although absolute values differ for the specific motors applied to commutator components the principles assumed for these curves are generally applicable. Typical minimum motor operating speeds to achieve

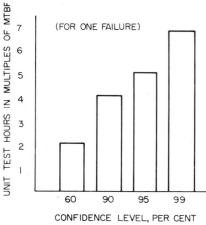

Fig. 72. Test hours required for a single failure to achieve a given value of MTBF.

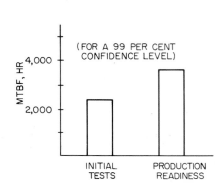

Fig. 73. Typical reliability-improvement program. Test program: 10 units, 30,000 test hours, 5-pole configuration, 250 contacts/sec per pole commutation rate.

reasonable torque and design flexibility in gearing are 5,500 rpm for d-c motors and 6,000 rpm for a-c types. Rotor speed stability with changes in rotor torque is, unfortunately, inversely proportional to the square of the gear ratio employed. A compromise must therefore be reached between such stability and motor reliability. The d-c brushless stepping motors are generally operated at lower equivalent speeds, giving some improvement over brushless a-c motors on this basis.

It is apparent from observations over the years that commutators exhibit a definite failure-time characteristics as shown in Fig. 75. This may be assumed to be a plot of failure rate as a function of time from initial assembly. It can be seen that an initial period exists during which a slight increase in failures is ob-

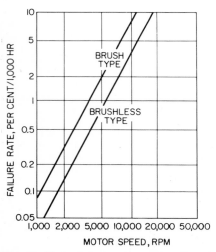

Fig. 74. Motor-failure rate as a function of speed.

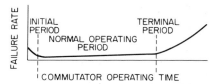

Fig. 75. Failure-rate pattern for electromechanical commutators.

served because of initial defects. These are apparent in the first tests to which the device is subjected, after which a normal operating life span can be assumed. Following this period an increase in failure rate occurs marking the end of life. Most well-developed complex devices exhibit a similar characteristic curve. The exact shape of the curve and time values corresponding to changes in slope depends on many design factors. The conclusions drawn from empirical testing, regardless of the number of samples and test hours employed, must be consistent with this type of experience curve.

Evidence of Reliability

Reliability of the commutator more so than of other components of an instrumentation system is subject to proof. In accord with accepted methods, proof of reliability is determined in two ways, first by the analytical (or objective) method by comparison

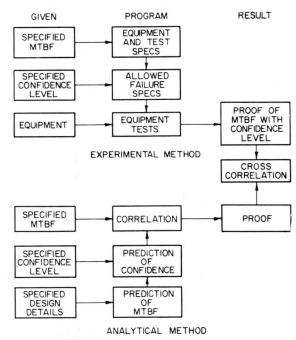

FIG. 76. Reliability proof methods for commutators.

of elements to be employed with the history of those previously evaluated and second by the experimental (or subjective) method by actual operation and failure observations. The main elements of these methods are shown diagrammatically in Fig. 76.

The experimental method has been found to be the most practical and realistic method for this type of component. For this reason only years of testing experience can provide completely satisfactory proof of reliability on long-life commutators. The reliability level is generally determined for the normal operating period following initial testing. The objective sought is an MTBF with the desired confidence level. MTBF is experimentally determined by the observation of equipment failures over a reasonable test time employing an adequate number of samples as already described above. The confidence level provides a measure of the assurance that such an MTBF will result.

Analytical methods are more frequently used for reliability predictions as applied to electronic commutators discussed below. Such an analysis is based on a knowledge of the failure rates for the individual components used in the particular design of interest. In some instances crosscorrelation can be made with empirical data, as also indicated on the diagram.

15 Electronic Commutation Devices

Although electronic commutation schemes have been in use for many years, the relatively recent development of improved solid-state elements has made them much more practical for airborne use. This is particularly true in terms of size, power, and weight, factors of great interest in aerospace telemetry. This section describes some of the principles and techniques employed in practical equipment for electronic commutation. It also suggests limitations and future possibilities. Although mechanical commutation has, in the past, dominated the industry whenever sampling rates would permit, electronic commutators have increasingly been used in their place for certain applications. This is particularly true at the higher sampling rates, for limited numbers of channels, and where cost is not an important factor. Commuta-

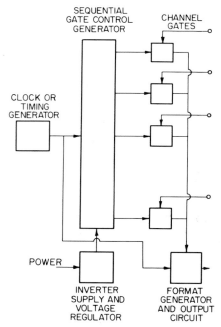

FIG. 77. Block diagram of basic electronic commutator.

tion at the lower signal levels using solid-state devices is more complex and has certain limitations in size and in electrical characteristics. Much progress has already been made, however, with respect to these limitations, and their use in telemetry is now chiefly limited by cost and reliability. A general comparison between electronic and mechanical types is made in Sec. 16.

Electronic commutators are generally made up of a few principal parts: the timing clock (or generator), the signal (or channel) gates, and the sequential gate control. In addition, an output circuit usually provides the format and synchronizing means required. Reference should be made to the block diagram of Fig. 77. The timing clock may be any one of a number of commonly known generators such as an oscillator followed by appropriate pulse-forming networks, or a stable multivibrator. The sequential gate control generator may be based on one of many electronic circuits which provide a sequential stepping or commutating action such as a binary chain, a thyratron chain, a ring counter, or a delay-line chain. The channel gates may be any electronic gating device capable of gating the signal to be sampled under the control of the sequence generator having the proper response and transfer characteristics. Diodes, transistors, and tubes have all been commonly used in gating circuits.

The following illustrates a few of the common gate-control generator types. Refer to Fig. 78 for applicable block diagrams.

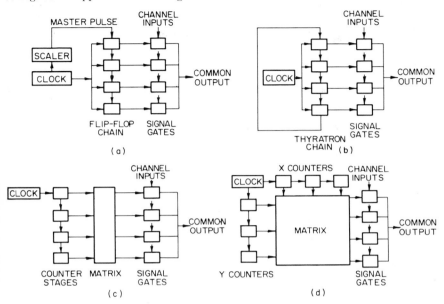

Fig. 78. Some basic electronic commutator schemes. (a) Broken-ring commutator. (b) closed-ring commutator. (c) Single counter-matrix commutator. (d) Dual counter-matrix commutator.

Broken-ring Commutator

In this system the basic time clock steps the elements of the sequential pulse generator, which may typically be a chain of bistable flip-flops.

The chain is actually driven by two related sets of pulses. A set of master (or framing) pulses is derived by counting down from the channel pulse rate. The master pulse turns the first element on. When the first element is turned off by the next occuring channel pulse, an interchannel master pulse, produced by element 1, turns on element 2. This action continues until the end of the chain. The stepping action then stops until it is again initiated by the next occurrence of a master pulse. The gate control is therefore based on a broken-ring operation as distinguished from a continuous-ring counter.

Closed-ring Commutator

A second type of sequential gate control is produced by forming a continuous ring in which the last stage actuates the first, the stepping again being synchronously controlled by the timing clock. Only one of the elements is on at a time. Thyratron devices and particularly their solid-state equivalents have been effectively applied to this use.

Counter-matrix Commutator

A third type of electronic commutator is based on the counter-matrix technique. A common form of this technique is one in which a clock is used to drive a set of binary counters. The counter waveforms are combined in such a manner as to produce gate-control pulses in the desired sequence. The number of channels generally corresponds to the overall scale of the counter. The frame rate is equal to the lowest frequency of the counter set. Refer to Fig. 79 showing a possible matrix configuration for a simple 16-channel electronic commutator. Gate-control signals may be derived by means of the diode matrix as shown or by a well-known resistance matrix from the primary waveform shown in Fig. 80. The clock waveform, having approximately half the period of the highest frequency shown, is sometimes used to gate out a 50 per cent duty cycle.

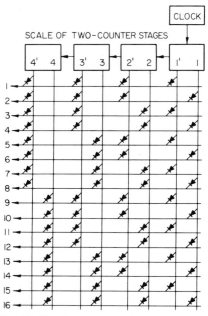

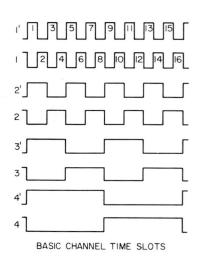

FIG. 79. Simplified counter-matrix arrangement for a 16-channel gate-control generator.

FIG. 80. Waveform timing relationships for 16-channel gate-control generator.

The number of channels in this example is the fourth power of 2, making a simple symmetrical matrix. Odd counter scales may also be used such as 3 and 5, producing unsymmetrical waveforms. Considerable work has been done to minimize the number of diode elements required. Combinations have been found to employ the smallest number of diodes using subbus or submatrix techniques. As the number of channels becomes larger the required number of matrix diodes approaches two per channel.

A second type of counter-matrix arrangement commonly used is that known as the dual (or XY) counter matrix. In this case two synchronous counter sets are employed, the coincident outputs of which can provide a sequence of pulses by use of a similar diode matrix. A common arrangement of this type is represented in Fig. 81 for a 12-channel commutator.

Commutator Signal Gates

One of the most critical parts of electronic commutators is the type of gating element employed. There are several well-known forms of electronic gates such as used early in the computing field. These have been refined and combined with appropriate drive methods for accurate data switching. A gate may involve one or more transistors, diodes, or tubes. Single elements or back-to-back pairs are often used for high-level single-ended sampling. For low-level differential circuits, a pair of transistors is generally used for each of the balanced differential inputs.

It is clear that the simple gates involve the smallest number of elements and hence generally make the most compact, reliable, and least expensive commutators. On the other hand, a compromise is often necessary to achieve a given input or transfer characteristics. Several important considerations should be kept in mind. In solid-state circuitry, simple gates usually operate with small currents flowing in the circuits which includes the sources. This is commonly called back current, which must be minimized by one of several techniques wherever the source impedance is great enough to produce an IR drop in excess of that allowed in the circuit.

The application of gate-control pulses must have a minimum influence on the signal level. Transformer coupling to the gates has certain desirable features in this respect but is bulky and makes the circuitry dependent on the gate-control waveform. D-C coupling, used in many modern commutators, provides uniform operation from d-c to relatively high channel rates but generally results in a current path through the source.

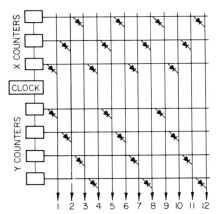

FIG. 81. Simplified X-Y (dual) counter-matrix 12-channel gate-control generator.

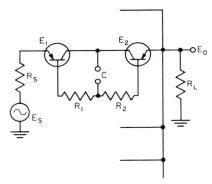

FIG. 82. Schematic of dual transistor gate.

D-C coupling arranged to avoid d-c path through the source reduces the back-current effects but becomes more bulky and expensive.

Various feedback techniques provide improved gate characteristics, such as higher input impedance, greater linearity, and improved temperature performance.

Fundamentally all common solid-state gates involve the superposition of signal currents and currents resulting from the gate control. They attempt to balance all nonsignal potentials produced by the gating action. They use back-to-back elements for this purpose, which also makes the circuit normally nonconducting. Reference is made to Fig. 82 showing a complete gate circuit of this type. The control signal C, produced by the sequential gate-control generator, causes the two transistors to conduct with the base-emitter current limited and balanced by the base resistances. These currents are adjusted so that the voltage drops across the two elements are equal and opposite in value. The principal objective is to cause each element to act as an ideal low resistance when conducting (low saturation resistance) and a very high resistance when not conducting (high leakage resistance). For very low-level signals, requiring optimum balance, selection of transistors must also consider variations over the specified temperature range. The circuit shown assumes ideal transformer application of the control pulse. However, any impedance between the control pulse sources and signal ground in practice may cause a portion of the control signal to flow through the source. Additional currents comprising the sum of currents flowing through the leakage resistance of all the other gates also flows through the source. This current is called back current and changes widely from off to on positions and over the temperature range. Back-current contribution to error is largely dependent on R_s. It is frequently measured dynamically by substituting a 10-kilohm resistor for the signal source, paralleling all inputs, and making a scope observation of voltage across the resistor.

The matched transistors shown in these gating circuits are p-n-p types, arranged to conduct when the base potentials is more negative than the collector and open when the relative potentials are reversed. This so-called inverted use of the transistor exhibits considerably lower collector-to-emitter offset voltage compared with standard transistor forward use, allowing improved matching of the pairs.

Recent developments of integrated pairs (sometimes called integrated choppers) have greatly improved the matching problem. Integrated transistors are constructed around a single silicon crystal which assures identical characteristics for both elements of the pair. Present integrated pairs, however, exhibit considerably greater saturation resistance than pairs of separate elements of high quality.

The offset voltages do not match precisely in practice. The main offset arises from the base current flowing through the so-called transfer resistance of the element. This resistance is the ratio of the incremental values of offset voltage and base current and is determined by the exact transistor structure. The transfer resistance of an integrated pair is approximately an order of magnitude lower than that of the individual transistor pair. Stability with age and temperature is of great concern in commutator design. The more stable planar transistors are often used for the very low levels after undergoing a temperature-cycling and aging process.

Leakage resistance of the transistor pairs was indicated as a factor affecting gate performance. Currents flowing through the resistances of all open gates combine to flow through the closed gate and hence the source. The total crosstalk effect of the gates is therefore dependent upon leakage resistance and the channel resistance. Arrangements of gates in groups of subgates provides increased isolation and considerably reduces leakage effects. The grouping of gates also makes it possible to simplify the total circuitry and matching problems. One transistor may be used to balance as many as five or ten other transistors, the latter being sequentially switched on and thereby paired with the single element.

Reference is made to Fig. 83, which shows a wide variety of gating techniques. a through f are generally considered to be high-level gates, whereas g to h may be used for low-level sampling in the millivolt full-scale range. S represents the transducer signals, C represents corresponding gate-control pulses, and E the common output. A and B designate the legs of the balanced circuitry. a shows the basic schematic for an early electronic tube gate using twin-triode cathode followers. This type of gate was used extensively in early aircraft telemetry equipment. b shows a common diode-pair gate in which the diodes are normally biased off except at the time of occurrence of the gate-control pulse. c shows a conventional transformer-coupled transistor-type gate using back-to-back transistors such as that discussed above. d represents a very simple and little-used single-diode gate which uses a resistive element in place of one of the diodes shown in b. e is a common four-diode (quad or bridge-type) gate which requires a balanced drive circuit. f is a simple dual-transistor circuit similar to c but employs d-c coupling. The base and collector are supplied from a relatively constant-current source. g is identical to c in concept but is double-ended for differential low-level use. h is similar to f but arranged for differential measurement. The particular form shown also makes use of a single pair of balancing transistors to service a group of channels. Active constant-current elements are shown supplying the collectors. Provision for balancing and temperature compensation not indicated here is made for most gates of this type. Extensive selection and matching is generally necessary only for extreme low levels.

A more complete functional schematic of a practical form of high-level commutator is shown in Fig. 84. A common channel gate is used to balance a group of single transistor gates. Constant-current sources are used for operating the gating elements.

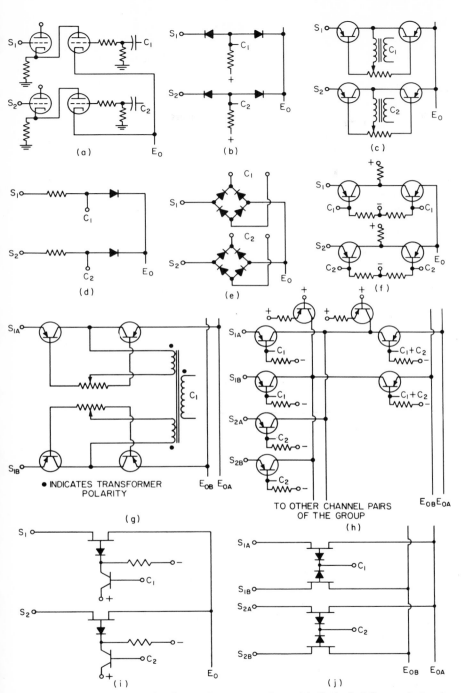

FIG. 83. Common forms of signal gates in commutation. (a) Cathode-follower electronic tube gate. (b) Balanced diode gate. (c) Transformer-coupled balanced transistor gate. (d) Simple single-diode gate. (e) Diode bridge gate. (f) Direct-coupled dual transistor gate. (g) Low-level differential gate with transformer coupling. (h) Low-level differential gates with common elements. (i) High-level single-ended FET gate. (j) Low-level differential FET gate.

4-83

D-C coupling is used for the sequential gate-control generator. Provision is made for simple temperature compensation and regenerative feedback is used to increase the input impedance.

A more recent form of solid-state gate is that which employs the field-effect transistor (FET). Of the solid-state devices, this type of transistor acts most like a vacuum-tube triode, since only one kind of charged carrier is involved in its operation. The field-effect transistor may be viewed as one in which current flows from what is termed its source end to the drain end through a channel whose effective width is controlled by the instantaneous bias on control gates situated about the channel. Because of this unique structure, this transistor offers certain desirable characteristics for signal-gating purposes. In particular, it offers low offset voltages, considerably lower back current, and may be operated by much simpler gate control circuitry. Since operation does not depend upon minority carriers, resistance to radiation is relatively good. In addition, transistors of this type are inherently free of certain common sources of noise found in other types.

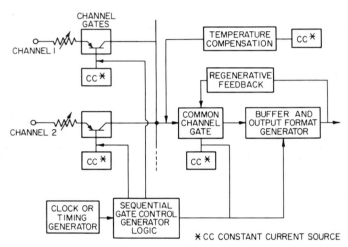

Fig. 84. Functional schematic showing principal portion of typical high-level solid-state commutator.

Despite these major advantages, both FET and standard-type transistors are widely used in airborne commutation at the present time. There are some particular deficiencies in the field-effects transistors which can be overcome with more complex circuitry. One disadvantage is the fact that this type of transistor normally conducts when the equipment power is turned off, resulting in a low-impedance loading effect on the signal sources. In addition, transistor failure often causes a similar loading effect, a condition which is highly undesirable for most airborne applications.

As in the case of the standard transistor gate mentioned above, a detailed analysis of the electrical operation of the FET gate may be found in many other publications and cannot be included within the scope of this handbook. Two representative forms of FET gates are shown, however, in Fig. 83i and j for high-level single-ended and for low-level differential applications, respectively. It should be noted that the metal oxide form of FET, commonly called the MOSFET, is particularly suited for use in data signal multiplex. In general, the trend in airborne telemetry equipment is toward circuit integration in whole or in part. Although considerable progress has been made in the application of microelectric techniques to this type of equipment, it will be some time before such equipment is capable of complete integration with all the desired characteristics and in a cost range suitable for common usage.

The Potentiometer Effect

The potentiometer effect is obtained when the output voltage of a commutator becomes strongly dependent upon the value of the channel source impedances. Refer to Fig. 85. If R_L is much larger than the source impedances, this effect is small. If source impedance is constant and the same for all channels (impedances may be padded to make them all equal to the largest), all channels may be calibrated by use of any one as a reference even where R_L is not extremely large. In the case of varying source impedance such as the potentiometer pickup shown, the impedance is zero at the extremes and a maximum of $R_2/4$ at the center of movement. This merely introduces a known nonlinearity in the calibration. In general, the effect can be avoided if the impedance differences in any channel (or from channel to channel) are small compared with R_1. The effective constant resistance of the switch is represented by R, which in some instances should be taken into account.

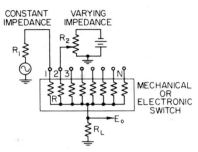

FIG. 85. The potentiometer effect in commutators.

16 Comparison of Basic Commutator Types

Much has been argued for and against each of the two basic commutation approaches, electronic and mechanical. As already stated, each has its advantages and a detailed critical appraisal is possible only for a given set of exact requirements. In a general way, however, the relative advantages of each for telemetry applications may be reviewed as follows:

Relative Merits of Mechanical Commutators

1. Small number of parts or elements
2. Relatively inexpensive for many channels
3. Direct contact, not temperature-dependent
4. Low contact resistance
5. Small in size, weight, and power
6. Channel-to-channel equivalence
7. No back-current effects
8. Not affected by radiation

Relative Merits of Electronic Commutators

1. Capable of relatively high sampling rates
2. Allows random programming
3. Has no switching inertia
4. Relatively accurate in timing and in synchronization
5. Wide flexibility in packaging its elements
6. Employs commonly known and tested elements
7. Highly reliable for small numbers of channels
8. No wear effects

Because of their basic simplicity, reliability considerations appear to favor electromechanical commutators for large channel numbers. In practice, however, reliability of both types of devices has proved to be more dependent on control of quality in fabrication than on limits in design and in technology. These factors were discussed in a previous section dealing with commutator reliability (Sec. 14). Since the choice of commutator is easily made when the requirements are completely specified it is necessary that the behavior of both types be well established. It is actually the limitation in knowledge of behavior which complicates the task of making such an optimum choice. A true indication of behavior can only be obtained empirically, using the various methods already discussed above. It is important, however, to program evaluation testing to take maximum advantage of the newest and best techniques available for both types without undue delay.

For a more detailed comparison of certain performance characteristics, values for both types are given quantitatively in Table 8. This tabulation gives commonly accepted or frequently measured values of salient parameters. Ranges are given in some cases since exact values depend on the particular design and how it is applied.

Table 8. Quantitative Comparison of Salient Commutator Characteristics

Measurement	Electromechanical	Electronic solid state
Channel equivalence, μv	10–30	20–100
Differential noise, μv	10–200	25–175
Contact resistance, ohms	½–10	10–250
Nonlinearity	Negligible	Negligible
MTBF (high confidence level), hr	500–4,000	1,000–40,000
Max sampling rate, channels/sec	5,000	100,000
Insulation or leakage resistance, megohms	100–1,000	500–1,000
Sampling-rate stability, %	1–10	0.1–0.3

There are several aspects of the comparison worth specific comment. It is assumed that voltage errors smaller than 5 μv are negligible. Noise is primarily of interest in low-level measuring systems where the circuits are differential. Noise is peak-to-peak and includes the induced potentials from sources within the commutator package. Contact-potential differences from channel to channel in mechanical commutators are assumed to be equivalent to offset in electronic types. Noise and offset variations with commutator size and sampling rate are not specifically accounted for. MTBF values for electromechanical commutators are based primarily on experimental data, whereas values for electronic multiplexers are largely analytical. The test data and reliability proof information are relatively limited for the electronic types.

17 Decommutation Devices

It is the intent in this section to consider decommutation only insofar as it concerns techniques involved in commutation. Decommutation has already been described as the process of separating channels contained in the composite waveform of a time-division signal. It is generally thought of as being the inverse of commutation but results directly in separate PAM signals for each channel. The separated PAM signals must then be appropriately interpolated in order to obtain the original form of intelligence.

Decommutation is accomplished with a wide variety of techniques using circuitry similar to that employed in the commutation process. The essential elements of a complete system are shown in Fig. 86. The sync selector circuitry extracts the information contained in the signal which marks channel and frame occurrence. These two sets of derived sync pulses then operate the decommutator in the same manner already described for electronic commutation. After separation, the individual channels are applied to pulse-stretching and interpolating circuitry. A buffer is used to provide a relatively low channel output impedance suitable for driving recording, display, and other terminal instruments.

Commercial equipment for electronic decommutation has been developed in compact and modular form. For standard telemetry applications it is supplied for use with

PAM or PDM signals and for all standard frame rates and channel numbers. Other time-division systems such as PCM and PPM all employ decommutators usually designed specially for the particular system. Equipment is well developed, reliable in operation, and available with commonly desired levels of accuracy and performance.

Standard equipment is designed for operation with both PAM and PDM waveforms by minor front-panel adjustments. Provision is also made in many types of equipment for automatic zero and sensitivity control. This means that zero and full-scale reference signals may be applied to two channels at the transmitting end. An electronic servo in the decommutator equipment then operates on the entire waveform to maintain these fixed channel values through the entire decommutation process; zero shift and sensitivity variations are therefore corrected in all channels.

All decommutators tend to provide for operation with deteriorated input signals. Typical of such degrading effects on the composite signal are the superposition of random noise, the addition of impulse noise, and the severe restriction of transmission bandwidth. In addition, certain defects may cause waveform deterioration such as missing channels or bridging adjacent channels. Suitable auxiliary circuitry is capable of providing reliable operation over a reasonable range of such degraded signals in most equipment.

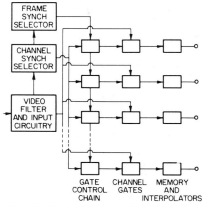

FIG. 86. Basic electronic decommutator.

Since cross strapping or paralleling of contacts in the frame is often done to increase the sampling rate of a particular measurement provision is made for corresponding patching of gate pulses in the decommutator.

Synchronization

It is not within the scope of this section to make an appraisal of circuit techniques for decommutator synchronization. Several important aspects should, however, be noted. It is necessary to detect the synchronization intelligence with the highest possible signal-to-noise ratio. This is often done for the case of channel sync pulses by passing the received PAM signal through a limiter and narrow filter. Both open-loop and closed-loop phase-locked methods are then used to drive the sequencer. The latter method allows a smaller effective bandwidth with varying commutation rate and hence higher effective signal-to-noise performance. Similarly frame-synchronization selection must be as noise-free and reliable as selection of the other intelligence. "Flywheel" and similar circuitry have been used with success to improve performance, especially in the presence of impulse noise.

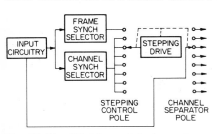

FIG. 87. Simplified form of electromechanical decommutator.

Electromechanical Decommutation

Although decommutators have in the past been largely electronic, the advent of the stepping commutator makes possible reliable synchronized decommutation for limited channel rates. The schematic of such a system is shown in Fig. 87 in simplified form. The principles involved are the same as those applied to the electronic case. As in the case of commutation the reliability of mechanical decommutators may be greater than their electronic counterparts for relatively large numbers of channels. There are obvious advantages in size, cost, and power consumption.

18. Commutation Terminology

Terminology is carefully treated in all sections of this chapter in order to provide the reader with an understanding required for communicating this technology. A glossary of *telemetry* terms will be found in Chap. 5. The following supplementary terms are listed for the convenience of the reader:

Accuracy—Commutator accuracy is the transfer accuracy of the switch as a whole. This is expressed in per cent of full-scale voltage either as the arithmetical sum of the maximum amplitude errors or as the rms error produced by all effects.

Adjacent-channel Crosstalk—Crosstalk generally produced by bandwidth restriction in time division.

Aliasing Effect—An apparent downward shift in the frequency spectrum produced by the sampling process.

Aperture Effect—Effect introduced by use of a finite rather than instantaneous sample aperture in the commutation process.

Blanking Level—Early term used to designate the level resulting from gating or clamping a pulse train between channel pulses.

Channel Equivalence—The extent of likeness in amplitude characteristics for all the channels. If semiconductor devices are used for channel gates, the output levels of the individual channels may be slightly different in each channel. The equivalence may be expressed in per cent of full scale, or in terms of absolute value referred to either input or output. Channel differences may change with the signal level and with temperature in most simple gating circuits.

Channel Interval—Duration of the time slot allotted for a given channel of a time-division system. For mechanical commutators this refers to the geometric channel period as opposed to the actual period, which varies from channel to channel.

Channel Rate—Generally used to specify the number of channels per second of a system or the system pulse-repetition rate. This rate is also called the commutation rate.

Clock—Pulse generator from which basic timing is derived for a system or for one of its component parts.

Commutation—In telemetry the process of sequential sampling of a multiplicity of data signal sources.

Commutation Rate—The number of channels per second resulting from commutation. The basic channel rate, equal to the product of the number of channels and the frame rate.

Contact Resistance—The effective resistance of a contacting segment of a commutator over the duration of contact.

Crosstalk—The change in amplitude on any channel due to the change in amplitude of one or more other reference channels. This value may be expressed in per cent or in decibels.

Decommutator—Device for separating, selecting, or distributing commutated signals. Generally the second of a commutator pair caused to operate in phase and frequency synchronism with the first.

Duty Cycle—Channel duty cycle is defined as the ratio of the channel on time and the channel period.

Dwell Period—Term used interchangeably with channel on time.

Edge Noise—Variations and uncertainty in the occurrence of the leading and trailing edges in time.

Efficiency of Conversion—The conversion efficiency refers to the gain or loss through the switch. It is defined as the ratio of the nominal incremental output amplitude to the corresponding modulating signal amplitude.

Frame—Interval between two samples of any channel of a normal time-division system. Generally one rotation of the commutator. Sometimes called a frame cycle.

Instantaneous Sampling—Sampling involving an aperture of very small duration compared with the frame period of a time-division process.

Interpolation—The process of reconstructing or recovering the original intelligence signal from its sampled forms.

Interval Sampling—Sampling involving an aperture of relatively long duration compared with the intelligence periods under observation.

Nonlinearity—Commutator nonlinearity is the maximum deviation from the best straight line determined by the method of least squares. This deviation applies to all channels taken together. If the individual channel linearity is to be separately considered the term channel linearity should be used.

Off Current in Source—This is defined as the current flowing through the source (back current) during the period in which its corresponding channel gate is turned off, that is, disconnected from the load. If this current flows in the same direction as that produced by a positive signal applied to the switch, it is called a positive current.

On Current in the Source—The on current refers to the current (back current) flowing during the time that the gate is turned on, that is, connected to the load.

Paralleling Channels—It is generally known that more than one contact of a particular contact plate may be connected together to provide two or more samples for each rotation of a mechanical switch. Similarly, in the electronic case it is possible to provide means for achieving more than one sample per frame. The technique for paralleling channels is dependent on the gating mechanism employed. In some instances it is possible to connect the inputs directly together as in the case of the mechanical switch. In other cases it is necessary to provide separate terminals in the logic circuitry for making appropriate selections for paralleling channels.

Pedestal—The pedestal is defined as a fixed level in the output signal with reference to some extreme signal value to which the channel amplitude is limited. In standard PAM signals for telemetering a pedestal is used for synchronizing ground decommutation equipment.

Sampling Principle—This principle or theorem states that at least two samples per cycle of intelligence are required to recover the original intelligence completely by minimum-bandwidth interpolation techniques.

Sampling Rate—The sampling rate is the number of times a particular channel is sampled in 1 sec. This generally corresponds to the frame rate, which is the number of times per second that all channels of the switch are sampled.

Stepping Commutator—A multicontact commutator in which the channel-to-channel switching is stepped in accordance with a control pulse.

Subcommutation—Further commutation of signals to be applied to a channel of the main or primary commutator. Sometimes called secondary commutation. Generally in synchronism with the primary commutator.

Supercommutation—Sometimes used to indicate more than one sample of a signal per commutator cycle. Often called "cross patching" or "paralleling."

Synch Selector—Circuitry for detecting and regenerating synchronization pulses which may be contained in a multiplex signal.

Timing Generator—The timing generator or clock is the basic timing device used to effect the switching operation. The period of the clock generally corresponds to the channel period.

Zero Signal Offset—The absolute value of the output level obtained with zero input signal. This may refer to a nominal or average value where all channels are considered together, the tolerance or peak-to-peak value being taken into account by the expression of equivalence.

BIBLIOGRAPHY FOR TELEMETRY COMMUTATION

Alpert, N., J. Luongo, and W. Wiener: 32 Channel High-speed Commutator, *Electronics*, vol. 23, pp. 94–97, November, 1950.
Backus, A. S.: Commutator Assembly, U.S. Patent 2,707,731.
Beaumont, J. O.: Electronic High Speed Multiplexing System, U.S. Patent 2,850,725.
Bolie, V. W.: Distributing Delay Line Using Non-linear Parameters, U.S. Patent 2,800,596.

Bowers, J. O., and W. L. Elden: The Application of Junction Type Field-effect Transistors to High-level Time Division Multiplexing, *Natl. Telemetering Conf. Record*, Houston, 1965.

Brinster, J. F., and E. Garretson: Specification and Design of Mechanical Sampling Devices Relative to Telemetering System Requirements, *IRE Proc. Natl. Symp. Telemetering*, 5.4, April, 1957.

Brinster, J. F.: "Electronic and Electromechanical Sampling Devices for Multichannel Instrumentation," General Devices, Inc., 1958.

Brinster, J. F.: "New Electromechanical and All Solid State Commutation Devices for Multichannel Telemetry," General Devices, Inc., 1963.

Brinster, J. F., and W. C. Johnson: Pulse Sequence Generator, U.S. Patent 2,861,202, 1958.

Brinster, J. F., and E. B. Garretson: Commutator Brush Assembly, U.S. Patent 2,948,795, 1960.

Brinster, J. F., and E. B. Garretson, Subminiature Multi-signal Mechanical Commutator, U.S. Patent 2,948,715, 1960.

Brinster, J. F., and W. C. Johnson: Multiple Output Sequential Signal Source, U.S. Patent 3,052,871, 1962.

Brinster, J. F.: Transistorized Motor Voltage Regulator, U.S. Patent 3,214,668, 1965.

Brinster, J. F., and W. C. Johnson: Multiple Signal Source, U.S. Patent 3,209,264, 1965.

Campbell, J. C.: Simplified Industrial Telemetering, Hayden Book Company, New York, 1965.

Chester, W. H., and W. P. Klemens: Decommutating Pulse Telemetry, *Electronics*, vol. 25, pp. 140–143, August, 1952.

Clark, W. L.: Automatic Frequency Phase Control of Television Sweep Circuits, *Proc. IRE*, vol. 37, pp. 497–500, May, 1949.

Coblenz, A., and H. L. Owens: Switching Circuits Using the Transistor, *Electronics*, vol. 26, p. 186–191, December, 1953.

Corry, J. D., and P. Garner: A Micro-PCM Encoder and Multiplexer, *Natl. Telemetering Conf. Record*, Houston, 1965.

Davis, W. R.: High Speed Mercury Jet Commutating Switch, 1955 *Natl. Telemetering Conf. Record*, pp. 105–111.

Den Hertog, M.: Electrical Communication Systems, U.S. Patent 2,761,903.

Den Hertog, M.: Group Selection Control Circuit, U.S. Patent 2,561,051.

Desch, J. R.: Electronic Counter, U.S. Patent 2,644,110.

DuBois, R. O., Jr.: Miniaturized Airborne Electronic Commutator, 1955 *Natl. Telemetering Conf. Record*, pp. 112–117.

Ely, W. B.: A Low Level Commutator with Field Effect Signal Gates, General Devices, Inc., 1965.

Electrical Contacts 1962, 1963, 1965, *Proc. Eng. Seminar Elec. Contacts*, University of Maine, College of Technology.

Field, O.S., et al.: Selective Signaling System, U.S. Patent 2,448,487.

Fink, D. G.: Synchronization in Color Television, *Electronics*, vol. 25, pp. 146–150, October, 1952.

Finkel, L.: A New Miniature Automatic Pulse Demultiplex Set, *IRE Proc. Natl. Symp. Telemetering*, 10.3, 1958.

Foss, R. C.: The Telemetry of Low Level D.C. Signals, *Proc. Intern. Telemetering Conf.*, London, 1963.

Foster, Le Roy E.: Telemetry Systems, John Wiley & Sons, Inc., 1965.

Francis, J. P.: A Low Level, High Speed Sampling System, *IRE Proc. Natl. Symp. Telemetering*, 1.3, April, 1957.

Gerring, F. H.: Commutating Switch Development for Critical Applications, *IRE Proc. Natl. Symp. Telemetering*, 5.3, April, 1957.

Gohorel, F. P.: Control or Signaling System, U.S. Patent 2,512,639.

Greig, D. D., J. J. Glauber, and S. Moskowitz: The Cyclophon: A Multipurpose Electronic Commutator Tube, *Proc. IRE*, vol. 35, pp. 1251–1257, 1947.

House, C. B., and R. L. Van Allen: Commutation and Nondestructive Read-out of Magnetic Memory Cores in Earth Satellite, *Natl. Telemetering Conf. Record*, 1-A-2, 1957.

Hunt, C. E., Jr.: Phasing Clutch for Facsimile Receivers, U.S. Patent 2,629,777.

Hussey, L. W.: Diode Gate, U.S. Patent 2,636,133.

Hellimer, A. J., and P. G. Pardey: The Design of Electro-mechanical Multiplexing Switches for Airborne Applications, *Proc. Intern. Telemetry Conf.*, London, 1963.

Inose, H., and M. Jakagi: A Method of Frame Separation in the Time-division-multiplexed Delta Modulation Telemetry, *Proc. Intern. Telemetry Conf.*, London, 1963.

Johnson, W. C., and J. F. Brinster: Multisignal Sampling Circuit, U.S. Patent 2,958,857, 1960.

Johnson, W. C.: Diode Switching Circuitry Using Constant Current Sources, U.S. Patent 3,126,488, 1964.

Kalbfell, D. C.: Low Level Magnetic Commutator, *IRE Proc. Natl. Symp. Telemetering*, 10.4, 1958.
Katz, L.: Determination of Visual Interpolation Errors in the Plotting of Curves from Commutated Data, *Trans. IRE*, TRC-1, pp. 15–24, February, 1955.
King, J. F., J. Elwell, and A. McCalmant: A Miniature Mechanical Ultra Noise Commutator, *Proc. Aerospace Instr. Symposium*, Dallas, April, 1961.
Koenig, W., Jr.: Signal Transmission System, U.S. Patent 2,629,779.
Kranzler, M. M.: Designer's Dilemma—Electronic or Mechanical Multiplexing, *IRE Proc. Natl. Symp. Telemetering*, 8.1, 1958.
Logan, C. A.: Miniature, Low Level Commutator, *IRE Proc. Natl. Symp. Telemetering*, 10.5, 1958.
Luhn, H. P.: Distributor and Method for Making the Same, U.S. Patent 2,649,513.
Mallets, E. S., R. E. Perkins, and H. W. P. Knapp: A Telemetry Data Processing Equipment, *Proc. Intern. Telemetry Conf.*, London, 1963.
Mallinckrodt, A. J., and R. M. Stewart: Data Smoothing Techniques—Sampled Data, *Rept.* 1026-P2, Mar. 21, 1955, Contract No. AF 04(611)-648.
Manley, J. C.: Electronic Counter, U.S. Patent 2,646,534.
Moody, W. N.: A Comparison of Low Level Commutators, *Natl. Telemetering Conf. Record*, Albuquerque, 1963.
Moss, H.: The Magnetron Beam Switching Tube, *Natl. Telemetering Conf. Record*, pp. 97–104, 1955.
Nichols, M. H., and J. F. Brinster: Multi-signal Transmission System, U.S. Patent 2,444,950.
Nichols, M. H., and L. L. Rauch: "Radio Telemetry," John Wiley & Sons, Inc., 1954.
North, M. E., and J. B. Crank: A High Level Magnetic Core Commutator, *IRE Proc. Natl. Symp. Telemetering*, 10.1, 1958.
Nyman, A.: Facsimile Framing System,U.S. Patent 2,632,810.
Perkins, L. C., et al.: Repeating Multi-pole Selector Relays, U.S. Patent 2,795,773.
Proc. Intern. Res. Symp. Elec. Contact Phenomena, University of Maine, College of Technology, November, 1961.
Phillips, J. L.: A Multiple-input Differential Amplifier—A Novel Approach to Multiplexing, *Natl. Telemetry Conf. Record*, Albuquerque, 1963.
Rack, A. J.: Electronic Switch, U.S. Patent 2,657,318.
Rauch, L. L.: Electronic Commutation for Telemetering, *Electronics*, vol. 20, pp. 114–120, February, 1947.
Rauch, L. L.: Electronic Commutation of Strain Gages for Telemetering, *Proc. Soc. Exptl. Stress Anal.*, yol. 5, pp. 111–121, 1947.
Rauch, L. L.: Multi-signal Transmission, U.S. Patent 2,445,840.
Reeves, A. H.: Telecommunication System, U.S. Patent 2,631,194.
Reynolds, F. N.: An Improved FM/FM Decommutator Ground Station, *Conf. Record IRE*, 1953 National Convention, pt. 1, pp. 73–76.
Richman, D.: Frame Synchronization for Color Television, *Electronics*, vol. 25, pp. 146–150, October, 1952.
Rochester, N.: Signal Transmission Network, U.S. Patent 2,570,716.
Rochester, N.: Crystal Matrix, U.S. Patent 2,476,066.
St. John, D. E.: High-speed Commutator, U.S. Patent 2,658,142.
Segerstrom, C. A.: Electronic Commutated Channel Separators, U.S. Patent 2,799,727.
Schoenwetter, H. K.: A New Approach to Solid-state Commutator Design, General Devices, Inc., 1965.
Shandelman, F., A. Hartung, and H. Golden: Low Level Commutation System for Telemetry Application, *IRE Proc. Natl. Symp. Telemetering*, 8.6, April, 1957.
Skellet, A. M.: Magnetically Focused Radial Beam Vacuum Tube, *Bell System Tech. J.*, vol. 23, pp. 190–202, April, 1944.
Skellet, A. M.: Electrostatically Focused Radial-beam Tube, *Proc. IRE*, vol. 36, pp. 1354–1357, 1948.
Slavin, P.: An Electronic Commutator, *IRE Proc. Natl. Symp. Telemetering*, 6.4, April, 1957.
Smith, F. D., and M. M. Kranzler: Data-scanning Semi-conductors or Metallic Conductors? *Natl. Telemetering Conf. Record*, Albuquerque, 1963.
Spielberg, A.: Parity Generator, U.S. Patent 2,674,727.
Stewart, R. M.: Statistical Design and Evaluation of Filters for the Restoration of Sampled Data, *Proc. IRE*, vol. 44, pp. 253–257, February, 1956.
Taylor, V. L.: Optimum PCM Synchronization, *Natl. Telemetering Conf. Record*, Houston, 1965.
Tolson, W. A.: Multiplexing Commutators, U.S. Patent 2,833,862.
Wendt, K. R., and G. L. Frendendall: Automatic Frequency and Phase Control of Synchronization in Television Receivers, *Proc. IRE*, vol. 31, pp. 7–15, January, 1943.

Wiener, N.: "The Extrapolation, Interpolation, and Smoothing of Stationary Time Series," John Wiley & Sons, Inc., New York, 1949.

Young, W. R., Jr.: Switching Network Using Diodes and Transformers, U.S. Patent 2,817,079.

RADIO LINKS

19 R-F Link System Design

General

In a radio-telemetry system, the r-f link design determines the range for a required accuracy of the transmitted data. Design of the link involves a series of trade-offs to derive system parameters that will permit satisfactory output signal-to-noise ratio at the desired range.

To limit the scope of this discussion it is assumed that the modulation method is selected for reasons other than range performance and therefore the required signal-to-noise ratio at the receiver output is fixed. Some modulation schemes such as FMFB (frequency modulation with feedback) and PCM exchange spectral occupancy for increased signal-to-noise ratio. These schemes may be resorted to when the r-f link becomes the limiting factor in system design.

The designer of an r-f link requires a complete knowledge of the total mission requirement and all the system constraints. In a typical limiting case, the receiving-station hardware is defined, modulation method has been selected, vehicle antennas exist, and all that remains is choosing an adequate transmitter output power. At the other end of the spectrum are design problems in which all the parameters are open to choice and the only limitations are natural effects and the state of the art in hardware development. In the past very few years, the approach to r-f link design has changed from the use of simple considerations with large margins for safety[1] to rather exact calculations involving radio astronomy, physics of the atmosphere, precise orbit predictions, etc.[2] The introduction of very low noise receivers and the need for efficient design of space probes are responsible.

Receiver Power

The objective of r-f link design is to have a satisfactory probability of an adequate-strength signal at the receiver so that when receiver noise and external noise are added, the signal-to-noise ratio at the receiver output permits the required data accuracy.

The free-space path attenuation, or range-squared attenuation, with loss correction determines the effective range of a telemetry system as shown in Eq. (15), which describes the received power as a function of system parameters:

$$P_r = G_t P_t A_r / 4\pi R^2 L \qquad (15)$$

where P_r = receiver available power, watts
 G_t = transmitting-antenna gain
 P_t = transmitter power, watts
 A_r = receiving-antenna effective capture area, length-squared units
 R = range, length units used for A_r
 L = losses (power ratio) absorption, etc.

It is to be noted that Eq. (15) is independent of frequency. In the practical sense, this is not true since the path attenuation is dependent on frequency and the effective receiving-antenna aperture is a function of frequency due to mechanical limitations. Atmospheric attenuation starts to become a factor at frequencies of 3 gc and above, as discussed in Sec. 20.

Anomalous propagation conditions due to variations in the refractive index of air at varying altitudes result in fades or ducting. These effects constitute a departure from free-space propagation conditions. Their effect is definitely dependent upon the choice of carrier frequency.

Multipath effects (the combining of radio-frequency energies received from the same source over different paths) is frequency-dependent since the coefficient of reflection of earth surfaces and atmospheric strata is a function of carrier frequency and multipath effects are dependent upon the path difference in wavelengths of the carrier frequency.

Receiver Sensitivity

The sensitivity of a receiver will be defined as that input power level necessary to produce a required signal-to-noise ratio in the noise-limiting r-f bandwidth. This power level is dependent upon the noise introduced into the antenna, losses in transmission lines, and the antenna and the receiver noise.

With the introduction of receivers generating less noise than that contributed externally, the concept of effective receiver temperature T_e has become the accepted way of establishing receiver sensitivity. Noise temperature is related to power by the Boltzmann equation

$$N = kT_eB \qquad (16)$$

where N = noise power, watts
k = 1.38 × 10⁻²³ joules/°K
T_e = effective receiver noise temperature, °K
B = noise bandwidth, cps

The noise bandwidth of a receiver is defined here as

$$B = \int_0^\infty A(f)\,df \qquad (17)$$

where $A(f)$ = frequency-dependent gain

If the gain is normalized and taken as unity at midband the above expression gives the equivalent rectangular passband of a receiver. It must be noted that this bandwidth is not necessarily that seen by the signal information components. It is the predetection bandwidth (the intermediate-frequency bandwidth in superheterodyne receivers) and for calculation of the postdetection signal-to-noise ratio must be modified by the detector characteristics and the postdetection bandwidth.

Receiver sensitivity is

$$\text{Sensitivity} = N(C/N) \qquad (18)$$

where C/N = carrier-to-noise power ratio in bandwidth B for the required output S/N ratio
$N = kT_eB$

It remains to determine T_e for the receiving system. Figure 88 shows that the noise temperature is the combination of the antenna temperature T_A, temperature due to losses T_L, and the noise temperature of the receiver referred to its input T_R.

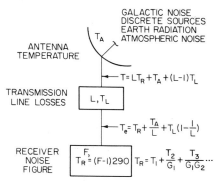

FIG. 88. Receiving-system noise temperatures.

$$T_e = T_R + T_A/L + T_L(1 - 1/L) \qquad (19)$$

where T_A = effective antenna temperature considering all external noise contributions (assuming no ohmic losses in the antenna)
T_R = receiver noise temperature referred to its input
T_L = temperature of losses between the antenna and the receiver (normally 290°K)
L = power-loss ratio between antenna and receiver

The antenna temperature may be expressed as the temperature of a resistor equal to the radiation resistance which produces a noise power equal to the received noise power. In the case of directional antennas with noise contributions of various intensities in the main lobe, side lobes, etc., the effective antenna temperature is

$$T_A = (1/4\pi) \oint G(\theta,\phi) T(\theta,\phi) \, d\omega \qquad (20)$$

where $G(\theta,\phi)$ = directional antenna gain
$T(\theta,\phi)$ = temperature in a given direction
$d\omega$ = an increment of solid angle

This three-dimensional integral is generally replaced by a summation of temperature contributions from a number of small solid angle regions depending upon the antenna pattern and the distribution of noise sources.[3,4]

The noise temperature of a single-response receiver referred to its input is[5]

$$T_R = (F - 1)290 \qquad (21)$$

where F = noise figure (power ratio)

Choice of Frequency

Equation (15) shows that the power delivered to a receiver is frequency-independent provided a fixed gain antenna is used at one end of the link and a fixed-aperture antenna at the other.

Practically all the significant parameters in determination of range are influenced by the selection of link frequency. The variation of link parameters with frequency is discussed in Sec. 20.

General-purpose telemetry at goverment test ranges is limited to the frequency assignments of document IRIG 106-60 dated June, 1962 (see Chap. 5). These assignments are abstracted in Table 9.

Table 9. United States Military Telemetry Frequencies

Frequency band	Channel spacing	Transmitter frequency tolerance, %	Receiver frequency tolerance, %
216.5–224.5 (noninterference)	500 kc	±0.01	±0.005
225–259.7 (protected)	500 kc min (44 channels)	±0.01	±0.005
1,435–1,435 Mc (manned aircraft)	1 Mc	±0.005	±0.001
1,486–1,535 Mc (missiles and space)	1 Mc	±0.005	±0.001
2,200–2,300 Mc	1 Mc	±0.005	±0.001

In satellite and space-probe systems, frequencies assigned to space communications by the International Telecommunication Union (Chap. 5) are generally used. The 1963 list of frequency bands is given in Table 10.

Table 10. International Space-communications Frequencies, Megacycles

30.005– 30.010	5,925– 6,425
136 – 137	7,250– 7,300
143.6 – 143.65	7,300– 7,750
267 – 273	7,900– 7,975
401 – 402	7,975– 8,025
1,427 –1,429	8,025– 8,500
1,525 –1,540	15,762–15,768
1,700 –1,710	18,030–18,036
2,290 –2,300	31,000–31,300
3,400 –4,200	31,500–31,800
5,250 –5,255	31,800–32,300
5,670 –5,725	34,200–35,200

20 Link Parameters

Level Diagram

The interrelation of link parameters can be expressed in tabular form, but use of a level diagram permits easy visualization and helps prevent omissions or errors of sign during link calculations. Figure 89 is a convenient form. The abcissa is plotted in decibels relative to 1 mw, but any relative power scale is satisfactory.

Reading from left to right, the first plateau is the output power of the transmitter in dbm. Step 1 represents the transmission-line loss between the transmitter and the

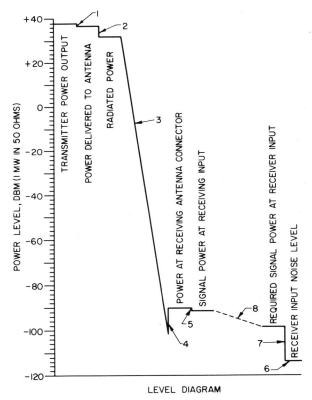

FIG. 89. Link power-level diagram.

antenna. Step 2 represents the transmitting-antenna gain which may, of course, be either positive or negative. In the example, a loss is shown to account for unfavorable aspects. Step 3 represents the transmission-path loss between isotropic antennas, and step 4 receiving antenna gain. Step 5 represents transmission-line loss between the receiving antenna and the actual receiver. The following plateau then indicates the power available at the input to the receiver.

Starting from the right-hand corner of the level diagram, the plateau at 6 represents the receiver noise level in dbm, which is a function of the receiver noise bandwidth and the effective noise temperature T_e.

Step 7 represents the signal-to-noise required at the input to the receiver for a specified and required signal-to-noise ratio in the output channel. The next plateau then

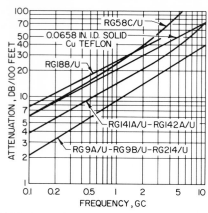

Fig. 90. Vehicle transmitter to antenna r-f transmission-line loss.

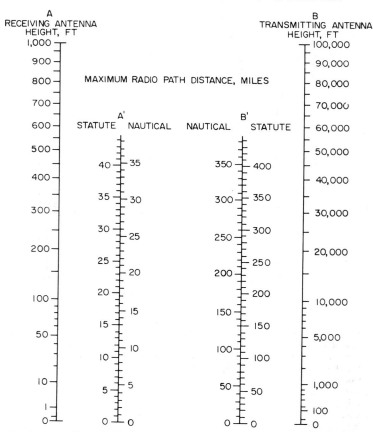

Fig. 91. Maximum radio path distances, air-to-ground links. Use A and A' scales with 0 of B scale as pivot. Use B and B' scales with 0 of A scale as pivot. Add reading of A' and B' scales.

indicates the required signal input level to the receiver. Step 8 represents the margin or safety factor required to account for unfavorable propagation conditions or degradation in system performance. A level diagram of the type shown in Fig. 89 may be entered at any point and is particularly useful in system calculations since its use requires that all aspects of the problem be considered.

Transmission-line Loss

Since the configuration of most vehicle telemetry installations requires flexible coaxial cable between the transmitter and antenna, the power loss in the cable must be considered. Figure 90 shows line loss as a function of frequency for various common cables.

Vehicle Antenna Gain

Section 21 treats vehicle antennas in detail. Generally, omnidirectional coverage is the goal for unstabilized vehicles; if this were achieved, the antenna would be isotropic and the gain would be 0 db. In the practical case, the vehicle antenna will exhibit some directivity and the relative gain will be greater in some directions with nulls at others. Gain is best specified as the lowest which is likely to be encountered due to changes in aspect of the test vehicle. In practice, this quantity may vary widely, say from plus 10 db for a positioned directive antenna to minus 20 db for some aspects of a flush-mounted antenna.

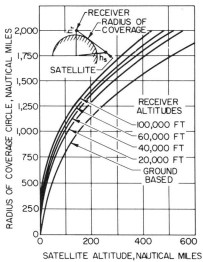

FIG. 92. Maximum line-of-sight distances, satellite to ground.

Path Loss

While transhorizon (scatter) propagation or ionospheric reflection propagation may be used for radio telemetry, most systems are designed so that the transmission path lies in the interference region; that is, the transmitter and receiver are in "radio line of sight." For air-to-ground links, Fig. 91 shows the effective distance to the radio horizon considering a smooth earth. The approximate equation solved by the nomogram is

$$d = (2h)^{1/2} \tag{22}$$

where d = distance, statute miles
h = antenna height, ft

Figure 92 shows the same information for satellite links. The graph is a plot of the equation

$$R_c = r \left(\cos^{-1} \frac{r}{r + h_s} + \cos^{-1} \frac{r}{r + h_r} \right) \tag{23}$$

where R_c = radius of coverage circle
r = radius of earth
h_s = satellite altitude
h_r = receiver altitude

Because of atmospheric refraction, the actual distance to the radio horizon is slightly longer than predicted by geometry, but this effect is usually negligible.

Provided the geometry of the path is correct, the free-space path loss may be taken from the nomograms of Figs. 93, 94, and 95 depending upon the range. These nomograms solve the equation for free-space path loss in decibels:

$$10 \log (P_t/P_r) = 36.58 + 20 \log f + 20 \log d \qquad (24)$$

where P_t = transmitted power
P_r = received power
f = frequency, Mc
d = distance, statute miles

The path attenuation is increased over the free-space loss by various absorption contributions. Additionally, cross-polarization effects may be considered as part of

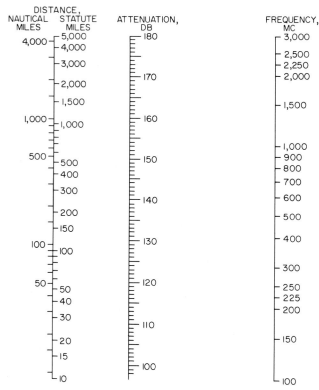

Fig. 93. Nomogram for determination of free-space path attenuation between isotropic antennas, short range.

the path loss. If both transmitting and receiving antennas are linearly polarized, the loss in decibels due to misalignment of the polarization by $\theta°$ in decibels is

$$L_{db} = 20 \log \cos \theta \qquad (25)$$

This loss is 3 db when θ is 45° and is very significant when θ is in the vicinity of 90°. For relay links within the atmosphere, linear polarization at both ends may be used.

Experimental vehicle testing, however, generally requires link operation at any vehicle attitude so that common practice is to use a linearly polarized transmitting antenna and a circularly polarized receiving antenna. For a given receiving-antenna aperature, this results in a 3-db added path loss. For space-to-ground links, the polarization of an electromagnetic wave passing through the ionosphere is changed by the effect known as Faraday rotation. A linearly polarized wave is split into two elliptically polarized waves traveling at different velocities through the magnetoionic path so that the resultant polarization at the receiving antenna is effectively rotated

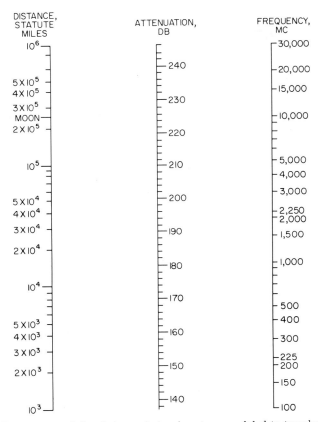

FIG. 94. Free-space path loss between isotropic antennas, global to translunar ranges.

from that transmitted. The magnitude of the rotation is plotted as a function of frequency and elevation angle in Fig. 96. For frequencies above 100 Mc, ionospheric attenuation is negligible so the remaining sources of attenuation are water in the atmosphere and attenuation due to the atmosphere itself. The attenuation per mile vs. frequency for various atmospheric conditions is plotted in Figs. 97 and 98. The data are plotted from the best available sources, but for critical applications the exact situation should be considered. References 6 and 7 give a good account of available information and its limitations. For specific applications, attenuation due to plasma sheaths in reentry and losses occurring at missile-staging times must be included. Much of the source material is classified but Refs. 12, 13, and 14 provide useful design information.

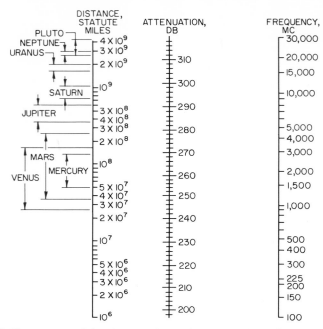

Fig. 95. Free-space path loss between isotropic antennas, interplanetary ranges.

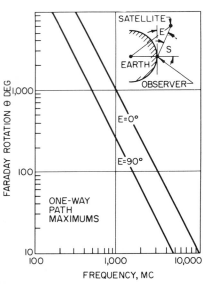

Fig. 96. Faraday rotation vs. frequency and elevation angle.

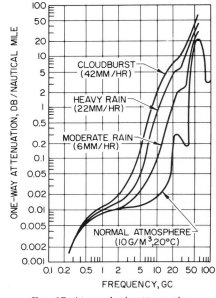

Fig. 97. Atmospheric attenuation.

RADIO LINKS 4-101

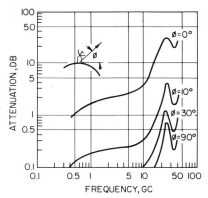

FIG. 98. Total one-way attenuation through atmosphere. (*After Hogg and Mumford Ref. 7.*)

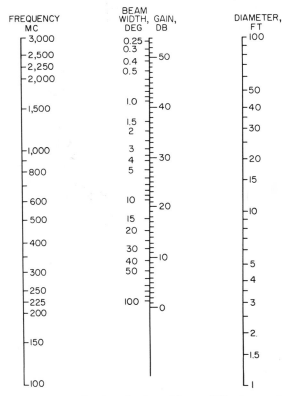

FIG. 99. Nomogram for determination of gain and beamwidth of a parabolic antenna.

Ground-antenna Gain

To be consistent with the other information in this section, receiving-antenna gain must be plotted in decibels relative to a linearly polarized isotropic antenna. Figure 99 is a nomogram for the determination of the gain of a parabolic antenna, with a

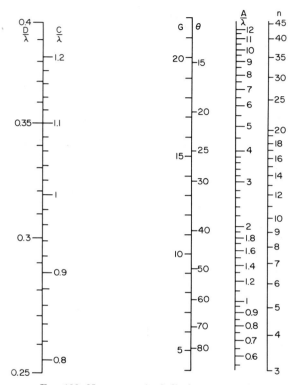

FIG. 100. Nomogram for helical antenna gain.

circularly polarized feed, as a function of frequency and physical size. The nomogram solves the equation

$$G = 20 \log f + 20 \log D - 49.6 \tag{26}$$

where G = gain, db, relative to a linearly polarized isotropic vehicle antenna (circular feed in parabola)
f = frequency, Mc
D = diameter, ft

An illumination factor of 0.54 is assumed. With matching polarization the gain would be 3 db higher. Figure 100 is a nomogram for determination of the gain of a helical antenna radiating in the axial mode based on the equations of Kraus.[8,9] The equation and symbols used in its construction are

$$G = 8.76 + 10 \log [(C/\lambda)^2 nS/\lambda] \tag{27}$$

where G = gain, db, relative to a linearly polarized isotropic antenna
 C = helix circumference
 λ = free-space wavelength at operating frequency in same units as measurements
 n = number of turns
 S = spacing between turns = C tan pitch angle
 A = axial length — nS
 D = helix diameter
 θ = half-power beamwidth, deg

The equation holds for a pitch angle of 12.5°, a ground plane diameter of at least 0.8λ, and conductor diameters of approximately 0.02λ. These two antenna types, the helical and parabolic, are representative of the more popular telemetering receiving

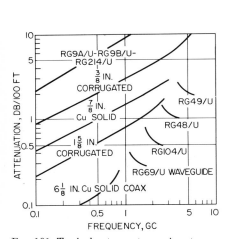

Fig. 101. Typical antenna-to-receiver transmissions-line loss.

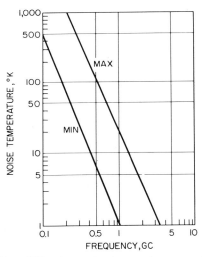

Fig. 102. Approximate maximum and minimum sky temperatures as function of frequency.

antennas. In most link-design problems, the actual antenna gain is known, but caution must be exercised in accounting for the reference to which the gain is quoted and in considering the polarization problem. If polarization diversity combining or polarization tracking is to be employed, the gain figure should be increased by the equipment performance figure.

Receiver Transmission-line Loss

The first receiver amplification stage is rarely mounted immediately adjacent to the exciting element or feed of a receiving antenna; so carrier attenuation due to transmission-line losses must be accounted for. Because weight and flexibility are not usually problems, solid coaxial lines or waveguide are generally used. Figure 101 plots the attenuation of a few of such transmission lines vs. frequency.

Receiver Noise

The noise level at the input to the receiver is found by adding the various noise-temperature contributions and converting to the equivalent noise power as shown by Eq. (16). The antenna noise T_A is the summation of galactic noise, noise from discrete sources (sun, hot stars, etc.), atmospheric noise, and earth radiation. Figure 102 plots the maximum and average noise temperatures vs. frequency for galactic noise. While this may suffice for some applications, critical designs require the plotting of a vehicle path on radio maps of the sky background such as those of Figs. 103 and 104. Noise inputs due to discrete sources such as the sun, moon, and hot stars must occasionally be considered when very narrow beamwidth antennas are used.

Figure 105 is a plot of antenna temperature due to the quiet sun when a $\frac{1}{2}°$ beamwidth antenna is pointed at the solar disk. The figure is a plot of Eq. (10) of Ref. 7. For antennas pointed at the sun having beamwidths greater than $\frac{1}{2}°$, a usable approximation of antenna temperature due to the sun T_{as} is

$$T_{as} = T_{sun}(0.5/\theta)^2 \tag{28}$$

where θ = antenna 3-db beamwidth, deg

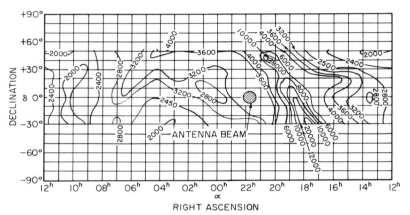

64-MC MAP (AFTER HEY, PARSONS AND PHILIPS)

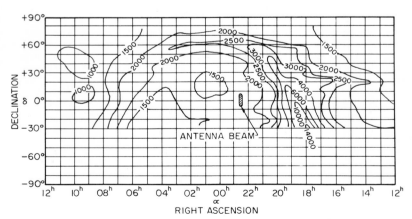

81-MC MAP (AFTER BALDWIN)

Fig. 103. Radio maps of the sky background—64, 81, 100, 160 Mc. The contours give the *Distribution of Cosmic Radio Background Radiation, Proc. IRE, vol. 46, pp. 208–215,*

For the more general case, noise contribution from either the main beam or side lobes may be determined from

$$T_{as} = T_{sun}[(0.5)^2/42{,}000]\, G \tag{29}$$

where G = antenna gain in the portion of the pattern pointed at the sun (power ratio to linearly polarized isotropic antenna)

The moon also subtends an angle of approximately $\frac{1}{2}°$; so Eqs. (28) and (29) may be used with a moon temperature of 240°K maximum (frequency-independent) to calculate its noise contribution.

Figure 106 by Millman[10] plots flux density vs. frequency for some intense point sources. The contribution of each point T_{ap} may be evaluated from

$$T_{ap} = S\lambda^2 G/8\pi K \tag{30}$$

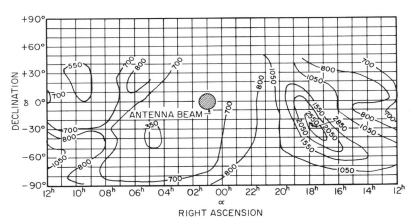

100-MC MAP (AFTER BOLTON AND WESTFOLD)

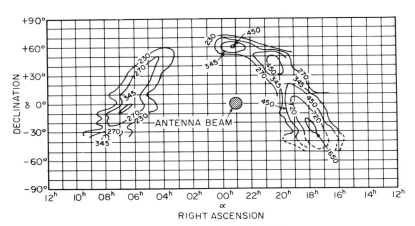

160-MC MAP (AFTER REBER)

absolute brightness temperature of the radio sky in degrees Kelvin. (*From H. C. Ko, The January, 1958.*)

where S = flux density, watts/sqm/cps
λ = wavelength, m
K = Boltzmann's constant 1.38×10^{-23} joules/°K
G = gain in direction of source for linearly polarized antennas (power ratio)

For perfect polarization diversity, the number obtained from Eq. (30) must be doubled since the flux density figures are based on random polarization.

The sky noise temperature due to oxygen and water vapor as given by Hogg and Mumford[7] is plotted in Fig. 107.

The contribution due to earth radiation into side lobes and back lobes of a receiving antenna is the most difficult to evaluate. Forward and Richey[4] present rather simplified sample calculations while Giddis[11] presents calculations involving computer simulation. For rough system calculations, the curves of Giddis (Fig. 108) are most useful. They plot measured antenna temperatures as a function of elevation angle for various antennas. The contributions due to receiver noise figure and line losses have been subtracted. Extrapolation from these curves to the case under consideration

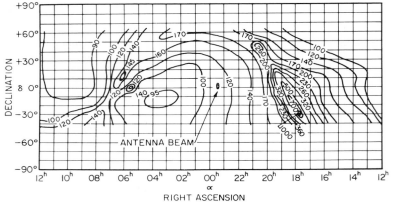

250-MC MAP (AFTER KO AND KRAUS)

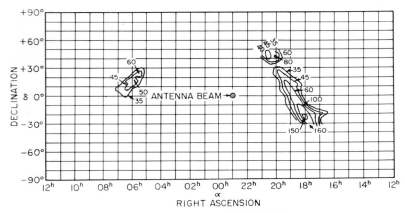

480-MC MAP (AFTER REBER)

Fig. 104. Radio maps of the sky background—250, 480, 600, and 910 Mc. The contours *Ko, The Distribution of Cosmic Radio Background Radiation, Proc. IRE, vol. 46, pp. 208–215,*

RADIO LINKS   4-107

can be used either as the basis for a system calculation or as check on figures obtained by summation as in Eq. (20).

In planning r-f links, the receiver noise bandwidth is given by Eq. (17) and the passband required is determined by the signal spectral occupancy, transmitter and receiver frequency tolerances, instability, and Doppler shift. For systems using receivers with automatic frequency control (AFC) the last two factors are compensated for, but the general case requires their consideration. The transmitter-receiver frequency uncertainty is the sum of their peak uncertainties and for nonrelativistic velocities the Doppler shift is approximately

$$f_d = (V/C) f_t \tag{31}$$

where f_d = Doppler frequency
      V = relative radial velocity
      C = velocity of light
      f_t = transmitted frequency

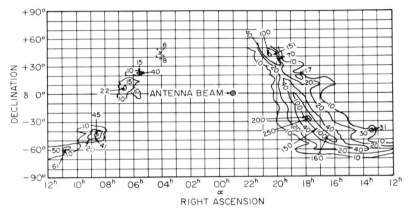

600-MC MAP (AFTER PIDDINGTON AND TRENT)

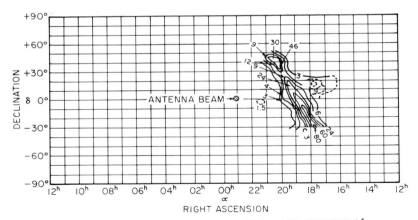

910-MC MAP (AFTER DENISSE, LEROUX AND STEINBERG)

give the absolute brightness temperature of the radio sky in degrees Kelvin. (*From H. C. January, 1958.*)

For the special case in which the transmitter is always moving away (space probe, missile tracked from pad, etc.) Eq. (31) is correct, but for airplanes and satellites the shift may be twice that given by (31). For the circular-orbit satellite case normalized Doppler shift vs. elevation angle above the horizon is plotted in Fig. 109. It must be noted that the shift given is for either an approaching or receding satellite, and generally twice the value at the lowest elevation angle of interest should be used.

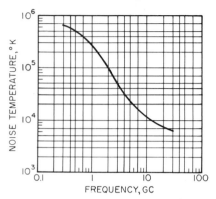

Fig. 105. Noise temperature of "quiet" sun for ½ beam-width antenna.

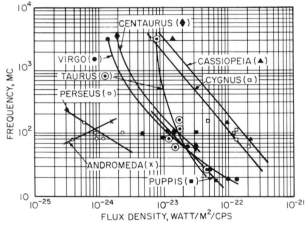

Fig. 106. Flux density vs. frequency for intense celestial radio noise emitters. (*Courtesy of G. H. Millman, Ref. 10.*)

An example of the calculation of receiver noise level using Eq. (16) is given here for a receiver bandwidth of 1 Mc and a T_e of 290°K. The factor $K = -198.6$ dbm/cps, $T_e = +24.6$ db, and 1 Mc is 60 db above 1 cps; so

$$N = KT_eB = 198.6 + 24.6 + 60 = -114 \text{ dbm}$$

The bandwidth B of 1 Mc includes all the frequency allowances just mentioned, including Doppler, and T_e all the noise-temperature contributions.

For some system calculations, receiver noise level may be obtained from Fig. 110 directly. The nomogram assumes an antenna temperature of 290°K, and for noise figures above 3 db and frequencies above 200 Mc the results will be resonably accurate and safe. This simplified approach is generally adequate for air-to-ground links.

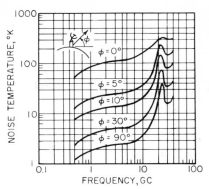

FIG. 107. Sky noise temperature due to oxygen and water vapor. (*After Hogg and Mumford, Ref. 7.*)

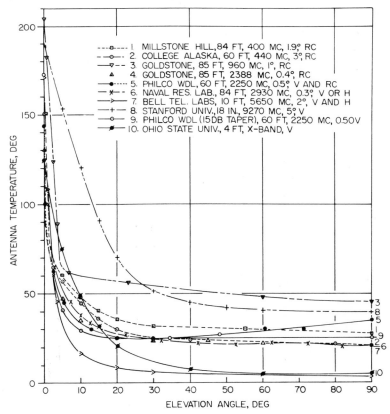

FIG. 108. Measured antenna temperatures as function of elevation angle for various large antenna installations. (*Courtesy of Albert R. Giddis, Ref. 11.*)

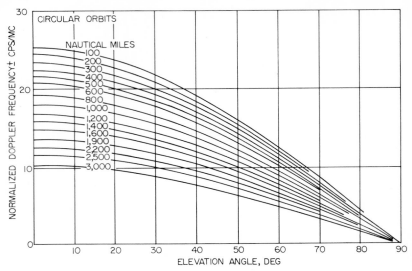

Fig. 109. Normalized Doppler shift vs. elevation angle above horizon.

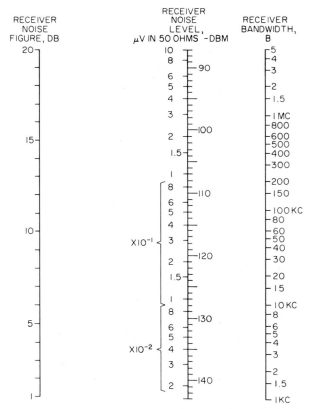

Fig. 110. Nomogram for determination of receiver noise level.

REFERENCES

1. H. Scharla-Nielsen, Radio Frequency Link Design for Telemetering, *1956 Natl. Telemetering Conf. Record.*
2. R. C. Hansen and R. G. Stephenson, Communications at Megamile Ranges, *J. Brit. IRE*, vol. 22, pp. 329–345, October, 1961.
3. M. L. Livingston, The Effect of Antenna Characteristics on Antenna Noise Temperature and System SNR, *IRE Trans. Space Electron. Telemetry*, vol. SET-7, pp. 71–79, September, 1961.
4. R. L. Forward and F. Richey, Effects of External Noise on Radar Performance, *Microwave J.* vol. 3, pp. 73–80, December, 1960.
5. IRE Standards on Electron Tubes: Definition of Terms, 1962 (62 IRE 7.S2).
6. E. S. Rosenblum, Atmospheric Absorption of 10–400 KMPS Radiation, Summary and Bibliography to 1961, *Microwave J.* vol. 4, pp. 91–96, March, 1961.
7. D. C. Hogg and W. W. Mumford, The Effective Noise Temperature of the Sky, *Microwave J.* vol. 3, pp. 80–84, March, 1960.
8. J. D. Kraus, Helical Beam Antenna Design Techniques, *Communications*, vol. 29, p. 6, September, 1949.
9. H. Scharla-Nielsen and P. H. Moore, Helical Antenna Chart, *Electronics*, vol. 33, p. 180, Mar. 11, 1960.
10. G. H. Millman, Cosmic Noise Limits Long Range Radar, *Space/Aeronautics*, January, 1961, pp. 124–129.
11. A. R. Giddis, The Influence of Natural Noise upon Antenna System Performance, *Trans. IEEE PGCS*, March, 1964, pp. 148–158.
12. W. C. Taylor, Analysis and Prediction of Radio Signal Interference Effects Due to Ionized Layer around a Reentry Vehicle, *IRE Proc. 1959 Natl. Symp. Space Electronics Telemetry*, paper 3.2, Sept. 28–30, 1959.
13. W. H. Drake and F. S. Howell, Radio Frequency Propagation to and from ICBM's and IRBM's, *IRE Proc. 1959 Natl. Symp. Space Electronics Telemetry*, paper 3.3, Sept. 28–30, 1959.
14. D. A. Jackman, C. R. Mullin, and R. H. Levy, The Reentry Plasma Sheath, *Space/Aeronautics*, vol. 41, p. 53, May, 1964.

VEHICULAR ANTENNAS

21 Systems Considerations for Antenna Design

The function of a vehicle antenna in a telemetry or remote-control r-f link is to provide adequate pattern coverage so that continuous reliable communications are maintained with the vehicle. Some of the major design considerations for vehicle antennas are:

1. Relative spatial orientation of the vehicle to the antenna at the other end of the r-f link
2. Size and shape of the vehicle and its relative size in terms of wavelengths
3. Environment in which the antenna must function (e.g., outer space, Mach 2 aircraft, etc.)
4. Allowable size and weight
5. Assigned frequencies
6. Mission profile of the vehicle

In this section a vehicle refers primarily to missiles, rockets, satellites, manned aircraft, and drones.

Many vehicles, such as tactical missiles, do not include telemetry as part of their primary mission profile. The telemetry is added for the test phases only, and it is obvious that neither the vehicle design nor the flight profile will be necessarily optimized for the telemetry. It is particularly critical, however, that the telemetry and remote control function at the extremes of orientation or even at orientations not contemplated for the test. These considerations typically point to vehicle-antenna systems that ideally provide isotropic or omnidirectional pattern coverage and no polarization restrictions.

Frequency allocations for telemetry and remote control are based primarily on availability and test-instrumentation standardization and not on considerations of optimum antenna performance.

Because of the factors described above, a new set of antennas is commonly designed for each new vehicle. The design of vehicle antennas is usually a compromise between desired electrical performance and the mechanical configuration that will be compatible with the vehicle. An attempt is made to compensate for nonoptimum antenna performance by increasing the system margin in other areas. The elimination of pattern nulls to minimize the probability of drop-outs and subsequent loss of data or control is frequently the major design objective of vehicle antennas.

There are some vehicles wherein the fundamental mission is the recovery of scientific data which are telemetered to the ground. Considerably more freedom of design is permitted, in this instance, because the structure and flight profile are governed by considerations of data collection and recovery.

The following sections contain a survey of the various types of antennas that are in common use on vehicles and describes briefly their characteristics. Some of the problems peculiar to vehicular antennas are discussed. Emphasis is placed on the relatively simple low- or medium-gain vehicle antennas rather than the elaborate high-gain types. These latter systems resemble ground systems, and a complete description is beyond the scope of this discussion.

22 Discussion of Antenna Characteristics and Terms

Pattern

The radiation pattern of an antenna or antenna system describes the spatial distribution of radiated energy. Some of the terms frequently used to describe antenna patterns are:

Isotropic—Having the same properties in all directions
Omnidirectional (in one plane)—Having radiation in all directions
Unidirectional—Lobe in one hemisphere
Bidirectional—Symmetrical lobes in both hemispheres
Omnidirectional—Figure of revolution with principal lobe on or near plane perpendicular to axis of rotation
Pencil Beam—Narrow lobe essentially a figure of revolution
Fan Beam—Narrow lobe in one plane and wide lobe in orthogonal plane
Shaped Beam—Designed shape in particular plane such as cosecant pattern

Linearly polarized antenna patterns are usually measured in the E plane (plane parallel to the electric field vector) and the H plane (plane perpendicular to the plane of the electric field vector).

In some instances vertical plane and horizontal plane or elevation and azimuth planes are used as the reference for the patterns.

Patterns are frequently described in terms of their half-power beamwidth, the angle at which the directivity (or gain) is down 3 db from the maximum. Antenna patterns are usually plotted in terms of voltage, but they may also be in power or decibels and should be so labeled.

Antenna Gain

Antenna gain is related to directivity. The *directivity* of an antenna D may be defined as the ratio of the radiated power density in a specified direction to the average power density radiated in all directions. *Maximum directivity* occurs in the direction of maximum radiation.

Radiation efficiency factor k is the ratio of the power radiated to the power input at some reference point.

The *gain* (or power gain) G of an antenna in a specified direction is defined by the equation

$$G(\theta,\phi) = \frac{\text{radiated power density in }(\theta,\phi)\text{ direction}}{\text{radiated power density of a reference antenna with the same power input}}$$

Thus

$$G(\theta,\phi) = kD(\theta,\phi) \tag{32}$$

when an isotropic source with 100 per cent radiation efficiency is taken as a reference source. If the test antenna is assumed to have 100 per cent efficiency,

$$G(\theta,\phi) = D(\theta,\phi) \tag{33}$$

Gain is normally expressed as a power ratio and can be expressed in decibels; thus

$$G(\text{db}) = 10\log_{10} G$$

If direction is not specified, gain usually refers to *maximum gain*. When specifying the gain of an antenna one should define the reference and the polarization of the reference.

The maximum gain of a medium- or high-gain antenna can be approximated in terms of its beamwidth by the equation

$$G = 30{,}000/\theta_E\theta_H \tag{34}$$

where θ_E and θ_H = half-power beamwidths in orthogonal planes, deg.

Polarization

The *polarization* of an antenna refers to the direction of the electric field vector in the far field, or radiated field. Actually the far field of an antenna has a spherical phase front but over a small surface a large distance from the antenna, it may be considered a plane wave. A *linearly polarized* plane wave consists of an electric and magnetic traveling wave with their vectors at right angles to each other such that if the E vector were rotated into the H vector a right-hand screw would move in the direction of propagation. In mks units $E = 120\pi H$. Figure 111a and b illustrates a *horizontally polarized* and a *vertically polarized* plane wave traveling to the right. A wave is *circularly polarized* if, at a given point in space, the electric field vector maintains a constant length and rotates with time. If an observer faces the source of the wave (wave approaching) and the electric field vector rotates counterclockwise in the plane of the observer, this is known as *right-circular polarization* by the IRE Standards. A clockwise rotation wave approaching is *left-circular polarization*. A right-circular wave is shown in Fig. 111c. A right-hand helix transmits and receives right-circular polarization by the IRE definition. If the magnitudes of the two orthogonal components of a wave are unequal or if they are not in phase quadrature at a point in space, the polarization becomes *elliptical*. The projection of the terminus of the electric vector on a plane perpendicular to the axis of propagation describes an ellipse. The voltage ratio of the major axis to the minor axis is defined as the *axial ratio*.

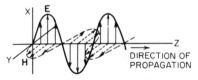
(a) VERTICALLY POLARIZED WAVE

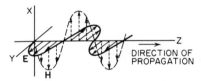
(b) HORIZONTALLY POLARIZED WAVE

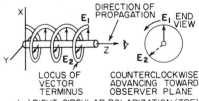
(c) RIGHT-CIRCULAR POLARIZATION (IRE)

FIG. 111. Types of polarized plane waves.

The loss associated with the transmission and reception of signals when the transmitting and receiving antennas do not have identical polarizations is shown in Table 11. Identical polarizations are used as a reference.

Table 11. Polarization Loss, Decibels

Transmit	Receive			
	Horizontal	Vertical	Right circular	Left circular
Horizontal...	0	$-\infty$	-3	-3
Vertical...	$-\infty$	0	-3	-3
Right circular...	-3	-3	0	$-\infty$
Left circular...	-3	-3	$-\infty$	0

It is obvious that opposite polarizations should be avoided to avoid nulls. There are two commonly used methods of reducing the probability of polarization drop-outs in an r-f link when the vehicle aspect is variable. One method is to make one of the antennas circularly polarized and the other linear. The second method is to employ polarization diversity on one antenna and select the larger of two oppositely polarized signals or combine the signals to obtain best signal-to-noise ratio.

Bandwidth

The bandwidth of an antenna is difficult to define in simple terms. For a particular application the usable bandwidth of an antenna may be limited by impedance characteristics; i.e., the VSWR exceeds the limits into which the terminal equipment can operate satisfactorily, or the patterns may deteriorate until acceptable pattern coverage is no longer obtained. In the case of circularly polarized antennas the bandwidth may be limited by the permissible axial ratio of the system. In a system specification the bandwidth usually refers to the frequency band over which the antenna must meet all specifications.

In addition to the dependence upon various characteristics the definition of bandwidth is relative. For example, an antenna used with a high-power transmitter may require a VSWR of less than 1.3:1 but the same antenna would be satisfactory with a receiver if the VSWR were less than 3:1.

A new class of antennas known as frequency-independent antennas have recently made their appearance. These include the equiangular spirals and the log-periodic antennas.

23 Effect of Vehicle on Antenna Patterns

Most vehicles employing telemetry antennas are metallic (good conductors) and irregular in shape. They may contain wings, fins, paddlewheel solar cells, etc. In many instances these irregularities impose boundary conditions that greatly distort the antenna pattern coverage from what would be obtained if the antenna were in free space or mounted on an infinite flat ground plane. As a result, the actual antenna pattern of the antenna mounted on a vehicle is frequently difficult, if not impossible, to calculate with any degree of accuracy, and pattern measurement must be made. If the vehicle is large in terms of wavelengths severe lobing is common with "holes" developing in the pattern coverage. Figure 112 illustrates the pattern distortion of a simple stub antenna pattern when the stub is mounted on the underside of an aircraft.

AZIMUTH　　ELEVATION

FIG. 112. Typical voltage patterns of a 400-Mc stub on bottom of fuselage.

Even the pattern and impedance characteristics of a stub on a flat circular ground plane vary considerably as a function of the diameter of the ground plane.[1] Figure 113 illustrates a typical pattern variation as a function of ground-plane diameter.

In many vehicle-antenna applications, it is desirable to provide antenna pattern coverage near a ground plane that is large compared with the antenna dimensions and height of the antenna above the ground plane. Also, it is frequently desirable to provide horizontal and/or vertical and/or circular polarization coverage. Since Maxwell's equations and the boundary conditions must apply, there are certain natural limitations to the pattern coverage that can be realized in practice. It is the purpose of this section to point out some of these limitations and indicate a method of calculating the performance of some types of antenna systems.

Assume an isotropic point-source antenna is located above an infinite ground plane. The far field patterns can be calculated in the region above the ground

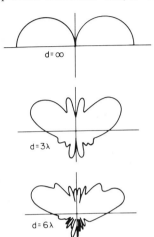
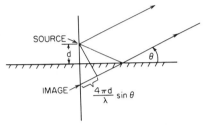

Fig. 113. Voltage patterns of a stub in a circular ground plane showing effect of ground-plane size.

Fig. 114. Geometry for far-field calculations.

plane by calculating the field produced by the source and its image. If the pattern of the source is only omnidirectional in one plane and symmetrical about a plane through the source parallel to the ground plane, the pattern can be calculated by multiplying the element pattern by the array pattern of the isotropic source and its image.

Assume the isotropic source is a distance d above the ground plane as shown in Fig. 114. The group pattern is given as follows:

For horizontal polarization

$$E_H = \sin\left[(2\pi d/\lambda) \sin \theta\right] \tag{35}$$

For vertical polarization

$$E_V = \cos\left[(2\pi d/\lambda) \sin \theta\right] \tag{36}$$

where E_H and E_V are normalized and d and λ are in the same units.

These curves are plotted for $d = \lambda/4$, $\lambda/2$, and $3\lambda/4$ in Fig. 115. Figure 116 illustrates the reduction in gain in decibels as a function of angle.

Several conclusions are obvious from these curves:

With horizontal polarization

1. It is impossible to obtain good horizontal polarization coverage at very small angles (near the surface) because of the shorting effect of the conducting boundary.
2. The more removed the source is from the ground plane, the better is the pattern coverage at very small angles. This coverage is obtained, however, at the expense of lobing (deep nulls) at greater angles.

With vertical polarization

1. Maximum coverage is at the horizon. (It should be pointed out that when the ground plane is not infinite this is not true and the field at small angles becomes a rather complicated function of ground-plane size and shape.)
2. As the source is removed from the ground plane, the small-angle lobe becomes more narrow and lobing occurs at higher angles.

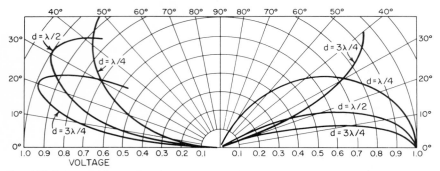

FIG. 115. Pattern of isotropic point sources over infinite ground plane. Horizontal polarization: $E_H = \sin\left(\frac{2\pi d}{\lambda} \sin \theta\right)$; vertical polarization: $E_V = \cos\left(\frac{2\pi d}{\lambda} \sin \theta\right)$; d = height above ground plane.

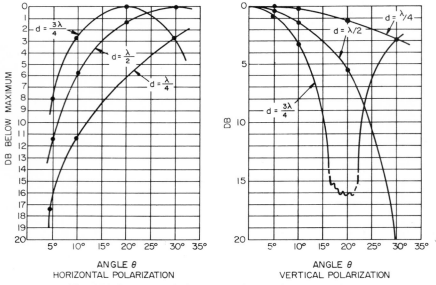

FIG. 116. Pattern variations caused by infinite ground plane.

Suppose optimum omnidirectional pattern coverage is desired between the ground plane and 30° with both polarizations. This might represent a microwave antenna in the bottom of an aircraft. Let the elements have doughnut-shaped patterns that provide maximum radiation in the region of plus or minus 30°. It is apparent from the curves that the optimum height for the element phase center is not the same for

Table 12. Characteristics of Specific Antenna Types

Antenna type	Pattern				Polarization			Bandwidth				Approx max gain, db, with respect to an isotropic source with same polarization	Comments
	Iso-tropic	Uni-direc-tional	Bi-direc-tional	Omni-direc-tional (in one plane)	Vari-able	Lin-ear	Cir-cular	Vari-able	1.1:1	1.3:1	2.0:1	10:1		
Whip, spike				x		x			x				2–4	Simple, lightweight
Blade, sleeve-stub				x		x			x				2–4	Rugged, streamlined
Isolated dipole				x		x				x			2	Simple
Turnstile (isolated)				x		x				x			0	Simple, lightweight
Swastika				x			x						2–4	
Loop-Vee				x			x						2–4	
Planar spiral (first mode)		x*	x				x				x		2–4 (4–8)*	Small protrusion
Planar spiral (second mode)				x			x					x	2–4*	Can be flush-mounted
Conical spiral (first mode)		x					x					x	4–8	Can be flush-mounted
Conical spiral (second mode)				x			x					x	2–4	
Scimitar					x	x		x						Rugged, streamlined
Log-periodic		x				x						x	6–12	Can be recessed and flush mounted
Axial-mode helix		x					x					x	6–15	
Slot antenna (broadband)		x				x					x		3–4	Flush-mounted
Annular slot			x			x					x		3–4	Flush-mounted
Poly rod	x	x				x	x	x		x			10–20	
Horn (pyramidal or conical)		x				x	x	x			x		6–15	Depends upon design

* With back cavity.

both polarizations. It is approximately $3\lambda/4$ for horizontal and $3\lambda/8$ for the vertical polarization. Other values produce nulls or poor coverage in the zone of interest. If the two sources were so phased as to produce circular polarization in the horizontal plane in the absence of the ground plane, the axial ratio would be degraded particularly at the edges of the zone of interest.

The above calculations and examples were for a single frequency. It is obvious that variations in axial ratio and pattern coverage at the edge of the zone will occur if the frequency is varied over a wide range.

If space and size are no problem, one can go to highly directive antenna systems with large protrusion or very large surface areas to improve performance. A second solution is to use multiple antennas to fill the pattern nulls caused by vehicle shadowing. This procedure is frequently used in missiles and satellites by placing an array of three or four antennas around the periphery of the vehicle to provide omnidirectional coverage.

To achieve additional gain and avoid pattern nulls with multiple antennas with a two-way link, separate receivers can be used with each antenna. The receiver outputs are compared and the one with the maximum output is selected for reception. The transmitter can be switched to the antenna receiving the greatest signal, thus effectively increasing gain toward the other ground receiver and conserving power.

When the dimensions of the vehicle are comparable with a wavelength, it is sometimes possible to utilize a part of the structure such as a fin, wing, or stabilizer for the antenna. Tail-cap and wingtip antennas on airplanes are examples of this technique. Another approach is to feed one half of the structure against the other so that it behaves like a dipole. In most practical cases where the size of the vehicle is small in terms of wavelengths, the antenna can usually be matched in terms of impedance, it does radiate, and the patterns are acceptable although far from ideal.

24 Specific Antenna Types

Table 12 lists many of the practical types of vehicle antennas and summarizes their typical characteristics. The table is intended as an aid to an engineer in determining what specific antenna types might be applicable to his requirements and as a quick reference for comparing different antenna types. The numbers in the characteristic columns represent typical values that may be achieved in practice and do not necessarily reflect the worst or the best possible designs.

Monopoles

If the size of vehicle is of the order of one-half wavelength or larger, one of the simplest and most widely used antenna systems is a monopole which protrudes from

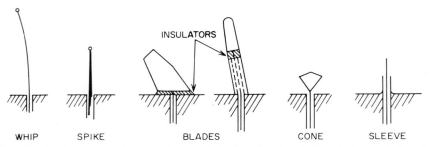

FIG. 117. Various types of monopoles and stubs (details of matching devices not shown).

the surface of the vehicle and usually utilizes the vehicle as a ground plane. Blades, spikes, whips, and stubs are all examples of this type of antenna. In general, they are lightweight and can be made flexible, folded, or streamlined.

A number of antenna systems following in this category are illustrated in Fig. 117.

The Cylindrical Quarter-wave Monopole. Figure 118 shows a sketch of a simple cylindrical monopole and the measured series-impedance characteristic for various length-to-diameter ratios as a function of the length L in wavelengths or L/λ.[2] When L/λ is approximately 0.25 the impedance is pure resistance and approximately 35 ohms. It should be noted that the "fatter" the stub the more constant is its impedance characteristic in the vicinity of the quarter-wave resonance and the more "broadband" it is in terms of impedance. For the quarter-wave stub over an infinite

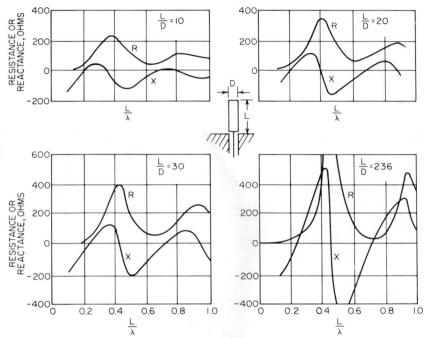

Fig. 118. Measured resistance and reactance of cylindrical monopole for various L/D ratios. (From Radio Research Laboratory Staff, "Very High Frequency Techniques," vol. 1, p. 110, McGraw-Hill Book Company, New York, 1947.)

ground plane the voltage pattern in the upper half plane is omnidirectional and given by the equation

$$F(\theta) = \frac{2 \cos (\pi/2 \cos \theta)}{\sin \theta} \approx \sin \theta \qquad (37)$$

where θ = angle from axis of antenna ($0 < \theta < \pi/2$).

With an infinite ground plane, maximum directivity occurs at the ground plane and it is 3 db higher than a half-wave dipole or 5.1 db. With finite ground planes the corresponding gain is typically about 2 db lower and the maximum at the pattern lobe is above the ground plane from 30 to 60° as shown in Fig. 113. Polarization is linear and in a plane perpendicular to the ground plane.

Whip Antenna. A whip antenna is usually a long thin flexible monopole that is much less than a quarter wavelength long. They are frequently used on vehicles for transmission and reception of signals below 20 Mc. Whips, short in terms of wavelengths, are characterized by an input impedance which has a small radiation-resistance component and a relatively large capacitive reactance. To obtain a reasonable efficiency the capacitive reactance must be tuned out by means of an inductance or loading coil but the overall efficiency is low because of resistance losses—typically in the order of a few per cent. Of course, if the length approaches a quarter wavelength the efficiency improves.

For higher frequencies whips are sometimes used that are one-half wavelength or more in length, and with proper design, good efficiency and greater directivity in the plane perpendicular to the whip axis can be obtained.

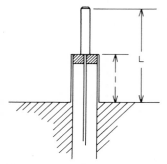

FIG. 119. Cylindrical sleeve stub.

A similar form of quarter-wave monopole is a thin tapered conducting element protruding from the ground plane called a spike. These are lightweight, rugged, easy to install, and make convenient vehicle antennas in the 100- to 400-Mc frequency range. Typical VSWR impedance characteristics of commercial units with respect to a 50-ohm line are

Frequency range	VSWR
1.07:1	1.5:1
1.09:1	2:1
1.56:1	5:1

Sleeve-stub Antenna. A sleeve antenna is one that utilizes the outside of a cylindrical conductor as a radiating element and the inside as the outer conductor of the coaxial feed line feeding the antenna. The sleeve antenna permits rugged mechanical construction because the sleeve can be rigidly attached to the vehicle surface or ground plane. Also, the sleeve antenna possesses broadband impedance characteristics that compare favorably with other antennas of comparable size.

A typical cylindrical sleeve-stub antenna is shown in Fig. 119, and considerable design latitude is permitted in selection of the parameters shown. Design values for the center of the design frequency bands are approximately $L = 2l = 0.34\lambda$ [2].

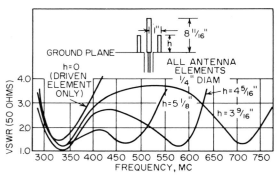

FIG. 120. Design data for open-sleeve antenna. (*From H. Jasik, "Antenna Engineering Handbook," pp. 27–34, Fig. 27-40, McGraw-Hill Book Company, New York, 1961.*)

Figure 120 shows some design data for an open-sleeve stub.[3] Typical measured patterns for a sleeve dipole show that satisfactory patterns and a VSWR of less than 2:1 over a bandwidth greater than 2:1 can be achieved.

It is obvious that sleeve-type monopoles can be incorporated into the structure of a vehicle in appendages such as wingtips or tail structures.

Half-Wave Dipoles

The half-wavelength dipole is frequently a convenient antenna to use on vehicles because of its simplicity. A variety of basic dipole types are illustrated in Fig. 121. The thin dipole has a driving-point impedance of about 73 ohms and a relatively narrow bandwidth. Various methods of making the elements "fatter" (e.g., larger diameter-to-length ratio, biconical, bowties, etc.) all tend to increase the impedance bandwidth of the antenna. The sleeve dipole, like the sleeve monopole, has excellent broadband characteristics.

Turnstile Antenna

The basic turnstile antenna element simply consists of two half-wave dipoles at right angles to each other and fed 90° out of phase as shown in Fig. 122. In a sense, the isolated turnstile element is one of the few antennas that produces omnidirectional pattern coverage with no nulls. The radiation normal to the plane of the elements is circularly polarized. In the plane of the elements it is linearly polarized and the pattern is nearly a circle. Between these two regions it is elliptically polarized. There are no deep nulls in the pattern coverage provided the second antenna has the proper polarization to match the turnstile.

The impedance bandwidth of a turnstile can be increased beyond that of the dipole elements by proper feeding. If the feed line connects to one dipole and the second dipole is connected to the first through a quarter-wave section, an impedance compensation takes place. At frequencies off resonance where the dipoles are reactive the transforming action of the quarter-wave section reverses the sign of the susceptance, and the susceptances tend to cancel at the feed point, thus improving the match. This impedance match can be maintained over a bandwidth many times that of the element alone. It should be pointed out, however, that the

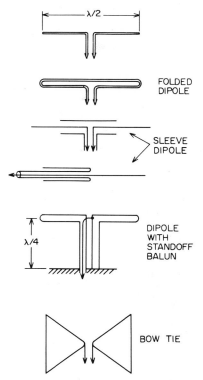

FIG. 121. Various types of dipole antennas.

transforming action of the quarter-wave section tends to unbalance the conductance terms at the driving point when off resonance and the power division to the two dipoles is not equal. This effect produces an elliptical antenna pattern in the plane of the dipoles and increases the axial ratio of the signal normal to the plane of the elements.

There are many variations of the turnstile. Elements may be stacked to yield relatively high-gain omnidirectional pattern coverage with linear polarization. Also, a turnstile element can be mounted over a ground plane to produce a unidirectional pattern with circular polarization.

Another conceivable turnstile arrangement contains three elements 120° apart mechanically and fed 120° apart electrically.

With missiles and satellites it is possible to build turnstile elements that fold during launch and spring into position when the vehicle is out of the atmosphere. A pair of crossed slots fed 90° out of phase is essentially a turnstile as is an array of three or four properly spaced and fed slots around the periphery of a cylindrical vehicle.

The Swastika Antenna

The swastika antenna, shown in Fig. 123, is an omnidirectional (in one plane) circularly polarized antenna consisting of four half-wave dipoles arranged in a square and each tilted at 30° with respect to the horizontal.[4] Each dipole is fed 180° out of phase with respect to the one on the opposite side of the square, and they are approximately one-half wavelength apart. The pattern is doughnut-shaped with a null straight up. The combination of element spacing and tilting the elements gives rise to circular polarization in the horizontal plane.

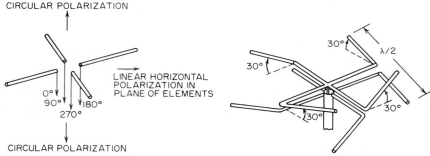

FIG. 122. Turnstile antenna showing relative phasing of feed lines.

FIG. 123. Swastika antenna, omnidirectional circularly polarized.

The impedance bandwidth is essentially limited by the variation in dipole impedance and the matching section with frequency, but a 1.3:1 bandwidth or more is feasible.

The Loop-Vee® Antenna*

The "Loop-Vee" antenna is a broadband antenna (2:1) that produces a circularly polarized pattern.[5] It is very efficient and is capable of handling high average powers. It may be fed by a coaxial line or waveguide by means of a waveguide-to-coax transition.

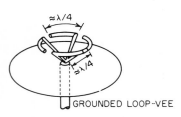

FIG. 124. Two types of Loop-Vee antennas.

There are several versions of the Loop-Vee, two of which are shown in Fig. 124. The grounded Loop-Vee is designed to provide omnidirectional pattern coverage above the ground plane with a minimum protrusion from it. The straight center elements are slightly over a quarter wavelength long at the center design frequency and each segment of the circle that forms the top portion is approximately one-quarter wavelength long. The optimum cone angle is slightly greater than 90°.

The pattern coverage of the grounded Loop-Vee is almost identical to that produced by a quarter-wave stub in conjunction with a current loop above the ground plane. Near the center of the frequency band the polarization is nearly circular in the direction of the main lobe, but it favors vertical at each end of the band.

The azimuth patterns are very nearly perfect circles for all polarizations.

The balanced Loop-Vee simply contains a second element similar to the single element of the grounded antenna that is connected to the outer conductor of the coax. The patterns are figures of revolution centered in a plane perpendicular to the axis of the antenna. Typical elevation patterns near the center frequency and VSWR are shown in Fig. 125.

* Registered U.S. Patent Office.

VEHICULAR ANTENNAS 4-123

The impedance characteristics of the Loop-Vee are similar to a discone or biconical antenna. The grounded version will match a 50-ohm line with a VSWR of less than 2:1 over a bandwidth of 2:1 or more. The impedance of the balance version is approx-

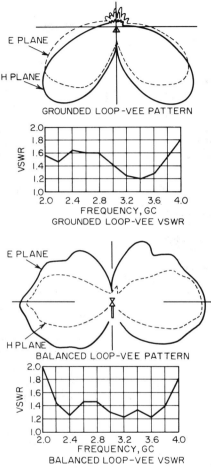

FIG. 125. Pattern and VSWR characteristics of Loop-Vee antenna. (*Courtesy of Radiation, Inc.*)

imately 100 ohms, but it can be matched to 50 ohms over a 2:1 band with a VSWR of less than 2:1 by means of a quarter-wave transformer section.

Planar Spiral Antennas

The spiral antenna has found considerable use for applications requiring broadband frequency coverage, circular polarization, and flush mounting.[6,7] With some designs the efficiency tends to be relatively low when compared with some other types, but spirals are relatively small, compact, and can be fabricated by etched-circuit techniques.

Two basic spiral configurations are in common use, the equiangular spiral and the Archimedean spiral. Mathematically, they are defined by

$$\rho = ke^{a\phi} \quad \text{equiangular spiral} \tag{38}$$
$$\text{and} \quad \rho = k\phi \quad \text{Archimedean spiral} \tag{39}$$

where ρ = radius vector from origin to a point on the curve
k = const
ϕ = angle of rotation

A sketch of each type is shown in Figs. 126 and 127.

Fig. 126. Equiangular spiral-slot antenna.

Fig. 127. Archimedean-spiral antenna.

Theoretically, an equiangular spiral will have pattern and impedance characteristics that are independent of frequency, but end effects, back cavities, element dimensions, etc., limit the theoretical bandwidth. 2:1 bandwidths are easy to achieve and 10:1 or more is entirely feasible.

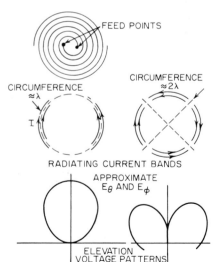

Fig. 128. Sketch showing band of current and radiation patterns for first-mode and second-mode spirals.

There are two principal modes of spiral operation, the first mode which has a main lobe broadside to the plane of the spiral, and the second mode which has a null in the broadside direction. The mode of operation depends upon the number of elements in the spiral and the relative phase of element excitation. The isolated spiral is bidirectional, but the addition of a back cavity, of course, produces a unidirectional pattern.

A qualitative explanation of the patterns obtained with a first-mode two-conductor Archimedean spiral can be obtained by considering a traveling wave progressing along the spiral from the feed point. In general, the currents in adjacent elements are more or less random in direction and hence they do not radiate, but at a radius at which the circumference of the spiral is a multiple of a wavelength, the current in adjacent turns tends to be in phase. Thus, a "band of in-phase current" is produced by adjacent elements and radiation occurs. Since this band is produced by traveling waves, it progresses around the spiral at a rate such that the polarization of the radiated field changes 90° for each quarter cycle of the excitation frequency. This fact accounts for the circular polarization of the spiral. The current band is illustrated in Fig. 128.

The "band of current" occurs when the circumference is one wavelength with a first-mode spiral and two wavelengths with a second-mode spiral.

Figure 129 shows an Archimedean-spiral antenna with back cavity. Typical characteristics of this first-mode type of flush-mounted spiral are:

Frequency	2–4 gc
VSWR	Less than 2:1 with respect to 50 ohms
Beamwidth	55–60° in both planes
Gain	5 to 8 db (relative to isotropic CP source)
Size	4½ in. in diameter
Polarization	Right-circular (IRE)
Peak power	2 kw max
Average power	100 watts max
Weight	5 oz

The first-mode spiral is fed at the center with the two arms 180° out of phase. The second-mode spiral has the arms joined at the center and they are fed in-phase with an

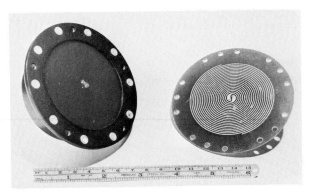

FIG. 129. A complete spiral antenna and a printed spiral. (*Courtesy of Sanders Associates, Inc.*)

unbalanced transmission line. The shield of the coax is connected to the back-cavity. The second-mode spiral produces a circularly polarized pattern with a null normal to the plane of the spiral as shown in Fig. 128.

The equiangular spiral antenna is very similar to the Archimedean spiral except that the larger conducting arms and fewer turns usually produce better efficiency and higher power-handling ability. On the other hand, they are slightly larger for a given frequency range and there is more variation in the shape of the elevation patterns as a function of azimuth.

First-mode spirals can be fed by means of a so-called "infinite balun" which is made by bringing the feed-line coax to the feed point with the outer conductor as part of the conducting arm. At the middle (feed point) the center conductor of the coax is connected across the gap between the arms to the second arm.

A novel antenna using two spiral antennas mounted on a spherical surface was used for broadband operation on the Transit satellite.[8] This antenna worked over a 4:1 frequency band and produced circular polarization of opposite sense off the poles and linear polarization off the equator.

Conical Spirals

The planar spiral is essentially a bidirectional device unless it has a back cavity which suppresses the radiation in one direction. By constructing an equiangular spiral, which is projected on a cone, a conical spiral can be obtained which has unidirectional radiating properties.[9] If the half cone angle is 30° or less, the back lobe is practically eliminated and a unidirectional pattern is obtained with first-mode operation.

Fig. 130. Scimitar antenna.

Scimitar

A scimitar antenna is a degenerate form of the equiangular spiral.[22] Its radiating element is similar to one arm of an equiangular spiral mounted at right angles to a ground plane and fed at the point as shown in Fig. 130. It is highly efficient broad-band antenna (better than 10:1) with high power-handling ability. It can be flush-mounted in a cavity or can be mounted external to the vehicle. Its shape is well adapted to rigid mounting and streamlining or it can be incorporated into the surface or appendages of an airframe. Arrays of scimitars are possible for particular types of pattern coverage.

The patterns of the scimitar vary with design and frequency, and they also vary considerably with polarization. In a very general way the patterns resemble those of a vertical stub near the azimuth plane (ground plane). In the elevation plane they resemble the patterns of a dipole parallel to the ground plane in the plane of the scimitar.

Log-periodic Antennas

Like the equiangular spiral antennas, the family of log-periodic antennas are essentially frequency-independent antennas and are capable of practical bandwidths of 10:1 or more with relatively little change in patterns and good VSWR.[11,3]

A log-periodic structure has a geometry that repeats periodically so that the electrical properties of the antenna also repeat periodically with the logarithm of the frequency. If the variation in pattern and impedance is small over a cycle frequency independence is achieved.

One basic form of log-periodic structure is made of two-toothed flat metal sheets intersecting at an angle ψ as shown in Fig. 131. The teeth may be trapezoidal, triangular, circular sectors, or other shape.

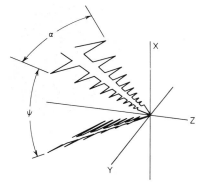

Fig. 131. Flat-plate triangular tooth log-periodic antenna. Polarization is in Y-Z plane and radiation is in the Z direction.

On each plate the teeth are complementary about the center line and the plates are arranged so the teeth are complementary on the two halves. When the two plates are fed at the tip a unidirection pattern is radiated off the point or vertex of the structure. It may be fed at the vertex with a balanced transmission line or by means of a so-called infinite balun which is a coax going up one element (with outer conductor bonded electrically to the conducting surface) and the center conductor crossing to the other plate.

A typical E-plane and H-plane pattern for a triangular-tooth log-periodic antenna is shown in Fig. 132.

A second type of log-periodic antenna is the dipole type shown in Fig. 133. This antenna was designed for a 1- to 2-gc band bandwidth with a VSWR less than 2:1 over the band.

Log-periodic antennas may be made to produce circular polarization on axis by placing two antennas orthogonal to each other and feeding them 90° out of phase. One novel method of achieving the 90° phase shift across a wide band is to locate the orthogonal elements between the original elements which have an inherent 180° phase shift between them.

A qualitative understanding of operation of the dipole log-periodic array can be had by considering a traveling wave along the feed line. When the elements are short compared with a dipole they simply represent a capacitive load on the line. The element that is one-half wave long has a low impedance and is effectively the driven element. The element to the rear of the driven element acts as a reflector and the element in front as a director. The radiation pattern is similar to a three-element yagi antenna.

The gain of the conventional unidirectional log-periodic antennas is typically about 6 to 12 db.

Inflatable log-periodic antennas, suitable for space vehicles, have been demonstrated. Flush-mounted log-periodic antennas have also been reported as well as log-periodic antennas which work with their ground-plane image.

Slot Antennas

One of the most commonly used antennas with missiles and aircraft is the slot antenna.[2,3] Slot antennas are particularly useful because they can be flush-mounted and they utilize the skin of the aircraft as part of the antenna.

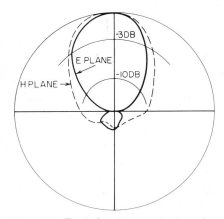

Fig. 132. Typical patterns of triangular-tooth log-periodic antenna.

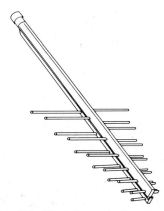

Fig. 133. Dipole log-periodic antenna.

The simplest form of slot is a half-wavelength resonant slot in an infinite ground plane as shown in Fig. 134. The radiation pattern of the slot is polarized in a plane across the narrow dimension of the slot. In the plane of the electric polarization, the pattern is a circle, and in the plane normal to the polarization plane (H plane), the pattern is the same as a half-wave dipole. With a cavity behind the slot, the coverage becomes unidirectional.

In practice, the finite size of the ground plane produces considerable distortion of the pattern in the plane of polarization as shown in Fig. 134.

One simple form of slot antenna is made by terminating a waveguide in a ground plane as shown in Fig. 135. The VSWR is less than 2:1 over a frequency band from the cutoff of the guide to twice the cutoff frequency if the thickness of the guide is at least one-third the width. Bandwidth is essentially limited by the coax-to-waveguide transition needed to feed the guide.

One type of broadband slot antenna which utilizes a bar transition for the coax-to-waveguide transition is shown in Fig. 136. This antenna has a slot about 0.62λ long and 0.2λ wide and a VSWR of less than 2:1 over approximately a 2:1 bandwidth.

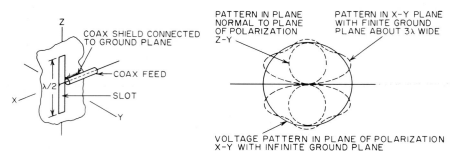

FIG. 134. Figure showing half-wave resonant slot and associated patterns. With back cavity pattern becomes unidirectional.

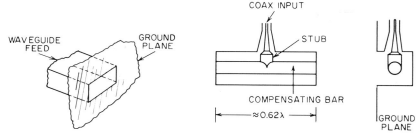

FIG. 135. Waveguide fed slot in ground plane.

FIG. 136. Broadband slot antenna.

The physical size of the slots can be reduced by loading with dielectric such as ceramic or Teflon, or by shaping the cavity and slot like a ridge or double-ridge waveguide. Most of these techniques tend to reduce the bandwidth of the antenna because the smaller slot is a poor match to space and the effective Q of the back cavity increases. Many of the loaded types must be tuned to a specific frequency in a given range to reduce the VSWR.

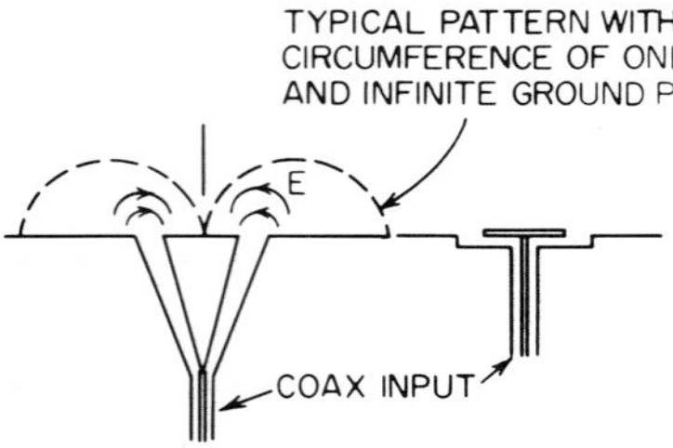

Annular-slot Antenna

One form of flush-mounted antenna that produces radiation patterns very similar to the stub monopole is the annular-slot antenna.[3] The radiating aperture is simply a circular slot in the conducting plane which is excited in-phase similar to the TEM mode in a coaxial line. The slot can be fed by means of a tapered coaxial section or a radial transmission line as shown in Fig. 137. To match the impedance of the slot to space efficiently, the circumference of the slot must approach one wavelength. Various impedance-matching systems are used in the radial section. A VSWR of under 2:1 can be obtained over the 225- to 400-Mc band with an antenna that is about 24 in. in diameter and 4.5 in. deep.

FIG. 137. Two basic forms of annular-slot antennas.

Axial-mode Helix

The axial-mode helix is a simple broadband end-fire antenna that produces essentially a circularly polarized radiation pattern on axis.[12] There are many variations of the helix, and much has appeared in the literature on helix design. A design nomograph for helix design is given on page 4–102.

A typical helix antenna has a bandwidth (with good patterns) of about 1.7:1 and an input impedance between 100 and 150 ohms which can be matched to less than 2:1 with respect to a 50-ohm line by means of a quarter-wave transformer. Typical design parameters are

     Circumference................... λ
     Pitch angle...................... 13°
     Conductor diameter.............. 0.02λ
     Ground-plane diameter........... 0.8λ

The number of turns depends upon the required beamwidth and gain, but from three to eight turns are common. Three turns will produce approximately a 60° beamwidth and a gain of about 8 db above a circularly polarized isotropic source. With eight turns the beamwidth is about 40° and the gain is increased to 12 db.

A short helix of about two turns can be mounted in a recessed cylindrical cavity to provide flush mounting on a ground plane. If the cavity is approximately 0.6λ in diameter the performance characteristics are very similar to the isolated helix. This is a commonly used type of flush-mounted circularly polarized antenna for vehicles.

Helices can be arrayed in groups of four or more to produce higher gain when this is needed. If the required gain exceeds about 15 db an array is usually better mechanically than a longer single element.

Surface-wave Antennas

A surface-wave antenna utilizes a slow-wave structure into which a surface wave is launched.[3] The launcher may be a horn, dipole, waveguide, or other coupling device. Typically, about 70 per cent of the total power is coupled into the surface wave and the rest is radiated directly by the launcher. The slow-wave structure may be pins, corrugated surface, disks, dielectric material, etc. The surface wave travels parallel to the surface of the structure and decays normal to the surface. At the end of the structure the wave illuminates the terminal aperture, a planar region of essentially uniform phase front perpendicular to the axis of the structure passing through the termination. The size of the terminal aperture depends upon the length of the structure and the attenuation of the surface wave along the structure.

The phase velocity of the surface wave along the antenna is the major design parameter in surface-wave antennas, and terminal phase difference between the surface wave and the free space wave is very important. For very short antennas the optimum terminal phase shift is about 60°, increases to about 120° for antennas between four and eight wavelengths long, and approaches 180° for longer antennas.

Surface-wave antennas may be designed for maximum gain or minimum side-lobe levels. A design for maximum gain with a length l between 3λ and 8λ has a gain given approximately by

$$G = 10l/\lambda \qquad (40)$$

and a half-power beamwidth of

$$\text{BW} = 55\sqrt{\lambda/l} \qquad (41)$$

The side-lobe level is about 11 db depending upon the type of structure. The bandwidth is about 1.3:1 for reasonably constant gain.

The gain of a surface-wave antenna increases as l/λ, whereas the gain of broadside apertures such as a parabola increases directly the area or $(\text{diam}/\lambda)^2$. For mechanical reasons surface-wave antennas seldom exceed 10λ in length, or 20 db gain unless they are arrayed.

When the surface-wave structure is designed for low side lobes rather than maximum gain it is possible to reduce side-lobe levels to -15 to -20 db with a reduction in gain of about 1.5 db. It is also possible to design for broad pattern bandwidth and achieve a 2:1 pattern bandwidth with a loss of about 2 db in gain.

The low silhouette of surface-wave antennas makes them desirable for flush or near-flush mounting on vehicles such as high performance aircraft and missiles. These antennas frequently employ a waveguide launcher that directs a wave parallel or nearly parallel to the conducting surface. The slow wave structure is a tapered dielectric slab or corrugated surface. A moderately directive pencil beam is obtained that has its lobe directed above the conducting surface. Polarization is usually normal to the conducting surface, but there are also designs for parallel polarization.

One antenna of this type is the flared slot antenna shown in Fig. 138.[13] This particular antenna produced good patterns over a 4:1 frequency band from about 2.4 to 10 gc. Typical patterns taken near the center of the frequency band are shown in Fig. 139. The flared-slot antenna described above had polarization normal to the ground plane, but a similar antenna with parallel polarization is described in the same reference.

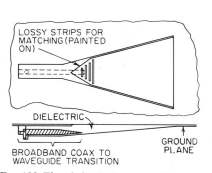

FIG. 138. Flared-slot flush-mounted antenna.

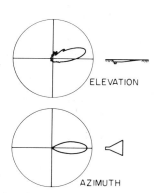

FIG. 139. Typical patterns of flared-slot antenna.

Dielectric-cylinder and Dielectric-tube Antennas

A dielectric-cylinder or -tube antenna is an end-fire surface-wave device made from a long tapered dielectric section. It may be a figure of revolution, in which case it can be linearly polarized or circularly polarized.[14,15] Plastics, foams, and ceramics can be used for the radiating element. The most common type is called a polyrod and the usual construction contains a dielectric that is linearly tapered over slightly more than half its length from 0.5λ diameter to 0.3λ with the remainder uniform. They are usually from three to nine wavelengths long and can provide a gain of 10 to 20 db with side lobes of about -13 db. Bandwidths of the order of 1.1:1 can be obtained with a VSWR less than 1.3:1.

Dielectric tubes and cylinders can be arrayed to increase gain, and the isolation between elements can be very high (30 db or more) if the spacing exceeds one-half wavelength.

Horn Antenna

A rectangular or circular electromagnetic horn can be used to provide unidirectional pattern coverage with either linear or circular polarization. The horn can be flush-mounted behind the skin of the vehicle, but the large physical size (length) usually precludes its use for high-gain applications. Circular polarization can be produced by exciting a square waveguide with a signal polarized at 45° and placing a quarter-wave plate between the feed point and the aperture. The bandwidth of a horn is essentially limited by the waveguide and transition feeding it. By using ridge waveguides and other broadbanding techniques the bandwidth can be increased.

25 Calculation and Display of Vehicle R-F Link Coverage

The Inter-Range Instrumentation Group (IRIG) (see Chapter 5) has prepared a working document (IRIG Document No. 102-61) entitled "IRIG *Standard Coordinate System and Data Format for Antenna Patterns.*" This document establishes a common language for different groups by defining a vehicle-antenna coordinate system and antenna pattern formats recommended for use at the National and Service Ranges.

The recommended consolidated coordinate system for relating the vehicle antenna pattern to the roll, yaw, and pitch axes of the vehicle are shown in Fig. 140. The combined vehicle and trajectory coordinate system are shown in Fig. 141.

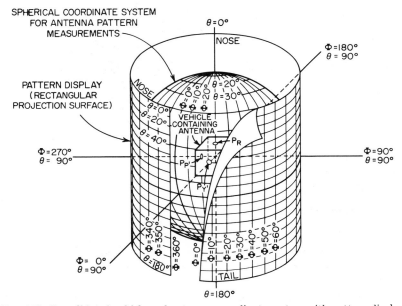

Fig. 140. Consolidated vehicle and antennas coordinate system with pattern display.

The vehicle nose is at $\theta = 0°$, with θ of the slant range vector measured toward the tail. When the vehicle is viewed from its nose, Φ is measured in a counterclockwise direction from the negative portion of the yaw axis to a projection of the slant range vector onto the roll plane, a plane perpendicular to the roll axis (Fig. 141).

This consolidated coordinate system enables one to determine the direction from the vehicle to any instrumentation site in terms of θ and Φ. With antenna patterns referenced to the same coordinate system, expected signal strength can be calculated for any instrumentation site, whether it be ground-based, airborne, or other.

A sample of a contour antenna pattern plotted on a cylindrical equal-spaced projection is shown in Fig. 142.

Sample Analysis of Contour Antenna Patterns

The purpose of this section is to give a simple example of the use of contour antenna patterns. This example describes an analysis of a linearly polarized transmitting antenna carried by a ballistic missile. Signals from this antenna are received at the ground station by a linearly polarized tracking antenna. These conditions are applicable to Azusa and most radar systems utilizing beacons.

Analyses and computer routines for calculating angles Φ and θ, polarization angles, and slant ranges are outside the scope of this handbook and are not presented here. Computer routines for calculating these values are available and are currently being used by the NRD and NASA.

Analysis. From the missileborne contour antenna pattern and other parameters, (1) transmitted power, (2) frequency, (3) ground antenna gain, (4) polarizations of both antennas, and (5) missile trajectory, the power received by a ground station can

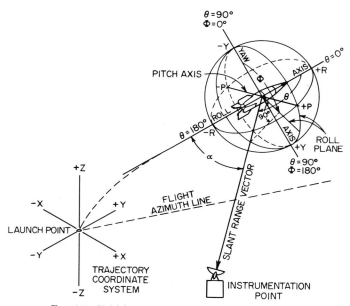

Fig. 141. Vehicle and trajectory coordinate system.

be readily calculated. The equation from which received power level can be obtained is expressed in decibel form as

$$\text{dbm}_r = \text{dbm}_t + G_g + G_m(\Phi,\theta) + 37.9 - 20 \log f_{\text{Mc}} - 20 \log d_{\text{ft}} + P \quad (42)$$

where dbm_r = received power level
dbm_t = transmitted power level
G_g = ground antenna gain, db
$G_m(\Phi,\theta)$ = missile antenna gain, db (directional gain, a function of aspect angles Φ and θ)
f_{Mc} = frequency, Mc/sec
d_{ft} = slant range, ft
P = polarization factor, db

This equation does not include effects of multipath or atmospheric and plasma absorption that may be encountered in the propagation path.

Transmitted Power and Frequency. The transmitted power level and frequency are assumed to be constant.

Missile Antenna Gain. The angles at which the vehicle's transmitting antenna views the ground site can be calculated if the vehicle trajectory (position and velocity

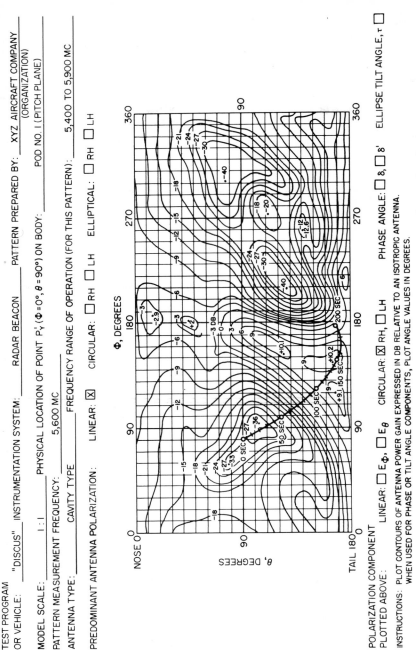

Fig. 142. Sample of contour antenna pattern on the cylindrical equal-spaced projection.

components) and attitude (pitch, yaw, and roll) parameters are known. These angles are called aspect angles Φ and θ. When Φ and θ are known, directional vehicle antenna gain $G_m(\Phi,\theta)$ can be obtained from the contour antenna pattern plot illustrated in Fig. 142. Calculated values of Φ and θ are plotted on the antenna pattern illustrated in Fig. 142. Directional gain values for any time during flight can be plotted and read from this pattern.

Ground Antenna Gain. If the ground antenna has tracking capability or is directed toward the missile at all times, G_g may be considered nearly constant. On the other hand, if the ground antenna is fixed, the missile moves through the ground antenna pattern, and a contour pattern may be needed for the ground antenna as well as for the missileborne antenna. This is not always necessary, since many fixed ground antennas produce a relatively well-known pattern, in which case a theoretical pattern is adequate.

Slant Range. Slant range can be easily calculated from vehicle position data.

Polarization Factor. Polarization factor P is a function of polarization angle difference β between the transmitted r-f wave and receiving antenna. For a linearly polarized antenna receiving a linearly polarized wave, the polarization factor is a function of $\cos^2 \beta$ or

$$P = 20 \log (\cos \beta) \quad \text{db} \tag{43}$$

and Eq. (42) becomes

$$\text{dbm}_r = \text{dbm}_t + G_g + G_m(\Phi,\theta) + 37.9 - 20 \log f_{\text{Mc}} - 20 \log d_{\text{ft}} + 20 \log (\cos \beta) \tag{44}$$

If the direction of the missileborne antenna polarization vector is known in reference to the vehicle's coordinate system, angle β can be calculated. The pattern shown in Fig. 142 represents the right-hand (RH) circularly polarized component of the wave. Assuming that the wave polarization is precisely linear, this component contains only one-half the total power; the remainder of the power is contained in the left-hand (LH) circularly polarized component. Thus, to represent the total power in the wave, 3 db is added to G_{RH}, the RH component gain.

The received power level may be calculated from information obtained from the contour plot (Fig. 142) and the results of the above calculations for a given time t of the missile flight.

These calculations can be repeated at desired intervals during missile flight; thus, expected power levels can be determined as a function of flight time. Signal drop-outs are readily determined by comparing these levels with the minimum power level required at the receiver antenna terminals.

REFERENCES

1. A. S. Meier, Measured Impedance of Vertical Antennas over Finite Ground Planes, *Ohio State Univ. Research Foundation Rept.* 233-3, Oct. 1, 1946.
2. Radio Research Laboratory Staff, "Very High Frequency Techniques," vol. 1, McGraw-Hill Book Company, New York, 1947.
3. H. Jasik, "Antenna Engineering Handbook," McGraw-Hill Book Company, New York, 1961.
4. G. H. Brown, and O. M. Woodward. Jr., Circularly-polarized Omnidirectional Antenna, *RCA Rev.*, vol. 8, p. 259, June, 1947.
5. L. P. Tuttle, Jr., and J. L. Moore, Design Parameters for the Loop-Vee Antenna, L. E. Williams, U.S. Patent 2,964,747.
6. J. A. Kaiser, The Archimedean Two-wire Spiral Antenna, *IRE Trans. Antennas Propagation*, vol. AP-8, no. 3, pp. 312–323, May, 1960.
7. J. D. Dyson, The Equiangular Spiral Antenna, *IRE Trans. Antennas Propagation*, vol. AP-7, no. 2, April, 1959.
8. H. B. Riblet, A Broadband Spherical Satellite Antenna, *Proc. IRE*, vol. 48, no. 4, pp. 631–635, April, 1960.
9. J. D. Dyson and P. E. Mayes, New Circularly-polarized Frequency Independent Antennas with Conical Beam or Omnidirectional Patterns, *IRE Trans. Antennas Propagation*, vol. AP-9, no. 4, pp. 334–342, July, 1961.

10. P. J. Klass, Airborne Spiral Antenna Minimize Drag, *Aviation Week*, July 14, 1958, pp. 75–79.
11. D. E. Isbell, Log Periodic Dipoles, *IRE Trans. Antennas Propagation*, vol. AP-8, no. 3, pp. 260–267, May, 1960.
12. J. D. Kraus, "Antennas," McGraw-Hill Book Company, New York, 1950.
13. J. W. Eberle, C. A. Levis, and D. McCoy, The Flared Slot: A Moderately Directive Flush-mounted Broadband Antenna, *IRE Trans. Antennas Propagation*, vol. AP-8, no. 5, pp. 461–466, September, 1960.
14. G. E. Mueller and W. A. Tyrrell, Polyrod Antennas, *Bell System Tech. J.* vol. 26, no. 4, pp. 837–851, October, 1947.
15. G. K. Ikrath and W. Schneider, Glass-fiber Array Produces Many Beams, *Electronics*, Nov. 1, 1963, pp. 28–29.

Chapter 5

STANDARDS FOR TELEMETRY

TELEMETRY STANDARDS.................................... 5-1
RADIO-FREQUENCY COORDINATION AND ALLOCATION... 5-4
  1 General ITU Rules for the Assignment and Use of Frequencies. 5-8
  2 ITU Categories of Services and Allocations.................. 5-9
  3 ITU Terms and Definitions................................. 5-9
  4 ITU Designation of Emissions.............................. 5-14
  5 Special Rules Governing the Aeronautical and Space Services. 5-23
IRIG TELEMETRY STANDARDS, DOCUMENT 106-60....... 5-25

TELEMETRY STANDARDS

Standardizing activity has been carried on in the field of *wired* telemetering in the American Institute of Electrical Engineers. Several documents have been published, such as "Telemetering, Supervisory Control, and Associated Circuits," AIEE Special Publications, 1932, 1948, 1959. This work has resulted in standards and definitions which are included in American Standards Association Standards Publications C 42.30 on Instruments and C 42.65 on Communications.

A most extensive work in standardization in the *radio* telemetry field has been undertaken by IRIG,[1] the Inter-Range Instrumentation Group. This activity was started by the Commanders of the Guided Missile Ranges in 1952 at their second Range Commanders' Conference. In 1957, the IRIG was expanded to include all seven existing ranges. These ranges are: (1) The Atlantic Missile Range (AMR), (2) Pacific Missile Range (PMR), (3) The White Sands Missile Range (WSMR), (4) The Naval Ordnance Test Station (China Lake, Calif.), (5) The Naval Ordnance Missile Test Facility, (6) The Air Force Missile Development Center located at WMSR (these constitute the Integrated Range Mission), and (7) The Air Proving Ground Center (Eglin Air Force Base, Florida).

The charter of the Inter-Range Instrumentation Group includes the following responsibilities:

1. To interchange information on common problems concerning instrumentation
2. To recommend to the range commanders the standardization of instrumentation systems and equipment techniques, methods, procedures, coordinate systems, etc.
3. To facilitate the interchange of instrumentation systems and personnel between the ranges
4. To facilitate the joint development and procurement of instrumentation and data-reduction systems
5. To recommend to the range commanders technical and management problems for test-range instrumentation

The IRIG consists of a steering committee which is limited to one member from each member range which is assisted by management representatives from the Range Management Agency. These agencies are the Air Research and Development Command, the Bureau of Aeronautics, the Bureau of Ordnance, and the Office of Chief of Ordnance. There is in addition an honorary or advisory associate member from each of the following: the Office of the Director of Defense for Research and Engineering, the Office of the Secretary of Defense (Advanced Research Projects Agency); Headquarters, U.S. Air Force, Wright Air Development Center (now Aeronautics Systems Division, U.S. Air Force); the National Aeronautics and Space Agency; the National Bureau of Standards; and Sandia Corporation. The steering committee is assisted by a secretary who is in charge of the IRIG secretariat.

Under the steering committee are several working groups. The membership of these groups is limited to three military or civil service employees from each member range organization. A working group may have a nonvoting associate member from any United States government agency having problems common to those of the working group or from a contractor organization employed by the United States government in the operation of a member range. There are 10 IRIG working groups, but this number is subject to change. The membership of an average working group is about 11 members and 9 associates. The 10 working groups are:

1. Data Reduction and Computing Working Group (DR and CWG)
2. Electromagnetic Propagation Working Group (EPWG)
3. Electronic Trajectory Measurements Working Group (ETMWG)
4. Frequency Coordination Working Group (FCWG)
5. Geodetic Working Group (GWG)
6. Meterological Working Group (MWG)
7. Optical Systems Working Group (OSWG)
8. Photo Processing Working Group (PPWG)
9. Telecommunications Working Group (TCWG)
10. Telemetry Working Group (TWG)

The steering committee meets three times a year to exchange information, to establish guidance of the working groups, to act on recommendations addressed to the steering committee by the working groups, and to make recommendations to its parent group, the Range Commanders' Conference. Three is established at White Sands Missile Range, New Mexico, a permanent IRIG secretariat to serve as a central point for steering committee correspondence and IRIG record keeping. Requests for information on the IRIG or on the IRIG documents that are available from IRIG should be addressed to the Secretary, Inter-Range Instrumentation Group, White Sands Missile Range, New Mexico. The working groups vary their procedures. They meet from one to four times a year.

The overall effectiveness of the member ranges as a national resource is enhanced by the IRIG through standardization, coordination of joint procurement, and exchange of equipment for the services. Standardization includes establishment of standard nomenclature, of uniform practices and policies, of common procurement specifications, etc. This function requires the preparation of a formal IRIG recommendation document having the approval of the steering committee and bearing an IRIG number. Such a document is not a binding agreement but can be used as the basis for individual action by any member range or the basis for cooperative action such as a joint development or procurement program. In a rare instance where an action would require the official concurrence and cooperation of all the member ranges the IRIG steering committee addresses the recommendation to its parent group, the Range Commanders' Conference.

Although several documents have been published by the IRIG, one of the most important is IRIG Document 106-60 entitled "Telemetry Standards." This document is of such importance that it is presented in this chapter (pages 5-25 to 5-67). It represents a comprehensive guide to telemetry design requirements available today. It should be clear that these specifications are mandatory only at goverment test sites. Since this represents the overwhelming utilization of radio telemetry and associated equipment, the IRIG standards largely shape the practice in the industry. IRIG

106-60 is a combination of several previously standardized IRIG documents on telemetry and magnetic recording, but it is expanded to include a standard on PAM-FM telemetry, a statement of frequency utilization parameters, and a glossary of terms.

This standard is largely the work of the Telemetry Working Group of IRIG.[2] It is of importance to understand the philosophy upon which IRIG based its standardization work. One objective is to reduce expense by providing range facilities which may serve as many different programs as possible. Another reason is to make it possible, from time to time, to combine the facilities of several ranges in carrying out a test involving the geographical areas of the several ranges. Since it is not the intention of the IRIG to stifle development effort on new systems and equipment, the standards are usually prepared to define, for example, the nature of the transmitted signals and not the performance hardware, at either the transmitting or receiving ends of the system.

All the personnel of the TWG is derived from government. Liaison is established with industry in the development of new standards or significant changes in the existing standards.

Once an IRIG working group has agreed on a standard, the document is submitted to the IRIG steering committee for approval; an IRIG number is assigned; and the document is printed, distributed widely, filed with a government technical information-disseminating agency, and stocked by the IRIG secretariat to meet requests for copies.

The TWG initiated its standardization work by revising the Telemetry Standards initially promulgated by the Research and Development Board (RDB) in 1948. This revision was made as IRIG Recommendation 102-55, dated July 13, 1955. An IRIG Standard, Telemetry Standard for Guided Missiles, IRIG Document 103-56, was issued on Oct. 9, 1956, which covered the FM-FM and PDM-FM systems. In 1959, they added new data on radio frequencies and data on automatic drift and gain-correction signals. In March, 1959, IRIG published the initial Standard for Pulse Coded Modulation Telemetry, IRIG 102-59. This and revised Standards on Magnetic Tape Recorders and Reproducer Standards, which were initially issued as IRIG 101-57, dated Feb. 14, 1957, are included in the new standard referred to above, namely, IRIG 106-60, which includes a new standard on PAM-FM.

Of particular importance has been the work on telemetry radio-frequency allocations which is being done in coordination with the IRIG frequency-coordinating working group.[2] More details on this subject are presented in the next section.

As previously pointed out, the TWG of the IRIG really represents the government's interest in telemetry. To represent the interests of all telemetry users and manufacturers in the matter of telemetry standards, the Telemetry Standards Coordination Committee was formed by the National Telemetering Conference in September, 1960. It is intended to serve as a focal point to represent the member societies of the National Telemetering Conference and serves to receive, coordinate, and disseminate information and to recommend and endorse standards, methods, and procedures to users, manufacturers, and supporting agencies. The member societies of the NTC are:

1. American Institute of Aeronautics and Astronautics
2. Institute of Electrical and Electronic Engineers
3. Instrument Society of America

(Refer to the 1962 National Telemetering Conference Proceedings for a history of the TSCC, its charter, organization, bylaws, and first annual report.)

The TSCC has already made a number of recommendations to the TWG on the PAM-FM standards. The TSCC has formed the following subcommittees:

1. Frequency Division Subcommittee
2. Time Division Subcommittee
3. Deep Space and Satellite Subcommittee
4. Radio Frequency Standards and Allocations Subcommittee

In the field of frequency division, the committee is considering the extension of FM standards to include higher-frequency subcarriers. In time division, the committee has recommended against the use of a digital synchronization as permitted in Sec. 4.5

of the PAM-FM standards. The Deep Space and Satellite Subcommittee is considering standardization of the FM system for smaller satellites.

A further development in telemetry standardization has been the formation of Panel 59-A by the Aerospace Research and Testing Committee (ARTC) of the Aerospace Industries Association. The major objectives of the panel's activities are to determine the industry's present and future requirements for telemetry systems and to encourage the development of systems in conformance with these requirements. This committee is establishing a set of Integrated Telemetry Standards, which is intended to point the way toward advances in complete telemetry systems technology. They are not intended to replace the IRIG standards, but to set up more definitive design criteria and test procedures.

In particular, these standards and recommendations cover the design criteria of a Constant Bandwidth Sub-Carrier FM-FM or PM system (Chap. 6) and provide design objectives for size of telemetry transmitters.

IRIG Standard 106-60 on Telemetry Systems, issued June, 1962, is presented in its entirety at the end of the chapter (see page 5-25). This standard is, of course, subject to change, and the latest issue should be obtained from the Secretariat mentioned above.

RADIO-FREQUENCY COORDINATION AND ALLOCATION

The expansion in the use of radio frequency for space research, space communications, telemetry and telecommand has increased the potential usage of the radio-frequency spectrum. The telemetering frequency allocations used at the various ranges are derived through government sources. However, it is important to understand both the national and international procedures involved in frequency allocation in order to establish an appropriate operating frequency for a given communications purpose. The United States authority for granting frequency allocations for civilian purposes is vested in the FCC, whereas all governmental uses of the radio-frequency spectrum are provided for by the IRAC (the Inter-Department Radio Advisory Committee) subject to the approval of the Office of Civil and Defense Mobilization. These two activities coordinate their work.

The above agencies have authority for frequency allocation and administration only within the national boundaries of the United States. It is the policy of the United States to seek international agreement concerning the allocations of frequencies that cannot be confined to national boundaries. This is negotiated through the State Department Office of Transport and Communications. The principal agency involved is the *International Telecommunications Union (ITU)*, Geneva, Switzerland. This union publishes radio regulations which are promulgated by the Administrative Radio Conference. These conferences are held from time to time as the state of the art in radio communications demands.

Once these regulations are adopted, they have the force of a treaty. Previous regulations were adopted in 1959. A branch of the ITU is the International Radio Consultative Committee (CCIR). This committee is charged with studying matters of significance to the radio-communications field. Study Group IV, for example, is charged with the problems of space communications. The CCIR meets in plenary session once every 3 years. Because of the rapid changes in the communications state of the art, an emergency meeting of the Administrative Radio Conference was convened in October, 1963. Significant changes in the radio regulations were agreed to at this conference and are effective after January 1, 1965. These changes provide for both space radio communications and radio astronomy, the latter having been inadequately provided for in the 1959 radio regulations. Table 1 is a summary of the allocations of radio spectra for space communications and radio astronomy which was agreed upon at the 1963 Extraordinary Administrative Radio Conference.

The significance of these allocations is that the use of the services indicated must be protected from interference by any other users. The primary user has priority over the secondary user. The regions referred to are three in number and are defined pre-

RADIO-FREQUENCY COORDINATION AND ALLOCATION

Table 1. Frequency Allocations for Space, etc.

Frequency, kc/sec	Service	Category	Region*
2,500	Radio astronomy	FN	2, 3
5,000	Radio astronomy	FN	
10,000	Radio astronomy	FN	
10,003–10,005	Space research	S, FN	
15,000	Radio astronomy	FN	
15,762–15,768	Space research	S	
18,030–18,032	Space research	S	
19,990–20,010	Space research	S	
20,000	Radio astronomy	FN	
20.007 ± 3 kc	Space research (distress channel)	FN	
25,000	Radio astronomy	FN	

Frequency, mc/sec	Service	Category	Region*
30.005– 30.010	Space-research space (satellite identification)	P, S	
37.75 – 38.25	Radio astronomy	S	
39.986– 40.002	Space research	S	
40.68 ± 0.25	Scientific, industrial	FN	
73 – 74.6	Radio astronomy	P	2, EX
79.75 – 80.25	Radio astronomy	FN	3
117.975–132	Aeronautical mobile (R) (to be used with communications satellites)	P	
132 –136	Aeronautical mobile (R) (to be used with communications satellites)	P	1
136 –137	Space research (telemetering and tracking)	P	2
		P, S	1, 3
137 –138	Meteorological satellite space (telemetering and tracking)	P, S	
143.6 –143.65	Space research (telemetering and tracking)		
149.9 –150.05	Radio navigation satellite	P	
150 –153	Radio astronomy	FN	1
183.1 –184.1	Space research	FN	
267 –272	Space (telemetering)	S	
272 –273	Space (telemetering)	P, S	
322 –329	Radio astronomy	FN	
399.90 –400.05	Radio-navigation satellite	P	
400.05 –401	Meteorological satellite (maintenance telemetering)	P, S	
	Space research (telemetry and tracking)		
401 –402	Space (telemetering)	Shared	
404 –410	Radio astronomy	FN	2
406 –410	Radio astronomy	FN	1, 3
449.75 –450.25	Space telecommand	FN	
460 –470	Meteorological satellite	S	
606 –614	Radio astronomy	FN	1
608 –614	Radio astronomy	FN	2
610 –614	Radio astronomy	FN	3
900 –960	Space research (experimental)	FN, S	
1,400 –1,427	Radio astronomy	P	
1,427 –1,429	Space (telecommand)	P, S	
1,525 –1,535	Space (telemetering)	P, S	1, 3
		P	2
1,535 –1,540	Space (telemetering)	P	

Table 1. Frequency Allocations for Space, etc. (*Continued*)

Frequency mc/sec	Service	Category	Region*
1,540 – 1,660	Aeronautical radio-navigation	P	
1,660 – 1,664.4	Meteorological satellites	P, S	EX
1,664.4– 1,668.4	Meteorological satellites	P, S	EX
1,664.4– 1,668.4	Radio astronomy	S	EX
1,668.4– 1,670	Meteorological satellites	P, S	EX
1,690 – 1,700	Meteorological satellites	P, S	EX
1,700 – 1,710	Space research	P	2
		P, S	1, 3
2,110 – 2,120	Space research (deep-space telecommand)	FN	
2,290 – 2,300	Space research (telemetry and tracking in deep space)	P	2, EX
2,290 – 2,300	Space research (telemetry and tracking in deep space)	P	EX
2,690 – 2,700	Radio astronomy	P	EX
3,165 – 3,195	Radio astronomy	FN	EX
3,400 – 3,600	Communications satellite (satellite to earth) includes associated telemetry and tracking	P, S	2, 3, EX
3,500 – 3,700	Communications satellite (satellite to earth) includes associated telemetry and tracking	P, S	2, 3, EX
3,700 – 4,200	Communications satellite (satellite to earth) and associated telemetry and tracking	P, S	2, 3, EX
3,600 – 4,200	Communications satellite (satellite to earth) and associated telemetry and tracking	P, S	1
4,200 – 4,400	Aeronautical navigation	P	EX
4,400 – 4,700	Communications satellite (earth to satellite) and associated telecommand	P, S	
4,800 – 4,810	Radio astronomy	FN	EX
4,990 – 5,000	Radio astronomy	P, S	1, 3
		P	2
5,000 – 5,255	Space research	S	EX
5,670 – 5,725	Space research (deep space)	S	EX
5,725 – 5,850	Communications satellite (earth to satellite) and telecommand	P, S	1, EX
5,800 – 5,815	Radio astronomy	FN	
5,850 – 5,925	Communications satellite (earth to satellite) and telecommand	P, S	1, 3
5,925 – 6,325	Communications satellite (earth to satellite) and telecommand	P, S	
7,120 – 7,130	Space (telecommand)	FN	
7,200 – 7,250	Meteorological satellite (including tracking and telemetry)	FN	
7,250 – 7,300	Communications satellite (satellite to earth)	P, S	
7,300 – 7,750	Meteorological satellite (including tracking and telemetry)	FN	
		P, S	
7,300 – 7,750	Communications satellite (satellite to earth)		
7,250 – 7,750	Passive communications satellite systems	FN	
7,900 – 7,975	Communications satellite (earth to satellite) and telecommand	P, S	
7,975 – 8,025	Communications satellite (earth to satellite) and telecommand	P	EX
8,025 – 8,400	Communications satellite (earth to satellite) and telecommand	P, S	EX
8,400 – 8,500	Space research	P	S, EX
		P, S	1, 3, EX
8,680 – 8,700	Radio astronomy	FN	EX
9,975 –10,025	Meteorological satellite (weather radar)	FN	

Table 1. Frequency Allocations for Space, etc. (*Continued*)

Frequency, gc/sec	Service	Category	Region*
10.68–10.7	Radio astronomy	P	EX
14.3 –14.4	Radio-navigation satellite	P	
15.25–15.35	Space research	P	EX
15.35–15.4	Radio astronomy	P	EX
15.4 –15.7	Aeronautical radio navigation	P	EX
19.3 –19.4	Radio astronomy	P	EX
31 –31.3	Space research	S	EX
31.3 –31.5	Radio astronomy	P	EX
31.5 –31.8	Space research	P	EX
31.8 –32.3	Space research	S	EX
33 –33.4	Radio astronomy	P, S	1
33.4 –34	Radio astronomy	FN	
33.4 –33.5	Meteorological satellite (weather radar)	FN	
34.2 –35.2	Space research	Shared	

EX—Some countries are excepted, usually Soviet bloc.
FN—Footnote to the Frequency Allocation Table. These are modifications governing the allocation. The primary and secondary services must be protected as stipulated in the footnote.
P—Primary service.
S—Secondary service.
P, S—Primary service, shared with other service.
* Worldwide unless otherwise designated. For explanation of numbers, see pp. 5-4, 5-7.

cisely by the Radio Regulations (see Chart 1). Region 1 embraces Europe and Africa; Region 2, North and South America and Hawaii; and Region 3 Australia-Asia. As Table 1 indicates, the regulations permit different usage occasionally in different regions. Also different uses are permitted on a national basis or in areas less than regional in extent in some cases.

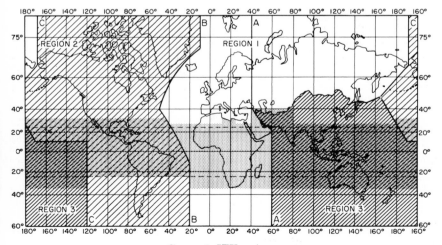

Chart 1. ITU regions.

The Amended Radio Regulations recognize the following new services:

1. Space service
2. Broadcasting-satellite service
3. Communications-satellite service
4. Meteorological-satellite service
5. Radio-navigation-satellite service
6. Space-research service

These services are recognized in the same sense as fixed and mobile services.

A point influencing the allocations for space service is that telecommand and telemetry channels should be spaced within 8 to 15 per cent frequency, one from the other. This is to permit sufficient separation of spacecraft receiver and transmitter frequencies and yet sufficiently close spacing to permit a single *antenna* in the space vehicle to be used for both the channels.

The Satellite Communications Service needs wide bands for trade-off of bandwidth for power. The allocations also recognize the need for and practicality of sharing the microwave frequencies between satellite communications services and ground services.

The international frequency channels are not directly authoritative for range and launch communications as long as wholly located within the United States. These allocations are obtainable from United States governmental agencies under the jurisdiction of IRAC. Frequencies assigned exclusively for government use are the ones preferred for this use since FCC regulation does not apply. These frequencies are monitored by range frequency coordinators appointed by the range commanders. Frequency coordinators have been established at each of the major ranges. In the case of the Air Force Missile Test Center, the area is enclosed within a 200-mile radius of the headquarters building. The frequency-band assignments of the missile ranges are reflected in the IRIG Standards. Where international operations take place, as in case of space telemetry and space communications, the international channels are to be respected.

Extracts from the amended ITU Radio Regulations are provided here. Included are:

1. ITU Rules for the Assignment and Use of Frequencies
2. ITU Categories of Services and Allocations
3. ITU Terms and Definitions
4. ITU Designation of Emissions
5. Special Rules Governing the Aeronautical and Space Services

1 General ITU Rules for the Assignment and Use of Frequencies

The Members and Associate Members of the Union agree that in assigning frequencies to stations which are capable of causing harmful interference to the services rendered by the stations of another country, such assignments are to be made in accordance with the Table of Frequency Allocations and other provisions of these Regulations.

Any new assignment or any change of frequency or other basic characteristic of an existing assignment shall be made in such a way as to avoid causing harmful interference to services rendered by stations using frequencies assigned in accordance with Table 1 and the other provisions of these Regulations, the characteristics of which assignments are recorded in the Master International Frequency Register.

Administrations of the Members and Associate Members of the Union shall not assign to a station any frequency in derogation of either the Table of Frequency Allocations given in this Chapter or the other provisions of these Regulations, except on the express condition that harmful interference shall not be caused to services carried on by stations operating in accordance with the provisions of the Convention and of these Regulations.

The frequency assigned to a station of a given service shall be separated from the limits of the band allocated to this service in such a way that taking account of the frequency band assigned to a station, no harmful interference is caused to services to which frequency bands immediately adjoining are allocated.

For the purpose of resolving cases of harmful interference, the radio astronomy service shall be treated as a radiocommunication service. However, protection from services in other bands shall be afforded the radio astronomy service only to the extent that such services are afforded protection from each other.

Where, in adjacent Regions or sub-Regions, a band of frequencies is allocated to different services of the same category the basic principle is the equality of right to operate. Accordingly, the stations of each service in one Region or sub-Region must operate so as not to cause harmful interference to services in the other Regions or sub-Regions.

ITU Regions and Areas

For the allocation of frequencies the world has been subdivided into three Regions; see Chart 1.

2 ITU Categories of Services and Allocations

Primary Services, Permitted Services, and Secondary Services

Permitted and primary services have equal rights, except that, in the preparation of frequency plans, the primary service, as compared with the permitted service, shall have prior choice of frequencies.

Stations of a secondary service:

(a) shall not cause harmful interference to stations of primary or permitted services to which frequencies are already assigned or to which frequencies may be assigned at a later date;

(b) cannot claim protection from harmful interference from stations of a primary or permitted service to which frequencies are already assigned or may be assigned at a later date;

(c) can claim protection, however, from harmful interference from stations of the same or other secondary service(s) to which frequencies may be assigned at a later date.

Where a band is indicated in a footnote to the Table as allocated to a service "on a secondary basis" in an area smaller than a Region, or in a particular country, this is a secondary service.

Where a band is indicated in a footnote to the Table as allocated to a service "on a primary basis," or "on a permitted basis" in an area smaller than a Region, or in a particular country, this is a primary service or a permitted service only in that area or country.

Additional Services

Where a band is indicated in a footnote to the Table as "also allocated" to a service in an area smaller than a Region, or in a particular country, this is an "additional" service, i.e., a service which is added in this area or in this country to the service or services which are indicated in the Table.

If the footnote does not include any restriction on additional service apart from the restriction to operate only in a particular area or country, stations of this service shall have equality.

If restrictions are imposed on an additional service in addition to the restriction to operate only in a particular area or country, this is indicated in the footnote to the Table.

Alternative Allocations

Where a band is indicated in a footnote to the Table as "allocated" to one or more services in an area smaller than a Region, or in a particular country, this is an "alternative" allocation, i.e., an allocation which replaces, in this area or in this country, the allocation indicated in the Table.

If the footnote does not include any restriction on stations of the services concerned, apart from the restriction to operate only in a particular area or country, these stations shall have an equality of right to operate with stations of the primary service or services.

If restrictions are imposed on stations of a service to which an alternative allocation is made, in addition to the restriction to operate only in a particular country or area, this is indicated in the footnote.

Miscellaneous Provisions

Where it is indicated in these Regulations that a service may operate in a specific frequency band subject to not causing harmful interference, this means also that this service cannot claim protection from harmful interference caused by other services to which the band is allocated.

3 ITU Terms and Definitions

For the purposes of ITU Regulations, the following terms shall have the meanings defined below. These terms and definitions do not, however, necessarily apply for other purposes.

General Terms

Telecommunication: Any transmission, emission or reception of signs, signals, writing, images and sounds or intelligence of any nature by wire, radio, visual or other electromagnetic systems.

General Network of Telecommunication Channels: The whole of the existing telecommunication channels open to public correspondence, with the exception of the telecommunication channels of the mobile service.

Simplex Operation: Operating method in which transmission is made possible alternately in each direction, for example, by means of manual control.*

Duplex Operation: Operating method in which transmission is possible simultaneously in both directions.*

Semi-duplex Operation: Operating method which is simplex at one end of the circuit and duplex at the other.*

Radio Waves (or Hertzian Waves): Electromagnetic waves of frequencies lower than 3,000 Gc/s, propagated in space without artificial guide.

Radio: A general term applied to the use of radio waves.

Radiocommunication: Telecommunication by means of radio waves.

Telegraphy: A system of telecommunication which is concerned in any process providing transmission and reproduction at a distance of documentary matter, such as written or printed matter or fixed images, or the reproduction at a distance of any kind of information in such a form. The foregoing definition appears in the Convention, but, for the purposes of these Regulations, telegraphy shall mean, unless otherwise specified, "A system of telecommunication for the transmission of written matter by the use of a signal code."

Frequency-Shift Telegraphy: Telegraphy by frequency modulation in which the telegraph signal shifts the frequency of the carrier between predetermined values. There is phase continuity during the shift from one frequency to the other.

Four-Frequency Diplex Telegraphy: Frequency-shift telegraphy in which each of the four possible signal combinations corresponding to two telegraph channels is represented by a separate frequency.

Telegram: Written matter intended to be transmitted by telegraphy for delivery to an addressee; this term also includes radio-telegram unless otherwise specified. In this definition the term Telegraphy has the meaning defined in the Convention.

Radiotelegram: Telegram originating in or intended for a mobile station transmitted, on all or part of its route, over the radio-communication channels of a mobile service.

Telemetering: The use of telecommunication for automatically indicating or recording measurements at a distance from the measuring instrument.

Radiotelemetering: Telemetering by means of radio waves.

Telephony: A system of telecommunication set up for the transmission of speech or, in some cases, other sounds.

Radiotelephone Call: A telephone call, originating in or intended for a mobile station, transmitted on all or part of its route over the radiocommunication channels of a mobile service.

Television: A system of telecommunication for the transmission of transient images of fixed or moving objects.

Facsimile: A system of telecommunication for the transmission of fixed images, with or without half-tones, with a view to their reproduction in a permanent form.

Radio Systems, Services and Stations

Station: One or more transmitters or receivers or a combination of transmitters and receivers, including the accessory equipment, necessary at one location for carrying on a radiocommunication service. Each station shall be classified by the service in which it operates permanently or temporarily.

Fixed Service: A service of radiocommunication between specified fixed points.

Fixed Station: A station in the fixed service.

Aeronautical Fixed Service: A fixed service intended for the transmission of information relating to air navigation, preparation for safety of flight.

Aeronautical Fixed Station: A station in the aeronautical fixed service.

Tropospheric Scatter: The propagation of radio waves by scattering as a result of irregularities or discontinuities in the physical properties of the troposphere.

Ionospheric Scatter: The propagation of radio waves by scattering as a result of irregularities or discontinuities in the ionization of the ionosphere.

---

* In general, duplex and semi-duplex operation require two frequencies in radio-communication; simplex may use either one or two.

RADIO-FREQUENCY COORDINATION AND ALLOCATION   5-11

Broadcasting Service: A radiocommunication service in which the transmissions are intended for direct reception by the general public. This service may include sound transmissions, television transmissions or other types of transmissions.

Broadcasting Station: A station in the broadcasting service.

Mobile Service: A service of radiocommunication between mobile and land stations, or between mobile stations.

Land Station: A station in the mobile service not intended to be used while in motion.

Mobile Station: A station in the service intended to be used while in motion or during halts at unspecified points.

Aeronautical Mobile Service: A mobile service between aeronautical stations and aircraft stations, or between aircraft stations, in which survival craft stations may also participate.

Aeronautical Station: A land station in the aeronautical mobile service. In certain instances an aeronautical station may be placed on board a ship or an earth satellite.

Aircraft Station: A mobile station in the aeronautical mobile service on board an aircraft or an air-space vehicle.

Maritime Mobile Service: A mobile service between coast stations and ship stations, or between ship stations, in which survival craft stations may also participate.

Port Operations Service: A maritime mobile service in or near a port, between coast stations and ship stations, or between ship stations, in which messages are restricted to those relating to the movement and safety of ships and, in emergency, to the safety of persons.

Coast Station: A land station in the maritime mobile service.

Ship Station: A mobile station in the maritime mobile service located on board a vessel, other than a survival craft, which is not permanently moored.

Ship's Emergency Transmitter: A ship's transmitter to be used exclusively on a distress frequency for distress, urgency or safety purposes.

Survival Craft Station: A mobile station in the maritime or aeronautical mobile service intended solely for survival purposes and located on any lifeboat, life-raft or other survival equipment.

Land Mobile Service: A mobile service between base stations and land mobile stations, or between land mobile stations.

Base Station: A land station in the land mobile service carrying on a service with land mobile stations.

Land Mobile Station: A mobile station in the land mobile service capable of surface movement within the geographical limits of a country or continent.

Radiodetermination: The determination of position, or the obtaining of information relating to position, by means of the propagation properties of radio waves.

Radiodetermination Service: A service involving the use of radiodetermination.

Radiodetermination Station: A station in the radiodetermination service.

Radionavigation: Radiodetermination used for the purposes of navigation, including obstruction warning.

Radionavigation Service: A radiodetermination service involving the use of radionavigation.

Radionavigation Land Station: A station in the radionavigation service not intended to be used while in motion.

Radionavigation Mobile Station: A station in the radionavigation service intended to be used while in motion or during halts at unspecified points.

Aeronautical Radionavigation Service: A radionavigation service intended for the benefit of aircraft.

Maritime Radionavigation Service: A radionavigation service intended for the benefit of ships.

Radiolocation: Radiodetermination used for purposes other than those of radionavigation.

Radiolocation Service: A radiodetermination service involving the use of radiolocation.

Radiolocation Land Station: A station in the radiolocation service not intended to be used while in motion.

Radiolocation Mobile Station: A station in the radiolocation service intended to be used while in motion or during halts at unspecified points.

Radar: A radiodetermination system based on the comparison of reference signals with radio signals reflected, or re-transmitted, from the position to be determined.

Primary Radar: A radiodetermination system based on the comparison of reference signals with radio signals reflected from the position to be determined.

Secondary Radar: A radiodetermination system based on the comparison of reference with radio signals re-transmitted from the position to be determined.

Instrument Landing System (*ILS*): A radionavigation system which provides aircraft with horizontal and vertical guidance just before and during landing and, at certain fixed points, indicates the distance to the reference point of landing.

Instrument Landing System Localizer: A system of horizontal guidance embodied in the instrument landing system which indicates the horizontal deviation of the aircraft from its optimum path of descent along the axis of the runway.

Instrument Landing System Glide Path: A system of vertical guidance embodied in the instrument landing system which indicates the vertical deviation of the aircraft from its optimum path of descent.

Marker Beacon: A transmitter in the aeronautical radionavigation service which radiates vertically a distinctive pattern for providing position information to aircraft.

Radio Altimeter: A radionavigation equipment, on board an aircraft, which makes use of the reflection of radio waves from the ground to determine the height of the aircraft above the ground.

Radio Direction-Finding: Radiodetermination using the reception of radio waves for the purpose of determining the direction of a station or object.

Radio Direction-Finding Station: A radiodetermination station using radio direction-finding.

Radiobeacon Station: A station in the radionavigation service the emissions of which are intended to enable a mobile station to determine its bearing or direction in relation to the radiobeacon station.

Safety Service: A radiocommunication service used permanently or temporarily for the safeguarding of human life and property.

Earth Station: A station in the earth-space service located either on the earth's surface or on an object which is limited to flight between points on the earth's surface.

Radio Astronomy: Astronomy based on the reception of radio waves of cosmic origin.

Radio Astronomy Service: A service involving the use of radio astronomy.

Radio Astronomy Station: A station in the radio astronomy service.

Meteorological Aids Service: A radiocommunication service used for meteorological, including hydrological, observations and exploration.

Radiosonde: An automatic radio transmitter in the meteorological aids service usually carried on an aircraft, free balloon, kite or parachute, and which transmits meteorological data.

Amateur Service: A service of self-training, intercommunication and technical investigations carried on by amateurs, that is, by duly authorized persons interested in radio technique solely with a personal aim and without pecuniary interest.

Amateur Station: A station in the amateur service.

Standard Frequency Service: A radiocommunication service for scientific, technical and other purposes, providing the transmission of specified frequencies of stated high precision, intended for general reception.

Standard Frequency Station: A station in the standard frequency service.

Time Signal Service: A radiocommunication service for the transmission of time signals of stated high precision, intended for general reception.

Experimental Station: A station utilizing radio waves in experiments with a view to the development of science or technique. This definition does not include amateur stations.

Special Service: A radiocommunication service, not otherwise defined in this Article, carried on exclusively for specific needs of general utility, and not open to public correspondence.

Terrestrial Service: Any radio service defined in these Regulations, other than a space service or the radio astronomy service.

Terrestrial Station: A station in a terrestrial service.

Space Systems, Services and Stations

Space Service: A radiocommunication service between earth stations and space stations; or between space stations; or between earth stations when the signals are re-transmitted by space stations, or transmitted by reflection from objects in space, excluding reflection or scattering by the ionosphere or within the earth's atmosphere.

Earth Station: A station in the space service located either on the earth's surface, including on board a ship, or on board an aircraft.

Space Station: A station in the space service located on an object which is beyond, is intended to go beyond, or has been beyond, the major portion of the earth's atmosphere.

Space System: Any group of co-operating earth and space stations, providing a given space service and which, in certain cases, may use objects in space for the reflection of the radiocommunication signals.

Communication-Satellite Service: A space service between earth stations, when using active or passive satellites for the exchange of communications of the fixed or mobile service; or between an earth station and stations on active satellites for the exchange of communications of the mobile service, with a view to their re-transmission to or from stations in the mobile service.

Communication-Satellite Earth Station: An earth station in the communication-satellite service.

Communication-Satellite Space Station: A space station in the communication-satellite service, on an earth satellite.
Active Satellite: An earth satellite carrying a station intended to transmit or re-transmit radiocommunication signals.
Passive Satellite: An earth satellite intended to transmit radiocommunication signals by reflection.
Satellite System: Any group of co-operating stations providing a given space service and including one or more active or passive satellites.
Space Research Service: A space service in which spacecraft or other objects in space are used for scientific or technological research purposes.
Space Research Earth Station: An earth station in the space research service.
Space Research Space Station: A space station in the space research service.
Broadcasting-Satellite Service: A space service in which signals transmitted or re-transmitted by space stations, or transmitted by reflection from objects in orbit around the Earth are intended for direct reception by the general public.
Radionavigation-Satellite Service: A service using space stations on earth satellites for the purpose of radionavigation, including, in certain cases, transmission or re-transmission of supplementary information necessary for the operation of the radionavigation system.
Radionavigation-Satellite Earth Station: An earth station in the radionavigation-satellite service.
Radionavigation-Satellite Space Station: A space station in the radionavigation-satellite service, on an earth satellite.
Meteorological-Satellite Service: A space service in which the results of meteorological observations, made by instruments on earth satellites, are transmitted to earth stations by space stations on these satellites.
Meteorological-Satellite Earth Station: An earth station in the meteorological-satellite service.
Meteorological-Satellite Space Station: A space station in the meteorological-satellite service, on an earth satellite.
Space Telemetering: The use of telemetering for the transmission from a space station of results of measurements made in a spacecraft, including those relating to the functioning of the spacecraft.
Maintenance Space Telemetering: Space telemetering relating exclusively to the electrical and mechanical condition of a spacecraft and its equipment together with the condition of the environment of the spacecraft.
Altitude of the Perigee: Altitude above the surface of the Earth of the point on a closed orbit where a satellite is at its minimum distance from the center of the Earth.
Stationary Satellite: A satellite, the circular orbit of which lies in the plane of the earth's equator and which turns about the polar axis of the Earth in the same direction and with the same period as those of the earth's rotation.
Spacecraft: Any type of space vehicle, including an earth satellite or a deep-space probe, whether manned or unmanned.

Technical Characteristics

Assigned Frequency: The center of the frequency band assigned to a station.
Characteristic Frequency: A frequency which can be easily identified and measured in a given emission.
Reference Frequency: A frequency having a fixed and specified position with respect to the assigned frequency. The displacement of this frequency with respect to the assigned frequency has the same absolute value and sign that the displacement of the characteristic frequency has with respect to the center of the frequency band occupied by the emission.
Frequency Tolerance: The maximum permissible departure by the center frequency of the frequency band occupied by an emission from the assigned frequency or, by the characteristic frequency of an emission from the reference frequency. The frequency tolerance is expressed in parts in 10^6 or in cycles per second.
Assigned Frequency Band: The frequency band the center of which coincides with the frequency assigned to the station and the width of which equals the necessary bandwidth plus twice the absolute value of the frequency tolerance.
Occupied Bandwidth: The frequency bandwidth such that, below its lower and above its upper frequency limits, the mean powers radiated are each equal to 0.5% of the total mean power radiated by a given emission. In some cases, for example, multi-channel frequency-division systems, the percentage of 0.5% may lead to certain difficulties in the practical application of the definitions of occupied and necessary bandwidth; in such cases a different percentage may prove useful.
Necessary Bandwidth: For a given class of emission, the minimum value of the occupied bandwidth sufficient to ensure the transmission of information at the rate and with the

quality required for the system employed, under specified conditions. Emissions useful for the good functioning of the receiving equipment as, for example, the emission corresponding to the carrier of reduced carrier systems, shall be included in the necessary bandwidth.

Spurious Emission: Emission on a frequency or frequencies which are outside the necessary band, and the level of which may be reduced without affecting the corresponding transmission of information. Spurious emissions include harmonic emissions, parasitic emissions and intermodulation products, but exclude emissions in the immediate vicinity of the necessary band, which are a result of the modulation process for the transmission of information.

Harmful Interference: Any emission, radiation or induction which endangers the functioning of a radionavigation service or of other safety services or seriously degrades, obstructs or repeatedly interrupts a radiocommunication service operating in accordance with these Regulations.

Power: Whenever the power of a radio transmitter, etc., is referred to, it shall be expressed in one of the following forms: peak envelope power (P_p); mean power (P_m); carrier power (P_c).

For different classes of emissions, the relationships between peak envelope power, mean power and carrier power, under the conditions of normal operation and of no modulation, are contained in Recommendations of the CCIR, which may be used as a guide.

Peak Envelope Power of a Radio Transmitter: The average power supplied to the antenna transmission line by a transmitter during one radio frequency cycle at the highest crest of the modulation envelope, taken under conditions of normal operation.

Mean Power of a Radio Transmitter: The power supplied to the antenna transmission line by a transmitter during normal operation, averaged over a time sufficiently long compared with the period of the lowest frequency encountered in the modulation. A time of $1/10$ second during which the mean power is greatest will be selected normally.

Carrier Power of a Radio Transmitter: The average power supplied to the antenna transmission line by a transmitter during one radio frequency cycle under conditions of no modulation. This definition does not apply to pulse modulated emissions.

Effective Radiated Power: The power supplied to the antenna multiplied by the relative gain of the antenna in a given direction.

Gain of an Antenna: The ratio of the power required at the input of a reference antenna to the power supplied to the input of the given antenna to produce, in a given direction, the same field at the same distance. When not specified otherwise, the figure expressing the gain of an antenna refers to the gain in the direction of the radiation main lobe. In services using scattering modes of propagation the full gain of an antenna may not be realizable in practice and the apparent gain may vary with time.

Isotropic or Absolute Gain of an Antenna: The gain (G_{is}) of an antenna in a given direction when the reference antenna is an isotropic antenna isolated in space.

Relative Gain of an Antenna: The gain (G_d) of an antenna in a given direction when the reference antenna is a half-wave loss free dipole isolated in space and the equatorial plane of which contains the given direction.

Gain Referred to a Short Vertical Antenna: The gain (G_v) of an antenna in a given direction when the reference antenna is a perfect vertical antenna, much shorter than one quarter of the wavelength, placed on the surface of a perfectly conducting plane earth.

Antenna Directivity Diagram: A curve representing, in polar or cartesian co-ordinates, a quantity proportional to the gain of an antenna in the various directions in a particular plane or cone.

4 ITU Designation of Emissions

Emissions are designated according to their classification and their necessary bandwidth.

Classification

Emissions are classified and symbolized according to the following characteristics:*
(1) type of modulation of main carrier; (2) type of transmission; (3) supplementary characteristics.

(1) *Types of modulation of main carrier:*   *Symbol*
    (a) Amplitude   A
    (b) Frequency (or phase)   F
    (c) Pulse   P

(2) *Types of transmission:*
    (a) Absence of any modulation intended to carry information   0
    (b) Telegraphy without the use of a modulating audio frequency   1

* As an exception to the provisions of (1) to (3), damped waves are designated by B.

RADIO-FREQUENCY COORDINATION AND ALLOCATION 5-15

(c)	Telegraphy by the on-off keying of a modulating audio frequency or audio frequencies, or by the on-off keying of the modulated emission (special case: an unkeyed modulated emission)	2
(d)	Telephony (including sound broadcasting)	3
(e)	Facsimile (with modulation of main carrier either directly or by a frequency modulated sub-carrier)	4
(f)	Television (vision only)	5
(g)	Four-frequency diplex telegraphy	6
(h)	Multichannel voice-frequency telegraphy	7
(i)	Cases not covered by the above	9

(3) *Supplementary characteristics:*

(a)	Double sideband	(none)
(b)	Single sideband:	
	Reduced carrier	A
	Full carrier	H
	Suppressed carrier	J
(c)	Two independent sidebands	B
(d)	Vestigial sideband	C
(e)	Pulse:	
	Amplitude modulated	D
	Width (or duration) modulated	E
	Phase (or position) modulated	F
	Code modulated	G

Bandwidths

Whenever the full designation of an emission is necessary, the symbol for that emission, as given above, shall be preceded by a number indicating in kilocycles per second the necessary bandwidth of the emission. Bandwidths shall generally be expressed to a maximum of three significant figures, the third figure being almost always a nought or a five.

Table 2. Nomenclature of the Frequency and Wavelength Bands Used in Radiocommunication

The radio spectrum shall be subdivided into nine frequency bands, which shall be designated by progressive whole numbers in accordance with this table. Frequencies shall be expressed: in kilocycles per second (kc/s) up to and including 3,000 kc/s; in megacycles per second (Mc/s) thereafter up to and including 3,000 Mc/s; in gigacycles per second (Gc/s) thereafter up to and including 3,000 Gc/s.

Band number	Frequency range (lower limit exclusive, upper limit inclusive)	Corresponding metric subdivision
4	3 to 30 kc/s (kHz)	Myriametric waves
5	30 to 300 kc/s (kHz)	Kilometric waves
6	300 to 3,000 kc/s (kHz)	Hectometric waves
7	3 to 30 Mc/s (MHz)	Decametric waves
8	30 to 300 Mc/s (MHz)	Metric waves
9	300 to 3,000 Mc/s (MHz)	Decimetric waves
10	3 to 30 Gc/s (GHz)	Centimetric waves
11	30 to 300 Gc/s (GHz)	Millimetric waves
12	300 to 3,000 Gc/s (GHz) or 3 Tc/s (THz)	Decimillimetric waves

Note 1: "Band Number N" extends from 0.3×10^N to 3×10^N c/s (Hz).
Note 2: Abbreviations:
    c/s = cycles per second,   Hz = hertz
    k = kilo (10^3), M = mega (10^6), G = giga (10^9), T = tera (10^{12})
Note 3: Abbreviations for adjectival band designations:
    Band 4 = VLF   Band 8 = VHF
    Band 5 = LF   Band 9 = UHF
    Band 6 = MF   Band 10 = SHF
    Band 7 = HF   Band 11 = EHF

Table 3. Examples of Necessary Bandwidths and Designations of Emissions

The necessary bandwidth may be determined by one of the following methods: (a) use of the formulae included in this table which also gives examples of necessary bandwidths and designation of corresponding emissions; (b) computation in accordance with CCIR Recommendations; (c) measurement, in cases not covered by (a) or (b) above.

The value so determined should be used when the full designation of an emission is required.

However, the necessary bandwidth so determined is not the only characteristic of an emission to be considered in evaluating the interference that may be caused by that emission.

In the formulation of the Table, the following terms have been employed:

B_n = Necessary bandwidth in cycles per second.
B = Telegraph speed in bauds.
N = Maximum possible number of black plus white elements to be transmitted per second, in facsimile and television.
M = Maximum modulation frequency in cycles per second.
C = Sub-carrier frequency in cycles per second.
D = Half the difference between the maximum and minimum values of the instantaneous frequency. Instantaneous frequency is the rate of change of phase.
t = Pulse duration in seconds.
K = An overall numerical factor which varies according to the emission and which depends upon the allowable signal distortion.

Description and class of emission	Necessary bandwidth in cycles per second	Examples — Details	Designation of emission
		I. Amplitude Modulation	
Continuous wave telegraphy, A1	$B_n = BK$ $K = 5$ for fading circuits $K = 3$ for non-fading circuits	Morse code at 25 words per minute, $B = 20$, $K = 5$; Bandwidth: 100 c/s. Four-channel time-division multiplex, 7-unit code, 42.5 bauds per channel, $B = 170$, $K = 5$; Bandwidth: 850 c/s.	0.1A1 0.85A1
Telegraphy modulated by an audio frequency, A2	$B_n = BK + 2M$ $K = 5$ for fading circuits $K = 3$ for non-fading circuits	Morse code at 25 words per minute, $B = 20$, $M = 1,000$, $K = 5$; Bandwidth: 2,100 c/s.	2.1A2
Telephony, A3	$B_n = M$ for single sideband $B_n = 2M$ for double sideband	Double sideband telephony, $M = 3,000$; Bandwidth: 6,000 c/s. Single sideband telephony, reduced carrier, $M = 3,000$; Bandwidth: 3,000 c/s. Telephony, two independent sidebands, $M = 3,000$; Bandwidth: 6,000 c/s.	6A3 3A3A 6A3B
Sound broadcasting, A3	$B_n = 2M$ M may vary between 4,000 and 10,000 depending on the quality desired.	Speech and music, $M = 4,000$; Bandwidth: 8,000 c/s.	8A3

RADIO-FREQUENCY COORDINATION AND ALLOCATION 5-17

Table 3. Examples of Necessary Bandwidths and Designations of Emissions (*Continued*)

Description and class of emission	Necessary bandwidth in cycles per second	Examples	
		Details	Designation of emission
Facsimile, carrier modulated by tone and by keying, A4	$B_n = KN + 2M$ $K = 1.5$	The total number of picture elements (black plus white) transmitted per second is equal to the circumference of the cylinder multiplied *by* the number of lines per unit length *and by* the speed of rotation of the cylinder in revolutions per second. Diameter of cylinder = 70 mm, number of lines per mm = 5, speed of rotation = 1 r.p.s., $N = 1{,}100$, $M = 1{,}900$; Bandwidth: 5,450 c/s.	5.45A4
Television (vision sound), A5 and F3	Refer to relevant CCIR documents for the bandwidths of the commonly used television systems.	Number of lines = 625; Number of lines per second = 15,625; Video bandwidth: 5 Mc/s; Total vision bandwidth: 6.25 Mc/s.; FM sound bandwidth including guard bands: 0.75 Mc/s; Total bandwidth: 7 Mc/s.	6,250A5C 750F3

II. Frequency Modulation

Frequency-shift telegraphy, F1	$B_n = 2.6D + 0.55B$ for $1.5 < \dfrac{2D}{B} < 5.5$ $B_n = 2.1D + 1.9B$ for $5.5 \leqslant \dfrac{2D}{B} \leqslant 20$	Four-channel time-division multiplex with 7-unit code, 42.5 bauds per channel, $B = 170$, $D = 200$; $\dfrac{2D}{B} = 2.35$, therefore the first formula in Column 2 applies; Bandwidth: 613 c/s.	0.6F1
Commercial telephony, F3	$B_n = 2M + 2DK$ K is normally 1 but under certain conditions a higher value may be necessary.	For an average case of commercial telephony, $D = 15{,}000$, $M = 3{,}000$; Bandwidth: 36,000 c/s.	36F3
Sound broadcasting, F3	$B_n = 2M + 2DK$	$D = 75{,}000$, $M = 15{,}000$ and assuming $K = 1$; Bandwidth: 180,000 c/s.	180F3

Table 3. Examples of Necessary Bandwidths and Designations of Emissions (*Continued*)

Description and class of emission	Necessary bandwidth in cycles per second	Examples	
		Details	Designation of emission
Facsimile, F4	$B_n = KN + 2M + 2D$ $K = 1.5$	(See facsimile, amplitude modulation.) Diameter of cylinder = 70 mm, number of lines per mm = 5, speed of rotation = 1 r.p.s., $N = 1,100$, $M = 1,900$, $D = 10,000$; Bandwidth: 25,450 c/s.	25.5F4
Four-frequency diplex telegraphy, F6	If the channels are not synchronized, $B_n = 2.6D + 2.75B$ where B is the speed of the higher speed channel. If the channels are synchronized the bandwidth is as for F1, B being the speed of either channel.	Four-frequency diplex system with 400 c/s spacing between frequencies, channels not synchronized, 170 bauds keying in each channel, $D = 600$, $B = 170$; Bandwidth: 2,027 c/s.	2.05F6

III. Pulse Modulation

Unmodulated pulse, P0	$B_n = \dfrac{2K}{t}$ K depends upon the ratio of pulse duration to pulse rise time. Its value usually falls between 1 and 10 and in many cases it does not need to exceed 6.	$t = 3 \times 10^{-6}$, $K = 6$; Bandwidth: 4×10^6 c/s.	4,000 P0
Modulated pulse, P2 or P3	The bandwidth depends on the particular types of modulation used, many of these being still in the development stage.	—	—

Table 4. ITU Frequency Tolerances*

1. Frequency tolerance is defined in Sec. 3, page 5-13, and is expressed in parts in 10^6 or, in some cases, in cycles per second.
2. The power shown for the various categories of stations is the mean power defined in Sec. 3, page 5-14.

Frequency bands (lower limit exclusive, upper limit inclusive) and categories of stations	Tolerances applicable until 1st January, 1966† to transmitters in use and to those to be installed before 1st January, 1964	Tolerances applicable to new transmitters installed after 1st January, 1964 and to all transmitters after 1st January, 1966†
Band: 10 to 535 kc/s		
1. *Fixed Stations:*		
10 to 50 kc/s	1,000	1,000
50 to 535 kc/s	200	200
2. *Land Stations:*		
(a) Coast Stations:		
Power 200 W or less	500	500
Power above 200 W	200	200
(b) Aeronautical Stations	200*	100*
3. *Mobile Stations:*		
(a) Ship Stations	1,000 (a)	1,000 (a)
(b) Ship's Emergency Transmitters	5,000	5,000
(c) Survival Craft Stations	5,000	5,000
(d) Aircraft Stations	500	500
4. *Radiodetermination Stations*	200*	100*
5. *Broadcasting Stations*	20 c/s	10 c/s
Band: 535 to 1,605 kc/s		
Broadcasting Stations	20 c/s	10 c/s (b)
Band: 1,605 to 4,000 kc/s		
1. *Fixed Stations:*		
Power 200 W or less	100	100
Power above 200 W	50	50
2. *Land Stations:*		
Power 200 W or less	100	100
Power above 200 W	50	50
3. *Mobile Stations:*		
(a) Ship Stations	200	200
(b) Survival Craft Stations	. . . . . .	300
(c) Aircraft Stations	200*	100*
(d) Land Mobile Stations	200	200
4. *Radiodetermination Stations:*		
Power 200 W or less	100	100
Power above 200 W	50	50
5. *Broadcasting Stations*	50	20
Band: 4 to 29.7 Mc/s		
1. *Fixed Stations:*		
Power 500 W or less	100	50
Power above 500 W	30	15
2. *Land Stations:*		
(a) Coast Stations:		
Power 500 W or less	50	50
Power above 500 W and less than or equal to 5 kW	50*	30*
Power above 5 kW	50	15
(b) Aeronautical Stations:		
Power 500 W or less	100	100
Power above 500 W	50	50

Table 4. ITU Frequency Tolerances* (Continued)

Frequency bands (lower limit exclusive, upper limit inclusive) and categories of stations	Tolerances applicable until 1st January, 1966* to transmitters in use and to those to be installed before 1st January, 1964		Tolerances applicable to new transmitters installed after 1st January, 1964 and to all transmitters after 1st January, 1966*	
(c) Base Stations:				
Power 500 W or less	100		100	
Power above 500 W	50		50	
3. *Mobile Stations:*				
(a) Ship Stations:				
(1) Class A1 emission	200		200	
(2) Emission other than Class A1:				
Power 50 W or less	50	(c)	50	(c)
Power above 50 W	50		50	
(b) Survival Craft Stations	200		200	
(c) Aircraft Stations	200*		100*	
(d) Land Mobile Stations	200		200	
4. *Broadcasting Stations*	30		15	
Band: 29.7 to 100 Mc/s				
1. *Fixed Stations:*				
Power 200 W or less	200*		50*	
Power above 200 W	200		30	
2. *Land Stations:*				
Power 15 W or less	200		50	
Power above 15 W	200		20	
3. *Mobile Stations:*				
Power 5 W or less	200		100	
Power above 5 W	200		50	
4. *Radiodetermination Stations*	200		200	
5. *Broadcasting Stations (other than television):*				
Power 50 W or less	50		50	
Power above 50 W	30		20	
6. *Broadcasting Stations (television sound and vision):*				
Power 50 W or less	100		100	
Power above 50 W	30		1,000 c/s	
Band: 100 to 470 Mc/s				
1. *Fixed Stations:*				
Power 50 W or less	100*		50*	
Power above 50 W	100*		20*	
2. *Land Stations:*				
(a) Coast Stations	100		20	
(b) Aeronautical Stations	100		50	
(c) Base Stations:				
Power 5 W or less	100		50	
Power above 5 W	100		20	
3. *Mobile Stations:*				
(a) Ship Stations and Survival Craft Stations:				
In the band 156–174 Mc/s:	100		20	
Outside this band	100	(d)	50	(d)
(b) Aircraft Stations	100		50	
(c) Land Mobile Stations:				
Power 5 W or less	100		50	
Power above 5 W	100		20	

Table 4. ITU Frequency Tolerances* (Continued)

Frequency bands (lower limit exclusive, upper limit inclusive) and categories of stations	Tolerances applicable until 1st January, 1966† to transmitters in use and to those to be installed before 1st January, 1964		Tolerances applicable to new transmitters installed after 1st January, 1964 and to all transmitters after 1st January, 1966†	
4. *Radiodetermination Stations*	200*		50*	(d)(e)
5. *Broadcasting Stations* (other than television)	30		20	
6. *Broadcasting Stations* (television sound and vision):				
Power 100 W or less	100		100	
Power above 100 W	30		1,000 c/s	
Band: 470 to 2,450 Mc/s				
1. *Fixed Stations:*				
Power 100 W or less	7,500		300	(f)
Power above 100 W	7,500		100	(g)
2. *Land Stations*	7,500		300	
3. *Mobile Stations*	7,500		300	
4. *Radiodetermination Stations*	7,500	(e)	500	(e)
5. *Broadcasting Stations* (other than television)	7,500		100	
6. *Broadcasting Stations* (television, sound and vision) in the band 470–960 Mc/s:				
Power 100 W or less	7,500		100	
Power above 100 W	7,500		1,000 c/s	
Band: 2,450 to 10,500 Mc/s				
1. *Fixed Stations:*				
Power 100 W or less	7,500		300	(f)
Power above 100 W	7,500		100	(g)
2. *Land Stations*	7,500		300	
3. *Mobile Stations*	7,500		300	
4. *Radiodetermination Stations:*	7,500	(e)	2,000	(e)
Band: 10.5 to 40 Gc/s				
1. *Fixed Stations*	—		500	
2. *Radiodetermination Stations*	—		7,500	(e)

* Certain services may need tighter tolerances for technical and operational reasons.

† 1st January, 1970, in the case of all tolerances marked with an asterisk.

(a) At the present time some administrations permit ship transmitters fulfilling the role of standby to a main transmitter not only for distress but also for traffic purposes to operate with a tolerance of 5,000. These administrations should make every effort to ensure that by 1st January, 1966, all ship transmitters operating in the band 10–535 kc/s, other than ship's emergency transmitters, have a frequency tolerance of 1,000.

(b) In the area covered by the North American Regional Broadcasting Agreement (NARBA) the tolerance of 20 c/s may continue to be applied.

(c) For ship transmitters, of power 50W or less, using frequencies below 13 Mc/s in tropical regions, the tolerance of 50 can be increased to 200 since these transmitters are sometimes used in such regions in the same circumstances as those of the band 1,605–4,000 kc/s.

(d) This tolerance is not applicable to survival craft stations operating on the frequency 243 Mc/s.

(e) Where specific frequencies are not assigned to radar stations, the bandwidth occupied by the emissions of such stations shall be maintained wholly within the band allocated to the service and the indicated tolerance does not apply.

(f) For transmitters using time division multiplex the tolerance of 300 may be increased to 500.

(g) This tolerance applies only to such emissions for which the necessary bandwidth does not exceed 3,000 kc/s; for larger bandwidth emissions a tolerance of 300 applies.

Table 5. ITU Table of Tolerances for the Levels of Spurious Emissions

1. The following table indicates the tolerances which shall apply to the mean power of any spurious emission supplied by a transmitter to the antenna transmission line.
2. Furthermore, spurious radiation from any part of the installation other than the antenna system, i.e., the antenna and its transmission line, shall not have an effect greater than would occur if this antenna system were supplied with the maximum permissible power at that spurious emission frequency.
3. These tolerances shall not, however, apply to ship's emergency transmitters or survival craft stations.
4. For technical or operational reasons, specific services may demand tolerances tighter than those specified in the table.
5. The final date by which all equipment shall meet the tolerances specified in Column B is 1st January, 1970. Nevertheless, all administrations recognize the urgent need to implement Column B tolerances for all equipment at the earliest possible dates and will endeavor to ensure that necessary changes are made to all transmitters under their jurisdiction well before this date and wherever possible by 1st January, 1966.
6. No tolerance is specified for transmitters operating on fundamental frequencies above 235 Mc/s. For these transmitters the levels of spurious emissions shall be as low as practicable.

Fundamental frequency band	The mean power of any spurious emission supplied to the antenna transmission line shall not exceed the values specified as tolerances in Columns A and B below	
	A	B
	Tolerances applicable until 1st January, 1970, to transmitters now in use and to those installed before 1st January, 1964	Tolerances applicable to transmitters installed after 1st January, 1964, and to all transmitters after 1st January, 1970
Below 30 Mc/s	40 decibels below the mean power of the fundamental without exceeding the power of 200 milliwatts	40 decibels below the mean power of the fundamental without exceeding the power of 50 milliwatts[1] [2] [3]
30 Mc/s to 235 Mc/s: For transmitters having mean power: Greater than 25 watts		60 decibels below the mean power of the fundamental without exceeding 1 milliwatt[4]
25 watts or less		40 decibels below the mean power of the fundamental without exceeding 25 microwatts and without the necessity for reducing this value below 10 microwatts[4]

[1] For transmitters of mean power exceeding 50 kilowatts and which operate below 30 Mc/s over a frequency range approaching an octave or more, a reduction below 50 milliwatts is not mandatory, but a minimum attenuation of 60 decibels shall be provided and every effort should be made to keep within the 50 milliwatts limit.

[2] For hand-portable equipment of mean power less than 5 watts which operates in the frequency band below 30 Mc/s, the attenuation shall be at least 30 decibels, but every effort should be made to attain 40 decibels attenuation.

[3] For mobile transmitters which operate below 30 Mc/s any spurious emission shall be at least 40 decibels below the fundamental without exceeding the value of 200 milliwatts, but every effort should be made to keep within the 50 milliwatts limit wherever practicable.

[4] For frequency modulated maritime mobile radio-telephone equipment which operates above 30 Mc/s, the mean power of any spurious emission falling in any other international maritime mobile channel, due to products of modulation, shall not exceed a limit of 10 microwatts and the mean power of any other spurious emission on any discrete frequency within the international maritime mobile band shall not exceed a limit of 2.5 microwatts. Where, exceptionally, transmitters of mean power above 20 watts are employed, these limits may be increased in proportion to the mean power of the transmitter.

5 Special Rules Governing the Aeronautical and Space Services

Aeronautical Mobile Service

Frequencies in any band allocated to the aeronautical mobile (R) service are reserved for communications between any aircraft and those aeronautical stations primarily concerned with the safety and regularity of flight along national or international civil air routes.

Frequencies in any band allocated to the aeronautical mobile (OR) service are reserved for communications between any aircraft and aeronautical stations other than those primarily concerned with flight along national or international civil air routes.

Administrations shall not permit public correspondence in the frequency bands allocated exclusively to the aeronautical mobile service, unless permitted by special aeronautical regulations adopted by a Conference of the Union to which all interested Members and Associate Members of the Union are invited. Such regulations shall recognize the absolute priority of safety and control messages.

Terrestrial Services Sharing Frequency Bands with Space Services between 1 Gc/s and 10 Gc/s

Choice of Sites and Frequencies. Sites and frequencies for terrestrial stations, operating in frequency bands shared with equal rights between terrestrial and space services, shall be selected having regard to the relevant recommendations of the CCIR with respect to geographical separation from earth stations.

Power Limits. (1) The maximum effective radiated power of the transmitter and associated antenna, of a station in the fixed or mobile service, shall not exceed +55 dbW.

(2) The power delivered by a transmitter to the antenna of a station in the fixed or mobile service shall not exceed +13 dbW.

(3) The limits given in 470B and 470C apply in the following frequency bands allocated to reception by space stations in the communication-satellite service, where these are shared with equal rights with the fixed or mobile service:

    5,800–5,850 Mc/s (for the countries mentioned in 390)
    5,850–5,925 Mc/s (Regions 1 and 3)
    5,925–6,425 Mc/s
    7,900–8,100 Mc/s

Space Services Sharing Frequency Bands with Terrestrial Services between 1 Gc/s and 10 Gc/s

Choice of Sites and Frequencies. Sites and frequencies for earth stations, operating in frequency bands shared with equal rights between terrestrial and space services, shall be selected having regard to the relevant recommendations of the CCIR with respect to geographical separation from terrestrial stations.

Power Limits. (1) Earth Stations in the Communication-Satellite Service.

(2) The mean effective radiated power transmitted by an earth station in any direction in the horizontal plane* shall not exceed +55 dbW in any 4 kc/s band, except that it may be increased subject to the provisions of (2) and (3). However, in no case shall it exceed a value of +65 dbW in any 4 kc/s band.

(3) In any direction where the distance from an earth station to the boundary of the territory of another administration exceeds 400 km, the limit of +55dbW in any 4 kc/s band may be increased in that direction by 2 db for each 100 km in excess of 400 km.

(4) The limit of +55 dbW in any 4 kc/s band may be exceeded by agreement between the administrations concerned or affected.

(5) The limits given in (2) apply in the following frequency bands allocated to transmission by earth stations in the communication-satellite service, where these are shared with equal rights with the fixed or mobile service:

    4,400–4,700 Mc/s
    5,800–5,850 Mc/s (for certain countries)
    5,850–5,925 Mc/s (Regions 1 and 3)
    5,925–6,425 Mc/s
    7,900–8,400 Mc/s

Minimum Angle of Elevation. (1) Earth Stations in the Communication-Satellite Service.

*For the purpose of this Regulation, the effective radiated power transmitted in the horizontal plane shall be taken to mean the effective radiated power actually transmitted towards the horizon, reduced by the site-shielding factor that may be applicable.

(2) Earth station antennas shall not be employed for transmission at elevation angles less than 3 degrees, measured from the horizontal plane to the central axis of the main lobe, except when agreed to by the administrations concerned or affected.

(3) The limit given in (2) applies in the following frequency bands allocated to transmission by earth stations in the communication-satellite service, where these are shared with equal rights with the fixed or mobile service:

> 4,400–4,700 Mc/s
> 5,800–5,850 Mc/s (for certain countries)
> 5,850–5,925 Mc/s (Regions 1 and 3)
> 5,925–6,425 Mc/s
> 7,250–7,750 Mc/s
> 7,900–8,400 Mc/s

Power Flux Density Limits. (1) Communication-Satellite Space Stations.

(a) The total power flux density at the earth's surface, produced by an emission from a communication-satellite space station, or reflected from a passive communication satellite, where wide-deviation frequency (or phase) modulation is used, shall in no case exceed -130 dbW/m^2 for all angles of arrival. In addition, such signals shall if necessary be continuously modulated by a suitable waveform, so that the power flux density shall in no case exceed -149 dbW/m^2 in any 4 kc/s band for all angles of arrival.

(b) The power flux density at the earth's surface, produced by an emission from a communication-satellite space station, or reflected from a passive communication satellite, where modulation other than wide-deviation frequency (or phase) modulation is used, shall in no case exceed -152 dbW/m^2 in any 4 kc/s band for all angles of arrival.

(c) The limits given in a and b apply in the following frequency bands allocated to transmission by space stations in the communication-satellite service, where these are shared with equal rights with the fixed or mobile services:

> 3,400–4,200 Mc/s
> 7,250–7,750 Mc/s

(2) Meteorological-Satellite Space Stations*

(a) The power flux density at the earth's surface, produced by an emission from a meteorological-satellite space station, where wide-deviation frequency (or phase) modulation is used, shall in no case exceed -130 dbW/m^2 for all angles of arrival. In addition, such signals shall if necessary be continuously modulated by a suitable waveform, so that the power flux density shall in no case exceed -149 dbW/m^2 in any 4 kc/s band for all angles of arrival.

(b) The power flux density at the earth's surface, produced by an emission from a meteorological-satellite space station, where modulation other than wide-deviation frequency (or phase) modulation is used, shall in no case exceed -152 dbW/m^2 in any 4 kc/s band for all angles of arrival.

(c) The limits given in a and b apply in the following frequency bands allocated to transmissions by space stations in the meteorological-satellite service, shared with equal rights with the fixed or mobile service:

> 1,660–1,670 Mc/s
> 1,690–1,700 Mc/s
> 7,200–7,250 Mc/s
> 7,300–7,750 Mc/s

The limits given in a and b also apply in the band 1,770–1,790 Mc/s, although the meteorological-satellite service is a secondary service in this band.

Space Services

Cessation of Emissions. Space stations shall be made capable of ceasing radio emissions by the use of appropriate devices† that will ensure definite cessation of emissions.

REFERENCES

1. B. W. Pike, IRIG Inter Range Instrumentation Range Group History Function and Status, 1959, *IRE Trans. Space Electron. Telemetry*, vol. SET-6, p. 59, March, 1960.
2. B. W. Pike, Telemetry Work Group TWG of the Inter-Range Instrumentation Group, IRIG, *IRE Trans. Space Electron. Telemetry*, vol. 6, p. 61, March, 1960.

* In view of the absence of any CCIR Recommendations relative to sharing between the meteorological-satellite service and other services, power flux density levels applicable to communication-satellite space stations are extended to meteorological-satellite space stations.

† Battery life, timing devices, ground command, etc.

IRIG DOCUMENT No. 106-60

TELEMETRY STANDARDS

Revised June 1962

Prepared by
Telemetry Working Group
of the
Inter-Range Instrumentation Group

Incorporating
"Frequency Standards for Telemetry"
prepared by the
Frequency Coordination Working Group
of the
Inter-Range Instrumentation Group

Document approved August 1962

This document was reissued March 1963
Additional copies may be obtained from:
Secretariat
Range Commanders Conference
ATTN: STEWS-ROD-RCC
White Sands Missile Range, New Mexico

or

Armed Services Technical Information Agency
Arlington Hall Station, Arlington 12, Virginia

Foreword

A standard in the field of telemetry for guided missiles was established in 1948 by the Research and Development Board and was thereafter revised and extended as the result of periodic reviews by that agency. The last official RDB revision of the standard was published as MTRI 204/6, dated 8 November 1951.*

Since then, the Inter-Range Instrumentation Group (IRIG) has prepared new standards in telemetry. The Steering Committee representing IRIG and the Department of Defense test ranges assigned the task of promulgating new or revised telemetry standards to the Telemetry Working Group (TWG). This publication, the first revision of *IRIG Telemetry Standards* dated November 1960, contains the current combined standards and supersedes the following IRIG documents:

101-55	Testing for Speed Errors in Instrumentation Type Magnetic Tape Recorders
102-55	Telemetry Standards for Guided Missiles
103-56	Revised Telemetry Standards for Guided Missiles
101-57	Magnetic Recorder/Reproducer Standards
102-59	Standards for Pulse Code Modulation (PCM) Telemetry
101-60	Magnetic Recorder/Reproducer Standards
106-60	IRIG Telemetry Standards (November 1960)

* In 1951, the RDB was succeeded by the Office of the Assistant Secretary of Defense (Research and Development).

These telemetry standards were established to further the compatibility of airborne transmitting equipment and ground receiving and data-handling equipment at the test ranges. To this end, the IRIG Steering Committee recommends that telemetry equipment at the test ranges conform to the standards presented here.

The quality of terminal equipment in general will be raised by concentrating development on a minimum of system types. Research should be continued, however, on telemetry systems that may offer substantial improvements over those described in this document.

Agencies proposing to use equipment that deviates from these standards should be required to show that their proposed action is both technically necessary and economically feasible.

To ensure that the standards remain current, the Telemetry Working Group will review them at each meeting and, if necessary, will revise them annually.

Now being considered for inclusion in the IRIG telemetry standards are the following: a revision of the FM/FM section to extend subcarrier bands to higher frequencies, standards for video recorder/reproducers, and recommended techniques for predetection recording.

1 Radio Frequencies

Detailed information on frequency usage is found in IRIG Recommendation No. 101-59 Revised, *Frequency Standards for Telemetry*, which is included in these standards as Appendix I.

2 FM/FM or FM/PM Standards

2.1 General

These telemetry systems are of the frequency-division multiplex type. That is, a radio-frequency carrier is modulated by a group of subcarriers, each of a different frequency. The subcarriers are frequency-modulated in a manner determined by the intelligence to be

Table I. Subcarrier Bands

Band	Center frequency (cps)	Lower limit* (cps)	Upper limit* (cps)	Maximum deviation (per cent)	Frequency response** (cps)
1	400	370	430	±7.5	6.0
2	560	518	602	"	8.4
3	730	675	785	"	11
4	960	888	1,032	"	14
5	1,300	1,202	1,399	"	20
6	1,700	1,572	1,828	"	25
7	2,300	2,127	2,473	"	35
8	3,000	2,775	3,225	"	45
9	3,900	3,607	4,193	"	59
10	5,400	4,995	5,805	"	81
11	7,350	6,799	7,901	"	110
12	10,500	9,712	11,288	"	160
13	14,500	13,412	15,588	"	220
14	22,000	20,350	23,650	"	330
15	30,000	27,750	32,250	"	450
16	40,000	37,000	43,000	"	600
17	52,500	48,562	56,438	"	790
18	70,000	64,750	75,250	"	1,050
A***	22,000	18,700	25,300	±15	660
B	30,000	25,500	34,500	"	900
C	40,000	34,000	46,000	"	1,200
D	52,500	44,625	60,375	"	1,600
E	70,000	59,500	80,500	"	2,100

Notes: * Rounded off to nearest cycle.
** The frequency response given is based on maximum deviation and a deviation ratio of 5 (see discussion in paragraph 2.2.2).
*** Bands A through E are optional and may be used by omitting adjacent bands as shown in the following table. In the process of recording the foregoing subcarriers on magnetic tape at a receiving station, provision may also be made to record a tape-speed-control tone and tape-speed-error-compensation signals, as specified in section 6 of these standards.

Band used	Omit bands—
A	13, 15 and B
B	14, 16, A and C
C	15, 17, B and D
D	16, 18, C and E
E	17 and D

transmitted. One or more of the subcarriers may be modulated by a time-division multiplex scheme (commutation) in order to increase considerably the number of individual data channels available in the system. The modulation of the radio-frequency carrier may be by either of two methods—frequency modulation (FM) or phase modulation (PM).

2.2 Subcarrier Bands

Eighteen standard subcarrier-band center frequencies, with accompanying information on frequency deviation and nominal intelligence frequency response, are specified in Table I. It is intended that the standard FM/FM receiving stations at the test ranges be capable of simultaneously demodulating a minimum of any 12 of these subcarrier signals. The nominal frequency response listed for each band is computed on a basis of maximum deviation and a deviation ratio of 5, and it is intended that the standard receiving station be capable of demodulating data with these frequency responses. It should be remembered, however, that the actual frequency response obtainable depends on many things, such as the actual deviation used, the characteristics of the filters, etc. The primary reason for specifying a frequency response is to ensure that elements in the receiving station, such as filters and recording oscillographs, provide the frequency responses shown in Table I.

2.2.1 While deviation ratios of 5 are recommended, ratios as low as 1 or less may be used, but in that case low signal-to-noise ratios, possibly increased harmonic distortion and cross-talk must be expected.

2.2.2 The 18 bands were chosen to make the best use of present equipment and the frequency spectrum. There is a ratio of approximately 1.3:1 between center frequencies of adjacent bands, except between 14.5 kilocycles and 22 kilocycles, where a larger gap was left to provide for compensation tone in magnetic-tape recording. The deviation has been kept at ± 7.5 per cent for all bands, with the option of ± 15 per cent deviation on the five higher bands to provide for transmission of higher frequency data. When this option is exercised on any of these five bands, certain adjacent bands cannot be used (see footnote to Table I).

2.2.3 It is likely that certain applications will make amplitude pre-emphasis of some subcarrier signals desirable, and it is recommended that the ground equipment capable of accommodating this pre-emphasized signal. A de-emphasis capability of up to 9 db per octave may be required.

2.3 *Automatic Correction of Subcarrier Zero and Sensitivity Drift*

2.3.1 *General:* In some cases it is found necessary to automatically correct for subcarrier zero and sensitivity drift during the course of a test. To provide for such corrections, calibration signals are applied to the subcarrier oscillators, which must have such correction by an in-flight calibrator. Also, a signal is required to arm and actuate the automatic correction equipment in the receiving or data-playback station. When employed, automatic correction command and calibration signals shall conform to the standards set forth in section 2.3.2, which follows.

2.3.2 *Automatic Correction Command:* A standard IRIG ± 7.5 per cent deviation subcarrier band, multiplexed with the data subcarriers, shall be employed to transmit the correction commands. The command subcarrier which has a lower frequency than any of the data subcarriers, shall be modulated as follows:

  2.3.2.1 *Command Sequence:* The command sequence shall be: "data," "correct for zero drift," "correct for sensitivity drift," "data."

  2.3.2.2 *Command Subcarrier Modulation:*

   2.3.2.2.1 The "data" command is indicated by the command subcarrier operating at its nominal center frequency $\pm 0.75\%$ to f_c.

   2.3.2.2.2 The "correct for zero drift" command is indicated by a displacement of the command subcarrier upward in frequency to f_c plus (6.75% f_c $\pm$ 0.75% f_c). This command shall occupy 50 per cent of the total calibration time interval.

   2.3.2.2.3 The "correct for sensitivity drift" command is indicated by a displacement of the command subcarrier downward in frequency to f_c minus (6.75% f_c $\pm$ 0.75% f_c). This command shall occupy 50 per cent of the total calibration time interval.

2.3.3 *Data Subcarrier Calibration:*

  2.3.3.1 *Calibration Sequence:* The calibration sequence shall be: "data," "center frequency," "80 per cent of design deviation," "data."

  2.3.3.2 *Subcarrier Modulation* (See Figure 1):

   2.3.3.2.1 The "data" position is the subcarrier connected to its normal data source (transducer, commutator, etc.).

FM/FM OR FM/PM STANDARDS

2.3.3.2.2 The "center frequency" position is the subcarrier connected to a signal source that would result in the nominal subcarrier center frequency if no zero or sensitivity drift has occurred. The subcarrier shall remain at this position for 50 per cent of the calibration interval.

2.3.3.2.3 In the "80 per cent of design deviation" position, the subcarrier is connected to a signal source that would result in a frequency equal to 80 per cent of design deviation if no zero or sensitivity drift has occurred. The subcarrier shall remain at this position for 50 per cent of the calibration interval.

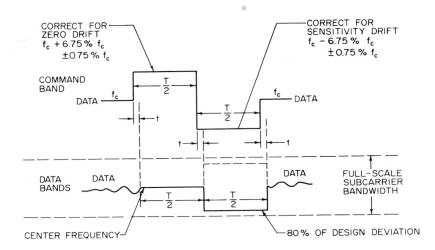

FIG. 1. Automatic zero and sensitivity drift calibration, command and data channel signals.

2.3.3.3 *Phasing of Calibration Signal:* The data subcarrier calibration signals shall lag the command signals by 200 milliseconds, $-0 + 50$ milliseconds (see Figure 1).

2.3.4 *Correction Capability:* Automatic correction equipment shall be capable of correcting zero and sensitivity drift errors of up to ± 10 per cent of full-scale subcarrier bandwidth per calibrate cycle.

2.3.4.1 *Calibration Duration:* For maximum (± 10 per cent) zero and sensitivity drift correction, the calibration interval shall be 5 seconds. Where maximum corrections per calibrate cycle are not required, the calibration interval may be correspondingly reduced.

2.4 *PAM/FM/FM Commutation*

Commutation (time-division multiplexing) may be used in one or more subcarrier bands. A nearly limitless variety of commutation schemes could be devised, but a few relatively simple methods will satisfy most telemetry needs. The specifications listed below for commutation were chosen to give maximum flexibility consistent with presently available equipment and techniques, and it is intended that, in order to limit the varieties that must be handled at test ranges, the following restrictions on commutation be observed:

2.4.1 The total number of samples per frame (number of segments of a mechanical commutator) and the frame rates shall be one of the combinations shown in Table II. If a higher commutation rate is required for certain information, two or more samples per frame (equally spaced in time) can be used to represent one telemetered function at the expense of the total number of information channels. This process is referred to as cross-strapping or supercommutation.

2.4.2 The commutation pattern in the subcarrier frequency vs. time domain shall be as shown in Fig. 2.

2.4.3 A frame-synchronizing pulse of full-scale amplitude and duration equal to two "on" periods plus one "off" period shall be provided once every frame, as shown in Figure 2.

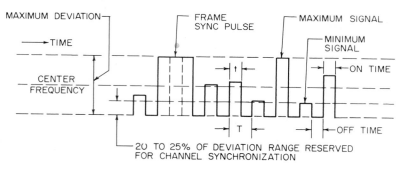

Fig. 2. PAM pulse train waveform.

2.4.4 The commutator speed (or frame rate) shall not vary more than +5.0 per cent to −15 per cent from the nominal values given in Table II.

2.4.5 The duty cycle shall be 40 per cent to 65 per cent.

2.4.6 A channel synchronization pedestal is required for automatic decommutation (see Figure 2).

Table II. Commutation Specification for Automatic Decommutation

No. of samples per frame*	Frame rate (frames per sec)	Commutation rate** (samples per sec)	Lowest recommended subcarrier bands (cps)
18	5	90	14,500
18	10	180	22,000 ($\pm 15\%$) or 30,000 ($\pm 7.5\%$)
18	25	450	30,000 ($\pm 15\%$) or 70,000 ($\pm 7.5\%$)
30	2.5	75	10,500
30	5	150	22,000 ($\pm 7.5\%$)
30	10	300	22,000 ($\pm 15\%$) or 40,000 ($\pm 7.5\%$)
30	20	600	40,000 ($\pm 15\%$)
30	30	900	70,000 ($\pm 15\%$)

Notes: * The number of samples per frame available to carry information is 2 less than the number indicated, because the equivalent of 2 samples is used in generating the frame-synchronizing pulse.

** Frame rate times number of samples per frame.

2.5 In-flight Zero and Full-scale Calibration

On all pulse-amplitude-modulation (PAM) commutators, channels one and two, following the synchronizing pulse, are recommended for zero and full-scale calibration, respectively.

3 PDM/FM or PDM/PM or PDM/FM/FM Standard

3.1 General

The pulse-duration-modulation (PDM) systems are intended for use where a time-division multiplex system can meet the bulk of the telemetry requirements of a given application. A relatively large number of information channels can be accommodated, but with a relatively low frequency response capability in comparison with the subcarrier channels of the FM/FM system.

3.2 PDM/FM or PDM/PM†

The following are the specifications for the PDM signal:

Number of samples per frame*	30	45	60	90
Frame rate (frames per second)	30	20	15	10
Commutation rate (samples per second)**	900	900	900	900

Notes: * The number of samples per frame available to carry information is two less than the number indicated, because the equivalent of two samples is used in generating the frame-synchronizing pulse.
    ** Frame rate times number of samples per frame.

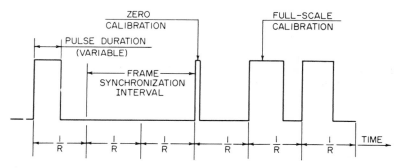

FIG. 3. PDM pulse train waveform.

The amplitude of the measurands being transmitted in each channel shall determine the duration of the corresponding pulses. The relation between measurands and pulse duration should, in general, be linear.

Minimum pulse duration (zero level information)	90 ± 30 microseconds
Maximum pulse duration (maximum level information)	700 ± 50 microseconds
Pulse rise and decay time (measured between 10% and 90% levels)	10 to 20 microseconds (constant to ±1 microsecond for a given transmitting set)

† Use of PDM/PM requires the use of equipment to reshape the receiver output signal for proper operation of many display scopes, decommutators and tape recorder/reproducers. PDM/FM is recommended over PDM/PM.

3.2.1 The time interval between the leading edges of successive pulses within a frame shall be uniform from interval to interval within ±25 microseconds. This time interval shall have a nominal period equal to 1 divided by the total sampling rate.

3.2.2 The commutator speed or frame rate shall not vary more than plus 5.0 per cent to minus 15.0 per cent from nominal.

A frame-synchronizing interval equal to two successive pulse time intervals shall exist in the train of pulses transmitted, to be used for synchronization of the commutator and the decommutator. A representation of the pulse train waveform is shown in Figure 3.

3.3 PDM/FM/FM

PDM systems may also be employed on the ±15 per cent deviation channels of the standard FM/FM multiplex systems. When so used, they are designated PDM/FM/FM telemetry. It should be recognized that this application of PDM is wasteful of bandwidth and that it places three wide-band-modulation systems in cascade. Gaussian-type-output low-pass filters should be used at the subcarrier discriminator outputs for this application.

3.3.1 *Subcarrier Channels:* The recommended subcarrier channels for this application are bands B, C, D or E. Operating criteria for use of these specific bands are shown in Table III. Satisfactory performance is contingent upon the use of optimum-output low-pass filters.

Table III. PDM Modulation of FM/FM Subcarrier Channels

Samples per second	Channel allocation	FM/FM channel (kcps)	Deviation utilized (per cent)	Recommended value of minimum pulse length (microseconds)
900	B	30.0	±7.5	200 + 30 − 0
900	C	40.0	±7.5	170 + 30 − 0
900	D	52.5	±7.5	150 + 30 − 0
900	E	70.0	±7.5	110 + 30 − 0

Note: Institute of Radio Engineers, "The Transmission of Pulse Width Modulated Signals over Restricted Bandwidth Systems," *IRE Transaction on Telemetry and Remote Control*, Vol. TRC-3, No. 1, April 1957.

3.3.2 *Time-interval Variation Between Leading Edges of Successive Pulses:* Section 3.2.1 shall apply.

3.3.3 *Commutation Speed:* Section 3.2.2 shall apply.

3.4 In-flight Zero and Full-scale Calibration

On all PDM commutators, channels 1 and 2, following the synchronizing pulse, are recommended for zero and full-scale calibration, respectively.

4 PAM/FM Standards

4.1 *General*

Pulse amplitude modulation (PAM) data specified in these standards shall be transmitted as time-division multiplexed analog sample pulses.

This standard defines recommended operating parameters and design practices insuring efficient and reliable performance, for the implementation of Pulse Amplitude Modulation (PAM) telemetry systems. Reasonable latitude is provided for moderate variation about theoretical optimal configurations, where such is justified by practical considerations. Attention is also directed to Standards for PAM/FM/FM Systems in Section 2.4.

4.2 *IF Bandwidth and Transmitter Deviation**

4.2.1 Selection of the design IF bandwidth (exclusive of the band shift factor) to be used should be guided by the permissible RMS error at minimum required receiver power.

* See Appendix IV, Section 1, for additional information.

See Figure 4. A total sampling rate shall be used which lies in the range of ½–¼ times the actual receiver IF bandwidth (3 db points) divided by the IF band shift factor (1.5–3.3).

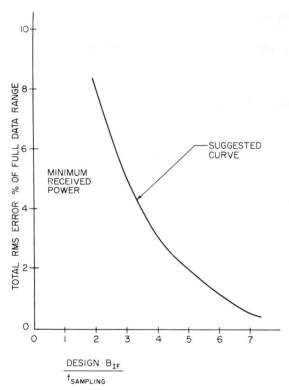

Fig. 4. $\dfrac{\text{Design IF bandwidth}}{\text{Sampling frequency}}$ vs. error at minimum required receiver power

4.2.2 The peak-to-peak frequency deviation shall not exceed 0.75 times the design IF bandwidth selected in accordance with paragraph 4.2.1.

4.2.3 The actual receiver IF bandwidth shall be between three (3) and twenty-three (23) times the total sampling rate. This includes a system IF band shift factor (1.5–3.3). Selection of receiver intermediate frequency (IF) bandwidth (3 db points) shall be from the set of discrete bandwidths listed in Table IV.

Table IV. List of Receiver IF Bandwidths (3-db points; cps)

12,500*
25,000*
50,000*
100,000
300,000
500,000
1,000,000**
1,500,000**

* For use only under special conditions. See Appendix II, Section 1.2.
** For use in the 1435–1535 Mc and 2200–2300 Mc telemetry frequency bands only.

4.3 Sampling Rate Stability

The long term stability of the sampling rate shall be one per cent or better of the specified nominal sampling rate measured over any consecutive N samples, where N is the number of samples closest to 1000 within an integral number of frames. Within any period of N consecutive samples, no data pulse shall deviate from its average position, measured over that period, by more than five per cent of a PAM sample period (T). See Figure 5(a).

4.4 Frame and Pulse Structure

4.4.1 The number of primary channels, including those assigned to synchronization signals shall not exceed 130 (frame length).

4.4.2 The sample pulses generated for transmission shall have a duty cycle of either 50 per cent ±5 per cent or 100 per cent ±5 per cent. See Figure 5(a), 5(b), 5(c) and 5(d).*

4.5 Synchronization

Each frame shall be identified uniquely by the transmission of a pulse amplitude assigned to synchronization,† or by means of a binary code word‡ as illustrated in Figures 5(a), 5(b), 5(c) and 5(d). The maximum length of the code word when used shall not exceed 33 times the period of time defined by the duration of the individual pulses comprising the

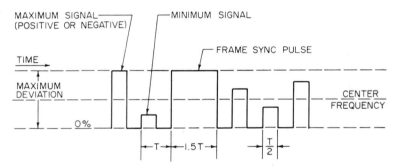

20 TO 25% OF DEVIATION IS RESERVED FOR CHANNEL SYNCHRONIZATION.

CHANNEL SYNCHRONIZATION IS RECOMMENDED BUT NOT REQUIRED.

IF THE 20 TO 25% DEVIATION IS NOT USED FOR SYNCHRONIZATION, MINIMUM SIGNAL WOULD CORRESPOND TO THE OPPOSITE DEVIATION EXTREME FROM MAXIMUM SIGNAL.

FIG. 5a. PAM pulse train waveform, conventional frame synchronization 50 per cent duty cycle.

code word. The time duration of the pulses comprising the synchronizing code word shall evenly divide the duration of the one or more PAM sample pulse periods occupied by the code word. The minimum duration of the pulses comprising the code word is defined as 1/M times the data sampling period, where M is the largest integer not in excess of the design IF bandwidth divided by the total sampling rate.

4.6 Supercommutation and Subcommutation

4.6.1 Supercommutation and subcommutation are acceptable methods of exchanging the number of measurements and sampling rates. A unique synchronization pattern shall be used to indicate the beginning of the longest subcommutated frame. All other sub-commutated frames with shorter lengths shall be integral sub-multiples of the longest

---

* See Appendix IV, Section 2, for additional information.
† See Appendix IV, Section 3, for additional information.
‡ See Appendix IV, Section 3, for discussion of binary code words.

PAM/FM STANDARDS 5-35

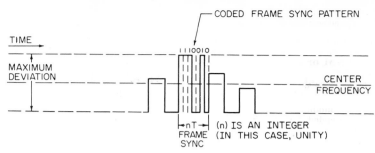

Fig. 5b. PAM pulse train waveform, coded frame synchronization 50 per cent duty cycle.

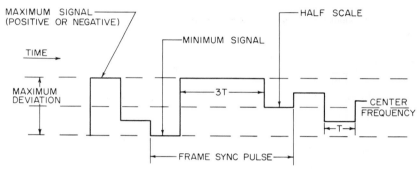

Fig. 5c. of IRIG Document 160-60 corrected PAM pulse train waveform, conventional frame synchronization 100 per cent duty cycle.

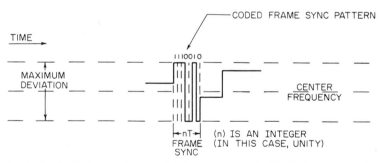

Fig. 5d. PAM pulse train waveform, coded frame synchronization 100 per cent duty cycle.

frame and shall have a fixed and known relative phase with respect to the longest subcommutated frame.

4.6.2 The longest subcommutated frame shall not exceed 130 channels, including the channels used for synchronization.

4.7 Premodulation Filtering

Premodulation filtering shall be used in the transmitter to obtain an overall radiated spectrum which does not exceed the system design IF bandwidth. The amplitude of the power spectrum of the radiated signal outside the design IF bandwidth shall not exceed one per cent of the unmodulated carrier amplitude.*,†

* See Appendix IV, Section 4, for recommended filter characteristic.

† The requirements of the "Frequency Standards for Telemetry" shall be met.

4.8 *RF Carrier Modulation*

Frequency modulation of the radio frequency carrier shall be used. Transmission of frame and channel synchronizing code pulses when used shall be such that the carrier is deviated to the higher frequency deviation limit to transmit a "one" and to the lower frequency deviation limit to transmit a "zero." Once a frequency deviation limit is reached for either a "one" or a "zero," the resulting frequency shall remain essentially constant for consecutive like bits. See Figure 5(b) and 5(d).

5 PCM/FM or PCM/PM Standards

5.1 *General**

Pulse-code-modulation (PCM) data specified in these standards shall be transmitted as serial-binary-coded, time-division multiplexed samples.

5.2 *Bit Rate vs. Receiver IF Bandwidth (3-db points)*

Selections of bit rates and corresponding receiver intermediate-frequency (IF) bandwidth shall be made from those listed in Table V. Only those discrete receiver IF bandwidths listed shall be used (optional below 12,500 cps). The selections in Table V have been made on the consideration that automatic tracking of radio-frequency (RF) carrier drift or shift will be used in the receiver.

Table V. PCM Bit Rate and Receiver IF Bandwidth (3-db points)

System type	Bit rate (bits per sec)*	Receiver IF bandwidth (cps)
A	8,000 and lower**	12,500 (and as required for lower bit rates)
B	8,000 to 65,000	25,000 — 50,000 — 100,000
C	50,000 to 330,000	100,000 — 300,000 — 500,000
D	320,000 to 800,000	500,000 — 1,000,000*** — 1,500,000***

In the 225- to 260-mc band, bit rates that require bandwidth in excess of those prescribed in section 2.1.2.1 of Appendix I shall not be used.

Notes: * See restrictions imposed by section 6, "Magnetic Tape Recorder/Reproducer Standards."
 ** Systems in this category may require special consideration and nonstandard hardware; these systems will be handled separately, as required by the test ranges concerned.
 *** For use only in 1435- to 1535-mc and 2200- to 2300-mc telemetry bands.

It is recommended that, for practical design considerations, a bit rate equal to the receiver IF bandwidth (3-db points), divided by a factor ranging from 1.5 to 3.3, be used. The bandwidth/bit-rate relationships in Table V were selected on this basis.†

5.3 *Bit-rate Stability*

The change in bit rate shall not exceed 1.0 per cent of the nominal bit rate. In addition, spurious time displacement of bit phase in an interval "T" relative to the frequency and phase established by an average over the preceding interval "T" shall not exceed 0.1 bit period. The interval "T" shall be taken as 10 times the maximum period between assured bit transitions. Such transitions or changes in state may be provided by appropriate parity, fixed programming, the guarantee that all data will not simultaneously go to zero or full scale, etc.

The allowable change in bit rate given above accommodates the use of magnetic drums. When drums are not involved, crystal-controlled clock frequency is recommended.

5.4 *Word and Frame Structure*

The number of bits per frame shall not exceed 2,048, including those used for frame synchronization. The frame length selected for a particular mission‡ shall be kept con-

---

* See Appendix II for additional PCM information and recommendations.
† See Appendix II, section 1, for additional information.
‡ See Appendix II, section 2, Word and Frame Structure.

stant. Word length for any given word position can range from 6 to 64 bits but shall be kept constant for that position for a particular mission.[2] It is recommended that an odd parity bit be included where a higher order of confidence in bit transmission is desired.

5.5 Synchronization

Frames shall be identified by a single frame-synchronization word, which shall be limited to a maximum length of 33 adjacent bits.*

5.6 Supercommutation and Subcommutation

5.6.1 Supercommutation and subcommutation are acceptable methods of exchanging the number of measurements and sampling rate. A selected coded word shall be used to indicate the beginning of the longest subcommutated frame. All other subcommutated frames with shorter lengths, shall be submultiples of the longest frame and shall have a fixed and known relative phase with respect to the longest subcommutated frame.

5.6.2 The longest subcommutated frame shall not exceed 130 channels, including the channel used for synchronization.

5.7 Premodulation Filtering

Filtering shall be used before the transmitter modulator.†

5.8 RF Carrier Modulation

5.8.1 The RF carrier modulation method shall be either FM or PM.

5.8.2 Frequency modulation of the carrier, when used, shall be of the type in which the carrier is deviated to the higher frequency deviation limit to transmit a "one" and to the lower frequency deviation limit to transmit a "zero." Once a frequency deviation limit is reached for either a "one" or a "zero," the resulting frequency remain essentially constant for consecutive like bits.

5.8.3 Phase modulation of the carrier, when used, shall be of the type in which the carrier is deviated to the leading phase-deviation limit to transmit a "one" and to the lagging phase-deviation limit to transmit a "zero." Once a phase-deviation limit is reached for either a "one" or a "zero," the resulting phase remains essentially constant for consecutive like bits.

6 Magnetic-tape Recorder/Reproducer Standards

6.1 Scope

These standards define terminology and specify the configuration and operating characteristics of magnetic-tape recording/reproducing devices used for telemetry and airborne-data-collection applications at the missile ranges. Also included are standards applying to magnetic tapes used by magnetic-tape recording-reproducing devices.

Although primarily for use in telemetry data recording and reproducing, these standards are also intended to serve as a guide in the procurement of airborne magnetic-tape-recording equipment. Compatibility of airborne recording equipment is desirable so that standard reproducing equipment on the ground can be used for playback.

Because the magnetic tape is the only common element between the recording and reproducing devices, the configuration of the devices is referenced, where applicable, to the magnetic tape.

The standard speeds for instrumentation type magnetic tape recorder/reproducers are 1⅞, 3¾, 7½, 15, 30, 60 and 120 inches per second. Though no specific recorder may be required to operate at all speeds, all such recorders, suitably configured, must be capable of performing at these speeds in accordance with the specifications presented herein. The number and/or combinations of speeds to be provided will be delineated by the application involved.

6.2 Requirements

Magnetic-tape recording and reproduction of telemetry signals are used throughout all the missile ranges. For maximum utilization of the various telemetry-recording activi-

---

* See Appendix II, section 3, for suggested PCM synchronization patterns.

† See Appendix II, section 4, for recommended filter characteristics.

5-38 STANDARDS FOR TELEMETRY

ties' potential, it is desirable that recording information and equipment be interchangeable to the maximum degree. These standards are to be used to that end. It is quite possible that, at any one establishment—or even during a single operation, one of the several methods of information storage set forth here may be used, or any combination may be used simultaneously.

6.3 Direct Recording

Direct recording is used for FM/FM-derived telemetry data and also for such applications as airborne recording, in which the data have not been telemetered but are of the same general form as FM/FM telemetered data would be at the output of the ground receiver. Standards for this type of telemetry data are set forth in section 2 of this document.

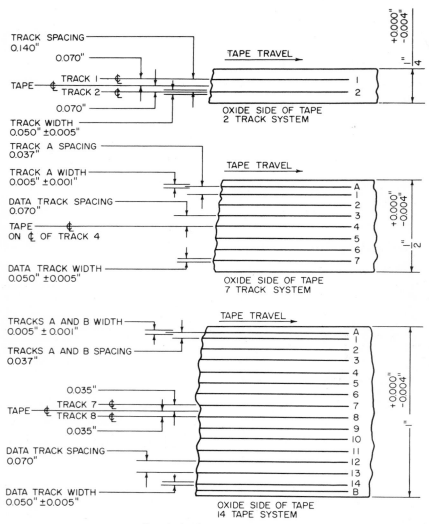

Fig. 6. Analog tape geometry.

MAGNETIC-TAPE RECORDER/REPRODUCER STANDARDS

6.3.1 *Tape:*
   6.3.1.1 *Tape Widths:* The standard tape widths are $\frac{1}{4}$ inch, $\frac{1}{2}$ inch and 1 inch, with tolerances on all widths of plus 0.000 inch, minus 0.004 inch. The preferred tape width for radio telemetry is $\frac{1}{2}$ inch, and this should be used whenever it is compatible with program requirements.
   6.3.1.2 *Tape Thickness:* The over-all thickness (base plus coating) of the tape used for telemetry applications will be within the limits 1.0 mil and 2.2 mils, depending on the type of base and coating used. Federal specification MIL-T-21029A (Ships), "Military Specification Tapes, Sound Recording, Telemetering Instrumentation, Magnetic, Oxide Coated," sets forth the requirements for instrumentation-quality magnetic tape as a function of tape thickness. A standard recorder/reproducer must be capable of handling or utilizing any of the tapes qualified under this specification without degrading the characteristics of the specified recorder/reproducer system.
   6.3.1.3 *Track Geometry:* (See Figure 6.)
      6.3.1.3.1 *Track Width:* The track width for multiple-track recording shall be 0.050 ± 0.005 inch. Track width is defined as the physical width of the head that would be used to record or reproduce any given track, although the actual width of the recorded track may be somewhat greater because of the magnetic fringing effect around each record head.
      6.3.1.3.2 *Track Spacing:* Tracks shall be spaced 0.070 inch center-to-center across the tape and, as a group, shall be centered on the width of the tape. Therefore, the preferred width of tape ($\frac{1}{2}$ inch) would contain seven tracks, with one track located at the center of the tape.
      6.3.1.3.3 *Track Numbering:* The tracks on a tape shall be numbered consecutively, starting with track number 1, from top to bottom when viewing the oxide-coated side of a tape with the earlier portion of the recorded signal to the observer's right. Annotation tracks, if employed, shall be located on the edge of the tape, with track A adjacent to track 1 and track B adjacent to the highest numbered track.
   6.3.2 *Head and Head-stack Configuration:* (See Figure 7.)
      6.3.2.1 *Head Placement:* The standard placement is to locate the heads (both record and playback) for alternate tracks in separate head stacks. Thus, to record in all tracks of a standard-width tape, two record-head stacks will be used; to reproduce all tracks of a standard-width tape, two playback-head stacks will be used.
      6.3.2.2 *Head-stack Placement:* The two stacks of a pair (record or reproduce) shall be mounted in such a manner that the centerlines through the head gaps of each stack are parallel and spaced 1.500 ± 0.001 inches apart as measured along the tape path.
      6.3.2.3 *Head-stack Numbering:* Head stack number 1 of a pair of stacks (record or reproduce) is the first stack over which an element of tape passes when moving in the normal record or reproduce direction.
      6.3.2.4 *Head and Stack Numbering:* Heads (both record and reproduce) shall be numbered to correspond to the track on the magnetic tape which they normally record or reproduce. Stack number 1 of a pair will contain all odd-numbered heads, while stack number 2 will contain all even-numbered heads. Where only a single stack is needed, stack number 1 shall be used.
      6.3.2.5 *Individual Gap Azimuth Alignment:* The alignment of individual gaps within a head stack shall be within ±1 minute of arc, referenced to a straight line that is perpendicular to the direction of tape travel and in the plane of the tape.
      6.3.2.6 *Head-stack Tilt:* The plane tangent to the front surface of the head stack at the centerline of the head gaps shall be perpendicular to the head-mounting plate within ±3 minutes of arc.
      6.3.2.7 *Gap Scatter:* Gap scatter shall be 0.0001 inch or less.
      6.3.2.8 *Head Location:* Any head in a stack shall be located within ±0.002 inch of the nominal position required to match the track location set forth in paragraph 6.3.1.3.
      6.3.2.9 *Head Interchangeability:* Where rapid interchangeability of heads is specified, the method of head mounting, locating and securing shall ensure that all alignment and location requirements are satisfied without shimming or mechanical adjustment.
      6.3.2.10 *Annotation Heads:* Optional tracks A and B shall be used for voice annotation, location and identification only. Paragraphs 6.3.2, 6.3.2.5 and 6.3.2.7 shall not apply to these optional heads.
   6.3.3 *Head Polarity:*
      6.3.3.1 *Record Head:* Each record-head winding shall be connected to its respective amplifier in such a manner that a positive-going pulse with respect to system ground, at the amplifier input, will result in the generation of a specific magnetic pattern on a segment of tape passing the record head in the normal direction of tape motion. The resulting magnetic pattern shall consist of a polarity sequence of south-north-north-south.

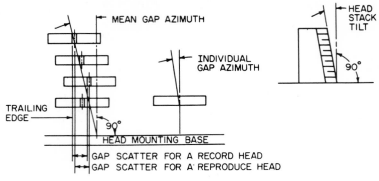
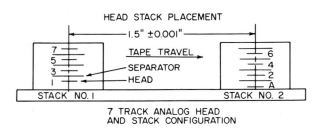

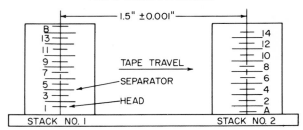

FIG. 7. Analog head configuration.

6.3.3.2 *Reproduce Head:* Each reproduce-head winding shall be connected to its respective amplifier in such a manner that a segment of tape exhibiting a south-north-north-south magnetic pattern will produce a positive-going pulse, with respect to system ground, at the output of the reproduce amplifier.

6.3.4 *Tape Guiding:* The tape guiding that determines the position of the tape relative to the head stacks shall in no way contribute to the deterioration of machine performance and shall not permit vertical movement of the tape relative to the head-mounting plate and/or misalignment of the tracks to exceed ±0.003 inch.

6.3.5 *Record/Reproduce Parameters:*

6.3.5.1 *Bias:* The high-frequency bias signal shall be of a frequency greater than three times the highest data frequency for which the machine is designed and shall not be less than 50,000 cps.

MAGNETIC-TAPE RECORDER/REPRODUCER STANDARDS

6.3.5.2 *Frequency Response:* The nominal frequency or pass band of direct-recorded data is a function of tape speed, as given in Table VI. (Performance characteristics set forth in 6.3.5 do not apply to optional tracks A and B.)

6.3.5.3 *Tape Speed:* The nominal tape speeds shall be those listed in Table VI.

6.3.5.4 *Record Amplifier:*

6.3.5.4.1 *Input Impedance:* Input impedance shall be 20,000 ohms minimum, with or without meter.

6.3.5.4.2 *Nominal Input Level:* This level shall be 1.0 rms.

6.3.5.4.3 *Transfer Characteristics:* The record amplifier shall provide a transfer characteristic (determined by a direct gap flux measurement) that is basically a constant current vs. frequency characteristic, upon which is superimposed a pre-emphasis characteristic to correct for loss of record-head efficiency with frequency.

Table VI. Direct-record Parameters

Tape speed (ips)	±3 db pass band (cps)	Record bias set frequency (cps)	Record level set frequency (cps)
60	100 – 100,000	20,000 ± 10%	1000 ± 10%
30	100 – 50,000	10,000 ± 10%	1000 ± 10%
15	100 – 25,000	10,000 ± 10%	1000 ± 10%
7½	50 – 12,000	500 ± 10%	500 ± 10%
3¾	50 – 6,000	500 ± 10%	500 ± 10%
1⅞	50 – 3,000	500 ± 10%	500 ± 10%

6.3.5.4.4 *Record Bias Setting:* For optimum record characteristics, the amplitude of the bias current shall be adjusted for maximum reproduced signal while recording signals of the frequency set forth in Table VI.

6.3.5.4.5 *Record-level Setting:* The level of recording shall be set at a value that yields 1 per cent third-harmonic-signal content on playback. This level shall be set while recording a signal of nominal input level and of the frequency indicated in Table VI.

6.3.6 *Speed Control and Compensation:*

6.3.6.1 *Speed Control:*

6.3.6.1.1 *Speed-control Signal:* The speed-control signal is an AM signal with the following characteristics:

Subcarrier frequency	17.0 kc ± 0.5%
Modulating frequency	60 cps ± 0.02%
Percentage modulation	45 to 55%
Operating level	10 db ± 0.5 db, below normal record level

6.3.6.1.2 *Record Speed:* All tape shall be recorded at a tape speed within ±0.5 per cent of the nominal standard speed.

6.3.6.1.3 *Playback Speed:* Tape-playback speed without external speed control shall be within ±0.5 per cent of the nominal standard speed. With external speed control, the tape-playback speed shall be within ±0.25 per cent of the record speed.

6.3.6.2 *Compensation:* Compensation signals to be used for correction of tape-speed-error effects are a function of tape speed as follows:

Tape speed (ips)	Compensation tone frequency
15	25 kc ± 0.01%
30	50 kc ± 0.01%
60	100 kc ± 0.01%

6.3.7 *Reproduce Amplifier:*

6.3.7.1 *Output Impedance:* Output impedance shall be 100 ohms maximum within the pass bands specified in Table VI.

6.3.7.2 *Nominal Output Level:* This level shall be 1.0 volt rms.

6.3.7.3 *Transfer Characteristics:* The reproduce amplifier shall provide signal equalization as a function of frequency considering the nature of the recorded signal as set forth in paragraph 6.3.5.4.3, head-to-tape, tape and tape-to-head transfer characteristics, which will provide the over-all recorder/reproducer system frequency response within the passband requirements set forth in Table VI.

6.4 PDM Recording

PDM recording is done by differentiating the input duration-modulated rectangular waveform and driving the record head with the resulting positive and negative spikes which correspond in time to the leading and trailing edges of the input pulses. The tape is thereby magnetically marked in such a manner that the pulses during the reproduce process may be used to trigger pulse-reconstruction circuitry. Although recorded PDM data may be reproduced through a direct-recorded data-reproduce amplifier and pulse reconstruction performed later, the PDM reproduce amplifier reconstructs the original duration-modulated rectangular waveform.

   6.4.1 *Tape:* Standards for tape used in PDM recording are the same as for direct recording (section 6.3.1).

   6.4.2 *Head and Head-stack Configuration:* Standards for PDM recording are the same as for direct recording (section 6.3.2).

   6.4.3 *Head Polarity:* Standards for PDM recording are the same as for direct recording (section 6.3.3).

   6.4.4 *Tape Guiding:* Standards for PDM recording are the same as for direct recording (section 6.3.4).

   6.4.5 *Record/Reproduce Performance Parameters:* The record/reproduce system shall be capable of recording and subsequently reproducing and reconstructing pulses whose minimum duration as a function of tape speed is given in Table VII. It shall be capable of recording and subsequently reproducing and reconstructing pulses with time errors, as a function of tape speed, not to exceed the values given in Table VII. The maximum pulse jitter measured at the half-amplitude point of the leading edge of the pulse shall not exceed 2 microseconds.

Table VII. PDM Record Parameters

Trade speed (ips)	Minimum pulse duration (microseconds)	Accuracy (microseconds)
60	75	±2
30	75	±2
15	100	±3

   6.4.6 *Record Amplifier:*
      6.4.6.1 *Input Impedance:* Input impedance shall be 20,000 ohms minimum.
      6.4.6.2 *Normal Input Level:* This level shall be 1.0 volt, peak-to-peak.
      6.4.6.3 *Transfer Characteristic:* The record amplifier shall drive the record head with a pulse signal that is obtained by differentiation of the input duration-modulated rectangular wave pulse train. The time constant of the differentiation shall be 10 microseconds.

   6.4.7 *Reproduce Amplifier:*
      6.4.7.1 *Function:* The PDM reproduce amplifier will amplify the pulse output of the reproduce head and reconstruct the basic duration-modulated rectangular pulse wave train.
      6.4.7.2 *Output Impedance:* Output impedance shall be 100 ohms maximum.
      6.4.7.3 *Nominal Output Level:* This level shall be 20 volts, peak-to-peak, across 1000 ohms resistance.
      6.4.7.4 *Pulse Rise Time:* Rise and decay time of the output rectangular pulses shall be less than 2 microseconds from 10 to 90 per cent amplitude levels.
      6.4.7.5 *Missing Pulse Protection:* The reproduce amplifier shall incorporate circuitry to detect defective pulses during the reproduce process and provide automatic resetting to preclude loss of subsequent data.

6.5 PCM Recording

PCM data may be recorded in several ways. The signal to be recorded (either in the air or on the ground at the output of the receiver) will be in serial form. The serial data may be recorded directly in serial form, utilizing a direct-recording technique or saturation-

MAGNETIC-TAPE RECORDER/REPRODUCER STANDARDS   5-43

recording techniques with adequate bandwidth, or it may be converted into a parallel form and recorded in parallel on a multitrack tape recorder.

Another method of recording PCM data is by predetection recording. Because serial or predetection recording of PCM data involves no special techniques beyond those required for recording any other type of data, this section deals specifically with the standards for the recording of PCM data on tape in parallel form.

There are two standard systems—a 16-track system and a 31-track system. The 31-track system consists of interleaved 16-track and 15-track stacks. The two stacks are employed as independent record/reproduce systems. Track spacing and location of tracks 1 through 16 in the 31-track system are identical to the 16-track system. Additional optional tracks A and B located beyond tracks 1 and 16 may be used. Performance standards specified herein shall not apply to the optional tracks.

6.5.1   *Tape:*

    6.5.1.1   *Tape Width:* The standard tape width is a nominal 1 inch. Standard tape width shall be 0.997 ± 0.001 inch.

    6.5.1.2   *Tape Type:* Tape for PCM use shall have physical and electrical characteristics equal to or better than the polyester tape.

    6.5.1.3   *Tape Thickness:* Paragraph 6.3.1.2 shall apply.

6.5.2   *Track Geometry:* (See Figure 8.)

    6.5.2.1   *Track Width:*

        6.5.2.1.1   *Sixteen-track System:* Track width for 16-track systems shall be 0.025 ± 0.002 inch. Track width is defined as the physical width of the head that would be used to record or reproduce any given track, although the actual width of the recorded track may be somewhat greater because of the magnetic fringing effect around each record head. Track width for optional tracks A and B for the 16-track system shall be 0.010 ± 0.002 inch.

        6.5.2.1.2   *Thirty-one-track System:* Track width for 31-track systems shall be 0.020 ± 0.001 inch. Track width is defined as the physical width of the head that would be used to record or reproduce any given track, although the actual width of the recorded track may be somewhat greater because of the magnetic fringing effect around each record head. Optional tracks A and B, when employed, shall also be 0.020 ± 0.002 inch in width.

    6.5.2.2   *Track Spacing:*

        6.5.2.2.1   *Sixteen-track System:* Spacing between track centers on 16-track systems shall be 0.060 inch. Optional tracks A and B shall be centered 0.035 inch from the centerlines of tracks 1 and 16, respectively.

        6.5.2.2.2   *Thirty-one-track System:* Spacing between track centers on 31-track systems shall be 0.030 inch.

    6.5.2.3   *Track Location:*

        6.5.2.3.1   *Sixteen-track System:* On 16-track systems, the center of the tape shall be centered between tracks 8 and 9.

        6.5.2.3.2   *Thirty-one-track System:* On 31-track systems, the center of the tape shall be centered on the centerline of track 24.

    6.5.2.4   *Track Numbering:* (See Figure 8.)

        6.5.2.4.1   *Sixteen-track System:* For 16-track systems, paragraph 6.3.1.3.3 shall apply.

        6.5.2.4.2   *Thirty-one-track System:* Paragraph 6.3.1.3.3 shall apply, except that the numbering from top to bottom shall be A (optional), 1, 17, 2, 18, 3, 19, . . . 31, 16, B (optional).

6.5.3   *Head and Head-stack Configuration:* (See Figure 9.)

    6.5.3.1   *Head-stack Placement* (31-Track System): Paragraph 6.3.2.2 shall apply.

    6.5.3.2   *Head-stack Numbering* (31-Track System): Paragraph 6.3.2.3 shall apply.

    6.5.3.3   *Head and Stack Numbering:* Heads shall be numbered to correspond to the track on the tape that they normally record or reproduce. For 31-track systems, stack number 1 of a pair wil contain heads numbered 1 through 16, and stack number 2 will contain heads numbered 17 through 31 and, optionally, tracks A and B.

    6.5.3.4   *Individual Gap Azimuth Alignment:* Paragraph 6.3.2.5 shall apply.

    6.5.3.5   *Mean Gap Azimuth Alignment:* Mean gap azimuth error shall not exceed ±⅓ minute of arc.

    6.5.3.6   *Head-stack Tilt:* Paragraph 6.3.2.6 shall apply.

    6.5.3.7   *Gap Scatter:* Paragraph 6.3.2.7 shall apply.

    6.5.3.8   *Head Location in Stack:* The location of any head in a stack shall be within ±0.001 inch, nonaccumulative, of the nominal position required to match the track location, as set forth in paragraphs 6.5.2.1.1, 6.5.2.1.2, 6.5.2.2.1 and 6.5.2.2.2.

6.5.4   *Head Polarity:* Section 6.3.3 shall apply.

6.5.5   *Tape Guiding:* Tape guiding, which determines the position of the tape relative to the head stacks, shall in no way contribute to the deterioration of machine performance and shall not permit transverse movement of the tape relative to the head-mounting plate and/or misalignment of the tracks to exceed 0.0025 inch, including tape slitting tolerance.

6.5.6 *Tape Speeds:* The standard speeds for instrumentation type magnetic tape recorder/reproducers shall be employed.

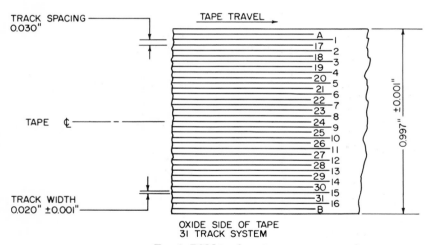

FIG. 8. PCM track system.

6.5.7 *Bit-packing Density:* The playback device shall be capable of playing back data recorded at bit-packing densities of up to at least 1,000 bits per linear inch per track.* The nominal maximum bit-packing density at the test ranges shall be 1,000 bits per linear inch per track.

6.5.8 *Cross-talk and Transverse Sensitivity:* Cross-talk between any two channels and transverse sensitivity between any two tracks shall be less than 25 db for 16-track systems and 20 db for 31-track systems.

*When recording at 120 inches per second the minimum bit-packing density capability must be at east 600 bits per linear inch per track.

MAGNETIC-TAPE RECORDER/REPRODUCER STANDARDS 5-45

6.5.9 *Record/Reproduce Reliability:* The maximum allowable error shall be 1 bit in 100,000.

6.5.10 *Skew and Differential Flutter:* This shall not exceed 125 microinches, peak-to-peak.

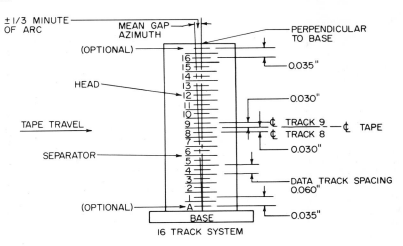

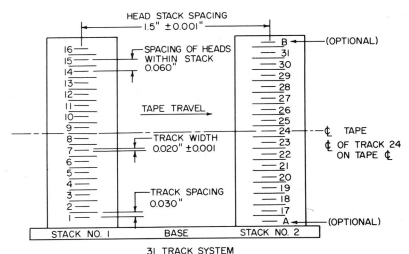

FIG. 9. PCM head configuration.

6.5.11 *Type of Recording:* Non-return-to-zero (NRZ) recording shall be employed, wherein a change in magnetization of the tape from saturation level of one polarity to saturation level of the opposite polarity is used to indicate the digit "one," and no change in magnetization during a bit interval indicates a "zero." Recorder/reproducer electronics shall be designed to meet the requirements of paragraphs 6.5.13.3 and 6.5.14.3.

6.5.12 *Timing:* Track 16 shall be reserved for range timing.

6.5.13 *Recorder Input Characteristics:*

    6.5.13.1 *Input Impedance:* This shall be 20,000 ohms minimum.

6.5.13.2 *Input Voltage:* This shall be 2 to 20 volts plus or minus, balanced or unbalanced and polarity-selectable.
  6.5.13.3 *Input Format:* This shall be parallel input, non-return-to-zero.
 6.5.14 *Output Characteristics:*
  6.5.14.1 *Reproduce Output Format:* This shall be parallel output, non-return-to-zero. Reproducer output shall compensate for all recorder/reproducer-induced time displacement errors to within 5 per cent of the word interval, or 1.6 microseconds, whichever is greater.
  6.5.14.2 *Output Impedance:* This shall be 100 ohms maximum.
  6.5.14.3 *Output Voltage:* This shall be 20 volts, peak-to-peak, minimum; one polarity for "one," opposite polarity for "zero," selectable polarity.
 6.6.1 *Tape:*

6.6 Single Carrier FM (Not Including Video)

 6.6.1.1 *Tape Width:* Paragraph 6.3.1.1 shall apply.
 6.6.1.2 *Tape Thickness:* Paragraph 6.3.1.2 shall apply.
 6.6.1.3 *Track Geometry:*
  6.6.1.3.1 *Track Width:* Paragraph 6.3.1.3.1 shall apply.
  6.6.1.3.2 *Track Spacing:* Paragraph 6.3.1.3.2 shall apply.
  6.6.1.3.3 *Track Numbering:* Paragraph 6.3.1.3.3 shall apply.
 6.6.2 *Head and Head-stack Configuration:*
  6.6.2.1 *Head Placement:* Paragraph 6.3.2.1 shall apply.
  6.6.2.2 *Head-stack Placement:* Paragraph 6.3.2.2 shall apply.
  6.6.2.3 *Head-stack Numbering:* Paragraph 6.3.2.3 shall apply.
  6.6.2.4 *Head and Stack Numbering:* Paragraph 6.3.2.4 shall apply.
  6.6.2.5 *Individual Gap Azimuth Alignment:* Paragraph 6.3.2.5 shall apply.
  6.6.2.6 *Head-stack Tilt:* Paragraph 6.3.2.6 shall apply.
  6.6.2.7 *Gap Scatter:* Paragraph 6.3.2.7 shall apply.
  6.6.2.8 *Head Location:* Paragraph 6.3.2.8 shall apply.
  6.6.2.9 *Head Interchangeability:* Paragraph 6.3.2.9 shall apply.
 6.6.3 *Tape Guiding:* Paragraph 6.3.4.1 shall apply.
 6.6.4 *Tape Speeds:* For carrier deviation limits ($\pm 40\%$), see Table VIII.

Table VIII. Carrier Deviation Limits*

Tape speed (ips)	Carrier center frequency (cps)	Carrier plus deviation (cps)	Carrier minus deviation (cps)
1⅞	1,688	2,363	1,012
3¾	3,375	4,725	2,025
7½	6,750	9,450	4,050
15	13,500	18,900	8,100
30	27,000	37,800	16,200
60	54,000	75,600	32,400
60** – 120	108,000	151,200	64,800

Note: * Input ± 1.4 volts dc nominal; deviation direction vs. voltage not specified.
 ** Magnetic tape recorders configured with high performance heads are capable of achieving this performance at 60 i. p. s.

 6.6.5 *General Requirements:*
  6.6.5.1 *Frequency Drift:* Maximum center frequency drift during any 15 minute period after 30 minutes warm-up shall not exceed 0.5 per cent of peak-to-peak deviation.
  6.6.5.2 *Modulation Frequency:* Maximum modulation shall be 20 per cent of the carrier center frequency.

6.7 Video Recording*

6.8 Predetection Video Recording†

6.9 Tape Reels, Reel-centering and Hold-down Device

6.9.1 *Tape Reels:* The standard magnetic-tape reel sizes are 10½ and 14 inches in diameter. The preferred machine shall handle standard flanged and flangeless reel sizes. Dimensions and tolerances or reel flanges and/or hubs shall be as specified by MIL-T-21029A (Ships), except as specified herein. The hub shall be at least as thick as the maximum thickness of the reel.

6.9.2 *Tape-reel Positioning:* The tape reels shall be positioned in conformance with the requirements of paragraph 6.3.4.

6.9.3 *Tape-reel-centering and Hold-down device:* In the engaged position, the reel-centering and hold-down device shall not permit motion of the reel relative to the reel spindle in any plane. Hold-down eccentricity shall not exceed 0.01 inch.

6.9.4 *Tape Wind:* This shall conform with MIL-T-21029A (Ships).

Appendix I

IRIG Frequency Standards for Telemetry

IRIG Recommendation No. 101-59, Revised
Approved 22 January 1959
Revision 1 September 1960
Revision 2 September 1961

Prepared by the
Frequency Coordination Working Group
Inter-Range Instrumentation Group
September 1961

1 Introduction

The parameters and criteria given here were devised by the Frequency Coordination Working Group of the Inter-Range Instrumentation Group, with the assistance of members of the Telemetry Working Group and development groups of the three Military Services and aircraft industries. The purpose of these parameters is to provide development and coordination agencies with design specifications on which to base equipment development and modification in an effort to ensure interference-free operation for all concerned and efficient utilization of the telemetry radio-frequency spectrum.

It has long been recognized that the frequency spectrum is a limited entity, a resource that must be conserved. It has been further recognized that frequency utilization is a system problem; the transmitter-receiver link must be considered as a system. Efficiency of spectrum utilization should be a goal; susceptibility to interference should be minimized.

The wasteful use of the spectrum by any system using electromagnetic radiation and reception can have far-reaching effects in many phases of military and civil activities. It is firmly believed that, unless the basic philosophy of spectrum conservation is recognized and applied by all agencies in the electronics field (designers, manufacturers, testers and users), serious consequences are inevitable.

It is emphasized that these parameters and criteria have been devised for application at military test ranges where congestion of portions of the usable frequency spectrum is a severe problem. It is hoped that, where applicable, these same principles will be applied to other fields outside the scope of instrumentation systems.

---

* Video and predetection recording techniques have developed along lines not anticipated when these sections were prepared for previous editions. Accordingly, they were deleted from this issue. These techniques have not been used sufficiently to make it possible to prepare definitive standards at this time. However, the Telemetry Working Group will frequently review these areas and prepare standards as soon as the state of the art has been sufficiently stabilized.

† *Ibid.*

2 Frequency Parameters and Criteria for Design of Telemetry Transmitter and Receiver Systems

2.1 Frequency Band 216 to 260 Megacycles

216 to 225 Mc: Channel spacing is based on 0.5 Mc separation on the integral and one-half-megacycle channels. Assignments are made on a noninterference basis to established services.

225 to 260 Mc: A total of 44 (500 kc) channels are allocated on a protected basis until 1 January 1970.

Efficiency of Spectrum Usage (216 to 260 Mc Band)

2.1.1 *Transmitter Systems (FM/FM; PDM/FM; PAM/FM;* and *PCM/FM):*
    2.1.1.1 *Maximum RF Deviation:* This shall be ±125 kc. (Optimum deviation for PDM/FM systems is ±60 to ±90 kc.)
    2.1.1.2 *Transmitter Frequency Tolerance:* The transmitted RF carrier, including drift and all other variables, will be within 0.01 per cent of the assigned carrier frequency.
    2.1.1.3 *Bandwidth:* The bandwidth of the modulated carrier shall not exceed 500 kc. Carrier components appearing outside to 500 kc bandwidth must meet the limits for spurious and harmonic emissions, as stated in paragraph 2.1.1.5.1 below.
    2.1.1.4 *Power:* The power shall be 100 watts maximum, never more than absolutely necessary.
    2.1.1.5 *Spurious and Harmonic Emissions:*
        2.1.1.5.1 Spurious and harmonic emissions from the transmitting-antenna system are of primary importance insofar as these criteria are concerned. Spurious and harmonic outputs, antenna-conducted (i.e., measured in antenna transmission line) as well as antenna-radiated (i.e., measured in free space), shall be limited to the values derived from the formula:

$$\text{db (down from carrier)} = 55 + 10 \log_{10} P_t,$$

where P_t is the measured power output in watts. Measurements to determine relative levels of RF power shall be made under the following conditions:

(1) Transmitter to be operated into a matched, shielded dummy load, with a suitable coupling device inserted in the antenna cable to sample the transmitter RF output. As an alternative, the actual antenna can be substituted for the dummy load, provisions being made to remove the field strength meter from the influence of signals radiated from the antenna.

(2) Transmitter to be tested under conditions of zero and full normal modulation.

(3) Commercial, Category Class "I" Field Strength Measuring Equipment, as listed in current MIL-I-26600, will be used.

        2.1.1.5.2 Spurious, harmonic and fundamental signals conducted by power leads or radiated directly from equipment units or cable (except antenna) shall be within the limits specified in current MIL-I-26600.

    2.1.1.6 *Flexibility of Operation:* The system shall be capable of operating on any of the following frequencies without design modification (all given in megacycles):*

216.5	223.0	228.2	237.8	248.6
217.0	223.5	229.9	240.2	249.1
217.5	224.0	230.4	241.5	249.9
218.0	224.5	230.9	242.0	250.7
218.5		231.4	243.8	251.5
219.0		231.9	244.3	252.4
219.5		232.4	244.8	253.1
220.0	225.0	232.9	245.3	253.8
220.5	225.7	234.0	245.8	255.1
221.0	226.2	235.0	246.3	256.2
221.5	226.7	235.5	246.8	257.3
222.0	227.2	236.2	247.3	258.5
222.5	227.7	237.0	247.8	259.7

* All telemetry assignments within the 225 to 260 Mc band shall conform with these assignments. No change in assignments in the 216 to 225 Mc band is contemplated. However, it should be kept in mind that telemetry assignments in the 216 to 225 Mc band are on the basis of noninterference with other established users.

APPENDIX I

2.1.2 *Receiver Systems (FM/FM, PDM/FM; PAM/FM; and PCM/FM):*
    2.1.2.1 *Maximum Bandwidth Between 60 db Points:* This shall be 600 kc.
    2.1.2.2 *Receiver Stability:* This shall be 0.005 per cent.
    2.1.2.3 *Spurious Receiver Responses:* These shall be more than 60 db below fundamental frequency response.
    2.1.2.4 *Spurious Emissions:* Oscillator energy, either radiated from the unit or antenna-conducted, shall be within the limits specified in current MIL-I-26600.
    2.1.2.5 *Flexibility of Operation:* The system shall operate on any of the frequencies listed under paragraph 2.1.1.6 without design modification.

2.2 Frequency Band 1435 to 1535 Megacycles

Channel spacing of the 1435 to 1535 Mc band should be in increments of 1 Mc.
The 1435 to 1485 Mc portion of the band should be reserved primarily for use in connection with aeronautical flight testing of manned aircraft.
The 1486 to 1535 Mc portion of the band should be reserved primarily for use in connection with aeronautical flight testing of missiles and space vehicles.

Efficiency of Spectrum Usage

2.2.1 *Transmitter Systems (1435 to 1535 Mc Band).*
    2.2.1.1 *Transmitter Frequency Tolerance:* The transmitter RF carrier, including drift and all other variables, shall be within 0.005 per cent of the assigned carrier frequency.
    2.2.1.2 *Power:* The power shall be as dictated by the intended use, never more than absolutely necessary.
    2.2.1.3 *Spurious and Harmonic Emissions:* Spurious and harmonic emissions from the transmitting-antenna system are of primary importance insofar as these criteria are concerned. Spurious and harmonic outputs, antenna-conducted (i.e., measured in antenna transmission line) as well as antenna-radiated (i.e., measured in free space), shall be limited to the values derived from the formula:

$$\text{db (down from carrier)} = 55 + 10 \log_{10} P_t,$$

where P_t is the measured power output in watts.
    2.2.1.4 *Spurious, Harmonic and Fundamental Emissions:* Such signals conducted by power leads or radiated directly from equipment units or cables (except antenna) shall be within the limits specified in current MIL-I-26660.
    2.2.1.5 *Measurements to Determine Relative Levels of Spurious and Harmonic Emissions:* These measurements shall be made under the following conditions:
    (1) Transmitter to be operated into a matched, shielded dummy load, with a suitable coupling device inserted in the antenna cable to sample the transmitter RF output. As an alternative, the actual antenna can be substituted for the dummy load, provisions being made to remove the field strength meter from the influence of signals radiated from the antenna.
    (2) Transmitter to be tested under conditions of zero and full normal modulation.
    (3) Commercial, Category Class "I" Field Strength Measuring Equipment, as listed in current MIL-I-2660, will be used.
    2.2.1.6 *Flexibility of Operation:* The RF transmitter shall be capable of operating throughout the entire frequency band 1435 to 1535 Mc without design modification.
2.2.2 *Receiver Systems (1435 to 1535 Mc Band):*
    2.2.2.1 *Receiver Stability:* This shall be 0.001 per cent.
    2.2.2.2 *Spurious Receiver Responses:* These shall be more than 60 db below fundamental frequencies.
    2.2.2.3 *Spurious Emissions:* Oscillator energy, either radiated from the unit or antenna-conducted, shall be within the limits specified in current MIL-I-26600.
    2.2.2.4 *Flexibility of Operation:* The system shall be tunable over the entire 1435 to 1535 Mc band without design modification and with variable bandwidth selection.
2.2.3 *Bandwidths:*
    2.2.3.1 In specifying bandwidths, the transmitter and receiver shall be considered as a system. The designer should be required to adhere to rigid engineering design practices to conserve the frequency spectrum. Each system should be subjected to a critical review with respect to the amount of information contained in a given bandwidth vs. type of modulation. The designer should be required to demonstrate and prove the system design in order to justify the use of the frequency spectrum.
    2.2.3.2 As a general guideline, it is anticipated that, for a deviation of ± 125 kc, a maximum of 1 Mc bandwidth as reference to the 60 db points will be permitted. For a

wide-band system with a deviation of ±1.4 Mc, a maximum of 10 Mc as reference to the 60 db points will be permitted. Bandwidth requirements for PCM shall be determined in accordance with criteria set forth in the *IRIG Telemetry Standards*, No. 106-60, section 5. Bandwidth for telemetry systems in excess of 10 Mc as reference to the 60 db points shall not be used. Bandwidth requirements for transmission of video (television) shall be considered on the basis of the individual case.

2.3 Frequency Band 2200 to 2300 Megacycles

Channel spacing of the 2200 to 2300 Mc band shall be in increments of 1 Mc.

Efficiency of Spectrum Usage

2.3.1 *Transmitter Systems (2200 to 2300 Mc):*
    2.3.1.1 *Transmitter Frequency Tolerance:* The transmitted RF carrier, including drift and all other variables, shall be within 0.005 per cent of the assigned carrier frequency.
    2.3.1.2 *Power:* The power shall be as dictated by the intended use, never more than absolutely necessary.
    2.3.1.3 *Spurious and Harmonic Emissions:* Spurious and harmonic emissions from the transmitting-antenna system are of primary importance insofar as these criteria are concerned. Spurious and harmonic outputs, antenna-conducted (i.e., measured in antenna transmission line) as well as antenna-radiated (i.e., measured in free space), shall be limited to the values derived from the formula:

$$\text{db (down from carrier)} = 55 + 10 \log_{10} P_t,$$

where P_t is the measured power output in watts.
    2.3.1.4 *Spurious, Harmonic and Fundamental Emissions:* Such signals conducted by power leads or radiated directly from equipment units or cable (except antenna) shall be within the limits specified in current MIL-I-26600.
    2.3.1.5 *Measurements To Determine Relative Levels of Spurious and Harmonic Emissions:* These measurements shall be made under the following conditions:
(1) Transmitter to be operated into a matched, shielded dummy load with a suitable coupling device inserted in the antenna cable to sample the transmitter RF output. As an alternative, the actual antenna can be substituted for the dummy load, provisions being made to remove the field strength meter from the influence of signals radiated from the antenna.
(2) Transmitter to be tested under conditions of zero and full normal modulation.
(3) Commercial, Category Class "I" Field Strength Measuring Equipment, as listed in current MIL-I-26600, will be used.
    2.3.1.6 *Flexibility of Operation:* The RF transmitter shall be capable of operating throughout the entire frequency band 2200 to 2300 Mc without design modification.

2.3.2 *Receiver Systems (2200 to 2300 Mc):*
    2.3.2.1 *Receiver Stability:* This shall be 0.001 per cent.
    2.3.2.2 *Spurious Receiver Response:* This shall be more than 60 db below fundamental frequencies.
    2.3.2.3 *Spurious Emissions:* Oscillator energy, either radiated from the unit or antenna-conducted, shall be within the limits specified in current MIL-I-26600.
    2.3.2.4 *Flexibility of Operation:* The system shall be tunable over the entire 2200 to 2300 Mc band without design modification and with variable bandwidth selection.

2.3.3 *Bandwidths:*
    2.3.3.1 In specifying bandwidths, the transmitter and receiver shall be considered as a system. The designer should be required to adhere to rigid engineering design practices to conserve the frequency spectrum. Each system should be subjected to a critical review with respect to the amount of information contained in a given bandwidth vs. type of modulation. The designer should be required to demonstrate and prove the system design in order to justify the use of the frequency spectrum.
    2.3.3.2 As a general guideline, it is anticipated that, for a deviation of ±125 kc, a maximum of 1 Mc bandwidth as reference to the 60 db points will be permitted. For a wide-band system with a deviation of 1.4 Mc, a maximum of 10 Mc as referenced to the 60 db points will be permitted. Bandwidth requirements for PCM shall be determined in accordance with criteria set forth in the IRIG *Telemetry Standards*, No. 106-60 (Rev. 1), section 5. Bandwidth in excess of 10 Mc for telemetry systems as referenced to the 60 db points shall not be used. Bandwidth requirements for transmission of video (television) shall be considered on the basis of the individual case.

Appendix II

PCM Standards
Additional Information and Recommendations

1 Bit Rate vs. Receiver IF Bandwidth (3-db Points)

1.1 For reference purposes, in a well-designed system, a receiver IF signal-to-noise ratio (power) of approximately 15 db will result in a bit error probability of about 1 bit in 10^6. A 2 db change (increase or decrease) in this signal-to-noise ratio will result in an order-of-magnitude change (10^7 or 10^5 from 10^6, respectively) in the bit error probability.
1.2 It should be recognized that the range of factors 1.5 to 3.3 recommended in section 5.2 of the *IRIG Telemetry Standards* may result in a compatibility problem when using current FM receivers for standard IRIG FM/FM and PDM/FM systems, as well as PCM/FM systems designed in accordance with the standard given here. Modifications to video amplifier stages and other circuitry may be required.

2 Word and Frame Structure

2.1 The assignment of word positions to convey special information in designated frames on a programmed basis is acceptable. The substituted words, including the necessary identifier and padding bits, shall exactly match the replaced word or words in total number of bits.

3 Suggested PCM Synchronization Patterns

3.1 It is suggested that an n-bit frame-synchronization pattern be selected under the criterion that the probability of displacement of the pattern by ±1 bit be minimized, at the same time restricting the probability of pattern displacement by 2 to (n − 1) bits below a prescribed maximum. A 31-bit synchronization pattern satisfying this criterion is:

0101011010100101101001101010111

3.2 An analysis leading to the selection of this criterion and resulting synchronization patterns is presented in Technical Memorandum 73-53, "Synchronization Methods for PCM Telemetry," Naval Ordnance Laboratory, Corona, California.
3.3 Other synchronization patterns will be suggested when their suitability has been established.
3.4 It is recommended that a flexible shift-register type pattern recognizer be used to accommodate possible variations in patterns.

4 Premodulation Filtering

4.1 For a well-designed system, it is recommended that a premodulation low-pass filter with the following characteristics be used:

(1) Cutoff frequency (3 db) equal to the nominal bit rate.
(2) Maximally linear phase response.
(3) Final slope of 36 db per octave.

Appendix III

Recommended Methods for Testing
Magnetic-tape Recorder/Reproducers

1 Method for Testing for Speed Errors

1.1 *Scope*

This appendix defines procedures only. The actual performance data will be entered by the tester on the Standard Performance Sheet shown in Figure 1.

STANDARD PERFORMANCE SHEET

Data entered in this form is the result of the average of five tests made using five randomly chosen reels of magnetic recording tape.

Type _____ Lot No. _____

Manufactured by _____

on a reel size of _____ -inch by _____ feet in length.

Average absolute speed error is _____ % of recorded frequency.

Instantaneous relative reproduce speed error is _____ % of recorded frequency.

Wow-and-flutter error is:

_____ % peak-to-peak of recorded frequency at 0 – 450 cps bandpass

_____ % peak-to-peak of recorded frequency at 0 – 2100 cps bandpass

_____ % peak-to-peak of recorded frequency at 0 – 10,000 cps bandpass

Recorder Tested:

Model _____ Manufacturer _____

Tape Speed _____ in/sec. Number of tracks _____

Track Tested _____

Auxiliary Equipment used in Test:

Test Equipment Used:	Manufacturer:	Model:	Serial:
Audio Oscillator			
Frequency Discriminator			
Vacuum Tube Voltmeter			
Frequency Meter			
Oscilloscope			

FIG. 1. Standard performance sheet.

Since tape and recorder performance are interrelated, this method attempts to ensure that the effects of the tape on the recorder measurements are minimized by the test procedure.

1.2 Test Definitions

Speed errors—all distortions in recording and playback that are exhibited as apparent frequency changes.

APPENDIX III

Absolute speed—(for this appendix) the velocity at which a 1-inch unit of tape moves by the heads (expressed in inches per second).

Instantaneous speed error (without high-frequency effects)—the instantaneous speed at which the recorder is reproducing a tape with respect to the speed at which the tape was recorded, with all speed effects that occur at rates in excess of 8 cps removed.

Speed errors (wow-and-flutter effects)—the cumulative sum of record and reproduce speed errors caused by instantaneous mechanical characteristics of the recorder when recording and reproducing are not performed simultaneously.

1.3 Test Conditions

1.3.1 *General:* Each of the error measurements must use a complete reel of tape of the largest size the recorder normally accommodates. If extension arms are used to obtain greater capacity, this must be stated and a test made on each alteration. The tape used shall be of random choice from commercially available stock, and the type and manufacturer shall be stated. A complete set of measurements shall be made on one reel of tape and then repeated on a new reel of tape until five complete tests have been run. The results of these measurements, when averaged, give the final results to be entered in the Standard Performance Sheet. All the described tests are made with the recorder adjusted as stipulated by the recorder manufacturer, and the recorder shall not be readjusted during the series of tests.

1.3.2 *Calibration:* The following procedure determines the standard calibration that is used to arrive at the percentage error figures:

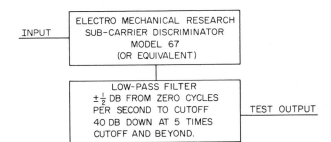

Fig. 2. Frequency discriminator.

Connect the audio oscillator directly to the frequency discriminator input (see Fig. 2) and adjust the audio oscillator for the discriminator center frequency while monitoring it with a frequency meter. Display the discriminator output on the direct-coupled oscilloscope, and observe the amount by which the oscilloscope trace is displaced as the oscillator is readjusted to produce a definite discriminator deviation, e.g., $+1$ per cent of $+10$ per cent of center frequency. The per cent deviation of the discriminator causing this displacement is called the calibrated deviation.

1.4 Test Procedures

1.4.1 *Absolute Speed:* Absolute speed will be measured in the following manner:

A precision 60 cps source (0.02 per cent or greater accuracy) shall be used to generate a 60 cps square wave. The square wave shall be differentiated and recorded on the tape at a recording speed of 60 inches per second (see Figure 3). This signal shall be recorded

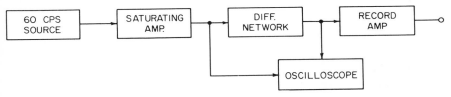

Fig. 3. Absolute speed test setup.

at a minimum of three places (beginning, center and end) on a full reel of tape. The three recorded sections of tape will be removed and "developed" with an iron-oxide solution. The spacing between five consecutive recorded spikes will be measured on a universal Telereader or equivalent reader. (See Figure 4.)

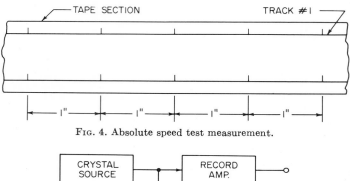

Fig. 4. Absolute speed test measurement.

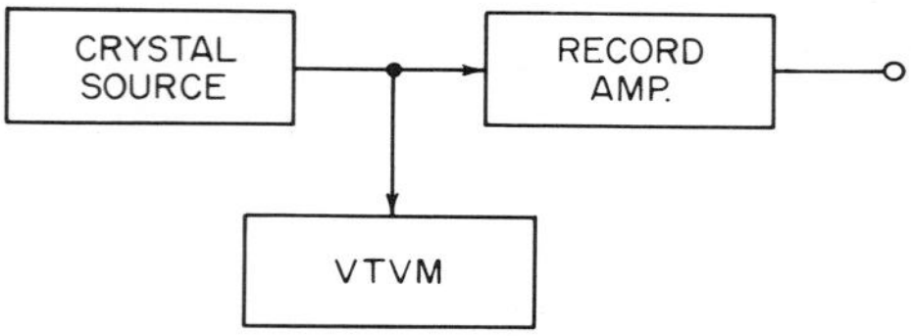

Fig. 5. Record setup for measuring instantaneous speed error.

1.4.2 *Instantaneous Speed Error:* Instantaneous speed error (without high-frequency effects) will be measured in the following manner:

A complete reel of tape will be recorded with a constant frequency (see Figure 5) (at normal record level), as specified in Table I.

Table I. Frequency Recommendations for Measurement of Instantaneous Speed Errors

Tape speed (ips)	Frequency (kc)
60	50
30	25
15	10
7½	5.4
3¾	2.3
1⅞	1.3

All frequencies will be crystal-derived of 0.02-per cent (or greater) stability.

The tape will be reproduced with the recorder output signal supplying a discriminator (EMR Model 67F or equivalent). (See Figure 6.) The discriminator will use a low-pass

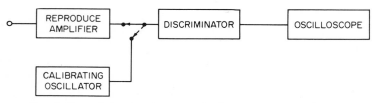

Fig. 6. Reproduce setup for measuring instantaneous speed error.

APPENDIX III

filter of 8-cps cutoff frequency. The discriminator output will be connected to a dc oscilloscope (Tektronix Model 535 or equivalent). The discriminator and oscilloscope will be calibrated as a unit so that a given frequency deviation, e.g., +1 per cent of +10 per cent of center frequency will produce an oscilloscope deflection of +1 centimeter. The oscilloscope sweep rate will be adjusted to 5 seconds per centimeter for visual observation or for photographing speed variations (see Figure 6).

Both of these tests must be performed before the recorder speed performance can be determined.

1.4.3 *Speed Errors:* Speed errors (wow and flutter) will be measured in the following manner:

A complete reel of tape will be recorded (all channels) with a constant 50-kcps frequency (at normal record level) for tape speeds of 60 ips and 30 ips (see Figure 7). For measurements at other tape speeds, the frequencies outlined in Table I are recommended. All reference frequency sources will be crystal-derived of 0.02-per cent (or greater) stability.

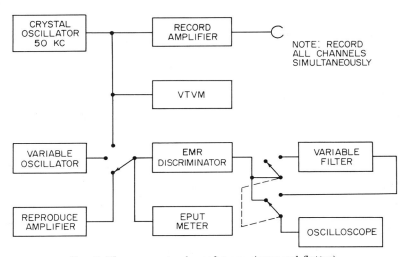

FIG. 7. Measurements of speed errors (wow and flutter).

The recorded tape will be reproduced with the recorder output signal (from the desired channel) supplying a discriminator (EMR 67F or equivalent). The discriminator output will be connected to a dc oscilloscope (Tektronix Model 535 or equivalent). The discriminator and oscilloscope will be calibrated as a unit (see section 1.3.2) so that a given frequency deviation, e.g., +1 per cent of +10 per cent of center frequency will produce an oscilloscope deflection of +1 centimeter (or a convenient deflection for measurement being made). The oscilloscope sweep rate will be adjusted to 5 second per centimeter (minimum) in order to display both low-frequency (wow) and high-frequency (flutter) effects. The discriminator low-pass filter will be selected to cover the desired areas of interest. Recommended low-pass filters are:

0 to 450 cps
0 to 2,100 cps
0 to 10,000 cps

Various sections of the speed-error (wow-and-flutter) spectrum can be examined by using a variable filter (SKL or equivalent) supplied by the discriminator output, with a discriminator 0 to 10,000 cps low-pass filter. When using such an active variable filter, the filter gain must be unity in order to preserve the calibration, and it must be recognized that the dc to 20 cps (approximate) bandpass is also eliminated because of the ac coupling used in the filter.

1.5 *Equipment*

Test equipment other than that specified below may be used, if it will produce equivalent results.

> Audio oscillator: Hewlett Packard Model 200-C
> Frequency discriminator: Electro-Mechanical Research (EMR) subcarrier discriminator Model 67F with low-pass filter, as shown in Figure 2.
> Vacuum-tube voltmeter: Ballantine Model 320
> Frequency meter: Berkeley Model 554 or Hewlett Packard 524
> Oscilloscope: Tektronix Model 535

1.6 *Statement of Measurements*

Test result shall be tabulated on the Standard Performance Sheet (Figure 1). The test results shall be followed by a statement of the auxiliary equipment used when the measurements were made; for example, an external means was used to control the speed; extension arms for 4800-foot tape reels were used, etc.

2 Method for Electronic Test of Mechanical Tolerances

It is suggested that the mechanical tolerances of a record/reproduce system can be checked electronically. The test is made by simultaneously recording on all tracks at 15 inches per second and noting, during reproduction at 15 inches per second, the time displacement of the signal on each track when referenced to any one track (the chosen reference track).

When the recovered signals are viewed on an oscilloscope, the gap scatter appears as a fixed time displacement, while other sources of error such as jitter and skew appear as constantly varying displacements that can be averaged out by the operator. This measurement shall be made while running the tape through in the normal manner and reversed (upside-down and backward) at 15 inches per second. The fixed time displacement for these tests should not exceed 30 microseconds.

Appendix IV

PAM Standards
Additional Information and Recommendations

1 IF Bandwidth and Transmitter Deviation

1.1 The appropriate receiver final IF bandwidth and transmitter deviation depend primarily on the total sampling rate (including provision for synchronization) and system noise and distortion tolerances.

1.2 Transmitter and receiver instabilities (refer to Appendix I for frequency tolerances) may cause frequency drifts as great as $\pm 39,000$ cps in the VHF band. Instabilities of this magnitude preclude the use of the three lower receiver IF bandwidths, 12,500; 25,000; and 50,000 cps unless special techniques, such as automatic frequency control capable of accommodating PAM waveforms, are applied.

2 Frame and Pulse Structure

2.1 It is recommended that ground system equipment be capable of decommutating both 50 per cent and 100 per cent duty cycle pulses, and that new systems applications recognize that 100 per cent duty cycle pulses improve the radio frequency (RF) spectrum utilization.

3 Synchronization

3.1 The amplitude type synchronization format for the 100 per cent duty cycle pulse structure provides: (1) A synchronizing pulse at full scale amplitude and duration 3 times

APPENDIX V 5-57

that of individual channel pulses, giving good discrimination between synchronizing and channel pulses; (2) specific modulation levels at the beginning and end of the synchronization period (zero and half-scale levels); (3) assured full scale signal at least once per frame, thus permitting use of full scale clamp circuitry. These levels, zero, half scale, and full scale permit automatic calibration correction in the receiving stations.

3.2 Binary code patterns should be selected that have a high probability of recognition in a random PAM pulse train. It is pointed out that the amplitude synchronization formats described in figures 5(a) and 5(b) of the PAM standards can be considered as specific binary codes. (The 100 per cent duty cycle case may be 0111.)

3.3 It is recommended that a flexible shift-register type pattern recognizer be used to accommodate possible variation in binary code patterns.

4 Premodulation Filtering

4.1 It is recommended that the premodulation passband exhibit a final attenuation slope of 36 db per octave beyond the bandwidth specified in Section 4.7 of the PAM/FM Standards.

Appendix V

Glossary of Telemetry Terms

Acceleration error: The maximum difference at room conditions, at any measurand value within the specified range, between output readings taken before and during the application of specified constant acceleration along specified axes. See Transverse Sensitivity when applied to acceleration transducers.

Acceleration error band: The error band applicable when constant accelerations within a specified range of amplitudes are applied to a transducer along specified axes at room conditions.

Accuracy: (1) Freedom from mistakes or errors. A measure of conformity to a specified value. (2) Transducer: The ratio of the error to the full-scale output (expressed as "within ±—— per cent of full scale output") or the ratio of the error to the output, expressed in per cent.

Active leg: An electrical element within a transducer which changes its electrical characteristics as a function of the application of the measurand.

Altitude: The vertical distance above a stated reference level.

Ambient conditions: The conditions (pressure, temperature, etc.) of the medium surrounding the transducer.

Ambient pressure error: The maximum change in output, at any measurand value within the specified range, when the ambient pressure surrounding the transducer is changed from room conditions to specified extremes.

Ambient pressure error band: The error band applicable when the transducer operates over a specified range of ambient pressures.

Analog to digital conversion: A process by which a sample of analog information is transformed into a digital code.

Analog to digital converter: (Also called Digitizer, ADC, and Encoder.) A device which will convert an analog voltage sample to an equivalent digital code of some finite resolution.

Analog voltage: A voltage that varies in a continuous fashion in accordance with the magnitude of a measured variable.

Angular velocity: The time rate of change of angular displacement expressed in radians per second and generally designated by the Greek letter omega "ω."

Attitude: The relative orientation of a vehicle or object represented by its angles of inclination to three orthogonal reference axis.

Attitude error: The error due to orientation of the transducer.

Automatic zero and full-scale calibration correction: Zero and sensitivity stabilization by utilization of electronic servos for continuous comparison of demodulated "zero" and full-scale signals with "zero" and full-scale reference voltages.

Band: A bounded continuous portion of a frequency spectrum.

Bar-graph monitoring oscilloscope: Oscilloscope for observation of commutated signals appearing as a series of bars with lengths proportional to channel modulation. The same oscilloscope is commonly used for set-up and trouble-shooting observations.

Barker code: A binary code suitable for PCM frame synchronization having optimal correlation properties with the unique property, when decoded, of relative immunity to phase displacement by random pulses immediately adjacent to the pattern, and relative immunity to phase displacement by error in the transmitter code.

The Barker codes are:

                        3-bit: 110
                        7-bit: 1110010
                        11-bit: 11100010010

Reference: Communication Theory; Willis Jackson; Academic Press Inc., London 1953; Papers read at a Symposium Applications of Communication Theory held at Institute of Electrical Engineers, London, September 22–23, 1952.

Bias set frequency: In direct magnetic tape recording, a specified recording frequency employed during the adjustment of bias level of optimum record performance. (Not the frequency of the bias.)

Binary number system: A number system which uses two symbols (usually denoted by "0" and "1") and therefore has 2 as its base, just as the decimal system uses ten symbols (0, 1, 2 . . . 9) and has the base ten.

Bit: A unit of information which is carried by an identifiable character and which can exist in either of two states. (An abbreviation of binary digit.)

Bit rate: The frequency derived from the period of time required to transmit one bit.

Blanking level: Level of multiplexed signal between channel pulses.

Bonded: The adhesive method used to provide mechanical coupling of transducer sensing elements, usually strain gages, to the point at which measurements are made.

Breakdown voltage rating: The d-c or sinusoidal a-c voltage stated in a specification which can be applied across specified insulated portions of a transducer without causing arcing or conduction above a specified current value across the insulating material. Time duration of application, ambient conditions, and a-c frequency must be specified.

Burst pressure rating: The pressure stated in a specification which must be applied to the sensing element of a transducer without rupture of either the sensing element or the transducer case as specified. The minimum number of applications and time duration of each application must be specified.

Calibration: The process of obtaining the data for a calibration record.

Calibration curve: A record of the measured relationship of the transducer output to the applied measurand over the transducer range.

Calibration cycle: The application of known values of measurand, and subsequent recording, of corresponding output readings over the range of a transducer, in an ascending and descending direction.

Calibration record: A record of the measured relationship (e.g., table or graph) of the transducer output to the applied measurand over the transducer range.

Carrier: A wave suitable for being modulated to transmit intelligence. The modulation represents the information; the original wave is used only as a "carrier" of the modulation.

Carrier frequency: In a periodic carrier, the reciprocal of its period. In a PCM system the carrier frequency is the midpoint between the deviation limits.

Case pressure: The preferred terms are *burst pressure*, *proof pressure*, or *reference pressure*, as required.

Channel interval: Time allocated to a channel including ON and OFF times.

Channel pulse synchronization: Synchronization of local channel rate oscillator by comparison and phase-lock with separated channel synchronizing pulses.

Channel sampling rate: Number of times per second individual channels are sampled. Note: Use of "channel sampling rate" to designate commutation rate, or commutator switching rate, is not recommended because of likelihood of confusion in terms.

Channel, subcarrier: The channel required to convey telemetric information involving a subcarrier band.

Channel synchronizing pulse separator: A device for separating channel synchronizing pulses from commutated signals.

Channel translator: A device which converts individual separated channel pulses or signals derived therefrom to analog form for subsequent monitoring and/or recording. Alternate names for channel translator are channel demodulator, channel decoder, information gate.

Clock pulse: A pulse used for timing purposes. In PCM Systems, a timing pulse which occurs at the bit repetition rate.

Closed loop telemetry: (1) A telemetry system which is used as the indicating portion of a remote control system. (2) A system used to check out test vehicle and/or telemetry performance without radiation of r-f energy.

Code: A system of characters and rules for representing information.

APPENDIX V 5-59

Combined error: A term used to specify the largest error of an instrument in the presence of adding or interacting parameters. Generally applied to the largest error due to the combined effect of nonlinearity and hysteresis.

Commutation: Sequential sampling, on a repetitive time sharing basis, of multiple data sources for transmitting and/or recording on a single channel.

Commutation duty cycle: Channel dwell period expressed as per cent of channel interval.

Commutation frame period: Time required for sequential sampling of all input signals. This period would correspond to one revolution of a single multicontact rotary switch.

Commutation rate: Number of commutator inputs sampled per second.

Commutator: A device used to accomplish time division multiplexing by repetitive sequential switching.

Commutator channel dwell period: Channel ON time.

Commutator segment: One of the stationary contacts of a mechanical commutator.

Compensation: Provision of a supplemental device or special materials to counteract known sources of error.

Compensation signals: A compensation signal is a signal recorded on the tape, along with the data and in the same track as the data, which is used during the play-back of data to electrically correct for the effects of tape speed errors.

Conduction error: The error in a temperature transducer due to heat conduction between the sensing element and the mounting to the transducer.

Continuous rating: The rating application to operation for a specified uninterrupted length of time.

Critical damping: This term is defined under "Damping."

Cross-axis acceleration: The preferred term is *transverse acceleration*.

Cross sensitivity, cross-axis sensitivity: The preferred term is *transverse sensitivity*.

Cross talk: Interference in a given transmitting or recording channel which has its origin in another channel. Often used as equivalent to transverse sensitivity.

Cycle: A cycle is defined as a completed scheduled sequence of events, i.e., a frame cycle would be the sequence of samples from all of the prime channels.

Damping: (1) Refers to the resistance, friction or similar cause that diminishes the amplitude of an oscillation with each successive cycle. (2) Transducer: The energy dissipating characteristic which, together with natural frequency, determines the upper limit of frequency response and the response-time characteristics of a transducer. Note 1: In response to a step change of the measurand an underdamped (periodic) system oscillates about its final steady value before coming to rest at that value; an overdamped (aperiodic) system comes to rest without overshoot; and a critically damped system is at the point of change between the underdamped and overdamped conditions. Note 2: Viscous damping uses the viscosity of fluids (liquids or gases) to effect damping. Note 3: Magnetic damping uses the relative motion of electric currents and magnetic fields to effect damping.

Damping factor: The ratio of any one amplitude and the next succeeding it in the same sense or direction, when energy is not supplied on each cycle. In second-order systems with single degree of freedom the decrement is constant. The amplitude decays as $e^{-\delta t}$

where: t—time
 δ—logarithmic decrement

Data assimilator: A device which synchronizes the flow of data between digital systems whose flow rates and internal timing are independent, different and/or asynchronous.

Dead volume: The total volume of the pressure port cavity of a transducer with room barometric pressure applied.

Decommutator: Equipment for separation, demodulation or demultiplexing commutated signals.

Dependent linearity: A manner of expressing non-linearity errors as deviation from a desired straight line of fixed slope and/or position.

Design IF bandwidth: The information bandwidth required for a specified PAM system performance.

Deviation: In frequency modulation, the peak difference between the instantaneous frequency of the modulated wave and the carrier center frequency.

Deviation ratio: Deviation Ratio is given by $m \frac{\Delta f}{f_{max}}$ where Δf is the maximum frequency difference between the modulated carrier and the unmodulated carrier and f_{max} is the maximum modulation frequency.

Deviation ratio (PCM/FM): In PCM systems, the ratio of the peak-to-peak carrier deviation to the bit rate.

Differential flutter: Speed change errors occurring at different magnitudes, frequencies or phase across the width of a magnetic tape.

Digital: Expressing value in terms of numbers. Measurable in discrete, discontinuous steps.

Digital magnetic tape recording: The method of recording binary coded information using two discrete flux levels.

Digital output: Transducer output that represents the magnitude of the measurand in the form of a series of discrete quantities coded in a system of notation, distinguished from analog output.

Digital resolution: The value of the least significant digit in a digitally coded representation.

Digitizer: A device which converts analog data into numbers expressed in digits in a system of notation.

Direct recording, magnetic tape: The method of recording using high frequency bias in which the input (electrical) signal is delivered to the recording head unaltered in form.

Direct writing recorder: A strip chart recorder which produces a readable record without further processing.

Discriminator, FM: A device which converts variations in frequency to proportional variations in voltage or current.

Discriminator tuning unit: A device which tunes the discriminator to a particular subcarrier.

Displacement: The change in position of a body or point with respect to a specified reference point. When no reference frame is specified, a reference frame fixed with respect to the earth is assumed.

Dithering: The application of intermittent or periodic acceleration forces sufficient to minimize the effect of static friction within the transducer, without introducing other errors.

Diversity reception: The use of several receivers or receiver front ends to improve signal reception. Diversity in space, phasing, polarization, etc. of antennas feeding the receivers is employed.

Double amplitude: In the field of vibratory acceleration the term *double amplitude* is employed to indicate the peak-to-peak value.

Dropout: Any discrete variation in signal level during the reproduction of recorded data which results in a data reduction error.

Dynamic calibration: A calibration during which the measurand varies and time in a known manner and the output is recorded as a function of time.

Dynamic response: The preferred term is *frequency response.*

Dynamic test: A test performed on accelerometers by means of which information is gathered pertaining to the over-all behavior, frequency response and/or natural frequency of the device.

End device, end instrument: The preferred term is *transducer.*

End points: The outputs at the upper and lower limits of the specified transducer range.

Environmental conditions: Specified external conditions (shock, vibration, temperature, etc.) to which a transducer may be exposed during shipping, storage, handling, and operation and which may adversely affect its performance or reliability.

Error: The algebraic difference between the indicated value and the true value of the measurand. Usually expressed in per cent of the full scale output.

Excursion: The application of measurand in a controlled manner in one direction only, whether it be increasing or decreasing. Ordinarily the term implies application of stimulus over the entire range of the transducer.

False pulse generator: Unit which supplies substitute pulses for missing PAM or PDM channel pulse.

FM: Frequency modulation.

FM/AM: Amplitude modulation of a carrier by subcarrier(s) which is (are) frequency modulated by information.

FM discriminator (subcarrier): A device which converts frequency variations to proportional variations in the amplitude of an electrical signal. Discriminators may be of several basic types, such as: Pulse Averaging, Foster Seely, Ratio Detector, Phase-Lock Correlation Detector.

FM/FM: Frequency modulation of a carrier by subcarrier(s) which is (are) frequency modulated by information.

FM/PM: Phase modulation of a carrier by subcarrier(s) which is (are) frequency modulated by information.

Frame: In time division multiplexing, one complete commutator revolution. In PCM systems, an integral number of words which includes a single synchronizing signal.

Frame frequency: Same as *frame rate.*

Frame pulse synchronization: Synchronization of local channel rate oscillator by comparison and phase-lock with separated frame synchronizing pulses.

Frame rate: The frequency derived from the period of one frame.

APPENDIX V

Frame synchronization signal: In PAM, uniquely coded pulses or interval to mark start of commutation frame period. In PCM, any signal used to identify a frame of data.

Frame synchronizing pulse separator: Unit for separating frame synchronizing pulses or intervals from commutated signals.

Free-running local synchronizer oscillator: Free-running oscillator in the decommutator normally triggered by separated channel synchronizing pulses. It supplies substitute pulses for missing channel pulses.

Frequency division multiplex: A system for the transmission of information about two or more quantities (measurands) over a common channel by dividing the available frequency bands. Amplitude, frequency or phase modulation of the subcarriers may be employed.

Frequency-modulated output: An output which is obtained in the form of a deviation from a center frequency, where the deviation is proportional to the applied stimulus.

Frequency output: An output in the form of frequency which is a function of the applied measurand (e.g., angular speed and flow rate).

Frequency, resonant: The measurand frequency at which a transducer responds with maximum output amplitude: for subsidiary resonance peaks use "resonances".

Frequency shift keying (FSK): Modulation accomplished by switching from one discrete frequency to another discrete frequency.

Friction error: The maximum change in output, at any measurand value within the specified range, before and after minimizing friction within the transducer.

Friction-free calibration: Calibration under conditions minimizing the effect of static friction often obtained by dithering.

Full excursion: The application of measurand, in a controlled manner, over the entire range of a transducer.

Gap azimuth alignment: The azimuth is the alignment of the line through the gaps relative to a line perpendicular to the precision-milled mounting pads in a plane parallel to the surface of the tape.

Gap scatter: Gap scatter is defined as the distance which includes the trailing edges of the gaps for record headstack and the center lines of the gaps for reproduce headstack.

Ground gating: Conversion of PAM signals at telemeter ground station to 50% duty cycle signals.

Harmonic content: The distortion in a transducer sinusoidal output, in the form of harmonics other than the fundamental component, usually expressed as a percentage of rms output.

Head stack: A group of two or more heads mounted in a single unit for the purpose of obtaining multiple track recording or reproduction.

High frequency bias: A sinusoidal signal which is mixed with the data signal during the direct record process on magnetic tape for the purpose of increasing the linearity and dynamic range of the recorded signal. The bias frequency is usually 3 to 4 times the highest information frequency which is to be recorded.

High velocity noise: The noise in wire-wound potentiometric transducers which appears as a series of momentary open circuits when the slider bounces along the coil if moved too quickly.

Hysteresis: The maximum difference in output, at any given measurand value within the specified range, when the value is approached first with increasing and then decreasing measurand. Hysteresis is expressed in per cent of full scale output, during any one calibration cycle. Friction dithering is specified.

IF band shift factor: A factor by which the design IF bandwidth in PAM telemetry is multiplied to produce a receiver IF bandwidth that is sufficiently wide to allow for doppler shifts, and receiver and transmitter drifts.

IF bandwidth: For telemetry receivers, the *post-conversion bandwidth*.

Inactive leg: An electrical element within a transducer which *does not* change its electrical characteristics as a function of the applied stimulus. Specifically applies to elements which are employed to complete a Wheatstone bridge in certain transducers.

Individual gap azimuth: In a magnetic recorder or reproduce stack, the angle of an individual gap relative to a line perpendicular to the precision milled mounting pads in a plane parallel to the surface of the tape.

Information gate: A device which, when triggered, allows information pulses to pass.

Inhibitor gate: A device which, when triggered, prevents information pulses from passing.

Input impedance: The impedance (presented to the excitation source) measured across the excitation terminals of a transducer.

Input recorder: Any device which makes a record of an input electrical signal.

Instability: It is preferred that this term not be used. See *stability*.

Insulation resistance: (1) An electrical measure of the insulation, at a specified voltage, between given components. Usually expressed in megohms. (2) Transducers: The resistance measured between specified insulated portions of a transducer when a specified d-c voltage is applied.

Integrating accelerometer: A transducer designed to measure velocity and/or distance by means of time integration of acceleration.

Intermittent rating: The rating applicable to operation over a specified number of time intervals of specified duration; the length of time between these time intervals must also be specified.

Intermodulation: Modulation of the components of a complex wave by each other, producing new waves whose frequencies are equal to the sums and differences of integral multiples of the component frequencies of the original complex wave.

Internal impedance: The preferred term is source impedance.

Internal pressure: The preferred terms are burst pressure, proof pressure or reference pressure.

Interval calibration: The preferred term is *step calibration*.

Jerk: The time rate of change of acceleration. Expressed in feet/sec^3, cm/sec^3, or g/sec.

Keyer: A PAM to PDM converter.

Leakage rate: The rate at which a specified fluid applied to the sensing element at a specified pressure is permitted to leak out of the case.

Least average deviation: A method of calculating the best fit straight line for which the average residuals are minimized.

Life cycling: The minimum number of full range excursions or specified partial range excursions over which a transducer will operate without changing its performance beyond specified tolerances.

Life, operating: The minimum length of time over which the specified continuous and intermittent rating of a transducer applies without change in transducer performance beyond specified tolerances.

Life, storage: The minimum length of time over which a transducer can be exposed to specified environmental storage conditions without changing its performance beyond specified tolerances.

Linearity: The similarity of a calibration curve to a specified straight line. Linearity is expressed as the maximum deviation of any calibration point from the corresponding point on a specified straight line, during any one calibration cycle. It is expressed as "within ± -per cent of full scale output."

Linearity, end point: Linearity referred to a straight line between the end points.

Linearity, independent: Linearity referred to the best fit straight line.

Linearity, least squares: Linearity referred to a straight line for which the sum of the squares of the residuals is minimized.

Linearity, terminal: A special form of theoretical slope linearity for which the theoretical end points are 0 and 100% of both measurand and output.

Linearity, theoretical slope: Linearity referred to a straight line between the theoretical end points.

Line pressure: The preferred term is *reference Pressure*.

Load impedance: The impedance presented to the output terminals of a transducer by the associated external circuitry.

Magnetic recorder/reproducer: A machine which converts electrical data signals to magnetic patterns on a magnetic tape during a recording process and/or converts the remanent magnetic patterns on a magnetic tape to electrical data signals during a reproducing process.

Measurand: (1) A physical or electrical quantity, property or condition which is to be measured. (2) Transducers: A physical or electrical quantity, property or condition which is measured. Referred to transducers this term is preferred to "entity," "excitation," "parameter to be measured," "physical phenomenon," "stimulus," and "variable."

Microlock: A trade name applied to a satellite telemetry system which uses phaselock techniques in the ground receiving equipment.

Minitrack: A trade name applied to a satellite tracking system which uses a miniature pulse type telemeter and a precise directional antenna system with phase comparison tracking techniques.

Modulation: The process of impressing information on a carrier for transmission.

    AM = Amplitude Modulation
    PM = Phase Modulation
    FM = Frequency Modulation

Modulation index: In angle modulation with a sinusoidal modulating wave, the modulation index is the ratio of the frequency deviation to the frequency of the modulating wave.

Most favorable straight line: A line from which non-linearity deviations are minimized for the largest number of points. Often used as synonymous with *best fit straight line*.

APPENDIX V

Mounting error: The error resulting from mechanical deformation of the transducer caused by mounting the transducer and making all measurand and electrical connection.

Multicoupler: A device for connecting several receivers to one antenna and properly matching the impedances of the receivers and the antenna.

Multiplexing: The simultaneous transmission of two or more signals within a single channel. The three basic methods of multiplexing involve the separation of signals by time division, frequency division and phase division.

Nominal bit rate: Bit rate established as a specific system design center.

Nominal range: See *rated range*.

Non-linearity: It is preferred that this term not be used. See *linearity*.

Non-repeatability: The preferred term is *repeatability*.

Normal record level: Normal record level is the level of record head current required to produce 1% third harmonic distortion of the reproduced signal at the Record Level Set Frequency when the distortion is a function of magnetic tape saturation and is not a function of electronic circuitry.

Octave: The interval between two frequencies having a ratio of 2:1.

Operating temperature: The range of temperatures, in which a transducer is expected to operate within specified limits of error.

Oscillograph: A strip chart recorder that employs light sensitive paper and mirror type galvanometers to produce traces.

Output: The electrical signal from a system or device which is a function of the applied measurand or input.

Output impedance: The preferred term is *source* impedance.

Overload: The maximum magnitude of measurand that can be applied to a transducer without causing a change in performance beyond specified tolerances.

Overrange: The preferred term is *overload*.

Overshoot: In an underdamped transducer, the amount of output measured beyond the final steady output value, in response to a step change in the measurand, expressed in per cent of the equivalent step change in output.

PAM: Pulse Amplitude Modulation.

PAM/FM: Frequency modulation of a carrier by pulses which are amplitude modulated by information.

PAM/FM/FM: Frequency modulation of a carrier by subcarrier(s) which is (are) modulated by pulses which are amplitude modulated by information.

PAM signal integrator: Unit which integrates individual PAM pulses for fixed period. Integration is usually delayed to avoid transient effects of leading edge of pulses.

Parallel recording, magnetic tape: The technique of simultaneously energizing heads in a head stack to record an ordered set of bits.

Parameter (to be measured): The preferred term is *measurand*.

Parity: A symmetry property of a wave function; the parity is 1 (or even) if the wave function is unchanged by an inversion (reflection in the origin) of the coordinate system; and -1 (or odd) if the wave function is changed only in sign.

Parity bit: A bit added to a binary code group which is used to indicate whether or not the number of recorded "1's" or "0's" is even or odd.

Parity check: A self-checking code employing binary digits in which the total number of 1's (or 0's) in each permissible code expression is always even or always odd. A check may be made for either even parity or odd parity.

PCM: Pulse Code Modulation-Pulse modulation of a carrier by coded information. In PCM telemetry, information transmission by means of a code representing a finite number of values of the information at the time of sampling.

PCM/FM: Frequency modulation of a carrier by pulse code modulated information.

PCM/FM/FM: Frequency modulation of a carrier by subcarrier(s) which is (are) frequency modulated by pulse code modulated information.

PCM/PM: Phase modulation of a carrier by pulse code modulation information.

PDM (PWM): Pulse Duration Modulation (Pulse Width Modulation).

PDM/FM: Frequency modulation of a carrier by pulses which are modulated in duration by information.

PDM/FM/FM: Frequency modulation of a carrier by subcarrier(s) which is (are) frequency modulated by pulses which are duration modulated by information.

PDM/PM: Phase modulation of a carrier by pulses which are duration modulated by information.

PDM recording: The method of recording Pulse Duration Modulated telemetry data in which the signal delivered to the recording head is the differential of the input signal.

PDM signal integrator: Unit which integrates individual PDM pulses of fixed amplitude.

Peak amplitude: The maximum deviation of a phenomenon from its average or mean position. When applied to vibration, same as single amplitude.

Peak-to-peak: (1) The maximum algebraic difference between two or more stimuli or signals. (2) Transducers: The preferred term is *double amplitude.*

Pedestal, PAM: An arbitrary minimum signal value assigned to provide for channel synchronization and decommutation.

Phase-lock loop: An electronic servo system used either as a tracking filter or as a frequency discriminator.

Phase-locked local channel rate oscillator: Local oscillator is maintained at channel switching frequency by servo control. Error voltage is derived from separated frame and/or channel synchronizing pulses when compared in frequency and phase with signals generated by local ring counter chain. The ring counter chain is triggered by the local oscillator in the absence of normal channel synchronizing pulses.

Pickup: The preferred term is *transducer.*

Point-based linearity: A manner of expressing non-linearity as deviation from a straight line which passes through a given point or points.

Post-conversion bandwidth: In a telemetry receiver, the bandwidth presented to the detector.

Pot: In general usage, a contraction of *potentiometer.* When applied to instrumentation, contraction of *potentiometric transducer.*

PPM: Pulse Position Modulation.

PPM/AM: Amplitude modulation of a carrier by pulses which are position modulated by information.

Primary calibration: Calibration in which the transducer output is observed, or recorded, while a direct known stimulus is applied under controlled conditions.

Primary element, primary detector: The preferred term is *sensing element.*

Primary standard: A unit directly defined and established by some authority, against which all secondary standards are calibrated.

Prime channels: The channels which are sequentially sampled by the basic commutator of the system.

Prime frame: A group of words resulting from a complete sampling of the prime channel.

Proof pressure: The maximum pressure which may be applied to the sensing element of a transducer without changing the transducer performance beyond specified tolerances.

Pulse position modulator: A device which converts analog information to variations in pulse position.

Pulse sample-and-hold circuit: A circuit which holds final amplitude of an integrated pulse until the final amplitude of the succeeding integrated pulse is reached. A less desirable sample-and-hold circuit resets after each hold period to a fixed level before integration of succeeding pulse.

Quantization: The process of converting from continuous values of information to a finite number of discrete values.

Quantization error: The difference between actual values of information and the corresponding discrete values resulting from quantization.

Quantization noise: Inherent noise resulting from quantization.

Radio telemetry: Telemetry in which an r-f link is used as a portion of the transmission path.

Random vibration: Non-periodic vibration, described only in statistical terms, most commonly taken to mean vibration characterized by an amplitude distribution which follows the normal error curve (Gaussian distribution).

Ratio calibration: A method by which potentiometric transducers may be calibrated, in which the value of the measurand is expressed in terms of decimal fractions representing the ratio of output resistance to total resistance.

Read: See *reproduce, magnetic tape.*

Record, magnetic tape: The process by which an electromagnetic transducer (record head) and associated electronic circuitry convert electrical data to a magnetic flux pattern on a magnetic tape.

Record head: An electromagnetic transducer used during the record process for inducing magnetic patterns into the magnetic tape.

Recovery time: The time interval, after a specified overload, after which a transducer again performs within its specified tolerances.

Reliability: A measure of the probability that a system or device will continue to perform within specified limits of error for a specified length of time under specified conditions.

Reproduce (playback), magnetic tape: The process by which an electromagnetic transducer (reproduce head) and associated electronic circuitry convert the magnetic flux pattern on a magnetic tape to an electrical signal containing the recorded information.

Reproduce head: An electromagnetic transducer which converts the remanent flux patterns in a magnetic tape into electrical signals during the reproduce process.

Reproducibility: The preferred term is *repeatability.*

APPENDIX V 5-65

Resonances: Amplified vibrations of transducer components, within narrow frequency bands, as vibration in specified ranges of frequencies and amplitudes is applied along specified transducer axes.

Resonant frequency: The measurand frequency at which a transducer responds with maximum output amplitude.

Response (transducer): A quantitative expression of the output of a transducer as a function of the input, under conditions which must be explicitly stated.

Ripple: The rms a-c component of a transducer's d-c output voltage expressed in per cent of the average value of the total output voltage.

Rise time: The length of time for the output of a transducer to rise from a small specified percentage of its final value of a large specified percentage of its final value.

Room conditions: Conditions for conducting operational tests shall be as follows:
 (a) Temperature: $25° \pm 10°C$ ($77° \pm 18°F$).
 (b) Relative Humidity: 90 per cent or less.
 (c) Barometric pressure: 26 to 32 inches Hg.

Selectivity: The degree of falling off in response of a resonant device with departure from resonance.

Self heating: Internal heating resulting from electrical energy dissipated within the transducer.

Sensing element: Where applicable that part of the transducer which responds directly to the measurand. This term is preferred to "primary element," "primary detector."

Sensitivity set: A permanent change in sensitivity attributable to any cause, such as over-ranging, shock, aging, etc.

Serial recording, PCM: The technique of recording a train of bits on a single magnetic tape track.

Signal separation filter: A bandpass filter which selects the desired subcarrier channel from the FM composite.

Single carrier FM recording: The method of magnetic tape recording in which the input signal is frequency modulated onto a carrier and the carrier is recorded on a single track at saturation and without bias.

Skew: Tape motion characterized by an angular velocity between the gap center line and a line perpendicular to the tape center line.

Source impedance: The impedance across the output terminals of a transducer presented by the transducer to the associated external circuitry.

Stability: The ability of a system to retain its specified performance.

Standard: A value or a criterion that has been established by authority, custom, or agreement, to serve as a model or rule in the measurement of a quantity or in the establishment of a practice or procedure.

Standard PAM (or PDM) signals: Signals with format conforming to IRIG standards.

Static calibration: A calibration performed under room conditions by application of the measurand to the transducer in discrete amplitude intervals.

Static error band: The error band applicable at room conditions and in the absence of any vibration, shock or acceleration.

Static test: A measurement taken under conditions where neither the stimulus nor the environmental conditions fluctuate.

Strain error: The error resulting from a strain imposed on a surface to which the transducer is mounted.

Strip chart recorder: A recorder whose record is a time display of traces on a continuous chart.

Subcarrier: A carrier which is applied as a modulating wave to modulate another carrier or an intermediate subcarrier.

Subcarrier band: A band associated with a given subcarrier and specified in terms of maximum subcarrier deviation.

Subcarrier composite: Two or more subcarriers combined in a frequency division multiplexing scheme.

Subcarrier oscillator: In a telemetry system, the oscillator which is directly modulated by the measurand, or by the equivalent of the measurand in terms of changes in the transfer elements of a transducer.

Subchannel: The route required to convey the magnitude of a single subcommutated measurand.

Subcommutation: Commutation of additional channels with output applied to individual channel of the primary commutator. Subcommutation is synchronous if its rate is a submultiple of that of the primary commutator. Unique identification must be provided for the subcommutation frame pulse.

Subcommutation frame: In PCM systems, a recurring integral number of subcommutator words which includes a single subcommutation frame synchronization word. The number of subcommutator words in a subcommutation frame is equal to an integral number of primary commutator frames. The length of a subcommutation frame is equal to the total number of words or bits generated as a direct output of the subcommutator.

Supercommutation: Commutation at higher rate by connection of single data input source to equally spaced contacts of the commutator (crosspatching). Corresponding crosspatching is required at the decommutator.

Synchronizing pulse selector: Unit for separating synchronizing pulses from commutated pulse trains.

Tape speed errors: Any variation of the tape speed from its nominal speed over the record or reproduce head regardless of cause.

Tape speed error compensation: The process of correcting for tape speed errors electrically.

Tapping: The preferred term is *dithering*.

Telemetering: Measurements accomplished with the aid of intermediate means which allows perception, recording or interpretation of data at a distance from a primary sensor. The widely employed interpretation of telemetering restricting its significant to data transmitted by means of electromagnetic propagation is more properly called Radio Telemetry.

Telemetry: The science of measuring a quantity or quantities, transmitting the results to a distant station, and there interpreting, indicating and/or recording the quantities measured.

Temperature effect: The difference between the output at room temperature and at any other specified temperature at any one value of the stimulus within the range of the measuring device.

Temperature error: The maximum change in output, at any measurand value within the specified range, when the transducer temperature is changed from room temperature to specified temperature extremes.

Temperature error band: The error band applicable over stated environmental temperature limits.

Temperature gradient error: The transient deviation in output of a transducer at a given measurand value when the ambient temperature of the measurand temperature changes at a specified rate between specified magnitudes.

Theoretical curve: The specified relationship (table, graph, or equation) of the transducer output to its applied measurand over the range.

Theoretical end points: The specified points between which the theoretical curve is established and to which no end point tolerances apply.

Thermal coefficient of resistance: The relative change in resistance of a conductor or semiconductor for each unit change in temperature, expressed in ohms per degree F or C.

Thermal coefficient of sensitivity: The change in full scale output due to the effects of temperature only. Usually expressed in percentage of the full scale output at room temperature per unit or interval, change in temperature.

Thermal compensation: A method employed to reduce or eliminate the thermal effects on one or more of the performance parameters of a transducer.

Thermal zero shift: The preferred term is *zero shift*.

Thermistor: A resistor whose value varies with temperature in a definite desired manner.

Thermocouple: A transducer which depends on the production of an emf in two dissimilar metals as a function of the temperature or temperature change.

Threshold: The point at which an effect is first produced, observable, or otherwise sensed.

Time constant: The length of time required for the output of a transducer to rise to 63% of its final value as a result of a step change of measurand.

Time division multiplex: A system for the transmission of information about two or more quantities (measurands) over a common channel by dividing available time intervals among the measurands to form a composite pulse train. Information may be transmitted by variation of pulse duration, pulse amplitude, pulse position, or by a pulse code. Abbreviations used are PDM, PAM, PPM and PCM respectively.

Timing signal: Any signal recorded simultaneously with data to provide a time index.

Torque error: The preferred term is *mounting error*.

Traceability: The relation of a transducer calibration, through a step-by-step process, to an instrument or group of instruments calibrated and certified by the National Bureau of Standards.

Track: A portion of a magnetic tape whose width and position on the magnetic tape is specified. A track extends throughout the entire length of a reel of tape and always exists regardless of its state of magnetization.

Tracking filter: A bandpass filter whose center frequency follows the average frequency of the input signal.

Transducer, bi-directional: A transducer capable of measuring stimulus in both a positive and a negative direction from a reference zero or rest position.

APPENDIX V   5–67

Transducer, bonded: A transducer which employs the bonded strain gage principle of transduction.

Transverse recording, magnetic tape: The technique of recording with rotating heads which are oriented perpendicularly to the edge and the surface of the magnetic tape.

Transverse response: The preferred term is *transverse sensitivity*.

Transverse sensitivity: (1) Tape recording: Susceptibility of a track to interference from flux patterns generated by adjacent record heads.   (2) Transducers: The maximum sensitivity of a transducer to a specified value of transverse acceleration or other transverse measurand.

Variable: The preferred term is *measurand*.

Variable-erase recording: The method of recording on magnetic tape by selective erasure of a prerecorded signal.

Vibration: Motion due to a continuous change in the magnitude of a given force which reverses its direction with time.   Vibration is generally interpreted as symmetrical or nonsymmetrical fluctuations in the rate at which acceleration is applied to an object.

Vibration error: The maximum change in output at room conditions, at any measurand value within the specified range, when vibration levels of specified amplitude and range of frequencies are applied to the transducer along specified axes other than the normal sensing axes.

Vibration error band: The error band applicable when vibration levels with a specified range of frequencies and amplitudes are applied to a transducer along specified axes.

Video recording (magnetic tape): The methods of recording information having a bandwidth in excess of 500 kilocycles on a single track.

Viscous damping: Included under *Damping*.

Voltage breakdown test: A test whereby a specified voltage is applied, between given points in a transducer, circuit, or device to determine whether breakdown occurs at said voltage.

Voltage ratio: The ratio of output voltage to excitation voltage, expressed in per cent.

Wire link telemetry: Also called "hard wire" telemetry or "line" telemetry in which no radio frequency link is used.

Word: An ordered set of digits processed as a unit.

Word rate: The frequency derived from the elapsed period between the beginning of transmission of one word and the beginning of transmission of the next word.

Wow and flutter: Terms derived from disc and motion picture sound recording often used in reference to speed change errors in magnetic recording.

Zeroing: A deliberate translation of recorded data, or data reduction equipment output, to a position selected as zero reference.

Zero pedestal: Amplitude level of PAM channel signal corresponding to "zero" or minimum channel signal.   See *Pedestal, PAM*.

Chapter 6

FM-FM TELEMETRY SYSTEMS

C. D. ALBRIGHT, W. P. McGARRY, O. J. OTT, and R. A. RUNYAN, *Data-Control Systems, Inc., Danbury, Conn.* (*System Characteristics, Receivers, Subcarrier Oscillators, Subcarrier Discriminators, Frequency Translation*)

D. B. LEESON, *Applied Technology, Inc., Palo Alto, Calif.* (*Transmitters*)

SYSTEM CHARACTERISTICS

1	FM-FM System Functions	6–2
2	System Configurations	6–3
3	Theory of FM Systems	6–7
4	FM Multiplexing Principles	6–13
5	Sources of Error	6–14

TRANSMITTERS

6	Characterization of FM Telemetry Transmitters	6–20
7	Functional Blocks	6–22
8	Mechanical and Thermal Design	6–27

RECEIVERS

9	Functional Description of Telemetry Receivers	6–27
10	R-F Amplifiers, Converters, and Filters	6–28
11	Receiver Demodulators and Output-signal Processing	6–31
12	Predetection Recording	6–32
13	Typical Receiver Performance	6–36

SUBCARRIER OSCILLATORS

14	Types and Functions	6–36
15	Oscillator Performance Parameters	6–38

SUBCARRIER DISCRIMINATORS

16	Function and Design Requirements	6–40
17	Bandpass Input Filters	6–41
18	Limiter and Demodulator	6–43
19	Output-signal Processing	6–51
20	Tape-speed Compensation and Automatic Standardization	6–52

FREQUENCY TRANSLATION

21	System Considerations	6–55
22	Translation Techniques	6–55
23	Applications and Typical Performance	6–59
24	SS-FM Telemetry	6–61

SYSTEM CHARACTERISTICS

1 FM-FM System Functions

The FM-FM telemetry-system block diagram in Fig. 1 provides a simple, flexible technique for transmitting a number of analog channels via a common r-f link with good high-frequency response and reasonable accuracy. Since the data values are encoded in the form of subcarrier frequencies, the system is insensitive to level changes either in the r-f transmission path or in magnetic-tape-recorder outputs, and frequency response to direct current can be obtained. In addition, the nature of the FM system allows the designer considerable latitude in exchanging spectrum utilization and data bandwidth for noise immunity. Since a frequency-division multiplexing format is

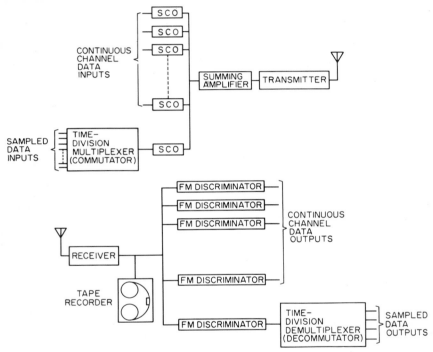

FIG. 1. Typical FM-FM system block diagram.

employed, the system enjoys the additional advantage of simplicity in adding or deletion of data channels, as it is merely necessary to add or remove a modular subcarrier oscillator of the appropriate frequency to revise the system data capacity.

Standards have been set up to provide for a set of frequencies having fixed percentage deviations and other standards providing multiplexed channels of identical bandwidth have been proposed, thereby making a highly flexible and versatile but thoroughly proved series of channel configurations available to the telemetry-system designer. This standardization also results in the easy availability of components made in volume production by a number of manufacturers. Proportional- and constant-bandwidth standard channel assignments are discussed in Sec. 2.

FM-FM systems are normally employed when these performance levels are desired:

    Accuracy.............. 1–5 %
    Data bandwidth....... D-C to approx 10 kc max
    No. of channels........ Less than 25

SYSTEM CHARACTERISTICS 6–3

They find widest application in medium-accuracy wideband applications such as vibration telemetry, and in areas where high-frequency signal components might cause aliasing errors in time-shared systems.

Another widespread use of FM-FM telemetry systems consists of transmitting a time-division multiplexed signal over one of the FM subcarrier channels. They are also noted for higher reliability because failure or degradation of one data channel does not normally affect the accuracy of the remaining channels.

Every FM-FM telemetry system performs the following functions:

1. It generates a multiplex of subcarriers of various frequencies ranging from several hundred cps to thousands or hundreds of thousands of cps.

2. It encodes the information to be transmitted in the form of frequency modulation of the individual subcarriers.

3. It transmits the FM multiplex to the receiving station via an FM radio-frequency carrier.

4. It separates and demodulates the subcarriers forming the multiplex.

An FM-FM telemetry system may be briefly described as a frequency-division multiplex of frequency-modulated subcarriers transmitted via a frequency-modulated r-f carrier. Of course, many other functions than those listed above may be performed by a typical FM-FM system. For example, one or more of the subcarriers may be modulated by the output of a commutator sampling a number of data points rather than one per subcarrier channel, or magnetic recording of the multiplex may be performed before or after transmission. In addition, such functions as display and processing for computer entry might properly be considered part of a telemetry system, but this chapter will be restricted to a discussion of the characteristics of the encoding, multiplexing, separation, and demodulation equipment. R-F transmission and reception are also discussed in this chapter.

The block diagram shown in Fig. 1 will be of help in considering the FM-FM system functions and the influence which each of the system components has on overall accuracy. An immediate observation which may be drawn from the block diagram is that if the subcarrier oscillators do not have a linear frequency vs. stimulus curve, or if the demodulators do not have a linear output vs. frequency curve, the channel linearity will be adversely affected. A principal function of the SCO and discriminator is to provide linear transformation of data input to output. In addition, the drift and gain stability of the subcarrier oscillators and discriminators have a direct effect on system accuracy unless automatic stabilization subsystems are employed. The requirements of linearity, gain stability, and low drift also apply to commutation and decommutation equipment; so what we might call the FM-FM system *static* characteristics have been defined. Not so the performance of an FM-FM system in handling *dynamic* signals, a function for which it is especially well suited.

The task of evaluating the performance of an FM-FM system in transmitting *rapidly varying signals* is far more difficult and often much more important than the determination of system linearity and drift. For one thing, the static characteristics of an FM-FM system can be evaluated on a "per channel" basis, while the measurement of dynamic performance invariably involves testing groups of channels and so the number of parameters to be considered is vastly increased. The sources of *dynamic error* can, however, be grouped into four categories: *interchannel crosstalk* due to harmonic and intermodulation distortion products in the multiplex, *noise* arising from the transmission-system *thermal fluctuations*, distortion arising from *phase nonlinearity in filters*, and errors introduced by *tape-recorder speed fluctuations*.

Each of these sources of error is discussed in the subsequent paragraphs; and with prudent concern for the factors limiting system performance, the designer can arrive at an optimal combination of channel assignments, modulation indices, and component specifications to satisfy the program requirements confronting him.

2 System Configurations

Each set of telemetry requirements can best be served by a unique system configuration specifically tailored to accommodate the data channels in question, but compatibility with existing ground-station equipment prohibits custom system design except under very unusual circumstances. However, the system designer has available to him two types of standard systems offering a variety of bandwidth capabilities. These are the proportional-bandwidth and constant-bandwidth subcarrier channel systems. Proportional bandwidth means that the subcarrier deviation ratio is a fixed percentage of the subcarrier center frequency. Constant bandwidth implies that the subcarrier deviation is a fixed frequency, regardless of subcarrier center frequency.

The IRIG standards for frequency division multiplexing presented in Chap. 5 (Sec. 2 of the IRIG Standards document) cover only proportional-bandwidth subcarrier channels and are 18 in number. These have been in use for many years. The need for greater channel capacity and more flexible operation has called for proposed revisions of these standards to include more proportional-bandwidth channels and the standardization of constant-bandwidth channels.

The original IRIG FM-FM standard system configuration plus proposed additions are shown in Table 1. These standards provide for a graduated series of channels, ranging from a ±30 cps deviation, 6 cps data channel at 400 cps center frequency, to the proposed channel H, which allows 5,000 cps data bandwidth on a 165-kc subcarrier

Table 1. Proportional-bandwidth Subcarrier Channels (IRIG)

Channel	Center frequency, cps	Lower limit,* cps	Upper limit,* cps	Max deviation, %	Frequency response,† cps
1	400	370	430	±7.5	6.0
2	560	518	602	±7.5	8.4
3	730	675	785	±7.5	11
4	960	888	1,032	±7.5	14
5	1,300	1,202	1,399	±7.5	20
6	1,700	1,572	1,828	±7.5	25
7	2,300	2,127	2,473	±7.5	35
8	3,000	2,775	3,225	±7.5	45
9	3,900	3,607	4,193	±7.5	59
10	5,400	4,995	5,805	±7.5	81
11	7,350	6,799	7,901	±7.5	110
12	10,500	9,712	11,288	±7.5	160
13	14,500	13,412	15,588	±7.5	220
14	22,000	20,350	23,650	±7.5	330
15	30,000	27,750	32,250	±7.5	450
16	40,000	37,000	43,000	±7.5	600
17	52,500	48,562	56,438	±7.5	790
18	70,000	64,750	75,250	±7.5	1,050
19	93,000	86,025	99,975	±7.5	1,400
20	124,000	114,700	133,300	±7.5	1,900
21	165,000	152,625	177,375	±7.5	2,500
A‡	22,000	18,700	25,300	±15	660
B	30,000	25,500	34,500	±15	900
C	40,000	34,000	46,000	±15	1,200
D	52,500	44,625	60,375	±15	1,600
E	70,000	59,500	80,500	±15	2,100
F	93,000	79,050	106,950	±15	2,800
G	124,000	105,400	142,600	±15	3,700
H	165,000	140,250	189,750	±15	5,000

* Rounded off to nearest cycle.
† The frequency response given is based on maximum deviation ratio of 5.
‡ Channels A through H are optional and may be used by omitting adjacent lettered and numbered channels in Table 2. In the process of recording the foregoing subcarriers on magnetic tape at a receiving station, provision may also be made to record a tape-speed-control tone and tape-speed-error-compensation signals.

SYSTEM CHARACTERISTICS 6–5

deviated ±25 kc. Thus, given a schedule of channel bandwidth requirements, it becomes a fairly straightforward process to match these needs with the appropriate IRIG band capacities. Vibration and other wideband data are transmitted via the letter and higher number channels, and narrowband signals are fed to low-frequency subcarrier oscillators.

As flexible and useful as the IRIG proportional bands are, there are many instances in which their progressive channel capacities do not provide a satisfactory distribution of channel bandwidths, particularly when a large number of vibration transducer outputs must be transmitted, or when close time correlation between channels is desired.

The constant-bandwidth standards shown in Table 2 were adopted by the Aircraft Industries Association to supplement IRIG standard systems in this situation. They

Table 2. Constant-bandwidth Subcarrier Channels (AIA)

Channel	Center frequency, kc	Deviation, kc	Frequency response, kc	
			$m = 2$	$m = 5$
1C	12.5	±2	1	0.4
2C	20.8	±2	1	0.4
3C	29.2	±2	1	0.4
4C	37.5	±2	1	0.4
5C	45.8	±2	1	0.4
6C	54.2	±2	1	0.4
7C	62.5	±2	1	0.4
8C	70.8	±2	1	0.4
9C	79.2	±2	1	0.4
10C	87.5	±2	1	0.4
11C	95.8	±2	1	0.4
12C	104.2	±2	1	0.4
13C	112.5	±2	1	0.4
14C	120.8	±2	1	0.4
15C	129.2	±2	1	0.4
16C	137.5	±2	1	0.4
17C	145.8	±2	1	0.4
18C	154.2	±2	1	0.4
19C	162.5	±2	1	0.4
20C	170.8	±2	1	0.4
21C	179.2	±2	1	0.4
22C	187.5	±2	1	0.4
2CW	20.8	±4	2	0.8
4CW	37.5	±4	2	0.8
6CW	54.2	±4	2	0.8
8CW	70.8	±4	2	0.8
10CW	87.5	±4	2	0.8
12CW	104.2	±4	2	0.8
14CW	120.8	±4	2	0.8
16CW	137.5	±4	2	0.8
18CW	154.2	±4	2	0.8
20CW	170.8	±4	2	0.8
22CW	187.5	±4	2	0.8

The number of these channels that may be transmitted over r-f link simultaneously will be limited by the available transmitter deviation.

The channels with ±4-kc deviation will be designated with the suffix W

The even-numbered wideband channels, 2CW through 22CW, are optional and may be used by omitting the adjacent odd-numbered narrowband channels.

provide up to 22 channels of 1 kc bandwidth or 11 channels of 2 kc bandwidth for a total base-band spectrum occupancy of less than 200 kc. The double-bandwidth channels, denoted by the suffix CW, can be interspersed with standard-bandwidth channels in any desired combination by omitting the adjacent narrowband channels.

The proposed IRIG constant-bandwidth standard subcarrier channels are given in Table 3. Combinations of proportional- and constant-bandwidth subcarriers may be used, provided proper precautions are taken to avoid crosstalk.

It should be observed that the nominal frequency responses have been stated in terms of a modulation index of 5. Higher frequency response may be obtained at

Table 3. Constant-bandwidth subcarrier channels (IRIG)*

Deviation = ±2 kc/sec Nominal frequency response = 0.4 kc/sec Maximum frequency response = 2 kc/sec†		Deviation = ±4 kc/sec Nominal frequency response = 0.8 kc/sec Maximum frequency response = 4 kc/sec†		Deviation = ±8 kc/sec Nominal frequency response = 1.6 kc/sec Maximum frequency response = 8 kc/sec†	
Channel	Center frequency, kc/sec	Channel	Center frequency, kc/sec	Channel	Center frequency, kc/sec
1A	16				
2A	24				
3A	32	3B	32		
4A	40				
5A	48	5B	48		
6A	56				
7A	64	7B	64	7C	64
8A	72				
9A	80	9B	80		
10A	88				
11A	96	11B	96	11C	96
12A	104				
13A	112	13B	112		
14A	120				
15A	128	15B	128	15C	128
16A‡	136				
17A‡	144	17B‡	144		
18A‡	152				
19A‡	160	19B‡	160	19C‡	160
20A‡	168				
21A‡	176	21B‡	176		

* Proposed in 1966.

† The indicated maximum frequency response is based upon the maximum theoretical response that can be obtained in a bandwidth between deviation limits specified for the channel.

‡ Recommended for use in uhf transmission systems only.

modulation indices as low as 1; however, data quality suffers. It is important to consider carefully the laws governing frequency modulation and frequency division before assembling a multiplex. These are presented below in Secs. 3 to 5.

It must be stressed that, when a system is designed around the standards shown in Tables 1, 2, and 3, the design engineer has available to him a variety of production-proved components from a number of competent vendors. Compared with developing a new channel configuration, the use of standard systems results in far fewer engineering problems and offers substantial economies because of the large volume of production in which the standard system components are made.

3. Theory of FM Systems

Although a rigorous derivation of FM theory is beyond the scope of this chapter, an attempt will be made to develop the basic concepts in an intuitive manner and outline the manner in which effects such as crosstalk, capture, and thresholding occur.

We begin by comparing the spectra of amplitude-modulated and angle-modulated waves, and find (Fig. 2a) that for small levels of modulation, the spectra bear a strong resemblance even though the appearance of the two signals is quite different (Fig. 2b). The explanation for the similarity of the spectra and the dissimilarity of the waveforms can be obtained from the phasor diagram in Fig. 2c in which it is seen that in each case, the signals consist of a carrier ω_c, an upper sideband $\omega_c + \omega_m$, and a lower sideband

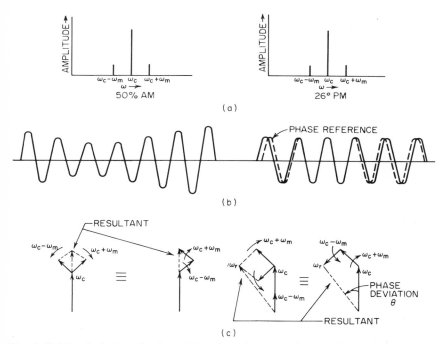

FIG. 2. Sideband relationships in modulated carriers. (a) Spectra of modulated carriers. (b) Waveforms of modulated carriers. (c) Phasor diagrams of modulated carriers.

$\omega_c - \omega_m$, but the phasing of the sidebands is different. In the amplitude-modulated case, the phasing of the sidebands is such that their horizontal components cancel and their vertical components sum, thereby periodically increasing and decreasing the amplitude of the resultant but not varying its phase, while in the angle-modulated case the sidebands are phased in such a manner that their vertical components cancel and their horizontal components reinforce, thereby periodically varying the apparent phase of the carrier.

If, in the amplitude-modulated case, we increase the sideband level, we reach a point at which the composite signal amplitude reaches zero at the valley point of the modulating cycle and twice the unmodulated level at the peak point of the modulating cycle. At this 100 per cent modulation level, the amplitude of each of the sidebands is equal to half the carrier amplitude and the power in each sideband is one-quarter of the carrier power.

Referring to Fig. 3a, it is evident that increasing the modulation level of a phase-modulated system is not such a simple matter. Here two problems become evident. First, the amplitude of the resultant signal does not remain constant as it should in a true phase-modulated system—it increases sharply at the positive and negative peaks of deviation; so we observe spurious amplitude modulation at twice the modulating frequency. More significantly, an examination of the phasor diagram (Fig. 3a) reveals the modulation process is grossly nonlinear when large phase deviations are to be produced by a single pair of sidebands. If a phase excursion of 90° is desired, the required sideband amplitudes will be infinite. This situation can be alleviated and the

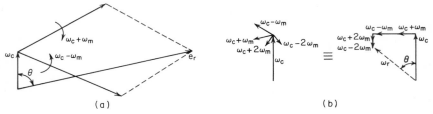

Fig. 3. (a) Large-amplitude angle modulation with one pair of sidebands. (b) Contribution of second-order sidebands to phase modulation.

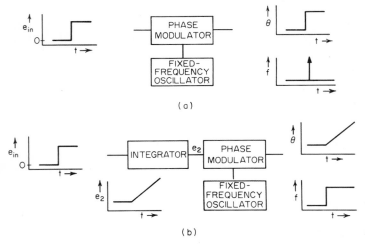

Fig. 4. Relation between phase and frequency modulation. (a) Phase modulator. (b) Frequency modulation produced by phase modulator and integrator.

resultant phasor kept much more nearly constant in amplitude if a second set of sidebands, spaced away from the carrier by twice the modulating frequency, are added (Fig. 3b). To maintain amplitude constancy and linearity of the deviation, increasing sideband orders are required as the amount of phase deviation is increased.

The relationship between phase and frequency modulation is easily ascertained by examining Fig. 4. Part a of this figure shows a phase-modulator response to an input step while in part b, an integrator has been placed between the input and modulator, converting the step to a ramp. The output phase now continues to advance in response to an input step, and the performance of the integrator and phase-modulator combination is indistinguishable from that of a frequency modulator. In practice, the block

SYSTEM CHARACTERISTICS 6–9

diagram of Fig. 4b is not often used to generate FM carriers since a variety of simpler and more practical circuit configurations are available.

A number of terms are of special importance in discussing FM theory. These are

ω_c = carrier frequency
ω_m = intelligence frequency, or modulating frequency
$\Delta\omega_c$ = peak carrier deviation
M = modulation index = $\Delta\omega_c/\omega_m$

Using these terms, we can write the expression for a frequency-modulated wave

$$e = E_{max} \sin (\omega_c + \Delta\omega_c \sin \omega_m t)t \tag{1}$$

where e = instantaneous voltage
E_{max} = peak voltage

Sidebands of an FM Carrier

The spectrum of this frequency-modulated wave can be analyzed and is found to consist of a number of components:

	Frequency	Amplitude
Carrier term................	ω	$E_{max}J_0(M)$
First-order sidebands........	$(\omega_c + \omega_m), (\omega_c - \omega_m)$	$E_{max}J_1(M)$
Second-order sidebands.......	$(\omega_c + 2\omega_m), (\omega_c - 2\omega_m)$	$E_{max}J_2(M)$
Third-order sidebands........	$(\omega_c + 3\omega_m), (\omega_c - 3\omega_m)$	$E_{max}J_3(M)$
nth-order sidebands..........	$(\omega_c + n\omega_m), (\omega_c - n\omega_m)$	$E_{max}J_n(M)$

A plot of the values of the Bessel function $J(M)$ of various orders for values of the argument M is shown in Fig. 5. (See Appendix, Fig. A.10-3 or 4, for a more extensive

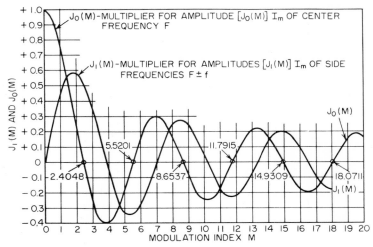

Fig. 5. Bessel curves by means of which the magnitude or the amplitude of center frequency F and of respective side frequencies $F \pm f$ can be computed. (*From Hund, "Frequency Modulation," McGraw-Hill Book Company, New York, 1942.*)

plot.) Spectra for values of M of 5, 2, and 1 are shown in Fig. 6. The data transmitted by means of a frequency-modulated carrier are perturbed if the relative amplitudes of the carrier and sidebands are altered. No simple analytical technique permits accurate prediction of the distortion introduced by altering the relative amplitudes or phases of the various sidebands which make up the signal. The effects of normal telemetry bandpass filters on transmitted data are described in Sec. 17.

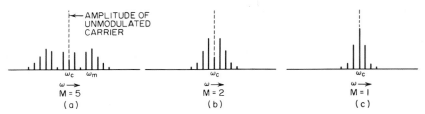

Fig. 6. FM spectra for various modulation indices. (a) Modulation index = 5.0. (b) Modulation index = 2.0. (c) Modulation index = 1.0.

Phase Modulation of One Carrier by Another

If a carrier applied to the input of an FM demodulator has another carrier superimposed on it, an output whose fundamental frequency equals the difference frequency between the two carriers appears. An analysis of this effect is relatively simple and provides a basis for an analysis of crosstalk between multiplexed channels, permits calculation of wideband improvement ratio in an FM system, and provides an explanation of the "capture" effect in FM demodulators.

Figure 7 shows a vector diagram of a carrier e_c with an interfering carrier e_n superimposed to form the resultant e_r. If $e_n \ll e_c$, the angle Θ of the e_r relative to e_c is

$$\Theta = (e_n/e_c) \sin (\omega_n - \omega_c)t \qquad (2)$$

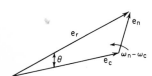

Fig. 7. Superposition of carrier voltage e_c of frequency ω_c and interference e_n at frequency ω_n.

The frequency of the resultant e_r is given by ω_r

$$\omega_r = \omega_c + d\Theta/dt$$
$$= \omega_c + (\omega_n - \omega_c)(e_n/e_c) \cos (\omega_n - \omega_c)t \qquad (3)$$

The second term represents the apparent frequency modulation of the carrier e_c which results from adding e_n to e_c.

It is important to note that the apparent modulation occurs at the frequency difference between the two carriers and that the peak deviation is

$$(\omega_n - \omega_c)(e_n/e_c) \qquad (4)$$

and the effect of e_n (noise) increases linearly with frequency offset. This gives rise to the familiar triangular shape of the FM noise spectrum.

Noise Improvement in FM Systems

The noise present at the output of an FM demodulator when white noise is superimposed on the carrier at the input to the demodulator can be calculated in a similar fashion. Define the following terms:

$$2\Delta\omega_{max} = \text{channel width, radians/sec}$$
$$P_n = \text{power spectral density} = \frac{(\text{rms volts})^2}{\text{radians/sec}}$$
$$e_c = \text{rms carrier voltage}$$

For white noise, P_n is constant across the channel passband and the *input* signal-to-noise ratio is

$$S/N = \frac{e_c}{(P_n 2 \Delta\omega_{max})^{1/2}} \tag{5}$$

The rms deviation of the carrier resulting from the noise in infinitesimal passbands at frequencies ω above and below the carrier frequency is [from Eq. (4)]

rms deviation at noise modulation frequency $\omega = [(P_n)^{1/2}/e_c]$ (6)

rms deviation at output of demodulator in frequency range from 0 to max modulation frequency
$$= \sqrt{\int_0^{\Delta\omega_{max}/M} 2\{[(P_n)^{1/2}/e_c]\omega\}^2 \, d\omega}$$
$$= [(2P_n/3e_c^2)(\Delta\omega^3_{max}/M^3)]^{1/2} \tag{7}$$

The wideband improvement ratio is defined as

$$\frac{S/N \text{ output}}{S/N \text{ input}}$$

$$S/N \text{ output} = \sqrt{3M^3} \, (S/N \text{ input}) \tag{8}$$

If a given intelligence frequency must be transmitted through a fixed noise level, the power spectral density P_n is unchanged and the input signal-to-noise ratio deteriorates 3 db every time $\Delta\omega_{max}$ is doubled (M doubled) but the output S/N ratio is improved by 6 db since M appears to the $3/2$ power in the improvement expression. Thus $M^3 = 8$ or 9 db; this less 3 db input loss in S/N input results in $9 - 3 = 6$ db improvement. This expression holds only above threshold (input $S/N \geq 10$ db) so that if unlimited bandwidth is available, the widest deviation which yields better than 10 db signal-to-noise ratio should be used to achieve the highest output signal-to-noise ratio. This conclusion is not valid for coherent detectors (phase lock and frequency feedback) (see Chap. 9, Secs. 8 and 9). The resulting output power spectral density has the shape of Fig. 8, and it is

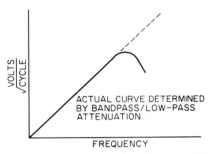

FIG. 8. Output power spectral density, FM demodulator, above threshold.

from this curve that the commonly used preemphasis schedules arise. If we wish to obtain a uniform signal-to-noise ratio in all subcarrier channels, it is evident that the relative levels of the subcarriers must be adjusted to compensate for both the nonuniform noise density and whatever variation in bandwidth the subcarriers may have. The development of a preemphasis schedule is discussed in more detail in Sec. 5.

Capture Effect

In an earlier part of this section the phase modulation which results from superposition of two carriers at the input to a limiter has been analyzed. In that analysis, it is assumed that the interfering carrier is small relative to the signal carrier, and the apparent modulation of the larger carrier at the difference frequency between the carriers is calculated. The capture effect is concerned with the case where the interfering carrier approaches the amplitude of the desired carrier at the input to the limiter. As the level of this interfering carrier approaches that of the signal, there is a level (dependent on the design parameters of the demodulator) where the modulation of the interfering carrier appears in the output of the demodulator.

Figure 9 shows the vector combination of two nearly equal carriers at two instants. In a, the vectors are opposed and the vector resultant e_r is a minimum but the rate of change of phase differs from the carrier frequency ω_c by the maximum amount.

In b, the vectors add and the frequency deviation of the resultant is much less than in a. The linear velocity at the tip of the resultant vector is the same in both

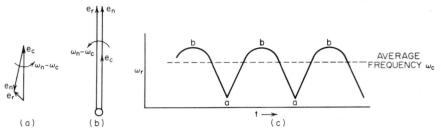

FIG. 9. Instantaneous frequency of resultant of two carriers of nearly equal amplitude superimposed at input to limiter. (a) Minimum instantaneous frequency. (b) Maximum instantaneous frequency. (c) Instantaneous frequency vs. time.

cases but the angular velocity is inversely proportional to the length of the vectors. A plot of instantaneous frequency ω_r vs. time is shown in c. The dotted line is the value of ω_c and the areas under the instantaneous ω_r curve above and below are equal. The period of the deviation of ω_r is the difference of the two carrier frequencies. For the adjacent-channel case, this represents a frequency that exceeds the base band of the channel which can be rejected with a filter. If the demodulator used cannot linearly accommodate the full range of frequencies that results from superposition of the carriers, clipping of an instantaneous frequency peak results in displacement of the average frequency ω_c and modulation of ω_n appears in the output of the ω_c demodulator. To improve the tolerance of a demodulator to large interfering carriers, a large linear range of demodulator bandwidth is desirable.

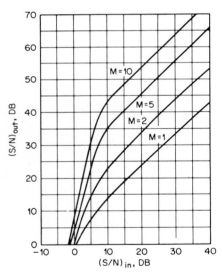

FIG. 10. $(S/N)_{out}$ vs. $(S/N)_{in}$ for pulse-averaging discriminator with signal-suppression effects included.

Thresholding and Signal Suppression

The wideband improvement characteristics of an FM system are derived above. It is assumed in this derivation that the signal carrier is substantially larger than the interference. When the interference is white noise rather than a fixed carrier, the validity of this assumption is statistical rather than absolute. For an input signal-to-noise ratio of 10 db, the noise is larger than the signal 0.1 per cent of the time. Stumpers[1] has analyzed this effect in detail and Martin[2] has modified Stumpers's conclusions to include the effect of a bandpass limiter analyzed by Davenport[3] to yield the curves of Fig. 10. These curves show a degradation of output S/N which increases abruptly for input S/N ratios below about 10 db, the "threshold" of an FM demodulator.

A related effect also analyzed by Stumpers is the signal-suppression effect shown in the curve of Fig. 11 in which the presence of noise at the input to an FM demodulator is shown to decrease the amplitude of the intelligence component at the demodulator output. This effect can be considered an aspect of the "capture" effect in an FM demodulator. The noise at the input to a demodulator near threshold exceeds the signal for a portion of the time which increases as the signal-to-noise ratio decreases. The bandpass filtering at the demodulator input results in an average frequency of the noise near the center of the passband; hence when the noise "captures" the demodulator, the output tends toward the band center, resulting in a reduction in the apparent amplitude of the intelligence.

4 FM Multiplexing Principles

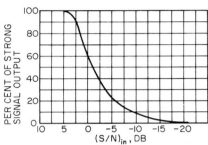

Fig. 11. Signal suppression in threshold region.

The design of a multiplex configuration is influenced by three major considerations: realizability of practical filters, subcarrier oscillator/discriminator stability, and tape-recorder-speed fluctuations. At percentage deviations in excess of 20 per cent, the design of bandpass filters having acceptable phase linearity and attenuation characteristics becomes increasingly difficult, while at percentage deviations below 2 per cent, the stability of bandpass-filter components under temperature variations and handling becomes more critical. At small deviations, the drift of the subcarrier oscillator and discriminator also become more critical in determining the system static accuracy, and tape-recorder-speed errors introduce proportionately greater spurious modulation of the subcarrier. In fact, as the percentage deviation of a subcarrier is reduced, there comes a point at which tape-speed errors introduce so much spurious subcarrier deviation that it is no longer possible to design a filter which will pass the extremes of subcarrier frequency shift while satisfactorily attenuating the adjacent channels. In this situation, it is necessary to frequency-translate the entire group of small percentage-deviation subcarriers down to lower center frequencies, using a translation reference which has a corresponding flutter. In this way, the percentage deviations will be increased and the flutter components can be kept from driving the subcarriers out of the filter passbands.

Using high-performance equipment, therefore, it is possible directly to generate, filter, and demodulate subcarrier multiplexes having percentage deviations ranging from approximately 20 to approximately 2 per cent. When extension of the system capability to add more narrow-deviation channels is desired, group translation of subcarriers having deviations of less than 2 per cent is indicated.

The next consideration in subcarrier multiplex layout is the effect of channel width/guard band distribution on crosstalk. In most modern FM demodulators, Gaussian (linear-phase) bandpass filters are used to minimize data distortion. (See Secs. 17 and 19.)

The low-pass filter used depends upon the type of data being gathered; but the system is usually most susceptible to "beats" with the adjacent channels when Gaussian output filters are used with a cutoff frequency corresponding to a modulation index of 1.0 for full bandwidth deviation. For this reason, a general analysis of system crosstalk for this case is useful in determining the limitations of a given set of multiplexing standards.

Define the channel bandwidth as $2\Delta f_{max}$ and the guard band as $K(2\Delta f_{max})$ and assume that Gaussian filters are used where

$$L = 3(\Delta f/\Delta f_{max})^2 \qquad (9)$$

where L = loss, db, of a frequency Δf from band center

Δf_{max} = peak channel deviation

The worst-case analysis is based on a channel carrier at the band edge with the smallest difference frequency to the adjacent band edge and an interfering carrier of equal amplitude at the adjacent band edge. The peak apparent deviation of the resulting demodulator "beat" is given by the frequency difference between the two carriers multiplied by the ratio of their amplitudes at the output of the bandpass filter. This is given by

$$20 \log 2K - 3(2K + 1)^2 - 3(2MK)^2 + 3 = \text{peak-to-peak amplitude, db, relative to full bandwidth} \quad (10)$$

where K = ratio of guard to channel bandwidth as above.

The first term of Eq. (10) is the ratio of the frequency difference to half bandwidth in decibels; the second term is the attenuation of the Gaussian bandpass filter; the third term is the attenuation of the beat frequency by the Gaussian output filter whose cutoff frequency corresponds to a modulation index M; and the constant 3 is the attenuation of the signal carrier by the bandpass filter.

A plot of this relationship is shown in Fig. 12. It should be remembered that the filters assumed in this analysis are ideal Gaussian filters. In practice, filters with a small number of poles may produce crosstalk in excess of that shown in the figure and filters having a cutoff characteristic sharper than Gaussian will reduce the crosstalk level.

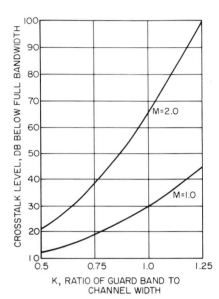

FIG. 12. Crosstalk between FM channels for carriers located at adjacent band edges. Bandpass filters, Gaussian, 3 db down at band edges. Low-pass filters, Gaussian, 3 db down at cutoff. No preemphasis.

5 Sources of Error

Inaccuracies in a data channel can arise from defects internal to the channel, such as subcarrier oscillator nonlinearity, or the errors may result from the fact that the data channel shares the frequency spectrum with other channels. This section is concerned with the latter class of problem.

Data errors arising from frequency multiplexing do not affect the static characteristics of the channel—zero drift and linearity are not influenced by inclusion of the subcarrier in a multiplex. However, the *dynamic* performance of an FM subcarrier channel can be substantially impaired by spurious frequency components falling in or near its band. These can result from spurious outputs from other subcarrier oscillators, adjacent-channel interference, intermodulation and harmonic distortion generated in the transmission link, noise, tape-recorder nonlinearity, and imperfection in other circuitry through which the multiplex passes.

Spurious Outputs of Components

Subcarrier oscillator harmonic distortion can be a significant source of noise in multiplexed FM systems. Although virtually all oscillators have output filters which are designed to minimize harmonic outputs which might disturb higher-frequency bands, the high component density in airborne subcarrier oscillators can result in capacitive coupling of high-frequency components into the multiplex in spite of the output filter. Another source of noise in a multiplex is data feedthrough in high-

SYSTEM CHARACTERISTICS   6–15

frequency oscillators. For example, when IRIG band E is used for voice transmission, any component of the voice frequency signal in the band E subcarrier output will corrupt all the lower channels occupying that spectrum range. Another case of spurious signal generation which has been observed is the nonlinear interaction of high-frequency signals with a tape-recorder bias carrier which can result in the down conversion of these signals into the data spectrum.

Although even the finest FM-FM system components will generate spurious outputs when overdriven or otherwise misapplied, a first step in any system design must be to specify suitable component performance when operating under the expected conditions.

Adjacent-channel Interference

All the standard-frequency multiplexing-system channel assignments have been made in a manner which ensures that adjacent-channel interference will not be a major factor if the systems are properly adjusted and modulation indexes of 2 or more are employed. In other words, if the demodulator low-pass filter bandwidth is made equal to or less than half the maximum channel deviation, interference beats with adjacent channels will be attenuated to less than 0.5 per cent of full-scale output.

Intermodulation Products

In any FM-FM data system which has adequate signal strength to produce receiver quieting, intermodulation products are the greatest source of dynamic error. The multiplex may be passed through a summing amplifier, transmitter, receiver, recorder, and various other amplifiers before separation and demodulation of the subcarriers and the transfer function of none of these components is perfectly linear even when operated within its normal range. When overdriven, each of these devices becomes grossly nonlinear and generates new frequency components related to harmonics, sums, differences, and higher-order cross products of the input frequencies. In fact, the total number of spurious frequency components which can be generated from a 10-channel multiplex subjected to a transfer nonlinearity is so great that the resulting noise background can be treated as wideband random noise. The spurious components generated by a nonlinear transfer function can be calculated by first plotting the input-to-output transfer-characteristic curve, approximating the curve by a polynomial in the input variable, and finally multiplying the input complex wave by the polynomial to determine the output wave which can then be analyzed to find the spurious components.

As an example, consider a link which has overall independent nonlinearity of 2 per cent, consisting entirely of a second-order characteristic superimposed on a linear transfer function.

$$e_2(t) = e_1(t) + 0.04 e_1^2(t) \qquad -1 < e_1 < +1 \qquad (11)$$

Let us make the simplifying assumptions that considerable preemphasis is used, so that only the two highest-frequency channels need be considered, and that the two channels considered have equal amplitudes, and total two-thirds of the output. The input waveform can be represented by

$$e_1(t) = \frac{1}{3} \sin(\omega_1 t) + \frac{1}{3} \sin(\omega_2 t) \qquad (12)$$

and the output waveform contains, in addition to the input, terms of the form

$$(0.04/9),\ \sin^2 \omega_1(t) + 2 \sin \omega_1(t) \sin \omega_2(t) + \sin^2 \omega_2(t) \qquad (13)$$

Since we are considering distortion produced by high-frequency bands, we are justified in dropping the harmonic terms and considering only the intermodulation term e_{im}.

$$e_{im} = 0.04 \frac{2}{9} (\sin \omega_1 t \sin \omega_2 t)$$

$$= (0.04/9) \cos(\omega_1 - \omega_2)t - (0.04/9) \cos(\omega_1 + \omega_2)t \qquad (14)$$

The sum-frequency term in this expression is outside the frequency range of interest and can be discarded, but the difference-frequency term can easily lie in a lower-frequency (and, therefore, lower-level) subcarrier band. The difference frequency resulting from intermodulation between bands 17 and 18 can fall as low as band 12 (10.5 kc), and since this band is often transmitted as much as 18 db below the highest band, the intermodulation products can introduce a beat into band 12 of as much as 4 per cent of full scale for a modulation index of 1.

While the example given above is obviously oversimplified, it is designed to show that the use of extreme preemphasis to obtain a uniform noise susceptibility in all bands can intensify other sources of error.

Oversimplification of the calculation, as can occur if a small number of subcarriers is used, or if they are unmodulated, can lead to the fallacious conclusion that channel assignments can be chosen in such a manner that spurious components will fall in the unoccupied guard bands between channels. For example, consider the "nearly perfect" IRIG band numbers 8, 9, 10, and 11. For these channels at center frequency, spurious responses arising from second- and third-order nonlinearities appear in the guard bands, but it will be found that as soon as the subcarriers are shifted away from band center by an input stimulus, the spurious terms move into the subcarrier bands.

The main result of performing these calculations will be to sharpen the system designer's dislike for nonlinearities anywhere in the multiplex flow path and encourage him to devote the necessary effort to uncovering and removing these sources of distortion rather than trying to devise multiplex configurations which are unaffected by transfer nonlinearities.

A major source of distortion arises in the r-f link. To minimize system noise, it is desirable to operate with the largest carrier deviation permitted under IRIG standards but the multiplex composite signal occasionally peaks to sizable levels and during these intervals may drive the transmitter beyond its linear-deviation range or move the receiver intermediate frequency into a region of filter-phase nonlinearity, thereby substantially increasing the system distortion. To predict the range of peak levels which will be encountered in a multiplex of uniform spectral density, one can turn to probability tables. For example, given the rms voltage of a random wave, one finds that the peak exceeds 2.6 times the rms value less than 1 per cent of the time and exceeds 3.3 times the rms value less than 0.1 per cent of the time. Analytical solutions to the multiplex peaking problem are complicated by the fact that in practical systems, preemphasis of high-frequency subcarriers is employed to provide uniform signal-to-noise ratios in all the subcarrier channels and consequently the spectrum of the multiplex is far from uniform.

In operational practice, deviation levels are best set by attenuating the transmitter output to the minimum expected level and adjusting the r-f carrier deviation until the intermodulation noise equals the system thermal noise. This provides best overall signal-to-noise ratio for the weakest signal case. Although the system can be further improved by modifying the r-f carrier deviation as the vehicle range changes, the complexity of this method has ruled it out except for unusual circumstances.

R-F Link Noise and Preemphasis Schedules

Noise in the r-f link is the major concern when weak signals are being received. The receiver output noise above threshold (see Sec. 11) has a triangular spectrum, concentrating the noise power in the higher subcarrier channels, while below r-f signal-to-noise ratios of 10 db, the r-f link is said to be thresholding and the subcarrier signal-to-noise ratios decline rapidly. The spectrum of the noise tends to be more nearly uniform under these conditions, and the increase in noise level is accompanied by suppression of the subcarrier levels below r-f signal-to-noise ratios of approximately 7 db. In the overwhelming majority of FM-FM systems, operation at r-f signal-to-noise ratios below threshold is not regarded as productive of useful data and the subcarrier preemphasis schedules are designed for operation at r-f signal-to-noise ratios above 10 db. Arriving at the preemphasis schedule consists of assigning deviation levels to each channel after calculating the noise in each channel so that

the assigned deviation makes the subcarrier demodulator output signal-to-noise ratio the same for all channels.
 The calculation first assumes that the thermal and shot noise density in the r-f spectrum is constant. Thus, the predetection noise voltage in a given band is proportional to the square root of the bandwidth. In addition, the assumption is made that the system modulation index is below 1, so that the predetection spectrum can be represented by a carrier and set of first-order sidebands corresponding to the subcarriers.
 Noise in any subcarrier sideband "slot" phase modulates the carrier and the rms value of this noise is proportional to the square root of the sideband slot. To maintain a uniform signal-to-noise ratio in each band, therefore, the sideband amplitude-to-rms-noise ratio must be constant. This discussion has been couched in terms of phase modulation; and when a conversion to frequency-modulation nomenclature is made, using the results of Sec. 3, it becomes apparent that the r-f carrier *deviation* assigned to a subcarrier is proportional to both the square root of subcarrier bandwidth and the subcarrier center frequency.
 This argument explains the amplitude vs. frequency relationship which forms the basis for IRIG and constant-bandwidth telemetry preemphasis schedules. The receiver output noise power in each channel is proportional to center frequency to make constant-bandwidth-channel signal-to-noise ratios uniform. The IRIG bandwidths which are proportional to subcarrier center frequency require that the constant-bandwidth preemphasis relationship be multiplied by the square root of subcarrier frequency, resulting in a three-halves power IRIG proportional-bandwidth preemphasis schedule.
 The practical technique for assigning r-f carrier deviations to the various subcarriers combines this theoretical noise susceptibility relationship with considerations involving

Table 4. Example of Preemphasis Scheduling

IRIG/AIA band	CF, kc	$\dfrac{wt}{f_1 B_1}\; f_n B_n$	Min. deviation, kc	Final deviation per channel
1	0.4	1.0	2	2.0
2	0.56	1.7	2	2.0
3	0.73	2.5	2	2.0
4	0.96	3.7	2	2.0
5	1.3	5.9	2	2.0
6	1.7	8.8	2	2.0
7	2.3	14	2	2.0
8	3.0	20.5	2	2.0
9	3.9	30	2	2.0
10	5.4	50	2	2.0
11	7.35	80		1.34 (increase to 2 kc)
Remaining deviation divided among these channels				
12	10.5	135		2.25
13	14.5	220		3.7
14	22.0	410		6.9
3C	29.2	600		10.0
4C	37.5	755		12.6
5C	45.8	940		15.7
6C	54.2	1,100		18.4
8CW	70.8	2,050		34.3
		6,290		125.19
Peak carrier deviation.....................				125 kc
Combined min. deviations				20 kc
Remaining carrier deviation..............				105 kc

intermodulation distortion and transmitter and receiver microphonism. First, the channels to be employed are listed, along with the produce of their center frequency and square root of bandwidth divided by the product of center frequency and square root of bandwidth of the lowest channel. Also tabulated is the minimum deviation allocated to any channel.

In the example given in Table 4, a minimum deviation of 2 kc is assigned to the lower 10 IRIG channels, leaving 105 kc of deviation to be divided among the remaining IRIG and constant-bandwidth channels on a preemphasized basis. Distributing the remaining deviation in this manner leaves channel 11 with only 1.34 kc of r-f carrier deviation; so that deviation allocated to this channel is arbitrarily increased to 2.0 kc and the resulting overmodulation of the r-f carrier is alleviated by very slightly reducing the composite modulation level at the transmitter.

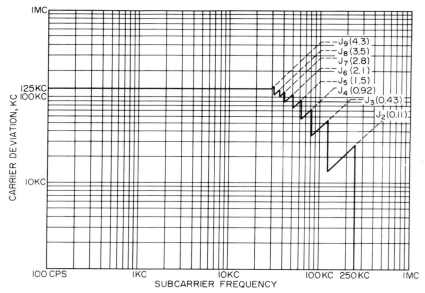

FIG. 13. Carrier deviation vs. subcarrier frequency. NOTES: 1. For a vhf carrier, the bounded ordinate is the maximum deviation that can be assigned to any single subcarrier channel. 2. The dashed lines of constant M (modulation index) indicate where the sidebands J_2 and J_9 are outside the assigned ± 250-kc band and violate the requirement of 55 db below the undeviated carrier level.

IRIG standards require that power outside a ± 250-kc channel bandwidth be at least 55 db below the undeviated carrier level. Although normal operation of both proportional- and constant-bandwidth FM-FM multiplexes does not violate this rule, large deviations assigned to high-frequency subcarriers can result in excessive sideband power output. Figure 13 presents a convenient way of estimating conformity of sideband power to IRIG standards. For example, when IRIG channel 17 is allocated a deviation of more than 73 kc, $M = 1.5$, the fifth-order sidebands will contain excessive power, and when IRIG channel 18 is given more than 56 kc of r-f carrier deviation, the fourth-order sideband power will be excessive. Lines are plotted for fixed modulation index M and sideband order $J_n(M)$ which show those deviations of subcarrier center frequencies which will cause sideband power to exceed the permissible limits. As increasing use of high-frequency constant-bandwidth and extended IRIG bands is made, closer watch over the sideband power distribution will be required.

SYSTEM CHARACTERISTICS 6–19

For subcarrier frequencies other than those shown, the maximum permissible deviation can be determined from a table of Bessel functions.*

Effect of Tape-speed Variation

The FM multiplex is generally recorded on magnetic tape, and when the subcarriers are reproduced, the speed errors during record and playback cause corresponding changes in the zero, sensitivity, and time base of the data channel. To derive these relationships, consider the output E_o of a discriminator with center frequency f_o and input frequency f_{in}. Then E_o is given by

$$E_o = K[(f_{in} - f_o)/f_o] \qquad (15)$$

(see Sec. 19). That is, when $f_{in} = f_o + \Delta f_{max}$ at high band edge

$$E_{o\;max} = K(\Delta f_{max}/f_o) \qquad (16)$$

If a fractional tape-speed error $\epsilon(t)$ is introduced, the input frequency changes from f_{in} to

$$f_{in}' = (1 + \epsilon)f_{in} = (1 + \epsilon)f_o + (1 + \epsilon)\Delta f \qquad (17)$$

This expression can be related to a corresponding output E_o'. Then the relative error can be found from

$$(E_o - E_o')/E_{o\;max} = (\epsilon_o - \epsilon)\Delta f/f_{max} \qquad (18)$$

An examination of this relative-error expression shows a channel's per cent *zero error* is proportional to the magnitude of the speed variation and inversely proportional to the percentage deviation of the channel. The *sensitivity error* is proportional to the magnitude of the speed variation but not to the percentage deviation of the channel. Consequently, tape-speed variations will cause the same sensitivity error in all channels but the zero error will be inversely proportional to the percentage deviation of the channel. Compensation systems to correct for these errors are described in Sec. 19.

The third effect of tape-speed variation is the *time-base error*. When tape is pulled at the correct and constant rate, the length S pulled in an amount of time t is simply $S = Vt$. If the speed varies some small fractional amount $\epsilon(t)$ about the correct speed V_o, then the amount of tape pulled will be

$$S + \Delta S = V_o \int_0^t [1 + \epsilon(t)]\,dt \qquad (19)$$

The error in the length $\Delta S = V_o \int_0^t [1 + \epsilon(t)]\,dt - V_o \int_0^t dt$ will be given by

$$\Delta S = V_o \int_0^t \epsilon(t)\,dt \qquad (20)$$

Since the varying velocity component is small compared with the average, the time-base error can be expressed as

$$\Delta t = \Delta S/V_o = \int_0^t \epsilon(t)\,dt \qquad (21)$$

As an example, when the speed error $\epsilon(t)$ is of the form $\epsilon \sin 2\pi ft$, the instantaneous time-base error will be given by

$$\Delta t = \epsilon \int_0^t (\sin 2\pi ft)\,dt = (\epsilon T/2\pi)(1 - \cos 2\pi ft) \qquad (22)$$

This expression states that the time-base error will, on the average, be zero but will have peak errors proportional to the period and magnitude of the frequency components of the speed variation.

* See Sec. A.10-14 of the Appendix.

TRANSMITTERS

6 Characterization of FM Telemetry Transmitters

The purpose of the telemetry transmitter is to provide a modulated r-f (radio-frequency) signal of sufficient power and stability to accomplish telemetering of data to a remote receiving location. The equipment to be discussed here is provided with angle-modulation capabilities. The angle modulation is generated as either frequency modulation or phase modulation or, in some cases, an advantageous combination of both. There is, of course, no inherent difference between FM and PM except the behavior of modulation index with modulating frequency.

Since the circuit realizations for generating FM in a simple manner are somewhat different from the simplest phase modulators, a distinction between FM and PM transmitters is often made; FM transmitters generally have d-c or very low audio-frequency-modulation capabilities as these are inherent in the modulation scheme.

The principal characteristics of a telemetry transmitter are determined by the parameters of the telemetry situation itself. Since telemetry systems are often employed to transmit data which are not accessible because of environmental reasons, telemetry transmitters must themselves withstand extremes of operating environment. Telemetry is seldom the most important part of an overall system; hence the volume, weight, and power allocated are invariably a minimum. A telemetry transmitter must generally be a miniature, lightweight, efficient unit providing only as much power as is required for the expected range and data bandwidth. Loss of data or control can be tremendously expensive—this gives rise to a requirement for exceptional reliability and insensitivity to environment.

This combination of constraints points to the use of all-solid-state design, and the present design trend is to make use of semiconductor circuitry wherever the appropriate electrical characteristics are available. Equipment employing vacuum tubes has achieved some outstanding contemporary successes, and there are a number of high-power applications which can be satisfied only by thermionic devices.

Table 5. Typical Telemetry-transmitter Functional Blocks

Element	Frequency	Power level
Modulators:		
FM	1 Mc–2.3 gc	1–10 mw
PM	1 Mc–2.3 gc	1–100 mw
Oscillators:		
Quartz crystal	1–200 Mc	10 mw max
LC and cavity	100 Mc–2.3 gc	1 mw–10 watts
Amplifiers:		
Video	D-C to 50 Mc	
R-F	1 Mc–2.3 gc	1 mw–100 watts
Frequency multipliers:		
Active	1 Mc–2.3 gc	
Passive (varactor)	50 Mc–2.3 gc	10 mw–100 watts
Power-supply accessories:		
Regulators	D-C	
RFI filters	D-C to 10 gc	
D-C to d-c converters	D-C	

TRANSMITTERS

The transmitter must generate a stable carrier and provide significant amounts of power in the chosen r-f bands. Power levels vary from a few milliwatts for very short range systems to hundreds of watts for long-range or wideband systems. Frequency stability depends upon application, but a fractional stability of ± 0.003 per cent over all conditions is a common requirement. The principal telemetry bands are 215 to 260, 1,435 to 1,535, and 2,200 to 2,300 Mc, as well as some specialized space-telemetry allocations at 136 to 137, 400 to 401, and 960 Mc. Equipment in these frequency ranges makes common use of quartz-crystal oscillators as the stable source of output frequency. Crystal resonators are limited by physical size to the frequency range below 200 Mc and to power levels below a few milliwatts.

This limitation coupled with typical output-frequency and power specifications gives rise to requirements for amplification and frequency multiplication.

The major functional blocks found in telemetry transmitters are listed in Table 5. Detailed discussion of circuit characteristics is found in subsequent sections.

Table 6. Typical Telemetry-transmitter Specification Points

	Typical figure
R-F output port:	
Frequency	2,250,000 Mc
Stability, accuracy	$\pm 0.003\%$
Adjustment capability	Replaceable oscillator
Power output	5 watts min
Spurious output	-87 db with respect to carrier
Load impedance or VSWR	2:1 all phases
Modulation:	
Deviation	± 125 Mc
FM or PM	FM
Linearity or intermodulation	2% linearity
Video bandwidth	250 kc
Modulation input port:	
Impedance	>75 kilohms
Video frequencies	250 cps–250 kc
Video level	0–5 volts
Power input port:	
Voltage	$+28$ volts d-c
Current	750 ma max
Regulation	± 2 volts
D-C or a-c	D-C
Transients and RFI	100 mv
Environment (storage and operating):	
Temperature	0–50°C
Pressure or altitude	10^{-5} mm Hg
Vibration	20–2,000 cps 5 g's rms
Shock	10 g's each axis
Acceleration	40 g's one axis
Humidity or spray	95% for 24 hr
Corrosive atmosphere	Salt spray
Explosive atmosphere	—
Radiation	Nominal
Magnetic or electric field	—
Meteorite flux	—
RFI	Per MIL-I-6181D
Warmup time	100 msec
Mechanical:	
Configuration	Rectangular box
Dimensions and volume	2 by 4 by 10 in.
Weight	35 oz
Connectors	OSM/BRM
Mounting	Flush to plate
Cooling	Conduction
Finish	Gold
Marking	—
Reliability and quality control	Q.C. per NPC-200, Rel. 98% for 10^4 hr

A telemetry transmitter is in essence a three-port device, with inputs for modulation and power and an output for modulated r-f energy. The transmitter can be characterized by the electrical characteristics of each of the three ports, and by its environmental capabilities and mechanical construction.

A complete specification for a telemetry transmitter will contain information pertaining to most of the points of Table 6, as well as additional data relating to any unusual parameters for a particular application.

The typical figures do not represent any particular transmitter but are included for the purpose of establishing some scale factors for a typical space-borne S-band unit. Design improvements are aimed primarily at size, weight, efficiency, and reliability.

7 Functional Blocks

The requirements outlined in the previous section are met in practice by a transmitter design which combines the functional elements of Table 5. It is necessary to consider each in its turn with comments on the general state of the art of each area. Basic to most areas is application of sound r-f techniques such as minimum parasitic reactance, low inductance grounding, shielding, and separation of radio frequency and direct current.

Lumped-element techniques are common at vhf and below and are finding increasing application up to S band as the general size of active elements and circuits is reduced. Coaxial or other structures are satisfactory above 1,000 Mc, and hybrid circuits employing both lumped and distributed elements bridge the once difficult gap between vhf and microwave frequencies.

Modulators

The modulator is a principal component of an FM telemetry transmitter, and a number of approaches have been developed. The angle modulation must impress the information on the carrier without deteriorating frequency stability or introducing amplitude modulation.

Table 7 shows the principal angle-modulation methods with comments on advantages and disadvantages. The various types of modulators are identified by the diagrams shown in Fig. 14 *a* through *g*.

Table 7. Angle-modulation Schemes

	Advantages	Problem areas
PM:		
Varactor phase modulators	Size, linearity	Cost, losses
Ferrite phase modulators	Linearity	Weight, power, cost
Transistor or vacuum tube	Simplicity	Limited deviation
FM:		
VCO	Linear	Poor stability
Offset-VCO	Linear	Spurious
Discriminator-stabilized VCO	Low spurious, linear	Complex, loss of lock problem
Voltage-controlled crystal oscillator (VCXO)	Simplicity, size, spurious-free	Modulation singularities
VCXO + phase modulation	Same as VCXO	Crossover matching

Of the phase-modulator schemes, the varactor phase modulator appears most generally useful. High-pass, low-pass, and bridge-T all-pass configurations have been used, and the high-pass configuration shown in Fig. 14*a* yields best linearity and least incidental AM.

The various FM modulators have individual strong points: the VCO and offset-VCO methods provide outstanding range and linearity at a cost of poor stability; the discriminator-stabilized VCO provides both good linearity and stability at a price of large size and complexity; the VCXO methods provide good stability but require understanding and careful control of quartz-crystal parameters. The discriminator-stabilized VCO is presently in widespread use, but increased quartz-crystal capabilities point to the use of VCXO techniques in future miniature equipment.

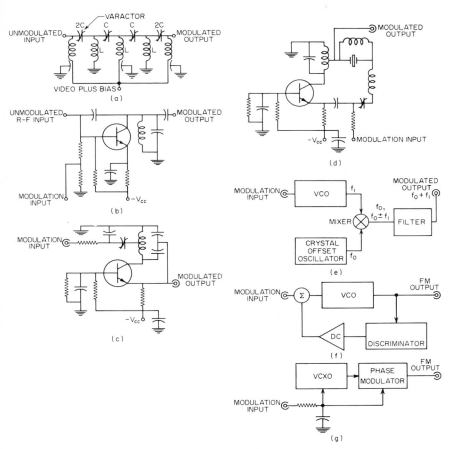

Fig. 14. Various FM, PM modulator circuit configurations. (a) High-pass varactor phase modulator. (b) Transistor phase modulator. (c) Voltage-controlled LC oscillator (VCO). (d) Voltage-controlled crystal oscillator (VCXO). (e) Offset VCO. (f) Discriminator-stabilized VCO. (g) VCXO + phase modulator.

Quartz Crystals

A brief discussion of quartz crystals as they apply to FM telemetry applications is in order here. The principal crystal characteristics affecting FM telemetry systems are

1. Long-term stability, aging and temperature stability
2. Short-term stability in the presence of vibration
3. Motional and spurious parameters affecting VCXO operation

Aging ranges from 10^{-5} per year for solder-seal rugged-mount units through 10^{-6} per year for ribbon-mount cold-weld units to better than 10^{-7} per year for highly specialized units for 2.5- or 5- Mc frequency standards.

Short-term stability under vibration is largely affected by the crystal mount and its resonances. The vibration sensitivity of quartz itself is $\Delta f/f \cong 2 \times 10^{-9}$ per g. Short-term variations range from this low value for ribbon-mount units to 2×10^{-7} per g for ordinary wire-mount crystals. Typical resonant frequencies are 400 cps in HC-6, 700 cps in HC-18, and 3,000 cps in ribbon-mount TO-5 enclosures. There is often a trade-off between long- and short-term stability.

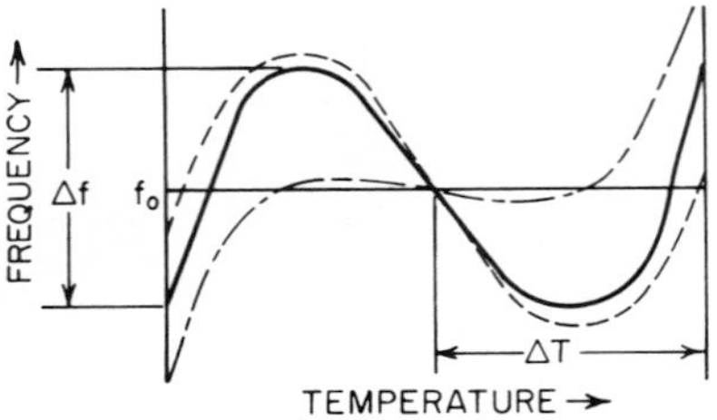

Crystal requirements for VCXO operation are (1) maximum ratio of motional to electrical to allow wide slewing range and (2) minimum spurious resonances in the slewing or modulation range to avoid frequency hopping or modulation singularities. Circuit requirements often favor a high value of electrical capacitance, which is at odds with the spurious requirement. Present understanding of anharmonic-overtone spurious responses is such as to allow design and fabrication of fundamental and over-tone crystals which are essentially spurious-free.

FIG. 15. Temperature behavior of AT-cut quartz-crystal resonator. (Solid line—optimum cut angle for temperature range $26°C \pm \Delta T$. Broken lines—deviation from optimum cut angle.)

Oscillators employing quartz crystals as the reference element can approach the stability of the crystal itself. Temperature behavior for an AT-cut crystal is cubic in form, as shown in Fig. 15. The range of frequency variation for the optimum angle cut for a certain temperature range is given by $\Delta f/f = 5 \times 10^{-11}(\Delta T)^3$, where ΔT is the maximum temperature excursion from the crossover point, usually 26°C. This value can be approached within a factor of 10 for reasonable production-angle tolerances. Compensation techniques have been developed which allow frequency variations of less than $\pm 5 \times 10^{-7}$ from -40 to $+60°C$. Compensation holds considerable promise for future applications.

Oscillators

Transistors are available in all frequency ranges and power levels appropriate to crystal-oscillator use, and vacuum-tube oscillators are used only for thermal environments outside the capabilities of semiconductors. A number of equivalent oscillator circuits are in use. Typical circuits are shown in Fig. 16. The inductors shown in broken lines across the crystal are required for overtone operation—they are chosen to

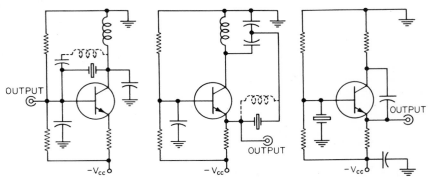

FIG. 16. Typical transistor crystal-oscillator circuits.

have reactance equal to that of the electrical capacitance of the crystal at the desired overtone frequency.

High-power transistor and vacuum-tube LC and cavity oscillators can be constructed over almost the entire frequency and power range of interest in telemetry. These oscillators suffer from poor long-term frequency stability, but this can be provided by locking to a crystal oscillator or discriminator reference. There are an increasing number of situations in which a high-power oscillator locked to a reference provides the smallest, simplest, and most efficient transmitter. Oscillator power and frequency capabilities are quite similar to amplifier capabilities outlined in the following paragraphs.

Amplifiers

The combined requirements for frequency stability and output power lead to consideration of r-f power amplifiers for transmitter use. Video amplifiers are required to provide appropriate modulation sensitivity and bandwidth.

Almost all video requirements are satisfied by field-effect or bipolar transistor circuits. Design of wideband, small-signal circuits is classical and need not be treated deeply here. Use of conventional feedback techniques provides a wide range of capabilities in impedance levels and bandwidth. Distortion is a major problem, as intermodulation of various subcarriers is deleterious to system operation. Integrated circuitry is finding increased application in video-frequency adders and amplifiers.

Power amplifiers are evaluated on the basis of power gain, efficiency, output level, size, and reliability. Both transistors and vacuum tubes are employed as active elements, and rapid development in both areas precludes clear-cut choice in many circumstances. For power levels below a few watts at frequencies below 1 gc, transistors are satisfactory in most telemetry application. Similarly, thermionic devices are outstanding at power levels of tens of watts or more and at frequencies above 2 to 3 gc.

Class C operation for high efficiency is common in high-power stages. Design of vacuum-tube class C amplifiers is a well-established discipline which is covered in the literature. Principal design points are attention to biasing for desired gain and current-pulse duration, and to output circuit matching and Q. Traveling-wave devices are most appropriate for very high microwave frequencies and are found in specialized telemetry equipment.

Transistor output stages are generally operated near the limits of transistor capabilities. For this reason, design is quite empirical, with emphasis on input and output impedance matching and variation of emitter d-c impedance to optimize efficiency or gain.

Frequency Multipliers

In situations where multiplication of signal frequency is required, active or passive multipliers are available. Class C stages can be operated in classical fashion as relatively inefficient active multipliers. Varactor (nonlinear-reactance) multipliers provide a highly efficient means of frequency multiplication at higher power levels, and these devices are used to augment transistor power-frequency capabilities in present solid-state designs. Combined transistor amplifier-varactor multiplier operation using the nonlinear collector-base capacitance of a vhf or uhf power transistor shows promise.

The design of a class C frequency multiplier is well documented. It is based upon obtaining fundamental-frequency current pulse to drive an output tank resonated at the desired harmonic. Multipliers of this type are relatively inefficient and follow a $1/N^2$ efficiency behavior for Nth-harmonic output compared with fundamental output of the same device.

Varactor multipliers employ nonlinear-capacitance diodes to provide highly efficient multiplication. Several low-order stages may be cascaded to provide high-order multiplication. Efficiency is limited only by varactor and circuit parasitic losses and may approach 90 to 95 per cent for vhf doublers. Power-handling capability extends from hundreds of watts at 100 Mc to tens of watts at 2 gc.

A principal shortcoming of varactor circuits has been a tendency toward spurious oscillations associated with the parametric nature of the circuit. The causes and cures for this undesired behavior are becoming better known, and it is presently possible to provide varactor multipliers which are essentially spurious-free over a wide range of environments and load conditions.

Comparison of Output Devices

The three output devices considered here are listed in Table 8 with notes on their advantages and disadvantages. Vacuum-tube output stages are capable of providing

Table 8. Telemetry-transmitter Output Devices

	Advantages	Disadvantages
Vacuum-tube amplifier:		
Triode...............	Cost, frequency-power range	Fragile; limited cathode life
TWT.................	High gain	Life; large size
Transistor amplifier.......	Reliable, rugged, miniature	Limited frequency-power range
Varactor multiplier........	Reliable, rugged, miniature	Limited bandwidth, potential instability

as much power as would ordinarily be required in a telemetry transmitter. Because of the penalties of size, weight, and reliability incurred in the choice of thermionic devices, the attainable and expected frequency-power capabilities of transistors and varactor multipliers are of interest.

Representative characteristics for a particular transistor type are shown in Fig. 17a. Transistor available output power drops asymptotically at 6 db/octave relative to some low-frequency value. The figure shows the increased power available through

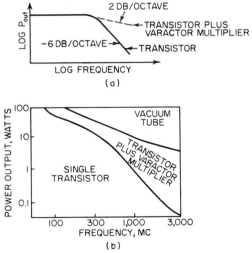

Fig. 17. (a) Power-frequency behavior for a transistor. (b) Frequency-power capabilities.

the use of varactor-multiplier technique, since efficiencies can be 2 db/octave or better. This behavior is typical for any transistor choice.

Power-frequency capabilities are summarized in Fig. 17b. The improvement rate for semiconductor devices is approximately 3 db in power-frequency product per year.

Power-supply Accessories

The primary power available to a telemetry transmitter is generally not well regulated or filtered. The voltage level may not be appropriate for all the transmitter circuitry. For these reasons, a typical transmitter contains power-supply regulators, RFI filters, and d-c to d-c converters.

This type of circuitry is ideally suited to solid-state realization. Regulators may take a number of canonical forms with characteristics such as regulation and efficiency chosen to suit the particular applications.

The requirement for a d-c to d-c converter arises in situations where the high-power elements of a transmitter cannot be operated directly from regulated primary power. Efficiency of solid-state d-c to d-c converters ranges from almost unity down to 30 to 40 per cent in high-power applications requiring a large voltage transformation. Vacuum-tube circuits generally require the use of a d-c to d-c converter.

8 Mechanical and Thermal Design

Telemetry transmitters require packaging suitable both for radio-frequency equipment and for wide environment applications. At vhf frequencies no less than at microwave frequencies, the physical details of component position, local fields and currents, interloop and intermode coupling, etc., are important design parameters.

Miniature physical designs have been developed which allow proper r-f layout while permitting attention to mechanical and thermal requirements. Separation of r-f and d-c circuits is typical wherever possible. Low-density foam potting is used to provide mechanical integrity in the presence of vibration.

Conductive cooling is most common, and the thermal impedance from a heat source to the external environment is an important parameter. Heat-conductive insulating materials such as beryllium oxide and boron nitride are in common use.

The emphasis on miniature design has led to the use of lumped-circuit techniques well up into the microwave region. Strip-line techniques have shown promise, and future work will include the use of thin film for r-f circuitry. Logic and video application of solid-state circuitry is almost universal, and the possibility of total application of monolithic or hybrid integrated techniques is very strong.

Case design is adapted to the particular requirements of an application. Increased use of cast construction incorporating all interstage shielding promises a decrease in the application of modular circuitry for future designs. The elimination of superfluous interfaces and connectors is dictated by reliability requirements.

RECEIVERS

9 Functional Description of Telemetry Receivers

Although the overwhelming majority of telemetry channels to date have been transmitted via FM carriers lying in the vhf band, modern systems are employing an increasing variety of transmission frequencies and formats. This trend has caused a major departure from previous "FM radio" design practices toward the concept of a flexible group of signal-processing modules which may be conveniently rearranged and supplemented to cope with unprecedented input signals. Present requirements entail handling r-f carriers ranging from less than 100 to more than 2,000 Mc, while the range of future carrier frequencies must be considered virtually limitless.

However requirements of modulation format and carrier frequency may change, the basic functions of a telemetry receiver will remain the following:

1. Select the desired r-f carrier from the multiplex of strong and weak signals arriving at the antenna.
2. Amplify the selected carrier to a usable level.
3. Demodulate the signal to retrieve the transmitted information.

Thus a design based on the double-conversion superheterodyne-receiver block diagram shown in Fig. 18 will be suitable for any received signal configuration.

Selection of the desired channel is here accomplished in several steps, rather than at the input terminals of the receiver as would be preferred since the design of a tunable narrowband filter at radio frequencies presents an insoluble insertion loss and tracking problem. So the preselector is designed to reject interfering signals lying more than a few per cent away from the desired carrier frequency while offering minimum attenuation of the received signal. After this broad but vitally important tuning function has been performed, an active element may be introduced to amplify the received signal to a level which minimizes the degradation caused by the relatively noisy converter.

Further amplification and band limiting is performed by the first intermediate-frequency amplifier, and since the overall gain of the receiver must be approximately 100 db, a second intermediate-frequency amplifier is also employed to minimize the

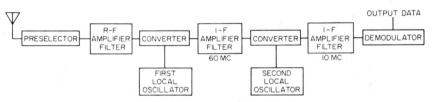

Fig. 18. Block diagram, typical telemetry receiver.

stability problems encountered in high-gain tuned amplifiers. The use of a second, lower intermediate frequency also simplifies the design of narrowband filters.

The final r-f processing element in the receiver is generally a demodulator which generates an estimate of the transmitted base band waveform. This demodulator can range in complexity from a simple AM detector to FM detectors of the Foster-Seeley, phase-lock, or frequency-feedback varieties, depending on the modulation format employed.

In some instances, the demodulator is replaced or supplemented by an additional converter which heterodynes the carrier to a frequency suitable for storage on a wideband tape recorder prior to demodulation. This predetection recording technique allows storage of certain signals, such as PAM-FM and PCM-FM, whose base-band formats cannot be satisfactorily tape-recorded because of recorder nonlinearity and lack of low-frequency response.

Keeping in mind that the function of the telemetry receiver is to provide the best possible representation of the transmitted data under a wide range of received signal strengths and in the presence of interfering signals, the succeeding paragraphs will examine the characteristics of each of the signal-processing components used in a typical receiver to determine what effect each component has on the accuracy of the output signal. Finally, the typical performance of a modern receiver will be described.

10 R-F Amplifiers, Converters, and Filters

The input circuitry of a receiver must cope with an array of radiation ranging from automobile ignition impulses to the transmissions of powerful radar and commercial broadcast stations in addition to the relatively low power telemetry signals. Although the system designer can often reduce the level of undesired signals by means of a directional receiving antenna, in most systems the presence of strong interference must

still be presumed. There are three common ways in which out-of-band components can enter the frequency range of the desired signal:
  1. They may lie at the intermediate frequency of the receiver and simply pass through the input circuitry and converter.
  2. They may lie at the "image" frequency. For example, if a received signal of 250 Mc is translated to a 30-Mc intermediate frequency by heterodyning against a 280-Mc local oscillator, then a 310-Mc carrier will also appear at the 30-Mc intermediate frequency after passing through the converter.
  3. Two undesired components may generate a spurious signal lying within the band of interest by intermodulating with one another in one of the nonlinear elements of the receiver input circuitry, either the r-f amplifier or converter.[5] Typically, this type of interference might occur in a receiver tuned to 250 Mc when unwanted input signals at 40 and 290 Mc also appear at the input. If the unwanted signals are not sufficiently attenuated prior to the first nonlinear element, a second-order distortion component at 290 Mc $-$ 40 Mc $=$ 250 Mc will be generated to interfere with the desired signal.

Consequently, it is important that a selective filter be placed before the first nonlinear element if data errors due to interference are to be avoided. This filter also serves to attenuate local oscillator radiation which might otherwise cause the receiver to interfere with other nearby systems. However, the use of input filtering necessarily introduces some attenuation of the desired signal, and any preselector design represents a compromise between selectivity and insertion-loss characteristics.

Another problem encountered in preselector design is that of "windows" in the stop band. In cavity preselectors, such spurious passbands can arise at the third, fifth, seventh, and all other odd harmonics of the desired frequency, and other types of preselectors based on lumped-constant design can also have undesired regions of low attenuation. For this reason, the spurious responses of a receiver should be specified over the entire range of frequencies which might interfere with the desired signal band.

Up to this point, the advantages of selectivity prior to the first nonlinear element have been stressed, but with selectivity comes the problem of tracking between the preselector and local oscillator in a tunable receiver. If the preselector and local oscillator do not track properly throughout the tuning range, the receiver noise factor will be a function of the channel being received, and noise factors measured at one point in the tuning range will not be duplicated elsewhere in the band.

Since frequency converters have poorer noise performance characteristics than linear amplifiers, good practice requires placing an r-f amplifier between the preselector and mixer where frequency limitations permit. Since the r-f amplifier noise factor is better than that of a frequency converter, the overall receiver noise factor will be improved according to the relationship

$$N_r = N_1 + (N_2 - 1)/G_1 + (N_3 - 1)/G_1G_2 \cdots \qquad (23)$$

where N_r = receiver noise factor
$N_1, N_2, N_3 \cdots$ = noise factor of individual stages
$G_1, G_2, G_3 \cdots$ = power gain of individual stages
The noise factor can be related to noise figure N.F. by the relationship

$$\text{N.F.} = 10 \log_{10} N \qquad (24)$$

In stressing the need for reduction of the receiver bandwidth at the earliest possible point in the signal path, it has been assumed that the restricted converter dynamic range determines the receiver interference sensitivity. However, the r-f and i-f amplifiers must also be designed with care to maximize their dynamic range, or serious spurious response can be generated in these circuits.

While problems of linearity have been described as part of the receiver-design problem, the systems designer should keep in mind that severe distortion can also occur because of nonlinearities in broadband antenna preamplifiers, active multicouplers, etc. Many ultra-low-noise antenna preamplifiers such as tunnel-diode amplifiers have very restricted dynamic ranges.

In telemetry receivers operating at frequencies below 1 gc, best practice calls for r-f amplification prior to the first frequency conversion because an improvement in noise figure and strong signal performance and a reduction in local oscillator radiation can be obtained in this manner. Nonetheless, the converter noise and linearity characteristics are an important factor in determining receiver performance. Enough low-noise-power gain must be provided prior to the converters to minimize the receiver noise figure, but this power gain must be controlled to prevent overloading of the frequency converter by strong signals.

Virtually all telemetry receivers contain provisions for crystal-controlled or continuously tunable (VFO) mode of operation. Although telemetry applications do not usually place extreme requirements on the accuracy of the first local oscillator frequency, spectral purity is essential to prevent false modulation of the first intermediate-frequency signal and spurious response to out-of-band signals. The first local oscillator signal is normally generated at frequencies in the neighborhood of 80 Mc where stable harmonic modes of crystal frequencies can easily be employed, and multiplied to the appropriate injection frequency. At the injection point, the power level of the local oscillator must remain reasonably constant over the receiver tuning range since excessive noise can result from either insufficient local oscillator injection level or excessive local oscillator input causing heating of the converter components. The system designer can assure satisfactory performance in this area by specifying a maximum receiver noise figure over the tuning range rather than a spot or typical noise figure. In some systems, short-term frequency stability of the local oscillators may be significant. Typical short-term stabilities which can be achieved in modern receivers range below 1 ppm in a 1-cps to 5-kc bandwidth in VFO mode of operation.

The first intermediate frequency is chosen to allow adequate image rejection by the r-f preselector and is normally 30 or 60 Mc, a low first intermediate frequency minimizing the noise contribution of the first i-f amplifier, and a high intermediate frequency minimizing the receiver image response, particularly at microwave frequencies. The first intermediate frequency is too high to allow overall bandwidth determination in telemetry receivers; so the filtering performed is simply designed to protect the second converter from image responses.

Since the converter dynamic range is normally a limiting factor in receiver performance, it is desirable to have as little power gain as possible prior to the final conversion. Thus an ideal receiver would be designed with just enough r-f amplification to mask the converter noise contributions and the first i-f section would consist of a lossless filter. However, the design of a second i-f tuned amplifier having tightly controlled phase and amplitude characteristics together with 60 or 70 db of power gain presents a formidable challenge, and common practice allocates at least part of the overall power gain to the first i-f amplifier.

Since the second oscillator does not undergo multiplication prior to injection and has a smaller tuning range than the first local oscillator, obtaining the necessary frequency stability and spectral purity presents fewer design problems than are encountered in the first local oscillator. However, a complicating factor in the design of the second local oscillator is the requirement for automatic frequency control and fine tuning, as the circuitry used to implement these functions can introduce substantial temperature coefficients and cause spurious hum and noise modulation of the oscillator frequency.

The requirements placed on the second frequency converter are comparable with those places on the first frequency converter insofar as distortion is concerned, but since considerable power gain is normally provided prior to this unit, the noise contribution of the second converter is usually negligible. This is the case only when proper AGC system performance is provided since an AGC which reduces the gain of the front end faster than the gain of later stages can substantially increase the effective noise contribution of the second converter.

The remaining predetection power gain is apportioned to the second i-f amplifier. Before this section, or integrated with it, is the final predetection bandwidth-determining network. Separating the final i-f filter from the amplifier makes the receiver bandwidth less dependent on signal strength and the tolerances associated with active elements while integration of the filter with the amplifier provides production economies. This is particularly true of solid-state receivers.

When dsb (double-sideband) AM signals are being processed, the i-f bandwidth need only be sufficient to accommodate the sidebands corresponding to the highest received frequency, that is, twice that frequency. However, when FM signals are being received, other factors must be considered in establishing the i-f filter characteristics. The passband of the filter must be sufficient to accommodate the peak-to-peak deviation of the carrier or the first-order sidebands of the highest modulation frequency, whichever is greater. In addition, the filter must have a linear phase characteristic to minimize dynamic distortion of the transmitted intelligence.

In many receiver systems, the legitimate concern for selectivity has led to specification of i-f filter 6- to 60-db attenuation-bandwidth ratios of 3 or less, with little or no emphasis on phase linearity. General telemetry-receiver design practice has been based on using cascaded Butterworth two-pole filters similar to those used in AM communications receivers, but since Butterworth and Chebyshev filters introduce dynamic distortion when used in FM-FM systems, these configurations can be a cause of substantial errors in received data. To minimize the distortion caused by i-f filters, it has become common practice to use a filter bandwidth approximately twice that of the incoming signal, even though the resultant increase in system bandwidth raises the threshold level. This technique improves distortion because the phase nonlinearity of a Butterworth filter is most severe near the band edges.

It is not difficult to show that the distortion produced by phase nonlinearity is proportional to both the time-delay variation in the filter and the modulation frequency.[3] Consequently, the system designer must specify the means of measurement of distortion as well as the tolerable distortion levels. A convenient technique for evaluating receiver distortion consists of using a two-tone modulation format, such as IRIG channels 17 and 18, combined to provide full-scale modulation of the transmitter and measuring the difference frequency produced by transmitter and receiver nonlinearities. The distortion term can then be compared with the expected levels of nearby subcarriers to predict the effect on system accuracy. It should be noted that most commercial test equipment is not sufficiently linear to serve in place of the transmitter.

11 Receiver Demodulators and Output-signal Processing

Two primary characteristics—symmetry and dynamic range—determine the effect of the limiter on the overall receiver performance. If a satisfactory capture ratio is to be achieved, limiting must be symmetrical about the composite signal baseline. If this requirement is not met, a sharp reduction of limiter output will result when the interfering signal is phased oppositely to the desired carrier, and in addition to the inevitable difference frequency beats in the demodulator output, suppression of the desired signal will result (Fig. 19).

The demodulators employed to extract the base-band signal from the modulated carrier can be grouped into two categories: broadband detectors, such as the Foster-Seeley and related discriminators, and bandwidth compressive circuits, among which the most common varieties are the phase-locked and frequency-compressive feedback demodulators. The latter circuits improve the system threshold performance by their ability to reject certain types of signals which represent unlikely modulation of the carrier, and in order to improve noise rejection, certain a priori knowledge of the modulation format must be built into them. This information usually relates to the maximum deviation rate impressed on the carrier, and thus bandwidth-compressive demodulators are effective only in extending the system threshold when signals having high modulation indexes are received. Inasmuch as normal telemetry formats employ modulation indexes of 1 or less, these demodulator circuits have not found widespread use. Additional information regarding the characteristics of phase-locked demodulators may be found in Chap. 9 and Sec. 18.

Broadband discriminators, such as the Foster-Seeley detector (Fig. 20) and its variations, provide a signal output proportional to the instantaneous frequency of the i-f carrier. To provide an accurate representation of the system modulation format, the demodulator must be linear over the full range of limiter output frequencies. It should be noted that when in-band interference is present, the limiter substantially broadens the bandwidth of the composite signal applied to the demodulator and, therefore, linearity over a frequency range substantially greater than the i-f filter bandwidth is required.[4]

In addition to reconstituting the base-band signal, the demodulator provides tuning information, and if minimum distortion and best weak-signal performance are to be achieved, the center frequency of the demodulator must be well matched to the center of the i-f passband.

Following demodulation, the resulting base-band signal, often referred to as the video signal, is band-limited, amplified, and processed to generate tuning, deviation, and output-level indications. Since a considerable number of telemetry formats utilize direct pulse modulation of the r-f carrier, system response to direct current is highly

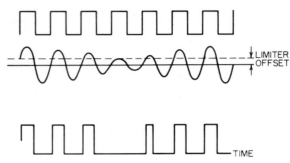

Fig. 19. Limiter output error due to lack of symmetry.

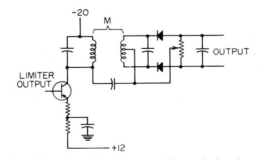

Fig. 20. Simplified schematic diagram of Foster-Seeley demodulator.

desirable to maintain flexibility and incidentally provide an output for Doppler measurements. The pulse-modulation formats also make the use of deviation meters calibrated in terms of peak-to-peak modulation rather than average modulation desirable. By the same token, a tuning meter displaying mean of peaks of modulation is preferable to one showing average carrier-frequency error. Accurate and stable meter calibrations are essential to allow accurate tuning to the center of the i-f passband where best weak-signal and distortion performance will be achieved. In addition, AFC circuitry, if used, should be carefully aligned to ensure a return to the center of the i-f bandpass.

12 Predetection Recording

Predetection recording is rapidly supplanting postdetection recording as a primary data-acquisition tool because of its superior transfer linearity, gain stability, and, more importantly, its d-c response and relative immunity from drop-outs. These latter characteristics allow recording of nonsymmetrical waveforms such as PAM and PCM signals without undesirable baseline shift as the average duty cycle of the pulse train changes. Perhaps the most important reason for the favor which predetection recording has found lies in its versatility, since the ranges are faced with such a variety of modulation formats that they have been forced to seek a method for recording which, while not optimum for any signal, at least provides reasonable accuracy and is applicable to all the formats normally encountered. Another advantage obtained by the use of this technique is the possibility of tailoring the predetection filter to the specific signal being analyzed. Unfortunately, this advantage is seldom exploited since it is somewhat more time-consuming and requires a fair degree of skill on the part of the people engaged in recording the data.

Another claim sometimes made for predetection recording is that it is inherently a more simple acquisition technique, but even a quick look at some representative systems suffices to show that this simplicity is more philosophical than real.

Choice of I-F Bandwidth and Recording Center Frequency

Influence of Image Responses. Obviously the i-f bandwidth selected for predetection recording must pass the full carrier deviation extremes and at least the first-order modulation sidebands. Inasmuch as predetection filtering can be performed during playback, the temptation may be to employ a very broad i-f filter during recording,

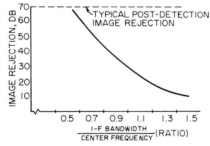

FIG. 21. Image rejection of typical predetection recording system. As the recorder bandwidth is utilized more completely, a decrease in image rejection results.

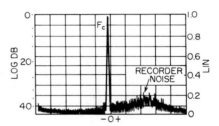

FIG. 22. Tape-recorder noise spectrum. Spectral analysis of the output of a 1.5-Mc tape recorder reproducing a 900-kc unmodulated carrier.

thereby making sure that the filter characteristics do not cause distortion or loss of data. However, this solution ignores the greater susceptibility of these systems to image responses. Where a typical telemetry-receiver image response may be 70 db or more below the desired response, the image response of most predetection systems is passed over in silence. Using six-pole linear phase filters to minimize system distortion and operating at moderate i-f bandwidths, typically two-thirds of the recording center frequency, an image response only 55 db below the desired response will be obtained, while if the recorded bandwidth is increased to equal the center frequency, the image response will rise to only 30 db below the desired response (see Fig. 21). This, then, tends to make the choice of a high recording center frequency seem desirable.

Influence of Recorder Noise and Flutter. Other factors, however, act to favor the use of a low center frequency. The first of these is the tape-recorder noise spectrum (see Fig. 22) which results in some degradation in the overall signal-to-noise ratio as the carrier frequency is raised. In addition, raising the recorded center frequency decreases the percentage deviation of the carrier, thereby increasing the effect of recorder wow and flutter. The magnitude of the error introduced by recorder-speed errors may be directly calculated by taking the ratio of carrier deviation introduced by the recorder-speed errors to the expected deviation introduced by the desired signal, keeping in mind that the recorder-speed error is introduced during record and playback and, again, if dubbing is performed. Since typical recorders have flutter totaling approximately 0.2 per cent at 120 in./sec, a rough guide as to the need for tape-speed compensation is as follows:

1.. If the expected carrier deviation is less than ±10 per cent and the modulation index is 1 or more
2. If the expected deviation is less than ±5 per cent

It should be noted, however, that flutter effects will be more severe at lower tape speeds, typically reaching 0.5 per cent at 15 in./sec. Although tape-speed compensation can reduce the apparent deviation introduced by recorder-speed errors, it cannot

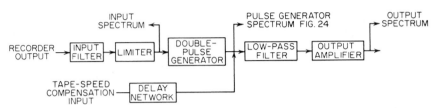

FIG. 23. Wideband FM demodulator block diagram. The demodulator shown operates directly at the recorded carrier frequency.

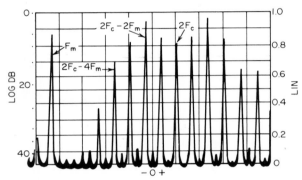

FIG. 24. Double-pulse-generator spectrum. Carrier has been doubled and a modulation component has been generated.

correct for time-base errors. Low-inertia servo-controlled tape recorders can perform both the functions of time-base correction and low-frequency flutter suppression, and developments in the performance of electrically variable delay lines indicate that extremely stable time-base performance can be achieved.

Influence of Spurious Responses. In the direct-playback demodulation system, which is favored by many system designers because of its simplicity and generally superior distortion characteristics, a relationship exists between permissible maximum modulating frequency and the carrier deviation limits. This relationship may be ascertained with the help of Fig. 23, the demodulator block diagram, and Fig. 24, the spectral response at the low-pass filter input when a 900-kc carrier having 300-kc deviation is modulated by a 200-kc signal. It can be seen that in consisting of a pulse at each input axis crossing, the output pulse train has components at the modulating frequency, twice the carrier frequency, and at frequencies integral multiples of the modulating frequency above and below the doubled carrier. When these sidebands move down into the data passband because of an increase in the modulating frequency, they are termed spurious responses and, together with the recorder bandwidth, place a limit on the intelligence passband of the system. The method of calculation of these sideband magnitudes consists of first computing the modulation index for the modulation frequency and deviation in question (remembering that since the pulse generator doubles the carrier frequency, it also doubles its deviation) and then finding the magnitude of the sideband from a table of Bessel functions (see Appendix Sec. A.10-14). Once the amplitude of the sideband relative to the carrier has been determined, it is compared with the modulation which produced it by using an equation from the Fourier analysis of the pulse train

$$\frac{e_c}{e_m} = 1.27/D \qquad (25)$$

where D = fractional deviation
e_m = signal amplitude
e_c = the unmodulated carrier, assuming a 50 per cent pulse duty cycle at band center

The simple procedure described above applies to all FM systems, be they conventional 40 per cent deviation systems, wideband FM and predetection recording, or video recorders employing FM carriers.

Predetection Filtering

Since wideband FM recordings can be made under ideal signal conditions, provisions for predetection filters are not normally made in the wideband FM demodulators. This is most unfortunate, because these demodulators are well suited to the task of reducing predetection recorded data as well as the wideband FM signals, and the quality of predetection recorded data can often be substantially improved by using the appropriate bandpass filter prior to demodulation.

When the carrier modulation consists of a multiplex of subcarriers and the interference consists simply of thermal and receiver noise, a linear-phase bandpass filter will generally provide threshold extensions in the neighborhood of 3 to 4 db, but when the noise is in the form of multipath interference, better results may be obtained when using a more flat-topped filter, even at the expense of higher distortion caused by phase nonlinearity in the filter. The improved multipath performance obtained from a flat-topped filter is due to the fact that when the desired carrier moves to the edge of a conventional filter passband, its amplitude is reduced 3 db relative to the band center value and if the spurious path signal happens to be at band center, the apparent capture ratio of the system is degraded by a corresponding amount.

If noisy PCM-FM signals having a bit rate comparable with or less than the deviations are being demodulated, an appreciable reduction in bit error rate can be made by using an input having a double-humped amplitude characteristic, the points of maximum response corresponding to the carrier excursions. The bandwidth of each hump must be sufficient to pass the bit rate employed, and when the bit rate approaches the carrier deviation, it becomes impossible to introduce much attenuation in the center of the filter passband, thereby diminishing the utility of this technique. However, when bit rates in the vicinity of half the carrier deviation are encountered, threshold extensions of 3 or more db can be obtained using these filters.

Additionally, when a defective recording is made because of improper setup procedures, strong spurious signals are often recorded together with the desired carrier. On several occasions, it has been found possible to retrieve data which would normally have been lost by using appropriate high-pass, bandpass, or low-pass predetection filters.

System Setup Procedures

A great number of noisy or unusable predetection recording and wideband FM tapes are due to improper system setup, most often at the recorder. This problem is in great measure caused by the fact that adjustment procedures supplied by the manufacturers of tape recorders are generally designed to permit demonstration of compliance with recorder specifications rather than maximize the signal-to-noise ratio of the actual data being recorded. For example, if the received signal does not have frequency components lying near the upper frequency-response limits of the recorder, a substantial improvement in data accuracy can often be achieved by simply increasing the bias and record levels. On the other hand, if a tape-speed reference carrier is being multiplexed with the input signal, a slightly reduced record level plus higher bias amplitude will generally give improved results.

It is impossible to prescribe a setup procedure which is optimum for all recorders and all formats, but at least some suggestions for monitoring the performance level can be given. First, a means for simulating the expected signal during system setup is essential if good-quality data are to be obtained. This need not, for example, be a complete PCM test generator feeding a test r-f link, but it might consist of a pulse- or square-wave-modulation format driving an r-f signal generator while the operator adjusts i-f bandwidth, record level, and bias amplitude to optimize the system rise time and signal-to-noise ratio. The use of meters has never been found satisfactory in setting up either a predetection recording or wideband FM system. Oscilloscope monitoring of the demodulated playback signal has proved most effective in evaluating the performance of these systems, but other instruments such as carrier telephone test sets might be worthy of consideration. In spite of the great efforts which have been made to mechanize and automate data acquisition and reduction systems, it is essential to remember that the quality of data obtained from these systems still depends very heavily on the skills of the operator.

13 Typical Receiver Performance

The following abbreviated specifications are representative of the performance levels achieved by a typical modern receiver:

1. *Preselector Bandwidth.* 215- to 260-Mc band, 7 Mc maximum; 2,200- to 2,300-Mc band, 15 Mc maximum (determines the sensitivity of the receiver to out-of-band overload signals)
2. *Input VSWR.* 3.5:1 maximum (determines the quality of impedance matching to the antenna)
3. *Input Impedance.* 50 ohms
4. *Noise Figure.* 215- to 260-Mc band, 6 db maximum; 2,200- to 2,300-Mc band, 9 db maximum (determines the weak-signal performance of the receiver)
5. *RFI Characteristics.* Meets requirements of MIL-I-26600 and MIL-I-6181D
6. *Image Rejection.* 60 db
7. *Data Distortion.* Intermodulation distortion between any two IRIG subcarriers less than 0.5 per cent when using an i-f filter having a bandwidth equal to the peak-to-peak carrier deviation
8. *I-F Bandwidth Range.* 100 kc to 2.5 Mc standard. Narrower bandwidth available with third conversion
9. *Video-frequency Response.* Switchable 20 kc to 1 Mc linear phase

SUBCARRIER OSCILLATORS

14 Types and Functions

The conversion of physical parameters to frequency-modulated subcarriers has generally been accomplished by one of two principal techniques. Either the force, displacement, temperature, etc., to be telemetered has been converted to a d-c volt or millivolt level signal and applied to the input of a voltage-controlled oscillator (VCO) or the transducer and oscillator circuitry are combined in a manner which permits generation of a frequency-modulated subcarrier without the intermediate conversion to a d-c signal. Examples of the latter technique are found in strain-gage oscillators and reactance-modulated oscillators.

It would appear that the intermediate conversion of the parameter under study to d-c voltage simply adds an opportunity for increased system error, but in fact, a marked trend to this approach has developed in recent years. The larger and more complex instrumentation systems which have come into being have made separate calibration of transducers and oscillators highly desirable and since subcarrier channel assignments are frequently changed from one flight to another in a program, the use of a transducer-controlled oscillator which must be calibrated together with the transducer it will be used with presents a considerable logistic and operational problem. Although certain of the transducer-modulated oscillators, such as strain-gage oscillators, need not be calibrated in conjunction with a specific transducer, the operational problems associated with capacitively balanced cable connections between the oscillator and transducer have tended to limit the application of these units.

Another factor which has served to encourage transducing physical parameters to voltage prior to the conversion to FM subcarriers has been the increasing availability of high-performance millivolt-controlled subcarrier oscillators suitable for use with

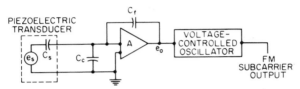

Fig. 25. Charge-controlled oscillator with piezoelectric transducer. e_s = piezoelectric transducer voltage; C_s = piezoelectric transducer effective series capacitance; C_c = cable capacity, typically 0.005 μf; C_f = feedback capacitor; $e_s C_s = Q_{in}$ (input charge); $e_o C_f = -Q_{in}$; $e_o = -e_s(C_s/C_f)$ (output voltage of amplifier A); A = high-gain inverting amplifier.

strain-gage transducers. With these units, the connecting leads between oscillator and strain-gage transducer carry d-c modulation signals rather than subcarriers and so need not be capacitively balanced. In addition, since millivolt-controlled subcarrier oscillators are generally supplied with floating input circuits, common d-c excitation to all strain-gage transducers can generally be employed in a carefully designed system.

Another family of subcarrier oscillators, specifically designed for capacitive transducers, such as piezoelectric vibration pickups, known as charge-controlled oscillators, is also finding widespread acceptance in FM-FM telemetry systems. The circuit normally employed for this purpose normally consists of a conventional multivibrator voltage-controlled oscillator preceded by a charge-to-voltage converter. As shown in Fig. 25, the input circuitry produces a node at the charge converter input, thereby drastically reducing the shunting effect of cable capacity and eliminating the need for high-input-impedance buffers located at the transducer where severe environments are often encountered.

The circuits which have been employed to convert analog voltages to FM subcarriers are too numerous to even list, let alone describe in detail, but two main categories stand out. The first of these consists of sinusoidal oscillators in which the input signal is made to modulate one of the reactive elements controlling the oscillator frequency. While these units eliminate the need for an output filter many designs suffer from poor deviation linearity, and circuits which operate with sufficiently high resonant-circuit Q to give an acceptably low output distortion level often produce severe amplitude modulation when deviated rapidly. A second common configuration which is now preferred in most designs because of its simplicity and low cost consists of a free-running multivibrator which is directly modulated by the input signal, followed by a buffer amplifier and filter to provide a sinusoidal output. A typical schematic diagram of such an oscillator is shown in Fig. 26. This circuit can be made to provide modulation linearity in the neighborhood of 0.1 per cent and does not

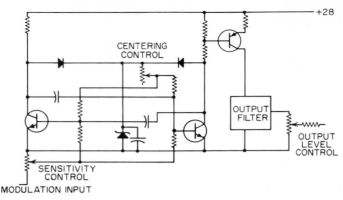

FIG. 26. Typical airborne subcarrier oscillator.

suffer from amplitude modulation when deviated rapidly. (See Chap. 10 for other subcarrier-oscillator circuits details.)

15 Oscillator Performance Parameters

Since the overwhelming majority of telemetry channels consists of charge-, volt-, and millivolt-controlled oscillators, let us consider the effects on system performance of these critical elements. We can conveniently group these effects into four general categories:

1. "Per-channel" static characteristics, such as linearity, drift, and environmental sensitivity
2. "Per-channel" dynamic characteristics, such as intelligence distortion produced by time-delay skew, and channel bandwidth reduction caused by oscillator frequency-response limitations
3. System crosstalk errors in adjacent and higher channels resulting from excessive deviation, harmonic output, and sideband output from the oscillators
4. System crosstalk effects in lower channels resulting from intelligence feedthrough terms in the oscillator outputs

Although the static characteristics and environmental sensitivity of volt- and millivolt-controlled oscillators can be measured by means of eput meters, precision d-c voltage sources, and test chambers, the use of a precision subcarrier discriminator in the transfer function-display setup shown in Fig. 27 can sharply reduce the test time and data-processing effort required to evaluate VCO performance. The test configuration shown allows the systems designer to display the oscillator frequency error vs. position in the band as a Lissajous pattern on a cathode-ray oscilloscope. Here the sinusoidal test signal is first split into two phases to facilitate testing oscillators having either modulation sense, and the oscillator being evaluated is modulated from a source impedance simulating the operational case. The resulting FM output

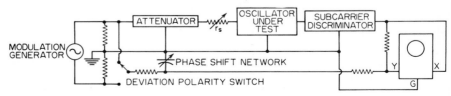

FIG. 27. Typical voltage-controlled oscillator test setup.

is demodulated by a precision subcarrier discriminator which, in the absence of oscillator drift and nonlinearity, produces an exact replica of the test signal, delayed by the various filters in the test setup. Display of the error is accomplished by subtracting the RC delayed and inverted test signal and applying the difference signal to the oscilloscope vertical input. The discriminator output is applied to the oscilloscope horizontal input to provide a Lissajous pattern allowing quick and convenient measurement of zero shift, sensitivity shift, nonlinearity, and source-loading effects. A perfectly linear oscillator exactly calibrated produces a straight horizontal line on the oscilloscope face, while zero errors result in vertical displacement of the pattern, sensitivity errors are indicated by tilting of the pattern away from the horizontal, phasing errors produce an ellipse in place of the single line, and nonlinearity produces a curved trace.

Two additional measurements of "per-channel" oscillator performance are facilitated by the availability of a subcarrier discriminator: common-mode rejection in oscillators having floating inputs and dynamic intelligence distortion in oscillators which employ bandpass output filters. In this latter area, the systems designer is faced with an unavoidable performance trade-off since highly selective bandpass output filters are desirable as a means of protecting other channels from harmonic output, intelligence feedthrough, interference due to overmodulation, and crosstalk produced by excessive sideband output when high-frequency modulating signals are handled, while the use of a steep-skirted output filter can introduce severe high-frequency intelligence distortion due to nonlinearity of the filter phase vs. frequency characteristic. Worst-case time-delay distortion is generally observed with a discriminator low-pass filter corresponding to unity modulation index and a modulating frequency equal to one-half or one-third of low-pass cutoff. If it is not certain that the discriminator input filter used to make the distortion tests has a linear phase characteristic, it should be bypassed since separation of the tested channel from a multiplex is not necessary.

Obviously, while the transfer-function-display system makes evaluation of the "per-channel" characteristics a fairly straightforward operation, the effect of the oscillator's spurious outputs on other data channels cannot be measured in this way. To determine crosstalk effects due to harmonic output, intelligence feedthrough, and sideband output, a wave analyzer or preferably a spectrum display is essential. When measuring the spurious outputs of a subcarrier oscillator, data should be taken at the output-level setting which will be used operationally, or if this level is not known at the time of the tests, at a variety of output-level settings since the spurious output cf a subcarrier oscillator is often a function of the adjustment of this control.

Subcarrier harmonic-distortion measurements should be made with the oscillator statically deviated to both band edges as well as at band center since many of the filters employed to reduce distortion in oscillator circuits have highly variable attenuation in the stop band. When the intelligence feedthrough and sideband output components of oscillator outputs are being evaluated, the highest modulation frequency expected should be employed since this generally provides worst-case data. Excessive sideband output is most generally a problem when commutated inputs having steep transients are being accommodated. In this case, the designer must trade off the crosstalk introduced in nearby channels against increased rise time caused by premodulation filtering.

Once data on spurious oscillator outputs have been assembled, the crosstalk effect can readily be calculated using the equation

$$\text{Per cent error} = R(S/D)a_1 a_2 \qquad (26)$$

where R = ratio of harmonic output to expected amplitude of affected channel
D = full-scale deviation of affected channel, cps
S = separation of harmonic signal from subcarrier, cps, normally taken when separation is most nearly equal to perturbed channel demodulator low-pass filter cutoff for worst-case analysis
a_1 = attenuation of interfering signal relative to affected channel subcarrier, fractional

a_2 = attenuation of output beat frequency by demodulator low-pass filter, fractional (beat frequency equals difference of interfering and desired subcarrier frequencies)

Using the test techniques outlined on the previous page, the designer will be able to evaluate these characteristics of volt- and millivolt-controlled oscillator performance:

Deviation linearity and environmental sensitivity
Dynamic distortion and common-mode rejection
Crosstalk effects on other channels

Many other factors must be considered in selecting a subcarrier oscillator for a specific application—in some systems accuracy may be sacrificed to obtain minimum size or power consumption, while in others the ever-present factor of cost will be of unusual significance.

SUBCARRIER DISCRIMINATORS

16 Function and Design Requirements

A subcarrier discriminator, as shown in Fig. 28, is a complete single-channel FM demodulator. It normally includes a bandpass input filter, limiter, demodulator, and

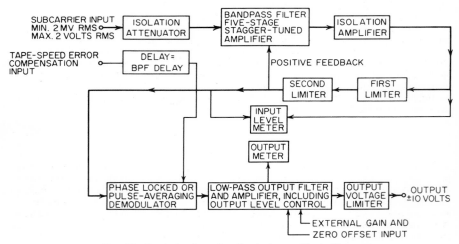

Fig. 28. Typical subcarrier discriminator block diagram.

low-pass output filter, plus a number of peripheral functions such as tape-speed compensation and metering. The demodulation section is either "pulse-averaging" or "phase-locked" in current practice, although many other types have been used. Provisions for tape-speed-error compensation are normally included.

Most subcarrier discriminators are designed to permit adaptation to a range of carrier frequencies and deviations with plug-in frequency-determining units. To permit this, the circuits used are wideband amplifiers and trigger circuits with as few circuit elements as possible involved in frequency changes. A typical unit will operate over a carrier-frequency ratio of 1,000:1. Performance levels achieved in modern discriminators are so high that discriminator deficiencies contribute only slightly to errors in an FM-FM telemetry system.

For telemetry utilizing normal IRIG standards where modulation indices of 5 or less are used, phase-locked or pulse-averaging demodulators are used interchangeably. For quasi-static data, a substantial improvement in performance with input S/N ratios poorer than about 10 db can be achieved with phase-locked types. The behavior of these circuits is described in Sec. 18.

The filter must select the desired band with maximum rejection of other channels while introducing the smallest dynamic distortion of the signal information due to time-delay variations. Following this, the limiter/discriminator demodulates the FM subcarrier with minimum drift and nonlinearity. The capture ratio of the demodulation is also determined in this section. Following demodulation, the output is filtered to remove undesired noise and subcarrier components and amplified to a suitable output level. The output amplifier must provide considerable current, typically 100 ma at full-scale load, to drive high-frequency galvanometers or other low-impedance loads with linearity and drift characteristics which do not materially degrade the overall demodulator performance.

In addition, most subcarrier demodulators provide inputs for tape-speed-error compensation and remote gain and zero correction, and include circuits for metering of subcarrier level and subcarrier position in the selected band.

17 Bandpass Input Filters

The bandpass filter serves to select the carrier to be demodulated from a multiplex. The design of this filter represents a compromise between selectivity (to minimize crosstalk) and the bandwidth phase characteristics required to minimize distortion of the channel intelligence.

In frequency-modulated systems, the intelligence transmitted is contained in the amplitude and phase of the various orders of sidebands relative to the carrier. A frequency-selective filter inevitably alters the relative amplitude and phase of these components of the signal and hence may introduce considerable distortion of the transmitted intelligence. An exact analysis of the effect of a frequency-selective network upon the intelligence present in a frequency-modulated signal is difficult and cumbersome. Van der Pol[3] has analyzed this problem for a "quasi-stationary" case. He finds that modulation $g(t)$ is modified by a network with phase characteristic $\phi(\omega)$ so that the intelligence at the output of the network becomes

$$g[t + \phi'(\omega)] + \Delta\omega g(t)g'(t)\phi''(\omega) \qquad (27)$$

The first term shows the intelligence as retarded by the envelope delay $\phi'(\omega) = d\phi/d\omega$; the second term represents the distortion introduced by the filter. The distortion is proportional to the deviation, the rate of change of deviation, and the second derivative of phase shift with respect to frequency. The quasi-stationary analysis shows the distortion to depend only upon the phase characteristic of the network and not upon the amplitude. Experiments show that this analysis ia a good approximation for cases where the carrier and its significant sidebands lie within a portion of the filter phase characteristics that can be represented by a second-order transfer function. As the spectrum of the carrier occupies most of the filter passband the quasi-stationary approximation fails to describe the effect of the filter upon the channel intelligence.

In practice, the effect of a given filter upon channel intelligence can usually be measured more easily than it can be calculated. The general type of filter used is "Gaussian" in character, but it is important to recognize that the linear phase character of a Gaussian low-pass filter is not preserved by the low-pass to bandpass transformation of the low-pass filter poles. Practical bandpass filters can be designed analytically with a computer or experimentally. The attenuation and envelope delay characteristics of a typical discrimination bandpass filter are shown in Fig. 29.

In many telemetry discriminators, the bandpass filter used is responsible for the major part of the harmonic and intermodulation distortion of the demodulated intelligence. For a discriminator operated with a given output filter cutoff frequency, the worst harmonic distortion is usually observed for the case of full bandwidth deviation at a modulation frequency of either one-half or one-third of the low-pass filter cutoff frequency. For higher modulation frequencies, the harmonics generated are eliminated from the output by the low-pass filter.

In bandpass filters where the envelope delay is to be controlled precisely, the filters used generally employ transistor isolation of the poles from each other to facilitate adjustment. A typical circuit employed in such a filter is shown in Fig. 30. In the

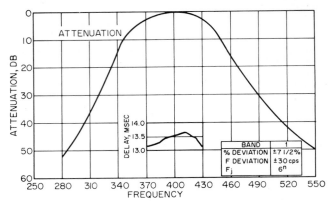

Fig. 29. Typical discriminator bandpass-filter delay and attenuation characteristics.

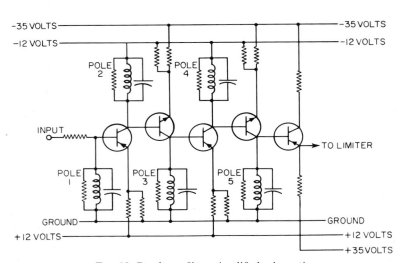

Fig. 30. Bandpass filter, simplified schematic.

arrangement shown, the first filter pole is driven directly through a resistor from the input; this arrangement will accommodate a multiplex of carriers of high amplitude since the carriers that are to be rejected are reduced in amplitude prior to the first transistor element. This arrangement is susceptible to change in the filter characteristic as a function of the internal impedance of the signal source. A buffer amplifier before the filter eliminates this source of error, but the amplifier must accommodate the full amplitude of the multiplex of carriers without introducing significant distortion.

18 Limiter and Demodulator

Following separation from the multiplex, the subcarrier is processed by an amplifier/limiter which provides a fixed-level square wave to the subsequent demodulation circuitry. Although a variety of circuits have been employed for this purpose, the most effective technique employs diode clamping networks which do not permit the amplifiers associated with the limiter to depart from linear operation, to either saturation or cutoff. Thus, there is no bias change with input signal level and symmetrical limiting is assured over the full 60 db or more of input signal range which must be accommodated.

A variety of circuit configurations has been developed to demodulate FM subcarriers, beginning with low-frequency versions of the well-known Foster-Seeley discriminator and including pulse averaging, double pulse averaging, and phase-locked detectors. The great majority of subcarrier demodulators presently in use employ one of the two latter techniques, and since the other demodulation methods listed have no inherent advantages, this discussion will be restricted to a description and comparison of double-pulse-averaging and phase-locked discriminators.

Analytically, the double-pulse-averaging demodulator shown in Fig. 31 is the simpler device since it simply generates a fixed-area pulse for each limiter axis crossing. An increase in the subcarrier frequency raises the number of pulse generator

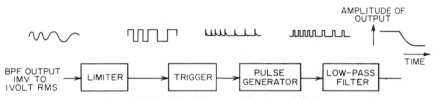

Fig. 31. Typical pulse-averaging demodulator.

outputs per second, increasing the average input level to the low-pass filter. Thus, the low-pass filter output, after high-frequency ripple and noise have been attenuated, is proportional to the subcarrier input frequency. It should be noted that the limiter, trigger, and pulse-generation circuitry can be combined in many ways, and the block diagram given in Fig. 31 is for illustrative purposes only. Numerous generations of subcarrier discriminators have resulted in pulse-generator circuitry capable of operation at carrier frequencies in excess of 10 Mc and having output temperature coefficients of less than 50 ppm/degree centigrade when operated at frequencies below 200 kc. Higher-frequency operation entails some deterioration in temperature coefficient since transistor storage times and circuit stray-capacity temperature sensitivity begin to influence the overall discriminator stability. The double-pulse-averaging circuit described above provides an exact measure of the frequency output of the limiter each half-cycle of the subcarrier. It has no provisions for ignoring improbable shifts in subcarrier frequency. Therefore, its threshold characteristics follow the curve of Fig. 32. A more complete description of the threshold characteristics will be found in Sec. 3.

The other subcarrier demodulator circuit in common use employs the phase-locked principle illustrated in Fig. 33a. Here the limiter output is compared in phase with the output of a voltage-controlled oscillator which is quiescently displaced 90° from the input signal. Departure from the quadrature-phase relationship produces a d-c output component from the phase-sensitive demodulator which, after processing by the loop filter, is applied to the oscillator modulation input, readjusting its operating frequency in a manner which tends to reduce the phase error. The behavior of the phase-locked loop in response to modulation of the subcarrier input is determined by the loop filter, which determines the maximum rate at which the voltage-controlled oscillator frequency may be changed before the departure from quadrature, or phase error, reaches ±90°, after which point the discriminator is said to be out of lock. It is the ability of the phase-locked discriminator to lose lock when improbable limiter output frequencies are received, which allows extension of the FM threshold.

From the discussion of FM theory in Sec. 3, it is seen that noise may be represented as random frequency components randomly amplitude-modulated, and therefore, the signal will be suppressed occasionally by superimposed noise. Interfering noise components having amplitudes smaller than the desired signal will cause phase modulation

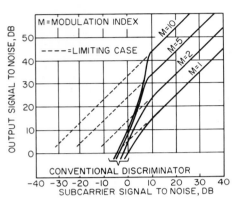

FIG. 32. Thresholding in FM discriminator.

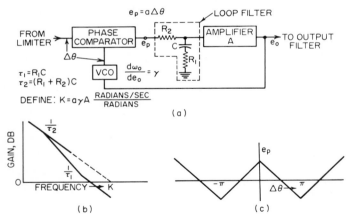

FIG. 33. (a) Phase-locked loop demodulator. (b) Bode diagram of loop. (c) Phase-comparator characteristic for linear loop.

of the limiter output which cannot exceed ±90° (see Fig. 9) while, if the noise amplitude exceeds that of the signal, very large phase excursions will occur. The threshold extension capability of the phase-locked demodulator arises from the fact that phase excursions in excess of 90° from the quiescent operating point do not result in proportionally greater input to the loop filter, thus making it possible for the demodulator to ignore improbable frequency excursions.

The basic advantage of a phase-locked loop discriminator lies in its application to threshold signals. In conventional, that is, noncoherent, FM demodulation methods, the threshold of the demodulator is independent of the modulation index used, that is, independent of the data to be transmitted.

In the phase-locked loop, this is not true; if it is known in advance that the intelligence transmitted in an FM channel is relatively slowly varying compared with the upper limit of the channel capacity, it is possible to choose the phase-locked-loop parameters in such a way that a substantial improvement in output signal-to-noise ratio can be achieved. This improvement must be based upon advance knowledge of the character of the intelligence to be transmitted so that an appropriate choice of loop parameters can be made. It will be shown that, if an unfortunate choice of loop parameters is made, it is possible that a phase-locked-loop will actually cause a deterioration in the quality of the demodulated data rather than an improvement.

The most fundamental problem in the use of the phase-locked loop is how to recognize the limitations imposed upon the loop by the modulation applied to the input of the FM channel. The phase-locked-loop parameters must be chosen for intelligence actually transmitted rather than for the intelligence of interest.

A further problem in the use of phase-locked-loops has been the application of this technique to wide-deviation FM channels. While it has been possible to build phase-locked demodulators that would accommodate ± 40 per cent deviation FM channels, it has been difficult to accommodate step-function modulation of more than a fraction of the bandwidth of these demodulators. However, it should be stated that this is not a fundamental limitation of the phase-locked loop and that a loop with the proper design characteristics can be used for demodulating ± 40 per cent channels with full-bandwidth step functions. As the loop cutoff frequency used in the phase-locked loop approaches the frequency which corresponds to the half bandwidth of the channel in question, the characteristics of the phase-locked discriminator approach those of the conventional pulse-averaging or Foster-Seeley types of demodulators.

Figure 33b shows a Bode diagram of the loop. The dotted line, which has a 6 db/octave slope and intercepts the 0-db gain curve at a frequency of K radians/sec, represents the basic integrator characteristic of the loop itself; that is, the loop is phase-locked and control of the phase of the FM oscillator applied to the phase comparator represents a 6 db/octave response with respect to frequency. The two breaks in the solid loop-gain curve shown at $1/\tau_1$ and $1/\tau_2$ represent the pole and zero introduced by the loop filter. The loop filter shown in Fig. 33a consists of resistors R_1 and R_2 and the capacitor C. τ_1 and τ_2 are defined in Fig. 33a. The Bode diagram shown is based upon the assumption that the loop is a linear feedback system. The output voltage vs. phase difference characteristic of the phase comparator is shown in Fig. 33c. This phase comparator produces a linear relationship between the output voltage e_p and the phase difference $\Delta\phi$. It should be noted that this phase characteristic is linear only over a range of phase difference of π radians.

To understand the modulation limits that the characteristics of this loop impose, it is important to obtain a mental picture of the operation of the loop. The VCO shown as part of the loop must produce a frequency ω_o which corresponds precisely to the input frequency in the static case and which deviates from the input frequency in the dynamic case by the loop frequency-response characteristics. As we apply an input signal which is frequency-modulated to the demodulator, an examination of the block diagram will show that since we expect a nearly constant deviation of the frequency ω_o as the modulation frequency is varied out to the loop cutoff frequency ω_n, the voltage input to the VCO must remain nearly constant over this range of intelligence frequency. By the same token, the voltage across the series combination of R_1 and C in the loop filter must also remain constant. It follows then that action of the feedback loop causes the output voltage of the phase comparator e_p to increase as the frequency of the applied intelligence increases. Since the phase comparator is linear only over a limited range of phase difference and it can produce only a specific maximum voltage e_p, it follows then that as the modulation amplitude or frequency increases, the phase error required may exceed the linear range of the phase comparator. When this occurs, the demodulator is said to lose lock.

It is important to recognize in examining this problem that this is a fundamental property of the loop and cannot be changed by altering the characteristics of the phase comparator or changing the value of R_2. For instance, if we decide to avoid this problem by increasing the maximum value of the voltage e_p that can be obtained from the phase comparator, we find that in so doing, we must increase the time constant τ_2 by increasing either the resistor R_2 or capacitor C, and we find that to achieve the same natural frequency in the loop, we exceed the capability of the phase comparator at the same intelligence frequency and deviation as formerly, assuming that the static gain of the loop is high (the usual case).

The various considerations that enter into the permissible modulation have been shown in detail by Martin.[2] To illustrate the modulation capability of a loop, Fig. 34 shows the limiting modulation capability of a loop in terms of its natural frequency ω_n. The particular loop chosen for the analysis has a damping factor of 0.7. In this curve, the abscissa is the modulation frequency relative to the loop natural frequency ω_n and the ordinate is the permissible deviation $\Delta\omega$ in decibels above $\Delta\omega = \omega_n$. An examination of this curve shows that the minimum permissible deviation occurs at a frequency slightly in excess of the cutoff frequency (which is equivalent to natural frequency ω_n) and that modulation capability increases either side of this limiting frequency. Below the cutoff frequency ω_n, this curve is a 6 db/octave slope which decreases with increasing frequency. Noting that the permissible deviation at the natural frequency ω_n is 4 db greater than the value of ω_n, we derive the relationship

$$\Delta\omega = 1.6\omega_n(\omega_n/\omega_m) \qquad (28)$$

where ω_m is defined as the intelligence frequency. Defining the modulation index in the usual way, we define $M = \Delta\omega/\omega_m$. From this we get

$$\Delta\omega/\omega_m = M = 1.6(\omega_n/\omega_m)^2 \qquad (29)$$

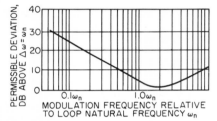

FIG. 34. Limiting modulation capability of a phase-locked loop in terms of natural frequency ω_n. Damping factor = 0.7.

It follows that the modulation index as defined in terms of the deviation $\Delta\omega$ and the intelligence frequency ω_m is proportional to the square of the ratio of the loop cutoff ω_n to the intelligence frequency ω_m. This is an important *basic* relationship used to determine the lowest loop filter cutoff that can be used. It is calculated on the basis that the maximum permissible phase difference ($\pm\pi/2$ radians) will be utilized in accommodating the channel intelligence. In practice, if there is *any noise whatever* or the static phase error is appreciable, a somewhat lesser value of intelligence frequency or deviation must be used for a given loop cutoff frequency. An attempt to force the loop to follow intelligence frequencies beyond its capability results in loss of lock and virtually complete loss of the data, as will be illustrated later in this section. Therefore it is reasonable to state that the ratio of loop cutoff ω_n to the maximum intelligence frequency ω_m must equal the square root of the modulation index M, where M is defined as the ratio of the deviation $\Delta\omega$ to the intelligence frequency ω_m.

Gruen[1] obtains the phase transfer function of the loop

$$\theta_2/\theta_1\,(j\omega) = \frac{1 + j2\zeta(\omega/\omega_n)(1 - \omega_n/2\zeta K)}{1 + j2\zeta(\omega/\omega_n) - (\omega/\omega_n)^2}$$

and inserts this expression into the equation for noise bandwidth

$$B_0 = \int_{-\infty}^{+\infty} |\theta_2/\theta_1(j\omega)|^2\,d\omega$$

to obtain

$$B_0 = \omega_n \int_{-\infty}^{+\infty} \frac{1 + 4\zeta^2(\omega/\omega_n)^2[1 - (\omega_n/2\zeta K)]^2}{1 - (2 - 4\zeta^2)(\omega/\omega_n)^2 + (\omega/\omega_n)^4}$$

For the practical case where K, the static gain, is very large, and ζ, the damping coefficient, is equal to $1/\sqrt{2}$, the integral is of the form

$$B_0 = \omega_n \int_{-\infty}^{+\infty} \frac{1 + 2x^2}{1 + x^4} dx$$

where $x = \omega/\omega_n$. The roots of the integrand lie at $\pm \sqrt{j}$ and $\pm j\sqrt{j}$, and the residues can easily be computed to be

$$\pm \frac{1 + 2j}{4j \sqrt{j}} \quad \text{and} \quad \pm \frac{2 + j}{4j \sqrt{j}}$$

The theory of complex variables then permits the evaluation of the integral

$$B_0 = \omega_n \int_{-\infty}^{+\infty} \frac{1 + 2x^2}{1 + x^4} dx = \omega_n 2\pi j \left(\frac{1 - 2j}{4j \sqrt{j}} + \frac{2 + j}{4j \sqrt{j}} \right)$$

Taking the absolute value, we obtain

$$B_0 = (3\sqrt{2}/2)\pi\omega_n$$

that is, the noise bandwidth of the loop is approximately seven times the natural frequency.

The modulation index was previously related to the cutoff frequency ω_n and the intelligence frequency by

$$M = (\omega_n/\omega_m)^2$$

From this the noise bandwidth B_0 can be related to the modulation index thus

$$B_0 = (3\pi/\sqrt{2}) \sqrt{M} \, \omega_m$$

The deviation $\Delta\omega = M\omega_m$ so that

$$B_0 = (3\pi/\sqrt{2})(\Delta\omega/\sqrt{M})$$
$$= (3\pi/2\sqrt{2})(F/\sqrt{M}) \quad (30)$$

where F = channel width = $2\Delta\omega_{max}$.

Thus when signal formats having low modulation indexes are being handled, the system noise bandwidth is determined by the bandpass filter prior to the demodulator rather than the loop noise bandwidth, and no improvement in output

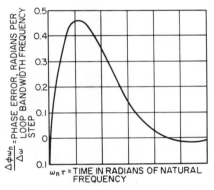

FIG. 35. Transient phase error in phase-locked loops for step frequency change. $\Delta\omega$ = step change in frequency, radians/sec; ω_n = natural frequency of phase-locked loop; $\Delta\phi$ = transient phase error, radians.

signal-to-noise ratio results from using a phase-locked loop. A noticeable threshold extension is provided by phase-locked demodulation when $B_0/F < 1$. This is so when

$$M > (3\pi/2\sqrt{2})^2$$

or when the modulation index is greater than 11.

The limitations that a phase-locked loop imposes on the step-function inputs to a data channel can be understood with reference to Fig. 35, which shows the waveform of transient phase error resulting from a frequency step function applied to the input of the phase-locked loop. The peak phase error is equal to 0.46 radian per loop bandwidth-frequency step. From this, it is readily shown that a full-bandwidth step function can be tolerated for a loop filter whose cutoff frequency corresponds to a modulation index of 1.7. This, however, provides no phase margin to accommodate static phase error or noise.

Examples

Figure 36 shows a waveform photograph that compares the output of a pulse-averaging discriminator with a phase-locked discriminator with the same input signal applied to both; the input signal to both discriminators consists of a modulated FM carrier with superimposed noise. The loop filter used in this case is the loop filter that corresponds to a modulation index of 1, and an examination of the photographs show that there is little to choose between the two types of demodulators in the case.

Figure 37 contains three photographs that compare the outputs of a pulse-averaging and a phase-locked discriminator with identical input signals applied. The loop used in this case is that which corresponds to a cutoff frequency for a modulation index of 20. Figure 37a shows the situation that prevails when the modulation is full-bandwidth at an intelligence frequency of 1 cycle. It should be noted that the

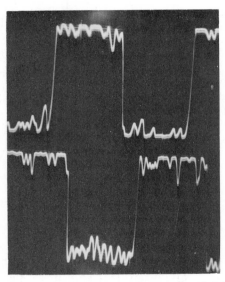

FIG. 36. Comparison of phase-locked and pulse-averaging discriminator.

distinction between these two photographs does not result from the simple filtering action of the loop since the loop cutoff frequency in both cases is far beyond the intelligence frequency applied. Another interesting thing which can be observed from this photograph is the signal-suppression effect; in the absence of noise, the waveforms at the outputs of the two discriminators would have the same amplitude. The pulse-averaging discriminator (the lower waveform in the photograph) appears to have a greatly reduced amplitude of fundamental as compared with the phase-locked loop. The reduction of signal suppression is a basic property of the phase-locked loop.

Figure 37b shows the pulse-averaging discriminator and the phase-locked loop with identical input signals consisting of a carrier modulated at approximately 8 cps with superimposed noise (superimposed noise is less than the case shown in Fig. 37a). The phase-locked loop in this instance is clearly superior to the pulse-averaging unit, but the output levels of fundamental frequency differ only slightly.

Figure 37c represents again the loop filter setup for a cutoff modulation index of 20. The intelligence frequency now is 12 cps instead of 8 and there is some noise superimposed on the input signal. It can be noted that the intelligence output of the phase-locked loop has virtually disappeared; that is, we have exceeded the modulation

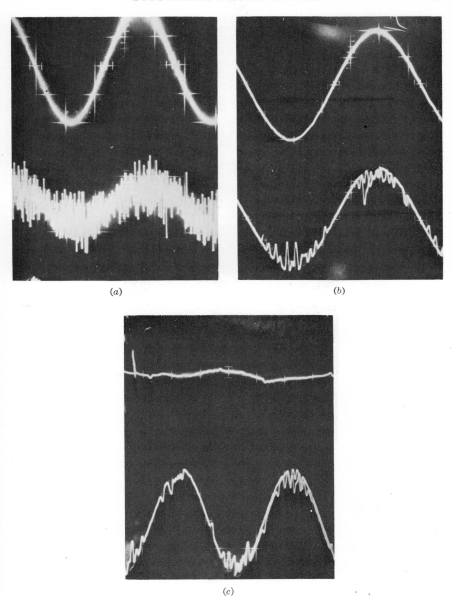

Fig. 37. (a) 22 kc ± 15 per cent discriminator comparison, intelligence 1 cps full bandwidth; $M_{co} = 20$; $M = 400$. (b) 22 kc ± 15 per cent discriminator comparison, intelligence = 8 cps full bandwidth; $M_{co} = 20$; $M = 50$. (c) 22 kc ± 15 per cent discriminator comparison, intelligence = 12 cps full bandwidth; $M_{co} = 20$; $M = 36$. M_{co} is modulation index for cutoff frequency ω_n.

capability of the loop and the phase-locked loop's output is inferior to that of the pulse-averaging discriminator. This hazard is inherent in phase-locked-loop technique since an attempt to narrow the bandwidth beyond the permissible value will cause the loop to lose lock and make the output useless.

Figure 38 illustrates the point that the phase-locked loop must be chosen to accommodate the modulation actually present in the data channel rather than the portion of the intelligence which is of interest. The topmost waveform in the figure shows the intelligence output of the phase-locked discriminator demodulating channel* A in which approximately equal amplitudes of 5-cps intelligence and 165-cps intelligence are superimposed to modulate the channel to its capacity. The phase-locked loop used in this case is set up for a modulation index of 1 and the low-pass filter used has a 600-cps cutoff. The middle waveform shown represents the output of the same

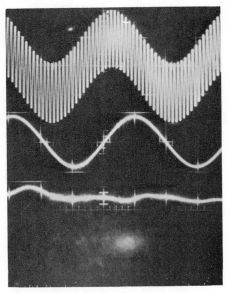

FIG. 38. Output of phase-locked discriminator with superimposed 5-cps and 165-cps intelligence for various loop filter cutoff frequencies.

channel with an 8.4-cps output filter on the basis that only the components of intelligence below 8.4 cps are of interest. The loop filter used in this case is still set up for a modulation index of 1.

Examining this situation where we are attempting to extract the portion of the data below 8.4 cps and recognizing that the channel in question is a channel* A, we find that the information-modulation index is approximately 400. Going back to our rule developed earlier that the modulation index defined in terms of the loop cutoff is the square of the modulation index defined in terms of intelligence, we conclude that a loop filter with a 165-cps cutoff can be used. This corresponds to a loop modulation index of 20. The bottom waveform in this figure shows the results of reducing the loop cutoff frequency to correspond to a modulation index of 20. It is easily seen from this waveform that the data in this channel are now essentially worthless, and it follows that the choice of loop filter made in demodulating a channel must be made not on the basis of the intelligence of interest but on the basis of the modulation actually present in the channel.

Further information on phase-locked-loop theory will be found in Chap. 9.

* IRIG proportional.

19 Output-signal Processing

The demodulator output, consisting of double-frequency subcarrier, noise, and a lesser signal must be filtered to suppress unwanted components and amplified to a usable output level. These functions are usually performed by a sophisticated filter/amplifier combination which, by means of a variety of plug-in networks, can have its high-frequency response cutoff varied over a range from a few cycles per second to approximately 100 kc. Although it would seem that separation of the filtering and amplification functions would simplify the design, a look at the component sizes in filters having low-pass cutoff frequencies corresponding to the lower IRIG bands quickly reveals why the Rauch filter[6] configuration has gained widespread acceptance. The use of an operational amplifier in conjunction with RC networks (see Fig. 39)

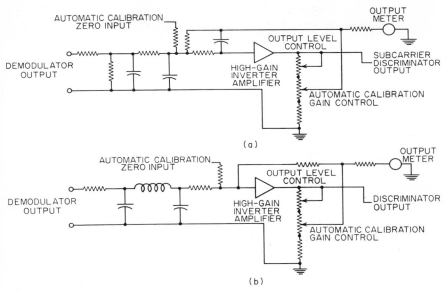

Fig. 39. Subcarrier discriminator output filter and amplifier. (a) Typical circuit configuration. (b) Equivalent circuit.

permits synthesis of a wide range of output-filter response characteristics without concern for practical inductor design problems. In practice, the majority of filter response characteristics employed are designed either for maximally flat frequency response (Butterworth) or maximally linear phase (Gaussian) characteristic. Some manufacturers offer filter characteristics which combine flat time-delay passband response with Butterworth stop-band performance. The frequency and transient response characteristics of the Butterworth and Gaussian filters are shown in Figs. 40 and 41. In selecting a linear-phase filter cutoff frequency for pulse reproduction, it is convenient to remember that the rise time of the filter is very nearly 0.35 of a cycle of the cutoff frequency. Besides forming part of the low-pass filter, the output amplifier incorporates provisions for remote zero and gain adjustments so that automatic calibration for drift and sensitivity can be included. The amplifier drives an output meter to show the position of the subcarrier in the band. This meter indication must not be affected by changes in output-level setting.

Additional design requirements are placed on this amplifier by the need for high-power outputs—typically 100 ma and 10 volts at band edge—combined with source impedance of less than 1 ohm and an ability to drive large cable capacities. The additional requirement that it operate at high-source-impedance levels in order to minimize the size of capacitors used in the low-pass filter often results in choice of a chopper-stabilized configuration to reduce drift, especially in units developed prior to the availability of field-effects transistors.

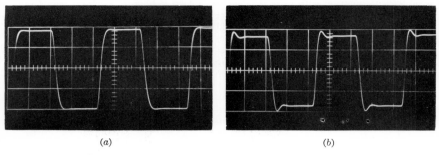

(a)   (b)

Fig. 40. Three-pole filter responses. (a) Butterworth frequency and transient response. (b) Gaussian frequency and transient response.

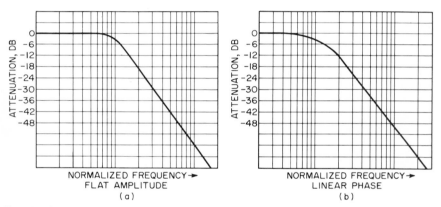

Fig. 41. (a) Amplitude-response constant-amplitude (Butterworth) low-pass filter. (b) Amplitude-response linear-phase (Gaussian) low-pass filter.

20 Tape-speed Compensation and Automatic Standardization

An important factor in the success of FM multiplex systems for data acquisition is the capability of the playback equipment for tape-speed-error compensation. The block diagram of a typical tape-speed-compensation system is shown in Fig. 42. Along with the subcarrier multiplex, a crystal-controlled reference frequency is recorded, usually above the highest subcarrier frequency. During playback, the reference frequency is demodulated and any variation of its frequency is assumed to be evidence of tape-speed fluctuation. Thus the reference-discriminator output may be used to retune the individual data discriminators, canceling the apparent modulation of a subcarrier produced by wow and flutter.

In principle, the system is simple and easily implemented, and the differences between various types of equipment used in tape-speed compensation relate more to the manufacturing tolerances held in various components than to differences in operating principle. For example, the provision of phase and amplitude controls on the individual compensation-signal-processing circuits allows the manufacturer to loosen tolerances on absolute delay in the subcarrier-tuning units, on the reference-discriminator gain, and on the compensation sensitivity of the data discriminators. On the other hand, equipment in which adjustments are unnecessary is simpler to set up and use but normally costs somewhat more to manufacture.

As shown, the compensation-signal input is applied in such a manner that the pulse-generator or phase-locked VCO is actually *retuned;* that is, the zero and slope of the data discriminator are both corrected by the compensation signal. This technique supplants earlier systems in which the compensation signals were simply subtracted from the discriminator output at the low-pass filter input, and which corrected for zero shifts but not sensitivity errors arising from tape-speed fluctuations. The

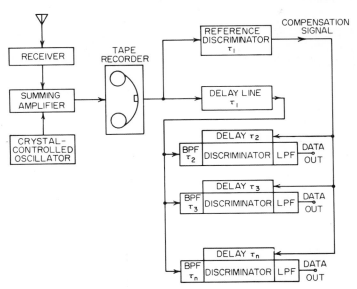

Fig. 42. Block diagram, tape-speed-compensation system.

subtractive compensation method may be analyzed by first writing the discriminator output in the presence of a tape-speed error:

$$E_o = K \frac{(1 + \epsilon)(\omega_o + \Delta\omega) - \omega_o}{\omega_o} \quad (31)$$

where E_o = discriminator output, volts
ϵ = fractional tape-speed error
$\Delta\omega$ = intelligence deviation of carrier from band center
ω_o = band center frequency
K = constant

Since there is no modulation on the reference carrier, the reference-discriminator output is given by

$$E_{or} = \epsilon K$$

Subtracting the reference-discriminator output from the data-discriminator output, we have

$$E_o - E_{or} = K(1 + \epsilon)(\Delta\omega/\omega_o) \quad (32)$$

and find that compensation is perfect at band center only. As the subcarrier moves toward either band edge, the maximum improvement attainable is a factor of 13.33 for a channel in which the deviation is ±7.5 per cent.

If, on the other hand, the compensation signal is used to retune the discriminator, the term ω_o in Eq. (31) is replaced by

$$\omega_{oc} = \omega_o(1 + \epsilon) \tag{33}$$

where ω_{oc} = effective center frequency of compensated discriminator
Making this substitution, we find that

$$E_{oc} = K \frac{(1 + \epsilon)(\omega_o + \Delta\omega) - \omega_o (1 + \epsilon)}{\omega_o(1 + \epsilon)} \tag{34}$$

where the term $(1 + \epsilon)$ can be divided from both numerator and denominator, showing that the effect of speed change ϵ has been eliminated regardless of the position of the subcarrier in the band.

Another useful function which is performed by most subcarrier discriminators consists of changing output level and zero in response to remote commands. This allows

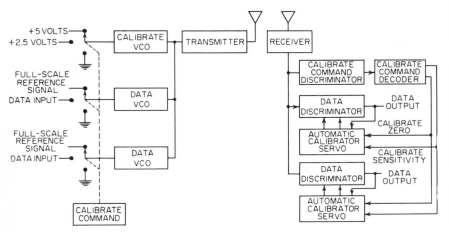

FIG. 43. Simplified block diagram—automatic calibration system.

implementation of an automatic standardization system such as that shown in Fig. 43.

Here, upon command, all data subcarrier-oscillator inputs are switched to a zero reference and time is allowed for the ground-based automatic calibration system to adjust the subcarrier discriminator outputs automatically to zero. Then the data-oscillator inputs are switched to a full-scale reference signal, following which the discriminator output levels are servo-adjusted to predetermined values. Command signals are normally transmitted via the lowest bandwidth subcarrier. The adjustment ranges usually encompass approximately ±10 per cent of full-scale output, and speed of response is largely determined by the settling time of the lowest-frequency cutoff data-discriminator filter. A complete standardization cycle can normally be completed in less than 5 sec.

While this system can reduce system gain and sensitivity errors by an order of magnitude or more, it has no effect on linearity or dynamic effects such as noise, crosstalk, and dynamic distortion. These data errors are usually comparable with or greater than drift in gain and zero, especially when higher-performance transistorized subcarrier oscillators and discriminators are used. For this reason, plus the fact that the autocalibration cycle necessarily interrupts data transmission, automatic calibration systems are not so commonly used as are tape-speed-compensation systems.

FREQUENCY TRANSLATION

When the available transmission or storage bandwidth is subdivided into a large number of extremely narrow sections, FM subcarrier multiplex systems begin to suffer in accuracy because of the small percentage deviation allocated to the higher-frequency channels, as described in Sec. 4. One method which has been explored to increase the number of narrowband channels which may be transmitted over a given link is to use "triple FM," that is, to modulate the wider-band IRIG subcarrier oscillators with groups of low-frequency narrowband subcarriers. The communication efficiency of this multiplex configuration is low, and two important alternative systems both based on AM frequency translation have been developed to provide increased numbers of channels with better overall efficiency.

The most direct and efficient of these systems, known as SS-FM telemetry, simply heterodynes the actual data signal into a new frequency band, while suppressing both the reference carrier and one sideband resulting from the translation process. This technique provides excellent bandwidth utilization and is frequently used to transmit vibration data or other signals which do not require frequency response to direct current, phase linearity, or tight time correlation.

A second method, offering frequency response to direct current and good time correlation and phase linearity, consists of heterodyning subcarriers, either singly or in groups, into new slots in the transmission spectrum. The transmitted subcarriers may thus have very small percentage deviations but may be reconverted to signals suitable for demodulation by heterodyning against a reference frequency transmitted along with the subcarrier multiplex. The bandwidth utilization of this system is intermediate between that of the triple FM system and SS-FM, and its performance under poor signal-to-noise conditions is also superior to SS-FM and somewhat poorer than triple FM.

21 System Considerations

In all systems which translate data into new spectral regions, a fundamental requirement is that a reference signal must be retained in order to perform the inverse translation. The bandwidth allocated to the reference is determined by the time-base perturbation introduced by tape-recorder flutter, Doppler effect, etc. The power allocated to the reference is determined by both the time-base perturbation and the noise power in the reference band introduced by either thermal and shot noise or noise generated by the transmission link nonlinearity acting on the signal multiplex.

In the triple FM system, the reference consists of the wideband IRIG channel carrier on which the low-frequency subcarriers are impressed as sidebands, and the low-frequency carriers are obtained by demodulating the wideband channel. Part of the reason for the greater efficiency of the SS-FM system lies in the fact that a single reference is transmitted for all data channels and its amplitude is held to a level consistent with both system accuracy and transmission efficiency.

The FM frequency-translation system transmits a common reference for all data channels in a manner similar to that employed in the SS-FM system, but a basic difference between these systems lies in the fact that the base-band signal, or data, is transmitted as sidebands about a subcarrier and errors in reference processing do not affect the frequency of the output data but only the gain and zero level of the data channel. The price paid for this reduced sensitivity to reference processing errors is, of course, reduced transmission efficiency.

Other systems considerations which apply to all the frequency-translation telemetry links are the sources of error in any frequency-division multiplex system such as translator distortion and link nonlinearity thermal noise discussed in Sec. 4.

22 Translation Techniques

An ideal frequency-translation device will permit the translation of a portion of the spectrum to some other frequency range without altering the interrelationship between the various carriers that make up the original spectrum and without generating any spurious carriers which lie in the translated spectrum. If the portion of spectrum to be translated includes a group of carriers with frequencies f_a, f_b, f_c, through f_n, and these carriers are to be translated against a reference frequency f_r, an ideal translation device will produce at its output the frequencies $(f_r + f_a)$, $(f_r - f_a)$, $(f_r + f_b)$, $(f_r - f_b)$, $(f_r + f_c)$, $(f_r - f_c)$, etc. The output of the translation device should not contain the original inputs f_a, f_b, f_c, through f_n, nor should it contain the reference frequency f_r. In addition, no components which represent intermodulation products of the original group of carriers supplied to the input should be present. That is, there should be no terms of the type $(f_r + f_a + f_b)$.

These requirements describe a four-quadrant high-frequency multiplier. In practice, making such a multiplier is a difficult circuit problem. A considerable simplification can be made if instead of using a reference frequency which is a sinusoidal carrier at the reference frequency f_r, a square wave of fundamental frequency f_r is used as the reference. If this is done, the problem of building a multiplier is simplified since a switching circuit can be used to simulate a true multiplier.

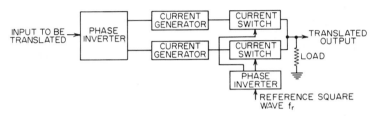

Fig. 44. Block diagram of switch-type translator.

Fig. 45. Frequency-translation spectra. (a) Spectrum prior to translation. (b) Switch-type frequency-translator output.

A block diagram of a switch-type translator is shown in Fig. 44. The group of carriers which is the input to the translation system is supplied to a phase inverter which produces two equal outputs which are opposite in phase. These two outputs are supplied to current generators which generate currents which contain the original spectrum components; these currents are then supplied in turn to a pair of current switches. The current switches are driven in phase opposition from the reference carrier f_r, and the outputs of the two current switches are superimposed to provide the translated output. This type of translator is not the only type that will produce the desired results but represents one which can be readily implemented to achieve a high level of performance.

The spectra associated with the translation system are shown in Fig. 45. Figure 45a shows the portion of spectrum to be translated prior to translation. Figure 45b shows the reference carrier f_r which is shown as a dotted spectrum line and the upper and lower sidebands that represent the translated original spectrum. In addition, another group of sidebands occurs about $3f_r$, which is the third harmonic of the reference frequency. Still another group would occur about $5f_r$, etc. Since the sidebands that lie about the odd harmonics of the reference carrier lie much farther from the desired-portion spectrum than the sideband which must be eliminated, their presence does not complicate the filter design. As an example of the level of performance readily achieved with a switching modulator, it has been found that a modulator, can be designed in which the "feedthrough" of the original input frequencies is down at least 40 db from the sidebands that are desired, where the reference carrier is down 40 db, and where the intermodulation-product sidebands between the various carriers that made up the original multiplex can be down in excess of 50 db. A circuit with this level of performance can be designed to require no adjustments or component selection and can achieve the performance specified for frequencies as high as several megacycles.

In the general case where the recording or transmission spectrum is to be utilized as fully as possible, it is desirable to single-sideband the output of the translator since the two sidebands are redundant. A filter is required to achieve this and the requirements imposed on the filter represent a fundamental limitation of frequency-translation systems. Initially, it is important to recognize the requirements imposed on such a filter. A study of the effect of interfering carriers on an FM system shows that the superposition of a single carrier on the spectrum that lies within the band of a single data channel will have its maximum effect if the spurious carrier is separated from the desired carrier by a frequency which is equal to the intelligence cutoff frequency for the channel. Under these circumstances, the interference level can be calculated approximately by simply taking the level of the interfering carrier relative to the unmodulated carrier and reducing this relative level by a factor equal to the data-channel modulation index. For instance, in an IRIG channel E in which the carrier is at center frequency (70 kc), a spurious carrier located at 80 kc would represent a band-edge carrier. If this interfering carrier were 1 per cent of the normal level of the channel E carrier, the result at the output of the discriminator would be a 10-kc spurious intelligence which had an amplitude which was 1 per cent of bandwidth peak-to-peak. If the interfering carrier had been at 72 kc in a channel E which is set up for a modulation index of 5 and the level of the spurious carrier was still 1 per cent, the resulting output beat would be 0.2 per cent of bandwidth peak-to-peak beat.

In a translation system in which the sideband to be eliminated by the filter occupies a portion of the spectrum in which it is desired to locate additional channels, it is therefore necessary to achieve between 30 and 50 db of attenuation of the undesired sideband. If this rejection of the undesired sideband is to be achieved without degradation of the performance of the system, it is safest to assume that linear-phase filters (Bessel polynomial) will be used in the process (this is not to state that a satisfactory system cannot be made with filters that are sharper than the linear-phase ones are). For the conservative case where linear-phase filters are used, it is possible to analyze this system. Defining the frequency f_h as the high-frequency end of the spectrum to be translated and f_l as the frequency which represents the low end of the spectrum to be translated, we define a frequency Δf which is the half bandwidth of the bandpass filter used in rejecting the undesired sideband. This frequency $\Delta f = (f_h - f_l)/2$. Defining another frequency difference $\Delta f'$ as the frequency difference from the filter band center to the adjacent band edge of the rejected sideband, the frequency $\Delta f' = \Delta f + 2f_l$. From these two expressions, we can obtain the ratio of $\Delta f'$ to Δf in terms of the ratio f_h/f_l. This becomes

$$\frac{\Delta f'}{\Delta f} = \frac{(f_h/f_l) + 3}{(f_h/f_l) - 1} \tag{35}$$

The dimensionless ratio $\Delta f'/\Delta f$ represents the abscissa against which normalized filter-selectivity curves are usually plotted. By referring to page 204 of Ref. 9 we find general curves for linear-phase filters plotted in terms of this ratio.

Using the data obtained from Ref. 9 and replotting them in other terms, we find in Fig. 46 the requirements for a number of filter poles plotted against the dimensionless ratio f_h/f_l. The parameters for the various curves shown are 30, 40, and 50 db of rejection of the undesired sideband in the translated system. An examination of these curves shows that a practical frequency ratio for f_h/f_l ranges approximately from 1½ to 2½ for linear-phase filters, and that the filter problem increases in severity as the ratio f_h/f_l increases.

To recapitulate the requirements on which this analysis is based, it is assumed that the portion of the spectrum rejected is to be occupied by intelligence carriers. If this is not so, the requirements for filters may be greatly reduced or in some instances eliminated. In this filter analysis, it has been assumed that there is no reference-carrier feedthrough in the output of the balanced modulator. In practice, this is usually not true; however, the reference carrier is a stable frequency. This means that if the amount of reference feedthrough is objectionable, it is a simple matter to use a "trap" to eliminate the feedthrough term; this can be done with a high-Q zero so that the effect on the phase of the multiplex will be small.

In many cases it is desired to reverse the heterodyne process used in the encoding end of the system to permit the demodulation of the data. This requires that a reference carrier for the heterodyne circuit be included in the inputs supplied to the data-playback equipment. Where a radio link is being used, it is possible to substitute a stable oscillator at the receiving end of the system for the stable oscillator used in heterodyning the inputs at the encoding end. The consequence of this approach to reconstituting the translation reference is that a difference in frequency between the record and playback references appears as a frequency error in all the translated channels. If the oscillator stability which can be achieved under the particular circumstances is such that the possible difference in frequency between the record and playback references represents an acceptable frequency error in the carriers, this is a practical method for providing a playback reference. It has the advantage that the reference does not require any of the transmitter deviation and hence does not reduce the signal-to-noise ratio of the data channels. In cases where the acceptable frequency shift of the carriers is much smaller than the tolerances that can readily be achieved by substitution of one stable oscillator for another, it is necessary to transmit the reference frequency over the r-f link used. Since this reference carrier contains no modulation, it is possible to transmit it at a vestigial level and from the output of the radio receiver to reconstitute the carrier by using a narrow-phase lock-tracking filter. This method has been used for translating an IRIG multiplex against a 200-kc reference where $6/10$ cycle variation in the 200-kc reference oscillator represents a 1 per cent error in the data transmitted on channel 1. A frequency tolerance this small is not a practical one for airborne reference oscillators; hence the reference carrier had to be transmitted.

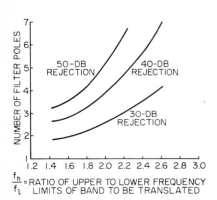

$\dfrac{f_h}{f_l}$ = RATIO OF UPPER TO LOWER FREQUENCY LIMITS OF BAND TO BE TRANSLATED

FIG. 46. Flat time-delay filters required for specified rejection of undesired sideband in frequency translation.

Where the data after translation are recorded on tape, rather than transmitted, the problem is more complex. Since the reference channel now must be used not only to transmit the basic reference but also to permit automatic readjustment of the system for variations in tape speed during recording and during playback, the bandwidth of the filter used in selecting the reference carrier from the tape must be sufficient to permit the passage of sidebands which represent the frequency modulation imposed by the tape recorder out to the highest flutter frequencies for which compensation is required in the system. This means that the reference-channel bandwidth is approximately equal to that of one of the data channels, and the reference channel then is treated as another data channel.

When the reference channel is selected out with the filter whose output is supplied to the frequency translator, the multiplexed carriers that form the input to the frequency translator must be delayed by a time which equals the time delay that the selective reference filter introduces, if the flutter of the difference frequency is to be correct. If a very narrow filter is used for selecting the translation reference, it follows that a delay line or filter with a considerable delay must then be used on the various data channels to achieve time correlation of the inputs to the translator.

Many such systems can be designed to require delay lines which are quite impractical. In some types of systems, it becomes advantageous to record the frequency-translated multiplex in a double-sideband mode rather than to use a single-sideband technique in spite of the fact that this represents wasteful utilization of the recorder spectrum available. The use of a double-sideband approach simplifies the recovery of the reference carrier since it becomes possible then to use a bandpass filter which is much wider than that required in the single-sideband case as the presence of symmetrically located sidebands eliminates the apparent frequency modulation of the reference by the data channels. In the double-sideband case, a broad bandpass filter followed by a limiter can be used to recover the reference carrier. This technique has been used in recording several IRIG multiplexes on a wideband tape recorder.

23 Applications and Typical Performance

Most frequency-translation systems are used in applications where a large number of data channels must be accommodated on a single transmission link or tape track, so

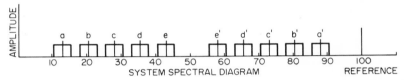

FIG. 47. Typical constant-bandwidth telemetry system employing frequency translation. Reference frequency: 100 kc

Five direct channels: 12.5 kc ± 2.0 kc; 20.0 kc ± 2.0 kc; 27.5 kc ± 2.0 kc; 35.0 kc ± 2.0 kc; 42.5 kc ± 2.0 kc.

Five translated channels: 57.5 kc ± 2.0 kc; 65.0 kc ± 2.0 kc; 72.5 kc ± 2.0 kc; 80.0 kc ± 2.0 kc; 87.5 kc ± 2.0 kc.

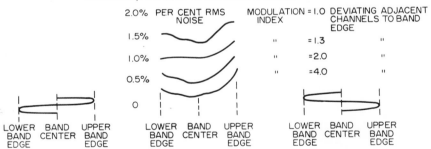

FIG. 48. Modulation index = 1.0 system. Output low-pass filter: five-pole linear phase.

that the percentage bandwidths of individual channels become small. If, for example, the percentage bandwidth of a data subcarrier lies in the range of 1 per cent or less, drift of subcarrier oscillator and demodulator becomes prohibitive and recorder flutter can drive the subcarrier out of the demodulator bandpass filter range, causing loss of data even though the discriminator section of the demodulator may be adequately tape-speed-compensated. Under these conditions, frequency translation is an indispensable tool in the hands of a system designer.

Typical performance of a system such as that specified in Fig. 47 can be ascertained from Figs. 48, 49, and 50, where measured values of adjacent-channel interference are

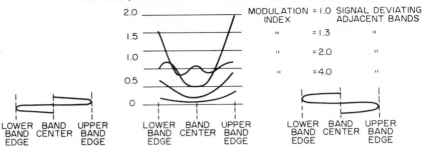

Fig. 49. Measurement of adjacent-channel interference between constant-bandwidth channels. Modulation index = 1.0 system. Output low-pass filter: five-pole constant amplitude. Guard bands = $7/8$ data band.

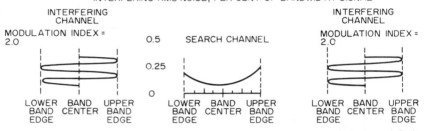

Fig. 50. Measurement of adjacent-channel interference of a constant-bandwidth modulation index = 2.0 system. Modulation index = 2.0 signals on adjacent channels. Output low-pass filter: five-pole linear phase. Guard bandwidth = $7/8$ channel bandwidth.

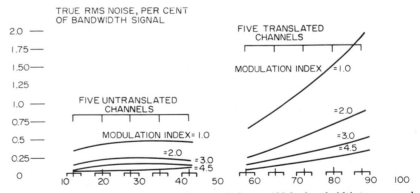

Fig. 51. Study of noise in 10 FM channels recorded on a 100-kc bandwidth tape recorder played back with tape-speed compensation.

given. Figure 48 shows the rms noise in a linear-phase channel with a modulation index = 1.0 when the adjacent channels are deviated band edge to band edge by various sine waves whose frequencies ranged from a modulation index of 1.0 to a modulation index of 4.0. It can be seen that this worst-case noise is less than 2 per cent of a full-band edge-to-band edge signal. Figure 49 shows how much the noise can be reduced by changing from a linear-phase to a constant-amplitude low-pass filter. Figure 50 gives similar data for a system with a modulation index = 2.0. In normal applications, this is the widest-bandwidth system recommended since other sources of system noise such as recorder distortion act to increase the error level somewhat above the crosstalk values given in this series of figures.

The overall performance which constant-bandwidth frequency-translation systems can achieve including tape-recorder effects is given in Fig. 51. Here a complete 10-channel multiplex with five channels translated was recorded, reproduced, and demodulated. The demodulator outputs were monitored to determine the noise in each channel as a function of its frequency and modulation index. It can be seen that for modulation indexes of 2.0 or more, rms noise levels of 1.0 per cent or better are readily attainable. Although no preemphasis was employed in the test system, it is obvious that redistribution of the subcarrier levels to correspond more closely to the recorder noise-power spectral-density curve would permit further improvement in system performance.

24 SS-FM Telemetry

In many applications where d-c response and phase linearity are not required, the SS-FM telemetry system shown in block-diagram form in Fig. 52 is utilized. This

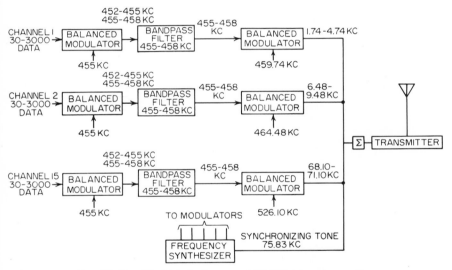

FIG. 52. Simplified block diagram of SS-FM telemetry system.

transmission technique, which omits the preliminary encoding of data channels in the form of FM subcarriers and simply translates them to new slots in the frequency spectrum, offers considerably better spectrum utilization and communication efficiency. The data channels are translated to their locations in the transmission multiplex by first converting to a 455- to 458-kc band, sharply rejecting the unwanted 452- to 455-kc sideband, and balancing the unwanted 455-kc carrier, and then performing a second conversion to the output frequency. The highly selective filters needed to suppress the unwanted sideband resulting from the first conversion are responsible for some phase distortion in the transmitted signals, but since most vibration data are subjected to spectral analysis, phase linearity is not normally of paramount concern.

In performing the inverse translation to base band at the receiving station, the pilot carrier is used to generate a series of heterodyne reference signals which are then used to convert each channel of the multiplex to a base-band signal. It should be noted that the distortion produced in FM data channels by phase nonlinearity does not appear in SS-FM since none of the channels consists of a carrier plus sidebands as is the case in FM multiplexing. By the same token, since the reference carrier transmitted with the multiplex is required to establish the frequency of the reconverted base-band signals, the time delay of reference and data channels must be carefully matched if tape-recorder flutter is not to produce dynamic output-frequency errors.

Since the FM wideband improvement factor is not obtained in SS-FM telemetry, a somewhat better received-signal strength—usually in the neighborhood of 10 db—is required for satisfactory data accuracy, and the system is somewhat more sensitive to crosstalk resulting from nonlinearities in the transmission link and tape recorder. For this reason, best results in recording an SS-FM multiplex are usually obtained using wideband FM or predetection recording techniques.

Properly implemented, SS-FM is capable of providing 15 data channels, each having response from 30 cps to 3 kc in a transmission band between 1.5 and 76 kc with 1 db channel frequency-response ripple.

REFERENCES

1. Wolf J. Gruen, Theory of AFC Synchronization *Proc. IRE*, August, 1953.
2. Benn D. Martin, Threshold Improvement in an FM Subcarrier System, *IRE Trans. Space Electron. Telemetry*, March, 1960.
3. Van der Pol, Balth., The Fundamental Principles of Frequency Modulation, *Proc. IEE*, (*London*) vol. 93, pt. III, pp. 153–158, 1946.
4. Arguimbau and Stuart, "Frequency Modulation," Chap. 5, John Wiley & Sons, Inc., New York, 1956.
5. F. E. Terman, "Radio Engineers' Handbook," pp. 645–647, McGraw-Hill Book Company, New York, 1943.
6. M. W. Nichols and L. L. Rauch, "Radio Telemetry," John Wiley & Sons, Inc., New York, 1956.
7. F. L. H. M. Stumpers, Theory of Frequency Modulation Noise, *Proc. IRE*, September, 1948.
8. W. B. Davenport, Jr., Signal-to-Noise Ratios in Band-pass Limiters, *J. Appl. Phys.*, vol. 24, June, 1953.
9. IT&T, "Reference Data for Radio Engineers," 4th ed., pp. 1118–1121, 1956.

Chapter 7

PDM TELEMETRY SYSTEMS

JOHN F. BRINSTER, *General Devices, Inc., Princeton, N.J.*

1. Introduction to PDM 7-1
2. Noise in PDM Radio Transmission 7-4
3. PDM on an FM Carrier 7-5
4. PDM-FM Bandwidth Requirements 7-6
5. Comparison with Other Multiplex Modulation Methods 7-7
6. PDM on an FM Subcarrier 7-8
7. Significant Relationships for PDM-FM 7-9
8. PDM-FM Standards 7-9

PDM EQUIPMENT

9. PDM Airborne (or Transmitting) Equipment 7-11
10. The FM Radio Link for PDM 7-22
11. PDM Ground-station Equipment 7-27

SOURCES OF ERROR IN PDM SYSTEMS

12. Impulse Noise 7-31
13. Random Noise 7-32
14. Synchronization Errors 7-32
15. Crosstalk in PDM 7-32
16. Sampling and Interpolation Errors 7-33
17. Commutator Noise 7-36
18. Other Sources of Error 7-36

STORAGE OF PDM DATA

19. PDM Recording on Magnetic Tape 7-36
20. The PDM Data Logger 7-41

1 Introduction to PDM

Pulse-duration telemetry is one of the commonest forms of telemetry used in recent years. It is generally considered to be an instrumentation method in which data to be transmitted are used to vary the time duration of a series of pulses in some predetermined manner. The pulses are then transmitted to a remote point by means of wires or by a radio link (or stored on a magnetic-tape link) and the time durations of the received pulses are measured to obtain a form of the original data. Accepted terminology for this modulation method is "PDM," an abbreviation of "pulse-duration modulation." The designation "PWM," meaning "pulse-width modulation," is an older form of terminology.

Many different forms of PDM telemetry systems are in use today in a wide variety of applications. The simplest of these is probably a single-channel commercial telemetering system for wire transmission in which the duration of the transmitted pulse is determined electromechanically by means of a rotating cam. The most complex systems, on the other hand, are multichannel radio systems involving a large number of data channels, measuring both high and low signal levels, with several main channels further subcommutated to increase the system capacity for special requirements. This

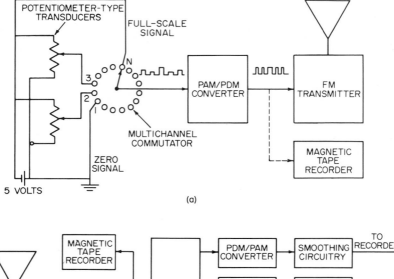

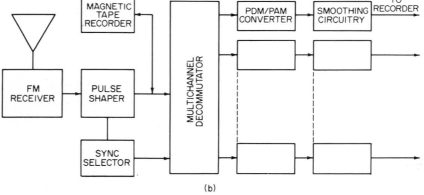

Fig. 1. Simplified form of standard PDM-FM radio-telemetry system. (a) Basic PDM-FM airborne transmitting system. (b) Basic PDM-FM ground-station receiving system.

chapter will deal primarily with practical multichannel radio telemetering using PDM techniques with emphasis on standard systems. Detailed theoretical considerations and broad data-handling techniques related to PDM telemetry are more specifically treated in other chapters of this handbook. For these and other peripheral aspects of this subject only brief references will be made to orient the reader.

Consider the simplified multichannel PDM radio-telemetering system represented by the block diagram of Fig. 1a. The transducers shown, by way of example, are of the well-known potentiometer type, described elsewhere. They are connected to a voltage source in such a manner as to produce an output signal in the range of 0 to 5 volts linearly related to their shaft positions. The voltage outputs from all the transducers are sampled in sequence by an appropriate commutator or multicontact switch producing a relatively short amplitude sample of each one. Referring to the representative waveform shown in Fig. 2a, one rotation of the commutator constitutes a single *frame*. The terms *frame rate* (in frames per second) and *sampling rate* (in samples per second per channel) are equivalent terms and are used interchangeably. It is implied that sampling rate refers to that of each channel. The term *channel rate* (in channels per second) is this rate multiplied by the number of channels. This is sometimes called the *pulse-repetition rate* (in pulses per second) or *commutation rate* (in samples per second).

The sequence of amplitude samples produces a "PAM" or "pulse-amplitude modulation" signal. (The chapter of the handbook dealing with commutation provides a detailed description of PAM signals and how they are derived for telemetry.) The

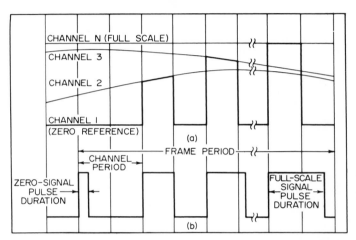

Fig. 2. Basic waveforms in PDM system. (a) PAM waveform produced by commutator. (b) PDM waveform produced by PAM-PDM converter.

duration of a sample is called the channel "on time," and the time interval from a reference point of a given sample to the corresponding point of the next sample is called the *channel period*. To obtain PDM the resulting PAM signal is commonly applied to a PDM converter which transforms the samples one by one in sequence to pulse-duration samples whose durations are proportional to their corresponding amplitude samples such as shown in Fig. 2b. In radio telemetry the PDM signal is then applied to a radio transmitter for transmission to the remote receiving station or stations. In other nonradio instrumentation applications the PDM signal is stored on magnetic tape for subsequent playback.

When the PDM signal appears at the output of the radio receiver it may be considerably deteriorated by the radio-transmission process. The signal is therefore always regenerated by means of pulse-shaping circuitry. Referring to the block diagram of Fig. 1b, it may then be stored at this point on magnetic tape or directly selected (decommutated) in real time to create the original number of independent channels in their proper sequence and converted back to a PAM form. Smoothing or integrating circuitry is then used to provide the original channel data often finally recorded on direct-writing instruments. The decommutator, like the commutator, is assumed to be a multichannel sampling device operating in both speed and phase synchronism. As will be explained below in more detail, the decommutator at the receiving end is always electronic in nature whereas the transmitting commutator is often of the mechanical motor-driven rotary-switch type.

The frame-synchronization process is accomplished by introducing a framing or master pulse in the transmitted wave train which can be distinguished electrically from the channel pulses. Channel (pulse-by-pulse) synchronization is accomplished by using the pulse-repetition rate as a timing means for the decommutator.

Although PDM may be produced in several forms, the most common form is that in which the leading edge of the PDM pulse is fixed in time and the trailing edge is varied in accord with the modulating intelligence. The instantaneous value of a sample may correspond to the time occurrence of the leading edge, the trailing edge, or some other reference point of the resultant PDM signal. This depends upon the nature of the converter employed and whether or not intermediate storage techniques are involved. Common PAM-PDM converter types are discussed below.

2 Noise in PDM Radio Transmission

Before discussing actual equipment some mention should be made of fundamental considerations which allow comparison with other possible methods of multiplex modulation. The general theory of multiplex-modulation methods and comparative analysis techniques is discussed elsewhere and is treated more fully in various references of the accompanying bibliography. A nonmathematical discussion of the most salient considerations is included here for PDM.

Assuming ideal operation of the telemetering equipment, it is first desirable to look briefly at the radio-transmission problem for PDM. The main difference between wired and radio links is that the former is linear but has limited bandwidth whereas the latter is capable of a rather large bandwidth but limited with respect to its linearity capabilities. Perhaps the most important difference, however, in determining limits of performance is that the radio link introduces well-known random noise together with the desired signal in the input of the radio receiver. This noise is particularly a function of the receiver bandwidth.

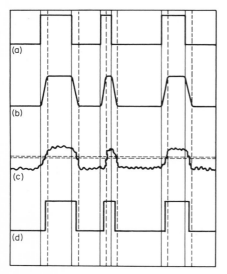

FIG. 3. PDM waveform after radio transmission. (*a*) Ideal PDM signal waveform. (*b*) Sloping PDM signal (with limited high-frequency components). (*c*) PDM signal in presence of noise (showing a 50 per cent clipping level). (*d*) PDM signal after regeneration.

Referring to Fig. 3, *a* represents an ideal PDM signal which is normally assumed in circuit discussions. *b* represents the signal after it is transmitted through a typical radio link with a relatively good signal-to-noise ratio. We assume here that the bandwidth is restricted to the extent required to minimize fluctuation noise but that it is sufficient to pass a reasonable number of the high-frequency components which make up the edges of the PDM signal. *c* represents the condition where the r-f signal-to-noise ratio is low. In this case it is quite apparent that in regeneration by a well-known slicing technique noise adds to the signal with its finite slopes contributing to timing errors and therefore to the system measuring errors. Slicing is therefore done at the 50 per cent level corresponding to the point of greatest slope. It is also apparent that if amplitude slicing is accomplished at the 50 per cent level to regenerate the original PDM signal, it is necessary for the amplitude of the demodulated video signal to be at least twice that of the noise measured in the video circuit. This threshold condition is called the *video-improvement threshold*.

To demonstrate the effect of amplitude noise on PDM consider the PDM signal whose rise time is approximately 5 per cent of the duration corresponding to full-scale modulation (typical of PDM practice after transmission). It is clear from Fig. 4 that a 10 per cent amplitude change of the leading edge would produce a change of only ½ per cent in the PDM signal. The total PDM noise is actually considered to be the square root of the sum of the squares of the noise on the two pulse edges. If the noise contributions of both edges were equal, the total noise would therefore be the value of either multiplied by $\sqrt{2}$.

It is important to differentiate between fluctuation noise (also called random or white noise) and other major types of noise interference, the most common of which is considered to be *impulse noise*. Whereas fluctuation noise has a random character, impulse noise is often manmade and consists of discrete pulses with definitive characteristics. In general, the condition for optimum system performance in the presence of one type of noise is different for the other type. To obtain optimum performance for fluctuation noise, for example, the system bandwidth is limited to the greatest possible degree consistent with the type of carrier and its modulation. On the other hand, to reduce the effects of impulse noise it is necessary to be able to distinguish the impulses from the information pulses. In PDM it is apparent that this can be done only if the shortest information pulses have greater time durations than the noise impulses. To accomplish this with increased system bandwidth the threshold signal levels are generally increased and the signal-to-noise performance for fluctuation noise is reduced. These relationships are considered below in more detail.

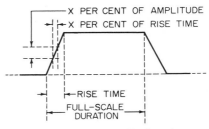

FIG. 4. The effect of amplitude noise on a PDM signal.

3 PDM on an FM Carrier

Pulse-duration modulation applied to a frequency-modulated radio link, designated by the abbreviation PDM-FM, has been evolved over the years as one of the best practical methods of multiplex transmission available in the form of standard equipment for a large number of similar channels of limited frequency response. This method not only has theoretical advantages relative to alternative methods but has certain practical equipment advantages which are also discussed in this chapter.

Frequency modulation differs in its noise characteristics from the well-known AM case in that the noise per unit bandwidth in the FM receiver output is proportional to frequency, constituting the so-called *triangular noise spectrum*. In the case of a frequency-modulated carrier it is also well known that when the signal strength falls below a certain level, the signal-to-noise ratio of the output of the system deteriorates rapidly. This point is defined as the *FM carrier-improvement threshold* and occurs when the amplitude of the carrier is approximately equal to the noise peaks (peak value is generally considered to be approximately four times the rms value of such noise). The signal-to-noise ratio at the output of the receiver for a given carrier signal-to-noise ratio is a function of the FM *deviation ratio*. This is defined as the ratio of the frequency deviation for full modulation and the maximum modulating frequency (which, for analysis purposes, may be termed *video bandwidth*). Where the deviation ratio is greater than unity the signal-to-noise ratio above threshold is considerably greater than for an equivalent amplitude-modulation system. Below threshold, on the other hand, the signal-to-noise ratio may be considerably worse for frequency modulation since the FM carrier occupies a somewhat greater bandwidth to transmit the same video signal. Figure 5 indicates the nature of the FM threshold effect compared with amplitude modulation for fluctuation noise. Several important points not apparent from these curves should be noted. The threshold level is higher for higher deviation ratios where wider bandwidths are involved. The carrier threshold is actually proportional to the square root of the bandwidth. This curve for rms noise in the output of the receiver shows a relatively slow drop-off. In contrast, the signal-to-peak-noise ratio in the output of the receiver drops off somewhat sooner and much more sharply as the carrier is reduced. This difference is said to become more pronounced as the deviation ratio is reduced.

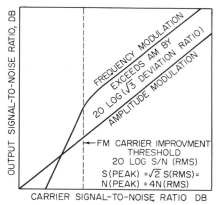

Fig. 5. Frequency-modulation threshold.

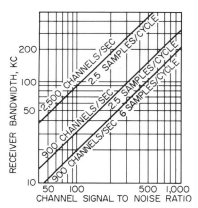

Fig. 6. Threshold bandwidth requirements for PDM-FM.

4 PDM-FM Bandwidth Requirements

The PAM to PDM converter of the system provides a keying signal for the frequency-modulated transmitter. The frequency may be assumed to shift first in one direction and then in the other from some nominal center frequency. This shift is the FM deviation of the carrier. If the bandwidth is greater than twice the frequency deviation, the output of the FM receiver has about the same character as would an equivalent AM receiver. As the bandwidth is compressed and becomes closer to twice the deviation, the pulse wave train at the FM receiver output exhibits the well-known ringing effect which may reduce the PDM rise time and increase crosstalk effects between one channel and the next. It is therefore desirable to restrict the deviation to a value somewhat lower than half the bandwidth.

Video bandwidth requirements (determined by the channel rate for a given crosstalk condition) demand a greater FM deviation for higher S/N ratios. An analysis of PDM-FM therefore shows a direct relationship between the channel signal-to-noise ratio attainable after decommutation and detection under threshold conditions and the r-f bandwidth required. Figure 6 shows this bandwidth relationship for two channel rates and channel filter types. Two curves are shown for the common channel rate of 900 channels per second where 6 samples and 2.5 samples are used per cycle of highest intelligence frequency in the channel. Another curve involves 2,500 channels per second using 2.5 samples per cycle. The theoretical bandwidth required at threshold is proportional to the number of channels per second but inversely proportional to the square root of the number of samples per cycle.

Several papers in recent years have suggested the possibilities of achieving S/N performance in systems of the PDM-FM type by further video bandwidth conservation. For the upper region of practical channel signal-to-noise ratios, one method given was that of narrowing the video bandwidth to the limit and using wider carrier deviations. This technique decreases the slopes until the PDM modulation range is severely restricted by crosstalk specifications, claiming that the video noise is reduced more rapidly.

As already defined above, there are two thresholds in a PDM-FM system, namely,

COMPARISON WITH OTHER MULTIPLEX MODULATION METHODS   7-7

the *carrier threshold* and the *video threshold*. Both thresholds are proportional to the square root of the video bandwidth. The video threshold, however, is also inversely proportional to the deviation ratio. If the optimum condition for operation of a system is determined to be that at which the two thresholds occur at the same time, it is possible to adjust the deviation ratio accordingly. As an example, the deviation ratio necessary to provide this condition is somewhat greater than one-half for an ideal PDM-FM system where the r-f bandwidth is considered to be approximately twice the video bandwidth and approximately 3.5 times the deviation. For deviation ratios greater than this value, the carrier threshold is higher than the video threshold. This means that the threshold criterion would theoretically be met for the video signal for a longer time than for the carrier, as the carrier signal strength is decreased.

It can be shown that, for a given r-f bandwidth and for a given crosstalk figure, there is a value of video bandwidth which will maximize the theoretical performance for PDM-FM. This results from the ability to optimize the balance between FM noise and video noise improvement. Maximizing the wideband gain provides an optimum value for full-scale PDM (maximum duty cycle). For this duty-cycle value one may then determine the video bandwidth required for a given crosstalk condition. Reference should be made to a later section on crosstalk and particularly to Fig. 31 showing crosstalk values for various video bandwidths and channel rates.

5 Comparison with Other Multiplex Modulation Methods

In the past several years analytical methods for multiplex communication have been applied to radio telemetry to give the designer a basis for optimizing his data-transmission systems. The problems in telemetry, however, are sufficiently different from the problems in voice communication to demand special consideration. Particularly, the needs for high accuracy in the transmission process and for d-c response in the data channels become major points of difference.

The primary interests, in comparing multiplex radio-transmission methods of this type, lie in those factors which determine the minimum transmitted power required to provide the desired signal-to-noise ratio in the data channels at the output of the receiving equipment. (For a valid comparisons in most vehicular applications, power must ultimately be computed in terms of the total power required at the transmitting end.) It is also sufficient for comparison purposes to deal with the relative carrier signal strength or carrier power at the input of the receiving system. The threshold signal level is generally determined for each modulation method on a relative basis for the required signal-to noise ratio in the channel output. The relative power required is the square of the threshold-signal voltage ratio.

The PDM-FM system is one of the many so-called *wideband modulation systems*. Its performance relative to the well-known AM system is conveniently expressed by a ratio called the *wideband gain* of that system. This gain is defined as the ratio of the output signal-to-noise ratio of the wideband system and the output signal-to-noise ratio of the equivalent AM reference system, both operating above threshold with the same carreir power at the receiver input. This concept plays an important role in the understanding and comparison of modulation methods.

Two other wideband FM multiplex systems of practical interest in modern radio telemetry are the FM-FM subcarrier systems and the PCM-FM digital systems described elsewhere. The FM-FM system is a frequency-division multiplex system in which a group of frequency-modulated subcarrier oscillators are applied to a frequency-modulated carrier. The PCM system is a digital system in which the data are coded into pulse groups after commutation and applied to an FM carrier.

Figure 7 shows the result of one theoretical comparison of PDM-FM made with particular forms of these two systems assuming similar characteristics. Relative power is given as a function of the per cent of noise to be expected in channel output relative to the data signal. Each system here was assumed to have 10 channels with 100 cps response. These curves show that the hypothetical FM-FM system requires about two times the average transmitted power to achieve a 2 per cent noise value at threshold. On the other hand, the ideal PCM system of comparable intelligence capacity with an FM carrier requires approximately one-half the average power needed for the PDM-FM system to achieve the same signal-to-noise performance. There are many variations possible in these relationships taking into account all practical factors, component limitations, and special techniques applicable to a particular modulation method when considered in terms of physical equipment. For example, it may be more advantageous to use two or three times the transmitted power in a well-designed compact PDM system than to use a much more complex, larger, and heavier type of PCM equipment requiring more total power. For information relating to a more detailed comparison, the reader is referred to Chapters 2 and 9 of this handbook.

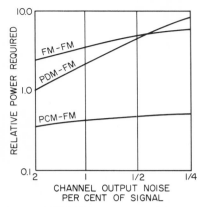

Fig. 7. PDM-FM compared with other wideband FM systems.

Some mention should be made of the phase-modulated carrier for PDM transmission. Since the equivalent frequency modulation of a phase-modulated transmitter is the rate of change of phase, the FM receiver output is a differentiated form of the modulating PDM waveform. PM improvement is somewhat less than FM improvement resulting in a higher required threshold power for PDM-PM. In addition, there are certain practical considerations which favor use of an FM transmitter relating to the detection of the differentiated waveform in the presence of impulse noise encountered in practice.

6 PDM on an FM Subcarrier

PDM has often been used in place of PAM as a means for increasing the capacity of the standard FM-FM system when many like channels of low-frequency response are required. This constitutes a form of a hybrid triple modulation system called PDM-FM-FM, although the latter designation more accurately applies to the case where only one subcarrier is employed and is modulated by the PDM time-division system.

Some improvements over the corresponding PAM techniques have been reported based on an analysis comparing the two modulation methods in practical detail.

Table 1. PDM Values for Modulating FM Subcarriers

IRIG band	SCO center frequency, kc	FM deviation, %	PDM channel rate	PDM zero and full scale, μsec	Channel filter cutoff, kc
10	5.4	±7½	55	2,200–11,000	0.40
11	7.35	±7½	75	1,600– 8,000	0.55
12	10.5	+7½	112	1,050– 5,500	0.79
13	14.5	±7½	150	800– 4,000	1.1
14	22.0	±7½	225	530– 2,600	1.9
15	30	±7½	300	380– 1,900	2.25
16	40	±7½	400	285– 1,450	3.0
17	52.5	±7½	500	240– 1,200	3.9
18	70	±7½	550	220– 1,100	5.25
A	22.0	±15	450	260– 1,300	3.8
B	30	±15	550	210– 1,050	4.5
C	40	±15	700	165– 840	6.0
D	52.5	±15	800	140– 730	7.9
E	70	±15	900	130– 660	10.5

PDM-FM STANDARDS   7-9

Advantages claimed are based on an improvement factor achieved by working in time rather than in amplitude and on practical advantages in equipment design considering crosstalk, hum, distortion, drift, etc.

The bandwidths available for this purpose are quite obviously limited. Some recommended values of pulse-width ranges for standard IRIG subcarrier bands are shown in Table 1 together with suggested cutoff values for Gaussian-type low-pass channel output filters. In other respects, PDM equipment for subcarrier modulation is analogous to that used for the transmission of PDM on the main carrier.

It is apparent from the noise considerations that when only one PDM wave train is present in the system and subcarriers are needed for no other purpose, it is better to apply the PDM signal directly to the transmitter rather than to employ an intermediate subcarrier. The use of the subcarrier merely enlarges the bandwidth requirements unnecessarily, increasing the power requirement by several times to achieve the same performance.

7 Significant Relationships for PDM-FM

This section is primarily intended to provide information of a general and practical nature rather than to supply derivations of mathematical relationships among the various system parameters. The relationships shown in Table 2, however (written out because applicable symbols are not completely standardized), are fundamental in analyzing PDM-FM and comparing it with other multiplex modulation methods.

8 PDM-FM Standards

Before discussing existing types of PDM equipment, it is of interest to review some of the technical standards set forth in the industry. The application of PDM-FM equipment has been strongly influenced by standards developed some years ago based largely upon mechanical commutation limitations. It is reasonable to assume that PDM systems of the future will be extended to include much higher information capacity with many improvements in performance. The following is a condensed version of PDM-FM standards.

For greater detail the reader is referred to Chapter 5 of this handbook which contains the most recent industry standards for PDM-FM systems.

Pulse-duration-modulation Specifications*

The following are the specifications for the pulse-duration-modulated signal:

Number of samples/frame*............	30	45	60	90
Frame rate, frames/sec.................	30	20	15	10
Commutation rate (samples/sec)†.......	900	900	900	900

* The number of samples per frame available to carry information is two less than the number indicated because the equivalent of two samples is used in generating the frame-synchronizing pulse.

† Commutation rate is equal to frame rate multiplied by the number of samples per frame.

The commutator speed (or frame rate) shall not vary more than plus 5 to minus 15 per cent from nominal.

Frame synchronization of the receiving station shall be provided for by leaving a longer than normal gap time in the train of pulses transmitted.

This gap shall be the same as that normally occupied by two successive data channel pulses. A representation of the pulse-train waveform is shown in Fig. 8.

The information being transmitted in each channel shall determine the duration of

* These systems have sometimes been designated as PWM/FM.

Table 2. PDM-FM Relationships

1. Deviation ratio $= \dfrac{\text{carrier deviation}}{\text{video bandwidth}}$

2. Carrier bandwidth = const (video bandwidth)

   The constant factor involved here is often considered to be at least 2 in order to achieve an adequate step-function response. It is always assumed that the video low-pass filter at the receiver discriminator output has a cutoff value equal to the video bandwidth

3. Carrier bandwidth = const (carrier deviation)

   When the constant value is reduced to a value approaching 2, the step function causes an increased ringing effect (resulting in a frequency equal to the carrier deviation). Although the video filter serves to reduce the ringing effects, practical systems employ a somewhat larger bandwidth

4. Channel bandwidth $= \dfrac{\text{sampling rate}}{\text{number of samples per data cycle}}$

   The number of samples per data cycle which has the minimum theoretical value of 2 is frequently given the value of 2.5. Practical systems, however, often involve 6 or more samples per data cycle. The term "channel bandwidth" is synonymous with channel frequency response

5. Wideband gain $= \dfrac{\text{channel } S/N \text{ ratio}}{S/N \text{ ratio of equivalent AM radio link}}$

   Wideband gain is a convenient term used in multiplex theory for quantitatively comparing one modulation method with another. This is done by relating the operation of one of its channels to an equivalent single-channel AM system of well-known character operating above threshold, both having the same carrier power and noise per unit bandwidth in their carrier channels

6. Wideband gain for PDM-FM

   $= \dfrac{\text{const (carrier deviation)(max channel duty cycle)}}{\text{(number of channels)(video bandwidth)}^{1/2}\text{(sampling rate)}^{1/2}}$

   $= \dfrac{\text{const (video bandwidth)}^{1/2}\text{(max channel duty cycle)}}{\text{(number of channels)(sampling rate)}^{1/2}}$

   Wideband gain is proportional to the square root of the video bandwidth for a constant frequency deviation. The carrier threshold is also proportional to the square root of the video bandwidth. The wideband gain therefore cannot be increased without a proportionate increase in the carrier threshold level

7. PDM-FM channel signal-to-noise ratio at threshold

   $= \dfrac{\text{const (video bandwidth)(max channel duty cycle)}}{\text{(number of channels)(sampling rate)}^{1/2}\text{(channel bandwidth)}^{1/2}}$

   $= \dfrac{\text{const (carrier deviation)(number of samples per cycle)}^{1/2}\text{(max channel duty cycle)}}{\text{(number of channels)(sampling rate)}}$

   The channel signal-to-noise ratio is obtained by multiplying the wideband gain by the S/N ratio of the AM reference system. This ratio at threshold is obtained by using the threshold signal value for the system which is proportional to the square root of video bandwidth

   The channel signal-to-noise ratio at threshold is proportional to the carrier deviation employed and inversely proportional to the commutation rate (see Fig. 6)

   The signal-to-noise ratio is directly proportional to the carrier amplitude above the threshold value. Note that for a given signal-to-noise ratio the threshold carrier power is proportional to the number of channels

---

the corresponding pulses. The relation between information magnitude and pulse duration should, in general, be linear.

Min pulse duration (zero-level information), μsec............ 90 ± 30
Max pulse duration (max level information), μsec............ 700 ± 50
Pulse rise nad decay time (measured between 10 and 90 % levels),
  μsec.. 10–20
  (constant to 3 μsec for a given transmitting set)

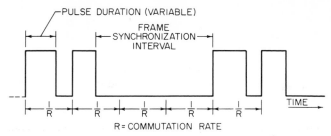

Fig. 8. Standard IRIG format and framing method for PDM.

The time interval between the leading edges of successive pulses within a frame shall be uniform from interval to interval and shall be constant within plus or minus 25 μsec.* This time interval shall have a nominal period equal to 1 divided by the total sampling rate.

Radio-frequency Carrier Specifications

The following are the specifications for the radio-frequency carrier and its modulation:
Radio Frequency. 216 to 235 Mc/sec. The specific frequency assignment in this band shall be obtained by negotiation with the pertinent test range. Because up to approximately 0.2 Mc/sec bandwidth may be occupied by a PDM-FM telemetry signal, appropriate spacing between adjacent r-f channels must be provided. The extent of the guard bands required will be determined by the operating conditions existing at the individual test ranges.
Stability. The r-f carrier frequency shall be stable to within plus or minus 0.01 per cent.
Radio-frequency Deviation. 25 to 45 kc/sec.
Power. 100 watts maximum, depending on distance and propagation problems; no more should be used than that necessary to achieve reliable transmission.
Spurious-signal Radiation. The radiated power of harmonics and all other spurious signals shall be 60 db or greater below the power level of the fundamental.
Polarization. If a circularly polarized transmitting antenna is used on a missile or aircraft, the signal transmitted to the rear, or downward, shall be right-hand polarized (by IRE† definition). Any use of circularly polarized transmitting antennas should be coordinated in advance with the test range involved in order to ensure compatibility with receiving antennas.

PDM EQUIPMENT

9 PDM Airborne (or Transmitting) Equipment

That portion of radio-telemetry equipment which collects and transmits the data has come to be known as the *airborne equipment* since it is most often installed in instrumented flight vehicles. More generally it is called the *transmitting equipment* since it is installed at the transmitting end of the remote measuring system. In contrast, the equipment used at the receiving end (the r-f receiver, decommutation equipment, etc.) is often located at a fixed point on the ground and is therefore commonly termed the *ground station*. The ground-station equipment as a whole is alternatively termed the *receiving equipment*.

* In practice, channel period scatter is allowed to exceed 150 μsec since this does not affect measuring accuracy and is more typical of electromechanical equipment.
† Now IEEE.

The PDM Converter

A typical PDM system was shown in Fig. 1 in block-diagram form. This figure included both transmitting and receiving equipment. The portion of most interest in our discussion of airborne equipment is the PAM-PDM converter, which is also called the *keyer*. The term "keyer" used in the radio-telemetry field is a carry-over from the concept of frequency-shift keying used in the radio-communication field. Equipment of this type involved a manual or automatic keying switch to shift the frequency of the carrier from one fixed value to another.

The early *converter* was ordinarily a multivibrator-type circuit with various degrees of complexity. In one form the circuit is triggered by a so-called timing pulse just after the channel amplitude sample has been applied to the particular circuit element whose potential controls its period. The interval between the time of triggering and the time of reverting back to its original state is therefore a function of the channel signal amplitude. The converter design and its timing are dependent upon the type of commutation device employed. Newer more stable designs are based on the principle of a ramp level comparator. The following discussion on converter types treats the application of both mechanical and electronic commutation devices to PDM.

The PDM waveform produced by the converter should ideally have uniform periods. This means that the time interval from one leading edge to the next is always the same. The difference between extremes is called *channel period scatter*.

Similarly the pulse width corresponding to a given converter input signal may vary between two extremes. This spread is termed *pulse-width scatter*. The term *pulse-width jitter* is also sometimes used but refers more properly to a description of the phenomenon than to its measurement.

In general, channel period scatter results from nonuniformity of time spacing in the pulse train which initiates the PAM-PDM conversion. Pulse-width scatter, on the other hand, is generally a property of the converter itself, assuming a stable PAM input signal.

PDM Converters with Electronic Commutation

One of the most successful converters used for telemetry today is the all-silicon solid-state type used in conjunction with electronic commutation. This type of equipment has obvious advantages of long life at high sampling rates, low power requirements, and high reliability. It is not only well suited for aircraft telemetry and ground equipment in which long maintenance-free performance is desired but also for instrumentation requirements in small space vehicles in which low power and weight are desired, retaining the feature of high reliability.

The features attainable with new electronic commutation techniques have been described in some detail in Chap. 4. It particularly compared electronic and mechanical commutation. Among the advantages of electronic commutation for PDM equipment is the possibility of equipment packaging in almost any physical shape and in any practical number of subassemblies. Improved insensitivity to shock and vibration has also made it possible to fabricate solid-state equipment having high reliability under extreme environments.

System-synchronization problems are somewhat reduced where electronic commutation is employed since the frame rate and the channel period are kept more uniform than is possible with mechanical devices under varying conditions generally encountered. The problem of channel period scatter primarily caused by edge noise effects of the commutator timing mechanism has been substantially eliminated. The pulse-width scatter at the output of the converter is largely a function of the circuit design. Equipment is now available with uniformity far below the tolerable errors for most requirements.

As discussed more fully in the section on commutation, electronic commutation has been made practical for both high- and low-level signals. The high-level switching system employs diode or transistor gates whereas the most successful low-level solid-state switches employ a combination of matched transistors in a special double-pole differential gating arrangement. The low-level electronic switch is followed by a specially designed low-level amplifier, the latest type of which is completely solid-state in design. Signal levels produced by the commutator or by the commutator amplifier combination are adjusted to operate the converters with a signal varying over the range of 0 and 5 volts. The amplifier gain is generally adjusted to provide a compatible output level in this range regardless of the input-level range.

PDM Converters with Mechanical Commutation

Historically the first PDM equipment for telemetry used primitive mechanical commutation having service-free operation for only a matter of minutes. In the past decade the technology has been constantly improved to a point where a few thousand service-free hours are attainable from well-constructed commutators.

The reader is again referred to the section on mechanical commutation for a more complete treatment of commutator characteristics which influence their use in PDM equipment. It may be sufficient here to be aware of several major characteristics which affect converter designs. These may be summarized as follows:

1. Mechanical commutation methods, particularly at the higher switching rates, produce amplitude samples whose leading and trailing edges may not be accurately defined (or stabilized) in time. These edge variations are generally termed "edge noise."

2. The contact resistance may vary appreciably over the "on time" for a particular sample. This is often called the "on-time noise" but has no relation to "generated noise" which may be troublesome only at extremely low signal levels. "On-time noise" may include an effect called "contact bounce" which is a condition of infinite contact resistance at one or more points across the "on time."

3. The duty-cycle requirements of the commutator must be related to the converter design since in some instances the channel "on time" must be as great as the maximum range of PDM. A memory circuit is often used in place of a long duty cycle directly out of the commutator.

4. Timing variations due to motor-speed changes or irregularly spaced contacts influence the design of the timing-signal generator and synchronization with the sampling process.

Modern mechanical commutators have not only provided long service-free life and reliability but also much improved characteristics for PDM. Timing accuracy and speed control have been among the most important improvements. The latter is the result of much improved mechanical drive systems such as the speed-sensing regulator described in the section on commutation of Chap. 4.

Timing Signals

Regardless of the converter design it must function in synchronism with the commutator. Where an electronic commutator is used preceding the converter, the timing arrangements consist of relatively simple electronic circuits.

Refer to Fig. 9 showing the basic waveform timing for the PDM converter. *a* shows a 100 per cent duty cycle typical output from the electronic commutator. *b* shows the type of PAM output obtained from a mechanical commutator. The timing pulse shown in *c*, derived from the clock or from the mechanical commutator, is used to initiate the PDM operation. It is generally delayed as shown relative to the leading edges of the PAM signal to the extent required to avoid edge effects.

Timing arrangements for a mechanically commutated system are derived from either an auxiliary magnetic trigger wheel or an auxiliary set of contacts on the commutator switch. Both methods produce a series of trigger pulses which are arranged to occur just after the region of leading-edge noise on the channel sample. Representative waveforms for common timing methods are shown in Fig. 10. *a* shows typical magnetic trigger pulses and *b* those following amplification and shaping. A technique of double differentiation is often used here. *c* shows a typical waveform as produced by a contact pole of the switch. *d* is again the final converter timing pulse

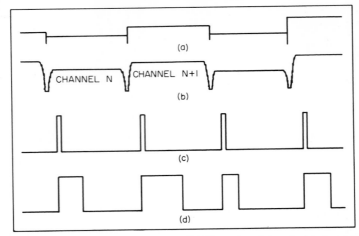

FIG. 9. Typical converter waveforms. (a) PAM produced by electronic commutator. (b) PAM (with edge noise) produced by mechanical commutator. (c) Trigger or timing pulses applied to converter. (d) Output of PDM converter.

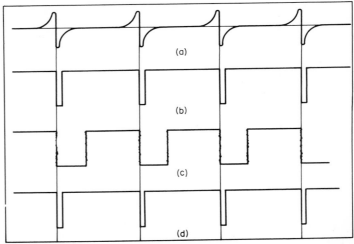

FIG. 10. Converter timing methods. (a) Waveform out of magnetic pickup. (b) Negative trigger pulses after amplification and shaping. (c) Timing pulses output after integration and slicing. (d) Negative trigger pulses after differentation and clipping.

after integration and shaping. The waveform does not show the factors affecting timing as discussed above.

The magnetic trigger wheel can be used only where sufficient peripheral speed is involved. A ring of magnetic material with small projections (or slots) is attached to the rotor so that a changing magnetic field induces a voltage pulse for each projection in an associated miniature magnetic pickup. One pulse is usually produced for each active channel of the system. The pulses are then shaped and applied to the converter. In practice, it is often possible to utilize a simple trigger of this type in all the standard configurations. If, however, the mechanical commutator is expected to be subjected to accelerational vibration in excess of 10 to 15 g's, it is advisable to limit the use of the magnetic pickup to the very high sampling rates. This limitation arises because the pickup voltage is proportional to the sampling rate in its amplitude whereas the unwanted voltage induced as a result of vibration remains constant for a particular vibration frequency and displacement.

Where a contact trigger is used, an additional pole of the commutator is generally connected to provide the required number of timing pulses. Generally because of edge noise related to the make and break of contact, it is necessary to provide additional circuitry capable of producing a noise-free timing pulse. In practice this is done by integrating the signal as produced by the mechanical commutator brush, essentially removing the high-frequency noise components, and then slicing the resultant waveform to provide reasonably jitter-free pulses, one for each information channel. It should be noted that channel period scatter (variation) does not directly affect accuracy in the PDM system.

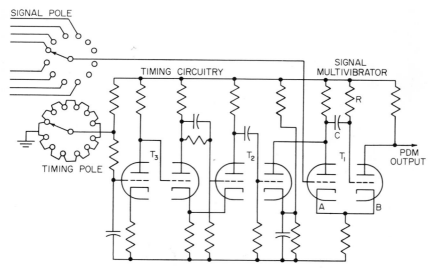

Fig. 11. Schematic of simple PDM converter.

Converter Circuitry

The converter or keyer accepts the PAM samples from the commutator and produces a pulse corresponding to each applied sample whose duration is proportional to the amplitude. A simplified circuit for a relatively simple converter of very early design employing contact triggering is shown in Fig. 11. Circuit operation is as follows: The output of the sampling switch is applied to the grid of T 1 A. A timing pulse, which is phased to occur approximately 0.1 msec later, is applied to grid T 1 B through a buffer stage consisting of T 2. T 1 B is normally turned on since its grid is returned to the positive plate supply through a high resistance. A negative pulse from the timing circuit turns off T 1 B, which drops the potential on the cathode of T 1 A, allowing the latter to conduct. The signal on the grid of T 1 A therefore determines the potential drop across its plate load resistance. This potential drop for the full range of input signal is always low enough to maintain cutoff for T 1 B, until the capacitor c discharges sufficiently through the resistance R to raise the grid of T 1 B above its cutoff voltage. As soon as this point is reached, T 1 A is cut off again in the regenerative operation of the multivibrator, reestablishing a stable condition. It is clear that the duration over which T 1 B will remain off is proportional to the magni-

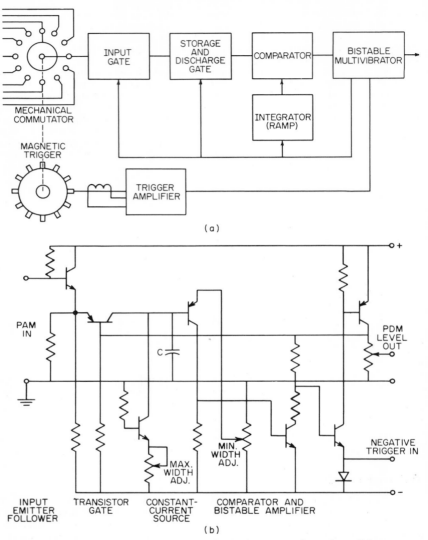

Fig. 12. PAM-PDM converter for electromechanical commutation using solid-state components. (a) Block diagram. (b) Simplified schematic diagram.

tude of the signal applied to T 1 A. The linearity of conversion from amplitude to duration is quite high since only a small portion of the normal RC curve is actually used.

As mentioned above, the mechanical commutator, when not operating properly, produces an effect of increasing dynamic contact resistance. If one were to examine the "on" time of a poor commutator output signal in some detail, one might find large negative pulses corresponding to periodic increases in contact resistance.

If this increase is significant to the total resistance of the commutator circuit, it is obvious that erroneous operation will result. To eliminate or to reduce this effect in early equipment the converter circuitry was equipped with a storage device which in effect remembers the level of lowest contact resistance for each amplitude sample.

It is possible to improve the PDM converter design by using a simple comparator and sawtooth ramp providing greater stability and linearity. This more complex circuit form becomes quite practical when fabricated with solid-state elements. A block diagram of this type of converter is shown in Fig. 12a. Operation is initiated when trigger pulses from the pickup on the motor-driven commutator actuate the bistable multivibrator which in turn actuates the integrator (or ramp generator) and the input gate. When the amplitude of the ramp reaches the amplitude of the stored sample, as detected by the comparator, the multivibrator is reset to its original state. The time difference is therefore a direct measure of the amplitude of the applied signal sample.

Recent designs, employing somewhat more simplified circuitry retain most of the advantages of this form of converter. They are compact and lightweight and use extremely small amounts of power. Advances in commutator design obviate the need for a magnetic trigger. Figure 12b is a simplified schematic diagram of a solid-state PAM-to-PDM converter. The analog PAM signals are applied to the input emitter follower charging capacitor C through the transistor gate. After appropriate stabilization of the capacitor voltage, a trigger or timing signal is applied as shown; this opens the gate and allows the capacitor to discharge linearly through the constant-current circuit. Discharge takes place to a predetermined level, at which time the comparator amplifier produces a signal returning the output to its starting level. The duration between the occurrence of the trigger and the operation of the comparator is clearly a function of the capacitor voltage and hence the input signal level. The resultant PDM signal may be adjusted in its extremes of duration and amplitude as indicated. Amplitude limiting of the input signal, usually required, is not shown here.

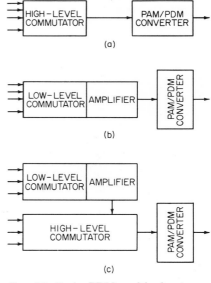

Fig. 13. Basic PDM multicoder types. (a) High-level multicoder. (b) Low-level multicoder. (c) Mixed high- and low-level multicoder.

PDM Multicoder Types

The term PDM multicoder is commonly used to describe the combination of commutator and converter. It is used loosely to describe the complete PDM airborne package including inverters and voltage-regulating circuitry. The following is a brief discussion of the various types of multicoders available today. Reference should be made to Fig. 13, indicating their major differences in block-diagram form.

High-level PDM Multicoders

Although electronic commutation has occasionally replaced mechanical commutation for high-level (0- to 5-volt signals) sampling for PDM, mention should be made of simple and inexpensive forms of electromechanical PDM multicoders. The simplest multicoder of this type, as indicated above, was originally made up of a mechanical commutator with a single information pole and a suitable timing mechanism packaged together with a simple tube-type converter to produce the PDM signal. Although this combination has been used extensively in past years, problems of unreliability, instability, and service life invariably encountered in its use have contributed widely to reduced confidence in PDM-type equipment. In applications for missile instrumentation involving short operating periods this type equipment has been more successful than in aircraft applications where operation is required for relatively long periods of time on a repetitive basis.

The section of Chap. 4 of the handbook dealing with commutation has described some of the specific commutator improvements which have been realized in recent years. Mechanical commutators for this purpose now have few limitations. In addition, certain features in the electronic circuitry described above have made it possible to employ electromechanical commutators successfully even after the commutator performance has appreciably deteriorated. Figure 14a shows a typical modern PDM electromechanical multicoder for high-level operation. This is the so called plug-in design intended for mounting from below. The unit measures approximately 3 by 1¾ by 1½ in. and has a capacity of up to 45 nonshorting channels. Output format is according to IRIG standards. Life expectancy exceeds 1,000 hr.

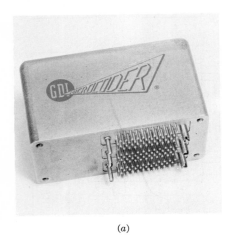

(a)   (b)

Fig. 14. Typical high-level PDM multicoders. (a) Electromechanical-type multicoder. (b) Solid-state-electronic-type multicoder.

High-level all-solid-state multicoders of modern design are also much improved in life and reliability. The principal problems encountered in this type of product have been in fabrication of smaller and smaller package sizes using high-density circuit techniques. As opposed to the simple mechanical types the solid-state form requires a large number of separate elements contributing to decreased reliability. The cost of solid-state multicoders is generally greater than that of the electromechanical types. Figure 14b shows a typical high-level PDM solid-state multicoder. This unit employs high-density packaging of discrete elements arranged in modular blocks for increased reliability. For 45-channel operation the power requirement is approximately 3 watts and dimensions are 4 by 2⅞ by 2⅜ in. More recent designs employing microelectronic circuitry, in whole or in part, are considerably smaller but more expensive and hence cannot be justified for the majority of PDM telemetry applications.

Low-level PDM Multicoders

Because of limitations previously encountered in the technology of low-level (of the order of 10 mv full scale) signal commutation by electronic means, the electromechanical commutator proved to be the only practical method of handling small-signal multiplex requirements. Not only were there limitations due to offset potential but also severe limitation in reliability because of the high parts count. More recently, however, improvements in solid-state switching of low-level signals have made it possible to compete more favorably with the mechanical switch for certain applications.

Low-level multicoders consist principally of three major subassemblies, namely, the low-level commutator, the common low-level amplifier, and the PDM converter. The overall package used for low-level signals is designed quite differently from the high-level multicoder package. The channel inputs are double-ended balanced inputs in which the circuitry is arranged to favor common-mode noise rejection. Shielding, both magnetically and electrostatically, is specifically arranged to minimize induced noise effects. The power supply, in general, is better regulated and filtered than in the case of high-level multicoders.

One of the most significant components in the low-level multicoder is the amplifier which raises the level of the input signal to the standard working range of 0 to 5 volts.

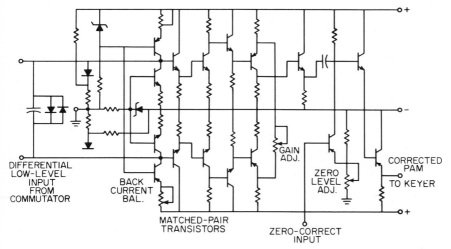

Fig. 15. Simplified schematic of transistorized low-level drift-stabilized d-c amplifier.

Amplifiers for this purpose generally require d-c response since the normal PAM wave train produced by the commutator must carry the lowest-frequency component appearing in any of the channels. The same result can be acheived with a specially designed a-c amplifier having the feature of d-c restoration. This type of amplifier has been used successfully with certain limitations. A third type of amplifier employing a chopper and using the well-known carrier principle has also been used successfully.

Reference should be made to Fig. 15, which shows a simplified schematic diagram of a switch-stabilized d-c amplifier employing matched-pair transistors. This amplifier is drift-compensated using a technique well-known in the analog-computer art. Once or more in each frame the amplifier input is short-circuited and its output simultaneously driven to zero through an appropriate electronic switch circuit. The sensitivity and response of the amplifier are adjusted to meet the requirements of the overall measuring system. This form of circuitry is shown functionally in Fig. 16 for the electromechanical multicoder case. The output level is set either by electronic gating techniques or by use of a third pole on the commutator designed for that purpose.

This type of equipment made up of a miniaturized electromechanical commutator and the most modern solid-state circuitry is far superior in overall performance to multicoders available in the past. Increased MTBF and low cost are significant considerations.

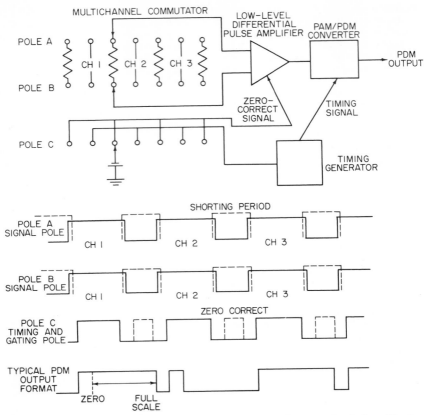

Fig. 16. Timing relationships for typical low-level PDM electromechanical multicoder.

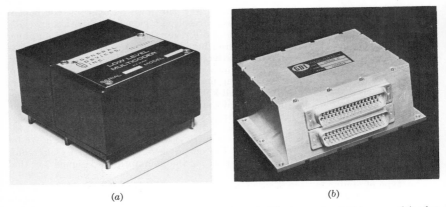

Fig. 17. Typical low-level PDM multicoders. (a) Electromechanical-type multicoder. (b) Electronic-solid-state-type multicoder.

Figure 17 shows typical low-level PDM multicoders. *a* is a 90-channel electromechanical type having separate packages for the mechanical commutator and electronics. A smaller version, of single-package construction, includes the electronics in the commutator case itself. *b* is a 30-channel completely electronic solid-state type employing back-to-back transistorized gates as described above. Economy of parts and power drain is achieved by use of a technique in which a single balancing pair (half gate) is used for many channels. The same unit is available using field-effects transistors with improved performance and higher reliability. Refer to pages 4-82 to 4-84 for technical description.

Mixed High- and Low-level PDM Multicoders

A third type of PDM multicoder involving either electromechanical or electronic commutation is that in which both low- and high-level signals can be handled in any combination. This type of system is generally more complex and more expensive than those described above. Although the mixed signals can be handled in a number of different ways, one common method involves an additional high-level signal pole on the commutator synchronized with the low-level sampling and capable of connecting to either the amplifier output or the high-level signal input as desired.

The Universal Multicoder

A multicoder of increasing interest is the universal multicoder. It has acquired this name because it is capable of operation over a wide range of sampling rates and numbers of channels. In addition to the normal PDM output it has a record amplifier output and a simultaneous PAM output. The most popular of this type is one which provides standard IRIG configurations for both PAM and PDM outputs including master pulses and pedestals. The unit measures 4 by 4 by 5 in., weighs approximately 2 lb, and operates from a source of about 3 watts of power at 28 volts.

Multicoders for FM Subcarriers

Any of the multicoder types described above may be used on the higher standard FM subcarrier channels. Special multicoders with nonstandard durations may also be used as explained in an earlier section (see Table 1).

The most practical forms of PDM multicoders for this use are those specially packaged to fit into the subcarrier rack together with the subcarrier oscillators and their associated electronics.

Special Multicoder Forms

Another form of low-level PDM equipment, which departs somewhat from the standard techniques, involves the use of light beams in conjunction with galvanometer elements. A mechanical schematic of this equipment is shown in Fig. 18. The switch

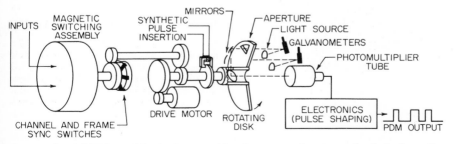

FIG. 18. Special PDM multicoder for use with galvanometer inputs, mechanical schematic.

assembly consists of double-pole single-throw switching units operating in sequence by a rotating magnet. By this means six different sets, of 15 galvanometers per set, are sampled in sequence, providing a total of 90 channels. Light beams from all 15 galvanometers of each set are directed at a rotating disk which contains an aperture. As the opening moves across the light path of a given galvanometer, light is admitted to a photomultiplier tube. The time duration of the pulse thus obtained is directly proportional to the position of the galvanometer beam which in turn is proportional to the current input to the galvanometer. The pulse output from the photomultiplier tube is reshaped to provide a suitable PDM waveform. The framing pulse for synchronization is obtained by the use of a synchronization switch which effectively blanks out two data channels. This equipment was developed to provide a means for converting galvanometer-type measurements directly into a PDM signal.

Calibration and Signal Conditioning

Calibration and signal-conditioning techniques for PDM are similar in most respects to those for other multichannel telemetry systems and are described in several other portions of this handbook. The standard PDM system is a relatively low capacity system and may readily employ automatic calibrators synchronized with the channel commutation in such a way as to obtain one of several preselected signal reference levels for a duration corresponding to one or more frames. These reference levels can be used in providing higher accuracies in manual and semiautomatic data-reduction processes. The following section of this chapter describes how the standard zero and full-scale reference channels carried in the frame may be also used for automatic correction of zero and full-scale conditions in all PDM channels through the use of special ground equipment. The reader is also referred to a following section dealing with PDM errors.

Signal-conditioning systems have been used in various degrees of complexity where the channel inputs demanded some normalization to the standard 0- to 5-volt high-level range. The signal-conditioning package becomes a significant part of the total system where the types of measurements are widely different. Single-channel d-c amplifiers have been used extensively in the past to raise the signal level to the standard value where the number required relative to the total number of channels has been small. Mixed high- and low-level equipment described above now provides additional flexibility in this regard together with increased economy.

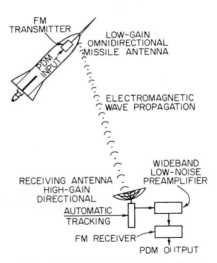

FIG. 19. The basic FM link for PDM.

10 The FM Radio Link for PDM

Since the general subject of radio links for telemetry is treated elsewhere in this publication, only a brief discussion of the problems peculiar to PDM will be included here. The basic components of the complete radio link are shown in Fig. 19. These include the PDM transmitter, the transmitting antenna, the receiving antenna, and the r-f receiving system. In radio telemetry for vehicle instrumentation the characteristics of the electronic equipment including a radio link are largely controlled by the IRIG standards mentioned above. Except for special-purpose equipment occasionally required, the application of these instruments is in accord with these standards.

Conventional telemetering links for FM-FM telemetering are sometimes used to handle PDM data. For some applications compensation is desirable to minimize tilt due to poor response at low frequencies. Since the signal is regenerated in PDM, considerable tolerance on rise time and tilt is usually acceptable.

It is preferable to utilize an FM radio link having d-c response from the modulator input to the receiver output. Under these conditions the center frequency corresponds to the 50 per cent or nominal slicing level of the wave train. If, on the other hand, the link lacks d-c response in the modulator the effective center frequency of the link will adjust to the average of the waveform. The output-signal center frequency would vary considerably as the channels are simultaneously varied from zero to full-scale condition. It would appear that a somewhat greater bandwidth is required under such conditions or the modulation must be restricted to a lesser portion of total normally available range of the link in order to provide for this shift. Slicing at the 50 per cent level can be fixed by clamping the signal at the receiver output.

An early FM modulator designed specifically for PDM is shown in Fig. 20. In this system PDM pulses applied to the input flip the multivibrator from one stable position to another, thus switching the potential across the modulator varicap or diode, between

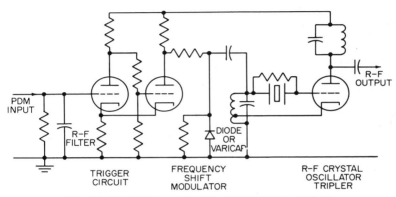

FIG. 20. Modulation method for PDM-FM transmitter.

two predetermined levels. Two distinct carrier frequencies are thus produced corresponding to the two levels, providing frequency-shift operation. Variations in pulse amplitude do not affect the operation of the modulator if the pulse amplitude is sufficiently greater than required to trigger the multivibrator.

This type of modulator provides the desired d-c response. The frequency corresponding to the 50 per cent level does not shift as the PDM duty cycle is varied from one extreme to the other. However, in crystal-stabilized transmitters which employ an FM discriminator in a closed-loop AFC, special provision must be made to eliminate the effects of duty cycle on the control signal.

A typical FM receiver for PDM telemetry is shown schematically in Fig. 21. This is a crystal-controlled FM receiver operating in the range of 215 to 260 Mc. It is a double superheterodyne type with a grounded-grid input amplifier. The noise figure claimed for this equipment is less than 8 db using the well-known 417A type tube. The first local oscillator is crystal-controlled with high stability whereas the second is of the Hartley type with vernier adjustment from the panel for exact tuning. A choice of two second i-f amplifiers is provided: one which has a 500-kc bandwidth, 60 db down at 500 kc from center frequency; the other having a bandwidth of 100 kc with attenuation better than 60 db at 250 kc for center frequency. The latter amplifier is intended for PDM-FM operation with 50-kc FM deviation.

The noise figure is a well-known measure of receiver performance, indicating the amount of noise introduced in the signal channel by the front-end circuitry of the receiver itself. It is defined specifically as the ratio of the signal-to-noise ratio at the

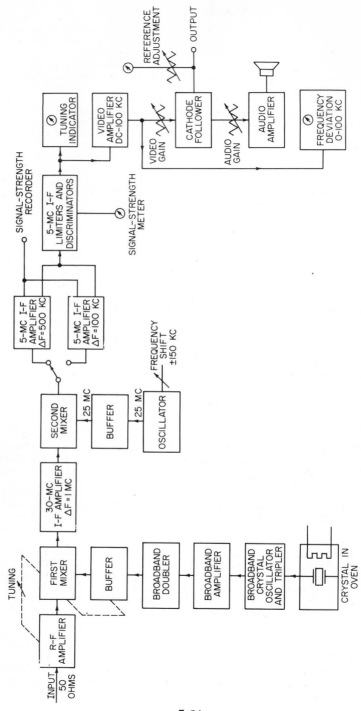

Fig. 21. Block diagram of a typical FM receiver for PDM.

receiver input to the signal-to-noise ratio at its output. For example, a receiver which has a noise figure of 7 db introduces noise power in the signal channel to the extent of five times the noise power at its input. Similarly, one which has a noise figure of 3 db produces twice the noise power of its input. Telemetry receivers are now available with a noise figure of better than 8 db. However, when used properly with special r-f preamplifiers, a noise-figure improvement of as much as 3 to 5 db is claimed. Such preamplifiers are designed for mounting at the antenna location for optimum performance.

Of interest to the practical engineer is the relationship between the r-f carrier level at the receiver input required to achieve the assumed FM threshold conditions, and the r-f bandwidth. This relationship is shown in Fig. 22 for a practical range of noise figures.

Considerably further improved forms of receivers have been developed for critical applications utilizing correlation-detection and other techniques. One common improved type employs a phase-locked principle in which the i-f signal is generated locally. Using a series of relatively complex circuits, including a phase-locked circuit with very narrow bandwidth, the local signal is compared with the incoming signal and made to simulate it as closely as possible. The advantages are obviously that much higher effective signal-to-noise performance can be attained. The minimum usable signal-to-noise power may approach unity with a well-designed receiving system of this type. Disadvantages are in the complexity of equipment design and in certain limitations in use. The small bandwidths employed may be in conflict with stability of the oscillators of the system, with the possible Doppler shifts present in space applications, and with response times required in the use of search equipment. A block diagram showing the elements of a typical phase-locked type receiver used in PDM-FM is shown in Fig. 23. Refer to Chap. 6, Secs. 11 and 18.

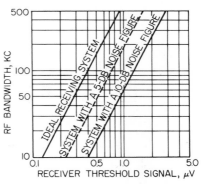

Fig. 22. FM receiver input and bandwidth relationship.

Except under ideal free-space conditions, the problem of accurately calculating the expected performance of a radio link can be rather complex. In an actual case one can usually determine only the approximate average transmitted power required for a given set of conditions by taking into account present-day knowledge of antenna designs, polarization effects, ionization effects, signal reflection, major noise sources, etc. Much useful experimental information is available today on these subjects. Detailed considerations of antenna design, of propagation problems, and of radio-link equipment are taken up more specifically for telemetry in other chapters of the handbook.

As a typical telemetry example consider the problem of PDM transmission from a satellite approximately 1,000 miles distant: ($D = 5.28 \times 10^6$ ft). Considering only space loss $4\pi D^2$ one may compute the approximate required transmitted power for a given receiving system. The minimum acceptable signal power P_r at the receiver input is given by the relationship shown below. The receiver noise figure (NF) is assumed to be 5 (7db). The bandwidth (BW) is assumed to be in the region of 80 kc. The minimum usable signal-to-noise power ratio R_m is assumed to be 8 (see Fig. 5). The value of kT is assumed to be 4×10^{-21} watts/cycle-sec.

$$P_r = kT(\text{BW})(\text{NF})R_m$$

The telemetry-transmitting antenna on the vehicle would have a gain of approximately 1, since it is generally desirable to minimize directional effects with respect to vehicle orientation. On the other hand, the telemetry-receiving antenna can generally

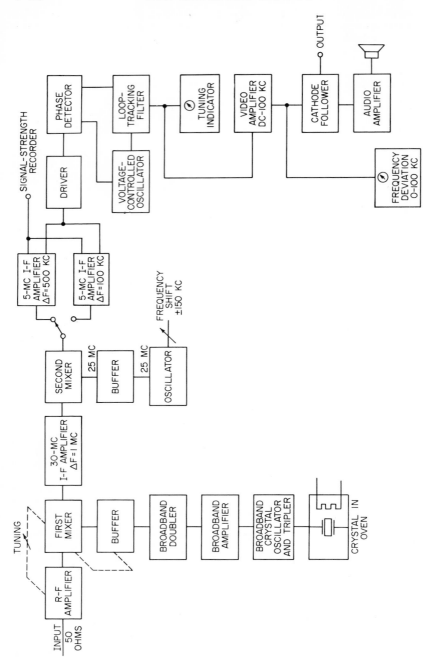

Fig. 23. Block diagram of phase-locked-type FM receiver for PDM.

be made quite large and directional even at the assigned range with an effective capture area A of approximately 50 sq ft. The ratio of required transmitter power to minimum acceptable received power for the link would then be derived by the relationship

$$P_t/P_r = 4\pi D^2/A$$

From these relationships the value of the required transmitter power is therefore computed to be approximately $\frac{1}{10}$ watt. Looking toward future space-data communication needs, it is quite practical to build large directional receiving antennas with a capture area of 2×10^5 sq ft. In the above example this would allow a range twenty times greater for the same power. The use of phase-locked techniques discussed above suggests an improvement in required power of approximately eight times.

Most radio-telemetry equipment in the past has operated in the r-f range between 216 and 235 Mc. In recent years it has appeared desirable to find other bands for use for telemetry transmission, preferably at the higher frequencies. A band in the region of 2,200 Mc was therefore allotted, and both transmitting and receiving equipment was developed for use at this higher frequency. One type of FM equipment is, for example, tunable from 2,150 to 2,300 Mc, which is approximately a factor of 10 higher in frequency from the well-known telemetry band. This higher frequency not only supplies more of the spectrum to work with, minimizing problems of interference at various proving grounds, but also facilitates greater bandwidths for higher-capacity systems. Although the transmitting antenna is simplified and the ground antenna of a given size provides a greater gain at these higher frequencies, more accurate vehicle tracking is necessary to benefit by these advantages.

11 PDM Ground-station Equipment

The essential parts of a simple ground station were shown in an earlier block diagram (Fig. 1). The actual receiving equipment for PDM, however, may take one of several

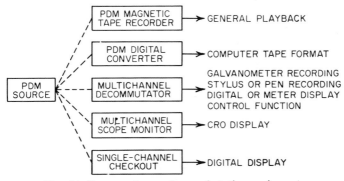

Fig. 24. Common forms of ground-station equipment.

forms, as shown in the block diagram of Fig. 24. The decommutation equipment, sometimes called the channel separator, and a PDM magnetic-tape recorder are contained in most of the major installations. This section will deal primarily with the decommutator and its closely associated equipment.

The decommutator equipment consists of the following major functional elements:

1. The master or framing pulse-synch selector
2. The channel pulse-synch selector
3. The sequential pulse generator
4. The gating circuitry
5. The PDM-PAM converters
6. The channel output circuitry

Reference should be made to the functional block diagram of Fig. 25, which defines a typical decommutator. The input to the commutator, as explained above, may be derived from one of a number of sources, namely, the receiver output, the output of the magnetic-tape playback equipment, the output of an FM subcarrier channel discriminator, or the output of a PDM signal simulator.

The PDM signal is first regenerated and shaped by appropriate circuitry and the channel pulse-synchronizing circuit derives a set of trigger pulses from the leading edges. These pulses are then commonly used to drive a free-running multivibrator type clock which maintains accurate channel synchronism and provides a flywheel effect if the signal should be momentarily obscured by noise or by drop-outs. Under these circumstances the clock multivibrator continues to operate at the rate for which it was last adjusted until the signal reappears.

The frame synchronizing pulses (or master pulses), on the other hand, are derived by a duration-discriminating circuit which detects the occurrence of the absence of two

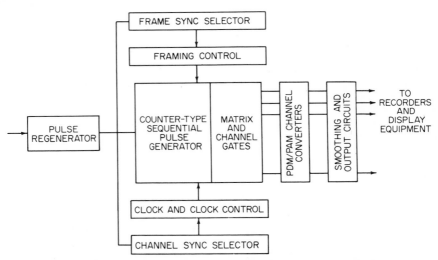

Fig. 25. Functional block diagram of typical PDM decommutator.

sequential channel pulses. A common principle of a master pulse-synch selection for PDM is demonstrated by the RC circuit shown in Fig. 26. Capacitor C charges through a suitable resistance R until discharged by the constant-duration negative pulses from the trigger multivibrator and cathode follower. It is evident that the amplitude of the voltage across C increases to a considerably greater value where the two adjacent pulses are omitted in the wave train. A suitable comparator or amplitude selector may then be used to provide a framing pulse for operating the ground station. This type of circuitry can be arranged to provide reliable operation over a relatively wide tolerance in waveform timing. As mentioned elsewhere, the master pulse and the PDM wave train must be selected with minimum bandwidth for minimum noise influence. Circuitry for flywheel gating and for artificially inserting missing channel pulses is not shown since it varies widely in various equipments.

The two sets of pulses (channel and framing pulses) are used together to drive a counter-matrix circuit which provides the sequential pulses for channel gate control. False frame triggering is avoided by a continuous comparison of the framing pulse with the state of the counter, using a circuit which always tends to maintain the proper frame phasing. The output of the gates is generally in the form of a separated single-channel PDM signal. This is then converted to a useful amplitude form by one of a number of processes. The most common one involves a sawtooth whose amplitude

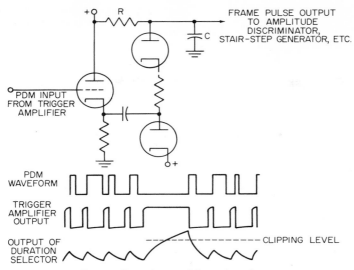

FIG. 26. Duration-sensitive pulse selector.

builds up and remains at a value proportional to the pulse duration. The smoothing circuit and the output circuitry which follow are relatively standard. In general, the final output impedance is arranged to operate recorders and other output equipment through interconnecting cables and patch panels.

Automatic-zero and Full-scale Correction

As previously mentioned, most PDM instrumentation installations involve the use of two or more standard reference signals carried on individual channels. It is very

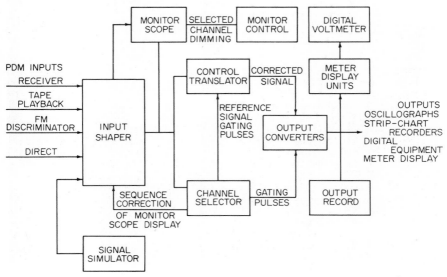

FIG. 27. Functional block diagram of decommutator with automatic correction.

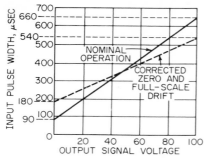

FIG. 28. PDM translator characteristic.

common to apply zero signal level to channel 1 and full-scale signal level to an additional channel in the frame. Some users employ additional channels for intermediate reference levels such as levels of 10, 50, and 90 per cent of full scale. Occasionally automatic calibrators are employed to provide these reference steps continuously in one particular channel.

Several forms of ground-station equipment utilize the zero and full-scale reference signals to adjust zero level and channel sensitivity automatically in servo fashion. The block diagram of Fig. 27 represents a system with this automatic feature. The key to automatic correction is a special type of PDM-PAM converter circuit called the *translator*. This converter is a form of a Miller integrator in which the slope and zero position of its transfer characteristics are both adjustable by varying the potentials of the translator elements. The characteristic of such a translator is shown in Fig. 28 showing the relationship between the input pulse width and the amplitude of the output voltage. A control unit made up of two additional translators (corresponding to the zero and full-scale channels) supplies the control signals to the control elements of all the translators. Within the operating limits of the equipment an electronic servo thereby maintains the two control translators at zero and at a fixed potential corresponding to full scale regardless of variations of the reference channels. Thus all channels maintain the same zero to full-scale levels all corresponding to the signals which appear in the two reference channels. The

FIG. 29. Complete PDM ground station.

SOURCES OF ERROR IN PDM SYSTEMS

7–31

output range most commonly used is 100 volts, allowing the digital voltmeter readout to provide conveniently a percentage of full-scale reading. Solid-state decommutation equipment operates at much lower equivalent potentials.

Figure 29 shows a complete PDM receiving station installed at a government base for missile and aircraft testing. This station has a complete facility for real-time or magnetic-tape playback with direct-writing output recorders.

Figure 30 shows a typical bar-graph monitor scope for PDM. This monitor console may be used for real-time monitoring or for monitoring stored PDM data in the playback process.

FIG. 30. PDM bar-graph monitor console.

SOURCES OF ERROR IN PDM SYSTEMS

The following items contribute most heavily to errors which may be experienced in PDM telemetry equipment:

1. Impulse noise
2. Random noise
3. Synchronization errors
4. Crosstalk
5. Sampling and interpolation errors
6. Commutator noise
7. Other sources of error

These sources of measuring errors are described briefly in the following paragraphs.

12 Impulse Noise

Impulse noise may occur in the r-f system, as already described above, or induced in the circuits at an undesirable point. This type of noise may contribute to an amplitude error where the information is in PAM form or to a timing error after conversion to PDM. In either case it contributes to the total system error. A common source of impulse noise within the PDM equipment itself is that produced by poorly designed mechanical commutators driven by d-c motors. In this case, the motor brush noise may not be sufficiently isolated from the critical circuitry by line filters or properly electrostatically and magnetically shielded. Another source of impulse noise within the equipment is that due to the transistorized inverter-type power supply which produces sharp pulses corresponding to the high-frequency components resulting from switching of the primary voltage supply. Here again the effects can be generally eliminated by proper circuit isolation, filtering, and shielding. Inverters are also available which do not have the high-frequency switching transients.

Low-level PDM multicoder equipment requires special consideration in eliminating the effects of impulse noise inside and outside of the PDM equipment. For most occasions a balanced system with an amplifier having a reasonable a-c common-mode rejection characteristic is sufficient. Specific care must be taken where a storage system is employed to avoid storing to the level of the noise rather than to the desired signal level. In compact equipment induced noise in the conductors may be eliminated by judicious use of magnetic and electrostatic shielding.

13 Random Noise

The basic effects of fluctuation noise introduced in the radio link have been discussed above. The type of analysis described provides the parameters of the PDM system required for achieving a given signal-to-noise ratio in each channel at threshold. Any increase in signal strength over the threshold value at the receiver input further reduces errors from this source.

14 Synchronization Errors

Most signal-noise analyses for PDM multiplex systems, for simplicity, assume that the leading edge of the channel pulse is a noise-free time reference. In actual practice both edges might be expected to have about the same amount of noise. The total channel noise is determined by the square root of the sum of the squares of the noise due to each of the channel edges.

As in all time-division systems, one must be concerned particularly with the basic frame synchronization since a reduced signal-to-noise ratio may cause the loss of frame synchronization before the loss of a reasonable degree of the channel intelligence. This again depends upon how the framing pulse is selected and on the complexity of synchronizing equipment. It is possible to build elaborate electronic circuitry containing flywheel techniques to optimize this characteristic. It is possible, but not common, to create all the channel references from the master pulse in actual equipment. If the master pulse, and hence the related channel references, are to have minimum noise, selection must be made with the smallest possible bandwidth. It should be pointed out that the more complex circuitry is of importance in systems in which radio-link operation is expected to be marginal in this respect. In many early telemetering systems for missile and aircraft testing, more than adequate power was available for the transmission ranges involved. Similarly, more than adequate signal-to-noise ratio was obtained in applying PDM to a subcarrier channel of an FM-FM system. In present-day long-range missile telemetry, however, particularly that intended for space transmission, it is becoming more important to achieve the optimum performance in terms of the radio-link characteristics.

15 Crosstalk in PDM

Apart from crosstalk effects due to common circuit elements, leakage, and circuit intercapacity in practical equipment, PDM crosstalk occurs only by overlapping of pulses from one channel into the following group of adjacent channels. As in PAM systems, crosstalk is dependent on the bandwidth restrictions, that is, on the ratio of

SOURCES OF ERROR IN PDM SYSTEMS 7-33

the cutoff of the video low-pass filter to the product of the number of channels and their sampling rate. Unlike PAM, however, the crosstalk effect in PDM is the variation of the pulse edges in time due to related amplitude disturbance. The amplitude disturbance (as indicated in Fig. 4) is divided by the slope of the pulse to provide position error.

Only a fraction of the total channel time is used for full-scale PDM modulation, as explained elsewhere in this chapter. The percentage of crosstalk due to bandwidth restriction is plotted in Fig. 31, for a maximum pulse modulation of 0.8 times the channel period (a duty cycle of 0.8). The bandwidth restriction is given in terms of the ratio of the video filter cutoff frequency to the product of the number of PDM channels and the sampling rate per channel.

16 Sampling and Interpolation Errors

In early PDM systems it was common practice to present the composite signal obtained from the FM receiver output on the face of a special cathode-ray oscilloscope for direct observation or for moving-film recording. Several different forms of raster were produced and photographed by means of a continuous-film camera. The resulting records were either read manually by means of a simple projector or read automatically by devices specially constructed for this purpose. Two common forms were the so-called "dots" and the "lines" presentations as shown in Figs. 32 and 33. The "dots" represent the edges of the PDM pulse whereas the "lines" represent the actual durations of the pulses. Both are clearly a measure of the data applied to the channels.

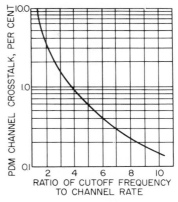

Fig. 31. Crosstalk in PDM.

Only the "lines" display is used at this time but more for quick-look purposes rather than for recording. For example, a flight-test analyst in direct radio communication with the test pilot may observe results from instrumented aircraft in real time. In some cases "dots" or "lines" presentations have been made after channel decommutation. This type of recording (or monitoring) is called recording without interpolation

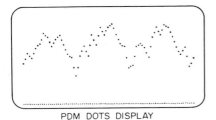
PDM DOTS DISPLAY
Fig. 32. Oscilloscope display of PDM.

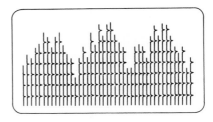
Fig. 33. Oscilloscope display of PDM. PDM lines or bar-graph display.

(or without channel filtering). The final information is recovered by drawing a smooth curve through the dots (or through the extremities of the lines in the lines presentation).

It has been shown that in instantaneous sampling processes where the samples corresponding to a given channel are separated, displayed, and optically integrated, more samples are required per intelligence cycle. If one were using the data system primarily for quick display or monitoring purposes, accuracy requirements would be limited and 6 to 7 samples per cycle would probably be satisfactory. On the other hand, if an accuracy of the order of 1 or 2 per cent is required in the system, it is estimated that 8 or 9 samples per cycle would be required for manual reconstruction of the original data from such recorded samples.

The wideband gain of a system without interpolation is reduced by a factor of approximately 2.5. In addition, since about twice as many samples per cycle are required than are necessary where a low-pass filter is used, the wideband gain is effectively reduced by about five times. The threshold signal level for a constant channel bandwidth would be correspondingly higher.

The sampling principle has been discussed in Chaps. 2 and 4 of the handbook. It simply states that more than 2 samples per cycle of intelligence are required in order to reproduce the original intelligence signal by interpolation using an ideal low-pass filter. Stated in another way, it means that the maximum channel frequency response is approximately equal to half the sampling rate. Practical multichannel telemetering systems, as has already been shown, have effective channel responses of between 0.1 and 0.2 times the sampling rate.

Following the process of channel decommutation and reconversion to PAM, as previously described, the well-known process of pulse stretching is often used. This term refers to a technique for holding the amplitude level for each sample of the decommutated signal until the next sample establishes a new level which is similarly held constant, etc. The resulting waveform is shown in Fig. 34.

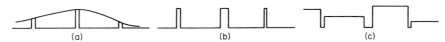

Fig. 34. PDM output waveform. (*a*) Channel signal and corresponding PAM samples. (*b*) PDM channel after decommutation. (*c*) PDM channel output after PAM conversion and stretching.

The output of the pulse stretcher is generally filtered to produce a relatively smooth curve for recording. In fact, the recorder element itself, having limited frequency response, is sometimes used as the filter. As was implied above, an ideal filter with a cutoff at one-half the sampling rate is not possible in practice. A large number of filters even approximating such a characteristic, one for each channel, would occupy an unreasonable volume. Although some systems have employed improved filter techniques, such as by the use of active filters or electromechanical filters to achieve a cutoff corresponding to 3 to 4 samples per cycle, most of the accepted PDM equipment appears to involve at least 6 samples per cycle.

The principal advantage in pulse lengthening is that of increasing the energy in the channel to a level great enough to avoid amplification. The gain is appreciable relative to the condition where PAM samples of their original duty cycle are involved. It should be pointed out, however, that the pulse-stretching process introduces considerable distortion in the signal which is a function of the intelligence frequency applied to the channel. Figure 35 shows how the amplitude error actually increases with the modulating frequency in a channel under these conditions. It is apparent that even where 6 samples per cycle are used, there would be a measuring error as great as 5 per cent without amplitude compensation. Fortunately, simple compensation means can be applied and have been used to correct this error to some extent in a practical manner.

A PAM signal when passed through a low-pass filter under certain conditions will result in the full reproduction of the original signal sample. Similarly a PDM signal after decommutation can be used directly in the process of interpolation. The major differences in the spectra of the two signals is that PDM signals contain components in the form $af \pm bf$, whereas PAM signals have components of the form $af \pm f$ where a and b are integers, F is the sampling rate, and f is the modulating intelligence frequency (see Fig. 36). It is apparent from Fig. 36a that in the case of PAM an ideal low-pass filter can be used with no distortion, if the intelligence frequency is less than the filter cutoff frequency, which is in turn less than one-half of the sampling rate. For decommutated PDM shown in Fig. 36b, on the other hand, it can be shown that a similar low-pass filter technique can be used with a negligible distortion error only if the duty cycle is less than 3 per cent. This fact must be kept in mind in the process of paralleling channels to achieve a higher sampling rate.

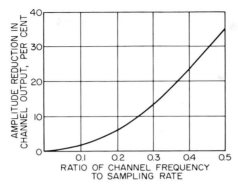

Fig. 35. Frequency error in interpolation.

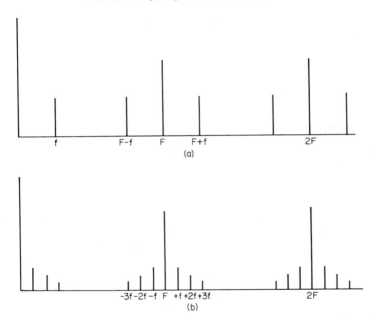

Fig. 36. Modulation spectra. (a) PAM modulation spectrum (single channel). (b) PDM modulation spectrum (single channel).

It is important to restate the major source of error in commutation of data whose character is not generally known. If the sampled data have frequency components which are higher than one-half of the channel sampling rate, obvious errors will appear in the output of the system. In this case the frequency of the channel signal appearing at the output of the system will be the difference between its actual frequency and the sampling rate.

17 Commutator Noise

Noise in PDM systems may arise from faulty or noisy operation of the commutation device. In the case of electronic commutation, amplitude noise may occur through the influence of interacting circuits such as counter matrixes or power supplies. In addition, in low-level electronic systems the inherent noise of the low-level switch must be considered. If the switch is composed of solid-state elements, it is likely that some low-frequency noise will be present particularly when operating at higher sensitivities. Also, as pointed out elsewhere in this handbook, electronic commutation devices if not properly designed are likely to differ slightly in transfer characteristics from channel to channel. This error must also be taken into account in the overall system appraisal. Amplitude errors of this type are directly converted into timing errors having the same proportions in which they originally appear. This is in contrast to the effect produced by amplitude noise on the signal after it is converted to PDM form, as discussed in detail in an earlier section of this chapter.

Where mechanical commutators are used, the two major electrical noise influences apart from pickup are generated or contact noise and variation in contact resistance, commonly termed dynamic contact resistance. The value of generated noise in well-designed commutators is quite small and can be neglected except where very low level signals are being commutated. The effects of dynamic contact resistance, however, are often significant and depend directly on the circuit parameters and converter type employed. As explained in detail in the section on commutation, the dynamic contact resistance can take any one of a variety of forms. The extreme conditions are large spikes of high resistance on the one hand, and resistance variations of low amplitude with the character of random noise on the other hand. Regardless of its form, the influence on subsequent circuitry is in the ratio of the effective contact-resistance value to the resistance value of the total commutator including source and load. The manner in which noise affects the converter is a function of its design.

18 Other Sources of Error

There are obviously many possible sources of errors in a complicated telemetry system involving both airborne and ground-station components. Normal accuracies in the d-c component of the channel signal are in the region of from $\frac{1}{4}$ to 5 per cent depending upon the complexity of equipment employed. With modern equipment one can expect to be able to achieve accuracies of better than 1 per cent in a properly adjusted PDM system. In order to accomplish this accuracy all components must be functioning well within design limits. With care in calibration, it is quite possible to obtain increased measuring accuracies of the order of $\frac{1}{4}$ per cent. Accuracy can be expressed for given equipment only under a given set of conditions including power-supply variations, environmental conditions, operating life, etc. The type of error must also be specified. Some applications required a knowledge of possible extremes of errors for the entire system whereas others can tolerate a root-mean-square expression of error probability.

The reader is again referred to the section on commutation (Chap. 4) for further information on errors relating to that element of the system.

STORAGE OF PDM DATA

19 PDM Recording on Magnetic Tape

The above section included a discussion of display methods involving film recording of PDM data. Several years ago it was realized that a new medium of storage and reproduction was required for data handling in multiplex telemetry. There were specific needs for local recording in airborne vehicles for subsequent laboratory playback and processing and for field recording at ground-station installations where equipment was not adequate (or not planned) for complete data processing. Magnetic-tape techniques already developed for audio use were extended to make it possible to deal with telemetry video signals such as PDM. The diagram of Fig. 37 represents the magnetic-tape link sometimes used in place of the radio link.

The following description of the magnetic-tape mechanism is essential to an understanding of recording of PDM signals: A magnetic-tape recorder for instrumentation consists basically of (1) the record and playback magnetic heads capable of converting electrical signals into suitable magnetic patterns on the tape and vice versa, (2) a tape transport capable of moving the flexible magnetic tape smoothly across the heads, and (3) suitable electronic conversion equipment.

The recording head itself consists of an electrical winding on a small core having an air gap at the point of tape contact. The tape is basically a plastic material with fine particles of magnetic material dispersed uniformly along its surface. The magnetic material therefore becomes a significant part of the magnetic circuit of the head. The signal current in the head winding produces a proportional magnetic flux in its core.

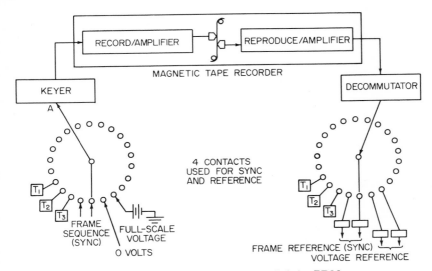

FIG. 37. Schematic of magnetic-tape link for PDM.

If the tape is moving smoothly over the head, the portion of the material in the gap remains in a state of magnetization the magnitude of which is, in turn, proportional to the flux. The magnetization therefore retained by an element of the tape is that present at the instant that particular element passes out of the influence of the gap.

For playback, a similar magnetic head is used. The magnetic field present on the tape causes a corresponding flux to flow in the core. The magnitude of this flux is a function of the average field in the gap due to the tape magnetism. The voltage generated in the winding of the pickup is proportional to the *rate of change* of the flux in the magnetic circuit. The output voltage is therefore frequency-dependent and varies in direct proportion to the frequency of the signal recorded.

A significant factor in PDM recording is the so-called gap effect. If the wavelength of magnetization on the tape is physically short compared with the gap length, the average output is small. When the wavelength is equal to the gap, the output is obviously zero. The high-frequency response of a tape system is therefore limited. The wavelength for a given frequency can be increased by proportionally increasing the tape speed at the expense of possible equipment wear. The high-frequency response can also be increased by decreasing the gap size but at the expense of voltage output. It is interesting to note that a practical tape system can be made to have a relatively uniform response from 50 to 100,000 cps at a speed of 60 in./sec. At 15 in./sec the high-frequency response is correspondingly lowered to 25,000 cps.

In PDM recording, the rectangular waveform output from the PAM-PDM converter (or from the receiving-station equipment) is modified to its differentiated form using a record amplifier circuit. Reference should be made to the record amplifier

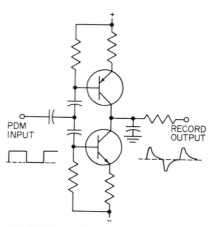

Fig. 38. PDM record amplifier-circuit schematic.

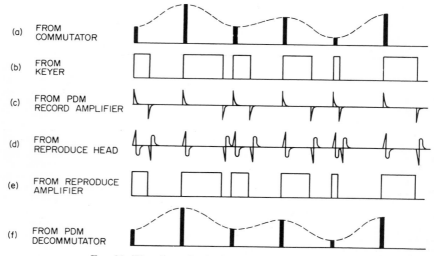

Fig. 39. Waveforms in the PDM magnetic-tape link.

circuit of Fig. 38. Waveforms for the magnetic-tape link are shown in Fig. 39. The differentiated unipolar PDM signal provides a set of positive and negative pulses corresponding to its leading and trailing edges. The output of the record head as described above is the rate of change of the applied voltage waveform. In practice this waveform is applied to a reshaping circuit (often in the form of a multivibrator) which is part of the so-called "reproduce electronics" for the tape system.

Tape-speed variations are particularly important for PDM telemetry recording since measurements are dependent upon time. Long-term variations in speed called "quasi-static errors," such as occur over many frame periods, may be handled by use of automatic corrective circuits in the recording equipment or in other related ground-station apparatus. This type of error was considered above in connection with automatic correction circuitry in PDM ground-station equipment. Changes in tape speed within a frame are called "dynamic errors" and are eliminated primarily by proper equipment design.

A brief mention should be made of major sources of error in tape equipment which may be due to the transport mechanics. Refer to Fig. 40. The supply reel feeds the tape over an inertia roller which exists for the purpose of smoothing out variations in tape speed. The capstan drives the tape at a constant velocity, being driven by a constant-speed synchronous motor. The take-up arm is intended to take up any slack in the tape. The most critical element in uniform speed control is the capstan. To maintain a timing accuracy of 0.25 per cent in record and playback, the tolerance of the capstan diameter must be better than 0.125 per cent. Any eccentricity on the pulleys or in the bearings of shafts tends to produce speed variations in the tape. Speed errors are also introduced by variations in tape tension. The take-up arm senses variation in tension and actuates a suitable holdback. Drive motor-speed variations such as may be involved in "cogging" or "hunting" may cause appreciable time errors.

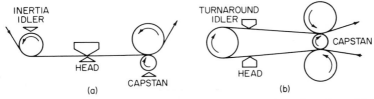

Fig. 40. Mechanical schematic of tape-drive methods. (a) Open-loop tape drive. (b) Closed-loop tape drive.

Most of these errors are expressed by the terms "flutter" and "wow." The standard method of measuring flutter and wow for instrumentation is to record an unmodulated 54-kc carrier whose center frequency corresponds to a tape speed of 60 in./sec. The peak-to-peak frequency modulation produced by wow and flutter is measured on an oscillograph. The frequency content of this modulation is measured on a cumulative basis by using a variable-cutoff low-pass filter whose cutoff may be raised to 10 kc.

Another significant timing error arises from the vibration of the unsupported loop existing between the capstan and the head which represents a distributed mass under tension. This vibration is generated by scraping effects commonly called "stiction." The resulting vibration is known as the "violin-string effect." This effect is considerably reduced in the so-called closed-loop-type drive in which the length of the unsupported tape is smaller. A physical distortion in the tape may also cause a major error in PDM recording. Mylar tapes are much less sensitive to these effects than are the acetate tapes.

Quasi-static errors in the time base are overcome where necessary by use of a servo-speed control system. This is accomplished by recording an accurate reference signal on one track of the tape. On playback this signal is compared with a similar signal of constant frequency with the difference in phase applied as an error signal to the capstan motor through an appropriate amplifier in such a way that the playback signal will always be exactly the frequency which was recorded. In practice the available long-term accuracy from this system can be as high as 20 ppm. This system may also provide short-term corrections giving an accuracy better than 1 msec for variations up to about ½ cps.

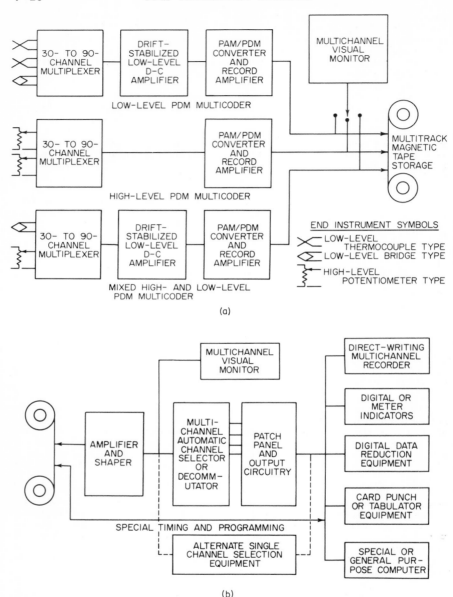

FIG. 41. Block diagram of PDM data logger. (a) Block diagram of typical data logger (multichannel data acquisition and storage equipment). (b) Block diagram of typical playback and analysis equipment.

Amplitude instability is a term used for describing instantaneous lapses or reductions in signal level (called "drop-outs") which are caused primarily by the surface conditions of the tape. The most serious defect is due to the occurrence of nodules of oxide particle clusters. These clusters as well as other foreign particles cause the tape to be momentarily lifted away from the head. Whereas this type of drop-out is relatively unimportant for audio recording, it is completely intolerable beyond a specific magnitude for a given PDM system.

The *dynamic range* of the magnetic-tape recorder is normally defined as the ratio of the maximum signal which can be recorded within a given distortion limit and the minimum signal which can be recorded within the limits of the inherent noise level. The dynamic range, in practice, may be as high as 55 db for approximately 20 per cent distortion (magnetic saturation) whereas a range of 35 db may be obtained for a 1 per cent distortion requirement. PDM recording has the advantage over frequency-division recording that the distortion level is of little concern.

20 The PDM Data Logger

The PDM data logger is a device for data acquisition and playback using PDM recording on magnetic tape. In its simplest form the acquisition equipment consists of one of the many types of multicoders mentioned above and a suitable magnetic-tape recorder. The playback equipment, as in the case of the ground equipment discussed above, can take a variety of forms. The most frequently used forms are represented by the block diagram of Fig. 41. Applications range from laboratory and test-stand use to crash and operational recording for aircraft.

BIBLIOGRAPHY ON PDM AND PDM-FM

The information included on PDM above is intended to provide general information on how it functions, the forms in which it is used, and the essential points of limitation. For those whose interest extends to detail circuitry or to more theoretical information, it is suggested that reference be made to the various instruction manuals prepared by the equipment manufacturers or to the publications noted in the following bibliography.

Adelson, S. L.: U.S. Patent 2,623,112.
Brinster, J. F.: "A Survey Report on Telemetry," Applied Physics Laboratory, Johns Hopkins University, P.O. 13272, May, 1947.
Brinster, J. F.: Standard Pulse Width Radio Telemetering, *Natl. Telemetering Conf. Record*, 1953, pp. 27–34.
Brinster, J. F., et al.: U.S. Patents 2,853,235, 2,765,211, 2,843,840.
Brinster, J. F.: "New Electromechanical and All Solid State Commutation Devices for Multichannel Telemetry," General Devices, Inc., September, 1963.
Brinster, J. F.: "Electronic and Electromechanical Sampling Devices for Multichannel Instrumentation," General Devices, Inc., 1958.
Bryan, F. E.: A Complete Telemeter System for the Flight Testing of Aircraft, *IRE Professional Group on Instrumentation, Data Handling Systems Symp. Proc.*, PGI-2, June, 1953, p. 88.
Bryan, F. E.: A Survey of Data Processing Equipment of PWM-FM Telemetry, *Natl. Telemetering Conf. Record*, 1954, pp. 169–180.
Chambers, F. T.: A Novel Technique for Direct Digital Conversion of Pulse Width Multiplexed Data, *IRE Proc. Natl. Symp. Telemetering*, 1958, paper 1.1.
Corrington, M. S.: Frequency Modulation Distortion Caused by Multipath Transmission, *Proc. IRE*, vol. 33, pp. 878–891, 1945.
Corrington, M. S.: Variation of Bandwidth with Modulation Index in FM, *Proc. IRE*, vol. 35, pp. 1013–1020, 1947.
Crosby, M. G.: Frequency Modulation Noise Characteristics, *Proc. IRE*, vol. 25, pp. 472–514, 1937.
Davey and Matte: Frequency Shift Telegraphy, *Bell System Tech. J.*, April, 1948.
Donath, E., et al.: U.S. Patent 2,753,547.
Friis, H. T.: Noise Figure of Radio Receivers, *Proc. IRE*, vol. 32, pp. 419–422, 1944.
Gerlich, A. A., and D. S. Schover: Pulse-width Discriminator, *Electronics*, vol. 24, pp. 105–107, June, 1951.

Goldberg, H. S., and C. Pilnick: Precision PDM/FM Telemetering System, *IRE Proc. Natl. Symp. Telemetering*, 1958, paper 12.1.

Heffernan, H. J.: The Transmission of Pulse Width Modulated Signals over Restricted Bandwidth Systems, *IRE Proc. Natl. Symp. Telemetering*, April, 1957, paper 2.2.

Hill, H. M.: Miniature Airborne Telemetering System, *Tele-Tech*, vol. 11, pp. 68–72, 84–96, December, 1952.

Kaplan, H.: Solid State Pulse Width Modulator, *IRE Proc. Natl. Symp. Telemetering*, April, 1957, paper 3.3.

Knowles, W. S.: U.S.Patent 2,753,546.

Kretzmer, E. R.: Distortion in Pulse Duration Modulation, *Proc. IRE*, vol. 35, pp. 1230–1235, 1947.

Kuehn, R. L., and W. T. Johnson: PDM-PAM Conversion System, *IRE Proc. Natl. Symp. Telemetering*, April, 1957, paper 8.5.

Landon, V. D.: Impulse Noise in FM Reception, *Electronics*, vol. 14, pp. 26–30, February, 1941.

Landon, V. D.: Theoretical Analysis of Various Systems of Multiplex Transmission, *RCA Rev.*, vol. 9, pp. 433–482, 1948.

Miller, E. C.: Frequency Discriminator for Narrow-band FM, *Electronics*, February, 1943, pp. 128–129.

Nichols and Rauch: "Radio Telemetry," 2d ed., John Wiley & Sons, Inc., New York, 1956.

Riblat, H. B., and J. P. Randolph: "Satellite Telemetry and Data Processing," International Telemetering Conference, London, September, 1963.

Riedel, J. A.: A Transistorized Pulse Width Keyer, *IRE Proc. Natl. Symp. Telemetering*, April, 1957, paper 6.5.

Rock, F. E.: PDM/FM/FM on Lower Subcarrier Frequencies, *IRE Proc. Natl. Symp. Telemetering*, 1958, paper 12.2.

Salinger, H.: Transients in Frequency Modulation, *Proc. IRE*, vol. 30, p. 378, 1942.

Sargeant, P. N.: "A Simple Supervisory Encoder Using Pulse-length Modulation," International Telemetering Conference, London, September, 1963.

Smith, J. H.: Design Problems of a PDM/FM Transistorized Telemeter System, *Natl. Telemetering Conf. Record*, 1957.

Smith, J. H.: Transistor Circuits Applied to Telemetering, *IRE Proc. Natl. Symp. Telemetering*, April, 1957, paper 3.1.

Spearow, R. G., and G. Helms: A High Performance Transistorized Pulse Duration Modulator, *Natl. Telemetering Conf. Record*, 1957, paper IV-B-4.

Uglow, K. M.: Noise and Bandwidth in PDM/FM Radio Telemetering, *IRE Proc. Natl. Symp. Telemetering*, April, 1957, paper 7.5.

Weber, P. J.: "The Tape Recorder as an Instrumentation Device," Ampex Corporation, 1958.

White, R. W.: A PDM-FM Telemeter for Low Level DC Inputs, *Natl. Telemetering Conf. Record*, 1955, pp. 78–81.

Zadeh, L. A.: Optimum Non-linear Filters for the Extraction and Detection of Signals, *IRE Natl. Conv. Record*, 1953, pt. 8.

Chapter 8

PCM TELEMETRY SYSTEMS

GEORGE P. SARRAFIAN (*Introduction*), LEO A. CHAMBERLIN (*Data-transmission-system Design*), GEORGE E. GOODE (*Data Recovery*), ALAN L. McBRIDE (*RF Link, System Error Analysis*), *Texas Instruments Incorporated, Dallas, Tex.*

INTRODUCTION

1	Introduction to PCM	8–2
2	PCM System-design Principles	8–2
3	System Advantages and Applications	8–3

DATA-TRANSMISSION-SYSTEM DESIGN

4	Signal Conditioning	8–5
5	Multiplexers	8–11
6	Programmers	8–15
7	Encoders	8–23
8	Data-processing Design Considerations	8–27
9	Digital Modulation Techniques	8–28

DATA RECOVERY

10	Data-recovery-system Considerations	8–30
11	Bit and Group Synchronization	8–31
12	Data Regeneration, Processing, and Display	8–43

R-F LINK

13	General Considerations	8–48
14	PCM Data Modulation and Transmission	8–49
15	PCM Data Reception and Demodulation	8–52

SYSTEM ERROR ANALYSIS

16	Sampling	8–56
17	Quantization Errors	8–58
18	Systematic Errors	8–58
19	Random Errors	8–59
20	Example of Error Calculations	8–60

1 Introduction to PCM

Pulse-code-modulation (PCM) telemetry systems have come into wide usage in the last few years, particularly for missile and space applications. For applications requiring high accuracy or the sampling of large numbers of channels of varying characteristics, PCM systems offer certain net advantages over competing modulation systems.

Basically, a PCM system is a time-multiplexed sampled-data system in which the values of the input channels sampled are expressed in digital (usually binary) form. In principle, any form of digital code may be used to represent the channel values; however, the two most common codes probably are the straight binary code, used with electronic analog-to-digital converters, and the Gray code, used with shaft-position digital encoders.

The modulating signal in a binary PCM system is thus a sequence of 1's and 0's, which are grouped in binary "words" describing the value of the particular channel sampled at the instant of sampling. The two-level analog signal corresponding to the 1's and 0's may then be used to modulate a radio-frequency carrier signal in any of the standard ways: AM, FM, PM, etc. Depending upon the modulation technique selected, the modulation system would then be labeled a PCM-AM, PCM-FM, or PCM-PM system. On some occasions it may be desirable to provide two parallel digital channels; in this case two stages of modulation may be used with the two bilevel signals modulating subcarriers and the subcarriers in turn modulating the main carrier. If phase modulation were used in such a system, the modulation would be termed PCM-PM-PM.

The primary system advantages of PCM systems include capability for handling both analog and digital signals, flexibility as to numbers of channels and channel sampling rates, capability of transmitting data of high accuracy with little or no degradation in the r-f link, and generally superior characteristics of information efficiency and noise immunity in the r-f link. Offsetting these advantages is the fact that the complexity of a PCM system varies little with the number of channels, and thus although the complexity of a PCM system compares quite favorably when the number of input channels is large (≥ 50), for systems with small numbers of input channels and moderate accuracy requirements, analog systems may be desirable for reasons of simplicity.

2 PCM System-design Principles

Figure 1 indicates the basic elements of a typical PCM system. Analog signals are fed through signal-conditioning circuitry into a multiplexer, which is actually a series of on-off switches with inputs fed by the analog data channels and with the outputs of the switches tied to a common point. The multiplexer is controlled by the programmer in such a way as to ensure that the input channels are sampled one at a time in sequence. The sequence may be simple but more often is complex, allowing some channels to be sampled more frequently than others. The resulting analog signal at the output of the multiplexer is fed to an amplifying unit which may contain a "sample-and-hold" circuit, designed to keep the analog value of the channel being sampled at a constant level during the sampling interval, even though the value of the channel itself may be varying during that interval. This provision reduces errors in the digital encoding process for rapidly varying analog signal-digital errors which under some conditions may exceed the value of the analog level change causing those errors.

Next, the signal is fed to an analog-to-digital converter, which is programmed to make one digital conversion per sampling interval. The output of the analog-to-digital converter is a sequence of binary digits, represented in the physical system by a two-level analog signal. This signal is fed to a set of digital gating circuits to be mixed with digital data from a synchronization generator and often also digital input data from off-on switches, computer digital registers, etc. In a system with both analog and digital input data it is obviously true that the analog inputs would all be suppressed, generally as a result of having all the multiplexer switches in the closed positions.

The PCM system programmer programs all timing and control functions in the system. The programmer includes a clock, countdown logic plus decoding logic for the command outputs to the multiplexers, sample-and-hold circuit, analog-to-digital converter, and digital gating circuits. Also, as shown in Fig. 1, it also includes a synchronization generator which generally provides a pseudorandom binary sequence with a bit rate which is an integral multiple of the analog-to-digital converter. The synchronization data are normally mixed with the output of the analog-to-digital converter but may alternatively be transmitted along a parallel channel to another modulator.

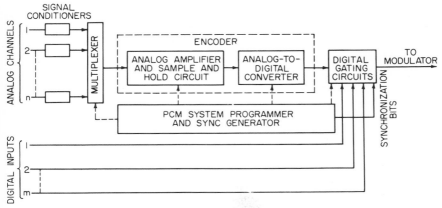

Fig. 1. Basic elements of a PCM system.

Of course, the functions described above are merely the rudimentary ones performed by any PCM system. Many additional functional blocks may be provided such as core buffers and magnetic-tape transports to record the output data, provision for storing alternate formats within the PCM system programmer, and parity coding and error-correcting coding circuits at the output of the analog-to-digital converter. But the basic elements of any PCM system are the ones shown on the diagram.

3 System Advantages and Applications

Pulse-code-modulation systems have been known for many years to provide superior flexibility, communications efficiency, and performance in the presence of noise.[1] However, some time elapsed before PCM systems came into wide use, primarily because the newness and slightly greater complexity of such systems delayed their usage until applications emerged which clearly demanded PCM with its unique advantages. The first extensive operational use of PCM telemetry systems occurred on the Polaris, Minuteman, and Titan missile-test programs.[2] In the missile tests, it was necessary to transmit with high accuracy the contents of certain registers in the guidance computer; consequently a digital data-transmission system was highly desirable. Given the requirement for digital transmission on the r-f link, it logically followed that it was desirable to digitize the other (analog) information required concerning missile status and transmit all the data in digital form. Currently PCM systems predominate where mixed digital and analog data must be transmitted (a category covering most missile and space programs) and where the higher communications efficiency of PCM is a strong advantage, as in deep-space communications. Some specific advantages of PCM systems are summarized below.

Higher Communications Efficiency

The *efficiency* of a given modulation technique in a communications system is best measured by a figure of merit defined as the received energy required per bit of *informa-*

tion (in the Shannon sense) which is received at the demodulator. The actual value of this figure of merit for any modulation system, of course, depends upon the tolerable error in a given case; in a digital system, this can be represented either by an allowable bit error rate, by setting an arbitrary maximum on the root-mean-square error of the received sampled-data sequence, or by other methods. Sanders[3] has compared the communications efficiency achieved with various modulation techniques on the basis of each of the above criteria. The modulation techniques compared include coherent and noncoherent AM, FM, FM-FM, FM with phase-locked loop, and the class of orthogonal systems. The orthogonal systems (with coherent detection) may be shown to approach the theoretical Shannon communications-efficiency limit with increasing symbol complexity. Simple PCM-PM with the carrier shifted $\pm 45°$ by the 1 and 0's is an example of a two-symbol orthogonal system. A commonly used technique involves phase modulation over a $\pm 90°$ range and performs with slightly higher efficiency. This technique may be considered the degenerate case ($n = 1$) of a class of biorthogonal systems with slightly higher communications efficiency than the orthogonal systems.[4] In the simple two-level uncoded case, such systems perform with greater communications efficiency than do the AM and FM systems mentioned above. Furthermore, PCM techniques offer the simplest route to the mechanization of complex-symbol orthogonal and biorthogonal systems with communications efficiencies approaching within a very few decibels of Shannon limit. The simplest procedure for accomplishing this (described elsewhere in this handbook) involves the generation of orthogonal or biorthogonal binary code sets and using the resulting codes to phase-modulate the carrier over a $\pm 90°$ range.

Noise Immunity and Communications Reliability

A direct result of the greater communications efficiency of coherent PCM systems is that, with a fixed available transmitter power, the performance of such systems will be degraded less by noise in the data link than the performance of the commonly used analog systems. In addition the digital nature of the data being transmitted makes it easy to insert parity checks for error detection; such a feature provides both a warning when data are being lost and reassurance when the link is operating properly. In addition, error-correcting codes may be employed if desired. Finally, the "go no-go" nature of the demodulation process makes PCM relatively insensitive to minor non-linearities in the transmitter and receiving-system circuitry that might cause distortion in an analog system.

Capability for Handling Both Analog and Digital Inputs

As described earlier, many current programs require transmission of information which is inherently digital in form. Common examples are the contents of computer registers, and event counters, readout of shaft position from a shaft-angle encoder, etc. In these situations, the usage of analog techniques would be awkward and unwise and would not provide the accuracy required in most cases. Once the decision is made to send the digital data via a PCM link, it is natural to encode the analog data digitally also so that all data can be transmitted on a single link.

System Flexibility

Pulse-code-modulation systems are extremely adaptable to changes in numbers of channels, absolute sampling rates, and relative sampling rates between channels. Systems can be easily mechanized which provide (for example) 7-bit (1 part in 128) accuracy for some analog channels, 10-bit accuracy for others (1 part in 1,024), and 24-bit accuracy (1 part in 16,000,000) for digital-computer registers. The same basic system design may be used to transmit at an information rate varying by several orders of magnitude. Digital systems often offer the only practical way of transmitting at extremely low rates. Finally, PCM systems allow the use of flexible storage means (magnetic tape, core buffers, etc.) to allow matching of data-collection to data-transmission rates over many orders of magnitude.

DATA-TRANSMISSION-SYSTEM DESIGN 8-5

Compatibility with Digital Computers for Ground Data Processing

Typically, large amounts of data are generated during a missile-test flight or during the operational lifetime of an orbiting satellite. Furthermore, there are generally quite a few different missile and space programs active during the same time period, each with its own data requirements. Considerations of economy demand flexible ground data-handling equipment which can accept data at the varying rates, accuracies, etc., characteristic of the different programs rather than highly specialized equipment for each program. In addition it is often desirable to provide the capability for automated data editing, scale shifting, linearization, etc. These functions are most flexibly and efficiently performed by computers and other digital equipment. These systems of course operate on the data in digital form, and though the data may be converted on the ground from analog to digital before processing it simplifies things greatly if the data are digital as received.

DATA-TRANSMISSION-SYSTEM DESIGN

4 Signal Conditioning

Signal conditioning is defined as any operation performed on a transducer signal, with the exception of multiplexing, occurring between the transducer and the encoder. By far the most common form of signal conditioning is simple linear amplification, although other operations such as logarithmic amplification and frequency and amplitude demodulation are at times necessary. Level adjustment is necessary to bring all signals to a common range so that a single encoder may be used. Also, low-level signals must be amplified to a level suitable for encoding. In conjunction with amplification, the bandwidth of the signal generally must be limited to minimize the effects of noise on the signal lines and to minimize sampling errors.

Signal conditioning may be categorized by the characteristics of the signal being operated on and whether the signal conditioner is used by only one data-input channel or is time-shared among several channels.

Using the above criteria, we shall categorize signal conditioners as

Low- or high-level
Single-ended or differential
Pre- or postmultiplexer

Following the standards set for FM-FM telemetry, "high-level" generally connotes a signal whose range of values is 0 to 5 volts. Such signals are typically derived from transducers such as thermistors or from monitoring signals within some electronic equipment. Low-level signals are typified by the outputs of strain-gage or thermocouple instrumentation. These signals typically vary from 0 volts to a full-scale value lying between 10 and 50 mv. These definitions are purely arbitrary, and there certainly is a class of signals with between 50 mv and 5 volts full-scale outputs. The techniques listed below can certainly be used in this midvalue region; however, it is usually found that commercially available transducers can be made to fit in one of the categories described above, since a transducer capable of from 1 to 3 volts output can usually be modified for 5 volts full scale, while strain-gage and similar transducers are inherently limited to much smaller full-scale outputs of the order of tens of millivolts.

The second criterion classifies signals as single-ended or differential. Most low-level signals are derived from bridge configurations of strain gages, thermocouples, etc., and are almost always differential while high-level signals are more often than not single-ended in nature.

The last classification depends upon the physical location of the signal-conditioning amplifiers. Figure 2 illustrates two possible configurations. Figure 2a shows one signal-conditioning amplifier per data input with the amplifiers preceding the multiplexer; on the other hand, Fig. 2b illustrates a PCM system with one amplifier time-shared among all inputs. The advantages and disadvantages of each scheme are apparent. The premultiplexing scheme allows tailoring the signal conditioner for each channel and alleviates the necessity of multiplexing the low-voltage levels, while the postmultiplexer single amplifier greatly reduces the amount of hardware required, offering a sizable reduction in size, weight, and power consumption. However, far more stringent requirements are placed on the multiplexer and all channels receive the same conditioning. Also, the requirements for the amplifier are made far more stringent by forcing it to accommodate the wide-bandwidth pulse-amplitude-modulated signal from the multiplexer rather than the relatively low bandwidth signal from the transducer.

Historically, the premultiplexing technique was the first used, following experience gained with FM-FM, PAM, and PDM telemetry systems, and when low-level signals were being commutated, limitations caused by multiplexer design. As commutator technology improved and switches capable of multiplexing millivolt-level signals became available, the postmultiplexer conditioner was widely adapted. The recent advances in semiconductor-device technology have made available integrated-circuit amplifiers in packages requiring no more space than one or two conventional transistors. Hence the size and weight incurred by the premultiplexer conditioner (but not power consumption) have been greatly reduced and the individual-amplifier-per-channel technique is again being used.[5]

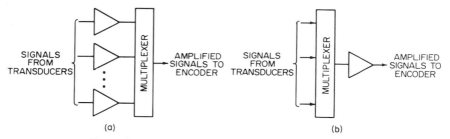

FIG. 2. Pre- and postmultiplexed signal conditioning.

In missile and satellite telemetry, reliability has become at least as important a design criterion as performance; so some mention should be made of the relative merits of the two schemes. Even granting that the time-shared amplifier is considerably more complex than any single-channel amplifier, the sheer reduction in number of components for the single amplifier results in a higher probability of retrieving 100 per cent of all data. However, the probability of retrieving most, but not all, of the data is usually higher for multiple amplifiers since failure of one amplifier results in loss of only one data channel. Exact comparisons between the two methods depend upon the number of channels, parts count of the amplifiers, and failure rate of parts, and also the effect of increased power requirements for the one-amplifier-per-channel technique must be considered. Hence the technique to be used must be determined for each individual design.

Finally, it should be noted that combinations of techniques, or the use of more than one amplifier following the multiplexer may be required. Frequently two amplifiers follow the multiplexer, one for low-level and one for high-level inputs.

In addition to amplification (or attenuation) each input should be filtered to bandwidth-limit the signal. Sampling theory requires that the signal be sampled at a rate at least twice as high as the highest frequency present in the signal. If frequencies higher than one-half the sampling rate are present, encoding errors will occur. The filtering operation must take place before the multiplexer. Often the filter is placed in a common enclosure with the multiplexer switch. A filter which has a flat amplitude response to a frequency just above the highest frequency of interest and a rapid cutoff thereafter is desired. Because of the size and weight limitations, RC filters of one or two sections are commonly used. Filters derived from modern network-synthesis techniques, such as the Butterworth polynominal filter, provide flatter response in the passband and sharper cutoff at the expense of size and weight. Filters of both the simple RC and the LC type may be designed directly from published tables.[6,7] In this case another advantage of the premultiplexer amplifier is apparent as both the gain and bandwidth of the individual amplifiers may be selected. This is most advantageous for channels sampled at a slow rate, where the size of passive filter components becomes very large.

The parameters affecting signal-conditioner design for any signal-conditioning technique include

1. Sufficiently high input impedance to prevent error caused by loading the transducer
2. Gain accuracy and stability
3. Low noise generation within the signal conditioner
4. Prevention of crosstalk

In addition, if the input signal is differential, the signal conditioner will be required to provide common-mode rejection.

The need for the first three requirements listed above is obvious. The techniques for high- and low-level signals are usually basically the same, with the amount of gain being the differentiating factor, the most widely used amplifier configuration being a differential amplifier, with one input grounded in the case of single-ended inputs. Good noise performance and very high input impedance can be obtained by use of field-effect transistors as the input pair, although some sacrifice in bandwidth can be expected. New field-effect devices appear to be overcoming this disadvantage.

An amplifier preceding the multiplexer must have d-c response if the signal is very slowly varying; however, for a postmultiplexer amplifier, d-c response is not mandatory as the signal presented at this amplifier input is a pulse-amplitude-modulated wave train. Thus, this amplifier may use a-c coupling although the response required will be near d-c as will be shown below. Also, chopper-carrier-demodulator and chopper-stabilized amplifiers are suitable, with the restriction noted below.

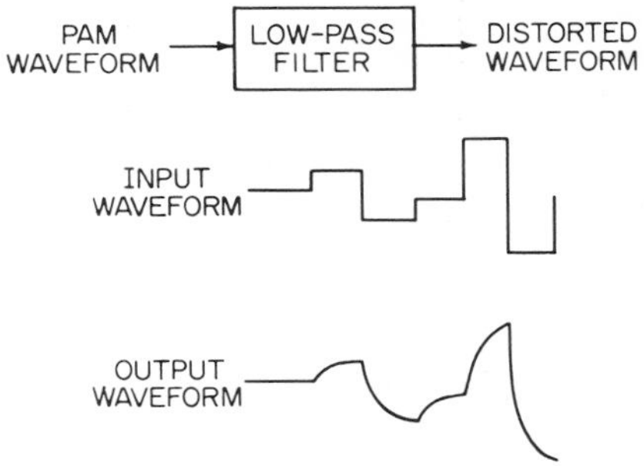

The requirements for high input impedance, common-mode rejection, and prevention of crosstalk are often conflicting and are of prime importance in the design of telemetry systems. Crosstalk is defined as the error produced in any given channel caused by the signals on all other channels. Prior to multiplexing, crosstalk is caused by inductive and capacitive coupling between wiring carrying the signals. It is minimized by use of shielded cables, and careful layout of wiring, particularly in the multiplexer.

Crosstalk introduced after multiplexing is, assuming a linear system, due almost completely to the finite bandwidth of that portion of the telemetry system lying between the multiplexer and encoder and thus is a far greater problem in systems using a time-shared amplifier.

FIG. 3. "Crosstalk" errors caused by finite system bandwidth.

To a good approximation, crosstalk error may be analyzed by replacing the portion of the system between multiplexer and encoder by a low-pass filter and, if the system does not have d-c response, a high-pass filter.[8] Figure 3 shows a low-pass filter excited by a PAM wave train. The waveforms (greatly exaggerated) show the effect of finite bandwidth.

The error is most obvious when observing a data sample following one of higher amplitude. Crosstalk becomes an even more severe problem when system nonlinearities are considered. Consider that low-level signals are frequently derived from bridge transducer configurations. A conservatively designed telemetry system should suffer no degradation on other channels due to a failure of one transducer. A common failure of bridge-type transducers is an open or shorted arm, in which case the transducer excitation voltage (on the order of 10 or more volts) is suddenly impressed on the input of an amplifier which normally accommodates millivolt-level signals. If the amplifier contains energy-storage devices, particularly transformers, the device will be saturated and several sample times may be required to discharge the stored energy. The problem is present even if transformers and large capacitors are eliminated since the active elements of the amplifier will be driven into saturation.

Common-mode rejection is a requirement for differential signals only. Common-mode voltages arise from two sources. Consider the bridge transducer depicted in Fig. 4. The differential signal between points a and b is on the order of millivolts

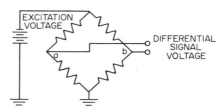

FIG. 4. Bridge transducer showing source of common-mode voltage.

while a voltage of approximately one-half the excitation voltage is present on both leads. The second source of common-mode voltage is pickup of stray signals, particularly power frequencies, and also potential differences in "ground" at various points within a system. Aircraft and missile "ground" points at physically separated points within the airframe may vary with respect to each other by several volts.

The ability of a signal conditioner to reject common-mode signals is governed by the input impedance of the system (from input to common-mode ground) and the degree of balance of impedances.

For a linear system, the ability of a system to reject common-mode voltage is a function of the impedance of the system to the common-mode ground. Consider the circuit shown in Fig. 5a. The transducer is represented by two generators with an impedance in each transducer lead. In general, the impedance in each load will differ by some amount R. For resistance-bridge transducers, such as strain-gage bridges, there will be unbalance in the transducer for all values of strain except one. The common-mode source is represented as a generator. The system is represented as an ideal differential amplifier, i.e., one whose input impedance is infinite and whose output is proportional to the difference in its input. The resistor-capacitor networks shown simulate the actual amplifier input impedance.

By application of superposition, we may consider the effect of the common-mode source independent of the signal. By further application of superposition to the transducer circuit, we may examine the effect of the common-mode signal in a circuit which neglects the transducer unbalance and one which includes only the unbalance. These two circuits are shown in Fig. 5b and c, respectively. Consider the balanced circuit. Assuming R_{in_1} and R_{in_2} to be equal and C_{in_1} and C_{in_2} also equal, there is no error produced by the common-mode signal as the common-mode currents I_{cm_1} and I_{cm_2} are equal, and hence the voltages developed across R_{in_1} and R_{in_2} are equal.

Consider now the circuit of Fig. 5c. The impedances in each branch of the circuit differ by R, and hence the common-mode currents differ. This results in a difference in voltages across R_{in_1} and R_{iu_2} and thus results in an error. Notice also that the input capacitance causes the common-mode error to increase with the frequency of the common-mode signal.

The higher the input impedance of the system, measured from signal input to common-mode ground, the better the common-mode rejection will be. From Fig. 5a, it

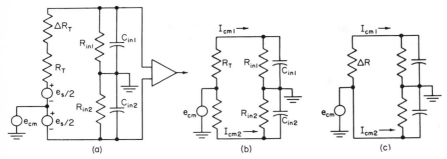

FIG. 5. Circuits for the analysis of common-mode error. (*a*) Generalized circuit for common-mode analysis. (*b*) The balanced input circuit. (*c*) The unbalanced input circuit.

is also obvious that it is necessary that the input impedance be both large and balanced. Any unbalance will produce an error in exactly the same manner as transducer unbalance. The desire for high and balanced input resistances imposes requirements primarily on the amplifier and multiplexer switch designs, while capacitive balance is dependent both on these factors and on the mechanical layout of the signal-conditioner wiring and of the wiring between the transducer and signal conditioner.

Various circuit configurations have been used for signal conditioners and each has advantages and disadvantages, with none being superior in all cases. A few of these circuits will be discussed below.

D-C Differential Amplifier

The direct-coupled differential amplifier,[9] operating differentially or with one input grounded for single-ended operation, is possibly the most straightforward approach. Input impedance is raised in this type of amplifier by feedback, including positive current feedback to supply base current to the input-transistor pair. The principal difficulty with amplifiers of this type is d-c or zero drift; however, since the amplifier is time-shared, the drift-correction circuit shown in Fig. 6 can be used to minimize drift. This amplifier is a-c coupled with d-c restoration between samples. Advances in device technology, particularly the use of field-effect transistors, have permitted the design of amplifiers of this type having input impedance measured in megohms.

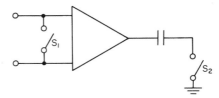

FIG. 6. D-C restoration to eliminate drift.

Transformer-coupled Amplifiers

Transformer-coupling the input to the signal-conditioning amplifier (this includes chopper-carrier-demodulator amplifiers) provides excellent d-c common-mode rejection. Since the amplifier must have near-d-c response, the primary inductance of the transformer must be large, thus increasing size and weight.

The high primary inductance necessary requires a relatively large number of turns and results in considerable interwinding capacitance, thus adversely affecting a-c common-mode rejection. The iron losses of the transformers limit input impedance to a few hundred kilohms, and overload recovery is poor because of saturation of the transformer.

Floating Amplifiers

One solution to the common-mode problem is to float the amplifier as seen in Fig. 7. When the multiplexer switch S_1 closes, S_2 is closed, allowing the capacitor to charge to the sample value. After changing, S_2 is opened and S_3 closed, making the sampled value available to the encoder.

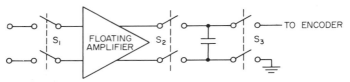

Fig. 7. Floating amplifier.

This method requires a separate power supply for the amplifier and increased switching circuitry. The common-mode rejection obtainable is limited by the impedance of the amplifier to ground, which is predominately through the power supply. The d-c impedance is high, usually being the insulation resistance of the power-supply transformer, but a-c impedance is lowered by the primary-secondary capacitance of the power-supply transformer.

Bridge Amplifiers

Besides the more straightforward techniques described above, many ingenious techniques have been devised to meet the amplifier requirements. One in particular is very successful. This is the potentiometric-bridge amplifier shown in Fig. 8.

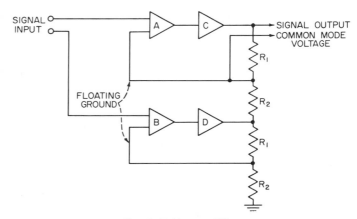

Fig. 8. Bridge amplifier.

Amplifiers A and B are floating while C and D are conventional amplifiers. After the amplifier output has settled, the input impedance is very high, as the potential from amplifier input to floating ground is near zero and thus little input current is required. Another feature of this amplifier is that the voltage at point a is the common-mode voltage. An amplifier that senses the common-mode voltage provides a means of further improving common-mode rejection, as will be discussed in Art. 5.

5 Multiplexers

The multiplexer, or commutator, of a PCM system samples the data inputs to the telemetry system in some predetermined order. The signals being multiplexed may be analog voltages to be encoded by the analog-to-digital converters or digital signals which will be inserted in the output data train. We shall first be concerned with switching analog voltages.

The multiplexer is essentially a multipole switch or group of switches. For purposes of this discussion, the logic required for sequencing switch closure will be discussed in the section on programmers to follow. This section will primarily discuss types of switches suitable for use in multiplexers.

Before discussing actual switches, it is worthwhile to compare any physical switch with an ideal switch. Figure 9 shows an ideal switch and the equivalent circuit of a

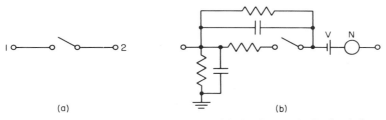

FIG. 9. Equivalent circuit of switch. (a) Ideal switch. (b) Real switch.

physical switch. For the ideal switch, an infinite impedance will be measured between points 1 and 2 with the switch open, zero with it closed. The impedance from points 1 or 2 to ground is infinite. By contrast, the real switch will have a finite open impedance and a nonzero closed impedance and there will be both measurable resistance and capacitance to ground. In addition, an offset voltage V and a noise voltage N can be expected.

Switches may be broadly categorized as electromechanical or electronic. Let us consider first the electromechanical switches.

Rotating Commutators

The earliest form of multiplexer switch was the rotating commutator. This switch consists of a mechanically driven rotor and a number of contacts arranged in a circle about the rotor axis. A second set of poles are all connected to a common voltage source and are sequentially contacted by a second contact on the wiper. This second set of contacts is used to furnish pulses in synchronism with the data samples used to trigger sequential operations in the encoder. The wiper is driven by an electric motor, usually of the hysteresis synchronous type. Speed regulation is obtained by careful frequency control of the motor power voltage. The rotating commutator is one of the most highly developed switching mechanisms, as it enjoyed wide use in earlier time-division multiplex systems (PAM and PDM) and the experience gained from d-c motor commutators is directly applicable. It has been well covered in the literature.[10]

As far as open and closed impedance and impedance to ground are concerned, the rotating commutator is a very good approximation to an ideal switch; however, the noise generated by the sliding contacts has limited its usefulness as a low-level switch. An even more serious disadvantage is the limited lifetime of the equipment due to mechanical wear and its low reliability compared with some electronic switches. Maximum sampling rates are lower than those obtainable with electronic switches though high enough for many applications. Power consumption is rather high, several watts being required to drive the motor. Attempts have been made to eliminate the drawbacks of this type of commutator caused by the sliding wiper-pole contact by replacing the metal wiper with a jet of mercury or other conducting liquid.[11] A rotating nozzle directs the stream of fluid against poles which physically are pins arranged around the circumference of a circle whose center is the axis of rotation of the nozzle.

A variation of the rotating commutator is the stepper switch, which, in principle, utilizes the basic commutating scheme of contacts and a wiper described above, except that a stepper motor replaces the synchronous motor. The advantage to be gained is that the sampling rate can be controlled by logic circuitry within the telemetry programmer since the motor advances one position each time it is pulsed.

Relay Multiplexers

The relay is a very common electromechanical switch and has found wide use in telemetry commutators. A great advantage of the relay, as well as the electronic switches described below, is that they can be opened and closed in any arbitrary sequence, determined solely by the logic in the system programmer. A relay switch, like the rotating commutator, gives an excellent approximation to an ideal switch insofar as open and closed impedance is concerned, and since sliding contact is avoided, noise is considerably reduced. Very low offset voltages are incurred.

The common moving-armature relay is suitable for low-speed switching and relatively small-sized low-cost units are available. Because of the mass of moving elements, switching speed is relatively slow and the relays are sensitive to vibration. Lifetime is also limited. The moving mass also gives rise to contact bounce, further limiting sampling rate since sufficient time must be allowed between switch closure and encoding for switching transients to die out.

A serious drawback to the use of these relays is the relatively large currents required for operation, necessitating heavy driver circuits, power supplies, etc. The undesirable power requirement can be greatly reduced by the use of magnetic latching relays which require power only while the relay changes states. Also, the latching relay, being a bistable device, serves as a memory element capable of performing logic functions. By providing the latching relay with two sets of transfer contacts, it is possible to build a commutator in which both the sequencing logic and the data multiplexing are performed by a number of relays equal to the number of data-input channels, with no other components required.[12]

The reed relay or switch (Fig. 10) possesses all the advantages noted for the moving-armature relay and is capable of higher switching speeds because of the low mass of the moving elements. This switch is composed of two metal reeds enclosed in a sealed glass enclosure. The reeds are of a soft magnetic material, and when a magnetic field is applied by a solenoid as shown, the reed ends are attracted to each other, causing switch closure. Since the contacts are enclosed in the glass envelope, very little degradation due to contact contamination is experienced. The low mass, coupled with the magnetic attraction of the reeds, lessens contact bounce and reduces sensitivity to vibration. Lifetimes of several million operations are obtainable.

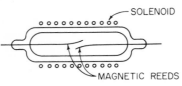

FIG. 10. Reed relay switch.

Diode Switches

The unidirectional current-flow characteristic of the diode may be exploited to make a switch. The earliest electronic switches were diode switches, and although vacuum diodes were used in some applications, tubes found relatively little use in telemetry equipments. In general, either semiconductor or vacuum diodes can be used for any of the circuits discussed here; however, semiconductors are assumed for purposes of explanation.

Since diode switches have been in use for several years, many circuits have evolved and several have been described in standard texts.[18] A representative circuit will be shown here.

Figure 11 shows a typical diode switch. When point 3 is made more positive than the bias $+V_1$, diodes CR5 and CR6 are back-biased. Assuming the absolute magnitude of the bias voltages V_1 is sufficiently high to keep the current flow in the diodes CR1 through CR4 in the forward direction, the equivalent circuit of Fig. 11b applies,

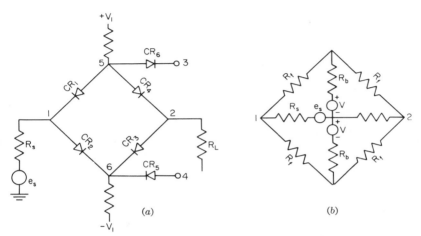

FIG. 11. Diode switch. (a) Switch schematic. (b) Equivalent circuit.

where the R_f's are the forward resistances of the diodes. If the diodes are matched to have the same forward resistances, a balanced-bridge configuration is created and the voltages at points 1 and 2 are equal. It may also be seen that, because of the balanced bridges, bias current does not flow through the transducer impedance R_s. When the polarity of the control voltages at points 3 and 4 is reversed, the bridge diodes are all back-biased and the switch is in effect opened.

There are many variations of this basic switch. Diodes CR5 and CR6 and the bias supply can be eliminated if carefully balanced control voltages are used. Likewise a transformer secondary may be placed across points 5 and 6 and the switch may be closed by pulsing the transformer. This technique does not positively back-bias the diodes, so the "off" impedance of the switch will be somewhat lower than for the other cases described above.

The primary sources of error in this type of switch are the impossibility of perfectly matching the switch diodes and the nonlinear voltage-current relationship for the diodes. The imperfect match may be compensated for by placing a potentiometer between CR1 and CR2 or CR4 and CR5. The terminals of the resistance element are connected to the diodes and the control voltage is supplied through the wiper. The effect of diode nonlinearity is not easily compensated for. If the signal voltage is other than zero, the current through the branch composed of CR1 and CR3 differs from that through CR2 and CR4. Because of the nonlinear diode characteristics the forward resistances of the diodes are no longer equal and an error is generated. No linear compensation is possible; however, the error is minimized by making the bias voltage and hence the current through the diodes as high as practical, thus reducing the relative variation of diode current due to the signal.

The diode switch shown here suffers one further disadvantage which for many applications is quite serious. While control current does not flow back through the transducer, signal current does flow through the control circuit. Consider again the equivalent circuit and assume ideal diodes, i.e., let $R_f = 0$. The impedance seen looking into the switch is then the parallel combination of the bias-current-limiting resistors R_b and the local resistor. Thus the impedance looking into the switch is

$$R_{in} = R_L R_b / (2R_L + R_b)$$

The bias-current resistors cannot be made excessively large—certainly not so large as R_L, which is the imput impedance of the encoder or the signal-conditioning amplifier. This low impedance, coupled with the nonlinearities of the diodes and difficulties in matching, causes the diode switch to be less suitable for low-level differential multiplexing than the transistor switch, which will be described in the following section.

Transistor Switches

The basic transistor analog gate, or switch, is shown in Fig. 12a. By saturating both transistors, a conduction path is established from point a to b. Ignoring the driving source, the equivalent circuit for the closed switch is shown in Fig. 12b. The closed impedance is the sum of the saturation resistances of the two transistors ($2R_c$'s). The voltages measured from collector to emitter of the saturated transistors tend to cancel as the transistors are connected collector to collector. By careful matching of transistors, total offset voltages of less than 20 μv are obtainable. For high-level applications where several millivolts offset do not cause intolerable errors, only one of the transistors may be required. The driving source must provide current to saturate the transistors and in addition isolation must be provided; i.e., the impedance to ground seen looking into the switch must be high. A typical driver is shown with the basic switch in Fig. 12c. Isolation is provided by a transformer and transistors Q_1 and Q_2 are turned on when Q_3 is turned on, causing a current to flow in the transformer primary and inducing a current pulse in the secondary. Note that the switch transistor drive is applied base to collector; i.e., the transistor is operated in the inverted configuration. This results in lower collector-emitter voltages than in the case of base-emitter drive.

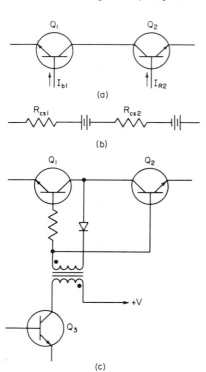

FIG. 12. Transistor multiplexer switch. (a) Basic switch. (b) Equivalent circuit. (c) Typical driver.

Multiplexer Configuration

After selection of a switch type, the multiplexer itself may be designed. If electromechanical switches are used, the multiplexer is basically just one switch per input (two switches or double poles per switch for differential inputs). When semiconductor electronic switches are used and particularly when differential low-level signals are being switched, further refinement in design may be required. Recall that the resistance to ground seen looking into one switch is quite high but not infinite and also that a nonzero capacitance to ground is observed and in addition that a reverse leakage current will flow when the switch is in the open state. If many switches are in parallel as shown in Fig. 13a, the impedance looking into the one closed switch will be the impedance of all switches in parallel and the leakage current of all switches will be fed back to the transducer connected to the one closed switch. The impedance looking into the one closed switch can be increased and the leakage current reduced by the supermultiplexing technique shown in Fig. 13b. The switches are arranged in groups and the outputs of the groups are in turn multiplexed onto the common data bus. From an impedance standpoint, the number of switches in parallel is then the number in each group plus the number of supermultiplexer switches or

$$S_p = S_t/N_g + N_g$$

where S_p = number of switches in parallel
S_t = total number of switches
N_g = number of groups and hence number of supermultiplexer switches

To find the minimum values of S_p, we take the derivative of the above equation with respect to N_g and equate to zero. The optimum number of groups is

$$N_g = S_t^{1/2}$$

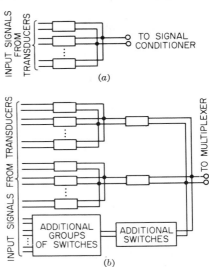

Fig. 13. Supermultiplexing.

The common-mode rejection of the multiplexer for low-level signals in a system employing postmultiplexer amplification can be improved by effectively nullifying the capacitance measured from the signal path to ground. This can be accomplished by connecting the shields of all leads between the transducers and the multiplexer switches and driving them at the common-mode voltage of the one channel being sampled. The shields of course are left floating at the transducer ends and the signal path is thus surrounded by an electrostatic shield at the common-mode voltage.

Digital Multiplexing

In addition to analog voltages, a telemetry system may be required to multiplex data already in digital form into its data output. Examples of such data may be digital words from a missile-guidance computer or data from an inherently digital output transducer, such as a shaft-position encoder.

Unless the telemetry system and the source supplying the digital data are operating in synchronism, buffering is required. The data from the buffer are gated into the telemetry-system output by replacing an analog switch with a digital gate. Upon command of the telemetry-system programmer the gate is opened, and by means of clock pulses supplied by the programmer the data are mixed with the other telemetry-system output data. In some cases, often involving shaft-position encoders, the programmer may interrogate the data source at will, but the entire output word, i.e., all binary bits of the data, must be accepted simultaneously on parallel lines. In this case, the telemetry system must provide a register to accept the parallel word and then shift the data serially out of the register.

6 Programmers

The programmer generates all timing information for the system and performs any other digital or logic functions that may be required. The largest bulk of the programming problem involves sequencing the closure of the multiplexer switches (unless a rotating commutator is used).

Although we shall be discussing electronic or relay-type commutators, the commutator sequencing process is more easily seen by imagining the commutator to be a rotating device. Figure 14a shows a simple rotating commutator in which each data channel, identified by numbers on the diagram, is sequentially sampled. The format of the commutator output may be represented by the diagram in Fig. 14b. The order of switch closure is obtained by reading from top to bottom of the column.

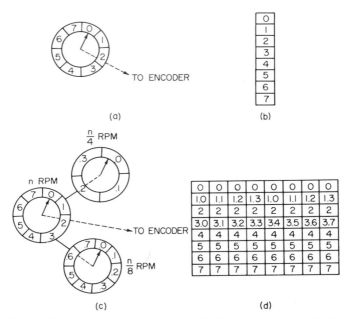

FIG. 14. Format diagrams. (a) Commutator. (b) Format diagram. (c) Commutator with subcommutator. (d) Format diagram.

In most applications, it will be found that sampling all channels at the same rate is not desirable and is indeed extremely wasteful in terms of bandwidth of the system output. This is because, in a system operating at a standard sampling rate, the rate is of necessity set by the highest expected frequency on any input channel. By employing the subcommutation scheme illustrated in Fig. 14c and d, all channels need not be sampled at the same rate. To achieve subcommutation, additional commutators are attached to the prime commutator. In the numbering scheme on the format drawing the number to the left of the decimal point is the switch number of the prime commutator, while that to the right is the subcommutator switch number. The format drawing is read as before from top to bottom and after reading one column proceeding to the next one at the right. A vertical column of the format drawing will be referred to as a prime frame and the block of data sampled before a subcommutator samples the same channel twice as a subframe. The format shown has subframes at one-fourth and one-eighth the prime-frame rate.

Note that, when using discrete switches, there is no direct equivalent to the separate commutation of the rotating system. There is simply a group of switches which close in a fixed order to give the same output format.

The sequencing of the switch closures and other timing functions may be performed in a variety of ways, with the most common being the use of counting and decoding circuits. Figure 15 illustrates a typical programmer and shows its relation to the remainder of the system.

The timing source for the system is a crystal or other stable oscillator operating at a binary multiple of the system-output bit rate. The output frequency of the oscillator is successively divided by two in stages of the binary counter. Frequency higher than the bit rate may be provided if needed in the system. The bit-rate frequency signal shown in Fig. 15 drives the bits per word counter whose output states may be decoded as required for synchronization purposes.

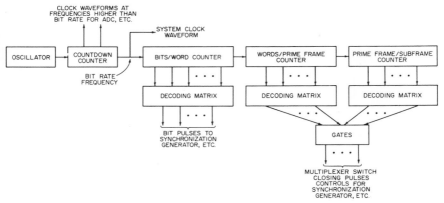

FIG. 15. Typical programmer.

The output of the bits per word counter then drives the word (or channel) per prime-frame counter, which in turn drives the prime frame per subframe counters. By decoding the state of these counters, pulses to close the multiplexer switches are generated.

The designer has a wide degree of freedom in selecting counter and decoder combinations to perform a given task. The number of circuit elements and hence equipment reliability and cost are greatly affected by the choice made, and thus a brief discussion of counter types is warranted.

Ripple-through Counter

The simple ripple-through counter shown in Fig. 16a provides a means of counting using few elements. With no additional logic elements, the counter will generate 2^n states and recycle as shown by the sequence table accompanying the logic diagram. This type of counter circuit is well suited to counting down the primary oscillator frequency to the bit rate. However, its usefulness for a counter whose states are to be decoded is limited. The difficulty arises from the ripple-through property. Consider the change of state from 1111 to 0000. Since, because of switching-time delays, the change of state of each flip-flop does not occur simultaneously, the intermediate states of 0111, 0011, 0001, are observed for brief instants. Each of the extraneous intermediate states is a permissible state of the counter; so any gate which decodes one of the intermediate states may produce a brief output during the logic-propagation time, causing improper closure of a multiplexer switch.

Synchronous Binary Counter

The undesired propagation-delay effect of the ripple-through counter can be alleviated by use of the synchronous binary counter of Fig. 16b. As shown here, the counter has four stages and system states, and the output states represented the binary numbers 0 through 15, occurring in sequential order. All flip-flops are clocked by the same

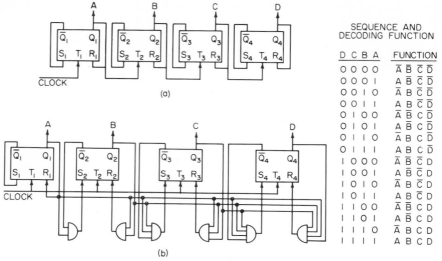

FIG. 16. Binary counter circuits. (a) Ripple-through counter. (b) Synchronous counter.

clock frequency and all state changes occur simultaneously. Using clocked reset-set (RS) flip-flops the logic equation for a counter of n stages, generating 2^n states is

$$S_n = \bar{Q}_n Q_{n-1} Q_{n-2} \cdots Q_1$$
$$R_n = Q_n Q_{n-1} Q_{n-2} \cdots Q_1$$

By modifying the interstage logic, the counter can be made to count to any number less than 2^n and recycle.

Shift-register Counters

While the synchronous binary counter eliminates the transient-error possibility of the ripple-through counter, it will be noted that considerably more logic circuitry is required for the synchronous counter. The synchronous counter is made to count in a specified sequence; specifically, the output states represent the binary numbers, 0 through 2^{n-1}, occurring in numerical order. Some reduction in logic complexity might be obtained if the sequence of states follows some other order.

Indeed, a whole class of counters based on logic feedback applied to shift registers exists which can result in reduction of circuitry. Several examples of feedback counters will be presented below. We may tentatively classify these counters as maximum- (or near-maximum-logic counters) or minimum-logic counters. We have seen that it is possible to obtain 2^n states from a counter with n flip-flop or memory elements. Such counters are of the minimum-logic type. At the other extreme, a maximum-logic counter is one in which there is one flip-flop for each output state.

Although at first it might appear that maximum-logic counters are wasteful of circuit elements, it will be shown later that, although the counters uses excessive elements, the decoding networks are simplified and the total system complexity may be reduced.

Ring Counter

An n-stage flip-flop register with output of the last stage fed back to the first makes a simple counter. If, by means of additional logic circuitry, binary zeros are fed to the input stage until the register is completely filled with zeros, and then a single binary 1 is inserted, the 1 will, at each clock pulse, be propagated to the next stage of the counter and, after propagating through the entire length of the counter, will be returned to the first stage. This is a maximum-logic counter; however, note that no decoding is necessary as each state of the counter is represented by there being a binary 1 in one and only one flip-flop. It should be noted that this type of counter is easily implemented with a moving armature or magnetic latching relays and that if relays with two sets of transfer contacts are used the relay serves as both a counter element and multiplexer switch.

The danger in using such a counter is that, because of induced noise, power-supply transients, or other cause, more than one binary 1 will be introduced into the counter. The extra 1 or 1's will continue to be propagated, causing erroneous operation. This is a property of all non-minimum-logic counters. Since they do not utilize all 2^n possible states obtainable from the n memory elements of the counter, the danger

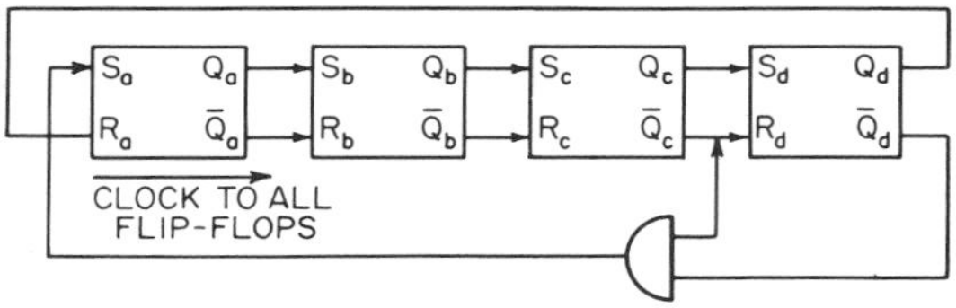

SEQUENCE TABLE

STATE	A	B	C	D	DECODING FUNCTION	
0	0	0	0	0	$\bar{A}$	$\bar{D}$
1	1	0	0	0	A	$\bar{B}$
2	1	1	0	0	B	$\bar{C}$
3	1	1	1	0	C	$\bar{D}$
4	1	1	1	1	D	A
5	0	1	1	1	$\bar{A}$	B
6	0	0	1	1	$\bar{B}$	C
7	0	0	0	1	$\bar{C}$	D

FIG. 17. Shift-register counter.

exists that they will enter an unintended sequence if because of a malfunction the counter elements ever assume a state not in the sequence. When using non-minimum-logic counters precaution must be taken to avoid erroneous sequences. For the ring counter, one method of detecting an erroneous condition is to use $n - 1$ two-input AND gates with the state of the last flip-flop in the shift register fed to all gates. The second input of each gate is from another stage of the register. The output of all the $n - 1$ gates is then combined in an $n - 1$ input OR gate. Whenever the OR gate has a true output more than one binary 1 is in the register and feedback to the first stage is made to shift zeros into the register until a single binary 1 is left in the register.

A second near-maximum-logic counter and its sequence are shown in Fig. 17. This counter produces $2n$ output states from n elements, and the decoding for each state involves only two terms, hence requiring a two-input AND gate for decoding. Note that no gating logic is required in the feedback path, except for prevention of erroneous states. If the true output of the last stage is applied to the reset input of the first stage and the false output to the set input the sequence is obtained. The additional gates prevent the erroneous sequences.

An important minimum-logic class of shift-register counter is the maximum-length-sequence generator. Such counters derive the feedback signal by summing modulo 2 the outputs of various states. (Modulo 2 summing of a pair of signals A and B is given by the logic equation sum = $A\bar{B} + \bar{A}B$. A logic element performing this function is often called a half adder or EXCLUSIVE OR gate.) A shift register of n elements connected to generate a maximum-length sequence produces $2^n - 1$ states. The condition of all zeros is prohibited. Figure 18 illustrates a four-stage maximum-length-sequence generator and lists the 15 output states.

Maximum-length counters of $2^n - 1$ states can be designed for any length of register, and tables are available listing the feedback connections.[13] A short listing for counters of length from 3 to 12 stages is presented in Table 1, which is read as follows: The flip-flops are labeled beginning with the letter A. Under the labeled columns, a 0 or 1 will be found for each flip-flop in the counter. The set term for the input flip-flop

Table 1. Feedback Connections for Maximum-length-sequence Generators

	A	B	C	D	E	F	G	H	I	J	K	L
3	0	1	1									
3	1	0	1									
4	0	0	1	1								
4	1	0	0	1								
5	0	0	1	0	1							
5	0	1	0	0	1							
6	0	0	0	0	1	1						
6	1	0	0	0	0	1						
7	0	0	0	0	0	1	1					
7	1	0	0	0	0	0	1					
8	0	0	0	1	1	1	0	1				
8	0	1	1	1	0	0	0	1				
9	0	0	0	0	1	0	0	0	1			
9	0	0	0	1	0	0	0	0	1			
10	0	0	0	0	0	0	1	0	0	1		
10	0	0	1	0	0	0	0	0	0	1		
11	0	0	0	0	0	0	0	1	0	0	1	
11	0	1	0	0	0	0	0	0	0	0	1	
12	0	0	0	0	0	1	0	1	0	0	1	1
12	1	0	0	1	0	1	0	0	0	0	0	1

is then the sum modulo 2 of the outputs of the flip-flops under which a 1 is entered. For example, the feedback for the four-stage counter just described is

$$S_A = C \oplus D$$
$$R_A = \bar{S}_A$$

or

$$S_A = C\bar{D} + \bar{C}D$$

All counters of this class have one forbidden state, all zeros. Note that, for the four-stage counter, should the register ever have the state 0000, the feedback will always cause S_A to be zero and R_A to be 1, and hence the register will continue to be filled with zeros. Such a state could be assumed when power is first turned on or as a result of noise within the circuits. Erroneous operation can be eliminated by detecting the forbidden state and causing the term S_A to be 1 when it occurs. Thus the complete expression for the input term of the four-stage counter is

$$S_A = C\bar{D} + \bar{C}D + \overline{ABCD}$$
$$R_A = \bar{S}_A$$

Decoding of this class of counter is the same as for binary counters, and thus considerable saving in circuitry is obtainable when compared with the binary counter since most gating is eliminated.

Of course, it is rare that one would need only counters with $2^n - 1$ states. Maximum-length counters can be modified to generate a sequence of less than maximum length. Any number of states less than $2^n - 1$ can be obtained. Refer again to the list of states for the four-stage counter. Suppose a counter of 9 states is desired. Consider state 6, which is 0100. If, on the next clock pulse, flip-flop A had been set to 1 instead of 0, the counter would jump to state 13, which is 1010. This jump would skip states 7 through 12, resulting in a 9-state counter. The feedback to give the 9-state counter is then

$$S_A = \underbrace{C\bar{D} + \bar{C}D}_{\substack{\text{Max-length}\\\text{feedback}}} + \underbrace{\bar{A}B\overline{CD}}_{\substack{\text{Jump}\\\text{term}}} + \underbrace{\overline{ABCD}}_{\substack{\text{Prohibited}\\\text{state}}}$$

$$R_A = \bar{S}_A$$

By examining the list of states in Fig. 18, it will be found that any length of counter could have been designed up to and including 15 states. For any sequence generator of n elements, a sequence of length $2^n - 1$ or less may be generated by the jump method. Unfortunately, no general method for finding the states to be jumped has been found and the designer must determine the jump feedback term for a given counter by listing all possible states and finding the proper jump. Table 2 lists the feedback term for counters of 3 through 6 stages.* The column marked P contains a 1 if the feedback is applied to the S term or reads 0 if applied to a reset term. The two examples below illustrate the design process.

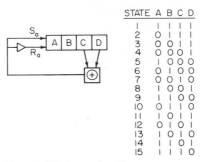

FIG. 18. Maximum-length-sequence generator.

STATE	A	B	C	D
1	1	1	1	1
2	0	1	1	1
3	0	0	1	1
4	0	0	0	1
5	1	0	0	0
6	0	1	0	0
7	0	0	1	0
8	1	0	0	1
9	1	1	0	0
10	0	1	1	0
11	1	0	1	1
12	0	1	0	1
13	1	0	1	0
14	1	1	0	1
15	1	1	1	0

1. Design of a 26-state counter using the jump technique. Since 26 lies between 2^4 and 2^5, a 5-stage shift register is required. The maximum-length feedback term obtained from Table 1 is

$$S_A = C \oplus E = \bar{C}D + C\bar{D}$$

Table 2. Jump Terms for Maximum-length-sequence Counters

Cycle length	A	B	C	D	E	F	P	Cycle length	A	B	C	D	E	F	P
1	X							33	1	1	0	0	1	0	0
2	X							34	0	0	1	0	0	0	1
3								35	0	1	0	1	1	1	1
4	0	1	1				1	36	0	0	1	1	0	1	0
5	1	0	0				1	37	1	1	1	1	0	1	0
6	1	1	0				0	38	0	1	0	1	0	1	0
7								39	0	1	1	1	1	1	1
8	0	0	1	0			0	40	0	1	1	0	0	1	0
9	0	1	0	0			1	41	1	0	1	0	0	0	1
10	1	1	0	0			1	42	1	0	1	1	1	1	1
11	0	0	1	1			1	43	0	0	0	1	0	0	1
12	1	0	0	0			1	44	1	1	0	0	0	0	1
13	1	0	1	1			1	45	1	1	0	1	0	0	1
14	1	1	1	0			0	46	1	0	0	1	0	1	0
15								47	1	0	0	1	1	1	1
16	0	0	0	1			0	48	1	1	1	0	1	1	1
17	1	0	0	0	1		0	49	1	0	0	0	1	1	1
18	0	0	0	1	1		0	50	0	1	1	1	0	0	1
19	0	1	1	0	1		1	51	0	0	0	1	1	1	0
20	1	1	1	0	0		0	52	1	0	1	1	0	1	0
21	1	0	1	0	1		1	53	0	1	0	0	0	0	1
22	0	1	1	1	1		1	54	1	1	1	0	0	1	0
23	1	1	0	0	1		0	55	0	0	1	0	1	0	0
24	0	0	1	1	0		0	56	0	1	1	0	1	0	0
25	1	0	0	1	0		1	57	0	0	0	0	1	1	1
26	0	0	1	0	1		1	58	1	0	0	0	0	0	1
27	1	0	1	1	0		0	59	1	1	0	1	1	0	0
28	0	1	0	0	0		1	60	0	1	0	0	1	1	1
29	0	1	0	1	0		0	61	1	0	1	0	1	1	1
30	1	1	1	1	0		0	62	1	1	1	1	1	0	0
31								63							
32	0	0	0	0	1		0								

* The tabulated data presented here are from unpublished notes of W. J. Watson, used in a logic design course taught by Mr. Watson at Texas Instruments.

The term to prevent the prohibited state is

$$S_A = \bar{A}\bar{B}\bar{C}\bar{D}\bar{E}$$

From Table 2 it is noted that the entry under P is a 1, indicating that the jump term is an additional set term and is

$$S_A = \bar{A}\bar{B}C\bar{D}E$$

The unsimplified logic equations for the counter are

$$S_A = C\bar{D} + \bar{C}D + \overline{ABCDE} + \bar{A}\bar{B}C\bar{D}E$$
$$R_A = \bar{S}_A$$

To implement the counter the above expression could be simplified, but as the last two terms are states of the counter which are decoded for use in the remainder of the system, the unsimplified equation is often more useful.

2. Design of a 27-state counter. The maximum-length and prohibited feedback terms are the same as for a 26-state counter; however, the entry under P in Table 2 indicates the jump term is applied to the R term (i.e., the jump is obtained by causing the initial flip-flop to reset to a 0 where the maximum-length sequence caused it to set to 1). The feedback term is

$$R_A = A\bar{B}C\bar{D}\bar{E}$$

The equations for the maximum-length counter are

$$S_A = C\bar{E} + \bar{C}E + \overline{ABCDE}$$
$$R_A = \bar{S}_A$$

or $S_A = \bar{R}_A$

$$R_A = C\bar{E} + \bar{C}E + \overline{ABCDE}$$

Introducing the jump term, the unsimplified equations are

$$R_A = C\bar{E} + \bar{C}E + \overline{ABCDE} + A\bar{B}C D\bar{E}$$
$$S_A = \bar{R}_A$$

Cascading Counters

To produce a counter of a given number of states, two or more counters may be used. Figure 19a shows two 10-stage ring counters in series. Using the decoding shown, a 100-state counter is derived. This technique is also applicable to generating the subcommutated-channel switch-closure pulses. In this case, counter A is the prime channel counter and B is the subcommutated channel counter. If more than one subcommutation rate is employed, counter B may be replaced by one counter for each rate and only the states required for closing switches are decoded. The same technique is applicable to any of the counters discussed. Each individual counter is decoded and then the respective outputs are gated together.

Figure 19b illustrates a second method of obtaining a long sequence from two shorter counters. If the total number of states desired can be factored into two relatively prime numbers, i.e., numbers which have no common divisor other than 1, two counters of the factored lengths can generate the sequence, by gating together all possible combinations of the counter output states. Thus, the 9- and 11-state counter shown in Fig. 19b generates a 99-state sequence.

Other Programmer Functions

In addition to generating the timing information for the telemetry, the programmer performs any other digital functions which may be required. The most important of these is generating the synchronization code. Unique codes must be inserted in the serial output-data stream so that the channel format may be identified at the receiving station. They are inserted in the place of data channels. The synchronization codes

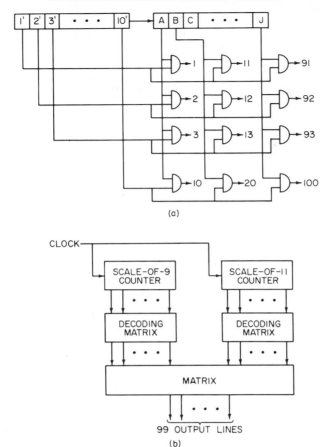

FIG. 19. Cascaded counters. (a) Cascaded ring counter. (b) Parallel relatively prime counters.

can be generalized serially by combining the pulses from the bit per word counter with the pulses from the prime-frame and subframe counters.

7 Encoders

Many techniques have been devised for encoding analog voltages into digital words. A mere cataloging of the various analog-to-digital conversion techniques would require more space than is available here, and description of their operation is, and already has been, the subject of complete volumes.[14]

Instead of attempting to describe all techniques in use, this discussion will be limited to a detailed description of two of the most widely used techniques plus a brief discussion of a few other types.

Ramp Encoder

Figure 20 illustrates an analog-to-digital converter that has found wide use and requires relatively few precision components. A command to encode (from the pro-

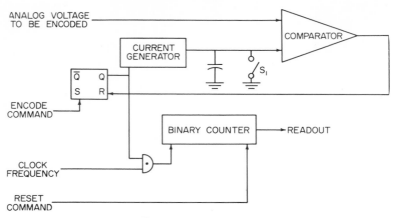

Fig. 20. Ramp encoder.

grammer) sets the control flip-flop, which in turn causes a current generator to charge a capacitor and which also gates clock pulses to a binary counter (see Sec. 6). If the current generator provides a current of constant magnitude, the voltage across the capacitor will increase as a linear ramp. The ramp voltage and voltage to be encoded are compared by a comparator, and when the ramp and input voltages are of equal magnitude, the comparator changes state, causing the control flip-flop to be reset and the counter stopped. The binary number contained in the counter is then proportional to the magnitude of the analog input. Another pulse from the programmer causes the capacitor to be discharged through a switch. After reading the contents of the counter into a register, the counter is reset and ready for the next encoding cycle. The principal sources of error in the encoder are nonlinearities in the ramp and variations in the slope of the ramp. The biggest deviation from linearity usually occurs at the initiation of the ramp. An improvement can be made by causing switch S_1 to discharge the capacitor to a negative voltage. When an encoding cycle is begun the generator is turned on although the counter is not turned on. A second comparator detects when the ramp voltage equals zero and causes the counter to be turned on. Thereafter the operation is the same as described above.

This modification improves linearity but has no effect on the magnitude of the slope. To provide control over slope, another comparator can be added which compares the ramp voltage to a precision reference voltage near full scale. If this technique is used neither the current source nor the counter is turned off when the ramp voltage equals the analog input, but the contents of the counter are nondestructively read out into another register when coincidence is reached. After coincidence is reached, the process continues until the second comparator determines that the ramp voltage is equal to the precision reference. The counter should then contain a predetermined number corresponding to the encoded value of the reference voltage. Any error occurring may be used to generate a signal to correct the magnitude of the current charging the capacitor.

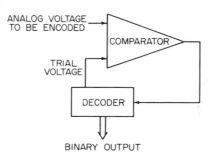

Fig. 21. Feedback encoder.

DATA-TRANSMISSION-SYSTEM DESIGN

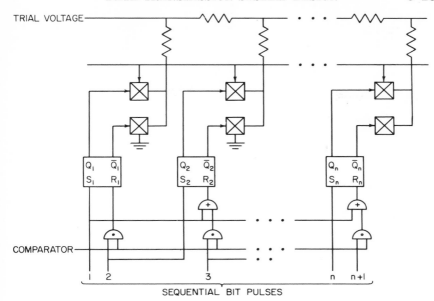

FIG. 22. Feedback-encoder logic diagram.

Feedback Encoder

The feedback, or successive-approximation encoder, operates by successively comparing the voltage to be encoded with a series of trial voltages and changing the trial voltage in accordance with decisions as to whether the trial voltage was greater or less than the analog input.[15] The first voltage with which the input is compared is one-half of the maximum or full-scale value. If the input exceeds the trial voltage, the trial voltage is increased by one-fourth the full-scale value, to three-fourths full scale, and another comparison is made. If the first trial voltage had exceeded the input, the one-half full-scale value would have been removed and the one-fourth value inserted in its place. The third trial involves adding one-eighth of the full scale, the fourth one-sixteenth, etc., until the input is encoded to the desired precision. Figure 21 shows a typical encoder. The trial voltages are generated by a decoder (so named because it decodes a digital word into an analog voltage). The decoder in Fig. 22 is a flip-flop register which controls, through switches, a resistor ladder. For the first trial all switches except the one controlled by flip-flop 1 connect the ladder to ground while the first switch connects the ladder to the reference voltage, which is the full-scale

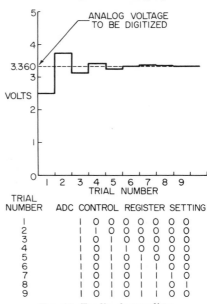

FIG. 23. Feedback encoding.

value. The first trial is performed by comparing the voltage generated by the decoder with the input. On the second clock pulse of the encoding cycle, the second switch is closed to the reference voltage. The first switch is left connected to the reference if the first trial indicated the input exceeded the trial, or it will be switched back to ground if the results of the first trial were reversed. Figure 23 illustrates the encoding of an input of 3.360 volts by an encoder whose full-scale value is 5.000 volts.

The error introduced by quantizing, i.e., the amount by which the trial voltage may differ from the input (assuming perfect circuits), is voltage equal to the value of the least significant (last) bit of the encoded word. This is easily seen since an input whose analog value corresponds to just slightly less than the least significant bit will be encoded as 000 . . . 0. Note that the error is always positive so that error equals +1 bit, −0. The error can be made symmetrical, i.e., ±½ bit, by offsetting all trial voltages by a voltage equal to the value of one-half of the least significant bit. This is easily done by adding one more section to the decoder ladder. No switch is required since this section is permanently connected to the reference.

The accuracy of encoding is dependent upon the accuracy with which the trial voltages may be generated and the stability of the comparator. The trial-voltage error is dependent on the tolerance and stability of the resistors used in the ladder, the reference and offset voltages, and the closed resistance of the switches. The effect of reference-voltage drift can be eliminated if all transducers can be excited by the encoder reference voltage. The stability of the comparator can be improved by modifying the encoder in such a way that the trial voltages are of a polarity opposite to the inputs. The input and trial voltage may then be added in a resistive summing network and applied to one input to the comparator while the other input to the comparator is held at ground potential.

The maximum operating speed of the encoder is determined by the speed with which the ladder switches may be opened and closed and by the time required for the comparator decision. Clock rates of 500,000 bits/sec are common with over 1,000,000 bits/sec possible using transistor switches in the ladder. By using high-speed gallium arsenide diodes as the switches, clock rates as high as 12,000,000 bits/sec have been obtained.[16] The encoder just described is a "linear" encoder, in that the conversion between the binary number obtained from the encoding to the actual value of the input involves only multiplication by a constant, if quantizing error is ignored. (Strictly speaking, any analog-to-digital conversion is nonlinear since an infinite number of input states generate a finite number of output states.) If the trial-voltage generator does not provide an output linearly proportional to the binary input, nonlinear encoding can be obtained. Decoders with logarithmic and other characteristics have been reported[17,18] and, while finding little or no use in general-purpose telemetry appear quite useful for encoding voice data.

Other Encoder Techniques

The ramp and feedback encoders described above are the most widely used in telemetry, but many others have been developed and reported. A paper has described an encoder built entirely of cascaded identical networks, each network being composed only of operational amplifiers. This encoder was built using integrated semiconductor networks for all components.

An encoder using only one operational amplifier and one comparator has been reported.[19] This encoder, by means of switches and capacitors, performs repeated trials on the input by successively subtracting trial voltages and storing remainders between trials. One of the earliest and still possibly the fastest encoder uses a coding tube.[20] This tube is a modified cathode-ray tube which has a screen made of a pattern representing coded outputs. The voltage to be encoded is applied to one set of deflection plates of the tube and the output is obtained from light sensors monitoring the illuminated pattern on the screen.

Sample-and-Hold Circuits

All the encoders described require some time to encode the input. If the input can change during the encoding cycle, erroneous encoding will result. The effect of encoding time can be overcome with the sample-and-hold circuit illustrated in Fig. 24. The switch S_1 closes at the time of the closure of the multiplexer switch or slightly afterward, allowing the buffer amplifier to charge the memory capacitor. At some later point in time before the time of the opening of the multiplexer switch, S_1 is opened and the voltage to be encoded is held on the capacitor. Actually, S_1 should close slightly after and open slightly before the multiplexer switch so no errors are introduced by multiplexer or amplifier transients. The output impedance of the buffer amplifier should be as low as possible to allow the capacitor to be charged rapidly to the signal voltage and to allow the capacitor voltage to follow closely the variations in

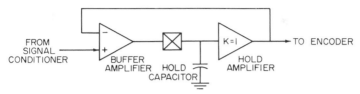

Fig. 24. Sample-and-hold circuit.

input signal while S_1 is closed. The stored value in this case is very nearly the value of the input at the instant when the switch was opened; so the sample-and-hold circuit becomes a very good approximation to an ideal impulse sampler.

To prevent error being introduced by discharge of the memory capacitor, the input impedance of the hold amplifier should be as high as possible.

8 Data-processing Design Considerations

Thus far, the PCM telemetry system has been described as only a data-gathering and -encoding device. This, with a few exceptions, is the manner in which it is most often used. Data are gathered in some remote place and transmitted to a centralized receiving station where they are processed to extract the useful information. For most industrial applications, as well as aircraft, missile, and near-earth orbital applications, this is the only practical method as it allows time sharing of expensive data-reduction equipment. Further, in airborne applications, as much equipment should be on the ground as possible, where size and weight are of little consequence. In the area of aircraft, missile, and near-earth orbital applications, a vast majority of the telemetered data are of no real-time importance. The data are recorded as received and reduced at some later time. Here again data-processing equipment should be on the ground, where failure of the processor results only in a delay in obtaining information rather than an experiment ruined because some equipment in the telemetry system failed.

A different situation prevails in deep-space probes. The transmitter power must be made as small as possible to minimize transmitter and power-supply size and weight. The real expense of inefficient transmission is that worthwhile hardware for experiments is often replaced by batteries and transmitters.

A cursory examination of telemetry data shows that a tremendous amount of redundant data is telemetered. Parameters such as power-supply voltages, temperature of an equipment, or stress on a structural member are expected to have some value. We have no interest in these data unless they deviate from an expected value. Further, even experimental data will undergo periods of no change.

At times we are not interested in each piece of data detected by a transducer but in some information computed from these data. Some reduction in data rate, and hence required bandwidth and power, is often obtainable by computing the quantity desired before transmission.

The subject of data processing in the telemetry system is far too broad to attempt coverage of any extent here. In fact, it does not lend itself to coverage in a general manner as the reduction technique depends strongly on the application. There are, however, a few papers of a general nature in the literature.[21]

Very broadly, we can consider two types of data processing which might be applied in a telemetry-system processing—those which theoretically preserve all information obtained from the transducer and those which do not preserve all information.

The first process, preserving information, involves primarily removing redundancy from the data. In the earlier discussion of programming and multiplexing, a step in this direction was made when subcommutation was discussed. Data channels are sampled at a rate compatible with maximum data frequency. Even when tailoring sampling rate to the maximum expected frequency, redundancy in the telemetered data must exist, since the sampling rate was chosen for maximum rate of change of input, yet the input will most probably not always be changing at maximum rate. Two possible solutions exist to removing this redundancy. The system may be made adaptive, in that it may contain computational capability to monitor the actual rate of change of data and adjust sampling rates to be compatible. Alternately, new sampling rates may be determined at the receiving site and format changes commanded by another link.

The redundancy may also be removed while maintaining constant sampling rates by storing each data sample and then comparing each sample with the previous sample and transmitting the data only if a change occurs. The output data rate is set proportional to the average expected rate at which data channels will be changing. A buffer must be added to accept the excess data when more than the average number of channels are changing and reading the data out during times of less than average activity.

The second class of data processing listed above involves losing some data. In many instances, it is not necessary to reproduce the telemetry system's input signal completely at the receiving site. For example, consider the data obtained from nuclear-particle detectors, radiation detectors, or meteorite detectors aboard a space probe. A typical detector, or group of detectors, would provide signals proportional to the energy of the impinging particle or photon. The energy levels could be quantized and counters provided in the telemetry system to accumulate the number of bits of particles in each energy band during the same time span. At the end of the interval, the system relays the content of each counter to the receiving station. Certainly, some information has been lost—the receiver cannot determine the energy level of any particular particle, nor when the hit occurred. Such information would be available if the sensor output had been continuously telemetered; however, in experiments of this type, the statistical distribution of the data may be the only information desired.

The processes described above require considerable equipment and are justified only when the weight of the additional telemetry equipment is far less than the weight saved in transmitter and power-supply equipment. One process is available, however, which may require little additional data equipment. Again recall the advantages of using several sampling rates in one telemetry system and also recall that, in a PCM system, bandwidth is proportional to both sampling rate and accuracy, as greater accuracy requires greater word length. A system with many input channels will very likely be sampling data from sources of widely varying accuracy. Consequently, only a very few channels require high accuracy. Considerable savings in bandwidth are obtainable by varying word length. Those channels whose data are derived from very accurate transducers utilize the full capability of the encoder, which might be 10 or more bits per word, while data from less accurate devices would be transmitted as shorter words. This increases the complexity of the programmer as the number of bits per word must be programmed; however, the reduction of bandwidth often saves sufficient weight in the transmitter and related hardware to justify the increased programmer complexity. Even the weight savings may not be important when other data links of fixed bandwidth, such as telephone lines, are employed. The objective in such a case may be to reduce bandwidth, reduce the number of channels, or seek a new data link.

9 Digital Modulation Techniques

After the data have been multiplexed and encoded and synchronization words inserted, they must be put in a form for transmission. Final transmission is accomplished after a second modulation process in which the pulse-code-modulated signal in turn modulates the carrier via FM, PM, or AM techniques. It is also possible to have a multichannel PCM system in which each of the pulse trains modulates a subcarrier which in turn modulates the carrier signal. In this section we shall consider only the ways in which the digital information at the output of the system may be represented before the FM, AM, or PM process.

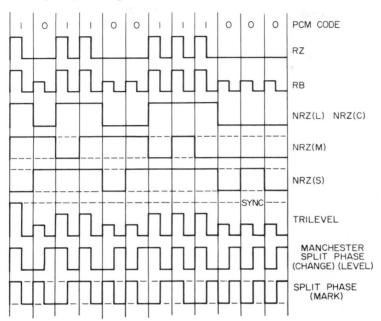

FIG. 25. Methods of binary data representation.

A number of different coding techniques are available (Fig. 25).

RB—The return-to-bias method of representing binary data which entails three fixed envelope levels, the highest (generally) for a 1, the middle level for a 0, and the lowest or bias level at which the envelope always remains during the latter half of each bit period.

RZ—The return-to-zero method of representing binary data where a 1 is indicated by a change of envelope to the 1 level for one-half the bit interval after which the signal returns to the reference level for the remaining half of the bit interval; a zero is indicated by no change in the signal (that is, it remains at the reference level).

NRZ(L)—The non-return-to-zero representation of a bit wherein a bit pulse remains in one of its two level states for the entire bit interval [also denoted by *NRZ(C)*].

NRZ(M)—A similar coding method wherein a level change is used to indicate a "mark" or 1.

NRZ(S)—Same as above, except that the level shift is used to indicate a "space" or 0.

Tri-level—Similar to *RB* except that four fixed envelope levels are required, the highest for word sync, second highest for a 1, second lowest for a 0, and lowest for the bias level at which the envelope always remains for the latter half of each bit period.

Split-phase (Manchester)—A method whereby a 1 is represented in the first half of the bit interval with a 1 level, then shifted to a data 0 level for the latter half; and a data 0 is indicated by the reverse representation. The split-phase (mark) method represents a "mark" or 1 with a 180° change in phase (with reference to the previous phase) at the beginning of the bit period.

DATA RECOVERY

10 Data-recovery-system Considerations

Before discussing the details of the bit, word, and frame-synchronization processes, it would be well to consider the general data-handling requirements of the ground station. In a typical telemetry system, a fixed number of bits constitutes a syllable or a word; and a fixed number of words constitutes a data frame. Ordinarily a word would represent one digitized sample from a data channel and throughout one frame each data channel would occur at least once. The frame is repeated continuously at a fixed rate for the duration of the transmission unless there is some change in the multiplexer which would result in a different frame format.

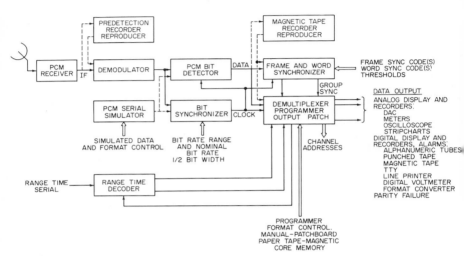

FIG. 26. PCM data-recovery system.

Some channels will provide a data sample exactly once per frame, whereas other data channels will provide samples of data at an integral number of times, equally spaced throughout the frame. Since the format of the data remains fixed throughout a transmission and is known at the receiving station, the problem of the data-recovery system is to regenerate or detect the incoming data bits, determine the location of the first bit in the frame, then by means of a counter with programmed control to decommutate or demultiplex each of the data words to its appropriate address (display, recorder, format converter, or printer) as desired by the operator. Specifically, the order of operations which must be executed by the data-recovery system (Fig. 26) for it to perform its function is

1. Recover bit or clock timing—lock onto the incoming bit frequency and phase of the data envelope
2. Detect and regenerate the data, then translate them to the logic levels of the demultiplexer and processor
3. Acquire frame and then word synchronization (or the converse)
4. Commence demultiplexing, recording, and processing of the received data

Data cannot be regenerated or detected accurately and reliably unless accurate bit frequency and phase information are available in the bit-detection and data-regeneration circuit. Similarly, it is impossible to demultiplex and process data until group synchronization, both frame and word sync, has been acquired which enables the programmed counting circuits in the demultiplexer to be reset at the proper time.

In the material which follows different approaches are presented for implementing the four steps mentioned above. Generally a method is discussed in more detail where it is considered to be the prevalent or the most illustrative technique that is in use.

11 Bit and Group Synchronization

Bit Synchronization

Fundamental to all PCM telemetry systems is the requirement of the ground station for recovering bit timing, i.e., bit synchronization. In short, before PCM data may be detected with a high degree of accuracy and thereafter demultiplexed, the receiver must be able to determine the beginning and the end of each bit interval so that it may sample the input waveform at the optimum time (middle of the bit interval for sampling detection) or begin and end its integration at the correct time (for an integrating detector).

In modern PCM telemetry systems there are at least three basic methods of deriving bit timing or bit synchronization at the receiver. They are

1. Derivation from a primary or a secondary time standard—in applications where both the transmitter and receiver stations may be slaved to a master timing system.

2. Utilization of a separate synchronization channel or signal.

3. Derivation of bit timing implicit in the modulated carrier. This technique is by far the most common in PCM systems with r-f as opposed to wire links. Such systems are said to be self-timed and take advantage of the 1-0 and 0-1 transitions in the data stream. There are at least three ways in which this can be done: (a) The bit transitions (or zero crossings) are conditioned by a slicing circuit, then differentiated and rectified to drive a high-Q resonant circuit which will free-wheel or coast over nontransition periods. This method has been applied to a number of wire-line data-transmission systems.[22] (b) A second method involves the utilization of the bit transitions by means of a slicing circuit, differentiator, and rectifier in conjunction with a one-shot multivibrator, phase discriminator, loop filter, and VCO to obtain frequency and phase of the clock rate by means of a phase-locked loop (PLL). It should be noted that the first and second methods mentioned above receive no bit-timing information during long sequences of 1's or 0's (*NRZ* data). (c) A third method involves setting the bit period equal to an integral number of carrier cycles; therefore, by using a phase-locked loop on the carrier it is a relatively easy matter to provide bit frequency by counting down to the bit rate and then to determine the correct bit phase by use of the bit transitions mentioned above.

A number of basic assumptions are made in the discussion of bit synchronization to follow. Briefly, these assumptions are:

1. The use of *NRZ* modulation.

2. Equal probability of a data 1 and a data 0—with the restriction that at least one data transition will occur every 64 bits (see IRIG Telemetry Standards, Chap. 5). In some applications, an odd parity check is used to ensure at least one bit transition per word, which would provide a higher minimum transition density for most systems; however, this method is effective only when the word length (including parity bit) is an even integer.

3. The received video pulses are assumed to be of the form $[(\sin x)/x]^2$, which is the impulse response of an ideal low-pass filter with flat response throughout the passband and no phase delay. No spectral component at the bit frequency is present; consequently, some nonlinear process is needed (e.g., differentiation and rectification) to introduce a frequency component at the bit rate (see Bennett[22]).

4. The noise in the received signal is additive band-limited white Gaussian noise and introduces a time or phase jitter on the detected zero crossings.

5. Video input-filter response is constant from direct current to 0.5 to $0.8R$, where R is the bit rate.

Basic Phase-locked-loop Considerations

As Bennett has shown, the quasi-periodic zero crossings (or transitions) of an NRZ sequence of bits may be used to control a pulse generator which drives a circuit tuned to the bit rate. The axis crossings of the tuned circuit will be evenly spaced but the amplitude will vary in accordance with the density of input pulses. Since noise is generally present on the NRZ data the output of the tuned circuit would contain an rms phase jitter, the amplitude varying additionally because of the noise fluctuations. Such a method of implementing a bit-timing-recovery circuit may be used only in cases where the clock source and channel are relatively stable. In most missile and space-vehicle applications, some means must be employed to track the incoming bit rate over a range which may be as much as ± 5 or 10 per cent of the nominal bit rate. Hence a phase-locked loop is employed which in essence may be thought of as a relatively narrowband high-Q filter whose center frequency follows or tracks the rate of the received bit transition pulses. In general, a phase-locked loop is used to allow a local oscillator to follow the phase and frequency of a continuous incoming signal. In the present application, however, a phase-locked loop would not have a continuous signal but a random input as a function of the bit transitions or zero crossings of the received data; consequently, this latter application works more like a pulsed error servosystem with an input or an error signal occurring intermittently.

For the purpose of the discussion to follow the main components of a phase-locked loop are grouped into the following blocks: (1) signal preconditioning, (2) phase detection, (3) loop compensation, and (4) voltage-controlled oscillator. These major blocks will be discussed with reference to a typical implementation in a bit synchronizer indicating the main characteristics of each. (For further information concerning design and performance of phase-locked loops in general the reader is referred to the papers by Jaffe, Rechtin, Gilchriest, and Weaver,[23,24,25,26] and to Chap. 9, Secs. 16 to 18.

Signal Preconditioning

Generally, the clock-timing-system phase-locked loop is a part of what is called the input section of a telemetry-receiving station. Such an input section would have two primary functions: to remove noise from the raw data entering the ground station and to provide a bit-rate clock. Figure 27 is a block diagram of a typical input section for a telemetry-receiving ground station. The input signal comes from a telemetry ground-station receiver, a serial code simulator, or a serial tape recorder. The data train enters the input section through the input attenuator which is designed to bring the input level to that appropriate for a given unit in the ground station. The signal from the input attenuator is fed to the video amplifier before going to the input low-pass filter. The amplified video then goes to the input low-pass filter, which has been designed to remove wideband noise. In typical applications, this low-pass filter would be adjustable to handle the full range of bit rates from the lowest to the highest. Ordinarily, the cutoff frequency for the filter should be from 0.5 to 0.8 of the bit rate.[27,28]

From the low-pass filter, the signal is fed to the level sampler and slicing circuits. The level sampler has been designed to sample the positive and the negative pulses and provide a signal to the AGC-ADCL (automatic d-c level) amplifier and the appropriate slice levels to the slicers. The AGC-ADCL amplifier functions as both an automatic gain control for the video amplifier and a d-c restoration circuit to keep the video-amplifier output symmetrical around zero volts. The latter is required since shifts in the mean signal level will occur because of duty-factor variation implicit in a long series of 1's or 0's. Although a-c coupling is desired for both stability and gain where impedance transformations can be used, this form of coupling causes the voltage level to decay back to the coupling reference-potential level and thereby reduces the signal-to-noise margin. The AGC-ADCL amplifier feeds the difference of the average peak levels in the filter output back to the video amplifier as AGC and feeds the sum of the average peak levels back to the video amplifier as ADCL correction.

The output of the low-pass filter is also fed to the slicer which serves to reshape the received data, remove low-amplitude noise, and provide uniform rise and fall times independent of those in the received envelope. This is done by slicing the incoming waveform at the horizontal symmetry axis, i.e., halfway between positive and negative peaks, as determined by the level sampler. This slicing circuit serves as a sampling detector, and its output is used to set or reset the data flip-flop at the appropriate clock time. In addition, the output of the slicer is fed to the differentiator and rectifier. Figure 28a shows the typical waveforms of the signals into the slicer and out to the data flip-flop and to the differentiator-rectifier.

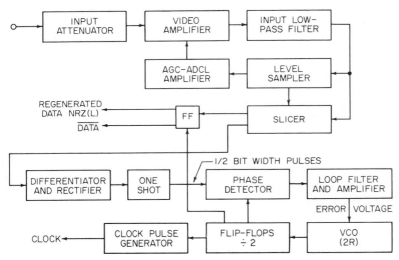

FIG. 27. Input section, telemetry-receiving station.

The differentiator and rectifier circuit performs the required nonlinear operation on the input signal to provide the discrete power spectrum containing the appropriate frequencies for use in the phase-locked loop. The output of the differentiator and rectifier circuit is a series of positive-going spikes that are used to trigger the one-shot multivibrator which produces pulses adjusted to one-half the bit width. This may be seen by referring to Fig. 28b.

Phase Detection

The half-bit-width pulse (and complement) is fed to the phase detector, which compares its phase with that of the square wave (and complement) produced by the VCO. If the VCO is at the exact frequency and correct phase, then the mean voltage level out of the phase detector will be zero and will provide no error voltage for the VCO. If, on the other hand, a frequency or phase error is present, the output of the phase detector will be asymmetrical, resulting in a d-c level being generated by the phase detector, then filtered by the loop filter, and thereby used to control the VCO to bring it back to the correct frequency and phase. Figure 28c shows typical signals in the bit-synchronization and data-regenerator system. Notice that the phase-detector output consists of a symmetrical signal which has a mean value of the bias level. However, when the VCO is off in frequency or phase, the output of the phase detector is asymmetrical as shown in Fig. 28d and thereby, when integrated, provides a d-c level on top of the steady-state bias of the phase detector. The loop filter senses this level, which is proportional to the frequency error, and uses it to control the VCO. In Fig. 28d the incoming-signal bit rate was 20 per cent lower than the center frequency or rest frequency of the VCO; therefore, the VCO is controlled in such a way as to decrease its frequency. If the incoming bit rate had been 20 per cent higher than the rest frequency of the VCO, then the asymmetry and the waveform would be reversed as compared with that in the diagram.

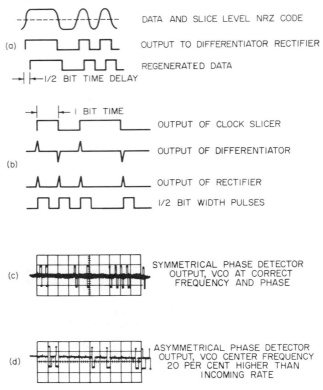

Fig. 28. Telemetry-receiving station input section waveforms. (a) At input and output slicer. (b) Pulse regeneration (to phase detector). (c) Symmetrical phase detector output. (d) Asymmetrical phase detector output.

With regard to implementation of the phase detector, a variety of techniques are available, although most are variations of the split-gate phase detector used in tracking-radar systems.[29] Sometimes called an early gate–late gate phase detector it consists in its simplest form of two logic AND gates, one being used to AND gate the VCO waveform with the leading edge of the one-shot square wave, while the other is used to AND gate the 180° phase shifted VCO output with the trailing edge of the one-shot multivibrator complement. The outputs of the two AND gates are OR gated (summed) together.

As is characteristic of most phase detectors and phase-locked loops, the phase of the VCO must be less than $+90°$ or greater than $-90°$ in order for the system to be pulled into synchronization. A phase difference between the VCO phase detector of less than $-90°$ or greater than $+90°$ tends to produce an error signal which will make the phase error even greater between the VCO and the incoming *NRZ* data envelope.

Loop Compensation

For proper operation, the bit-synchronizer phase-locked loop requires both a loop filter and a loop video amplifier. The loop filter is merely a low-pass filter designed to pass the d-c level generated by the phase detector to control the VCO. It has a very low cutoff frequency and its output is a slowly varying d-c level. An optimum filter would be one that is designed so that the noise interference is minimized while maintaining some desired minimum transient phase error between the actual output and the desired output from the loop. The first question that arises with reference to the loop filter is: "What should its bandwidth be?" A loop filter with a relatively narrow bandwidth is one that would tend to have a memory for coasting over relatively long periods of no error-signal input and would tend to maximize noise rejection. However, two disadvantages of a relatively narrow bandwidth filter are that it could not track changes so readily as a relatively wideband loop filter and acquisition would be more difficult because of the narrower capture range of the loop. The capture range is defined as the frequency difference between the VCO and the incoming bit rate over which the system will pull into synchronism.

If, on the other hand, the filter is a relatively wideband filter, then its memory for coasting over long periods without an error-signal input would be relatively short and its noise rejection is relatively poor. Bit-rate changes can be tracked more readily and acquisition is faster since the capture range is greater. Thus it is apparent that in an optimum system, the bandwidth of the loop filter should be relatively wide for acquisition since this would tend to minimize the acquisition time and should be narrow after the phase-locked loop has locked in order to minimize the effect of noise and provide a minimum mean-squared phase error in the operation of the system. Although some phase-locked loops are designed with a variable-loop filter to take advantage of the wideband and narrowband filter characteristics, most bit synchronizers that have been developed to date have a single-loop filter bandwidth for both acquisition and lock so that a compromise has to be made for both the acquisition and lock modes of the loop. If the assumed requirement of rms phase jitter is less than 8° (0.14 radian) with a 3-db video signal-to-noise ratio, then it can be shown that the ratio of loop bandwidth BW_L to bit rate R should be

$$BW_L/R \approx 0.0015$$

for a typical phase-locked loop.[30] This would limit the tracking of the bit rate of change to $0.0015R$ cps.

The loop amplifier shown in Fig. 27 is designed to provide the desired loop gain. The loop gain required for the system is a linear inverse function of the bit-transition density—in other words, it decreases linearly with an increase in density of bit transitions. Also, the wider the required hold-in range, the higher the gain in the loop. Hold-in range is defined as the frequency difference between the rest frequency of the VCO and the incoming bit rate over which the system will maintain synchronism. In addition, a more rapid response to a change in bit rate implies the need for higher gain in the loop. On the other hand, too much gain could result in an instability and too little gain would imply that the system would be sluggish in responding to changes and in acquiring synchronism.

Voltage-controlled Oscillator

In the block diagram of Fig. 27, the voltage-controlled oscillator consists of a free-running RC multivibrator operating at twice the incoming bit rate with its frequency controlled by modulating a constant-current source that changes the charging time of the cross-coupling capacitors. The output of the multivibrator is used to drive two flip-flop circuits both of which provide a square wave at the bit rate with the output on both sides of both flip-flops; thus four phases at the clock rate are available for use in the phase detector, the clock data-regenerator flip-flop, and for control of the system clock pulse generator. This may be seen in Fig. 29, which represents a master phase diagram of a typical bit synchronizer showing the four phases of the output flip-flops from the VCO (A, B, C, and D). The symmetrical detector output indicates that the VCO is operating at the correct frequency and phase.

An LC voltage-control oscillator may be used with control by some type of saturable reactor.

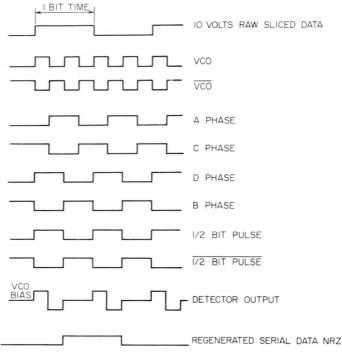

Fig. 29. Master phase diagram—*NRZ* mode.

Conclusion

Operationally, there are four characteristics that affect the input signal to the phase-locked loop: (1) the randomly intermittent input of one-zero transitions which provide an intermittent error signal; (2) instabilities in the clock source at the transmitter; (3) changes in the propagation-path length between the transmitter and receiver; and (4) time jitter on the input signal due to noise.

All the above factors influence the operation of the phase-locked loop and produce some rms phase error in the output of the loop. However, if the signal-to-noise ratio is greater than 3 to 5 db, the effect on bit error rate is negligible.[31]

Because of the nonstationarity of the input signal and the nonlinearity of the phase-locked loop during acquisition (the most critical period) no satisfactory analysis has been made which accurately describes the operation of the system; however, simplifying assumptions have been made on the nature of the input signal which indicate that the loop as described operates in a nearly linear fashion after lock-in.

Group Synchronization

It is assumed that bit synchronization has been acquired and that data are being regenerated at the appropriate logic levels for use throughout the data-recovery system. The next step, therefore, is to acquire group synchronization in order that demultiplexing or decommutation of the received telemetry data may be commenced. Although there are numerous approaches to recovering group synchronization, only three basic ones will be discussed: (1) acquisition of word sync as a prerequisite for acquiring frame sync; (2) acquisition of frame sync as a prerequisite for determining word sync; (3) complementary or integrated frame- and word-sync acquisition.

DATA RECOVERY 8-37

A brief survey discussion is given on each of these approaches with considerable detail given on the problem of frame-sync acquisition with a table of some preferred frame-synchronization codes.

Word-sync Acquisition

Although the acquisition of word sync is not sufficient for decommutation, it is very helpful in the acquisition of frame synchronization since it enables the frame-sync circuits to be gated at word-sync time and therefore would minimize the probability of a false frame-sync indication due to random coincidence. In recent years, there has been a tendency to omit word-synchronization codes in telemetry formats because of the relatively high bandwidth requirement. Nevertheless, in some applications where word-sync coding would enhance the operation of the system or simplify the requirements at the receiving station the use of word-sync codes may be justified.

There are two principal methods of acquiring word synchronization. The most common is one in which one or more bits are used in each word at a given location so that the receiver merely has to scan for these bits until the word-sync pattern continually reoccurs at the word rate. In general, the fewer the bits used for the word-sync pattern, the longer the acquisition process since the probability of a shorter pattern, say a single 1 or a 0 at the end of the word, is likely to occur in the middle of the word for several words consecutively. During the acquisition of word sync, circuitry must be used to keep track of the number of times the word-sync pattern has been detected in a given trial bit position in the word, and this must be compared with the minimum requirement to satisfy a given level of acquisition confidence. After a given interval of time at one bit position if the word-sync test has not been satisfied, then the circuit slips to the next bit position for a new trial. The process is repeated until the correct bit position for word sync has been determined and verified.

In applications where the demultiplexer contains a relatively long serial shift register, advantage may be taken of it to detect in-parallel word-sync codes in multiple adjacent word positions. For example, suppose the input serial shift register contained 32 stages and the data word length were 8 bits; then the word-sync detector could look for four word-sync codes simultaneously in one bit position, which may be sufficient for one trial at that bit position. This speeds the word-sync-acquisition process. Tests have been made with word-synchronization circuits operating in such a fashion and have shown that, even under poor signal-to-noise ratios, word sync was acquired within 1,000 bits, or well within the period of a typical missile or spacecraft telemetry frame.[32]

The other method of acquiring word synchronization is to utilize the parity-check bit which occurs at the end of each word and constrains it to either an even or odd number of 1's. The receiver determines word synchronization by trying out different bit positions for the parity condition; if the position being checked is not the correct parity position, the parity check will fail on the average one-half the time. Circuitry must be provided to reject a bit position being tried if it does not satisfy some minimum criterion so that the next bit position may be tested. When at the correct position, the circuit must indicate that the sync criterion has been satisfied, i.e., a very high percentage of satisfactory parity checks have been made over the interval of the trial.

The primary advantage of this method is that it makes use of the parity bits which have been inserted in the data to indicate their quality after being received; hence word sync is provided at no extra cost in bandwidth. However, this method is primarily useful when the signal-to-noise ratio is relatively high (i.e., bit error rate 10^{-2} or better) since marginal signal conditions entail much longer acquisition times because of the relatively high percentage of parity failures.

In general, it may be said that the constraints put on the telemetry data by error-detection and error-correction codes may be similarly used to acquire word synchronization, and several such systems have been proposed.[33]

As previously mentioned, once word sync has been acquired, the frame-sync-detector circuit may be gated at word-sync time in order to minimize the probability of a false sync indication. Furthermore, knowledge of word sync implies that the frame-sync code may be detected with a serial device (with reduced hardware requirements) as readily as with a parallel device.

Frame-sync Acquisition

From the system standpoint, the design of a frame-synchronization system involves three main steps:

1. The selection of a frame-sync code for a given format and system
2. The determination of a means of detecting the selected code at the receiving station
3. The logic design of the acquisition circuit

In the selection of a frame-sync code, a decision must be made as to its length and as to its composition of 1's and 0's. In regard to its length, it may be seen that the problem of determining the location of the beginning of the frame is essentially one of positioning or labeling one bit out of M bits in the frame. Thus information theory indicates that the number of bits required to specify the beginning of the frame is

$$L = (\log_2 M)$$

where the parentheses indicate the next largest integral value. This, of course, is the minimum number of bits required in the case of an error-free channel and would be required only one time since, after once being acquired, frame sync will be maintained as long as bit sync is maintained. However, the error-free case is unrealistic because the frame-sync code is subject to degradation due to a noisy channel, particularly when the signal-to-noise ratio is marginal. In addition, the problem is further complicated by the fact that the frame-sync code has to be multiplexed with the other PCM data; therefore, the frame-sync detector has no means of discriminating data from the frame-sync code except by appearance alone. In most applications, the frame-sync code will appear in the data with approximately the expected probability for random data, 2^{-n}, where n is the length of the code (assuming no errors are allowed). Thus the two problems of a noisy channel and the probability for random data having the appearance of the frame-sync code indicate the need for a frame-sync code with a length somewhere between two and five times L in order to achieve rapid frame-sync acquisition. The specific value selected for a system would depend upon the worst-case operating conditions of signal-to-noise ratio and the desired mean acquisition time. In many missile and spacecraft telemetry formats, a factor of three times L is found to be typical. This redundancy factor in the frame-sync code performs two important functions: namely, it enables the code to correct for errors which may occur in it because of the channel noise, and it reduces significantly the probability that the sync code would appear in random data and produce a false sync indication during acquisition, thus prolonging the acquisition process.

Still another problem is that of a false sync indication being produced while part, but not all, of the sync code has been received. The simplest example is that of an all 1's sync code where all but the last bit of code has been received. If the data bit received just prior to the first bit of the frame-sync code happens to be a 1 then the sync-code detector will produce a false or premature sync indication. Any code having an inherent periodicity (all 1's, all 0's, alternating 1's and 0's, etc.) will have this problem since the data bits just prior to the code will, with a certain probability, conform to the periodicity and cause a detection ambiguity. In general, codes which have no inherent periodicities are preferred not only because they minimize the probability of premature sync indications but also because they maximize the tolerance for errors in the code throughout the acquisition process.[34]

The simplest means of checking for periodicities in a code is to compute its cyclic autocorrelation functions $\Phi_c(t)$. Let $a_0, a_1, \ldots, a_{n-1}$ be an n-bit code where the $a_i = \pm 1$, then

$$\Phi_c(t) = \sum_{i=0}^{n-1} a_i a_{i+t}$$

DATA RECOVERY

with $i + t$ reduced modulo n as necessary. A simple method of computing this function is illustrated below for the 7-bit code 1110010:

t	0	1	2	3	.	.	.						
Code plus one repetition	1	1	1	0	0	1	0	1	1	1	0	0	1	0
Code to be shifted		1	1	1	0	0	1	0						

If the shifting code in its above location ($t = 1$) is compared with the corresponding bits above it, it will be seen that there are 3 bits alike and 4 different; therefore, $\Phi_c(1) = 3 - 4 = -1$. This need only be done for $t = 1, 2,$ and 3 since the function has a period of 7 and is symmetrical about the value $\Phi_c(0) = +7$.

The cyclic autocorrelation function has a maximum value of n for $t = 0$; any other relative maxima in the plot of the function are indications of periodicities in the code and result in detection ambiguities. Hence codes that have ideal (or nearly ideal) cyclic autocorrelation functions, i.e., possess only one maximum, minimize detection ambiguity and maximize the tolerance for errors in the code. Such codes are usually referred to as pseudorandom codes and consist of two main categories: (1) maximal length shift-register generator codes and (2) Legendre codes. A number of these preferred codes are shown in Table 3. An ideal cyclic autocorrelation function has the value of n for $t = 0$ and -1 for the values of $t = 1, 2, \ldots, n - 1$. Figure 30 is a plot of the cyclic autocorrelation function of a pseudorandom code.

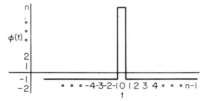

Fig. 30. Cyclic autocorrelation function of a pseudorandom code.

Pseudorandom codes are limited to values of $n = 2^k - 1$ (for maximal-length codes) and to values of $n = 4m - 1$ where m takes on any positive integer value such that n is a prime (for Legendre codes). There will be occasions where a pseudorandom code will not be of the appropriate length for use in a particular requirement. In general, the length of the code is chosen to be equal to the word length or a multiple thereof which more than likely will not be exactly equal to the length of a pseudorandom code (where n has the value of 7, 11, 15, 19, 23, 31, 43, and 47). Therefore, some means must be devised to determine preferred codes which are not of the length of a pseudorandom code. Often, a pseudorandom code may be augmented or truncated as required to meet a frame-sync-code requirement. However, the cyclic autocorrelation function of such a modified code should be checked to determine if it has any undesirable peaks or side lobes near the main peak of the function.

One limitation of the cyclic autocorrelation function in checking the desirability of the code is that it gives no indication as to which of the n cyclic permutations of a basic code is the best to use. Consequently, the truncated autocorrelation function $\Phi(k)$, in conjunction with the sample variance S^2, as defined below, serves as an excellent measure for determining the most desirable cyclic permutation of a given code:

$$\Phi(k) = \sum_{i=0}^{k-1} \overline{(a_i \oplus a_{i+n-k})}$$

where $\oplus$ is mod 2 addition

$$S^2 = 1/(n-1) \sum_{k=1}^{n-1} [\Phi(k)/k - \tfrac{1}{2}]^2$$

Table 3

n	Code	Remarks
	Pseudorandom codes, n = 7 to 31	
7	1110010	SRG (3, 1, 0)*
11	11100010010	Legendre
15	110010101111000	SRG (4, 1, 0)
19	1100100100001111101	Legendre
23	11110010011010010000101	Legendre
31	1111100110100100001010111011000	SRG (5, 2, 0)
	Other miscellaneous codes, n = 27 to 43	
27	101110100010011100000 (27 bits)	Truncated pseudo-random, SRG (5, 3, 2, 1, 0)
31	1111010100101011010101100110010	IRIG, Doc. 106-60 Recommended Code for Maximizing Sync Retention (3-bit Aperture)
43	1110111100011100010110101101101010010100100	Legendre

*This notation indicates that a simple three-stage shift-register generator (SRG) will produce the code by adding the third- and first-stage outputs together, mod 2, and feeding the result to the input of the first stage at the next clock time.[35]

For a given code, the permutation which has the smallest S^2 will tend to have the minimum probability of a false sync indication.

It is pointed out that the complement of a code, its time inverse, and the complement of the time inverse all contain the same statistical properties as the basic code word. Therefore, once a good code is found it automatically provides three additional good codes for use in subframe synchronization, etc.

The second aspect of the frame-synchronization problem is that of determining a means of detecting the frame-sync code.

Frame-sync Acquisition

Assuming that no word synchronization is available, the problem of frame-sync acquisition becomes one of determining at some place in the frame the location of the first bit. Thus the frame-sync-code detector must scan all data bits since it has no way of determining the location or the start of each word; this implies that there is a need for parallel means of detecting the frame-sync code and preferably a device which would have some type of adjustable threshold so that errors may be tolerated in the

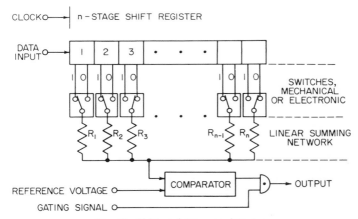

FIG. 31. Shift-register sync detector.

received code. Under very good channel conditions, however, a simple AND gate may be used to detect the frame-sync code. But if even one error is to be tolerated in the received code and a parallel means of detection is employed, the simplest method so far proposed, which is in operational use today, consists of a shift register of length at least equal to that of the number of bits in the code which feeds in parallel to a linear summing network in such a way that the code to be detected may be programmed by means of appropriate switches or logic. The block diagram shown in Fig. 31 is typical of that used in practice today. The threshold of the detector may be adjusted by several methods such as a variable-level Schmitt trigger or by leaving the Schmitt trigger at a fixed trigger level and feeding in offset voltages into the linear summing network to take care of the error tolerance desired.

In essence, such a shift-register code recognizer serves as a crosscorrelator and utilizes an m-out-of-n threshold logic circuit for a decision as to when the code, with acceptable number of errors, is in place on the shift register. It is readily apparent that this same shift register may be used to drive several or a number of linear summing networks each of which is programmed to detect a specified code or cluster of codes. As was mentioned previously, such a device may be easily used for the detection of word-sync codes if provision is made for tying the inputs to the linear summing network from the unused stages on the shift register to the appropriate voltage level. In the case of a digital correlator which uses manually controlled toggle switches for programming in the desired code, a center-off position may be used to satisfy this requirement.

In implementing the acquisition circuitry for the frame-sync system, there is a requirement for a sequential circuit to keep track of the present state or mode of the synchronizer. There are three basic modes in the frame-synchronizer circuit: (1) acquisition, (2) check or verification, and (3) lock. Ordinarily, when the system is in the acquisition mode it is scanning all bit positions until a sync indication is produced by the frame-sync correlator, at which time the acquisition sequential circuit would change to the verification mode. Since a false sync indication can be produced by a data word or combination of data words which pass the frame-sync-code test, some means is needed to evaluate especially the first received sync indication. On the other hand, after the system has locked into frame sync there is a definite probability that a frame-sync indication will be missed because of an excessive number of errors in the received code or a signal drop-out; consequently, there must be a built-in memory capability in the acquisition circuitry to allow the system to coast through short signal perturbations. As would be expected, the response of the system to the sync indications produced by the frame-sync correlator should be a function of the past synchronization history. Therefore, initially (before any codes arrive) no past history is available and the first sync indication received is assumed to be true until the screening process proves otherwise. Gating circuits are employed so that the correlator output will be observed only at the time that the next sync indication is expected and thereby a time-phase test is performed on the sync indications.

In a typical system, when a sync indication is first produced, at least one more indication is required during the next two sampling periods (at precisely one and two frames later) before the initial indication is considered true and lock-in is indicated. The lock-in condition would be maintained satisfactorily as long as no more than one or two sync indications are missed. In a universal system, both the verification-mode criteria and the lock-in-mode criteria may be programmed by the operator through the appropriate controls. For example, in a given application, the lock-in condition may be considered satisfactory as long as no more than three frame-sync indications are missed over an interval of eight consecutive frames of data. The adjustment of these parameters would be a function of the operating conditions of the system, in particular the signal-to-noise ratio under marginal conditions and the typical frequency and average intervals of signal drop-outs.

In many applications there are still other adjustments which can be made to the frame-sync-acquisition circuitry to improve its performance in maintaining lock on a given signal. Specifically, the error tolerance may be increased to some maximum allowable value as determined by the cyclic autocorrelation function of the code.[34] Further, since signal drop-outs may cause the bit synchronizer to slip one or more bits under a no-signal condition, it would be desirable to increase the aperture of the gating circuits which are used to verify the continued detection of the frame-sync code. In this manner, a slippage in the bit synchronizer would be detected and corrected by the frame synchronizer at the first instant when the code satisfies the threshold test of the correlator, and the loss of data to the multiplexer and data-processing system would thereby be minimized. If the system operated under the strategy where two, three, or even more frame-sync indications have to be missed prior to returning to the acquisition mode, it is apparent that at least four or five frames of data would be demultiplexed incorrectly before frame sync could be reacquired.

In an attempt to optimize the maintenance of lock-in by using a 3- or 5-bit aperture to verify sync detection, codes have been proposed which have a relatively high negative correlation close to the main spike of the code and therefore allow for a relatively larger number of errors in the frame-sync-correlator circuit.[36] However, this will tend to sacrifice performance of the code during acquisition because the relatively high negative-correlation levels of the code near the main spike tend to cause relatively high positive spikes in the code (or undesirable maxima) which will tend to lower the error tolerance during acquisition. If the error tolerance is not reduced, the probability of a premature sync indication while part of the code is on the correlator is increased, which

DATA RECOVERY 8-43

increases the mean acquisition time of the system. Consequently, the designer of the data-recovery system must decide where his emphasis must be placed in attempting to optimize his system for its given application or applications. In a number of flexible data-recovery systems that are operational, the operator has the option of programming the frame-synchronizer circuit to optimize it for a particular application.

Integrated Word- and Frame-sync Acquisition

In an application where the telemetry format contains both word- and frame-sync codes, a more effective means of group synchronization is one in which complementary action between the otherwise independent recognition circuits is employed. Although frame-sync lock-in implies word-sync lock-in, the reverse is not true. If word sync is acquired first, it may be used to decrease the time for acquiring frame sync by checking each frame-sync indication for compatibility (time phase) with the word-sync timing.

In such formats word-sync lock-in will generally occur before frame sync since the word-sync code appears repeatedly throughout the frame. Similarly, in formats containing more than one frame- or subframe-sync code, the subframe-sync code occurring most frequently should be used for acquisition.

This method of compatibility testing between word- and frame-sync indications to decrease sync acquisition time can be extended to the lock-in mode to decrease further the probability of needless sync drop-outs. In essence, the method here is an inhibiting action that each sync detector imposes on the other. Under certain conditions, word sync might fail its lock-in test but would be prevented from returning to the search mode by the frame-sync lock-in condition, which may be satisfactory; and vice versa.

An extension of the technique of compatibility testing is that of giving one of the sync circuits a priority over the other. This would be a function of the relative effectiveness of the two codes being used, since in a given application a format might contain a relatively strong word-sync code and yet a relatively weak frame-sync code or vice versa. In such a case, the code lock-in condition having the higher confidence would be required before the other code would be allowed to begin its acquisition process. In brief, the operator would have the option of giving either of the codes a dominant role in acquisition or neither, depending upon the conditions.

12 PCM Decommutation, Processing, and Display

Since PCM time-division-multiplexed telemetry consists of a series of bits grouped into words, generally with each word coming from a separate channel in the multiplexer and a fixed number of words composing a frame, the basic problem of the decommutator is to take the incoming data bits and segment them into words and to route each of the words (or channels) to the appropriate output device. Nearly every different telemetry application requires a unique format. Consequently, to handle a large variety of telemetry formats requires considerable built-in flexibility such as the typical one to be described.

The primary functional parts of such a typical PCM decommutator are (1) a serial-to-parallel converter; (2) bit, word, and frame counters; (3) a patch or programming board; (4) a multiplicity of output-device logic gates; and (5) a variety of output buffer registers and digital-to-analog converters (DAC), all of which are used to drive various processing, monitoring, and display devices.

The processing function, as distinguished from the decommutation function, will be defined here to mean the preparation of the raw data which have been decommutated, or "stripped out," for human or computer analysis and reduction. In particular, the processing function includes converting and editing the data and giving them a format as required by the respective output equipments such as recorders (both analog and digital), printers (both teletype and line), and display devices. In many telemetry applications the telemetry encoder must be made as simple as possible to reduce weight and to minimize power consumption; consequently, the data stripped out at the receiving station by the decommutator often must be converted and/or given a format into the appropriate code and/or format before they can be displayed or analyzed. For example, numerical data may be received in the form of straight binary (SB) and must be converted to binary-coded decimal (BCD) before they may be displayed visually in decimal form. Similarly, data being recorded for use in a computer for data analysis and reduction must be recorded in blocks according to a specified format as required by the computer program.

Editing of raw PCM data includes the selection (via gating) of specified channels for routing to the appropriate output and display devices, time-code searching as a means of starting and stopping selected output devices, and the reduction of sampling rates on highly redundant channels.

In contrast to the non-real-time data analysis and reduction which is performed on much of the telemetry information received from missile and aircraft tests and from satellite and spacecraft scientific sensors, some telemetered engineering data channels as well as data-recovery-system operational status must be monitored in real time in order that the operator may know almost instantaneously the performance of these critical channels and the station's operation. In referring to telemetered data which are monitored in real time the term often used is "quick-look." One of the simplest examples of a monitor for the data-recovery-system operation is a parity-check circuit which indicates when parity has failed or, better, counts the number of parity failures over a fixed large sample of received words and thereby gives the operator a quantitative measure of the quality of the raw data received. A number of different analog displays such as oscillographs or strip-chart recorders may be used for preliminary evaluation of the data quality while certain channels monitored in real time may provide alarm indications so that the operator may take the appropriate corrective action, if such is possible, by means of the command or control equipment.

In many, if not most, PCM missile and space-telemetry data-recovery systems there are insufficient output devices to handle simultaneously all the channels in the telemetry format. Hence the telemetry data must be recorded so that at a later time channels which were not stripped out, converted, or given a format on the original run may be set up for processing on subsequent runs. As shown in Fig. 26, the data-recovery system may include a predetection recorder/reproducer which records the relatively wideband i-f signal that goes into the demodulator or detector. Alternatively or simultaneously incoming data may be recorded on a standard magnetic-tape recorder/reproducer which takes as its input the regenerated binary data at the output of the PCM detector. Even if the data-recovery system could process all the channels simultaneously it would still be the requirement in many applications to record the incoming data in case the demultiplexer or processor were to have a failure while an operation was in progress. Ordinarily, however, after each successful run of the received data through the decommutator and processor, the subsequent runs would entail a change in the patchboard (for manual systems) or the program (for stored program-control systems) in order that the unprocessed channels may be stripped out and routed to the appropriate output devices.

The following section places primary emphasis upon a patch-programmable decommutator since an understanding of the fundamentals of its operation enables one to appreciate some of the newer trends in modern high-speed, large-capacity and versatile PCM telemetry-data-recovery systems.

Fundamentals of Decommutation

In a PCM time-division-multiplex telemetry format after group sync has been acquired the first requirement is to convert the received serial stream of bits to parallel words synchronized with the encoder at the transmitter. The second requirement is to time-gate the parallel words to the required output destination by means of decommutator counters, output-device logic gates, and programming patchboard. In reality, the parallel data are not switched but the clock or gating signal to each output buffer or DAC is controlled as necessary to receive the parallel data. Ancillary requirements are generation and checking of parity on the incoming data and the provision of channel and subchannel identification numbers for use in such output

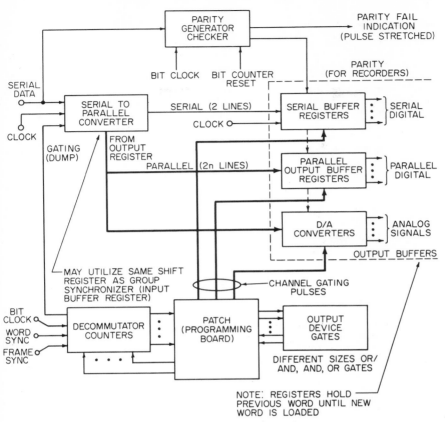

Fig. 32. PCM decommutator block diagram.

devices as a computer formatter. A block diagram of the main functional parts of a typical PCM decommutator are shown in Fig. 32. A discussion of each of the blocks will indicate its operation.

Serial-to-Parallel Converter

The inputs to the serial-to-parallel converter include the serial telemetry data, the bit clock, and a word-gating signal. For an n-bit word the output consists of $2n$ lines which are routed to the parallel output buffer registers and to the DAC's. In addition, a serial data line is routed over to the serial buffer registers. In simplest form, a serial-to-parallel converter consists of a serial register of sufficient length to handle the largest data word. And gates on the Q and $\bar{Q}$ outputs of each stage may be used to gate the data in parallel off the serial shift register onto the parallel data lines. In some applications, there will be a requirement for an additional static register of the same length to buffer the data from the serial shift register prior to unloading it to the output buffers. Since the data words will generally be of shorter length than the frame-synchronization code, it will be found that in many systems in the same input serial shift register used in the data-recovery system for driving the frame-synchronization correlator may also be used in the serial-to-parallel converter.

Decommutator Counters

In most decommutators there is a requirement for at least two counters: a bits-per-word counter and a words-per-frame counter. In other decommutators there is a further requirement for counters to index the number of prime frames within a master or major frame. The interconnection of these counters in relation to the patchboard may be seen by referring to Fig. 33. A bit clock pulse from the bit synchronizer advances the bits-per-word counter, which has previously been synchronized or reset at the proper time in accordance with the acquired word sync. A counter matrix is

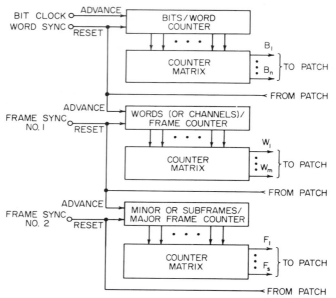

Fig. 33. Decommutator counters.

driven by the bits-per-word counter to produce an output line to the patchboard to connect the desired matrix output line back to the reset line of the counter and also to provide the advance pulse for the words (or channels-per-frame counter). The most frequently occurring frame-sync code, after acquisition, is used to reset the words-per-frame counter, which in turn drives a counter matrix whose outputs W_1 through W_m go to the appropriate jack locations on the patchboard. In addition a jumper wire from the patchboard is connected from the desired counter state for resetting the word counter and for use in advancing the minor-frames-per-major-frame counter. In some decommutators there may be more than one frame counter depending upon the subcommutation scheme and the format; in such implementations each of these counters requires a correct reset pulse from the appropriate frame synchronizer once it has locked in.

Patchboard and Output-device Gates

The patchboard provides a relatively simple and highly flexible means of performing two functions: (1) connecting the desired counter-matrix output lines to the inputs of the required output-device gates, which provides the channel gating pulses needed by the output buffer devices for stripping out the various channels; and (2) interconnecting the channel gating pulses produced by the output-device gates to the selected output buffer gating lines. In short, the patchboard receives inputs from the decommutated counter-matrix outputs and from the output-device gates outputs. It provides outputs for resetting the decommutator counters, inputs for the output-device gates, and inputs to each one of the output buffer gating lines. This is illustrated in Fig. 34a, where the patching is shown for a supercommutated channel. Here the format consists of 50 words making up a primary frame and four primary frames making up a major frame. A supercommutated channel appears on words 10, 20, 30, 40, and 50; and on all the prime frames $F_1, \ldots, F_4$. The output-device gates consist of different sizes of OR and AND gates as determined by functional requirements. In this case, a four-input and a five-input OR gate are connected by the patchboard to a two-input AND gate, and the output of the latter goes in jumpers on the patchboard to DAC No. 3 clock line.

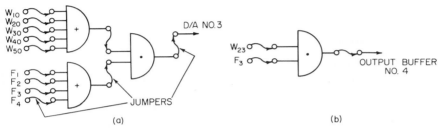

FIG. 34. Patchboard interconnections. (a) Supercommutation on channels W_{10}, W_{20}, W_{30}, W_{40}, W_{50} to D/A 3. (b) Subcommutation on W_{23}, F_3 to parallel output buffer 4.

A simpler example is that of the decommutation of word 23 in the third prime frame. Figure 34b indicates that the counter-matrix outputs W_{23} and F_3 are patched as inputs to a two-input AND gate, the output of which is patched to output buffer No. 4 clock line.

Further gating may be required on the clock lines to ensure that the data are transferred from the serial-to-parallel converter to the required output buffer within one bit period or within any other system time restriction. In some applications a high-speed direct-coupled transistor AND gate is cascaded on the clock gating signals coming into each of the output buffers to accomplish the data transfer within the time limitation required.

Output Buffers

Three types of output buffers are shown in the block diagram of the decommutator. The serial buffer registers consist of a group of single flip-flops used to transfer data serially from the serial-to-parallel converter to a serial output device such as a tape recorder. Each of the parallel output buffer registers consists of an n-stage static register which receives parallel words from the serial-to-parallel converter and holds them statically for a sample period of the respective channel. Each of the DAC's contains its own n-bit static register for holding the data word for the duration of the sample period. The output of a DAC may be fed through an impedance-matching and deflection-control attenuator to an oscillograph, bar graph, strip-chart recorder, analog computer, or merely a meter. A polarity-reversing switch may be employed on the DAC to control direction of deflection of a galvanometer.

Processing and Display

Requirements for editing, conversion, and choosing a format vary widely with data-recovery systems, applications, and data-reduction facilities. Consequently, no attempt is made to generalize or to point out a predominant requirement in conversion or choosing a format. The proliferation of codes and formats continues; however, it may be said that code conversion and choice of format tend to be problems that are more straightforward than those of synchronization and decommutation, especially in view of the increased use of magnetic-core memories in code conversion and computer formatting. Indeed, magnetic-code memories are being used increasingly in stored-program decommutators to facilitate changes in the decommutator operation for reruns and for different formats.

Two other ancillary functions required for processing and display are (1) parity generation and checking and (2) range-time decoding. Parity generation is a simple matter for data received serially since a flip-flop may be used after being reset at word-sync time to toggle on each of the 1's in the data word and thereby provide the even or odd parity bit required for comparison with the parity bit in the data (if there is one) or to provide parity for use in digital recording devices.

In order that data and events received by the data-recovery system may be located in time some type of time information is recorded along with the data. At test ranges and space telecommunication centers, time codes are provided for this use and must be suitably decoded for recording along with the decommutated data. Thus selection and elimination of received data may be accomplished for data reduction by utilization of the recorded time information. AND gates with selectable inputs and associated detection logic may be used to detect the time-coded signals and thereby control the various operations of the data-reduction system such as turning on recorders and starting the computer format.

The output data from the decommutator may be recorded digitally in parallel or serially depending upon requirements and the equipment available. Detailed discussions of magnetic-tape-recording characteristics and specifications are given in Chap. 10.

Trends

One of the principal trends in decommutation has been the use of stored-program-control decommutators built around magnetic-core memories. Typically, a format is punched on a paper tape and fed into a paper-tape reader for loading of the magnetic core of memories. Therefore, changes in the demultiplexer program may be easily made. The paper tape may be made up from a manual punch or even by a computer. Channel and subchannel addresses as well as identification tags and word length are stored for decommutation and routing of each of the data words. Usually, a stored-program-controlled demultiplexer provides a common-language-format output consisting of 12 bits for the received data, 11 bits for channel address, 11 bits for subchannel address, 4 bits for word length, 1 bit for continued word tag, and 1 bit for parity. However, it should be stressed that this increasing trend toward automation and common-language output is prevalent primarily at the larger range complexes.

R-F LINK

13 General Considerations

Pulse-code modulation is defined in the IRIG standards as "pulse modulation of a carrier by coded information. In PCM telemetry, information is transmitted by means of the code representing a finite number of values of the information at the time of assembling." Because of the reliability and convenience of bistable circuits, such as flip-flops and magnetic-core memories, the binary digital code is usually used in pulse-code modulation. This code can be transmitted as the presence or absence of the pulse in a sequence of N pulses. This technique is sometimes referred to as PCM by impulse. Another technique widely used in space communication is the non-return-to-zero codes (PCM-NRZ) where the coded binary information assumes either a pulse or minus voltage depending on whether the information is a logic 0 or 1. M-ary PCM systems have been extensively analyzed in the literature,[36a] but because they are more difficult to implement than the binary system, we shall devote our attention mainly to the binary mode.

A system block diagram of a PCM telemetry link having an analog source of data and an analog data sink is shown in Fig. 35. In some PCM telemetry systems, the source of data is digital and the received data are to be used in digital form. If this is the case, the portion of the system block diagram from the carrier modulator to the carrier demodulator is approximately all that is needed in this digital-to-digital link (Fig. 36).

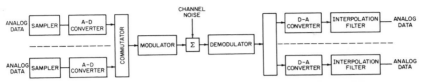

Fig. 35. A block diagram of an analog-to-analog PCM telemetry link.

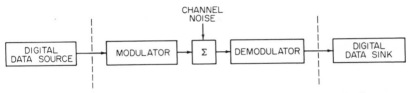

Fig. 36. A block diagram of a digital-to-digital PCM telemetry link.

A generic performance index widely used in analog-to-analog telemetry links is the mean-square error, which is the sum of the square of all the error occurring in the overall link. Some of these errors occur in the digital modulation and demodulation portion of the PCM link, and the remainder occur in the quantization and interpolation portion of the system. In Secs. 14 and 15, we shall review the modulation and demodulation process and the digital errors that occur in this subsystem. In the remaining section of this chapter, the overall PCM telemetry link will be reviewed.

14 PCM Data Modulation and Transmission

In specifying multiplexing and modulation it is customary to work from the individual input channels toward the carrier. The channels that have been sampled, quantized, and commutated (time-multiplexed) into a single series of binary digits are usually specified as PCM binary data. This PCM signal is then used to modulate or key the carrier. For example, PCM-FM means that the binary PCM data will frequency-modulate the carrier. Similarly, PCM-PM means that the PCM data will phase-modulate the carrier.

Another terminology that is often used is that of keying, such as phase-shift keying (PSK), amplitude-shift keying (ASK), and frequency-shift keying (FSK). There is no universally accepted terminology for specifying the carrier-modulation technique; however, AM, FM, and PM usually refer to the binary PCM data modulating a single carrier oscillator.

On the other hand, PSK may refer to switching between the output and a delayed output of the same oscillator, and it usually refers to biphase modulation with a phase excursion of π radians; FSK generally implies switching between two randomly phased oscillators; and ASK almost universally means switching the output of an oscillator on and off. To eliminate any misunderstanding, Table 4 lists the implied meaning of these modulation and keying acronyms; these definitions will be used in this section.

Table 4

Acronym	Meaning
ASK	Switching the output of a single oscillator on and off
PSK	Switching the phase of an oscillator between zero and π radians. Special case of PCM-PM
FSK	Switching between two randomly phased oscillators
PCM-FM	Frequency-modulating an oscillator with binary PCM-NRZ data. No discontinuities in the output at bit transitions
PCM-AM	Amplitude modulation of oscillator by PCM data. 100% modulation infers ASK
PCM-PM	Phase modulation of an oscillator by PCM data; not necessarily π-radian excursions

A look at the output-frequency spectrum for these keying and modulation techniques follows.

Modulator Output-frequency Spectrum

Amplitude Modulation (PCM-AM and ASK). The frequency spectrum for both these amplitude-modulation techniques consists of a spectral spike at the carrier frequency and a continuous ($\sin^2 x/x^2$) term centered at $\pm f_c$. For 100 per cent modulation, PCM-AM will be identical to our definition of ASK. A plot of the spectrum for a 100 per cent amplitude-modulated oscillator is shown in Fig. 37. Table 5 gives the equation for the spectral density of a 100 per cent modulated ASK signal.

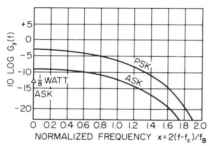

FIG. 37. Power spectrum $G_y(f)$ for PCM-AM (ASK) and PCM-PM (PSK) modulation techniques.

Phase Modulation (PCM-PM and PSK). The spectrum for PCM-PM will be similar to that for PAM-AM, i.e., a continuous ($\sin^2 x/x^2$) term plus spectral spikes at the carrier. This is true as long as the total phase excursion of the carrier is not equal to π radians. When this occurs, the spectral spikes disappear leaving only a continuous spectrum; this situation of a total phase excursion of π radians is called PSK. The spectrum for a PSK modulated sine wave is shown plotted in Fig. 37, and the equation for the spectral density of a PCM-PM signal is given in Table 5.

Frequency Modulation (PCM-FM and FSK). Pelchat[37] has obtained the spectrum for a PCM-NRZ binary signal frequency-modulating an oscillator (Fig. 38). Note that the shape of the spectrum is predominantly controlled by the modulation index or deviation ratio of the oscillator. Also, this is a continuous spectrum without spectral spikes unless the total carrier-frequency excursion $2 \Delta f$ is an integral of the bit rate f_b. When this occurs, spectral spikes appear.

In PCM-FM, the output time signal $y(t)$ is a continuous wave with no discontinuity occurring at the bit transition. This is not the case, however, for FSK since discontinuities can occur as a result of switching between two randomly phased oscillators. Titsworth and Welch[38] have determined the spectrum for this type of binary FSK signal when the two frequencies have an integral relationship to the bit rate f_b. The equation for this expression is given in Table 5. In the spectral-density equation, if the frequency separation between the two oscillator frequencies is large enough, $G_y(f)$, the spectral density, can be approximated by the sum of two $\sin^2 x/x^2$ terms centered at their respective frequencies. For small separation between the frequencies, the spectrum will have the appearance of the PCM-FM spectrum for $D = 0.8$.

Table 5. Equations for Power Spectra of Different PCM Modulation Techniques

Definitions:

f_c = carrier frequency
f_B = bit rate
$G_y(f)$ = output spectral density*
$y(t)$ = output signal

$$x = \frac{2(f - f_c)}{f_B}$$

$$F\left(x; \frac{f_B}{f_c}\right) = \frac{x^2(f_B/f_c)^2 + 4x(f_B/f_c) + 8}{x^2(f_B/f_c)^2 + 8x(f_B/f_c) + 16}$$

PCM-AM and ASK:

$$y(t) = m(t)A\cos\omega_c t \qquad m(t) = +b \text{ or } +a \qquad \text{where } b > a$$

$$G_y(f)_{PCM\text{-}AM} = \frac{A^2}{2f_B}\left(\frac{b-a}{2}\right)^2 \left(\frac{\sin \pi x/2}{\pi x/2}\right)^2 F\left(x; \frac{f_B}{f_c}\right)$$
$$+ A^2 \frac{[(b-a)/2]^2}{4}\left\{\delta(x) + \delta\left[x + 4\left(\frac{f_c}{f_B}\right)\right]\right\}$$

$a = 0, b = 1, A = \sqrt{2}$

$$G_y(f)_{ASK} = \frac{1}{4f_B}\left(\frac{\sin \pi x/2}{\pi x/2}\right)^2 F\left(x; \frac{f_B}{f_c}\right) + \frac{1}{8}\delta(x) + \frac{1}{8}\delta\left(x + 4\frac{f_c}{f_B}\right)$$

PCM-PM and PSK:

$$y(t) = A\cos[\omega_c t + \phi(t)] \qquad \phi(t) = \pm a\pi$$

$$G_y(f)_{PCM\text{-}PM} = \frac{A^2 \sin^2 a\pi}{2f_B}\left(\frac{\sin \pi x/2}{\pi x/2}\right)^2 F\left(x; \frac{f_B}{f_c}\right) + \frac{A^2 \cos^2 a\pi}{4}\left[\delta(x) + \delta\left(x + 4\frac{f_c}{f_B}\right)\right]$$

$a = \frac{1}{2}, A = \sqrt{2}$

$$G_y(f)_{PSK} = \frac{1}{f_B}\left(\frac{\sin \pi x/2}{\pi x/2}\right)^2 F\left(x; \frac{f_B}{f_c}\right)$$

PCM-FM and FSK:

a. PCM-FM:

$$y(t) = \sqrt{2}\cos\left[\omega_c t + \Delta\omega \int_{t_1}^{t} V(x)\,dx + \phi\right]$$

where $\Delta\omega = 2\pi \Delta f$ = constant which fixes the degree of modulation
$V(x)$ = PCM-NRZ data
$D = 2\Delta f/f_B$ = deviation ratio
ϕ = arbitrary carrier phase

when $|\cos \pi D| < 1.0|$,

$$G_y(f) = \frac{4}{f_B}\left[\frac{D}{\pi(D^2 - x^2)}\right]^2 \frac{(\cos \pi D - \cos \pi x)^2}{1 - 2\cos \pi D \cos \pi x + \cos^2 \pi D}$$

when $|\cos \pi D| = 1.0$,

$$G_y(f) = \frac{1}{4}\delta(x + D) + \frac{1}{4}\delta(x - D) + \frac{2}{f_B}\left[\frac{D}{\pi(D^2 - x^2)^2}\right]^2 (1 - \cos \pi D \cos \pi x)$$

b. FSK:

$$y_i(t) = A\sin(\omega_i t + \phi_i)$$

where ϕ_i = uniformly distributed between 0 and 2π
$\omega_i = n_i \pi f_B$, n_i is integral

$$G_y(f)_{FSK} = \frac{A^2}{4f_B}\left[\left(\frac{\sin x_1\pi/2}{x_1\pi/2}\right)^2 F\left(x_1; \frac{f_B}{f_i}\right) + \left(\frac{\sin x_2\pi/2}{x_2\pi/2}\right)^2 F\left(x_2; \frac{f_B}{f_2}\right)\right]$$

$$x_1 = \frac{2(f - f_1)}{f_B}, \quad x_2 = \frac{2(f - f_2)}{f_B}$$

---

* Spectral density is frequently designated by Φ instead of G.

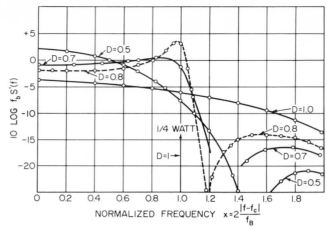

Fig. 38. Power spectrum of PCM-FM with deviation ratio as a parameter. (1) Total power of 1 watt. (2) $S'(f)$ is in watts per cps. (3) No premodulation filtering. (4) Modulating signal—random binary NRZ waveform with bit rate f_B.

15 PCM Data Reception and Demodulation

There are several metrics that could be used to specify the performance of the PCM binary modulation and demodulation subsystem, such as probability of bit error, information efficiency, and minimum energy per bit. However, the probability of bit-error metric P_e is the simplest and most frequently used when describing the performance of the binary channel. In a binary symmetric channel, which we shall assume, most of the other performance metrics can be obtained in a straightforward manner from the probability of bit error. For this reason, we shall use P_e to judge the performance of our modulation and demodulation techniques.

Hancock and Sheppard[39] have defined this probability of bit error P_e as

$$P_e = P(0_t)P(1_r/0_t) + P(1_t)P(0_r/1_t)$$

where $P(0_t)$ = probability of transmitting a logic 0
$P(1_t)$ = probability of transmitting a logic 1
$P(1_r/0_t)$ = probability of deciding a 1 when a 0 was transmitted
$P(0_r/1_t)$ = probability of deciding a 0 when a 1 was transmitted

They point out that a binary symmetrical system is one in which the transitional probabilities of error are equal, i.e.,

$$P(0_r/1_t) = P(1_r/0_t)$$

Therefore, since for a binary PCM

$$P(0_t) + P(1_t) = 1$$

it follows that

$$P_e = P(0_r/1_t) = P(1_r/0_t)$$

Thus, determining the probability of bit error reduces to determining the transitional probabilities.

These transitional probabilities are defined in the following manner: Assume that a signal $y(t)$, the transmitted signal $x(t)$ plus noise $n(t)$, is received. If the transmitted signal was $X_1(t)$ we need to determine $p(y/X_1)$, the conditional probability density function that Y was received knowing that X_1 was sent. Similarly, when the transmitted signal is $X_0(t)$, we must determine $P(Y/X_0)$. These conditional density functions are usually very complex and difficult to derive. Fortunately, though, most that are used for detection processes of interest are available in the literature.

If we assume that $P(Y/X_1)$ and $P(Y/X_2)$ can be represented by the curves shown in Fig. 39, then the transitional probability or the probability of error is

$$P_e = \int_{-\infty}^{\Delta} P(Y/X_1)\,dY = \int_{\Delta}^{\infty} P(Y/X_0)\,dy$$

These conditional probability density functions depend on (1) the type of channel, (2) the detection process, and (3) the type of interference.

In all our detection processes discussed in this section, we shall assume a binary symmetrical channel and white Gaussian noise interference. For these conditions, it has been shown[40] that the optimum detection process, i.e., the one which yields the minimum P_e, is the coherent matched filter or correlator detector. Although there is much inconsistency in the literature, a demodulation technique requiring precise knowledge of the phase is referred to as "coherent" demodulation; demodulation without this phase information is often called "noncoherent" demodulation. This is a classification which we adhere to for the demodulation process discussed here.

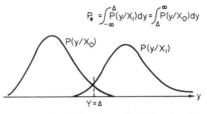

Fig. 39. Conditional probability density functions where $Y = \Delta$ is the threshold setting.

There are numerous detection schemes which can be employed to detect the binary information of the modulated signal. Simplified block diagrams of some of the most commonly used detection schemes are shown in Table 6. Included with the block diagrams are the equations for the probability of bit error for the binary symmetrical channel perturbed by Gaussian noise and also references to literature sources that indicate how P_e was derived.

For the case of coherent demodulation, the probability of bit error is derived as a function of the average signal energy per bit and the noise density per cps of bandwidth. For two possible transmitted signals $X_1(t)$ and $X_0(t)$, the average energy per bit is defined as

$$E = \tfrac{1}{2} \int_0^{T_B} [X_0^2(t) + X_1^2(t)]\,dt$$

where T_B is the bit duration. The noise power density per cps of bandwidth N_0 is defined on a one-sided-spectrum basis; i.e., it is the density of the noise power when all the noise that enters the system is assumed to be from positive-frequency components.

The probability of bit error for the noncoherent detection system is usually derived on a signal-power-to-noise-power ratio. Therefore, to be able to compare coherent and noncoherent demodulation processes on an equal basis we must convert the signal-power-to-noise-power ratio to energy-per-noise-density ratio. To do this, define the signal power as S and the noise power as N or $N_0 W$ where W is the bandwidth of the system. The signal-to-noise ratio S/N can then be expressed as a function of the average energy per noise power density as

$$S/N = S/N_0 W = ST_B/N_0 T_B W = (E/N_0)(1/T_B W)$$

If we define the bandwidth of our system from the output spectrum of the modulation process discussed earlier, as the frequency separation between the first two nulls in the $\sin^2 x/x^2$, then $T_B W = 2$, and

$$E/N_0 = 2(S/N)$$

Thus, using our definition of the system bandwidth we can relate the signal-to-noise-power ratio to the energy-to-noise-density ratio.

Table 6. Block Diagram of the Most Commonly Used PCM Detection Schemes

CURVES A AND D

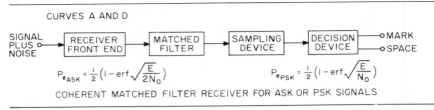

$P_{e_{ASK}} = \frac{1}{2}\left(1-\text{erf}\sqrt{\frac{E}{2N_0}}\right)$ $P_{e_{PSK}} = \frac{1}{2}\left(1-\text{erf}\sqrt{\frac{E}{N_0}}\right)$

COHERENT MATCHED FILTER RECEIVER FOR ASK OR PSK SIGNALS

CURVE B

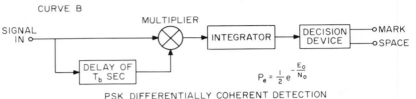

$P_e = \frac{1}{2} e^{-\frac{E_0}{N_0}}$

PSK DIFFERENTIALLY COHERENT DETECTION

CURVE C

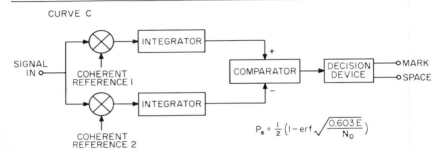

$P_e = \frac{1}{2}\left(1-\text{erf}\sqrt{\frac{0.603 E}{N_0}}\right)$

PCM-FM COHERENT MATCHED FILTER

CURVE D

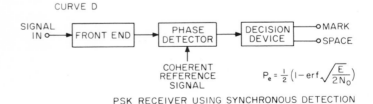

$P_e = \frac{1}{2}\left(1-\text{erf}\sqrt{\frac{E}{2N_0}}\right)$

PSK RECEIVER USING SYNCHRONOUS DETECTION

Table 6. Block Diagram of the Most Commonly Used PCM Detection Schemes (*Continued*)

CURVE E

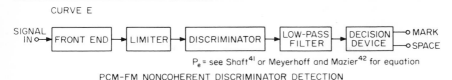

P_e = see Shaft[41] or Meyerhoff and Mazier[42] for equation

PCM-FM NONCOHERENT DISCRIMINATOR DETECTION

CURVE F

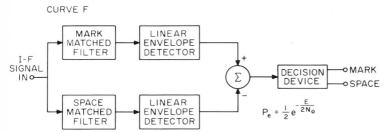

$P_e = \frac{1}{2} e^{-\frac{E}{2N_0}}$

NONCOHERENT MATCHED FILTER RECEIVER FOR FSK SIGNAL

CURVE F

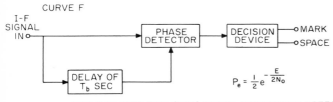

$P_e = \frac{1}{2} e^{-\frac{E}{2N_0}}$

PSK RECEIVER EMPLOYING PHASE-COMPARISON DETECTION

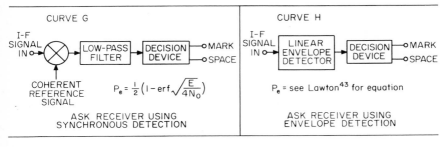

CURVE G

$P_e = \frac{1}{2}\left(1 - \mathrm{erf}\sqrt{\frac{E}{4N_0}}\right)$

ASK RECEIVER USING SYNCHRONOUS DETECTION

CURVE H

P_e = see Lawton[43] for equation

ASK RECEIVER USING ENVELOPE DETECTION

From the probability-of-bit-error curves plotted in Fig. 40, we see that the PSK coherent matched-filter detection scheme gives the minimum probability of bit error for the specified channel constraints. Although this detection scheme is optimum from the probability-of-bit-error standpoint, it is probably one of the most difficult to implement since precise phase knowledge of the incoming signal must be known. On

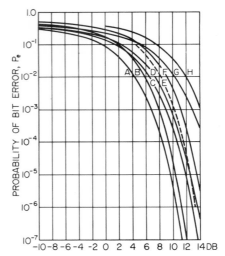

FIG. 40. Probability of bit error P_e as a function of the input signal-to-noise ratio of a binary symmetrical channel for different detection schemes. Curve A: PSK—coherent. B: PSK—differential coherent. C: PCM-FM matched filter coherent, PSK synchronous. D: FSK matched filter coherent, ASK matched filter coherent. E: PCM-FM discriminator detection. F: FSK matched filter noncoherent, PSK phase comparison. G: ASK synchronous detection. H: ASK linear envelope detection.

the other hand, the PCM-FM discriminator detection scheme, curve E, is approximately 3.5 db worse than the PSK coherent system; but only bit synchronization is needed in this detection process and precise knowledge of the phase is not required. This is one of the reasons why PCM-FM discriminator detection is frequently used in telemetry links.

SYSTEM ERROR ANALYSIS

16 Sampling

If we have an information source which is known to have no frequency components higher than W cps the signal can be reconstructed from samples of the original signal if the sampling rate f_s is greater than twice the highest-frequency component, i.e., $f_s \geq 2W$. Harman[44] uses the convolution integral to demonstrate this result:

Ideally, the sampling operation can be represented as multiplication of the signal by a periodic train of impulses; thus, define the sampling function $s(t)$ as

$$s(t) = \sum_n \delta(t - mTs)$$

for which the Fourier transform $F_s(f)$ is

$$F_s(f) = \sum_m 1/T_s \, \delta(f - m/T_s)$$

If $x_s(t)$ is the sampled signal, then

$$x_s(t) = X(t)S(t) = \sum_n X(mT_s)\delta(t - nT_s)$$

This multiplication in the time domain is equivalent to convolving in the frequency domain; thus, the Fourier integral of $x_s(t)$ is

$$F_{xs}(f) = \int_{-\infty}^{\infty} F_x(f')F_s(f - f')\,df'$$

Physically, this convolution has the effect of repeatedly locating the signal frequency spectrum $F_x(f)$ of the signal $x(t)$ along the frequency axis, centering it at the harmonics of the sampling rate. This is illustrated in Fig. 41. Note that as long as the sampling rate is greater than twice the bandwidth of the signal, these repeated spectra do not overlap. If this sampled signal is now filtered by the low-pass filter whose transfer function is $H(f)$, as shown in Fig. 41, the Fourier transform of the output of the filter will be identical to the Fourier transform of the input

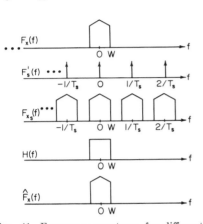

FIG. 41. Frequency spectrum for different portions of an ideal sampled data system.

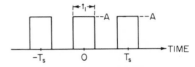

FIG. 42. The nonideal sample function $S(t)$.

signal. Therefore, for ideal impulse sampling pulses and an ideal low-pass filter, the sampled signal can be reconstructed to be exactly the original input signal.

For the nonideal case where the sampling pulse now has a nonzero width as shown in Fig. 42, Childers[45] shows that the frequency spectrum of the signal is no longer spikes occurring at the harmonics of the sampling rate as shown in Fig. 41 but is in fact represented by repeated $\sin x/x$ type curves centered at these harmonics; the power density $G_s(f)$ of this signal is

$$G_s(f) = A^2 \left(\frac{t_1}{T_s}\right)^2 \sum_{n=-\infty}^{\infty} \left[\frac{\sin(\pi n t_1/T_s)}{\pi n t_1/T_s}\right]^2 \delta\left(f - \frac{n}{T_s}\right)$$

where the parameters in the equation can be defined from Fig. 42. Note that the spectrum does approach the train of delta functions shown in Fig. 41 when $A = 1/t_1$ and t_1 is allowed to approach zero. With this nonideal sampling function, data-aliasing errors occur because of interference with the desired data spectrum by the sampled-data-spectrum sidebands which will be within the bandpass of the signal-reconstruction filter (interpolation filter).

In PCM telemetry, this sampled signal is usually boxcarred, i.e., put into a holding circuit where the sample amplitude is held fixed until the next sample arrives. The resulting stair-step function approximating the analog signal is called a 100 per cent duty cycle PAM waveform. Childers[45] shows that the power spectrum for this waveform is

$$G(f) = (1/T_s)|B(f)|^2[(\bar{a}^2 - \bar{a}^2) + \bar{a}^2/T_s \Sigma \delta(f - n/T_s)]$$

where $B(f) = T_s^2[\sin(\pi f T_s)/\pi f T_s]^2$
T_s = sampling period = $1/f_s$
f_s = sample frequency
a = random variable, the pulse amplitude
$\overline{a^2}$ = mean-square value of a
$\bar{a}^2$ = square of the mean value of a

Note that when the mean value of the signal amplitude ($\bar{a}$) is zero, the spectral spikes disappear, leaving for the power spectrum the familiar ($\sin^2 x/x^2$) curve. Also note that the possibility of aliasing errors has been removed.

17 Quantization Errors

In PCM telemetry, only certain discrete levels of the analog information can be transmitted by the code because, in the process of coding, the signal must be "quantized" or "digitalized" to the nearest discrete value; this process introduces errors into the telemetry link called quantization errors.

If the analog signal is to be quantized into L evenly spaced levels for a total amplitude excursion of $\pm V$, then the quantized signal may contain positive or negative errors up to $\pm V/L$. Furthermore, if the levels are closely spaced compared with the rms value of the signal, it is usually assumed that the probability is the same for the original analog signal to occur at any location at the interval. Thus the probability for an error y at any level is uniformly distributed for $(-V/L < y < +V/L)$, and the mean-square value of the quantization error σ_q^2 is

$$\sigma_q^2 = \int_{-V/L}^{+V/L} y^2 p(y)\, dy = (V^2/L^2)(\tfrac{1}{3})$$

where $\bar{y} = 0$.

The quantization error is independent of any other error occurring in a system, and its mean-square value can therefore be added directly to the other mean-square errors to obtain the total error of the system. However, the quantization error may cause a much greater degradation in the usefulness of the output analog signal than that caused by other errors occurring in the system. It is sometimes necessary to add a weighting factor to the mean-square quantization error when it is to be added to other mean-square errors occurring in the system.

18 Systematic Errors

The error in the output analog signal caused by the sampling and interpolation filter is generally classified as the systematic error. Childers[45] has analyzed the mean-square value of this error for a 100 per cent duty cycle PAM signal by using the simplified model shown in Fig. 43. If for the PCM telemetry link, we assume that the only

Fig. 43. Model for systematic error determination.

errors are interpolation errors, then Fig. 43 can also be used to analyze the interpolation error in the PCM link. The error function $\epsilon(t)$ is defined from Fig. 43 as

$$\epsilon(t) = f(t) - d(t - t_0)$$

where t_0 is the time delay in the link and $f(t)$ and $d(t)$ subject to the constraint

$$\overline{f^2(t)} = \overline{d^2(t - t_0)}$$

The normalized mean-square interpolation error E^2 is defined as

$$E^2 = \sigma_\epsilon^2/\sigma_d^2$$

where $\sigma_\epsilon^2 = \int_{-\infty}^{\infty} G_\epsilon(f)\,df$

$\sigma_d^2 = \int_{-\infty}^{\infty} G_d(f)\,df$

The power density of the error $G_\epsilon(f)$ is given by

$$G_\epsilon(f) = \lim_{T\to\infty} (1/T)|F_\epsilon(f)|^2$$

where $F(f)$ = Fourier transform of (t).

In Sec. 16, the equation for the power density $G_o(f)$ of the output of the ideal sample-and-hold circuit was given as

$$G_o(f) = \sigma_d^2 T_s [\sin(\pi f T_s)/\pi f T_s]^2$$

where $\sigma d^2 = \overline{a^2}$, the mean square of the zero mean signal amplitude. The output power density from the interpolation filter $G_f(f)$ is then

$$G_f(f) = G_o(f)|I(f)|^2$$

where $I(f)$ is the transfer function of the interpolation filter. By using these results of the $\sigma d^2 = \sigma f^2$ constraint, Childers derives the nomalized mean-square error E^2 as

$$E^2 = 2\left(1 - \frac{1}{\sigma_d}\frac{\int_{-\infty}^{\infty}[\sin(\pi f T_s)/\pi f T_s]|I(f)|\,|F_d(f)|\,df}{\left\{\int_{-\infty}^{\infty}[\sin(\pi f T_s)/\pi f T_s]^2|I(f)|^2\,df\right\}^{1/2}}\right)$$

where $F_d(f)$ is the Fourier transform of the data signal $d(t)$. Thus we see that we must have spectral knowledge of the input analog signal in order to predict the normalized mean-square error. Recognizing this situation, Childers[45] has classified most telemetry data into five types of spectrum and created simple models for the spectrum. However, even with these models for the spectrum of the data, calculating the interpolation normalized mean-square error is no easy task.

19 Random Errors

The errors occurring in the reconstructed analog signal caused by the fluctuating noise in the channel are usually referred to as random errors. In Sec. 15, we discussed the probability of bit error for the binary symmetrical channel. This probability of bit error along with the probability distribution of an amplitude of a signal can be used to determine the portion of a total mean-square error caused by the noise in the channel.

Viterbi[46] has demonstrated how to calculate this mean-square error by using the probability of bit error P_e and the signal-amplitude probability density function. In his discussion, he assumes that the source amplitude is uniformly distributed between $-V$ and $+V$ volts and that the bit-error probability P_e is independent of the location of the bit within the word structure. Using these basic assumptions, Viterbi derives an expression for the expected squared error σ_N^2 due to channel noise:

$$\sigma_N^2 = [4V^2(L^2 - 1)P_e/3L^2]$$

Since the random errors and the quantization error are independent, the total mean-square error from these sources is

$$\sigma_N^2 + \sigma_q^2 = [4V^2(L^2 - 1)P_e/3L^2] + V^2/3L^2$$

where σ_q^2 is the quantization mean-square error and L is the number of levels.

It should be noted that this expression for the mean-square error is valid only for those sources in which the amplitude of the source is uniformly distributed between upper and lower limits. Viterbi[46] indicates, however, that the results could be extended to any amplitude-limited source by using his outlined procedure.

Sometimes, error-correcting codes are used to decrease the probability of word error P_w by adding controlled redundancy to the sequence of binary bits. If this is done, the probability of bit error defined from the binary detection circuit will be used in a decoding scheme to predict the probability of word error. When error-correcting codes are used, if the information rate is to remain constant the noise bandwidth of the system must necessarily increase, thereby letting more noise into the detection process and consequently increasing the probability of bit error. Usually, codes are designed to combat different types of interference, such as burst noise and impulse noise. Therefore, they can actually increase the overall performance of the telemetry link for these channel constraints. A more complete treatment of channel coding is given in Ref. 13.

20 Example of Error Calculations

When all errors are considered, the signal w that arrives at the input to the interpolation filter will be the sum of three random variables

$$w = x + y + z$$

where x is the data signal amplitude, y is a quantization error, and z is the error in reconstructing the sample amplitude from the incoming binary data. If we assume that the only effect the errors have on the signal arriving at the interpolation filter is a possibility of changing the amplitude of each bit transition in the staircase waveform, then the spectrum of the incoming signal $G_g(f)$ defined in Art. 18 is only scaled in amplitude. The equation for the normalized mean-square interpolation error E^2, which was discussed in Art. 18, is a function of the amplitude of the power density for the incoming signal. This amplitude of the power density is simply the variance of the data signal times the sample time T_s. To determine the normalized mean-square error when all errors are included, we simply replace σ_d^2 with the variance for the sum of the errors σ_w^2. If the amplitude random variable x as well as the quantization error y and the reconstruction error z have zero mean, and since the quantization error y is independent of both x and z, the variance for w is

$$\sigma_w^2 = \sigma_x^2 + \sigma_y^2 + \sigma_z^2 + 2E(x,z)$$

where $E(x,z)$ = joint expectation of x and z
$\sigma_y^2 = \sigma_q^2$, mean-square quantization error
$\sigma_x^2 = \sigma_d^2$, mean-square value of data

Unfortunately, the joint expectation $E(x,z)$

$$E(x,z) = \int_{-\infty}^{+\infty} \int_{-\infty}^{\infty} xz p(x,z) \, dx \, dz$$

where $p(x,z)$, the joint density of x and z, does not readily yield to a closed-form analytical solution. However, for very low probability of bit error P_e, and where the number of quantization levels is not too great,

$$\sigma_w^2 \approx \sigma_x^2 + \sigma_y^2 = \sigma_d^2 + \sigma_q^2$$

and the total normalized mean-square error E_T^2 defined from Art. 18 reduces to

$$E_T^2 = 2\left(1 - \frac{1}{\sqrt{\sigma_d^2 + \sigma_q^2}} \frac{\int_{-\infty}^{\infty} [\sin(\pi f T_s)/\pi f T_s]|I(f)|\,|F_d(f)|\,df}{\left\{\int_{-\infty}^{\infty} [\sin(\pi f T_s)/\pi f T_s]^2 |I(f)|^2 \, df\right\}^{\frac{1}{2}}}\right)$$

Childers[6] shows that the rms error is minimized when

$$K|I(f)| = \frac{\sigma_d}{\sigma_d^2 + \sigma_q^2} \frac{1}{\sqrt{T_s}} \frac{|F_d(f)|}{\sin(\pi f T_s)/\pi f T_s}$$

where K is a gain constant associated with the interpolation filter to satisfy the constraint that

$$\sigma_d{}^2 + \sigma_f{}^2$$

where $\sigma_f{}^2$ is the variance of the output signal from the interpolation filter. Therefore, if we can determine the Fourier transform $F_d(f)$ of the data signal we are in a position to determine the optimum interpolation filter, i.e., the interpolation filter that minimizes the rms error. With the determination of $I(f)$ and knowing the Fourier transform of the signal data, we can then evaluate the total normalized mean-square error. For most nontrivial cases, the evaluation of this mean-square error will most likely have to be done by computer techniques.

REFERENCES

1. B. M. Oliver, J. R. Pierce, and C. E. Shannon, The Philosophy of PCM, *Proc. IRE*, vol. 36, no. 11, pp. 1324–1331, November, 1948.
2. J. J. Stiltz, "Aerospace Telemetry," Prentice-Hall, Inc., Englewood Cliffs, N.J., 1961.
3. R. W. Sanders, Communication Efficiency Comparison of Several Communication Systems, *Proc. IRE*, vol. 48, no. 4, pp. 575–588, April, 1960.
4. A. J. Viterbi, On Coded Phase Coherent Communications, *Trans. IRE*, PGSET-7, March, 1961.
5. F. A. Galindo and G. Antle, "An Integrated Circuit PCM Telemetry Encoder," presented at 1963 Western Electronic Show and Convention.
6. "Reference Data for Radio Engineers," 4th ed., International Telephone and Telegraph Corporation, New York, 1963.
7. P. R. Goffe, "Simplified Modern Filter Design," John F. Ryder Publisher, Inc., New York, 1963.
8. Harold M. Sharke, Dependency of Cross-talk on Upper and Lower Cutoff Frequencies in PAM Time Multiplexed Transmission Paths, *IRE Trans. Commun. Systems*, September, 1962, p. 268.
9. Texas Instruments, Inc., "Transistor Circuit Design," McGraw-Hill Book Company, New York, 1963.
10. M. V. Kiebert, "Basic Design of Commutating Devices," *Trans. IRE Professional Group Radio Telemetry Remote Control*, vol. PGR TRC-1, p. 7, August, 1954.
11. S. Davis, High Speed Mercury Jet Commutating Switch, *Natl. Telemetry Conf. Record*, 1955.
12. J. F. Meyer, Low Speed Time Multiplexing with Magnetic Latching Relays, *IRE Trans. Space Electron. Telemetry*, vol. SET-7, p. 34, June, 1961.
13. W. R. Peterson, "Error Correcting Codes," The M.I.T. Press, Cambridge, Mass., 1961.
14. A. K. Susskind, "Analog-to-Digital Conversion Techniques," The Technology Press of the Massachusetts Institute of Technology, Cambridge, Mass., 1957.
15. B. D. Smith, Coding by Feedback Methods, *Proc. IRE*, vol. 41, no. 6, p. 1053, August, 1953.
16. E. F. Kovenic, A High Accuracy Nine Bit Digital-to-Analog Converter, *IEEE Trans. Commun. Electron.*, March, 1964, p. 185.
17. H. Keneko and T. Sekimuto, Logarithmic PCM Encoding without Diode Compandok, *IEEE Trans. Commun. Systems*, vol. CS-11, p. 296, September, 1963.
18. J. C. H. Davis, A PCM Logarithmic Encoder for a Multichannel TDM System, *Proc. IEEE*, vol. 109, p. 109, November, 1962.
19. B. K. Smith and R. H. Kelzene, A Universal Serial Recirculating Analog-to-Digital Converter for Aerospace Applications. *IRE 1963 Intern. Conf. Record*.
20. R. W. Sears, Electron Beam Deflection Tube for Pulse Code Modulation, *Bell System Tech. J.*, vol. 27, pp. 44–57, 1948.
21. H. Blasbalg and R. Van Blerkom, Message Compression, *IRE Trans. Space Electron. Telemetry*, September, 1962, p. 228.
22. W. R. Bennett, Statistics of Regenerative Digital Transmission, *Bell System Tech. J.*, vol. 37, pp. 1501–1542, November, 1958.
23. R. Jaffe and E. Rechtin, Design and Performance of Phase-Lock Circuits Capable of Near Optimum Performance over a Wide Range of Input Signal and Noise Levels, *IRE Trans. Inform. Theory*, vol. IT-1, pp. 66–76, March, 1955.
24. E. Rechtin, Design of Phase-lock Oscillator Circuits, *Section Rept.* 8-566, Feb. 7, 1957, Jet Propulsion Laboratory, California Institute of Technology, Pasadena.

25. C. E. Gilchriest, Application of the Phase-locked Loop to Telemetry as a Discriminator or Tracking Filter, *IRE Trans. Telemetry Remote Control*, vol. TRC-4, pp. 20–35, June, 1958.
26. C. S. Weaver, A New Approach to the Linear Design and Analysis of Phase-locked Loops, *IRE Trans. Space Electron. Telemetry*, vol. SET-5, pp. 166–178, December, 1959.
27. E. R. Hill, Techniques for Synchronizing Pulse-Code-Modulated Telemetry, *Proc.* 1963 *Natl. Telemetering Conf.* May 20–22, Albuquerque.
28. Wolf J. Gruen, Theory of AFC Synchronization, *Proc. IRE*, August, 1953, pp. 1043–1048.
29. D. D. McRae and F. A. Perkins, "PCM Synchronization and Demultiplexing Study—Final Report," George C. Marshal Space Flight Center, Contract NAS 8-5111, prepared by Radiation Inc., June 14, 1963.
30. Aeronautical Systems Division, AF Avionics Laboratory, Electronics Warfare Division, Wright-Patterson Air Force Base, Ohio, Report ASD-TDR-62-1073, "Study on Telemetry and Range Safety Techniques for Advanced Aerospace Vehicles," Final Report 73, March, 1963.
31. F. A. Perkins, The Effect of Bit and Group Synchronization on the Reception of PCM/FM Telemetry Signals, *Proc. Natl. Telemetry Conf.*, June, 1964, Los Angeles.
32. J. L. Phillips and G. E. Goode, Correlation Detection and Sequential Testing for PCM Group Synchronization, *Proc.* 1962 *Natl. Telemetering Conf.*, vol. 1, Session V, May 23–25, Washington, D.C.
33. M. S. Maxwell and R. L. Kutz, An Efficient PCM Error Correction and Synchronization Code, 1963 *National Space Electronics Symposium*, *PTGSET Record*, Miami Beach, Oct. 1–4, 1963.
34. G. E. Goode and J. L. Phillips, Optimum PCM Frame Synchronization Codes and Correlation Detection, *Proc.* 1961 *Natl. Telemetering Conf.*, May 22–24, Chicago.
35. T. G. Birdsall and M. P. Ristenbatt, Introduction to Linear Shift-register Generated Sequences, *Tech. Rept.* 90, University of Michigan Research Institute, October, 1958 (ASTIA AD 225-380).
36. E. R. Hill and J. L. Weblemoe, Synchronization Methods for PCM Telemetry, *Proc.* 1961 *Natl. Telemetering Conf.*, May 22–24, Chicago.
36a. H. Zabronsky, Statistical Properties of M-ary Frequency-shift-keyed and Phase-shift-keyed Modulated Carrier, *RCA Rev.*, vol. 22, p. 431, September, 1961.
37. M. G. Pelchat, The Autocorrelation Function and Power Spectrum of PCM-FM with Random Binary Modulating Waveforms, *Trans. IRE*, PTGSET-10, no. 1, p. 39, March, 1964.
38. R. C. Titsworth and L. R. Welch, Modulation by Random and Pseudo-random Sequences, *Progress Rept.* 20-387, Jet Propulsion Laboratory, California Institute of Technology, June 12, 1959, ORDCIT Project Contract DA-04-495-Ord. 18, Ordnance Corps.
39. J. C. Hancock and E. M. Sheppard, "Information Transfer Efficiency of Wide Band Communication Systems," Part 1, Information Efficiency of Binary Communication Systems, ASTIA Document AD 284450, July, 1962, prepared for the Electromagnetic Warfare and Communications Laboratories, Aeronautical Systems Division, Air Force Command, Wright-Patterson Air Force Base, Ohio, under contract AF33(616)-8283.
40. Elie J. Baghdady (ed.), "Lectures on Communication System Theory," McGraw-Hill Book Company, New York, 1960.
41. Paul D. Shaft, Error Rate of PCM-FM Using Discriminator Detection, *Trans. IRE*, PTGSET-9, no. 4, p. 131, December, 1963.
42. Alan A. Meyerhoff and William H. Mazier, Optimum Binary FM Reception Using Discriminator Detection and I-F Shaping, *RCA Rev.*, vol. 22, pp. 698–728, December, 1961.
43. John G. Lawton, Comparison of Binary Data Transmission Systems, *Proc. 2d Natl. Conf. Military Electron.*, Washington, D.C., June 16–18, 1958, pp. 54–61.
44. Willis A. Harman, "Principles of the Statistical Theory of Communication," p. 28, McGraw-Hill Book Company, New York, 1963.
45. D. G. Childers, Study and Experimental Investigation of Sampling Rates and Aliasing in Time Division Telemetry Systems, *Trans. IRE*, PTGSET, vol. SET-8, no. 4, pp. 267–283, December, 1962.
46. Andrew Viterbi, Lower Bounds on Maximum Signal-to-Noise Ratios for Digital Communication over the Gaussian Channel, *Trans. IRE*, PTGCS, vol. CS-12, pp. 10–17 March, 1964.

Chapter 9

PAM AND ADVANCED TELEMETRY SYSTEMS

DANIEL HOCHMAN, *Lockheed Missiles and Space Company, Sunnyvale, Calif.* (*PAM-FM Telemetry*)

ELLIOT L. GRUENBERG, *International Business Machines Corporation, Yorktown Heights, N.Y.* (*High-efficiency Telemetry Systems; Phase-locked-loop Systems*)

PAM-FM TELEMETRY

1	Introduction..	9–2
2	Communication Properties of the PAM-FM System...........	9–2
3	Principles of Pulse-amplitude Modulation....................	9–5
4	Elements of a PAM-FM Telemeter..........................	9–9
5	Typical Application..	9–11
6	Ground Station..	9–13
7	PACM Telemetry System..................................	9–18

HIGH-EFFICIENCY TELEMETRY SYSTEMS

8	Theoretical Limits of Communication Efficiency...............	9–30
9	Comparison of Discrete and Equivalent Discrete Modulation Systems...	9–31
10	PCM Systems...	9–32
11	Multisymbol Discrete Systems..............................	9–34
12	Quantized Pulse-position Modulation (QPPM)................	9–34
13	Coded Phase-coherent Systems.............................	9–39
14	PFM Telemetry...	9–48
15	Sampled Analog Transmission..............................	9–54

PHASE-LOCKED-LOOP SYSTEMS

16	Phase-locked-loop Operation................................	9–60
17	Linear Model..	9–61
18	Nonlinear Model—Noiseless Case...........................	9–70
19	Nonlinear Model—Noisy Signals............................	9–77

In this chapter will be presented several systems which have been developed somewhat later than those previously discussed. The basic objective of these systems has been to maximize communication efficiency in some manner. The PAM system has been standardized by the IRIG. It has been shown to be advantageous where lower-accuracy data are sufficient. The chief disadvantage of the system is the high power per sample it requires, although its bandwidth requirement is the least of any radio-telemetry system. The PAM system is discussed in Secs. 1 to 6.

PAM-FM performance is best where lower-accuracy data are required whereas PCM-FM outperform other standard telemetry systems for higher-precision data. It is possible to combine both systems into one. Such a system, called PACM, is discussed in Sec. 7 together with an evaluation of the standard telemetry system from the viewpoint of maintaining overall *analog* data *accuracy* for the telemetry user.

Because of the limitations of power in missiles and space vehicles considerable demand has existed for telemetering systems which can minimize such requirements. It is possible to show from Shannon's theorem derived in Chap. 2, Sec. 15 that the ideal communications system will improve in efficiency as more bandwidth is used per information bit. Several wideband high-efficiency systems also having this property will be discussed (Secs. 11 to 14). These depend on matched filters of high time-bandwidth product and include Digilock, PFM, and PPM. It is of interest to note that not all systems using large bandwidths per information bit improve in efficiency. In particular this is true of FM systems because of the threshold effect, as shown in Sec. 8.

Phase-locked loop technique has been used as a component of high-efficiency telemetry systems. It is another approach to a matched-filter system. It is a fundamental component to any coherent system since it can be made to track the Doppler-shifted carrier. Principles of design of such filters are therefore presented in this chapter.

PAM-FM TELEMETRY

List of Symbols for Secs. 1 to 7

B	Receiver bandwidth
B_i	Total information bandwidth
S	RMS carrier voltage
f_s	Sampling rate
f_i	Highest-frequency component input signal
k_i	RMS noise voltage per unit root bandwidth
$s(t)$	Sampling function
$\delta(t - \tau)$	Impulse at $t = \tau$
ω_c	Filter cutoff frequency (angular)

1 Introduction

Following the initial applications of PAM-FM telemetry in the late 1940s, the use of this system in the United States was discontinued as a result of FM-FM standardization. However, the recent rise in data-handling requirements occasioned by the expansion of missile and space programs instigated renewed efforts on telemetry optimization in which the PAM-FM system received increased attention.

This section should familiarize the reader with the PAM-FM telemeter as presently employed. The basic properties of the PAM-FM system, the criteria for selecting it for certain applications, and the principles of the system will be discussed, together with the description of a typical application.

2 Communication Properties of the PAM-FM System

The problem of transmitting several intelligence signals simultaneously by either wire or radio is usually solved by means of multiplexing. In frequency-division multiplex, each intelligence signal modulates a separate subcarrier. The subcarriers are of different frequencies and are spaced in the spectrum to avoid overlapping of the sidebands produced in the modulation process. The modulation is AM, PM, or FM, depending on the requirements and constraints present in the particular case. In time-division multiplex, each intelligence signal modulates a separate pulse subcarrier. The pulses of the subcarriers are spaced in time (electrical angle) in such a way that the summation of all the subcarriers is a pulse train with a repetition rate equal to the sum of the repetition rates of the individual subcarriers. The modulation process may involve changes in pulse amplitude, phase, duration, or a combination thereof. The pulse-modulation multiplexing methods most often used in telemetry practice are pulse-amplitude modulation (PAM), pulse-duration modulation (PDM), pulse-position modulation (PPM), and pulse-code modulation (PCM).

For radio transmission, it is necessary to modulate the r-f carrier with the multiplexed intelligence signals. As a result, a double modulation process is performed: first, in multiplexing the signals; second, in modulating the radio-frequency carrier. It is possible to extend this procedure further and arrive at triple modulation, quadruple modulation, etc. Since there is no theoretical limit to the number of steps taken between the intelligence signal and the final transmission, the number of possibilities is endless. In practice, however, the process is usually limited to two or three modulation steps. In view of the different characteristics of various multiplexing and modulation methods, it is readily recognized that the overall system performance will largely depend on the choice of the multiplexing and modulation schemes.

Considering the multiplicity of choices available, the selection of an optimum system based on theoretical considerations is a prohibitive task, indeed, unless some practical constraints are imposed on the parameters of the systems under consideration. For telemetry applications, the following constraints seem realistic:

1. The data-transmission system is limited to two modulation processes.
2. Single-sideband multiplexing is not considered suitable for general application because of poor low-frequency-response characteristics of practical systems.
3. From previous experience, most practical telemetry systems are FM-FM, PDM-FM, PPM-FM, PCM-FM, and PAM-FM.

Based on these constraints, an analysis of communication properties of various telemetry systems can be performed[1,2,3,4] and the systems ordered in terms of their behavior relative to the selected parameters. A summary of the results obtained follows.

Signal-to-Noise Improvement (Wideband Gain)

The wideband gain is defined as the ratio of signal-to-random-noise ratios comparing a single channel of a given multiplex system to a one-channel AM system of equal intelligence bandwidth and equal carrier power. The results indicate that, of those considered, the best systems, in order of merit, are PAM-FM, PWM-FM, and PPM-AM. Systems involving PCM are not taken into account here, as the signal-to-noise ratio for these systems is theoretically infinite (excluding quantization noise) if the signal amplitude is above threshold.

The Threshold of Various Systems

The threshold of any system is the received average power required to realize the wideband gain. Generally, pulse systems have lower threshold levels than other systems.

Signal-to-Noise Ratio on Impulse Noise

On impulse noise of less than signal amplitude, the wideband gain is the same as for random noise. On strong-impulse noise, the wideband gain for frequency-division multiplex systems and for PAM-FM is approximately the same as for random noise. For other pulse systems, the S/N ratio becomes unity or worse.

Interchannel Cross-modulation

In frequency-division systems, cross-modulation is caused by nonlinearity in amplifiers and by envelope nonlinearity in modulators and demodulators. In time-division systems, cross-modulation is caused by a slight overlapping of adjacent pulses because of insufficient bandwidth, or by overmodulation in PDM or PPM. The actual signal-to-cross-modulation ratio obtained in each system is more a function of the skill and techniques employed than a function of the system.

Interchannel Crosstalk

In frequency-division systems, the crosstalk is governed by subcarrier spacings and the design of the frequency-selective filters. In time-division systems, proper shieldings should entirely eliminate crosstalk, other than that caused by cross-modulation.

Requirements on Transmitter Power and Link Bandwidth

In considering these requirements, certain assumptions have to be made relative to system accuracy. It is readily understood that the permissible system error affects the signal-to-noise ratio requirement at the output which, in turn, is a function of the transmitter power and the radio-link bandwidth. The accuracy requirements in telemetry systems vary with application and within a single telemeter may vary from channel to channel.

Statistical studies conducted on this subject show that it is reasonable, for the purpose of system comparison, to assume that a 2 per cent accuracy will satisfy a great majority of applications. Distinctly separate are those applications in which accuracies of 0.1 to 1 per cent are required. In Ref. 4, a comparison was made between telemetry systems with the information accuracy fixed at 2 per cent. The results are shown in Fig. 1. The definitions of the symbols are as follows:

B Receiver bandwidth
B_i Total information bandwidth
k_1 RMS noise voltage per unit root bandwidth
S RMS carrier voltage

The ordinate in Fig. 1 is a ratio of the signal power to the noise power within the information bandwidth, i.e., is representative of the required transmitter power. The chart shows that for an accuracy of 2 per cent, the rating of the systems is PAM-FM, PCM-FM, PDM-FM, and FM-FM, in that order. It is also evident from Fig. 1 that for a given transmitter power output PAM-FM will require the minimum link bandwidth.

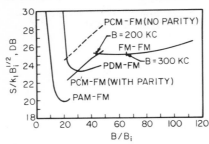

Fig. 1. System-comparison chart for 2 per cent error.

The weighting of the systems in terms of their efficiency changes as the accuracy requirements become more stringent. Generally speaking, because of various design problems in analog systems, their capability reaches a limit at accuracies of 1 per cent. From this point on, digital systems become more efficient since the only limitation on error is quantization noise. As shown in Fig. 2, PCM-FM becomes compatible with PAM-FM at 1 per cent accuracy and, from the standpoint of transmitter power and bandwidth, is superior to the PAM-FM system where greater accuracies are required. For comparison purposes, the curves have been normalized with respect to the sampling rate f_s.

Other Considerations

With regard to mutual interference between systems operating on the same assigned frequency, a disadvantage exists in the amplitude modulation of the r-f carrier, since

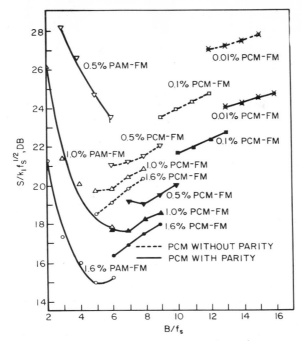

Fig. 2. PAM-FM and PCM-FM performance for various accuracies.

such systems are susceptible to the serious fixed-pitch beat-note interference. The interference between systems operating on adjacent frequency assignments is generally associated with two effects: the sideband splash in the adjacent channel, and imperfect receiver selectivity. Except for cases where the interference may exceed the threshold level, the susceptibility to sideband splash places the systems in the same order of merit as signal-to-random-noise-ratio considerations. In these cases, the pulse systems have the advantage insofar as they have a high crest factor. In susceptibility to adjacent-channel interference due to imperfect selectivity, the systems utilizing phase modulation of the r-f carrier have a definite advantage. The problems connected with propagation and selective fading are basically dependent on the operating frequencies of the r-f carrier and enter the multiplexing problem only to the extent of threshold consideration.

From the foregoing, it can be concluded that
1. PAM-FM is superior to other telemetry systems where accuracy requirements do not exceed 1 per cent maximum allowable error.
2. For accuracies of better than 1 per cent, pulse-code modulation is mandatory.
Consistent with these conclusions, PAM-FM systems could be used in the majority of telemetry applications. The flexibility of time-division multiplex permits a large range of variation in the number of data channels and in the assignments of frequency response per channel.

3 Principles of Pulse-amplitude Modulation

Sampling Theorem

Oliver, Pierce, and Shannon[5] have shown that a function of time $f(t)$ having a Fourier transform $F(\omega)$, which contains no frequency component greater than f_i is uniquely determined by the values of $f(t)$ at any set of sampling points spaced $1/2f_i$ sec apart. This sampling theorem is mathematically expressed by

$$f(t) = \sum_{n=-\infty}^{\infty} f\left(\frac{n}{2f_i}\right) \frac{\sin (2\pi f_i t - n\pi)}{2\pi f_i t - n\pi} \qquad (1)$$

In other words, $f(t)$ may be thought of as a sum of a series of elementary functions of the form $\sin x/x$ centered at the sampling points, each having a peak value equal to $f(t)$ at the corresponding sampling point. To reconstruct $f(t)$, all that is needed is to generate a series of $\sin x/x$ pulses proportional to the samples and then add the ensemble. This can be accomplished by a system as shown in Fig. 3. Here $f(t)$ is a band-limited low-pass function of which a Fourier transform exists, and which has a highest-frequency component f_i. The sampler is a switch which closes periodically at intervals $1/2f_i$ sec apart and possesses an infinitesimal contact dwell time. Its function can be described by

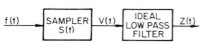

Fig. 3. Sampling and smoothing system.

$$s(t) = \sum_{n=-\infty}^{\infty} \delta(t - n/2f_i) \qquad (2)$$

where n is an integer.

The output of the sampler $v(t)$ represents a series of samples of the input function $f(t)$ conforming to the first term of the sampling-theorem equation. The second term is represented by the ideal low-pass filter with a cutoff frequency ω_c whose weighting function has the familiar $\sin x/x$ form. The output of the low-pass filter $Z(t)$ is an exact replica of $f(t)$ except for the time delay involved in the process of sampling and interpolation.

Feldman and Bennett[6] further discussed the subject of minimum sampling rate and evolved a sampling theorem for low-pass and bandpass functions. They showed that

$$f_s = 2B_i(1 + k/m) \qquad (3)$$

in which f_s = minimum sampling frequency
B_i = bandwidth of the sampled function
f_2 = highest-frequency component of the sampled function
m = largest integer not exceeding f_2/B_i
k = $f_2/B_i - m$

The value of k varies between zero and unity. When the band is located between adjacent multiples of B_i, the value of k is zero and $f_s = 2B_i$ no matter how high the frequency range of the signal may be. As k increases from zero to unity, the sampling rate increases from $2B_i$ to $2B_i(1 - 1/m)$. The curve of minimum sampling rate vs. the highest frequency in a band of constant width becomes a series of sawteeth of successively decreasing height, as shown in Fig. 4.

It is important to note that at the minimum sampling rate the highest-frequency component is not precisely specified since, with a sampling rate of $2f$, a sine wave of frequency f will contribute two fixed sample values half a wavelength apart. Therefore, in order to specify fully a low-pass function whose highest-frequency component is f_i, the sampling rate must be at least a very small amount greater than $2f_i$.

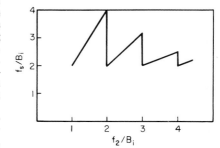

Fig. 4. Minimum sampling rate.

Spectrum

In our previous considerations, the sampling function was assumed to be a series of impulses, spaced in time by $1/2f_i$. The spectrum of such a function is periodic (the period being f_s) and extends to infinitely high frequencies. The amplitudes of the individual spectral components remain the same for each period. In practice, however, the sampling function consists of a series of rectangular pulses of appreciable duration. The composite spectrum of such output is shown in Fig. 5. It contains the spectrum of the function, together with higher-frequency terms, spaced periodically at intervals as shown in Fig. 5. As a result of the appreciable width of the sampling pulse, the magnitudes of the periodic terms die off with a $\sin x/x$ envelope depending on the sample duration. For narrower pulse widths, the rate of decay decreases, and in the limit, as the pulse width approaches zero, the spectrum approaches a uniform-amplitude distribution.

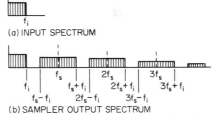

Fig. 5. Spectrum of a sampled function.

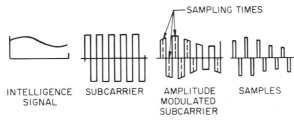

Fig. 6. Plus-and-minus sampling.

The spectrum of the sampler output includes the lowest frequencies present in the input signal. Therefore, if the input signal contains a d-c component, a corresponding d-c component will appear in the output of the sampler. Where a radio link is used to transmit such a signal, the presence of direct current in the modulating signal may pose difficult problems as regards the stability and frequency response of the communications link. In order to avoid this difficulty, a form of sampling can be used whereby the original signal is first amplitude-modulated on a subcarrier. The resultant signal is then sampled at the appropriate sampling rate. It can be readily visualized, that if the subcarrier frequency is selected to be an odd multiple of half the sampling frequency, the samples will fall alternately during positive and negative half-cycles of the amplitude-modulated subcarrier, hence the term "plus-and-minus sampling." Figure 6 illustrates the plus-and-minus sampling process described above. The spectrum of the sampler output for the case where the subcarrier frequency is equal to one-half the sampling frequency is shown in Fig. 7.

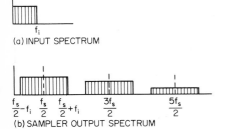

Fig. 7. Spectrum of a plus-and-minus sampled function.

Aliasing Error

A practical implementation of a sampled-data system in strict accordance with the sampling theorem leads to two basic problems:

1. Implied in the sampling theorem is a statement that the function to be sampled contains frequency components up to a certain maximum frequency f_i beyond which all spectrum components are equal to zero. This behavior is not encountered in practice where the input function is either band-limited by a physically realizable filter or is known, from experience, to contain very little energy beyond a certain specified maximum frequency.

2. The ideal low-pass filter used for interpolation is not physically realizable and instead an interpolation filter is used with some finite cutoff rate.

Recognizing these two restrictions, the sampling rate in any particular system will exceed the theoretical $2f_i$ by a factor which is dictated by the "a priori" knowledge of the spectrum of the input function, the selection of the cutoff rate in the presampling and interpolation filters, and the allowable magnitude of the aliasing error selected by the system design. The origin of the aliasing error becomes evident from the examination of the spectrum of a sampled function (see Fig. 8).

Since the input function contains spectrum components extending beyond one-half of the sampling rate, the lower sideband of the modulation process centered around the sampling frequency will extend into the frequency region which lies within the pass-band of the interpolation filter and

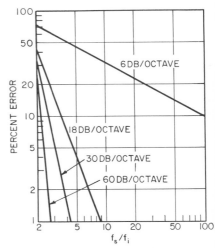

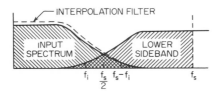

Fig. 8. Aliasing error.

Fig. 9. Aliasing error for various presampling filters.

will cause the interpolated output to differ from the input to the sampler by more than the gain factor. Several analyses of this problem have been performed [7,8,9] with the result shown in Fig. 9. It can be seen that the ratio of the sampling rate to input bandwidth for a given maximum aliasing error decreases with increasing attenuation in the presampling filter. In practical applications where an overall-accuracy objective is of the order of 2 per cent, the selection of 60 db/octave input filter and a 2.5:1 ratio between the sampling rate and the filter break frequency is satisfactory to keep the aliasing-error contribution within allowable limits.

Band Limiting and Interchannel Cross-modulation

As shown before, the spectrum of the sampler output contains frequency components which extend far beyond the bandwidth of the input function. In the interpolation process, it is not necessary to consider any frequency components higher than one-half the sampling rate. It is therefore desirable to band-limit the spectrum of the sampler prior to radio transmission in order not to occupy more bandwidth than necessary for the data transmission. The consequence of band limiting is spreading out of the sample pulse which, in turn, causes interchannel cross-modulation. This statement has the implication that a minimum-phase filtering network such as an RC or RLC is used for the purpose of band limiting. Filters of this nature are characterized by an infinite memory; i.e., it takes an infinite amount of time until the output of the filter reaches the steady-state value of a step input. This characteristic causes the subsequent channels in a PAM multiplexer to be affected by the voltage amplitude in all preceding channels; i.e., interchannel cross-modulation exists. In practice, an assignment of the maximum value for interchannel modulation can be made, based on which the design of the band-limiting filter can be accomplished. For example, using an RC filter, it is possible to select a configuration (number of stages and cutoff frequency) which will result in a maximum cross-modulation of less than a certain specified value. If an RLC filter is selected, it is possible to control the response of the filter in such a way that zero crossings of its output will occur at particular times such that the cross-modulation effects will be negligible. In these cases, however, it is necessary to start the sampling process with infinitesimally short samples and by a holding process ensure that a flat-top pulse is presented to the input of the filter.

There is another class of filters composed of non-minimum-phase networks which exhibit a finite memory; i.e., the output will reach the peak amplitude of the input step function in a fixed controllable time without any additional overshoot or ringing. Because of their complexity, such filters are used primarily in ground installations and are not considered for application in vehicle-borne equipment where the band-limiting process must first occur.

4 Elements of a PAM-FM Telemeter

The airborne package of a PAM-FM telemeter consists of a PAM multiplexer and a frequency-modulated transmitter. The PAM multiplexer is the device which accomplishes the sampling of a large number of inputs in time division and presents at its output a pulse train representing the sampled amplitudes of all the inputs in a periodic time sequence. The duty cycle of the output waveform is selected by the system designer to best suit the facilities available in the ground station. In practice, the duty cycle is either 50 or 100 per cent depending on the ground-station system. The apparent bandwidth advantage of the 100 per cent duty cycle has to be examined from a standpoint of interchannel cross-modulation caused by the selection of the band-limiting filter.

In general, the PAM multiplexer consists of a sequencing circuit and a sampling circuit. The sequencing circuit generates a sequence of control pulses which cause the appropriate sampling gates to open at designated times. Typically, the circuit consists of an oscillator followed by a binary frequency divider and a logic matrix, or an oscillator followed by a closed ring counter (or a matrix of ring counters).

The most important component of the multiplexer is the sampling gate. It must exhibit adequate sensitivity to low-level inputs and should have a constant gain over the whole range of input to 0.1 per cent. It must not feed back current in excess of a predetermined value to the data source, and it must generate a minimum amount of noise and d-c offset. A simple form of a sampling gate is shown in Fig. 10a. To open the gate, current sufficient to saturate the transistor is drawn from the base. To close the gate, the transistor base is raised to a potential more positive than that at the input or output. A buffer amplifier is placed in front of the switching transistor in order to prevent base current coming from the gate-control signal from flowing through the signal source and also to raise the input impedance (approximately 500 kilohms). The output of several gates may be connected together by a common bus.

Another form of an analog gate is shown in Fig. 10b. Two complementary transistors are formed into a bridgelike configuration. The circuit is somewhat similar to the familiar diode gate, and its operation is as follows: $R1$ and $R2$ are chosen such that in the gate-closed condition the base of $Q1$ is negative, cutting it off. $R3$ and $R4$ are chosen similarly such that in the gate-closed condition the base of $Q2$ is positive, cutting it off.

To open the gate, a positive pulse is applied to $C1$, and simultaneously a negative pulse is applied to $C2$, causing $CR1$ and $CR2$ to block. The current through $R2$ then flows into the base of $Q1$, saturating it, and similarly the current through $R4$ causes $Q2$ to saturate. With $Q1$ and $Q2$ saturated, the impedance between gate input and gate output changes from a high value to a low value. At this point, a further restriction may be seen on the values of $R2$ and $R4$, in that the current through $R2$ should equal the current through $R4$.

With the gate closed, input impedance is determined by the cutoff characteristics of the two transistors and is typically several megohms. With the gate open, the input impedance is the parallel combination of $R2$, $R4$, and the load impedance. Series gate impedance is 10 to 100 ohms.

By using complementary transistors, n-p-n-p-n-p, errors induced by one transistor (leakage current, voltage offset collector-to-emitter) are in the opposite direction to the errors induced by the other transistor and tend to cancel. However, the greatest

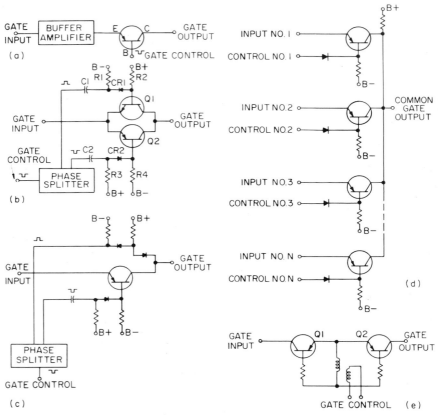

Fig. 10. Sampling gates.

advantage of this gate over the gate shown in Fig. 10a is that the base current of one transistor may be balanced against the base current of the other transistor so that no current from the gate control need flow through the signal source, and hence no buffer amplifier is needed.

Considering the practical characteristics of available switching transistors, the p-n-p transistors appear to be better than n-p-n. Hence the results of using one p-n-p transistor as compared with using a p-n-p-n-p-n combination are often as good or better. In Fig. 10c, the n-p-n transistor has been replaced with a diode, and the operation is the same as in Fig. 10b.

The significance of the gate in Fig. 10c is that it points the way to a simplicity of design when a group of signals are to be sampled sequentially, one at a time with a common output. Under these conditions, as each gate is opened it passes current from B+ to the common gate output through a resistor and diode. Since one and only one gate is always open, one common resistor between B+ and the common gate output can be used to carry this current for all the gates. The phase splitter becomes unnecessary. A group of such simplified gates is shown in Fig. 10d. One transistor, one resistor, and one diode are required per gate plus one resistor per group of gates. To operate the gates, a negative voltage is applied to the control input of one gate, reverse-biasing the diode. Current flows from B+ through the common resistor to this one gate, through the collector junction of the transistor and on to B− through a resistor. A positive voltage is applied to all the other gate-control inputs, cutting off all the other transistors. The negative control signal may be applied to the gates in any sequence that may be desired so long as one and only one gate is opened at a time.

This gate represents an optimized form of a direct-coupled transistor gate. For slow-sampling-rate applications, a gate can be held open indefinitely. For fast-sampling-rate applications, the speed is limited only by the response of the diode and transistor.

Another sampling-gate circuit exhibiting high input impedance is shown in Fig. 10e. To open the gate, a pulse is applied through the transformer to saturate both transistors. In the closed condition, $Q1$ blocks passage of a signal when the input is negative with respect to the output, and $Q2$ blocks when the input is positive. The gate is simple and accurate but is restricted by sample frequency and duration limitations because of the transformer characteristics.

Because of the varying degree of difficulty involved in designing circuits for PAM multiplexers, they are normally divided into two categories—high level and low level. The high-level multiplexer is normally designed for a 0- to 5-volt input, whereas the low-level multiplexer may have a full-scale range of 20 to 50 mv. In the low-level case, the multiplexer should provide capability to handle differential signals with a high degree of common-mode ejection.

Where desirable, the functions of the common-output band-limiting and duty-cycle conversion may be incorporated in the multiplexer unit, although in recent designs these are handled in a separate unit commonly called the *programmer*. This unit is used to coordinate the operation of several multiplexers of a larger PAM-FM system and to some extent may also assume the function of the pulse sequencer.

The transmitter used in a PAM-FM telemeter is designed to provide a r-f power output as dictated by the transmission-link considerations. The important modulation characteristics are the linearity of the frequency deviation with respect to signal amplitude and the frequency response of the modulation network which should preferably extend to direct current at its low limit.

5 Typical Application

To illustrate a typical application of a PAM-FM application, consider the instrumentation assignment in Table 1.

Table 1. Instrumentation Schedule

Item	No. of data channels	Signal amplitude	Required accuracy, %	Channel bandwidth	Channel sampling rate, samples/sec
1	3	0–5 volts	10	2 kc	5,000
2	2	0–5 volts	5	1 kc	2,500
3	5	0–5 volts	2	100 cps	312.5
4	28	0–5 volts	2	25 cps	78
5	55	0–50 mv	1	5 cps	39
6	115	0–50 mv	2	5 cps	19.5
7	110	0–5 volts	2	1 cps	19.5

Total number of channels: 318.
Total intelligence bandwidth: 10,160 cps.
Total sampling rate: 40 kc.

The implementation of a vehicle system to transmit the data shown in Table 1 is depicted in block-diagram form in Fig. 11. The system is centered around a 16-channel main multiplexer in which each channel is sampled at a rate of 2,500 per second. Channels 1 and 11 are used for synchronization purposes. Since item 1 requires 2-kc channel bandwidth, channels 2 and 10, 4 and 12, and 5 and 13 are used in pairs (supercommutation), thus providing 5,000 samples per second for each of the 2-kc inputs. In employing supercommutation, it is important to select channels in a manner which will provide uniform time separation between data samples; e.g., in

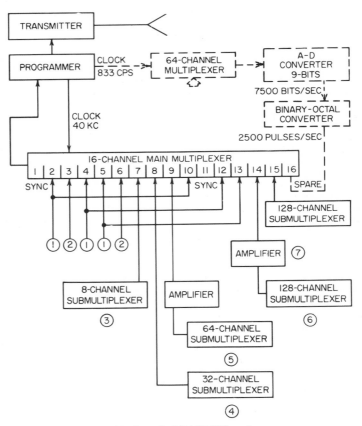

Fig. 11. A typical PAM-FM system.

the 16-channel main multiplexer, a separation of eight channels should exist between the supercommutating channels.

Item 2 is handled conveniently by channels 3 and 6, respectively, since 2,500 samples per second are sufficient to satisfy the bandwidth and accuracy requirements of these measurements. The transmission of item 3 is accommodated by an eight-channel submultiplexer feeding channel 7 of the main multiplexer. Similarly, items 4 and 7 are handled by submultiplexing channels 8 and 15, using a 32-channel and a 128-channel submultiplexer, respectively. In the case of the low-level inputs, items 5 and 6, the appropriate low-level multiplexers are each followed by a common bus amplifier (gain: 100) which, in turn, feed channels 9 and 14 of the main multiplexer. The effective sampling rate per channel is higher than required to satisfy the accuracy demands on these data.

In each of the submultiplexers, sufficient spare capacity exists to include calibration pulses where desired. The design of the system is such that one set of calibrating pulses will effectively calibrate the entire system.

As seen in Fig. 11, channel 16 is available as a spare which could be used for some additional data. For example, it could be used to transmit data requiring high accuracy, say 0.2 per cent, by adding the components of the system shown in dotted lines. Here a 64-channel multiplexer could provide 13 samples per second in each of the channels to sample slowly varying data inputs. The analog-to-digital converter would encode each of the samples to 9 bits, resulting in a binary digital output of 7,500 bits/sec. This output could be converted by the binary-to-octal converter into 2,500 pulses per second, each pulse assuming one of eight discrete amplitudes. This, in effect, provides for transmission of PCM data at a rate of 3 bits per PAM

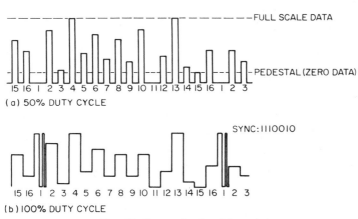

Fig. 12. Composite signal format.

pulse. The performance of the overall PAM-FM system, of course, permits recognition of the discrete eight levels. A subsequent conversion to binary data, if desired, would be accomplished in the ground station.

The programmer as shown in Fig. 11 has the function of providing the proper clock pulses to the PAM multiplexers and converting the composite waveform of the 16-channel multiplexer into the format selected for transmission. For example, two possible formats are shown in Fig. 12. In Fig. 12a, data are transmitted as a pulse train of 50 per cent duty cycle. All data are placed on a pedestal of approximately 20 per cent of the full-scale data level. The synchronization pulses are characterized by the absence of data and pedestal. Figure 12b shows a 100 per cent duty-cycle waveform in which no pedestal is used. The synchronization interval is identified by a digital code word inserted at the proper time.

6 Ground Station

General Description

The ground station of a PAM-FM telemetry system should offer the necessary flexibility to accommodate the wide variety of signal formats made possible by the vehicle sampled-data systems. Figure 13 shows the equipment layout and may be used as a guide for a brief survey of station functions.

Rack 1 contains a wideband magnetic-tape recorder especially adapted for predetection recording, using two wideband tracks on a reel of 2-in.-wide video tape. Also recorded longitudinally on the tape are two low-frequency tracks (response to 15 kc) for voice annotation and timing signals, respectively. The equipment is designed to accept either a 5-Mc or a 30-Mc frequency-modulated carrier with a deviation of ± 1 Mc. Because recording is performed prior to second detection, the recorder is insensitive to modulation signal.

Rack 2 contains the r-f detection equipment, which accepts the r-f carrier from the antenna or r-f preamplifier and in turn provides the 5-Mc carrier to the magnetic-tape recorder. In addition, this equipment detects the signal for real-time processing. In the recorder playback mode, the r-f unit accepts the 5-Mc carrier reproduced from the tape and detects it for later, off-line data analysis. The detector may be a general-purpose device or may be optimized for a specific signal format.

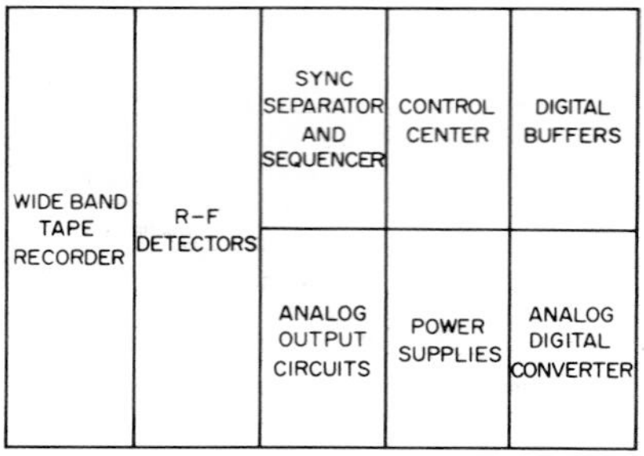

WIDE BAND TAPE RECORDER	R-F DETECTORS	SYNC SEPARATOR AND SEQUENCER	CONTROL CENTER	DIGITAL BUFFERS
		ANALOG OUTPUT CIRCUITS	POWER SUPPLIES	ANALOG DIGITAL CONVERTER

FIG. 13. PAM-FM ground-station layout.

Rack 3 contains the sync separation and sequencing circuitry. This equipment reconstructs the clock rate of the incoming signal and performs the necessary logic to identify the frame- and subframe-sync patterns. The racks may also contain a limited number of analog output circuits to provide demultiplexed data from the pulse-amplitude-modulated samples.

Rack 4 contains power supplies for the entire equipment and a control center for station operation and monitoring. All critical signals in the station are brought to selector switches at the monitor panel to permit checking station operation before, during, and after an actual mission. This capability is of particular value for rapid isolation of possible malfunctions during a mission and quick replacement of the faulty component. It is also a considerable help in off-line trouble shooting and routine maintenance.

Rack 5 contains the necessary digital circuitry for converting any or all of the PAM data into digital format for further processing by digital computers.

The following describes in greater detail the various functions performed in the ground station.

Data Recording

A primary function of nearly all ground stations is to provide permanent storage of all data for future off-line processing and study. To be most efficient and reliable, the storage device should be electrically as close as possible to the receiving antenna and should be insensitive to the type of data being stored. For example, pulse rates, number of channels, sync format, and data format should have no effect upon the reliability and accuracy of data storage. Long recording times may be required as the systems may operate in high-orbiting vehicles or in deep-space probes.

While the above requirements establish certain desired characteristics of the data-storage system, other characteristics of the data links themselves must be established before the preferred storage device can be selected. A first consideration is that the systems will utilize a frequency- or phase-modulated carrier. Secondly, because of the growing desire to transfer all telemetry operations to higher frequencies, a capability must be available to handle transmission bandwidths considerably greater than the 500 kc utilized in the present telemetry bands.

Combining the foregoing considerations leads to the selection of predetection recording utilizing a wideband magnetic-tape recorder as the preferred method for data storage. As the data are recorded prior to detection, the multiplexed data format does not affect the storage. In the reproduce mode, the frequency-modulated carrier is fed into the r-f detector and handled just as a real-time signal off the air.

Real-time Data Handling

The received signal from the antenna is supplied to the detection equipment where it is first converted down to the 5-Mc range. This is done in one or more heterodyne stages depending upon the transmission frequency being used. The 5-Mc signal is then fed directly to the wideband recorder for storage as described above. The signal may also be fed to a data detector at those stations requiring real-time readout of any of the data carried by the link. The commonly used detectors are standard, general-purpose frequency discriminators. However, in special cases optimum detectors may be employed for certain data formats, utilizing all the a priori knowledge of the signal that will help recover data under adverse conditions.

The primary function of the sync separator and sequencer is to identify and sort out each data sample and deliver it to the appropriate processing or display device. Its operation depends to some degree on the format of the multiplexed PAM signal. For the format shown in Fig. 12a (used extensively in the Lockheed system), the sync separator and sequencer is as shown in Fig. 14.

The input amplifier (block A) is used primarily as a buffer to match the input line. It provides adequate drive for the circuits to follow (blocks B and E).

In cases where optimized detectors are utilized, additional data filtering is not required. In the more general case, the detector will be quite broadband to permit adequate detection of high-sampling-rate systems. To realize the maximum-output signal-to-noise ratio in the ground station, the video bandwidth must be restricted to reject noise components which are outside the information bandwidth. The characteristics of the noise filter (block B) are determined by the sampling rate, the overall system accuracy, and the allowable crosstalk between successive samples. For low-speed low-accuracy systems, a simple RC filter similar to that used in the vehicle and having a cutoff at 2.2 times the sampling rate will suffice. For higher-speed systems with greater accuracy requirements, a finite memory filter is employed which restricts the effective video bandwidth to 1.7 times the sampling rate, with minimum crosstalk introduced into successive samples.

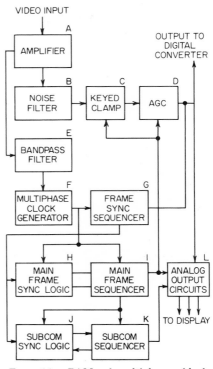

Fig. 14. PAM demultiplexer—block diagram.

The purpose of the keyed clamp (block C) is to remove the pedestal from the data pulse train. As received, the signal is proportional to a 6.5-volt peak-to-peak signal of which 5 volts is data and 1.5 volts is the pedestal, as described before. The clamp is keyed to operate during a zero data calibrate and thereby in conjunction with the AGC to keep the output data within the limits of the zero and full-scale calibrates received from the vehicle.

The automatic gain control (block D) is provided to compensate for gain variations within the data link and to normalize the gain corresponding to a 5-volt peak-to-peak data input to the vehicle multiplexer. The operation of the AGC is predicated upon having a full-scale calibrate pulse present in the composite pulse train. Under the control of a keying pulse from the sequencer, the AGC circuit samples the calibrate signal and compares it with a fixed reference. The error signal, if any, adjusts the system gain to reduce the error to zero.

The AGC has only a limited range of control, approximately ± 25 per cent of nominal gain. This is adequate to compensate for the gain variations normally expected but will not permit abnormal conditions to drive the system gain to zero or to saturation. This feature also makes it possible to synchronize the incoming signal and thereby locate the calibrate pulse, regardless of the gain status of the AGC loop. In special cases, it is possible to disable the AGC loop and operate the system at a fixed gain.

This completes the path of the composite sampled-data pulse train. Succeeding operations require separating the composite signal into individual channels.

The bandpass filter (block E) is a narrowband filter used to separate out the clock frequency while rejecting the noise and modulation components. At the same time, the high Q of this filter provides the memory required to retain the clock frequency through periods of interference or of signal fades. The memory of the high-Q circuit is adequate for most transmission conditions and is comparable with that obtainable from a phase-locked system or other clock regenerators that must be able to track small deviations of the vehicle clock.

The multiphase clock generator (block F) accepts the output of the bandpass filter, compares it with the data pulse train, and adjusts a variable-delay element to phase the clock properly with respect to the data. It also provides a variety of outputs having fixed time relationships to permit keying subsequent circuits at the proper times. These outputs are used to control all logical functions within the station.

The frame-sync separator (block G) receives an input from the filtered gain-stabilized output of the AGC circuit. This input is clipped and shaped to provide a pulse for each pulse present in the data pulse train. This is then combined with the reconstructed clock from the multiphase clock generator in a gate that provides an output only when a data pulse is missing. Both main frame and subframe synchronization is identified by the complete absence of data or pedestal during a selected channel time. Therefore, the output of the frame-sync separator is a train of pulses generated when there was no pulse present in the incoming signal. These are potential main frame- or subframe sync pulses which must be further analyzed by subsequent logic circuitry to identify true sync and reject any noise or other false sync indications.

The main frame-sync logic (block H) is based upon the use of a 16-bit shift register so that sync pulses with any spacing up to 16 channels may be identified. The train of potential sync pulses generated by the frame-sync separator is applied to the input of the shift register while the register is clocked by the proper output of the multiphase clock generator. The train of potential sync pulses is also applied to an input of an AND gate. The other input to the AND gate is taken from the appropriate stage of the shift register. The stage selected is determined by the spacing of the main frame-sync pulses. For example, in Fig. 11 the synchronization periods are assigned to channels 1 and 11. In this case, the output of the sixth stage of the shift register would be applied to the AND gate. When the pulse from channel 11 is delayed exactly six clock pulses, it will arrive at the AND gate in coincidence with the channel 1 pulse and a main frame-sync pulse will be generated. This main frame-sync pulse is then used to reset all sequencer logic to channel 1 and thereby establishes the true identity of each channel.

As is shown in the block diagram, there is a signal fed back from the sequencer to the sync logic. This feedback provides added noise immunity and works as follows: A sync gate follows the shift-register logic which is keyed by an output from the main frame sequencer to open when a main frame-sync pulse is expected (i.e., when channel 1 time is due). The main frame-sync pulse itself closes this gate until the channel 16 pulse opens it again. Thus, throughout the entire main frame, noise pulse cannot cause false resetting of the sequencer and only true frame sync is accepted. However, if for any reason the frame-sync pulse does not occur to close the sync gate, the gate remains open, ready to accept a pulse when it does appear.

This operation of the sync gate may also be used to operate an alarm circuit to indicate if the system is in sync. When the sync gate is opening and closing regularly, with only a one-channel open time, the system is in sync. When the gate remains open, the system may not be truly synchronized and the sync search mode is in process.

The main frame sequencer (block I) controls all the output circuits by generating keying pulses in proper time sequence to identify and separate each main channel. It consists of two 16-bit shift registers, each connected in a ring but with one operating at one-sixteenth the rate of the other. Outputs from each ring are combined in an AND gate to generate a keying pulse for any desired channel from 1 to 256.

The keying pulses from the main frame sequencer control the AGC and keyed clamp, the output circuits, and all subcommutate sync logic and sequencers. These control signals are routed to the proper destinations by means of the patch panel.

The subcommutate sync logic (block J) is identical to the main frame-sync logic except that it receives its timing signal from the appropriate output of the main frame sequencer. This signal performs two functions. It opens a gate allowing an input into the shift register only when the proper channel is present at the output of the frame-sync separator, and it clocks this input down the shift register at the frame rate instead of the composite clock rate. The same sync pattern used on the main channel can be used on the subcommutate channels.

Since the equipment may employ several asynchronous subcommutate channels, each with its own sequencer and sync pattern, a separate set of sync logic should be provided for each independent subcommutate channel.

The subcommutate sequencer (block K) is essentially similar to the main frame sequencer and requires no further discussion.

In cases where a 100 per cent duty-cycle waveform is transmitted (as shown in Fig. 12b), the synchronization circuits will involve a synchronous clock reference oscillator and a matched filter for synchronization-pattern recognition. The degree of complexity of such a system will vary, of course, with the sync code assignment. Some experimental systems have been built employing this technique. Indications are that this method of synchronization will provide improved performance over other systems.

In all cases where immediate analog display of data is required, a sufficient number of output circuits should be provided (block L). There are two types of output circuits. The most frequently used is a sample-and-hold circuit, which is keyed to sample the data pulse at its peak and then to hold that value until the next sampling period. This provides an essentially d-c output that is a step-function approximation of the data signal. It is used on low-frequency data where amplitude rather than true wave shape is most important, and usually where the sampling rate is much higher than the maximum data frequency. The sampling switch is bilateral and can charge or discharge the holding capacitor so that it is possible to hold a d-c level without returning to zero. The second type of output circuit is used whenever a smooth data output is required. This circuit consists of an analog data gate, keyed by an output from the sequencer, to pass the desired data sample; an interpolation filter; and an amplifier, the output of which is a smoothed replica of the input data.

The digital data-handling equipment, contained in rack 5 of Fig. 13, performs a number of functions when digital operations are involved. For example, if any or all of the data is transmitted as PCM, the appropriate bits must be reassembled into a word describing the data sample. If the PCM data are transmitted in octal code, three eight-level pulses describe the value of a data sample with 0.2 per cent resolution. It may be desirable to convert the octal code into straight binary for subsequent processing. Therefore, under the control of the sync separator and sequencer, the octal-coded pulses are sent to an octal-to-binary converter and then to a buffer memory until the required number of bits have been accumulated to represent the data sample fully. The buffer is then read out into the desired storage or display equipment as a complete data sample. In case the PCM data are transmitted in binary form, the octal-to-binary conversion is omitted and the bits are fed directly into the buffer for assembly into a complete word.

In the case where digital processing is to be used for all data whether transmitted as PCM or PAM, an additional step is required for the PAM data. This step is accomplished by an analog-to-digital converter also contained in rack 5. Thus there is available from the ground station all data in digital format ready for further processing. Analog outputs from desired channels may also be available either in addition to or in place of the digital outputs. Thus complete flexibility is provided according to the requirements of a particular ground-station or data-acquisition program.

Acknowledgment

The author gratefully acknowledges the contribution of C. H. Burley on sampling gate circuits and C. M. Kortman on PAM-FM ground stations.

7 PACM Telemetry System*

A study was made for the U.S. Army Signal Research and Development Laboratory in 1959 by the Aeronutronics Department of the Ford Motor Company (now Philco Division)[10] with the object of recommending improved telemetry systems or standards for use on the missile test ranges. A users' survey was made to establish the expected volume of measurements, the data rates, and the required accuracy. This of course reflected current practice to a large extent. The characteristics of four types of telemetry systems, PAM-FM, PCM-FM, PDM-FM, and FM-FM, were carefully evaluated, both analytically and experimentally, from the viewpoint of communications efficiency, spectrum occupancy, effect of interference, information capacity, and overall accuracy. The report recommended that a hybrid system of telemetry called PACM be standardized, which is a combination of PAM and PCM. Only FM carrier modulation was considered at length in this study. This recommendation stemmed from an attempt to meet all user requirements in one system. These were found to extend over a wide range of accuracies and data rates.

The study is valuable because it investigated the *error* performance of the standard telemetry systems and experimentally justified the selection of important telemetry *system* parameters, such as q, the allowable samples per cycle of information bandwidth f_D, the rms frequency-deviation ratio, and B/B_i, the ratio of r-f bandwidth (receiver) to information bandwidth. Such work is more fundamental than the concept of PACM itself and can be used by telemetry-system designers where any of the four systems may find application. For this reason this section will present first the major points from this study of importance to telemetry-system design and then a brief description of the PACM system.

Analog data were the major concern of the study in keeping with the users' requirements. Some requests for digital information were emerging at the time, as, for example, words from the missile-borne digital computer; however, these were not considered. Hence system analog rms error was chosen as the basis for evaluating telemetry-system performance after consideration of other alternatives.

The greater flexibility of time-division multiplexing was noted as compared with the standard FM-FM. The latter was considered adequate for a compact 10- to 12-channel system of 1 or 2 per cent accuracy. In this case it is simple and reliable. However, it cannot expand in capacity without undue increase in bandwidth, easily meet the better accuracy requirements of 0.01 or 0.1 per cent, or conveniently trade channels for bandwidth as can be easily done in time division by supercommutation. In addition unequal time delays occur in the standard FM-FM system. Hence the

---

* The section is by E. L. Gruenberg. It is based upon work done by Aeronutronics, Newport Beach, Calif., for USASRDL, on Contract DA-36-039-SL-73182 and reported in Ref. 10.

time-division systems were found superior for future growth. PCM is particularly attractive for its flexibility; however, PAM proves more useful for low-accuracy data, requiring less bandwidth and power than PCM. See Fig. 15, which is drawn for a 2 per cent error. This fact led naturally to the thought that a hybrid PCM and PAM system could give best performance for both high- and low-accuracy inputs. The ability to combine the two modulations into one system with minimum design problems was discussed fully in the report.

The standardization of PACM has not occurred. The widespread use of PCM and PAM as separately required may account for this, since in fact there is little actual difference between the combined system and a PAM and PCM system designed so

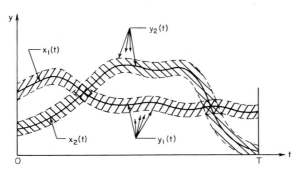

Fig. 15. Illustration of two possible signals from an ergodic ensemble of duration T and frequency band limit B_i. Transmitted signals are $x_i(t)$, received signals $y_i(t)$.

that they may be synchronously used. This will be more apparent when the design of the PACM system is discussed in more detail.

User Requirements

The requirements for instrumentation of a typical missile were found to consist of

1. A large number of channels for low-frequency data at 0.1 to 1 per cent accuracy
2. A medium number of channels for 1 to 2 per cent data with a frequency response of 100 to 200 cps
3. A few channels to accommodate high-frequency (up to 2,000 cps) data

The average missile required a telemetry capacity of 75,000 bits/sec; however, a telemetry system for 75 per cent of the missile types must have a capacity of 200,000 bits/sec. Large missiles require a telemetry system with a capacity of 1,000,000 or more bits/sec and more than 400 channels.

The data are substantially analog in nature with a substantial d-c component. Otherwise they can be assumed to have a flat spectrum to $f = B_i$ and a sharp rolloff above this frequency. Peak amplitudes can be restrained by limiters.

Performance Criteria

Two possible approaches to the evaluation of telemetry systems may be considered. They are:
1. How much information is effectively transmitted over the available bandwidth with the available signal power?
2. How efficiently is the telemetry facility being used to transmit data within the desired *accuracy* limitations?

9–20 PAM AND ADVANCED TELEMETRY SYSTEMS

The former approach concerns itself with determination of equivocation or loss of information (Chap. 2) in transmission, since the information rate H is

$$H = 2B_i[H(x) - H_y(x)] \tag{4}$$

where $H(x)$ is the entropy of the source and $H_y(x)$ is the equivocation or uncertainty in received signals caused by noise.

$$H_y(x) = -\int_{-\infty}^{\infty} P(y) \int_{-\infty}^{\infty} P_y(x) \log P_y(x) \, dx \, dy \tag{5}$$

where $P(y)$ is the probability that the received signal is $y(t)$ regardless of the signal that is sent and $P_y(x)$ is the probability that signal x_i was sent when signal y_i is received. This is presented by the shaded area in Fig. 15, shown for two transmitted signals $x_1(t)$ and $x_2(t)$. Any received signal $y_i(t)$ fitting within the shaded area might have been originated by the transmitted signal $x_i(t)$ owing to the smearing caused by the noise. In the case of analog telemetry signals this expression may be simplified by the assumption that $P_y(t)$ is independent of y or $P(y)$; then

$$H_y(x) = -\int_{-\infty}^{\infty} P_y(z) \log P_y(z) \, dz \tag{6}$$

where $z = y - x$, the linear error. A distribution for $P_y(x)$ has to be assumed which can be easily manipulated but is also physically realistic. Also z is small for additive noise or $z \ll A =$ the maximum signal. This permits ignoring variations in $P_y(z)$ for limit values of y and hence permits the use of Eq. (6) in the case of symmetrical additive noise.

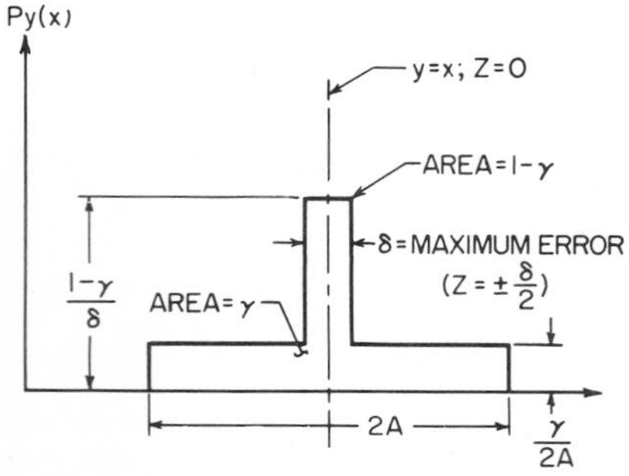

For the case of $P_y(z)$ assumed to be Gaussian (this means a Gaussian noise distribution) and for $P(x)$ uniformly distributed, that is, a uniform peak limited signal, the resulting information rate is

$$H = B_i(-\log 2\pi\sigma_n{}^2 - 1) \quad \text{nepits/sec} \tag{7}$$

where $\sigma_n = \sigma/2A$ and σ is the variance of the error. A is the maximum amplitude. Hence σ_n is the *normalized rms error*.

FIG. 16. Possible distribution of $P_y(x)$ for intermittent break in operation.

PCM is a digital system. The information rate is thus readily computed from Eq. (5). $P(x)$ is determined by assuming equal probability for 0 and 1 and $P_y(x)$ is obtained directly from the bit-error probability e. $1 - e$ and e are the respective probabilities $P_y(x)$ for the two signals. Then, for $e \ll 1$,

$$H = Nf_s(1 + e \log e) \quad \text{bits/sec} \tag{8}$$

where $N =$ number of bits per sample and $f_s =$ number of samples per second. The corresponding PCM analog-error equation is

$$\epsilon_{rms} = \sqrt{\frac{e}{3} \frac{2^N + 1}{2^N - 1}} \tag{9}$$

A third case which might be considered is where an analog signal is transmitted but there is a small probability that the channel is broken. During the break the signal may assume any value within $\pm A$ with a small probability e. The distribution for $P_y(x)$ is shown in Fig. 16. It is assumed that the maximum error in the absence of the break is $z = \delta/2$. The area γ is the total probability of the break. Hence $1 - \gamma$ must be the probability of uninterrupted signal. $e = \gamma/zA$ is the magnitude of the probability distribution of a signal being a particular value A during intermittency. Then for uniform signal probability distribution $P(x)$

$$H = B_i[\log 2A - \gamma \log (2A/\gamma) + (1 - \gamma) \log [(1 - \gamma)/\delta]] \tag{10}$$

and the error is

$$\sigma_e{}^2 = \frac{2}{3} \gamma A^2 \tag{11}$$

The expressions [Eqs. (7) through (11)] cover a sufficiently wide range of practical situations to examine the relationships of information rate and normalized error. This has been done in Fig. 17. The intermittent channel is plotted for the very low noise peak-to-peak signal value of 10^{-5} when there is no break in the channel (this closely equals e).

The figure shows that there is no direct relationship between information rate and normalized error that holds for all cases. Thus it is necessary to consider the use of the second approach to a performance criterion, namely, the transmission of data to a specified accuracy. This approach permits the telemetry designer to better match the telemetry to the system user's requirements.

The user has indicated two extreme types of data:

1. Low-frequency high-precision data in which there is a *low* probability of unforeseen transient phenomena during test
2. High-frequency low-precision data with considerable probability of new transient phenomena

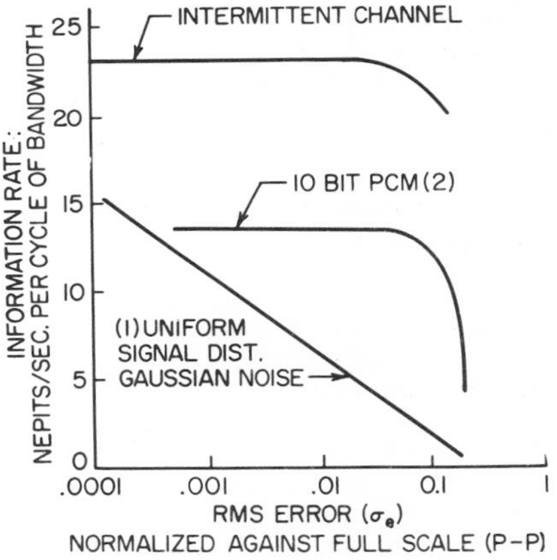

Fig. 17. Information rates vs. normalized rms error.

The error-moment criterion seems best suited to cope with this situation. This provides a means of deciding between a reasonable transient variation in signal as opposed to an actual error. The maximum sample-to-sample variation possible δ_{ss} for a *band-limited* signal is

$$\delta_{ss} = 2A \sin (\pi/q) \tag{12}$$

where A is maximum data amplitude and q = samples per cycle. When a variation exceeds this amplitude it is likely to be error with probability *increasing* as the variation amplitude. Since this is not infallibly noise a safety factor of 2 or 3 may be used. Below this value the errors are indistinguishable and the larger of these are the most important.

Equation (12) recognizes the reality of the *finite bandwidth* of the channel as a basis for distinguishing signal from noise. Hence it is possible to vary q, the samples per cycle, with the type of data, using a smaller value for type 1 data than for type 2 data for the same bandwidth.

An additional determinant of the value of q to be used in a telemetry system comes from the need to minimize aliasing as a source of error in sampled-data systems. Aliasing results when a sampling rate f_s is used which is insufficient with respect to the highest message frequency. When this occurs the sampling system may receive "alien" information (noise) centered in bands around multiples of the sampling frequency f_s. Even though the message is restricted to frequencies below a sharp data corner f_m, the noise is not and has finite power above and below $nf_s \pm f_m$.

Table 2 lists the values of q for various total errors.

Table 2. Estimated Samples per cps q for Various Accuracies

E, % rms normalized to output peak to peak.	2	0.1	0.01
q, samples/cps.	4	10	20
"Noise" spectrum rolloff, db/octave.	18	30	30

Assumptions:
 1. D-C signal component $\cong$ a-c component.
 2. Error due to filter distortion is 1 per cent or less for cutoff frequency = twice f_m; the data "corner" frequency.
 3. RMS error in interpolation is $\frac{1}{2}E$ or less for high-accuracy data where computers would be used.
 4. Message spectrum is filtered by ideal rectangular filter cornered at f_m so that message power above f_m is zero.
 5. The noise-spectrum power rolloff is at the rate given in the third line and refers to frequencies above and below $nf_s \pm f_m$.

The considerations there substantiate an evaluation of the system on the basis of rms error E in combination with bandwidth at baseband B_i. Performance at r-f carrier band is best evaluated by S/N and receiver bandwidth B, or signal S to noise intensity k_i and B. Knowledge of two of these parameters does not determine the others but cross plots are used to establish working values. The user survey for missile-test-range requirements and the need for spectrum conservation indicate the following procedure:
 1. Fix the rms error E and B_i.
 2. Optimize each system for minimum bandwidth B while operating at FM carrier threshold.

Since all systems considered in this analysis are FM carrier systems, minimum bandwidth and minimum signal occur simultaneously, at threshold. Hence this procedure will result in optimum performance for the practical systems considered. It should be noted that these FM systems do *not* trade bandwidth for signal-to-noise improvement as do the advanced communications systems of Sec. 11 and as does the ideal communications channel of Chap. 2, Sec. 15, which obeys Shannon's law. The latter requires the signal to be transmitted with Gaussian statistics, a hard-to attain ideal. Hence they are not of the highest achievable *communications* efficiency but do make optimum use of available bandwidth at low S/N for meeting given analog *error* requirements.

Optimizing Performance on Error Basis

Characteristic curves of error performance of S/N vs. receiver bandwidth B per unit information bandwidth B_i for FM systems are shown in Fig. 18. For the larger B/B_i ratios fluctuation noise dominates and causes the threshold because the lower B/B_i distortion and crosstalk transcends the fluctuation. Hence, the typical constant-error curve must change from a vertical to a horizontal line with increasing bandwidth.

Since the noise is proportional to bandwidth B, a more basic parameter is S/k_1 or signal-to-noise intensity ratio. Error curves plotted against this parameter vs. B/B_i will appear as in Fig. 19. This chart plots most conveniently system performance against external (or input-output) parameters only. These parameters include accuracy, required signal-to-noise intensity, and bandwidth. This permits optimization of the internal telemetry parameters.

The clear-cut minimum point of Fig. 19 is, of course, the desired design objective. An understanding of the factors causing it is important. As mentioned in connection with Fig. 18, there are two basic error sources in the telemetry system. They are

1. E_F, the fluctuation noise error
2. E_D, the error due to the distortion noise, which includes interpulse crosstalk in time multiplex, FM sideband distortion in FM-FM, and any other noises caused by band limiting.

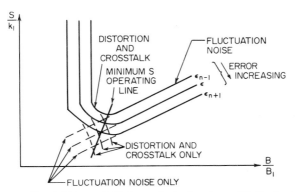

FIG. 18. Family of S/k_1 vs. B/B_i for optimized system with fixed errors ϵ_n.

If circuit nonlinearity errors are neglected then

$$E^2 = E_F^2 + E_D^2 \qquad (13)$$

since these two types of errors should be statistically independent. It is apparent from Fig. 19 that increase of signal power can be useful in reducing error only where the distortion error E_D is small, since E_D is mainly a function of bandwidth, whereas

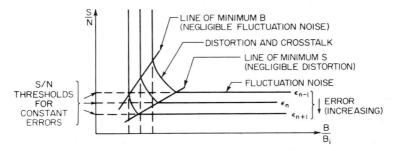

FIG. 19. S/N vs. B/B_i for optimized system with fixed errors ϵ_n.

E_F is a function of both signals and bandwidth. It is clear that the optimum point of operation is not much above the bandwidth B at which distortion error becomes important with decreasing B.

Table 3 lists the nature and origin of errors arising in telemetry systems. Some error sources are common to all systems whereas others, like quantization, are peculiar to a specific system. Details of the computation of the latter errors are presented in previous chapters (6, 7, 8) and Secs. 1 to 6.

Table 3. Telemetry-system Error Sources

1. All systems
    - *a.* Aliasing and interpolation........ A function of the *data* source; determines q
    - *b.* Fluctuation noise............... Arises from the communications link
2. PCM
    - *a.* Quantization................... Depends upon precision of encoding—the major error source
    - *b.* Interpulse crosstalk............ Pulse spillover
    - *c.* Distortion..................... Pulse distortion *b* and *c* are minor in PCM
3. PAM
    - *a.* Crosstalk..................... From pulse spillover
    - *b.* Pulse distortion................ Change of pulse shape from filtering. These are major error sources in PAM
4. PDM
    - *a.* Crosstalk..................... Between the adjacent pulses
    - *b.* Distortion.................... Change of pulse shape affects the susceptibility of fluctuation noise
5. FM-FM
    - *a.* Distortion of carrier sidebands.... Occurs in r-f and i-f
    - *b.* Overmodulation drop-out noise... Excess carrier-frequency deviation causes noise during limiter drop-out (normally part of *a*)
    - *c.* Subcarrier crosstalk............ Between adjacent subcarrier channels
    - *d.* Subcarrier distortion........... Occurs in subcarrier bandpass filter or the output low-pass filter

Crossover of Analog and Digital Systems

It has just been shown that optimum performance for FM systems centers at the point where $E_D \cong E_F$. E_D may also be called the system residual errors since they are relatively fixed once the telemetry-system design (FM-FM, PDM-FM, etc.) has been established. In the case of the PCM system, this error is mainly quantization noise the rms value of which is

$$E_q = (1/\sqrt{12})(1/2^N) \tag{14}$$

where N is the number of bits per sample or word used to designate the sample value to a given precision. The variations of this error with bandwidth follow directly from the fact that

$$N = B/f_s \tag{15}$$

where B is the receiver bandwidth and f_s is the sampling rate. A plot of log E_q with B/f_s must show a linear improvement with bandwidth. This has been done in Fig. 20, where it is the curve marked PCM-FM (no parity).

In contrast, the residual errors of the analog systems, both time- and frequency-division types, do not improve markedly with additional bandwidth per sample. Actual experimental values are shown in Fig. 20 as well as theoretical estimates. However, the analog-system performance might have been improved at lower errors since the experimental systems were optimized for 2 per cent accuracy.

Thus typically the digital PCM system crosses over and outperforms the analog systems where *higher-precision* data are required. In this case the digital system becomes superior to the nearest analog competitor, PAM, at a B/f_s of 7.5. A further comparison of PAM-FM and PCM-FM systems on the basis of signal-to-noise and total system accuracy has already been shown in Fig. 2. The signal-to-noise has been normalized to sampling rate f_s. Here the equivalence occurs at 1 per cent overall system accuracy and a bandwidth per sample of 6 for the case of the PCM system *using parity*. Aliasing and interpolation errors are not included. The parity bit improves the PCM system S/N performance at the crossover point in Fig. 2 far more than it worsens the residual error by being withheld from use to quantize the data as shown in Fig. 20. At this point the major error for the PCM system is the quantization noise.

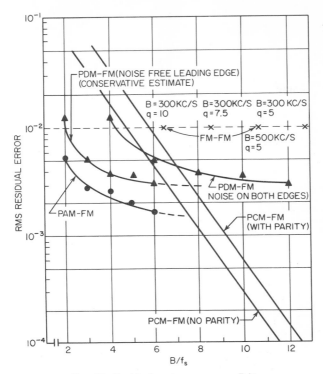

Fig. 20. Residual system error vs. B/f_s.

Spectrum Occupancy

Practical systems must be evaluated not only on a basis of minimum power and receiver bandwidth needed for optimum operation of the particular channel but also on a basis of minimum interference with adjacent channels and susceptibility to interference from such channels. These factors and transmitter drift rates determine channel allocation. The points of minimum power for each system from Fig. 1 are shown in Table 4 in terms of the normalized band occupancy B_0/B_i along with the channel allocation factor $I_0 = B_0/B$. In the tabulation both B_0/B_i and I_0 are

Table 4. Band-occupancy Comparison and Channel Allocation

System	B_0/B_i		$I_o = B_o/B$		B/B_i	$S/k_1 B_i^{1/2}$, db
	$I_p = 0$ db	$I_p = 60$ db	$I_p = 0$ db	$I_p = 60$ db		
PAM-FM	18	56	0.9	2.8	20	19.9
PCM-FM (with parity)	24	65	1.0	2.7	24	22.3
PDM-FM	36	80	1.3	2.8	28	23.2
FM-FM ($B = 200$ kc/sec)	34	120	0.81	2.8	42	25.3
FM-FM ($B = 300$ kc/sec)	47	170	0.75	2.6	64	25.0

shown for two relative power advantages for the interfering carrier I_p, 0 and 60 db. The 60-db figure is considered conservative but practical for actual use in determining channel assignments.

The data of the tabulation are based upon a total rms system error of 2 per cent of full signal range. B_0 was determined by permitting 0.5 per cent error due to an interfering adjacent band signal. For the time-division systems residual errors and fluctuation noise errors were allowed to total 1.6 per cent leaving a permissible error of 1.2 per cent for aliasing and interpolation with an assumed q of 4.

It is seen from a comparison of the systems on the basis of band occupancy with $I_p = 60$ db that they maintain their relative positions shown in Fig. 2 on the basis of receiver bandwidth B. The band-occupancy and allocation factors for the PACM-FM system may be taken to be those for the pure PCM-FM system. These factors will then become increasingly conservative as the portion of PAM in the hybrid system increases.

PACM-FM System Design

As suggested by Fig. 2, a telemetry system employing both PCM-FM and PAM-FM could accommodate a wide range of data accuracies by use of various pulse widths

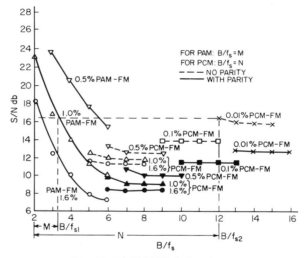

Fig. 21. PACM-FM design chart.

for PAM and word lengths for PCM. The design of such a system is facilitated by redrawing the chart using S/N instead of the normalized parameter $S/k_1 f_s^{1/2}$. Figure 21 is such a diagram. Horizontal lines represent systems having equal power requirements, and intersection of such a line with a given system-error curve provides the value of bandwidth per sample B/f_s required to obtain the stated performance for either modulation; $M = B/f_{s1}$ is the value for PAM-FM and $N = B/f_{s2}$ is that for PCM-FM. Thus, for the horizontal line drawn on Fig. 21 a combined system could consist of a PAM-FM system of 1.0 per cent accuracy having an $M = 3.2$ and a PCM-FM system (no parity) of 0.01 per cent having an $N = 12$. The received signal-to-noise ratio would be 16.2 db. Since a common bandwidth B is used, it is determined by the bit rate f_B of the PCM system. N is then the number of PCM bits per word (sample) and $M = f_s/f_{s1}$. Then for 100 per cent duty cycle PAM and NRZ (non-return-to-zero) PCM, which are recommended,

$$M = \tau_1/\tau_2 \tag{16}$$

where τ_1 is PAM pulse period and τ_2 is PCM *bit* period. In this case a PAM pulse period at least three times and possibly four times the PCM bit rate is indicated; the chart shows that somewhat better PAM accuracy will result from the choice of a longer PAM pulse period. The design of the PACM system to meet the specified accuracies has now been established with one possible precaution; that is, the q = samples per cycle will be chosen high enough so that aliasing and interpolation errors will be negligible (see page 9-22); otherwise such errors must be subtracted properly from the design accuracies and the revised-accuracy objectives must be used in the design.

To accommodate a given information bandwidth consisting of a number of primary PAM and PCM channels n and n_2, we proceed as follows:
The frame rate must be

$$F = 1/(n_1\tau_1 + n_2 N_2 \tau_2) \qquad (17)$$

where $n = n_1 + n_2$ = total number of primary channels. From Eqs. (16) and (17) and using $\tau_2 = 1/B$,

$$F = B/(n_1 M + n_2 N) \qquad (18)$$

The system sampling rate is determined from summing the individual data-channel information bandwidths. B_{ij} each multiplied by the appropriate factor q_j to provide the proper samples per cps

$$f_s = \sum_{j=1}^{j} B_{ij} q_j \qquad (19)$$

where j is total number of data channels, but the frame rate F must be equal to $f_s n$. Then from Eqs. (19) and (20),

$$B = [(n_1/n)M + (n_2/n)N]\Sigma B_{ij} q_j \qquad (20)$$

This bandwidth will have to be increased to accommodate factors such as drifts and mistuning and B adjusted accordingly. Allowance for synchronization will normally increase B by 1 to 3 per cent. Synchronization provisions are equivalent to those used in PCM systems (Chap. 8).

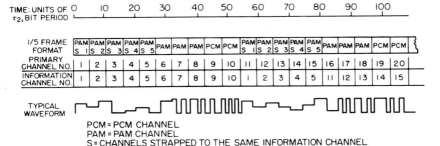

Fig. 22. Example of partial PACM frame format and pulse waveform.

The division of channels between PAM and PCM, that is, the determination of n and n_2, is established on a trial-and-error basis and is dependent upon the division of accuracy requirements in the data channels. This process is facilitated, however, by using a larger rather than smaller total number of primary channels n, because supercommutation is easier to handle than subcommutation. Practical problems such as synchronization limits n to a maximum of 100 to 130 (IRIG Standards, Chap. 5, call for a maximum of 130 words per frame for PCM). This completes the design procedure of a PACM system.

A PACM word format which would result is shown in Fig. 22.

A block diagram of the vehicle-borne PACM-FM equipment is shown in Fig. 23. Functionally the system is similar to PCM-FM, the major exception being the encoder bypass switch which directs the PAM signals directly to the modulator, bypassing the analog-to-digital encoder. The programmer must necessarily be more complex

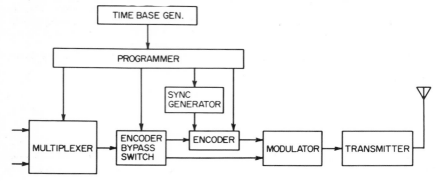

Fig. 23. Block diagram—PACM airborne system.

to take care of the variable word lengths, sampling rate, and intermixed operation. It controls all the timing and logical operations, to bring about the final pulse wavetrain format of Fig. 22. It generates the following signals:

Function	*Device controlled*
Stepping pulses	Multiplexer
Stepping pulses	Submultiplexer
Sample command pulses	Sample-and-hold circuit
PAM, PCM selection pulses	Encoder bypass switch
Sequential PCM clock pulses T_j	Encoder
Parity interrogate pulse	Encoder
Reset pulse	Encoder
Sync word 1's	Encoder
Sync word 0's	Encoder
Inhibit pulses	Encoder

The programmer is controlled only by the time-base generator. A stability of 1 part in 10^5 obtainable from a small crystal oscillator should be sufficient.

The ground station is shown in block form in Fig. 24. An r-f filter is used to prevent front-end overload from adjacent-strong-channel interference. The unique functional requirement is the separation of the PAM and PCM signals. The programmer performs the timing and logical operations to

1. Separate the PAM and PCM words within the frame
2. Synchronize the detectors
3. Synchronize the decommutator
4. Perform parity check

The video bandwidth is established by a low-pass filter with cutoff at one-half the bit rate. This filtered video is applied to the synchronizer to establish bit and frame sync.

The PAM and PCM detectors should be of the synchronous integrating type which has superior performance by about 1 db.

Conclusions

The work, which incidentally substantiated the concept of a combined PAM-PCM-FM telemetry system as workable, represented an extensive evaluation of the standard systems. The evaluation method developed was oriented to the criterion of *accuracy* of information data transmitted rather than to optimizing the transmission of information. This approach focuses upon the data *source* and optimizes the system to transmit source data to a prescribed accuracy. A different approach is to optimize the *link* so as to transmit a bit of information with lowest possible energy. This presents a different parameter β, energy per bit per unit noise intensity, which is used

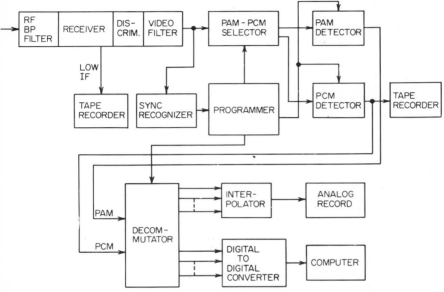

Fig. 24. Block diagram ground system.

in Sec. 8 to compare several modulation systems for their communication efficiency. This criterion focuses upon the *link* efficiency and does not directly guarantee a prescribed-accuracy performance of the telemetry system.

REFERENCES

1. M. H. Nichols and L. L. Rauch, "Radio Telemetry," John Wiley & Sons, Inc., New York, 1956.
2. V. D. Landon, Theoretical Analysis of Various Systems of Multiplex Transmission, *RCA Rev.*, vol. 9, 1948.
3. H. S. McGaughan, "Performance Characteristics of Time and Frequency Multiplexed Telemetering Systems," Navy Ordnance Laboratory, TM 13-15.
4. Aeronutronics Systems, Inc., "Telemetry System Study, Final Report," ASI Publication U-743, Newport Beach, Calif., Dec. 18, 1959.
5. B. M. Oliver, J. R. Pierce, and C. E. Shannon, The Philosophy of PCM, *Proc. IRE*, November, 1948.
6. Feldman and Bennett, Bandwidth and Transmission Performance, *Bell System Tech. J.*, July, 1949.
7. J. J. Spilker, Theoretical Bounds on the Performance of Sampled Data Communications Systems, *IRE Trans. Circuit Theory*, September, 1960.
8. R. M. Stewart, Statistical Design and Evaluation of Filters for the Restoration of Sampled Data, *Proc. IRE*, February, 1956.
9. H. W. Bode and E. E. Shannon, A Simplified Derivation of Linear Least Squares Smoothing and Prediction Theory, *Proc. IRE*, April, 1950.

10. Aeronutronics Dept., Philco Div., Ford Motor Co., Newport Beach, Calif., "Telemetry Systems Study," Aeronutronics Publication U-743, Dec. 18, 1959. Three volumes of which vol. I is referenced in this section and is ASTIA Doc. 234958, vol. II is 234959, and vol. III is 234960.

HIGH-EFFICIENCY TELEMETRY SYSTEMS

8 Theoretical Limits of Communication Efficiency

A fundamental consideration in selecting a modulation scheme for telemetry service is optimum immunity to noise for a given transmitted power. This is of particular importance in missile and space applications where weight and power aboard the space vehicle are limited. This implies that the least possible signal power be sent for a given permissible error rate P_e. This subject has been treated by Jacobs and Sanders.[2]

Shannon's expression for channel capacity, given in Chap. 2, Sec. 15, states the maximum rate of transmission of information with arbitrarily small error in the presence of white Gaussian noise:

$$C = B \log_2(1 + S/N) \tag{21}$$

But for Gaussian noise

$$N = N_0 B$$

where N_0 = noise spectral density, watts/Mc
H = information rate, bits/sec = C (ideal maximum)
S = EH where E is energy per bit = signal power

Then

$$\frac{S}{N} = \frac{EH}{N_0 B} = \frac{E/N_0}{B/H} = \frac{\beta}{\alpha} \tag{22}$$

where α = cps per bit/sec or, simply, cycles/bit

$\beta = E/N_0$ is a more fundamental measure of communication efficiency than signal-to-noise, since it represents the contrast of energy per transmitted information bit to the spectral noise *density* or noise intensity. It is only this contrast ratio that determines the detectability of a transmitted signal. Furthermore, for low signal-to-noise ratios (21) becomes

$$C = B (\log_2 e) S/N = S/N_0 (\log_2 e) \tag{23}$$

which shows that the ideal transmission rate *at low signal-to-noise* depends only on the noise *intensity* and not on the channel bandwidth (for white Gaussian noise.)

From Eqs. (21) and (22) can be derived

$$\frac{E}{N_0} = \frac{2^{H/B} - 1}{H/B} \quad \text{or} \quad \beta = (1/\alpha)(2^{1/\alpha} - 1) \tag{24}$$

The curve is shown in Fig. 25 as the theoretical limit. It represents the ultimate efficiency of communication. It is worth repeating that the theory states that the error rate can approach zero above this limit. However, as this limit is approached, the message must be encoded in a more complex manner. The complexity increases in an exponential manner as the error rate is reduced. All practical systems therefore do have finite error rates and higher β or poorer efficiencies than the ideal receivers. β is a means of characterizing the relative noise immunity of a communications systems as the term is used by Kotel'nikov.[3]

Figure 25 shows that β should ideally fall as the transmission-channel bandwidth per information bit is increased. It can be shown that the limit for infinite α or bandwidth is $\log_e$ or 0.693. This limit is very closely approached for $B/H = 10$.

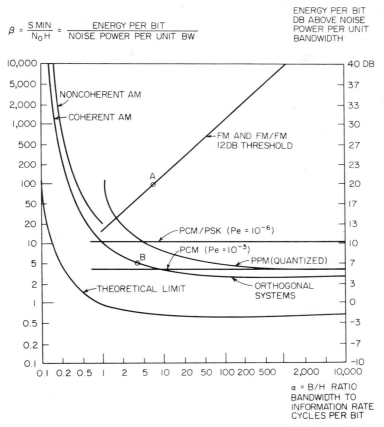

Fig. 25. Communication efficiency of various systems and the ideal in terms of β = energy per bit per unit noise intensity vs. α = bandwidth cycles per bit. The bit-error rate is $P_e = 10^{-6}$ where not otherwise stated.

9 Comparison of Discrete and Equivalent Discrete Modulation Systems

In addition to the performance of the ideal system, the efficiencies of practical modulation systems are also plotted on Fig. 25. There is an important assumption in these plots, namely, that the information input signal has been digitally encoded before being fed into the transmission channel. The performance shown for each system is for an output error rate per bit $P_e = 10^{-6}$, which is caused by band-limited white Gaussian noise. The channel bears no responsibility for any quantization error which results when a signal is to represent an analog or continuously variable signal. The channel discussed in Fig. 25 is *digital* or *discrete*, whereas a channel for transmitting analog data is called a *continuous* channel. Sanders[2] deals with this point in some detail. A *continuous* channel may be made to communicate digital data by quantizing the channel. AM and FM modulations systems are normally continuous but are considered to be quantized in this manner in Fig. 25. The data should be quantized such that errors would have equal effect regardless of the bit position in which they occur in a sample; hence a Gray code or equivalent should be used. This would be an equivalent discrete communications system.

It is more difficult to make comparisons between the transmission of *analog* data over *continuous* channels and *analog* communicated over *discrete* channels. This will be dealt with in Sec. 15. It is important to clearly distinguish between this case and the simpler comparison being made here. The problem of the transmission of analog data requires that a relationship between quantization error and error due to noise in the transmission system be established.

The derivation of the efficiencies of a number of discrete communications systems will be considered below.

10 PCM Systems

In PCM systems the information is encoded in binary form. Waveforms $a_1(t)$ and $a_2(t)$ may be used to represent the 0's and 1's, respectively. $a_1(t)$ and $a_2(t)$ may be made unequal or $a_2(t) = 0$. The most efficient system is to set $a_2(t) = a_1(t)$. Kotel'nikov,[3] page 39, has shown that in this last case, the probability of mistaking either a 1 for a 0 or a 0 for a 1 is

$$P_e = \Phi(-z) = 1/\sqrt{2\pi} \int_{-\infty}^{z} e^{-x^2/2} \, dx \qquad (25)$$

where $z = \sqrt{2E/N_0} = \sqrt{2\beta}$
P_e = error rate per bit
$\Phi(-z)$ = normal distribution function. (The Appendix, Sec. A.14-3, describes its properties)

Because values of $\Phi(-z)$ for large values of z are of interest, a special plot of this function is presented in Fig. 26. Equation (25) may also be expressed as (Chap. 1, Sec. 14)

$$P_e = \Phi(-z) = \tfrac{1}{2} \operatorname{erfc} (z/\sqrt{2}) = \tfrac{1}{2} \operatorname{erfc} \sqrt{\beta/2} \qquad (26)$$

For the case where $a_2(t) = 0$ mentioned above

$$z = \sqrt{E/2N_0}$$

or four times the signal energy is required for a given error rate, which is 3 db poorer performance.

Of particular note is the fact that P_e depends only upon the ratio of signal energy to per unit spectral noise power and does not depend on the wave shape of $a_1(t)$ and $a_2(t)$, provided only that the noise is white Gaussian, or at least flat over the bandwidth of the signal. This means that the error rate is independent of signal bandwidth.

In accordance with the sampling theorem, the bit rate H cannot be greater than $2B$, or $H/B = 2$ and $B/H = 0.5$. It follows that the PCM/PSK system is independent of B/H and its performance can be represented for a given error rate as a horizontal line starting from $B/H = 0.5$ for increasing B/H. Two such lines are shown in Fig. 25, one for $P_e = 10^{-3}$ and another for $P_e = 10^{-6}$, the values of β being derived from Fig. 26.

Thus no improvement is effected by increasing the bandwidth per information bit in a PCM system. It is possible to improve performance somewhat by using redundancy codes, such as devised by Hamming and Slepian (Chap. 2, Sec. 20), to correct for the bit errors. The excess bandwidth is used for transmitting the redundant bits. The improvement is slight, as shown in Fig. 27.

The detection process implied in Kotel'nikov's derivation [Eq. (25)] is a correlation detector. This is a form of maximum-likelihood estimator[4] which uses the criterion that a 0 was sent if

$$\int_0^T a_1(t)y(t) \, dt = \int_0^T a_2(t)y(t) \, dt \qquad (27)$$

otherwise a 1 was sent. $y(t)$ is the received signal plus noise and $a_1(t)$ and $a_2(t)$ are the signals for 0 and 1, respectively. Equation (27) implies the presence at the receiver of replicas of the wave shapes $a_1(t)$ and $a_2(t)$ to be sent. Also implied is that the correlation function of the noise should approach an impulse function. This condition is met if the signal center frequency is much larger than the signal bandwidth and the noise bandwidth at the detector is larger than the signal bandwidth.

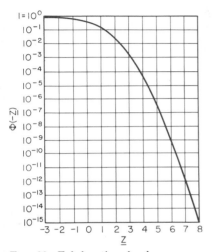

Fig. 26. Erf function for large arguments. (*After T. A. Kotel'nikov, "Theory of Optimum Noise Immunity," p. 12, McGraw-Hill Book Company, New York, 1960.*)

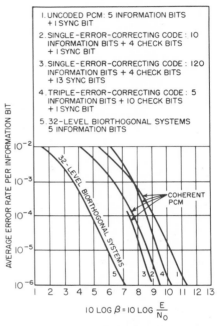

Fig. 27. Relative efficiencies of error-correcting systems.

Other methods of detecting PCM signals are not so efficient as the correlation detector. In Fig. 28 are shown the efficiencies of several PCM detection schemes. The

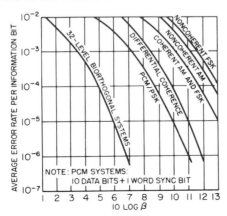

Fig. 28. Relative communication efficiencies of various PCM techniques.

curve marked PCM/PSK represents the performance of the correlation detector. Work by Shaft[5] indicates that PCM/FSK is only 1 db rather than 3 db poorer than PCM/PSK, as shown in Fig. 28.

11 Multisymbol Discrete Systems

More generalized discrete systems exist which improve in performance as more bandwidth is used per information bit. Such systems encode information in discrete codes which are equivalent to n bits per symbol (or word). Reception is based upon detection of a complete symbol among a set of 2^n symbols. Only the case of equiprobable symbols has been treated to any extent, because of severe analytic difficulties.

Kotel'nikov[3] has analyzed the optimum detection of this kind of signal. Such a system involves correlation of the received signal with replicas of the set of possible symbols. Other methods are possible which may be easier to instrument, for example, a PPM system in which the appearance of a pulse at a specific time position in a time interval can be used as a multisymbol discrete system. Here the detection consists of observance of the signal above a threshold at specific time intervals. Jacobs[1] has discussed such a system, which is presented in Sec. 12. The analysis, which is based on the work of Golay, shows performance about 1.5 db poorer than optimum detection at $B/H = 100$.

Two other implementations of such systems have been developed and used in telemetry. These are the coded phase-coherent system and the PFM system. These will be discussed in Secs. 13 and 14. In the coded phase-coherent system the information is encoded in one of 2^n discrete symbols consisting of n binary bits each. Each bit is either an advanced (+) or a retarded (−) phase of a carrier. In the PFM system, the information is encoded as a frequency of duration equivalent to n bits long. The frequencies for the different symbols are separated by a minimum frequency of $1/T$ or $1/nt$ and are detected by a filter bank.

All the systems discussed here are *orthogonal* systems (Chap. 2); that is, the crosscorrelation coefficient between the message symbols (words) is zero. In some cases biorthogonal systems may be used, in which case the crosscorrelation coefficient between the words is either 0 or −1. The three systems to be described below may be considered the analogs of time modulation, coded modulation, and frequency modulation in more conventional communications systems.

12 Quantized Pulse-position Modulation (QPPM)

It was noted by Golay[6] in 1949 that pulse-position modulation (PPM) can approach the minimum energy per bit given by Eq. (24) as closely as desired and with arbitrarily small error rate at sufficiently low bits per cycle.

Consider the sequence of band-limited functions

$$a_k(t) = \sqrt{2E_sB}\,\frac{\sin 2\pi B(t - k/2B)}{2\pi B(t - k/2B)} \qquad k = 0, 1, 2, \ldots, n - 1 \qquad (28)$$

This sequence of displaced time functions is orthogonal; that is,

$$\int_{-\infty}^{\infty} dt\, a_j(t)a_k(t) = \begin{cases} E_s \text{ for } j = k \\ 0 \text{ for } j \neq k \end{cases} \qquad (29)$$

where E_s is the energy contained in each signal.

For each time interval T

$$T = n/2B \qquad (30)$$

one of these n functions is transmitted, which corresponds to quantized (n-valued) PPM. There are $\log_2 n$ bits per symbol; so that the information rate is

$$H = \log_2 n/T = 2B(\log_2 n)/n \qquad (31)$$

and the energy per bit is
$$E = E_s/\log_2 n \tag{32}$$

Optimum detection[3] involves the correlation of the received signal with each of the n orthogonal waveforms, with decision determined by the largest output. The determination of the error probability for this case requires the evaluation of difficult integrals. This can be done by approximate methods or digital computers. Sanders[2] and Viterbi[7] have made such evaluations, the latter using a 704 computer. However, threshold detection of PPM is only slightly poorer than optimum (Fig. 25) and has been analyzed by Golay. This analysis follows, as presented by Jacobs.[1] To determine which of the n time functions has been transmitted, the received signal is sampled at time $k/2B$, $k = 0, 1, 2, \ldots, n - 1$, and each of these samples is compared with a threshold level. An error occurs if the noise of any of the $(n - 1)$ "incorrect" sampling points exceeds the threshold, or if a negative noise spike at the "correct" point causes the signal to go below threshold. The error probability is then as given by Golay.[6]

$$P_e \cong (n - 1)\, \text{erfc}\,(\alpha u) + \text{erfc}\,[(1 - \alpha)u] \tag{33}$$

where α = ratio of the threshold to the peak signal amplitude
u = ratio of the peak signal amplitude to the rms noise

$$u = \sqrt{2E_s/N_0} \tag{34}$$

(As before, white Gaussian noise with a spectral density N_0 is assumed.) It is also assumed in Eq. (33) that the error probability is small enough so that the probability of simultaneous occurrence of multiple sources of error is negligible.

If the threshold is chosen to minimize P_e, it follows from Eq. (33) that

$$\alpha = \tfrac{1}{2} + [\log_e(n - 1)]/u^2 \tag{35}$$

Let
$$u = 1 - [2 \log_e(n - 1)]/u^2 \tag{36}$$

It follows from Eqs. (36), (33), and (32) that

$$E/N_0 = [1/(1 - \varepsilon)] \log_e 2 \tag{37}$$

so that as ε approaches zero, E/N_0 approaches the theoretical minimum. To show that this is consistent with an arbitrarily small error rate, it follows from Eqs. (32) and (36) that

$$P_e \cong [2/(2 - \varepsilon)]\, \text{erfc}\,(\varepsilon u/2) \tag{38}$$

so that for arbitrarily small ε, u may be chosen sufficiently large to make P_e as small as desired.

This is essentially the result obtained by Golay. We wish to consider the quantitative implications. For a given H/B and E/N_0, P_e is calculated as follows:
From Eq. (31), n is determined by H/B. That is,

$$H/B = 2 \log_2 n/n$$

From Eqs. (32) and (33), u is determined by E/N_0 and n. Thus

$$u = \sqrt{2E \log n/N_0}$$

From Eq. (36), ε is determined by n and u.
From Eq. (38), P_e is determined by ε and u.
The result of these calculations is shown in Fig. 29, where error probability is given as a function of E/N_0 with H/B as a parameter. It is seen that error probability diminishes exponentially with increasing E/N_0, and that the rate of decrease becomes larger as H/B diminishes.

In Fig. 25 E/N_0 is given as a function of B/H for fixed error probability. The performance of PPM, PCM-PSK, and the theoretical limit are shown in this figure. It is seen that as B/H decreases, the energy per bit for PPM does indeed decrease and approach the theoretical limit. For example, at $B/H = 10^4$, to achieve an error probability of 10^{-6} requires an energy per bit 8 db above the theoretical minimum, as compared with 12.5 db for conventional PSK.

In order to reduce the energy per bit even further, *coherent* PPM may be used in which the transmitted waveforms are $\pm a_k(t)$, where the a_k are given by Eq. (28). The sequence $\pm a_k$ is said to be *biorthogonal*. If there are $n/2$ pulse positions, the total

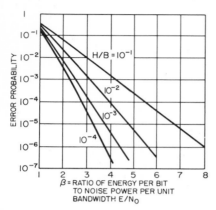

FIG. 29. Error probability for multiposition PPM.

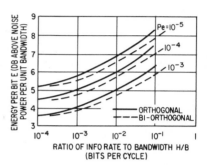

FIG. 30. Comparison of orthogonal and biorthogonal pulse-position modulation.

number of bits per symbol is the same as before. However, the time per symbol is halved, so that the information rate is doubled:

$$T = n/4B \qquad (39)$$
$$H = [4B(\log_2 n)]/n \qquad (40)$$

Equations (32) through (38) are unchanged, so that, for fixed energy per bit and error probability, coherent PPM gives twice the information rate of the noncoherent system. The two are compared in Fig. 30. For fixed information rate, the coherent system requires about 0.4 db less energy per bit at $H/B \cong 10^{-2}$. The improvement diminishes as H/B decreases.

Thus, in one sense, coherent PPM offers an appreciable advantage in that it permits the information rate to be doubled without increasing bandwidth or energy per bit. On the other hand, the same increase in information rate may be obtained by increasing the transmitter power by less than 0.5 db in the noncoherent system. For a given power, the coherent system gives only a 5 per cent increase in range.

Time-Bandwidth Product TB

More complex coding and modulation schemes may be devised to approach the theoretical limit more closely. Ultimately the problem becomes not what one can accomplish in principle, but rather what one can accomplish in practice. In considering practical limitations, a parameter of particular importance is the *time-bandwidth* product TB where T is the integration time (memory) and B the bandwidth of the signal processor. For an n-symbol orthogonal system the $TB = n/2$. This follows directly from Eq. (30). Using this in Eq. (31) yields

$$H/B = (\log_2 TB)/TB \qquad (41)$$

In Fig. 31, the time-bandwidth product is shown as a function of H/B for orthogonal PPM. For $H/B = 10^{-2}$ the required time-bandwidth product is approximately 1,000.

Signal-to-noise ratio, energy-per-bit-to-noise-density ratio, and time-bandwidth product can be related by Eqs. (22) and (41). Thus

$$S/N = (E/N_0)(H/B) = (E/N_0)[(\log_2 2TB)/TB] \qquad (42)$$

For a given E/N_0, the required signal-to-noise ratio may be made arbitrarily small by making the time-bandwidth product sufficiently large.

A strong impetus for the development of large time-bandwidth processors has been provided by the requirements of high-resolution radar. Good range resolution requires large bandwidths. Similarly, long pulses are desired in order to utilize available power efficiently and/or obtain Doppler resolution.

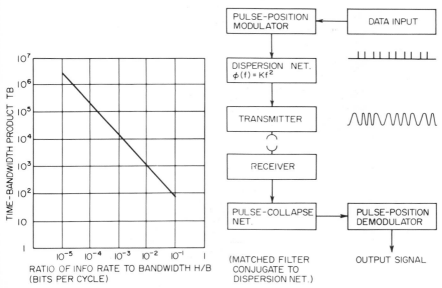

FIG. 31. Time-bandwidth product required for multiposition PPM.

FIG. 32. Pulse-position-modulation systems with linear FM pulse dispersion.

Radar signal processing, in which there are many resolvable cells in position and velocity, is directly applicable to communication systems in which the alphabet size (number of distinguishable signal functions) is large.

An example of large time-bandwidth signal processing developed for radar applications is a technique known as "chirp." In chirp, the transmitted signal is generated by passing a short pulse through a dispersive network which spreads the signal in time and imparts a frequency modulation to it. Time-bandwidth products in the hundreds may be obtained. The signal is then modulated up to the carrier frequency and amplified for transmission. At the receiver, the inverse process is applied. The spread-out signal is coupled into a collapsing network which concentrates the signal energy into a narrow pulse. In this way, multiposition PPM may be used without excessive peak-power requirements. The combination of QPPM and pulse-compression techniques is illustrated by the block diagram in Fig. 32.

Virtually any storage device can be used as the basis of a signal processor. In addition to lumped-constant all-pass filter networks, the following devices might be made the basis of a processing system: quartz delay lines, wire delay lines, magnetic drums, magnetic tapes, thermoplastic tapes, and waveguides. The processing may also be done entirely in a high-speed digital computer.

Figure 33 presents a summary of the time delays and operating bandwidths for a number of devices, most of which are now in use in signal processors.* The examples shown are representative, and the illustration should not be taken as definite or complete. Techniques at the upper left of the figure are applicable to detection of extremely weak signals when long integration times are employed. At the lower right are the extremely wideband techniques which permit high data rates at extremely low signal-

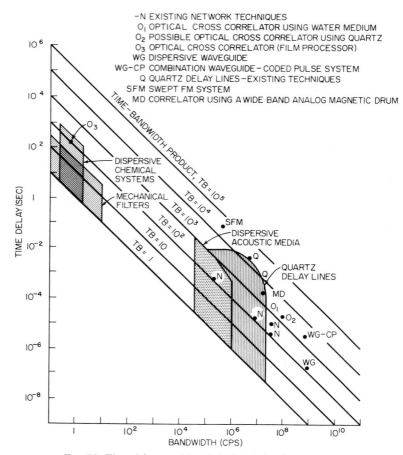

FIG. 33. Time delays and bandwidths of signal processors.

to-noise ratios. We have seen, however, that reduction in bits per cycle below 10^{-2} gives only a small reduction in the required energy per bit. Consequently, the use of processors with time-bandwidth products in excess of 1,000 would seem to offer little advantage except in a strong interference environment. The investigation of more general modulation and coding techniques might also modify this conclusion.

There are other considerations affecting the use of large TB such as receiver gain stability, phase control, and time synchronization.

Large time-bandwidth products are required in systems with low input-signal-to-noise ratios and high processing gains. Analog systems, such as wide-deviation FM,

* Compiled by C. W. Hoover, Jr., of Bell Telephone Laboratories, Incorporated.

exhibit input *thresholds* below which this processing gain is not realized. Techniques to lower the threshold have been developed, notably wide-deviation FM with feedback, but there still exists a signal-to-noise ratio below which the signal rather than the noise is suppressed.

Similar catastrophic effects do not explicitly appear in analyses of digital systems. However, in these analyses, it is usually assumed that all elements of the system are completely known, with the exception of additive noise, the statistics of which are completely known. As processing gains are increased, system performance depends critically on the precise knowledge of system parameters. For example, if long time integration is used to detect a weak signal in overriding noise, it is necessary that the signal be greater than the *uncertainty in the noise*.

Optimum detection of digital signals requires *precise time synchronization* at the receiver.* For example, in the analysis of PPM, use was made of the fact that the signal functions are identically zero at all but one sampling point. Lack of synchronization invalidates this assumption and gives rise to "crosstalk" between signal functions. It then becomes pertinent to ask: How is synchronization obtained, what is the effect of noise on maintaining synchronization, and what is the effect of lack of synchronization on error performance?

The above considerations imply that, at very low signal-to-noise ratios, noise may no longer be considered *as a simple additive effect*. More detailed analysis is required to evaluate error performance properly in this region.

We may conclude that there is no upper limit on time-bandwidth products and consequent processing gains which may be achieved in principle. However:

1. It is generally difficult to obtain devices with time-bandwidth products much greater than 1,000.
2. Larger time-bandwidth products impose severe stability requirements on system performance and tend to invalidate a simple model in which noise is considered only as an additive effect.

13 Coded Phase-coherent Systems

Another instrumentation of a multisymbol high-efficiency communication system has been analyzed by Viterbi.[7] The following discussion is based on his work. In

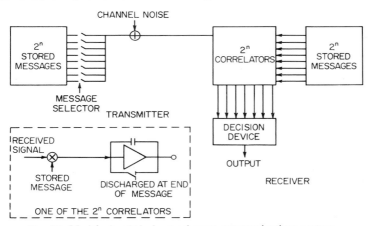

FIG. 34. Model of coded phase-coherent communications system.

this case information is encoded into words n bits long which require nT sec to transmit. As shown in Fig. 34, the transmitter is equipped with 2^n messages the encoding of

* If the information is conveyed by differences between consecutive pulses, absolute synchronization is not required, but the required signal power is somewhat larger.

which has been previously standardized. The messages which are sent are selected from the store and are transmitted as phase shift of the carrier frequency of π radians for the 1 bit and 0 shift for the 0 bits. This is more clearly shown in Fig. 35. Replicas of the messages stored at the transmitter are stored at the receiver. Reception of messages consists of correlating the incoming message with the 2^n stored replicas, for a

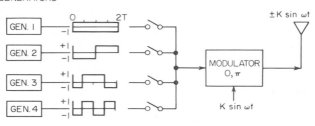

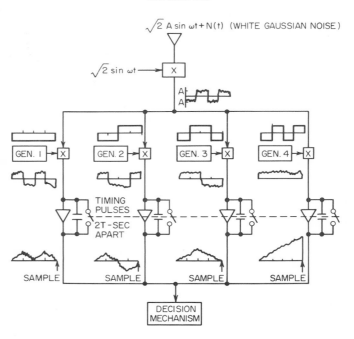

Fig. 35. Coded phase-coherent system for transmission of 2 bits per word.

period of time nT, then sampling the output of the correlators at the end of the interval nT, establishing which correlator has maximum output, and selecting the waveform corresponding to this correlator as the correct message.

Fano[8] has shown that this is the ideal receiver for detecting multisymbol signals in white Gaussian noise. The process just described is maximum-likelihood detection and was shown by Middleton and Van Meter[9] to minimize the probability of error where all the signals are equally likely and contain equal energy. The similarity to the process analyzed by Kotel'nikov[3] should be noted.

HIGH-EFFICIENCY TELEMETRY SYSTEMS

Thus the output of the kth correlator, which corresponds to the kth word x_k, is

$$\int_0^{nT} x_k(t) y(t) \, dt \tag{43}$$

where $y(t) = x_m(t) + N(t)$, $x_m(t)$ is the received signal, and $N(t)$ is the channel noise. If the 2^n words were a priori all equally likely to be transmitted with equal energy, i.e.,

$$P(x_i) = P(x_j) \quad \text{and} \quad \int_0^{nT} x_i^2(t) \, dt = \int_0^{nT} x_j^2(t) \, dt \tag{44}$$

for all i and j, then the conditional probability that x_k was sent, given that y was received, is proportional to the exponential of the output of the kth correlator.

$$P(x_k|y) \sim \exp \int_0^{nT} x_k(t) y(t) \, dt \tag{45}$$

It follows intuitively that in order to achieve low error probabilities, the waveforms should be as unlike as possible, such that in a noisy channel there will be the least possible chance to make the wrong selection of the word transmitted. More precisely, the crosscorrelation coefficients among all pairs of words,

$$\rho = \frac{\int_0^{nT} x_i(t) x_j(t) \, dt}{\left[\int_0^{nT} x_i^2(t) \, dt \int_0^{nT} x_j^2(t) \, dt \right]^{1/2}} \tag{46}$$

should be made as low as possible. The least possible value of ρ is -1. However, this value can be achieved only when the number of words in the set is two ($n = 1$). In this case, if $x_1(t) = -x_2(t)$, $\rho = -1$, and the words are said to be antipodal. In general, it is possible to make all the crosscorrelation coefficients equal to zero. The set of words is then said to be *orthogonal*. Actually, it is possible to obtain sets of words for which some or all of the crosscorrelations are negative.

Figure 35 represents an example of a binary-coded phase-coherent system. The term "phase-coherent" refers not only to the coherence between the transmitted carrier and the locally generated carrier but also to that between the transmitted and locally generated code words.

Blocks of two bits of information are transmitted by selecting one of a set of four binary code words. This set is orthogonal, since the words switch between $+1$ and -1, and it is easily verified that

$$\int_0^{2T} x_i(t) x_j(t) \, dt = 0 \tag{47}$$

for $i \neq j$.

Phase modulation of $K \sin \omega t$ by π radians when the word is at the -1 level is equivalent to amplitude modulation of the carrier by $+1$'s and -1's. At the receiver, the noisy signal is demodulated and fed to the four correlators. Only the low-frequency component of these inputs is shown in Fig. 35. Actually, the component centered at a frequency of 2ω radians/sec is eliminated by the integrator provided ω is a multiple of $\pi/2nT$, where n is the number of bits per word.

Because the code words are orthogonal, the outputs of all correlators, except the one corresponding to the word sent, are zero in the absence of noise. If this were not the case, the noise-free output of the ith correlator would be proportional to

$$\int_0^{2T} x_i(t) x_j(t) \, dt$$

when the jth word was sent.

It should be noted that multiplication of the additive noise by the locally generated words does not alter its white Gaussian statistics, since multiplying successive uncorrelated samples of noise arbitrarily by $+1$ and -1 does not alter the first-order distribution, nor does it render them correlated.

In general, if n bits are transmitted as one word, the integrating time is nT. The integrate-and-discharge filter is assumed to produce an attenuation of $1/nT$. Thus the signal will produce an output at time nT of $e(nT) = A$, the rms carrier input volts, provided that ωnT is a multiple of $\pi/2$. The channel noise is white Gaussian with spectral density $N/2B$. (This input spectral density would produce a power of N watts at the output of a bandpass filter of bandwidth B.) The variance at time nT at the output of the integrate-and-discharge filter is

$$\sigma^2 = \left\{ E[1/(nT)^2] \int_0^{nT} \sqrt{2}\, N(t) \sin \omega t\, dt \int_0^{nT} \sqrt{2}\, N(u) \sin \omega u\, du \right\}$$
$$= 1/(nT)^2 \int_0^{nT} \int_0^{nT} E[N(t)N(u)] 2 \sin \omega t \sin \omega u\, dt\, du \tag{48}$$

Since the noise is white with density $N/2B$,

$$E[N(t)N(u)] = (N/2B)(t - u) \tag{49}$$

Therefore,

$$\sigma^2 = N/2B(nT)^2 \int_0^{nT} 2 \sin^2 \omega t\, dt = N/2BnT' \tag{50}$$

provided that ωnT is a multiple of $\pi/2$. The ratio of peak output signal to the noise standard deviation is

$$\frac{e(nT)}{\sigma} = \frac{A}{(N/2BnT)^{1/2}} = \left(\frac{2A^2 nT}{N/B} \right)^{1/2} = \left(\frac{2SnT}{N/B} \right)^{1/2} \tag{51}$$

where $S = A^2$ is the received signal power. Since T is the transmission time per bit, the ratio

$$\frac{ST}{N/B} = \frac{\text{(received signal energy)}/\text{bit}}{\text{(noise power)}/\text{(unit bandwidth)}} = \beta$$

or
$$e(nT)/\sigma = \sqrt{2n\beta} \tag{52}$$

β is the same basic parameter for communications system which is discussed in Sec. 8, Eq. (22). Equation (52) shows that only increasing the energy per bit or n can improve the output signal-to-noise ratio for a given channel.

In orthogonally coded systems the noise components of the correlator outputs are mutually independent. This follows from the fact that for orthogonal codes there must be exactly as many subintervals during which different code words have opposite signs as subintervals during which they have equal signs. Hence ρ_N, the crosscorrelation between the noise components of any two channels which had been multiplied by such codes, will be zero after a period of nT.

Coding

The coded phase-coherent system makes use of binary-coded words for the multisymbol transmission. These codes should be judged for their ability to minimize error in decoding the entire word. They should therefore have a uniformly low crosscorrelation coefficient. Certain error-correcting codes[10,11] have such useful properties. These codes were devised to add redundancy to binary words so as to provide for correction of error after transmission. See Chap. 2, Sec. 20. Incorrect individual bits in a word are detected by logical operations on the redundant code pattern and individually corrected. On the other hand, the redundancy is used in the coded phase-coherent system to improve the detection of the entire word.

The construction and properties of such codes will be discussed here. They include the orthogonal, biorthogonal (Reed-Muller), and shift-register codes.

Orthogonal Codes. A set of orthogonal codes has the property that the crosscorrelation coefficients among all pairs in the set are zero. That is, for the code words

$$\{x_1, x_2, \ldots, x_k\} \quad \text{and} \quad \{y_1, y_2, \ldots, y_k\}$$

(where the x_i's and y_i's can take on the values $+1$ or -1), the sum of the products of corresponding symbols

$$\sum_{i=1}^{k} x_i y_i = 0$$

It is sometimes more convenient to write the codes using the symbols 0 and 1 rather than ± 1. The orthogonal property can then be stated as follows: Two code words are orthogonal if the number of symbol positions in which they are similar equals the number in which they are dissimilar.

Sets of orthogonal codes can be constructed in a multitude of ways, since the 2^n elements of any basis of a 2^n-dimensional vector space over the field of two elements can be made orthogonal to one another.[12] A simple inductive construction of a set of orthogonal codes follows.

A single bit of information may be sent by selecting from a set of two orthogonal code words of two symbols each:

$$\begin{matrix} 00 \\ 01 \end{matrix}$$

Two bits might be sent by using the code word set

$$\begin{matrix} 0000 \\ 0101 \\ 0011 \\ 0110 \end{matrix}$$

It should be noted that this set can be constructed by extending the set for 1 bit both horizontally and vertically. The lower right-hand square is filled by the complements of these words. A code set for 3 bits may be constructed by extending the set for two:

$$\begin{matrix} 0000 & 0000 \\ 0101 & 0101 \\ 0011 & 0011 \\ 0110 & 0110 \\ 0000 & 1111 \\ 0101 & 1010 \\ 0011 & 1100 \\ 0110 & 1001 \end{matrix}$$

An appreciation of the use of these codes can be derived by calculating the correlation coefficients for a number of cases. First consider the transmission of 1-bit messages,

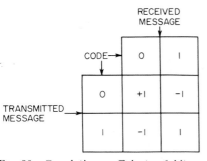
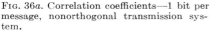

Fig. 36a. Correlation coefficients—1 bit per message, nonorthogonal transmission system.

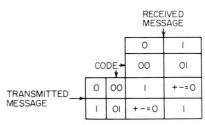

Fig. 36b. Same for orthogonal transmission system, 1 bit per message.

not coded orthogonally but sent in the base-band code as either 0 or 1. The result is shown in Fig. 36a. The coefficients are either positive for true message or negative for false message. In contrast to this, Fig. 36b shows the same 1-bit message orthogonally encoded, that is, a 0 is 00 and 1 is 01. Here the false messages yield correlation coefficients of 0 while the true messages still result in a unity correlation coefficient.

In Fig. 37 are shown correlation coefficients for 1-, 2-, and 3-bit messages coded orthogonally in the codes just explained. The correlation coefficient continues to be zero for all false messages.

Biorthogonal Codes. These codes were first discovered by Muller and Reed. They can be generated by taking a set of orthogonal code words and adding to it the complements of each word. Thus biorthogonal codes are really two sets of orthogonal

FIG. 37. Correlation coefficients—orthogonally coded 1-, 2-, and 3-bit messages.

codes which are mutually orthogonal, except that each code word in one set has its complement (or antipode) in the other set. A biorthogonal or Reed-Muller code for 4 bits can be constructed from the preceding orthogonal code for 3 bits:

```
00000000    11111111
01010101    10101010
00110011    11001100
01100110    10011001
00001111    11110000
01011010    10100101
00111100    11000011
01101001    10010110
```

Figure 38 shows the corresponding correlation coefficients for this biorthogonal code. It is seen that some unwanted messages have negative correlation coefficients, but most have zero correlation coefficients. This causes no difficulty in decoding.

One advantage of this set over the corresponding orthogonal set is that it requires one-half as many symbols per code word. Thus the bandwidth required it transmit the same number of bits per second is cut in half. Also, the average crosscorrelation coefficient among all the codes in a set of 2^n words is $-1/(2^n - 1)$, as will now be shown. There are in all $(2^n - 1)2^{n-1}$ pairs. The crosscorrelations are -1 for 2^{n-1} pairs, and zero for all the rest. Thus the average correlation is

$$\frac{(-1)2^{n-1}}{(2^n - 1)2^{n-1}} = -\frac{1}{2^n - 1}$$

Sets of biorthogonal codes have equal numbers of 0's and 1's. This is a favorable property since, if all words are equally likely, this assures that the modulating signal will have zero mean; hence all the power in the carrier will be modulated.

Shift-register Codes. It is known that shift registers with linear modulo-2 feedback logic produce codes which have two-level autocorrelation functions. If the register has length n and the code is maximal length, $2^n - 1$, the lower level will be $-1/(2^n - 1)$ (see Fig. 39). Thus a set of $2^n - 1$ codes with a uniform negative crosscorrelation coefficient can be constructed by taking all shifted replicas of one maximal-length shift-register sequence. For example, a set of seven code words can be generated by taking all possible shifts of the sequence from a three-stage shift register

MESSAGE NO. T↓	R→ CODE	0 0000 0000	1 0110 0101	2 0011 0011	3 0110 0110	4 0000 1111	5 0101 1010	6 0011 1100	7 0110 1001	8 1001 0110	9 1100 0011	10 1010 0101	11 1111 0000	12 1001 1001	13 1100 1100	14 1010 1010	15 1111 1111
0	00000000	+1	0	0	0	0	0	0	0	0	0	0	0	0	0	0	−1
1	01010101	0	+0	0	0	0	0	0	0	0	0	0	0	0	0	−1	0
2	00110011	0	0	+1	0	0	0	0	0	0	0	0	0	0	−1	0	0
3	01100110	0	0	0	+1	0	0	0	0	0	0	0	0	−1	0	0	0
4	00001111	0	0	0	0	+1	0	0	0	0	0	0	−1	0	0	0	0
5	01011010	0	0	0	0	0	+1	0	0	0	0	−1	0	0	0	0	0
6	00111100	0	0	0	0	0	0	+1	0	0	−1	0	0	0	0	0	0
7	01101001	0	0	0	0	0	0	0	+1	−1	0	0	0	0	0	0	0
8	10010110	0	0	0	0	0	0	0	−1	+1	0	0	0	0	0	0	0
9	11000011	0	0	0	0	0	0	−1	0	0	+1	0	0	0	0	0	0
10	10100101	0	0	0	0	0	−1	0	0	0	0	+1	0	0	0	0	0
11	11110000	0	0	0	0	−1	0	0	0	0	0	0	+1	0	0	0	0
12	10011001	0	0	0	−1	0	0	0	0	0	0	0	0	+1	0	0	0
13	11001100	0	0	−1	0	0	0	0	0	0	0	0	0	0	+1	0	0
14	10101010	0	−1	0	0	0	0	0	0	0	0	0	0	0	0	+1	0
15	11111111	−1	0	0	0	0	0	0	0	0	0	0	0	0	0	0	+1

Fig. 38. Correlation coefficients—biorthogonal system $n = 16$ message symbols, transmitting 4 bits of information per symbol (word).

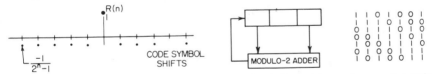

Fig. 39. Autocorrelation function of shift-register code.

Fig. 40. Shift-register and generated code.

with linear logic, as shown in Fig. 40. The eighth code word in this figure is the 0 vector (0000000). The crosscorrelation coefficient among all possible pairs is $-1/(2^n - 1)$.

Shift registers can be used to generate orthogonal or biorthogonal codes quite simply. For example, if a zero is added to every word of the set of Fig. 40 and to the 0 vector, a set of eight orthogonal code words is obtained. By taking the complemented output of the shift register, the complementary orthogonal set is also obtained:

```
01101001    10010110
01110100    10001011
00111010    11000101
00011101    11100010
01001110    10110001
00100111    11011000
01010011    10101100
00000000    00000000
```

Note that this is not the same biorthogonal set as that generated above. This example can be generalized to any number of bits.

Optimal Decision and Probability of Error

Orthogonal Codes. The typical receiver for coded phase-coherent communication was shown in Fig. 35. The outputs of the correlators are fed into a device which determines the waveform most probably sent. If the a priori probabilities of the various code words are all equal, the disturbance is white Gaussian noise, and the energy in all transmitted words is the same, Eq. (46) indicates that the word which was most probably transmitted is that which corresponds to the maximum correlator output.

The probability that the word which was sent will be chosen correctly is equal to the probability that the output of all the other correlators will be smaller than the output of the given correlator. Assume that in the absence of noise, the output of the correlator corresponding to the word sent is A and that the standard deviation of the output noise of any correlator is σ. For a set of 2^n code words, the probability that the correct one will be chosen is

$$P_c(n) = \int_0^\infty p(x_i) \, dx_i \, P(y_1, y_2, \ldots, y_j \ldots y_{2^n-1} < x_i)$$

$$= \int_0^\infty p(x_i) \, dx_i \prod_{j=1}^{2^n-1} P(y_j < x_i) \tag{53}$$

where $p(x_i)$ is the probability density of the output of the correct correlator, and

$$P(y_j < x_i) = \int_{-\infty}^{x_i} p(y_j) \, dy_j$$

is the probability that the output of the jth incorrect correlator will be less than the correct correlator output. The second equality of (53) holds because the correlator noise outputs are mutually independent as shown above.

Then, for the given parameter,

$$P_c(n) = \int_{-\infty}^\infty \frac{e^{-(x-A)^2/2\sigma^2}}{\sqrt{2\pi}\sigma} \, dx \left(\int_{-\infty}^x \frac{e^{-y^2/2\sigma^2}}{\sqrt{2}\sigma} \, dy \right)^{2^n-1} \tag{54}$$

Making the substitutions $z = y/\sigma$ and $u = (x - A)/\sigma$ yields

$$P_c(n) = \int_{-\infty}^\infty \frac{e^{-u^2/2}}{\sqrt{2\pi}} \, du \left[\int_{-\infty}^{u+(A/\sigma)} \frac{e^{-z^2/2}}{\sqrt{2\pi}} \, dz \right]^{2^n-1} \tag{55}$$

The probability that a word is in error is

$$P_w(n) = 1 - P_c(n)$$

It was shown in (52) that the ratio of the output from the correct correlator to the standard deviation of the noise is

$$\frac{A}{\sigma} = \left(\frac{2nST}{N/B} \right)^{1/2}$$

Then

$$P_w(n) = 1 - \int_{-\infty}^\infty \frac{e^{-u^2/2}}{\sqrt{2\pi}} \, du \left\{ \int_{-\infty}^{u+[2nST/(N/B)]^{1/2}} \frac{e^{-z^2/2}}{\sqrt{2\pi}} \, dz \right\}^{2^n-1} \tag{56}$$

This integral cannot generally be evaluated analytically. However, numerical integration by an IBM 704 computer yielded the results of Fig. 41 for code words containing up to 20 bits of information.

Equation (56) may be evaluated analytically when n tends to infinity. The result is also shown in Fig. 41.

Biorthogonal Codes. To demodulate a set of 2 biorthogonal code words carrying n bits, only 2^{n-1} correlators are required. This is due to the fact that in the absence of noise any one correlator will produce a positive voltage $+A$ at time nT for one code word, a negative voltage $-A$ for its complement, and zero voltage for all the rest. Thus only one orthogonal code set need be generated at the receiver. The first step in the decision process is to establish whether the voltage at time nT at the output of a given correlator is positive or negative; therefore, the situation is the same as for orthogonal codes, and the optimal decision in the presence of white Gaussian noise is to choose the one corresponding to the greatest output.

The correct word will be selected if the absolute values of the outputs of all the other correlators are less than that of the given one, and furthermore, if the output of the

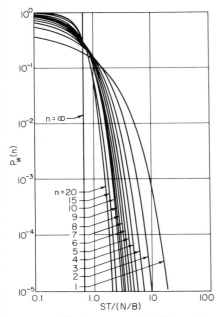

FIG. 41. Word-error probability—orthogonal codes.

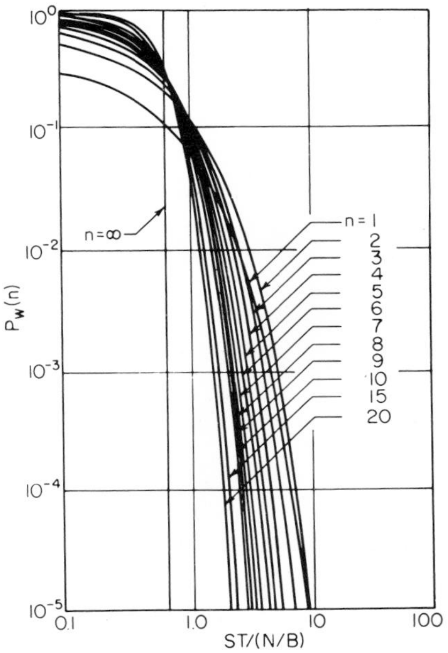

FIG. 42. Word-error probability—biorthogonal codes.

correct correlator is of the right sign. Without loss of generality, assume that a word has been sent which produces a voltage $+A$ at time nT on correlator x_i. The probability that it will be selected by the decision process is

$$P_c(n) = \int_0^\infty p(x_i)\, dx_i \left[\prod_{j=1}^{2^n-1} p(|y_j| < |x_i|) \right] \tag{57}$$

where

$$P(|y_j| < |x_i|) = \int_{-x}^{x} p(y_j)\, dy_j$$

This expression is valid because the noise outputs of the correlators are independent since the noise components are again multiplied by orthogonal words. Then, in terms of the Gaussian probability densities,

$$P_c(n) = \int_0^\infty \frac{e^{-(x-A)^2/2\sigma^2}}{\sqrt{2\pi}\sigma}\, dx \left[\int_{-x}^{x} \frac{e^{-y^2/2\sigma^2}}{\sqrt{2\pi}\sigma}\, dy \right]^{2^{n-1}-1}$$

Making the substitutions,
$$v = (x - A)/\sigma \quad \text{and} \quad z = y/\sigma$$
and recalling from Eq. (52) that
$$\frac{A}{\sigma} = \left(\frac{2nST}{N/B}\right)^{1/2}$$
the word-error probability for biorthogonal coding is

$$P_w(n) = 1 - P_c(n) = 1 - \int_{-[2nST/(N/B)]^{1/2}}^{\infty} \frac{e^{-v^2/2}}{\sqrt{2\pi}} dv$$
$$\times \int_{-\{v+[2nST/(N/B)]^{1/2}\}}^{v+[2nST/(N/B)]^{1/2}} \frac{e^{-z^2/2}}{\sqrt{2\pi}} dz^{2^{n-1}-1} \quad (58)$$

This expression was also evaluated, using an IBM 704, for various values of n and the results plotted in Fig. 42. Its limit as n approaches infinity can be computed in the same way as the orthogonal codes, and the result is the same.

Comparison of Coded and Uncoded Word-error Probabilities

If a single bit were to be sent using a biorthogonal code, the code set would degenerate to two words of one symbol each. This is the special case of communication with two antipodal signals (such as $+1$ and -1). In this situation, to which we shall refer as uncoded, the probability that each bit is in error is obtained by letting $n = 1$ in (58)

$$P_B = 1 - \int_{-[2ST/(N/B)]^{1/2}}^{\infty} \frac{e^{-v^2/2}}{\sqrt{2\pi}} dv$$
$$= \int_{-\infty}^{-[2ST/(N/B)]^{1/2}} \frac{e^{-v^2/2}}{\sqrt{2\pi}} dv \quad (59)$$

If it is desired to transmit an n-bit word by sending one bit at a time by means of antipodal signals, the probability that the word will be received in error is one minus the product of the probabilities that each bit will be detected correctly. Thus

$$P_w(n) = 1 - (1 - P_B)^n \quad (60)$$

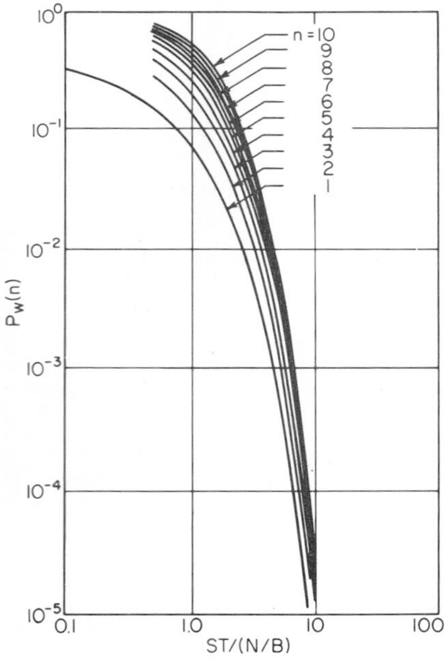

This expression is plotted in Fig. 43. For the sake of comparison, Fig. 44a and b shows the word-error probabilities for coded and uncoded transmission, and as might be expected, the two coding schemes produce almost identical results

FIG. 43. Word-error probability—uncoded.

for large 2^n. Also, the improvement due to coding for $n = 10$ is almost twice as great as for $n = 5$.

14 PFM Telemetry

This method of multisymbol (or multiword) signal transmission relies entirely on frequency encoding and discrimination of the words. Such a system was used on the

HIGH-EFFICIENCY TELEMETRY SYSTEMS 9-49

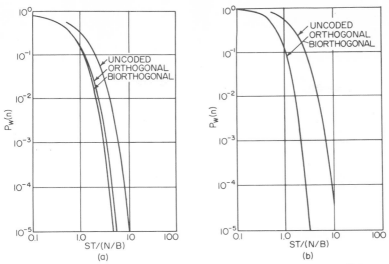

Fig. 44a. Comparison of coded and uncoded word-error probabilities, $n = 5$. b. Same, $n = 10$.

	SIGNAL FREQUENCY, KC									
	5.0	5.1	5.2	5.3	— — — — —	14.7	14.8	14.9	15.0
5.0	100	0	0	0	0	0	0	0	0	0
5.1	0	102	0	0	0	0	0	0	0	0
5.2	0	0	104	0	0	0	0	0	0	0
5.3	0	0	0	106	0	0	0	0	0	0
	0	0	0	0		0	0	0	0	0
	0	0	0	0	0		0	0	0	0
	0	0	0	0	0	0	0	0	0	0
14.7	0	0	0	0	0	0	294	0	0	0
14.8	0	0	0	0	0	0	0	296	0	0
14.9	0	0	0	0	0	0	0	0	298	0
15.0	0	0	0	0	0	0	0	0	0	300

(CORRELATOR FREQUENCY, KC)

Fig. 45. Pulse-frequency-modulation correlation table.

Vanguard III satellite (1959 η1) and subsequently on several other small scientific-satellite probes. Rochelle,[13] who has analyzed this system in detail, has pointed out that PFM can be considered also as a special case of coded pulse-coherent modulation where the code is restricted to sequences of alternate 0's and 1's only. Thus only one code is selected from orthogonal codes of a given word length of the kind discussed in the preceding section. This results in the correlation table of Fig. 45.

Here the number 100 represents 100 correlations of 50 0's and 50 1's, 102 represents 51 0's and 51 1's, and so on. The restriction in admissible codes does not affect the system performance. It permits the use of simpler hardware for the encoding and decoding.

Two sine-periodic waves are orthogonal to each other if the following condition is met:

$$\int_0^{nT} \sin \omega_m t \sin [(\omega_m + 2n\nu/nT)t + \phi]\, dt = 0 \tag{61}$$

where ω_m = carrier frequency of first (reference) signal
$\nu = \pm 1, \pm 2, \pm 3$, etc.
ϕ = initial relative phase of the second signal
n = message bits per word
T = duration of message *bit*
nT = duration of *word* (symbol) = T_0

The allowable frequency spacing between adjacent signals is then

$$\Delta f = 1/nT = 1/T_0 \tag{62}$$

provided ω is a *multiple* of π/nT, if we permit ϕ any value. If ϕ is *restricted* to either 0 or π then

$$\Delta f = 1/2nT = 1/2T_0 \tag{63}$$

When divided by nT (normalized) Eq. (61) is the crosscorrelation function of two sine functions for zero time lag ($\tau = 0$). This is plotted in Fig. 46 for the case of

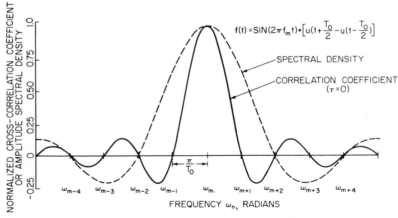

FIG. 46. Crosscorrelation function and power spectrum of pulsed sine wave.

$\phi = 0$. It is compared with the spectrum of an impulse sine wave of the same duration T_0. The first 0 crossing occurs at one-half the frequency width of the impulse wave. Thus the influence of the phase ϕ is very important, since if it is controlled the spectrum spacing between words is cut in half. It can be controlled by modulating the words all in the same manner. It is to be recalled that sampling of the output of the system is done at the end of interval $T_0 = nT$ as in Sec. 13.

The minimal-bandwidth requirement for PFM is thus readily determined as

$$B = M/2T_0 \tag{64}$$

where M is the number of possible words into which messages may be encoded. The system is quite flexible as additional words can be readily added by adding bandwidth. The bandwidth of the coded phase-coherent system can be determined from Eq. (61) for comparison. In this case, nT must be replaced by an interval equal to the duration of the smallest coded subinterval (see Fig. 37). This is $nT/2^n$ since there must be 2^n subintervals in an orthogonal word containing n bits per message. In the minimal case

$$B = 2^n/2nT \tag{65}$$

But M in (64) is 2^n where n is the bit value of the word and $T_0 = nT$. Thus PFM and CPC modulation use equivalent message bandwidths. However, CPC requires the use of a higher-order code as the bits per message (sample) increases whereas PFM requires merely the addition of another frequency.

The implementation of the PFM system has been facilitated by the development of a digitally timed oscillator in which, for example, 3 bits from a register is used to set up a single frequency as shown in Fig. 47a. The transmitted signals are shown in Fig. 47b. Eight possible signal frequencies are used.

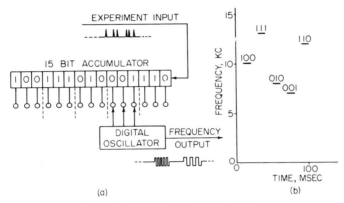

FIG. 47. Digital data-readout system.

The transmission format is shown in Fig. 48. Time frames and channel sampling times are set up by sync signals operating in the channel 0 interval. The precision of the measurements is determined by the number of available frequencies, in this case eight. Analog measurements can be made by converting to a frequency within the transmission band. Quantization need not be done at transmission; it can be done at the receiver with some degradation of performance.

Reception is optimally performed by the use of a bank of *matched filters* which are matched to the expected fixed frequencies. Unmatched filters might also be used. They have the advantage of simplicity and permit better detection of signals not precisely centered in the frequency band. This may result from drift in the oscillator on the reception of analog signals. The disadvantage of unmatched filters is a degradation is signal-to-noise performance. Figure 49 shows an arrangement of filters for reception and a coding of the received signal into digital form. Three adjacent filters are sampled on a "greatest of" basis; that is, the encoding of the digital signal output is based on the largest output of each of the three filters at the sampling time.

Rochelle[13] has computed the word-error probability vs. energy per bit per unit noise intensity $[\beta = ST/N_0;$ see Eq. (22) above] for both matched and unmatched filter reception. This has been done in a manner directly comparable with the work of Viterbi given in Fig. 41 presented in Sec. 13. The probability density functions for signal and signal plus noise differ between matched and unmatched filters. Unmatched

9-52 PAM AND ADVANCED TELEMETRY SYSTEMS

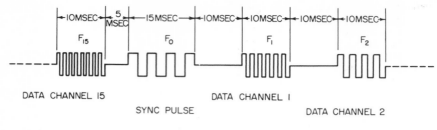

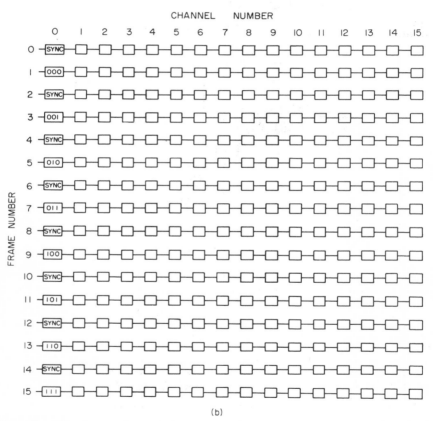

FIG. 48. Pulse-frequency-modulation format. (a) Telemetry frame. (b) Sequence format.

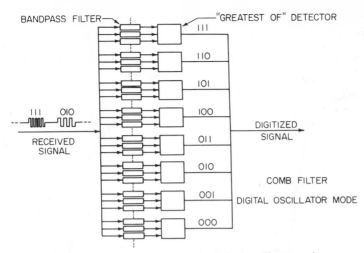

FIG. 49. Contiguous filter bank in digital-oscillator mode.

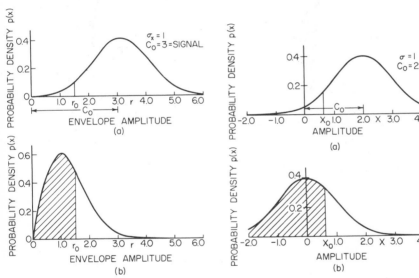

FIG. 50. Probability density functions. (a) Signal plus Rayleigh noise. (b) Rayleigh noise.

FIG. 51. Probability density functions. (a) Signal plus Gaussian noise. (b) Gaussian noise.

filters result in Rayleigh noise output statistics as shown in Fig. 50, whereas matched filters result in the Gaussian probability density functions of Fig. 51. The expression for $P(e)$ for the *unmatched* filter is given by the following equation:

$$P_e(N) = 1 - \int_0^\infty \left[(1/\sigma_x^2) \int_0^{r_0} re^{-r^2/2\sigma_x^2} \, dr \right]^{N-1} (r_0/\sigma_x^2)$$
$$e^{-(r_0^2+C_0^2)/2\sigma_x^2} (1/2\pi) \int_0^{2\pi} e^{r_0 C_0 \cos(\theta-\theta_0)/\sigma_x^2} \, d\theta \, dr_0 \quad (66)$$

where r, θ represents an envelope space of signals of arbitrary magnitude and phase angle. C_0 represents a signal of magnitude C_0 and arbitrary phase angle θ_0. r_0 is the threshold value of r. N is equal to 2^n; σ_x is the in-phase variance of the noise. Figure 52 shows the word-error probability computed from Eq. (66) by the use of an IBM 7090 computer. The curves have been plotted in terms of $n = \log_2 N$ so that they are directly comparable with those in Sec. 13.

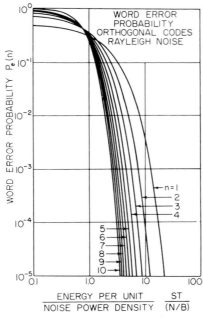

FIG. 52. Word-error probability curves for Rayleigh noise.

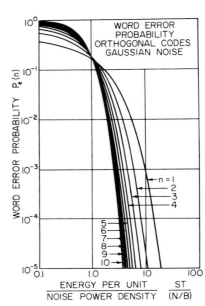

FIG. 53. Word-error probability curves for Gaussian noise.

For the *matched* filter the expression for $P_e(N)$ is

$$P_e(N) = 1 - \int_{-\infty}^{\infty} \left[(1/\sqrt{2\pi\sigma_m}) \int_{-\infty}^{x_0} e^{-x^2/2\sigma_m^2} \, dx \right]^{N-1} (1/\sqrt{2\pi\sigma_m}) e^{-(x_0-C_0)^2/2\sigma_m^2} \, dx_0 \tag{67}$$

Here σ_m is the noise variance at output of matched filter, x_0 is the threshold, and C_0 is the carrier amplitude. The results of integrating this expression on a 7090 computer are shown in Fig. 53. This figure is directly comparable with Fig. 41.

15 Sampled Analog Transmission

Analog data may be transmitted by means of the discrete transmission systems just previously discussed, provided that the data are sampled and coded into symbols (words). Comparative results of such transmission and those of a continuous (analog) transmission system differ considerably depending upon whether equal information rates or equal signal-to-noise ratio is used as the basis.

Sanders[2] has made comparisons both ways, and the results are presented here in a somewhat different fashion.

Figure 54a shows the analog transmission system and Fig. 54b the comparable discrete transmission system. The discrete channel includes a sample quantizer which converts the analog signal to time samples of a specific number of levels L. The levels

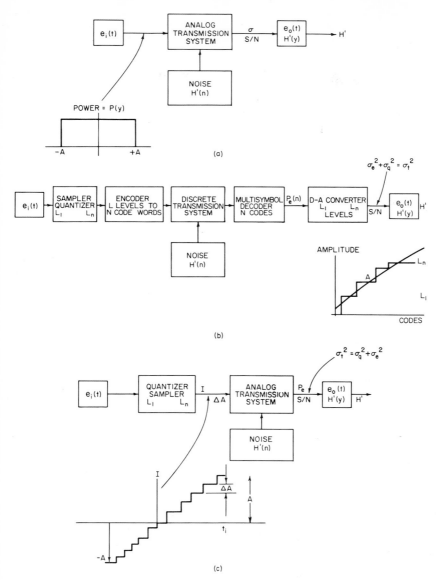

Fig. 54. (a) Analog data inputs and outputs—analog transmission system. (b) Analog data (inputs and outputs)—*discrete* transmission. (c) Sampled *analog* data (inputs and outputs)—*analog* transmission.

are converted into discrete code words of one of the types discussed in Secs. 12 to 14 for transmission to the discrete channel. At reception a multisymbol decoder identifies the code word with a probability of error P_e. The received code identifies a level L in the digital-to-analog converter. This level is used as the equivalent analog output signal $e_o(t)$. This signal is in error by an amount having two components. One is the error, the probability of which is $P_e(n)$ in specifying the correct code word at the output of the discrete transmission link proper; the other is caused by the discrete number of levels used, that is, the error due to quantization. In the general case the total variance σ_t caused by these errors is

$$\sigma_t^2 = \sigma_q^2 + \sigma_e^2 \tag{68}$$

σ_q = variance of quantization errors
    = $(\Delta A)^2/12$ where ΔA is difference in level amplitudes
σ_e = variance caused by random noise in transmission channel

$$\sigma_e = (\Delta A)^2 \frac{P_e(n)L(L+1)}{6} \tag{69}$$

Equation (69) assumes that the probability of transmission of any level is equal and that the probability of error is distributed equally over all levels.

The system is usually designed for a reasonable relationship between the errors so that

$$\sigma_e^2 = K\sigma_q^2 \tag{70}$$

The probability of error P_e, of a sample (word) is then, using Eqs. (68) to (70),

$$P_e = \frac{K}{2L(L+1)} \tag{71}$$

This is equivalent to $P_e(n)$ of Secs. 12 to 14.

The S/N of the discrete channel can also be derived from the above. The signal in terms of ΔA is, assuming all levels equally probable,

$$S = [(\Delta A)^2/12](L^2 - 1) \tag{72}$$

The noise power N is equal to the total variance σ_t^2 which is [from (68) and (70)]

$$\sigma_t^2 = (\Delta A^2/12)(1 + k)^{\frac{1}{2}} \tag{73}$$

The signal-to-noise ratio is therefore

$$S/N = (L^2 - 1)/(1 + k) \tag{74}$$

Sanders[2] has computed values of $P_e(n)$ and S/N for several levels L for a *discrete* channel for $k = 1$, the case where $\sigma_q^2 = \sigma_e^2$. These are shown in the third and fourth columns of Table 5.

The information per sample (word) at the output is H' and is reduced by the action of the noise source $H'(n)$ from the information available in the source, $\log_2 L$. The information loss is, of course, the equivocation.

The output information rate per sample is

$$H' = H'(y) - H'(n) \tag{75}$$

where $H'(y)$ = entropy of the received signal including *noise*
$H'(n)$ = entropy of the perturbing noise (independent of the transmitted signal)
For the *discrete* channel the information rate per sample is

$$H' = \log_2 L - P_e \log_2 (L - 1) + P_e \log P_e + (1 - P_e) \log (1 - P_e) \tag{76}$$

or using the value of P_e in Eq. (71)

$$H' = \log_2 L - \frac{k}{2L(L+1)} \log_2 (L-1)(2L^2 + 2L - k) \\ + \log_2 (2L^2 + 2L - k) - \log_2 (2L^2 + 2L) \tag{77}$$

This equation for $k = 1$ was used to determine the information rates in the second column of Table 5.

Table 5

L signal levels	Digital channel				Analog channel	
	H' information rate, bits per sample	P_e	$10 \log_{10} S/N$, db		P_e	$10 \log S/N$, db
4	1.791	0.025	8.75		0.0830	11.05
8	2.921	0.00694	14.98		0.0558	18.58
16	3.973	0.00184	21.06		0.0452	25.21
32	4.992	4.73×10^{-4}	27.09		0.0414	31.46
64	5.998	1.20×10^{-4}	33.11		0.0399	37.58
128	6.999	3.03×10^{-5}	29.14		0.0393	43.64
256	8.000	7.60×10^{-6}	45.16		0.0390	49.68
512	9.000	1.90×10^{-6}	51.18		0.0389	55.71
1,024	10.000	4.76×10^{-7}	57.20		0.0388	61.74

Conditions: H' for discrete channel equals H' for continuous channel. Discrete channel is equal-increment quantized. RMS quantization noise = rms error due to uniform error probability P_e.
For analog channel, transmitted signal is uniformly distributed over an interval. Noise in channel is uniformly distributed.
P_e = error rate per sample if transmitting L-level digital data.

The comparative information rate for the analog channel was calculated assuming that the transmitted signal is uniformly distributed over an interval $-A/2$ to $+A/2$ (see Fig. 54a) and that the noise is normally distributed with zero mean and variance σ^2. Then

$$P(y) = \frac{1}{A}\left[\Phi\left(\frac{2y+A}{2}\right) - \Phi\left(\frac{2y-A}{2}\right)\right] \tag{78}$$

$$P(n) = (1/\sigma\sqrt{2\pi})e^{-n^2/2\sigma^2} \tag{79}$$

where Φ is the error function.
The value of H' is difficult to calculate except for S/N ratio greater than 5. In this case

$$H' \cong \log_2 A - \log_2 \sigma \sqrt{2\pi e} + 2.6062 \, (\sigma/A)$$
$$\cong \tfrac{1}{2} \log S/N + 0.7523 \sqrt{N/S} - 0.2546 \tag{80}$$

where $S = A^2/12$
$N = \sigma^2$

Values for the S/N ratio for the analog channel corresponding to the same information rates previously computed for the discrete case are shown in the sixth column of Table 5.

In order to compare the channels on the basis of error rate it is necessary to quantize the inputs to the analog link into specific levels L of separation ΔA as shown in Fig. 54c. The noise, of variance σ^2, is again assumed to be Gaussian and to be added linearly. Then

$$P_e(i) = 2\Phi(-\Delta A/2\sigma) \quad i = 2, 3, \ldots, L-1$$

and
$$P_e(i) = \Phi(-\Delta A/2\sigma) \quad i = 1, L$$

Assuming all levels equally likely

$$P_e = \frac{2(L-1)}{L}\Phi\left(-\frac{\Delta A}{2\sigma}\right)$$

but the signal S (assuming 0 mean value) is

$$S = [(\Delta A)^2/12](L^2 - 1)$$

or

$$S/N = [(L^2 - 1)/3](\Delta A/2\sigma)^2 \tag{81}$$

Hence

$$P_e = \frac{2(L-1)}{L} \Phi\left(-\sqrt{\frac{S}{N}\frac{3}{L^2-1}}\right) \tag{82}$$

These values are given in the fifth column of Table 5. The analog error rates are considerably higher than those of the discrete channel. In the case of the continuous channel, however, the quantization is such that the error causes a selection of an adjacent value to the correct one and hence only a small error is made. This is not true of the discrete case. Here noise error can cause a completely erroneous word (sample) to be selected, unrelated to the true value. Hence discrete channels are characterized by infrequent significant errors and analog channels by relatively frequent small errors.

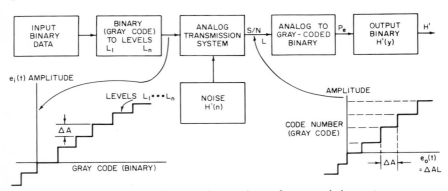

Fig. 55. *Discrete* data (input and output)—*analog* transmission system.

It is of interest for completeness to examine the case in which *digital* information is transmitted over a continuous channel. This is shown in Fig. 55. If the same assumptions are made about the signal and error distributions as for Fig. 54c, Eqs. (81) and (82) hold for the P_e and S/N of this channel. However, for minimum error

Table 6. Transmission of Binary Information over an Analog Channel of Various Levels L

L	P_e	$10 \log_{10} S/N$	H' bits per sample
2	10^{-6}	13.54	1.000
4	2×10^{-6}	20.42	2.000
8	3×10^{-6}	26.56	3.000
16	4×10^{-6}	32.55	4.000
32	5×10^{-6}	38.50	5.000
64	6×10^{-6}	44.46	6.000
128	7×10^{-6}	50.42	7.000
256	8×10^{-6}	56.39	8.000
512	9×10^{-6}	62.36	9.000
1,024	10×10^{-6}	68.34	10.000

Bit error rate is $P_e' = 10^{-6}$.

the binary information must be encoded in Gray code before encoding to the quantized level so that adjacent levels differ by only 1 bit position.

The error per bit is then

$$P_e' = P_e/\log_2 L \tag{83}$$

for small P_e. Hence

$$P_e' = \frac{2(L-1)}{L \log_2 L} \Phi\left(-\sqrt{\frac{S}{N} \frac{3}{L^2 - 1}}\right) \tag{84}$$

Values of analog signal-to-noise ratios required for a bit error rate of 10^{-6} are shown for various numbers of levels in Table 6.

REFERENCES

1. Ira Jacobs, "Weak Signal Processing Techniques," pp. 415–423, Space Research II, H. B. Kallman et al. (eds.), North Holland Publishing Company, Amsterdam, 1961.
2. Ray W. Sanders, Communication Efficiency Comparison of Several Communication Systems, *Proc. IRE*, 1960, pp. 575–588.
3. V. A. Kotel'nikov, "The Theory of Optimum Noise Immunity," McGraw-Hill Book Company, New York, 1960.
4. W. B. Davenport, Jr. and W. L. Root, "Introduction to Random Signals and Noise," Chap. 14, pp. 343–345, McGraw-Hill Book Company, New York, 1958.
5. Paul D. Shaft, Error Rate of PCM-FM Using Discriminator Detection, *IEEE Trans. Space Electron. Telemetry*, vol. SET-9, pp. 131–137, 1963.
6. M. J. E. Golay, *Proc. IRE*, 1949, p. 1031.
7. A. J. Viterbi, On Coded Phase-coherent Communications, *IRE Trans. Space Electron. Telemetry*, vol. SET-7, no. 1, pp. 3–14, March, 1961.
8. R. M. Fano, "Communication in the Presence of Additive Gaussian Noise," pp. 169–182, Communication Theory, W. Jackson (ed.), New York, Academic Press Inc., 1953.
9. D. Middleton and D. Van Meter, Detection and Extraction of Signals in Noise from the Point of View of Statistical Decision Theory, *J. Soc. Ind. Appl. Math.*, vol. 3, pt. I, pp. 192–253, December, 1955; vol. 4, pt. II, p. 86, June, 1956.
10. I. S. Reed, A Class of Multiple-error-correcting Codes and the Decoding Scheme, *IRE Trans. Inform. Theory*, PGIT-4, p. 38, September, 1954.
11. J. H. Green, Jr., and R. L. San Soucie, An Error-correcting Encoder and Decoder of High Efficiency, *Proc. IRE*, vol. 46, no. 10, pp. 1741–1744, October, 1958.
12. G. Birkhoff and S. MacLane, "A Survey of Modern Algebra," The Macmillan Company, New York, 1953.
13. Robert W. Rochelle, Pulse-frequency-modulation, *NASA Tech. Rept.* TR R-189, January, 1964. (Available from Office of Technical Services, Washington, D.C.)

PHASE-LOCKED-LOOP SYSTEMS

The phase-locked loop is a key element of many high-efficiency communication systems. It is used in two fundamentally different ways:

1. To track carrier signals which move in frequency with tuning drifts and Doppler shifts.
2. To follow information-bearing modulated frequency or phase changes so as to improve the signal-to-noise performance of communication systems in cases where the carrier bandwidth is greater than the information bandwidth.

In some cases both techniques are used in the same system, for example, where a pilot carrier is tracked (method 1) and a modulated subcarrier is demodulated (method 2). In the second use the phase-locked loop is a matched filter because it creates a replica of the received signal modulation to compare coherently with the modulation to be received. The earliest work in this connection was that of Chaffee[1] on FM feedback. The theory of operation of the loop as a demodulator was first presented by Jaffe and Rechtin.[4] To do so, however, a linearized model was used. Linear models were used by Gruen[2] and Richman[3] for TV sync problems resembling method 1.

To study problems such as phase-locking time and threshold more exact models are required. These have been developed by Viterbi,[11] Develet,[14] and others. No general exact method exists for calculating the pull-in of signals in noise. However, Viterbi[15] has presented an exact solution for limited cases which provides insight as to the nature of loop dynamics.

The range of application of the phase-locked system includes demodulation to improve signal-to-noise, synchronization for frame time and bit rate, Doppler tracking to improve signal-to-noise and to measure velocity, and range tracking (delay-lock discriminator).[21,22]

We shall derive the tracking characteristics from the exact model assuming no noise and develop the operation under noise conditions from the linear model. Then the operation as a demodulator shall be presented. Finally methods of dealing with noisy signals using the more exact nonlinear-loop model will be discussed.

16 Phase-locked-loop Operation

Figure 56 shows the basic configuration of a phase-locked loop. Here it is assumed that there is no noise at the input, only signal e_i. The input signal is a sinusoid which

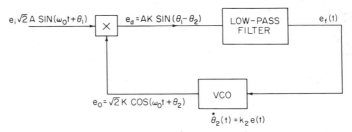

FIG. 56. Basic function diagram of a phase-locked loop.

is phase- or frequency-modulated with either intelligence or Doppler, or is arbitrarily changed by tuning.

$$e_i = \sqrt{2}\, A \sin[\omega_0 t + \theta_1(t)] \tag{85}$$

where A^2 is the received signal power and $\theta_1(t)$ represents the phase or frequency modulation in question. The output of the voltage-controlled oscillator (VCO) is a pure sinusoid of arbitrary phase. It can be expressed as

$$e_o = \sqrt{2}\, K \cos[\omega_0 t + \theta_2(t)] \tag{86}$$

K being rms amplitude. The output of the multiplier is then

$$\begin{aligned} e_d = e_i e_o &= 2AK \sin[\omega_0 t + \theta_1(t)] \cos[\omega_0 t + \theta_2(t)] \\ &= AK \sin[\theta_1(t) - \theta_2(t)] + \sin[2\omega_0 t + \theta_1(t) + \theta_2(t)] \end{aligned} \tag{87}$$

The low-pass filter of the loop effectively discards the higher-frequency term. The filter operates on $AK \sin \theta_1(t) - \theta_2(t)$ to produce the signal e_f of which the instantaneous frequency of the output of the VCO (voltage-controlled oscillator) is related to its input by

$$\dot{\theta}_2(t) = K e_f \tag{88}$$

For $e_f = 0$ let $\dot{\theta}_2(t) = \omega\theta$. Then

$$\theta_2(t) = \int e_f(t) \tag{89}$$

The block diagram can be redrawn to emphasize the operation in terms of phase error ϕ, which is $\theta_1 - \theta_2$ and which is driven toward null by the loop action. This is shown in Fig. 57 in operational notation. This system differs from the familiar servo loop in that there is a nonlinear transfer function $Y = \sin X$ in the loop. This gives rise to fundamental difficulties in describing the action of the loop. If $\phi < 1$ radian, $\sin \phi = \phi$ and with this linear approximation the system may be analyzed using linear servo theory. In actual practice it is necessary to pull in signals differing from ω_0 by large phases and to track signals in noise. Techniques discussed later in this section have been developed to analyze these situations but they are difficult and are sometimes restricted in applicability.

It is common practice to refer to first-, second-, and higher-order loops. This refers to the number of integrations in the loop, as in servosystems (Chap. 15). The VCO is inherently one integration as noted in Fig. 57; the filter can supply additional integrations. Higher-order loops than second have inherent stability problems and are not generally used. The second-order loop is particularly useful because it provides 0 phase error when the incoming signal is a fixed frequency or, equivalently, changes phase at a constant rate.

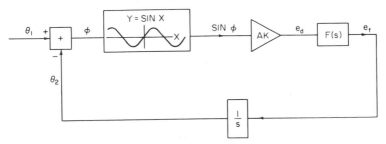

FIG. 57. Exact model of phase-locked loop.

The analysis of loop operation becomes progressively more difficult under the following conditions:

1. Small initial separation of incoming signal phase and VCO phase ($\phi \ll 1$ radian)
2. Motion of incoming signal in frequency
3. Presence of noise with signal
4. Wide ($\phi > \pi/2$ radians) separation of phase of incoming signal from VCO
5. Wide initial separation (large ϕ) and presence of noise

The first three cases can be handled with the linear approximate model for both tracking and modulation following applications. The fourth case requires the use of the exact model and may be treated by phase-plane analysis. The last situation can be analyzed only by using the exact model for the first-order loop and approximate methods for second and higher order. Most situations can be covered by use of the linear model by imposing design constraints on the loop. Accordingly we shall present this method in greatest detail and then provide an acquaintance with the methods of nonlinear analysis.

17 Linear Model

The linear model may be used for both noise and noiseless conditions and for tracking and modulation applications, the only restriction being that the phase error ϕ be constrained below approximately 1 radian; otherwise dropout will take place. In this case $\sin \phi$ in Fig. 57 is replaced by ϕ in Fig. 59. In order to justify the use of the linear model under noisy input conditions it must be assumed that the propagation of signal phase modulation and that of random noise through the network are independent; this is equivalent to stating that the superposition principle holds for linear systems. In order for this assumption to be at least approximately fulfilled, as well as for the loop to remain near the phase-locked condition, the total phase error, which consists of $\theta_1 - \theta_2$, the error in the absence of noise, and ν, the phase jitter on the VCO output due to noise, must be at all times small compared with 1 radian. The best justification for the validity of the assumption of linearity in the vicinity of phase lock is experimental evidence.

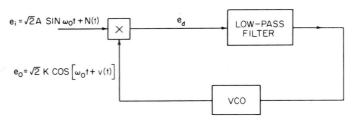

Fig. 58. Phase-locked loop—functional block diagram, with noise input only.

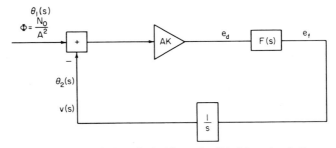

Fig. 59. Linear model phase-locked-loop simplified functional diagram.

Based on the superposition assumption the operation of the loop can be examined when only noise is applied to the input and no signal. Signal and noise will then be additive if the linear model proves valid for noise. The signals are shown in Fig. 58.

The input signal is
$$e_i = \sqrt{2}\, A \sin \omega_0 t + N(t) \tag{90}$$

where $N(t)$ represents additive noise. The VCO output is
$$\begin{aligned} e_o &= \sqrt{2}\, K \cos [\omega_0 t + \nu(t)] \\ &= \sqrt{2}\, K \cos [\omega_0 t - \nu(t) \sin \nu_0 t] \end{aligned} \tag{91}$$

where the phase noise jitter $\nu(t)$ is assumed small compared with 1 radian. The phase error signal is the product of these two signals
$$e_d = e_i e_o \approx AK[\sin 2\omega_0 t - (\nu t)(1 - \cos 2\omega_0 t)] \\ + \sqrt{2}\, KN(t)[\cos \omega_0 t - \nu(t) \sin \omega_0 t] \tag{92}$$

This expression can be simplified according to a further approximation based on the following observation. The loop filter will not pass the double-frequency terms, and $\nu(t)$ is a slowly varying function at all times small compared with 1 radian. Thus it is possible to discard the double-frequency terms and consider $\nu(t) \sin \omega_0 t$ negligible compared with $\cos \omega_0 t$:
$$\begin{aligned} e_d(t) &\approx -AK\nu(t) + \sqrt{2}\, K \cos \omega_0 t N(t) \\ &= AK[(\sqrt{2}/A)(\cos \omega_0 t)N(t) - v(t)] \end{aligned} \tag{93}$$

The second term within the brackets is simply the low-frequency VCO phase noise. In the first term $N(t)$ is taken to be white noise of spectral density N_0. Multiplication of white noise by $(\sqrt{2}/A) \cos \omega_0 t$ produces a signal whose autocorrelation function is a product of the autocorrelation functions of the two signals. This means that the first term of Eq. (93) must be white noise of spectral density N_0/A^2. It follows that e_d is KA times the difference between white noise of spectral density N_0/A^2 and the VCO phase jitter $\nu(t)$. Hence the linear model of Fig. 59 is a valid representation of the loop performance in the vicinity of phase lock with a noisy input signal. Note that in the linear model the multiplier is replaced by an adder and linear servo concepts apply. The symbol Φ is generally used to represent the noise spectral density normalized by the signal power N_0/A^2.

The propagation of the input phase modulation $\theta_1(t)$ through the linear model of the loop is readily computed by Laplace-transform methods. The VCO output phase due to $\theta_1(t)$ only is derived from Fig. 59 from

$$[\theta_1(s) - \theta_2(s)]AKF(s)(1/s) = \theta_2(s) \tag{94}$$

and is

$$\theta_2(s) = \frac{AKF(s)}{s + AKF(s)} \theta_1(s) = H(s)\theta_1(s) \tag{95}$$

where $H(s)$ is the closed-loop gain of the linear feedback loop. The VCO phase jitter $\nu(t)$, which is assumed to be a stationary random signal, since $N(t)$ is stationary, can be computed by transform methods from the input noise. However, since the latter is a random signal for which only statistical information is known, only the statistical parameters of $\nu(t)$ can be obtained. The simplest parameter to compute is the variance of the noise jitter

$$\sigma_\nu^2 = \Phi \int_{-j\infty}^{j\infty} \frac{|H(s)|^2}{2\pi j} ds \tag{96}$$

The integral

$$\int_0^{j\infty} \frac{|H(s)|^2}{2\pi j} ds \tag{97}$$

is defined as the loop noise bandwidth B_L since an ideal low-pass filter of bandwidth B_L cps with a white-noise input of spectral density N_0 will have the same amount of noise power at its output as the variance of loop-phase noise. B_L is a key parameter in linear-model phase-locked-loop design.

If the phase modulation $\theta_1(t)$ is a random signal also (which is the case in telemetry), the *total phase error*, which is the sum of the phase error due to modulation and that due to noise, $\theta_1(t) - \theta_2(t) - \nu(t)$, has a variance or mean-square value whose dimension is power. This is the *mean-square phase error* and can be expressed in terms of the time functions as

$$\overline{e_d^2} = \lim_{T \to \infty} (1/2T) \left\{ \int_{-T}^{T} [\theta_1(t) - \theta_2(t)]^2 dt + \int_{-T}^{T} \nu^2(t) dt \right\} \tag{98}$$

The cross term of $(\theta_1 - \theta_2)$ and ν is zero because the modulation and noise are independent. Parseval's theorem states that Eq. (98) can be expressed in terms of the transforms:

$$\overline{e_d^2} = (1/2\pi j) \left[\int_{-j\infty}^{j\infty} |\theta_1(s) - \theta_2(s)|^2 ds - \int_{-j\infty}^{j\infty} |\nu(s)|^2 ds \right] \tag{99}$$

The second integral is the variance of σ_{ν^2} given by Eq. (96). Hence the mean-square error can be expressed as the sum

$$\overline{e_d^2} = (1/2\pi j) \int_{-j\infty}^{j\infty} |\theta_1(s)|^2 |1 - H(s)|^2 ds + \sigma_\nu^2 \tag{100}$$

where

$$\sigma_\nu^2 = \Phi \int_{-j\infty}^{j\infty} \frac{|H(s)|^2}{2\pi j} ds = 2\Phi B_L \tag{101}$$

In Eq. (98) the first term is the mean-square error due to modulation and the second term σ_ν^2 is the phase noise power.

It is also convenient at times to consider the peak phase error; this is

$$E_d = |\theta_1 - \theta_2|_{max} + |\nu|_{max} \tag{102}$$

Signal Tracking—Linear Model

The linear-model analysis may be directly applied to tracking signals in noise. This problem is common in space telemetry for these reasons:

1. It is necessary, to conserve power, to restrict receiver bandwidth as closely as possible to base band. Here it is desired to follow the received signal to reduce allowance for Doppler shifts and tuning instability.

2. It is often desirable, again to save transmitted power, to have the local receiver oscillator track a pilot unmodulated subcarrier of a complex signal with a phase-locked loop and to provide for demodulation of a different subcarrier.

3. Tracking Doppler frequency changes yields direct information as to space-vehicle radial velocity. This measurement, combined with others, can be used to establish position and velocity of the vehicle. One use of this information is to order high-gain antennas toward such vehicles to improve ground-to-space link communications efficiency.

There are two cases of practical importance:

1. The incoming signal is a constant frequency shift from initial VCO center frequency. Then

$$\theta_1(t) = \Omega t \quad \text{or} \quad \theta_1(s) = \Omega/s^2 \tag{103}$$

2. The incoming signal frequency varies with time as when a vehicle accelerates radially with respect to the receiver. Here

$$\theta_1(t) = \tfrac{1}{2} D t^2 \tag{104}$$

Consider the constant-frequency case first. The loop filter, acting as a low-pass filter, will normally have a transfer function

$$F(s) = 1 + a/s \tag{105}$$

As previously mentioned, this will make the linear model equivalent to a second-order servo loop. From Eq. (95) the closed-loop transfer function becomes

$$H(s) = \frac{AK(s + a)}{s^2 + AK(s + a)} \tag{106}$$

The parameters AK and a can be related to those commonly used in servomechanisms to express natural frequency and damping constants:

$$aAK = \omega_n^2 \quad \text{and} \quad a = \omega_n/2\zeta \tag{107}$$

Then

$$H(s) = \frac{2\zeta\omega_n s + \omega_n^2}{s^2 + 2\zeta\omega_n s + \omega_n^2} \tag{108}$$

The phase error due to modulation is

$$\theta_2(s) - \theta_1(s) = \frac{\Omega[1 - H(s)]}{s^2} = \frac{\Omega}{s^2 + 2\zeta\omega_n s + \omega_n^2} \tag{109}$$

The steady-state error is zero, since by the final-value theorem of Laplace transforms

$$\lim_{t \to \infty} [\theta_2(t) - \theta_1(t)] = \lim_{s \to 0} s[\theta_2(s) - \theta_1(s)] = 0 \tag{110}$$

Thus in the absence of noise the loop would achieve perfect phase lock. The noise, however, produces a phase noise variance given by Eq. (101).

$$\sigma_\nu^2 = \Phi \int_{-j\infty}^{j\infty} \frac{|H(s)|^2}{2\pi j} ds = 2\Phi B_L \qquad (111)$$

Using the form of $H(s)$ given by Eq. (101) yields

$$B_L = \frac{\omega_n(1 + 4\zeta^2)}{8\zeta} \qquad (112)$$

The phase jitter can be determined from these equations for any second-order loop. The optimum filter, however, requires a criterion for minimizing noise. Jaffe and Rechtin[4] used the constraint that

$$\int_0^\infty [\theta_1(t) - \theta_2(t)]^2 dt \qquad (113)$$

be equal to a constant when determining the optimum filter to minimize σ_ν^2. They have also shown that a second-order loop is optimum with respect to this criterion.

A case of greater importance than that of achieving phase lock or synchronization with a received signal of constant frequency is that of the tracking signal which is *accelerating* radially with respect to the receiver. Then the Doppler frequency is $(\alpha t/c)f_0$ cps, where α is the acceleration. Hence the received signal phase

where
and
$$\begin{aligned}\theta_1(t) &= \tfrac{1}{2} D t^2 \quad \text{radians} \\ D &= (2\pi\alpha/c)f_0 \\ \theta_1(s) &= D/s^3 \end{aligned} \qquad (114)$$

If the second-order phase-locked loop is used, the phase error is given by

$$\theta_2(s) - \theta_1(s) = \frac{D}{s^3}[1 - H(s)] = \frac{D}{s(s^2 + 2\zeta\omega_n s + \omega_n^2)} \qquad (115)$$

and the steady-state error is

$$\lim_{t \to \infty} \theta_2(t) - \theta_1(t) = D/\omega_n^2 \qquad (116)$$

It is fundamental in control theory that in order to obtain a zero steady-state error for a $1/s^3$ input three integrators must be contained in the loop. (The VCO accounts for one, and the other two must be in the loop filter.) As would be expected, the optimizing procedure of Jaffe and Rechtin arrives at a *third-order loop* for these conditions. Practical considerations render it undesirable to increase the complexity of the loop filter. For example, three integrators in the loop render the system conditionally stable. That is, if the gain of the system falls below a given level, the system will oscillate. Since the loop gain depends on the signal power A^2, this can be a serious difficulty.

Linear Model—Modulation Following

In modulation following, the phase-locked loop is required to reproduce frequency or phase modulation which is on the incoming signal carrier. Much of the theory developed for the case of tracking can be applied. The statistics of the modulation and noise interference may differ in many respects, but for linear analysis it is only important that the modulation spectrum is much narrower than the noise.

It is appropriate to consider the phase-locked loop demodulator as a matched filter. In such a filter the incoming signal is crosscorrelated with a replica of the signal modulation. The advantage is that the detection bandwidth can be limited to the modulation bandwidth. Consider an FM signal (similar analysis holds for phase modulation) which can be written as

$$\sqrt{2} \, A \sin [\omega_0 t + \theta_1(t)] \quad (117)$$

where
$$\theta_1(t) = K_M \int e(t) \, dt$$

and K_M is the modulation index and $e(t)$ the data signal. Following the matched-filter concept, the receiver must contain a VCO tunable over the frequency range of the transmitter oscillator. If a reasonable replica of the data signal, which will be designated $e_f(t)$ (Fig. 56), could be reconstructed, application of this signal to the receiver VCO input would produce at its output

$$\sqrt{2} \, K \cos [\omega_0 t + \theta_2(t) + \psi] \quad (118)$$

where
$$\theta_2(t) = \int e_f(t) \, dt$$

and ψ is an arbitrary initial phase. It is clear that multiplying together the received signal with its local replica produces a term whose low-frequency component is proportional to the sine of the difference between the transmitted phase and the locally reconstructed replica,

$$\sin [\theta_1(t) - \theta_2(t)]$$

If this signal is filtered and used to modify the modulation replica signal $e_f(t)$, the result is the phase-locked loop of Fig. 60. Hence the loop has been shown to be a coherent

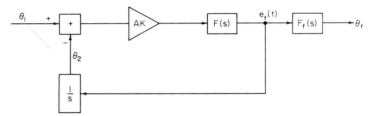

Fig. 60. Linear block diagram—phase-locked-loop demodulator.

frequency demodulator producing a replica of the data signal at the input of the receiver VCO.

For the phase-locked loop to function properly as a frequency discriminator, the phase error must be maintained small compared with 1 radian. Since both the modulation and the interference are random in this case, the only measure of phase error which can be easily calculated is the mean-square error

$$\overline{e_d^2} = \lim_{T \to \infty} (1/2T) \int_{-T}^{T} [\theta_1(t) - \theta_2(t)]^2 + \nu^2(t) \, dt$$

$$= (1/2\pi j) \int_{-j\infty}^{j\infty} [|\theta_1(s)|^2 |1 - H(s)|^2 + \Phi |H(s)|^2] \, ds \quad (119)$$

If the phase-locked loop is to produce a good replica of the modulating signal, it must maintain the mean-square error considerably below 1 sq radian.

The criterion for optimizing the loop is therefore the minimization of e_d^2. This was first done by Lehan and Parks[6] for the case in which the data signal $e(t)$ has a spectrum of the form $2a/(\omega^2 + a^2)$ and hence a 3-db radian frequency at a. The result of the Wiener optimization technique which minimizes e_d^2 is that the optimum loop filter has transfer function

$$F(s) = \frac{\alpha s + \beta}{s + a} \quad (120)$$

where α and β are constants on A^2 and N_0, the signal power and noise spectral density for which the demodulator was to be optimized, and K_M, the modulation index. This the loop filter is a one-stage RC low-pass filter.

The signal $e_f(t)$ is not the best filtered replica of the data signal $e(t)$. To obtain this the loop at $e_f(t)$ must be followed by a low-pass filter $F_f(s)$ as shown in Fig. 60. The criterion for optimizing $F_f(s)$ is that of minimizing the mean-square error between the filtered $e_f(t)$, which is $\theta_f(t)$, and the derivative of $\theta_1(t)$, which is proportional to $e(t)$. This will result in a transfer function for $F_f(s) = \alpha/s + \delta$ where the constants depend on parameters similar to those determining α and β in Eq. (120). The low-pass filter requires only a single resistor and capacitor. The calculations, however, are involved.

An example of the design of a phase-locked FM demodulator will be given following Viterbi[8] using the linear-model analysis developed here. For a particular maximum signal-to-noise ratio the loop filter can be designed to keep either the mean-square error [Eq. (100)] or the peak error [Eq. (102)] at a minimum. If the former is used the Wiener optimization technique will evolve a transfer function of the form in Eq. (120). However, a simpler design procedure can be developed using the peak error criterion, and it will be used here.

The modulating frequency will be assumed never to exceed a maximum time rate of change of $\dot{f}$ radians/sec². Therefore, the peak phase input is $\tfrac{1}{2}\dot{f}t^2$, the transform of which is $\dot{f}/s^3$. This calls for a second-order loop (Figs. 59 and 60) for the error to be finite. The loop filter must provide one integration and a minimum of one zero in the left-hand plane (Chap. 15). These considerations establish the simplest form of the loop filter as

$$F(s) = (1 + bs)/s \tag{121}$$

The phase transfer function $H(s)$ can be written as either Eq. (106) or (108), the latter being in terms of natural frequencies and damping constants. It has been established experimentally that a satisfactory balance between the transient error and the output noise occurs when the loop damping constant $\zeta = 0.707$. From Eqs. (107) and (112) and noting that $b = 1/a$ in Eq. (105)

$$B_L = \tfrac{3}{4}\omega_n/\sqrt{2} = \tfrac{3}{4}\sqrt{AK}/\sqrt{2}$$
$$\omega_n = \tfrac{4}{3}\sqrt{2}\,B_L \tag{122}$$

Equation (108) can be rewritten for this damping factor in terms of the loop noise bandwidth:

$$H(s) = \frac{\tfrac{8}{3}B_L s + \tfrac{32}{9}B_L^2}{s^2 + \tfrac{8}{3}B_L s + \tfrac{32}{9}B_L^2} \tag{123}$$

From Eq. (107) we find that, again noting that $b = 1/a$

$$b = \sqrt{2}/\omega_n = 3/4B_L \tag{124}$$

and the loop filter becomes

$$F(s) = \frac{1 + (3/4B_L)s}{s} \tag{125}$$

The transform of the phase error is

$$[\theta_1(s) - \theta_2(s)] = \theta_1(s)[1 - H(s)] \tag{126}$$

When the frequency ramp is applied at the input

$$[\theta_1(s) - \theta_2(s)] = \frac{\dot{f}}{s^3} \frac{s^2}{s^2 + \tfrac{8}{3}B_L s + \tfrac{32}{9}B_L^2} \tag{127}$$

By the final-value theorem of Laplace transforms [Eq. (110)] the steady-state error is

$$(\theta_{1ss} - \theta_{2ss}) = 9\dot{f}/32B_L^2 \tag{128}$$

Because of the heavy damping the peak phase transient error is only slightly higher than the steady-state error. We shall consider these equivalent.

How closely loop output e_f follows the input frequency θ_1 is established by

$$\frac{e_f(s)}{\theta_1(s)} = sH(s) = s\frac{1 + (3/4B_L)s}{1 + (3/4B_L)s + (9/32B_L{}^2)s^2} \qquad (129)$$

In the absence of noise e_f is a filtered replica of the derivative of the input. This transfer function does not approach zero for infinite frequencies and therefore does not filter wideband noise. An output filter is required for this, as in Fig. 60. In this case the transfer function is the inverse of the numerator in Eq. (129) and the discriminator transfer function is

$$\frac{e_f'}{\theta_1} = \frac{s}{1 + (3/4B_L)s + (9/32B_L{}^2)s^2} \qquad (130)$$

The resulting configuration of the discriminator is given in Fig. 61. The total output noise is

$$\sigma_f{}^2 = \frac{\Phi}{2\pi}\int_{-\infty}^{\infty}\left|\frac{e_f'}{\theta_1}(\omega)\right|^2 d\omega = 64\!/\!27\Phi B_L{}^3 \qquad \text{radians/sec}^2 \qquad (131)$$

If a limiter is used before the phase-locked-loop demodulator, noisy signals will have the effect of reducing the available signal power A^2 and hence the loop noise bandwidth B_L. Hence B_{L_0} the loop noise bandwidth corresponding to minimum signal-to-noise conditions, taking into account the suppression factor, must be used in the design.

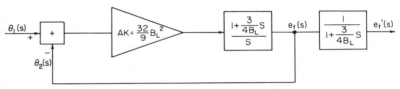

Fig. 61. Phase-locked-loop FM demodulator—linear model design to minimize peak phase errors.

The peak error must be restricted to prevent loop drop-out. Using Eq. (102) and assuming peak error is equal to the steady-state error of Eq. (128) and the noise to be normally distributed and that it be acceptable that the phase noise never exceeds peak for 95 per cent of the time, we obtain

$$\begin{aligned}9\dot{f}/32B_{L_0}{}^2 + 2\sigma_\nu &\leq \text{acceptable phase limit}\\ 9\dot{f}/32B_{L_0}{}^2 + 2\sqrt{2\Phi B_{L_0}} &\leq \pi/2 \text{ radian}\end{aligned} \qquad (132)$$

Equation (132) implies that the noise is normally distributed. A more conservative approach would be to use $3\sigma_\nu$, which would drastically reduce the probability of drop-out.

The minimum acceptable output signal-to-noise ratio is set by σ_f of Eq. (131) in contrast with the smallest acceptable output signal level. In the case of pictures this is set by the number of gray levels desired, in data by the minimum acceptable accuracy. If Δ is the maximum modulating frequency deviation, then

$$d = \Delta/\sigma_f \qquad (133)$$

is minimum data accuracy, and from Eq. (131)

$$\Delta = (8/3\sqrt{3})\, d\Phi^{1/2}\, B_{L_0}{}^{3/2} \qquad \text{radians/sec} \qquad (134)$$

The maximum information rate is

$$H = \dot{f}/\Delta \tag{135}$$

From Eq. (132) and (134)

$$H = (1/d\sqrt{3})(2\pi B_{L_0}^{1/2}/\Phi^{1/2} - 8\sqrt{2}\,B_{L_0}) \tag{136}$$

Figure 62 shows the information rate for $\Phi = 0.01$.
This expression has a maximum with respect to B_{L_0} at

$$B_{L_0} = \pi^2/128\Phi \tag{137}$$

for which the maximum information rate is

$$H_{max} = (\pi^2/16)\sqrt{2/3}\,(1/d\Phi) \cong 1/2d\Phi \tag{138}$$

The design procedure may now be summarized:

1. Establish the expected noise spectral density Φ from communication-link parameters (the range equation).
2. Establish the data accuracy d (or dynamic range).
3. Determine H_{max} and B_{L_0} from Eqs. (138) and (137).
4. Calculate Δ, the required deviation, from Eq. (134).

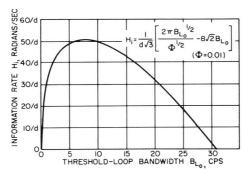

Fig. 62. Information rate as a function of threshold-loop bandwidth for phase-locked frequency-modulation discriminator.

Sanders[13] has analyzed the communications efficiency for a modulation following second-order phase-locked loop such as has just been discussed. He used the same linear model. From the definition of β [Eq. (22), Sec. 8] and loop noise bandwidth the following expression B_L can be derived:

$$\beta = (1/2\sigma_p^2 H')(B_L/B_d) \tag{139}$$

where B_d is the data bandwidth, H' is the number of bits per cycle, and σ_p^2 is the variance of the phase error due to noise. When the 3-db passband of the system is made equal to the data bandwidth

$$B_L/B_d = 3\pi/2\sqrt{2} \tag{140}$$

As has been already discussed, the drop-out probability of the loop depends directly on this variance plus the transient error due to modulation. The transient error is approximately

$$2\Delta/\pi B_d \tag{141}$$

Sanders shows that the *lower bound* of this probability P_p is

$$P_p > \frac{\pi\{\Phi[-(\pi/2\sigma_p)(1 - 4\Delta/\pi^2 B_d)]\}^2}{(4/\sigma_p)(\Delta/B_d)\phi\{(\pi/2\sigma_p)[1 - (4/\pi^2)(\Delta/B_d)]\}} \quad (142)$$

where $\quad \Phi(z) = \mathrm{erf}(z) \quad$ and $\quad \phi(z) = (1/\sqrt{2}\,\pi)e^{-z^2/2} \quad (143)$

This is the probability of the total phase error exceeding $\pm\pi/2$ radians. The output signal-to-noise S_o/N_o ratio determines the information rate per sample H' per Table 5.

Sanders shows that

$$S_o/N_o = (1/\sigma_p^2)(\Delta/B_d)^2 \quad (144)$$

or the output S_o/N_o is determined entirely by the product of the transient and noise errors. β can now be determined for a given S_o/N_o or information rate per sample H' and drop-out probability P_p by first establishing σ_p and Δ/B_d from Eqs. (142) and (144). Then β is directly calculated from Eq. (139), which can be rewritten

$$\beta = (3\pi/4\sqrt{2})(1/\sigma_p^2 H') \quad (145)$$

For $P_p = 10^{-4}$, $H' = 5$, $S_o/N_o = 2{,}000$, and $\beta = 254$.

This large value is caused by the large product of σ and the transient error, signifying too much carrier power, and is due to the restriction on the peak allowable phase error. Sanders shows that if this phase error is allowed to vary over a wider range and if the post phase-locked-loop filtering [$F(s)$] is revised, β can be reduced to the order of 10. The threshold performance of the loop on the basis of Develet's quasi-linear analysis, which does not require a restriction on the phase errors and thus permits a truer assessment of the loop in the presence of noise, will be shown in Sec. 19.

18 Nonlinear Model—Noiseless Case

Linear-model analysis of the phase-locked loop does not give a complete understanding of the lock-on time of the loop, particularly when initial separation of the signal from loop center frequency is greater than 1 radian. Lock-on can be studied using phase-plane techniques[11] when the noise input can be considered negligible, or at least 10 db below signal power level. Then the exact model (Fig. 57) of the phase-locked loop may be used. The operational equation becomes

$$\phi = \theta_1 - [F(s)/s]AK \sin \phi \quad (146)$$

For the first-order loop the loop filter has no storage and $F(s)$ may be set equal to unity.

Equation (146) then becomes

$$s\phi = s\theta_1 - AK \sin \phi \quad (147)$$

which is the operational equivalent of the first-order differential equation

$$\dot{\phi} = d\phi/dt = \Omega - AK \sin \phi \quad (148)$$

where Ω, as before, represents the initial frequency offset and is equal to $\dot{\phi}_s - \omega_0$, a constant. Here $\dot{\phi}_s$ is the reference signal frequency and ω_0 is the VCO center frequency.

Figure 63 shows $d\phi/dt$ plotted against ϕ. If the VCO is initially at its center frequency $\dot{\phi}(0) = \Omega$ and $\phi(0) = \pm n\pi$ (where n is an integer). If the frequency error $\dot{\phi}$ is positive, the phase error ϕ tends to increase; ϕ decreases for negative $\dot{\phi}$. If n is an *even* integer and $\Omega < AK$ the system will travel along the sinusoidal trajectory of Fig. 63 until it reaches the ϕ axis at $\phi = \sin^{-1}(\Omega/AK) + n\pi$. This is a stable point; $\dot{\phi}$ cannot be negative because ϕ would then tend to decrease and return the system to the ϕ axis. If n is an odd integer, the system will go through a larger part of the sinusoidal trajectory until it reaches a stable point at $(n + 1)\pi + \sin^{-1}(\Omega/AK)$. If $\Omega > AK$, however, the ϕ trajectory never crosses the ϕ axis and a stable or phase-lock point is never reached.

From the above the maximum pull-in frequency range of the loop is

$$\Omega_{max} = AK \quad \text{radians/sec} \tag{149}$$

For the first-order loop AK is also equal to the closed-loop 3-db bandwidth. When phase lock does occur the steady-state phase error is

$$\phi_{ss} = \sin^{-1}(\Omega/AK) \tag{150}$$

Pull-in time is derivable from Eq. (148) in the following way:

$$dt/d\phi = 1/(\Omega - AK \sin \phi) \tag{151}$$

$$t = \int_{\phi \text{ initial}}^{\phi \text{ final}} \frac{d\phi}{\Omega - AK \sin \phi} \tag{152}$$

If ϕ final is taken as $\sin^{-1}(\Omega/AK) + n\pi$ the denominator becomes zero and the pull-in time becomes *infinite*. This is the theoretically correct value for pull-in to steady-state error for a first-order loop. A more realistic value of t can be obtained by using a slightly smaller value than the steady error and integrating Eq. (152).

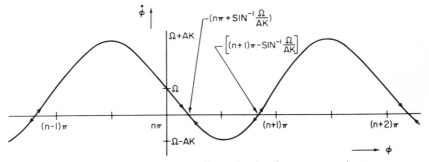

FIG. 63. First-order loop pull-in behavior (in even integer).

Summarizing, the first-order loop will pull-in within one cycle provided the frequency offset does not exceed Eq. (149) (or the closed-loop bandwidth); the lock-on time depends on the initial phase error.

The first-order loop is restricted in application because of its limited pull-in range and because it cannot be used to track linearly varying frequencies with narrow bandwidths. As previously discussed, the second-order loop is generally used for such purposes; its loop filter characteristic is

$$F(s) = (s + a)/s \tag{153}$$

This equation assumes the filter contains a perfect integrator. Equation (146) then becomes, in operational terms,

$$s^2\phi + (s + a)AK \sin \phi = s^2\theta_1 \tag{154}$$

or as a function of time

$$d^2\phi/dt^2 + AK[\cos \phi \, (d\phi/dt) + a \sin \phi] = d^2\theta_1/dt^2 \tag{155}$$

For a constant-frequency incoming (reference) signal the second derivative of its phase is zero. Equation (155) then becomes a homogeneous second-order nonlinear equation in ϕ. Making the substitutions,

$$\omega_n^2 = aAK \quad AK = 2\zeta\omega_n \tag{156}$$

which are similar to Eq. (107)]. Eq. (155) becomes

$$d^2\phi/dt^2 + 2\zeta\omega_n \cos \phi \, (d\phi/dt) + \omega^2 \sin \phi = 0 \tag{157}$$

The substitutions make the analysis compatible with linear servomechanism concepts. ω_n is the undamped natural frequency and ζ is the damping factor of such a servo loop. Equation (157) may be normalized on a time basis by making the substitution $t = \tau/2\zeta\omega_n$; this implies $d\phi/d\tau = (1/2\zeta\omega_n)(d\phi/dt)$. Then Eq. (157) becomes

$$d^2\phi/d\tau^2 + \cos\phi\,(d\phi/d\tau) + (1/4\zeta^2)\sin\phi = 0$$
$$\ddot\phi + \dot\phi\cos\phi + (1/4\zeta^2)\sin\phi = 0 \tag{158}$$

In this way ω_n has been eliminated as a parameter by normalization. Equation (158) is solvable graphically by phase-plane techniques developed for nonlinear servomechanism problems (see also Chap. 15). The normalized phase-plane plot, based on Eq. (158) and presented in Fig. 64, shows the phase error ϕ as the abscissa and its normalized derivative, the frequency error, as the ordinate. Figure 64 is shown for the case where $AK/a = 2$. Then from Eq. (156)

$$2\zeta\omega_n = 2a$$
$$\omega_n^2 = 2a^2$$
$$2\zeta(\sqrt{2}\,a) = 2a \tag{159}$$
$$\zeta = 1/\sqrt{2} = 0.707$$

The plot is actually periodic with respect to the abscissa; so only one period ($-\pi \leq \phi \leq \pi$) need be shown. The curves represent the system trajectories. That is, if initially there are given phase and frequency errors ϕ and $\dot\phi$, respectively, the system follows the trajectory starting at this point and proceeds to the right if $\dot\phi > 0$ and to the left if $\dot\phi < 0$. It is clear from Fig. 64 that the loop in question will always proceed to zero phase and frequency errors no matter what the initial errors may be. The stable point in the phase-plane plot is referred to as the phase-locked condition. However, if the initial frequency error is large, the system must pass through a trajectory of several cycles in order to pull into phase lock. Time can also be read from a phase-plane plot since, as before [Eq. (151)],

$$\dot\phi = d\phi/dt$$
$$t = \int_{\phi\,\text{initial}}^{\phi\,\text{final}} d\phi/\dot\phi \tag{160}$$

For large initial frequency errors the pull-in time is considerable. However, the preceding establishes that for the case where the incoming signal frequency is constant a second-order phase-locked loop will always pull in in the absence of noise. In other words the pull-in range is infinite provided the loop filter is a perfect integrator.

An approximate expression for the pull-in time has been derived by Viterbi.[11] It holds best for large initial frequency errors:

$$t_{lock} = \Omega^2/2\zeta\omega_n^3 \tag{161}$$

where both Ω and ω_n have the dimensions of radians per second. When $\zeta = 0.707$,

$$\omega_n = (4\sqrt{2}/3)B_L \tag{162}$$

from Eq. (112) and the pull-in time becomes

$$t_{lock} = 27\Omega^2/256 B_L^3 \tag{163}$$

The time required to achieve phase lock is proportional to the square of the initial frequency error and inversely proportional to the cube of the loop natural frequency or loop noise bandwidth. This shows clearly why a narrowband loop must have a small initial frequency error if it is to achieve lock in a reasonable time.

The dynamics of the loop in the case of a constant rate of change of reference frequency (corresponding to vehicle acceleration) is of extreme interest in tracking appli-

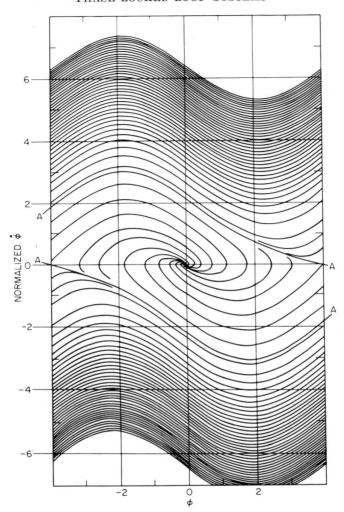

FIG. 64. Phase-plane plot of second-order loop—constant incoming (reference) signal frequency.

cations. Equation (157) becomes

$$d^2\phi/dt^2 + 2\zeta\omega_n \cos\phi(d\phi/dt) + \omega_n^2 \sin\phi = D \qquad (164)$$

where
$$D = (2\pi\alpha/c)f_0 = \Delta f_0 \qquad (165)$$

for Doppler situations where α is the acceleration, c is the velocity of light, and f_0 is the carrier frequency. D represents also the modulation slope of the modulation-following case. Normalized equation (159) becomes, in this case,

$$\ddot\phi + \dot\phi \cos\phi + (1/4\zeta^2)\sin\phi = D/4\zeta^2\omega_n^2 \qquad (166)$$

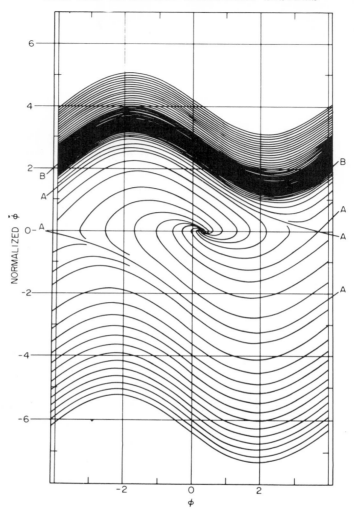

Fig. 65. Phase-plane plot, second-order phase-locked-loop, constant moderate rate of frequency change.

The same normalized phase-plane technique can be used as for the constant-frequency case. Figure 65 has been plotted for

$$\zeta = 1/\sqrt{2} \qquad D/\omega_n^2 = \tfrac{1}{4} \tag{167}$$

The abscissa represents phase error ϕ, and the ordinate as before represents frequency error divided by $2\zeta\omega_n$. The important features of the nonlinear behavior are seen in this plot. First of all, the behavior is periodic in ϕ, as before, and hence only one period need be shown. Second, the line B-B, commonly known as an unstable limit cycle, separates trackable and nontrackable signals. If the initial frequency and phase errors place the system above this line, the frequency error tends to increase and phase lock is never achieved. If the initial point is below the line, the system does achieve phase lock. However, the stable point does not occur at zero phase error as in Fig. 64 but rather at

$$\phi = \sin^{-1}(D/\omega_n^2) \qquad (168)$$

In this case the steady-state phase error is small, since

$$D/\omega_n^2 = \tfrac{1}{4} \qquad (169)$$

This confirms the analysis using the linear model. In Eq. (116) it was found that the steady-state phase error for a second-order loop tracking a phase

$$\theta_1 = \tfrac{1}{2} D t^2 \qquad (170)$$

is D/ω_n^2, which is a small overestimate of the correct value of $\sin^{-1}(D/\omega_n^2)$ for small values of the argument.

The physical situation is related to the phase-plane plot in the following manner: Positive initial-frequency errors put the initial state of the system in the upper or positive half plane. Since the Doppler frequency rate or modulation slope is positive, this means that the frequency to be tracked is moving linearly away from the initial VCO frequency; and if the initial-frequency error was too great, the loop cannot track the signal. On the other hand, if the initial-frequency error is negative, positive Doppler rate brings the frequency to be tracked toward the initial VCO frequency, making phase lock easier to attain. In this particular case, it is seen from Fig. 65 that if the initial-frequency error is negative, phase lock is assured.

However, when the Doppler rate is increased for a given loop so that the ratio D/ω_n^2 increases, the situation is less favorable. Figure 66 shows the case for

$$D/\omega_n^2 = \sqrt{3}/2 \qquad (171)$$

The steady-state error and hence the stable point are moved to

$$\sin^{-1}(\sqrt{3}/2) = \pi/3 \text{ radians} \qquad (172)$$

Also, *positive* initial-frequency errors usually render the signal untrackable. A difficulty is apparent in Fig. 66 also for negative initial-frequency errors. As the positive Doppler rate brings the frequency to be tracked toward the VCO frequency, it may occur that the state of the system enters a strip (A-A) in Fig. 66 from which it is swept into the positive-frequency-error region rather than into phase lock. Physically, this is explained by the fact that, if the Doppler rate sweeps the signal toward the VCO at too high a rate, the loop may be too sluggish or narrowband to catch and hold onto the signal and may let it slip by.

The important quantitative results which are obtained from the phase-plane analysis[11] are that the signal can never be tracked (even if the initial-frequency error is zero) when

$$D > \omega_n^2 \qquad (173)$$

This is explained by the fact that, when this condition holds, there is no stable point in the phase-plane plot. On the other hand, the signal can always be tracked if the initial-frequency error is negative or slightly positive provided

$$D < \omega_n^2/2 \qquad (174)$$

The average value of the unstable limit cycle of Fig. 65 which separates the positive initial-frequency errors which can be tracked from those which cannot is given approximately by

$$\Omega_{av} = \zeta \omega_n^3/D \qquad (175)$$

Linear analysis shows that a third-order loop should achieve better results in Doppler tracking, since the steady-state error is zero and the optimum linear synthesis pro-

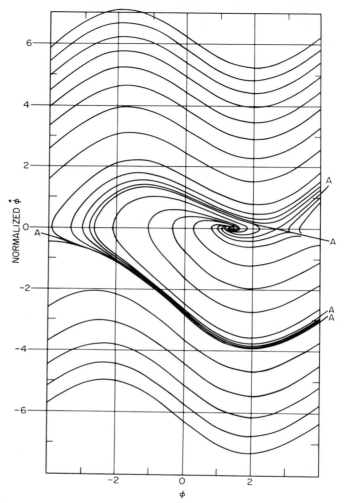

Fig. 66. Phase-plane plot, second-order loop, constant faster rate of frequency change than Fig. 65.

cedure arrives at such a configuration. This is brought out also by the nonlinear noiseless analysis, and the limits for stability can also be determined.

Swept-frequency Technique

When the receiver attempts to acquire the signal for the first time, the initial frequency may be known only approximately, say within 100 cps. If the loop bandwidth B_L is only 5 cps, the phase-lock time as given by Eq. (163) would be about 5.5 min, even in the absence of noise. This is an extreme amount of time for most applications, and in the presence of noise phase lock might never occur. In this case the technique often used is to sweep the VCO frequency at a constant rate over the possible range of the signal frequency. This rate appears in the system equation in the same way that the Doppler rate appears. Thus information on the maximum VCO sweep rate to be used to achieve phase lock can be obtained from the above results.

Imperfect Integrator

In the previous linear and nonlinear analyses of the *second*-order phase-locked loop, the filter function $F(s)$ was assumed to be

$$F(s) = (s + a)/s$$

see Eqs. (105) and (153). If a passive RC filter is used or if the integrator in the loop is not perfect, the loop filter transfer function becomes

$$F(s) = \frac{s + a}{s + \alpha} \tag{176}$$

This case has been analyzed by Gruen.[2] Use of Eq. (176) in the loop operational equation 94 and using the normalizing process of Eq. (156) will result in the parallel expression to Eq. (157):

$$\ddot{\phi} + (\alpha/2\zeta\omega_n + \cos\phi)\dot{\phi} + (1/4\zeta^2)\sin\phi = \alpha\Omega/4\zeta^2\omega_n^2 \tag{177}$$

Here $\Omega = \dot{\phi} - \omega_0$ as before is the frequency of offset. This is the expression for the case of a *constant signal frequency* offset. Phase-plane analysis of Eq. (177) shows that for

$$\Omega > \omega_n^2/\alpha \tag{178}$$

there is no stable point and the system cannot achieve phase lock. This is the upper bound on the initial-frequency error. However, for lesser Ω, pull-in also may not occur under certain conditions, as shown by Viterbi.[11] By analog-computer plots he determined that if the VCO is initially at ω_0, for $\zeta = 0.707$ and $\alpha/2\zeta\omega_n = 0.1$ the limit for pull-in lies in the region

$$3\sqrt{2}\,(\Omega/\omega_n) < \tfrac{7}{2}\sqrt{2} \tag{179}$$

An approximate analytic expression for the pull-in range was also derived by Viterbi. It is

$$\Omega < 2\omega_n\sqrt{\zeta\omega_n/\alpha + 1} \tag{180}$$

The approximation depends on the assumption that the frequency error $d\phi/dt$ is large compared with ω_n.

19 Nonlinear Model—Noisy Signals

Solution of the phase-locked loop with noise and for phase errors greater than 1 radian requires the use of nonlinear analytical techniques. Three approaches to this problem have been made:

1. The use of Fokker-Planck diffusion equations to obtain an exact solution[15]
2. Obtaining an approximate solution by Volterra functional-expansion technique[16]
3. An approximation approach using Booton's quasi-linearization method[14]

The first method, first employed for phase-locked loops by Tikhonov,[18] and discussed in detail by Viterbi,[15] provides an exact solution to the first-order loop, but only approximate expressions have been obtained for the stationary probability density of the second-order loop. The Fokker-Planck equations for higher-order loops cannot yet be solved nor can the expression for a modulated signal being followed by any order loop. Approximate expressions, to any desired accuracy, may be obtained by the second approach. In this method the sinusoidal nonlinearity in the error channel is replaced by a series expansion and a series of functional equations are solved by the Volterra functional calculus. Expansions much above the fifth power result in considerable complexity of the expressions to be used, however. Van Trees[16] has applied the method to the phase-locked loop, both modulated and unmodulated, and to first- and second-order loops.

The third method, developed by Develet,[14] approximates loop operations by replacing the sinusoidal nonlinearity with statistical gain under the assumption that the input signal is approximately *Gaussian* statistical in nature. This method remains simple for all loop filters and even for modulated signals. It is particularly useful in establishing the threshold value of input signal-to-noise for modulated signals.

Fokker-Planck Method

Figure 67 shows a form of the block diagram of the phase-locked loop useful for analysis of the nonlinear case with noise, which was first used by Develet. The phase error can be derived in a similar manner to Eq. (146) and is, in operational form,

$$\phi(t) = \theta_1(t) - [KF(s)/s][A \sin \phi + n'(t)] \tag{181}$$

As before, the double carrier-frequency terms have been assumed to be completely eliminated by the combination of the filter $F(s)$ and the VCO. It can be shown[15] that the statistics of the noise $n'(t)$ are the same as the original noise $n(t)$ regardless of the nature of the input signal.

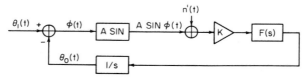

FIG. 67. Model of the phase-locked loop.

In the Fokker-Planck approach the phase error is considered a Markov process. A Markov process is one which can be completely described by the statistical parameters of the incremental change of position only as a function of the present position. For the first-order loop, the only one which has been solved by this method, Eq. (181) becomes

$$\dot{\phi}(t) = (\omega - \omega_0) - K[A \sin \phi(t) + n'(t)] \tag{182}$$

The incremental phase is, considering that $\phi(t)$ is a continuous process,

$$\Delta\phi = \int_t^{t+\Delta t} \dot{\phi}(t)\, dt = (\omega - \omega_0)\, \Delta t - (AK \sin \phi)\, \Delta t - K \int_t^{t+\Delta t} n'(u)\, du \tag{183}$$

For a given position ϕ, ϕ is a Gaussian variable with mean

$$\overline{\Delta\phi} = [(\omega - \omega_0) - AK \sin \phi]\, \Delta t \tag{184}$$

and variance

$$\sigma_{\Delta\phi}^2 = \overline{(\Delta\phi)^2} - \overline{(\Delta\phi)}^2$$
$$= K^2(N_0/2)\, \Delta t \tag{185}$$

With the knowledge of the statistical parameters of the increment we may proceed to obtain $p(\phi,t)$. It was shown by Uhlenbeck and Ornstein[19] that for a continuous Markov process described by a first-order differential equation with a white Gaussian input, the instantaneous probability density $p(\phi,t)$ must satisfy the partial differential equation

$$\partial p(\phi,t)/\partial t = -(\partial/\partial\phi)[A(\phi)p(\phi,t)] + \tfrac{1}{2}(\partial^2/\partial\phi^2)[B(\phi)p(\phi,t)] \tag{186}$$

with the appropriate initial condition, where

$$A(\phi) = \lim_{\Delta t \to 0} (1/\Delta t)\, \overline{\Delta\phi}$$

$$B(\phi) = \lim_{\Delta t \to 0} (1/\Delta t)\, \overline{(\Delta\phi)^2}$$

provided
$$\lim_{\Delta t \to 0} (1/\Delta t)\overline{(\Delta \phi)^n} = 0 \quad \text{for } n > 2$$

Equation (186) is known as the Fokker-Planck equation or the diffusion equation because it is a generalization of the equation for heat diffusion. From (184) and (185) we obtain for the first-order loop

$$A(\phi) = (\omega - \omega_0) - AK \sin \phi$$
$$B(\phi) = K^2 N_0/2$$

and inserting the coefficients into (186) we obtain

$$\partial p/\partial t = (\partial/\partial \phi)[(AK \sin \phi + \omega_0 - \omega)p] + (K^2 N_0/4)(\partial^2 p/\partial \phi^2) \quad (187)$$

Viterbi solves this equation with two simplifications. First he recognizes that the probability $p(\phi,t)$ may be replaced by probability

$$P(\phi,t) = \sum_{n=-\infty}^{\infty} P(\phi + 2\pi n, t) \quad (188)$$

to take advantage of the periodic nature of ϕ. The form of Eq. (187) is essentially unchanged and may be solved after the interval of one period of $\phi(-\pi \leq \phi \leq \pi)$ with the initial condition

$$P(\phi,0) = \delta(\phi - \phi_0)$$
$$-\pi \leq \phi \leq \pi \quad (189)$$

the boundary condition

$$P(\pi,t) = P(-\pi,t) \quad \text{for all } t \quad (190)$$

and the normalizing condition

$$\int_{-\pi}^{\pi} P(\phi,t) = d\phi = 1 \quad \text{for all } t \quad (191)$$

The other simplification is that only the steady-state value of P is obtained

$$P(\phi) = \lim_{t \to \infty} P(\phi,t) \quad (192)$$

By definition, the steady-state distribution is stationary.

In this case the partial differential equation becomes an ordinary differential equation. In the case when $\omega = \omega_0$, that is, when the frequency of the received signal is determined beforehand and the VCO quiescent frequency is tuned to this frequency so that the problem consists only of acquiring and tracking phase

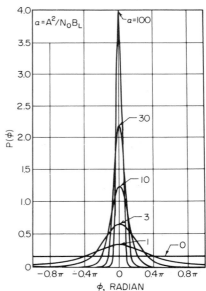

Fig. 68. First-order loop steady-state probability densities for $\omega = \omega_0$.

$$P(\phi) = \frac{\exp(\alpha \cos \phi)}{2\pi I_0(\alpha)} \quad -\pi < \phi < \pi \quad (193)$$

where
$$\alpha = \frac{4A}{KN_0} = \frac{A^2}{N_0(AK/4)}$$
$$= A^2/N_0 B_L \quad (194)$$

where A^2 is the received signal power and B_L, loop noise bandwidth. α is actually the S/N ratio in the bandwidth of the loop.

The probability of Eq. (193) is plotted in Fig. 68. For large α (high S/N ratio)

$P(\phi)$ approaches Gaussian, which is the result which would be obtained for the linear model.

The cumulative steady-state probability distribution

$$\text{Prob } (|\phi| < \phi_1) = \int_{-\phi_1}^{\phi_1} P(\phi) \, d\phi \qquad 0 < \phi_1 < \pi \tag{195}$$

indicates the percentage of time during which the absolute value of the loop phase error ϕ is less than a given magnitude ϕ_1. For the case where $\omega = \omega_0$ it is

$$\text{Prob } (|\phi| < \phi_1) = 2 \int_0^{\phi_1} P(\phi) d(\phi)$$

$$= \frac{\phi_1}{\pi} + \frac{2}{\pi} \sum_{n=1}^{\infty} \frac{I_n(\alpha) \sin n\phi_1}{n I_0(\alpha)} \tag{196}$$

for $\qquad 0 < \phi_1 < \pi \quad$ and $\quad \omega = \omega_0$

This is shown in Fig. 69.

The variance of ϕ is

$$(\sigma_\phi)^2 = \int_{-\pi}^{\pi} \phi^2 \exp (\alpha \cos \phi) \, d\phi / 2\pi I_0(\alpha)$$

$$= [1/2\pi I_0(\alpha)] \int_{-\pi}^{\pi} \phi^2 \left[I_0(\alpha) + 2 \sum_{n=1}^{\infty} I_n(\alpha) \cos n\phi \right] d\phi$$

$$= \frac{\pi^2}{3} + 4 \sum_{n=1}^{\infty} \frac{(-1)^n I_n(\alpha)}{n^2 I_0(\alpha)} \tag{197}$$

The variance obtained from the linear model is simply $1/\alpha$. Both these variances are plotted in Fig. 70.

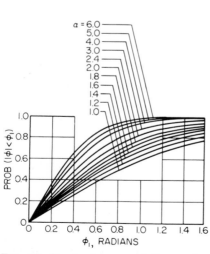

Fig. 69. Steady-state cumulative probability distributions of first-order loop for $\omega = \omega_0$.

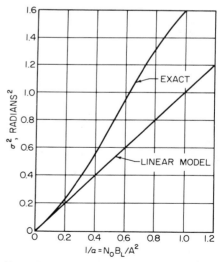

Fig. 70. Variance of phase error for first-order loop where $\omega = \omega_0$.

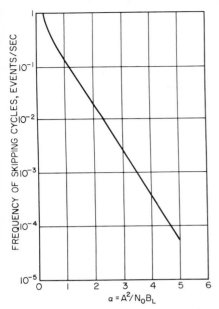

FIG. 71. Frequency of skipping cycles normalized by loop bandwidth for first-order loop where $\omega = \omega_0$.

Viterbi also derived an expression from the frequency of skipping full cycles for the first-order loop. This parameter affects the accuracy of range (distance) measurement derived from integrating Doppler frequencies.

Frequency of skipping cycles
$$= (2B_L)/\pi^2 \alpha I_0^2(\alpha) \quad (198)$$

This parameter normalized by B_L is shown as a function of α in Fig. 71. For large signal-to-noise ratio α,

$$I_0(\alpha) \sim (e^\alpha)/(2\pi\alpha)^{1/2}$$

so that for large α,

Frequency of slipping cycles
$$\simeq [(4B_L)/\pi] e^{-2\alpha} \quad (199)$$

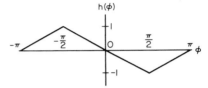

FIG. 72. Triangular error function.

In the second-order loop $F(s) = (s + a)/s$ for the case of greatest interest, where a is the constant of the integrator. The expression for steady-state probability density of ϕ, $P(\phi)$, can be generalized to the case of the second-order loop for large α thus

$$P(\phi) \simeq \frac{\exp(\alpha' \cos \phi)}{2\pi I_0(\alpha')} \quad \text{for large } \alpha' \quad (200)$$

where the effective signal-to-noise ratio α' is given by

$$\alpha' = (A^2)/[N_0(AK + a)/(4)] \quad (201)$$

If we let $B_L = (AK + a)/4$ this is the same expression as that for the first-order loop with $\omega = \omega_0$. As would be expected, this expression for loop bandwidth for the second-order loop is the same as that obtained from the linear model of the loop.

On the other hand, if $a \ll AK$ the expression (200) becomes the same as for the first-order loop equation (193).

The results obtained for the sinusoidal signals can be generalized for square-wave signals. Such a system is the delay-lock tracker[21,22] in which a comparison is made of coded binary input signals with the contents of a linear shift register.

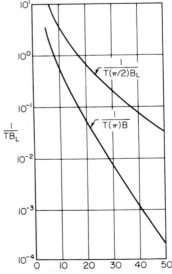

FIG. 73. Inverse-mean times to loss of lock and to first passage from linear region (normalized by loop bandwidth).

The triangular-shaped error signal which results in these cases is shown in Fig. 72. The times to loss of lock $T(\pi)$ and of first passage from the linear region $T(\pi/2)$ are of special importance in this case because there is no restoring force as in the sinusoidal case after exceeding these points. These times are plotted in Fig. 73.

Volterra Functional Technique

Van Trees[16] has applied the Volterra functional-expansion technique to the analysis of the nonlinear phase-locked loop. Such functionals are generalizations of the con-

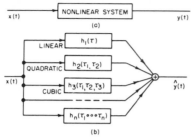

FIG. 74. Functional expansion of nonlinear system.

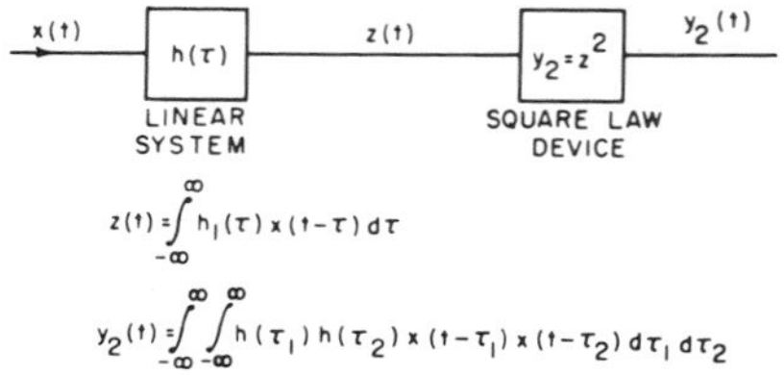

FIG. 75. A simple quadratic system.

volution integral used in linear-system analysis. As shown in Fig. 74, the nonlinear system at a may be represented by convolution integrals of several orders summed together as shown in b. If the system consists of a linear system and a square-law device the resulting expressions are shown in Fig. 75. The two-dimensional *kernel* which results is

$$h_2(\tau_1, \tau_2) = h(\tau_1) h(\tau_2) \tag{202}$$

If $h(\tau)$ were a simple RC filter the kernel is shown in Fig. 76. In the linear case the output is

$$y_1(t) = \int_{-\infty}^{0} h_1(\tau) x(t - \tau)\, d\tau = \int_{-\infty}^{\infty} h_1(t - \tau) x(\tau)\, d\tau \tag{203}$$

In the cubic case

$$y_3(t) = \int_{-\infty}^{\infty} \int_{-\infty}^{\infty} \int_{-\infty}^{\infty} h_3(\tau_1, \tau_2, \tau_3) x(t - \tau_1) x(t - \tau_2) \tag{204}$$

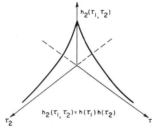

FIG. 76. A two-dimensional kernel. (*From Van Trees*.[16])

The total output $\hat{y}(t)$ is, in general, an infinite sum of $y_i(t)$:

$$\hat{y}(t) = \sum_{i=1}^{\infty} y_i(t) \tag{205}$$

Application of the method to phase-locked loops may be understood in connection with the model shown in Fig. 67, but consider the filter function $F(s) = 1$ (the first-order loop). The differential equation is

$$d\phi(t)/dt + AK \sin \phi(t) = \theta_1(t) - AKn(t) \equiv X(t) \tag{206}$$

where $X(t)$ is defined as an equivalent input. The behavior of the error signal $\phi(t)$ is that which must be determined. Let

$$\phi(t) = \phi_1(t) + \phi_2(t) + \cdots = \sum_{i=1}^{\infty} \phi_i(t) \tag{207}$$

Here $\phi_1(t)$ is the output of a linear system as shown in Fig. 74, $\phi_2(t)$ that of a second-order (quadric system) having kernel $h_2(\tau_1,\tau_2)$, and so forth. The analytic form of the various kernels is found by substituting Eq. (207) into Eq. (206) and sorting the terms according to the order in which they involve $X(t)$. Expanding $\sin \phi(t)$ and performing the substitution gives

$$[\phi_1(t) + \phi_2(t) + \phi_3(t) + \cdots] + AK\{(\phi_1(t) + \phi_2(t) + \phi_3(t) + \cdots)$$
$$- \frac{1}{3!}[\phi_1(t) + \phi_2(t) + \phi_3(t) + \cdots]^3 + \frac{1}{5!}[\phi_1(t) + \phi_2(t) + \phi_3(t) + \cdots]^5$$
$$+ \cdots \} = x(t) \quad (208)$$

All terms of equal order in Eq. (208) are then equated, giving the following set of equations, keeping in mind that $x(t)$ is itself first-order, being an input.

(a) $\dot{\phi}_1(t) + AK\phi_1(t) = x(t)$
(b) $\dot{\phi}_2(t) + AK\phi_2(t) = 0$
(c) $\dot{\phi}_3(t) + AK\phi_3(t) = (AK/3)\phi_1^3(t)$ \quad (209)
(d) $\dot{\phi}_4(t) + AK\phi_4(t) = (AK/2)\phi_1^2(t)\phi_2(t)$
(e) $\dot{\phi}_5(t) + AK\phi_5(t) = (AK/2)\phi_1^2(t) - (AK/5)\phi_1^5(t)$

These equations may be solved in sequence, the *first*-order equation being solved first and its solution used in the solution of the higher-order equations. The first-order solution is

$$\phi_1(t) = \int_0^\infty e^{-AK\tau} x(t-\tau)\, d\tau$$
$$= \int_0^\infty h_1(\tau) x(t-\tau)\, d\tau \quad (210)$$

where we define

$$h_1(\tau) \equiv e^{-AK\tau} \quad \tau \geq 0$$
$$\equiv 0 \quad \tau < 0 \quad (211)$$

The *second*-order equation solution must be

$$\phi_2(t) = 0 \quad (212)$$

Hence all higher-order *even* terms are zero. The *third*-order solution of Eq. (209c) is

$$\phi_3(t) = \int_0^\infty h_1(\tau)(AK/3!)\phi_1^3(t-\tau)\, d\tau \quad (213)$$

To write $\phi_3(t)$, as a function of $x(t)$, Eq. (210) is substituted for $\phi_1(t)$ in the last expression; this yields

$$\phi_3(t) = \int_0^\infty d\tau \int_0^\infty d\tau_1 \int_0^\infty d\tau_2 \int_0^\infty d\tau_3 [(AK/3!) h_1(\tau) h_1(\tau_1) h_1(\tau_2) h_1(\tau_3) x(t-\tau-\tau_1)$$
$$x(t-\tau-\tau_2) x(t-\tau-\tau_3)] \quad (214)$$

In this form, it is easy to see that $\phi_3(t)$ depends in a third-order manner on $x(t)$. From Eq. (206), we know that we want a third-order relationship of the form

$$\phi_3(t) = \int_0^\infty d\tau_1 \int_0^\infty d\tau_2 \int_0^\infty d\tau_3 h_3(\tau_1,\tau_2,\tau_3) x(t-\tau_1) x(t-\tau_2) x(t-\tau_3) \quad (215)$$

Similarly,

$$\phi_5(t) = \int_0^\infty h_1(\tau)[(AK/2)\phi_1^2(t-\tau)\phi_3(t-\tau) - (AK/5!)\phi_1^5(t-\tau)]\, d\tau \quad (216)$$

Expressing in terms of $x(t)$ only, we have

$$\phi_5(t) = (A^2K^2/12) \int_0^\infty d\tau_1 \int_0^\infty d\tau_2 \int_0^\infty \cdots \int_0^\infty d\tau_7 [h_1(\tau_1)h_1(\tau_2) \cdots h_1(\tau_7)]$$
$$x(t - \tau_1 - \tau_2)x(t - \tau_1 - \tau_3)x(t - \tau_1 - \tau_4 - \tau_5)x(t - \tau_1 - \tau_4 - \tau_6)$$
$$x(t - \tau_1 - \tau_4 - \tau_7) + (AK/5!) \int_0^\infty d\tau_1 \int_0^\infty d\tau_2 \cdots \int_0^\infty d\tau_6 h_1(\tau_1) \cdots h_1(\tau_6)$$
$$x(t - \tau_1 - \tau_2)x(t - \tau_1 - \tau_3)x(t - \tau_1 - \tau_4)x(t - \tau_1 - \tau_5)x(t - _1 - \tau_6)] \quad (217)$$

Once again, the fifth-order relationship between $\phi_5(t)$ and $x(t)$ is clear. Equation (217) is equivalent to the form

$$\phi_5(t) = \int_0^\infty d\tau_1 \int_0^\infty d\tau_2 \cdots \int_0^\infty d\tau_5 h_5(\tau_1, \tau_2, \tau_3, \tau_4, \tau_5,)$$
$$x(t - \tau_1)x(t - \tau_2)x(t - \tau_3)x(t - \tau_4)x(t - \tau_5) \quad (218)$$

Higher-order kernels can be found in a similar manner. Observe that the kernels do not depend on the nature of the input. Thus once the functional relations are derived they are a property of the system. Now assume the series in Eq. (207) converges. We have an explicit representation for $\phi(t)$.

The convergence problem is not trivial. However, most convergence proofs[23] are involved and give conservative regions of convergence. Our primary concern is how well a small number of kernels approximate the actual system behavior. Evaluation of the output of a kernel of higher order than five is too tedious to be of any practical value. It does not appear possible to obtain a closed-form solution in any very interesting cases. In general, the functional solution should be regarded as a method of obtaining approximate answers over a reasonable range of signal and noise levels.

For the above reasons only approximate representations of $\phi(t)$ are used, the most exact one used being fifth-order, thus

$$\phi(t) \cong \phi_{(5)}(t) = \phi_1(t) + \phi_3(t) + \phi_5(t) \quad (219)$$

Case I: First-order Loop; Constant-frequency Input

The parameter of greatest interest in phase-locked loops is the variance

$$\sigma^2 = \langle \phi^2(t) \rangle$$

which Van Trees approximates by $\langle \phi^2_{(5)}(t) \rangle$ in the case of the first-order loop. Consider the case of a constant-frequency input, the frequency of which is the same as the VCO center frequency; then $\theta_1(t) = 0$ and

$$x(t) = \theta(t) - AKn(t) = AKn(t) \quad (220)$$

Let $n(t)$ be a sample function from a white Gaussian random process with a correlation function

$$R_N(\tau) = (N_0/2A^2)\delta(\tau) \quad (221)$$

The constant in the correlation function corresponds to a phase-locked loop whose received signal is a sine wave of rms value A corrupted by white Gaussian noise with a double-sided spectral height $N_0/2$.

$x(t)$ of Eq. (220) is used in Eqs. (210), (214), and (216) to represent $\phi_1(t)$, $\phi_3(t)$, and $\phi_5(t)$, respectively. The variance $\langle \phi^2_{(5)}(t) \rangle$ is determined from the double sum

$$\langle \phi_{(5)}^2(t) \rangle = \sum_{i=1,3,5} \sum_{j=1,3,5} \langle \phi_i(t) \phi_j(t) \rangle \quad (222)$$

This results in the need to evaluate terms such as $\langle \phi_i(t) \rangle$, $\langle \phi_i(t) \phi_3(t) \rangle$, $\langle \phi_5^2(t) \rangle$, etc. When this is done Eq. (221) becomes important. Van Trees (Ref. 16, Appendix I) has performed these evaluations.

A fundamental quantity in the solution is

$$KN_0/4A = 1/\alpha = N_0 B_L/A^2 \quad (223)$$

which is the inverse of the loop signal-to-noise ratio previously given in Eq. (194). If we retain only terms in our answer which are third power or less in $1/\alpha$, we have

$$\langle \phi^2{}_{(5)}(t) \rangle \cong 1/\alpha + \tfrac{1}{2}(1/\alpha)^2 + \tfrac{13}{24}(1/\alpha)^3 \quad (224)$$

The variance as a function of $1/\alpha$ is plotted in Fig. 77 as "Van Trees's Model." Comparisons with the results of the other methods and the linear model which are also shown in this figure are discussed in a later paragraph.

Techniques have been developed for expediting the evaluation of Eqs. (210), (214), (216), etc., which resemble the Laplace-transform method used in linear-system analysis. These are summarized in Ref. 23 and are worked out in more detail in George.[24] For example, the first-order kernel transform is $H_1(s) = 1/(S + AK)$, corresponding to (210). The reader is referred to these sources for the development of the functional transform methods.

Case II: Second-order Loop, Constant-frequency Input

The method has also been applied by Van Trees to the case of the *second-order* loop, *constant-frequency* input. The variances which result are

Linear Approximation

$$\sigma^2 = \langle \phi_1{}^2(t) \rangle = 1/\alpha \quad (225)$$

Second-order Approximation

$$\sigma^2 = \langle \phi^2{}_{(3)}(t) \rangle = 1/\alpha + \tfrac{2}{3}(1/\alpha)^2 \quad (226)$$

where α has the same meaning as in Eq. (223) and is the coherent S/N ratio in the loop bandwidth. The loop bandwidth B_L, however, is modified by the presence of the loop filter $F(s)$. Figure 77 presents the results of Eqs. (225) and (226).

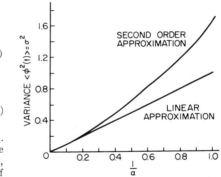

FIG. 77. Mean-square error in second-order loop.

Case III: First-order Loop, Random-frequency Modulation

In this case $\theta_1(t) \neq 0$, as in the previous cases, and $x(t)$ becomes

$$x(t) = \theta(t) + G(p)n(t) \quad (227)$$

$x(t)$ can be regarded as being made up of two independent nondeterministic Gaussian processes. Then

$$x(t) = x_1(t) + x_2(t) \quad (228)$$

Both these terms can be considered to have originated from independent white Gaussian sources $Z_1(t)$ and $Z_2(t)$ having equal correlation functions

$$R_{z_1}(\tau) = R_{z_2}(\tau) = \delta(\tau) \quad (229)$$

$x_1(t)$, the signal term, is derived from $Z_1(t)$ through a linear filtering process which shapes the spectrum according to the message. $x_1(t)$ becomes nonwhite, though still Gaussian. $x_2(t)$ remains white Gaussian, although its amplitude varies independently of $x_1(t)$. The modulation process need only be characterized by its correlation function, which is

$$R_{\theta_1}(\tau) = a^2 E^{-\omega_m(t)} \qquad (230)$$

and the signal spectral power is

$$S_{\theta_1}(j\omega) = a^2 \frac{2\omega_m}{\omega^2 + \omega_m^2} \qquad (231)$$

where a^2 is the mean-square amplitude of instantaneous frequency and ω_m is the effective bandwidth of the modulation process.

The linear approximation of the variance is

$$\langle \phi_1^2(t) \rangle = \mathfrak{P} \frac{1}{r(1+r)} + \mathfrak{Q}\tau \qquad (232)$$

where we have defined

$\mathfrak{P} = a^2/\omega_m^2$: $\dfrac{\text{mean-square value of instantaneous frequency}}{\text{square of effective bandwidth of process}}$

$\mathfrak{Q} = (N_0/2A^2)(\omega_m/2)$: coherent noise-to-signal ratio in modulation bandwidth

$r = B_L/\omega_m$: ratio of loop bandwidth to modulation-process bandwidth

$\mathfrak{P}$ corresponds to the square of the modulation index. r_{min}, the value of r which minimizes $\langle \phi_1^2(t) \rangle$, is only a function of $\mathfrak{P}/\mathfrak{Q}$ and is given in Fig. 78.

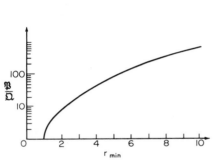

FIG. 78. Ratio of bandwidths for minimum error (linear approximation).

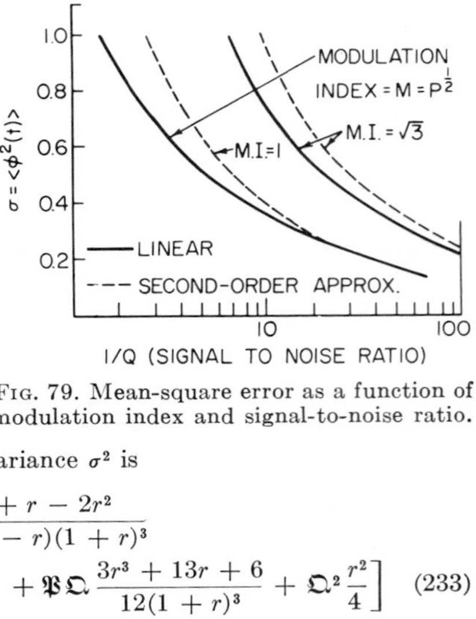

FIG. 79. Mean-square error as a function of modulation index and signal-to-noise ratio.

The second-order approximation to the variance σ^2 is

$$\langle \phi_{(2)}^2(t) \rangle = \mathfrak{P} \frac{1}{r(1+r)} + \mathfrak{Q}r + 2 \left[\mathfrak{P}^2 \frac{1 + r - 2r^2}{4r^2(1-r)(1+r)^3} \right.$$
$$\left. + \mathfrak{P}\mathfrak{Q} \frac{3r^3 + 13r + 6}{12(1+r)^3} + \mathfrak{Q}^2 \frac{r^2}{4} \right] \qquad (233)$$

or $\quad \langle \phi_{(2)}^2(t) \rangle = \mathfrak{Q} \left[\dfrac{\mathfrak{P}}{\mathfrak{Q}} \dfrac{1}{r(1+r)} + r \right] + 2\mathfrak{Q}^2 \left[\dfrac{\mathfrak{P}^2}{\mathfrak{Q}^2} f_1(r) + \dfrac{\mathfrak{P}}{\mathfrak{Q}} f_2(r) + \dfrac{r^2}{4} \right] \qquad (234)$

This can be interpreted most conveniently by assuming that we use $r = r_{min}$ from Fig. 78.

Both the linear and second-order approximation to the variance are given in Fig. 79 as a function of $\mathfrak{Q}$.

Equivalent-gain Model

Referring back to the model of the phase-locked loop in Fig. 67, it is possible to replace the nonlinear element $A \sin [\ ]$ with an equivalent gain K_A. Such a procedure for handling nonlinear control systems was developed by Booton[20,25] and has been applied to analysis of the nonlinear phase-locked loop by Develet.[14] This technique essentially determines the average gain of the nonlinear device under the expected operating conditions. K_A may be found by utilizing an averaging procedure which is a slight variation of Booton's equation (25) thus:

$$K_A = \int_{-\infty}^{\infty} g'(x) p_1(x) \, dx \qquad (235)$$

where K_A = equivalent element gain
$g'(x) = A \cos x$
$p_1(x)$ = probability density of $\phi(t)$, which must be Gaussian to conform to Booton's criteria

Let $\langle \phi^2(t) \rangle = \sigma^2$. Substituting this into Eq. (235) and integrating yields

$$K_A = A \exp(-\sigma^2/2) \qquad (236)$$

The quasi-linear receiver representation obtained by linearizing the $A \sin [\ ]$ transfer function is now shown in Fig. 80.

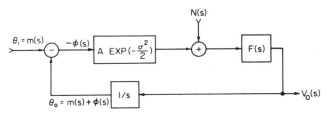

FIG. 80. Quasi-linear phase-locked receiver.

Denoting the signal and noise one-sided power spectral densities of $m(s)$ and $N(s)$ as $\Phi_m(\omega)$ radians^2/cps and $\Phi_n(\omega)$ watts/cps, respectively, it is a simple matter to show

$$\sigma^2 = \overbrace{\int_0^\infty \Phi_m(\omega) \left| 1 - \frac{\theta_0}{\theta_1}(\omega) \right|^2 df}^{\text{Modulation error}} + \overbrace{\int_0^\infty \frac{\exp(\sigma^2) \Phi_n(\omega)}{2S} \left| \frac{\theta_0}{\theta_1}(\omega) \right|^2 df}^{\text{Noise error}} \qquad (237)$$

where
$$\frac{\theta_0}{\theta_1}(\omega) = \frac{A \exp -(\sigma^2/2) F(s)/s}{1 + A \exp(-\sigma^2/2) F(s)/s} \qquad (238)$$

and S is the received signal power $A^2/2$. Equation (237) implies a threshold criterion for the receiver model of Fig. 80 as a consequence of the $\exp[\ ]$ term. The received signal power can be determined from Eq. (237) as follows:

$$S = \frac{(\exp \sigma^2/2) \int_0^\infty \Phi_n(\omega)(\theta_0/\theta_1)(\omega) \Big|^2 df}{\sigma^2 - \int_0^\infty \Phi_m(\omega)|1 - (\theta_0/\theta_1)(\omega)|^2 df} \qquad (239)$$

The maximum receiver sensitivity is established when the loop is adjusted for minimum S. This is done by optimizing $(\theta_0/\theta_1)(\omega)$ for specific σ, $\Phi_m(\omega)$ and $\Phi_n(\omega)$. Let

$$\theta_0/\theta_1(\omega) = A(\omega) E^{j\phi(\omega)} \qquad (240)$$

For minimum S, $j\phi(\omega)$ should be zero. Then it can be shown that maximum receiver sensitivity occurs for

$$A(\omega)\Big|_{opt} = \frac{\Phi_m(\omega)}{\Phi_m(\omega) + \dfrac{\Phi_n(\omega) \exp(\sigma^2)}{2S_{min}}} \qquad (241)$$

The same optimization occurs for minimizing σ^2. This means that the same method of optimizing the loop transfer function as employed by Wiener for *linear servosystems* may be employed for the *nonlinear* case, except that the noise power spectral density is replaced by

$$\frac{\Phi_n(\omega) \exp(\sigma^2)}{2S_{min}} \qquad (242)$$

Then the Wiener-Hopf solutions may be directly extended to this nonlinear problem. The expressions so far presented may be employed for any kind of noise input. However, if the noise spectrum is white Gaussian a particularly useful application of linear theory can be made; that is, the expression for the optimum realizable transfer function derived by Yovits and Jackson[26] can be used. For white Gaussian noise,

$$\Phi_n(\omega) = 2\Phi_i \qquad (243)$$

where Φ_i is the one-sided predetection power spectral density of the receiver. Equation (242) becomes

$$\frac{\Phi_i \exp(\sigma^2)}{S_i} \qquad (244)$$

The optimum linear realizable filter referred to above[26] is

$$\left| 1 - \frac{\theta_0}{\theta_1}(\omega) \right|^2_{opt} = \frac{\Phi_n}{\Phi_n + \Phi_m(\omega)} \qquad (245)$$

The corresponding minimum following error is

$$\sigma^2_{min} = \Phi_n \int_0^\infty \log_\epsilon \left[1 + \frac{\Phi_m(\omega)}{\Phi_n} \right] df \qquad (246)$$

For the quasi-linear case Φ_n is replaced by Eq. (242), as explained above. The following fundamental relations are thereby obtained:

$$\left| 1 - \frac{\theta_0}{\theta_1}(\omega) \right|^2_{opt} = \frac{\Phi_i \exp(\sigma^2)}{\Phi_i \exp(\sigma^2) + S\Phi_m(\omega)} \qquad (247)$$

$$\sigma^2_{min} = \frac{\Phi_i \exp(\sigma^2)}{S} \int_0^\infty \log_\epsilon \left[1 + \frac{S\Phi_m(\omega)}{\Phi_i \exp(\sigma^2)} \right] df \qquad (248)$$

This result shows in a very general way that for a fixed output variance the input signal power S can become too small relative to noise density to give a useful output. The value of S below which the solution for σ^2_{min} ceases to exist is the threshold for phase-lock demodulation obtained from the quasi-linear model.

The result of Eq. (248) enables one to calculate the variance of the loop when θ contains complex modulation Φ_m for both optimum transfer function and any desired filter characteristic. This is a decided advantage over the other approaches to nonlinear-loop analysis. In particular, Develet has analyzed the following cases:

1. Reception of a band-limited phase-encoded white Gaussian signal spectrum with the *optimal* transfer function
2. Reception of a band-limited phase-encoded white Gaussian signal spectrum with a *second-order* transfer function

The first case is important to illustrate optimal communication of information by use of a phase-locked receiver.

The second situation is of practical interest since a second-order loop is easily realized and is amenable to measurements verifying the theory which has been developed.

Optimal Receiver; Band-limited Phase-encoded White Gaussian Signal

The signal power spectral density is be given by

$$\Phi_m(\omega) = \Phi_m \qquad 0 \leq f \leq f_m \qquad (249)$$
$$\Phi_m(\omega) = 0 \qquad f_m < f$$

Integration of (248) yields

$$\sigma^2 = \frac{\Phi_i \exp(\sigma^2)}{S} f_m \log_\epsilon \left| 1 + \frac{\Phi_m S}{\Phi_i \exp(\sigma^2)} \right| \qquad (250)$$

Since $\Phi_i \exp(\sigma^2/S)$ is the one-sided phase noise power spectral density in the receiver output and $\Phi_m f_m$ is the mean-square signal power in the receiver output, one may rewrite (250) in a form relating input and output signal-to-noise power ratios,

$$(S/N)_i = [\exp(\sigma^2)/2\sigma^2] \log_\epsilon [1 + (S/N)_o] \qquad (251)$$

where $(S/N)_i$ = input signal-to-noise power ratio referred to twice the information bandwidth
$(S/N)_o = 2(S/N)_i [\sigma_m^2/\exp(\sigma^2)]$
= output signal-to-noise power ratio referred to the information bandwidth
$\sigma_m^2 = \Phi_m f_m$ = modulation index, radians

Equation (251) has a minimum value for $(S/N)_i$ at $\sigma = 1.0$ for a fixed system output quality $(S/N)_o$. Substitution of $\sigma = 1.0$ radian in (251) yields the threshold result depicted in Fig. 81.

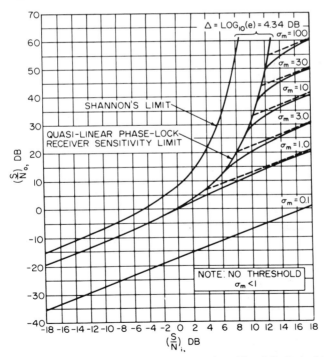

Fig. 81. Quasi-linear receiver performance for the situation of band-limited white Gaussian phase-encoded signals with the optimal transfer function.

The curves above threshold represent system output quality vs. input signal-to-noise power ratio, and asymptotically converge on the conventional high $(S/N)_i$ relation. These are constructed with the modulation index $\sigma_m = \sqrt{\Phi_m f_m}$ as a parameter. Since

$$\left(\frac{S}{N}\right)_o = \frac{2\sigma_m^2 (S/N)_i}{\exp(\sigma^2)} \qquad (252)$$

and σ is a function of $(S/N)_i$ governed by (251), a simultaneous solution of this relation and (251) yields σ and hence $(S/N)_o$ vs. $(S/N)_i$. It is interesting to note the curvature approaching threshold. Note also for $\sigma_m < 1.0$ the quasi-linear model yields no threshold.

The threshold criterion depicted by Fig. 81 is significantly below a standard FM or PM discriminator, as would be expected since more optimal demodulation is employed. It is, however, higher than that predicted by use of Shannon's results by a value of 4.34 db.

Second-order Loop Receiver; Band-limited Phase-encoded White Gaussian Signal

The optimal receiver just discussed is an ideal one which requires an infinite number of elements to realize, as has been previously discussed. The second-order loop is typical of the practical loop receiver of high performance.

The signal power spectral density is the same as that of Eq. (249). However, Eq. (248) cannot be used to calculate variance since the receiver is not optimal. Instead Eq. (237) must be used. $\theta_0/\theta_1(\omega)$ is the loop transfer function which has been derived for the second-order loop as Eq. (109) above. The following simplifying assumption is made:

$$2\pi f_m / \omega_n \ll 1 \qquad (253)$$

which restricts the validity of the results to be obtained to the region of high $(S/N)_o$. The variance for the second-order receiver is then

$$\sigma^2 = \frac{(2\pi)^4 \Phi_m f_m^5 \exp(\sigma^2)}{5\omega_n^4} + \frac{\Phi_i \exp(\sigma^2)[1 + 4\zeta_0^2 \exp(-\sigma^2/2)]}{8S\zeta_0} \omega_{n0} \qquad (254)$$

Assuming a fixed damping ζ_0 and Φ_i, Φ_m and f_m, Eq. (254) may be solved for S and minimized with respect to ω_{n0}. Thus maximum receiver sensitivity is achieved. Performing this manipulation one obtains as the minimum receiver input signal-to-noise power ratio, defined as in the optimal receiver [for Eq. (251)],

$$\left(\frac{S}{N}\right)_i = \frac{S}{2\Phi_i f_m} = \frac{\exp(\frac{6}{5}\sigma^2)}{2\sigma^2} \left\{\frac{5\pi[1 + 4\zeta_0^2 \exp(-\sigma^2/2)]}{16\zeta_0}\right\}^{4/5} \left(\frac{S}{N}\right)_o^{1/5} \qquad (255)$$

where the relation for signal-to-noise output power ratio is given by the same relation as in the optimal receiver, i.e.,

$$\left(\frac{S}{N}\right)_o = \frac{2\sigma_m^2 (S/N)_i}{\exp(\sigma^2)} \qquad (256)$$

As in (251), (255) has a minimum value at a particular value of σ which depends on system output quality $(S/N)_o$. In this case it also depends on receiver damping ratio ζ_0. Only one particular value of $\zeta_0 = 1/\sqrt{2}$ is used. For this damping the value of σ which yields minimum $(S/N)_i$, and hence minimum receiver threshold, is 1.01 radians. Substitution of $\sigma = 1.01$ radians into (255) yields the following threshold relation:

$$(S/N)_i = 4.08(S/N)_o^{1/5} \qquad (257)$$

Figure 82 graphs (255). Note that only values for $(S/N_o) > 20$ db are considered. This is consistent with the approximation made above [Eq. (253)].

As with optimal receiver, the curves above threshold represent system output quality vs. signal-to-noise power ratio input asymptotically converging on the conventional high $(S/N)_i$ relation. These are constructed with the modulation index σ_m as a parameter. Loop error σ is eliminated for purposes of the graphs in Fig. 82 by simultaneous solution of (255) and (256).

Shannon's lower limit and the optimal phase-locked receiver sensitivity are shown for comparison. It is interesting to note that for output signal-to-noise ratios of

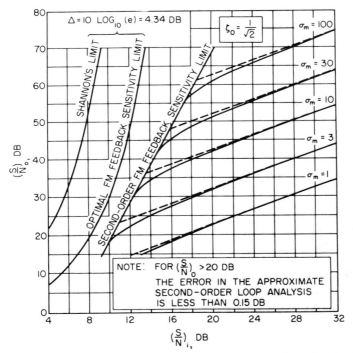

FIG. 82. Quasi-linear receiver performance for the situation of band-limited white Gaussian phase-encoded signals with a second-order transfer function.

practical interest (≈ 20 to 40 db) the second-order receiver is only 2 to 3 db poorer than optimal receiver or 6 to 7 db poorer than Shannon's limit.

It is important to note that this threshold analysis was based on a mean-square signal-to-noise, and loop error criterion. Threshold is defined as the input signal-to-noise power ratio at which, for a given quality constraint, the loop error becomes unbounded. If, however, short-term statistics of the receiver output are important to the observer, an entirely new criterion may require development. In any event, this analysis should present the lower bound on sensitivity, because with any new criterion a bounded mean-square loop error will certainly be a prerequisite.

Comparisons of the Methods

Viterbi[15] has made a comparison of the three techniques for the case of the first-order loop. The parameter compared is the variance as a function of noise-to-signal ratio $1/\alpha$ when the noise is confined to the loop noise bandwidth. The results are shown in

Fig. 83. The linear model is valid to 20 per cent for $1/\alpha = 0.25$ or $\alpha = 4$ or a signal-to-noise ratio of 6 db. The variance shown for the Van Trees model is based on Eq. (224) and is the result obtainable by calculating the first five terms of the Volterra functional expansion of the variance and using the first three terms of the power series.

Using the quasi-linearization approach of Develet the variance of the phase error for the first-order loop is

$$\sigma^2 = (1/\alpha) \exp(\sigma^2/2) \tag{258}$$

The solution of this transcendental equation yields the value of the variance shown in Fig. 82. The maximum of $\sigma^2 \exp(-\sigma^2/2)$ is $2/e$ so that there can be no solution for $\alpha < e/2$, which means that the validity threshold of this model can be no lower than this value.

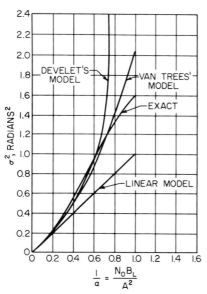
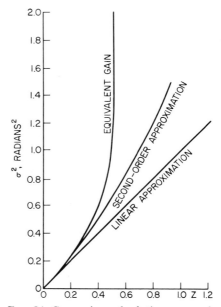

Fig. 83. Comparison of variance for first-order loop with results of approximate models.

Fig. 84. Comparison of solutions—second-order loop.

From Fig. 83 we note that the error in the Develet model is less than 10 per cent for $1/\alpha = 0.65$ or $\alpha = 1.54$, while the Van Trees approximation involving the first five Volterra kernels yields results of this accuracy for $1/\alpha = 0.80$ or $\alpha = 1.25$. Of course, with sufficient effort one can compute arbitrarily many terms of the Volterra series and consequently obtain arbitrarily many terms of the power-series expansion of σ^2, thus extending the validity threshold of the model as far as may be desired. However, for higher-order loops, Van Tree's method becomes exceedingly complex and tedious, while Develet's method remains simple for all loop filters and even for modulated signals. In fact, using this method he has obtained fairly general results on the threshold of the phase-locked loop as a frequency-modulation discriminator.

Van Trees made a comparison of the Volterra expansion and equivalent-gain method for the *second*-order loop. The results are shown in Fig. 84. Equation (226) is used for the second-order approximation. The equation derived for the equivalent-gain case was

$$\frac{1}{\alpha} = \frac{\tfrac{3}{2}\sigma^2}{e^{+\sigma^2/2} + \tfrac{1}{2}e^{+\sigma^2}} \tag{259}$$

For this particular system, the results agree within 25 per cent for $1/\alpha < 0.4$, $(S/N > 2.5)$. The maximum value of $1/\alpha$ which satisfies (259) is approximately 0.52.

FM with Frequency-feedback Systems

FM with frequency feedback is another technique for achieving high-sensitivity performance of receivers employed for space communications. This system was originated by Chaffee.[1] Several analyses have been made, notably by Enloe[27,28] and Cahn,[29] which make use of linearized feedback theory based on a twin-threshold concept. Develet,[17] using the quasi-linear (equivalent-gain) approach presented in a previous paragraph, has been able to show that the phase-locked-loop receiver and the FM with frequency feedback provide identical results when designed to operate at maximum sensitivity.

Figure 85 shows a block diagram of the FM feedback receiver. The principal difference from the phase-locked loop is the use of a discriminator in the loop so that the VCO receives a voltage proportional to instantaneous frequency rather than phase. The VCO output frequency mixes with the incoming signal and noise to provide an intermediate frequency ω_{IF} entirely within the discriminator frequency range. The loop compensation filter $F(s)$ must be carefully chosen so that a maximum-sensitivity design results. A narrow BPF (bandpass filter) is used. The noise bandwidth of the BPF in the intermediate frequency is small enough to assure linear operation of the frequency discriminator well above threshold.

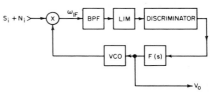

FIG. 85. FM feedback receiver.

Develet shows that the discriminator output is

$$\dot{\alpha} = \dot{\phi}(t) + \frac{\dot{y}(t)}{A \exp(-\sigma^2/2)} \qquad (260)$$

where $\dot{\alpha}$ = instantaneous frequency of the BPF filter vector
A = carrier amplitude at the BPF output
$\dot{y}(t)$ = quadrature noise amplitude of the BPF output
$\dot{\phi}(t)$ = instantaneous frequency error signal at the BPF output

The resulting quasi-linear model is given in Fig. 86a and is compared with the phase-locked-loop receiver at b. $s/[1 + 2s/\omega_0]$ represents the action of the discriminator and the bandpass filter.

As a result of the $\exp(-\sigma^2/2)$ in the noise term, the quasi-linear model yields a precise threshold criterion. In order to determine this threshold, the close-loop transfer function is defined as

$$\frac{\theta_o}{\theta_i}(s) = \frac{F(s)/1 + 2s/\omega_0)}{1 + F(s)/(1 + 2s/\omega_0)} \qquad (261)$$

The mean-square error of this receiver may be determined from Eq. (237) just the same as in the phase-locked case. All the equations of the section on equivalent gain apply. It can be concluded, therefore, that *within the approximation of quasi-linearization* the performance of an FM feedback receiver with a narrowband i-f filter is identical to that of a phase-lock receiver when each has the same closed-loop transfer function θ_o/θ_i.

It is interesting to compare the models for the two implementations. Note that either a decrease in signal strength or an increase in loop error tends to enhance the equivalent noise input in FM feedback, while the loop parameters remain fixed. This is a direct result of the limiter which precedes discrimination in the frequency-feedback demodulator of Fig. 85. On the other hand, in the phase-locked receiver the input noise remains fixed but the loop gain changes with either signal strength or loop error. This latter characteristic of phase-locked receivers has long been known.

Phase-locked loops are usually optimally designed for a particular set of conditions, e.g., noise, loop error, and signal strength at threshold. Parameter variation at other conditions is accepted as characteristic of the device. The tendency for parameters to remain fixed in the FM feedback receiver of Fig. 85 may be advantageous in certain situations. The detailed differences of the two models given in Fig. 86 in no way influence the ultimate sensitivity of an optimally designed receiver. Configurations 86a and b are equivalent servomechanisms, if $F(s)$ and $F'(s)$ are selected to make θ_o/θ_i identical in both cases.

It can be questioned whether or not threshold could be further enhanced by increasing the bandwidth of the i-f filter; this would allow for greater modulation error. For *nonoptimum* loop design, threshold is enhanced. However, if one strives for an optimum design [transfer function given by (250)], the threshold value for the loop error σ invariably lies in the vicinity of 1 radian. Since modulation error is only a small fraction of the total error σ, especially at high indexes, it is clear that little is to be gained by special accommodations for a negligible quantity. In nonoptimum designs, however, the increased modulation error which results will require a wider i-f filter to achieve maximum sensitivity.

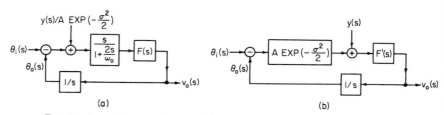

FIG. 86. Quasi-linear receiver models. (a) FM feedback. (b) Phase-lock.

Only the situation of Gaussian signals and noise was considered. Considering that the modulation error becomes vanishingly small at high indices, generalization to arbitrary signal probability distributions may be valid as long as the noise remains Gaussian.

A significant tactic in this analysis was the introduction of the narrowband i-f filter to eliminate irrelevant open-loop threshold considerations. This approach deviates from other design techniques.[27,28] These must consider open-loop threshold, since the Bode filter[27,28] and a constrained compression optimization[29] do not collapse modulation error to zero at high indices as does the design procedure explained here. Cahn[29] demonstrates that an FM feedback receiver influenced by open-loop threshold, becomes inferior to the phase-locked loop at large modulation indices. On the other hand, by following the design procedure set forth here, it was shown that the FM feedback receiver in its most sensitive form performs identically to the phase-locked receiver. The ultimate sensitivity is within 4.34 db of the theoretical limit. This 4.34-db degradation holds regardless of modulation index. Finally, it is observed, as first noted by Chaffee[1] in the original work on this device, that the limiter in Fig. 85 is unessential. Postulating a very narrow i-f filter, the fluctuations in the amplitude of the filter response are negligible and require no limiting to reduce their effect on the discriminator output. Of course, without the limiter, parameter variation will occur with signal strength and loop error as in the phase-locked loop.

Actual side-by-side comparisons are required for conclusive proof of the advantage of *nonoptimum* FM feedback design vs. *nonoptimum* phase-locked receiver design.

The results presented here show that a choice between the FM feedback or the phase-locked approach to receiver design should be dictated by hardware complexity rather than by theoretical performance.

Maximum-frequency Tracking Sensitivity

In a coherent communication system, oscillator (clock) stability is of great importance especially when accurate Doppler measurements or low information rates are to be conveyed through the link.

The ability to track such weak signals is strongly affected by the spectral shape of the oscillator signal. In general, however, this shape is not known. The effect may be calculated in the specific instance when the resulting random process imposed on the signal phase is caused by thermal noise. In this case, Edson[30] has shown that the frequency modulation $\dot{m}(t)$ has a white power spectral density. Develet[31] then obtained by simple integration and Fourier transformation the one-sided phase power spectral density Φ_m given by

$$\Phi_m = 2/\tau_c \omega^2 \qquad \text{radians}^2/\text{cps} \qquad (262)$$

where τ_c = coherence time of the oscillator system, the time in seconds it takes the phase drift to build up to 1 radian rms
ω = radian frequency, radians/sec
This power spectral density can be identified with $m(t)$, a random-walk process.

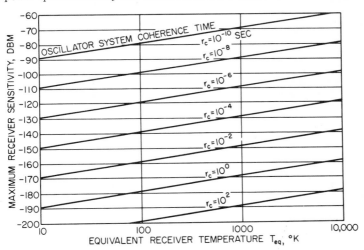

Fig. 87. Receiver sensitivity vs. equivalent receiver temperature.

We wish to determine the maximum receiver sensitivity in the presence of additive white Gaussian noise for the received signal which has a phase power spectral density governed by (262). Substitution of (262) in (250) for the optimum receiver yields

$$\sigma^2 = \frac{\Phi_i \exp(\sigma^2)}{S} \int_0^\infty \log_\epsilon \left[1 + \frac{2S}{\omega^2 \tau_c \Phi_i \exp(\sigma^2)} \right] df \qquad (263)$$

Integration gives the simple result

$$S = (\Phi_i/2\tau_c)[\exp(\sigma^2)/\sigma^4] \qquad (264)$$

Equation (264) may be differentiated with respect to σ to find the minimum value of S. Thus

$$S\bigg|_{min} = (\Phi_i/\tau_c)(e^2/8) \qquad (265)$$

for which value $\sigma = \sigma_{min}$ and is $\sigma_{min} = \sqrt{2}$ radian.
Considering that the receiver noise density is given by Boltzmann's constant times the equivalent receiver temperature, (265) may be restated as

$$S\bigg|_{min} = (KT_{eq}/\tau_c)(e^2/8) \qquad (266)$$

where $K = 1.38 \times 10^{-23}$, joules/°K
T_{eq} = equivalent receiver temperature, °K

Figure 87 plots phase-locked receiver sensitivity as a function of T_{eq} with τ_c as a parameter. It represents a fundamental sensitivity limitation for phase-locked reception given by the quasi-linear receiver model, for reception of signals disturbed by a random-walk-phase process such as an oscillator whose frequency is modulated by thermal noise.

REFERENCES

1. J. G. Chaffee, The Application of Negative Feedback to Frequency Modulation Systems, *Proc. IRE*, May, 1939, pp. 317–331.
2. W. J. Gruen, Theory of AFC Synchronization, *Proc. IRE*, vol. 41, pp. 1043–1048, August, 1953.
3. D. Richman, Color Carrier Reference Phase Synchronization Accuracy in NTSC Color TV, *Proc. IRE*, vol. 42, pp. 106–153, January, 1954.
4. R. Jaffe and E. Rechtin, Design and Performance of Phase-lock Circuits Capable of Near-optimum Performance over a Wide Range of Input Signal and Noise Levels, *IRE Trans. Inform. Theory*, vol. IT-1, pp. 66–72, March, 1955.
5. S. G. Margolis, The Response of a Phase-locked Loop to a Sinusoid Plus Noise, *IRE Trans. Inform. Theory*, vol. IT-3, pp. 136–143, June, 1957.
6. F. W. Lehan and R. J. Parks, Optimum Demodulation, *IRE Conv. Record*, pt. 8, PGIT, pp. 101–103, March, 1953.
7. C. E. Gilchriest, Application of the Phase-locked Loop to Telemetry as a Discriminator or Tracking Filter, *Trans. IRE Telemetry Remote Control*, vol. TRC-4, pp. 20–35, June, 1958.
8. A. J. Viterbi, System Design Criteria for Space Television, *J. Brit. IRE*, vol. 19, pp. 561–570, September, 1959.
9. R. L. Choate, Analysis of a Phase-modulation Communication System, *IRE Trans. Commun. Systems*, vol. CS-8, pp. 211–227, December, 1960.
10. M. Easterling, A Long-range Precision Ranging System, *JPL Tech. Rept.* 32-80, presented at URSI meeting, Washington, D.C., May, 1961.
11. A. J. Viterbi, Acquisition and Tracking Behavior of Phase-locked Loops, *Proc. Symp. Active Networks and Feedback Systems*, Polytechnic Institute of Brooklyn, April, 1960.
12. A. J. Viterbi, Phase-lock-loop Systems, in A. V. Balakrishnan (ed.), "Space Communications," Chap. 8, pp. 123–141, McGraw-Hill Book Company, New York, 1963.
13. R. W. Sanders, Communication Efficiency Comparison of Several Communication Systems, *Proc. IRE*, 1960, pp. 575–588.
14. J. A. Develet, Jr., A Threshold Criterion for Phase-lock Demodulation, *Proc. IEEE*, vol. 51, pp. 349–356, February, 1963, and correction, p. 580, April, 1963.
15. A. J. Viterbi, Phase-locked Loop Dynamics in the Presence of Noise by Fokker-Planck Techniques, *Proc. IEEE*, vol. 51, pp. 173–175, December, 1963.
16. H. L. Van Trees, Functional Techniques for the Analysis of the Non-linear Behavior of Phase-locked Loops, *Proc. IEEE*, vol. 52, pp. 894–910, August, 1964.
17. J. A. Develet, Jr., Statistical Design and Performance of High-sensitivity Frequency-feedback Receivers, *IEEE Trans. Military Electron.*, vol. 7, pp. 281–284, October, 1963.
18. V. I. Tikhonov, The Effects of Noise on Phase-lock Oscillation Operation, *Automatika i Telemakhanika*, vol. 22, no. 9, 1959.
19. G. E. Uhlenbeck and L. S. Ornstein, On the Theory of Brownian Motion, *Phys. Rev.*, vol. 36, pp. 823–841, September, 1930.
20. R. C. Booton, Jr., The Analysis of Nonlinear Control Systems with Random Inputs, *Proc. Symp. Nonlinear circuit Analysis*, Polytechnic Institute of Brooklyn, pp. 369–391, April, 1953.
21. J. J. Spilker, Jr., and D. T. Magill, The Delay-lock Discriminator—an Optimum/Tracking Device, *Proc. IRE*, vol. 49, pp. 1403–1416, September, 1961.
22. J. J. Spilker, Jr., Delay-lock Tracking of Binary Signals, *IEEE Trans. Space Electron. Telemetry*, vol. SET-9, pp. 1–8, March, 1963.
23. N. Wiener, Non Linear Problems in Random Theory, The M.I.T. Press, Cambridge, Mass., 1959.
24. P. A. George, Continuous Nonlinear Systems, MIT Research Laboratory of Electronics, Cambridge, Mass., *Rept.* 353, July, 1959.
25. R. C. Booton, Jr., Nonlinear Control Systems with Statistical Inputs, *MIT, Rept.* 61, pp. 1–35, Mar. 1, 1952.
26. M. C. Yovits and J. L. Jackson: Linear Filter Optimization with Game Theory Considerations, 1955 *IRE Natl. Conv. Record*, pt. 4, pp. 195–196.

27. L. H. Enloe, Decreasing the Threshold in FM by Frequency-feedback, *Proc. IRE*, vol. 50, pp. 18–30, January, 1962.
28. L. H. Enloe, The Synthesis of Frequency-feedback Demodulators, *Proc. Natl. Electron. Conf.*, vol. 18, pp. 477–497, October, 1962.
29. C. R. Cahn, Optimum Performance of Phase-lock and Frequency-feedback Demodulators for FM, Magnavox Research Labs, Torrance, Calif.
30. W. A. Edson, Noise in Oscillators, *Proc. IRE*, vol. 48, pp. 1454–1466, August, 1960.
31. J. A. Develet, Jr., Fundamental Sensitivity Limitations for Second Order Phase-lock Receivers, presented at URSI Spring Meeting, Washington, D.C., May 4, 1961, *STL Tech. Note* 8616-0002-NU-000, June 1, 1961.

Chapter 10

TELEMETRY-SYSTEM COMPONENT DESIGN

EDWARD Y. POLITI, *Solid State Electronics Corp., Sepulveda, Calif.* (*Environmental Problems; Transistor Characteristics; Transistorized Telemetering Circuits; Miniaturization Techniques*)

NORMAN D. WHEELER, *Spacecraft Department, Missile and Space Division, General Electric Company, Oklahoma City, Okla.* (*Power Sources for Remote Use*)

ENVIRONMENTAL PROBLEMS

1	Introduction	10-3
2	Aircraft and Missile Environments	10-4
3	Environmental Specifications	10-4
4	Outer-space Environment	10-9

TRANSISTOR CHARACTERISTICS

5	Equivalent Circuit Parameters	10-18
6	Transistor Equivalent Circuits	10-20
7	Hybrid Parameters	10-24
8	Shunt-feedback Amplifier Stabilization	10-25
9	Series-feedback Amplifier Stabilization	10-32
10	External Amplifier Stabilization	10-33
11	Self-stabilization of Low-frequency Small-signal Transistor Parameters	10-37

TRANSISTORIZED TELEMETERING CIRCUITS

12	FM-FM Circuits and Systems	10-42
13	Resistance-controlled Oscillator	10-42
14	Inductance-controlled Oscillator	10-43
15	Voltage-controlled Oscillator	10-46
16	Current-controlled Oscillator	10-52
17	Summing Amplifier	10-52
18	R-F Transmitter	10-53
19	Electronic Time Multiplexer	10-55
20	PDM Keyer	10-61
21	Solid-state PCM System	10-64

10-1

MINIATURIZATION TECHNIQUES

22	Introduction	10-69
23	Efficient Circuit Design	10-69
24	Small Components	10-71
25	Modularization and the Micromodule	10-72
26	Integrated Circuits	10-73
27	Printed Circuits	10-77
28	Encapsulation	10-78

POWER SOURCES FOR REMOTE USE

29	Introduction	10-79
30	Electrochemical Cells	10-79
31	Solar Cells	10-92
32	Thermoelectric Power Generation	10-96
33	Thermionic Cells	10-98
34	Nuclear Cells	10-99
35	Fuel Cells	10-100
36	Battery-selection Considerations	10-107

The essence of the systems discussed in this handbook is that some part of the system must be located at a remote location. Interest in the remote location in most cases stems from the need to avoid a hostile environment, at the same time accomplishing the necessary measurement or control. Examples readily prove the point:

1. Aircraft test flights
2. Missile remote control
3. Nuclear-reactor test
4. Space-satellite monitoring

The most rigid requirements are imposed on the remote system components which must operate in airborne, missile, or space environments. In actual fact many hostile environments exist in terrestrial locations such as desert, mountainous, or seaborne situations. Expansion in the art of telemetry and telecontrol should lead to many more terrestrial applications as techniques are developed for the more difficult situations.

In addition to the purely environmental aspect of the problem, other special requirements dictate the design of these remote components. These are size, weight, power availability, and equipment life or reliability. These considerations arise for air and space applications from the penalties on performance which result from an excess of the first three and a lack of the fourth. In general, the necessary performance of the electric circuits must be developed in opposition to these factors; better electrical performance tends to require more unfavorable figures in all these factors for a given state of the art in electrical technology. Vibration and shock environments impose an additional motive for reducing size and weight, since this is a direct method of improving ability of the electric circuits to withstand these disturbances.[1]

Other unusual environments are encountered for measurements internal to human and animal bodies and internal to rotating machinery such as large motors and generators. Here severe restrictions are imposed on size of transducers as well as transmitting units to fit into very small tubes and cavities.

The special qualities of telemetry-component circuit design stem from the characteristic demands on the remote components. The environmental problems posed will be better defined by a summary of Air Force specifications governing missile and aircraft and a review of the space environmental factors. The latter are subject to constant revision as further probes add to our knowledge of the radiations and particles in space.

In the majority of cases transistorized circuits meet the severe requirements imposed for the telemetry functions to be performed. Transistors provide miniaturized circuits which are able to perform all functions from amplification to low-power r-f radiation up to the uhf telemetry band. They are particularly useful for the distinctively telemetry functions such as parameter-controlled oscillators, FM modulators, switches, pulse modulators, and PCM encoders, also remote-control functions such as receivers, AGC, AFC, discriminators, servo-control amplifiers. They are best applied below 300 Mc. Other solid-state components are developing which can be used at microwave frequencies between 300 and 10,000 Mc. These include the tunnel (Esaki) diode and the varactor. The former can be used as oscillator, switch, or r-f amplifiers and the latter as switch or frequency multipliers. Microwave components based on traveling-wave amplification have been developed with sufficient compactness and ruggedness for operation up to 10,000 Mc. These components provide power outputs in the order of 10 watts over a wide frequency range (10 per cent of carrier frequency) and are to be used where higher microwave transmitted power is necessary in space applications than can be provided by solid-state components. Because the transistor is the solid-state component most applicable to the strictly telemetry and telecontrol functions, this chapter will confine itself to a discussion of transistor circuits.

Miniaturization by transistors has gone a long way to reduce the acceleration errors resulting from resonances in the telemetry system such as existed in vacuum-tube structures[1] by permitting much higher structural natural frequencies. The telemetry-circuit designer is thus forced to be familiar with the principles of transistor-circuit design for application to specific telemetry problems. A section of this chapter sets forth the characteristics of transistors and the various equivalent circuit approaches used to schematicize transistor performance. A major concern with transistors expected to operate over the wide temperature range in the telemetry environments is the stabilization of performance. Several methods are discussed in this chapter. Specific transistorized designs of important functional components of the standard telemetry systems are presented in a separate major section of this chapter.

The various approaches to physical miniaturization of circuits are reviewed in a section on miniaturization techniques where packaging techniques, printed circuits, and modularization approaches are discussed.

Electrical equipment at remote locations is apt to require unusual power sources. A survey of such sources, including chemical, solar, nuclear, and fuel cell, is provided which shows important characteristics of the sources such as voltage, current, and energy per unit weight.

ENVIRONMENTAL PROBLEMS

1 Introduction

Further advances toward miniaturization of electronic components, circuits, and systems will continually contribute toward increased reliability within the imposition of increasingly severe environmental conditions. Telemetering systems and components must be capable of reliable operation throughout any expected environmental spectrum if data integrity is to be maintained. Knowledgeable environmental specifications are set forth as a result of analysis and experience. Since previous data may be meager, interpolation and extrapolation techniques must also be used.

Aircraft, guided-missile, and space-vehicle telemetering requirements generally necessitate a high degree of simulation and operational testing utilizing statistical techniques to determine and verify the environmental reliability of components and systems. Once a determination has been made of the most unfavorable environmental conditions to be expected, a suitable safety factor must be included to provide for tolerance variations in production and operation.

2 Aircraft and Missile Environments

In general, severest environmental operational conditions are specified where military service requirements are concerned. Adverse conditions for military equipment include the extreme environments encountered in combat, tactical, and surveillance situations together with the conditions met in transport and storage. It is not unreasonable to expect telemetering systems and components to withstand environments to a greater degree than other accompanying equipment in order to preserve the accuracy of and confidence in the data it is designed to acquire.

Aircraft and missile environments vary with the specific usage and design. Flight tests are necessary to determine the range of environmental stresses at particular locations within the vehicle. Tests and computations are made to determine operation under the severest expected conditions. Reliance is placed on the telemetering system to measure the environment while monitoring the vehicle and other associated equipment.

3 Environmental Specifications

Current environmental specifications for electronic equipment used in military aircraft and guided-missile systems are referenced[1] in Table 1. Specification MIL-E-4970 is applicable to ground-support communications while MIL-E-5272 covers airborne communications.

Table 1. U.S. Air Force Environmental Specifications for Electronic Equipment

Category of the environment	Ground-support communications		Aircraft communications	Guided-missile communications
	Sheltered	Unsheltered		
Acceleration............	....	....	Mil-E-5272	Mil-E-5272
Altitude...............	Mil-E-4970	Mil-E-4970		
Explosion proof........	....	....	Mil-E-5272	Mil-E-5272
Fungus................	Mil-E-4970	Mil-E-4970	Mil-E-5272, Procedure I	Mil-E-5272, Procedure I
Humidity..............	Mil-E-4970	Mil-E-4970	Mil-E-5272, Procedure I	Mil-E-5272, Procedure I
Rain..................	....	Mil-E-4970		
Salt spray.............	....	Mil-E-4970		
Sand and dust.........	....	Mil-E-4970	Mil-E-5272	Mil-E-5272
Shock.................	Mil-E-4970	Mil-E-4970	Mil-E-5272	Mil-E-5272
Temperature, high.....	Mil-E-4970, Procedure II	Mil-E-4970, Procedure I	Mil-E-5272, Procedure I	Mil-E-5272, Procedure I
Temperature, low......	Mil-E-4970, Procedure III	Mil-E-4970, Procedure I	Mil-E-5272, Procedure II	Mil-E-5272, Procedure II
Temperature, shock....	....	....	....	Mil-E-5272, Procedure I
Temperature, altitude..	....	....	Mil-E-5272, Procedure I	Mil-E-5272, Procedure I
Vibration, acoustical...	For appropriate test procedure contact WCLOD-4, Wright-Patterson AFB, Ohio			
Vibration, mechanical..	Mil-E-4970	Mil-E-4970	Mil-E-5272	Mil-E-5272

The requirements of MIL-E-5272 are more severe than those of MIL-E-4970. The four military specifications MIL-E-5272 (USAF), MIL-E-5400 (ASG), MIL-T-5422 (ASG), and MIL-STD-202 are widely used for telemetering applications. Military environmental test procedures called out by these specifications[2] are summarized below. MIL-E-5400 covers design and manufacturing requirements for electronic equipment utilized in piloted aircraft. MIL-T-5422 is supplemental to MIL-E-5400 and is concerned with accelerated and simulated environments. MIL-STD-202 is a general specification that relates to electric and electronic circuit elements. MIL-STD-810 establishes uniform methods for environmental tests of aerospace and ground equipment and is intended to reconcile the differences among the above specifications.

Typical Environmental Test Specifications

Typical test procedures called for by military specifications are summarized below. This information follows closely a table prepared by Ref. 2. These requirements are subject to change but are representative of tests for missile and aircraft electronic equipment. The actual specifications must be consulted for latest official requirements.

In the following please note that the specifications are denoted as follows:

USAF = MIL-E-5272B (USAF)
E = MIL-E-5400B (ASG)
T = MIL-T-5422C (ASG)
STD = MIL-STD-202

Hot Test. *USAF.* 71°C, less than 5 per cent relative humidity, test unit at 71°C. Duration: 50 hr (results compared with room-condition operation).
E. To +260°C, depending on class of equipment and duty. Temperatures constant or varying at 1°C/sec.
T. +85°C nonoperating. +71°C operating and testing. Duration: Part of complex cycling.
STD. +85°C, 30 min cycles, nonoperating.
Cold Test. *USAF, Procedure I.* −54°C, test at −54°C. Duration: Stabilization (results compared with room-condition operation).
USAF, Procedure II. −62°C, 48 hr, check deterioration −54°C, 24 hr or stabilization whichever longer. Test at −54°C. Results compared with room-condition operation; check for deterioration.
E. To −54°C, operating, −62°C nonoperating. Long exposure to temperature extremes and temperature shock.
T. −62°C nonoperating. −54°C operating and testing. Duration: Part of complex cycling.
STD. −55°C, 30 min cycles, nonoperating.
Humidity. *USAF, Procedure I.* Temperature +71°C, relative humidity 95 per cent nonoperating, 240 hr of cycling, 10 cycles, compared with room-condition operation; check for deterioration.
USAF, Procedure II. Temperature +49°C, relative humidity 95 per cent, nonoperating. Duration: 360 hr, results compared with room-condition operation; check for deterioration.
E. Humidities to 100 per cent including condensation conditions. During all operating and nonoperating conditions.
T. Temperature +71°C, relative humidity 95 per cent, nonoperating. Duration: 240 hr, part of complex cycling.
STD. Temperature +40°C, relative humidity 95 per cent nonoperating. Duration: 250 hr.
Altitude. *USAF, Procedure I.* Pressure, 23.98 in. (Hg abs), temperature −55°C, test and operate under these conditions. Duration: Per detail specifications. Results compared with room-condition operation.
USAF, Procedure II. Pressure 3.44 in. (Hg abs), temperature −54°C, test and operate under these conditions. Duration per detail specifications. Results compared with room-condition operation.
USAF, Procedure III. Conditions specified in detail specifications.
USAF, Procedure IV. Conditions specified in detail specifications.

USAF, Procedure V. Pressure 1.32 in. (Hg abs), temperature −54°C, test and operate under these conditions. Duration: Per detail specifications. Results compared with room-condition operation.

E. From sea level to 80,000 ft, operating and nonoperating, depending on class of equipment. May vary at ½ in. Hg per second.

T. Pressure 3.44 in. (Hg abs), temperature −55°C; operate and test under these conditions. Duration: Part of complex cycling procedure.

STD. Pressure 3.44 in. (Hg abs), temperature not specified. Duration: Per detail specifications. Dielectric breakdown per detail specifications while at this reduced pressure.

Salt Spray. *USAF, T, STD.* Temperature +35°C, relative humidity greater than 85 per cent, salt solution 20 per cent ± 2 per cent, 99.8 per cent pure salt. Duration longer than 50 hr, nonoperating; results compared with room-condition operation. Salt deposits may be removed prior to operation.

E. Operate and nonoperate in salt-sea atmosphere.

Fungus. *USAF, T.* Temperature +30°C ± 2°C, relative humidity 95 per per ± 5 per cent. Fungus: five types. Duration: 28 days, nonoperating check for deterioration.

E. Operate and nonoperate when exposed to fungus encountered in tropical climate.

STD. None.

Sand and Dust. *USAF, Procedure I, STD.* Density: 1 to 5 g/cu ft. Velocity: 2,300 ± 500 ft/min, relative humidity never more than 30 per cent. Duration: 6 hr at 25°C, 6 hr at 71°C. Results compared with room-condition operation. Check for deterioration.

USAF, Procedure II, STD. Conditions same as Procedure I. Duration 8 hr at 25°C and test at 71°C deleted.

E. Operate and nonoperate on exposure to sand and dust as encountered in desert areas.

T. Duration: 6 hr at 71°C. Nonoperating.

Immersion. *USAF.* Liquid: water or equivalent. Temperature: room. Pressure: 1 in. (Hg abs). Examine for leaks. Increased to 2½ in.

E. None.

T. None.

STD. Liquid: saturated salt solution or water. Temperature +65°C. Examine and test.

Acceleration. *USAF.* Acceleration: per detail specifications. Duration: 1-min tests per detail specifications during acceleration.

E. None.

T. None.

STD. None.

Temperature and Altitude. *USAF, T.* Complex cycling of conditions at combinations of pressures and temperatures.

E. Temperature and altitude requirements call for temperature to 260°C at sea level depending on class of equipment.

STD. None.

Explosion. *USAF, Procedure I, T.* Temperature +52 to 71°C. Unit to operate in explosive mixture of butane and air at altitudes up to 40,000 ft. Different fuel mixtures required.

USAF, Procedure II, T. Unit to be filled with explosive mixture and placed in chamber with explosive mixture. The mixture in the unit is ignited at equivalent altitudes up to 5,000 ft.

E. No ignition shall be caused when equipment is operated in an explosive mixture.

Shock. *USAF, Procedure I, STD.* Shock test per detail specifications, operating. Equipment, Jan. 5, 1944.

USAF, Procedure II, T. 1. Equipment, MIL-S-4456. Shock: 15 g, 11 ± 1 msec. Duration: 18 shocks, 6 in each axis. 2. Shock mount with a dummy load 30 g 11 ± 1 msec. Duration: 6 shocks.

USAF, Procedure III. Per detail specifications. Machine per MIL-S-901. High impact.

*E.*1. With vibration isolators in place, 18 shocks of 15 g on 3 axes, 11 $\pm$ 1 msec.
2. 12 shocks of 30 g along 3 axes, 11 $\pm$ 1 msec. Bending and distortion permitted.

Vibration. *USAF, Procedure I.* For assembly mounting on aircraft. Resonance search: 5 to 500 cps, 10 g maximum; cycling 10 to 55 cps, 10 g maximum.

USAF, Procedure II. For assembly mounting directly to reciprocating engines. Resonance search: 5 to 500 cps; 20 g maximum. Vibrate at resonant frequencies.

USAF, Procedure III. For assembly mounting on vibration-isolated portions of aircraft. Resonance search: 5 to 55 cps; 15 g maximum. Vibrate at resonant frequencies.

USAF, Procedure IV. For panel-mounted instruments, resonance search: none. Cycling: 5 to 50 cps at amplitude determined by detail specifications.

USAF, Procedure V. Same as Procedure IV except amplitude is 0.018- to 0.020-in. double amplitude. To test for faulty construction in panel-mounted instruments.

USAF, Procedure VI. For rotary equipment. The equipment to be vibrated torsionally through 3° at frequencies from 10 to 40 cps.

USAF, Procedure VII. For alternators and generators, this test is a resonance search from 100 to 500 cps. At maximum acceleration of 10 g's.

USAF, Procedure VIII. For alternators and generators. This test is a functional weakness test for power equipment mounted directly on aircraft engines.

USAF, Procedure IX. For testing efficiency of vibration isolators. Test: from 5 to 55 cps. Double amplitude at 0.036 in.

USAF, Procedure X. For constructional details. Cycling: 6 hr, 10 to 55 cps. 0.03-in. double amplitude. The unit shall also be vibrated from 20 to 200 cps at 10 g maximum acceleration.

USAF, Procedure XI, T. Cycling with shock mounts 10 to 55 cps at 0.06-in. double amplitude. 4½ hr in 3 directions. Cycling without shock mounts for 6 hr at 5 to 500 cps. Double amplitude of 0.01-in. or 2 g's, whichever is limiting factor.

STD. Duration: 2½ or 5 hr. Frequency: 10 to 55 cps. Double amplitude 0.06-in.

Description of Tests and Typical Failures

A brief description of current environmental tests is provided below. The purpose of these tests is to simulate actual environmental conditions so as to reveal troublesome component and system weaknesses. Considerable experience has already been gathered regarding characteristic failures. These are discussed in connection with each type of test as a clue to improvement in equipment design.

Temperature. Constantly higher temperatures are being encountered by missiles and space vehicles. Very large chambers are used for testing of complete systems.

At high temperatures metals lose design strength, insulations weaken, lubricants fail, dimensions change, semiconductors fail to operate, capacitor life is reduced and bearings and lubricants freeze. Moisture condensation can freeze bearings and create noise on potentiometer sliders. Temperature shock can destroy finishes and crack seals. Wide temperature ranges are a principal problem in the design of temperature-compensated equipment. Temperature failures constitute a principal source of environmental failure.

Shock. Shocks are imposed by hard landings, transportation over rough terrain, weapon firing, rapid accelerations, and a variety of other causes. Shock tests are carried out using standard test equipment as described in particular government specifications. Missile applications have increased the need for devices producing a precise and reproducible shock pattern. Shock-testing equipment may be based on impact with a swinging hammer under precisely defined conditions, dropping from a prescribed height onto a bed of sand, acceleration in a pneumatic cylinder, dropping on a platform which is stopped by a lead pellet, etc. Crystal-type accelerometers having an extremely rapid time response are used to calibrate these machines.

Shock failures usually take the form of deformed or broken parts. Fastener failures are common. Sometimes an entire framework of a subsystem may be deformed or fractured. Brush contacts may open up. Tube filaments may open. The other electrodes of a vacuum tube may short circuit or become structurally weakened so as to generate microphonics. In general, smaller components having little mass are far less sensitive to shock.

Vibration. This is one of the more severe environmental problems encountered in current system requirements. Early military specifications calling for vibration testing from 5 to 55 cps are being superseded by more stringent requirements covering the range to 2,000 cps. For the lower frequency a straightforward mechanical shake table is employed, energized by an adjustable-speed motor through a driving linkage. Cam arrangements can be incorporated for cyclical speed variations over the band. Amplitude control is also required. Higher frequencies are covered by loudspeaker type devices with the shake table attached to the "voice coil." Power is derived from a variable-frequency supply, which may be an alternator driven by an adjustable-speed motor or a power amplifier energized by a variable-frequency oscillation.

Vibration tests introduce electrical contact failures in potentiometers, relays, and devices with brush contacts, fatigue failures at flanges, and other stressed portions of cantilevered members, loosening of fasteners, intermittent contact of tube elements, failure of solder joints, wear and misalignment in shafts and bearings, opening up of electrical connections, breaking of lead connections, and the greatest variety of other mishaps. As with shock tests, smaller encapsulated units are most resistant to vibration. Resonances should be avoided where possible. Frames and shells should be sufficiently stiff to avoid stiffness without increasing mass and lowering resonant frequency. Shock mounts are widely used, although on occasion they can accentuate a dangerous resonance. Silicone rubber may be poured or tucked into the corners of subassemblies to provide a damping support. Shock mounts are of no value in the face of vibration induced by acoustic noise.

Humidity. Telemetering equipment must be designed and tested to assure reliable functioning following storage or during operation under high-humidity conditions. Humidity-generating equipment may be controlled by heaters immersed in distilled water or by spraying water against rod or strip heaters. Humidity may generally be varied from 5 to 95 per cent.

To varying degrees humidity causes corrosion, dimensional changes, and reduction of insulation resistance due to moisture absorption. Water vapor is injurious to electrical contacts and can create short or open circuits through corrosive processes. Accumulation of moisture in hydraulic lines can be especially serious. Humidity followed by a sudden drop in temperature can cause ice formation with occasional disastrous results.

Salt Spray. To simulate marine environment tests are performed by pumping a standard salt solution through a nozzle as a fog in a salt-spray chamber. Baffles prevent direct infringement of salt particles on test samples. Washing away excess salt is often permitted before retesting units having rotating shafts.

Salt spray causes severe corrosion due to electrolytic action and may result in binding of rotating members, malfunctioning of electrical systems, and chemical attack on certain lubricants. Suitable design techniques and finishes have minimized the difficulty of meeting salt-spray requirements.

Explosion. Explosion tests are designed to determine whether equipment operating in an explosive atmosphere will cause an explosion, or whether an externally caused explosion will fragmentize the equipment. Explosion tests are conducted at high ambient temperatures. Lean, intermediate, and rich fuel mixtures are prescribed.

Explosion tests are of extreme importance because of the probable disastrous consequences of this type of failure. Sparks at contacts, commutating and slip rings, or even sparks generated by mechanical impact of certain materials may initiate an explosion. Cutout switches, circuit breakers, and relays can be offenders. Arcing due to insulation failure may cause an explosion. Construction which permits destructive fragments to be scattered by an explosion can cause damage over a wide area.

Sand and Dust. Sand and dust conditions are encountered primarily in dry desertlike areas. Tests are performed in special chambers comprising a blower, ducts, and heaters to generate specified ambient temperatures. Military specifications define the size and composition, and velocities of the particles used in test.

Sand and dust are particularly injurious to ball bearings of rotating components and devices having gearing or other moving parts. Unsealed contacts on relays, slip rings, and commutators, and potentiometer contacts are sensitive to minute dust particles. The best design protection is to provide appropriate seals and dust covers. The contact pressures required to maintain good performance in the presence of shock and vibration are not usually in themselves adequate for dust particles. Hermetic sealing or encapsulation may be necessary for dust protection of certain types of components.

Fungus. High temperatures and humidities in tropical areas promote the growth of fungus which consumes materials and damages finishes, causing a great variety of failures. It is remarkable how many engineering materials represent "good eating" for fungus. Fungus growth may be countered by use of only fungus-resistant materials, which contain no nutrients, or by a suitable protective coating of nonnutrient materials. Equipment is generally designed directly to resist fungus, and fungus tests are therefore frequently waived.

Fungus tests are carried out by spraying the equipment with a water suspension of fungus spores, and then storing at 30°C and 95 per cent relative humidity for 28 days.

Fungus consumes many of the standard construction materials. Because fungus-resistant materials or coatings are readily available and widely used, the problem of fungus resistance is not a severe one. Failure is usually associated with some overlooked design detail which may be readily corrected. Fungus failure in its milder forms can obscure nameplates or readout points and ruin finishes, while in its more severe forms, it can completely wreck the proper operation of the equipment.

Sunshine. Sunshine tests are accomplished by exposure to lamps of spectral distribution simulating strong sunshine. Radiant energy is applied of 100 to 120 watts/sq ft in specified wavelengths while temperature is held at 113°F.

Sunshine may cause fading of colors, darkening of glass, and reduced legibility of markings, meters, and dials. Semiconductors, particularly the photosensitive variety, may be influenced by the action of sunshine.

Rain. The equipment is subjected to the equivalent of heavy rain, small droplets of water being applied, rather than a fine mist as in the humidity test. The equipment is positioned to simulate installation conditions. Water penetration and resulting damage are similar to effects of humidity, although less difficult to meet from a design point of view. Water may enter principally through openings provided for ventilation.

Advanced-project Environmental-test Requirements

Typical military aircraft and guided-missile environmental specifications applicable to telemetering equipment for advanced projects are shown[1] in Table 2. This tabulation establishes design criteria for future aircraft and missiles approaching altitudes up to 600 miles. Examination of these requirements shows challenging temperature problems extending from -54 to $+1093$°C. Intense acoustical vibration environments extend from 30 to 10,000 cps at levels up to 183 db.

References (Art. 1-3)

1. W. W. MacDonald, Electronics Handbook for Design Engineers, *Electronics*, p. R-6, mid-month, June, 1958.
2. Chart published in Servomechanisms, Inc., *Eng. Rev.*, January, 1957.
3. V. J. Junker, The Evolution of USAF Environmental Testing, Flight Dynamics Laboratory, Rand TD, AF Systems Command, Wright-Patterson Air Force Base, Ohio, *Tech. Rept.* AFFDL-TR-65-197, October, 1965.

4 Outer-space Environment

Future telemetering requirements will find challenging applications in the probing and exploration of outer space, first by unmanned vehicles, finally followed by manned

Table 2. Air Force Environmental Design Specifications*

	Aircraft							
	Present						Estimated	
	Test A	Test B	Test C	Test D	Test E	Test F	Test G	Test H
Altitude, ft†	30,000 0–30,000	50,000 0–50,000	60,000 0–60,000	70,000 0–70,000	80,000 0–80,000	90,000 0–90,000	100,000 0–100,000	150,000 0–150,000
Shock, min..	20	15	15	15	5	5	5	5
Temperature, °C:								
Max..	71, 1 hr	95, 10 min 71, 1 hr	125, 10 min 95, 1 hr	150, 10 min 95, 4 hr	260, 10 min 125, 4 hr	315, 10 min 260, 1 hr	400, 10 min 375, 1 hr	500, 10 min 400, 1 hr
Military emergency..	54, 4 hr	54, 4 hr	71, 4 hr	95, 4 hr	125, 4 hr	260, 4 hr	260, 4 hr	375, 4 hr
Cruise..								
Temperature, °C..	−54 to 71	−54 to 95	−54 to 150	−54 to 150	−54 to 268	−54 to 375	−54 to 400	−54 to 500
Shock, min..	20	15	15	10	10	5	5	5
Noise, acoustical, 37.5–9,600 cps‡								
External, db..	130	150	145–165	148–168	152–170	155–170	160–170	160–170
Internal, db..	110–120	115–135	130–150	133–163	137–157	140–160	145–165	145–165
Vibration, cps..	10–55	5–55	5–55	5–55	5–55	55–1,000 to 2,000 cps 10–20 g		
	0.06 in. da§	0.06 in. da§ 56–500 cps‡ 10 g	0.06 in. da§ 56–500 cps‡ 10 g	0.06 in. da§ 56–500 cps‡ 10 g	0.06 in. da§ 56–500 cps‡ 10 g			

	Missiles			
	Present		Estimated	
	Test A	Test B	Test C	Test D
Altitude, ft..	100,000 0–50,000	0–3,000,000	3,000,000 0–3,000,000	0–3,000,000
Altitude, ft..				
Shock..	10 sec	20 min	20 min	20 min
Shock, g..	50, 11 ± 1 msec	50, 11 ± 1 msec	50, 11 ± 1 msec	50, 11 ± 1 msec
Temperature, °C..	−54 to 93	−54 to 371	−54 to 538	−54 to 1093
Temperature, °C..	−54 to 93	−54 to 371	−54 to 538	−54 to 1093
Shock, min..	5	5	5	5
Noise, acoustical, 37.5–9,600 cps				
External, db..	150–170	150–170	153–170	138–170
Internal, db..	130–155	130–150	133–158	138–163
Vibration, cps..	5–55 cps 56–2,000 cps			Complex wave

* Summary values only. Refer to MIL-STD-810.
† ARDC Model Atmosphere, 1959.
‡ By octaves.
§ da = double amplitude.

space flights. The environments of interest will be concerned with the following space domains:

1. Terrestrial
2. Hyperatmospheric
3. Lunar
4. Interplanetary
5. Interstellar
6. Intergalactic

Exploratory telemetering equipment will have to be designed to operate and report on relatively unknown environments. These environments include

1. Atmosphere
2. Vacuum
3. Radiation
4. Temperature
5. Auroras
6. Solid-particle collisions

Of particular concern will be the induced environments due to interaction between the vehicle and its surrounding environment.

Atmosphere

Atmospheric effects are of concern during vehicle exit and reentry. During exit, atmospheric turbulence will result in some vehicle perturbations. Vehicle entry into the atmosphere at relatively high speeds results in aerodynamic heating of leading edges. The heating severity is more pronounced as the density of the atmosphere and relative velocity of the missile increase. In addition, entry at high speeds causes dissociation, ionization, and recombination of the atmospheric gases immediately surrounding the missile. An ion sheath forms about the vehicle, causing the gas molecules to undergo a change in conductivity and dielectric constant according to the equations[2]

$$\sigma = \frac{E_0 \omega_{cr}^2 \nu}{\nu^2 + \omega^2} \tag{1}$$

$$E = E_0 \left(1 - \frac{\omega_{cr}^2}{\nu^2 + \omega^2}\right) \tag{2}$$

where E_0 = free-space dielectric constant
ω = incident radiation frequency, radians
ν = electron collision frequency
ω_{cr} = critical frequency, radians

Ionization attenuates the propagation of electromagnetic energy and seriously limits radio transmissions to and from the vehicle. This takes the form of modified antenna impedance, receiver noise level, and voltage breakdown.

Natural ionized strata exist in the earth's upper atmosphere, termed the E and F layers (Table 3), which cause radio signals of different frequencies to be reflected

Table 3. Upper-atmosphere Ionized Layers

Region	Altitude, ft	Ion density, N/cu cm	Relative ion density, %
E	330,000	1.5×10^5	0.000001
F_1	650,000	2.5×10^5	0.01
F_2	1,000,000	1.5×10^6	1.0

out of time phase. Recent radio measurements revealed two E layers between 300,000 and 430,000 ft having a relatively low ion concentration ranging from 10^4 to 1.5×10^5 electrons/cu cm. As a result, the E-layer concentrations are not so serious. The F_1 and F_2 layers at 650,000 and 1,000,000 ft, however, are sufficiently high in concentration to affect satellite transmissions below 100 mc. The electron density is related to the frequency according to $N = 1.24 f^2 \times 10^{16}$.

Water-vapor content in the atmosphere may under certain conditions form ice crystals at 80,000 and 250,000 ft. Though highly improbable, impact with ice crystals on reentry can result in some vehicle abrasion.

Probable gaseous atmospheric content of the solar system's planets is shown in Table 4.

Table 4. Gaseous Atmospheric Content for Solar System's Planets

Planet	Probable atmosphere
Mercury	None
Venus	CO_2, H_2O
Earth	H_2O, CO_2, N_2, O_2, O_3, A, Ne, He, Kr, Xe
Mars	H_2O, CO_2, N_2
Jupiter	CH_4, NH_3, H_2, He
Saturn	CH_4, NH_3, H_2, He
Uranus	CH_4, NH_3, H_2, He
Neptune	CH_4, NH_3, H_2, He
Pluto	None
Moon	None

Vacuum

The vacuum environment increases as a vehicle egresses from a planet having an atmosphere and will approach a near total vacuum in outer space as far as is known. Typical values of earth air pressure vs. altitude are given[3] in Table 5.

Table 5. Atmospheric Pressure vs. Altitude

Altitude, ft	Pressure, psi
50,000	1.7
100,000	0.16
200,000	0.465×10^{-2}
300,000	0.256×10^{-4}
1,000,000	0.258×10^{-9}

Effects that may become apparent in a vacuum environment include

1. Low-temperature boiling
2. Electric arcing and corona discharge
3. No air damping of vibration
4. Outgassing
5. Explosive decompression
6. No convection heating or cooling

Radiation

Outside the earth's atmospheric shield, space vehicles will be subjected to cosmic and heat radiation. Cosmic radiation manifests itself in the form of high-speed particles. Approximately 80 per cent of these particles are protons, the remainder consisting of subatomic particles. Space exploration to date has uncovered two concentric toroidally shaped radiation belts (the Van Allen belts) situated around the earth (Fig. 1). These high-intensity belts are presumed to consist of charged particles such as protons and electrons and are maintained by the earth's geomagnetic field. The shape of these regions has not been exactly established. It is not clear, for example, that the inner and outer belts are different; they may actually be part of one large belt. Concentric contours indicated particle counts per second. Each count indicates about 1.6 particles per square centimeter. The inner and outer belts of highest radiation are located at about one and three earth radii, respectively, from earth center. Maximum radiation intensity is approximately 40,000 particles per square centimeter per second. The energy level of the proton particles ranges from 10 to 100 mev, 1 to 10 mev for electron particles. Electronic components subjected to cosmic radiation require adequate shielding depending on intensity and time of exposure. Effective shielding can be provided with the proper thickness of metallic enclosure, although weight considerations may require other solutions.

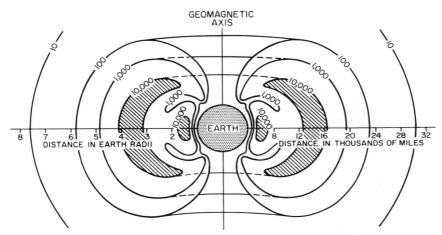

Fig. 1. Earth-radiation particle-count contours, counts per second.

Solar energy radiates virtually unabated through the space void until intercepted by the earth's atmosphere, where certain portions of the solar light spectrum are selectively attenuated through absorption. Solar-energy wavelength distribution at the earth's surface is shown in Table 6. Wavelengths less than 2,800 angstroms are entirely absorbed. About 50 per cent of the ultraviolet radiation between 2,800 and 3,800 angstroms is absorbed.

Table 6. Solar-energy Distribution

Type	Wavelength, angstroms	%
Far ultraviolet	1– 2,000	0.2
Near ultraviolet	2,000– 3,800	7.5
Visible	3,800– 7,000	41
Infrared	7,100– 10,000	22
Infrared	10,000– 20,000	23
Infrared	20,000–100,000	6

Short wavelengths of less than 100 angstroms, such as X rays, are not readily absorbed. Ten per cent of 1-angstrom light penetrates the atmosphere to 200,000 ft of altitude. About 50 per cent of most wavelengths greater than 3,100 angstroms finally reaches the earth's surface. Solar-energy levels for certain wavelength ranges are given in Table 7.

Satellites at altitudes exceeding 200 miles would be subject to all the sun's radiation. Most radiation wavelengths will be completely absorbed by sufficient thickness of metal, including X rays. Aluminum satellite shells having a 0.04-in. thickness would absorb in excess of 90 per cent of all solar radiation. Greater absorption can be obtained from heavier metals such as steel.

Advanced aircraft and space vehicles will utilize nuclear propulsive means which will require the operation of electronic equipment in reactor radiative environments. Reactor radiation is in the form of charged particles such as electrons and protons, or uncharged particles which include gamma rays as well as slow and fast neutrons. The

Table 7. Solar-energy Levels for Various Wavelength Ranges

Wavelength range, angstroms	Energy, 10^{-6} watts/sq m
1,040–1,240	400
1,230–1,340	200
1,180–1,300	100
5–10	0.1
1,216, Lyman-Alpha	20–2,000

uncharged particles have the most effect in causing atomic dislocations. Fast neutrons are 150 times more effective than the slow neutrons, which are about 15 times more effective than gamma rays. Radiation studies have been performed on resistors, capacitors, vacuum tubes, transistors, semiconductors, and magnetic and other materials. Transistors, diodes, and other semiconductors are highly susceptible to radiation damage, photocells being the most adversely affected. Resistors are able to withstand radiation with little tolerance variation, the wirewound types being very slightly affected compared with the composition carbon types. Ceramic, mica, and

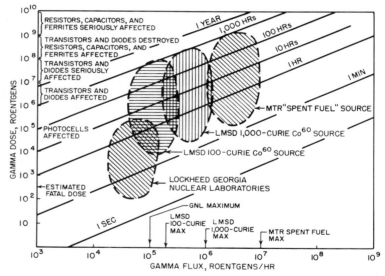

FIG. 2. Flux-dose plot for gamma radiation.

glass dielectric capacitors are slightly affected when compared with plastic and paper types. Vacuum tubes are generally able to withstand radiation exposure, although certain types are affected because of the materials and structure used. For example, gas-filled tubes generally experience an increase in ionization voltage with time.

Exposure of electronic components to radiation is greatly dependent on the reactor or isotope source. A convenient method[6] of describing these radiation environments has been devised and denoted as the flux-dose plot. Three such plots for gamma and slow- and fast-neutron radiation are shown in Figs. 2, 3, and 4. Flux vs. dose is

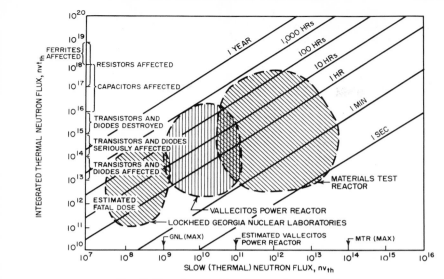

Fig. 3. Flux-dose plot for slow-neutron radiation.

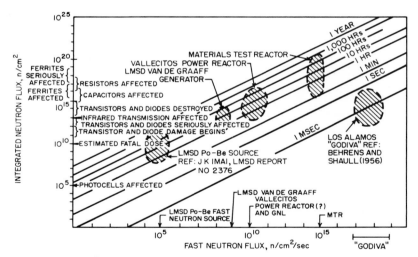

Fig. 4. Flux-dose plot for fast-neutron radiation.

plotted with exposure time as a parameter. Shadowed areas are for specific radiation sources where lifetime is limited. Radiation dose is in terms of particles per unit area, the roentgen being the dose for gamma rays. Radiation flux is given in terms of dose per unit area. Another radiation unit utilized is the product of the number of particles per unit volume and their velocity, called *nvt* for neutrons.

The use of radiation shielding to protect electronic equipment will require study for each particular case. If personnel shielding is required, there will be no radiation problem for electronic components in the same environment. For space vehicles, shielding adds a weight disadvantage.

Light shielding may prove to be optimum. For example, slow radiation particles cause more damage to transistors than the faster particles. Therefore, less shielding than the maximum for perfect shielding whould be the most desirable in many cases.

Solar-heat radiation outside any atmosphere will be of serious concern in the design of space vehicles and equipment. Reference[4] to Table 8 gives the mean distance

Table 8. Planetary Distances to Sun and Maximum Surface Temperature

Planet	Approx mean distance to sun ($\times 10^6$)	Max surface temp, °C
Mercury	36	410
Venus	67	427
Earth	93	60
Mars	142	30
Jupiter	484	−129
Saturn	887	−152
Uranus	1,790	−168
Neptune	2,800	−200
Pluto	3,680	−223
Moon	93	−152

to the sun for each of the solar system's planets, together with the maximum surface temperature.

Without the protective shield of the atmosphere, a black or gray body at the same distance from the sun as the earth would reach temperatures approaching 400°K. Temperatures of various black[4] or gray bodies in space in terms of planetary distance from the sun are given in Fig. 5 for various shapes.

The coolest shape is that of the shadowed cone with the base facing toward the sun. Expected temperature limits[4] for some electronic components are listed in Table 9.

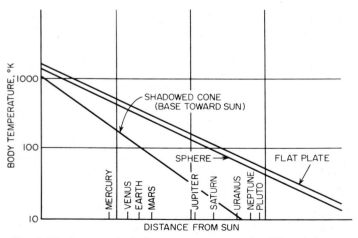

FIG. 5. Black or gray body in-space temperature for different shapes.

ENVIRONMENTAL PROBLEMS

Table 9. Temperature Limits of Electronic Components, °C

Batteries	225	A-C generators	250
Bearings	540	A-C servomotors	500
Capacitors	500	Insulated wire	500
Circuits	550	Potentiometers	200
Relays	125	Radiation-detection tubes	100
Selsyns	450	Semiconductor silicon	200
Transformers	500	Semiconductor germanium	75
Tubes	500	Semiconductor selenium	125

Auroras

Illuminations of the night sky are characteristic of the aurora phenomena.[1] Auroras give the appearance of a glow, arc, corona, or ray lasting from a fraction of a second up to half an hour. Source of the auroras is attributed primarily to an electrically charged current of corpuscular radiation emitted from the sun. Previous rocket probes have identified the radiation as X rays having energies from 10 to 100 kev and intensities from 1,000 to 100,000 photons/sq cm/sec.

The auroras occur most frequently at 70 degrees of geomagnetic latitude and extend from altitudes of 300,000 to 600,000 ft. During magnetic storms the auroras may stretch out to 3,000,000 ft at reduced latitudes. Polar-traversing satellites at altitudes below 200 miles may pass through the auroras. At values of 30,000 ev, the X-ray wavelength is 0.4 angstrom. The intensity-penetration equation is $I = I_0 E^{-kx}$. k for aluminum and iron is 4 and 85, respectively, which results in 50 per cent penetration values of 0.062 and 0.012 in., respectively.

Solid Particles

Space vehicles will be subject to sporadic collisions with solid particles. A meteor is a visible reaction between a high-velocity solid particle (meteoroid) and the atmosphere. Such a reaction results in frictional heating to a degree where light is emitted and ionization occurs. Meteorites are particles of sufficiently large dimensions to penetrate to the surface of the earth. Micrometeorites are particles having a large surface-to-mass ratio such that high-velocity reaction with the atmosphere results in quick radiation of the energy. This effectively stops the particle without serious physical change or ablation. Original particles are considered to be in hyperbolic solar orbit.[5]

Radar observations have shown particle velocities outside the atmosphere of 135,000 ft/sec, the escape velocity from the sun to the earth. Maximum meteor speeds in the atmosphere are 230,000 ft/sec; minimum speeds are about 36,000 ft/sec. Meteors exist at altitudes from 100,000 to 500,000 ft and disintegrate from 50,000 to 250,000 ft. Micrometeorites eventually reach the earth in the form of fine dust particles. Rocket experiments at speeds of Mach 1 have indicated approximately 17 particle impacts per minute.

Estimates are that 1,000 tons of space-particle debris fall to earth each day. Table 10 presents meteor data[1] such as mass, brightness, size, number, and electron line density of the trail. Ninety-nine per cent of the total mass consists of particles less than 40 but more than 0.4 microns in diameter. Smaller particles are ejected from the solar system by radiation pressure due to sunlight.

Space vehicles can undergo penetration from large particles and erosion from smaller particles. One-microgram particles will penetrate aluminum to a depth of 1 mm. Meteoric dust having a time density of 10^{11} atoms/sq m/sec, each with an assumed energy of 420 ev, would erode an area of 4×10^{-4} sq in. to a depth of 1 mil in 1 hr at a maximum rate.

Table 10. Meteor Data

Types of particles	Mass, g	Visual magnitude	Radius	Number swept up by earth each day	Electrons per meter of trail length
Particles reaching ground.........	10^4	-12.5	8 cm	10	
Particles totally disintegrated in the upper atmosphere	10^3	-10.0	4 cm	10^2	
	10^2	-7.5	2 cm	10^3	
	10	-5.0	0.8 cm	10^4	10^{18}
	1	-2.5	0.4 cm	10^5	10^{17}
	10^{-1}	0.0	0.2 cm	10^6	10^{16}
	10^{-2}	2.5	0.08 cm	10^7	10^{15}
	10^{-3}	5.0	0.04 cm	10^8	10^{14}
(Approximate limit of radar measurements)	10^{-4}	7.5	0.02 cm	10^9	10^{13}
	10^{-5}	10.0	0.008 cm	10^{10}	10^{12}
These react with atmosphere but do not produce enough light or ions to be detected	10^{-6}	12.5	40 microns		
	10^{-7}	15.0	20 microns		
	10^{-8}	17.5	8 microns		
Micrometeorites...............	10^{-9}	20.0	4 microns		
	10^{-10}	22.5	2 microns		
	10^{-11}	25.0	0.8 micron		
	10^{-12}	27.5	0.4 micron		
Particles removed from the solar system by radiation pressure	10^{-13}	30.0	0.2 micron		

References (Art. 4)

1. R. A. Di Taranto and J. J. Lamb, The Space Environment, *Elec. Mfg.*, October, 1958, pp. 54–56.
2. W. Sisco and J. M. Fiskin, Shock Ionization Changes EM Propagation Characteristics, *Space/Aeronautics*, March, 1959, p. 66.
3. R. A. Minzer and W. S. Ripley, ARDC Model Atmosphere, 1956, *Geophysics Research Directorate*, AFCRC TN-56-204, 1956.
4. Haig A. Manoogian, The Challenge of Space, *Electronics*, Apr. 24, 1959, pp. 67–68.
5. L. La Paz, Meteoroids, Meteorites and Hyperbolic Meteoric Velocities, "Physics and Medicine of the Upper Atmosphere," The University of New Mexico Press, 1952.
6. T. R. Nisbet, How Radiation Affects Electronic Equipment, *Electron. Equipment Eng.*, May, 1959, p. 36.
7. Staff Report, Radiation Effects Data Sheets, *Elec. Mfg.*, February, 1959, p. 112.

TRANSISTOR CHARACTERISTICS

5 Equivalent Circuit Parameters

In developing transistorized circuitry, the designer must of necessity be concerned with the problem of selecting the most appropriate equivalent representation of the transistor for analysis purposes. A very large number of transistor equivalent circuits have been presented in the literature.[1,2,3,4,5] For any particular design problem the choice of equivalent transistor circuit to be used is not always easily determined. Familiarity with as many equivalents as possible is advantageous so that the most logical choice can be made.

From a practical standpoint, it is desirable to go from the circuit concept through the analysis to the final optimum design in the shortest possible time. To this end a great deal of consideration will be given to the hybrid low-frequency small-signal parameters, because the common-base hybrid parameters are most frequently specified by the transistor-device manufacturers. Adoption of these parameters is based on the simplicity and accuracy of measurement utilizing ordinary laboratory equipment.

Small-signal low-frequency impedance parameters are presented first, since electronic designers have a greater intuitive appreciation of these. The relationship between the transistor equivalent impedance parameters and the hybrid parameters will be shown. The hybrid parameters for the common-emitter and common-collector connections will be presented in terms of the common-base hybrid parameters, in order to simplify conversion to quantitative analysis. Transistor operation will also be described in terms of (1) input impedance, (2) output impedance, (3) voltage gain, (4) current gain, and (5) power gain. Exact equations will be presented, together with their approximations. In most instances, the approximate equations will be satisfactory, but their validity should be established by the designer for any particular case.

The transistor can be considered a two-port four-terminal active device as shown in Fig. 6.

FIG. 6. Two-port active network.

This is a classical representation familiar to network analysts, and it permits the application of a great deal of previous work in networks to transistor circuit design.

Two equations are required to relate the input and output currents in terms of four independent parameters. Once these parameters are determined the four-terminal network is completely characterized. There are six possible sets of these equations. However, only three such sets are in general use. They are given below in both equation and matrix form.

In terms of z, or open-circuit impedance parameters,

$$\text{Equation} \qquad\qquad \text{Matrix}$$
$$v_1 = z_{11}i_1 + z_{12}i_2 \qquad \begin{bmatrix} v_1 \\ v_2 \end{bmatrix} = \begin{bmatrix} z_{11} & z_{12} \\ z_{21} & z_{22} \end{bmatrix} \begin{bmatrix} i_1 \\ i_2 \end{bmatrix} \qquad (3)$$
$$v_2 = z_{21}i_1 + z_{22}i_2$$

In terms of y or short-circuit admittance parameters,

$$i_1 = y_{11}v_1 + y_{12}v_2 \qquad \begin{bmatrix} i_1 \\ i_2 \end{bmatrix} = \begin{bmatrix} y_{11} & y_{12} \\ y_{21} & y_{22} \end{bmatrix} \begin{bmatrix} v_1 \\ v_2 \end{bmatrix} \qquad (4)$$
$$i_2 = y_{21}v_1 + y_{22}v_2$$

In terms of h or hybrid parameters,

$$v_1 = h_{11}i_1 + h_{12}v_2 \qquad \begin{bmatrix} v_1 \\ i_2 \end{bmatrix} = \begin{bmatrix} h_{11} & h_{12} \\ h_{21} & h_{22} \end{bmatrix} \begin{bmatrix} i_1 \\ v_2 \end{bmatrix} \qquad (5)$$
$$i_2 = h_{21}i_1 + h_{22}v_2$$

These network parameters are determined from these equations under short- or open-circuit conditions.

Thus:

From Eq. (3), output open-circuited, $i_2 = 0$,

$$z_{11} = v_1/i_1 = \text{input impedance}$$
$$z_{21} = v_2/i_1 = \text{forward transfer impedance}$$

From Eq. (3), input open-circuited, $i = 0$,

$$z_{12} = v_1/i_2 = \text{reverse transfer impedance}$$
$$z_{22} = v_2/i_2 = \text{output impedance}$$

From Eq. (4), output short-circuited, $v_2 = 0$,

$$y_{11} = i_1/v_1 = \text{input admittance}$$
$$y_{21} = i_1/v_2 = \text{forward transfer admittance}$$

From Eq. (4), input short-circuited, $v_1 = 0$,

$$y_{12} = i_1/v_2 = \text{reverse transfer admittance}$$
$$y_{22} = i_2/v_2 = \text{output admittance}$$

From Eq. (5), output short-circuited, $v_2 = 0$,

$$h_{11} = v_1/i_1 = 1/z_{11} = \text{input impedance}$$
$$h_{21} = i_2/i_1 = \alpha = \text{forward current amplification}$$

From Eq. (5), input open-circuited, $i_1 = 0$,

$$h_{12} = v_1/v_2 = \mu = \text{reverse voltage amplification}$$
$$h_{22} = i_2/v_2 = 1/z_{22} = \text{output admittance}$$

The matrix form of Eqs. (3), (4), and (5) is useful when combining networks. It is very useful in determining the properties of terminated networks.

Table 11 shows the interrelations among the matrices of Eqs. (3), (4), and (5).

Table 11. z, y, and h Matrix Interrelations

From To	(z)		(y)		(h)	
(z)	z_{11}	z_{12}	$\dfrac{y_{22}}{\Delta^y}$	$\dfrac{-y_{12}}{\Delta^y}$	$\dfrac{\Delta^h}{h_{22}}$	$\dfrac{h_{12}}{h_{22}}$
	z_{21}	z_{22}	$\dfrac{-y_{21}}{\Delta^y}$	$\dfrac{y_{11}}{\Delta^y}$	$\dfrac{-h_{21}}{h_{22}}$	$\dfrac{1}{h_{22}}$
(y)	$\dfrac{z_{22}}{\Delta^z}$	$\dfrac{-z_{21}}{\Delta^z}$	y_{11}	y_{12}	$\dfrac{1}{h_{11}}$	$\dfrac{-h_{12}}{h_{11}}$
	$\dfrac{-z_{21}}{\Delta^z}$	$\dfrac{z_{11}}{\Delta^z}$	y_{21}	y_{22}	$\dfrac{h_{21}}{h_{11}}$	$\dfrac{\Delta^h}{h_{11}}$
(h)	$\dfrac{\Delta^z}{z_{22}}$	$\dfrac{z_{12}}{z_{22}}$	$\dfrac{1}{y_{11}}$	$\dfrac{-y_{12}}{y_{11}}$	h_{11}	h_{12}
	$\dfrac{-z_{21}}{z_{22}}$	$\dfrac{1}{z_{22}}$	$\dfrac{y_{21}}{y_{11}}$	$\dfrac{\Delta^y}{y_{11}}$	h_{21}	h_{22}

$\Delta^z = z_{11}z_{22} - z_{12}z_{21} = z$ determinant.
$\Delta^y = y$ determinant.
$\Delta^h = h$ determinant.

6 Transistor Equivalent Circuits

The three basic transistor connections used are common-emitter, common-base, and common-collector, corresponding to the transistor terminal that is common to both the input and output. These three connections are shown in Fig. 7 and can be considered somewhat analogous to the vacuum-triode connections of common-cathode, common-grid, and common-plate.

Table 12, from Ref. 1, summarizes the general characteristics to be expected from the transistor for the three types of connections.

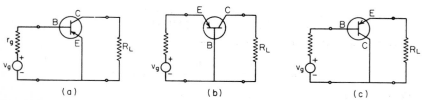

Fig. 7. Three possible transistor connections. (a) Common emitter. (b) Common base. (c) Common collector.

TRANSISTOR CHARACTERISTICS

Table 12

Common-emitter (grounded-cathode)*	Common-base (grounded-grid)*	Common-collector (grounded-plate)*
Large current gain	Approximately unity current gain	Large current gain
Large voltage gain	Large voltage gain	Approximately unity voltage gain
Highest power gain	Intermediate power gain	Lowest power gain
Low input resistance	Very low input resistance	High input resistance
High output resistance	Very high output resistance	Low output resistance

* Tube equivalents in parentheses.

The physical equivalent circuits useful for most low-frequency analysis are shown in Fig. 8 for the common-emitter, -base, and -emitter connections. These circuits are in terms of the r parameters as defined below, which represent resistances which are assumed to exist in the transistor.

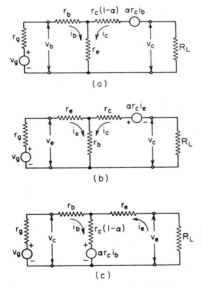

Fig. 8. Physical or T equivalent transistor circuits. (a) Common emitter. (b) Common base. (c) Common collector.

r_b = ohmic base resistance which ranges from tens to hundreds of ohms

r_e = *incremental* emitter resistance in ohms $\cong 26/I_E$ at room temperature, where I_E, the static emitter current, is in milliamperes (the actual relationship is kT/qI_E where k is Boltzmann's constant, T is temperature in degrees Kelvin, and q is electron charge)

r_c = *incremental* collector resistance, typically several megohms, since the collector is a reverse-biased diode

i_b, i_e, and i_c are the base, emitter, and collector *incremental* currents, respectively

$\alpha = i_c/i_e$ = short-circuit current-amplification factor

$\beta = i_c/i_b = \alpha/(1-\alpha)$

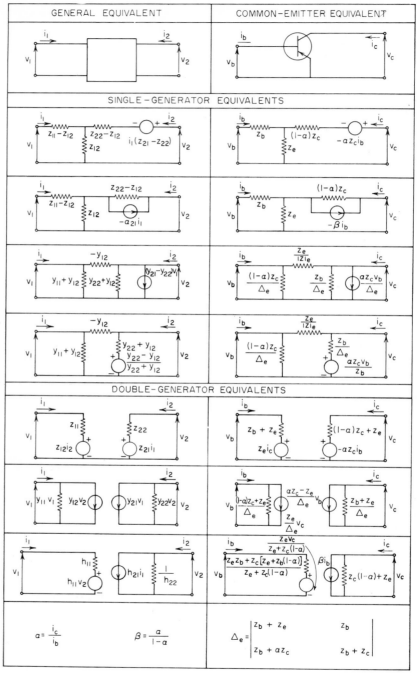

Fig. 9. Equivalent transistor circuits.

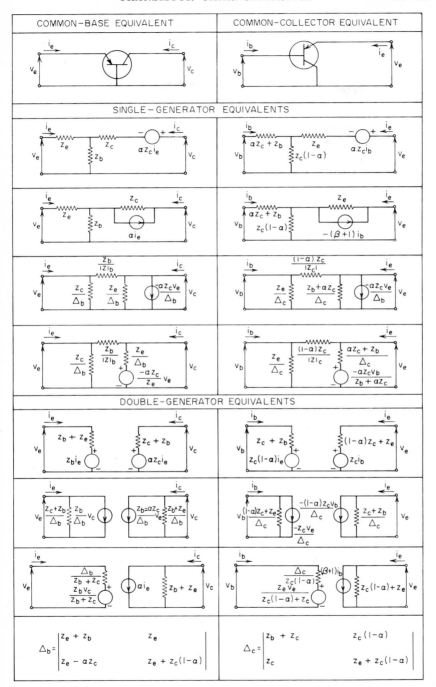

Fig. 9. Equivalent transistor circuits. (*Continued*)

Because the equivalent circuit is in the form of a T, these representations have been called T equivalent circuits and will be referred to in this fashion below. It is important to keep in mind the fact that these circuits relate *incremental* currents and voltages. Bias currents are required to maintain the proper quiescent operating points. Fixed bias is not practical with transistors, and self-regulating schemes must be used. Refer to Ref. 5, pages 105 to 109, for detailed approaches to this problem.

The equivalent circuits of Fig. 9 show transformations among the general equivalent circuits in terms of the parameters of Eqs. (3), (4), and (5). These are shown in the left-hand column. The relationship between the T equivalent-circuit single-generator representation and the z parameter double-generator circuit may be seen by comparing the second row with the fifth row. The circuits for common emitter, base, and collector are shown in each row. z_b, z_e, z_c are equivalent to r_b, r_e, and r_c in Fig. 8.

Subscript Notation. An alternative subscript notation has become common.

$11 = i = $ input
$12 = r = $ reverse
$21 = f = $ forward
$22 = o = $ output

Whether emitter, base, or collector configuration is meant is also indicated by subscript. As an example, Eq. (5) is rewritten for the common-base connection.

$$\begin{aligned} v_e &= h_{ib}i_e + h_{rb}V_c \\ i_c &= h_{fb}i_e + h_{ob}V_c \end{aligned} \qquad \begin{bmatrix} V_e \\ i_c \end{bmatrix} = \begin{bmatrix} h_{ib} & h_{rb} \\ h_{fb} & h_{ob} \end{bmatrix} \begin{bmatrix} i_e \\ V_c \end{bmatrix} \tag{6}$$

7 Hybrid Parameters

The hybrid or h parameters have been widely adopted for describing transistor characteristics because of their adaptability to precise measurement with common electronic laboratory instruments and unsophisticated methods. At present the h parameters are most often specified by transistor manufacturers. Developing a familiarity with these allows for more rapid quantitative analysis.

The h parameters are obtainable graphically from characteristic curves as shown in Figs. 10, 11, 12, and 13.

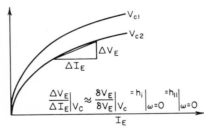

FIG. 10. Graphical determination of $h_i \big|_{\omega=0}$.

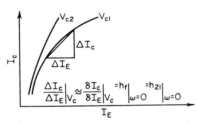

FIG. 11. Graphical determination of $h_f \big|_{\omega=0}$.

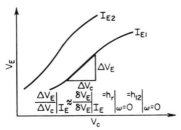

FIG. 12. Graphical determination of $h_r \big|_{\omega=0}$.

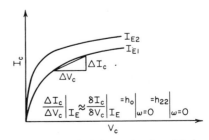

FIG. 13. Graphical determination of $h_o \big|_{\omega=0}$.

These figures show determination of the parameters by a change in current or voltage, which can be applied as slowly as desired; hence they represent the values at low-frequency conditions ($\omega = 0$). The measurements are made in accordance with the definitions in Sec. 5. However, there is a different set of h parameters for each of the transistor connections.

For example, the common-base h parameters are defined as follows:

$h_{ib} = h_{11b}$ = common-base input impedance with output short-circuited
$h_{rb} = h_{12b}$ = common-base reverse voltage amplification with input open-circuited
$h_{fb} = h_{21b}$ = common-base forward current amplification with output short-circuited
$h_{ob} = h_{22b}$ = common-base output admittance with input open-circuited

Relationship between the h and r parameters are given in Tables 13 and 14.

There is occasional use for y parameters in connection with high-frequency problems. These are related to the h and z parameters as shown in Table 11.

The input and output impedance, voltage, current, and power gains in terms of h and r parameters are given in Table 15.*

Input and output impedances, voltage, and current gains in terms of z and y parameters are set forth in Table 16.*

Typical values of h parameters are shown in Table 17.

8 Shunt-feedback Amplifier Stabilization

A general current-feedback equation can be derived by considering Fig. 14. The current gain is

$$A_i = \frac{i_o}{i_i} = \frac{A_i - BA_i}{i - \gamma A_i} = \frac{A(1 - B)}{1 - \gamma A} \qquad (7)$$

To achieve amplifier independence of transistor parameters the gain-feedback factor should be large such that

$$\gamma A \gg 1$$

The current gain becomes

$$A_i = (B - 1)/\gamma \qquad (8)$$

If feedback is obtained through a series resistor (shunt feedback), as shown in Fig. 15, then $B = \gamma$ and

$$A_i = (B - 1)/B \qquad (9)$$

For stability, B should be made a function of stable passive elements. If the input impedance of the feedback amplifier is considered negligible, then

$$B = r/(r + R_L) \qquad (10)$$

or from an admittance standpoint,

$$B = y/(y + Y_L) \qquad (11)$$

* Load and generator impedances, admittances, and resistances, etc. are designated here, and generally, by capital letters, thus Z_L, R_L, R_G. Subscript L generally designates the load.

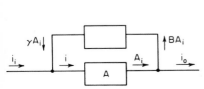

Fig. 14. Transistor current feedback.

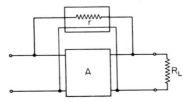

Fig. 15. Shunt resistance feedback.

Table 13. Common-base and Common-collector h Parameters*

h parameter	Common-emitter	Common-collector	T equivalent circuit
h_{ib}	$\dfrac{h_{ie}}{(1+h_{fe})(1-h_{re})+h_{ie}h_{oe}} \cong \dfrac{h_{ie}}{1+h_{fe}}$	$\dfrac{h_{ic}}{h_{ic}h_{oc}-h_{fc}h_{rc}} \cong -\dfrac{h_{ic}}{h_{fc}}$	$r_e + (1-\alpha)r_b$
h_{rb}	$\dfrac{h_{ie}h_{oe}-h_{re}(1+h_{fe})}{(1+h_{fe})(1-h_{re})+h_{ie}h_{oe}} \cong \dfrac{h_{ie}h_{oe}}{1+h_{fe}}-h_{re}$	$\dfrac{h_{fc}(1-h_{rc})+h_{ic}h_{oc}}{h_{ic}h_{oc}-h_{fc}h_{rc}} \cong h_{re}-1-\dfrac{h_{ic}h_{oc}}{h_{fc}}$	$\dfrac{r_b}{r_c+r_b} \cong \dfrac{r_b}{r_c}$
h_{fb}	$\dfrac{-h_{fe}(1-h_{re})-h_{ie}h_{oe}}{(1+h_{fe})(1-h_{re})+h_{ie}h_{oe}} \cong -\dfrac{h_{fe}}{1+h_{fe}}$	$\dfrac{h_{re}(1+h_{fc})-h_{ic}h_{oc}}{h_{ic}h_{oc}-h_{fc}h_{rc}} \cong -\dfrac{1+h_{fc}}{h_{fc}}$	$-\alpha$
h_{ob}	$\dfrac{h_{oe}}{(1+h_{fe})(1-h_{re})+h_{ie}h_{oe}} \cong \dfrac{h_{oe}}{1+h_{fe}}$	$\dfrac{h_{oc}}{h_{ic}h_{oc}-h_{fc}h_{rc}} \cong \dfrac{h_{oc}}{h_{fc}}$	$\dfrac{1}{r_c+r_b} \cong \dfrac{1}{r_c}$

h parameter	Common-emitter	Common-base	T equivalent circuit
h_{ic}	h_{ie}	$\dfrac{h_{ib}}{(1+h_{fb})(1-h_{rb})+h_{ob}h_{ib}} \cong \dfrac{h_{ib}}{1+h_{fb}}$	$r_b + \dfrac{r_e r_c}{r_e+r_c-\alpha r_c} \cong r_b + \dfrac{r_e}{1-\alpha}$
h_{rc}	$1-h_{re}$	$\dfrac{1+h_{fb}}{(1+h_{fb})(1-h_{rb})+h_{ob}h_{ib}} \cong 1$	$\dfrac{r_c-\alpha r_c}{r_e+r_c-\alpha r_c} \cong 1-\dfrac{r_e}{(1-\alpha)r_c}$
h_{fc}	$-(1+h_{fe})$	$\dfrac{h_{rb}-1}{(1+h_{fb})(1-h_{rb})+h_{ob}h_{ib}} \cong -\dfrac{1}{1+h_{fb}}$	$-\dfrac{r_c}{r_e+r_c-\alpha r_c} \cong \dfrac{-1}{1-\alpha}$
h_{oc}	h_{oe}	$\dfrac{h_{ob}}{(1+h_{fb})(1-h_{rb})+h_{ob}h_{ib}} \cong \dfrac{h_{ob}}{1+h_{fb}}$	$\dfrac{1}{v_e+r_c-\alpha r_c} \cong \dfrac{1}{(1-\alpha)r_c}$

* From Ref. 5.

Table 14. Common-emitter h Parameters and Transistor T Parameters*

h parameter	Common-base	Common-collector	T equivalent circuit
h_{ie}	$\dfrac{h_{ib}}{(1+h_{fb})(1-h_{rb})+h_{ob}h_{ib}} \cong \dfrac{h_{ib}}{1+h_{fb}}$	h_{ic}	$r_b + \dfrac{r_e r_c}{r_e + r_c - \alpha r_c} \cong r_b + \dfrac{r_e}{1-\alpha}$
h_{re}	$\dfrac{h_{ib}h_{ob} - h_{rb}(1+h_{fb})}{(1+h_{fb})(1-h_{rb})+h_{ob}h_{ib}} \cong \dfrac{h_{ib}h_{ob}}{1+h_{fb}} - h_{rb}$	$1 - h_{rc}$	$\dfrac{r_e}{r_e + r_c - \alpha r_c} \cong \dfrac{r_e}{(1-\alpha)r_c}$
h_{fe}	$\dfrac{-h_{fb}(1-h_{rb}) - h_{ob}h_{ib}}{(1+h_{fb})(1-h_{rb})+h_{ob}h_{ib}} \cong \dfrac{-h_{fb}}{1+h_{fb}}$	$-(1+h_{fc})$	$\dfrac{\alpha r_c - r_e}{r_e + r_c - \alpha r_c} \cong \dfrac{\alpha}{1-\alpha}$
h_{oe}	$\dfrac{h_{ob}}{(1+h_{fb})(1-h_{rb})+h_{ob}h_{ib}} \cong \dfrac{h_{ob}}{1+h_{fb}}$	h_{oc}	$\dfrac{1}{r_e + r_c - \alpha r_c} \cong \dfrac{1}{(1-\alpha)r_c}$

T parameter	Common-emitter	Common-base	Common-collector
α	$\dfrac{h_{fe}(1-h_{re})+h_{ie}h_{oe}}{(1+h_{fe})(1-h_{re})+h_{ie}h_{oe}} \cong \dfrac{h_{fe}}{1+h_{fe}}$	$-h_{fb}$	$\dfrac{h_{ic}h_{oc} - h_{rc}(1+h_{fc})}{h_{ic}h_{oc} - h_{fc}h_{rc}} \cong \dfrac{1+h_{fc}}{h_{fc}}$
r_c	$\dfrac{h_{fe}+1}{h_{oe}}$	$\dfrac{1-h_{rb}}{h_{ob}}$	$-\dfrac{h_{fc}}{h_{oc}}$
r_e	$\dfrac{h_{re}}{h_{oe}}$	$h_{ib} - (1+h_{fb})\dfrac{h_{rb}}{h_{ob}}$	$\dfrac{1-h_{rc}}{h_{oc}}$
r_b	$h_{ie} - \dfrac{h_{re}(1+h_{fe})}{h_{oe}}$	$\dfrac{h_{rb}}{h_{ob}}$	$h_{ic} + \dfrac{h_{fc}(1-h_{rc})}{h_{oc}}$
a	$\dfrac{h_{fe}+h_{re}}{1+h_{fe}}$	$-\dfrac{h_{fb}+h_{rb}}{1-h_{rb}}$	$\dfrac{h_{fc}+h_{rc}}{h_{fc}}$

* From Ref. 5.

Table 15. Transistor Impedance and Gain in Terms of h and T Parameters*

h parameter	Input impedance	Output impedance
Common-base T equivalent circuit	$Z_i = \dfrac{v_i}{i_i} = h_i - \dfrac{h_p h_r Z_L}{1 + h_o Z_L}$ $r_c + r_b + \dfrac{r_c - \alpha r_c + R_L}{r_c + r_b + R_L} \cong r_e + r_b(1-\alpha)$	$Z_o = \dfrac{v_o}{i_o} = \dfrac{1}{h_o - \dfrac{h_f h_r}{h_i + Z_g}}$ $r_c + r_b\left(1 - \dfrac{\alpha r_c + r_b}{r_e + r_b + R_g}\right) \cong r_c$
Common-emitter T equivalent circuit	$r_b + \dfrac{r_e(r_c + R_L)}{r_c - \alpha r_c + r_e + R_L} \cong r_b + \dfrac{r_e}{1-\alpha}$	$r_c - \alpha r_c + r_e\left(1 + \dfrac{\alpha r_c - r_e}{r_e + r_b + R_g}\right) \cong \dfrac{r_c}{1-\alpha}$
Common-collector T equivalent circuit	$r_b + \dfrac{r_c(r_e + R_E)}{r_c + \alpha r_c + r_e + R_L} \cong r_b + \dfrac{r_e + R_L}{1-\alpha}$	$r_e + (r_b + R_g)\dfrac{r_c - \alpha r_c}{r_c + r_b + R_g}$
h parameter where Z_g and Z_L are pure resistance	Insertion power gain $\left(\dfrac{\text{power into load}}{\text{power generator would deliver directly}}\right)$ $G_i = \dfrac{h_f{}^2(R_g + R_L)^2}{[(h_i + R_g)(1 + h_o R_L) - h_f h_r R_L]^2}$	Transducer power gain $\left(\dfrac{\text{power into load}}{\text{maximum available generator power}}\right)$ $G_t = \dfrac{4h_f{}^2 R_g R_L}{[(h_i + R_g)(1 + h_o R_L) - h_f h_r R_L]^2}$

	Current gain	Voltage gain
h parameter	$A_i = \dfrac{i_o}{i_i} = \dfrac{h_f}{1 + h_o Z_L}$	$A_v = \dfrac{v_o}{v_i} = \dfrac{1}{h_r - \dfrac{h_i}{Z_L}\dfrac{1 + h_o Z_L}{h_f}}$
Common-base T equivalent circuit	$\dfrac{\alpha r_c + r_b}{r_c + r_b + R_L} \cong \alpha$	$\dfrac{(\alpha r_c + r_b) R_L}{r_e(r_c + r_b + R_L) + r_b(r_c - \alpha r_c + R_L)} \cong \dfrac{\alpha R_L}{r_e + r_b(1-\alpha)}$
Common-emitter T equivalent circuit	$\dfrac{-(\alpha r_c - r_e)}{r_c - \alpha r_c + r_e + R_L} \cong \dfrac{\alpha}{1-\alpha}$	$\dfrac{-(\alpha r_c - r_e) R_L}{r_e(r_c + R_L) + r_b(r_c - \alpha r_c + r_e + R_L)} \cong -\dfrac{\alpha R_L}{r_e + r_b(1-\alpha)}$
Common-collector T equivalent circuit	$\dfrac{r_c}{r_c - \alpha r_c + r_e + R_L} \cong \dfrac{1}{1-\alpha}$	$\dfrac{r_c R_L}{r_c(r_e + R_L) + r_b(r_c - \alpha r_c + r_e + R_L)} \cong \dfrac{1}{1 + r_e + r_b \dfrac{1-\alpha}{R_L}}$
	Available power gain $\left(\dfrac{\text{maximum available output power}}{\text{maximum available generator power}}\right)$	Operating power gain $\left(\dfrac{\text{power into load}}{\text{power into transistor}}\right)$
h parameter where Z_g and Z_L are pure resistance	$G_a = \dfrac{h_f^2 R_g}{(h_i + R_g)[h_o(h_i + R_g) - h_f h_r]}$	$G_1 = A_v A_i = \dfrac{v_o i_o}{v_i i_i} = \dfrac{h_f}{1 + h_o R_L} h_r - \dfrac{h_i}{R_L}\dfrac{1 + h_o R_L}{h_p}$

* From Ref. 5.

Table 16. Transistor Impedance and Gain in Terms of z and y Parameters*

Parameter	Input impedance	Output impedance	Voltage gain	Current gain
z	$\dfrac{\Delta z + z_{11} Z_L}{z_{22} + Z_L}$	$\dfrac{\Delta z + z_{22} Z_g}{z_{11} + Z_g}$	$\dfrac{\Delta z + z_{11} Z_L}{z_{21} Z_L}$	$\dfrac{z_{22} + Z_L}{-z_{21}}$
y	$\dfrac{y_{22} + Y_L}{\Delta y + y_{11} Y_L}$	$\dfrac{y_{11} + Y_g}{\Delta y + y_{22} Y_g}$	$-\dfrac{y_{21}}{y_{22} + Y_L}$	$\dfrac{y_{21} Y_L}{\Delta y + y_{11} Y_L}$

$\Delta z = z_{11} z_{22} - z_{12} z_{21}.$
$\Delta y = y_{11} y_{22} - y_{12} y_{21}.$
* From Ref. 5.

Table 17*

	Common-emitter	Common-base	Common-collector
h_i	2,000 ohms	39 ohms	2,000 ohms
h_r	600×10^{-6}	380×10^{-6}	1
h_f	50	0.98	-51
h_o	25 μmhos	25 μmhos	25 μmhos

* From Ref. 1, chap. 11.

A transistor shunt-feedback circuit as shown in Fig. 16 can be treated as a pair of four-terminal networks. The forward and feedback networks are connected in parallel and therefore should be treated with admittance parameters which can be added. The admittance matrix equation for the feedback amplifier is

$$\begin{bmatrix} i_1 \\ i_2 \end{bmatrix} = \begin{bmatrix} y_{11} & y_{12} \\ y_{21} & y_{22} \end{bmatrix} \begin{bmatrix} e_1 \\ e_2 \end{bmatrix} \quad (12)$$

For the forward portion of the amplifier the admittance matrix is

$$\begin{bmatrix} i_{1t} \\ i_{2t} \end{bmatrix} = \frac{1}{\Delta} \begin{bmatrix} r_e + r_c(1 - \alpha) & r_e \\ \alpha r_c - r_e & r_e + r_b \end{bmatrix} \begin{bmatrix} e_1 \\ e_2 \end{bmatrix} \quad (13)$$

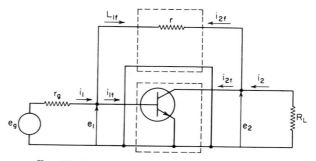

Fig. 16. Transistor amplifier with shunt feedback.

where $\Delta = r_e r_b + r_c(r_e + r_b(1 - \alpha))$. Terms have been defined in connection with Fig. 8. The admittance matrix equation for the feedback network is

$$\begin{bmatrix} i_{1f} \\ i_{2f} \end{bmatrix} = \frac{1}{r}\begin{bmatrix} 1 & -1 \\ -1 & 1 \end{bmatrix}\begin{bmatrix} e_1 \\ e_2 \end{bmatrix} \tag{14}$$

Matrix addition of Eqs. (13) and (14) results in

$$\begin{bmatrix} i_1 \\ i_2 \end{bmatrix} = \begin{bmatrix} \dfrac{r_e + r_c(1-\alpha)}{\Delta} + \dfrac{1}{r} & \dfrac{r_e}{\Delta} - \dfrac{1}{r} \\ \dfrac{\alpha r_c - r_e}{\Delta} - \dfrac{1}{r} & \dfrac{r_e + r_b}{\Delta} + \dfrac{1}{r} \end{bmatrix}\begin{bmatrix} e_1 \\ e_2 \end{bmatrix} \tag{15}$$

The input impedance is

$$z_i = \frac{y_{22} + Y_L}{\Delta_y + y_{11}Y_L} \tag{16}$$

Substituting from Eq. (15), the input impedance of the feedback amplifier is

$$z_i = \frac{(r_e + r_b)/\Delta + 1/r + 1/R_L}{\Delta_y + \{1/r + [r_e + r_c(1-\alpha)]/\Delta\}(1/R_L)} \tag{17}$$

$$= \frac{rR_L(r_e + r_b) + \Delta(r + r_e)}{\Delta + r[r_e + r_c(1-\alpha)] + rR_L + R_L(r_b + r_c)} \tag{18}$$

The output impedance is

$$z_o = \frac{y_{11} + y_g}{\Delta_y + y_{22}y_g} \tag{19}$$

Substituting from Eq. (15),

$$z_o = \frac{[r_e + r_c(1-\alpha)]/\Delta + 1/r + 1/r_g}{\Delta_y + (1/r_g)[(r_e + r_b)/\Delta + 1/r]} \tag{20}$$

$$= \frac{rr_g[r_e + r_c(1-\alpha)] + \Delta(r + r_g)}{\Delta + r(r_e + r_b) + r_g(r + r_b + r_c)} \tag{21}$$

The voltage gain becomes

$$A_v = -y_{21}/(y_{22} + Y_L) \tag{22}$$

Substituting from Eq. (15),

$$A_v = \frac{(-\alpha r_c + r_e)/\Delta + 1/r}{(r_e + r_b)/\Delta + 1/r + 1/R_L} \tag{23}$$

$$= \frac{(-\alpha r_c + r_e + \Delta)R_L}{rR_L(r_e + r_b) + \Delta(R_L + r)} \tag{24}$$

The current gain from Table 15 is

$$A_i = y_{21}Y_L/(\Delta_y + y_{11}Y_L) \tag{25}$$

Substituting from Eq. (15),

$$A_i = \frac{-i/r + (\alpha r_c - r_e)/\Delta}{\Delta_y + \{[r_e + r_c(1-\alpha)]/\Delta + 1/r\}} \tag{26}$$

$$= \frac{\Delta R_L - (\alpha r_c - r_e)rR_L}{\Delta_y \Delta rR_L + [r_e + r_c(1-\alpha)]r + \Delta} \tag{27}$$

$$G = A_v A_i \tag{28}$$

The admittance parameters in terms of h parameters are

$$y_{11} = 1/h_{11} \qquad y_{12} = -h_{12}/h_{11} \qquad y_{21} = h_{21}/h_{11} \qquad y_{22} = \Delta h/h_{11}$$

With shunt feedback, the amplifier characteristics in terms of hybrid parameters are

$$Z_i = \frac{\Delta_h + h_{11}(r + R_L)}{rR_L h_{22} + R_L(1 - h_{12} + h_{21} + \Delta_h) + h_{11} + r} \quad (29)$$

$$Z_o = \frac{1 + h_{11}(r + r_g)}{rr_g h_{22} + r_g(1 - h_{12} + h_{21} + \Delta_h) + h_{11} + r} \quad (30)$$

$$\left. \begin{array}{l} A_v = \dfrac{R_L(h_{11} - h_{21}r)}{rR_L \Delta_h + (R_L + r)h_{11}} \\[6pt] A_i = \dfrac{R_L(h_{11} - h_{21}r)}{rR_L h_{22} + R_L(1 - h_{12} + h_{21} + \Delta_h) + h_{11} + r} \\[6pt] G = A_v A_i \end{array} \right\} \quad (31)$$

If values for the h parameters are $h_{11} = 50$ ohms, $h_{12} = 150 \times 10^{-6}$, $h_{21} = -0.95$, $h_{22} = 0.6$ micromho, and if $R_L < 25$ kilohms, $r < 10R_L$, $r_g < 500$ ohms, then the negative-feedback amplifier characteristics simplify to

$$Z_i \simeq h_{11} \quad (32)$$
$$Z_o \simeq r_g \quad (33)$$
$$A_v \simeq h_{21} R_L / h_{11} \quad (34)$$
$$A_i \simeq h_{21} \quad (35)$$
$$G \simeq h_{21}^2 R_L / h_{11} \quad (36)$$

For the assumptions made, the foregoing equations give some indication of the stabilizing effect of shunt negative feedback. The dependence of amplifier characteristics on the transistor has been fairly well restricted to the two parameters h_{11} and h_{21}, both of which are fairly stable as a function of temperature when compared with the other h parameters.

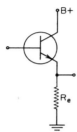

FIG. 17. Transistor amplifier with series feedback.

9 Series-feedback Amplifier Stabilization

Series feedback in a transistor circuit, as shown in Fig. 17, is similar to cathode degeneration in a vacuum-tube circuit. This is known as a common-collector connection and is characterized by relatively high input impedance and low output impedance. Like the cathode follower, the voltage gain is less than unity. Characteristics for the series-feedback amplifier are as follows:

$$z_i = r_b + r_c \frac{r_e + R_e}{r_c(1 - \alpha) + r_e + R_e} \quad (37)$$

$$z_o = r_e + r_c(1 - \alpha) \frac{r_g + r_b}{r_g + r_b + r_c} \quad (38)$$

$$A_v = \frac{r_c R_e}{r_b[r_c(1 - \alpha) + r_e + R_e] + r_c(r_e + R_e)} \quad (39)$$

$$A_i = \frac{r_c}{r_c(1 - \alpha) + r_e + R_e} \quad (40)$$

$$G = \frac{r_c^2 R_e}{[r_c(1 - \alpha) + r_e + R_e] + \{r_b[r_c(1 - \alpha) + r_e + R_e] + r_c(r_e + R_e)\}} \quad (41)$$

Assuming $r_e \ll r_c(1 - \alpha)$; $r_b \ll r_c$; $r_e \ll R_e \ll r_c(1 - \alpha)$; $r_g \ll r_c$, the series-feedback amplifier characteristics simplify to

$$Z_i = r_b + R_e/(1 - \alpha) \quad (42)$$
$$Z_o = R_e + (r_b + r_g)(1 - \alpha) \quad (43)$$
$$A_v = \frac{R_e}{r_b(1 - \alpha) + R_e} \quad (44)$$
$$A_i = \frac{1}{1 - \alpha} \quad (45)$$
$$G = \frac{R_e}{r_b(1 - \alpha)^2 + R_e(1 - \alpha)} \quad (46)$$

10. External Amplifier Stabilization

The problem of stabilizing transistor amplifiers with respect to temperature has been analyzed[1] with temperature incremental techniques similar to those used in the "small-signal analysis" of transistor parameters. Transistor d-c characteristics for the common-base connection are

$$I_e = f_1(V_{eb}, V_{cb})$$
$$I_c = f_2(V_{eb}, V_{cb}) \qquad (47)$$

Partial differentiation of (47) with respect to absolute temperature results in

$$dI_e/dT = \partial f_1/\partial T + (\partial f_1/\partial V_{eb})(dV_{eb}/dT) + (\partial f_1/\partial V_{cb})(dV_{cb}/dT)$$
$$dI_c/dT = \partial f_2/\partial T + (\partial f_2/\partial V_{eb})(dV_{eb}/dT) + (\partial f_2/\partial V_{cb})(dV_{cb}/dT) \qquad (48)$$

Substitute for the coefficients

$$\partial f_1/\partial V_{eb} = g_{ee} \qquad \partial f_1/\partial V_{cb} = g_{ec}$$
$$2f_2/2V_{eb} = g_{ce} \qquad 2f_2/2V_{cb} = g_{cc}$$

and let primes indicate temperature derivatives so that (48) becomes more simply

$$I_e' = I_{et}' + g_{ee}V_{eb}' + g_{ec}V_{cb}'$$
$$I_c' = I_{ct}' + g_{ce}V_{eb}' + g_{cc}V_{cb}' \qquad (49)$$

The incremental temperature equivalent circuit for the common-base connection is shown in Fig. 18.

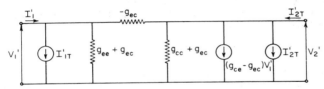

FIG. 18. Transistor common-base incremental temperature equivalent circuit.

Shockley, Sparks, and Teal have generated the following temperature-dependent quiescent relationships for the common-base connection:

$$I_e = (G_{ee}/\Lambda)(E^{\Lambda V_{eb}} - 1) + (G_{ec}/\Lambda)(E^{\Lambda V_{cb}} - 1) \qquad (50)$$
$$I_c = (G_{ce}/\Lambda)(E^{\Lambda V_{eb}} - 1) + (G_{cc}/\Lambda)(E^{\Lambda V_{cb}} - 1) \qquad (51)$$

where $\Lambda = q/kT$ for p-n-p, $\Lambda = -q/kT$ for n-p-n
k = Boltzmann's constant
q = electronic charge

The conductance coefficients are related to the intrinsic-type, p-type, and n-type conductivities such that

where $\quad G_{xy} \propto \sigma_i^2/\sigma_p \quad$ for p-n-p $\quad G_{xy} \propto \sigma_i^2/\sigma_n \quad$ for n-p-n
and $\qquad\qquad\qquad\qquad \sigma_i^2 \propto E^{-E/kT^2}$
so that $\qquad\qquad\qquad \sigma_n \text{ or } \sigma_p \propto T^{-3/2}$
$\qquad\qquad\qquad\qquad G_{xy} \propto E^{E_g/kT} T^{3/2}$

where E_g is the energy gap between the valence and conduction band.

The percentage change of the G coefficients as a function of temperature is expressed by

$$(dG_{xy}/dT)(1/G_{xy}) = (1/T)(E_g/kT + 3/2) \simeq 10 \text{ per cent per °K for germanium,}$$
$$15 \text{ per cent per °K for silicon} \qquad (52)$$

The common-emitter connection temperature-incremental equations can be obtained from the associated d-c characteristic. This results in

$$I_b = (1/\Lambda)(G_{ee} + G_{ec} + G_{ce} + G_{cc}) - (1/\Lambda)[G_{ee} + G_{ce} + (G_{ec} + G_{cc})E^{\Delta V_{ce}}]E^{-\Delta V_{be}}$$
$$I_c = (-1/\Lambda)(G_{ee} + G_{cc}) + (1/\Lambda)(G_{ce} + G_{cc}E^{\Delta V_{ce}})E^{-\Delta V_{be}} \tag{53}$$

The corresponding temperature-differential equations are

$$I_b' = I_b T' + g_{bb}V_{be}' + g_{bc}V_{ce}'$$
$$I_c' = I_c T' + g_{cb}V_{be}' + g_{cc}V_{ce}' \tag{54}$$

where $g_{bb} = g_{ee} + g_{ec} + g_{ce} + g_{cc}$
$g_{bc} = -g_{ec} - g_{cc}$
$g_{cb} = -g_{ce} - g_{cc}$

The incremental temperature equivalent circuit for the common-emitter connection is shown in Fig. 19.

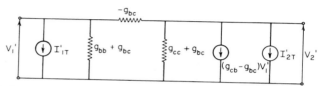

Fig. 19. Transistor common-emitter incremental temperature equivalent circuit.

Temperature considerations so far have been for the intrinsically perfect transistor. Extrinsic effects, which include base spreading resistance and collector leakage resistance and collector leakage resistance, should also be taken into account. This can be accomplished by substituting

$$V_{eb} = V_{eb}' + I_b r_{bb}$$
$$V_{cb} = V_{cb}' + I_b r_{bb} \tag{55}$$

in Eq. (51).

Circuitry external to the transistor will affect the temperature dependence of the transistor. This is especially true of the d-c circuitry. The transistor nodal equations are

$$I_1 = g_{11}V_1 + g_{12}V_2 + \sum_i G_i E_i$$
$$I_2 = g_{21}V_1 + g_{22}V_2 + \sum_j G_j E_j \tag{56}$$

from which the incremental-temperature equations are found to be

$$I_1' = g_{11}V_1' + g_{12}V_2' + \theta_1$$
$$I_2' = g_{21}V_1' + g_{22}V_2' + \theta_2 \tag{57}$$

If the network is constant as a function of temperature, then $\theta_1 = \theta_2 = 0$. If thermistors are used, θ_1 and θ_2 will vary conductance with temperature according to the relationship

$$G_T = G_{T_0}E^{B/T_0}E^{-B/T}$$

where G_{T_0} is the conductance at $T = T_0$ and B is a semiconductor-material constant. G_T, the thermistor conductance, changes with temperature according to

$$G_T' = BG_T/T_2$$

The complete thermal behavior and equivalent thermal equivalent circuit can be derived from Eq. (56). For a common-emitter transistor amplifier where $I_1' = I_b'$, $I_2' = -I_c'$, $V_1' = V_{be}'$, and $V_2' = V_{ce}'$, the representative thermal equations are found from (54) and (56) to be

$$-I_{bT}' - \theta_1 = (g_{bb} + g_{11})V_{be}' + (g_{bc} + g_{12})V_{ce}'$$
$$-I_{cT}' - \theta_2 = (g_{cb} + g_{21})V_{be}' + (g_{cc} + g_{22})V_{ce}' \qquad (58)$$

Solving for the temperature-dependent voltages results in

$$V_{be}' = \frac{(g_{bc} + g_{12})(I_{cT}' + \theta_2) - (g_{cc} + g_{22})(I_{bT}' + \theta_1)}{(g_{bb} + g_{11})(g_{cc} + g_{22}) - (g_{bc} + g_{12})(g_b + g_{21})}$$
$$V_{ce}' = \frac{(g_{cb} + g_{21})(I_{bT}' + \theta_1) - (g_{bb} + g_{11})(I_{cT}' + \theta_2)}{(g_{bb} + g_{11})(g_{cc} + g_{22}) - (g_{bc} + g_{12})(g_{cb} + g_{21})} \qquad (59)$$

Solutions for the four parameters g_{11}, g_{12}, g_{21}, and g_{22} may be gleaned from the simultaneous solution of (56) and (59).

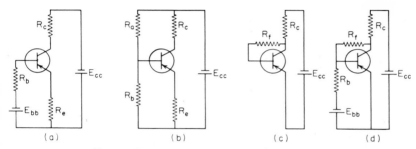

FIG. 20. Some common-emitter biasing networks.

Many transistor biasing networks are possible. Some of the more usual common-emitter biasing circuits are shown in Fig. 20. Analysis of Fig. 20a results in the conductance matrix in terms of the resistance network.

$$\begin{bmatrix} g_{11} & g_{12} \\ g_{21} & g_{22} \end{bmatrix} = \frac{1}{\Delta_R} \begin{bmatrix} R_e + R_c & -R_e \\ -R_e & R_e + R_b \end{bmatrix} \qquad (60)$$

where
$$\Delta_R = (R_e + R_b)(R_e + R_c) - R_e^2$$

Following substitution of (60) into (59) there ensues the temperature-incremental equations for this particular bias network in the form

$$V_{be}' = \frac{(g_{bc} - R_e/\Delta_R)I_{cT}' - [g_{cc} + (R_e + R_b)/\Delta_R]I_{bT}'}{[g_{bb} + (R_e + R_c)/\Delta_R][g_{cc} + (R_e + R_b)/\Delta_R] - (g_{bc} - R_e/\Delta_R)(g_{cb} - R_e/\Delta_R)}$$
$$V_{ce}' = \frac{(g_{cb} - R_e/\Delta_R)I_{bT}' - [g_{bb} + (R_e + R_b)/\Delta_R]I_{cT}'}{[g_{bb} + (R_e + R_c)/\Delta_R][g_{cc} + (R_e + R_b)/\Delta_R](-g_{bc} - R_e/\Delta_R)(g_{cb} - R_e/\Delta R)} \qquad (61)$$

Figure 21 is an example of experimental data-verifying equations (61) taken from an amplifier circuit using the connection of Fig. 20a. Note the linear variation of V_{be}, V_{ce}, I_b, and I_c as a function of temperature.

Similarly for Fig. 20d,

$$\begin{bmatrix} g_{11} & g_{12} \\ g_{21} & g_{22} \end{bmatrix} \begin{bmatrix} G_b + G_f & -G_f \\ -G_f & G_c + G_f \end{bmatrix} \qquad (62)$$

where $G_b = 1/R_b$, $G_f = 1/R_f$, $G_c = 1/R_f$. Then

$$V_{be}' = \frac{(g_{bc} - G_f)I_{cT}' - (g_{cc} + G_c + G_f)I_{bT}'}{(g_{bb} + G_b + G_f)(g_{cc} + G_c + G_f) - (g_{bc} - G_f)(g_{cb} - G_f)}$$
$$V_{ce}' = \frac{(g_{cb} - G_f)I_{bT}' - (g_{bb} + G_b + G_f)I_{cT}'}{(g_{bb} + G_b + G_f)(g_{cc} + G_c + G_f) - (g_{bc} - G_f)(g_{cb} - G_f)} \qquad (63)$$

The following example illustrates the manner in which the foregoing incremental equations can be utilized for the stabilization of transistor amplifier characteristics with variations in temperature.

Consider the common-emitter amplifier of Fig. 22, wherein a thermistor (temperature-variable resistor) is used as a temperature-sensitive conductance element G_T to

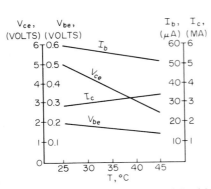

FIG. 21. Variation of V_{be}, V_{ce}, I_c, and I_b with temperature.

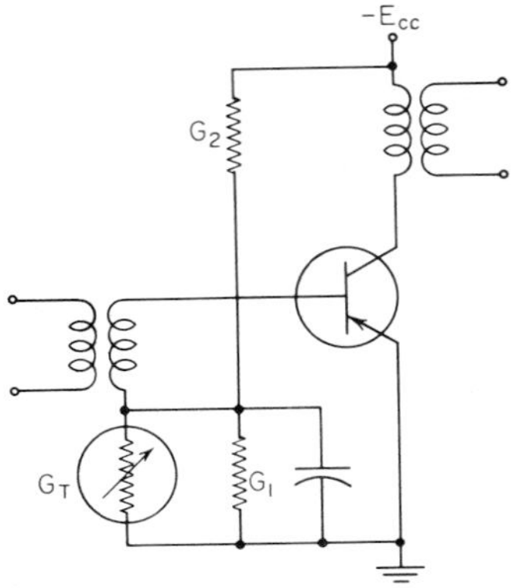

FIG. 22. Thermistor compensated common-emitter amplifier.

stabilize I_c with temperature. Biasing is also a function of conductances G_1 and G_2. Substituting in Eqs. (56) and (57) we obtain

$$-I_b = (G_T + G_1 + G_2)V_{be} - G_2 E_{cc} \tag{64}$$
$$-I_b' = (G_T + G_1 + G_2)V_{be}' + G_T'V_{be} \tag{65}$$

Since $V_{ce} + E_{cc} = $ constant, then $V_{ce}' = 0$. From (64) and (65)

$$V_{be}' = \frac{-I_{bT}' - G_T'V_{be}}{g_{bb} + G_T + G_1 + G_2} \tag{66}$$

$$I_c' = I_{cT}' + g_{cb}V_{be}' \tag{67}$$

Since we wish to stabilize I_c with temperature, $I_c' = 0$; then

$$I_{cT}' = -g_{cb}V_{be}'$$

$$I_{cT}' = -g_{cb}\frac{-I_{bT}' - G_T'V_{be}}{g_{bb} + G_T + G_1 + G_2} \tag{68}$$

Substituting (64) into (68), we obtain

$$G_2 = \left(g_{cb}\frac{I_{bT}' + G_T'V_{be}}{I_{cT}'} - g_{bb}\right)\frac{V_{be}}{E_{cc}} + \frac{I_b}{E_{cc}} \tag{69}$$

All the conductance parameters to stabilize I_c with temperature can then be calculated.

References

1. L. P. Hunter et al., "Handbook of Semiconductor Electronics," Sections 11, 12, 2d ed., McGraw-Hill Book Company, New York, 1962.
2. Richard F. Shea et al., "Principles of Transistor Circuits," Chaps. 3, 9, John Wiley & Sons, Inc., New York, 1953.
3. A. W. Lo et al., "Transistor Electronics," Chap. 2, Prentice-Hall, Inc., Englewood Cliffs, N.J., 1955.

4. Richard F. Shea et al., "Transistor Circuit Engineering," Chap. 2, John Wiley & Sons, Inc., New York, 1957.
5. Texas Instruments, Inc., "Transistor Circuit Design," Chaps. 6, 7, McGraw-Hill Book Company, New York, 1963.
6. L. M. Vallese, Temperature Stabilization of Transistor Amplifiers, *Commun. Electron.*, vol. 75, pp. 379–384, 1956.
7. A. W. Lo et al., "Transistor Electronics," p. 304, Prentice-Hall, Inc., Englewood Cliffs, N.J., 1955.

11 Self-stabilization of Low-frequency Small-signal Transistor Parameters

The following analysis is concerned with the stabilization of the low-frequency small-signal parameters of a transistor over a wide temperature range. The analysis is based on the transistor bias voltages V_E and V_c shown in Fig. 23, where V_E = base-to-emitter d-c voltage and V_c = base-to-collector d-c voltage.

The low-frequency small-signal characteristics of a transistor are completely described by the four conductance parameters g_{11}, g_{12}, g_{21}, g_{22}, which can be defined as follows:

$$\begin{aligned} g_{11} &= \frac{\partial I_e}{\partial V_E}\bigg|_{V_c = \text{const}} \\ g_{21} &= \frac{\partial I_c}{\partial V_E}\bigg|_{V_c = \text{const}} \\ g_{12} &= \frac{\partial I_e}{\partial V_c}\bigg|_{\substack{V_E = \text{const} \\ W = \text{const}}} \\ g_{22} &= \frac{\partial I_c}{\partial V_c}\bigg|_{\substack{V_E = \text{const} \\ W = \text{const}}} \end{aligned} \quad (70)$$

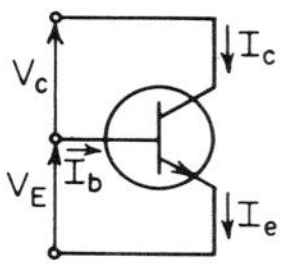

Fig. 23. Transistor circuit.

where I_e = d-c emitter current
I = d-c collector current
V_E = d-c voltage measured from emitter to base
V_c = d-c voltage measured from collector to base
W = effective base width

The following approximations have been shown in the literature:[1]

$$\begin{aligned} g_{11} &\simeq KT^{3\!/\!2} \exp \frac{qV_E - E_{ge}}{kT} \\ g_{21} &\simeq -KT^{3\!/\!2} \exp \frac{qV_E - E_{gc}}{kT} \\ g_{12} &\simeq -KT^{3\!/\!2} \exp \frac{qV_c - E_{ge}}{kT} \\ g_{22} &\simeq KT^{3\!/\!2} \exp \frac{qV_c - E_{gc}}{kT} \end{aligned} \quad (71)$$

where T = temperature
q = electronic charge
E_{ge} = emitter band-gap energy
E_{gc} = collector band-gap energy
k = Boltzmann's constant
$K = b/(1 + b)^2 W$ = const
$b = \mu_n/\mu_p$ = const
μ_n = electron mobility
μ_p = hole mobility

To stabilize g_{xy} with temperature it is required that

$$dg_{xy}/dT = 0 \quad (72)$$

Carrying out the differentiation results in the equations of parameter stability. For g_{11} and g_{21} to be stable with temperature

$$V_E = K_1 - K_3 T \tag{73}$$

Similarly, to stabilize g_{22} and g_{12} with temperature

$$V_c = K_2 - K_3 T \tag{74}$$

where $K_1 = E_{ge}/q = \text{const}$
$K_2 = E_{gc}/q = \text{const}$
$K_3 = 3k/2q = \text{const}$

K_1, K_2, and K_3 are all constant terms which may vary in value for different transistors and junctions because of crystal impurities and leakage effects. Therefore, K_3 should not be considered as always being equal in Eqs. (73) and (74), because of physical differences in the emitter and collector junctions of commercial transistors. The foregoing has established that the transistor d-c bias voltages V_E and V_C must vary linearly with temperature according to Eqs. (73) and (74) to stabilize the transistor conductance parameters g_{xy}.

The conductance parameters can be expressed in terms of resistance, hybrid, or any other type of parameters. In other words, stabilization of all parameters of one type infers that all parameters of any other type are also stabilized. The resistance r and hybrid h parameters can be transformed from the conductance g parameters in the following way:

$$\begin{aligned}
r_{11} &= g_{22}/\Delta_g \\
r_{12} &= -g_{12}/\Delta_g \\
r_{21} &= -g_{21}/\Delta_g \\
r_{22} &= g_{11}/g \\
h_{11} &= 1/g_{11} \\
h_{12} &= -g_{12}/g_{11} \\
h_{21} &= g_{21}/g_{11} \\
h_{22} &= \Delta_g/g_{11} \\
\Delta_g &= g_{11}g_{22} - g_{12}g_{21}
\end{aligned} \tag{75}$$

Since

$$\begin{aligned}
g_{11} &\propto g_{21} \\
g_{22} &\propto g_{12}
\end{aligned} \tag{76}$$

the following proportionalities are obtained from Eq. (75):

$$\begin{aligned}
r_{11}, r_{12} &\propto 1/g_{11} \\
r_{21}, r_{22} &\propto 1/g_{22} \\
h_{11} &\propto 1/g_{11} \\
h_{12} &\propto g_{22}/g_{11} \\
h_{21} &= \text{const} \\
h_{22} &\propto g_{22}
\end{aligned} \tag{77}$$

The stabilization of g_{11} and g_{22} should mean, therefore, that all g, r, and h parameters are stabilized. In other words, with the assumptions made, the stabilization of the small-signal low-frequency parameters g_{11} and g_{22} should stabilize all types of small-signal low-frequency parameters associated with transistors.

Figure 24 shows the bias network that can be used to stabilize the small-signal low-frequency parameters of a junction transistor.

Proper ratios of R_1/R_2 and R_3/R_4 are determined by two factors. First, a desired d-c operating point is required and second, the emitter and collector voltages must be made to satisfy Eqs. (73) and (74) as a function of temperature. The two requirements must be satisfied simultaneously, but if the temperature-stabilization consideration is paramount, then the d-c operating point will depend on the stabilization network.

The emitter-to-base and collector-to-base d-c bias voltages will depend on the transistor temperature characteristic which is, in turn, controlled by the resistor bias network. The d-c circuit is qualitatively that of Fig. 25.

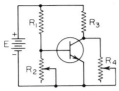

Fig. 24. Transistor d-c bias control circuit.

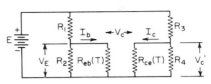

Fig. 25. D-C bias circuit with temperature-sensitive transistor characteristics.

Resistors R_1 and R_3 act as series dropping resistors. R_2 and R_4 are used as current bleeders. $R_{eb}(T)$ is the equivalent d-c resistance from emitter to base and is a function of temperature. $R_{ce}(T)$ is the equivalent d-c resistance from collector to emitter and is also a function of temperature. It is considered that

$$V_E \ll V_C$$
therefore $$V_C' \simeq V_C$$

The equivalent transistor resistances $R_{eb}(T)$ and $R_{ce}(T)$ are dependent on V_E and V_C, respectively, as well as temperature. Because of poor tolerance control of transistor parameters in manufacture in the present state of the art, it is necessary to find the proper bias-resistance values for R_2 and R_4 by trial-and-error test methods. The idea is to make the transistor self-biased so as to satisfy Eqs. (73) and (74).

Experimental verification of the dependence of h_{11} ($= 1/g_{11}$) on V_e and V_c on I_B and I_E is shown in Fig. 26. The solid-line plots for each variable are for the case where h_{11} is unstabilized as a function of temperature. The dashed-line curves are for the case where h_{11} is temperature-stable. Study of these curves shows a change in slope or linearity of V_c and V_E when h_{11} is temperature-sensitive. On the other hand, linear variation of V_E and V_C with temperature results in a constant condition for h_{11}. Note the linearization of I_B and I_E as well, which is to be expected if V_E and V_C are linearized.

The following method can be used to temperature-stabilize the transistor conductance parameters. (Note: The stabilization of g_{11} and g_{22} can be attempted one at a time or at the same time, depending on requirements, available equipment, and convenience.)

1. Choose a d-c operating point for the transistor.
2. Calculate values of R_1, R_2, R_3, and R_4 so that the transistor will operate at the chosen d-c operating point.
3. Insert the transistor into a variable-temperature chamber and measure the conductance parameter g_{11} and g_{22} at various temperatures T over the desired temperature range.
4. Note whether g_{11} vs. T and g_{22} vs. T are within the desired tolerance. If they are, then the desired temperature stabilization is accomplished. If not, note whether the coefficient of g-parameter drift is positive or negative with temperature. Then decrease R_2 and R_4 and repeat the procedure from step 3.
5. If the g parameters increase their coefficient of temperature drift, the resistance R_2 and R_4 should be increased. If the g parameters decrease their coefficient of temperature drift, the adjustment of R_2 and R_4 is in the proper direction. This adjustment, followed by a temperature test, can then be repeated until the desired temperature stabilization is realized.
6. If the coefficient of temperature drift of g parameters cannot be stabilized to the desired tolerance by any adjustment of R_2 and R_4, then R_1 and R_3 should be decreased and the procedure repeated from step 3. If the drift coefficient of the g parameter

10-40 TELEMETRY-SYSTEM COMPONENT DESIGN

increases, then R_1 and R_3 should be increased and the procedure repeated from step 3.

The stabilizing network and procedure described are intended to be the simplest possible to achieve fairly good stability of transistor parameters over a wide temperature range without sacrificing much voltage or power gain. A small number of components are used. Thus the values of resistance have been set so that the transistor acts as a self-biasing device to stabilize its small-signal low-frequency parameters

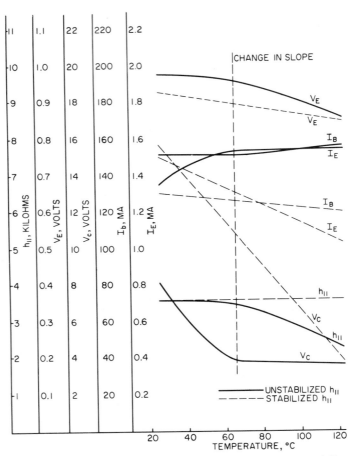

FIG. 26. Temperature dependence of h_{11} as a function of V_E and V_C.

as a function of temperature. Temperature-sensitive resistors such as thermistors and sensistors can also be used in the bias control network. However, the effects of thermal lag between the transistor and the temperature-sensing element may not be satisfactory if rapid temperature transients are experienced. On the other hand, transistor self-stabilization becomes evident on demand.

Indirect methods can be utilized to stabilize the conductance parameters. For example, the frequency of an oscillator may depend on one or more of the conductance parameters. The foregoing method of stabilization can still be used except that frequency measurements can be substituted for conductance-parameter measurements. This would be a simplification since the frequency can be monitored by an electronic digital counter instead of having to read two voltmeters for conductance measurements. The frequency reading would be a direct indication of the stability whereas the ratio of the voltmeter readings would have to be computed to provide conductance measurements.

Other indirect methods can also be used. For example, the amplitude of a transistor amplifier that would be a function of transistor conductance parameters might be measured. Indirect methods require accurate analysis to determine and verify dependent parameters.

A secondary effect is also present that can be used for temperature-stabilizing the input conductance as a function of temperature. This effect is due to the forward-biased base-to-emitter diode of the transistor that acts as a rectifying element in a-c circuits (see Fig. 27). The magnitude of this effect becomes greater as the a-c

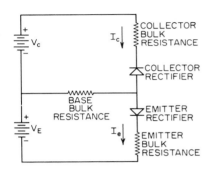
Fig. 27. Diode equivalent transistor circuit.

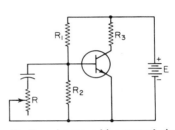
Fig. 28. Transistor a-c bias control circuit.

signal amplitude is increased. As a result, the emitter bias V_E can also be controlled by changing a passive parameter that affects the voltage amplitude in an a-c part of the circuit. For example, the resistor R in Fig. 28 can be adjusted to stabilize the effective input conductance.

Since this type of stabilization depends on relatively large signals applied between the base and emitter, small-signal theory cannot be used and small-signal parameter measurements will not give accurate results. Therefore, temperature stabilization of the input impedance of the transistor is more readily obtained if the measurements are made on a circuit connected according to its intended application. The resistance R can be placed in any convenient part of an a-c circuit. For a sine-wave signal it has been noted that this type of temperature stabilization of the input impedance of a transistor is most easily accomplished when the amplitude of the sine wave is such that it appears to be on the verge of, or just slightly, clipping when viewed with an oscilloscope.

The procedure to be followed when using this type of stabilization is very similar to the one previously outlined except that an effective conductance, frequency, or amplitude, depending on the application, is monitored as a function of temperature, rather than the small-signal conductance parameters.

References

1. Lo, Endres, Zawels, Waldhauer, and Cheng, "Transistor Electronics," pp. 279, 281, 286, 304, 305, Prentice-Hall, Inc., Englewood Cliffs, N.J., 1955.
2. Richard F. Shea, "Principles of Transistor Circuits," p. 102, John Wiley & Sons, Inc., New York, 1953.

TRANSISTORIZED TELEMETERING CIRCUITS

12 FM-FM Circuits and Systems

Telemetering programs require advances in the art so as to provide reliable instrumentation through critical environments. The introduction of solid-state devices, integrated circuits, and molecular electronics to the development of new telemetering systems is providing increased reliability and information capacity. Important advantages which have been realized include high resistance to shock and vibration, reduced power requirements, and a high degree of miniaturization.

A typical FM-FM telemetering system is shown in block form in Fig. 29. Variable resistance, induction, and voltage-controlled oscillators are used in conjunction with

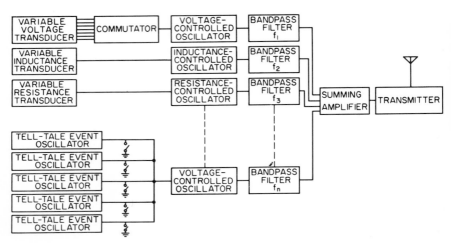

Fig. 29. Block diagram of transistorized FM-FM telemetering system.

their respective transducers and signal-conditioning circuits to provide most of the instrumentation generally required. Variable-inductance transducers are available for instrumentation of pressure, acceleration, and position. The voltage-controlled oscillator is a high-sensitivity unit which can be modulated with voltages as low as those normally encountered with thermocouple measurements. The variable-resistance oscillator is designed to operate with potentiometer-type transducers or any variable-resistance device. Tone oscillators may be utilized to function as telltale discrete-event indicators.

13 Resistance-controlled Oscillator

The variable-resistance controlled oscillator in Fig. 30 utilizes a resistance-capacitance phase-shift network wherein three or more L sections are cascade-coupled from the output of a transistor amplifier back to its input.

The amplifier and RC network each contribute 180° of phase shift, thereby satisfying the condition that the closed-loop gain be unity for oscillations to be maintained. A resistance leg of the feedback network, such as R_2, may be that of the variable-resistance transducer to be used. Neglecting transistor loading on the network, the oscillating frequency for an n-section phase-shift oscillator is

$$\omega = (R_1 C_1 R_2 C_2 \cdots R_n C_n)^{-1/n}$$

This relationship holds if the source impedance is zero and the load impedance infinite. The resultant curve for this type of oscillator is shown in Fig. 31.

An interesting application of this particular oscillator is for a solid-state missile-roll indicator, wherein a light-sensitive semiconductor diode is utilized as a variable-resistance element. The sun therefore becomes a source of reference from which missile roll rate can be determined. A typical application of a missile-roll indicator

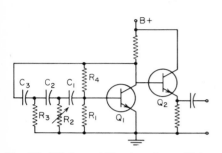

Fig. 30. RC phase-shift subcarrier oscillator with emitter-follower output.

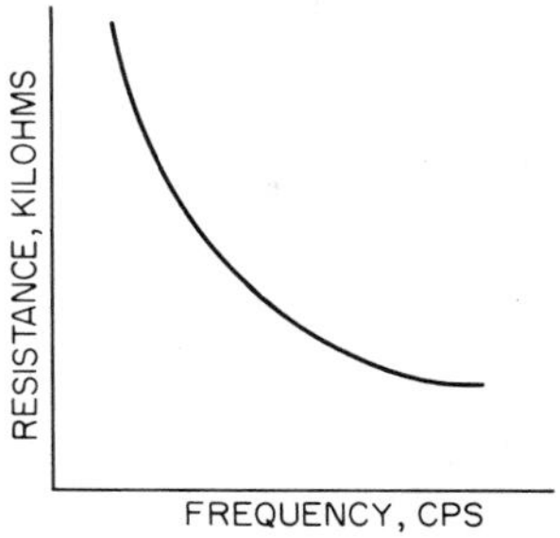

Fig. 31. Resistance vs. frequency for RC oscillator.

positioned on the outer shell of a ballistic missile may be seen in Fig. 32. The lens is quartz to resist high temperatures due to frictional skin heating during atmospheric reentry. All electronic components including the diode are embedded in an epoxy resin.

14 Inductance-controlled Oscillator

Analysis shows that inductance-controlled transistor oscillators of the Colpitts type can be designed so as to have minimum dependence upon transistor parameters which have considerable variation over an extended temperature range.[1] Negative feedback is helpful in decreasing transistor effects on oscillator drift, but another

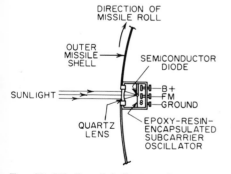

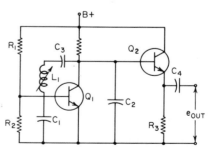

Fig. 32. Missile roll indicator using sun as reference.

Fig. 33. Inductance-controlled subcarrier oscillator.

means of stabilization is sometimes more effective if simplicity and a minimum number of transistor stages are to be utilized.

The Colpitts-type oscillator, shown in Fig. 33, appears preferable to the Hartley type under the consideration that it allows for simple adjustment of the feedback ratio to maintain sinusoidal oscillation. It is accomplished by capacitance tapping of the tank circuit as opposed to inductance tapping. This affords a design freedom for obtaining optimum operation with variable-inductance transducers having different characteristics. Referring to Fig. 33, transistor $Q1$ is the active element of the oscillator. $Q2$ provides a low-impedance emitter-follower output and is direct-coupled from $Q1$. The oscillator is tuned to a specific frequency by means of capacitor $C2$. Resistor $R2$ is adjusted to provide the required balance between gain, bias, and frequency stability.

An analysis of this oscillator for the high-Q coils is available in the literature.[2] It is necessary to extend the analysis to the low-Q case since variable-inductance transducers available for telemetering purposes generally have a relatively low L/R ratio. A comparison of the high- and low-Q cases shows different oscillating and stability criteria. The a-c equivalent circuit for the transistorized Colpitts oscillator using a low-Q inductor is shown in Fig. 34. For oscillations to occur the loop gain is unity and the current feedback ratio is

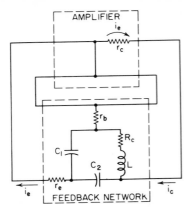

Fig. 34. Transistor equivalent of Colpitts oscillator.

$$\frac{i_c}{i_e} = \mu \approx \frac{L/C_1 + R_c/j\omega C_1 + r_b(j\omega L + R_c + 1/j\omega C_1 + 1/j\omega C_2)}{L/C_1 + R_c/j\omega C_1 - 1/\omega^2 C_1 C_2 + (r_e + r_b)(j\omega L + R_c + 1/j\omega C_1 + 1/j\omega C_2)} \quad (78)$$

where R_c is the effective coil resistance.

At low frequencies (78) can be considered real. Removal of the imaginary term of (78) results in the following expression for oscillating frequency:

$$f = \frac{1}{2\pi}\left(\frac{C_1 + C_2}{LC_1C_2} + \frac{R_c g_{11}}{LC_1}\right)^{1/2} \quad (79)$$

g_{11} is the transistor short-circuit input conductance for which the approximate expression is

$$g_{11} \approx \frac{1 - a}{r_c + r_b(1 - a)} \quad (80)$$

Oscillator stability as a function of transistor parameters can be expressed as a stability coefficient S and defined as a ratio of percentage frequency change to a percentage change of g_{11}, so that

$$S = \frac{d\omega/\omega}{dg_{11}/g_{11}} = \frac{1}{2[(C_1 + C_2)/g_{11}R_cC_2 + 1]} \quad (81)$$

To improve oscillator stability as a function of g_{11}, which is temperature- and bias-dependent, the magnitudes of g_{11}, R_c, and $C_2(C_1 + C_2)$ should be reduced as much as possible. It is also true from Eq. (79) that least dependence on g_{11} is obtained when the ratio L/R_c is highest or when the Q of the coil is increased.

Frequency deviation of the oscillator over any particular band requires that

$$\Delta L_0/L_0 = 2\Delta f_0/f_0 \quad (82)$$

where L_0 and f_0 are the mid-band inductance and frequency, respectively. A typical calibration curve is shown in Fig. 35 for a subcarrier oscillator of this type. Figure 36 represents the degree of frequency stability attainable as a function of temperature and supply voltage independently. Figure 37 is a plot of bandwidth sensitivity

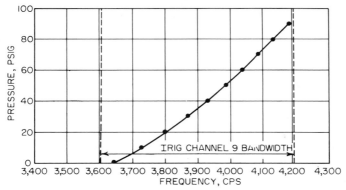

Fig. 35. Transistorized inductance-controlled oscillator; pressure vs. frequency. Supply voltage = 28 volts d-c. Temperature = 25°C. $f_0 = 3{,}900$ cps.

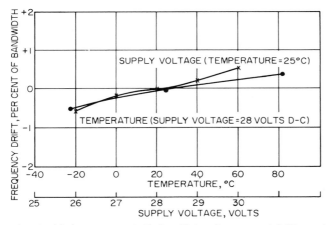

Fig. 36. Transistorized inductance-controlled oscillator; frequency stability vs. temperature and supply voltage. $f_0 = 3{,}900$ cps.

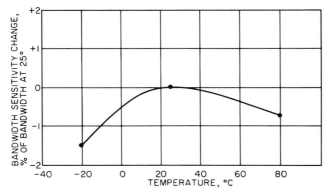

Fig. 37. Transistorized inductance-controlled oscillator; sensitivity stability vs. temperature. Supply voltage = 28 volts d-c. $f_0 = 3{,}900$ cps.

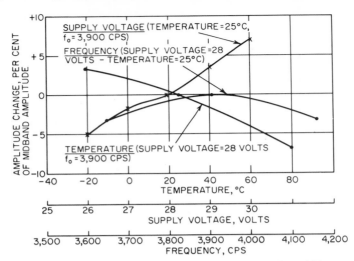

Fig. 38. Transistorized inductance-controlled oscillator; amplitude stability vs. temperature, supply voltage, and frequency.

change vs. temperature, and Fig. 38 shows amplitude stability as a function of temperature, supply voltage, and frequency.

References

1. Edward Y. Politi, Progress Report on a Solid State FM-FM Telemetering System; IRE Trans. Telemetry Remote Control; Proc. 1957 Natl. Symp. Telemetering, April, 1957, p. 3.6.
2. Richard F. Shea et al., "Principles of Transistor Circuits," p. 279, John Wiley & Sons. Inc., New York, 1953.

15 Voltage-controlled Oscillators

The voltage-controlled oscillator (VCO) is a convenient and widely used type of oscillator for the conversion of analog signals from the amplitude domain (amplitude modulation) to the frequency domain (frequency modulation). A large number of transducers are available that convert various types of stimuli directly to a voltage. These stimuli can be in the form of temperature, pressure, force, acceleration, vibration, motion, position, light, or electromagnetic and radioactive fields. Table 18 shows various stimuli and the associated transducer types utilized for remote measurements. All these devices can be used to convert a stimulus to an equivalent voltage. Another advantage of the VCO is the comparatively high degree of simple conversion linearity that can be achieved. As a practical result of its great versatility, the VCO is very often applied in FM telemeter systems.

The problems associated with VCO design are mainly concerned with stability, sensitivity, linearity, and frequency response. A simple multivibrator-type VCO[1] together with its waveforms is shown in Fig. 39a and b. To ensure stability and repeatability as a function of temperature and supply voltage, this circuit was chosen to minimize variations in transistor parameters due to the base-to-emitter forward voltage drop. The choice of transistor is based on a stable high-speed switching device having low leakage, low noise generation, and a stable well-packaged semiconductor chip. To reduce variability all other components in the circuit such as resistors, capacitors, and diodes should be chosen for stability and low noise generation. This consideration also applies to the power supply and packaging design.

Referring to Fig. 39b, when the base voltage of $Q1$ reaches V_x, thereby causing collector conduction, the base of $Q2$ is driven to cutoff. The base of $Q2$ has a high resistance when cutoff and the capacitor C_b at the collector of $Q1$ discharges through R_b to V_s until the base of $Q2$ reaches V_x, causing $Q2$ to conduct. This process repeats regeneratively and periodically so that an oscillating condition exists.

Table 18. Stimuli and Transducers

Types of stimuli	Associated transducer types
Heat	Thermocouples, resistance thermometers, thermistors
Pressure	Diaphragm, twisted bourdon tube, bellows, in combination with strain gages, potentiometers, piezoelectric crystals, vibrating string
Acceleration	Cantilevered beams in combination with strain gages, piezoelectric crystals
Vibration	Cantilevered beams in combination with strain gages, piezoelectric crystals, photoelectric cells
Force	Strain gages, piezoelectric crystals
Motion	Gyros, potentiometers, synchros, piezoelectric crystals, photoelectric cells, variable differential transformers, tachometers
Position	Gyros, potentiometers, synchros, photoelectric cells, variable differential transformers
Light	Photoelectric cells, solar cells, light spectrometers
Vacuum	Pirani gages
Electromagnetic fields	Inductors, transformers, Hall-effect devices
Radiation fields	Geiger and scintillation counters, mass spectrometers
Particle impacts	Particle counters, Geiger and scintillation counters, mass spectrometers

The ratio of R_c to R_b is determined by the minimum transistor d-c current gain H_{FE} and the d-c saturated collector current I_{cs}, such that

$$I_b = H_{FE} I_{cs}$$

For adequate margin

$$R_b = 0.5 H_{FE} R_c$$

The frequency of operation is determined by the switching time

$$t = R_b C_b \log_e \frac{V_s + V_c - V_{be}}{V_s - V_x}$$

where V_c = collector voltage swing from cutoff to saturation
V_{be} = base-to-emitter forward voltage drop
V_s = supply voltage
V_x = switching-point voltage

Transistor parameters that can affect the frequency through discharge of C_b include V_x, V_{bc}, I_{co} (base-to-collector cutoff leakage current), and I_{eo} (base-to-emitter cutoff leakage current). The percentage frequency effect due to leakage current is given by

$$\text{Per cent frequency shift} = R_b \frac{I_{co} + I_{eo}}{V_s} \times 100 \text{ per cent}$$

Junction and zener diodes have temperature coefficients very similar to the temperature characteristics of the transistor base-to-emitter junction as shown in Fig. 39 c, d, e, and f. Diodes are placed in the collector circuit of the multivibrator circuit to compensate for the transistor V_{bc} drift.

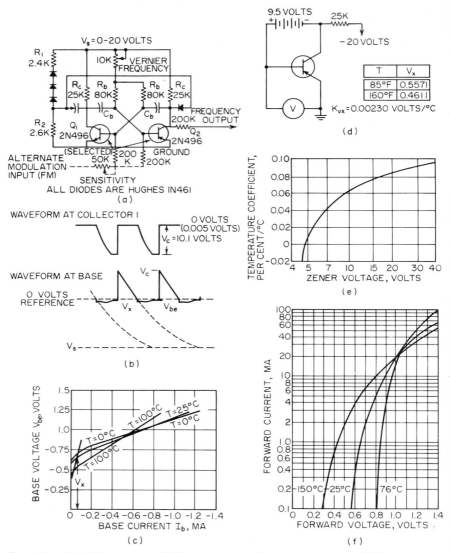

Fig. 39. (a) VCO Schematic. (b) Waveforms. (c) Input characteristic, 2N496 grounded emitter. $V_{CE} = 10$ volts. (d) Circuit to determine switching-point temperature coefficient. (e) Temperature coefficient vs. zener voltage. (f) IN461 diode forward bias voltage vs. current.

Temperature differences between the various components in the circuit can be minimized by providing a common high-conductivity heat sink for all or most components. The temperature coefficient of the multivibrator period is derived in Ref. 1 as

$$T_c = K_{tc} + \frac{V_1}{V_1 + V_2}\frac{K_1}{V_1} - \frac{K_2}{V_2}\frac{1}{\log_e(1 + V_1/V_2)}$$

where $K_1 = K_{vc} + K_{vx} - K_{vbe}$
$V_1 = V_c - V_{be} + V_x$
$K_2 = K_{vs} - K_{vx}$
$V_2 = V_s - V_x$
$K_{tc} = K_r + K_c$
K_{vc} = injected temperature coefficient, volts/°C, at V_c (collector limit voltage)
K_{vbe} = temperature coefficient, volts/°C, on V_{be} (base-to-emitter diode forward bias drop for collector-current saturation)
K_{vx} = temperature coefficient, volts/°C, on V_x (switching-point voltage at base)
K_{vs} = temperature coefficient, volts/°C, on V_s (supply voltage for base-current saturation)
K_{tc} = temperature coefficient, ppm/°C, of time-constant resistor and capacitor neglecting second-order temperature terms

Typical values of temperature coefficients for K_{vc}, K_{vbe}, and K_{vx} have been determined as $+0.000154$ volt/°C, -0.00172 volt/°C, and -0.0023 volt/°C.

If it is assumed that $K_{vc} = K_{vs} = 0$ and typical measured values are used as follows:

$K_1 = K_{vx} - K_{vbe} = -0.00230 - (-0.00172)$
$\quad = 0.00058$ volt/°C
$V_1 = V_c - V_{be} + V_x$
$\quad \simeq 10$ volts (since $V_c \gg -V_{be} + V_x$)
$K_2 = K_{vx} = -0.0023$
$\quad = 0.00230$ volt/°C
$V_2 = V_s - V_x$
$\quad = 20$ volts (since $V \gg -V_x$)
$K_{tc} = K_c + K_r = (120 \pm 5) + (\pm 20)$
$\quad = 120 \pm 25$ ppm/°C nominal
$\quad = 0.000120$ parts/°C nominal

then
$$T_c = 0.000120 + \frac{10}{10 + 20}\left(\frac{-0.00058}{10} - \frac{+0.0023}{20}\right)\frac{1}{0.405}$$
$\quad = -22.5$ ppm/°C nominally

Final temperature stabilization of the multivibrator frequency may be accomplished by the collector limit circuit, which can be readjusted following actual temperature tests.

Diodes may be added or deleted or the values of R_1 and R_2 varied to modify the multivibrator frequency vs. temperature coefficient. The temperature coefficient of R_b and C_b can be changed by proper selection of component types.

Other types of VCO's include the phase-shift and variable-reactance types.

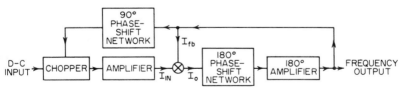

Fig. 40. Phase-shift-type VCO block diagram.

The phase-shift VCO can be frequency-deviated by changing either resistance, inductance, capacitance, or voltage so as to effect out-of-phase vector summing of voltage or current.

The variable-reactance type of VCO introduces a quadrature-current component into a basic LC oscillator in order to modulate the frequency. The out-of-phase current is a function of the input current to the reactance modulator.

A phase-shift[2] VCO designed for high sensitivity, adequate for thermocouple instrumentation, is shown in the block diagram of Fig. 40. The phase-shift oscillator

(PSO) contains an *RLC* feedback network which provides the necessary 180° phase shift at the oscillating frequency. The PSO output undergoes a 90° phase shift and is then used to drive a transistor chopper. The chopper output amplitude is proportional to the d-c input and always 90° out of phase with the PSO output. The output of the chopper is amplified and summed into the PSO so as to shift its frequency. The theory of frequency shift in the oscillator can be explained by referring to the block and vector diagrams of Fig. 41.

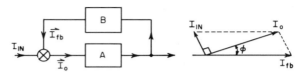

FIG. 41. Phase-shift oscillator and summing-point voltage vectors.

Assuming that the phase shift through the amplifier A is constant, the oscillator phase changes are attributed to the phase-shift network β and the phasing between the vectors I_{fb} and I_0. The sum of all phase shifts around the oscillator loop must instantaneously be 0° according to the Barkhausen principle. As a result, the exchange of phasing occurs between the network β and the vectors I_{fb} and I_0. Since β is a passive network, its phase-shift contribution must be initiated by a simultaneous frequency change. A variation in the magnitude of I_{in} results in a corresponding change in the phasing angle ϕ between I_0 and I_{fb}. In order to maintain a 0° phase shift around the closed loop of the oscillator, the phase shift of β must undergo a change by an amount $-\phi$, and it follows that this will cause the frequency of the oscillator to deviate. As a final result, the oscillator frequency is proportional to the d-c input.

Figure 42 is a schematic of the transistorized phase-shift VCO.

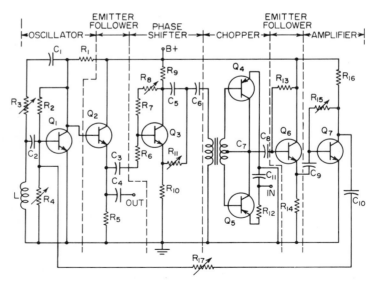

FIG. 42. Low-level phase-shift VCO.

Transistor $Q1$ comprises an RLC phase-shift oscillator, wherein 180° of phasing are obtained through $R3$, $R4$, $C1$, $C2$, and $L1$. $Q2$ is an emitter-follower stage which acts to isolate the oscillator from the following phase splitter, $Q3$, which couples to a 90° phase-shift network consisting of $R11$ and $C5$. The output of the phase shifter is transformer-coupled to drive the inverted chopper consisting of $Q4$ and $Q5$. The low-level d-c input is introduced at the chopper where it is converted to an a-c signal. $Q6$ is an emitter follower for impedance matching to the a-c amplifier $Q7$. The output of $Q7$ supplies the current I_{in} which is fed back to the oscillator $Q1$. $R17$ is a sensitivity control for I_{in}. $R3$ will adjust the frequency

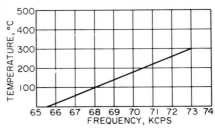

Fig. 43. Phase-shift VCO thermocouple temperature vs. frequency.

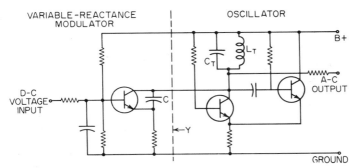

Fig. 44. Variable-reactance VCO.

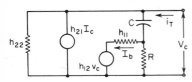

Fig. 45. Equivalent-reactance modulator circuit.

of oscillation. The oscillator bias control $R4$ can be adjusted for desired gain and stability. The amplifiers are designed with a-c and d-c negative feedback for improved stability and reduced dependence on transistor parameters. A typical VCO frequency-deviation curve is shown in Fig. 43. Stimulus is derived from a chromel-alumel thermocouple. The linearity is within 1 per cent of a best straight line.

A VCO of the variable-reactance type[3] can be seen in Fig. 44. The reactance modulator supplies a quadrature current to the oscillator which appears useful as a capacitance across the tank circuit. The equivalent modulator circuit is shown in Fig. 45. The effective variable capacity may be found from the equivalent circuit and the ratio i_T/V_c. This results in a susceptance

$$b = \omega C(1 + h_{21}K)$$
$$K = R/(R + h_{11})$$

where

The transistor parameters are dependent on emitter current, which is in turn controlled by the input signal voltage.

References

1. Warren E. Wilk and Wendell B. Sander, Development of a Transistorized Voltage Controllable Frequency Source, *IRE Wescon Conv. Record*, vol. 2, pt. 5, pp. 86–97, August, 1958.

2. Edward Y. Politi, Progress Report on a Solid State FM-FM Telemetering System, *IRE Trans. Telemetry Remote Control, Proc. Natl. Symp. Telemetering*, vol. PGTRC-3, no. 1, p. 3.6.1, 1957.
3. C. B. McCampbell, R. H. Gablehouse, P. F. Scheele, and R. P. Mathews, Transistors Applied to an Operational FM-FM Telemetry System, *Natl. Telemetering Conf. Rept.*, 1956, p. VII, 4-1.
4. F. M. Riddle, A Temperature-stable Transistor VCO, *IRE Trans. Telemetry Remote Control*, vol. PGTRC-2, p. 11, November, 1954.

16 Current-controlled Oscillator

An interesting variation of a variable-reactance CCO[1] (current-controlled oscillator) is one wherein the out-of-phase current is introduced over some controlled period

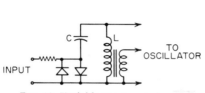

Fig. 46. Variable-reactance circuit.

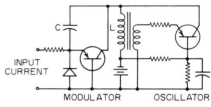

Fig. 47. Current-controlled modulator and subcarrier oscillator circuit.

of time. The circuit of Fig. 46 shows that this may be accomplished by a pair of diodes connected back to back. The proportion of a-c current passed by the diodes is controlled by the d-c input. The a-c current in the tuned loop is 90° out of phase with the voltage across the parallel-tuned circuit. The diodes introduce a negligible amount of resistance into the loop, so as not to affect the tuned circuit Q appreciably. To increase sensitivity, one of the diodes may be replaced by the equivalent emitter-to-base connection of a transistor. In this manner, advantage can be taken of the available transistor power gain by connecting the collector to supply reactive current to the tuned circuit as shown in Fig. 47.

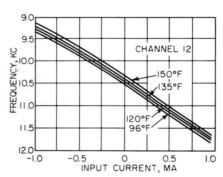

Fig. 48. Current-controlled oscillator frequency vs. current with temperature as a parameter.

The input voltage controls the fraction of each cycle wherein quadrature current is introduced into the LC tank circuit. As a result, the oscillator frequency will deviate as a function of the input voltage or current. A typical calibration curve for this type of oscillator is shown in Fig. 48, with temperature as a parameter.

17 Summing Amplifier

An FM telemetering system employs bandpass filters at the outputs of all subcarrier oscillators. To obtain maximum advantage of the filter frequency characteristics, adequate isolation must be provided for the filters. This is accomplished by inserting sufficient resistance in series with the input and output terminals. The attendant signal attenuation that results requires an amplifier that sums all subcarrier frequencies into a composite signal. Figure 49 shows the manner in which n subcarrier channels are connected at the input to a summing amplifier. The amplifier utilizes a high-

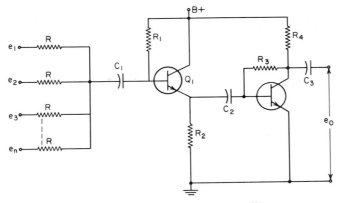

Fig. 49. Transistorized summing amplifier.

impedance common-collector input stage that is capacitor-coupled to a common-emitter amplifier. Sufficient a-c and d-c negative feedback are provided to ensure stable operation over a wide temperature range. The frequency-attenuation characteristics of the summing amplifier are given in Fig. 50. The curves indicate that the response is essentially flat from 50 cps to 15 kc/sec and within 3 db from 16 cps to 100 kc/sec.

18 R-F Transmitter

A principal source of malfunction of airborne telemetering transmitters is attributable to kinetically induced frequency modulation due to shock, vibration, and acceleration. Since the transmitter is a series link in the telemetering chain, its faulty operation will affect all information channels. At high radio frequencies the mode of operation is affected to a great degree by component spacing and the immediate surrounding structure. Assuming that structural and component stability can be maintained

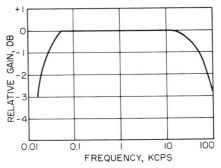

Fig. 50. Summing-amplifier attenuation characteristic.

by careful and considered design, the vacuum tube remains as the weakest component because of its relatively fragile filamentary structure. From this standpoint, it would, of course, be desirable to substitute solid-state devices in place of vacuum tubes. The development of high-frequency transistors has progressed sufficiently to permit power generation in the order of watts at frequencies below 300 Mc.

FM-FM telemetering transmitters are generally of two types, self-excited and crystal-controlled. Self-excited transmitters have the attributes of wideband frequency modulation and low distortion but generally suffer from microphonic noise and excessive frequency drift. Transmitters that are crystal-controlled are relatively drift-free with respect to frequency and exhibit excellent resistance to low-level microphonics; however, the very stability of the crystal does not allow for wideband modulation without excessive distortion.

Shown in Fig. 51 is a block diagram of a transistorized transmitter that combines self-excitation with crystal control to minimize the disadvantages of each when used individually. Mixed frequency-modulated signals ranging from 400 cps to 70 kc/sec are fed to the reactance modulator. The modulator converts voltage level to frequency deviation about a 5-Mc center frequency. This signal is mixed with the output of a 23.75-Mc crystal-controlled oscillator, after which the two frequencies are summed to provide 28.75 Mc. Following some amplification, three stages of frequency doubling multiply the frequency to 230 Mc. A final stage of amplification provides 2 watts of r-f power output.

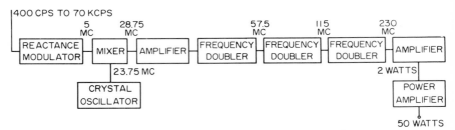

Fig. 51. FM transmitter block diagram.

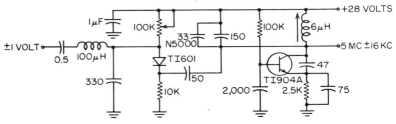

Fig. 52. Reactance modulator.

Figure 52 shows the transistorized reactance modulator. The tank circuit of the oscillator is shunted by a diode circuit which draws reactive current proportional to the input modulating voltage. The modulator will respond to inputs from 100 cycles to 100 kc within 2 db. A ±1-volt signal will cause a frequency deviation of ±16 kc about the center frequency of 5 Mc. Frequency stability of the modulator as a function of temperature is shown in Fig. 53. The crystal-controlled oscillator of Fig. 54 supplies the reference frequency of 23.75 Mc. A tetrode transistor is utilized with the crystal connected between the collector and emitter. A parallel-tuned circuit at the collector increases the efficiency of the oscillator.

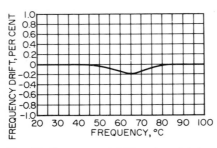

Fig. 53. Frequency stability of modulator.

The remainder of the transmitter consists of the vacuum-tube stages shown in Fig. 55. The first four stages are cathode-biased and plate-tuned. The final stage utilizes a grounded-grid connection to minimize the effects of interelectrode capacity and to improve gain at very high frequencies.

The vhf power amplifier[1] increases the power level from 2 to 50 watts. A beam-tetrode power-amplifier tube is used having a capacitive input impedance of 50 ohms. Inductance $L1$ is used to balance the capacitive reactance. The output of the amplifier consists of a quarter-wave transmission line. The actual length of the transmission line is shortened by means of an annular ring placed about the plate of the tube which acts as a capacitance $C1$. This ring is movable in a longitudinal direction, thereby serving as a coarse tuning adjustment. The transmission line is also folded back on itself, which effectively shortens it even further. A rotatable loop near the end of the line provides for fine tuning. Another loop, similarly placed and mounted to a coaxial connector, provides the final output of the power amplifier.

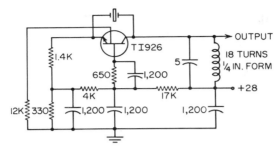

Fig. 54. Crystal-controlled tetrode oscillator.

Operation at high altitudes is extended by the addition of a Teflon ring about the surface of $C1$. This minimizes the possibility of flashover and enhances cooling of the transmitter tube's cooling fins.

Typical power output of the amplifier exceeds 50 watts, a nominal power gain of 14 db. A typical power amplifier measures 6 in. in length and 3 in. in diameter.

19 Electronic Time Multiplexer

The development of solid-state switching devices such as the semiconductor diode and transistor has encouraged their application to electronic time multiplexers as a substitute for functionally equivalent mechanical units. Advantages of the solid-state commutators include long life; immunity to shock, vibration, and acceleration; elimination of contact noise, bounce, and pitting; as well as greater power efficiency. The major disadvantages to be reckoned with are higher equivalent contact resistance and leakage current, feedthrough of the switching drive, temperature sensitivity, and complexity.

The electronic multiplexer to be described[2] functions according to the standard IRIG (Inter-Range Instrumentation Group) commutation pattern presented in Fig. 56.

This commutator comprises 150 active channels sampled at a rate of 150 samples per second, or 10 samples per second for each channel. Sampling rates as high as 1,000 samples per second have been utilized.

The following specifications are provided for:

1. Zero to five volts full-scale inputs
2. Linearity ±0.4 per cent of full scale
3. Interchannel uniformity within ±0.4 per cent of full scale for 100-kilohm signal source impedance
4. Zero and sensitivity drift within ±0.5 per cent of full scale
5. ±20 per cent change in supply voltage causes negligible change in output
6. Input impedance of 0.5 megohm with 1-megohm load resistance
7. Sampling-rate stability within 1 per cent up to 70°C
8. Contact voltage less than 50 mv at full scale
9. Built-in limiting to prevent transmitter overmodulation
10. Power required approximately 1 watt

A simplified functional block diagram for accomplishing this type of sequential switching is shown in Fig. 57. A detailed block diagram is referenced in Fig. 58. The

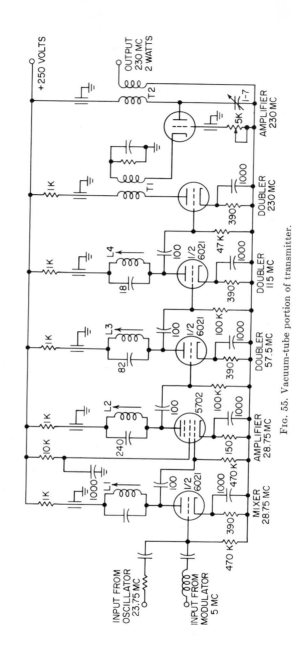

Fig. 55. Vacuum-tube portion of transmitter.

TRANSISTORIZED TELEMETERING CIRCUITS   10–57

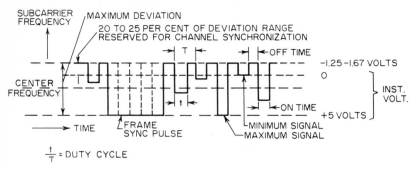

Fig. 56. Standard IRIG commutation signal.

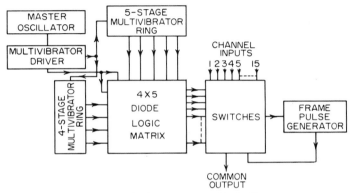

Fig. 57. Simplified commutator block diagram.

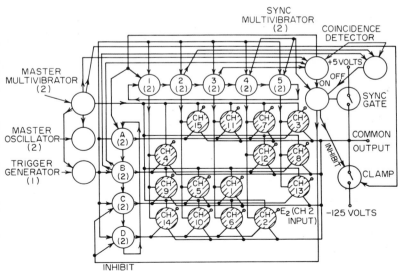

Fig. 58. Detailed commutator block diagram.

switching time of the multiplexer is controlled by a sine-wave master oscillator triggering a drive multivibrator. The drive multivibrator simultaneously triggers both a five-stage and a four-stage multivibrator closed-ring stepping chain. Both rings are connected to a rectangular 4 × 5 diode logic matrix so that coincident pulses from each ring sequentially gate each of 20 switches. Eighteen of the 20 matrix elements are utilized; 15 are active channels and 3 are used for frame synchronization. Sampling is at a rate of 10 per second, so that the two-ring stepping chain is triggered 180 times per second. The frame-synchronizing pulse has an amplitude of 5 volts and a duration 2.5 times greater than each channel sample. Reset of the stepping rings is executed during the period of frame synchronization to inhibit possible interference with active channels.

The basic commutator channel switch in Fig. 59 is comprised of eight silicon diodes and two resistors. If a positive voltage is injected at point A and a negative voltage at point B, the four matched ring diodes $D1$, $D2$, $D3$, and $D4$ conduct such that E_o is approximately equal to E_i. The switch is cut off by applying a negative voltage at point F and a positive voltage at point G.

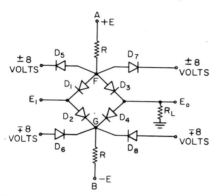

Fig. 59. Basic commutator channel switch.   Fig. 60. Channel gating multivibrator.

Matching of forward voltage drops across the diodes within 2 per cent results in less than ±10 mv differential between the input and output. The output voltage is limited to $ER_L/R + R_L$ and can be used to prevent overmodulation of the transmitter or other equipment.

Gating control for the switch is obtained by means of a combination p-n-p and n-p-n multivibrator as shown in Fig. 60. For the ON condition, both transistors are conducting, and for the OFF condition, both transistors are nonconducting. The gating signals are obtained from points A and B. The coupling of these multivibrators into a ring is shown in Fig. 61. A differentiated square-wave drive for each succeeding stage of the ring is obtained from point G of the preceding station. Silicon breakdown diodes are used to provide precise switching voltages, thereby greatly improving the reliability of operation.

Referring to Figs. 60 and 61, the common-emitter busses are approximately at zero potential if

$$R_4 = \frac{R_2(R_1 + R_3)}{R_1 + R_2 + R_3}$$

The common-emitter resistances allow just a single multivibrator to be in the ON state, thereby providing economy of power.

A four- and five-stage multivibrator ring, respectively, drives a 4 × 5 rectangular-switch matrix. Coincident pulses from a single ON multivibrator in each ring will open a particular switch. Triggering each ring simultaneously will sequentially step the sampling switch.

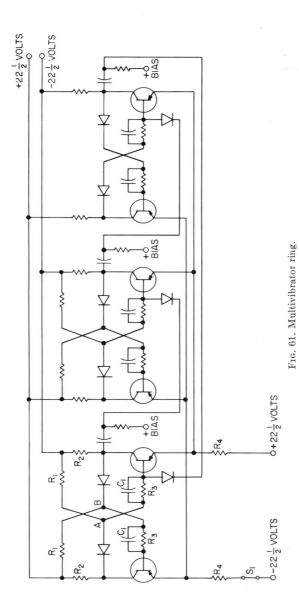

Fig. 61. Multivibrator ring.

10-59

Reset of the commutator is obtained by two of the 20 matrix coincidence detectors. A 5-volt synchronizing pulse utilizes 2½ of 3 available channel intervals and is gated by means of a bistable multivibrator. A constant commutating rate is provided by the master oscillator of Fig. 62. This constitutes a regenerative feedback amplifier.

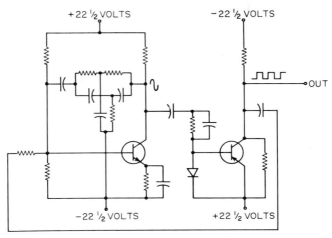

Fig. 62. Parallel-T master oscillator.

Feedback is obtained through a parallel-T network. Maximum regenerative feedback is obtained at the frequency to which the T network is tuned. It is essentially at this frequency that the amplifier oscillates. Frequency stability is mainly a function of passive network parameters and is virtually independent of transistor parameters. The ring stepping chains are triggered 180 times per second.

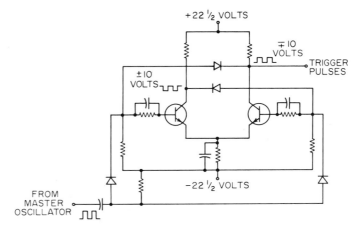

Fig. 63. Multivibrator driver.

IRIG specifications require that the duration of each information pulse be within the bounds of 47 to 53 per cent of the nominal channel period.

The required 50 per cent duty cycle for each channel is maintained by gating each channel off for half the channel period. This is achieved with the multivibrator driver shown in Fig. 63. An additional pair of diodes per channel is also required to inhibit each half period. The multivibrator is driven by the master oscillator at 360 cps. Two *n-p-n* transistors are used in a conventional multivibrator configuration except for the silicon breakdown diodes used as coupling elements. The synchronizing multivibrator is similar to the drive multivibrator. Waveforms for principal parts of the system are shown in Fig. 64.

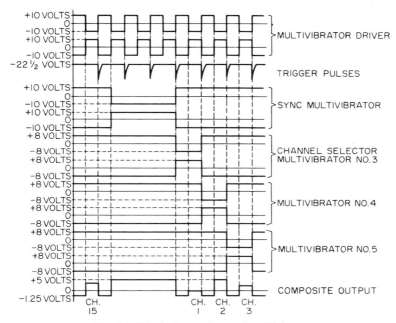

Fig. 64. Principal waveforms of multiplexer.

References

1. F. M. Riddle, A Temperature-stable Transistor VCO, *IRE Trans. Telemetry Remote Control*, vol. PGTRC-2, p. 11, November, 1954.
2. J. M. Sacks and E. R. Hill, Transistorized Time Multiplexer for Telemetering, *IRE Trans. Telemetry Remote Control*, vol. PGTRC-3, p. 26, May, 1957.

20 PDM Keyer

The keyer functions to convert sampled voltages to pulses of time duration proportional to the amplitude of the samples. Keyer design problems are generally concerned with conversion linearity, high input impedance, and minimum crosstalk and jitter.

A block diagram for a transistorized pulse-duration modulation (PDM) is shown in Fig. 65. A comparison principle is utilized wherein the sampled input signal is com-

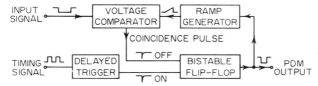

Fig. 65. PDM keyer block diagram.

pared with a generated linearly increasing sawtooth or ramp signal. The output pulse starts when the ramp commences and ends when the ramp level equals the sampled input signal level. As a result, the output pulse width of the bistable flip-flop is directly proportional to the input signal amplitude.

The keyer utilizes a trigger circuit comprising an amplifier and integrator as shown in Fig. 66. This circuit provides a leading edge for the trigger that is more reliable than that available from the commutator time contacts. Commutator stray capacitances and cable capacitance cause noise pickup in high-impedance circuits; commutator contact bounce and noise under severe environmental conditions can cause false keying.

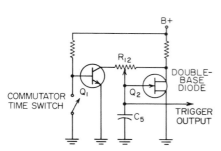

FIG. 66. Trigger delay circuit.

The operation of the trigger delay circuit depends on the position of the commutator switch. When open, transistor $Q1$ is conducting heavily and the collector potential is near zero. When closed, $Q1$ stops conducting, causing the emitter voltage of the double-base diode $Q2$ to rise exponentially with time constant $R12\ C5$. This voltage increases until the emitter resistance of the double-base diode goes into its negative-resistance region, causing the emitter voltage to decrease sharply to almost zero. This sudden change is differentiated and is used as the trigger pulse. Once triggering occurs, the double-base diode is insensitive to further triggering due to noise. Its regenerative action allows it to be in only either of two states, conducting or nonconducting.

The trigger delay is adjustable with resistor $R1$, thereby providing optimum synchronization between signal and trigger without mechanical adjustment of commutator brushes.

The voltage comparator utilizes the astable p-n-p-n-p-n flip-flop of Fig. 67. A negative signal input puts $Q3$ in a position to conduct. When the ramp voltage increases to a level proportional to the signal input amplitude, the regenerative feedback from $Q4$ through $C6$ takes effect, producing a negative coincidence pulse with a fast rise time.

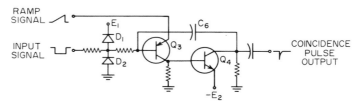

FIG. 67. Regenerative voltage comparator.

The bistable flip-flop of Fig. 68 functions to drive the ramp generator and supplies the PDM keyer output. The flip-flop is conventional except for the cross-coupling transistor $Q7$. This stage minimizes capacitive loading on the collector of $Q6$, thereby allowing a fast rise time. This has the effect of decreasing keying jitter. The coupling transistor also provides a low-impedance output for the keyer.

Clamping diodes $D5$ and $D6$ at the flip-flop collectors prevent saturation and determine collector voltages so that the output amplitude is independent of transistor parameters. Ambiguity of triggering is inhibited by diode gating of the "on" and "off" trigger pulses by means of diodes $D3$ and $D4$.

Ramp generation is provided by the circuit of Fig. 69. Considering switch S to be

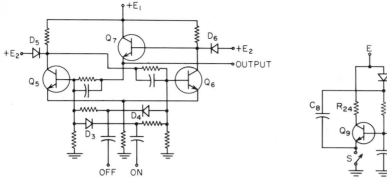

Fig. 68. Keyer bistable flip-flop.

Fig. 69. Constant-current ramp generator.

open, the condenser $C9$ charges through resistor R. The emitter resistance is infinite as the ramp increases. This is a constant-current condition, and the voltage gain of $Q9$ approaches unity. Closing switches S causes $C8$ to discharge through the base of $Q9$. Resistor $R24$ limits the collector current to prevent saturation when the ramp is generated. The mechanical switch S is replaced by a transistor switch, but the operation is essentially as explained. The switch is driven by the bistable flip-flop.

Figure 70 is an overall schematic of the described PDM keyer.

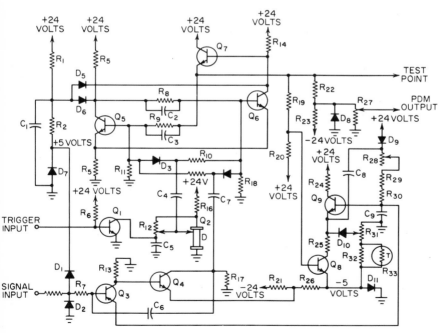

Fig. 70. Transistorized PDM keyer.

References

1. D. A. Williams, A Stable Transistorized PDM Keyer, *IRE Wescon Conv. Record*, pt 5, Instrumentation, Telemetry and Remote Control, 1957.

21 Solid-state PCM System

Recently, pulse-code-modulation techniques have received increased consideration and effort for telemetering purposes. Benefits derived from this flexible system include a favorable exchange between bandwidth and signal-to-noise ratio for a specified information capacity; i.e., $I = W \log_2 [1 + (S/N^2)]$. Also, the signal-to-noise ratio of the PCM signal is independent of the carrier signal-to-noise ratio if the carrier is above threshold.[1] That is, no distortion of the signal will be evident if the pulse or absence of it is detected. Noise in a PCM system is denoted as "quantization noise." It is introduced because of the ambiguity of signals that are smaller than the smallest quanta to which the incoming analog signals are compared.

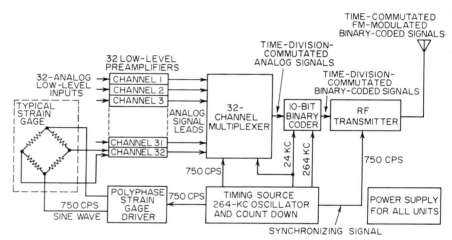

Fig. 71. Block diagram of PCM telemetering system.

PCM transmissions are in a favorable form for introduction into a digital system or computer. PCM systems developed so far are relatively complex, but the advantages are noteworthy, especially for large data samplings, over extended distances and periods of time.

Figure 71 is a block diagram of a transistorized PCM system[2] utilizing solid-state components exclusively with the exception of r-f vacuum tubes utilized in the transmitter. This particular system accepts up to 32 low-level analog data inputs, each sampled by a multiplexer at a rate of 750 per second, or a total of 24,000 samples per second. Each sample is transformed into a 10-bit binary code which then modulates the r-f transmitter. The information capacity, therefore, is 240,000 bits/sec. At 750 samples per second, each channel has an information bandwidth of 375 cps. Paralleling all channels would provide an information bandwidth of 12 kc/sec.

Various types of transducers may be driven in synchronism with the sampling. Preamplifiers are utilized where required for low-level input signals.

Figure 72 is a schematic of the preamplifier which utilizes four transistors connected in a repetitive direct-coupled common-emitter common-collector connection.

Each common-emitter stage contains emitter degeneration for improving stability and linearity. An overall negative-feedback network is also utilized for increasing stability. This network may be adjusted by means of a potentiometer to modify the preamplifier sensitivity over a 100 to 1 range. At maximum sensitivity the input signal levels extend from 0 to 2.5 mv to provide an output of over 5 volts. The input impedance is approximately 100 kilohms.

Outputs of the respective preamplifiers are fed to an electronic time-division multiplexer which commutates all outputs onto a single bus. Thirty-two sequentially

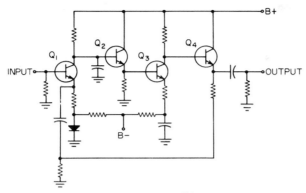

Fig. 72. Preamplifier.

gated diode bridges, such as are shown in Fig. 73, are utilized for switching. The sequential gating generator provides the trigger that drives the bridge so as to present a low switching impedance path to the analog input during the gated period. The current through the diode bridge charges a capacitor which holds the bridge in a nonconducting state until the appearance of the succeeding pulse.

The basic analog-to-digital conversion is accomplished by a 10-bit coder, using the "half-split" sampling method, a feedback sampling technique. The coder block diagram is shown in Fig. 74. The coder functions to convert each analog input into a 10-bit binary code at a rate of 24,000 samples per second. An eleventh bit is used for reset purposes, giving the coder a bit rate of 264,000 per second.

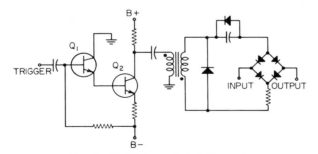

Fig. 73. Multiplexer diode bridge gate.

In the "half-split" sampling method, the analog input samples are sequentially compared with generated precision binary weighted voltages. The initial comparison is accomplished at one-half of the full-scale input range. If the input signal is greater than this 50 per cent reference, it is rejected; if smaller, it is retained. The coder generates a binary 1 for greater signals and a binary 0 for smaller signals. If the decision is "greater than 50 per cent," the succeeding comparison is made at 75 per cent of full scale. The 50 per cent voltage reference level is retained and an additional

25 per cent is applied, or a total of 75 per cent. A decision on the first comparison of "less than 50 per cent" will cause the second comparison to be made at a generated 25 per cent level, where the original 50 per cent level is dropped. These comparisons are continued along this "half-split" principle, until a 10-bit binary word is formed. Quantization of the input analog signals, therefore, can be as low as 1 part in 1024.

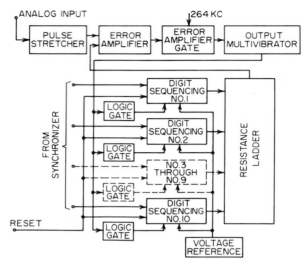

Fig. 74. Coder block diagram.

The coder synchronizing pulses are generated by means shown in the block diagram of Fig. 75. Precise timing is obtained from a 264-kc crystal-controlled oscillator which drives a bistable multivibrator binary frequency-divider chain and a diode matrix. The multivibrator chain is composed of four flip-flops in tandem, which are reset every eleventh cycle. The diode matrix in conjunction with the 264-kc signal

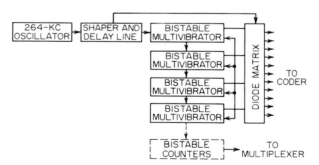

Fig. 75. Synchronizer block diagram.

is also driven by the divider chain to provide 11 parallel sequential outputs at a 24-kc rate. The matrix parallel output pulses are utilized to drive flip-flops, which in turn are used to gate precision-weighted binary currents for use in the comparison circuit.

The pulse stretcher is shown in Fig. 76. The analog input signal is coupled through a buffer; and input amplifier to a diode gate. The gate is controlled by a transformer-coupled bistable multivibrator, driven by the system synchronizing pulses. After the gate is opened, a charged capacitor maintains this condition until the succeeding sample is taken. The output amplifier acts as an impedance-matching stage between the pulse stretcher and the coder. This amplifier has an input impedance of 50 megohms and an output impedance lower than 1 ohm.

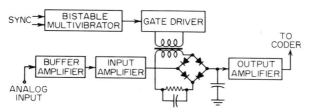

Fig. 76. Pulse-stretcher block diagram.

Digit sequencing is controlled by the circuit of Fig. 77. The synchronizer provides sequential pulses at a 24-kc rate to a regenerative pulse amplifier through a diode logic matrix. The amplifier output forces a bistable multivibrator to the "on" state, which opens a diode gate. It is through this gate that the binary precision-weighted currents are provided to the resistance ladder network. The ladder is grounded when the gate is off. When the gate is on, a precision current is generated into the ladder from a reference voltage supply through a precision resistor. Ten identical currents are sequentially applied to the ladder at the 24-kc rate. The currents are

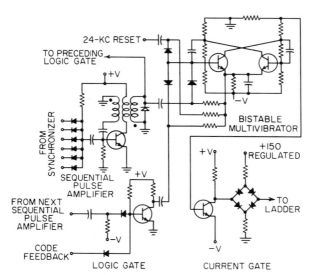

Fig. 77. Digit-sequencing circuit.

progressively divided in the ladder so as to form 10 precision binary comparison voltage levels which are then fed to the error amplifier.

The 150-volt regulated reference supply is shown in Fig. 78. Stable referencing is obtained from a zener diode.

The d-c error amplifier, shown in Fig. 79, receives the analog input voltage and compares it with the sequentially generated binary weighted voltages. Differentials of voltage exceeding plus or minus 5 mv will respectively saturate or cut off the amplifier. This output is fed to a gate that is sampled 264 times per second. The gate output drives a readout flip-flop. If the analog signal exceeds the comparison voltage, a "yes" bit is generated; if less, a "no" bit occurs. If a "no" bit is coincident

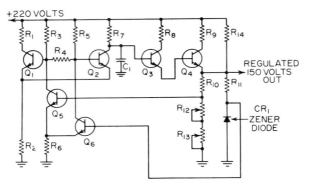

FIG. 78. 150-volt regulated reference supply.

with the succeeding sequential pulse, the binary comparison voltage is eliminated, and the following comparison is made. A "yes" bit retains the reference voltage for the next comparison. As previously explained, these comparisons are accomplished by the "half-split" sampling method.

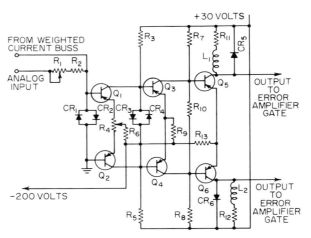

FIG. 79. Error amplifier.

Following 10 comparisons, the digit-sequencing flip-flops are reset to their original state and coincidentally the precision binary currents are also removed. This is done at a 24-kc rate.

The degree of error expectancy for the system from the standpoint of pulse-code-modulation theory is shown in Table 19.

Table 19. PCM Error Expectancy vs. Signal-to-Noise Ratio

Signal-to-noise ratio, db	Probability of error	This is about one error every
13.3	10^{-2}	378 μsec
17.4	10^{-4}	37.8 msec
19.6	10^{-6}	3.78 sec
21.0	10^{-8}	7.56 min
22.0	10^{-10}	9 hr
23.0	10^{-12}	1 month

References

1. M. H. Nichols and L. L. Rauch, "Radio Telemetry," 2d ed., John Wiley & Sons, Inc., New York, 1956.
2. R. E. Marquand and W. T. Eddins, A Transistorized PCM Telemeter for Extended Environments; 1957 *IRE Wescon Conv. Record*, pt. 5, Instrumentation Telemetry and Remote Control.

MINIATURIZATION TECHNIQUES

22 Introduction

Most telemetering-system designers have been faced with increased requirements of miniaturization. Aircraft and missile telemetering necessities have generated a considerable amount of effort in this direction. Satellites and space vehicles require further advanced concepts. The development of solid-state electronic devices has provided a considerable impetus toward microminiaturization. Eventual component densities of one hundred million per cubic foot are contemplated.

Advantages associated with miniaturization include improved reliability and the conservation of space, weight, and power. Reduced size and weight simplify mechanical design and increase resistance to adverse environmental conditions. Higher-density packaging allows for added control and instrumentation of aircraft, missiles, and spacecraft, together with increased payloads.

The elements of miniaturization techniques utilizing available components include

1. Efficient circuit design
2. Small components
3. Modular construction
4. Integrated circuits
5. Printed circuits
6. Encapsulation
7. Special materials

New microminiaturization techniques are also being developed which will be discussed.

23 Efficient Circuit Design

Miniaturization begins with efficient design where unnecessary components, circuits, and materials are eliminated. Simple, straightforward circuits are generally the most advantageous. Designs should be optimized so as to provide minimum losses and maximum gain consistent with stability. Decreasing the number of different power supplies required will result in significantly smaller overall system size. New components should be investigated to possibly replace those which are conventional, but large. Many electromechanical devices have their analog in purely electronic devices. For example, mechanical tuning condensers can sometimes be replaced by voltage-controlled semiconductor capacitors as shown in Fig. 80.

Transistorized choppers can be substituted for electromechanical choppers, as illustrated in Fig. 81. For low-level signals, the transistor chopper is conservative of power and is able to switch at relatively high frequencies independent of inertia effects. Similarly, electromechanical vibrators can be replaced by a transistorized d-c to a-c converter as in Fig. 82.

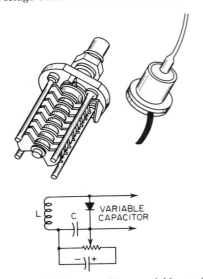

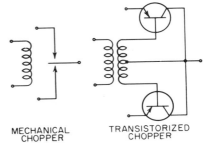

FIG. 80. Miniature voltage variable semiconductor capacitors can replace larger mechanical condensers.

FIG. 81. Transistor chopper analog of electromechanical chopper.

Research in the field of semiconductors has resulted in the development of a multitude of miniature components.

Active vacuum-tube elements such as the triode amplifier, diode rectifier, diode

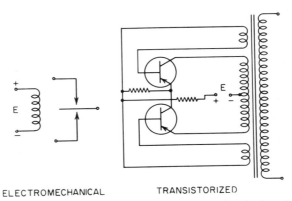

FIG. 82. Electromechanical vibrator replaced by transistorized equivalent.

detector, thyratron, and voltage regulator have their equivalents in respective semiconductor components, such as the transistor, semiconductor diode rectifier and detector, and the zener diode.

24. Small Components

A primary consideration in miniaturization is the selection of the smallest available components in accordance with operational requirements. Many new components, such as resistors, are miniaturized by a scaling down and derating of larger existing components. Significant capacitor miniaturization has been realized through the development of new dielectric materials and improved processing methods.

The utilization of tantalum has resulted in large capacitance in a small space (approximately 20,000 µf-volts/cu in.) as well as long operating and shelf life over extended temperature ranges. In addition d-c leakage and power factor are generally low and satisfactory for most applications. Solid tantalum capacitors, as opposed to liquid and semiliquid electrolytes, have higher stability and improved seal characteristics, together with increased operating and shelf life. Test data as a function of temperature and life are shown in Figs. 83 through 87. Table 20 provides life-test results.

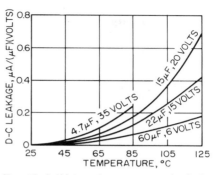

Fig. 83. Solid tantalum capacitor d-c leakage vs. temperature for a series of ratings.

Many new plastic-film capacitors have been developed through the introduction of such dielectric materials as polystyrene, polyethylene, Mylar, Teflon, and cellulose acetate. Highest operating temperatures are obtained with Teflon, where the d-c conductivity and dissipation factor remain low up to 200°C. However, Teflon

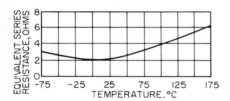

Fig. 84. Equivalent series resistance vs. temperature for 6-µf 10-volt solid tantalum capacitors.

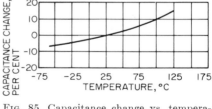

Fig. 85. Capacitance change vs. temperature for 6-µf 10-volt solid tantalum capacitors.

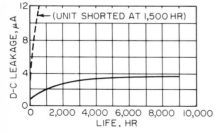

Fig. 86. Solid tantalum capacitor d-c leakage vs. life.

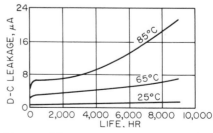

Fig. 87. Direct-current leakage at various temperatures vs. life for solid tantalum capacitors.

Table 20. Solid Tantalum Capacitor Life-Test Results

Rating, volts	Quantity tested	No. of failures	Total test time, hr	Time of failure, hr
10 μf, 20	6	1	9,000	4,500
10 μf, 20	6	1	9,000	1,500
6 μf, 10	6	0	9,000	
100 μf, 8	16	0	9,000	
10 μf, 20	5	1	9,000	6,000
15 μf, 15	12	1	215	215
4 μf, 35	10	5	215	24–215
6 μf, 30	6	0	9,500	
20 μf, 35	6	2	1,000	390
4.7 μf, 35	12	1	3,000	250
4 μf, 35	10	2	3,000	12
4 μf, 50	10	4	2,700	100–2,700

capacitors are more expensive and larger than other film capacitors. Mylar film capacitors are usable up to 130°C. Their higher dielectric constants permit a greater degree of miniaturization. Metallizing of film capacitors will also contribute to further miniaturization. Other film types are generally unstable at higher temperatures.

25 Modularization and the Micromodule

Advantages of modularization for telemetering equipment include

1. Standardization
2. Interchangeability
3. Rapid maintenance
4. Volumetric efficiency

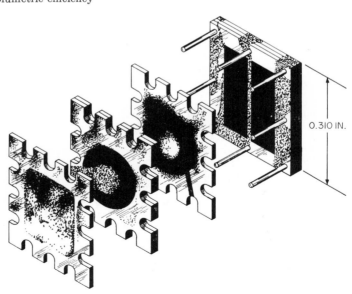

FIG. 88. Microelemental wafers are stacked and interconnected to form a micromodule.

Considerations of modularization will generally consider necessity, practicality, and cost. Fairly long term telemetering programs requiring large amounts of data acquisition, where schedules are tight and changes frequent, would warrant a high degree of modularization.

The modular element is generally standardized with respect to size, mounting, connections, and outer covering. This allows for accurate prediction of the volume and weight of the telemetering-system package for any particular program.

Because of the continuing demand for increasing amounts of data from missiles and space vehicles, it has not been possible to standardize on modules over extended periods of time. New miniaturization techniques and components are perennially evolving to aid telemeter designers toward smaller modules.

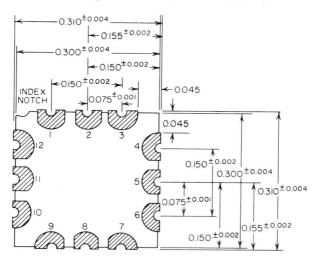

Fig. 89. Standard 12-terminal dimensions for micromodule wafer.

Two new microminiaturization techniques are being developed. One method, denoted as the wafer micromodule,[2] will place individual components on small plates which may then be combined to form complete circuits. A second method, the integrated micromodule, will place an entire circuit or system on an extremely small plate. The wafer micromodule consists of microelemental plates that are stacked and interconnected according to a circuit function, as shown in Fig. 88.

Basic subassemblies are formed and encapsulated to make the micromodule, which can, in turn, be assembled into larger systems.

The standardized 12-terminal wafer dimensions of Fig. 89 are utilized in all tools, molds, and jigs. This simplified design facilitates automatic fabrication and assembly.

A typical circuit module-connection schedule is shown in Fig. 90, for a high-frequency i-f amplifier. Interconnections are shown for flat wafers containing resistors, capacitors, a choke, transformer, and a transistor.

It is expected that micromodule packaging densities will average to approximately 250,00 parts per cubic foot.

26 Integrated Circuits

The integrated micromodule combines passive and active components within a common semiconductor unit. Controlled masking, etching, and diffusion methods are utilized. Resistors are made by forming ohmic contacts to n- or p-type semiconductor material. An indication of the physical configuration can be obtained from Fig. 91. Capacitors are formed by relatively large area p-n junctions; distributed

10–74 TELEMETRY-SYSTEM COMPONENT DESIGN

capacitors, by combining capacitors and resistor elements; transistors and diodes by diffused-base techniques.

An example of an integrated silicon semiconductor micromodule is shown in Fig. 92. This unit makes up a phase-shift oscillator utilizing a transistor and a phase-shift

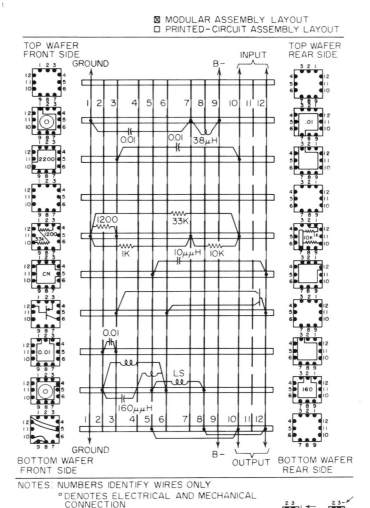

Fig. 90. Modular connection schedule for i-f amplifier.

network, consisting of resistance and distributed capacity. The dimensions of this unit are ¼ by ⅛ by ¹⁄₃₂ in. Typical component densities of this type could reach 30 million per cubic foot.

Extension of this technique toward complete subsystems would further increase component densities up to 10^8 parts per cubic inch. For example, an integrated micromodular shift register can be formed from the basic thyristor of Fig. 93a. This

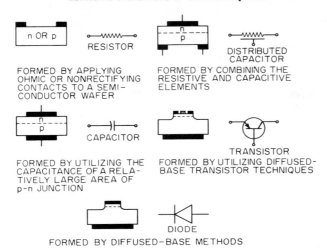

FIG. 91. Passive and active circuit components formed from semiconductor materials.

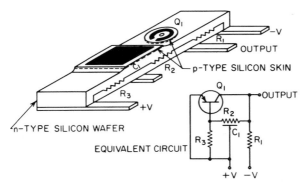

FIG. 92. Integrated semiconductor micromodule for a phase-shift oscillator having dimensions of $\frac{1}{4}$ by $\frac{1}{8}$ by $\frac{1}{32}$ in.

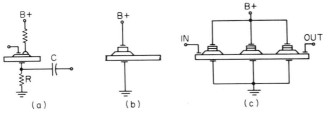

FIG. 93. Shift-register development from (a) conventional negative-resistance device like thyristor to (b) integrated micromodule and then to (c) continuously repeating structure to form an integrated shift register.

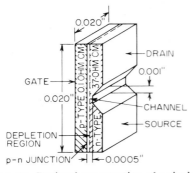
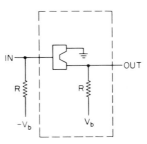

FIG. 94. Regional construction of unipolar transistor.

FIG. 95. Basic biasing circuit for unipolar (field-effect) transistor.

device has a negative-resistance region which makes it usable as a bistable digital element. The conventional RC coupling circuit function is replaced by a germanium delay line. Resistors are likewise replaced by n- or p-type germanium to form the integrated module of Fig. 93b. A shift pulse is applied to the register which propagates minority carriers from the ON stage down along the line. Following the transit time, minority carriers are collected at the end of the line where they are available for triggering the following stage ON. The entire shift register may be constructed by a continuously repeating structure to form an integrated shift register, as shown in Fig. 93c.

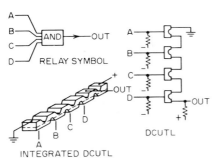

FIG. 96. Unipolar transistor AND gate.

The unipolar (field-effect) transistor may be especially applicable to switching logic for lightweight telemetering systems. This device is constructed as shown in Fig. 94. In essence, this device is analogous to a voltage-controlled relay with a high-frequency cutoff up to 10 Mc. The application of reverse bias at the input causes the depletion layer of the p-n junction to increase, which, in turn, narrows the channel region. This transition takes the form of an effective increase in channel

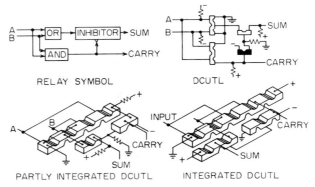

FIG. 97. Unipolar transistor half-adder.

resistance to more than 10 megohms. In contrast, a decrease in reverse bias increases the channel width, reducing its resistance to approximately 5,000 ohms. The dynamic resistance characteristic is highly nonlinear, thereby assuring positive switching action. Isolation between the control point and the relay is in the order of 100 megohms for silicon. Basic biasing for a unipolar transistor can be seen in Fig. 95.

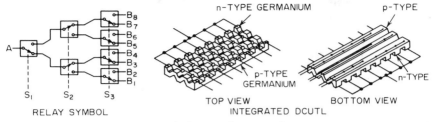

Fig. 98. Unipolar transistor transfer tree.

Examples of an AND gate, a half-adder, and a transfer tree are shown in Figs. 96, 97, and 98. These circuits employ direct-coupled unipolar transistor logic (DCVTL).

27 Printed Circuits

Printed-circuit techniques and their refinements are assuming increased importance in extending the art of microminiaturization. Vacuum deposition methods of metals and semiconductors are being studied and developed using photolithography. Photoengraving processes are being utilized to produce extremely fine lines for intercomponent connections. Lines as fine as 1 mil[4] wide are reproducible.

The great promise of printed circuits is the eventual microminiaturization and the mechanization of production methods for complete electronic systems. However, great difficulties still exist in devising smaller printed-circuit coils and transformers for use at audio frequencies. Much progress is also required in developing magnetic films for vacuum deposition.

The modern printed circuit embodies the following components in combination:

1. Dielectrics
2. Conductors
3. Resistors
4. Capacitors
5. Inductors
6. Transistors and diodes

The dielectric should essentially act as a matrix for the circuit, serving mutually to isolate or insulate from each other the various electronic circuit components which are utilized in any particular design. The dielectric should also function to protect the circuit from physical damage.

Printed-circuit conductors for microminiature circuits of deposited metal are generally formed by some means of metal deposition onto a suitable dielectric. Common processing methods include screen printing, spraying, and vacuum deposition techniques. Component leads can be connected to printed conductors by soldering, welding, plating, or conductive cements.

Screen printing of conductive patterns onto dielectric plates is accomplished by hand-operated or automatic machinery. Fully automated screen printing requires machinery for conveying, feeding, indexing, holding, and removing various sizes and shapes of printed-circuit plates. For microminiature printed circuits, the smaller structures will assume more fragility, thereby requiring delicate handling.

Spray-printing methods utilize a mask having the desired pattern. This is placed over the dielectric plate and a silver paint is sprayed over the masked plate. Removal of the mask leaves the desired printed circuit. It is difficult to obtain sharp patterns when using this method. As a result, it is not expected to find extensive application for microminiature circuits.

The vacuum deposition of conductive material is finding increased interest in microminiaturization since this technique tends to produce finer lines under a controlled environment. This method requires that the desired metal be placed within a vacuum chamber, together with masked dielectric plates having the desired pattern. After the chamber is evacuated, the metal is heated until vaporization occurs. The metallic vapors disperse and are evenly deposited on the prepared plates. Then the vacuum is eliminated, after which the patterned plates are removed. With this method 10-mil-wide lines having a spacing of 4 mils have been produced using electroetched vacuum-deposited palladium.

Printed resistors can be applied to circuits by means of tape, screen, or pen printing, injection molding, and vacuum evaporation.

Printed capacitors can be formed by screening silver electrodes on either side of a thin wafer having a high dielectric constant. These wafers can be mechanically and electrically fastened to a circuit with conductive cement.

Printed inductors have been made practical only for values in the range 0.1 to $1\mu h$. Although great strides have been made in developing small windings, the inductive component is much larger than any other electronic component. In microminiaturization, the approach taken is to substitute RC circuitry in lieu of inductive elements wherever practical.

Diodes and transistors can be formed by controlled masking, etching, and diffusion techniques.

28 Encapsulation

Encapsulating techniques have been advantageously employed in furthering the art of miniaturization. The conditions of operation and environment determine the type of encapsulent or coating material to be used.

Previously, protection against mechanical shock, vibration, or acceleration within electronic packages was rendered by mounting, clamping, and supporting hardware. Protection against humidity generally required some means of gasketing. Intricate designs were required which generally defeated all attempts toward miniaturization. The advent of the modern encapsulent has allowed close-proximity stacking of components, alleviating any requirement for supporting structure or further protection against humidity.

Outstanding among the potting components for miniaturization purposes are the epoxy resins, which have the advantages of

1. Excellent dielectric properties
2. Dimensional stability
3. Excellent adhesion
4. Bonding strength
5. Hardness
6. Resistance to impact
7. Minimum water absorption

Careful encapsulating procedures must be implemented to minimize certain disadvantages in utilizing the epoxy resins such as

1. Short pot life
2. Low viscosity
3. Exothermic reaction
4. Residual stresses
5. Moisture entrapment

The main objections toward the epoxy encapsulents are the additional weight and the impractical aspect of repairing damaged units. Foam encapsulents are used when minimized weight is required. Air pockets within the foam result in improved dielectric constant and also provide additional heat insulation. Foam encapsulents are more difficult to apply because of reduced viscosity. They also lack the rigidity of epoxy resins after curing.

References

1. Albert Lunchick, Characteristics of Solid Tantalum Capacitors, *Elec. Mfg.*, June, 1959.
2. Paul G. Jacobs, Micro-module Design Progress, *Elec. Mfg.*, March, 1959, p. 79.
3. J. T. Wallmark and S. M. Marcus, Semiconductor Devices for Microminiaturization, *Electronics*, June 26, 1959, p. 35.
4. E. F. Horsey and L. D. Shergolis, "Symposium on Microminiaturization of Electronic Assemblies," p. 1, Hayden Publishing Company, New York, 1958.

POWER SOURCES FOR REMOTE USE

29 Introduction

There are many possible sources of electrical power for use at remote locations. The selection of the optimum power source requires consideration of the logistics, economics, and physical environment of the application. For example, for huge blocks of power required at remote locations on the surface of the earth, the engine-driven mechanical generator would most likely be optimum provided an adequate supply of fuel could be maintained. This solution is not too practical for underwater or space applications. For some applications, a combination of different types of power sources might be required. Such is the case for long-term orbiting satellites where rechargeable batteries are utilized to supply power during periods when solar cells are inactive. When active, the solar cells, in turn, supply sufficient power both to meet load requirements and to recharge the battery. Each situation must be examined in the light of its own particular requirements.

The information in this section includes a compilation of data provided by manufacturers and various other sources on primary, secondary, reserve, thermal, thermionic, ferroelectric, nuclear, and solar cells. Mechanical converters, such as engine-driven generator sets, and converters which manipulate electrical energy, such as rectifiers and inverters, have not been included. The information given should provide guide lines and aid in evaluating various types of "batteries" as applied to a given set of design parameters.

30 Electrochemical Cells†

Electrochemical cells usually approach the essential characteristics desired of a power source for remote use; i.e., they are reliable, inexpensive, lightweight, and compact. The following paragraphs present a brief discussion of electrochemical-cell properties to aid in the understanding of basic cell actions.

Cell actions are basically oxidation-reduction reactions. Any oxidation-reduction reaction can be broken into "half-cell reactions" or "couples" that indicate the overall transfer of electrons from the reductant to the oxidant. Table 21 shows several half-cell reactions and their resultant voltages. It should be noted that the hydrogen-gas/hydrogen-ion couple is arbitrarily defined as zero and used as the standard reference couple for the potentials of all other couples. This is necessary since the absolute potential of a couple cannot be readily determined. The sign convention is that adopted by the American Chemical Society; i.e., with the couple written with the electrons on the right-hand side of the equation, a positive value for $E°$ will mean that the reduced form of the couple is a better reducing agent than hydrogen gas. Table 22 is an electromotive-force chart showing the voltage obtainable from various couples from the most active positive couple to the most active negative couple.

† Portions of this section appear in "Survey of Electrochemical Batteries" by Norman D. Wheeler, *Electro Technology*, June, 1963, pp. 68–73. Copyright, 1963, by C-M Technical Publications Corp., New York.

Table 21. Half-cell Reactions and Potentials at Standard Conditions

Couple	$E°$, volts	Couple	$E°$, volts
$Na \rightarrow Na^+ + e$	2.712	$H_2 \rightarrow 2H^+ + 2e$	0.000
$Mg \rightarrow Mg^{++} + 2e$	2.34	$Cu \rightarrow Cu^+ + e$	-0.522
$Mg + 2OH^- \rightarrow Mg(OH)_2 + 2e$	2.67	$Cu^+ \rightarrow Cu^{++} + e$	-0.167
$Al \rightarrow Al^{+++} + 3e$	1.67	$Cu + Cl^- \rightarrow CuCl + e$	-0.124
$Al + 3OH^- \rightarrow Al(OH)_3 + 3e$	2.31	$Cu + 2OH^- \rightarrow CuO + H_2O + 2e$	0.258
$Al + 4OH^- \rightarrow AlO_2^- + H_2O + 3e$	2.35	$2Hg \rightarrow Hg_2^{++} + 2e$	-0.799
$Zn \rightarrow Zn^{++} + 2e$	0.762	$Hg + 2OH^- \rightarrow HgO + H_2O + 2e$	-0.098
$Zn + 2OH^- \rightarrow Zn(OH)_2 + 2e$	1.248	$Ag \rightarrow Ag^+ + e$	-0.800
$Zn + 3OH^- \rightarrow HZnO_2^- + H_2O + 2e$	0.72	$Ag + Cl^- \rightarrow AgCl + e$	-0.222
$Fe \rightarrow Fe^{++} + 2e$	0.44	$Ag + 2OH^- \rightarrow AgO + H_2O + 2e$	-0.457
$Fe^{++} \rightarrow Fe^{+++} + e$	-0.771	$Mn(OH)_3 + OH^- \rightarrow MnO_2 + 2H_2O + e$	-0.8
$Fe + 2OH^- \rightarrow Fe(OH)_2 + 2e$	0.877	$Mn^{++} + 2H_2O \rightarrow MnO_2 + 4H^+ + 2e$	-1.28
$Cd \rightarrow Cd^{++} + 2e$	0.402	$MnO_2 + 4OH^- \rightarrow MnO_4^- + 2H_2O + 3e$	-0.57
$Cd + 2OH^- \rightarrow Cd(OH)_2 + 2e$	0.815	$MnO_2 + H_2O \rightarrow MnO_4^- + 4H^+ + 3e$	-1.67
$Ni \rightarrow Ni^{++} + 2e$	0.25	$4OH^- \rightarrow O_2 + 2H_2O + 4e$	-0.401
$Ni(OH)_2 + 2OH^- \rightarrow NiO_2 + 2H_2O + 2e$	-0.49	$2H_2O \rightarrow O_2 + 4H^+ + 4e$	-1.229
$Pb \rightarrow Pb^{++} + 2e$	0.126	$2H_2O \rightarrow H_2O_2 + 2H^+ + 2e$	-1.77
$Pb + 2OH^- \rightarrow PbO + H_2O + 2e$	0.578	$H_2O_2 \rightarrow O_2 + 2H^+ + 2e$	-0.682
$Pb^{++} + 2H_2O \rightarrow PbO_2 + 4H^+ + 2e$	-1.456	$2Cl^- \rightarrow Cl_2 + 2e$	-1.358
$PbSO_4 + 2H_2O \rightarrow PbO_2 + 4H^+ + SO_4^{--} + 2e$	-1.685		

The power obtainable from any "practical" cell is always less than the theoretical power. Losses associated with cell operation are conventionally divided into two classes: self-discharge and overvoltage.

Self-discharge

Theoretical voltages as given on the half-cell-reaction charts are rarely, if ever, realized because of the self-discharge effects (local action) of the electrodes. Minute impurities in the electrode material react with pure electrode material in the presence of the electrolyte. For example, high-purity magnesium (99.8 per cent Mg) in a 3 per cent solution of salt has a theoretical potential of 2.67 volts; however, because of self-discharge effects, the half-cell potential is observed as 1.4 volts. Self-discharge effects occur at both electrodes, though predominantly at the anode.

Overvoltage

Overvoltage effects are the result of three basic parameters: electrolytic resistance, concentration changes, and polarization.

Electrolytic resistance occurs since current in a cell is carried by ion movement. The ion-migration velocity is dependent upon the potential gradients and ion size, which produce an IR drop representing a resistive load. The electrolytic resistance is a function of the composition of the electrolyte and the geometry of the cell. Table 23 shows representative values of some types of electrolytes.

Concentration-change effects occur because the ion-migration rates are finite. Thus, when current flows in a cell, an ion-concentration gradient exists between the bulk of the electrolyte and the electrolyte in direct contact with an electrode. These effects not only will reduce the cell output voltage but will also create the need for an increased charging voltage. The loss in efficiency is directly proportional to the current flow rate.

Table 22. Selections from Electromotive-force Table*

Positive electrode (Volts, Oxidizing ↑ / Reducing ↓)	Half-cell reaction	Equivalent weight
3 — Fluorine	$\tfrac{1}{2}F_2 + e \to F^-$	19.0
2 — Lead dioxide	$PbO_2 + 4H^+ + 2e \to Pb^{++} + 2H_2O$	217.6†
— Chlorine	$\tfrac{1}{2}Cl_2 + e \to Cl^-$	35.5
— Bromine	$\tfrac{1}{2}Br_2 + e \to Br^-$	79.9
1 — Iodine	$\tfrac{1}{2}I_2 + e \to I^-$	127
— Cupric ion	$\tfrac{1}{2}Cu^{++} + e \to \tfrac{1}{2}Cu$	31.8
0 — Hydrogen	$\tfrac{1}{2}H_2 \to H^+ + e$	1
— Lead	$\tfrac{1}{2}Pb \to \tfrac{1}{2}Pb^{++} + e$	103.6
— Nickel	$\tfrac{1}{2}Ni \to Ni^{++} + e$	29.4
— Iron	$\tfrac{1}{2}Fe \to Fe^{++} + e$	27.9
— Zinc	$\tfrac{1}{2}Zn \to \tfrac{1}{2}Zn^{++} + e$	32.7
1 — Manganese	$\tfrac{1}{2}Mn \to \tfrac{1}{2}Mn^{++} + e$	27.47
— Aluminum	$\tfrac{1}{3}Al \to \tfrac{1}{3}Al^{+++} + e$	9.0
2 — Magnesium	$\tfrac{1}{2}Mg \to \tfrac{1}{2}Mg^{++} + e$	12.16
— Sodium	$Na \to Na^+ + e$	23
— Calcium	$\tfrac{1}{2}Ca \to \tfrac{1}{2}Ca^{++} + e$	20.04
3 — Lithium (Negative electrode)	$Li \to Li^+ + e$	6.94

* As given in E. C. Pitzer, Battery Capabilities, *TIS Rept.* 59GL1, General Electric Co. General Engineering Laboratory, Schenectady, N.Y.

† Assuming $4H^+$ are furnished by $2H_2SO_4$.

Table 23. Resistance of Typical Aqueous Electrolytes

Electrolyte types	Concentration (normality)	Temp., °C	Resistance range,* ohm-cm
Strong acids	2–6	20	1.3–2
Strong bases	6	20	3–5
Highly ionized salts	2	20	6–8
Weakly ionized salts	2	20	12–40
Sea water	...	20	18
Sea water	...	0	30

* Resistance for metallic copper is 1.7×10^{-6} ohm/cm.

Polarization effects are the result of reaction rates at the electrode surfaces, since the chemical reactions within a cell are typically a step reaction and one or more of the reactions may be slow. Thus, if the reducing action leads the oxidizing action within the cell, gas is liberated. These polarization effects can be quite sizable, introducing not only voltage-drop problems but also problems associated with hermetic sealing and the presence of a volatile gas.

Electrochemical cells can be classified into three major types: primary, secondary, and reserve (or delayed-action) cells.

Primary Cells

Primary cells are those cells in which the chemical reaction which releases the electrical energy is irreversible from a practical standpoint. There are both dry and wet primary cells.

Dry Primary Cells. *Leclanche.* The Leclanche dry cell is perhaps the best-known cell in common use today. It is widely used in flashlights and other such equipment. This type of cell was originally described by Georges Leclanche in 1868 and has undergone many improvements since that time. Basically this cell consists of a nearly pure (99.99 per cent) zinc negative terminal, a carbon positive terminal, and a mixture of ammonium chloride, manganese dioxide, acetylene black, zinc chloride, chrome inhibitor, and water. The mixture acts as a depolarizing agent to reduce the formation of hydrogen bubbles on the positive electrode as discharge takes place. Improvements which have been made include leakproofing, longer shelf life, pepped-up depolarizers, improved insulation, and miniaturization. In addition, attempts have been made recently to substitute magnesium or aluminum for the zinc in this cell. Work on the aluminum electrode has shown some promise, while the magnesium electrode cell is presently in production. The latter shows good promise of supplementing zinc electrode cells. It should be noted, however, that magnesium cells have two disadvantages which must be resolved: (1) their voltage is too high to work with existing light bulbs now on the market, and (2) a time delay is exhibited by magnesium cells; up to a minute may be required for a current to flow after the switch is closed.

Mercury Cells. The development of the mercury dry cell has been acclaimed as the most significant development in the dry-cell field within the past two decades. The mercury cell was invented by Samuel Ruben and is manufactured principally by the P. R. Mallory Co.

The depolarizing cathode consists of mercuric oxide to which graphite and, in some cases, electrolytic manganese dioxide is added. The graphite reduces internal cell resistance by increasing the depolarizer conductivity, while manganese dioxide provides added oxygen-giving material. The anode is of pressed amalgamated zinc particles, while the electrolyte is a concentrated aqueous solution of potassium hydroxide saturated with zincate. The cell container usually is of steel.

The main advantage of a mercury cell is its unusually high ratio of energy to size and weight. For example, mercury cells can yield as much as three to four times the energy of other types of batteries. The main disadvantage seems to be in the area of cost, because a watthour costs about $1.50 for small-capacity cells.

Alkaline. The alkaline dry cell has experienced rapid progress in the last few years because of military needs for small batteries. This cell is similar to the Leclanche cell except that an electrolyte of potassium hydroxide is substituted for the ammonium chloride. The construction of this cell might be likened to two steel bottle caps. The anode is housed and in contact with one bottle cap; the cathode is housed in another and the housings are crimped together. The electrolyte is in a gelatinous medium contained between the electrodes. Mercuric oxide mixed with manganese dioxide is used in some cells to form the cathode.

The Dichromate Cell. The dichromate cell is similar to the alkaline cell described above except that potassium dichromate is used as a depolarizer. This cell is suitable for very-low-current-drain applications such as a grid-bias cell.

The Vanadium Pentoxide Cell. This type of cell is usually a button-type cell similar to the two above-mentioned. It uses vanadium pentoxide as a depolarizer, cadmium as the anode, and a weak acid as the electrolyte. The electromotive force of this cell is reportedly 1.02 to 1.04 volts. Exactly similar in construction but employing zinc instead of cadmium and an electrolyte of ammonium glycol-borate is another vanadium pentoxide cell which has an electromotive force of 1.2 volts. Both types are quite small and are suitable for sources of potential rather than current. They are generally used as grid-bias cells. Reportedly, the life of these cells may be 10 years or more with only a negligible change in the emf.

The Air Cell. The air dry cell is generally used in such applications as flashlights and hearing aids. It has a zinc anode and an electrolyte of sodium or potassium hydroxide. The electrolyte is nonspillable, being retained in gelatinized starch. The cathodic depolarizer is atmospheric oxygen which is absorbed on a surface of a special water-repellent, porous carbon. Air dry cells are intended for service where current drains are approximately 20 to 30 ma. They have relatively long life but cannot be used in environments where the temperature drops below 15°C.

Silver Chloride Dry Cell. This dry cell employs silver chloride as the depolarizer. It finds application in medical apparatus, blasting galvanometers, and applications requiring a source of constant potential at very low current rates. For best service the current should not exceed 10 ma continuous discharge; however, 300 to 500 ma may be drawn from the cells for very short intervals. This type of cell has exceptionally good shelf-life characteristics, has an emf of about 1 volt, and weighs about 45 g.

The Titanium-alloy Dry Cell. The titanium-alloy dry cell is like the mercury cell except that it utilizes a titanium alloy instead of zinc as the anode. This cell is primarily used in applications where high temperatures are encountered. At 170°F, for example, its operating voltage is about 1 volt. At this high temperature the cell gases at a very low rate but has the disadvantage of having a high temperature coefficient; i.e., its working voltage changes 2 to 3 mv/°F.

The Indium Dry Cell. The indium dry cell consists of an anode of indium-bismuth alloy, a cathode of mercury oxide, and an electrolyte of potassium hydroxide in a gel. It is generally made in a miniature size, about that of a dime, and its widest use at present is for wrist watches. The cell delivers power evenly at extremely low drains in the order of one-millionth watt.

Zamboni Piles. Zamboni piles are dry batteries consisting of cells stacked in series to give a compact source of high voltage with low current drains. There are several types of these piles. One type consists of thin wafers of paper with aluminum molecularly distilled on one side and manganese dioxide spread on the other. With proper insulation, these wafers of paper can be stacked one on another to give a high-voltage battery. For example, a battery which will have an open-circuit voltage of 100 volts can be made $\frac{1}{2}$ in. in diameter and 1 in. long. However, this battery would sustain only a few millimicroamperes of current. The second type of design uses sheets or foils of magnesium rather than the molecularly distilled aluminum as anodes. The main drawback of the Zamboni pile is its shelf life which, in general, rarely extends over 1 year.

Low-temperature Cells. Two general types of low-temperature cells have been investigated. One uses inorganic chemicals, the other organic salts. The Germans are said to have used magnesium chloride in place of ammonium chloride in cells designed for low temperatures during World War II. Reportedly, at the relatively light loads of 10 to 12 μa for a B supply, and 100 μa for A supplies, these batteries were used at temperatures as low as $-30°C$. Zinc chloride may be used in larger proportion than in ordinary cells; however, its resistivity is high, and consequently it is generally used in combination with other chlorides such as calcium chloride or lithium chloride. These latter two cells can be used to temperatures as low as $-40°C$. A three-salt (four components, including water) cell has also been investigated. This would be a calcium chloride–zinc chloride–ammonium chloride cell. This cell is usually used to $-30°C$. As a general rule, the starch-flour separator or gelling agent originally employed was not satisfactory at low temperatures. It was found that the locust-bean gum was a substitute for the ordinary cereal paste. The locust-bean gum, however, was not compatible with the methylamine hydrochloride which has also been used as an electrolyte in a low-temperature cell. This last-mentioned electrolyte is generally stabilized by adding ammonium chloride to the solution and is in general less satisfactory than the calcium or lithium chloride cells.

Wet Primary Cells. Wet primary cells find restricted use at the present time except possibly for the newly developed fuel cells. Wet primary cells may be listed as follows:

The Lalande or Caustic Soda Cell. This cell is made with anodes of zinc, an electrolyte of sodium hydroxide, and cathodes of copper oxide. It has a working voltage of 0.6 volt and is best suited for long service at low current drains. This cell is used widely in rural areas by the railroads to activate signals and similar equipment.

The Air Cell. The air cell is similar to the Lalande cell except that atmospheric oxygen is absorbed on water-repellent porous carbon and used as the depolarizer instead of copper oxide. The air cell has an open-circuit voltage of about 1.45 volts. This cell is used for long service at low current drains.

The LeCarbone Cell. The LeCarbone cell is a special air cell which contains manganese dioxide.

The Gravity or Daniel Cell. The gravity cell is practically obsolete. This cell uses copper sulfate and zinc sulfate electrolytes and is generally operated on closed circuit at all times. It, too, is used for long-term low-current drains. It was the predecessor of the Lalande cell in railroad service.

Standard Cell. Standard cells are not used as power sources since they can supply only very minute currents. In general, standard cells have extremely long life, stable voltages, and low temperature coefficients and are used as standard references for electromotive force.

Silver-Zinc Cell. Recent development of silver-zinc cells have permitted a degree of recharging. These cells are therefore discussed under Secondary Cells.

Some of the more common primary cells are compared in Table 24 in terms of open-circuit voltage and capacity.

Secondary Cells

Secondary cells are more commonly known as storage batteries or as accumulators. In general, storage batteries are used in applications requiring high current rates and where recharging facilities are available.

The Lead-Acid Cell. The lead-acid cell is probably the best known of the common storage batteries. The negative plate is nearly always a pasted or Faure plate, that is, one in which the grid or framework on the plate is pasted with a blend of lead oxide which is then converted electrolytically to spongy lead. The positive plate may be a similar pasted construction, where high discharge rates are required. For lower discharge rates, the construction may be quite different. For example, the positive plate could be of a Plante or formed-plate construction; that is, the active material is formed from the lead plate itself by electrolytic action. Another type of anode would be a tubular or perforated plate, a type most commonly used in

Table 24. Primary-cell Characteristics

Cell	Materials	Open-circuit voltage, volts	Capacity*				Watt-hour per dollar†	Remarks
			Watt-hr/lb	Watt-hr/cu in.	Amp-hr/lb			
Leclanche	$Zn/NH_4Cl/MnO_2$	1.5	2–30	0.2–2.5	1.5–20	2–50	Operable from 30–90°F
Mercury	$Zn/KOH/HgO$	1.34	51.4	4–8	38.4	0.7–14	Operable from +50 to 160°F
Alkaline	$Zn/NaOH/MnO_2$	1.46	47	2.2	41	33	Operable at temp 40–90°F under further development
Oxygen-depolarized Lalande	$Zn/NaOH\text{-}Ca(OH)_2/O_2$	1.4	50	1.5		75	Storage and feeding of O_2 not solved
Hydrogen peroxide depolarized Lalande	$Zn/NaOH\text{-}Ca(OH)_2/O_2$ from H_2O_2	1.4	56	2.1		75	Storage, feeding, decomposition of H_2O_2 not solved
Lalande (wet)	$Zn/NaOH/CuO$	0.65	19.6	0.94		57	Very reliable but bulky
Magnesium	$Mg/MgBr/MnO_2$	2.0	28–44	3.3		28	Still in development. May have delay in reaching cell voltage at room temp
Zinc chloride	$Zn/ZnCl/Cl$	1.7	20–40	1–2.2		75	"Dry cell" connected to Cl_2 cylinder, Cl_2 feed not solved. Not designed for low-drain operation
Neutralization	$NaOH/H_2SO_4$	0.25	28	2.6		23	Uses ion barrier membranes
Magnesium-bismuth	Mg/Bi_2O_3	1.6	47	3.3			Sketchy information available
Organic depolarized (dry cell)	$Zn/C_6H_4(NO_2)_2$	1.5	103	6.6		76	Questionable data
Sodium amalgam-oxygen	$Na/25\% NaOH/O_2$	1.85	190	3.3			Questionable data

* Based on large capacity. Figures dependent upon operating conditions.
† Variable depending upon size and operation.

motive-power batteries. It has been found that the tubular form of construction can be simulated using pasted positive plates which are encased in a number of foils, generally of fiber glass. The electrolyte is usually a dilute solution of sulfuric acid.

The ratio of power obtained on discharge to power required for charging is a great advantage of the lead-acid cells when compared with other couples such as nickel-iron, nickel-cadmium, or zinc-silver. This, together with its lower initial cost, makes the lead-acid cell preferable in many applications where economy of operation is of importance.

The Nickel-Cadmium Cell. Nickel-cadmium alkaline type cells were developed in Sweden about 1900. They have been used widely in the United States since 1930. Nickel-cadmium cells come in two distinct forms; the older form is the Jungner or pocket-plate type of cell, and the other is the Durac or impregnated sintered-plate type which was developed in Germany during World War II. Sintered plates enjoy a great advantage over the tubular pocket plates in that they can be made thinner and thus occupy less space. Further, owing to their larger surface area, sintered plates can sustain higher current drains. It should be noted, however, that relative merits of the two types depend upon the current-drain rate. The higher the current-drain rate, the better the advantage of a sintered-plate battery. Sintered-plate batteries also are advantageous for those applications which require "hermetically" sealed units needing no vent caps or double chambers. The main disadvantage of the sintered-plate

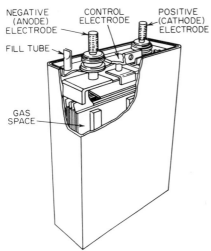

FIG. 99. Typical sealed cell with control electrode. (*Courtesy of General Electric Co.*)

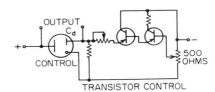

FIG. 100. Typical transistor control circuit for use with control electrode cells. (*Courtesy of General Electric Co.*)

battery is that it has not yet had sufficient experience time to prove its reliability as has the tubular-type pocket-plate battery. The electrolyte used with the nickel-cadmium battery is potassium hydroxide.

An innovation introduced by the Battery Products Section of the General Electric Company in 1963 is the sealed nickel-cadmium cell with an auxiliary electrode for positive control and determination of full charge of the cell. The additional electrode operates as an oxygen electrode, promoting recombination of the gas evolved during battery charge and overcharge by cathodic reaction of oxygen with water to provide hydroxyl ions at usable voltages and current densities. Current flows from the auxiliary "control electrode" through some type of external circuit to the cadmium electrode. Figure 99 shows a typical sealed cell with control electrode. Figure 100 shows one type of external charging circuit which may be used with this cell.

With the control electrode, the effects of pressure built up during charge can be greatly reduced since the oxygen reaction takes place quickly. Thus, internal cell pressures can generally be kept below 1 atm, with the result that lighter-weight plastic cases may be used which permit great savings in battery size. The control electrode also permits increased charge rates (up to $2c$,* or nearly fifteen times the

* c is 1-hr discharge rate.

rate of conventional cells) which may permit increased discharge depths in many applications. Figure 101 shows some typical charge-retention characteristics of a General Electric control-electrode cell.

The Nickel-Iron, or Edison-type Cell. The nickel-iron battery is generally made as a pocket-type cell in which the positive plates are filled with nickelous hydroxide and nickel and the negative plates are filled with a finely divided mixture of metallic iron, ferrous oxide, and mercuric oxide. The electrolyte is a solution of potassium hydroxide in water to which a small amount of lithium hydroxide has been added.

The advantage of the nickel-iron battery is its ability to withstand complete discharge and short circuiting. The main disadvantage of the nickel-iron battery is that the battery gases throughout nearly the entire period of charge and is subject to a rather high rate of loss of charge immediately after the charging period. The efficiency of the Edison cell on an ampere-hours basis is quite low.

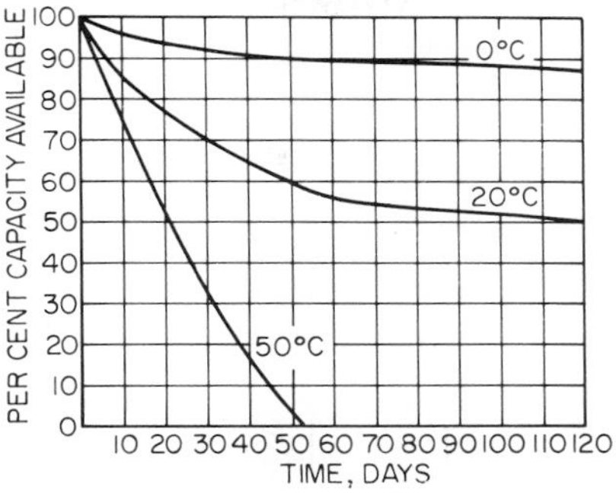

Silver Oxide Cell. Silver oxide cells employ zinc and silver oxide as electrodes in a solution of potassium hydroxide, saturated when in the discharged condition with zinc hydroxide. Of particular interest because of its large output per unit weight and unit volume, this type of battery cannot be hermetically sealed because of the evolution of oxygen. Another drawback is that the zinc-silver battery has poor retention of charge when stored. The zinc-silver battery has a limited cycle life unless the batteries are discharged and charged very carefully at very low rates. The ampere-hour efficiency of the zinc-silver cell is 80 per cent. The zinc-silver battery is especially suited for very high current drains.

Fig. 101. Typical charge-retention characteristics of nickel-cadmium control electrode cell. (*Courtesy of General Electric Co.*)

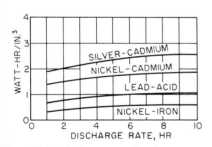

Fig. 102. Typical energy-unit volume curves for various battery systems. (*Courtesy of Yardney Electric Corp.*)

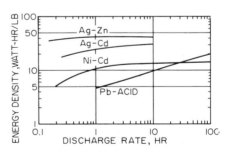

Fig. 103. Energy density vs. drain rate for various types of secondary batteries. (*Courtesy of General Electric Co.*)

Silver cells can be operated in series or parallel provided that they are of the same type and size, and provided that some necessary precautions are taken. For example, before assembly into a battery, silver cells should be completely discharged to 0 volts at a rate equal to the 1-hr rate or less. Series-connected cells are charged on a constant-current basis. Parallel-connected cells should not be charged while so connected.

The Cadmium-Silver Cell. Cadmium-silver batteries are presently under investigation. These cells are similar in construction to the zinc-silver batteries except

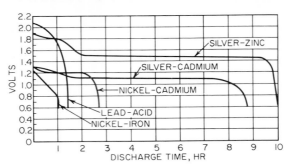

Fig. 104. Typical discharge characteristics of various battery systems of equal weight discharging under the same conditions. (*Courtesy of Yardney Electric Corp.*)

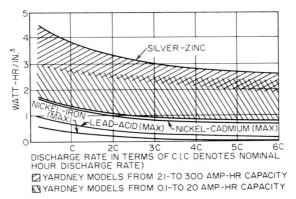

Fig. 105. Typical energy/unit volume curves for various battery systems. (*Courtesy of Yardney Electric Corp.*)

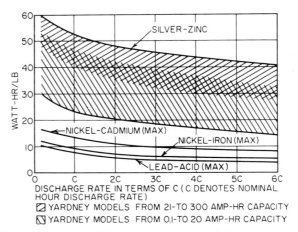

Fig. 106. Typical energy/unit weight curves for various battery systems. (*Courtesy of Yardney Electric Corp.*)

Table 25. Secondary-cell Characteristics

Characteristics	Lead-acid	Nickel-cadmium (pocket plate)	Nickel-cadmium (sintered plate)	Edison (nickel-iron)	Silver-zinc	Silver-cadmium
Materials	$Pb/H_2SO_4/PbO_2$	$Cd/KOH/NiO_2$	$Cd/KOH/NiO_2$	$Fe/KOH/NiO_2$	$Zn/KOH/AgO$†	$Cd/KOH/AgO$†
Open-circuit voltage of fully charged cell	2.1	1.3	1.3	1.4	1.86	1.4
Nominal voltage	2.0	1.2	1.2	1.2	1.5	1.1
Average discharge voltage at 1-hr rate:*						
At 80°F	1.86	1.05	1.18	0.85	1.43	0.99
At 0°F	1.84	1.00	1.15	....	1.27	0.95
At −40°F	1.72	0.75	1.08	....	....	0.77
Voltage range for constant-current charge	2–2.6	1.3–1.7	1.3–1.7	1.5–1.8	1.6–2.1	1.2–1.7
Constant-current charging at 80°F	10-hr rate to constant voltage	5-hr rate for 7 hr	5-hr rate for 7 hr	5-hr rate for 7 hr	20-hr rate to 2.1 volts	20-hr rate to 1.7 volts
Time to 50% capacity retention:						
80°F	55 days	300 days	300 days	25 days	Estimated over 2 years	Over 2 years
125°F	7 days	17 days	17 days	....	115 days	115 days (est)
160°F	¾ day	4 days	4 days	....	58 days	58 days
Cycle life	200–2,000	Over 2,000	Over 2,000	Over 2,000	10–400	300–1,000
Watt-hr/lb‡	7–26	6–10	10–15	11–14	25–56	11–38
Watt-hr/cu in.‡	0.45–2.75	0.4–0.7	1.08	0.98–1.16	1.3–3.8	0.67–2.9
Watt-hr/dollar‡	45	9.6 (est.)	8.5 (est.)	9.5	2.8	2.8
Amp-hr/lb‡	4–13	4–8	8–12	9–12	16–37	10–35
Major advantages	Low cost, general availability, good cycle life, high voltage per cell, good capacity, life and charge retention	Excellent cycle life, reliable with care, rugged	Excellent cycle life, reliable with care, good low-temperature charge/discharge performance, rugged, can be hermetically sealed	Excellent cycle life, reliable, extremely rugged, not damaged by overcharge or overdischarge	Excellent energy output per unit weight and volume, excellent performance at high discharge rates	Good energy output per unit weight and volume, good cycle life

10-89

Table 25. Secondary-cell Characteristics (*Continued*)

Characteristics	Lead-acid	Nickel-cadmium (pocket plate)	Nickel-cadmium (sintered plate)	Edison (nickel-iron)	Silver-zinc	Silver-cadmium
Major disadvantages......	Sulfates on discharged stand, cannot be charged at sub-zero temperatures, cannot be hermetically sealed	High cost, poor high rate and low-temperature performance	High cost	High cost, poor charge retention, poor low-temperature performance	High cost, poor cycle life, poor low-temperature performance	High cost

* Based on 20 per cent drop in voltage during the discharge period.
† Some cells use silver peroxide Ag_2O_2.
‡ Figures dependent upon capacity. Figures based on large capacity at 80°F operation.

POWER SOURCES FOR REMOTE USE   10-91

that the electrodes are cadmium and silver in a solution of potassium hydroxide. The cycle life of these cadmium-silver batteries is considerably better than that of the zinc-silver battery. The cadmium-silver battery apparently has good energy output per unit weight and volume. Its major disadvantage is its high cost.

Table 26. Secondary-cell Temperature-discharge Characteristics

	Temp, °F	Discharge rate, hr	Lead-acid	Nickel-cadmium (pocket plate)	Nickel-cadmium (sintered plate)	Edison (nickel-iron)	Silver-zinc	Silver-cadmium
Watt-hr/lb (based on an allowable voltage drop of 20%)	80	5	11.4	6	15	15	52.8	24.6
		1	8.4	4.8	11.6	9.5	45.8	19.8
		0.25	6.8	2.9	9.9	.....	39	15.6
	0	5	5.4	4.2	11.5	.....	39.4	20.9
		1	3.9	3.1	9.9	.....	33	16.6
		0.25	2.9	1.7	8.7	.....	24.6	12.1
	-40	5	3.7	3.2	9	.....	.....	10.7
		1	2.5	2.4	8.4	.....	.....	7.7
		0.25	1.5	0.9	5.3	.....	.....	
Watt-hr/cu in. (based on an allowable voltage drop of 20%)	80	5	0.83	0.47	1.08	1.16	3.36	1.57
		1	0.61	0.38	0.84	0.73	2.92	1.26
		0.25	0.49	0.23	0.72	.....	2.48	0.99
	0	5	0.39	0.33	0.83	.....	2.51	1.33
		1	0.28	0.24	0.72	.....	2.1	1.06
		0.25	0.21	0.13	0.63	.....	1.67	0.77
	-40	5	0.27	0.25	0.65	.....	.....	0.68
		1	0.18	0.19	0.61	.....	.....	0.49
		0.25	0.11	0.07	0.38	.....	.....	

Figures 102 through 106 give comparative data for some types of secondary cells as functions of operating conditions. Just as with primary batteries, no single secondary battery can provide all the necessary attributes to satisfy all conditions. Tables 25 and 26 compare some of the characteristics of various secondary cells.

Reserve or Delayed-action Batteries

Reserve or delayed-action batteries are batteries that are activated at the time of use or shortly before the battery is to be placed in service. The common reserve batteries, with means of activation, are as follows:

Type	Activation	Type	Activation
Silver chloride............	Sea water	Zinc-lead.................	Mechanical
Cuprous chloride..........	Sea water	Zinc-silver...............	Mechanical
Ammonia.................	Gas	Lead-acid................	Mechanical
Chlorine.................	Gas	Perchloric acid...........	Mechanical
Thermal.................	Heat	Fluoroboric acid..........	Mechanical
Solar...................	Light	Fluorosilicic acid.........	Mechanical
Sodium..................	Fuel	Dry frozen...............	Alternating current
Cadmium-lead............	Mechanical		

The water-activated types both use magnesium as anodes. They are made operative by dunking in sea water or, with less degree of efficiency, in pure water. These batteries, which are used in weather forecasting and oceanographic equipment, show an initial sharp rise to peak voltage, and become warm during operation. Typical weight and size curves are given in Fig. 107. The chlorine and sodium amalgam cells are described in the fuel-cell section of this report. Indeed, fuel cells in general may be considered reserve batteries. The thermal batteries are activated when the electrolyte is heated above its melting point; then the ions are free to move and carry current, and the electrode processes take place. Generally, quick activation is obtained. One type of experimental thermal cell is made with magnesium anodes, cathodes of magnesium dioxide housed in carbon, and an electrolyte of solid anhydrous sodium hydroxide. These cells generally will sustain a very small current drain. More details are given in another section on pp. 10-96 to 10-99. Solar cells are likewise discussed in a separate section. With the exception of the dry frozen type of cell, the other cells listed are activated mechanically. For example, a capsule containing the liquid electrolyte is punctured, allowing the electrolyte to be introduced into the cells either by gravity flow or by forced flow. The cadmium-lead, zinc-lead, and lead-acid cells are activated with aqueous solutions of sulfuric acid. The zinc-silver cell is activated by an aqueous solution of potassium hydroxide. Since the shelf life of these batteries when activated is limited, they are most generally used only for special purposes. The perchloric acid, fluoroboric acid, and fluorosilicic acid batteries have good low-temperature characteristics. These batteries utilize lead anodes and lead dioxide cathodes, and are activated by the addition of aqueous solutions of the particular acid. Since the electrode reaction products are soluble in the electrolytes, these cells have a short storage life. In the dry frozen battery, the entire cell is maintained at a very low temperature, thus limiting its activity and preserving its energy for use when the temperature is raised by an alternating current applied to the terminals of the battery. A blocking capacitor is used in the circuit to prevent the cell from discharging through the external circuit. Table 27 gives some of the characteristics of reserve-type cells.

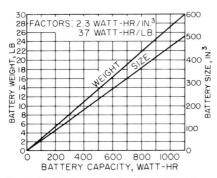

FIG. 107. Water-activated batteries (silver chloride type).

31 Solar Cells

Photovoltaic or solar cells are being used increasingly in satellites and space vehicles as power sources. These cells have limited application at the earth's surface because of the large volume requirements as well as overall cost. This type of battery must be considered only as a reserve supply, since the battery ceases to provide power in the absence of light.

Two types of photovoltaic cells are in common use today. One is the silicon photovoltaic cell, which uses a boron-diffused silicon p-n junction device developed by the Bell Telephone Laboratories, and the other is the selenium photovoltaic cell. Although other metals (such as tellurium) exhibit photovoltaic characteristics, they are not widely used in the construction of photovoltaic batteries. Although selenium has been used in many photovoltaic devices such as light meters and in other photocell applications, its typical current sensitivity in the linear operating range is 3.6 ma/sq. in./ft-c compared with 27.5 ma/sq in./ft-c illumination for a silicon photocell. Because of this higher linear operating range, silicon cells are the most commonly used elements in solar batteries.

With the sun at the zenith on a clear day, the solar radiant power density at sea level is about 100 mw/sq cm. (Silicon cells that will give approximately 8 to 10 mw/sq cm under these lighting conditions have been constructed.) More energy is available outside the atmosphere of the earth, since the radiant energy from the sun is as much as 1.4 times that found on the surface of the earth. (The power-density figure most frequently used for the space above the atmosphere is 120 mw/sq cm.)

Table 27. Reserve-cell Characteristics

Cell	Materials	Open-circuit voltage, volts	Capacity* Watt-hr/lb	Capacity* Watt-hr/cu in.	Watt-hr/dollar*	Remarks
Silver-sea	Mg/sea water/ AgCl-K$_2$S$_2$O$_8$	1.50	20-80	Up to 7.8	16.3	Life uncertain. High fabrication costs
Copper-sea	Mg/sea water/ Cu$_2$Cl$_2$	1.3	20-38	1.1-1.35	22.8	Life uncertain. High fabrication costs
Magnesium-sea	Mg/sea water	0.5	190	6.6	190	Cathode sealing is main problem
Concentration	Saturated brine/ sea water	0.05	5.7	0.26	38	Uses ion barrier membranes
Magnesium-hydrochloric acid	Mg/sea water/ HCL	0.5	44	2.2	125	Hydrochloric acid feed is main problem
Zinc-silver oxide	Zn/KOH/AgO	1.86	25-56	1.3-3.8	2.8	Same as secondary cell with electrolyte being added at time of use
Lead-lead dioxide	Pb/H$_2$SO$_4$/PbO$_2$	2.1	7-26	2-2.75	45	"Dry charge" storage battery
Zinc-lead dioxide		2.1-2.4	26	3.7		Complete data not reported
Cadmium-lead dioxide		1.8-2.1	14	1.06		Complete data not reported
Magnesium–organic N-halogen		1.9-2.3	Up to 61	2.24		Complete data not reported
Zinc–silver chloride	Zn/H$_2$O/AgCl	1	40	3		Complete data not reported.
Ammonia (Corson)		2-volt and 210-volt batteries tested, per cell value not given.	0.33-1.65 design goal 13	0.04-0.18 design goal 0.8		Activation time 0.1-1.0 sec. Very short life (approx. 30 min). Temp range: −65 to +165°F

* Based on large-capacity cells.

The power converted by a silicon cell depends on

1. The temperature of operation. The parameters of most solar cells are specified at 30°C, and an efficiency factor is given for temperature varying from this standard. In the usual case, the loss of the efficiency with increased temperature is at the rate of about 0.6 per cent/°C above 30°C. Although efficiency increases as the temperature is lowered, the rate of change in maximum output decreases below 10°C and is, for most cases, negligible below −100°C. Figures 108 and 109 show the voltage current characteristics for a 1- by 2-cm cell as a function of cell temperature.

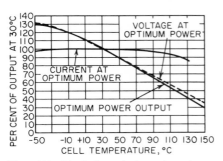

Fig. 108. Typical variation of optimum power output with temperature (along with current at optimum power, and voltage at optimum power). (*Courtesy of International Rectifier Corporation.*)

2. The amount of incident radiant power. The variation in output voltage at maximum power as a function of incident radiation is small and nearly linear above a level of approximately 40 mw/cm at sea level. Below this radiation level, the voltage decreases more rapidly with decreasing radiation.

3. The spectral distribution of radiation. The silicon solar cell has its highest efficiency at a wavelength of approximately 0.8 micron. It can, however, convert radiant energy into electrical energy for radiation at wavelengths from about 0.4 to 1.1 microns.

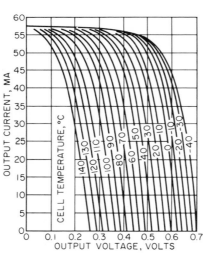

Fig. 109. Current-voltage characteristic for 1- by 2-cm cell at 1,400 watts/sq m solar irradiation with cell temperature as parameter (10 per cent efficient cell). (*Courtesy of International Rectifier Corporation.*)

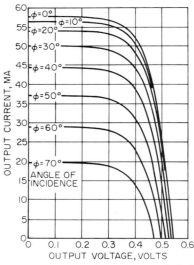

Fig. 110. Current-voltage characteristic for 1- by 2-cm cell at 30°C cell temperature for 1,400 watts/sq m irradiation at normal incidence with angle of incidence as parameter (10 per cent efficient cell). (*Courtesy of International Rectifier Corporation.*)

4. A function of the angle of incidence. Figure 110 shows a current-voltage characteristic for a silicon cell as a function of the angle of incidence. In general, the power converted is a cosine function of the angle of incidence.

POWER SOURCES FOR REMOTE USE 10-95

Although cells will operate as units, shingled modules have been found to increase the overall efficiency in the construction of a battery. Figures 111 and 112 show representative constructions for silicon solar cells.

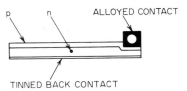

FIG. 111. Cross section of a solar cell. Outline dimensions: 1 by 2 cm. Active area: 1.8 sq cm. Efficiency of conversion: 7, 8, 9, 10 per cent. Weight: 0.3 g.

FIG. 112. Diagram showing shingle construction of cells. Outline dimensions: 2 by 4.6 cm. Active area: 9.0 sq cm. Efficiency of conversion: 7, 8, 9 per cent. Weight: 1.5 g average.

Table 28 shows some characteristics for typical silicon cells. One square foot of silicon solar cell will provide approximately 8 watts of power and weigh approximately 2.5 lb. In the application of solar cells, the area is of greater importance than the volume, since the thickness of the cell is nearly negligible when compared with the area required.

Table 28. Unmounted Cell Types—Commercial Grades*,†
Typical electrical characteristics for maximum power transfer at 30°C cell temperature

Overall dimensions, cm × cm	Nominal active area, sq cm	Avg conversion efficiency, %	Incident energy 100 mw/sq cm (1 sun)		
			Output power, mw	Approx output voltage, volts	Approx output current, ma
1 × 2	1.75	4	7.0	0.35–0.4	17.5
1 × 2	1.75	6	10.5	0.4	26
1 × 2	1.75	8	14.0	0.4	35
½ × 2	0.75	4	3.0	0.35–0.4	7.5
½ × 2	0.75	6	4.5	0.4	11.2
½ × 2	0.75	8	6.0	0.4	15
½ × 1	0.37	4	1.5	0.35–0.4	3.8
½ × 1	0.37	6	2.25	0.4	5.6
½ × 1	0.37	8	3.0	0.4	7.5

Notes:
1. Series-parallel configurations may be used to obtain desired power-supply rating. (a) For series connection, cells may be shingled together. (b) Cells connected in series must be exposed to uniform incident radiation for optimum power output.
2. Under dark conditions, the cells may be isolated from the storage battery or other load with a silicon blocking diode. This introduces a voltage drop of approximately 0.7 volt at 30°C.
* Courtesy of International Rectifier Corporation.
† For military and other special applications requiring an optimum power-to-weight ratio, cells with conversion efficiencies to 10 per cent are available.

32 Thermoelectric Power Generation

The phenomena of thermoelectricity have been understood for many years. The Seebeck and Peltier effects were discovered over 100 years ago. The recent breakthroughs in semiconductor materials have led to more practical possibilities for employing this type of technique to power generation.

Semiconductors of the silicon and germanium types are not generally suitable since they have the disadvantage of a temperature limitation due to "intrinsic" conduction. Should the temperature of these semiconductors be raised above that at which intrinsic conduction occurs, the resistivity is no longer controllable by adding various "impurities" but has a low value entirely dependent upon the temperature change. Although this low value of resistivity is favorable, the value of the Seebeck

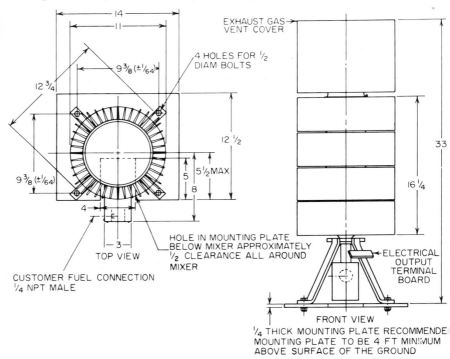

FIG. 113. Outline drawing for Westinghouse 50-watt type FTE-12 thermoelectric generator. (*Courtesy of Westinghouse Electronic Corp.*)

coefficient tends to diminish to zero, since the Seebeck effects of electrons (i.e., n-type conduction) and holes (p-type conduction) tend to cancel one another in most cases in which intrinsic conduction takes place. This property eliminates many semiconductors from operation at temperatures above 600°C. Even though many insulating crystals do not show this intrinsic conductivity at very high temperatures, their resistivity is much too high to yield a high figure of merit as a thermoelectric material. Development of the rare-earth compounds will no doubt be exploited since these materials hold good potential for high-efficiency thermoelectric conversion. Also under development are mixed-valence-type materials such as samarium sulfide which will be used at temperatures above 600°C.

Recent research on the thermoelectric properties of tertiary (three-element) compounds has shown that this class of materials may include some semiconductors

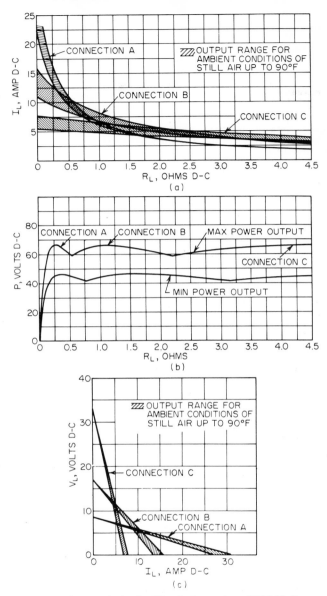

Fig. 114. Performance characteristics for Westinghouse type FTE-12 thermoelectric generator. (a) Load current vs. load resistance thermoelectric generator type FTE-12 part no. 914F627-1. (b) Load power vs. load resistance thermoelectric generator type FTE-12 part no. 914F627-1. (c) Load voltage vs. load current thermoelectric generator type FTE-12 part no. 914F627-1. (*Courtesy of Westinghouse Electric Corp.*)

which could be useful for power generation at intermediate temperatures (175 to 525°C). Tertiary compounds having a cubic structure—rock salts, zinc blendes, and fluorides—have been prepared, and some of their thermoelectric properties have been determined. Of particular interest has been the rock-salt structure silver-antimony-tellurium ($AgSbTe_2$). This tertiary compound has been found to have a low lattice thermal conductivity of approximately 0.0063 watt/cm/deg, an estimated band gap of less than 0.2 ev, a thermoelectric power of greater than 200 μv/deg, and a low electrical resistivity in the order of 10^{-3} ohm-cm. Alloys of this silver-antimony-tellurium tertiary compound with germanium tellurium appear promising for operation at temperatures in the 25 to 175°C area.

Figure 113 shows a sketch of a 50-watt thermoelectric generator as built by the Aerospace Electrical Division of the Westinghouse Electric Corporation. This device reportedly has a weight of 65 lb and operates at approximately 300°F utilizing propane, butane, or natural gas fuels to provide an input of 15,000 Btu/hr. Figure 114 shows performance curves obtained for one of these devices.

In general, the conversion efficiency of thermoelectric devices increases as the temperature of operation is increased. Temperatures in the region around 600°C have been used on one design to obtain an efficiency of about 3 per cent. These temperatures may be attained by the use of a solar concentrator (either a lens system or a mirror system—the latter being more efficient by elimination of the absorption of energy by the lens media). These solar devices have many problems such as curvature accuracy in mirrors, homogeneity in lens, and precise orientation of the collector to permit interception of the sun's rays at the best angle.

Radioisotope thermoelectric generators were first used for space power on the Transit 4A satellite launched June 28, 1961. Depending upon the selection of radioisotope-thermoelectric material combinations, the specific weight for a generator output in the 300- to 400-watt range varies from 260 to 770 lb/kw.

33 Thermionic Cells

Thermionic cells depend upon the Edison effect to generate electricity. That is, an emitting surface is heated until electrons are "boiled off." The "boiled-off" electrons are collected at a cathode surface. The emitting surface is a high-work-function surface and the cathode surface is a low-work-function surface. The voltage output is a function of the difference between the two work functions and may be as high as 1 volt.

The theoretical maximum efficiency is approximately 50 per cent. The highest efficiencies are attainable at higher temperatures.

Solar thermionic devices utilizing a silver-cadmium battery to supply the power requirements during the dark part of a 300-nautical-mile orbit are estimated to have a specific weight of 250 lb/kw for a generator in the 300- to 400-watt range.

Based upon a generator using curium-244 designed by the General Electric Spacecraft Department, the specific weight for a radioisotope thermionic generator with an output in the 300- to 400-watt range varies from 125 to 250 lb/kw depending upon safety requirements imposed by the radioisotope.

The Special Purpose Nuclear Systems Operation of the General Electric Company has been conducting extensive investigations on cesium-vapor refractory-metal thermionic cells which use the thermal energy released by a nuclear reactor utilizing uranium dioxide fuel pellets. This source of power, which operates with emitters in the 1500 to 1800°C region, will provide large blocks of power (tens of kilowatts to megawatts) and could find many uses in space or terrestrial applications. Single-cell converters have been operated out-of-pile with outputs exceeding 30 watts/sq cm and efficiencies of 20 per cent. Out-of-pile experimental converter life has been demonstrated in the thousands of hours. In-pile tests using nuclear fission have demonstrated life in the hundreds of hours. Utilizing demonstrated performance of single cells, system designs have been suggested which could be developed to provide a specific weight of 10 lb/kwe* in the 1-mwe* size.

---

* "kwe" and "mwe" stand for "kilowatt electric" and "megawatt electric" as opposed to "kw" and "mw" of heat.

34 Nuclear Cells

Nuclear batteries convert radioactive energy of particles or rays ejected from the atomic nucleus into electrical energy. There are seven general types of nuclear batteries:

Beta Current

The best-known type is a strontium-90 battery. This battery consists of elements of strontium-90 embedded in gold foil (which acts as the emitter electrode), an aluminum rod (which acts as the collector electrode), and polystyrene (which acts as the insulating medium). Although the emitter and collector may be separated by a vacuum, a plastic is generally used to absorb the slow-moving electrons that are reflected from the collector electrode and which tend to reduce the efficiency of the cell. Although the voltage across the electrodes may become as great as 10,000 volts, the maximum current obtainable is very minute. The beta particles, having an energy which is large compared with the terminal voltage, account for the fact that the field produced by the voltage has little influence on the electron current and thus, the cell acts like a constant-current generator. A typical beta cell 1 in. high, $3/8$ in. in diameter, and weighing only 5 oz has been built using 2 millicuries of strontium-90. This cell can provide a maximum current of only 50 μa and has capacitance of 10 pf. Summarizing, the beta-current cells are expensive, have no current multiplication or amplification, operate efficiently at high voltages, produce a very low current, and are temperature-independent.

Contact Potential

The contact-potential-type nuclear battery utilizes current multiplication. The most commonly used radioisotope for this battery is tritium. In this battery two dissimilar electrodes are used with a space between them filled with either a radioactive substance or an irradiated gas. The basic principle of operation is the contact-potential-difference principle, with the space between the electrodes being filled with electrons, ions, or both, produced by the radioactivity present. In contrast to the beta-current type of nuclear battery, the contact-potential-type nuclear battery has low voltage (in the order of 1 volt) but can sustain high current drains because of the avalanche of the electrons produced.

p-n Junction Nuclear

The p-n-junction-type battery consists of a p-n junction activated by a radioactive source. This cell, like the contact-potential type, has current multiplication and operates at low voltages. Strontium-90 has been used in the construction of this cell. The major drawback of this type of nuclear battery is the fact that radiation from the sotope damages the silicon crystal, and hence the battery has a life of only a few weeks.

Thermojunction

The thermojunction battery relies on the heat released from the radioactive material. This heat is absorbed by a sensitive thermopile placed near the source of radioactivity. Most generally, the hot junction is operated at high temperatures to minimize voltage fluctuations. A cell of this type has been built using 150 curies of polonium-210, which delivers 0.75 volt and 21 ma and weighs only 31 g. Polonium-210 has a half-life of only 138 days and is quite expensive. (The cell described costs approximately $375,000.) Larger longer-lived thermojunction batteries have been built and are in use on satellite systems.

Photojunction Nuclear

In the photojunction nuclear battery, the radioactive energy is first converted to light in a phosphor. The light in turn liberates electrons from a photojunction such as might be found in a solar battery. At present the efficiency of this type of cell is quite low since the efficiency of converting the radioactivity into light is approximately 30 per cent and the efficiency of converting the light into electrical energy (through the use of a solar cell) may be only 8 to 10 per cent. A wafer-thin cell of the photojunction type which is less than 1 cm in diameter has recently been announced. This cell utilizes promethium-147 as a radioactive source and is stated to give 20 mw of power and to operate better at $-200°F$ than at room temperature. The main application for this cell has been for use in electric wrist watches.

One disadvantage of the photojunction nuclear battery is that the radioactivity present may cause deterioration of the photojunction material, resulting in limited battery life.

Photoelectric Nuclear

The photoelectric nuclear battery is similar to the photojunction type except that the light given off at the phosphor liberates electrons from a photoelectric source. The efficiency of this cell is quite low because of the low efficiency of the photoelectric cell.

Secondary-emission Nuclear

The secondary-emission nuclear battery makes use of a sensitive surface which when bombarded with beta particles releases electrons. These electrons are then collected on an insensitive surface. The maximum voltage obtainable is approximately 20 volts corresponding to the maximum energy of the secondary electrons. As yet not much work has been done with this type of cell.

The nuclear batteries listed above will be used mainly in those applications requiring low current drains over extended periods of time and/or those applications requiring operation at low temperatures such as may be found in Arctic regions. Table 29 shows the characteristics of a few nuclear batteries.

35 Fuel Cells

Although not yet economically competitive for the short-term power requirement, the fuel cell is increasing in importance because of its long-term power-to-weight ratio. This device is not new, having been discussed as early as 1842 by W. R. Grove. Although the problem of converting the chemical energy of fuels directly into electric energy has for a long time challenged the best efforts of many electrochemists, more research has been done in the past 15 years than in the previous 150. The impetus given to fuel cells has been largely the need in space vehicles for a reliable, lightweight, compact source of energy with high current density, an inherently long shelf life, and a long operating life.

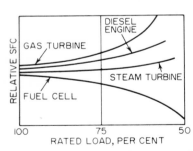

Fig. 115. Fuel consumption vs. rated load.

In many ways a fuel cell is like a common galvanic cell. A fuel cell consists of an anode at which oxidation takes place, a cathode at which an oxidizing agent is reduced, and an electrolytic medium. The main distinctions are these: (1) batteries store in themselves the chemical energy to be converted to electrical energy while a fuel cell converts fuel brought to it, and (2) the substances batteries consume in anodes (for example, zinc or lead) are not conventional fuels, as is the case with fuel cells.

A fuel cell is a type of energy converter rather than an energy source. Consequently, fuel cells cannot provide electricity in great surges as can be accomplished by batteries; the fuel and oxidizing agent must be brought to the cell for conversion

Table 29. Characteristics of Nuclear Batteries*

Battery type	Constant-current charging			Contact-potential difference	Junction	Photojunction	Thermojunction
Radioactive material	Sr^{90}	H^3	Kr^{85}	H^3	Sr^{90}	Pm^{147}	Po^{210}
Half-life	25 years	12 years	10 years	12 years	25 years	2.6 years	138 days
Quantity	10 Mc	1 curie	1 curie	1.5 Mc/cell	50 Mc	4.5 curies	3,000 curies
Size	1 cu in.	1 cu in.	5 cu in.	1 cu in.	....	0.2 × 0.7 in. diam	5.5 × 4.75 in. diam
Weight	6 oz	1 oz	14 oz	1.5 oz	....	0.6 oz less shielding	5 lb
Current amp	10^{-12}	6×10^{-10}	10^{-9}	10^{-10}	5×10^{-6}		
Voltage	14 kv	1 kv	1 kv	100 volts (66 cells)	0.2 volt	0.25–1 volt, 20 μw	5 watts
Development status	Sr batteries in production; prototypes of H, Kr batteries under test			Development complete, but not in production	Development complete, but not in production	Development complete	Prototypes completed larger units being investigated
Manufacturers	Radiation Research Corp.; Patterson Moos Div., Universal Winding Co.			Tracerlab, Inc.	RCA	Elgin National Watch Co.	Mound Laboratory; Martin Company

* Courtesy of *Electronics*.

at some maximum finite rate. Thus, fuel cells provide the greatest advantage over batteries when loads are moderate over a continuous long period.

An important characteristic of a fuel cell is that the cell is more efficient at partial load than at full load. This is contrary to the operation of conventional generating devices such as the engine generator or turbine generator (see Fig. 115). Curves show relative efficiency of various power sources as a function of less than full-load demand.

Conversion of chemical energy found in a fuel into electrical energy by electrochemical means, as in a fuel cell, does not involve a conversion of heat into work as is the case when converting fuel energy into electrical energy by means of a heat cycle. The normal "heat-engine-cycle" conversions of fuel energy involve the following steps:

1. Oxidation of the fuel to obtain heat
2. Use of the heat to obtain a high-velocity fluid (usually steam)
3. Use of the high-velocity fluid to drive a turbine
4. Use of the rotary motion of the turbine to drive an electromagnetic generator

The energy conversion accomplished in a fuel cell (an electrochemical device) eliminates three of these steps; the oxidizing action involves the release of ions which are directly utilized to supply an electric current. Thus the process occurring in a fuel cell is essentially isothermal and escapes the Carnot limitations. Avoiding the Carnot limitations is the primary reason for the present interest in fuel cells.

Under good operating conditions, the overall efficiency of a modern-day power-generating plant employing a heat-cycle-type conversion might be 40 per cent. The maximum efficiency attainable using a fuel cell is at least 70 per cent. No moving parts are required for the operation of most fuel cells with a resultant long life with a minimum of maintenance problems. (Some types of fuel cells, however, do require auxiliary pumps on the plumbing and controls to either circulate an electrolyte or heat and pressurize the system. These devices would require a certain amount of surveillance and maintenance.)

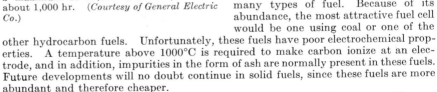

FIG. 116. Fuel-cell volume comparison. Fuel cells pack more energy in less volume for many applications. High energy density exceeds batteries even for relatively short operating periods of about 10 hr and reaches a maximum of 27,000 watt-hour/cu ft at about 1,000 hr. (*Courtesy of General Electric Co.*)

Fuel cells have been attempted for many types of fuel. Because of its abundance, the most attractive fuel cell would be one using coal or one of the other hydrocarbon fuels. Unfortunately, these fuels have poor electrochemical properties. A temperature above 1000°C is required to make carbon ionize at an electrode, and in addition, impurities in the form of ash are normally present in these fuels. Future developments will no doubt continue in solid fuels, since these fuels are more abundant and therefore cheaper.

Figures 116 and 117 show how ion-exchange-membrane fuel cells compare with more conventional power sources.

One of the major problems associated with most fuel cells is that the reaction between the gas, electrolyte, and the electrode can take place only at a three-phase boundary. Reaction products tend to dilute the electrolyte or flood the electrode, thus interrupting the reaction.

Electrode design, then, is of paramount importance and is one of the most serious problems in designing a fuel cell. Fuel-cell electrodes should have high electrical conductivity, possess catalytic properties to increase reaction rates, and have a stable pore-size distribution. This distribution maintains the electrolyte-gas interface at the proper position within the electrode over extended periods of time with a wide range of load currents. Finding an electrode which meets these requirements but has great corrosion resistance is no easy task.

Fuel-cell electrolytes must be chosen on the basis that there be no change in composition or amount of electrolyte with cell operation. That is, ions consumed at one electrode must be replaced by reactions at the opposite electrode.

At least two types of fuel cells avoid the three-phase boundary problem; however, they do suffer certain disadvantages. One type of fuel cell is the high-temperature fuel cell, which uses molten salts as electrolytes. The disadvantage, of course, in having to operate at the high temperature is that greater erosion and other problems

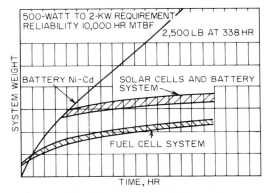

FIG. 117. Power-supply-system weight comparison. Comparison assumes (1) system is utilized for a manner orbital vehicle requiring a water consumption of 0.5 lb/hr, (2) 100 per cent basic system redundancy for reliability, (3) fuel-cell curves based on supercritical hydrogen and oxygen storage, and (4) solar cells which are used for recharging are in dark 50 per cent of the time. (*Courtesy of General Electric Co.*)

can take place. The second type is a so-called Redox cell. Here, an intermediate substance in aqueous solution is reduced by the fuel and then circulated over an inert electrode where it is oxidized, releasing electrons to the electrode. Similarly, an intermediate oxidant is oxidized by the oxygen and circulated to an inert electrode, and there reduced. The disadvantage of the Redox cell is that auxiliary pumps are required to circulate the electrolyte.

Operation

Figure 118 is a basic diagram of a typical fuel cell using an acid electrolyte. In this fuel cell, hydrogen reacts at the anode to give up electrons. Hydrogen ions are set free. The electrons then travel the external circuit through the load to the cathode, where they combine with the oxygen and hydrogen ions that move through the electrolyte, completing the circuit and producing water. It can be seen that both the material and the electrical charge balances are maintained by migration of both electrons and ions;

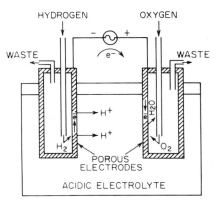

FIG. 118. Typical hydrox fuel cell.

interrupting the electron flow in the external circuit or the ion flow in the internal circuit will interrupt the power to be derived from a fuel cell.

Types of Fuel Cells

There are three major types of fuel cells currently under development. They are (1) the high-temperature low-pressure cell, (2) the moderate-temperature high-pressure cell, and (3) the low-temperature low-pressure cell. Serious technical difficulties to be expected in the high-temperature high-pressure cells have precluded further work for this type of cell.

Some of the basic construction of fuel cells may be illustrated by four fuel cells that have recently emerged from development activity: (1) the Bacon or hydrox cell, (2) the Redox cell, (3) the hydrocarbon cell, and (4) the ion-exchange-membrane fuel cell.

The hydrox fuel cell shown in Fig. 118 utilizes hydrogen and oxygen in the fuel-cell action. Hydrogen is supplied from tanks containing compressed hydrogen (or

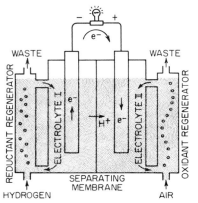

FIG. 119. Redox fuel cell.

hydride fuels such as calcium or lithium hydride, which are decomposed to hydrogen with water). Oxygen is either supplied in its pure state or taken from the air. This is a moderate-temperature cell that uses either acid or alkaline electrolytes and depends upon nickel or carbon electrodes of controlled porosity to get a sufficiently large effective reaction area. These solid electrodes help maintain stable reaction sites, lessening the possibility of electrode drowning by the electrolyte or the water produced by the cell action. The reaction rate of hydrox cells can be increased by raising the cell temperature considerably. However, to ensure adequate partial pressures of the fuel and oxidant the per cell pressure must also be increased. Another problem is the fact that the electrolyte is gradually poisoned by the carbon dioxide and other impurities in the hydrogen and in the air.

Even though high-temperature high-pressure hydrox cells have been operated successfully at 700 psi with high reaction rates, the theoretical maximum efficiency of the hydrogen-oxygen system decreases at elevated temperatures. This factor, plus the problems of pressurization of the air used, limits the utility of this particular approach.

The Redox cell is a hydrogen-oxygen cell that avoids the difficulties associated with gaseous-fuel electrodes by separating the fuel reaction from the electrochemical reaction. It is, rather, a two-step process in which the fuel and oxidant consumed in the cell are not reacted at the electrodes. As shown in Fig. 119, the Redox cell consists of two electrolytes separated mechanically by a membrane. Intermediate reactions take place on each side of the cell. A reducing reaction takes place between the gasified fuel and a liquid electrolyte on one side, and an oxidizing reaction takes place between the air and a liquid electrolyte on the other side. The liquid electrolyte in each case contacts both the solid electrode and the membrane. Thus the electron and ion balance in the system is maintained by the migration of the hydrogen ions across the separating membrane, closing the internal circuit. The overall action within the cell is still oxidation of the fuel gas, but the various reactions are carried on independently and under optimum conditions, using the auxiliary apparatus. Auxiliary equipment is usually necessary to assure good circulation of the electrolytes.

The advantages of the Redox cell are lower internal resistance losses, lower polarization losses, potentially higher current densities, and the ability to operate on relatively impure hydrogen gas. These advantages may offset the inherently lower efficiency found with this design.

The hydrocarbon fuel cell, one form of which is shown schematically in Fig. 120, is generally a high-temperature cell that operates above 500°C and uses molten carbonates as its electrolyte. The electrolytes are generally held in a sponge matrix of magnesium oxide. The metallic electrodes are in direct contact with the electrolyte matrix.

The usual fuel for a hydrocarbon cell is gasoline or some other hydrocarbon, which is "cracked" to produce hydrogen and carbon monoxide. This combination of gases is diffused into the cell at one electrode. The combination then reacts with ions in the carbonate to form water and carbon dioxide while releasing electrons to the electrode. Oxygen or air at the other electrode picks up the electrons to produce ions that migrate through the electrolyte to complete the internal circuit.

This type of cell has advantages in that it may be possible to use hydrocarbon gases such as propane, with a suitable catalyst, in a low-temperature version of the cell. This would permit wide application where hydrogen may be undesirable

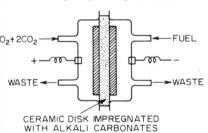

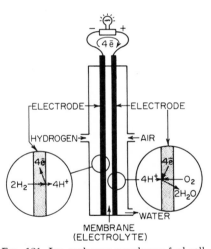

Fig. 120. Schematic drawing of ceramic matrix-molten alkali carbonate fuel cell.

Fig. 121. Ion-exchange-membrane fuel cell.

as a fuel. This particular fuel may also have a greater efficiency potential than any other, depending upon the chemical energy of the fuel reactions. However, these cells have not as yet reached the stage of practical application.

The ion-exchange-membrane fuel cell shown schematically in Fig. 121 is a General Electric development. This cell solves the three-phase contact-stability problem by eliminating the requirements for a liquid electrolyte. A plastic membrane holds a solid electrolyte which permits the hydrogen ions to migrate from one electrode to another.

Hydrogen and oxygen enter chambers on opposite sides of the ion-permeable membrane and penetrate the porous electrodes, effecting a contact on the surface of the membrane. On the hydrogen side, the electrons are given up, collected in the electrode, and passed through the electrical load to the other side of the cell. They are then combined, in the presence of oxygen, with the hydrogen ions, which travel

Table 30. Fuel-cell Characteristics

Type	Fuel	Oxidant	Electrode	Electrolyte	Operating temp	Operating pressure	Open-circuit voltage Theoretical	Open-circuit voltage Measured	Current density, amp/sq ft	Output, watt-hr/lb	Output Cu ft /kw	Output Lb /kw	% efficiency thermal to electric	Remarks
Ion-exchange membrane	Hydrogen	Oxygen or air	Activated metal	(Solid) ion-exchange membrane	50°F above ambient −65 to +165°F	Atmospheric	1.1	1 0.8 volt at 10 ma	22 at 0.8 volt	100 (measured)	3.5–5	250–500	60 at 15–20 amp/sq ft	National Carbon Co. and General Electric Co. Direct Energy Conversion operation developing. Used on Gemini project
Redox	Liquefied fuel	Oxygen or air	Porous metal	Liquids	70–85°C	Approx atmospheric	~1	....	200	1,200 with air 1,600 with pure oxygen	5	50–75	....	Units have been operated for 3-month periods without interruption. Has definite weight and size advantage over hydrogen-oxygen type cells. GE and Lockheed
Carbox	HCO-petroleum hydrocarbons (kerosene)	Oxygen or air	Porous metal	Fused carbonate	500–800°C	Atmospheric	~1	0.7	60	....	....	....	64	Being developed by Patterson-Moos and Pittsburgh Cons. Coal Co.
Hydrox	Hydrogen	Oxygen	Porous metal	....	200–250°C, 400–500°F	10–55 atm, 400–600 psi	1.1	1.0	Up to 1,000	....	0.25–1.2	40–90	....	1,500-hr life possible. Work by Patterson-Moos
Thermal regenerative	Hydrogen	Group I metals	Fuel metal and nickel	Fused group I chlorides	608 or 1004°F	200–500 mm Hg abs	0.75	0.72	245 at 0.72 volt with lithium	~5	....	....	Carnot: 40	Operated nonregeneratively for 2 weeks. Operating temp depends upon fuel (MSA Res. Corp.)
Solar regenerative	Nitric oxide	Chlorine	Carbon-disk	Liquid nitrosyl chloride	70°F	15 psig	0.21	0.21	2 at 0.1 volt	....	....	....	....	Development by Sundstrand Aviation
Low temp-pressure	Hydrogen	Oxygen	Specially processed carbon	12 molar solution of KOH	70–150°F	1–5 atm	1.2	1.12	100 at 0.95 volt and 140°F and oxygen at 5 atm	1,620	3.5–5	250–500	75	Being developed by Union Carbide Consumer Products Co.

through the solid electrolyte to the other side of the membrane. Water droplets which form as a result of this reaction on the open-air side are then drained off. Since there is no liquid electrolyte there is no problem of diluting the electrolyte or in removing the water. As a matter of fact, since this cell is easily reversible, the water may be collected and electrolyzed to produce fuel and oxidant for later use in much the same manner as a storage battery might be charged. With this feature, if the charging current were to be interrupted, the cell would instantly commence operation in its normal mode of furnishing electrical energy. Thus it can be seen that this cell could be used for emergency power or in some application where electrical energy might be made available for regenerating the system.

Figures 116, 117, and 122 show typical fuel-cell performance characteristics for an ion-exchange-membrane type of cell.

Table 30 shows some of the characteristics of present-day fuel cells, including their construction, efficiency, power output, and present developers. The performance data as given for the various fuel cells must be loosely interpreted, not held as absolute values. The reason for this is that nearly all cells listed are under continuous investigation with hope for cost improvement. In addition, the performance of most cells can be varied over wide ranges by changes in operating temperature, pressure, or current density.

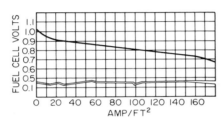

FIG. 122. Ion-exchange-membrane fuel cell polarization curve. (*Courtesy of General Electric Co.*)

36 Battery-selection Considerations

To select the proper battery for a given application, the following data should be determined:

1. Power-output requirements (watt-hours)
2. Current-drain characteristics (e.g., continuous or surge, maxima and minima)
3. Operating voltage, including tolerable variances (e.g., open-circuit voltage, starting voltage under load, and end-point voltage. The end-point voltage is the closed-circuit voltage below which the equipment will not operate)
4. Discharge schedule. (For intermittent duty these data should indicate the period or periods each day during which the battery must furnish current. Surge-current characteristics should also be specified. The schedule should also indicate the desired battery life.)
5. Type of load (e.g., plate supply, resistive load, etc.)
6. Size, shape, and weight restrictions
7. Storage and operating environmental conditions and the duration of exposure
8. Storage periods prior to use
9. Availability of heater power

For rechargeable batteries the following should also be supplied:

1. The number of charge/discharge cycles required
2. The type of battery charging available for maintenance as well as during operation

The requirements of a particular application will dictate the amount of power required. The type of load determines the voltage required although it should be remembered that it is sometimes desirable to utilize d-c–to–d-c inverters to obtain higher voltages from low-voltage batteries. If the power and voltage are known, the current is readily determined, provided the overall power requirements do not include intermittent surge loads. These surges are added to the average current drain to obtain the ampere rating required from the battery. (Rest periods between surges are desirable, since they aid in moderating the battery operating temperature and in permitting depolarizing action to be more effective.) If the required current rating is known, the optimum battery composed of various series and parallel connections of similar cells can be estimated.

The number of cells in a battery bank is determined by dividing the required voltage by the output voltage of an individual cell. The number of banks is determined by dividing the required current by the output current of an individual cell. The total number of cells in the battery is then the number of cells per bank times the number of banks required.

References

1. K. Henney, C. Walsh, and H. Mileaf, "Electronic Components Handbook," vol. 2, McGraw-Hill Book Company, New York, 1958.
2. Interservice Group for Flight Vehicle Power, *Project Briefs*, Power Information Center, University of Pennsylvania, Philadelphia, Pa.
3. G. W. Vinal, "Primary Batteries," John Wiley & Sons, Inc., New York, 1950.
4. G. W. Vinal, "Storage Batteries," 4th ed., John Wiley & Sons, Inc., New York, 1955.
5. "Burgess Engineering Manual," Burgess Battery Co., 1960.
6. P. J. Rappaport, Performance Ratings of Secondary Batteries, *Electronics*, vol. 33, no. 8, pp. 60–62, Feb. 19, 1960.
7. E. C. Pitzer, "Battery Capabilities," General Electric Co., General Engineering Laboratory, Schenectady, N.Y., *Rept.* 59GL1, 1959.
8. W. J. Hamer, Modern Batteries, *IRE Trans. Component Pts.*, vol. CP4, pp. 86–96, 1957.
9. Electronic Design Staff, Chemical and Solar Power Sources, *Electronic Design*, vol. 9, no. 4, pp. 51–73, Feb. 15, 1961.
10. D. Linden and A. F. Daniel, New Batteries for the Space Age, *Electronics*, vol. 31, no. 29, pp. 59–65, July 18, 1958.
11. D. L. Douglas, Advances in Basic Sciences—Part I, Fuel Cells, *Elec. Eng.*, vol. 78. no. 9, pp. 906–910, September, 1959.
12. D. Linden and A. F. Daniel, New Power Sources for Space-age Electronics, *Electronics* vol. 32, no. 12, pp. 43–47, Mar. 20, 1959.
13. "Some Plain Talk about Fuel Cells," General Electric Co., Direct Energy Conversion Operation, Pamphlet GED-4111B.
14. Accessory Components, *Space/Aeronautics*, "R & D Handbook," 1960–1961, pp. G 20–21.
15. W. Luft, "A Method for Calculating the Number of Silicon Solar Cells Required for Power Supply of Satellites and Space Vehicles," International Rectifier Corporation Pamphlet.
16. Marvin P. Eisen, "Fuel Cell System Analysis for Space Power Applications," General Electric Co., Direct Energy Conversion Operation, Pamphlet DE-81, August, 1963.
17. J. L. Schanz and E. K. Bullock, "Gemini Fuel Cell Power Source—First Spacecraft Application," General Electric Company, Direct Energy Conversion Operation, Pamphlet DE-35, 1962.

Chapter 11

DATA-HANDLING EQUIPMENT

JOHN E. GAFFNEY, JR., *Federal Systems Division, International Business Machines Corporation, Rockville, Md.* (*Introduction, Data-acquisition Equipment, Analog-to-Digital Converters, Recorders, Data Processors, Data Presentation*)

CHARLES H. DOERSAM, JR., *Polytechnic Institute of Brooklyn, Brooklyn, N.Y.* (*Automatic Checkout Equipment*)

INTRODUCTION

1	Telemetry-system Functions	11-2
2	Scope	11-3
3	Date Processors	11-4
4	Programming	11-4
5	Data Acquisition	11-5
6	Magnetic Tapes—Storage Devices	11-5
7	Analog-to-Digital Converters—Data Presentation	11-6

DATA-ACQUISITION EQUIPMENT

8	Data-acquisition Components and Use	11-6
9	Inputs and Signal Conditioning	11-9
10	Electrical Analog Signal Conditioning, Multiplexing, and Noise	11-10
11	Analog Signal Degradation	11-11
12	Data-selection Control and Programmer	11-11
13	Calibration Techniques	11-12
14	Data-acquisition-system Outputs	11-13

ANALOG-TO-DIGITAL CONVERTERS

15	Analog-to-Digital Converter Functions	11-13
16	ADC Characteristics and Error Sources	11-14
17	Voltage-to-Number ADC's	11-16
18	Position-type ADC's	11-19
19	Time-and-Frequency-to-Number Conversion	11-22

RECORDERS

20	Application of Magnetic-tape Recording	11-24
21	Characteristics of Storage Devices	11-24
22	Recording Principles	11-25
23	Physics of Recording	11-27
24	Tape Transports	11-28
25	Digital Recording—Computer Magnetic Tapes	11-30

26	Computer Tape-recording Techniques........................	11–31
27	Telemetry Data-Recording Techniques......................	11–33
28	Predetection Recording..	11–33
29	Skew, Wow, and Flutter.......................................	11–34

DATA PROCESSORS

30	Data-processor Applications.................................	11–34
31	Computer Fundamentals and Classification.................	11–35
32	Binary Computer Operation..................................	11–38
33	Operation of the Arithmetic and Logic Unit................	11–38
34	Control-logic Operation......................................	11–42
35	More Sophisticated ALU Operations.......................	11–44
36	Input-Output Control...	11–45
37	Input-Output Data-channel Operation.....................	11–46
38	Input-Output Data-channel Control of a Data-acquisition Subsystem...	11–47
39	Programming..	11–48

DATA PRESENTATION

40	Data Presentation in Telemetry.............................	11–48
41	Data-presentation Devices...................................	11–50
42	Symbol-display Techniques..................................	11–53
43	Image-formation Displays....................................	11–53
44	Data-presentation-equipment Operation....................	11–54

AUTOMATIC CHECKOUT EQUIPMENT

45	Introduction..	11–56
46	Function...	11–57
47	Types of Automatic Checkout Equipment..................	11–59
48	Data Transfer...	11–69
49	Desirable Improvements.....................................	11–72

INTRODUCTION

1 Telemetry-system Functions

The essence of a telemetry system is the gathering of data about another system, which is either under test or in action. The chief task executed by the data-handling components of such a system is the refinement of the data gathered so as to be comprehensible to an engineer or other person using it. This task may be subdivided into several subtasks. The principal subtask of data refinement is the display of the data to the engineer in a form useful to him. There are two forms of display of interest in this connection: "quick-look" and that of processed data. The former provides a means by which the status of an experiment or test can be ascertained *during* its course, as opposed to after it has been completed. Such timely information can have a substantial impact upon the continuation of a test. When sensed by the engineer, it can enable him to stop the test, for example, if the "quick-look" data indicate to him that continuation of the test would be dangerous and/or would not yield the data desired.

The principal direction of information flow in a telemetering system is that from the data source to the recording, processing, and display units. Parallel flow of data to a storage medium, for archival purposes, is occasioned in many applications of telemetering systems. Such a repository is frequently termed a "journal." The human analog of this function is that of an individual's daily recording of the events

INTRODUCTION

of his life in a diary. In some telemetering applications, in particular those involved in testing and checkout, such as those for various systems guidance, power, etc., of a rocket vehicle, the function of stimulation or probing the device or system being tested is exercised. In an installation of this nature, a sequence of signals (the "stimuli") is applied in some prescribed order to the device undergoing the test. The device's responses to these signals, in terms of certain measures desired by the test engineer, are telemetered to the processing units of the telemetering system. Figure 1 is a schematic diagram of a "typical" telemetering or instrumentation system. The interrelationships of the various components of the system are illustrated. The components relating to the stimulation function are set apart from those principally associated with the measurement function by means of dashed lines.

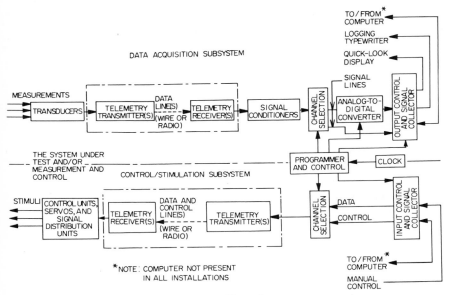

FIG. 1. Telemetry data-handling subsystem components.

The stimulation function, portrayed in Fig. 1, is not present in all telemetry systems. The connections to an on-line computer which *could directly* program and operate the subsystem components are indicated. In an actual system which provides both the data-acquisition and control/stimulation functions, separate channel-selection equipment may be provided for each of these functions, as in the "sample" system in the figure. One unit may satisfy the requirements for both functions, however.

2 Scope

This chapter describes the equipment and the means by which it is controlled or sequenced (including programming) to operate, to acquire the telemetry data, to process them, and ultimately to present the result of the processing as accurate and useful information, in a form most suitable for the experimenter and design engineer. The units which provide these functions may be functionally grouped into the categories of data acquisition, data processing, and data presentation or display. Together, they comprise the data-handling portion of a telemetry subsystem.

Little material in this chapter is devoted to describing the stimulus function. Chapter 12 of this handbook discusses the process of reducing the data to the proper form. Other chapters describe details of various telemetry systems, FM-FM, AM-FM, PCM, etc., the theory behind them, and the means to accomplish the transmission of these data to the data-handling subsystem (the subject of this chapter).

The principal objective of this chapter is to provide the reader with an understanding of the equipment used in the data-handling function both for its operation and in connection with the data-reduction function to enable him to select and apply the equipment to telemetry data-reduction jobs. The material of this chapter should provide the reader with the insight to describe the functions to be met by telemetry data-handling subsystems. There is a fairly wide choice of possible equipment to achieve the functions of this subsystem.

The choice among computers depends on such considerations as memory capacity, access time of fast memory, instruction repertoire, and programming language(s) available. A basic knowledge of programming is necessary in order to appreciate the magnitude of the tasks required of the data processor in order to estimate run times and to achieve a proper balance between machine operating time and programming time, as well as between machine capabilities and programming complexity.

An optimum installation is realized through an appreciation of the trade-offs between programming requirements and machine complexity. Not infrequently, the total cost of programming, including both the initial expenditure and that required for updating the routines, comprises a substantial portion of the overall system cost. In many purely data-processing applications, such as for the maintenance of company records, programming accounts for 50 per cent or more of the total system cost. Clearly, an appreciation of the programming and equipment requirements is mandatory for those who are responsible for selecting the initial system-design approach as well as for monitoring its realization.

3 Data Processors

A principal section of this chapter is devoted to a description of data processors. The data processor is responsible for the actual reduction of the data gathered by the data-acquisition equipment. Further, it controls this equipment either directly or indirectly, depending upon the installation. The discussion in this chapter is devoted strictly to binary, digital, stored-program calculators. Neither analog machines nor digital differential analyzers are considered. Further, certain machines are discussed which rely on a program or sequence of instructions provided by a recirculating paper tape or plugboard. Such machines are frequently used in automatic testing applications. Such units also find wider application in those installations in which only limited processing (such as linearizing and square rooting) are required. Such units are sometimes used in conjunction with a stored-program processor in a hybrid system, each unit performing the part of the data selection, processing, and programming for storage and display to which it is most suited.

Only binary machines are described in this chapter. The term "binary" in this instance means that all data, numbers, and description of computers and system operations are represented by numbers specified by a sequence of digits which can assume one of two values, 1 or 0, equivalent linguistically to a "yes" or "no."

4 Programming

A computer program, in its simplest form, is a sequence of orders, which is essentially a description of how the processor is to solve the problem given to it by the user. The point of view most frequently assumed in describing a program imparts the qualities and function of a control device to it. This "device" (the program) causes the calculator to perform a desired task. Considerable effort has gone into designing programming languages which make the task description easier for the human programmer to write and for the machine to follow. Thus these languages are more descriptive than functional.

Sometimes programs are written to describe a desired end objective, with "suggestions" on means of accomplishing it. This technique is distinguished from the more common one, in which the detailed steps to achieve the objective are provided. Sophisticated programs of this nature are frequently termed "heuristic." They have found little or no application to telemetry data reduction thus far. Their chief avenue of advance has been in stimulating developments in learning theory, including pattern classification and recognition. Computers so programmed have been used for playing various games such as checkers and chess. The chief quality of such programs is the ability to make a procedural judgment (of however limited a scope), in concert with a partial ability (in some cases) of being able to define the problem more clearly. The capacity will undoubtedly find eventual application in telemetry-system control and data reduction. Even today, claims of a learning or responsive capability have been made for some telemetry systems, ascribing "adaptive" capacities to them because they are able to vary their input scanning sequence and processing bandwidth in response to varying requirements imposed upon the processing of the data being collected.

It is necessary to view the program and the data processor as equal partners in the useful application of a telemetry system.

5 Data Acquisition

Data-acquisition equipment is that which is required to select and convert electrical signals directly into oscillograph records, that used for the direct recording of raw data, and that which provides direct entry into the digital data-processing equipment. With the exception of predetection recording, this chapter considers the processing of video or baseband data only.

Common data-presentation equipment includes oscilloscopes, X-Y plotters, TV tubes, card punches, strip printers, typewriters, and Nixie-type indicator tubes. These units may be connected either directly to data sources, or indirectly to them via data processors.

In the purely DA type of system, the equipment which performs the functions of data-point selection and recording is physically separated from the data processor which performs the data-reduction functions. Installations of this nature exist at many test stands. One data processor, of sufficient capacity may be employed to reduce the data written on computer compatible magnetic tapes by equipment at several such test stands. In general, these tapes are (subsequently) physically transported to the centrally located data-processing center for reduction. A variation on this scheme is that in which the properly edited data from one or more test locations may be transmitted by wire or radio link (the former the more likely) to a central data-reduction facility. One processor suitable for use in this situation is the IBM 7700.

6 Magnetic Tapes—Storage Devices

A magnetic tape is one of the chief types of large-volume storage as well as being an input-output medium for digital computers. A computer uses its tapes for storage of programs, constants, and other such information relevant to its processing activity. In addition, a tape may be prepared for a data processor by another machine and then used as a means for entering data into it for processing.

Magnetic disk files are also extensively used for high-volume computer storage. They are more efficient than tape for certain purposes, as a particular item of data stored on a disk can be accessed directly without requiring searching through unwanted information, as is the case with magnetic tape. A disk-file unit operates mechanically similarly to a juke box, some types of which it physically resembles.

Another type of computer storage unit in relatively wide use is the magnetic drum. It affords the relatively rapid, direct-access capability which can be obtained with a disk file, but without many of the physical appendages required for that device's use. Hence it *can* be more compact than that unit.

Up to the present time, neither the disk file nor the drum has found much use in telemetry data-recording systems. One reason for this is the very extensive use and familiarization with magnetic-tape units in telemetry engineering circles. Another reason is that the recording portion of the drum- and disk-file equipment on which the data are recorded is not removable and hence is *not* readily transportable from one location and machine to another as *is* the case with tapes. A relatively new disk-file unit, available with certain IBM data processors, uses removable disk packs, however. Eventually, such equipment may be applied to telemetry data-reduction facilities.

Both magnetic-drum- and disk-file units have been used by computers for communication to the "outside world." Also, they have been used for communication *between* duplexed computer units in the same installation. Magnetic drums are used for the SAGE Air Defense Control Computers as buffer units for radar and other data which define the status of the environment under the control of the SAGE computer. The drums are used to match the asynchronously arriving data with the operating cycles of the computer.

7 Analog-to-Digital Converters—Data Presentation

This chapter describes both analog-to-digital and digital-to-analog converters. The former units are widely used in telemetry data-reduction systems to convert FM (analog) signals to digitally coded representation suitable for machine processing. Digital-to-analog converters are used for the conversion of a computer-language (digital) representation of various quantities of interest to the experimenter to signals capable of actuating X-Y plotters, TV tubes, operating meters, etc. It is wise, from the point of view of good design, to provide such analog information directly to the plotter, or other output device used, if possible. Such connection minimizes the overall error in the representation of the measure. There are two conversions, analog-to-digital and digital-to-analog, each of which introduces errors in the final display. Such a "direct" display affords a type of "quick-look" which is required in certain experimental situations.

The possible errors in ultimate (output) analog representation that could arise, as has been indicated, could be large. A computer can operate upon these raw input data to reduce the error in that which is displayed to an experimenter or system operator. Even simple editing is helpful in this regard.

The subsequent sections of this chapter provide more detail on the equipment described in this introduction.

DATA-ACQUISITION EQUIPMENT

8 Data-acquisition Components and Use

The data-acquisition equipment is that portion of the telemetry data-handling equipment which conditions, selects, and converts incoming data to a form proper for processing and/or presentation by the other portions of the telemetry data-handling configuration. Figure 2 illustrates the principal units of a data-acquisition subsystem of a telemetry data-handling system. The units are input signal conditioners; data (signal) selector—frequently called a commutator or a multiplexer; data-selector control and programmer, converter calibration; and output. Every data-acquisition subsystem does not necessarily include all these units. The particular arrangement of units illustrated will not be that found in all cases. However, the configuration shown may be considered as typical.

The input portion of the data-acquisition equipment collects electrical, recorded, and manual input signals accepted by the data-acquisition subsystem. The signal conditioner units put all (or most in some cases) of the inputs on a common base (electrical) suitable for their acceptance by the other units of the system. For example, a typical data-acquisition installation may accept both d-c analog signals in the 0- to 100-mv range and FM signals. Both classes of signals must be transformed into equivalent signals suitable for the same analog-to-digital converter. The conditioner equipment could include FM discriminators. The conditioners must compensate for noise on the signal lines.

In some instances, the signal-conditioning equipment performs simple arithmetic transformations on one or more of the inputs. One such transformation is a scale change. The transformation in this case is $D_o = D_i S + Z$, where D_i is the input signal, S is the scale change, Z is the zero offset (see the sections on Analog-to-Digital Converters), and D_o is the output of the conditioner. The amplification (or attenuation) function S is frequently performed by an amplifier, often a feedback amplifier. Because of the difficulty in designing a stable d-c amplifier, this function may be performed alternatively by an a-c pulse amplifier (in conjunction with other components). The amplifier would be in the "front end" of the ADC, isolated from the signal lines themselves, in this case.

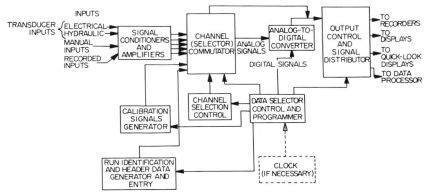

Fig. 2. Data-acquisition subsystem (schematic diagram).

The channel selector is that unit which switches the proper data channel into the data-acquisition subsystem for processing. The particular sequence of channels selected is determined and controlled by the data-selector control and programmer unit. In the simplest case, all the input points, 1 to n, are selected successively on a repetitive basis. The initiation of each such scan or "sweep" of data may be accomplished manually (by an operator pressing an activating button) or automatically by either a clock pulse or by the occurrence of a particular condition in the system generating the data. An additional possibility is initiation by the data processor to which the data-acquisition equipment transmits information.

The data-selector control and programmer direct the operation of the entire data-acquisition subsystem, in a manner analogous to that of the control-logic section of a data processor (Sec. 34). This unit may include a clock as a part of its mechanism. The programmer, or operation sequencer, may be implemented by a manually set plugboard, magnetic-core or card capacitor ("read-only") memory unit, or other such device. Also, this function can be assumed by a data processor if the data-acquisition subsystem is attached to one.

The converter unit consists of an analog-to-digital converter as a principal component if the output of the data-acquisition subsystem is a digital sequence of numbers (generally binary) equivalent to the analog voltages fed to the data-acquisition subsystem. Because of the importance of analog-to-digital converters in telemetry data-handling systems, several sections of this chapter are devoted to their operation. Only the interrelations of the ADC to the other elements of a data-acquisition subsystem will be described here. Some other topics, such as a discussion of noise and compensation for it, are also discussed here because of their relevance to the subject of data-acquisition subsystems as a whole. The sections on analog-to-digital converters describe the several basic types of implementation and relate the principles illustrated to the operation of FM discriminators, PDM-to-digital conversion, and other principal telemetry signal-reduction techniques.

In some data-acquisition subsystems, various elementary arithmetic functions are performed upon the numerical outputs of the ADC's. These functions include scale changes and squaring, among others. The functions performed by the converter may differ from channel to channel. Generally, the object of these functions is to convert the raw data to engineering units. The programmer unit would select the function corresponding to each given channel. As an illustration of these ideas, suppose that the ADC produces an output that consists of a number in the range 0 to 800. The signal on one input channel might be derived from an FM discriminator. The measurand represented could be pressure in the range 50 to 1,000 psi. A second input could correspond to another telemetry channel, which transmits a signal representative of a temperature in the range 100 to 1000°F, obtained from a thermocouple. Clearly, different transformations are required to generate the value of each signal in terms of the ADC output in the range cited. A further degree of sophistication in preprocessing is to calibrate each ADC output in engineering units, prior to its transmission to the output section of the data-acquisition subsystem. Typically, this would be done only if the data generated by the data-acquisition equipment were to be displayed *directly* for use without additional processing in a computer. In telemetry data-handling systems in which the data-acquisition output is processed by a computer prior to the use of these data, the calibration function described is most likely to be performed in the data processor. The calibration function typically includes reading a channel, the value of which is known, once during every scan of the input channels. The digitized value of this measurand, obtained from the ADC, is compared with the value it "should" have. The difference is the error in the reading. This error value may then be added algebraically to all readings taken since the last reading of the calibration channel. Calibration, as described, may be used to compensate for long-term drift of the subsystem's analog-to-digital converter, among other such error-contributing factors. A standard voltage is used for ADC calibration. It may be supplied by a standard cell (battery) connected to one or more input channels of the data-acquisition subsystem.

In addition to the important one cited, there are other types of calibration which are particularly significant in telemetry systems. For example, a PDM signal channel will typically include two calibration pulses per frame. The one, of duration t_h sec, corresponds to 100 per cent of (full) scale, while the other, of duration t_L sec, corresponds to the 0 per cent value of the scale of measurand(s) values transmitted by the channel.

An FM telemetering channel might transmit signal bursts at frequencies f_h and f_L, corresponding to the full (100 per cent) and low (0 per cent) scale values of the measurand(s) conveyed by that channel.

The "output" portion of the data-acquisition equipment consists of various display, recording, and data-processor interface media and their controls. All data-acquisition subsystems have at least one such class of outputs. The general class of "output" data includes the three principal categories of data: raw data, preprocessed (e.g., scaled) data, and data sufficiently processed and calibrated to be of immediate engineering use. The first type of data is generally developed by a data-acquisition subsystem which feeds a data processor. The third class of output is that which is developed by data-acquisition equipment which is not associated with an additional (data-processing) machine. "Quick-look" outputs may be available with any one of the three configurations. Generally, they are found in systems of the second type. A "quick-look" display affords the system operator or director/evaluator the opportunity of assessing the progress of the test and/or the processing of the data developed by the test. The data are processed by the telemetry data-handling equipment. Among other reasons, a "quick-look" capability is valuable because through it a faulty test may be spotted before it is completed, thus saving a considerable amount of money in many cases. In others, this facility enables the system (test) director/evaluator the opportunity to make a preliminary assessment of the test, without waiting for the final results from the data processor. This capability can be an asset, particularly in those situations in which the data processing is done "off-line" on a non-real-time basis. In such a situation, there may be a considerable delay before final results are available to the engineering staff which has the responsibility for evaluating them.

9 Inputs and Signal Conditioning

The types of inputs accepted by data-acquisition equipment may be categorized in two ways: first with respect to the source of the information, and second with respect to requirements imposed upon the source devices. Principal characteristics of the several types of noise which can be expected to corrupt input signals are cited below.

The are three categories of inputs which a data-acquisition equipment accepts. They are (according to source) instrumentation, manual, and data-acquisition system (self) generated.

Instrumentation

Signals of the "instrumentation" category are those telemetered to the data-acquisition equipment, such as those generated during the course of a test. These inputs are either electrical or hydraulic. If hydraulic, they are converted to the standard form required for the conversion equipment (e.g., to a d-c voltage in the range 1 to 10 volts). Electrical inputs include the outputs of telemetry receivers and FM discriminators. The operation of these devices is not considered in this section. The signals may be analog or digital. A common analog signal is the d-c voltage output of an FM discriminator. A common digital signal is the binary voltage output, or pulse sequence, derived from a PCM receiver. The electrical signals are either real-time, such as from telemetry receivers, or recorded, read from a magnetic tape. One type of data from a magnetic tape is that which is recorded during a test which occurred prior to the processing of the tape. A data-acquisition subsystem might be used in such an instance to convert an FM-analog tape to a binary-digital one, properly formatted for use by a digital computer. Because of their central role in telemetry data-handling systems, magnetic recorders are described in considerable detail subsequently in this chapter. Thus, the instrumentation category includes all the measured inputs. As stated previously, all signals of this category either originate in electrical form or can be converted to such form. If they did not originate as such, they will be transduced appropriately to yield the electrical signals requisite for processing by the system. Further, processing of the signals may be required in the signal-conditioning circuits to convert the signals to the system standard, say d-c voltage in the range of 1 to 5 volts, in which the intelligence is conveyed by the amplitude of the signals. For example, pneumatic signals, conveyed by air under pressure in the range of 3 to 15 psi would be so converted. The electrical signals may be a-c or d-c. If a-c, they would be frequency, phase, or amplitude-modulated. If d-c, they would be amplitude or duration-modulated.

Manual Inputs

The category manual inputs includes all those signals developed by the system operator or other responsible individuals concerned with the use of the telemetry data-handling equipment. There are two principal types of manually entered data. The first type consists of information about the course of the experiment itself, which supplements that obtained automatically with the instrumentation equipment. An example of such data would be the temperature of a rocket's exhaust, "estimated" by a human observer using an optical pyrometer as an aid. Information of this class includes experimental measurements not readily available with automatically operated (conventional) instruments. These measurements have a strong subjective element in their determination.

The other class of manual inputs includes those which specify *conditions* of an experiment, as distinguished from the *data* derived from it. Important items in this class include those items of data commonly found in the "header" portion of a telemetry tape. Typically, they are date, time, the identification number of the experiment, the run number, etc.

Information may be manually entered into a data-acquisition subsystem in a variety of ways. The more common ones are dials and on-off switches. More sophisticated means are available. For example, the data-acquisition equipment might include a visual display on a TV viewer on which are portrayed the values of certain measurands and/or other indices through reference to which the course of the experiment may be evaluated. The operator could be equipped with a light wand, which when pointed toward a certain portion of the image would cause equivalent information to be entered into the data-acquisition subsystem. Similarly, the operator could manually trace a curve, which might then be smoothed by the processor and displayed *by it* on the TV viewer. This technique could be used for the entrance of functions for which analytic expressions are not available to the system. Manual input devices may be connected to the data-acquisition equipment via the channel-selector equipment employed for the entrance of instrumentation data. Alternately, special channels which bypass the instrumentation/selection equipment may be provided for the entrance of manual inputs. The activation of a manual-input device may interrupt the "normal" sequence of instrumentation channel selection.

System-generated Inputs

All system-generated inputs treated in this chapter are assumed to be electrical in nature. Such inputs include clocking data, some types of calibration signals (e.g., standard voltage applied to analog-to-digital converter), and certain signals generated by the various operational units of the data-acquisition subsystem. The last category includes "progress report" data, which defines the current operational status of these units. The data-acquisition equipment may include a clock which initiates a scan of the instrumentation inputs at predetermined intervals. Also, the "time," the value of the clock register at a particular moment, may be recorded as part of the "header data" on a tape unit which is the output device of a data-acquisition subsystem.

10 Electrical Analog Signal Conditioning, Multiplexing, and Noise

The electrical signals accepted by the data-acquisition system's analog-to-digital converter may either be high- or low-level. A high-level signal is one whose amplitude is in the range of 1 to 10 volts, approximately. A low-level signal is one in the millivolt range, typically in the range of 0 to 50 mv, more or less. Some ADC's have been designed to handle signals exhibiting variation with a full-scale swing of 0 to 10 mv. Such units have not been made on a true production basis because of difficulties caused by corruption of the signals by noise. Generally, ADC's are designed to operate on signal voltages of a higher level. Consequently, lower-level signals are frequently amplified prior to being fed to the ADC. If the rate of scanning input signals is fairly high, on the order of 1,000 or more data points, or channels, per second, then each low-level channel *may* need to have its own amplifier. This is due in part to the difficulties which have been encountered in designing low-level high-speed multiplexers suitable for large-scale production. Laboratory-type models of such equipment have been built, however. Another problem, which tends to preclude the use of one amplifier in common among several high-speed switched low-level signals, is that of amplifier settling time. This is the minimum amount of time between which samples of signal voltages may be applied to the amplifier and be properly operated upon to raise the level of the input signal to the level required.

Because of the need for various rates of switching from one data channel to another in a given telemetry data-handling installation, the multiplexing, or switching function, may be done in tandem. Multiplexers are frequently termed "commutators" in telemetry-system terminology. "Commutation" may be defined as that process by which a plurality of individual information channels are sampled periodically and combined into one channel for processing. All the analog telemetry data channels which are the inputs of a data-acquisition equipment configuration must be blended together in an orderly way to be transmitted to the analog-to-digital converter(s) of the system.

The process of providing several data channels from one commutator channel is termed "subcommutation." This is simply the use of several multiplexers in tandem, making a hierarchy of units which operate at successively lower switching rates. Note that this equipment arrangement is directed toward requirements of successive scans, executed repetitively. The configuration for switching among different telemetry channels in a random, or nonsequential, manner need not involve a succession of subcommutators, since an individual channel is selected by a programmer, which connects it to the ADC (if an analog signal) at just the proper time.

11 Analog Signal Degradation

The degrees of accuracy, resolution, and other criteria of system performance achieved are dependent upon the manner in which signal-generation equipment is connected to the data-acquisition system multiplexing equipment. Suppose each signal uses two leads, neither of which is at the ground potential of the data-acquisition equipment. Interference may be either transverse (normal mode), (across the terminals), or longitudinal (common mode), appearing between ground and both signal lines. The noise voltage, whether normal or common mode, may be classified as spike, a-c, or crosstalk. The first type is that generated by switch closures, especially ones in which energy, in an amount considerably above the level of that in the signal, is suddenly released. A-C noise is that resulting from coupling to power (50- to 60-cycle) or a-c control (400-cycle) lines. Crosstalk noise has the same statistical description as the actual signal, since it results from the cross coupling of one or more information-bearing channels with the particular one selected at a certain moment. This type of interference is the most difficult to compensate for.

From the point of view of the electric circuits involved, interference is caused by conductive, electrostatic, or electromagnetic induction, ground loops, back circuits, and/or thermal or junction potentials.

Some of the remedies for interference are the use of shielded line between signal source and input section of the data-acquisition equipment, grounded at one end only; employing floating differential input amplifiers, when possible; twisting signal lines; and balancing the signal lines (see Chap. 4).

12 Data-selection Control and Programmer

This unit directs the operation of the complete data-acquisition subsystem. It selects which input-data channel is to be connected to the data-acquisition equipment. The programmer includes some type of storage facility, in which the requisite sequence of data-point addresses is recorded. Such a facility provides either fixed- or variable-sequence data-point selection. A plugboard is commonly used for fixed-sequence selection, especially in data-acquisition subsystems which are not linked to the data-processor portion of the telemetry data-handling system. Many modern systems use other means such as endless loops of paper or magnetic tape for this purpose. Also, read-only memory storage units, such as a card-capacitor memory or a magnetic-core storage unit operated in the read-only mode, are used for this purpose.

Variable-sequence point selection is accomplished under the control of the data processor which is the "host" to the data-acquisition equipment. Material in the section of this chapter on data processors illustrates a way in which a computer organized in a manner similar to the IBM 7090 could select input-data channels according to a sequence stored in its magnetic-core (fast) storage unit. The sequence in this case could be changed at the volition of the system operator, as well as under the control of the data processor's stored program, in response to the occurrence of a particular condition, such as that of one particular measurement's exceeding a certain prescribed bound.

The data-acquisition subsystem's programmer unit may be used to vary the scale of the analog-to-digital converter (ADC) of the subsystem. Further, it might direct the other simple arithmetic operations to be performed upon the data.

In its role as the controller of the data-acquisition subsystem, the programmer meshes the operation of the outputs of the system with the scanning and preprocessing of the data derived from them. If an output is the interface to the data processor of the telemetry data-handling system, then the programmer will also serve to mesh the operations of the data-acquisition equipment with the data processor. Other output units such as magnetic-tape handlers may be employed. Such devices have their own logic to perform such functions as providing the sequence of signals required to start, stop, or rewind the tape. The data-acquisition subsystem programmer provides the signals to initiate the execution of these operations by the tape handler's own logic unit. Thus the programmer of a data-acquisition subsystem performs many of the functions of the control logic of a data processor and thus in many cases is similar to it in design, although generally not in size or complexity.

13 Calibration Techniques

In order to realize the maximum degree of accuracy attainable with a telemetry data-handling system or any other measurement system, calibration of the system as a whole, as well as of various of the units comprising it, is required. There are two types of calibration. The first is that performed prior to the system's becoming operational as well as at certain intervals after it has been completed. The second is that done during the operation of the system. One might term the former classification "fixed calibration" and the later "variable calibration." The meaning of "calibration" can be made clear through reference to the transfer function of the system or component to be calibrated. The transfer function of a unit is the mathematical expression, which prescribes its output, corresponding to a given input. The function takes scale and power-level changes into account, as well as phase shift, or delay, encountered in the unit. The function, within the ranges of interest of the variables, is single-valued. Outside this range it is probably not single-valued, however, either with respect to there being two different inputs which could cause the same output, or the reverse, in which one input could result in either of two outputs. Let the transfer function be of the form $y = F(x)$, where y is the output and x is the input. $F(x)$ is generally expressed as the quotient of two polynomials in x. The preoperational calibration procedure consists of obtaining the function $F(x)$. In some cases an analytic expression may be determined, while in others it may not. In the latter situation, an *approximation* to $F(x)$ is developed by applying various values of x to the system and determining the corresponding ones for y. In some cases, more sophisticated probing signals are applied to the input, from which an expression for $F(x)$ can be evaluated. Simple examples of such probing functions are the ramp function and the impulse function.

The *actual* response function [e.g., $F(x)$] of a unit or system will generally vary from that specified by the function $y = F(x)$ during the operation of the system. This effect is to be distinguished from that of output uncertainty Δy corresponding to an input uncertainty of Δx, such that $(y_i + \Delta y_i) = F(x_i + \Delta x_i)$ at the ith time or measurement value. The actual response will be $y_{i,j} + \delta y_{i,j} = F_{i,j}(x_i)$, corresponding to an input of X_i, at a time or during an interval j. This means that the actual transfer function departs from the ideal and hence must be calculated. Thus $F_{ij}(X_i)$ is a stochastic function, the mean of which is $F(X)$.

Practically, the unit or system can be calibrated during the operational period in various ways depending upon the configuration of the system itself. In the case of a PDM telemetry channel, calibration pulses corresponding to the lower and upper scale values would be transmitted periodically. In the case of an amplitude-modulated channel, calibration signals spanning the range of signal amplitudes would be transmitted periodically, at preestablished time intervals, between actual measured signal values. Signals corresponding to intermediate values of the signal, as well as the low and high limits, may be transmitted as well. More material relevant to calibration is found in the sections on analog-to-digital converters later in this chapter.

14 Data-acquisition-system Outputs

A data-acquisition system can have three different types of outputs, displays, recorders, and direct connection to the data processor. "Quick-look" displays are included in the first category, along with plotters and other such devices which provide data on the full range of the telemetered data. The category "displays" includes TV screens, printers, and plotters. The section on data presentation describes these units in greater detail. Any one or combinations of all three classes of outputs can be present in any one system.

ANALOG-TO-DIGITAL CONVERTERS

15 Analog-to-Digital Converter Functions

Analog-to-digital converters (ADC's) play a central role in telemetry data-reduction systems. The previous section dealt with the data-acquisition subsystem (DAS) of such systems. This section describes the principles of operation of the chief types of ADC's, used as components in the DAS's. ADC's are found in two parts of telemetry systems, either immediately adjacent to the data source or at some distance from it. An example of the first area of application is the use of an ADC in a vehicle (such as a rocket) under test to convert continuously varying measurements such as temperatures into PCM signals which are transmitted to the telemetry data-handling equipment for reduction. An instance of the second application area is the use of an ADC to digitize analog data recorded on tape after it is first converted to an AM signal from an FM representation by a discriminator.

An analog-to-digital converter is a device which transforms the value of a measurand capable of assuming a continuous range of values, to a number equivalent to its current value. There are three principal types of ADC's. They convert voltage (or current), position, or time and frequency (considered as equivalent) into a number which is equivalent to the (original) quantity converted. An analog quantity is one which at any time assumes one of a continuum of values, generally between two specific numbers. A digital quantity, on the other hand, can assume one of only a finite number of values.

A specific number may be assigned on a one-for-one basis to each value which a measurand can assume. The region over which a measurement can vary is commonly termed its "range." This region may be defined by stating the two end-of-scale (low and high) values. The algebraic difference between these two values is termed the "span." A measured quantity may be represented by a number having any number of binary (or decimal) places. Any nonfractional *binary* number having n places may be represented by the expression $a_n 2^{n-1} + a_{n-1} 2^{n-2} + \cdots + a_i 2^{i-1} + \cdots + a_2 2^1 + a_1 2^0 = a_n a_{n-1}, \ldots, a_2 a_1$. Each of the a_i's can be either 1 or 0. Similarly, any nonfractional *decimal* number having n places can be expressed as $a_n 10^{n-1} + a_{n-1} 10^{n-2} + \cdots + a_i 10^{i-1} + \cdots + a_2 10^1 + a_1 10^0 = a_n a_{n-1}, \ldots, a_2 a_1$. Each digit a_i can assume any integer value between 0 and 9. A number representing a particular quantity can be written to any specific number of places. The number of places used implies that the quantity can be distinguished to the equivalent degree of granularity or precision. This may not be the case for a particular measurement, however. For example, a number representing a certain measurement might have 6 decimal places, implying that the value of the quantity can be ascertained to 1 part in 1,000,000 or 0.0001 per cent. This is the "precision" or "resolution" of this value. "The resolution" or, better stated, the "dead band" is defined as the range through which the measured quantity can be varied without initiating a response in the measuring system. Thus, if the resolution to which a particular measurement can be made corresponds to a smaller number of places than is used, then a smaller number of places should have been used. This is the number of significant digits. If the smallest variation in the measurand that can be discerned is 1/1,000 of the scale, then the number representing it can be stated to three significant figures, or decimal places. Any digits beyond that are purely random and have no significance. In many systems, the scale selected depends upon the actual range of magnitude of the particular measurement.

As an example, suppose the range of a given measurand were 0 to 1,000,000, determined to a precision of 1 part in 1,000, or 0.1 per cent. The range over which the measurand could vary is the sequence 000000, 000001, . . . , 999,999, 1,000,000. Note that, although 6 decimal places are required to represent the number, the three least significant positions are meaningless throughout the range in this example. In a situation like this, especially during a calculation employing this number, these three places would generally *not be explicitly* carried along in the computation. Their presence would be implied. This is analogous to the way in which balance sheets of large corporations are written for distribution to the stockholders. The figures are typically cited in thousands of dollars. Thus, the three least significant places (always taken to 0's) are inferred by those who read the reports.

Another term of importance in defining the capability of a measurement system is "accuracy." The accuracy to which a measurand can be represented is defined as the quantity which specifies the limit or maximum value of error throughout the scale in which this measurand exists. The "error" is the difference between the indicated value of the measurement and its "true" quantity. The "true" value is that which the measurement system would indicate, were it perfect and were there no noise to corrupt the signal representing the measurand. Note that although a number may be known to a *precision* of 0.01 per cent, for example, it may be known *only to an accuracy* of 0.1 per cent. This means that the measurement system affords a granularity of 0.01 per cent of the span. In the illustration, various perturbations superimposed upon the measurement make determination of its value more closely than ± 0.1 per cent impossible, *even though its apparent* value can be stated to 0.01 per cent.

The measurement problem described is analogous to that involved in the reception of a signal S, combined with additive noise N. The receiver responds to S, where $S = S + N$. Even if S is determinable to 0.01 per cent, for example, S itself may be known to only ± 0.1 per cent if N is known to be within that percentage of S. Generally the accuracy of a measured quantity is stated as plus or minus a numerical quantity, stated after its nominal value. Thus, in the example cited above, the maximum value of the scale which could be stated to precision is 1,000,000 or "1,000 thousands." When accuracy is taken into account, however, the maximum value might be stated as $(1,000,000 \pm 10,000)$ or, stated in terms of percentage, as $(1,000,000 \pm 1 \text{ per cent})$.

16 ADC Characteristics and Error Sources

Typically, an ADC does not operate upon the (analog) measurement in the form in which it is developed. Rather, the original medium of the measurement may be altered to the one acceptable by the ADC mechanism itself. For example, if the ADC processes only d-c signals in the range of ± 50 mv, then a measurand, say a current, which can assume any one of a range of values between 2 and 20 ma, for example, would have to be converted to a voltage within the required range.

A device which can make conversions from one medium to another or from one range to another within the same medium is termed a "transducer." The response of the transducer, the ADC, and the other units comprising the analog train of a data-acquisition system ideally operate in a linear manner. This means that any input signal x is transformed to another having a value y, in accordance with the equation $y = Sx + z$, within the limits of the analog subsystem. The y is the *number* corresponding to the analog value of the measurand x, which is typically a voltage. S is the scale factor. It accounts for *both* magnitude and dimensional changes, with respect to the pair x and y. Z is the zero offset, which ideally has zero value. Figure 3 illustrates the two ideal responses of analog subsystems, including analog-to-digital converters. Phase shifts caused by reactances, both mechanical or electrical, are not considered. However, they are present in every system, including an "ideal" one which employs filters in its signal-conditioning portion. Further, delays (which could be lumped together with phase shifts) arising because of the finite time required for conversion in the ADC proper are excluded from the present treatment.

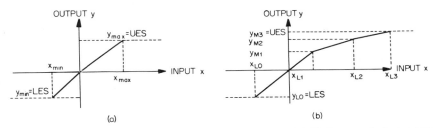

FIG. 3. Ideal transducer and/or analog system responses. (a) Uniform scale factor over range. (b) Several scale factors over range.

Figure 3a illustrates the response of an ideal unit which has the same scale factor over its entire range, from the lower end-scale (LES) value, to the upper end-scale (UES) value. Figure 3b illustrates the other principal type of response, in which there are several different scale factors over the range of operation of the system. The latter type of response, less common than the other, is useful in those circumstances in which a measurement can assume a wide range of values, while only a particular portion of the total range must be available with a relatively high degree of sensitivity. Both responses shown in Fig. 3 illustrate subsystems in which negative excursions of the measurement may occur. This is not the case for all measurements, of course. An analog subsystem exhibiting a response only to positive signals can accommodate negative excursions as well if the expedient of sensing and reversing the polarity of the input signal, if it be negative, is taken. This is the case with most voltage-tape ADC's.

There are four principal types of departures from the ideal response to be expected in the practical situation. They are zero uncertainty, scale-factor aberrations, variation in the instantaneous slope, and hysteresis. The first three of these factors are illustrated in Fig. 4. Descriptions of each of them are provided below and are made with reference to the figure.

    a. Zero uncertainty relates to the actual location of the unit's Z or zero offset, which may be at any value between 0^1 and 0^2 (in the figure).

    b. Scale factor. For any point p_i the scale factor S_i is equal to y_i/x_i, which geometrically may be represented as a line extending from the unit's zero to this point. The band lying between lines L_1 and L_2 established the limits of scale-factor uncertainty for the full range of the unit.

    c. Instantaneous slope. For any point p_k the slope $\bar{S}_k$ is equal to $\Delta y_k/\Delta x_k$. This factor defines the magnitude of the incremental departure in X from the value S_k required to produce an arbitrary output Δy_k. This factor is related to the "dead band," or "resolution," defined earlier in this section.

    d. Hysteresis. The phenomenon that excursions in the input x generally do *not* yield the same value of y for a particular value of x, upon application of x successively to the system. This factor relates to repeatability or reproducibility, a very important factor in any statement of the quality of an analog-to-digital converter.

Repeatability or reproducibility is defined as the ability to generate the same value of output (within the resolution of the system) for applications of the same value of input signal (on the same input channel) under identical operational conditions at different times. Variations in repeatability are attributable in large measure to the drift in the response of the ADC unit. Both long- and short-term drift are significant in determining a unit's repeatability. If an ADC has a low short-term drift, a particular input may be sampled repeatedly, at relative closely spaced intervals of time, and be converted to the same number (within the range of the short-term drift specified). If the unit exhibits a good long-term drift, then readings of the same input with the same value will be converted to the same value.

A useful ADC must exhibit a minimum value of drift during at least one complete scan of the input channels which it samples. Significant variation during that interval would render an ADC virtually useless, since there would be no way of calibrating it.

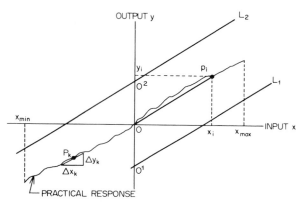

Fig. 4. Practical transducer and/or analog system responses (for uniform scale factor, ideal).

It would be impossible to reconstruct the actual values of the sampled signals. In a typical data-acquisition or other system using an ADC, a calibrated, or precisely known, signal is connected to one channel of the scanner and is read and converted by the ADC at least once during every scan of the input channels. The converted value of the calibration signal is compared with what it should be. Thus a corrective value can be determined which can be applied to the other signals converted by the ADC during that scan to correct them. This correctable error is one type of bias error associated with the operation of an ADC. Random errors are also significant. They are caused principally by the impression of noise on the input signals and power-supply voltage fluctuations.

17 Voltage-to-Number ADC's

This type of ADC converts amplitude-varying input signals to numerical values. A voltage ADC also may convert *a current* signal to a number by first developing a voltage across a highly stable resistor having an accurately known resistance.

There are two types of voltage analog-to-digital converters. They are the "successive-comparison" and the "ramp" types. Both of them generate voltages against which the (unknown) input value is compared. A number equivalent to each possible value of the input voltage can be generated by the ADC, within its resolution capability. Thus, when equivalence between the input and the ADC-generated, or "yardstick," voltage is attained, the conversion is complete, and the numeric representation of the input signal is the digital value in the ADC's output register. Equivalence between the two voltages must be achieved to within $\pm \frac{1}{2}$ of the least significant bit of the binary representation of the input signal.

Ramp Comparison

The yardstick voltage is a repetitively generated "sawtooth" (Fig. 5). During each ADC sampling period equal to or greater than the period ΔT of one sawtooth, the time t_i required for the sawtooth voltage to be equal to the value of the input signal V_i is representative of the value of that signal. Conventionally, a counter is used to develop the numerical equivalent of the sample voltage. A pulse generator is gated to the counter at the beginning of each sampling period, incrementing it until the value of the ramp equals that of the input. Then the number in the counter is read out, and the counter is reset. Then the process is begun anew for the next

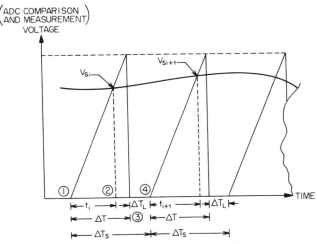

FIG. 5. Measurand sampling procedure.

sample. The sampling period of the ADC must include the time required to read out the counter and reset it to zero, unless two counters are used alternately for successive samples. This would enable the number corresponding to input i to be read out while the conversion of input $(i + 1)$ is under way. Figure 5 illustrates the sampling procedure, and Fig. 6 illustrates a simplified schematic of a ramp-type ADC which is used to effect it. The components labeled as F/F in the figure are "flip-flops," or switches, which can be put into either of two positions. The position at any time corresponds to which one of the two leads was last actuated by a pulse. The flip-flop will remain in this position until a pulse corresponding to the other position occurs. A "gate" is defined as a switch that permits a signal to pass as long as a required control lead is active. The "programmer" is a simple counter, caused to advance and make active a succession of leads corresponding to sequential steps in the sampling period. If it is desired to digitize a given input (continuous) signal current without loss of information, then ΔT_S must be equal to or less than $1/2f_{max}$, where f_{max} is that frequency component of the signal beyond which a power-spectrum analysis of the signal would show little energy content. If the ADC converts different inputs successively, ΔT_S, sometimes termed the "dwell time" or "aperture," should be as short as possible in order to assure that the signal value varies little during the time it is being digitized. This is the pragmatic equivalent of the reasoning above, given relative to the power spectrum of an input signal.

PAM as well as continuous AM signals can be directly converted by this equipment.

Note that FM telemetry signals, having passed through a discriminator, can likewise be converted.

Converters operating at rates in the range of 2,500 volts/sec are commerically available.

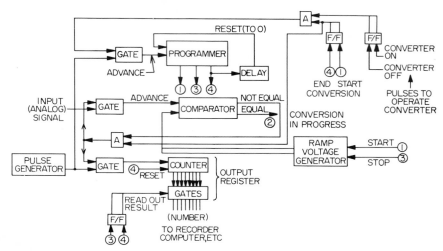

Fig. 6. Simplified ramp analog-to-digital converter.

Successive Comparison

Figure 7 shows a converter which operates on this principle. The principal difference between this device and the ramp converter lies in the way the "yardstick" or comparison voltage is generated. In this unit, the voltage is built up successively, through a prescribed sequence, until equality between this test voltage and the sample is achieved. The heart of the unit is a ladder network or an equivalent capable of developing a finite set of voltages over the range of the inputs sampled. To understand the ADC's operation, assume that the input voltage is to be quantified into 1 of 8 values, from 0 to 7. In binary notation, this is the range from 000 to 111. Four resistors, the values of which are weighted in the ratio 4:2:1, are switched in and out of the circuit path through the ladder, in accordance with the sequence shown in Fig. 8. At each stage in the sequence, the sample voltage V_{si} is compared with the comparison voltage V_C to determine whether it is greater or less than V_c. Examination of Fig. 8 indicates that a maximum of 4 "yes" or "no" binary decisions (tests) are required to quantize V_{si} in this case. The first test indicates whether V_{si} exceeds the capacity of the ADC, or the maximum value of the signal value convertible.

Figure 7 illustrates the relationship of the ladder network to the other principal components of the comparison-type ADC. The capacitor is alternately charged with V_{si} and the current value of V_c. A switch, called a chopper, driven by means not shown in the figure, alternately switches the capacitor to each voltage source. The pulse amplifier amplifies the pulse, negative or positive, according to the polarity of the difference between V_{si} and V_c. Corresponding to the occurrence of each such pulse, trial values for V_{si}, as shown in Fig. 8, are generated. Finally, the binary representation of V_{si}, within the resolution and accuracy of the ADC, is available in the output register. This register is comprised of flip-flops, or equivalent two-state devices.

ANALOG-TO-DIGITAL CONVERTERS   11–19

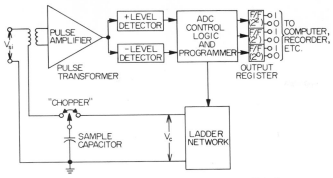

Fig. 7. Simplified successive comparison analog-to-digital converter.

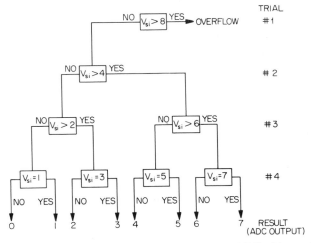

Fig. 8. Binary voltage comparison (for successive comparison ADC with maximum output of 7).

Numerical performance figures for some typical units are presented below.

One such unit can convert 30 voltage signals lying in the range 0 to 50 mv to 3 decimal digits per second. This converter exhibits the following performance:

Accuracy	±0.25% of full scale
Linearity	±0.25% of full scale
Dead band (resolution)	5 μv
Zero drift (long-term)	±0.02% per month
Zero offset	15 ± 20 μv
Ambient operating temperature	40–95°F
Ambient operating humidity	10–80% relative humidity
D-C common-mode rejection	400,000:1
A-C common-mode rejection	200,000:1
Max safe input voltage	200 volts peak

Another converter unit accepts signals in the range 0 to 20 volts. It can convert 20,000 of such signals to 14-bit representation per second. This unit is characterized by the following performance:

Accuracy	±0.10% of full scale ± 1 bit
Dead band (resolution)	1 mv
Ambient operating temperature	60–120°F
Ambient operating humidity	20–90% relative humidity

Designing an ADC which converts low-voltage signals, measured in millivolts, at a high rate is quite difficult. Therefore, production converters which must convert thousands of signals per second are built to accept signals at higher levels, measured in volts.

18 Position-type ADC's

Position-to-number ADC's convert an analog quantity, represented as an angular position of a shaft, to a number. The present material assumes the number is a binary one. Of course, ADC's are not restricted to binary output. Conversion to decimal or other digital representation may be done very conveniently. The input to the ADC, the measurement itself, may be the position of a linkage or other mechanism which is resolved into a rotation of a code disk. Alternatively, the measurement might be a d-c voltage, which is converted to a binary number through the use of a servomechanism employing a code disk.

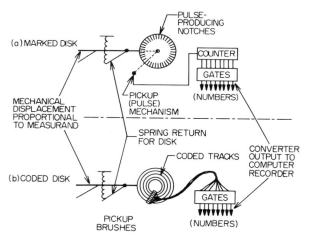

Fig. 9. Angle-to-code position ADC's.

The simplest type of a position-to-digital converter is a counter which registers the number of revolutions of a shaft. A variation of such a mechanism would be one in which only one revolution would be proportional to the full range of values possible for that measurement. The 360° (or other lesser value) of rotation would be indicated in machine-sensible marks on the disk itself, such that the number of marks counted, from some zero or reference point to the point actually reached on the disk, will be proportional to the value of the measurement.

Figure 9 portrays the two principal designs of angle-to-code ADC's. Figure 9a illustrates the one type which uses an incrementally marked disk. Rotation of this disk causes one or more marks to be sensed and pulses to be generated which are added in the counter shown. Figure 9b pictures one of the other principal types of angle-type encoders. This one uses a disk which is encoded such that its angular displacement from some fixed reference may be sensed from it directly. A principal feature of design of such code disks is the type of encoding employed. The means selected for sensing the codes is a significant feature, as well.

Figure 10 illustrates a sample disk encoding, and the form which a disk might assume. The table of codes in the figures shows both the ordinary binary and a reflected

DECIMAL	BINARY	(GRAY) REFLECTED BINARY	REFLECTED BINARY COMMUTATOR CODING			
0	0000	0000				
1	0001	0001				
2	0010	0011				
3	0011	0010				
4	0100	0110				
5	0101	0111				
6	0110	0101				
7	0111	0100				
8	1000	1100				
9	1001	1101				
10	1010	1111				

SAMPLE TABLE OF DISK ENCODINGS

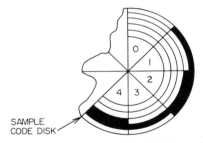

SAMPLE CODE DISK

FIG. 10. Sample table of disk encodings.

binary, or Gray, code preferred for use with disk encoders. The Gray code is so constructed that only one binary digit changes in a progression from one resolution element to the next. This is not the case with ordinary binary representation, however. Several binary digits can change in a progression from one resolvable element to the next in that code. A severe ambiguity problem can arise because of misalignment of readout units if a binary code is used whereas, if reflected binary encoding is used, such a misalignment would be expected to result in a resolution error in the low-order bits only.

The principal means of sensing code disks are electrical with contact brushes, magnetic with no-contact sensing, and optical. In the first case, brushes are in direct contact with the code disk. A circuit is closed between each brush and the track corresponding to it, if conductive material is present at the point of contact. Alternatively, the tracks may be bipolar magnetized sectionally, to the degree of resolution of the disk. The magnetization represents the presence or absence of a bit.

Optical sensing can be afforded in two ways. In the one, a light is projected on the disk, which passes through it or not, depending upon whether the particular segment is opaque or not, corresponding to the presence or absence of a "bit" at that position.

Code disks which use contact brush sensing have been employed which provide a resolution of 1 part in 2^{14}. A current optical system employing separate readout units for coarse and fine sensing resolves 1 part in $2^{19}(-524,288)$.

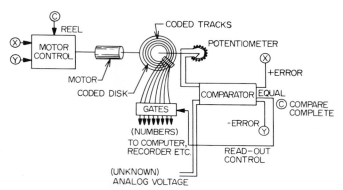

FIG. 11. Voltage-to-code position ADC.

Figure 11 shows a representative system which uses a code disk in an ADC which accepts analog voltages for conversion. The device illustrated compares the input voltage with that controlled by a potentiometer mounted on the same shaft as the coded disk. The difference between these voltages is applied to control a motor which displaces the shaft from its current angular position in such a direction as to null the difference in voltages. When the null is reached, the position of the coded disk is read out. The number represents the amplitude of input (measurement) voltage.

19 Time-and-Frequency-to-Number Conversion

Time-and-frequency-to-number converters employ counters as their principal elements. A time interval of unknown duration can be readily converted into a number by counting the number of pulses produced by a pulse source of known frequency, having a stability within the degree required to attain the prescribed accuracy of the converter overall. Thus, at the beginning of the sampling interval, the oscillator (or clock) pulses will be gated to a counter. The connection will be maintained until the end of the interval. At this time, the reading in the counter will be transferred to the unit's output, data processor, recorder, or display unit. The resolution of this type of ADC can be enhanced through the use of two oscillators which differ in frequency to a slight extent. The oscillators might be such that m pulse periods of one equal $m + 1$ pulse periods of the other. If a particular interval of time were found lying between m and $m + 1$ pulse periods, then the vernier scale afforded by the oscillator which produces m pulse periods in that time could be used to specify the time.

A frequency-to-number converter produces a number equivalent to the frequency of

ANALOG-TO-DIGITAL CONVERTERS   11–23

the input voltage during a particular interval. The accuracy possible with such a unit is directly related to the degree to which constancy of frequency is maintained during that interval. Many converters of this type operate by counting the number of cycles of the input signal which occur during the measurement interval. Generally, the input will pass through a limiter and then be differentiated. The limiter removes variations in the amplitude of (clips) the signal. The differentiation function converts the clipped signal voltage into a sequence of pulses. If only positive differentiation is used, there will be one pulse per cycle. If both positive and negative differentiation are used, the times of occurrence of both the beginning and mid-points of each cycle are available. Variations in design of the frequency-to-number converter are possible. Either one or both types of pulses may be used to advance the ADC's counter at a rate proportional to the frequency of the signal, or two times its value in the latter case. The accuracy of such a converter may be related to the number of cycles over which a count is obtained. The minimum number of cycles over which the count must be made is one. If there is noise imposed upon the signal, in terms of frequency jitter, the number of cycles over which the count must be made increases to attain a desired amount of accuracy. The signal is integrated or smoothed by this action. The number of cycles over which a count must be taken limits the rate at which information can be transmitted by the varying frequency signal. To illustrate this point, suppose the frequency can assume any one of 10 values. Then, the equivalent of one decimal number can be conveyed by the signal during each interval over which a count is made. If two converters were to operate upon the signal, and one required an interval having two times the duration of that required by the other, its conversion rate would be twice that of the other. At the end of a conversion period, the number in the counter is transferred to a data processor, recorder, or display device, as in the case of the time-to-frequency converter.

Figures 6, 10, and 11 which illustrate ramp-type voltage-to-number and position-to-number converters, respectively, show counters used in the ways described for the time and frequency to number converters described in this section.

Time-and-frequency-to-number conversion can be applied to converting PDM and FM signals to PCM format. A PDM pulse may be converted to PCM format for recording on digital tape by gating a clock (oscillator) to a counter for the duration of the pulse and counting and recording the number of pulses that occur during the interval. PDM signals may be converted to digits by first converting them to PAM representation and then using a voltage-to-number converter as described earlier in this section. The PDM-to-PAM converter could employ a ramp voltage generator such as the one in the sawtooth voltage generator of Fig. 6. The generator would be gated on at the beginning of the pulse (PDM) and would stop its increase in voltage at the end of the pulse. The total voltage would be maintained on a sample-and-hold circuit, consisting principally of a capacitor charged by the voltage generated by the ramp. This would be the PAM representation of the PAM (pulse) input. The voltage on the capacitor would be converted to a number by any one of the means described.

Earlier it was stated that many frequency-to-number converters use counters as their principal components. Another principle, used in some FM detectors, can be used also. This is the conversion of the FM signal to a sequence of constant-width pulses, the intervals between which vary in proportion to the frequency of the signal. The output of this converter is a d-c voltage, proportional to the frequency of the signal, which may be applied to a voltage-to-number converter of one of the types described. The unit limits or clips and squares the FM sinusoid, producing a two-valued signal the zero crossings of which occur at the same time as the sinusoid from which it is derived. The output of the limiter is then differentiated. The sequence of spike pulses thus developed is applied to a monostable multivibrator, which produces one pulse of duration t_1 sec each time a triggering pulse is applied to it. The interval of time between the beginning of one output of the multivibrator and the next is equal to the period T of the FM signal, the instantaneous frequency of which is f, where $T = 1/f$.

If the output pulse of the multivibrator has a magnitude of E volts, then the d-c component of the waveform during the entire interval T is given by $E_{dc} = E(t_1/T)$. If $t_1 = 1/f_0$ then $E_{dc} = E(1/f_0)/(1/f) = E(f/f_0)$.

RECORDERS

20. Application of Magnetic-tape Recording

The widespread application of magnetic-tape units in telemetry data-handling systems requires that the telemetry-system user and design engineer understand their use in connection with two principal functions. The first is the recording of such data at the telemetry center, and the generation of either "analog" tapes or PCM tapes. The second area of application is digital (binary) recording and recording for the purposes of computer input and output as well as for large-volume storage for the processor. Some significant characteristics differentiate these two functional areas of application of magnetic-tape technology. The tape transports used in instrumentation recording are designed to operate in a continuously running mode. The units used by the computers as well as others employed in the generation of digital tapes suitable for their use have the ability to start and stop rapidly and to run for brief periods for the writing or readout of relatively small groupings of information.

Unquestionably, magnetic-tape-recording equipment offers great advantages relative to alternative storage means for both the acquisition and reduction of telemetry data. Its use enables various tests to be "redone" in order that their progress may be critically examined. In addition, the data gathered during a test or run may be processed in more than one way with various processing techniques to obtain more than one type of result.

For telemetry installations some of the more important characteristics which should be considered in selecting any particular equipment are the rates of data flow required, the information density necessary, the need for a synchronization track, the amount of circuitry required for positioning the tape and the transfer of information to and from it, and the reliability of the tape units.

The sections on recorders of this chapter define the principles of magnetic recording as well as specifying the specific attributes of telemetry as opposed to computer data recording. Further, the differences between saturated and nonsaturated recordings are pointed out.

21. Characteristics of Storage Devices

There are three types of storage devices, categorized in terms of the manner in which information may be loaded into or read out of them. They are random access, cyclic, and linear devices. A magnetic-core memory of a computer is a random-access device. The various registers comprising it may be uniquely addressed, with no variation in the amount of time required to enter information or retrieve it from any register in the memory unit. Cyclic stores, such as magnetic drums and disk files, although having uniquely addressable registers (basic subdivisions of information-recording space), exhibit a variance among registers with respect to the delay between the time a particular register is specified or selected and the time information transfer into or out of it can be effected. In linear stores, such as paper cards or magnetic tape, a particular register can be accessed, in general, only after other registers or storage units which precede it in its sequence have been accessed. When magnetic tape is used, various storage locations must be passed over before a specified unit may be accessed. When information is to be recorded on a tape, the tape is advanced until a point is reached at which either there is no information currently recorded or that which is recorded is no longer desired to be retained. In the latter case, it is simply written over. In some units, previously recorded information must be erased positively; this process is termed "degaussing." It is effected by the application of an a-c voltage to a writing head which produces an alternating flux which tends to randomly orient the magnetic dipoles in the recording surface.

The information recorded upon a tape can be attributed an "address," or location specification for readout, in several ways. These ways are: identify all or a portion (a "tag") of the data in the desired location; identify a given value of a "key," such as a time code, with the desired data; or simply count the number of items of information between the specified location relative to a certain known position. In the case of digital-computer tapes, the information is divided first into "files," then into "records," and ultimately into "words." A word is comprised by the group of bits at a particular longitudinal (in the direction of the length of the tape) position. Second, such words comprise a "record." One record is separated from another by a blank segment, termed a "record gap." One file is separated from another by either another blank segment of tape and/or a special character, called a "tape mark."

Whenever a segment of information is to be located by the "tag" or "key" method, the desired value of the tag or key is set up in a register, and the tape is advanced until the portion on which the desired value is sensed is read out.

22 Recording Principles

Information is stored upon the surface of a magnetic tape by appropriately magnetizing the recording surface which is sprayed, plated, dipped, or otherwise caused to adhere to the basic body of the tape, which is an appropriate plastic material, such as acetate or Mylar (Mylar is a registered trademark of the Du Pont Company). The recording surfaces used are iron oxide or electrodeposited cobalt-nickel plating. The thickness of the coating ranges from about 0.4 to 3 mils. The highest-quality tape is comprised of a Mylar base (1.5 mils thick) coated with a red oxide layer of recording material of 0.5 mil thickness.

Three possible directions of surface magnetization exist: longitudinal, transverse, and vertical. Longitudinally recorded information is written in the direction, call it Y, of the tape movement. Transversely recording techniques produce magnetization in the plane of the surface, normal to the direction of motion of the tape, call it Z. In vertical recording (in the Z direction), the portion of the surface magnetized is in a direction vertical to the plane of motion of the tape. Longitudinal recording is universally employed in today's magnetic data-recording systems. Figure 12 illustrates the three modes of recording.

A ring structure, as illustrated in Fig. 12, is quite generally used for both read and write heads. This structure has a gap in it, placed close to the tape's surface. Coils of wire are placed on the ring for writing and reading data and for erasing the tape. When a magnetomotive force is effected in the ring, caused by a flow of current in the writing coil, lines of flux are induced in magnetic material of the ring. The flux tends to pass through the surface magnetic material of the tape rather than across the air gap. The result is that a localized area of that surface is magnetized. If the surface is visualized to consist of dipoles, the equivalent of very minute bar

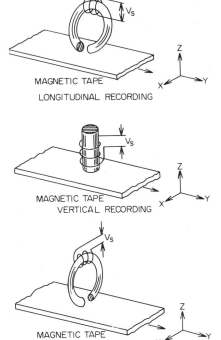

Fig. 12. Magnetic-recording modes.

magnets, the magnetization action may be viewed as an orientation of them in a preferred direction, directly (not necessarily linearly) in proportion to the amount of flux passed through them. This orientation process may be applied to dipoles which are essentially randomly oriented. This is not necessarily so in all instances, however, especially in the case of digital (binary) recordings in which the local area

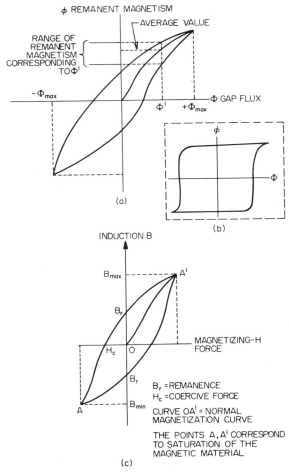

Fig. 13. Hysteresis loops. (a) Remanent magnetism of magnetic tape. (b) Square loop. (c) Hysteresis loop induction vs. magnetizing force.

of the magnetic material is put into one of two saturated states. The application of an additional magnetizing force in the same direction will not augment the flux density. Application of magnetizing in the opposing sense, will, however, vary the flux density. Figure 13b illustrates the type of hysteresis loop desired for saturation-type recording. It is highly nonlinear and is called "square-loop" material for evident reasons.

Material used for magnetic-tape recording must be such as to retain the degree of flux density imposed by the writing coil, even after the magnetizing current in the coil ceases to flow. Note the contrast to both "permanent" magnetic materials, which when once magnetized in a certain orientation remain that way more or less indefinitely, and those materials used for the "cores" of electromagnets. The latter type of material exhibits a high flux density when the existing coil around it is activated. When the current in the coil ceases to flow, however, the core material reverts to a state of low residual or remanent flux density.

Materials which are used for the base of magnetic tapes must exhibit very low hydroscopic properties and a minimal tendency to stretch or shrink regardless of direction in response to variations in temperatures. Further, they should exhibit little variation in physical structure, such as being brittle, in response to variations in temperature and humidity.

Information recorded on tape in the manner described above may be nondestructively read out from it simply by passing a (read) head over the magnetized areas. The head used can be either the same head by which it was recorded, or one similar to it. Recalling that the magnetized areas are like true bar magnets, it is evident that lines of flux will connect both ends through the air. As the alternative path, the base of the tape, is material with an extremely low permeability it has a high resistance to lines of magnetic flux. The passage of the head over the magnetized area affords the lines of flux a path of extremely low reluctance relative to the air path. The variation in the flux, caused by the relative motion of the head and the tape, causes an open-circuit voltage to be induced in a coil on the head. The coil from which the induced voltage is sensed is the read coil.

Some recorders have a coil especially provided for an erase function. A section of tape may be erased by passing it under the head while an alternating current is feeding the erase coil. The alternating current need not be purely sinusoidal but, in fact, could be a random, noiselike voltage, exhibiting a roughly equal number of positive and negative excursions. The effect of applying the flux induced by such a voltage to the recording medium is to randomly orient the various domains of magnetic dipoles which comprise it. An a-c voltage of a particular frequency may be applied to one track of the tape during the course of recording pure data for the purpose of later using a signal derived from it for synchronization of data readout and for compensation of anomalous variations in tape speed.

23 Physics of Recording

Magnetic recording relies upon the relative motion between a "permanent" magnetic storage surface and a transducer, capable of inducing flux in that surface or having a current induced in it by passing through lines of flux from the magnetized elements on the recording surface. The path of the relative motion described is called a track. Generally, parallel recording is used in which several such tracks are recorded simultaneously. Serial recording has been employed to some degree in certain applications of telemetry data recording, also.

There are two principal types of mechanical-recording arrangements found in practice, "contact" and "noncontact" recording. The former method involves an actual physical contact between head and tape. This type of operation affords the maximum degree of recording resolution and potential recording density. Audio and video magnetic-tape recorders use this principle to attain the greatest recording bandwidth possible. Unfortunately, the maximum tape speed is rather limited to minimize the wear (frictional) caused by the contact. "Noncontact" recording is preferred for data recording because of the potentially greater recording speeds possible, and higher reliability attainable if contact recording is used.

Contact recording may be considered a particular case of noncontact recording in which the spacing between head and recording surface has been reduced to zero. How closely regions of opposite magnetization may be placed without nullifying each other's effects defines a fundamental limitation of magnetic recording. This may be termed "writing definition." Essentially, it is the length along the track over which a saturation change in magnetization can be effected. "Reading resolution" refers to the segment of tape extending on both sides of the head within which a step change in the magnetization previously registered induces a perceptible voltage during reading, there being a relative motion between head and tape. In the common situation in which the same head is used for both reading and writing, whenever conflicts in construction arise, the requirements of enhancing the reading resolution should predominate.

Information desired to be recorded is first converted to variations in the amplitude of the current supplied to the head's writing coil, producing a changing gap flux. The corresponding value of remanent magnetization is recorded on the tape as it passes the head gap.

In the case of nonsaturating recording, as for PAM-PM, audio, and video, the value of the remanent magnetization is the analog of the original information. The equations $\Phi = kNI$ and $d\Phi/dt = kN(dI/dt)$ describe the relationship between the gap flux and the writing coil current. We note that Φ = gap flux, k = a constant, N = number of turns on the coil, I = current in the coil, t = time, and y = longitudinal distance along the track. The implied equivalence of the distance and the time derivative is to be noted, since the time of a change in flux and the position y along the track at which it is recorded are equivalent in the case of a magnetic tape. Although the amount of flux passing through the portion of the tape to be magnetized is directly proportional to Φ, the flux in the gap, the amount of *magnetization* transferred to the tape, as well as that to the recording surface after the removal of Φ, called the remanent magnetization, is not directly proportional to Φ because of hysteresis, as illustrated in Figs. 12 and 13.

Hysteresis, having to do with the history of the magnetic-storage material, may be defined as "the lag of induction with respect to the magnetizing force." Referring to Fig. 13, one can observe that variations of gap flux from $-\Phi_{max}$ to $+\Phi_{max}$ have an average value of zero. Hence, if a current to produce such a flux is applied to the writing head, the effect will be to neutralize the recording medium. If a flux corresponding to Φ' is applied, the remanence will be nonzero and will be fairly closely proportional to Φ'. This affords a physical basis for the writing of analog information on the tape. Introduction of an alternating binary flux, say from $-\Phi_{max}$ to $+\Phi_{max}$, as shown in Fig. 13, will result in the variations of the remanent magnetism along the tape (in the y direction) being quite well proportional (linear) to the variations of the original information-bearing signal with time. This is the basis for analog data recording. This technique is used for direct telemetry recording, such as for FM-borne signals.

The readout head does not respond to the amount of magnetic intensity on the tape. Rather, it responds to *variations* in that intensity, the flux induced in the head being proportioned to the variation. Thus the readout voltage E is related to the rate of change of flux Φ along the tape by the equation

$$E = KN(d\Phi/dt)$$

As the voltage in the readout head is proportional to the time rate of change of the flux, the head is frequency-sensitive. It develops a voltage that increases by about 6 db / octave with frequency increase. In order to make the ultimate reproduction as faithful to the original input as possible, "equalization" is required in the reproduce amplifier.

There are two choices for effecting the equalization process required. The one would boost the amplifier output in the low-frequency range. The second would attenuate it in the higher-frequency range.

24 Tape Transports

The term tape transport refers to the equipment employed to hold the reels of tape and to move tape past the writing, reading, and recording heads. In the case of intermittent-motion units, the reel units holding the tape are designed to maintain a certain degree of slack between them and the tape drive units in order that the tape may be accelerated and decelerated at a sufficiently high rate. This type of arrangement places the burden of rapid accelerations and decelerations upon the slack length of tape, rather than upon the tape-laden reels, which have a relatively high inertia as compared with the length of tape passing through the heads between them.

For directly recording instrumentation data, transports without the dual control of slack tape and reels may be used. Figure 14 illustrates the three types of tape

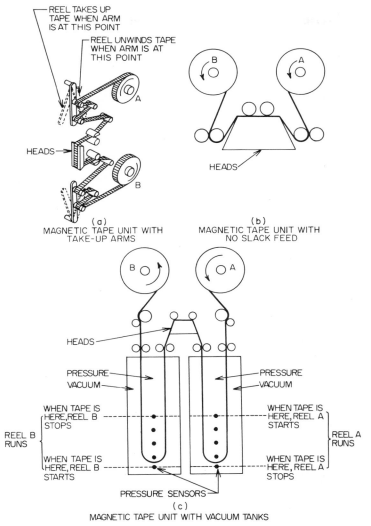

FIG. 14. (a) Magnetic-tape unit with take-up arms. (b) Magnetic-tape unit with vacuum tanks. (c) Magnetic-tape unit, no slack feed.

transports. The first type is one having a movable take-up or follower. They are spring-loaded and designed to maintain a desirable tape tension, say $1\frac{1}{2}$ oz or so. As the tape is moved, the resulting positional shifts of the follower arms are sensed, and signals are generated to activate the reel motors to reestablish the desired degree of tension.

The other type of transport used for applications in which intermittent tape motion is required relies upon the use of two vacuum surge tanks, one corresponding to the feed reel and the other to the take-up reel. In each tank, a pressure differential is maintained. Air pressure is maintained above the tape, while a (relative) vacuum is maintained below it. As the tape is moved past the head(s), the corresponding movement of the portions of tape in the tanks is sensed, by either vacuum or photoelectric means. Signals to control the tape reels are developed in response to the sensing of these conditions which cause the reels to rotate in one or the other direction to reestablish the specified degree of slack.

25 Digital Recording—Computer Magnetic Tapes

There are several ways in which information may be represented on computer magnetic tape. All of them employ saturation-recording means. In some schemes, both the two possible directions of saturation as well as the zero-magnetization situation are employed. Some of them are self-clocking, while others require a special timing or synchronization track. For a given track, the flow rate of information to or from it is given by the expression (bpi) = v(bps), where (bpi) stands for the storage density in bits per inch, and v is the relative velocity of the storage surface and the head, while (bps) is the information flow rate in bits per inch. Table 1 summarizes some of the currently available maximum combinations of values of the items cited.

Table 1

bpi	v, in./sec	bps
200	75	15,000
556	75	41,700
200	112.5	22,500
556	112.5	62,500
800	112.5	90,000

Commonly, digital-computer tapes have had 7 tracks. A group of 7 bits at corresponding positions on the tracks can be called a "character." The information in the character may be conveyed by 5 or 6 bits, with another used for an odd or even "parity" check in the information binary bits. One of the tracks may be used for clocking if the information is not self-clocking. Some computers, such as those in the IBM System 360 series, have 9 track tapes, 8 for data and 1 for parity. This is the case for *NRZ* (non-return-to-zero) recording (see below, and Fig. 15), which also happens to be the IRIG standard for PCM recording. If odd parity checking is used, a 1 bit is added, if required, to make the sum of all the bits at the particular (longitudinal or y) position on the track an odd number. The opposite is the case with "even" parity.

Groups of characters may be combined into a larger element of information termed a "record." Records sequentially recorded upon the same tape are separated by lengths of blank tape ¾ in. long on IBM tape units, for example. Further, a sequence of records may be combined into a "file." In some units, an "end-of-file" condition is indicated by an end-of-record gap and/or by a special character, termed a "tape mark." In other units, an elongated gap, about 3¾ in. long, with or without (in certain cases) a "tape mark," indicates the end-of-file condition.

An entire record generally has a parity check made on it, derived in the same manner as that for an individual character, except that the check is made on all the bits in a track.

An additional check is provided by certain tape units, in order to reduce errors due to the not infrequently found imperfections in the tapes used by data-processing centers. A two-gap (dual) head is used while writing. This equipment enables the information to be read out into a compare register immediately after it has been written. If there is a disagreement, an indication is made. Similar indications exist for the failure of the character and longitudinal redundancy-check characters described above. In most cases, the processor program will be able to take cognizance of such a failure and perform some corrective action.

When estimates are made of the capacity of tapes for storage purposes, care should be taken to assure that the space required for the various interrecord and end-of-file gaps has been taken into account. Generally, these gaps reduce the *actual* storage capacity of a tape a not inconsiderable amount from that quoted in many advertising brochures. The maximum data-transfer rates typically quoted are not usually realized operationally because of the presence of gaps and/or the fact that transfer between the computer's internal (working) memory and the tape cannot proceed at that rate because of other demands placed upon the computer.

Unless specially adapted to read gapless magnetic tapes, data processors require that the tapes which they read be written with gaps. The gaps are used for control and data-flow purposes, as has been described. Consequently, installations which record telemetry data directly on magnetic tapes in a computer-compatible form must employ a buffer unit, typically a magnetic-core device, placed between the source of digitized data and the tape unit. Conventional "telemetry recording tapes" record in a gapless continuous mode, however, as has been indicated.

26 Computer Tape-recording Techniques

Recording techniques refer to the ways of writing and reading data, the process by which trains of bits (0's and 1's) are translated into and out of a waveform generated by the writing-current pattern. The most important of these techniques are return-to-zero (RZ), non-return-to-zero (NRZ), modified or non-return-to-zero on 1 ($NRZI$), and the Manchester, or phase-modulation method. Each of these methods is described below. The appropriate parts (a, b, c, d) of Fig. 15 may be consulted as an aid to understanding this material.

a. RZ

Undoubtedly the most straightforward method of recording digital (binary) data is that in which a current pulse of one direction is applied when a 1 is to be recorded and a pulse of another direction is applied when a 0 is to be recorded. Note that the current returns to zero between such pulses. Note that, because a pulse is present every "bit time," no synchronizing signal is required to extract the data recorded on the tape. In fact, a timing signal can be provided simply by inverting all negative pulses and combining them with the positive ones, together with shaping if necessary.

b. NRZ

Several methods within this category exist. They are all characterized by the fact that the current is switched to record information but that it does not remain at zero for a significant amount of time. The conventional schemes use continuous-current waveforms which endure throughout each bit period, in one direction when representing a 1, and in the other when representing a 0. Consequently, if a string of 1's or 0's occurs in the data being recorded, the polarity of the writing current, and hence the flux, remains the same throughout the period that those data are being recorded.

This type of recording offers the advantage over RZ recording of greater bit density, and the erasure of data previously recorded on the tape may be more nearly completely realized.

The disadvantages of NRZ are a lack of self-clocking capability and a higher head duty-cycle requirement than RZ recording. Further, it suffers from the fact that if a single pulse which happens to be the first of a sequence of its kind is missed, the remaining members of the sequence will be read erroneously.

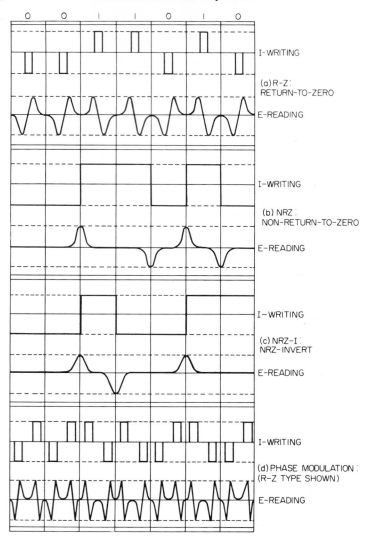

Fig. 15. Types of binary magnetic recording.

c. NRZI

An important variation on the *NRZ* method is called the change on 1, or *NRZ* Invert method. With this method, the current is inverted each time a 1 is to be recorded. When several channels are recorded in parallel, the bits comprising the character can be so encoded that each character contains at least one 1. Consequently, the presence of a 1 at each character position, longitudinally sequenced along the tape, can be used to derive a self-clocking feature of the same nature as is obtainable from data recorded in an *RZ* mode.

RECORDERS   11-33

d. Phase-modulation Method (RZ Type)

With this method, when a 1 is recorded, current is sent through the head first in the positive direction and then in the negative direction. Each "half-bit" pulse occupies one-half of the digital interval. A 0 is recorded with the opposite sequence. This method offers a self-clocking feature. Note that NRZ techniques may be used to record data in the "PM" manner.

27 Telemetry Data-recording Techniques

This section describes the means for recording pulse-duration (or width) modulated (PDM or PWM) signals. Also, the means for recording pulse-code-modulated (PCM) signals is reviewed. The methods employed differ little in essentials from those described for computer tape units.

The recording of FM, PAM-PM, and other amplitude analog signals is described fully in the earlier section of this chapter which deals with recording principles and hence is covered no further here.

PDM

Pulse-duration signals (of constant current or voltage amplitude) convey information in terms of their (time length) duration, typically varying from about 60 to 750 μsec, corresponding to a full-scale range in the measurement to which a signal corresponds. Information corresponding to a train of pulses of this type is recorded on a single track. The data recorded are obtained by differentiating each pulse of the train. This produces a positive "spike" followed by a negative spike, separated by an amount of time equal to the duration of the data-bearing PDM pulse. Upon readout, the sequence of spikes may be readily reconstructed into PDM signals, simply by using them to gate a voltage or current having a constant amplitude of the level desired for the PDM pulse train on and off.

Several PDM pulse trains may be recorded on different tracks of the same tape if desired. One track can be used for framing or other timing signals, as needed in the particular installation.

PCM

The IRIG (Inter-Range Instrumentation Group) standards require that PCM data be recorded employing the NRZ method. The principle of NRZ is described in the previous section. In the case of recorders for PCM which are directly coupled to the data source, it is possible that the data will be written without the gaps described in connection with computer tapes.

28 Predetection Recording

Predetection recording signifies the recording of the telemetry data prior to the demodulation of the r-f carrier with which it is associated. Frequency conversion, limiting, and other such operations are distinguished from demodulation in this regard. The versatility of predetection recording has greatly simplified the instrumentation of many telemetry systems, especially of the aerospace field, in which signals are conveyed by a wide variety of types of lines, including FM-FM, PCM-FM, and PAM-FM. The predetection-recording equipment may be used for all these forms. The processes of demodulation and demultiplexing can be done at a central data-processing center. The use of predetection-recording equipment at the telemetry-receiver site can substantially reduce the expenditure involved in the operation of the site. This is due to the fact that different telemetry links can be introduced without requiring a corresponding increase in the instrumentation requirements for these links that would otherwise be necessary.

29 Skew, Wow, and Flutter

Skew

In a multichannel tape, the "skew" is the measure of angular departure of the physical orientation of the head assembly from the normal to the direction of the tape travel. Skew must be minimized so that corresponding segments of the tape may have data entered into and abstracted from them without error due to this cause.

Wow and Flutter

"Wow" and "flutter" are the terms used to describe the variations in tape speed which can cause nonlinearity in the time-distance relation involved in the recording process. The former term refers to low-frequency variations, while the latter refers to the higher-frequency variations. The values for wow and flutter generally quoted on specifications for recorders refer to the maximum as the peak-to-peak magnitudes of these variations. Compensation for these speed variations is very important, especially in the case of predetection and analog recording. The scheme usually involves the recording of a precision reference-frequency standard with the information. Upon reproduction, the reference frequency is compared with respect to phase or frequency of the standard used during recording. Variations in the phase or frequency of the information recorded are used by a servo to produce an error voltage which will vary the speed of the motor driving the reproduce unit accordingly, thus compensating for any undesired variation in the tape speed which was introduced at the time the information was recorded.

DATA PROCESSORS

30 Data-processor Applications

Various types of data processors are used in telemetry data-handling systems. Frequently the terms "computer" and "calculator" are used synonymously with "data processor." Sometimes a distinction is drawn between them, however. Frequently the former terms connote the execution of a sequence of arithmetic operations, such as might be required to calculate the trajectory of a missile tested on the Atlantic Missile Range. In distinction to this the latter term denotes a broader application of this equipment, to control the acquisition of the run data, to process it, and finally to record it in permanent form (if desired), and to display all or part of it to various personnel as required.

As pointed out earlier in this chapter, the sequencing function may be performed with a plugboard, tape-controlled sequencer, or a special- or general-purpose data processor. An excellent example of this latter type of equipment is "the digital data processor" built by personnel of the U.S. Naval Ordnance Test Station at China Lake, Calif. This device is "actuated by instructions from an internally stored program" to operate a *randomly addressable* mutiplexing unit which replaces a fixed-rate commutator. The output of the machine is a magnetic tape, written by an IBM 729 tape unit. The tape is processed subsequently (on an off-line basis) by an IBM 7090 at another location at the test station.

Data processors may be used either directly in the telemetry-recording process or indirectly subsequent to the collection of data. Any data processor or simple calculator must be programmed. A program is merely a sequence of instructions, available to the machine, which *explicity* describes *every step* which it must take to perform the tasks assigned to it. There are two types of programming systems used by processors: external, and internal or "stored." The former category includes plugboards (such as are used to advance the commutator in certain telemetry data-acquisition systems), and tape (generally paper), used to a large extent in automatic testers, which are used to check out circuits on an assembly line as well as missile

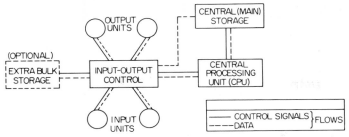

Fig. 16. Data processor—functional diagram.

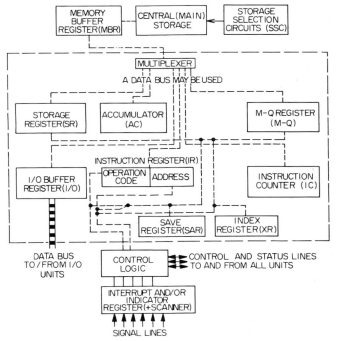

Fig. 17. Sample arithmetic and logic unit detail.

subsystems, while the vehicle is on the launch pad or in an experimental test cell. The latter category, the one which this section describes, includes all modern data-processing machines. The instructions comprising the program are stored in a magnetic storage unit, to which the processor's arithmetic and logical unit (see Figs. 16 and 17) have access at electronic speeds, in the range of 1 to 2 μsec in some currently available machines.

31 Computer Fundamentals and Classification

The performance of the functions provided by a data processor should be interpreted as the joint responsibility of the equipment comprising it together with the program controlling it. Indeed, these two segments of a data-processing system, the equipment and the program, are frequently referred to as "hardware" and "software," respectively. The subject of programming is treated separately as the next major section of this chapter. Further mention of programming is made in this section only to the extent necessary to clarify the equipment description which is provided.

A data processor (DP) consists of four principal components: the central processing unit (CPU), the central (internal or main) storage, the input-output control, and the input and output units. The CPU is comprised by two subdivisions of equipment, the arithmetic and logic unit (ALU) and the control logic (CU). The CPU is the "heart" of the DP. It consists of the ALU executing all arithmetic and logical operations prescribed by the program, and the CU which decodes all the instructions comprising the program. The CU translates them into a sequence of voltage pulses (short duration) and gates (relatively long duration). The pulse and gates assume one of two values. Similarly, all instructions, data, control signals, etc., are represented in the machine by two-valued, or binary, voltages. Other possibilities, such as ternary and decimal representation, exist but have received attention of primarily a theoretical nature.

There are two categories of input and output units, those which the computer uses for storage and those through which it communicates with its environment. The former category of equipment includes supplementary core storage units, magnetic drums, and magnetic tapes. The latter category consists of card printers and readers, paper-tape punches and readers, typewriters, TV displays, and switches and lights. Also included in this category could be the data-acquisition equipment, including multiplexers or commutators, analog-to-digital converters, etc., included in a telemetry data-handling configuration. Further, equipment by which the data processor could be connected to telegraph and other long-distance data links is of this nature. Some data processors have been designed to switch data links and to serve as a means of interconnecting such links with other data processors programmed to insert, process, and develop the data. These processors are called "message exchanges."

Several formats for computer control have been used by the various manufacturers. Since the scope of the section is, of necessity, limited, the material presented assumes that the processors described use "single-address" instructions. The majority of instructions consist of at least two parts, the "operation" part, which designates the elementary function prescribed, such as add, substract, shift left and right, compare, write and read (a tape, for example), plus the "address" or "operand" part. The operand portion of the instruction designates the location in core storage to or from which information is to be moved, which tape unit data are to be read or sent (and written on the tape in the unit), as well as other prescriptions of possible action which the machine might take, such as shifting the number in the accumulator register (see Fig. 17). Sometimes, an instruction contains information in addition to the operation and address portions. One or several bits (binary units of information) grouped into one or several "tags" may be used to modify the operation prescribed, in terms of associating it with an "index register," etc.

Some computers have instructions of various lengths, in terms of bits or of characters comprising them. In the case of arithmetic operations in a single-address machine, the locations of one of the operands as well as the result of the calculation are implicit. In the case of an addition, for example, the following sequence of instructions might be used:

Instruction number	Instruction location	Instructions	
		Operation	Location to which instruction refers
1	100	CLEAR and ADD	1000
2	101	ADD	200
3	102	STORE	500

Instruction 1, located at the position 100, in (fast-access) storage, causes the number at location 1000, also in central storage, to replace the number in the accumulator register (AC). Instruction 2 causes the number of location 200 to be added to the number in the AC, the result replacing the number loaded into the AC by the previous instruction. Instruction 3 causes the number in the AC to be placed in location 500 of the central storage, destroying what was there previously.

Several things of importance relating to the operation of the data processor can be noted through a brief examination of this table of three instructions. The *instructions* themselves are written in the fast storage in *identically* the same fashion as the data upon which they operate. The instructions are taken from sequential positions in storage. The "instruction counter" (IC), see Fig. 17, specifies the address of the instruction to be executed. The IC can both hold a given number and be incremented to the next successive integer. Thus it is both a "register" and a "counter," a device to hold a "number." The IC will advance from one address to another. It could be so incremented through each address in the core memory. Eventually the program contains either a HALT instruction which precludes this or a TRANSFER instruction which simply causes the next instruction to be taken from a position in core memory which is *not* the location at which the TRANSFER instruction is located, augmented by 1, as is normally the case. Note that both the ADD and the STORE instructions explicitly refer to a location in core memory and implicitly refer to the accumulator register. A little reflection on these instructions indicates that they could be modified (and indeed are in many programs) in exactly the same way as data.

The term "register" has already been mentioned on several occasions in this section and will be referred to again subsequently. "Registers" are employed in virtually every aspect of a data processor's operation. A register is a device which may receive, retain, and transfer information as directed by the machine's control circuits. The term register refers to both devices comprised of active bistable elements, called "flip-flops," as well as positions in core storage. Those comprised by active devices are often referred to as "live" registers.

Data processors are also described in terms of the way in which information is organized within them with respect to both operating upon data (and instructions) and the way in which data are stored. Machine operation and data storage may be done in either serial or parallel fashion. The distinction is particularly important with respect to the way the processor performs arithmetic. In a serial machine, numbers to be added are considered one (decimal or binary) position at a time. In a parallel machine, addition is performed on complete words. Essentially, the largest number which the machine can represent is that which can be contained in one word. A serial machine can potentially represent a number of any length, by simply using as many characters as necessary.

Another point of interest is the way in which numbers are represented or encoded in a machine. The three most common ways are binary, binary-coded-decimal (BCD), and 2-out-of-5. Consider a serial machine using 6-bit characters. If binary representation is used, 2^6 or 64 possibilities can be represented. If one bit is used for parity checking, 32 possible characters can be uniquely represented. If the 2-out-of-5 scheme is used then $C_2^5 = \dfrac{5!}{(5-2)!2!} = 10$ values can be represented. Machines representing numbers in this way are customarily called decimal machines. The "extra" bit in this case can be used for various control purposes, among which is an indication that two particular characters should be operated upon in a pairwise fashion as a single character. A character pair could be used to represent both alphabetic and numerical data, since it offers 10×10 or 100 possible representations in all. Characters of 6 bits can represent 64 possibilities if binary representation is used or 40 possibilities if binary-coded-decimal representation is employed. In this case, 4 of the 6 bits will be used to represent 10 possibilities, and the other 2 will be used as "zone bits," identifying how the other 4 bits should be interpreted. The IBM 704, 709, and 7090 employ words of 36 bits in length. Arithmetic is performed on these words as a whole. From the point of view of storage, however, these words are frequently used to represent 6 BCD characters. The magnetic tapes on these machines can be written in either binary or BCD format.

32. Binary Computer Operation

The remainder of the sections of this chapter devoted to data processors describe the operation of a binary, parallel (or word-oriented) computer. Although *single-address* instructions are assumed in the material, other formats are cited here to emphasize that alternatives do exist and in fact are implemented in some machines. A two-address instruction contains an operation part as well as one or two data addresses. In the one case, the address of the operand to which the operation is to be referred, together with the address, or location in storage where the next instruction is to be located, is represented by an instruction. An instruction having this format is often referred to as a "one-plus-one" instruction. Machines using this format differ in two important respects from the single-address system. No instruction counter is required, and no unconditional transfer instruction is necessary. This format is used by the IBM 650, among other machines. It is particularly suited to a machine of that type which employs a magnetic drum for central storage, as a drum is a "cyclic" storage device. By properly locating instructions on the drum, "minimum-access programming" can be employed. This means that a minimum of time will be lost waiting for the next instruction to come under the drum's reading heads.

The other type of two-address instruction specifies the locations of the two operands of an arithmetic instruction. They are the locations of the result of the operation designated by the operation portion of the instruction and the location of one of the operands. The location of the other operand is the processor's accumulator. A three-address format of an instruction designating an arithmetic operation would include an operation portion, as well as the addresses of both operands and the location where the result is to be placed. A machine, no longer manufactured, which uses this format is the "ELECOM 100."

33. Operation of the Arithmetic and Logic Unit

This section describes the operation of the arithmetic and logic unit of a specimen single-address, binary, synchronous computer. The term "synchronous" in this instance means that all operations of a computer take place at well-defined fixed intervals of time. The execution of a particular instruction is divided into one "instruction (I) cycle" and one or more "execution (E) cycles." Although different amounts of time are required for their execution, the length of the I and E cycles of which they are comprised are equal. In an "asynchronous" machine, the execution

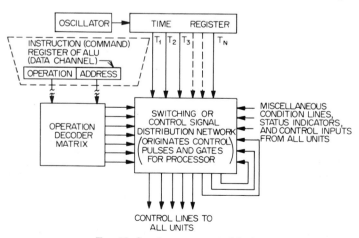

Fig. 18. Synchronous control logic.

of an instruction would proceed as fast as possible. Although this section deals with synchronous machine operation, a comparison between the control logic for both synchronous and asynchronous operation is made. Figures 18 and 19 show the organization for the synchronous and asynchronous approaches, respectively.

There are many variations in the ways in which a synchronous single-address stored-program data-processing machine can be organized. The machine described here is a hypothetical one, which incorporates the features of modern machines. Some of the more significant variations in the structure described are mentioned as appropriate.

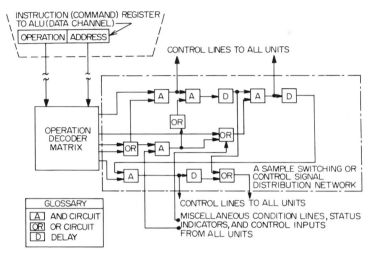

FIG. 19. Asynchronous control logic.

Figure 17 shows the ALU, with nine "live" registers, one of which, the instruction counter (IC), can be incremented as well as hold numbers. The other registers in the ALU are the memory buffer register (MBR); the storage register (SR); the accumulator (AC), which can add and perform a bit-by-bit comparison of two numbers; the multiplier-quotient (M-Q); the input-output (I-O) buffer register; the instruction register (IR) (subdivided into an operation and an address portion); the instruction counter (IC); one index register (XR); and one save (SAR) register. In some machines, one or several of these registers may be provided by using one or more locations in the central storage, which is (are) addressed automatically by the control logic when needed in the execution of an instruction.

Descriptions of Machine Cycles

Instruction (I) Cycle. I time begins when the IC transfers the address of the instruction to the storage-selection circuits (SSC), causing the instruction to be transferred to the IR via the MBR. The operation part of the instruction is placed in the operation part of the IR, from which it is directly routed to the operation decoder of the control logic. Similarly, the address portion of the instruction is placed in the address part of the IR. This tells the machine the location of what it is to work with, which is generally the location of an operand in storage for an arithmetic operation the address of an input or output device to or from which data are to be transferred. Some machines are constructed which use variations on this procedure. Variation is particularly evident with regard to the determination of the *actual* address of the operand upon which the computer acts. The most significant of these modifications to the "normal" procedure described here are *indexing* and *indirect addressing*. These items are considered in a subsequent section.

Execute (E) Cycles. Usually, I time (as described above) is followed by one or more E times or cycles. One exception to this would be the execution of a REWIND operation for a magnetic tape in certain computers such as the IBM 704. The instruction which accomplishes this objective might be written as REW A, in which REW designates the operation REWIND, while A, the address portion of the instruction, designates the particular tape.

In order to clarify the operations which are usually executed during the one or more E cycles of an instruction, we now describe the operation of the specimen machine considered here, with particular reference to Fig. 17. E time starts with the removal from storage of the operand whose address is currently in the address portion of the IR. Sometimes, a data movement of this kind (internal to the processor) is termed a "fetch" operation. This operation is accomplished by the address portion of the IR activating the storage-selection circuits (SSC) and the subsequent readout of the operand to the MBR, with its being immediately transferred either to the SR or to the AC.

The SR is used during multiplication for storage of the multiplicand, and during division for the storage of the divisor. The use of the SR during these operations obviates the need for repeated reference to storage during their course. The multiplexor is used considerably in these operations. It is essentially a switch which routes information transferred into the storage (via the MBR) and into each of the live registers in the ALU, under the direction of the control logic. Direct data transfers between some of the registers are also possible, such as between the AC and the M-Q. Alternative to the use of the multiplexor, a bus system could be employed, which would be nothing more than an "information highway" to which all the live registers could gain access, their connection to the bus being by switches or gates operated by the control logic. Such a bus could be used for transfer both to and from the MBR and the other registers, as well as between certain registers directly.

The SR is not logically required for the processing of data within the ALU. Strictly speaking, the MBR could serve just as well. However, the availability of the SR enables the MBR to be used to gain access to storage simultaneously with the execution of an arithmetic or logical instruction by the ALU. This capability is required if asynchronous input-output data operations are to be possible.

Continuing with the description of the operation of the ALU during arithmetic operations, we note that the two numbers involved are sent successively to the ALU where they are combined as prescribed. During a CLEAR and ADD instruction (described previously), the operand is routed from the MBR to the AC, just after the AC has been reset—made to hold a zero value. These operations might occur simultaneously, the contents of the MBR destroying those of the AC. Execution of the E cycles of the ADD instruction causes the contents of the MBR to be transferred to the SR, freeing the MBR for the reception of other words. The connections of the flip-flops which constitute the AC are altered so that the SR may be gated to the AC and the sum of the contents of those two registers developed in the AC.

Multiplication requires the execution of two instructions, plus a third (and possibly a fourth) to store the result in the desired location in central storage. The required sequence of instructions might be as indicated in the table:

Instruction number	Instruction	
	Operation	Address of operand
i	LOAD M-Q	200
$(i+1)$	MULTIPLY	170
$(i+2)$	STORE M-Q	205

The numbers i, $i+1$, $i+2$ (generalized values) are successively placed in the

DATA PROCESSORS 11–41

instruction counter. These numbers are the addresses in storage at which the instructions to which they correspond are stored.

The instruction LOAD M-Q 200 requires one I and one E cycle for completion. During I time, storage position i is accessed, and its contents (LOAD M-Q 200) are placed in the IR. The latter causes the contents of location 200 to be placed in the M-Q register. This quantity is the multiplier.

The instruction MULTIPLY 170, requires one I and a plurality of E cycles for its execution, depending upon the number of significant binary digits in the multiplier. The first E cycle causes the contents of location 170 to be placed in the SR and the AC to be reset. The quantity placed in the SR is the multiplicand. At this time, the multiplication process starts. It consists of repetitively adding the contents of the SR to the AC, and shifting to the right one position after each addition is performed. The product is thereby successively built up and is shifted into the M-Q register during its formulation. At the conclusion of the operation, the double-length product is available in the AC and M-Q registers, the high-order bits in the AC. If both portions of the product are to be kept, two data-movement instructions are required. One could be STORE 204, which would transfer the contents of the AC to location 204, and the other STORE M-Q 205, which would transfer the contents of the M-Q (the low-order bits of the product) to location 205.

In order to clarify the foregoing description, let us consider a numerical example:

$$
\begin{array}{lr}
\text{Multiplicand (in SR)} & -1111 \\
\text{Multiplier (initially in M-Q)} & -1101 \\
\text{Partial products:} & \\
\quad \text{PA} & 1111 \\
\quad \text{PB} & 0000 \\
\quad \text{PC} & 1111 \\
\quad \text{PD} & \underline{1111} \\
\quad \text{Product} & 11000011
\end{array}
$$

The contents of the SR, the AC, and the M-Q registers are tabulated below as they appear during successive intervals (E cycles and parts thereof) in the execution of the MULTIPLY instruction. Note how the right-hand—least significant—bit of the con-

Time interval	$C(SR)_i$	Contents of AC at time $i = C(AC)_i =$	$C(AC)_i$ Carry bits	$C(AC)_i$ Body	M-Q, $C(M\text{-}Q)_i$
0	1111	Cleared = $C(AC)_0$	00	0000	1101
1	1111	PA + $C(AC)$ = $C(AC)$	00	1111	1101
2	1111	$C(AC)_1$ shifted	00	0111	1110
3	1111	$C(AC)_2$ + PB	00	0111	1110
4	1111	$C(AC)_3$ shifted	00	0011	1111
5	1111	$C(AC)_4$ + PC	01	0010	0111
6	1111	$C(AC)_5$ shifted	00	1001	0111
7	1111	$C(AC)_6$ + PD	01	1000	0111
8	1111	$C(AC)_7$ shifted	00	1100	0011

tents of the M-Q, at any time interval i, $C(M\text{-}Q)_i$, controls the development of partial products which are either 0 or equal to the multiplicand $C(SR)_i$ according to whether the bit is a 0 or a 1. Further, note that two "carry bit" positions are provided in the AC. When a STORE X instruction is executed, these bits would not be transferred to location X. They are useful, however, as seen during the intermediate steps in the formation of the product, as well as after it is formed, should the value of the product exceed the capacity of the AC and the M-Q in combination.

Division is performed in the specimen machine as the reverse of multiplication.

The divider is loaded into the AC or the AC and M-Q in combination, if required. The divisor is placed in the SR. The quotient is developed in the M-Q and the remainder in the AC.

Some computers use programmed algorithms to do division and successive additions to perform multiplication. It is also possible to use a table look-up capability for these functions as well as for doing addition and subtraction. The IBM 1620 employs two "table areas" of its memory to which the machine refers automatically during arithmetic operations. The idea is to add successively the digits comprising two multiple-digit numbers. Such a system has an advantage over fixed-circuit adders, for example, if the computer is to be used for arithmetic of different radices, from time to time. For example, a machine performing calculations in the familiar decimal system might be used for octal (radix or base 8) arithmetic. In this type of scheme, 7 is the greatest-order digit available for each order (position). The number 8_{10} (eight in the decimal system), would be written as 10_8 in the octal, or "base (radix) eight" system.

All arithmetic operations can be reduced to sequences of additions or subtractions. Division can be performed by successive subtraction of the divisor from the dividend. Subtraction and addition are complementary functions, which can be exchanged for each other by taking the complement of the result of either.

34 Control-logic Operation

We have examined the ALU portion of the central processing unit. The various operations of which it is capable are initiated by the instructions stored in the central storage, as has been described. These instructions control the operation of the input and output equipment as well. As was indicated in the description of the performance of arithmetic operations, the execution of the instructions required can be subdivided into many suboperations. Major suboperations are the I and E cycles for each instruction. Many logical functions, such as the movement of information and the clearing of a register, must be performed during each cycle. The control logic directs the performance of each such suboperation.

A CL can be synchronous or asynchronous. This discussion devotes principal attention to the synchronous model. Reference should be made to Fig. 18. The principal components of a synchronous control-logic unit are the operation-decoder matrix, the switching or distribution network, and the oscillator and time register. The IR is gated to the matrix, which has an output wire for each possible instruction. Each output wire is gated to the switching network for interpretation. In addition to the operation-selection lines from the decoder matrix, the network has other inputs available to it. They include status lines from the various units, registers and entire devices controlled by the processor, as well as various switches controllable by the system operator. Included in this category of inputs would be leads from counters which monitor the flow of data between the processor and input and output equipment. A time register, incremented by an oscillator, sequentially activates one line after another, corresponding to the passage of time intervals of prescribed length, into which the I and E cycles may be subdivided. In addition to the components shown in the figure, a counter is required to control the number of E cycles of instructions executed. In one implementation, this counter is loaded at the beginning of the execution period of the instruction with the number of E cycles required for the prosecution of that instruction. The counter "counts down" to 0. When 0 is reached, the execution of the instruction is terminated. The control logic transforms the occurrence of the number representing an order given to the computer (instruction) into the activation of a sequence of control lines connected to the units of the data processor and its peripheral equipment. The sequence of and the particular combination of lines activated may vary, depending upon the status of the various condition lines.

It is of interest to note that as a programmer transforms an arithmetic task and/or logical problem into a sequence of instructions for the computer to execute, so each instruction is transformed by the control logic into a sequence of micro-operations for the various flip-flops, counters, and other devices constituting the machine to perform. The logical equivalence of these two stages in the processor's performance of a task has influenced the design of some processors. Machines have been designed which allow variation of the connections of their switching and distribution networks. In some cases, the program itself can gain access to the network, even during the execution of a program. Such operations are within the category of "microprogramming." In other instances, the network is such that some or all of its connections can be altered externally, in a very simple manner.

Switching networks are generally composed of various logic elements of the same type used for the other parts of the calculation. They can be composed, in part, of arrangements of bits stored in a core storage unit especially constructed for the purpose. A microprogrammed computer might be implemented with such a unit.

The IC (Instruction Counter) normally advances incrementally, from one number i to its successor $(i + 1)$. Each number corresponds to the address in central storage of the instruction which is to be executed at that time. At various points in the course of a program, the next instruction will not be the successor to the one currently being executed. At such points, the program is said to "jump" or transfer to another location. Thus a portion of the sequence of numbers in the IC might be " . . . 100, 101, 102, 605, 606 . . . ." Note the break in the sequence, and the transfer from location 102 to 605. There are four principal types of program transfers. They are shown below, classified by the type of signal to which they respond, and the logic of their response mechanism:

In response to	Conditional	Unconditional
Programmed signals.......	1	3
Interrupt signals..........	2	4

A transfer can be initiated by the execution of an unconditional transfer instruction, which branches the program (such as from location 102 to 605), as shown previously, automatically upon its execution. Alternatively, it can be effected upon the execution of a conditional transfer instruction if certain requirements specified in the instruction are satisfied. Conditional branches are generally based upon the result of a test of the sign of the number in the accumulator, the availability of a certain input or output permit for use, etc. A sample instruction of this type is TRANSFER IF POSITIVE, X, stored at location i in central storage. In this case, if the contents of the accumulator is a positive number, the program will branch to location X to obtain its next instruction, rather than taking it sequentially from location $i + 1$. An example of an unconditional branch, located at some position j is TRANSFER TO X. The execution of this instruction causes the IC to be loaded with the number X, and the instruction located at that position to be executed. Either data or an instruction can be at any storage location. Whenever the IC contains the address of that location, its contents will be interpreted to be an instruction and will be executed if possible. Particular attention is made to their use in the operation of telemetry data-acquisition equipment.

The other major type of program transfer is that caused by the occurrence of an event, within or without the data processor, which need not to be tested for by the execution of an instruction, as is the case with programmed branching. This capability, not found in all computers, is called "interrupt" and is sometimes referred to as "trapping" or other similar names. The occurrence of a specified "event," as indicated by the activation of a particular line to the data processor's control logic, causes the program to branch to an instruction, not in the numerical sequence of the program. Some machines can respond to several such events by branching to a location corresponding to a particular instruction. This capability can be valuable, since it frees the (main) program from periodically testing various indications with transfer instructions to determine whether specified actions have been accomplished. An application of this is in the use of an analog-to-digital converter (ADC) associated with the conversion of FM telemetry signals to digital data ingestible by the computer. The ADC in a telemetry data-handling system might be connected to an interrupt line which it would activate upon the completion of a conversion.

The process of interrupt is logically equivalent to the execution of a transfer instruction, with the exception of the fact that interrupt occurs at an unscheduled time while the occurrence of a transfer is predetermined. Some processors are provided with an instruction which precludes the occurrence of interrupt or locks it out. A comparison instruction would be available in such a machine to reactivate the interrupt mechanism when desired. When a program transfer, particularly a conditional one, occurs, provision must be made for the return of the progression of the program in the sequence from which it branched (was diverted). At the minimum, this means saving the value of the instruction counter (or its value +1) at the time at which the transfer was effected. Frequently, other items in addition must be preserved at this time, relating to the status of the data processor. A list of such items might include the status of the overflow, or (extra) carry bit(s) in the accumulator register, the status of interrupt lines (if there are any), etc. If the transfer is executed under the aegis of one or a series of instructions, the need to save these items of system-status data can be met by including appropriate instructions to place these values in the storage. If the transfer occurs because of one interrupt, that is, without the "knowledge of the program," such status information could be lost, since it may not be possible to preserve the "return address" (the last value of the IC or that value augmented by 1), etc., by programming. Consequently, it must be done automatically. As shown in Fig. 17, a special save register could be used for this purpose. At the occurrence of an interrupt, the contents of the IC (possibly plus 1) would be loaded into that register. Further, the status of various machine indications, as described above, would be loaded into it. Alternative to the use of a "live register," one or more locations in storage could be used for this purpose. As with the use of a "live" register, all the transfers of information and concomitant exercise of control lines would be directed by the control logic.

Attendant to the use of an interrupt scheme as outlined here, care must be exercised by proper programming so that the "return addresses," the machine status, etc., are not lost through the occurrence of an interrupt or an interrupt-originated program execution.

35 More Sophisticated ALU Operations

The arithmetic and logical operations of the data processor, described above, can be incorporated into a machine having greater flexibility if some additional capabilities are provided. Two of these are indexing and indirect addressing. Both of them are means by which some relatively sophisticated processing can be accomplished with relatively simple programming. They both refer to ways in which the actual address of the operand of an instruction may be modified to be different from that written in the instruction.

First, let us consider indexing. A computer may have one or more index registers [see the index register (XR) in Fig. 17]. An index register is a register which can hold a number which can be incremented (or decremented in some cases). There is a considerable range in the number of XR's available in today's data processors, from none to 100 or so. The XR's may be simulated by programmed reference to prescribed locations in storage for instruction modification as opposed to employing a "live" register for each, such as that shown in Fig. 17. Index registers are used for a variety of purposes, the chief among them being a way to vary the address to which an instruction applies, either without rewriting the program or by providing a multitude of contingent branches to various "copies" of the same instructions or group of instructions which differ only in with regard to the addresses of their operands. Another principal application of an index register is in counting. An example of this use is to tally the number of times a particular sequence of instructions, called a "subroutine," is executed.

When an instruction is indexed, the address of its operand is not used directly. Instead, an "effective address" is computed by subtracting the value of the number in the XR from the one in the address (operand) portion of the instruction. This operation may be modified somewhat by requiring that the number in the XR be decremented each time it is accessed. This capability can be used to advantage in the case in which a sequence of instructions is repeated. Each time the indexed instruction in the sequence is executed, for example, its "effective" address is augmented. A conditional transfer instruction applied to the index register (XR) can be used to determine whether the number in the XR is equal to a specified number located in a prescribed location in storage. An instruction of this type located at position j would be TRANSFER TO X IF XR = 5 and would be written as TRANS X, XR, 5. Thus, if the number in the index register were numerically equal to 5, the program would take its next instruction from location X. If this condition were not met, however, the next instruction, at position $j + 1$, would be executed.

Indirect addressing is another significant capability, available in most of the medium- and large-sized data processors. When an instruction is indirectly addressed, its address portion refers to the location in storage in which (the number equal to) the address of the operand is located.

36 Input-Output Control

Common data processors are constructed in such a way that their ALU's can perform only one operation at a time. Parallel and other multiprocessors have been constructed, but their use is not yet widespread. When the sequential modus operandi is extended to include input and output operations, the effective speed of the ALU that would otherwise be possible without active I-O, can be considerably reduced from that value. Many machines, including relatively large-scale ones like the IBM 704, allow the execution of only one operation at a time. That is, the transfer of information between the storage and an input or output unit, such as a magnetic tape, cannot take place simultaneously with the execution of an instruction in the arithmetic and logic unit. Because the speeds of the processor storage and most I-O units do not match, it is necessary to interpose a unit, called a buffer, which is nothing more than a memory unit, between them. The IBM 704 uses its M-Q register for this purpose. Its M-Q register serves the arithmetic functions ascribed to the M-Q register of the specimen machine shown in Fig. 17, as well as those of the I-O register shown. If data are to be transferred into the storage of an IBM 704 from a magnetic tape attached to it, a sequence of instructions, such as indicated below, is required:

1. READ TAPE X
2. COPY M
3. COPY M + 1

The execution of the READ instruction selects magnetic-tape unit X, readies its circuits to read the information on the tape, and starts the tape moving at the proper speed. After a delay of a certain amount of time, sufficient for these things to be accomplished, the COPY instructions can be executed, also with the proper amount of delay between them. The execution of instruction number 2 causes the group of characters passing under the tape read heads at that moment to be transferred to location M. The same thing is true with regard to position (M + 1), and so on, under the control of instruction 2, 3, . . . and so on. In an actual situation, only one COPY instruction would be written. It could be modified through the use of index operations to advance the "effective" address of the operand to successive storage locations.

There are two disadvantages with the conduction of I-O operations in a data processor of type described: an instruction must be executed for the transfer of each word of data, not allowing the processor to do other words; and the instructions must be scheduled to occur at exactly the right moments. If a COPY instruction is given too early, either the wrong data will be transmitted or the processor will be "hung up" (stopped). If the instruction is given too late, on the other hand, the data desired will be lost. Through judicious programming, the intervals between the executions of COPY instructions conceivably could be used for calculations, but only in a relatively inefficient manner, if at all, in most realistic cases.

There is another form of operating input and output equipment which can obviate these difficulties. The processor can be made to operate asynchronously with respect to the I-O equipment, as opposed to synchronously with it, as in the case of processors organized as described above. Many medium- and large-capacity processors, such as the IBM 7090, are organized so that the actual processing of data in the arithmetic and logical unit can occur simultaneously with the transfer of data between the central storage unit and the I-O units, after the operational sequence has been initiated. The equipment to provide this capability includes the units (or ones similar) of the type shown in Fig. 20, which may be called a data channel (DC). In addition, an "I-O

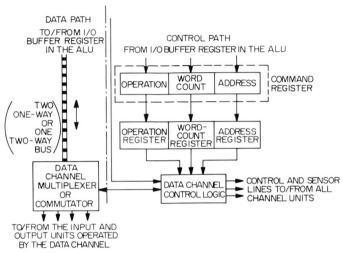

Fig. 20. Data channel (schematic diagram).

buffer register," shown in Fig. 17, is provided. The ALU and the DC are operated under the control of two different programs, although both of them are stored in the central storage. The operation of the DC is started by an instruction in the main program. Once started, however, the DC operates *independently of the main program* until it has exhausted its set of instructions, at which point it interrupts the main or processing program to alert it to this fact. In order to distinguish between processor and data-channel programs, the instructions of the latter are frequently referred to as "commands."

37 Input-Output Data-channel Operation

Figure 20 shows a simplified data-channel logic, similar in many respects to the central processing unit of the data processor itself. A data channel is incorporated in most large processors to facilitate I-O operations. The principal units of the data channel are the command register, the operation register, the word-count register, the address register, and the data-channel control logic. The DC control logic directs the DC, in a manner analogous to the way in which the data processor's control logic directs its operation. The DC control logic interprets each "command" held in the command register. Under its control, the CPU and the DC proceed independently. It is possible that the execution of a command or an instruction can be held up very briefly, only if both the DC and the CPU request the use of central storage at the same instant. Otherwise, the main program proceeds *as if there were no input-output operations under way.*

Input-output operations commence with the selection of a particular input or output device with a READ or a WRITE instruction, as described in connection with input-output operations in a machine like the IBM 704. From then on, however, the two procedures differ radically. The next set of commands for the data channel are fed to it as needed. In the 7090, these commands specify various operations with the I-O unit selected by the main program instructions, with respect to the transfer of words and/or groups of words called records, between storage and the device. The operations are conducted on the basis of a stated number of words to be transferred, entered in the word-count portion of the command, or in terms of the transfer of a complete record without regard to the number of words to be transferred. Also, some commands control the skipping over of a designated number of words by not allowing their transfer to or from storage. The address portion of the command designates the first address of the block of locations in storage, into or from which the transfer of data is to take place.

38 Input-Output Data-channel Control of a Data-acquisition Subsystem

A data channel can be applied to the control of a commutator and analog-to-digital converter in a telemetry data-reduction system controlled by a data processor. A hypothetical arrangement by which this might be done is specified below. Each successive word transferred to storage in this case would be the value of the signal in successive channels digitized by an analog-to-digital converter (ADC) provided, as shown in dashed lines in Fig. 20. The word count would correspond to the number of channels and the address part to the locations in storage to which the data would be sent. The use of one command in this way would enable all the channels served by the commutator to be served.

The commands described could also be used to avail nonsequential commutation equivalent to a plugboard used in many data-acquisition systems. Their use in this connection is illustrated below. The commands of the 7090 are used. They are:

      Command mnemonic*   *Function ("I-O" = "I-O operation")*
      1. IOCP   I-O, count, and proceed
      2. IOCPN   I-O, count, and proceed, but no data transferred
      3. IORP   I-O, record control, and proceed
      4. IOCD   I-O, count, and disconnect
* Abbreviation.

The format of the commands is | OP CODE | WORD COUNT | ADDRESS |. Commands are preceded by a SELECT instruction of the form | READC X |. The address X could be used to refer to the first channel, to which the READ C or READ COMMUTATOR instruction would refer. These instructions and commands are applied to a sample situation below:

Problem
a. Read channels 10, 11, . . . , 20 into storage positions 1000, 1001, . . . , 1010
b. Read channels 2, 3, 4, 5, . . . , 8, 9 into storage positions 600, 601, . . . , 607
c. Skip channels 10, 11, . . . , 20
d. Read channels 21, . . . , 30 into storage positions 650, . . . , 659

Instructions and Command Format

Number		
1	READ C 10	. . . .
2	IOCD 11	1000
3	READ C 2	. . . .
4	IOCP 8	600
5	IOCP N 11	. . . .
6	IOCD 10	650

Upon the completion of each IOCD command, the processor would be signaled, indicating the completion of the channel program. The processor would respond with the next instruction if there were one, such as number 3 in the example.

This section describes only one of many ways in which a data processor might accept telemetry data from a previously developed magnetic tape or in "real time" as the event which originated the data takes place. It demonstrates the concept of a "stored-program plugboard."

39 Programming

A good understanding of programming is essential to one's comprehension of the operation of modern stored-program automatic test and telemetry data-gathering systems, as well as of telemetry data-reduction systems. Very often the choice of computer is governed by conveniences in programming which reduce time to write programs. This is an important trade-off versus computer running time. The nature of the mathematical techniques involved in telemetry data reduction is shown in Chap. 12. Knowledge of the methodology of programming is of course a most desirable asset to anyone faced with the need to apply the equipment discussed in this chapter to telemetry problems. The subject of programming methods is too broad to be adequately treated in this handbook, however. The reader is directed to the numerous publications now available on the subject. In particular he is directed to Refs. 33, 34, 35, 36, 37, and 43

DATA PRESENTATION

40 Data Presentation in Telemetry

The balance of this chapter considers both the types of data displayed and the means for presenting it to the telemetry-data-system user. Information associated with the operation of a telemetry system may be presented in alphanumeric, graphical, or pictorial form. Meters, light-indicating plotters, strip charts, TV viewers, strip printers, output typewriters, and plotters are used to this end. The information may be presented concurrent with or subsequent to the occurrence of the activity (e.g., static rocket test) from which it is derived. The former class includes "quick-look" data, while the later includes the results of processing the telemetry data in a data processor. Frequently, such processing is done on an off-line basis, that is, after the test or run has been completed. Thus data-presentation equipment is used to indicate the telemetry data in both their raw and processed form. Also, this equipment is used in connection with operation of the telemetry data-handling equipment itself. Operational use of telemetry data includes monitoring the system's operations, indicating errors in its operation, and aiding in the servicing of improperly operating devices which are part of it. The equipment used to perform these functions is also used in the preoperational or system-setup phase. For example, the data processor's card punch or paper-tape punch is used in conjunction with various console units, switches, and indicator lights, in the debugging of the data processor's program. The "hard copy" outputs (e.g., the actual cards or tape) of the data-processing functions are frequently interpreted to aid the user. "Interpreted" means that the alphanumeric or special symbols to which given code characters in the copy correspond are printed adjacent to them for easy comprehension by the operator. He checks the cards or tape by reading the interpreted code.

Data-presentation equipment includes that operated by both the data-acquisition and the data-processor subsystems of a telemetry data-handling installation. A data processor operates standard computer output devices, such as paper tape or card punches, line or strip printers, output typewriters, and lights on its operating console. The coverage on data-presentation equipment of this chapter excludes such devices as magnetic-tape handlers which do not present information in a manner readily comprehensible to a human being from the category of data-presentation equipment. Magnetic tape is covered elsewhere in this chapter, however, in connection with its use as a storage device.

Certain devices such as paper-tape punches useful for telemetry-data presentation are employed with a data processor in other types of applications, such as payroll and inventory maintenance. When used in a telemetry system, the data processor may operate output devices such as plotters and graphical displays, in addition to those needed in the "standard applications" of data processors. Some standard output devices such as line printers may be used to present graphical or pictorial information. For example, the line printer could be used to make the equivalent of a two-tone picture by printing I's corresponding to dark points and O's corresponding to the lighter ones. Other pairs of characters, such as the letters X and O, could be employed alternatively for this function.

Certain telemetry data-handling systems include data processors operated on line, interposed between the data-acquisition equipment and the display equipment. The data processor accepts the run or preprocessed data, operates upon it, and converts it, with respect to form and content, for presentation to the personnel concerned. As an example, photographs from the Tiros weather satellite can be converted to a Mercator projection and then printed by the computer.

The significance of an on-line data processor to the display function in a telemetry system can be seen in the contrast between two types of remotely monitored gas-pipeline installations. In the one, various data, including intake and discharge pressures at the several pumping stations along the right of way, are transmitted to a central dispatching center and displayed there on meters mounted on wall panels in the system control room. Also, the values may be logged (printed) on an output typewriter provided for this purpose. The telemetry data-handling equipment, in this case, can be classified as a data-acquisition system, which has some preprocessing capability. A data processor may be added to the dispatching-office configuration. It may operate either off or on line. If on line, it will be directly connected to the data-acquisition, which may function as described above, directly controlling various display devices. There is an alternative on-line configuration toward which the current trend in the use of data-presentation equipment is directed. That trend is the one in which the processor operates all displays including ones which present run or preprocessed data. The computers would be programmed for the presentation of that level of data as well as for the processing and presentation of data in the (relatively) more complete form. The computer might operate the "system" types of displays, such as digital meters, which are used in telemetry systems which do not include a computer. The computer could use a different type of display unit, employing a television tube, in lieu of some or all of the meters. In this case, the numerical values otherwise shown on a meter would be displayed or printed on the screen of the TV view tube. Such an approach might provide a less expensive display system than the alternative of a multiplicity of meters. Thereby it could at least partially compensate for the cost of the computer. The TV view tube used for the display of alphanumerical information could present graphical data as well. Thus it could be used in place of a plotter.

Large visual displays, capable of presenting both alphanumerical and pictorial information, have been built using solid-state materials. Each such display would be made of numerous little cells, or resolution elements, each of which can produce illumination over a range of gray levels. The TV tube and the solid-state displays are illustrative of the sophistication possible in telemetry-data presentation through the use of a data processor. For additional information about pipeline and similar industrial telemetry systems, the reader should consult Chap. 13.

Data-input and -presentation processes are intimately associated in many instances. For example, the same typewriter can be used for both display of data to the operator and his entry of additional data or commands into the system. Another display system, in which certain classes of input and output are even more intimately associated, is that used in several military data systems. An image is displayed to an operator on a TV view screen. He can point to a character displayed or a blank area of the screen with a wand and therewith designate the position of a character or a symbol to be added or deleted from the display.

41 Data-presentation Devices

This section treats the more important data-presentation devices used in telemetry data-handling systems.

Paper-tape Punches

Paper-tape punches are used as output devices for a number of modern data processors, among them the IBM 1620. Information is recorded on a tape by mechanically punching holes in it. A hole can stand for a binary 1 and the lack of a hole for a binary 0. The converse is possible also. Tape punches are available which punch 5, 6, 7, and 8 channels (characters) on tapes of approximate width from 0.7 to 1 in. Each character is punched simultaneously at one lineal position along the tape. In addition to the punches in the information channels, holes are prepunched at regular intervals, linearly along the middle of the tape. They are the feed holes which are engaged by the sprocket (s) which move (s) the tape. Typically, wood-sulfide paper of good quality having a thickness on the order of 0.004 in. is used.

The tape may be punched with or without chad. With chadless punching, the punched chip of paper remains attached to the body of the tape. Tape may or may not be punched and interpreted simultaneously. The characters punched on market ticker tape are interpreted adjacent to the hole punched.

Paper-tape punches have been used for a long time in teletype service. Some teletypewriter exchanges employ the physical movement of a tape from a punch operated by a receiving line to a punch activating the transmitting line which terminates either in the next exchange along the route or at the destination of the message itself.

When specifying the use of either paper tape or cards, one must critically determine the range of humidity anticipated for the location of the punch or reader unit. Ideally, of course, the paper would not be hydroscopic to any degree. This is not possible, practically, unfortunately. The question of moisture absorption is more critical in the case of punched cards, since the knife-thin blades that move the cards are critically adjusted for use with cards having a narrow thickness tolerance.

There are two principal types of punching mechanisms. One employs solenoids to drive the punches directly. The other uses a piece executing either a reciprocating or a rotary motion which drives a particular set of punches which are otherwise positioned to be engaged with this mechanism. Paper-tape punches are available which operate at rates up to several hundred characters per second.

Printers

The several types of mechanical printers are line printers, strip printers, and typewriters. Also, there are several types of electronic printers. One uses an image tube and microfilm. The film can be developed and photographic prints made from it, thus providing a hard-copy output. A variation of this method employs an image tube which can be used in conjunction with any one of several *direct*-printing means, including electrostatic. Several types of special paper are available which are sensitive to all or part of the optical spectrum. When exposed, this paper produces a fixed image. The image may be fixed through self-induced, thermally induced, or other chemical reactions. Direct-printing electronic units suffer the disadvantage of being able to produce only one copy at a time. The hard copy may be reproduced by other means.

A line printer can produce an entire line during one of its cycles. There is a printing device at each character position on the line in this type of printer. Printing can be effected by a type wheel, an endless-type "chain," a rigid type bar, or a set of wires. In the first case, each of the type wheels is rotated to the position corresponding to the character to be printed at its position in the line. After the wheels have been so "set up," either they all are simultaneously moved and pressed against the paper, or the paper is moved against them.

DATA PRESENTATION 11–51

The "chain" printer operates in a similar fashion. It differs from the wheel printer described in that the symbols are mounted sequentially on a flexible tape or chain. The tape or chain is advanced to the proper position.

Bar printers use a rigid bar to mount the symbols which they print. A particular symbol is selected for printing by properly moving the bar up or down. Once in position, either the bar or the paper must be moved so that physical contact can be made between them. Single-character-at-a-time bar printers can be constructed also. They are similar to a typewriter in which only one character is printed at a time. In this type of bar printer, the bar moves laterally to the proper character position on the page.

A strip printer is one which prints characters sequentially along a paper tape. A stock-ticker or a teleprinter used for telegrams which consist of printed strips pasted onto a block are of this type.

There is another type of printer which is considered neither a line nor a strip printer. It uses a strip of paper wide enough to print several characters. A stenotyper operates in this way. Some industrial telemetry and control ("process-control") systems use these devices as annunciators which indicate and record in hard copy indications of an undesirable condition or of variations out of preset margins of some of the variables monitored by the system.

Typewriters are widely used as the output units of data loggers and data processors. The two principal mechanizations of the unit are the "standard" and the "ball" unit. In the first case, the proper type of bar is selected and pressed against the paper at the center of the printing bar of the machine. The platen moves laterally, right to left, across the machine successively, one character position at a time, until the entire line, or whatever portion of it is desired, has been printed. The "ball" typewriter, such as the IBM Selectric, employs a stationary platen. All the type symbols are on a ball, approximately the size of a golf ball. The ball is rotated on two of its axes about its center in order to position the symbol selected at a particular time to the orientation proper for pressing it against the paper and printing the symbol. The ball is moved laterally, left to right, across the page or the line, printing each character on the line successively. The effect is similar to that of the bar printer which uses only one bar, described above.

The wire printer uses a set of wires, arrayed within a rectangle, to develop a character. One unit uses a 5 by 7 array of wires. The proper wires are pressed against the paper (after having been inked) to form a character composed of dots; each dot is the imprint of one wire.

Punched-card Output

Card punches are the most extensively employed output devices used by data processors. Paper cards have been used for many years as the information-bearing means in electromechanical accounting machines. Their use in electronic data-processing installations is a natural evolution of this original, and still widely used, application. The use of cards in connection with the operation of electronic data-processing equipment makes that equipment compatible with electromechanical accounting equipment. Also, because the card is a unique physical entity, the information on it can be ordered in a variety of sequences with that or other cards. Generally, the card is considered by a data processor to be one *record* of information. A group of cards is correspondingly called a *file*.

Meters

Meters may be classified as analog or digital, in accordance with the manner in which they represent a measurement. A meter which uses a pointer to signify the value of the measurement on a scale, is an "analog" meter. A digital meter, on the other hand, is one which indicates the value of the measurement as a number. Common examples of the digital meter are automobile odometers and gas meters. Both are essentially counters. Most automobile speedometers are analog meters, on the other hand. Digital meters hold an advantage over analog meters in that they provide an exact value of the measurement, to the degree of accuracy possible. Conversely, obtaining an accurate reading from an analog meter typically requires interpolation between marks on the scale. Parallax error may be introduced at this point as well.

When included in a digital data-handling system, digital meters are particularly attractive. If a computer is to provide an output on a meter, a minimum of equipment will be required to translate the data as represented in the computer to a numerical display on a meter. The material in the section on Symbol-display Techniques provides more information on various means for presenting symbols, including both categories of meters as well as other mechanisms.

Plotters

Plotters are widely employed to display relationships between two variables. Use of these devices in telemetry data-handling systems has been extensive. Generally, plotter output in this case indicates the succession of values of a particular measurement, with respect to time. A substantial portion of the older telemetry data-handling systems employ manual techniques for data processing. The personnel responsible use the outputs of plotters to obtain these data. The more modern installations employ machine techniques to do the majority of the data processing. Even in these configurations, plotters are used fairly extensively for "quick-look" purposes to reveal trends in the data being gathered. Thus, in the older telemetry systems, the plotter outputs frequently constitute the record of the test or other activity monitored by the telemetry system. In the more modern installations, however, permanent records are retained in *machine-readable* form on magnetic paper tape, cards, or film. Plotters are the chief output devices for analog computers.

The present material describes (electromechanical) plotters, per se, as distinguished from other less specialized data-presentation equipment such as TV viewers and oscilloscopes which can be used for this purpose.

Plotters may be classified as analog or digital, in accordance with the means by which their (writing) styli are positioned during writing. In the analog field particularly, many forms of plotters are available.

A plotter may be used to display more than one graph. Typically, each graph represents the values of a measurand or some function of it at successive points in time. The paper on which the graph is drawn will advance at a steady rate, proportional to the time variable. The proportionality or scale factor will depend strongly upon the resolution required. The faster the paper moves, the greater the resolution possible. One measurand or function of it may be plotted with respect to another one, used in place of time as an independent variable. This can be difficult unless the "pseudo-independent" variable is nondecreasing, as is time. If not, the paper will have to be moved backward and forward. This is not possible with all plotters.

Plotters may be classified according to whether their paper feed is linear or rotary, in addition to the distinction between analog and digital writing means cited. Plotters used in aerospace and other applications to which this book is addressed use linear paper feed principally. However, certain plotters, extensively employed in industrial applications, employ rotary feed. The variations in the measurand are recorded on a disk of paper which rotates about a fixed point at an angular rate proportional to the passage of time.

Both analog or digital function values may be indicated as a sequence of discrete points or a continuous curve through them. In the discrete case, a symbol stamp prints one of a set of symbols on the plotter paper at each time point at a position corresponding to the value of the function at that point. Using this technique, several graphs may be plotted on the same paper simultaneously without loss of clarity. Each graph would consist of a sequence of the same symbol, such as □, X, or O. In this case, the same color of ink may be used.

A continuous-curve plot uses a pen which is in contact with the paper continuously. A plotter can have two or more pens so that two or more graphs may be drawn simulta-

DATA PRESENTATION   11-53

neously on the paper. To avoid ambiguities and confusion in using these graphs, each pen should use ink of a different color.

An output printer may be used by a computer in lieu of a digital plotter for certain applications. The printer's writing means, whether it be a bar, a wheel, or other device, is used in a manner equivalent to a symbol stamp. An output printer can be used, more generally, to display pictorial information as well.

There are several commonly employed means for positioning the plotter's recording head with respect to the paper. The means selected for a particular application is very significant in determining the frequency response of the device. Head control is accomplished by either a combination of paper movement and one-dimensional head movement or by two-dimensional head movement. The former method is the more common. However, there may be no actual head movement. Effective head movement can be achieved through the use of several fixed heads. The effect of their use is to produce a quantized curve.

Recorders based upon the principle of a moving-coil galvanometer are widely employed. These devices use styli which can be moved in one dimension. The stylus is part of a member which is responsive to varying voltages. It employs the galvanometer principle of operation.

42 Symbol-display Techniques

Symbol displays present information for immediate consumption only. This distinguishes them from plotters and printers which generate outputs in a form suitable for storage as well as for immediate use. Techniques for symbol presentation important to telemetry-system designers and users are selectively positioned symbols, including rotary-positioned devices (e.g., an odometer, tapes, and disks); selectively illuminated symbol systems including those in which one of a complete alphabet of symbols is selected, as well as those in which a selected symbol is formed (e.g., dot-formed characters); image-formation tubes; and solid-state image-formation units.

Selectively positioned displays include meters the pointers of which are directed by d'Arsonval movements, synchros, etc. Also included in this category are digital counter wheels like those used in odometers. Endless tapes, on which are printed the symbols selectable, are also used. They may be direct-reading or may be used in projection systems by shining a lamp upon a selected symbol, represented in a transparency on the tape. Disks on the periphery of which are printed the selectable symbols may be similarly used for direct-viewing and projection systems.

Selectively illuminated symbol systems include the projection-type instrumentation described as well as certain other techniques. Included are displays consisting of a sandwich of clear plastic sheets, each one of which has one symbol, which upon selection is edge-illuminated. Electronic tubes, which have grids, each one shaped in the form of a symbol, are also used. To select a particular symbol, the grid corresponding to it is activated. Systems in which the character is formed upon selection are also used. All alphabetic and numeric, as well as certain special symbols, can be formed by selectively illuminating combinations of lights in a planar array of 35 (5×7) lamps. Arrays having other width-to-height proportionality factors such as 6×8 have been used. Arrays of this nature are frequently used to display the contents of register in digital data-handling equipment. Simpler ones, such as N lamps for N bit positions, are used to display the contents of the principal computer registers to an operator. A particular lamp's being on indicates that there is a 1 bit in the register position to which it corresponds.

43 Image-formation Displays

There are two types of image formation or TV tube display systems. The one uses a standard tube the beam of which is deflected in such a way that a selected symbol is displayed on the face of the tube. A multiplicity of such symbols can be maintained on the view screen simultaneously, as far as the human receiver can detect. Circuitry in the display system deflects the beam in the proper sequence of positions to form each symbol presented to the observer. The material used for this target, on the face of the tube, must be chosen to provide the proper degree of persistence of the image. Tubes have been made which can provide image persistence of several hours. One tube, used to display radar data, retains an image until hot air is applied to its face to "erase" it.

The other principal type of image tube is one specially constructed for character displays which employ optical and electro-optical means to shape the beam to produce the image desired. These tubes use stencils in which each symbol of the alphabet it is possible to display are cut out. Images can be formed in two ways. In the one, the stencil is put inside the tube and completely illuminated by the beam. The electronic image is then positioned so that only the selected symbol can pass through an aperture in the neck of the tube from which it is deflected and positioned to the desired location on the face of the tube, using conventional techniques. The other type of tube uses a stencil external to the tube, in connection with a photoemissive cathode at the end of the tube. A lamp illuminates the stencil, casting the image upon the photoemissive cathode. The cathode generates an electronic image of the stencil, from which a given symbol is selected and its image positioned on the face of the tube in the manner previously described for the other tube.

Solid-state image-formation units have been constructed using photoemissive materials. Displays using this technique can be large in area and can display any type of image, including pictures and symbols. The unit of resolution of this type of display depends upon the size of the individual cells from which the display is constituted. Each cell responds to a voltage applied to it with an illumination having a gray scale having more than two values.

Large displays are used in various industrial telemetry and control systems, such as in pipeline control systems. The dispatching center of a pipeline typically has a display with a stylized map of the system, indicating principal paths along which oil is pumped. Data about the system, such as the pressures at the pumping stations, the positions of main valves, and other such information, from a knowledge of which the status of the system can be determined, are indicated in meters, lights mounted on the display. Data are shown in the context of their physical relationships among each other in this type of display.

44 Data-presentation-equipment Operation

The purpose of this section is to draw the reader's attention to the functions of control, sequencing, and buffering necessary to the operation of the displays. The units are controlled by the data-acquisition equipment or data processor to which they are attached, with respect to synchronization of the displays and the transfer of data to them. The observer controls a unit in the sense of turning it on or off, and with respect to making certain adjustments in its operation, such as adjusting the brightness of an image tube included in it. The operator may enter data into the system which will cause the information displayed to be varied accordingly.

Buffering of data transferred is generally required because of the differences in speeds and form between the display and its host (e.g., data-acquisition equipment or data processor).

The interface to the data-presentation equipment includes any necessary changes of voltage levels and interchange of control signals between the equipment and its host required to match them together successfully.

REFERENCES

1. "General Information Manual—Introduction to IBM Data Processing Systems," Form F22-6517-1, International Business Machines Corporation.
2. K. M. Roehr and R. D. Coleman, The Digital Data Processor for the Skytop Static Test Facility, *IRE Trans. Military Electron.*, October, 1961, pp. 300–306.
3. R. K. Richards, "Digital Computer Components and Circuits," D. Van Nostrand Company, Inc., Princeton, N.J., 1957.

REFERENCES

4. H. D. Huskey and G. A. Korn, "Computer Handbook," McGraw-Hill Book Company, New York, 1962.
5. E. M. Grabbe, S. Ramo, and D. E. Wooldridge (eds.), "Handbook of Automation, Computation, and Control," vol. 2, John Wiley & Sons, Inc., New York.
6. J. W. Dahnke, Computer-directed Checkout for NASA's Biggest Booster, *Control Eng.*, August, 1962, pp. 84–87.
7. R. H. Bhavani and K. Chen, Assigning Confidence Levels to Process Optimization, *Control Eng.*, August, 1962, pp. 75–78.
8. C. L. Dawes, "A Course in Electrical Engineering," 4th ed., vols. I, II, McGraw-Hill Book Company, New York, 1947, 1952.
9. Digits Can Lie, *Gen. Radio Experimenter*, December, 1962.
10. M. P. Pastel, Analyzing Random Signals in Linear Systems, *Control Eng.*, December, 1961, pp. 85–89.
11. E. S. Ida, Reducing Electrical Interference, *Control Eng.*, February, 1962, pp. 107–111.
12. E. J. Kompass, What about Digital Transducers? *Control Eng.*, July, 1958, pp. 94–99.
13. Is the Zero Output Really Zero? *Control Eng.*, December, 1959, pp. 95–97.
14a. H. A. Cook, Chopper Technology—Part I, *Autom. Control*, January, 1962, pp. 28–32.
14b. H. A. Cook, Chopper Technology—Part II, *Autom. Control*, February, 1962, pp. 55–62.
14c. H. A. Cook, Chopper Technology—Part III, *Autom. Control*, June, 1962, pp. 10–14.
15. R. C. Platzek, H. F. Lewis, and J. J. Mielke, High Speed A/D Conversion with Semiconductors, *Autom. Control*, August, 1961, pp. 34–41.
16. A. K. Susskind (ed.), "Notes on Analog-Digital Conversion," The M.I.T. Press, Cambridge, Mass., 1957.
17. R. K. Richards, "Arithmetic Operations in Digital Computers," D. Van Nostrand Company, Inc., Princeton, N.J., 1955.
18. P. A. Borden and W. J. Mayo-Wells, "Telemetering Systems," Reinhold Publishing Corporation, New York, 1959.
19. H. L. Stiltz (ed.), "Aerospace Telemetry," Prentice-Hall, Inc., Englewood Cliffs, N.J., 1959.
20. C. A. Walton, A Direct-reading Printed-circuit Commutator for Analog-to-Digital Data Conversion, *IBM J. Res. Develop.* July, 1958, pp. 179–192.
21. Optical Encoder Technology, *Computer Design*, January, 1964, pp. 28–30.
22. H. L. Funk, T. J. Harrison, and J. Jursik, Converter Digitizes Low Level Signals for Control Computers, *Automatic Control*, March, 1963, pp. 21–23.
23. A. G. Ratz, R. D. Lavin, and D. G. Lammers, Automatic Tape-editing Equipment, *IEEE Trans. Space Electron. Telemetry*, June, 1963, pp. 51–60.
24. "Fundamentals of Instrumentation for the Industries," Minneapolis-Honeywell Regulator Co., Industrial Division, Philadelphia, Pa.
25. V. N. Smith, Predicting and Evaluating the Performance of Analytical Instruments, *Control Eng.*, October, 1961, pp. 93–99.
26. R. L. Visser, Digital Tape Handling Systems, *Automatic Control*, June, 1963, pp. 27–40.
27. D. L. Reed, Computers and Programmers in Hybrid Checkout Systems, *Control Eng.*, April, 1963, pp. 79–82.
28. C. E. Bradshaw, Structural Testing with Hybrid Checkout Systems, *Control Eng.*, April, 1963, pp. 83–84.
29. J. Sweeney, Information Control Computers—Messages Pay for Data Links, *Control Eng.*, May, 1962, pp. 122–126.
30. R. E. Wright, How to Make Computer Compatible Data Tapes, *Control Eng.*, May, 1962, pp. 127–129.
31. "A Handbook of Time Code Formats," Astrodata, Inc.
32. Trends in Control Components and Instruments: Graphic Recorders, *Automatic Control*, January, 1962, pp. 59–71.
33. "IBM 7700 Data Acquisition System," Form A22-6798, International Business Machines Corporation.
34. "IBM 709-7090 Data Processing Systems, General Information Manual," Form D22-6508-2, International Business Machines Corporation.
35. "IBM 7080 Data Processing System, General Information Manual," Form D22-6512-1, International Business Machines Corporation.
36. "IBM 7070-7074 Data Processing Systems, General Information Manual," Form D22-7004-4, International Business Machines Corporation.
37. H. D. Leeds and G. M. Weinberg, "Computer Programming Fundamentals," McGraw-Hill Book Company, New York, 1961.
38. A. S. Hoagland, High-resolution Magnetic Recording Structures, *IBM J. Res. Develop.*, April, 1958, pp. 91–104.
39. J. Jeenel, Programs as a Tool for Research in Systems Organization, *IBM J. Res. Develop.*, April, 1958, pp. 105–122.

40. M. M. Astrahan, B. Housman, J. F. Jacobs, R. P. Mayer, and W. H. Thomas, Logical Design of the Digital Computer for the SAGE System, *IBM J. Res. Develop.*, January, 1957, pp. 76–83.
41. R. A. Doyle, R. A. Meyer, and R. P. Pedowitz, Automatic Failure Recovery in a Digital Data Processing System, *IBM J. Res. Develop.*, January, 1959, pp. 2–17.
42. IBM Commercial Translator, Form F28-8013, International Business Machines Corporation.
43. D. D. McCracken, "A Guide to Fortran Programming," John Wiley & Sons, Inc., New York, 1961.
44. J. B. Scarborough, "Numerical Mathematical Analysis," The Johns Hopkins Press, Baltimore, 1950.
45. C. Hastings, Jr., "Approximations for Digital Computers," Princeton University Press, Princeton, N.J., 1955.
46. J. A. Rajchman, G. R. Briggs, and A. W. Lo, Transfluxor Controlled Electroluminescent Display Panels, *Proc. IRE*, November, 1958, p. 1808.
47. E. A. Sack, P. N. Wolfe, and J. A. Asars, Construction and Performance of an ELF Display System, *Proc. IRE*, April, 1962, p. 432.
48. R. T. Loewe, R. L. Sisson, and P. Horowitz, Computer Generated Displays, *Proc. IRE*, January, 1961, p. 432.
49. J. E. Gaffney, Jr., Character Synthesizing Tube, U.S. Patent 2,978,608.

AUTOMATIC CHECKOUT EQUIPMENT

45 Introduction

Some of the most difficult aspects of automatic-checkout-equipment design concern problems of transferring information. In the checkout of a complicated weapon-system missile or a space vehicle, we must establish conditions in the system and measure the effects of these conditions. In some cases large numbers of connections to and from the many different parts of the system are involved. For safety considerations, the system frequently must be located at a considerable distance from the automatic checkout equipment. It is this part of the problem that requires telemetering. It is the purpose of this section to discuss the telemetering aspects of the automatic-checkout-equipment design problem.

Government agencies lack enough skilled personnel to maintain their complex systems in the necessary state of readiness. In many instances, the systems are so complicated that, even if the necessary skills were available in sufficient quantity, it would be difficult and sometimes impossible to perform the maintenance necessary within the times available. Thus a system may involve the simultaneous operation of so many varied subsystems or components that, if one of these fails, determining its cause of failure and repairing it would consume a considerable length of time. During this time, two or more other failures may occur; consequently, the system never would obtain the state of readiness necessary for a successful operation. In still other cases, such as long-range ballistic missiles, a state of alert to provide readiness for firing in periods somewhat shorter than the normal manual countdown may be desirable. For these reasons the automatic-checkout-equipment field has expanded considerably during the past several years.

Automatic checkout equipment provides a combination of the skills and the speed needed for rapid checkout compared with manual methods. In some instances, 8 h of manual checkout have been compressed to approximately 3 min by using automatic checkout equipment. Such automatic equipment takes many forms and has been applied to many problems. The primary application area is missiles and rockets. Other examples are to be found in the field of military weapons systems, which in turn divides into two parts: (1) fighting equipment and (2) military repair. First-line fighting equipment involves missiles, aircraft, radar, and information systems on board ship, in fact, any complicated military weapon system. The purpose of this checkout is to determine fighting readiness; consequently, it occurs in a short period usually just before the operational use of a weapon. In this application, it determines whether or not the system is in operational readiness, not what is wrong, if anything. If the fighting equipment is not operational, the checkout equipment determines which black box or plug-in assembly needs to be unplugged and replaced. In the military-repair application, the object of automatic checkout equipment is to analyze black boxes or parts of systems that have been returned to a central point for repair. The nature of the failure within the black box is detected so that a suitable repair may be made.

In both these military applications, note that the automatic checkout equipment provides the two necessary factors discussed above: *speed* and *skill*. The skill needed to repair and locate trouble is built into the automatic checkout equipment. Thus it may be operated by unskilled and untrained technicians on the flight line and at the repair point.

But even skilled technicians sometimes do not operate effectively when under high pressure, as may sometimes occur during battle situations. Consequently, even if the skilled manpower and time were available, they probably would not be effectively used under the pressures of battle, especially for repairs on such complicated systems, without automatic checkout equipment.

A second major area of checkout problems is commercial applications. Automatic checkout equipment is older than we sometimes realize. In fact, in the late eighteen hundreds, a machine was used to actuate typewriter mechanisms to determine whether the setting of the various typewriter linkages was correct and whether the machines were suitable for rapid typing. Electronic methods have become prevalent in our modern electronic fabrication factories. In the simplest form, this equipment checks cables or circuits. In a more elaborate form, it measures voltage, and in still more elaborate forms, it checks the dynamics of equipment as it performs in test situations. In the final instance, the equipment is operated in a dynamic environment duplicating that in which it is intended to operate ultimately.

Other, more widely based commercial applications are beginning to appear. One of these is determining flight readiness of large commercial airplanes. Such aircraft are profitable only during flight, and takeoff or en route failures that delay schedules result in lost revenue for the airline. Besides, the resulting inconvenience frequently causes loss of airline business. Methods for automatically checking readiness before aircraft takeoff are certain to play an important part in the future of the airline operation in the United States.

Thus, as we have indicated, the need for automatic checkout equipment to combine the knowledge and skills of a number of highly trained technicians into an automatic high-speed system is present and growing.

46 Function

Figure 21 illustrates the major information flow. The system to be checked may have many modes of operation, and after turning the system on, the operator selects one of these modes. Next, for proper operation, the system being checked may need to acquire signals from external sources. If these sources are not normally available when and where the system is to be checked, these signals must be provided by the automatic checkout equipment in a simulated form. Thus the automatic checkout equipment's first function is to control the operation of the system, the mode of system operation, and the simulated signals fed into the system.

The next function of the automatic checkout equipment is to measure the performance of the system. The information describing this performance, in turn, is returned to the automatic checkout system for evaluation.

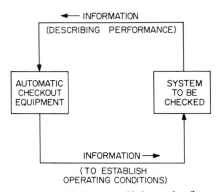

Fig. 21. Block diagram of information flow.

The third function of the automatic checkout system is to evaluate these measurements by comparing the results of the measurements with those expected for the conditions established in the system in step 1.

Figure 22 shows a somewhat expanded block-diagram version of Fig. 21. In establishing the operating conditions, signal generators supply steady-state signals, whereas signal simulators supply dynamic signals. These, in turn, must be fed into the proper parts of the system to be tested. At this point, the proper signal generator must be connected to the proper point in the system. Information describing the system's performance must, in turn, be measured from a number of different points. These points vary, depending upon what is to be measured and upon the operating conditions established by the automatic checkout equipment. Here, again, switching is required. A number of properly selected possibilities exist to simplify the telemetering problem.

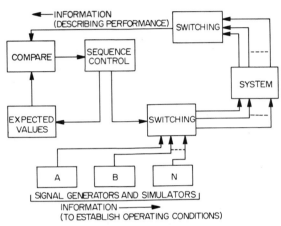

Fig. 22. Basic block diagram of automatic checkout equipment.

This information-flow problem is common to many telemetering and control problems and is a particularly important part of the design of automatic checkout equipment. Some systems require many connections, particularly those for the automatic checkout of electronic equipment in a military aircraft. One such system requires 1,152 connections to and from the automatic checkout equipment. This number appears to be impractically large, yet it is typical of the number of signals and amount of information to be telemetered to and from a complex system to accomplish automatic checkout.

At this point, and again referring to Fig. 22, note that several different alternatives to this massive information flow exist. First of all, if the system is to be checked under dynamic conditions, signal simulators will be required. But if it is sufficient to test the system under static or pseudo-static conditions, signal generators will be sufficient. Secondly, we must decide whether a large number of signals are to be injected at a large number of different points in the system or whether only a few signals are to be injected. Note that, if only a few signals are injected, many points will have to be measured to determine which unit within the system needs to be replaced. Conversely, if a large number of signals can be injected, then only a few points will need to be measured to determine which point within the system is at fault.

The difficulty of telemetering the information to and from the system should be studied with particular regard to minimizing complicated, simulated, and dynamic signals, which generally are more difficult to telemeter. Measured values are preferred, particularly if the measurements are in digital form. A number of alternatives should be studied to select the most efficient method of transferring the information

AUTOMATIC CHECKOUT EQUIPMENT   11-59

to and from the system. This problem of information transfer is the key to a successful automatic checkout system, and while it may appear that it is unimportant, the problem should be carefully studied before the designer progresses to further system implementation.

The remaining blocks in Fig. 22 contain some of the fundamental processes performed by the automatic checkout equipment. The measured values describing the performance of the system are compared in the comparator with the expected values, and the result of this comparison is forwarded to the sequencer and control, which, in turn, determines the conditions for the next test. Other control conditions needed by several other points of the system are furnished by the sequencer and control. For simplicity, they have been left out of the diagram but will be covered later.

In summary, automatic checkout equipment provides an artificial environment for the system with signal generators and simulated signals. Then, it controls the operation of the system in this artificial environment. Under these conditions, the automatic checkout equipment makes certain measurements and compares these with the expected performance of the equipment. In this manner, it determines whether the system is performing properly. Further, by injecting signals along the system or by measuring results at various points in the system, the automatic checkout equipment can locate areas of trouble between points and thus identifies faulty units. These units are normally referred to as line-replaceable units, or "LRU's." An LRU is a unit that can be replaced during an operational check, typically on the flight line. Locating and replacing faulty LRU's provides the operational readiness necessary for modern weapon-system success. This is the most important use for automatic checkout equipment.

47 Types of Automatic Checkout Equipment

There are many types of automatic checkout equipment. As we have seen, some forms of automatic checkout equipment have been used for many years. Only recently, however, has the name "automatic checkout equipment" become widely used, primarily because of the continuing growth and expansion of electronics. In the following discussion, the types of automatic checkout equipment are arranged in order of increasing complexity and are explained by means of illustrative examples.

Simple Sequencer

Example 1 is illustrated in Fig. 23. A simple sequencer is used to measure a series of voltages or circuits in a system, and the results of these measurements are fed to an indicator. The indicator is marked with an allowable range, which shows the limits of the go and no-go indication, and this is read manually. If the indication falls within the limits, the operator repeatedly advances the sequencer to the next step. The sequencer automatically puts in the proper multiplying resistors so that the meter will always read in the same range. This rather trivial example of automatic checkout has been used for some time, and it has such ramifications as automatic sequencing and other means for measuring sequential resistances through various combinations of circuits.

Frequently, a keyboard type of switch which the operator can set beforehand, is used in place of the sequencer. This type of switch is a good example of a simple sequence of tests. The operator performs the sequence and notes the results. If the results are within limits, he so indicates and puts the piece in another bin. Most of the conditions for automatic checkout are satisfied in that (1) the operator does not need to

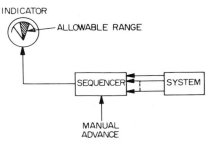

Fig. 23. Example 1. Simple sequencer.

understand what he is doing, and (2) there is a considerable saving of time. However, he does not, at this point and with this equipment, determine exactly what is wrong and how it should be fixed. Also, the sequence he follows is fixed, and no effort is made to vary his sequence as a function of the results obtained during the test. All these important factors will be explained in more detail when applied to more complicated systems.

Sequencer

Example 2 is illustrated in block-diagram form in Fig. 24. This system, once again, is designed to make a series of measurements in the system under test. For this purpose, a sequencer works through these measurements in a manner similar to that already described in Example 1. The difference, however, is that measurements are compared automatically; consequently, the speed of comparison can be considerably

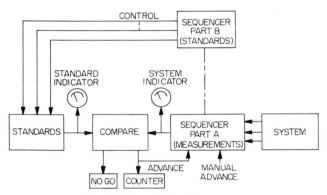

Fig. 24. Example 2. Sequencer.

greater than possible by the manual method shown in Fig. 23. To yield this speed, standard values are needed. These standard values, switched in synchronism with the tests, are controlled by a portion (part B) of the same sequencer, part A of which sequences the measurement through the system. As the test progress, each new standard value is compared with a result of the measurement made at that point. If this comparison shows a correct result or a "go" indication, it comes out as an advance signal, which steps the sequencer to the next test. If, however, the comparison is outside of limits and a "no-go" indicator light is indicated, the sequencer is not advanced. Instead, a "no-go" indicator light is lit. At this point, the operator makes a note that one of the tests has failed, together with the number of the test as shown on the counter. This counter is attached to the advance signal connection to the sequencer; consequently it counts up each of the tests, in sequence. Indicators may also be added to show the value of the standard voltage and the measured voltage in the comparator. These indicators would allow the operator the option of recording these two values at the same time. To continue the test sequence, the operator then pushes the manual advance button, which moves the sequencer to the next test. The advantage of such a system is that the speed is considerably higher than could be obtained with a human operator. Also, in the majority of cases, the operator has very little to do, and it is conceivable that he might be able to perform similar operations on a large number of separate setups simultaneously.

The checkout system just described measures a system in only one physical state; that is, while the system being checked may be in full operation, it nevertheless is in a static situation with respect to changes in operating modes and with respect to the presence of any artificial signals. Examples of such systems are large cables, power supplies, or output voltages in a system where measured values should remain constant for the duration of the test.

Automatic Sequenced Checkout

Example 3 differs from the semiautomatic systems which require the attention of the operator for the performance of the test. Entirely automatic checkout equipment is a matter of degree, but normally such equipment must provide signals to the equipment being checked as well as make automatic measurements on the equipment before it can be called truly automatic equipment. Figure 25 shows such a system.

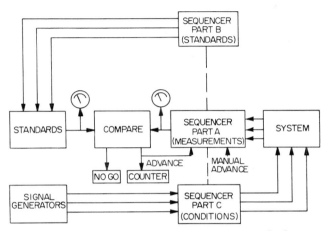

FIG. 25. Example 3. Block diagram of sequencer checkout.

Its major difference from the previous example results from an additional unit in the sequencer that controls the conditions. This unit controls the switching of the signal generators, the power switches, and other switching functions in such a way that the system can be put into any one of a number of different conditions. This automatic system can perform a series of different tests on a system without any manual intervention. Thus, if the system being tested has 10 or more modes of operation, each one of these modes can be checked and each in a number of different ways. Automatic checkout equipment becomes particularly applicable where a large number of operations must be repeated in order to achieve complete checkout. Thus even this simple system can expeditiously check out 100 different conditions in each of 10 modes, as might be the case in a rather complicated cable.

Programmed Automatic Sequenced Checkout

Example 4, the next step in increasing complexity and in corresponding gain in capability and flexibility of an automatic checkout system, is illustrated in Fig. 26. Figure 26 looks exactly like Fig. 25 except that the sequencer here has been replaced by a large block labeled "programmer." This difference is extremely significant. Whereas the sequence merely provides for connecting in sequence a number of wired-in fixed paths, the programmer has a considerable amount of additional flexibility. The saving in equipment and the many new functions provided can be illustrated by considering that the sequence of tests performed by a programmer is not a wired-in fixed sequence but instead may be controlled by punched paper tape, punched cards, magnetic tape, and other digital control means. Besides controlling the sequence of tests in a flexible manner, the programmer can also use common switching circuits.

The same circuits may be used in different tests in different points of the control problem. Referring again to Fig. 25, notice that there are three major switching functions: (1) the switching in of the proper standards, (2) the switching of the measured values, and (3) the switching of the conditions. A properly designed programmer can frequently use a smaller number of switching circuits in these three functions by changing the position of the switching circuits so that they may be used in two or more of the three functions. This form of multiplexing equipment is common to the digital-computer design art.

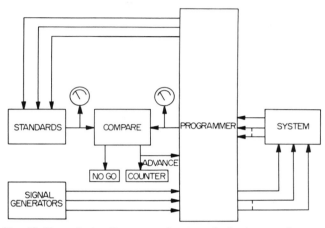

Fig. 26. Example 4. Programmed automatic checkout equipment.

Figure 27 illustrates the basis for the design of a typical programmer. The key element here is the box labeled "decoder," which consists of a switch that examines each group of signals sent its way from the punched paper tape as read by the reader. For illustration, the tape and reader are taken here as 8 binary levels, although any other number is equally good. The decoder function is to examine these eight bits and to determine whether or not they were intended for the tree associated with the decoder. The tree is merely a configuration of switching circuits that switch incoming signals to any one of a number of points 1 through N at the output. The path of incoming signals A to the outputs 1 through N is determined by control signals B. Thus, for a normal binary arrangement of three levels in the tree and three control pulses at B, any one of eight outputs may be selected. Once such a path is set up, the signals applied at A will go to the selected output. This will continue until the path is modified by again addressing the decoder with the proper code to change the setting control signals B.

The decoder decides whether or not the particular tree has been addressed by the eight-level tape. This is done by reserving several bits (for example, three levels) on the tape to correspond again to eight different trees or eight different parts of the system for addresses of trees or switching points. These switching points correspond in the previously described systems to the various points where switching is needed. These, in turn, were defined as switching of the measured values, switching of the signal generators, or condition of the standard. These three points of switching represent the rudimentary points. However, as it will be seen, there are many other points which it frequently becomes desirable to address and change. Having addressed the tree, the decoder takes the remaining bits of that particular message and sets up the control signals of the tree. These control signals are permanent and stay set until such time as a reset signal is received. The method of reset will be separately described later.

The system in Fig. 27 illustrates how eight bits from a paper-tape reader can be made to control a tree. It is a simple extension to show how this may be cascaded to control other trees, as shown in Fig. 28. Thus sequences of tree networks can be constructed of serial and parallel paths and connected to fit the routing needs of the particular problem. The decoder associated with each tree may be addressed in the

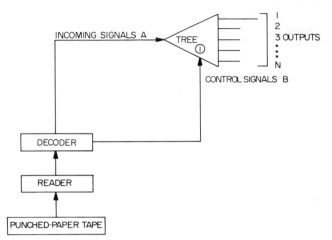

Fig. 27. Programmer elements.

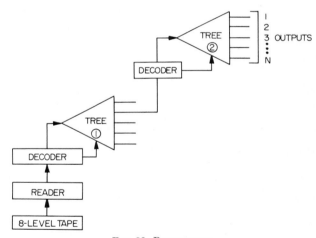

Fig. 28. Programmer.

same manner, and in this way the trees are set up. The tape controls the ability of the programmer to set any tree in a desired manner by means of the decoder. Thus numbers and control signals from the tape may be routed to any one of a large number of places in the system. For each three-level tree, eight different selections are possible. Each one of these may be used as a control signal to set up a path that actually controls the routing of a test signal or of a measured or standard value. As an alternative, each one of these selected paths may be used in turn to send a signal to another tree, which is used for additional branching.

In the serial cascading of trees, interrogating decoder codes may be duplicated, since the only decoders that will receive succeeding codes are those which lie along the path already established. This permits the control of a larger number of trees without the need for an individual unique code for each.

The sequence of operation of the entire programmer can be controlled in this fashion by the sequence of orders put on the punched paper tape. A typical sequence is shown in Table 2

Table 2. Typical Programmer Sequence

Step	Operation
1	Set system conditions
2	Set signal generators
3	Route measurements
4	Select and set standards
5	Select and adjust comparators
6	Advance or stop (go or no go)
7	Print out results
8	Manual or automatic reset

Step 1 is a setting of conditions. This is a general term to indicate that the paths to the various points where signals are to be injected into the system must be established. Step 2 is a setting of the signal generators. These generators may put out a variety of different signals. The programming sequence usually must use them in several places. Consequently, such parameters as frequency, attenuation, and similar controllable values must be set. Step 3 switches the measurements from the system under test to the proper points in the test system, which means that the paths must be established to connect the various points in the system under test to the proper comparator. So far, we have indicated that there was a single comparator. In actual practice, there may be several types of comparators, and each comparator may have several adjustments which are necessary to establish the proper range for each test that is to be performed. Step 4 is to switch in the proper standards to the other input of the comparator. This forms the reference on which the comparison is to be based. These standard signals will have several different forms of adjustment—frequency, attenuation, and modulation, which are necessary in order to establish the proper standards from which the comparison is to be made. Step 5 selects the proper comparator and determines the limits of the comparator. On certain tests, a ± 10 per cent variation may be permissible, while on others a level of ± 0.5 per cent or less may be necessary. Consequently, the range of allowable "go–no-go" limits must be set at this time as one of the input settings to the comparator.

In step 6 the comparison is actually made. This is the point at which timing must be introduced. Timing is particularly essential in the comparison operation because frequently a transient is present and will give an erroneous result if sufficient time is not permitted for the comparison to reach equilibrium. This time may be built into the response of the comparator, or it may be built into the timing of the overall system. As a result of the comparison, the system will either advance or stop depending upon whether the comparison was "go" or "no-go."

In step 7 the results of the test may be printed out. In general, it is not necessary to print out the results if they are proper, although a historical record may be required for some forms of test. In such a case, the printer may be addressed in the same way that the other parts of the system are addressed—by the programmer (consequently, whether or not the data are printed out at this time can be put under the control of the programmer by means of the punched paper tape). Step 8 is the final step: the "reset" to prepare for the next test. This may be either manual or automatic, depending on the conditions.

Timing

Consider the problem of timing the entire system. First of all, the tape reader is a device that provides 8 bits at a time (in the example selected). These 8 bits must be routed to the proper decoding assemblies. This is repeated until all or part of the sequence described in Table 2 is performed, after which the system is reset and advanced to the next test. There are a number of ways in which the system can be timed. Some of them are better than others.

One possible way to time is to slow down the tape reader to allow all other circuitry and all other functions ample execution time before the next order is read off the tape. This method, however, limits the operating speed of the system to the slowest operation in the entire system. Since there is a great difference in speed between different operations, this is a severe limitation. Also, in such an arrangement, the advance signal can occur before the operation is complete, and this occurrence would, in turn, cause considerable confusion.

These, then, are the two elements which any good timing system must fulfill: (1) the speed of the automatic checkout test sequence must be a function of the time to perform each test and must not be restricted by the slowest operation in the entire system and (2) before the next test operation starts, the previous test operation must be complete. From a theoretical point of view, then, the top speed of such a system is attained when the next operation takes place at the completion of the previous operation.

An improved method, which is the first step toward meeting these objectives, is to provide a sequencer for the tape reader to control the rate at which the tape is read. This sequencer normally is arranged to provide more than enough time for each of the orders to be performed. This time, in turn, is geared to the particular order so that greater or less time is provided depending on the order.

Finally, the third and best system for timing is to provide separate completion signals from each point in the system so that the tape reader will not advance until its previous order has been successfully completed. This method sometimes is more difficult to implement, but it results in a system with better performance in speed and accuracy than any other system.

Tape Programming

The functions of reset and advance to the next sequence of tests occur in the same manner as other timing functions. If we wish to control the reset or the advance to the next test by the tape, we can reserve a special order on the tape to designate the reset or advance to the next test routine. For an 8-level tape in the system, there are only 256 selectable possibilities. These must be allotted between addressing various points and information. Three bits are necessary if we want to address only 8 points. If, for example, we felt that 64 points were needed, we would use 6 of the available bits, and only 2 would be available for information. This may seem to be a rather severe limitation. However, it should be noted that the flexibility provided by a system of this type makes it, in fact, an infinitely expandable system because each successive set of 8 bits is routed through the sum of the previously established paths. Thus paths may be compounded virtually indefinitely. Each decoder examines each group of 8 bits that comes along to see whether the group is meant for it.

While, as we have stated, we could use part of the 8 bits to designate the address and the rest for the function, we may also use the bits in succeeding groups of 8 to designate the function or to control successive trees. Thus the first grouping would be used to perform a function at the selected output of the tree. This can be extended to more than one group. Groups of 8 bits do not have to be stored at a central distributing point at the tape reader, a common misunderstanding that has led in some instances to the use of such things as punched paper cards. While punched paper cards will indeed perform this function, it is unnecessary to have so many bits in permanent storage at the reader in order to set up a programmer. Modern high-speed punched paper tapes are particularly well suited to the flexible requirements of any modern programmer. With a design of this type, it is possible to make changes very readily in the field and to extend the problem of checkout in many directions without increasing noticeably the amount of equipment. Thus a series of orders on punched paper tape can control an automatic checkout system. This tape must be prepared accurately and carefully, as shown in Fig. 29. In step 1, operator A uses the keyboard to punch a tape. In step 2, a second operator, B, replaces operator A and performs the same function at a similar keyboard. The difference in step 2 is that the tape operator A prepared is read into a reader in synchronism with operator B. This reader reads operator A's work and compares it with operator B's work. Only if the two compare exactly is operator B permitted to punch a new tape. When the tapes are properly compared, the final verified tape is punched during step 2.

Although this last described system of automatic checkout can be programmed with a punched paper tape and can flexibly perform any one of quite a large number of automatic checkout functions, its embodiment in this description is strictly one of relays and punched paper tape. As a result, the information-transfer rate to and from the system under test is normally quite limited. Another limitation is in the sequence of tests, which is usually determined by the sequence coded onto the tape.

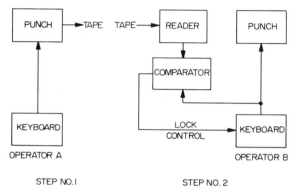

FIG. 29. Tape preparation.

In this respect the device is strictly a test sequencer that will automatically perform a series of tests and print out the results of these tests when they are "no-go" or indicate if the test sequence has been performed satisfactorily when the results are in the "go" category. The next example illustrates a system of increased complexity which does more than sequence the tests. The other limitation of the system just described is its slow speed because of its relay switching and sequential logic. Before we go to the next system, we shall note some improvements and variations that can be added to this system. Among these is the ability to perform more than one test at a time. Because of the way the system is designed, this is merely a question of what the man who plans the programming decides to do. Thus the man who prepares the paper tape may elect to initiate more than one test at one time, a particularly valuable decision when there are some long tests which require time to reach equilibrium. During this time, he may also want to start other tests in order to minimize the overall time required for the checkout.

Another simple change that can result in improved performance is that of modifying the sequence of test as the test progresses. This may include modifying the sequence as a function of the results of the previous tests. This change can be made by adding more than one tape reader, for example. Two tape readers would provide one tape reader with the master, or central control sequence, which determines the tests normally to be conducted if all tests are successful. If one test was a "no-go," it may be desirable to modify the routine to include tests which are necessary only if the previous test was "no-go." These tests will locate troubles in the system and define these troubles to the point of a removable unit. In this case, a second tape reader is provided which advances in "block synchronism" with the first tape reader. That is, for each major test of the first, or control tape, the second tape reader advances

AUTOMATIC CHECKOUT EQUIPMENT 11–67

to a block of words describing a subroutine for fault isolation. This routine is to be entered if the result of the main test is a "no-go" rather than a "go."

The extension of this to shift the control between more than two tape readers is apparent, and in fact three or four seems to be a practical quantity on some systems. It is interesting to note at this point that the problem of hunting for a subroutine on a long tape can be solved by the use of the high-speed slew-search mode which is presently available for tape readers. In some cases this can also be solved by duplicating the sequence repeatedly in such an order that most or all tape readers will read in the same, or forward, direction at all times.

Computer-controlled Automatic Checkout

Example 5 is the final example, and at this time it represents the state of the art in complexity of automatic checkout equipment. As we have seen, the limitations of the previous system were in these areas:

1. Speed
2. Ability to perform large numbers of tests simultaneously
3. Difficulty of modifying the tested sequence based on the results of previous tests
4. A point not previously mentioned: the desirability of drawing conclusions based on a number of tests

This last function is defined as "compound logic" and is useful for reducing the number of connections or telemetering channels provided to and from the system.

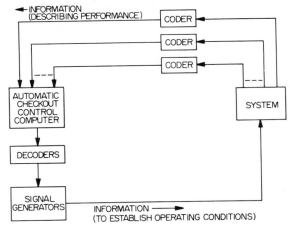

Fig. 30. Example 5. Automatic checkout equipment, computer-controlled.

Compound logic makes it possible to interpret a large series of tests by logical means and from this logic deduce the performance at points in between without specifically measuring at these points. This provides a valuable means of limiting the number of connections necessary to and from the system under test.

Figure 30 shows schematically how such a system is designed. The key to Fig. 30 is the word "computer," and it appears in the left-hand middle box labeled "automatic checkout control computer." What we have been trying to do, in making automatic checkout equipment, is a similar problem to that faced by digital-computer designers. The difference is that the digital computer operates at a considerably higher speed. The original computers were made with relays, but the day of the relay computer has long since passed. The digital-computer people have also learned to time-share a majority of the active elements in their computers, allowing a reduction in the overall number of components. The checkout systems described in the previous examples imply that the components are all wired-in and used for only one function, whereas in a digital-computer system the active elements may be used for one function during one part of the cycle and for a completely different function for another part of the cycle. In addition, the sequence of computer orders can commonly be modified by so-called branch orders. These are orders based on comparison of two numbers and a decision based on the nature of a number. Thus a typical branch order may be based on a comparison as to whether or not the sign is positive or negative. In the event that it is positive this might correspond to a "go" action, if negative to a "no-go" action.

In a digital computer, it is therefore quite easy to conceive of a system in which a number of tests are run and their meaning resolved by a logical analysis: "compound logic." These stored data, in turn, are used to build up a criterion as to whether or not the performance is satisfactory. This "learning" or accumulation of operation experience can be a very valuable asset to operational use. A typical example of this might be the performance of a radar where the transmitter power has fallen off—in fact may be below limits. But if the receiver sensitivity or some other compensating factor were such that the overall performance of the system was satisfactory, the radar could be left in operation. Consequently, it is quite important and necessary to be able to build up a criterion that indicates the overall system performance.

Another use of compound logic is its ability to accumulate the results of a number of tests and from the results of these tests make certain deductions as to what is wrong or whether or not anything is wrong. A further advantage of the flexibility provided by compound logic is the ability to take very large hops within the system, that is, for example, to check from one end to the other and, if the results are not proper, to decide how and where a division is to be made. If a check from one end to the other results in a "no-go," the next most logical point might be to check from zero to 50 per cent. If this section is found to be in order, the next hop might be between 50 and 75 percent, and so on. This type of branching enables a highly efficient checkout system to be made.

A further advantage of applying digital techniques to automatic checkout equipment is the inherent timing accuracy built into most digital computers. Thus it is possible in many instances to use the time base of the computer for measuring time intervals, a frequent parameter in modern weapons systems. A further advantage is the ability to initiate a large number of tests simultaneously and to work fast enough so that all tests are properly evaluated at the time they are concluded. Finally, the reason for going to the high speed of a digital computer is that, in some systems, the information rate is so high that it is impossible to measure it in any other manner without providing great quantities of auxiliary precalculated dynamic response data. In other words, the so-called standards become dynamic storage places instead of becoming merely standard cells of constant voltage. This results in considerable complexity which can only be practically solved by the utilization of modern digital-computer techniques.

Returning to Fig. 30, we find that a computerized system involves several other functions, but otherwise it is quite simple. The basic assumption is that all information is to be handled in digital form. Consequently, digital outputs must be provided from the system to be tested, a need met simply by digital pickups. At present, digital pickups are not widely used and are generally not available, although they will probably become common in future systems; consequently, Fig. 33 indicates the use of a coder in each of the information lines between the system and the automatic checkout computer. This coder is simply a device that converts an analog value, such as voltage, to a digital value so that it may be properly entered into the computer.

Conversely, the output of the computer is also in digital form, and decoders at this point are needed to control the signal generators properly. This arrangement sets up the conditions of the test systems.

AUTOMATIC CHECKOUT EQUIPMENT 11–69

48 Data Transfer

In the foregoing discussions, we have alluded to the volumes of data that need to be transferred to and from the system under test. Wire will serve this function, in the simpler systems, but as the complexity of the system increases, the volume of simple wires becomes prohibitive. Coaxial cables may also be necessary where information is of such a bandwidth as to require this special treatment. Such cables also fall victim to system complexity. In automatic checkout equipment for present systems, the total number of connections can exceed 1,000 and require a derrick for connection of cables to the system. This is obviously bordering on, if not already an impractical solution to a problem. We must therefore examine the problem to discover ways in which the number of wires may be reduced.

The engineer familiar with telemetering practices will discover that there are a number of similarities between the problem of telemetering to a remote location and that of transferring information in and out of a weapon system under test for use by an automatic checkout system. In fact, almost all the many telemetering techniques will be found to be very helpful. We have previously defined the type and form of information we are likely to encounter. There are two main categories:

Analog Signals

These may be in any one of a number of forms: d-c voltages, currents, frequencies, time differences, pulse amplitudes, pulse widths, etc. This information is more difficult to transmit to and from the system under test. Because its content is in an analog form, the means utilized for transmitting it must preserve the analog value. Consequently, the bandwidth of the channels normally required for this type of analog transmission becomes wider than if we used measured values.

Measured Values

In this form—pulse-code information—the information has been measured and is presented as a series of pulses representing binary digits, or bits. This method has the advantage of avoiding an analog value; thus the equipment needed to telemeter these data to the system under test or back out of the system is far simpler than the equipment needed for analog-information transmission. A further simplification results from the data's being measured at a discrete time as opposed to being transmitted continuously. The discrete time measurement enables a number of different values of various parameters to be transmitted over the same channel at different times (a form of time-division multiplexing). This time sharing of the channels results in fewer connections. In its simplest form, such a method takes the form of a distributor that commutates a number of different measurements into a single wire or channel. Such a distributor, or stepping switch, runs continually or it may be under the control of the automatic checkout equipment, enabling the equipment to select the proper connection at the proper time. It should be noted that the stepping switch or distributor in most applications will physically ride along with the tested system. In one typical problem, it developed that the saving in pounds by eliminating the wires and the plugs necessary to bring the connections to the skin of the aircraft, as opposed to the installation of this type of multiplexer in the system, favored the multiplexer by a considerable margin. This is a rather preliminary example of ways in which even the crudest methods common to telemetering can be of assistance in automatic checkout systems.

Information Feedback Control

In the next system classification, information communication is considerably more important. It envisions a system consisting of automatic checkout equipment that communicates with the equipment in the system under test over a common path.

This common path may be an r-f connection, a coaxial line, or other connection, usually with fairly wide bandwidth. Each parameter to be measured in the system is interrogated. When the proper interrogation code is transmitted to it, it will respond with the value of its parameter. Such a system enables the automatic checkout equipment to interrogate or sample the value of any parameter at any time and in any sequence and only with a single path. The previous distributor stepping-switch version, in general, is limited to stepping in a fixed sequence as a function of the manner in which the parameters are wired to the switch. This interrogation method enables the automatic checkout equipment to skip around through the 1,000 or more parameters in any order that happens to be the most logical for the particular difficulty being diagnosed. Once interrogated, a particular parameter will respond

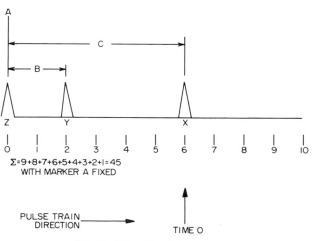

Fig. 31. Pulse-time interrogation.

with information describing its state or measured value. Two systems are particularly well suited to this type of response:

1. The first is a pulse-code system. A pulse-code system consists of two parts. First is the device that recognizes the various interrogations and second is the code generator. All parameters receive all interrogations in a single-channel system and consequently must respond only to the unique code which describes their particular parameter. There are two classifications of ways in which pulse code is used to interrogate a number of different parameters. First is a common tree arrangement in which the pulse code is used in conjunction with a matrix to select one of a large number of points. This would involve sharing the decoder among many points. The second method is to provide a decoder at each parameter. If the interrogation code is serial, this decoder can be provided in a manner similar to that described below as "pulse-time modulation." When the interrogation is parallel along a number of paths, a diode matrix that recognizes the presence or absence of 0's and 1's must be built. In the proper combinations, this matrix then goes into an AND gate, which provides a method of ensuring that all the conditions are met before an ACCEPT pulse is generated.

Having received an interrogation at the parameter desired, it is now necessary to produce the pulse code that describes the value of the parameter. Here again, there are two general methods by which this may be accomplished. First of all, a common coder may be shared among all the many variables. In this instance, the analog signal is furnished to the coder that produces the code and transmits it back to the automatic checkout system. The second method is preferred, although suitable equipment is not always presently available for its implementation. In this method, the parameter is coded at its pickup point by a pickup which generates a pulse code directly. This device will enjoy more common usage in the future.

2. The second system for implementing an interrogator responder is based on similar methods to those for the old and presently unclassified IFF, "identification friend or foe." This system, which utilizes a pulse-time method of coding, has many of the advantages of the pulse-code system. The disadvantages are, however, that it is fundamentally an analog method which, however, may be reclocked in order to preserve some of the quantitizing advantages of pulse code. Also the utilization of information capacity is not so good in the pulse-time system as in the pulse-code system. Pulse time, however, does have the advantage that, with the present state of the art, it is easier to implement with the available components and consequently is more widely used.

In a pulse-time system, the method for interrogating is illustrated in Fig. 31. A train of three pulses (for example) is transmitted from the automatic checkout equipment to all the parameters in the system under test. These parameters are all connected over a single coaxial line to the point where this code is generated so that they

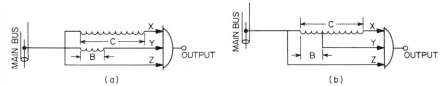

FIG. 32. Pulse-time decoder.

all receive the code. They all receive every code that is transmitted as well. If there are 10 possible positions where pulses can be transmitted, and if we are using a 3-pulse interrogation code consisting of a fixed marker pulse and two information pulses both free to occur in any of the ten positions, then there are 45 possible parameters which can be interrogated. $\Sigma = 9 + 8 + 7 + 6 + 5 + 4 + 3 + 2 + 1 = 45$.

Figure 32a shows the decoder schematically as three delay lines, each of a different length. The input of the three lines is common, and the pulses are all received into each input. Only for the unique combination of spacing which corresponds to the interrogated decoder will all three pulses arrive at the terminations of the three separate delay lines simultaneously. In this instance, the AND gate ands all these together and furnishes the output pulse, which indicates that this is the parameter which has been interrogated. In practice, three delay lines are not necessary, since one pulse need not be delayed at all, and the other two delays may be taken from a single tapped delay line (Fig. 32b). This makes a unit that can easily be packaged in a very small volume and contributes to the practicality of the pulse-time interrogation system, since the interrogation interpreter is actually very small compared with the usual length of wire necessary for such a connection.

The next part of the problem is making the interrogated parameter respond with its value. (Figure 33 shows one of the methods for doing this.) A sawtooth, or other reference, is transmitted on the common line, and at the instant that this value on the sawtooth is equal to the measured value, a pulse is generated which is transmitted on top of the sawtooth reference back to the automatic checkout equipment. The time between the initiation of the sawtooth and the arrival of the spike from the interrogated parameter is a direct measure of the value of that parameter in pulse-time form. Such a device is very simple and can easily be converted into pulse-code modulation at the receiving end without the need for a separate pulse coder in the equipment under test.

In making connections to a system under test, consideration should be given to the existing means for communicating with the system. Thus, if the nature of the system is such that it has an inherent r-f link connection, such as a radar, it may frequently be possible to connect into the system without the need for any wires at all. Such an r-f link, radar or otherwise, is a very useful means for getting in and out of the system. Some systems also contain a computer, and if so, this computer could be used as a place for controlling portions of the test or as temporary storage. The

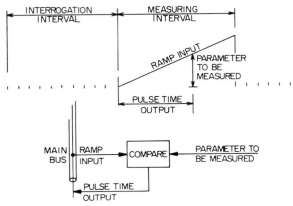

FIG. 33. Pulse-time responder.

computer could also assist in putting the information in a form that is readily communicable in and out of the system to the automatic checkout equipment.

49 Desirable Improvements

Computer

The most important development presently under way is a computer for automatic checkout equipment manipulation. Such a computer is characterized by the following:
  1. It must be extremely reliable, certainly more reliable than equipment being tested.
  2. It should be special-purpose such that the number of components and the simplicity of use are considerably improved over the devices presently being utilized.
  It is likely, therefore, that such a computer will be the outgrowth of efforts in two fields: (1) logic design, which will greatly simplify the logic makeup of the system, and (2) electronic-component development, which will provide more reliable elements for this machine.

Pickups

There is a great need for pickups that will develop pulse-code information directly from the parameter being measured without the need of time sharing a rather complicated coding device as in the present state of the art.

Improved Measuring

Methods of measuring data, such as by means of the pulse-time system described above and recently devised for this purpose, can stand considerable improvement. Measuring is the key to any data-transfer process as it is in the telemetering art. Devices to provide measured information in more useful forms continue to be needed.

AUTOMATIC CHECKOUT EQUIPMENT 11–73

Automatic Checkout Equipment

Automatic checkout equipment needs to operate more rapidly. If this need for rapid operation is extended to the point where the operation becomes extremely high speed, then the checkout process may be repeated frequently; that is, an entire system may be checked out not in 8 hours, 8 minutes, or 8 seconds, but in something more like a tenth or a hundredth of a second. In such a case, the process may be repeated immediately after being completed. The times selected may be arranged relative to the bandwidth of the monitored quantity so that "continuous" monitoring results. "Continuous monitoring" of this sort will tell immediately whether or not the system is functioning properly and is instantly ready to go. In fact, in a dynamic situation, where the parameters measured are changing rapidly, it may be desirable to compare these parameters with the values defined by limits to determine whether or not they are in fact staying within the desirable limits and whether the system is performing properly. Such a system has many ramifications, and it is interesting to note just a few at this point. One of the most interesting is the problem of not firing a missile that is not actually performing properly. In the limiting case, it would be quite desirable to check everything a fraction of a second before a rocket is launched. Thus it would be possible to assure that things were working at launch in the way they had been working several hours earlier.

A natural extension of this system would enable checks to be made during the flight of the missile. At the present time, missiles telemeter data back to the ground. The telemetering is preprogrammed, and the information which comes back is determined before the vehicle is launched. If it were possible to interrogate the operation of the system during the flight, the information brought back would be considerably more valuable because the automatic checkout equipment would be in a position to decide what needs to be measured and when it should be measured. Consequently, the limited capacity of the communication links transmitting the data to and from the system could be used more efficiently. This ability to decide what you want and when you want it by interrogation and control of response from the weapon system is one which should prove to be valuable. As a matter of fact, such an operation has taken place in some space satellites. A repair of Telstar, for example, has been effectuated in this manner.

Chapter 12

DATA REDUCTION

HARRY H. ROSEN, *Reentry Systems Department, General Electric Company, Philadelphia, Pa.,* **and Staff***

SCOPE AND FUNCTION

1. The Four Phases of Data Processing.................... 12–2
2. The Input-data Phase................................. 12–3
3. Data Handling and Conversion........................ 12–4
4. Computer Data Processing............................ 12–9
5. Data-processing Analysis............................. 12–10

DATA-PROCESSING SYSTEMS

6. The Systems Approach............................... 12–11
7. Preparatory Planning................................. 12–12
8. Manual and Semiautomatic Processes................. 12–14
9. Automatic Digitizing Systems........................ 12–18
10. Analog-to-Analog Conversion........................ 12–25
11. Final Presentation and Summary..................... 12–27

COMPUTER DATA PROCESSING

12. Introduction.. 12–28
13. Selecting the Computer.............................. 12–28
14. Numerical Techniques............................... 12–29
15. Programming Systems and Techniques................ 12–37
16. Presentation of Final Reports........................ 12–47
17. Data Computations.................................. 12–48
18. Special Computer Data-processing Applications....... 12–49

DATA-PROCESSING APPLICATION

19. The Telemetry Information-processing System........ 12–50
20. Flight-test Planning................................. 12–52
21. Data-reduction Requirements........................ 12–52
22. Establishing Data-reduction Requirements............ 12–54
23. Flight-test Program Planning........................ 12–56
24. Telemetry-data Defects.............................. 12–56
25. Remedial Measures.................................. 12–57
26. Elements of Integrated Data Processing.............. 12–57
27. Evaluating the Results............................... 12–59

* Staff members contributing to this chapter are J. W. King, M. A. Martin, L. A. Meeks, M. H. Slud, and R. Todd.

DATA REDUCTION

There is little disagreement as to the basic function of data processing, namely, that of *reducing* the measured or experimental *data* to meaningful terms. The normal process starts with a relatively large quantity of data and ends with a condensed version of what the test or experiment yielded in the way of information. Hence the term "data reduction" is frequently used to describe this process interchangeably with "data processing"; however, a number of different concepts exist as to the nature of data reduction in terms of its scope. In this section the broadest concept is used; that is, data reduction is the technique by which the *information* being sought by means of a test or an experiment is *processed* from raw data to final answer form. Thus data reduction, data processing, and information processing are all considered to be synonymous.

It is apparent that any attempt to describe fully all the problems and solutions involved in as complex a field as information processing is far beyond the scope of a handbook on telemetry and remote control. This chapter will limit itself to dealing with the various problems involved in designing, building, and operating a data-processing center where most of the data come from telemetry instrumentation systems used in missile flight testing. In the first part of this chapter, Scope and Function, the general scope, function, and language of data processing will be introduced. The second part, Data-processing Systems, describes systems of men and machines used to accomplish the various functions required in reducing telemetry data. The third, Computer Data Processing, describes the role of digital and analog computers in the processing of telemetry data. The fourth part, Data-processing Application, shows how the processing of telemetry data must be mathematically considered in order to produce valid and accurate results.

SCOPE AND FUNCTION

1 The Four Phases of Data Processing

The field of data processing is one of the newest and fastest-growing technologies. It suffers, however, the usual growing pains of youth—lack of definition. While there is no universal agreement as to the scope and function of data processing, there is little question as to the phases telemetry data must go through to fulfill the function of giving the tester or experimenter the answers he seeks. There are innumerable ways of organizing the work to be done in proceeding from raw telemetry data to the answers which the telemetry system was designed to yield. Many differences of opinion exist as to what portion of the process should be the responsibility of design engineering, instrumentation engineering, systems engineering, data processing and computation, or project engineering.

For the purpose of this handbook, it is of little importance how the responsibility is divided or centralized; instead, the entire process will be described, and the reader can select the organizational pattern best suited to his needs. First, it is essential that the language of the trade be understood and that the objectives of data processing be agreed to. To this end, the first part of the chapter will define those terms which are either new or have multiple interpretations. This part will also describe the work needed to accomplish the ultimate objective of providing accurate and useful information to the experimenter or tester. For the sake of simplicity, the process is arbitrarily divided into four phases, the input data, the data handling and conversion, the computation, and the application phases.

Figure 1 is a simplified block diagram of the flow of work to be done in arriving at final answers from raw telemetry and other test data. Although the diagram indicates four separate functions, in actual practice much overlap exists among the boxes. For example, computation is used in calibrating the data in the "data-handling and conversion phase," and it is certainly essential to error analysis in the analysis phase. Similarly, the output of computation is frequently converted to plots and tabulations for report or presentation purposes. The interaction or interdependence of the four phases will become more apparent as each is described in greater detail.

SCOPE AND FUNCTION 12-3

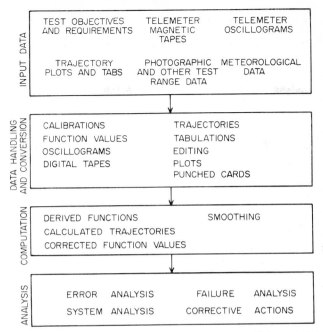

Fig. 1. Simplified flow chart for processing telemetry flight-test data.

2 The Input-data Phase

The forms and content of this phase are obviously dependent upon the test under consideration. But these forms and content are probably most important to the data processor in that they define the work to be done by him. It is here that he learns about the particular objectives of the test: the requirements for accuracies, frequency responses, sampling rates, rate of tests, and how much time he has to do his work; the number of telemeter channels, the type of telemetry, and what other data the test range will furnish him. It is here that data-processing personnel defines in great detail what is to be done to the data in order that they may meet the objectives of the test. This is the point at which data-processing specialists have the opportunity of demonstrating that the instrumentation systems, both airborne and ground, are adequate, inadequate, or overly redundant to meet the objectives of the test. For example, the selected sampling rate may be inadequate to furnish the desired accuracy for a given band of frequencies, the instrumentation of the test range may be such as to defy direct measurement of trajectory to the degree of accuracy required, or the sensors used to measure temperatures may be improperly located to give heat flow at designated rates. Perhaps these problems can and should be handled by the systems designer, but all too frequently he is not so well aware of the data-handling problem as is the data processor. In the past, wherever data processing has been omitted from system design considerations, either the desired data were not obtained or the instrumentation system had to be redesigned.

Assuming, then, that all test requirements are known and are resolved, it becomes essential to know the physical characteristics of the data which the data processor will receive from the test site. Again, the types, quantity, and formats of the data will depend upon the type and place of the test. It is virtually impossible to list all the possible data which a data-processing center will need to handle all types of telemetry effectively. Table 1 lists the major types of data and configurations that are currently available from missile test centers. It will be noted that much data are required that are not the outputs of telemeter systems. The supplementary data, such as trajectory and meteorological information, are normally essential to the full processing of telemetered information. Thus meteorological, trajectory, and telemetry data are normally combined to obtain various aerodynamic coefficients.

Table 1. General Input-data Types

Type of instrumentation	Recording medium	Data obtained	Processed output form	Remarks
From optical instrumentation systems:				
Metric photography	Film or plates	Space position, velocity, acceleration, attitude	Digital tapes, punched tapes or cards, tabulations, plots	Includes all fixed and tracking systems used for triangulation. Normally processed at test-range facilities. Raw-data film or readings can be obtained for specialized computations
Engineering sequential photography	Film	Events	Film, tabulations	Timing recorded on film for correlation. Instrument normally tracking telescope. Can be used for altitude measurements
Documentary photography	Film	Historical data	Film	Used for close-up purposes such as required for flame studies as well as historical
Geodetic........	Reports	Reference points and lines	Tabulations	Required for accurate frame of reference
From electronic ground instrumentation systems:				
Tracking........	Magnetic tape, film, punched cards, paper-tape reports, plots	Space coordinates, velocity, acceleration	Digital tapes, punched tapes and cards, tabulations, plots	Includes all types of electronic tracking systems such as radar, Dovap, Cotar, external guidance. Reduction usually accomplished by test center—providing processed data in specified forms
Telemetry.......	Magnetic tape	Missile and environmental functions	Magnetic-tape oscillograms	¼-, ½-, and 1-in. tape width with 2, 7, and 14 tracks, respectively. Recording speeds normally either 30 or 60 in./sec
	Oscillograph-paper strip charts	Missile functions	Annotated oscillograms	Used for "real-time" data applications as well as playback
	Reports	Calibration data	Tabulations, plots	Field calibrations—test range as well as contractor equipment
Meteorological...	Punched paper tape, reports	Atmospheric properties	Tabulations, plots	Rawinsonde, high-altitude rockets for soundings. Required for trajectory parameters

NOTE: In addition, there are many specialized systems such as splash nets for impact time and location, infrared measurements, ionospheric measurements, and r-f flame attenuation.

3 Data Handling and Conversion

A glance at Table 1 is sufficient to conclude that the data obtained from test ranges are seldom directly usable. Human beings are incapable of reading magnetic tapes, punch tapes, or even photographs without the aid of conversion devices. Data handling and conversion are the manipulation of raw data in any form through the use of hand measurement techniques, manual calculations, semiautomatic or automatic digitizing equipment, analog translation equipment, and analog or digital computers to convert raw test data into usable function values in the form of tabulations and plots. As the definition implies, this process is a basic function of any data-reduction center and one on which the work of subsequent processes is dependent. The required accuracy and time allotted to the operation usually determine the particular techniques or combinations of techniques to be used. Basically, there are three conversion processes: analog-to-analog, analog-to-digital, and digital-to-analog.

Analog-to-Analog Conversion

The analog-to-analog conversion process is the translation of analog data to another analog form, normally for the purpose of increasing or making possible human comprehension of the data. For example, a single oscillogram may have dozens of traces with many crossovers. Analog conversion will separate the various traces and will eliminate crossovers. In the case of magnetic tape there is little choice but to translate magnetic variations on the tape to varying lines on paper; the bulk of telemetry data is normally recorded on magnetic tapes. In the latter case, varying d-c voltages representing the physical phenomena being measured are retrieved through conversion equipment and converted to graphic form by oscillographs or such direct-writing devices as pen recorders. The number of traces per record is arbitrarily restricted by equipment capabilities and presentation requirements. Variable paper speeds allow for the expansion or contraction of the time base, and the variable zero offset and scale-factor settings of amplifiers provide some flexibility in placement and amplitude determination of each trace. Analog-to-analog outputs are used in several ways, and they represent various degrees of data reduction from the first step of a process to the end product of the process. Examples are shown in Fig. 2.

The "real-time" display of telemetered functions illustrates the use of analog outputs as a final process. At missile test sites this technique is frequently used to present an immediate display of a few important telemetered functions indicative of missile behavior and to provide decision-making information during critical portions of a flight; however, equipment and personnel limitations normally restrict to very few channels of information the direct recording of flight-test information in this manner. Although the process is performed at the test site, most of the elements of data handling and conversion are needed except that the output requirements are normally less accurate than those for engineering and design purposes.

A second and more widely used application of analog-to-analog conversion is the "quick-look" process. The first playback records of test results from telemeter magnetic tapes are required in the shortest possible time to review the overall behavior and success or failure of a test as well as to provide the necessary preview of the data before further processing. The importance of speed in this phase of the work cannot be overemphasized. This need is particularly true in research and development programs where the results of one test are awaited so that design changes can be incorporated in the next test. In the case of converting from telemeter magnetic tape to analog form, the output is an oscillogram of time-varying functions. Conversion of these traces to function value may be provided through the use of calibrated scales drawn either on the record or on some type of overlay. The problems associated with preparing calibrated scales will be discussed in some detail later.

Analog-to-Digital Conversions

The analog-to-digital process is a conversion of analog forms of data to digital forms through use of hand techniques or of semiautomatic or automatic equipment. The inputs to this process can be in a variety of forms: magnetic tapes, oscillogram records, strip charts, and photographs; outputs take the form of punched cards, punched tapes, digital magnetic tapes, or tabulations. For telemetry data, the desirable technique is one of using automatic equipment to permit speed, accuracy, and ease of handling; however, provisions must be made for semiautomatic methods to handle those portions of data which are not amenable to automatic digitizing, owing to such difficulties as excessive noise or poor recordings. In those cases where the digitizing workload is comparatively small or where time is not of great significance, semiautomatic or manual methods may be sufficient.

Many types of semiautomatic machines are available for digitizing film data, oscillograph records, or any type of graphic presentation. The output of these devices is usually in the forms of punched cards, punched paper tape, or automatically typed listings. Some machines include provisions for conversion to direct-function value with corrections for nonlinearity and zero offset; however, most devices used in this fashion convert graphic displacements linearly by means of arbitrary numerical count; conversion to corrected functional values is obtained by means of hand or automatic computations.

Several systems are available for converting telemeter data on magnetic tape to digital form on magnetic tape, on punched cards, or on punched tape. Individual systems are usually designed in such a manner as to permit the output format to be

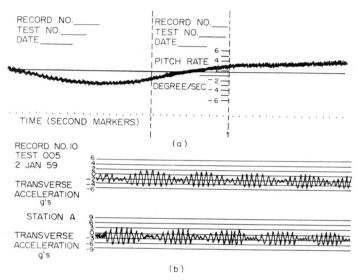

Fig. 2. Examples of annotated oscillograms. (a) Portion of oscillogram record showing transparent overlay. The advantage to this system is the movable overlay, providing identification and reading scale at any spot on the record. (b) Portion of oscillogram record with grid lines ruled directly on record. This technique is not used commonly because of excessive time of preparation.

compatible with high-speed computing devices. Calibrations are applied by means of a computer through table look-up or curve-fitting techniques or through analog correcting devices which precede the conversion to digital form.

The reasons for digitizing analog data are threefold: (1) If data are to be used as inputs to digital computation it is imperative that the data be in digital form. (2) Digitizing of data is used to condense long-time function histories to a format of reasonable length by selecting those points of a trace which are significant. (3) Where oscillograms consist of numerous traces with a multitude of indistinguishable crossovers, it may be necessary to digitize and replot each function to permit readability of the records. An example of this practice is shown in Fig. 3. Some data-reduction centers digitize all their data so that zero offset, scale factor, and nonlinearity corrections can be made by digital-computation means. The results are then replotted. The effectiveness and efficiency of going from analog to digital and back to analog form is questioned by many data processors; therefore, in many data-reduction centers, analog data are digitized only when required for computation purposes.

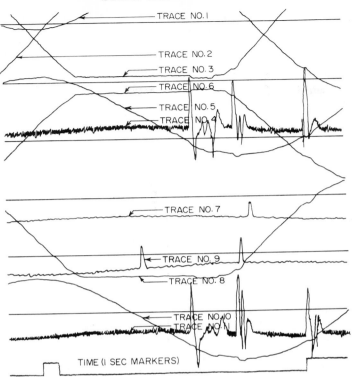

Fig. 3. Example of oscillogram with excessive crossovers. This type of record is extremely hard to follow, and such a setup is avoided whenever possible.

Digital-to-Analog Conversion

The digital-to-analog process, as the name implies, is the translation of digital data to specialized analog forms, normally in the form of plots. The conversion in this direction is completely flexible, since scaling and function value can be computed arbitrarily, and is limited in speed by the plotting equipment available. This latter limitation is a serious one in that present technology is such that the speed of plotting is far below the speed of the computer in providing points to be plotted. A number of developments, which show promise of overcoming this particular shortcoming, are going on at this time.

The need for plotting digital data is frequently questioned, particularly by mathematicians who understand numbers. It is agreed that the plotting of numerical values adds nothing to the accuracy of data; but the human mind has not developed to the stage where it can see relationships by means of numbers as well as it can by means of plots. Until such time that engineers develop the capability of visualizing numbers, the data-processing center will be required to present the answers it obtains by means of plots as well as by tabulations.

Another problem associated with producing plots at high rates of speed is that of annotating the output records. This bottleneck exists even where plots are produced automatically and rapidly as in the case of oscillograms. It should be borne in mind that in some instances the data processor has no other choice than to use digital-to-analog conversion in presenting his data—for example, pulse-code-modulated telemetry (PCM). In this case there is no way of obtaining "quick-look" records other than by means of a digital-to-analog conversion process.

In summary, the presentation of derived parameters and other digital computation results, in a manner conducive to rapid comprehension, nearly always returns the data processor to the problem of digital-to-analog conversion.

Calibration Data

The need for calibrations arises from the very nature of the telemetry system which converts measured functions to electrical analogs. The end instrument or transducer, which responds to the physical phenomenon being measured, translates its response into electrical parameters. As described in earlier sections of the handbook, these parameters are converted, sampled, monitored, and transmitted by various methods, depending upon the particular telemeter system being used. At the receiving end, the data are in some form of varying voltage which is normally recorded on magnetic tape. The basic function of data reduction is to reverse this process by converting these varying voltages back to function values which resemble the physical phenomena as closely as possible. Calibration is required, then, (1) to relate the physical unit measured to the voltage output so that conversion to function value can be accomplished and (2) to provide means for correcting for system errors such as drift and nonlinearities.

The most straightforward way to calibrate the airborne telemetry system would be by "end-to-end" measurements; i.e., stimulate the transducers by accurately controlled and measured environments over the complete range of the instruments, record the outputs through the entire system, and compare with the known inputs. This method is generally used wherever possible; however, the difficulty in creating, applying, and controlling the physical stimuli required, the fact that some sensors are designed for destruction during the environmental stresses to be measured, and the physical size of the item frequently prohibit the use of end-to-end calibration. Hence, the more practical approach is taken in that (1) the various end instruments are calibrated individually and curves provided and (2) simulated sensor output voltages are applied to the rest of the system, the outputs of which are compared with the known inputs, and a system calibration obtained. These two results are combined, yielding an "end-to-end" calibration.

In missile applications, further calibration data are obtained on the instrumentation system in three possible ways. The first is the preflight calibration procedure which introduces known electrical analogs of the transducers into the system just prior to the flight. The calibrations are recorded on magnetic tape to provide an overall check on the airborne telemetry system, less transducers. The second is the "in-flight" calibration, which again inserts reference voltages into the system by (1) interrupting transmission for short periods over the flight intervals or (2) providing continuous calibration pulses used chiefly in commutated data channels. Again, this is a means for providing a check on the airborne system, particularly for correcting system drift during the flight. The third technique employs the use of the actual flight data and is one which is utilized whether or not the first two techniques are available. This process involves the use of known check points along the trajectory to remove any bias errors. For example, pressure and temperature measurements on the launcher should read the current ambient conditions; at extreme altitudes, pressure is zero; in free flight, accelerometers should read zero; etc. The careful use of actual data does much to improve the calibration data obtained on the ground.

The data-reduction equipment is also calibrated as a system, except for tape playback units which are checked separately, by introducing known step functions to each channel. Automatic calibration equipment is normally utilized to provide accurate calibration steps of up to 11 points. Although current telemeter systems have become relatively stable and linear, increased accuracy requirements still make it mandatory to correct for as many system errors as practical.

4 Computer Data Processing

The data obtained from the data-handling and -conversion processes are fundamental: they are corrected so that system errors are reduced to acceptable limits, and they are expressed in the units of the particular measured phenomena. But this is not the end of the data-processing chain of events. In order to interpret results, it is usually necessary to provide further computations which introduce derived parameters and problem solutions based on telemetered data, and such appropriate supplementary data as meteorological soundings, survey data, and tracking data. Computer data processing comprises the mathematical services to edit and smooth basic data, to obtain parameters describing the physical environment which are either technically impossible and/or too expensive to measure directly, and to provide analytical solutions which will aid the experimenter or tester in evaluating test results and will guide him in making decisions.

The computations involved can, of course, be accomplished by hand calculations provided the number of points to be examined is small. In many test situations, such as the missile-test applications, the amount of data to be handled and the computations required are enormous. Hence the need for a high-speed digital computer is unquestioned if basic data handling and computations are to be completed so that results can be available within a reasonable length of time.

It can be shown easily that, in the case of computation, if hand calculations were performed to obtain derived parameters, much of the data would be wasted, since no one would have time to perform the work. For example, it is not unusual for an experienced person on a desk calculator to take one to two days or more to perform calculations for one point through a complex solution, such as a flow-field problem; looking at only 500 points could become an impossible task. In fact, it appears that even with a high-speed computer, good judgment must be exercised as to what computations should be made, since time in a large-scale computer can be saturated quickly in handling large amounts of data.

In addition to the straightforward computation processes concerning the derivation of various parameters, computation is used in editing and smoothing of raw data. Editing, particularly, is one of the most difficult problems in the realm of data processing and one, along with smoothing techniques, which is of major interest throughout all phases of the work. The need for the editing and smoothing process is inherent in measurement systems in that the data processor is required to deal with such problems as discontinuities of sampling systems, omissions caused by failures, antenna breakdown, signal dropouts, crosstalk, and noise of all types. The function of editing is that of systematically identifying, discarding, and replacing bad data points, and that of smoothing to remove random noise.

While it is true that digitized data can be plotted and smoothed by means of French curves and other manual methods, these methods are not repeatable or describable. To overcome these objections to hand-smoothing techniques, numerous methods, which use computational techniques to accomplish the same end result by definable and reproducible methods, have been developed. Another advantage of using computational techniques for accomplishing smoothing and editing is that the errors introduced by these techniques are definable and known. The problem is further compounded when data are missing, because missing data necessitate possible use of curve fits of unknown validity. Particular techniques will be discussed in more detail in Sec. 14.

The use of a computer requires that a library of general and specialized subroutines and programs be available to meet the needs as dictated by the requirement of advance studies as well as of flight-test work. In order to meet schedules, it is mandatory that the bulk of the programming be completed before the test program gets under way. Hence known requirements are collected, data-handling procedures and computer programs specified, and programming completed. The problems encountered in preprogramming and possible methods of approach in allowing for flexibility required, ensuring the compatibility of input-output, etc., are discussed in Sec. 15.

5 Data-processing Analysis

Data-processing analysis is the development and the systematic and detailed study of requirements, techniques, methods, and procedures to provide current and

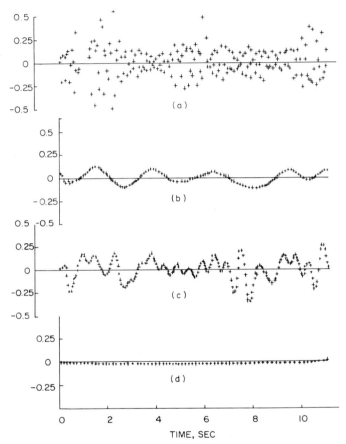

FIG. 4. (a) Raw data. (b) Digital filter, 0.2 to 0.5 cycles. (c) Digital filter, 0.2 to 2.0 cycles. (d) Polynomial fit second degree. The polynomial was fit over the entire span. The points are shifted from the axis slightly so that they are distinguishable.

advanced data-handling processes, problem solutions, and error estimates—all based on sound mathematical and physical criteria. Although the processes described thus far represent the bulk of the work with respect to handling, condensing, and performing computations on large amounts of data, the analysis function is the integrating element of data processing. In brief, the broad function of data-processing analysis is to describe and attest to the validity and accuracy of tests and the *information* yielded.

It is obvious, from the definition above, that the role of the analysis function in data processing is broad and that it encompasses all phases of the work. For example,

DATA-PROCESSING SYSTEMS 12–11

in the data-handling and -conversion area, analysis is required to determine adequacy of equipment—such as optimum filters—to reproduce the phenomena to be measured, the optimum sampling rates to be used based on a knowledge of further processes required, basic editing techniques, adequacy of calibration data and the application thereof, and error estimates as to the validity of the data as processed to that point.

Within the area of computer data reduction, the role of the analyst expands to that of detailed analysis of initial requirements to determine adequacy of measured data for computation of derived parameters, the editing and smoothing of data, the detailed specification for computer programs, the follow-through to ensure accuracy of computations, and estimates of error introduced through the various approximations necessary in the solution of complex problems on digital computers.

Although the data-handling and computation phases normally become somewhat routine in defined areas, the analysis function does not end here. The data results must be examined critically in terms of reasonableness and validity within the framework of the mathematical criteria established, and reports must be written to provide the data user with a systems analysis with respect to the data obtained, the accuracy and adequacy of the final presentation, together with recommendations as to possible improvements to be gained through new techniques or instrumentation changes, and/or deletions to improve basic measurements.

This responsibility for continued progress in new and improved techniques is a basic function of data-processing analysis. In addition, there can be numerous special studies required to obtain a more comprehensive picture of test results and the physical relationships involved. For example, Fig. 4b and c shows two of the curves resulting from a frequency analysis of a set of pressure-difference data obtained from actual flight conditions. Curve a is a plot of the raw data. The technique used was that of digital filtering; i.e., a mathematical model of a bandpass filter is constructed and a series of weights computed for various desired bandwidths. This material was analyzed to allow detection of dominant frequencies present to show correlation with body motions; but the example is used to show the use of filters for smoothing applications and to demonstrate the powerful techniques available to the data processor. The ever-present possibility of inadvertently distorting basic data is shown by curve d, which was obtained by fitting the data to a second-degree polynomial in a least-squares sense. It can be seen that actual trends could easily be masked by the indiscriminate choice of method. For discussion of techniques, see Sec. 14.

In summary, sound analysis is required in the selection and application of techniques to ensure that the user is provided with data with which he can confidently interpret test results. See the fourth part of this Chapter for further discussion of data-processing application analysis.

DATA-PROCESSING SYSTEMS

6 The Systems Approach

Note the definition of "data processing" used in the introduction to this chapter. To repeat, it is the system which provides the information, sought by means of a test, by manipulating raw data to final-answer form. The term "system" is used here in its broadest sense. It includes machines, personnel, procedures, forms, plans, and other elements required to complete the data-processing job. It would be foolhardy to attempt to describe *every* possible system for handling all combinations of input data and output information requirements; instead, a limited number of examples, which are deemed to be representative of telemetry and remote-control problems, will be presented. In the interest of economy of space and avoidance of duplication, specific equipments will not be described in great detail; this type of information is available in other sections of the handbook or may be obtained from the equipment manufacturers. However, equipment performance characteristics and sources for further information will be included wherever such information is required.

As indicated, the most desirable input forms to a data-processing center are those amenable to automatic handling. However, to reduce supplementary data and to render services to the engineer performing tests in which direct-recording techniques are the only means available, automatic and semiautomatic handling are usually required. The ideal procedure to ensure inputs compatible with data-reduction capabilities is for the engineer to meet with data-processing and instrumentation personnel prior to testing and clearly define (1) the purpose of the test, (2) what final answers are required to define the test results adequately, (3) what basic measurements are required to obtain final answers, (4) what instrumentation would best provide the basic measurements specified, (5) what schedules are required to meet deadlines for the overall test programs and (6) what recording medium would provide the basic data, and the flexibility in handling required, to meet schedules.

Data processing is not always represented in test planning. Indeed, planning this function is not always possible; the flexibility to handle a variety of input forms therefore becomes mandatory. The need for flexibility is generally met by providing three basic data-handling systems, i.e., manual or semiautomatic digital, automatic digital, and analog systems. Since the approaches to the work vary greatly, an effort will be made to discuss the processes within the area of the three basic systems in terms of three general considerations: preparatory planning, the process, and final presentation. The preparatory planning is common to all systems and will be discussed first.

7 Preparatory Planning

As mentioned previously, the ideal procedure in planning an experiment is one in which data processing is active in planning the test prior to the actual running of the experiment. Whether or not this activity can be realized, the first need in reducing data to information is the determination of requirements—what is the information to be gained from the test? A complete description of the test should be obtained whenever it is possible to do so; however, there will always be crash programs for which no time exists except to determine that data are needed. Included within the area of the test description should be performance details of the airborne hardware, test procedures, equipment used to record data, description of pretest calibration procedures, telemetry recording speeds, compensation and speed-lock frequencies, channel allocations, and data-reduction requirements. The description may be provided by the test engineer or it may be a product of data-reduction personnel assigned to the test.

Calibration Considerations

Just as the value of an instrument is dependent upon its calibration, the information obtained from the data-processing system is dependent upon calibration data for its usefulness in providing adequate and accurate information. In the planning stage it is essential to data-processing personnel to examine the calibrations available to them so that a determination can be made as to whether the accuracy requirements can be met under the given set of conditions; thus, a two-point calibration of an instrument, the linearity of which is unknown, can hardly be expected to yield data of 1 per cent accuracy. It is therefore essential that the tester, the instrumentation man, and the data processor recognize limitations imposed by the calibration data available. Depending upon the particular application, calibration data may be provided in various physical forms: tables, graphs, and voltages recorded on magnetic-tape or oscillograms, for example. In a similar fashion, the number of calibration points used may vary from two through eleven or more.

The adequacy of the calibration data provided must be determined by a thorough knowledge of the test, the data-reduction process, and the accuracy required of the final answer. It is normally a function of a data-processing group to accumulate, to convert, to present, to publish, and to distribute calibration data associated with the tests with which they are concerned. The data-processing function can perform a valuable service at this point in feeding back information to instrumentation, to quality control, and to other interested groups relative to the accuracy and adequacy of calibration data. This feedback frequently might include suggestions for reducing, as well as for increasing, the degree of testing and tolerance requirements.

Supplementary Data

Ideally, again, it would be desirable to have all required test data recorded automatically on one magnetic tape which then could be processed automatically to final-answer form. The prevailing situation, however, is radically different because of the limitations of telemetry instrumentation systems to provide all the data required. The causes for these limitations are numerous; for example, information bandwidths are restrictive, technical difficulties are encountered in economically accomplishing certain measurements by means of telemetry, weight and volume of airborne instrumentation items are limited, or expensive and adequate ground-based instrumentation systems are in place. These nontelemetry, or supplementary, data are those associated with such items as trajectories, timing signals, external environments, and other information without which a missile or aircraft flight test cannot be fully understood. In the preparatory-planning phase, details of the supplementary data are studied carefully to determine how they can be combined with telemetry data to best meet the objectives of the test. As an example, it is virtually impossible to analyze telemetry data without associating them with the data obtained from trajectory measurements. The data-processing system must therefore provide the means by which photographic, tracking, radiosonde, timing, and many other types of data can be welded together to perform analyses of tests in a rapid, accurate, and complete manner.

Data-formats Considerations

We assume that the information desired from the test and the measurements taken by the test are now defined. An important consideration is that of the formats to be used for both the input to and the output from the data-processing system. It has been found expeditious to define standard input-data formats for consistent use in the reduction process. For example, if punched cards are used, the definition of a fixed number of fields with a fixed number of digits, plus ample space for identification, is useful. A *field*—as related to data processing—is the assignment of specific character positions utilizing any recording media to allow standardization of input-output formats. Establishing standard formats for particular types of telemetry simplifies and speeds the automatic-digitizing process and the production of analog records. The use of standards also allows the programming of generalized computer programs to ease the data-handling problem. Similarly, the form of the final presentation of the data will influence the process, provided a choice exists. For example, if plots of function value vs. time are all that is required the analog process may be adequate. Or, using semiautomatic equipment, it may be possible to obtain automatic plots directly from the data readings. If further computation is required, the fastest method of digitizing should be used to provide computer inputs in a fixed format. With complete input and output format design accomplished, the data-processing system which can best accomplish the transformation can be selected.

Computer Considerations

The increase in complexity of modern military systems with parallel increase in telemetry complexity and utilization has made the need for high-speed computation a necessary element of the data-reduction process. The need for new or modified computer programs must be examined long before the tests are to take place because programming normally requires significantly long lead times. Although automatic programming techniques are becoming more available, a significant time is still required to code, to assemble, and to check out programs for actual use. There are numerous technical decisions to be made relative to the numerical techniques to be used for smoothing, integration, differentiation, accuracy checks, and for the solution of a variety of equations. It is essential that a large percentage of the required programs and subroutines be completed and fully checked out prior to the time that the first test is conducted; otherwise the delays encountered in turning out test results will reduce the effectiveness of the entire test program.

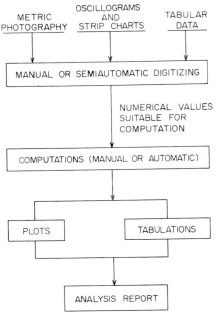

Fig. 5. Simplified flow of manual or semiautomatic data processing.

8 Manual and Semiautomatic Processes

Where the data reaching the data-processing center are on media other than magnetic tape, punched tape, punched cards, or other forms compatible with electronic computers, manual or semiautomatic methods are needed. Figure 5 illustrates the typical flow of this type of data in deriving answers from raw data. The figure is simplified in that many essential elements are omitted; for example, calibration data are assumed to be included as one of the three types of inputs, and data in proper digital format may be introduced into the computation from other sources. The manual methods may be disposed of directly by stating that these use people with rules, paper, and pencil to accomplish the same tasks which the semiautomatic machines, described below, do much more rapidly and accurately.

Semiautomatic processes are those techniques by which the editing, the selecting, and the reading of data are completed by skilled personnel using machine aids. The data are read by means of manually setting crosshair(s) on the selected point; the automatic process of the machine is the translation of the movement of the crosshairs into reader counts and the provision for readout via visual display for manual recording, or via typewriter, punched card, or tape for automatic recording. Such motion is translated into digital form by means of mechanical coupling of a shaft or position code disk. As the shaft rotates, electrical contacts on the disk are closed to construct the desired digital code. The process includes the direct manual measurement by use of fixed or variable scales to obtain digital data and, in some cases, is still used in current practice as the best method for particular applications. For example, it would be wasteful to use any type of equipment to digitize event data (normally step functions fixed in time and occurring only once) unless the particular data were read out as a by-product of the digitizing of other information required.

There are two broad categories of semiautomatic equipment in use: metric film readers and oscillogram readers. The two types are different in that film readers normally require projection systems for magnification and are two-axis (X-Y) readers; oscillogram readers, in general, have no magnification systems and are one-axis readers. There are exceptions, of course, and a single machine may have combined features. The choice as to which readers to obtain, with respect to type and resolution, should be determined by the particular requirements of the facility. The semiautomatic process is used to convert analog data in the form of oscillograms, strip charts, graphic data, and film of plate data to digits. The methods used are all basically the same while the type of equipment and the mechanics will certainly vary. To discuss the general processes, we must treat the two basic types separately.

Film Data

In handling film data, the film is first previewed carefully to (1) check the general quality of the film, i.e., contrast image quality and calibration shots (target boards, star shots, etc.); (2) ascertain if timing is recorded on film to assure consistency throughout the run for the purpose of ensuring adequate correlation; (3) determine whether total coverage is obtained together with history of missing data (this is important if the desired results require multicamera coverage in the solution); (4) in some cases, note which of the frames are to be read; (5) in all cases, make sure that identification is cross-checked and marked accordingly; (6) determine, to save wasted effort, which of the records or portions thereof are usable in the reduction; (7) preview timing records associated with the test, in particular where time is not recorded in the film and correlation must be established by shutter-return signals or by other techniques. There will be other details to be checked within any of the previous items peculiar to the type of film and source. The results of the preview not only are important to data reduction but are the basis for important feedback to the data-acquisition people for improvement of subsequent test coverage.

Once the film coverage is established, the actual reading process can proceed. The different reading equipment currently available is diverse, varying from simple preview projection mechanisms—for which calibrated overlays can be constructed and readings made directly—to general, all-purpose readers and specialized types for particular camera outputs. Unless the particular facility has reason to handle large amounts of film repetitively, most requirements can be met by selecting the most flexible machines possible by relying on the test range to reduce the bulk of the specialized photographic records. The important points to consider in film-reading equipment follow.

Film Size. Readers can be obtained that will accommodate the standard film sizes—i.e., 8, 16, 35, and 70 mm—either through interchangeable heads housing the optics and transport system or by design of the single film transport. In addition, the same machine may handle film plates up to 5 by 7 in. Preview equipment should be considered in the same way. Although the reader itself may be used for preview purposes (and is in many instances), it is convenient to have minimum preview equipment available in the form of still and motion projection, simple light boxes or equivalent, and machines built specifically for this purpose.

Film Alignment. The problem of film alignment during the reading process can be handled in two ways. First, some readers have adjustable film transports with control knobs within easy access of the operator. Hence, with a fixed crosshair system, the film reference fiducials can be aligned by slight adjustments of the transport as the reading progresses. The second and most commonly used type employs the fixed film transport. Since most film readers are fixed two-axis readers, this property can be used to read fiducial marks or any two known points, such as target boards, between which the horizontal angle is known. These readings are differenced and the angle computed. If the angle differs from that expected, the data readings are rotated to the desired axis. Since the process is included within the computer program, the rotation is automatic and represents little additional work. Indeed, even if hand calculations are necessary, film alignment in this way is much faster than attempting to keep the film perpendicular to the machine's horizontal axis.

Resolution. The accuracy achievable is dependent on the resolution and the repeatability of the particular machine. If the film quality were perfect, the resolution and repeatability of the reader could represent the absolute limit in accuracy obtainable. Perfection of film includes such characteristics as proper contrast, good and consistent image definition, and no distortion by lens, film, or other elements of the system. The resolution is defined in terms of reader counts and is usually associated with counts per centimeter or inch. Repeatability is the capability of the man-machine combination to read a point, move off the point, and then reread the same point within a specified count limit, assuming no operator error. The repeatability of current equipment varies from one count on up. However, one count can represent 1 micron, such as on the fine ballistic plate readers (lower limit for screw-driven machines), to 10 microns as associated with certain powered comparators, and on up to thousandths or hundredths of an inch. An application of the interferometer principle applied to ballistic plate reads coordinates in terms of light fringes yielding extreme precision. Resolution is also a function of magnification. However, the limit here is normally the film because the image quality deteriorates rapidly and "washes out," owing chiefly to grain size of the emulsion.

Human Engineering. The film-reading job is a fatiguing one. Studies have shown that, even under the best conditions, reading accuracy and efficiency of the operator fall off rapidly after 4 hours of continuous operation; hence the construction of the reader is an important part in minimizing operator fatigue. Such features as the angle of the projection surface and the operator position to minimize parallax are extremely important; location of keyboard or dials for manual insertion of data such as time, frame count, data or identification, light intensity, operator controls, and readability of display should be considered.

Output. The output of the equipment should be automatic, and it should be recorded on a medium suitable for minimum handling to yield compatibility with the rest of the data-processing facility. Again, the use of standard output formats can be used to advantage.

Oscillogram Data

As in the film-handling process, the preview of the records in an important step in the procedure and is accomplished in this case to review record quality, determine that calibration data are present, identify the traces frequently throughout the record, indicate the particular data to be read, review the timing trace for consistency, determine total coverage, and indicate any portions to be omitted because of excess noise or drop-outs.

The preview of oscillogram or strip-chart records is implied by use of standard preview machines. No recommendation is offered, since the current machines are all comparatively the same.

Paper Size. The first consideration in specifying a reader is the size to be handled; i.e., provision for widths up to 12 in. and 450-ft lengths is required. Most of the current machines are adequate in this respect.

Alignment. In setting up the record to be digitized, the alignment of the paper on the reading surface must be precise. Here, the alignment is accomplished by obtaining the same count value for a reference line at opposite ends of the record. This process is not stringent, since most readers accommodate a minimum of 2 to 3 ft of record before advancing to the next portion. Although errors due to poor alignment are small, consistency is not obtained unless reasonable care is exercised.

Resolution. Unlike film readers, most oscillogram readers contain no projection systems—hence no magnification. Back lighting of the reading surface is the only aid increasing the legibility of the record. The resolution of the equipment depends on the type. One type is the fixed-scale reader. Representative resolution of this type is 300 counts per inch with a repeatability of one count. The variable-scale reader actually has infinite resolution, achieved by the capability for setting up the machine so that the vertical "crosshair" can be fixed at any desired slope. Hence the limitations are (1) the ability to read the intersection of two lines as the angle between them becomes smaller and (2) the limitation of the machine in terms of the number of digits to be handled. For this type of machine, 1,000 counts per inch is the best currently achievable with an expected repeatability of one count.

Reading Process. The reading process differs from that of film basically in that the amplitude only is read so that the independent variable, such as time or frame number, has to be inserted indirectly. In some cases, the reader can be programmed to advance an initial setting by unity, automatically reading out the point number

DATA-PROCESSING SYSTEMS   12–17

(time) if readings are taken at one per second. In other cases the data must be added manually.

Calibration. The use of calibration data varies in that there may be several types to be applied through the run. The first of these is the airborne sensor calibrations, represented by a single curve for each sensor or type of sensor. This calibration may be applied directly to the readings by constructing an overlay which replaces the linear reading arm. This process is used only under ideal conditions. Normally, the nonlinearities are removed in the computational process.

The second type of calibration is that applied to the ground conversion system in the form of a step function. The steps are read and can be checked for linearity separately or left as a normal input to the computing process.

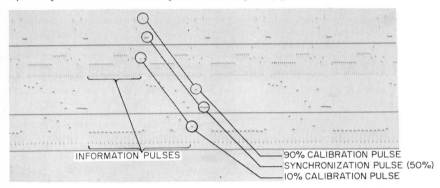

FIG. 6. Examples of commutated wave trains. The upper trace is from a 30 × 5 commutator, the lower trace from a 30 × 2.5. (The first number indicates number of segments used, and the second the speed of the commutator in revolutions per second.)

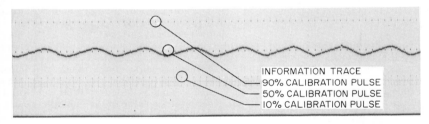

FIG. 7. A portion of an oscillogram record illustrating a decommutated trace with calibrations pulse presented as the same galvanometer.

The third type is the preflight calibration (pretest) to yield last-minute information as to the status of the system (sans sensors). Here again, the values are represented as step functions and are used to correct nonlinearities of the airborne system. If preflight steps are not recorded, the latest ground-test data obtainable are used.

The fourth type of calibration is the in-flight data. This can be in the form of steps, as in the previous case, or in the form of continuous calibration presented as pulses of the commutated data. Figure 6 shows an example of a wave train with calibration pulses indicated. Figure 7 shows a portion of a record illustrating a decommutated segment with calibration data presented on the same galvanometer to minimize error. Figure 8 shows an actual in-flight calibration "break." The term "break" is used since the in-flight calibration is achieved by interrupting or breaking the circuit carrying the actual information and inserting the rotary stepping switch output at regular intervals; the technique is required only on continuous channels.

The last technique, as mentioned in the introduction, is the use of actual data as calibration points to remove known bias errors. The check points required are numerous and are determined by careful examination of the physical conditions existing during the test. For example, an excellent check on pressure data is obtained by reading a point at launch where ambient conditions are accurately known, and another point is obtained by comparing readings anywhere out of the atmosphere (vacuum). The calibration curve can be shifted to coincide with the two measured points; it constitutes a valid means for obtaining refined measurements.

With the calibration types reviewed, the general methods for reading can be stated.

1. With a fixed-scale reader, the best technique is to read data points referenced to a static trace (the edge of the paper if necessary) or any other static point. The readings can then be differenced to obtain amplitudes. This method requires that all the calibration information be read and corrections applied during the computational process, either by hand or by machine.

2. The variable-scale machine can be used in the same way as in method 1. However, more flexibility is possible in that the reader can be set up to read directly referenced to particular calibration pulses or step functions. Thus, if calibration pulses on commutated data are at 0 and 100 per cent, the reader could be set up to read 0 to 1,000 counts corresponding to 0 to 100 per cent, or minimum and maximum function values, respectively. This operation assumes that the calibration pulses are stable. This allows output in function value directly. However, the chief advantage of a variable-scale reader is that resolution can be adjusted to requirements. As familiarity is gained by operators, some improvement in techniques and speed can be realized with semiautomatic equipment. The chief disadvantages are slow reading speeds of large amounts of data and the deterioration in accuracy due to human-operator fatigue.

FIG. 8. Portion of oscillogram record showing a continuous trace with a typical in-flight calibration where transmission is interrupted for short periods regularly over the flight interval. The five steps normally represent 0, 25, 50, 75, and 100 per cent information levels.

Final Presentation. Final presentation of semiautomatic output normally represents intermediate steps in the overall process. The major reason for digitizing is to obtain a data form compatible for further digital computation. Although the process is used as a means for condensing the results of long tests into plots of time vs. function as final outputs, the final format is usually designed to be compatible with the input capability of the particular computer available to the reduction center. This does not affect the capability of direct automatic plotting since, on most machines, only two fields can be plotted simultaneously, and the position of the fields is not significant. Plotting the results of further computations will normally require rescaling so that needed manipulation of formats can be achieved during this process.

The chief element in considering final presentation is the determination beforehand of the customer requirements. Much time can be consumed in having to redo portions of completed work owing to an inadequate understanding of what the particular responsible engineer wants to present. Although the rework of data is a necessary evil at times, experience has shown that a well-planned effort can keep this expensive activity to a minimun.

In summary, the semiautomatic process is an *important* one for filling in the gaps which automatic equipment cannot handle.

9 Automatic Digitizing Systems

For conversion of analog data to digital form, there are a variety of choices in equipment and accuracies. The simplest, least accurate, and least expensive is that just discussed, i.e., the semiautomatic system. Limited by human-operator speed and eye precision to about five data points per minute and 500 to 800 counts per linear inch, these machines are adequate for a minimal system. With advent of high-speed analog-to-digital electronic converters, it is possible to convert to digital form early in most playback processes and to avoid most of the inaccuracies inherent in analog-system conversions.

There are several basic techniques presently used in the conversion of electrical analog voltage into digital voltage representation. The choice as to which technique is to be used is dependent on the form of the electrical analog inputs and the intended use of the digital output. One of the earliest concepts, which is used extensively on semiautomatic equipment, is that of continuous-balance electromechanical servosystems. These devices are usually based upon a resistance-bridge network with mechanical servomechanisms reestablishing null conditions caused by variations in the electrical analog input and at the same time providing the digital output. Because of the comparatively slow response time of servomechanisms, the "sampling rate" of these devices is limited to less than 20 per second and ordinarily to 1 or 2 samples per second.

If the electrical analog input is of a frequency-modulated form or can be conveniently converted to such a form, electronic cycle-counting devices may be used to convert to digital output. The chief advantage to this type of conversion is the extreme precision obtainable at moderate sample speeds. This advantage of time-domain conversion is lost if conversion from an amplitude-proportional electrical analog to a frequency-proportional signal is required. In addition, the conversion speed is limited by the counting rate of present-day electronic counters. The sampling rates for conversion devices of this type range from 100 to 1,200 samples per second.

Several methods of analog-to-digital conversion, using voltage-comparison techniques, have been developed. Accuracy limitations result because of the necessity to make an analog feedback comparison with the input analog voltage sooner or later. Stability of comparison circuits becomes difficult to push beyond 12 binary bits. Anodige and "sawtooth" comparison coders operate in the following basic manner: Initially a counter is reset to zero. When a start pulse is given, a master oscillator of fixed frequency is gated into the counter and a "staircase" ramp (in case of anodige) or a smooth "sawtooth" ramp (in case of "sawtooth") is also initiated, and rises from zero. When the ramp amplitude matches the amplitude of the input analog voltage, the master oscillator frequency is gated off. Thus the counter which has been counting cycles of the fixed frequency now contains a count proportional to the input analog voltage. Suitable electronic gating provides the digital readout at the proper time. A variation of the counter technique is to use a "reversible" counter with analog feedback in such a way that the counter continuously adjusts its count contents to maintain zero voltage difference between input volts and feedback volts.

Another type of analog-to-digital converter uses "successive approximation" and works like an automatic self-balancing bridge. When a start pulse is given, a train of switching pulses causes a switch to insert a fixed current proportional to its bit position into a ladder network. At the switching intervals a comparison is made between input and feedback from the network to determine whether there is too little or too much. If too little the 1 remains; if too much the 1 is returned to 0. Controlled by the pulse train, the converter proceeds through all bits from most significant bit first to least significant bit last; conversion is complete when the least bit has been examined.

All the above types of voltage-to-digital converters have been successfully used and are commercially available in transistorized or tube versions. The chief advantage of intermediate conversion to voltage technique is that resolution (sufficient for a large area of physical measurements encountered) can be attained simultaneously with high sampling rates. Sampling rates of 40,000 samples per second with resolutions of 1 part in 1,023 are not unusual.

One of the greatest users of automatic analog-to-digital conversion systems requiring high resolution and high sampling rates is those groups interested in reduction and handling of FM telemetry data. There are several points at which conversion of telemetry data to digital form can take place. By referring to the simplified block diagram (Fig. 9) it can be seen that the selection of where to interrupt the analog system for conversion to digits is a compromise choice between the desired sampling rates, output device, and precision desired.

The last practical point for automatic digitalization is that point just prior to the analog recorders (i.e., immediately following the discriminators, continuous FM, or immediately following the time-multiplex decommutators, PAM or PDM). Most systems start digital conversion following the FM discriminators and realize 0.5 per cent full scale precision or better. By addition of specialized synchronization circuits for PAM and PWM type telemetry signals, the recognition of synchronization pulses and data pulses allows the optimum sampling rate to be determined by the signal itself. The primary reason for using the lowest possible sampling rate for any

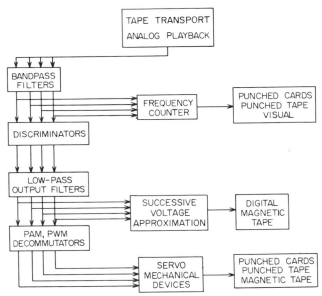

FIG. 9. Block diagram, illustrating the various points a signal can be picked off for digitizing process.

signal can be attributed to the compelling desire to reduce running time on digital computers. If greater precision is desired and economy is not a limit, special bandpass filters can be used and the digitalization can take place entirely in the time domain. Care must be taken not to violate sampling theory in any case. The ultimate practical limit in such systems is approximately 0.1 per cent reproducibility with analog recorded "flutter" noise being the limitation.

The format of the outputs of the analog-to-digital conversion devices is usually devised to assure compatibility with the available digital computing equipment. In most cases, a digital magnetic tape provides the most easily manipulated output medium for analog-to-digital converters. The use of digital tapes also provides an effective buffering system to assure optimum scheduling of both the available computing facilities and the analog-to-digital conversion equipment. Once conversion to digital format has been accomplished in a medium which is compatible with the computer to be used, the linearization, decommutation, editing, and derivation of test parameters can proceed in the variety of ways available to digital techniques. The biggest problem then facing the data handler after computation is that of presentation.

Preparation for Automatic Digitization

Although a large-scale electronic digital computer is capable of handling very large quantities of data it can be saturated easily if it is required to handle all the telemetry data obtained from a long test. As an example, it is not unusual for a flight test to yield as many as 10 magnetic tapes each of which contains 200 data channels for 15 min of recording. Even at a low rate of sampling, say 10 samples per second, this amounts to 10 tapes $\times$ 200 channels $\times$ 900 sec $\times$ 10 samples per second, or 18 million data points. Fortunately there is little real need to digitize whole tapes of telemetry information. Instead, the contents of the telemetry tapes are first presented in oscillogram form so that the data can be scanned and those portions of interest marked accordingly. Within the portions of interest, the records are examined to obtain such information as approximate subcarrier deviations, the quality of the time code, drop-outs, and approximate function values. In addition, the oscillogram will show those areas where automatic digitization will prove difficult because of excessive noise. This preview simplifies checking the results of the automatic digitization.

There is a need for providing at least minimum tape duplicating and time-code generating capability so that a good time code, compatible with the facility time-code translator, can be recorded on the information tape. The duplicating capability is normally a necessary element of a data-processing facility. Although the original recording may be good, experience has shown that timing signals are troublesome. The need for a well-defined time code is twofold. First the time correlation of sampled data is required to a reasonable degree of confidence—in most cases to 1 msec. Although the dubbing of a secondary time code can cause some distortion, because of a variation in tape speed, the ability to correlate functions accurately far outweighs such errors. Since most automatic digitizing systems include time-code translators which permit the selection of automatic start and stop times, the need for a good time code is apparent. The second need, then, is to facilitate the use of the equipment in choosing the intervals to be digitized so that the computer facility is relieved of excessive conversion time. Additionally, in reviewing and editing data, the analyst can spend his time more profitably by restricting himself to the selected areas of interest. Finally, the process allows the compilation of telemetry data from multiple tape coverage without excessive overlap.

The automatic digitizing system is a complex one and must be treated as such. Hence the operator-maintenance personnel must be highly qualified technicians in order to maintain ready operational status of equipment. In addition, the training must include the ability to recognize when the equipment is performing correctly during the data runs. Otherwise excess conversion time will be used on the computer in handling bad data. It is important then to (1) develop and sustain a complete preventive-maintenance program (experience throughout the industry shows that performance of equipment is related directly to the quality and especially to the regularity of such a program), (2) develop diagnostic-type runs for aid in trouble shooting as well as for use in the preventive-maintenance program, (3) develop and constantly improve setup and operational procedures consistent with the capabilities of the equipment, and (4) devise and maintain high standards for the calibration of the system.

Additional Equipment Factors

In preceding paragraphs much attention has been paid to the selection of the analog-to-digital converter. In operating practice, other elements give cause for more concern. The analog-to-digital converter can easily be the most precise and reliable element in the system. The reader's attention is directed now to considerations of magnetic-tape properties, characteristics of tape-handling mechanisms, and FM discriminators. The most commonly forgotten characteristic of magnetic tape is its dimensional stability under changing temperature and relative humidity conditions, as shown in Table 2. Precise time correlation can be affected for long time intervals over extended tape lengths recorded under field conditions. If yield strength

Table 2

	Acetate base	Mylar base
Temp. elongation fraction (per °F rise in range 30–130).	3×10^{-5}	2×10^{-5}
Relative humidity elongation factor (per % relative humidity rise in range 20–80).....................	15×10^{-5}	1.1×10^{-5}

is exceeded, stretch in excess of 1 per cent in very short lengths is possible but not so likely to go unnoticed as the temperature and humidity effects. Reproducible amplitude magnetization (B-H curve) along the tape is rarely attainable to better than 10 or 20 per cent. For this reason alone, all high-quality instrumentation systems designers employing magnetic tape for analog recording will adopt some form of frequency modulation. Frequency modulation imposes stringent requirements for speed stability on tape-handling mechanisms. The better machines guarantee 0.4 per cent peak-to-peak "wow and flutter" cumulative over the band from direct current to above 2,000 cps. "Wow and flutter" mean apparent frequency deviation peaks expressed as a per cent of speed and are derived from their source within the mechanism. Lower frequencies—wow frequencies—arise primarily from eccentricity of the drive capstan cylinder and bearing runout. Higher frequencies—flutter frequencies—arise from the "violin string" length of tape unsupported between head and pinch roller and roughness of tape and head surfaces. Other sources are due to dynamic tape stretch and roughness of tape and head surfaces. The effects of "wow and flutter" can be corrected only partially within the tape mechanism. Nevertheless, the "wow" correction obtained is very valuable and is sufficient for wideband (± 40 per cent deviation) systems. The most usual correction method involves first prerecording a frequency containing the desired capstan drive frequency as an amplitude modulation. Within the playback machine a phase-locking servocontrol minimizes difference between the prerecorded tone and a local precision standard frequency. Standard systems correct up to 0.1 cps variations in this manner; special machines are in use which provide significant improvement up to 20 or 25 cps.

In any event for narrowband FM ($\pm 7\frac{1}{2}$ per cent or ± 15 per cent deviation) telemeter systems, the higher flutter frequencies must be removed electronically by retuning the individual discriminators. For this purpose, a high-frequency tone such as 100 kc is prerecorded and multiplexed on the same track as data (skewing effects require use of a tone on each track). A reference discriminator is used in the playback process and puts out a difference voltage corresponding to the difference between the frequencies of the prerecorded tone and a stable local reference oscillator. The difference signal thus derived is used to retune the data discriminators. At first glance one would be tempted to drop the wow compensation if he has flutter compensation. However, the flutter compensation corrects instantaneous amplitude only; the time interval could still be wrong unless the tape speed is accurately controlled. Therefore, good practice requires inclusion of both "wow" and "flutter" compensation. Some manufacturers claim 40 to 1 or better improvement for their various flutter-compensation circuits. A more practical value is about 20 to 1, which can be maintained without excessive use of elaborate calibration equipment. The importance of flutter is seen by the following relation:

$$\text{Per cent flutter} \times \frac{100}{\text{per cent deviation}} = \text{per cent flutter noise in data band}$$

FM discriminators commercially available have easily demonstrated linearity of 0.1 per cent of best straight line and drift figures less than 0.1 per cent per hour. Analog multiplexers also yield 0.1 per cent figures if carefully designed. It will be noted that the dynamic accuracies of the system have not been discussed, primarily because adequate measurement methods are not currently available. However, instrumentation engineers agree that static accuracy values are still usable under dynamic conditions and take great pains to remain within the confines imposed. Experience indicates that the input bandpass filter discriminator low-pass output filter combination must provide a minimum of 40 db margin between adjacent subcarrier bands and approximately 50 to 60 db amplitude range of input signal level.

The most satisfactory equipment calibrations are obtained by insertion of known standard static frequencies as inputs to the system. As an overall system calibration, a few standard prerecorded tapes are quite satisfactory, particularly for equipment-readiness checks, and can be used in conjunction with data runs to eliminate much unnecessary "knob twisting." Provision should be made for observation of intermediate points between elements such as recorder amplifier output, bandpass filter, discriminator, and various trigger and synchronization points. Benefits to be derived from these additional test points are increased operator confidence and speed-up of trouble shooting.

One more warning—digital representation necessarily involves quantization and sampling. Sampling is *not* filtering; it is frequency folding. It is suggested that the reader refer to Chap. 2 of the handbook. The point is made again at risk of repetition because the digitizing process is not generally so well understood as analog processes. One cannot reduce the sampling rate without first reducing the input analog bandwidth. For this reason inclusion of adjustable cutoff frequency or plug-in low-pass filters will be found most useful in adjusting sample rate.

The Analog-to-Digital Process

The setup and operation of an automatic analog-to-digital system will be different for each type and installation; therefore, it is of little use to attempt a detailed discussion. As indicated in Fig. 10, the output of such a system is a magnetic tape containing digital information normally in binary form with format suitable for direct input to the particular computer employed by the data-processing center. It is possible that an additional translation may be required to convert from one computer format to another; such equipment is available. The next logical step is the conversion of the data to either function value or straight counts, so that the data can be checked. Again, the analog-to-digital system may include the capability for tabulated readout, or the computer facility may be considered a part of the system. Conversion, using a computer, is discussed in detail under Computer Data Processing.

After obtaining the printout of the data, which have been calibrated and converted to function value, the results must be checked carefully in terms of the known physical conditions inherent in the test. Even though the equipment setup may be satisfactory, the possibility of a bias being introduced is always present. A method of correcting for such biases is the use of known conditions to correct the data before further computation, that is, the use of the test data to refine the calibration points. Several examples have been mentioned previously, and the point is reiterated here to stress the importance of adjusting the data in a valid way to achieve the best obtainable results.

The most troublesome aspects in reviewing the raw data and preparing the inputs for further computational processes are (1) how can a large amount of digital data be checked carefully without consuming excess time and (2) what techniques can be employed for editing and smoothing of data including removal of obviously bad points as well as fill-in of missing data? The first problem is usually answered by using high-speed plotting techniques achieved through whatever equipment is available. One way to obtain high-speed plots is through the use of high-speed printers associated with computer peripheral equipment. Programs have been developed which utilize the print positions as plotting positions. By restricting the printout to time vs. one function, a tabulation and corresponding plot can be obtained. This type is particularly helpful in reviewing data. Another approach is the use of high-speed mechanisms specifically designed as plotters or printer-plotter combinations. Several prototypes have been built to date with varying degrees of success. However, progress has been and will continue to be made since the trend is toward the handling or more and more information requiring increased plotting and printing capability. At any rate, high speed is achieved by simultaneous multiple plotting and through the use of high-response media. The techniques for handling the data up to the point of recording are basically the same. The writing methods used are numerous, such as the use of a cathode-ray tube and associated photographic equipment, electromechanical mechanisms, xerographic, hot needle, etc. The recording medium may be electrostatic or electrolytic types of paper or photographic film as the method requires. Plotting speeds currently available vary from 100 points per minute to

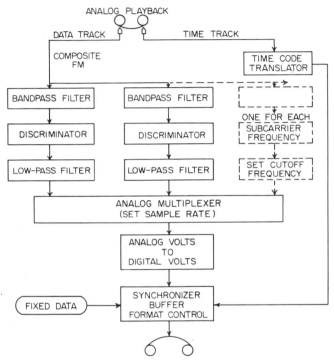

FIG. 10. Block diagram, illustrating equipment alignment for conversion of continuous FM data.

5,000 points per second. The printer-plotter combinations are particularly valuable since plots can be annotated automatically; annotation can become a true bottleneck if the plotting load is great.

The second problem, that of actually editing the data for bad points or missing portions, is necessarily handled on the computer for large masses of data and even though techniques are developed to do the bulk of the work, a great deal of hand work still may be required. In fact, it has been found necessary on some types of rough and noisy data to fair a curve through the measured data manually and to reread that curve to achieve the job of editing. This method is most undesirable, yet it may be the only way available to accomplish the required editing. One may ask, why is editing necessary? The problem arises out of the nature of telemetry or any measurement system; measurement systems are not precise. Characteristic noise is present as well as possible bias. In the computational process, the normal requirement is for smooth data since any noise present is amplified by the numerical

DATA-PROCESSING SYSTEMS   **12-25**

procedure. The smoothing process requires that data scatter be limited for the same reason; i.e., bad points could be weighted too heavily, thus distorting the output.

In summary, automatic digitizing systems provide a means for fast and accurate digitizing of large amounts of data. The capability of present systems is dependent on design, and more so on the quality of telemetry tapes received. Performance of automatic systems for this application should be considered worthwhile if 80 per cent of the overall digitizing load is achieved.

10 Analog-to-Analog Conversion

Analog-to-analog systems, as stated previously, are utilized for real-time display, quick-look reduction, or as the final process as determined by particular requirements. Analog systems are the backbone of the telemetry reduction center because of the flexibility achievable and because of the simple fact that one generally needs to see a visual representation of the information before proceeding to the expense of further processing. Flexibility is achieved through an optimum combination of men and machines. Although the following discussion will stress the machine process, there is little that can be done without good personnel with the special skills required to maintain and operate the equipment properly, set up procedures, and perform analyses.

Preparation for Analog Conversion

The preparatory aspects stated previously apply generally to the analog process. In addition to those steps which apply, some particular considerations are (1) The identification of the tape as to the recording format is mandatory, i.e., track assignments for voice, time, reference frequencies, and data. If IRIG (Inter-Range Instrumentation Group) standards (see Chap. 5) are used, the tape could be "searched" and the information obtained, but the process is slow and costly as well as uncertain. (2) In addition, accurate identification is required for channel and commutator-segment assignments so that the traces can be properly identified as well as provide information as to frequency response required to reproduce faithfully the phenomena being measured. (3) If time is critical, the record formats can be established so that the switching assignments can be made and equipment setup can proceed before receipt of the tape. If the airborne assignments are relatively stable, standard formats can be utilized. (4) Preventive-maintenance schedules should be established to ensure operational status of equipment. (5) Within the realm of planning is the critique of the previous reduction in order to improve current procedures and scheduling. This is particularly important where schedules are tight, such as always prevails in the quick-look application.

The Analog-to-Analog Process

The process involves the separation of the recorded complex wave into its component bands by use of bandpass filters, feeding each of these outputs to a tuned discriminator, smoothing the demodulated signal with a low-pass output filter, decommutating the time-multiplexed signals if required, and finally driving a recording device with the resultant output voltage (see Fig. 11). This overall process is the same for all applications of the system. The degree to which the equipment setup is checked, tuned and retuned, the tape run and rerun, and calibrations applied determines the accuracy obtained. Hence let us discuss the two extremes, i.e., the real-time display and the quick-look vs. the final reduction.

Real-time display and quick-look are mentioned together since the use is primarily the same; only lapsed time distinguishes the two applications. Real-time display is used in missile applications to present a few critical telemetry functions to the test conductor so that decisions can be made prior to and during the test. The technique also yields immediate information for determining overall success or failure. Since the information is needed in real time (little or no delay between presentation and actual phenomena), pen recorders are normally used. The recording paper with such mechanisms generally has grid lines preprinted so that the measurements presented can be calibrated to read in function value directly. It can be suspected that accuracy is probably on the order of 10 to 15 per cent but the primary purpose of the monitoring is to identify catastrophic changes—a purpose well answered by the method. Of course, real-time display of much more complex processes can be presented. A good example is that of impact prediction where tracking data are received, fed to an analog-to-digital converter, then to a digital computer, and finally to a digital-to-analog converter to be presented as a plot. The plot yields continuous

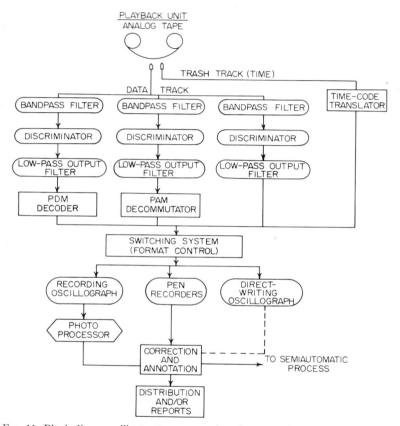

Fig. 11. Block diagram, illustrating a general analog-to-analog conversion system.

information giving the impact point of the missile if the power was shut down as of that moment. Although there is an obvious delay before presentation, on the order of 2 sec, it is essentially real time. The quick-look application yields information of the same nature—after the fact. The design engineers and other interested personnel at the home site are always anxious to receive results of test performed; the analog system at the data-reduction center serves this function. Upon receipt of the telemetry tape from the test site, an effort is made to provide "quick-look" records within as short a time span as possible. Here the process becomes more complex. Modern missile techniques, in the research and development stage, require the monitoring of hundreds of measurements. If the process is to be really "quick-look," recording capability must be adequate. Even when functions of little interest are omitted, the number of functions can still be numerous. Hence it is not unusual to find 15 to

30 multichannel recorders for simultaneously recording all data of interest. A data-processing center which requires repetitious setup and rerunning of the tape, because recording facilities are inadequate, will consume much time and would certainly not qualify in the quick-look category. It is necessary to provide a means for reading the records to obtain function values for preliminary analysis. There are many methods possible, with the determining factor being that of speed; the two illustrations in Fig. 2 are representative. The fastest method is the use of the overlay where all the information can be preprinted with the scales and identification being the only items to be completed after the test. "Quick-look" techniques yield accuracies on the order of 5 to 10 per cent under large-volume and high-pressure conditions.

The only difference between "quick-look" and the use of analog records for final reduction is the attempt at better accuracies, if required. Although the equipment setup procedures could be more exhaustive, the best procedure is one of standardizing the operation such that all the elements required are provided for each run. Then, only those variable steps which can be accomplished with precision in the time allotted are performed. Assuming good maintenance throughout, for each run, the procedure may be outlined as follows: (1) An equipment calibration run is made. This is usually performed through the use of an automatic calibrator, with the number of points varying from 3 through 11. The calibrator puts out precise frequency steps which are recorded as step functions on the data records. These steps yield linearity and bias correction for the equipment. (2) A standard format is used where one restricts the number of traces per record and reduces the total deflection to eliminate trace crossover, and adjusts paper speeds for minimum lengths. The oscillograph recorder setup is the time-consuming operation, but it is possible to obtain subsystems for automatic zero-offset and scale-factor settings to speed this process.

(3) The setup for the rest of the system will vary according to how much equipment is available. In Fig. 11, it will be noted that various paths are possible. Although it is desirable to have sufficient equipment to allow reduction of all information in one pass, multiple running is often necessary. (4) Development of the records is accomplished on standard photographic processing equipment. (5) The annotation process consists of labeling the records for precise identification of time and traces, and applying one of the various calibration methods to provide easily interpreted scales reading function values. Hence, if a standard operating procedure is used, the annotation process becomes the area of greatest concern as to how detailed and precise the final output will be. The quick-look records could be published with overlays prepared using a nominal calibration. More precise overlays can be constructed, and the process varies from one of strictly manual methods to that of reading the various equipment calibration steps and the in-flight calibration (if present) semiautomatically, combining sensor calibrations via a computer, and constructing an overlay on an automatic plotter. The better techniques can yield accuracies of approximately 2 to 3 per cent. By accuracy, it is meant that the data-reduction process does not degrade the data any more than by the amount stated.

It is emphasized that the use of standard procedures is only one way to operate. Even then, there are usually numerous special problems requiring special handling such as the requirement to display data recorded using nonstandard telemetry frequencies, transient analysis, and vibration analysis—to mention just a few. With the number of nonstandard tests usually associated with a large research and development program, the best system to develop is one allowing flexibility and enough redundancy to provide the means to make simultaneous runs. Redundancy is a necessary concession for overcoming equipment-reliability problems in large systems where tight schedules must be met. Recording characteristics affecting the conversion system are discussed under the automatic analog-to-digital systems and also apply here.

11   Final Presentation and Summary

Although the data-reduction processes were separated into three primary systems, it can be readily noted that no such division can truly exist. The processes are normally combinations of all the facilities available. In fact, Figs. 10 and 11 indicate clearly that the automatic analog-to-digital and the analog-to-analog systems need not be separate systems even in terms of the hardware, since they can both use identical elements.

In presenting the mass of data obtained from flight testing and associated in-house tests, the primary purpose is to provide the engineers with all the information needed in as condensed and meaningful a form as possible. There is no one answer. However, final reports consist of portions of outputs from all the equipments discussed, including further computation and analysis results. The job is enormous in scope; improvements and new approaches are needed in all phases of the work. This is the function of the man in the "men and machines" combination—an element not likely to be replaced for some time.

COMPUTER DATA PROCESSING

12 Introduction

It is assumed here that the reader has some familiarity with analog and digital computers, which have become common tools for the engineer.[1] In the discussion below, emphasis will be on digital computers used as general-purpose machines, that is, machines able to perform a large variety of numerical and logical operations. However, mention will be made occasionally of certain analog computers which are much more efficient to solve special problems.

Basically any of the elementary computations performed by a digital computer can be performed by a human being; but the repetition of these operations a considerable number of times is practically beyond the limit of human possibilities when available time, cost, and reliability of results are taken into consideration. The digital computer has then two fundamental uses; namely, it permits the automatic handling of large amounts of data and it permits a large amount of computations reliably and in a short time. Therefore, it must be considered as a powerful tool for data processing. However, its usefulness depends upon the intrinsic capability of the computer, and the experience and ingenuity of the personnel devising the data-handling system and selecting the numerical techniques to accomplish the calculations.

Since the selection of a computer is strictly a decision on which off-the-shelf pieces of equipment should be purchased, and since the techniques used will either make or break an installation, only a short portion of this discussion will be devoted to the choice of a computer and the bulk of this part of the chapter will be on the techniques and problems involved in using a computer for data processing. The part is therefore divided into three segments namely, Selecting the Computer, Numerical Techniques, and Organization of Computer Data Processing.

13 Selecting the Computer

The choice of the digital computer is obviously one of the far-reaching decisions in computer data processing. A digital computer usually requires a considerable amount of time for its construction and installation, and always represents a large investment in equipment and in labor (site preparation, training of programmers, library of computer programs). Budgetary considerations are often the determining factor in the choice of a digital computer, but the selection of an inadequate computer has dire consequences; requirements of data processing will not be met.

For data processing of missile flight-test data, a large-scale computer is necessary for which the *minimum* requirements can be specified as follows: at least 8,000 words of high-speed memory, 6 to 10 magnetic-tape units, and adequate peripheral facilities for conversion of media (tapes, punched cards), printing, and plotting. Most machines in this class work in binary arithmetic and have a floating-point capability, contributing to speed of operation and ease of programming, respectively. Such a computer used under these conditions usually handles numbers with at least a mantissa of eight significant decimal digits and a characteristic of $10^{\pm 40}$ and performs from 40,000 to 200,000 additions or subtractions per second or 4,000 to 25,000 multiplications or divisions per second. However, the speed of the arithmetic operations is not the only element to consider in the choice of a digital computer. The logic of the computer is extremely important and questions like those which follow should be asked. Is the computer able to perform various input and output operations at the same time as computations are performed? Are the provisions for repeating instructions satisfactory? Can accumulated products be accomplished?

The technology of computer design is improving continuously; hence it is necessary to follow this progress very closely, either to increase the capabilities of the present installation by a suitable modification or substitution in the internal or peripheral equipment or, more drastically, to replace the present installation by one which is more advanced in speed, accuracy, capacity, reliability, and flexibility. These modifications or replacements are usually complex and require a considerable amount of time to accomplish; hence they must be considered very carefully and as early as possible.

14 Numerical Techniques

To solve a data-processing problem with a digital computer, the following steps are required: (1) Determination of a numerical algorithm, that is, a method of solution of the problem, regardless of the means of implementation. During this step, approximations in the definition of the problem are made if necessary. (2) Numerical analysis to determine the sequence of arithmetic and logical operations which can be programmed in the computer. Since very often there are several methods which can be used, the analysis should include, when possible, an evaluation of the truncation and round-off errors as well as an estimate of the speed of the computation and of the capacity of the computer memory required. (3) Programming and checkout of the program before the actual problem is put on the computer.

In the field of scientific and engineering problem solution, digital computers and the numerical analysis taking advantage of the power of computers have developed extensively in relatively recent years. However, numerical analysis in the computer data-processing field is even more recent, and this science has developed in various directions according to the immediate needs of the data processor or the programmer. One of the problems in data processing is the noise which affects all data. Techniques which would be satisfactory with perfect data may provide meaningless results if the noise level is too high. On the other hand, the quality of the data may sometimes be improved by appropriate techniques to a point where it is possible to combine several pieces of data to derive new data. Study of this two-way interaction between errors in the data and computer data processing is an important part of the error analysis, which is discussed in detail on pages 12-60 and 12-61.

Another characteristic problem with which the numerical analyst is faced in solving data-processing problems is the overabundance of test data in which he has limited confidence. Despite the limited accuracy of these data, they are frequently used as boundary conditions. Since the complexity of a problem increases with the number of boundary conditions, there is a tendency to write computer data-processing programs as simply as possible by the use of a minimum of boundary conditions. Investigations are made to see how closely other boundary conditions are satisfied; then minor changes are made in the original boundary conditions to see if these changes improve or deteriorate the solution with respect to the other boundary conditions. When such a parametric study is possible, it is a powerful tool to investigate the sensitivity of the solution to changes in some data. This method of parametric study, judiciously applied, provides a means for solving complex problems by successive approximations.

For that purpose, the digital computer is a very valuable tool. Once a program is written, it is immediately available at any time, and the same problem with various input data can be run again and again without loss of time. On the contrary, an analog computer always requires an appreciable, sometimes extensive time for setting the conditions of the problem.

The computer data processor recognizes that it is desirable to have programs ready in the shortest possible time after a test takes place, yet he recognizes that exact solutions are virtually impossible because of the prevalence of noisy data. As a result, programs are usually written in a simplified manner first and then refinements, such as the correction of terms in an equation or the computation of additional results, are added. Occasionally, entirely different methods of solution are attempted as the interpretation of the results obtained from successive flight tests detects systematic biases in the computation.

From the discussion above, it can be seen that many numerical techniques have to be developed for computer data processing. It is expected that progress will be slow, because of the complexity of the investigations. Effectively, methods have to be developed usually for perfect data first; then they have to be adapted for systematic or random noise. Numerous experiments have to be made with digital computers, including many analyses of results which often result in many changes in computer programs. However, the reward is great when a new technique improves the accuracy of the results derived from the same initial data, or when the accuracy of the initial data themselves can be improved.

It is not intended to review here the various methods of numerical analysis, which are treated in many books.[2] However, a few important problems of computer data processing will be considered, and mention of some of the methods currently used to solve them will be made. A typical case of data processing, the determination of a missile trajectory, will be then discussed in some detail to illustrate the use of various methods to solve the problems involved.

Editing

Telemetry data are usually sampled data, and it sometimes happens that some data points are missing or deviate from the correct values by intolerable amounts. Points in this last category are often called "bad points." It is the purpose of the editing operation to remove bad points and usually to replace them with points which do not depart too much from the correct values. Editing involves a very complex process: first, assumptions must be made as to what type of function the sequence of data points is supposed to represent; then the reasonable departure from the values of this function must be estimated for each point. This complex process results in a nonlinear operation on the data points which is extremely difficult to process on a computer because of the number of parameters involved. To the author's knowledge, there is no general editing technique which has been programmed on a digital computer. The reason is that no set of rules has been defined which can apply to all cases without the operator's intervention.

Editing, performed by eye, usually gives satisfactory results. This editing can be a simple inspection of a listing or plot of the data points. If the function is a slowly varying one, bad points can be detected very easily and replaced, for instance, by an interpolated value from the two nearest good data points. Such a process can be programmed on a digital computer. The real difficulty is for data points showing considerable changes in the slope of the function representing their variation, such as data points of oscillating functions. Under these conditions, it is difficult, if not impossible, to discriminate between good and bad points by simple inspection of a listing; consequently the data points must be plotted and a curve fitting made by eye, with or without the aid of drafting devices like French curves. In this case, both editing and smoothing operations are performed simultaneously; the interpretation of this process is discussed below, under Smoothing and Curve Fitting. The principal problem with the "eyeball" technique is that repeatability is extremely poor, especially on noisy data. Even if the same person makes several tries, his lines do not usually coincide; and different individuals attempting this method show that accuracy of the data is very questionable. Therefore, any automatic method, where the steps have reason behind them, is better than sketching in the data. There are other operations which are sometimes classified as editing; but since they involve either interpolation or frequency analysis, they are discussed under these headings.

COMPUTER DATA PROCESSING   12-31

Smoothing and Curve Fitting

If the data points are sampled data, it is impossible to detect any frequency higher than half the sampling frequency. In practice, the smoothing operation consists in leaving only frequencies much lower than this upper bound. The purpose of smoothing a set of data usually is to replace each data point by another point lying on a smooth function through the raw data or to find the values of the parameters defining the smooth function, such as the coefficients of a least-squares polynomial. Smoothing, with the usual editing as a preliminary step, is one of the most important operations of data processing. The purpose is to reduce the noise in the data since all subsequent operations of data processing are performed with smoothed data. It is then of prime importance that the smoothing remove the noise with minimum alteration of the function that the available data are representing.

If the form of the exact function the data are supposed to represent is known, the parameters of this function can be determined by least-squares or other techniques and an estimate of the accuracy of the coefficients of the function can be made, provided the accuracy of the data points is known. The same technique applies if a specified form of function is known to represent the variation shown by the data points accurately enough. However, the exact or approximate function which can be used for that purpose is seldom known. The choice of any function is then left completely to the discretion of the data processor. There are several techniques commonly used which will be described briefly.

Polynomial Smoothing. Smoothing by polynomials consists in passing a polynomial, usually by the least-squares method, of a selected degree through a certain number of consecutive data points. The degree of the polynomial must be as low as possible to avoid multiple points of inflexion; however, it must be of high enough degree to follow the data points closely enough. "Goodness of fit" may be estimated, for instance, by the root mean square of the residuals between the data points and the computed points.[3] Orthogonal polynomials can also be used, the coefficients being obtained independently by integration.[4]

Digital Filters. Smoothing can be obtained by trigonometric polynomials also, a certain number of data points being represented by a sum of frequencies each of which is a multiple of the frequency corresponding to the time interval between the extreme points of these selected data points. Smoothing is obtained by using only a few of the lower frequencies of the trigonometric polynomial. See also Appendix Sec. A.16.

The concept of frequency selection used in the trigonometric polynomials can be extended. In the case of sampled data, this leads to the notion of digital filters which perform on numerical data operations similar to those performed by electrical filters on electrical quantities. Digital filtering consists basically of taking a weighted average of a certain number of successive data points and replacing one of these data points, usually the central one, by this weighted average. The weights used define characteristics of the filter, which by analogy with the electrical filters, can be called "gain" and "phase shift." Since the weights can be applied to the points following, as well as to those preceding the data point to be operated on, the digital filters do not have the limitations of physically realizable filters. For instance, it is possible to design digital filters with exactly zero phase shift by making the weights for the points equidistant from the data point to be operated on equal in value.

The result of fitting a polynomial by least-squares through a certain number of sampled data, and computing the value of the polynomial for the central point, is a linear combination of all the sampled data points. Hence a polynomial least-squares fit of sampled data can be obtained directly by the equivalent digital filters; in other words, least-squares polynomial fitting of sampled data is a particular case of a digital filter. It has zero phase shift, and a gain curve which depends upon the degree of the polynomial and the number of data points.

A digital filter is not restricted to having its gain curve depend upon the number of data points. If a gain curve is approximated in a satisfactory manner (for instance, with an absolute error less than 0.01 if the maximum gain is 1) with a certain number of weights, it can be approximated also (with even a smaller error) with a larger number of weights. The art of digital filter design consists of approximating desired gain curves closely enough with a minimum number of weights, in order to reduce the computing time of the weighted average and to provide a larger number of outputs for a given number of data points.

Selecting the Filter. The problem of the choice of the degree of the least-squares polynomial is a particular case of a more general one: how to select the gain curve of a digital filter. In other words, what transmission is allowed for each of the frequencies contained in the data points? This is a very difficult question, and the quality of the answer depends upon the knowledge the data processor has of the phenomenon represented by the data points.

If the frequency spectra of the signal and of the noise are known, it is possible to design a digital filter which minimizes the root mean square of the differences between

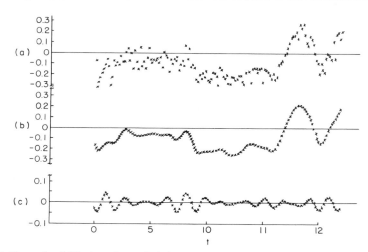

FIG. 12. Example of filtering of sampled data. (a) Raw data. (b) Output of a low-pass filter ($f < 0.10 f_s$). (c) Output of a bandpass filter ($0.05 f_s < f < 0.10 f_s$). fs = sampling frequency. f = frequency passed by filter.

the sampled data points and the "filtered" points, i.e., the outputs of the digital filter. The gain of such a filter is near unity for the frequencies where the signal-to-noise ratio is large, and near zero for the frequencies where it is small. This is the optimum filter treated by Wiener; however, very often the noise spectrum is unknown. Under these conditions, the proper choice of the digital filter depends upon the knowledge of the phenomenon possessed by the data processor. Ordinarily he is able to obtain from the engineer the limit frequencies between which all the frequencies of interest will lie. He then selects his filter accordingly. The experience of the author has proved that a filter having a gain of nearly unity for all frequencies of interest, and zero for all frequencies outside the zone of interest, usually gives satisfactory results. To make this transition between gain unity and gain zero as steep as possible, a large number of weights are required, with 101 weights commonly used.

The filtering of a large set of data is made by moving the set of weights along the data points. Figure 12 shows a set of sampled raw data, the output of a low-pass filter which essentially does not pass any frequency above a tenth of the sampling frequency, and the output of a bandpass filter which essentially passes only frequencies between one-twentieth and one-tenth of the sampling frequency. In both cases, a filter with 101 weights was used. For details of digital filtering techniques the reader is referred to Ref. 5.

Interpolation and Extrapolation*

The linear interpolation method, which consists of passing a straight line between two consecutive data points, is extensively used in data processing for several reasons: (1) it is easy to program on a digital computer, (2) the interpolation is fast, and (3) it permits close following of any desired shape of function represented by the data points if, within the desired accuracy, the variation between two successive points can be assumed to be linear. For instance, linear interpolation is commonly used for table look-up and calibration; thus, when the original data appear as a curve, discrete data points are selected with variable increments of the argument to satisfy this condition of approximate linearity between successive points. Linear interpolation is frequently used to produce data corresponding to times different from those corresponding to the original data. This is used, for instance, to change the sampling frequency of data; as when data computed every fifth of a second must be replaced by data for every twentieth of a second. Time correlation and adjustment also require interpolation as in the case where times corresponding to different sets of data may be different and the data may have to be adjusted to the same time basis.

It is very important to remember that linear interpolation is justified only when low frequencies (compared with the sampling frequency) are involved between successive points. Consequently, linear interpolation should be done on smoothed data, not on raw data. However, interpolation involving only two successive points need not be linear. For such cases, many types of interpolation are available, such as Newton's, Gauss's, and Lagrange's.[6] As an example, logarithmic interpolation is often used to compute atmospheric pressure between pressures known at two specified altitudes, while atmospheric temperature between these two same altitudes is obtained by linear interpolation. Also interpolation is frequently accomplished by fitting a polynomial of selected degree to a certain number of successive data points; this fit may be exact or approximate if it results from a smoothing operation. Extrapolation is a much more uncertain operation than interpolation and should be avoided as much as possible. However, extrapolation is frequently necessary in data processing, as it is often the only way of estimating the values of some functions. For instance, radiosonde data for atmospheric pressure and temperature are available up to a certain altitude. Since trajectory determination should take into account the values of these quantities for higher altitudes, an extrapolation is made, based on the data at the maximum measured altitude and the best available physical constants about the behavior of the atmosphere above that altitude. Many processes used on data can be interpreted as extrapolation operations: the determination of a complete trajectory from initial conditions only is nothing other than an extensive extrapolation, justified only if the numerous assumptions made are reasonable. Wiener's method of extrapolating by taking into account the frequency spectra of signal and noise could be used in data processing.[7]

Differentiation*

To obtain the derivative at one data point of a function known only at a finite number of discrete points, a method of curve fitting can be used and the function approximating the points can then be differentiated. When the data points are sampled data, the derivative of a polynomial fitted by least squares is a linear combination of the values of the successive data points; hence this reappears as a particular case of the digital filter. The phase shift of a digital filter used as differentiator must be 90°. This can be obtained without error by selecting weights of equal absolute value, but opposite signs, for points symmetrical with respect to the central point for which the derivative is desired. The gain of an ideal differentiating filter is proportional to the frequency.

Two techniques can be used to perform differentiation of sampled data with digital filters: One consists of first smoothing the data points, then differentiating the smoothed

*See also Appendix Sec. A.16.

data; the second operation requires only a few weights (7 to 11 are sufficient). The other technique consists of using a single digital filter to perform the operations of smoothing and differentiating simultaneously. The weights of a filter which yields, in one pass, an output equal to the resulting output of two filters used in succession can be obtained in a straightforward manner when the weights of the successive filters are known. The first method is usually preferred, because both the smoothed data and the derivatives of the smoothed data are generally of interest.

It is well known that noise in data increases when differentiation of these data is performed; a perfect differentiation has high gain for high frequencies, where the noise is predominant. Hence the derivative obtained by a filtering technique is extremely sensitive to the gain curve of the filter used to smooth the data. Great care must be taken, then, in the selection of the smoothing filter, if derivatives are to be computed later on.

In ordinary differential equations, the derivatives may sometimes be obtained without differentiation. For instance, in a point mass trajectory in a vacuum around the earth, the second derivatives of the variables representing the coordinates of the moving point are equal to the gravity components at that point and thus are explicit functions of the coordinates of the point.

Integration*

To obtain the definite integral between two points of a function known only at a finite number of discrete points, the method of curve fitting can be used, if the function which is fitted can be integrated.

When the data points are sampled data, filtering techniques are commonly used in one of two ways: (1) The filter gives the value of the definite integral for an interval of the argument equal to twice the sampling interval. The integral is for the intervals on both sides of the central point. If the number of weights is three, the filter is Simpson's rule, which is used extensively for integrating data. Values of the summation are obtained at every other data point, since two intervals are necessary for each integration. (2) The filter gives the value of the definite integral for an interval of the argument equal to the sampling interval. In this case, it is advisable to have an even number of weights, and the integral is for the interval between the two central weights. If the number of weights is two, the filter is the trapezoidal rule, which is not used too often in data processing because its gain is satisfactory only for very low frequencies.

Frequency Analysis*

Frequency analysis is a very important operation of data processing, as it provides information about the significant frequencies contained in the data and is helpful in the interpretation of data and the selection of smoothing filters. Spectrum analysis on sampled data assumes that in the data points there is no frequency higher than half the sampling frequency; otherwise the sampling would not permit discrimination between these higher frequencies and frequencies lower than half the sampling frequency. Caution should be observed in interpreting the two-per-cycle sampling rate referred to here and previously. This is the theoretical minimum sampling rate; in actual practice it is advisable to use 5 to 10 samples per cycle of the highest frequency of interest.

If only the power spectrum is desired, autocorrelation techniques can be used, and these are fast (Ref. 8 and Chap. 2). Crosscorrelation is useful when the relationship between two functions is of interest (Ref. 9 and Chap. 2).

The amplitude spectrum could be determined exactly only if an infinite number of data points were available. The larger the number of data points, the finer the spectrum analysis can be made. For instance, it would be theoretically possible to make a very fine spectrum analysis of the output of a vibration sensor, using all the data covering a full flight test, but this would be almost completely useless. What is

*See also Appendix Sec. A.16.

needed in practice is information about the spectrum during various critical portions of the flight; but the smaller the number of data points used, the cruder the spectrum analysis. Consequently, there must be a compromise in order to operate on enough data points to have a spectrum of sufficient resolution, and yet not too many points so that changes in spectrum composition can be detected. One hundred and one points can often be considered as a satisfactory number of data points.

Filters in Spectrum Analysis. Filtering techniques can be used advantageously to make amplitude spectrum analyses. Filters are selected with very narrow bandwidths around their nominal frequencies. A pair of these filters is designed for each nominal frequency to be investigated; one filter provides a zero phase shift, the other a 90° phase shift.

A single filter can be used if the spectrum analysis is made for numerous successive points; the output is the same as the output of an ordinary filter except that it appears approximately as a sine wave of variable amplitude. The operation is repeated with a filter for each frequency of interest.

Ordinarily both filters corresponding to the same nominal frequency are used consecutively and the square root of the sum of the squares of the outputs for the central point is computed. This is proportional to the spectrum amplitude for that frequency if the following restrictions are satisfied: (1) the bandwidth of the filters is very small, (2) the gain curves of the two filters corresponding to the same nominal frequency are practically identical, and (3) the spectrum of the data within the bandwidth of the filters has constant amplitude and constant phase angle so that there is no isolated frequency within the bandwidth i.e., that there is no single predominant frequency within the bandwidth to cause a "spike" in the spectrum. By repeating the operation with a pair of filters for each frequency of interest and the same group of data points, the amplitude spectrum at the time corresponding to the central point of these data can be determined. Fourier analysis applied to sampled data may be considered as a particular case of this filtering technique, the frequencies used being multiples of the sampling frequency divided by the number of time intervals between the extreme data points.

Analog Spectrum Analysis. Frequency analysis by digital techniques requires a considerable amount of computer time but provides accurate information relatively easily*. Faster and less expensive methods are often desired, especially when the limitation of the number of data points available prevents an accurate spectrum analysis. In this case analog devices, which do not require digitalization of data, can be used advantageously. An analog spectrum analyzer has been devised which uses a fixed number of bandpass filters which accept the input signal for a predetermined amount of time, the outputs of which go through a squaring and integrating circuit. The output of this analyzer gives a signal proportional to the power spectral density for the nominal frequency of each bandpass filter.

Statistical Analysis

Statistical analysis is another important element of data processing. The most common operation is the determination of the mean and standard deviation of a certain number of selected consecutive data points. This operation is repeated for all data points by moving the selected data points. The determination of the mean can be interpreted as a particular case of filtering. Basically such an operation is a frequency analysis for zero frequency and consequently has the characteristics mentioned earlier; namely, the larger the number of data points operated on at one time, the less variation there is in the value of the trend, but the more lag there is in the detection of the trend. Besides these statistical routines, many special statistical analyses can be performed by the data processor to evaluate the quality of the data and make appropriate error analyses.

Computation of Derived Data by Explicit Expressions or Algebraic Equations

There are many cases where formulas are available to compute new quantities from original data or from previously derived data. The form of such equations

* The Cooley-Tukey algorithm has recently reduced the computer time considerably.

may be of a large variety of types which may require numerical analyses to convert them into expressions that the computer can accept. The nature of the quantities entering into these formulas may be quite varied; in particular, the quantities may be the results of differentiation, integration, or other data-processing operations.

Sometimes the data to be computed cannot be expressed as explicit functions of other data, but relations between several unknown quantities are available. A system of such relations may be sufficient to determine the unknowns. When such a system includes nonlinear equations, it may be very difficult to solve for explicit relations, and solutions by successive approximations are used whenever possible. These approximations are often obtained by solving a system of linear equations, for which many methods are available.[10]

Ordinary Differential Equations

When the initial conditions are known, the integration of ordinary differential equations can be obtained by many methods. A popular method for data processing is the Runge-Kutta method, especially in the particular form known as Gill's method.[11]

The accuracy of that method is very great, the truncation error being proportional to the fifth power of the incremental interval used in the integration. A *truncation error* is an error due to the replacement of exact by approximate equations for the convenience of numerical solutions. Also, no past history is required, thus making it easy to change the incremental interval of integration to match particular requirements. In this method, it is necessary to compute approximations to the derivatives four times per incremental step. In spite of this increased computing time, this method is very suitable for data processing because of its great flexibility; changes can be made in the logic of the computer program without need for providing past history.

One question which appears in any solution of ordinary differential equations is the choice of the size of the incremental step. This choice depends upon the conditions of the problem, and it is frequent that in the course of the same problem several different sizes of the increment have to be used in different portions of the computation. Effectively, to reduce the computing time and the roundoff errors, the size of the interval should be as large as possible, but a larger size increases the truncation error. This requires a compromise, and an optimum one is usually reached only after sufficient experience with the particular problem has been acquired. A very convenient method of developing an appreciation of the effect of the size of the interval consists of running the same program for the same problem with various sizes of the interval. The range of sizes for which the results are practically independent of the size of the interval defines the optimum compromise; for smaller sizes, the roundoff errors are predominant; for larger sizes, the truncation errors are predominant.

Partial Differential Equations

Partial differential equations are of great importance in data processing. For instance, problems related to heat transfer, heat conduction, and diffusion lead to such equations. The solution of these equations is usually quite involved and requires use of many simultaneous data. Different methods are employed, ordinarily, for elliptic, hyperbolic, and parabolic partial differential equations.[12] The computer program for any of these problems is always a very complex one, and the numerical analyst must very carefully evaluate the various methods available and any possible adjustment which might reduce the computer time to solve the problem or increase the accuracy of the solution. Moving boundaries, usually encountered in data-processing investigations, add to the complexity of the problem.

An Example: Determination of a Particle Trajectory

To show how various types of computations are involved successively or simultaneously in the same data-processing problem, an example will be treated in some detail, namely, the determination of a particle trajectory. Basically, this is a problem involving solution of a system of ordinary differential equations; however, many other operations are also involved.

First, an adequate mathematical model has to be established, including proper representation of the earth, and selection of proper functions for the acceleration-of-gravity components and forces acting on the point mass. A computer program has to be prepared, able to accept data on atmospheric characteristics (pressure, temperature, wind components) and initial conditions (three components of the position vector and three components of the velocity vector).

Careful attention must be given to providing the best possible values for these six initial conditions. The atmospheric conditions are usually obtained by interpolation or extrapolation of a data table. The data for this table may be observed either at the time of the flight or from some standard for the geographic area involved. Also, both linear and exponential techniques are used for temperature and pressure, respectively. Usually the six components of position and velocity obtained from tracking data are given for various successive points of the trajectory, but each piece of data is altered by some unknown noise. Data from any of these points could be used as initial conditions, but advantage should be taken of the availability of a certain number of successive points on the same trajectory. One method consists of independently smoothing each of the six quantities; the smoothing can be performed efficiently by a filter passing only the very low frequencies (such a filter can be designated a "low-pass filter"). However, the number of weights in that filter cannot be too large because of the relatively few data points available. The result of this smoothing is filtered initial conditions for a small number of successive points. Usually trajectories obtained from several of these successive points differ very little. Another method of smoothing the initial conditions consists of using the equations of motion to optimize a trajectory between the various data points available, using some selected criterion of good fit, for instance, the maximum likelihood[13] or the minimization of the sum of the squares of the distances between tracked points and computed points on the trajectory. This method is much more involved than the first one but judiciously used may provide more accurate results.

If only these six quantities are available, the trajectory is computed from these adjusted data and all additional information about atmospheric conditions which have been obtained during the flight test are included. But usually other trajectory data are known, such as the time of impact, values of acceleration at some positions, or downrange tracking data. Since, ideally, data processing should make use of all the available information, the additional data are evaluated to determine their quality; then the differences between them and the computed data are analyzed so that their significance can be assessed. If there is a definite discrepancy, changes in initial conditions have to be considered or changes in some parameters of the program have to be contemplated. It is part of the art of data processing to determine the proper course of action to investigate the cause of a discrepancy and the means of minimizing it, if possible.

In that manner, a trajectory is obtained which is the best available with the data-processing techniques used. As these techniques improve, a better trajectory may be obtained if sufficient information is provided. It must be added that sometimes the task of the data processor is almost hopeless, as an example, when a single set of initial conditions is available and no additional information can be secured. This kind of thing does happen! In the processing of flight-test data, the determination of the best trajectory is fundamental, as many subsequent computations to determine aerodynamic and thermodynamic properties depend upon quantities obtained from the trajectory program.

15 Programming Systems and Techniques

Telemetry is usually used where the system being tested will not be recovered after a test, as in the case of missiles. Also, it is used because little is known about the pressures, temperatures, and other environmental conditions in which the item being tested must survive, and information is required for as long as the test continues. The design of a weapon system is based, primarily, on projected theories derived from analytical studies using the most modern and advanced techniques and information available. The major purpose of testing is to check the assumptions that were made and the analytical techniques that were used in arriving at the design so that further extrapolations on the theories can be made, if the assumptions were correct.

Obviously, the verification of assumptions and techniques requires much more than can be obtained through mere measurement of parameters. For example, a flight test of a modern missile can yield many hundred thousands of data-seconds which must be analyzed to determine whether the assumed environment and laws of nature and their assumed effect on the system's performance were correct. It must be borne in mind, however, that the methods of analysis will probably be as much in the realm of new technology as the item being tested, and that any fallacious assumption can cause a serious difference between observed and predicted results. This difference, in turn, will require an adjustment of the assumed theory. It should also be evident that extensive computations are required to complete this type of analysis.

Organizing for Computer Data Processing

Since it is usually necessary to complete a thorough analysis within an extremely short time of the test, it is essential that computer programs to accomplish these analyses be established long before the actual tests take place. However, the programs must be sufficiently flexible to permit last-minute changes to be made to adapt them to the particular results obtained from the test. It should be emphasized that the short time permitted for the reduction of test results requires that no extensive changes be necessary to any of the major analytical programs.

One method of permitting this flexibility without sacrificing the long time delays required for reprogramming is to establish a modular approach to data reduction. With this concept, data are kept in some standard format. These processes are programmed as extremely flexible subroutines (sometimes called "modules" because they are elements of a modular system) and stored as a master file on one of the tapes of the computer system for rapid access when needed. Throughout this portion of the discussion the terms "subroute," "module," and "program" will be used synonymously to mean an extremely flexible set of computer instructions which are set up at "computation" time to accomplish a specific task by supplying numerous parameters required to define this task. It is assumed that "static" subroutines for transcendental functions, square roots, etc., will be considered as a part of the module coding, and the above terms do not infer this simple and more standard type.

Types of Modules

The subroutines required for data reduction can be divided, roughly, into two general categories: (1) data-preparation operations (editing, smoothing, interpolation, differentiation, etc.) and (2) solution of problem-oriented equations, such as computation of derived data by explicit expressions, solutions of systems of algebraic equations, and solution of ordinary and partial differential equations. These equation solutions are specifically designed to obtain only one or a limited amount of answers and would include computing trajectories for known initial and terminal conditions, converting vibration data to force and acceleration values, etc. The calculations performed will depend upon the item being tested and will not be treated here, except to emphasize the great need for flexibility. Flexibility can be incorporated into modules of the system by, wherever possible, treating the known variables of the system such as the number of sensors used and the boundary conditions as parameters to be supplied just before running on the computer. Also, since these equations are generally prepared by the design engineer or systems analyst, they are often based extensively on extrapolated theories. Therefore, it is advisable to have a responsible individual indicate where logical breaks exist so that the discrete portions can be programmed as separate modules. In this manner, only small parts of the analysis need to be changed, reducing the need for a complete system checkout after every modification.

The data-preparation modules can be further subdivided into the classes of data-handling operations and general data computations. In the data-handling area are input programs, rearranging (permuting) programs, and output programs, while the general data computations include such processes as calibrating and filtering, or smoothing.

Figure 13 is an example of how a thermodynamics analysis can be flow-charted and broken into modules. Each block represents a subroutine through which the data

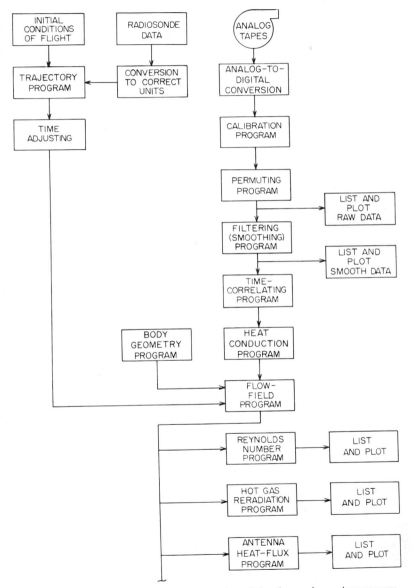

FIG. 13. A simplified flow chart of one portion of the thermodynamics program.

(results from the preceding block) must pass. The figure demonstrates the subdivision of the analysis into smaller portions (programs for antenna heat flux, hot gas, and reradiation, etc.) because modification to each part is possible when the data are evaluated.

Standardization: Key to Compatibility

Several of these data-preparation programs (calibration, permuting, filtering, etc.) will be discussed in greater detail in the ensuing paragraphs, but the need for compatibility among these subroutines is so important it must be stressed again. Assurance that various items of hardware will work together is usually of great concern, at least to the manufacturer, but the compatibility of the second half of the computational system, the organization of the programs, is often overlooked.

Since the order in which a set of subroutines will be used to reduce data is often unknown until the actual data are received, the system must allow for variety of possible sequences. In particular, the format of data on the tape should be similar for all stages of processing, and any variations should be decided upon in advance so that these variations can be provided for in the program.

One generally accepted method of setting up data files is to write one item (all the information for a given time interval, position on an airframe, etc.) per record for a file and begin the file with a definitive record which will explain the arrangement of the data within the items. For example, the first field of the definitive record can contain the number of fields in the data records, the second field can tell the number of records in the file, the third can be the identification of the file, etc. With this method, the data in the definitive record can be used for preliminary setup of each subroutine, and the subroutines can be used in any order. This procedure is sometimes called an executive program.

A second approach to standardization for compatibility is the use of floating-point numbers for data in order to avoid the necessity of scaling problems. Most large-scale computers have a built-in floating point as either a standard or optional item, and this type of logic is almost a necessity for data reduction. With this logic in the machine, each field of the above "standard" file can be a computer word—a floating-point number—and the programs accept this format, which adds to the standardization.

To these remarks on standardization for compatibility, a word of caution be added about the design of a computer system for handling telemetry data. Care should be exercised in evaluating what portions of an analysis should be placed on a computer in order to utilize this expensive machinery efficiently. While as much time can be spent getting data out of a computer as processing them within the machine, some operations (sorting, permuting fields, selecting, etc.) are much more economical and practical on standard card-handling equipment, and if a series of consecutive operations do these tasks, it may save money to remove that portion from the machine. Where elapsed time is important, the time delay should also be considered in deciding how much of a specific problem should be done off line.

Time: The Independent Variable

In most data-reduction applications, time is considered as the independent variable and all other data are sampled at discrete intervals of time. Time is usually written on a telemetry tape at the ground receiving station in one of several code forms.

Example a in Fig. 14 shows a 1-pps 13-bit 24-hr code. The reference pulses are marked, and this particular code is read as follows:

Reference........	Wide pulse
Bits 1, 2.........	15- and 30-sec markers, respectively
Bits 3–8.........	Binary code, min
Bits 9–13........	Binary code, hr
Bit 14...........	Filler
Reference........	Wide pulse

COMPUTER DATA PROCESSING 12–41

The time is for the leading edge of the preceding reference marker, and 15 sec is required for one complete readout.

Example *b* in Fig. 14 shows a 100-pps 10-bit, straight binary code, frequency-modulated on a 1-kc carrier. The code group appears once each second, and the time presented is for the leading edge of the preceding reference marker.

The simplest problem involving time is the analog-to-digital conversion of commutated telemetry data. The sampled data do not necessarily occur at uniform time intervals since commutators may speed up or slow down during a test, thereby changing the sampling rate. Since these telemetry data will probably be matched to either calculated values for fixed intervals or other data at variable intervals, an interpolation program will be required to obtain values of the dependent variables at fixed intervals of time.

A second problem occurs when data from more than one receiving site are to be combined and the recorded time signals are not identical. For instance, one site may record time relative to a zero time of event and recycle every 15 min, while a

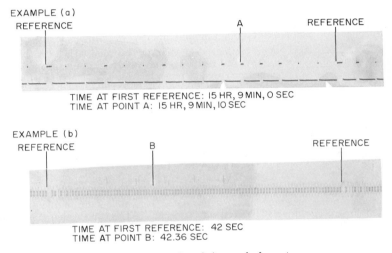

FIG. 14. Examples of time-code format.

second location may be writing the output of WWV (EST) on its tape, and other sites may be using the output of local time-signal generators to write the time channel on their tapes. Therefore, a program is necessary to prepare an output tape with a "universal" time code and interval, and to insert patches of data from various input tapes on this new time base.

A third program which has proved useful is one which will add a given amount (positive or negative) to an existing time code on a data tape. This permits final adjustment of time when the certain key events indicate that one time is transposed slightly from another.

Finally, this discussion of time-code problems would not be complete without a suggestion on the method of writing time on a computer tape when a binary-logic computer is to be utilized; namely, it is best to keep the units as integers. That is, if thousandths of a second are to be the time bases, keep the time so that integral numbers of milliseconds are recorded on the tape. The reason for this suggestion is that most decimal fractions do not convert exactly to a binary counterpart and an accumulation of the small remainder prohibits later time comparisons. However, all integers convert precisely; and since a match on time (the independent variable) is often required to ensure valid results, the two identical times on different input tapes will correspond and not cause alarms or fallacious stops in the programs.* An alternate technique is to allow a tolerance on the match of times from different sets of data. This method does require considerable extra programming, however, and still does not eliminate the cause of the disagreement.

If some such set of "time-adjusting" modules are available to make the independent variable the primary chain of continuity which binds the other pieces of data together, the backbone of the system has been established. To strengthen the framework further, data-permuting and data-patching programs are required. An explanation of these types follows.

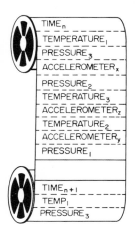

Fig. 15. Example of data records on a magnetic tape.

A Word-permuting Program

The order in which the various items of data are collected is not a serious concern to the instrumentation engineer, who is interested more in the orderly arrangement of the wires in the test vehicle rather than logical grouping of the final data. Therefore, the consecutive data samples within a record of raw data on magnetic tape may be as shown in Fig. 15. Usually, it is desirable to have the samples grouped by type, by logical location, or by some other definable key. In order to rearrange the words within a record, combine data from several tapes for the same instant, or to add or delete specific types of data, a permutation program must be written. At first glance, this appears to be a very easy program to produce, but the logic required to make it a useful, extremely flexible part of the data-reduction system will be challenging enough for an experienced programmer. However, this program is a "must" for any extensive system, in order to prepare the correct sequence of data for any time interval.

The Patching Program

As the permuting program arranges the data "perpendicular" to the time flow, so the patching program sets the data up "parallel" to the time scale. If complete coverage (from start to finish) of an event is required and only portions are available on any one tape, a program must be written to combine, or patch, the several segments into one complete flow of data, throughout the period of activity. This also can be an extremely complex program if different time intervals are permitted for each data sample, more than two tapes are combined onto one output tape, and time transposition is included. However, as with the permuting program, one or more programs of this type are necessary in order to manipulate large masses of data properly. Permuting and patching programs are also referred to as "collating, merging, and sorting" routines.

Intermediate Results

One may question the ability of an individual to design these manipulations correctly, since one erroneous parameter could cause trouble. This question is extremely valid, and for this purpose intermediate output programs are required. It is not economically feasible to allow complete reduction to take place on high-speed digital computers unless proper care has been taken to prevent erroneous data from entering the lengthy arithmetic computations. Even after data handling has

* In converting from decimal to binary, a small remainder occurs on most decimal fractions, and an accumulation of this remainder for many intervals causes disagreement in results at a later time. Since all decimal integers convert to an exact binary counterpart, no accumulation of roundoff occurs and a precise match can be obtained anywhere throughout the data.

entered into the realm of engineering computations it is extremely vital that the engineers predict, by looking at intermediate results, if their initial parameters were correct. In cross and intermediate checking, the computer programs should relieve personnel as much as possible of this tedious task. However, the option for producing quick-look material must be included in every good reduction system.

To offer this flexibility to the engineers, two methods may be used to produce the desired results. One is the printing of tabulated listings and the other is the plotting of graphs. The preference for either type must be left to the discretion of the engineer, since his application may require the accuracy of the numerical answers, or trends as shown by a graph.

```
1
1                       154
059001 5B 99A00059
0   TIME     ALTITUDE    PRESSURE   TEMP.    DENSITY         V(S)      VISCOSITY
273661.00    1111540.       0.      1851.    0.1916E-13     2110.      0.8974E-06
275661.00    1127929.       0.      1871.    0.1754E-13     2121.      0.9032E-06
277661.00    1144252.       0.      1891.    0.1608E-13     2132.      0.9089E-06
279661.00    1160510.       0.      1910.    0.1476E-13     2144.      0.9146E-06
281661.00    1176701.       0.      1930.    0.1357E-13     2154.      0.9202E-06
283661.00    1192827.       0.      1949.    0.1249E-13     2165.      0.9258E-06
285661.00    1208887.       0.      1969.    0.1151E-13     2176.      0.9313E-06
287661.00    1224883.       0.      1988.    0.1062E-13     2187.      0.9367E-06
289661.00    1240812.       0.      2007.    0.9807E-14     2197.      0.9421E-06
291661.00    1256675.       0.      2026.    0.9070E-14     2207.      0.9474E-06
293661.00    1272474.       0.      2045.    0.8398E-14     2218.      0.9527E-06
295661.00    1288207.       0.      2064.    0.7785E-14     2228.      0.9579E-06
297661.00    1303874.       0.      2083.    0.7225E-14     2238.      0.9630E-06
299661.00    1319477.       0.      2101.    0.6712E-14     2248.      0.9681E-06
301661.00    1335014.       0.      2120.    0.6242E-14     2258.      0.9732E-06
303661.00    1350487.       0.      2138.    0.5812E-14     2268.      0.9782E-06
305661.00    1365894.       0.      2156.    0.5416E-14     2277.      0.9831E-06
307661.00    1381235.       0.      2175.    0.5052E-14     2287.      0.9880E-06
309661.00    1396513.       0.      2193.    0.4717E-14     2296.      0.9928E-06
311661.00    1411725.       0.      2211.    0.4409E-14     2306.      0.9976E-06
313661.00    1426873.       0.      2229.    0.4124E-14     2315.      0.1002E-05
315661.00    1441955.       0.      2246.    0.3862E-14     2324.      0.1007E-05
317661.00    1456973.       0.      2264.    0.3619E-14     2333.      0.1012E-05
319661.00    1471926.       0.      2282.    0.3394E-14     2343.      0.1016E-05
321661.00    1486815.       0.      2299.    0.3186E-14     2351.      0.1021E-05
323661.00    1501639.       0.      2317.    0.2994E-14     2360.      0.1025E-05
325661.00    1516398.       0.      2334.    0.2815E-14     2369.      0.1030E-05
327661.00    1531094.       0.      2351.    0.2649E-14     2378.      0.1034E-05
329661.00    1545724.       0.      2368.    0.2495E-14     2386.      0.1039E-05
331661.00    1560291.       0.      2385.    0.2352E-14     2395.      0.1043E-05
333661.00    1574793.       0.      2402.    0.2218E-14     2404.      0.1047E-05
335661.00    1589231.       0.      2419.    0.2094E-14     2412.      0.1052E-05
337661.00    1603604.       0.      2436.    0.1978E-14     2420.      0.1056E-05
339661.00    1617913.       0.      2452.    0.1870E-14     2428.      0.1060E-05
341661.00    1632159.       0.      2469.    0.1768E-14     2437.      0.1064E-05
343661.00    1646341.       0.      2485.    0.1674E-14     2445.      0.1068E-05
345661.00    1660458.       0.      2501.    0.1586E-14     2453.      0.1072E-05
347661.00    1674511.       0.      2518.    0.1503E-14     2461.      0.1076E-05
349661.00    1688501.       0.      2534.    0.1426E-14     2469.      0.1080E-05
351661.00    1702426.       0.      2550.    0.1353E-14     2476.      0.1084E-05
353661.00    1716289.       0.      2566.    0.1285E-14     2484.      0.1088E-05
355661.00    1730087.       0.      2582.    0.1221E-14     2492.      0.1092E-05
357661.00    1743821.       0.      2597.    0.1161E-14     2499.      0.1096E-05
359661.00    1757492.       0.      2613.    0.1105E-14     2507.      0.1100E-05
360661.00    1764304.       0.      2621.    0.1078E-14     2511.      0.1102E-05
361161.00    1767703.       0.      2625.    0.1065E-14     2512.      0.1103E-05
381161.00    1900448.       0.      2625.    0.1065E-14     2512.      0.1103E-05
381161.00    1900448.       0.      2625.    0.1065E-14     2512.      0.1103E-05
277661.00    1144252.       0.      1891.    0.1608E-13     2132.      0.9089E-06
279661.00    1160509.       0.      1910.    0.1476E-13     2144.      0.9146E-06
281661.00    1176701.       0.      1930.    0.1357E-13     2154.      0.9202E-06
283661.00    1192827.       0.      1949.    0.1249E-13     2165.      0.9258E-06
285661.00    1208887.       0.      1969.    0.1151E-13     2176.      0.9313E-06
287661.00    1224883.       0.      1988.    0.1062E-13     2187.      0.9367E-06
289661.00    1240812.       0.      2007.    0.9807E-14     2197.      0.9421E-06
291661.00    1256675.       0.      2026.    0.9070E-14     2207.      0.9474E-06
```

Fig. 16. Sample page of high-speed listing for quick-look purposes.

Printed Quick-look

Two basic criteria should be considered in the preparation of "quick-look" material: speed, which should be maximized, and cost, which should be minimized. Since most large-scale reduction is done on binary machines, the conversion routines for printing intermediate results should be chosen carefully, sacrificing "fancy" output

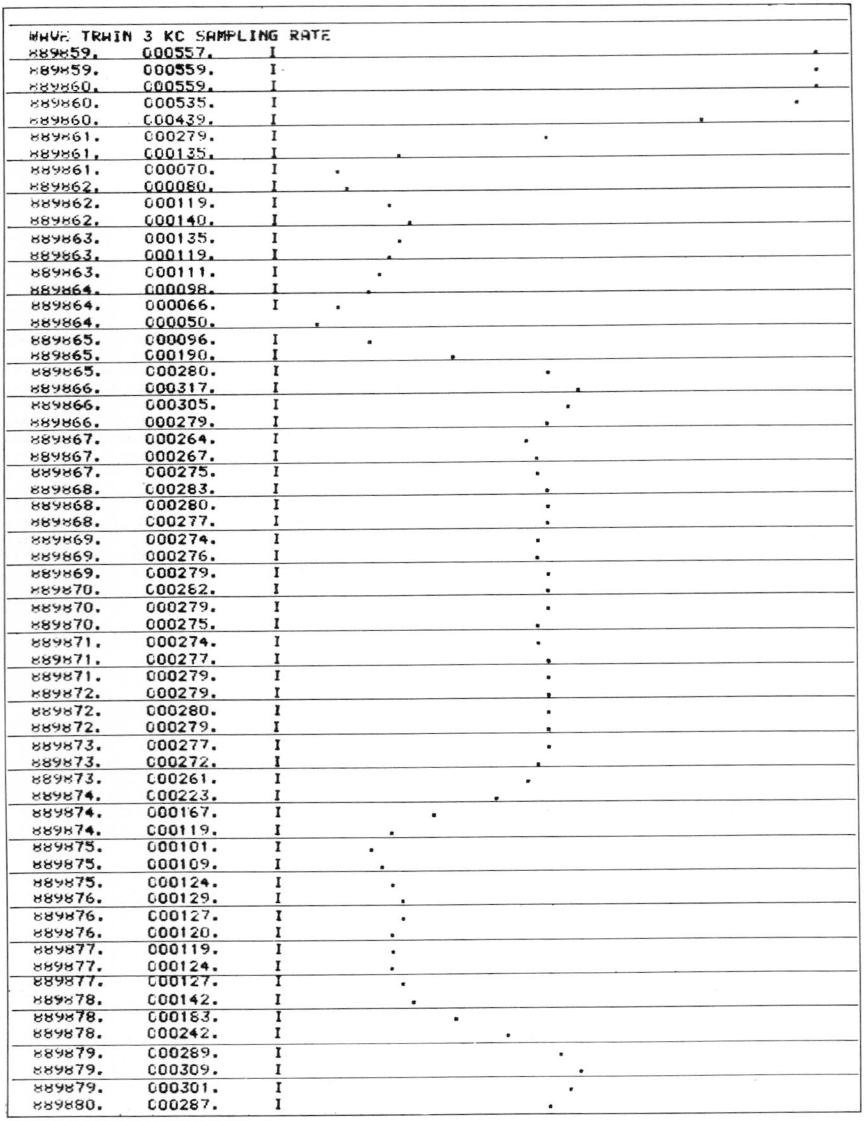

```
WAVE TRAIN 3 KC SAMPLING RATE
889859.   000557.   I                              .
889859.   000559.   I                               .
889860.   000559.   I                              .
889860.   000535.   I                                      .
889860.   000439.   I                                              .
889861.   000279.   I                 .
889861.   000135.   I         .
889861.   000070.   I  .
889862.   000080.   I    .
889862.   000119.   I       .
889862.   000140.   I          .
889863.   000135.   I         .
889863.   000119.   I        .
889863.   000111.   I       .
889864.   000098.   I      .
889864.   000066.   I   .
889864.   000050.   .
889865.   000096.   I     .
889865.   000190.   I            .
889865.   000280.   I                   .
889866.   000317.   I                       .
889866.   000305.   I                      .
889866.   000279.   I                  .
889867.   000264.   I                 .
889867.   000267.   I                 .
889867.   000275.   I                  .
889868.   000283.   I                   .
889868.   000280.   I                  .
889868.   000277.   I                  .
889869.   000274.   I                  .
889869.   000276.   I                  .
889869.   000279.   I                  .
889870.   000282.   I                   .
889870.   000279.   I                  .
889870.   000275.   I                  .
889871.   000274.   I                  .
889871.   000277.   I                  .
889871.   000279.   I                  .
889872.   000279.   I                  .
889872.   000280.   I                  .
889872.   000279.   I                  .
889873.   000277.   I                  .
889873.   000272.   I                 .
889873.   000261.   I                .
889874.   000223.   I             .
889874.   000167.   I         .
889874.   000119.   I       .
889875.   000101.   I     .
889875.   000109.   I     .
889875.   000124.   I       .
889876.   000129.   I        .
889876.   000127.   I        .
889876.   000120.   I       .
889877.   000119.   I       .
889877.   000124.   I       .
889877.   000127.   I        .
889878.   000142.   I         .
889878.   000183.   I            .
889878.   000242.   I              .
889879.   000289.   I                   .
889879.   000309.   I                      .
889879.   000301.   I                     .
889880.   000287.   I                    .
```

FIG. 17. Sample plot from peripheral printer.

with elaborate annotation. This procedure along with other possible programming short cuts could save a large amount of conversion time as compared with using a standard routine obtained through a computer user's distribution agency. Page headings on the first page only, with the remaining output completely numeric, will usually satisfy quick-look requirements. Figure 16 is an example of this type of listing.

Plotting Techniques

The problems in plotting are much the same as they are in printing. If neatly annotated plots are prepared, cost and time that cannot be afforded will be unneces-

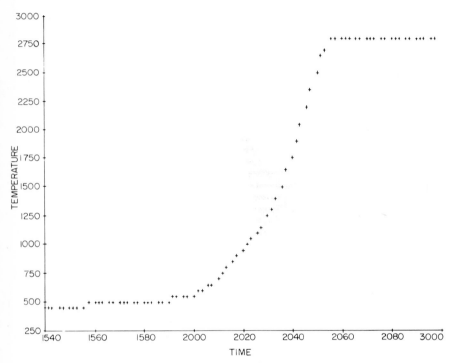

Fig. 18. Sample of one-page plotting program output.

sarily consumed. One existing method of obtaining quick-look plots is by utilizing a peripheral printer associated with the computer. An example of a plot prepared on such a device is shown in Fig. 17. Annotated one-page graphs can also be produced by this method as in Fig. 18.

Work is being done in this field to build hardware which will be capable of extremely fast high-density plotting. This type of plotting, although it imposes some undesirable restrictions, provides a fast method of preparing plots. The speed of such plotting more than makes up for the disadvantages. It is not unreasonable to expect up to 5,000 points per second from such a machine. Present "X-Y" type plotters, although much neater and more flexible, can come nowhere near such a rate. This type of plotter would be ideal for a quick-look at edited or filtered data.

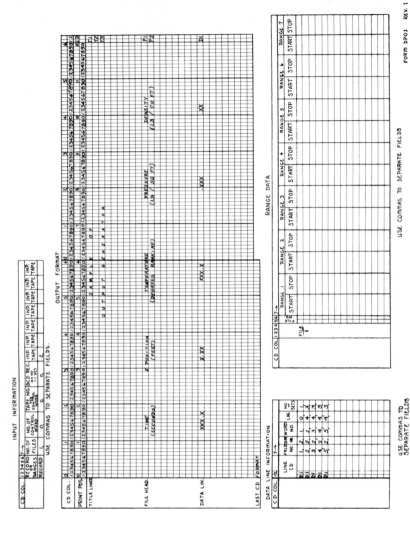

FIG. 19. Form for preparing cards for an output generator.

16 Presentation of Final Reports

In considering a second type of output, the preparation of data for final reports, much of the discussion above is completely reversed. In the presentation of final output and results, neatness and accuracy are of utmost importance and cannot be sacrificed to relatively small savings in computer time. The two mediums basically used for type of output are again printing and plotting. In printing, however, report generators or similar output-conversion routines are the most feasible and afford the user excellent flexibility for varying output formats to include titles, headings, security classifications, scaling, page number, etc. These output generators usually require the user to prepare a form which "displays" the format of the output and also serves,

| | | SAMPLE OF | | |
| | | OUTPUT GENERATOR | | |
TIME (SECONDS)	X POSITION (FEET)	TEMPERATURE (DEGREES RANKINE)	PRESSURE (LB / SQ FT)	DENSITY (LB / CU FT)
122.5	3.05	619.5	0.399	0.09
	3.76	403.8	0.399	0.16
	4.47	558.2	0.088	0.02
122.6	3.05	622.6	0.458	0.10
	3.76	406.5	0.458	0.18
	4.47	561.1	0.101	0.03
122.7	3.05	628.6	0.606	0.14
	3.76	411.8	0.606	0.24
	4.47	566.5	0.136	0.03
122.8	3.05	633.1	0.754	0.17
	3.76	415.8	0.754	0.29
	4.47	570.4	0.166	0.04
122.9	3.05	641.5	1.170	0.26
	3.76	423.5	1.170	0.45
	4.47	557.5	0.259	0.06
123.0	3.05	649.4	1.850	0.41
	3.76	430.9	1.850	0.71
	4.47	583.5	0.409	0.10
123.1	3.05	655.6	2.760	0.61
	3.76	436.8	2.760	1.00
	4.47	587.7	0.608	0.15
123.2	3.05	659.2	3.530	0.78
	3.76	440.3	3.530	1.30
	4.47	589.7	0.779	0.19
123.3	3.05	662.8	4.530	1.00
	3.76	443.7	4.530	1.70
	4.47	591.6	0.999	0.25
123.4	3.05	671.0	8.260	1.80
	3.76	451.4	8.260	2.90
	4.47	594.5	1.820	0.46

Fig. 20. Sample of output from a report generator.

in a secondary capacity, as the specification for preparing the program required by the computer. Figure 19 is an example of such a form. Usually, a somewhat greater amount of computer time is required to prepare for printing because proper spacing is important for neatness of the final report, but since such routines are not used too frequently during a reduction process, the time lost is insignificant. Figure 20 is an example of the output of this type of generator.

When considering plotting for final reports, usually the "X-Y" type is the best. These will produce neat and accurate plots. Care should be given to the selection of such plotters since, even though this plotting is not done very often during a reduction process, if much final plotting is to be done, an excessive amount of time will be consumed to finish the reports. X-Y plotters plotting at a rate of 100 points a minute or less can become a bottleneck. Also in the selection of X-Y plotters, annotation capabilities should be considered. If the plotter is not capable of annotation, a tedious task is placed on personnel. Figure 21 is an example of a plot from a typical X-Y plotter.

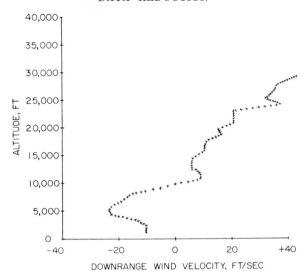

Fig. 21. Sample of plot from an X-Y plotter.

17 Data Computations

Thus far the discussion has involved the data-handling type of subroutine, and with the exception of the discussion on time and conversion for printing and plotting, no real computations have been performed on the data. As previously mentioned, the mere measurement and presentation of data is not sufficient to check the complex theories employed in the design of an item which would require telemetry for its testing. Therefore, we proceed to the programs which will process the data by performing computations. These programs will include conversion of raw data counts from the digitized tape (previously obtained by digitizing the telemetry analog tape) to calibrated function values, editing techniques to remove obviously bad data, and smoothing methods which utilize digital filters.

Calibration of Digitized Telemetry Data

To explain how the information recorded on the telemetry tapes is restored and put into a form suitable for further computations, let us assume that an analog-to-digital conversion system has produced magnetic tapes which now contain digital information that originally had been obtained by one of the three commonly used telemetry systems, PAM, PDM, or FM.

The resulting magnetic tape from the analog-to-digital conversion equipment now contains the binary-encoded and, perhaps, packed (data with more than one data value in any one computer word) "counts" which represent the actual voltage readings of the transducers or sensors as recorded at the ground station. In addition the digital tape contains other pertinent information such as time or some other independent variable to identify the frame, some type of file identification for later computer checking, a record count, sampling rate, etc. The latter type of data may be inserted manually or automatically, at the time that the analog information is being digitized.

Computer Checks on the Digitizer

In order to ensure that the analog-to-digital conversion equipment is functioning properly, a complete check of the digital tapes is made via a computer program to provide a decimal print-out of the count values for the functions, the status of "error-bit" indicators, and the time difference between successive records. This permits rapid scanning of the print-out and will indicate where "data drop-out" or missing information has occurred. Not only does this program serve as a tool to monitor the proper operation of the analog-to-digital conversion equipment and thus aid in its maintenance, but as is readily realized, it eliminates wasting computer time by the processing of erroneous or improperly digitized data.

Calibration Considerations

When the decimal print-out indicates a tape is suitable for processing, it is ready for conversion by an "analog-to-digital tape-conversion program" which will arrive at the temperatures, pressures, altitudes, etc., that the transducers sensed and relayed to the telemetry data-transmission system in flight. Note that the information content and, therefore, the actual value of the variable being monitored is dependent upon its voltage as a proportional part of the calibrated full-scale voltage which was applied to that sensor at the time its value was being read for conversion and subsequent data transmission. With this thought in mind, it is seen that a fluctuation in the full-scale voltage—unless likewise recorded—will result in false conclusions if based on a constant full-scale voltage value. For this reason, many systems include a sampling of certain percentages of these full-scale values, e.g., 10, 50, and 90 per cent of full-scale values, in order to enable proper calibration when the binary counts representing these voltage values are interpreted. Another important point is that, for each sensor, actual values of the variable to be measured are recorded vs. its percentage full-scale reading prior to flight testing, and these then constitute the unconverted calibration curves so necessary for proper conversion of the digitized telemetry data.

With these point-calibration curves and the other required parameters for an analog-to-digital binary tape in the machine's memory at the same time, the conversion process can begin. To be certain that the tape being processed is the correct one, a check of the file identification is made with the condition that inequality results in an error print-out and a stoppage of the machine. Other conditions are also tested as the program progresses, and depending on their seriousness or noteworthiness, alarms are printed out with or without machine stoppage. The program can be prepared to continue if the check does not indicate a critical condition.

After all necessary time corrections are made and a check on the time lapse between records (time for one revolution of the commutator) is made and accepted, any "unpacking" or separating of the binary data samples must be accomplished before processing continues. A check is made at this point to determine if data samples have been flagged with error bits, denoting a bad analog tape or erroneous conversion to the digital format. Characteristic bit patterns can be inserted for these "bad" data samples for later interrogation and action by succeeding programs or the immediate action of interpolation or extrapolation may be initiated and a record of such action made to aid in later decisions.

For each cycle of the commutator a table can be set up of known voltage divisions vs. the analog-to-digital conversion counts read on the commutator pins representing these divisions. The counts for a good data sample can use this table for its cycle to determine the proportion of full-scale voltage which had been read by the represented sensor. This per cent full-scale value obtained can then be used as a search argument in the sensor's calibration table of actual function value vs. per cent full-scale and linear interpolation or extrapolation employed to obtain the correct value. Knowing the time of the first sample and the time of the last one, intermediate times can be assigned to each data sample and the completely converted group of functions written on a magnetic tape in a standard format suitable for reading and computation by the other subprograms of this very versatile and flexible "modular-type" system.

18 Special Computer Data-processing Applications

Although most computing machines have included in their hardware the capability of floating-point arithmetic, all the computations are performed in fixed-point arithmetic with some small saving in machine time. Other means of saving machine time or reducing the amount of rerunning time can be accomplished by having provisions in such a program for converting only a specified number of functions of all those available, for starting and stopping at predetermined times and thereby overlooking insignificant sections of data, for manufacturing time and data when data drop-out occurs, for multiple processing option, for automatic specification of tape units to be used, for any rewinding to be done on same, etc.

One additional use for the digital computer in the data-reduction field is the replacing of semiautomatic techniques when automatic systems cannot be employed because of poor quality of the telemetry data. This substitution of a general-purpose computer for a specialized piece of electronic equipment is not recommended for general use, but it can be invaluable as backup in emergencies.

For example, a PAM decommutator usually requires a sharp rise at the leading edge of any pulse in voltage in order to trigger the sampling mechanism. Two problems can occur: (1) noisy data may trigger this mechanism too often or (2) an elongated rise time will not activate the mechanism when it is required. In either case, erroneous data will be sent to the computer for processing and will require special techniques for extracting the data. One solution to this problem is to digitize the entire wave train at a high sampling rate, as if it were a continuous signal, and let the computer scan these values and select the values corresponding to the tops of the pulses. This technique can also be useful to decommutate the output of nonstandard commutators digitally.

Another example of special use of a computer is to perform a spectrum analysis by utilizing digital filters on a section of data from an unknown waveform. This method, although much slower, can be used in place of automatic analog filters for narrower bands than possible otherwise, and also to extend the range of frequencies which can be handled. This topic of digital techniques for spectrum analysis has been treated in much greater detail elsewhere in this chapter.

DATA-PROCESSING APPLICATION

19 The Telemetry Information-processing System

Our primary concern here is with the data processing of missile flight-test telemetry. This implies no real restriction of scope, for all the problems of data processing are reflected here, combined with the peculiar difficulties and defects of telemetered data as well as the imperative need for the very high-speed processing of enormous quantities of data. This compels us to give primary consideration to automatization through the use of large-scale computers and their associated equipment. That is to say that the manual methods, and methods using relatively simple equipment, must be slighted or overlooked. The restriction implied is actually more apparent than real, however. As the speed and capacity—and cost—of data-processing equipment increase, more and more careful consideration must be given to every detail of the preparation and manipulation of data, so that it may fairly be said that the introduction of the new methods has enforced a habit of closer attention to the merits of alternative procedures and meticulous attention to computational detail. This has had the effect of raising the level of appreciation of the numerical art at an unprecedented rate. It seems reasonable, therefore, to assume that to achieve understanding of data processing on a large scale includes understanding of the problems on any scale.

An aspect of the transition to the wholesale processing of huge masses of data has been the emergence of a high degree of specialization of personnel. This is inevitable because of the manifest advantages it brings. The work could probably not be done without it. The price that must be paid is the usual one, that the individual has such limited scope that in a very real sense he does not know what he is doing. Since, however, the need for an integrated effort consistent with a unified outlook is not primarily a matter of good taste or good business, but an absolute necessity enjoined

DATA-PROCESSING APPLICATION 12–51

by the nature of physical experimentation, it follows that the quality and validity of the result turned out by the data-processing facility are directly dependent upon the state of balance, harmony, and mutual appropriateness among (1) the data-reduction requirements, (2) the state of the data-processing art, (3) the capacity of available data-processing equipment, and (4) the capabilities of the people concerned in the data-processing and flight-test program, both individually and collectively. These interactions are so important that an uncompensated limitation in any one of them becomes an absolute limitation upon the overall capability.

It has become customary to say that large-scale data processing demands a high degree of integration and that it should be imbued with the "systems" point of view. Failure in this regard may well be prejudicial to the success of missile flight testing (depending, of course, upon the character of the flight-test objectives). There is a certain danger of misconception, however. It is not the mass of data, the complexity of the machinery, or the size of the data-processing staff which demand integration. *It is the nature of physical experimentation which demands it.* The highest degree of integration is to be found in the smallest-scale experiment, conceived and conducted by one man, who also reduces the data, analyzes and interprets them, then prepares his own report of the results and conclusions and stakes his reputation on it. As the experimental enterprise grows to the scale of the flight tests for missile development, there is increasing difficulty in keeping the various aspects of the effort in consonance. Failure in this regard threatens failure to meet the objectives of the flight-test program. Large-scale data processing shows the need for integrated systems in the seriousness of the consequences of the lack of sufficient integration.

A balance must be struck between the actual practice of data processing and the availability of relevant knowledge. That this may not be easy to achieve can be seen by glancing at Fig. 22, which illustrates the various essential components and their interrelationships. It must be made clear at once that only the logical parts and connections are shown here. It is not suggested that this is a flow chart for flight-test data-processing organization at any installations now in existence. This is not to say, of course, that it could not be.

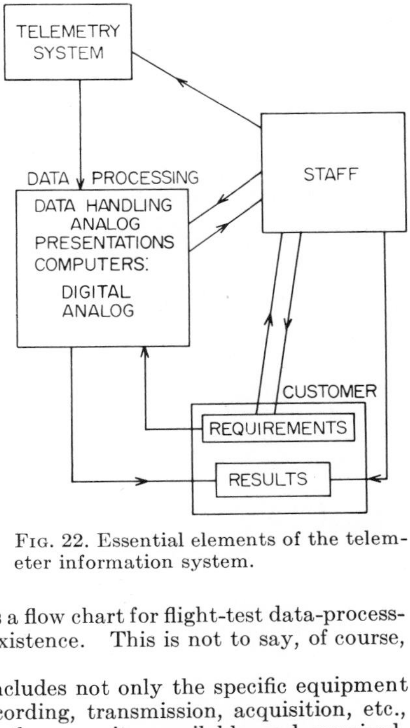

Fig. 22. Essential elements of the telemeter information system.

The "telemetry system" block in Fig. 22 includes not only the specific equipment components for measurement, conversion, recording, transmission, acquisition, etc., but also the instrumentation specifications, the capacity available and required, priorities, sampling rates, and all other features concerned with obtaining the actual observations of the physical phenomena of the flight. The "data-processing" block is intended to include not only all the data-processing equipment but also the people directly involved in carrying out the data-processing operations described in Chap. 12. It also is intended to include such intangibles as the limitations of the equipment as computer speed and data-handling capacity; logical flexibility; computer programs and techniques which compensate for specific defects in the data; and the degree to which the installation can cooperate to meet special requests. The "customer" is the ultimate source of objectives, goals, requirements, and the final recipient of results (so far as "data processing" is concerned). Most in need of elucidation is the block labeled "staff." This block is intended to comprehend the following functions: planning of flight tests, formulating flight-test objectives, proposing new problems for flight test, devising computational methods, adapting requirements to the available people and machines, interpretations and studies on the computer, collateral technical studies, appraisal of the value and effect of additional or alternative equipment, etc. It is evident that each installation has its own peculiar organizational arrangements and that these various essential "staff" functions may be performed by data-processing personnel or specialists having no organizational ties with data processing. Whatever the arrangements, however, it is clear that the functions are indispensable. It is also clear that if these essential functions are distributed among autonomous groups, one to design missiles, another to specify the instrumentation, still another to plan flight tests, and so on, that it would be only by singular good fortune that the outcome of the flight tests would be as successful as they might be. The various functions are so obviously interrelated in the most intimate way that a material change in any one of them is likely to have the most serious effects on all the rest. Nevertheless, it cannot be said that unified and coordinated planning of flight tests, in which all these aspects are in a planned balance, is the rule rather than the exception.

20 Flight-test Planning

The flight testing of a missile involves a program of flights each of which occupies a definite place in the entire program. The flight tests are designed to meet three kinds of objectives: (1) to provide information which will be of use in correcting and improving the design of the missile; (2) to provide information believed to be of value in understanding or designing missiles in general, not necessarily only the specific one under test; and (3) to demonstrate to the customer that the missile performs in a prescribed manner to a successful degree. It is obvious that any two or all three of these types of objectives may be represented in any single flight and that certain of the flights will have objectives of the third type.

That a flight-test program cannot hope to succeed unless it is planned to meet a detailed and specific list of objectives is evident. Consequently there is always a program plan book which lists them. In establishing objectives and the tests and procedures for meeting them the prime role of data reduction is often overlooked, with the result that after the flight is over and the data submitted for final processing it not infrequently turns out that the information required cannot be extracted because the data collected, or the procedures specified, are incapable of providing the results required. Flight-test planning which does not take the fullest account of the procedures of data reduction from the very outset, even in the establishment of objectives, runs grave risks of failure.

Examples of this sort of oversight are abundant: commutated pressure measurements at a hopelessly inadequate sampling rate; the specification of an instrumentation system without giving consideration to whether it was even theoretically possible to infer the information required from the data collected; a fundamental objective of an entire series of flight tests could not be met because no satisfactory computational method existed to determine the quantities needed accurately enough; and so on.

From what has been said, it is clear that once the principal objectives have been established, it is necessary to determine what observations to make in flight, i.e., what measurements to make, what instruments to use in making them, how to handle the outputs of the transducers (continuous channel or commutated, what frequency, what bandwidth, what sampling rate), in what form the results should be presented, and the precise manner of converting the received signals into the required presentation. The data-reduction facility has contact with these decisions through the data-reduction requirements with which it is presented.

21 Data-reduction Requirements

Data-reduction requirements are specifications of the data about physical or other quantities which it is the object of the flight testing to obtain, and of the manner in which these data are to be presented.

DATA-PROCESSING APPLICATION 12–53

These requirements are prepared by the various engineering groups which require flight-test data in order to be able to discharge their responsibilities in design, development, reliability, or other. The following are some of the more common types of requirement:

Events

These are separate occurrences, happening at one or more instants in the course of a flight test. For any of a variety of reasons it may be necessary to know whether particular phenomena relating to the action of a system being studied by telemetry have occurred in the manner intended and at the proper time. Such events include the chain of settings and adjustments involved in the process of arming and fuzing, the time of occurrence of a specified value of acceleration, whether and when a nose cone became separated from the missile airframe, whether a structural or other component failed, etc. They may sometimes be inferred from discontinuities in oscillograph records; sometimes they can only be inferred by computation.

Time Histories of Directly Measured Quantities

These include such items as temperatures, pressures, accelerations, and angular rates. The flight-test specifications contain descriptions of the instruments to be used in making the measurements, with their frequency response; and system tests are specified for determining system response. This sort of result may be presented as an oscillograph record, a tabulation, or a plot. The results presented, however, would have to be obtained by converting the (calibrated) signal values into function values, i.e., to the numbers representing the physical quantities measured in customary units, as computed from given formulas. In some cases, these function values would not be presented until after they had been subjected to such operations as filtering, smoothing, and editing. It is evident that this type of requirement would almost demand the use of computing machinery.

Time Histories of Derived Quantities

Most flight-test data requirements are for derived quantities. This means simply that the variables measured in the telemeter are not precisely those about which information is needed but, it is hoped, contain the required information implicitly in a form in which it can be inferred by the procedures presented with the requirements. The following possibilities arise most commonly:

1. Functional Transformations of the Observed Data. The time history required is obtained from the "function values" of the observed time history (or from a time history derived from function values) by substitution in a formula which is a part of the data-reduction requirement and which may involve simultaneous data from other time histories. For example, translational acceleration time histories are obtained by substituting telemetered accelerometer readings in formulas which also involve telemetered rate-gyro readings.

2. Use of Tables, Graphs, Etc. The time history required is obtained from observed or derived time histories by looking up values in tables, graphs, nomograms, and the like. When the tables, etc., can be represented by empirical or theoretical formulas, this case reduces to the previous one. Otherwise, the distinction between the two cases is a significant one from the point of view of practical data processing. A particularly clear illustration of this distinction is that between using a graph and a formula. The formula can be computed automatically in a computer; the graph, on the other hand, must be read by somebody, at best with manually operated reading equipment, much more slowly, and with more errors.

3. Time Histories Obtained by Formal Data-processing Operations. It is often intended that the required data be subjected to subsequent analysis, either qualitative or quantitative; or that they will be used in a process involving further mathematical manipulation. In such cases we commonly find that quantities like those classified above under items 1 and 2 either are not in a form suitable for further computation or are not precisely the quantities which have to be used. In such cases, therefore, it is necessary to have recourse to operations such as computing moving averages, smoothing, filtering, trend analysis, curve fitting (perhaps by least squares), frequency analysis, numerical differentiation or integration, interpolation, extrapolation, or combining alternate records.

4. Solution of Functional Relations. The quantities required may not be amenable to measurement or obtainable by formal processing of the data or by substitution in explicit formulas. They may, however, be implicit in functional relationships constructed with quantities obtained by telemetry, and inferrable from them by analytical or numerical procedures. Such relations could, for example, be a set of algebraic equations obtained from the equations of motion of a missile when the telemeter is used to measure the variables of the motion. These equations could then be solved for the stability coefficients. The relations could assume the form of a system of ordinary differential equations with prescribed initial conditions. This, indeed, is exactly what is used for determining a nose-cone trajectory, the initial conditions being the conditions at separation of the nose cone from the airframe. An even more elaborate example would be a partial differential equation which is to be solved for prescribed boundary conditions. As an example of this we may mention the determination of temperature and heat-flux time histories obtained by solving the equation of heat conduction for boundary conditions including a temperature time history measured by telemetry.

5. Formal Derivations. There are usually numerous quantities of considerable interest which are more or less readily computable once we are in possession of certain other variables which must be determined by more elaborate means, such, for example, as those in the preceding paragraph. For example, only a simple formula must be evaluated in order to obtain the dynamic pressure, once the velocity has been determined by the rather formidable computation of the trajectory.

6. Statistical Quantities. It follows from their definition that statistical quantities are not observable directly but must somehow be inferred from masses of data. The degree of difficulty encountered in making such inferences is quite variable. It is common to require the determination of mean values or measures of variability such as the standard deviation or probable error. Such calculations are simple, but tedious. It is a bit more difficult to carry out a spectrum analysis in the presence of noise. More difficult still would be an estimation of the degree of confidence to be placed in the results issued by the data-processing facility in meeting a flight-test data-reduction requirement.

Miscellaneous Associated Requirements

These must also be satisfied—such as the determination of launch conditions or separation conditions; atmospheric data, including wind and pressure measurements at all accessible altitudes at the launch site and downrange sites, as well as the forecast values for the flight.

22 Establishing Data-reduction Requirements

It may be noted that in every case these requirements are such that the operations involved could be performed manually. As a matter of fact, that is how airplane flight-test telemetered data were actually processed in the past. It is evident, however, that the quantity of data transmitted by a missile telemeter in a flight test is simply too large to be processed by any means other than automatic machinery. More than that, certain of the requirements imposed on a flight-test program, especially those classified as under items 3 and 4, could occupy an army of human computers for many months. This would normally be intolerable, with respect to both cost and time. Obviously such demands were never placed upon airplane flight-test telemetry-data processing.

At this point it is again appropriate to look back to Fig. 22 and the discussion

DATA-PROCESSING APPLICATION 12-55

accompanying it. It was indicated there that the various blocks, disjointed as they might look at first glance, were actually most intimately bound together. Now once more consider a missile or nose cone which has been designed and built; instrumentation has been procured and put into it; and the customer has provided his requirements in the form of objectives to be met. Let us sketch briefly how a set of data-reduction requirements can be established. Naturally, only a few illustrative parameters can be considered here without going completely out of bounds. They should be sufficient to show what is involved in representative cases.

As a first illustration, suppose that we are dealing with a nose cone containing an ejectable data capsule. It is intended to eject the capsule at a specified altitude by means of a JATO rocket. The nose cone contains no instrumentation capable of measuring ambient air pressures, so that—let us suppose—the only way to determine within the nose cone when to eject the capsule is to use knowledge of the trajectory of the nose cone, namely, to let ejection be triggered when the nose-cone acceleration assumes a specified value. The recording equipment in the capsule would also be turned on at a specified acceleration. The flight-test program would therefore include measures to determine whether the recorder was turned on when it ought to have been; whether the JATO was ignited at the proper altitude; whether it actually went off; whether the capsule was actually ejected at the proper altitude and fell clear; and whether the capsule actually recorded the data it was supposed to record. It is clear that in this case most of the occurrences of interest are "events," so that instrumentation and data channels must be provided. For them, the task imposed on data processing is relatively light; it is sufficient to scan the analog quick-look records and check and note the times of occurrence of the various events. To determine how well the data capsule as a recording device functions, it is, of course, necessary to recover the capsule, examine the tape it contains, and run it through the computer programs and quick-look procedure in order to evaluate the results. This may be regarded as the simplest case, and relatively little consultation and interaction among groups is required in order to complete the job properly.

Now consider something more demanding. Suppose it is required to determine the attitude of a nose cone during reentry and that for some reason pressure probes are regarded as not available. It is a basic objective to find out what the aerodynamic performance of the vehicle is in various phases of the reentry and to demonstrate ultimately what the performance of the vehicle is in various phases of the reentry, and to demonstrate ultimately that the performance is satisfactory for the completion of its designed mission. It is not difficult to find some excellent questions. In the first place, and most obvious: What physical phenomena will be used to detect the nose-cone attitude? In one actual case it was decided to use instrumentation consisting of three rate gyroscopes rotating around the principal axes of the nose cone and three accelerometers whose lines of action were to be respectively the axes of the nose cone. This meant that the "staff" in this case consisted of aerodynamicists to make this decision and to devise a method of inferring the nose-cone attitude from the readings they would get from these instruments. It also involved instrumentation engineers to select the proper instruments, to decide whether to use continuous or commutated channels and in what frequency range, and how to fit the equipment required into the telemeter. It was necessary to present the data-processing facility with at least the theoretical foundation of the method of deducing the nose-cone attitude, and for the data-processing personnel to determine either by themselves or in consultation with engineers what the quality and character of the received data would be and how the received data should be prepared for computation. Quick-look procedures had to be devised; digital-computer programs had to be worked out, coded, debugged. In this case, therefore, we see that almost every type of data-reduction requirement had to be met. Time histories of all six instruments had to be telemetered, edited, smoothed, calibrated, time-correlated, and their function values computed. Supplementary information in the form of specification of trim had to be supplied and utilized in conjunction with the rate-gyro data to provide corrected acceleration histories. A system of ordinary differential equations had to be solved in the digital computer, the rate-gyro histories being used in the variable coefficients of the differential equations, and the corrected accelerometer readings being used to provide boundary conditions. It then emerged that attitude information was valuable to other than aerodynamicists, so that some of the data-processing "staff" began to investigate the possibility of applying the attitude information in the study of other parameters. Moreover, results from the earliest flight tests already indicated that the instrumentation was only marginally capable of providing the data with sufficient precision, while the numerical procedures which had been selected were unduly sensitive to certain errors, and that these difficulties had to be circumvented.

This is an excellent example of a complex problem involving people of rather diverse background. It also shows clearly how the mere possibility of satisfying even a fundamental objective may hinge upon the feasibility of carrying out an operation in data reduction.

23 Flight-test Program Planning

It has been shown that the developmental flight testing of a missile involves an interrelated set of flights, each with its own place in the entire program. Furthermore, in view of the uncertainty of success in testing new missiles sufficiently, many redundant flights must be scheduled to compensate for unsatisfactory flights. Rational planning of such a program is impossible unless all the separate parts, including data processing, are properly integrated. In planning the data processing of flight-test data, the following considerations must be taken into account:

1. The objectives of the testing program
2. The objectives of each flight in the program, and how it fits into the program as a whole
3. How fast the results of each test must be ready
4. How complex or difficult the required data-processing operations happen to be
5. The accuracy to be provided in the results
6. The characteristics and capabilities of the data-processing equipment available for the job, especially the capacity of the large-scale digital computer

Clearly these factors are closely related. For example, with given machinery and schedules, there is a limitation on practical precision requirements, a precision which may be rapidly altered for the better by an improvement in data-processing methods or by improvements in the machinery; or changed in the opposite direction if the schedule is accelerated. Also to be considered, of course, are such factors as sensor errors, calibration errors, drift, sampling rates, and the like which influence attainable accuracy.

24 Telemetry-data Defects

Before one can proceed very far toward evaluating a telemetry-data-processing program, it is important to underline the consequences of the use of telemetry. The intrusion of telemetry into the processes of experimental measurement and observation means that it is inconvenient to make the required observations and that we may consequently anticipate obstacles on the way toward satisfactory measurements. As soon as it is recalled that we are only performing a task which would offer relatively little difficulty if performed in the normal manner on the ground, we can see at once what should be the difficulties introduced by telemetry, simply because of inaccessibility and irremediability. These difficulties are presented in outline form as follows:

1. Impairment of the measurement because of
    a. Distortion in the airborne and/or ground electronic equipment
    b. Distortion in the transmission link
    c. Noise arising at every point of the process, and the effects of modulation by noise
    d. Crosstalk
    e. Failure to transmit proper signals during periods when attenuation is excessive

DATA-PROCESSING APPLICATION 12–57

through ionization, unfavorable aspect of the antenna pattern, etc.—all of which appear as data drop-outs
   f. Total or partial failure of any of the radio elements to function within specifications—data drop-outs
   g. The limitations of the processes of recording on magnetic tape
   h. Synchronization difficulties connected with analog-to-digital conversion
   i. Equipment incompatibilities, naturally to be expected in a large system using independently designed components of diverse origin
2. Errors and uncertainties in time because
   a. Any of the defects considered in item 1 may occur in the time code and are particularly destructive there—drop-outs, extra bits, etc.
   b. There may be no time code on the signal—for example, because it was received on an aircraft which has no time-code generator
   c. There may be no time code because the signal came from the playback recorder, which transmits its entire message backward, so that time may be associated only later in the course of data processing
   d. Different tapes of the same flight, received at distinct points, may not have perfectly compatible time codes
3. Compromises with respect to quality and design of the sensing used in the telemeter, where considerations of size or weight or resistance to shock may have greater influence than the nature of the measurement to be made and the preferred manner of making it
4. Limitations on vehicle payload or space, limitations of telemeter power or capacity, or simple oversights which may occur in a complex enterprise on a tight schedule all may lead to marginal or insufficient quantities of data, leading to difficulty or impossibility of obtaining required data, without hope of correction because no more flight tests are planned with the same instrumentation
5. Inexpediency of direct measurement of the quantity of interest, so that a more or less elaborate computer program and analysis may be required before a particular data requirement can be met
6. The fact that flight tests are scheduled far in advance, and a vehicle or its instrumentation cannot be modified in the light of the results of previous flights, except in cases of great weight and with much difficulty. Since some failures always occur, the later phases of the data processing may be made difficult or impossible by the absence of irreplaceable data which should have been obtained in an earlier flight

25 Remedial Measures

This partial listing of data defects is enough to indicate the magnitude of the problem of data reduction in unusually bad cases. Some of these problems arise simply from the fact that a complex remote system must be used to make the observations and that failures are inevitable. There is little that can be done about this, except perhaps to improve the various components as improvements occur and to raise the proficiency of all operators.

On the other hand, many of these defects are entirely remediable. For example, use of recoverable data capsules instead of direct transmission reduces the hazard of complete loss of data during periods of high attenuation. The noise which impairs the signals may be effectively removed by autocorrelation, filtering, or other techniques. Expedients have been devised for dubbing in time codes. There is a steady advance in the quality and design of sensory elements. Even the planning of flight-test programs is being carefully and critically reviewed, so that the maximum benefit may be derived from the program.

26 Elements of Integrated Data Processing

The quality of the results of data processing will, in the last analysis, depend upon the success with which the elements described in the following paragraphs are achieved and integrated.

The bulk of the data to be processed. This is influenced by available instrumentation and telemeter capacity, as well as by the speed and capacity of the data-handling equipment on the ground. In turn, it imposes demands upon the speed and capacity of the computing equipment and its peripheral equipment which must be met if the the data requirement is to be met. It is here that the capacity of the computer determines which objectives of the flight test are within reach.

The complexity of the procedures to be followed—whether simple records and tabulations will suffice or it is necessary to perform extensive analyses and intermediate processing. Here again we see how computer capacity may play a limiting role in setting reasonable objectives and data-reduction requirements.

The accuracy requirements and the degree of confidence demanded of the results. The ultimate in what is achievable is obviously within reach only if there is a proper match between the computer speed and capacity, the ingenuity of the instrumentation engineers and the availability of suitable components, the specialists who require the data, the skill of the numerical analysts in choosing procedures which allow the required accuracy and also permit the analysts to determine the accuracy limits, etc.

The quality and availability of essential auxiliary information, such as indications of results obtained in other flight tests which offer information of relevance to the present one, and indications of subsequent tests being contemplated at the present time.

The state of the data-processing art. Large-scale data processing at high speed is quite recent. Numerical analysis has been the serious preoccupation of large numbers of mathematicians only within the past few years. Progress in the development of mechanical aids to both is very rapid. Consequently, numerical methods have been developed in a most irregular, even haphazard manner, according to the interests or purposes of the individual mathematician, as influenced by the machines with which he is obliged to work. The rapidity with which the new and reliable computers were developed is itself an embarrassment, for the classical methods and judgments of numerical analysts have to be completely reconsidered in the light of the new capabilities. This field is growing rapidly, therefore, but unsystematically. What is practical numerically is not readily visible to the nonspecialist, and the situation changes so rapidly as to challenge the specialist. Evidently, therefore, the realizability of flight-test objectives and success in meeting data-reduction requirements depend heavily upon (1) whether, or to what extent, the mathematical methods required for the solution of data-processing problems of interest for the flight test have been developed to the point where they will give practical results; (2) whether, or to what extent, the art of numerical analysis and computational technique are capable of obtaining practical solutions and of making useful estimates of the errors incurred; and (3) whether the state of development of the technique of computer application is adequate to the demands which are made.

Naturally, considerations of this sort may be more or less ignored, provided that only the simplest requirements for data reduction are made. However, it is difficult to be consistently successful in being simple enough. Moreover, the full possibilities of extracting information from the data that can be made available by flight test cannot then be realized.

The state of instrument and computational technology. It is difficult to keep abreast, of the many and various improvements in computers, particularly special-purpose computers; in data-handling equipment; and in instrumentation for measurement in the missile, both available and in process of development—as may be seen elsewhere in this handbook. When planning flight tests, however, we find that improvements in equipment per se do not necessarily imply increased capacity for extracting information by telemetry. Even from the purely technical side we must consider the entire chain which involves (1) available airborne instrumentation—considering the accuracy of transducers and freedom from drift; instruments which contain special-purpose computers so that some results are computed before transmission, etc.; (2) telemeter capacity and flexibility; (3) component reliability and accuracy, both airborne and on the ground; and (4) types of computers available, their capacities and speeds, and their peripheral equipment. It is quite apparent that in such a chain of instrumentation and other equipment, the capacity of the whole is determined by the least of them. It is equally important to know what this capacity is and to adapt the other aspects of the program to it.

The composition, capabilities, and versatility of the data-processing staff. This factor, the most difficult to measure, is in many ways the dominant one. It determines what the customer may have, or perhaps a little more accurately, it determines what he cannot have. The limitations of the staff in any of the three preceding divisions entail a corresponding limitation upon what the customer may reasonably expect to get. On the other hand, to the extent that this staff is conversant with the level of theoretical knowledge and with the best of existent techniques in using all the mechanical aids to data processing that modern technology can furnish, to that extent the greatest value can be extracted from the telemetered data.

Before this listing was presented, we remarked that the quality of the reports on flight tests depends upon the success with which all such factors cooperate. This obvious point is being stressed because it is so readily violated, as experience richly illustrates. For example, it is the rule, not the exception, for the data-reduction requirements to be drawn up by a number of separate groups, working with little or no intercommunication. Quite commonly, by the time that the data-processing group is made aware of what will be required of it, it is likely to be too late for it to be able to influence the flight-test program to any very significant extent. It is most unusual for engineers or their scientific associates to have more than the most superficial acquaintance with the field of practical numerical computation, and they are likely to have none at all with the capabilities and limitations of modern computing equipment. The result is that impossible demands are sometimes made while, at the same time, opportunities for securing useful information from the flight tests, with little or no modification of schedules or instrumentation, are likely to be overlooked. This only illustrates the remark made in the introduction, that large-scale high-speed data processing of telemetry data must be regarded as an activity for a closely integrated system, the need for integration growing with the size of the data-processing installation, and that this is so because the nature of physical observation and experiment make it so.

Thus, it follows that when a data-reduction requirement is presented, setting forth what the user of telemetry wants, the data-processing group must determine for itself, with whatever help it can get (1) exactly how the data are to be prepared for further processing; (2) exactly how the prepared data will be processed to provide the required result; (3) how the results obtained will be checked and evaluated; and (4) what deficiencies it has discovered in its ability to provide the results required; in what direction methodological research should point; what useful results it can provide which have not been requested; and how it can report the results it has obtained in the most useful manner. Normally, these procedures will be set up by data-processing personnel with little or no significant guidance or assistance from those who present the requirements for data. This is a condition which is likely to persist, and even develop further, with the further development and specialization of data-processing techniques.

27 Evaluating the Results

It would seem to be distinctly beneficial if some knowledge of the potentialities of data-processing technique were to become a part of the background of every user of telemetry data. Since, however, this is not likely to happen rapidly, it should become a recognized function of data processing to furnish a critical evaluation of the results it presents. Such an evaluation should have the following three aspects:

1. Validity of Results

This may be considered under the following headings:

Does the quality or the nature of the data justify the anticipated degree of confidence in the result to be provided? For example, if an oscillation is to be measured in a commutated channel, it should be determined whether the sampling rate is adequate and what degree of uncertainty or imprecision is caused by the noise on the signal. Again, if a quantity is to be presented with an assigned precision, it should be checked whether the instrumentation of the numerical procedures is such that this precision is a realistic demand.

Is the prescribed data-processing procedure adequate to provide the results anticipated? For example, suppose that a received noisy signal is to be smoothed and differentiated before the next step in processing it. If its further application is such as to be sensitive to the value of the derivative, a very poor result should be expected. In such a case, the user will either have to reconcile himself to poor results or find an alternative procedure which will not be subject to the same defect. The detection of such situations should be regarded as a major function of a data-processing facility.

Is the prescribed form of presentation of results adequate to show what is desired? In simple cases, for example, a graph may be plotted on an inappropriate scale, or the wrong kind of plot may be requested. Such problems are easy to solve. On the other hand, the variable chosen for presentation may be inappropriate in that it conceals the phenomenon of interest. This is a far less obvious sort of difficulty, and it may not be at all easy to detect or remedy.

Can the result required be validly inferred from the type of data presented? Unfortunately, it is by no means rare to find that the data obtained by telemetry are simply not able to provide the information for which the flight test was undertaken. In commutated channels, this could happen because the sampling rate is too low. Another case is that the transducer is not sensitive or accurate enough to provide measurements of accuracy sufficient for the requirements. Still another, far less obvious example can arise when the data from several channels must be used together to obtain the required result, and it is found that the relative phase shifts in the several channels are of the same order as the phenomenon to be observed, so that it is unlikely that anything can be found out. An even more difficult case has arisen in the following manner: It is found after the telemetered data have begun to be processed that the quantity desired cannot be derived from the data collected, because the data obtained are either insufficient or partly inappropriate; it is a case of theoretical impossibility.

Does the procedure prescribed really lead to the conclusion required (in terms of the particular theoretical structure supposed to be governing the phenomenon)? If it does not, then of course a blunder has been committed. This happens. It seems reasonable that somewhere in a data-processing group there should be somebody to examine the logic of the work to be done. Bad logic cannot be justified any more than bad arithmetic, its near cousin. Under this heading, in brief, we determine whether what is being done possesses validity or not. If it should appear that what is to be done is valid and that the result obtained will actually be what is intended, then we apply a second type of criterion:

2. Level of Confidence in the Results

It is not quite rational, though customary, to present the results of measurement or of computation applied to the results of measurement, without indicating the size of the cloud of uncertainty surrounding each one. At the same time it must be admitted that the analysis of the errors arising from computation is poorly developed, as a rule. Errors arising from rounding off, errors arising in the solution of differential equations, errors arising from truncation combined with the other two kinds easily can lead to error-estimation problems which are beyond the present power of mathematics to solve. The following list presents a few examples; it has no systematic pretensions.

Obtain the probable error, standard deviation, or other index of precision. This is an absolute minimum and is likely to be requested in most cases.

Is the data-processing procedure stable? That is, when results are obtained from the given data, is the result effectively unique? It is possible that the method used in processing the data is such that considerable uncertainties are introduced into the result, so that slight variations in procedure may lead to appreciable discrepancies. This must be suspected in iterative procedures until it has been demonstrated that it does not occur. Such unstable methods must be detected and replaced by others. On the other hand, a method which is generally satisfactory may become involved in difficulties in special cases, as when one attempts to solve poorly conditioned systems of linear algebraic equations. Difficulties of this sort are a common occurrence, but there are no general rules for suspecting or detecting them. Another closely related difficulty, sometimes hard to detect, is the existence of singular cases for which the method fails. Many of these bad cases can be detected in a competent preliminary analysis. If the precaution of such an analysis is not taken, the difficulty will arise in the course of the data processing. In any case, data processing will be involved in it and should have the responsibility for avoiding this sort of error or oversight.

Does the theory provide a stable solution? Although the relevant theory, and the equations derived from it, may have been shown to possess unique solutions analytically, it is yet a not uncommon occurrence to find in the course of numerical solution that a unique and determinate solution is hard to come by. This could happen because solutions are unstable; i.e., small differences in conditions can lead to large differences in solutions.

The extent to which *independent checking* is possible; the possibility of *obtaining alternate results for comparison.* It is not uncommon to find that although an algorithm can be found for obtaining a numerical solution to a problem, it is not known how to check the result. Computations of this sort must be always suspect, and the suspicion (and the basis for it) should be furnished with the solution.

Inherent uncertainties in at least one of the procedures used in the course of preparing the data or manipulating them. For example, it may be difficult or impossible to estimate the types and magnitudes of the errors committed when an empirical time history is differentiated or smoothed. Such uncertainties may propagate as further uncertainties throughout subsequent operations, perhaps to vitiate the final results. It is the responsibility of the data-processing facility to detect and, if possible, estimate, perhaps nullify, the effects of such errors.

Inherent uncertainties in the raw data. Such uncertainties may arise from the inaccuracies of measurement and would therefore provide a source of error which must be estimated and followed as it propagates through subsequent operations. On the other hand, errors may be introduced by the data-handling procedures, with an effect very similar to errors of measurement. As a data-processing procedure proceeds, it may reduce the level of uncertainty, thus actually nullifying the effects of errors in the data. This is mentioned to indicate that there is really no simple way to dispose of the problems of error analysis. As a complicated example, suppose that a data-processing problem included the solution of a partial differential equation with boundary conditions which must be obtained by measurement and may contain the kinds of errors we have discussed. It could easily prove to be impossible, in the present state of mathematical knowledge, to provide any kind of estimate of the effect of such errors upon the solution obtained numerically.

After the legitimacy of data-reduction procedures has been established and, where possible, estimates of the errors in results have been furnished, there remains a third kind of evaluation which can be provided by a properly trained and organized data-processing facility, namely,

3. Feedback

As a result of the critical examination that that would have been made in the course of the evaluation of the flight-test telemetry-data-processing program, the data-processing facility should be in a position to offer constructive suggestions of the following kinds:

Suggestions of further information that could be derived from the same data. Even when such further information was not needed in connection with the flight test in which the data were obtained, it could serve as a point of departure for new investigations.

Indication of data-reduction requirements that cannot be met because of the nature of the data (poor quality, insufficient in amount, wrong kind).

Indication of the extent to which compromises made in the physical description of the phenomenon being measured, or of the system being used to measure them, cause the intentions of the flight test to be fulfilled or not to the requisite degree of validity and precision.

Indication of data requirements that could be met if appropriate data were supplied by flight test.

Proposals of alternate methods of handling the data processing in the interest of obtaining a more complete picture.

REFERENCES

1. (a) D. D. McCracken, "Digital Computer Programming," John Wiley & Sons, Inc., New York, 1957. (b) G. A. Korn and T. M. Korn, "Electronic Analog and Hybrid Computers," McGraw-Hill Book Company, New York, 1964. (c) H. A. Huskey and G. A. Korn, "Computer Handbook" McGraw-Hill Book Company, New York, 1961.
2. (a) F. B. Hildebrand, "Introduction to Numerical Analysis," McGraw-Hill Book Company, New York, 1956. (b) W. E. Milne, "Numerical Calculus," Princeton University Press, Princeton, N.J., 1949. (c) J. B. Scarborough, "Numerical Mathematical Analysis," The Johns Hopkins Press, Baltimore, 1955. (d) C. Lanczos, "Applied Analysis," Prentice-Hall, Inc., Englewood Cliffs, N.J., 1956. (e) E. Whittaker and G. Robinson, "The Calculus of Observations," 4th ed., Blackie & Son, Ltd., Glasgow, 1944.
3. (a) E. Whittaker and G. Robinson. (b) W. E. Deming and R. T. Birge, "Statistical Theory of Errors," U.S. Department of Agriculture Graduate School, Washington, 1938. (c) W. E. Deming, "Some Notes on Least Squares," U.S. Department of Agriculture Graduate School, Washington, 1938.
4. (a) G. Szego, "Orthogonal Polynomials," American Mathematical Society Colloquium Publications, vol. 23, 1939. (b) A. C. Aitken, On the Graduation of Data by the Orthogonal Polynomials of Least Squares, *Proc. Roy. Soc. Edinburgh*, vol. 53.
5. (a) M. A. Martin, "Frequency Domain Applications in Data Processing," Technical Information Series 57SD340, General Electric Co., Missile and Space Vehicle Dept., Philadelphia, Pa., May, 1957. (b) M. A. Martin, Frequency Domain Applications to Data Processing, *IRE Trans. Space Electron. Telemetry*, vol. SET-5, pp. 33–41, March, 1959.
6. Ref. 2a–e.
7. N. Wiener, "Extrapolation, Interpolation, and Smoothing of Stationary Time Series," John Wiley & Sons, Inc., New York, May, 1950.
8 and 9. R. B. Blackman and J. W. Tukey, The Measurement of Power Spectra from the Point of View of Communications Engineering—Parts I and II, *Bell System Tech. J.*, vol. 37, nos. 1, 2, January, February, 1958.
10. (a) Ref. 2d. (b) Alston S. Householder, "Principles of Numerical Analysis," McGraw-Hill Book Company, New York, 1953. (c) E. Bodewig, "Matrix Calculus," Interscience Publishers, Inc., New York, 1956. (d) V. N. Faddeeva, "Numerical Methods in Linear Algebra," Dover Publications, Inc., New York, 1958.
11. S. Gill, A Procedure for the Step-by-step Integration of Ordinary Differential Equations in an Automatic Digital Computing Machine, *Proc. Cambridge Phil. Soc.*, vol. 47, p. 96, 1951.
12. (a) Ref. 2a–e. (b) W. E. Milne, "Numerical Solution of Differential Equations," John Wiley & Sons, Inc., New York, 1953. (c) L. Collatz, "Numerischer Behandlung der Differentialgleichungen," 2d ed. Springer-Verlag OHG, Berlin, 1951, 1955. (d) K. S. Kunz, "Numerical Analysis," McGraw-Hill Book Company, New York, 1957.
13. I. Shapiro, "The Prediction of Ballistic Missile Trajectories from Radar Observations," McGraw-Hill Book Company, New York, 1958.

Chapter 13

INDUSTRIAL TELEMETRY AND REMOTE CONTROL

JOHN E. GAFFNEY, JR., *Federal Systems Division, International Business Machines Corporation, Rockville, Md.*

REMOTE-CONTROL-SYSTEM PRINCIPLES

1	Telemetering Systems	13-1
2	Control Systems	13-3
3	Computers and Communications	13-7
4	Multiunit Systems	13-9

COMMUNICATIONS CONFIGURATIONS

5	Data-link Components	13-11
6	Data-link Configurations	13-13
7	Communications-system Configurations	13-15

COMMUNICATIONS SYSTEMS

8	Telemetering Systems	13-20
9	Modulation Systems	13-21
10	Sampling	13-23
11	Information Theory	13-24
12	Pulse-code Modulation	13-26
13	Codes and Transmission Security	13-28
14	Multiplexing and Timing	13-30
15	Types of Signal Transmission	13-32
16	Current and Voltage Telemetering Systems	13-32
17	Frequency Telemetering Systems	13-35
18	Position Telemetering Systems	13-37

COMMUNICATION-BASED PROCESSING CONTROL SYSTEMS

19	Pipelines	13-39
20	Operational Security	13-39
21	System Components	13-41
22	Pipeline Control	13-42
23	Power-system Control	13-44

REMOTE-CONTROL-SYSTEM PRINCIPLES

1 Telemetering Systems

The essential object of a telemetering system is to gather data, translate them into a form suitable for transmission, transmit them to a remote location, and there display them in a form useful for human interpretation. "Telemetering" means "measurement at a distance." The location at which the display of data is effected may be several feet or several thousands of miles from the source of the information. Any telemetering system consists of three groupings of equipment, a transmitter, a transmission medium, and a receiver or display unit. Depending upon the particular configuration, the greatest source of problems encountered in the development of the system will be found in one, or each, of these sections. The chief function executed by a telemetering system, then, is to extend the link between the meter indicating the present condition of a given variable and that variable. This is done for the sake of convenience and efficiency. This concept has been in existence and has been effected practically for more than forty years. Commonly, such systems have been classified in terms of the manner in which the electrical analog of the measured variable is conveyed to the display device. These categories are (1) current, (2) voltage, (3) frequency, (4) position, and (5) impulse.

Current, voltage, and frequency systems are representative of the most historical configurations, although they are far from being outdated at the present time. Position- and impulse-based systems are more recent arrivals in this field. A multitude of experience lies behind them, however. Some early measurement-transmission schemes involve the use of pulses for conveying the data to the display device. Today's more sophisticated schemes transmit data encoded in impulse-coded form. Varieties of the basic impulse scheme are meeting with ever-increasing popularity because of their adaptibility to a variety of transmission schemes and because of their suitability to inclusion in systems employing digital computers. By "encoding" in impulse systems is meant the choice of the one of the several possible schemes of representation possible for the measurand. That is, in impulse-transmission systems (particularly in PCM-pulse code), the value of the variable is *not* given in a form analogous to its primary (the originally measured) representation. Thus, in such systems, another dimension of design is made available to the system engineer by which he may be able to counteract the effects of errors (corruption added to the data during their transmission) and by which he may better avail himself of the information through-put capacity of the channel portion of his system. The parameter of specific interest in this regard is the channel bandwidth. As an example, consider a system in which the measurand can assume one of four values. Only three binary digits are required to assume the one out of "five conditions the channel can assume (the four values of the measurand and the "no traffic" status of the link). Using the least expensive telegraph channel available, one having a bandwidth of 15 cps, at *least* five such measurements could be transmitted per second. The theoretical maximum for such transmission is 10 such measurements per second. On the other hand, in a system in which tones were representative of the measurand, a bandwidth of 100 cps or more might be required.

Minimization of the effects of errors is of crucial importance to the designer. This is so, since the basic security of the data in his system can be affected by error-producing conditions. In those systems in which modification (control) of the measured configuration is done with communication channels similar to the telemetering link, security of control is of crucial significance. More on this will be said later.

In those systems in which an electrical analog of the measurand is transmitted, two rather than just one measurements must be made, compounding the possibility for introduction of error into the displayed image of the variable. These measurements take place at the location of the sensor (the original measurement) and at the display unit. There are varying conditions of the data link between the transmitter and the receiver which may cause erroneous transmission. Consequently, any action which can be taken to nullify the effects of parameter variation is generally employed, such as balancing the instruments, for example. Current systems are insensitive to voltage drops in the transmission line due to line resistance and other causes. Ideal voltage systems draw no current. Practical ones draw little current and are consequently more conservative in their energy requirements than are current systems. Clearly, statements can be made favoring either one of these "dual" systems. Similar comparisons can be made among facilities in which other transmission schemes are employed.

Current, voltage, and frequency systems have had the most extensive development of all telemetering systems employing hard (solid) conductors. Their use has been most extensive in those configurations in which the distance between transmitter and receiver is minimum. Indeed, in systems in which the distances are very small, these "traditional" systems frequently prove to be the least expensive. However, in those configurations in which there is a considerable physical separation between the transmitter and the receiver, the current and similar systems become quite unattractive. In these cases, the equipment requirements can become considerable. Reshaping, amplifying, and filtering circuits become a necessity. Further, in really extensive systems, construction of the communications links by the concern which owns the telemetering system loses its appeal. The availability of links on a lease or, in certain cases, intermittent rental basis in such instances is desirable. Rental service is equivalent to dialing another party on the telephone. The relatively low cost associated with such configurations is due principally to the minimal requirements placed on them. A telegraph channel is a good example of such a link. In one type of channel, the presence or absence of a voltage on the line is the only indicator of intelligence required. Voltage variations of the signal relative to the nominal signal values are of minimal importance. Sophisticated versions of such simple on-off signaling are available at present. Elaborate multiplexing techniques have been constructed. Nevertheless, the art behind the simplest telegraph setup is a well-developed one. The repeaters and other equipment constituting a telegraph system are frequently built from highly reliable, low-cost relays.

Remote-measurement schemes can be classified in the broadest sense as to whether they are incremental or continuous, that is, whether the transmitted image of the measurand can assume one of a finite or one of an infinite number of values. Another classification, not quite overlapping with this one but more generally accepted, is that of analog vs. digital signaling systems. In analog systems, some quality of the transmitted signal is directly analogous to the measurand. Strictly speaking, certain of the impulse systems, for example, pulse-amplitude (PAM) and pulse-position (PPM) installations, are of this category. Digital systems are those in which the primary measurement is encoded into some chosen on-off format and which need *not* be measured *again* prior to display. Furthermore, in a digital system, the transmitted intelligence is a number. While we generally work with systems in which the radix of the number scheme representing the transmitted numbers is two (an on-off or "binary" system), such a restriction is not necessary. For example, a digital system has been postulated in which the transmitted intelligence is encoded in terms of four basic symbols (to use coding terminology). This system is particularly suitable to implementation in a digital subset in which the data are conveyed as the phase of an a-c signal.

2 Control Systems

The original concept of a telemetering system, as stated above, was to reproduce measured quantities as accurately as possible at a distance from the source of these data. Subsequently, the conception was broadened to include activation of devices from a remote location. In a typical automated system, switches to open valves, etc., are located at some location which is chosen to be central to a plurality of such controlled devices. The central location might be the control board of a plant, with the devices within several hundred feet of the control panel's location. Or the controlled devices might correspond to the remotely operated pumping stations in a gas pipeline, coordinated from a central location deemed most advantageous from the point of view of the company's management. In many cases, remote measurement is of less interest than remote control. Indeed the advent of the digital computer with its ability to be connected with digital data links which rely on a standard data or message format has offered a new dimension of expansion for process-control systems. The functions of a telemetering system have been expanded to include automatic modification of measured data, generation of control "messages," the operation of a communications link, and the transmission of "messages" to operate controllers remotely located from the computer. Thus there appear to be two *basic* classes of communications-based process "control" systems, telemetering systems and "teleoperating" systems. In the former, *measurement* is the essential purpose for the configuration. In the latter, manipulation of a selected (process) environment is the primary objective. This difference implies that one must take cognizance of the fact that a different philosophy of design is implied in the two types of systems. Such factors as error control and time delay acquire crucial significance in such systems. Systems in which both control and measurement are exercised over some distance may be termed "teleoperating systems." The concept of teleoperating systems may be extended to cover manual inputs and operator displays at the points of process control. Certain factory data-collection systems and/or components of those systems may be included in the teleoperating-system category. However, little specific mention will be made of the input-output equipment for such systems in this chapter. The emphasis of the present material is placed upon direct measurement and, where applicable, control.

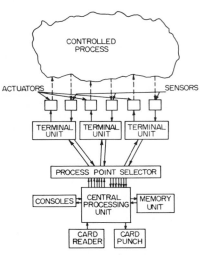

Fig. 1. Basic ("local") process-control computer system.

The general classification of "teleoperating" systems is further subdivisible into two categories. In the roughest sense, they represent closed-loop vs. open-loop systems. The closed-loop-system configuration is derived, figuratively speaking, by separating the arithmetic and logical processing unit of the computer from the process terminal equipment. This basic closed-loop arrangement is the "usual" process-control computer system. In this configuration, the complete control system is located at the process site. Such an installation can be termed a "local process-control computer." Its configuration is detailed in Fig. 1.

The open-loop system basically is composed of a central computer operating in conjunction with an array of remote "terminal" units. The system can be comprised of several "local process-control computers" plus a central processing unit, as illustrated in Figure 2a. Such a system is employed in those instances in which each "local" unit requires a relatively high measure of control. In this situation, the principal function of the central machine is one of coordinating the remote machines, integrating them *overall* with respect to the *system's* performance objective. Such an objective is through-put maximization, allocation of facilities, etc. With this approach, individual controller settings and measurements of process variables are not transmitted between the central and its satellites. Rather, they are derived by the local equipment.

Alternatively, the open-loop system can be comprised of a centrally located computer as well as an array of remotely located installations which can exhibit only "fixed-logic" activity. That is, they scan points at prestated intervals and execute various sequencing operations in a precisely predesignated manner. Their functions are "built in" either in terms of hardware or in terms of program tapes. etc. The flexi-

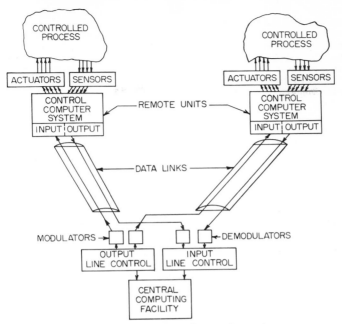

Fig. 2a. Multiple-control computer system.

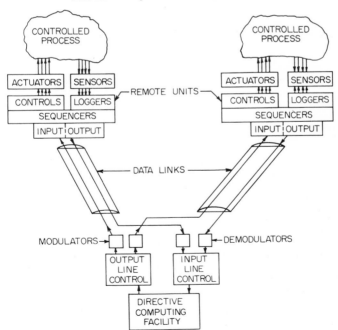

Fig. 2b. Multiple low-level control computer system.

bility of response exhibited by either a human operator or a stored-program computer is not expected of such equipment. The layout described is detailed in Fig. 2b.

A typical application for this type of system is one in which the principal economic justification for the computer lies in its use for business-type calculations. Here, computations directly involved with controlling a process, a pumping station, etc., require but a minimal segment of the computer's time. In such installations, the ability of the computer to exercise control of a process provides a portion of the economic justification for the central computing installation itself. However, it must

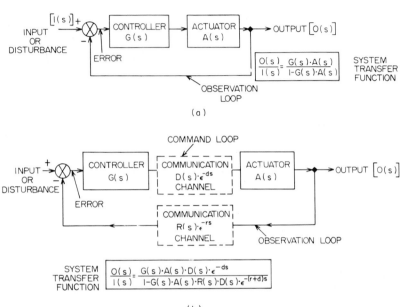

Fig. 3. Servomechanisms. (a) Simple servo. (b) Simple servo with communications loop. 1. $D(s)$ and $R(s)$ are the transfer functions of the command and observation loops, respectively. 2. d and r are the (time) delays induced in the command and observation loops, respectively.

be recognized when approaching the system design of such an installation that the computer might well have been installed without regard to the existence of the processes in the configuration with which it must work.

The closed-loop system is more nearly equivalent to a regular servo. Indeed, the effect of the data link can be taken into account by the inclusion of both noise and delay in the forward and the feedback loops. See Fig. 3 for diagrams of a simple servo and this servo "modified" to take transmission delay into account. The equations accompanying the diagrams indicate the effect of this modification. Chapter 15 discusses design principles of such feedback control systems. More nearly instantaneous information transmission is required for closed-loop than for open-loop systems. Consequently the encoding scheme employed develops a crucial significance in the closed-loop system. An encoding scheme, not well chosen to match the communication channel to the terminal equipment so as to minimize the effect of noise, can produce a correspondingly greater *time lag* in attaining a successful transmission than a code, the structure of which is well tailored to the conditions of transmission. Note how the possibility for instability arises here. If a "message" from the central to a remotely located controller is delayed, effectively a phase shift is introduced. A propagation delay of 5.05 μsec/mile exists in free-space transmission links. Corruption, causing retransmission or equivocation of the messages transmitted on such lines, can only result as an *additional* delay, or equivalently, a phase shift. The amount of such basic delay, caused by the transmission of information, plus the time required for its processing at both ends of the line, constitutes the principal determinant of whether open-loop teleoperating control is a realistic possibility in any given instance. Speed-up and/or other compensation is applicable in many cases to offset the deleterious effects possible because of the communications system.

The principal task of the communications equipment in the systems described is to connect the central computer with the remote equipment, enabling these principal components of the configuration to "converse." The communications control equipment at either end of the data lines serves a multiplexing function, principally. If the equipment at the remote station(s) were contiguous with that at the central, multiple lines would be employed for the transmission of signals between these sections. Many broadband, e.g., frequency-multiplexed teleoperating systems, essentially maintain this philosophy in their operation. That is, a multiplicity of channels are used for communication between the central and the remote stations. For example, one channel is used to operate one valve, while another channel is used to operate another valve, and so on. Multiplexing techniques permit the use of one channel of *sufficient capacity* for the transmission of several signals simultaneously. While this technique is superior to using multiple conductors, it is not maximally efficient. The poor economy of such systems has been recognized by the manufacturers who provide automation equipment for installations in which communications plays an important role.

3 Computers and Communications

Clearly, the use of simplified communications facilities becomes *the* goal in the types of configurations described here. Single-line systems become the objective. Reflection on the inefficiency of using one (multiplexed) channel for *each* individual signal led system designers to the concept of using a *message* structure for the transmissions between the central and the remote stations. Digitally encoding such messages appeared to be the wisest objective. Such encoding provides maximum consistency with the aims of minimizing the channel bandwidth requirements and providing the simplest hardware to effect the operation demanded of the remote station.

Utilizing the message-structure system, the "central" or master control station generates a message which is dispatched to the unit (terminal) addressed by it. Such a message designates a controller setting, an on-off position for a switch, a request for a meter reading, etc. The message is essentially a command by which the master orders the slave to perform one (or several) designated operations. Equipment at the slave location responds to the command interpreting it in terms of control lines to be activated, etc.

The intelligence of the master station resides in a digital computer in increasing numbers of such installations. The digital computer is a device suitable for performing both the logic and the arithmetic calculations required of a control system. It is comprised of binary logical circuits, grouped to perform designated functions. A computer has five principal sections:

1. Input
2. Output
3. Memory
4. Control
5. Internal communications

A typical machine configuration is shown in Fig. 4. A computer would be a completely useless device without the "orders" which provide it a complete sequence for executing the functions necessary for it to demonstrate a required behavior—perform a prescribed task. The assembly of orders "telling" a computer how to perform a certain task is called a "program." A program is contained in the computer's memory, into which it must be entered prior to the machine's commencing to work on the job. The individual orders comprising the program are commonly called "instructions." A computer normally executes the instructions of the program in its memory, one at a time, in sequence. In some computers, the sequence can be altered during the operation, corresponding to the introduction of new information. The *combination* of the computer and its program constitutes a logical decision-making and computational device suitable for providing the intelligence required of the master station in a multiple-unit configuration. One can envision the program of the "central's" computer as composed of two types of instructions, operations internal to the computer and those designating individual operations or sequences of operations of equipment external to the computer. Consider the "commands" or messages which prescribe the actions of the remote units to be of this latter nature. These "commands" are intermingled with "instructions" to form a program. Whenever such a command is issued (generated by the computer and transmitted to the remote station addressed by

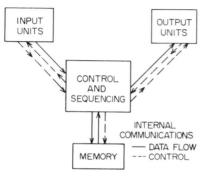

Fig. 4. Digital computer schematic.

it), the computer *itself* continues its operation. The command message is interpreted upon its arrival at the remote station. This action results in the contact (on-off status) points at that location being scanned, digitized, and transmitted back to the central, for its processing, for example. In Fig. 5 is shown a sample message of this type.

The structure and operation of a command are not strictly as has been described. Actually the message sent would be developed by a group of instructions, called a subroutine, which would be solicited by the computer-program combination during the course of its activity. A "subroutine" by definition is a group of computer instructions assembled to enable the computer to execute a particular function such as the computation of a square root.

The command message transmitted to the remote station is a macro-order for that station. Its content varies depending upon the specifics of the system requirements. At the least, it contains an order code, one or more addresses, and possibly data. Further, error-control characters might be included as part of the message.

ERROR CONTROL	CONTROLLER SETTING	CONTROLLER ADDRESS	ORDER CODE	ADDRESS OF REMOTE UNIT

Fig. 5. Control message.

These error-control characters could be the "longitudinal redundancy check" characters, for example. The binary digits, or bits, constituting such a character are 0 or 1 as required to make the sum of the bits comprising the message odd or even (as required by the parity scheme). A message for prescribing a set point for a remote controller is shown in Fig. 5.

Figure 6 shows a message whose characters are transmitted in odd-parity form. This macro-order is interpreted by the equipment at the remote station to execute the function stipulated in the order-code portion of the command message. The individual logical elements of the remote station are activated accordingly. Thus the logical ability of the computer is combined with the logical ability of the remote station. Most of today's systems do not rely upon *on-line* computers for their control functions. Rather, the abilities of a human operator are combined with those of a "fixed-logic" device to execute the central functions. However, computers will be seen more and more in such roles.

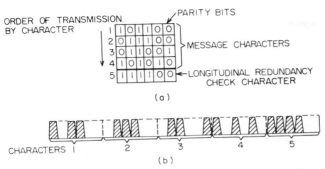

FIG. 6. Checked message. (*a*) Odd-parity checked message. (*b*) Signal profile of message.

Generally, the logical ability of the remote station is minimal. Even in remotely operated pipeline pumping stations, the equipment directly responding to remotely generated commands is simple in nature. The start-stop equipment which they *call into operation* actually to sequence the station "through its paces" is generally far more complex. Such equipment is customarily "fixed-logic." It has been generally implemented by relay techniques but transistorized equipment is being used for this purpose on a larger scale.

4 Multiunit Systems

The concept and the practice of telemetering have been discussed. It has been demonstrated that the possibilities afforded by the central display of remotely derived data lead naturally into the concept of teleoperating systems. Teleoperating systems employ the centrally available data to determine the status of the complete system and to modify this indicated status. The modification is performed at the operating points—the remote, slave stations—of the systems by commands issued from the central point. By recognizing that the most current data pertinent to the configuration's performance are available in one location, one can readily conclude the intriguing possibilities thereby afforded for noncontrol functions. The centralization of the control function can conveniently lead to consolidation with the accounting function of the overall business operation of which the controlled process is *the* (or an) operating part. In a bulk (petroleum products) pipeline, for example, one of the "remote" functions could be an unattended customer facility. The customer would drive his tank truck up to the tank holding the product he wants. Upon his insertion of an identification card into an automatic reading mechanism, a message could be transmitted to the central computer where his "right to purchase" would be checked. Then, an "O.K. to load" message would be dispatched to the remote equipment, enabling the customer to load. The tank levels before and after the loading would be dispatched to the central, thus providing it with sufficient information to adjust its *centrally* maintained inventory. At any time, the appropriate supervisory or management personnel could communicate with the computer by means of manual input-output equipment, to determine the level of the inventory for the system overall, or any tank in particular.

The typewriter is the most common means for an operator to communicate with a computer. The input-output typewriter looks like any common office typewriter, with a few extra controls added to it. Such a device intrinsically need not be immediately adjacent to the computer with which it converses. In fact, it can be remote from the computer, connected to it via a communications channel of exactly the same nature as that linking the remote-control station and the master station. In Fig. 7 is a schematic diagram of a central with inquiry stations. This configuration is illustrative of the fact that automatic control and normal accounting functions can meet in the centralized communications-based process-control system. The side benefit of "closing the management loop" is added to enhancing the purely mechanical aspect of making system control a reality. Management response time and directive errors due to the unavailability of sufficient information are minimized. Perhaps the chief role which communications, principally digital communications, has in the type of design described above is to enable the more nearly complete *integration* of the operational mechanism owned by a company with its operations overall.

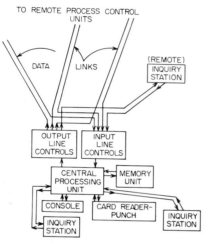

FIG. 7. Teleoperating central.

There are two approaches to both the synthesis and the analysis of a multiunit system. It can be viewed either as a dynamic data-handling system or else as a (closed-loop) control system. That is, the system's operational and developmental problems can be interpreted as problems either of control or of data manipulation *depending on the orientation* of the group responsible for its development. The man with a communications or a computer background will think in terms of items such as channel bandwidth, information theory, and message structures. The concept of a phase shift in the feedback loop of the "servo" consisting of remote actuator, communications link, and computer are not of so great interest to him as to the control-oriented man. Mutual understanding and interchange of information among those working on the evolvement of a complex system exemplified by automated pipeline or electric-power-distribution systems is very necessary. Implementation of a successfully functioning system demands such cooperation.

Automatic data-collection systems for industrial installations have been developed. In one of them, workers' badges can be remotely sensed and centrally verified against centrally stored criteria. Systems of this nature can accept cards on which are recorded numbers relating to the extent of completion of a given task, in a job shop, for instance. Installations like these can be classified as industrial control systems along with systems controlling switches and valves. In the two examples cited immediately above, the emphasis is on information handling exclusively. No criteria for evaluating servo performance are directly applicable to their analyses, for example. We can categorize them as closed-loop or open-loop systems, however. The criterion in this instance is whether the centralized operations on the remotely derived data result in subsequent actions by equipment at the remote locations. If they do, the system is closed-loop, while if they do not, it is open-loop.

It has been noted that industrial control systems in concept can be oriented either toward data handling or toward being an "extended servo" system. In many cases, the system exhibits characteristics of *both* categories of system. Future developments in this field will be facilitated by recognizing the cross fertilization possible between the (two) philosophies as crystallized in each design approach.

COMMUNICATIONS CONFIGURATIONS

5 Data-link Components

All telemetering and teleoperating systems are communications-based configurations. This means that a crucial determinant of their overall performance is the behavior of the data links interconnecting their component parts. In this section, the equipment common to the several types of data network configurations suitable for use in telemetering and teleoperating systems are discussed. The networks of the common carriers are particularly significant, because of their increasing use in such systems. The use of pulse-modulated (digital) signals and the increasingly large geographic extent of today's configurations have led to the use of such networks. Many companies owning control facilities prefer to own the communication networks necessary for their operation, however. Generally, cost considerations dictate whether leased or private facilities are used.

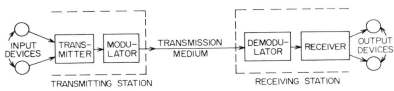

FIG. 8. Two-station communications system.

Communication networks consist of modulation equipment, repeaters, and the data links themselves. Figure 8 shows a typical two-station communications system consisting of transmitter and modulator, transmission medium, demodulator, and receiver. The transmitter could be a remote telemetering pickup, which develops a reading which is to be transmitted to another (central) location for display or, in general, processing there. The reading is suitably converted for transmission over the given medium by the modulating equipment, such conversion being an unambiguous representation by information-bearing signals appropriate to that medium. The demodulating equipment executes essentially the converse operation upon the information-bearing signals received by it, transforming them into inputs suitable for the receiver. The repeaters regenerate the modulated signal as it passes from the transmitting station to the receiving station. The input to the repeater is a corrupted version of the original signal as put onto the transmission medium by the modulator. The output of the repeater is a "regenerated" version of this signal, one having characteristics *identical* (at least ideally) to those of the original signal. The operation is effected with regard to such factors as the specified timing, the wave (or pulse) form, and its magnitude. Repeaters may be required because of the basic nature of the transmission medium. The signals suffer phase distortion, amplitude distortion, etc. The transfer function of any physically realizable transmission system is of the form $T = A(\omega)\epsilon^{j\theta(\omega)}$. $A(\omega)$ is called the amplitude function, and $\theta(\omega)$ is called the phase function. Distortionless transmission requires that the amplitude function must be a constant or invariant with respect to the frequency of the signal. Further, the phase function must be proportional to the signal frequency for distortionless transmission to be achieved. Unfortunately, all media exhibit nonlinearities in these functions. Figure 9a shows both amplitude and phase functions for an "ideal" low-pass filter, one which passes all signals (considered to be sinusoids) in the range of frequency from 0 to ω_0 radians/sec. Such a filter represents a good model of the typical transmission link. A less than ideal filter would have characteristics like those shown in Fig. 9b, in which nonlinearities are found even in the operating range 0 to ω_0 radians/sec.

A Fourier analysis of any given complex waveform decomposes it into a sequence of harmonics, or sinusoids, varying among each other in respect to both amplitude and frequency. According to the considerations given above, such a waveform will suffer distortion when passing through a transmission line. The process of counteracting such distortion is commonly termed "equalization." The ideal equalizer has a transfer function which is the *converse* of that of the stretch of line between it and the transmitter (or the previous equalizer if there is a plurality of them on the line).

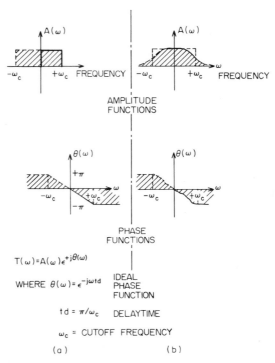

FIG. 9. (*a*) "Ideal" low-pass filter. (*b*) "Nonideal" low-pass filter.

The effect of the equalizer, consequently, is to nullify the effect of the line. Essentially, the equalizer network is such that the product of its response and that of the system (the preceding line segment) are almost constant over the spectrum of frequencies. Equalization of telephone lines has been done principally with regard to the amplitude factor. Phase correction has been of lesser concern, because the human ear is much less responsive to this transmission impairment than it is to amplitude distortion. Digital signals, however, are subject to this class of distortion. Consequently, attention has to be given to it when facilities designed for the transmission of voice signals are considered for use by digital signals, such as those employed in some telemetering systems.

A Fourier analysis of the pulse shown in Fig. 10*a* demonstrates that an infinite bandwidth would be required of a transmission system to convey such a pulse *exactly* from a transmitter to a receiver. Because such a requirement is an impossibility to achieve as has been described, the pulse in a real system will look more like the form shown in Fig. 10*b*. The ideal low-pass filter produces an output having the analytic form called the sine integral, a function which is tabulated. Realistic systems will produce outputs having even less fidelity to the original signal than this.

Because pulse modulation and pulse-code or impulse systems have come to be the most widely used modulation schemes in modern telemetering practice, the above points are of considerable significance. Pulse modulation is distinct from pulse-code modulation in that it is the classification for *unquantized* pulse signaling. Impulse systems have gained precedence over the older, essentially *unmodulated* systems such as voltage and current in modern practice because of their minimized subjection to

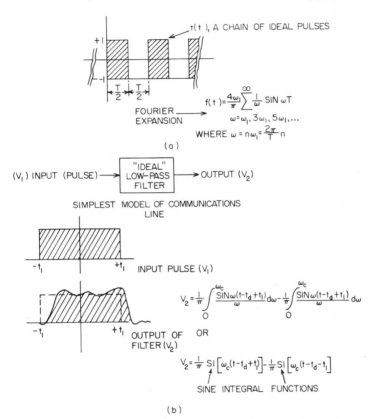

FIG. 10. (a) Ideal pulse train. (b) Nonideal pulse transmitted.

error, their suitability for long-distance transmission, and their suitability for use in systems in which digital computing equipment is included. In modern practice, the pulses may be converted to changes in either frequency (FSK) or frequency shift keying or phase (PM, or phase modulation) for transmission purposes.

6 Data-link Configurations

The devices which constitute a transmission link have been described. A transmission network is comprised by a plurality of such links. In the following, descriptions of several different network configurations important to control systems are given. While the details for the most part assume digital, or impulse-signaling, techniques, the overall philosophy is also representative of analog-signaling techniques.

The data links used in these systems commonly run through switching centers. The same centers serve to cross-connect several links of a network at junction points. The lines may be owned either privately or by a common carrier. In the latter case, two types of service are available to the customer, lease or individual call connection. A user can lease a line between any two locations that he wishes, stipulating the characteristics which he requires for that line to be suitable for his system. Depending upon his intended usage of the facility, amount of data per stipulated time interval, whether or not *all* transmissions take place between the stipulated station *and* the same *other* station, etc., the leasing scheme might not be considered most economical to him. In this case, he might choose to connect the given station to his system only periodically. Several systems are currently either available or are under development which permit this. They vary with regard to the rates at which they pass data as well as with regard to how they set up a connection. On one system, an ordinary telephone call is placed, and when it is satisfactorily completed, the data-transmission system is called into action upon the operator's depression of a button. Another system affords a completely automatic connection procedure, in which *no* operator is required. The station transmitting equipment contains an apparatus which automatically makes the connection to the other station with which the originating station is to "converse." This process is equivalent in its essentials to placing a telephone call. The transmitting equipment notifies the switching center of which it is a subscriber that it wishes service. Upon receiving an "O.K. to connect" signal, it transmits a code starting with the address of the desired station. When the connection is completed, the station originating the connection is given an "O.K. to transmit" signal. Then the message is transmitted.

This system is similar to that used in teletype or telegraph transmission. In such a system, every station has an address. When one station wishes to send a message to another station either on "its" line or in another network but with which it nevertheless can be connected, it merely transmits the appropriate address characters. That such a process can be made automatic is easy to understand.

Teletype and telegraph networks are suitable for certain telemetering and teleoperating configurations. Telegraph networks are in very common use in automated gas pipelines, for example. The section on Pipeline Control describes the communications configuration for this type of system, giving details of the equipment. The automatic connection of stations for digital transmission is naturally suited to the capabilities of the digital computer. Dial pulses or other code groups suitable for the activation of network exchanges can be readily produced by a computer's program. The facilities required for connecting a computer to communications links of interest to the control field are available today.

Let the connection centers or exchanges found in the communications networks be considered. Both automatic and manual centers exist. The automatic centers are not only in principle the same as dial-telephone offices. In some centers, the same switching equipment is used *both* for subscriber telephone calls and for digital data transmission. Manual switching centers are still used in teletype and telegraph networks. However, their use is decreasing with the increasing use of automatic (dial-up) systems. In manual centers, the digital data are received and recorded on paper tape by more or less standard receiving equipment. A routing clerk then reads the address of the message's destination and *physically* transports it to an appropriate transmitter which reads the tape and retransmits the message. Such a procedure can be repeated several times for one message in different switching centers along its route. These centers are being replaced because of the delays involved and the increasingly difficult tasks of manually handling the messages. By using leased lines or dial-up facilities, messages between two selected stations can be transmitted without most of this switching delay. The use of leased facilities is definitely recommended where *time* is of any significance. This is certainly the case in many of the teleoperating systems of interest. Thus, "torn-tape" or other manual systems like those just described are of little interest for control systems.

A telemetering system can employ a completely private communications network, with all components including the switching centers operated entirely by the concern which owns the configuration. Sometimes, a system employed by a pipeline company to measure temperatures, pressures, etc., and to operate pumps, valves, and other equipment at locations along its right of way installs wires (or microwave equipment) along its right of way for the communications network when it installs such equipment. Such might be the case, for example, if the right of way were to depart significantly from the routings of any common carrier's trunk lines. The most common example of extensive private-line facilities is found in the railroads. Also, America's major airlines utilize several different types of leased networks. They are used to transmit reservations information between the stations of a passenger's itinerary, as well as for other purposes of an information-handling nature. In this case, all the facilities are leased on a more or less "permanent" basis.

The similarity of the airlines' and the railroads' configurations to those suitable for an extensive industrial control system such as either an electric-power-distribution system or an automated pipeline is to be emphasized. Though the latter category is one in which control of an operation is the end result, while the former is one in which the movement of business-type information is of principal concern, the principal aim of the communication network in each case is *identical*. This aim is the movement of data arranged in a group called "messages" with a minimum amount of inaccuracy and with a maximum of processing simplicity. *Processing* refers principally to the encoding, decoding, and manipulation of the data at either end of the line required to transform them into operations upon the controlled system to make the system respond to the directive or "command" signals imposed upon it in a minimum amount of time. Problems relating exclusively to the communications network of a teleoperating system should be considered as *communications* problems *only* and *not* as *control* problems. Obviously the communications segment of the control system should be viewed as being a segment of a control system. The maximum delays acceptable to permit the system to operate safely and effectively will help to determine the types of equipment and lines used as well as the coding and the structure of the messages to be employed.

7 Communication-systems Configurations

The profiles of some of the commonly employed line configurations are described in this section. Some of the links may be solid conductors, others radio or microwave. Both full- and half-duplex lines are available. A full-duplex line permits simultaneous transmission in both directions, while a half-duplex line permits only one-way conversation. The majority of long-distance (as distinguished from local lines from a central office) lines are one-way in character. This is because the amplifiers (particularly the electronic) as well as other equipment in the links are unilateral devices. Consequently, a full-duplex line is the *basic* arrangement, with a *modification required* to make it act like a half-duplex line as far as the user is concerned. The modification includes equipment to connect the customer alternatively from one line to another upon reception of a signal from his equipment. The networks in which we are interested are built up on segments like that shown in Fig. 8. They are categorized in terms of two "dimensions":

1. Whether they are hub (or wagon wheel) or mesh, overall
2. Whether they are serial or hub or in groupings of individual links

Figure 11a shows a "wagon wheel" network. The central is the hub, and the individual stations are on spokes which lead out from the hub. There may or may not be more than one station on each spoke. Figure 11b shows a mesh network. Although there are a plurality of lines emanating from the central in this case as well as in the wagon wheel, a given station is not identified uniquely with a specific trunk line. This implies that the existence of alternative paths through the network between a station and the central are possible. Only information transmission between the central and either one (or a group of) station(s) and not between two stations is considered here. The possibility of choice among alternative paths presents a realistic picture of a common carrier's facilities. In this system, only the fact of a connection between two desired points is guaranteed and *not* the specific path between them. For a given amount of traffic between the stations and the central, the individual segments can have a lesser data-transmission capacity than in the hub system. Furthermore, the reliability of the system is increased because of the availability of alternative connections. A mesh system operates in one of two ways; either a connection is set up for *each* "conversation" between the central and a station or else a "normal" path exists but an alternative can be selected in an emergency. When operating in the latter mode, it behaves like the hub system.

In the basic hub system, there is only one station on any line. With a leased-wire system, the addressing of a message is utterly simple. The central merely selects the proper line and executes the transmission. With several stations on one

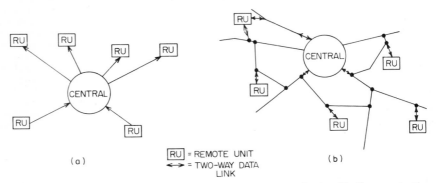

Fig. 11. (*a*) Communication-network configuration; wagon wheel. (*b*) Communication-network configuration; mesh.

line, the central must specify both the line and the station. The outgoing line is a "party" line, effectively, with each station "listening" for a message bearing its address.

The mesh configuration requires the most general form of message addressing. Each station has its own address. The proper routing through the network junction points is made by the switching centers at those points. The output lines (from the central) cannot be "party" lines in this case.

Inputs (messages from the stations to the central) present a different addressing requirement. A station having a unique leased or otherwise always "open" line can transmit at any time—if the central has a storage to hold the message in the event that it cannot immediately accommodate it. If several stations share the line, only one can transmit at a time (assuming no multiplexing). Therefore, before transmitting its message, the station must "know" if the line is available. There are two principal schemes to assure the nonsimultaneity of transmission of input: the preemptive and the "polled" systems. With the preemptive scheme, whenever any given station "gets the line," it automatically locks out or otherwise precludes transmission by the other stations on the line until this station has completed its transmission. Alternatively remotely initiated transmissions can be controlled more directly from the central. In such a system, a station desiring to transmit a message activates an indicator, whenever this message is available. Although the message will not necessarily be dispatched as soon as it is produced, it can be constructed "at any time" if sufficient storage is available at the station to hold it until the line becomes available *for* that station. Whenever it wants messages, the central addresses the stations individually, in some sequence. Whenever a station wants the line, this status will be available to the central by virtue of the "request" setting of its indicator. When the station receives the "poll" message addressed to it, it responds by transmitting the data it has developed for the central.

Not all configurations operate in one of the manners described. However, the

COMMUNICATIONS CONFIGURATIONS 13–17

concepts presented are generally correct and are illustrative of the problems encountered in the specification of data-communication systems. Not all of them allow a (remote) station to initiate a transmission to the central. It might have had to be requested by the central. Also, if a conversational (full-duplex) link between a station and a link has been set up, message transmission in both directions can take place. That is, the establishment of an "open" link in one direction establishes it in the other also.

The characteristics of the individual links are explored in the following section. The gross distinction between hub and mesh for data networks has been made. The differences between serial and hub subnetworks are elucidated below.

This material assumes that the lines require synchronization to be effected by the modulator and demodulator units which are commonly referred to as subsets or modems.

There are two types of serial subnetworks, the regenerative, as illustrated in Fig. 12*a*, and the bridged, as in Fig. 12*b*. In the regenerative system, all the stations are connected in series with regenerative repeaters at each station. Transmission over a line of this sort occurs in a series of point-to-point jumps. At each station between the transmitting station and the central, both demodulation and modulation occur. The transmitted message, when received by the station, is demodulated and is reshaped into either two discrete voltage levels on one conductor or the energization of either one of two conductors. This is done according to whether the binary digit is a 1 or a 0, a "mark" or a "space." These signals are applied to similar line(s) on the input of the modulator unit found on the *outgoing* line of this station. This is the place at which message signals developed by equipment at this station enter into the communications facilities for transmission to the central.

Synchronization of all the subsets in the stations between the transmitting station and the central has to be effected in the regenerative system. The methods used vary with the system employed. For example, if the common (Baudot) telegraph signaling scheme is employed, synchronization is automatically accomplished with the transmission of each character by the insertion of "start" and "stop" binary digits preceding and trailing the character, respectively. The receiving (demodulating) subset has a commutator which is caused to keep "in step" with a similar device in the transmitter. Start-stop systems are in widespread use because of their association with telegraph equipment.

There is a variety of other subsets built to operate with binary coding schemes other than Baudot. In general, they require two types of synchronization, character and bit. By character synchronization is meant the process by which the receiver is made to recognize the occurrence of the beginning (or end as the case may be) of a given fixed-length grouping of pulses. Such a grouping is a character. The receiver (demodulator) must know when to sample the line. For systems using marks and spaces which are transmitted as signals of equal length, this process is considerably simplified over cases in which the signals are not of equal duration. Bit synchronization is the process by which the transmitter times the receiver so that it will sample the line at the proper instants. If a 0 is conveyed as a unique voltage, the receiver must be set to recognize that two 0 bits have been transmitted after a certain period of time has passed, for example. In some transmission schemes, the indication of a mark or a space (a 1 or a 0) is given by the phase of the sinusoidal carrier signal on the transmission line. In this case, locking the demodulator (receiver) into phase with the modulator (transmitter) is necessary. Bit synchronization in a binary transmission scheme is the process by which the receiver is retimed to recognize the *transition* from a 0 to a 1 and vice versa, *as well as* to *when* and for what duration of time to do this. In the phase-modulated system, the receiver *may* have been 180° out of phase and consequently be interpreting a 0 as a 1. Let the use of 7-bit characters in the transmission be assumed for the moment. Then, the synchronization sequences in Fig. 13 are applicable. Figure 13*a* shows bit synchronization which consists of an alternating sequence of 0's and 1's. The alternations lock in the local oscillator of the receiving subset to either their frequency or a multiple of it. The bit-synchronization pattern can be said to be producing a regular pattern of repetitive pulsations, applying them to the oscillator. The pattern is transmitted for a time sufficiently extensive to ensure that the required locking in has taken place.

A sequence of transmitted characters suitable for character synchronization are shown in Fig. 13*b*. They *are* transmitted subsequent to the bit-synchronization

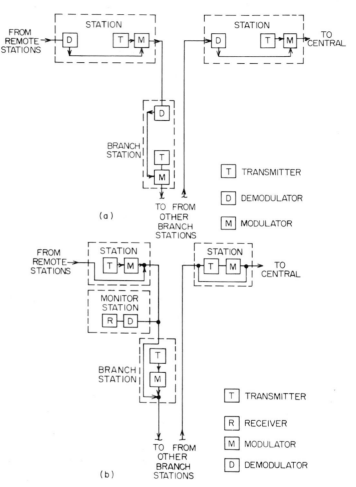

FIG. 12. (*a*) Regenerative serial subnetwork; only input-to-central shown. (*b*) Bridged serial subnetwork; only input-to-central shown.

group. At this stage, 1's can be improperly recognized as 0's, and the process will still work. The transmission of the first character of the sequence (all 1's) tells the receiver to await the arrival of the next character of the group. If it is *not* the (expected) synchronization character, the character-synchronization process does not occur. The six 1's followed by the *transition* to the 0 causes two things to occur; this 0 is recognized as a 0, and the receiver is reset so that it will interpret the very next binary digit as the first digit of the next character. These characters have to be considered illegal—not to be transmitted—*except* for this use. Alternative means of synchronization are possible for the binary-coded configuration described. They include the use of other (different) synchronization characters. Further, they use a separate channel for timing.

The synchronization of transmission facilities employing one of the other pulse or impulse systems is roughly similar to what is described above.

The pulse-position and pulse-duration systems commonly use marks or frame pulses. The position (or the width) of the pulse within that frame comprises the signal. The pulse-amplitude-modulation scheme uses either a pulse of a different amplitude or a different width such as the above to indicate the extent of a frame. Our discussion of communications networks does not implicitly assume the pulse-code scheme about which most of the detail above is concerned. However, because of their increasing application and in order to simplify the discussion, emphasis is placed on pulse-code systems where detail is required.

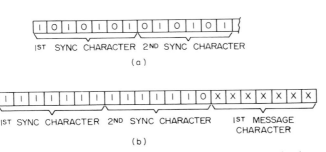

FIG. 13. (a) Bit-synchronization sequence. (b) Character-synchronization sequence.

The serial bridged system has all the stations connected in series as with the regenerative system. It has the modulators bridged across the line, however. Transmission over this line from the remote station to the central is done directly, without regeneration at every station. Realistically, regeneration may be required at selected points along the line in order to maintain a certain level of signal quality. This configuration is illustrated in Fig. 12b. Monitoring such a line is a relatively simple operation, since the only addition to do it is a demodulator connected to an additional bridge point in the line.

The hub system is the other principal class of communications subnetwork. A typical layout for it is pictured in Fig. 14. While the individual spokes of the hub could operate in a regenerative mode, more generally a bridged arrangement is used. The hub subnetwork is virtually the same as the serial bridged network except that messages not originating at a station on a spoke are not required to pass through stations on that spoke in their passage to the central. Essentially, the spoke, or subnetwork "branch," plugs into subsets bridged on the main line (to the central).

Both the serial and the hub stations require synchronization and interlock provisions to prohibit simultaneous use of a given line. If a given communication network actually employs the switching facilities of a common carrier, the interlocking could be done with this equipment. This equipment could both provide the storage necessary for holding a message until a connection to the central can be made, as well as making the connection through the usual or alternative point as required. However, in a real-time system, especially one in which minimum cost is important, centrally controlled transmission by such schemes as polling would probably be used. The individual subnetworks in the configurations described in this section might use wire, radio, or microwave transmission facilities. The equipment at each station presently labeled "subset" or "modulator and demodulator" becomes labeled "transmitter" and "receiver" when radio or microwave links are used, and thus acquires additional responsibilities for the system's operation. Microwave equipment has found some use in the pipeline industry for telemetering. If the interest is expanded to encompass remote, automatic missile-guidance systems, then wireless links become of importance. In such "command" systems, information about a missile's attitude, speed, etc., is telemetered to the ground station where it is digested by a computer which develops the appropriate control signals. They are telemetered to the missile, causing it to alter its direction, change its speed, etc. This is the prime example of a "tight" communications-based control system.

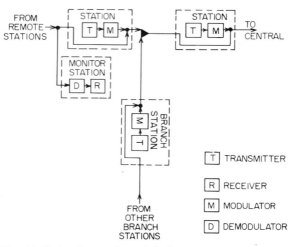

FIG. 14. Hub subnetwork. (Only input-to-central shown.)

The descriptions of this section, regardless of specific detail, do present quite a fair picture of the configurations employed for today's (land) communications networks. The problems of synchronization and so forth described appear no matter what hardware is used to construct the system.

COMMUNICATIONS SYSTEMS

8 Telemetering Systems

There are various ways in which telemetering systems may be categorized. The methods of classification most pertinent to the communications aspect of such systems are important to the discussion of this section. However, other methods do exist. Indeed, telemetering systems can be classified in terms of the nature of the measurand and mode of display to the observer (the end product of the remote measuring system), the principles of operation of the telemetering system as a whole, etc. In terms of the communications channel, the conventional system of classification is that adopted by the American Standards Association and is based upon the nature of the electrical variable employed as signaling. These systems are

1. Current
2. Voltage
3. Frequency
4. Position
5. Impulse

This categorization does not correspond exactly to the division of signaling methods in terms of the types of modulation used to convey the information in the transmitting channel. Modulation in the broadest sense means the particular method of conveying the message intelligence in a uniquely and as nearly unambiguously specified manner as possible. Modulation schemes are most commonly classified as being of one of the following types:

1. Amplitude modulation
2. Angle modulation
3. Pulse modulation
4. Pulse-code modulation

The thing common to *all* these schemes is that some alteration of the original form of the intelligence to be conveyed in the data link is effected by the modulation process. The principal purpose of this alteration is to adapt the means of expressing this intelligence to one most suitable for transmission by a given channel. This is to be done with a minimum distortion and with a maximum of economy. The process of modulation is executed roughly as follows: The information to be transmitted is made available as an electrical variable which is impressed upon or mixed with a "carrier," another electrical signal. The carrier, when in its unmodulated state, displays a constancy with regard to all its measurable characteristics. The intelligence signal acts upon the carrier in such a way to alter one (or possibly more) of these characteristics in a manner proportional to this signal.

9 Modulation Systems

In both amplitude- and angle-modulation schemes, the carrier is a sinusoidal wave of the form

$$C = A \sin(\omega t + \phi) \qquad (1)$$

With amplitude modulation, A, the amplitude, is varied as a function of time. The function *is* the signal which is to be transmitted. In angle modulation, the instantaneous phase angle of the carrier is varied in one of two ways. If the instantaneous *frequency* $\omega t + \phi$ is varied, the modulation is designated "frequency modulation." If the phase angle ϕ is varied, the modulation is described as "phase modulation." Frequency and phase modulation should be viewed as being really the same type of modulation. This is so because of the nature of a sine wave. If the frequency parameter is varied, the phase is of necessity also varied. Also, any change in phase necessarily involves a change in frequency. This is so, since the angular frequency ω is mathematically identical to the time rate of change of the phase angle ϕ, which is $d\phi/dt$.

The modulating signal, the signal bearing the intelligence to be transmitted, may or may not be a continuous function of time. This depends strictly on the nature of the signal produced by the transducer (measuring and converting device) which it is the purpose of the communications configuration to telemeter. Some pickups develop continuous and others discrete signals with respect to time. The nature of the signal should have *nothing* to do with the type of modulation used in the telemetering link if the system is well designed. The modulation scheme should be selected *only* according to the requirements of maximally efficient signaling by the given channel.

In all the varieties of both pulse-modulation and pulse-code schemes, the modulating signal is first sampled before it is imposed upon the carrier. Thus *samples* of the intelligence are telemetered to the display equipment. Whether the intelligence is developed (at the source, pickup, or other device) as a discrete entity or not, it is transmitted as such. The pulse-modulation systems available are called pulse-position (PPM), pulse-duration (PDM), and pulse-amplitude (PAM). In these schemes, the information is sampled at prescribed intervals and is imposed upon the pulse carrier, such that some quality of the pulses which constitute the carrier is varied in a fashion analogous to the variation of the intelligence signal. In a PPM system, the *position* of a pulse having constant width and amplitude is the analog. The position can be measured with respect to marks or frame pulses. See Fig. 15a for an example of a PPM transmitted waveform. PDM systems employ pulses whose widths correspond to the value of the variable at each sampled instant. All the modulated pulses in a PDM system have both a constant amplitude and a constant position relative to the appropriate marks or frame pulses. Figure 15b pictures a signal modulated by PDM. PPM and PDM systems are very widely employed in modern land-based telemetering system. Much less frequently used is PAM equipment. In PAM, the amplitude of standard-width pulses occurring at the sampling instants of the data signal constitute the modulated waveform which is conveyed to the remote display device

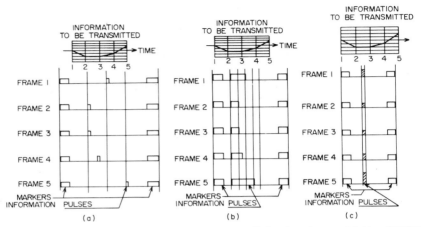

Fig. 15. (a) Pulse-position modulation (PPM). (b) Pulse-duration modulation (PDM), trailing-edge modulation. (c) Pulse-amplitude modulation.

over the transmission line as shown in Fig. 15c. A PAM signal is probably the most easily generated of the three pulse analog systems. A PAM signal in its simplest form is really only the intelligence derived from the pickup which is periodically gated onto the line for transmission. Figure 16 shows a much simplified gating circuit suitable for its generation.

The three types of pulse-modulation systems are analog in their usual application. However, the amplitudes which the pulses in a PAM system can assume can be quantized. In PAM, the system in the strictest sense is a digital one. Similarly, in PDM and PPM systems, the variable can be quantized, and the system could consequently be termed digital. This is not the accepted classification, however. In PAM, the indication of the variable is an absolute one, while in PDM and PPM systems, this is not the case. The amplitude of the pulse in PAM is the analog of the measured variable and is referenced to the no-signal (no-pulse) condition. PDM and PPM systems require marks or frame pulses against which the comparison of the information-bearing pulse is made in order to "decode" (interrupt) them. These two systems lend themselves very neatly to use in time-division multiplexing. In such a multiplexed system, there is a plurality of intelligence pulses between the markers (standard pulses) of each frame, as shown in Fig. 17. Pulse 1 corresponds to the information being conveyed from pickup 1, pulse 2 corresponds to the information being transmitted from pickup 2, and so on. Time multiplexing is possible only when the information being transmitted from each pickup is sampled, and the band of the (single) transmitting channel is wide enough to accommodate the required number of samples for each subchannel being conveyed by that band. We say that the one information

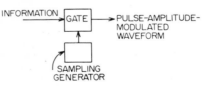

Fig. 16. PAM generation circuit.

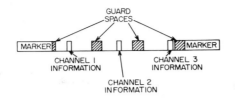

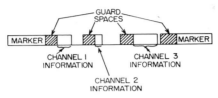

Fig. 17. Time-division-multiplexing systems.

channel is subdivided into several subchannels to which each information pulse in a frame corresponds.

10 Sampling

Implicit in both pulse modulation and pulse-code modulation, which together comprise the category of impulse transmission systems, is the concept of sampling. Considerable effort has been devoted to the subject of sampling by many workers in the field during the past twenty years. Much of the basic work was done by C. E. Shannon, thinking in terms of more efficient usage of communication systems. This work has led to many interesting developments in the art of telephony. For example, the Bell Telephone Laboratories have developed systems, very economical in bandwidth, in which voice (amplitude-modulated) signals are sampled and then the sampled values are encoded as *numbers* for transmission. This work is obviously directly related to the material of the present section. Other work by people such as J. R. Ragazzini and L. A. Zadeh has resulted in the development of mathematical tools for the analysis and synthesis of sampled-data control systems. Indeed, an entire linear-transform calculus dealing with so-called "Z transforms" which are the logical extension of the calculus of Laplace, Fourier, and other *continuous* transforms has been evolved. This mathematics is used to quantify the design, performance, and analysis of equipment which processes either continuous information discretely and/or discretely represented information. An entire analytic procedure for dealing with the spectrum of discretely operating control systems from simple servomechanisms to extremely complex devices like computers has been developed. This procedure might be termed the analog of the "traditional" theory built up over the years for continuous systems. In the discontinuous domain criteria for stability and other important aspects of system performance have been developed for the analysis of sampled systems. This body of theory is very important for evaluating communications-based control systems. The simplest of such systems are those used for telemetering—for the remote display of data. The use of impulse techniques for the transmission of measurands in such configurations has been described in brief in the foregoing material. The considerations required in developing telemetering systems employing sampling techniques employ more the communications-oriented techniques of Shannon than the more control-oriented techniques developed by Ragazzini, etc. However, the analysis of teleoperating systems requires the use of sampled-data control techniques. This is so, since teleoperating systems, particularly the tightly coupled ones, act as closed-loop control systems. In both the forward and the feedback loops of these systems, the signals are discrete. Furthermore, if the control or intelligence position (the "central") of these systems is a digital computer, these techniques are the most suitable for application there. In the application of sampled-data techniques to communications-based control systems, the fact of the necessary merger of communications theory and practice with control theory and practice is seen. Regardless of whether the control engineer or the communications engineer is considering a data-transmission link, both the *philosophy* of approach taken and the techniques employed for solution are virtually coincidental. Only the terminology used is in variance. (Refer to Chap. 15 for Z transform theory.)

11 Information Theory

In both (sampled and continuous) signaling and information-processing systems, the aims of conveying all the information content of the signal as accurately and as quickly as possible are paramount. The word "information" has been used fairly indiscriminately in our work thus far, with only an intuitive definition achieved. Let us look at this term more carefully at this point. There are two principal meanings to the work "information." They are "selective information" and "semantic information." "Selective" implies a *choice* of one (or more) of *a number of alternative* possibilities. "Semantic" implies a *meaning* attached to what is conveyed from one place to another. When the *semantic content* of a message is questioned, what is asked is, "What was said by that message?" In fact, the choice among alternatives—every possible sentence that could be written in English, in an English-language message consisting of one sentence, is virtually infinite. To "guess" what a given sentence will be prior to its reception is very difficult (assuming *completely independent* sentences—those with no interredundancy), if not completely impossible in any realistic sense.

Essentially, all the body of work now generally termed "information theory" which has been developed deals with *selective* information. It is based on an arbitrarily selected, but *precise*, mathematical *measure* of the *degree of uncertainty* involved in a choice among a range of alternative messages which can be received by a *given* receiver working with a *given* transmission medium and transmitter combination. If the probabilities of messages 1,2,3, . . . , N are $P_1, P_2, P_3, \ldots, P_N$, respectively, then the average *selective* information which can be conveyed by these message taken as whole is

$$H = -\sum_{i=1}^{N} P_N \log_2 P_N \qquad (2)$$

If the probabilities of transmission of all the messages are equal then a *guess* about which one will be transmitted is the most difficult, since the choice among the messages that could be sent is the widest. Consequently, a *maximum* of knowledge will be gained when the message is received. Far more knowledge is gained in this case than if the *probability* of a given message's being transmitted was very high *compared* with the probabilities of transmission of the other possible alternatives. The amount of information H in Eq. (2) is the *weighted average* of the knowledge conveyed by each possible message when *it* is transmitted. The process of information transmission, as far as selective information is concerned, can be likened to gambling. One gains (learns) the most when one makes the right guess under the condition that the guessing is the most difficult. That is the case when the probabilities of occurrence of the various alternatives are the most nearly equal and, consequently, the choice is the most difficult.

The concept of the "measure of information" is based on the premise that if the receiver had perfect knowledge—could predict in advance—of which message it was going to receive, nothing would be learned at the receiver upon reception of the message. Whether or not the message really "says anything"—has a meaning—is irrelevant to the selective information theory, which comprises most modern thinking on the subject. C. E. Shannon demonstrated that any given channel used for information transmission has a theoretical capacity for the transmission of information. The capacity is equal to the maximum value of a function called the "mutual information." This maximization is effected with respect to all the variable parameters of importance to the system consisting of transmitter, channel, and receiver. The average mutual information

$$I = \sum_i \sum_j p(t_i, R_j) \log_2 \frac{p(t_i/Rj)}{p(t_i)} \qquad (3)$$

where t_i is the ith transmitted message; R_j is the jth message received; $p(t_i/R_j)$ is the probability that t_i is the message that was transmitted, given that R_j is the message received; and $p(t_i)$ is the probability that message t_i is transmitted.

This function is derived, based on the following argument: The a priori knowledge that t_i was the message transmitted is $p(t_i)$. The a posteriori knowledge of the receiver is based upon the fact of its reception of message R_j. If R_j is the received message that corresponds to the transmitted message t_i, then it is desirable that $p(t_i/R_j)$ be maximized. This means that the corruption of the transmitted signal by the channel must be minimized. This a posteriori knowledge is mathematically equal to the conditional probability that t_i was transmitted, given that the particular message R_j is received. Thus the gain of information at the receiver about conditions at the transmitter (as inscribed in the message) is equal to the ratio of the natural logarithms of the final and initial values of the uncertainties concerning the transmitted message. One should note that the logarithm of the ratio of two quantities is equal to the difference between the logarithms of these two quantities taken separately. If one thinks about selective information in these terms, the basis of the mathematical definition seems intuitively reasonable.

The channel capacity derived by Shannon, $C = \max I$, is realistically unattainable. However, it can serve as a valuable guide to the systems engineer to help him test various hypotheses relating to encoding messages most efficiently in order to *match* them to the capabilities of the channel. Information theory presents a method for quantifying the error-making structure of a transmission channel. The method, at least in some cases, enables one to compute the probability that a given symbol developed by the transmitter will be corrupted by the channel into another symbol and thus will be received (erroneously) as this symbol.

The unit of information is the "bit." It is the amount of information associated with a choice between two messages (or symbols) having an equal probability of occurrence. Recalling Eq. (2),

$$H = P_1 \log_2 P_1 + P_2 \log_2 P_2 = \tfrac{1}{2} \log_2 \tfrac{1}{2} + \tfrac{1}{2} \log_2 \tfrac{1}{2} = 1 \text{ bit}$$

There is a very prevalent confusion, due to imprecise definitions, between the term "binary digit" and "bit." *Strictly* speaking, a binary digit is either one of the two symbols (0 or 1, + or −) used in the binary (radix 2) numbering system. In a binary signaling system, all messages are encoded and transmitted in terms of two symbols only. Each of the binary digits which comprise these messages *most likely* does *not* convey one bit's worth of information. One must be very careful, consequently, when dealing with binary systems, to avoid the many possible pitfalls attributable to the imprecision with which these definitions customarily are made. This author suggests the use of the term "binit" in place of "binary digit," letting the term "bit" be resolved solely as the dimension of information.

Sampling is very important to the impulse-modulation schemes of data transmission. Its importance relates to the reliable and accurate functioning of such systems. The sampling theorem specifies the least number of discrete values of any function of time required for its subsequent *unique* reconstruction. Mathematically, the sampling theorem is

Let $f(t)$ be a function possessing a *band-limited* Fourier transform $F(j\omega)$.

Then $f(t)$ is *completely* determined for *all* values of time t by a knowledge of its values at times equal to $n\pi/\omega_0$, for all $n = 1, 2, \ldots, \infty$. ω_0 is the limit of the band of the Fourier integral of the function. That is, $F(j\omega) = 0$ for $|W| > |\omega|_0$.

The sampled time function can be reconstructed as a continuous function

$$f(t) = \sum_{n=-\infty}^{\infty} f\left(\frac{n\pi}{\omega_0}\right) \frac{\sin W_0(t - n\pi/\omega_0)}{W_0(t - n\pi/\omega_0)} \qquad (4)$$

F. M. Reza has pointed out an interesting example of the sampling theorem to verify its significance and meaning in a practical circumstance. The ideal low-pass filter has the Fourier transform

$$F(j\omega) = 0 \qquad |\omega| \geq \omega_0$$
$$|F(j\omega)| = k \qquad |\omega| \leq \omega_0$$

The corresponding time function, or dual, is $f(t) = 2\omega_0 k \sin W_0 t / W_0 t$.

Let a *band-limited* voltage $V(t)$, that is, one whose sinusoidal components are limited in frequency, be applied to a low-pass filter. If this voltage is quantized at the appropriate instants, $(\pi n/W_0)$ $(n = 0, 1, \ldots, 0)$, and these voltages are successively fed to the filter, its output will be

$$V(t) = \sum_{n=-\infty}^{\infty} V(\pi n/W_0) \frac{\sin W_0(t - \pi n/W_0)}{W_0(t - \pi n/W_0)}$$

These quasi-theoretical considerations have been used as the basis of deriving a practically useful version of the sampling theorem. It states that if a continuous function of time is sampled at a constant rate slightly in excess of twice the highest-frequency component (based on a Fourier breakdown into harmonics), then the function can be reconstructed (at a remote location) exactly. The only penalty for transmitting a sampled version of a continuous signal may be delay and an amplitude change. Both these factors can be readily compensated for in the communication system overall, however,

For the purposes of sampling, let time be divided into equal segments of T sec. T must be less than half the period of the highest *significant* frequency component of the signal. One sample is taken from each such interval in any manner. Then, knowledge of the sampling instants plus knowledge of the instantaneous values of the magnitude of the signal ensures that the function can be reconstructed exactly. In other words, this knowledge is equivalent to perfect knowledge. The highest *significant* frequency was referred to above as being important. It should be recognized that many signals of practical importance will not be precisely band-limited in the sense expressed by the sampling theorem in which the Fourier transform of the signal must have no content beyond a certain frequency. A signal can be *approximately* band-limited, in that the amplitude of its components at frequencies greater than some frequency ω_0 are very small. Such a signal will suffer a minor distortion upon reproduction if it is sampled only at the rate of $2\omega_0$. However, this distortion will be very small. Therefore, it will be of little consequence to the receiver.

Nyquist has shown that the sampling of the function need *not* be done at equally spaced points. If the function is *approximately* time-limited, that is, it exists for a certain period only, then $2\omega_0 T$ samples are sufficient to define the function in a virtually unique manner. Here, a contradiction exists between the mathematics of the theory and the literal interpretation of this statement. A function can *not* be both band-limited (frequency) and time-limited. This fact can be shown by analysis of the *dual* Fourier transforms of a sampled function, one in the time domain and the other in the frequency domain. *Practically*, this factor is not very bothersome according to the same argument given above with regard to the approximately band-limited function. See also Chap. 2 for additional discussion of information theory and sampling.

12 Pulse-code Modulation

Pulse-code modulation has acquired great importance in the communications field. All data systems of the telegraph and teletype variety use coded representations of numbers and other symbols which they transmit. If a signal is sampled at the proper rate, as described in the foregoing section, then these samples can be encoded to any degree of accuracy and transmitted to a remote location for interpretation there. Pulse-code modulation is not modulation in the same sense as the analog schemes: amplitude, angle, and pulse. That is, a relatively straightforward variation of an intelligence-bearing signal is not directly applied to a carrier signal for transmission. Rather, the actual *values* of the signal, taken at the appropriate sampling instants, are encoded and transmitted over the data link.

Important to discussing pulse-code schemes are certain definitions. An encoding scheme is an algorithm for representing only one (of a finite discrete set) value as a particular arrangement of another prescribed set of symbols. This latter set constitutes the code elements, or alphabet of the scheme. In a binary scheme, there are two such elements, commonly represented as a 0 or a 1 in the literature. A teriary scheme employs three symbols. In general, an n-ary scheme utilizes arrangements of n symbols to encode messages. There is a wide variety of codes currently in use, some of which will be described in detail. A code character, the arrangement of alphabetic symbols of the given scheme, represents a given value. In the simplest binary scheme, four of these elements are required to represent one decimal value. Generally, they are given the weights 8, 4, 2, and 1. For example, since $9 = 2 \times 4 + 1$, it is coded as 1001, since the binary digits of this character have the values 8, 4, 2, and 1 in that order. Thus the 10 decimal digits are encoded in this scheme as in Table 1.

Table 1. Binary-coded Decimals

VALUE OF BINITS	8	4	2	1
DECIMAL				
0	0	0	0	0
1	0	0	0	1
2	0	0	1	0
3	0	0	1	1
4	0	1	0	0
5	0	1	0	1
6	0	1	1	0
7	0	1	1	1
8	1	0	0	0
9	1	0	0	1
10	1	0	1	0

Quite often, such an unsophisticated scheme is not sufficient. Communication links afford a very significant source of errors in telemetering and teleoperating systems. Security of data, particularly in a link which is part of a control system, is very important. Consequently, one of a range of error-correcting codes is frequently chosen for data transmission in these systems. The chief purpose of choosing an encoding scheme is to utilize the communication system in the most efficient manner. In a sense, the choice of an encoding scheme for any given installation is made on the basis of its compensating for the error-producing capabilities of the system, as well as according to the statistics of the message ensemble from which the transmissions are chosen.

Pulse-code-modulation systems have gained great importance for several reasons. First, quantized samples of a signal may approximate any continuous function to within a maximum of one-half of a quantum step. Second, the techniques for the handling of quantized values have become highly developed over the last several years.

Pulse-code systems are most frequently referred to as digital systems. Further, as binary systems are virtually the only class of digital systems of interest today, the terms binary and digital have become virtually interchangeable. It is very interesting to note that while (theoretically) an increasingly greater information rate can be achieved by increasing the number of quantization levels, the maximum reliability of the transmitted information is achieved in a binary system. The greater the interference or noise level, the more pronounced is this effect. See also Chap. 8.

13 Codes and Transmission Security

Several types of noise are of interest in a digital (which we equate with binary) system. They are single-frequency noise, spike (or single-impulse) noise, Gaussian noise, and burst noise. Single-frequency noise is undoubtedly the simplest type to nullify. Its effect can be eliminated by the use of a simple single-frequency filter. This noise is introduced by a data link's being in close proximity to a power line, for example. Spikes are frequently encountered with electrical equipment in general. They are induced by switch closures, for example. Such natural phenomena as lightning are frequently responsible for this type of noise. The extent of the spectrum of the frequencies included within a spike is inversely proportional to the duration of the spike. Theoretically, an impulse has no duration in time, and an infinite range of frequencies. Gaussian noise is generally taken to be *white* Gaussian noise. It has a normal (bell-shaped) distribution of amplitudes in time, and a uniform distribution of power over the range of frequencies. Burst noise is very important to data-transmission links, particularly on lines which were developed for use in voice transmission. Recent experimentation by the American Telephone and Telegraph Company (A.T.&T.), the International Business Machines Corporation (IBM), and others has indicated that errors, in general, do not exist singly on such lines. Commonly, they occur in bursts of multiple errors. Work by a variety of people has been devoted toward the development of codes which automatically correct this type of error. Most of this work is very recent.

Probably the majority of the work on error-correcting codes has stemmed from a famous paper by Hamming published in 1950. He demonstrated a technique for constructing codes with the ability to detect and correct any number (within reason) of errors. Others, such as Huffman, have made very valuable contributions to the art. (See Chap. 2.) Chief among the difficulties in implementing error-detecting and -correcting codes is the practical one of paying the cost of the equipment required for working with them. All the burst-correcting schemes developed to date predicate their implementation on sometimes rather elaborate techniques using shift registers. Depending upon the error statistics of a given data link or system, it may be more practical (efficient and inexpensive) to rely upon error detection in combination with retransmission of incorrectly received information than to use an error-correcting code.

An interesting scheme was described several years ago for the improvement (with regard to errors) of binary transmission systems. Two varieties of the system were developed, and experimentally operated. In the one, the receiver could indicate three conditions: binary 1, binary 0, or "uncertain." In the other, four conditions could be indicated: binary 1, *probably* binary 1, binary 0, and *probably* binary 0. The single-null scheme is capable, under the most favorable conditions, of achieving about one-half of the possible (theoretical) improvement over "regular" binary transmission, measured with respect to the extreme of having an infinite number of gradations of signal be intelligible to the receiver. The chief advantage of a system of this type is that fairly efficient systems can be designed to facilitate retransmission of binits received in error. This is so, since in most cases these errors will be located by the receiver's having indicated a null.

Because of the great importance of data security in a communications-based control system, fairly elaborate schemes of "answer back" are frequently employed. "Answer back" is the term applied to any scheme in which a remote station "replies" to an order given to it from the system's master station before executing that order. The General Electric Company's "Code Selector Supervisory Equipment" is an example of this type of system. To operate a remote unit with this system, the operator depresses a button selecting the equipment. It replies to this selection by turning on an indicator showing that it is "ready." Then the operator activates the desired control operation by depressing the desired "control" button. This system closes the loop completely in order to validate the selected operation. One might term such a loop a "virtual" closed loop, since it exists only as a means of validating a control selection. It does not provide "closed-loop control" in the accepted sense of the word.

Such an elaborate scheme to assure security will frequently not be required in a teleoperating system. Reasonable though not necessarily exact transmission of control data such as "orders" from the "master" to the "slave" station may be sufficient for this purpose. If the addressed controller at the remote station is set to *reject* settings transmitted to it that are out of limits, safe control is assured. Another method, simple to achieve, is to transmit every control message twice. If there is

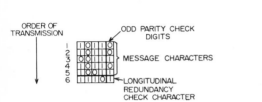

Fig. 18. Parity-checked coded message.

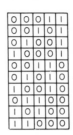

Fig. 19. Biquinary code characters.

a disagreement (or an unreconstructable erroneous character) then transmission will be requested of the master. If the remote station has the logical ability of a stored-program digital computer, fairly elaborate "reasonableness" or validity checks can be made on all control messages from the central. Generally, however, this is not the case. Probably, the best system for the money is to have the remote unit reject all messages received in error and signal the "master" that a retransmission is required in those cases. This scheme minimizes the memory requirement of the slave station.

The control-system designer should not take a pessimistic view, as could be implied by accepting complicated error-control procedures as being required in all cases. Many relatively simple error-checking procedures have been devised based on simple codes, which have been found to be reasonably well suited to safe data transmission.

The most basic error-detecting code employs the concept of parity. Each code character of a message has an extra binary digit appended to it which is a 1 or 0 according to whether the arithmetic sum of the binits of the character are odd or even. For an odd-parity code, the parity digit is chosen to make the sum of all the digits of the character an odd number. This check is called a "vertical check." A similar check can be made on a group of characters. This check is called a "longitudinal-redundancy" check. Figure 18 illustrates this concept for a code having four-digit characters. This code is important since, with the parity bit included, each character is five binary digits. Telegraph and teletype equipment uses a five-binit code called Baudot. The checked code described can be used in teletype networks. This is very important, since teletype channels are used very frequently in remote-control systems. See also Chap. 2.

Another five-digit code of importance is the two-out-of-five, or biquinary, code. The possible characters of this code are shown in Fig. 19. It is a good code for certain applications, since the presence of more or less than two 1 binits in a character indicates an error (or several).

Work has been done using data-transmission equipment using a four-out-of-eight code. In this code, four binary digits of the eight of a character must be binary 1's. IBM's 068 Card Transceiver uses this code for transmission. It has been found highly successful, having produced few errors. IBM has built a system consisting of an analog-to-digital converter and transceiver equipment which has transmitted process data to a card punch 400 miles from the pickup. This is a good example of a digital telemetering system. Closed-loop control is not part of this system, but a control line could be provided to make closed-loop operation possible.

It is very useful to have a transmission code which is convertible into a computer code by simple hardware rather than programmed "looking up" of a required character in a table. This factor will become increasingly more of a requirement as computers become more typical in remote-control and telemetering systems.

14 Multiplexing and Timing

Frequently, several unique signals are transmitted on one band. This process is called multiplexing. There are two principal types of multiplexing, frequency-division and time-division. Time-division is essentially a serial scheme, in which the information, whether coded or not, is transmitted in time sequence with regard to the different subchannels; whereas, in frequency-multiplexed schemes, the subchannels

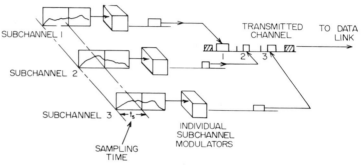

Fig. 20. Time-division-multiplexing PDM system.

occupy various subdivisions of the overall channel. Consequently, the several unique information sequences are transmitted in parallel. The subchannels are separated by "guard spaces," or unused portions of the frequency band, to minimize interchannel interference. Figures 20 and 21 illustrate the concepts of time- and frequency-division multiplexing.

Timing or synchronization is a very important aspect of data transmission. Both the pulse-analog schemes, today chiefly PPM and PDM, and the pulse-code schemes require synchronization. PPM and PDM generally use "frame" or marker pulses. The receiver, by "locking in" to these pulses, can reset its gating circuitry appropriately so that the individual subchannels' information is distributed to the proper equipment there. In these two systems, which are time-multiplexed, "commutation" plays an important role. One can visualize the commutating switch as rotating, connecting to one contact after another in sequence. The idea of operating two commutating switches in synchronism so as to connect two corresponding circuits simultaneously at either end of a transmission line has sparked inventive talent for many years. Probably, the first work along these lines is found in the early multiplexed telegraphy schemes. In fact, the terms "commutator" and "commutation" are in general use in telegraphy circles today.

In most telegraph systems, the synchronization is done on every character. A Baudot-code character consists of the binary digits which represent the information preceded by a synchronizing digit (start) and followed by another synchronizing digit (stop). The lead digit has the same time duration as the information digits. The trailing digit is longer, however, being 1.42 or 1.5 times as long as the information digit. The practice of using pulses of different length arose at least partially because of the ease with which relay equipment used for both the terminals and repeater equipment can differentiate between pulses of different duration, if the difference is sufficiently great. Relay equipment is still used very extensively in this type of installation; the newer equipment is "solid-state" (transistorized) generally, however.

If commutators are viewed as rotating devices, the following can be said about them. Means must be provided to assure that the commutators in the transmitter and the receiver rotate at the same velocity, thus switching at the same rate. Rotation is a valid concept for use in describing commutation equipment, even in the case of purely electronic devices. The idea of rotating vectors or "phasors" is one in common use by electrical engineers. Further means must be provided to restore the synchronous operation of the transmission circuit in the event of momentary interruption, noise bursts, power loss, and simple drift of the commutation equipment itself. These requirements are satisfied by the use of a synchronizing pulse, differing in some readily distinguishable respect from the signal pulses on the line. This difference may be in polarity (in certain voltage systems), duration (as in telegraph systems), or frequency. The use of one channel of a multiple-channel system for the transmission of synchronizing signals is common in frequency-multiplexed systems. Should the

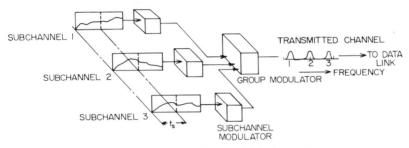

FIG. 21. Frequency-division-multiplexing system.

commutators fall out of step, a correction action is immediately executed at the receiver.

Digital systems in general require two types of synchronization, bit and character synchronization. Character synchronization is the process by which the receiver is made to recognize the occurrence of the beginning (or end as the case may be) of a given group of pulses—*a character*. Bit (binary digit) synchronization is the process by which the receiver is retimed to recognize the transition from a 0 to a 1 and vice versa *as* such a transition and not its opposite. It can mean the process of "looking at"—gating—the line at the proper intervals of time. The definition of bit synchronization used depends upon the particular data modem (subset) employed. The first definition applies in particular to those transmission systems in which phase reversals stand for alternations. The receiver can get out of phase in such systems and consequently, it may misinterpret at 0 as a 1, or vice versa. If seven-digit (binary) characters are used, a most common scheme for data-processing machines the synchronization sequences shown in Fig. 13 can be used.

For bit synchronization, the sequence 10101010 . . . serves to "lock in" the local oscillator of the receiving subset with the one in the transmitting subset. In the character-synchronization sequence, the fact that thirteen 1 bits in sequence have been received indicates that the next digit *is* a 0 (even if it is interpreted as a 1) *and* that the very next digit is the first digit of a (message) character. Thus the receiver will now recognize subsequent digits as what they are, whether they be 0's or 1's, and the commutator (character bit counter) will be placed into synchronism.

There is a variety of synchronization schemes corresponding to the varieties of data modems. The description above was of only one such system. The principles used in this scheme are found in virtually *any* synchronization system.

15 Types of Signal Transmission

A variety of data-transmission systems are available, suitable for use in telemetering and industrial control systems. A word will be said here about the nature of the transmitted signal. In modern telegraphy circuits, frequency-shift-keying (FSK) equipment has become quite prevalent. Two tones are used, one standing for "mark" and the other for both "space" and "no signal." A similar scheme is used for both tone-signal dialing and one type of "Data-Phone" service offered by A.T.&T. Both these systems use five tones. The data system uses these tones in a two-out-of-five code. The receiver has to recognize which two tones comprise the signal. In this system, synchronization is a virtually insignificant consideration. Each time another signal is received, it is (asynchronously) interpreted. This tone system has the disadvantage of being an unconservative user of bandwidth since data could be transmitted at rates up to several thousand bands on one voice channel, depending upon the transmission schemes employed.

A very interesting modem uses multiple phases of the same sinusoidal carrier wave to transmit data. Consider the phasor diagram in Fig. 22. This figure shows four different vectors, each of which can represent two binary digits (referred to as a "di-bit") at one time. Various manufacturers have developed different subsets using varieties of this scheme. The Collins Radio Company has done extensive work on this type of equipment, having developed the "Kineplex" system. There is a fairly wide variety of this type of equipment on the market today. None of it is ideal. However, the rapid progress made in recent years leads one to believe that more nearly ideal solutions to the major problems plaguing data-transmission facilities—that of errors—will be solved relatively soon.

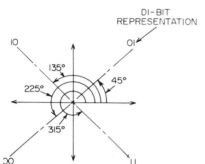

Fig. 22. Phasor diagram.

The majority of the discussion on data transmission in the section has been developed in terms of a relatively modern approach to communications in telemetering and industrial control systems. By "modern" is meant that a more general, a more communications-oriented approach has been taken with regard to the transmission-facilities position of a control system than the tradition-oriented presentation of such facilities as current, voltage, frequency, position, or inpulse systems. It is suggested that a classification in terms of the type of modulation used would make the telemetering art more nearly conformant with the terminology in general use in the rest of the communications field. Another important dimension of classification is whether the system is analog or digital. The original restriction of telemetering systems to the unsophisticated use of transmission circuits using simple metallic conductors greatly hampered progress in the field. The availability of electronic systems such as carrier and other means to afford multiplexing, thus eliminating the characteristic wire conductors as imposing limitations in a direct manner on signal transmission, has released the telemetering art toward achieving great levels of advancement.

16 Current and Voltage Telemetering Systems

"Current" and "voltage" telemetering systems are virtually the same in principle. In the current system, the value of a current in the circuit is measured at a point remote from the source of the current, which is the pickup which reads the measurand and develops the current as its analog. In a voltage system, the analog quantity is the voltage which is developed by the pickup. As is well known, the means for measuring potential differences and current flows are almost identical. A voltmeter is in essentials an ammeter which is shunted by a resistance. Consequently, it draws little current by being connected across two lines whose difference in potential it is measuring. The parallel combination of shunt resistance and internal voltmeter resistance must be very high to minimize the amount of current which it draws. An ammeter passes all the current in a circuit. It must be a low-resistance device so as to draw as little current from the circuit as possible. The ideal voltage system works so that virtually no line current is drawn, and consequently, the problem of (resistance) line loss is negated. The transmission circuit should look like an infinite-impedance source to the (voltage) measuring instrument at the display location.

In a current system, the line current must be maintained at measurable levels such that a range of current is available to provide the complete scale for the measurement at the receiving (display) location of the line. The transmission system should look like a very low impedance to the measuring device. As line conditions vary, more (or less) current may have to be made available to the circuit to compensate for these variations.

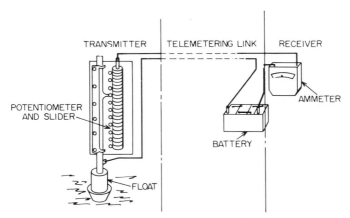

Fig. 23. Current telemetering system.

Null systems, based on one of the electric-bridge principles, such as that of the Wheatstone bridge, are commonly employed in both current and voltage telemetering systems. It is possible to design a current system so that no line current actually flows at the time of measurement (the instant of balance). It is a moot point as to whether such a system is voltage or current.

The principal components of both current and voltage telemetering systems are a pickup, or measuring instrument, a line suitable for conveying a desired electrical variable, and either an ammeter or a voltmeter. The latter device is one of a class called "displays" in telemetering parlance. Note that *two* measurements are made in both these systems, by both pickup and receiver. The basic design for this type of system is shown in Fig. 23.

Variations of these schemes called "conductance" and "resistance" systems have been developed which tend to minimize variations in supply voltage, line conditions, etc. In the conductance system, the ratio of current to voltage is developed by means of energizing an instrument having both a fixed and a variable winding. One could term this a resistance system equally well. A variation of this theme uses a resistance, the value of which is varied according to the value of the measurand. The receiver of this system is comprised by a bridge circuit, of which one arm is the pickup and the line.

Systems predicated on balance arrangements occur as both current and voltage systems. In their operation, a suitable variable is developed by the pickup which acts upon the balancing device in such a way that an electrical quantity (an adjustable current or voltage) is generated so as to nullify the effect of the measured quantity upon the balancing device. This quantity constitutes the variable transmitted and depends upon the equipment used. The electrical quantity which is proportional to the *opposing* balancing force is transmitted to the remote indicator. This type of system tends to nullify the effects of unwanted parameter variations such as varying supply voltages.

An instrument frequently used as the primary means of measurement in systems of this type takes the form of a Kelvin balance. This device is electromechanical and can be very elegantly embodied. It consists of a balance arm, on each end of which is a coil. Corresponding to each one of these movable coils is one (or two, depending on the design) stationary coil. The mutual induction between each pair of coils produces a force. A balance is obtained with this instrument in identically the same manner as a laboratory scale when the (mechanical) forces on either end of the beam are equal (actually, when the associated moments about the beam's fulcrum are

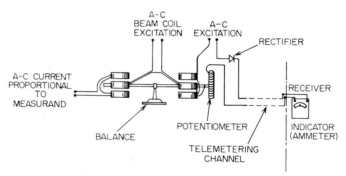

Fig. 24. Balance telemetering system.

equal). Note that this principle could be used as the basis of a weighing device in which an object to be weighed could be placed on one end of the beam, and the electrical (coil) equipment on the other. In this case, the amount of current supplied to the balancing coil would be indicative of the weight of the object.

A telemetering system could be readily constructed employing the principles embodied in this device. A constant current would be supplied to the beam coils. A current proportional to the measurand is supplied to one of the stationary coils. A current, supplied by a servo circuit, is applied to the other (at the opposite end of the beam) coil, causing it to react with its counterpart (coil) on the beam in the manner of creating a "restoring" force. When "perfect" balance (within the capabilities of the device) is achieved, the current generating this restoring force is proportional to the measurand. An embodiment of this device is pictured in Fig. 24. This (restoring) current may be transmitted to some (relatively) remote location for measurement and interpretation there.

Sometimes a given telemetering system will be comprised by both current- and voltage-type equipment. A good example of this is found in certain thermocouple heat-measuring systems. Thermocouples produce voltages in the millivolt range. By means of fairly expensive special wires, these signals can be transmitted a reasonable distance—several thousand feet as a maximum. Equipment is available today which will convert these voltage signals into current (high-impedance) signals more suitable for relatively distant transmission. This type of equipment could be significant to a process-control computer system in which the pickups are fairly widely (up to several miles) removed, or scattered, with respect to the control computer.

Current, voltage, and "resistance" or "conductance" transmission concepts are not directly suitable for use in microwave or radio (non-solid-conductor) systems. This is self-evident, since a variable current or voltage cannot be transmitted over a radio beam. Thus, if it is desired to transmit either a variable voltage or current, the signal variations occurring on either of these media first must be converted into an amplitude, a frequency (or phase) of a carrier wave (of suitable radio frequency). As has been stated, both the current and voltage systems were designed for relatively "local" transmission. These systems became available considerably prior to the electronic systems which use the carrier techniques of modern telephone practice for *both* voice and coded-data service. They can, however, be readily integrated with these "modern" systems when required.

17 Frequency Telemetering Systems

The so-called "frequency" systems of telemetering have a considerable history. Though today's electrical and electronic engineers tend to think in terms of frequencies in the order of those in the range produced by vacuum tube and other *all electronic* techniques, the word "frequency" originally denoted a system for use in the power-frequency range. Such a system consists of electromechanical equipment such as some special variable-speed alternators capable of generating power of up to several hundred cycles per second oscillation frequency. It is interesting to note that the first transatlantic radio transmitter used alternators to generate very long wave r-f (radio-frequency) signals for transmission of Morse code (CW—"continuous-wave") signals.

A very ingenious use of cascaded motor-generator sets was made in a relatively early installation to generate a load indication in a power-generating system. The speed and hence the frequency of the output (a-c) current of such a motor-generator set was controlled by a servo actuated by a wattmeter. The currents developed by each such motor-generator set were electrically (vectorially) added together to form a single signal, the frequency of which was varied to be proportional to the total system power load.

The other extreme of "frequency" telemetering systems is found in today's frequency-modulation schemes in which signals in the multimegacycle range are employed. Radio telemetering systems, such as those used for missile checkout and missile-firing-range telemetering, use various types of FM systems almost exclusively. FM-FM and PDM-FM systems are the most common of the frequency systems. The simplest type of FM-FM systems use each frequency-modulated channel as an analog means for transmitting the status of a given pickup. In PDM-FM system, the measurand is represented in the pulse-duration form, which modulates an r-f carrier in the manner of appropriately varying the frequency. That is, the information signal is in the form of a sequence of pulses occurring at constant intervals. Each pulse's duration is proportional to the value of the measurand as sampled during that interval. These *pulse-modulated* signals are then impressed upon an r-f carrier and vary *its* frequency according to the duration of the pulses (sampled values of the measurand).

Multimodal or "multidimensional" modulation techniques such as described here are quite common. Theoretically, as many types and levels of modulation as desired can be executed upon any given intelligence signal. For example, a continuously available measurand first could be amplitude-modulated and then frequency-modulated subsequently. Many of the various carrier systems in use in modern telephony employ these techniques. First, a typical system works as follows: The voice modulates a d-c "carrier," which in turn modulates a higher-frequency carrier. The resultant signal is impressed upon a multifrequency channel for transmission by a solid conductor. Then, this signal is impressed upon a microwave-frequency carrier for wireless transmission. This sequence of "enmodulating" events has become quite typical for voice transmission. The techniques are quite applicable for data transmission as well.

There are two principal categories of frequency telemetering systems. There is one in which the signals are in the power-frequency spectrum, up to several kilocycles as absolute maximum, but which are generally only in the range of a few hundred cycles. Also, there are other configurations which are virtually unlimited as to possible maximum frequency. They cover the range from radio frequency up through the megacycle and multimegacycle band, to the limit of microwave data-transmission facilities.

The low-frequency systems are used with solid-conductor facilities, while the high-frequency systems utilize either solid conductor or wireless (typically microwave) equipment in their operation, depending upon the range and the use of the system.

The simplest type of frequency telemetering transmitter for use in the power-frequency range employs an oscillator, the tank circuit of which is varied in accordance with a shaft position which is indicative of the values of the measurand. This causes the oscillator to produce a signal of variable frequency proportional to this position. Either a capacitor or an inductor serves as the variable element of the tank circuit. The oscillator output may be used directly, or else it can be beaten against a signal of standard, constant frequency, to produce a "beat" proportional to the angular position of an indicator, the setting of a knob, etc.

A variable, not directly (primarily) measured by a device of the "frequency" type, may be transmitted as a frequency signal. A simple telemetering transmitter of this type is one in which a voltage analog of the measurand is developed which is applied to an oscillator to vary its frequency. The oscillator in this instance may be of either the continuous-frequency (sinusoidal) or the relaxation (pulse) type. In the latter, the "triggering level" of a multivibrator is set according to the value of the measurand. More or fewer pulses are generated according to the value of this voltage. The *number* of standard amplitude and width pulses generated per unit time corresponds to the value of the measurand.

Systems of the latter type, while classified as frequency systems, differ extensively from continuous-wave (sinusoidal) frequency systems which are actually angle-modulated systems. The varying-repetition pulse-rate scheme could also be very validly classified as an impulse system, since the telemetered information is conveyed by pulses. There is at least one other well-known and important pulse-type scheme which is equally a candidate for a dual classification. This is the so-called "frequency-code" system in which a prescribed number of different tones are simultaneously transmitted, usually over a voice-width channel. This combination encodes the measurand. The code used may be any one of several, depending upon the individual system requirements. The two-out-of-five or the four-out-of-eight codes, both containing a certain degree of error-detecting ability (as has been discussed in the section on codes) are good candidates. A code wheel with various apertures can be used to generate the digital representation of an angular position, corresponding to the measurand to be transmitted. It selects the combination of tones required to represent the character in the transmission system. A light impinging upon the disk will or will not activate several photocells, depending upon the angular position of this code wheel. The photocells in turn activate oscillators attached to them. The outputs of these (independently operating) oscillators are electrically added and transmitted to the receiver where they are decoded and converted into a suitable display. The decoding technique employed is very similar to that used for encoding. The bandwidth required for this type of system is not nearly so small as that required for a pulse-code-modulation system. Nevertheless, it is eminently well-suited to certain applications. Bendix markets a system of this type. A parallel-tone signaling scheme of the same nature as has been described is used in the telephone industry for one system of dialing. A.T.&T. has available a dial-up data-transmission scheme called "Data-Phone" which operates in a very similar manner to the parallel-tone telemetering system that is described in this section.

A number of different mechanisms have been developed for receiving frequency telemetered signals and making them ready for display. They comprise two principal categories: frequency meters and amplitude receivers. The frequency meters are of the same construction as used in power work to measure frequency variations. They are calibrated in either absolute values or deviations from the nominal value. These devices are suitable for use in low-frequency (up to several hundred cycles per second) systems. Becoming increasingly more widely employed in the industry are receivers which convert the varying frequency signal into a *constant frequency* the *amplitude* of which is proportional to the frequency of the received signal. A standard amplitude-indicating instrument (an ammeter or voltmeter) is employed as the display device at the receiving station. It is calibrated to read the value of the measurand directly.

Frequency-modulation systems have acquired great importance in modern telemetering practice. They are of particular interest in radio systems. Probably, the chief areas of their employment are for missile tracking and checkout. The FM systems in use are multichannel configurations, for the most part. Each channel conveys the value of a single variable. Undoubtedly, one important reason for their wide use is the relatively low degree of interchannel (cross) modulation enjoyed by these systems. Such distortion can become very critical in those installations in which there are a large number of channels.

18 Position Telemetering Systems

Another one of the basic classes of telemetering systems according to the "classical" system of categorization is the "position" type of system. Both d-c and a-c positioning systems are in current use. In both, the fact of *simultaneous* transmission of two or more electrical variables (current or voltage) to indicate the measurand is more important than the nature of the electrical signals. The interrelationship between the several (principally two) transmitted signals in terms of the ratio of one of their more significant aspects represents the measurand. The ratio may be in terms of amplitude, angle (frequency or phase), etc. Any factor which generally designates one of the types of modulation systems which are available can be likewise employed in the implementation of a "position" system.

The transmitting equipment develops the ratio of the two quantities according to the value of the measured variable. The values of the *several* signals representing this ratio are transmitted according to the techniques appropriate to the method of modulation used for this signal—just as for a current, voltage, or frequency system. No unique transmission techniques are employed in position systems relative to those in the other (single "channel") types of transmission systems.

The receiver is an instrument which converts the transmitted signal to the position of an indicator, for example. In some systems, this is accomplished by there being a torque developed proportional to the ratio between the two signals. Systems using synchros are of this type. Generally little torque is developed directly by the receiving device. Consequently, if any significant amount of "work" is to be done by the indicator, torque multiplication will be required. "Synchro" is the commonly used designation for an a-c positioning motor. The name is usually extended to include d-c devices, however. This device and others employing various modifications which represent simple modifications of its principles are probably the best known of the "positioning" systems. The a-c synchro is very similar in electrical structure to the synchronous motor. Units having identical electrical and mechanical structures are coupled together to form a telemetering system transmitter and receiver. If the single-phase (rotor) windings are excited from the same source, and the polyphase (stator) windings are not excited but are connected together, the rotors will assume identical angular orientations. Customarily, the rotor windings are *not* fed from a common source, since *any* a-c source will be satisfactory for transmitter and receiver if the phases of the supplies differ by only a small amount. Consequently only three conductors are generally needed to connect the transmitter to the receiver in a synchro system. Figure 25 pictures a synchro system.

Some d-c ratio devices have come to be termed "synchros" also, probably because they use a rotatable magnetic field in an electric-motor-like structure. In this type of system, the transmitter takes the form of a rheostat which may be continuous. The receiver consists of coils with suitable cores which develop a field the angular alignment of which is directly correspondent to the tapped position of the rheostat. Within the coils of the receiver is mounted a rotatable magnet—typically a permanent one—to which is attached an indicating pointer. The magnet aligns itself with the electrical position of the field developed in the stationary coils, thus causing the pointer to indicate properly the value of the measured variable—as it is transmitted.

In Fig. 26 is shown a generalized version of a system of this type. The configuration in the diagram can operate as an a-c device also.

Other than the motor-like systems described, there is another general category of positioning system in common use. This is the bridge system. The bridge may be either self-balancing or require manual balancing. In principle, two sides of the

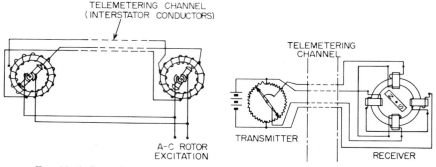

Fig. 25. A-C synchro system.   Fig. 26. D-C synchro system.

bridge are provided by the transmitter and are linked to the other sides by the transmission facilities. Figure 27 shows the basic principle of this type of system which is comprised by two slide wires, one at the transmitter and the other at the receiver. Each slide wire represents two resistors of the bridge. The transmitter slider is positioned according to the value of the measurand. The receiver slider is positioned either manually or by a servomotor so as to drive the bridge's imbalance,

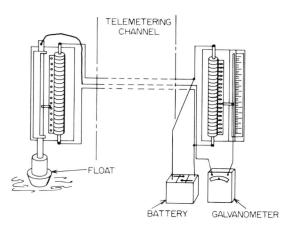

Fig. 27. Bridge telemetering system.

indicated by a meter at the receiver, to zero. The simplified circuit shown in Fig. 27 stands for either a d-c or an a-c system.

Other, more complicated bridge structures can be built particularly if a-c excitation is employed. Variable inductances, capacitors, and/or combinations of both with resistors are used in some equipment configurations. Also, there are some positioning devices which it is difficult to classify as either motor or bridge systems.

The transmitter is structured like part of a bridge, while the receiver operates using magnetic forces as described.

COMMUNICATION-BASED PROCESSING CONTROL SYSTEMS

19 Pipelines

Gas and bulk (petroleum products) distribution systems as well as power-distribution systems represent prime examples of communication-based process-control systems. These configurations might be termed "dispersed transportation" systems, since the control problems associated with their operation are ones principally related to product flow, whether this flow be of a physical quantity, such as oil, or of electricity. Automation in these systems has generally not yet reached the maximum level—that in which a computer performs the centralized task of data gathering, and the development and dispatch of control settings. Computers, both general- and special-purpose, have come to be virtually a required tool for off-line operations in such installations, however. They serve an auxiliary function in that they compute load distributions to satisfy varying load requirements, process centrally logged data, develop operational settings for remote stations (to be manually dispatched to them), etc. In common automatic-control parlance, these systems which employ computers for on-line decision making are termed "closed-loop." Those using computational facilities in an off-line manner, on the other hand, are designated "operator-guide."

Many companies are currently engaged in the manufacture of both systems and components related to this field. For example, Union Switch and Signal, a division of the Westinghouse Air Brake Company, produces various systems under the title of "Centralized Transport Control." These systems comprise equipment to facilitate a completely centralized monitoring operation and control of remote pumping stations, in an extended pipeline system. They make possible the operation of an entire pipeline from *one* location.

In 1958, Union Switch completed such an installation for the Columbia Gulf Transmission Company, in which a dispatcher in Nashville oversees the operation of a gas transmission which extends 841 miles. With this system five compressors (pumping stations) and several gas-delivery points are centrally operated. The General Electric Company markets its "Code Selector Supervisory Equipment" systems for the remote control of a-c switching centers in power-distribution systems, hydrostations, pipeline pumping stations, and other units exhibiting similar control requirements. General Electric produces a high-speed digital, scan logging system to be associated with this supervisory system. This logging system is called the "APRIL" system, this name standing for "Automatically Programmed Remote Indication Logging" (system). Many other companies, such as the Radio Corporation of America, the North Electric Company, and the Westinghouse Electric Manufacturing Company, both manufacture and market these systems. Much of the equipment currently installed for supervisory control and centralized telemetering is implemented by tubes and/or relays. However, solid-state systems employing transistors, magnetic materials, etc., are becoming increasingly more significant in the market. Various factors, including reliability, are responsible for this.

20 Operational Security

Security is the accepted term which denotes the code techniques—principally for digital signaling—which are included in the encoding procedure, to minimize, and hopefully completely preclude, ambiguity due to noise on the line. It may be recalled that error-detecting and -correcting codes and associated operational features are described in the sections on codes and modulation. There are several different methods to achieve security which are incorporated in the supervisory equipment marketed. Among these methods are

1. Answer back
2. Checking codes—including redundancy methods, parity checking, self-correcting structures, etc.
3. Transmission of a definite number of pulses in a data group
4. One or more pulses in a group distinguishable from the others in the group

Answer back is a technique by which the remote station echoes the control signal transmitted to it from the master station *back* to it for (automatic) comparison there prior to its being executed. This scheme determines the accuracy of the transmission of the data or the command message in a manner quite satisfactory for many installations; with answer back, the operator selects the point (and station to be operated) to be operated, a command message is dispatched to the remote location, and an "O.K." message (the answer back) is transmitted back to the master, confirming the connection. Then, the "execute" signal is dispatched to the remote location, causing the designated operation to be initiated. The "execute" is issued automatically in some systems, while in others the operator must actuate an appropriate switch to activate the operation.

Many varieties of checking codes exist. Two of these are the four-out-of-eight and the two-out-of-five binary codes, both of which have found fairly widespread use in digital communications systems. The more simple codes such as pure binary and binary-coded decimal have been elaborated upon by various parity-checking techniques to detect and in some cases to correct one or more errors in either one or a group of code characters. Parity checking is also accomplished on a group of characters, those comprising a given command message, for example. Parity, it may be recalled, may be odd or even. It denotes adding the proper one or several binary digits to the group of binary digits designating the transmitted intelligence to make the arithmetic sum of the digits odd or even, as the case may be. Various methods of introducing redundancy into the character group forming a message have been developed. They go from the extreme of repeating a message twice in its entirety to the relatively simple expedient of transmitting certain key characters. Some systems of the redundancy category impinge quite closely on the realm of answer back.

Many systems employ message groups of constant length. These fixed (constant) word- or group-length schemes operate either in terms of there being a preestablished *number* of pulses in every message or else in terms of the length of time required for a transmission, this *time* being required to be a constant in every case. Messages transmitted with a fixed-pulse criterion are very easy to check.

Another common method used to attain security is to require that one or more pulses in a message (character group) be longer or otherwise distinguishable from the information-bearing pulses transmitted on the system's data links. Teletype systems employ longer pulses for the stop bit (a synchronizing digit) than for information bits, for example. Some other systems use a hierarchy of several (commonly two) pulse amplitudes in message transmission. The security guarantee can be provided under these conditions by requiring a certain number of pulses of each possible amplitude.

The means of security in use in today's equipment in many cases evolved more from good engineering than from a scientific and/or theoretical evaluation of the various transmission media used in these systems. Modern consciousness of experimental techniques and the tools of information theory are altering this situation rapidly, however. Interestingly, the "tried-and-true" security techniques described here have met with excellent success in practice, several of them over many years of rugged use in a wide variety of applications.

Associated with the several means of message-transmission security described is the operation called "station check." This is one of the elements of supervisory control. A supervisory station check is the automatic selection of each (control and/or sensing) point in a station in some prescribed sequence. Customarily, it is done in numerical order, upon a single initiating action at the central, or master, station. In some installations, every point in the entire system is selected for indication of status at the central in this manner. This operation enables the dispatcher (or operator) to check out the system for possible transmission errors. This testing procedure is invoked during these times in which control or data communications between central and the remote station are not needed. This procedure is an obvious aid in preventive maintenance.

21 System Components

Several codes are in use for the various centralized transportation control systems, as is evident from the preceding discussion. There is a similar lack of common definition of the various parts of and significant concepts related to these systems. The Instrument Society of America (ISA) recognized this state of affairs and formed a commitee which drew up a "recommended practice" for Data Transmission for Centralized Operation of Petroleum Systems. The material of this section is based upon the Recommended Practice for such data transmission. The parlance used to describe the centralized supervisory and logging functions in a petroleum system is valid for gas-transmission systems and to a large extent for power-distribution systems as well.

The basic parts of a (communication-based) supervisory control system are the master station, the several remote stations, and the interconnection (data) channels which link these equipment groupings together. A master station is the central station from which the remotely located units or other such equipment are operated by supervisory control and/or received telemetered indications of the status of equipment at these remote locations. In semiautomatic systems, an operator digests these telemetered readings and executes control of the system by varying the system parameters which are determinable from his location. In addition, he varies these parameters according to other factors such as delivering a desired product to a prescribed takeoff or stripping (product-unloading) station along the length of the pipeline. A completely automatic system operates in virtually the same manner with the exception that a computer performs many of the functions of the operator. It does *not replace* him, however. While the computer cannot do much more than a person can do, it can do it far more efficiently and with far fewer errors that a person can. The computer operates tirelessly, day and night. The incorporation of a computer into a centrally operated distribution system enables the fullest exploitation of the extensive telemetering facilities currently available in these systems. It provides an optimum response to data relating to the loads at various pumping stations in the system as well as the total product inventory of the system, specified by the amount of product(s) in the tanks at the stripping locations. The accuracy and speed of a computer make possible the integrated operation of the several pumping stations in the system. The pumping stations of a pipeline *cannot* be operated independently if the pipeline is to work efficiently and in many cases, safely. Pipeline pressures, temperatures, and product flow rates are important in these configurations from the point of view of *immediate* evaluations of the system's performance and status. Levels of the storage tanks at both the terminals and the takeoff stations of the line as well as data on customer demand provide some of the necessary vital information required to schedule movements of gas in the system, or the even more complicated problem of scheduling the flows of various grades of product in a bulk line.

The stations at the various pumping sites in a gas pipeline and at the switching stations in a power-distribution network are called "remote stations." There is equipment at these locations to operate the pumps and otherwise activate various other operational equipments. Further, the pickups used to sense the status of this equipment are located there. The values obtained by these pickups are telemetered to the master station for processing there. The remote station's process-operating equipment is controlled from this "central" station. Two types of control for this remote operating equipment exist. They are remote supervisory control and remote reset automatic control. With the former, essentially all steps of the logical sequence implemented in exercising the station in startup and shutdown, for example, are derived from the master station. With the latter type of control, however, the operational devices at the site are caused to change according to some function of automatically introduced information about the process and a logical, computer-type program. The program can be altered by remote control. Remote stations can be virtually autonomous in their operation in that they are automatically cycled. Such a station is one in which the operation of one or more of its constituent devices such as the pumping unit executes certain operational sequences in concord with a prescribed schedule. "Start up" and "shut down" are examples of such sequences. This sequencing equipment is analogous to, and in some cases, similar to, the program-execution circuitry of a computer in which the operation-code portion of an instruction is translated into the activation of a certain group of control leads according to some predesignated sequence. Any given operation code initiates the activation of a given set of leads in some such sequence unique to that operation code.

Many remote stations are virtually completely automatic in their operation. They customarily have quite elaborate safety provisions to safeguard their operation. This is particularly so in pipeline pumping stations, some of which have triple-level protection against an unsafe low pump suction pressure, for example. These particular devices provide a high degree of assurance that cavitation of a pump will not take place. An automatic station is one that is unattended (usually) which commences operating according to an automatic sequence, and generally which ceases operation in accordance with another, similar automatic sequence. The deactivation takes place under predetermined conditions, generally. An automatic station may be placed into or removed from service in response to supervisory control, varying load conditions, failure of a built-in safety check, a given time, as well as many other conditions.

22 Pipeline Control

Both gas and bulk pipelines are ideal candidates for complete automation since the development of the semiautomatically operated pipeline has great experience behind it. The first unattended pump station was built over thirty years ago. Remotely controlled pipelines utilizing the techniques and equipment described in the material above are relatively commonplace today. Information about the status of the system's pumping stations, stripping stations (product-takeoff facilities), and storage tanks in the case of the bulk line is telemetered to the dispatcher who is located at the line's master station. He operates controls to exercise these remote facilities in accordance with the *current* status information presented to him on the telemeter displays. He operates the system in accord with the load pattern facing his system. Centralized dispatching makes possible a more precise movement of product to the customer more safely than is possible without the centralized availability of information about the system's status. The next big step is to close the loop with a computer, transferring the decision-making functions of the dispatcher over to the computer. Up to this time, pipeline companies have used computers for off-line (noncontrol) computations. The dispatcher might frequently use the computer to help him schedule product movements in his line. Then, he applies these *results to aid* him in determining the settings for the control points in the remote stations of his line. The computer would operate upon a mathematical model programmed into it. The steps required to connect the computer directly to the controls and the telemetering displays are straightforward. Pipeline companies use computers for the normal accounting-type data-processing jobs of payroll, bill accounting, and so forth. They will also be expected to fill the post of the dispatcher, interpret incoming (telemetered) data at the central and establish corrective action at any segment of the pipeline's control, these operations being consistent with the overall operational objectives of the pipeline.

Figure 28 illustrates the basic features of the automatic pipeline. Centralized pipeline control is validated by the fact that experience with semiautomatic pipeline complexes indicates that either the system dispatcher or the control computer (in the case of the completely automated system) requires information from all the pumping and takeoff stations in order to interpret the data from any one station properly.

Several principal operations are performed by the remote stations of a bulk pipeline. They are

COMMUNICATION-BASED PROCESSING CONTROL SYSTEMS   13-43

1. Product tracking, including the monitoring of line conditions and detection of product interface
2. Line balance control—the adjustment of pumps and valves
3. Product removal (sometimes called stripping)

Just as the movement of freight in a railroad system must be accurately tabulated, so must the precise location of all product loads in the pipeline system be centrally available. The suction and discharge pressure at each pump must be available.

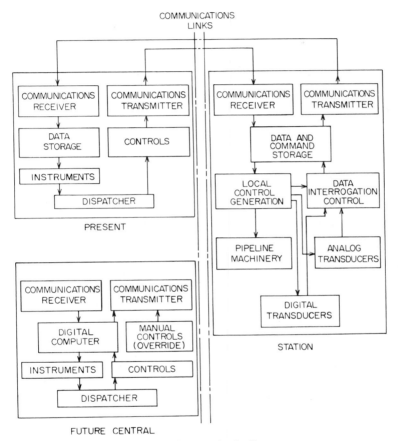

Fig. 28. Automatic pipeline.

In addition, the temperature and viscosity of the product in the line at that point must be made available at the dispatcher's location. Detection of the interface between two different product batches is quite important. Automatic detection can be combined logically with a preestablished condition transmitted from the central to connect the proper tank automatically to the line at the proper time. Accurately monitoring the level of each tank in the system is of crucial importance in successfully operating the pipeline systems as an integrated whole.

Line balancing, the proper control of the pumps at each station in relation to the data telemetered to the dispatcher about the system conditions, both normal such as during unloading, and abnormal such as a complete pumping-station failure, is a principal task performed by the central dispatcher. It is essential that proper pressures be maintained at each pumping location in order that the lines do not burst, the pumps are not cavitated, and no line section suffers a "water hammer," the unwanted and generally severely harmful oscillation of the fluid in the line. Figure 29 details a generalized remote pipeline station, illustrating the interrelationship of these various functions to each other.

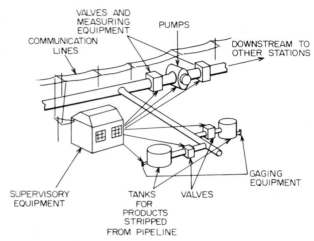

Fig. 29. Remote station on pipeline.

Depending upon the particulars of the application, not all the measurements at each remote station important to the system's operation are always simultaneously available at the dispatcher's location. The dispatcher's display panel may have fewer instruments than the total number of variables required for this operation of the system. By means of selectors, the chief dispatcher can connect desired telemetering channels to the instrument banks in turn, so that he can read *all* the pertinent measurands in turn.

23 Power-system Control

The use of specialized computation equipment, coupled with elaborate telemetering schemes, has come to be of considerable importance to the power industry. The high quality of virtually uninterrupted service would not be available today were it not for the "computer control" employed in modern automatic dispatching systems in nearly all the multitude of power companies of the United States whose stations, gas-fired, oil-fired, coal-fired, and water-powered, are interconnected through tie lines. Automatic dispatch provides

1. Automatically regulated frequency
2 Metered (and regulated) power interchange among interconnected power producers
3. Economic allocation of available generating capabilities for a given operating area

The power interchange between generation areas is set both by contract and by the emergency conditions that periodically arise such as unscheduled operating-station shutdown or main feeder-line breakage. Equipment is available to implement automatically both area interchange and intra-area allocation of generating capacity. Two of the principal manufacturers of this equipment are the General Electric Company and the Leeds and Northrup Company.

The basic controlled variable of a power-generating system is the speed of each generator. The frequency of the generated current is proportional to the revolutions per minute of the generator. A generator reacts to load variations (differing current requirements) as follows: The rotor tends to slow down in response to a load increase, while *tending* to exhibit the opposite behavior in response to a load decrease. Each generator of a station has a speed-level changer which consists of a controller and a motor to adjust the machine's governor. The relationship between speed and load can be adjusted so that the generator's speed may be the same for different loads. These devices constitute a servosystem to which are sent speed up and slow-down signals developed by the system computer equipment under the supervision of the dispatcher. These signals are developed as a function of several factors relating to the output of *the* particular generator. Among these factors are

1. Area interchange requirement
2. Frequency requirement
3. Time-error correction
4. Unit incremental cost

The area interchange requirement is set as a schedule which can be entered into the generator requirement equipment as either a "remembered" value in the computing device or else manually available through the dispatcher. The difference between the metered interchange value and the schedule corresponds to the interchange error which is fed into the generator speed-control servo. Frequency requirements take two forms, the necessity for 60 cps generation (within a stipulated allowable tolerance) and the adjustment of integrated frequency or time. This latter factor relates to the need for reducing any accumulated system error—as would be indicated by a customer's electric clock—being reduced to zero over a prescribed period of time, generally a day. The instantaneous frequency error is developed as the difference between the output of a frequency standard, such as a crystal oscillator, and the measured system frequency. Economic considerations are of extreme importance in the operation of a multiunit power system. For maximum economy, the incremental cost of power for each unit should be equal, neglecting transmission loss. The values on the cost curve can be available either through manual entry into the system by the dispatcher or else by means of a hardware storage capability. This economic information is combined with the other control signals described and is fed to the generator speed controller.

The need for precalculated generation schedules can be removed through the use of generation-scheduling computers. Principally, such a computer is an analog machine which calculates the allocation of generating facilities based on a manually entered value of incremental cost. Telemetered values of power flows through the tie lines comprise the basic inputs to the computer. The analog computer can be taken out of the control loops and employed for other activities, such as simulation studies related to future system requirements.

The digital computer, with its almost unlimited operational versatility due to the use of *readily alterable* programs, should find increasing use in power-control systems. The multifunctional use of a single large computer in both direct power-system control and the normal off-line run problems of the business aspects of the company should become increasingly more important.

Local dispatching conducted with the type of equipment described has resulted from the rapid expansion of today's alternating-current facilities to serve large geographic areas. Through the extensive tie-line facilities, the capabilities of individual generators have come to be used more efficiently. Peaking of the load requirements tend to be smoothed out in these systems, since locally available surplus can be used elsewhere *when* it is needed. The centralized distribution control ensures coordinated power generation in response to the observed power requirements of each customer area. Centralized operation serves the purposes of enhanced reliability of service, improved quality of service, and maximum economy of power production.

REFERENCES

1. Perry A. Borden and Wilfrid J. Mayo-Wells, "Telemetering Systems," Reinhold Publishing Corporation, New York, 1959.
2. Elliot L. Gruenberg, Telecontrol, *IRE Trans. Telemetry Remote Control*, May, 1957.
3. Alexander A. McKenzie and Haig A. Manoogian, Telemetering, Electronic Data Transmission, *Electronics*, April, 1956.
4. *IRE Professional Group Trans. Space Electron. Telemetry*, vol. TRC-4, September, 1958.
5. *IRE Professional Group Trans. Space Electron. Telemetry*, vol. TRC-5, September, 1959.
6. Zeev Bonenm, The Influence of Coding on Closed Loop Remote Control Systems, *IRE Trans. Telemetering Remote Control*, May, 1957.
7. Harold S. Black, "Modulation Theory," D. Van Nostrand Company, Inc., Princeton, N.J., 1953.
8. R. L. Carbrey, Pulse Code Modulation Terminal and Repeater Methods, *Electron. Design*, July 6, 1960.
9. Peter R. Brograton, Application of Tone Equipment to Data Transmission, *Pipeline Eng.*, May, 1959.
10. F. J. Bloom et al., Improvement of Binary Transmission by Null-zone Reception, *Proc. IRE*, July, 1957.
11. R. W. Hamming, Error Detecting and Error Correcting Codes, *Bell System Tech. J.*, vol. 29, 1950.
12. A. A. Alexander et al., Capabilities of the Telephone Network for Data Transmission, *Bell System Tech. J.*, vol. 29, no. 3, May, 1960.
13. C. M. Melas, A New Group of Codes for Correction of Dependent Errors in Data Transmission, *IBM J. Res. Develop.*, January, 1960.
14. J. E. Meggitt, Error Correcting Codes for Correcting Bursts of Errors, *IBM J. Res. Develop.*, July, 1960.
15. D. W. Hagelbarger, Recurrent Codes Easily Mechanized Burst Correcting Binary Codes, *Bell System Tech. J.*, July, 1959.
16. Leo Rosen, Characteristics of Digital Codes, *Control Eng.*, December, 1959.
17. Montgomery Phister, Jr., "Logical Design of Digital Computers," John Wiley & Sons, Inc., New York, 1958.
18. R. K. Richards, "Arithmetic Operations in Digital Computer," D. Van Nostrand Company, Inc., Princeton, N.J., 1955.
19. J. R. Ragazzini and G. Franklin, "Sampled-data Control Systems," McGraw-Hill Book Company, New York, 1958.
20. L. A. Zadeh and J. R. Ragazzini, The Analysis of Sampled-data Systems, *Trans. AIEE*, vol. 71, pt. II, pp. 225–239, 1952.
21. C. E. Shannon and W. Weaver, "The Mathematical Theory of Communications," University of Illinois Press, Urbana, Ill., 1949.
22. F. M. Reza, "Probability and Information Theory," class notes E.E. 370–371, Syracuse University.
23. Max T. Nigh, The Automatic Pipe Line, *Oil Gas J.*, July 8, 1957.
24. G. L. Maciula, Centralized Monitoring with Digital Telemetering, *Pipeline Eng.*, May, 1959.
25. Leon K. Kirchmayer, "Economic Control of Interconnected Systems," John Wiley & Company, Inc., New York, 1959.
26. Bernhardt G. A. Skrotzki, "Electric System Operation," McGraw-Hill Book Company, New York, 1959.
27. Leon K. Kirchmeyer, "Economic Operation of Power Systems," John Wiley & Company, Inc., New York, 1958.
28. J. E. Gaffney, Jr., "Digital Communications in Process Control," ISA Conference Paper 115-NY60, 1960.
29. Fazlollah M. Reza, "An Introduction to Information Theory," McGraw-Hill Book Company, New York, 1961.
30. International Symposium on Data Transmission (at the) Delft Institute of Technology, Delft, Netherlands, Sept. 19–21, 1960, *IRE Trans. Commun. Systems*, vol. C5-9, no. 1, March, 1961.
31. Edward Bedrosian, The Analytical Signal Representation of Modulated Waveforms, *Proc. IRE*, October, 1962.

Chapter 14

SPACE SYSTEMS AND TELEMETRY

STAFF, *TRW Space Systems, James B. Kendrick, Ed., Redondo Beach, Calif.* *(Space Systems)*

MORTON J. STOLLER, *Formerly NASA, Washington, D.C.* *(Satellite Telemetry)*

ELLIOT L. GRUENBERG, *International Business Machines Corporation, Yorktown Heights, N.Y.* *(Space Telemetry)*

INTRODUCTION

SPACE SYSTEMS

1	Introduction...	14-2
2	Glossary of Astronautical Terms...........................	14-3
3	Space Missions...	14-10
4	Astrodynamics..	14-15
5	Attitude Control of Spacecraft............................	14-33
6	Propulsion...	14-39
7	Example of Space-mission Planning........................	14-49
8	Astronomy..	14-52
9	Environmental Constraints................................	14-58
10	Communications Constraints...............................	14-69

SATELLITE TELEMETRY

11	Introduction...	14-73
12	Constraints on Satellite Telemetry and Data-recovery Systems	14-73
13	Past Telemetry and Data-recovery Systems.................	14-75
14	Data-handling Considerations in Satellite Experiments....	14-83
15	Future Satellite Telemetry Systems.......................	14-88
16	Orbiting Geophysical Observatory Specifications for Data-handling Devices...	14-91

SPACE TELEMETRY

17	Mission Constraints.......................................	14-95
18	Communications System Design............................	14-95
19	Modulation System...	14-97
20	The Synchronization Process..............................	14-98
21	System Characteristics...................................	14-101

INTRODUCTION

This chapter presents applications of telemetry during vehicle deployment in near-earth missions and in deep-space probes. Besides dealing with such vehicles during flight, we shall also consider telemetry's role in these missions during the rocket-launch phases of flight.

In supporting space missions with telemetry applications, the engineer must be familiar with a wide range of disciplines, including astrodynamics (spacecraft motion in the solar system), propulsion, stabilization and control, structural design, astronomy, power systems, and thermal control besides a requisite knowledge of communications and data management. The telemetry engineer must, of course, rely upon specialists in most of these disciplines for detailed knowledge, with the exception of the latter two, in which the telemetry engineer should himself be a specialist.

During space missions, the telemetry engineer must assure the safe and reliable acquisition of two types of data:

1. How the spacecraft's components are performing in executing the designed mission, sometimes referred to as housekeeping functions
2. Scientific or other observed data required by the mission objective or planned experiments

In this data-acquisition process, the telemetry engineer cannot be expected to replace the scientific specialist, but he must understand mission objectives so that he can estimate data rates needed to provide efficient communications and data-gathering equipment. The telemetry engineer contributes to the total process of system design not only his understanding of how communications can contribute to a space mission but also his knowledge of communications limitations.

In Chap. 10 we covered circuit-design requirements associated with missile telemetry. Much of this information is also applicable to space telemetry, but with the added requirement of designing components to operate reliably for operational periods of months and years. In all applications of telemetry aboard rocket-launched vehicles, moreover, components must be able to withstand the short, but severe, launch period of shock and vibration.

To provide the telemetry engineer the "feel" we believe necessary to him in designing telemetry systems for space missions, we shall precede the discussions of satellite telemetry and space telemetry with summary space-system fundamentals such as mission planning, astronautics, propulsion, and astronomy.

SPACE SYSTEMS

1 Introduction

The telemetry-system designer has an interest in many of the questions of concern to the space-system designer. These are
 1. The objectives of the space system.
 2. The mission profile of the space vehicle.
 3. The environment of the vehicle or what the space system will encounter at any point in its voyage.
 4. What energy (cost) is required to get to any point in the path of the mission, including changes in path. A subsidiary question is what types of rockets and propulsion are available to provide this capability.
 5. Where the space vehicle is with respect to bases or other vehicles with which it is required to communicate, and at what time and vehicle attitudes may it be able to communicate.

Often parametric information is used to trade off one possible system equipment parameter against others, so as to optimize system performance with respect to a criterion such as weight or cost. The telemetering engineer is frequently called upon

SPACE SYSTEMS **14–3**

to provide information for such trade-offs. In the immediately subsequent sections there has been assembled such system information as will be of general usefulness for space-system planning. It has been largely derived from *Space Data*, a publication of TRW Space Technology Laboratories, Redondo Beach, Calif. This information is not intended as complete space-system engineering information. The reader is referred to the references for further data.

To facilitate understanding of the space-system data in this chapter a glossary of astronautical terms is presented here. Terms more specifically associated with telemetry are presented in the glossary in Chap. 5.

2 Glossary of Astronautical Terms*

Ablation—The removal of surface material from a body by vaporization, melting, or other process, due to aerodynamic effects while moving at high speed through a planetary atmosphere.

Abort—To cancel or cut short flight after it has been launched.

Acquisition—The process of locating the orbit of a satellite or trajectory of a space probe by properly pointing an antenna or telescope, to allow gathering of tracking or telemetry data.

Actinic—Pertaining to electromagnetic radiation capable of initiating photochemical reactions, as in photography or the fading of pigments.

Active—Transmitting a signal, as "active satellite," in contrast to "passive."

Aerobiology—Study of the distribution of living organisms freely suspended in the atmosphere.

Aerodynamic Heating—Surface heating of a body caused by air friction and compression processes on passage of air or other gases over the body; significant chiefly at high speeds.

Aerodynamics—Science of motion through air and other gaseous media, and the forces acting on bodies moving through such fluids.

Aerodynamic Vehicle—A device, such as an airplane or glider, capable of flight only within a sensible atmosphere—aerodynamic forces serving to maintain flight.

Aeroelasticity—Study of the effect of aerodynamic forces on elastic bodies.

Aeroembolism—Formation or liberation of gases in the blood vessels of the body, brought on by a change from a relatively high atmospheric pressure to a lower one. The result of such gas liberation is characterized by neuralgic pains, cramps, and swelling—sometimes death. Also called "decompression sickness."

Aerolite—A meteorite composed principally of stony material of low density.

Aeronomy—Study of the upper regions of the atmosphere where physical and chemical reactions due to solar radiation take place.

Aerospace (from *aero*nautics and *space*)—Earth's atmospheric envelope and the space beyond, domain of operations for airborne vehicles, rocket vehicles, and spacecraft.

Agravic—Of or pertaining to a condition of weightlessness.

Air Shower—A grouping of cosmic-ray particles observed in the atmosphere.

Albedo—The albedo of a celestial body is the ratio of the total amount of sunlight reflected from the body in all directions, to the amount that falls upon the body.

Alpha Particle—A positively charged particle emitted from the nuclei of certain atoms during radioactive disintegration. The alpha particle has an atomic weight of 4 and a positive charge equal in magnitude to 2 electronic charges; hence it is essentially a helium nucleus.

Altitude—The altitude of a celestial body is its elevation angle above the horizon measured on the vertical circle passing through the body.

Ambient—Condition of the environment surrounding an aircraft or other body in motion, but undisturbed or unaffected by it, as in "ambient air," or "ambient temperature."

Angstrom—A unit of length, used chiefly in expressing short wavelengths. One meter equals 10^{10} angstroms.

Anomalisitic Year—The period between successive perihelion passages of earth, namely, 365.2596 mean solar days. It differs from the sidereal year (365.2422 days) because of earth's annual perihelion advance due to the perturbative effects of the other planets.

Antigravity—Resulting effect of a hypothetical field, capable of canceling the local gravitational field of the earth or other body.

Aphelion—The point which is farthest from the sun on the orbit of a celestial object orbiting the sun.

* From *Space Data* by Staff of TRW Space Technology Laboratories with permission.

Apofocus (apoapsis)—On the orbit of an object, the point which is farthest from the body orbited.

Apogee—On the earth orbit of an object, the point farthest from the earth.

Areo—Combining form of *Ares* (Mars) as in "areography" (geography of Mars).

Argument of Latitude—The angle measured in the orbit plane in the direction of motion from the ascending node to the object in orbit. It is numerically equal to the sum of the argument of perifocus and the true anomaly.

Argument of Perifocus (ω)—The angle measured in the orbit plane in the direction of motion from the ascending node to perifocus.

Artificial Gravity—A simulated gravity field established within a space vehicle, as by rotating the spacecraft or a portion thereof, the centrifugal force generated being similar to the force of gravity.

Ascending Node—The point at which an object's orbit crosses the reference plane (usually the ecliptic) from south to north.

Asteroid—One of the many small celestial bodies revolving around the sun, mostly between the orbits of Mars and Jupiter. Also called "planetoid," "minor planet."

Astro—A prefix pertaining to the stellar universe, sometimes used in reference to the space domian, as in *astro*nautics.

Astroballistics—Study of the phenomena arising from the motion of a solid through a gas at speeds high enough to cause ablation; for example, the interaction of a meteoroid with the atmosphere.

Astrodynamics—The practical application of celestial mechanics, astroballistics, propulsion theory, and allied fields to the problem of planning and directing the trajectories of space vehicles.

Astronaut—A person who occupies a space vehicle. Specifically, one of the test pilots selected to participate in a manned-space flight program.

Astronautics—The art, skill, or activity of operating space vehicles. In a broader sense, the science of space flight.

Astronomical Horizon—The circle of revolution normal to the local vertical, dividing the celestial sphere into two hemispheres.

Astronomical Latitude (ϕ_a)—The angle between the local vertical and the plane of the equator, measured north or south along the meridian passing through the point.

Astronomical Unit (AU)—Semimajor axis of earth's orbit about the sun.

Atmosphere—The envelope of air surrounding the earth; also the body of gases surrounding any planet or other celestial body.

Attitude—The position or orientation of an aircraft, spacecraft, etc., either in motion or at rest, as determined by the relationship between its axes and some other frame of reference such as the horizon.

Aurora—The sporadic visible emission from the upper atmosphere over middle and high latitudes. Also called "northern lights."

Autumnal Equinox (−)—The point and time (Sept. 23) of the sun's crossing the celestial equator southward (see Vernal Equinox).

Azimuth (A)—The azimuth of a celestial object is the angle of the horizon measured clockwise from the north point to the vertical circle passing through the object.

Ballistic Coefficient ($W/C_D A$)—A design parameter indicating the relative magnitude of inertial and aerodynamic effects, used in performance analysis of objects which move through the atmosphere.

Ballistic Trajectory—The trajectory followed by a body under the action of gravitational forces and the resistance of the medium through which it passes. (A rocket vehicle without lifting surfaces will describe a ballistic trajectory after its engines are shut off.)

Binary Notation—A system of positional notation in which the digits are coefficients of powers of the base 2, in the same way as the digits in the conventional decimal system are coefficients of powers of the base 10.

Blackout—A fadeout of radio communications due to environmental factors such as ionospheric disturbances or a plasma sheath surrounding a reentry vehicle.

Boiloff—The vaporization of a cold propellant, such as liquid oxygen or liquid hydrogen, as the temperature of the propellant mass rises in response to an external heat source, as in a rocket being readied for launch.

Booster—Short for "booster engine" or "booster rocket."

Braking Ellipses—A series of ellipses (decreasing in size because of aerodynamic drag in the perifocal region) which might be described by a spacecraft on entering a planetary atmosphere.

Bremsstrahlung—Electromagnetic radiation produced by the rapid change in velocity of an electron or another fast, charged particle as it approaches an atomic nucleus and is deflected by it.

Burnout Angle—The angle between the local vertical and the velocity vector at termination of thrust.

Burnout Velocity—Velocity attained by a launch vehicle at the termination of thrust.

Capsule—A small, sealed, pressurized cabin with an internal environment capable of supporting human or animal life during extremely high-altitude flight, space flight, or emergency escape. Also refers to a recoverable instrumentation package.

Cavitation—The turbulent formation of bubbles in a fluid, occurring whenever the static pressure at any point in the flow falls below the fluid vapor pressure.

Celestial Axis—The axis of the celestial sphere is the earth's axis of rotation, extending indefinitely.

Celestial Longitude (λ)—An angle measured on the ecliptic eastward from the vernal equinox to the great circle perpendicular to the ecliptic passing through the object.

Celestial Meridian—A great circle passing through the zenith and the north celestial pole. See Fig. 47.

Celestial Sphere—To an observer on earth, the sky seems to be a hemisphere with the earth at its center. This visible hemisphere, together with the part of the sky below the horizon, forms the celestial sphere. All objects are seen in projection against the celestial sphere.

Centrifuge—An apparatus rotating about a central axis, used to create an acceleration field at the periphery, similar to the prolonged accelerations encountered in high-performance aircraft, rockets, and spacecraft.

Circular Velocity—Velocity necessary for maintaining a circular orbit.

Cislunar—Region in which lunar gravitational effects become significant or dominant over those of earth.

Civil Day—A solar day beginning at midnight. The civil day may be based on either apparent solar time or mean solar time; it begins 12 hr earlier than the astronomical day of the same data.

Coherent—Continuous, as opposed to pulsed or discontinuous, as "a laser producing coherent light."

Command—A signal that initiates or triggers an action in the receiving device.

Communications Satellite (Comsat)—A satellite designed to reflect or relay radio or other communication waves.

Conjunction—Defined as the time when the celestial longitude of the sun and the planet are the same (though generally their celestial latitudes differ). An inferior planet may be at either "superior" or "inferior" conjunction, while a superior planet may be only at "superior" conjunction.

Coordinate Systems—The "topocentric," whose reference plane is the observer's local horizon; the "geocentric," which considers the earth as the center of reference; the "heliocentric," whose plane of reference is the ecliptic and has the sun as center of origin; and the "galactocentric," which uses the celestial position of the Milky Way galaxy as the reference.

Cospar—Abbreviation for "Committee on Space Research," international council of scientific unions for coordinating space data of international significance.

Cryogenics—The science of producing and maintaining very low temperatures; for example, from $-50°F$ to absolute zero.

Day—The time required for earth to complete one revolution with respect to a reference point (i.e., a mean solar day with reference to the mean sun or a sidereal day with reference to the vernal equinox). 1 sidereal day = 0.99726957 mean solar day; 1 mean solar day = 1.0027391 sidereal days. (See Sidereal Time.)

Declination—Declination of a celestial object is its angular distance from the celestial equator (see Fig. 47), measured north or south along the hour circle through the object.

Direct Motion—Real or apparent eastward motion. Opposite of retrograde motion.

Doppler Shift—The change in frequency with which energy reaches a receiver when the source of radiation (or a reflector of the radiation) and the receiver are in motion relative to each other. The Doppler shift is used in many tracking and navigation systems.

Eccentric Anomaly (E)—The eccentric anomaly of a point on an ellipse is the angle measured at the center of the ellipse from the major axis to the outward projection (normal to the major axis) of that point on the bounding circle (having a radius equal to the semimajor axis). The given point on the ellipse and the corresponding point on the circle have the same x coordinates.

Eccentricity (e)—Ratio of distance between foci (of ellipse) to length of major axis.

Ecliptic—The plane of the earth's orbit around the sun, inclined to the earth's equator by about $23°27'$.

Ecological System—A habitable environment; occurring naturally as on the surface of the earth, or created artificially as in a manned space vehicle; in which man, animals, or other organisms can live in mutual relationship with each other.

Electric Propulsion—The generation of thrust by acceleration of a propellant with some electrical device, such as an arc jet, ion engine, or magnetohydrodynamic accelerator.

Electromagnetic Radiation—Energy (propagated through space or through material media) in the form of an advancing disturbance in electrical and magnetic fields existing in space or in the media. Also called simply "radiation."

Ephemeris (pl. Ephemerides)—A tabular statement of the positions of objects in space at specified intervals of time. A standard yearly reference used by the United States is "The American Ephemeris and Nautical Almanac," issued in Great Britain as "The Astronomical Ephemeris."[6]

Epoch—An instant of time or a date selected as point of reference.

Equinox—A point of intersection of the celestial equator and the ecliptic. (See Vernal and Autumnal Equinox.)

Escape Velocity—Velocity required to escape from the gravitational attraction of a central body.

Exchange Ratio—Differential effect of one parameter on another, used for optimization of design characteristics.

Fixed Satellite—An earth satellite that orbits from west to east at such a speed as to remain constantly over a given place on the earth's equator.

Galactic Coordinates—The Milky Way galaxy is believed to have the appearance of a "flying saucer," with our solar system located approximately two-thirds of the way (10,000 parsecs) from the center to the rim. Because of its shape, the Milky Way appears to divide the sky quite neatly into two hemispheres, one centered upon the point $\alpha = 12^h40^m$, $\delta = +28°$, the other centered upon the opposite point at $\alpha = 0^h40^m$, $\delta = -28°$. These points are designated respectively as the north and south "galactic poles"—the axis between these poles passing through the center of the Milky Way. The great circle normal to the polar axis is called the "galactic plane" which bisects the Milky Way galaxy. Because of its broader encompassment it is often convenient to refer stellar positions to galactic coordinates. "Galactic latitude" is measured north and south from the galactic plane; "galactic longitude" is measured eastward from the point at $\alpha = 18^h40^m$, $\delta = 0°$, where the galactic plane crosses the celestial equator.

Geo—A prefix meaning "earth," as in "geology" and "geophysics."

Geocentric Latitude (ϕ')—Geocentric latitude of a given point on earth is the angle, as seen from the geometric center of the earth, between the given point and the equator, measured north or south along the meridian of the point.

Geodetic Latitude (ϕ)—Geodetic latitude of a given point on earth is the angle between the equatorial plane and the normal to the geoid at that point.

Giga—A prefix meaning multiplied by one billion, as in "gigacycles."

Gravitation—The acceleration produced by the mutual attraction of two masses, directed along the line joining their centers of mass, and of magnitude inversely proportional to the square of the distance between the two centers of mass, and the product of their masses.

Guidance—The process of directing the movements of an astronautical vehicle or spacecraft, with particular reference to the selection of a flight path or trajectory.

Heat Sink—A material capable of absorbing heat; a device utilizing such material for the thermal protection of a spacecraft or reentry vehicle.

Heliocentric—Orbiting about the sun as central body.

Hour Angle (H)—See Fig. 47.

Hour Circle—(See Fig. 47.) An imaginary great circle on the celestial sphere passing through both celestial poles, analogous to the meridian of longitude on the terrestrial sphere.

Human Engineering—The art or science of designing, building, or equipping mechanical devices or artificial environments suitable to the anthropometric, physiological, or psychological requirements of the men who will use them.

Hyperbolic Velocity—Greater than escape velocity.

Hypergolic—Self-igniting, with reference to combinations of chemical fuels and oxidizers, and their ability to ignite when brought together.

Impact Area—The area in which a rocket strikes the earth's surface.

Inclination (i)—The angle by which the orbital plane of an object in space is inclined to the plane of reference (usually the equator in geocentric work, or the ecliptic in heliocentric work).

Injection—The process of putting an artificial satellite into orbit. Also the time of such action.

Ion—An atom or molecular group of atoms having an electric charge. Sometimes also a free electron or other charged subatomic particle.

Ionosphere—The part of the earth's outer atmosphere where ions and electrons are present in quantities sufficient to affect the propagation of radio waves.

Kepler's Laws—An astronomical theory advanced by Johannes Kepler (1571–1630) on

the basis of extensive mathematical analysis of the copious records of observations made by Tycho Brahe, summarized in the following three laws:
1. The orbit of any planet is an ellipse with the sun at one focus.
2. The line joining sun and planet will sweep over equal areas in equal intervals of time.
3. The ratio of the squares of the periods of any two planets is equal to the ratio of the cubes of their mean distances from the sun.

Launch Complex—Entire area of launch-site facilities, including blockhouse, launch pad, gantry, etc.

Launch Vehicle—Any device that propels and guides a spacecraft into orbit about the earth or into a trajectory to another celestial body. Often called "booster."

Launch Window—An interval of time during which a rocket can be launched to accomplish a particular purpose; as "liftoff occurred 5 min after the beginning of the 82-min launch window." May also refer to launch angle, velocity, etc.

Libration—A real or apparent oscillatory motion like that of a balance before coming to rest; such as the apparent oscillation of the moon. Libration *points* refer to unique positions of gravitational balance, measured from a celestial body and its satellite, at which points smaller objects can orbit in formation with the other two bodies, Trojan group.

Light-year—The distance traveled by light in 1 year; about 63,000 AU.

Line of Nodes—The intersection of an orbit plane and a reference plane.

Longitude, Celestial—Celestial longitude of an object is the angle measured in the reference plane eastward from the vernal equinox to the hour circle passing through the object.

Longitude, Geographic—Geographic longitude of any given point is the angle (measured east or west) from the prime meridian (Greenwich) to the meridian through the given point.

Mach Number—(After Ernst Mach, Austrian scientist, 1838–1916.) A number expressing the ratio of the speed of a body with respect to the speed of sound in the medium surrounding the body; the speed represented by this number.

Magnetic Storm—A worldwide disturbance of the earth's magnetic field.

Magnitude—Magnitude of a star is a convenient index of brightness.

Mass Ratio—The ratio of weight before and after consumption of propellant.

Mean Anomaly (M)—The angle through which an orbiting body would move in a specified period of time if it moved at its mean angular rate.

Mean Distance of Planets—Semimajor axis of planetary orbits around the sun.

Mean Motion (n)—The average angular rate of an object in orbit.

Mega—A prefix meaning multiplied by one million as in "megacycles."

Meridian of Longitude (or simply meridian)—An imaginary great circle on the surface of earth, running from pole to pole.

Meteor—The light phenomenon which results from the entry into the earth's atmosphere of a solid particle from space; more generally, any physical object or phenomenon associated with such an event.

Meteoroid—Small particles moving in space.

Missile—Any object thrown, dropped, fired, launched, or otherwise projected with the purpose of striking a target. Short for "ballistic missile," "guided missile."

Module—A self-contained unit of a launch vehicle or spacecraft, serving as a building block for the overall structure. The module is usually designated by its primary function as "command module," "lunar landing module," etc. A one-package assembly of functionally associated electronic parts; usually a plug-in unit.

Nadir—The point on the celestial sphere directly opposite the zenith, i.e., the point directly under the observer's feet.

Newton's Laws of Motion—A set of three fundamental postulates forming the basis of the mechanics of rigid bodies, formulated by Sir Isaac Newton in 1687.
The first law is concerned with the principle of inertia, stating that if a body in motion is not acted upon by an external force, its momentum remains constant (law of conservation of momentum). The second law asserts that the rate of change of momentum of a body is proportional to the force acting upon the body and is in the direction of the applied force. A familiar statement of this law is the equation $F = ma$, where F is the vector sum of the applied forces, m the mass, and a the vector acceleration of the body. The third law is the principle of action and reaction, stating that for every force acting upon a body, there exists a corresponding force of the same magnitude exerted by the body in the opposite direction.

Nodes (ascending and descending)—Points of intersection of an orbit with the reference plane (usually ecliptic or equator). (See Ascending Node.)

Noise—Any undesired sound. By extension, noise is any unwanted disturbance within a useful frequency band, such as undesired electric waves in a transmission channel or device. When caused by natural electrical discharges in the atmosphere, noise may be called "static."

Nuclear Fuel—Fissionable material of reasonably long life, used or usable in producing energy in a nuclear reactor.

Nuclear Reactor—An apparatus in which nuclear fission may be sustained in a self-supporting chain reaction. Commonly called "reactor."

Nutation—A small perturbation in the motion of precession; a libratory motion of a spinning body like the nodding of a top.

Obliquity of the Ecliptic—The angular inclination (about 23°27′) of earth's equatorial plane to the ecliptic.

Orbital Elements—The six basic quantities which serve to describe any planet's orbit completely. They are

- a—semimajor axis of the orbit
- e—eccentricity of the orbit
- i—inclination of the orbit
- Ω—longitude of the ascending node
- ω—argument of perifocus
- T—time of perifocal passage

Parameter (P)—Distance from focus of conic to point on conic, measured perpendicular to the major axis.

Parsec—A unit of measurement for interstellar space; namely, the distance at which one astronomical unit subtends one second of arc; 3.26 light-years.

Payload Ratio—Ratio of launch weight to payload weight. Inverse of payload fraction. Payload is that portion of the launch weight over and above the weight of propellant and hardware necessary for the operation of the vehicle during its flight.

Perifocus (Periapsis)—An orbiting body's point of nearest approach to the surface of the central body.

Perigee—On the earth orbit of an object, the point nearest to the earth.

Perihelion—The point of nearest approach to the sun of a solar orbit

Perturbation—A disturbance in the regular motion of a celestial body, as the result of an additional force to those causing the regular motion.

Photon—According to the quantum theory of radiation, the elementary quantity or "quantum," of radiant energy.

Plasma—An electrically conductive gas comprised of neutral particles, ionized particles, and free electrons, but which, taken as a whole, is electrically neutral.

Precession—The reactive motion of a gyroscopic object in a direction normal to that of the disturbing input, and to the angular-momentum vector.

Precession of the Equinoxes—A slow, circular movement (i.e., $50^s.2$ per year) of the earth's axis of rotation about the poles of the ecliptic, due primarily to an action of the moon, secondarily to the sun. As a result, the plane of the earth's equator remains inclined 23°27′ to the plane of its orbit; but the points of intersection, the equinoxes, are continually shifting. (See Zodiac.)

Prime Meridian—The meridian which passes through Greenwich, England, adopted as a standard of east-west reference.

Prime Vertical—The vertical circle that is at right angles to the celestial meridian and therefore intersects the horizon at its east and west points.

Probe—Any device inserted in an environment for the purpose of obtaining information about the environment. Specifically, an instrumented vehicle moving through the upper atmosphere or space, or landing upon another celestial body in order to obtain information about the specific environment.

Propellant—Fuel and oxidizer used for rocket propulsion.

Proton—A positively charged subatomic particle having a charge equal to the negative charge of the electron, but of 1,837 times the mass; a constituent of all atomic nuclei.

Radiation, Space—Energy transmission from sources such as the sun and other cosmic bodies. Short for "electromagnetic radiation" "nuclear radiation."

Reaction Engine—An engine that develops thrust by the ejection of a substance from it, such as a jet or stream of gases created by the burning of fuel within the engine.

Recovery—The procedure or action that occurs when the whole of a satellite, or a section, instrumentation package, or other part of a rocket vehicle is recovered after a launch; the result of this procedure.

Reentry—The event occurring when a spacecraft or other object returns to the sensible atmosphere after being rocketed to altitudes above the sensible atmosphere; the action involved in this event.

Regenerative Cooling—The cooling of a part of an engine by the propellant being delivered to the combustion chamber; specifically, the cooling of a rocket-engine combustion chamber or nozzle by circulating the fuel or oxidizer, or both, around the part to be cooled.

Reliability—Fractional probability of accomplishing all the functions required for success of a given task or mission, within a specified time.
Rendezvous—The bringing together of two or more spacecraft in orbit at a preconceived time and place; orbital rendezvous.
Retrograde Motion—Real or apparent westward motion. Opposite of direct motion.
Retrorocket—A rocket fitted on or in a spacecraft, satellite, or the like to produce thrust opposed to forward motion.
Right Ascension (α)—Right ascension of a celestial object is its angular position measured eastward along the celestial equator from vernal equinox to the hour circle of the object. See Fig. 47.
Rocket Vehicle—One or more stages of propulsion, for acceleration of a payload by means of a thrust acting independently of the medium in which it operates.
Satellite—An attendant body that revolves about another (primary) body; especially in the solar system, a secondary body, or moon, that revolves about a planet. A manmade object revolving about a celestial body, such as a spacecraft orbiting about the earth.
Scale Height—A measure of the altitude variation between density and/or temperature in an atmosphere. For example, $\rho/\rho_0 = \exp(-\beta h)$, $1/\beta$ is the scale height, ρ/ρ_0 the density ratio, and h the altitude.
Selenocentric—Orbiting about the moon (or a moon) as central body.
Semilatus Rectum—(See Parameter.)
Sidereal Time—Time reckoned by reference to the stars. It is determined by using the vernal equinox as the index; thus the hour angle of the vernal equinox is equal to the local sidereal time. The sidereal day is shorter than the mean solar day by $3^m 56^s$. (See Day.)
Solar Constant—The rate at which solar radiation is received on a surface perpendicular to the incident radiation and at the earth's mean distance from the sun, but outside the earth's atmosphere. $G = 442$ Btu/hr/ft^2 or 1.94 cal/min/cm^2.
Solar Flare—A bright eruption from the sun's chromosphere, characterized by an increase in electromagnetic radiation.
Solstice—The point and time when the sun is at its northernmost (summer solstice, June 22) and southernmost (winter solstice, Dec. 22) point on the ecliptic.
Space—The part of the universe lying outside the limits of the earth's atmosphere. More generally, the infinite domain in which all celestial bodies move, including the earth.
Spacecraft—Devices, manned and unmanned, that are designed to be placed into an orbit about the earth or into a trajectory to another celestial body.
Space Power System—Spacecraft energy source.
Specific Impulse—Impulse content per unit weight of propellant.
Stage, Rocket—One element of rocket vehicle, consisting of a propulsion system, propellant tanks, propellant, and interconnections to adjacent stages.
Stoichiometric—Of a combustible mixture, having the exact proportions required for complete combustion.
Storable Propellant—Rocket propellant (usually liquid) capable of being stored for prolonged periods of time.
Structure Ratio—The ratio of structural weight to stage weight (less payload).
Superinsulation—Alternate layers of reflective material and insulation in near vacuum, assembled around an object to inhibit heat flux to or from the object.
Synchronous Satellite—A satellite orbiting the earth at period equal to, or multiples of, the earth's rotational period; i.e., making one, two, three, etc., orbits in a 24-hr period.
Synodic Month—The time between two successive conjunctions or oppositions of the moon. It averages $29^d.53059$.
Thrust—Propelling force developed by an aircraft engine or a rocket engine. In rocket engines, thrust is the product of propellant mass flow rate and exhaust velocity relative to the vehicle.
Tracking—The process of following the movement of a satellite or rocket by radar, radio, and/or photographic observations, generally for the purpose of recording its trajectory or for improving the reception of signals from the body.
Transfer Maneuver—Velocity change in orbit.
Transportation Cost—The unit cost (in dollars per pound of payload) of the launch-vehicle system necessary to elevate and accelerate a spacecraft to a desired objective, i.e., to earth orbit, lunar or interplanetary landing, etc.
True Anomaly (v)—An angle in the orbit plane measured from the perifocus to the object in the direction of motion of the object.
Upper Transit—The time or place when an object crosses the observer's local celestial meridian at maximum elevation.
Vernal Equinox (γ)—The point and time (Mar. 21) of the sun's crossing the celestial equator northward. It is derived from the Latin "equinoctium" meaning equal night, because at such times day and night are of equal length.

Vertical Circle—A great circle drawn from the zenith at right angles to the horizon and passing through a given celestial point.

X Ray—Electromagnetic radiation of very short wavelength, within the range between gamma rays and ultraviolet radiation. Also called "X radiation," "Roentgen ray." X rays penetrate various thicknesses of solids, and act upon photographic plates in the same manner as light. Secondary X rays are produced whenever X rays are absorbed by a substance; in the case of absorption by a gas, the result is ionization.

Zenith—The point on the celestial sphere toward which the local vertical is directed, i.e., the point directly overhead. See Fig. 47.

Zodiac—A belt of sky extending about 9° to each side of the ecliptic. Since ancient times, the zodiac has been sectioned at intervals of 30° along the ecliptic, each of these sections being designated by a "sign of the zodiac." Each sign bears the name of the constellation which occupied it in the second century B.C. Precession of the equinoxes has since shifted each constellation forward one sign; thus, while the sun is said to enter Aries at the vernal equinox, it is then in the constellation of Pisces. The annual revolution of earth causes the sun to appear to enter a different constellation of the zodiac each month. The 12 constellations are Aries (Ram), Gemini (Twins), Leo (Lion), Libra (Scales), Sagittarius (Archer), Aquarius (Water bearer), Taurus (Bull), Cancer (Crab), Virgo (Maiden), Scorpius (Scorpion), Capricornus (Sea goat), Pisces (Fish).

Zodiacal Light—A faint glow extending around the entire zodiac but showing most prominently in the neighborhood of the sun. This real but inconspicuous glow is explained as sunlight reflected from a great number of particles of meteoritic size believed to be present in or near the plane of the ecliptic in the planetoid belt.

3 Space Missions

Telemetry requirements arise out of the functional requirements generated by the mission objectives. These mission objectives are worked out by system-engineering specialists charged with the job of fulfilling functional needs. These overall functional needs might be (1) *scientific*, such as determining the meteoroid distribution in the space between earth and the moon or the determination of extent of radiation belts; (2) *planetary exploration*—establishing the nature of planetary and lunar bodies; (3) establishing better understanding of *resources of the earth* from orbital operations; (4) a continuing mission, such as *meteorology*, for the purpose of continuously obtaining data on the earth's atmosphere; (5) an example of a more continuing mission is a *communication satellite*, sometimes called an *applications* mission, for the continuous purpose of relaying information from and to different points on earth.

The above list provides examples of mission objectives. In most of these cases the telemetry system is not primary to the system mission. However, telemetry must support these missions in the two ways previously mentioned; that is, either to monitor the performance or to transmit collected data. This function is required regardless of whether the mission requires a man or is unmanned. In the unmanned case, however, the telemetry system in most cases represents almost the entire communications link.

This section provides basic information about space missions, including space missions already planned, aerospace transportation costs, trajectory considerations, and mapping coverage.

The latter is presented because mapping coverage is a fundamental consideration to exploratory-type missions which bulk heavily among space missions, especially those which rely heavily on telemetry.

Other factors influencing space-systems design and planning are astronautics and space environmental conditions. These are treated in later sections (4 through 13).

Table 1 summarizes manned-space programs according to regime and purposeful intention. Table 2 presents unmanned missions currently planned or executed.

Aerospace Transportation Costs

The cost of space missions is largely determined by the cost of launch vehicles required to accelerate and elevate the spacecraft to their mission velocities and altitudes. Since the boosters are presently nonrecoverable, the transportation costs per pound of payload to orbit with the current catalog of launch vehicles are rela-

Table 1. Manned-spacecraft Programs*

Near-earth Missions

X-15—Hypersonic flight in upper atmosphere with rocket-powered aircraft
Mercury—Fundamental manned-space-flight technology and operational capability (program terminated after 2 suborbital and 4 orbital flights)
Gemini (2-man spacecraft)—Demonstrate feasibility of long-duration (14-day) flight, rendezvous and docking procedure, postdocking maneuvers, controlled reentry and landing, selected DOD and scientific space experiments
   Flight 1 (unmanned orbital)—Structural integrity, launch guidance system, checkout launch complex 19, and exercise Gemini launch team
   Flight 2 (unmanned suborbital)—Flight safety, reentry heat protection, subsystems and recovery operations
   Flight 3 (manned, orbital)—2-man performance, controlled reentry, ground-systems support, and spacecraft readiness for longer missions
   Flight 4 (4-day, manned, orbital)—Effects of prolonged space flight on the astronauts, DOD experiments, preliminary extravehicular activity, including cabin depressurization and hatch opening
   Flight 5 (7-day, manned, orbital)—Long-duration capability, DOD and scientific experiments, extravehicular activity, and rendezvous evaluation exercises with the rendezvous pod
   Remaining flights will involve interrelated duration tests and rendezvous
Gemini Target (modified Atlas D-Agena)—Rendezvous and docking procedures

Lunar Missions

Apollo—Land men on the moon and return them safely to earth in this decade. Spacecraft includes command, service, and LEM modules. Lunar orbital rendezvous approach to be used, with limited surface exploration, and return to earth
Apollo Earth-orbit Tests (development flight tests in earth orbit)
   Unmanned Orbital Flights—Verify Saturn 1B launch vehicle, including guidance and control operation and structural integrity; spacecraft and adapter integrity, firing and restarting of the spacecraft engines, recovery of the spacecraft command module, and capability of guidance and control system to perform entry at earth orbital speeds
   Manned Orbital Flights (*Phase I*)—Equipment test reliability, rendezvous, and docking of command and service module with unmanned LEM. (*Phase II*)—To perfect rendezvous and docking operations
Apollo Circumlunar Tests—Test manned spacecraft in lunar trajectory
Apollo Landing Mission—Land 2 men on moon, explore immediate surface (24- to 48-hr period), and return men and surface samples to earth
LEM (Lunar Excursion Module—2 stages: soft lander and ascent engine)—Ferry astronauts from lunar orbit to surface and back
LLV (Lunar Logistics Vehicles)—Deliver support payloads for lunar exploration
ALSS (Apollo Logistic Support System)—Initial phases of advanced manned lunar exploration
LESA (Lunar Exploration System for Apollo)—Provide maximum support for the manned lunar exploration mission within capabilities of Apollo (provide shelter and 1,500 miles of surface mobility for 3 men during a 90-day stay)
Lunar Base—Extend lunar exploration capabilities (concepts include modular structures, preassembled vehicle shells, permanent installations, roving vehicles, and communications facilities)

Orbital Laboratories

MOL (Manned Orbiting Laboratory) (2-man spacecraft utilizing Gemini B)—Explore man's contribution to systems of potential military value in space, experiments in military hardware and sensors, economic feasibility of manned space laboratory
Extended Apollo—Provide a minimum orbital laboratory capability with 2 men for earth-orbit experimentation; AORL (Apollo Orbital Research Laboratory)—to provide laboratory capability for 2 to 6 men; EALM (Extended Apollo Laboratory Module)—3 to 12 men, up to 2 years' duration
MORL (Manned Orbital Research Laboratory) (2- to 6-man crew)—Provide laboratory facilities with orbit stay capability greater than 1 year
LORL (Large Orbital Research Laboratory) (20- to 30-man crew)—Advanced concepts, including "cartwheel" design, artificial-gravity methods, nuclear-power sources, orbital capability of 1 to 5 years

* From Ref. 12.

Table 2. Precursor Missions*

Planetary Exploration

Advanced Mariner B—Drop instrument capsules on Mars and Venus
Advanced Pioneer—Study solar phenomena and look at Mercury
Cometary Probe—Approach and measure nature of comets
Interstellar Probe—Obtain data on nearest stars
Jupiter Reconnaissance—Advanced concept of reconnaissance of Jupiter and possibly asteroid belt in favorable 1971–1973 period.
Mars Landing—Advanced concept of early manned Mars spacecraft for flyby and landing missions for 1970s and 1980s.
Venus/Mars Reconnaissance—Advanced concepts, based on study of earliest possibilities for manned planetary missions
Mariner C—Mars flyby and TV reconnaissance operations
Out-of-Ecliptic Probe—Space environment away from ecliptic plane
Outer Solar System Probe—Asteroid belt and Jupiter probe
Pioneer—Solar effect on space environment at 0.8 to 1.2 AU from sun
Ranger—Photograph lunar surface
Surveyor—Explore lunar surface; analyze samples
Voyager—Orbit and land on Venus and/or Mars

Space Sciences—Unmanned

Advanced Meteoroid Satellite—Test meteoroid hazards in cislunar space
Air Density Injun—Measure upper atmospheric density in polar regions
AOSO (Advanced OSO)—Monitor individual solar flares with improved telemetry, volume, and weight capability
Arents—Measure environmental conditions and radiation in synchronous orbit
Atmospheric Structures Monitor—Upper-atmosphere composition, density, pressure, and temperature
Biosatellites—Measure biological effects of long-term space flights
Energetic Particles Satellites—Measure radiation from high-altitude nuclear blasts
FIRE—Determine effects of high-speed reentry (Apollo)
Greb (SR Series)—Study solar and cosmic radiation
Helios—Study particle radiation as well as electromagnetic radiation
IMP (Interplanetary Monitoring Platform)—Study Apollo radiation hazards
Lofti—Test propagation of very low frequency radio signals in ionosphere
Meteoroid Detection Satellites—Determine micrometeoroid hazard
OAO—Map heavens in ultraviolet spectrum (orbiting telescope)
OGO—Study radiation belts, magnetic fields, and ionospheric phenomena
OSO—Study solar ultraviolet and X radiation
Polar Ionospheric Beacon—Determine factors that disrupt communications
Saturn Meteoroid Satellite—Support design of long-term Apollo spacecraft
Sounding Rockets—Gather data at altitudes between 40 and 160 miles
Topside Sounder—Measure electron distribution and density of ionosphere
Traac—Test gravity-gradient stabilization system for Transit satellite
TRS—Map earth radiation belt

* From Ref. 12.

tively high, as shown in Table 3. The economic advantage of recoverable boosters is so great (Fig. 1) that the eventual development of such devices is expected to occur, but the time schedule is uncertain. Possibly the aerospace airplane will be the first truly recoverable booster (see Sec. 6 and Fig. 42).

Table 3

Vehicle	Payload to orbit	Approx cost per launch
Saturn 5	6 men + RV† + 200,000 lb	$150 × 10^6$
Saturn 1B	6 men + RV† + 10,000 lb	25
Titan IIIC	Approx 28,000 lb	25
Titan II—Gemini	2 men + RV†	10
Titan II Delivery	No men + 6,000 lb	10
Titan II + Centaur	12,000 lb	20

†Reentry vehicle.

SPACE SYSTEMS

The cost per launch will drop with repetitive launches of the same vehicle as the reliability improves and as a function of learning (Fig. 2).

Cost-effectiveness-reliability Analysis

Figure of merit, or measure of desirability of a system, is $M = M(P,C)$ where P represents the overall effectiveness (or total performance with respect to one or more specified standards of quality) and C = total cost of the system. The effect of a small change to the reference system defined by P, C, M will then be determined as the sum of a performance effect and a cost effect.

$$\Delta M = \frac{C}{\alpha} \frac{\Delta P}{P} - \Delta C \quad \text{where } \alpha = \left(\frac{\Delta P/P}{\Delta C/C}\right)_{M \text{ const}} \tag{1}$$

The term α denotes the break-even ratio ($\Delta M = 0$) between a per cent change in performance and a per cent change in cost, which can generally be taken as unity.

Unit Effectiveness. The unit performance p represents a measure of "quantity" at the specified value of "quality" for any particular mission. For example, with performance specified and fixed, p = constant, $\Delta p/p = 0$. On the other hand, with velocity specified and fixed, payload W may represent the performance of interest; then $p = KW$, $\Delta p/p = \Delta W/W$. As a third example, a long-term operating system may involve the expected lifetime T and efficiency ϵ; and the performance is denoted as $p = K\epsilon T$; $\Delta p/p = \Delta \epsilon/\epsilon + \Delta T/T$.

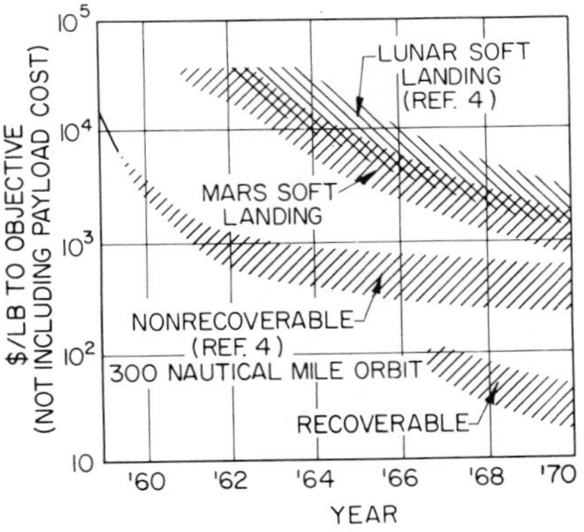

Multiple Missions. The incremental figure of merit ΔM, with appropriate break-even ratio α, provides the means to evaluate quickly the effect of changes in performance and cost on the overall cost effectiveness. For a multimission program, the terms (P, C, M) are evaluated for the ith mission, and the overall ΔM_T for a system change is given as the summation of effects like those of Eq. (1) for each individual mission.

FIG. 1. Cost per pound delivered to objective. (*From Space Data, courtesy of TRW Space Technology Laboratories, Redondo Beach, Calif.*)

$$\Delta M_T = \sum_i \Delta M_i \tag{2}$$

Cost Reliability. In order to accomplish ν successful missions with a system reliability R, it is necessary to schedule $N = \nu/R$ flights. Let p denote the unit performance associated with a single successful mission; then $P = \nu p = NRp$. The total cost C of N flights, where C_d denotes the nonrecurring cost and C_u the unit cost per flight, will be

$$C = C_d + NC_u = C_d + (\nu/R)C_u$$

then $\quad \Delta M = N[(C/\alpha N)(\Delta p/p) + \beta_R(\Delta R/R) - \Delta C_d/N - \Delta C_u] \tag{3}$

where $\quad \beta_R = (1/\alpha)(C_u + C_d/N) \quad$ for N held constant

$\quad\quad\quad \beta_R = C_u \quad\quad\quad\quad\quad\quad$ for ν held constant

In this manner, the proper weighting factors are established for including the effect of reliability (with performance and cost) in a routine analysis procedure involving N flights.

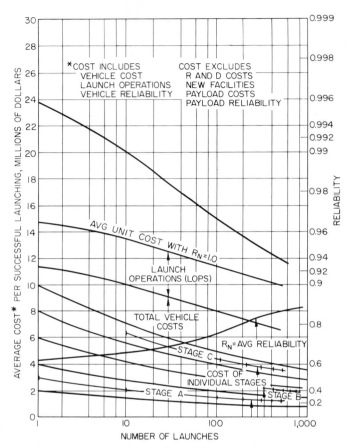

FIG. 2. Cost per launch as function of learning. The costs of individual stages A, B, and C are plotted on their learning curves. The total cost per successful launching is then (stage A + stage B + stage C + LOPS)/R_N. (*From Space Data, courtesy of TRW Space Technology Laboratories, Redondo Beach, Calif.*)

Effect of Reliability on Overall System Cost

A large percentage of the cost of space programs is frequently expended in overcoming the effects of unreliability of essential elements of the system. The launch-vehicle assembly, guidance and control system, payload components, human errors in design and operation—any of these may be responsible for the loss of a mission, making it necessary to refinance and reprogram the flight. It is therefore important that overall system reliability be properly evaluated in planning the design and development procedure, to minimize the loss of funds, facilities, and manpower due to unsatisfactory performance.

Figure 3 illustrates the effect of overall system reliability on a series of 100 space launches, nominally costing about $20 million each, including launch vehicles, payloads, and associated operating costs. The figure on the left shows the characteristic growth of system reliability with number of launches. Three specific examples (*A*,

B, and C) represent systems with inherently different complexity. During the course of the program, an improvement in reliability ΔR costs only $35 million for the reliable system A, increasing to $500 million for system C. Independently of the development cost, the slope of the overall program cost shown in the right-hand figure indicates that a 10 per cent improvement in reliability will result in a $250 million saving in the cost of launch vehicles and payloads for the program of 100 launchings. It is therefore desirable that a development program ensure that every element of the system which

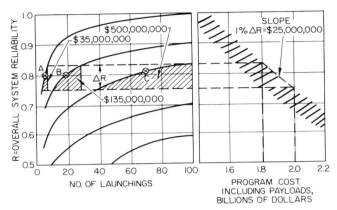

Fig. 3. System costs savings as function of reliability. (*From Space Data, courtesy of TRW Space Technology Laboratories, Redondo Beach, Calif.*)

could possibly have an effect on the reliability be perfected to a degree commensurate with the economics of the rest of the program (i.e., no weak links in the reliability chain for lack of development funds).

Orbital Mapping

The systematic mapping of the moon and planets is recognized as an essential precursor mission to obtain data on potential landing sites and to discover regions of special interest for more detailed scientific investigations. Data on mapping coverage are shown in Fig. 4. Resolution ρ (i.e., size of objects discernible) is equal to $(h/f)(1/l)$, where h = altitude, f = focal length, l = line pairs per unit length of film. For aerial-mapping film, $l \geq 50$ line pairs per millimeter. For photoscanning, it is customary to at least double the number of lines; i.e., $l \geq 100$ line pairs per millimeter. The number of bits per frame to be transmitted is then al^2, where a is the film area in each frame. For improved definition and contrast, it may be necessary to increase the number of bits by a factor of 5 (32 gray levels = 2^5 with analog data handling). For example, a lunar-mapping satellite at 100 NM, f = 6 in., l = 50 line pairs per millimeter, 70 × 70 mm frame will obtain a resolution of ρ = 80 ft and will require $(70 \times 100)^2 \times 5 = 250 \times 10^6$ bits/frame. The frame rate with 10 per cent overlap is about 48 sec/frame; so the bit rate for continuous transmission (no appreciable film storage) is about 5×10^6 bits/sec.

4 Astrodynamics

Astrodynamics includes all those considerations of space flight which require an understanding of celestial mechanics and propulsion for planning and directing the trajectories of space vehicles. As such it is the point of departure for system analysis of a projected space mission.

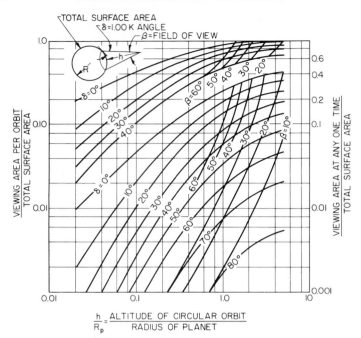

Fig. 4. Mapping coverage per view and per orbit as function of fractional altitude. (*From Space Data, courtesy of TRW Space Technology Laboratories, Redondo Beach, Calif.*)

Lunar and Interplanetary Trajectory Considerations

The initial planning of a lunar or interplanetary flight involves the selection of a favorable departure date, to minimize the distance and time in transit, and the formulation of a nominal trajectory by which the specified launch vehicle will accomplish the flight. Figures 5 and 6 show the timing and distances involved in inter-

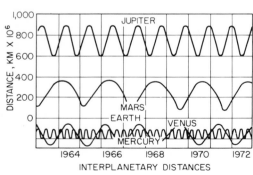

Fig. 5. Interplanetary distances. (*From Space Data, courtesy of TRW Space Technology Laboratories, Redondo Beach, Calif.*)

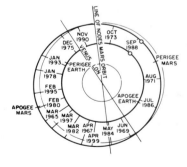

Fig. 6. Earth-Mars, oppositions. (*From Space Data, courtesy of TRW Space Technology Laboratories, Redondo Beach, Calif.*)

planetary flight. The nominal trajectory involves characteristics of the launching, midcourse, and terminal operations, to ensure the vehicle's passage through the "launch window," across space to the "approach window" and then to the target. The trajectory analysis is normally conducted by computer-machine solution of the N-body problem for the nominal trajectory and variations.

The launch site is a moving platform (fixed to moving, rotating earth) having a certain latitude, rate of rotation, and axis tilt to the ecliptic. Special machine runs are made to concentrate attention upon the effects of launch-vehicle propulsion characteristics, weight, atmospheric drag, etc., and their influence on the establishment of the launch window—permissible variations in burnout velocity, altitude, azimuth, elevation angle, and lift-off time.

The destination of the flight is a prescribed spot on a moving target whose orbital characteristics are specified. In approaching the target, the necessary velocity, grazing angle, and altitude need to be attained to permit a spot landing. The effects of gravity, atmosphere, entry technique, axis tilt, and rotation are considered in machine calculations of the approach phase, in order that the timing and direction of deceleration thrust can be applied correctly. The approach window is thus established.

The midcourse trajectory is determined by the burnout velocity vector and by the various sources of gravitational attraction such as earth, moon, and sun. Errors accrued during the launching phase are observed and the necessary corrections applied during the midcourse phase to attain the desired approach window. Beyond a certain point in the trajectory (the null point) the gravitational attraction of the objective will exceed that of earth.

The following sections present the fundamental laws governing the orbits and trajectory of bodies in space, modes of changing orbit in space, and rendezvous maneuvers. A series of orbital mechanical charts, derived from the principles of conservation of energy and of momentum (see Fig. 11), are presented which can be used for determining many orbit characteristics of a small body operating in the vicinity of a central body. Such characteristics as burnout velocity and angle to reach specific altitude, and injection velocity to circularize orbit vs. altitude, can be determined. Charts are provided for earth, moon, Mars, Venus, and the sun as central body.

Further subsections include data useful for near-earth astronautics (such as estimating near-earth orbital lifetimes and suborbital ranges and maximum altitudes) and planetary encounter and entry, including earth reentry. Other astronautical topics are attitude control and propulsion, which are treated in Secs. 5 and 6.

Orbit Equations

The differential equations of motion of the spacecraft in orbit are as follows:

$$\ddot{r} - r\dot{\theta}^2 = -\mu/r^2 \qquad (4)$$
$$r\ddot{\theta} + 2\dot{r}\dot{\theta} = 0 \qquad (5)$$

μ is the gravitational parameter of central body at focus. Other variables are defined in Fig. 7. Solution of these equations simultaneously gives $1/r = (\mu/C^2) + D \cos \theta$, where $C = r_p V_p$ and $D = (1/r_p) - \mu/(r_p V_p)^2$. Maximum (burnout) velocity V_p occurs at perifocus. The eccentricity $\epsilon = DC^2/\mu = (r_p V^2_p/\mu) - 1$ then determines whether the orbit is elliptic, parabolic, or hyperbolic (see Fig. 7).

Earth Orbital Motion and Maneuvers

The period of elliptical or circular orbits around the *earth* follows Kepler's third law (see Glossary) and is expressed in hours:

$$T = 1.41 \ (a/R)^{3/2}$$

where a, as defined in Fig. 7, is one-half the major axis of the elliptical orbit and R is the average radius of the earth. For circular orbits $a = r$ and

$$r = R + h, \ R = 3{,}441.66 \ \text{NM} = 20.92 \times 10^6 \ \text{ft}$$

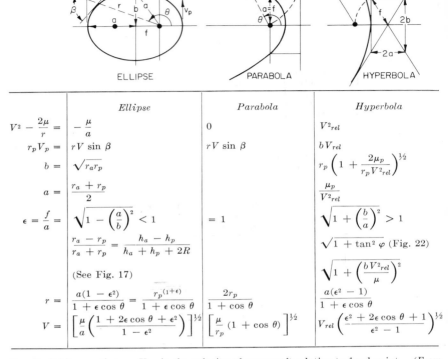

FIG. 7. Orbit equations. V_{rel} is the velocity of spacecraft relative to focal point. (*From Space Data, courtesy of TRW Space Technology Laboratories, Redondo Beach, Calif.*)

where h is the altitude of the orbit. The circular orbital velocity V_c varies inversely as the square root of r, so that the velocity V_c, at altitude h, is related to V_0, the circular orbital velocity at an altitude of 100 NM by

$$V_c = V_0[(R + 100)/r]^{1/2}$$

Coplanar transfers to higher orbits require an increase in velocity V_1 over initial circular velocity V_0. Such transfers are between orbits in the same plane. The transfer is effected by an elliptical orbit (Hohmann transfer ellipse). At the apogee of the ellipse the spacecraft has velocity V_a, which is insufficient to maintain a circular orbit at altitude h. The velocity increment V_2 is required to inject the spacecraft into circular orbit at this altitude, with final circular velocity V_c. Figure 8 shows a typical sequence for coplanar transfer and the velocity increments required.

Note the following relations:

$V_0 + V_1 =$ total velocity required to reach altitude h from $h = 100$ NM.
$V_a + V_2 = V_c =$ total velocity required for injection into circular orbit at altitude h.
$V_1 + V_2 =$ total velocity *increments* required to change from a 100-NM orbit to a coplanar one at altitude h.

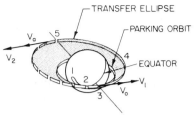

Fig. 8. Typical sequence of events for coplanar transfer. 1. Launch vehicle into desired orbit plane. 2. Inject into 100-NM parking orbit at velocity V_0. 3. Add transfer velocity V_1 when crossing latitude of desired perigee. 4. Coast to altitude h along transfer ellipse. 5. Reach apogee with velocity V_a; add V_2, sufficient to circularize the orbit at altitude h. (*From Space Data, courtesy of TRW Space Technology Laboratories, Redondo Beach, Calif.*)

Table 4 presents velocity values for earth orbital motion and coplanar transfers from a 100-NM orbit for orbits of specific altitude or period.

Table 4. Earth Orbital Characteristics
Coplanar Transfer from Circular Earth Orbit at 100-NM Altitude to Circular Orbit at Altitude h

Period, days	Altitude h, NM	Circular orbit velocity, ft/sec	V_1, ft/sec	V_2, ft/sec	$V_1 + V_2$, ft/sec
0.06124984	100	25,567 = V_0	0	0	0
0.06386214	200	25,213	177	176	353
0.06651065	300	24,874	349	344	693
$\frac{1}{14}$	482	24,290	646	630	1,276
$\frac{1}{12}$	906	23,073	1,275	1,211	2,486
$\frac{1}{10}$	1,469	21,713	1,993	1,836	3,829
$\frac{1}{8}$	2,256	20,156	2,707	2,583	5,290
$\frac{1}{6}$	3,461	18,313	3,827	3,232	7,059
$\frac{1}{4}$	5,603	15,998	4,994	3,997	8,991
$\frac{1}{2}$	10,917	12,697	6,817	4,710	11,523
1	19,351	10,078	8,071	4,851	12,922
2	32,740	7,999	8,941	4,621	13,562
10	101,684	4,644	9,996	3,494	13,490
28 (moon)	208,000	3,310	10,226	2,710	12,936
365.25	1,161,210	1,409	10,535	1,300	11,835

Figure 9 provides a chart showing similar data to Table 4, but it permits determining V_1, V_2, and V_c for intermediate altitudes and periods. In addition, the chart may be used for determining the period T for a circular orbit of given altitude. Curves are also provided for total velocity increment $V_1 + V_2$ and total velocity $V_0 + V_1 + V_2$ required to achieve a circular orbit at altitude h through Hohmann transfers.

The time required for coplanar transfer T, which is the half period of the transfer ellipse from the 100-NM parking orbit, may be read for any given final altitude, h, by using the time scale at the top of the chart. Rendezvous operations may be figured with the aid of the small insert diagram. If the target is in circular orbit of altitude h, a chaser in parking orbit may rendezvous with it at apogee if it is launched into transfer orbit when the target is θ_t degrees before apogee, where $\theta_t = 2\pi \frac{\tau}{T}$.

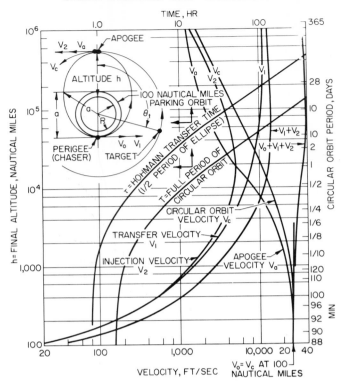

Fig. 9. Earth orbital maneuvers—coplanar.[10] Velocity increments for coplanar transfer from circular earth orbit at 100-NM altitude to circular orbit at altitude h (see Table 4). (*From Space Data, courtesy of TRW Space Technology Laboratories, Redondo Beach, Calif.*)

Additional energy (velocity increment) is required if transfer is to be effected to an orbit in a different plane. As shown in Fig. 10, orbital plane change can be done in several ways:
1. Low altitude: Add transfer vector V_{r1} instead of coplanar transfer vector V_1.
2. Apogee altitude: Add injection velocity V_{r2} instead of coplanar transfer vector V_1.
3. Both altitudes: Add smaller lateral components.

The chart in Fig. 10 shows as solid lines the *total* velocity increment V_{r1} required to change orbits from 100-NM altitude to altitude h *and* to change the plane of the orbit by δ_1 degrees. The dashed curves show the *total* velocity increment V_{r2} needed to change orbital plan by δ_2 degrees at altitude h *and* to inject into a circular orbit at that altitude. The corresponding period of the circular orbit is also given. The velocity $V_{r1} + V_{r2}$ for combined maneuvers is minimized by making initial dogleg δ_1 in the range $0 < \delta_1 < 5°$ for all values of $\delta_1 + \delta_2$.

Orbital Mechanics

An orbiting spacecraft moves in accordance with the principles of the *conservation of energy* and the *conservation of angular momentum*. Charts showing the resulting behavior of the orbit parameters can be used for the analysis of all unperturbed orbits, as well as for cases involving energy changes.

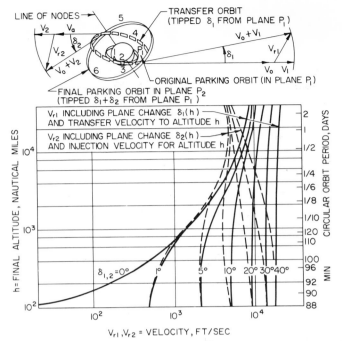

Fig. 10. Noncoplanar transfer and rendezvous maneuvers[10] (from 100-NM circular orbit). *(From Space Data, courtesy of TRW Space Technology Laboratories, Redondo Beach, Calif.)*

Constant-energy curves (the sum of kinetic and potential energy) of an object with respect to a central body of much greater mass are drawn to show various levels of energy, according to the following equation:

$$E = \tfrac{1}{2}mV^2 + m[(-\mu)/r] = \text{const} \qquad E^* = C_3 = 2E/m = V^2 - 2\mu/r \qquad (6)$$

where E = total energy of object
m = mass of object in orbit
V = speed of object in orbit
r = radius from focus to object
R = radius of spherical central body
V_0 = speed at $h = 0$
μ = gravitational parameter of central body = $g_0 R^2$; g_0 is the gravitational acceleration at $h = 0$ (g_0 is negative since it acts in opposite direction to $+r$)

Rearranging Eq. (6) and letting $E^* = 2E/m$ gives

$$\frac{r}{R} = 1 + \frac{h}{R} = \frac{2(-\mu)/R}{E^* - V^2} = \frac{-V_{e0}^2}{-V_{e0}^2 + V_0^2 - V^2} = \frac{1}{1 - (V_0/V_{e0})^2 + (V/V_{e0})^2} \qquad (7)$$

where h is the altitude of the object above the reference surface of radius R and $V_{e0} = (2\mu/R)^{1/2}$ is the surface escape velocity. Moving upward along the constant-energy curve from the initial conditions of altitude and velocity (burnout is assumed to occur at $h = 0$, velocity V_0, zero drag) the values at other altitudes can be determined by use of Eq. (7), as shown in Fig. 11.

Constant-angular-momentum lines are represented by plotting the horizontal component of velocity $V \sin \beta$, as it varies with radius. β is the angle measured from the local vertical to the orbit velocity vector V.

$$\text{Angular momentum per unit mass} = RV_0 \sin \beta_0 = rV \sin \beta = \text{const} \quad (8)$$
$$\text{Horizontal component of velocity} = V \sin \beta = (V_0 \sin \beta_0)(R/r)$$

The constant-angular-momentum lines, when plotted on the log-log scale as shown, are parallel lines of unit slope, intercepting the base line at $V_0 \sin \beta_0$. The values V_0 and β_0 represent the resultant velocity and angle when $r = R$ ($h = 0$). Moving upward along the constant-momentum line, the horizontal component of orbit velocity can be readily determined as a function of altitude by means of Eq. (8).

Apogee. It is noted that for given burnout conditions of V_0 and β_0, the maximum altitude or apogee is the intersection of the energy curve with that of angular momentum, i.e., the point at which the resultant velocity and the horizontal component thereof are equal. Such points are indicated by the short tick marks across the curves, as shown by the locus of apogees for $\beta_0 = 30°$ in Fig. 11. From the divergence of the energy and momentum curves for values of $V_0 > V_{e0}$, it is evident that no apogee exists for launch conditions in this (hyperbolic) region.

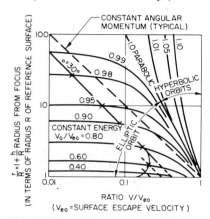

FIG. 11. Generalized chart of orbit characteristics relating radius from orbit focus—*altitude* (as fraction of radius of central body) to *velocity* as fraction of surface escape velocity—$(1 + h/R$ vs. $V/V_{e0})$—indicating paths of constant energy and constant momentum. (From *Space Data*, courtesy of *TRW Space Technology Laboratories, Redondo Beach, Calif.*)

The following additional factors should be noted on the energy-momentum chart to aid in its interpretation:

1. The line of escape velocity (parabolic) as a function of altitude, obtained from Eq. (7) when $V_0 = V_{e0}$, marks the boundary between elliptic orbits (to the left) and hyperbolic orbits (to the right); $V_e = \sqrt{2\mu/r}$.

2. The line of circular velocity as a function of altitude is drawn parallel to the escape velocity; $V_{c0} = V_{e0}/\sqrt{2}$. (See Fig. 12.) $V_{c0} = V_c$ in Figs. 9 and 10.

3. Curves are plotted to show the variation in apogee altitude as a function of burnout angle $\beta_0 = 15°$, $30°$, $45°$, and $90°$. Note that apogee points for $\beta_0 = 0$ are at the far left, off the chart; while for $\beta_0 = 90°$, apogee points occur only for $V_{c0} < V_0 < V_{e0}$. (See Fig. 12.)

Some of the orbit characteristics which can be found from the energy-momentum charts are as follows:

1. Burnout velocity and angle to reach specified altitude
2. Velocity and angle vs. altitude in coasting orbits
3. Transfer velocity required to reach higher apogee
4. Injection velocity to circularize orbit vs. altitude
5. Midcourse corrections to maintain nominal trajectory
6. Exchange ratios between various orbit parameters

For each of the central bodies of the solar system, orbit-characteristics curves of this type can be constructed as described above. The corresponding charts for earth, moon, Mars, Venus, and sun are presented in Figs. 12 through 16.

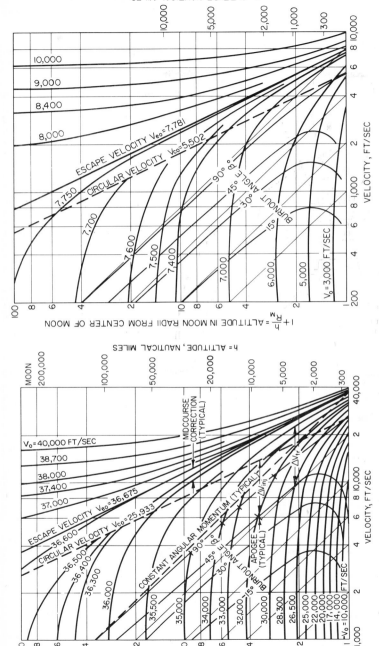

Fig. 13. *Lunar orbits*—altitude vs. velocity and launch angle (Fig. 11 applied to moon). Period of lunar orbit = $1.81(a/R_M)^{3/2}$ hr, where a = semimajor axis of orbit, R_M = radius of moon. (*From Space Data, courtesy of TRW Space Technology Laboratories, Redondo Beach, Calif.*)

Fig. 12. *Earth orbits*—altitude vs. velocity and launch angle (Fig. 11 applied to earth). See Fig. 9 for other earth-orbital data. (*From Space Data, courtesy of TRW Space Technology Laboratories, Redondo Beach, Calif.*)

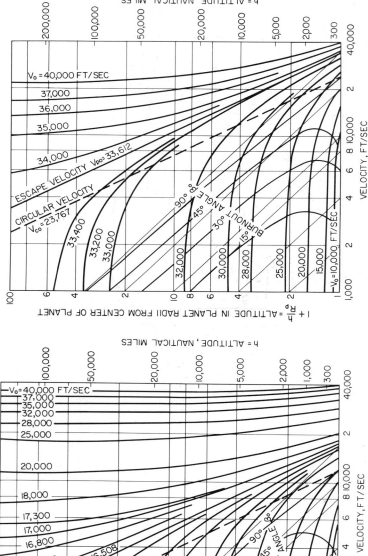

Fig. 15. Venus orbits—altitude vs. velocity and launch angle (Fig. 11 for Venus). Period of Venus orbits = $1.48(a/R_v)^{3/2}$ hr, where a = semimajor axis of orbit, R_v = radius of Venus. (*From Space Data, courtesy of TRW Space Technology Laboratories, Redondo Beach, Calif.*)

Fig. 14. Mars orbits—altitude vs. velocity and launch angle (Fig. 11 for Mars). Period of Mars orbits = $1.68(a/R_m)^{3/2}$ hr, where a = semimajor axis of orbit, R_m = radius of Mars. (*From Space Data, courtesy of TRW Space Technology Laboratories, Redondo Beach, Calif.*)

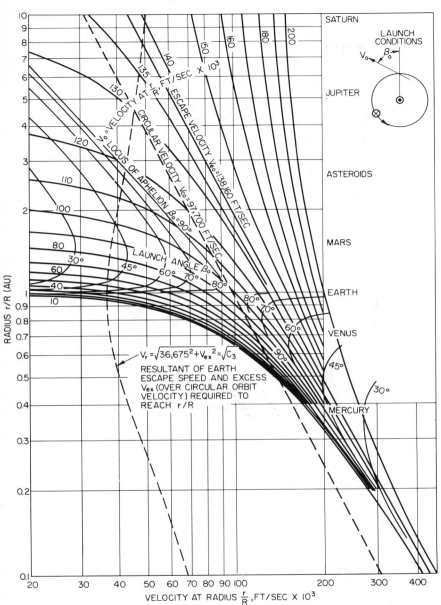

Fig. 16. Solar orbit characteristics—orbital radius in fractional AU vs. velocity. Period of solar orbits (circular or elliptical) = $a^{3/2}$ (years), where a = semimajor axis of orbit, AU. (*From Space Data, courtesy of TRW Space Technology Laboratories, Redondo Beach, Calif.*)

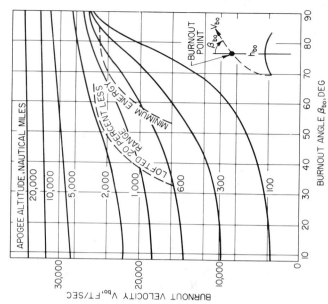

Fig. 18. Apogee altitude vs. burnout velocity and angle. $h_{bo} = 400{,}000$ ft, zero drag. (*From Space Data, courtesy of TRW Space Technology Laboratories, Redondo Beach, Calif.*)

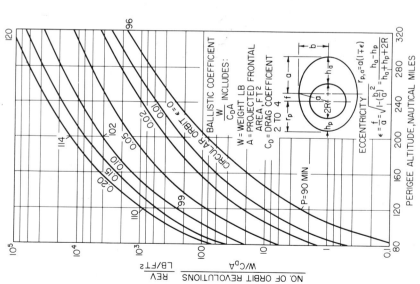

Fig. 17. Earth-orbital lifetime as function of ballistic coefficient and orbital parameters. (*From Space Data, courtesy of TRW Space Technology Laboratories, Redondo Beach, Calif.*)

Near-earth (and Planet) Astronautics

Near-earth orbits are affected by atmospheric drag, which limits orbital lifetime. The ballistic coefficient is the major determinant. Lifetimes may be estimated from Fig. 17.

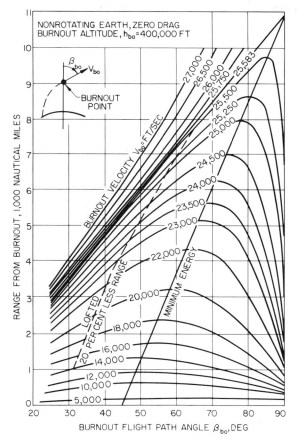

Fig. 19. Impact range from burnout as function of burnout angle and velocity. The chart above shows the ballistic range for the conditions of zero drag and nonrotating earth. It is very useful in preliminary range safety analysis as an indication of the effect of burnout velocity and flight-path angle on the impact range. (*From Space Data, courtesy of TRW Space Technology Laboratories, Redondo Beach, Calif.*)

Suborbital trajectories are used for missile and space-vehicle testing and for sounding rockets used for upper-atmospheric research. Figures 18 and 19 may be used to estimate maximum altitude and range from burnout characteristics.

For reasons of safety, firing angles at missile ranges are restricted. Figures 20 and 21 show permissible firing sectors at the Atlantic and Pacific Missile Ranges.

Planetary Encounter and Reentry Dynamics. During the approach of the spacecraft to a planet or moon, but before the gravitational effect (varies as μ/r) becomes appreciable, observations will be made to determine the range and range rate

(V_{rel}) of the planet. The approach angle of the spacecraft will then be adjusted to give the desired offset value b at encounter.

   a. Trajectory Modulation during Flyby. In the absence of any energy dissipation, the hyperbolic trajectory of the spacecraft with respect to the planet follows the principles of conservation of energy and angular momentum (Fig. 22a).

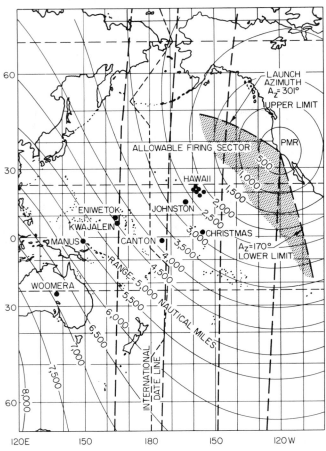

Fig. 20. Pacific missile range sector. Dash lines show typical ground tracks for southward launch from PMR. (*From Space Data, courtesy of TRW Space Technology Laboratories, Redondo Beach, Calif.*)

Constant energy:
$$V^2{}_{rel} = V_p{}^2 - 2\mu/r_p$$

Constant angular momentum:
$$r_p V_p = b V_{rel}$$

The approach offset b is then as follows:
$$b = r_p \sqrt{1 + 2\mu/r_p V^2{}_{rel}}$$

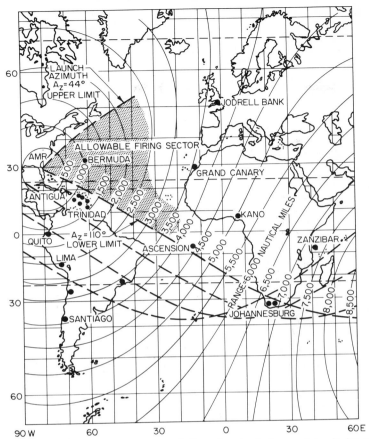

Fig. 21. Atlantic missile range sector. Dash lines show typical ground tracks for eastward launch from AMR. (*From Space Data, courtesy of TRW Space Technology Laboratories, Redondo Beach, Calif.*)

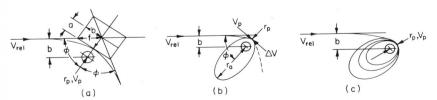

Fig. 22. Planetary encounter of spacecraft. (a) Trajectory modulation during flyby. (b) Transfer from hyperbolic to elliptic orbit by propulsive acceleration. (c) Capture by atmospheric deceleration during flyby. (*From Space Data, courtesy of TRW Space Technology Laboratories, Redondo Beach, Calif.*)

The eccentricity $e = f/a = \sec \varphi = \sqrt{1 + \tan^2 \varphi}$.

It is then determined that
$$e = \sqrt{1 + (bV^2_{rel}/\mu)^2}$$
$$\tan \varphi = b/a = bV^2_{rel}/\mu$$

b. Transfer from Hyperbolic to Elliptic Orbit by Propulsive Deceleration. Capture of the spacecraft by the gravitational field of the planet is accomplished by applying ΔV at perifocus (for maximum energy dissipation) (Fig. 22b). The value of ΔV must be sufficient to reduce the spacecraft velocity below escape velocity $V_e = \sqrt{2\mu/r_p}$; i.e., $\Delta V > V_p - V_e$. Substituting for V_p and V_{rel} from the energy and momentum equations above, the required condition for capture is expressed as follows:

$$\frac{\Delta V}{V_e} > \sqrt{\frac{r_p V^2_{rel}}{2\mu} + 1} - 1 = \frac{1}{\sqrt{2}}\left[1 + \sqrt{1 + \left(\frac{bV^2_{rel}}{\mu}\right)^2}\right]^{\frac{1}{2}} - 1 = \left[1 - \left(\frac{r_p}{b}\right)^2\right]^{-\frac{1}{2}} - 1$$

See energy charts (Figs. 11 to 15) for proper value of ΔV to produce desired eccentricity ϵ of elliptic orbit, where $e = (r_a - r_p)/(r_a + r_p)$.

c. Capture by Atmospheric Deceleration during Flyby. Another mode of capture is to adjust the approach offset b so the first pass of the spacecraft through the planetary atmosphere will produce sufficient aerodynamic drag to decrease the velocity by an amount similar to the case above (Fig. 22c). The reentry charts (Figs. 23, 24) may be used to determine the proper approach conditions for capture during the first pass, followed by a series of braking ellipses.

Entry into Planetary Atmosphere[13]

Gravitational attraction causes the velocity of a space vehicle to increase as it approaches a planet or other celestial body. In the absence of an atmosphere (i.e., with zero drag) an object falling from rest at an infinite distance will accelerate to an impact velocity equal to the surface escape velocity

$$V_{e0} = (2\mu/R)^{\frac{1}{2}}$$

When the atmospheric density ρ becomes appreciable, the aerodynamic drag force will cause the object to decelerate at a rate of

$$\frac{\rho g_0 V^2/2}{W/C_D A}$$

until the velocity reaches its terminal value (where drag equals weight), at which time the deceleration becomes very small. (See Fig. 23.) β = scale factor for altitude density variation (see Fig. 24).

Nonlifting Entry. For nonlifting entry the velocity during the high-speed portion of the flight can be given approximately by

$$V/V_E = \exp(g_c/g_p)(p/2\Delta \sin \gamma_E)$$
$$= \exp(g_c\rho/2\beta\Delta \sin \gamma_E)$$

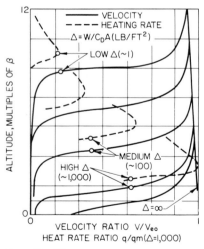

Fig. 23. Planetary entry parameters. Heat-rate ratio $q/qm\ (=1,000)$. (*From Space Data, courtesy of TRW Space Technology Laboratories, Redondo Beach, Calif.*)

where V = velocity, ft/sec; p = atmospheric pressure, lb/ft²; ρ = atmospheric density, slug/ft³; $\Delta = W/C_D A$, ballistic coefficient, lb/ft²; γ = flight-path angle measured positive up from the local horizontal; g = acceleration of gravity, ft/sec²; g_c = conversion constant, ft/sec².

Subscripts p and E refer to planet and entry, respectively. This formula is relatively good for entry angles steeper than about $\gamma_E = -10°$. The maximum deceleration for the same trajectory is

$$G_{max} = -\beta V_E^2 \sin \gamma_E / 2eg_c \quad \text{(in } g\text{'s)}$$

A nomograph giving the solution to this formula for entry to the atmospheres of earth, Mars, and Venus is given in Fig. 24.

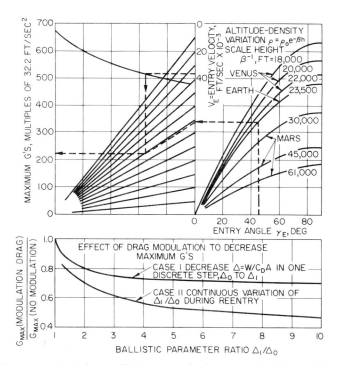

Fig. 24. Maximum g's during nonlifting atmospheric entry. Initial conditions of entry velocity V_E and angle γ_E are taken at an altitude h of approximately $13/\beta$, denoting the outer edge of the sensible atmosphere, for example, 13 (23,500) $\doteq$ 300,000 ft for earth entry. (*From Space Data, courtesy of TRW Space Technology Laboratories, Redondo Beach, Calif.*)

The convective heating rate at the stagnation point on the spherical segment at the nose of a vehicle during entry can be given by the expression

$$q = 17(\rho/R)^{1/2}(V/1{,}000)^3 \quad \text{Btu/ft}^2/\text{sec}$$

where ρ is the atmospheric density in slugs/ft^3, V the velocity in ft/sec; and R the nose radius in feet. The maximum heating rate at the stagnation point is given by

$$q_{max} = 1.05[\beta\Delta \sin(-\gamma_E/R)]^{1/2}(V_E/1{,}000)^3 \quad \text{Btu/ft}^2/\text{sec}$$

See Fig. 23 for plot of q/q_{max}. The total heating for the stagnation point down to impact for this trajectory is given approximately by

$$Q = 0.0053[\Delta/R\beta \sin(-\gamma_E)]^{1/2}(V_E/1{,}000)^2 \quad \text{Btu/ft}^2$$

These heating formulas are for air and should be increased about 10 per cent for an atmosphere which is primarily carbon dioxide.

Lifting Entry. Lifting entry permits the spacecraft to descend more gradually through the atmosphere, with correspondingly lower resultant deceleration, as shown in Fig. 25, for lifting entry to earth at the velocities noted. In addition to the lower

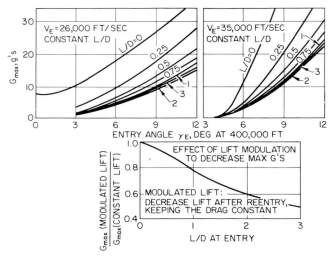

Fig. 25. Maximum g's during lifting entry. (*From Space Data, courtesy of TRW Space Technology Laboratories, Redondo Beach, Calif.*)

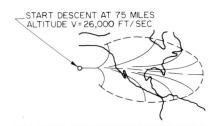

Fig. 26. Reentry "footprint." (*From Space Data, courtesy of TRW Space Technology Laboratories, Redondo Beach, Calif.*)

deceleration, lifting entry permits the spacecraft to maneuver laterally to a desired landing site, a feature not attainable by the zero-lift entry vehicle. The "footprint" of the accessible landing area is shown in Fig. 26.

Deboosting from Earth Orbits

In order to find the atmospheric-entry conditions for deboosting from a circular earth orbit, Fig. 27 can be used. The firing angle for the deboost rockets which will give minimum range and minimum range error due to error in the firing angle is given in curve a as a function of ΔV and orbital altitude. The range error due to an error in ΔV is given in curve b where the range is to impact (for vacuum conditions).

Using the firing angle given, the entry angle at an altitude of 400,000 ft is then given as a function of orbital altitude and deboost velocity.

5 Attitude Control of Spacecraft[11]

Thrust Vector Control. During periods of thrust application, orientation of the spacecraft is generally maintained along a desired trajectory by thrust vector control (TVC), which provides the most economical means of attitude control during such

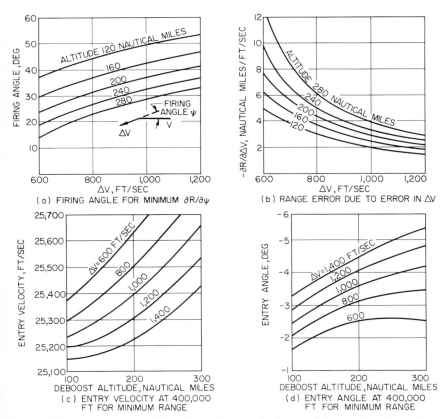

FIG. 27. Planetary entry conditions from deboost velocity and firing angle. (*From Space Data, courtesy of TRW Space Technology Laboratories, Redondo Beach, Calif.*)

periods. Appropriate servomechanisms are designed to actuate the TVC in response to prescribed (or sensor determined) attitude programs.

Starting Transient. The TVC mechanism should react promptly to the maximum starting transient imposed by thrust misalignment about the center of gravity, to avoid undue deviation of the attitude from the prescribed program. The response magnitude (thrust moment ÷ inertia) determines the type of servomechanism best suited to control the initial transient and the subsequent attitude program of the spacecraft.

Cutoff Transient. The residual angular velocity at thrust cutoff produces an initial condition for design of the attitude-control system to be employed during the coasting period. Here again, it is desirable for the control system to overcome the initial angular velocity before excessive displacement occurs.

Coasting Conditions. During coasting periods, the relatively inert spacecraft may require attitude control with respect to the sun, earth, or other desired direction (Fig. 28). Various systems of forces and moments may produce influences either beneficial or detrimental to the desired attitude control, as discussed in the following paragraphs.

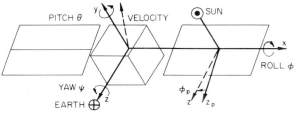
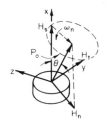

FIG. 28. Fully oriented satellite. (*From Space Data, courtesy of TRW Space Technology Laboratories, Redondo Beach, Calif.*)

FIG. 29. Spin-stabilized body. (*From Space Data, courtesy of TRW Space Technology Laboratories, Redondo Beach, Calif.*)

Spin Stabilization. A body of revolution with spin velocity p_0 (Fig. 29) about either maximum or minimum principal axes of inertia will maintain its orientation in inertial space, in the absence of external moments. If the total angular momentum vector H_s initially coincides with the spin axis, and an impulse angular momentum H_n is added normal to the spin axis, the body spin axis will then nutate about the new total angular momentum vector H_t, at a frequency of $\Omega = p_0 J_x/J_z$ and half-cone amplitude $\Omega = J_z \omega_n / J_x p_0$ where $\omega_n = H_n/J_z$. When $J_x > J_z$, passive damping can be used to dissipate the nutational energy, or it can be removed by applying another moment impulse, equal to the first and also normal to the spin axis, after the body has rotated through an angle $\phi = \pi J_z/(J_x - J_z)$ radian. This double-impulse method will have caused the z axis to precess through the angle 2θ. In this manner, the spin-stabilized body can be reoriented to any desired attitude, by the application of one or more small moment impulses, properly timed with respect to the angular position.

Sensor Accuracy. Accuracies of typical attitude-control sensors (with their fields of view, in parentheses) are shown in Fig. 30.

Solar-radiation Torque. Bombardment of the various surfaces of the satellite by photons emanating from the sun will create forces whose magnitude and direction are determined by the reflective properties of the surfaces. If the center of radiation

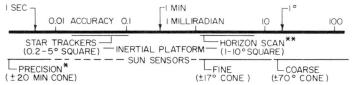

FIG. 30. Accuracy of sensors. (*From Space Data, courtesy of TRW Space Technology Laboratories, Redondo Beach, Calif.*)

pressure is not coincident with the vehicle center of mass, then a torque will act on the vehicle, sometimes with sufficient magnitude to affect the control-system design.

The radiation power in the vicinity of the earth is 1.94 cal/cm²/min corresponding to a pressure of 9.4×10^{-8} lb/ft² for complete absorption. In preliminary design it is necessary to calculate an upper bound for the radiation-torque effect, to determine the cumulative momentum-storage requirements for control-system design.

Gravity Gradient. The earth's gravitational potential varies with altitude. For this reason the center of gravity and the center of mass of a satellite are not exactly coincident. Unless the force of gravity, applied at the center of gravity, acts along a line passing through the center of mass, the resulting torque will tend to rotate the satellite. If the latter is properly configured, this torque can be usefully employed for orientation with respect to the earth's gravitational field.

The gravity force acting on a unit mass can be expressed in terms of the gradient of the gravitational potential as $F = -\bar{\nabla}(-GM/r)$, where r is the magnitude of the radius vector $\bar{r}$ from the earth's center to the unit mass; GM is the gravitational constant (approximately 1.4×10^{16} ft³/sec²); and $-GM/r$ is the gravitational potential, a representation sufficiently accurate for the purpose here. The torque $\bar{M}_g$ tending to rotate the satellite about its center of mass is then

$$\bar{M}_g = -\int \bar{p} \times \bar{\nabla}(-GM/r)\, dm$$

where $\bar{p}$ is a radius vector from the satellite center of mass to the differential mass dm.

Earth's Magnetic Field. For satellite altitudes greater than 100 miles above the surface, the earth's magnetic field is approximated by a simple magnetic dipole at the center of the earth, with field intensity and resulting torque as shown in Figs. 31 and 32. The axis of this dipole is skewed at an angle of approximately 18° with

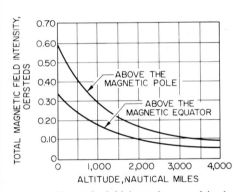

Fig. 31. Magnetic field intensity vs. altitude. (*From Space Data, courtesy of TRW Space Technology Laboratories, Redondo Beach, Calif.*)

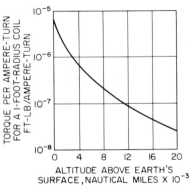

Fig. 32. Torque per ampere-turn vs. altitude. Maximum torque per ampere-turn 1-ft-radius coil over magnetic equator. (*From Space Data, courtesy of TRW Space Technology Laboratories, Redondo Beach, Calif.*)

respect to the earth's spin axis. The north magnetic pole is at approximately 70°N latitude, 97°W longitude. The south magnetic pole is at approximately 72.5°S latitude and 155°E longitude. This means that the axis of the dipole, and hence the field, precesses around the earth's spin axis. The significance of this precession is that the effects of the magnetic field will be the same during successive orbits (or multiple orbits, depending on the satellite period) only when these orbits are synchronous with respect to the earth's period.

A current-carrying coil in the magnetic field tends to assume a position to include the largest possible flux through it in a positive sense. That is, the force $\bar{F}$ on an element of wire in a flux field $\bar{B}$ is given by $\bar{F} = i\, dl \times \bar{B}$. A coil whose center lies along the spacecraft z axis would cause the spacecraft to experience a torque M_m in dyne-centimeters of

$$\bar{M}_m = (\pi r^2{}_c i/10) n (\bar{u}_x B_y + \bar{u}_y B_x)$$

where $\bar{u}_x$ and $\bar{u}_y$ are unit vectors along x and y spacecraft axes, respectively; r_c is the radius of the coil in centimeters; i is the current in amperes; B is the flux density in gauss; and n is the number of turns.

Aerodynamic Torques. The earth satellite in a low-altitude orbit, or with a low-altitude perigee, will experience aerodynamic forces during its lifetime. The nature of these forces, depending on the orientation of the vehicle and its altitude, may be of consequence and may require an expenditure of a significant amount of control-system momentum. In fact, for some extremely low altitude satellites, certain proposals have been made to use the aerodynamic torques produced by a large rudder or trailing drag device for control torques.

"Hyperthermal free molecular flow" theory is generally used to obtain the shearing and normal stresses on the various flat surfaces of a space vehicle. The equations which describe the pressure P and the shearing stress τ are $P = 2(2 - \sigma') q \sin^2 \beta$ and $\tau = 2\sigma q \sin \beta \cos \beta$, where the pressure P acts normal to the surface. The angle of attack β may be expressed in terms of the yawing angle ψ for the body surface. Hence $\beta = \psi$ and $\beta = 90° - \psi$ for surfaces whose normals are along the y and x body axes, respectively.

The two quantities σ and σ', respectively, defined as the surface-reflection coefficients for tangential and normal momentum exchange, have a significant influence on the magnitude of the pressure and shearing stress. The nature of the molecular reemission, and hence the value of σ and σ', are functions of the type of surface material, the velocity angle of incidence, and the wall temperature. The value of σ and σ' can vary between 0 and 1. The few engineering measurements made on typical surfaces indicate values of σ' between 0.8 and 1.0. However, for low angles of attack it appears that the characteristic of reemission may be altered sufficiently to cause considerable deviations of σ' from these values. The quantity σ has not yet been measured experimentally; however, the values of σ and σ' should not differ greatly. Analyses conducted at Space Technology Laboratories have assumed a value of about 0.8 for σ and σ'.

Internal Moving Parts. When considering the attitude control of a space vehicle the small torques produced by the motion of parts or personnel within the spacecraft cannot be ignored. In fact, internal moving parts (reaction wheels) can be used to considerable advantage to provide momentum storage to resist and to damp vehicle motions. While such devices are normally considered as part of the attitude-control system, the designer must bear in mind that noncontrol rotating systems will also cause torques which must be overcome by control torques. It is best to avoid, through proper orientation and/or counterbalancing, the large momentum reactions imparted to the spacecraft when such devices accelerate, as well as the gyroscopic effects resulting from such devices when the spacecraft itself rotates.

The effect of reaction wheels is reviewed briefly by noting that the time rate of change of wheel momentum is equal to the torque. Generally speaking, the attitude-error signal is used with filtering to control the wheel speed; hence, for a small pointing error there will be a nonzero momentum change. Integral control can be used to alleviate this problem in the steady state. Since most momentum devices have limited storage capability, they must be used in conjunction with the mass expulsion devices to permit frequent restoration of the momentum storage capability. See Fig. 33.

Mass Expulsion. The attitude control of a spacecraft with initial rates and/or external torques can be achieved simply by use of mass expulsion jets. The actuation of such devices can be controlled by the output of a sensor to change the angular position of the spacecraft within the prescribed error limit (or to precess a spin-stabilized spacecraft). The mass expulsion system may produce either a steady torque

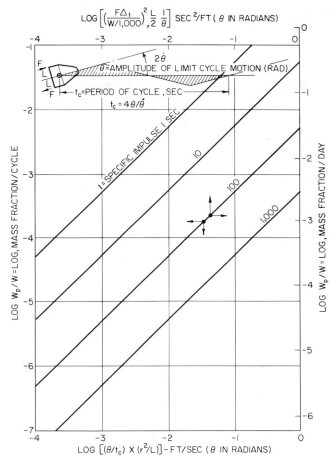

Fig. 33. Attitude control by mass expulsion. (*From Space Data, courtesy of TRW Space Technology Laboratories, Redondo Beach, Calif.*)

$$\text{Change in angular impulse per cycle} = 4J\dot\theta\, \Delta tL = W_p IL$$

$$\text{Mass fraction per cycle} = \frac{W_p}{W} = \frac{4}{I}\frac{F\,\Delta t}{W} = \frac{16r^2\theta}{gI_{sp}Lt_c}$$

$$\text{Mass fraction per day} = \frac{86{,}500}{10^6}\left(\frac{F}{W/1{,}000}\right)^2 \left(\frac{Lg}{r^2}\right)\left(\frac{\Delta t^2}{\theta I_{sp}}\right)$$

where W_p = propellant weight, lb
W = spacecraft weight, lb
r = radius of gyration, ft
L = radius of jet centerline from center of gravity
g = 32 ft/sec^2
θ = amplitude of motion, rad
t_c = period of motion, sec
F = thrust of reaction jets, lb
Δt = on-time of jets, sec

proportional to the error signal ($F = K_\theta$), or a quantized torsional impulse ($2LF\Delta t$, where $\Delta t \geq 0.01$ sec) adequate to reverse the direction of rotation.

Typical values of specific impulse I for various propellants	Steady thrust I (vac), sec	Pulsed operation		Thrust range, lb
		Δt, sec	I (vac), sec	
Stored cold gas, N_2........	65	0.01	40	0.0– 50
Monopropellant, H_2O_2.....	150	0.10	40	1.0–100
Bipropellant, N_2O_4/MMH..	280	0.01	75–150	0.2–100

A more efficient "dual mode" of operation for long-duration missions is to expend electrical power (which can be replenished easily with solar energy) to control body-fixed reaction wheels, in response to fluctuations in angular momentum, and to expel

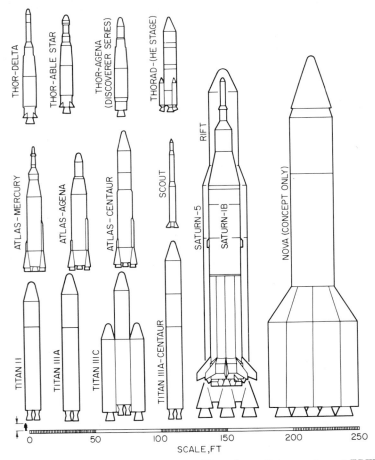

Fig. 34. Rocket vehicles for space projects. (*From Space Data, courtesy of TRW Space Technology Laboratories, Redondo Beach, Calif.*)

mass through reaction jets only when reenergizing the reaction wheels. The expenditure of propellant is only a fraction as great for the dual mode as for the on-off pneumatic mode shown in Fig. 33; however, the added weight of reaction wheels may be significant.

6 Propulsion

The prime method of propelling vehicles into space and to objectives in space is by liquid- and solid-fuel rockets. Methods under development are nuclear, which will be able to provide much higher terminal velocities; ion, which can provide small thrusts, but at very high specific impulse and for indefinite periods of time which is ideal for long space missions; and aerospace airplane, which is intended to save the huge cost of rocket fuel during the launch phase by using atmospheric air instead.

This section will present data on existing rocket vehicles (See Fig. 34)*, the fundamental rocket equation, trade-off exchange ratios, method of calculating thrust, and the burnout-velocity capability of existing single and multistage rockets. In essence, the data presented here show what is required in practice to achieve the velocities at the altitudes or orbits required by the astronautical considerations presented in Sec. 4. Section 7 gives an example of a planned mission taking both propulsion and astrodynamics into account.

Rocket Equations

The propulsion equations derive from Newton's laws applied to a body in motion maintaining a constant thrust by ejection of mass through a nozzle.[21] Thus

$$F = -ma = -dmc/dt = -c(dm/dt) \qquad (9)$$

where F = thrust, lb
m = mass
a = acceleration
c = exhaust velocity of the *expelled mass* (note this is *not* the forward velocity V of the vehicle)
dm/dt = rate of discharge of mass
$= (1/g)(dw/dt) = \dot{W}/g$
$\dot{W}$ = propellant flow rate
g = acceleration of gravity

$$F = (c/g)\dot{W} = I\dot{W} \qquad (10)$$

where I = specific impulse or pounds of thrust per pound per second of fuel flow.

The *forward velocity* V of the rocket vehicle is then, assuming vacuum trajectory, from (9) and (10),

$$(W/g)(dV/dt) = I\dot{W}$$
$$dV/dt = Ig(\dot{W}/W)$$
$$V = \int_0^{t_p} Ig \frac{\dot{W}}{W_0 - \dot{W}t} dt \qquad (11)$$

where W_0 = launch weight
t_p = burnout time
Let r = mass ratio = $W_0/(W = \dot{w}t_p)$
$= \dfrac{\text{launch weight}}{\text{burnout weight}}$

Then (11), when integrated, becomes

$$V = Ig \ln r \qquad (12)$$

This is plotted in Fig. 35. Table 7 provides typical values of I.

Exchange ratios, or partial derivatives, derived from this equation are given in Fig. 36. These permit determination of effect on velocity V of small variations in specific impulse, launch weight, or mass ratio.

* See also Tables 5 and 6.

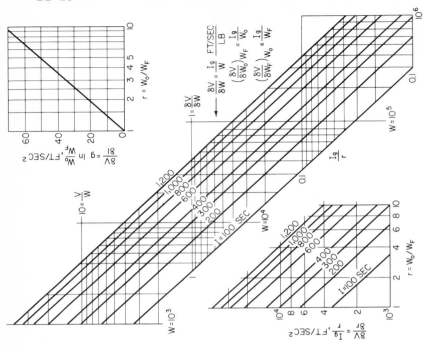

Fig. 36. Exchange ratios derived from rocket-velocity equation. Exchange ratios, or partial derivatives of the velocity equation, are used to determine the effect on velocity V of small variations in specific impulse I, or increments of initial weight W_0, or mass ratio r. (From Space Data, courtesy of TRW Space Technology Laboratories, Redondo Beach, Calif.)

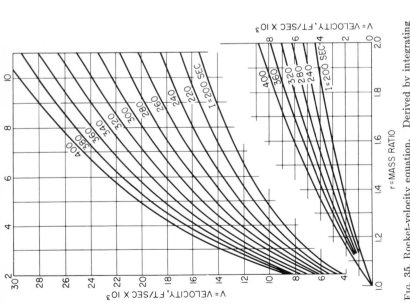

Fig. 35. Rocket-velocity equation. Derived by integrating equation $F = m\,dV/dt$, where F = thrust = constant; m = mass = $(W_0 - \dot{w}t)/g$; W_0 = launch weight; $\dot{w}$ = propellant flow weight; $F/\dot{w} = I$ = specific impulse. (From Space Data, courtesy of TRW Space Technology Laboratories, Redondo Beach, Calif.)

Table 5. Liquid Stages and Vehicles

Vehicle stages	Stage	Contractor	Engines per stage	Propulsion Engine	Propulsion Propellants	Thrust, lb
Atlas......	1½	G/D Astro	2 + 1	R/D MA-2	LO_2/RP	367,000
Thor......	1	Douglas	1	R/D MB-3	LO_2/RP	165,000
Titan II....	1	Martin	2	AGC XLR87-AJ5	N_2O_4/N_2H_4 UDMH	430,000
	2		1	AGC XLR91-AJ5	N_2O_4/N_2H_4 UDMH	100,000
Titan III...	1	Martin	2	UTC 5-Segment	120 $N_2O_4/$ N_2H_4 UDMH solid	1,000,000
	2			See Titan II stage 1		
	3			See Titan II stage 2		
	4		2		N_2O_4/N_2H_4 UDMH	16,000
Ablestar....		Aerojet	1	Aerojet AJ10-104	IRFNA/ UDMH	7,870
Agena......		Lockheed	1	Bell 8096	IRFNA/ UDMH	16,000
Centaur....		G/D Astro	2	P/W RL10-A3	LO_2/LH_2	30,000
Delta......	1	Douglas	1	AGC AJ10-118	1 WFNA/ UDMH	7,700
	2		1	ABL 258	Solid	5,400
Saturn						
S-I......		Chrysler	8	R/D H-1	LO_2/RP	1,504,000
S-IC.....		Boeing	5	R/D F-1	LO_2/RP	7,500,000
S-II.....		NAA	5	R/D J-2	LO_2/LH_2	1,000,000
S-IV.....		Douglas	6	P/W RL10-A3	LO_2/LH_2	90,000
S-IVB...		Douglas	1	R/D J-2	LO_2/LH_2	200,000
S-V......		G/D Astro	2	See Centaur		
Nova						
N-1......			8	R/D F-1	LO_2/RP	12,000,000
N-2......			4	AGC M-1	LO_2/LH_2	5,000,000
N-3......			1	R/D J-2	LO_2/LH_2	200,000

Launch vehicles	Performance,* lb To orbit	Performance,* lb To escape	Programs
Atlas..................	2,700	......	Mercury
Atlas-Agena............	5,000	750	Samos, Midas, Saint, Bambi, OAO OSO-II, Mariner A, Ranger, OGO
Atlas-Centaur..........	9,000	2,300	Mariner B, Surveyor
Thor-Ablestar..........	900	......	Transit, Courier, ANNA
Thor-Agena............	1,600	......	Discoverer, Nimbus, Echo, OGO
Thor-Delta............	750	140	OSO-I, Explorer, Syncom, Telstar, Tiros, Relay
TAT..................	1,800	......	Secor
TAT-Agena............	Class 1,000	Class 140	Biosatellite, Commercial Comsat, Pioneer
TAD..................			
Atlas/Fire.............	4,000	850	Reentry Test Vehicle
Titan II..............	6,000	......	Gemini
Titan II-Agena.........	Class	Class	
Titan II-Centaur.......	Class	Class	
Titan IIIC............	28,500	8,000	
Saturn 1 (S-I, S-IV)....	20,000	5,000	Prospector, Voyager
Saturn IB (S-IB, S-IVB)..	30,000	13,000	Apollo A
Saturn 5 (S-IC, S-II, S-IVB)...	220,000	90,000	Apollo B, Apollo C
Nova.................	400,000	200,000	Might be replaced by larger version, to place either 500,000 or 1,000,000 lb in orbit

* Performance data from various unclassified periodicals.

Table 6. Solid Stages and Vehicles

Stages	Contractor	Engine	Thrust
Aerojet Jr.	Aerojet	....	57,000
Algol	Aerojet	33KS-120,000	120,000
Algol 2	Aerojet	....	150,000
Alcore	Aerojet	30KS-8000	8,000
Altair	ABL	X 248	3,100
Altair 2	ABL	X 258	5,800
Antares	ABL	X 254	14,100
Antares 2	ABL	X 259	21,300
Castor	Thiokol	TX-33-35	64,000
Cetus	NOTS	100A	900
Recruit	Thiokol	TE 29-1	36,000
Titan III Booster	UTC	UA 1205	1,000,000

Vehicles	Stage no.	Stage name	Performance	Contractor
Blue Scout 1 (XRM89)	1	Algol	100-lb probe to 1,610 miles	Aeronutronic
	2	Castor		
	3	Antares		
Blue Scout 2 (XRM90)	1	Algol	100-lb probe to 2,395 miles	Aeronutronic
	2	Castor		
	3	Antares		
	4	Altair		
Blue Scout Jr. (XRM91)	1	Castor	100-lb probe to 4,550 miles	Aeronutronic
	2	Antares		
	3	Alcore		
	4	Cetus		
Little Joe I		4x Castor 4x Recruit	Short-range Mercury recovery tests. Staggered ignition, 2 Castor, 4 Recruits, 2 Castor	NAA
Little Joe II	1	7x Algol	80,000 lb to 400,000 ft	GD/A
RAM	1	Castor 1	Proposed extension of Scout	
	2	Antares		
	3	Alcore		
Scout (XRM90)	1	Algol or Algol 2	150 to 260 lb in 100-NM orbit with 4 stages	LTV
	2			
	3	Antares or Antares 2	50 lb to 1,200-NM altitude	
	4	Altair or Altair 2		
	5	Cetus		

The effective exhaust velocity c, already indicated in Eq. (9), is useful in describing the overall performance of a rocket using a particular propellant and is directly related to performance in the combustion chamber and expansion through the nozzle. It is related to thrust coefficient C_f as follows:

$$c = C_f C^* \tag{13}$$

where C^* is the characteristic velocity. The thrust coefficient describes the amplification of thrust resulting from the expansion of gases in the nozzle. C_f is obtained from measured data from the test stand and design data and is

$$C_f = F/p_c A_t \tag{14}$$

where p_c = chamber pressure, lb/sq in.
A_t = throat area of nozzle, sq in.
Expressions (13) and (14) can be used with (9) and (10) to establish rocket thrust from a number of different aspects.

$$F = C_f p_c A_t = I\dot{W} = C_f C^*(\dot{W}/g) \tag{15}$$

Table 7. Rocket-propellant Characteristics
Typical Liquid-propellant Characteristics

Oxidizer	Fuel	Mixture ratio $r = O/F$	T_c, °F	I = specific impulse, sec S. L. Opt. ϵ p_c = 300 psi	$\epsilon = 40$ vac	Bulk density, g/cu cm	Sp. ht. ratio $\gamma = c_p/c_v$
ClF_3^s	NH_3^{*3}	3.93	4980	240	314	1.34	1.32
ClF_3	$N_2H_4^{*s}$	2.77	6400	258	341	1.45	1.23
F_2^c	NH_3^*	3.3	7280	311	417	1.2	
F_2	$N_2H_4^*$	2.4	7320	315	411	1.4	1.33
F_2	H_2^{*c}	12.0	7220	355	475	0.6	1.31
F_2	MMH^{*s}	2.4	7040	298	409	1.12	
90% $H_2O_2^s$	$N_2H_4^*$	2.04	4710	253	337	1.24	1.25
$N_2O_4^s$	50/50*s	2.1	5590	257	338	1.21	1.19
N_2O_4	MMH*	2.3	5420	250	339	1.23	1.19
N_2O_4	$UDMH^{*s}$	2.8	5280	250	337	1.18	1.23
N_2O_4	$N_2H_4^*$	1.25	5080	255	341	1.23	1.27
O_2^c	H_2	5.0	5340	347	456	0.33	1.26
O_2	JP-4^s	2.4	5740	263	349	1.00	1.24
O_2	N_2H_4	0.83	5730	274	363	1.06	1.23
O_2	RP-1^s	3.5	5830	261	355	1.05	
O_3^c	H_2	3.5	5030	375	497	0.28	1.21
IRFNA^s	$N_2H_4^*$	1.4	4800	246	326	1.27	
IRFNA	UDMH*	3.0	5340	246	346	1.26	1.21
Monopropellant	H_2O_2	.....	1380	133	174	1.39	1.27
Monopropellant	N_2H_4	.....	1850	190	230	1.00	
Monopropellant. B-120^s	Cavea	.....	5200	222	290	1.48	
Nuclear H_2^c	Vacuum	Liquid density: 0.076 at 25°R; 0.070 at 37°R; 0.060 at 49°R					
p_c = 1 atm	I sec	615	675	740	800	875	990
$\epsilon = 40$	T_c, F°	2000	2500	3000	3500	4000	4500

* Hypergolic; s = storable; c = cryogenic.

Typical Solid-propellant Characteristics

	S. L. I at 1,000 psi Opt. expansion, sec	Density, lb/cu ft	T_c, °F	$\gamma = c_p/c_v$
Ammonium nitrate composite	175	97	2200	1.28
Double-base NC/NG	225	102	4900	1.22
Polysulfide-ammonium perchlorate composite	208	104	3700	1.22
Polyurethane-ammonium perchlorate composite, aluminized	234–242	107–110	4700–5800	1.18
PBAA-ammonium perchlorate composite, aluminized	238–245	107–110	4900–5500	1.18
Double-base-ammonium perchlorate composite, aluminized	247	110	6300	1.16

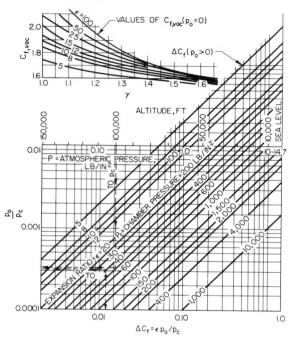

FIG. 37. Rocket nozzle characteristics. *(From Space Data, courtesy of TRW Space Technology Laboratories, Redondo Beach, Calif.)*

Rocket thrust can be expressed in several ways as follows:

$$\text{Thrust } F = C_f p_c A_t = I \dot{W} = C_f C^* \dot{W}/g$$

Thrust coefficient $C_f(p_c, p_e, p_0, \gamma, \epsilon) = C_{f vac} - \Delta C_f$ (see figures below)
where p_c, p_e, p_0 = chamber, exit, and ambient pressures, respectively
ϵ = nozzle expansion ratio = A_t/A_e
A_t/A_e = throat area and exit, area, respectively
C^* = characteristic exhaust velocity of the propellant at the specified mixture ratio (nearly independent of chamber pressure p_c)
γ = ratio of specific heats of exhaust products
$\dot{W}$ = weight per second of propellant flow

Thrust coefficient C_f depends on several parameters of the chamber and the propellant. These are presented in Fig. 37.

Multistaging makes possible attainment of greater velocities for practical mass ratios. Design curves are given in Fig. 38, to permit estimate of payload of both single-stage and multistage rockets for specific mass, structure, and payload ratios (see Fig. 38 for definition). Figure 39 shows launch weight, payload weight vs. burnout velocity (and hence orbit attained) for various knows and theoretical combinations of vehicles. Figure 40 may be used for more refined estimates of performance, based on specific impulses and structure ratio.

Other Forms of Payload Delivery

Two methods of potentially reducing cost to deliver payload in space will be treated here. They are

1. Orbital assembly
2. Aerospace airplane

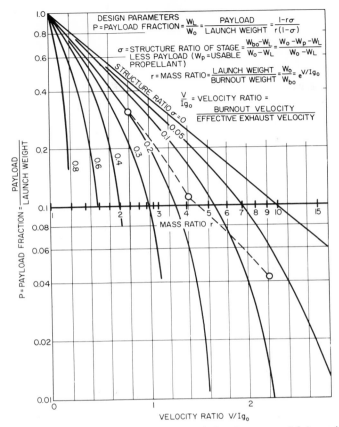

Fig. 38. Rocket-vehicle design chart. The chart below is very useful for quickly determining the payload of a one-stage or multistage rocket vehicle; i.e., payload = pW_0 = payload fraction × launch weight. For multistage vehicles, each stage is first considered to originate from $V = 0$ on the σ contours, after which the upper stages are translated to their proper positions, as shown by the dotted lines. Note similarity of these curves to those of payload vs. velocity in Fig. 40. (*From Space Data, courtesy of TRW Space Technology Laboratories, Redondo Beach, Calif.*)

Spacecraft can be launched directly to their space objectives, by utilizing the available launch vehicles previously described; or the individual payload modules can be assembled in orbit to make up a larger spacecraft for more difficult missions. By means of multiple launches and orbital rendezvous and assembly, it is possible to attain the payload requirements for many interesting missions prior to the availability of the larger boosters necessary to accomplish the missions directly with one launching. The spacecraft modules which may be needed for orbital assembly are classified as manned modules with reentry capsules, tanker modules to increase the velocity capability of the vehicle to which they are attached, and nuclear-propulsion modules for the higher-velocity objectives (see Fig. 41). Some of the space missions attainable with this type of operation are described in Table 8.

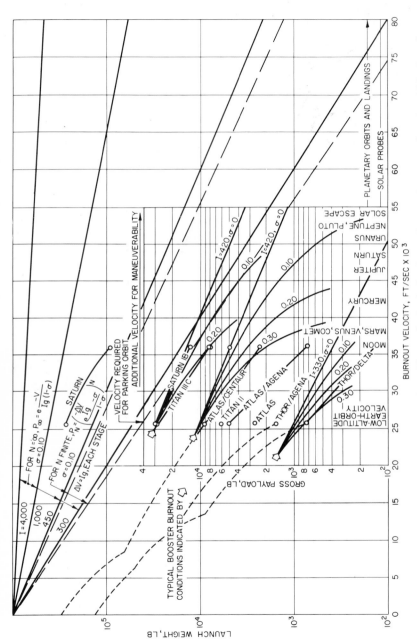

Fig. 39. Space payload capabilities of launch vehicles. The payload-velocity data points (0) for current launch vehicles have been taken from unclassified sources without modification. The performance of additional stages can be obtained by use of the rocket-vehicle design chart in Fig. 40. Examples of this method of analysis are given below by the "comet-tail" curves marked with the indices (⇨). (*From Space Data, courtesy of TRW Space Technology Laboratories, Redondo Beach, Calif.*)

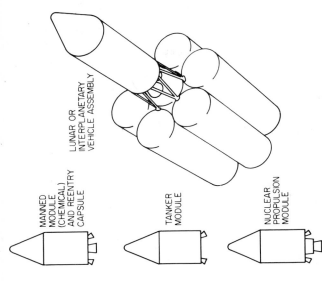

Fig. 41. Orbital assembly of space modules. (*From Space Data, courtesy of TRW Space Technology Laboratories, Redondo Beach, Calif.*)

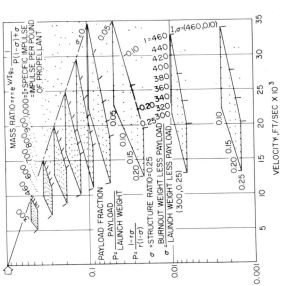

Fig. 40. Rocket-vehicle performance. The reader can plot payload-velocity curves for any combination of specific impulse I and structure ratio σ by interconnecting the corresponding dots in the table below. If such curves are traced on vellum, they can be transferred to Fig. 39 for performance analysis of upper stages. By placing the index point ⇧ of the vellum over the burnout condition for the preceding stage, one may read off the corresponding values of payload vs. velocity. (*From Space Data, courtesy of TRW Space Technology Laboratories, Redondo Beach, Calif.*)

Table 8. Orbital Assembly for Space Missions

Mission	Velocity, ft/sec × 10⁻³	Payload ratio P*	Payload, ton*	No. launches per mission
Interplanetary:				
1. Mars and Venus landing and return, manned	57–87	18–250	8 Venus (N) 8 Mars	6–7
2. Mars and Venus orbit and return, manned	52–59	11–21	4–10	2
3. Mars and Venus orbit and return, unmanned	52–59	11–21	2–5	1
4. Escape from solar system	54	13	6	2
5. Mars and Venus landing	38	3.3	8	1
6. Mercury or Jupiter				
Flyby	43–45	6	16	2
Orbit	66	40	2–8(N)	2
7. Sun impact	104	400	1(N)	2
8. Explore comet head or tail	38–42	4.5	11	1
9. Mars and Venus orbit	45–49	6–9	6–9	1
10. Mars and Venus impact	38	3.3	8	1
11. Mars and Venus flyby (no heat shield required)	38	3.3	16	1
Lunar:				
1. Lunar colony	45	6.0	50	6
2. Lunar landing and return, manned	54	13	8	2
3. Lunar fly around and return, manned	38	3.3	16	1
4. Lunar landing and return	54	13	4	1
5. Lunar landing	45	6.0	9	1
6. Lunar orbit	40	3.7	14	1
Earth:				
1. Manned launch and refuel platform	25	1	50	1
2. Manned laboratory and reconnaissance	25	1	300	6

The space missions shown in this table assume a basic launch vehicle capable of placing 100,000 lb into a 300-NM earth orbit. Various numbers of modules are then assembled in orbit prior to departure for their space objectives. The spacecraft design parameters are $I = 420$ sec, $\sigma = 0.12$, except in a few cases designated by (N), where nuclear propulsion was assumed, $I = 750$ sec.

* Payload ratio = $\dfrac{\text{weight assembled in earth orbit}}{\text{payload carried to space objective (or returned to earth)}}$

Aerospace Airplane

The expanding space program cannot indefinitely accept the losses associated with each flight of the large and costly chemical rocket boosters currently planned. There is an urgent need for more economical transport modes to orbit. One of the more promising systems is the aerospace airplane concept (Fig. 42), which utilizes winged aircraft with air-breathing engines for the recoverable booster stage, and chemical- or nuclear-rocket final stages at altitudes and velocities beyond the capability of the air-breathing system.

Performance. The performance characteristics of various aerospace engine systems are shown in Fig. 43, as the specific impulse of those systems vs. vehicle Mach number. The corresponding engine weights are combined with other overall system weights in an estimation of the payload/initial gross weight of aerospace airplane systems designed to place a payload in a 300-NM orbit (Fig. 42a). In spite of the high engine system weight, it is shown that large gains are attainable with the aerospace airplane. The relatively low payload to orbit of the conventional rocket launch vehicle is due to its having to accelerate the mass of its own inefficiently used propellant. A comparison of the relative performance of rocket and air-breathing systems (Fig. 42c) shows that 37 per cent of the launch weight of a hydrogen-fuel rocket is used to accelerate the vehicle to 20 per cent of its orbital speed, compared with only 3 per cent for the air-breathing engine.

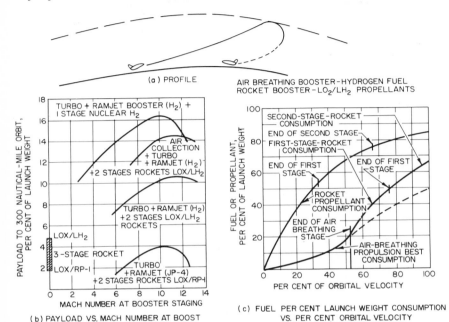

FIG. 42. Aerospace-airplane concept. Horizontal takeoff and landing one or more stages to orbit full recovery of vehicles. (a) Profile. (b) Payload vs. Mach number at boost. (c) Fuel (per cent launch weight) consumption vs. per cent orbital velocity. (*From Space Data, courtesy of TRW Space Technology Laboratories, Redondo Beach, Calif.*)

System Considerations. The potential performance gains and recoverability of the air-breathing boosters must be weighed against the following considerations in assessing the relative advantage with respect to a pure rocket system:

Added structural weight vs. higher specific impulse
Smaller payload to orbit vs. multiple flights and orbital assembly
Engine operation and staging at Mach numbers of 5 to 8
    Variable inlet geometry required means added weight
    Airframe aerodynamic heating requires special heat exchangers
    Engines need intercoolers and liquid-cooled turboblades

7 Example of Space-mission Planning

The astronautical planning of a lunar landing from Atlantic Missile Range launching will serve to illustrate the use of the data of Secs. 4 and 6. See also Fig. 44.

Nominal Trajectory. The nominal trajectory will be planned for a 50-hr transit, with departure date selected to intercept the moon near its monthly perigee position. The moon travels at a mean velocity of 3,350 ft/sec (27.322 days per revolution) in

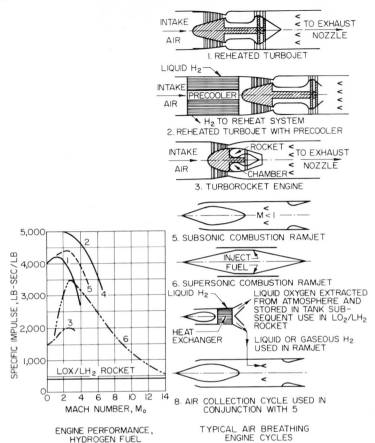

Fig. 43. Typical aerospace engine systems. (*From Space Data, courtesy of TRW Space Technology Laboratories, Redondo Beach, Calif.*)

1. Reheated turbojet or reheated bypass turbojet
2. Precooler with reheated turbojet or reheated bypass turbojet
3. Turborocket
4. Turborocket with precooler
5. Ramjet with subsonic combustion
6. Ramjet with supersonic combustion
7. Variations of turbomachinery coupled with ramjets
8. Air-collection cycle used in conjunction with (*a*) hyperjets (LACE), (*b*) turbomachinery and ramjet (Super LACE)
9. Combination of the above air-breathing systems with nuclear-reactor heat in place of chemical-reaction heat

a plane inclined 5°9′ to the ecliptic. For a 50-hr transit time and a coast-orbit altitude $h = 100$ NM, the nominal burnout conditions are selected from Fig. 44 as $V_{b0} = 36,090$ ft/sec and $\phi = 166°$. The space trajectory will be highly eccentric, since the velocity is only slightly below parabolic.

Injection Point. The lunar vehicle remains in the parking orbit for a short time θ_{coast} before reaching the injection point, where the transfer velocity ΔV_1 is applied—serving to increase the apogee and cause the trajectory to intercept the moon. The nominal launch azimuth of 90° from the latitude of AMR (28.45°) determines the inclination angle $I = 28.45°$ of the coast orbit, which with the declination of the moon $\delta_m = 10°$ will yield (by solving spherical triangles) a coast angle $\theta_{coast} = 126°$.

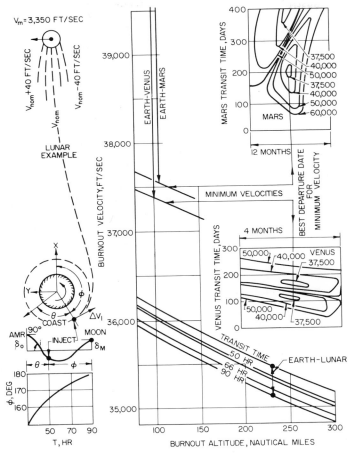

FIG. 44. Mission velocity requirements for Moon, Mars, and Venus. (*From Space Data, courtesy of TRW Space Technology Laboratories, Redondo Beach, Calif.*)

Lift-off Time. The lift-off time for the 50-hr transit is computed as $(50/24 \times 27.322) = 0.0763$ revolution of the moon, or 1.83 earth-hr, subtracted from the nominal value of 13.47 hr before moonrise (for $\phi + \theta_{coast} - 90° = 202°$); giving the lift-off times as 11.64 hr before moonrise.

Error Analysis. The nominal trajectory is used as a basis for trade-offs and error analysis to perfect the efficiency and reliability of the proposed lunar flight, and to specify the final mission more completely. Some of the trade-off values to be considered are as follows, the miss distance being denoted by M:

Initial guidance error: $\partial M / \partial \epsilon = 207$ NM/milliradian (error at 207,000-NM mean distance to moon, due to 1-milliradian displacement of injection velocity). $M = R_{moon}$ for $\epsilon = 0.26°$.

Lift-off timing error: $\partial M/\partial t = 93$ NM/min (error due to movement of target and aiming point of launcher). $M = R_{moon}$ for $\Delta t = 10$ min.

Velocity error: $\partial M/\partial V = 25$ NM/ft/sec (variation from nominal 50-hr transit time and movement of target). $M = R_{moon}$ for $\Delta V = 40$ ft/sec.

Midcourse and terminal velocity corrections which can be applied are (see Fig. 36)

$\partial V/\partial W = 1$ to 10 ft/sec/lb; for $W = 10^3$ to 10^4 lb, $I = 300$ sec
$\partial V/\partial I = 50$ ft/sec/sec; for $r = 5$
$\partial V/\partial r = 2{,}000$ ft/sec/unit; for $r = 5$, $I = 300$ sec

The launch window can then be specified as a combination of permissible variations from the nominal trajectory, which can be compensated by midcourse corrections to attain the approach window. Likewise, the approach-window variations can be compensated by terminal corrections to attain the specified landing conditions. Mission velocity requirements for Mars and Venus can also be determined from Fig. 44.

8 Astronomy

Astronomy is the science concerned with the definition and understanding of celestial phenomena. As such, astronomical data are essential to any space operation, providing information for both mission objectives and as to environmental constraints on the mission profile, the forces to be encountered, the distances, the temperatures, etc. Some of the most basic astronomical data are therefore presented in this section. Further environmental constraints are presented in Sec. 9.

The Solar System (See Fig. 45)

The sun acts as a gigantic thermonuclear reactor[1] in the center of the solar system. Its gravitational pull holds the bodies of the system in orbit. Its radiation energy flux into space is so abundant that the interception of the tiny fraction of one-millionth

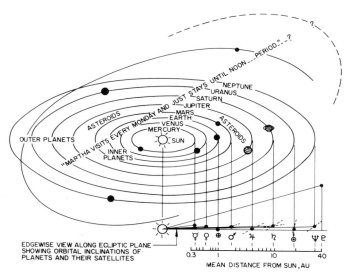

Fig. 45. The solar system. (*From Space Data, courtesy of TRW Space Technology Laboratories, Redondo Beach, Calif.*)

of a per cent of the total solar output suffices to maintain all chemical processes on the plants, including life on earth and possibly also on Mars and Venus (although the form of life may be considerably different from earth forms because of the extremes of atmosphere and environment).

All the planetary orbits can be bounded by a thin disk, having a diameter of 80 astronomical units or 1/800 light-year (7.4 billion miles). This small subassembly of matter and space is part of the great Milky Way, a vast, lens-shaped, spiral galaxy, extending 80,000 light-years in diameter by 16,000 light-years in thickness at the center. The sun is located far out from the galactic center (two-thirds of the radius) but is only 45 light-years from the principal galactic plane.

The solar system consists of the sun as central body (with 98 per cent of the total mass), plus its 9 planets, their 31 known satellites, tens of thousands of asteroids, and countless numbers of smaller objects, normally unobserved, which occasionally signal their presence as comets, meteors, or meteorites. All these bodies qualify as members of the solar system, revolving in elliptical orbits about the sun, because their energy level is below the parabolic limit required for escape from the solar gravitational field.

Some of the characteristics of the solar family will be described in sufficient detail to permit the consideration of rocket-propelled flights from earth to these objectives. Orbital properties, masses, gravitational attractions, surface escape velocities, and other physical characteristics will be presented as accurately as possible from the background of current knowledge of such phenomena.

Table 9. Basic Characteristics of the Solar System

Body	Symbol	Semi-major axis to sun, AU*	Period earth-years ($\oplus = 1$)	Mean diam ($\oplus = 1$)	Mass ($\oplus = 1$)	No. of natural satellites	Surface escape velocity ($\oplus = 1$)
Sun	☉	......	......	109.0	3×10^5	...	55.0
Mercury	☿	0.387	0.241	0.38	0.054	0	0.371
Venus	♀	0.723	0.616	0.97	0.815	0	0.915
Earth	⊕	1.00	1.00	1.00	1.00	1	1.00
Mars	♂	1.52	1.88	0.52	0.108	2	0.449
Jupiter	♃	5.20	11.9	11.0	318.0	12	5.38
Saturn	♄	9.54	29.5	9.03	95.1	9	3.26
Uranus	♅	19.2	84.0	3.72	14.5	5	1.97
Neptune	♆	30.1	165.0	3.38	17.0	2	2.24
Pluto	♇	39.5	248.0	1.02?	0.8	0	0.85?
Earth's moon	☾	......	0.075	0.27	0.012	0	0.212

* 1 AU = 92,959,670 miles.

The Sun—☉²

The sun, a relatively small member of the galactic family, is typical of the most numerous class of stars, with spectral designation *dG2* and surface temperature of around 6000°K. It is heated to incandescence by thermonuclear reactions in its deep interior, where the theoretical temperature is near 14,000,000°K. The sun gives off radiation from 30 m wavelength r-f to 10-angstrom X rays. More than 95 per cent of the radiation is between 2,900 and 25,000 angstroms. The mean distance of the sun from earth is used as a unit of measurement in the solar system—the astronomical unit or AU. Recent determinations of the value of the AU have been made by radar observations of the planet Venus.

Table 10. Principal Physical Characteristics of the Sun

Semimajor axis of earth orbit: 1 AU = $(1.49599 \pm 0.00004) \times 10^8$ km
Apparent angular diameter: $32'.25$
Diameter: 109.2 ($\oplus$ = 1); 751,200 NM; $(1.396 \pm 0.002) \times 10^6$ km
Volume: 1.302×10^6 ($\oplus$ = 1)
Mass: 332,952 ($\oplus$ = 1); 2.19×10^{27} tons; $(1.987 \pm 0.002) \times 10^{33}$ g
Mean density: 0.2554 ($\oplus$ = 1); 88 lb/cu ft; 1.410 ± 0.002 g/cu cm
Surface gravity: 28 ($\oplus$ = 1); $(2.738 \pm 0.003) \times 10^4$ cm/sec^2
Total energy output: $(3.86 \pm 0.03) \times 10^{33}$ erg/sec
Energy flux at surface of sun: $(6.34 \pm 0.07) \times 10^{10}$ erg/(cm^2) (sec)
Energy flux at 1 AU (solar constant): 1.97 cal/cm^2/min
Effective surface temperature: $5780° \pm 50°$K
Stellar magnitude (photovisual): -26.73 ± 0.03
Absolute magnitude (photovisual): $+4.84 \pm 0.03$
Inclination of axis of rotation to ecliptic: 7°
Period of rotation: About 27 days. The sun does not rotate as a solid body; it exhibits a systematic increase in period from 25 days at the equator to 31 days at the poles.
Gravitational parameter: GM^* = $\mu = 1.32715$ (10^{11}) km^3/sec^2
$\mu = 2.08929$ (10^{10}) NM^3/sec^2
$\mu = 4.68679$ (10^{21}) ft^3/sec^2
$\mu = 2.959122083$ (10^{-4}) AU^3/day^2
$\sqrt{GM} = 0.01720209895$ AU$^{3/2}$/day

$^*G = 6.6695 \times 10^{-8}$ (cgs units) = universal gravitational constant; M = mass, and $V_{escape} = (2\mu/a)^{1/2}$ where a = heliocentric radius (see Fig. 46b)
= 138,160 ft/sec at a = 1 AU

The Moon—☾

Earth's only natural satellite, the moon, is seemingly devoid of atmosphere and physical activity. However, the lunar topography bears evidence of volcanic and meteoric evolution of past eras, as indicated by the disruption of relatively smooth plains (the maria and terrae, ranging in size from 200 to 500 miles) by craters of all sizes up to 150 miles diameter, mountains rising to 25,000-ft altitude, ridges and rays extending hundreds of miles in length, and many smaller rills and faults. Superimposed over these gross features is the (much hypothesized) surface covering, generally believed to consist of finely pulverized material of unknown thickness (millimeters to meters) caused by internal eruptions, the impact of meteors, and eons of bombardment by solar and cosmic radiation particles, and by micrometeorites. Analysis of the selenographs obtained by Ranger 7 provides valuable information on lunar topography for the design of future landing craft and the planning of exploratory missions.

Table 11.

Physical Characteristics

Apparent angular diameter: $31'4''.6$
Diameter: 0.2725 ($\oplus$ = 1); 1,877 NM; 2,160 miles; 3,476 km
Mean density: 0.604 ($\oplus$ = 1); 3.33 g/cu cm
Mass: 0.012299 ($\oplus$ = 1); 7.35×10^{25} g
Mean surface gravity: 0.165 ($\oplus$ = 1); 5.31 ft/sec^2; 162 cm/sec^2
Surface escape velocity: 7,750 ft/sec; 1.27 NM/sec; 2.38 km/sec
Surface temp: $-58°$C to $+130°$C (sun side); $-150°$C (dark side)
Albedo: 0.07 (see Glossary, Sec. 2)

Orbital Characteristics

Semimajor axis of lunar orbit: 207,559 NM; 384,400 km; 238,855 miles
Orbital eccentricity: 0.0549
Mean orbital velocity about earth: 0.55 NM/sec; 1.022 km/sec
Sidereal period (about earth): 27.322 days
Inclination of orbital plane to ecliptic: 5°9'
Period of axial rotation: 27.322 days
Gravitational parameter: $\mu = 1.7313 \times 10^{14}$ ft^3/sec^2

The Inner Planets, Mercury, Venus, Earth, Mars

☿ Mercury, the smallest of the planets (3,008 miles diameter) and nearest to the sun (36×10^6 miles mean distance), has a sidereal period of 88 days (4+ revolutions per year). Its orbit eccentricity (0.206) and inclination to the ecliptic (7°) are the highest of all the planets except Pluto. It presents the same face to the sun at all times, the average surface temperature being 340°C on the bright side, and only a few degrees above absolute zero on the dark side. Mercury's appearance through a telescope is generally similar to that of the moon seen with the naked eye. Also like the moon, it has very little atmosphere, possibly only a small amount of heavy gas like argon.

♀ Venus, the brightest of the planets, is about the same diameter as earth and travels in a near circular orbit at 67×10^6 miles from the sun with a period of 225 days. Venus is surrounded by a dense, turbulent, cloudy atmosphere which obscures its surface; hence its period of rotation is not accurately known but is believed to be about 30 days. The upper atmosphere is predominantly CO_2; its temperature is about $-38°C$, and the surface temperature is found to be approximately 430°C (Mariner II). It is uncertain whether the surface is completely dry and dusty, or covered with water, but the majority of data now point to the former.

⊕ Earth is the only part of the universe known to present an intermingling of water, air, and land, making possible a life zone accommodated to these elements. The earth moves about the sun in a slightly elliptical orbit (eccentricity 0.0167) at a mean distance of 93×10^6 miles (1 AU). The period is 365.24 days. The axis of the earth is tilted 23°27' to the plane of its revolution (ecliptic). The earth's only natural satellite is the moon, at a mean distance of 238,000 miles, period $27^d7^h43^m$, orbit eccentricity 0.0549, inclination to ecliptic 5°9'.

♂ Mars, the bright red planet, tinted by its oxide-covered desert regions, ranges in distance from earth (and stellar magnitude) between 35×10^6 miles (-2.9) and 250.10^6 miles ($+1.5$). It is believed to have a tenuous atmosphere less than 10 per cent as dense as that of earth, consisting mainly of CO_2 and N_2, with <0.1 per cent O_2. Its surface temperature varies between -120 and $+30°C$ with diurnal variations of about 60°C. Polar caps of water ice (not CO_2) covering 4×10^6 sq miles are clearly visible. Clouds of blue and white haze (ice crystals) are observed, and yellow dust storms with wind velocities up to 60 mph. Seasonal color variations, recurring in certain areas and along definite lines, are believed to indicate the presence of some form of life, perhaps moss or lichens. The so-called canals, however, have not been confirmed by modern photographic methods.

The Outer Planets, Jupiter, Saturn, Uranus, Neptune, Pluto

♃ Jupiter, the largest of the planets (86,860 miles in diameter), revolves around the sun in 11.86 years at a mean distance of 5.2 AU. It has 12 satellites, of which 4 are comparable in size with the moon. Its atmosphere consists of clouds or bands of hydrogen, helium, methane, ammonia, and other gases. Its period of rotation is about 10 hr. The temperature at the cloud layer is about 130°K. Surface markings, such as the great Red Spot, have been observed for periods of many decades. The surface temperature is found to be as high as 600°K by thermal radio emission, and radio noise at 15-m wavelength seems to indicate the presence of hot energy sources under a solid surface.

♄ Saturn, the second largest planet (71,500 miles in diameter), has an orbital period of 29.5 years around the sun, at a mean distance of 9.5 AU. The mean density is 0.7, lowest of all the planets. In addition to its remarkable engirdling system of thin rings, which is composed of a dense swarm of small solid bodies, it has nine other satellites beyond the outer edge of the rings. The atmosphere above the dense cloud layer consists of methane, hydrogen, helium, and ammonia, the latter probably in the form of tiny ice crystals, which constitute the visible clouds. The temperature is about 120°K. The largest satellite (Titan) is 3,100 miles in diameter and is the only satellite known to have an atmosphere.

Table 12. Physical Characteristics of the Planets*

	☿ Mercury	♀ Venus	⊕ Earth	♂ Mars	♃ Jupiter	♄ Saturn	⛢ Uranus	♆ Neptune	♇ Pluto
No. of natural satellites	0	0	1	2	12	9	5	2	
Apparent equatorial angular diam, sec	4.6–12.7	9.9–64.5		3.5–25.1	30.8–50.0	14.9–20.6	3.4–4.2	2.2–2.4	0.4–0.6
Equatorial radius:									
(⊙ = 1)	0.00346	0.00873	0.00915	0.00488	0.102	0.0865	0.0337	0.0320	0.010
(⊕ = 1)	0.379	0.956	1.00	0.535	11.14	9.47	3.69	3.50	1.1
km	2.42^{3}	6.10^{3}	6.37816^{3}	3.41^{3}	7.14^{4}	6.04^{4}	2.35^{4}	2.23^{4}	7^{3}
miles	1.50^{3}	3.79^{3}	3.96320^{3}	2.12^{3}	4.43^{4}	3.75^{4}	1.46^{4}	1.39^{4}	4^{3}
NM	1.31^{3}	3.29^{3}	3.44393^{3}	1.84^{3}	3.85^{4}	3.26^{4}	1.27^{4}	1.21^{4}	4^{3}
Oblateness:									
f	0	0	1/298.3	1/192	1/16.1	1/10.4	1/16	1/50	0
J_2	0	0	1.082^{-3}	1.92^{-3}	1.47^{-2}	1.67^{-2}	1.5^{-2}	5^{-3}	0
Volume (⊕ = 1)	0.054	0.87	1.00	0.153	1,400	850	50	43	1.3
Mean density:									
(⊙ = 1)	4.06	3.68	3.94	2.78	0.89	0.44	1.14	1.6	2.4
(⊕ = 1)	1.03	0.934	1.00	0.705	0.227	0.112	0.290	0.40	0.6
g/cu cm	5.7	5.16	5.52	3.89	1.25	0.62	1.60	2.2	3.3
lb/cu ft	355	322	344.6	243	78	39	100	138	200
Mass (including satellites):									
(⊙ = 1)	1.64^{-7}	2.4477^{-6}	3.04039^{-6}	3.236^{-7}	9.5475^{-4}	2.857^{-4}	4.360^{-5}	5^{-5}	$(2.5 \pm 0.3)^{-6}$
(⊕ = 1)	0.0546	0.81498	1.01230	0.1077	317.89	95.12	14.52	17	0.8 ± 0.1
$GM = \mu = g_0R^2$ (see Fig. 11) $\mu = aV_e^2/2$:									
km³/sec²	2.18^{4}	3.2485^{5}	3.98604^{5}	4.293^{4}	1.2671^{8}	3.7927^{7}	5.788^{6}	6.8^{6}	3.2^{5}
mile³/sec²	5.23^{3}	7.7936^{4}	9.56302^{4}	1.0299^{4}	3.0399	9.098^{6}	1.388^{6}	1.6^{6}	7.7^{4}
ft³/sec²	7.70^{14}	1.1472^{16}	1.40766^{16}	1.516^{15}	4.4747^{18}	1.339^{18}	2.044^{17}	2.4^{17}	1.1^{16}
AU³/day²	4.85^{-11}	7.2430^{-10}	8.88757^{-10}	9.576^{-11}	2.8252^{-7}	8.454^{-8}	1.290^{-8}	1.5^{-8}	7.4^{-10}
Equatorial surface gravity y_e:									
(⊕ = 1)	0.380	0.893	1.00	0.377	2.54	1.06	1.07	1.4	0.7
cm/sec²	372	873	978.031	369	2,490	1,040	1,050	1,400	700
Albedo (see Glossary, Sec. 2)	0.06	0.76	0.39	0.15	0.51	0.50	0.66	0.62	0.16
Max surface temp, °F	750	210	140	90	−200	−240	−270	−330	−370

NOTE: Superscripts denote exponents of 10; e.g., $4.6^{-8} = 4.6 \times 10^{-8}$.
$G = 6.6695 \times 10^{-8}$ (cgs units) = universal gravitational constant.
* From Ref. 3.

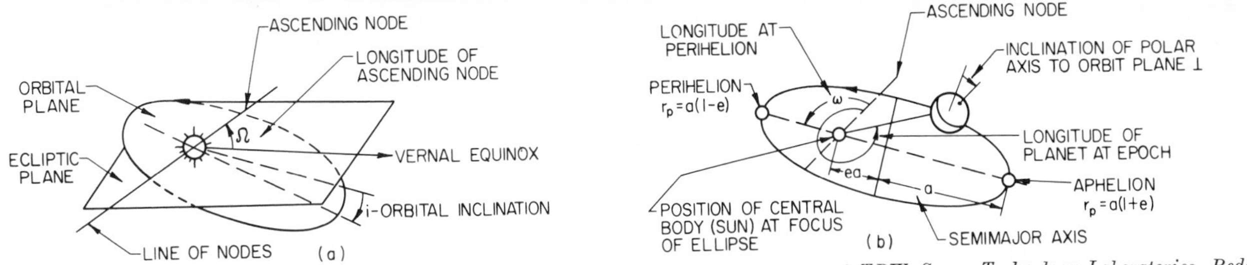

FIG. 46. Defining diagrams for planetary orbital characteristics. *(From Space Data, courtesy of TRW Space Technology Laboratories, Redondo Beach, Calif.)*

Table 13. Orbital Characteristics of the Planets*

	☿ Mercury	♀ Venus	⊕ Earth	♂ Mars	♃ Jupiter	♄ Saturn	⛢ Uranus	♆ Neptune	♇ Pluto
Semimajor axis a, AU	0.387099	0.723332	1.000000	1.523691	5.202803	9.538843	19.181973	30.057707	39.51774
Perihelion distance, AU	0.307501	0.718418	0.983207	1.381431	4.950829	9.007604	18.276561	29.800040	29.691899
Aphelion distance, AU	0.466697	0.728264	1.016727	1.665951	5.454777	10.07008	20.087385	30.315374	49.343581
Orbital eccentricity $e = f/a$	0.205627	0.006791	0.0167272	0.093370	0.048490	0.051618	0.0443132	0.0073350	0.2481112
Mean orbital velocity									
(⊕ = 1)	1.607271	1.175794	1.00	0.806855	0.438411	0.323782	0.2283249	0.1823988	0.1590757
km/sec	47.90	35.05	29.77	24.02	13.05	9.64	6.797	5.43	4.73
NM/sec	25.87	18.92	16.08	12.97	7.047	5.20	3.6705	2.93	2.55
ft/sec	157,186.0	114,958.0	97,702.1	78,805.7	42,817.6	31,595.2	22,302.0	17,802.7	15,560
Sidereal mean daily motion, in.	14,732.4	5,767.670	3,548.193	1,886.519	299.128	120.455	42.235	21.532	14.283
Period of revolution (⊕ = 1)	0.2411	0.6156	1.00	1.8822	11.86	29.46	84.0	164.8	247.7
Orbital inclination i to ecliptic, deg.	7.00402	3.39425	0	1.84992	1.30618	2.48715	0.77220	1.77320	17.16908
Inclination of equatorial plane to orbit, deg.			23.44436	25.2	3.115	26.745	97.983	29	
Mean longitude of ascending node Ω, deg.	47.87873	76.33611		49.26308	100.0708	113.3360	73.6925	131.4056	109.8975
Mean longitude of perihelion ω, deg (epoch Oct. 28, 1961)	76.86145	131.0339	102.2697	335.3563	13.2459	89.6225	172.5428	25.3654	224.5272
Mean longitude of planet at epoch, deg.	60.24552	160.9267	35.6137	236.8616	310.4313	299.6033	147.05343	221.01954	159.46964
Axial rotational period	$88^d.0$	150^d to 280^d	$23^h56.07^m$	$24^h37.38^m$	9^h53^m	10^h26^m	10^h42^m	15^h48^m	
Escape velocity $(2\mu/R)^{1/2}$, ft/sec	13,600	33,500	36,675	16,500	197,500	119,500	72,500	82,400	31,300

* From Ref. 3.

♅ Uranus, the third largest planet, is 19 AU from the sun, about 29,500 miles in diameter, and barely visible to the naked eye. Its sidereal period is 84 years, and it rotates on its axis in 10.7 hr. Its equatorial plane, and the orbits of its four largest satellites, are nearly perpendicular to the ecliptic. The cloudy atmosphere appears blue-greenish in color and occasionally forms into faint bands or belts parallel to the equator; the constituents are similar to Saturn's. A temperature of 100°K prevails at the cloud surface.

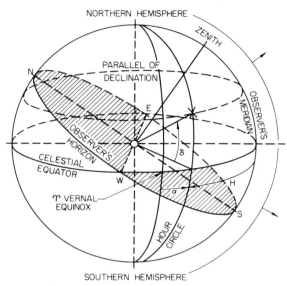

Fig. 47. Position on celestial sphere. Right ascension α, measured (eastward) along celestial equator from vernal equinox to hour circle of *. Declination δ, measured N or S along hour circle from celestial equation to *. Hour angle H, measured (westward) along celestial equator from observer's meridian to hour circle of *. (*From Space Data, courtesy of TRW Space Technology Laboratories, Redondo Beach, Calif.*)

♆ Neptune (26,800 miles in diameter) travels around the sun at 30 AU in a period of 165 years. Its discovery in 1846, as the result of computations by Leverrier and Adams, was considered a great triumph of mathematical astronomy. Like Uranus, its greenish atmosphere consists of methane, hydrogen, helium, and ammonia (solid state).

♇ Pluto, the most remote planet known (40 AU), is about the same size as earth and has a sidereal period of 248 years. Its orbital eccentricity (0.25) and inclination to the ecliptic (17°9′) are higher than any other planet. Its existence was predicted by Lowell in 1905, and confirmed photographically in 1925 by Tombaugh. The surface gravity is believed sufficient to retain an atmosphere of heavy gases.

9 Environmental Constraints

Space-system operations are limited by the environmental hazards, the physical conditions to which the systems are exposed during a space mission, the requirements of supporting man in space, and the availability of power in those missions. This section presents essential data on the geophysical environment, including the very important radiation effects, solar flares, heat transfer and thermal equilibrium in space (Fig. 55), manned-space-flight data, and space power systems (Fig. 57). Additional information on environment and power supplies will be found in Chap. 10.

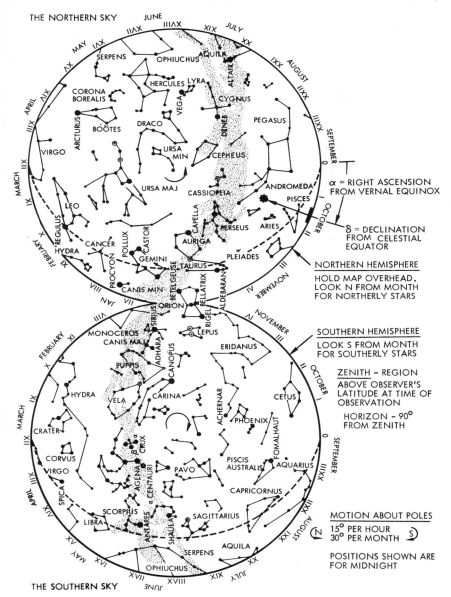

Fig. 48. Map of stars and constellations. (*From Space Data, courtesy of TRW Space Technology Laboratories, Redondo Beach, Calif.*)

Geophysical Environment

Studies of the space environment in the vicinity of earth up to 20,000-NM altitude show that the following geophysical effects are of prime importance in the design of spacecraft: gravitation, atmosphere, radiation, and micrometeorites.

Gravitational effects for operations near the earth or other celestial bodies are normally represented by the point-source gravitational parameter μ, from which the velocity required for circular and escape orbits can be computed as $(\mu/r)^{1/2}$ and

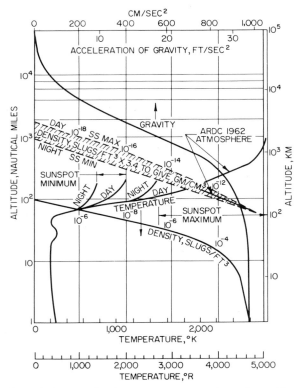

Fig. 49. Altitude variations of density, temperature, and gravity. (*From Space Data, courtesy of TRW Space Technology Laboratories, Redondo Beach, Calif.*)

$(2\mu/r)^{1/2}$, respectively (r = distance from the gravitational source). For near-earth orbits of long duration, and for precise space-trajectory calculations, it is also necessary to account for the nonspherical shape of the earth, which causes deviations to occur from the original orbit plane. Gravitational effects are overcome by propulsion and attitude-control devices, suitable for the original boost velocity, for midcourse corrections, and for reentry deboost. Guidance and control are also necessary to facilitate the application of propulsive forces in the proper direction for accurate trajectory shaping.

Atmospheric characteristics of the earth and other celestial bodies include the variation of density and pressure with altitude due to gravitational effects; the temperature resulting from solar radiation and atmospheric convection; and the chemical constituents of the atmosphere. Provisions necessary to overcome these atmospheric effects are pressurized cabins with adequate oxygen supply, adequate bearings and lubricants, aerodynamic drag and lift devices for reentry, heat shielding, dielectric protection for electronic apparatus, etc.

Radiation effects include the temperature and pressure from high-energy sources like the sun, with allowance for distance, eclipse, and anomalies from sunspot activity. Radiation dosage from solar and cosmic electrons and protons is also of prime concern. These effects require the provision of adequate temperature control for solar radiation

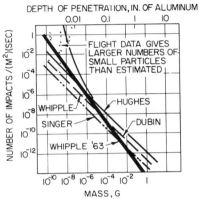

Fig. 50. Depth of penetration of micrometeoroids. (*From Space Data, courtesy of TRW Space Technology Laboratories, Redondo Beach, Calif.*)

Depth of penetration d

$$d \sim (E/pH_F)^{1/3}$$

where E = kinetic energy of particle
 p = density of spacecraft material
 H_F = heat of fusion of spacecraft material

(and the absence thereof during eclipse periods); shielding to protect against injury or damage to personnel or apparatus from bombardment of radiation particles; and attitude controls suitable for overcoming solar pressure, gravity gradient, and other small torques.

Micrometeorites are prevalent in sufficient quantities that their size, momentum, frequency, and penetration effects need to be considered. Provisions for protection against damage due to normal or abnormal micrometeorite density include reserve wall thickness and puncture-sealing provisions.

Figure 50 shows depth of penetration as a function of number of impacts per square meter per second which penetrate a specific thickness of aluminum, as given by a number of investigators. Depth of penetration may also be predicted by the expression given in the figure. See also Table 14.

Other effects such as the geomagnetic field and its variations with electromagnetic disturbances are presently believed to be of lesser importance in spacecraft design, except for special purposes. A few pages of geophysical data are provided as an aid to the analysis of space missions (see Figs. 51 to 54 and Tables 15 to 19).

Solar-flare Data[9]

Cosmic rays are primarily protons with energies ranging from less than 10 Mev up to several bev. During solar flares the intensity of cosmic-ray flux may increase, for intervals of 1 or 2 days, to many thousand times greater than the normal galactic flux intensity of 1.5 to 4 particles/sq cm/sec.

The integrated yearly dose of 5 to 12 roentgens then results from the 0.5 to 1.2×10^8 particles/sq cm (with an estimated specific ionization of three times the minimum flux

Table 14. Probability of Micrometeoroid Penetration*
(Body of 1 sq m Cross Section)

Meteor visual magnitude	Kinetic energy, ergs	Mass, g	Particle radius, mm at 4 g/cu cm	Probable no. of encounters in 24 hr in 2 years		Penetration of aluminum, mm
0	1.0^{13}†	1.25	4.6	3.6^{-9}	2.6^{-6}	109
1	4.0^{12}†	0.5	3.4	9.3^{-9}	6.8^{-6}	80
2	1.6^{12}†	0.198	2.5	23.1^{-9}	16.8^{-6}	59
3	0.63^{12}	0.079	1.8	6.0^{-8}	4.4^{-5}	43
4	0.25^{12}	0.031	1.4	14.7^{-8}	10.7^{-5}	32
5	0.10^{12}	0.012	1.0	3.6^{-7}	2.6^{-4}	23
6	40.0^{9}	5.0^{-3}	0.74	9.3^{-7}	6.8^{-4}	17
7	16.0^{9}	2.0^{-3}	0.54	23.1^{-7}	16.8^{-4}	13
8	6.3^{9}	0.79^{-3}	0.40	6.0^{-6}	4.4^{-3}	9.3
9	2.5^{9}	0.31^{-3}	0.29	14.7^{-6}	10.7^{-3}	6.9
10	1.0^{9}	0.12^{-3}	0.22	3.6^{-5}	2.6^{-2}	5.1
11	4.0^{8}	50.0^{-6}	0.16	9.3^{-5}	6.8^{-2}	3.7 (0.145 in.)
12	1.6^{8}	20.0^{-6}	0.12	23.1^{-5}	16.8^{-2}	2.7 (0.106 in.)
13	6.3^{7}	7.9^{-6}	0.086	6.0^{-4}	4.4^{-1}	2.0 (0.079 in.)
14	2.5^{7}	3.1^{-6}	0.063 (0.0025 in.)	1.5^{-3}	1.1	1.5 (0.059 in.)
15	1.0^{7}	1.2^{-6}	0.046 (0.0018 in.)	3.6^{-3}	2.6	1.1 (0.043 in.)

* From Ref. 1.
† NOTE: Superscripts are exponents of 10; i.e., $4.0^{12} = 4.0 \times 10^{12}$.

Table 15. Annual Radiation Dosage for Satellites in Various Earth Orbits*
Number of impacts/sq cm/year for orbits shown by roman numerals in Fig. 51

Orbit no.	Electrons							Protons						
	Energy exceeding values in mev line below													
	0.045	0.11	0.20	0.50	0.80	1.60	5.00	1	4	10	20	40	80	100
I	3.4^{13}	1.2^{13}	8.0^{12}	2.5^{12}		2.5^{10}	1.2^{8}	8.4^{12}	1.6^{11}	1.6^{10}	2.8^{9}	5.0^{8}		5.0^{7}
II	3.0^{14}	1.1^{14}		2.2^{13}		2.2^{11}	1.1^{9}	8.0^{12}	1.6^{11}	1.6^{10}	2.8^{9}	5.0^{8}		5.0^{7}
III	7.0^{13}	2.6^{13}		5.1^{12}		5.1^{10}	2.6^{8}	5.7^{12}	1.6^{11}	1.6^{10}	2.8^{9}	5.0^{8}		5.0^{7}
IV	6.3^{15}	2.3^{15}		4.7^{14}		4.7^{12}	2.3^{10}	6.2^{14}	3.1^{11}	2.0^{8}				
V	6.4^{15}	2.3^{15}		4.7^{14}		4.7^{12}	2.3^{10}	5.0^{14}	1.2^{10}					
VI	1.9^{15}	7.0^{14}		1.0^{14}		1.4^{12}	7.0^{9}	6.0^{13}	3.0^{8}					
VII	3.1^{14}		6.6^{13}	2.2^{13}	6.0^{12}	2.2^{11}		3.0^{12}	3.3^{8}	3.3^{8}	3.3^{8}	2.7^{8}		1.3^{8}
VIII	5.2^{14}	1.9^{14}		4.1^{13}		3.9^{11}					2.3^{12}	1.4^{12}	1.0^{12}	6.2^{11}
IX	1.3^{16}	4.6^{15}		1.0^{15}		9.4^{12}					8.3^{9}	4.0^{9}	2.7^{9}	1.5^{9}
X	1.6^{12}	5.8^{11}		1.2^{11}		1.2^{11}					1.0^{9}	7.0^{8}		1.0^{8}
XI	2.0^{11}	7.3^{10}		1.5^{10}				5.0^{12}	1.6^{11}	1.6^{10}	2.8^{9}			5.0^{7}
XII	4.1^{14}	1.5^{14}		3.2^{13}							5.1^{11}	2.6^{11}	1.3^{11}	
XIII	3.0^{12}	1.0^{12}		2.0^{11}							1.0^{10}	7.0^{9}	4.0^{9}	

NOTES: Superscripts are exponents of 10; i.e., $7.0^{9} = 7.0 \times 10^{9}$.
The above estimates should be regarded as accurate to an order of magnitude, because of the lack of data in certain regions of space.
* From Ref. 8.

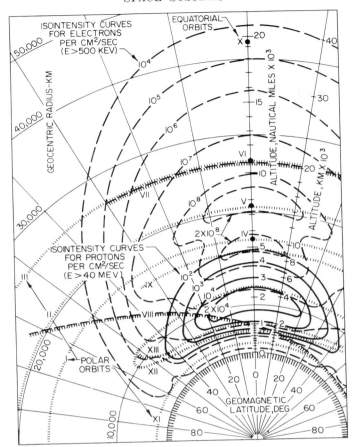

Fig. 51. Electron and proton intensity with typical earth orbits. (*From Space Data, courtesy of TRW Space Technology Laboratories, Redondo Beach, Calif.*)

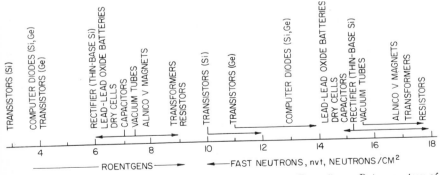

Fig. 52. Radiation thresholds for electronic components. (*From Space Data, courtesy of TRW Space Technology Laboratories, Redondo Beach, Calif.*)

Table 16. Radiation Damage to Solar Cells*
Based on data provided in Table 15
Time, years,† to reach 25% degradation in silicon p-n cells by the electron and proton components of the radiation environment for various orbits and shield thicknesses L
(1.35×10^{10} defects/sq cm = 25% degradation)

L, mils	Orbit						
	I	II	III	IV	V	VIII	IX
4	0.35	0.34	0.26	0.038	0.038	0.0004	0.019
10	1.73	1.52	0.8	0.061	0.061	0.0022	0.029
30	10.4	7.95	3.14	0.218	0.218	0.012	0.11
65	25.5	23	13.5	1.12	1.12	0.31	0.56
100	37.5	37.5	33.8	12.2	12.2	0.45	4.1

L, mils	Orbit					
	VI	VII	X	XI	XII	XIII
4	0.146	0.8	26.5	0.355	0.193	0.096
10	0.199	1.43	27.6	1.73	0.010	0.48
30	0.71	4.5	30.0	10.4	0.056	2.88
60	4.2	29.2	31.4	25.4	0.14	7.1
100	45	312	31.4	37.5	0.202	10.4

† n-p cells are more resistant to radiation damage than p-n cells, as follows:

$$\frac{\text{Lifetime of } n\text{-}p \text{ cells}}{\text{Lifetime of } p\text{-}n \text{ cells}} = \begin{cases} 3 \text{ for protons} \\ 10 \text{ for electrons up to about 8 mev} \\ 30 \text{ for electrons up to about 1 mev} \end{cases}$$

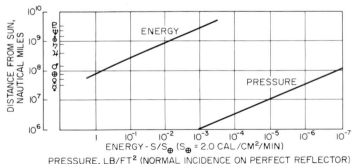

Fig. 53. Solar-radiation effects.[1] (*From Space Data, courtesy of TRW Space Technology Laboratories, Redondo Beach, Calif.*)

intensity). During the period 1956 to 1961, the annual dose was increased by solar flares by a factor of 2.5 for particles above 100 mev and by a factor of 15 for particles above 30 mev, as shown in Table 18. The yearly integrated intensities are listed by year, and also the most active solar flare observed during each year. Such data are expected to be useful as design requirements for manned spacecraft, as well as for critical subsystems and unmanned spacecraft.

Table 17. Radiation Effects on Man

Definitions

One r (roentgen) is the quantity of gamma or X radiation that produces an energy absorption of 83 ergs/g of dry air.

One *rep* (roentgen equivalent—physical) is the quantity of radiation that produces an energy absorption of 93 ergs/g of aqueous tissue.

One *rad* (radiation absorbed dose) is required to deposit 100 ergs/g in any material by any kind of radiation.

One *rem* (roentgen equivalent per man) is the unit of particulate radiation that produces tissue damage in man.

The conversion factor from rad to rem is the "relative biological effectiveness," i.e., dose in rem = dose in rad × RBE.*

Exposure-tolerance Values for Man, Whole-body Radiation Doses

0.001 rem/day	Natural background radiation
0.01 rem/day	Permissible dose range 1957
0.1 rem/day	Permissible dose range 1930–1950
1 rem/day	Debilitation 3–6 months; death 3–6 years (projected from animal data)
10 rem/day	Debilitation 3–6 weeks; death 3–6 years (projected from animal data)
100 rem, 1 day; 150 rem, 1 week; 300 rem, 1 month	Survivable emergency exposure dose but permitting no further exposure for life
25 rem	Single emergency exposure
100 rem	20-year career allowance
500 rem	Maximum permissible 20-year career allowance

Conversion: rad to rem*
Based on most detrimental chronic biological effects for continuous low dose exposures

Type of radiation	RBE
X rays	1
Gamma rays	1
Beta particles, 1.0 mev	1
Beta particles, 0.1 mev	1.08
Neutrons, thermal	2.8
Neutrons, 0.0001 mev	2.2
Neutrons, 0.005 mev	2.4
Neutrons, 0.02 mev	5
Neutrons, 0.5 mev	10.2
Neutrons, 1.0 mev	10.5
Neutrons, 10 mev	6.4
Protons > 100 mev	1–2
Protons, 1 mev	8.5
Protons, 0.1 mev	10
Alpha particles, 5 mev	15
Alpha particles, 1 mev	20

* Example for total dose: For a given exposure time, a dose of 0.2 rad of γ radiation, plus 0.04 rad of thermal neutrons, gives a total dose of (0.2 × 1 RBE) + (0.04 × 2.8 RBE) = 0.312 rem.

Recommended Maximum Weekly Dosage

Radiation	Rems/week					
	Skin		Lens of eye	Gonads	Blood-forming organs	Intermediate tissue (0.07–5.0 cm depth)
	Total body	Appendages				
X or γ rays < 3 mev	0.45	1.5	0.45	0.3	0.4	0.4–0.45
Electrons or β	0.6	1.5	0.3	0.3	0.3	0.3–0.6
Protons	0.6	1.5	0.3	0.3	0.3	0.3–0.6
Fast neutrons	0.3–0.6	0.75–1.5	0.3	0.3	0.3	0.3–0.6
Thermal neutrons	0.5	1.2	0.3	0.1	0.17	0.17–0.5
Alpha particles (α)	1.5	1.5	0.3	0.3	0.3	0.3–1.5
Heavy nuclei (O, N, C) (locally generated)	1.5	1.5	0.3	0.3	0.3	0.3–1.5

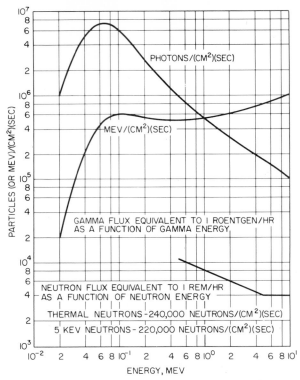

Fig. 54. Conversion of flux to dose.[18] The figure below shows the number of gammas/sq cm/sec or mev/sq cm/sec that is equivalent to 1 roentgen/hr. (Accepted laboratory tolerance is about 300 milliroentgens per 48-hr week.) The neutron flux that is equivalent to 1 rem/hr is also given as a function of neutron energy. (*From Space Data, courtesy of TRW Space Technology Laboratories, Redondo Beach, Calif.*)

Table 18a. Yearly Integrated Intensities Compared for Solar Protons and Galactic Cosmic Rays

Year	No. of events	Solar-proton integrated intensity (10^7 particles/sq cm)		Galactic cosmic-ray integrated intensity (10^7 particles/sq cm)
		>30 mev	>100 mev	
1956	2	70	32	10
1957	4 or 5	40	1	7
1958	6	70	1	6
1959	4	360	34	6
1960	8	200	50	8
1961	5	32	6	10
Total......	30	772	124	47

SPACE SYSTEMS

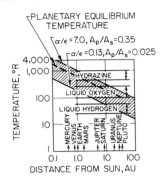

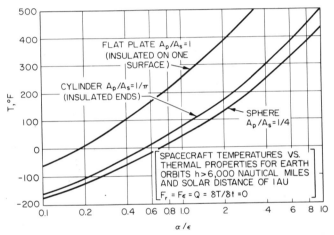

Fig. 55. Equilibrium temperatures in space. (*From Space Data, courtesy of TRW Space Technology Laboratories, Redondo Beach, Calif.*)

The heat-balance equation is as follows:

$$\alpha G A_p + \alpha E_r F_r A_s + \epsilon E_\epsilon F_\epsilon A_s + Q = \epsilon \sigma T^4 A_s + W C_p (\partial T / \partial t)$$

α/ϵ = absorptivity to solar spectrum/emissivity

        0.15/0.85 to 0.25/0.90 for white paint
        0.08/0.04 to 0.35/0.10 for aluminum
        0.66/0.05 for 2024 Al alloy, as received
        0.77/0.07 for beryllium foil
        0.23/0.02 for gold deposited on resin over metal
        0.25/0.03 for gold on aluminum
        0.20/0.04 for gold on fiberglass

G = solar radiation (at earth) = 442 Btu/hr/sq ft
E_r = earth-reflected solar radiation = 0 − 177
E_ϵ = earth-radiated infrared = 66–72 Btu/hr/sq ft
F_ϵ = configuration factor for emitted radiation
F_r = configuration factor for reflected radiation
A_p = projected area to solar radiation, sq ft
A_s = radiating surface area, sq ft
T = absolute temperature of surface, °R
W = weight of heat sink, lb
C_p = specific heat of material, Btu/lb °F
σ = Stefan-Boltzmann constant = 0.1714×10^{-8} Btu/(hr)(sq ft)(°R)[4]
Q = internally generated heat, Btu/hr

Table 18b. Major Solar-proton Outbursts during Years 1956-1961

Date	Time of optical max	Time of r-f max	Max r-f signal 3-10 gc (10^{-19} watts/ sq m/cps)	Onset + rise time, hr		Decay time, hr		Peak intensity (10^5 protons/ sq cm/sec)	
				>30 mev	>100 mev	>30 mev	>100 mev	>30 mev	>100 mev
2/23/56	0340	0341	20	6-8	3-4	30	16	6.2	5
1/20/57	1120	....	....	.....	.....	.....	.....	2-3	0.15
7/7/58	0115	0112	2	.....	.....	32	16-20	1.5-2	0.08
7/14/59	0349	0352	6.3	16-20	12-18	18	9-12	10-12	1.2
11/12/60	1329	1329	10	12-16	8-10	18-24	14-18	12	2.5
7/18/61	1010	....	5	6-10	2-3	24	12	2.5	0.6

Temperature in Space

Figure 55 provides information for estimating equilibrium temperatures of certain shaped bodies in space.

The heat loss in Btu/hr through a surface of thickness X is computed as follows:

$$\text{Btu/hr} = KAT/X = \text{watts}/3.413$$

where K = thermal conductivity, Btu/hr/sq ft/°F/ft
A = cross-sectional area of heat path, sq ft
T = temperature difference across heat path, °F
X = length of heat path, ft

Table 19. Summary of Thermal-conductivity Values

Material	Pressure	Temp, °R	Density ρ, lb/cu ft	Thermal conductivity K, Btu/hr/ft²/°F
Aluminum..........................	.....	Room	200	130.0
Magnesium (alloys).................	.....	Room	110	35.0 (to 100)
Stainless steel.....................	.....	Room	500	10.0
Magnesium oxide...................	.....	Room	220	1.0
Solid plastic: Teflon, Nylon, polyethylene.........................	.....	Room	100	0.15
Foam plastic:				
Polycel...........................	1 atm	40-500	2	0.005
Styrofoam.......................	1 atm	40-500	2	0.010
Santocel A.......................	1 atm	160-520	6	0.014
	0.1μ	160-520	6	0.0012
Perlite...........................	0.1μ	36-520	8	0.00072
Fiberglas AA.....................	1 atm	140-535	15	0.0125
	10 μ	140-535	15	0.00031
	10 μ	140-525	4	0.0005
Superinsulation:* Multiple reflective layers SI-4 Alt. Al foil and paper	1 atm	36-520	3-5	0.020
NRC-2 crinkled aluminized Mylar	0.1μ	36-520	3-5	0.000030

* See Fig. 56.

Table 20. Physiological Design Criteria for Manned Space Flight[a]

Condition	Opt value	Conventional limits	Extreme limits (g)
Relative humidity, %	50	30 70	dna
Temp, °F, effective	68	66–71 66–71	dna
Temp, °F, dry bulb	72	72–78 68–74	dna
Atmosphere:			
Total cabin pressure, psia			
Standard air	14.7	10.1–14.7	5.4–14.7
100% oxygen[h]	dna	3.5– 5	2.7– 8.3
Oxygen partial pressure, mm Hg			
In standard air (21% O_2)	160	110–160	59–160
All oxygen[a]	dna	180–260	141–430
Nitrogen partial pressure, mm Hg	593	0–580	0–619
Water partial pressure, mm Hg	9.6[b]	5–15[c]	dna
CO_2 partial pressure, mm Hg	0.3	0–8[h]	0–23[h]
Acceleration tolerance (g's):[d,e]			
Eyeballs up	dna	0– –1	–3
Eyeballs down	1	0–4	Grayout 3–4
			Blackout 5–6
			Unconscious 6.5–7
Eyeballs in	1	0– 10	12–14 for 120–180 sec
Eyeballs out	dna	0– –1	–5 for 15 sec
Human whole-body tolerance to sinusoidal vibrations:[f]			
Short time	dna	dna	3.8 g_{max} at 1 cps
			2.2 g_{max} at 6 cps
			6.7 g_{max} at 15 cps
1 min	dna	dna	2.9 g_{max} at 1 cps
			1.0 g_{max} at 6 cps
			2.9 g_{max} at 15 cps
3 min	dna	dna	2.3 g_{max} at 1 cps
			0.6 g_{max} at 6 cps
			1.6 g_{max} at 10 cps
			No data at 15 cps
Where A = amplitude, microns; f = frequency, cps	....	0–0.2 A_{max}	$A_{max} = 10^5/f^2$
Acoustic tolerance, db	....	0–80	135$_{max}$
Ionizing radiation tolerances (see Table 17)			

* There are exceptions and additional conditions to the above data.[16,17]
[a] Same as total cabin pressure with 100% O_2.
[b] At 74°F and 50% relative humidity.
[c] At 74°F and 30 to 70% relative humidity.
[d] Tolerance to acceleration is in reality a function of rate of onset, duration, and time spent at zero g.
dna = does not apply.
[e] Long-term effects of zero g not established.
[f] Resonance of the thoracic-abdominal system limits human tolerance at 4 to 8 cps. If the accelerations g are increased, or if the time period for the 3-min levels is increased, tissue-organ systems will be damaged.
[h] Time-dependent.

Physiological Constraints

Tables 20 and 21 summarize the principal factors affecting manned operations in space.

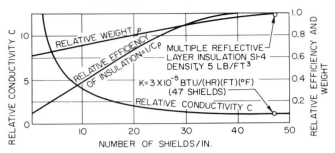

FIG. 56. Superinsulation efficiency and thermal conductivity vs. number of layers per inch. (*From Space Data, courtesy of TRW Space Technology Laboratories, Redondo Beach, Calif.*)

Table 21. Life Support and Environmental Control

	Lb/man-day
O_2 consumption, on 3,000-kg-cal diet	2.0
CO_2 production, on 3,000-kg-cal diet	2.25
Water intake:*	
To reconstitute dried food, avg	2.12
For drinking, avg	3.68
Metabolic H_2O	0.86
Wash water	1.5
Dried food (0% H_2O), avg 3,000-kg-cal diet	1.3

Metabolic heat output, Btu/hr:	Avg	Routine duties
Shirtsleeve (in cabin)		
Sensible	315	354
Latent	173	312
Mercury pressure suit (in cabin)		
Sensible	156	133
Latent	332	533
Apollo suit (hard work on moon), total metabolic heat		600–2,000

	Avg lb/man-day
Feces water	0.25
Feces solids	0.11
Urine water	4.21
H_2O condensate, respiration and perspiration	2.2
Growth—hair, nails, etc	0.1

* Total amount is dependent on electrolyte management and environmental conditions. Also, depending on the per cent of water in the dried food, the ratio of water for drinking to water for reconstitution will vary.

Space Power Supplies

Figure 57 provides application data for primary power sources. Refer to pp. 10-79 through 10-108 for more detailed information.

10 Communications Constraints

A number of special considerations apply to communications from space vehicles and from points in space. It is usually necessary to track such vehicles by some electronic trajectory-measurement system (Chap. 3). This tracking is expedited by the use of a beacon, useful also as a command receiver. Thus unified tracking command and telemetry systems have evolved, particularly in the S-band microwave region. These systems, called unified S band, use a particular frequency allocation for all these functions.

For near-earth orbits communication with ground stations is necessarily discontinuous as overfly time is limited by the line of sight between ground station and orbiting vehicle. Such contacts must be planned from orbital ephemeris data and the minutes of transmission time available estimated. Such work is most conveniently done by means of a digital computer. The computer can take into account the effect of the oblate earth and atmospheric drag on the ephemeris of the vehicle. The elevation-angle coverage of the ground-station antenna is taken to be 5° from the horizon and above to minimize atmospheric effects and interference from ground stations.

A network of stations may be used to receive data accumulated during an orbit. In such a case it is necessary to reduce the total overfly time per orbit, or day, of the stations by the *redundant* time when two or more stations may be in common view of the space station and hence receive the same data.

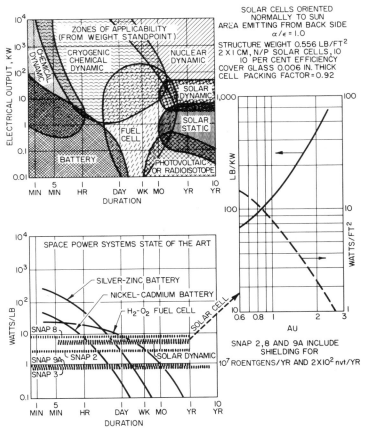

FIG. 57. Space power systems. (*From Space Data, courtesy of TRW Space Technology Laboratories, Redondo Beach, Calif.*)

The required inputs to the program for computing the total time of communication are the station locations, orbital inclination, and orbital altitude.

Laser Communications

Lasers make use of the phenomenon of light amplification by stimulated emission. They are attractive for communication in space because this form of light-energy generation may be coherently emitted, thus providing a highly directive beam, typically in the order of 1 milliradian. This narrow beam represents extremely high gain compared with microwave systems and energy concentrated in only the desired direction. Because of the short wavelength of light the aperture required for the energy concentration is conveniently small. These considerations indicate most usefulness of lasers will be to long-distant points in space having low relative motion to each other.

To exploit the laser the following problems must be solved:

1. A highly precise attitude or directional control must be perfected so that a user can derive benefit from the highly directional beam.
2. Methods must be found to limit and stabilize the bandwidth of laser receivers, so as to exclude all but the desired information bandwidth.
3. Higher carrier-wave power sources are required.

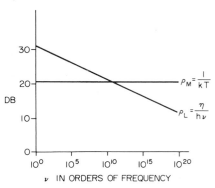

FIG. 58. Comparison of quantum signal-to-noise ratio with thermal signal-to-noise as function of frequency (assuming received power per unit bandwidth is 1 watt per cycle).

The advantage of the laser is partially offset by the quantum "noise," which is relatively larger at light frequencies than the thermal noise is at microwaves. Thus the signal-to-noise ratio of a laser is, neglecting strong background light,

$$\rho_L = \eta P_r/h\nu B \qquad (16)$$

where ρ_L = signal-to-noise ratio (laser)
η = quantum efficiency of detecting device ($= 0.004$ to 0.1 for photoemissive-type detectors such as photomultipliers and 0.5 for photodiodes)
P_r = received power
B = effective bandwidth
h = Planck's constant = 6.6×10^{-34} joule-sec
ν = light frequency, cps

This may be compared with the signal-to-noise ratio for microwave receivers

$$\rho_m = P_r/kTB \qquad (17)$$

where k = Boltzmann's constant = 1.34×10^{-23} joule/°K
T = receiver antenna temperature, °K

If a nominal value for quantum efficiency of 0.01 is used and T is taken as 300°, a comparative plot can be made of the two signal-to-noise ratios. This is shown in Fig. 58.

REFERENCES

1. H. S. Seifert (ed.), "Space Technology," John Wiley & Sons, Inc., New York, 1959.
2. "The McGraw-Hill Encyclopedia of Science and Technology," McGraw-Hill Book Company, New York, 1960.
3. E. C. Levy (ed.), "SRL Design Manual," Systems Design and Analysis Department, Space Technology Laboratories, 1961.
4. H. H. Koelle (ed.), "Handbook of Astronautical Engineering," McGraw-Hill Book Company, New York, 1961.
5. J. C. Lair, Mission to a Comet, STL Rept. 9844-0023-MU-R01, July 11, 1962.
6. "American Ephemeris and Nautical Almanac," Government Printing Office, Washington, D.C., 1961.

7. Robert H. Baker, "Astronomy," 7th ed., D. Van Nostrand Company, Inc., Princeton, N.J., 1959.
8. A. Rosen, "Comsat Flight Environment," Contract SD-117, Space Technology Laboratories, 1962.
9. W. R. Webber, Solar Flare Proton Data, *Nucleonics*, August, 1963.
10. R. W. Wolverton (ed.), "Flight Performance Handbook for Orbital Operations," Space Technology Laboratories, John Wiley & Sons, Inc., New York, 1961.
11. R. K. Whitford, Design of Attitude Control Systems for Earth Satellites, *Control Eng.*, February, 1962; April, 1962.
12. "1965 NASA Authorization," hearings relative to H.R. 9641, Government Printing Office, Washington, D.C., February, March, 1964.
13. J. F. White (ed.), "Flight Performance Handbook for Powered Flight Operations," Space Technology Laboratories, John Wiley & Sons, Inc., New York, 1962.
14. R. J. Lane, Recoverable Air-breathing Boosters for Space Vehicles, *J. Roy. Aeronaut. Soc.* vol. 66, pp. 373–376, June, 1962.
15. S D. Ellis and R. H. Ellis, Jr., Dosimetry Conversion Factors, *Nucleonics*, November, 1960.
16. "Life Sciences Data Book," Directorate of Aerospace Medicine, Office of Manned Space Flight, NASA Hq., June, 1962.
17. H. L. Mandelstam and A. W. Optican, First Generation Manned Spacecraft Environmental Control System, *Space Technol. Lab. Rept.* 9990-6397-RU-000, March, 1963.
18. J. Moteff, "Miscellaneous Data for Shielding Calculations," APEX 176, AND Department, General Electric Corp., Dec. 1, 1964.
19. Carsbie C. Adams, "Space Flight," Chap. 3, McGraw-Hill Book Company, New York, 1958.

SATELLITE TELEMETRY*

11 Introduction

This section considers some of the controlling considerations in the design of telemetry and data-acquisition systems for scientific and unmanned observatory satellites. This is followed by a brief review of the data-acquisition systems which have seen the most use in the United States program. Some comments are then offered on system concepts and component developments which are affecting the telemetry and data-acquisition situation. The requirements set for the telemetry and data-recovery systems for a typical small satellite for the Scout vehicle and for the large Orbiting Geophysical Observatory (OGO) satellites are presented and analyzed. The material presented does not cover the detailed design, in the electronic engineering sense, of signal-conditioning devices, radio-frequency links, etc., for the telemetry but is limited to a review of existing systems developments and the factors affecting the choice of techniques for processing the information collected by the sensors aboard a spacecraft.

12 Constraints on Satellite Telemetry and Data-recovery Systems

The engineering of the early U.S.-IGY systems[1–15] was very much affected by the difficulty of meeting the scientific experimenters' needs for information transfer, within the restricted limits of satellite weight and power-supply capacity, which were themselves dictated by the booster vehicle's performance capability. Performance constraints have not left us, nor are they ever likely to, as we can expect that the

---

* This chapter subsection is taken, with some revision, from a paper of this title presented at the Second International Space Science Symposium, Florence, 1961, and published in Space Research II, edited by H. C. Van De Hulst, C. DeJager, and A. F. Moore, and published by North Holland Publishing Company, Amsterdam, 1961. Permission for the use of the material is gratefully acknowledged. The author wishes to acknowledge the important part played in the formulation of many of the concepts discussed here by many members of the staff of the NASA Goddard Space Flight Center and the NASA Headquarters. In particular, the assistance of R. W. Rochelle, C. Creveling, C. Stout, P. T. Cole, and M. J. Aucremanne in the preparation of this material has been most valuable.

desire to collect additional scientific data will always outrun the ability to engineer a system that will collect and transmit it. For this reason, opportunity certainly still exists for the development of better data-transmission techniques and for the application of new theoretical developments relating to information collection and processing. A major characteristic of the problem is the low signal-to-noise ratio normally encountered in satellite data transmissions. In this respect, the system problems are appreciably different from those which faced the designer working on aeronautical and missile test telemetry. However, with the increase in permissible payload weight it is becoming possible to relax some of the constraints; a typical example of the capability to carry a larger power supply is the growth in transmitter-radiated power from the time of Explorer I (60 mw for 2 weeks and 10 mw for several months) to the more recent systems using solar cells to recharge storage batteries (Explorer VI, 5 watts on command for several hours per day or an equivalent of about 500 mw continuously; Explorer VII, 600 mw continuously).

Weight and power constraints were not the only factors dictating system design for the earlier satellites. As it was necessary to use the data-transmission link as a component of the position-locating system, appropriate compromises in system design were made for the first Explorers and Vanguards. A specific example was the original Minitrack requirement that the radio-frequency carrier, which was to be used for tracking, have no modulation components at 500 cps or at its harmonics up to and including 2,500 cps.

It was apparent at the initiation of the earth-satellite program that it would be impractical to provide worldwide receiver coverage for continuous data transmission by telemetry. A great deal of ingenuity has accordingly gone into the design of the components and systems which make it possible to collect scientific data for one or more orbits and which then permit command control of the data transfer to the ground recorders.

Not the least of the constraints on the satellite systems designer are the demands for reliability and long life in orbit and for stability of performance in the environment encountered in space. These demands affect every detail of the design of satellite-borne equipments and lead to the selection of those components, circuits, and modulation schemes which will be least affected by any changes in component operating characteristics that may occur during the life of the satellite.

Finally, it is necessary for the designer to select a system for data handling, which will acquire and process the data with sufficient speed, as well as precision, for the experimenter to make effective use of the results in planning his next experiment. This requirement grows in importance as the complexity of experiments increases. A satisfactorily performing satellite can be tracked fairly easily by the observing networks and the precision orbital elements and ephemerides produced in a few weeks. At the same time, the number of telemetry tapes can run into the thousands and the number of data samples into the hundreds of millions. For example, in the 2 months of useful life of Explorer VIII, the Minitrack network collected from over 300 million data points. A careful experimental design and a plan for data handling which adopts automatic processing as a technique to recover the bulk of the useful information is needed in every case. Any such detailed plan must, however, retain for the experimenter an ability to detect an unusual or unexpected event. The automatic processor should not throw away such data as if they were noise on the desired signal.

Without some form of automatic processing, many experiments which are designed to monitor specific physical quantities for spatial and temporal fluctuations will overwhelm the experimenter with masses of indigestible data. The properly planned data-processing system serves a primary function in that it makes it practical to plan on collecting survey data throughout the life of a satellite. The design of the total data-handling system must therefore consider the condensed form in which the final presentation of the survey data will be made, as well as the possibility of having to display almost every bit of information collected when a search is being made for a previously unanticipated or undetected phenomenon.

13 Past Telemetry and Data-recovery Systems

As Williams[16] has noted, the United States satellites which have been orbited up to this time have generally used one of three telemetry systems. These were (1) systems based on the use of several frequency-modulated subcarriers which simultaneously modulated a radio-frequency carrier, (2) systems using bursts of frequency-modulated subcarrier energy switched in sequence to the transmitter modulator and varied in burst length and spacing to transfer additional information, and (3) the pulse-code-modulation system in which quantized data are transmitted by a sequence of coded groups of pulses.

The initial United States efforts in the design of telemetry systems for scientific satellites resulted in the systems used in Explorers I, III, and IV and in those used in Vanguards II and III. The early Explorer systems[2,10,11,12,13,14,15,16,17,18] were

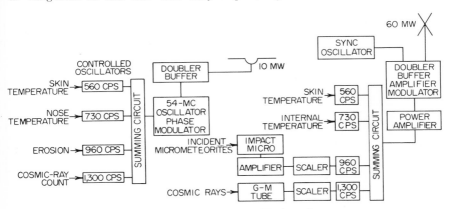

FIG. 59. Block diagram of Explorer I.

designed and fabricated at the State University of Iowa and at the Jet Propulsion Laboratory of the California Institute of Technology. The Vanguard II unit[19,20,21,22] was designed and fabricated at the U.S. Army Signal Research and Development Laboratory Vanguard III[23,24,25] was designed and fabricated by Naval Research Laboratory personnel working with Project Vanguard. The work was initiated when the Vanguard project was under the direction of the U.S. Naval Research Laboratory and was completed after the project's transfer to NASA. The Explorer VI system[26,27,28,29,30,31,32] was designed and fabricated by the Space Technology Laboratories under NASA and U.S. Air Force direction. Explorer VII[33,34,35,36,37] was designed and built by personnel of the Army Ballistic Missile Agency (now the NASA Marshall Space Flight Center). Explorer VIII's telemetry[38] was designed and built by the NASA Goddard Space Flight Center and was integrated into the satellite payload by the Marshall Space Flight Center.

The systems arrived at by the Vanguard, Jet Propulsion Laboratory, and Space Technology Laboratories groups were distinctly different in their method of attack on the problem of transmitting scientific data from an orbiting satellite to the network of grounds stations. Figure 59 is a block diagram of the telemetry system used in Explorer I.[10,11,12,13,14,15,16,17,18] There were two transmitters in the satellite, each with its own power supply and antenna, in order to obtain positive assurance of tracking and data transmission through redundancy. While the low-power (10-mw) transmitter was phase-modulated and the high-power (60-mw) transmitter was amplitude-modulated, both were modulated by four frequency-modulated subcarriers. Data-recovery provisions were quite simple: Microlock receivers[2,12] were used and the detected outputs recorded; the magnetic-tape recordings were made by a number of stations cooperating in the program; these records had a channel carrying timing pulses recorded simultaneously with the incoming telemetry; the tapes were mailed to central processing locations for reduction; both phase-locked and conventional discriminators and oscillographic recorders were used to decode and display the data for analysis.[12,18] The satellite-borne equipment carried on the basic concept of simultaneous transmission of several frequency-modulated subcarriers which had previously been used in rocket and aircraft telemetry systems and the ground equipment introduced the use of the phase-locked receiver for the detection of satellite transmissions under unfavorable signal-to-noise ratio conditions.

Explorer III[11,13,14,15,18] introduced the use of the satellite-borne magnetic-tape recorder and command receiver. These units had originally been planned for use in one of the Vanguard satellites but the earlier availability of the Explorer launching vehicle led to their use in Explorer III.

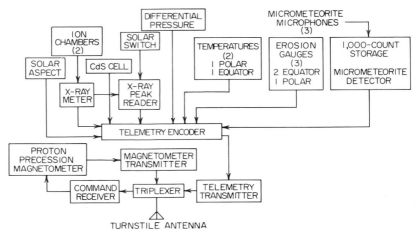

FIG. 60. Block diagram of Vanguard III.

Explorer IV telemetry[39] was quite similar to that of Explorer I. No tape recorder was included.

Vanguard II did not use the telemetry system designed by the Vanguard project personnel. The system devised by the Signal Corps staff was one using amplitude modulation of the radio-frequency carrier. The modulating signal was a vestigial-sideband frequency-modulated subcarrier signal which was played back from the tape recorder on command.[21,22] There was a second transmitter on Vanguard II to provide a continuous signal for tracking purposes. The shift in carrier frequency resulting from the use of a temperature-sensitive crystal control unit for this transmitter made it possible to monitor the satellite's temperature. The Vanguard II telemetry system operated quite successfully during the life of the satellite but because of its specialized applicability has not been used in other projects.

Figure 60 is a block diagram for the telemetry used in Vanguard III. This was the first time the basic Vanguard system was used in an orbital vehicle. Here again, two transmitters were used: the unit that was assigned primarily for tracking was also used as a link for certain items of instrumentation; the second transmitter, which radiated only on command, was amplitude-modulated with the signal from the proton precession magnetometer. The low-powered tracking transmitter was modulated with a sequence of pulses of variable length and spacing, each pulse consisting of a burst of audio frequency. The audio frequency was variable and was directly related to the outputs of the sensors for the measurement of solar X-ray radiation and to the cumulative reading of the micrometeorite counter. The basic design concepts for the Vanguard telemetering units are discussed in Refs. 4, 5, 6, 7, 8, 9, and 15. The Vanguard III satellite used the solar radiation and micrometeorite instrumentation system described in these references with the addition of a proton precession magnetometer and a transmitter for the magnetometer data. Static storage devices (magnetic cores) were used for the first time in this satellite.

The Vanguard telemetry system was designed to function as an integral element of the Minitrack system, and as such it was designed to conform to the constraints of that system's interferometer tracking units. To avoid interference with the Minitrack instrumentation it was necessary, as noted previously, to avoid modulation components at 500 cps and its lower harmonics, up to and including 2,500 cps. (The

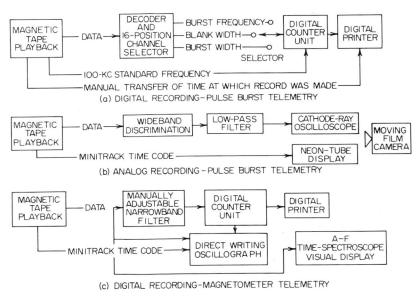

FIG. 61. Block diagram of Vanguard III data processing.

revised Minitrack system, operating in the 136 Mc/sec band, has changed this constraint to the avoidance of modulation components at 100 cps and its lower harmonics up to and including 500 cps.) Tapes recorded in the field at the various stations of the Minitrack network were returned to a central data-processing point. Ground facilities for handling the telemetry tapes were set up at the central point. The experiments in Vanguard III were separable into three categories: (1) the magnetometer which utilized its own transmitter and data link, (2) the solar-radiation experiments, and (3) the remainder of the sensors shown in Fig. 60. As two different sponsoring groups were responsible for the two sets of sensors, which were on the same transmitter, it was decided to carry out only the first steps in the data processing at the central point before distributing the partially reduced data to the experiments. Figure 61 shows in simplified form the three independent processing arrangements used for the Vanguard III tapes. In Fig. 61a is shown the digital recording scheme used for the reduction of the data from the micrometeorite experiments, the temperatures, and the differential pressure. The analog recording process (Fig. 61b) which prepared a time-history record of the frequency variations occurring in each burst, was used for examination of the solar-radiation data. (Unfortunately, the ion chambers were frequently saturated by bremsstrahlung resulting from the Van Allen belt radiation flux. As this reduced the percentage of time that useful data were being collected, more elaborate procedures for reduction of these data were not pursued.)

The varying frequency signals from the proton precession magnetometer were recovered after tape playback by using a narrowband tunable filter, which was manually adjustable, to separate the usable signal from the noise. The output of the filter was connected to a digital counter and printer. The time signals, the analog representation of the data, and a synchronizing pulse from the digital frequency recorder were simultaneously recorded on a direct-writing oscillograph. This made it possible to derive the exact time, and later the exact location, at which the magnetometer

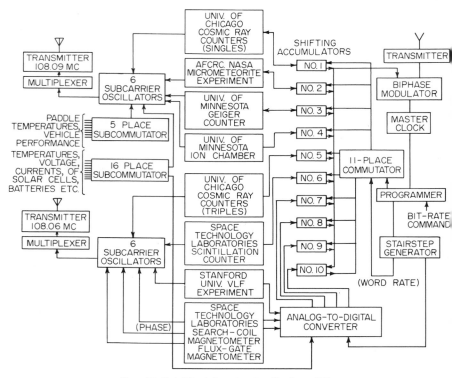

Fig. 62. Satellite telemetry for Explorer VI.

data were recorded. Figure 61c also indicates the use of an audio-frequency time-history spectrograph, sometimes called a sonograph. It was noticed in reducing the magnetometer readings that whistler-like signals had been recorded a number of times. The spectrograph was used to study these signals. The detection of these signals is an example of the discovery of the unexpected in what was planned as a relatively routine experiment. A fully automatic, preprogrammed data-processing arrangement would probably not have revealed these signals.

Explorer VI, the "paddle-wheel" satellite, was the first to use a digitally coded data-transmission system. The satellite carried three transmitters; two of these were of the relatively common frequency-modulated subcarrier type and the third was pulse-code-modulated. The general systems arrangement is shown in Fig. 62. The major conceptual change in this system was that all data would be digitally encoded and transmitted over the radio-frequency channel not as an amplitude, a pulse width, or a frequency variation, but as a specified sequence of pulses. Thus the transmission was analogous to the transfer of information in an electronic digital computing machine of the serial type. This meant that the continuous output analog signals were first sampled and then converted to digitally coded pulse trains, which then modulated the transmitter. Counting devices could be coupled directly to the digital coder without having to go through the analog digital converter. At the ground stations, it was possible to feed the detected output of the receiver to a decoding device which then provided control signals for a conventional teletype paper-tape perforator. The data, supplemented by the appropriate identity and time information, were transmitted over the usual teletype communication links to the central data-processing center maintained at Los Angeles by the Space Technology Laboratories. This mode of handling the data was reserved for the limited quantity of information whose importance warranted the use of the relatively expensive teletype communications channels. Data which could be processed with less urgency were recorded in the field on punched paper tape and on magnetic tape, and returned by mail to the central data-processing location. Here a decoder was used to prepare

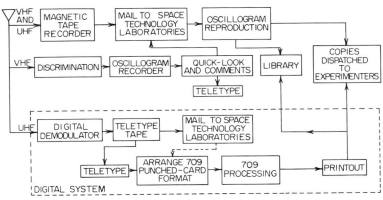

Fig. 63. Explorer VI data-acquisition and recording system.

punched cards which were used as input to a large-capacity digital computer which carried out the first phase of the data reduction in preparation for the data analysis by the scientist who designed the experiment (Fig. 63).[26,27,28,29,30,31,32]

Explorer VII utilized the familiar encoding scheme of frequency-modulated subcarriers, amplitude-modulating the telemetry's radio-frequency carrier. Among the outputs to be telemetered were the states of several counters. A conversion device for digital-to-analog conversion was included, so that the counter data could be handled by the electronic multiplexer along with other analog data. This multiplexer was used to couple a group of sensors to a single voltage-controlled subcarrier oscillator. A mechanically driven commutating switch was included as part of the University of Wisconsin experiment. The outputs of this commutator were also coupled to a voltage-controlled oscillator. One of the subcarrier oscillators was controlled directly by the outputs of two binary scalers.

A second, short-lived transmitter was provided for tracking. This second transmitter originally was to have been unmodulated but was converted to a telemeter by the addition of a group of sensors, a multiplexer, a resistance-controlled oscillator, and a phase modulator. The modulation characteristics were carefully selected so as not to interfere with the Minitrack interferometer receivers. The block diagram of the satellite system is given in Fig. 64.[37]

Data acquisition and reduction for Explorer VII has been continuing since its launch in October, 1959. A timer which was to have turned off the transmitter about 13 months after launch did not operate, and some of the telemetry stations have continued to collect data on a limited schedule. Some valuable correlations with regard to solar-flare radiations have been obtained in the period that has elapsed since the transmitter should have been cut off. The Minitrack network, supple-

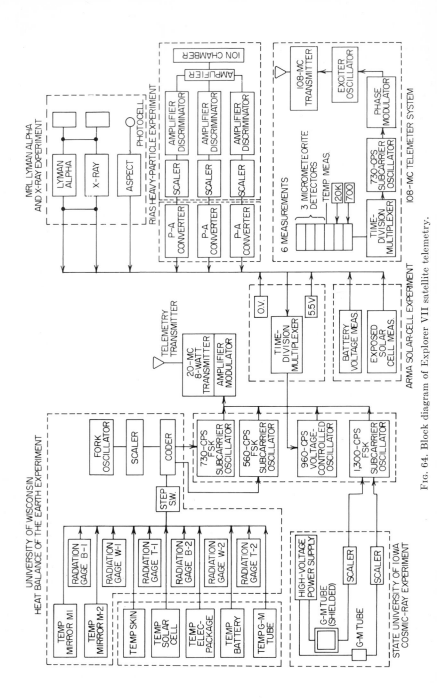

Fig. 64. Block diagram of Explorer VII satellite telemetry.

mented by a few special telemetry stations at other locations, made magnetic-tape recordings of the data. These were returned to the NASA Goddard Space Flight Center for indexing, cataloging, and distribution to the tape processors. Central data processing was not used for Explorer VII. The tapes were duplicated so that extra copies would be available for parallel processing by several groups. Certain channels of the data were recorded on direct-writing oscillographs which were connected to the outputs of standard subcarrier discriminators. Solar radiation in the ultraviolet and X-ray wavelengths and the data from the heavy cosmic-ray primaries experiment were recorded and read from the oscillograph with conventional digitizing oscillograph record readers. The sets of duplicate tapes were furnished to Drs.

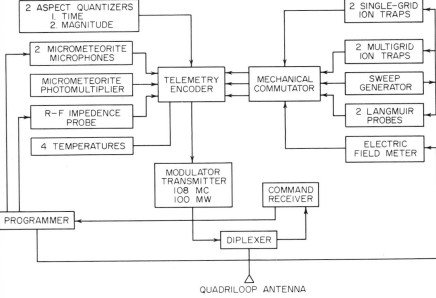

FIG. 65. Block diagram of Explorer VIII.

Suomi and Van Allen who had developed tape-processing devices particularly adapted to their own needs, and carried out the data processing for their experiments.

Explorer VIII was launched into orbit on Nov. 3, 1960, carrying a number of sensors for the direct measurement of ionospheric parameters.[38] A block diagram of the Explorer VIII telemetry is shown in Fig. 65. This system was built of the basic subsystems developed for Vanguard, but its arrangement included one important modification. The reduction of the data from Vanguard III had shown that the use of the nonsynchronous channel switching scheme (a result of the variable-length frequency burst and spacing) imposed rather more of a burden on the decommutating portion of the data-processing system than had been anticipated, when low signal-to-noise tapes were being reduced. As a result, in designing Explorer VIII, almost all the frequency bursts were held to a fixed length, making the system quasi-synchronous. It was not possible to eliminate entirely the variable-blank-length modulation because of the need for the large number of independent data samples and 4 of the 16 blank lengths did vary in accordance with inputs from internal temperatures. As these vary quite slowly, even this limited change was found to improve the "lock-on" of the decoding equipment when signal-to-noise ratios were low.

Data processing for Explorer VIII has been proceeding with equipment of the type indicated in Fig. 66. After playback the frequency bursts are passed through a comb filter which considerably improves the system's ability to handle noisy records. There are quantizers to convert frequency, pulse width, and blank width to a digital format and a multiplexer and programmer to control the way the data are presented to the computer format-control buffer. This last assembly accepts digitized data, time signals, and manually set identity data, and prepares from these, in a computer notation, magnetic tapes ready for the next step in the processing procedure. These tapes may then be used to prepare listings with a digital printer or they may be converted in a computer operation, by appropriate routines, for linearization, scaling, or reordering for later computations.

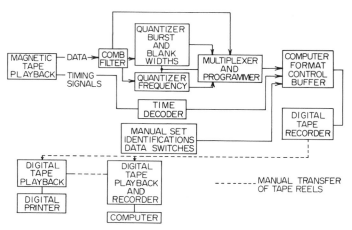

Fig. 66. Block diagram of Explorer VIII data processor.

Let us review some of the characteristics of the systems that have been mentioned. As often happens, each of the design engineering groups arrived at a somewhat different solution to the basic problem. It is obvious that in each case the designer tried to make the maximum use of the technological capabilities as they were then known. Thus with the Microlock system, correlation-detector theory was applied and advantage was taken of the phase-locked technique to recover a weak radio-frequency signal in the presence of noise. It was then possible to use a minimum-weight transmitter. The radio-frequency bandwidth was held to a minimum by the choice of the lowest-frequency subcarriers which could readily accommodate the amount of information to be transmitted.

The Vanguard system was designed to transmit as much information as could be effectively transferred by a low-powered transmitter from a relatively large number of independent sensors in the satellite. The typical Vanguard design was set up to handle as many as 48 channels of information on a single radio-frequency carrier, and as a result the bandwidth associated with the transmission was appreciably greater than that of the Microlock system.

The performance of the frequency-modulated subcarrier Microlock telemetry system in Explorer I indicated that its design had been excellent, and similar designs were therefore used in the later satellites designed by the Jet Propulsion Laboratory-Army Ballistic Missile Agency groups (Explorer III, Explorer IV, and Explorer VII). Similarly, as the performance of the telemetry on Vanguard III indicated that the basic design was satisfactory, a modification of the design was later used in Explorer VIII.

As has already been noted, a magnetic-tape recorder and command receiver were used for the first time in the Explorer III satellite, where they immediately proved their effectiveness as part of the data-collection system. Another type of control subsystem was used for the first time in Vanguard II where an internal switching device, which was responsive to the day or night condition it detected, was used to control the operation of the tape recorder carried in the satellite. The introduction of the analog-digital converter and the pulse-code-modulation scheme in Explorer VI made it possible to translate data directly at the receiving station to a tape format for computer entry. Some of the systems mentioned have used multiplexers or commutating devices to sample in sequence those data channels which could be telemetered at relatively low information-transfer rates.

14 Data-handling Considerations in Satellite Experiments

Telemetry and data-recovery systems for scientific satellites have already taken the first steps on the road to the incorporation of automatic data processing as an integral part of the system design. An outstanding example of this was the Explorer VI system. Figure 67 outlines in a functional block diagram the flow of data and control information in a typical satellite experiment.

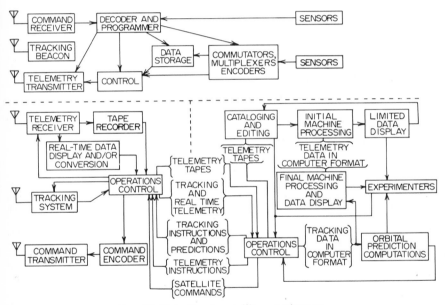

Fig. 67. Data flow in satellite experiments.

The upper portion of the figure shows the major elements in the satellite, the lower left those at a typical field location, and the lower right the functional units at the central location to which the field stations report. The satellite itself incorporates a command receiver and a programmer. The programmer utilizes as additional control inputs, data from special sensors, and timing and control pulses from the telemetry commutating and encoding units. With these items of input information to operate on, the programmer's internal preset logic controls the data format leaving the commutators, multiplexers and encoders, the mode of operation of the data-storage unit, and the selection of either the data storage or the direct sensor output for the telemetry transmitter.

At the field station the telemetered data are received, tape-recorded, and returned by mail to the central processing location. At the time the data are received, they may also be partially decoded in a real-time display or conversion device, so as to provide information from which commands for satellite operations in the next few days can be derived. This type of data, as well as that from the tracking system, is generally returned to the central operations control location by teletype or an equivalent communications link.

The central operations control distributes the incoming data to the groups responsible for telemetry data processing and orbital computations. Preliminary information on the progress of the satellite operation is also available to the experimenters as the orbital predictions are computed, and as the telemetry data go through their initial processing after the tapes are cataloged and edited. A limited amount of the data may be worked up to the point of charts to guide the experimenters and data processors in the establishment of routine procedures they wish to use both for final processing and for initial editing. The orbital predictions are returned to the field locations through operations control to ensure continued acquisition of data. The experimenters provide information to the operations control for commands to the satellite after reviewing the available telemetry data. Final data processing and display takes place after the data received at central control are converted into an appropriate computer-notation format by preliminary processing.

In such a pattern as this, there is an obvious complexity of paths for information flow. Some of the channels are electrical in nature and some are channels of communication between experimenters, data processors, station controllers, and operators and the other individuals who are concerned with the project. Some of the channels are therefore not in simultaneous operation and the "phase lags" that occur may cause oscillations to build up around the feedback loops. It is important to provide low-resistance, short delay paths for communications between the experimenters and the operations group. It pays to spend a good deal of effort on expediting the flow of information between the human control elements in the system in order to ensure a productive operation.

Signal-to-Noise Ratio and Encoding

As satellite telemetry experience accumulated, it became apparent that the satellite systems design groups may have attempted to transfer too large a volume of data over the very low powered radio links with which they were working. Obviously, it was the designer's hope that satisfactory data recovery could be accomplished by careful processing of the magnetic-tape recordings which were made in the field. This position is now considered to be somewhat undesirable for two reasons: First, if the signal-to-noise ratio at the time the tape is recorded is not fairly satisfactory, no simple process will derive adequate data from the record; secondly, if the signal-to-noise ratio at the antenna of the receiving stations is consistently low, there is a burdensome continuing pressure on the staff at these stations to maintain the ground receiving facilities at maximum performance. If an improvement can be realized in the average signal-to-noise ratio at the field-station receiver terminals, then the drops in the field equipment's performance which sometimes creep in can be considered as deficiencies but not as catastrophes.

There are many points within a given system at which technical changes can produce the desired improvement in signal-to-noise ratio. However, one of the first things that is apparent in attempting to improve this ratio is that the telemetry and data-recovery problem is best handled as a system design problem. Data processing starts with the scientific sensor and continues from its terminals, through the telemetry signal conditioners and encoders, the data-transmission link, the ground decoding units, and the computers and does not stop until the information displayed on a printing or plotting device in the ground data-processing center is analyzed by those who designed the experiment. While the sensor units and final data analysis for satellite-borne experiments may be undertaken by independent scientific investigators, the telemetry, data link, receivers and ground processing equipment can best be handled by a unified system attack. Such an attack, taking into account the technological state of the art, while still considering carefully the needs of the

SATELLITE TELEMETRY

scientific experimenters will produce the best system design. Obviously, there must be a balance between the experimenter's desire to transmit a large volume of information from the satellite, and the telemetry designer's desire to maximize the signal-to-noise ratio. Trying to transmit too much data will only deteriorate much of it to the point that only those few satellite passages which are almost directly over the receiving station will be usable. Experience has indicated that it is very seldom that the experimenter will be satisfied with just close-in passes over the stations of the Minitrack network. The experimenter, and often the data-processing engineer, tries to reduce data from all records noted as having a reasonably adequate radio-frequency signal level. As all marginal tape recordings are sent in to the central data-processing point along with those with high signal-to-noise ratios, the question arises of how much effort should be spent on recovering data when the recording is of marginal quality. This becomes quite important when automatic decoding and processing is introduced. It then is necessary to edit the records manually to remove those sections which will be processed incorrectly because of known or readily recognizable abnormalities in the record, or which will be a waste of time to run because they have little or no useful data recorded. As an alternate it is possible to devise highly reliable, mechanized elements (these may be computer programs[32]) based on an adequate set of logical rules, in order to identify those portions of the data which have a sufficiently adequate signal-to-noise ratio for proper operation of the automatic processor.

At this point the selection of the modulation technique for the telemetry radio link and the encoding process for the information must be considered. There have been a number of studies in this area[40,41,42,43] directed to the specific problems of rocket and aircraft testing, and space systems telemetry. These studies generally conclude that the use of pulse-code modulation is desirable. With the use of such an encoding technique, it is possible to realize a more reliable information transfer with a lower signal-to-noise ratio at the receiver. An encoding and modulation scheme which has a sharp signal-to-noise threshold, in the sense that if the ratio drops below a specified level data processing abruptly becomes completely useless, would be very desirable. Advantage could be taken of such a system to minimize the effort spent on attempts to reduce recordings of marginal quality.

Pulse-code modulation is directly adaptable to digital data handling at the ground stations and at the central processing point. Thus it was possible at the time of the Explorer VI satellite and later the Pioneer V space probe operations to prepare punched tapes at the data-acquisition stations, to transmit the data by teletype to the central digital computer, to incorporate calibration data in the program of the computer, and to prepare tabulations of the data for the experiments.[30,31,32] This type of modulation requires that all data be quantized before transmission. Quantization may be a continuous operation, but the nature of the modulation also requires that a sampling process be applied. This usually is periodic in nature. The sampling frequency then sets an upper limit on the spectral band within which the data can be expected to be transmitted without the occurrence of "aliasing" of the higher-frequency components into the desired low-frequency band. A discussion of this problem is given in Ref. 44.

Data Storage within the Satellite

A very important role will be played, in the satellites now being designed, by devices for internal data storage, programmed operations, and external command. As has already been noted, Explorer III introduced the use of the satellite magnetic-tape recorder. The tape recorder has become a vital component in the design of satellite data systems, as the recorder makes it possible to acquire data over those orbits or parts of orbits during which the satellite is not within the range of a telemetry receiving station. The inclusion of a recorder or an equivalent data-storage device has come to be almost routine in satellite systems design.

Several tape-recorder designs have been qualified for satellite use. Recorders have been used in Tiros[45] and in Project Score[46] as well as in Explorer III and Vanguard

II. The Tiros recorders for the cloud-cover television system run at constant speed during the recording and playback sequences. Some of the recorders used in scientific satellites have been designed to record continuously at a low tape speed during the period in which direct telemetry is precluded. The recorded information is played back at a much higher speed when in the vicinity of the receiving station. This change in tape speed presents a difficult mechanical problem to the recorder designers and it is to their credit that reliable units with tape-speed ratios as high as 50:1 have been developed. The performance specifications for the Scout magnetic-tape recorder are given in Table 22.

Table 22. Performance Specifications for Magnetic-Tape Recorders for Use in the U.S.-U.K. Scout Satellite

1. Record for about 100 min with record capabilities from 100 to 500 cps.
2. Play back entire recorded period in 2 min.
3. Dynamic signal-to-noise ratio of at least 30 db.
4. Flutter shall be no greater than 1 per cent peak-to-peak for any one discrete flutter frequency component or for any 200-cps bandwidth notch between 4,900 and 25,000 cps during playback.
5. Amplitude fluctuations shall be no greater than a 10 per cent variation at the highest recordable frequency.
6. Power consumption shall not exceed 0.7 watt during record phase and 1 watt during playback phase.
7. Operational voltage shall be 12 ± 2 per cent volts supplied by solar cells and nickel cadmium or silver cadmium battery.
8. The recorder shall remain continuously in the record mode until it is switched into playback by a command receiver.
9. The recorder shall include an internal timer to determine the playback time to within 1 sec.
10. The recorder record to playback tape speed ratio shall be 1:48 with a tolerance of ± 0.5 per cent.
11. The recorder's long-term speed stability shall be within ± 1 per cent over the entire temperature range.
12. The overall dimensions of the recorder shall be 7 in. in diameter and not more than ½ in. high.
13. The shell of the recorder shall be pressure-sealed and shall withstand an external pressure of 10^{-5} mm Hg for a period of 1 year.

Magnetic-tape recorders are not the only storage devices which are expected to be used in the next few years. For the larger satellites, which will use digital telemetry and control devices, magnetic-core storage may be used. Much as in the development of digital computers, the desire for a capability for random access to the storage device has led to consideration of cores. The Electrostatic tape storage system[47] is another method of bulk storage.

Some flexibility in storage control can be obtained just with satellite-borne elements. The storage and switching circuitry used in the Vanguard III solar radiation experiment is an example. The night-day switch in Vanguard II is another example. However, once the decision is made to use a tape recorder, a ground-to-satellite command link is practically a necessity.

If a command receiver is available, and a suitable multichannel decoding device is incorporated in the satellite, it becomes possible to take the first step to remote operation of the instrumentation. The Tiros satellites[48,49,50,51,52,53] carried multichannel command receivers. Other types of multichannel command receivers have been developed for satellites now being built.

There is another mode of control for storage and telemetry devices which it is believed will come to play a larger part in the planning of future space-science experiments. Reference is made here to detectors which are appropriately responsive to changes in significant physical quantities. For example, the onset of a solar flare should in itself be enough to switch the satellite's recording equipment from a quiescent state or one of low rate of sampling of the solar radiation, to a continuous high-speed sampling of the events occurring during the flare. Increased effort on the design of *automatic monitoring* equipment of this sort is most desirable. Far more efficient use of the data channel capacity can be expected if such devices are developed and brought into regular use in those satellite programs which are intended to monitor solar, stellar, or geophysical phenomena.

Multiplexing or commutating devices are specific forms of what may be called "programmers" for the data-handling system. In the sense used here, a "programmer" is a device which performs a specified set of operations in a predetermined sequence after the initiating order is given. Thus, either a motor-driven commutator switch or an electronic multiplexer are programmers, which run continuously, generating a cyclic switching pattern. Far more complex programmers are incorporated in the large observatory satellites. The sequence of operations, their timing and the internal satellite conditions under which the first two factors are to be modified will be variable in the Orbiting Astronomical Observatory. All these factors will be resettable from the central ground control point by the transmission of commands over the ground-to-satellite data-transfer link. The Tiros satellites carried fairly complex programmers of this general type which could be reset from the ground station.[50,51]

Data Presentations

Three categories of data presentation can be distinguished. The first is one in real time which is used to close a command loop. This is one of the data displays which is required in the operation of the large observatory satellites. It furnishes the satellite controller the information needed to command new or corrective operations within the satellite. This telemetered data display provides the feedback information which the satellite operator uses to make his decisions regarding the control of the pointing system of the orbiting astronomical observatory, the selection of an appropriate instrument, and switching from a direct data transfer by telemetry mode to a tape-recording mode.

When satellites are put into orbit to observe continuously those scientific phenomena which can only be monitored from an orbiting vehicle, fast response displays for monitoring purposes are also needed. Solar-flare, magnetic-field, and cosmic-radiation monitors come immediately to mind.

A second basic type of telemeter data presentation is a medium-speed readout and display system for engineering performance data. This type of data does not necessarily result in the immediate generation of commands in response to a change in the display, but the data have to be fairly readily available to the engineer controlling the operation, so that he may monitor the condition of the batteries, the temperature of the structure, the position of movable components, etc. Using the information presented in this display it is possible to take appropriate corrective action to prevent a system failure as a result of continued abnormal operation.

The third category includes the normal readout for the scientific experiments carried by the satellite. In many cases, the time it takes to reduce and display these data can be quite reasonable, since extensive analysis and correlation of the data with ground-based observations and with other satellite observations may be required after the data acquisition is completed. These operations are in themselves time-consuming, and extreme speed in preparing the data for such operations is not required.

It is appropriate to consider at this point another aspect of the data-presentation planning with respect to the needs of the scientific experimenter. In many cases the experimenter is more interested in examining directly the variations of a given phenomenon with respect to the geographic or geomagnetic location than he is in having two independent records, one of the variations with time of the location of the satellite and the other the time history of the magnitude of the physical quantity that was under study. Accordingly, the data-processing system should be capable of merging the coordinate information collected by the tracking network with the reduced data from the telemetry network so as to prepare either tabulations or graphs, as required, of physical quantities measured as functions of both location and time.

The Explorer VI system,[26,27,28,29,30,31,32] for example, arranged to first provide tabulations of experimental results as a function of time. Following this, there was a calibration and trajectory correlation routine for the computer, which used the decoded data from the first computer operation to produce final listings. In the second series of operations, the data were divided in accordance with the institution originating the experiment, merged with the trajectory information, and printed out separately. This print-out showed each experimental value, the corresponding time, and the corresponding position of the payload. Calibration curves were stored in the computer so as to convert raw reduced data from the telemetry to physical quantities.

A review of some of the papers in the *Journal of Geophysical Research*, which present data collected in earth-satellite experiments, indicates that time histories of the variable under study, for use in correlations with other variables, and contour plots of geographic variations are the most often used data presentations. Tabulations are generally of maxima and minima as functions of time or location, or counts of the number of significant events. Computer routines may be developed for the transfer of reduced data to digital plotting devices for automatic preparation of the data in the desired graphic format. For example, the preparation of contour maps showing magnetic field or energetic particle flux intensities as measured in geophysical experiments, and sky maps showing star locations and their intensities in the ultraviolet for astronomical experiments can be taken as typical objectives.

15 Future Satellite Telemetry Systems

Unmanned satellites tend to fall into two major classes: those smaller spacecraft for vehicles of the size of Scout, which will continue to require the compactness of design and economy in power usage which characterized the earlier satellites; and the large observatory satellites, which will permit the designer some freedom for the incorporation of internal data-processing devices.

A review was made in 1959–1960 by the Goddard Space Flight Center of the satellite telemetry systems which had been used and which were then under active development. An evaluation was made of the applicability of these systems to the projects which the center was scheduled to undertake, so as to arrive at a systems choice for immediate application and future development.

Satellites for Scout Class Vehicles

For the smaller, Scout-launched satellites, it has been decided to continue for the present with the development of the Vanguard modulation system in its modified form. The major modification in the design concept from that used in the original Vanguard telemetry is the elimination of the variable pulse length and spacing from the encoded information wave. Figure 68 shows in (*a*) the original Vanguard wave train and in (*b*) the later wave train. The elimination of the variable burst length and blank length makes the system one in which each of a group of sensors produces a frequency-modulated subcarrier signal. These are then sampled in sequence for transmission over the radio-frequency link. The system is then synchronous in concept, as the time required to sample each of the channels and complete the framing sequence is constant. This should make the demodulation of the information somewhat easier as a locked oscillator sampling detector can be introduced in the ground processor. The system is for use in the joint U.S.-U.K. Scout satellite project. A block diagram of the telemetry system for this satellite is given in Fig. 69. Although this is a system for a relatively small satellite it contains a tape recorder and command receiver in addition to the telemetry system. To provide the signal encoding which is required for the low-speed tape recording, a low-speed encoder is provided as a part of the telemetry unit. The various components of the telemetry are built up into standardized subsystems which are mounted as required on cards of a standardized mechanical design. In this way, changes in details of the encoder can be effected in subsequent satellites by the assembly of the necessary subsystems in the required configuration. Listed in Table 23 are some of the characteristics of this telemeter. The NASA Minitrack facilities will be used for the reception of the signals from the satellite. The data-handling system planned for this unit will be similar to that used for Explorer VIII.

Observatory Satellites

If we turn to the large scientific satellites we find a need for a multiplicity of telemetry systems on a single satellite. There are certain functions, such as the measurement of component temperature, the mechanical position of a controlled element,

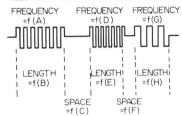

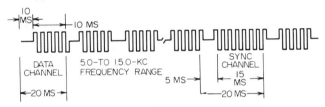

Fig. 68. Wave shapes, pulse-frequency encoding.

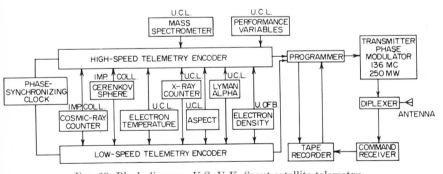

Fig. 69. Block diagram, U.S.-U.K. Scout satellite telemetry.

the charging current to the batteries in the power supply and the like, which we have come to call familiarly "housekeeping" functions, which a relatively low information-transfer rate will cover quite adequately. There are also certain scientific experiments whose data-acquisition needs can be handled by a telemetry system with a low data-transfer rate. These experiments can be combined with the "housekeeping" on a link which requires only a narrow bandwidth in the radio-frequency spectrum. There will still be a large number of experiments which will require a relatively rapid data-transfer rate; for these a high-capacity wideband telemetry system will be needed. Finally, we find in the large satellites that a need also exists for a very wideband or video type of telemetry for the transfer of signals from imaging devices such as television-type tubes. The Tiros satellites[46,47,48,49,50,51] were among the first to use several telemetry channels in this way.

Table 23. Characteristics of the U.S.-U.K. Scout Satellite Data System

Number of channels
    High-speed encoder: 44
    Low-speed encoder: 25
Bit rates
    High-speed encoder (real-time): 300 bits/sec
    Low-speed encoder (tape-recorded):
        Record: 5.5 bits/sec
        Playback: 260 bits/sec
Channel frequencies
    High-speed encoder:
        Burst frequency range: 5 to 15 kc/sec
        Burst duration: 10 msec
    Low-speed encoder:
        Burst frequency range: 104–312 cps
        Burst duration: 0.48 sec
    Playback from tape recorder:
        Burst frequency range: 5 to 14 kc/s
        Burst duration: 10 msec
Synchronization
    High-speed encoder:
        One burst (1.5 times length of other bursts) in every 16-burst frame. Burst frequency alternates from 4.5 kc/sec to a progressive step of an 8-level digital oscillator. Synchronizing information from two adjacent frames identifies location within the 16-frame sequence
    Low-speed encoder:
        Record—One burst (1.5 times length of other bursts) in each 16-burst frame is a 93.7-cps (4.5/48 kc/sec) tone. Two frames constitute a sequence
        Playback—One burst (1.5 times length of other bursts) in each 16-burst frame is a 4.5 kc/sec tone. Two frames constitute a sequence
Transmitter
    Frequency: 136 Mc/sec approx
    Power output: 250 mw approx
    Modulation: Phase, excursion ± 50°
    Expected range: 3,218 km (2,000 miles) approx
Tape recorder
    Records low-speed encoder throughout orbit
    Record time: Up to 100 min approximately
    Playback time: 2 min approx
    Ratio playback to record speed: 48:1
    Playback initiated by ground command; input to modulator simul-switched from high-speed encoder to tape-recorder playback
    Tape-speed compensation: By fork resonator frequency of 320.83 cps recorded between bursts. Produces 15.4 kc/sec tone between bursts on playback

NASA has therefore undertaken, in developing its large satellites, to specify that flexible data handling and telemetry be included. Section 16 abstracts the paragraphs relating to these systems from the specifications prepared by the Goddard Space Flight Center for the Orbiting Geophysical Observatories. Briefly the specifications call for a narrowband, a wideband, and a special-purpose (magnetometer) telemeter capability, which is to be supplemented by data storage, and a command system capable of modifying the format of the data-transmission sequence. It should be noted that for the large satellites pulse-code modulation has been specified for both the narrow- and wideband telemeters. This has been done to ease the transfer of the data from the receiving-station tape recordings to the tape format needed for the computer processing. A block diagram of the system designed by the Space Technology Laboratories, Inc., is shown in Fig. 70.

The Orbiting Astronomical Observatory (OAO) specifications were quite similar, in that they required the inclusion of narrow- and wideband telemetry, the latter of which is to transmit either digital or wideband analog (video) signals. The OAO also includes a fairly elaborate on-board data-storage and programming system, which is intended to give the ground control station flexible control of the orbiting instrumentation. This satellite also has a pair of low-power unmodulated radio tracking

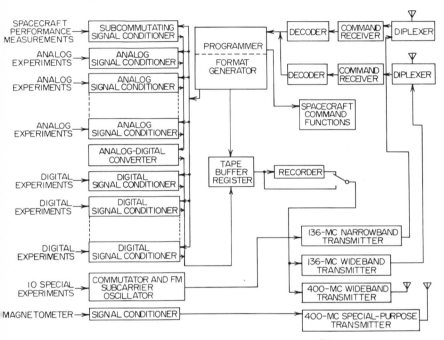

FIG. 70. Block diagram, telemetry system, OGO.

beacons to act as sources for the Minitrack tracking network. The Nimbus satellite uses multiple telemetry channels with all "housekeeping" and low-data-rate information transmitted on a pulse-code-modulation link. Television pictures and other data will use an FM-FM system.

16 Orbiting Geophysical Observatory Specifications For Data-handling Devices

Telemetry

There shall be three telemetry systems, operated on command, utilizing assigned NASA frequencies, and conforming to NASA/GSFC standards where applicable. Conservative design practices shall be employed in determining system characteristics and the predetection carrier-to-noise ratio must be at least $+15$ db under the most unfavorable operating conditions. Advantage shall be taken of the earth-pointing stabilization system to provide a directive 400 Mc/sec antenna in addition to the omnidirectional antennas for the wideband telemetry. All telemetry reception, except as necessary for the contractor to perform his assigned prelaunch functions, will be performed by NASA/GSFC. All telemetry systems must be compatible with existing or programmed equipment at the Minitrack stations; however, if significant advantages accrue from the use of incompatible methods of carrier modulation and new receiving equipment, then it should be so specified in an alternate proposal. Redundancy should be used where possible to increase reliability. The telemetry system shall include the following:

Wideband Telemetry. One wideband PCM system, operating with a carrier frequency of 400.800 Mc/sec and 136 to 137 Mc/sec shall be provided to telemeter the bulk of the experimental data and measurements necessary to determine spacecraft performance. The bit rate shall be variable upon command from the ground or the programmer and shall have a maximum of 200,000 bits/sec. Capability shall be provided for telemetering only the experiments that are energized as well as a separate multiplexer for the environmental and performance measurements necessary to evaluate spacecraft operation. Readouts from data storage shall also be telemetered over this system. The 136 Mc/sec transmitter shall be used whenever limitations in ground equipment preclude the use of the 400 Mc/sec link.

Narrowband Telemetry. One narrowband telemetry and tracking system, operating within the 136 to 137 Mc/sec frequency band, shall be provided. Approximately ten experiments will be telemetered on this system, and the bit rate shall be approximately 50 bits/sec. This should be a simple system capable of reception by scientific institutions over the world and should not require high-gain antennas, sophisticated receivers, and decommutation equipment. One such form of modulation that might meet these requirements would be FM-AM. This system will also be used for tracking, and consideration shall be given to proposing a locked Doppler system compatible with the GSFC Tracking Ground Station, in order to enhance the tracking accuracies at extreme range. The r-f power in the carrier shall be at least 1 watt and no modulation components shall fall within the band ± 1 kc/sec from the carrier in order to prevent interference with tracking.

Special-purpose Telemetry. One directly modulated transmitter operating at 400.300 Mc/sec shall be provided for telemetering data from the magnetometer and other selected experiments. The signal from the magnetometer will be a sinusoid ranging in frequency from 5 cps to 270 kc/sec. This frequency decreases with range approximately as the ratio A^3/R^3 where A is the earth's radius and R is the range from the center of the earth. The maximum frequency is at perigee. This transmitter may also be used for tracking at apogee and must have adequate power in the carrier for this function. The modulation proposed must restrict the spectrum to the first pair of sidebands in order not to interfere with the wideband telemetry and to permit synchronous detection.

Data Storage

The minimum requirement for data storage is 500 bits/sec for up to 3 hr, and the desired capacity is 1,000 bits/sec for up to 8 hr. Nondestructive readout is required. It is expected that most of the experiments will be operating at a low bit rate and will be continuously stored. However, these experiments may be interrupted at regular intervals by a few experiments operating at higher bit rates. Data storage shall be read out upon ground command and the read-in shall be controlled by the programmer. Synchronization and spacecraft time shall also be stored in order to aid in data acquisition and reduction. A clock, stable to 1 part in 10^6 for the lifetime of the spacecraft, shall be provided to supply timing.

Operation of the Spacecraft Shall Be Controlled by Ground Command and the Programmer as Described Below

Command System. A digital command system that is compatible with the NASA/GSFC command transmitter shall be provided. The predetection carrier-to-noise ratio shall be at least $+20$ db under the most unfavorable operating conditions and the system shall be as secure as feasible consistent with reliability considerations.

The number of commands shall be a minimum consistent with the experiments and spacecraft subsystem requirements. At least two command receivers are required and a backup decoder shall also be provided.

Programmer. The program shall be established prior to launch and shall provide the capability for turning on and off five experiments in increments of 5 min throughout orbit. The programmer may also be used for other spacecraft functions and must be able to be reset by ground command.

Antennas

Omnidirectional antenna systems are required for the command system at 100 Mc/sec (exact frequency classified) and the telemetry and tracking systems at 137 and 400 Mc/sec. A directive antenna is also required at 400 Mc/sec for the wideband telemetry. The omnidirectional antennas shall have no nulls greater than -3 db within the downward-facing hemisphere. Consideration must be given to the effects of polarization on the ground receiving systems, and polarization diversity is required for the command receivers.

NASA/GSFC Standards for PCM Telemetry

Code. A non-return-to-zero (*NRZ*) binary code shall be used. Time sharing shall be the multiplexing mode.

Bit Rate. The bit rate shall not be lower than 1 bit/sec and not higher than 200,000 bits/sec. The rate shall be constant for as many frames as possible and shall never change more than 0.25 bit/frame.

Word Structure. One word shall consist of up to 16 binary digits including word synchronization. One of the bits may be parity pulse. This parity pulse, if included, shall be odd parity at the end of the word. The word-synchronization pulse at the beginning of the word shall be opposite polarity from the frame synchronization. The most significant bit shall be transmitted as the first following the synchronization pulse.

Frame Synchronization. One word shall be used for frame synchronization. The word shall be all 1's. All 0's are permissible if required for compelling reasons.

Words per Frame. The maximum number of words per frame shall be 128 including the synchronization word.

Subcommutation. Subcommutation (or supercommutation) of channels is permissible and shall be at a subharmonic (or harmonic) of the frame rate. Subcommutation synchronization shall be provided by using the first word following the frame-synchronization word to indicate the forthcoming subcommutator segment number.

REFERENCES

1. J. Kaplan et al., *Proc. IRE*, vol. 44, pp. 741–767, 1956.
2. H. L. Richter, Jr., et al., 1957 *Natl. Telemetering Conf. Rept.* (sponsored by AIEE, IAS, ISA), paper V-A-1 (May, 1957). Also in *Jet Propulsion*, vol. 28, pp. 532–540, 1958.
3. H. Friedman, *Elec. Eng.*, vol. 76, pp. 470–474, 1957.
4. W. Matthews, *Elec. Eng.*, vol. 76, pp. 562–567, 1957. Also in 1957 *Natl. Telemetering Conf. Rept.* (sponsored by AIEE, IAS, ISA), paper I-A-I (May, 1957).
5. J. T. Mengel, *Elec. Eng.*, vol. 76, pp. 666–672, 1957.
6. W. Matthews, *Elec. Eng.*, vol. 76, pp. 976–981, 1957.
7. R. W. Rochelle, *Elec. Eng.*, vol. 76, pp. 1062–1065, 1957. Also in 1957 *Natl. Telemetering Conf. Rept.* (sponsored by AIEE, IAS, ISA), paper I-A-3 (May, 1957).
8. D. H. Schaefer, *Elec. Eng.*, vol. 77, pp. 52–56. Also in 1957 *Natl. Telemetering Conf. Rept.* (sponsored by AIEE, IAS, ISA), paper I-A-4 (May, 1957).
9. W. Matthews et al., *Electronics*, vol. 31, pp. 56–66, 1958.
10. Explorer I, *Jet Propulsion Lab. External Publ.* 461, 1958.
11. W. Pilkington, *Jet Propulsion Lab. External Publ.* 483, 1958.
12. J. Koukol, *Jet Propulsion Lab. External Publ.* 487, 1948.

13. W. K. Victor, *Jet Propulsion Lab. External Publ.* 491, 1958.
14. H. L. Richter, Jr., *Jet Propulsion Lab. External Publ.* 523, presented at CSAGI meeting July 30–Aug. 9, 1958.
15. L. V. Berkner (ed.), "IGY Manual on Rockets and Satellites," vol. 6, pts. I–V, pp. 276–442, 477–487, 1958.
16. W. E. Williams, Jr., *Proc. IRE*, vol. 48, pp. 685–690, 1960.
17. H. L. Richter, Jr., et al., *Electronics*, vol. 32, pp. 39–43, 1959.
18. W. K. Victor et al., *IRE Trans. Military Electron.*, MIL-4, pp. 78–85, 1960.
19. W. G. Stroud et al., *Ann. Intern. Geophys. Yr.* 6, pt. 3, pp. 340–345, 1958.
20. R. A. Hanel and R. A. Stampfl, *IRE Natl. Conv. Record*, pt. 5, pp. 136–141, 1958.
21. R. Hanel et al., *Electronics*, vol. 32, no. 18, pp. 44–49, 1959.
22. R. A. Hanel et al., *IRE Trans. Military Electron.*, MIL-4, pp. 245–247, 1960.
23. *IGY Bull.* 28, October, 1959. Also *Trans. Am. Geophys. Un.*, vol. 40, pp. 384–388, 1959
24. J. P. Heppner et al., Space Research, *Proc. First Intern. Space Science Symp.*, Nice 1960, H. Kallmann-Bijl (ed.) (North Holland Publishing Company, Amsterdam, 1960), pp. 982–999.
25. H. E. Lagow and W. M. Alexander, Space Research, *Proc. First Intern. Space Scienc Symp.*, Nice, 1960, H. Kallmann-Bijl (ed.) (North Holland Publishing Company Amsterdam, 1960), pp. 1033–1041.
26. Applied Physics Department, Space Technology Laboratories, Inc., *IRE Trans Military Electron.*, MIL-3, pp. 129–143, 1959.
27. J. E. Taber, *IRE Trans. Military Electron.*, MIL-3, pp. 143–149, 1959.
28. R. E. Gottfried, "Explorer VI Digital Telemetry, Telebit," Space Technology Laboratories, Inc., Feb. 3, 1960.
29. G. E. Mueller, *Astronautics*, vol. 5, pp. 26–27, 88, 90, 92, 94, 96, 1960.
30. "Project Able-3, Final Mission Report," Space Technology Laboratories, Inc., vol. 2 August, 1960.
31. E. W. Greenstadt, *IRE Trans. Space Electron. Telemetry*, set 6, pp. 122–129, September-December, 1960.
32. J. M. Seehof et al., "Processing and Presentation of Digital Data from Outer Space II Explorer VI, 'Paddlewheel' Satellite," Space Technology Laboratories, Inc., no date
33. *IGY Bull.* 29, November, 1959. Also *Trans. Am. Geophys. Un.*, vol. 40, pp. 401–405 1959.
34. A. W. Thompson, *IRE Trans. Military Electron.*, MIL-4, pp. 93–98, 1960.
35. J. Boehm, *IRE Trans. Military Electron.*, MIL-4, pp. 86–92, 1960.
36. G. B. Heller, *IRE Trans. Military Electron.*, MIL-4, pp. 98–112, 1960.
37. O. B. King, *IRE Trans. Fifth Natl. Symp.*, 1960, *Space Electron. Telemetry*, paper 4–2 September, 1960.
38. *IGY Bull.* 42, December, 1960. Also *Trans. Am. Geophys. Un.*, vol. 41, pp. 726–729 1960.
39. *IGY Bull.* 15, September, 1958. Also *Trans. Am. Geophys. Un.*, vol. 39, pp. 1002–1003 1958.
40. R. W. Sanders, *Proc. IRE*, vol. 48, pp. 575–588, 1960.
41. R. W. Rochelle, *Proc. IRE*, vol. 48, pp. 691–693, 1960.
42. Z. Jelonek, *IEE J.*, vol. 94, pt. 3A, 1957.
43. "Telemetry System Study," Final Report, Aeronautronic Publication U-743, vols I, II, III, Dec. 18, 1959.
44. R. B. Blackman and J. W. Tukey, *Bell System Tech. J.* no. 1, pp. 185–282, January 1958; no. 2, pp. 485–569, March, 1958.
45. J. A. Zenel, *J. SMPTE*, vol. 69, pp. 818–820, 1960.
46. S. P. Brown and G. F. Senn, *Proc. IRE*, vol. 48, pp. 624–630, 1960.
47. E. C. Hutter, J. A. Inslee, and T. H. Moore, *J. SMPTE*, vol. 69, pp. 32–35, 1960.
48. *IGY Bull.* 35, May, 1960. Also *Trans. Am. Geophys. Un.*, vol. 21, pp. 379–384, 1960
49. S. Sternberg and W. G. Stroud, *Astronautics*, vol. 5, pp. 32–34, 84–86, 1960.
50. J. D. Freeman, *Astronautics*, vol. 5, pp. 35, 78–79, 1960.
51. E. A. Goldberg and V. D. Landon, *Astronautics*, vol. 5, pp. 36–37, 98–99, 1960.
52. M. H. Mesner, *IRE Trans. Fifth Natl. Symp.*, 1960, *Space Electron. Telemetry*, pape 4–1, September, 1960.
53. *IGY Bull.* 43, January, 1961. Also *Trans. Am. Geophys. Un.*, vol. 42, no. 1, tentativ pp. 107–111, March, 1961.

SPACE TELEMETRY

17 Mission Constraints

Space telemetry is distinguished from satellite telemetry in that space telemetry is related to missions to the moon and beyond. The constraints placed upon the communications system of such vehicles are severe. Payload weight is limited to the extreme. At the same time, mission lifetime requirements are long. For example, the Mars II probe to Venus in 1962 required 4 months whereas Mariner Mars 1964 had an 8-month mission. Thus, reliability requirements are very high.

One of the few parameters which may be exploited to assist communications for space probes is antenna gain, both on the spacecraft and on earth. Gains of 53 db at 1,000 to 2,000 Mc have been achieved at the Deep Space Instrumentation Facility. Higher gains are precluded by deformations in the large antenna structures. Gain can also be used in the space vehicle, since it remains in relatively stable orientation to earth. The limitation here is the attitude-control stability, which determines pointing accuracy.

In spite of these aids, communication capacity is restricted, and efficient methods of modulation and demodulation, such as those discussed in Chap. 9, are called for. Data rates on the order of 10 bits/sec have been achieved from Mars distance, but 2,000 bits/sec or more are planned for Voyager missions in the 1970s. At the same time, the demand is for visual data requiring at least 200,000 bits/frame. Thus, data rates are not limited by propagation time (8 min from Mars) but by channel capacity. Much progress needs to be made if the element of remote control, such as planetary soft landing, for example, is added to the requirements.

18 Communications System Design

There are compelling reasons to combine the functions of *telemetry, tracking,* and *command* into an integrated communications system for space vehicles. The tracking subsystem provides data for position location of the vehicle, both in distance and angle, which is necessary for mission control, scientific observation and, not least, to point the directive antennas required for achieving data bandwidth. In fact, integrated systems are used for near-earth space missions, such as manned space flights, where bandwidth and data requirements have dictated the use of microwave frequencies and highly directive receiving antennas. The "Integrated S-band System" is such an approach. The command subsystem is the upward link for communicating control orders to the space vehicle to change its functioning during a mission. The telemetry subsystem is the downward link, providing both the engineering data (housekeeping or maintenance data about the vehicle) and the scientific data which is the observed information gathered by the probing vehicle.

The major difference between the telemetry and command systems from a communication-system design point of view is the tolerable error rate. Acceptable telemetry can be obtained at error rates as poor as 1×10^{-2} errors per bit, but command errors at a rate greater than 1×10^{-5} errors per bit would be intolerable to the command system.

Taken together, the telemetry, tracking, and command subsystem form one kind of remote control system (see Chap. 15), but the great time lags preclude very intricate control. For example, the time for a round-trip distance equal to one astronomical unit, which is one earth-sun distance, is 8 minutes. However, such remotely actuated operations as change of track, and combined local and remote control for landing maneuvers, mission mode changing, and even remote repair operations have been successfully executed.

The integrated tracking and telemetry system requires Doppler measurements to provide accurate velocity information. This means that the spacecraft must provide a trackable CW signal without suppression by the telemetry modulation. This restriction rules out many wideband modulation schemes and has led to the use of a subcarrier for the telemetry data and synchronization signals which place negligible energy into the tracking band of the carrier-following automatic phase-locked receiver.

The free-space path loss is the dominant restriction to the communication data rate in deep-space telemetry. In Chap. 4, Sec. 20, this loss between isotropic transmitter and receiver is given in decibels, as

$$10 \log (P_t/P_r) = 36.58 + 20 \log f + 20 \log d \tag{18}$$

where P_t = transmitted power
P_r = received power
f = frequency, Mc
d = distance, statute miles

If d is in kilometers, 4.1 db must be subtracted from Eq. (18).

Figures 94 and 95 of Chap. 4 provide nomograms for calculating this loss. A few representative values are given in Table 24 for 2,300 Mc.

Table 24. Free-space Path Loss in Solar System

	db
Moon	210
Sun	270
Venus	255–272
Mars	252–270
Jupiter	275–278
Pluto	295

The signal-to-noise capability per cps of bandwidth for either the up or down link is given by

$$S/(N/B) = P + G_T + G_R - \Phi_K - L_S - L_m - L_T \tag{19}$$

where P = transmitted power, dbm
Φ_K = receiving system noise spectral density, dbm/cps
L_S = space loss, db
G_T = transmit antenna gain, db
G_R = receive antenna gain, db
L_m = miscellaneous losses, db
L_T = total negative system losses, db
$S/(N/B)$ = normalized received SNR, db-cps

This expression can be used to determine either carrier or sideband power requirements by using the appropriate value for P.

Table 25 gives the set of parameters for the Mariner 1964 mission to Mars.

These figures may be used to calculate an example of deep-space telemetry-data rate capability using Eq. (19). The ground system parameters are those of the Deep Space Instrumentation Facility which is intended to interface all space missions, while the spacecraft figures are those of the Mariner craft. The telemetry link is comprised of the spacecraft transmitter and the DSIF receiver. For the Mars range of 2.2×10^8 km the space loss is 266.5 db. If $L_S + L_T$ total 8.0 db, $S/(N/B)$ is approximately 26.8 db − cps.

The PCM data-rate capability depends upon the bit error probability permissible. Figure 26 in Chap. 9 shows that $(S/N/B)T = \beta$ is 6.8 db for P_e per bit of 1×10^{-3}. $1/T = R$ is then $10 \log^{-1} (26.8 - 6.8) = 100$ bits/sec. Similarly, the up or command link, when working with the low-gain omnidirectional spacecraft antenna (emergency mode) would achieve a bit rate of 3 bits/sec for a bit error probability of 1×10^{-5}. This example suffices to show the order of magnitude of the transmission rates in deep-space missions. Improvements in antenna gains, power levels, and noise temperatures will serve to better this performance in the future, the most likely areas being spacecraft noise temperature and antenna gain. The latter requires improved methods of pointing the spacecraft antenna beam. Thus data rates will

SPACE TELEMETRY 14-97

range between 5 and 5,000 bits/sec and command rates between 1 and 20 bits/sec. The upper limit on a data rate is determined by the DSIF bandwidth.

Table 25. Space Mission Communication System Parameters
(a) Ground Station—DSIF

	Transmit	Receive
Frequency	2,116 Mc	2,298 Mc
Antenna size*	85 ft	85 ft
Gain	51.0 db	53.0 db
Noise temperature (maser front end)	—	55° K
Φ_K	—	−185.1 dbm/cps
Transmit power total	+70 dbm	

(b) Space Station
(Mariner 1964 Mission)

	Transmit	Receive
Frequency	2,298 Mc	2,116 Mc
Antenna size*	46.0 × 21.2 in.	46.0 × 21.2 in.
Gain	23.2 db	21.8 db
Noise Temperature	—	2,700°K
Φ_K	—	−166.6 dbm/cps
Power	+40 dbm	

* High-gain mode. Lower-gain antenna used in near-earth phases of mission.

19 Modulation System

Discrete rather than analog communications systems have been chosen for deep-space work as well as for the satellite telemetry systems just previously described. One of the reasons for this choice is the greater compatibility of a discrete system with digital computers which are used ultimately to process the derived data. Secondly, a discrete system is an extremely convenient way of instrumenting the command link. Third, as discussed in Chap. 9, Secs. 8 and 13, coded-phase coherent sequences provide highly efficient transmission. The theory of this modulation system is presented in Sec. 13 of Chap. 9. It requires a minimum of β energy per bit per unit noise power to modulate the subcarrier and β drops with greater bandwidth per data bit.

The so-called binary-symmetric channel (Chap. 2), for the conveyance of information in the state of one of two possible orthogonal conditions, is the most efficient digital means of communication (Chap. 9). Furthermore, it has been shown that biphase modulation, or the transmission of one of two mutually opposed phase states (i.e., 0 and 180°, or ±90°) of a carrier frequency, is the simplest and perhaps best mechanization of such an orthogonal system.[5] This modulation technique is often called phase-shift keying, or "PSK" modulation.

If the information state (i.e., 1 or 0) in such a system is conveyed in a single-phase state, the system is often referred to as an uncoded binary system, and the detection technique required on the ground is called "bit-by-bit" detection. By using a combination of two phase states, literally a coded phase word, to represent a single information state, the modulation efficiency of the system can be increased (see also Chap. 9).[6] The coded systems are also referred to as "redundant," since more than one phase state is used to represent a single information state, and the corresponding detection technique in the ground system is often called "word" detection.

14-98 SPACE SYSTEMS AND TELEMETRY

To demodulate a biphase-modulated carrier frequency the receiving station must know the exact phase of the unmodulated carrier. Such a demodulation technique is most often referred to as "coherent" demodulation; demodulation without this phase information is most often referred to as "noncoherent."

Following demodulation, the noise-corrupted binary waveform must be detected, and the optimum detector for the case of systems corrupted by white Gaussian noise is the so-called "matched filter." This detector will always maximize the output signal-to-noise ratio; it is characterized by an impulse response that is the inverse time function of the waveform to be detected. Matched-filter detection requires a knowledge of the phase of the transmitted bits or, equivalently, the time at which a change of phase could occur if two successive bits differ. A number of techniques can be used to mechanize coherent demodulation, but the techniques that readily yield bit synchronization under conditions of low bit rates, and therefore low signal to-noise ratios, are few. This bit-synchronization problem must be solved before near-optimum use of PSK techniques can be achieved.

A functional diagram of coherent demodulation and matched-filter detection is shown in Fig. 71. The matched filter shown is the "integrate-and-dump" type, and it may be mechanized exactly as shown with a sufficiently stable operational amplifier. To implement it, we need to decide whether the matched-filter output sample is less than or greater than zero. Note that the symbol ⊗ stands for the function of coherent demodulation, or multiplication in this case but is also equivalent to biphase or "balanced," modulation in some instances.

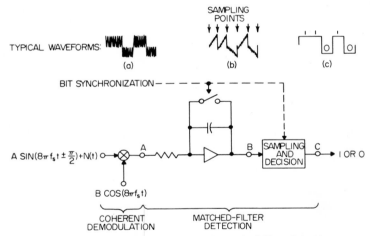

Fig. 71. Coherent demodulation and matched-filter detection.

20 The Synchronization Process

The synchronization process, used in both the command and telemetry system is a unique combination of the basic properties of phase-coherent, or "phase-locked," loops and pseudo-random binary sequences, commonly called pseudo-noise (PN) sequences. A typical PN sequence and its fundamental properties are presented in Fig. 72. These binary codes are inherently odd in length and are best characterized by their two-level autocorrelation function. Simply stated, this means that, as a PN sequence slides discretely by a duplicate of itself, it looks consistently unlike this duplicate until the two sequences are in perfect alignment, at which time, of course, they are identical. The autocorrelation function of such a sequence can be obtained by comparing adjacent bits in the two identical but out-of-phase sequences, and then plotting the following function:

$$\frac{(\text{Number of similar bits}) - (\text{number of dissimilar bits})}{(\text{Total number of bits in sequence})}$$

as a function of the relative displacement between the two identical sequences. This has been done in Fig. 72 for a PN sequence of length 7. Since the sequences are inherently odd in length there will always be one more bit of one state than the other, and the absolute value of the numerator in the function above will therefore be unity until such time as the sequences are in alignment (see also Chap. 9, Sec. 13).

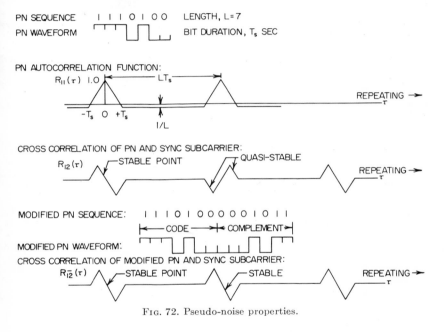

FIG. 72. Pseudo-noise properties.

How these pseudo-noise sequences are used in the basic synchronization system is shown in Fig. 73. At the transmitting end a stable frequency source, sometimes called the system "clock," is frequency-multiplied by 2 to drive the PN code generator at a bit rate equal to twice the clock frequency. The resulting PN sequence then biphase-modulates the original clock frequency, essentially making it a carrier, or subcarrier, for the PN synchronization code.

An identical PN generator exists at the receiving end of the link to be synchronized, and for the moment let it be assumed that this generator is precisely synchronized and aligned with its mate at the transmitting end. If, under these conditions, the received biphase-modulated sync subcarrier is passed through a second biphase modulator driven by the synchronized PN code generator, the original unmodulated clock frequency will appear at the modulator output. In other words, the second biphase modulator essentially "unmodulates" the received signal. If this recovered subcarrier frequency is now made to synchronize, or "lock" a phase-coherent loop, the loop oscillator output can be frequency-multiplied by 2 and the resulting signal used to drive the PN generator. As long, then, as the two PN code generators remain in step, the system will be precisely synchronized.

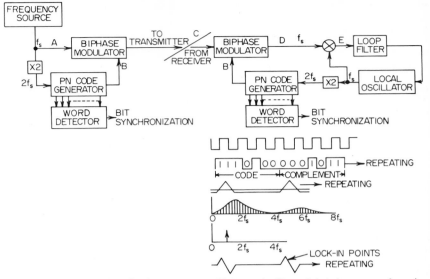

FIG. 73. Basic synchronization system. Notes: *A*. Unmodulated sync subcarrier. *B*. Typical pseudo-noise waveform. PN autocorrelation function. *C*. Modulated sync subcarrier spectrum. *D*. Recovered sync subcarrier. *E*. Loop correlation function.

A word detector completes the transmitting system. This detector is used to recognize that the code generator has made a complete cycle and produces a pulse to indicate this fact. An identical word detector used in conjunction with the PN generator in the receiving system would then produce pulses coincident with those in the transmitting system, and a unique synchronization signal would be available.

Let us now consider the process of achieving synchronization in such a system when it is initially "out of lock." Figure 74 shows the power-density envelope, or spectrum, of the PN-modulated sync subcarrier appearing at the input to the receiving system biphase modulator. Since the PN code generators are out of step, the signal appearing at the output of the receiving-system biphase modulator will look, for all practical purposes, like random noise. The phase-coherent loop will therefore have no

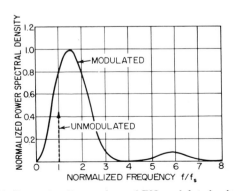

FIG. 74. Power-density envelope of PN-modulated subcarrier.

SPACE TELEMETRY 14-101

discrete frequency upon which to lock, and the system will remain unsynchronized as long as the local oscillator in the loop remains static at the clock frequency. In this state, the two PN generators are driven at the same bit rate but are out of phase by some number of bits. If the frequency of the loop oscillator is now made to differ slightly from the transmitted clock frequency, the PN generators will be driven at slightly different rates, and the second sequence will essentially slide by the first at a rate proportional to the relative frequency difference between the two oscillators. Assuming this rate to be small compared with the response time of the loop, the system will automatically become synchronized when the two PN sequences once again come into alignment. If the relative rate at which the sequences slide past each other is too large, the loop will not be able to respond when the reconstructed subcarrier clock appears, and the system will never synchronize.

An important characteristic of any servo is its error function, or the curve which relates the loop control signal to a given difference between its input and output. For the case of the normal phase-coherent loop configuration, the error function is simply the crosscorrelation function of the input frequency and local oscillator signal, a simple sine function, and stable over limits of $\pm 90°$ about any zero.

The loop error, or correlation, function for the synchronization system just described is the crosscorrelation function of the PN sequence and the local oscillator signal and is shown in Fig. 72. Because of the odd length of the code sequence, the function has a stable point, at which the loop can lock only every other cycle of the PN sequence, and quasi-stable points in between. Laboratory tests have verified that the loop will lock at these quasi-stable points and thereby create a situation of conditional stability.

However the PN sequence can be modified to consist of the original code and its complement, as shown in Figs. 72 and 73, and the crosscorrelation function will become even, with stable lock-in points for each cycle of the sequence or its complement. The only revision necessary in Fig. 73 is the mental note that the "PN code generator" now produces a cyclic output which is the original code followed by its complement (see Note B of Fig. 73).

The system described thus far can provide a unique synchronization pulse rate, which is a function of the length of the pseudo-noise code and sync subcarrier clock frequency. A coherent reference at f_s derives automatically from the locked oscillation.

An absolute threshold of 9.0 db in $2B_L$ where B_L in the sync loop bandwidth has been experimentally established. It is possible to use a separate data and sync channel. The data channel can be modulated on a subcarrier at $4f_s$, which is not occupied by sync modulation as shown in Fig. 74. A power division between data and synchronism is then necessary. This method, however, enhances the phase jitter on the subcarrier by a factor of 4. The sync power must be raised by a factor of 9, assuming a tolerance of 15° phase jitter must be maintained on the data reference.

It has been found possible to place the data and sync on the same channel subcarrier, f_s, by modulo 2 addition (half adder). A special circuit, called a single-channel detector, has been developed to demodulate the composite signal. Data and reference signal f_s are derived by multiplying the input signal $\pm PN \oplus 2f_s$ by PN^* and PN in separate channels[4]. (Here PN^* stands for complement of pseudo-random sequence.)

21 System Characteristics

The Mariner coded phase-coherent telemetry system bit and word formation is shown in Fig. 75. A PN sequence is 63 bits long. Word detectors establish each ninth PN pulse, forming 7 word bits. Figure 76 provides a resume of the waveform and bit timing of the telemetry system. The highest rate is 2,400 cps, which is derived from the spacecraft power supply. Table 26 summarizes the parameters of the telemetry and data storage systems.

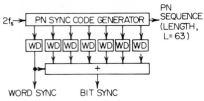

Fig. 75. Telemetry bit- and word-sync mechanization.

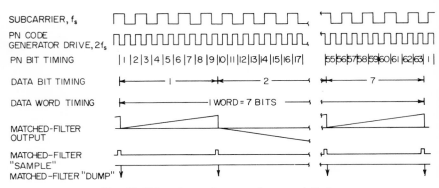

Fig. 76. Telemetry-system waveforms and timing.

Table 26
Mariner Telemetry Parameters

Engineering measurements, No.	90 + 4 events
Scientific data, type	video
Word length, data, bits	7 bits
Word length, PN bits	63 bits
Transmission rate	8⅓, 33⅓ bits/sec
Sync channel threshold	11 db-cps
Word error probability	1 in 18
Bit error probability	5×10^{-3}

Data Handling

Storage	5.24×10^6 bits
Record rate	10,700 bits/sec
Playback rate	8⅓ bits/sec
Tracks	2

The command system is similar in design to the telemetry system but operates with 26 bit words at a single data rate of 1 bit/sec. The system provides discrete commands for actuating switches, and quantitative commands which require a magnitude as well as address for execution. Bit error probability in a completely executed quantitative command is less than 2×10^{-4}. The probability of a false-command execution is less than 2×10^{-9}. The probability of execution in one attempt is 0.5 to 0.7.

Figure 77 shows a block diagram of the complete communications system for the Mars mission. The range and Doppler measurements use the carrier whereas the command and telemetry system use a subcarrier. The diagram stresses the nature of information flow and synchronization requirements. The ground system of 77b corresponds to the DSIF. The tracking system is the TRAC(E) system mentioned on page 3-48.

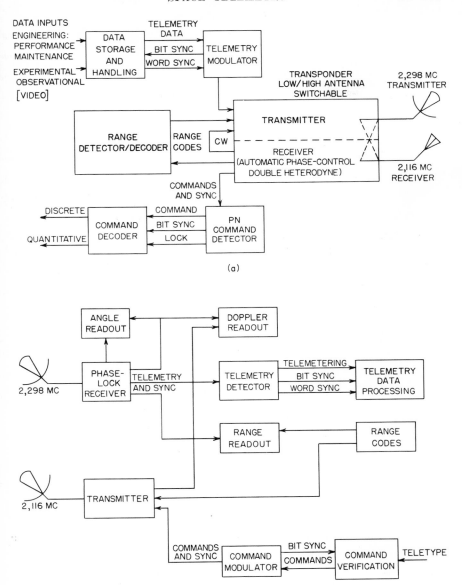

Fig. 77. Space communication system: (a) spacecraft, (b) ground.

REFERENCES

1. B. D. Martin, The Mariner Planetary Communication System Design, *Jet Propulsion Lab. Tech. Rept.* 32-85 (Rev. 1), Pasadena, Calif., May 15, 1961.
2. R. P. Mathison, Mariner Mars 1964 Telemetry and Command System, *Jet Propulsion Lab. Tech. Rept.* 32-684, Pasadena, Calif.,
3. R. P. Mathison, Constraints in Space Telecommunications Systems, *Astronautics*, pp. 40-50, May, 1962.
4. J. C. Springett, Telemetry and Command Techniques for Planetary Spacecraft, *Jet Propulsion Lab. Tech. Rept.* 32-495, Pasadena, Calif., January 15, 1965.
5. J. C. Springett, Command Techniques for the Remote Control of Interplanetary Spacecraft, *Jet Propulsion Lab. Tech. Rept.* 32-314, Pasadena, Calif., August 1, 1962.
6. R. C. Tausworthe, Theory and Practical Design of Phase-locked receivers, *Jet Propulsion Lab. Tech. Rept.* 32-819, vol. 1, Pasadena, Calif., February 15, 1966.
7. R. C. Titsworth, The Role of Pseudorandom Codes in Communications, *Jet Propulsion Lab. Tech. Rept.* 33-185, Pasadena, Calif., August 3, 1964.
8. R. C. Titsworth, Optimal Ranging Codes, *Trans. IEEE* PTG-SET, vol. SET 10, No. 1, March, 1964.
9. F. M. Riddle, Communications with Deep Space Vehicles, *Proceedings of the 1962 National Telemetering Conference*, Washington, D.C., vol. 1, p. 8-1.
10. H. D. Becker and J. G. Lawton, Theoretical Comparison of Binary Data Transmission Systems, *Cornell Aeronaut. Lab. Rept.* CA-1172-S-1, May, 1958.
11. L. Baumert, M. Easterling, S. W. Golomb, and A. Viterbi, Coding Theory and Its Applications to Communications Systems, *Jet Propulsion Lab. Tech. Rept.* 32-67 Pasadena, Calif., March 31, 1961.
12. H. Richter, R. Stevens, and W. Sampson, Microlock: A Minimum Weight Instrumentation System for a Satellite, *Jet Propulsion*, August, 1958.
13. M. Brockman, H. Buchanan, R. Choate, and L. Malling, Extra-terrestrial Radio Tracking and Communication, *Proc. IRE*, vol. 48, no. 4, April, 1960.
14. J. P. Fearey, J. R. Hall, B. J. Ostermier, and N. A. Renzetti, "Radio Tracking Techniques and Performance of the United States Deep Space Instrumentation Facility,' paper presented at the COSPAR symposium, Florence, Italy, April, 1961.
15. "System Capabilities and Development Schedule of the Deep Space Instrumentation Facility," Technical Memorandum 33-83, Jet Propulsion Lab., Pasadena, Calif., April 24, 1964.
16. Staff, Jet Propulsion Lab., "Mariner: Mission to Venus," McGraw-Hill Book Company New York, 1963.

Chapter 15

REMOTE CONTROL

JOHN G. TRUXAL, *Polytechnic Institute of Brooklyn, Brooklyn, N.Y. (Principles of Vehicular Guidance and Control)*

MARTIN L. SHOOMAN, *Polytechnic Institute of Brooklyn, Farmingdale, N.Y. (Feedback Systems)*

WILLIAM B. BLESSER, *Polytechnic Institute of Brooklyn, Brooklyn, N.Y. (Components)*

JOHN W. CLARK, *Formerly Hughes Aircraft Co., Fullerton, Calif. (Remote Handling)*

INTRODUCTION TO REMOTE CONTROL*

1. Remote Control Defined.................................. 15-2
2. Scope of the Chapter.................................... 15-3

PRINCIPLES OF VEHICULAR GUIDANCE AND CONTROL

3. Guidance Fundamentals.................................. 15-4
4. Guidance Systems...................................... 15-7
5. Maneuver Strategy..................................... 15-14

FEEDBACK SYSTEMS

6. Feedback Theory....................................... 15-20
7. Stability... 15-27
8. Performance Evaluation................................ 15-36
9. Sampled-data Systems.................................. 15-48
10. Nonlinear Systems.................................... 15-55
11. Simulation and Machine Computation................... 15-63
12. Computers in Control................................. 15-72
13. Pure Time Delay...................................... 15-81

* By Elliot L. Gruenberg.

COMPONENTS

14	Accelerometers	15-88
15	Gyroscopic Instruments	15-99
16	Stable Platforms	15-124
17	Prime Movers	15-127

REMOTE HANDLING

18	Introduction	15-137
19	The Manipulatory Subsystem	15-144
20	The Locomotion Subsystem	15-149
21	The Sensory Subsystem	15-152
22	The Command and Data Link	15-158
23	The Control Console and Power Subsystem	15-164
24	Areas for Future Research	15-167

INTRODUCTION TO REMOTE CONTROL

1 Remote Control Defined

Remote control is a relatively new art; hence the concept is not so standardized as that of telemetering. An early conception of remote control was the bridge telegraph system between a ship's bridge and engine room. This system required human intervention to read signals and to activate the necessary control valves. Later in process control these were actuated remotely. Telemetered data were used to establish the need for valve control and the extent of control. A later manifestation has been pilotless aircraft, where the pilot has been replaced by remote control techniques for performing dangerous tests. Perhaps the most dramatic form of remote control is the guided missile, where remote control requires a system of navigation to assist in guidance of the vehicle.

We shall define remote control to include any system of control which requires a definite communication system to control action at a distance from the control point. This definition seems to rule out a completely self-contained system, such as an inertially guided missile, since such a missile is really not under control of a control center once launched. Such a missile is controlled only by commands which have been preprogrammed before flight and local corrections sensed and used by the missile itself. However, in many cases, correction of the program is required after flight, which requires a remote-control operation within our definition.

A remote-control system as defined here is a closed-loop system which must consist of at least the following elements:

1. Sensors of information
2. A transmission system to communicate information to a *remote* control point
3. A control point which includes a human or automatic decision-making system
4. Devices to translate information into appropriate control signals
5. Links to communicate information to actuators at remote action points
6. Actuators operated by control signals to effect responsive controlled operation

Systems of this sort are sometimes called telecontrol systems.[1] They are classified in some cases as real-time or non-real-time systems, particularly if a computer is involved at the control point in 3 above. This classification is based on the consideration as to whether or not the responsive action must occur within the control response time of the system. A remote-control system could be composed of the elements

INTRODUCTION TO REMOTE CONTROL

listed above without, however, requiring that action occur immediately upon receipt of the sensed information. Rather, a considerable delay might be tolerated while long assessments or correlations are made of this information at the control point, or command center, as it would be called in this case. Action might be delayed for long periods of time or may be suspended entirely. Many supervisory or monitoring systems operate in this fashion. These systems are non-real-time systems.

The type of remote control which concerns us here falls into the real-time category. The fundamental form of the real-time system is one which presupposes a state of equilibrium in the system and which acts to restore that equilibrium. In fact, equilibrium is the underlying criterion of system action. Various references are set up which define the desired state at which equilibrium is desired; these can be complex and interrelated; in the case of remote-control systems, they are usually remote from the control center. Deviations from these references are sensed and actuators act to restore these deviations in accordance with the logic and plan of the system.

Such systems as these constitute *feedback systems* and require an understanding of the principles governing their design. Of particular importance is the question of stability. A distinctive feature of remote-control systems is the longer time delay between disturbing signal and restoring action which can exaggerate the stability problem. This will be discussed below in the section on Feedback Systems.

Many remote-control systems rely on discontinuous signals or sampled data. In the case of the real-time remote-control signals to be discussed in this chapter the sampling rate is less than the shortest response time of the system.

In remote-control systems involving the guidance of vehicles, the reference system, usually derived from a system of navigation, is complex and requires special understanding.

A development in the field of remote control in quite a different direction than guidance is that of remote handling. These systems have been developed to perform tasks in hostile environments normally performed by human hands. The actuators for this are quite complex, being able to take on more than seven degrees of freedom. This has forced development in the direction of projecting the mental powers of the human to the remote location by using visual and other clues as to the state of the remote situation.

Remote-control systems cover a wide field of application. Examples are

1. Pipeline and utility control
2. Pilotless aircraft and drones
3. Guided missiles
4. Space-vehicle control
   Maneuver
   Operation
   Rendezvous
5. Remote observatory, such as orbital observatory for
   Astronomy
   Geophysics
   Weather
   Lunar exploration
6. Remote repair or assembly operations
   Space
   Undersea
   Nuclear environment

2 Scope of the Chapter

A coordinated treatment of remote control, including both the communications and control aspects, is rare in the literature. Remote control is treated briefly by Nichols and Rauch,[2] whereas the communications theory of telemetry is treated comprehensively. An attempt at a coordinated approach has been made recently by a Russian author.[3] In the past, most remote-control situations could be treated by keeping control theory and communications separate. As *real*-time remote-control systems become more common, operating over longer distances, the need for a more integrated approach should become more apparent.

This chapter will emphasize control theory. It will present the design principles and methods of analyzing control-system problems involved in remote control. The methods for providing coordinate reference signals and schemes of guidance for remote vehicles will be presented in the next section. The principles of design of feedback systems and ensuring stable operation will be next discussed, as well as special methods of handling sampled-data systems. Of importance to remote control is the treatment of pure time delay. Methods of using computers in remote-control systems or for system simulation are detailed.

Components of particular use to remote systems are those for sensing acceleration and angular position. The theory of operation of these devices is given in a section on Components, together with a similar discussion on prime movers useful in actuators. Other types of information sensors are discussed in Chap. 3.

A final section is devoted to the present practice and design fundamentals of remote-handling systems. As will be seen, this art has developed along considerably different lines from guidance in a number of respects.

Communications effects on remote-control systems, such as noise, data rate, modulation, and bandwidth restrictions, can be studied by applying the principles of the previous chapters to a functional description of the communications links and then using the design procedures outlined in this chapter.

REFERENCES

1. E. L. Gruenberg, Telecontrol, *IRE Trans. Telemetry Remote Control*, vol. TRC-3, pp. 5–8, May, 1957.
2. M. H. Nichols and L. L. Rauch, "Radio Telemetry," 2d ed., pp. 359–361, John Wiley & Sons, Inc., New York, 1960.
3. V. N. Tipugin and V. A. Veytsel, "Radio Control," Moscow, 1962. Translation available from Office of Technical Services, U.S. Dept. of Commerce.

PRINCIPLES OF VEHICULAR GUIDANCE AND CONTROL

3 Guidance Fundamentals

A transportation vehicle can be divided into two basic functional divisions, motive power and guidance control. The following sections will deal with the functions of the guidance and control sections of modern vehicles.

A most important use of feedback-control techniques is in the guidance and control of vehicular motion. Currently the vehicles of primary importance are the railroad train, automobile, submarine, aircraft, low-altitude missiles, high-flying missiles, and satellites. Certainly space ships will be a reality in the near future and, if one possesses foresight and imagination, earth vehicles which tunnel under terra firma like gigantic moles. The motion of all these devices must be controlled in a specific manner by a human being or inanimate pilot for successful operation.

Coordinate Axes

Motion of a rigid body is governed by three translational equations and three rotational ones, yielding six degrees of freedom. An orthogonal-axis reference system[1] is shown in Fig. 1, where the long dimension of the body lies along the x axis (longitudinal axis). The y axis is called the lateral axis. Rotations about the y axis are called pitch, about the z axis yaw, and about the x axis roll. These are denoted respectively as θ, ψ, and ϕ. In an automobile, "nose-diving" during braking or "tail-sitting" under acceleration are pitch motions, sway when rounding a turn is roll,

PRINCIPLES OF VEHICULAR GUIDANCE AND CONTROL

and the motion of skids on icy road surfaces a combination of yaw and sideslip (lateral translational motion).

If the suspension system is tight and road friction in control, the quantities θ, ϕ, ψ, z, y, and their derivatives are zero. In driving along a straight road, the velocity vector is along the x axis, and motions of the steering wheel tend to redirect the velocity vector, producing a yaw reaction torque. To a first approximation the effect of the road surface and gravity is to limit motion to the earth's tangential plane.

In a low-speed aircraft, the wing surfaces and the tail surface are large, and addition of a simple control system provides good stabilization in roll and pitch. Thus the system can be approximately described by expressions involving only θ, x, y, and z. Unfortunately such simplifications are not easily made in many cases and the equations of motion contain all six degrees of freedom.

Terminology

Telemetry is most frequently used in the control of air and space vehicles, and the major part of this chapter is concerned with problems and techniques in this area. In this field the terminology, a hybrid of missile and aircraft terms, is far from standardized.[2] Some of the most important terms and their general meanings are discussed below:

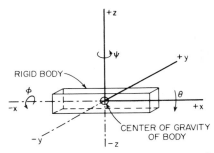

FIG. 1. Reference system for a six-degree-of-freedom rigid body.

Stabilization—Design of a system so that it yields satisfactory natural and driven response modes (balance and equilibrium)

Navigation—Determination of translational and rotational positions, velocities, and accelerations of a device relative to a fixed or moving coordinate system (sensation and measurement)

Guidance—Generation and utilization of steering signals to determine the flight path of a vehicle (thinking and computation)

Control—Operation on the steering signals in a manner that forces the vehicle to follow a given flight path (motion and activation)

Fire control—A specialized form of guidance describing the management of a missile attacking a target, or the aiming of guns at a target

The control problems of missiles and aircraft vary with the size and type of the vehicle in question.[3] Missiles may be roughly categorized in four classes:

Short-range—Range 1 to 30 miles; speed supersonic up to Mach 5; maneuverability as high as 10 to 15 g forces. Primary considerations—rapid, accurate control under large loads. Secondary importance—drag and efficiency

Medium-range—Range 100 to several hundred miles. Appreciable cruising at constant altitude. Important considerations—low drag, large lift-to-drag ratio, high accuracy

Long-range—Range several hundred to thousands of miles. Primary consideration—low drag and large lift-to-drag ratio

Ballistic missiles—Short-, medium-, and long-range. Trajectory mainly determined by the initial velocity and gravity. Important—frictional heating on reentry. Secondary—aerodynamic forces, forces of drag, lift, and control moments

Missiles are also classified according to their mission:

SS—Surface-to-surface missiles
SA—Surface-to-air missiles
AS—Air-to-surface missiles
AA—Air-to-air missiles

Aerodynamic Forces

Aircraft and missiles at low altitudes experience large aerodynamic forces and moments.[4,5] These forces dominate the control problems of atmospheric vehicles. The atmosphere may be considered as extending from sea level to 150,000 ft, at which altitude the air density is 13 per cent of its value at sea level. Calculation of aerodynamic forces is very difficult, and experimental wind-tunnel measurements are generally used.

In straight and level flight, the rearward, aerodynamic drag on an aircraft is balanced by the forward engine thrust, and the weight of the aircraft is balanced by the lift. The total drag is the sum of profile drag due to air flow over the fuselage and induced drag caused by tilting the lift vector backward. The effective lift acts at a point

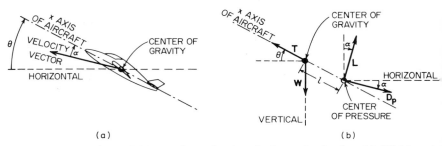

FIG. 2. Angle notation and force configuration for a body moving in air. (a) Flight-angle notation. x axis, velocity vector, and horizontal all lie in the xz plane. Cf. Fig. 1. $\theta \equiv$ pitch angle. $\alpha \equiv$ angle of attack. (b) Force configuration for small α and θ. $\mathbf{T} \equiv$ engine thrust; $\mathbf{D}_p \equiv$ profile drag; $\mathbf{W} \equiv$ gravity force; $\mathbf{L} \equiv$ lift force.

called the center of pressure (Fig. 2). If the angles θ and α are small as in the case of normal flight, expressions for the lift, drag, and pitching moment are

$$L = qSC_L$$
$$D = qSC_D \quad (1)$$
$$M = qSlC_M$$

where q = dynamic pressure = $\frac{1}{2}\rho v^2$
p = air density
v = relative velocity
S = effective area (wings and/or fuselage)
C_L = dimensionless lift coefficient
C_D = dimensionless drag coefficient
C_M = dimensionless pitch-moment coefficient
l = distance between center of gravity and center of pressure

For supersonic flight, the lift and drag coefficients for idealized configurations are shown in Table 1. For practical configurations, order-of-magnitude results should be expected.

Since the center of gravity (c.g.) and center of pressure (c.p.) are not coincident, an unbalanced torque exists. If the c.p. is behind the c.g., a deflection of a tail-control surface (elevator trim moment) provides a stable control configuration. If the c.p. is in front of the c.g., elevator trim establishes an unstable configuration and a nose-control surface (canard) or automatic control system is needed to establish a stable equilibrium.

Whenever possible, aerodynamic control surfaces are used because of their moderate input-power requirements. When the airspeed and air density are too small to provide sufficient aerodynamic force, reaction controls are used. Reaction controls change the velocity vector by changing the direction or magnitude of the thrust

PRINCIPLES OF VEHICULAR GUIDANCE AND CONTROL   15-7

Table 1. Idealized Lift and Drag Coefficients

Coefficient	Cross section	Expression	Limitations
Lift coefficient (wing)	t – Thickness; c – Chord	$C_L = \dfrac{4a}{\sqrt{M^2-1}}$	Length of wing must be much larger than the width (chord).
Profile Drag coefficient (wing)	(Same as above)	$C_{D_p} = \dfrac{5\left(\frac{t}{c}\right)^2}{\sqrt{M^2-1}}$	Length of wing must be much larger than the width (chord).
Profile Drag coefficient (blunt body of revolution)	(blunt body, diameter d, length l)	$C = \pi^2 \left(\dfrac{d}{l}\right)^2$	Holds for a blunt body with maximum cross section at base.
Induced drag coefficient (wing)	Induced drag, angle a, lift L, Center of pressure	$C_{D_i} = a C_L$	Holds for small a. Total drag $= C_D = C_{D_p} + C_{D_i}$

force. This can be accomplished by rotating the tail pipe or the entire engine of a jet or rocket motor, by movement of a deflecting vane in the exhaust stream, or by auxiliary jets, rockets, air nozzles, or ionic propulsion units.

4 Guidance Systems

The function of a guidance system is to provide steering signals to control the flight path of a vehicle. The necessary information can be provided in several ways, by three techniques of major importance. Information can be transmitted from a remote source in the form of steering signals or a beam that the missile must follow. Steering commands can be stored in the missile prior to launch, and "readout" of the stored commands can be actuated by appropriate measurements made within the missile during the flight. If some form of energy is either transmitted to or reflected from the target, the radiated or reflected energy can be used as a homing signal by the seeking missile.

The guidance function can often be chronologically separated into three parts: initial, midcourse, and terminal guidance. Many systems use different guidance techniques for each phase of missile guidance.

Remote Source of Steering Information

Command Guidance. A command guidance (command control) system (Fig. 3) utilizes a remote installation to determine target and missile position, compute necessary steering signals, and transmit these signals to the missile.

The missile range and angle and the target range and angle are measured by separate tracking systems and fed into the computer. The computer must calculate the missile and target positions, generate steering signals on the basis of a preprogrammed attack philosophy, and convert these signals into the missile-axes coordinate system. The computer output is transmitted via a data link to the missile.

Many variations of this scheme exist.[6] In some cases, the missile flight path can be satisfactorily predicted using launch data, wind information, previous commands, and the equations of motion of the missile. In this case, only a single target-tracking radar is used and the necessary missile flight path is "programmed" in the computer. A similar situation exists in guiding a rocket to the moon. Astronomers have precisely determined the motion of the moon relative to the earth, and this information can be stored in the computer. Thus, only a missile-track radar would be necessary. Target or missile position might be determined by measurements made within the missile or target and transmitted to the computer. An example might be the guidance of

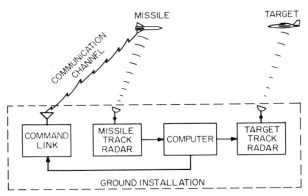

Fig. 3. A command guidance system. (*From Ref. 6, p. 563.*)

a supply rocket to a space station which is continuously transmitting its position back to the earth. Tracking need not be restricted to radar, since optical tracking or other energy-sensing means can be employed.

The advantages of command guidance lie in its flexibility. Since the majority of the equipment is situated at the fixed base, the tracking, transmission, and computing equipment can be flexible, accurate, and long-range. Some disadvantages are distance and line-of-sight limitations (apparent in intercontinental and interplanetary flight), restrictions in traffic handling to a single missile and target for each command unit, and vulnerability to countermeasures because of the radio data link. Wire transmission of steering signals eliminates the latter problem but is feasible only in very short-range applications such as an antitank missile.

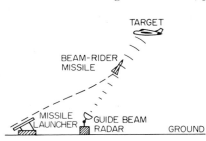

Fig. 4. A beam-rider system. (*From Ref. 6, p. 575.*)

Beam Rider. The basic configuration of a beam-rider guidance system is shown in Fig. 4. The radar unit tracks the missile and also produces a guide beam. The missile is launched into the beam and an internal steering system causes the missile to ride up the center of the beam to the target.[7]

The beam must contain information on the magnitude of displacement from the center line as well as the direction in which the missile must turn to follow the center line. One technique for producing such a beam is to use a conical scan. A conventional radar beam is rotated by a small angle δ from the line of sight (ground-to-target) and then nutated about the center line (Fig. 5). The signal strength is proportional to the displacement r_m from the center of the beam, and the phase of the signal indicates the rotation θ_m about the circumference of the beam. Thus r_m is contained in the amplitude modulation of the beam and θ_m is carried in the phase modulation of the beam.

PRINCIPLES OF VEHICULAR GUIDANCE AND CONTROL

The block diagram of a typical beam-rider missile receiver is shown in Fig. 6. The reference signal is transmitted with the radar beam to establish a phase reference. At launch the missile velocity is too small to provide aerodynamic control and the missile follows a ballistic trajectory. If the missile flight path crosses the beam at too large an angle, the beam may not capture the missile. In this case the missile is lost and the mission aborted. To avoid such a difficulty the launcher is pointed along the beam to provide initial ballistic guidance while beam riding provides midcourse and terminal guidance.

As an alternative, a broad beam is sometimes used for acquisition and a narrow one for guidance. The beamwidth of a radar antenna narrows as the ratio of diameter to wavelength increases.

The missile must be stabilized in roll; otherwise a rotation of 90° from the proper orientation would reverse the roles of pitch and yaw controls, causing the missile to diverge from (rather than converge to) the beam center line. A more sophisticated approach utilizes missile roll rather than avoiding it. Since missile roll does produce a desirable gyroscopic stabilization along the longitudinal axis, the missile can be permitted to roll and proper steering signals can be computed as a function of the roll angle ϕ.

If the beam radar is ground-based, power and antenna size are flexible. When the radar is airborne, as in an air-to-air mission, size and weight become of prime importance.

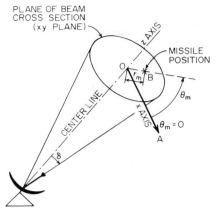

Fig. 5. A conical scan system. Note:
1. Line OA in cross-section plane.
2. Line OB in cross-section plane.
3. Lines OA and OB determine cross-section plane which is normal to the center line.
4. Angular rotation of beam from center line is δ.
5. Polar coordinates (in cross-section plane) of the missile are r_m, θ_m.

The missile receiver is made relatively simple and reliable. More than one missile may be guided along the same beam. Some of the disadvantages of beam guidance are similar to those of command guidance: line-of-sight requirement and countermeasure vulnerability. In addition the initial-capture problem is present.

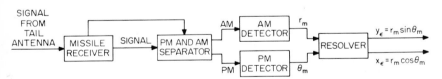

Fig. 6. A beam-rider missile receiver.

Programmed Steering Command

A vehicle can be guided along a flight path on the basis of its position point by point with respect to certain fixed reference axes, as well as the launch and target position relative to these axes. Initially, the positions of launcher and target are known, and the flight program is set from these data. Position, velocity, and acceleration measurements are made during flight with respect to the reference frame and compared with programmed values to obtain error signals which are used as steering commands.

Dead Reckoning. Dead-reckoning navigation predicts present position by noting the direction and amount of progress from a known initial position. This is the simplest means of guidance available to a human navigator. A pilot flying a light airplane from New York City to Chicago might use dead reckoning. Taking off from Long Island, he would note the time he passed over New York City and set his bearing toward Chicago using a magnetic compass and a map. If he were cruising at constant airspeed and altitude and knew the direction and magnitude of the wind when he took off, a simple vector addition would tell him his approximate ground

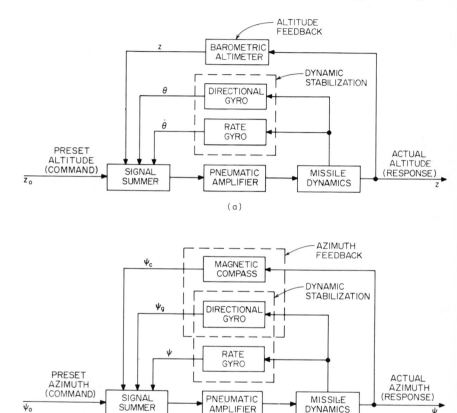

FIG. 7. Altitude and yaw control for a V-1 missile. (a) Altitude-control system. (b) Yaw-control system.

position after 1 hr of cruising. His estimates could be corrected whenever he passed over a recognizable landmark listed on the map.[8]

A similar programmed guidance system was used in the German V-1 missile.[9] The pitch- and yaw-control systems are shown in Fig. 7. The altimeter was used to hold a constant altitude. The rate and directional gyroscopes stabilized the missile dynamics. The directional gyroscope was aligned with the missile roll axis and the missile was launched in the proper direction. The gyroscopic inertia would tend to keep the missile on course, except for gyro drift errors. The magnetic compass helped to reduce the drift error in azimuth. The air mileage was measured by a windmill

PRINCIPLES OF VEHICULAR GUIDANCE AND CONTROL 15-11

airspeed indicator driving a counter through a large reduction-gear ratio. The control actuators and amplifiers were pneumatic and operated from compressed-air bottles inside the missile. The initial programming consisted of manually setting the magnetic compass, the altitude, and the range. When the preset target mileage was reached, the counter actuated a switch, locking the control surfaces and forcing the missile into a ballistic dive at the target.

The advantages of dead-reckoning guidance are simplicity, insensitivity to countermeasures, and lack of complex land-based equipment. Some disadvantages are accuracy problems and limitation to fixed targets. (Only if the target motion can be precisely predicted before launch can the missile be programmed to fly to the collision rendezvous point in space.)

Inertial-guidance Systems. Inertial measuring instruments are used within a guided missile to determine its position in inertial space.[10] Position is measured using a free or integrating rate gyro, angular velocities with a rate gyro, linear or incremental angular accelerations with an accelerometer, altitude with a barometric or radar altimeter, and the local gravity with pendulums or bubble levels. Integrating and differentiating circuits can be used to obtain derivative and integral information from one basic measurement.

Ideally, an inertial system navigates with respect to a set of reference axes stabilized in inertial space. A stable platform uses rate gyros and accelerometers to establish such an inertial reference frame. To simplify navigation between two points on the earth, a terrestrial reference system is used. The reference frame can be slaved to the earth using the terrestrial magnetic or gravitational field. Generally the axes are slaved to the direction of the local gravity using a pendulum. This forms an xy reference plane tangent to the earth and a z axis aligned with the local gravity vector. Knowledge of the coordinates of the launch site and the target with respect to this reference frame permits computation of a missile flight path which is stored in the missile as the flight program.

A long-range ballistic missile can utilize an inertial-guidance system. An initial trajectory plane is defined, passing through the earth's center, the launch site, and the target. The missile is stabilized in roll and yaw to keep its flight path within the trajectory plane. Lateral acceleration may be used to control any sideslip caused by wind or thrust misalignment. Pitch angle is programmed as a function of time to obtain the desired flight path in the trajectory plane. The longitudinal acceleration can be integrated yielding a velocity signal, which is used for motor cutoff, after which the missile follows a ballistic trajectory to the target. A typical system is shown in Fig. 8.

A more sophisticated means of effecting motor cutoff is to compute the range miss distance, assuming the missile will stray a bit from the optimum flight path. At the point of motor shutoff, the remaining flight path can be computed in terms of the existing positions and velocities. The range error ΔR due to nonoptimum position and velocity can be approximately computed by expanding the range equations in a Taylor series about the desired range R_d and retaining first-order terms.

$$\Delta R = R - R_d = (\partial R/\partial x)(x - x_d) + (\partial R/\partial y)(y - y_d) + (\partial R/\partial \dot{x})(\dot{x} - \dot{x}_d) \\ + (\partial R/\partial \dot{y})(\dot{y} - \dot{y}_d) \quad (2)$$

A motor-shutoff computer is designed on the basis of Eq. (2) which turns off the motor when ΔR goes to zero.

An inertial reference system need not be slaved to a terrestrial reference if the drift errors of the components are tolerable and if navigation with respect to arbitrary reference axes is practical. The earth reference need not be gravitational, and magnetic, optical, radio, or infrared properties of the earth may be used. In fact celestial references can be used in conjunction with an inertial system. With the use of two optical star trackers, tabulated astronomical data, and a computer, an inertial reference system can be slaved to an astronomical reference frame.

The limitations and advantages of dead-reckoning equipment also apply to inertial systems; however, the latter yield better accuracy, are bulkier, and are more complex than simple systems of the type used in the V-1 missile. Gravitational slaving of a stable platform introduces additional problems. Gravity is nonuniform over the earth's surface, and the gravity-indicating pendulum is sensitive to the earth's rotations, causing a false indication of the vertical. The former can be alleviated using a gravity computer and the latter by careful feedback adjustment of the pendulum period to 84 min (Schuler tuning), so that the pendulum angle changes at the same rate as the earth's normal.

Modern techniques of integrated circuitry have permitted substantial reductions in the size and weight of inertial navigation computers. Many systems now solve the exact equations of motion governed by Kepler's laws at each point along the trajectory, yielding a more flexible and accurate system than the Taylor series expansion technique.[11,12]

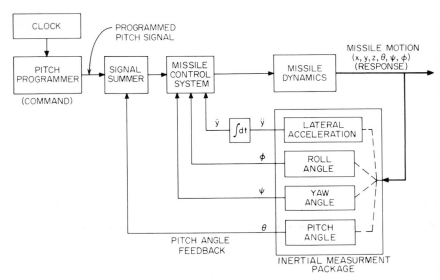

Fig. 8. An inertial-guidance system for a ballistic missile.

Radio Navigation. Radio navigation is a technique used to establish a radio reference grid on the earth's surface using two or more transmitters.[13] Loran, Decca and Gee are commercial names for radio-navigation systems. The two most popular pattern geometries are the hyperbolic and circular reference grids, shown in Fig. 9. In hyperbolic navigation, the missile flight path lies on a hyperbola if $(r_2 - r_1)$, the difference in range between transmitters A and B, is constant. A signal transmitted from the missile is received by ground stations A and B, which trigger transmitters located at A and B (slightly different frequencies are used to avoid confusion). The signals are received at the missile and the time difference is compared.

If the time difference exceeds or falls short of the preset time τ_n, the missile steers left or right back on course. The transmission frequencies and the geometry of the hyperbolic path passing through the target determine τ_n. The missile is held at a constant preset altitude with an altimeter or an altitude-hold system. The range to the target is determined by comparing the transmission-reception lag from one of the ground stations with a stored range time t_r. The missile then begins terminal guidance or dives at the target.

When circular navigation is used, the missile remains at a constant distance r_1 from transmitter A. As in hyperbolic navigation, the ground transmitters are triggered by signals from the missile. The missile is stabilized in altitude and is kept on a circular flight path by steering the missile so that the transmission-reception lag of transmitter A is constant. The time lag of the signal from transmitter B is used to determine range to the target.

Radio-navigation systems have high traffic-handling capacities and require relatively simple missile and ground-station equipment. The target location with respect to the ground stations must be accurately known. Also, the enemy can locate the ground stations and missile by detecting the transmitted signals. Of course, the latter difficulty is irrelevant if the missile is used for nonmilitary purposes, e.g., transporting mail or cargo.

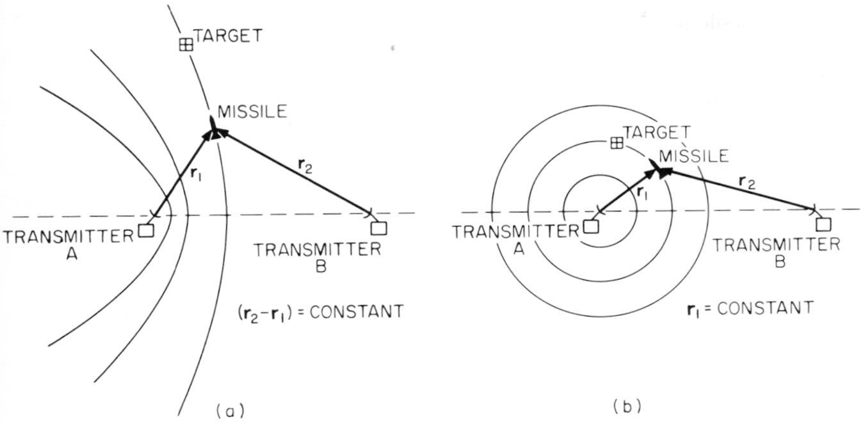

FIG. 9. Geometry of hyperbolic and circular navigation. (a) Hyperbolic geometry. (b) Circular geometry.

Homing Systems

Homing systems generate steering signals from information received at the missile from the target.[14] An active homing system beams a signal at the target and generates steering commands from the reflected signal. In a semiactive homing system, a remote transmitter bounces signals off the target to the missile. A passive homing system receives energy emanating from the target. The homing radiation may be radio, heat, light, or sound signals.

The particular missile application, surface to surface, surface to air, etc., and the form of the energy determine the details of the homing system. For example, in an active radio-homing system, the missile antenna is stabilized and tracks the target. If the missile computer steers the missile at the target, the missile roll axis tends to align with the antenna; if another attack philosophy is used, the missile axis and antenna axis are skew. With an active infrared system, a conical scan similar to that used with beam riders can be used. A passive homing system may use a receiving antenna on either side of the missile and determine the direction of the target by comparing phase differences much as the human ear does.

In an active system both transmitting and receiving equipment must be contained in the missile, with a resultant weight penalty as well as vulnerability to enemy detection. When a tracking signal is continuously transmitted and received in the missile, it is difficult to separate direct reception of the transmitted signal from reception of the reflected signal. Pulse transmission is often used to alleviate this problem. When the missile and target approach each other, the reflected signal strength increases greatly and the system may saturate.

The semiactive system achieves decreased size and weight and increased range by moving the transmitter from the missile to a remote location where powerful transmission is feasible. The remote transmitter now reveals its position rather than that of the missile. The transmitter is now responsible for, and better able to thwart countermeasures. Saturation presents less of a problem than in active systems.

Passive homing depends on target energy emission in a form, and of sufficient amplitude, for adequate detection at the missile. Certain characteristics can be used to locate direction and shape. The amplitude, phase, and frequency of the received radiation can be interpreted to yield distinguishing characteristics of the target. The complexity of a passive homing guidance system depends on the signal strength available and the ease with which tracking and range information can be obtained. Often considerable detailed information about the average characteristics of the target radiation must be known. Both sonic and infrared emission of an aircraft, for example, have directional patterns much like antennas. A sonic system also suffers from a virtual-image effect due to the low speed of sound and the high speed of missile and target. If the signal source is of high intensity and tracking simple, the equipment can be made small, simple, reliable and difficult to fool. System accuracy is highly dependent on the specific target, and on the energy form employed. This mode of guidance might be useful to the first visitors on the moon to guide subsequent supply rockets to them. A solar-powered radio beacon might be used to provide the homing signal.

5 Maneuver Strategy

An important aspect of the guidance function is the plan of attack or attack philosophy used in hunting down a moving target.[15] If a defense weapon has relatively good maneuver and speed capabilities and is launched when an attack weapon is far from the target, any reasonable attack plan will probably suffice. If initial detection of the attacker is at short range and the defense is the inferior in maneuver and speed characteristics, there is little hope for interception no matter how shrewdly the defender maneuvers. Of course, the area between the two extremes where both offense and defense are capable, fast, and maneuverable contains most of the practical problems.

The maneuver strategy of a missile depends in part on the form of guidance employed. A beam-rider missile must follow the line of sight to the target. In a command guidance system, the ground-based computer is free to generate any appropriate strategy. Navigation of a scientific satellite transmitting television pictures of the earth would require a different procedure. In the latter case the flight path as well as the orientation of the vehicle (attitude) must be controlled.

The quantities of interest are the missile flight path, time to intercept, missile turning rate, and missile accelerations. The flight path and intercept time determine whether the tactical mission is accomplished; the turning rate determines the control moments and forces necessary, and the acceleration determines the stresses applied to the missile. Large or infinite missile turn rates and tangential accelerations need not cause immediate discard of the proposed system if they exist for only short periods of time. Limiting means can be built in to protect the missile, and the concomitant errors tolerated if small. The various attack philosophies are briefly discussed in the material that follows. Detailed studies of specific attack plans appear in the literature. Since the differential equations of motion are, in general, nonlinear and transcendental, approximate techniques or computer solutions are generally employed.

Line-of-sight Course

A beam-rider missile follows a line-of-sight path, since its steady-state *velocity* vector is along the line joining the *beam origin* and the *target*. A typical flight path is shown in Fig. 10a. The target is assumed to be flying at a constant altitude and rate. The launch angle depends on the target altitude and the initial range R_0. Th

PRINCIPLES OF VEHICULAR GUIDANCE AND CONTROL 15-15

intercept time t_i can be determined from the target velocity V_T and distance of travel from R_0 to collision range R_c.

$$t_i = \frac{(R_0 - R_c)}{V_T} \tag{3}$$

The missile turning rate is always finite if the missile velocity is larger then the target velocity; i.e., $p = (V_M/V_T) > 1$. Since the magnitude of the tangential acceleration is equal to the missile speed multiplied by the turning rate, the acceleration is finite whenever the turning rate is finite. As the target and missile approach, there is a rapid increase in turn rate and a similar change in acceleration. One of these phenomena may make this attack philosophy impractical in certain applications.

Pure Pursuit

In a pure-pursuit course, the missile *velocity* vector is directed along the line of sight from the *missile* to the *target*. Initially the pure-pursuit course and the line-of-sight course are similar, since the missile is close to the launch site. As the missile moves away from the launch site, they begin to differ. The line-of-sight course can be likened to a stationary hunter pointing his gun at a moving rabbit, while a pursuit course would be the path taken by the hunter's dog chasing the rabbit. A typical flight path for a pure-pursuit attack of a constant-altitude constant-speed target is shown in Fig. 10b.

For a pure-pursuit course, an intercept always occurs if missile velocity exceeds target velocity, $p > 1$. It can be shown that for outgoing as well as incoming targets, the missile ends up in a head-on chase or in a tail chase. For $p \geq 2$, the maximum turn rate occurs at closing; moreover, for $p > 2$ it becomes infinite, and for $p = 2$, the turn rate is finite and proportional to V_T/R_0. (R_0 is the initial separation between missile and target.) For $1 < p < 2$ the maximum turn rate is proportional to V_T/R_0 and occurs somewhere between launch and intercept. The acceleration (for all ranges of p) is proportional to the missile velocity and turn rate as before.

Lead Pursuit

A lead-pursuit (deviated-pursuit) course maintains the *angle* between the *missile velocity* vector and the line of sight from *missile* to *target* constant. This angle is called the lead angle δ. When the lead angle is zero, a pure-pursuit course is obtained. A common example of lead pursuit might be a boy chasing his dog. The chase might begin as a pure-pursuit course but would probably become a lead pursuit when the boy tried to "head off" the dog. An example of a lead-pursuit flight path is shown in Fig. 10c.

Detailed analysis shows that if $p^2 \sin^2 \delta > 1$, the missile spirals about the target, and no interception occurs. When $p^2 \sin^2 \delta \leq 1$, interception occurs. The turning rate becomes infinite for

$$2 < \frac{p \cos \delta}{\sqrt{1 - p^2 \sin^2 \delta}} \tag{4}$$

The turn rate is always finite for

$$2 \geq \frac{p \cos \delta}{\sqrt{1 - p^2 \sin^2 \delta}} \tag{5}$$

As before, acceleration is given by the product of missile velocity and turn rate.

Constant-bearing Course

In a constant-bearing course, the line of sight from the *missile* to the *target* is maintained at a *constant direction* in space. This means that if the target flies a straight course, the missile does also. This situation results in a collision course,

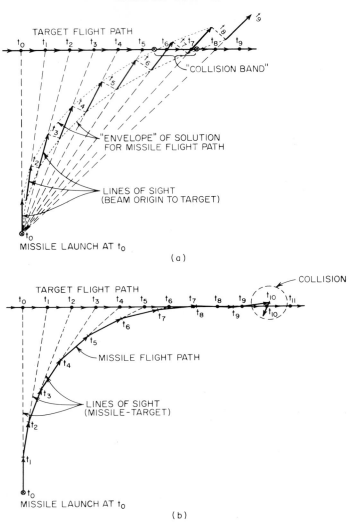

Fig. 10. Typical missile and target flight paths illustrating various attack philosophies. (a) Line-of-sight course. Graphical solution for $p \equiv V_m/V_T = 1.5$. Collision occurs between $t_{5.5}$ and $t_{7.2}$ sec. (b) Pure-pursuit path. Graphical solution for $p \equiv V_m/V_T = 1.5$. Collision occurs at approximately t_{10} sec.

where the missile and target collide at the point of intersection of the two velocity vectors. A typical example is shown in Fig. 10d.

The flight path is the same as a lead-pursuit course when

$$\delta = \gamma_M = \sin^{-1}[\sin(\gamma_T/p)]$$

No interception occurs when $\cos \gamma_M \leq \cos(\gamma_T/p)$. When $\cos \gamma_M > \cos(\gamma_T/p)$, collision occurs, and the flight time is given by $T_f = R_0/(V_M \cos \gamma_M - V_T \cos \gamma_T)$. From the definition of the constant-bearing course, the angular rate and tangential

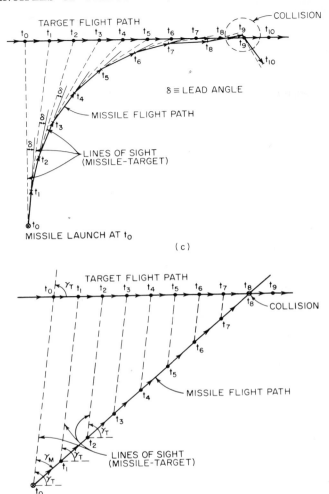

Fig. 10 (*Continued*). Typical missile and target flight paths illustrating various attack philosophies. (c) Lead-pursuit course. Graphical solution for $p \equiv V_m/V_T = 1.5$, $\delta = 5°$. Collision occurs at approximately t_9 sec. (d) Constant-bearing course. $\gamma_T \equiv$ constant bearing. Graphical solution for $p \equiv V_m/V_T = 1.5$. $\gamma_m = 42°$. Collision occurs at t_8 sec.

acceleration are zero for a nonmaneuvering target. This latter attribute makes this a highly desirable attack strategy.

Proportional Navigation

In a proportional-navigation course, the *rate of change* of missile *heading* is proportional to the *rate of rotation* of the line of sight from *missile* to *target*. A typical proportional-navigation course is shown in Fig. 10e. The governing constraint is is that $\dot{\phi}_m = a\dot{\phi}_T$. The constant a is called the navigation constant (navigational correction). Integration of the above relation yields $\phi_m = a\phi_T + \phi_0$, where ϕ_0 is a constant of integration. The availability of two design parameters a and ϕ_0 results in greater flexibility for a proportional-navigation course than for the other maneuver strategies discussed. The latter asset is of even greater advantage if a maneuvering target is to be encountered.

A proportional navigation course becomes a pure-pursuit course for the specific adjustment $a = 1$ and $\phi_0 = 0$. When $a = 1$ and ϕ_0 is fixed at some constant bearing, a lead-pursuit course is obtained.

Closed-form solutions are obtainable for the particular case when $a = 2$, and appear in the literature. For arbitrary a, numerical methods can be used. As before, acceleration is proportional to the product of missile velocity and turn rate.

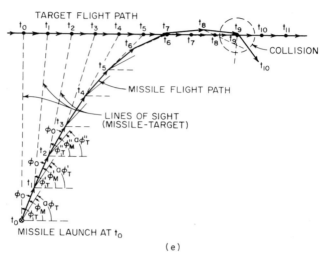

(e)

Fig. 10 (*Continued.*). Typical missile and target flight paths illustrating various attack philosophies. (e) Proportional-navigation course. ϕ_T = target bearing. ϕ_m = missile bearing. ϕ_0 = constant bearing. a = navigation constant. $\phi_m = a\phi_T + \phi_0$. Graphical solution for $a = \frac{2}{3}$, $\phi_0 = 10°$. Collision occurs at approximately t_9 sec.

Attitude Control

The task that a vehicle must perform is a major factor in determining the problems encountered and the basic control-equipment configurations that are to be used. Control of a "space-mail" missile during a flight from New York City to London differs from the control functions necessary to guide a passenger rocket between the same two points.

The mail missile could be launched from the ground or from an aircraft. It could descend via a parachute, be caught in a landing net, or use any other feasible means to terminate its voyage. During its flight, it would be of minor concern if the missile tumbled or gyrated in some other manner. The important consideration would be the *path* taken by the *center of gravity* of the vehicle. (Actually the two end points are the items of primary importance.) This category of control problems could be called flight-path control.

Control of a passenger vehicle between the same points would be more complex. The rocket ascent and descent would be governed by the limits of passenger safety and comfort. The missile orientation during flight (rotational position, velocity, and acceleration) would be important in determining satisfactory performance. Thus the *orientation* of the vehicle with respect to a set of *coordinate axes* (fixed on the missile, another body, or in inertial space) is of major importance. Control problems in this field are known as attitude control.[16] A combination of attitude and flight-path control is necessary in many practical problems.

A contemporary example of interest in the attitude-control field is the control of a photographic satellite. Such a vehicle must be stabilized so that its optical system points at the object to be photographed. If the resulting picture is to be sent back to earth via radio means, the transmitting antenna on the satellite has to point in the general direction of the receiving station. Inertial, terrestial, or celestial navigation might be used to stabilize the vehicle properly. Control moments can be applied using compressed air, rocket, or ionic reaction controls.

At present, the field of attitude control is developing and basic philosophy is evolving. Discussions of specific attitude-control problems are quite prevalent in the missile- and satellite-control literature.[17]

REFERENCES

Coordinate Axes
1. Herbert Goldstein, "Classical Dynamics," pp. 107, 108, Addison-Wesley Publishing Company, Inc., Reading, Mass.

Terminology
2. Allen E. Puckett and Simon Ramo, "Guided Missile Engineering," Chap. 7, McGraw-Hill Book Company, New York, 1958.
3. Richard B. Dow, "Fundamentals of Advanced Missiles," Chaps. 8, 9, John Wiley & Sons, Inc., New York, 1958.

Aerodynamic Forces
4. O. G. Sutton, "The Science of Flight," Penguin Books, Inc., Baltimore.
5. Ref. 2, chap. 2.

Command Guidance
6. Arthur S. Locke, "Guidance," chaps. 2, 16, D. Van Nostrand Company, Inc., Princeton, N.J., 1955.

Beam Rider
7. Ref. 6.

Dead Reckoning
8. Ref. 2, chap. 7.
9. Ref. 6, chap. 2.

Internal-guidance Systems
10. William T. Russell, Inertial Guidance for Rocket-propelled Missiles, *Jet Propulsion*, January, 1958, pp. 17–24.
11. C. F. O'Donnell, "Inertial Navigation," McGraw-Hill Book Company, New York, 1964.
12. C. T. Leondes, "Guidance and Control of Aerospace Vehicles," McGraw-Hill Book Company, New York, 1963.

Radio Navigation
13. Ref. 6, chap. 16.

Homing Systems
14. Ref. 6, chap. 16.

Maneuver Strategy
15. Ref. 6, chap. 12.

Attitude Control
16. R. E. Roberson, Orbital Behavior of Earth Satellites, *J. Franklin Inst.*, vol. 264, pp. 181–202, September, October, 1957.

FEEDBACK SYSTEMS

6 Feedback Theory

Feedback is a descriptive term embracing a group of system analysis and design philosophies.[1,2] Feedback may occur in both passive and active systems. It may appear as an inherent system characteristic or be purposely introduced to yield special effects. Some systems appear to contain feedback when viewed in one manner but the effect appears to vanish if another approach is used. The distinguishing characteristic of the feedback is that some function of the system output is returned to the input in such a manner as to affect the system response. The simple resistive voltage divider of Fig. 11a can be described in feedback terms.

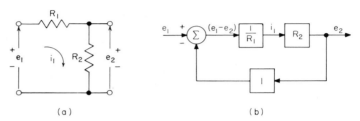

Fig. 11. Circuit and block diagram of a resistive voltage divider. (a) Circuit diagram. (b) Block diagram.

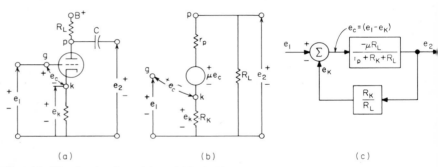

Fig. 12. Circuit, signal model, and block diagram for a triode amplifier. (a) Circuit. (b) Signal model. (c) Block diagram.

The equations for the network of Fig. 11a are, by inspection,

$$(e_1 - e_2) = R_1 i_1 \qquad (6)$$
$$e_2 = R_2 i_1 \qquad (7)$$

The block diagram of Fig. 11b contains the same information as Eqs. (6) and (7). The advantage of the block-diagram approach is that it transforms a set of equations into a functional model yielding topological insight into the cause-and-effect relationships governing the system: the reference signal e_1 excites the system; the controlled variable e_2 is subtracted from e_1 forming the actuating signal $(e_1 - e_2)$; the actuating signal multiplied by $1/R_1$ yields the loop current i_1; the loop current multiplied by R_2 is the system output e_2. Thus a simple voltage divider can exhibit feedback effects if the analysis is approached in an appropriate manner.

The circuit-diagram signal models of a triode amplifier are shown in Fig. 12. If the amplifier is biased class A and the capacitor C is a short circuit at all frequencies of interest, the voltage across the cathode resistor provides degeneration (a feedback effect reducing the incremental voltage gain), since the actuating signal e_c is equal to the input signal e_1 diminished by the *subtractive* effect of e_K. The circuit equations are

$$(e_1 - e_K) = e_c \tag{8}$$

$$e_2 = \frac{-\mu R_L}{r_p + R_K + R_L} e_c \tag{9}$$

$$e_K = \frac{-\mu R_K}{r_p + R_K + R_L} e_c = \frac{R_K}{R_L} e_2 \tag{10}$$

Equations (8), (9), and (10) yield the block diagram of Fig. 12c.

Transfer Functions and Functional Models

The preceding section illustrates the manner in which a block-diagram model can be substituted for a set of algebraic equations. In order that a similar representation may be used for a set of integrodifferential equations, the operations of time integration and differentiation must be modeled. There are two common approaches that are used. Integration and differentiation operators may be defined, or the integrodifferential equations may be Laplace-transformed to yield an associated set of algebraic equations involving s, the complex frequency.[3]

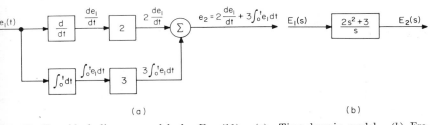

FIG. 13. Two block-diagram models for Eq. (11). (a) Time-domain model. (b) Frequency-domain model.

Example. Voltages $e_1(t)$ and $e_2(t)$ are related by the differential equation

$$de_2(t)/dt = 2[d^2 e_1(t)/dt^2] + 3e_1(t) \tag{11}$$

There is no energy storage in the system at $t = 0$ (zero initial conditions). Integrating both sides of Eq. (11) yields

$$e_2(t) = 2[de_1(t)/dt] + 3 \int_0^t e_1(t)\, dt \tag{12}$$

A block-diagram model corresponding to Eq. (12) is shown in Fig. 13a. The Laplace transform of Eq. (11) gives

$$sE_2(s) = 2s^2 E_1(s) + 3E_1(s) \tag{13}*$$

Solving for $E_2(s)/E_1(s)$ gives the system function or transfer function.

$$\frac{E_2(s)}{E_1(s)} = \frac{2s^2 + 3}{s} \tag{14}$$

The block-diagram model associated with Eq. (14) is shown in Fig. 13b.

* Equation (13) is correct only if the initial conditions are zero. If $e_1(t)$ and $e_2(t)$ are nonzero at $t = 0$, the values $e_1(0)$ and $e_2(0)$ appear as separate inputs in the block diagram of Fig. 13a, and a transfer function is not so simply defined.

Feedback-model Manipulation

In the study of a physical system, many different sets of system equations can be written. Each set is valid as long as the equations are independent and equal in number to the number of system unknowns. Different sets of equations produce different, but equivalent, block diagrams. One set of equations can be transformed into another by using ordinary algebraic manipulations.[4]

A set of a topological relations for manipulation of block diagrams is shown in Table 2.

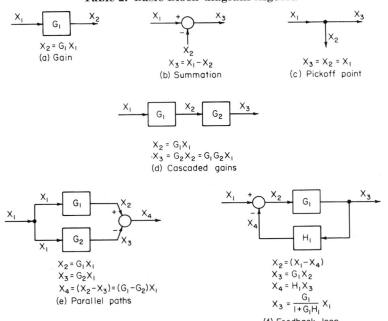

Table 2. Basic Block-diagram Algebra

Application of relations (d) and (f) of Table 2 to Fig. 12b results in the familiar voltage-divider relationship

$$e_2 = \frac{R_2}{R_1 + R_2} e_1 \qquad (15)$$

Applying the same manipulations to Fig. 12c yields

$$e_2 = \frac{-\mu R_L}{r_p + R_L + (\mu + 1)R_K} e_1 \qquad (16)$$

Equation (17) is the familiar expression for the gain of a grounded-cathode triode amplifier with unbypassed cathode resistor.*

Another functional model useful in feedback analysis is the signal flow graph.[4,5,6] A flow graph may be viewed as an evolution of a block diagram so that the specific transfer functions are pictorially deemphasized and the topology of the system clearly revealed. Table 2 is transformed into signal-flow-graph notation in Table 3. The

* The preceding section and the bulk of this chapter have been benefited by the suggestions and comments of Irwin I. Sterman.

FEEDBACK SYSTEMS

Table 3. Signal-flow-graph Algebra

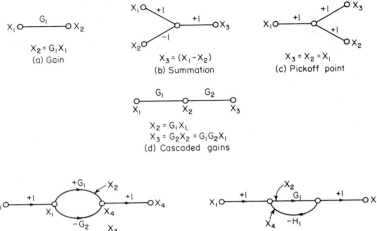

flow graph consists of nodes and branches. A signal flowing through a branch is multiplied by the gain of the branch. Signals entering a node (arrows point toward the node) add at the node. All signals leaving a node have the same value, which is equal to the algebraic sum of all the signals that enter the node.

Many theorems and manipulating rules have been developed for signal flow graphs to simplify calculation and reveal the nature of the feedback effects (most of these are, of course, equally valid when applied to block diagrams). In order to illustrate the use of signal flow graphs, and in addition to stress an important principle, the "loading effect," the following example is discussed.

Example. In Table 4, step 1, a simple low-pass filter is shown. The system equations are easily formulated if the capacitor is represented by its complex impedance $1/c_1 s$, step 2. (This last step is essentially replacement of the differential equations by $\mathcal{L}$ equations with zero initial conditions.) Step 3 consists of a choice of input and output nodes and at least some of the intermediate nodes. The branches of the graph are drawn between the nodes in step 4. The gain of the graph is calculated in step 5, yielding the system transfer function. A circuit model of a two-stage filter is shown in Fig. 14a. It is incorrect to assume that the transfer function of the two-stage filter is the product of two factors similar to those in Table 4, step 5, i.e.,

$$\frac{E_4(s)}{E_0(s)} \neq \frac{1/R_1 C_2 s}{1 + 1/R_1 C_2 s} \frac{1/R_3 C_4 s}{1 + 1/R_3 C_4 s}$$

The second stage of the filter loads the first, that is, current i_3 affects voltage e_2. The true expression is given by solving the flow graph shown in Fig. 14b. By using more general flow-graph methods, the transfer function can be shown to be

$$\frac{E_4(s)}{E_0(s)} = \frac{(1/R_1 C_2 s)(1/R_3 C_4 s)}{1 + 1/R_1 C_2 s + [1/R_3 C_2 s] + 1/R_3 C_4 s + 1/R_1 R_3 C_2 C_4 s^2} \qquad (17)$$

Table 4. Evolution of a Signal Flow Graph

Step	Method	Results	Comments
1. Circuit models	Draw a circuit model for the system in question. Clearly label input and output and intermediate variables.	Circuit diagram with R_1, C, input $e_1(t)$, current $i_1(t)$, output $e_2(t)$	Circuit model approximation for electrical, mechanical, thermal, pneumatic, hydraulic, etc., lumped-parameter systems. Other models may be used or the process may start at step 2.
2. Circuit equations	Write down an appropriate set of independent $\mathcal{L}$ equations relating input to output.	$I_1(s) = \dfrac{E_1(s) - E_2(s)}{R}$ $E_2(s) = \dfrac{1}{C_1 s} I_1(s)$	Use of the complex impedance of the elements is a useful short cut. Initial conditions must be zero. [See footnote Eq. (13).] Since many sets of equations can be written, many flow graphs are valid for the same system.
3. Choose node variables	Choose an input and an output node. Include intermediate nodes that may be necessary.	Nodes: $E_1(s)$ Input, $E_2(s)$ Output, $I_1(s)$	Sometimes it is not obvious what intermediate variables will be necessary and it is best to proceed, including intermediate variables as their need arises.
4. Cause and effect	Start at input and work from left to right in a cause and effect manner.	A. $I_1(s) = \dfrac{E_1(s)}{R_1} - \dfrac{E_2(s)}{R_2}$ B. $E_2(s) = \dfrac{1}{C_1 s} I_1(s)$	In complex systems it is best to proceed slowly, step by step.
5. Gain computation	Find gain from E_1 to E_2 by using relations (f) and (e) of Table 2.	$\dfrac{E_2(s)}{E_1(s)} = \dfrac{1}{R_1} \times \dfrac{\dfrac{1}{C_1 s}}{1 + \dfrac{1}{R_1 C_1 s}} = \dfrac{1}{R_1 C_1 s + 1}$	This is the transfer function of a low-pass filter.

The bracketed term in the denominator of Eq. (17) represents the "loading" of the second stage on the first.

Uses of Feedback

The more interesting applications of feedback theory deal with the intentional introduction of feedback into control systems. In control work, feedback is use primarily to change the system sensitivity to or to alter the output time response to certain transient inputs.

The sensitivity[8,9] of a transmission (transfer function) T to a change in an internal parameter K is denoted by the symbol S_K^T and is equal to the percentage change in T divided by the percentage change in K, for small changes in K. Thus sensitivity

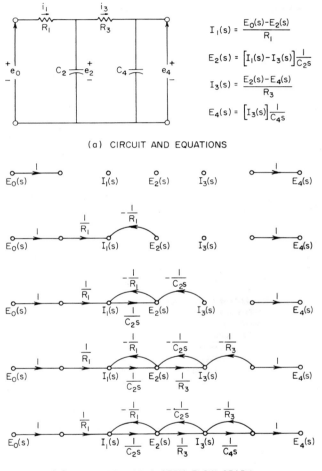

(a) CIRCUIT AND EQUATIONS

(b) EVOLUTION OF SYSTEM FLOW GRAPH

FIG. 14. Circuit and flow graph models for a two-stage filter.

is a measure of the effect on the output of drift or change in value of an internal system component.

$$S_K^T = \frac{dT/T}{dK/K} = \frac{\partial T}{\partial K}\frac{K}{T} \qquad (18)$$

As an illustration of the manner in which feedback affects sensitivity, S_K^T is calculated for the three systems in Fig. 15. The transmission T is C/R. The gain G is assumed

15-26   REMOTE CONTROL

to vary, $G = K$.   Application of Table 2 (a) and (f) yields, respectively,

$$T_a = G \quad (19a)$$
$$T_b = G/(1 + GH) \quad (19b)$$
$$T_c = G/(1 - GH) \quad (19c)$$

Taking partial derivatives of the above transmission with respect to G yields

$$\partial T_a/\partial G = 1 \quad (20a)$$
$$\partial T_b/\partial G = 1/(1 + GH)^2 \quad (20b)$$
$$\partial T_c/\partial G = 1/(1 - GH)^2 \quad (20c)$$

Substitution of Eqs. (19) and (20) into (18) yields

$$S_G^{T_a} = 1 \quad (21a)$$
$$S_G^{T_b} = 1/(1 + GH) \quad (21b)$$
$$S_G^{T_c} = 1/(1 - GH) \quad (21c)$$

If the system in question is a feedback amplifier with $G = 100$, $H = 1/120$, the sensitivities and gains of the three configurations of Fig. 15 are

$$S_G^{T_a} = 1 \quad (22a)$$
$$S_G^{T_b} = 6/11 \quad (22b)$$
$$S_G^{T_c} = 6 \quad (22c)$$
$$T_a = 100 \quad (23a)$$
$$T_b = 54.5 \quad (23b)$$
$$T_c = 600 \quad (23c)$$

Negative feedback decreases the system sensitivity (generally a beneficial effect) at the expense of a loss in system gain. Positive feedback degrades the system b

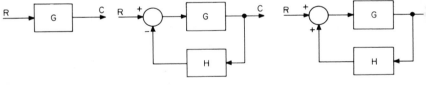

(a) NO FEEDBACK   (b) NEGATIVE FEEDBACK   (c) POSITIVE FEEDBACK

FIG. 15. Three feedback systems. ($R \equiv$ reference, input; $C \equiv$ controlled variabl output.)

increasing the sensitivity sixfold but also increases the system gain by a factor of (frequently an advantageous effect). A trade-off between sensitivity and gain ca be effected by adjustment of the polarity and magnitude of the feedback. Often combination of positive and negative feedback is useful in a complex system. A an illustration of the manner in which feedback affects system transient response the following example is cited. A system with an input $r(t)$, an output $c(t)$, and ϵ intermediate variable $e(t)$ is described by the following differential equations:

$$e(t) = c''(t) + ac'(t) + bc(t)$$
$$e(t) = r(t) - kc'(t) \quad (24)$$
$$c(0) = c'(0) = 0$$

The Laplace transforms of Eqs. (24) are

$$E(s) = (s^2 + as + b) \times c(s)$$
$$E(s) = R(s) - ks \times c(s) \quad (25)$$

A block diagram for Eq. (25) is shown in Fig. 16. (If the initial conditions are nonzero, they appear as additional signal inputs in the block diagram.)
Solution of the block diagram of Fig. 16 gives

$$C(s) = \frac{1}{s^2 + (k+a)s + b} \times R(s) \qquad (26)$$

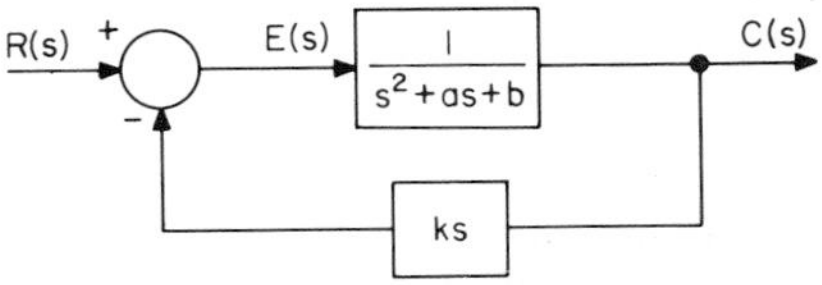

FIG. 16. Block diagram for the system described by Eqs. (25).

The unit step response of the system $R(s) = 1/s$ is given by the Laplace-transform pair shown below. [It should be noted that Eqs. (26) and (27) differ by the numerator constant scale factor ω_n^2.]

$$\mathcal{L}^{-1} \frac{\omega_n^2}{s(s^2 + 2\zeta\omega_n s + \omega_n^2)} = 1 - \frac{\epsilon^{-\zeta\omega_n t}}{\sqrt{1-\zeta^2}} \sin(\omega_d t + \phi) \qquad (27)$$

where $\omega_d = \omega_n\sqrt{1-\zeta^2} \qquad \phi = \sin^{-1}\sqrt{1-\zeta^2}$

A normalized graph of Eq. (27) appears in Fig. 17. Feedback in the system in Fig. 16 does not affect the value of ω_n, since it is independent of k. The damping ratio ζ is $(a+k)/2b$. If positive rather than negative feedback were used, $\zeta = (a-k)/2b$. Thus, by proper choice of the polarity and magnitude of the feedback gain k, the damping ratio of the system and consequently its step response can be varied over a wide range.

The sections that follow describe methods for studying the feedback effect and the manner in which it controls system gain, sensitivity, and transient response in a wide variety of control systems.

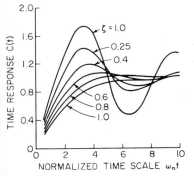

FIG. 17. Time-normalized step response of a second-order system. $c(t) = \mathcal{L}^{-1}\{\omega_n^2/(s^2 + 2\zeta\omega_n s + \omega_n^2)\}$. (Taken from page 3-55, Fig. 3.54, of Ref. 9.)
$c(t) = 1 - (e^{\zeta\omega_n t}/\sqrt{1-\zeta^2})\sin(\omega_d t + \phi)$
$\phi = \sin^{-1}\sqrt{1-\zeta^2} = \tan^{-1}\omega_d/\omega_n$
$\omega_d = \omega_n\sqrt{1-\zeta^2}$

7 Stability

Definition

Stability might be described intuitively as the property of systems which tends to return them to rest after being disturbed. Furthermore, such systems should yield reasonable and predictable responses for reasonable excitations. The former property is an equilibrium condition and the latter a boundedness characteristic. Either concept can be used to formulate a definition of stability. A more rigorous definition that is based on boundedness states

Definition 1. A system is stable if its response is bounded when the system is excited by all possible bounded inputs.

A working definition of stability for linear rational systems (systems for which the transfer functions are expressible as a ratio of polynomials in s) is

Definition 2. A linear rational system is stable if the impulse response of the system contains no terms with amplitudes that increase without limit as time increases. Thus an impulse response that decays with increasing time is stable.

The borderline case, comprising systems that yield a constant-amplitude impulse response, may be defined as stable or unstable depending on personal preference. Definition 2 calls the borderline case stable, while Definition 1 places it in the unstable category.

These stability definitions may be compared with the definitions of system equilibrium in a gravitational (potential) field. A cone resting on its base or a sphere in a concave potential well is said to be in stable equilibrium; if the cone rests on its vertex with its axis oblique or the sphere is on a convex surface, the system is in unstable equilibrium; when the cone rests on its vertex with its axis normal or the sphere is on a horizontal surface, the system possesses neutral equilibrium. In these cases, discussion centers about the configuration yielding minimum potential energy, and the curvature of the equipotential surfaces.

Stability in the s Plane

The complex frequency plane can be used to good advantage in testing the stability of a linear rational system. (A rational system is described by a set of linear constant-coefficient differential equations.) The general form of a rational transfer function is shown in polynomial and factored form in Eq. (28).

$$\frac{C(s)}{R(s)} = T(s) = \frac{a_m s^m + a_{m-1} s^{m-1} + \cdots a_0}{b_n s^n + b_{n-1} s^{n-1} + \cdots b_0} = \frac{(s - s_1)(s - s_3) \cdots}{(s - s_2)(s - s_4) \cdots} \quad (28)$$

The frequency s is a complex quantity with real part σ and imaginary part $j\omega$. The roots of the numerator polynomial are called zeros, and those of the denominator are called poles. If the input is a unit impulse, $R(s) = 1$, the output $C(s)$ can be expanded in partial fractions, yielding

$$C(s) = A/(s - s_2) + B/(s - s_4) + C/(s - s_6) + \cdots \quad (29)$$

The time function associated with Eq. (29) is

$$c(t) = A\epsilon^{s_2 t} + B\epsilon^{s_4 t} + C\epsilon^{s_6 t} + \cdots \quad (30)$$

If the pole position s_2 is real and negative, the first term of Eq. (30) is a decaying exponential; if real and positive a growing exponential. If $s_2 = 0$, the term is a

Table 5. Typical System Dynamics

Transfer function	Pole-zero pattern	Unit-impulse response
$G(s) = \dfrac{A}{s(s+a)}$	poles at 0 and $-a$ on real axis	$g(t)$ rises from 0 to A/a with time constant $1/a$
$G(s) = \dfrac{A(s+a)}{s(s+b)}$	poles at 0 and $-b$, zero at $-a$ on real axis	$g(t)$ rises to $A\dfrac{a}{b}$, asymptote A
$G(s) = \dfrac{\omega_n^2}{s^2 + 2\zeta\omega_n s + \omega_n^2}$	complex conjugate poles at angle $\cos^{-1}\zeta$, radius ω_n	$g(t)$ damped oscillation, shown for $\zeta = 0.8$ and $\zeta = 0.5$
$G(s) = \dfrac{s+2}{(s^2 + 2s + 10)(s+5)}$	poles at -5, $-1 \pm j3$; zero at -2	$g(t)$ damped oscillation, period $\pi/3$, time constant = 1 sec

FEEDBACK SYSTEMS

step starting at $t = 0$. When s_2 is pure imaginary, there must be another root s_2' which is its conjugate, two roots combining to form a sinusoid. When s_2 is a complex number, another root $s_2{}^*$, its complex conjugate, must exist, the two roots combining to form a sinusoidal time function with an amplitude that varies exponentially. If the real part of the complex root is positive, the amplitude grows exponentially with time; if negative, the amplitude decays exponentially with time. The conclusion is that the *character* of the time function depends only on the nature of the system *poles*. The *magnitudes* [A, B, C, etc., in Eq. (30)] of the time-response components depend on the nature of both the *poles and zeros* of the system. Thus the nature of the poles of a system reveals its stability.

Definition 2 can therefore be reworded in terms of the poles and zeros of the system.

Definition 3. A system is stable if it has no poles with positive real parts or multiple-order pure imaginary poles. This is often stated as: A stable system has no poles in the right half s plane, and only simple poles on the $j\omega$ axis. The poles and zeros of several representative transfer functions, together with their associated impulse responses, are shown in Table 5.

To prove system stability, it is necessary that all the poles satisfy the stability requirements. To prove instability, it is sufficient that one pole violate the stability requirements.*

Routh-Hurwitz Criteria

As long as the system transfer function appears in the factored form shown in Eq. (28), system stability can easily be tested by inspecting the pole locations in an s-plane plot of the system poles and zeros. When $T(s)$ appears in polynomial form, and when it is impractical or impossible to factor the denominator, other stability criteria are used.

Both the Routh test and the Hurwitz test are algebraic procedures for detecting the presence of right-half-plane roots of a polynomial.[10] These tests permit investigation of the roots of the denominator polynomial, i.e., the system poles. Certain preliminary observations can be made concerning the roots of a polynomial by inspection of the coefficients of the polynomial. The denominator polynomial of Eq. (28), often called the characteristic equation, is repeated below.

$$b_n s^n + b_{n-1} s^{n-1} + \cdots b_0 = 0 \qquad (31)$$

Preliminary Observations

1. For the polynomial to have no right-half-plane roots, the algebraic signs of all the b_n coefficients must be the same (all positive or all negative). The number of alternations of signs indicates the number of right-half-plane roots.

Example

[$5s^3 + 2s^2 + s + 1$]   Satisfies *necessary* condition stated above. Must be
(+, +, +, +)   tested further
[$2s^4 - 3s^3 + s^2 + 2s + 1$] Two sign inversions indicate that the polynomial has two
(+, -, +, +, +)   right-half-plane roots. No further testing necessary

2. If all the odd or even coefficients are zero, all the roots are pure imaginary and of course occur in conjugate pairs.

Example

[$3s^3 + s$]   Roots at $s = 0$, $\pm(1/\sqrt{3})j$. No further testing necessary
[$s^6 + 2s^4 + 3s^2 + 5$] Can be written in the form [$(s^2 + \omega_1{}^2)(s^2 + \omega_2{}^2)(s^2 + \omega_3{}^2)$].
                          Test to see if $\omega_1 = \omega_2$, $\omega_1 = \omega_3$, or $\omega_2 = \omega_3$.

3. If any combination of terms is missing, other than those described in observation 2 above, the polynomial has at least one right-half-plane root.

* This assumes that any common factors in the numerator and denominator have been canceled first.

Example

$[s^4 + 2s^3 + s + 2]$ Polynomial has at least one right-half-plane root. No further testing necessary

4. If any roots are known, they can be factored out of the expression, and the remaining reduced polynomial treated. Roots at $s = 0$ are easy to detect.

Example

$[3s^5 + s^4 + 2s^3 + 4s^2 + s]$ One root at $s = 0$; test further
$[2s^4 + 10s^3 + 6s^2]$ Two roots at $s = 0$; since this indicates instability, no need for further testing

If the preliminary observations indicate further testing is necessary, either the Routh or Hurwitz tests can be used.

The *Routh* test is based on a table of terms, the Routh table, constructed from the polynomial coefficients. Construction of the Routh table is begun by forming two rows of terms from the polynomial coefficients. (Any zero roots are removed beforehand as in observation 4.) The first row starts with the first coefficient of the polynomial b_n and the second row starts with second coefficient b_{n-1}. One row contains even coefficients and the other odd coefficients (see Table 6). The other rows in the

Table 6. General Form of the Routh Table

Polynomial coefficients
$\begin{cases} b_n & b_{n-2} & b_{n-4} & \cdots \\ b_{n-1} & b_{n-3} & b_{n-5} & \cdots \end{cases}$

Computed terms
$\begin{cases} c_k & c_{k-1} & c_{k-2} & \cdots \\ d_k & d_{k-1} & d_{k-2} & \cdots \\ \cdots \end{cases}$

$$c_k = \frac{-\begin{vmatrix} b_n & b_{n-2} \\ b_{n-1} & b_{n-3} \end{vmatrix}}{b_{n-1}} = \frac{b_{n-1}b_{n-2} - b_n b_{n-3}}{b_{n-1}}$$

$$c_k = \frac{-\begin{vmatrix} b_n & b_{n-4} \\ b_{n-1} & b_{n-5} \end{vmatrix}}{b_{n-1}} = \frac{b_{n-1}b_{n-4} - b_n b_{n-5}}{b_{n-1}}$$

Routh table are computed from the terms in the two preceding rows as indicated in Table 6. Each row contains an equal or lesser number of terms than the preceding row; thus the table terminates with a single-term row.

Example

$$b_5 s^5 + b_4 s^4 + b_3 s^3 + b_2 s^2 + b_1 s + b_0 \tag{32}$$

Routh Table for Polynomial of Eq. (32)

Coefficients of polynomial in question $\begin{cases} b_5 & b_3 & b_1 \\ b_4 & b_2 & b_0 \end{cases}$

$$c_1 = \frac{-\begin{vmatrix} b_5 & b_3 \\ b_4 & b_2 \end{vmatrix}}{b_4}$$

Computed terms $\begin{cases} c_1 & c_2 \\ d_1 & d_2 \\ e_1 \end{cases}$

$$c_2 = \frac{-\begin{vmatrix} b_5 & b_1 \\ b_4 & b_0 \end{vmatrix}}{b_4}$$

$$d_1 = \frac{-\begin{vmatrix} b_4 & b_2 \\ c_1 & c_2 \end{vmatrix}}{c_1}$$

$$d_2 = \frac{-\begin{vmatrix} b_4 & b_0 \\ c_1 & 0 \end{vmatrix}}{c_1}$$

$$e_1 = \frac{-\begin{vmatrix} c_1 & c_2 \\ d_1 & d_2 \end{vmatrix}}{d_1}$$

etc.

FEEDBACK SYSTEMS   15–31

For the polynomial to have no right-half-plane roots, it is *necessary* that all terms in the left-hand column of the Routh table have the same sign. If all the terms do not have the same sign, the number of right-half-plane roots equals the number of sign alternations. (For computational convenience, any row of the Routh table may be multiplied or divided by any positive number without altering the criterion.)

Example

$s^5 + s^4 + 3s^3 + 4s^2 + s + 2$
Routh table
1   3   1
1   4   2
−1 −1
3   2
−⅓
2

Four right-half-plane roots since there are four alternations of sign

$s^3 + 3s^2 + 2s + k$
Routh table
1   2
3   k
$\dfrac{6-k}{3}$
k

For stability: $\dfrac{6-k}{3} > 0,\ k > 0$

$\therefore 0 < k < 6$

It should be emphasized that zero alternations of sign in the left-hand column of the Routh table is *not a sufficient* criterion for stability. The *necessary* conditions discussed in preliminary observation 1 must also be considered. The Routh criterion can be stated concisely as: The *sufficient* condition for a polynomial to have no roots in the right half plane is: All the polynomial coefficients have the same sign, and all the left-column terms in the Routh table have the same sign. These conditions are not all *necessary* however.

Example

$a_4 s^4 + a_3 s^3 + a_2 s^2 + a_1 s + a_0 = 0$
Routh table
a_4 a_2 a_0
a_3 a_1
b_1 b_2
d_1

$b_1 = \dfrac{-\begin{vmatrix} a_4 & a_2 \\ a_3 & a_1 \end{vmatrix}}{a_3} = \dfrac{(a_2 a_3 - a_1 a_4)}{a_3}$

$b_2 = \dfrac{-\begin{vmatrix} a_4 & a_0 \\ a_3 & 0 \end{vmatrix}}{a_3} = a_0$

$d_1 = \dfrac{-\begin{vmatrix} a_3 & a_1 \\ b_1 & b_2 \end{vmatrix}}{b_1} = \dfrac{(a_1 b_1 - a_3 b_2)}{b_1}$

The *sufficient* conditions for this example are that

$a_4 > 0,\ a_3 > 0,\ a_2 > 0,\ a_1 > 0,\ a_0 > 0,$ [polynomial constraints] (33)
$a_4 > 0,\ a_3 > 0,\ b_1 > 0,\ d_1 > 0$ [Routh table] (34)

The first two inequalities in Eq. (33) are repeated in Eq. (34). (This will always be so.) From $b_1 > 0$,

$$(a_2 a_3 - a_1 a_4) > 0 \tag{35}$$

From $d_1 > 0$, $(a_1 b_1 - a_3 b_2) > 0$, which reduces to

$$(a_2 a_3 - a_1 a_4) > (a_3)^2 a_0 / a_1 \tag{36}$$

Since a_0 and a_1 are positive, Eq. (36) includes Eq. (35).* Therefore, if $d_1 > 0$ then $b_1 > 0$. The minimal set of sufficient conditions (the *necessary* and *sufficient* con-

---

* The reader is reminded that for real numbers A, B, and C, where A can have any value, if $A > B$ and $B > C$, then $A > C$; if $A > B$, then $(A + C) > (B + C)$ for *all* C; if $A > B$ and $C > 0$, then $AC > BC$; if $A > B$ and $C < 0$ then $AC < BC$; $A^2 > 0$ for *all* A.

ditions) for stability in this example are

$$a_4 > 0,\ a_3 > 0,\ a_2 > 0,\ a_1 > 0,\ a_0 > 0,\ (a_2 a_3 - a_1 a_4) > (a_3)^2 a_0/a_1 \quad (37)$$

The *Hurwitz* test is based on the properties of real-coefficient polynomials and continued fraction expansions. In order to apply the Hurwitz stability criterion, the polynomial $P(s)$ in Eq. (31) is expressed in terms of an even part $m(s)$ and an odd part $n(s)$.

$$P(s) = m(s) + n(s) \quad (38)$$

$m(s)$ contains only even powers of s and $n(s)$ only odd powers. An auxiliary rational function $\psi(s)$ is formed by taking the ratio of even to odd parts of $P(s)$

$$\psi(s) = m(s)/n(s) \quad (39)$$

[The procedures that follow could equally well be applied to

$$\psi'(s) = 1/\psi(s) = n(s)/m(s)]$$

The function $\psi(s)$ is expanded in a continued fraction expansion

$$\psi(s) = \alpha_1 s + \cfrac{1}{\alpha_2 s + \cfrac{1}{\alpha_3 s + \cfrac{1}{\alpha_4 s + \cdots}}} \quad (40)$$

If $P(s)$ is to have only left-half-plane roots, all the coefficients α_n in the continued fraction expansion must be positive, in which case the later is called a finite, Stieltjes continued fraction. (Strictly speaking, if the Hurwitz test fails, a root could be on the $j\omega$ axis as well as in the right half plane.)

Example

$$a_4 s^4 + a_3 s^3 + a_2 s^2 + a_1 s + a_0 = 0$$

$$\psi(s) = \frac{a_4 s^4 + a_2 s^2 + a_0}{a_3 s^3 + a_1 s}$$

$$a_3 s^3 + a_1 s \overline{\smash{\big)}\, a_4 s^4 + a_2 s^2 + a_0} \quad \boxed{(a_4/a_3) = \alpha_1}$$

$$\underline{a_4 s^4 + (a_1 a_4/a_3)s^2}$$

$$(a_2 - a_1 a_4/a_3)s^2 + a_0$$

$$\psi(s) = \alpha_1 s + \cfrac{1}{\cfrac{a_3 s^3 + a_1 s}{(a_2 - a_1 a_4/a_3)s^2 + a_0}}$$

$$(a_2 - a_1\alpha_1)s^2 + a_0 \overline{\smash{\big)}\, a_3 s^3 + a_1 s} \quad \boxed{\dfrac{a_3}{a_2 - a_1\alpha_1} = \alpha_2}$$

$$\underline{a_3 s^3 + \dfrac{a_0 a_3}{(a_2 - a_1\alpha_1)} s}$$

$$\left[a_1 - \dfrac{a_0 a_3}{(a_2 - a_1\alpha_1)}\right] s$$

$$\psi(s) = \alpha_1 s + \cfrac{1}{\alpha_2 s + \cfrac{1}{\cfrac{(a_2 - a_1\alpha_1)s^2 + a_0}{(a_1 - a_0\alpha_2)s}}}$$

FEEDBACK SYSTEMS

$$\psi(s) = \alpha_1 s + \cfrac{1}{\alpha_2 s + \cfrac{1}{\alpha_3 s + \cfrac{1}{\cfrac{(a_1 - a_0\alpha_2)}{a_0}s}}}$$

$$(a_1 - a_0\alpha_2)s \overline{\smash{\big)}\,(a_2 - a_1\alpha_1)s^2 + a_0} \atop \underline{(a_2 - a_1\alpha_1)s^2} \atop a_0$$

with quotient $\left(\dfrac{a_2 - a_1\alpha_1}{a_1 - a_0\alpha_2}\right)s$

$$\boxed{\dfrac{(a_2 - a_1\alpha_1)}{(a_1 - a_0\alpha_2)} = \alpha_3}$$

$$\boxed{\dfrac{(a_1 - a_0\alpha_2)}{a_0} = \alpha_4}$$

For the system to have only left-half-plane roots, it is sufficient that

$$\alpha_1 > 0, \; \alpha_2 > 0, \; \alpha_3 > 0, \; \alpha_4 > 0$$
$$a_0 > 0, \; a_1 > 0, \; a_2 > 0, \; a_3 > 0, \; a_4 > 0$$

Manipulation of the above inequalities results in exactly the same set of necessary and sufficient conditions as those derived from the Routh test, Eq. (37).

For high-order polynomials, both the Routh and the Hurwitz criteria involve considerable computational labor. If the polynomial to be tested contains literal coefficients, the Routh test is usually simpler to perform; if the coefficients are numerical, then the Hurwitz test is generally easier.

The Nyquist Criterion

The Routh-Hurwitz criteria[11] allow investigation of stability of a system through its characteristic equations but develop no information about system behavior other than the presence or absence of unbounded response roots. The Nyquist criterion provides a method of graphically depicting the regions of stable or unstable system operation. This method can also be used to describe relative system stability, and is discussed in Sec. 8. The Nyquist criterion is also applicable to problems for which a transfer function is not available but for which experimental measurements describing the system behavior have been made.

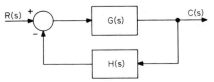

Fig. 18. Basic closed-loop control system.

$$T(s) = \frac{C(s)}{R(s)} = \frac{G(s)}{1 + G(s)H(s)}$$

The ideas embodied in the Nyquist criterion are applicable in general to polynomials, but the results are generally applied to the study of the basic closed-loop control system shown in Fig. 18.

In general, $G(s)$ will represent the open-loop transfer function of the device or process of interest, and $1 + G(s)H(s)$ the feedback effect, inserted in the problem to improve operation. $T(s)$ is called the closed-loop transfer function. The poles of the closed-loop transfer function are given by the roots of the equation $1 + G(s)H(s) = 0$, or equivalently by values of s for which $G(s)H(s) = -1$. Conformal mapping techniques and residue theorems available from complex variable theory can be utilized to simplify the investigation of the roots of this equation. In the mapping process, the right half of the s plane is enclosed by the contour shown in Fig. 19. The mapping of the s plane contour produces a corresponding locus in the $G(s)H(s)$ plane. Each value of s on the contour in the s plane corresponds to a single point on the locus in the $G(s)H(s)$ plane. The contour consists of three parts, the negative imaginary axis ($\sigma = 0$; $-\infty \leq \omega \leq 0$), the positive imaginary axis ($\sigma = 0$; $0 \leq \omega \leq +\infty$), and the arbitrarily large semicircle $|s| \to \infty$; $+\pi/2 \geq \underline{/s} \geq -\pi/2$). Because most $G(s)H(s)$ functions for physical systems behave as $(1/s)^n$ ($n = 1, 2, 3, \ldots$, etc.), for large s, the semicircular portion of the contour maps into a portion of the locus encircling the origin, with a *vanishingly small* radius and a rotation equal to $n\pi$ radians. This portion of the locus is often ignored in practical usage. The positive imaginary portion of the contour maps into the d-e-f portion of the locus as shown by the solid line. The mapping of the negative imaginary portion of the contour can be drawn by inspection, since it is the mirror image of the positive contour and is shown as the dotted curve a-b-c in Fig. 19. Thus the locus is determined by the behavior of $G(s)H(s)$ along the positive imaginary axis in the s plane. This is the frequency response of $G(s)H(s)$, and the locus can be sketched quickly by calculation of the magnitude and phase of $G(j\omega)H(j\omega)$ for a few representative frequencies or inspection of experimental gain-phase data. The frequency ω appears as a parameter along the locus.

The Nyquist theorem states that the net number of positive encirclements of the -1 point (the critical point) by the $G(s)H(s)$ contour is equal to the number of zeros minus the number of poles of the function $1 + G(s)H(s)$ in the right half plane. The right-half-plane *zeros of* $1 + G(s)H(s)$ *are right-half-plane poles of* $T(s)$. The right half-plane poles of $1 + G(s)H(s)$ are equal to the right-half-plane poles of $G(s)H(s)$.

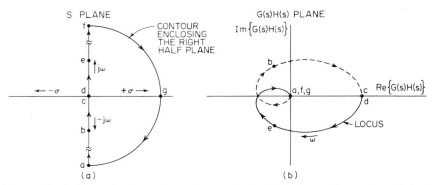

FIG. 19. Typical mapping from the s plane to the $G(s)$ $H(s)$ plane. Locus shown is for a $G(s)$ $H(s)$ of the form $\dfrac{K}{(s + \alpha)(s + \beta)(s + \gamma)}$. (a) s-plane contour. (b) $G(s)$ $H(s)$-plane locus.

Thus the number of right-half-plane poles of $G(s)H(s)$ plus the number of positive encirclements of the critical point in the $G(s)H(s)$ plane determines the number of right-half-plane zeroes of $1 + G(s)H(s)$.

$$\mathcal{P}_{rhp} = Z_{rhp} = N_{crit} + P_{rhp} \qquad (41)$$

where $\mathcal{P}_{rhp}$ = number of right-half-plane poles of $T(s)$
Z_{rhp} = number of right-half-plane zeros of $1 + G(s)H(s)$
N_{crit} = algebraic number of positive encirclements of the -1 point by the locus in the $G(s)H(s)$ plane
P_{rhp} = number of right-half-plane poles of $G(s)H(s)$

If Z_{rhp} is zero, the closed-loop system is stable; if Z_{rhp} is a positive integer, the system is unstable. Z_{rhp} can *not* be negative, and such a result represents an error in applying the criterion. It is a well-known fact that certain systems that are unstable in the open-loop configuration can be stabilized by closing a feedback loop around the system. In terms of Eq. (41), there must be one net negative encirclement of the critical point for each right-half-plane pole of $G(s)H(s)$. Several representative open-loop transfer functions and their Nyquist loci are shown in Table 7. When a pole occurs on the $j\omega$ axis in the s plane, a detour in the mapping contour must be taken around it, as shown in the third and fifth examples. This detour in the s plane maps into a large semicircle in the $G(s)H(s)$ plane, closing the locus through infinity.

As an example of the use of the Nyquist criterion in testing system stability Example 4 from Table 7 is discussed below.

Table 7. Nyquist Loci of Several Representative Systems

	$G(s)H(s)$	s-plane contour	$G(s)H(s)$-locus	Stability
1.	$\dfrac{1}{s+1}$		Im{GH}, Re{GH}	$P_{r.h.p.} = 0$ $N_{crit} = 0$ $\mathcal{P}_{r.h.p.} = Z_{r.h.p.} = 0$ System – <u>stable</u>
2.	$\dfrac{1}{s^2+as+b}$			$P_{r.h.p.} = 0$ $N_{crit.} = 0$ $\mathcal{P}_{r.h.p.} = Z_{r.h.p.} = 0$ System – <u>stable.</u>
3.	$\dfrac{1}{s(s+1)}$ †			$P_{r.h.p} = 0$ $N_{crit} = 0$ $\mathcal{P}_{r.h.p.} = Z_{r.h.p.} = 0$ System – <u>stable.</u>
4.	$\dfrac{K\omega_n^2}{s(s^2+2\zeta\omega_n s+\omega_n^2)}$ †		Region II, Region III, Region I	Region I: $P_{r.h.p.} = 0$; $N_{crit.} = +2$ $\boxed{2\zeta \leq K \leq +\infty}$ $\mathcal{P}_{r.h.p.} = Z_{r.h.p.} = 2$ System <u>unstable</u>, oscillations occur at frequency ω_1 given by $\angle(s^2+2\zeta\omega_n s+\omega_n^2) = +90°$, i.e. $\omega_1 = \omega_n$ --- Region II: $P_{r.h.p.} = 0$; $N_{crit.} = 0$ $\boxed{0<K<2\zeta}$ $\mathcal{P}_{r.h.p.} = Z_{r.h.p.} = 0$ System <u>stable</u>. $0 \leq K < 2\zeta$ --- Region III: $P_{r.h.p.} = 0$; $N_{crit.} = +1$ $\boxed{-\infty \leq K \leq 0}$ $\mathcal{P}_{r.h.p.} = Z_{r.h.p.} = 1$ System <u>unstable.</u> Making K negatives correspond to switching the critical point to +1
5.	$\dfrac{K}{s(s+a)(s-a)}$ †		Region I, Region III, Region II	Region I: $P_{r.h.p.} = 1$; $N_{crit.} = +1$ $\boxed{2a \leq K \leq +\infty}$ $\mathcal{P}_{r.h.p.} = Z_{r.h.p.} = 1+1 = 2$ System <u>unstable.</u> Frequency of oscillation ω_1, given by $\angle(s+a)(s-a) = 90°$, ie. $\omega_1 = a$ --- Region II: $P_{r.h.p.} = 1$; $N_{crit.} = -1$ $\boxed{0<K<2a}$ $\mathcal{P}_{r.h.p.} = Z_{r.h.p.} = 1-1 = 0$ System <u>stable.</u> --- Region III: $P_{r.h.p.} = 1$; $N_{crit.} = 0$ $\boxed{-\infty \leq K \leq 0}$ $\mathcal{P}_{r.h.p.} = Z_{r.h.p.} = 1$ System <u>unstable.</u>

† When a system has poles on the $j\omega$ axis, the contour in the s plane must skirt the singularity with a small semicircular detour. This is illustrated in systems 3, 4, 5 by the pole at the origin. Using an argument similar to that used to explain the mapping of the large semicircular portion of the s-plane contour, it can be shown that a *vanishingly small* detour (about an nth-order pole on the $j\omega$ axis) in the contour maps into an *infinitely large* portion [of the $G(s)H(s)$ locus] with a rotation of $n\pi$ radians. (Counterclockwise rotation of the contour corresponds to clockwise rotation of the locus, and infinitesimally speaking, a "right turn" encountered along the contour must correspond to a "right turn" along the locus.)

System stability depends on the values of the parameters K, ζ, and ω_n. The locus crosses the negative real axis when the phase shift is $-180°$. This occurs at the natural resonant frequency of the quadratic term $\omega_1 = \omega_n$. The $|GH(\omega_n)| = K/2\zeta$. If $0 \leq K/2\zeta < 1$, the critical point is in region I and the system is stable. When $1 \leq K/2\zeta \leq +\infty$, the critical point is within region II and the system is unstable. The system is unstable for all negative K.

The Nyquist test depends on a knowledge of the magnitude and phase of $G(j\omega)H(j\omega)$ vs. frequency. This can be determined quite rapidly, even for complex systems, using the gain-phase approximations that are discussed in the next section.

8 Performance Evaluation

Absolute stability is generally the first criterion of satisfactory system performance. In addition, restrictions on system transient response frequency are usually specified. The latter is often called adjustment of the "relative stability" of the system.

Error Constants

One means of specifying performance is in terms of steady-state system errors. In the unity feedback system in Fig. 20, the difference between system input and output is the error $e(t)$. The error is given by

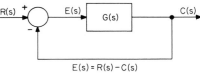

FIG. 20. A unity feedback-control system.

$$E(s) = R(s) - C(s) = R(s)/[1 + G(s)] \quad (42)$$

The steady-state value of the error $e_{ss}(t)$ is given by the final-value thorem

$$e_{ss}(t) = \lim_{t \to \infty} \{e(t)\} = \lim_{s \to 0} \{sE(s)\} \quad (43)*$$

Applying this theorem to Eq. (42) yields

$$e_{ss}(t) = \lim_{s \to 0} \left\{ \frac{sR(s)}{1 + G(s)} \right\} \quad (44)$$

The steady-state errors for step, ramp, and acceleration inputs are shown in Table 8, where the position, velocity, and acceleration error constants are, respectively,

Table 8. Control-system Errors

Input	Steady-state errors
Step input $r = A$, $t > 0$	$e_{ss} = \dfrac{A}{1 + K_p}$
Ramp input $r = At$, $t > 0$	$e_{ss} = \dfrac{A}{K_v}$
Acceleration input $r = \tfrac{1}{2}At^2$, $t > 0$	$e_{ss} = \dfrac{A}{K_a}$

$$K_p = \lim_{s \to 0} \{G(s)\}$$
$$K_v = \lim_{s \to 0} \{sG(s)\} \quad (45)$$
$$K_a = \lim_{s \to 0} \{s^2G(s)\}$$

Control systems are sometimes classified according to their steady-state response. The definitions of type 0, I, II systems, and the associated error constants, are shown in Table 9.

* This is not applicable to systems in which $E(s)$ has poles in the right half s plane.

Table 9. Control-system Classification

System type	Description	K_p	K_v	K_a
0	A *step* input *commands a step* output	K_p	0	0
I	A *step* input *commands a ramp* output	∞	K_v	0
II	A *step* input *commands an acceleration* output	∞	∞	K_a

System Design in the Time Domain

A control system can be designed in the time domain to yield a certain transient response. Consider the position-control system shown in Fig. 21. The block diagram represents a d-c motor with a potentiometer providing position feedback and a tachometer velocity feedback; $r(t)$ is the electrical input to the system, and $c(t)$ is the motor shaft position. The system transfer function is

$$\frac{C(s)}{R(s)} = \frac{1}{s^2 + (a + K_T)s + K_p} \quad (46)$$

The step response of this system is given by Eq. (27), where $\omega_n = \sqrt{K_p}$ and

$$\zeta = (a + K_T)/(2\sqrt{K_p})$$

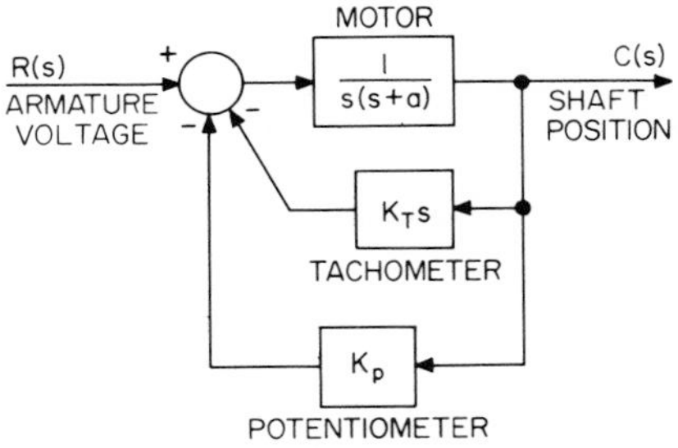

The resonant frequency ω_n can be adjusted by choosing an appropriate value for K_p. The damping ratio is adjusted by varying K_T. Experience has shown that values of ζ ranging from 0.3 to 0.7 are satisfactory in a large variety of systems. The percentage overshoot and damping ratio are related as shown in Fig. 22. Systems are sometimes designed for a particular value of overshoot, chosen through experience and a knowledge of system specifications. The design technique used in this example is applicable only in simple cases. If $G(s)$ were more complex, the transfer function $C(s)/R(s)$ could not be factored because of the literal coefficients. Direct design in the time domain is, in general, feasible only when the denominator of the closed-

Fig. 21. A typical position-control system.

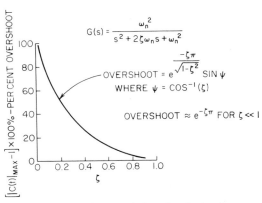

Fig. 22. Relationship between overshoot and damping factor for a second-order system.

loop transfer function can be factored, placing the system poles in evidence. All the techniques that are discussed subsequently predict the closed-loop transfer function $T(s)$ in terms of the open-loop transfer functions $G(s)H(s)$, since the latter is more easily dealt with.

Root-locus Technique

The root-locus technique[12] is a direct descendant of the pole-zero test for stability that was discussed in Sec. 7.

The closed-loop transfer function of the system in Fig. 23 is given by

$$T(s) = \frac{C(s)}{R(s)} = \frac{KG(s)}{1 + KG(s)H(s)} \qquad (47)$$

The poles of the closed-loop response are the zeros (roots) of the denominator polynominal $1 + KG(s)H(s) = 0$. This constraint can also be written in the form

$$|G(s)H(s)| = 1/K \qquad (48)$$
$$<G(s)H(s) = n\pi \quad (n = 1, 3, 5, \ldots \text{ etc.}) \qquad (49)$$

The heart of the root-locus technique lies in a set of rules for rapidly sketching, in the s plane, the locus of the roots of the denominator of Eq. (47) as the system gain K is varied from 0 to ∞. These rules can be developed by consideration of Eq. (49). The following is a condensed set of rules for sketching the root locus of a system.

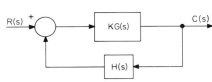

Fig. 23. Block diagram of a feedback-control system with adjustable loop gain K.

Rules for Root-locus Sketching

1. The root loci are continuous curves, symmetrical about the real axis, starting at the poles ($K = 0$) and terminating on the zeros ($K = \infty$) of $G(s)H(s)$. (Zeros at infinity must be considered.)

2. The locus exists along all portions of the real axis for which the number of real-axis poles and zeros to the right is an odd number. [Imaginary, complex, or infinite poles or zeros of $G(s)H(s)$ are not counted.]

3. As K becomes large, the loci approach straight-line asymptotes which intersect the imaginary axis at angles ϕ_a, where

$$\phi_a = n\pi/(\#P - \#Z)$$

where $n = 1, 3, 5, \ldots$
$\#P$ = number of finite poles of GH
$\#Z$ = number of finite zeros of GH

4. The asymptotes intersect at a point on the real axis called the centroid, given by

$$\text{Centroid} = \frac{\Sigma(\text{real parts of poles}) - \Sigma(\text{real parts of zeros})}{\#P - \#Z}$$

(The real part of a complex conjugate pole *pair* $s = -\sigma \pm j\omega$ is -2σ.)

Examples

(a) $\qquad\qquad H(s) = 1, \; KG(s) = K(s+1)/(s-1)$

The root locus is shown in Fig. 24a. The locus starts at the pole and ends at the zero. An odd number of real poles and zeros (i.e., one) exists to the right between the pole and zero; thus this portion of the real axis is part of the locus.

(b) $\qquad\qquad H(s) = 1/s, \; KG(s) = K/[(s+1)(s+2)]$

The locus is shown in Fig. 24b. The locus exists on the real axis between 0 and -1, and between -2 and $-\infty$. Loci start at the poles and terminate on the zeros (There are three zeros at $s = \infty$.) The asymptotes intersect the real axis at angles

given by
$$\phi_a = \frac{n\pi}{(3) - (0)} = \frac{\pi}{3}, \frac{3\pi}{3}, \frac{5\pi}{3}$$

The centroid of the asymptotes is located at
$$\text{Centroid} = \frac{(0 - 1 - 2)}{(3) - (0)} = -1$$

5. If $1 + KG(s)H(s)$ is written in polynomial form $b_n s^n + b_{n-1} s^{n-1} + \cdots + b_0$ the product of the roots equals $(-1)^n b_0$ and the sum of the roots $-(b_{n-1})$. This

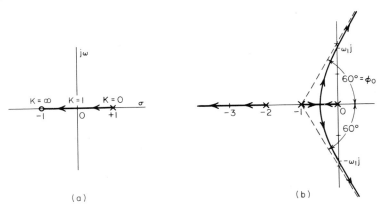

Fig. 24. Root-locus sketches for two systems. (a) Root locus of Example (a). (b) Root locus of Example (b).

fundamental theorem of algebra is valuable in determining intermediate values of K along the locus.

Example. For Example (a),
$$1 + KG(s)H(s) = 1 + K(s + 1)/(s + a)$$
When the root is at the origin, $(s = 0)$, $(K - 1) = 0$, and $K = 1$.
For Example (b),
$$1 + KG(s)H(s) = 1 + \frac{K}{s(s + 1)(s + 2)}$$
The denominator polynomial of the closed-loop transfer function is therefore
$$s^3 + 3s^2 + 2s + K$$
The sum of the roots is -3 and the product $-K$. The two imaginary roots $s = \pm j\omega_1$ sum to zero, and the real root (on the real-axis loci) must be at -3. The product of the roots is $(+j\omega_1)(-j\omega_1)(-3) = (-K)$. The value of K that produced borderline stability (roots on the imaginary axis) can be found by using the Routh test, giving $K = 9$. The corresponding value of frequency is $\omega_1 = \sqrt{3}$.

Additional rules for sketching root loci can be found in the literature. Other points on the locus may be found by graphical techniques or by solution of the denominator polynomial for a few well-chosen values of K. As an aid in drawing root loci, a spirule (combination protractor-slide rule), is commercially available for summing angles and multiplying magnitudes.

Choice of an appropriate value of K depends on the location of all the closed-loop poles. Poles located at large radial distance from the origin yield rapidly decaying transients (high-frequency effects). If a single pair of complex poles lies close to the origin, it will dominate the transient response, and the system performance can be adjusted in terms of ζ and ω_n for this dominant pole pair. This technique is appropriate if the remaining poles are at least five times farther away from the origin than the dominant pole pair. When this approximation cannot be made, a bit of experimentation and experience are necessary. The utility of the root-locus technique rests on the fact that the poles of the closed-loop transfer function are determined by considering the poles and zeros of the open-loop function $KG(s)H(s)$, and the rapidity with which the loci can be sketched.

Frequency-response and Compensation Techniques

Another family of techniques useful in system analysis and design considers the frequency response of a system.[13] The magnitude and phase of a system transfer function when the system is driven by sinusoidal inputs $s = j\omega$ are related to the system time response. The high-frequency response characteristics determine the initial transient behavior while low-frequency properties affect the steady-state time response. Frequency-response techniques will be developed by considering a second-order system, as was done for the time-domain approach, discussed above.

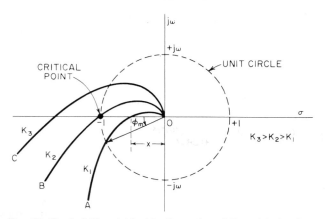

Fig. 25. Typical Nyquist loci in the region of the critical point.

The stability criterion can be expressed in terms of the gain and phase of $KG(s)H(s)$; cf. Eqs. (48) and (49). A portion of the Nyquist loci of a typical system* for three different values of K is shown in Fig. 25. Locus A is stable, locus C unstable, and locus B on the verge between stability and instability. There are two common ways of expressing the margin by which locus A misses encircling the critical point, i.e., the stability margin. Locus A crosses the negative real axis at the point $(-\sigma) = (-x)$. The ratio $1/x$ is called the gain margin and is usually expressed in decibels, abbreviated db. The gain margin in decibels $= 20 \log_{10} (1/x)$.† The difference between the phase of $KG(s)H(s)$ and $-180°$ when the locus crosses the unit circle is called the phase margin ϕ_n.

A "rule-of-thumb" value for satisfactory transient response is a gain margin equal to 2; i.e., 6 db. A satisfactory choice for ϕ_n can be made by considering a second-order system. Referring to Fig. 18, for $H(s) = 1$ and $KG(s) = \omega_n^2/[s(s + 2\zeta\omega_n)]$, the phase margin and damping ratio are related as shown in Fig. 26. Systems are frequently adjusted for phase margins from 35 to 55°. Use of gain margin may not be applicable or very meaningful for higher than second-order systems. The phase-margin criterion is a bit more successful in a variety of systems than the gain-margin

* See System 2 in Table 7.

† The quantity A expressed in decibels is defined as $20 \log_{10} (A)$. Other logarithmic units often employed in gain-phase plotting are decilogs $= 10 \log_{10} (A)$ and loru $= \log_{10} (A)$.

criterion. *Only for second-order systems are gain and phase margin directly related to each other and to system transient response.* In higher than second-order systems, phase margin becomes only a qualitative measure of system performance, unless system response is predominantly second-order as in the case of a dominant pole pair.

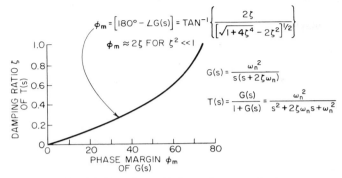

FIG. 26. Damping ratio ζ vs. phase margin ϕ_m.

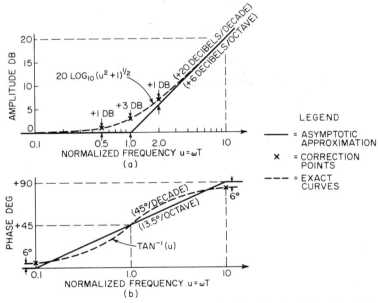

FIG. 27. Gain-phase curves and asymptotic approximations for the factor $(ju + 1)$, where $u = \omega T$. (a) Gain vs. frequency. (b) Phase vs. frequency.

The use of frequency-response techniques is facilitated through Bode plots, a technique that allows rapid sketching, by means of asymptotic straight-line approximations, of the gain and phase characteristics of rational systems. The real poles and zeros of a transfer function are of the form $(s + a)$, and complex roots and their conjugates combine to give factors $(s^2 + 2\zeta\omega_n s + \omega_n^2)$. All rational transfer functions can be written as a quotient of products of these two basic factors. Approximations for the gain and phase of the complex quantity $(j\omega T + 1)$ are shown in Fig. 27. The frequency scale is logarithmic for both the gain and phase plots. The amplitude scale is in decibels and the phase angle in degrees. The asymptotic approximation to the gain curve is a straight line, horizontal before the break frequency (corner frequency) $u = 1.0$, and rising at a rate of 20 db/decade (6 db/octave) beyond (a decade is a fac-

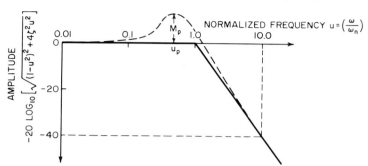

Fig. 28. Gain vs. normalized frequency for a second-order factor, $1/[(ju)^2 + 2\zeta(ju) + 1]$, where $u = (\omega/\omega_n)$. (Dotted curve is sketch of exact curve, while solid lines are the asymptotic approximation.)

tor of 10 in the frequency scale and an octave a factor of 2). At the break frequency the actual gain curve is 3 db above the asymptote; at 1 octave above and below the corner frequency, the actual curve is 1 db above the asymptotic approximation. The phase asymptote is horizontal up to a decade below the break frequency, and then rises with a slope of $+45°$/decade, passing through $+45°$ at the break frequency. Beyond $u = 10$, the phase shift remains constant at $+90°$. The phase approximation shown is within $6°$ throughout the range. The curves shown are for a *zero*, i.e., a *numerator factor* $(T_s + 1)$. If the factor were a *pole*, $1/(T_s + 1)$, the curves would be identical in appearance except that the gain would be in *negative* decibels and the phase shift in *negative* degrees.

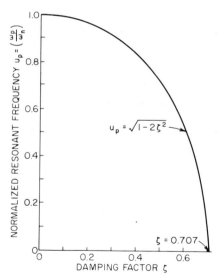

Fig. 29. Normalized resonant frequency of a second-order system $[1/(s^2 + 2\zeta\omega_n s + \omega_n^2)]$ vs. damping factor.

A second-order factor presents a bit more trouble since there is an additional variable parameter ζ. For $\zeta > 1$, the quadratic has two negative real roots which can be handled as above. For $0 < \zeta < 1$, the common range of interest, the gain and phase curves are difficult to sketch accurately using straight-line approximations. An approximate gain-frequency plot is shown in Fig. 28.

The solid-line asymptote is horizontal up to the natural resonant frequency of the system, $u = 1$, and has a slope of -40 db/decade(-12 db/octave) beyond. M_p is the height of the resonant peak of the system. When additional accuracy is desired, the peak is sketched by reading values of U_p and M_p from Figs. 29 and 30. Unfortunately, asymptotic plots for the phase angle of a second-order system are inaccurate. At low frequencies, the phase shift is zero; at ω_n it is $-90°$, and at high frequencies the phase approaches $-180°$. The slope of the phase curve increases as ζ decreases. Exact gain and phase plots for a second-order system are shown in Figs. 31 and 32. These curves, especially the phase curves, are generally used when accurate results are required. The use of Bode plots and gain and phase margin is illustrated in the following example.

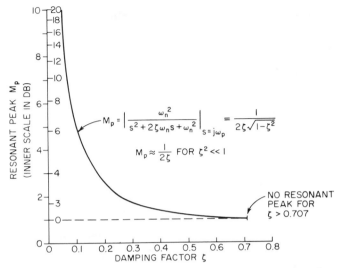

Fig. 30. Resonant peak of a second-order system $[1/(s^2 + 2\zeta\omega_n s + \omega_n^2)]$ vs. damping factor.

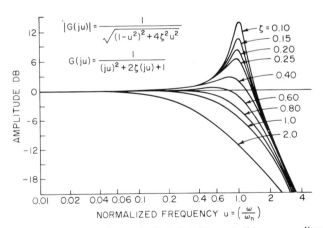

Fig. 31. Amplitude of a second-order factor $[1/(s^2 + 2\zeta\omega_n s + 1)]$ vs. normalized frequency. (*Taken from Ref. 4, p. 128, Fig. 5.3.*)

Example. A hypothetical-pitch-axis autopilot for an aircraft is shown in Fig. 33. The rate gyro is introduced in order to alter the open-loop step response, which is unsatisfactory because of the light damping and low resonant frequency of the aerodynamic transfer function. Suppose that experience and experimentation have shown that pitch autopilots perform satisfactorily if $0.4 < \zeta < 1.0$ and 2 radians/sec

$< \omega_n < 5$ radians/sec. The problem is to determine a value of K that satisfies these requirements.

The loop gain is

$$KG(s)H(s) = \frac{(KK_sK_{el}K_{ap}K_g)s}{s(T_ss+1)(s^2+2\zeta\omega_n s+\omega_n^2)} = \frac{K}{(0.1s+1)(s^2+s+6.25)} \quad (50)$$

The gain and phase characteristics of $G(S)H(s)$ are sketched in Fig. 34. Note how the logarithmic amplitudes and the phase angles are combined through addition.

Inspection of Fig. 34 reveals that the phase of the system changes rapidly over the range $0 \leq \omega \leq 1$, passing through 180° at approximately $\omega = 7$ radians/sec

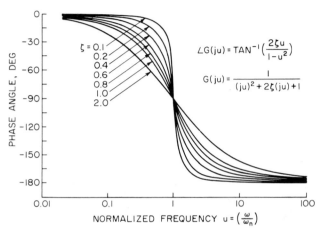

Fig. 32. Phase of a second-order factor $[1/(s^2 + 2\zeta\omega_n s + 1)]$ vs. normalized frequency (*Taken from Ref. 4, p. 128, Fig. 5.3.*)

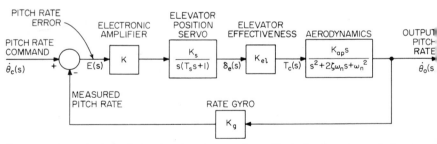

Fig. 33. A hypothetical pitch axis, aircraft autopilot. Typical values (numerical value of quantities denoted with + were chosen for computational convenience to be unity and therefore need not be representative).

$\delta_e(s) \equiv$ elevator deflection (range of -10 to $+20°$)
$T_c(s) \equiv$ control torque (newton-meters)
$T_s(s) \equiv$ servo time constant, 0.1 sec
$\zeta \equiv$ aerodynamic damping ratio, 0.2 (dimensionless ratio)
$\omega_n \equiv$ aerodynamic resonant frequency, 2.5 radians/sec
$K_g \equiv$ rate gyro sensitivity, 1 volt/(radian/second)$^+$
$K_s \equiv$ servo gain, 1 radian/volt$^+$
$K_{el} \equiv$ elevator effectiveness, 1 (newton-meters)/radian$^+$
$K_{ap} \equiv$ aerodynamic gain, 1 (radians/sec)/(newton-meter)$^+$
$K \equiv$ adjustable gain of electronic amplifier, design parameter to be determined

FEEDBACK SYSTEMS

This rapid phase change will cause difficulties in stabilizing the system. As a first approach to system design, the gain K is chosen to yield approximately 45° phase margin $[<G(s)H(s) = -135°]$. This occurs at $\omega \approx 3$ radians/sec, and the system gain at this frequency is approximately unity. Therefore, K should be chosen equal

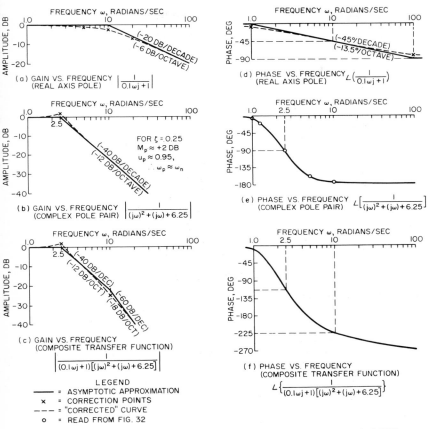

Fig. 34. Gain-phase diagrams for $G(s) H(s) = 1/[(0.1s + 1)(s^2 + s + 6.25)]$.

to 1.0. Computation of the closed-loop gain, and subsequent factoring, yields

$$T(s) = \frac{KG(s)}{1 + KG(s)H(s)} = \frac{1}{[(0.1s + 1)(s^2 + s + 6.25)] + 1}$$
$$= \frac{10}{(s + 10.1)(s^2 + 0.9s + 7.2)} \quad (51)$$

The natural frequency of the quadratic term has been raised slightly to $\omega = 2.68$, but the damping factor has been lowered to 0.17. This result could have been predicted more precisely if the gain-phase diagrams were accurately drawn. (It must be remembered that gain-phase techniques are only *approximately* correlated with transient response in systems that are higher than second-order.) However, it should be clear that other choices of system gain in the range $\frac{1}{2} < K < 2$ would be little better. In order to change the system to conform with design objectives, either the component transfer functions or the feedback configuration is modified. This technique is known as compensation of the system.

Compensation through altering the component transfer functions is called cascade compensation. This is effected by inserting compensating elements in cascade in the forward path of the system. In this example, cascade compensation can be achieved by adding an appropriate RC filter network to the electronic amplifier. The network could be chosen to yield appreciable positive phase shift in the range $1 \leq \omega \leq 10$ to compensate for the excessive negative phase shift of the quadratic term. Efficient utilization of cascade compensation becomes a matter of practice, experience, and trial and error. Cascade compensation is probably the simplest and most widely used method of compensation. It yields good results in simple systems and is amply documented by examples and descriptions in the control literature.

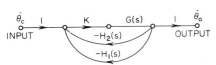

Fig. 35. Proportional and derivative feedback.

$$G(s) = \frac{1}{(0.1s + 1)(s^2 + s + 6.25)}$$

$H_1(s) = K_g$
$H_2(s) = K_a s$

If the feedback arrangement is changed to provide better performance, the process is called feedback compensation. As was demonstrated above in this section, proportional feedback (potentiometer) around a second-order system directly controls ω_n and indirectly ζ. If derivative feedback (tachometer) is used in addition, it can be used to adjust ζ independent of ω_n.* The resultant system is shown in flow-graph form (for convenience, and to accentuate the structural aspects) in Fig. 35. In order to analyze this system, the configuration $KG(s)H_2(s)$ can be simplified as shown in Fig. 36a and b. The bandwidth of the elevator servo chosen is only 10 radians/sec. The phase shift contributed by this component at low frequencies is troublesome. Since high-performance position systems are readily available, the bandwidth will be raised to 50 radians/sec.

$$\frac{K_s}{s(T_s s + 1)} = \frac{1}{s(0.02s + 1)} \approx \frac{1}{s}$$
for $\omega < 5$ radians/sec (52)

This approximation is justifiable since the magnitude of a first-order factor is approximately unity a decade below the break, and the phase shift is only 6°. The transfer function of the inner loop is given by

$$\frac{KG(s)}{1 + KG(s)H_2(s)}$$

$$= \frac{1}{s^2 + (1 + KK_{ac})s + 6.25} \quad (53)$$

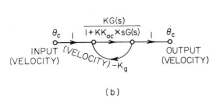

Fig. 36. Reduction of an inner loop. (a) Two-loop structure. (b) Reduced inner loop.

Choosing $KK_{ac} = 4$ yields a pair of complex poles $s^2 + 5s + 6.25$ with a damping ratio of 1.0. Inspection of Fig. 26 reveals that 55° of phase margin corresponds to a damping ratio of about 0.5. From Figs. 31 and 32 one can see that, for a damping ratio of 1.0, $\phi_m = 55°$ occurs at $u = 2$, and the gain at this frequency is -13 db (magnitude = $\frac{1}{4}$). The loop gain KK_g can be chosen equal to 4.0, and the closed loop system will have a damping ratio of approximately 0.5 and a resonant frequency

* Since the controlled variable (output) in this problem is θ_0, *proportional feedback* is supplied by a rate *gyro*, and *derivative feedback* is supplied by an angular accelerometer.

of about 5 radians/sec. In order to check the validity of the approximations made, the closed-loop transfer function is computed for $K = 1$, $K_{ac} = 4$, $K_g = 4$.

$$\frac{\theta_0(s)}{\theta_c(s)} = \frac{\dfrac{KG(s)}{1 + KG(s)H_2(s)}}{1 + \dfrac{KG(s)H_1(s)}{1 + KG(s)H_2(s)}} = \frac{KG(s)}{1 + KG(s)[H_1(s) + H_2(s)]}$$

$$\frac{\theta_0(s)}{\theta_c(s)} = \frac{\dfrac{1}{(0.02s + 1)(s^2 + s + 6.25)}}{1 + \dfrac{4s + 4}{(0.02s + 1)(s^2 + s + 6.25)}} = \frac{1}{(s + 45.6)(s^2 + 5.36s + 11.4)} \quad (54)$$

The system has a damping ratio $\zeta = 0.78$ and a resonant frequency $\omega_n = 3.4$ radians/sec. Of course, other combinations of tachometer and gyro gains would result in systems with equivalent ζ and ω_n. Other specifications and criteria would be required to choose the best configuration.

Error Criteria

A comparison of the advantages and drawbacks of the performance criteria that have been discussed appears in Table 10. The limitations of these criteria have encouraged the development of other measures of system performance.[14,15]

Table 10. Evaluation of Performance Criteria

Performance criterion	Advantages	Drawbacks
Error constants	Easy to evaluate; yields figure of merit; classifies the system	Yields no information about transient response
Time-domain design	Gives complete response of system excited by a periodic input	Response must be sketched or evaluated on a computer and results evaluated visually; too complex for higher-order problems; no figure of merit
Root locus	Relatively quick and simple even for higher-order problems; eliminates much of the labor associated with time-domain design	Same disadvantages as time-domain design; simplification possible when a dominant pole pair exists
Gain-phase techniques	Relatively simple; can be used with experimental information; yields gain- and phase-margin figures of merit; transcendental terms can be handled	Gain and phase margin have little meaning in higher-order systems

The difference between system output and input is called the system error $e(t)$ and is a natural measure of system performance. An ideal system might be defined as one which had zero system error for all time when excited by a particular class of input functions. A practical approach is to use some sort of weighted time average of the system error response to a particular test input, usually a step, as a measure of system performance. Several possible error criteria are discussed in Table 11. Appreciable use of these criteria in analysis and synthesis of control systems is found in the literature. The most common is the mean-squared error criterion. This choice represents a reasonable compromise between computational difficulty and the power of the criterion as a figure of merit. An expression for mean-squared error in terms of the residues of $E(s)E(-s)$ can be derived using Fourier transform and contour integration theorems.

In many problems it is more significant to describe the system in terms of the statistical nature of the input and output signals than to use the signal waveforms. Time domain correlation functions and Fourier-transform power-density spectra are used to summarize the statistical properties of the transmission. The mean-squared error between an ideal statistical output and the actual statistical output in the presence of corrupting noise can be minimized. When the resultant transfer function for the system is chosen on this basis, a statistically "optimum" system is produced.

When error criteria become too complex for manual computation, digital or analog machine computation can be used. Analog integrators and nonlinear components can be used to evaluate the error criteria of Table 11. Statistical signals, with specified power-density spectra, can be generated using random-noise generators and filter networks.

Table 11. Error Criteria

Type	Usefulness
1. Mean system error criterion $\frac{1}{\tau}\int_0^\tau e(t)\,dt$	Algebraic cancellation of positive and negative areas may give misleading results; computationally feasible
2. Mean-squared error criterion $\frac{1}{\tau}\int_0^\tau [e(t)]^2\,dt$	Weights initial and final values equally; thus initially large errors in step response receive too much weighting; computationally feasible
3. Mean absolute error criterion $\frac{1}{\tau}\int_0^\tau |e(t)|\,dt$	Same comment as 2, except computationally difficult
4. Mean time-weighted squared error criterion: $\frac{1}{\tau}\int_0^\tau t[e(t)]^2\,dt$	Gives greater weight to errors for large t than initial errors. Computationally more difficult than 2
5. Mean time-weighted absolute error criterion: $\frac{1}{\tau}\int_0^\tau t|e(t)|\,dt$	Same comment as 4, except computation even more difficult

None of the performance criteria of Tables 10 and 11 is without blemish nor does any of these dominate the group with regard to its advantages. Proper choice of a performance criterion may depend on experience or contemplation of the problem at the outset.

9 Sampled-data Systems

Sampled signals occur in systems containing scan radars, digital computers, and telemetering. Sampled signals are number sequences existing at discrete intervals of time known as the sampling instants.[16,17,18,19,20]

A model of the sampling process using an impulse sampler is shown in Fig. 37. This model is valid whenever the system constants are much longer than the pulse

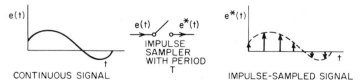

FIG. 37. An impulse sampler model.

FEEDBACK SYSTEMS

width of a physical sampler. The continuous signal input is $e(t)$ and the sampled output $e^*(t)$.

Sampling Described in the Time and Frequency Domain

The sampled output $e^*(t)$ can be written as

$$e^*(t) = e(t)i(t) \tag{55}$$

where $i(t)$ is a train of unit impulses with a period of t sec. Equation (55) can also be written as

$$e^*(t) = \sum_{n=0}^{\infty} e(nt)u_0(t - nT) \tag{56}$$

where $u_0(t - nT)$ is a unit impulse occurring at $t = nT$. The Laplace transform of Eq. (56) is

$$E^*(s) = 1/T \sum_{n=0}^{+\infty} e(nT)\epsilon^{-nTs} \tag{57}$$

Expanding Eq. (57) in a complex Fourier series, and taking the Laplace transform, yields a different but equivalent form*

$$E^*(s) = 1/T \sum_{n=-\infty}^{+\infty} E(s + jn\omega_S) \tag{58}$$

where $\omega_S = 2\pi/T$.

The Sampling Theorem

Equation (58) can be interpreted in terms of modulation sidebands. For a continuous signal whose amplitude spectrum is shown in Fig. 38a, the spectrum of the sampled signal appears in Fig. 38b.

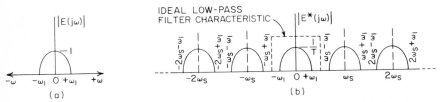

Fig. 38. Amplitude spectra of a continuous and a sampled waveform. (a) Continuous signal. (b) Sampled spectrum.

The impulse sampler reproduces the spectrum of the continuous time function (fundamental signal) as well as sidebands (complementary signals) centered about integral multiples of the sampling frequency. If the sidebands do not overlap the fundamental signal and the sampled spectrum is passed through a properly chosen ideal low-pass filter, the continuous signal is recovered without distortion. The sampling theorem (Shannon-Nyquist theorem) states: *If a continuous function with a bandwidth-limited spectrum* ($-\omega_1 \leq \omega \leq \omega_1$) *is sampled, the original function can be*

---

* Actually, Eq. (57) is valid in general, whereas Eq. (58) is in error if $e(t)$ has a nonzero initial value. For generality, Eq. (58) should be written

$$E^*(s) = 1/T \sum_{n=-\infty}^{+\infty} E(s + jn\omega_S) + \tfrac{1}{2}e(0^+)$$

recovered without distortion by passing the sampled waveform through an ideal low-pass filter of bandwidth $\omega_1 \leq \omega_c \leq (2\omega_S - \omega_1)$, where the sampling frequency $\omega_S \geq 2\omega_1$.

Since an ideal low-pass filter is not physically realizable, various filter characteristics (hold circuits) are used as approximations. The most commonly used signal-recovery circuit, the zero-order hold, maintains the level (clamps) of the sampled signal between sampling periods, as shown in Fig. 39. The low-frequency transmission of this filter, $0 \leq \omega \leq \omega_0$, approximates the ideal low-pass filter.

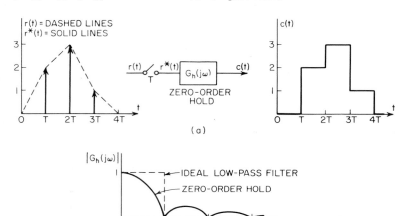

Fig. 39. Characteristics of a zero-order hold. (a) Input-output waveforms. (b) Frequency response.

The z Transform

The Laplace transform can be used in the analysis of sampled-data systems; however, the expressions are complex and difficult to interpret. The z transform (also called sampled transform or pulse transform) is introduced to facilitate analysis and design. The z transform describes the behavior of the sampled time function only at the sampling instants. To compute the z transform, the sampled time function is computed [Eq. (56)]; the Laplace transform found [Eq. (57)]; and finally the substitution $z = \epsilon^{+st}$ is made. (Some authors use $z = \epsilon^{-st}$. Either is correct, as long as one is consistent.) Thus the z transform of $e(t)$ is

$$Z[e(t)] = E(z) = \sum_{n=0}^{\infty} e(nT)z^{-n} \tag{59}$$

The z transform of a unit step occurring at $t = 0$ will be computed as an example.

Example 1. $e(t) = u_{-1}(t)$

$$e(nT) = \left\{ \begin{array}{ll} 1 & \text{for } n \geq 0 \\ 0 & \text{for } n < 0 \end{array} \right\} \tag{60}$$

Substituting Eq. (60) in Eq. (59) yields

$$E(z) = 1 + z^{-1} + z^{-2} + \cdots + z^{-n} \tag{61}$$

The power series in Eq. (61) can be written in closed form as $1/(1 - z^{-1})$. Therefore,

$$Z\{u_{-1}(t)\} = z/(z - 1) \tag{62}$$

Example 2. Similarly, for an exponential time function $e(t) = \epsilon^{-at}$,

$$e(nT) = \epsilon^{-anT}$$
$$E(z) = 1 + \epsilon^{-aT}z^{-1} + \epsilon^{-2aT}z^{-2} + \cdots + \epsilon^{-naT}z^{-n} = 1/(1 - \epsilon^{-aT}z^{-1})$$
$$\mathcal{Z}\{\epsilon^{-at}\} = z/(z - \epsilon^{-aT}) \tag{63}$$

The z transform can be found directly from the Laplace transform in a simple manner by writing Laplace transform as a partial-fraction expansion.

Example. The z transform of the time function that possesses a Laplace transform $a/[s(s + a)]$ is to be found symbolically,

$$\mathcal{Z}\{\mathcal{L}^{-1}[a/s(s+a)]\}, \text{ or simply } \mathcal{Z}\{a/s(s+a)\})$$
$$E(s) = a/s(s+a) = 1/s - 1/(s+a)$$
$$\mathcal{Z}\{E(s)\} = \mathcal{Z}\{1/s\} - \mathcal{Z}\{1/(s+a)\}$$

Since $1/s$ and $[1/(s + a)]$ are the Laplace transforms of $u_{-1}(t)$ and ϵ^{-at}, respectively, substitution of Eqs. (62) and (63) in the above equation gives

$$\mathcal{Z}\left\{\frac{a}{s(s+a)}\right\} = \frac{z(1 - \epsilon^{-aT})}{(z-1)(z - \epsilon^{-aT})} \tag{64}$$

A short list of the z transforms of common time functions appears in Table 12.

Table 12. z Transforms of Common Time Functions

Time function	Laplace transform	z transform
1. Unit impulse $u_0(t)$	1	Not defined*
2. Unit pulse $[u_{-1}(t) - u_{-1}(t - \alpha)], 0 < \alpha < T$	$\dfrac{1}{s} - \dfrac{\epsilon^{-\alpha s}}{s}$	1
3. Unit step $u_{-1}(t)$	$\dfrac{1}{s}$	$\dfrac{z}{z-1}$
4. Exponential ϵ^{-at}	$\dfrac{1}{s+a}$	$\dfrac{z}{z - \epsilon^{-aT}}$
5. Unit ramp t, $[u_{-2}(t)]$	$\dfrac{1}{s^2}$	$\dfrac{Tz}{(z-1)^2}$
6. Unit acceleration $\tfrac{1}{2}t^2$, $[u_{-3}(t)]$	$\dfrac{1}{s^3}$	$\dfrac{1}{2}T^2 \dfrac{z(z+1)}{(z-1)^3}$
7. Sine wave $\sin(at)$	$\dfrac{a}{s^2 + a^2}$	$\dfrac{z \sin(aT)}{z^2 - 2z \cos(aT) + 1}$
8. Translation theorem $e(t + T)$	$\epsilon^{-Ts}E(s)$	$\dfrac{E(z)}{z}$

* See Ref. 17, sampled-data chapter.

A pulse transfer function can be defined for sampled signals in the same manner as an ordinary transfer function is for continuous signals. A system with a continuous input appears in Fig. 40a, and the same system with a sampled input is shown in Fig. 40b.

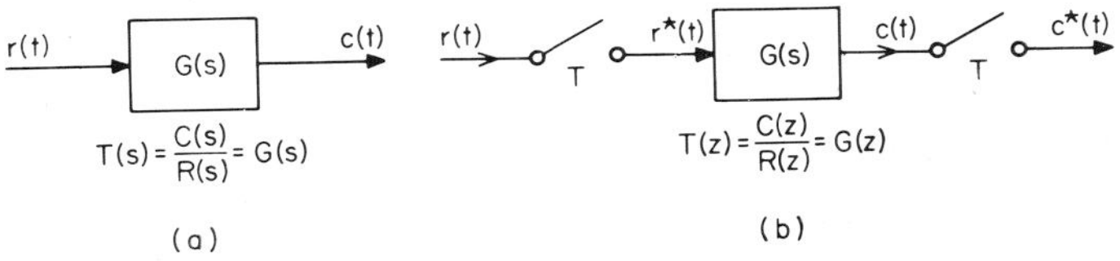

$$r(t) \rightarrow \boxed{G(s)} \rightarrow c(t)$$
$$T(s) = \frac{C(s)}{R(s)} = G(s)$$
(a)

$$r(t) \rightarrow \diagup_T \rightarrow r^*(t) \rightarrow \boxed{G(s)} \rightarrow c(t) \rightarrow \diagup_T \rightarrow c^*(t)$$
$$T(z) = \frac{C(z)}{R(z)} = G(z)$$
(b)

Fig. 40. Continuous and pulse transfer functions. (Input and output sampler are synchronized in the sampled system shown.) (a) Continuous system. (b) Sampled system.

15-52 REMOTE CONTROL

One of the advantages of the z transform is the ease with which the inverse transform is found. The z transform is expanded in a power series in z by dividing the denominator into the numerator. Each term in the series is a delayed impulse, occurring at one of the sampling instants. There is no information contained in the z transform about the behavior of the time function between sampling instants.

Example. $E(z) = z/(z - 0.5)$; find $Z^{-1}\{E(z)\}$.

$$
\begin{array}{r}
1 + 0.5z^{-1} + 0.25z^{-2} + 0.125z^{-3} + \cdots \\
z - 0.5 \overline{\smash{)}z} \\
\underline{z - 0.5} \\
+ 0.5 \\
\underline{0.5 - 0.25z^{-1}} \\
0.25z^{-1} \\
\underline{0.25z^{-1} - 0.125z^{-2}} \\
0.125z^{-2}
\end{array}
$$

$$E(z) = 1 + 0.5z^{-1} + 0.25z^{-2} + 0.125z^{-3} + \cdots \tag{65}$$

Reversing the steps used in finding $E(z)$ from $e(t)$, $E^*(s)$ is found from $E(z)$ and $e^*(t)$ from $E^*(s)$. The *continuous* time function $e(t)$ is not recovered!

$$
\begin{aligned}
E^*(s) &= 1 + 0.5\epsilon^{-st} + 0.25\epsilon^{-2st} + \cdots \\
e^*(t) &= u_0 + 0.5u_0(t - T) + 0.25u_0(t - 2T) + \cdots
\end{aligned}
\tag{66}
$$

The sample of values of Eq. (66) is plotted in Fig. 41.

When the z transform is a rational function, its inverse can be found in closed form by expanding $E(z)/z$ in partial fractions, multiplying by z, and then finding the inverse transform of each term in the series by successive application of transform pair 4 in Table 12.

Example. $E(z) = \dfrac{0.5z}{(z-1)(z-0.5)}$; find $Z^{-1}\{E(z)\}$.

$$\frac{E(z)}{z} = \frac{0.5}{(z-1)(z-0.5)} = \frac{1}{z-1} - \frac{1}{z-0.5}$$

$$E(z) = \frac{z}{z-1} - \frac{z}{z-0.5}$$

Application of transform 4 of Table 12 for $\epsilon^{-aT} = 0.5$ gives

$$e^*(t) = (1 - \epsilon^{-0.692t/T})i(t) \tag{67}$$

where $i(t)$ is unit impulse train as in Eq. (55).

The Pulse Transfer Function

The pulse transfer function of interconnected networks and samplers is easily derived. Referring to Fig. 42a

$$E(s) = G_1(s)R^*(s) \tag{68}$$
$$E^*(s) = [G_1(s)R^*(s)]^* = G_1^*(s)R^*(s) \tag{69}$$
$$C(s) = G_2(s)E^*(s) \tag{70}$$
$$C^*(s) = [G_2(s)E^*(s)]^* = G_2^*(s)E^*(s) \tag{71}$$

Substituting Eq. (69) in Eq. (71),

$$C^*(s) = G_2^*(s)G_1^*(s)R^*(s) \tag{72}$$

Thus the pulse transfer function for the system of Fig. 42a is

$$C(z)/R(z) = G_1(z)G_2(z) = Z\{G_1(s)\} \times Z\{G_2(s)\} \tag{73}$$

In a similar manner, for the system of Fig. 42b,

$$C(z)/R(z) = Z\{G_1(s)G_2(s)\} = G_1G_2(z) \tag{74}$$

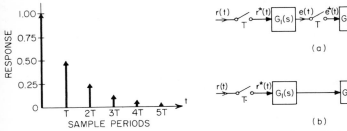

Fig. 41. Sampled time functions, Eq. (66).

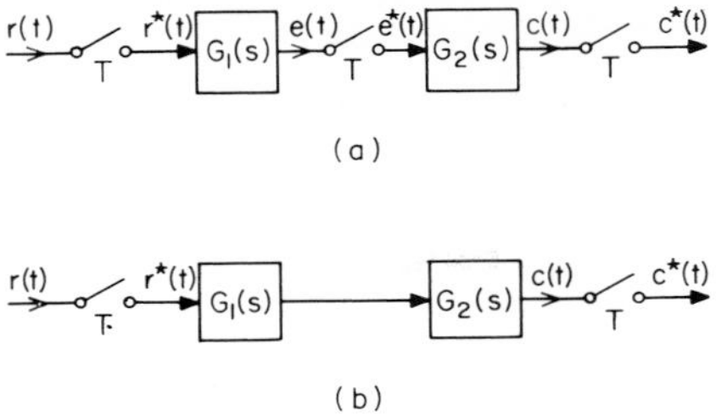

Fig. 42. Two tandem combinations of networks and samplers. (a) Networks isolated by a sampler. (b) Only input and output samplers.

The pulse transfer functions of the two systems appear similar but are actually quite different. The z transform relations for three feedback-control systems are shown in Table 13.

Table 13. Output Transforms for Typical Sampled-data Systems

System	Z Transform of output, C(Z)
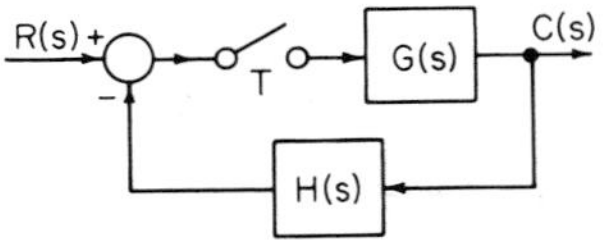	$\dfrac{G(Z)R(Z)}{1+HG(Z)}$
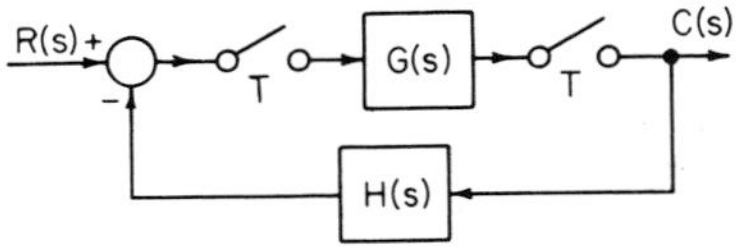	$\dfrac{G(Z)R(Z)}{1+H(Z)G(Z)}$
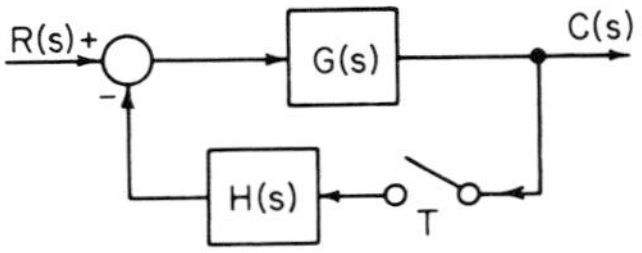	$\dfrac{RG(Z)}{1+HG(Z)}$

Stability of Pulse Transfer Functions

It can be shown that the stability requirements on the roots of $T(s)$ in the s plane ($s = \sigma + j\omega$) have counterpart requirements on the roots of $T(z)$ in the z plane ($z = \text{Re}\{z\} + j\,\text{Im}\{z\}$).

These requirements are summarized in Table 14. Thus the poles and zeros of $T(z)$ can be plotted in the z plane stability determined by the pole positions with respect to the unit circle.

The Routh test can be applied in modified form to pulse transfer functions. The transformation

$$z = \frac{(\lambda + 1)}{(\lambda - 1)}$$

Table 14. Stability Criteria in the s and z Planes

Region in s plane	System behavior	Region in z plane
$\sigma < 0$ Left half plane	Stable	$\|z\| < 1$ Interior of unit circle
$\sigma > 0$ Right half plane	Unstable	$\|z\| > 1$ Outside the unit circle
$\sigma = 0$ Imaginary axis	1. First-order poles yield sustained oscillations (if ω is zero, step response occurs) 2. Higher-order poles yield growing oscillations (if ω is zero, the response is a singularity function, i.e., step, ramp, etc.)	$\|z\| = 1$ Circumference of unit circle

is made, and the criterion applied in the normal manner to the polynomial in λ.

Example. Test the stability of $T(z) = \dfrac{N(z)}{D(z)} = \dfrac{z^2 + 4z + 3}{z^3 + 2z^2 - 0.5z - 1}$. Substituting for z in terms of λ in the denominator polynomial yields

$$D(\lambda) = \left(\frac{\lambda + 1}{\lambda - 1}\right)^3 + 2\left(\frac{\lambda + 1}{\lambda - 1}\right)^2 - 0.5\left(\frac{\lambda + 1}{\lambda - 1}\right) - 1$$

$$D(\lambda) = \frac{3\lambda^3 + 17\lambda^2 - 3\lambda - 1}{2(\lambda - 1)^3} \tag{75}$$

The numerator of Eq. (75) will be examined by Routh's criteria.
Routh table

$$\begin{array}{rr} 3 & -3 \\ 17 & -1 \\ \hline \text{Inversion } -48 & \\ 17 & \end{array}$$

There is one sign inversion, and the system has one unstable root.

The Nyquist stability criterion can also be developed for z transforms by mapping the exterior of the unit circle into the $G(z)H(z)$ plane. This is analogous to mapping the right half s plane into the $G(s)H(s)$ plane. The net number of encirclements of the critical point in the $G(z)H(z)$ plane determines system stability.

Design Methods

The root-locus technique can be adapted to pulse transfer functions by plotting the poles and zeros of $GH(z)$ or $G(z)H(z)$ (depending on the system) in the z plane. The same rules for rapidly sketching the root loci in the s plane are valid in the z plane. Consider the sampled-data feedback-control system of Fig. 43. If $G(s) = aK/(s + a)$, the pulse transfer function of the system is given by

$$T(z) = \frac{C(z)}{R(z)} = \frac{G'(z)}{1 + G'(z)}$$

where $\quad G'(z) = \mathcal{Z}\left\{\dfrac{(1 - \epsilon^{-Ts})}{s} G(s)\right\} = (1 - z^{-1})\mathcal{Z}\left\{\dfrac{aK}{s(s + a)}\right\}$

If $\epsilon^{-aT} = 0.5$,

$$G'(z) = \frac{(z - 1)}{z} \frac{0.5zK}{(z - 1)(z - 0.5)} = \frac{0.5K}{z - 0.5} \tag{76}$$

The root locus of $1 + G'(z)$ appears in Fig. 44.

The denominator polynomial of $T(z)$ is $[z + 0.5(K - 1)]$. The system becomes unstable when the locus crosses the unit circle at $z = -1$. This corresponds to $K = 3$ and places the system on the verge of instability.

Gain-phase techniques are also applicable to sampled-data systems. The Laplace transfer function of the system in Fig. 43 can be written as

$$T^*(s) = \frac{C^*(s)}{R^*(s)} = \frac{G'^*(s)}{1 + G'^*(s)}$$

The gain and phase of $G'^*(j\omega)$ are the quantities of interest. These can be calculated using Eq. (58).

$$G'^*(j\omega) = 1/T \sum_{n=-\infty}^{+\infty} G'(j\omega + jn\omega_S) \tag{77}$$

Unfortunately, the laborsaving feature of the gain-phase methods, the use of asymptotic approximations, is no longer applicable. In a continuous system, the gain and

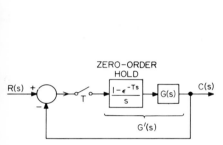

Fig. 43. Closed-loop sampled-data systems   Fig. 44. Root locus of $1 + G'(z)$.

phase can be computed directly for a few well-chosen values of ω. If this technique is applied to Eq. (77) each computation, for a particular value of $j\omega$, involves an infinite series. Fortunately, in many cases the series converges rapidly and only the first few terms need be retained.

As an example, consider the case where $G(s) = K/[s(s + 1)]$ and $T = 1$ sec ($\omega_S = 2\pi/T = 6.28$ radians/sec). For $\omega = 1$, Eq. (77) gives

$$G'^*(j) = G'(j) + G'(7.28j) + G'(-5.28j) + \cdots$$

Computing each term in the series yields

$$G'^*(j) = 0.678\underline{/-163°} + 0.0024\underline{/-380°} + 0.0063\underline{/-321°} + \cdots \tag{78}$$

In this example, only the first term of the series need be retained. Once appropriate values of $|G'^*(j\omega)|$ and $\angle G'^*(j\omega)$ have been computed, the design proceeds as with continuous systems.

The techniques of cascade compensation previously discussed are also applicable to sampled-data systems. Great flexibility is possible since a digital controller (special- or general-purpose digital computer) can be used as the compensating element.

10 Nonlinear Systems

The continuous systems discussed in Sec. 7 are linear and are described by a set of constant-coefficient linear differential equations. [The a's in Eq. (79) are independent of y or t.]

$$a_n(d^n y/dt^n) + a_{n-1}(d^{n-1}y/dt^{n-1}) + \cdots + a_0 y = f(t) \tag{79}$$

Laplace-transform methods are valid for systems of this type. The sampled systems discussed in Sec. 9 are governed by time-varying coefficient, linear differential equations or constant-coefficient, linear difference equations. Both the Laplace transforms and the more convenient z transform are applicable in the analysis of sampled systems. There exists a large class of systems containing most practical problems which are described by nonlinear differential or difference equations.[21,22,23] In continuous systems, nonlinear effects can be due to saturation, dead zone, backlash, or temperature effects in physical components; intentionally introduced nonlinear effects; or inherently nonlinear components such as relays, position limiters, or on-off type components. Nonlinear difference equations occur in sampled systems containing the aforementioned nonlinearities as well as whenever the sampled signal is quantized in amplitude. Linear-transform methods are not applicable in nonlinear systems.

Properties of Nonlinear Systems

Nonlinear systems possess some interesting properties not found in linear systems, some of which may be advantageous and some detrimental to system performances depending on the application in question. A most fundamental property of nonlinear systems is the failure of the principle of superposition. If the system response is $c_1(t)$, when excited by $r_1(t)$, and $c_2(t)$ when excited by $r_2(t)$, then the response to an excitation $r_1(t) + r_2(t)$ need not be $c_1(t) + c_2(t)$. The character of nonlinear system response depends not only on the present input but also on the past history. In nonlinear systems, limit cycles may occur, where the output is an oscillation of fixed amplitude and frequency but nonsinusoidal waveform. If a linear system is excited by sinusoids of frequency ω_1 and ω_2, the response is composed of two sinusoids of frequency ω_1 and ω_2; where the system transfer function determines only the output amplitudes and phases. In a nonlinear system, the above-mentioned input produces output frequency components of $(m\omega_1 + n\omega_2)$, where m and n assume all possible integer values including zero.

Linearization

When the nonlinearities in a system do not predominate, the system can be approximated by linearization techniques. If the nonlinearity is presented in graphical form, the system can be represented by appropriate linear models. Examples of this are the linearization of torque-speed curves for a two-phase motor and equivalent circuit models of vacuum tubes. If the system is described by a nonlinear differential equation, it can be linearized by expanding the nonlinear terms in a power series and retaining only first-order terms. An alternate procedure is illustrated below.

Example. The following equation is to be linearized:

$$\frac{d^3y}{dt^3} + \frac{dy}{dt}\left(\frac{d^2y}{dt^2}\right)^2 + y^2 \frac{dy}{dt} + y = f(t) \tag{80}$$

$$y(0) = 1 \quad \dot{y}(0) = 2 \quad \ddot{y}(0) = 4 \quad f(0) = 1$$

Assume $y(t) = y(0) + \xi(t)$ where $y(0)$ is the initial value of y and ξ a small increment about $y(0)$, similarly for the other derivatives.

$$\begin{aligned} y &= y(0) + \xi \\ \dot{y} &= \dot{y}(0) + \dot{\xi} \\ \ddot{y} &= \ddot{y}(0) + \ddot{\xi} \\ \dddot{y} &= \dddot{y}(0) + \dddot{\xi} \end{aligned} \tag{81}$$

Substitution of Eqs. (81) in (80) yields

$$[\dddot{y}(0) + \dddot{\xi}] + [\dot{y}(0) + \dot{\xi}] + [\ddot{y}(0) + \ddot{\xi}]^2 + [y(0) + \xi]^2[\dot{y}(0) + \dot{\xi}] + [y(0) + \xi] = f(t) \tag{82}$$

Equation (82) is expanded and the higher-order terms are struck from the expression, yielding a linearized equation. Substitution of the given values of $y(0)$, $y'(0)$, and

FEEDBACK SYSTEMS

$y''(0)$ in the original differential equation, Eq. (80), allows one to solve for the initial value of the third derivative $y'''(0)$. The resulting linearized equation is

$$\dddot{\xi} + 16\ddot{\xi} + 17\dot{\xi} + 5\xi - 18 = f(t) \quad (83)$$

Equation (83) is now linear and can be solved by conventional methods for $\xi(t)$ and combined with Eqs. (81) for the complete solution.

Numerical Techniques

Numerical techniques are a powerful tool in the analysis of nonlinear systems.[24,25] The control engineer is fortunate that this topic has interested many mathematicians and that the field is well developed. A simple nonlinear equation will be solved numerically below as an illustration of the concepts and techniques involved.

The nonlinear equation written below will be solved numerically.

$$\begin{align} dy/dt + 2y^2 &= 0 \\ y(0) &= 1 \end{align} \quad (84)$$

The terms in Eq. (84) are approximated in Eq. (85).

$$\frac{dy}{dt} \approx \frac{y[nT] - y[(n-1)T]}{T} \quad (85)$$

$$y^2 \approx y[nT]y[(n-1)T]$$

Substitution of Eqs. (85) in Eq. (84) yields

$$y(nT) = \frac{y[(n-1)T]}{1 + 2Ty[(n-1)T]} \quad (86)$$

Equation (86) is evaluated in Table 15 for $T = \tfrac{1}{2}$ sec.

Table 15. Evaluation of Eq. (86)

n	nT	$y[(n-1)T]$	$y[nT]$
0	0	0	0
1	½	1	½
2	1	½	⅓
3	1½	⅓	¼
4	2	¼	⅕
.	.	.	.
.	.	.	.
.	.	.	.
n	$n/2$	$1/n$	$\dfrac{1}{n}\left(1 + \dfrac{1}{n}\right)^{-1} = 1/(n+1)$

Describing-function Method

When the system nonlinearities are severe, a linearized solution may no longer be indicative of system behavior and specialized nonlinear techniques must be used. Certain nonlinear differential equations have been solved by analytical techniques, and their solutions appear in the mathematical literature. The phase-plane method is a graphical method that yields exact solutions for second-order nonlinear differential equations that are homogeneous or are excited by driving functions that are replaceable by initial conditions. In the describing-function approach, one expands the sinusoidal response of a nonlinear element in a Fourier series and approximates the transmittance of the nonlinearity by the fundamental term in the series expansion. The describing-function technique is applicable to a wide variety of nonlinearities and is described below.

FIG. 45. A nonlinear system component.

If a nonlinear element is excited by a sinusoid of frequency ω_1, and the output of this element is periodic and can be represented by a Fourier series with no d-c term and with a fundamental frequency ω_1, a describing function exists. The describing function approximates the output with the fundamental term of the series expansion. The magnitude of the describing function is equal to the ratio of magnitudes of the fundamental output component to the input and the phase shift is equal to the phase of the fundamental output term. From Fig. 45,

When $x(t) = X \sin \omega_0 t$
$$y(t) = f(x,t)x(t) \tag{87}$$

$$y(t) = \sum_{n=0}^{\infty} Y_n \sin (n\omega_0 t + \theta_n) \tag{88}$$

The d-c component y_0 is zero, and the describing function $N(x,\omega)$ is given by

$$N(x,\omega) = Y_1/X \underline{/\theta_1} \tag{89}$$

Choice of this form for the describing function automatically minimizes the mean-squared error (difference between complete and approximate output), since this is a fundamental property of Fourier coefficients. The validity of the approximation

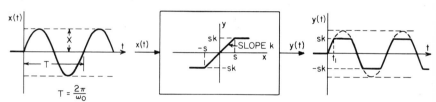

FIG. 46. Saturation nonlinearity.

depends on the rapidity with which the Fourier-series coefficients decrease, and the relative attenuation of the transfer function of the closed loop containing the nonlinear element. The smoother the output waveform $y(t)$, the more rapidly the y_n's decrease. If the loop gain is low-pass in nature with sharp cutoff above ω_0, the describing-function approximation will be nearly exact.

Consider the saturation characteristic shown in Fig. 46. The output waveform is an odd function and possesses symmetry about $\pi/2$; therefore, the Fourier-series representation contains only odd-harmonic sine terms. The coefficients are given by

$$Y_n = 4/\pi \int_0^{\pi/2} y(t) \sin (n\omega_0 t) d(\omega t) \tag{90}$$

where $y(t) = kX \sin (\omega_0 t) \quad 0 \leq t \leq t_1$
$\qquad\qquad sk \qquad\qquad t_1 < t < [(\pi/\omega_0) - t_1]$ (91)

Substitution of Eq. (91) in Eq. (88) and simplification yields

$$Y_1 = 2kX/\pi(t_1 - \sin 2t_1/2)$$
$$t_1 = \sin^{-1} (s/X) \tag{92}$$

Thus the describing function [Eq. (89)] is of the form

$$N = Y_{1/X} = (2k/\pi)(t_1 - \sin 2t_1/2) \quad (93)$$

where $t_1 = \sin^{-1}(s/X)$.

Note that Eq. (93) has zero imaginary part; therefore, the describing function for saturation has no phase shift. This describing function is plotted in Fig. 47, where it is normalized with respect to the linear gain k. If desired, the next largest harmonic can be computed (in this case the third) and compared with the fundamental to estimate accuracy. Describing functions for common nonlinearities are shown in Table 16. The describing functions of the first three nonlinearities in Table 16 exhibit no phase shift, while in the case of backlash, both the amplitude and phase of the

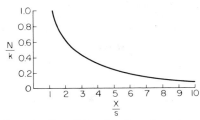

Fig. 47. Describing function for saturation. (Taken from Ref. 1, Fig. 10-12.)

Table 16. Describing Functions for Common Nonlinearities

Type	Characteristic and output for sinusoidal input $x(t) = X \sin(\omega t)$		Describing function
Saturation	Slope k, $\pm s$, $\pm sk$	$y(t)$: $X, sk, -sk, -X$; $t_1, T/2, T$	$\frac{N}{k} = \frac{2}{\pi}(t_1 + \frac{\sin 2t_1}{2})$; $t_1 = \sin^{-1}(s/X)$
Dead zone (threshold)	Slope k, $\pm D$	$y(t)$: $X, (X-D), -(X-D), -X$; $t_1, T/2, T$	$\begin{cases} 0; & X/D < 1 \\ \frac{N}{k} = \frac{2}{\pi}(\frac{\pi}{2} - t_1 - \frac{\sin 2t_1}{2}); & X/D > 1 \end{cases}$ $t_1 = \sin^{-1}(D/X)$
Saturation and dead zone	$k(s-D)$, $-k(s-D)$, Slope k, $\pm s, \pm D$	$y(t)$: $X, k(s-D), -k(s-D), -X$; $t_1, t_2, T/2, T$	No saturation, $s/D = \infty$, $s/D = 2$; $\frac{N}{k} = \frac{2}{\pi}(t_2 - t_1 + \frac{\sin 2t_2 - \sin 2t_1}{2})$ $t_1 = \sin^{-1}(D/X);\ t_2 = \sin^{-1}(s/X)$
Backlash	$(X - \frac{a}{2})$, $-\frac{a}{2}$, Slope $= 1$, $\frac{a}{2}$, $-(X-\frac{a}{2})$	$y(t)$: $(X-\frac{a}{2})$, $-(X-\frac{a}{2})$; $\frac{\pi}{2}, T/2, \frac{3\pi}{2}, T$, $(\pi - \omega t_1)$	Amplitude, Phase, $65°$, $0°$

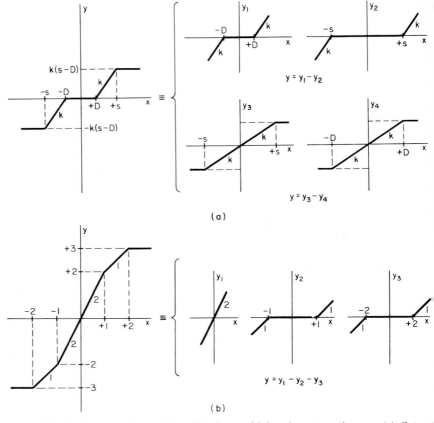

Fig. 48. Synthesis of complex nonlinearities by combining elementary forms. (a) Saturation and dead zone. (b) Piecewise linear characteristic.

output vary with the input. Thus the describing function for backlash will have a phase shift depending on the input. Other nonlinear characteristics can be generated from those shown by proper combination. Thus the saturation and dead zone

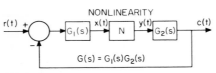

Fig. 49. System with a nonlinearity in the forward path.

describing function is *not* the sum of the describing function for saturation and dead zone, but is the difference of two dead-zone characteristics or two saturation characteristics (see Fig. 48). In general, describing functions are complex quantities, and when they are combined they must be treated as such.

The system shown in Fig. 49 possesses a nonlinearity in the forward path.

Since the describing function serves as a transfer characteristic for the nonlinearity, the stability of the system depends on the zeros of $1 + NG(j\omega)$. This expression has meaning only for $s = j\omega$, since the describing function is based on sinusoidal signals. Thus the system becomes unstable for values of frequency and amplitude that satisfy

$$-N = 1/G(j\omega) \quad (94)$$
$$G(j\omega) = -1/N \quad (95)$$

Since it seems more natural to deal with $G(j\omega)$ rather than its reciprocal, Eq. (95) is generally used (both forms are found in the literature). A convenient means of obtaining graphical solutions of Eq. (95) is to combine the Bode plots of gain vs. ω and phase vs. ω into one gain-phase plot, where gain is plotted vs. phase with ω as a parameter along the curve. This curve is known as the frequency locus, since it traces a path in the gain-phase plane with ω as the parameter. The negative reciprocal of the describing function $-1/N$ is plotted on the same axis with amplitude as a parameter and is called the amplitude locus. Solutions of Eq. (95) are given by intersections of the two curves. At the intersection points the frequency and magnitude of oscillation are read respectively from the frequency and amplitude loci. Typical situations are shown in Fig. 50a,b,c for a certain $G(j\omega)$ and a phase-independent describing function. In Fig. 50a, the two loci do not intersect; therefore, the system is stable. Since the describing function is a frequency-response description, gain and phase margin have some meaning, and the system of Fig. 50a can be said to have $\sim +10$ db gain margin. Attempts to relate this gain margin to the degree of

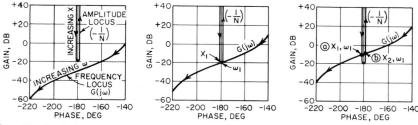

FIG. 50. Typical amplitude and frequency loci. (a) Stable system. (b) Limit cycle. (c) Stable and unstable limit cycles.

damping of the transient response of the system are unsuccessful since the describing function has meaning only for inputs that are approximately sinusoidal. If the gain of $G(j\omega)$ is increased by 10 db, the loci of Fig. 50b are obtained. In the latter case, a limit cycle is excited, and the system oscillates at $\sim\omega_1$ radians/sec, and with amplitude $|x| = X_1$. Thus, referring to Fig. 49, $y(t) \approx N(X_1)X_1 \sin(\omega_1 t)$, and $x(t) \approx C_0 \sin(\omega_1 t + \theta)$; where $C_0 = X_1 N(X_1)|G(j\omega_1)|$ and $\theta = \underline{/G}(j\omega_1)$. The output waveform is only approximately sinusoidal. Since the system is nonlinear the existence of this oscillation depends on input amplitude X. If, because of some signal input $r(t)$ the input to the nonlinearity x becomes equal to X_1, the system will oscillate as described above. Suppose the input became $(X_1 - \epsilon)$, i.e., slightly smaller than X_1. This corresponds to a slight movement up along the left flank of the amplitude locus, giving a slight increase in $|1/N|$, or a decrease in $|N|$. Decreasing $|N|$ tends to decrease the loop gain, which decreases X_1 and the oscillation decays. Similarly, increasing the input to $(X_1 + \epsilon)$ increases $|1/N|$ and decreases $|N|$, tending to decrease the loop gain, restoring the constant-amplitude oscillations. For inputs in the range $0 \le x < X_1$ and $X_1 < x < \infty$, no limit cycle exists. For $x = X_1$ a limit cycle is present. If the limit cycle is excited, and the signal input subsequently removed, oscillations die out and the limit cycle is unstable. In Fig. 50c two intersections a and b are present. By the same reasoning as used above, the limit cycle of point a is unstable and that of point b stable. For signal inputs $0 \le x < X_1$ no limit cycle is excited. For the case $x = X_1$ limit cycle a is excited, but upon removal of the input, the oscillations either decay to zero or grow until the stable limit cycle is excited. Application of signal input in the range $X_1 < x < \infty$, and subsequent removal of the signal excites the stable limit cycle b.

Since the gain-phase plot of $-1/N$ is convenient to use in stability studies, describing functions are often plotted directly in this form.

The describing function for a relay with different pickup and drop-out levels is plotted in this manner in Fig. 51.

In some cases proportional control of a system variable is impractical or undesirable and on-off type control (also called bang-bang, relay, or contactor control) is used. The most common example of this is the use of a thermostat in controlling house temperature via a home heating system. If the system were proportional, the furnace would run continuously, with variable Btu output controlled by a temperature error signal. Actually on-off control is used, and the furnace turns full on when the error

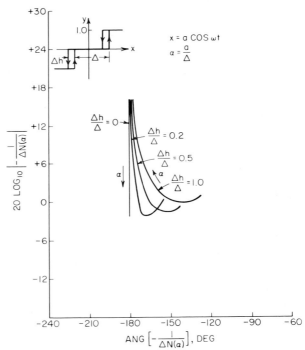

Fig. 51. Gain-phase plot of $(-1/N)$ for a relay. (*Taken from E. C. Johnson, thesis Servomechanisms Laboratory, M.I.T., Jan. 12, 1953.*)

signal exceeds a certain value and shuts off when the error is reduced below that value. This system is nonlinear and stable limit cycles may exist. If such oscillations do exist and are of small amplitude, the performance may be perfectly acceptable. For example, an oscillation amplitude of 1°F in a heating system may be acceptable if the occupants of the house are unable to detect such a small change.

On-off control finds extensive use in space vehicles. A missile may be controlled in roll using compressed-air nozzles that produce clockwise and counterclockwise roll torques, by venting blasts from an air-storage reservoir.

In general on-off control is advantageous when control must take place at high power levels or when a minimum cost, weight, and/or size must be achieved in the control equipment.

FEEDBACK SYSTEMS 15-63

11 Simulation and Machine Computation

The term simulation implies the use of one device to study the behavior of another, because the former is simpler, cheaper, more convenient to use, or readily available. If direct study of a mechanical device is not feasible, an investigation of a substantially equivalent electrical, thermal, pneumatic, hydraulic, or optical device might be advantageous. If in designing a high-fidelity audio amplifier, an engineer cannot adequately describe the system by paper-and-pencil analysis, an actual laboratory prototype can easily be built and laboratory tests made on the device. The cost, size, and time required make this approach attractive. If the computation of film thickness and indices of refraction for antireflection coatings on an optical system become difficult, an equivalent electrical model using wave filters or transmission lines might be used. When the system in question approaches the complexity, cost, and size of a guided missile, simulation becomes a veritable necessity. Analog simulation replaces the system under study with another more convenient one governed by the same integrodifferential equations. A device that uses this technique to solve a specific problem is called a simulator. In general, a simulator approximates only one specific problem or class of problems. A collection of basic building blocks designed to be used as a general-purpose simulator for a wide variety of problems is called an analog computer. Rather than simulating a system per se, an analog computer simulates mathematical operations which can, in turn, be used to simulate the differential equations describing system operation.

Analogy and Simulation

Analogy is the basic principle used in simulation. One device is called the analog of another (actually they are mutual analogs) if the dependent and independent variables governing operation of the two devices are related by the same set of integrodifferential equations. The most convenient analogs to deal with are those for two-terminal lumped-circuit elements. A number of common mechanical-electrical analogs appear in Table 17. The analogs of rotational mechanical quantities can be found from Table 17 if one makes the following substitutions: torque for force, radian velocity for linear velocity, rotational viscous damping for translational viscous damping, torsional spring constant for linear spring constant, inertia for mass, and gear ratio for lever-arm ratio.

An example of the use of analogous mechanical and electrical circuit models is shown in Table 18. Part a of Table 18 depicts a shaft that transmits the power developed by an electric motor, via a belt drive to an inertia load (flywheel). The shaft is supported by two sleeve bearings that contribute viscous friction. In part b a lumped, rotational, mechanical circuit diagram is shown. Part c shows an electrical analog of part b. If the plane of the belt is not perpendicular to the axis of the shafts, longitudinal as well as circumferential forces are produced. The shaft is spring-loaded by K_1 to prevent lateral motion. If the shaft is rigid in translation and longitudinal friction is negligible, the model of part d holds for translational motion.

If the designer of the mechanical system wished to study experimentally the vibration modes of the shaft, it would be more convenient to construct a simple electrical analog in the laboratory. Rather than changing inertias and friction coefficients, variable capacitors and resistors could be used. A laboratory oscillator could be used to generate the excitation, and an oscilloscope to view and measure the response.

Analog Computers

The power of an analog computer as a research tool depends on the flexibility and ease with which it can be used to simulate a physical system.[26,27,28] An analog computer designed for solving linear, constant-coefficient differential equations must be able to perform the operations of differentiation and integration, addition and subtraction, and multiplication by a constant. The computer must also introduce initial conditions, inject driving functions, and display responses.

Early analog computers used mechanical components to perform these operations. Variables were represented by shaft rotations, integration performed with ball-and-disk integrators, summation effected using differential gears, and multiplication by

Table 17. Common Mechanical-electrical Analogs

Quantity	Mechanical model	Electrical model
Through variable source	Force $f(t)$	Current $i(t)$
Across variable source	Velocity $v(t)$	Electric potential Voltage $e(t)$
Energy dissipation element	Dashpot $f_B(t) = B v_B(t)$; $v_B(t) = \frac{1}{B} f_B(t)$ $B \sim \frac{1}{R}$	Resistor $i_R(t) = \frac{1}{R} e_R(t)$; $e_R(t) = R i(t)$ $R \sim \frac{1}{B}$
Potential energy storage element	Spring $f_K = K \int v_K \, dt$; $v_K = \frac{1}{K} \frac{df_K}{dt}$ $K \sim \frac{1}{L}$	Inductor $i_L = \frac{1}{L} \int e_L \, dt$; $e_L = L \frac{di_L}{dt}$ $L \sim \frac{1}{K}$
Kinetic energy storage element	Mass / Inertial reference $f_m = M \frac{dv_m}{dt}$; $v_m = \frac{1}{M} \int f_m \, dt$ $M \sim C$	Capacitance $i_c = \frac{c \, de_c}{dt}$; $e_c = \frac{1}{C} \int i_c \, dt$ $C \sim M$
Energy coupling element	Lever $f_2 = -(\frac{l_1}{l_2}) f_1$; $v_2 = (\frac{l_2}{l_1}) v_1$ $(\frac{l_1}{l_2}) \sim (\frac{n_1}{n_2})$	Transformer $i_2 = -(\frac{n_1}{n_2}) i_1$; $e_2 = (\frac{n_2}{n_1}) e_1$ $(\frac{n_2}{n_1}) \sim (\frac{l_2}{l_1})$

a constant realized through use of gear ratios. Modern analog computers use electrical components.

The heart of an electric analog computer is a high-gain electronic amplifier (operational amplifier) shown in Fig. 52. Resistor R_i is an input resistance and R_f feedback resistance. The currents are related by $i_G = i_1 + i_f$. The voltage gain of an operational amplifier is 10^4 to 10^6 or higher. The maximum output voltage

FEEDBACK SYSTEMS 15-65

e_2 for linear operation is about 100 volts.* For the values $A = 10^6$ and $e_2 = 100$ volts, $e_g = 10^{-4}$. Therefore, to a first approximation $e_g \approx 0$ and $i_1 = e_1/R_i$, $i_f = e_2/R_f$. The current i_g is the grid current of a vacuum-tube operating class A and is essentially zero. If the active elements are transistors, the circuit is *designed* to have very high

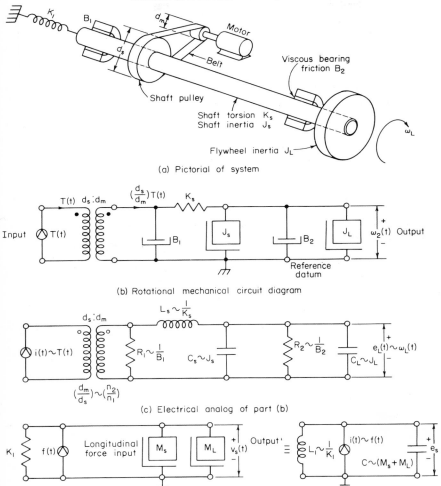

Table 18. Analogous Mechanical and Electrical Circuit Models of a Mechanical System

(a) Pictorial of system

(b) Rotational mechanical circuit diagram

(c) Electrical analog of part (b)

(d) Translational model and its analog

input impedance and essentially the same performance as vacuum-tube operational amplifiers. Solving for e_2,

$$e_2(t) = -(R_f/R_i)e_1(t) \qquad (96)$$

Thus the circuit of Fig. 52 multiplies the excitation by the constant $-R_f/R_i$. In

*Transistor analog computers operate at lower voltage levels such as 10 to 30 volts.

the circuit of Fig. 53, R_i and R_f are replaced by impedances. The amplifier is symbolized by a triangle, and the common ground is deleted for simplicity. The transfer function is

$$E_2(s)/E_1(s) = -Z_f(s)/Z_i(s) \qquad (97)$$

By proper choice of $Z_f(s)$ and $Z_i(s)$ a variety of transfer functions can be synthesized. For any desired transfer function, there are of course many choices for $Z_f(s)$ and $Z_i(s)$. Choosing $Z_f(s) = 1/C_f s$ and $Z_i(s) = R_1$ results in an integrator. Interchange of the resistor and capacitor produces a differentiator. Summation can be obtained

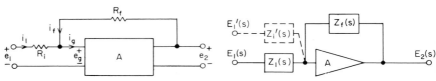

Fig. 52. Basic operational amplifier.   Fig. 53. Generalized operational amplifier

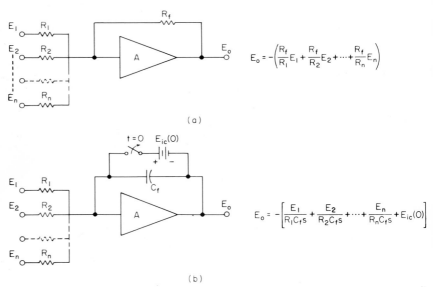

Fig. 54. Basic analog computing components. (a) Amplifier-summer. (b) Integrator-summer.

by adding input impedances and signals at the grid as shown in the dotted branch of Fig. 53. The output is then

$$E_2(s) = -\{[Z_f(s)/Z_i(s)]E_1(s) + [Z_f(s)/Z_i'(s)]E_1'(s)\} \qquad (98)$$

Since the grid of the operational amplifier is virtually at zero potential, the output voltage is equal to the voltage across Z_f. In the case of an integrator, Z_f is a capacitor, and the initial value of the output is the initial voltage across the capacitor. This allows the introduction of initial conditions. Practical realization of a differentiator is difficult because of noise and saturation problems; therefore, the differential equation is generally manipulated so that only integrations are necessary.

FEEDBACK SYSTEMS 15-67

The two basic components, the amplifier-summer and integrator-summer, are shown in Fig. 54. Because of various practical considerations, resistances between 0.1 and 10 megohms, capacitances between 0.1 and 10 μf, and output voltages in the range ± 100 volts are commonly used.

An example of the use of an electronic analog computer in solving a differential equation is illustrated in the example below.

$$d^3y/dt^3 + 5(d^2y/dt^2) + 2(dy/dt) + 0.1y = f(t) \tag{99}$$
$$\ddot{y}(0) = +2 \quad \dot{y}(0) = 0 \quad y(0) = -1$$

The equation is first solved for its highest derivative

$$\dddot{y} = -5\ddot{y} - 2\dot{y} - 0.1y + f(t) \tag{100}$$

An interesting "circular-reasoning" procedure is used to devise the computer diagram. The left side of Eq. (100) is considered "given" and the right side of the equation "computed" from it, using amplifiers and integrators. The given and computed quantities are then equated to satisfy Eq. (100).

Assuming that the highest derivative ($\dddot{y}$) is known, lesser derivatives ($\ddot{y}$, $\dot{y}$, y) can be found by successive integration as shown in Fig. 55a. As a next step toward synthesis of the right side of Eq. (100), each derivative is given its proper algebraic sign using unity-gain amplifiers (called sign inverters) as shown in Fig. 55b. The computed function is formed using an amplifier-summer as shown in Fig. 55c. The computer diagram is closed using an inverter as shown in Fig. 55d. A flow-diagram representation is shown in Fig. 55e. The primitive diagram corresponds to Fig. 55d. The simplified computer diagram, using an integrator-summer rather then an amplifier-summer, is shown in Fig. 55f.* The initial conditions are also introduced in the later diagram. The output can be taken at α, if the negative sign is acceptable; if not the addition of an inverter yields a positive output at β.

The common driving functions used in physical problems are steps and ramps. A step can be generated using a d-c voltage and a switch, and a ramp by applying a step to an integrator, as shown in Fig. 56. Other input functions can be generated by auxiliary simulation of a differential equation that has a homogeneous or driven response equal to the input function in question.

Simulation of Block Diagrams

When a control system is to be simulated from its block diagram, the differential equations governing the block diagram can be written and the simulation procedure just described can be used. A more convenient technique is generally used if the system block diagram is available. Rather than simulating the differential equation, the poles and zeros of the system are introduced by using operational amplifiers with appropriate networks for Z_i and Z_f. A number of useful combinations appear in Table 19.

An example of direct analog-computer simulation of a block diagram is shown in Fig. 57. The block diagram in Fig. 57a is rearranged in flow-diagram form in Fig. 57b. A computer realization of the flow diagram appears in Fig. 57c. Of course, many equivalent flow diagrams can be written for any block diagram and, therefore, many equivalent computer diagrams exist.

Simulation of Nonlinear Systems

Transcendental equations (linear differential equations with time-varying coefficients) can be simulated on an analog computer. Such simulations involve the

---

* The simplified computer diagram is generally preferred, since fewer computing components are used. One minor disadvantage is that $\dddot{y}$ is no longer an explicit signal available for display but is contained "inside" the integrator-summer.

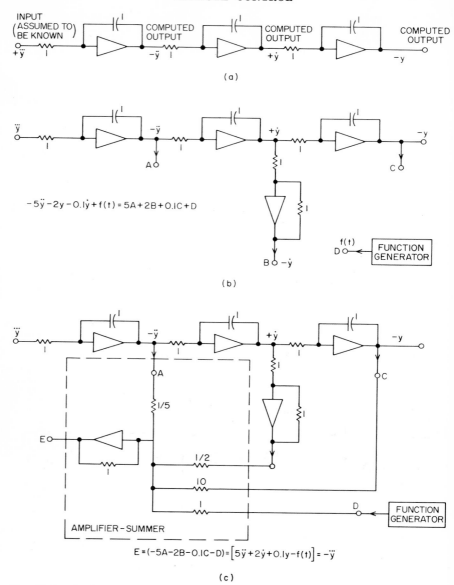

Fig. 55. Synthesis of a computer diagram that solves Eq. (99). All resistor values are in megohms and capacitor values in microfarads. (a) Computation of all necessary time derivatives of y, from "given" quantity. (b) Formation of the computed function components. (c) Formulation of computed function.

FEEDBACK SYSTEMS

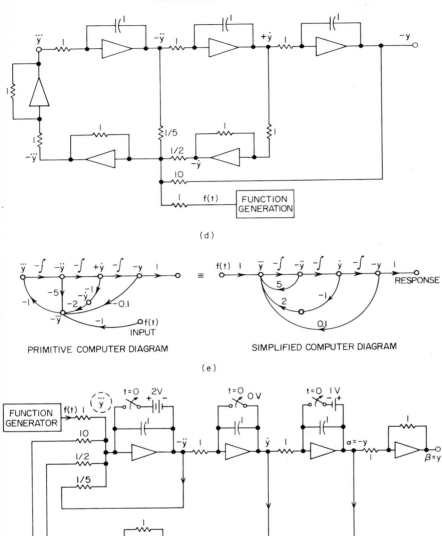

FIG. 55. (*Continued*). (*d*) Primitive computer diagram. (*e*) Flow-diagram representation of computer diagram. (*f*) Simplified computer diagram. Initial condition switches open synchronously when solution starts at $t = 0$.

transcendental operations of sin, cos, tan, transportation lag, etc. Simulation of nonlinear equations involves nonlinear operations such as multiplication, division, powers and roots, saturation, and dead zone. Transcendental and nonlinear opera-

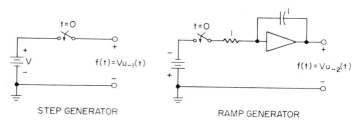

FIG. 56. Step and ramp generation.

tions are generally performed with auxiliary equipment or specially modified operational amplifiers. Examples of dead zone and saturation nonlinearities simulated with biased diodes and operational amplifiers appear in Fig. 58.

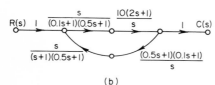

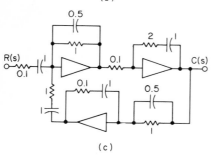

FIG. 57. Analog-computer simulation of a block diagram. (a) Block diagram of a typical control system. (b) One possible flow diagram corresponding to the block diagram of (a). (c) Computer realization of (b). All values in megohms and microfarads.

Digital Computation

Use of a digital computer to calculate system response differs from analog simulation.[29,30] In using the digital computer, a difference equation that approximates the system differential equation is used as the point of departure. A computer program for solving the difference equation is then written. The program must be converted into computer language and then read into the computer. The size, cost, and availability of digital computers tend to limit their application to problems complex enough to tax the computer, thereby fully utilizing its potential. The digital computer can handle larger and more complex problems than an analog computer. However, for medium-sized problems it is more expensive and harder to use in trial-and-error solutions, since it does not usually operate in real time. An analog-computer study is generally performed by the engineer concerned with the design of the system or by one of his associates. Modifications and changes in the system can be easily evaluated while the program is in progress. With a digital computer, the engineer poses the problem to the programmer, who writes the program and changes it into machine language. Changes in the system are difficult, and any modifications that might be necessary are best included at the outset. Of course, the digital computer still retains its great capacity and usefulness in solving difficult system-optimization or design equations and is finding increased application as the level of sophistication of control problems increases.

Table 19. Some Useful Transfer Functions Simulated with Operational Amplifiers

Network	Transfer function $E_2(s)/E_1(s)$
1. $E_1 - Z_i -$ [amp with Z_f feedback] $- E_2$	$-\dfrac{Z_f}{Z_i}$
2. $E_1 - R_1 -$ [amp with $R_2 \parallel C$ feedback] $- E_2$	$-\dfrac{(R_2/R_1)}{(R_2 C s + 1)}$
3. $E_1 - R_1 - C -$ [amp with R_2 feedback] $- E_2$	$-\dfrac{(R_2 C)s}{(R_1 C s + 1)}$
4. $E_1 - R_1 -$ [amp with R_2 in series with C feedback] $- E_2$	$\dfrac{-(R_2 C s + 1)}{(R_1 C)s}$
5. $E_1 - C_1 -$ [amp with R_2 in series with C_2 feedback] $- E_2$	$-(R_1 C_1 s + 1)$
6. $E_1 - R_1 - C_1 -$ [amp with R_2 in series with C_2 feedback] $- E_2$	$-\left(\dfrac{C_1}{C_2}\right)\dfrac{(R_2 C_2 s + 1)}{(R_1 C_1 s + 1)}$
7. $E_1 - R_1 - C -$ [amp with $R_2 \parallel C_2$ feedback] $- E_2$	$\dfrac{-(R_2 C_1)s}{(R_1 C_1 s + 1)(R_2 C_2 s + 1)}$
8. $E_1 -$ [$C_1 \parallel R_1$] $-$ [amp with R_2 in series with C_2 feedback] $- E_2$	$\dfrac{-(R_1 C_1 s + 1)(R_2 C_2 s + 1)}{(R_1 C_2)s}$

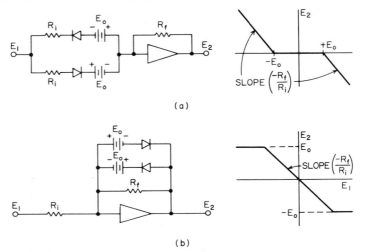

Fig. 58. Simulation of dead zone and saturation. (a) Dead zone. (b) Saturation.

12 Computers in Control

A control system is inherently a computing device. The error signal in a negative-feedback loop is computed as the difference between the input and feedback signal. The system performing this calculation might be called a computer; however, this term is generally reserved for a section of a system wherein a large number of computations are performed at one time. The computations can be effected using a special-purpose analog or digital computer or, if warranted, a general-purpose computer programmed for the specific computations.[31,32,33]

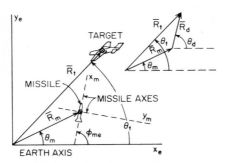

Fig. 59. Geometry of a pursuit course fire-control problem.

Analog Computation

As an example a hypothetical fire-control problem is considered in which a missile follows a pursuit course toward a target. To simplify the problem, the missile and target flight paths are assumed to be coplanar. When on a pursuit course the missile is always directed along the line of sight to the target (see Fig. 59). The ground installation contains a target-track radar, a missile-track radar, a computer, and a data link. The target-track radar determines $|R_t|$ and θ_t, the range and elevation angle of the target in earth coordinates. The missile-track radar determines $|R_m|$, θ_m, and ϕ_{me}, the angle that the missile velocity vector makes with the earth axes. $\bar{R}_d$ is the desired direction the missile should take to head at the target. The error in elevation angle $(\theta_d - \phi_{me})$ is the command that must be computed and telemetered to the missile. In terms of earth-axis components,

$$\bar{R}_t = |R_t|(\cos \theta_t \bar{\imath} + \sin \theta_t \bar{\jmath}) \qquad (101)$$

$$\bar{R}_m = |R_m|(\cos \theta_m \bar{\imath} + \sin \theta_m \bar{\jmath}) \qquad (102)$$

$$\bar{R}_d = \bar{R}_t - \bar{R}_m = (|R_t| \cos \theta_t - |R_m| \cos \theta_m)\bar{\imath} + (|R_t| \sin \theta_t - |R_m| \sin \theta_m)\bar{\jmath} \qquad (103)$$

$$\tan \theta_d = \frac{|R_t| \sin \theta_t - |R_m| \sin \theta_m}{|R_t| \cos \theta_t - |R_m| \cos \theta_m} = \frac{y_d}{x_d} \qquad (104)$$

Assuming that the interception of the target always occurs from below, the y component of $\bar{R}_d$ is always positive; this means that the angle θ_d is always in the first or second quadrant. If interception occurs before the target passes over the radar, the X component of $\bar{R}_d$ is positive. Under these assumptions, $0 \leq \theta_d \leq \pi/2$. Solution of Eq. (104) for the command signal is shown operationally in Fig. 60. Physical realization of these operations is of considerable interest.

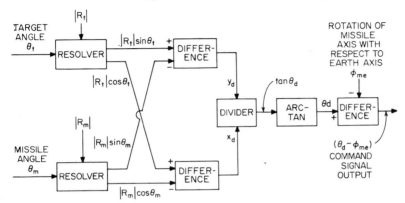

FIG. 60. Operational diagram of a missile command computer.

θ_t, $|R_t|$ = signal inputs from target-track radar
θ_m, $|R_m|$, ϕ_{me} = signal inputs from missile-track radar

Transcendental and Nonlinear Operations

The more challenging control problems generally involve transcendental and/or nonlinear computations. Because of their individuality and importance, several practical realizations of these operations are discussed below.

Resolver. A resolver is a device that converts polar coordinates R and θ into rectangular coordinates x and y. A popular means of performing this and other operations uses a motor-generator servo (Fig. 61). Simplification of the block diagram yields a transfer function $C(s)/R(s) = K_m/s[s + (a + K_mK_T)]$. When a high-quality servomotor is used, the break frequency $\omega_b = a$ is 30 radians/sec or higher. By adjusting the term K_mK_T, the time constant of the motor with tachometer feedback can be moved to high frequencies. With this gain adjustment, the transfer function can be approximated by K'/s over a frequency range $0 < \omega < (a + K_mK_T)$ where $K' = K_m/(a + K_mK_T)$. A resolver using a position servo of this type is shown in Fig. 62. The input voltage θ_i drives the motor-generator unit, positioning the motor shaft angle θ_0. The shaft θ_d drives three potentiometers (dotted lines), one with a linear winding, one with a sinusoidal winding, and the third with a cosinusoidal winding. The linear potentiometer is excited with a d-c voltage providing position feedback and the transfer function is $\theta_0(s)/\theta_i(s) = K'/(s + K')$. Using the expressions of Table 8, the steady-state position error is found to be zero. Furthermore, for a frequency range $0 < \omega \ll K'$, $\theta_0 \approx \theta_i$, the output of the sine and cosine potentiometers yields the rectangular components $r \cos \theta_i$ and $r \sin \theta_i$.

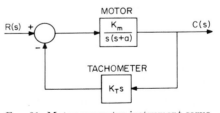

FIG. 61. Motor-generator instrument servo.

Difference Amplifier. The sum of two signals can be computed using an operational summer-amplifier. When taking the difference of two quantities, an inverter can be used first to change the sign of one of the quantities. When accuracy limitations permit, specialized circuitry less complex then standard operational amplifiers can be used. An electronic difference-amplifier circuit is often used.

Divider. The most common divider circuit shown in Fig. 63 uses an instrument servo. The input b excites the position servo, driving the shaft to the steady-state position $\theta_0 \approx \theta_i = b$. The second potentiometer is excited by the output of the

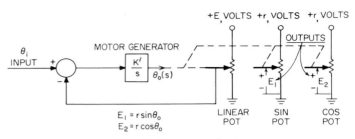

Fig. 62. A servo resolver.

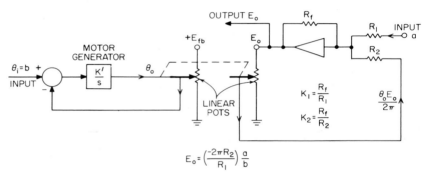

Fig. 63. Servo divider circuit.

operational amplifier. The output voltage at the center arm of this potentiometer is $\theta_0 E_0/2\pi$. The amplifier output voltage is given by

$$E_0 = -(R_f/R_2)(\theta_0 E_0/2\pi) - (R_f/R_1)a \tag{105}$$

Solving for E_0 yields

$$E_0 = -(R_f/R_1)a/1 + (R_f/2\pi R_2)\theta_0 \tag{106}$$

As before, $\theta_0 \approx \theta_i = b$ over a range of frequencies $0 < \omega \ll K'$. By adjusting $(R_f b/2\pi R_2) \gg 1$, the output voltage becomes

$$E_0 = (-2\pi R_2/R_1)(a/b) \tag{107}$$

When b becomes very small so that $(R_f b/2\pi R_2) \le 1$, Eq. (107) is no longer valid and the system no longer divides. As $b \to 0$, the denominator of Eq. (106) becomes unity, yielding an incorrect result. The coefficient $(R_f/2\pi R_2)$ is made large enough so that this happens only for very small b. It is not surprising that a physical division circuit breaks down when the divisor approaches zero.

Arctangent Generator. Physical realization of the arctan operation required in Fig. 60 illustrates a technique useful in the generation of specialized functions. The circuit of Fig. 64 solves the equation $\tan \theta_o = (\sin \theta_o / \cos \theta_o) = \theta_i$. The output

θ_o is

$$\theta_o = (K'/s)E \tag{108}$$
$$E = \theta_i - \sin\theta_o/\cos\theta_o = \theta_i - \tan\theta_o \tag{109}$$

Substitution of Eq. (109) in (108) yields

$$\theta_o + (K'/s)\tan\theta_o = (K'/s)\theta_i \tag{110}$$

The steady-state value of Eq. (110) is found by multiplying both sides of the equation by s and letting $s \to 0$. This yields the required relation $\theta_o = \tan^{-1}\theta_i$. The range

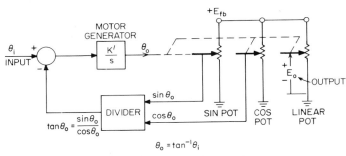

Fig. 64. An arctangent computer.

of frequencies over which this device approximates the arctangent function is given by the restriction $(K'/s)\tan\theta_o \gg \theta_o$. This leads to an operating bandwidth

$$K'|\tan\theta_o/\theta_o| \gg \omega$$

Feedback Function Generation. The technique used to devise an arctangent generator can be adapted to generate many nonlinear and transcendental functions. In general the function $\theta_o = f(\theta_i)$ is to be generated, where the inverse relation $\theta_i = g(\theta_o)$ holds. Figure 65 presents a flow graph illustrating the general approach. This approach is used to generate the functions shown in Table 20.

Other important function-generating techniques utilize cathode-ray tubes, cams, and biased diodes. The function can be plotted on a black paper mask and placed on a cathode-ray-tube where an electron beam follows the edge of the mask. The required scan potentials generate the function in question. The function may also be plotted on the surface of a cam and a motion transducer used to "read out" the function. A very flexible technique is to approximate the function using piecewise linear analysis. The piecewise linear function is then synthesized using diodes, bias sources, and operational amplifiers.

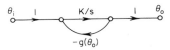

Fig. 65. General approach to feedback function generation.

$\theta_o = (K/s)\theta_i - (K/s)g(\theta_o)$
$\theta_i \approx g(\theta_o)$, i.e., $\theta_o \approx f(\theta_i)$ for $|(K/s)g(\theta_o)| \gg \theta_o$

Multipliers. Perhaps the most frequently used nonlinear computing component is a multiplier. Of the many multiplying devices used, the commonest is a servo-multiplier. If the input θ_i to the motor generator of Fig. 62 is a, and an additional linear potentiometer excited by a voltage b is placed on the shaft, the signal at the center arm of this potentiometer is $E_3 = ab$. A diode squarer and "quarter-square" multiplier are shown in Fig. 66. This circuit is common in current transistor computers.

The analog realization of Fig. 60 results in a moderately complex system. If the problem were analyzed in three dimensions (as it should be), the computer would be considerably more complex. Also other guidance philosophies might prove more

Table 20. Feedback Function Generators

Function	Block diagram	System equations
1. Arc sin	θ_i Input → (+/−) → K'/s → θ_o Output; feedback via Resolver with $\sin\theta_o$	$\theta_o = \dfrac{K}{s}\theta_i - \dfrac{K}{s}\sin\theta_o$ $\theta_o \approx \sin^{-1}\theta_i$ for $\begin{cases}\omega \ll \dfrac{2\sqrt{2}}{\pi}K \\ -\pi/4 < \theta_o < +\pi/4\end{cases}$
2. Arc cos	θ_i Input → (+/−) → K'/s → θ_o Output; feedback via Resolver with $\cos\theta_o$	$\theta_o = \dfrac{K}{s}\theta_i - \dfrac{K}{s}\cos\theta_o$ $\theta_o \approx \cos^{-1}\theta_i$ for $\begin{cases}\omega \ll \dfrac{2\sqrt{2}}{\pi}K \\ -\pi/4 < \theta_o < +\pi/4\end{cases}$
3. Square root	θ_i Input → (+/−) → K'/s → θ_o Output; feedback via Squarer with θ_o^2	$\theta_o = \dfrac{K}{s}\theta_i - \dfrac{K}{s}\theta_o^2$ $\theta_o \approx \sqrt{\theta_i}$ for $K|\theta_o| \gg \omega$
4. Divider	$\theta_i = a$ Input → (+/−) → K'/s → θ_o Output; feedback via Multiplier with $b\theta_o$, b	$\theta_o = \dfrac{K}{s}\theta_i - \dfrac{K}{s}b\theta_o$ $\theta_o \approx \dfrac{a}{b}$ for $K|b| \gg \omega$

intricate. At a certain level of system complexity, use of a full-scale digital computer becomes competitive with analog equipment with regard to cost, weight, size, etc.

Digital Computation

Digital computation in a control system may not imply a full-scale digital computer. There are many control applications involving discrete computations that are small-scale in nature and are performed using special-purpose digital circuitry. Some of the pertinent aspects of digital design are illustrated in the two design problems discussed in the sections that follow.

Digital Controller. The use of a digital controller as a compensation element in a sampled data system was mentioned in Sec. 9. A typical sampled-data system is shown in Fig. 67. The transfer function of the closed-loop control system is given by

$$\frac{C(z)}{R(z)} = \frac{D(z)\mathcal{Z}[G(s)]}{1 + D(z)\mathcal{Z}[G(s)]} \tag{111}$$

$$\mathcal{Z}[G(s)] = \mathcal{Z}\frac{1 - \epsilon^{-Ts}}{s(s+1)(s+2)} = \left(1 - \frac{1}{z}\right)\mathcal{Z}\left[\frac{\frac{1}{2}}{s} - \frac{1}{s+1} + \frac{\frac{1}{2}}{s+2}\right] \tag{112}$$

$$\mathcal{Z}[G(s)] = \frac{\frac{1}{8}(z + \frac{1}{2})}{(z - \frac{1}{2})(z - \frac{1}{4})} \tag{113}$$

An ideal transfer function for many control purposes would be $[C(z)/R(z)] = 1$. This would mean that the output, at the sampling instants, is identical with the input

Unfortunately, this is too stringent a requirement. A "second-best" choice would be if the output duplicated the input after a delay of one sample period. This type of system is known as a prototype system and has a transfer function $[C(z)/R(z)] = z^{-1}$.

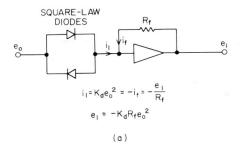

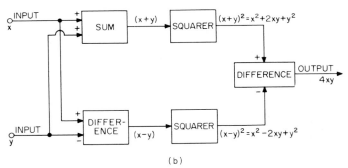

FIG. 66. A diode squarer and a quarter-square multiplying circuit. (a) A diode-squaring circuit. (b) A "quarter-square" multiplier.

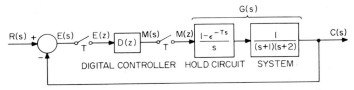

FIG. 67. A sampled-data control system with a digital controller.

Substitution of this condition and Eq. (113) in Eq. (11) determines the pulse transfer function of the digital controller necessary for prototype response.

$$D(z) = \frac{M(z)}{E(z)} = \frac{8z^2 - 6z + 1}{z^2 - \frac{1}{2}z + \frac{1}{2}} \qquad (114)$$

A pulse transfer function is physically realizable if the denominator polynomial is of equal or higher order than the numerator. Thus Eq. (114) is physically realizable. The difference equation associated with Eq. (114) appears in Eq. (115).

$$M_n = \frac{1}{2}M_{n-1} + \frac{1}{2}M_{n-2} + 8E_n - 6E_{n-1} + E_{n-2} \qquad (115)$$

The special-purpose digital controller described by Eq. (115) is shown in operational form in Fig. 68. The system shown contains delays, scale factors, and an adder. It might be noted in passing that a prototype sampled-data system is not a panacea. If the sampling period T is long, the delay in system response is too long and the system is too slow. For T very short, the system supposedly responds very quickly, but the signals within the loop become very large, saturating the components. Optimum response (which may or may not surpass that of other design techniques) occurs for some intermediate value of T.

The addition shown in Fig. 68 can be carried out using an operational amplifier-summer. Wherever sign changes or multiplying constants greater than unity are necessary, operational amplifiers can be used. For multiplying constants less than unity, potentiometers or voltage dividers can be used. Realization of a time delay requires more ingenuity. Time delays in the order of milli- and microseconds can be realized using lossless or distortionless electrical transmission lines, electric-wave filters, or pulse-circuit techniques. Pneumatic or hydraulic transmission lines and lumped-component filters allow longer delays. A common technique is to record the signal on magnetic tape with a recording head and to read out the signal with a

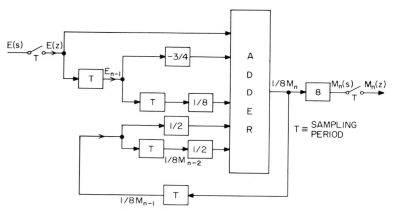

FIG. 68. A digital controller.

playback head. The displacement between the two heads and the tape speed determines the time delay. Another method is to drive a two-deck wafer switch with a constant-speed servo (synchronous motor, or d-c motor with velocity feedback). The signal is read in on the center arm of the first deck, which has a capacitor between each terminal and ground. Each terminal on the second deck is connected to the corresponding terminal on the first deck. The signal is read out on the center arm of the second deck, which lags the center arm of the first deck by an angle θ. The time delay is proportional to θ and the speed of rotation.

Digital Logic. The heart of specialized digital circuitry is digital logic. Digital logic is based on Boolean algebra, the mathematics describing a system with discrete states. In a digital system only two discrete states are possible; these are designated 0 (zero) and 1 (unity). Zero denotes absence and unity denotes presence of a quantity. The two most common logical operations are the AND operation (intersection) and the OR operation (union). An AND operation, written $AB = C, (A \cap B = C)$, means that C occurs only if A and B both occur. $A + B = C, (A \cup B = C)$ is an OR operation, meaning that C occurs if A or B or both A and B occur. Both these operations are explained in the Venn diagrams of Fig. 69 and the "truth tables" shown in Fig. 70.

Proper combination of these and other logical operations results in synthesis of complex logical operations. The error-sensing system described below is an example.*

The control system shown in Fig. 71 is designed to minimize the mean time-weighted

* This system was originally suggested to the author by Prof. L. Braun.

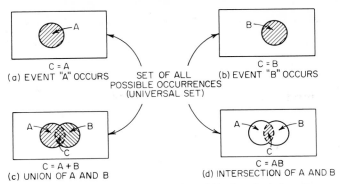

FIG. 69. Venn-diagram representation of Boolean logic functions.

AB = C		
A	B	C
0	0	0
0	1	0
1	0	0
1	1	1

(a) "AND" OPERATION

A+B = C		
A	B	C
0	0	0
0	1	1
1	0	1
1	1	1

(b) "OR" OPERATION

FIG. 70. Truth tables for OR and AND operations.

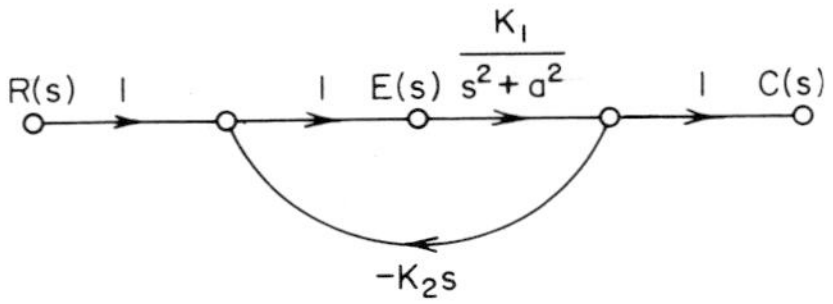

FIG. 71. Second-order control system.

absolute-error criterion (MTWAE) (cf. Table 11, Error Criteria). This means that $\int_0^\infty t|e(t)|\,dt$ is to be minimized when the system is driven by a step input. The transfer function of the closed-loop system is given by

$$\frac{C(s)}{R(s)} = \frac{K_1/(s^2+1)}{1 + K_1 K_2 s/(s^2+1)}$$

$$= \frac{K_1}{s^2 + K_1 K_2 s + 1} \quad (116)$$

By comparing Eq. (110) with Eq. (27), it is seen that $\omega_n = 1$ and $\zeta = (K_1 K_2)/2$. A plot of the MTWAE criterion vs. ζ is shown in Fig. 72.

Since the minimum value of MTWAE occurs for a $\zeta = 0.7$, adjustment of

$$K_1 K_2 = 1.4$$

would be optimum. Unfortunately K_2 varies over the range $2 \le 10 \le 50$ because

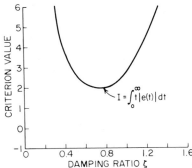

FIG. 72. Error criteria as a function of damping ratio. (*Taken from Ref. 4.*)

of environmental factors. (Such large variations in dynamics are not uncommon in the process-control industry where a chemical being mixed, pumped, or stirred changes viscosity; also large variations in aircraft transfer functions occur when an aircraft goes from sea level to high altitudes.) In order to keep the loop gain constant, the feedback gain K_2 is varied as gain K_1 varies, by adjusting a potentiometer with a servomotor. An analog-computation circuit is used to compute $I = \int_0^\infty t|e|\,dt$;

Table 21. Truth Table for Sensing Minimum of Curve in Fig. 72

ΔI	$\Delta \zeta$	$\dfrac{\Delta I}{\Delta \zeta}$	Operating condition	Proper action
0	0	0	Neither K_1 nor K_2 is changing	Off
1	0	0	Logically impossible	
1	1	0	Logically impossible	
1	1	1	Both K_1 and K_2 are changing	Drive motor
0	1	1	Logically impossible	
0	0	1	Neither K_1 nor K_2 is changing	Off
0	1	0	System at minimum of curve	Off
1	0	1	Input amplitude is changing; K_1 is constant	Off

Table 22. Basic Logic Circuits

Function	Truth table	Relay circuit	Electronic circuit
AND gate	$xy = w$ — truth table: (0,0)=0, (0,1)=0, (1,0)=0, (1,1)=1		
OR gate	$(x+y) = w$ — truth table: (0,0)=0, (0,1)=1, (1,0)=1, (1,1)=1		
NOT gate	$\bar{x} = w$ — truth table: 0→1, 1→0		
Comparison	$w = \begin{cases} 1; & x > y \\ 0; & x < y \end{cases}$ $w' = \begin{cases} 0; & x > y \\ 1; & x < y \end{cases}$		

the gain K_1 is proportional to an external parameter which can be measured. By measuring ΔI, $\Delta \zeta$, and $\Delta I / \Delta \zeta$, digital logic can be used to sense the minimum of the curve. A "truth table" describing the possible combinations is shown in Table 21.

Letting $A \equiv \Delta I$, $B \equiv \Delta \zeta$, and $C \equiv \Delta I / \Delta \zeta$, the logic expression corresponding to the truth table is written $ABC = D$. This means that D (motor-drive indication) is present only when A and B and C are all present. The motor-drive direction is determined by the polarity of ΔI, which indicates whether ζ is too small or too large. Logical elements, or as they are sometimes called "gates," can be physically realized with vacuum tubes, transistors, relays, diodes, or saturable magnetic cores. Some illustrative circuits are shown in Table 22.

An operational diagram for the error-sensing system described is shown in Fig. 73.

A simple relay and diode realization of the digital-logic portion of the diagram shown in Fig. 73 is given in Fig. 74. The positive and negative motor drives are easily realized as shown.

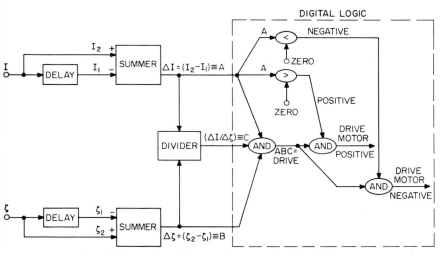

Fig. 73. Operational diagram of an error-sensing detector.

This example was chosen to illustrate the fact that digital principles have many applications in the control field that utilize equipment far simpler than a large-scale digital computer and possess many important features. Such devices as relays, thyratrons, diodes, switching transistors, controlled rectifiers, and many others allow very large amounts of power to be controlled by comparatively small devices. For example, an ideal diode (switch) absorbs no power since no reverse current flows in the off state and no forward voltage drop is present in the on state.

13 Pure Time Delay

The classical and well-established theory for the analysis and design of feedback-control systems is predicated on the assumption that the various system components are described by transfer functions which are ratios of polynomials in the complex frequency s (or by the corresponding linear, constant-coefficient differential equations). The additions of sampling or nonlinearity or time variation of the parameters represent three significant extensions of this classical theory. The presence of transcendental factors (e.g., ϵ^{-Ts} or $\epsilon^{-a\sqrt{s}}$) in the transfer function provides an additional dimension to control theory.

If the system contains a pure time delay of T sec, the transfer function includes the multiplicative factor ϵ^{-Ts} or, if $s = j\omega$, $\epsilon^{-j\omega T}$. Such a transfer characteristic possesses, for sinusoidal frequencies, a magnitude of unity and a phase lag of ωT radians:

$$G_d(j\omega) = \epsilon^{-j\omega T} = 1/\!-\!\omega T \tag{117}$$

In an open-loop system, the presence of the $G_d(s)$ factor causes no difficulty, since the time function corresponding to $G(s)\epsilon^{-Ts}$ is just the time function for $G(s)$ with an

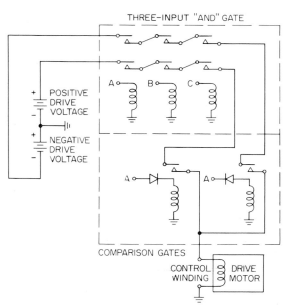

Fig. 74. A relay and diode circuit realizing the digital logic shown in Fig. 73.

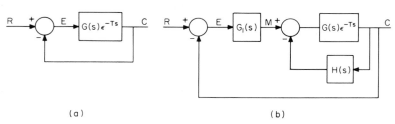

(a)   (b)

Fig. 75. Feedback systems with time delay. (a) Single-loop system. (b) Multiloop system.

extra time delay of T sec. When, however, feedback exists around the delay factor (Fig. 75), the analysis and design difficulties are much greater than for the corresponding system without delay, since the phase shift corresponding to the delay affects the system stability characteristics.

For example, the dynamic characteristics of Fig. 75a depend on the gain and phase of the forward transfer function $G(s)\epsilon^{-Ts}$ or the root loci associated with this factor. The complication is even greater in the case of Fig. 75b: here the design problem of selecting $G_1(s)$ to assure satisfactory system performance is confused by the nature of the overall forward transfer function

$$G_f(s) = \frac{G_1(s)G(s)\epsilon^{-Ts}}{1 + G(s)H(s)\epsilon^{-Ts}} \tag{118}$$

$G_f(s)$ is the ratio of two transcendental functions involving both polynomials and exponentials in s.

Sources of Time Delay

The pure time delay (also termed the transportation lag) represented by ϵ^{-Ts} arises whenever the system includes a transportation element or a communication link. For example, in a typical process-control system or automatic manufacturing process, the material undergoing processing must be moved from point to point within the plant between successive dynamical operations. Alternatively, in a control system involving a radio or telemetry link, the finite velocity of propagation of the signal energy results in a time delay which can be significant because of either the distances involved (in space guidance, for example) or the low speeds of propagation (in pipeline flow control, where the electrical signals are transmitted along wires).

In addition, a variety of electronic and data-processing equipment is described approximately in terms of a pure time delay. Demodulators, modulators, and amplifiers involving the modulation process (such as magnetic amplifiers) typically possess response times which are nearly independent of the signal level. For example, the response time may be one-half of the period of the carrier signal over the entire usable range of the control signal; in such a case, the element is approximately described by the linear model

$$G(s) = \epsilon^{-(T/4)s} \tag{119}$$

where T is the carrier period.

Furthermore, control systems of reasonable complexity are increasingly employing special-purpose computers for the controller elements. When the computer involves digital techniques (either as a digital computer or as a hybrid analog-digital device), computation time is customarily represented as by a pure delay which is an average of the delays for the various pertinent operations. Even when the controller is an analog computer, analysis and design are frequently simplified if the high-order transfer function is replaced by a low-order approximating transfer function coupled with pure delay. Finally, when the control system involves a human operator, the inherent delay of human response (usually increasing upward from 0.2 sec with complex tasks, aged people, etc.) must be included in the model of system performance.

Thus time delay is characteristic of telemetry and remote-control applications of feedback principles because of both the propagation delays and the delays associated with the complexity of modern control systems.

Analysis and Design Techniques with Delay

The block diagram of a typical simple system including a transportation lag is shown in Fig. 75a. The overall system function is

$$\frac{C(s)}{R(s)} = \frac{G(s)e^{-Ts}}{1 + G(s)e^{-Ts}} \tag{120}$$

As a result of the numerator exponential, there is a direct lag of T sec between input and output. In addition, the closed-loop performance of the system is affected by the lag because of the factor e^{-Ts} in the denominator. For example, the stability of such a system is modified by the presence of this factor.

In any analytical stability analysis, the transcendental transfer function has classically been considered by approximating the exponential by a rational algebraic function, after which the usual Routh criterion, root-locus methods, etc., can be applied. A number of approximations have been used.

1. The exponential function can be expressed by the limit

$$e^{-Ts} = \lim_{n \to \infty} \left(\frac{1}{1 + Ts/n}\right)^n \tag{121}$$

If a finite value of n is used, the exponential function is approximated by a pole of order n located at $-n/T$ on the negative real axis in the s plane. For example, an n of 3 yields

$$e^{-Ts} \cong \left(\frac{1}{1 + Ts/3}\right)^3 \tag{122}$$

The corresponding impulse response of the actual function and the approximation are sketched in Fig. 76. The approximation is not particularly good, with the maximum value of the impulse response occurring at $2T/3$.

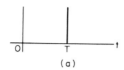
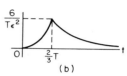

Fig. 76. Impulse responses. (a) Ideal response. (b) Response of approximation.

2. The exponential function can be approximated by the first few terms of the Maclaurin series. The series for either the positive or the negative exponential may be used:

$$e^{-Ts} = 1 - Ts + \frac{T^2s^2}{2!} - \frac{T^3s^3}{3!} + \cdots \tag{123}$$

$$e^{-Ts} = \frac{1}{1 + Ts + T^2s^2/2! + T^3s^3/3! + \cdots} \tag{124}$$

3. A somewhat better approximation can be achieved without an increase in the complexity of the analytical expression for the open-loop transfer function by the use of a rational algebraic function with both numerator and denominator different from unity. The Pade table for e^{-Ts} furnishes a particularly simple algebraic function.[34] The Pade approximation is the rational algebraic function, with numerator polynomial of degree n and denominator of degree m, such that the maximum number of terms in the Maclaurin expansion of the approximating function agree with similar terms in the expansion of the exponential function. In other words, if e^{-Ts} is to be approximated by the ratio of cubic to quadratic polynomials, there are six coefficients which can be selected arbitrarily:

$$e^{-Ts} \cong \frac{1 + a_1s + a_2s^2 + a_3s^3}{b_0 + b_1s + b_2s^2} \tag{125}$$

These six coefficients can be chosen such that at least the first six terms are equal in the two Maclaurin expansions. In this specific example, the appropriate rational algebraic function is

$$e^{-Ts} \cong \frac{1 - \tfrac{3}{5}Ts + \tfrac{3}{20}T^2s^2 - \tfrac{1}{60}T^3s^3}{1 + \tfrac{2}{5}Ts + \tfrac{1}{20}T^2s^2} \tag{126}$$

There are two primary disadvantages of an analytical approach as outlined:
1. The required accuracy of approximation can be determined only after the approximate expression has been used in the analysis.
2. Any satisfactory approximation is ordinarily moderately complicated. The degree of the polynomials involved in the stability analysis rapidly becomes onerous, particularly if the transfer function of the rest of the system is reasonably complex.

Because of these disadvantages, a graphical procedure ordinarily provides a much simpler attack on problems involving transportation lags. The transfer function $e^{-j\omega T}$ is readily interpreted in terms of either the Nyquist diagram or the Bode plots without the necessity of any approximation.

In either the Nyquist diagram or the logarithmic plots, multiplication of a transfer function by $e^{-j\omega T}$ represents merely a phase shift varying linearly with frequency. In terms of the Nyquist diagram, each point on the diagram is rotated through an angle of $-\omega T$ radians, where ω is the angular frequency corresponding to the point on the original locus. Figure 77 shows the Nyquist diagram of the simple system with

$$G(s) = K/[s(s + 1)]$$

Inclusion of the e^{-Ts} factor (with T equal to 0.5 and 1 sec) changes the diagram to the forms shown in curves b and c, respectively. The plot spirals inward toward the origin as the frequency increases toward infinity since the added phase shift, $-\omega T$ radians, increases without bound. The form of the diagram indicates that, as either

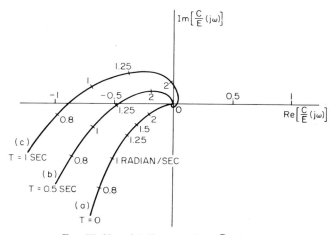

Fig. 77. Nyquist diagrams for $\epsilon^{-Ts}/s(s+1)$.

the gain K or the transportation lag is increased, more and more poles of the closed-loop system function move into the right half plane, as evidenced by the increase in the number of encirclements of the -1 point in the $(C/E)(j\omega)$ plane.

The logarithmic gain and phase plots permit even simpler analysis of the closed-loop system with a transportation lag. The gain curve is unchanged by the introduction of the lag factor, but the phase lag is increased proportional to frequency. Figure 78 presents the change caused by the introduction of the e^{-Ts} factor in the $K/[s(s+1)]$ transfer function considered previously. The gain K allowable if the system is to be stable is rapidly determined. The gain of the $1/[s(s+1)]$ plot at the frequency at which the phase shift of the total open-loop transfer function is $-180°$ is -1.1 db if T is 1 sec. K can then be as high as $+1.1$ db, or 1.14.

The very desirable characteristic of this approach through logarithmic plots to the problem of a transportation lag is that the situation is not particularly complicated by increased complexity in the rest of the open-loop transfer function. The problem with a transportation lag is essentially no more difficult than the analysis of the same system without the lag.

Attempts to apply root-locus techniques to systems with transportation lags generate very real difficulties similar to those experienced with sampled-data systems. When the control system includes a sampler as well as the time delay, analysis is simple if the time delay is an integral multiple of the sampling period (or if we can find a hypothetical sampling period which is a low-order submultiple of both the sampling period and the time delay). Under such circumstances the delay is represented by a z transfer function of the form $1/z^n$, and analysis proceeds in the same way as with the conventional sampled-data system. The simplicity of the sampled-data system can be exploited in the analysis of continuous systems if we purposely introduce a hypothetical sampler, with a sampling frequency sufficiently high to avoid significant distortion of the output (i.e., f_S should be chosen eight times the system bandwidth or significantly higher than the significant signal energy).

When the system with time delay also includes a nonlinearity,[36] the conventional analysis techniques for nonlinear systems are directly applicable, with only slightly more complexity in the determination of phase portraits for low-order systems.

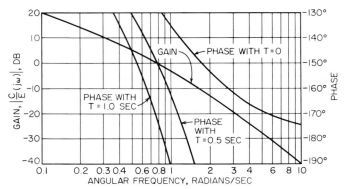

Fig. 78. Gain and phase curves for $\epsilon^{-Ts}/s(s+1)$.

Simulation

Since time delay conventionally arises in the characterization of complex control systems, the primary interest in analysis and design resides in analog-computer simulations of the delay.[37,38] An appropriate basis for simulation depends markedly upon the magnitude of the time delay and the bandwidth required for the device representing the delay. If the delay is less than a second, and the bandwidth is less than the order of magnitude of the reciprocal of the delay time, electric networks or computer programs based upon the algebraic transfer-function approximations described above are useful.

For longer delays or greater bandwidths, the signal can be recorded on magnetic tape or on a magnetic drum. The delayed signal is then obtained from a subsequent readback of the recorded signal, with the delay time adjustable (in the case of tape recording) simply by control of the mechanical position of the recording and reading heads. The primary disadvantage of the recording system is the requirement that the signal be modulated, with the associated problems of controlling distortion. Alternate techniques for realizing long time delays in an analog-computer simulation involve plotting and tracing (electrically or photoelectrically).

REFERENCES

Feedback

1. John G. Truxal, "Automatic Feedback Control System Synthesis," chap. 2, McGraw-Hill Book Company, New York, 1955.
2. H. W. Bode, "Network Analysis and Feedback Amplifier Design," D. Van Nostrand Company, Inc., Princeton, N.J., 1945.

Transfer Functions and Functional Models
3. John A. Aseltine, "Transform Method in Linear System Analysis," McGraw-Hill Book Company, New York, 1958.

Feedback Model Manipulation
4. C. J. Savant, Jr., "Basic Feedback Control System Design," chap. 1, McGraw-Hill Book Company, New York, 1958.
5. Sam J. Mason and Henry J. Zimmermann, "Electronic Circuits, Signals, and Systems," John Wiley & Sons, Inc., New York, 1960.
6. Ref. 1.
7. John N. Warfield, "Electronic Analog Computers," chap. 4, Prentice-Hall, Inc., Englewood Cliffs, N.J., 1959.

Uses of Feedback
8. Ref. 1.
9. John G. Truxal, "Control Engineers' Handbook," pp. 4-43 to 4-60, McGraw-Hill Book Company, New York, 1958.

Routh-Hurwitz Criteria
10. E. A. Guillemin, "The Mathematics of Circuit Analysis," chap. VI, art. 26, John Wiley & Sons, Inc., New York, 1949.

Nyquist Criterion
11. H. Chestnut and R. W. Mayer, "Servomechanisms and Regulating System Design," vol. I, John Wiley & Sons, Inc., New York, 1951.

Root-locus Technique
12. Ref. 1, chap. 4.

Frequency Response and Compensation Techniques
13. Ref. 4, chaps. 5, 6.

Error Criteria
14. D. Graham and R. C. Lathrop, Optimum Transient Response: Criteria and Standard Forms, *Trans. AIEE*, pt. 2, vol. 72, pp. 273–285, November, 1953.
15. G. Newton, L. Gould, and J. Kaiser, "Analytical Design of Linear Feedback Controls," John Wiley & Sons, Inc., New York, 1957.

Sampled-data Systems
16. Refs. 1 and 9.
17. Eli Mishkin and Ludwig Braun, "Adaptive Control Systems," McGraw-Hill Book Company, New York, 1961.
18. John R. Ragazzini and Gene F. Franklin, "Sampled-data Control Systems," McGraw-Hill Book Company, New York, 1958.
19. E. I. Jury, "Sampled-data Control Systems," John Wiley & Sons, Inc., New York, 1958.
20. J. T. Tou, "Digital and Sampled-data Control Systems," McGraw-Hill Book Company, New York, 1959.

Nonlinear Systems
21. Refs. 1 and 9.
22. Ref. 4.
23. Ref. 11, vol. II.
24. F. Hildebrand, "Advanced Calculus for Engineers," Prentice-Hall, Inc., Englewood Cliffs, N.J., 1950.
25. F. Hildebrand, "Methods of Applied Mathematics," Prentice-Hall, Inc., 1952.

Analog Computers
26. Ref. 7.
27. C. L. Johnson, "Analog Computer Techniques," 2d ed., McGraw-Hill Book Company, New York, 1963.
28. Paynter, "Palimpsest on the Electronic Analog Art," Geo. Philbrick Researches, Inc., Boston, 1955.

Digital Computation
29. Ref. 9.
30. R. Richards, "Arithmetic Operations in Digital Computers," D. Van Nostrand Company, Inc., Princeton, N.J., 1955.

Computers in Control
31. Ref. 9.
32. Electronic Computers, *IRE Natl. Conv. Record*, pt. 4, New York, 1958.
33. Electronic Computers, *IRE Wescon Conv. Record*, pt. 4, 1958.

Time Lags
34. Ref. 1, pp. 546–557.
35. Y. Chu, Feedback Control Systems with Dead Time Lag or Distributed Lag by Root locus Method, *AIEE Trans.*, vol. 71II, pp. 291–296, November 1952.
36. J. C. Gille, P. Decaulne, and M. Pelegrin, "Methodes modernes d'étude des systemes asservis," pp. 228–229, Dunod, Paris, 1960.
37. N. R. Scott, "Analog and Digital Computer Technology," pp. 112–118, McGraw-Hill Book Company, New York, 1960.
38. J. M. L. Janssen, Discontinuous Low-frequency Delay Line with Continuously Variable Delay, *Nature*, Jan. 26, 1952, p. 148.

COMPONENTS

14 Accelerometers

Modern guidance systems generally include and sometimes rely entirely on device which can determine vehicle acceleration without depending upon external measurements. The basic device in such a self-contained system is the accelerometer—an instrument which senses vehicle accelerations by measuring acceleration reaction forces. The energy to drive the accelerometer is derived entirely from the motion of the vehicle so that no external sources (such as radiation, electrical or otherwise are necessary. The acceleration intelligence thus derived can be used to measure vehicle position or can be used in the guidance-control system to provide so-called second-derivative damping. When used to provide positional data, accelerometer must have a high degree of accuracy as the positional data are derived from a continuous double integration of the accelerometer signal. Any small accelerometer error thus causes a positional error which builds up according to the second power of time.

In the following paragraphs, the basic theory of accelerometers is discussed. Some basic types of accelerometers and construction features are briefly described. Special types of accelerometers are then mentioned. The final paragraphs deal with responses performance specifications, and accelerometer testing.

Elementary Theory

The basic accelerometer consists of a simple mass which is constrained to move linearly and which is coupled to the vehicle framework through an elastic member (see Fig. 79). The theory of operation is briefly explained as follows: When the vehicle and accelerometer mass are both traveling at the same constant velocity (no acceleration), there is no relative motion between them and the mass experiences no unbalanced force (the spring forces balance each other). When the vehicle accelerates, its velocity increases. The mass, however, tends to maintain its constant velocity by virtue of its inertia. There is, therefore, a relative motion between the vehicle and mass. The springs are thus deflected and an unbalanced force is applied to the mass. This unbalanced force on the mass causes it to accelerate. Ultimately (assuming for simplicity a constant vehicle acceleration) a steady state is reached where the spring force produces an acceleration of the mass equal to the vehicle acceleration. By measuring this accelerating force on the known mass, the vehicle acceleration can be determined.

From Newton's first law,

$$a = f/m \qquad (127)$$

where f = accelerating force exerted by the spring
m = accelerometer mass
a = vehicle acceleration

For a linear spring, the spring force is proportional to the spring deflection:

$$f = Kx \qquad (128)$$

where K = spring constant
x = spring deflection

Equation (128) inserted in Eq. (127) yields

$$a = (K/m)x \qquad (129)$$

Thus the vehicle acceleration can be determined if the spring deflection is measured, i.e., if the relative displacement between the vehicle and mass is measured. In order that the instrument may respond faithfully to rapid changes in vehicle accelerations, the spring should be rather stiff (large K) and the mass small. The accelerometer mass can then be rapidly accelerated so as to match the vehicle acceleration. With a stiff spring and small mass, however, the spring deflection will be rather small. For very accurate acceleration measurements, very sensitive displacement-measuring devices are therefore required. Some common means used to pick up these small displacements are mentioned below in the discussion of accelerometer construction.

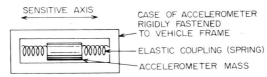

Fig. 79. Schematic diagram of accelerometer.

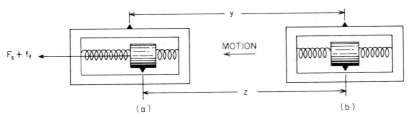

Fig. 80. Forces on accelerometer mass. (a) After vehicle acceleration (b) Before vehicle acceleration.

While the above discussion does explain the basic operating mechanism of the accelerometer, it is an oversimplified treatment. A more thorough analysis requires that the dynamic differential equation be written for the accelerometer mass. This equation may be derived as follows:* By d'Alembert's principle,

$$\Sigma f - \text{inertial force} = 0 \qquad (130)$$

where Σf is the summation of externally applied forces on the accelerometer mass. The external forces acting on the mass (refer to Fig. 80) are the spring force f_s and the friction force f_f. The inertial force of the mass is $m\ddot{z}$ where z is the displacement of the mass referred to inertial space and the dots indicate the second time derivative of z. Therefore,

$$f_s + f_f - m\ddot{z} = 0 \qquad (131)$$

If the spring is linear, the spring force may be represented by

$$f_s = Kx \qquad (132)$$

where x is the *relative* displacement between vehicle and mass. If viscous friction

* A more complete treatment can be found in almost any good vibrations text, e.g., Ref. 1 or 2.

is assumed, then the frictional force is directly proportional to the *relative* velocity between the vehicle and mass. If B is the viscous-friction proportionality constant then

$$f_f = B\dot{x} \tag{133}$$

Relations (132) and (133) when used in (131) yield

$$Kx + B\dot{x} - m\ddot{z} = 0 \tag{134}$$

Finally, if the relative-displacement equation

$$y - z = x \text{ or } y - x = z \tag{135}$$

is used to replace z in Eq. (134), the result is

$$m\ddot{x} + B\dot{x} + Kx = m\ddot{y} \tag{136}$$

or

$$\ddot{x} + (B/m)\dot{x} + (K/m)x = \ddot{y} \tag{137}$$

where $\ddot{y}$ is the vehicle acceleration and all other terms are as previously defined. Equation (129) previously derived is actually the steady-state solution of the above differential equation (137) when $\ddot{y}$ is a constant. Thus, when $\ddot{y}$ equals a and the steady state is considered, the time derivatives in Eq. (137) drop to zero and Eq. (137) reduces to Eq. (129).

The transfer function of the accelerometer can be obtained directly from Eq. (137). If zero initial conditions are assumed then,

$$\frac{x}{\ddot{y}}(s) = \frac{1}{s^2 + (B/m)s + K/m} \tag{138}$$

Ideally, the relative displacement x of the accelerometer mass should at all times be proportional to the input acceleration. Equation (137) [or Eq. (138)], however, indicates that x will not respond instantaneously to give a reading for $\ddot{y}$. Instead, the relative displacement of the mass, may undergo a sustained transient oscillation depending upon the value of B/m. To examine the situation quantitatively the roots of the characteristic equation [of Eq. (137)] or the poles of Eq. (138) are examined

$$s^2 + (B/)ms + K/m = 0 \tag{139}$$

$$s = -B/2m \pm \sqrt{(B/2m)^2 - K/m} \tag{140}$$

The x response will be oscillatory if the roots (or poles) are complex. The critical value for B/m therefore is

$$B/m = 2\sqrt{K/m} \tag{141}$$

$$B/2\sqrt{Km} = 1 \tag{142}$$

The ratio $B/2\sqrt{Km}$ is often called the damping factor ζ of the system. From the above discussion, if $\zeta > 1$ there will be no oscillation; if $\zeta < 1$ there will be an oscillation. It might appear the $\zeta > 1$ would be desirable to preclude the possibility of oscillation. However, with such heavy damping the instrument will be sluggish and will take a long time to reach a final value. On the other hand, if ζ is too low (say below 0.1) the transient oscillation will persist for a long time and again the instrument will not settle down to a final reading in a reasonable time.* The optimum value for ζ depends, in general, upon the input acceleration, but in most cases the acceleration is not known. A compromise value for ζ must therefore ordinarily be used. As an overall practical value, ζ is generally adjusted to about 0.7.

To adjust ζ, the viscous friction in the system (i.e., the viscous damping coefficient B) is generally varied. Variable friction is incorporated into the instrument by both electrical and mechanical means. A more complete description of how variable damping is achieved will be deferred until accelerometer construction is discussed

* The accelerometer response is that of a general second order system. A brief discussion of second-order systems is given in a later section.

Construction Features

Elastic Coupling between Mass and Frame. In many accelerometer designs the elastic coupling is made to serve at least two functions. The prime function is to provide a linear force to accelerate the mass. As a secondary function, the elastic member is ordinarily designed so as also to guide the mass, i.e., to constrain the mass to move only along a linear path. The most common means of coupling is through specially designed mechanical springs some of which are shown in Fig. 81a, b, and c. The compliance of these springs along the sensitive axis is much less than that along any other axis. The mass can therefore move only along the sensitive axis; in any other direction, movement is prevented by the stiffness of the springs. By such design there is no need to guide the mass in bearings. The possibility of inaccuracies due to nonlinear friction is thus reduced. In some cases cantilever and bow-spring suspensions are used primarily to guide the mass rather than act as elastic suspensions. The mass is then elastically coupled to the frame through some other means. Statham

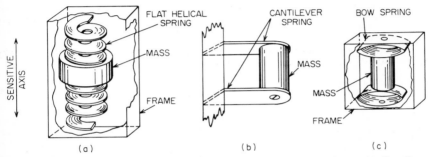

FIG. 81. Elastic suspensions in accelerometer. (a) Flat helical spring. (b) Cantilever suspension. (c) Bow spring.

Laboratories, for example, suspends the mass by strain-gage wires (see Fig. 82). The elasticity of the wires serves as an elastic suspension and at the same time the wires can be used to transduce the displacements of the mass into an electrical signal. Strain-gage transducers will again be mentioned when displacement-measuring devices are discussed.

While magnetic and electric fields have been used in the laboratory for suspending masses [3] this writer knows of no company which has utilized this effect in a commercial accelerometer. The force-balance accelerometer does use a magnetic field instead of a spring to provide the accelerating force. However, the principle of operation in this case is such that it could not properly be called a magnetic suspension. The force-balance accelerometer (sometimes referred to as servo-operated or feedback accelerometers) involves a closed-loop feedback system and is separately discussed later.

Displacement Transducers. In most applications, information obtained from accelerometer measurements is alternately fed into electrical computing equipment—integrators, analog computers, etc. The intelligence derived from the accelerometer is thus most conveniently handled if it is in the form of an electrical signal. In most accelerometer designs, therefore, some displacement-to-voltage transducer is generally included. The most familiar transducer of this kind is the simple potentiometer. A schematic arrangement of an accelerometer with a potentiometer included is shown in Fig. 83. Depending upon the accuracy required, the potentiometer can be of wirewound or slide-wire type. The latter will in general be more accurate when small accelerations are to be measured, as there is essentially no problem of resolution. In either case, however, the problem of stiction is present because of mechanical contact between slider and wire. Stiction is a nonlinear friction force and, if appreciable, will impair the accuracy of instrument and also increase the threshold of the unit (minimum input signal needed to produce a detectable output).

Instead of a potentiometer, a differential transformer is often used to transduce mechanical motion into electrical voltage. This device operates through magnetic rather than mechanical coupling so that stiction is virtually eliminated. Figure 84 shows a schematic diagram of an accelerometer which includes a differential transformer as a transducer. The primary coil C of the transformer is energized with alternating current. When the core is centered, equal and opposite voltages are induced in coils A and B (these are oppositely wound). When the core is centered, therefore, the voltages across both coils cancel and no signal appears across the output. When the core is moved off center (because of, say, an applied acceleration)

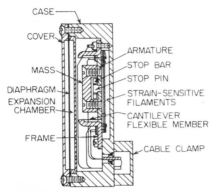

Fig. 82. Model A5A accelerometer, Statham Laboratories.

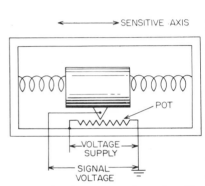

Fig. 83. Accelerometer schematic with a potentiometer for displacement to voltage transduction.

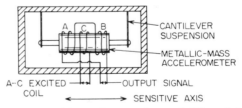

Fig. 84. Accelerometer with cantilever springs and differential transformer.

the system becomes unbalanced and a signal appears across the output terminals. If the transformer is designed properly, the output voltage will vary linearly with the core displacement over a limited range. The output voltage is thus essentially a measure of the applied acceleration. In that there is no mechanical contact between the core and coils, stiction and sliding friction are eliminated. The operating threshold of the accelerometer in this case is limited by electrical noise.

Another type of stictionless transducer often used in accelerometers is the strain gage. Figure 85a shows a schematic diagram of a bonded-type strain gage. The gage consists of filamentary wires cemented to a paper or cardboard back. A force applied across the paper (as, for example, F-F in Fig. 85a) stretches the wire filaments so that the wire length increases while the wire diameter decreases. These changes in physical dimensions are reflected by a change in electrical resistance. For small wire strains, the per cent resistance change is proportional to the strain. Figure 85 shows how a bonded strain gage may be used as a transducer in an accelerometer

The strain gage is cemented to the cantilever support near the base so that strains in the cantilever leaf are transmitted directly to the gage. When the accelerometer mass is moved (as when an acceleration signal is applied) the cantilever spring is deflected and a strain is produced on the surface of the cantilever spring which is proportional to the deflection. This strain is transmitted to the gage so that the gage resistance is changed. This resistance change is sensed by a Wheatstone bridge. When properly designed, the bridge voltage output will be proportional to the core displacement and so the bridge voltage can be used to measure the applied acceleration.

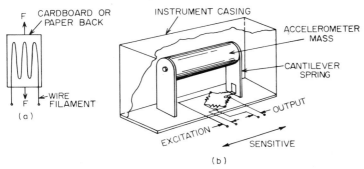

Fig. 85. Accelerometer with strain-gage transducers. (a) Bonded strain gage. (b) Accelerometer with bonded-strain-gage transducer.

To increase the sensitivity of the system the Wheatstone bridge can include two strain gages which are placed on opposite faces of the cantilever spring.

Strain gages are temperature-sensitive so that erroneous results are possible unless temperature regulation or compensation is used. Regulation has the advantage in that variations in damping will also be eliminated under constant-temperature conditions. However, compensation is also frequently used and is accomplished simply by using an unstressed strain gage in one of the legs of the bridge. Resistance changes due to temperature in one leg are thus compensated for by comparable resistance changes in another leg. Temperature changes alone will therefore not unbalance the bridge.

When bonded strain gages are used, the gage is used only as a detection device. In some accelerometer designs unbonded strain gages are used. The gages then serve as both an elastic suspension and a detection device. Unbonded strain gages are simply taut filamentary wires strung between two posts. The posts are arranged so that one post can move relative to the other. Figure 86 shows an accelerometer in which unbonded strain gages are used. The strain-gage wires couple the accelerometer mass to the vehicle frame and serve as a stiff spring as well as a transducer. The cantilever springs are included merely to guide the mass. The compliance of the cantilever springs along the sensitive axis is very much less than that of the wire filaments. The voltage output is again obtained by using the gage wires as legs in a Wheatstone bridge.

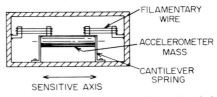

Fig. 86. Accelerometer with unbonded strain gages.

The transducers mentioned above are those which are most commonly used in present-day accelerometers. Other types used are reluctance and capacitance transducers. In these, core displacements cause inductive or capacitive changes which can be picked up by a bridge. Piezoelectric crystals and magnetostrictive materials have been used in some designs (the former more than the latter), but these are generally for special applications.

In general, the signal level of the potentiometer output is high enough to be used directly. This is not ordinarily true of the differential transformer output or the output of the bridge–strain gage combination. When these transducers are used, amplification is required to raise the signal level.

Damping. As pointed out in a previous section, damping in the accelerometer should to some extent be controlled in order that instrument response may be rapid and accurate. For many applications, however, damping is not so critical as may be imagined. A variation in damping factor from about 0.1 to 0.8 can often be tolerated without severely affecting accuracy. When there are no stringent requirements on the damping factor, air metered by an orifice is often used to provide the necessary damping. Figure 87 schematically shows one possible arrangement for an air-damped system. The size of the orifice can be varied to adjust the damping. It would be expected that the damping obtained in this manner would be nonlinear (i.e., the friction force is not linearly related to velocity) and variable. The claims of one manufacturer, however, are that an air system provides a more stable damping factor as the characteristics of air are not severely altered by temperature changes.

In order to achieve a more linear damping coefficient, the accelerometer is often filled with an organo-silicone oil instead of air. Figure 88 is a schematic diagram of

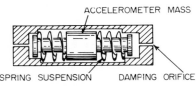

Fig. 87. Air-damped accelerometer.

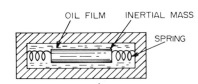

Fig. 88. Accelerometer with oil damping.

an accelerometer using oil damping. The space between the accelerometer mass and the instrument case is generally very narrow (in the order of thousandths of an inch) so that an oil film essentially surrounds the mass. The basic premise is that Newton's viscous-shear law for fluids is applicable, i.e., the shear stress generated in an oil film is proportional to the velocity gradient across the oil film. For the construction shown in Fig. 88 the shear area of the oil and the film thickness are essentially constant so that the shear force opposing the motion of the mass is proportional to the velocity of the mass. If the shear force in the oil were the only velocity-generated force opposing the mass, then linear damping would be closely approximated. However motion of the mass is also opposed by pressure gradients in the oil produced by the motion of the mass so that truly viscous friction is not realized. The pressure effect can be reduced although not eliminated by using a hollow cylinder on a thin flat plate as the accelerometer mass.

The viscous frictional force that can be derived from an oil is directly proportional to the viscosity of the oil. Unfortunately, oil viscosity is highly temperature sensitive. In order to obtain a constant damping coefficient B, the oil temperature must be controlled or some means of compensation must be used. Temperature control can be simply arranged by enclosing the accelerometer in a heating jacket and controlling the jacket with a thermostat. Compensation devices for controlling B are many and varied. Most involve some means to vary the oil film thickness such that the film thickness is decreased when the temperature is increased. In one design advantage is taken of the dissimilar expansion rates of two different metals. If the core (inertial mass) expands at a faster rate than the instrument case, the clearance between the two decreases with increasing temperature. Hence the film thickness decreases, thus offsetting the lowered oil viscosity. Other designs involve an air-bellows arrangement whereby the air expansion in the bellows causes a decrease in the clearance between the core and the casing. In general, compensation devices are hard to adjust and are somewhat complicated. Temperature control ordinarily proves to be more practicable.

Viscous frictional damping can also be achieved through electrical rather than mechanical means. Such damping is often called "eddy-current damping." The "viscous" force in this case results from the interaction between two magnetic fields, one fixed field and the other generated by eddy currents. Figure 89 shows a schematic diagram of an accelerometer which employs eddy-current damping. The accelerometer mass is guided so that it (or fins on it) passes between the pole pieces of magnets. When the mass moves in the magnetic field eddy currents are generated within the mass. These currents give rise to a magnetic field in opposition to the exciting field. There is thus generated a repulsive force (force-opposing motion) on the mass. The greater the velocity of the mass the greater are the eddy currents generated and the greater is the repulsive force. The relation between the force generated and the velocity of mass is very nearly linear, and hence the force can be considered a viscous-friction force.

While eddy-current damping is also temperature-sensitive, it is not so highly sensitive as is oil damping. The system is, however, somewhat bulky it requires rather large magnetic fields to get a ζ of about 0.5. Both permanent and electromagnets are used to provide the necessary fields. Permanent magnets are more efficient when space and weight are considerations but the field strengths available in commercial magnets are in general not adequate to provide sufficient damping.

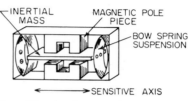

Fig. 89. Accelerometer with eddy-current damping.

Force-balance Accelerometers

Force-balance accelerometers were first used by the Germans in their V-2 rocket. They were developed to overcome some of the inherent inaccuracies of the simple spring-mass system. The spring, which is a large source of errors (spring nonlin-

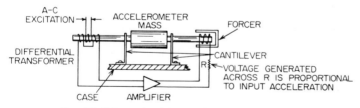

Fig. 90. Schematic of force-balance accelerometer.

arities, hysteresis, etc.), is removed and instead the accelerating force is produced by an electrical "forcer." Figure 90 shows a schematic drawing of the system. The cantilever springs have a low spring constant and are merely to guide the mass. The mass drives the core of a differential transformed transducer on the one side; on the other side of the mass is the core (a permanent magnet) of an electrical forcer. The forcer is simply a device which generates a mechanical force proportional to an electric current. Movement of the case with respect to the inertial mass (due to an acceleration signal) causes an unbalanced signal across the differential transformer. This signal voltage is amplified and causes a current signal to flow through the coil of the forcer. The forcer generates a force which drives the mass and accelerates it. Ultimately, the generated force is just sufficient to give the mass the same acceleration as the case. This force is now measurable as it is proportional to the current in the forcer. Measurement of this current (or its effect—as the voltage drop it causes across a resistor) is thus a measure of the input acceleration.

A more quantitative treatment of the force-balance accelerometer requires that the differential equation for the system be derived. The analysis follows the same outline as that used for the spring-mass accelerometer. The forces on the mass are frictional and that due to the forcer. The sum of these forces equated to the inertial force gives

$$m\ddot{z} = B\dot{x} + f_{f0} = m(\ddot{y} - \ddot{x}) \qquad (143)$$

where f_{f0} is the force due to the forcer and the other terms are as defined for Eq. (131). The electrical forcer delivers a force proportional to the relative displacement* (i.e., $f_{f0} = K_c x$). Therefore,

$$m\ddot{y} = m\ddot{x} + B\dot{x} + K_c x \qquad (144)$$

A comparison of Eq. (144) and Eq. (136) indicates, as previously stated, the mechanical spring has been replaced by an "electrical spring" of spring constant K_c (where K_c is the product of the transducer gain, the amplifier gain, and the forcer gain). The electrical spring constant is however, conveniently variable (e.g., by varying the amplifier gain) and can be made very high so that the natural frequency of the system can be made advantageously high.

The transfer function of this system has already been derived [see Eq. (138)]. A more complete discussion of the response of a second-order system is presented in a later section.

Equation (144) indicates that a measure of x could be used to measure $\ddot{y}$ (the case acceleration). As in the spring mass, however, a transduced variable is used to measure $\ddot{y}$. An examination of Fig. 90 shows that the measured variable is the voltage across a resistor R in the amplifier output. This voltage output is iR (where i is the forcer current). Use of the proportionality equations given in the preceding footnote shows that this voltage can now be expressed as

$$e_R = K_a K_t R x \qquad (145)$$

where K_t is the transducer gain and K_a is the amplifier gain.

Pendulous Accelerometers

The accelerometer described in the foregoing pages all employed inertial masses which were constrained to move linearly. Instruments are also available in which the accelerometer mass is pendulous, i.e., the mass is suspended from a pivot point. The mass in this case moves in a circular arc rather than a straight line. The elastic coupling is through a torsional elastic member, sometimes in the form of a spring, sometimes in the form of a torsion bar. Figure 91a shows an arrangement where a torsion spring is used; in Fig. 91b a torsion bar is employed.

---

* Actually, the relative displacement causes a proportional voltage across the transformer

$$e_t = K_t x$$

This voltage drives an amplifier which delivers an output current proportional to the input voltage

$$i = K_A e_t$$

The current from the amplifier drives the forcer whose output is

$$f_f = K_f i$$

when these last equations are combined, the result is

$$F_f = K_f K_A K_t x = K_c x$$

COMPONENTS

The differential equations which govern this system are again derived by applying Newton's first law. With reference to Fig. 92a, at $t = 0$ an acceleration is applied to the instrument case. Because of the inertia of the mass, it lags the motion of the pivot as shown in Fig. 92b. The torsion spring therefore undergoes a twist. Spring torque and friction torque are thus generated around the pivot point. These torques

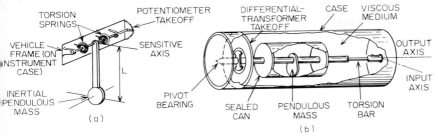

Fig. 91. Pendulous accelerometer. (a) Schematic of pendulous accelerometer. (b) Practical realization of pendulous accelerometer.

are equivalent to the inertial torque which must exist on the mass. This inertial torque can be evaluated as follows: The inertial force on the mass is

$$F_i = m\ddot{z} \tag{146}$$

But
$$z = y - x \tag{147}$$

Therefore,
$$F_i = m\ddot{y} - m\ddot{x} \tag{148}$$

If, now, the angular deflection of the mass is small, then

$$x \approx L\theta \tag{149}$$

and
$$\tau_i \approx F_i L \tag{150}$$

Therefore,
$$\tau_i \approx (m\ddot{y} - m\ddot{x})L = m\ddot{y}L - mL^2\ddot{\theta} \tag{151}$$

Since this inertial torque is equivalent to the applied torques

$$m\ddot{y}L - mL^2\ddot{\theta} = B\dot{\theta} + K\theta \tag{152}$$

or
$$m\ddot{y}L = mL^2\ddot{\theta} + B\dot{\theta} + K\theta \tag{153}$$

where now $\ddot{y}$ is the applied acceleration, θ is the resultant angular displacement, m is the mass of the pendulum, and L is the pendulum length. (Note: mL^2 is the moment of inertia of the pendulous mass around the pivot point.) B and K are the torsional friction and torsional spring constants, respectively.

Equation (153) is of the same general form as Eq. (136). In this case, however, the dependent variable is θ, an angular displacement rather than the translational displacement x. In this case, therefore, θ (an angular displacement) is used to measure the input translational acceleration $\ddot{y}$. As in the "translational" accelerometers, a heavy spring and a light mass should be used to provide accuracy and rapid response. Deflections are thus correspondingly small and sensitive transduction is still required. The advantage of the pendulous accelerometers is the single-point suspension (pivot) which permits easier control of nonlinear friction effects. In fact, by use of the flotation principle,* pivot-point friction is practically eliminated. Figure 91b shows an arrangement in which the flota-

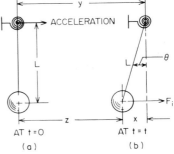

Fig. 92. Inertial reaction of a pendulum.

---
* Used also in gyroscopic instruments.

tion principle is employed. The pendulous mass is enclosed in a hermetically sealed can. The sealed can is then floated in a fluid contained in an outer casing. The densities of the fluid and of the sealed can are adjusted to be approximately equal. The fluid thus essentially supports the full weight of the can. The function of the pivot bearing is therefore no longer for support (i.e., it is no longer subjected to a radial load). Instead, the bearing now serves only to keep the sealed can centered. The frictional torques generated are in this way minimized. Furthermore, in that the sealed can now rotates in a fluid bath, viscous damping is effectively incorporated. The fluid also provides a cushioning effect to protect the instrument from shock or vibration.

Displacement transducers for pendulous accelerometers are of the same general nature as those used in translational designs. Potentiometric takeoffs, strain gages, etc., are some of the means employed. Most common, however, is some variation of the differential transformer (Microsyn is one trade name, though this would more appropriately be called a variable-reluctance device) designed for rotary operation. Basically these devices deliver an output a-c voltage with an amplitude proportional to the input angular displacement of a rotor. A fuller description of this unit is deferred until gyro transducers are discussed.

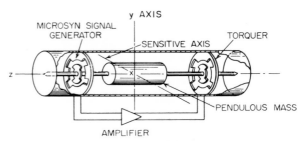

Fig. 93. Force-balance pendulous accelerometer.

The force-balance principle is also employed in pendulous accelerometers. The mechanical torsion spring is removed and replaced by an electrical "torquer." The torquer delivers an output torque proportional to an input current. In other particulars the unit operates in much the same manner as the translational force-balance accelerometer. A schematic diagram of a force-balance pendulous accelerometer is shown in Fig. 93.

Another type of pendulous accelerometer which is often used employs a gyroscopic element. The instrument is in fact called a pendulous gyroscopic accelerometer (PGA). Discussion of this device will, however, be deferred until gyroscopic devices are discussed.

Response of the Accelerometer

The transfer function of the accelerometer given by Eq. (138) indicates that the instrument is a second-order system. The various responses of a second-order system are well known and are available in most texts concerned with control systems or servomechanisms.* Below (Figs. 94a and 94b) are shown the unit-step and impulse responses of a second-order system. Figure 94c shows the frequency response for such a system. From the frequency-response curves, it is noted that the gain is roughly constant for the largest bandwidth when ζ is between about 0.5 and 0.8. In accelerometer design, this would mean that when the ζ of the system is between these limits, x will most nearly measure $\ddot{y}$ for the largest range of input frequencies. For these same values of ζ, it is noted from the step- and impulse-response curves that

* See, for example, Refs. 4 and 5; Ref. 5 contains an excellent plot of the frequency response characteristics included in a pocket inside the cover. (See also pp. 15-41 to 15-44.)

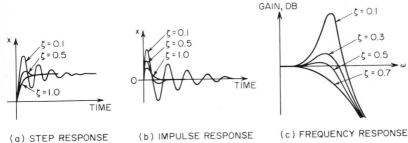

(a) STEP RESPONSE   (b) IMPULSE RESPONSE   (c) FREQUENCY RESPONSE

FIG. 94. Responses for a system with transfer function $\omega_n^2/s^2 + 2\zeta\omega_n + \omega_n^2$.

qualitatively the system responds with fair rapidity and yet does not have an extended transient.

15 Gyroscopic Instruments

In its most elementary form, the gyro is simply a spinning wheel whose spin axis can move relative to some reference mount. The basic constituent of all gyroscopic instruments is thus a rotating wheel (or rotor) with special mounting arrangements to permit limited or complete freedom of the spin axis (see Fig. 95).

Gyroscopic instruments were virtually unheard of until about the beginning of the twentieth century. Prior to that time the gyro was considered only a fascinating toy. Foucault, in the eighteenth century, did use a "spinning wheel" to demonstrate the rotation of earth. However, such practical demonstrations were few and far between until Sperry in the United States and Anshultz-Kaemfe in Germany used a "spinning wheel" to construct some practical navigational instruments. Since that time a tremendous amount of research and development has gone into the field of gyroscopic devices. Today some of the most accurate and reliable navigational devices are those which employ the gyro spinning wheel. The great developments in missile guidance and control are due in great measure to the accuracy and dependability of these gyroscopic instruments.

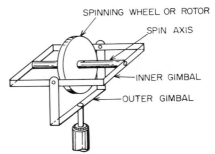

FIG. 95. Schematic of freely mounted spinning wheel.

Gyro devices are not in general "on-the-shelf items," i.e., there are relatively few "standard designs." In most cases gyros are tailored to do special jobs, each job requiring a modification of an existing design, and often an entirely new design must be developed. Because of this variability, many different classifications have developed in the gyro field. In some literature one finds gyros classified according to use. Thus one hears of

1. Instrument gyros—used primarily for measurements and indicating purposes (artificial horizons, turn-bank indicators, gyrocompass, etc.).

2. Control gyros—used to generate signals which in turn control other devices. In some cases such gyros are called master gyros.

3. Stabilizing gyros—These are large units used to generate torques for stabilization purposes. (Sperry attempted to use such a gyro for ship stabilization with limited success; the often discussed monorail railroad would require such a stabilizing device. The stabilizing gyros are often slaved to a master-control gyro.)

Recently gyros have been classified according to both basic construction and function. In terms of construction, the freedom afforded the spin axis of the gyro rotor is often used to separate gyros into two main categories:

1. Single-degree-of-freedom gyros (sometimes called "captive" gyros)—Fig. 96a
2. Two-degree-of-freedom gyros—Fig. 96b

In a single-degree-of-freedom gyro, only one coordinate is necessary to locate the spin axis with reference to the instrument mount; in a two-degree-of-freedom unit, two coordinates are necessary. These terms will be amplified as particular gyros are discussed. These two main categories are further subdivided according to the

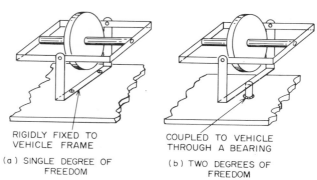

RIGIDLY FIXED TO VEHICLE FRAME

(a) SINGLE DEGREE OF FREEDOM

COUPLED TO VEHICLE THROUGH A BEARING

(b) TWO DEGREES OF FREEDOM

FIG. 96. Single- and two-degree-of-freedom gyros. (a) Single degree of freedom. (b) Two degrees of freedom.

application of signals or the restraints used on the movement of the spin axis. Thus in single-degree-of-freedom gyros, the spin axis may have

    a. Elastic restraint (rate gyros which measure input angular velocities)
    b. Viscous restraint (integrating rate gyros which measure input angular displacements)
    c. Combined elastic and viscous restraints (turn-bank instruments in aircraft)
    d. No restraints (double integrating gyros)

Two-degree-of-freedom gyros may have their spin axes

    a. Completely free except for unavoidable frictional restraints (used to establish reference directions with respect to inertial space)
    b. Supplied with torques for correction or measurement purposes (directional or vertical gyros to maintain directions referenced to the earth, pendulous gyro accelerometers)

The problem of classification will become somewhat more understandable as the types of gyro instruments are examined.

Gyro Rotors. The type and use of the gyro instrument notwithstanding, all gyro devices have one feature in common, namely, a spinning rotor. The operation of the gyro depends wholly on the angular momentum* provided by this spinning wheel. Theoretically, the higher the angular momentum, the better the accuracy of the instrument. However, because of practical considerations in instrument gyros (drift-rate performance† for one thing) exceptionally high angular momentum does

---

* H (angular momentum) = J (rotor moments of inertia)$|$ W (rotor angular velocity). Gyros are often rated by the rotor angular momentum. The common range of values is about 10^3 to about 10^6 g-cm^2 (sometimes rated in dyne-cm-sec).

† Drift rate, when applied to a two-degree-of-freedom gyro, is a measure of the angular rate at which the spin axis moves away from a desired reference direction. In some appli-

not appear to be advantageous. One manufacturer of instrument gyros claims that optimum drift performance is achieved when the angular momentum is about 10^6 or 10^7 g-cm^2/sec. In order to attain momenta of such an order of magnitude in the small space ordinarily allowed for instrument gyros, high spin velocities are necessary. Rotors are generally designed to give the highest moment of inertia feasible in the available space;* however, a spin velocity is then also needed to boost the momentum to the desired level.

There are today three common methods employed to give the rotor a high spin velocity:

1. Pneumatic drives
2. Electric drives
3. Spring drives

Though pneumatic drives are to a large extent being replaced by electrical drives, there are still some cases where the air-driven gyro is quite practical (some torpedoes and aircraft instruments, for example, still use air-driven gyros). The rotor for an air-driven gyro instrument is constructed with blades around the periphery much as a turbine wheel would be made. An air blast passing across the blades spins the rotor. Spin velocities up to about 8,000 rpm are possible with such an arrangement. In most cases the rotor is in a sealed compartment which is continuously evacuated by a vacuum pump. Atmospheric air entering the compartment from a jet passes air across the vanes, and this air jet drives the rotor. Such an arrangement lends itself to the use of pneumatic control devices.

Electrically driven gyros date back to the latter portion of the nineteenth century. These gyros did not, however, enjoy great popularity until the early part of the twentieth century when the "inside-out" electric motor was developed. The "inside-out" motor is designed so that the rotor is built around the stator. The rotating member thus has the larger diameter. This provides a rotor with a large moment of inertia as required for effective gyro operation. Spin velocities of 24,000 rpm are not uncommon with electrically driven rotors. The type of motor to be used to drive the rotor depends upon the application for which the unit is to be used. For very high accuracy and long operating time, a constant rotor speed is essential and synchronous or multiphase-hysteresis motors are used.† For less severe applications, induction motors are acceptable; 400 and 60 cps a-c, two- and three-phase motors are the most popular types. D-C motors are used in some applications but the need for brushes precludes their use for high-accuracy units. Brush wear causes unbalance and electrical noise.

Spring power for rotor drives has found only limited use. Design difficulties and short operating times have not permitted widespread use of the spring-driven rotor. The Waltham Precision Instrument Co., however, has recently developed a spring-powered gyro to be used in antitank missiles. The unit employs a helical spring instead of a flat coil spring. The spring drives a gear which in turn drives the rotor. The gyro is caged‡ until the rotor reaches its operating speed (approximately 7,000

---

cations a drift rate of 1°/hr is tolerable; for good accuracy, however, 0.1°/hr is necessary.

When applied to a single-degree-of-freedom gyro, drift rate has a different significance; it is a measure of the instrument-reading error due to frictional and vibratory effects.

These definitions for drift rate will become more meaningful during the discussion of the single- and two-degree gyros.

* The design is generally such that the major weight of the rotor is concentrated at the rim of the wheel. This gives the highest J within a given volume. The rotor material is ordinarily a high-density material with reasonable strength such as brass or bronze. Laminated steel is used when electric drives are used to minimize eddy-current losses.

† Constant spin velocity is generally more important in single-degree gyros as the measurements depend on the spin velocity.

‡ That is, the spin axis is held mechanically so that it cannot move relative to the instrument mount. The rotor is free to spin, however.

rpm) and is then automatically released. Operating time is only for a few minutes During this time the rotor coasts at an approximately constant speed.

The high spin velocities employed in gyros make it important that extreme care be exercised in the balance of the rotor. Any small rotor unbalance will cause very large centrifugal forces to exist (the centrifugal force is proportional to the square of the spin velocity). Such forces will cause vibrations, drift, and large bearing loads to be generated. The rotor must therefore be statically and dynamically balanced (balanced when not running and then again when at operating speed). As an order of magnitude, in precision applications, rotor eccentricities should be kept less than a microinch. It is of interest to note that in previous years, accuracy (drift performance for one thing) of gyro instruments was limited by frictional effects. However, more recently, the effect of friction has been successfully reduced to such an extent that now mass unbalance appears to be the limiting factor on instrument performance.

Bearings. The problem of friction in gyro instruments still exists. The quest for better bearings, better pivot design, etc., still goes on. The ball bearings used in the gyro (pivot bearings between supporting frames or gimbals and bearings to support the rotor) are about the finest that can be made. Bearing races are ground to a superfine finish; the balls in the bearing are carefully hardened and checked for roundness. The bearings are assembled in an air-conditioned room in which the air is lint-free and the workers wear special caps and gowns to reduce the dust level. After

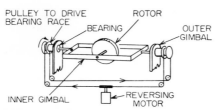
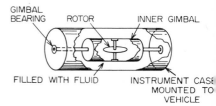

FIG. 97. Principle of the Sperry "Rotorace."

FIG. 98. Gyroscopic instrument in which flotation is employed.

assembly the bearings are themselves sealed or are used in hermetically sealed compartments of the instruments. In spite of the elaborate precautions, the level of Coulomb friction and stiction is high enough so that additional schemes are necessary to reduce the frictional effects further. One approach to reducing frictional effects was developed by Sperry only a few years ago. The trade name for the system is "Rotorace." The principle involved is alternate rotation of the bearing outer (or inner) races in opposite directions (see Fig. 97). The claims are that by such rotation random drift errors due to friction are averaged out and so minimized that bearing loads are more equitably distributed and mechanical irregularities are averaged out. According to laboratory reports such a scheme is said to provide an improvement in drift rate in the order of 1,200 per cent.

The Rotorace system described above is applicable to both single- and two-degree-of-freedom gyros. However, in a single-degree-of-freedom gyro, the flotation principle is used almost exclusively. The flotation principle was previously described in the discussion of the pendulous accelerometer. As applied to the gyro instruments, the sealed can, instead of containing the pendulous mass, now contains the spinning rotor. The flotation of the can in a viscous fluid takes the radial load off the bearing, and so reduces the bearing friction and wear (Fig. 98).

The most critical bearings in gyro instruments at the present time appear to be supporting the rotor. Here the friction does not cause an operating problem* but

---

* The gimbal bearings rotate rather infrequently and at relatively low speeds. Stiction in the bearing is therefore an operating difficulty. Rotorace purportedly eliminates this problem. The gimbal bearings experience relatively few cycles of operation so that wear is not a major consideration.

rather one of longevity. The high speed of the rotor causes a severe wear problem so that after a relatively short period of operation, wear in the bearings causes dynamic unbalance. This unbalance in turn causes vibrations, tendency to drift, and further wear. Because of this wear the life of the instrument is sorely limited.

Bearing development has now about reached the point of diminishing returns. Much work and money must be invested to realize even small gains. For this reason many companies have investigated and tried other schemes for supporting the rotating members of the gyro instruments. The air bearing has met with some success and is being used by some companies* in some of their gyro designs. The air bearing essentially floats the gimbal shafts of the gyro on a cushion of compressed air. The claims are such a design reduces the friction level to an almost nonmeasurable value (about one-millionth of the friction encountered in a good-quality ball bearing).

While air bearings can be used to reduce pivot friction, they cannot be used for rotors which are enclosed in a sealed can.† For rotor bearings, electrostatic bearings have been considered. In this instance, the rotor shaft would be supported by electrostatic levitation. While the literature discusses this possibility, the writer knows of no company which has yet constructed a commercial unit.

Single-degree-of-freedom Gyro Instruments

As mentioned in the previous section, gyros can be grouped into two general categories depending upon the freedom afforded the spin axis. If the position of the spin

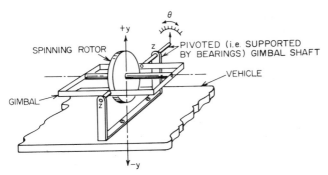

FIG. 99. Single-degree-of-freedom gyro.

axis (with respect to the vehicle) can be described by one coordinate, the gyro is said to have a single degree of freedom. Figure 99 illustrates a gyro with a single degree of freedom.

The spinning rotor is mounted in a frame commonly called a gimbal. The gimbal is pivoted in bearing mounts so that it can move with respect to the vehicle. The gyro has a single degree of freedom as the spin axis of the rotor is located with respect to the vehicle when angle θ (pointer on gimbal shaft) is specified.

Basic Theory. The easiest way to derive the general equations governing the single-degree-of-freedom gyro is to use a vector notation.[6] In most applications, however, the gyro motion is restricted so that a completely general set of equations is not needed. Here, therefore, a rather simplified treatment is presented. The results of

* Ford Instrument Co. is an example of one company which has been using air bearings in their work with the Army Ballistic Missile Agency. Eclipse-Pioneer Division of Bendix Aviation is another company which has worked with air-bearing gyros.

† Minneapolis-Honeywell is presently working on a hydrodynamic gas bearing to be used for the rotor. As far as this writer knows, however, this is still in the design stage and has not yet been translated into commercial hardware (Carl Green, "Gas Bearings," Minneapolis-Honeywell Aero Document R-ED 29004-1B).

this analysis, while not completely general, are adequate to describe gyroscopic applications.

The basic governing equation is Newton's first law as applied to rotating bodies:

$$\tau = d(I_x\dot\psi)/dt = I_x(d\dot\psi/dt) \quad \text{(when } I_x \text{ is time-independent)} \tag{154}$$

where I_x, $\dot\psi$, and τ are moment of inertia, absolute angular velocity,* and torque, respectively. Equation (154) states that any change in angular velocity (i.e., any angular acceleration) must be accompanied by a torque. The torque and angular acceleration must be similarly directed.

Equation (154) is really a vector equation so that $d\dot\psi/dt$ must be interpreted as a vector change in angular velocity where the change may be one of magnitude, direction, or both.

Examine now Fig. 100 where a spinning rotor is shown with the spin axis along the xx axis. For the spin direction shown, the spin vector $\dot\psi$ is directed to the right in accordance with the right-hand screw convention. Axes yy and zz are constructed so that these and the spin axis form a mutually perpendicular system. Assume now that the rotor is given an angular velocity $\dot\phi$ around the yy axis in the positive y direction (i.e., the $\dot\phi$ vector is directed in the positive y direction). The spin axis now moves from aa to bb; while now $\dot\psi$ may remain constant in magnitude, the direction of $\dot\psi$ has changed. This angular-velocity change $\Delta\dot\psi$ is shown as line $\bar{a}b$ in Fig. 100. The change in $\dot\psi$ can be expressed as a function of $\dot\phi$ if, for a small increment of time Δt, the line $\bar{a}b$ is considered equivalent to the arc $\widehat{ab'}$ (subtending angle $\dot\phi\,\Delta t$). Then,

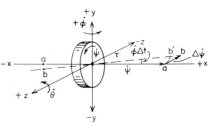

FIG. 100. Vector relations for gyro rotor.

$$\Delta\dot\psi = \widehat{ab'} = (\dot\psi)(\dot\phi\,\Delta t) \tag{155}$$
$$\text{or} \quad \Delta\dot\psi/\Delta t = \dot\psi\dot\phi \tag{156}$$

In the limit the changes become differentials and

$$d\dot\psi/dt = \dot\psi\dot\phi \tag{157}$$

Equation (157) implies that there is an angular acceleration $d\dot\psi/dt$ of the rotor. This acceleration vector is directed perpendicularly to the xx axis (in the limit the angle between $\dot\psi$ and $\Delta\dot\psi$ approaches 90°) and in the negative z direction (see Fig. 100). Equation (154), however, states that if such an acceleration exists it must be accompanied by a torque in the same direction. By Eqs. (154) and (157), therefore,

$$\tau_{zg} = -I_x\dot\psi\dot\phi \tag{158}$$

where τ_{zg} is considered to be the gyroscopic torque in the z direction which accompanies the angular acceleration $d\dot\psi/dt$ $(=\ddot\psi)$ caused by the precessional velocity $\dot\phi$.

Equation (158) can be interpreted from two points of view:

1. A torque applied along the positive z axis will cause a negative precessional velocity $\dot\phi$ around the y axis
2. A positive angular velocity $\dot\phi$ applied to the rotor around the y axis will give rise to a gyroscopic torque acting in the negative z direction.

Both viewpoints have been used to construct gyroscopic instruments. The integrating accelerometer (presented in a later discussion), for example, utilizes the viewpoint exemplified by 1 while the "rate" gyro (see the discussion following) employs the viewpoint exemplified by 2.

It should be noted here that Eq. (158) is applicable only if mutual perpendicularity is maintained between the three significant axes. However, in most applications rotor

* That is, angular velocity referred to inertial space.

COMPONENTS

movement is restricted so that the required perpendicularity is maintained at least approximately.

Rate Gyro. The rate gyro is an instrument which responds to an angular input velocity with an angular output displacement. The output displacement thus measures the input velocity. A schematic diagram of a rate gyro is shown in Fig. 101. The operation of the instrument can be very easily explained if transient effects are for the moment disregarded. Assume then for simplicity that a constant angular velocity is applied to the case around the yy axis. When the steady state is reached a gyroscopic torque will exist around the zz axis as defined by Eq. (158). This torque is balanced by a countertorque produced by a torsion spring* so that in the steady state there is no motion around the zz axis. At the balance condition

$$\tau_s = I_x \dot\psi \dot\phi \tag{159}$$

or

$$\dot\phi = \tau_s / I_x \dot\psi \tag{160}$$

where τ_s is the spring torque, I_x and $\dot\psi$ are as previously defined. If now I_x and $\dot\psi$ are

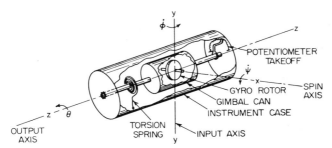

Fig. 101. Schematic diagram of a rate gyro.

known constants then τ_s can be used to measure $\dot\phi$. τ_s can be determined if the spring torque and its deflection are linearly related, i.e.,

$$\tau_s = K_t \theta_{g/c} \tag{161}$$

where K_t is the torsional spring constant and $\theta_{g/c}$ is the angular deflection of the gyro gimbal with respect to the instrument case around the z axis. Combination of Eqs. (161) and (162) results in

$$\dot\phi = (K_t / I_x \dot\psi) \theta_{g/c} \tag{162}$$

Equation (162) implies that the relative angular displacement around the z axis ($\theta_{g/c}$) can be used to measure the input angular velocity $\dot\phi$.

Use of Eq. (159) in the above discussion automatically assumes that the significant axes remain mutually perpendicular during operation. In order to ensure that this condition is at least approximately satisfied, the torsion spring used in the rate gyro must be very stiff. The corresponding deflections $\theta_{g/c}$ will therefore be very small—and very sensitive measuring devices will thus be required if any reasonable velocity measurements are to be made. Angular-measurement devices will again be mentioned during a discussion of construction details.

The above analysis has been simplified to illustrate the basic operating principle of the rate gyro. A more complete analysis will require that the differential equation around the zz axis be determined. This equation can be simply derived if all effects

---

* Compare with a linear accelerometer which responds to input accelerations with output displacements. Both instruments, linear accelerometer and rate gyro, employ the same basic principle. In the first case a spring is used to supply the accelerating force; in the second case a spring is used to supply the gyroscopic torque. In either case the spring deflection is used to measure the input variable.

around the z axis are accounted for. Thus, if the gyro has a precessional velocity $\dot{\phi}$ (see Fig. 101), a gyroscopic torque will exist around the z axis. This unbalanced torque will tend to drive the gyro (and its gimbal) around the z axis; i.e., this torque will tend to cause an angular acceleration around the z axis, tend to cause a spring displacement, and tend to overcome any friction opposing motion. Therefore,

$$\text{Gyro torque} = \text{inertial torque} + \text{spring torque} + \text{frictional torque} \tag{163}$$
$$I_x \psi_g \dot{\phi}_g = I_z \ddot{\theta}_g + K_z \theta_{g/c} + B_z \dot{\theta}_{g/c} \tag{164}$$

where I_z and θ_g are the moment of inertia and angular displacement,* respectively, around the z axis. The spring torque and friction torque are due to the relative motion between the gyro gimbal and instrument case so that $\theta_{g/c}$ is a relative angular displacement between the gimbal and case. The relative and absolute motions are related by

$$\theta_{g/c} = \theta_g - \theta_c \tag{165}$$

where θ_c is the absolute angular displacement of the case around the z axis and θ_g the

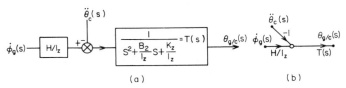

FIG. 102. Diagrams for transfer function of rate gyro. (a) Block diagram. (b) Signal flow diagram.

absolute gimbal displacement around the z axis. The relation expressed by Eq. (165) when employed in Eq. (164) results in

$$I_x \psi_g \dot{\phi}_g = I_z \ddot{\theta}_{g/c} + B_z \dot{\theta}_{g/c} + K_z \theta_{g/c} + I_z \ddot{\theta}_c \tag{166}$$
or
$$I_x \psi_g \dot{\phi}_g - I_z \ddot{\theta}_c = I_z \ddot{\theta}_{g/c} + B_z \dot{\theta}_{g/c} + K_z \theta_{g/c} \tag{167}$$

If now $I_x \psi_g$ (the angular momentum around the z axis) is replaced by H and the significant variables written as a function of s, Eq. (167) can be rewritten as

$$H \dot{\phi}_g(s) - I_z \ddot{\theta}_c(s) = \theta_{g/c}(s)(s^2 I_z + B_z s + K_z) \tag{168}$$

wherefore
$$\frac{H \dot{\phi}_g(s) - I_z \ddot{\theta}_c(s)}{I_z[s^2 + (B_z/I_z)s + K_z/I_z]} = \frac{(H/I_z)\dot{\phi}_g(s) - \ddot{\theta}_c(s)}{s^2 + (B_z/I_z)s + K_z/I_z} = \theta_{g/c}(s) \tag{169}$$

Equation (169) is often represented by a block diagram such as shown in Fig. 102a or the signal flow diagram shown in Fig. 102b. If for the moment the case acceleration term $\ddot{\theta}_c$ is neglected, then Eq. (169) (or the diagrams in Fig. 102) show that the the transmittance from $\dot{\phi}(s)$ to $\theta_{g/c}(s)$ can be written as

$$\theta_{g/c} = \dot{\phi}_g \frac{H/I_z}{s^2 + (B_z/I_z)s + K_z/I_z} \tag{170}$$

The relative angular displacement $\theta_{g/c}$ has, therefore, a second-order response (as does the accelerometer described in previous pages). The response curves for $\theta_{g/c}$ are thus the same as for any second-order system. The natural frequency of the system is

$$\omega_n = \sqrt{K_z/I_z} \tag{171}$$

and the damping factor is defined by

$$2\zeta \omega_n = B_z/I_z \tag{172}$$

* Angular displacement referred to inertial space. Single and double dots above the variables indicate first and second derivatives, respectively.

Equation (170) shows that for large ω_n ($= \sqrt{K_z/I_z}$)* the relative angular displacement measures the input angular velocity to within a multiplicative constant. That is,

$$\theta_{g/c}\omega_n{}^2(I_z/H) = \theta_{g/c}(K_z/H) \approx \dot{\phi}_g \qquad (173)$$

Equation (173) can now be compared with the result previously derived when transient effects were neglected [see Eq. (162)].

As in any second-order system, the "best" value for ζ depends to a large extent upon the nature of the input signal (in this case $\dot{\phi}_g$). In general, however, the signal is not known beforehand so that a compromise ζ of about 0.5 to 0.7 is used for most purposes.

The case acceleration term $\ddot{\theta}_c$ represents an input disturbance and will again be mentioned later when errors are discussed.

The schematic diagram of Fig. 101 roughly shows the general physical arrangement of parts found in most rate gyros. The overall dimensions of the unit do not appear to be limited as to a maximum value. A diameter of approximately 2 to 3 in. and an overall length of about 4 to 5 in. seems to be fairly common. Such units weigh roughly 3 lb. As to minimum dimensions, the U.S. Time Corporation claims its rate

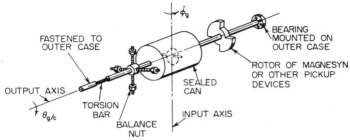

FIG. 103. Schematic diagram of rate gyro with torsion bar and a variable-reluctance takeoff.

gyro to be "the smallest, lightest, and most rugged production unit" available. The overall dimensions of this unit are approximately $2 \cdot 3\!/\!16$ in. long by $1 \cdot 5\!/\!16$ in. diameter with an overall weight of approximately $3\!\frac{1}{2}$ oz.

For accuracy and extended life, the gyro rotor is ordinarily mounted in a sealed can (the can thus serves as an inner gimbal). The can is filled with helium (or a mixture of helium and nitrogen)† to provide for a nonoxidizing atmosphere and yet allow reasonable heat conductivity. In order to balance the can and rotor after sealing, a set of balancing nuts is often included in the shaft supporting the rotor-can assembly (see Fig. 103).

While a spring could be used to provide for elastic restraint (as shown in Fig. 103), a more practical arrangement is the use of a torsion bar (as shown in Fig. 103). The torsion bar is merely a "necked-down" section on the main shaft (the shaft which supports the sealed can). One end of the main shaft is anchored to the outer instrument casing‡ while the other end is supported in a bearing mounted in the outer casing. An angular-velocity input around the input axis causes a rotation of the shaft around the output axis. The torsion bar twists and by so doing provides countertorque to stop rotation. The angle of twist which now measures the input velocity [see Eq. (162)] is picked up by an angular measuring device.

* ω_n approximately three times the maximum input frequency.
† Nitrogen is included to increase the dielectric strength of helium. This reduces the possibility of arc-over in rotor-motor leads.
‡ The Fairchild Control Corporation supports the torsion bar in what they call a "spider network." This spider network is fastened to the instrument casing. The claim is that the spider network takes over the job of supporting the rotor can, thus relieving the torsion bar of this load.

The potentiometer as a pickoff device (see Fig. 101) is sometimes used (though not ordinarily in rate gyros) when measurements are not severely critical and when the contingent frictional effects are not objectionable. However, when the requirements become more severe, devices with greater accuracy and less frictional loading must be employed. Of the various devices available, the variable-reluctance devices* appear to be the most favored; in particular the Microsyn seems to be most often used. Figure 104 is a schematic diagram of a Microsyn signal generator. The generator delivers a voltage output across the secondary terminals proportional to the input angular displacement of the rotor. The windings on the stator are such that the secondary voltages at alternate poles are series-aiding and at adjacent poles are series-opposing. When the rotor is centered, an equal amount of flux threads through each pole. The secondary voltages are therefore all equal in magnitude but cancel each other around the closed loop [Eq. (174)]. The total output across the secondary is therefore zero.

$$v_1 - v_2 + v_3 - v_4 = 0 \qquad (174)$$

When the rotor is moved, the reluctance of the magnetic circuit changes and one pair of opposite poles carries more flux lines while the flux lines through the other pair are reduced. The voltage at one pair of opposite poles is therefore increased a small amount while the voltage at the other pair of poles is reduced by a small amount. The total charge at the output, therefore, is four times the voltage change at each pole [Eq. (175)].

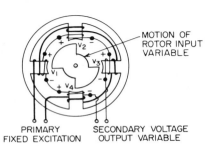

PRIMARY FIXED EXCITATION   SECONDARY VOLTAGE OUTPUT VARIABLE

Fig. 104. Schematic of Microsyn generator.

$$(v_1 + \Delta v) - (v_2 - \Delta v) + (v_3 + \Delta v) - (v_4 - \Delta v) = 4\Delta v \qquad (175)$$

For small rotations (less than 10°) the output voltage is closely proportional to the input angular displacement (at the maximum angular displacement the Microsyn output can be made to be linear to within 0.1 per cent).

The flotation principle, as described earlier in a discussion of the pendulous accelerometer, is also used in gyro instruments. In fact, the principle was first mentioned by Lord Kelvin in connection with a study he was making on gyroscopes. Others after him also toyed with the idea. However, the first real progress in this direction was due to C. S. Draper at M.I.T. His successful development of the integrating gyro (to be discussed later) is in good measure due to flotation.

As mentioned previously, flotation is realized by simply filling the instrument casing with a fluid so that the gimbal can is supported by fluid buoyancy. Ideally the buoyancy effect is adjusted so that the main shaft bearings are not required to support any radial load. The function of the bearing is thus reduced to alignment and centering. Flotation is said thus to reduce frictional effects by a factor of 1,000.

Various commercial fluids are now available specifically made for filling gyro instruments. One such fluid, manufactured by the Hooker Electrochemical Co., is sold under the trade name of Florolube. The density of this fluid is about twice that of water, and it can be obtained in various viscosities. Another such fluid recently developed by Sperry is called Gyrolube. Many of the fluids available are solids† (or semisolids) at room temperature. In order for them to be used in instruments, therefore, the instrument must be heated and kept at constant temperature to maintain constant fluid characteristics.

Aside from providing buoyancy, the flotation fluid also serves a "cushion" for the

---

* The Magnesyn, Inductosyn, Microsyn, Telegon, etc. A rather complete description of most of these units can be found in Ref. 7.

† Gyrolube, however, retains its fluidity even at relatively low temperatures.

gimbal can* and provides a viscous-damping medium. As noted previously, the response of the rate gyro is that of a second-order system [see Eq. (170)]. With bearing friction reduced to a minimum and if no damping medium were present, the instrument would have an extended transient oscillatory response. The damping fluid is thus necessary to preclude the possibility of sustained oscillations. The fluid viscosity is ordinarily chosen so that the instrument will operate with a damping factor of about 0.5 to 0.7.

The viscous friction provided for by the fluid is due to the shear stresses set up in the oil when the gimbal can moves relative to the core. The friction coefficient [B_z in Eq. (164)] varies directly with viscosity and shear area. If the shear area is kept constant, then the viscosity must also be kept constant if the damping factor is to remain fixed. Viscosity is, however, highly temperature-sensitive. It is necessary, therefore, that the instrument be temperature-controlled (when the shear area is fixed) if a constant damping factor is required. Control to within ±5°C is ordinarily sufficient for most purposes to ensure reasonably constant viscosity.† In that the heated fluid expands, accommodations must be included in the instrument to allow

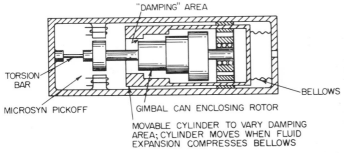

FIG. 105. A rate gyro with variable shear area to compensate for viscosity variations due to temperature changes.

for this expansion. For this purpose a bellows is often incorporated into the casing design. The bellows permits the casing to be completely filled (so that no air remains in the instrument casing) yet allows for fluid expansion to prevent fluid-pressure buildup.

While a relatively constant damping factor can be realized by temperature control, this method is sometimes considered objectionable as it requires a separate heating and control system. Another method which has been used is to permit the shear area to increase as the temperature increases. The increased shear area compensates to some extent the viscosity decrease which accompanies the increase in temperature. The value of the friction coefficient B_z is thus maintained at a relatively constant value. A schematic diagram of an instrument with variable shear area is shown in Fig. 105.

Errors in Rate Gyros. 1. Case acceleration: Equation (169) shows that the output relative displacement of rate gyros $\theta_{g/c}$ consists of two terms. The first of these is due to the input signal $\dot{\phi}_g$ (the angular input rate), which is the variable to be measured. The second term is due to an angular acceleration input, applied to the case, around the spin axis and represents the case acceleration error. The instrument cannot discriminate between $\dot{\phi}_g$ and $\ddot{\theta}_c$; it would interpret the case acceleration input as an input rate $\dot{\phi}_g$ of magnitude $\ddot{\theta}_c(I_z/K_z)$ and thus give an erroneous reading.

The output error due to case acceleration can be reduced if K_z is increased or I_z

---

* The fluid surrounding the gimbal can protects it from shock and spurious vibrations. Shock loads as high as 1,000 g's have been applied to fluid-filled instruments with little or no resultant damage.

† Heat is supplied through a simple coil heater built into or around the outer casing. Thermistors or bimetallic strips are used for control.

is decreased. The latter possibility must, however, be ruled out as it will require that the spin-axis inertia I_z also be reduced. Reduction of I_x would adversely affect the performance of the instrument. The spring constant K_z should therefore be increased to reduce the case acceleration error. High K_z is advantageous from another point of view: It increases the useful range of the instrument (i.e., increases the natural frequency). However, as K_z is increased, the magnitude of the output $\theta_{g/c}$ decreases. The maximum value of K_z is therefore limited by the ability of the pickoff device to detect small angular displacements.* The limit of K is thus to a great extent dependent on the resolving power of the displacement transducer used.

It is possible to compensate for the case acceleration error. The case acceleration is detected by some other instrument. This information is then sent through a device with the same transfer function as the gyro. The output is then combined with the gyro output to cancel the error (see the block diagram in Fig. 106). For the rate gyro, an extra block (with its associated errors) is required to effect a cancellation; this method is therefore not particularly effective for rate gyros. However, for integrating gyros no extra block is needed and error cancellation is feasible.

2. *Alignment and hysteresis errors:* In that the output of the rate gyro is delivered from a transducer, it is obvious that the transducer must be zeroed to the gyro output.

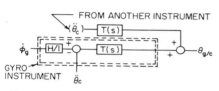

FIG. 106. Compensation of the rate gyro.

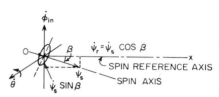

FIG. 107. Cross coupling between the spin and reference axes.

That is, the transducer must be aligned to the gyro so that the instrument will read zero when there is zero input. Initial alignment can be mechanically adjusted so that the error in reading is no more than 1 per cent. This residual error can then be nulled out by the use of compensating voltages. Because of mechanical hysteresis, however, some alignment error results during operation. After deflection, the torsion bar does not return to its original zero. The transducer and the torsion bar are thus misaligned and an alignment error results. The error due to the hysteresis effect, however, is small; the order of magnitude is about 0.01 per cent of full scale.

3. *Coupling errors:* The derivation of the equation for the rate gyro [Eq. (167)] depends upon mutual perpendicularity between the three significant axes. However, when the instrument operates (i.e., gives a reading) the spin axis and the input axis must become misaligned and some of the spin velocity becomes coupled with the input velocity to affect the output reading. To describe the coupling error quantitatively the following approximate analysis is presented:† If the spin axis is deflected a small angle β from the reference spin axis (Ox in Fig. 107) then the angular velocity around the input axis includes a component of the spin angular velocity.

$$\dot{\phi}_{total} = \dot{\phi}_{in} - \dot{\psi}_s \sin \beta \qquad (176)$$

Along the spin reference axis, the spin velocity is

$$\dot{\psi}_r = \dot{\psi}_s \cos \beta \qquad (177)$$

If β is small, the inertia around the spin reference axis is almost the same as the inertia around the actual spin axis (i.e., I_s). The torque generated around the z axis is

---

* This is essentially the resolution of the pickoff device. The resolution is the smallest input change applied to the transducer which will produce a measurable output.

† For a more exact presentation, see Ref. 6, p. 8-32. The approximate analysis presented above does, however, lead to correct results.

therefore
$$\tau = I_s \psi_s \cos \beta (\dot{\phi}_{in} - \dot{\psi}_s \sin \beta) \quad (178)$$
If there is no coupling between the axes, the torque generated around the z axis is
$$\tau = I_s \psi_s \phi_{in} \quad (179)$$
The generated torque error is therefore
$$\tau_\epsilon = I_s \psi_s \phi_{in}[1 - \cos \beta + (\psi_s/\phi_{in}) \sin \beta \cos \beta] \quad (180)$$
The corresponding angular error is
$$\theta_\epsilon = \tau_\epsilon/K = (I_s \psi_s \phi_{in}/K)\,[1 - \cos \beta + (\psi_s/\phi_{in}) \sin \beta \cos \beta] \quad (181)$$
This error is inherent with the instrument and cannot be eliminated. Compensation is not possible as it requires a knowledge of ϕ_{in}, which is the variable being measured. The error can be minimized by making the torsional-spring constant large. However, as mentioned during the discussion of case acceleration error, increases in K are limited by the sensitivity of the pickoff device. Compromises are therefore necessary.

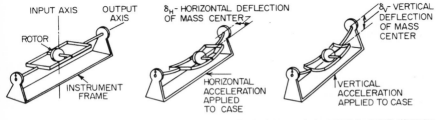

FIG. 108. Rotor deflections due to acceleration. (a) No acceleration applied. (b) Horizontal acceleration applied. (c) Vertical acceleration applied.

4. Errors due to vibratory effects:* a. Anisoelastic errors: Isoelasticity implies equal elastic suspension of the rotor in all directions. Thus, if a force (due, say, to an acceleration applied to the instrument case) is applied to a rotor which is isoelastically supported, the displacement of the center of gravity of the rotor will be the same amount and in line with the applied force (or acceleration) regardless of the direction of the force. An exaggerated version of how the mass center deflects when the instrument undergoes a linear acceleration is shown in Fig. 108. In Fig. 108a, there is no acceleration and the mass center is along the output axis. In Fig. 108b, the instrument is subjected to a horizontal acceleration. The elasticity of the output shaft and the "give" of the gimbal support both permit the mass center to move horizontally. The mass center and output axis now no longer coincide. The deflection of the mass center from its position along the output axis is designated as δ_H. In Fig. 108c, a vertical acceleration is applied, and again because of the "give" in the various supporting members, the mass center moves away from the output axis. The resultant vertical deflection is designated δ_V. If δ_V and δ_H are equal when the vertical and horizontal acceleration are equal, then the rotor may be considered to be supported isoelastically. If this condition is not fulfilled, then the support is anisoelastic (or monisoelastic, as it is often called). How an error torque is created when the support is anisoelastic is schematically shown in Fig. 109. Figure 109 shows end views of the output axis (i.e., the output axis is perpendicular to the page). In Fig. 109a, the rotor deflects isoelastically and the mass center remains along the line of the applied acceleration. The reaction force of the rotor thus acts

* A good simplified treatment of these errors can be found in Ref. 8.

through the center of rotation of the output axis. Hence the torque arm of the reaction force is zero and no torque is generated. In Fig. 109b, the rotor deflects anisoelastically; the mass center moves off the line of applied acceleration (because of unequal "give" of the supporting structure). The reaction force of the rotor no longer acts through the center of rotation and now has a torque arm δ. The torque thus generated causes an output deflection which is misinterpreted by the instrument as an angular velocity input $\dot{\phi}^*$. The anisoelastic torque can be semiquantitatively determined as follows: The reaction force is given by

$$F_r = ma \qquad (182)$$

The torque arm is proportional to the applied acceleration. That is,

$$\delta \propto a \qquad (183)$$
or
$$\delta = \mu a \qquad (184)$$

where μ is a proportionality constant. Therefore,

$$\tau = F_r \delta = \mu m a^2 \qquad (185)$$

The corresponding output-error deflection is

$$\theta = \tau/K_t = (\mu m/K_t)a^2 \qquad (186)$$

where K_t is the torsional-spring constant. The coefficient $\mu m/K_t$ is sometimes

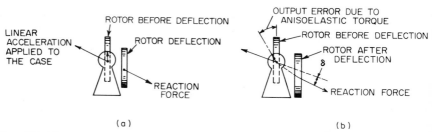

Fig. 109. Front view of rotor deflections. (a) Front view of isoelastic deflection. (b) Front view of anisoelastic deflection.

referred to as the anisoelastic coefficient and is used to give an indication of drift rate. The units of this coefficient are output angular displacement per squared lineal acceleration (output degrees/g^2 where g is acceleration in terms of gravity). Since the output angular deflection is equivalent to an input angular rate, the units of the coefficient are often given as degrees/hr/g^2. To be acceptable, a gyro should have an anisoelastic coefficient in the order of $0.05°/\text{hr}/g^2$.

Equation (186) shows that an anisoelastic error will result when a constant or vibratory acceleration is applied as the error is proportional to the square of the acceleration. If, for example, the applied acceleration a is sinusoidal ($A \sin \omega t$), then θ is proportional to $\sin^2 \omega t$ and the average value of θ is not zero. Furthermore the magnitude of error due to an oscillatory acceleration will be frequency-sensitive. As the input frequency approaches the resonant frequencies along the spin and input axes, the error magnitude mounts. These resonant points should therefore be significantly higher than the maximum expected input frequency. Structural stiffness along these axes is therefore an important requirement.

Equal stiffness along the input and spin axes should, theoretically, eliminate the anisoelastic error. In actual practice, however, it may be found that isoelasticity is not the whole answer. It has been shown[9] that unequal damping along the spin

---

* In some literature this output is often called the *drift rate* due to anisoelastic error. The drift rate would be the equivalent angular-velocity input necessary to create the same output torque as does the anisoelastic torque.

and input axes also causes errors. (These errors are often also called anisoelastic errors.)

While the explanation of anisoelastic error was given with reference to the rate gyro, this type of error is not restricted to the rate gyro. This error is, in fact, present in all gyros (single- and two-degree gyros) which have nonrigid structural supports.

Another error which is due to the directional mechanical properties of the instrument is the anisoinertial error. This error is due to the difference in the moment of inertia around the input and spin axes. This error is, however, of lesser importance than the anisoelastic effects.

b. *Cylindrical error:* While anisoelastic error is caused by linear vibration of the input shaft, cylindrical error is caused by rotational motion of the input shaft. Both errors, however, are due, at least in part, to the nonrigidity of the supporting gimbal structure. To explain the mechanics of how the cylindrical error is produced, reference is made to Fig. 110. Figure 110a shows a schematic diagram of the instrument. The instrument is given a rotary motion so that all points on the unit describe a circle. Because of the centrifugal forces generated by the rotation, and the structural weakness of the supporting gimbals and shaft, there is a deflection of the gimbal structure (as shown by the phantom lines in Fig. 110a). Because of this deflection,

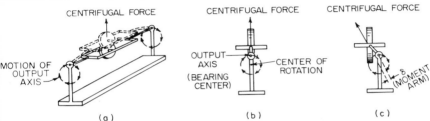

FIG. 110. Explanation of cylindrical error. (a) Cylindrical motion of instrument. (b) End view—deflection with no hysteresis. (c) Deflection with hysteresis.

the mass center no longer lies along the bearing center line. An end view of the situation is shown in Fig. 110b. If no frictional forces were present in the metal, then the mass center, bearing center, and center of rotation would all be collinear. Under these circumstances no error torques are produced around the output axis. However, because of internal and external frictional losses, the mass center lags the bearing center (as shown in Fig. 110c). When this condition exists, the centrifugal force acting on the mass center has a lever arm δ around the bearing center, and a torque around the output shaft results. As the lever arm varies during a single revolution of the output axis, the torque error also varies. However, since the mass center always lags the bearing center, this variable error has a nonzero average value (assuming that the cylindrical motion is in a constant direction). There is therefore an average error torque produced in the output because of the cylindrical motion of the output shaft and the hysteresis of the gimbal material. This error exists whether the gimbal structure is or is not isoelastic. If the structure is anisoelastic, the instantaneous cylindrical error is altered slightly but the average value remains virtually unaffected.

If the motion of the output shaft is not purely cylindrical but is, say, elliptical, then both an anisoelastic error and a cylindrical error result, causing a cumulative error which is roughly equal to the sum of the individual errors.

c. *Coning error:* The rate gyro should indicate any angular velocity applied to the input axis. If, therefore, the input axis of the instrument describes a cylinder or cone,* the instrument should indicate an input rate.

---

* For example, the input axis of a rate gyro will describe a cylinder if rigidly fastened to an airplane or missile which executes an unbanked turn. The axis will describe a cone if fastened to a vehicle which executes a banked turn. If the radius of the turn in either case is the same, the instrument will read the same angular rate in both cases.

It is possible to cause the input axis to describe a cone through vibrations applied to the instrument case. If the instrument input axis does describe a cone, the instrument output will indicate an input angular rate even though the vehicle which carries the instrument has had no angular motion. A conical error is then said to exist.

In order to explain the coning of the input axis, reference is made to Fig. 111. In Fig. 111a, the instrument case is subjected to an angular oscillation around an axis parallel to the spin axis. The input axis will now describe an arc as shown by aa. In Fig. 111b, the instrument is subjected to an angular vibration around an axis parallel to the input axis. This input causes the gyro rotor to have an oscillatory precessional motion so that the input axis now describes an arc bb as shown in the figure. If now the input motions of Fig. 111a and b are superimposed* then the composite motion of the input axis will be conical (see Fig. 111c). The instrument will therefore indicate an angular rate which is considered to be the conical-error drift rate.

Of the three vibration errors just discussed, the coning error is the only one which is independent of mechanical imperfections. Anisoinertial and cylindrical errors are due to frictional effects and imperfect mechanical properties. The conical error is, however, due to the geometry of the instrument and is inherent in the instrument

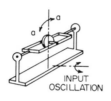
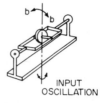
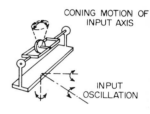

(a) OSCILLATION AROUND AXIS PARALLEL TO SPIN AXIS   (b) OSCILLATION AROUND INPUT AXIS   (c) OSCILLATION IN PLANE OF INPUT AND SPIN AXIS

FIG. 111. Coning of the input axis. (a) Oscillation around axis parallel to spin axis. (b) Oscillation around input axis. (c) Oscillation in plane of input and spin axis.

regardless of the materials used in its construction. To minimize this error, the instrument should be vibration-isolated. When it is mounted in a stable platform, this error is minimized.

5. *Drift errors:* The drift error or drift rate as applied to rate gyros is the output reading which would be obtained when no input signal is applied. From the previous discussion it is evident that torques can be generated around the output axis because of vibratory effects, for example. These torques will cause an output reading even though no input signal was applied. The equivalent input turning rate which will produce the same amount of output torque is the drift error caused by vibratory effects.

Torques which cause drift errors are often classified as follows:

a. *Acceleration-sensitive torques*—those which depend upon acceleration of the instrument. In this category are vibratory errors previously discussed.

b. *Acceleration-insensitive torques*—those present even when the instrument is standing still. In this category are the torques due to thermal-convection currents which tend to rotate the inner gimbal, electromagnetic torques stemming from the signal generators, and the torques due to the flexible leads feeding the gyro motor. To some extent these torques can be compensated for. Unfortunately, however, because of aging, cold storage, etc., the amount of compensation is not constant from day to day.

c. *Random torques*—The true cause of these torques is not precisely known, and they are sometimes called "uncertainty torques." These are the torques which

---

* If, for example, an oscillatory angular vibration has components along the spin axis and input axis.

remain after all other known effects have been accounted for. A plot of these torques with time looks like a noise function, and these torques are therefore considered to be due to random effects. The standard deviation of the noise vs. time plot serves to measure the spread of the random torques. An acceptable value for a gyro would be a torque standard deviation which is equivalent to an input rate of 0.005°/hr.

The Integrating Gyro. The integrating gyro is basically an instrument which responds to an angular input displacement with an output angular displacement. The output displacement is thus used as a measure of the input displacement. The reason for the name "integrating gyro" becomes apparent when one considers that the rate gyro measures angular velocity via an angular displacement whereas the integrating gyro measures the integral of angular velocity (i.e., angular displacement) via an output angular displacement.*

The basic difference between the rate and integrating gyros is the restraint used on the output axis. The rate gyro uses a spring restraint on the output axis whereas the integrating gyro uses viscous restraint. Figure 112 shows a simplified schematic of

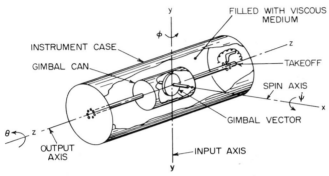

FIG. 112. Simplified schematic of an integrating gyro.

an integrating gyro. The only opposition to motion of the gimbal can is that of the fluid viscosity.

The equation which governs the operation of the integrating gyro is derived in exactly the same manner as employed to arrive at the rate gyro equation. Thus Eq. (163) is again used except that in this case the spring force is now zero. Equation (164) therefore becomes

$$I_x \dot{\psi}_g \dot{\phi}_g = I_z \ddot{\theta}_g + B_z \dot{\theta}_{g/c} \tag{187}$$

If the relative-angular-velocity equation [Eq. (165)] is again used and the results are rearranged Eq. (187) becomes

$$H \dot{\phi}_g - I_z \ddot{\theta}_c = I_z \ddot{\theta}_{g/c} + B_z \dot{\theta}_{g/c} \tag{188}$$

where all the terms are as previously defined in connection with the rate-gyro derivation. As in the rate-gyro discussion, if the case acceleration term $\ddot{\theta}_c$ is neglected and zero initial conditions are assumed then the transfer function for the integrating gyro becomes

$$\frac{\theta_{g/c}}{\phi_g} = \frac{H/I_z}{s + B_z/I_z} \tag{189}$$

* There is some confusion here as the literature often speaks of an "integrating gyro" used to measure angular rates. However, such "integrating gyros" have a feedback loop incorporated in the unit. This loop is to drive a torque, and its function is primarily to incorporate an electrical spring in the gyro (like the force-balance accelerometer). Since a simulated spring is being used such instruments might more appropriately be called torque-balance rate gyros. Such instruments will be more fully described later.

The integrating gyro thus has a simple first-order response characterized by a time constant of I_z/B_z. Equation (189) shows that if B_z/I_z is large (i.e., system time constant is small) then the relative displacement of the output axis with respect to the case is approximately proportional to the input displacement. That is,

$$\theta_{g/c} \approx (H/I_z)\phi_g \tag{190}$$

The successful development of integrating gyros as a practical instrument is in good measure due to the work of the M. I. T. Instrumentation Laboratory under the direction of Dr. C. Draper. The sealed rotor* can and the use of flotation evolved from work done by this laboratory. These ideas (sealing and flotation) have been

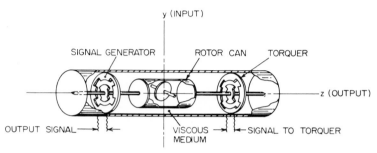

Fig. 113. Integrating gyro with torque generator.

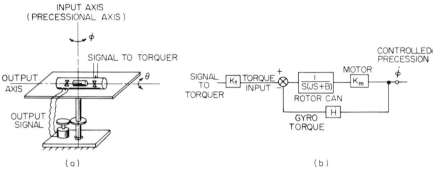

Fig. 114. Use of integrating gyro to index a table. (a) Schematic arrangement of parts. (b) Signal flow diagram.

incorporated in other gyro instruments, but they stem from work done in connection with the integrating gyro. As such, integrating gyros almost always use the floated-rotor construction and in most particulars the integrating gyros are constructed pretty much in the same manner as the rate gyro (except, of course, that no spring is included). Potentiometric takeoffs are, however, almost never used in integrating gyros. Instead, the Microsyn (or some comparable device) is used.

In many cases, the integrating gyro is equipped with a torquer mounted on the output axis (see Fig. 113). This torquer can be energized according to some predetermined program. The gyro can then be incorporated into a servo loop to control the turning speed of a table. Such a scheme is often used to Shuler-tune a platform (index a platform so that it always remains in a horizontal position) in inertial-guidance navigation systems. Figure 114 shows schematically how a table may be

---

* The use of a hermetically sealed rotor has given rise to the name "hermetically sealed integrating gyro," which is now often abbreviated to simply the HIG gyro.

indexed around the input axis of an integrating gyro. The gyro is mounted on the table and the torquer energized by an external signal. The torquer causes the rotor to rotate about the *output* axis, and this rotation is sensed by the takeoff device. The signal from the takeoff device excites a motor which drives the table (and gyro) around the *input* axis of the gyro. This precessional velocity around the input axis of the gyro generates a gyroscopic torque around the gyro output axis in a direction to null the torque signal from the torquer. Figure 114b shows a signal flow diagram of the operation.

When the output of the signal generator is directly fed back to the torquer the effect is to incorporate an electrical spring into the system. The result is then essentially a rate gyro. This instrument will be described later.

Errors. The discussion of errors in connection with the rate gyro applies in almost all particulars to the integrating gyro. There is, of course, no problem with hysteresis as there is no spring in the integrating gyro. In this connection, however, in that there is no spring, there is no automatic centering device. Therefore, by simple drifting or because of spurious signals the input and spin axes can easily become misaligned. The cross-coupling error under these circumstances becomes prohibitive and the instrument can no longer be used for angular measurements. For this reason the integrating gyro is seldom used to measure angular inputs directly but instead is used as an error sensor and nulling device. In this capacity the instrument is quite

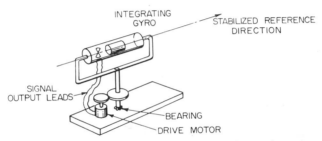

Fig. 115. Integrating gyro used to stablize a reference in space.

effective as its function is now merely to correct for any nonperpendicularity between input and spin axes rather than measure angular displacements. As an error sensor, the integrating gyro is particularly useful when it is desired to maintain a specific reference direction. Figure 115 shows how the instrument might be used in this capacity. The integrating gyro is freely mounted on a base and coupled to a drive motor. The drive motor is energized by signals originating from the instrument. When the instrument is deflected away from the reference direction, it detects the error and sends a signal (from the signal generator) to the motor. The motor drives the instrument back until the error is nulled.

Torque-balance Rate Gyros. As mentioned above, when the HIG unit is equipped with a torquer and when the torquer receives its signal from the signal generator, the integrating gyro effectively becomes a rate gyro. The instrument now measures angular velocity via an angular displacement. Actually, by connecting the torquer and signal generator in tandem, the effect is to incorporate an electrical spring into the system. That this is so can easily be demonstrated by an examination of the governing equations. Equation (164) can still be used but must now be modified to care for this new situation. It should now read

$$\text{Gyro torque} + \text{externally supplied torque} = \text{inertial torque} + \text{frictional torque} \quad (191)$$

The externally supplied torque is that delivered by the torquer. This torque is proportional to the signal from the signal generator which in turn is proportional to the output angular displacement. The output of the torquer may then be expressed as

$$\tau = -K_s K_a K_t \theta_{g/c} \tag{192}$$

where K_s, K_a, and K_t are the gains of the signal generator, amplifier, and torquer, respectively. The negative sign is used, as negative feedback will be used. That is, a positive-output angular displacement should generate a torque to oppose the output motion. When Eq. (192) is used in (191) the result is

$$I_x \dot{\psi}_g \dot{\phi}_g - K_s K_a K_t \theta_{g/c} = I_z \ddot{\theta}_g + B_z \dot{\theta}_{g/c} \tag{193}$$

where all terms are as previously defined in connection with the derivation of Eq. (164). If the acceleration of the case is presumed to be zero then Eq. (193) can be written

$$I_x \dot{\psi}_g \dot{\phi}_g = I_z \ddot{\theta}_{g/c} + B_z \dot{\theta}_{g/c} + K_s K_a K_t \theta_{g/c} \tag{194}$$

and finally

$$\theta_{g/c} = \dot{\phi}_g \frac{H/I_z}{s^2 + (B_z/I_z)s + K_s K_a K_t/I_z} \tag{195}$$

A comparison of Eq. (195) with Eq. (170) shows the two to be essentially the same. The only difference is that the mechanical spring constant of Eq. (170) (i.e., K_z) has been replaced by the product $K_s K_a K_t$. This product can therefore be considered as an electrical equivalent to the mechanical spring and has the advantage that it can be easily adjusted to suit the needs of a particular application.

Two-degree-of-freedom Gyros

A gyro rotor is considered to have two degrees of freedom when it is necessary to use two coordinates to locate the spin axis with respect to the carrying vehicle. In Fig. 116, for example, the spin axis can be located with respect to the vehicle if angles ϕ and θ are specified. This definition is not so rigorous as might be desired, for by this definition the gyro in Fig. 114 or 115 might also be called a two-degree-of-freedom unit.

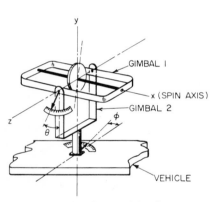

FIG. 116. Two-degree-of-freedom gyros.

In order to pin the definition down a bit more a two-degree-of-freedom gyro will be considered one in which the rotor spin axis can move with respect to the vehicle around two axes in an uncontrolled manner. Since in Figs. 114 and 115 the motion around the vertical axis is motor-controlled, the gyro could be considered to be one with a single degree of freedom. In Fig. 116 no such control exists and the gyro may be considered as one having two degrees of freedom. Actually a single-degree gyro can be made into a two-degree unit by mounting the single-degree unit so that it can move freely with respect to the vehicle. Two-degree gyros are, in fact, often called freely mounted gyros.

The prime use of the two-degree-of-freedom gyro is to establish a pair of reference axes in space. The high momentum of the spinning rotor tends to keep the rotor spin axis rigid in space. Thus in Fig. 116, if the x axis (spin axis) and the z axis were originally set to lie in a horizontal plane, the angle ϕ can be used to measure the turn of the vehicle around the y axis, while angle θ can be used to measure the angular displacement of the vehicle around the z axis.

The explanation of gyroscopic motion and rigidity on a mathematical basis is quite complex even when simplifying assumptions are made.* A qualitative expla-

---

* A complete derivation of the equations can be found in any good mechanics text. See Ref. 10, for example.

nation, however, can be given without too much difficulty. The experience gained in connection with the single-degree gyro can, in this instance, be used to good advantage.

It will be recalled that when a spinning rotor is given a precessional angular velocity around its y axis (see Figs. 100 and 117) a gyroscopic torque is generated around the z axis. Or, taking a reverse viewpoint, if a torque is exerted around the z axis, the gyro wheel will have a tendency to precess around the y axis. If, then, a torque around the z axis is the initial "signal" (or disturbance) input, the gyro wheel will not turn continuously around the z axis but will instead take up a precessional motion around the y axis. A pointer on the z axis (see Fig. 117a) therefore will retain its direction in space even when acted upon by external torques. The pointer P in Fig. 117a will thus remain essentially vertical as if it were rigid in the vertical direction even though it is subjected to disturbing torques. If now a free gyro were mounted in an aircraft and the aircraft made a turn around the z axis (see Fig. 117a), the degree of turn could be monitored by the pointer P mounted on the z axis. One might well ask why the system of Fig. 117a cannot be used to monitor the turn even if the rotor were nonspinning, i.e., if it were an inert mass. As a matter of fact, in the

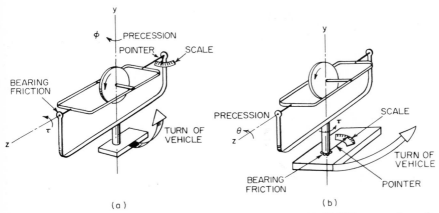

FIG. 117. Gyroscopic rigidity. (a) Rigidity around z axis. (b) Rigidity around y axis.

absence of friction, a simple inert mass would maintain its rigidity in space and could therefore be used to monitor turns. However, with friction present, frictional torques (from the bearings) are generated around the z axis when the aircraft turns around the z axis. These torques, small though they may be, would tend to move the pointer in the direction of the turn. The use of the spinning rotor inhibits this motion. It should be mentioned that, because of the relatively small torques involved, the precession around the y axis is for most practical purposes negligible over short periods of time. For longer periods corrections must be made.

A similar discussion could now be applied to motions around the y axis. If the craft executes a turn around the y axis, frictional torques are generated around the y axis. These torques, instead of causing motion around the y axis, cause a negligible precession around the z axis. The pointer on the y axis therefore remains essentially rigid in space and can thus be used to monitor a turn around the y axis.

Directional Gyros. A directional gyro is a two-degree-of-freedom gyro which has its spin axis set in a horizontal plane and which is used to detect or measure angular motions around the vertical. In brief, the directional gyro is used to measure or control vehicle heading.

A description of operation of this gyro has been essentially covered just above. The only difference between the directional gyro and the gyro previously discussed is the disposition of the spin axis. When the spin axis is located in any arbitrary direction, the gyro is called a free gyro. When the spin axis of the gyro is restricted to lie only in the horizontal plane (its direction in this plane can, however, be arbitrary), the gyro is called a directional gyro. The directional gyro therefore is used to establish an arbitrary reference in the horizontal plane. The pitch-control surfaces of an aircraft can now be slaved to the heading of the gyro, and the heading can in this manner be automatically maintained. For short-term flights, the gyro will maintain an accurate reference with no monitoring necessary. For long-term flights, corrections are necessary for both heading and resetting in the horizontal plane. The corrections are sometimes manually made by the pilot based upon his recognition of known references or information otherwise obtained. In some cases the corrections can be effected automatically by use of servo slaving to an earth reference or computer controls.

Gyro Compass. The gyro compass is a directional gyro in which the gyro spin axis is "slaved" to lie in a north-south direction. The spin of the earth relative to the gyro and the earth's gravitational field serve to orient the spin axis so that it

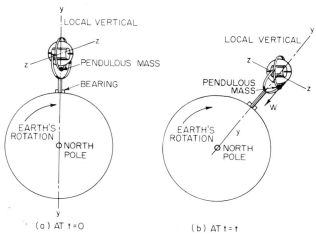

FIG. 118. Operation of gyro compass.

lies along or parallel to a meridian. The gyro compass thus indicates the geographic north-south direction rather than the magnetic north-south direction.

In order to understand how the compass operates an explanation of gyroscopic "apparent precession" is necessary. Apparent precession is the precession that an observer on earth would notice if he were watching a gyro mounted on the earth's surface. Since the gyro is rigid in space and the earth is turning, an earth-based observer would notice the motion of the gyro with respect to the earth. Foucault, in fact, used this scheme to demonstrate the earth's rotation. Reference to Fig. 118 will clarify the concept of apparent precession. In Fig. 118a is shown a top view of the earth. A gyro is mounted at the equator with its spin axis oriented in the east-west direction. Figure 118a shows the condition which exists at $t = 0$. At a later time $(t = t)$, the earth rotates to a new position and the situation is now as is shown in Fig. 118b. Because of gyroscopic rigidity, the spin axis retains its orientation in space. To an observer on earth, however, the gyro rotor appears to be turning about axis zz. The observer sees, therefore, an apparent precession with respect to himself.

If a simple two-degree gyro has no additional devices associated with it, it will have no north-seeking quality. However, if a mass is pendulously suspended on the inner gimbal of the gyro (shown in Fig. 118) an actual precession of the gyro is induced. The spin axis then oscillates in space until it finally settles down in a north-south plane.

To demonstrate how this motion develops, reference is again made to Fig. 118. In Fig. 118a, the pendulous mass is aligned with the local vertical; it has no moment arm around axis zz. After a small time t (Fig. 118b), because of apparent precession, the mass no longer lies along the local vertical. In this new position, the weight of the mass has a moment arm around axis zz. A torque is therefore exerted around axis zz and this torque causes the rotor actually to precess (i.e., move in inertial space) around the yy axis. The precession is such that the spin axis of the gyro turns toward alignment with the earth's spin axis. The precession continues even after alignment so that there is overshoot. The pendulous mass thus moves to the other side of the local vertical, where it exerts a reversing torque. However, the earth's rotation is such that now the pendulosity of the mass is reduced and the reverse precession is slower than the forward precession. Because of damping and the decreasing effect of the mass, the gyro finally settles down with the spin axis in the north-south plane and with the pendulous mass aligned with the local vertical.

Precession in the gyro compass is induced by an accelerating force acting on the pendulous mass. When the accelerating force is due only to gravity, the compass will precess so that it will ultimately indicate the geographic north-south direction. If, however, the instrument is subjected to other accelerations (e.g., centrifugal), the compass will give erroneous readings. For this reason, the gyro compass cannot

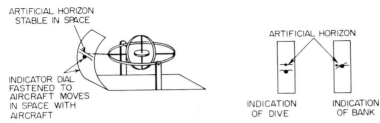

FIG. 119. Vertical gyro used to indicate bank and dive of aircraft.

be depended upon when the carrying vehicle undergoes high accelerations or violent maneuvers. The use of the gyro compass is thus limited to periods during which the vehicle is under steady flight conditions. In general, the gyro compass is found to be most effective in vessels which execute slow and deliberate maneuvers. Much work has been done to adapt the gyro compass so that it will be useful in aircraft even when subjected to acceleration disturbances. Special damping devices have been incorporated into the unit; feedback compensation has been employed. While many advances have been made in these directions, the sustained accuracy of the gyro compass in airborne vehicles is still fairly limited, and most applications of the instrument are limited to ships and other waterborne craft.

Vertical Gyro. The vertical gyro is a two-degree-of-freedom gyro in which the spin axis is oriented to seek the local vertical. A schematic view of a vertical gyro used as a bank-and-dive indicator is shown in Fig. 119. If the spin axis is maintained in a vertical direction, a line perpendicular to the axis indicates the horizon. A display of this artificial horizon on an instrument panel shows the aircraft's orientation with respect to the horizon wherefore the climb, dive, or bank of the aircraft can be visually indicated.

In order to locate the spin axis along the local vertical, the local vertical must first be defined. There are a number of definitions for the "local vertical."* A few of these are

1. The geocentric vertical—a line drawn from the center of the earth to a point on the earth's surface

* An excellent discussion of the vertical and the problems associated with indication of the vertical is given in Ref. 11.

2. The geographic vertical—a line perpendicular to the ellipsoidal approximation of the earth's surface

3. The "true" vertical—the direction taken by a plumb bob whose pivot point is fixed with respect to the earth

A diagrammatic representation of these definitions for the vertical is shown in Fig. 120.

The definition for the vertical most often used is that of the true vertical, i.e., the vertical indicated by a plumb bob. The difference between the true vertical and geographic vertical is often small enough to be negligible.* The differences are due primarily to geographical anomalies.

The use of a plumb bob would seem to be an obvious way to indicate the local vertical and is indeed used on structures fastened to the earth's surface. However,

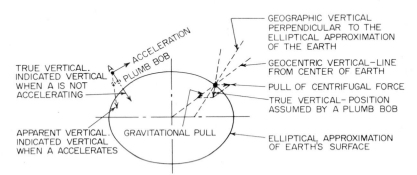

Fig. 120. Definitions of the vertical.

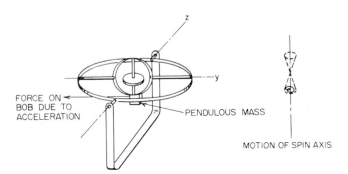

Fig. 121. Gyro with a pendulous bob.

on moving vehicles, the plumb bob experiences other accelerations besides gravity. These accelerations cause the bob to be in almost continuous oscillation. The period and magnitude of oscillation can be reduced if the inertia of the pendulous bob is increased, but practically speaking the bob inertia cannot be sufficiently increased to provide for efficient operation. In order to reduce the oscillations (and in a sense increase the effective inertia of the bob) a gyro-spinning rotor is used. In Fig. 121 is shown a pendulous bob mounted on a two-degree-of-freedom gyro. Any spurious acceleration which is now applied to the bob will cause the bob to move off the local vertical and at the same time torque the gyro. The torque around the z axis (see Fig. 121) causes a precession around the y axis. The spin axis is thus displaced small

* There are, however, differences up to 1 min of arc at some points on the earth. For accurate navigation these differences must be taken into account.

amounts in both the y and z directions and begins to trace out a conical volume in space. The axis of the cone is the average direction of the plumb bob. Because of friction the axis will slowly spiral in to assume finally again the average direction indicated by the plumb bob. The gyro has, in a sense, made the bob relatively insensitive to spurious accelerations. If the average value of the spurious accelerations is zero, then the spin axis will ultimately settle along the local vertical. In order to decrease the settling time of the gyro spin axis, the gyro is often torqued by other devices which are in turn actuated by the pendulous bob. Figure 122 shows one arrangement whereby the motion of the pendulum causes the gyro to be torqued by an air blast. As the plumb bob moves with the local vertical, the gyro is appropriately torqued so that the spin axis of the gyro follows the motion of the plumb bob. The high effective inertia of the gyro still integrates out the spurious acceleration effects so that, if the average applied acceleration is zero, the spin axis will closely follow the local vertical. There are other means used to torque the gyro.

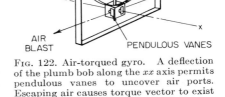

Fig. 122. Air-torqued gyro. A deflection of the plumb bob along the xx axis permits pendulous vanes to uncover air ports. Escaping air causes torque vector to exist in xx direction.

These include the use of electric motors, magnetic devices, friction, eddy currents, and ball-disk action.

The basic difficulty with the use of any of the gyro pendulous devices previously described is that the spin axis aligns itself with the direction indicated by the plumb bob. The plumb bob, however, is itself subject to accelerating forces and may not be aligned with the local vertical but instead may indicate what is called an "apparent vertical" (see Fig. 120). When the carrying vehicle accelerates, therefore, the spin axis of the gyro becomes aligned with the apparent vertical rather than the true local vertical. Over long periods of time, it is expected that continuous acceleration will not be applied and in fact that the overall average of acceleration will be zero. For long-term operation, therefore, the gyro pendulum can be used to indicate the local vertical. For accuracy during short-term operation however, the instrument is inadequate. When the accuracy requirements are stringent a single-axis HIG gyro with feedback is used to indicate the true local vertical. In this application, the gyro is precessed by a velocity signal obtained from an integrating accelerometer. As the carrying vehicle travels over the earth's surface, the local vertical "follows" the vehicle so that the angular velocity of the local vertical can be determined from the vehicle velocity (angular-velocity = vehicle velocity/radius of earth). By precessing the HIG gyro in accordance with the signal derived from the vehicle velocity, the angular velocity of the spin axis can be made to be equal to the angular velocity of the local vertical. In this manner, the spin axis remains aligned with the local vertical. When the feedback system is adjusted so that the spin axis follows the local vertical, the system is said to be Shuler-tuned.

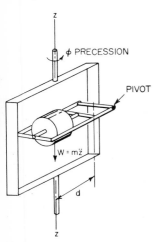

Fig. 123. Gyro integrating accelerometer.

Gyroscopic Integrating Accelerometer. The gyroscopic integrating accelerometer is a two-degree-of-freedom gyro which is used to measure *linear* accelerations. A schematic diagram of the instrument is shown in Fig. 123. In this instrument the rotor spin axis is supported only at one end; the weight of the gyro rotor thus causes a torque to be exerted around the yy axis. This torque causes the rotor to precess around the zz axis. The relations involved are

$$H\dot\phi = \tau = \text{torque} \qquad (196)$$
$$\tau = mgd \qquad (197)$$

where H and $\dot\phi$ are as previously defined, m is the rotor mass, g the acceleration of gravity, and d the torque arm. If $\dot\phi$ is determined from the above two equations the result is

$$\dot\phi = (md/H)g \qquad (198)$$

Thus the angular precession around the zz axis is a measure of gravitational acceleration along the zz axis. If now the acceleration $\ddot z$ is applied along the zz axis, the precessional velocity becomes

$$\dot\phi = (md/H)\ddot z \qquad (199)$$

wherefore $\dot\phi$ measures the total acceleration (including g) along the zz axis. Since g is constant a correction can be applied so that $\dot\phi$ could be made to indicate only those accelerations below or above the acceleration due to gravity.

16 Stable Platforms

A stable platform is simply a set of three orthogonal coordinate axes which has a fixed orientation with respect to inertial space (i.e., the fixed stars). The purpose of

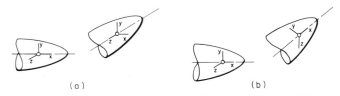

Fig. 124. Stable vs. nonstable coordinate systems. (a) Missile containing a coordinate system stabilized with respect to inertial space. (b) Missile containing a coordinate system fixed to the missile frame. Sometimes this system is called a no-gimbal or "strapped down" system.

the platform is to provide a reference mount for instruments and in particular for accelerometers. Accelerometers, as used in inertial-guidance systems, serve to measure vehicle accelerations from which vehicle position can be determined. In general, it is necessary to determine the vehicle vector acceleration so that acceleration measurements along three mutually perpendicular axes are necessary. Actually, any set of three orthogonal measurements would be sufficient to establish the vector acceleration. Thus if accelerometers are mounted on each of the x, y, z axes in Fig. 124a and b, the total vehicle acceleration can be determined with either arrangement. However, an overall simpler navigation system results if acceleration measurements are always made along the same reference line. In the system shown in Fig. 124a the accelerometer mounted in the x axis would always measure accelerations in the XX inertial direction. Similarly, the accelerometers on the y and z axis would measure accelerations in the YY and ZZ inertial directions, respectively. In order to arrange for the system shown in Fig. 124a, the reference axes x, y, z must be stable in inertial space and hence these axes must be independent of the motion of the carrying vehicle.

Realization of a Stable Platform with Single-degree Gyros

If frictionless bearings were feasible, then it should be possible, theoretically at least, to construct a stable platform by merely mounting an inertial mass in an appropriate gimbal system. The arrangement shown in Fig. 125, for example, would constitute a stable platform if the supporting bearings were frictionless. The accelerometers would tend to maintain a fixed orientation in space because of their inherent inertia.* Practically speaking, however, because of friction and inertial-reaction forces, the scheme proposed by Fig. 125 is not practicable.

In order to overcome the friction in the bearings and to counter inertial forces caused by accelerometers, energy from an external source must be supplied. This energy could come from a motor which would be called upon to deliver an output when spurious forces tend to disturb the reference heading. In Fig. 115, for example, a motor actuated by a single-degree integrating gyro serves to counter the disturbing forces which would tend to move the gyro output axis away from a set direction. The output axis of the gyro is therefore an axis which has been stabilized in space. In fact, now if three accelerometers were mounted to form an orthogonal triad as shown in Fig. 126 the accelerometers would be stabilized against vehicle motions around the yy axis. In order now to stabilize the triad against vehicle motions around the xx and zz axes, two additional angular-displacement sensors (gyros) and two additional correction motors are needed. The evolution of a completely stabilized system is shown in Fig. 126 with the complete system shown in Fig. 126c. The innermost gimbal, on which the gyros and accelerometers are mounted, is stabilized in space; it will retain its orientation in space in the face of disturbance caused by vehicle motions.

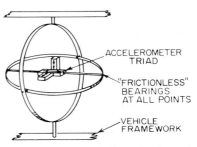

FIG. 125. Theoretical realization of stable reference system.

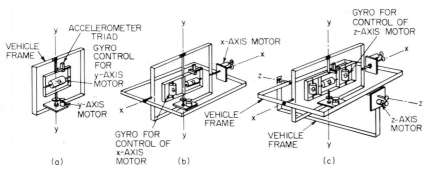

Fig. 126. Evolution of three-axis stabilized system. (a) Stabilization around y axis. (b) Stabilization around x and y axes. (c) Stabilization around x, y, and z axes.

Stabilization with Two-degree-of-freedom Gyros

It has been shown just above that a stable reference frame could be constructed if three correcting motors are actuated by three angular-displacement sensors. A two-degree-of-freedom gyro can sense displacements about two perpendicular axes. It is therefore possible to use two two-degree gyros to sense displacements about three mutually perpendicular axes, and these gyros can thus be used to stabilize a reference platform. Figure 127 shows a platform stabilized by two two-degree gyros. The diagram is self-explanatory.

* Actually even if the bearings were frictionless the arrangement of Fig. 125 would not constitute a stable platform. The inertial reaction of the accelerometer masses would exert a force on the gimbal system and tend to move it.

Both methods for stabilizing platforms have been used. The Redstone and Jupiter missiles, for example, use single-degree gyros, while the Atlas and Titan employ two-degree gyros. Each system has its individual merits. A platform stabilized by two-degree gyros requires only two such gyros; three gyros are needed if single-degree gyros are used. However, two-degree gyros need an extra pair of bearings to support the second gimbal. These two bearings are a source of friction and can cause inaccuracies. While it is difficult to note any trend, it seems that platforms stabilized by single-degree gyros are gaining favor.

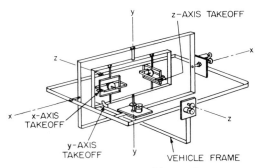

FIG. 127. Stabilized system with two-degree gyros.

Gimbal Systems

The platforms shown in Figs. 126 and 127 are suspended in what is known as an external three-gimbal system (see also Fig. 128a). In such a system, the instrument are all mounted within or internal to the gimbal rings. Three gimbals must be used in order to stabilize around three axes but three gimbals do not afford 360° freedom around all axes. In general, two axes will have 360° freedom but the third will b free to move only within ±85°. The limited motion around the third axis is due to mechanical constraints sometimes unavoidable and sometimes included to prevent gyro gimbal lock. In most missile applications, the limited movement around the third axis is not objectionable as the missile is not expected to undergo any complicated maneuvers. In aircraft and some ships, maneuvers are such that 360° motion around all axes is necessary. For such applications, a fourth and sometimes a fifth gimbal is added to afford greater flexibility.

An external gimbal does not represent the most efficient use of space and materials. In order to support the table and instruments, the gimbals must be large and should be sturdy to provide for rigidity. The amount of instrumentation that can be mounted on the table is limited by the volume available inside the gimbal system. As a means of rating a stable system, the weight ratio instrument weight/structure weight is often used. For an external gimbal system this ratio is about 1:4. In order to improve this weight ratio, the internal-gimbal system has been developed (see Fig. 128b). By such an arrangement, the weight ratio is improved to about 1:1.5. As a further advantage of the internal-gimbal system, the instruments ar now outside the gimbal system, affording easy access without any interference from the platform gimbals. This latter point is of some importance in star-tracking systems when an unobstructed view of the star reference is necessary. Internal gimbaling cannot, however, be used when full freedom about all axes is required This system can provide limited freedom around two axes and 360° freedom around only one axis. This restriction may not be critical for missiles but it does preclude its use in many carrying vehicles.

The ability of a platform to maintain its orientation in space is a gage of its effec tiveness. A 1° error in orientation can cause a navigational error of about 60 mile The rate at which a platform moves away from its set direction is known as th platform drift rate. This drift is due in part to the drift rate of the gyros and in part to the actual platform drift caused by vibration, nonperfect bearings, and other mechanical imperfections. Platform drift is, to a large extent, due to the nonisoelasticity of the structural members. The problem of nonisoelasticity is not so severe in internal-gimbal systems but does give difficulties in systems with external gimbaling.

Even if table vibration does not cause platform drift, it can cause errors in the instruments carried. Table vibrations around two axes can cause a rectification-type error in accelerometers known as a sculling error; the accelerometers show a reading even though the average acceleration is zero. To obviate this difficulty, platforms are often mounted on some sort of vibration-isolation device. Flotation of the whole platform has been contemplated, but this writer knows of no company which has as yet employed such a scheme.

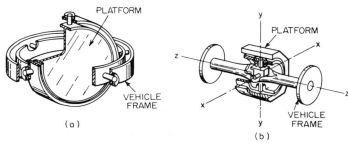

FIG. 128. Gimbal systems. (a) External gimbal system. (b) Internal gimbal system. (Drawings taken from F. K. Mueller, A New Look in Gimbal Systems, Missiles and Rockets, March, 1958.)

17 Prime Movers

The term prime mover as used in the following discussion includes those devices which are used to convert various forms of power into mechanical power. The mechanical power thus derived is used for the purposes of control—not indication.

At the present stage of development, there are three main classes of prime movers used in missiles and aircraft control systems:

Electric prime movers—devices which transform electric power into mechanical power. In this category are the electric motors, torquers, etc.

Hydraulic prime movers— devices which transform fluid power (as derived from the potential or kinetic energy of the fluid) into mechanical power. This category includes pistons or linear motors which produce linear motion and piston-, vane-, and gear-type motors which produce rotary motion.

Pneumatic or gas prime movers—devices which produce mechanical power from gas energy sources (i.e., from the potential, kinetic, or thermal energy contained within the gas). The pneumatic and gas devices are for the most part similar to the hydraulic devices. The main difference is the working medium.

Each of the classes of prime movers mentioned above has some attractive feature which makes it ideally suited for some application. No one type, however, can be singled out as being "the best all-purpose method" for power transmission. In general, a high torque-to-inertia ratio and power-to-mass ratio are considered to be desirable features in a prime mover. Hydraulic and pneumatic devices are particularly efficient along these lines. However, the added equipment which must be used in conjunction with these devices often precludes their use. Other factors such as operating climate and economy must also be considered before a decision is made as to which class of prime mover should be employed. While a hydraulic system may appear to be a good choice in one case, in other cases severe ambient conditions (e.g., subzero temperatures) may not permit the use of a hydraulic system and instead an electrical system may be employed.

Electric Prime Movers

The most popular prime mover being used today is the rotary electric motor. However, of the great host of available electric motors, relatively few types are useful in servo applications. Motors with toroidal-wound armatures, for example, are not often used in servosystems because of the high armature inertia. The drum-wound armature instead is used almost exclusively when a motor with a wound armature is required. In many cases, the servosystem requirements are such that only motors with a linear torque-speed curve can be used. Such a restriction limits the field so that relatively few types of motors can be used. In the following paragraphs some of the motors which are often considered for control applications are described.

D-C Motors. Of the available d-c motors, those with separate excitation seem to be the most popular for servo applications. Figure 129 shows diagrammatically the two types of d-c motors with separate excitation. In Fig. 129a is shown the field-controlled motor while in Fig. 129b an armature-controlled motor is shown. The

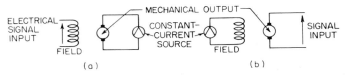

Fig. 129. Separately excited d-c motors. (a) Field-controlled motor. (b) Armature-controlled motor.

approximate characteristics and transfer function of each machine can be derived from the following equations:

$$e_f = L_f(di_f/dt) + i_f R_f \qquad (200)$$
$$e_a = L_a(di_a/dt) + i_a R_a + e_b \qquad (201)$$
$$e_b = K i_f \omega_a \qquad (202)$$
$$\tau_a = K_1 i_f i_a \qquad (203)$$
$$\tau_a - \tau_L = J_a \dot{\omega}_a + B_a \omega_a \qquad (204)$$

The electrical variables are voltage e and current i; τ and ω are the mechanical variables torque and angular velocity, respectively. L and R represent the electric circuit elements, inductance and resistance, while J and B are the mechanical-circuit elements of inertia and damping coefficient. The constants K and K_1 are the electromechanical coupling coefficients. If a consistent set of units is used, these constants will be equal, and they will be so considered in the following developments. The subscript f is applied to elements and variables associated with the field circuit while subscript a is applied to elements and variables associated with the armature circuit. The back emf in the armature and load torque are e_b and τ_L, respectively.

The characteristics of the armature-controlled motor can be derived from the above equations by considering the field current i_f a constant I_{f0}. All equations above are then linear and can be Laplace-transformed. A combination of the transformed equations then results in

$$(KI_{f0})E_a - \tau_L(s + R_a/L_a)L_a = [s^2 + s(B_a/J_a + R_a/L_a) + (KI_{f0})^2/J_a L_a \\ + B_a R_a/J_a L_a]J_a L_a \Omega_a \qquad (205)$$

In particular, if the transfer function between applied voltage and output angular velocity is required, τ_L in Eq. (205) can be considered zero and the result written

$$\frac{\Omega_a}{E_a} = \frac{(KI_{f0})/J_a L_a}{s^2 + s(B_a/J_a + R_a/L_a) + [(KI_{f0})^2/J_a L_a + B_a R_a/J_a L_a]} \qquad (206)$$

COMPONENTS

As a first-order approximation, the armature inductance is ordinarily small enough to be neglected. Under these conditions Eq. (206) reduces to

$$\frac{\Omega_a}{E_a} = \frac{(KI_{f0})/R_a}{sJ_a + B_a + (KI_{f0})^2/R_a} \qquad (207)$$

The total damping in the system is thus seen to consist of a mechanical as well as an electrical term. In fact, the electrical damping is in most cases the more significant term.

For the static characteristics (i.e., zero frequency behavior) of the armature-controlled motor, Eq. (205) is used with s reduced to zero. Then

$$\frac{KI_{f0}}{(KI_{f0})^2 + B_aR_a} E_a - \tau_L \frac{R_a}{(KI_{f0})^2 + B_aR_a} = \Omega_a \qquad (208)$$

Equation (208) is a family of straight lines as shown in Fig. 130a. These straight-line characteristics make the armature-controlled d-c motors particularly useful in many control applications.

The above analysis is also applicable to d-c motors which have a constant field flux as produced by a permanent magnet. The only changes that would be made

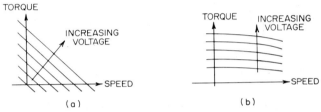

FIG. 130. Static characteristics of separately excited motors. (a) Armature-controlled d-c motors. (b) Field-controlled d-c motors.

would be in the constant term KI_{f0}. This would be replaced by another constant to represent the fixed field flux of the permanent magnet. Permanent-magnet motors have increased in popularity in the last few years because of this relative cheapness. However, the use of these motors is restricted to the fractional-horsepower range. The armature of these motors cannot be severely loaded as armature reaction can demagnetize the field.

The characteristics of the field-controlled motor can also be derived from the basic machine equations. In this case i_a is a constant I_{a0}. Equations (200), (203), and (204) can now be transformed and combined to give

$$\frac{KI_{a0}/L_f}{(s + R_f)/L_f} E_f - \tau_L = \left(s + \frac{B_a}{J_a}\right) J_a \Omega_a \qquad (209)$$

The transfer function Ω_a/E_f is now written

$$\frac{\Omega_a}{E_f} = \frac{KI_{a0}/L_fJ_a}{(s + R_f/L_f)(s + B_a/J_a)} \qquad (210)$$

In general, the field inductance L_a is never small enough to be neglected. The field-controlled motor thus has a significant electrical time constant and does not therefore respond so rapidly as the armature-controlled motor.

The static characteristics of the field-controlled motor are again determined by reducing s to zero in the basic equation [Eq. (209) in this case]. The result of this operation is

$$(KI_{a0}/R_f)E_f - \tau_L = B_a\Omega_a \qquad (211)$$

It is noted that the only damping which appears in the field-controlled motor is due to the mechanical circuit. This damping coefficient B_a is often small enough to be neglectable. Under these circumstances the right side of Eq. (211) reduces to zero and the static characteristic curves can be drawn as parallel horizontal lines as shown in Fig. 130b.

In order to operate the separately excited motors described above, constant-current sources must be available. When only small power is being handled, a pentode can be used to provide a constant-current supply. This arrangement is practicable where the power requirements are about 10 watts or less. Above 10 watts to about 500 watts, a ballast resistor can be used. This method is often used in field-controlled motors. In effect, all that is necessary is that a series resistor be included in the armature circuit. This resistor reduces the significance of the armature reaction so that the armature presents effectively constant resistance to the constant-voltage supply. The armature current thus remains relatively constant. Such an arrangement is somewhat inefficient as energy is lost across the ballast resistor. An alternate arrangement consists of driving the armature with an a-c supply through a bridge rectifier. A capacitor in the a-c line then serves as a ballast.[5] The energy lost across the ballast is relatively small. When the power requirements exceed 500 watts, a rotating amplifier such as the metadyne is often used. This unit employs armature reaction in such a fashion as to change a constant-voltage source into a constant-current source.*

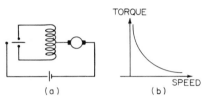

Fig. 131. Split-field series motor. (a) Schematic diagram. (b) Torque-speed characteristics.

It is somewhat difficult to compare the two types of separately excited motors discussed above without including the necessary related equipment. In order to judge fairly the relative merits of each, the excitations used should be included. Other things being equal, however, one might say that the armature-controlled motor will probably be the more responsive of the two as it has a relatively low time constant. Also, the armature-controlled motor is the more linear of the two as the field flux does not vary significantly because of armature reaction. On the other hand, the field-controlled motor can serve as a better power amplifier than the armature-controlled unit. The power needed to drive the field coils is appreciably less than that needed to drive the armature.

Another type of d-c motor often used in servo applications is the split-field series motor. The connections for this motor are shown in Fig. 131a. It is noted that the speed-torque characteristics for this motor (shown in Fig. 131b) are nonlinear. This motor is used therefore only where linearity is not a consideration. In general, this motor is for low-power installations where high starting torque is required. One of the conveniences of the motor is that it can be driven from a relay arrangement and can be reversed with relative ease. Alternatively it can be driven from a push-pull amplifier arrangement or with a thyratron switching circuit.[13] The equations for this motor are nonlinear and are rather involved.[14] They can be linearized by making first-order approximations but the results are only qualitatively useful.[12] This is especially true when the motor is relay-operated, for then the relay nonlinearity must be included. In general, when such a motor is to be used in a servosystem, experience and trial and error are the primary guides for the designer.

A-C Motors. The two-phase induction motor is, by far, the most popular a-c motor used in control systems. It is the simplest polyphase motor that can be used which has the features necessary to qualify it as a servomotor. The single-phase motor, while it is simpler than the two-phase unit, cannot be used for control as it is not self-starting and must run at synchronous speed. Polyphase motors with more

---

* A rather brief but complete treatment of rotating amplifiers, and in fact motors in general, can be found in Ref. 13.

than two phases can be used for control applications, but these provide unnecessary complications and the two-phase motor is favored.

The motor is powered by two supplies which are phase-displaced by 90°. The voltage of one of the phases is the reference voltage; the other is a variable control voltage. A schematic diagram of the general arrangement is shown in Fig. 132a. In most military applications, the supply voltages are from a 400-cps source. To permit tolerances, however, the motors are generally built to be able to operate effectively from a supply which may vary from 380 to 420 cps. In some applications, the supply frequency may have even greater variations. This situation would be encountered when the supply generators are powered by variable prime movers as an aircraft engine. For these situations, variable-frequency two-phase motors are available. There are, for example, motors built which can operate effectively in the range 40 to 450 cps, others which can operate in the range 300 to 2,000 cps. These motors are, however, less efficient than the fixed-frequency machines.

The two-phase motor is basically an induction motor, and the standard rotor designs for induction motors are ordinarily used. However, because of the low inertia requirement, the solid and squirrel-cage machines are often built with long narrow rotors. That is, the length of the rotor is much greater than its diameter. In this manner rotor inertia is kept low. For the squirrel-cage machine, the copper bars are

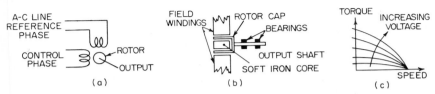

Fig. 132. Two-phase servomotor. (a) Circuit connections. (b) Drag-cup design. (c) Torque-speed curves.

generally skewed to prevent slot lock and cogging. When rotor inertia must be minimized, the drag-cup rotor design is often employed (see Fig. 132b). In this case the rotor is simply a hollow cylinder rotating around a soft-iron core. The latter is included merely to complete the magnetic circuit.

In general, the two-phase motor is used in applications where the power requirements are less than 100 watts. For larger installations, 1,000-watt motors are available, but beyond this point d-c motors are ordinarily employed as the a-c motor becomes grossly inefficient and overheating becomes a problem. In fact efficiency and power per unit pound are not the strong points of the two-phase machine. These motors can, however, be built with relatively low inertias and have high initial acceleration. These features make them useful in servo applications.

The torque-speed characteristics of the two-phase machine (see Fig. 132c) are approximately straight lines when the rotor resistance is high. The curves are not, however, parallel lines, and hence the motor does not provide for linear operation through a full speed range. For this reason, these motors are seldom operated beyond about one-third of synchronous speed. A theoretical derivation of these curves involves the use of induction-motor and rotating-field theory. The derivations are somewhat lengthy,* but results indicate that near stall the torque is very nearly proportional to the voltage across the control phase. These results are pretty well confirmed by actual measurements and further justify the use of these motors in the near-zero-speed range.

Experimental evidence indicates that the transfer function of the two-phase motor can be represented by

$$\frac{\Omega}{E} = \frac{K}{(sT_E + 1)(sT_M + 1)} \qquad (212)$$

* Derivations can be found in Refs. 12 and 13.

where Ω is the output angular velocity and E the control-phase voltage. The coefficient T_E and T_M are the electrical and mechanical time constants, respectively. For small-signal operation, the mechanical time constant can be determined from the characteristic curves. Its value is proportional to the slope of the curves at the operating point. That is,

$$T_M = J(d\Omega/d\tau) \qquad (213)$$

where J is the rotor inertia. The constant K can also be determined from the static curves. Its value is given by the horizontal distance between the curves at the operating point. That is,

$$K = \frac{d\Omega}{dE}\bigg|_{\tau = \text{const}} \qquad (214)$$

The electrical time constant must be evaluated from other information. As an approximation, L/R can be used where L and R are the effective rotor inductance and resistance, respectively, at the operating speed. These values would, however, have to be experimentally determined. Alternatively, the time constants could be derived from the frequency-response curves.

A-C motors have, in many applications, replaced the d-c motor. Probably the greatest advantage of the a-c motor is that it eliminates the need for commutators and brushes. The commutator and brushes of the d-c machine are a troublesome source of mechanical friction, electrical resistance, and electrical noise. As a further advantage of a-c motors, by their use, instrumentation and power supplies can be somewhat simplified. With an all a-c system, the induction-type instruments and transducers such as synchros and related devices can be used. Furthermore, a-c amplifiers can be employed, thus eliminating the problems of d-c amplification and the contingent drift problems. On the other side of the ledger, a-c motors cannot deliver the power, per given frame size, that the d-c motor can; nor is the a-c motor so efficient as the d-c counterpart. The d-c motor is more easily controlled and gives smoother operation. As in most engineering problems, the ultimate choice of which type to use depends on the particular problems on hand and the acceptable compromises which can be made.

Torquers. The motors described above are almost always high-speed units. That is, the armature or rotor speed is such that gear trains must be used to reduce speeds to acceptable levels. The torquer, on the other hand, is a device designed primarily to deliver torque. It operates at low or zero speed. The need for gear trains is thus eliminated and direct drives can be employed.

The torque-to-inertia ratio of a torquer is significantly higher than that of a comparable motor. As an example, a torquer capable of 60 oz-in. peak torque has a torque-to-inertia ratio of about ten times that of a motor which can deliver the same peak torque. In general, torquers, because of their relatively low inertia, have a faster response than conventional motors and they also consume less power per unit torque than do conventional motors.

Torquers are available from about 0.1 lb-ft (physical dimensions about 2 in. in diameter by 1/2 in. thick) to larger units capable of delivering as high as 3,000 lb-ft. Designs for 10,000 lb-ft units have been contemplated though not yet built.

Hydraulic Prime Movers

Two basic designs for prime movers are often used in hydraulic systems. Both are generally called "motors," but only the hydraulic rotary motor is a motor in the conventional sense; i.e., its output is rotary. The other hydraulic motor is a simple hydraulic cylinder (often also called an actuator). Here the output is rectilinear rather than rotational.

The simple cylinder (see Fig. 133a) needs little explanation. A piston guided in a cylindrical sleeve is free to move when there is a differential pressure across the piston head. The motor can be made single- or double-acting depending upon the valving arrangements used. A single-acting piston is driven in one direction only by hydraulic pressure. Reversing is accomplished through a spring drive or by the external load. The double-acting cylinder is driven in both directions by application of oil pressure on either side of the piston (see Fig. 133a). The maximum thrust available for a given set of piston dimensions is limited primarily by the pressure which can be delivered to the cylinder. In this instance it is not always the cylinder which limits the operating pressure. As often as not it is the oil-delivery lines which set the upper operating limit. Today, operating pressures of 3,000 psi are not uncommon. Higher-pressure systems are used but only in few isolated cases.

The cylinder with a simple three-valve arrangement (Fig. 133a) is not a follow-up system but more of an on-off system. Once the valve is energized (either mechanically or electrically) the piston will move and continue to do so until the valve is deenergized or the piston reaches the end of its stroke. In order to provide for a follow-up or an indexing system, the arrangement shown in Fig. 133b is often used. Here, the valving mechanism is integral with the cylinder; the piston is anchored and the cylinder moves. A small linear motion applied to the valve stem opens the ports to the cylinder. The cylinder moves until the ports are again closed. By such a

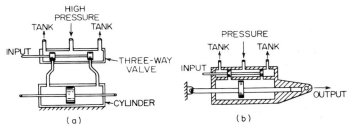

Fig. 133. Piston-valve arrangements. (a) Double-acting piston with three-way valve. (b) Force amplifier.

scheme, the cylinder is made to follow the valve-stem motion with a force amplification. That is, by proper design the valving mechanism can be actuated by ounces of force; the cylinder output can, however, be thousands of pounds. This device is, in fact, often called a hydraulic or force amplifier. There are other arrangements possible where valve and cylinder are coupled to give force amplification, some of which involve special feedback links. These devices are well described in almost all literature which includes mention of hydraulic mechanisms.* It is possible also to build linear actuators by using an electric motor coupled to a screw jack. It is of interest to note, however, that an electric actuator would weigh approximately twenty-five times a comparable hydraulic actuator having the same time constant and thrust.

Rotary hydraulic prime movers may be of the vane- or piston-type construction (see Fig. 134). There are other designs possible† besides the two aforementioned, but these appear to be the most versatile and hence are the most popular. The vane-type motor shown in Fig. 134a consists of a solid rotor with radial slots. Steel or brass vanes ride in these slots (only four vanes are schematically shown in Fig. 134a; actually there are many more) and wipe up against the outer casing which serves as a camming surface. The vanes are kept in contact with the cam surface by springs

* Refs. 5, 12, 13, and 15. All the above-mentioned references have a reasonably complete description of hydraulic components and first-order approximations to the various transfer functions. A very detailed treatment of these devices may be found in the last-named reference where not only are the transfer functions derived, but also analog-computer circuits are as shown when a more elaborate analysis yields nonlinear results.

† Gear-type motors have been attempted but these, because of friction problems, have proved thus far to be inadequate. Gear-type pumps are entirely feasible, however, and are in fact a popular type of power supply.

or by fluid pressure. High-pressure oil enters between the vanes. Since one vane is extended a greater amount than the one behind it, there is a greater force on the extended vane and the rotor turns, carrying the oil to the low-pressure port. Such a motor is of the fixed-displacement type. The rpm of the motor depends only on the quantity of oil delivered. A constant oil delivery will provide a constant output speed. The torque output is proportional to the differential pressure. The operating pressure of these motors is in the 2,000-psi range.

Shown in Fig. 134b and c are the two kinds of piston-type motors most often used today. The radial type shown in Fig. 134b is constructed with a solid rotor in which radial holes have been drilled. The pistons, which ride in these radial holes, are driven away from the center of rotation by fluid pressure which is introduced underneath the pistons through special porting arrangements. The upper end of the piston (which generally has a roller) is now cammed forward by a camming surface, thus giving the rotor its rotary motion. Like the vane-type motor, this radial-type piston motor shown in the diagram is a fixed-displacement unit. The rpm is constant for a

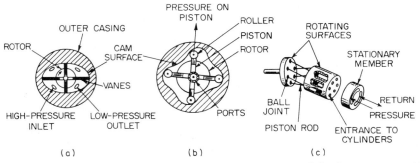

FIG. 134. Hydraulic motors. (a) Vane-type motor. (b) Radial piston motor. (c) Axial-piston motor.

constant-volume oil delivery. The operating pressure of the piston-type motors is higher than that of the vane-type unit. Pressures of 3,000 psi and better are possible with these motors.

The axial-piston motor shown in Fig. 134c* is probably the most versatile rotary motor of the three mentioned. The cylinder block shown is a rotating member as is the driven plate. Piston rods with ball joints butt up against the drive plate; the pistons themselves ride in axial holes drilled into the cylinder block. When the center lines of the cylinder block and valve plate are not in line with each other, the pistons around the periphery of the cylinder block have a variable stroke. Oil at high pressure is now introduced when the piston is closest to the stationary member. The force generated on the piston is transmitted to the drive plate through the ball joint. The drive plate is thus caused to rotate. In turn the plate drives the whole cylinder block around so that the piston moves around *with* the cylinder block until the oil is passed into the return line and the next cylinder aligns itself with the inlet port. The sequence then repeats itself for the next piston. Motors of this type are essentially variable-displacement devices. With a fixed input flow of oil, the output rpm can be varied by varying the angle between the drive-plate center line and cylinder-block center line. When these two center lines are collinear the pistons have no variable stroke and the output rpm is zero. In practice the angle between the center lines can be varied by ±20°. Reverse rotation is therefore possible with a fixed input flow. Motors of this type can be operated from pressure supplies up to and exceeding 3,000 psi.

It should be mentioned that any of the above-mentioned motors can be made to

* Some excellent pictorial representations of this and the other motors mentioned can be found in Ref. 15, chap. 4.

operate as pumps by merely driving the output shafts. In general, almost all pumps and motors (except possibly the gear-type units) can be used interchangeably except for minor adjustments. The vane- and radial-type pumps are in fact often made into variable-delivery pumps by permitting the rotor to run eccentrically with respect to the camming surface.

Hydraulic transmissions are available in which the pump and motor are built in one integral housing. This arrangement constitutes one type of hydraulic control system often called a pump-controlled transmission. Such a system ordinarily consists of a variable-stroke pump and a fixed-stroke motor. Motor control is effected by varying the pump delivery. This system is efficient but cannot be used when several independent motors are to be fed from the same pump supply. When such an arrangement is desired, a valve-controlled hydraulic system is usually employed. In this instance both the pump and motor are of the fixed-displacement type. Flow to the motor is varied through the use of valves such as variable-flow and variable-pressure valves. A schematic diagram of the two types of systems is shown in Fig. 135.

As yet, hydraulic transmissions are not nearly so popular as electric transmissions. Hydraulic systems are very susceptible to dirt, air entrainment, fire, and high-pressure

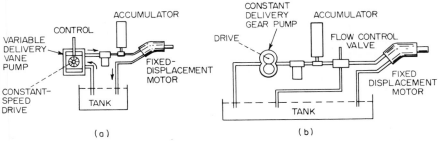

Fig. 135. Transmissions systems. (a) Pump-controlled system. (b) Valve-controlled system.

leaks. Some of these difficulties have been reduced but not eliminated. However, the operating characteristics of these transmissions often outweigh their disadvantages. The horsepower per pound of equipment for hydraulic systems is much higher than that of equivalent electrical systems. For linear motors, the weight advantage is in the ratio of 25:1 in favor of hydraulics; for rotary motors the advantage is much reduced and is only about 4:1, but this saving is significant, especially in aircraft. A hydraulic system is relatively a stiff system—the compressibility of oil, while not infinite, is significantly greater than the "give" of a magnetic field. As such, hydraulic time constants are much lower than electrical time constants for comparable equipment. When there is a choice to be had, electrical systems are still favored. In some cases, however, and these are occurring more and more frequently, no choice exists. Operating requirements are such that hydraulic systems must be used.

Pneumatic Prime Movers

As mentioned previously, the basic difference between the hydraulic and pneumatic systems is the working medium. The components of both systems, especially with regard to vane motors and valves, are almost identical. In some cases it is even possible to interchange components. The pneumatic diaphragm motor (Fig. 136) is probably unique to pneumatic systems, however. These motors are almost always used as valve positioners rather than prime movers, however. The stroke is small and they are seldom used with pressures above 15 psi. For heavier-duty work the pneumatic cylinder is used.

Pneumatic systems are almost always valve-controlled transmissions. The reason for this is that variable-delivery pumps are not commercially feasible. Pneumatic systems are thus powered, either from a constant-delivery pump or from a "charged bottle," i.e., a compressed-air tank. The bottle is in a sense a "pneumatic battery." In either case the only reasonable way to control the motor output is through valving.

Within recent years pneumatic systems have been growing in popularity and have in many instances replaced hydraulic systems, especially in aircraft systems. For aircraft, it is necessary that storage tanks be used to make the pneumatic system practicable. Until recently storage-tank capacities were relatively low. But now 3,000-psi storage systems have been built and tanks with higher capabilities are being contemplated. With the ability to operate at higher pressures, the size of pneumatic components was reduced until now, considering working medium, piping, supplies, and components, pneumatic systems are in some cases superior to hydraulic systems. Besides the comparable horsepower per pound feature, pneumatic systems offer other advantages. They are cleaner, relatively unaffected by temperature, and do not constitute a fire hazard. On the other hand, pneumatic systems are not so stiff as hydraulic systems. They therefore have lower resonances and longer time constants. The systems are not self-lubricating, and oil must be delivered to moving parts. Finally, hydraulic systems can have an overall efficiency of almost 60 per cent whereas the air system seldom has an efficiency exceeding 30 per cent.*

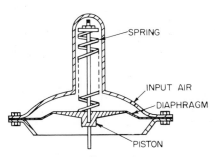

FIG. 136. Pneumatic motor.

REFERENCES

1. N. O. Myklestad, "Fundamentals of Vibration Analysis," McGraw-Hill Book Company, New York, 1956.
2. J. P. Den Hartog, "Mechanical Vibrations," 4th ed., McGraw-Hill Book Company, New York, 1956.
3. J. W. Beams, J. D. Ross, and J. F. Dillon, Magnetically Suspended Vacuum Type Ultra Centrifuge, *Rev. Sci. Inst.*, vol. 22, no. 2, February, 1951.
4. John G. Truxal, "Automatic Feedback Control System Synthesis," McGraw-Hill Book Company, New York, 1955.
5. T. C. Gille, M. J. Pélegrin, and P. Decaulne, "Feedback Control Systems," McGraw-Hill Book Company, New York, 1959.
6. W. R. Weems, "An Introduction to the Study of Gyroscopic Instruments," mimeographed notes, McGraw-Hill Book Company, New York, 1948.
7. I. A. Greenwood, J. Vance Holdam, Jr., and Duncan MacRae, Jr., "Electronic Instruments," McGraw-Hill Book Company, New York, 1948.
8. W. E. Fellows, "Vibration Effects on Gyroscopes and Accelerometers," Minneapolis-Honeywell Aero Document R-ED 29004-1A.
9. L. F. Warnok, "Frequency and Damping Dependence of Anisoelastic Torques in the M. I. G. Gyros," Minneapolis-Honeywell Aero Research, Nov. 4, 1957.
10. H. Goldstein, "Classical Mechanics," Addison-Wesley Publishing Company, Inc., Reading, Mass., 1956.
11. W. Wrigley, F. E. Houston, and H. R. Whitman, Indication of the Vertical from Moving Bases, *Mass. Inst. Technol. Instrumentation Lab. Rept.* R156, June, 1956.
12. J. G. Truxal, "Control Engineers' Handbook," chap. 12, McGraw-Hill Book Company, New York, 1958.
13. John E. Gibson and F. B. Tuteur, "Control System Components," p. 149, McGraw-Hill Book Company, New York, 1958.
14. Ref. 13, p. 215.
15. J. F. Blackburn, Gerhard Reethof, and J. Lowell Shearer, "Fluid Power Control," The M. I. T. Press, Cambridge, Mass., and John Wiley & Sons, Inc., New York, 1960.

* An excellent comparative analysis of hydraulic vs. pneumatic systems is found in Ref. 15, chap. 19. A weight comparison between electrical, hydraulic, and pneumatic systems is found in Ref. 13, Sec. 13.2.

REMOTE HANDLING

18 Introduction

This part of the handbook is concerned with remote control in the sense that this term is used in the nuclear and allied fields. This usage should be distinguished from that applicable to such devices as remotely operated valves as employed in the processing industries.

The communication link which is a necessary component of any remote-control system of the type to be discussed here is a special application for telemetry and promises to develop a variety of new and interesting applications for well-understood telemetry and multiplexing techniques.

The increasing development of hostile or hazardous environments for economic, military, or political purposes indicates a considerable expansion of the technology for operating in these environments in the next few years. The technology of hostile-environment operations is in its infancy, but the potentialities and the problems can be stated with some clarity. It is the purpose of this discussion to explore the current

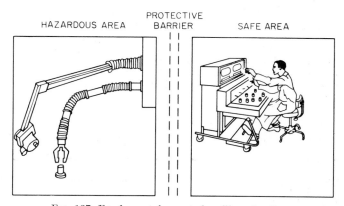

Fig. 137. Fundamental remote-handling situation.

state of the art, to define an analytical approach to the design of hostile-environment systems, and to outline the principal areas to which research effort may be effectively directed.

The discussion which follows is concerned with general system-design principles and accordingly will not cover specific equipment-design methods. The design methods and the structural mechanisms to be employed for hostile-environment remote-handling systems are not different from those applicable to any electronic or electromechanical system for the environment in question. On the other hand, we are depending upon those who specialize in the development of materials, components, and structures for hostile environments to make available to the remote-handling system engineer the materials and components from which he can assemble systems as required.

At present, hostile-environment system design is only beginning to be recognized as a technology in its own right. Like many other branches of engineering, it draws heavily upon previous experience in many fields. As experience is gained, it is confidently to be expected that remote-handling or hostile-environment technology will assume its place in engineering and as a rewarding subject for research.

In order to define more clearly the hostile-environment problem, consider Fig. 137. This schematic illustration shows an area which is hostile, i.e., unsafe for occupancy by personnel. Within this area is a task to be performed. This task may include any operation normally performed manually or with the aid of hand tools. Such operations as operating a wrench or screw driver, inserting or releasing electrical connectors, pouring liquids, or operating stopcocks or valves are but a few of the very large number of typical operations. In all cases these operations require manipulation of one or more objects, in other words, geometrical translation and/or rotation over at least a limited distance. Figure 137 shows schematically a mechanical "arm" to accomplish this manipulation and a mechanical "eye" to observe and monitor it.

A protective barrier separates the hostile area from a safe area in which a human operator may remain indefinitely. He is provided with a control console from which he directs the motions of the manipulating device.

Simple as it is, this situation illustrates all the essentials of any hostile-environment situation. For purposes of general analysis one is not concerned with the nature of

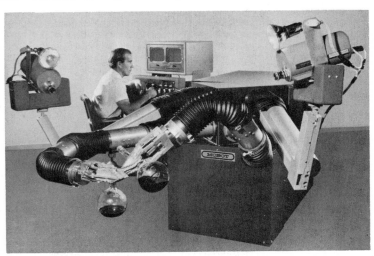

FIG. 138. A typical fully remote handling system. (*Hughes Aircraft Company photograph.*)

the hostile environment or of the protective barrier. One attempts to establish a body of theory with the aid of which systematic analysis can be made of any specific situation; this analysis will in turn result in the design of an optimal system to accomplish the required effort.

Figure 138 shows an example of a typical modern hostile-environment system, including all the features basic to such systems. This vehicle moves freely about on a three-wheel chassis and employs two multiply articulated arms for manipulation and two television cameras for vision. It is controlled by a three-conductor cable which may be, if required, a few thousand feet long. Its operator is provided with TV monitors, indicators for other sensory transducers on the vehicle, and a variety of control switches to direct all its motions. There is no communication between operator and vehicle other than that provided by the three-conductor interconnecting cable.

In the following discussion we shall analyze in more detail the several major subsystems which make up any remote-handling system and shall survey the current practice, as well as the future possibilities, of this new technology.

The Hostile Environment

Any environment into which it is undesirable or unsafe to introduce personnel may properly be referred to as "hostile." Specific examples of hostile environments now of serious economic and/or military importance are the nuclear environment, the undersea environment, and the space environment. Additional hostile environments may readily be added to this list, for example, elevated temperatures as in steel mills and the like, extremely low temperatures as in the Arctic, and subterranean situations as in mining, in which the principal hazard may be due to falling rock.

In all the cases noted, a man can enter the environment, if at all, for only a limited period of time. In some cases, protective clothing, such as deep-sea diving equipment, a space suit, or heat-resistant clothing as used by fire fighters, may be employed. Such protective clothing will increase the length of time which a man can remain in the environment but will decrease his ability to perform useful work by some quite measurable factor. Protective clothing can in fact be looked at as a means of placing the "safe environment" of Fig. 137 within the "hostile environment" while retaining the necessity for communicating between the two areas. Looked at in this way, a proper analysis of a hostile-environment situation will include among the possibilities the use of protective clothing or similar devices.

The "tasks" to be performed within the hostile environment encompass the complete gamut of work normally performed manually or with the aid of tools. The analysis of a specific hostile-environment situation properly starts with a clear definition of the task to be performed. One then analyzes a variety of alternate methods for performing these tasks and selects that system which will provide the most effective performance for minimum cost. In this way one attempts to avoid preconceived ideas concerning such irrelevant questions as "man versus machine" or "shielded vehicle versus remote-controlled vehicle" or a variety of related alternates which, being concerned with methods rather than results, do not always lead to factual analysis of the problem.

As a first step toward developing factual criteria for design of hostile-environment systems, the following classification into three groups is useful:

Group 1—Direct-vision Systems. Under this classification are considered all types of systems in which the physical separation between the hostile and safe environments is at most a few feet, and in which tools resembling tongs or forceps may extend the reach of the human arm. This classification also includes gloves or other flexible barriers which permit the indirect use of the operator's own fingers. Vision is accomplished through appropriate protective windows.

Group 2—Remote-control Devices. This group includes all mechanisms which are unrestricted in the separation between the safe and hostile regions, but in which all actions of the remote-handling system are directed in detail by a human operator. The system shown in Fig. 138 is an example of this type.

Group 3—Programmed Systems. Such systems are located fully within the hostile area and have no communication with the outside world. They are obviously applicable only when access to the hostile area is available at certain periods of time. They also are most applicable when the task to be performed is clearly predictable, so that a program to perform it need not be unreasonably complex. Many reactor-refueling devices come under this category, as do many of the first-generation instrumented probes for exploring space and the lunar surface.

Clearly, the boundaries between the three classes of remote handling noted above are not always clearly defined. On the contrary, sound system design may judiciously combine these in order to optimize the overall system. Specifically, a carefully thought out combination of a manual remote-controlled (group 2) system with a programmed system (group 3) in which the operator can select from a variety of predetermined programs will often far outperform a system confined to either of the two types of control.

Present Practice

As one might expect, the current technology and practice in hostile-environment operations is widely spread throughout industry and science. Perhaps the most dramatic of the hostile environments is the nuclear environment, and indeed one finds a variety of remote-handling gear has been developed and operated in connection with nuclear "hot labs" and reactors.

The simplest of such equipment is the so-called dry box developed from the familiar chemical dry box. The nuclear dry box employs flexible gloves for manipulating and a glass or plastic window for viewing. Quite elaborate ventilating systems and pass-through devices for introducing and removing articles into the dry box may be employed.

One should point out here that the technology indicated by the dry box has long been employed in both chemical and biological investigations. Wherever toxic, volatile, or explosive materials must be manipulated, and where the relatively limited protection afforded by the thin walls and the gloves is adequate, the dry-box technology is well proved and accepted.

It is possible that the development of electromechanical arms will make feasible the development of dry boxes without gloves. One or more miniature remotely controlled arms within a transparent box might permit more handling facility with less hazard and hence permit even wider use of the dry-box technology, which is of course a group 1 handling method.

Returning to nuclear practice, one often encounters situations in which the dry box is not adequate because of either extremely high radiation levels or extremely toxic as well as radioactive materials. Shielding must in many cases be from 3 to 6 ft of concrete to afford adequate personnel protection. Operations with radioactive sources of such intensity as to require these thick shielding walls are usually accomplished in the "hot cell," which has become a familiar feature in many nuclear laboratories. Figure 139 illustrates a typical hot cell employing heavy concrete shielding walls for operator protection, and employing equally heavy radiation-resistant glass windows for vision.

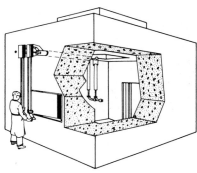

FIG. 139. A typical hot cell.

Manipulation is accomplished with the master-slave manipulator, as illustrated in the figure. This is a group 1 handling system in that it is inherently limited to distances between 6 and 10 ft and vision is accomplished through the radiation-resistant glass windows. The precision and dexterity of manipulation with a master-slave manipulator are of extremely high quality. The mechanisms described in Ref. 1 have been developed and improved over the years and are now widely accepted in nuclear and related work.

The hot cell as described above was developed primarily for chemical operations and for the handling of relatively small and light objects. When confronted with the necessity for working with large and heavy objects, and perhaps for transporting these over considerable distances, nuclear practice has typically employed very large shielded rooms in which electromechanical manipulating arms are employed, supplemented in some cases by master-slave manipulators. Vision is accomplished by a combination of radiation-shielding windows, periscopes, and television. Examples of such facilities are the FET (Flight Engine Test) facility at Arco, Idaho, which was operated for some years in connection with the nuclear-airplane program; the MAD (Manipulation and Disassembly) building, which is in increasing use in connection with the nuclear-rocket program; and the SERF (Sandia Engineering Reactor Facility), which is used for investigation of the effects of reactor radiation upon a great variety of electronic and electromechanical systems. The SERF employs a particularly effective combination of group 1 and group 2 handling systems, as described in Ref. 2. Reactor refueling is a task of increasing importance in connection with electric-power reactors. In general, this operation, which consists of the removal of a spent fuel element and its replacement by a fresh one, involves a completely predetermined sequence of mechanical motions. Most refueling machines have been purely mechanical in nature and can be considered as extremely complex cams. Thus they fall under the general heading of group 3, or programmed hostile-environment systems.

In the underwater environment, a lengthy history of protective devices can be studied. Present-day practice depends primarily upon the diving suit, which may

Fig. 140. The RUM (Remote Underwater Manipulator), a mobile underwater manipulating system. (*U.S. Navy photograph.*)

be a simple rubber fabric cover for relatively shallow water, ranging to heavily armored pressurized clothing capable of operation to depths of several hundred feet for brief periods. Obviously, a breathable atmosphere must be furnished to the diver and he must be protected as much as possible from hydrostatic pressure. Practically all undersea installations currently in use were installed with the aid of divers employing this familiar equipment.

For exploring the depths of the ocean beyond diver capability, a variety of diving bells and pressurized vessels have been successfully operated. Most spectacular of these is the Bathyscaphe,[3] with which the greatest penetration of the ocean's depths has been made. A scientist within a Bathyscaphe is a mere helpless observer of his surroundings; only recently are steps being taken to equip the Bathyscaphe with external manipulating and viewing equipment.

Fig. 141. The Solaris, a free-swimming, remotely controlled underwater grappling system. (*Courtesy of Vitro Corp.*)

The reader will note that both diving suits and Bathyscaphes are group 1 handling systems, in that the operator is limited by his own ability to reach and to see in his capability for accomplishing undersea tasks. An example of a group 2 system for undersea purposes is the RUM (Remote Underwater Manipulator) shown in Fig. 140. This vehicle, built upon a modified tank chassis, crawls upon the ocean bottom under commands transmitted through a cable 5 miles long. Vision is accomplished by both TV and sonar, and manipulation by a single planar arm mounted on the forward end of the vehicle. With the increasing pressure to develop the resources of the ocean, it is to be expected that a variety of advanced group 2 undersea vessels will be developed and operated in the near future.

Another example of a group 2 undersea system is "Solaris," shown in Fig. 141.

This is a free-swimming, cable-controlled device; its principal purpose is retrieval of experimental torpedoes by means of its integral grasping device.

The above brief survey of present practice omits any discussion of the space environment, since at present our accomplishment of actual tasks in space is limited to a few simple group 3 programmed instrumentation problems. It is, however, to be expected that the development of basic hostile-environment techniques will prove of increasing value as the exploration of space proceeds.

Analysis of the Basic Hostile-environment Problem

Consider the basic situation as shown in Fig. 137 from an operational rather than a mechanical-design viewpoint. If the environment were not hostile the man would enter the region and would approach the task and accomplish it using his muscular system for the physical motions, his sensory system to obtain information concerning his surroundings, and his nervous system to communicate between his intelligence and his senses and muscles. Also, of course, he uses a separate set of muscles to move himself about, as contrasted with those employed in manipulation.

In this method of analyzing the performance of a task, we note that the man is reduced to a few simple systems, as shown in Fig. 142. We note that by far the most difficult of the items shown in Fig. 142 to extend to a distance is the intelligence. The other subsystems, namely, the senses, the muscles, and the nerves can all be quite readily extended by electronic and electromechanical means to any arbitrary distance. We therefore locate the human intelligence in a safe region and provide it with an extension of the nervous system which connects with the senses and the muscles, which must of necessity be located in the hostile environment. Figure 143 shows this new analysis of the elements required.

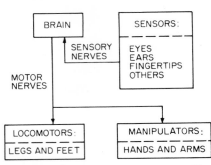

FIG. 142. Analysis of man as a handling system.

Properly engineered remote-handling systems provide for the operator complete information concerning the surroundings of the remote vehicle and complete ability to control the motions of the manipulating devices and the vehicle itself. In a very real sense, the operator of a system of this type loses temporarily all conscious awareness of his own physical situation and identifies himself with the remote vehicle. Subjectively, he is "where the vehicle is." In fact, until this subjective identification is accomplished, one does not have a really effective remote-handling-system operator.

This operational view of the basic hostile-environment problem, with its emphasis upon the "gestalt" which the operator gains from the remote-handling machine results from facing squarely the hostile-environment problem as noted above.

Experience to date clearly demonstrates that this gestalt can be accomplished with only a few hours of systematic practice. The psychological feasibility of this particular learning process had to be demonstrated by actual experience, since there was no past evidence upon which to evaluate it. This has now been accomplished and one can proceed with confidence to design equipments for the exploration of all hostile environments employing these principles.

Equipment Requirements of Hostile-environment Systems

This discussion, in contrast with the preceding one, will be concerned with analysis from an equipment viewpoint rather than an operational analysis of hostile-environment systems.

The basic subsystems required to accomplish the functions of a generalized hostile environment system are shown in Fig. 144. Here again we see the hostile environment and the safe environment. Contained within the hostile environment is a vehicle which must of necessity include within its structure a locomotion system to move the entire vehicle about, a sensory system to gather information concerning its surround-

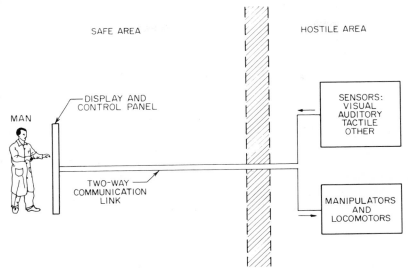

Fig. 143. Fundamental components of any remote-handling system.

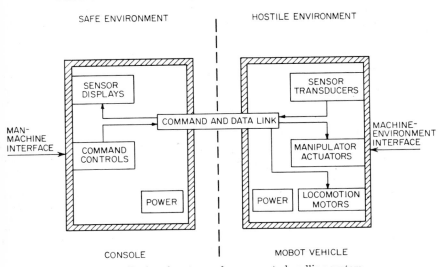

Fig. 144. Basic subsystems of any remote-handling system.

ings, and a manipulating system to perform whatever tasks may be required. In addition to these, a source of power must be available within the vehicle and a multiplexing system to distribute both incoming and outgoing information from the various subsystems.

A communication link must join the remote vehicle with the safe environment. This link must communicate in both directions. It is perhaps better referred to as "the command and data link," since commands travel from operator to vehicle and data from vehicle to operator over this link. Over short distances, this link may simply be a multiwire cable. Over longer distances it will be a two- or three-conductor cable, and over still longer distances a two-way radio system is suitable.

Within the safe environment, the equipment required consists of, again, a source of power and a multiplexing system. In addition, one must have the readouts or displays corresponding to the sensors in the sensory system on the vehicle, and there must be a variety of controls which actuate the physical motions of the vehicle. The connection between the human operator and the system involves only the displays of the sensory system and the actuating means for the physical motions. For convenience, these items are usually installed on a control console, which is the man-machine interface.

Any remote-handling system, whether simple or complex, must contain all the elements shown in Fig. 144, since these are basic to accomplishing hostile-environment operations. In approaching a new hostile-environment-system design problem one finds it extremely useful to employ this same subsystem analysis in order systematically to optimize the design of the system and the engineering of the specific subsystems.

The preceding general analysis of the basic hostile-environment situation enables one to define in terms of equipment six major subsystems which make up any hostile-environment system. These subsystems are

The manipulatory subsystem
The locomotion subsystem
The sensory subsystem
The communication and data link
The control console
The power subsystems

In approaching from the system viewpoint any specific hostile-environment problem one attempts first to state the requirements upon the overall system as clearly as possible; this is usually done in terms of weights to be lifted and the geometrical and velocity requirements upon their motions. Further constraints, such as fixed obstacles, nature of surface available for locomotion, and the like, should also be stated. Given these criteria, one then proceeds to design the six subsystems so that each will perform its required function and so that the total system will be an optimal one with due regard both to performance and to economics.

The sections which follow will examine in somewhat more detail the design methods applicable to each separate subsystem.

19 The Manipulatory Subsystem

In approaching an understanding of the manipulatory subsystem, it is first useful to separate the manipulating function as such from the simpler handling function. For purposes of this discussion, the term "manipulation" will refer to translation and/or rotation of objects with a minimum of six degrees of freedom. In other words, manipulation will be used to refer to the type of generalized unrestricted motions of which the human hand and arm are capable.

In contrast, the term "handling" will be used to refer to simple translations usually along predetermined geometrical coordinates. This handling function is typically performed by bridge cranes, forklift trucks, and other heavy equipmen familiar to the rigging industry.

In general, one does not manipulate heavy objects or handle light objects. The borderline between the two may be established somewhat arbitrarily somewhere between 25 and 100 lb. This limit is primarily dictated by the limitations of the human hand and arm, since many of the items to be worked with are designed for human handling and since both hand tools and heavy rigging equipment are engineered for a weight separation in this vicinity.

To restate the distinction between manipulation and handling, manipulation is the general motion of small objects; handling is the restricted motion of large objects. This distinction is of value in equipment design, since it enables one to make the fullest use of already existing equipment with a minimum of adaptation for remote control.

For the most part, the handling function as defined above can well be accomplished by remotizing* already available handling or rigging equipment; hence it will not be further discussed in this section. On the other hand, remote-manipulating equipment which should ideally approach the versatility of a human hand and arm requires development and study and will form the principal subject of the balance of this section.

Design Criteria for Manipulating Devices

As an attempt to arrive at quantitative design criteria for manipulative systems, one may start with the observation that, as a minimum, seven degrees of freedom are required for any such device. The seven degrees of freedom are three translational motions, three rotational motions, and one grasping motion. In an unobstructed situation in which only a single rigid body is to be manipulated, the seven degrees of freedom are quite adequate. Note that the translatory and rotatory motions both refer to the center of gravity of the object to be handled, not to any fixed point in the manipulatory system itself. The grasping motion is integral to the manipulator but is an essential function without which even the simplest system cannot perform.

In the simplest possible case then, one needs to define the maximum weight of the object to be handled, the distance to be translated, and the maximum translatory and rotary speeds required in order fully to define the handling requirement. These same criteria will also serve to evaluate specific handling "arms" for accomplishing a given task.

Such properties as jaw pressure and torque or speed limitations of specific structural portions of manipulators should be viewed as derivative properties in the light of which one can calculate the primary capability of the manipulator as defined in the preceding paragraph.

It is clear that, when one contemplates the fundamental geometrical problem of manipulation, which is to translate an object (which may or may not be rigid) from one point in space to another, that in general more than the minimum seven degrees of freedom will be required. Any additional constraint, consisting either of a fixed obstacle in the operating region or a geometrical constraint to travel along a predetermined path between the two end points, may require additional degrees of freedom. The structure of the manipulating arm itself may also impose constraints because of its bulk and geometry. While it is difficult to make general statements about these constraints, they can usually be quite clearly defined in particular instances, and an envelope of the space available to the manipulator can be computed. The necessary additional degrees of freedom required to conform with this availability envelope can then be determined.

By the same token, when one approaches the design of a general-purpose manipulating arm without regard to specific tasks, one attempts to design into it mechanical redundancy without unduly increasing the bulk of the arm itself. In other words, more than the minimal seven degrees of freedom are required, and in general one wishes to obtain within the working volume the largest possible freedom in direction of approach of each point and in trajectories available between different points.

In the manner briefly outlined above, workable design criteria and specifications for handling arms can be derived. These criteria will revolve around definition of the geometrical degrees of freedom in the arm in combination with the maximum weights which can be lifted and the maximum speeds of translation and rotation available both for the maximum weight and for objects of lesser weight or moment of inertia.

It should also be clear that overspecifying any of these properties may lead to ineffective performance of the complete system. The additional bulk required by a

*This term shall be used to mean rendering able to be used remotely.

system engineered for greater weight or speed than is actually needed may prove a severe operating limitation.

Current Arm Designs

Perhaps the most familiar current arm design is the "master-slave manipulator" widely used in nuclear hot cells. This arm, which is mechanically operated, accomplishes the minimal seven degrees of freedom by a combination of rotations about the upper mounting point of the tube which penetrates the shielding wall and of rotations about the so-called "wrist" which supports the grasping tong. The three translatory motions actually occur in arcs centered on the upper mounting point; the three rotary motions occur about the geometrical center of the wrist structure. The fact that the rotations are not centered in the object grasped and that the translations are not exactly linear has proved to be but little handicap in practice. There are of course

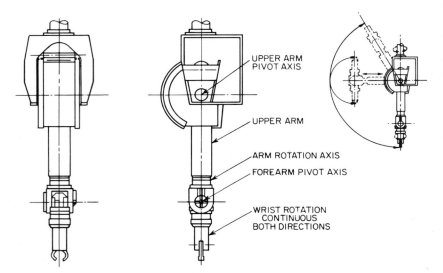

FIG. 145. A typical planar arm. (*Courtesy of General Mills, Inc.*)

numerous situations in which the vertical supporting tube for the tong interacts severely with fixed obstructions, limiting one's handling volume. Specifically, this arm is quite unable to reach under a shelf or tabletop. On the other hand, the speed and dexterity obtained as a result of the direct mechanical transmission of tactile information to the operator is of extremely high quality, as demonstrated by the wide and successful use of these master-slave manipulators.

Electronic master-slave manipulators employ bilateral servosystems instead of mechanical interconnections between master and slave. One such system was demonstrated at the 1958 Geneva Conference on Peaceful Uses of Atomic Energy.[4] An electronic master-slave manipulator employing force amplification is the "Handy-Man," developed for use in the nuclear-powered aircraft program. See Ref. 5 for a detailed description.

In order to handle items beyond the lifting ability of the master-slave manipulator, and in some cases to traverse greater distances, a variety of electromechanical arms have been developed by and for nuclear technology. These arms might be referred to as "planar arms," since in most cases their articulations confine them to a single plane. Figure 145 illustrates a typical arm of this class. When mounted upon a three-axis mount (usually adapted from conventional bridge-crane structures), these planar electromechanical arms can effectively cover a very large working volume, and if due regard in layout is given to their geometrical restrictions, very effective handling can be accomplished with them. Since they are electromechanical, they are almost unrestricted in the mass which can be handled; arms of this type have been developed capable of manipulating as much as 500 lb.

Still another basic arm type is shown in Fig. 146. This arm employs bidirectional articulations to give versatility in the direction of the approach of the operating point. An arm employing three such articulations can approach any point within its toroidal operating volume from any direction, thus permitting operation even in quite restricted situations. These same articulations permit the arm to depart from a single plane when required for operation among, for example, a variety of stationary pipes or supports. Figure 147 illustrates applications of this arm in a variety of typical restricted situations.

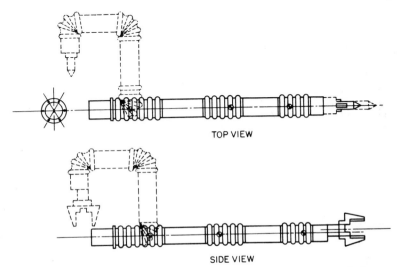

FIG. 146. An arm employing bidirectional articulations. (*Hughes Aircraft Company.*)

Numerous other electromechanical, electrohydraulic, and mechanical arms have been built and operated in the past few years. All are, however, in the general category of one of the types described above. There are few fields where ingenuity in mechanical design can be more rewarding; hence it is to be expected that the next years will see great activity in the design of improved remote-control arms which will very likely take shapes at present undreamed of, and which will in general not closely resemble either present mechanical arms or the human arm.

Manipulator Controls

The general subject of command and control systems will be explored more fully below in the discussion of the command and data link. It is, however, desirable at this point to explore briefly the requirements applicable to the control of the manipulating subsystem.

The fundamental decision concerning the control system involves a choice among position command, rate command, acceleration command, or combinations of these. These three nearly self-explanatory terms are defined below.

Position command refers to systems in which the controlling element (on the operator's console) is made to duplicate the position of the element controlled, which is within the hostile environment. In such systems the controlling element must be able to duplicate all the degrees of freedom of the control element. Further, either scaling factors must be employed, or it must also duplicate the physical forces upon the element controlled.

Rate-command systems are those in which the position of the element on the control desk determines the speed with which the controlled element moves. There is of course no simple relationship between the positions of the controlled and controlling elements.

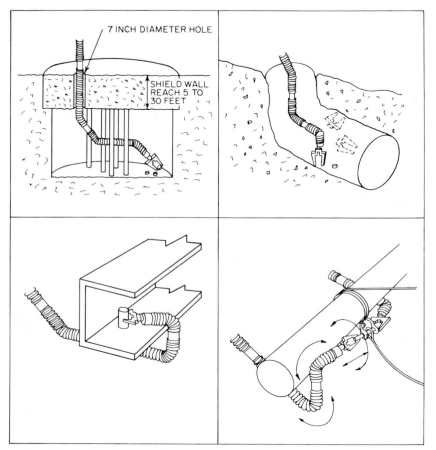

Fig. 147. Applications of bidirectional articulated arm to a variety of situations.

Acceleration-command systems are those in which the rate of change of velocity of the controlled element is determined by a movable element on the control console.

It is clearly possible to combine these in a variety of ways; effective system design will attempt to make such a combination in a manner most favorable to the particular tasks to be performed. It is noteworthy that position-command systems have the fundamental property that the operator has merely to move his hands and arms in exactly the manner desired in the remote location in order to have these motions duplicated within the hostile region. An example of such a system is the master-slave manipulator described in more detail under Present Practice. Such systems are quickly learned. In contrast, rate-command or acceleration-command systems require a certain training period before facility is obtained in their operation. In return for this, one gains complete freedom in determining both the configuration and the magnitude of the remote-controlled arms.

In overall system design, one must also consider the relative difficulty with which commands are transmitted over the command link. In general, bandwidth required is considerably greater for position-command systems than for rate- or acceleration-command systems. This point is explored more fully in the section on the command and data link.

The physical elements (switches, joysticks, wheels, foot pedals, etc.) mounted upon the control console, with which the operator commands the motions of the remote-controlled arm, are perhaps the most intricate portion of the control console. The relative merits and demerits of a number of alternate schemes for commanding complex remote-control arms are more fully explored in the section on the control console.

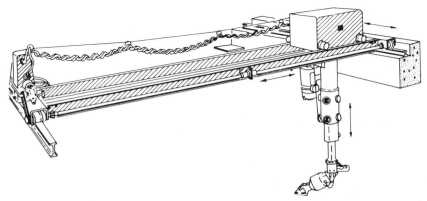

Fig. 148. Planar arm mounted on three-axis carriage, illustrating combination of locomotion and manipulating functions. (*Courtesy of General Mills, Inc.*)

20 The Locomotion Subsystem

The distinction between the manipulating and handling subsystem discussed in the preceding section and the locomotion subsystem is to some extent an arbitrary one. It is nevertheless useful for analytical purposes in permitting one to attack system design in an orderly manner. The distinction between the two subsystems is based upon distance to be traversed. In general, manipulating operations as defined above involve translations of, at most, a few feet. When translation of more than a few feet is required, it is usually more practical to design a separate mechanism for a pure translation (without manipulation) over the distance to be required, and to separate this locomotor mechanism completely from the "arms" which accomplish the manipulation. One example, which has become quite familiar in nuclear technology, of this separation is the planar arm mounted on a three-axis carriage, shown in Fig. 148. Here a conventional electromechanical planar arm, having a manipulating reach of 4 or 5 ft, is mounted upon a three-axis carriage which is almost unlimited in its traversing range. A few tens of feet in all three dimensions can quite readily be accomplished with such mechanisms, thus providing full coverage of a large nuclear hot cell or similar installation. Note the clear separation of the manipulating mechanisms or arm from the locomotor mechanism or traversing bridge. A more extreme example may involve the installation of handling arms upon a conventional road-mobile or off-road vehicle chassis, as shown in Fig. 149. With addition of a suitable command system and of TV cameras for viewing, many conventional vehicles can be remotized and can be driven, steered, and operated by a conventional radio link almost over any desired distance. Such vehicles can of course accomplish the locomotion function over any distance desired up to many miles. It is often desirable to add an intermediate locomotor function in the form of a simple extendable boom in order to permit translations of a few feet without the necessity of maneuvering the entire vehicle.

The attempt to design a single mechanical structure which can accomplish translations over many feet and also can accomplish delicate and precise manipulations usually results in mechanically unworkable structures. The bulk required adequately to support the considerable weights involved in a cantilever configuration over a distance of 15 or 20 ft usually results in a structure so heavy as to be incapable of rapid acceleration, and also incapable of entry into restricted regions.

The logic leading to the somewhat arbitrary separation between manipulation and locomotion is based upon an analysis very briefly summarized in the preceding paragraphs. As examples of the application of the analysis, the following sections discuss briefly the locomotion methods available to the remote-handling-system

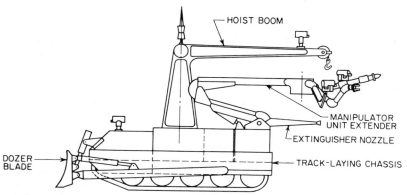

Fig. 149. A general-purpose off-road radio-controlled maintenance or rescue vehicle. (*Hughes Aircraft Company.*)

designer. One notes that many of these methods are well established for other applications, and commercial hardware can usually be employed or adapted for hostile-environment locomotion purposes.

Land Operations (*within Buildings*)

Locomotion within laboratory buildings will take full advantage of protection from the weather and of the frequent availability of smooth concrete surfaces. Simple hard-rubber-tired chassis permit great freedom of operation when unobstructed floor space is available. In many cases the floor space is obstructed, and three-axis bridge cranes mounted on overhead rails are to be preferred, giving access from above rather than below to the objects to be manipulated.

Consideration should also be given to operation within buildings where neither of the above methods is available. This can occur in excessively crowded conditions; it may also occur when because of accident or disaster the normal means of locomotion are not available and emergency vehicles must enter the building in spite of obstructions and debris on the floor and of unavailability of crane rails or other overhead means of locomotion. Development of locomotion methods which are able to climb over or crawl under obstacles, to climb stairways or ladders, or otherwise get about under difficult conditions, appears quite feasible but will require much experimental effort before it can be considered a proved capability.

REMOTE HANDLING

15-151

Land Operations (Outdoors)

For the most part, exterior remote-control vehicles can employ conventional rubber-tired or track-laying chassis, as noted in the introductory discussion. Many conventional vehicles can quite readily be remotized; i.e., their control systems for steering, throttling, and braking can be adapted to electrical actuators and these can be tied into an appropriate command system. Such vehicles need only be fitted with appropriately mounted TV systems for vision, and provide very satisfactory mobile operating chassis for outdoor remote operations.

The development of remote-handling technology promises a new departure in off-road vehicle capability which deserves brief mention. For negotiating extremely difficult terrain, a mechanism more like the feet and legs of animals may well outperform either wheels or track-laying systems. Even though attempts to develop "walking vehicles" in the past have not been conspicuously successful, electronic command systems may justify reconsideration of such vehicles. An articulated locomotion member which can step over or otherwise negotiate difficult obstacles may be assembled from manually remote-controlled articulated members. Exactly the same design principles which apply to remotely controlled arms enable one seriously to consider this possibility. In addition, the feasibility of preprogrammed operation, as discussed more in detail under Command Systems, relieves the operator of the necessity of carrying out in detail the actuation of each separate motion involved in walking or climbing. The operator can, however, regain detailed manual remote control when required for negotiation of unusually difficult obstacles. It is probable that the development of electronically commanded advanced locomotion systems will proceed first as a part of the program for exploration of the lunar surface, since one is here forced to cope with a surface whose nature is not accurately known. This in turn may lead to the development of superior off-road vehicles for use on earth in areas where more conventional vehicles are unable to operate.

Undersea Operations

In designing remote-handling systems for use beneath the ocean, one takes advantage of the dense medium and, when possible, designs systems which "swim" rather than crawl or walk on the ocean floor. Such systems, which somewhat resemble miniature unmanned submarines, may employ marine screws in a variety of ways to accomplish locomotion and attitude control. Figure 150 is a typical undersea handling system employing two independently mobile marine screws. This vehicle can hover, can swim vertically or horizontally or any combination thereof, and can control its attitude within wide limits.

For handling very heavy objects or for operations restricted to the ocean floor, it may be desirable to use any of the off-road land vehicles reengineered for the undersea environment in exactly the same manner as discussed in the preceding paragraphs. The most interesting feature, from the locomotion viewpoint, of the undersea environment is, however, the free-swimming type of vehicle with its ability to accomplish a variety of maneuvers in combination with a skillfully maneuvered surface or submarine controlling vessel.

Space Operations

The general consideration of locomotion in space naturally divides into separate consideration for lunar and planetary operations and for orbital operations. The former case differs from land operation as discussed above only in the severity of the environment which the equipment must survive, and in the lack of detailed knowledge concerning the terrain which must be traversed. With these rather important exceptions, one adapts proved terrestrial locomotion systems for lunar locomotion.

When operating in orbit, the most conspicuous point is the obvious lack of a gravitational reference and the corresponding need for providing an artificial reference for control of vehicle attitude. Locomotion is usually accomplished by auxiliary rockets or jets; the control may be from either a manned space ship or space station, or by satellite radio-relay command from earth. Figure 151 shows an artist's concept of an orbiting remote-controlled vehicle accomplishing a maintenance operation upon a satellite.

The lack of any fixed objects in the orbiting environment indicates the requirement for furnishing handholds or grasping devices upon any objects which must be manipulated. The handling system must be equipped with an additional set of grasping arms whose only purpose is to provide a firm operating base for exertion of torque or

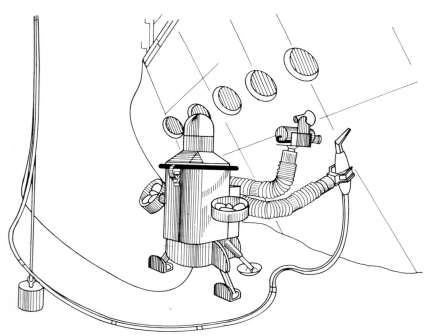

FIG. 150. A general-purpose underwater manipulating system as employed in a salvage operation. (*Hughes Aircraft Company*.)

thrust upon objects handled. An exactly similar requirement exists in connection with free-swimming undersea vehicles; the problem is solved in the same manner.

21 The Sensory Subsystem

The sensory subsystem, for purposes of hostile-environment-system analysis, includes all primary sensory transducers contained within the hostile environment itself. The transmission of the information from these transducers from the hostile to the safe environment is discussed under Command and Data Link, and the presentation of sensory information to the operator is discussed in under Control Console. The overall integration of these items into the total hostile-environment system is of course a part of the overall system-integration task.

The sensory subsystems to be provided on a remote vehicle correspond more or less to the senses of the human operator. It is accordingly perhaps logical to discuss these in their approximate order of importance, namely, vision first, since perhaps 80 per cent of our information concerning our surroundings is gained by means of vision, followed by tactile and auditory senses which in general are about equally useful. One considers finally the nonhuman senses which can readily be included in remote-handling systems.

It is in this area of hostile-environment-system design that one finds a most rewarding field for psychological and human-factors research. It has been demonstrated that the human mind has unsuspected capability for learning to react to stimuli quite different from those normally available to us.

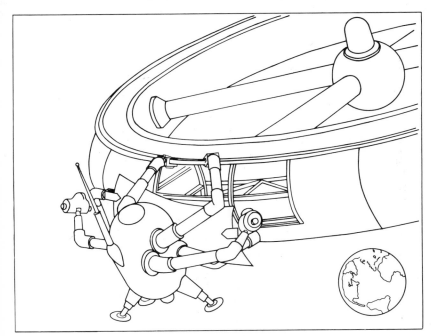

FIG. 151. Artist's concept of a general-purpose space handling system as employed for assembly of an orbiting vehicle.

Remote Vision

The obvious mechanism by which one accomplishes remote vision is the TV camera and its associated transmission link and display monitor. Conventional closed-circuit television systems are employed for this purpose; the present discussion will not enter into the mechanics of TV, since this is well understood, but will be confined to methods for employment of closed-circuit TV for remote vision.

The basic situation for remote viewing is seen in Fig. 152. Here one has an object to be grasped and a tong with which to grasp it. The viewing system must inform the operator of the three spatial coordinates $\Delta x, \Delta y, \Delta z$ which separate the two objects. This knowledge enables the operator properly to manipulate the arm controls so as to bring it into coincidence with the object to be grasped.

It is clear that a single TV camera cannot obtain this information, since a single TV camera presents a planar projection of the scene toward which it is directed   If the axes are appropriately oriented, a single TV camera can present, for example, Δx and Δz upon its monitor screen, but Δy, which is parallel to its line of sight, is not determined.

A second TV camera mounted at right angles to the first can determine Δy and Δz;

a comparison of the two pictures will give the required information with a certain amount of redundancy. It is thus geometrically evident that two single-channel closed-circuit TV cameras, so oriented as to observe the working area from mutually orthogonal directions, will present to the operator the necessary and sufficient information to accomplish his required task.

It is perhaps not evident but has been proved by experience that the interpretation of this type of TV picture is quite readily learned and that with practice the operator

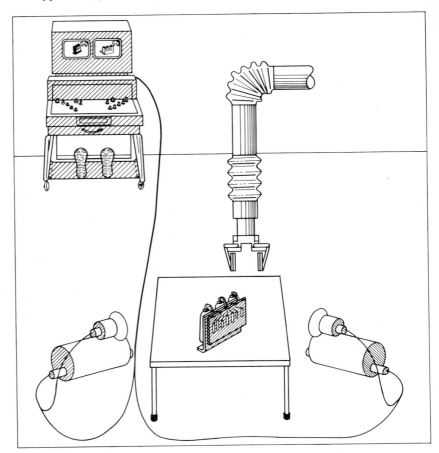

FIG. 152. Basic geometry of remote vision for manipulation.

obtains a subjective picture of the spatial orientation of objects observed, just as vivid as that obtained with his own eyes. This observation is the psychological fundamental upon which effective remote vision may be based.

Almost as important as the use of two orthogonally oriented cameras is the ability to move the cameras about. This camera motion not only permits one to see in the presence of fixed obstacles but permits one to obtain additional visual cues by means of parallax. This is particularly important if one is required to deal with objects of unknown size so that scale is not available as a distance cue.

Excellent work has been done with stereoscopic TV systems in which two separate pictures are presented to the operator's two eyes. Such systems, when properly adjusted, present startlingly clear three-dimensional pictures of the area viewed. In engineering hostile-environment systems, one must choose between the vividness of a stereo system as against the lower cost and superior reliability of the two-camera system.

The above comments are restricted to manipulation within restricted areas. For remotely controlling locomotion, stereoscopic vision is clearly unnecessary; note that it is not usually employed in manually driven automobiles, boats, or aircraft.

In both cases some means must be provided to make the operator aware of the position and orientation of his TV cameras. It is usually quite impossible to interpret the picture on a TV monitor unless one has prior knowledge of the "look angle" or direction in which the camera is pointed. Camera motions may then serve much the same purpose that head and shoulder motions do for direct visual observation.

Camera position can be presented to the operator in either or both of two distinct ways. First, the position of the camera mount may be telemetered and displayed to the operator; second, a third TV camera may be so mounted that it can observe the positions of the two close-in TV cameras. Practical experience indicates that a three-camera system in which the third camera is primarily intended for general observation of the working area, while the other two are employed for close-in detailed observation, is extremely effective.

It is clear that a variable-focal-length lens is extremely desirable in remote viewing. This can be accomplished either by a lens turret or by a lens of continuously variable focal length ("Zoomar"). Either of these devices enable the operator remotely to select a narrow-angle detailed view of a distant object or a wide-angle survey of the surroundings. They also permit, when required, the operator to present on the monitor screen a larger-than-life-size image of the object worked with.

When the operator is equipped with a full capability to move his TV cameras about and to adjust their focal length, no particular requirement for unusual picture definition is encountered. Conventional 525-line interlaced pictures are completely adequate for a wide range of handling operations. The normal range of contrast adjustment is also quite adequate.

No discussion of remote vision would be complete without mention of illumination. Controllable illumination is a great aid in remote operation. Not only does this ensure that the TV cameras are operating on the linear portion of their response curves, it enables the operator by throwing shadows in various directions to obtain additional cues concerning the position and orientation of objects handled. Ideally, a minimum of two light sources independently movable should be provided. Practically, a spotlight mounted on each TV camera and independently turned on and off can prove adequate in most cases.

In some cases the cost of the very wide bandwidth required for transmission of standard television necessitates reconsideration of the entire viewing subsystem. This is particularly true of space problems where distances become very great and the cost per kilocycle of communication bandwidth is extremely high. Low-frame-rate TV can reduce bandwidth required almost in proportion to the frame rate. The speed of operation is also reduced, although the information rate generated by conventional 30-frame TV is far beyond that actually required for manual operations. Reduction to 2 to 5 frames per second will only slightly reduce operating speed. Beyond this point, further reductions in frame rate will increase the time to perform any given operation in almost inverse proportion. These facts point out the method for optimizing the overall system in cases where cost of communication bandwidth is very high.

Hearing

The mechanics of accomplishing remote hearing are extremely simple, being no more than adaptation of conventional intercommunication systems. The bandwidth required is so limited as to be a minimal design constraint in almost all cases.

The principal value of auditory information in operation of a hostile-environment system rises from the ability of the operator to hear the motors, actuators, and other parts of the system perform their functions. This auditory information supplements the visual displays of the TV systems and facilitates precise and rapid remote operation. The "presence" added by the auditory channel is difficult to measure but is of definite value in making the total system an effective and comfortable one with which to work. One accordingly includes an auditory subsystem in hostile-environment systems whenever this is at all possible.

The application of stereophonic sound to remote-handling systems has been but little investigated. The use of two microphones and two speakers, exactly as in conventional high-fidelity systems for musical reproduction, can in principle give the operator a psychological presentation of the position or motion of any sound-producing objects in the vicinity of the remote vehicle. Systematic research will enable one to evaluate the value, if any, of this refinement to remote-handling systems.

The Tactile Sense

The sense of touch is a useful supplement to vision and hearing. Occasionally, when circumstances do not permit the use of vision, the tactile sense may be the only one available. This of course corresponds to the use of our own physiological senses when, because of darkness or for other reasons, it is necessary to use our tactile sense for accomplishing even rather complex manipulations.

One can distinguish at least three quite separate uses for the tactile sense, as follows:
1. The obtaining of information concerning texture and shape, commonly referred to as "feeling."
2. The obtaining of information concerning location of obstacles and/or items with which manipulations are to be accomplished; this is without regard to such subtleties as texture and is intended for manipulative purposes as opposed to information-gathering purposes.
3. The proprioceptive function, i.e., information concerning the configuration of the arms or other movable parts of the handling system.

These three principal tactile functions are completely separable; any or all of them may be employed to accomplish specific hostile-environment operations.

The accomplishment of such subtle tactile functions as texture, smoothness, warmth, or other surface properties is quite a difficult one. Adequate input transducers are not readily obtained, and this function would normally be employed only in unusual cases where this information was badly needed and where optical means could not be employed.

The much simpler function of identifying and locating objects can be accomplished by simple, readily available sensors. The familiar sensitive switch ("microswitch") connected to a suitable feeler-type actuator may be all that is required. Such switches can transmit information through the data link to inform the operator when contact is made with obstacles or with objects which one wishes to grasp. More subtly, pressure-sensitive transducers may be employed to provide quantitative information concerning contact forces existing within the system.

The information obtained from tactile sensors, such as those discussed in the preceding paragraph, has the interesting attribute that it is readily employed in connection with automatic or semiautomatic auxiliaries to the handling system. As discussed at more length in the section on the command system, feedback loops may be closed within the remote vehicle itself employing tactile sensory inputs. This type of information is much more usable for automation purposes than that obtained from TV systems, and hence is particularly adaptable to automation requirements.

The third principal tactile function, that of proprioception, may be accomplished in either or both of two distinct ways. The more obvious way is to imitate that employed in our own bodies and to provide position- or angular-indicating transducers on every moving part in the system. These transducers can continually transmit through the data link information concerning the position of each moving part which can be displayed in an appropriate manner to the operator.

·An alternate method is to employ the same vision system used for manipulation and locomotion. In a three-camera system, the third camera can be so mounted that with a wide-angle lens it can view the positions of both the handling arms and the TV arms. A glance by the operator at the third monitor screen shows him the positions of all the arms with which he is working and enables him to accomplish his manipulations with a minimum of hazard due to collision with obstacles by the "elbows" of the manipulating system.

In accomplishing rapid motions, the dynamics of the proprioceptive system play a vital role. Such actions as driving a nail with a hammer, twirling a hoop, or the like, involve, when performed manually, a closed feedback loop between the muscles and the proprioceptive input sensors. To accomplish such operations remotely may be done by simply extending the body's sensory system, as is done mechanically in the master-slave manipulators, and electronically in the "Handy-Man," both of which were described under Present Practice. If the dynamics of the transmission links between remote equipment and human operator are adequate, systems of this type permit one to accomplish complex and rapid motions. An alternative method is to close the feedback loop at the remote point electronically; this greatly reduces the demands upon the command and data link at the cost of complicating the equipment within the hostile environment. Some of the implications of this approach to the problem of operating dynamics are discussed further under the Command and Data Link.

Other Senses

There is no fundamental reason to restrict oneself to the human senses when designing a hostile-environment system. On the contrary, one should take the greatest possible advantage of technology to obtain information concerning any particular hostile environment. As an example, when working beneath the ocean it is highly desirable to employ sonar "vision" as a supplement to optical vision. Further, one might find it desirable to select quite a different color of light for undersea viewing from the blue-green which is most effective for our own eyes.

Such physical variables as temperature, nuclear-radiation level, hydrostatic pressure, and numerous others may be thought of as part of the sensory subsystem of a hostile-environment system. The measurement of these variables by suitable sensors and their display to the operator as a part of his control-console information can form a useful part of the total function of information gathering within the hostile environment.

This use of measurement of the physical state of the environment may be distinguished from the use of hostile-environment equipment to obtain scientific data concerning the environment. This is of course a tenuous distinction, and one upon which not too much emphasis should be placed. However, if one's objective is to obtain scientific information, one should view a remote-handling system as a mechanism with which to install, operate, and read scientific instruments. It is usually quite simple to distinguish these instruments, which may form the principal motivation for operating the entire system, from the sensors which are an integral part of the system itself and are essential for its successful operation.

One may well add to the complement of sensory equipment employed in hostile-environment systems navigation equipment, such as either gyrocompasses or magnetic compasses, as well as electronic navigation aids adapted from the familiar RDF, VOR, and DME of the aircraft industry. Finally, one often finds it desirable to include attitude-sensing instruments. These are particularly important in systems which do not have an inherent attitude reference; for example, both free-swimming underwater systems and space systems require, of necessity, three-axis attitude indication. Vehicles required to operate over extremely rough terrain may also benefit from a two-axis attitude display to enable the operator to avoid danger of overturning on steep slopes.

The above paragraphs can only suggest the wealth of sensory information available for use in connection with hostile-environment systems, and a few of the ways in which it can be employed in design of hostile-environment systems. Each requirement and each environment is different. We see in this area a field in which ingenuity and intelligent application of techniques and of equipment which have been developed and proved for other purposes can be most rewarding.

22 The Command and Data Link

As noted under the general discussion of hostile-environment systems, the command and data link is that subsystem which establishes communication between the mobile vehicle within the hostile environment and its human operator. It must of necessity transmit information in both directions; it may also be employed to transmit electrical and mechanical energy when its length is sufficiently short to make this feasible.

In its simplest possible form the command link may consist only of a multiwire electrical cable, as is typical of many electromechanical manipulative systems, or of a multiplicity of mechanical cables, as in the master-slave manipulator. The data link may be, in its simplest form, direct vision through a suitable window, perhaps supplemented by a tactile sense transmitted mechanically.

At the other extreme of sophistication, one may find completely isolated remote vehicles in which the only communication is by two-way radio, and in which all energy employed within the hostile environment is generated within the mobile vehicle itself. Such systems are of course almost unlimited in the possible separation between operator and vehicle. Only at astronomical, as opposed to terrestrial, distances do limitations to vehicle-operator separation become so severe as to require an additional increment of analysis in the overall system evaluation.

From the viewpoint of equipment design, it is usually desirable separately to consider the command link and the data link, since most communication media are undirectional.

The Command Link

As noted above, the command link must transmit from operator to vehicle all commands which direct the motions of the remote vehicle. The simplest method of accomplishing this function is clearly the multiconductor cable in which each conductor corresponds to a single degree of freedom or motion of the remote vehicle.

One can note immediately a hierarchy of increasing sophistication and complexity in command systems, as follows:

1. The multiwire system
2. The two-conductor multiplexing system
3. Multiplexing with radio transmission

Obviously, in analyzing specific hostile-environment systems one will select the command link which will meet the requirements in the simplest possible way.

In utilizing multiwire systems, one finds again a number of alternatives, each of which has numerous applications. One may actually transmit the power to perform each separate motion through its individual pair of conductors. This is frequently done in electromechanical systems of a few degrees of freedom. For systems of more degrees of freedom, where cable bulk becomes a problem, it is preferable to transmit control information through the multiconductor cable and to transmit the prime power separately. Either relays or magnetic or electronic amplifiers may be employed to utilize the command information to control much greater amounts of power. Sometimes variable command voltage may be transmitted through a multiwire cable to control, by means of analog amplifiers, either speed or force on the part of the manipulative or locomotive system. Multiwire systems are of necessity somewhat limited in the distance over which command can be accomplished. These limitations arise not only because of the bulk and cost of the cable but because of the increasing difficulty of accurately transmitting commands in the face of distortions and noise which may arise in a very long multiwire cable.

To minimize the cable problem, one may employ a multiplexing system in which the input commands are translated to a suitable code and transmitted through a single pair of conductors to the hostile environment, where they are decoded and employed to control the moving parts of the system. Figure 153 is a block diagram showing the essential elements which make up a system of this type.

Any of the coding systems familiar in telemetry may be employed for command coding. It is of interest to note some of the criteria and constraints applicable to command systems, since these differ somewhat from those applicable to conventional data transmission and telemetry as discussed elsewhere in this handbook.

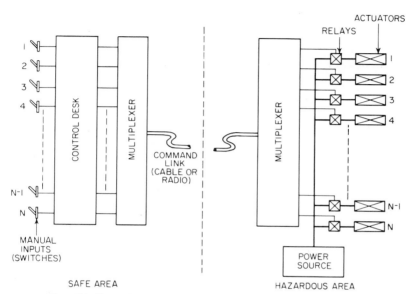

Fig. 153. Essential elements of a command multiplexing system.

The principal criteria applicable to the command subsystem for a hostile-environment system follow:
1. There must be no crosstalk between command channels.
2. The system must "fail safe," i.e., it must transmit either a correct command or none.
3. It must be economical of bandwidth.
4. It must be minimally affected by faults in the transmission medium.
5. It must be readily adaptable to either hard-wire or radio transmission.
6. Channels must be inexpensive.

One notes that once the point has been reached where multiwire cable is not satisfactory, the design of a command system for radio or for hard-wire link is subject to much the same criteria. In general, one should attempt to design the several components of the command subsystem so that they are usable with any suitable transmission medium.

The data rate required for manually commanded systems may be computed approximately by considering the fact that the speed of the entire system is determined by that of its operator. The minimum reaction time of the human mind is about 0.1 sec, indicating that there is no need for the system to respond much faster than this. This indicates that, for example, in a system with 50 degrees of freedom, if each degree of freedom is available to the operator within 0.1 sec he will be unable

15-160   REMOTE CONTROL

to distinguish the system response from one in which all degrees of freedom are accessible to him in parallel. This fact indicates the desirability of a time-division system in which each degree of freedom is made available on a time-sequential basis to its corresponding command channel. Each command channel transmits a digital pulse which indicates whether or not that particular channel should remain at rest or change its state. In the example noted, only 500 such pulses would be transmitted per second, and if each pulse were perfectly square and at 50 per cent duty cycle, somewhat less than 5,000 cycles bandwidth would be required for transmission.

As an interesting aside, one notes a rather unique attribute of systems having mechanically moving parts, namely, each moving element has three states; for

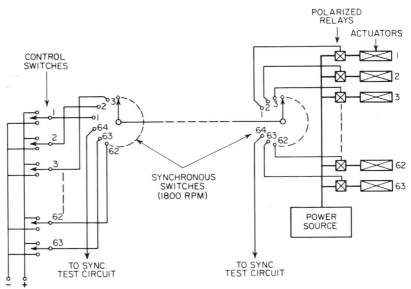

Fig. 154. Time-division-multiplex ternary-coded command system.

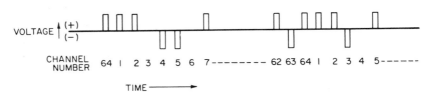

Fig. 155. A typical ternary-coded command sequence.

example, stationary, counterclockwise rotation, and clockwise rotation. This observation suggests the use of ternary as opposed to binary coding. One will recall from information theory that the ideal coding, from a viewpoint of overall economy of components as well as efficient bandwidth utilization, may be either binary or ternary. The binary coding so widely employed in computers is selected from practical considerations. These same practical considerations indicate some value to the employment of ternary coding in hostile-environment command systems. Figure 154 shows in block-diagram form a typical time-division multiplex ternary-coded command system, and Fig. 155 shows a typical ternary-coded command transmission. Such a system meets all the criteria outlined above. Like any digital system, it is relatively unaffected by fading, noise, or distortion in the transmission

REMOTE HANDLING   15–161

medium, and by this same token can readily be handled by repeaters which not only will improve the signal-to-noise ratio but will restore the command pulses to their original shapes.

The preceding discussion has not considered the choice among position-command, rate-command, and acceleration-command systems. These distinctions are basic to the overall stability and operability of the system, as discussed further under Control Console. The command link, as such, can transmit commands which control either position, rate, or acceleration; and indeed one may find all three employed in the same overall system.

It is most important in engineering the command system to obtain from those responsible for other portions of system design information concerning the number of commands required and the sampling interval applicable to each. These are the principal properties which determine the bandwidth required; this in turn will determine the power needed to transmit the command signals and (in radio systems) the dimensions and configurations of the antennas. Good command subsystem engineering ordinarily endeavors to minimize bandwidth and power required for commands; in some cases, subtle trade-offs are required between the requirements of the command subsystem and those of speed or response time applicable to the system as a whole.

The Data Link

The data link which transmits information from remote vehicle to operator is in no respect different from a conventional telemetry link. As noted in the discussion on the sensory subsystem, data are gathered from a variety of sensory transducers and suitably coded as inputs to the data link. Most of the data involved are very slowly varying and hence require very limited bandwidth. Obviously, the most demanding of the data to be transmitted are the television signals corresponding to the one or more TV cameras which may be employed. Here again, conventional methods and equipment for TV transmission may be employed.

The data link may be selected from the same three transmission media noted above under Command Link:

1. Multiwire
2. Two-conductor with multiplexing
3. Radio

For short distances, it may be desirable to employ a multiwire data link in which each separate sensory input is transmitted over a single pair of wires or coaxial cable. In such systems, the command link is ordinarily an additional group of conductors electrically separate from those of the data link. The many conductors may or may not be contained within a single cable for mechanical convenience.

When the multiwire system becomes unworkable, the next step is to employ a single two- or three-conductor hard-wire multiplexing system. It is usually desirable to employ the same pair of conductors for both the command and the data link, and to transmit video and audio by radio-frequency carrier. In a hard-wire system using mechanical rotary switches for multiplexing, some of the switch points may be employed for transmission of slowly varying analog data, since this system is bidirectional. Electric power may be transmitted on the same electrical conductors and separated from the command pulses and the analog data by suitable filters. It is sometimes desirable, however, to employ a third separate conductor to simplify the filtering problem.

In radio-transmission systems, the command link and the data link become completely separated and conventional telemetry transmitters may be employed for data-transmission purposes.

Programming and Computers

The preceding discussions have been concerned almost entirely with manually commanded systems in which every change of state within the system is initiated by operator decision and controlled by manual action of the operator through his control console. As noted in the introductory discussion, one may distinguish between such manually controlled systems and programmed systems, in which no communication with the exterior of the hostile environment is provided and in which all actions and decisions are predetermined. One can visualize a complete spectrum of systems ranging between these two extremes, and indeed sound hostile-environment-system engineering includes suitably combining fully manual remote operation with programmed operation. This subject is far too large to permit a complete discussion here; however, a few preliminary observations may be of interest.

As one example of a judicious use of programming, one can consider systems in which a small number of predetermined operations are anticipated. One can record the sequences of command signals initiating each operating sequence, and store these recordings within the remote vehicle. The operator then needs only to transmit a single coded command, "Execute sequence A," and thus initiate any one of a number of prerecorded sequences. This application of programming greatly reduces the demands upon the command link; it also relieves the operator of the necessity of manually initiating each detailed motion making up the motion sequence. This is accomplished at the cost of an appreciable increase in complexity on the part of the remote vehicle; this is a trade-off which must be made with great care.

A somewhat different application of electronic circuitry to reduce demands upon operator and command system has been made by Tomovich.[6] He has studied a grasping device in which the feedback loop is closed directly from a tactile sensor to the actuator without involvement of the command link. This system is the exact analog of the human reflex action and can be accomplished with quite simple equipment.

The comparison between "local" and "remote" feedback is shown schematically in Fig. 156. Remote feedback adds a sensor and a control element at the actuator. The entire sequence involved in grasping an object is initiated by a single command from the operator, in contrast to the remote-feedback, or manual-remote-control system in which each separate motion requires transmission of a separate command. The same is true of releasing an object. The reduction of requirements upon the command link resulting from use of local feedback is obvious. The reader will find it interesting to compare local feedback with the prerecorded sequence method discussed in the preceding paragraph; the hostile-environment-system designer selects appropriate combinations of these systems to accomplish specific system functions.

Still another combination between hostile-environment technology and computer technology is illustrated in Fig. 157. We see here the use of a typical high-speed high-performance computer to control a grasping device. This combination may be programmed in very complex ways, making full utilization of the enormous storage capacity and flexibility of modern computers. The combination of computer and handling system can execute search patterns and can make fairly subtle decisions, depending upon inputs received from tactile sensors.

These systems are group 3 systems as defined in Sec. 18. They do not require any direct contact with the world outside the hostile environment; i.e., no command link is required. The human intelligence intervenes in the system only when entering the program into the computer. In terms of Fig. 143, we have made it possible physically to remove the operator from the system completely; at the same time, of course, we lose the human ability to react to the unexpected which is inherent in group 2 systems.

Group 3 systems may be viewed as substituting the computer "brain" for the human brain in manually controlled systems and may be very useful for cases where the full capabilities of the human brain are not required. They can also be looked upon as a way to connect the computer output directly to something which does physical work. This is a computer application which seems to have been almost completely unexplored; so far, computer outputs have been restricted to printed pieces of paper or cathode-ray-tube displays which indirectly cause physical actions to be taken by people. We can now contemplate systems in which the computer itself directs physical motions without human intervention.

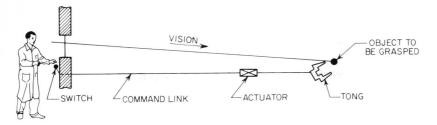

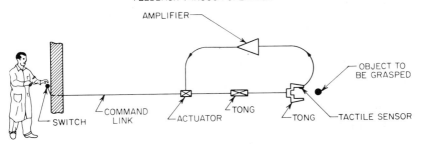

FIG. 156. Two-alternate feedback systems for grasping.

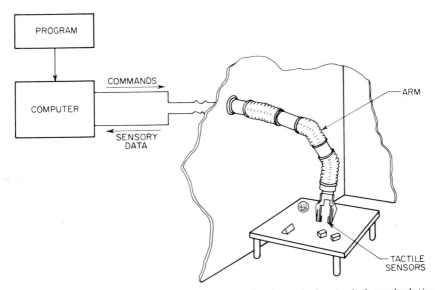

FIG. 157. Application of a general-purpose computer to control a typical manipulating system.

A final observation concerning application of computer technology to remote handling concerns the automation of repetitive operations. In much high-volume manufacturing, one finds personnel manually performing repetitive transfer operations which involve a small number of predetermined decisions. For example, a conveyor line is unloaded and objects contained thereon transferred to a different conveyor line. Electromechanical arms, controlled by simple stored motion-sequence-command codes, can accomplish a great variety of these industrial transfer operations. Their ability to select among a small number of predetermined motion sequences based upon information obtained from other equipment on the production line indicates a new potential flexibility in industrial automation.

23 The Control Console and Power Subsystem

As shown in the basic analysis of hostile-environment systems, any such system must include an interface at which the human operator interacts with the mechanical parts of the system. Figure 143 shows schematically the relationship between this interface and the several elements which make up the mechanical portions of any hostile-environment system.

For adequate analysis of these systems, one must consider the man who operates the system as one of its elements. His characteristics must be considered in exactly the same manner that "human factors" analyses are made in the design of, for example, aircraft cockpits.

Certain system design trade-offs are vitally dependent upon the characteristics of the human mind. The extent to which one may include programming or other automatic or semiautomatic features falls into this area. While the preceding section discussed some of the hardware aspects of automation in the system, the basic analytical choices as to how much, if any, of this type of automation is to be included must fall into this area of total system analysis.

Apart from its system-analysis aspect, the design of the control console itself is a difficult and challenging problem. This console forms one of the major subsystems; it is the only one with which a man normally comes in direct contact.

The control console must and should include the electric- or mechanical-circuit elements involved in the "safe-environment" portion of the system. Power-conversion equipment, multiplexing systems, and the like fall into this category, and while design-wise each is considered as a part of its appropriate subsystem, for packaging purposes these portions of these subsystems are best physically included in or adjacent to the control console.

Control-panel Configurations

All the foregoing discussion is intended to put in context the console itself, or more properly, the control panel, which is the physical man-machine interface. Figures 158 and 159 illustrate particular consoles in use today. They have been selected to illustrate roughly the very great range which can be covered. On the one hand we have a small hand-held box containing only a few switches. This is for a system in which vision is accomplished optically and there is no sensory system as such. On the other extreme, we see in Fig. 159 the console of a complete system, including controls and indicators for all the fundamental locomotor, manipulating, and sensory functions. This console employs a variety of multiposition switches to command the numerous degrees of freedom of the manipulating and locomotion systems. Foot treadles are used for speed control. Visual information is presented by television monitors located for comfortable operator viewing. A small loudspeaker provides an auditory channel.

A great variety of configurations for the operator's control and display console are in use. It will probably require an accumulation of a good deal of additional practical experience before an optimal solution can be achieved. At this point, it is even difficult to lay down clear guidelines. It is in general highly desirable, partic-

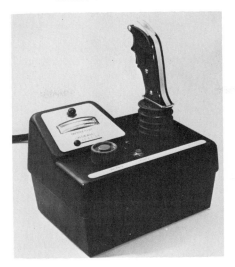

FIG. 158. A typical command system for a single arm. (*Courtesy of Lee Associates.*)

FIG. 159. A typical control console for a complete fully remote handling system. (*Hughes Aircraft Company photograph.*)

ularly for complex systems, to *build* and work with a small number of alternate control-panel configurations rather than to attempt on theoretical grounds to choose an ideal or optimal control-panel layout.

The philosophy applicable to control- and display-panel design can, however, be rather clearly stated. The principal requirement is to facilitate the interaction between the operator's mind (both conscious and subconscious) and the system itself. This means that all information displayed (visual, auditory, plus others as discussed in Sec. 21) must be presented in ways comfortable and natural to the operator. By the same token, the command elements which actuate physical motions in the remote system must be comfortable, natural, and easy to use.

These guidelines may seem obvious to the point of triteness. Nevertheless, they have frequently not been fully appreciated. In the same vein of laying down general philosophical guides, one notes that ease and rapidity of learning are usually not a primary requirement; one usually considers operator comfort and ability to operate with minimal fatigue for long periods of time as considerably more important than a minimal learning period. There is also no particular reason to retain an anthropomorphic configuration in control-panel layout, by the same token that one does not feel constrained to adopt anthropomorphic manipulating or locomotion geometries.

More detailed questions, such as the selection of a "joystick" type of control in contrast to one which uses a multiplicity of switches, can usually be evaluated for particular applications in the light of the general principles outlined above.

System-design Considerations

Basic questions of system design having to do, for example, with the choice between position command vs. rate command or acceleration command may also be evaluated in this light. This consideration, of course, is quite separate from the engineering consideration having to do with the relative quantity, size, and cost of the hardware required for each type of command system.

As an example of the contrast between position command and rate command, one notes that in the former type of system the operator's hands and arms must of necessity assume positions identical to those of the remote-manipulating elements. This has the advantage of fully utilizing the operator's proprioceptive and kinesthetic senses; this results in minimal learning time and also in ability to accomplish rapid motions such as twirling a hoop. The principal disadvantage of such systems is their inherent continual demands that the operator maintain his hands and arms in unnatural and uncomfortable positions. As a result, the position-command systems are in some cases extremely fatiguing to operate.

In contrast, the operator of a rate-command system can continually maintain his hands and arms in a relaxed and comfortable configuration employing either a multiple switch or a joystick type of control. This makes for operator comfort for extended periods of operation. The price paid for this is the need to learn a completely new set of reflexes relating the motion of the operator's hands and fingers on the controls to those of the remote element. Kinesthetic and proprioceptive information may be obtained via sensory channels, as discussed in Sec. 21, by means quite different from those employed in our own muscular and nervous systems. These new reflexes are rather easy to learn; there is indeed ample precedent for this type of learning, as illustrated by such machines as typewriters, calculating machines, and musical instruments. The process of learning any of these familiar devices actually consists of the creation of new reflex arcs completely comparable with those required for learning a complex remote-handling system.

In concluding this discussion, let us return briefly to the physiological or psychological view of the hostile-environment problem illustrated in Fig. 143. In this way of looking at a hostile-environment system, the man at the console plays the same role in the system that the human mind plays in "operating" the human body. Presumably, there is no physiological structure in the brain which corresponds to the control console. Nevertheless, there must be within the brain the means of assembling, analyzing, and interpreting sensory data and of activating the muscles which correspond functionally to the man at the console in the hostile-environment system.

The Power Subsystems

The need for prime power to operate the system is obvious. The purpose of setting the power subsystems out as separate elements is simply to systematize and clarify the task of the system designer.

The sources of power employed in remote-handling systems are in no wise different from those employed in other systems in the same environments. The hostile-environment-system designer selects power sources, which may be as simple as an electrical wall outlet or as exotic as a nuclear reactor or a bank of solar cells, from proved power-supply technology. The selection is based upon analysis of size, weight, and cost in comparison with the overall performance requirements imposed upon the system.

24 Areas for Future Research

The careful reader will have noted at numerous points during the preceding discussion areas where complete and precise knowledge is lacking. These are the areas where research effort can usefully be applied.

At the present time there is relatively little research in the field of remote handling, as such. This is largely because this field is only beginning to be recognized as a separate engineering discipline, and hence deserving of research support and research attention.

In spite of this fact, a considerable quantity of past and present research effort is applicable to hostile-environment technology, since this technology is an eclectic one and can benefit from work done primarily for other motivations.

Research areas of specific value to hostile-environment technology naturally fall into much the same classifications as those employed in the preceding sections for system analysis. It may, however, be of interest to group them somewhat differently in order to visualize those directions in which research of value to hostile-environment technology, as broadly defined, can be organized and directed. This is done in the following paragraphs.

Excluded from consideration under this heading is research in materials and components to survive in hostile environments. Such research is most logically considered a part of a different discipline, which may be called "environmental engineering." Specifically, the researches leading to the design and development of components which will operate normally in the nuclear environment, the undersea environment, or the space environment, are going forward in order to make possible the development of many types of equipment for these environments. The planner of a hostile-environment-equipment research program is probably justified in assuming that he will be able to obtain components as needed for the construction of equipments for the several hostile environments of both present and future interest.

Psychological Research

The development and use of fully remote handling machines presents new possibilities to the researcher in psychology, human factors, and related areas. These machines can be viewed as new tools with which to understand the possibilities of the human mind. Using a remote-handling system as a new, detached "body" controlled by the human mind, one can investigate the capabilities of the mind without being hampered by the limitations of the human body.

From an equipment-design viewpoint, psychological research is concerned primarily with the topics discussed in Sec. 23, in short, with learning how to design the entire system so that it is comfortable and easy to use for the operator and so that full account is taken in system design of both the capabilities and limitations of the human mind.

Electromechanical Research

Useful research can and should be done in electromechanical (or electrohydraulic or electropneumatic) devices to serve as the "muscles" for remote-handling systems. Devices of improved power-to-weight ratio, response speed, and the like are required. These "muscles" determine the size and capability of handling systems.

A related research area has to do with the kinematic and dynamic requirements upon actuators. Such studies would be concerned with realistic analysis of the manipulating function and its interpretation in terms of speed, torque, force, and the like, as applied to ideal rather than currently available actuators, manipulators, and handlers.

Still another interesting electromechanical research region has to do with methods of locomotion. Such investigations would be concerned with improved methods of steering and maneuvering in crowded quarters, as well as unconventional means of locomotion, such as walking vehicles, vehicles with multiple independently controlled wheels, and the like. This work should be closely related to the efforts now beginning to be undertaken in improving our understanding of the mechanical properties of soils and surfaces over which locomotion must be accomplished.

Electronic Research

A specific area of electronic research applicable to hostile-environment systems is multiplexing, where theoretical studies concerned with analyses of the fundamental requirements associated with data rates should be made. These should be paralleled with experimental investigations of a variety of means for multiplexing with both radio and cable communication links.

Completely separate from this, but in the electronic area, is further investigation of the interrelation between computers and handlers. As briefly discussed in Sec. 22, a simple computer contained within the remote system may act something like a "reflex arc" in physiology. At the other extreme of complexity, one may use a large high-speed computer to direct the actions of a correspondingly complex handling system.

Sensory Research

The several senses involved in hostile-environment systems were discussed in Sec. 21. Further research in this area should be concerned with investigations of unconventional ways of "seeing," employing one or more TV systems in methods more sophisticated than merely presenting a monitor to the observer.

The use of the tactile sense has been but little explored. Systematic investigations of a variety of means of conveying tactile, kinesthetic, and proprioceptive information may greatly improve the speed and simplicity of future remote-handling systems.

Similar comments apply to any and all additional sensory channels which one may wish to investigate. These researches would combine psychology, electronics, and physics in an interdisciplinary investigation which may carve out completely unexplored possibilities in advanced hostile-environment systems.

Future Potentialities

The foregoing discussion has been for the most part concerned with a discussion of applications for remote-handling technology which are now active or which will be active in the immediate future. It is of interest to close this discussion by a future-oriented look at these areas, with a few comments upon possibilities which can now clearly be discerned.

Simply to name the hostile environments serves to define in general the areas of future application. These are

Space
The ocean
The nuclear environment
Industrial hazards

Sufficient work has already been done clearly to demonstrate the technical feasibility of accomplishing a great variety of operations in all these environments; the problem is primarily an *economic* one. One cannot underestimate the human significance of these present and future developments. The increasing availability of hostile-environment systems promises ultimately to eliminate the necessity for men to be exposed to physical hazards or dangers in accomplishing their daily work. This is done without in any way displacing people from their employment, since each hostile-environment machine requires a man as its operator.

As a summary, a few typical tasks for each of the major hostile environments are noted below.

Space

Assembly, maintenance, and repair of orbiting systems
Orbital rendezvous
Development of lunar installations
Supplementary devices to manned lunar operations

The Ocean

Construction of permanent installations in the ocean
Development of oil fields beyond diver depth
Undersea mining
Salvage
Rescue (of submarine crews)

Nuclear

Scientific operations in "hot labs"
Fuel processing
Fuel-rod changing in reactors
Assembly and maintenance in nuclear-powered systems
Nuclear rescue

Industrial Hazards

Maintenance of high-temperature furnaces
Handling of high-temperature objects; for example, in enameling
Operations in mining, tunneling, etc., in which severe hazard from cave-in or falling rock is anticipated
Fire fighting

REFERENCES

1. L. G. Stang, Jr., "Hot Laboratory Equipment," 2d ed., Technical Information Service, Washington, D.C., April, 1958.
2. J. L. Colp, Hot Area Concept for Radiation Test Facilities, *Trans. Am. Nucl. Soc.*, vol. 3, no. 2, pp. 392–393, December, 1960.
3. Auguste Piccard, "Earth, Sky and Sea," Oxford University Press, Fair Lawn, N.J., 1956.
4. J. R. Burnett et al., *Proc. Intern. Conf. Peaceful Uses At. Energy*, vol. 14, Session 19C.3, P/69, 1956.
5. R. S. Mosher and W. B. Knowles, Operator-Machine Relationship in the Manipulator, *ASD Tech. Rept.* 61-430, pp. 173–186.
6. R. Tomovich, "Human Hand as a Feedback System," I Congress of IFAC, Moscow, 1960.

Chapter 16

SATELLITE-RELAY COMMUNICATIONS

SIDNEY SHAPIRO, *International Business Machines Corporation, Bethesda, Md.*

CONTENTS

1 Introduction... 16-1
2 Noise Considerations..................................... 16-3
3 Signal Power at Receiving Terminal....................... 16-5
4 Modulation Considerations................................ 16-16
5 Coverage Considerations.................................. 16-22
6 Operational System 16-29

1 Introduction

In 1945, Clark[1] first proposed using artificial satellites to relay signals from one point on the earth to another. Anticipating the objections of those who "consider the solution proposed too far-fetched to be taken seriously," he cited German V-2 rocket experience to predict that "extraterrestrial relays" could be established within "fifty to a hundred years." In fact, of course, development of communications satellites has been extraordinarily more rapid; within twenty years a variety of experimental communications satellites—Echo, Telstar, Relay, Syncom, and others—had already been demonstrated, and operational systems were in advanced stages of development.

As depicted in Fig. 1, communications satellites for two-way point-to-point communications may take a variety of forms. These may be classified in two different ways, according to the mode of relay, whether *passive* or *active;* and the altitude of the satellite. A *passive* communications satellite scatters incident energy received from the ground transmitting station, a portion of which is received by the ground receiving station. An *active* communications satellite contains amplifying equipment; it receives the transmitted signal, amplifies and transforms it, and reradiates it back to the receiving station.

A *medium-altitude* satellite orbits the earth at altitudes from 1,000 to 10,000 miles, corresponding to periods of about 2 to 10 hr, during which time the subsatellite point traces out a path which encircles the earth. Two-way communications between two points on the earth can be achieved during the time when the satellite is mutually visible to both points. This entails tracking the satellite with transmitting and receiving antennas. Generally, the period of mutual visibility is a small fraction of the orbital period, and the range of tracking angles is large.

A *stationary* satellite orbits the earth in the equatorial plane at an altitude of 19,320 nautical miles, with a period precisely equal to the 24-hr earth-rotation period. For such an orbit, the subsatellite point is fixed. Communications between any two points within the visible area of the satellite, about one-third of the earth, can thus be achieved continuously and, to a first order, without tracking motion by either transmitting or receiving antenna.

Passive satellite relays can take two forms: discrete reflectors, in which a small number of relatively large satellites are used; and distributed reflectors, in which a very large number of small, chafflike reflecting elements are used. In either case the concept is the same: microwave energy is directed at the orbiting scatterers by a narrow-beamwidth antenna and the scattered energy is received by high-gain receiving antennas and sensitive receivers.

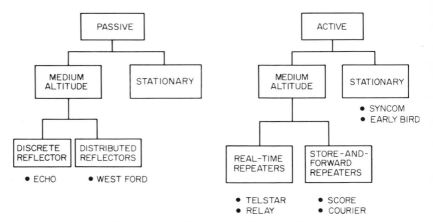

FIG. 1. Forms of communications satellites.

Active medium-altitude satellites also can take two forms, differentiated according to the time delay between transmission and reception of the signal. The time delay depends upon two factors: propagation delay, a function solely of the total distance from the transmitter to the satellite and back to the receiver; and processing delay within the satellite itself. If the satellite reradiates the signal essentially immediately, it is termed a real-time repeater. All passive satellites are real-time repeaters. If the satellite stores the received energy for retransmission at a later time when the satellite has moved to a different position, it is termed a delayed repeater or a store-and-forward satellite. Communication via a delayed-repeater communications satellite, a special class of active medium-altitude satellites, is not restricted to sites which are simultaneously visible to the satellite. For most common-carrier applications currently envisaged—including television and communications involving two-way conversation—real-time transmission is essential. For certain other applications, however, such as collecting data from a network of globally distributed meteorological sensors, for example, store-and-forward operation is admissible.

The capability of a communications-satellite system to relay information from a transmitting terminal to a receiving terminal is a complex function of many interrelated factors, concerning both the spacecraft and the ground-terminal facilities. Ideally, for each application, a system configuration is selected which minimizes the overall cost for a prescribed level of service. Practically, however, the optimization process is inexact and depends critically on parameters which cannot be accurately evaluated, as, for example, the expected operating lifetime of the satellite. Moreover, in so new a field as space communications, the "best" solution is dependent upon the development risks which one is willing to accept in utilizing untried advanced techniques. It is because of such uncertainties that the development of commercial and military operational satellite systems is being accomplished, not along any single approach, but rather in an evolutionary way, guided by experience obtained from a variety of experimental systems.

Although many of the important factors in overall system design are difficult to assess, the principles of operation of communications-satellite systems are well understood. These principles are reviewed in this chapter. Five principal aspects are discussed.

1. The factors which establish the noise power at the receiving terminal, with which the signal power received from the satellite competes to achieve intelligibility, are described in Sec. 2.

2. The relationship between the signal power at the receiving terminal and the principal parameters of the communications system are described in Sec. 3; the "range equations" are shown for passive discrete satellites, for systems employing belts of orbiting dipoles, and for active satellites.

3. Methods of modulation which achieve processing gain in the receiver (improvement in output signal-to-noise ratio over input signal-to-noise ratio) and which permit multiple access (the ability for more than one set of terminals to communicate simultaneously via the satellite) are described in Sec. 4.

4. Coverage considerations, which relate to the kinds of orbits and numbers of satellites required for a prescribed communication service, are described in Sec. 5.

5. The comparative advantages and limitations of the different kinds of communications-satellite system are summarized in Sec. 6.

2 Noise Considerations

The total noise power at the receiver of the ground receiving terminal, with which the signal power received from the satellite must compete to be intelligible, is comprised of externally generated noise induced into the receiving antenna by the sky, and by noise internally generated within the receiving equipment. Figure 2 shows the important noise sources.

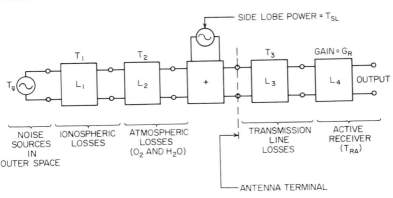

FIG. 2. Noise sources at receiving terminal. (*From Ref. 25.*)

If the receiving antenna were pointed at a blackbody radiator of absolute temperature T, it would receive a noise power

$$P_N = kTB \tag{1}$$

where k = Boltzmann's constant = 1.38×10^{-23} watt-sec/°K and B is the bandwidth of the antenna. The noise power contributed by the sky is conventionally expressed in terms of an equivalent noise temperature, as computed from Eq. (1). The variation of antenna noise temperature with frequency of operation is shown in Fig. 3. At frequencies below 1 gc, galaxy noise received from outer space is the major factor contributing to antenna temperature. At higher frequencies, absorption of radio waves by oxygen and water vapor in the earth's atmosphere causes the atmosphere to emit thermal radiation. This effect dominates in the range from 1 to 10 gc. Together with the cosmic noise at low frequencies, the atmospheric noise establishes the useful range of frequencies for communication satellites to be in the range from 1 to 10 gc.* As the noise depends on the total amount of oxygen and water vapor within the antenna beamwidth, the noise level varies with the elevation angle of the antenna above the horizon. The dependence of this noise component on frequency and on elevation angle is shown in Fig. 3.

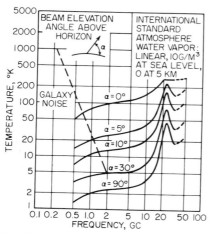

FIG. 3. Antenna temperatures due to oxygen, water vapor, and galaxy noise. (*From Ref. 2.*)

The temperatures in Fig. 3 are minimum values, which are achieved when the beamwidth intercepts only sky and atmosphere. The noise temperature at the antenna is increased when the antenna beamwidth intercepts "warmer" earth, sun, or discrete radio-noise sources such as stars or planets. If the beamwidth were narrow enough to "see" only the sun, the effective noise temperature at 3 cm would be about 6000°K. In general, if an antenna of beamwidth θ radians sees an object of diameter D at distance L, then the effective noise temperature will be

$$T_{eff} = T_{object}(D/\theta L) \quad \text{for } D \ll \theta L \tag{2}$$

If the antenna pointed at the earth, the noise temperature would be about 300°K. Accordingly, to preserve the low effective noise temperature of the "cold" sky, it is necessary to site the antenna carefully to limit noise pickup from the earth. This, in turn, generally requires that the antenna be operated at a minimum elevation angle above the horizon, of the order of a beamwidth. Further, care must be taken to limit the noise introduced through the antenna side lobes.

Losses in the antenna and feed structure reduce the signal and also add to the

---

*Formal international agreement on communications satellite frequency allocations was obtained at the International Telecommunications Union Geneva Space Radiocommunication Conference in October, 1963. The agreement set aside 2,850 Mc/sec for communications satellites as follows.[3]

a. Bands shared with other services:
  3,400–4,200 Mc/sec   Satellite-to-earth
  4,400–4,700   Earth-to-satellite
  5,725–5,850   Earth-to-satellite, only in region 1 (Europe, Africa, and Middle East)
  5,850–5,925   Earth-to-satellite, only in region 1 and region 3 (Asia and Australasia)
  5,925–6,425   Earth-to-satellite
  7,300–7,750   Satellite-to-earth
  7,900–7,975   Earth-to-satellite
  8,025–8,400   Earth-to-satellite

b. Bands for exclusive use:
  7,250–7,300   Satellite-to-earth
  7,975–8,025   Earth-to-satellite

For discussion of practical factors bearing on frequency allocations see W. H. Watkins.[4] (See also chap. 5.)

effective temperature. If the loss in the transmission circuit is L db, then the effective output temperature is related to the (lossless) input temperature T_{in} according to the relationship

$$T_0 = [(L-1)/L]T_L + (1/L)T_{in} \qquad (3)$$

where T_L is the ambient temperature of the transmission circuit. Small losses at room temperature contribute about 7°K for each 0.1 db of loss.

In addition to the noise present at the output of the receiving antenna and feed structure, noise is contributed by the subsequent receiving equipment. The thermal noise in the first stage of the amplifier effectively establishes the level of internally generated noise. Figure 4 shows the equivalent noise temperature of different types of commonly used front-end amplifiers. In communications-satellite systems using large antennas, the cost of the amplifier is a small part of the entire system, and the best maser amplifiers are generally used. In this case, the major component of the system noise is externally generated. However, for mobile stations using less complex parametric amplifiers, the contribution of the internally generated noise can be appreciable.

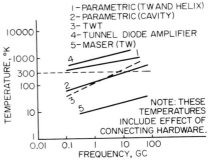

FIG. 4. Effective input noise temperatures of low-noise amplifiers. (*From Ref. 25.*)

3 Signal Power at Receiving Terminal

This section describes the dependence of the signal power received at the receiving terminal on the principal parameters of the communications-satellite system.

Discrete Passive Satellites[2]

For a communication link via a discrete passive reflecting satellite, the power received at the receiving antenna is given by the conventional range equation

$$P_R = (P_t G_t/4\pi R_{TS}^2)(\sigma G_s/4\pi R_{SR}^2)A_R L \qquad (4)$$

where P_T = power transmitted by the transmitting terminal
G_T = gain of the transmitting antenna
R_{TS} = distance from transmitter to satellite
σ = scattering cross section of the satellite reflector
G_S = gain of the reflector in the direction of the receiving terminal
R_{SR} = distance from the satellite to the receiving terminal
A_R = cross section of the receiving antenna
L = a loss factor to account for atmospheric attenuation, polarization, and practical system losses

Because of the dependence on R_{TS} and R_{SR}, the received power varies over a wide range as the satellite passes through the volume of mutual visibility, reaching a maximum when the satellite is directly above the midpoint between the two terminals. Minimum received power corresponds to the condition when R_{TS} and R_{SR} are equal to R_{max}, the distance from the terminal to the point in space at which the satellite rises over the horizon, corresponding to an antenna elevation angle of 0°. For a radio circuit, it is not feasible to operate with finite-beamwidth antennas pointing at the horizon. Receiving antennas must be restricted to a minimum elevation angle to prevent large amounts of noise and terrestrially generated interference from entering the receiver. Transmitting antennas are restricted to a minimum acceptable elevation angle (not necessarily the same angle as for the receiving antenna) to preclude large amounts of radiation which might interfere with other terrestrial communication services. In practice, since a terminal ordinarily receives and transmits essentially simultaneously, the minimum acceptable elevation angle is the larger of the two minimum angles. For communications-satellite systems, the minimum elevation angle is between 5 and 10°, depending on the antenna environment.

Table 1 shows R_{max}, the maximum slant range, as a function of orbital altitude for various values of minimum elevation angles. At altitudes above 100 nautical miles, the maximum slant range decreases approximately 60 nautical miles when the minimum beam elevation is increased 1°. Table 2 shows the corresponding maximum geocentric arc length between the subsatellite point and the communications terminal.

Table 1. Maximum Slant Range*

Satellite altitude h, nautical miles	Maximum slant range, nautical miles		
	$\alpha_{min} = 0°$	$\alpha_{min} = 5°$	$\alpha_{min} = 10°$
300	1,467	1,198	988
600	2,118	1,840	1,603
1,000	2,807	2,523	2,273
1,500	3,546	3,258	2,998
2,000	4,215	3,925	3,660
3,000	5,445	5,153	4,878
4,000	6,597	6,304	6,028
6,000	8,791	8,496	8,214
8,000	10,910	10,615	10,323
10,000	12,990	12,696	12,403

NOTE: Refractive bending not included.
* From Ref. 2.

Table 2. Maximum Arc Length between Subsatellite Point and Terminal*

Satellite altitude h, nautical miles	Maximum arc length, deg		
	$\alpha_{min} = 0°$	$\alpha_{min} = 5°$	$\alpha_{min} = 10°$
300	23.1	18.6	15.1
600	31.6	27.0	23.0
1,000	39.2	34.5	30.3
1,500	45.9	41.1	36.7
2,000	50.8	46.0	41.5
3,000	57.7	52.8	48.3
4,000	62.5	57.6	52.9
6,000	68.6	63.7	59.0
8,000	72.5	67.6	62.8
10,000	75.2	70.2	65.4

* From Ref. 2.

As an example of the magnitude of practical parameters of Eq. (4), Fig. 5 shows the minimum received power (corresponding to maximum terminal-to-satellite distance) plotted as a function of orbital altitude and ground antenna size for a representative circuit using a passive spherical reflector, such as the Echo I satellite.* It is apparent from Fig. 5 that, notwithstanding very large amounts of transmitted power and large antennas, the signal power at the receiver is very small.

* Echo I was made of 0.0005-in.-thick Mylar, coated with vapor-deposited aluminum. It measured 100 ft in diameter.

SIGNAL POWER AT RECEIVING TERMINAL

In general, the capability of a communications circuit to transmit information is described in terms of its output signal-to-noise ratio and its output bandwidth. The relationship between the output and input signal-to-noise ratios and bandwidths depends principally on the form of modulation used; this is described more fully in Sec. 4. As an approximate measure of the capacity of a passive communications-satellite system, the input bandwidth may be taken as characteristic. For the representative Echo I type passive communication link, Fig. 5 shows the usable input bandwidth as a function of orbital altitude. It is apparent that useful bandwidths (of the order of 10^4 cps) are achievable only at low orbital altitudes.

The limitation of the passive satellite is a consequence of the fact that only a small amount of power is reradiated from the satellite. For the example of Fig. 5, using a 40-ft transmitting antenna, the power radiated from the reflector at 1,000 nautical miles orbital altitude is less than 20 mw. Since it is entirely feasible to radiate tens of watts from an active satellite, passive satellites must achieve very significant increases in effective cross-section-to-weight ratio to be competitive with active systems. Several approaches which have been proposed for increasing the cross-section-to-weight ratio of passive satellites are discussed below.

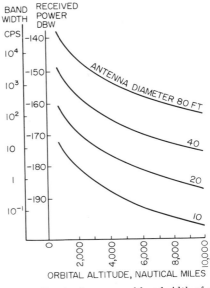

FIG. 5. Received power and bandwidth of a representative passive communications-satellite link. (*From Ref. 2.*) Reflector diameter, 100 ft. Transmitter power, 10 kw. Receiver temperature, 290°K. Signal-to-noise ratio, 20 db. Frequency, 6,000 Mc. Note: Curves apply for reflector location at maximum slant range from both terminals and antenna beam, elevation 5°.

Nonoriented Reflectors. An Echo-type spherical reflector wastefully radiates much of the reflected energy into space. Thus, one method of achieving an improvement in effective cross section is to use a reflector which preferentially reflects the incident radiation back to the earth, i.e., a reflector with gain ($G_S > 1$) regardless of orientation with respect to earth. A large number of reflector shapes which achieve gain have been suggested, including rough spheres, corner reflectors, saucers, and barrels.[5] Regardless of the detailed shape of the reflector, however, if communication is to be achieved from any point within the geographic "coverage" of the satellite, then a limit on satellite gain is imposed by the condition that the narrowest tolerable beamwidth of the reflection pattern be large enough to illuminate all points visible to the satellite.

Since the satellite is assumed to be unoriented, the satellite must, ideally, scatter incident energy into a cone having a vertex half-angle equal to the angle subtended at the satellite by the transmitting and receiving terminals. The half angle can be shown to be

$$\theta = 2 \tan^{-1} \left[\frac{r_0 \sin \dfrac{L}{2r_0}}{h + r_0 \left(1 - \cos \dfrac{L}{2r_0}\right)} \right] \tag{5}$$

where r_0 is the radius of the earth, h is the orbital altitude, and L is the great-circle distance between the terminals. At the limit of visibility, i.e., when the altitude is just

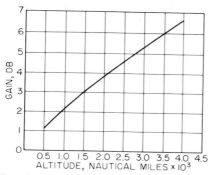

Fig. 6. Maximum gain of a nonoriented reflector.

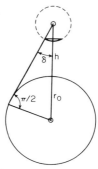

Fig. 7. Oriented spherical shell reflector. *(From Ref. 2.)*

adequate to "see" both terminals, the value of θ is

$$\theta_{max} = 2 \sin^{-1}\left(\frac{r_0}{r_0 + h}\right) \tag{5a}$$

The maximum gain of the antenna is thus

$$G_{max} = \frac{\text{area of unit sphere}}{\text{area of spherical cap of half-angle } \theta} = \frac{2}{1 - \cos \theta} \tag{6}$$

Figure 6 shows the maximum gain achievable as a function of altitude. For altitudes of practical interest, 1,000 to 10,000 miles, the gains are very modest.

Oriented Reflector. If the attitude of the satellite can be controlled so that the satellite always points along the local vertical, then it is possible to achieve gain and, at the same time, to reduce reflector weight. As an example, consider a spherical-segment reflector as shown in Fig. 7. The cross section of the segment as viewed from any point on the earth visible to the satellite is $\pi D^2/4$, as it would be for a full sphere. The weight of the segment, however, is only a fraction, $1/F$, that of the sphere, where

$$1/F = (1 - \cos \delta)/2 \tag{7}$$

and δ is the central half angle of the segment. Figure 8 shows the cross-section-to-weight ratio advantage achievable by such a reflector relative to a full sphere.

Operation of the spherical-segment reflector requires that the reflector rotate throughout the orbit to keep its axis pointing to the center of the earth. A simple method of accomplishing this attitude control makes use of the gravity gradient which exists across the dimensions of the satellite. The principle of gravity-gradient stabilization can be understood by considering a satellite consisting of two separated spheres

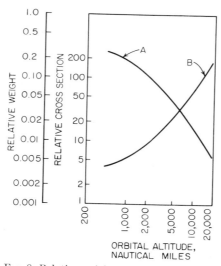

Fig. 8. Relative weight and cross section of spherical shell and spherical reflector. *(From Ref. 2.)* A = relative weight of oriented spherical shell and spherical reflector for same cross section. B = relative cross section of oriented spherical shell and spherical reflector for same weight.

joined by a long, rigid rod. In orbit about the earth, the dumbbell-like satellite experiences a balance between the centrifugal force and the earth's gravitational attraction applied at its center of gravity. Suppose the satellite is at some arbitrary inclination with respect to the local vertical. The gravitational force at the outer sphere is somewhat less than it is at the center of gravity of the satellite because it is slightly farther away. Similarly the gravitational force at the inner sphere is slightly greater than at the center. The unbalanced forces on the two spheres causes a small torque which tends to force the rod along the vertical. With the addition of a damping mechanism to damp out oscillations about the vertical, a purely passive means of attitude control is achieved. Stabilization by this means to a few degrees or less appears to be practicable.[6,7]

Mesh Spheres. Isotropic spheres made of wire mesh have been suggested[2] as another approach for achieving a higher cross-section-to-weight ratio for passive reflectors. Instead of covering the entire surface of a sphere with reflecting material, a high percentage (approximately 50 to 85 per cent) of the total cross section of the sphere can be achieved at greatly reduced weight using wire mesh. The mesh would be launched in folded form and would be deployed in orbit, possibly by centrifugal distension. The reflection coefficient of the mesh reflector varies with the mesh spacing and wire diameter. Table 3 compares the weight of wire-mesh reflectors (with best wire spacing) to that of Echo I-type reflectors made of ½-mil Mylar. In comparison with the Mylar reflectors, a maximum increase in cross-section-to-weight of about 25 can be achieved, for 1-mil wire at a frequency of 2,000 Mc.

Table 3. Weight, Lb, of Optimum Wire-mesh and Echo-type Reflectors*

Diameter, ft	½-mil Mylar	Frequency, 2,000 Mc			Frequency, 8,000 Mc		
		0.085λ spacing	0.10λ spacing	0.125λ spacing	0.105λ spacing	0.130λ spacing	0.170λ spacing
		1-mil wire†	3-mil wire	10-mil wire	1-mil wire	3-mil wire	10-mil wire
100	110	4.1	31	275	13	96	810
200	440	16	125	1,100	52	380	3,200
500	2,750	100	770	7,000	330	2,400	20,000
1,000	11,000	400	3,100	27,500	1,300	9,600	81,000

* From Ref. 2.
† Steel wire.

Distributed Passive Satellites

Distributed clouds of chafflike resonant-dipole reflectors have been suggested[5] as another means for increasing the cross-section-to-weight ratio of a passive satellite. The cross section of a half-wavelength dipole, averaged over all directions, is about $\lambda^2/6$. Its weight is $(\lambda/2)(a^2\pi\rho/4)$, where a is the dipole diameter and ρ is its density. The cross-section-to-weight ratio is thus $(4/3\pi)(\lambda^2/a^2\rho)$. The reflecting cross-section-to-weight ratio of copper dipoles of differing diameters is shown in Fig. 9 for frequencies of interest. Using 1-mil-diameter dipoles at 6000 Mc, a ratio of 1,750 sqm/lb is achieved; this is 250 times as high as achieved with an Echo-type balloon. Problems associated with use of dipole clouds concern the means by which they are deployed and their orbital lifetimes.

Although dipole clouds have not yet been tested, resonant dipoles deployed in orbital belts have been demonstrated in an experimental communications-satellite system called West Ford.[8] Figure 10 shows such a system. Signals are communicated from one point to another by the scatter action of a large number of dipoles in a volume of space V defined by the intersection of the receiving- and transmitting-antenna beam patterns with the belt. Each dipole behaves like an independent scatterer, and consequently the received signal at any given time is the sum of the signals scattered by a large number of dipoles.

As contrasted with the orbital clouds, a principal advantage of the orbital belt is the simplicity with which it can be established. Furthermore, for certain orbital configurations the belts provide a common volume of reflecting material which is stationary relative to the terminals; hence these belts provide continuous communications

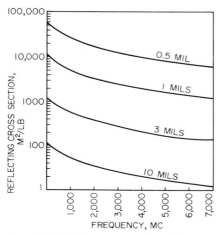

Fig. 9. Reflecting cross section of copper dipoles. (*From Ref. 2.*)

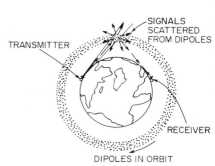

Fig. 10. Long-range communication by means of radio noise scattering from dipoles in orbit. (*From W. E. Morrow, Jr., and T. F. Rogers, The West Ford Experiment, Proc. IEEE, May, 1964, p. 462.*)

without the need for tracking antennas. It is interesting to note, however, that the cross-section-to-total-weight-in-orbit ratio for orbital-belt systems is less than that of an Echo-type reflector.

In general, power received by reflection from the volume of the belt illuminated by both the transmitting and receiving antennas is

$$P_R = \sum_{n=1}^{N} \frac{P_T G_{T_n} G_{R_n} \lambda^2 \sigma_n L}{(4\pi)^3 R_{T_n}^2 R_{R_n}^2} \tag{8}$$

where P_R = total power received, obtained by the incoherent addition of the contributions of each of the N dipoles within the common volume
G_{T_n} = gain of the transmitting antenna in the direction of the nth dipole
G_{R_n} = gain of the receiving antenna in the direction of the nth dipole
λ = wavelength
σ_n = scattering cross section of the nth dipole
R_{T_n} = distance from the transmitter to the nth dipole
R_{R_n} = distance from the receiver to the nth dipole
L = system loss factor
N = number of dipoles in the common sphere

Many gross properties of communications links using dipole belts may be deduced from simple geometric considerations and average values of dipole reflectivity. Lebow et al.[9] compute the overall path loss of a dipole belt channel from the bistatic radar equation. Designating transmitter and receiver parameters with the subscripts 1 and 2, respectively, the ratio of received power P_R to transmitted power P_T from a scattering volume with total scattering cross section (meters)² is given by

$$P_R = P_T G_1 G_2 \lambda^2 \sigma / (4\pi)^3 R_1^2 R_2^2 \qquad (9)$$

where the G's are the antenna gains, the R's are the slant ranges from the terminals to the common volume, and λ is the wavelength.

Calculation of received power is complicated by the fact that the scattering cross section σ presented by a dipole belt depends upon the geometry of the belt and the antenna beams. This may be expressed by the relation

$$\sigma = \bar{\sigma}_D N V / 2\pi (R_e + h) A = \sigma' V \qquad (10)$$

where a belt with cross-sectional area A in a circular orbit of radius

$R_e + h$ (R_e = earth radius)

containing N dipoles with average scattering cross section $\bar{\sigma}$ per dipole, is assumed. For random orientation of the dipoles, $\bar{\sigma}$ is approximately $\lambda^2/6$. Thus σ' is the average scattering cross-section density in the common volume V.

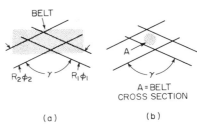

FIG. 11. Belt geometry. (a) Beam-limited case. (b) Belt-limited case. (From Ref. 9.)

Two simple geometric configurations are shown in Fig. 11. In Fig. 11a the intersection of the two beams is entirely contained within the belt. To make the common-volume estimates, it is assumed that each of the antenna beams is a parallelepiped with cross section a square of side $R\Phi$ where Φ is the 3-db beamwidth in radians.

In this beam-limited case, V is given by

$$V = (R_1 \Phi_1)^2 R_2 \Phi_2 / \sin \gamma \qquad (11)$$

where it is assumed that $R_1 \Phi_1 < R_2 \Phi_2$, and where γ is the angle between the beams. Relating the antenna beamwidth to its gain G by

$$G = \epsilon 4\pi / \Phi^2 \qquad (12)$$

where ϵ is a constant of proportionality (about 2,000), the common volume becomes

$$V = (R_1^2 R_2 / \sin \gamma)(4\pi)^{3/2} (\epsilon_1/G_1) \sqrt{\epsilon_2/G_2} \qquad (13)$$

The received power thus becomes

$$P_R = \frac{P_T \lambda^2}{(4\pi)^3} (4\pi)^{3/2} \sigma' \frac{\epsilon_1 \sqrt{G_2 \epsilon_2}}{R_2 \sin \gamma} \qquad (14)$$

In Fig. 11b, the belt is normal to the plane of the two beams and its cross-sectional area is entirely within the beam intersection. The common volume for the belt-limited condition is

$$V = A R_1 \Phi_1 \qquad (15)$$

resulting in

$$P_R = \frac{P_t \lambda^2}{(4\pi)^3} \sigma' (4\pi)^{1/2} \frac{A G_2 \sqrt{\epsilon_1 G_1}}{R_1 R_2^2} \qquad (16)$$

Lebow et al. note that "the received power relationship is much less sensitive to antenna gain and scatterer range than that for a single point target. This is due to the fact that, while each dipole acts as a point target, the number of dipoles in the common volume is proportional to the volume. Those geometric factors (G and R) which tend to increase the path loss per dipole tend to increase the number of dipoles being illuminated, the net result being the rather weak dependence given in Eq. (14). For the geometry of Fig. 11b, the effects of range and antenna gain are intermediate between that of the point target and the very extended target exemplified by the beam-limited case."

A typical path-loss ratio P_R/P_T representative of a West Ford type orbital belt that might be obtained 60 days after injection is about 10^{-21} or -210 db. Knowing the path-loss ratio, the received power-to-noise ratio can be computed as

$$P_R/N_o = (P_R/P_T)(P_T)(1/N_o) \qquad (17)$$

where N_o is the receiver noise power per cps of bandwidth (single-sided). For digital communications at a rate of R bits/sec, the energy-to-noise ratio per bit is

$$E/N_o = (P_R/N_o)(1/R) \qquad (18)$$

Assuming a 75°K receiver (corresponding to a noise of power N_o of -210 dbw) and 20 kw of transmitted power, $P_R/N_o = 2 \times 10^4$ or 43 db. This implies, as examples, an E/N_o per bit of 20 db at a data rate of 200 bits/sec or an E/N_o per bit of 10 db at a data rate of 2,000 bits/sec. As an indication of link capacities for an orbital-belt system, Table 4 shows typical parameters for an 8,000-Mc communications system for teletype and for voice service.

Table 4. Typical Parameters of West Ford Communication Links*

	Service	
	Teletype	Voice
Mass of belt in orbit, kg	100	1,000
Scattering cross section of dipoles per unit mass, sq m/kg	2,500	2,500
Altitude of belt, km	6,000	6,000
Ground-antenna size, m	5	10
Transmitter power, avg, kw	20	50
Receiver noise temp, °K	70	70
Capacity, bits/sec	Approx 100, i.e., 4 teletype circuits	Approx 50,000, i.e., 1 high-quality digital voice circuit

* From W. E. Morrow, Jr., and T. F. Rogers, The West Ford Experiment, *Proc. IEEE*, May, 1964, p. 466.

Active Satellites

In an active satellite, the signal is amplified within the satellite before it is relayed to the receiving terminal. It is convenient to speak of the "up link," from transmitting terminal to satellite, and the "down link," from satellite to receiving terminal. In practical systems, a different carrier frequency is used on the down link from that used on the up link to minimize isolation problems at the satellite. Frequently, too, the form of modulation is different for the two links.

Neglecting atmospheric attenuation and other system losses, the carrier power received at the satellite from the transmitting station is

$$P_{RS} = P_T G_T G_s \lambda_u^2 / (4\pi R_{TS})^2 \qquad (19)$$

where λ_u is the wavelength of the up link and the other symbols have the same meaning as in the previous sections. The signal-to-noise ratio at the satellite receiver is

$$(S/N)_\mu = [P_T G_T G_s \lambda_u^2 / (4\pi R_{TS})^2](1/kT_{eff}B) \qquad (20)$$

SIGNAL POWER AT RECEIVING TERMINAL

Similarly, the signal received by the ground receiving terminal is

$$P_{RG} = P_{TS}G_R G_S \lambda_d{}^2 / (4\pi R_{SR})^2 \qquad (21)$$

and the signal-to-noise at the receiving terminal is

$$(S/N)_d = [P_{TS}G_S G_R \lambda_d{}^2 / (4\pi R_{SR})^2](1/kT_{eff}B) \qquad (22)$$

Parameters of a representative active medium-altitude communications satellite, Telstar I, are shown in Table 5.

Table 5. Parameters of Telstar I Medium-altitude Active Satellite System*

Transmission
Signals handled:
 Television, 3-Mc bandwidth
 600 one-way telephone channels (simulated by noise)
 12 two-way telephone channels
Modulation, FM
R-F bandwidth:
 Ground station, 25 Mc
 Satellite, 50 Mc
Frequencies:
 Communication up, 6,389.58 Mc
 Communication down, 4,169.72 Mc
 Beacon, 4,079.73 Mc
 Telemetry and beacon, 136.05 Mc
 Command, about 123 Mc
Polarization:
 Microwave channels, circular
 VHF beacon, linear
 Command, circular

Satellite
Size, shape, and weight, 34-in. sphere, 170 lb
Orbit:
 Perigee 514.21 NM
 Apogee 3,051.37 NM
 Inclination 44.8°
Launch, Delta vehicle from Cape Kennedy
Repeater configuration, I-F type: amplifies, shifts frequency, does not alter modulation
Communications antennas, approximately isotropic, circularly polarized
R-F power output, 2 watts
Power plant:
 Silicon n-on-p solar cells, shielded
 Ni-Cd storage battery
Stabilization:
 Spin with axis normal to plane of ecliptic
 Magnetic torquing coil control

Ground station (Andover)
Communications antenna:
 3,600-sq-ft horn reflector
 Inflated radome
 Pointing by tape drive, autotrack, slave to precision tracker
Communications transmitter, 2 kw
Communications receiver, maser input, frequency-compression demodulator
Noise temperature, 32°K at zenith

 * From D. F. Hoth, E. F. O'Neill, and I. Welber, The *Telstar* Satellite System, *Bell System Tech. J.*, July, 1963, p. 770.

The critical path in active communications-satellite systems is the down link in which the satellite transmitter power and gain, as compared with the ground transmitter, are severely limited. At present, satellite power amplifiers use traveling-wave tubes and are limited to an output level of 10 to 50 watts; it is likely, however, that TWT's and solid-state devices with outputs of kilowatts can be developed.

With regard to satellite antenna gain, a number of significant new developments can be expected to become operational. To date only low-gain, essentially omnidirectional, antennas have been used on medium-altitude systems. For synchronous satellites, however, directional antennas, as indicated by Eq. (5), become worthwhile. To achieve this gain fine attitude control is normally required, to assure that the narrow beamwidth intersects the earth. Potentially, the simplest attitude-stabilization technique is passive gravity-gradient stabilization; this technique is under development. For current synchronous satellite systems, consideration has been given to full three-axes control systems and to spin stabilization. It has been found that the latter, notwithstanding the problem of despinning the beam, is considerably less complex. A spin-stabilized antenna system employing a 16-element phased

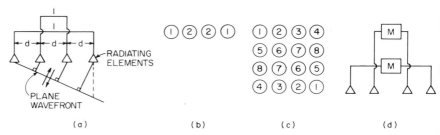

FIG. 12. Van Atta antenna operating principles. (a) Four-element passive array. (b) Linear array connection diagram. (c) Planar array connection diagram. (d) Semiactive Van Atta array. (*From Ref. 12.*)

array is currently being developed[10]. The elements are located on a circle about the spin axis, and the r-f energy is phased to each element so as to reinforce them in a single direction, toward the earth. As the spacecraft spins, the basic phasing pattern progresses in the opposite direction at exactly the same rate, so that a pencil beam, approximately 17° wide, is constantly directed at the earth.

Recent developments have indicated the possibility of using phased-array antennas for achieving directivity on unstabilized satellites, even those at medium altitudes. These phased-array techniques are based on the so-called Van Atta principle,[11] illustrated in Fig. 12a. Signals received from a distant transmitter by each element of a phased array differ in phase by an amount which depends on the geometry of the array and the angle of incidence which the array makes with a line connecting the antenna to the transmitter. If the elements of the array are equally spaced, then the incremental phase difference is

$$\Delta\phi = (2\pi d/\lambda) \sin \alpha \tag{23}$$

where d is the separation between elements, λ is the wavelength, and α is the angle of incidence. For suitable interconnections of elements in a linear array, the phase differences can be canceled out with the result that a reflected wave is returned to the illuminating source, in a manner similar to that of a corner reflector. The Van Atta principle can be extended to a surface array, by interconnecting conjugate elements of the array, as shown in Fig. 12c. The reflection property can thus be provided for both angular dimensions.

A so-called quasi-passive communications satellite employing these techniques has been proposed.[12] The satellite employs an antenna of the Van Atta type in which solid-state modulating devices are inserted in each of the paths connecting conjugate array elements. The transmitting terminal sends a modulated signal to the satellite from any direction within the field of view of the array, which can be made wide enough to encompass the earth and to permit variations in attitude of the satellite. Single elements of the array are used to receive the signals from the ground, and simple receivers are used to detect the signal. The output of these receivers then modulates the elements in the Van Atta array. The receiving terminal radiates the array with a high-powered source of r-f energy. The array reflects a portion of irradiating energy, now modulated with the information to be carried, back to the receiving terminal.

Except for the minute power required to modulate the interelement switches, the quasi-passive satellite requires no on-board power, the retransmitted power being provided by the receiving terminal itself. The satellite overcomes the limited sensitivity of conventional passive satellites through the high directivity of the antenna.

For transmission of digital data, varactor diodes are applicable as modulating components.

For transmission of analog information, wideband FM can be used, using a modification of the Van Atta array shown in Fig. 13.[13] Instead of interconnections between conjugate elements, image-frequency converters are placed at each array element. The image-frequency converters are connected by equal-length transmission lines to a local oscillator.

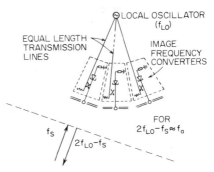

Fig. 13. Retrodirective array with image-frequency converters. (*From Ref. 13.*)

The local oscillator frequency f_{LO} is chosen to be approximately equal to the signal frequency f_S. The beat between the second harmonic of f_{LO} and the signal frequency produces an image frequency which is radiated from each of the antenna elements. The image frequency has a phase which is inverted with respect to f_S. Hence the radiated frequency has the correct phasing for a retrodirective wave. The system is particularly convenient for FM by modulating the local oscillator. The image converters amplify and provide a frequency offset for the reflected signal, since f_{LO} is not made exactly identical to f_S. The system provides a means for reducing the ground-transmitter power by providing both active amplifier gain and antenna-directivity passive gain in the satellite.

The configuration of Fig. 13 makes it convenient to use nonplanar arrays since there is no restriction on the placement of the array elements so long as the feed lines from the local oscillator are equal. It is convenient to construct a spherical array using this configuration.

Cutler et al[14] have proposed a similar Van Atta type phased-array antenna system in which a pilot tone transmitted by the ground receiving station causes the satellite retransmission to be beamed to the receiving ground station. Figure 14 shows a

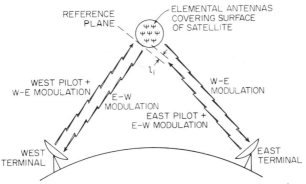

Fig. 14. Self-steering phased-array two-way satellite repeater configuration. (*From Ref. 14.*)

suggested system configuration. The satellite repeater consists of a large number of antenna elements distributed over its surface. Transmission from the west terminal consists of the normal west-east modulation and a separate carrier designated the "west pilot tone." Similarly, transmission from the east terminal consists of the normal east-west modulation and an "east pilot tone." The action of the repeater image-converter circuitry causes the west-east modulation to be amplified and retransmitted toward the east terminal; similarly the east-west modulation is retransmitted to the west terminal. Since the antenna beam is automatically steered to the receiver regardless of its location, the antenna can be as large as desired, independent of changing satellite position and attitude.

Most of the various phased-array techniques involve a large number of low-power solid-state elemental repeater amplifiers. In addition to the flexible performance characteristics described above, added reliability is provided by the many parallel paths through the repeater; failure of individual units, rather than causing total failure, only slightly degrades performance.

4 Modulation Considerations

The method of modulating the communications channel affects the ability to trade bandwidth for power, the vulnerability of the channel to interference and jamming, and the capability of the satellite repeater to handle simultaneous signals from more than one pair of terminals.

The Shannon formula relating the information capacity of a channel C in bits per second to the bandwidth W in cycles per second and signal-to-noise ratio

$$C = W \log_2 (1 + S/N) \qquad (24)$$

indicates the trade-off which may be effected between transmitter power and bandwidth. The equation shows that a given channel capacity can be achieved with less signal power using a wide bandwidth than a narrow bandwidth. For communications-satellite systems in which received power is severely limited (by R^4 attenuation for a passive systems, and by inherent limitations on satellite output power for the down link in active systems), wide-bandwidth modulation techniques are indicated.

Wide-bandwidth systems are also less susceptible to interference. For each sideband component, there is a certain probability, established by the level of noise and interfering signal, that noise will destructively interfere with it. If the destructive interference is statistically independent of sideband frequency components, the probability that all sidebands are destructively interfered with decreases as the number of sidebands increase.

Methods of Modulation

Two wide-bandwidth modulation techniques, analog wideband FM and digital PCM, have received extensive consideration for communications satellites. The following paragraphs describe their characteristics and compare them with narrow-band modulation methods.

Wideband FM. In frequency modulation the instantaneous frequency of the r-f carrier is varied in accordance with the amplitude of the message signal. If the bandwidth of the message is b and the maximum deviation of the carrier is $\pm \Delta f$, then an index of modulation m is defined as

$$m = \Delta f / b \qquad (25)$$

The spectrum of the modulated carrier is a complex function of m, consisting of a carrier component and a large number of sidebands. It can be shown that the r-f bandwidth B of the modulated signal is approximately

$$B = 2(\Delta f + b) = 2(m + 1)b \qquad (26)$$

On reception, the FM receiver recovers the message (of bandwidth b) from the wide-bandwidth signal. This is accomplished with an accompanying processing gain. If the signal-to-noise ratio at the input to the receiver is $(S/N)_i$, as computed by the range equations described in Sec. 3, then the signal-to-noise ratio at the output of the receiver $(S/N)_o$ is increased, according to the relationship

$$(S/N)_o = (S/N)_i 3m^2(m + 1) \qquad (27)$$

As the information capacity of any channel is a function of the output signal-to-noise ratio and output bandwidth, a prescribed capacity can be achieved at a reduction in transmitted power (proportional to S_i) by increasing the modulation index. It would appear from Eq. (27) that the transmitted signal could be reduced to an arbitrarily low level by using higher and higher values of m. Such is not the case. On the one hand, the value of m in any practical system is limited by the availability

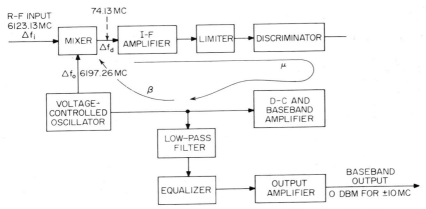

FIG. 15. Block diagram of FMFB receiver. (*From Ref. 15.*)

of r-f spectrum. In addition, a fundamental limitation to the ability to reduce the transmitted power is established by a threshold effect which occurs in all wideband systems; i.e., the processing gain is achieved only when the input signal-to-noise ratio exceeds a certain level, characteristic of the particular modulation method. When the input signal-to-noise level falls below the threshold, the performance of the system degrades sharply. For wideband FM the threshold is a function of modulation index; for large index ($m > 4$), it occurs when the carrier-to-noise level falls below about 13 db. For $m = 1$ the threshold occurs at about 2 db.

Wideband FM systems have been used extensively for the power-limited down link of experimental communications satellites. Giger and Chaffee have shown that the limitations imposed by the threshold effect can be reduced by the use of an FM feedback (FMFB) receiver.[15] The basic elements of the FMFB receiver used at the Telstar receiving terminal are shown in Fig. 15. The incoming r-f signal is combined in a mixer with the output of a voltage-controlled oscillator (VCO) to produce an intermediate-frequency signal. The frequency modulation which has been imparted to this signal, after going through an i-f amplifier and limiter, is then demodulated by the discriminator. The recovered baseband signals are amplified in the d-c-and-baseband amplifier and are fed back to the VCO in such a way as to reduce significantly the instantaneous frequency difference between the two signals going into the mixer. If this reduction is substantial, the resulting FM wave will occupy a much narrower band than the incoming r-f signal. This makes it possible to restrict the bandwidth of the i-f amplifier. In a sense, the feedback receiver can be considered as a "tracking" filter whose bandwidth is substantially narrower than that required to transmit the complete incoming FM wave; being narrower, the receiver is more immune to incoming noise.[12]

The processing gain of the FMFB receiver is identical to the standard FM system when the input signal-to-noise ratio is above the usual FM threshold. The advantage of the FMFB receiver is its ability to lower the threshold; threshold tests on Telstar have shown improvements of about 4 to 5 db.

PCM. In a pulse-code-modulation system, the analog baseband message is periodically sampled and converted to digital form. In binary PCM, the sampled value is transmitted as a series of 1, 0 pulses. The message bandwidth and the degree of fidelity to which the sampled value of signal is quantized determine the r-f bandwidth. Thus,

$$B = kf_m s \tag{28}$$

where f_m is the highest frequency in the message, s is the number of quantization levels, and k is a constant which, theoretically, can be as low as 2 but in practice is about 2.5. If the input signal power at the receiver is greater than the input noise power in the bandwidth B, then the 1, 0 binary pulses can be recognized with very low error rate. Figure 16 shows the error rate for a binary PCM system as a function of input-signal-to-noise ratio. The S/N at which the error rate increases significantly with decrease in input signal-to-noise ratio, about 18 db, establishes the system threshold. When the input signal-to-noise exceeds the threshold, the output signal-to-noise is determined by the quantization error, the error introduced by the nonzero width of the quantization intervals. This error is equivalent to a signal-to-noise ratio. It can be shown that the equivalent signal-to-noise ratio increases exponentially with the number of quantization levels. However, as given by Eq. (28), the bandwidth required increases only linearly with the number of quantization levels.

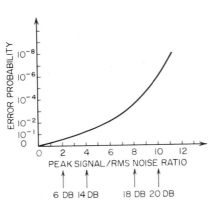

FIG. 16. Probability of error in transmitting binary digits. (*From Mischa Schwartz, "Information Transmission, Modulation, and Noise," McGraw-Hill Book Company, New York, 1959.*)

Accordingly, the output signal-to-noise ratio increases *exponentially* with bandwidth and is thus theoretically even more efficient than wide-bandwidth FM.

PCM has another important advantage for communication relays; because the signal is reconstituted at each repeater point, *noise* effects are not *cumulative*. This is more important, however, for multiple-hop systems than it is for single-hop satellite links.

Comparison with Narrowband Modulation Methods. In a practical communications-satellite system the ability to trade bandwidth for power by selection of modulation methods is limited by the availability of spectrum. Thus, for example, for a link carrying 600 simultaneous voice messages with a baseband bandwidth of about 2.5 Mc, it may not be practicable to use FM with a modulation index greater than $m = 2$, corresponding to an r-f bandwidth of 15 Mc.

Where spectrum conservation is critical, narrowband modulation techniques are more suitable than either FM or PCM. The simplest modulation technique, in concept, is single-sideband (SSB), which uses the minimum r-f bandwidth, equal to the base bandwidth. Conventional double-side band AM requires twice the bandwidth of SSB. Although AM is less complex in operation, this advantage is not sufficiently important to make it attractive for communications-satellite systems.

Wright[16] has compared SSB, PCM, and FM systems for a typical 600-telephone channel (each 4 kc) communications-satellite circuit. His conclusions, summarized in Table 6, indicate that:

1. SSB modulation is most economical of r-f bandwidth but requires the largest amount of peak power. Furthermore, it requires especially linear circuitry to prevent crosstalk between channels. It is suitable for the up link in active systems.

2. For the power-limited down link, wideband modulation techniques are more appropriate. Wideband FM is a well-tried technique for multichannel telephony and TV. PCM techniques appear to be about 3 to 5 db more efficient than wideband FM in more nearly approaching the ideal Shannon limit. Furthermore, PCM, being a regenerative system, avoids cumulative noise and distortion as it is relayed through the repeater.

3. Both FM and PCM suffer from threshold effects. In FM, since the signal-to-noise ratio at the output of the receiver is proportional to the signal-to-noise ratio at the input, extra transmitter power provided as margin for deterioration of components provides increased intelligibility; in PCM, the output signal-to-noise ratio is relatively unaffected by increase in power level so long as the input signal-to-noise ratio exceeds the threshold.

Table 6. Power, Bandwidth and Output Signal-to-Noise Ratios for Different Modulation Systems*

Modulation system	Channel output S/N, db, for 600 channels, at transmitter power levels of			Min bandwidth occupancy, Mc/s
	0.5 watt	1 watt	2 watts	
SSB................................	30	33	36	2.5
PCM,† 1/2t bandwidth..............	32.5	62	123	17.5
PCM,† 1/t bandwidth...............	13.5	28.5	59	35
FM, 50 Mc/sec bandwidth with feedback	41 (at threshold)	44	47	50
FM, 100 Mc/sec bandwidth with feedback	Below threshold	51	54	100

* From Ref. 16.
† PCM figures are for thermal error noise only. They do not include the quantization noise appropriate to a 7-bit system. t is bit duration.

The choice of modulation technique depends on the availability of power in the satellite, the allocation of spectrum and practical "hardware" questions, principally concerning the satellite repeater. The "modulation-handling" characteristics of the principal types of active satellite repeaters are shown in Table 7. In addition, the choice will be influenced by techniques devised for achieving multiple access.

Multiple Access

Multiple access refers to the capability of a satellite repeater simultaneously to receive and transmit signals from more than one ground terminal. In its simplest form, a satellite which provides a duplex voice circuit between two terminals may be said to possess a limited multiple-access capability; more generally, however, the term is used to express the ability of a common repeater to provide simultaneous service among many ground terminals. This capability is desirable, if not essential, for practicable communications-satellite systems to reduce the prohibitive expense of orbiting separate satellites for each terminal pair.

Satellite reception of a transmission containing several multiplexed communication channels from a single ground terminal does not constitute multiple access. However, a multiple-access repeater may multiplex signals received from a number of ground terminals for retransmission, provided that the various signals can eventually be separated at the intended receiving terminals.

Table 7. Characteristics of Principal Types of Satellite Repeaters*

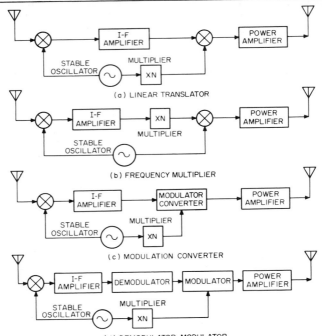

The *linear translator* simply performs a frequency translation on the incoming signal spectrum before retransmission. The translator may or may not include limiting, depending on whether angle modulation is employed and whether only one signal is present in the channel. The linear translator accepts a wide range of modulation types, especially in its nonlimiting form. Its noise performance is also relatively good in that satellite and ground contributions are simply additive.

The *frequency-multiplier repeater* reduces the received signal to an intermediate frequency and then multiplies it to the transmitted frequency. The frequency deviation is also multiplied by the same factor and the r-f bandwidth occupancy increased. The frequency multiplier is therefore useful in matching a high-power narrowband ground transmitter (e.g., a klystron) to a low-power wideband satellite transmitter (e.g., a traveling-wave tube). A high-power ground transmitter is required because the noise contribution of the satellite receiver is increased by the square of the multiplication ratio. Also the multiplier is suitable only for FM or PM signals.

The *modulation converter* is so named because it changes the input modulation form. It has been proposed specifically for converting single-sideband (SSB) transmissions to FM. The conversion can be accomplished by inserting at the satellite receiver a strong local carrier along with the stack of SSB telephone channels. The weak frequency modulation of the carrier that results is then enhanced (after limiting) by frequency multiplication. Both amplitude and frequency stability of the inserted carrier are vital. The former determines the ultimate frequency occupied by the SSB signals. The excess bandwidth required because of frequency instability means a S/N impairment compared to the linear translator repeater.

The *demodulator-modulator repeater* reduces the information to baseband, operates on it, and then allows it to modulate the satellite transmitter. Having been designed for one type of modulation, this repeater is inflexible in changing to another type. The ability to process the baseband signals may be important in the processing equipment but will, in general, be an additional burden on the satellite.

* From "Space Data," TRW Space Technology Laboratories, Redondo Beach, Calif., 1965.

MODULATION CONSIDERATIONS

A passive satellite, as a linear element, is well suited for multiple access. Different terminal pairs can simultaneously use the satellite by operating at different frequencies or by otherwise coding individual transmissions. In principle, similar operation can be achieved with an active satellite containing a wideband linear amplifier. In practice, however, repeater equipment difficulties constrain the multiple-access capability. These limitations have prompted extensive theoretical and experimental investigation into multiple-access techniques.

The requirements for multiple-access capability depend on the following factors according to Sawitz.[17]

Number of Ground Terminals. The larger the number of stations which are within the area of mutual visibility of the satellite, the greater the potential requirement for simultaneous communications through a single repeater. If the communications-satellite system employs a relatively small number of "gateways" —ground terminals through which all messages for a large surrounding area are routed—then the multiple-access requirements are reduced. Commercial satellite systems may conceivably take this form. However, nationalism and other factors may result in a commerical system in which many terminals are concentrated in a relatively small geographic area, for example, in England, France, Germany, and Sweden. For such a configuration there will be considerable overlap time when a single satellite is in common view of the United States and these stations. Similarly, for a military communications system or for civilian systems operating in areas in which ground-ground communications between gateways and users are poorly developed, the number of stations within view of a single repeater may be large. The multiple-access problem is complicated because of the differences in characteristics of ground terminals. If the stations to be simultaneously served include both high-power/high-gain transmitting stations and low-power/low-gain stations, the multiple-access repeater must process the two signals in such a way that the stronger does not suppress or interfere with the weaker. A desirable property of the repeater is the ability to allocate its capacity in such a way as to equalize the signal-to-noise ratios received at the various receiving terminals. Thus, in dividing its capacity between the two signals, the repeater should allocate output power unequally. The problem becomes even more complex when the variation in range between the satellites and stations is considered.

Number and Orbits of Satellites. The requirements for multiple access increase with satellite altitude, as the area of mutual visibility increases. It is particularly critical for a 24-hr satellite. For medium-altitude systems, the requirement decreases as the number of satellites is increased. If the number of satellites is such that each of the required simultaneous transmissions can be sent via a separate satellite, there will be no need for multiple access. If the number of stations within the area of mutual visibility is large, it is generally more economical to incorporate multiple access into the satellites. However, if the potential requirements for multiple access are small, then, recognizing that a multiple-access capability may increase the cost of the satellite and may reduce its useful life in orbit, it may in fact be more economical to launch more satellites.

Use Factor and Traffic Distribution. The basic problem of multiple access in active satellites is efficient sharing of a common repeater capacity. If the total capacity of a repeater is fully utilized by one ground transmitter, then multiple access is impossible. Defining use factor as the fraction of repeater capacity actually used by a particular transmitter, multiple access can be achieved only in a system in which individual transmitters have use factors less than 1. When the traffic requests of a system with a multiple-access capability temporarily exceed the repeater capacity, the use factors of the individual transmitters may be restricted and queuing procedures with orders of preference may be used until traffic has been reduced to repeater capacity. If the system does not have a multiple-access capability, then only queuing procedures are possible, and flexibility of control is lost.

Correlation of Use Factors. If the individual use factors of the stations for which multiple access is provided are uncorrelated, then large statistical deviations from the average can be tolerated since the fluctuations will tend to be compensating. This condition, however, is not likely to be met. In times of crises or emergencies the use factors of all users may be expected to increase simultaneously. This factor tends to militate against multiple-access systems. If the system is designed to handle the above-average use factors (as it must for military systems), it will of necessity have an inefficient channel utilization during average or below-average use periods. When the system is used appreciably below capacity, the requirements for multiple access are small because the probability of seeing an "empty" usable satellite is high. In times of crisis when the use factors increase for all stations, the system is used to capacity, and any multiple-access capability is likely to reduce the overall capacity of the system (because of equipment complexities) precisely when it is needed most. Under these circumstances, a queuing procedure might be preferable to a multiple-access capacity.

Both frequency-division (FDM) and time-division-multiplex (TDM) have been considered for achieving multiple access. In principle, a FDM system can be visualized in which each transmitter transmits on a separate frequency, each frequency corresponding to a particular transmitter-receiver combination. The signals would be received by the repeater and amplified by a wideband linear amplifier prior to retransmission. The difficulties with this method arise from the bandwidth required and the cross-modulation and suppression effects which take place when a number of signals are simultaneously processed by a practical, not perfectly linear, TWT. This method has been used in Telstar for achieving duplex service between two terminals. Separate carrier frequencies are used in each direction; voice signals are impressed on the carriers using wideband frequency modulation.

For multiple-access service in the larger sense, a more efficient method of frequency division is achieved using single-sideband modulation on the up link. This method has been proposed by Hughes Aircraft Corp. for use in commercial 24-hr satellites. Each of the different stations which use the repeater is assigned a portion of the spectrum. The absence of channel carrier frequencies permits simple linear summation in the satellite. The signals are amplified at the repeater and are frequency-modulated on the down-link carrier. Separation of signals is achieved at the ground stations by filters.

In a time-division (TDM) system, the voice signals are digitized prior to transmission. Each station may operate on a common frequency but they are all synchronized to a common time base. By interlacing their transmissions appropriately only one signal at a time is amplified at the repeater. This reduces the constraints on linearity and bandwidth of the repeater. Wright and Jolliffe[18] have shown that practical communications-satellite TDM systems can be achieved. W. Glomb of I.T.T. Federal Laboratories has studied such a system in detail, and has devised a method that relaxes the synchronizing timing accuracy required to the order of 1 msec.

In both FDM and TDM systems, the motion of the satellite relative to the terminals requires that guard bands be established to provide for Doppler shifts (for FDM) and transmission-delay variation (for TDM). The reduction in communication capacity corresponding to the guard bands is part of the cost of the multiple-access capability. Both FDM and TDM systems can be made to be adaptive—that is, their frequency or time allocations can be changed from a rigid schedule—by use of a central communications-satellite control facility, which senses allocation requirements.

Multiple-access capability independent of assigned allocations can be achieved using so-called random-access modulation methods. A variety of such techniques have been developed. Magnuski[19] describes the fundamentals of random access and its application to communications satellites.

5 Coverage Considerations

Real-time two-way communications via satellite can be effected only when the satellite is mutually visible to the transmitting and receiving terminals. For a stationary satellite, the coverage characteristics can be summarized as follows:

COVERAGE CONSIDERATIONS

A single stationary satellite provides coverage over an area of about 60×10^6 square miles, or approximately one-third of the earth. Three satellites provide global coverage, except for relatively small areas near the poles.

Communications service is continuous in time; i.e., there are no outage periods when the satellite is not visible to all transmitting and receiving terminals.

Communications service is achieved without wide-angle tracking by the transmitting and receiving antennas.

For a chaff belt system, at least two belts—one in polar and one in equatorial orbit—are required for global coverage. In principle, fixed antennas can be used at terminals which communicate via the equatorial belt and via that part of the polar belt directly over the poles, since the volume of reflectors common to the transmitting and receiving antennas is fixed relative to the terminals. In practice, limited tracking may be required to account for slight relative motion of the belts.

For a medium-altitude satellite system, the communications coverage characteristics cannot be stated so succinctly. Unlike the stationary and orbital belt systems, wide-angle tracking of the satellites is required. In general the area of coverage achieved by a single (or a small number of satellites) is very limited. In addition, the communications service is subject to frequent interruptions as the satellites pass out of view of the earth-rotating transmitting or receiving terminals. Accordingly, medium-altitude systems generally require a relatively large number of satellites.

A prescribed probability of coverage over a specified network of terminals can be achieved by a variety of orbital configurations involving differing numbers of satellites at different altitudes and in one, two, or more orbital planes. Furthermore, the satellites may be disposed within the orbital planes in various controlled patterns or in random patterns. As a consequence of the number and complexity of the factors involved, there does not exist any direct procedure for synthesizing the least-cost configuration. There is, however, no fundamental difficulty in determining the coverage attainable with any postulated configuration; indeed, this method is used for selecting and validating preferred configurations. It consists of the following steps:

1. Given a postulated configuration of satellites, the ephemeris of each satellite is accurately computed, using techniques of orbital mechanics.
2. The track of the subsatellite point of each satellite is projected forward in time for the life of the system (or until an inherent periodicity is observed).
3. For each required communication circuit, the occurrence and duration of outages are noted.
4. This process is repeated for each postulated configuration; the performance and overall cost of each configuration are compared and a preferred configuration selected.

The process of extrapolating the orbital positions of the satellites is numerically very involved and is usually accomplished by a digital computer. A large library of "trajectory" programs has been developed for this purpose.

Although all practical configurations must ultimately be validated by this exact procedure, it is useful to consider approximate methods for arriving at potentially acceptable orbital configurations. A method described by Reiger[2] which is particularly helpful in gaining insight into the coverage problem is presented below.

Consider first a single satellite in circular orbit at altitude h, as shown in Fig. 17. The satellite is visible from terminal T when the subsatellite point is within a spherical surface centered at T and bounded by circle A. The area of this surface depends on the orbital altitude and the minimum acceptable elevation angle α for the terminal antenna. It is

$$\text{Area of visibility} = 2\pi r_0^2 (1 - \cos \phi) \tag{29}$$

where ϕ, the half angle of the cone subtended by circle A at the center of the earth, is

$$\phi = \cos^{-1}[r_0/(r_0 + h)] - \alpha$$

Figure 18 shows the area of visibility, expressed as a percentage of the total earth area, as a function of altitude and minimum elevation angle. *If the orbit is such that* the *probability density of the ground track of the satellite is uniform over the entire surface of the earth, then the area of visibility, expressed as a percentage of total earth area, also represents the probability that the satellite is visible by terminal T at any specified instant.*

Similarly, a satellite at altitude h is visible from terminal R if the subsatellite point is within a spherical surface of identical area centered about R. For the satellite to be visible simultaneously by both terminals T and R, the subsatellite point must fall within the area common to both surfaces. This area is defined as the *area of mutual*

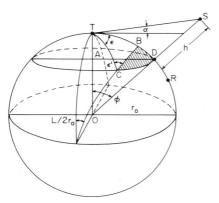

Fig. 17. Geometry of satellite visibility. (*From Ref. 2.*)

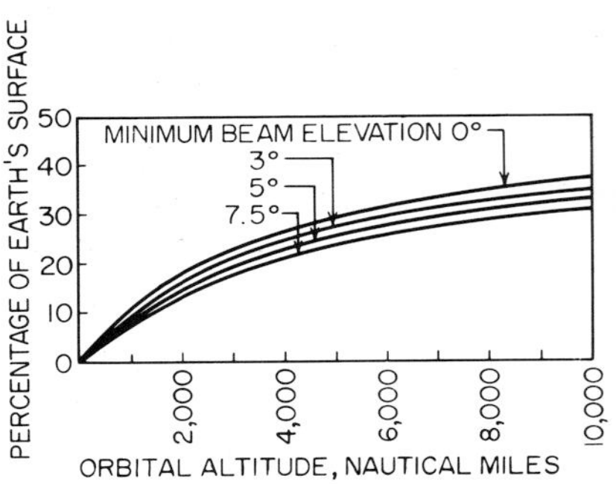

Fig. 18. Percentage of earth's surface from which a satellite is visible. (*From Ref. 2.*)

visibility: the area on the earth's surface within which the subsatellite point is located when the satellite is simultaneously visible from two terminals. The area of mutual visibility (AMV) is determined by the orbital altitude, the great-circle distance between the two terminals L, and the minimum elevation angle. It is

$$\text{AMV} = 4\pi r_0^2 p \tag{30}$$

where
$$p = \frac{1}{\pi}\left[\frac{\pi}{2} - \sin^{-1}\frac{\sin (L/2r_0)}{\sin \phi} - \cos \phi \cos^{-1}\frac{\tan (L/2r_0)}{\tan \phi}\right]$$

is the area of mutual visibility expressed as a fraction of the total areas of the earth. If the probability density of the subsatellite point over the entire surface of the earth is uniform, then p is the probability that the satellite is mutually visible by T and R at any instant. Similarly, if N satellites at altitude h are in orbit about the earth such that the probability density of the ground tracks is uniform over the entire surface, then the number of satellites which are mutually visible by T and R at any instant is pN.

Figure 19 shows p as a function of altitude and circuit length, the great-circle distance between terminals. It is apparent that the area of visibility decreases sharply with circuit length; for each orbital altitude there is a maximum length beyond which communications cannot be accomplished without intermediate relay points. The area of mutual visibility is the area in common of two overlapping circles on the earth's surface; its shape is roughly elliptical, centered on the midpoint between the two terminals. The "minor axis" is along the great circle connecting the terminals; the "major axis" is along the great circle through the midpoint, which is perpendicular to the great circle containing the terminals. These "dimensions," shown in Fig. 20, are useful for estimating the duration of mutual visibility—that is, the length of time a subsatellite point remains within the area of mutual visibility. Unfortunately there does not appear to be any simple way of expressing the duration in terms of the relevant parameters. The maximum possible duration of visibility is achieved

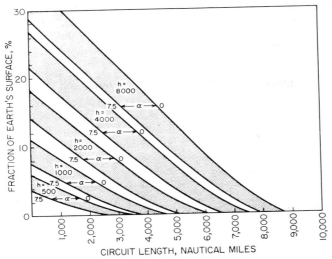

Fig. 19. Area of mutual visibility. h = satellite altitude, nautical miles. α = minimum elevation of antenna beam above horizon, degrees. (*From Ref. 2.*)

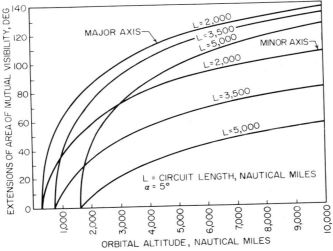

Fig. 20. Principal extensions of area of mutual visibility. 5° beam elevation. L = circuit length, nautical miles. $\alpha = 5°$. (*From Ref. 2.*)

when the subsatellite's track is along the "major axis" and can be found by dividing the major-axis dimension by the angular orbital velocity. Figure 21 shows the maximum period of mutual visibility plotted as a function of altitude and circuit length. To obtain the actual duration of visibility, one must plot individual circles of visibility about T and R, and trace the subsatellite path through the common region.

The concept of area of mutual visibility is fundamental to estimating the number of satellites required to provide a prescribed probability of service for a given circuit. Clearly, the minimum number of satellites is required for an orbital configuration in which one and only one satellite is mutually visible at any time. Such a configuration provides continuous coverage, without any outages, and requires that the satellites be launched into and maintained in a particular, controlled pattern. Its major disadvantage is the complexity of the equipment required to establish and to maintain the pattern in the face of orbital perturbations. To obviate these problems, a "random" orbit configuration can be used. A truly random system is one in which the probability density of the subsatellite point is uniformly distributed over the entire earth. The random configuration can take many forms: the satellites can be in one, two, or more planes; they can have the same or differing inclinations; they can be equally spaced within a plane or not. In all cases, the number of satellites required is greater than for the controlled configuration, and, unlike the controlled configuration, the probability of service is not continuous but can only approach 100 per cent.

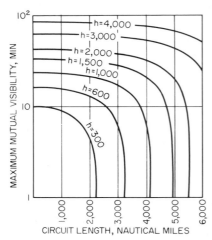

FIG. 21. Maximum mutual satellite visibility. h = orbital altitude, nautical miles. $\alpha = 5°$. (*From Ref. 2.*)

An approximate estimate of the number of random satellites required for a given probability of service can be made as follows: If there are N satellites in truly random orbit and the area of mutual visibility expressed as a fraction of the total earth's area is p, then the average number of satellites that are mutually visible by the two terminals at any instant of time is pN and the probability that a particular satellite is not mutually visible at that time is $1 - p$. The probability that no satellite is visible is

$$Q = (1 - p)^N \tag{31}$$

This expression is also the fraction of time during which no satellite is mutually visible, i.e., the fraction of time for which an outage exists. This fraction can be made arbitrarily small by increasing N;

$$N = \frac{\log Q}{\log (1 - p)} \tag{32}$$

This relation is plotted in Fig. 22.

As an example of use of random orbits, consider first a 2,000-nautical-mile altitude equatorial-orbit satellite configuration designed to provide service between two points located on the equator 2,400 nautical miles apart. Assume the minimum elevation angle is 5°. From Fig. 19, the fraction of earth's surface mutually visible is 7.5 per cent. The minor axis of the area of mutual visibility is along the equator and is, from Fig. 20, about 52°. The minimum number of satellites, in a controlled orbit, is thus about 7. This configuration would provide 100 per cent service. If 90 per cent service were to be provided by a set of random satellites meeting the assumptions specified in the preceding paragraph, the number of satellites required, from Fig. 22, would be about 30.

In actual practice, one would not employ a completely random system—whose subsatellite location probability density was uniform over the entire earth—to provide coverage between two points on the equator. Rather, a "less random" system, yielding a higher probability density for the equatorial regions, would be used. If instead of neglecting the actual inclination of the orbits and the location of the terminals. as was done in the preceding example, these factors are taken into account, then under certain simplifying assumptions a better estimate of the required number of

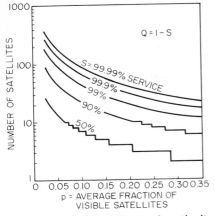

Fig. 22. Number of satellites and continuity of service. $Q = 1 - S$. (*From Ref. 2.*)

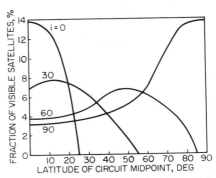

Fig. 23. Average fraction of visible satellites, AMV 5 per cent. i = orbit inclination, degrees. (*From Ref. 2.*)

satellites is obtained. Figures 23, 24, and 25 show the result of such analyses; the average fraction of visible satellites is plotted for various values of fractional area of mutual visibility, orbital inclination, and latitude of circuit midpoint. Inspection of these curves indicates that

While both polar and equatorial orbits favor circuit midpoints having a latitude near the orbital-inclination angle, inclined orbits (approximately 50 to 60°) provide the most uniform coverage over the entire surface of the earth, and the best coverage in the temperate zones.

Polar orbits are preferable if a considerable fraction of the circuits is near polar regions; polar orbits are only slightly inferior to inclined orbits for circuits in the temperate zones.

For high orbital altitudes and circuits in the temperate zone, the orbital inclination has very little influence on system performance.

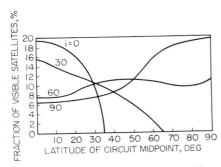

Fig. 24. Average fraction of visible satellites, AMV 10 per cent. i = orbit inclination, degrees. (*From Ref. 2.*)

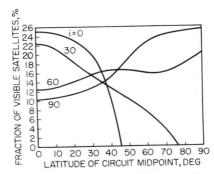

Fig. 25. Average fraction of visible satellites, AMV 15 per cent. i = orbit inclination, degrees. (*From Ref. 2.*)

Considering again the previous equatorial-orbit example, if a random satellite system is used in which all the satellites are launched with an inclination of 0°, then the fraction of visible satellites is (from Figs. 23 and 24) 17 per cent, rather than 7.5 per cent Thus, from Fig. 22, the number of satellites required is about 12.

As a second example of the use of Reiger's method for determining appropriate coverage, consider a circuit from New York to Paris, a length of 3,200 nautical miles. The latitude of the midpoint is approximately 43°. For a satellite system of 6,000 nautical miles altitude, the area of mutual visibility (from Fig. 20) is about 16 per cent.

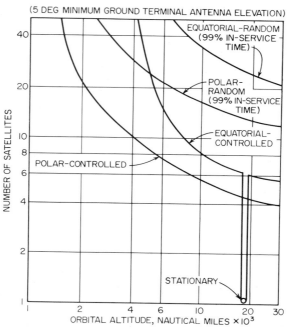

Fig. 26. Satellites required for New York to Paris circuit. (*From S. H. Reiger, Commercial Satellite Systems, Astronaut. Aerospace Eng., September, 1963, p. 29.*)

From Fig. 25, the average fraction of visible satellites is about 18 per cent for polar orbits. Thus, from Fig. 22, the average number of satellites for 99 per cent service is about 21.

Figure 26 shows the number of satellites plotted as a function of altitude and inclination; it also shows the reduction in number of satellites achievable with controlled orbits. Table 8 shows the continuity of service achievable with satellites in random 6,000-nautical-mile polar orbits. For a 21-satellite configuration, the average time between outages is 26 hr. The average time between outages of duration exceeding 15 min is 3 days, and the average time between outages of duration exceeding 30 min is 9 days. The mean duration of outages is 15 min.

It should be noted that notwithstanding the random nature of the configuration, the ephemeris of each satellite is accurately known. The occurrence and duration of each outage can therefore be calculated in advance and the activity of the circuit can be programmed accordingly; for example, messages to be transmitted during the outage can be routed by other means such as cable, or they can be handled like an ordinary telephone call that is held up when a busy signal is received.

Table 8. Continuity of Service with Satellites Randomly in 6,000 NM Polar Orbit*

Number of satellites†	In-service time, %‡	Two or more satellites visible, %	Expected interval between outages exceeding			Mean duration of outages
			0	15 min	30 min	
6	73.7	34.5	3.2 hr	4 hr	5 hr	50 min
9	86.5	56.4	4.2 hr	6 hr	10 hr	33 min
12	93.1	72.5	6 hr	11 hr	20 hr	25 min
15	96.5	83.3	9 hr	20 hr	42 hr	20 min
18	98.2	90.0	15 hr	38 hr	93 hr	17 min
21	99.1	94.2	26 hr	3 days	9 days	15 min
24	99.5	96.7	43 hr	6 days	21 days	12^m 30^s
27	99.8	98.1	3 days	12 days	49 days	11^m 10^s
30	99.9	98.9	6 days	25 days	120 days	10^m
33	99.94	99.4	10 days	52 days	−300 days	9^m 05^s
36	99.97	99.7	18 days	110 days	−1,200 days	8^m 20^s

* From S. H. Reiger, Commercial Satellite Systems, *Astronaut. Aerospace Eng.*, September, 1963, p. 28.
† The numbers are shown as multiples of three, because it is expected that three satellites can be orbited simultaneously by the same launch vehicle.
‡ One or more satellites simultaneously visible from both terminal stations. In-service time has been calculated for terminal stations located near New York and Paris, about 40 to 50° north latitude. For links of the same distance, outages would be slightly more frequent toward the equator and less frequent toward the pole.

6 Operational Systems

Two operational communications-satellite systems are presently being developed in the United States: a commercial system being developed by the Communications Satellite Corporation, and a military system being developed by the Department of Defense. The conceptual design differences between these systems are discussed by Pritchard and McGregor.[20] As even the abbreviated discussion presented here of the problems associated with communications satellites has established, the factors affecting the selection of operational communications-satellite systems are diverse and complex. Further, the selection is beset by a wide range of intricate nontechnological problems involving international agreements and basic ownership and management policies. No adequate discussion of the interrelationship of these factors can be given here. Rather, the discussion in this section is restricted to summarizing the relative advantages and limitations of passive and active systems and of active medium-altitude and stationary systems.

Comparison of Active and Passive Systems

The principal considerations affecting the choice between an active and a passive satellite system are summarized as follows:

Usable Lifetime of Satellite. Significantly higher with a passive satellite.
Multiple-access Capability. A passive satellite, as an essentially linear device over a wide bandwidth, has virtually unlimited multiple-access capability. Achieving a multiple-access capability with an active satellite, on the other hand, is a difficult problem.
Adaptability to Modulation Methods. As a linear device, a passive satellite can accommodate any form of modulation. An active satellite must be designed to

accommodate a fixed modulation method which cannot then be changed during the lifetime of the satellite without special design.

Ability to Share Spectrum with Terrestrial Services. Significantly less for passive satellite systems: on the average the transmitting terminal of a passive system requires about 30 db more power than that of an active system. Power levels for passive systems of the order of 500 kw are required. These high power levels require substantially greater isolation to prevent mutual interference with ground services.

Cost. In general, the spacecraft portion of a passive system is less costly than the corresponding active spacecraft. On the other hand, the ground terminal for a passive system is more expensive than for the active system. Figure 27 shows the result of a typical comparative cost analysis between an active satellite system and a passive

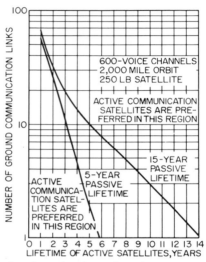

Fig. 27. Cost comparison of active and passive systems. (*From Ref. 2.*)

satellite system providing the same level of service. The illustration system provides 600 voice channels. The orbiting subsystem is assumed to consist of 30 satellites in 6 equally spaced belts at an altitude of 2,000 nautical miles. Both active and passive satellites are assumed to weigh 250 lb each. An Atlas-Agena booster capable of placing five randomly dispersed satellites into orbit is assumed. The active satellite is based on Telstar and Relay technology; the passive satellite is considered to be made of Echo-I material. The savings resulting from the longer lifetime in orbit of the passive satellite can be used to obtain a larger ground terminal transmitter-antenna combination. The choice between system depends upon the assumed lifetime of the satellites and the number of ground links. Figure 27 which shows the regions in which each type of system is preferred indicates the superiority of active satellites for most practical systems.

Comparison of Active Medium-altitude and Stationary Systems

Based on mid-1960s technology the superiority of active over passive satellite systems is generally recognized. The choice between medium-altitude and stationary satellites is, however, less clear. The advantages of the stationary satellite are:

Fewer number of satellites required to obtain coverage over habitable areas of earth. Ideally three satellites provide total coverage except for polar regions.

Continuous (no-outage) coverage.

OPERATIONAL SYSTEMS

Lower costs of ground terminals, as a consequence of the reduced tracking requirement. Since the satellite is (ideally) stationary, very large fixed antennas can be used.

Offsetting these advantages are the following factors:

The cost of launching the stationary satellite exceeds that of a medium-altitude satellite.

Since fewer satellites are used, the consequences of a failure in any satellite become more important. In a practical stationary satellite system, six or nine satellites, rather than three, are required to provide for failure of a satellite. In contrast, in a medium-altitude system (of say 21 satellites) failure of a single satellite causes only minor degradation in service. Thus, a lower percentage of spare satellites is required.

The stationary satellite is more complex, presenting more difficult design problems with respect to stabilization, position keeping, and multiple-access capability. In general, the reliability of the stationary satellite will be less than that of a medium-altitude satellite.

The longer transmission delays with stationary satellites present special problems for two-way voice and error-corrected digital data.

Coverage cannot be obtained in polar regions.

Whereas reduction in launching cost and improvement in reliability may confidently be expected, the last two factors are fundamental to stationary satellites. For certain applications, principally military, inability to provide coverage to polar regions may be very important. For all applications, the problems introduced by the long transmission delay critically affect the usefulness of the stationary satellite.

The transmission delay of the earth-to-satellite-to-earth path for a stationary satellite has a maximum value of about 275 msec; considering additional delay within the repeater, the total delay is of the order of 300 msec. For multiple satellite transmission paths, the delay is of the order of 600 msec. For service requiring only one-way transmission, such as television and uncorrected digital data, the transmission delay is unobjectionable. However, for two-way transmissions—principally telephone—the long delay degrades service by inserting an unnatural pause between question and answer. Pearson[21] describes the confusion which may result:

Consider, for example, what happens when one subscriber attempts to break into the other's speech. This is a regretable but common habit and normally is of no consequence. However, with a long transmission time, even if subscriber A stops speaking as soon as he hears B's interruption, B will get the impression that A is ignoring his interruption and will probably stop speaking. Then when B does hear A stop and starts to talk again, A has probably resumed because he thinks that B has given up.

Another way in which delay affects speech occurs when A stops speaking expecting B to reply, but gets apparently no reply and so resumes his speech, thereby discouraging B who has already started to talk. Then A hears B begin to reply and stops talking only to find that B has broken off. This start-stop procedure can go on for quite a while, with the result that the speakers are . . . confused . . . .

The transmission delay also exacerbates the normal echo problem. In practical telephone circuits, echoes are produced as a result of imperfect impedance matching. Typically the echoes are between -20 and -30 db; they are heard by both speaker and listener, but since the transmission time is short (under 100 msec by CCIR standards) they are unobjectionable. When the delay is of the order of 300 msec, however, the speaker can clearly distinguish between the echo and his current speech. This effect has been found to be psychologically disturbing. Reducing the level of the echo represents one approach toward alleviating the problem; generally, however, echo suppressors operate by inserting a high loss in one direction of transmission while speech is passing in the other direction. This limits the ability of the listener to interrupt the speaker without mutilating the received signal.

It is not clear whether the effects of transmission delay will prove so serious as to limit the use of stationary satellites. The effects, largely psychological in nature, are difficult to assess. Intensive testing programs are currently under way to determine users' adaptability to the inherent problem.[21] In addition, improved echo suppressors for use with communications satellites are being developed. It appears that acceptable solutions to the delay problem will be found.

Although the literature on communications satellites is of very recent origin (the first technical paper after Clark's original description was published by J. R. Pierce in 1955[22]), it is already extensive. The major papers, from which the material presented in this chapter was drawn, are given in the References below. The author is particularly indebted to S. H. Reiger and the Rand Corp. for permission to draw extensively from the significant contributions he has made to the communications-satellite art.

REFERENCES

1. A. C. Clark, Extra-terrestrial Relays, *Wireless World*, October, 1945, pp. 305–308.
2. S. H. Reiger, A Study of Passive Communication Satellites, *Rand Corp. Rept.* R-415-NASA, February, 1963.
3. Radio Regulations, Geneva 1959 Partial Divisions, Treaties and Other International Acts Series 5603, *Government Printing Office*, Washington, D.C.
4. W. H. Watkins, Preliminary Views of the U.S.A. for Frequency Allocations for Space Communication, in G. M. Brown (ed.), "Space Radio Communication," pp. 114–131, Elsevier Publishing Company, Amsterdam, 1962.
5. J. R. Pierce and C. C. Cutler, Interplanetary Communications, in "Advances in Space Science," pp. 55–109, Academic Press Inc., New York, 1959.
6. Richard B. Kershner, Gravity-gradient Stabilization of Satellites, *Astronaut. Aerospace Eng.*, September, 1963, pp. 18–22.
7. B. Paul, T. W. West, and E. Y. Yu, A Passive Gravitational Altitude Control System for Satellites, *Bell System Tech. J.*, vol. 42, pp. 2195–2283.
8. W. E. Morrow, Communication by Orbiting Dipoles, in G. M. Brown (ed.), "Space Radio Communication," pp. 492–521, Elsevier Publishing Company, Amsterdam, 1962.
9. I. L. Lebow et al., The West Ford Belt as a Communications Medium, *Proc. IEEE*, May, 1964, pp. 543–563.
10. H. A. Rosen, Directive Array for a Spinning Vehicle, *Proc. 1962 Natl. Telemetering Conf.*, Paper 1–4, May, 1962.
11. L. C. Van Atta, Electromagnetic Reflector, U.S. Patent 2,908,202, October, 1960.
12. E. L. Gruenberg and C. M. Johnson, Satellite Communications Relay System Using a Retro-directive Space Antenna, *IEEE Trans. Antennas Propagation*, March, 1964.
13. E. M. Rutz-Philipp and E. Kramer, An FM Modulator with Gain for a Space Array, *IEEE Trans. Microwave Theory Techniques*, September, 1963.
14. C. C. Cutler, R. Kompfner, and L. C. Tillotson, A Self-steering Array Repeater, *Bell System Tech. J.*, vol. 44, pp. 2013–2032, September, 1963.
15. A. J. Giger and J. G. Chaffee, The FM Demodulator with Negative Feedback, *Bell System Tech. J.*, vol. 42, pp. 1109–1135, July, 1963.
16. W. L. Wright, Choice of Optimum Modulation Method in Active Satellite Communication Systems, in G. M. Brown (ed.), "Space Radio Communication," pp. 409–428, Elsevier Publishing Company, Amsterdam, 1962.
17. Peter H. Sawitz, Requirements for Multiple Access in Satellite Communication Systems, *Inst. Defense Analysis Rept.*, July, 1963.
18. W. L. Wright and S. Jolliffe, Optimum System Engineering for Satellite Communication Links with Special Reference to the Choice of Modulation Method, *J. Brit. Inst. Radio Engrs.*, vol. 23, no. 5, pp. 381–391, 1962.
19. H. Magnuski, RADAS, Random Access Discrete Address System, *IRE Natl. Symp. Space Electronics Telemetry*, Paper 1.2, 1962.
20. W. L. Pritchard and N. McGregor, "Design Differences between Military and Commercial Communications Satellites," AIAA Paper 64-416, 1964.
21. K. W. Pearson, Transmission Problems of Satellite Communication with Special Reference to Transmission Time, in G. M. Brown (ed.), "Space Radio Communication," pp. 577–585, Elsevier Publishing Company, Amsterdam, 1962.
22. J. R. Pierce, Orbital Radio Relays, *Jet Propulsion*, vol. 25, pp. 153–157, April, 1955.
23. J. R. Pierce and R. Kompfner, Transoceanic Communications by Means of Satellites, *Proc. IRE*, March, 1959, pp. 372–380.
24. A. C. Clarke, The World of the Communications Satellite, *Astronaut. Aerospace Eng.*, February, 1964, pp. 45–48.
25. L. A. de Rosa and E. W. Keller, Potential Application of Recent Advances in Communications Technology, *AIEE 5th Natl. Symp. Global Communications*, Chicago, May 22–24, 1961.

Appendix

MATHEMATICAL FORMULAS, THEOREMS, AND DEFINITIONS*

GRANINO A. KORN, *Professor of Electrical Engineering, The University of Arizona;* and **THERESA M. KORN**.

Glossary of Symbols		A-1
A.1	Powers, Roots, Logarithms, and Decibels	A-4
A.2	Determinants and Linear Equations	A-7
A.3	Matrices	A-9
A.4	Boolean Algebras	A-11
A.5	Differential Calculus	A-13
A.6	Integration	A-16
A.7	Trigonometric Functions	A-18
A.8	Other Elementary Transcendental Functions	A-22
A.9	Progressions and Series	A-24
A.10	Special Functions	A-25
A.11	Step Functions and Symbolic Impulse Functions	A-32
A.12	Fourier and Laplace Transforms	A-35
A.13	Permutations, Combinations, and Related Topics	A-41
A.14	Probability Distributions	A-43
A.15	Mathematical Statistics	A-50
A.16	Numerical Calculations and Finite Differences	A-61
Numerical Tables		A-70

GLOSSARY OF SYMBOLS

The symbols and notations used in this mathematical section were chosen so as to permit reference to most standard textbooks while still maintaining consistency throughout the section. This glossary lists generally useful symbols whose definitions may not appear in their immediate context.

* The material in this section was taken by the publisher's kind permission from G. A. Korn and T. M. Korn, "Mathematical Handbook for Scientists and Engineers," McGraw-Hill Book Company, New York, 1961, which may be consulted for a fuller treatment of the topics presented here. Fourier analysis and the description of random processes is further treated in Chap. 2 (Sampling and Handling of Information, by G. R. Cooper).

A-2 MATHEMATICAL FORMULAS, THEOREMS, AND DEFINITIONS

Scalars and Matrices

$\alpha, \beta, \ldots$ represent scalar (numerical) quantities, α^* is the complex conjugate of α, and $|\alpha|$ is the absolute value of α.

$A, B, \ldots$ represent matrices, most frequently square matrices, with $A \equiv [a_{ik}]$

$A^* \equiv [a_{ik}{}^*]$, complex conjugate of A
$\tilde{A} \equiv [a_{ki}]$, transpose of A
$A\dagger \equiv [a_{ki}{}^*]$, hermitian conjugate of A
$x \equiv \{\xi_i\} \equiv \begin{bmatrix} \xi_1 \\ \xi_2 \\ \ldots \end{bmatrix}$, column matrix
$\tilde{x} \equiv \{\xi_i\} \equiv (\xi_1, \xi_2, \ldots)$, row matrix

E   shift operator
∇ forward-difference operator
Δ backward-difference operator
δ central-difference operator
μ central-mean operator

Expected Values Mean Values and Averages

$E\{x\} = \xi$ expected value, ensemble average
$<x>$ t average (a random variable)
$\bar{x} = \dfrac{1}{n}(x_1 + x_2 + \cdots + x_n)$ statistical sample average (a random variable)

General

Mean $\{x\}$	mean value over a group
$\arcsin z, \arccos z, \arctan z$	inverse trigonometric functions
$\arg z$	argument of z
Ci x	cosine integral
$\cos z$	cosine function
$\cosh z$	hyperbolic cosine
$\cosh^{-1} z$	inverse hyperbolic cosine
$\det [a_{ik}]$	determinant
$\mathrm{erf}\, x$	error function
$\mathrm{erfc}\, x$	complementary error function
$g_{xy}(\omega), G_{xy}(\omega)$	spectral densities
$g_{xy}(\omega), G_{xy}(\omega)$	ensemble spectral densities
$h_j{}^{(1)}(z), h_j{}^{(2)}(z)$	spherical Bessel functions of the third kind
$H_n(z)$	Hermite polynomials
$H_m{}^{(1)}(z), H_m{}^{(2)}(z)$	Hankel functions
$i = \sqrt{-1}$	unit imaginary number
$I_m(z)$	modified Bessel function
$I_z(p, q)$	incomplete beta-function ratio
Im z	imaginary part of z
inf x	greatest lower bound
$j_j(z)$	spherical Bessel function of the first kind
$J_m(z)$	Bessel function of the first kind
$K_m(z)$	modified Hankel function
$L_n(z)$	Laguerre polynomial
$L_n{}^m$	associated Laguerre polynomial or generalized Laguerre function
$L_2 \equiv L_2(V)$	class of quadratically integrable function
li (z)	logarithmic integral
lim z	limit
l.i.m. x	limit-in-mean
$\log_a z$	logarithm
max x, min x	maximum and minimum values
$n_j(z)$	spherical Bessel function of the second kind

GLOSSARY OF SYMBOLS A–3

$N_m(z)$	Neumann's Bessel function of the second kind
$P_n(z)$	Legendre's polynomial of the first kind
$P_j{}^m(z)$	associated Legendre "polynomial" of the first kind
$Q_n(z)$	Legendre function of the second kind
$R_{xy}(\tau)$, $R_{xy}(t_1, t_2)$	t correlation functions
$R_{xy}(\tau)$, $R_{xy}(t_1, t_2)$	ensemble correlation functions
Re z	real part of z
Res$_f\, a$	residue of $f(z)$ at $z = a$
$S_k{}^{(n)}$	Stirling numbers
Sgn x or sgn x	sign function
Si(x)	sine integral
sin z	sine function
sinh z	hyperbolic sine
sinh$^{-1} z$	inverse hyperbolic sine
sup x	least upper bound
tan z	tangent function
tanh z	hyperbolic tangent
tanh$^{-1} z$	inverse hyperbolic tangent
Tr $[a_{ik}]$	trace
$Z_m(z)$	cylinder function
$B(p, q)$	beta function
$B_z(p, q)$	incomplete beta function
$\Gamma(z)$	gamma function
$\Gamma_z(p)$	incomplete gamma function
$\gamma_{xy}(\nu)$, $\Gamma_{xy}(\nu)$	spectral densities
$\delta(x)$, $\delta_+(x)$, $\delta_-(x)$	impulse functions
$\delta(x, \xi)$	multidimensional delta function
Δy_k	forward difference
∇y_k	backward difference
δy_k	central difference
$\mathfrak{F}[f(t)] \equiv \mathfrak{F}[f(t); \nu]$	Fourier transform
$\mathfrak{F}_C[f(t)]$, $\mathfrak{F}_S[f(t)]$	Fourier cosine and sine transforms
$\mathcal{L}[f(t)] \equiv \mathcal{L}[f(t); s] \equiv F(s)$	Laplace transform
$\mathcal{V}$	a vector space
$n!$	factorial
$\binom{x}{n}$	binomial coefficient
$\dfrac{\partial(y_1, y_2, \ldots, y_n)}{\partial(x_1, x_2, \ldots, x_n)}$	Jacobian
$\cap$	cap
$\cup$	cup
$*$	convolution symbol
$\sum_{k=m}^{n}$	summation
$\prod_{k=m}^{n}$	product
$=$	equality symbol
$\equiv$	identity symbol
$\triangleq$	identity by definition
$\approx$	approximate equality
$\simeq$	asymptotically equal
$\sim$	asymptotically proportional
$<, >, \leq, \geq$	inequality, inclusion
$\subset, \supset$	inclusion
$\in$	element of

A–4 MATHEMATICAL FORMULAS, THEOREMS, AND DEFINITIONS

$\exists$ such that
D, V domain, region
S surface, boundary surface or hypersurface
C curve, boundary curve

A.1 POWERS, ROOTS, LOGARITHMS, AND DECIBELS

A.1-1. Powers and Roots. For $q \neq 0$

$$\left. \begin{array}{c} \dfrac{a^p}{a^q} = a^{p-q} \qquad a^{\frac{p}{q}} = \sqrt[q]{a^p} = (\sqrt[q]{a})^p \qquad \sqrt[p]{\sqrt[q]{a}} = \sqrt[pq]{a} \\[6pt] (ab)^p = a^p b^p \qquad \sqrt[p]{ab} = \sqrt[p]{a}\,\sqrt[p]{b} \\[6pt] \left(\dfrac{a}{b}\right)^p = \dfrac{a^p}{b^p} \qquad \sqrt[p]{\dfrac{a}{b}} = \dfrac{\sqrt[p]{a}}{\sqrt[p]{b}} \qquad (b \neq 0) \end{array} \right\} \qquad (\text{A.1-1})$$

A.1-2. Formulas for Rationalizing the Denominators of Fractions

$$\frac{a}{\sqrt{b}} = \frac{a}{b}\sqrt{b} \qquad \frac{a}{\sqrt[n]{b}} = \frac{a}{b}\sqrt[n]{b^{n-1}} \qquad (\text{A.1-2})$$

$$\frac{a}{\sqrt{b} \pm \sqrt{c}} = \frac{a}{b - c}(\sqrt{b} \mp \sqrt{c}) \qquad \frac{a}{\sqrt{b} + \sqrt{c}} = \frac{a}{b^2 - c}\sqrt{(b^2 - c)(b - \sqrt{c})} \qquad (\text{A.1-3})$$

A.1-3. Logarithms. The **logarithm** $x = \log_c a$ to the **base** $c > 0$ ($c \neq 1$) of the number (numerus) $a > 0$ may be defined as the solution of the equation

$$c^x = a \qquad \text{or} \qquad c^{\log_c a} = a \qquad (\text{A.1-4})$$

$$\left. \begin{array}{c} \log_c c = 1 \qquad \log_c c^p = p \qquad \log_c 1 = 0 \\ \log_c (ab) = \log_c a + \log_c b \qquad (\text{LOGARITHMIC PROPERTY}) \\ \log_c \left(\dfrac{a}{b}\right) = \log_c a - \log_c b \\ \log_c (a^p) = p \log_c a \qquad \log_c (\sqrt[p]{a}) = \dfrac{1}{p} \log_c a \end{array} \right\} \qquad (\text{A.1-5})$$

$$\log_{c'} a = \log_c a \, \log_{c'} c = \frac{\log_c a}{\log_c c'} \qquad \log_{c'} c = \frac{1}{\log_c c'} \qquad (\text{CHANGE OF BASE}) \qquad (\text{A.1-6})$$

Table A.1-1. Conversion of Logarithms to Different Base

Multiply → by to obtain ↓	$\log_2 a$	$\log_e a$	$\log_{10} a$	db $10 \log_{10} a$
$\log_2 a$	1	1.44268	3.32192	0.33219
$\log_e a$	0.693154	1	2.30529	0.23053
$\log_{10} a$	0.30103	0.43429	1	0.1
db	3.0103	4.3429	10	1

POWERS, ROOTS, LOGARITHMS, AND DECIBELS   A-5

Table A.1-2. Decibel-conversion Table*

Power ratio	Voltage or current ratio	—Db+	Voltage or current ratio	Power ratio	Power ratio	Voltage or current ratio	—Db+	Voltage or current ratio	Power ratio
10^{-1}		10		10	0.316	0.562	5.0	1.78	3.16
10^{-2}	10^{-1}	20	10	10^{2}	0.309	0.556	5.1	1.80	3.24
10^{-3}		30		10^{3}	0.302	0.550	5.2	1.82	3.31
10^{-4}	10^{-2}	40	10^{2}	10^{4}	0.295	0.543	5.3	1.84	3.39
10^{-5}		50		10^{5}	0.288	0.537	5.4	1.86	3.47
10^{-6}	10^{-3}	60	10^{3}	10^{6}	0.282	0.530	5.5	1.88	3.55
10^{-7}		70		10^{7}	0.275	0.525	5.6	1.91	3.63
10^{-8}	10^{-4}	80	10^{4}	10^{8}	0.269	0.519	5.7	1.93	3.72
10^{-9}		90		10^{9}	0.263	0.513	5.8	1.95	3.80
10^{-10}	10^{-5}	100	10^{5}	10^{10}	0.257	0.507	5.9	1.97	3.89
1.000	1.000	0	1.00	1.00	0.251	0.501	6.0	2.00	3.98
0.977	0.989	0.1	1.01	1.02	0.246	0.496	6.1	2.02	4.07
0.955	0.977	0.2	1.02	1.05	0.240	0.490	6.2	2.04	4.17
0.933	0.966	0.3	1.04	1.07	0.234	0.484	6.3	2.07	4.27
0.912	0.955	0.4	1.05	1.10	0.229	0.479	6.4	2.09	4.37
0.891	0.944	0.5	1.06	1.12	0.224	0.473	6.5	2.11	4.47
0.871	0.933	0.6	1.07	1.15	0.219	0.468	6.6	2.14	4.57
0.851	0.923	0.7	1.08	1.18	0.214	0.462	6.7	2.16	4.68
0.832	0.912	0.8	1.10	1.20	0.209	0.457	6.8	2.19	4.79
0.813	0.902	0.9	1.11	1.23	0.204	0.452	6.9	2.21	4.90
0.794	0.891	1.0	1.12	1.26	0.200	0.447	7.0	2.24	5.01
0.776	0.881	1.1	1.14	1.29	0.195	0.442	7.1	2.27	5.13
0.759	0.871	1.2	1.15	1.32	0.191	0.437	7.2	2.29	5.25
0.741	0.861	1.3	1.16	1.35	0.186	0.432	7.3	2.32	5.37
0.724	0.851	1.4	1.18	1.38	0.182	0.427	7.4	2.34	5.50
0.708	0.841	1.5	1.19	1.41	0.178	0.422	7.5	2.37	5.62
0.692	0.832	1.6	1.20	1.45	0.174	0.417	7.6	2.40	5.75
0.676	0.822	1.7	1.22	1.48	0.170	0.412	7.7	2.43	5.89
0.661	0.813	1.8	1.23	1.51	0.166	0.407	7.8	2.46	6.03
0.646	0.804	1.9	1.25	1.55	0.162	0.403	7.9	2.48	6.17
0.631	0.794	2.0	1.26	1.59	0.159	0.398	8.0	2.51	6.31
0.617	0.785	2.1	1.27	1.62	0.155	0.394	8.1	2.54	6.46
0.603	0.776	2.2	1.29	1.66	0.151	0.389	8.2	2.57	6.61
0.589	0.767	2.3	1.30	1.70	0.148	0.385	8.3	2.60	6.76
0.575	0.759	2.4	1.32	1.74	0.145	0.380	8.4	2.63	6.92
0.562	0.750	2.5	1.33	1.78	0.141	0.376	8.5	2.66	7.08
0.550	0.741	2.6	1.35	1.82	0.138	0.372	8.6	2.69	7.24
0.537	0.733	2.7	1.37	1.86	0.135	0.367	8.7	2.72	7.41
0.525	0.724	2.8	1.38	1.91	0.132	0.363	8.8	2.75	7.59
0.513	0.716	2.9	1.40	1.95	0.129	0.359	8.9	2.79	7.76
0.501	0.708	3.0	1.41	2.00	0.126	0.355	9.0	2.82	7.94
0.490	0.700	3.1	1.43	2.04	0.123	0.351	9.1	2.85	8.13
0.479	0.692	3.2	1.45	2.09	0.120	0.347	9.2	2.88	8.32
0.468	0.684	3.3	1.46	2.14	0.118	0.343	9.3	2.92	8.51
0.457	0.676	3.4	1.48	2.19	0.115	0.339	9.4	2.95	8.71
0.447	0.668	3.5	1.50	2.24	0.112	0.335	9.5	2.99	8.91
0.437	0.661	3.6	1.51	2.29	0.110	0.331	9.6	3.02	9.12
0.427	0.653	3.7	1.53	2.34	0.107	0.327	9.7	3.06	9.33
0.417	0.646	3.8	1.55	2.40	0.105	0.324	9.8	3.09	9.55
0.407	0.638	3.9	1.57	2.46	0.102	0.320	9.9	3.13	9.77
0.398	0.631	4.0	1.59	2.51	0.1000	0.316	10.0	3.16	10.00
0.389	0.624	4.1	1.60	2.57	0.0977	0.313	10.1	3.20	10.23
0.380	0.617	4.2	1.62	2.63	0.0955	0.309	10.2	3.24	10.47
0.372	0.610	4.3	1.64	2.69	0.0933	0.306	10.3	3.27	10.72
0.363	0.603	4.4	1.66	2.75	0.0912	0.302	10.4	3.31	10.96
0.355	0.596	4.5	1.68	2.81	0.0891	0.299	10.5	3.35	11.22
0.347	0.589	4.6	1.70	2.88	0.0871	0.295	10.6	3.39	11.48
0.339	0.582	4.7	1.72	2.95	0.0851	0.292	10.7	3.43	11.75
0.331	0.575	4.8	1.74	3.02	0.0832	0.288	10.8	3.47	12.02
0.324	0.569	4.9	1.76	3.09	0.0813	0.285	10.9	3.51	12.30

* From National Association of Broadcasters, Inc., "NAB Engineering Handbook," 5th ed., McGraw-Hill Book Company, New York, 1960.

A-6 MATHEMATICAL FORMULAS, THEOREMS, AND DEFINITIONS

Of particular interest are the "common" logarithms to the base 10 and the *natural* (*Napierian*) *logarithms* to the base

$$e = \lim_{n \to \infty} \left(1 + \frac{1}{n}\right)^n = 2.71828182 \cdots$$

e is a transcendental number. $\log_e a$ may be written $\ln a$, $\log a$, or log nat s. $\text{Log}_{10} a$ is sometimes written $\log a$.

Table A.1-2. Decibel-conversion Table (*Continued*)

Power ratio	Voltage or current ratio	—Db+	Voltage or current ratio	Power ratio	Power ratio	Voltage or current ratio	—Db+	Voltage or current ratio	Power ratio
0.0794	0.282	11.0	3.55	12.59	0.0251	0.159	16.0	6.31	39.81
0.0776	0.279	11.1	3.59	12.88	0.0246	0.157	16.1	6.38	40.74
0.0759	0.275	11.2	3.63	13.18	0.0240	0.155	16.2	6.46	41.69
0.0741	0.272	11.3	3.67	13.49	0.0234	0.153	16.3	6.53	42.66
0.0724	0.269	11.4	3.72	13.80	0.0229	0.151	16.4	6.61	43.65
0.0708	0.266	11.5	3.76	14.13	0.0224	0.150	16.5	6.68	44.67
0.0691	0.263	11.6	3.80	14.45	0.0219	0.148	16.6	6.76	45.71
0.0676	0.260	11.7	3.85	14.79	0.0214	0.146	16.7	6.84	46.77
0.0661	0.257	11.8	3.89	15.14	0.0209	0.145	16.8	6.92	47.86
0.0646	0.254	11.9	3.94	15.49	0.0204	0.143	16.9	7.00	48.98
0.0631	0.251	12.0	3.98	15.85	0.0200	0.141	17.0	7.08	50.12
0.0617	0.248	12.1	4.03	16.22	0.0195	0.140	17.1	7.16	51.29
0.0603	0.246	12.2	4.07	16.60	0.0191	0.138	17.2	7.24	52.48
0.0589	0.243	12.3	4.12	16.98	0.0186	0.137	17.3	7.33	53.70
0.0575	0.240	12.4	4.17	17.38	0.0182	0.135	17.4	7.41	54.95
0.0562	0.237	12.5	4.22	17.78	0.0178	0.133	17.5	7.50	56.23
0.0550	0.234	12.6	4.27	18.20	0.0174	0.132	17.6	7.59	57.54
0.0537	0.232	12.7	4.32	18.62	0.0170	0.130	17.7	7.67	58.88
0.0525	0.229	12.8	4.37	19.05	0.0166	0.129	17.8	7.76	60.26
0.0513	0.227	12.9	4.42	19.50	0.0162	0.127	17.9	7.85	61.66
0.0501	0.224	13.0	4.47	19.95	0.0159	0.126	18.0	7.94	63.10
0.0490	0.221	13.1	4.52	20.42	0.0155	0.125	18.1	8.04	64.57
0.0479	0.219	13.2	4.57	20.89	0.0151	0.123	18.2	8.13	66.07
0.0468	0.216	13.3	4.62	21.38	0.0148	0.122	18.3	8.22	67.61
0.0457	0.214	13.4	4.68	21.88	0.0145	0.120	18.4	8.32	69.18
0.0447	0.211	13.5	4.73	22.39	0.0141	0.119	18.5	8.41	70.79
0.0437	0.209	13.6	4.79	22.91	0.0138	0.118	18.6	8.51	72.44
0.0427	0.207	13.7	4.84	23.44	0.0135	0.116	18.7	8.61	74.13
0.0417	0.204	13.8	4.90	23.99	0.0132	0.115	18.8	8.71	75.86
0.0407	0.202	13.9	4.96	24.55	0.0129	0.114	18.9	8.81	77.62
0.0398	0.200	14.0	5.01	25.12	0.0126	0.112	19.0	8.91	79.43
0.0389	0.197	14.1	5.07	25.70	0.0123	0.111	19.1	9.02	81.28
0.0380	0.195	14.2	5.13	26.30	0.0120	0.110	19.2	9.12	83.18
0.0372	0.193	14.3	5.19	26.92	0.0118	0.108	19.3	9.23	85.11
0.0363	0.191	14.4	5.25	27.54	0.0115	0.107	19.4	9.33	87.10
0.0355	0.188	14.5	5.31	28.18	0.0112	0.106	19.5	9.44	89.13
0.0347	0.186	14.6	5.37	28.84	0.0110	0.105	19.6	9.55	91.20
0.0339	0.184	14.7	5.43	29.51	0.0107	0.104	19.7	9.66	93.33
0.0331	0.182	14.8	5.50	30.20	0.0105	0.102	19.8	9.77	95.50
0.0324	0.180	14.9	5.56	30.90	0.0102	0.101	19.9	9.89	97.72
0.0316	0.178	15.0	5.62	31.62	0.0100	0.100	20.0	10.00	100.00
0.0309	0.176	15.1	5.69	32.36					
0.0302	0.174	15.2	5.75	33.11					
0.0295	0.172	15.3	5.82	33.88					
0.0288	0.170	15.4	5.89	34.67					
0.0282	0.168	15.5	5.96	35.48					
0.0275	0.166	15.6	6.03	36.31					
0.0269	0.164	15.7	6.10	37.15					
0.0263	0.162	15.8	6.17	38.02					
0.0257	0.160	15.9	6.24	38.90					

A.2 DETERMINANTS AND LINEAR EQUATIONS

A.2-1. Definition. The **determinant**

$$D = \det[a_{ik}] = \begin{vmatrix} a_{11} & a_{12} & \cdots & a_{1n} \\ a_{21} & a_{22} & \cdots & a_{2n} \\ \cdots & \cdots & \cdots & \cdots \\ a_{n1} & a_{n2} & \cdots & a_{nn} \end{vmatrix} \quad (A.2\text{-}1)$$

of the square array (matrix) of n^2 (real or complex) numbers (**elements**) a_{ik} is the sum of the $n!$ terms $(-1)^r a_{1k_1} a_{2k_2} \cdots a_{nk_n}$ each corresponding to one of the $n!$ different ordered sets $k_1, k_2, \ldots, k_n$ obtained by r interchanges of elements from the set $1, 2, \ldots, n$. The number n is the **order** of the determinant (1).

A.2-2. Minors and Cofactors. Expansion in Terms of Cofactors. The (complementary) **minor** D_{ik} of the element a_{ik} in the nth-order determinant obtained from (1) on erasing the ith row and the kth column. The **cofactor** A_{ik} of the element a_{ik} is the coefficient of a_{ik} in the expansion of D, or

$$A_{ik} = (-1)^{i+k} D_{ik} = \frac{\partial D}{\partial a_{ik}} \quad (A.2\text{-}2)$$

A determinant D may be represented in terms of the elements and cofactors of any one row or column as follows:

$$\boxed{\begin{array}{c} D = \det[a_{ik}] = \sum_{i=1}^{n} a_{ij} A_{ij} = \sum_{k=1}^{n} a_{jk} A_{jk} \\ (j = 1, 2, \ldots, n) \qquad \text{(SIMPLE LAPLACE DEVELOPMENT)} \end{array}} \quad (A.2\text{-}3)$$

Note also that

$$\sum_{i=1}^{n} a_{ij} A_{ih} = \sum_{k=1}^{n} a_{jk} A_{hk} = 0 \qquad (j \neq h) \quad (A.2\text{-}4)$$

A.2-3. Examples: Second- and Third-order Determinants

$$\begin{vmatrix} a_{11} & a_{12} \\ a_{21} & a_{22} \end{vmatrix} = a_{11} a_{22} - a_{21} a_{12} \quad (A.2\text{-}5)$$

$$\begin{vmatrix} a_{11} & a_{12} & a_{13} \\ a_{21} & a_{22} & a_{23} \\ a_{31} & a_{32} & a_{33} \end{vmatrix} = a_{11} a_{22} a_{33} - a_{11} a_{23} a_{32} + a_{12} a_{23} a_{31} - a_{13} a_{22} a_{31} + a_{13} a_{21} a_{32} - a_{12} a_{21} a_{33}$$
$$= a_{11}(a_{22} a_{33} - a_{32} a_{23}) - a_{21}(a_{12} a_{33} - a_{32} a_{13}) + a_{31}(a_{12} a_{23} - a_{22} a_{13})$$
$$= a_{11}(a_{22} a_{33} - a_{32} a_{23}) - a_{12}(a_{21} a_{33} - a_{31} a_{23}) + a_{13}(a_{21} a_{32} - a_{31} a_{22})$$
$$\text{etc.} \quad (A.2\text{-}6)$$

A.2-4. Miscellaneous Theorems. (a) *The value D of a determinant (1) is not changed by any of the following operations:*

1. *The rows are written as columns, and the columns as rows* [interchange of i and k in Eq. (1)].
2. *An even number of interchanges of any two rows or two columns.*
3. *Addition of the elements of any row (or column), all multiplied, if desired, by the same parameter α, to the respective corresponding elements of another row (or column, respectively).*

EXAMPLES:

$$\begin{vmatrix} a_{11} & a_{12} & \cdots & a_{1n} \\ a_{21} & a_{22} & \cdots & a_{2n} \\ \hdotsfor{4} \\ a_{n1} & a_{n2} & \cdots & a_{nn} \end{vmatrix} = \begin{vmatrix} a_{11} & a_{21} & \cdots & a_{n1} \\ a_{12} & a_{22} & \cdots & a_{n2} \\ \hdotsfor{4} \\ a_{1n} & a_{2n} & \cdots & a_{nn} \end{vmatrix}$$

$$= \begin{vmatrix} a_{11} + \alpha a_{12} & a_{12} & \cdots & a_{1n} \\ a_{21} + \alpha a_{22} & a_{22} & \cdots & a_{2n} \\ \hdotsfor{4} \\ a_{n1} + \alpha a_{n2} & a_{n2} & \cdots & a_{nn} \end{vmatrix} \quad (A.2\text{-}7)$$

(**b**) *An odd number of interchanges of any two rows or two columns is equivalent to multiplication of the determinant by* -1.

(**c**) *Multiplication of all the elements of any one row or column by a factor* α *is equivalent to multiplication of the determinant by* α.

(**d**) *If the elements of the jth row (or column) of an nth-order determinant D are represented as sums* $\sum_{r=1}^{m} c_{r1}, \sum_{r=1}^{m} c_{r2}, \ldots, \sum_{r=1}^{m} c_{rn}$, *$D$ is equal to the sum* $\sum_{r=1}^{m} D_r$ *of m n^{th}-order determinants D_r. The elements of each D_r are identical with those of D, except for the elements of the jth row (or column, respectively), which are $c_{r1}, c_{r2}, \ldots, c_{rn}$.*

EXAMPLE:

$$\begin{vmatrix} a_{11} + b_{11} & a_{12} + b_{12} & \cdots & a_{1n} + b_{1n} \\ a_{21} & a_{22} & & a_{2n} \\ \hdotsfor{4} \\ a_{n1} & a_{n2} & & a_{nn} \end{vmatrix}$$

$$= \begin{vmatrix} a_{11} & a_{12} & \cdots & a_{1n} \\ a_{21} & a_{22} & \cdots & a_{2n} \\ \hdotsfor{4} \\ a_{n1} & a_{n2} & \cdots & a_{nn} \end{vmatrix} + \begin{vmatrix} b_{11} & b_{12} & \cdots & b_{1n} \\ a_{21} & a_{22} & \cdots & a_{2n} \\ \hdotsfor{4} \\ a_{n1} & a_{n2} & \cdots & a_{nn} \end{vmatrix} \quad (A.2\text{-}8)$$

(**e**) *A determinant is equal to zero if*
1. *All elements of any row or column are zero.*
2. *Corresponding elements of any two rows or columns are equal, or proportional with the same proportionality factor.*

A.2-5. Multiplication of Determinants. The product of two n^{th}-order determinants $\det [a_{ik}]$ and $\det [b_{ik}]$ is

$$\det [a_{ik}] \det [b_{ik}] = \det \Big[\sum_{j=1}^{n} a_{ij} b_{jk} \Big] = \det \Big[\sum_{j=1}^{n} a_{ji} b_{kj} \Big]$$

$$= \det \Big[\sum_{j=1}^{n} a_{ij} b_{kj} \Big] = \det \Big[\sum_{j=1}^{n} a_{ji} b_{jk} \Big] \quad (A.2\text{-}9)$$

A.2-6. Simultaneous Linear Equations: Cramer's Rule. Consider a set (system) of n linear equations in n unknowns $x_1, x_2, \ldots, x_n$

$$\begin{aligned} a_{11}x_1 + a_{12}x_2 + \cdots + a_{1n}x_n &= b_1 \\ a_{21}x_1 + a_{22}x_2 + \cdots + a_{2n}x_n &= b_2 \\ &\hdotsfor{1} \\ a_{n1}x_1 + a_{n2}x_2 + \cdots + a_{nn}x_n &= b_n \end{aligned} \quad \text{or} \quad \sum_{k=1}^{n} a_{ik}x_k = b_i \quad (i = 1, 2, \ldots, n) \quad (A.2\text{-}10)$$

such that at least one of the absolute terms b_i is different from zero. If the **system determinant**

$$D = \det [a_{ik}] = \begin{vmatrix} a_{11} & a_{12} & \cdots & a_{1n} \\ a_{21} & a_{22} & \cdots & a_{2n} \\ \vdots & & & \\ a_{n1} & a_{n2} & \cdots & a_{nn} \end{vmatrix} \quad (A.2\text{-}11)$$

differs from zero, the system (10) has the unique solution

$$x_k = \frac{D_k}{D} \qquad (k = 1, 2, \ldots, n) \qquad (\text{Cramer's rule}) \qquad (A.2\text{-}12)$$

where D_k is the determinant obtained on replacing the respective elements $a_{1k}, a_{2k}, \ldots, a_{nk}$ in the kth column of D by $b_1, b_2, \ldots, b_n$, or

$$D_k = \sum_{i=1}^{n} A_{ik} b_i \qquad (k = 1, 2, \ldots, n) \qquad (A.2\text{-}13)$$

where A_{ik} is the cofactor (Sec. A.2-2) of a_{ik} in the determinant D.

A.3 MATRICES

Matrix techniques permit a simplified representation of various mathematical and physical operations in terms of numerical operations on matrix elements.

A.3-1. Rectangular Matrices. An array

$$A \equiv \begin{bmatrix} a_{11} & a_{12} & \cdots & a_{1n} \\ a_{21} & a_{22} & \cdots & a_{2n} \\ \vdots & & & \\ a_{m1} & a_{m2} & \cdots & a_{mn} \end{bmatrix} \equiv [a_{ik}]$$

of real or complex numbers ("scalars") a_{ik} is called a (**rectangular**) $m \times n$ **matrix** over the field F whenever one of the "matrix operations" defined in Sec. A.3-2 is to be used. The elements a_{ik} are called **matrix elements**; the matrix element a_{ik} is situated in the ith **row** and in the kth **column** of the matrix (1). m is the number of rows, and n is the number of columns.

$n \times 1$ matrices are **column matrices**, and $1 \times n$ matrices are **row matrices**. The following notation will be used:

$$\begin{bmatrix} \xi_1 \\ \xi_2 \\ \vdots \\ \xi_n \end{bmatrix} \equiv \{\xi_i\} \equiv x \qquad [\xi_1 \xi_2 \cdots \xi_n] \equiv [\xi_k] \equiv \tilde{x}$$

An $n \times n$ matrix is called a **square matrix** of **order** n. A square matrix $A \equiv [a_{ik}]$ is

Triangular (superdiagonal) if and only if $i > k$ implies $a_{ik} = 0$
Strictly triangular if and only if $i \geq k$ implies $a_{ik} = 0$
Diagonal if and only if $i \neq k$ implies $a_{ik} = 0$
Monomial if and only if each row and column has one and only one element different from zero

A.3-2. Basic Operations. Operations on matrices are defined in terms of operations on the matrix elements.

1. Two $m \times n$ matrices $A \equiv [a_{ik}]$ and $B \equiv [b_{ik}]$ are **equal** ($A = B$) if and only if $a_{ik} = b_{ik}$ for all i, k.
2. The **sum of two** $m \times n$ **matrices** $A \equiv [a_{ik}]$ and $B \equiv [b_{ik}]$ is the $m \times n$ matrix
$$A + B \equiv [a_{ik}] + [b_{ik}] \equiv [a_{ik} + b_{ik}]$$
3. The **product of the** $m \times n$ **matrix** $A \equiv [a_{ik}]$ **by the scalar** α is the $m \times n$ matrix
$$\alpha A \equiv \alpha[a_{ik}] \equiv [\alpha a_{ik}]$$
4. The **product of the** $m \times n$ **matrix** $A \equiv [a_{ij}]$ **and the** $n \times r$ **matrix** $B \equiv [b_{jk}]$ is the $m \times r$ matrix
$$AB \equiv [a_{ij}][b_{jk}] \equiv \left[\sum_{j=1}^{n} a_{ij} b_{jk} \right]$$
5. A (necessarily square) matrix A is **nonsingular** (**regular**) if and only if it has a (necessarily unique) bounded **multiplicative inverse** or **reciprocal** A^{-1} defined by
$$AA^{-1} = A^{-1}A = I$$
Otherwise A is a **singular** matrix.
A finite $n \times n$ matrix $A \equiv [a_{ik}]$ is nonsingular if and only if
$$\det(A) \equiv \det[a_{ik}] \neq 0$$
In this case A^{-1} is the $n \times n$ matrix
$$A^{-1} \equiv [a_{ik}]^{-1} \equiv \left[\frac{A_{ki}}{\det[a_{ik}]} \right]$$
where A_{ik} is the cofactor of the element a_{ik} in the determinant $\det[a_{ik}]$.

Products and reciprocals of nonsingular matrices are nonsingular; if A and B are nonsingular, and $\alpha \neq 0$,

$$\left. \begin{array}{cc} (AB)^{-1} = B^{-1}A^{-1} & (\alpha A)^{-1} = \alpha^{-1}A^{-1} \\ (A^{-1})^{-1} = A & \end{array} \right\} \quad \text{(A.3-1)}$$

Note

$$\left. \begin{array}{ll} A + B = B + A & A + (B + C) = (A + B) + C \\ \alpha(\beta A) = (\alpha \beta)A & \alpha(AB) = (\alpha A)B = A(\alpha B) \\ & A(BC) = (AB)C \\ \alpha(A + B) = \alpha A + \alpha B & (\alpha + \beta)A = \alpha A + \beta A \\ A(B + C) = AB + AC & (B + C)A = BA + CA \end{array} \right\} \quad \text{(A.3-2)}$$

A.3-3. Identities and Inverses. Note the following definitions:

1. The $m \times n$ **null matrix** (**additive identity**) $[0]$ is the $m \times n$ matrix all of whose elements are equal to zero. Then
$$A + [0] = A \qquad 0A = [0]$$
$$[0]B = C[0] = [0]$$
where A is any $m \times n$ matrix, B is any matrix having n rows, and C is any matrix having m columns.
2. The **additive inverse** (**negative**) $-A$ of the $m \times n$ matrix $A \equiv [a_{ik}]$ is the $m \times n$ matrix
$$-A \equiv (-1)A \equiv [-a_{ik}]$$
with $A + (-A) = A - A = [0]$.

3. The **identity matrix** (**unit matrix, multiplicative identity**) I **of order** n is the $n \times n$ diagonal matrix with unit diagonal elements:

$$I \equiv [\delta_k^i] \quad \text{where} \quad \delta_k^i = \begin{cases} 0 \text{ if } i \neq k \\ 1 \text{ if } i = k \end{cases}$$

Then
$$IB = B \qquad CI = C$$

where B is any matrix having n rows, and C is any matrix having n columns; and for any $n \times n$ matrix A
$$IA = AI = A$$

A.3-4. Rank, Trace, and Determinant of a Matrix. The **rank** of a given matrix is the largest number r such that at least one r^{th}-order determinant formed from the matrix by deleting rows and/or columns is different from zero. An $m \times n$ matrix A is nonsingular if and only if $m = n = r$, i.e., if and only if A is square and $\det(A) \neq 0$.

The **trace** (**spur**) of an $n \times n$ matrix $A \equiv [a_{ik}]$ is the sum

$$\text{Tr}(A) = \sum_{i=1}^{n} a_{ii}$$

of the diagonal terms. *For finite matrices* A, B

$$\left.\begin{array}{l} \text{Tr}(A + B) = \text{Tr}(A) + \text{Tr}(B) \\ \text{Tr}(BA) = \text{Tr}(AB) \end{array}\quad \begin{array}{l} \text{Tr}(\alpha A) = \alpha \text{ Tr}(A) \\ \text{Tr}(AB - BA) = 0 \end{array}\right\} \quad (A.3\text{-}3)$$

$$\det(AB) = \det(BA) = \det(A)\det(B) \qquad (A.3\text{-}4)$$

A.4 BOOLEAN ALGEBRAS

A.4-1. Boolean Algebras. A **Boolean algebra** is a class $\mathcal{S}$ of objects A, B, C, . . . admitting two binary operations, denoted as (*logical*) *addition and multiplication*, with the following properties:

(**a**) For all A, B, C in $\mathcal{S}$
1. $\mathcal{S}$ contains $A + B$ and AB (CLOSURE)
2. $A + B = B + A$
   $AB = BA$ (COMMUTATIVE LAWS)
3. $A + (B + C) = (A + B) + C$
   $A(BC) = (AB)C$ (ASSOCIATIVE LAWS)
4. $A(B + C) = AB + AC$
   $A + BC = (A + B)(A + C)$ (DISTRIBUTIVE LAWS)
5. $A + A = AA = A$ (IDEMPOTENT PROPERTIES)
6. $A + B = B$ if and only if $AB = A$ (CONSISTENCY PROPERTY)

(**b**) In addition,

7. $\mathcal{S}$ contains elements I and 0 such that, for every A in $\mathcal{S}$,
$$A + 0 = A \qquad AI = A$$
$$A0 = 0 \qquad A + I = I$$

8. For every element A, $\mathcal{S}$ contains an element $\tilde{A}$ (**complement** of A, also written $\bar{A}$ or $I - A$) such that
$$A + \tilde{A} = I \qquad A\tilde{A} = 0$$

In every Boolean algebra

$$A(A + B) \equiv A + AB \equiv A \qquad \text{(LAWS OF ABSORPTION)} \quad (A.4\text{-}1)$$

$$\left. \begin{array}{l} \widetilde{(A + B)} \equiv \tilde{A}\tilde{B} \\ \widetilde{(AB)} \equiv \tilde{A} + \tilde{B} \end{array} \right\} \qquad \text{(DUALIZATION)} \quad (A.4\text{-}2)$$

$$\begin{array}{ll} \tilde{\tilde{A}} \equiv A & \tilde{I} = 0 \qquad \tilde{0} = I \\ A + \tilde{A}B \equiv A + B & AB + AC + B\tilde{C} \equiv AC + B\tilde{C} \end{array} \qquad \begin{array}{l} (A.4\text{-}3) \\ (A.4\text{-}4) \end{array}$$

A.4-2. Boolean Functions. Reduction to Canonical Form. Given n Boolean variables $X_1, X_2, \ldots, X_n$, each of which can equal any element of a given Boolean algebra, a **Boolean function**

$$Y = F(X_1, X_2, \ldots, X_n)$$

is an expression built up from $X_1, X_2, \ldots, X_n$ through addition, multiplication, and complementation.

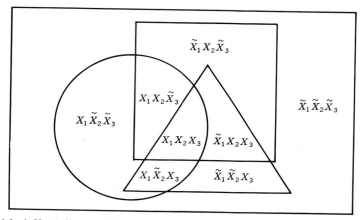

Fig. A.4-1. A Venn diagram (Euler diagram). Diagrams of this type illustrate relations in an algebra of classes. If the rectangle, circle, square, and triangle are respectively labeled by I, X_1, X_2, X_3, the diagram shows that every Boolean function of X_1, X_2, X_3 can be represented as a union of minimal polynomials in X_1, X_2, X_3. Note that there are $2^3 = 8$ different minimal polynomials. (*From G. A. Korn and T. M. Korn, "Mathematical Handbook for Scientists and Engineers," McGraw-Hill Book Company, New York, 1961.*)

In every Boolean algebra there exist exactly $2^{(2^n)}$ different Boolean functions of n variables. Every Boolean function is either identically equal to 0 or can be expressed uniquely as a sum of **minimal polynomials** $Z_1 Z_2 \ldots Z_n$, where Z_i is either X_i or $\tilde{X}_i$ (canonical form of a Boolean function; see Fig. A.4-1 for a geometrical illustration). A given Boolean function $Y = F(X_1, X_2, \ldots, X_n)$ may be reduced to canonical form as follows:

1. Use Eq. (2) to expand complements of sums and products.
2. Reduce $F(X_1, X_2, \ldots, X_n)$ to a sum of products with the aid of the first distributive law.
3. Simplify the resulting expression with the aid of the identities $X_i X_i \equiv X_i$, $X_i \tilde{X}_i \equiv 0$, and Eq. (4).
4. If a term f does not contain one of the variables, say X_i, rewrite f as $fX_i + f\tilde{X}_i$.

In many applications (e.g., design of switching circuits) it may be advantageous to omit step 4 and to continue step 3 so as to simplify each term of the expansion as much as possible.

A.5 DIFFERENTIAL CALCULUS

A.5-1. Differentiation Rules. Table A.5-1 summarizes the most important differentiation rules. The formulas of Table A.5-1a and b apply to *partial differentiation* if $\partial/\partial x_k$ is substituted for d/dx in each case. Thus, if $u_i = u_i(x_1, x_2, \ldots, x_n)$ $(i = 1, 2, \ldots, m)$

$$\frac{\partial}{\partial x_k} f(u_1, u_2, \ldots, u_m) = \sum_{i=1}^{m} \frac{\partial f}{\partial u_i} \frac{\partial u_i}{\partial x_k} \qquad (k = 1, 2, \ldots, n)$$

Table A.5-1. Differentiation Rules

(a) Basic Rules

$$\frac{d}{dx} f[u_1(x), u_2(x), \ldots, u_m(x)] = \frac{\partial f}{\partial u_1} \frac{du_1}{dx} + \frac{\partial f}{\partial u_2} \frac{du_2}{dx} + \cdots + \frac{\partial f}{\partial u_m} \frac{du_m}{dx}$$

$$\frac{d}{dx} f[u(x)] = \frac{df}{du} \frac{du}{dx} \qquad \frac{d^2}{dx^2} f[u(x)] = \frac{d^2 f}{du^2} \left(\frac{du}{dx}\right)^2 + \frac{df}{du} \frac{d^2 u}{dx^2}$$

(b) Sums, Products, and Quotients. Logarithmic Differentiation

$$\frac{d}{dx}[u(x) + v(x)] = \frac{du}{dx} + \frac{dv}{dx} \qquad \frac{d}{dx}[\alpha u(x)] = \alpha \frac{du}{dx}$$

$$\frac{d}{dx}[u(x)v(x)] = v\frac{du}{dx} + u\frac{dv}{dx} \qquad \frac{d}{dx}\left[\frac{u(x)}{v(x)}\right] = \frac{1}{v^2}\left(v\frac{du}{dx} - u\frac{dv}{dx}\right) \; [v(x) \neq 0]$$

$$\frac{d}{dx} \log_e y(x) = \frac{y'(x)}{y(x)} \qquad \text{[LOGARITHMIC DERIVATIVE OF } y(x)\text{]}$$

NOTE: To differentiate functions of the form $y = \dfrac{u_1(x) u_2(x) \cdots}{v_1(x) v_2(x) \cdots}$, it may be convenient to find the logarithmic derivative first.

$$\frac{d^r}{dx^r}(\alpha u + \beta v) = \alpha \frac{d^r u}{dx^r} + \beta \frac{d^r v}{dx^r} \qquad \frac{d^r}{dx^r}(uv) = \sum_{k=0}^{r} \binom{r}{k} \frac{d^{r-k} u}{dx^{r-k}} \frac{d^k v}{dx^k}$$

(c) Inverse Function Given. If $y = y(x)$ has the unique inverse function $x = x(y)$, and $dx/dy \neq 0$,

$$\frac{dy}{dx} = \left(\frac{dx}{dy}\right)^{-1} \qquad \frac{d^2 y}{dx^2} = -\frac{d^2 x}{dy^2} \bigg/ \left(\frac{dx}{dy}\right)^3$$

(d) Implicit Functions. If $y = y(x)$ is given implicitly in terms of a suitably differentiable relation $F(x, y) = 0$, where $F_y \neq 0$,

$$\frac{dy}{dx} = -\frac{F_x}{F_y} \qquad \frac{d^2 y}{dx^2} = -\frac{1}{F_y^3}(F_{xx} F_y^2 - 2 F_{xy} F_x F_y + F_{yy} F_x^2)$$

(e) Function Given in Terms of a Parameter t. Given $x = x(t)$, $y = y(t)$ and $\dot{x}(t) \equiv \dfrac{dx}{dt} \neq 0$, $\dot{y}(t) \equiv \dfrac{dy}{dt}$, $\ddot{x}(t) \equiv \dfrac{d^2 x}{dt^2}$, $\ddot{y}(t) \equiv \dfrac{d^2 y}{dt^2}$,

$$\frac{dy}{dx} = \frac{\dot{y}(t)}{\dot{x}(t)} \qquad \frac{d^2 y}{dx^2} = \frac{\dot{x}(t) \ddot{y}(t) - \ddot{x}(t) \dot{y}(t)}{[\dot{x}(t)]^3}$$

Multiplication of each formula of Table A.5-1a and b by dx or dx^r yields an analogous rule for *total differentiation;* thus

$$d(u + v) = du + dv \qquad d(uv) = v\,du + u\,dv \qquad (A.5\text{-}1)$$

Table A.5-2. Derivatives of Frequently Used Functions

(a)

$f(x)$	$f'(x)$	$f^{(r)}(x)$
x^a	ax^{a-1}	$a(a-1)(a-2)\cdots(a-r+1)x^{a-r}$
e^x	e^x	e^x
a^x	$a^x \log_e a$	$a^x (\log_e a)^r$
$\log_e x$	$\dfrac{1}{x}$	$(-1)^{r-1}(r-1)!\,\dfrac{1}{x^r}$
$\log_a x$	$\dfrac{1}{x}\log_a e$	$(-1)^{r-1}(r-1)!\,\dfrac{1}{x^r}\log_a e$
$\sin x$	$\cos x$	$\sin\left(x + \dfrac{\pi r}{2}\right)$
$\cos x$	$-\sin x$	$\cos\left(x + \dfrac{\pi r}{2}\right)$

(b)

$f(x)$	$f'(x)$	$f(x)$	$f'(x)$
$\tan x$	$\dfrac{1}{\cos^2 x}$	$\arcsin x$	$\dfrac{1}{\sqrt{1-x^2}}$
$\cot x$	$-\dfrac{1}{\sin^2 x}$	$\arccos x$	$-\dfrac{1}{\sqrt{1-x^2}}$
$\sec x$	$\dfrac{\sin x}{\cos^2 x}$	$\arctan x$	$\dfrac{1}{1+x^2}$
$\operatorname{cosec} x$	$-\dfrac{\cos x}{\sin^2 x}$	$\operatorname{arccot} x$	$-\dfrac{1}{1+x^2}$
$\sinh x$	$\cosh x$	$\sinh^{-1} x$	$\dfrac{1}{\sqrt{x^2+1}}$
$\cosh x$	$\sinh x$	$\cosh^{-1} x$	$\dfrac{1}{\sqrt{x^2-1}}$
$\tanh x$	$\dfrac{1}{\cosh^2 x}$	$\tanh^{-1} x$	$\dfrac{1}{1-x^2}$
$\coth x$	$-\dfrac{1}{\sinh^2 x}$	$\coth^{-1} x$	$\dfrac{1}{1-x^2}$
$\operatorname{vers} x$	$\sin x$	x^x	$x^x(1 + \log_e x)$

A.5-2. Jacobians and Functional Dependence. A set of *transformation equations*

$$y_i = y_i(x_1, x_2, \ldots, x_n) \qquad (i = 1, 2, \ldots, n) \qquad (A.5\text{-}2)$$

DIFFERENTIAL CALCULUS   A–15

define a reciprocal one-to-one correspondence between sets $(x_1, x_2, \ldots, x_n)$ and $(y_1, y_2, \ldots, y_n)$ throughout any open region of "points" $(x_1, x_2, \ldots, x_n)$ where the functions (2) are single-valued and continuously differentiable, and where the **Jacobian** or **functional determinant**

$$\frac{\partial(y_1, y_2, \ldots, y_n)}{\partial(x_1, x_2, \ldots, x_n)} \equiv \det\left[\frac{\partial y_i}{\partial x_k}\right] \tag{A.5-3}$$

is different from zero.

A.5-3. Values of Indeterminate Forms (see Table A.5-3 for examples). (a) Functions $f(x)$ of the form $u(x)/v(x)$, $u(x)v(x)$, $[u(x)]^{v(x)}$, and $u(x) - v(x)$ are not defined for $x = a$ if $f(a)$ takes the form $0/0$, ∞/∞, $0 \cdot \infty$, 0^0, ∞^0, 1^∞, or $\infty - \infty$; but $\lim_{x \to a} f(x)$ may exist. In such cases it is often desirable to *define* $f(a) = \lim_{x \to a} f(x)$.

(b) **Treatment of 0/0 and** ∞/∞. Let $u(a) = v(a) = 0$. *If there exists a neighborhood of $x = a$ such that* (1) $v'(x) \neq 0$, *except for $x = a$, and* (2) $u'(x)$ *and* $v'(x)$ *exist and do not vanish simultaneously, then*

$$\lim_{x \to a} \frac{u(x)}{v(x)} = \lim_{x \to a} \frac{u'(x)}{v'(x)} \tag{A.5-4}$$

whenever the limit on the right exists (L'Hôpital's *Rule*).

Let $\lim_{x \to a} u(x) = \lim_{x \to a} v(x) = \infty$. *If there exists a neighborhood of $x = a$ such that $x \neq a$ implies* (1) $u(x) \neq 0$, $v(x) \neq 0$, *and* (2) $u'(x)$ *and* $v'(x)$ *exist and do not vanish simultaneously, then Eq. (A.5-4) holds whenever the limit on the right exists*.

If $u'(x)/v'(x)$ is itself an indeterminate form, the above method may be applied to $u'(x)/v'(x)$ in turn, so that

$$\lim_{x \to a} \frac{u(x)}{v(x)} = \lim_{x \to a} \frac{u'(x)}{v'(x)} = \lim_{x \to a} \frac{u''(x)}{v''(x)} \tag{A.5-5}$$

If necessary, this process may be continued.

(c) **Treatment of** $0 \cdot \infty$, 0^0, ∞^0, 1^∞, **and** $\infty - \infty$. $u(x)v(x)$, $[u(x)]^{v(x)}$, and $u(x) - v(x)$ can often be reduced to the form $\varphi(x)/\psi(x)$ with the aid of one of the following relations:

$$\left.\begin{array}{c} u(x)v(x) \equiv \dfrac{u(x)}{1/v(x)} \equiv \dfrac{v(x)}{1/u(x)} \\[6pt] [u(x)]^{v(x)} \equiv e^{g(x)} \left[g(x) \equiv \dfrac{\log_e u(x)}{1/v(x)} \equiv \dfrac{v(x)}{1/\log_e u(x)} \right] \\[6pt] u(x) - v(x) \equiv \dfrac{\dfrac{1}{v(x)} - \dfrac{1}{u(x)}}{\dfrac{1}{u(x)} \dfrac{1}{v(x)}} \equiv \log_e g(x) \left[g(x) \equiv \dfrac{e^{u(x)}}{e^{v(x)}} \right] \end{array}\right\} \tag{A.5-6}$$

(d) It is often helpful to write $\lim_{x \to a} f(x) = \lim_{\Delta x \to 0} f(a + \Delta x)$ and to isolate terms of the order of Δx by algebraic manipulation or by a Taylor-series expansion.

A.5-4. One-sided (Unilateral) Limits. The methods of Sec. A.5-3 are readily modified to apply to the *one-sided limits* $\lim_{x \to a-0} f(x)$ and $\lim_{x \to a+0} f(x)$. To find $\lim_{x \to \infty} f(x)$, use $\lim_{x \to \infty} f(x) = \lim_{y \to 0+0} f(1/y)$.

A function $f(x)$ of a real variable x has the (*necessarily finite and unique*) **right-hand limit** $\lim_{x \to a+0} f(x) \equiv \lim_{x \to a+} f(x) \equiv f(a + 0) = L_+$ at $x = a$ if and only if for each positive real number ϵ there exists a real number $\delta > 0$ such that $0 < x - a < \delta$ implies that $f(x)$ is defined, and $|f(x) - L_+| < \epsilon$. $f(x)$ has the **left-hand limit** $\lim_{x \to a-0} f(x) \equiv \lim_{x \to a-} f(x) \equiv f(a - 0) = L_-$ at $x = a$ if and only if for each positive

Table A.5-3. Some Frequently Used Limits (Values of Indeterminate Forms)

$$\lim_{n \to \infty} \left(1 + \frac{1}{n}\right)^n = e \approx 2.71828 \quad (n = 1, 2, \ldots) \quad \lim_{x \to 0} (1 + x)^{\frac{1}{x}} = e$$

$$\lim_{x \to 0} \frac{c^x - 1}{x} = \log_e c \quad\quad\quad \lim_{x \to 0} x^x = 1$$

$$\lim_{x \to 0} \frac{\sin x}{x} = \lim_{x \to 0} \frac{\tan x}{x} = \lim_{x \to 0} \frac{\sinh x}{x} = \lim_{x \to 0} \frac{\tanh x}{x} = 1$$

$$\lim_{x \to 0} \frac{\sin \omega x}{x} = \omega \quad (-\infty < \omega < \infty)$$

$$\lim_{x \to 0} x^a \log_e x = \lim_{x \to \infty} x^{-a} \log_e x = \lim_{x \to \infty} x^a e^{-x} = 0 \quad (a > 0)$$

real number ϵ there exists a real number $\delta > 0$ such that $0 < a - x < \delta$ implies that $f(x)$ is defined, and $|f(x) - L_-| < \epsilon$. If $\lim_{x \to a} f(x)$ exists, then

$$\lim_{x \to a+0} f(x) = \lim_{x \to a-0} f(x) = \lim_{x \to a} f(x)$$

Conversely, $\lim_{x \to a-0} f(x) = \lim_{x \to a+0} f(x)$ implies the existence of $\lim_{x \to a} f(x)$.

A.6 INTEGRATION

A.6-1. The Fundamental Theorem of the Integral Calculus. *If $f(x)$ is single-valued, bounded, and integrable on $[a, b]$, and there exists a function $F(x)$ such that*

Table A.6-1. Properties of Integrals

(a) Elementary Properties. *If the integrals exist,*

$$\int_a^b f(x) \, dx = -\int_b^a f(x) \, dx \quad \int_a^b f(x) \, dx = \int_a^c f(x) \, dx + \int_c^b f(x) \, dx$$

$$\int_a^b [u(x) + v(x)] \, dx = \int_a^b u(x) \, dx + \int_a^b v(x) \, dx \quad \int_a^b \alpha u(x) \, dx = \alpha \int_a^b u(x) \, dx$$

(b) Integration by Parts. *If $u(x)$ and $v(x)$ are differentiable for $a \leq x \leq b$, and if the integrals exist,*

$$\int_a^b u(x) v'(x) \, dx = u(x) v(x) \Big]_a^b - \int_a^b v(x) u'(x) \, dx$$

or

$$\int_a^b u \, dv = uv \Big]_a^b - \int_a^b v \, du$$

(c) Change of Variable (Integration by Substitution). *If $u = u(x)$ and its inverse function $x = x(u)$ are single-valued and continuously differentiable for $a \leq x \leq b$, and if the integral exists,*

$$\int_a^b f(x) \, dx = \int_{u(a)}^{u(b)} f[(x(u)] \frac{dx}{du} \, du = \int_{u(a)}^{u(b)} f[x(u)] \left(\frac{du}{dx}\right)^{-1} du$$

(d) Differentiation with Respect to a Parameter. *If $f(x, \lambda)$, $u(\lambda)$, and $v(\lambda)$ are continuously differentiable with respect to λ,*

$$\frac{\partial}{\partial \lambda} \int_a^b f(x, \lambda) \, dx = \int_a^b \frac{\partial}{\partial \lambda} f(x, \lambda) \, dx$$

$F'(x) = f(x)$ for $a \leq x \leq b$, then

$$\int_a^x f(\xi)\, d\xi = F(x) \Big]_a^x = F(x) - F(a) \qquad (a \leq x \leq b) \qquad (A.6\text{-}1)$$

A.6-2. Integration Methods. (a) **Integration** is the operation yielding a (definite or indefinite) integral of a given integrand $f(x)$. Definite integrals may be calculated directly as limits of sums (numerical integration,) or by the calculus of residues; more frequently, one attempts to find an indefinite integral which may be inserted into Eq. (1). To obtain an indefinite integral, one must reduce the given integrand $f(x)$ to a sum of known derivatives with the aid of the "integration rules" listed in Table A.6-1.

The remainder of this section deals with integration methods applicable to special types of integrands. **Comprehensive tables of definite and indefinite integrals are presented in Korn and Korn, "Mathematical Handbook," Appendix E.**

(b) **Integration of Polynomials.**

$$\int (a_n + a_{n-1}x + a_{n-2}x^2 + \cdots + a_0 x^n)\, dx \equiv a_n x + \frac{1}{2} a_{n-1} x^2 + \frac{1}{3} a_{n-2} x^3 + \cdots$$
$$+ \frac{1}{n+1} a_0 x^{n+1} + C \qquad (A.6\text{-}2)$$

(c) **Integration of Rational Functions.** Every rational integrand can be reduced to the sum of a polynomial and a set of *partial fractions*. The partial-fraction terms are integrated successively with the aid of the following formulas:

$$\left.\begin{aligned}
\int \frac{dx}{(x - x_1)^m} &\equiv \begin{cases} -\dfrac{1}{(m-1)(x-x_1)^{m-1}} + C & (m \neq 1) \\ \log_e (x - x_1) + C & (m = 1) \end{cases} \\
\int \frac{dx}{[(x-a)^2 + \omega^2]} &\equiv \frac{1}{\omega} \arctan \frac{x - a}{\omega} + C \\
\int \frac{dx}{[(x-a)^2 + \omega^2]^{m+1}} &\equiv \frac{x - a}{2m\omega^2 [(x-a)^2 + \omega^2]^m} \\
&\quad + \frac{2m - 1}{2m\omega^2} \int \frac{dx}{[(x-a)^2 + \omega^2]^m} \\
\int \frac{x\, dx}{[(x-a)^2 + \omega^2]^{m+1}} &\equiv \frac{a(x-a) - \omega^2}{2m\omega^2 [(x-a)^2 + \omega^2]^m} \\
&\quad + \frac{(2m - 1)a}{2m\omega^2} \int \frac{dx}{[(x-a)^2 + \omega^2]^m}
\end{aligned}\right\} \qquad (A.6\text{-}3)$$

(d) **Integrands which Can Be Reduced to Rational Functions by a Change of Variables.**

1. If the integrand $f(x)$ is a rational function of $\sin x$ and $\cos x$, introduce

$$u = \tan (x/2)$$

so that

$$\sin x = \frac{2u}{1 + u^2} \qquad \cos x = \frac{1 - u^2}{1 + u^2} \qquad dx = \frac{2\, du}{1 + u^2} \qquad (A.6\text{-}4)$$

2. If the integrand $f(x)$ is a rational function of $\sinh x$ and $\cosh x$, introduce

$$u = \tanh (x/2)$$

so that

$$\sinh x = \frac{2u}{1 - u^2} \qquad \cosh x = \frac{1 + u^2}{1 - u^2} \qquad dx = \frac{2\, du}{1 - u^2} \qquad (A.6\text{-}5)$$

NOTE: If $f(x)$ is a rational function of $\sin^2 x$, $\cos^2 x$, $\sin x \cos x$, and $\tan x$ (or of the corresponding hyperbolic functions), one simplifies the calculation by first introducing $v = x/2$, so that $u = \tan v$ (or $u = \tanh v$).

3. If the integrand $f(x)$ is a rational function of x and either $\sqrt{1 - x^2}$ or $\sqrt{x^2 - 1}$, reduce the problem to case 1 or 2 by the respective substitutions $x = \cos v$ or $x = \cosh v$.

4. If the integrand $f(x)$ is a rational function of x and $\sqrt{x^2 - 1}$, introduce

$$u = x + \sqrt{x^2 + 1}$$

so that

$$x = \frac{1}{2}\left(u - \frac{1}{u}\right) \qquad \sqrt{x^2 + 1} = \frac{1}{2}\left(u + \frac{1}{u}\right) \qquad dx = \frac{1}{2}\left(1 + \frac{1}{u^2}\right) du \quad \text{(A.6-6)}$$

5. If the integrand $f(x)$ is a rational function of x and $\sqrt{ax^2 + bx + c}$, reduce the problem to case 3 ($b^2 - 4ac < 0$) or to case 4 ($b^2 - 4ac > 0$) through the substitution

$$v = \frac{2ax + b}{\sqrt{|4ac - b^2|}} \qquad x = \frac{v\sqrt{|4ac - b^2|} - b}{2a} \qquad \text{(A.6-7)}$$

6. If the integrand $f(x)$ is a rational function of x and $u = \sqrt{\dfrac{ax + b}{cx + d}}$, introduce u as a new variable.

7. If the integrand $f(x)$ is a rational function of x, $\sqrt{ax + b}$, and $\sqrt{cx + d}$, introduce $u = \sqrt{ax + b}$ as a new variable.

Many other substitution methods apply in special cases. Note that the integrals may not be real for all values of x.

(e) Integrands of the form $x^n e^{ax}$, $x^n \log_e x$, $x^n \sin x$, $x^n \cos x$ ($n \neq -1$); $\sin^m x \cos^n x$ ($n + m \neq 0$); $e^{ax} \sin^n x$, $e^{ax} \cos^n x$ yield to repeated *integration by parts*.

(f) Many integrals *cannot* be expressed as finite sums involving only algebraic, exponential, and trigonometric functions and their inverses. One may then expand the integrand as an infinite series, or one resorts to numerical integration.

A.6-3. Elliptic Integrals. If $f(x)$ is a rational function of x and $\sqrt{a_0 x^4 + a_1 x^3 + a_2 x^2 + a_3 x + a_4}$, $\int_a^b f(x)\, dx$

is called an **elliptic integral**; one may except the trivial case that the equation

$$a_0 x^4 + a_1 x^3 + a_2 x^2 + a_3 x + a_4 = 0$$

has multiple roots. *Every elliptic integral can be reduced to a weighted sum of elementary functions and normal elliptic integrals* available in tabular form.

A.7 TRIGONOMETRIC FUNCTIONS

A.7-1. Special Function Values and Relations between Functions of Different Arguments (Fig. A.7-1)

$$\left.\begin{array}{ll} \sin z = \cos\left(\dfrac{\pi}{2} - z\right) & \cos z = \sin\left(\dfrac{\pi}{2} - z\right) \\[2mm] \tan z = \cot\left(\dfrac{\pi}{2} - z\right) & \cot z = \tan\left(\dfrac{\pi}{2} - z\right) \end{array}\right\} \quad \text{(A.7-1)}$$

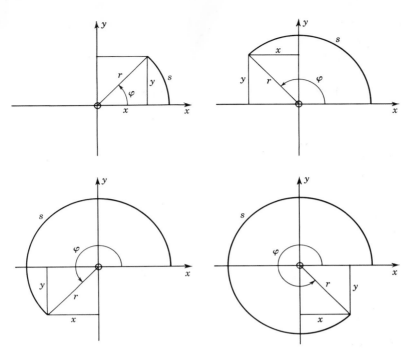

FIG. A.7-1. Definitions of circular measure and trigonometric functions for a given angle φ:

$$\varphi = \frac{s}{r} \text{ (radians)}$$

$$\sin \varphi = \frac{y}{r} \qquad \cos \varphi = \frac{x}{r}$$

$$\tan \varphi = \frac{y}{x} \qquad \cot \varphi = \frac{x}{y}$$

Table A.7-1. Special Values of Trigonometric Functions

A (degrees)	0° 360°	30°	45°	60°	90°	180°	270°
A (radians)	0	$\frac{\pi}{6}$	$\frac{\pi}{4}$	$\frac{\pi}{3}$	$\frac{\pi}{2}$	π	$\frac{3\pi}{2}$
$\sin A$	0	$\frac{1}{2}$	$\frac{1}{\sqrt{2}}$	$\frac{1}{2}\sqrt{3}$	1	0	-1
$\cos A$	1	$\frac{1}{2}\sqrt{3}$	$\frac{1}{\sqrt{2}}$	$\frac{1}{2}$	0	-1	0
$\tan A$	0	$\frac{1}{\sqrt{3}}$	1	$\sqrt{3}$	$\pm \infty$	0	$\pm \infty$
$\cot A$	$\pm \infty$	$\sqrt{3}$	1	$\frac{1}{\sqrt{3}}$	0	$\pm \infty$	0

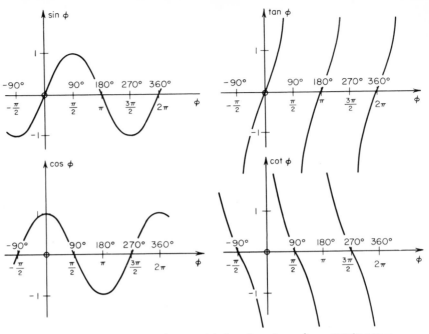

Fig. A.7-2. Plots of the trigonometric functions for real arguments $z = \varphi$.

Table A.7-2. Relations between Trigonometric Functions of Different Arguments

	$-A$	$90° \pm A$	$180° \pm A$	$270° \pm A$	$n360° \pm A$
sin	$-\sin A$	$\cos A$	$\mp \sin A$	$-\cos A$	$\pm \sin A$
cos	$\cos A$	$\mp \sin A$	$-\cos A$	$\pm \sin A$	$\cos A$
tan	$-\tan A$	$\mp \cot A$	$\pm \tan A$	$\mp \cot A$	$\pm \tan A$
cot	$-\cot A$	$\mp \tan A$	$\pm \cot A$	$\mp \tan A$	$\pm \cot A$

A.7-2. Relations between the Trigonometric Functions. The basic relations

$$\boxed{\begin{aligned} \sin^2 z + \cos^2 z &= 1 \\ \frac{\sin z}{\cos z} &= \tan z = \frac{1}{\cot z} \end{aligned}}$$
(A.7-2)

yield

$$\begin{aligned}
\sin z &= \sqrt{1 - \cos^2 z} = \frac{\tan y}{\sqrt{1 + \tan^2 z}} = \frac{1}{\sqrt{1 + \cot^2 z}} \\
\cos z &= \sqrt{1 - \sin^2 z} = \frac{1}{\sqrt{1 + \tan^2 z}} = \frac{\cot z}{\sqrt{1 + \cot^2 z}} \\
\tan z &= \frac{\sin z}{\sqrt{1 - \sin^2 z}} = \frac{\sqrt{1 - \cos^2 z}}{\cos z} = \frac{1}{\cot z} \\
\cot z &= \frac{\sqrt{1 - \sin^2 z}}{\sin z} = \frac{\cos z}{\sqrt{1 - \cos^2 z}} = \frac{1}{\tan z}
\end{aligned} \quad \text{(A.7-3)}$$

A.7-3. Addition Formulas and Multiple-angle Formulas.

The basic relation

$$\sin(A + B) = \sin A \cos B + \sin B \cos A \quad \text{(A.7-4)}$$

yields

$$\begin{aligned}
\sin(A \pm B) &= \sin A \cos B \pm \cos A \sin B \\
\cos(A \pm B) &= \cos A \cos B \mp \sin A \sin B \\
\tan(A \pm B) &= \frac{\tan A \pm \tan B}{1 \mp \tan A \tan B} \\
\cot(A \pm B) &= \frac{\cot A \cot B \mp 1}{\cot A + \cot B}
\end{aligned} \quad \text{(A.7-5)}$$

$$\left.\begin{aligned}
\sin 2A &= 2 \sin A \cos A \\
\cos 2A &= \cos^2 A - \sin^2 A = 2\cos^2 A - 1 = 1 - 2\sin^2 A
\end{aligned}\right\} \quad \text{(A.7-6)}$$

$$\left.\begin{aligned}
a \sin A + b \cos A &= r \sin(A + B) = r \cos(90° - A - B) \\
r &= +\sqrt{a^2 + b^2} \qquad \tan B = \frac{b}{a}
\end{aligned}\right\} \quad \text{(A.7-7)}$$

$$\left.\begin{aligned}
\sin A \pm \sin B &= 2 \sin \frac{A \pm B}{2} \cos \frac{A \mp B}{2} \\
\cos A + \cos B &= 2 \cos \frac{A + B}{2} \cos \frac{A - B}{2} \\
\cos A - \cos B &= -2 \sin \frac{A + B}{2} \sin \frac{A - B}{2} \\
\tan A \pm \tan B &= \frac{\sin(A \pm B)}{\cos A \cos B} \\
\cot A \pm \cot B &= \frac{\sin(B \pm A)}{\sin A \sin B}
\end{aligned}\right\} \quad \text{(A.7-8)}$$

$$\left.\begin{aligned}
2 \cos A \cos B &= \cos(A - B) + \cos(A + B) \\
2 \sin A \sin B &= \cos(A - B) - \cos(A + B) \\
2 \sin A \cos B &= \sin(A - B) + \sin(A + B) \\
2 \cos^2 A &= 1 + \cos 2A \\
2 \sin^2 A &= 1 - \cos 2A
\end{aligned}\right\} \quad \text{(A.7-9)}$$

$$\left.\begin{aligned}
\sin nA &= \binom{n}{1} \cos^{n-1} A \sin A - \binom{n}{3} \cos^{n-3} A \sin^3 A \\
&\quad + \binom{n}{5} \cos^{n-5} A \sin^5 A \mp \cdots \\
\cos nA &= \cos^n A - \binom{n}{2} \cos^{n-2} A \sin^2 A + \binom{n}{4} \cos^{n-4} A \sin^4 A \mp \cdots
\end{aligned}\right\} \quad \text{(A.7-10)}$$

If n is an odd integer,

$$\left.\begin{aligned}
\sin^n z &= \left(\frac{1}{2i}\right)^{n-1}\left[\sin nz - \binom{n}{1}\sin(n-2)z + \binom{n}{2}\sin(n-4)z \right.\\
&\qquad\left. - \binom{n}{3}\sin(n-6)z + \cdots (-1)^{\frac{n-1}{2}}\binom{n}{\frac{n-1}{2}}\sin z\right] \\
\cos^n z &= \left(\frac{1}{2}\right)^{n-1}\left[\cos nz + \binom{n}{1}\cos(n-2)z + \binom{n}{2}\cos(n-4)z \right.\\
&\qquad\left. + \cdots + \binom{n}{\frac{n-1}{2}}\cos z\right]
\end{aligned}\right\} \quad (A.7\text{-}11)$$

If n is an even integer,

$$\left.\begin{aligned}
\sin^n z &= \frac{(-1)^{\frac{n}{2}}}{2^{n-1}}\left[\cos nz - \binom{n}{1}\cos(n-2)z + \binom{n}{2}\cos(n-4)z \right.\\
&\qquad\left. - \cdots + (-1)^{\frac{n-2}{2}}\binom{n}{\frac{n-2}{2}}\cos 2z\right] + \binom{n}{\frac{n}{2}}\frac{1}{2^n} \\
\cos^n z &= \left(\frac{1}{2}\right)^{n-1}\left[\cos nz + \binom{n}{1}\cos(n-2)z + \binom{n}{2}\cos(n-4)z \right.\\
&\qquad\left. + \cdots + \binom{n}{\frac{n-2}{2}}\cos 2z\right] + \binom{n}{\frac{n}{2}}\frac{1}{2^n}
\end{aligned}\right\} \quad (A.7\text{-}12)$$

A.7-4. The Inverse Trigonometric Functions (Fig. A.7-3). The **inverse trigonometric functions** $w = \arcsin z$, $w = \arccos z$, $w = \arctan z$, $w = \text{arccot } z$ are respectively defined by

$$z = \sin w \qquad z = \cos w \qquad z = \tan w \qquad z = \cot w \qquad (A.7\text{-}13)$$

A.8 OTHER ELEMENTARY TRANSCENDENTAL FUNCTIONS

A.8-1. Hyperbolic Functions

$$\left.\begin{aligned}
\sinh z &= \frac{e^z - e^{-z}}{2} \qquad \cosh z = \frac{e^z + e^{-z}}{2} \\
\tanh z &= \frac{\sinh z}{\cosh z} \qquad \coth z = \frac{\cosh z}{\sinh z}
\end{aligned}\right\} \quad (A.8\text{-}1)$$

$$\left.\begin{aligned}
\sinh z &= \sqrt{\cosh^2 z - 1} = \frac{\tanh z}{\sqrt{1 - \tanh^2 z}} = \frac{1}{\sqrt{\coth^2 z - 1}} \\
\cosh z &= \sqrt{1 + \sinh^2 z} = \frac{1}{\sqrt{1 - \tanh^2 z}} = \frac{\coth z}{\sqrt{\coth^2 z - 1}} \\
\tanh z &= \frac{\sinh z}{\sqrt{1 + \sinh^2 z}} = \frac{\sqrt{\cosh^2 z - 1}}{\cosh z} = \frac{1}{\coth z} \\
\coth z &= \frac{\sqrt{1 + \sinh^2 z}}{\sinh z} = \frac{\cosh z}{\sqrt{\cosh^2 z - 1}} = \frac{1}{\tanh z}
\end{aligned}\right\} \quad (A.8\text{-}2)$$

OTHER ELEMENTARY TRANSCENDENTAL FUNCTIONS   A-23

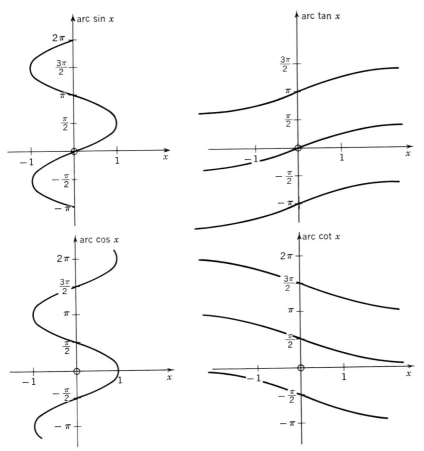

Fig. A.7-3. Plots of the inverse trigonometric functions.

A.8-2. Relations between Exponential, Trigonometric, and Hyperbolic Functions

$$e^{iz} = \cos z + i \sin z$$
$$\cos z = \frac{e^{iz} + e^{-iz}}{2} \qquad \sin z = \frac{e^{iz} - e^{-iz}}{2i} \tag{A.8-3}$$

$$e^{-iz} = \cos z - i \sin z \tag{A.8-4}$$
$$e^{z} = \cosh z + \sinh z \qquad e^{-z} = \cosh z - \sinh z \tag{A.8-5}$$
$$\cosh z = \frac{e^{z} + e^{-z}}{2} \qquad \sinh z = \frac{e^{z} - e^{-z}}{2} \tag{A.8-6}$$

$$\left. \begin{array}{ll} \cos z = \cosh iz & \cosh z = \cos iz \\ \sin z = -i \sinh iz & \sinh z = -i \sin iz \\ \tan z = -i \tanh iz & \tanh z = -i \tan iz \\ \cot z = i \coth iz & \coth z = i \cot iz \end{array} \right\} \tag{A.8-7}$$

A.8-3. Decomposition of the Logarithm

$$\log_e z = \log_e |z| + i \arg(z) \qquad \text{(A.8-8)}$$

$$\left. \begin{array}{l} \log_e (ix) = \log_e x + (2n + \tfrac{1}{2})\pi i \\ \log_e (-x) = \log_e x + (2n + 1)\pi i \end{array} \right\} \quad (n = 0, \pm 1, \pm 2, \ldots) \qquad \text{(A.8-9)}$$

A.9 PROGRESSIONS AND SERIES

A.9-1. Arithmetic Progression. If a_0 is the first term and d is the *common difference* between successive terms a_j, then

$$a_j = a_0 + jd \quad (j = 0, 1, 2, \ldots)$$

$$s_n = \sum_{j=0}^{n} a_j = \frac{n+1}{2}(2a_0 + nd) = \frac{n+1}{2}(a_0 + a_n) \qquad \text{(A.9-1)}$$

A.9-2. Geometric Progression. If a_0 is the first term and r is the *common ratio* of successive terms, then

$$a_j = a_0 r^j \quad (j = 0, 1, 2, \ldots)$$

$$s_n = \sum_{j=0}^{n} a_j = \sum_{j=0}^{n} a_0 r^j = a_0 \frac{1 - r^{n+1}}{1 - r} = \frac{a_0 - a_n r}{1 - r} \qquad \text{(A.9-2)}$$

A.9-3. The Binomial Theorem

$$\left. \begin{array}{l} (a \pm b)^2 = a^2 \pm 2ab + b^2 \\ (a \pm b)^3 = a^3 \pm 3a^2 b + 3ab^2 \pm b^3 \\ (a \pm b)^4 = a^4 \pm 4a^3 b + 6a^2 b^2 \pm 4ab^3 + b^4 \\ \cdots \cdots \cdots \cdots \cdots \cdots \cdots \cdots \cdots \cdots \end{array} \right\} \qquad \text{(A.9-3)}$$

$$\left. \begin{array}{l} (a + b)^n = \sum_{j=0}^{n} \binom{n}{j} a^{n-j} b^j \quad (n = 1, 2, \ldots) \\ \text{with } \binom{n}{j} = \frac{n!}{j!(n-j)!} \\ \qquad (j = 0, 1, 2, \ldots \leq n = 0, 1, 2, \ldots) \end{array} \right\} \qquad \text{(A.9-4)}$$

A.9-4. Taylor's Expansion and MacLaurin's Series. Given a function $f(x)$ such that all derivatives $f^{(k)}(x)$ exist and $\lim_{n \to \infty} R_n(x) = 0$ for $a \leq x < b$,

$$f(x) = \sum_{k=0}^{\infty} \frac{1}{k!} f^{(k)}(a)(x-a)^k \qquad (a \leq x < b) \qquad \text{(A.9-5)}$$

and the series converges uniformly to $f(x)$ for $a \leq x < b$ [*Taylor-series expansion of $f(x)$ about $x = a$*].

For $a = 0$, Taylor's series reduces to *MacLaurin's series* $\sum_{k=0}^{\infty} \frac{1}{k!} f^{(k)}(0) x^k$.

A.9-5. Power-series Expansions

$$\frac{1}{1-z} = 1 + z + z^2 + \cdots \qquad (|z| < 1; \text{ GEOMETRIC SERIES}) \qquad (A.9\text{-}6)$$

$$(1+z)^p = 1 + \binom{p}{1}z + \binom{p}{2}z^2 + \cdots \qquad (|z| < 1; \text{ BINOMIAL SERIES}) \qquad (A.9\text{-}7)$$

$$e^z = 1 + z + \frac{z^2}{2!} + \frac{z^3}{3!} + \cdots \qquad (z \neq \infty) \qquad (A.9\text{-}8)$$

$$\sin z = z - \frac{z^3}{3!} + \frac{z^5}{5!} \mp \cdots \qquad \cos z = 1 - \frac{z^2}{2!} + \frac{z^4}{4!} \mp \cdots \qquad (z \neq \infty) \quad (A.9\text{-}9)$$

$$\sinh z = z + \frac{z^3}{3!} + \frac{z^5}{5!} + \cdots \qquad \cosh z = 1 + \frac{z^2}{2!} + \frac{z^4}{4!} + \cdots \qquad (z \neq \infty)$$

$$(A.9\text{-}10)$$

$$\log_e (1+z) = z - \frac{z^2}{2} + \frac{z^3}{3} - \frac{z^4}{4} \pm \cdots \qquad (|z| < 1) \qquad (A.9\text{-}11)$$

$$\left. \begin{array}{l} \arcsin z = z + \dfrac{1}{2} \cdot \dfrac{z^3}{3} + \dfrac{1}{2} \cdot \dfrac{3}{4} \cdot \dfrac{z^5}{5} + \dfrac{1}{2} \cdot \dfrac{3}{4} \cdot \dfrac{5}{6} \cdot \dfrac{z^7}{7} + \cdots \\[4pt] \sinh^{-1} z = z - \dfrac{1}{2} \cdot \dfrac{z^3}{3} + \dfrac{1}{2} \cdot \dfrac{3}{4} \cdot \dfrac{z^5}{5} - \dfrac{1}{2} \cdot \dfrac{3}{4} \cdot \dfrac{5}{6} \cdot \dfrac{z^7}{7} \pm \cdots \end{array} \right\} \; (|z| < 1) \quad (A.9\text{-}12)$$

$$\left. \begin{array}{l} \arctan z = z - \dfrac{z^3}{3} + \dfrac{z^5}{5} \mp \cdots \\[4pt] \tanh^{-1} z = \dfrac{1}{2} \log_e \dfrac{1+z}{1-z} = z + \dfrac{z^3}{3} + \dfrac{z^5}{5} + \cdots \end{array} \right\} \; (|z| < 1) \quad (A.9\text{-}13)$$

A.10 SPECIAL FUNCTIONS

A.10-1. The Sinc Function (Fig. A.10-1)

$$\operatorname{sinc} z \equiv \frac{\sin \pi z}{\pi z} \qquad (A.10\text{-}1)$$

$\operatorname{sinc} z = 0$ for $z = n\pi$ ($n = \pm 1, \pm 2, \ldots$); $\operatorname{sinc} 0 = 1$. Note

$$\int_{-\infty}^{\infty} \operatorname{sinc}(t-n) \operatorname{sinc}(t-m)\, dt = \begin{cases} 0 & (m \neq n) \\ 1 & (m = n) \end{cases} \quad (m, nz, 0, \pm 1, \pm 2, \ldots)$$

$$(A.10\text{-}2)$$

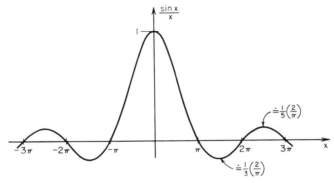

Fig. A.10-1. $\sin x/x$ vs. x. (From M. Schwartz, "Information Transmission, Modulation, and Noise," McGraw-Hill Book Company, New York, 1959.)

A.10-2. Sine and Cosine Integrals.
One defines

$$\operatorname{Si}(x) \equiv \int_0^x \frac{\sin x}{x} dx \equiv \frac{\pi}{2} - \int_x^\infty \frac{\sin x}{x} dx \equiv \frac{\pi}{2} + \operatorname{si}(x)$$

$$= x - \frac{1}{3!}\frac{x^3}{3} + \frac{1}{5!}\frac{x^5}{5} \mp \cdots \qquad \text{(SINE INTEGRAL)} \quad (A.10\text{-}3)$$

$$\operatorname{Ci}(x) \equiv -\int_x^\infty \frac{\cos x}{x} dx = C + \log_e x - \int_0^x \frac{1 - \cos x}{x} dx$$

$$= C + \log_e x - \frac{1}{2!}\frac{x^2}{2} + \frac{1}{4!}\frac{x^4}{4} \mp \cdots \qquad (x > 0)$$

$$\text{(COSINE INTEGRAL)} \quad (A.10\text{-}4)$$

where $C \approx 0.577216$ is the Euler-Mascheroni constant defined in Sec. A.10-4.

A.10-3. The Error Function

$$\operatorname{erf} x \equiv \frac{2}{\sqrt{\pi}} \int_0^x e^{-x^2} dx = \frac{2}{\sqrt{\pi}} \left(x - \frac{x^3}{3} + \frac{1}{2!}\frac{x^5}{5} - \frac{1}{3!}\frac{x^7}{7} \pm \cdots \right)$$

$$\text{(ERROR FUNCTION)} \quad (A.10\text{-}5)$$

The function

$$\operatorname{erfc} z = 1 - \operatorname{erf} z = \frac{2}{\sqrt{\pi}} \int_z^\infty e^{-\zeta^2} d\zeta \qquad (A.10\text{-}6)$$

is known as the **complementary error function**.

A.10-4. The Gamma Function.
(a) **Integral Representation.** The **gamma function** $\Gamma(z)$ is most frequently defined by

$$\Gamma(z) = \int_0^\infty e^{-t} t^{z-1} dt \qquad [\operatorname{Re}(z) > 0] \qquad \text{(EULER'S INTEGRAL OF THE SECOND KIND)}$$

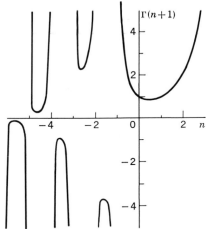

FIG. A.10-2. $\Gamma(n+1)$ vs. n for real n. Note $\Gamma(n+1) = n!$ for $n = 0, 1, 2, \ldots$, and the alternating maxima and minima given approximately by $\Gamma(1.462) = 0.886$, $\Gamma(-0.5040) = -3.545$, $\Gamma(-1.573) = 2.302$, $\Gamma(-2.611) = -0.888$, ...

Figure A.10-2 shows a graph of $\Gamma(x)$ vs. x for real x. Note

$$\left. \begin{array}{c} \Gamma(\tfrac{1}{2}) = \sqrt{\pi} \qquad \Gamma(1) = 1 \\ \Gamma(n+1) = n! \qquad (n = 0, 1, 2, \ldots) \end{array} \right\} \qquad (A.10\text{-}7)$$

$$\Gamma(z+1) = z\Gamma(z) \qquad (A.10\text{-}8)$$

(b) Other Representations of $\Gamma(z)$.

$$\Gamma(z) = \lim_{n \to \infty} \frac{n!}{z(z+1)(z+2)\cdots(z+n-1)} n^{z-1} \quad \text{(Euler's definition)} \quad \text{(A.10-9)}$$

$$\frac{1}{\Gamma(z)} = ze^{Cz} \prod_{k=1}^{\infty} \left(1 + \frac{z}{k}\right) e^{-z/k} \quad \text{(Weierstrass's product representation)} \quad \text{(A.10-10)}$$

C is the **Euler-Mascheroni constant** defined by

$$C = \lim_{n \to \infty} \left(\sum_{k=1}^{n} \frac{1}{k} - \log_e n \right)$$
$$= -\int_0^\infty e^{-t} \log_e t \, dt = -\int_0^1 \log_e \left(\log_e \frac{1}{\tau}\right) d\tau$$
$$\approx 0.5772157 \quad \text{(A.10-11)}$$

A.10-5. Stirling's Expansions for $\Gamma(z)$ and $n!$

$$\Gamma(z) = e^{-z} z^{z-\frac{1}{2}} \sqrt{2\pi} \left[1 + \frac{1}{12z} + \frac{1}{288z^2} - \frac{139}{51840z^3} - \frac{571}{2488320z^4} + O(z^{-5}) \right] \quad (|\arg z| < \pi) \quad \text{(Stirling's series)} \quad \text{(A.10-12)}$$

Stirling's series is especially useful for large $|z|$; *for real positive z, the absolute value of the error is less than that of the last term used.* Note, in particular,

$$\lim_{n \to \infty} \frac{n!}{n^n e^{-n} \sqrt{2\pi n}} = 1 \quad \text{or} \quad n! \simeq n^n e^{-n} \sqrt{2\pi n}$$

$$\text{as } n \to \infty \quad \text{(Stirling's formula)} \quad \text{(A.10-13)}$$

The fractional error in Stirling's formula is less than 10 per cent for $n = 1$ and decreases as n increases; this asymptotic formula applies particularly to computations of the *ratio* of two factorials or gamma functions, since in such cases the *fractional* error is of paramount interest.

More specifically

$$n^n e^{-n} \sqrt{2\pi n} < n! < n^n \sqrt{2\pi n} \, e^{-n+1/12n} \quad \text{(A.10-14)}$$

$$n! \simeq n^n \sqrt{2\pi n} \exp\left(-n + \frac{1}{12n} - \frac{1}{360n^3} + \cdots\right) \quad \text{as } n \to \infty \quad \text{(A.10-15)}$$

A.10-6. Beta Functions.

The **(complete) beta function** is defined as

$$B(p, q) \equiv \frac{\Gamma(p)\Gamma(q)}{\Gamma(p+q)} \equiv B(q, p) \quad \text{(A.10-16)}$$

or by analytic continuation of

$$B(p, q) = \int_0^1 t^{p-1}(1-t)^{q-1} dt \quad [\text{Re}(p) > 0, \text{Re}(q) > 0]$$
$$\text{(Euler's integral of the first kind)} \quad \text{(A.10-17)}$$

$$B(p, q) = \int_0^\infty \frac{t^{p-1}}{(1+t)^{p+q}} dt = 2 \int_0^{\pi/2} \sin^{2p-1}\vartheta \cos^{2q-1}\vartheta \, d\vartheta \quad \text{(A.10-18)}$$

Note
$$B(p, q)B(p + q, r) = B(q, r)B(q + r, p) \tag{A.10-19}$$

$$\frac{1}{B(n, m)} = m \binom{n + m - 1}{n - 1} = n \binom{n + m - 1}{m - 1} \quad (n, m = 1, 2, \ldots) \tag{A.10-20}$$

A.10-7. Incomplete Gamma and Beta Functions. The **incomplete gamma function** $\Gamma_z(p)$ and the **incomplete beta function** $B_z(p, q)$ are respectively defined by analytic continuation of

$$\Gamma_z(p) = \int_0^z t^{p-1} e^{-t} \, dt \quad [\text{Re } (p) > 0] \tag{A.10-21}$$

$$B_z(p, q) = \int_0^z t^{p-1}(1 - t)^{q-1} \, dt \quad [\text{Re } (p) > 0, \text{ Re } (q) > 0; 0 \le z \le 1] \tag{A.10-22}$$

$I_z(p, q) \equiv B_z(p, q)/B(p, q)$ is called the **incomplete-beta-function ratio**.

A.10-8. Bessel Functions and Other Cylinder Functions. (a) A **cylinder function (circular-cylinder function) of order** m is a solution $w = Z_m(z)$ of the linear differential equation

$$\frac{d^2w}{dz^2} + \frac{1}{z}\frac{dw}{dz} + \left(1 - \frac{m^2}{z^2}\right) w = 0 \quad (\text{Bessel's differential equation}) \tag{A.10-23}$$

where m is any real number; one usually imposes the recurrence relations

$$Z_{m+1}(z) = \frac{2m}{z} Z_m(z) - Z_{m-1}(z) = \frac{m}{z} Z_m(z) - \frac{d}{dz} Z_m(z)$$

$$= -z^m \frac{d}{dz} [z^{-m} Z_m(z)] \tag{A.10-24}$$

as additional defining conditions. The functions $e^{\pm i(Kz \pm m\varphi)} Z_m(iKr')$ are solutions of Laplace's partial differential equation in cylindrical coordinates r', φ, z.

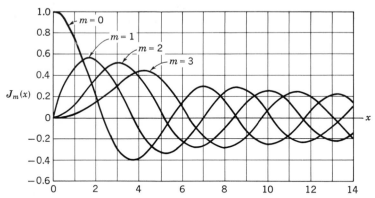

Fig. A.10-3. The Bessel functions $J_0(x)$, $J_1(x)$, $J_2(x)$, ... for real arguments. Note that $J_m(-x) = (-1)^m J_m(x)$.

(b) The most generally useful cylinder functions of order m satisfying the recurrence relations (24) are

$$J_m(z) = \left(\frac{z}{2}\right)^m \sum_{k=0}^{\infty} \frac{(-1)^k}{k!\Gamma(m + k + 1)} \left(\frac{z}{2}\right)^{2k} \quad (|\arg z| < \pi)$$

(Bessel functions of the first kind) (A.10-25)

$$N_m(z) \equiv \frac{1}{\sin m\pi} [J_m(z) \cos m\pi - J_{-m}(z)] \qquad (m \neq 0, \pm 1, \pm 2, \ldots)$$

$$N_m(z) = (-1)^{-m} N_{-m}(z) = \frac{2}{\pi} J_m(z) \left(\log_e \frac{z}{2} + C\right)$$

$$- \frac{1}{\pi} \left(\frac{z}{2}\right)^m \sum_{k=0}^{\infty} \frac{(-1)^k}{k!(m+k)!} \left(\frac{z}{2}\right)^{2k} \left(\sum_{j=1}^{k} \frac{1}{j} + \sum_{j=1}^{m+k} \frac{1}{j}\right)$$

$$- \frac{1}{\pi} \left(\frac{z}{2}\right)^{-m} \sum_{k=0}^{m-1} \frac{(m-k-1)!}{k!} \left(\frac{z}{2}\right)^{2k}$$

$$(m = 0, 1, 2, \ldots; \ |\arg z| < \pi) \qquad \text{(A.10-26)}$$

(Neumann's Bessel functions of the second kind)

$$H_m^{(1)}(z) \equiv J_m(z) + iN_m(z) \qquad H_m^{(2)}(z) \equiv J_m(z) - iN_m(z)$$

(Hankel functions of the first and second kind) (A.10-27)

The last three sums in Eq. (26) are given the value zero whenever the lower limit exceeds the upper limit; and $C \approx 0.577216$ is Euler's constant (11). Note that every function $N_m(z)$ has a singularity at the origin.

The Neumann functions $N_m(z)$ are sometimes denoted by $Y_m(z)$; some authors refer to them as *Weber's Bessel functions of the second kind*.

(c) Analytic Continuation. To obtain values of the cylinder functions for $|\arg z| > \pi$, use

$$\left. \begin{array}{l} J_m(e^{in\pi}z) = e^{imn\pi} J_m(z) \\ N_m(e^{in\pi}z) = e^{-imn\pi} N_m(z) + 2iJ_m(z) \sin mn\pi \cot m\pi \end{array} \right\} \qquad (n = 0, 1, 2, \ldots) \quad \text{(A.10-28)}$$

where one uses $\sin mn\pi \cot m\pi = (-1)^{mn} n$ for $m = \pm n$; and

$$\left. \begin{array}{l} H_m^{(1)}(e^{i\pi}z) = -e^{im\pi} H_m^{(2)}(z) = -H_{-m}^{(2)}(z) \\ H_m^{(2)}(e^{-i\pi}z) = -e^{im\pi} H_m^{(1)}(z) = -H_{-m}^{(1)}(z) \end{array} \right\} \qquad \text{(A.10-29)}$$

Note that *cylinder functions of integral order are single-valued integral functions.*

(d) Every cylinder function of order m can be expressed as a linear combination of $J_m(z)$ and $N_m(z)$ and as a linear combination of $H_m^{(1)}(z)$ and $H_m^{(2)}(z)$:

$$Z_m(z) = aJ_m(z) + bN_m(z) = \alpha H_m^{(1)}(z) + \beta H_m^{(2)}(z) \qquad \text{(A.10-30)}$$

(e) Cylinder functions with $m = \pm \frac{1}{2}, \pm \frac{3}{2}, \ldots$ can be written as elementary transcendental functions:

$$J_{1/2}(z) \equiv \sqrt{\frac{2}{\pi}} \frac{\sin z}{\sqrt{z}} \qquad J_{-1/2}(z) \equiv \sqrt{\frac{2}{\pi}} \frac{\cos z}{\sqrt{z}} \qquad \text{(A.10-31)}$$

$$J_{3/2}(z) \equiv \sqrt{\frac{2}{\pi}} \left(-\frac{\cos z}{\sqrt{z}} + \frac{\sin z}{z\sqrt{z}}\right) \qquad J_{-3/2}(z) \equiv \sqrt{\frac{2}{\pi}} \left(-\frac{\sin z}{\sqrt{z}} - \frac{\cos z}{z\sqrt{z}}\right) \qquad \text{(A.10-32)}$$

$$J_{k+1/2}(z) \equiv \sqrt{\frac{2}{\pi}} z^{k+1/2} \left(-\frac{1}{z} \frac{d}{dz}\right)^k \frac{\sin z}{z} \qquad (k = 1, 2, \ldots) \qquad \text{(A.10-33)}$$

$$H_{1/2}^{(1)}(z) \equiv \sqrt{\frac{2}{\pi}} \frac{1}{i} \frac{e^{iz}}{\sqrt{z}} \qquad H_{-1/2}^{(1)}(z) \equiv \sqrt{\frac{2}{\pi}} \frac{e^{iz}}{\sqrt{z}} \qquad \text{(A.10-34)}$$

$$H_{1/2}^{(2)}(z) \equiv -\sqrt{\frac{2}{\pi}} \frac{1}{i} \frac{e^{-iz}}{\sqrt{z}} \qquad H_{-1/2}^{(2)}z \equiv \sqrt{\frac{2}{\pi}} \frac{e^{-iz}}{\sqrt{z}} \qquad \text{(A.10-35)}$$

A.10-9. Integral Formulas

$$J_m(z) = \frac{1}{\pi} \int_0^\pi \cos(mt - z \sin t)\, dt \qquad (m = 0, 1, 2, \ldots)$$

(BESSEL'S INTEGRAL FORMULA) (A.10-36)

$$\left. \begin{array}{l} J_{2m}(z) = \dfrac{2}{\pi} \int_0^{\pi/2} \cos(z \sin t) \cos 2mt\, dt \\[2mm] J_{2m+1}(z) = \dfrac{2}{\pi} \int_0^{\pi/2} \sin(z \sin t) \sin(2m+1)t\, dt \end{array} \right\} \quad (m = 0, 1, 2, \ldots) \quad \text{(A.10-37)}$$

$$J_m(z) = \frac{1}{2\pi} \int_{-\pi}^{\pi} e^{iz\cos t} e^{im(t-\pi/2)}\, dt = \frac{(-i)^m}{\pi} \int_0^\pi e^{iz\cos t} \cos mt\, dt$$

$(m = 0, 1, 2, \ldots)$ (HANSEN'S INTEGRAL FORMULA) (A.10-38)

A.10-10. Series Expansions. Bessel functions of nonnegative integral order $m = 0, 1, 2, \ldots$ are single-valued integral functions of z. They may be "generated" as coefficients of the Fourier series

$$\left. \begin{array}{l} \cos(z \sin t) = J_0(z) + 2 \sum_{k=1}^\infty J_{2k}(z) \cos 2kt \\[2mm] \sin(z \sin t) = 2 \sum_{k=1}^\infty J_{2k-1}(z) \sin(2k-1)t \end{array} \right\} \quad \text{(A.10-39)}$$

$$e^{\pm iz \sin t} = \sum_{m=-\infty}^{\infty} J_m(z) e^{\pm imt} = J_0(z) + 2 \sum_{k=1}^\infty [J_{2k}(z) \cos 2kt \pm i J_{2k-1}(z) \sin(2k-1)t]$$

(JACOBI-ANGER FORMULA) (A.10-40)

A.10-11. Modified Bessel and Hankel Functions. The **modified cylinder functions of order** m are defined by

$$\left. \begin{array}{l} I_m(z) = i^{-m} J_m(iz) \qquad \text{(MODIFIED BESSEL FUNCTIONS)} \\[2mm] K_m(z) = \dfrac{\pi}{2} i^{m+1} H_m^{(1)}(iz) \qquad \text{(MODIFIED HANKEL FUNCTIONS)} \end{array} \right\} \quad \text{(A.10-41)}$$

The functions (41) are linearly independent solutions of the differential equation

$$\frac{d^2w}{dz^2} + \frac{1}{z}\frac{dw}{dz} - \left(1 + \frac{m^2}{z^2}\right) w = 0 \qquad \begin{array}{l} \text{(MODIFIED BESSEL'S} \\ \text{DIFFERENTIAL EQUATION)} \end{array} \quad \text{(A.10-42)}$$

and satisfy the recursion formulas

$$\left. \begin{array}{l} I_{m+1}(z) = I_{m-1}(z) - \dfrac{2m}{z} I_m(z) = 2 \dfrac{d}{dz} I_m(z) - I_{m-1}(z) \\[2mm] K_{m+1}(z) = K_{m-1}(z) + \dfrac{2m}{z} K_m(z) = -2 \dfrac{d}{dz} K_m(z) - K_{m-1}(z) \end{array} \right\} \quad \text{(A.10-43)}$$

$I_m(z)$ and $K_m(z)$ are real monotonic functions for $m = 0, \pm 1, \pm 2, \ldots$ and real z.

A.10-12. Spherical Bessel Functions. The **spherical Bessel functions of the first, second, third, and fourth kind**

$$\left. \begin{array}{ll} j_j(z) \equiv \sqrt{\dfrac{\pi}{2z}}\, J_{j+\frac{1}{2}}(z) & n_j(z) \equiv \sqrt{\dfrac{\pi}{2z}}\, N_{j+\frac{1}{2}}(z) \\[3mm] h_j^{(1)}(z) \equiv \sqrt{\dfrac{\pi}{2z}}\, H_{j+\frac{1}{2}}^{(1)}(z) & h_j^{(2)}(z) \equiv \sqrt{\dfrac{\pi}{2z}}\, H_{j+\frac{1}{2}}^{(2)}(z) \end{array} \right\} \quad \text{(A.10-44)}$$

SPECIAL FUNCTIONS   A–31

satisfy the differential equation

$$\frac{d^2w}{dz^2} + \frac{2}{z}\frac{dw}{dz^2} + \left[1 - \frac{j(j+1)}{dz}\right]w = 0 \qquad (A.10\text{-}45)$$

and the recursion formulas

$$w_{j+1}(z) = \frac{2j+1}{z}w_j(z) - w_{j-1}(z) = -z^j\frac{d}{dz}[z^{-j}w_j(z)] \qquad (A.10\text{-}46)$$

For integral values of j, the spherical Bessel functions are elementary transcendental functions:

$$\left.\begin{aligned} j_0(z) &\equiv \frac{\sin z}{z} & h_0^{(1)}(z) &\equiv -\frac{ie^{iz}}{z} & h_0^{(2)}(z) &\equiv \frac{ie^{-iz}}{z} \\ j_{-1}(z) &\equiv \frac{\cos z}{z} & h_{-1}^{(1)}(z) &\equiv \frac{e^{iz}}{z} & h_{-1}^{(2)}(z) &\equiv \frac{e^{-iz}}{z} \end{aligned}\right\} \qquad (A.10\text{-}47)$$

$$j_j(z) \equiv z^j\left(-\frac{1}{z}\frac{d}{dz}\right)^j\frac{\sin z}{z} \qquad n_j(z) \equiv (-1)^{j+1}j_{-j-1}(z) \qquad (j = 1, 2, \ldots) \qquad (A.10\text{-}48)$$

A.10-13. Asymptotic Expansion of Cylinder Functions and Spherical Bessel Functions for Large Absolute Values of z. As $z \to \infty$,

$$\left.\begin{aligned} J_m(z) &\simeq \sqrt{\frac{2}{\pi z}}\left[A_m(z)\cos\left(z - \frac{m\pi}{2} - \frac{\pi}{4}\right) - B_m(z)\sin\left(z - \frac{m\pi}{2} - \frac{\pi}{4}\right)\right] \\ N_m(z) &\simeq \sqrt{\frac{2}{\pi z}}\left[A_m(z)\sin\left(z - \frac{m\pi}{2} - \frac{\pi}{4}\right) + B_m(z)\cos\left(z - \frac{m\pi}{2} - \frac{\pi}{4}\right)\right] \end{aligned}\right\} \ (|\arg z| < \pi) \quad (A.10\text{-}49)$$

where $A_m(z)$ and $B_m(z)$ stand for the asymptotic series

$$\left.\begin{aligned} A_m(z) &= 1 - \frac{(4m^2-1)(4m^2-9)}{2!(8z)^2} \\ &\quad + \frac{(4m^2-1)(4m^2-9)(4m^2-25)(4m^2-49)}{4!(8z)^4} \mp \cdots \\ B_m(z) &= \frac{4m^2-1}{8z} - \frac{(4m^2-1)(4m^2-9)(4m^2-25)}{3!(8z)^3} \pm \cdots \end{aligned}\right\} \qquad (A.10\text{-}50)$$

A.10-14. Graphical Presentations of $J_m(x)$. Figure A.10-3 shows a plot of the Bessel functions $J_m(x)$ for real arguments and integral values of m, and Fig. A.10-4 presents a three-dimensional plot of the same function. For values of x and m larger than shown in these figures (approximately 10 to 15), the function may be represented by the trigonometric approximations (49) and (50).

A.10-15. Tables of Cylinder Functions. Four-figure tables of J_0, J_1; N_0, N_1; I_0, I_1; and K_0, K_1 for $x = 0$ to $x = 15$ will be found in Harold Etherington ["Nuclear Engineering Handbook," McGraw-Hill Book Company, New York, 1958 (also contains a short index to other numerical tables)]; G. A. Korn and T. M. Korn, ("Mathematical Handbook for Scientists and Engineers," McGraw-Hill Book Company, New York, 1961); and E. Jahnke and F. Emde ("Tables of Higher Functions," 6th ed., McGraw-Hill Book Company, New York, 1960). For additional references, see A. Fletcher, J. C. P. Miller, and L. Rosenhead ("Index of Mathematical Tables," McGraw-Hill Book Company, New York, 1946).

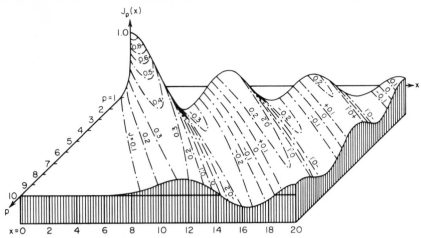

FIG. A.10-4. Bessel functions $J_p(x)$ vs. x and p. (*From E. Jahnke, and F. Emde, "Tables of Higher Functions," 6th ed., McGraw-Hill Book Company, New York, 1960.*)

A.11 STEP FUNCTIONS AND SYMBOLIC IMPULSE FUNCTIONS

A.11-1. Step Functions (see also Fig. A.11-1). (a) A **step function** of the real variable x is a function which changes its value only on a discrete set of discontinuities. The function values at the discontinuities may or may not be defined. The most frequently useful step functions are*

$$U(x) = \begin{cases} 0 \text{ for } x < 0 \\ \tfrac{1}{2} \text{ for } x = 0 \\ 1 \text{ for } x > 0 \end{cases} \quad \text{(SYMMETRICAL UNIT-STEP FUNCTION)} \quad (A.11\text{-}1)$$

$$U_-(x) = \begin{cases} 0 \text{ for } x < 0 \\ 1 \text{ for } x \geq 0 \end{cases}$$
$$U_+(x) = \begin{cases} 0 \text{ for } x < 0 \\ 1 \text{ for } x > 0 \end{cases} \quad \text{(ASYMMETRICAL UNIT-STEP FUNCTIONS)} \quad (A.11\text{-}2)$$

(b) **Approximation of Step Functions by Continuous Functions**

$$U(x) = \lim_{\alpha \to \infty} \left[\frac{1}{2} + \frac{1}{\pi} \arctan(\alpha x) \right] \quad (A.11\text{-}3)$$

$$U(x) = \lim_{\alpha \to \infty} \tfrac{1}{2}[\text{erf}(\alpha x) + 1] \quad (A.11\text{-}4)$$

$$U(x) = \lim_{\alpha \to \infty} \frac{1}{\pi} \int_{-\infty}^{\alpha x} \frac{\sin \tau}{\tau} d\tau \quad (A.11\text{-}5)$$

(c) **Fourier-integral Representations.** The complex contour integral $\dfrac{1}{2\pi i} \displaystyle\int_{-\infty}^{\infty} \dfrac{e^{i\omega t}}{\omega} d\omega$ is respectively equal to $U(t)$ or $-U(-t)$ if the integration contour passes *below* or *above* the origin. The Cauchy principal value of the integral equals $U(x) - \tfrac{1}{2}$. Note also

$$U(t) = \frac{1}{2\pi} \int_{-\infty}^{\infty} \frac{\sin \omega t}{\omega} d\omega + \frac{1}{2} \quad (A.11\text{-}6)$$

$$U(1 - t) = \frac{1}{\pi} \int_{-\infty}^{\infty} \frac{\sin \omega \cos \omega t}{\omega} d\omega \quad (t \geq 0) \quad (A.11\text{-}7)$$

* The notations employed to denote the various unit-step functions vary; use caution when referring to different texts.

STEP FUNCTIONS AND SYMBOLIC IMPULSE FUNCTIONS   A–33

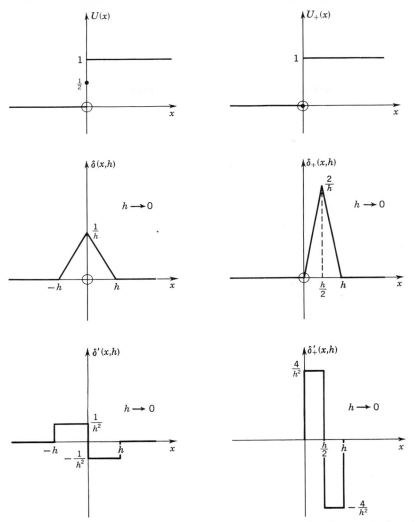

FIG. A.11-1. The unit-step functions $U(x)$ and $U_+(x)$ and approximations to the impulse functions $\delta(x)$, $\delta_+(x)$, $\delta'(x)$, and $\delta_+'(x)$.

A.11-2. The Symbolic Dirac Delta Function. The **symmetrical unit-impulse function** or **Dirac delta function** $\delta(x)$ of a real variable x is "defined" by

$$\int_a^b f(\xi)\delta(\xi - X)\,d\xi = \begin{cases} 0 \text{ if } X < a \text{ or } X > b \\ \tfrac{1}{2}f(X) \text{ if } X = a \text{ or } X = b \\ f(X) \text{ if } a < X < b \end{cases} \qquad (a < b) \qquad (A.11\text{-}8a)$$

where $f(x)$ is an arbitrary function continuous for $x = X$. More generally, one

A-34 MATHEMATICAL FORMULAS, THEOREMS, AND DEFINITIONS

"defines" $\delta(x)$ by

$$\int_a^b f(\xi)\delta(\xi - X)\,d\xi = \begin{cases} 0 \text{ if } X < a \text{ or } X > b \\ \tfrac{1}{2}f(X + 0) \text{ if } X = a \\ \tfrac{1}{2}f(X - 0) \text{ if } X = b \\ \tfrac{1}{2}[f(X - 0) + f(X + 0)] \text{ if } a < X < b \end{cases} \quad (a < b) \quad \text{(A.11-8}b\text{)}$$

where $f(x)$ is an arbitrary function of bounded variation in a neighborhood of $x = X$. $\delta(x)$ is *not* a true function, since the "definition" (8) implies the inconsistent relations

$$\delta(x) = 0 \quad (x \neq 0) \qquad \int_{-\infty}^{\infty} \delta(\xi)\,d\xi = 1 \qquad \text{(A.11-8}c\text{)}$$

$\delta(x)$ is a "symbolic function" permitting the formal representation of the functional identity transformation $f(\xi) \to f(x)$ as an integral transformation. The "formal" use of $\delta(x)$ furnishes a convenient notation permitting suggestive generalizations of many mathematical relations. Although no functions having the exact properties (8) exist, it is possible to "approximate" $\delta(x)$ by functions exhibiting the desired properties to any degree of approximation.

One can usually avoid the use of impulse functions by introducing Stieltjes integrals. It is possible to introduce a generalizing redefinition of the concepts of "function" and "differentiation" (*Schwarz's theory of distributions*). Otherwise, *mathematical arguments involving the use of impulse functions should be regarded as heuristic and require rigorous verification.*

A.11-3. "Derivatives" of Step Functions and Impulse Functions. Equations (8) and also the relation $\mathcal{L}[\delta(t - a)] = e^{-as} = s\mathcal{L}[U(t - a)] \; (a > 0)$ suggest the symbolic relationship

$$\delta(x) = \frac{d}{dx} U(x) \qquad \text{(A.11-9)}$$

The impulse functions $\delta'(x)$, $\delta''(x)$, . . . , $\delta^{(r)}(x)$ are "defined" by

$$\int_a^b f(\xi)\delta^{(r)}(\xi - X)\,d\xi = \begin{cases} 0 \text{ if } X < a \text{ or } X > b \\ \tfrac{1}{2}(-1)^r f^{(r)}(X + 0) \text{ if } x = a \\ \tfrac{1}{2}(-1)^r f^{(r)}(X - 0) \text{ if } x = b \\ \tfrac{1}{2}(-1)^r [f^{(r)}(X - 0) + f^{(r)}(X + 0)] \text{ if } a < X < b \end{cases}$$
$$(a < b) \quad \text{(A.11-10)}$$

for an arbitrary function $f(x)$ such that the unilateral limits $f^{(r)}(X - 0)$ and $f^{(r)}(X + 0)$ exist. The functions $\delta^{(r)}(\xi - X)$ are kernels of linear integral transformations representing repeated differentiations.

A.11-4. Approximation of Impulse Functions (see also Fig. 1.11-1). (*a*) **Continuously Differentiable Functions Approximating** $\delta(x)$. One can approximate $\delta(x)$ by the continuously differentiable functions

$$\delta(x, \alpha) = \frac{\alpha}{(\pi\alpha^2 x^2 + 1)} \quad \text{as } \alpha \to \infty \qquad \text{(A.11-11}a\text{)}$$

$$\delta(x, \alpha) = \frac{\alpha}{\sqrt{\pi}} e^{-\alpha^2 x^2} \quad \text{as } \alpha \to \infty \qquad \text{(A.11-11}b\text{)}$$

$$\delta(x, \alpha) = \frac{\alpha}{\pi} \frac{\sin \alpha x}{\alpha x} \quad \text{as } \alpha \to \infty \qquad \text{(A.11-11}c\text{)}$$

in the sense that $\lim_{\alpha \to \infty} \delta(x, \alpha) = 0$ $(x \neq 0)$, and

$$\lim_{\alpha \to \infty} \int_{-\infty}^{\infty} f(\xi)\delta(x - \xi, \alpha)\,d\xi = \tfrac{1}{2}[f(x - 0) + f(x + 0)]$$

wherever $f(x - 0)$ and $f(x + 0)$ exist; note also $\lim_{\alpha \to \infty} \int_{-\infty}^{\infty} \delta(\xi, \alpha)\,d\xi = 1$.

Integration of the approximating functions (11) yields the corresponding step-function approximations (3) and (4). $\mathcal{L}[\delta(x - a, \alpha)]$ $(a > 0)$ converges to

$$e^{-as} = \mathcal{L}[\delta(x - a)]$$

as $\alpha \to \infty$ for each function (11).

(b) **Discontinuous Functions Approximating** $\delta(x)$. $\delta(x)$ is often approximated by the central-difference coefficient

$$\delta(x, h) = \frac{U(x + h) - U(x - h)}{2h} \qquad \text{as } h \to 0 \qquad \text{(A.11-12)}$$

A.11-5. Fourier-integral Representations. Note the formal relations

$$\delta(x - X) = \frac{1}{2\pi} \int_{-\infty}^{\infty} e^{-i\omega X} e^{i\omega x}\,d\omega \qquad \text{(A.11-13)}$$

$$\delta^{(r)}(x - X) = \frac{1}{2\pi} \int_{-\infty}^{\infty} (i\omega)^r e^{-i\omega X} e^{i\omega x}\,d\omega \qquad \text{(A.11-14)}$$

$$\tfrac{1}{2}[\delta(x - X) + \delta(x + X)] = \frac{1}{\pi} \int_{0}^{\infty} \cos \omega X \cos \omega x\,d\omega \qquad \text{(A.11-15)}$$

A.11-6. Asymmetrical Impulse Functions. The **asymmetrical impulse functions** $\delta_+(x), \delta_+'(x), \ldots, \delta_+^{(r)}(x)$ are "defined" by

$$\int_{a+0}^{b} f(\xi)\delta_+(\xi - X)\,d\xi = \begin{Bmatrix} 0 & \text{if } X < a \text{ or } X \geq b \\ f(X + 0) & \text{if } a \leq X < b \end{Bmatrix} \quad (a < b) \quad \text{(A.11-16)}$$

$$\int_{a+0}^{b} f(\xi)\delta_+^{(r)}(\xi - X)\,d\xi = \begin{Bmatrix} 0 \text{ if } X < a \text{ or } X \geq b \\ (-1)^r f^{(r)}(X + 0) \text{ if } a \leq X < b \end{Bmatrix}$$

$$(a < b;\ r = 1, 2, \ldots) \quad \text{(A.11-17)}$$

One may write

$$\delta_+(x) \equiv 2\delta(x)U(x) \equiv \frac{d}{dx} U_+(x) \qquad \text{(A.11-18)}$$

One way to obtain approximation functions for $\delta_+(x)$ is to substitute the approximation function of Sec. A.11-4 into one of the relations (18), e.g.,

$$\delta_+(x, h) = \frac{U(x) - U(x - h)}{h} \qquad \text{as } h \to 0 \qquad \text{(A.11-19)}$$

A.12 FOURIER AND LAPLACE TRANSFORMS

A.12-1. Fourier Analysis (Harmonic Analysis) of Periodic Functions. Let $f(t)$ be a real periodic function with period T and such that $\int_{-T/2}^{T/2} |f(\tau)|\,d\tau$ exists. Then

$$f(t) = \tfrac{1}{2}a_0 + \sum_{k=1}^{\infty}(a_k \cos k\omega_0 t + b_k \sin k\omega_0 t)$$

$$= \sum_{k=-\infty}^{\infty} c_k e^{ik\omega_0 t} = c_0 + 2\sum_{k=1}^{\infty}|c_k|\cos(k\omega_0 t + \arg c_k)$$

$$a_k = \frac{2}{T}\int_{-T/2}^{T/2} f(\tau)\cos k\omega_0 \tau\, d\tau$$

$$b_k = \frac{2}{T}\int_{-T/2}^{T/2} f(\tau)\sin k\omega_0 \tau\, d\tau \qquad\qquad\text{(A.12-1)}$$

$$c_k = c_{-k}{}^* = \frac{1}{T}\int_{-T/2}^{T/2} f(\tau)e^{-ik\omega_0 \tau}\, d\tau$$

$$\left(\omega = \frac{2\pi}{T};\ k = 0, 1, 2, \ldots\right)$$

throughout every open interval where $f(t)$ is of bounded variation, if one defines $f(t) = \tfrac{1}{2}[f(t-0)+f(t+0)]$ at each discontinuity. $f(t)$ is thus expressed as the sum of

1. A **constant term** $a_0/2 = c_0$ [*average value of $f(t)$*], and
2. A set of *sinusoidal terms* (*sinusoidal components*) of respective **frequencies** $\nu_0 = 1/T$ (**fundamental** frequency), $2\nu_0 = 2/T$ (**2nd-harmonic** frequency),

$$3\nu_0 = \frac{3}{T}$$

(**3rd-harmonic** frequency), ...

The **kth-harmonic component** $2|c_k|\cos\left(k\dfrac{2\pi t}{T} + \arg c_k\right)$ has the frequency $k\nu_0 = k/T$, the **circular frequency** $k\omega_0 = 2\pi k\nu_0 = 2\pi k/T$, the **amplitude**

$$2|c_k| = +\sqrt{a_k{}^2 + b_k{}^2}$$

and the "phase angle" $\arg c_k = -\arctan(b_k/a_k)$.

A.12-2. The Laplace Transformation. The (one-sided) Laplace transformation

$$F(s) \equiv \mathcal{L}[f(t)] \equiv \int_0^{\infty} f(t)e^{-st}\, dt \equiv \lim_{\substack{a\to 0\\ b\to\infty}} \int_a^b f(t)e^{-st}\, dt \qquad (0 < a < b) \quad\text{(A.12-2)}$$

associates a unique **result or image function** $F(s)$ of the complex variable $s = \sigma + i\omega$ with every single-valued **object or original function** $f(t)$ (t real) such that the improper integral (2) exists. $F(s)$ is called the (**one-sided**) **Laplace transform** of $f(t)$.

The Laplace transform (2) exists for $\sigma \geq \sigma_0$, and the improper integral converges absolutely and uniformly to a function $F(s)$ analytic for $\sigma > \sigma_0$ if

$$\int_0^{\infty}|f(t)|e^{-\sigma t}\, dt = \lim_{\substack{a\to 0\\ b\to\infty}}\int_a^b |f(t)|e^{-\sigma t}\, dt \qquad (0 < a < b) \quad\text{(A.12-3)}$$

exists for $\sigma = \sigma_0$. The greatest lower bound σ_a of the real numbers σ_0 for which this is true is called the **abscissa of absolute convergence** of the Laplace transform $\mathcal{L}[f(t)]$. The region of definition of the analytic function

$$F(s) = \mathcal{L}[f(t)] \qquad (\sigma > \sigma_a)$$

can usually be extended by analytic continuation as to include the entire s plane with the exception of singular points situated to the left of the abscissa of absolute convergence.

Table A.12-1. Fourier Coefficients and Mean-square Values of Periodic Functions‡ $\left[s(x) \equiv \text{sinc } x \equiv \dfrac{\sin \pi x}{\pi x} \right]$

	Periodic function, $f(t) = f(t + T)$		Fourier coefficients (for phasing as shown in diagram)	Average value $\langle f \rangle = \dfrac{a_0}{2}$	Mean-square value $\langle f^2 \rangle$
1	Rectangular pulses	(rectangular pulse waveform, width T_0, height A)	$a_n = 2A \dfrac{T_0}{T} s\left(\dfrac{nT_0}{T}\right)$ $b_n = 0$	$A \dfrac{T_0}{T}$	$A^2 \dfrac{T_0}{T}$
2	Symmetrical triangular pulses	(triangular pulse waveform)	$a_n = A \dfrac{T_0}{T} s^2\left(\dfrac{nT_0}{2T}\right)$ $b_n = 0$	$A \dfrac{T_0}{2T}$	$A^2 \dfrac{T_0}{3T}$
3	Symmetrical trapezoidal pulses	(trapezoidal pulse waveform)	$a_n = 2A \dfrac{T_0 + T_1}{T} s\left(\dfrac{nT_1}{T}\right)$ $\cdot s\left[\dfrac{n(T_0 + T_1)}{T}\right]$ $b_n = 0$	$A \dfrac{T_0 + T_1}{T}$	$A^2 \dfrac{3T_0 + 2T_1}{3T}$
4	Half-sine pulses*†	(half-sine pulse waveform)	$a_n = A \dfrac{T_0}{T} \left\{ s\left[\dfrac{1}{2}\left(\dfrac{2nT_0}{T} - 1\right)\right] \right.$ $\left. + s\left[\dfrac{1}{2}\left(\dfrac{2nT_0}{T} + 1\right)\right] \right\}$ $b_n = 0$	$\dfrac{2}{\pi} A \dfrac{T_0}{T}$	$A^2 \dfrac{T_0}{2T}$
5	Clipped sinusoid $A = A_0\left(1 - \cos \dfrac{\pi T_0}{T}\right)$	(clipped sinusoid waveform, $2A_0$)	$a_n = \dfrac{A_0 T_0}{T} \left\{ s\left[(n-1)\dfrac{T_0}{T}\right] + \right.$ $\left. s\left[(n+1)\dfrac{T_0}{T}\right] - 2\cos\dfrac{\pi nT_0}{T} s\left(\dfrac{nT_0}{T}\right) \right\}$	$\dfrac{1}{\pi} A_0 \left(\sin\dfrac{\pi T_0}{T} - \dfrac{\pi T_0}{T}\cos\dfrac{\pi T_0}{T}\right)$	$\dfrac{1}{2\pi} A_0^2 \left(\dfrac{\pi T_0}{T} - \dfrac{3}{2}\sin\dfrac{2\pi T_0}{T}\right.$ $\left. + \dfrac{2\pi T_0}{T}\cos^2\dfrac{\pi T_0}{T}\right)$
6	Triangular waveform	(sawtooth/triangular waveform)	$a_n = 0$ $b_n = -\dfrac{A}{n\pi}\bigg\} n = 1, 2, \ldots$	$\dfrac{A}{2}$	$\dfrac{A^2}{3}$

* For $T_0 = \dfrac{T}{2} = \dfrac{\pi}{\omega}$, $f(t) = \dfrac{2}{\pi} A \left(\dfrac{1}{2} + \dfrac{\pi}{4}\cos \omega t + \dfrac{1}{3}\cos 2\omega t - \dfrac{1}{15}\cos 4\omega t + \dfrac{1}{35}\cos 6\omega t \pm \cdots\right)$ (HALF-WAVE RECTIFIED SINUSOID).

† For $T_0 = T = \dfrac{2\pi}{\omega}$, $f(t) = \dfrac{4}{\pi} A \left(\dfrac{1}{2} + \dfrac{1}{3}\cos 2\omega t - \dfrac{1}{15}\cos 4\omega t + \dfrac{1}{35}\cos 6\omega t \pm \cdots\right)$ (FULL-WAVE RECTIFIED SINUSOID).

‡ From G. A. Korn and T. M. Korn, "Mathematical Handbook for Scientists and Engineers," McGraw-Hill Book Company, New York, 1961.

Table A.12-2. Properties of Fourier Transforms*

Let
$$\mathcal{F}[f(t)] \equiv \int_{-\infty}^{\infty} f(t)e^{-2\pi i\nu t}\,dt = c(\nu) = \sqrt{2\pi}\,C(2\pi\nu)$$

$$f(t) = \int_{-\infty}^{\infty} c(\nu)e^{2\pi i\nu t}\,d\nu = \frac{1}{\sqrt{2\pi}} \int_{-\infty}^{\infty} C(\omega)e^{i\omega t}\,d\omega$$

$$\mathcal{F}_C[f(t)] \equiv 2 \int_{0}^{\infty} f(t)\cos 2\pi\nu t\,dt = c_C(\nu)$$

$$\mathcal{F}_S[f(t)] \equiv 2 \int_{0}^{\infty} f(t)\sin 2\pi\nu t\,dt = c_S(\nu)$$

and assume that the Fourier transforms in question exist.

(a) $\mathcal{F}[\alpha f_1(t) + \beta f_2(t)] = \alpha \mathcal{F}[f_1(t)] + \beta \mathcal{F}[f_2(t)]$ (LINEARITY)

$\mathcal{F}[f^*(t)] = c^*(-\nu)$

$\mathcal{F}[f(\alpha t)] = \dfrac{1}{\alpha} c\left(\dfrac{\nu}{\alpha}\right)$ (CHANGE OF SCALE, SIMILARITY THEOREM)

$\mathcal{F}[f(t + \tau)] = e^{2\pi i\nu\tau} c(\nu)$ (SHIFT THEOREM)

(b) $\mathcal{F}[f(t, \alpha)] \to \mathcal{F}[f(t)]$ as $\alpha \to a$ implies $f(t, \alpha) \to f(t)$ wherever $f(t)$ is continuous (*Continuity Theorem*). Analogous theorems apply to Fourier cosine and sine transforms.

(c) $\left. \begin{array}{l} \mathcal{F}[f_1(t)]\mathcal{F}[f_2(t)] = \mathcal{F}[f_1(t) * f_2(t)] \\ \mathcal{F}[f_1(t)f_2(t)] = \mathcal{F}[f_1(t)] * \mathcal{F}[f_2(t)] \end{array} \right\}$ (BOREL'S CONVOLUTION THEOREM)

where $f_1(t) * f_2(t) = \int_{-\infty}^{\infty} f_1(\tau)f_2(t-\tau)\,d\tau = \int_{-\infty}^{\infty} f_1(t-\tau)f_2(\tau)\,d\tau$

and $c_1(\nu) * c_2(\nu) = \int_{-\infty}^{\infty} c_1(\lambda)c_2(\nu-\lambda)\,d\lambda = \int_{-\infty}^{\infty} c_1(\nu-\lambda)c_2(\lambda)\,d\lambda$

$\left. \begin{array}{l} \mathcal{F}[f(t)e^{2\pi i\nu_0 t}] = c(\nu - \nu_0) \\ \mathcal{F}[f(t)\cos 2\pi\nu_0 t] = \tfrac{1}{2}[c(\nu - \nu_0) + c(\nu + \nu_0)] \\ \mathcal{F}[f(t)\sin 2\pi\nu_0 t] = \dfrac{1}{2i}[c(\nu - \nu_0) - c(\nu + \nu_0)] \end{array} \right\}$ (MODULATION THEOREM)

(d) $\mathcal{F}[f^{(r)}(t)] = (2\pi i\nu)^r \mathcal{F}[f(t)]$ ($r = 0, 1, 2, \ldots$) (DIFFERENTIATION THEOREM) provided that $f^{(r)}(t)$ exists for all t, and that all derivatives of lesser order vanish as $|t| \to \infty$.

(e) If $\int_{-\infty}^{\infty} |f_1(t)|^2\,dt$ and $\int_{-\infty}^{\infty} |f_2(t)|^2\,dt$ exist,

$\int_{-\infty}^{\infty} \mathcal{F}^*[f_1(t)]\mathcal{F}[f_2(t)]\,d\nu = \int_{-\infty}^{\infty} f_1^*(t)f_2(t)\,dt$ (PARSEVAL'S THEOREM)

* From G. A. Korn and T. M. Korn, "Mathematical Handbook for Scientists and Engineers," McGraw-Hill Book Company, New York, 1961.

If the **inverse Laplace transform** $f(t)$ corresponding to a given $F(s)$ exists, it is unique wherever it is continuous (*Lerch's theorem*). If the limit exists,

$$\frac{1}{2\pi i} \lim_{R \to \infty} \int_{\sigma_1 - iR}^{\sigma_1 + iR} F(s)e^{st}\,ds = \begin{cases} \tfrac{1}{2}[f(t-0) + f(t+0)] & \text{for } t > 0 \\ \tfrac{1}{2}f(0+0) & \text{for } t = 0 \\ 0 & \text{for } t < 0 \end{cases} \quad (\sigma_1 > \sigma_a) \quad \text{(A.12-4)}$$

where the integration contour lies to the right of all singularities of $F(s)$.

Table A.12-3. Theorems Relating Corresponding Operations on Object and Result Functions‡

The following theorems are valid whenever the Laplace transforms $F(s) = \mathcal{L}[f(t)]$ in question exist in the sense of absolute convergence

Theorem number	Operation	Object function	Result function
1	**Linearity** (α, β constant)	$\alpha f_1(t) + \beta f_2(t)$	$\alpha F_1(s) + \beta F_2(s)$
	Differentiation of object function*		
2a	... if $f'(t)$ exists for all $t > 0$	$f'(t)$	$sF(s) - f(0+0)$
2b	... if $f^{(r)}(t)$ exists for all $t > 0$	$f^{(r)}(t)$ ($r = 1, 2, \ldots$)	$s^r F(s) - s^{r-1}f(0+0) - s^{r-2}f'(0+0) - \cdots - f^{(r-1)}(0+0)$
2c	... if $f(t)$ is bounded for $t > 0$, and $f'(t)$ exists for $t > 0$ except for $t = t_1, t_2, \ldots$ where $f(t)$ has unilateral limits	$f'(t)$	$sF(s) - f(0+0) - \sum_k e^{-t_k s}[f(t_k+0) - f(t_k-0)]$
3	**Integration of object function** ... if $f'(t)$ exists for $t > 0$	$\int_0^t f(\tau)\, d\tau + C$	$\dfrac{F(s)}{s} + \dfrac{C}{s}$
4	**Change of scale**	$f(at)$ ($a > 0$)	$\dfrac{1}{a} F\left(\dfrac{s}{a}\right)$
5	**Translation (shift) of object function** ... if $f(t) = 0$ for $t \leq 0$	$f(t - b)$ ($b \geq 0$)	$e^{-bs}F(s)$
6	**Convolution of object functions**†	$f_1 * f_2 \equiv \int_0^\infty f_1(\tau) f_2(t - \tau)\, d\tau \equiv f_2 * f_1$	$F_1(s) F_2(s)$

Table A.12-3. Theorems Relating Corresponding Operations on Object and Result Functions‡ (*Continued*)

Theorem number	Operation	Object function	Result function
7	Corresponding limits of object and result function (continuity theorem; α is independent of t and s)	$\lim_{\alpha \to a} f(t, \alpha)$	$\lim_{\alpha \to a} F(s, \alpha)$
8a	Differentiation and integration with respect to a parameter α independent of t and s	$\dfrac{\partial}{\partial \alpha} f(t, \alpha)$	$\dfrac{\partial}{\partial \alpha} F(s, \alpha)$
8b		$\displaystyle\int_{a_1}^{a_2} f(t, \alpha) \, d\alpha$	$\displaystyle\int_{a_1}^{a_2} F(s, \alpha) \, d\alpha$
9a	Differentiation of result function	$-t f(t)$	$F'(s)$
9b		$(-1)^r t^r f(t)$	$F^{(r)}(s)$
10	Integration of result function (path of integration situated to the right of the abscissa of absolute convergence)	$\dfrac{1}{t} f(t)$	$\displaystyle\int_s^\infty F(s) \, ds$
11	Translation of result function	$e^{at} f(t)$	$F(s - a)$

* The abscissa of absolute convergence for $\mathcal{L}[f^{(r)}(t)]$ is 0 or σ_a, whichever is greater.

† The existence of $f_1 * f_2$ is assumed; absolute convergence of $\mathcal{L}[f_1(t)]$ and $\mathcal{L}[f_2(t)]$ is a sufficient condition for the absolute convergence of $\mathcal{L}[f_1 * f_2]$.

‡ From G. A. Korn and T. M. Korn, "Mathematical Handbook for Scientists and Engineers," McGraw-Hill Book Company, New York, 1961.

A.12-3. Solution of Linear Differential Equations with Constant Coefficients. (a) To solve a linear differential equation

$$\mathsf{L}y \equiv a_0 \frac{d^r y}{dt^r} + a_1 \frac{d^{r-1} y}{dt^{r-1}} + \cdots + a_r y = f(t) \quad (A.12\text{-}5)$$

with given initial values $y(0 + 0)$, $y'(0 + 0)$, $y''(0 + 0)$, ..., $y^{(r-1)}(0 + 0)$, apply the Laplace transformation to both sides, and let $\mathcal{L}[y(t)] \equiv Y(s)$, $\mathcal{L}[f(t)] \equiv F(s)$. The resulting linear *algebraic* equation (*subsidiary equation*)

$$\begin{aligned}(a_0 s^r + a_1 s^{r-1} + \cdots + a_r) Y(s) &= F(s) + G(s) \\ G(s) \equiv y(0 + 0)(a_0 s^{r-1} + a_1 s^{r-2} + \cdots &+ a_{r-1}) \\ + y'(0 + 0)(a_0 s^{r-2} + a_1 s^{r-3} + \cdots &+ a_{r-2}) \\ + \cdots & \\ + y^{(r-2)}(0 + 0)(a_0 s + a_1) + a_0 y^{(r-1)}&(0 + 0) \end{aligned} \quad (A.12\text{-}6)$$

is easily solved to yield the Laplace transform of the desired solution $y(t)$ in the form

$$Y(s) = \frac{F(s)}{a_0 s^r + a_1 s^{r-1} + \cdots + a_r} + \frac{G(s)}{a_0 s^r + a_1 s^{r-1} + \cdots + a_r} \quad (A.12\text{-}7)$$

Here the first term is the Laplace transform $Y_N(s)$ of the "normal response" $y_N(t)$ and the second term represents the effects of nonzero initial values of $y(t)$ and its derivatives. The solution $y(t)$ and $y_N(t)$ are found as inverse Laplace transforms by reference to tables.

(b) In the same manner, one applies the Laplace transformation to a system of linear differential equations

$$\varphi_{j1}\left(\frac{d}{dt}\right) y_1 + \varphi_{j2}\left(\frac{d}{dt}\right) y_2 + \cdots + \varphi_{jn}\left(\frac{d}{dt}\right) y_n = f_j(t)$$
$$(j = 1, 2, \ldots, n) \quad (A.12\text{-}8)$$

to obtain

$$\varphi_{j1}(s) Y_1(s) + \varphi_{j2}(s) Y_2(s) + \cdots + \varphi_{jn}(s) Y_n(s) = F_j(s) + G_j(s)$$
$$(j = 1, 2, \ldots, n) \quad (A.12\text{-}9)$$

where the functions $G_j(s)$ depend on the given initial conditions. The linear *algebraic* equations (8) are solved by Cramer's rule to yield the unknown solution transforms.

A.13 PERMUTATIONS, COMBINATIONS, AND RELATED TOPICS

Table A.13-1. Permutations and Partitions

1	Number of different orderings (**permutations**) of a set of n distinct objects	$n!$
2	i. Number of distinguishable sequences of N objects comprising $n \leq N$ indistinguishable objects of type 1 and $N - n$ indistinguishable objects of type 2, **or** ii. Number of distinguishable *partitions* of a set of N distinct objects into 2 classes of $n \leq N$ and $N - n$ objects, respectively	$\binom{N}{n} = \dfrac{N!}{(N-n)!n!}$ (**binomial coefficient**, Sec. 21.5-1)
3	i. Number of distinguishable sequences of $N = N_1 + N_2 + \cdots + N_r$ objects comprising N_1 indistinguishable objects of type 1, N_2 indistinguishable objects of type 2, ..., and N_r indistinguishable objects of type r, **or** ii. Number of distinguishable partitions of a set of $N = N_1 + N_2 + \cdots + N_r$ distinct objects into r classes of $N_1, N_2, \ldots, N_r$ objects, respectively	$\dfrac{N!}{N_1! N_2! \cdots N_r!}$ (**multinomial coefficient**)

Table A.13-2. Combinations and Samples
Each formula holds for $N < n$, $N = n$, and $N > n$

1	Number of distinguishable unordered **combinations** of N distinct types of objects taken n at a time: i. Each type of object may occur *at most once* in any combination (*combinations without repetition;* see also Table A.13-1, 2) ii. Each type of object may occur 0, 1, 2, ..., or n times in any combination (*combinations with repetition*) iii. Each type of object must occur *at least once* in each combination	$\binom{N}{n}$ $\binom{N+n-1}{n} = \binom{N+n-1}{N-1}$ $\binom{n-1}{N-1}$
2	Number of distinguishable **samples** (sequences, ordered sets, variations) of size n taken from a population of N distinct types of objects: i. Each type of object may occur *at most once* in any sample (*samples without replacement,* sequences without repetition) ii. Each type of object may occur 0, 1, 2, ..., or n times in any sample (*samples with replacement,* sequences with repetition)	$N(N-1) \cdots (N-n+1)$ $= \binom{N}{n} n!$ N^n

EXAMPLES: Given a set of $N = 3$ distinct types of elements a, b, c. For $n = 2$, there exist 3 *combinations without repetition* (ab, ac, bc); 6 *combinations with repetition* (aa, ab, ac, bb, bc, cc); 6 distinguishable *samples without replacement* (ab, ac, ba, bc, ca, cb); and 9 distinguishable *samples with replacement* ($aa, ab, ac, ba, bb, bc, ca, cb, cc$).

Table A.13-3. Occupancy of Cells or States
Each formula holds for $N < n$, $N = n$, and $N > n$

1	Number of distinguishable arrangements of n *indistinguishable* objects in N distinct cells (states): i. *No cell may contain more than one* object ii. Any cell may contain 0, 1, 2, ..., or n objects iii. Each cell must contain *at least one* object	$\binom{N}{n}$ $\binom{N+n-1}{n} = \binom{N+n-1}{N-1}$ $\binom{n-1}{N-1}$
2	Number of distinguishable arrangements of n *distinct* objects in N distinct cells: i. *No cell may contain more than one* object ii. Any cell may contain 0, 1, 2, ..., or n objects	$N(N-1) \cdots (N-n+1)$ $= \binom{N}{n} n!$ N^n

See also J. Riordan, "An Introduction to Combinatorial Analysis," John Wiley & Sons, Inc., New York, 1958.

A.14 PROBABILITY DISTRIBUTIONS

A.14-1. Discrete One-dimensional Probability Distributions. Tables A.14-1 to A.14-6 describe a number of discrete one-dimensional distributions of interest, for instance, in connection with sampling problems and games of chance. The generating function rather than the characteristic function or the moment-generating function is tabulated: the latter two functions are easily obtained from

$$\chi_x(q) \equiv \gamma_x(e^{iq}) \qquad M_x(s) \equiv \gamma_x(e^s) \qquad (A.14-1)$$

Table A.14-1. The Causal Distribution

$$p(x) = \delta_\xi{}^x = \begin{cases} 1 \text{ if } x = \xi \\ 0 \text{ if } x \neq \xi \end{cases} \qquad (x = 0, \pm 1, \pm 2, \ldots; \xi \text{ is a real integer})$$

$$E\{x\} = \xi \qquad \text{Var } \{x\} = 0 \qquad \gamma_x(s) \equiv s^\xi$$

Table A.14-2. The Hypergeometric Distribution

$$p(x) = \frac{\binom{N_1}{x}\binom{N - N_1}{n - x}}{\binom{N}{n}} \qquad (x = 0, 1, 2, \ldots, n; N \geq n \geq 0, \ N \geq N_1 = \vartheta N \geq 0)$$

$$E\{x\} = \frac{nN_1}{N} = n\vartheta \qquad \text{Var } \{x\} = \frac{nN_1(N - N_1)}{N^2}\left(1 - \frac{n-1}{N-1}\right)$$

$$= n\vartheta(1 - \vartheta)\left(1 - \frac{n-1}{N-1}\right)$$

Typical Interpretation. $p(x)$ is the probability that a random sample of size n *without replacement* contains exactly x objects of type 1, if the sample is taken from a population of N objects of which $N_1 = \vartheta N$ are of type 1.

Approximations. As $N \to \infty$ while n and $\vartheta = N_1/N$ remain fixed, the hypergeometric distribution approaches a *binomial distribution* (Table A.14-4; sampling with and without replacement becomes approximately equivalent if n/N is small). The binomial approximation is usually permissible if $n/N < 0.1$. The binomial distribution may, in turn, be suitable for approximation by a *normal distribution* (see Table A.14-4) or by a *Poisson distribution* (Table A.14-4).

Table A.14-3. The Poisson Distribution (Fig. A.14-1)

$$p(x) = e^{-\xi}\frac{\xi^x}{x!} \qquad (x = 0, 1, 2, \ldots; \xi > 0)$$

$$E\{x\} = \text{Var } \{x\} = \xi \qquad \gamma_x(s) \equiv e^{\xi s - \xi}$$

The Poisson distribution approximates a hypergeometric distribution (Table A.14-2) or a binomial distribution (Table A.14-4) as $\vartheta N \to \infty$, $n \to \infty$, $\vartheta/n \to 0$ in such a manner that ϑn has a finite limit ξ (*Law of Small Numbers*). The approximation is often useful for $\vartheta \leq 0.1$, $n\vartheta \geq 1$.

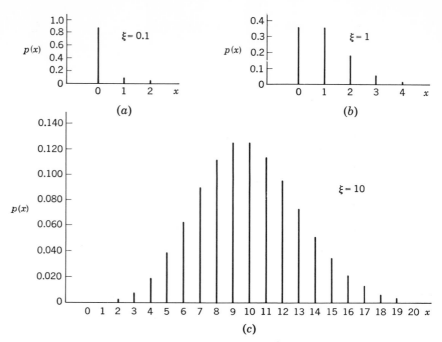

FIG. A.14-1. The Poisson distribution. (*From H. H. Goode, and R. E. Machol, "System Engineering," McGraw-Hill Book Company, New York, 1957.*)

Table A.14-4. The Binomial Distribution (Fig. A.14-2)

$$p(x) = \binom{n}{x} \vartheta^x (1-\vartheta)^{n-x} \qquad (x = 0, 1, 2, \ldots; 0 \leq \vartheta \leq 1)$$

$$E\{x\} = n\vartheta \qquad \text{Var}\{x\} = n\vartheta(1-\vartheta) \qquad \gamma_x(s) \equiv (\vartheta s + 1 - \vartheta)^n$$

Typical Interpretation. $p(x)$ is

1. The probability that a random sample of size n *with replacement* contains exactly x objects of type 1 if the sample is taken from a population of N objects of which ϑN are of type 1.
2. The probability of realizing an event ("success") exactly x times in n independent repeated trials (Bernoulli trials) such that the probability of success in each trial is ϑ.

Approximations. As $n \to \infty$, the binomial variable x is asymptotically normal with mean $n\vartheta = \xi$, and variance $n\vartheta(1-\vartheta) = \sigma^2$ (*De Moivre–Laplace Limit Theorem;* a special case of the Central Limit Theorem).

For $0 < \vartheta < 1$,

$$p(x) = \binom{n}{x} \vartheta^x (1-\vartheta)^{n-x} \simeq \varphi_u\left(\frac{x-\xi}{\sigma}\right)$$

$$\simeq \Phi_u\left(\frac{x + \frac{1}{2} - \xi}{\sigma}\right) - \Phi_u\left(\frac{x - \frac{1}{2} - \xi}{\sigma}\right) \qquad \text{as } \frac{(x-\xi)^3}{\sigma^4} \to 0$$

$$P[X_1 \leq x \leq X_2] \simeq \Phi_u\left(\frac{X_2 + \frac{1}{2} - \xi}{\sigma}\right) - \Phi_u\left(\frac{X_1 - \frac{1}{2} - \xi}{\sigma}\right)$$

$$\text{as } \frac{(X_1 - \xi)^3}{\sigma^4} \to 0, \frac{(X_2 - \xi)^3}{\sigma^4} \to 0$$

$$P\left[a \leq \frac{x - \xi}{\sigma} \leq b\right] \to \Phi_u(b) - \Phi_u(a) \qquad \text{as } n \to \infty \text{ for fixed } a, b$$

Approximations based on these relations are usually permissible for

$$\sigma^2 = n\vartheta(1 - \vartheta) \geq 9$$

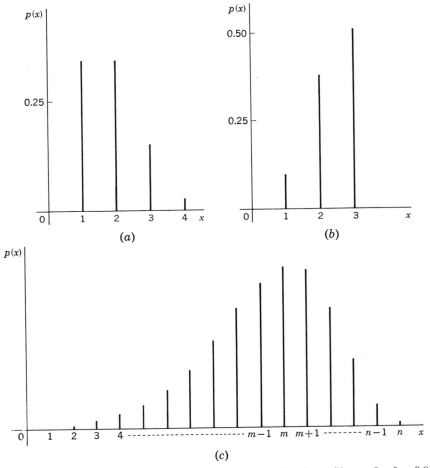

Fig. A.14-2. The binomial distribution. (a) $n = 4$, $\vartheta = 0.4$. (b) $n = 3$, $\vartheta = 0.8$. (c) $n = 16$, $\vartheta = 0.7$. m is the mode. (*From A. M. Mood, and F. A. Graybill, "Introduction to the Theory of Statistics," 2d ed., McGraw-Hill Book Company, New York, 1963.*)

Table A.14-5. The Geometric Distribution

$$p(x) = \vartheta(1 - \vartheta)^x \quad (x = 0, 1, 2, \ldots; 0 \leq \vartheta \leq 1)$$

$$E\{x\} = \frac{1 - \vartheta}{\vartheta} \quad \text{Var}\{x\} = \frac{1 - \vartheta}{\vartheta^2} \quad \gamma_x(s) \equiv \frac{\vartheta}{1 - (1 - \vartheta)s}$$

Typical Interpretation: $p(x)$ is the probability of realizing an event ("success") for the first time *after exactly* x Bernoulli trials with probability of success ϑ. $\Phi(x) = 1 - (1 - \vartheta)^{x+1}$ ($x = 0, 1, 2, \ldots$) is the probability that the first success occurs *after at most* x trials (see also Table A.14-6).

Table A.14-6. Pascal's Distribution

$$p(x) = \binom{m+x-1}{x} \vartheta^m (1-\vartheta)^x$$

$$(x = 0, 1, 2, \ldots; m = 0, 1, 2, \ldots; 0 \leq \vartheta \leq 1)$$

$$E\{x\} = m\frac{1-\vartheta}{\vartheta} \qquad \text{Var}\{x\} = m\frac{1-\vartheta}{\vartheta^2} \qquad \gamma_x(s) \equiv \left(\frac{\vartheta}{1-(1-\vartheta)s}\right)^m$$

Typical Interpretation: $p(x)$ is the probability of realizing an event ("success") for the mth time *after exactly* $m + x - 1$ Bernoulli trials with probability of success ϑ. $\Phi(x)$ is the probability that the mth success occurs *after at most* $m + x - 1$ trials. For $m = 1$ Pascal's distribution reduces to the geometric distribution (Table A.14-5).

A.14-2. Discrete Multidimensional Probability Distributions. (a) A **multinomial distribution** is described by

$$p(x_1, x_2, \ldots, x_n) = \frac{N!}{x_1! x_2! \cdots x_n!} \vartheta_1^{x_1} \vartheta_2^{x_2} \cdots \vartheta_n^{x_n}$$

$$(x_1, x_2, \ldots, x_n = 0, 1, 2, \ldots; x_1 + x_2 + \cdots + x_n = N) \quad \text{(A.14-2)}$$

where $\vartheta_1, \vartheta_2, \ldots, \vartheta_n$ are positive real numbers such that

$$\vartheta_1 + \vartheta_2 + \cdots + \vartheta_n = 1$$

Given an experiment having n mutually exclusive results $E_1, E_2, \ldots, E_n$ with respective probabilities $\vartheta_1, \vartheta_2, \ldots, \vartheta_n$ such that $\vartheta_1 + \vartheta_2 + \cdots + \vartheta_n = 1$, the expression (2) is the probability that the respective events $E_1, E_2, \ldots, E_n$ occur exactly $x_1, x_2, \ldots, x_n$ times in N independent repeated trials. In classical statistical mechanics, $x_1, x_2, \ldots, x_n$ are the occupation numbers of n independent states with respective a priori probabilities $\vartheta_1, \vartheta_2, \ldots, \vartheta_n$.

(b) A **multiple Poisson distribution** is described by

$$p(x_1, x_2, \ldots, x_n) = e^{-(\xi_1 + \xi_2 + \cdots + \xi_n)} \frac{\xi_1^{x_1} \xi_2^{x_2} \cdots \xi_n^{x_n}}{x_1! x_2! \cdots x_n!}$$

$$(x_1, x_2, \ldots, x_n = 0, 1, 2, \ldots; \xi_k > 0, k = 1, 2, \ldots, n) \quad \text{(A.14-3)}$$

A.14-3. Continuous Probability Distributions: The Normal (Gaussian) Distribution. A continuous random variable x is **normally distributed** (normal) **with mean** ξ **and variance** σ^2 [or **normal with parameters** ξ, σ^2; normal with parameters ξ, σ; normal (ξ, σ^2); normal (ξ, σ)] if

$$\varphi(X) \equiv \frac{1}{\sqrt{2\pi}\,\sigma} e^{-\frac{1}{2}\left(\frac{X-\xi}{\sigma}\right)^2} \equiv \frac{1}{\sigma}\varphi_u\left(\frac{X-\xi}{\sigma}\right) \quad \text{(A.14-4)}$$

$$\Phi(X) \equiv \frac{1}{\sqrt{2\pi}\,\sigma} \int_{-\infty}^{X} e^{-\frac{1}{2}\left(\frac{x-\xi}{\sigma}\right)^2} dx \equiv \Phi_u\left(\frac{X-\xi}{\sigma}\right)$$

$$\equiv \frac{1}{2}\left[1 + \text{erf}\left(\frac{1}{\sqrt{2}}\frac{X-\xi}{\sigma}\right)\right] \quad \text{(A.14-5)}$$

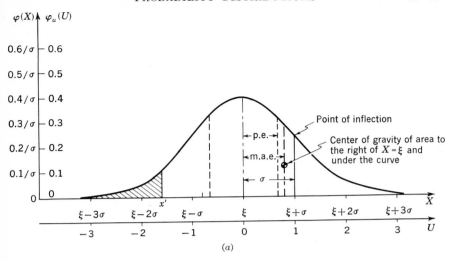

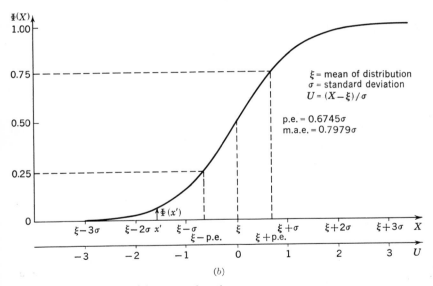

FIG. A.14-3. (a) The normal frequency function

$$\varphi(X) = \frac{1}{\sqrt{2\pi}\,\sigma} e^{-\frac{1}{2}\left(\frac{X-\xi}{\sigma}\right)^2} = \frac{1}{\sigma}\varphi_u(U) \quad \left(U = \frac{X-\xi}{\sigma}\right)$$

and (b) the normal distribution function

$$\Phi(X) = \frac{1}{\sqrt{2\pi}\,\sigma}\int_{-\infty}^{X} e^{-\frac{1}{2}\left(\frac{X-\xi}{\sigma}\right)^2} dx = \Phi_u(U) \quad \left(U = \frac{X-\xi}{\sigma}\right)$$

(From R. S. Burington and D. C. May, "Handbook of Probability and Statistics with Tables," McGraw-Hill Book Company, New York, 1953.)

A–48 MATHEMATICAL FORMULAS, THEOREMS, AND DEFINITIONS

The distribution of the **standardized normal variable (normal deviate)** $u = \dfrac{x - \xi}{\sigma}$ is given by

$$\varphi_u(U) \equiv \frac{1}{\sqrt{2\pi}} e^{-\frac{U^2}{2}} \quad \text{(NORMAL FREQUENCY FUNCTION)} \tag{A.14-6}$$

$$\Phi_u(U) \equiv \frac{1}{\sqrt{2\pi}} \int_{-\infty}^{U} e^{-\frac{U^2}{2}} du \equiv \frac{1}{2}\left[1 + \mathrm{erf}\left(\frac{U}{\sqrt{2}}\right)\right]$$

$$\text{(NORMAL DISTRIBUTION FUNCTION)} \tag{A.14-7}$$
$$E\{u\} = 0 \quad \mathrm{Var}\{u\} = 1 \tag{A.14-8}$$

(see also Fig. A.14-3). erf z is the frequently tabulated **error function (normal error integral, probability integral**; see also Sec. A.10)

$$\mathrm{erf}\, z \equiv -\mathrm{erf}(-z) \equiv \frac{2}{\sqrt{\pi}} \int_0^z e^{-\zeta^2}\, d\zeta \equiv 2\Phi_u(z\sqrt{2}) - 1 \tag{A.14-9}$$

$\varphi(X)$ has points of inflection for $X = \xi \pm \sigma$. Note

$$\chi_x(q) \equiv \exp\left(-\frac{\sigma^2}{2} q^2 + i\xi q\right) \tag{A.14-10}$$

A.14-4. Normal Random Variables: Distribution of Deviations from the Mean. (a) For any normal random variable x with mean ξ and variance σ^2,

$$P[a < x < b] = \Phi_u\left(\frac{b-\xi}{\sigma}\right) - \Phi_u\left(\frac{a-\xi}{\sigma}\right)$$

and, for $Y \geq 0$,

$$P[|x - \xi| < Y\sigma] = P[\xi - Y\sigma < x < \xi + Y\sigma]$$
$$= \frac{1}{\sqrt{2\pi}} \int_{-Y}^{Y} e^{-\frac{U^2}{2}} du = 2\Phi_u(Y) - 1 = \mathrm{erf}\left(\frac{Y}{\sqrt{2}}\right) = \Phi_{|u|}(Y)$$

$$P[|x - \xi| > Y\sigma] = \frac{1}{\sqrt{2\pi}} \int_Y^{\infty} e^{-\frac{U^2}{2}} du = 2[1 - \Phi_u(Y)] \tag{A.14-11}$$
$$= 1 - \mathrm{erf}\left(\frac{Y}{\sqrt{2}}\right) = 1 - \Phi_{|u|}(Y)$$

(b) The fractiles

$$|u|_P = u_{\frac{1+P}{2}} = |u|_{1-\alpha} = u_{1-\frac{\alpha}{2}} \tag{A.14-12}$$

defined by

$$P[|x - \xi| \leq |u|_P \sigma] = P = 1 - \alpha \tag{A.14-13}$$

are often referred to as *tolerance limits of the normal deviate* u or as α *values of the normal deviate*. Note

$|u|_{0.95} = u_{0.975} \approx 1.96 \qquad |u|_{0.99} = u_{0.995} \approx 2.58 \qquad |u|_{0.999} = u_{0.9995} \approx 3.29$

(c) Note the following measures of dispersion for normal distributions

The *mean deviation* (m.a.e) $E\{|x - \xi|\} = \sigma E\{|u|\} = \sqrt{\dfrac{2}{\pi}}\, \sigma \approx 0.798\sigma$

The *probable deviation* (p.e., median of $|x - \xi|$) $|u|_{\frac{1}{2}} = -u_{\frac{1}{4}} = u_{\frac{3}{4}} \approx 0.674\sigma$
One-half the half width $\sqrt{2 \log_e 2}\, \sigma \approx 1.177\sigma$

The lower and upper *quartiles* $x_{1/4} = \xi - u_{1/4}\sigma = \xi - |u|_{1/2}\sigma$

$$x_{3/4} = \xi + u_{3/4}\sigma = \xi + |u|_{1/2}\sigma$$

The *precision measure* $h = \dfrac{1}{\sqrt{2}\,\sigma}$

A.14-5. Miscellaneous Continuous One-dimensional Probability Distributions. Table A.14-7 describes a number of continuous one-dimensional probability distributions.

A.14-6. Two-dimensional Normal Distributions. (a) A **two-dimensional normal distribution** is a continuous probability distribution described by a frequency function of the form

$$\varphi(x_1, x_2) \equiv \frac{1}{2\pi\sigma_1\sigma_2\sqrt{1-\rho_{12}^2}}$$
$$\exp\left\{-\frac{1}{2(1-\rho_{12}^2)}\left[\left(\frac{x_1-\xi_1}{\sigma_1}\right)^2 \right.\right.$$
$$\left.\left. - 2\rho_{12}\frac{x_1-\xi_1}{\sigma_1}\frac{x_2-\xi_2}{\sigma_2} + \left(\frac{x_2-\xi_2}{\sigma_2}\right)^2\right]\right\}$$
$$(\sigma_1 > 0,\ \sigma_2 > 0,\ |\rho_{12}| \le 1) \quad \text{(A.14-14)}$$

The marginal distributions of x_1 and x_2 are both normal with respective mean values ξ_1, ξ_2 and variances σ_1^2, σ_2^2; ρ_{12} is the correlation coefficient of x_1 and x_2. The five parameters ξ_1, ξ_2, σ_1, σ_2, ρ_{12} define the distribution completely.

A.14-7. Circular Normal Distributions. Equation (14) represents a **circular normal distribution** with **dispersion** σ about the center of gravity (ξ_1, ξ_2) if and only if $\rho_{12} = 0$, $\sigma_1 = \sigma_2 = \sigma$. The contour ellipses $\varphi = \text{const.}$ become circles corresponding to fractiles of the **radial deviation (radial error)**

$$r = \sqrt{(x_1-\xi_1)^2 + (x_2-\xi_2)^2}$$

The distribution of r is given by

$$\varphi_r(r) = \frac{2r}{\sigma^2}\,\varphi_{\chi^2(2)}\left(\frac{r^2}{\sigma^2}\right) = \frac{r}{\sigma^2}e^{-\frac{r^2}{2\sigma^2}} \quad (r \ge 0) \quad \text{(Rayleigh distribution)} \quad \text{(A.14-15)}$$

(see also Table A.15-1).

Circular normal distributions are of particular interest in problems related to gunnery; *circular probability paper* shows contour circles for equal increments of $\Phi_r(R)$. Note

$$r_{1/2} = \sqrt{\chi^2_{1/2}(2)}\,\sigma \approx 1.1774\sigma \quad \text{(circular probable error, CPE, CEP)} \quad \text{(A.14-16)}$$

$$E\{r\} = \sqrt{\frac{\pi}{2}}\,\sigma \approx 1.2533\sigma \quad \text{(mean radial deviation, mean radial error)} \quad \text{(A.14-17)}$$

A.14-8. n-Dimensional Normal Distributions. The joint distribution of n random variables $x_1, x_2, \ldots, x_n$ is an **n-dimensional normal distribution** if and only if it is a continuous probability distribution having a frequency function of the form

$$\varphi(x_1, x_2, \ldots, x_n) \frac{1}{\sqrt{(2\pi)^n \det[\lambda_{jk}]}}$$
$$\exp\left[-\frac{1}{2}\sum_{j=1}^{n}\sum_{k=1}^{n}\Lambda_{jk}(x_j-\xi_j)(x_k-\xi_k)\right] \quad \text{(A.14-18)}$$

A-50 ATHEMATICAL FORMULAS, THEOREMS, AND DEFINITIONS

Table A.14-7. Continuous

No.	Distribution	Frequency function $\varphi(x)$	Distribution function $\Phi(x)$
1	Causal distribution	$\delta(x - \xi) = \begin{cases} 0 & (x \neq \xi) \\ \infty & (x = \xi) \end{cases}$	$U(x - \xi)$
2	Rectangular or uniform distribution	$\dfrac{1}{2\alpha}$ $(|x - \xi| < \alpha)$ 0 $(|x - \xi| > \alpha)$	0 $(x \leq \xi - \alpha)$ $\dfrac{1}{2\alpha}(x - \xi + \alpha)$ $(\xi - \alpha \leq x \leq \xi + \alpha)$ 1 $(x \geq \xi + \alpha)$
3	Cauchy's distribution	$\dfrac{1}{\pi\alpha} \dfrac{1}{1 + \left(\dfrac{x - \xi}{\alpha}\right)^2}$	$\dfrac{1}{2} + \dfrac{1}{\pi} \arctan \dfrac{x - \xi}{\alpha}$
4	Laplace's distribution	$\dfrac{1}{2\beta} e^{-\dfrac{|x - \xi|}{\beta}}$	$\tfrac{1}{2} e^{-\dfrac{|x - \xi|}{\beta}}$ $(x \leq \xi)$ $1 - \tfrac{1}{2} e^{-\dfrac{|x - \xi|}{\beta}}$ $(x \geq \xi)$
5	Beta distribution	0 $(x \leq 0, x \geq 1)$ $\dfrac{\Gamma(\alpha + \beta)}{\Gamma(\alpha)\Gamma(\beta)} x^{\alpha-1}(1 - x)^{\beta-1}$ $(0 < x < 1)$ $(\alpha > 0, \beta > 0)$	0 $(x \leq 0)$ $I_x(\alpha, \beta)$ $(0 \leq x \leq 1)$ 1 $(x \geq 1)$
6	Gamma distribution	0 $(x \leq 0)$ $\dfrac{1}{\beta^\alpha \Gamma(\alpha)} x^{\alpha-1} e^{-x/\beta}$ $(x > 0)$ $(\alpha > 0, \beta > 0)$	0 $(x \leq 0)$ $\dfrac{1}{\Gamma(\alpha)} \Gamma_{x/\beta}(\alpha)$ $(x \geq 0)$

Each normal distribution is completely defined by its center of gravity $(\xi_1, \xi_2, \ldots, \xi_n)$ *and its moment matrix* $[\lambda_{jk}] \equiv [\Lambda_{jk}]^{-1}$, *or by the corresponding variances and correlation coefficients.* The characteristic function is

$$\chi_x(q_1, q_2, \ldots, q_n) \equiv \exp\left[i \sum_{j=1}^{n} \xi_j q_j - \tfrac{1}{2} \sum_{j=1}^{n} \sum_{k=1}^{n} \lambda_{jk} q_j q_k\right] \quad (A.14\text{-}19)$$

Each marginal and conditional distribution derived from a normal distribution is normal. *n random variables* $x_1, x_2, \ldots, x_n$ *having a normal joint distribution are statistically independent if and only if they are uncorrelated.*

A.15 MATHEMATICAL STATISTICS

A.15-1. The Classical Probability Model: Random-sample Statistics. Concept of a Population (Universe). (a) In an important class of applications,

Probability Distributions*

Mean value $E\{x\}$	Variance Var $\{x\}$	Characteristic function $\chi_x(q)$	Remarks
ξ	0	$e^{i\xi q}$	x is almost always equal to ξ. Note that the rectangular, Cauchy, and Laplace distributions approximate a causal distribution as $\alpha \to 0$ or $\beta \to 0$
ξ	$\dfrac{\alpha^2}{3}$	$\dfrac{\sin(\alpha q)}{\alpha q} e^{i\xi q}$	x is uniformly distributed over the interval $(\xi - \alpha, \xi + \alpha)$
$E\{x\}$ and Var $\{x\}$ do not exist; the Cauchy principal value of $E\{x\}$ is ξ		$e^{i\xi q - \alpha|q|}$	Distribution of $x = \xi + \alpha \tan y$ if y is uniformly distributed between $y = -\pi/2$ and $y = \pi/2$ (rectangular distribution). Cauchy's distribution is symmetric about $x = \xi$. Half width and interquartile range are both equal to 2α
ξ	$2\beta^2$	$\dfrac{e^{i\xi q}}{1 + \beta^2 q^2}$	For $\xi = 0$, the characteristic function is proportional to the frequency function of a Cauchy distribution with $\alpha = 1/\beta$
$\dfrac{\alpha}{\alpha + \beta}$	$\dfrac{\alpha\beta}{(\alpha + \beta)^2(\alpha + \beta + 1)}$	$F(\alpha; \alpha + \beta; iq)$	$I_x(\alpha, \beta)$ is the incomplete beta-function ratio (Sec. A.10 6); unique mode $(\alpha - 1)/(\alpha + \beta - 2)$ for $\alpha > 1, \beta > 1$ $$\alpha_r = \frac{\Gamma(\alpha + r)\Gamma(\alpha + \beta)}{\Gamma(\alpha)\Gamma(\alpha + \beta + r)}$$
$\alpha\beta$	$\alpha\beta^2$	$(1 - i\beta q)^{-\alpha}$	$\Gamma_x(\alpha)$ is the incomplete gamma function (Sec. A.10-7)

* From G. A. Korn and T. M. Korn, "Mathematical Handbook for Scientists and Engineers," McGraw-Hill Book Company, New York, 1961.

a continuously variable physical quantity (observable) x is regarded as a one-dimensional random variable with the inferred or estimated probability density $\varphi(x)$. Each sample $(x_1, x_2, \ldots, x_n)$ of measurements of x is postulated to be the result of *n repeated independent measurements.* Hence $x_1, x_2, \ldots, x_n$ are *statistically independent random variables with identical probability density.* A sample $(x_1, x_2, \ldots, x_n)$ defined in this manner is called a **random sample of size** n and constitutes an n-dimensional random variable. The probability density in the n-dimensional sample space of "sample points" $(x_1, x_2, \ldots, x_n)$ is the **likelihood function**

$$L(x_1, x_2, \ldots, x_n) = \varphi(x_1)\varphi(x_2) \cdots \varphi(x_n) \qquad \text{(A.15-1)}$$

Statistical description and probability models apply to physical processes exhibiting the following empirical phenomenon. *Even though individual measurements of a physical quantity x cannot be predicted with sufficient accuracy, a suitably determined function $y = y(x_1, x_2, \ldots)$ of a set* (**sample**) *of repeated measurements $x_1, x_2, \ldots$ of x can often be predicted with substantially better accuracy,* and the prediction of y may still yield useful decisions. Such a function y of a set of sample values is called a **statistic,** and the incidence of increased predictability is known as *statistical regularity*.

Every random-sample statistic defined as a measurable function

$$y = y(x_1, x_2, \ldots, x_n)$$

of the sample values is a random variable whose probability distribution (**sampling distribution** of y) is uniquely determined by the likelihood function, and hence by the distribution of x.

As the size n of a random sample increases, many sample statistics converge in probability to corresponding parameters of the theoretical distribution of x; in particular, statistical relative frequencies converge in mean to the corresponding probabilities. Thus one considers each sample drawn from an infinite (theoretical) **population (universe, ensemble)** whose sample distribution is identical with the theoretical probability distribution of x. The probability distribution is then referred to as the **population distribution,** and its parameters are **population parameters.**

A.15-2. Relation between Probability Model and Reality: Estimation of Parameters. Statistical methods use empirical data (sample values) to infer specifications of a *probability model*, e.g., to estimate the probability density $\varphi(x)$ of a random variable x. An important application of such inferred models is to make *decisions* based on inferred probabilities of future events. In most applications, statistical relative frequencies are used directly only for rough *qualitative* (graphical) estimates of the population distribution. Instead, one infers (postulates) the general form of the theoretical distribution, say

$$\varphi = \varphi(x; \eta_1, \eta_2, \ldots)$$

where $\eta_1, \eta_2, \ldots$ are unknown *population parameters* to be estimated on the basis of the given random sample $(x_1, x_2, \ldots, x_n)$. In general, one attempts to estimate values of the parameters $\eta_1, \eta_2, \ldots$ "fitting" a given sample $(x_1, x_2, \ldots, x_n)$ by the empirical values of corresponding sample statistics $y_1(x_1, x_2, \ldots, x_n), y_2(x_1, x_2, \ldots, x_n), \ldots$ which measure analogous properties of the sample (e.g., sample average, sample variance). "Fitting" is interpreted subjectively and not necessarily uniquely; in particular, one prefers estimates $y(x_1, x_2, \ldots, x_n)$ which converge in probability to η as $n \to \infty$ (**consistent estimates**), whose expected value equals η (**unbiased estimates**), whose sampling distribution has a small variance, and/or which are easy to compute.

The *method of maximum likelihood* estimates each parameter η_k by a corresponding trial function $y_k(x_1, x_2, \ldots, x_n)$ chosen so that $L(x_1, x_2, \ldots, x_n; y_1, y_2, \ldots)$ is as large as possible for each sample $(x_1, x_2, \ldots, x_n)$. One attempts to obtain a set of m (joint) **maximum-likelihood estimates** $y_1(x_1, x_2, \ldots, x_n), y_2(x_1, x_2, \ldots, x_n), \ldots, y_m(x_1, x_2, \ldots, x_n)$ as nontrivial solutions of m equations

$$\frac{\partial}{\partial y_k} \log_e L(x_1, x_2, \ldots, x_n; y_1, y_2, \ldots, y_m) = 0$$

$(k = 1, 2, \ldots, m)$ (MAXIMUM-LIKELIHOOD EQUATIONS)   (A.15-2)

which constitute necessary conditions for a maximum of the likelihood function if the latter is suitably differentiable.

A.15-3. Random-sample Statistics. (a) The **statistical relative frequency** of the event E obtained from the given random sample is

$$h[E] = \frac{n_E}{n} \qquad (A.15\text{-}3)$$

where n is the size of the sample.

The random variable n_E has a binomial distribution (Table A.14-4) where

$$\vartheta = P[E] = \int_{S_E} d\Phi(x)$$

is the *probability* associated with the event E, and

$$E\{h[E]\} = P[E] \qquad \text{Var } \{h[E]\} = \frac{P[E]\{1 - P[E]\}}{n} \tag{A.15-4}$$

The statistical relative frequency $h[E]$ is an unbiased, consistent estimate of the corresponding probability $P[E]$; as $n \to \infty$, $h[E]$ is asymptotically normal with the parameters (4).
(**b**) Given a random sample $(x_1, x_2, \ldots, x_n)$, the **sample average** of x is

$$\bar{x} = \frac{1}{n}(x_1 + x_2 + \cdots + x_n) = \frac{1}{n}\sum_{k=1}^{n} x_k \tag{A.15-5}$$

In terms of the sample distribution over a set of class intervals centered at $x = X_1, X_2, \ldots, X_m$, $\bar{x}$ is approximated by

$$\bar{x}_G = \frac{1}{n}(n_1 X_1 + n_2 X_2 + \cdots + n_m X_m) = \frac{1}{n}\sum_{j=1}^{m} n_j X_j$$

(SAMPLE AVERAGE FROM GROUPED DATA) (A.15-6)

Note

$$E\{\bar{x}\} = \xi \qquad \text{Var } \{\bar{x}\} = \frac{\sigma^2}{n} \tag{A.15-7}$$

$\bar{x}$ *is an unbiased, consistent estimate of the population mean* $\xi = E\{x\}$; *if σ^2 exists, $\bar{x}$ is asymptotically normal with the parameters* (7) *as $n \to \infty$*.
(**c**) The **sample average of a function** $y(x)$ **of the random variable** x is

$$\bar{y} = \frac{1}{n}[y(x_1) + y(x_2) + \cdots + y(x_n)] = \frac{1}{n}\sum_{k=1}^{n} y(x_k) \tag{A.15-8}$$

(**d**) The **sample variances**

$$s^2 = \overline{(x - \bar{x})^2} = \frac{1}{n}\sum_{k=1}^{n}(x_k - \bar{x})^2 \tag{A.15-9}$$

$$S^2 = \frac{n}{n-1}s^2 = \frac{1}{n-1}\sum_{k=1}^{n}(x_k - \bar{x})^2 \tag{A.15-10}$$

are *measures of dispersion* of the sample distribution; s is called **sample standard deviation** or **sample dispersion**. Note

$$E\{s^2\} = \frac{n-1}{n}\sigma^2 \qquad E\{S^2\} = \sigma^2 \tag{A.15-11}$$

whenever the quantity on the right exists. *S^2 is an unbiased, consistent estimate of the population variance* $\sigma^2 = \text{Var } \{x\}$ *and is thus often more useful than s^2*.

A.15-4. Simplified Numerical Computation of Sample Averages and Variances. Corrections for Grouping. (a) For numerical computations, it is convenient to choose a computing origin X_0 near the center of the sample distribution ("guessed mean") and to compute

$$\bar{x} = X_0 + \frac{1}{n} \sum_{k=1}^{n} (x_k - X_0) \qquad (A.15\text{-}12)$$

or, for grouped data,

$$\bar{x}_G = X_0 + \frac{1}{n} \sum_{j=1}^{m} n_j (X_j - X_0) \qquad (A.15\text{-}13)$$

(b) The sample variances s^2 and S^2 may be computed from

$$s^2 = \frac{n-1}{n} S^2 = \frac{1}{n} \sum_{k=1}^{n} x_k^2 - \bar{x}^2 \qquad (A.15\text{-}14)$$

which is approximated for grouped data by

$$s_G^2 = \frac{n-1}{n} S_G^2 = \frac{1}{n} \sum_{j=1}^{m} n_j X_j^2 - \bar{x}_G^2 \qquad (A.15\text{-}15)$$

(c) Computations with grouped data are simplified if all class intervals are of equal length ΔX, and if one of the class-interval midpoints $X_j = X_0$ is taken to be the computing origin, so that

$$X_j = X_0 + Y_j \Delta X \qquad (A.15\text{-}16)$$

where the Y_j are "coded" class-interval centers which take integral values $0, \pm 1, \pm 2, \ldots$.

(d) **Sheppard's Correction for Grouping.** Let all class intervals be of equal length ΔX. Then, if the theoretical distribution of x has a high order of contact with the x axis at both "tails," one may improve the grouped-data approximation s_G^2 to the true sample variance s^2 by adding *Sheppard's correction* $-(\Delta X)^2/12$.

A.15-5. Samples Drawn from Normal Populations. For any sample of size n drawn from a normal population with mean ξ and variance σ^2.

1. $\dfrac{\bar{x} - \xi}{\sigma/\sqrt{n}}$ has a **standardized normal distribution** (u distribution, Sec. A.14-3).

2. $\dfrac{\bar{x} - \xi}{S/\sqrt{n}} = \dfrac{\bar{x} - \xi}{s/\sqrt{n-1}}$ (**Student's ratio**) has a t **distribution with** $n-1$ **degrees of freedom** (Table A.15-2).

3. $\dfrac{(n-1)S^2}{\sigma^2} = \dfrac{ns^2}{\sigma^2} = \dfrac{1}{\sigma^2} \sum_{k=1}^{n} (x_k - \bar{x})^2$ has a χ^2 distribution with $n-1$ degrees of freedom (Table A.15-1).

Table A.15-1. The χ^2 Distribution with m Degrees of Freedom (Fig. A.15-1)

(a) $\varphi_y(Y) \equiv \varphi\chi^2_{(m)}(Y) = \begin{cases} 0 & \text{for } Y < 0 \\ \dfrac{1}{\Gamma\left(\dfrac{m}{2}\right)\sqrt{2^m}} Y^{(m-2)/2} e^{-Y/2} & \text{for } Y > 0 \end{cases}$

(b) $E\{y\} = m$ \hspace{2em} Var $\{y\} = 2m$

Mode $\quad m - 2 \quad (m \geq 2)$ \hspace{2em} coefficient of skewness $2\sqrt{\dfrac{2}{m}}$

r^{th} moment about $y = 0$ \hspace{1em} $m(m+2) \cdots (m+2r-2)$ \hspace{2em} coefficient of excess $\dfrac{12}{m}$

characteristic function $(1 - 2iq)^{-m/2}$

(c) **Typical Interpretation.** Given any m statistically independent standardized normal variables $u_k = (x_k - \xi_k)/\sigma_k$, the sum $\chi^2 = \sum_{k=1}^{m} u_k^2$ has a χ^2 distribution with m degrees of freedom.

(d) The **fractiles** of y will be denoted by $\chi^2{}_P$ or $\chi^2{}_P(m)$; published tables frequently show $\chi^2{}_{1-\alpha}(m)$ vs. α.

(e) **Approximations.** As $m \to \infty$,
y is asymptotically normal with mean m and variance $2m$
y/m is asymptotically normal with mean 1 and variance $2/m$
$\sqrt{2y}$ is asymptotically normal with mean $\sqrt{2m-1}$ and variance 1
A useful approximation based on the last item listed above is

$$\chi^2{}_P(m) \approx \tfrac{1}{2}(\sqrt{2m-1} + u_P)^2 \quad (m > 30)$$

This approximation is worst if P is either small or large.

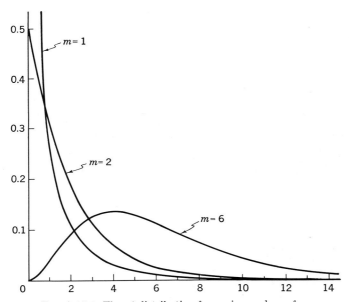

FIG. A.15-1. The χ^2 distribution for various values of m.

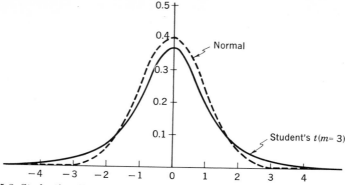

Fig. A.15-2. Student's t distribution compared with the standardized normal distribution.

Table A.15-2. Student's t Distribution with m Degrees of Freedom

(a) $\varphi_y(Y) \equiv \varphi_{t(m)}(Y) \equiv \dfrac{\Gamma\left(\dfrac{m+1}{2}\right)}{\Gamma\left(\dfrac{m}{2}\right)\sqrt{m\pi}} \left(1 + \dfrac{Y^2}{m}\right)^{-(m+1)/2}$

(b) $E\{y\} = 0 \quad (m > 1) \qquad \operatorname{Var}\{y\} = \dfrac{m}{m-2} \quad (m > 2)$

Mode $\quad 0 \qquad\qquad\qquad\qquad\qquad\qquad$ coefficient of skewness 0

$2r$th moment about $y = 0$: $\dfrac{1 \cdot 3 \cdots (2r-1)}{(m-2)(m-4)\cdots(m-2r)} m^r \qquad$ coefficient of excess $\dfrac{3(m-2)}{m-4}$

$\qquad\qquad\qquad\qquad (2r < m) \qquad\qquad\qquad\qquad\qquad\qquad (m > 4)$

(c) **Typical Interpretation.** y is distributed like the ratio

$$y = t = \dfrac{x_0}{\sqrt{\dfrac{1}{m}(x_1^2 + x_2^2 + \cdots + x_m^2)}}$$

where $x_0, x_1, x_2, \ldots, x_m$ are $m+1$ statistically independent normal variables, each having the mean 0 and the variance σ^2. Note that t is independent of σ^2.

(d) The **fractiles** y_P of y will be denoted by t_P; note $t_P = -t_{1-P}$. The distribution of $|y| = |t|$ is related to the t distribution by

$$P[|y| < Y] = P[-Y < y < Y] = \int_{-Y}^{Y} \varphi_{t(m)}(y)\, dy = 2\Phi_{t(m)}(Y) - 1 = \Phi_{|t(m)|}(Y)$$

$$P[|y| > Y] = 2\int_{Y}^{\infty} \varphi_{t(m)}(y)\, dy = 2[1 - \Phi_{t(m)}(Y)] = 1 - \Phi_{|t(m)|}(Y)$$

The fractiles

$$|t|_{1-\alpha} = t_{1-\alpha/2}$$

defined by $P[|y| > |t|_{1-\alpha}] = \alpha$ (α **values of** t) are often tabulated for use in statistical tests; note that some published tables denote $|t|_{1-\alpha}$ by t_α.

(e) **Approximations.** As $m \to \infty$, y is asymptotically normal with mean 0 and variance 1, so that $t_P \approx u_P$ and $|t|_{1-\alpha} = t_{1-\alpha/2} \approx |u|_{1-\alpha} = u_{1-\alpha/2}$ for $m > 30$.

MATHEMATICAL STATISTICS A–57

A.15-6. Statistical Hypotheses. Consider a "space" of samples (sample points) $(x_1, x_2, \ldots, x_n)$, where $x_1, x_2, \ldots, x_n$ are numerical random variables. Every self-consistent set of assumptions involving the joint distribution of $x_1, x_2, \ldots, x_n$ is a **statistical hypothesis**. A hypothesis H is a **simple** statistical hypothesis if it defines the probability distribution uniquely; otherwise it is a **composite** statistical hypothesis.

More specifically, let the joint distribution of $x_1, x_2, \ldots, x_n$ be defined by $\Phi(x_1, x_2, \ldots, x_n; \eta_1, \eta_2, \ldots)$, $\varphi(x_1, x_2, \ldots, x_n; \eta_1, \eta_2, \ldots)$, or $p(x_1, x_2, \ldots, x_n; \eta_1, \eta_2, \ldots)$, where $\eta_1, \eta_2, \ldots$ are parameters. Then a *simple statistical hypothesis* assigns definite values $\eta_{10}, \eta_{20}, \ldots$ to the respective parameters $\eta_1, \eta_2, \ldots$ ("point" in *parameter space*), whereas a *composite statistical hypothesis* confines the "points" $(\eta_1, \eta_2, \ldots)$ to a set or region in parameter space.

A.15-7. Fixed-sample Tests: Definitions. Given a fixed sample size n, a **test of the statistical hypothesis** H is a rule rejecting or accepting the hypothesis H on the basis of a test sample $(X_1, X_2, \ldots, X_n)$. Each test specifies a **critical set** (**critical region, rejection region**) S_C of "points" $(x_1, x_2, \ldots, x_n)$ such that H will be *rejected* if the test sample $(X_1, X_2, \ldots, X_n)$ belongs to the critical set; otherwise H is *accepted*.

Such rejection or acceptance does not constitute a *logical* disproof or proof, even if the sample is infinitely large. Four possible events arise:

1. H is *true* and is *accepted* by the test.
2. H is *false* and is *rejected* by the test.
3. H is *true* but is *rejected* by the test (**error of the first kind**).
4. H is *false* but is *accepted* by the test (**error of the second kind**).

For any set of true (actual) parameter values $\eta_1, \eta_2, \ldots$, the probability that a critical region S_C will *reject* the hypothesis tested is

$$\pi_{S_C}(\eta_1, \eta_2, \ldots) \equiv P[(x_1, x_2, \ldots, x_n) \in S_C; \eta_1, \eta_2, \ldots]$$
$$\equiv \int_{S_C} d\Phi(x_1, x_2, \ldots, x_n; \eta_1, \eta_2, \ldots)$$

(POWER FUNCTION OF THE CRITICAL REGION S_C)   (A.15-17)

A.15-8. Level of Significance. Neyman-Pearson Criteria for Choosing Tests of Simple Hypotheses. (a) It is, generally speaking, desirable to use a critical region S_C such that $\pi_{S_C}(\eta_1, \eta_2, \ldots)$ is small for parameter combinations $\eta_1, \eta_2, \ldots$ admitted by the hypothesis tested, and as large as possible for other parameter combinations. Given a critical region S_C used to test the *simple* hypothesis $H_0 \equiv [\eta_1 = \eta_{10}, \eta_2 = \eta_{20}, \ldots]$ ("null hypothesis"), let H_0 be *true*. Then the probability of falsely rejecting H_0 (error of the first kind) is $\pi_{S_C}(\eta_{10}, \eta_{20}, \ldots) = \alpha$. α is called the **level of significance** of the test.

(b) For each given sample size n and level of significance α

1. A **most powerful test** of the simple hypothesis $H_0 \equiv [\eta_1 = \eta_{10}, \eta_2 = \eta_{20}, \ldots]$ relative to the simple alternative $H_1 \equiv [\eta_1 = \eta_{11}, \eta_2 = \eta_{21}, \ldots]$ is defined by the critical region S_C which yields the largest value of $\pi_{S_C}(\eta_{11}, \eta_{12}, \ldots)$.
2. A **uniformly most powerful test** is most powerful relative to *every* admissible alternative hypothesis H_1; such a test does not always exist.
3. A test is **unbiased** if $\pi_{S_C}(\eta_{11}, \eta_{21}, \ldots) \geq \alpha$ for every alternative simple hypothesis H_1; otherwise the test is **biased**. A **most powerful unbiased test** relative to a given alternative H_1 and a **uniformly most powerful unbiased test** may be defined as above.

To construct the critical region S_C for a *most powerful test*, use all sample points $(x_1, x_2, \ldots, x_n)$ such that the **likelihood ratio** $\varphi(x_1, x_2, \ldots, x_n; \eta_{10}, \eta_{20}, \ldots)/\varphi(x_1, x_2, \ldots, x_n; \eta_{11}, \eta_{21}, \ldots)$ or $p(x_1, x_2, \ldots, x_n; \eta_{10}, \eta_{20}, \ldots)/p(x_1, x_2, \ldots, x_n; \eta_{11}, \eta_{21}, \ldots)$ is less than some fixed constant c; different values of c will yield "best" critical regions at different levels of significance α. *Uniformly most powerful*

Table A.15-3. Some Tests of Significance Relating

Tests are based on random samples. Obtain fractiles from published tables; the approximations given in Tables A.15-1 and A.15-2 apply if the sample size n is large.†

No.	Hypothesis to be tested	Test statistic $y = y(x_1, x_2, \ldots, x_n)$	Critical region Sc rejecting the hypothesis at the level of significance α ($\leq \alpha$ for composite hypotheses)	
1	$\xi = \xi_0$ (σ known)	$\dfrac{\bar{x} - \xi_0}{\sigma/\sqrt{n}}$	$\lvert y \rvert > \lvert u \rvert_{1-\alpha} = u_{1-\alpha/2}$	
2	$\xi \leq \xi_0$ (σ known)		$y > u_{1-\alpha}$	
3	$\xi \geq \xi_0$ (σ known)		$y < u_\alpha = -u_{1-\alpha}$	
4	$\xi = \xi_0$	$\dfrac{\bar{x} - \xi_0}{S/\sqrt{n}}$	$\lvert y \rvert > \lvert t \rvert_{1-\alpha} = t_{1-\alpha/2}$	$(m = n - 1)$
5	$\xi \leq \xi_0$		$y > t_{1-\alpha}$	$(m = n - 1)$
6	$\xi \geq \xi_0$		$y < t_\alpha = -t_{1-\alpha}$	$(m = n - 1)$
7	$\sigma^2 = \sigma_0^2$	$(n - 1)\dfrac{S^2}{\sigma_0^2}$	$y < \chi^2_{\alpha/2}$ $y > \chi^2_{1-\alpha/2}$	$(m = n - 1)$
8	$\sigma^2 \leq \sigma_0^2$		$y > \chi^2_{1-\alpha}$	$(m = n - 1)$
9	$\sigma^2 \geq \sigma_0^2$		$y < \chi^2_\alpha$	$(m = n - 1)$

tests are of particular interest if one desires to test H_0 against a *composite* alternative hypothesis. A *uniformly most powerful unbiased test* may exist even though no uniformly most powerful test exists.

A.15-9. Tests of Significance. Many applications require one to test a hypothetical *population property* specified in terms of a set of parameter values $\eta_1 = \eta_{10}$, $\eta_2 = \eta_{20}, \ldots$ against a corresponding *sample property* described by the respective estimates $y_1(x_1, x_2, \ldots, x_n)$, $y_2(x_1, x_2, \ldots, x_n), \ldots$ of $\eta_1, \eta_2, \ldots$ One attempts to construct a "test statistic"

$$y = y(x_1, x_2, \ldots, x_n; \eta_{10}, \eta_{20}, \ldots) \equiv g(y_1, y_2, \ldots; \eta_{10}, \eta_{20}, \ldots) \quad (A.15\text{-}18)$$

whose values measure a deviation or ratio comparing the sample property to the hypothetical population property for each test sample $(x_1, x_2, \ldots, x_n)$. The simple hypothesis $H_0 \equiv [\eta_1 = \eta_{10}, \eta_2 = \eta_{20}, \ldots]$ is then *rejected* at a given level of significance α (the "deviation" is *significant*) whenever the sample value of y falls *outside* an acceptance interval $[y_{P_1} \leq y \leq y_{P_2}]$ such that

$$P[y_{P_1} \leq y \leq y_{P_2}] = P_2 - P_1 = 1 - \alpha \quad (A.15\text{-}19)$$

Equation (19) specifies $y_{P_1} = y_{P_1}(\eta_{10}, \eta_{20}, \ldots)$ and $y_{P_2} = y_{P_2}(\eta_{10}, \eta_{20}, \ldots)$ as fractiles of the sampling distribution of $y(x_1, x_2, \ldots, x_n; \eta_{10}, \eta_{20}, \ldots)$. It is frequently to the Parameters ξ, σ^2 of a Normal Population*

Power function $\pi_{S_C}(\xi, \sigma)$	Remarks
$\Phi_u\left(u_{\frac{\alpha}{2}} - \dfrac{\xi - \xi_0}{\sigma/\sqrt{n}}\right) + \Phi_u\left(u_{\frac{\alpha}{2}} + \dfrac{\xi - \xi_0}{\sigma/\sqrt{n}}\right)$	Simple hypothesis. Uniformly most powerful *unbiased* test; no uniformly most powerful test exists
$\Phi_u\left(u_\alpha + \dfrac{|\xi - \xi_0|}{\sigma/\sqrt{n}}\right)$	Composite hypotheses. If the admissible hypotheses are restricted to $\xi \geq \xi_0$ for test 1, and to $\xi \leq \xi_0$ for test 2, either test is a uniformly most powerful test for the simple hypothesis $\xi = \xi_0$
Use Eq. (A.15-17)	Simple hypothesis; "two-tailed t test." CAUTION: many published tables denote $|t|_{1-\alpha}$ by t_α
	Composite hypotheses; "one-tailed t tests"
$P\left[\chi^2 < \dfrac{\sigma_0^2}{\sigma^2}\chi^2_{\frac{\alpha}{2}}\right] + P\left[\chi^2 > \dfrac{\sigma_0^2}{\sigma^2}\chi^2_{1-\alpha/2}\right]\ (m = n - 1)$	Simple hypothesis. No uniformly most powerful test exists
$P\left[\chi^2 > \dfrac{\sigma_0^2}{\sigma^2}\chi^2_{1-\alpha}\right]\ (m = n - 1)$	Composite hypotheses. If the admissible hypotheses are restricted to $\sigma^2 \geq \sigma_0^2$ for test 8, and to $\sigma^2 \leq \sigma_0^2$ for test 9, either test is a uniformly most powerful test for the simple hypothesis $\sigma^2 = \sigma_0^2$
$P\left[\chi^2 < \dfrac{\sigma_0^2}{\sigma^2}\chi^2_\alpha\right]\ (m = n - 1)$	

* From G. A. Korn and T. M. Korn, "Mathematical Handbook for Scientists and Engineers," McGraw-Hill Book Company, New York, 1961.
† Note that published tables often tabulate $\chi^2_{1-\alpha}$ rather than χ^2_α; check carefully on the notation used in each case.

possible to choose a test statistic y such that its fractiles y_P are independent of η_{10}, $\eta_{20}, \ldots$. Table A.15-3 lists a number of important examples.

A.15-10. Confidence Intervals Based on Tests of Significance. To find **confidence regions** relating values of one of the unknown parameters, say η_1, to the given sample value $Y = y(X_1, X_2, \ldots, X_n)$ of a suitable test statistic y, refer to Fig. A.15-3. Plot lower and upper acceptance limits (tolerance limits) $y_{P_1}(\eta_1)$ and $y_{P_2}(\eta_1)$ against η_1 for a given level of significance α. The intersections of these acceptance-limit curves with each line $y = Y$ define upper and lower **confidence limits** (**fiducial limits**) $\eta_1 = \gamma_2(Y)$, $\eta_1 = \gamma_1(Y)$ bounding a **confidence interval** $D\alpha \equiv [\gamma_1, \gamma_2]$ at the **confidence level** α. The confidence interval comprises those values of η_1 whose acceptance on the basis of the sample value $y = Y$ is associated with a probability $P[y_{P_1}(\eta_1) \leq Y \leq y_{P_2}(\eta_1)]$ at least equal to the **confidence coefficient** $1 - \alpha$.

A.15-11. The χ^2 Test for Goodness of Fit (see also Table A.15-1). (a) The χ^2 test checks the "fit" of the hypothetical probabilities $p_k = p[E_k]$ associated with r simple events $E_1, E_2, \ldots, E_r$ to their relative frequencies $h_k = h[E_k] = n_k/n$ in a sample of n independent observations. In many applications, each E_k is the event that some random variable x falls within a class interval, so that the test compares the hypothetical theoretical distribution of x with its empirical distribution.

The goodness of fit is measured by the test statistic

$$y = n \sum_{k=1}^{r} \frac{(h_k - p_k)^2}{p_k} = \sum_{k=1}^{r} \frac{(n_k - np_k)^2}{np_k} \quad (A.15\text{-}20)$$

y converges in probability to χ^2 with $m = r - 1$ degrees of freedom as $n \to \infty$. If all $np_k > 10$ (pool some class intervals if necessary), the resulting test *rejects* the hypothetical probabilities $p_1, p_2, \ldots, p_r$ at the level of significance α whenever the test sample yields $y > \chi_{1-\alpha}(m)$; for $m > 30$ one may replace the χ^2 distribution by a normal distribution with the indicated mean and variance.

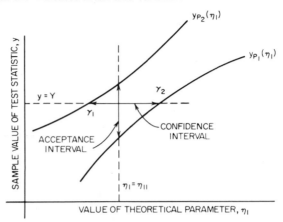

FIG. A.15-3. Definition of confidence intervals.

(b) The χ^2 Test with Estimated Parameters. If the hypothetical probabilities p_k depend on a set of q unknown population parameters $\eta_1, \eta_2, \ldots, \eta_q$, obtain their joint maximum-likelihood estimates from the given sample and insert the resulting values of $p_k = p_k(\eta_1, \eta_2, \ldots, \eta_q)$ into Eq. (20). *The test statistic y will then converge in probability to χ^2 with $m = r - q - 1$ degrees of freedom under very general conditions* (see below), *and the χ^2 test applies with $m = r - q - 1$.* Tests of this type check the applicability of a normal distribution, Poisson distribution, etc., with unspecified parameters.

Table A.15-4. Confidence Intervals for Normal Populations
Use the approximations given in Tables A.15-1 and A.15-2 for large n

No.	Parameter, η_1	Confidence interval $[\gamma_1 \leq \eta_1 \leq \gamma_2]$, confidence coefficient $1 - \alpha = P_2 - P_1$
1	ξ (σ^2 known)	$\bar{x} - u_{P_2} \dfrac{\sigma}{\sqrt{n}} \leq \xi \leq \bar{x} - u_{P_1} \dfrac{\sigma}{\sqrt{n}}$
2	ξ (σ^2 unknown)	$\bar{x} - t_{P_2} \dfrac{S}{\sqrt{n}} \leq \xi \leq \bar{x} - t_{P_1} \dfrac{S}{\sqrt{n}}$
3	σ^2	$S^2 \dfrac{n-1}{\chi^2_{P_2}} \leq \sigma^2 \leq S^2 \dfrac{n-1}{\chi^2_{P_1}}$

NUMERICAL CALCULATIONS AND FINITE DIFFERENCES   A–61

The theorem applies whenever the given functions $p_k(\eta_1, \eta_2, \ldots, \eta_q)$ satisfy the following conditions throughout a neighborhood of the joint maximum-likelihood "point" $(\eta_1, \eta_2, \ldots, \eta_q)$:

1. $\displaystyle\sum_{k=1}^{r} p_k(\eta_1, \eta_2, \ldots, \eta_q) = 1.$

2. The $p_k(\eta_1, \eta_2, \ldots, \eta_q)$ have a common positive lower bound, are twice continuously differentiable, and the matrix $[\partial p_k / \partial \eta_j]$ is of rank q.

A.16 NUMERICAL CALCULATIONS AND FINITE DIFFERENCES

A.16-1. Finite Differences. (a) Let $y = y(x)$ be a function of the real variable x. Given a set of equally spaced argument values $x_k = x_0 + k\,\Delta x$

$$(k = 0, \pm 1, \pm 2, \ldots; \Delta x = h > 0)$$

and a corresponding set or table of function values $y_k = y(x_k) = y(x_0 + k\,\Delta x)$, one defines the **forward differences**

$$\begin{aligned}
\Delta y_k &= y_{k+1} - y_k \quad \text{(FIRST-ORDER FORWARD DIFFERENCES)} \\
\Delta^2 y_k &= \Delta y_{k+1} - \Delta y_k = y_{k+2} - 2y_{k+1} + y_k \\
&\qquad\qquad \text{(SECOND-ORDER FORWARD DIFFERENCES)} \\
&\ldots\ldots\ldots\ldots\ldots\ldots\ldots\ldots\ldots\ldots\ldots\ldots \\
\Delta^r y_k &= \Delta^{r-1} y_{k+1} - \Delta^{r-1} y_k = \sum_{j=0}^{r} (-1)^j \binom{r}{j} y_{k+r-j} \\
&(r = 2, 3, \ldots)\,(r^{\text{th}}\text{-ORDER FORWARD DIFFERENCES}) \\
&\qquad\qquad (k = 0, \pm 1, \pm 2, \ldots)
\end{aligned} \quad \text{(A.16-1)}$$

and the **backward differences**

$$\left.\begin{aligned}
\nabla y_k &= y_k - y_{k-1} = \Delta y_{k-1} \\
\nabla^r y_k &= \nabla^{r-1} y_k - \nabla^{r-1} y_{k-1} = \Delta^r y_{k-r} \quad (r = 2, 3, \ldots) \\
&\qquad\qquad\qquad\qquad\qquad\qquad\qquad (k = 0, \pm 1, \pm 2, \ldots)
\end{aligned}\right\} \quad \text{(A.16-2)}$$

(b) Even though the function values $y_k = y(x_0 + k\,\Delta x)$ may not be known for half-integral values of k, one can calculate the **central differences**

$$\left.\begin{aligned}
\delta y_k &= y_{k+\frac{1}{2}} - y_{k-\frac{1}{2}} = \Delta y_{k-\frac{1}{2}} \\
\delta^r y_k &= \delta^{r-1} y_{k+\frac{1}{2}} - \delta^{r-1} y_{k-\frac{1}{2}} = \Delta^r y_{k-r/2}
\end{aligned} \quad (r = 2, 3, \ldots) \right\} \quad \text{(A.16-3)}$$

(c) Finite differences are conveniently tabulated in arrays like

$$\begin{array}{cccccc}
x_{-1} & y_{-1} & & & & \\
 & & \Delta y_{-1} & & & \\
x_0 & y_0 & & \Delta^2 y_{-1} & & \\
 & & \Delta y_0 & & \Delta^3 y_{-1} & \\
x_1 & y_1 & & \Delta^2 y_0 & & \Delta^4 y_{-1} \\
 & & \Delta y_1 & & \Delta^3 y_0 & \\
x_2 & y_2 & & \Delta^2 y_1 & & \\
 & & \Delta y_2 & & & \\
x_3 & y_3 & & & &
\end{array} \quad \text{(A.16-4)}$$

A.16-2. Operator Notation. (a) *Definitions.* Given a suitably defined function $y = y(x)$ of the real variable x and a fixed increment $\Delta x = h$ of x, one defines the **displacement operator** (**shift operator**) **E** by

$$\mathsf{E} y(x) \equiv y(x + \Delta x) \qquad \mathsf{E}^r y(x) \equiv y(x + r\,\Delta x) \quad \text{(A.16-5)}$$

where r is any real number.

The **difference operators** Δ, ∇, δ are defined by

$$\Delta y(x) \equiv y(x + \Delta x) - y(x) \quad \text{(FORWARD-DIFFERENCE OPERATOR)}$$
$$\nabla y(x) \equiv y(x) - y(x - \Delta x) \quad \text{(BACKWARD-DIFFERENCE OPERATOR)}$$
$$\delta y(x) \equiv y\left(x + \frac{\Delta x}{2}\right) - y\left(x - \frac{\Delta x}{2}\right) \quad \text{(CENTRAL-DIFFERENCE OPERATOR)} \quad (A.16\text{-}6)$$

(b) Operator Relations

$$\Delta = E - 1 = E\nabla = E^{1/2}\delta \qquad \nabla = 1 - E^{-1} = E^{-1}\Delta = E^{-1/2}\delta$$
$$\delta = E^{1/2} - E^{-1/2} = E^{-1/2}\Delta = E^{1/2}\nabla \qquad (A.16\text{-}7)$$
$$E = 1 + \Delta \qquad \Delta\nabla = \Delta\nabla = \delta^2$$

As an aid to memory, note

$$\Delta^r = (E - 1)^r = \sum_{j=0}^{r} (-1)^j \binom{r}{j} E^{r-i} \qquad (r = 1, 2, \ldots) \qquad (A.16\text{-}8)$$

A.16-3. General Formulas for Polynomial Interpolation (Argument Values Not Necessarily Equally Spaced). An nth-order **polynomial interpolation formula** approximates the function $y(x)$ by an nth-degree polynomial $Y(x)$ such that $Y(x_k) = y(x_k) = y_k$ for a given set of $n + 1$ argument values x_k.

(a) **Lagrange's Interpolation Formula.** Given $y_0 = y(x_0)$, $y_1 = y(x_1)$, $y_2 = y(x_2)$, ..., $y_n = y(x_n)$

$$Y(x) = \frac{(x - x_1)(x - x_2) \cdots (x - x_n)}{(x_0 - x_1)(x_0 - x_2) \cdots (x_0 - x_n)} y_0$$
$$+ \frac{(x - x_0)(x - x_2) \cdots (x - x_n)}{(x_1 - x_0)(x_1 - x_2) \cdots (x_1 - x_n)} y_1 + \cdots$$
$$+ \frac{(x - x_0)(x - x_1) \cdots (x - x_{n-1})}{(x_n - x_0)(x_n - x_1) \cdots (x_n - x_{n-1})} y_n$$

(LAGRANGE'S INTERPOLATION FORMULA)   (A.16-9)

(b) **Divided Differences and Newton's Interpolation Formula.** One defines the **divided differences**

$$\Delta_1(x_0, x_1) \equiv \frac{y_1 - y_0}{x_1 - x_0}$$
$$\Delta_r(x_0, x_1, x_2, \ldots, x_r)$$
$$\equiv \frac{\Delta_{r-1}(x_1, x_2, \ldots, x_r) - \Delta_{r-1}(x_0, x_1, \ldots, x_{r-1})}{x_r - x_0} \qquad (A.16\text{-}10)$$
$$(r = 2, 3, \ldots)$$

Then

$$Y(x) = y_0 + (x - x_0)\Delta_1(x_0, x_1)$$
$$+ (x - x_0)(x - x_1)\Delta_2(x_0, x_1, x_2) + \cdots$$
$$+ \left[\prod_{k=0}^{n-1}(x - x_k)\right]\Delta_n(x_0, x_1, x_2, \ldots, x_n)$$

(NEWTON'S INTERPOLATION FORMULA)   (A.16-11)

Unlike in Eq. (9), the addition of a new pair of values x_{n+1}, y_{n+1} requires merely the addition of an extra term. It is convenient to tabulate the divided differences (10) for use in Eq. (11) in the manner of Eq. (A.16-4).

NUMERICAL CALCULATIONS AND FINITE DIFFERENCES   A-63

(c) **The Remainder.** If $y(x)$ is suitably differentiable, the **remainder** (error) $R_{n+1}(x)$ involved in the use of any polynomial-interpolation formula based on the $n + 1$ function values $y_0, y_1, y_2, \ldots, y_n$ may be estimated from

$$|R_{n+1}(x)| \leq \frac{1}{(n+1)!} \max_{X \text{ in } I} |f^{(n+1)}(X)| \prod_{k=0}^{n} |x - x_k| \qquad (A.16\text{-}12)$$

where I is an interval containing $x_0, x_1, x_2, \ldots, x_n$, and x.

A.16-4. Interpolation Formulas for Equally Spaced Argument Values. Let $y_k = y(x_k)$, $x_k = x_0 + k\,\Delta x$ ($k = 0, \pm 1, \pm 2, \ldots$), where Δx is a fixed increment, and introduce the abbreviation $(x - x_0)/\Delta x = u$.

(a) **Newton-Gregory Interpolation Formulas.** Given $y_0, y_1, y_2, \ldots$ or $y_0, y_{-1}, y_{-2}, \ldots$,

$$\left. \begin{aligned} Y(x) &+ y_0 + \binom{u}{1} \Delta y_0 + \binom{u}{2} \Delta^2 y_0 + \cdots \\ Y(x) &= y_0 + \frac{u}{1!} \nabla y_0 + \frac{u(u+1)}{2!} \nabla^2 y_0 + \cdots \end{aligned} \right\} \begin{array}{c} \text{(Newton-Gregory} \\ \text{interpolation} \\ \text{formulas)} \end{array} \quad (A.16\text{-}13)$$

(b) **Symmetric Interpolation Formulas.** More frequently, one is given $y_0, y_1, y_2, \ldots$ and $y_{-1}, y_{-2}, \ldots$; Table A.16-1 lists the most useful interpolation formulas for this case. Note that Everett's and Steffensen's formulas are of particular interest for use with printed tables, since only even-order or only odd-order differences need be tabulated.

A.16-5. Numerical Harmonic Analysis. Given m function values $y(x_k) = y_k$ for $x_k = kT/m$ ($k = 0, 1, 2, \ldots, m - 1$), it is desired to approximate $y(x)$ within the interval $(0, T)$ by a trigonometric polynomial

$$Y(x) = \frac{1}{2} A_0 + \sum_{j=1}^{n} \left(A_j \cos j \frac{2\pi x}{T} + B_j \sin j \frac{2\pi x}{T} \right) \quad \left(n < \frac{m}{2} \right) \quad (A.16\text{-}14a)$$

so as to minimize the mean-square error $\sum_{k=0}^{m-1} [Y(x_k) - y_k]^2$. The required coefficients A_j, B_j are

$$A_j = \frac{2}{m} \sum_{k=0}^{m-1} y_k \cos j \frac{2\pi k}{m} \qquad B_j = \frac{2}{m} \sum_{k=0}^{m-1} y_k \sin j \frac{2\pi k}{m} \qquad \left(0 \leq j < \frac{m}{2} \right)$$

$$(A.16\text{-}14b)$$

The calculation of the sums (14b) is simplified whenever m is divisible by 4. Table A.16-2 shows a convenient computation scheme for $m = 12$.

If no harmonics higher than the third are required, note the simpler formulas

$$\begin{aligned} 6A_0 &= y_0 + y_1 + y_2 + \cdots + y_{11} & 6A_3 &= y_0 - y_2 + y_4 - y_6 + y_8 - y_{10} \\ 4A_2 &= y_0 - y_3 + y_6 - y_9 & 6B_3 &= y_1 - y_3 + y_5 - y_7 + y_9 - y_{11} \\ A_1 &= \tfrac{1}{2}(y_0 - y_6) + A_3 & B_1 &= \tfrac{1}{2}(y_3 - y_9) + B_3 \end{aligned}$$

$$(A.16\text{-}15)$$

and

$$4B_2 = y(T/8) - y(3T/8) + y(5T/8) - y(7T/8) \qquad (A.16\text{-}16)$$

The four additional function values required for Eq. (16) can often be read directly from a graph of $y(x)$ vs. x.

Table A.16-1. Symmetric Interpolation Formulas‡

One is given an odd number $n + 1 = 2m + 1$ of function values $y_k = y(x_0 + k\,\Delta x)$ $(k = 0, \pm 1, \pm 2, \ldots, \pm m)$, where Δx is a fixed increment; $u = (x - x_0)/\Delta x$

No.		Interpolation polynomial, $Y(x)$	Remainder, $R_{n+1}(x) \equiv R_{2m+1}(x)$ (ξ lies in the smallest interval containing x and every $x_0 + k\,\Delta x$ used)
1	Stirling's interpolation formula*	$y_0 + \sum\limits_{k=0}^{m-1} \binom{u+k}{2k+1} \dfrac{\delta^{2k+1}y_{-1/2} + \delta^{2k+1}y_{1/2}}{2} + \sum\limits_{k=1}^{m} \dfrac{u}{2k}\left[\binom{u+k-1}{2k-1}\right]\delta^{2k}y_0$	$\binom{u+m}{2m+1} y^{(2m+1)}(\xi)\,\Delta x^{2m+1}$
2	Bessel's interpolation formula†	$\dfrac{y_0 + y_1}{2} + \sum\limits_{k=0}^{m-1} \dfrac{u - 1/2}{2k+1}\binom{u+k-1}{2k}\delta^{2k+1}y_{1/2} + \sum\limits_{k=1}^{m-1}\binom{u+k-1}{2k}\dfrac{\delta^{2k}y_0 + \delta^{2k}y_1}{2}$	$\binom{u+m-1}{2m} y^{(2m)}(\xi)\,\Delta x^{2m}$
3	Everett's interpolation formula*	$(1-u)y_0 + uy_1 + \sum\limits_{k=1}^{m-1}\left\{\binom{u+k}{2k+1}\delta^{2k}y_1 - \binom{u+k-1}{2k+1}\delta^{2k}y_0\right\}$	$\binom{u+m-1}{2m} y^{(2m)}(\xi)\,\Delta x^{2m}$
4	Steffensen's interpolation formula	$y_0 + \sum\limits_{k=1}^{m}\left\{\binom{u+k}{2k}\delta^{2k-1}y_{1/2} - \binom{k-u}{2k}\delta^{2k-1}y_{-1/2}\right\}$	$\binom{u+m}{2m+1} y^{(2m+1)}(\xi)\,\Delta x^{2m+1}$

* Note that $\binom{u+k}{2k+1} = \dfrac{u(u^2 - 1^2)(u^2 - 2^2)\cdots(u^2 - k^2)}{(2k+1)!}$

† Bessel's modified formula

$$Y(x) = y_0 + u\,\delta y_{1/2} + \dfrac{u(u-1)}{2}\dfrac{\delta^2 y_0 + \delta^2 y_1}{2} + \dfrac{u(u - 1/2)(u-1)}{6}\delta^3 y_{1/2} + \dfrac{(u+1)u(u-1)}{24}\left(\dfrac{\delta^4 y_0 + \delta^4 y_1}{2} - \dfrac{13}{120}\delta^5 y_{1/2}\right)$$
$$+ \dfrac{(u+2)(u+1)u(u-1)}{24}\left(\dfrac{\delta^4 y_0 + \delta^4 y_1}{2} - \dfrac{191}{924}\dfrac{\delta^6 y_0 + \delta^6 y_1}{2}\right)$$

gives a simplified polynomial including the effect of sixth-order differences.

‡ From G. A. Korn and T. M. Korn, "Mathematical Handbook for Scientists and Engineers," McGraw-Hill Book Company, New York, 1961.

Table A.16-2. 12-ordinate Scheme for Harmonic Analysis*

$$y_k = y\left(\frac{kT}{12}\right); \text{ refer to Sec. A.16-5}$$

Line	How obtained											
1	Given function values	y_0	y_1	y_2	y_3	y_4	y_5	y_6				
2		$\ldots$	y_{11}	y_{10}	y_9	y_8	y_7	$\ldots$				
3	Sum of 1 and 2	s_0	s_1	s_2	s_3	s_4	s_5	s_6				
4	Difference of 1 and 2	$\ldots$	d_1	d_2	d_3	d_4	d_5	$\ldots$				
5	Copy	s_0	s_1	s_2	s_3							
6		s_6	s_5	s_4	$\ldots$							
7	Sum of 5 and 6	s'_0	s'_1	s'_2	s'_3							
8	Difference of 5 and 6	d'_0	d'_1	d'_2	d'_3							
		s''_0	s''_1	d'_0	s''_3							
9	Copy and multiply	s''_2	s''_3	$\ldots$	$\ldots$	d'_2	$-s'_3$	d'_0	s''_3	$\ldots$	$\ldots$	s''_1 s''_3
10		$\ldots$	$\ldots$	$\tfrac{1}{2}d'_2$	$\tfrac{\sqrt{3}}{2}d'_1$	$\ldots$	$\tfrac{1}{2}s'_1$	$\ldots$	$\tfrac{1}{2}s''_1$	$\tfrac{\sqrt{3}}{2}s''_2$	$\tfrac{\sqrt{3}}{2}d''_1$	$\ldots$ $\ldots$
11								s''_3				
12						$\ldots$	$-\tfrac{1}{2}s'_2$	$\ldots$	$\tfrac{\sqrt{3}}{2}s''_2$	$\tfrac{\sqrt{3}}{2}d''_1$	$\tfrac{\sqrt{3}}{2}d''_2$	$\ldots$ $\ldots$
13	Add each column (9 to 12)	S_1	T_1	S_2	T_2	S_3	T_3	S_4	T_4	S_5	T_5	S_6 T_6 S_7 T_7
14	$S_i + T_i$	$6A_0$	$6A_1$		$6A_2$		$6A_3$		$6B_1$	$6B_2$		$6B_3$
15	$S_i - T_i$	$12A_6$	$6A_5$		$6A_4$				$6B_5$	$6B_4$		

*From G. A. Korn and T. M. Korn, "Mathematical Handbook for Scientists and Engineers," McGraw-Hill Book Company, New York, 1961.

A.16-6. Numerical Integration Using Equally Spaced Argument Values. Newton-Cotes Quadrature Formulas.
Quadrature formulas of the closed Newton-Cotes type (Table A.16-3) use the approximation

$$\int_{x_0}^{x_0+n\,\Delta x} y(x)\,dx \approx a_0 y_0 + a_1 y_1 + a_2 y_2 + \cdots + a_n y_n$$

with
$$a_k = \frac{(-1)^{n-k}\,\Delta x}{k!(n-k)!}\int_0^n \frac{\lambda(\lambda-1)(\lambda-2)\cdots(\lambda-n)}{(\lambda-k)}\,d\lambda \quad\quad (\text{A.16-17})$$

where the $y_k = y(x_k)$ are given function values for $n+1$ equally spaced argument values $x_k = x_0 + k\,\Delta x$ ($k = 0, 1, 2, \ldots, n$); the resulting error vanishes if $y(x)$ is a polynomial of degree not greater than n. Instead of using values of $n > 6$, one adds m sums (17) of $n \leq 6$ terms for successive subintervals:

$$\int_{x_0}^{x_0+mn\,\Delta x} y(x)\,dx = \int_{x_0}^{x_0+n\,\Delta x} y(x)\,dx + \int_{x_0+n\,\Delta x}^{x_0+2n\,\Delta x} y(x)\,dx + \cdots \quad (\text{A.16-18})$$

Table A.16-3. Quadrature Formulas of the Closed Newton-Cotes Type
$[y_k = y(x_k) = y(x + k\,\Delta x),\ k = 0, 1, 2, \ldots, n]$

No.		$I' \approx \int_{x_0}^{x_0+n\,\Delta x} y(x)\,dx = I$ (add analogous expressions for m successive subintervals)	Error, $I - I'$ $(x_0 < \xi < x_0 + n\,\Delta x)$
1	Trapezoidal rule ($n = 1$)	$\dfrac{\Delta x}{2}(y_0 + y_1)$	$-\tfrac{1}{12}(n\,\Delta x)^3 y^{(2)}(\xi)$
2	Simpson's rule ($n = 2$)	$\dfrac{\Delta x}{3}(y_0 + 4y_1 + y_2)$	$-\dfrac{1}{90}\left(\dfrac{n\,\Delta x}{2}\right)^5 y^{(4)}(\xi)$
3	Weddle's rule ($n = 6$)*	$\tfrac{3}{10}\Delta x(y_0 + 5y_1 + y_2 + 6y_3 + y_4 + 5y_5 + y_6)$	$\dfrac{\vartheta}{212310}\left(\dfrac{n\,\Delta x}{2}\right)^7 y^{(6)}(\xi)$ $(0 < \vartheta < 1)$

* In Weddle's rule, the correct Newton-Cotes coefficient $^{41}\!/_{140}$ of $\Delta^6 y_0$ has been replaced by $^{3}\!/_{10}$.

A.16-7. Gauss and Chebychev Quadrature Formulas.
(a) Rewrite the given definite integral $\int_a^b y(x)\,dx$ as $\int_{-1}^{1} \eta(\xi)\,d\xi$ with the aid of the transformation

$$x = \frac{b-a}{2}\xi + \frac{a+b}{2} \qquad \eta(\xi) \equiv \frac{b-a}{2}y(x) \quad\quad (\text{A.16-19})$$

and approximate the latter integral by

$$\int_{-1}^{1} \eta(\xi)\,d\xi \approx \sum_{k=1}^{n} a_k \eta(\xi_k)$$

with
$$a_k = \frac{2}{(1-\xi_k^2)[P_n'(\xi_k)]^2} \quad (n = 1, 2, \ldots) \qquad \begin{array}{c}(\text{Gauss}\\ \text{quadrature}\\ \text{formula})\end{array} \quad (\text{A.16-20})$$

where the n argument values ξ_k are the n zeros of the nth-degree Legendre polynomial $P_n(\xi)$. Table A.16-4 lists the ξ_k and a_k for a number of values of n.

The error due to the use of the Gauss quadrature formula (20) is

$$E = \frac{(n!)^4(b-a)^{2n+1}}{(2n+1)[(2n!)]^3} y^{(2n)}(X) \qquad (a < X < b)$$

(b) A simpler class of quadrature formulas is obtained with the aid of the transformation (19) and an approximation of the form

$$\int_{-1}^{1} \eta(\xi)\, d\xi \approx \frac{2}{n}[\eta(\xi_1') + \eta(\xi_2') + \cdots + \eta(\xi_n')]$$

$(n = 2, 3, 4, 5, 6, 7, 9)$ (CHEBYCHEV QUADRATURE FORMULA)   (A.16-21)

Table A.16-4 lists the ξ_k' for a number of values of n. The use of equal weights minimizes the probable error if $y(x)$ is affected by normally distributed random errors. For $n = 3$, the error due to the use of the Chebychev quadrature formula (21) is

$$\frac{1}{360}\left(\frac{b-a}{2}\right)^5 y^{(4)}(X) \quad (a < X < b)$$

A.16-8. Numerical Differentiation. Numerical differentiation is subject to errors due to insufficient data, truncation, etc., and should be used with caution.

Table A.16-4. Abscissas and Weights for Gauss and Chebychev Quadrature Formulas

(a) Abscissas ξ_k and Weights a_k for the Gauss Quadrature Formula (20)

n	Abscissas	Weights
2	±0.577350	1
3	0	8/9
	±0.774597	5/9
4	±0.339981	0.652145
	±0.861136	0.347855
5	0	0.568889
	±0.538469	0.478629
	±0.906180	0.236927

(b) Abscissas ξ_k' for the Chebychev Quadrature Formula (21)

n	Abscissas	n	Abscissas
2	±0.577350	7	0
3	0		±0.323912
	±0.707107		±0.529657
4	±0.187592		±0.883862
	±0.794654	9	0
5	0		±0.167906
	±0.374541		±0.528762
	±0.832497		±0.601019
6	±0.266635		±0.911589
	±0.422519		
	±0.866247		

Note the following explicit *three-point differentiation formulas with error estimates*

$$\left. \begin{aligned} y_{-1}' &= \frac{1}{2\Delta x}(-3y_{-1} + 4y_0 - y_1) + \frac{\Delta x^2}{3} y'''(\xi) \\ y_0' &= \frac{1}{2\Delta x}(-y_{-1} + y_1) - \frac{\Delta x^2}{6} y'''(\xi) \\ y_1' &= \frac{1}{2\Delta x}(y_{-1} - 4y_0 + 3y_1) + \frac{\Delta x^2}{3} y'''(\xi) \end{aligned} \right\} \quad \text{(A.16-22)}$$

where $y_{-1} < \xi < y_1$.

A.16-9. Stepwise Solution of Initial-value Problems. The Runge-Kutta Method. To solve the first-order differential equation

$$y' = f(x, y) \quad \text{(A.16-23)}$$

for a given initial value $y(x_0) = y_0$, consider fixed increments $\Delta x = h$ of the independent variable x and use the notation $x_0 + k\,\Delta x = x_k$, $y(x_k) = y(x_0 + k\,\Delta x) = y_k$, $f(x_k, y_k) = f_k$ ($k = 0, \pm 1, \pm 2, \ldots$).

with

$$\left. \begin{aligned} y_{k+1} &= y_k + \tfrac{1}{6}(k_1 + 2k_2 + 2k_3 + k_4) \\ k_1 &= f_k\,\Delta x = f(x_k, y_k)\,\Delta x \\ k_2 &= f\left(x_k + \frac{\Delta x}{2},\, y_k + \frac{k_1}{2}\right)\Delta x \\ k_3 &= f\left(x_k + \frac{\Delta x}{2},\, y_k + \frac{k_2}{2}\right)\Delta x \\ k_4 &= f(x_{k+1},\, y_k + k_3)\,\Delta x \end{aligned} \right\} \quad \text{(A.16-24)}$$

A.16-10. Finite-difference Schemes for the Solution of Initial-value Problems. Each of the following *finite-difference schemes* for the numerical solution of the differential equation (A.16-23) approximates successive function values

$$y_{k+1} = y_k + \int_{x_k}^{x_{k+1}} f(x,y)\,dx = y_{k-1} + \int_{x_{k-1}}^{x_{k+1}} f(x,y)\,dx$$

by integration of an interpolation polynomial for $f(x, y)$ through three to five previously computed function values $f_i = f(x_i, y_i)$. The iterative methods of Secs. A.16-10d and e tend to reduce accumulation of errors, but the solution "stability" (sensitivity to small errors like round-off errors) should be separately investigated in each case.

(a) **Starting the Solution.** Given the initial value y_0, each finite-difference scheme requires one to compute the first three to five function values $y_1, y_2, \ldots$. This "starting solution" should be computed more accurately than the required solution by at least a factor of 10. If the Runge-Kutta method is used to start the solution, employ a step size $\Delta x = h$ smaller than that required for the subsequent difference scheme.

(b) **Extrapolation Schemes.** Given $y_{k-4}, y_{k-3}, y_{k-2}, y_{k-1}$, and y_k, tabulate f_k, ∇f_k, $\nabla^2 f_k$, $\nabla^3 f_k$, and $\nabla^4 f_k$ and obtain

$$y_{k+1} = y_k + (f_k + \tfrac{1}{2}\nabla f_k + \tfrac{5}{12}\nabla^2 f_k + \tfrac{3}{8}\nabla^3 f_k + \tfrac{251}{720}\nabla^4 f_k)\,\Delta x \quad \text{(A.16-25)}$$

or

$$y_{k+1} = y_{k-1} + (2f_k + \tfrac{1}{3}\nabla^2 f_k + \tfrac{1}{3}\nabla^3 f_k + \tfrac{29}{90}\nabla^4 f_k)\,\Delta x \quad \text{(A.16-26)}$$

The 4th-order term is usually omitted.

(c) **Interpolation-iteration Scheme.** As a first estimate, let $f_{k-1} \approx f_k$, $\nabla f_{k+1} \approx \nabla f_k$, $\nabla^2 f_{k+1} \approx \nabla^2 f_k$, ... Using this approximation, calculate

$$y_{k+1} = y_k + (f_{k+1} - \tfrac{1}{2}\nabla f_{k+1} - \tfrac{1}{12}\nabla^2 f_{k+1} - \tfrac{1}{24}\nabla^3 f_{k+1} - \tfrac{19}{720}\nabla^4 f_{k+1})\,\Delta x$$
(BASHFORTH-ADAMS FORMULA) (A.16-27)

Employ this new approximation to y_{k+1} to calculate improved values of f_{k+1}, f_{k+1}, $\nabla^2 f_{k+1}$, ..., and repeat the process until the difference between successive approximations to y_{k+1} is less than a few digits in the last decimal place required. The step size Δx should be chosen so that one or two iterations are sufficient.

(d) **Milne's Prediction-correction Scheme.** Use the simple approximation

$$y_{k+1} = y_{k-3} + \tfrac{4}{3}(2f_k - f_{k-1} + 2f_{k-2})\,\Delta x \quad (\text{"PREDICTOR"}) \quad (A.16\text{-}28a)$$

to obtain a trial value of f_{k+1}, and calculate

$$y_{k+1} = y_{k-1} + \tfrac{1}{3}(f_{k+1} + 4f_k + f_{k-1})\,\Delta x \quad (\text{"CORRECTOR"}) \quad (A.16\text{-}28b)$$

The step size Δx should be chosen so that the expressions (28a) and (28b) do not differ by more than 14 digits of the last decimal place required.

(e) **Change of Step Size.** To halve the step size, use the following simple interpolation formulas:

$$\left. \begin{array}{l} f_{k-\frac{1}{2}} = f_k - \tfrac{1}{2}\nabla f_k - \tfrac{1}{8}\nabla^2 f_k - \tfrac{1}{16}\nabla^3 f_k \\ f_{k+\frac{1}{2}} = f_k + \tfrac{1}{2}\nabla f_k + \tfrac{3}{8}\nabla^2 f_k + \tfrac{5}{16}\nabla^3 f_k \end{array} \right\} \quad (A.16\text{-}29)$$

REFERENCES

Eves and Newson, "Introduction to the Foundations and Fundamental Concepts of Mathematics," Rinehart & Company, Inc., New York, 1957.
Birkhoff and MacLane, "A Survey of Modern Algebra," rev. ed., The Macmillan Company, New York, 1953.
P. R. Halmos, "Finite-dimensional Vector Spaces," 2d ed., D. Van Nostrand Company, Inc., Princeton, N.J., 1958.
B. L. van der Waerden, "Modern Algebra," vols. 1 and 2, Frederick Ungar Publishing Co., New York, vol. 1, rev. ed., 1953; vol. 2, 1950.
Thrall and Tornheim, "Vector Spaces and Matrices," John Wiley & Sons, Inc., New York, 1957.
R. Courant, "Differential and Integral Calculus," vols. 1 and 2, Interscience Publishers, Inc., New York, vol. 1, 2d ed. rev., 1937; vol. 2, 1936.
K. Knopp, "Theory and Application of Infinite Series," 2d ed., Hafner Publishing Company, Inc., New York, 1948.
J. Todd, "Survey of Numerical Analysis," McGraw-Hill Book Company, New York, 1962.
K. S. Kunz, "Numerical Analysis," McGraw-Hill Book Company, New York, 1957.
D. J. Struik, "Differential Geometry," 2d ed., Addison-Wesley Publishing Company, Inc., Reading, Mass., 1961.
A. Erdélyi, "Higher Transcendental Functions," vols. 1 and 2 (Bateman Project), McGraw-Hill Book Company, New York, 1953.
N. W. McLachlan, "Bessel Functions for Engineers," Oxford University Press, Fair Lawn, N.J., 1946.
I. N. Sneddon, "The Special Functions of Physics and Chemistry," Oliver & Boyd Ltd., Edinburgh, 1956.
E. T. Whittaker and G. N. Watson, "Modern Analysis," The Macmillan Company, New York, 1943.
F. Oberhettinger and W. Magnus, "Formulas and Theorems for the Functions of Mathematical Physics," Chelsea Publishing Company, New York, 1954.
E. Yahnke and F. Emde, "Tables of Functions with Formulae and Curves," Dover Publications, Inc., New York, 1945.
W. B. Davenport, Jr. and W. L. Root, "Introduction to Random Signals and Noise," McGraw-Hill Book Company, New York, 1958.
D. Middleton, "An Introduction to Statistical Communication Theory," McGraw-Hill Book Company, New York, 1960.

R. S. Burington and D. C. May, "Handbook of Probability and Statistics with Tables," McGraw-Hill Book Company, New York, 1953.

W. J. Dixon and F. J. Massey, Jr., "An Introduction to Statistical Analysis," 2d ed., McGraw-Hill, New York, 1957.

A. Hald, "Statistical Theory with Engineering Applications," John Wiley & Sons, Inc., New York, 1952.

P. G. Hoel, "Introduction to Mathematical Statistics," John Wiley & Sons, Inc., New York, 1947.

A. M. Mood and F. A. Graybill, "Introduction to the Theory of Statistics," 2d ed., McGraw-Hill Book Company, New York, 1963.

NUMERICAL TABLES

For functions other than presented here refer to A. Fletcher, J. C. P. Miller, and L. Rosenlead, "An Index of Mathematical Tables," Scientific Computing Services Limited, London and McGraw-Hill Book Company, New York, 1946. This is a complete guide to all the important tables and contains over 2,000 references. Another useful source of tabulated numerical information is M. Abramowitz and I. A. Stogun, "Handbook of Mathematical Functions," National Bureau of Standards, Applied Mathematical Series 55, 1964. This contains much useful information on various mathematical formulas, as well as tabulated data.

Normal-distribution Areas*

Fractional parts of the total area (1.000) under the normal curve between the mean and a perpendicular erected at various numbers of standard deviations (x/σ) from the mean. To illustrate the use of the table, 39.065 per cent of the total area under the curve will lie between the mean and a perpendicular erected at a distance of 1.23σ from the mean.

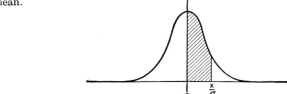

Each figure in the body of the table is preceded by a decimal point.

x/σ	0.00	0.01	0.02	0.03	0.04	0.05	0.06	0.07	0.08	0.09
0.0	00000	00399	00798	01197	01595	01994	02392	02790	03188	03586
0.1	03983	04380	04776	05172	05567	05962	06356	06749	07142	07535
0.2	07926	08317	08706	09095	09483	09871	10257	10642	11026	11409
0.3	11791	12172	12552	12930	13307	13683	14058	14431	14803	15173
0.4	15554	15910	16276	16640	17003	17364	17724	18082	18439	18793
0.5	19146	19497	19847	20194	20450	20884	21226	21566	21904	22240
0.6	22575	22907	23237	23565	23891	24215	24537	24857	25175	25490
0.7	25804	26115	26424	26730	27035	27337	27637	27935	28230	28524
0.8	28814	29103	29389	29673	29955	30234	30511	30785	31057	31327
0.9	31594	31859	32121	32381	32639	32894	33147	33398	33646	33891
1.0	34134	34375	34614	34850	35083	35313	35543	35769	35993	36214
1.1	36433	36650	36864	37076	37286	37493	37698	37900	38100	38298
1.2	38493	38686	38877	39065	39251	39435	39617	39796	39973	40147
1.3	40320	40490	40658	40824	40988	41149	41308	41466	41621	41774
1.4	41924	42073	42220	42364	42507	42647	42786	42922	43056	43189
1.5	43319	43448	43574	43699	43822	43943	44062	44179	44295	44408
1.6	44520	44630	44738	44845	44950	45053	45154	45254	45352	45449
1.7	45543	45637	45728	45818	45907	45994	46080	46164	46246	46327
1.8	46407	46485	46562	46638	46712	46784	46856	46926	46995	47062
1.9	47128	47193	47257	47320	47381	47441	47500	47558	47615	47670
2.0	47725	47778	47831	47882	47932	47982	48030	48077	48124	48169
2.1	48214	48257	48300	48341	48382	48422	48461	48500	48537	48574
2.2	48610	48645	48679	48713	48745	48778	48809	48840	48870	48899
2.3	48928	48956	48983	49010	49036	49061	49086	49111	49134	49158
2.4	49180	49202	49224	49245	49266	49286	49305	49324	49343	49361
2.5	49379	49396	49413	49430	49446	49461	49477	49492	49506	49520
2.6	49534	49547	49560	49573	49585	49598	49609	49621	49632	49643
2.7	49653	49664	49674	49683	49693	49702	49711	49720	49728	49736
2.8	49744	49752	49760	49767	49774	49781	49788	49795	49801	49807
2.9	49813	49819	49825	49831	49836	49841	49846	49851	49856	49861
3.0	49865									
3.5	4997674									
4.0	4999683									
4.5	4999966									
5.0	4999997133									

* This table was adapted, by permission, from F. C. Kent, *Elements of Statistics*, McGraw-Hill, New York, 1924.

Normal-curve Ordinates*

Ordinates (heights) of the unit normal curve. The height (y) at any number of standard deviations $\frac{x}{\sigma}$ from the mean is

$$y = 0.3989 e^{-\frac{1}{2}\left(\frac{x}{\sigma}\right)^2}$$

To obtain answers in units of particular problems, multiply these ordinates by $\frac{N_i}{\sigma}$ where N is the number of cases, i the class interval, and σ the standard deviation.

Each figure in the body of the table is preceded by a decimal point.

x/σ	0.00	0.01	0.02	0.03	0.04	0.05	0.06	0.07	0.08	0.09
0.0	39894	39892	39886	39876	39862	39844	39822	39797	39767	39733
0.1	39695	39654	39608	39559	39505	39448	39387	39322	39253	39181
0.2	39104	39024	38940	38853	38762	38667	38568	38466	38361	38251
0.3	38139	38023	37903	37780	37654	37524	37391	37255	37115	36973
0.4	36827	36678	36526	36371	36213	36053	35889	35723	35553	35381
0.5	35207	35029	34849	34667	34482	34294	34105	33912	33718	33521
0.6	33322	33121	32918	32713	32506	32297	32086	31874	31659	31443
0.7	31225	31006	30785	30563	30339	30114	29887	29658	29430	29200
0.8	28969	28737	28504	28269	28034	27798	27562	27324	27086	26848
0.9	26609	26369	26129	25888	25647	25406	25164	24923	24681	24439
1.0	24197	23955	23713	23471	23230	22988	22747	22506	22265	22025
1.1	21785	21546	21307	21069	20831	20594	20357	20121	19886	19652
1.2	19419	19186	18954	18724	18494	18265	18037	17810	17585	17360
1.3	17137	16915	16694	16474	16256	16038	15822	15608	15395	15183
1.4	14973	14764	14556	14350	14146	13943	13742	13542	13344	13147
1.5	12952	12758	12566	12376	12188	12001	11816	11632	11450	11270
1.6	11092	10915	10741	10567	10396	10226	10059	09893	09728	09566
1.7	09405	09246	09089	08933	08780	08628	08478	08329	08183	08038
1.8	07895	07754	07614	07477	07341	07206	07074	06943	06814	06687
1.9	06562	06438	06316	06195	06077	05959	05844	05730	05618	05508
2.0	05399	05292	05186	05082	04980	04879	04780	04682	04586	04491
2.1	04398	04307	04217	04128	04041	03955	03871	03788	03706	03626
2.2	03547	03470	03394	03319	03246	03174	03103	03034	02965	02898
2.3	02833	02768	02705	02643	02582	02522	02463	02406	02349	02294
2.4	02239	02186	02134	02083	02033	01984	01936	01888	01842	01797
2.5	01753	01709	01667	01625	01585	01545	01506	01468	01431	01394
2.6	01358	01323	01289	01256	01223	01191	01160	01130	01100	01071
2.7	01042	01014	00987	00961	00935	00909	00885	00861	00837	00814
2.8	00792	00770	00748	00727	00707	00687	00668	00649	00631	00613
2.9	00595	00578	00562	00545	00530	00514	00499	00485	00470	00457
3.0	00443									
3.5	0008727									
4.0	0001338									
4.5	0000160									
5.0	000001487									

* This table was adapted, by permission, from F. C. Kent, *Elements of Statistics*, McGraw-Hill, New York, 1924.

Distribution of t*

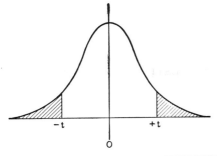

Values of t corresponding to certain selected probabilities (*i.e.*, tail areas under the curve). To illustrate: the probability is 0.05 that a sample with 20 degrees of freedom would have $t = 2.086$ or larger.

DF	Probability							
	0.80	0.40	0.20	0.10	0.05	0.02	0.01	0.001
1	0.325	1.376	3.078	6.314	12.706	31.821	63.657	636.619
2	0.289	1.061	1.886	2.920	4.303	6.965	9.925	31.598
3	0.277	0.978	1.638	2.353	3.182	4.541	5.841	12.941
4	0.271	0.941	1.533	2.132	2.776	3.747	4.604	8.610
5	0.267	0.920	1.476	2.015	2.571	3.365	4.032	6.859
6	0.265	0.906	1.440	1.943	2.447	3.143	3.707	5.959
7	0.263	0.896	1.415	1.895	2.365	2.998	3.499	5.405
8	0.262	0.889	1.397	1.860	2.306	2.896	3.355	5.041
9	0.261	0.883	1.383	1.833	2.262	2.821	3.250	4.781
10	0.260	0.879	1.372	1.812	2.228	2.764	3.169	4.587
11	0.260	0.876	1.363	1.796	2.201	2.718	3.106	4.437
12	0.259	0.873	1.356	1.782	2.179	2.681	3.055	4.318
13	0.259	0.870	1.350	1.771	2.160	2.650	3.012	4.221
14	0.258	0.868	1.345	1.761	2.145	2.624	2.977	4.140
15	0.258	0.866	1.341	1.753	2.131	2.602	2.947	4.073
16	0.258	0.865	1.337	1.746	2.120	2.583	2.921	4.015
17	0.257	0.863	1.333	1.740	2.110	2.567	2.898	3.965
18	0.257	0.862	1.330	1.734	2.101	2.552	2.878	3.922
19	0.257	0.861	1.328	1.729	2.093	2.539	2.861	3.883
20	0.257	0.860	1.325	1.725	2.086	2.528	2.845	3.850
21	0.257	0.859	1.323	1.721	2.080	2.518	2.831	3.819
22	0.256	0.858	1.321	1.717	2.074	2.508	2.819	3.792
23	0.256	0.858	1.319	1.714	2.069	2.500	2.807	3.767
24	0.256	0.857	1.318	1.711	2.064	2.492	2.797	3.745
25	0.256	0.856	1.316	1.708	2.060	2.485	2.787	3.725
26	0.256	0.856	1.315	1.706	2.056	2.479	2.779	3.707
27	0.256	0.855	1.314	1.703	2.052	2.473	2.771	3.690
28	0.256	0.855	1.313	1.701	2.048	2.467	2.763	3.674
29	0.256	0.854	1.311	1.699	2.045	2.462	2.756	3.659
30	0.256	0.854	1.310	1.697	2.042	2.457	2.750	3.646
40	0.255	0.851	1.303	1.684	2.021	2.423	2.704	3.551
60	0.254	0.848	1.296	1.671	2.000	2.390	2.660	3.460
120	0.254	0.845	1.289	1.658	1.980	2.358	2.617	3.373
∞	0.253	0.842	1.282	1.645	1.960	2.326	2.576	3.291

* This table is reproduced in abridged form from Table III of Fisher and Yates, *Statistical Tables for Biological, Agricultural, and Medical Research*, published by Oliver & Boyd, Ltd., Edinburgh, by permission of the authors and publishers.

Distribution of χ^2 *

Values of χ^2 corresponding to certain selected probabilities (*i.e.*, tail areas under the curve). To illustrate: the probability is 0.05 that a sample with 20 degrees of freedom, taken from a normal distribution, would have $\chi^2 = 31.410$ or larger.

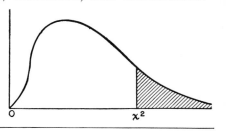

DF	Probability										
	0.99	0.98	0.95	0.90	0.80	0.20	0.10	0.05	0.02	0.01	0.001
1	0.0^{3}157	0.0^{3}628	0.00393	0.0158	0.0642	1.642	2.706	3.841	5.412	6.635	10.827
2	0.0201	0.0404	0.103	0.211	0.446	3.219	4.605	5.991	7.824	9.210	13.815
3	0.115	0.185	0.352	0.584	1.005	4.642	6.251	7.815	9.837	11.341	16.268
4	0.297	0.429	0.711	1.064	1.649	5.989	7.779	9.488	11.668	13.277	18.465
5	0.554	0.752	1.145	1.610	2.343	7.289	9.236	11.070	13.388	15.086	20.517
6	0.872	1.134	1.635	2.204	3.070	8.558	10.645	12.592	15.033	16.812	22.457
7	1.239	1.564	2.167	2.833	3.822	9.803	12.017	14.067	16.622	18.475	24.322
8	1.646	2.032	2.733	3.490	4.594	11.030	13.362	15.507	18.168	20.090	26.125
9	2.088	2.532	3.325	4.168	5.380	12.242	14.684	16.919	19.679	21.666	27.877
10	2.558	3.059	3.940	4.865	6.179	13.442	15.987	18.307	21.161	23.209	29.588
11	3.053	3.609	4.575	5.578	6.989	14.631	17.275	19.675	22.618	24.725	31.264
12	3.571	4.178	5.226	6.304	7.807	15.812	18.549	21.026	24.054	26.217	32.909
13	4.107	4.765	5.892	7.042	8.634	16.985	19.812	22.362	25.472	27.688	34.528
14	4.660	5.368	6.571	7.790	9.467	18.151	21.064	23.685	26.873	29.141	36.123
15	5.229	5.985	7.261	8.547	10.307	19.311	22.307	24.996	28.259	30.578	37.697
16	5.812	6.614	7.962	9.312	11.152	20.465	23.542	26.296	29.633	32.000	39.252
17	6.408	7.255	8.672	10.085	12.002	21.615	24.769	27.587	30.995	33.409	40.790
18	7.015	7.906	9.390	10.865	12.857	22.760	25.989	28.869	32.346	34.805	42.312
19	7.633	8.567	10.117	11.651	13.716	23.900	27.204	30.144	33.687	36.191	43.820
20	8.260	9.237	10.851	12.443	14.578	25.038	28.412	31.410	35.020	37.566	45.315
21	8.897	9.915	11.591	13.240	15.445	26.171	29.615	32.671	36.343	38.932	46.797
22	9.542	10.600	12.338	14.041	16.314	27.301	30.813	33.924	37.659	40.289	48.268
23	10.196	11.293	13.091	14.848	17.187	28.429	32.007	35.172	38.968	41.638	49.728
24	10.856	11.992	13.848	15.659	18.062	29.553	33.196	36.415	40.270	42.980	51.179
25	11.524	12.697	14.611	16.473	18.940	30.675	34.382	37.652	41.566	44.314	52.620
26	12.198	13.409	15.379	17.292	19.820	31.795	35.563	38.885	42.856	45.642	54.052
27	12.879	14.125	16.151	18.114	20.703	32.912	36.741	40.113	44.140	46.963	55.476
28	13.565	14.847	16.928	18.939	21.588	34.027	37.916	41.337	45.419	48.278	56.893
29	14.256	15.574	17.708	19.768	22.475	35.139	39.087	42.557	46.693	49.588	58.302
30	14.953	16.306	18.493	20.599	23.364	36.250	40.256	43.773	47.962	50.892	59.703

* This table is reproduced in abridged form from Table IV of Fisher and Yates, *Statistical Tables for Biological, Agricultural, and Medical Research*, published by Oliver & Boyd, Ltd., Edinburgh, by permission of the authors and publishers.

NUMERICAL TABLES — A-75

Sine Integral $Si(x)$*

$$Si(x) = \int_0^x \frac{\sin u}{u}\, du$$

x	$Si(x)$	(x)	$Si(x)$	x	$Si(x)$	x	$Si(x)$	x	$Si(x)$	x	$Si(x)$
0.0	0.00000	5.0	1.54993	10.0	1.65835	15.0	1.61819	20.0	1.54824	25.0	1.53148
0.1	0.09994	5.1	1.53125	10.1	1.65253	15.1	1.62226	20.1	1.55289	50.0	1.55162
0.2	0.19956	5.2	1.51367	10.2	1.64600	15.2	1.62575	20.2	1.55767		
0.3	0.29850	5.3	1.49732	10.3	1.63883	15.3	1.62865	20.3	1.56253		
0.4	0.39646	5.4	1.48230	10.4	1.63112	15.4	1.63093	20.4	1.56743		
0.5	0.49311	5.5	1.46872	10.5	1.62294	15.5	1.63258	20.5	1.57232		
0.6	0.58813	5.6	1.45667	10.6	1.61439	15.6	1.63359	20.6	1.57714		
0.7	0.68122	5.7	1.44620	10.7	1.60556	15.7	1.63396	20.7	1.58186		
0.8	0.77210	5.8	1.43736	10.8	1.59654	15.8	1.63370	20.8	1.58641		
0.9	0.86047	5.9	1.43018	10.9	1.58743	15.9	1.63280	20.9	1.59077		
1.0	0.94608	6.0	1.42469	11.0	1.57831	16.0	1.63130	21.0	1.59489		
1.1	1.02869	6.1	1.42087	11.1	1.56927	16.1	1.62921	21.1	1.59873		
1.2	1.10805	6.2	1.41871	11.2	1.56042	16.2	1.62657	21.2	1.60225		
1.3	1.18396	6.3	1.41817	11.3	1.55182	16.3	1.62339	21.3	1.60543		
1.4	1.25623	6.4	1.41922	11.4	1.54356	16.4	1.61973	21.4	1.60823		
1.5	1.32468	6.5	1.42179	11.5	1.53571	16.5	1.61563	21.5	1.61063		
1.6	1.38918	6.6	1.42582	11.6	1.52835	16.6	1.61112	21.6	1.61261		
1.7	1.44959	6.7	1.43121	11.7	1.52155	16.7	1.60627	21.7	1.61415		
1.8	1.50582	6.8	1.43787	11.8	1.51535	16.8	1.60111	21.8	1.61525		
1.9	1.55778	6.9	1.44570	11.9	1.50981	16.9	1.59572	21.9	1.61590		
2.0	1.60541	7.0	1.45460	12.0	1.50497	17.0	1.59014	22.0	1.61608		
2.1	1.64870	7.1	1.46443	12.1	1.50088	17.1	1.58443	22.1	1.61582		
2.2	1.68763	7.2	1.47509	12.2	1.49755	17.2	1.57865	22.2	1.61510		
2.3	1.72221	7.3	1.48644	12.3	1.49501	17.3	1.57285	22.3	1.61395		
2.4	1.75249	7.4	1.49834	12.4	1.49327	17.4	1.56711	22.4	1.61238		
2.5	1.77852	7.5	1.51068	12.5	1.49234	17.5	1.56146	22.5	1.61041		
2.6	1.80039	7.6	1.52331	12.6	1.49221	17.6	1.55598	22.6	1.60806		
2.7	1.81821	7.7	1.53611	12.7	1.49287	17.7	1.55070	22.7	1.60536		
2.8	1.83210	7.8	1.54894	12.8	1.49430	17.8	1.54568	22.8	1.60234		
2.9	1.84219	7.9	1.56167	12.9	1.49647	17.9	1.54097	22.9	1.59902		
3.0	1.84865	8.0	1.57419	13.0	1.49936	18.0	1.53661	23.0	1.59546		
3.1	1.85166	8.1	1.58637	13.1	1.50292	18.1	1.53264	23.1	1.59168		
3.2	1.85140	8.2	1.59810	13.2	1.50711	18.2	1.52909	23.2	1.58772		
3.3	1.84808	8.3	1.60928	13.3	1.51188	18.3	1.52600	23.3	1.58363		
3.4	1.84191	8.4	1.61981	13.4	1.51716	18.4	1.52339	23.4	1.57945		
3.5	1.83313	8.5	1.62960	13.5	1.52291	18.5	1.52128	23.5	1.57521		
3.6	1.82195	8.6	1.63857	13.6	1.52905	18.6	1.51969	23.6	1.57097		
3.7	1.80862	8.7	1.64665	13.7	1.53352	18.7	1.51863	23.7	1.56676		
3.8	1.79333	8.8	1.65379	13.8	1.54225	18.8	1.51810	23.8	1.56262		
3.9	1.77650	8.9	1.65993	13.9	1.54917	18.9	1.51810	23.9	1.55860		
4.0	1.75820	9.0	1.66504	14.0	1.55621	19.0	1.51863	24.0	1.55474		
4.1	1.73874	9.1	1.66908	14.1	1.56330	19.1	1.51967	24.1	1.55107		
4.2	1.71837	9.2	1.67205	14.2	1.57036	19.2	1.52122	24.2	1.54762		
4.3	1.69732	9.3	1.67393	14.3	1.57733	19.3	1.52324	24.3	1.54444		
4.4	1.67583	9.4	1.67473	14.4	1.58414	19.4	1.52572	24.4	1.54154		
4.5	1.65414	9.5	1.67446	14.5	1.59072	19.5	1.52863	24.5	1.53897		
4.6	1.63246	9.6	1.67316	14.6	1.59702	19.6	1.53192	24.6	1.53672		
4.7	1.61101	9.7	1.67084	14.7	1.60296	19.7	1.53357	24.7	1.53484		
4.8	1.58998	9.8	1.66757	14.8	1.60851	19.8	1.53954	24.8	1.53333		
4.9	1.56956	9.9	1.66338	14.9	1.61360	19.9	1.54378	24.9	1.53221		

* From P. O. Pedersen, *Radiation from a Vertical Antenna over Flat Perfectly Conducting Earth*, G. E. C. Gad, Copenhagen.

Entropy Tables*

$$q = 1 - p;\ H = -p \log_2 p - q \log_2 q$$

p	$-\log_2 p$	$-p \log_2 p$	H	$-q \log_2 q$	$-\log q$	q
.0001	13.287712	0.001329	0.001473	0.000144	0.000144	0.9999
.0002	12.287712	0.002458	0.002746	0.000289	0.000289	0.9998
.0003	11.702750	0.003511	0.003944	0.000433	0.000433	0.9997
.0004	11.287712	0.004515	0.005092	0.000577	0.000577	0.9996
.0005	10.965784	0.005483	0.006204	0.000721	0.000722	0.9995
.0006	10.702750	0.006422	0.007287	0.000865	0.000866	0.9994
.0007	10.480357	0.007336	0.008346	0.001010	0.001010	0.9993
.0008	10.287712	0.008230	0.009384	0.001154	0.001155	0.9992
.0009	10.117787	0.009106	0.010404	0.001298	0.001299	0.9991
.0010	9.965784	0.009966	0.011408	0.001442	0.001443	0.9990
.0011	9.828281	0.010811	0.012397	0.001586	0.001588	0.9989
.0012	9.702750	0.011643	0.013373	0.001730	0.001732	0.9988
.0013	9.587273	0.012463	0.014338	0.001874	0.001877	0.9987
.0014	9.480357	0.013273	0.015291	0.002018	0.002021	0.9986
.0015	9.380822	0.014071	0.016234	0.002162	0.002166	0.9985
.0016	9.287712	0.014860	0.017167	0.002306	0.002310	0.9984
.0017	9.200249	0.015640	0.018091	0.002450	0.002455	0.9983
.0018	9.117787	0.016412	0.019007	0.002595	0.002599	0.9982
.0019	9.039785	0.017176	0.019914	0.002739	0.002744	0.9981
.0020	8.965784	0.017932	0.020814	0.002882	0.002888	0.9980
.0021	8.895395	0.018680	0.021707	0.003026	0.003033	0.9979
.0022	8.828281	0.019422	0.022593	0.003170	0.003177	0.9978
.0023	8.764150	0.020158	0.023472	0.003314	0.003322	0.9977
.0024	8.702750	0.020887	0.024345	0.003458	0.003467	0.9976
.0025	8.643856	0.021610	0.025212	0.003602	0.003611	0.9975
.0026	8.587273	0.022327	0.026073	0.003746	0.003756	0.9974
.0027	8.532825	0.023039	0.026929	0.003890	0.003901	0.9973
.0028	8.480357	0.023745	0.027779	0.004034	0.004045	0.9972
.0029	8.429731	0.024446	0.028624	0.004178	0.004190	0.9971
.0030	8.380822	0.025142	0.029464	0.004322	0.004335	0.9970
.0031	8.333516	0.025834	0.030299	0.004465	0.004479	0.9969
.0032	8.287712	0.026521	0.031130	0.004609	0.004624	0.9968
.0033	8.243318	0.027203	0.031956	0.004753	0.004769	0.9967
.0034	8.200249	0.027881	0.032778	0.004897	0.004914	0.9966
.0035	8.158429	0.028555	0.033595	0.005041	0.005058	0.9965
.0036	8.117787	0.029224	0.034408	0.005184	0.005203	0.9964
.0037	8.078259	0.029890	0.035218	0.005328	0.005348	0.9963
.0038	8.039785	0.030551	0.036023	0.005472	0.005493	0.9962
.0039	8.002310	0.031209	0.036825	0.005616	0.005638	0.9961
.0040	7.965784	0.031863	0.037622	0.005759	0.005782	0.9960
.0041	7.930160	0.032514	0.038417	0.005903	0.005927	0.9959
.0042	7.895395	0.033161	0.039207	0.006047	0.006072	0.9958
.0043	7.861448	0.033804	0.039994	0.006190	0.006217	0.9957
.0044	7.828281	0.034444	0.040778	0.006334	0.006362	0.9956
.0045	7.795859	0.035081	0.041559	0.006477	0.006507	0.9955

* From Robert M. Fano, "Transmission of Information," The M.I.T. Press, Cambridge, Mass., 1961, by permission. Courtesy of Prof. W. W. Peterson.

NUMERICAL TABLES A-77

Entropy Tables (*Continued*)

p	$-\log_2 p$	$-p \log_2 p$	H	$-q \log_2 q$	$-\log q$	q
.0046	7.764150	0.035715	0.042336	0.006621	0.006652	0.9954
.0047	7.733123	0.036346	0.043110	0.006765	0.006797	0.9953
.0048	7.702750	0.036973	0.043881	0.006908	0.006942	0.9952
.0049	7.673002	0.037598	0.044650	0.007052	0.007087	0.9951
.0050	7.643856	0.038219	0.045415	0.007195	0.007232	0.9950
.0051	7.615287	0.038838	0.046177	0.007339	0.007377	0.9949
.0052	7.587273	0.039454	0.046936	0.007482	0.007522	0.9948
.0053	7.559792	0.040067	0.047693	0.007626	0.007667	0.9947
.0054	7.532825	0.040677	0.048447	0.007769	0.007812	0.9946
.0055	7.506353	0.041285	0.049198	0.007913	0.007957	0.9945
.0056	7.480357	0.041890	0.049946	0.008056	0.008102	0.9944
.0057	7.454822	0.042492	0.050692	0.008200	0.008247	0.9943
.0058	7.429731	0.043092	0.051436	0.008343	0.008392	0.9942
.0059	7.405069	0.043690	0.052177	0.008487	0.008537	0.9941
.0060	7.380822	0.044285	0.052915	0.008630	0.008682	0.9940
.0061	7.356975	0.044878	0.053651	0.008774	0.008827	0.9939
.0062	7.333516	0.045468	0.054385	0.008917	0.008973	0.9938
.0063	7.310432	0.046056	0.055116	0.009060	0.009118	0.9937
.0064	7.287712	0.046641	0.055845	0.009204	0.009263	0.9936
.0065	7.265345	0.047225	0.056572	0.009347	0.009408	0.9935
.0066	7.243318	0.047806	0.057296	0.009490	0.009553	0.9934
.0067	7.221623	0.048385	0.058018	0.009634	0.009699	0.9933
.0068	7.200249	0.048962	0.058739	0.009777	0.009844	0.9932
.0069	7.179188	0.049536	0.059457	0.009920	0.009989	0.9931
.0070	7.158429	0.050109	0.060172	0.010063	0.010134	0.9930
.0071	7.137965	0.050680	0.060886	0.010207	0.010280	0.9929
.0072	7.117787	0.051248	0.061598	0.010350	0.010425	0.9928
.0073	7.097888	0.051815	0.062308	0.010493	0.010570	0.9927
.0074	7.078259	0.052379	0.063015	0.010636	0.010716	0.9926
.0075	7.058894	0.052942	0.063721	0.010780	0.010861	0.9925
.0076	7.039785	0.053502	0.064425	0.010923	0.011006	0.9924
.0077	7.020926	0.054061	0.065127	0.011066	0.011152	0.9923
.0078	7.002310	0.054618	0.065827	0.011209	0.011297	0.9922
.0079	6.983932	0.055173	0.066525	0.011352	0.011443	0.9921
.0080	6.965784	0.055726	0.067222	0.011495	0.011588	0.9920
.0081	6.947862	0.056278	0.067916	0.011638	0.011733	0.9919
.0082	6.930160	0.056827	0.068609	0.011781	0.011879	0.9918
.0083	6.912673	0.057375	0.069300	0.011925	0.012024	0.9917
.0084	6.895395	0.057921	0.069989	0.012068	0.012170	0.9916
.0085	6.878321	0.058466	0.070676	0.012211	0.012315	0.9915
.0086	6.861448	0.059008	0.071362	0.012354	0.012461	0.9914
.0087	6.844769	0.059549	0.072046	0.012497	0.012606	0.9913
.0088	6.828281	0.060089	0.072729	0.012640	0.012752	0.9912
.0089	6.811979	0.060627	0.073409	0.012783	0.012897	0.9911
.0090	6.795859	0.061163	0.074088	0.012926	0.013043	0.9910
.0091	6.779918	0.061697	0.074766	0.013069	0.013189	0.9909
.0092	6.764150	0.062230	0.075442	0.013212	0.013334	0.9908
.0093	6.748554	0.062762	0.076116	0.013354	0.013480	0.9907
.0094	6.733123	0.063291	0.076789	0.013497	0.013625	0.9906
.0095	6.717857	0.063820	0.077460	0.013640	0.013771	0.9905

Entropy Tables (Continued)

p	$-\log_2 p$	$-p \log_2 p$	H	$-q \log_2 q$	$-\log q$	q
.0096	6.702750	0.064346	0.078130	0.013783	0.013917	0.9904
.0097	6.687800	0.064872	0.078798	0.013926	0.014062	0.9903
.0098	6.673002	0.065395	0.079464	0.014069	0.014208	0.9902
.0099	6.658356	0.065918	0.080129	0.014212	0.014354	0.9901
.0100	6.643856	0.066439	0.080793	0.014355	0.014500	0.9900
.0102	6.615287	0.067476	0.082116	0.014640	0.014791	0.9898
.0104	6.587273	0.068508	0.083433	0.014926	0.015083	0.9896
.0106	6.559792	0.069534	0.084745	0.015211	0.015374	0.9894
.0108	6.532825	0.070555	0.086051	0.015497	0.015666	0.9892
.0110	6.506353	0.071570	0.087352	0.015782	0.015958	0.9890
.0112	6.480357	0.072580	0.088647	0.016067	0.016249	0.9888
.0114	6.454822	0.073585	0.089938	0.016353	0.016541	0.9886
.0116	6.429731	0.074585	0.091223	0.016638	0.016833	0.9884
.0118	6.405069	0.075580	0.092503	0.016923	0.017125	0.9882
.0120	6.380822	0.076570	0.093778	0.017208	0.017417	0.9880
.0122	6.356975	0.077555	0.095048	0.017493	0.017709	0.9878
.0124	6.333516	0.078536	0.096314	0.017778	0.018001	0.9876
.0126	6.310432	0.079511	0.097574	0.018063	0.018293	0.9874
.0128	6.287712	0.080483	0.098831	0.018348	0.018586	0.9872
.0130	6.265345	0.081449	0.100082	0.018633	0.018878	0.9870
.0132	6.243318	0.082412	0.101329	0.018917	0.019170	0.9868
.0134	6.221623	0.083370	0.102572	0.019202	0.019463	0.9866
.0136	6.200249	0.084323	0.103810	0.019487	0.019755	0.9864
.0138	6.179188	0.085273	0.105044	0.019771	0.020048	0.9862
.0140	6.158429	0.086218	0.106274	0.020056	0.020340	0.9860
.0142	6.137965	0.087159	0.107499	0.020340	0.020633	0.9858
.0144	6.117787	0.088096	0.108721	0.020625	0.020926	0.9856
.0146	6.097888	0.089029	0.109938	0.020909	0.021219	0.9854
.0148	6.078259	0.089958	0.111151	0.021193	0.021511	0.9852
.0150	6.058894	0.090883	0.112361	0.021477	0.021804	0.9850
.0152	6.039785	0.091805	0.113566	0.021761	0.022097	0.9848
.0154	6.020926	0.092722	0.114768	0.022046	0.022390	0.9846
.0156	6.002310	0.093636	0.115966	0.022330	0.022683	0.9844
.0158	5.983932	0.094546	0.117160	0.022614	0.022977	0.9842
.0160	5.965784	0.095453	0.118350	0.022897	0.023270	0.9840
.0162	5.947862	0.096355	0.119537	0.023181	0.023563	0.9838
.0164	5.930160	0.097255	0.120720	0.023465	0.023856	0.9836
.0166	5.912673	0.098150	0.121899	0.023749	0.024150	0.9834
.0168	5.895395	0.099043	0.123075	0.024033	0.024443	0.9832
.0170	5.878321	0.099931	0.124248	0.024316	0.024737	0.9830
.0172	5.861448	0.100817	0.125417	0.024600	0.025030	0.9828
.0174	5.844769	0.101699	0.126582	0.024883	0.025324	0.9826
.0176	5.828281	0.102578	0.127744	0.025167	0.025618	0.9824
.0178	5.811979	0.103453	0.128903	0.025450	0.025911	0.9822
.0180	5.795859	0.104325	0.130059	0.025733	0.026205	0.9820
.0182	5.779918	0.105195	0.131211	0.026017	0.026499	0.9818
.0184	5.764150	0.106060	0.132360	0.026300	0.026793	0.9816
.0186	5.748554	0.106923	0.133506	0.026583	0.027087	0.9814
.0188	5.733123	0.107783	0.134649	0.026866	0.027381	0.9812
.0190	5.717857	0.108639	0.135788	0.027149	0.027675	0.9810

NUMERICAL TABLES

Entropy Tables (*Continued*)

p	$-\log_2 p$	$-p \log_2 p$	H	$-q \log_2 q$	$-\log q$	q
.0192	5.702750	0.109493	0.136925	0.027432	0.027969	0.9808
.0194	5.687800	0.110343	0.138058	0.027715	0.028263	0.9806
.0196	5.673002	0.111191	0.139189	0.027998	0.028558	0.9804
.0198	5.658356	0.112035	0.140316	0.028281	0.028852	0.9802
.0200	5.643856	0.112877	0.141441	0.028563	0.029146	0.9800
.0202	5.629501	0.113716	0.142562	0.028846	0.029441	0.9798
.0204	5.615287	0.114552	0.143681	0.029129	0.029735	0.9796
.0206	5.601212	0.115385	0.144796	0.029411	0.030030	0.9794
.0208	5.587273	0.116215	0.145909	0.029694	0.030325	0.9792
.0210	5.573467	0.117043	0.147019	0.029976	0.030619	0.9790
.0212	5.559792	0.117868	0.148126	0.030259	0.030914	0.9788
.0214	5.546245	0.118690	0.149231	0.030541	0.031209	0.9786
.0216	5.532825	0.119509	0.150332	0.030823	0.031504	0.9784
.0218	5.519528	0.120326	0.151431	0.031105	0.031799	0.9782
.0220	5.506353	0.121140	0.152527	0.031388	0.032094	0.9780
.0222	5.493297	0.121951	0.153621	0.031670	0.032389	0.9778
.0224	5.480357	0.122760	0.154712	0.031952	0.032684	0.9776
.0226	5.467533	0.123566	0.155800	0.032234	0.032979	0.9774
.0228	5.454822	0.124370	0.156886	0.032516	0.033274	0.9772
.0230	5.442222	0.125171	0.157969	0.032797	0.033570	0.9770
.0232	5.429731	0.125970	0.159049	0.033079	0.033865	0.9768
.0234	5.417348	0.126766	0.160127	0.033361	0.034160	0.9766
.0236	5.405069	0.127560	0.161202	0.033643	0.034456	0.9764
.0238	5.392895	0.128351	0.162275	0.033924	0.034751	0.9762
.0240	5.380822	0.129140	0.163346	0.034206	0.035047	0.9760
.0242	5.368849	0.129926	0.164413	0.034487	0.035343	0.9758
.0244	5.356975	0.130710	0.165479	0.034769	0.035638	0.9756
.0246	5.345198	0.131492	0.166542	0.035050	0.035934	0.9754
.0248	5.333516	0.132271	0.167603	0.035331	0.036230	0.9752
.0250	5.321928	0.133048	0.168661	0.035613	0.036526	0.9750
.0252	5.310432	0.133823	0.169717	0.035894	0.036822	0.9748
.0254	5.299028	0.134595	0.170770	0.036175	0.037118	0.9746
.0256	5.287712	0.135365	0.171822	0.036456	0.037414	0.9744
.0258	5.276485	0.136133	0.172870	0.036737	0.037710	0.9742
.0260	5.265345	0.136899	0.173917	0.037018	0.038006	0.9740
.0262	5.254289	0.137662	0.174961	0.037299	0.038303	0.9738
.0264	5.243318	0.138424	0.176004	0.037580	0.038599	0.9736
.0266	5.232430	0.139183	0.177043	0.037861	0.038895	0.9734
.0268	5.221623	0.139939	0.178081	0.038141	0.039192	0.9732
.0270	5.210897	0.140694	0.179116	0.038422	0.039488	0.9730
.0272	5.200249	0.141447	0.180149	0.038703	0.039785	0.9728
.0274	5.189680	0.142197	0.181180	0.038983	0.040081	0.9726
.0276	5.179188	0.142946	0.182209	0.039264	0.040378	0.9724
.0278	5.168771	0.143692	0.183236	0.039544	0.040675	0.9722
.0280	5.158429	0.144436	0.184261	0.039825	0.040972	0.9720
.0282	5.148161	0.145178	0.185283	0.040105	0.041269	0.9718
.0284	5.137965	0.145918	0.186303	0.040385	0.041566	0.9716
.0286	5.127841	0.146656	0.187322	0.040665	0.041863	0.9714
.0288	5.117787	0.147392	0.188338	0.040945	0.042160	0.9712
.0290	5.107803	0.148126	0.189352	0.041226	0.042457	0.9710

Entropy Tables (*Continued*)

p	$-\log_2 p$	$-p \log_2 p$	H	$-q \log_2 q$	$-\log q$	q
.0292	5.097888	0.148858	0.190364	0.041506	0.042754	0.9708
.0294	5.088040	0.149588	0.191374	0.041786	0.043051	0.9706
.0296	5.078259	0.150316	0.192382	0.042065	0.043349	0.9704
.0298	5.068544	0.151043	0.193388	0.042345	0.043646	0.9702
.0300	5.058894	0.151767	0.194392	0.042625	0.043943	0.9700
.0302	5.049308	0.152489	0.195394	0.042905	0.044241	0.9698
.0304	5.039785	0.153209	0.196394	0.043184	0.044538	0.9696
.0306	5.030325	0.153928	0.197392	0.043464	0.044836	0.9694
.0308	5.020926	0.154645	0.198388	0.043744	0.045134	0.9692
.0310	5.011588	0.155359	0.199382	0.044023	0.045431	0.9690
.0312	5.002310	0.156072	0.200375	0.044302	0.045729	0.9688
.0314	4.993092	0.156783	0.201365	0.044582	0.046027	0.9686
.0316	4.983932	0.157492	0.202353	0.044861	0.046325	0.9684
.0318	4.974829	0.158200	0.203340	0.045140	0.046623	0.9682
.0320	4.965784	0.158905	0.204325	0.045420	0.046921	0.9680
.0322	4.956795	0.159609	0.205307	0.045699	0.047219	0.9678
.0324	4.947862	0.160311	0.206288	0.045978	0.047517	0.9676
.0326	4.938984	0.161011	0.207268	0.046257	0.047816	0.9674
.0328	4.930160	0.161709	0.208245	0.046536	0.048114	0.9672
.0330	4.921390	0.162406	0.209220	0.046815	0.048412	0.9670
.0332	4.912673	0.163101	0.210194	0.047093	0.048711	0.9668
.0334	4.904008	0.163794	0.211166	0.047372	0.049009	0.9666
.0336	4.895395	0.164485	0.212136	0.047651	0.049308	0.9664
.0338	4.886833	0.165175	0.213104	0.047930	0.049606	0.9662
.0340	4.878321	0.165863	0.214071	0.048208	0.049905	0.9660
.0342	4.869860	0.166549	0.215036	0.048487	0.050204	0.9658
.0344	4.861448	0.167234	0.215999	0.048765	0.050502	0.9656
.0346	4.853084	0.167917	0.216960	0.049044	0.050801	0.9654
.0348	4.844769	0.168598	0.217920	0.049322	0.051100	0.9652
.0350	4.836501	0.169278	0.218878	0.049600	0.051399	0.9650
.0352	4.828281	0.169955	0.219834	0.049878	0.051698	0.9648
.0354	4.820107	0.170632	0.220788	0.050157	0.051997	0.9646
.0356	4.811979	0.171306	0.221741	0.050435	0.052296	0.9644
.0358	4.803897	0.171979	0.222692	0.050713	0.052596	0.9642
.0360	4.795859	0.172651	0.223642	0.050991	0.052895	0.9640
.0362	4.787866	0.173321	0.224589	0.051269	0.053194	0.9638
.0364	4.779918	0.173989	0.225536	0.051547	0.053494	0.9636
.0366	4.772013	0.174656	0.226480	0.051824	0.053793	0.9634
.0368	4.764150	0.175321	0.227423	0.052102	0.054093	0.9632
.9370	4.756331	0.175984	0.228364	0.052380	0.054392	0.9630
.0372	4.748554	0.176646	0.229304	0.052657	0.054692	0.9628
.0374	4.740818	0.177307	0.230242	0.052935	0.054992	0.9626
.0376	4.733123	0.177965	0.231178	0.053212	0.055291	0.9624
.0378	4.725470	0.178623	0.232113	0.053490	0.055591	0.9622
.0380	4.717857	0.179279	0.233046	0.053767	0.055891	0.9620
.0382	4.710284	0.179933	0.233977	0.054045	0.056191	0.9618
.0384	4.702750	0.180586	0.234908	0.054322	0.056491	0.9616
.0386	4.695255	0.181237	0.235836	0.054599	0.056791	0.9614
.0388	4.687800	0.181887	0.236763	0.054876	0.057091	0.9612
.0390	4.680382	0.182535	0.237688	0.055153	0.057392	0.9610

NUMERICAL TABLES A-81

Entropy Tables (*Continued*)

p	$-\log_2 p$	$-p \log_2 p$	H	$-q \log_2 q$	$-\log q$	q
.0392	4.673002	0.183182	0.238612	0.055430	0.057692	0.9608
.0394	4.665661	0.183827	0.239534	0.055707	0.057992	0.9606
.0396	4.658356	0.184471	0.240455	0.055984	0.058293	0.9604
.0398	4.651088	0.185113	0.241374	0.056261	0.058593	0.9602
.0400	0.643856	0.185754	0.242292	0.056538	0.058894	0.9600
.0402	4.636661	0.186394	0.243208	0.056815	0.059194	0.9598
.0404	4.629501	0.187032	0.244123	0.057091	0.059495	0.9596
.0406	4.622376	0.187668	0.245036	0.057368	0.059796	0.9594
.0408	4.615287	0.188304	0.245948	0.057644	0.060096	0.9592
.0410	4.608232	0.188938	0.246858	0.057921	0.060397	0.9590
.0412	4.601212	0.189570	0.247767	0.058197	0.060698	0.9588
.0414	4.594225	0.190201	0.248675	0.058474	0.060999	0.9586
.0416	4.587273	0.190831	0.249581	0.058750	0.061300	0.9584
.0418	4.580353	0.191459	0.250485	0.059026	0.061601	0.9582
.0420	4.573467	0.192086	0.251388	0.059303	0.061902	0.9580
.0422	4.566613	0.192711	0.252290	0.059579	0.062204	0.9578
.0424	4.559792	0.193335	0.253190	0.059855	0.062505	0.9576
.0426	4.553003	0.193958	0.254089	0.060131	0.062806	0.9574
.0428	4.546245	0.194579	0.254986	0.060407	0.063108	0.9572
.0430	4.539519	0.195199	0.255882	0.060683	0.063409	0.9570
.0432	4.532825	0.195818	0.256776	0.060958	0.063711	0.9568
.0434	4.526161	0.196435	0.257670	0.061234	0.064012	0.9566
.0436	4.519528	0.197051	0.258561	0.061510	0.064314	0.9564
.0438	4.512925	0.197666	0.259452	0.061786	0.064616	0.9562
.0440	4.506353	0.198280	0.260341	0.062061	0.064917	0.9560
.0442	4.499810	0.198892	0.261228	0.062337	0.065219	0.9558
.0444	4.493297	0.199502	0.262114	0.062612	0.065521	0.9556
.0446	4.486812	0.200112	0.262999	0.062887	0.065823	0.9554
.0448	4.480357	0.200720	0.263883	0.063163	0.066125	0.9552
.0450	4.473931	0.201327	0.264765	0.063438	0.066427	0.9550
.0452	4.467533	0.201933	0.265646	0.063713	0.066730	0.9548
.0454	4.461164	0.202537	0.266525	0.063989	0.067032	0.9546
.0456	4.454822	0.203140	0.267404	0.064264	0.067334	0.9544
.0458	4.448509	0.203742	0.268280	0.064539	0.067636	0.9542
.0460	4.442222	0.204342	0.269156	0.064814	0.067939	0.9540
.0462	4.435963	0.204942	0.270030	0.065089	0.068241	0.9538
.0464	4.429731	0.205540	0.270903	0.065363	0.068544	0.9536
.0466	4.423526	0.206136	0.271775	0.065638	0.068846	0.9534
.0468	4.417348	0.206732	0.272645	0.065913	0.069149	0.9532
.0470	4.411195	0.207326	0.273514	0.066188	0.069452	0.9530
.0472	4.405069	0.207919	0.274382	0.066462	0.069755	0.9528
.0474	4.398969	0.208511	0.275248	0.066737	0.070058	0.9526
.0476	4.392895	0.209102	0.276113	0.067011	0.070360	0.9524
.0478	4.386846	0.209691	0.276977	0.067286	0.070663	0.9522
.0480	4.380822	0.210279	0.277840	0.067560	0.070967	0.9520
.0482	4.374823	0.210866	0.278701	0.067834	0.071270	0.9518
.0484	4.368849	0.211452	0.279561	0.068109	0.071573	0.9516
.0486	4.362900	0.212037	0.280420	0.068383	0.071876	0.9514
.0488	4.356975	0.212620	0.281277	0.068657	0.072179	0.9512
.0490	4.351074	0.213203	0.282134	0.068931	0.072483	0.9510

Entropy Tables (Continued)

p	$-\log_2 p$	$-p \log_2 p$	H	$-q \log_2 q$	$-\log q$	q
.0492	4.345198	0.213784	0.282989	0.069205	0.072786	0.9508
.0494	4.339345	0.214364	0.283843	0.069479	0.073090	0.9506
.0496	4.333516	0.214942	0.284695	0.069753	0.073393	0.9504
.0498	4.327710	0.215520	0.285547	0.070027	0.073697	0.9502
.0500	4.321928	0.216096	0.286397	0.070301	0.074001	0.9500
.0502	4.316169	0.216672	0.287246	0.070574	0.074304	0.9498
.0504	4.310432	0.217246	0.288094	0.070848	0.074608	0.9496
.0506	4.304719	0.217819	0.288940	0.071121	0.074912	0.9494
.0508	4.299028	0.218391	0.289786	0.071395	0.075216	0.9492
.0510	4.293359	0.218961	0.290630	0.071668	0.075520	0.9490
.0512	4.287712	0.219531	0.291473	0.071942	0.075824	0.9488
.0514	4.282088	0.220099	0.292315	0.072215	0.076128	0.9486
.0516	4.276485	0.220667	0.293155	0.072489	0.076432	0.9484
.0518	4.270904	0.221233	0.293995	0.072762	0.076737	0.9482
.0520	4.265345	0.221798	0.294833	0.073035	0.077041	0.9480
.0522	4.259806	0.222362	0.295670	0.073308	0.077345	0.9478
.0524	4.254289	0.222925	0.296506	0.073581	0.077650	0.9476
.0526	4.248793	0.223487	0.297341	0.073854	0.077954	0.9474
.0528	4.243318	0.224047	0.298174	0.074127	0.078259	0.9472
.0530	4.237864	0.224607	0.299007	0.074400	0.078564	0.9470
.0532	4.232430	0.225165	0.299838	0.074673	0.078868	0.9468
.0534	4.227016	0.225723	0.300668	0.074945	0.079173	0.9466
.0536	4.221623	0.226279	0.301497	0.075218	0.079478	0.9464
.0538	4.216250	0.226834	0.302325	0.075491	0.079783	0.9462
.0540	4.210897	0.227388	0.303152	0.075763	0.080088	0.9460
.0542	4.205563	0.227942	0.303977	0.076036	0.080393	0.9458
.0544	4.200249	0.228494	0.304802	0.076308	0.080698	0.9456
.0546	4.194955	0.229045	0.305625	0.076580	0.081003	0.9454
.0548	4.189680	0.229594	0.306447	0.076853	0.081308	0.9452
.0550	4.184425	0.230143	0.307268	0.077125	0.081614	0.9450
.0552	4.179188	0.230691	0.308088	0.077397	0.081919	0.9448
.0554	4.173970	0.231238	0.308907	0.077669	0.082225	0.9446
.0556	4.168771	0.231784	0.309725	0.077941	0.082530	0.9444
.0558	4.163591	0.232328	0.310542	0.078213	0.082836	0.9442
.0560	4.158429	0.232872	0.311357	0.078485	0.083141	0.9440
.0562	4.153286	0.233415	0.312172	0.078757	0.083447	0.9438
.0564	4.148161	0.233956	0.312985	0.079029	0.083753	0.9436
.0566	4.143054	0.234497	0.313798	0.079301	0.084058	0.9434
.0568	4.137965	0.235036	0.314609	0.079572	0.084364	0.9432
.0570	4.132894	0.235575	0.315419	0.079844	0.084670	0.9430
.0572	4.127841	0.236113	0.316228	0.080116	0.084976	0.9428
.0574	4.122805	0.236649	0.317036	0.080387	0.085282	0.9426
.0576	4.117787	0.237185	0.317843	0.080659	0.085589	0.9424
.0578	4.112787	0.237719	0.318649	0.080930	0.085895	0.9422
.0580	4.107803	0.238253	0.319454	0.081201	0.086201	0.9420
.0582	4.102837	0.238785	0.320258	0.081473	0.086507	0.9418
.0584	4.097888	0.239317	0.321060	0.081744	0.086814	0.9416
.0586	4.092956	0.239847	0.321862	0.082015	0.087120	0.9414
.0588	4.088040	0.240377	0.322663	0.082286	0.087427	0.9412
.0590	4.083141	0.240905	0.323462	0.082557	0.087733	0.9410

NUMERICAL TABLES   A–83

Entropy Tables (*Continued*)

p	$-\log_2 p$	$-p \log_2 p$	H	$-q \log_2 q$	$-\log q$	q
.0592	4.078259	0.241433	0.324261	0.082828	0.088040	0.9408
.0594	4.073393	0.241960	0.325059	0.083099	0.088347	0.9406
.0596	4.068544	0.242485	0.325855	0.083370	0.088654	0.9404
.0598	4.063711	0.243010	0.326650	0.083641	0.088960	0.9402
.0600	4.058894	0.243534	0.327445	0.083911	0.089267	0.9400
.0605	4.046921	0.244839	0.329427	0.084588	0.090035	0.9395
.0610	4.035047	0.246138	0.331402	0.085264	0.090803	0.9390
.0615	4.023270	0.247431	0.333371	0.085940	0.091571	0.9385
.0620	4.011588	0.248718	0.335334	0.086615	0.092340	0.9380
.0625	4.000000	0.250000	0.337290	0.087290	0.093109	0.9375
.0630	3.988504	0.251276	0.339240	0.087965	0.093879	0.9370
.0635	3.977100	0.252546	0.341185	0.088639	0.094649	0.9365
.0640	3.965784	0.253810	0.343123	0.089313	0.095420	0.9360
.0645	3.954557	0.255069	0.345055	0.089986	0.096190	0.9355
.0650	3.943416	0.256322	0.346981	0.090659	0.096962	0.9350
.0655	3.932361	0.257570	0.348902	0.091332	0.097733	0.9345
.0660	3.921390	0.258812	0.350816	0.092004	0.098506	0.9340
.0665	3.910502	0.260048	0.352724	0.092676	0.099278	0.9335
.0670	3.899695	0.261280	0.354627	0.093348	0.100051	0.9330
.0675	3.888969	0.262505	0.356524	0.094019	0.100824	0.9325
.0680	3.878321	0.263726	0.358415	0.094689	0.101598	0.9320
.0685	3.867752	0.264941	0.360301	0.095360	0.102372	0.9315
.0690	3.857260	0.266151	0.362181	0.096030	0.103147	0.9310
.0695	3.846843	0.267356	0.364055	0.096699	0.103922	0.9305
.0700	3.836501	0.268555	0.365924	0.097369	0.104697	0.9300
.0705	3.826233	0.269749	0.367787	0.098037	0.105473	0.9295
.0710	3.816037	0.270939	0.369644	0.098706	0.106249	0.9290
.0715	3.805913	0.272123	0.371497	0.099374	0.107026	0.9285
.0720	3.795859	0.273302	0.373343	0.100041	0.107803	0.9280
.0725	3.785875	0.274476	0.375185	0.100709	0.108581	0.9275
.0730	3.775960	0.275645	0.377021	0.101376	0.109359	0.9270
.0735	3.766112	0.276809	0.378851	0.102042	0.110137	0.9265
.0740	3.756331	0.277968	0.380677	0.102708	0.110916	0.9260
.0745	3.746616	0.279123	0.382497	0.103374	0.111695	0.9255
.0750	3.736966	0.280272	0.384312	0.104039	0.112475	0.9250
.0755	3.727380	0.281417	0.386121	0.104704	0.113255	0.9245
.0760	3.717857	0.282557	0.387926	0.105369	0.114035	0.9240
.0765	3.708396	0.283692	0.389725	0.106033	0.114816	0.9235
.0770	3.698998	0.284823	0.391519	0.106696	0.115597	0.9230
.0775	3.689660	0.285949	0.393308	0.107360	0.116379	0.9225
.0780	3.680382	0.287070	0.395093	0.108023	0.117161	0.9220
.0785	3.671164	0.288186	0.396872	0.108685	0.117944	0.9215
.0790	3.662004	0.289298	0.398646	0.109348	0.118727	0.9210
.0795	3.652901	0.290406	0.400415	0.110009	0.119510	0.9205
.0800	3.643856	0.291508	0.402179	0.110671	0.120294	0.9200
.0805	3.634867	0.292607	0.403939	0.111332	0.121079	0.9195
.0810	3.625934	0.293701	0.405693	0.111992	0.121863	0.9190
.0815	3.617056	0.294790	0.407443	0.112653	0.122648	0.9185
.0820	3.608232	0.295875	0.409187	0.113312	0.123434	0.9180
.0825	3.599462	0.296956	0.410927	0.113972	0.124220	0.9175

A-84 MATHEMATICAL FORMULAS, THEOREMS, AND DEFINITIONS

Entropy Tables (*Continued*)

p	$-\log_2 p$	$-p \log_2 p$	H	$-q \log_2 q$	$-\log q$	q
.0830	3.590745	0.298032	0.412663	0.114631	0.125006	0.9170
.0835	3.582080	0.299104	0.414393	0.115289	0.125793	0.9165
.0840	3.573467	0.300171	0.416119	0.115948	0.126580	0.9160
.0845	3.564905	0.301234	0.417840	0.116606	0.127368	0.9155
.0850	3.556393	0.302293	0.419556	0.117263	0.128156	0.9150
.0855	3.547932	0.303348	0.421268	0.117920	0.128945	0.9145
.0860	3.539520	0.304399	0.422975	0.118577	0.129734	0.9140
.0865	3.531156	0.305445	0.424678	0.119233	0.130523	0.9135
.0870	3.522841	0.306487	0.426376	0.119889	0.131313	0.9130
.0875	3.514573	0.307525	0.428070	0.120544	0.132104	0.9125
.0880	3.506353	0.308559	0.429759	0.121200	0.132894	0.9120
.0885	3.498179	0.309589	0.431443	0.121854	0.133685	0.9115
.0890	3.490051	0.310615	0.433123	0.122509	0.134477	0.9110
.0895	3.481968	0.311636	0.434799	0.123162	0.135269	0.9105
.0900	3.473931	0.312654	0.436470	0.123816	0.136062	0.9100
.0905	3.465938	0.313667	0.438137	0.124469	0.136854	0.9095
.0910	3.457990	0.314677	0.439799	0.125122	0.137648	0.9090
.0915	3.450084	0.315683	0.441457	0.125774	0.138442	0.9085
.0920	3.442222	0.316684	0.443111	0.126426	0.139236	0.9080
.0925	3.434403	0.317682	0.444760	0.127078	0.140030	0.9075
.0930	3.426625	0.318676	0.446405	0.127729	0.140826	0.9070
.0935	3.418890	0.319666	0.448046	0.128379	0.141621	0.9065
.0940	3.411195	0.320652	0.449682	0.129030	0.142417	0.9060
.0945	3.403542	0.321635	0.451314	0.129680	0.143213	0.9055
.0950	3.395929	0.322613	0.452943	0.130329	0.144010	0.9050
.0955	3.388355	0.323588	0.454566	0.130978	0.144808	0.9045
.0960	3.380822	0.324559	0.456186	0.131627	0.145506	0.9040
.0965	3.373327	0.325526	0.457802	0.132276	0.146403	0.9035
.0970	3.365871	0.326490	0.459413	0.132923	0.147202	0.9030
.0975	3.358454	0.327449	0.461020	0.133571	0.148001	0.9025
.0980	3.351074	0.328405	0.462623	0.134218	0.148801	0.9020
.0985	3.343732	0.329358	0.464223	0.134865	0.149601	0.9015
.0990	3.336428	0.330306	0.465818	0.135511	0.150401	0.9010
.0995	3.329160	0.331251	0.467409	0.136157	0.151202	0.9005
.1000	3.321928	0.332193	0.468996	0.136803	0.152003	0.9000
.1005	3.314733	0.333131	0.470579	0.137448	0.152805	0.8995
.1010	3.307573	0.334065	0.472158	0.138093	0.153607	0.8990
.1015	3.300448	0.334996	0.473733	0.138737	0.154410	0.8985
.1020	3.293359	0.335923	0.475304	0.139381	0.155213	0.8980
.1025	3.286304	0.336846	0.476871	0.140024	0.156016	0.8975
.1030	3.279284	0.337766	0.478434	0.140668	0.156820	0.8970
.1035	3.272297	0.338683	0.479993	0.141310	0.157624	0.8965
.1040	3.265345	0.339596	0.481549	0.141953	0.158429	0.8960
.1045	3.258425	0.340505	0.483100	0.142595	0.159235	0.8955
.1050	3.251539	0.341412	0.484648	0.143236	0.160040	0.8950
.1055	3.244685	0.342314	0.486192	0.143877	0.160847	0.8945
.1060	3.237864	0.343214	0.487732	0.144518	0.161653	0.8940
.1065	3.231075	0.344109	0.489268	0.145158	0.162460	0.8935
.1070	3.224317	0.345002	0.490800	0.145798	0.163268	0.8930
.1075	3.217591	0.345891	0.492329	0.146438	0.164076	0.8925

NUMERICAL TABLES

Entropy Tables (*Continued*)

p	$-\log_2 p$	$-p \log_2 p$	H	$-q \log_2 q$	$-\log q$	q
.1080	3.210897	0.346777	0.493854	0.147077	0.164884	0.8920
.1085	3.204233	0.347659	0.495375	0.147716	0.165693	0.8915
.1090	3.197600	0.348538	0.496892	0.148354	0.166503	0.8910
.1095	3.190997	0.349414	0.498406	0.148992	0.167312	0.8905
.1100	3.184425	0.350287	0.499916	0.149629	0.168123	0.8900
.1105	3.177882	0.351156	0.501422	0.150266	0.168933	0.8895
.1110	3.171368	0.352022	0.502925	0.150903	0.169745	0.8890
.1115	3.164884	0.352885	0.504424	0.151539	0.170556	0.8885
.1120	3.158429	0.353744	0.505919	0.152175	0.171368	0.8880
.1125	3.152003	0.354600	0.507411	0.152811	0.172181	0.8875
.1130	3.145605	0.355453	0.508899	0.153446	0.172994	0.8870
.1135	3.139236	0.356303	0.510384	0.154080	0.173807	0.8865
.1140	3.132894	0.357150	0.511864	0.154715	0.174621	0.8860
.1145	3.126580	0.357993	0.513342	0.155348	0.175436	0.8855
.1150	3.120294	0.358834	0.514816	0.155982	0.176251	0.8850
.1155	3.114035	0.359671	0.516286	0.156615	0.177066	0.8845
.1160	3.107803	0.360505	0.517753	0.157247	0.177882	0.8840
.1165	3.101598	0.361336	0.519216	0.157880	0.178698	0.8835
.1170	3.095420	0.362164	0.520676	0.158511	0.179515	0.8830
.1175	3.089267	0.362989	0.522132	0.159143	0.180332	0.8825
.1180	3.083141	0.363811	0.523584	0.159774	0.181149	0.8820
.1185	3.077041	0.364629	0.525034	0.160404	0.181968	0.8815
.1190	3.070967	0.365445	0.526480	0.161035	0.182786	0.8810
.1195	3.064917	0.366258	0.527922	0.161664	0.183605	0.8805
.1200	3.058894	0.367067	0.529361	0.162294	0.184425	0.8800
.1205	3.052895	0.367874	0.530796	0.162923	0.185245	0.8795
.1210	3.046921	0.368677	0.532228	0.163551	0.186065	0.8790
.1215	3.040972	0.369478	0.533657	0.164179	0.186886	0.8785
.1220	3.035047	0.370276	0.535083	0.164807	0.187707	0.8780
.1225	3.029146	0.371070	0.536505	0.165434	0.188529	0.8775
.1230	3.023270	0.371862	0.537923	0.166061	0.189351	0.8770
.1235	3.017417	0.372651	0.539339	0.166688	0.190174	0.8765
.1240	3.011588	0.373437	0.540750	0.167314	0.190997	0.8760
.1245	3.005782	0.374220	0.542159	0.167939	0.191821	0.8755
.1250	3.000000	0.375000	0.543564	0.168564	0.192645	0.8750
.1255	2.994241	0.375777	0.544966	0.169189	0.193470	0.8745
.1260	2.988504	0.376552	0.546365	0.169814	0.194295	0.8740
.1265	2.982791	0.377323	0.547761	0.170438	0.195120	0.8735
.1270	2.977100	0.378092	0.549153	0.171061	0.195946	0.8730
.1275	2.971431	0.378857	0.550542	0.171684	0.196773	0.8725
.1280	2.965784	0.379620	0.551928	0.172307	0.197600	0.8720
.1285	2.960160	0.380381	0.553310	0.172929	0.198427	0.8715
.1290	2.954557	0.381138	0.554689	0.173551	0.199255	0.8710
.1295	2.948976	0.381892	0.556065	0.174173	0.200084	0.8705
.1300	2.943416	0.382644	0.557438	0.174794	0.200913	0.8700
.1305	2.937878	0.383393	0.558808	0.175415	0.201742	0.8695
.1310	2.932361	0.384139	0.560174	0.176035	0.202572	0.8690
.1315	2.926865	0.384883	0.561538	0.176655	0.203402	0.8685
.1320	2.921390	0.385623	0.562898	0.177274	0.204233	0.8680
.1325	2.915936	0.386361	0.564255	0.177893	0.205064	0.8675

Entropy Tables (Continued)

p	$-\log_2 p$	$-p \log_2 p$	H	$-q \log_2 q$	$-\log q$	q
.1330	2.910502	0.387097	0.565609	0.178512	0.205896	0.8670
.1335	2.905088	0.387829	0.566959	0.179130	0.206728	0.8665
.1340	2.899695	0.388559	0.568307	0.179748	0.207561	0.8660
.1345	2.894322	0.389286	0.569652	0.180365	0.208394	0.8655
.1350	2.888969	0.390011	0.570993	0.180982	0.209228	0.8650
.1355	2.883635	0.390733	0.572331	0.181599	0.210062	0.8645
.1360	2.878321	0.391452	0.573667	0.182215	0.210897	0.8640
.1365	2.873027	0.392168	0.574999	0.182830	0.211732	0.8635
.1370	2.867752	0.392882	0.576328	0.183446	0.212568	0.8630
.1375	2.862496	0.393593	0.577654	0.184061	0.213404	0.8625
.1380	2.857260	0.394302	0.578977	0.184675	0.214240	0.8620
.1385	2.852042	0.395008	0.580297	0.185289	0.215077	0.8615
.1390	2.846843	0.395711	0.581614	0.185903	0.215915	0.8610
.1395	2.841663	0.396412	0.582928	0.186516	0.216753	0.8605
.1400	2.836501	0.397110	0.584239	0.187129	0.217591	0.8600
.1405	2.831358	0.397806	0.585547	0.187741	0.218430	0.8595
.1410	2.826233	0.398499	0.586852	0.188353	0.219270	0.8590
.1415	2.821126	0.399189	0.588154	0.188964	0.220110	0.8585
.1420	2.816037	0.399877	0.589453	0.189575	0.220950	0.8580
.1425	2.810966	0.400563	0.590749	0.190186	0.221791	0.8575
.1430	2.805913	0.401246	0.592042	0.190796	0.222633	0.8570
.1435	2.800877	0.401926	0.593332	0.191406	0.223475	0.8565
.1440	2.795859	0.402604	0.594619	0.192016	0.224317	0.8560
.1445	2.790859	0.403279	0.595904	0.192625	0.225160	0.8555
.1450	2.785875	0.403952	0.597185	0.193233	0.226004	0.8550
.1455	2.780909	0.404622	0.598464	0.193841	0.226848	0.8545
.1460	2.775960	0.405290	0.599739	0.194449	0.227692	0.8540
.1465	2.771027	0.405956	0.601012	0.195056	0.228537	0.8535
.1470	2.766112	0.406618	0.602282	0.195663	0.229382	0.8530
.1475	2.761213	0.407279	0.603549	0.196270	0.230228	0.8525
.1480	2.756331	0.407937	0.604813	0.196876	0.231075	0.8520
.1485	2.751465	0.408593	0.606074	0.197481	0.231922	0.8515
.1490	2.746616	0.409246	0.607332	0.198086	0.232769	0.8510
.1495	2.741783	0.409896	0.608588	0.198691	0.233617	0.8505
.1500	2.736966	0.410545	0.609840	0.199295	0.234465	0.8500
.1505	2.732165	0.411191	0.611090	0.199899	0.235314	0.8495
.1510	2.727380	0.411834	0.612337	0.200503	0.236164	0.8490
.1515	2.722610	0.412475	0.613581	0.201106	0.237013	0.8485
.1520	2.717857	0.413114	0.614823	0.201709	0.237864	0.8480
.1525	2.713119	0.413751	0.616061	0.202311	0.238715	0.8475
.1530	2.708396	0.414385	0.617297	0.202913	0.239566	0.8470
.1535	2.703689	0.415016	0.618530	0.203514	0.240418	0.8465
.1540	2.698998	0.415646	0.619760	0.204115	0.241270	0.8460
.1545	2.694321	0.416273	0.620988	0.204715	0.242123	0.8455
.1550	2.689660	0.416897	0.622213	0.205315	0.242977	0.8450
.1555	2.685014	0.417520	0.623435	0.205915	0.243831	0.8445
.1560	2.680382	0.418140	0.624654	0.206514	0.244685	0.8440
.1565	2.675765	0.418757	0.625870	0.207113	0.245540	0.8435
.1570	2.671164	0.419373	0.627084	0.207711	0.246395	0.8430
.1575	2.666576	0.419986	0.628295	0.208309	0.247251	0.8425

Entropy Tables (Continued)

p	$-\log_2 p$	$-p \log_2 p$	H	$-q \log_2 q$	$-\log q$	q
.1580	2.662004	0.420597	0.629503	0.208907	0.248108	0.8420
.1585	2.657445	0.421205	0.630709	0.209504	0.248965	0.8415
.1590	2.652901	0.421811	0.631912	0.210101	0.249822	0.8410
.1595	2.648372	0.422415	0.633112	0.210697	0.250680	0.8405
.1600	2.643856	0.423017	0.634310	0.211293	0.251539	0.8400
.1605	2.639355	0.423616	0.635504	0.211888	0.252398	0.8395
.1610	2.634867	0.424214	0.636696	0.212483	0.253257	0.8390
.1615	2.630394	0.424809	0.637886	0.213077	0.254117	0.8385
.1620	2.625934	0.425401	0.639073	0.213671	0.254978	0.8380
.1625	2.621488	0.425992	0.640257	0.214265	0.255839	0.8375
.1630	2.617056	0.426580	0.641438	0.214858	0.256700	0.8370
.1635	2.612637	0.427166	0.642617	0.215451	0.257563	0.8365
.1640	2.608232	0.427750	0.643794	0.216043	0.258425	0.8360
.1645	2.603841	0.428332	0.644967	0.216635	0.259288	0.8355
.1650	2.599462	0.428911	0.646138	0.217227	0.260152	0.8350
.1655	2.595097	0.429489	0.647306	0.217818	0.261016	0.8345
.1660	2.590745	0.430064	0.648472	0.218409	0.261881	0.8340
.1665	2.586406	0.430637	0.649635	0.218999	0.262746	0.8335
.1670	2.582080	0.431207	0.650796	0.219588	0.263612	0.8330
.1675	2.577767	0.431776	0.651954	0.220178	0.264478	0.8325
.1680	2.573467	0.432342	0.653109	0.220767	0.265345	0.8320
.1685	2.569180	0.432907	0.654262	0.221355	0.266212	0.8315
.1690	2.564905	0.433469	0.655412	0.221943	0.267080	0.8310
.1695	2.560643	0.434029	0.656560	0.222531	0.267948	0.8305
.1700	2.556393	0.434587	0.657705	0.223118	0.268817	0.8300
.1705	2.552156	0.435143	0.658847	0.223705	0.269686	0.8295
.1710	2.547932	0.435696	0.659987	0.224291	0.270556	0.8290
.1715	2.543720	0.436248	0.661125	0.224877	0.271426	0.8285
.1720	2.539520	0.436797	0.662260	0.225462	0.272297	0.8280
.1725	2.535332	0.437345	0.663392	0.226047	0.273169	0.8275
.1730	2.531156	0.437890	0.664522	0.226632	0.274041	0.8270
.1735	2.526992	0.438433	0.665649	0.227216	0.274913	0.8265
.1740	2.522841	0.438974	0.666774	0.227799	0.275786	0.8260
.1745	2.518701	0.439513	0.667896	0.228383	0.276660	0.8255
.1750	2.514573	0.440050	0.669016	0.228966	0.277534	0.8250
.1755	2.510457	0.440585	0.670133	0.229548	0.278409	0.8245
.1760	2.506353	0.441118	0.671248	0.230130	0.279284	0.8240
.1765	2.502260	0.441649	0.672360	0.230711	0.280159	0.8235
.1770	2.498179	0.442178	0.673470	0.231292	0.281036	0.8230
.1775	2.494109	0.442704	0.674577	0.231873	0.281912	0.8225
.1780	2.490051	0.443229	0.675682	0.232453	0.282790	0.8220
.1785	2.486004	0.443752	0.676785	0.233033	0.283668	0.8215
.1790	2.481968	0.444272	0.677885	0.233612	0.284546	0.8210
.1795	2.477944	0.444791	0.678982	0.234191	0.285425	0.8205
.1800	2.473931	0.445308	0.680077	0.234769	0.286304	0.8200
.1805	2.469929	0.445822	0.681170	0.235347	0.287184	0.8195
.1810	2.465938	0.446335	0.682260	0.235925	0.288065	0.8190
.1815	2.461959	0.446845	0.683347	0.236502	0.288946	0.8185
.1820	2.457990	0.447354	0.684433	0.237079	0.289827	0.8180
.1825	2.454032	0.447861	0.685516	0.237655	0.290709	0.8175

A-88   MATHEMATICAL FORMULAS, THEOREMS, AND DEFINITIONS

Entropy Tables (*Continued*)

p	$-\log_2 p$	$-p \log_2 p$	H	$-q \log_2 q$	$-\log q$	q
.1830	2.450084	0.448365	0.686596	0.238231	0.291592	0.8170
.1835	2.446148	0.448868	0.687674	0.238806	0.292475	0.8165
.1840	2.442222	0.449369	0.688750	0.239381	0.293359	0.8160
.1845	2.438307	0.449868	0.689823	0.239955	0.294243	0.8155
.1850	2.434403	0.450365	0.690894	0.240529	0.295128	0.8150
.1855	2.430509	0.450859	0.691962	0.241103	0.296013	0.8145
.1860	2.426625	0.451352	0.693028	0.241676	0.296899	0.8140
.1865	2.422752	0.451843	0.694092	0.242249	0.297786	0.8135
.1870	2.418890	0.452332	0.695153	0.242821	0.298673	0.8130
.1875	2.415037	0.452820	0.696212	0.243393	0.299560	0.8125
.1880	2.411195	0.453305	0.697269	0.243964	0.300448	0.8120
.1885	2.407364	0.453788	0.698323	0.244535	0.301337	0.8115
.1890	2.403542	0.454269	0.699375	0.245105	0.302226	0.8110
.1895	2.399730	0.454749	0.700424	0.245675	0.303116	0.8105
.1900	2.395929	0.455226	0.701471	0.246245	0.304006	0.8100
.1905	2.392137	0.455702	0.702516	0.246814	0.304897	0.8095
.1910	2.388355	0.456176	0.703559	0.247383	0.305788	0.8090
.1915	2.384584	0.456648	0.704599	0.247951	0.306680	0.8085
.1920	2.380822	0.457118	0.705637	0.248519	0.307573	0.8080
.1925	2.377070	0.457586	0.706672	0.249086	0.308466	0.8075
.1930	2.373327	0.458052	0.707705	0.249653	0.309359	0.8070
.1935	2.369595	0.458517	0.708736	0.250219	0.310254	0.8065
.1940	2.365871	0.458979	0.709765	0.250785	0.311148	0.8060
.1945	2.362158	0.459440	0.710791	0.251351	0.312044	0.8055
.1950	2.358454	0.459899	0.711815	0.251916	0.312939	0.8050
.1955	2.354759	0.460355	0.712836	0.252481	0.313836	0.8045
.1960	2.351074	0.460811	0.713856	0.253045	0.314733	0.8040
.1965	2.347399	0.461264	0.714873	0.253609	0.315630	0.8035
.1970	2.343732	0.461715	0.715887	0.254172	0.316528	0.8030
.1975	2.340075	0.462165	0.716900	0.254735	0.317427	0.8025
.1980	2.336428	0.462613	0.717910	0.255297	0.318326	0.8020
.1985	2.332789	0.463059	0.718918	0.255859	0.319226	0.8015
.1990	2.329160	0.463503	0.719924	0.256421	0.320136	0.8010
.1995	2.325539	0.463945	0.720927	0.256982	0.321027	0.8005
.2000	2.321928	0.464386	0.721928	0.257542	0.321928	0.8000
.2010	2.314733	0.465261	0.723924	0.258662	0.323733	0.7990
.2020	2.307573	0.466130	0.725910	0.259780	0.325539	0.7980
.2030	2.300448	0.466991	0.727888	0.260897	0.327348	0.7970
.2040	2.293359	0.467845	0.729856	0.262011	0.329160	0.7960
.2050	2.286304	0.468692	0.731816	0.263124	0.330973	0.7950
.2060	2.279284	0.469532	0.733767	0.264235	0.332789	0.7940
.2070	2.272297	0.470366	0.735709	0.265344	0.334607	0.7930
.2080	2.265345	0.471192	0.737642	0.266451	0.336428	0.7920
.2090	2.258425	0.472011	0.739567	0.267556	0.338250	0.7910
.2100	2.251539	0.472823	0.741483	0.268660	0.340075	0.7900
.2110	2.244685	0.473629	0.743390	0.269761	0.341903	0.7890
.2120	2.237864	0.474427	0.745288	0.270861	0.343732	0.7880
.2130	2.231075	0.475219	0.747178	0.271959	0.345564	0.7870
.2140	2.224317	0.476004	0.749059	0.273055	0.347399	0.7860
.2150	2.217591	0.476782	0.750932	0.274150	0.349235	0.7850

NUMERICAL TABLES A-89

Entropy Tables (*Continued*)

p	$-\log_2 p$	$-p \log_2 p$	H	$-q \log_2 q$	$-\log q$	q
.2160	2.210897	0.477554	0.752796	0.275242	0.351074	0.7840
.2170	2.204233	0.478319	0.754652	0.276333	0.352916	0.7830
.2180	2.197600	0.479077	0.756499	0.277422	0.354759	0.7820
.2190	2.190997	0.479828	0.758337	0.278509	0.356606	0.7810
.2200	2.184425	0.480573	0.760167	0.279594	0.358454	0.7800
.2210	2.177882	0.481312	0.761989	0.280677	0.360305	0.7790
.2220	2.171368	0.482044	0.763803	0.281759	0.362158	0.7780
.2230	2.164884	0.482769	0.765608	0.282838	0.364013	0.7770
.2240	2.158429	0.483488	0.767404	0.283916	0.365871	0.7760
.2250	2.152003	0.484201	0.769193	0.284992	0.367732	0.7750
.2260	2.145605	0.484907	0.770973	0.286066	0.369595	0.7740
.2270	2.139236	0.485607	0.772745	0.287138	0.371460	0.7730
.2280	2.132894	0.486300	0.774509	0.288209	0.373327	0.7720
.2290	2.126580	0.486987	0.776264	0.289277	0.375197	0.7710
.2300	2.120294	0.487668	0.778011	0.290344	0.377070	0.7700
.2310	2.114035	0.488342	0.779750	0.291408	0.378944	0.7690
.2320	2.107803	0.489010	0.781481	0.292471	0.380822	0.7680
.2330	2.101598	0.489672	0.783204	0.293532	0.382702	0.7670
.2340	2.095420	0.490328	0.784919	0.294591	0.384584	0.7660
.2350	2.089267	0.490978	0.786626	0.295648	0.386468	0.7650
.2360	2.083141	0.491621	0.788325	0.296704	0.388355	0.7640
.2370	2.077041	0.492259	0.790016	0.297757	0.390245	0.7630
.2380	2.070967	0.492890	0.791698	0.298808	0.392137	0.7620
.2390	2.064917	0.493515	0.793373	0.299858	0.394032	0.7610
.2400	2.058894	0.494134	0.795040	0.300906	0.395929	0.7600
.2410	2.052895	0.494748	0.796699	0.301952	0.397828	0.7590
.2420	2.046921	0.495355	0.798350	0.302996	0.399730	0.7580
.2430	2.040972	0.495956	0.799994	0.304038	0.401635	0.7570
.2440	2.035047	0.496551	0.801629	0.305078	0.403542	0.7560
.2450	2.029146	0.497141	0.803257	0.306116	0.405451	0.7550
.2460	2.023270	0.497724	0.804876	0.307152	0.407364	0.7540
.2470	2.017417	0.498302	0.806488	0.308186	0.409278	0.7530
.2480	2.011588	0.498874	0.808093	0.309219	0.411195	0.7520
.2490	2.005782	0.499440	0.809689	0.310249	0.413115	0.7510
.2500	2.000000	0.500000	0.811278	0.311278	0.415037	0.7500
.2510	1.994241	0.500554	0.812859	0.312305	0.416962	0.7490
.2520	1.988504	0.501103	0.814433	0.313330	0.418890	0.7480
.2530	1.982791	0.501646	0.815998	0.314352	0.420820	0.7470
.2540	1.977100	0.502183	0.817557	0.315373	0.422752	0.7460
.2550	1.971431	0.502715	0.819107	0.316392	0.424688	0.7450
.2560	1.965784	0.503241	0.820650	0.317409	0.426625	0.7440
.2570	1.960160	0.503761	0.822185	0.318424	0.428566	0.7430
.2580	1.954557	0.504276	0.823713	0.319438	0.430509	0.7420
.2590	1.948976	0.504785	0.825234	0.320449	0.432455	0.7410
.2600	1.943416	0.505288	0.826746	0.321458	0.434403	0.7400
.2610	1.937878	0.505786	0.828252	0.322465	0.436354	0.7390
.2620	1.932361	0.506279	0.829749	0.323471	0.438307	0.7380
.2630	1.926865	0.506766	0.831240	0.324474	0.440263	0.7370
.2640	1.921390	0.507247	0.832723	0.325476	0.442222	0.7360
.2650	1.915936	0.507723	0.834198	0.326475	0.444184	0.7350

Entropy Tables (Continued)

p	$-\log_2 p$	$-p \log_2 p$	H	$-q \log_2 q$	$-\log q$	q
.2660	1.910502	0.508193	0.835666	0.327473	0.446148	0.7340
.2670	1.905088	0.508659	0.837127	0.328468	0.448115	0.7330
.2680	1.899695	0.509118	0.838580	0.329462	0.450084	0.7320
.2690	1.894322	0.509573	0.840026	0.330453	0.452057	0.7310
.2700	1.888969	0.510022	0.841465	0.331443	0.454032	0.7300
.2710	1.883635	0.510465	0.842896	0.332431	0.456009	0.7290
.2720	1.878321	0.510903	0.844320	0.333416	0.457990	0.7280
.2730	1.873027	0.511336	0.845737	0.334400	0.459973	0.7270
.2740	1.867752	0.511764	0.847146	0.335382	0.461959	0.7260
.2750	1.862496	0.512187	0.848548	0.336362	0.463947	0.7250
.2760	1.857260	0.512604	0.849943	0.337339	0.465938	0.7240
.2770	1.852042	0.513016	0.851331	0.338315	0.467932	0.7230
.2780	1.846843	0.513422	0.852711	0.339289	0.469929	0.7220
.2790	1.841663	0.513824	0.854085	0.340261	0.471929	0.7210
.2800	1.836501	0.514220	0.855451	0.341230	0.473931	0.7200
.2810	1.831358	0.514612	0.856810	0.342198	0.475936	0.7190
.2820	1.826233	0.514998	0.858162	0.343164	0.477944	0.7180
.2830	1.821126	0.515379	0.859506	0.344128	0.479955	0.7170
.2840	1.816037	0.515755	0.860844	0.345089	0.481968	0.7160
.2850	1.810966	0.516125	0.862175	0.346049	0.483985	0.7150
.2860	1.805913	0.516491	0.863498	0.347007	0.486004	0.7140
.2870	1.800877	0.516852	0.864814	0.347963	0.488026	0.7130
.2880	1.795859	0.517207	0.866124	0.348916	0.490051	0.7120
.2890	1.790859	0.517558	0.867426	0.349868	0.492079	0.7110
.2900	1.785875	0.517904	0.868721	0.350817	0.494109	0.7100
.2910	1.780909	0.518244	0.870009	0.351765	0.496142	0.7090
.2920	1.775960	0.518580	0.871291	0.352711	0.498179	0.7080
.2930	1.771027	0.518911	0.872565	0.353654	0.500218	0.7070
.2940	1.766112	0.519237	0.873832	0.354595	0.502260	0.7060
.2950	1.761213	0.519558	0.875093	0.355535	0.504305	0.7050
.2960	1.756331	0.519874	0.876346	0.356472	0.506353	0.7040
.2970	1.751465	0.520185	0.877593	0.357408	0.508403	0.7030
.2980	1.746616	0.520491	0.878832	0.358341	0.510457	0.7020
.2990	1.741783	0.520793	0.880065	0.359272	0.512514	0.7010
.3000	1.736966	0.521090	0.881291	0.360201	0.514573	0.7000
.3010	1.732165	0.521382	0.882510	0.361128	0.516636	0.6990
.3020	1.727380	0.521669	0.883722	0.362053	0.518701	0.6980
.3030	1.722610	0.521951	0.884927	0.362976	0.520769	0.6970
.3040	1.717857	0.522228	0.886126	0.363897	0.522841	0.6960
.3050	1.713119	0.522501	0.887317	0.364816	0.524915	0.6950
.3060	1.708396	0.522769	0.888502	0.365733	0.526992	0.6940
.3070	1.703689	0.523033	0.889680	0.366647	0.529073	0.6930
.3080	1.698998	0.523291	0.890851	0.367560	0.531156	0.6920
.3090	1.694321	0.523545	0.892016	0.368470	0.533242	0.6910
.3100	1.689660	0.523795	0.893173	0.369379	0.535332	0.6900
.3110	1.685014	0.524039	0.894324	0.370285	0.537424	0.6890
.3120	1.680382	0.524279	0.895469	0.371189	0.539520	0.6880
.3130	1.675765	0.524515	0.896606	0.372092	0.541618	0.6870
.3140	1.671164	0.524745	0.897737	0.372992	0.543720	0.6860
.3150	1.666576	0.524972	0.898861	0.373890	0.545824	0.6850

NUMERICAL TABLES

Entropy Tables (*Continued*)

p	$-\log_2 p$	$-p \log_2 p$	H	$-q \log_2 q$	$-\log q$	q
.3160	1.662004	0.525193	0.899978	0.374785	0.547932	0.6840
.3170	1.657445	0.525410	0.901089	0.375679	0.550043	0.6830
.3180	1.652901	0.525623	0.902193	0.376571	0.552156	0.6820
.3190	1.648372	0.525831	0.903291	0.377460	0.554273	0.6810
.3200	1.643856	0.526034	0.904381	0.378347	0.556393	0.6800
.3210	1.639355	0.526233	0.905466	0.379233	0.558517	0.6790
.3220	1.634867	0.526427	0.906543	0.380116	0.560643	0.6780
.3230	1.630394	0.526617	0.907614	0.380997	0.562772	0.6770
.3240	1.625934	0.526803	0.908678	0.381876	0.564905	0.6760
.3250	1.621488	0.526984	0.909736	0.382752	0.567041	0.6750
.3260	1.617056	0.527160	0.910787	0.383627	0.569179	0.6740
.3270	1.612637	0.527332	0.911832	0.384499	0.571322	0.6730
.3280	1.608232	0.527500	0.912870	0.385370	0.573467	0.6720
.3290	1.603841	0.527664	0.913901	0.386238	0.575615	0.6710
.3300	1.599462	0.527822	0.914926	0.387104	0.577767	0.6700
.3310	1.595097	0.527977	0.915945	0.387968	0.579922	0.6690
.3320	1.590745	0.528127	0.916957	0.388829	0.582080	0.6680
.3330	1.586406	0.528273	0.917962	0.389689	0.584241	0.6670
.3340	1.582080	0.528415	0.918961	0.390546	0.586406	0.6660
.3350	1.577767	0.528552	0.919953	0.391402	0.588574	0.6650
.3360	1.573467	0.528685	0.920939	0.392255	0.590745	0.6640
.3370	1.569180	0.528813	0.921919	0.393105	0.592919	0.6630
.3380	1.564905	0.528938	0.922892	0.393954	0.595097	0.6620
.3390	1.560643	0.529058	0.923859	0.394801	0.597278	0.6610
.3400	1.556393	0.529174	0.924819	0.395645	0.599462	0.6600
.3410	1.552156	0.529285	0.925772	0.396487	0.601650	0.6590
.3420	1.547932	0.529393	0.926720	0.397327	0.603840	0.6580
.3430	1.543720	0.529496	0.927661	0.398165	0.606035	0.6570
.3440	1.539520	0.529595	0.928595	0.399000	0.608232	0.6560
.3450	1.535332	0.529689	0.929523	0.399834	0.610433	0.6550
.3460	1.531156	0.529780	0.930445	0.400665	0.612637	0.6540
.3470	1.526992	0.529866	0.931360	0.401494	0.614845	0.6530
.3480	1.522841	0.529949	0.932269	0.402321	0.617056	0.6520
.3490	1.518701	0.530027	0.933172	0.403145	0.619271	0.6510
.3500	1.514573	0.530101	0.934068	0.403967	0.621488	0.6500
.3510	1.510457	0.530170	0.934958	0.404788	0.623710	0.6490
.3520	1.506353	0.530236	0.935842	0.405605	0.625934	0.6480
.3530	1.502260	0.530298	0.936719	0.406421	0.628162	0.6470
.3540	1.498179	0.530355	0.937590	0.407234	0.630394	0.6460
.3550	1.494109	0.530409	0.938454	0.408046	0.632629	0.6450
.3560	1.490051	0.530458	0.939313	0.408855	0.634867	0.6440
.3570	1.486004	0.530503	0.940165	0.409661	0.637109	0.6430
.3580	1.481969	0.530545	0.941010	0.410466	0.639355	0.6420
.3590	1.477944	0.530582	0.941850	0.411268	0.641604	0.6410
.3600	1.473931	0.530615	0.942683	0.412068	0.643856	0.6400
.3610	1.469929	0.530644	0.943510	0.412866	0.646112	0.6390
.3620	1.465938	0.530670	0.944331	0.413661	0.648372	0.6380
.3630	1.461959	0.530691	0.945145	0.414454	0.650635	0.6370
.3640	1.457990	0.530708	0.945953	0.415245	0.652901	0.6360
.3650	1.454032	0.530722	0.946755	0.416034	0.655171	0.6350

MATHEMATICAL FORMULAS, THEOREMS, AND DEFINITIONS

Entropy Tables (*Continued*)

p	$-\log_2 p$	$-p \log_2 p$	H	$-q \log_2 q$	$-\log q$	q
.3660	1.450084	0.530731	0.947551	0.416820	0.657445	0.6340
.3670	1.446148	0.530736	0.948341	0.417604	0.659723	0.6330
.3680	1.442222	0.530738	0.949124	0.418386	0.662004	0.6320
.3690	1.438307	0.530735	0.949901	0.419166	0.664288	0.6310
.3700	1.434403	0.530729	0.950672	0.419943	0.666576	0.6300
.3710	1.430509	0.530719	0.951437	0.420718	0.668868	0.6290
.3720	1.426625	0.530705	0.952195	0.421491	0.671164	0.6280
.3730	1.422752	0.530687	0.952948	0.422261	0.673463	0.6270
.3740	1.418890	0.530665	0.953694	0.423029	0.675765	0.6260
.3750	1.415037	0.530639	0.954434	0.423795	0.678072	0.6250
.3760	1.411195	0.530609	0.955168	0.424558	0.680382	0.6240
.3770	1.407364	0.530576	0.955896	0.425320	0.682696	0.6230
.3780	1.403542	0.530539	0.956617	0.426078	0.685014	0.6220
.3790	1.399730	0.530498	0.957333	0.426835	0.687335	0.6210
.3800	1.395929	0.530453	0.958042	0.427589	0.689660	0.6200
.3810	1.392137	0.530404	0.958745	0.428341	0.691989	0.6190
.3820	1.388355	0.530352	0.959442	0.429091	0.694321	0.6180
.3830	1.384584	0.530296	0.960133	0.429838	0.696658	0.6170
.3840	1.380822	0.530236	0.960818	0.430583	0.698998	0.6160
.3850	1.377070	0.530172	0.961497	0.431325	0.701342	0.6150
.3860	1.373327	0.530104	0.962170	0.432065	0.703689	0.6140
.3870	1.369595	0.530033	0.962836	0.432803	0.706041	0.6130
.3880	1.365871	0.529958	0.963497	0.433539	0.708396	0.6120
.3890	1.362158	0.529879	0.964151	0.434272	0.710756	0.6110
.3900	1.358454	0.529797	0.964800	0.435002	0.713119	0.6100
.3910	1.354759	0.529711	0.965442	0.435731	0.715486	0.6090
.3920	1.351074	0.529621	0.966078	0.436457	0.717857	0.6080
.3930	1.347399	0.529528	0.966708	0.437181	0.720232	0.6070
.3940	1.343732	0.529431	0.967332	0.437902	0.722610	0.6060
.3950	1.340075	0.529330	0.967951	0.438621	0.724993	0.6050
.3960	1.336428	0.529225	0.968563	0.439337	0.727380	0.6040
.3970	1.332789	0.529117	0.969169	0.440051	0.729770	0.6030
.3980	1.329160	0.529006	0.969769	0.440763	0.732165	0.6020
.3990	1.325539	0.528890	0.970363	0.441472	0.734563	0.6010
.4000	1.321928	0.528771	0.970951	0.442179	0.736966	0.6000
.4010	1.318326	0.528649	0.971533	0.442884	0.739372	0.5990
.4020	1.314733	0.528522	0.972108	0.443586	0.741783	0.5980
.4030	1.311148	0.528393	0.972678	0.444286	0.744197	0.5970
.4040	1.307573	0.528259	0.973242	0.444983	0.746616	0.5960
.4050	1.304006	0.528122	0.973800	0.445678	0.749038	0.5950
.4060	1.300448	0.527982	0.974352	0.446370	0.751465	0.5940
.4070	1.296899	0.527838	0.974898	0.447060	0.753896	0.5930
.4080	1.293359	0.527690	0.975438	0.447748	0.756331	0.5920
.4090	1.289827	0.527539	0.975972	0.448433	0.758770	0.5910
.4100	1.286304	0.527385	0.976500	0.449116	0.761213	0.5900
.4110	1.282790	0.527227	0.977023	0.449796	0.763660	0.5890
.4120	1.279284	0.527065	0.977539	0.450474	0.766112	0.5880
.4130	1.275786	0.526900	0.978049	0.451149	0.768568	0.5870
.4140	1.272297	0.526731	0.978553	0.451822	0.771027	0.5860
.4150	1.268817	0.526559	0.979051	0.452493	0.773491	0.5850

Entropy Tables (Continued)

p	$-\log_2 p$	$-p \log_2 p$	H	$-q \log_2 q$	$-\log q$	q
.4160	1.265345	0.526383	0.979544	0.453160	0.775960	0.5840
.4170	1.261881	0.526204	0.980030	0.453826	0.778432	0.5830
.4180	1.258425	0.526022	0.980511	0.454489	0.780909	0.5820
.4190	1.254978	0.525836	0.980985	0.455150	0.783390	0.5810
.4200	1.251539	0.525646	0.981454	0.455808	0.785875	0.5800
.4210	1.248108	0.525453	0.981917	0.456463	0.788365	0.5790
.4220	1.244685	0.525257	0.982373	0.457116	0.790859	0.5780
.4230	1.241270	0.525057	0.982824	0.457767	0.793357	0.5770
.4240	1.237864	0.524854	0.983269	0.458415	0.795859	0.5760
.4250	1.234465	0.524648	0.983708	0.459061	0.798366	0.5750
.4260	1.231075	0.524438	0.984141	0.459704	0.800877	0.5740
.4270	1.227692	0.524224	0.984569	0.460344	0.803393	0.5730
.4280	1.224317	0.524008	0.984990	0.460982	0.805913	0.5720
.4290	1.220950	0.523788	0.985405	0.461618	0.808437	0.5710
.4300	1.217591	0.523564	0.985815	0.462251	0.810966	0.5700
.4310	1.214240	0.523338	0.986219	0.462881	0.813499	0.5690
.4320	1.210897	0.523107	0.986617	0.463509	0.816037	0.5680
.4330	1.207561	0.522874	0.987008	0.464134	0.818579	0.5670
.4340	1.204233	0.522637	0.987394	0.464757	0.821126	0.5660
.4350	1.200913	0.522397	0.987775	0.465378	0.823677	0.5650
.4360	1.197600	0.522154	0.988149	0.465995	0.826233	0.5640
.4370	1.194295	0.521907	0.988517	0.466611	0.828793	0.5630
.4380	1.190997	0.521657	0.988880	0.467223	0.831358	0.5620
.4390	1.187707	0.521403	0.989237	0.467833	0.833927	0.5610
.4400	1.184425	0.521147	0.989588	0.468441	0.836501	0.5600
.4410	1.181149	0.520887	0.989933	0.469046	0.839080	0.5590
.4420	1.177882	0.520624	0.990272	0.469648	0.841663	0.5580
.4430	1.174621	0.520357	0.990605	0.470248	0.844251	0.5570
.4440	1.171368	0.520088	0.990932	0.470845	0.846843	0.5560
.4450	1.168123	0.519815	0.991254	0.471439	0.849440	0.5550
.4460	1.164884	0.519538	0.991570	0.472031	0.852042	0.5540
.4470	1.161653	0.519259	0.991880	0.472621	0.854649	0.5530
.4480	1.158429	0.518976	0.992184	0.473207	0.857260	0.5520
.4490	1.155213	0.518690	0.992482	0.473792	0.859876	0.5510
.4500	1.152003	0.518401	0.992774	0.474373	0.862496	0.5500
.4510	1.148801	0.518109	0.993061	0.474952	0.865122	0.5490
.4520	1.145605	0.517814	0.993342	0.475528	0.867752	0.5480
.4530	1.142417	0.517515	0.993617	0.476102	0.870387	0.5470
.4540	1.139236	0.517213	0.993886	0.476673	0.873027	0.5460
.4550	1.136062	0.516908	0.994149	0.477241	0.875672	0.5450
.4560	1.132894	0.516600	0.994407	0.477807	0.878321	0.5440
.4570	1.129734	0.516288	0.994658	0.478370	0.880976	0.5430
.4580	1.126580	0.515974	0.994904	0.478930	0.883635	0.5420
.4590	1.123434	0.515656	0.995144	0.479488	0.886299	0.5410
.4600	1.120294	0.515335	0.995378	0.480043	0.888969	0.5400
.4610	1.117161	0.515011	0.995607	0.480595	0.891643	0.5390
.4620	1.114035	0.514684	0.995829	0.481145	0.894322	0.5380
.4630	1.110916	0.514354	0.996046	0.481692	0.897006	0.5370
.4640	1.107803	0.514021	0.996257	0.482237	0.899695	0.5360
.4650	1.104697	0.513684	0.996462	0.482778	0.902389	0.5350

Entropy Tables (*Continued*)

p	$-\log_2 p$	$-p \log_2 p$	H	$-q \log_2 q$	$-\log q$	q
.4660	1.101598	0.513345	0.996662	0.483317	0.905088	0.5340
.4670	1.098506	0.513002	0.996856	0.483853	0.907793	0.5330
.4680	1.095420	0.512656	0.997043	0.484387	0.910502	0.5320
.4690	1.092340	0.512308	0.997225	0.484918	0.913216	0.5310
.4700	1.089267	0.511956	0.997402	0.485446	0.915936	0.5300
.4710	1.086201	0.511601	0.997572	0.485971	0.918660	0.5290
.4720	1.083141	0.511243	0.997737	0.486494	0.921390	0.5280
.4730	1.080088	0.510882	0.997896	0.487014	0.924125	0.5270
.4740	1.077041	0.510517	0.998049	0.487531	0.926865	0.5260
.4750	1.074001	0.510150	0.998196	0.488046	0.929611	0.5250
.4760	1.070967	0.509780	0.998337	0.488557	0.932361	0.5240
.4770	1.067939	0.509407	0.998473	0.489066	0.935117	0.5230
.4780	1.064917	0.509031	0.998603	0.489572	0.937878	0.5220
.4790	1.061902	0.508651	0.998727	0.490076	0.940645	0.5210
.4800	1.058894	0.508269	0.998846	0.490577	0.943416	0.5200
.4810	1.055891	0.507884	0.998958	0.491074	0.946194	0.5190
.4820	1.052895	0.507495	0.999065	0.491570	0.948976	0.5180
.4830	1.049905	0.507104	0.999166	0.492062	0.951764	0.5170
.4840	1.046921	0.506710	0.999261	0.492551	0.954557	0.5160
.4850	1.043943	0.506313	0.999351	0.493038	0.957356	0.5150
.4860	1.040972	0.505912	0.999434	0.493522	0.960160	0.5140
.4870	1.038006	0.505509	0.999512	0.494003	0.962969	0.5130
.4880	1.035047	0.505103	0.999584	0.494482	0.965784	0.5120
.4890	1.032094	0.504694	0.999651	0.494957	0.968605	0.5110
.4900	1.029146	0.504282	0.999711	0.495430	0.971431	0.5100
.4910	1.026205	0.503867	0.999766	0.495900	0.974262	0.5090
.4920	1.023270	0.503449	0.999815	0.496367	0.977100	0.5080
.4930	1.020340	0.503028	0.999859	0.496831	0.979942	0.5070
.4940	1.017417	0.502604	0.999896	0.497292	0.982791	0.5060
.4950	1.014500	0.502177	0.999928	0.497751	0.985645	0.5050
.4960	1.011588	0.501748	0.999954	0.498206	0.988504	0.5040
.4970	1.008682	0.501315	0.999974	0.498659	0.991370	0.5030
.4980	1.005782	0.500880	0.999988	0.499109	0.994241	0.5020
.4990	1.002888	0.500441	0.999997	0.499556	0.997117	0.5010
.5000	1.000000	0.500000	1.000000	0.500000	1.000000	0.5000

INDEX

Aberdeen Proving Ground, **3**-47
Ablation, **14**-3
Absolute gain, antenna, ITU definition of, **5**-14
Absolute speed, IRIG definition, **5**-53
Absolute speed test, recorder, IRIG recommended, **5**-53, **5**-54
A-C synchro system, **13**-37, **13**-38
A-C voltage converter, **4**-50, **4**-51
Acceleration, environmental test specification for, **10**-6
  of gravity, **14**-30
  transducer, **3**-17, **3**-21
Acceleration-command system, **15**-148
Accelerometer, **15**-88 to **15**-99
  air-damped, **15**-94
  compared to rate gyro, **15**-105
  construction of, **15**-91 to **15**-95
  damping in, **15**-94, **15**-95
  and elastic coupling, **15**-91
  electrically damped, **15**-95
  force-balance, **15**-95, **15**-96, **15**-98
  gyroscopic integrating, **15**-123, **15**-124
  integrating, for local vertical, **15**-123
  for missile attitude determination, **12**-55
  oil-damped, **15**-94
  pendulous, **15**-96 to **15**-98
  pendulous gyroscopic, **15**-98
  response of, **15**-98, **15**-99
  and steady state equation, **15**-89
  with strain gage, **15**-92, **15**-93
  suspensions in, **15**-91, **15**-96 to **15**-98
  theory of, **15**-88 to **15**-90
  transducer for, **15**-98
  transfer function of, **15**-90
  transient equation for, **15**-90
Access in storage device, **11**-24
Accumulator (AC), **10**-84 to **10**-92
  ALU, **11**-39
  (*See also* Secondary cell)
Accuracy, **3**-6
  of attitude sensor, **14**-34
  and bits, **4**-28, **4**-29
  and final analog processing, **12**-27
  and information rate, **4**-30
  of measurement, **11**-14
  missile position, **3**-42
  missile velocity, **3**-42
  overall system, **4**-39
  quick-look process for, **12**-27
  real-time process for, **12**-26
  and samples per cps, q, **9**-22
  as telemetry system criterion, **9**-19 to **9**-22
Accuracy requirements in telemetry system, **4**-9, **9**-4
Acetate base tape, **12**-22
Acquisition, frame synchronization, **8**-38 to **8**-43
  signal, swept frequency, phase-locked-loop, **9**-16
  word synchronization, **8**-37

Active homing system, **15**-13
Active satellite, **16**-12 to **16**-14
  ITU definition of, **5**-13
  (*See also* Communications satellite)
Adaptive telemetry, **8**-28
  systems for, **11**-4
ADC (*see* Analog-to-digital converter)
ADCL (automatic d-c level), PCM, **8**-32
Addition, modulo 2, **8**-19
Addressing, **11**-25, **11**-62 to **11**-64, **13**-16
  in checkout equipment, **11**-62 to **11**-64
  and industrial telemetry links, **13**-16
  in storage devices, **11**-25
Adhesives, strain-gage, **3**-35
Adiabatic frequency sweep, **2**-45
Admittance parameters, transistor, **10**-19, **10**-20
ADP (ammonium dihydrogen phosphate) transducer, **3**-30 to **3**-32
Advanced Research Projects Agency (ARPA), **5**-2
Advanced telemetry system, **9**-18 to **9**-60
Aero elasticity, **14**-3
Aerodynamic forces, **15**-6, **15**-7
Aerodynamic torque on satellite attitude, **14**-36
Aeronautical Mobile Service, special ITU rules, **5**-2
Aeronautics Systems Division (ASD), **5**-2
Aeronutronics Department of Ford Motor Company, **9**-18
Aerospace airplane, **14**-48, **14**-49
Aerospace Industries Association, **5**-4
Aerospace Research and Testing Committee, **5**-4
Aerospace transportation costs, **14**-10 to **14**-12
AGAVE (automatic gimbled-antenna vectoring equipment), **3**-53
Agravic, **14**-3
AIA (Aerospace Industries Association), **5**-4
AIAA (American Institute of Aeronautics and Astronautics), **5**-3
AIEE (American Institute of Electrical Engineers), **5**-1 to **5**-3
Air-bearing, gyro, **15**-103
Air cell, **10**-83
Air collection cycle in aerospace engine, **14**-49, **14**-50
Air-damping in accelerometer, **15**-94
Air Force Missile Development Center, **5**-1
Air Proving Ground Center, **5**-1
Aircraft recording, crash and operational, **7**-41
Airlines and data communication, **13**-15
AKT-14/UKR-7 Telemetry System, **4**-15 to **4**-17
Albedo, **14**-3, **14**-54 to **14**-56
Alcore, **14**-42
Algol, **14**-42

1

INDEX

Algorithm, **12**-29
  Cooley-Tukey, **12**-35
  for multiplication and division, **11**-42
Aliasing, **4**-8, **8**-57, **9**-7, **9**-8
  in commutation, **4**-56
  data, **8**-57
  effect on samples per cps, q, **9**-22
Alignment, **12**-15, **12**-16, **15**-110
  errors in, rate gyro, **15**-110
  film data reduction, **12**-15
  oscillogram data reduction, **12**-16
Alkaline cells, **10**-83 to **10**-85
Allocations, radio-frequency, **5**-5 to **5**-7
Allowable percentage deviation in FM-FM system, **6**-13
Alpha numeric, data presentation, **11**-37, **11**-48, **11**-53
Alphabet, encoding, **13**-27
Altair, **14**-42
Altitude, celestial body, **14**-3, **14**-8 to **14**-20, **14**-22 to **14**-26
  of communications satellite, **16**-6, **16**-7, **16**-23 to **16**-26
  environmental specifications for, **10**-5
ALU (*see* Arithmetic and logic unit)
AM (*see* Amplitude modulation)
Ambiguity, in commutation, **4**-56
  elimination of, and interferometers, **3**-46
  in trajectory measurement, **3**-45, **3**-46
AME (angle-measuring equipment), **3**-50
American Institute of Electrical Engineers (AIEE), **5**-1 to **5**-3
American Standards Association (ASA), **1**-6, **5**-1, **13**-20
  telemetry classification by, **13**-20
Amplification, **3**-25, **10**-20
  forward current, transistor, **10**-20
  photo conductive transducer, **3**-25
  reverse voltage, transistor, **10**-20
Amplifier, analog computer, **15**-64 to **15**-66
  bridge, **8**-10
  d-c, **4**-41 to **4**-44
  difference, **15**-73, **15**-74
  floating, **8**-10
  hydraulic, **15**-133
  operational, **15**-64 to **15**-66
  power, transistor, **6**-25
  transmitter, **6**-25, **6**-26
  vacuum tube, **6**-26
  settling time, in data acquisition system, **11**-10
  summing, **6**-2
  transformer-coupled, **8**-9
  TWT (travelling wave tube), **6**-26
Amplitude error, PDM, from pulse stretching, **7**-34, **7**-35
Amplitude instability, magnetic recorders, **7**-41
Amplitude locus, gain-phase plane, **15**-61
Amplitude modulation (AM), **2**-38, **5**-16, **5**-17, **9**-5, **16**-18
  carrier interference effects, **9**-5
  in communications satellite, **16**-18
  necessary bandwidth, ITU, **5**-16, **5**-17
Amplitude-shift keying, PCM, **8**-49, **8**-50
AMR (*see* Atlantic Missile Range)
AMV (area of mutual visibility), **16**-27, **16**-28
Analog computer, **12**-29, **15**-63 to **15**-70
  for data reduction, **12**-29
  differential equation solution, **15**-67 to **15**-69
  operational amplifier, **15**-64 to **15**-66

Analog gate, mechanical-electrical, **15**-63, **15**-64
  meters, **11**-51, **11**-52
PAM, **9**-9
  recording, IRIG standards for, **5**-38 to **5**-42
  magnetic, **11**-28
  signal conditioning, **11**-10
  transfer, in checkout equipment, **11**-69
Analog-to-analog conversion, **12**-5, **12**-25, **12**-28
Analog-to-digital conversion, **9**-18, **11**-68, **12**-5 to **12**-7, **12**-19, **12**-20
  in checkout equipment, **11**-68
  in data reduction systems, **12**-5 to **12**-7
  for FM telemetry data, **12**-19, **12**-20
  in PAM ground station, **9**-18
  semi-automatic, **12**-6
Analog-to-digital converter, **8**-2 to **8**-3, **8**-23 to **8**-27, **11**-6 to **11**-9, **11**-13 to **11**-24, **12**-19
  in data acquisition, **11**-7 to **11**-9, **11**-13 to **11**-24
  feedback type, **8**-25
  for PCM telemeter, **10**-65 to **10**-68
  position to number, **11**-19 to **11**-22
  ramp type, **8**-23
  voltage to number type, **11**-16, **11**-17
AND (logical operation), **15**-78, **15**-79
AND gate, **8**-34, **8**-41, **8**-48, **10**-76
  for frame-sync detection, **8**-41
  for phase detector, **8**-34
  for time-decoding, **8**-48
  unipolar transistor, **10**-76
AN/DKT-7 (XN-2), **4**-22
AN/FPS-16, **3**-53, **3**-54
Angle-measuring equipment (AME), **3**-50
Angle-modulation methods, **6**-22
Angle-to-code, analog-to-digital converter, **11**-20 to **11**-22
Angstrom, **14**-3, **14**-53
Angular displacement, accelerometer, **15**-97
  gyro measurement, **15**-100, **15**-115
  transducer, **3**-17, **3**-24
Angular momentum, gyro, **15**-100, **15**-101
Angular position, measurement requirements, **4**-9
Angular velocity, gyro measurement, **15**-100, **15**-105
Anisoelastic coefficient, **15**-112
Anisoelastic error, gyro, **15**-111
Anisoinertial error, gyro, **15**-113
Annotating, data record, **12**-7
Annotation processing, **12**-27
Annular-slot antenna, **4**-128
Anodige coder in analog-to-digital converter, **12**-19
Anomalies, vertical, **15**-122
Anomalistic year, **14**-3
Answer back, **13**-28, **13**-29
  security system, **13**-40
Antares, **14**-42
Antenna, **4**-20, **4**-24, **4**-111 to **4**-134, **14**-93
  annular-slot, **4**-128
  automatic tracking, **4**-24
  axial-mode helix, **4**-129
  bandwidth, **4**-114
  conical spiral, **4**-126
  directive, **4**-118
  gain of, **4**-112, **4**-113
  with ground plane, **4**-115, **4**-116
  horn, **4**-130
  log periodic, **4**-126, **4**-127
  Loop-Vee®, **4**-122, **4**-123
  OGO specification, **14**-93

INDEX

Antenna, pattern of, definitions of, **4**-112
    formats, IRIG, **4**-131 to **4**-134
    image and source, **4**-115
    vehicle effects, **4**-114 to **4**-118
  phased array, communications satellite, **16**-14
  placement, missile, **4**-20
  planar spiral, **4**-123 to **4**-125
  receiving, high-gain, **4**-20, **4**-21
  scimitar, **4**-126
  slot, **4**-127, **4**-128
  surface-wave, **4**-129, **4**-130
  swastika, **4**-121, **4**-122
  temperature of, **4**-94, **4**-104 to **4**-107
    communications satellite terminal, **16**-4
    effective, **4**-94
    from side lobe, **4**-107
    sun, **4**-104, **4**-105
  terminology to describe, **4**-112 to **4**-114
  turnstile, **4**-121
  vectoring system, automatic, **3**-53
  vehicular, **4**-111 to **4**-134
    design requirements of, **4**-111, **4**-112
  vehicular types of, **4**-117
  and VSWR, **4**-114
Antipodal, **9**-41
Aperture, analog-to-digital converter, **11**-17
Aperture effect, **4**-59
Aphelion, **14**-3, **14**-57
Apoapsis, **14**-4
Apofocus, **14**-4
Apogee, **14**-4, **14**-18, **14**-22, **14**-26
  suborbital trajectory, **14**-26
Apollo program, **14**-11
Approach offset, planetary flyby, **14**-30
Approach window, **14**-17, **14**-52
APRIL (automatically programmed remote indication logging), **13**-39
Arc cosine generator, **15**-76
Arc length between terminals in communications satellite, **16**-6
Arc sine generator, **15**-76
Archimedean-spiral antenna, **4**-124
Arctangent generator, **15**-74, **15**-75
ARDC 1962 atmosphere, **14**-60
Argument, of latitude, **14**-4
  of perifocus, **14**-4
Arithmetic and logic unit (ALU), **11**-35, **11**-36, **11**-38 to **11**-42, **11**-44, **11**-45
Arithmetic progression, **14**-24
Arm, bidirectional articulated, **15**-147
  designs of, remote handling, **15**-146
  planar, **15**-146, **15**-149
Armature-controlled d-c motors, **15**-128 to **15**-130
Army Ballistic Missile Agency, **14**-75
Array, retrodirective, **16**-14
  Van Atta, **16**-14
Articulated arm, **15**-138
ARTRAC (Advanced Range Tracking), **3**-56
ASA (*see* American Standards Association)
Ascending node, **14**-4, **14**-57
ASD (Aeronautics Systems Division), **5**-2
ASK (amplitude-shift keying), **8**-49, **8**-50
Aspect angle, missile antenna, **4**-134
Assigned frequency, ITU, **5**-13
Assignments, channel, commutator-segment, **12**-25
Asteroid, **14**-53
Astrodynamics, **14**-2, **14**-4, **14**-15
Astronautical terms, **14**-3 to **14**-10
Astronomical unit, **14**-4, **14**-53
Astronomy, **14**-52 to **14**-58
Asynchronous data processor, **11**-38

Atlantic Missile Range (AMR), **3**-47, **3**-52, **5**-1
  firing sector, **14**-27, **14**-29
Atlas, **14**-38, **14**-41, **14**-46
  stable platform, **15**-126
Atmosphere, **10**-11, **14**-61
  ARDC 1962, **14**-60
  density and drag of, **14**-30
  as environment, **10**-11
  planetary, gaseous content, **10**-12
  pressure of, vs. altitude, **14**-60
  and propagation loss, **4**-99, **4**-100
  variations of, with altitude, **14**-60
A.T.&T., data link noise tests by, **13**-28
  Data-Phone, **13**-32
Attitude, **14**-4
  control, **14**-33, **14**-34, **15**-18, **15**-19, **16**-9
    passive, **16**-9
    of spacecraft, **14**-33 to **14**-39
  determination, nose cone, **12**-55
  sensing, in remote-handling system, **15**-157
  sensor, accuracy, **14**-34
  torques, on satellite, **14**-34 to **14**-39
    from spacecraft internal moving parts, **14**-36
Auroras, **10**-17, **14**-4
Autocorrelation function, **2**-15
  cyclic, **8**-38, **8**-39
  of frame sync word, PCM, **8**-38, **8**-39
  PN sequence, **14**-98, **14**-99
  shift-register code, **9**-45
  truncated, **8**-39
Automatic calibration, **6**-54, **12**-8
Automatic checkout equipment, **11**-56 to **11**-73
  basic block diagram, **11**-58
Automatic d-c level control, PCM synchronizer, **8**-32
Automatic digitization, **12**-18 to **12**-24
  preparation, **12**-21
  sampling rates, **12**-18
Automatic dispatch, **13**-44
Automatic drift correction, subcarrier channels, **5**-28, **5**-29
Automatic frequency control, **6**-30, **6**-32
Automatic gain control, **6**-30
  PAM ground station, **9**-16
Automatic monitoring sensor, **14**-86, **14**-87
Automatic phase-locked receiver, **14**-96, **14**-103
Automatic processing of satellite experimental data, **14**-74
Automatically programmed remote indication logging (April), **13**-39
Automation, in pipeline systems, **13**-39
  and remote handling, **15**-164
Autopilot, control systems design, **15**-43, **15**-47
Average fraction of visible satellites, **11**-27
  (*See also* Mutual visibility, probability of)
Average signal power, AM, **2**-38
Average unmodulated carrier power, AM, **2**-38
Axial-mode helical antenna, **4**-129
Axial piston hydraulic motor, **15**-134
Axial ratio, polarization, **4**-113
Azimuth, **14**-4, **14**-28, **14**-29
AZUSA, **3**-52

B, coded phase-coherent system, **9**-42
  energy per bit, per unit noise intensity, **9**-29 to **9**-31, **9**-42, **9**-70, **14**-96, **14**-97

B, for phase locked discriminator, peak phase error constrained, 9-70
Back-bias, 8-13
Back current, 4-81
Backlash, **15-56**, **15-59**, **15-60**
Bacon fuel cell, **10-103**, **10-104**
Bad points, data, **12**-30
Bakelite strain gage, **3**-38
Balance, gyro rotor, **15**-102, **15**-103
Balance telemetering system, **13-33**, **13-34**
Balanced gate, **4**-82, **4**-83
Ballistic coefficient, **14**-4, **14**-26, **14**-29
Ballistic trajectory, **14**-4
Balun, **4**-121
Band, **5**-16
  ITU, **5**-15
  r-f, telemetry, **7**-27
Band occupancy, normalized, **9**-25, **9**-26
Band shift factor, **5**-32, **5**-33
Bandpass filter in data reduction, **12**-23
Bandwidth, **2**-50, **2**-51
  antenna, **4**-114
  per bit, α, **9**-30, **9**-31
    PCM, **9**-32
  coded phase-coherent system, **9**-51
  command link, remote handling, **15**-161
  and communications satellite, **16**-16
  designation of, ITU, **5**-15
  finite, effect of, on signal discrimination, **9**-22
  IF, telemetry receiver, IRIG recommendations for, **5**-32, **5**-33
  loop noise, **9**-63, **9**-65
  message, **2**-50
  necessary, ITU computation, **5**-16 to **5**-19
    ITU definition of, **5**-13
  noise, of receiver, **4**-93
  occupancy, **2**-51
  occupied, ITU definition, **5**-13
  PACM receiver, **9**-27
  passive communications satellite, **16**-7
  PCM communications satellite, **16**-18
  PDM r-f link requirements, **7**-23
  PFM requirements, **9**-50
  pulse signal modulation, **2**-50
  ratio of, r-f, to information, **9**-18
    and signal to loop bandwidth for minimum phase error variance, **9**-86
  receiver, **2**-51
  requirements vs. accuracy, in FM telemetry system, **9**-4
  signal, **2**-50
    AM, **2**-50
    FM, **2**-50
  television for remote handling, **15**-155
Bang-bang control, **15**-62
Bank-and-dive indicator, **15**-121, **15**-122
Bar printer, **11**-51
Bar-graph data presentation, PDM, **7**-33
Barium titanate transducer, **3**-30 to **3**-32
Barkhausen principle, **10**-50
Base resistance ohmic transistor, **10**-21
Base spreading resistance and transistor temperature effects, **10**-34
Baseband (see Bandwidth, message)
Bathyscaphe, **15**-141
Battery, selection considerations, **10**-107, **10**-108, **14**-71
  (See also Electrochemical cell; Primary cell; Secondary cell)
Baudot code, **13**-29, **13**-30
Baudot telegraph signaling, **13**-17
BCD (binary-coded-decimal) number representation, **11**-37
Beacon, radar, **3**-54, **3**-55

Beam rider guidance, **15**-8, **15**-9
Beamwidth, antenna and gain, **4**-113
Bearing system, gyros, **15**-102
Beat-Beat trajectory measurement system, **3**-53
Bendix Radio Corporation, **13**-35
Bennett, W. R., **9**-6
Bessel function, **2**-45, **6**-9, **6**-35, **A**-28, **A**-32
  first kind, order n, **2**-45
  first order, **6**-9
  integral formulas, **A**-30
  for large arguments, **A**-31
  series expansion, **A**-30
  zero order, **6**-9
Bessel's interpolation formula, **A**-64
Best chord, **3**-4
Beta current cell, **1**-99
Beta distribution, **A**-50, **A**-51
Beta function, **A**-27
Bias correction, data reduction, **12**-23
Bias error, **12**-30
Bias recording, **5**-41
Biasing, **10**-24
  temperature compensation, **10**-36
  transistor, **10**-24, **10**-35, **10**-36
  and transistor signal stabilization, **10**-38, **10**-39
  unipolar transistor, **10**-76, **10**-77
Biconical antenna, **4**-121
Bidirectional antenna pattern, **4**-112
Binary code, **8**-2
Binary-coded-decimal representation, **11**-37
Binary counter, **8**-18
Binary data processor, **11**-4
Binary data transmission, **8**-3, **9**-13
  by PAM system, **9**-13
Binary decimal conversion, **4**-28
Binary number representation, **4**-27
  in data processor, **11**-37, **11**-38
Binary-to-octal conversion, **9**-13
Binary PCM, **8**-48, **8**-49
Binary recording data representation, **11**-31 to **11**-33
Binary representation, **11**-14
Binary symmetric channel, **2**-26
  PCM, **8**-52, **8**-53, **8**-56
  in space communication, **14**-97
Binary telemetry system, **1**-8
Binit, **13**-25
Binomial distribution, **A**-44, **A**-45
Binomial theorem, **A**-24
Biorthogonal system, discrete communications, **9**-34
  PCM, **8**-4
  QPPM, **9**-36
Biphase modulation, **8**-49, **14**-97, **14**-98
Bipropellant for attitude control, **14**-38
Biquinary code, **13**-29
Bismuth, **3**-21
Bit-by-bit detection, **14**-97
Bit detector, **8**-30
Bit packing density, PCM, IRIG recording standard for, **5**-44
Bit rate, **5**-36, **5**-51, **8**-17, **9**-30, **14**-96
  generation, PCM, **8**-17
  PCM, IRIG standard, **5**-36
    receiver, bandwidth for, **5**-51
  and receiver, IF bandwidth, **5**-36
  in space telemetry, **14**-96
  stability, PCM, IRIG standard, **5**-36
  per unit bandwidth, **9**-30
Bit synchronization, **8**-30 to **8**-36
  at low bit rates, **14**-98
  sub set, **13**-17, **13**-19

INDEX

Bit synchronizer, **8**-30
   PCM performance limitations, **8**-36
Bit timing recovery, **8**-30 to **8**-32
   (*See also* Bit synchronization)
Bit transition density, effect of, on bit synchronization, **8**-35
Bits, **2**-22, **4**-28 to **4**-30, **13**-25
   per sample, PCM, **9**-24
Black body, in-space temperatures, **10**-16
Blackout, **14**-4
Block diagram, for feedback analysis, **15**-22
   and synchronism in automatic checkout equipment, **11**-66
Blue scout, **14**-42
Bode filter optimization, FM feedback, **9**-94
Bode plot, **15**-41 to **15**-48
   for control systems with delay, **15**-85
Boltzmann's constant, **1**-22, **2**-53, **4**-93, **4**-106 to **4**-108, **10**-33
Bonded strain gage, **3**-35
Boolean algebra, **15**-78, **A**-11, **A**-12
Booster, **14**-7
Booton's quasi-linearization method, phase-locked loop analysis, **9**-77, **9**-78, **9**-86 to **9**-96
Boundary conditions in data reduction, **12**-29
Bourdon tubes, **3**-33
"Bowtie" antenna, **9**-14
Braking ellipses, **14**-4
Branch, program, data processor, **11**-43
Branch orders in automatic checkout, **11**-68
Branch station, communication subnetwork, **13**-18
Break, calibration, **12**-17, **12**-18
Break-even ratio, **14**-13
Bremsstrahlung, **14**-4, **14**-77
Brewer, G. A., and strain gage, **3**-36
Bridge, and resistive transducer, **3**-35, **3**-38
   strain-gage, **3**-35, **3**-38
   for telemetering system for position telemetering, **13**-38, **13**-39
Bridged system, subnetwork, industrial telemetry, **13**-17
Broken ring commutator, **4**-79
Buffer, I-O, data processor, **11**-45, **11**-46
   output, PCM, **8**-47
   PCM input, **8**-15
Burnout angle, **14**-5, **14**-22 to **14**-25
   conditions, planned, mission, **14**-50, **14**-51
   velocity, **14**-5, **14**-17, **14**-21 to **14**-25, **14**-27, **14**-45 to **14**-47
Burst noise, **13**-28
Business control, **1**-6
Butterworth filter, **6**-31, **6**-51, **6**-52, **8**-6

Cadmium selenide phototransducer, **3**-26
Cadmium-silver cell, **10**-87, **10**-89, **10**-91
Cadmium sulfide phototransducer, **3**-26
Cahn, C. R., **9**-93
Calibration, **4**-5 to **4**-6, **4**-8
   of analog-to-digital converter, **11**-8
   automatic, **6**-54
      in FM-FM systems, **5**-28 to **5**-29
      in PPM systems, **7**-29 to **7**-30
   computer program, **12**-48 to **12**-49
   in data acquisition, **11**-8, **11**-12
   in data reduction, **12**-8
   data reduction equipment, **12**-23
   down range telemetry, **4**-5
   film data, **12**-15
   in-flight, PDM, IRIG standards of, **5**-32

Calibration, oscillogram data, **12**-17 to **12**-18
   PAM-FM-FM, IRIG recommendation, **5**-31
   PDM systems, **7**-22
   planning, **12**-22
   processing, analog-to-analog, **12**-27
   program module, **12**-39 to **12**-40
   of radio telemetry system, **4**-6, **4**-8
Camera location, in remote vision, **15**-155
Canonical form, Boolean functions, **A**-12
Capacitance transducer, **3**-9
   accelerometer, **15**-93
   sensitivity, **3**-10
Capacitor, **10**-90 to **10**-92
   miniaturization, **10**-71 to **10**-72
   plastic film, **10**-71
   printed circuit, **10**-78
   semiconductor, voltage variable, **10**-70
   tantalum, **10**-71 to **10**-72
Capacity, channel, **13**-2, **13**-25
   missile telemetry requirements, **9**-19
   Shannon limit, **9**-30
Capstan errors, **7**-39
Captive gyro (*see* Single-degree-of-freedom gyro)
Capture effect in FM system, **6**-11
Card capacitor memory, **11**-7, **11**-11
Card handling equipment for data reduction, **12**-40
Card punch, **11**-51
Carnot cycle limitations, **10**-102
Carrier bandwidth, PDM, **7**-10
Carrier frequency modulation, **1**-8
Carrier power, ITU defined, **5**-14
Carrier stability PDM-FM systems, **7**-11
Carrier threshold, PDM-FM, **7**-7
   (*See also* Improvement threshold)
Carrier-to-noise ratio, **4**-93
Cascade compensation, **15**-46, **15**-55
Cascading tree, checkout equipment, **11**-64
Case acceleration error, rate gyro, **15**-109 to **15**-110
Case design, **6**-27
Cast construction, **6**-27
Castor, **14**-42
Cauchy's distribution, **A**-50 to **A**-51
Causal probability distribution, **A**-43
Caustic soda cell, **10**-84
CCIR (Radio Consultative Committee), **5**-4
Celestial axis, **14**-7
Celestial mechanics, **14**-4, **14**-16 to **14**-30
Celestial sphere, **14**-5, **14**-58
Cell, electrochemical, **10**-79 to **10**-92
   alkaline, **10**-83, **10**-85
   dry primary, **10**-82 to **10**-84
   lead-acid, **10**-84, **10**-86, **10**-89 to **10**-90
   Leclanche, **10**-82
   losses, **10**-80
   low temperature, **10**-84
   mercury, **10**-82, **10**-85
   nickel-cadmium, **10**-86, **10**-89 to **10**-91
   primary, **10**-82 to **10**-85
   (*See also* Fuel cell; Nuclear cell; Secondary cell; Solar cell; Thermionic cell)
Cells of occupancy, **A**-42
Centaur, **14**-38, **14**-41, **14**-46
Central body in orbital motion, **14**-17 to **14**-22
Central-limit theorem, **2**-46
Central operations control, satellite experiments, **14**-84
Central processing unit, **11**-35, **11**-36, **13**-10
Central station communications, **13**-16
Centralized transport control, **13**-39

INDEX

Cessation, space emission, **5**-24
Cetus, **14**-42
Chad, punched tape, **11**-50
Chaffee, J. G., **9**-59, **9**-93, **16**-17
Chain printer, **11**-50 to **11**-51
Change on one, data recording, **11**-32
Channel(s), **9**-30 to **9**-32
  continuous, **9**-31 to **9**-32
  discrete, **9**-30
  PAM, maximum allowable IRIG, **5**-34
  telegraph, **13**-2
  telemetry, zero error, **6**-19
Channel allocation factor in FM carrier telemetry system, **9**-25
Channel assignments, data, **12**-25
Channel bandwidth, **7**-10
Channel capacity, **2**-24, **16**-16
  (*See also* Information capacity)
Channel equivalence voltage, commutator, **4**-86, **4**-88
Channel filtering, PDM, **7**-33 to **7**-34
Channel gate control, PDM, **7**-28
Channel gating multivibrator, **10**-58
Channel "on" time, **7**-3
Channel period scatter, **7**-12
Channel rate, **7**-3
Channel requirements for typical measurements, **4**-9
Channel selector, data acquisition, **11**-7
Channel separator, PDM, **7**-27 to **7**-29
Character, binary coding, **13**-17
Character synchronization, subset, **13**-17 to **13**-19
Characteristic velocity, **14**-42
Chebychev filter, **6**-31
Chebychev quadrature formulas, A-66, A-67
Check digit, **2**-37
Checking number, **2**-36
Checkout, preflight, **4**-4, **4**-6
  system, **11**-56 to **11**-57
Checkpoint, **12**-18
  in calibration, **12**-8
Chi squared (x^2) distribution, A-54, A-55
Chi squared test, statistical, A-59, A-60
Childers, D. G., **8**-58
Chirp, **9**-37
Chopper, transistor, **10**-70
Chopper amplifier, **4**-41 to **4**-44
Circuit, **10**-42 to **10**-69
  sample and hold, **8**-2, **8**-3
  signal conditioning, **4**-41, **8**-9 to **8**-11
  temperature equivalent incremental, transistor, **10**-33 to **10**-35
  transistor equivalent, **10**-20 to **10**-24
  transistorized telemetering, **10**-42 to **10**-69
Circuit analog, **15**-64
Circuit protection, **4**-38
Circuit transistor, physical equivalent, **10**-21
Circular orbital velocity, **14**-22 to **14**-26
Circular polarization, **4**-113
Circular probable error, A-49
Circular radio navigation, **15**-13
Clamp, keyed, PAM demultiplexer, **9**-15
Clock generator, PAM ground station, **9**-16
  in PN coded telemetry system, **14**-99, **14**-100
Closed-loop system, **13**-6, **13**-10
  pipeline control, **13**-39
  virtual, **13**-28 to **13**-29
Closed ring commutator, **4**-79
Coast attitude control, **14**-34
Coaxial cable, flexible, **4**-97
  transmission loss, **4**-96

Code, Baudot, **13**-29
  BCD, **13**-27
  binary-coded-decimal, **13**-27
  biorthogonal, **9**-44 to **9**-45
  biquinary, **13**-29
  checking, security system, **13**-40
  error checking, **13**-29 to **13**-30
    four-out-of-eight, **13**-29, **13**-30
    two-out-of-five, **13**-29
  error correcting, **2**-35, **2**-36, **13**-27
    group alphabets, **2**-36
    to improve PCM efficiency, **9**-32, **9**-33
    in PCM, **8**-60
    and phase-coherent codes, **9**-42
  Gray, **9**-31, **9**-58, **11**-21
  minimum redundancy, **2**-32
  multiple error correcting, **2**-36
  orthogonal, **9**-42 to **9**-44
    construction, **9**-43
  Reed-Muller, **9**-44
  reflected binary, **11**-21
    (*See also* Gray, *above*)
  shift-register, **9**-45
  standardizing, transportation control system, **13**-41
  transmission, of computer, convertibility, **13**-30
    for IBM Card Transceiver, **13**-29 to **13**-30
    for transmission security, **13**-28
  variable length of, **2**-34
Code disk, **11**-20 to **11**-22, **12**-14
Code generator, **9**-40
Code redundancy to improve PCM efficiency, **9**-32 to **9**-33
Code Selector Supervisory Equipment, **13**-28 to **13**-29
Code words in coded phase-coherent system, **9**-41
  (*See also* Symbols)
Coded phase-coherent system, **9**-39 to **9**-48
  binary, **9**-41
  two bit transmission, **9**-41
Coded phase-coherent telemetry, **9**-39 to **9**-48
  in space communication, **14**-97 to **14**-98
Coder, **10**-65, **10**-66, **12**-14
  analog-to-digital, **10**-65 to **10**-66
  shaft, **12**-14
Coding, **2**-32 to **2**-35, **9**-42 to **9**-45
  efficiency, **2**-34
  Hamming, **2**-35
  Huffman, **2**-34
  industrial telemetering, digital, **13**-7
  PCM data, **8**-29
  phase-coherent-system words, **9**-45
  PN (pseudo noise), **14**-98 to **14**-101
  remote handling commands, **15**-160
  Shannon-Fano, **2**-33
  for transmission, PCM, **8**-29
Cofactors, A-7
Cogging, **7**-39
Coherence, differential, in PCM system, **9**-33
Coherence time, oscillator, **9**-95
Coherent AM and FSK, efficiency of, for PCM, **9**-33
Coherent demodulation, **14**-98
  PCM, **8**-53
Cold test, environmental specification, **10**-5, **10**-7
Collapsing network, **9**-37
Collating program, **12**-42
Collector resistance, incremental, transistor, **10**-21
Collins Radio Company, **13**-32

INDEX

Colpitts oscillator, **10**-44
Columbia Gulf Transmission Company, **13**-39
Comb filter for telemetry signal enhancement, **14**-82
Combinations, **A**-42
Command, data processor, **11**-46
  space system, **14**-5
Command guidance, **15**-7, **15**-8
Command link, **1**-6
  remote handling of, **15**-144, **15**-158 to **15**-161
Command message, **13**-8
Command rates in space communications, **14**-97
Command receiver, **14**-76, **14**-82, **14**-86
  first satellite use of, **14**-76, **14**-82
  multi-channel, **14**-86
Command system, **14**-95, **14**-102 to **14**-103
  of OGO, **14**-90
  OGO specification for, **14**-93
Commercial equipment checkout, **11**-57
Common base, **10**-20, **10**-21
  h parameters, **10**-25
  transistor connection, **10**-20, **10**-21
Common carrier, **13**-13
Common collector, transistor connection, **10**-20, **10**-21
Common emitter, transistor connection, **10**-20, **10**-21
Common mode interface, and data acquisition system, **11**-11
Common-mode rejection, **8**-7 to **8**-9
  in chopper amplifier, **4**-42
  in multiplexers, **8**-15
  in subcarrier oscillator, **6**-39
Common-mode voltage, bridge type transducers, **8**-8, **8**-10
Communications, process control of, **13**-7 to **13**-9
Communications based control system, **13**-41 to **13**-42
Communications constraints, space missions, **14**-70 to **14**-71
Communications efficiency, **8**-3, **8**-4, **9**-69, **9**-70
  PCM, **8**-3 to **8**-4
  phase-locked discriminator, second order, **9**-69 to **9**-70
Communications link in control system, **15**-83
Communications link in remote handling, **15**-158 to **15**-161
Communications network configuration, industrial telemetry, **13**-15 to **13**-21
Communications properties, PAM-FM, **9**-2 to **9**-5
Communications satellite, **16**-1 to **16**-32
  active, **16**-12 to **16**-16
  cost comparisons, **16**-29 to **16**-32
  coverage, **16**-23 to **16**-29
  distributed passive, **16**-9
  earth coverage angle, **16**-7
  frequency allocations, **16**-4
  medium altitude, **16**-1
  modulation comparison, **16**-19
  modulation methods, **16**-16
  multiple access, **16**-21
  and operational systems, comparison of, **16**-29 to **16**-31
  orbital configuration, **16**-23
  passive, **16**-1 to **16**-2
  pulse-code modulation in, **16**-18
  quasi-passive, **16**-14 to **16**-15
  signal power at receiving terminal, **16**-5
Communications satellite, spectrum sharing with terrestrial services, **16**-30
  stationary, **16**-2
Communications Satellite Corporation, **16**-29
Communications satellite service, **5**-12
Communications satellite system, modified random orbit, **16**-26 to **16**-28
Communications system, **1**-2, **9**-31, **9**-39 to **9**-48, **14**-95 to **14**-97
  coded phase-coherent, **9**-39 to **9**-48
  definition of, **1**-2
  design for space telemetry, **14**-95 to **14**-97
  discrete, **9**-31
Commutation, **4**-53 to **4**-92
  comparison of types of, **4**-85 to **4**-86
  in data acquisition systems, **11**-11
  definitions of, **4**-88 to **4**-89
  electromechanical, **4**-60 to **4**-78
  electronic, **4**-78 to **4**-85, **7**-12
  high-level, **7**-12
  in industrial telemetry, **13**-30, **13**-31
  low-level, **7**-12 to **7**-13
  mechanical, **4**-60 to **4**-78, **7**-13
  multichannel, **4**-53
  PAM-FM-FM system, IRIG specification for, **5**-29 to **5**-30
  rate, PDM, **7**-3
  terminology, **4**-88 to **4**-89
Commutator, electronic, **10**-55 to **10**-61
  PDM, **7**-3
  rotating, **8**-11
  solid state, **10**-55 to **10**-66
Commutator noise, in PDM systems, **7**-36
Comparator, automatic checkout, **11**-59
  ramp level, **7**-12
Comparison, logic operation, **15**-80
  of modulation multiplex methods, **7**-7, **7**-8
Compatibility, program, **12**-40
Compensation, control system, **15**-46
  oil damping, **15**-94
  of rate gyro for case acceleration, **15**-110
  tape speed, **6**-52 to **6**-54
  variable shear area, **15**-109
  viscosity, for rate gyro, **15**-109
  wow and flutter, **4**-21, **6**-52, **12**-22
Complement, PN code, **14**-101
Completion signals for checkout timing, **11**-65
Components, telemetry, general requirements of, **10**-2 to **10**-3
Compound logic in automatic checkout, **11**-67 to **11**-68
Computer, **11**-4, **11**-34 to **11**-48
  analog, **15**-63 to **15**-70
  in control system, **15**-72 to **15**-80
  in data reduction, **12**-28 to **12**-29, **12**-49 to **12**-50
  digital, **11**-34 to **11**-48, **12**-28 to **12**-29, **13**-7 to **13**-8
  digital address format, **11**-36
  digital control format, **11**-36
  for generator scheduling, **13**-46
  local process control, **13**-4
  missile command, **15**-72 to **15**-73
  nonlinear analog, **15**-73 to **15**-76
  pipeline control functions, **13**-39
  process control, **13**-4, **13**-7 to **13**-9
  in remote handling, **15**-161 to **15**-164
  transcendental analog, **15**-73 to **15**-76
  (*See also* Data processor)
Computer controlled automatic checkout, **11**-67 to **11**-69
Computer data processing, **12**-9
Computer magnetic tape, **11**-30 to **11**-34

INDEX

Computer register display, **11**-53
Conditional probability density function, **2**-13
Conductance coefficients, transistor, **10**-33
Conductance telemetry system, **13**-33
Confidence level in data reduction results, **12**-60 to **12**-61
Conical spiral antenna, **4**-126
Coning error, rate gyro, **15**-113 to **15**-114
Connection(s), automatic, data link, **13**-14
 in automatic checkout, **11**-58
Constant-bearing course, **15**-15 to **15**-17
Constant current sources for d-c control motors, **15**-130
Constant energy trajectory, **14**-21 to **14**-26
Constant length data security code, **13**-40
Constant momentum trajectory, **14**-21 to **14**-26
Constants, physical, **1**-22
Constellations, map of, **14**-59
Constrained compression optimization in FM feedback receiver, **9**-94
Contact bounce, **7**-13, **8**-12
Contact potential nuclear cell, **10**-99
Contact recording, **11**-27
Contact resistance, **4**-36, **4**-71 to **4**-72, **7**-13, **7**-16 to **7**-17
 dynamic, **7**-36
Continuous channel, **2**-24
 capacity of, **2**-30 to **2**-31
Continuous checkout, **11**-73
Continuous probability distribution, A-46 to A-51
Contour antenna patterns, **4**-131 to **4**-134
Contour plot, scientific data presentation, **14**-88
Control, **15**-2 to **15**-136
 attitude, **15**-18 to **15**-19
 pipeline, **13**-42 to **13**-44
 power system, **13**-44
 process, **13**-4, **13**-6 to **13**-7
  closed-loop, **13**-4, **13**-6 to **13**-7
  multiplexing, **13**-7
  open-loop, **13**-4
 vehicular, **15**-4, **15**-19
 vehicular defined, **13**-5
Control console, remote handling, **15**-144, **15**-164, **15**-167
Control gyro, **15**-91
Control logic, **11**-36, **11**-42 to **11**-45
 asynchronous data processor, **11**-39
 synchronous data processor, **11**-38
Control medium for automatic sequenced checkout, **11**-61
Control message, **13**-4
Control system, **13**-1 to **13**-46, **15**-2, **15**-136
 communication based, **13**-1 to **13**-46
 computer, **15**-72 to **15**-80
 with delay, **15**-83 to **15**-86
  analysis and design, **15**-83 to **15**-86
  graphical design, **15**-85 to **15**-86
 design of, autopilot (example), **15**-43 to **15**-47
  frequency response, **15**-40, **15**-47
  gain-phase method, **15**-40 to **15**-48
  time domain, **15**-37 to **15**-40
 digital computation in, **15**-76 to **15**-81
 digital logic in, **15**-78 to **15**-81
 errors, **15**-36
 feedback effects on, **15**-25 to **15**-27
 industrial, **13**-3 to **13**-7
 optimization, **15**-70
 performance criteria comparison, **15**-47
 remote handling, **15**-147 to **15**-149, **15**-166
 simulation, **15**-67, **15**-70

Control system, stability, **15**-27, **15**-36
 time delay in, **15**-81 to **15**-86
 type 0, **15**-37
 type I, **15**-37
 type II, **15**-37
Control transfer, improper tape controlled checkout, **11**-67
Convair, **3**-52
Convergence, Volterra expansion, **9**-84
Conversion, analog-to-analog, **12**-5, **12**-25, **12**-28
 analog-to-digital (*see* Analog-to-digital conversion)
Conversion of base, logarithms, A-4
Conversion factors, **1**-11 to **1**-23
Converter, a-c voltage, **4**-50 to **4**-51
 analog-to-digital (*see* Analog-to-digital converter)
 d-c to a-c, transistorized, **10**-70
 digital-to-analog, **8**-43
 frequency, **4**-44 to **4**-48
 image frequency, in retrodirective satellite, **16**-15
 PDM, **7**-12
 receiver, design requirements, **6**-30
Convolution integral, **2**-19
 Volterra functional technique, **9**-82
Cooley-Tukey algorithm, **12**-35
Coordinate axis, body, **15**-4
Coordinate converter, polar to rectangular, **15**-73
Coordinate system, astronautic, **14**-5
 consolidated antenna and vehicle, **4**-131
Coplanar transfer, **14**-18 to **14**-20
Copper oxide photo-transducer, **3**-28
Corner reflectors, **16**-7, **16**-14
Correction, to mission trajectory, **14**-52
 wow and flutter, **12**-22
 zero and full scale, **7**-29 to **7**-30
Correlation, **2**-15 to **2**-16
 in data reduction, **12**-34
Correlation coefficient, biorthogonal coded message, **9**-45
 in coded phase-coherent system, **9**-41
 nonorthogonal message, **9**-43
 orthogonal coded message, **9**-43 to **9**-44
Correlation detector, **9**-38 to **9**-39, **14**-82
 PCM, **8**-53, **9**-32
Correlation functions of random process, **2**-15, **2**-16
Correlation tracking and ranging, **3**-50
 (*See also* COTAR)
Correlator, frame-sync, PCM, **8**-41 to **8**-42
Cosine integral, A-26
Cosmic radiation, **10**-12 to **10**-13
Cosmic rays, **14**-61, **14**-66, **14**-75
Cost comparisons in communications satellites system, **16**-30 to **16**-31
Cost effectiveness, **14**-13
Cost per launch, **14**-13 to **14**-14
COTAR (correlation tracking and ranging), **3**-50, **12**-4
Counter, bits-per-word, **8**-46
 cascaded, **8**-22
 forbidden state of, **8**-20
 frame, **8**-46
 jump feedback, **8**-20
 logic design, **8**-21 to **8**-22
 matrix, **8**-46
 matrix commutator, **4**-79 to **4**-80
 maximum length, **8**-19 to **8**-20
 maximum logic, **8**-18
 minimum logic, **8**-18
 ring, **8**-18 to **8**-19
 ripple through, **8**-17

INDEX

Counter, shift-register, **8**-18 to **8**-19
  synchronous binary, **8**-17 to **8**-18
  words-per-frame, **8**-45
Couples, electrochemical, **10**-79 to **10**-80
Coupling, energy, **15**-64
Coupling gyro error, **15**-110, **15**-117
Coupling stray circuit, **6**-27
Courier, **16**-2
Course, attack, **15**-14 to **15**-18
  constant bearing, **15**-15
  lead pursuit, **15**-15
  line-of-sight, **15**-14
  proportional navigation, **15**-17 to **15**-18
  pure pursuit, **15**-15
Coverage, communications satellite, **16**-22 to **16**-29
  (*See also* Mutual visibility)
  vehicle, of r-f link, **4**-131 to **4**-135
CPU (central processing unit), **11**-35, **11**-36, **13**-10
Cramer's rule, **A**-8 to **A**-9, **A**-41
Crest factor, **2**-53 to **2**-54
Criteria, performance, telemetry systems, **9**-19
Criterion, error moment, **9**-21
Criterion modification in computer-controlled checkout, **11**-68
Critical point, control system, **15**-40
Cross-coupling error, of integrating gyro, **15**-117
  (*See also* Coupling gyro error)
Cross-modulation, interchannel, **9**-3, **9**-4, **9**-8, **9**-9
  PAM-FM, **9**-3 to **9**-4
Cross-section-to-weight ratio, **16**-7, **16**-8, **16**-10
  and distributed communications satellite, **16**-10
Cross spectral density, **2**-17
Cross strapping, in decommutation, **4**-87
Crosscorrelation, **2**-16
  and PFM words (codes), **9**-49
  of pulsed sine wave, **9**-50
  and random process, function of, **2**-16
  ratio of, in phase-coherent correlator outputs, **9**-42
Crossover, of analog and digital telemetry, **9**-24
  processing, **12**-27
  in telemetry data, **12**-5, **12**-6, **12**-7
Crosstalk, **1**-9
  command link, **15**-159
  error from, **9**-23, **9**-29
  for FM-FM system, **6**-13 to **6**-14
  and frequency translation system, **6**-60 to **6**-61
  in gates, **4**-82
  from node response, **4**-58
  PAM-FM, **9**-4
  PAM signal, **4**-56 to **4**-58
  PCM, **8**-7
  PCM IRIG recording standard, **5**-94
  in PDM system, **7**-32 to **7**-33
  by limiting r-f bandwidth, **7**-6
  testing, in FM-FM system, **6**-38 to **6**-39
Crystal, quartz, **6**-23 to **6**-24
  at cut temperature characteristic, **6**-24
CU (*see* Control logic)
Cubic Corporation, **3**-50
Curie point, **3**-31
Current controlled oscillator, **10**-52
Current gain, A_i, transistor, **10**-29
Current telemetering systems, **13**-32 to **13**-35
Current telemetry, **13**-2 to **13**-3

Curve fitting, **12**-10 to **12**-11, **12**-31
Cutler, C. C., **16**-15
Cycles, data processor, **11**-39 to **11**-42
Cyclic storage, data processor, **11**-38
Cylinder function, **A**-28
Cylinder hydraulic prime mover, **15**-132 to **15**-133
Cylindrical equispace projection, antenna pattern display, **4**-133, **4**-136
Cylindrical error, rate gyro, **15**-113 to **15**-114
Cylindrical quarter-wave monopole, **4**-119

DAC (digital-to-analog converter), **8**-43
Damping, and accelerometer, **15**-90, **15**-94 to **15**-95
  and accelerometer response, **15**-98 to **15**-99
  control of, in accelerometer, **15**-90
  critical, **3**-7
  and d-c control motor, **15**-129 to **15**-130
  and transducer, **3**-7
Damping factor, phase-locked loop, second order, **9**-64 to **9**-65
  and phase margin, **15**-41
  in quasi linear loop receiver, **9**-90
  rate gyro, **15**-106 to **15**-107
  for second order system, **15**-37
Damping factor band resonant frequency, **15**-42
Damping ratio (*see* Damping factor)
Daniel cell, **10**-84
Dark current, **3**-25, **3**-27
Dark resistance, **3**-25
Data, **1**-4
Data accuracy, minimum, phase-locked frequency discriminator, **9**-68
Data-acquisition equipment, **11**-5 to **11**-13
Data analog, PAM display, **9**-17
Data analysis, **4**-11 to **4**-12
  post flight, use of radio telemetry, **4**-4
Data capsule, **12**-35, **12**-57
Data channel (DC), and data processor I-O unit, **11**-46
  used for telemetry commutation, **11**-47 to **11**-48
Data-collection system, industrial, **13**-10
Data compaction, **4**-10, **4**-34, **8**-28
  telemetry, **8**-28
Data compression (*see* Data compaction)
Data conversion, **12**-4 to **12**-5
  analog-to-analog, **12**-5
Data editing, **4**-6 to **4**-7, **8**-44
  PCM, **8**-44
Data event, **12**-4, **12**-14
Data flow, in satellite experiments, **14**-83
Data folding, **4**-8
  (*See also* Aliasing)
Data formats, in planning data processing, **12**-13
Data handling, **9**-15 to **9**-18, **12**-2 to **12**-9
  equipment for, **11**-1 to **11**-56
  in Mariner 1964 mission, **14**-102
  OGO specifications for, **14**-91 to **14**-93
  in PAM ground station, **9**-17 to **9**-18
  in PCM ground station, **8**-30 to **8**-48
  and program modules, **12**-39
  real-time, PAM, **9**-15 to **9**-18
  in satellite telemetry, **14**-83 to **14**-84
Data input, telemetry, **12**-3 to **12**-4
Data interdependency as data compression means, **4**-34
Data link, **13**-11 to **13**-20, **15**-114, **15**-161
  configurations, **13**-13 to **13**-15

INDEX

Data link, industrial, **13**-11 to **13**-20
  remote handling, **15**-144, **15**-161
Data logger, PDM, **7**-40 to **7**-41
Data modulation, PCM, **8**-40
Data-Phone, **13**-32
Data plotting, **4**-16 to **4**-18
Data presentation, and data input, **11**-49
  equipment for, **11**-5 to **11**-6
    operation and control of, **11**-34
  PDM, **7**-33
  and types of satellite, **14**-87
Data processing, and analysis for data reduction, **12**-10
  automatic, **4**-26
  film, **12**-15 to **12**-16
  information non-preserving, **8**-28
  information preserving, **8**-28
  of PFM telemetry, **14**-82
  systems of, for data reduction, **12**-11
  telemetry, **4**-10 to **4**-11, **8**-27 to **8**-28
  (*See also* Data reduction)
Data processor, **1**-8, **11**-4, **11**-34 to **11**-48
  as multiplexer, random access, **11**-34
Data rate, and remote handling commands, **15**-159 to **15**-160
  and space command system, **14**-46, **14**-102
  in space telemetry, **14**-95 to **14**-97, **14**-102
Data recording and pulsed telemetry signals, **11**-33
Data records, telemetry, **4**-9 to **4**-10
Data recovery, **8**-30 to **8**-48, **14**-71, **14**-75 to **14**-83
  PCM, **8**-30 to **8**-48
  systems, for Explorer I telemetry, **14**-71
  satellite, **14**-75 to **14**-83
Data reduction, **3**-2, **3**-4, **4**-6, **4**-7, **4**-10, **4**-26, **11**-5, **12**-1 to **12**-62
  automatic, **4**-26, **12**-18 to **12**-24
  and computing, working group, IRIG, **5**-2
  definition of, **12**-2
  and flight test system, **12**-54 to **12**-56
  manual, **12**-14
  in radio telemetry, **4**-6 to **4**-7
  requirements of, **12**-52 to **12**-54
  systems of, semi-automatic, **12**-14 to **12**-18
Data representation, binary, **8**-29
  digital, **8**-29
Data security, **13**-27 to **13**-28
Data selector control and programmer, **11**-7
  and data acquisition system, **11**-12 to **11**-13
Data storage, devices for, **11**-5, **11**-6
  in Mariner space mission, **14**-102
  OGO, **14**-90, **14**-91
  OGO specifications for, **14**-92
  predetection, **9**-14
  satellite, **14**-85 to **14**-87
Data transfer in checkout equipment, **11**-69
Data transmission, **1**-8
  noise in, **13**-28
  PCM, **8**-49
  sampled analog, **9**-54 to **9**-59
DC (*see* Data channel)
D-C amplifiers, **4**-41 to **4**-44
D-C coupling, gating circuit, **4**-81
D-C to D-C converter, **6**-27
D-C motors, control system, **15**-128
D-C offset, PAM gates, **9**-9
D-C response, and multicoders, **7**-19
  and PDM systems, **7**-23
D-C restoration, **8**-9
D-C step-function converter, **4**-50

D-C synchro systems, **13**-37 to **13**-38
D-C voltage monitor, **4**-49, **4**-50
Dead zone, **15**-56 to **15**-60
  and simulation, **15**-70, **15**-72
Deboosting from earth orbits, **14**-32 to **14**-33
Decca, **15**-12
Deceleration, maximum, at entry (or reentry), **14**-31 to **14**-32
Decibel conversion table, **A**-5 to **A**-6
Decimal data processor, **11**-37
Decision, optimal, signal reception, **9**-46 to **9**-48
Decision mechanism, detection, **9**-40
Declination, **14**-5
Decoder, in checkout equipment, **11**-62 to **11**-64
  multi-symbol, **9**-55 to **9**-56
  pulse-time interrogation, **11**-71
Decommutation, **4**-58
  devices for, **4**-86 to **4**-88
  PCM, **8**-43, to **8**-45
  PDM, **7**-27 to **7**-29, **7**-31
  programmable, PCM, **8**-44 to **8**-45
Decomposition of logarithm, **A**-24
Defects, telemetry data, in data reduction system, **12**-56 to **12**-57
Definitions, commutation, **4**-88 to **4**-89
  IRIG, **5**-57 to **5**-67
  ITU, **5**-9 to **5**-14
  telemetry, IRIG, **5**-57 to **5**-67
Degaussing, in magnetic records, **11**-24
Degrees of freedom, **A**-54, **A**-55, **A**-56
  of body, **15**-4 to **15**-5
  for manipulation, **15**-145
  and remote handling commands, **15**-160
  and remote handling control, **15**-164
Delay, from communication link, **15**-83
  from computers, **15**-83
  in control systems, **15**-81 to **15**-86
  from human operators, **15**-83
  simulation of, **15**-78, **15**-86
Delay lines, and bandwidths, **9**-38
  electrically variable, **6**-20, **6**-34
Delay lock, **9**-81
Delay time, in process control, **13**-6 to **13**-7
  transducer, **3**-8
Delayed action battery, **10**-91
Demodulation, direct-play-back system, **6**-34 to **6**-35
  PCM, **8**-52 to **8**-56
Demodulator, in control system, **15**-83
  frequency coherent, **9**-66
  frequency-compressive, **6**-31
  phase-locked, **4**-31, **6**-31, **9**-66 to **9**-67
  phase sensitive, **4**-48 to **4**-49
  receiver, **6**-31 to **6**-32
  wideband, **6**-34
Demodulator-modulator, communications satellite repeater, **16**-20
Demultiplexer, PAM ground station, **9**-15 to **9**-17
  r-f, **4**-20
Density, atmospheric, variation with attitude, **14**-60
Density function (*see* Probability density function)
Depolarizing, dry cell, **10**-82, **10**-84
Derivative, **A**-14
Derivative feedback, **15**-46
Derived data in data reduction, **12**-35
Derived quantities in flight testing, **12**-53 to **12**-54
Describing functions for nonlinear systems solutions, **15**-57

Despinning, antenna beam, **16**-14
Detection, coherent, **8**-53 to **8**-56
 data, PAM ground stations, **9**-15
 frame synchronization, **8**-41 to **8**-42
 greatest of, PFM, **9**-51, **9**-53
 maximum likelihood, **9**-40
 PCM signals, **8**-52 to **8**-56, **9**-34
 synchronous, for PCM, **8**-54 to **8**-55
Determinant, **A**-7 to **A**-9
Develet, J., **9**-60, **9**-78, **9**-93
Deviation, carrier, linear range, **6**-16
 frequency, **2**-38
  instantaneous, **2**-58
  maximum angular, **2**-37
 peak frequency, **2**-50
 permissible, phase-locked discriminator, **6**-46
 phase, **6**-7
  instantaneous, **2**-39
 ratio, **7**-10
 subcarrier, IRIG recommendations, **5**-27 to **5**-28
 RMS frequency, **2**-50
 subcarrier frequency, in multiplex design, **6**-17, **6**-18
Diagnostic program, computer, **12**-21
Dial-up data links, **13**-14
Diamagnetism, **3**-15
Di-bit, **13**-32
Dichromate cell, **10**-83
Dielectric-cylinder antenna, **4**-130
Dielectric-tube antenna, **4**-130
Difference amplifier, **15**-73 to **15**-74
Difference equation, **15**-70
Differential amplifier, direct-coupled, **8**-9
Differential calculus, **A**-13 to **A**-16
Differential equation, analog computer solution, **15**-67 to **15**-69
 in data reduction, **12**-36
 nonlinear first-order, **9**-70
 nonlinear second-order, **9**-71
 nonlinearized, **9**-72
Differential noise, **4**-86
Differential signal, **8**-5
Differential transducer, variable-reluctance, **6**-23 to **6**-24
Differential transformer, **3**-17
 accelerometer output, **15**-92
 frequency response, **3**-17
 magnetic shielding, **3**-17
 sensitivity, **3**-17
 temperature errors, **3**-18
Differentiation, in data reduction, **12**-33 to **12**-34
 numerical, **A**-67 to **A**-68
Differentiation rules, differential calculus, **A**-13
Differentiator rectifier for bit synchronizer, **8**-33
Diffused-base, active integrated circuits, **10**-74
Digit, binary, **13**-25
Digit sequencing circuit, a-d converter, **10**-67
Digital coding, in process control, **13**-7
Digital computer, for control system simulation, **15**-70
 for data reduction, **12**-28 to **12**-29
 and delay in control system, **15**-83
 special purpose, **15**-76 to **15**-81
Digital controller, **15**-76 to **15**-78
Digital data representation in magnetic recorder, **11**-30 to **11**-34
Digital data transmission (*see* Binary data transmission)

Digital filter, **12**-10, **12**-11, **12**-31, **12**-32, **12**-35
Digital logic, **15**-78 to **15**-81
Digital magnetic tape in data reduction systems, **12**-4, **12**-20
Digital meter, **11**-51, **11**-52
Digital modulation system, wideband, **4**-39
Digital recording, **11**-30 to **11**-34
Digital signal transfer in checkout equipment, **11**-69
Digital telemetry system, **1**-8
Digital-to-analog conversion, **12**-7
Digital-to-analog converter, **8**-43
Digital transducer, **3**-11
Digital transmission, telephone lines, **13**-12
Diode, and printed circuit, **10**-78
Diode gate, **4**-83
Diode matrix, **4**-80
Diode rectifier, semiconduction, **10**-70
Diode squarer multiplier, **15**-77
Diode switch, **8**-12 to **8**-13
Dipole, half-wave, **4**-120
Dipole cloud, **16**-9
Dirac delta function, **A**-33 to **A**-34
Direct-playback demodulation, **6**-34 to **6**-35
Direct printing, for data presentation, **11**-50
Direct recording, IRIG standards for, **5**-38 to **5**-42
Direct recording parameters, IRIG, **5**-41
Direct vision systems remote-handling, **15**-139
Direct cosine, **3**-46
Directional gyro, **15**-100, **15**-117, **15**-119 to **15**-121
 missile application of, **15**-10
Directivity, antenna, **4**-112
Directly measured quantities in flight testing, **12**-53
Discharge characteristics, storage battery, **10**-88, **10**-91
Discrete channel, **9**-31
 with noise, fundamental theorem for, **2**-28
Discrete command, **14**-102
Discrete communications system, multi-symbol, **9**-34
Discrete noiseless channel, **2**-24, **2**-25
 fundamental theorem for, **2**-25
Discriminator, **4**-50, **4**-51
 a-c voltage, **4**-50
 and errors, **12**-23
 FM, **6**-2
 for PCM, **8**-54 to **8**-56
 FM-FM, **4**-21
 Foster-Seeley, **6**-31, **6**-32
 frequency, coherent, **9**-66
 phase-locked, **4**-13, **4**-24, **4**-25
  peak phase error constrained, **9**-67 to **9**-69
 receiver, **6**-31, **6**-32
 reference, **6**-52, **6**-53
 subcarrier, **6**-40 to **6**-55
Disk file, data processor, **11**-6
Dispatching, centralized, pipeline control, **13**-42
Dispersed transportation system, **13**-39
Dispersion sample, **A**-53
Dispersive network, **9**-37
Displacement, in accelerometer, **15**-89
 angular, and accelerometer, **15**-97
 gyro measurement, **15**-115
 rate gyro output, **15**-105
 linear, gyro measurement, **15**-123, **15**-124
 transducer, **3**-3, **3**-10 to **3**-13, **3**-17, **3**-21 to **3**-24, **3**-33
 accelerometer, **15**-91 to **15**-94

Displacement, transducer, angular, **3**-34
   variable reluctance, **3**-23
Display, antenna and vehicle coordinates, **4**-131
   PAM ground station, **9**-17
   PCM telemetry data, **8**-44, **8**-47, **8**-48
   pipeline control, **13**-44
   system control and operation, **11**-54
Distance-measuring equipment (DME), **3**-49
Distances, interplanetary, **14**-16
   line-of-sight, aircraft and satellites, **4**-96, **4**-97
   radio transmission, air-to-ground, **4**-96
   satellite-to-ground, **4**-97
Distortion, dynamic, **6**-31
   due to subcarrier oscillator, **6**-38, **6**-39
   error from, **9**-23, **9**-24
     PDM, low-pass filter, **7**-35
   FM signal, by frequency selective network, **6**-41
   harmonic and intermodulation, in subcarrier discriminators, **6**-41, **6**-42
   PAM signal, from r-f channel, **4**-56, **4**-58
   phase nonlinearity, **6**-31
   by receiver, **6**-36
   time-delay skew, **6**-38 to **6**-41
   transmission line, **13**-11
Distribution system, gas and petroleum, **13**-39
Diversity polarization reception, **4**-7, **4**-114
Divider, analog, **15**-74, **15**-76
Division in data processor, **11**-41, **11**-42
DME (distance-measuring equipment), **3**-49
Documents, IRIG, where mandatory, **5**-2, **5**-3
DOETS (dual-object electronic tracking system), **3**-51
Domain, discontinuous, **13**-23
DOPLOC (Doppler phase lock), **3**-47
Doppler measurements and integrated tracking and telemetry system, **14**-95, **14**-102
Doppler phase-locked loop tracking, **9**-64
Doppler ranging (DORAN), **3**-49
Doppler shift, **14**-5
   effect of, on receiver bandwidth, **4**-107, **4**-108, **4**-110
Doppler tracking loop, phase-plane analysis of, **9**-72 to **9**-76
Doppler trajectory measurement, **3**-44, **3**-45
DORAN (Doppler ranging), **3**-49
Dosage, radiation, annual in earth orbits, **14**-62
   maximum for man, **14**-65
Double integrating gyro, **15**-100
Double pulse generator for FM system test, **6**-34, **6**-35
DOVAP (Doppler velocity and position), **3**-47, **12**-4
Down link, active communications satellite, **16**-12
Down-range telemetering, **4**-4, **4**-5
Drag, atmospheric, **14**-27
   and modulation, reentry, **14**-31
Drag coefficients, **15**-7
Draper, C. S., **13**-108, **13**-116
Drift, in analog-to-digital converter, **11**-16
   frequency of, and transmitter, **10**-53
Drift correction, **8**-9
   and subcarrier channel, **5**-28, **5**-29
Drift error, rate gyro, **15**-114, **15**-115
Drift rate, due to anisoelasticity, **15**-112
   of gyro, **15**-100 to **15**-102
   platform, **15**-127
Drift transducer, **3**-5

Drive, gyro, **15**-101
Driving function, analog-computer, **15**-67, **15**-70
Drones, **4**-2
Drop-out, data, **12**-49, **12**-57
   phase-locked-loop, **9**-68
   probability of, and phase-locked discriminator, **9**-20
Drop-out recorder, **7**-40
Drum, magnetic, **11**-3, **11**-6
Dry box, and nuclear hostile environment, **15**-140
Dry primary cells, **10**-82 to **10**-84
DSIF (Deep Space Instrumentation Facilities), **14**-93, **14**-96, **14**-97, **14**-102
Dual mode attitude control, **14**-38, **14**-39
Dubbing-in, time codes, **12**-57
Duplex operation, **5**-10
   voice circuit, **16**-19
Duty cycle, **7**-13
   and baseline shift, **6**-33
   electromechanical commutator, **4**-63
   PAM, **9**-9, **9**-11
   IRIG standards, **5**-34
   in PDM radio links, **7**-23
Dynamic signals for automatic check out, **11**-58
Dynamics, operating, in remote-handling system, **15**-157, **15**-159
Dynodes, **3**-28

e, natural logarithm base, **A**-6
E layer, ionized, **10**-11
Early Bird, **16**-2
Earth, **14**-53, **14**-55 to **14**-57
   and earth-Mars oppositions, **14**-17
   magnetic field of, **14**-35
   orbits around **14**-17 to **14**-21
   and altitude vs. object velocity and launch angle, **14**-23
   resources of, space survey of, **14**-10
EBU (efficiency of bandwidth utilization), **2**-54
Eccentric anomaly, **14**-5
Eccentricity, **14**-5, **14**-17, **14**-18
Echo, in communications satellite, **16**-31
Echo suppressor, **16**-31
Echo-type reflector, and passive satellite, **16**-9
Ecliptic, **14**-5, **14**-52, **14**-57
Ecological system, **14**-5
Eddy-current damping, accelerometers, **15**-95
Edison cell, **10**-87, **10**-89 to **10**-91
Edison effect, **10**-98
Editing, data, **4**-6, **4**-7, **12**-9, **12**-21, **12**-24, **12**-25, **12**-30
Effective exhaust velocity, **14**-39, **14**-42, **14**-45
Effective radiated power, ITU defined, **5**-14
Efficiency, of bandwidth utilization, **2**-54
   communications, **9**-22
   and telemetry links, **9**-29 to **9**-31
   of fuel cell, **10**-100, **10**-102
   of silicon solar cell, **10**-95
   telemetry application of, **4**-30
Efficiency factor, radiation, **4**-112
Eglin Test Range, **3**-51
EHF band, defined, **5**-15
Einstein's equation, **3**-25
Elastic coupling accelerometer, **13**-91
ELECOM 100, **11**-38
Electric prime mover, **15**-128 to **15**-132

INDEX

Electrical angle, potentiometer, 3-33
Electrical circuit equivalents, mechanical system, **15**-63, **15**-64
Electrical conversion factors, **1**-21 to **1**-23
Electrical damping of accelerometer, **15**-95
Electrical time constant, a-c servomotor, **15**-132
Electrochemical cell, **10**-79 to **10**-92
Electrodes in fuel cell, **10**-102
Electrolytic resistance, **10**-80
Electromagnetic transducer (*see* Magnetic transducer)
Electromechanical commutation, 4-60 to 4-78
Electromotive force, **10**-79, **10**-81
Electron intensity in earth orbits, **14**-63
Electronic commutation, 4-78 to 4-85
Electronic logic circuits, basic, **15**-80
Electronic printer, **11**-50
Electronic trajectory measurement, 3-42
  systems for, 3-46 to 3-53
  techniques for, 3-43 to 3-46
Electronic Trajectory Measurements Working Group (ETMWG), 5-2
Electronic transducer (mechanical-displacement-input-tube), 3-12
Electrostatic bearings, gyro, **15**-103
Electrostatic printer, **11**-50
Electrostatic tape storage, **14**-86
Elevation angle, communications satellite terminal, **16**-6
  minimum earth stations in space service, **5**-23, **5**-24
Elliptic integrals, **A**-18
Elliptical orbit, **14**-17, **14**-18, **14**-22
Elliptical partial differential equations, **12**-36
ELSSE (electronic sky screen equipment), 3-52 to 3-53
EMA (electronic missile acquisition) trajectory system, 3-53
Emissions, characteristics of, ITU definitions of, **5**-13, **5**-14
  ITU designation of, **5**-14 to **5**-19
  shutdown of, at space station, **5**-24
Emitter resistance, incremental, transistor, **10**-21
Encapsulation, **10**-78
Encoder, analog-to-digital, **8**-23 to **8**-27
  feedback, PCM, **8**-24, **8**-25
  PCM, **8**-23 to **8**-27
  ramp, PCM, **8**-24
Encoding, modulation, **1**-8
  systems of, for PCM transmission, **13**-27
End-to-end calibration, **12**-8
Energy, analogs, mechanical-electrical, **15**-64
  per bit, **8**-3, **8**-53
  per unit noise intensity, β, **9**-29, **9**-30
    in space communications, **14**-96, **14**-97
    and West Ford system, **16**-12
  body, in orbital motion, **14**-20 to **14**-22
  coupling, **15**-64
  dissipation of, **15**-64
  per unit volume, for battery systems, **10**-88
  per unit weight, for battery systems, **10**-88
Engine, aerospace, **14**-49
Engineering performance data presentation, **14**-87
Enloe, L. M., 9-93
Entropy, **2**-22
  of source, 9-20
Entropy power, **2**-29

Entropy tables, **A**-76 to **A**-94
Entry angle, **14**-30 to **14**-32
Envelope of random process, **2**-14
Environment, aircraft, **10**-4
  hostile, **15**-138, **15**-139
  missile, **10**-4
  outer-space, **10**-9 to **10**-18
  reactor radiative, **10**-14
  telemetry system, **10**-3 to **10**-18
  test specifications for, **10**-4 to **10**-10
Environmental conditions, in radio telemetry, 4-6
Environmental constraint, **14**-59 to **14**-70
Environmental control for man in space, **14**-70
Environmental effect on transducer, 4-8
Environmental specifications, **10**-4
Environmental tests, description of, **10**-7 to **10**-9
  requirements of, **10**-9, **10**-10
Ephemeris, **14**-6
  and communications satellite, **16**-23
Epoch, **14**-6
Epoxy resin encapsulent, **10**-78
Equalization in magnetic recording, **11**-28
Equalizer, communication link, **13**-12
Equations of motion for orbiting spacecraft, **14**-17, **14**-18
Equiangular spiral antenna, 4-114, 4-124
Equilibrium temperatures in space, **14**-67
Equinox, **14**-4
Equipment errors in data reduction, **12**-21 to **12**-23
Equivalent gain, 9-87
  and phase-locked loop analysis, 9-77, 9-78, 9-86 to 9-96
Equivocation, **2**-26, **2**-27, 9-20
  as function of error probability, 9-20
Erasing in magnetic records, **11**-27
Ergodic, 9-19
Ergodic process, **2**-10, **2**-11
Ergodicity, **2**-10
Error, acceleration insensitive, rate gyro, **15**-114
  acceleration sensitive, rate gyro, **15**-114
  aliasing, 9-7, 9-8
  alignment, rate gyro, **15**-110
  analog, for given PCM bit error probability, 9-20
  anisoelastic, gyro, **15**-111, **15**-112
  from band limiting, 9-23
  per bit, digital data over analog channel, 9-58, 9-59
  bit, equivalent, 4-28, 4-29
  case acceleration, rate gyro, **15**-109, **15**-110
  coning, rate gyro, **15**-113, **15**-114
  control system, **15**-36
  coupling, gyro, **15**-110, **15**-117
  from crosstalk, 9-23, 9-24
  cylindrical, rate gyro, **15**-113, **15**-114
  data conversion, **11**-6
  distortion noise, 9-23, 9-24
  dynamic, radar tracking, 3-55
  fixed, radar tracking, 3-55
  fluctuation noise, 9-23
  FM-FM system, **6**-19
  guidance, space mission, **14**-52
  hysteresis, rate gyro, **15**-110
  in IF bandwidth design, **5**-32, **5**-33
  in industrial telemetry, **13**-27
  and integrating gyro, **15**-117
  lag, radar tracking, 3-54
  lift-off timing, **14**-52
  linear, of analog telemetry signal, 9-20

INDEX

Error, mean square, **2**-7
  from noise, PCM, **8**-59, **8**-60
  phase, by Volterra expansion method, **9**-85
  normalized, and information rate, telemetry, **9**-21
  normalized *rms*, **9**-20
  PCM total normalized, mean square, **8**-60, **8**-61
  PDM, **7**-31 to **7**-36
    in PDM system, d-c level, **7**-36
    as wow and flutter, **7**-39
  quantization, **8**-26, **9**-23, **9**-24, **9**-31, **9**-32, **9**-56
    in PCM, **8**-58
  radar boresight, **3**-56
  radar tracking, **3**-54
  random, **3**-6
    in PCM, **8**-59, **8**-60
    torque, gyros, **15**-114
  rate gyro, **15**-109 to **15**-115
  recorder, instantaneous speed, **5**-54, **5**-55
    quasi-static, **7**-39
  residual, telemetry, **9**-24
  rms, interrupted analog signal, **9**-21
  sampling and interpolation, PCM, **8**-58, **8**-59
  sensitivity, **6**-19
  space mission, **14**-51, **14**-52
  stable platform, **15**-126, **15**-127
  steady-state, tracking phase-locked loop, **9**-65
  synchronization, PDM, **7**-32
  systematic, **3**-6
    in PCM, **8**-58
  tape-speed, **6**-19
  telemetry system, **9**-23, **9**-24
  time-base, **6**-19
  timing, in PDM recorders, **7**-39
  transaction, **12**-29
  transducer, **3**-6
  transient, phase-locked discriminator, **9**-69
  velocity, space mission, **14**-52
  vibrating, rate gyro, **15**-111
  zero, **6**-19
Error analysis, and data reduction, **12**-29
  of planned lunar mission, **14**-51, **14**-52
Error bit check, **12**-48
Error checking, **13**-8, **13**-9
  longitudinal redundancy, **13**-29
  two-out-of-five code, **13**-29
  by validity, **13**-29
  by vertical check, **13**-29
Error constraint, control system, **15**-36, **15**-37
Error control message, **13**-8
Error correcting codes, PCM, **9**-32, **9**-33
Error correction, **13**-27 to **13**-30
  implementing difficulties, **13**-28
  in PDM recorders, **7**-39
  by retransmission, **13**-28
  by validity checks, **13**-29
Error criteria, control system, **15**-48
Error detection by parity, **13**-24
Error function (erf), **9**-32, **9**-33, **9**-57, A-26, A-48
Error minimization, telemetry link, **13**-2
Error moment criterion, **9**-21
Error performance analysis in FM carrier telemetry system, **9**-18 to **9**-29
Error probability, word, PFM, **9**-47 to **9**-49, **9**-53, **9**-54

Error rate, **9**-30, **9**-31
  per bit, PCM, **9**-32
  QPPM, **9**-35
  comparative, discrete vs. analog channels, **9**-57, **9**-58
  in space command system, **14**-95
  in space telemetry, **14**-95, **14**-96
Error signal, phase coherent synchronization loop, **14**-101
Error sources, in analog-to-digital converter, **11**-14 to **11**-16
  in telemetry system, **9**-24
Escape velocity, surface, **14**-6, **14**-21 to **14**-26, **14**-30
  of planets and sun, **14**-53
Estimate, statistical, A-52
Etched-foil strain gage, **3**-35, **3**-37
Euler diagram, A-12
Euler-Mascherani constant, A-26, A-27
Everett, interpolation formula of, A-64
Events in flight testing, **12**-53
Exchange, data communication, **13**-14
Exchange ratio, **14**-6, **14**-22, **14**-39 to **14**-40
Execute (E) cycle, data processor, **11**-40 to **11**-47
Executive program, **12**-40
Exhaust velocity, propellant, **14**-39, **14**-42
Exit, atmospheric, **10**-11
Expectation, mathematical, **2**-13
Experiment, example of large scale, **12**-55, **12**-56
  preparatory planning for, **12**-12
Experimental data presentation, **14**-87
Experimentation, organizing, **12**-51
  (*See also* Flight test)
Explorer I, telemetry system block diagram of, **14**-25
Explorer III, telemetry system of, **14**-76
Explorer VI, data acquisition and recording system of, **14**-79
  telemetry and instrumentation of, **14**-78
Explorer VII, data processing system of, **14**-79, **14**-80
  telemetry and instrumentation system of, **14**-79
Explorer VIII, data processing of, **14**-82
  instrumentation and telemetry of, **14**-81
Explorer satellite, **14**-74, **14**-75
Explosion, test specification for, **10**-6, **10**-8
Exponential functions, **2**-4, A-23
Exponential transfer function, **15**-84
External gimbal system, **15**-126, **15**-127
Externally programmed processor, **11**-34
EXTRADOP (extended-range Doppler velocity and position), **3**-47
Extrapolation in data reduction, **12**-33

F layer, ionized, **10**-11
Fades, in radio-telemetry, **4**-7
  radio, in industrial telemetry, **4**-3
Failure rate pattern, mechanical commutators, **4**-76
Failures, typical, environmental, **10**-7 to **10**-9
  explosion, **10**-8
  fungus, **19**-9
  humidity, **10**-8
  rain, **10**-9
  salt spray, **10**-8
  sand and dust, **10**-8
  shock, **10**-7
  sunshine, **10**-9
  temperature, **10**-7
  vibration, **10**-8

INDEX

False-command probability, **14**-102
False keying, **10**-62
Fano, R., **9**-40
Faraday rotation, **4**-99, **4**-100
Faraday's induction law, **3**-11, **3**-20
Fast-neutron radiation, **10**-14, **10**-15
Faure cell, **10**-84
FCC (Federal Communications Commission), **5**-4
Federal Communications Commission (FCC), **5**-4
Feedback, **15**-20
  in control system, **15**-24 to **15**-27
  in gating circuit, **4**-81
  in remote handling, **15**-162
Feedback comparison, analog-to-digital converter, **12**-19
Feedback compensation, **15**-46
Feedback encoder, PCM, **8**-25
Feedback function generation, **15**-75, **15**-76
Feedback system, **1**-6, **15**-20, **15**-88
  block diagram analysis of, **15**-22
Feeling (tactile) in remote handling, **15**-156
Feldman, C. B., **9**-6
Ferrite phase modulator, **6**-22, **6**-23
Ferromagnetism, **3**-15
FET (field-effect transistor), **4**-83, **4**-84
  (*See also* Field-effect transistor)
FET (Flight Engine Test), at Arco, Idaho, **15**-140
Fetch operation, data processor, **11**-40
Fiducial marks, **12**-15
Field, data processing, **12**-13
Field-controlled d-c motors, **15**-128
Field-effect transistor, **10**-76, **10**-77
  for signal conditioning, **8**-7, **8**-9
  for signal gating, **4**-83, **4**-84
  (*See also* Unipolar transistor)
Fighting readiness check, **11**-56
Figure of merit, system, **14**-13
File of punched card information, **11**-51
Film data processing, **12**-15, **12**-16
Film reader, **12**-14 to **12**-16
Filter, band-limiting, **9**-8, **9**-9
  finite memory, **9**-9
  band pass, in FM-FM modulator, **4**-19, **4**-20
  Bessel polynomial, **6**-57
  Butterworth, **6**-31, **6**-51, **6**-52, **8**-6
  clock frequency, PAM demultiplexer, **9**-16
  comb, PFM, **9**-53
  constant amplitude, **6**-51, **6**-52
  (*See also* Filter, Butterworth)
  digital, **12**-31, **12**-32, **12**-55
  finite memory, PAM data detector, **9**-15
  Gaussian, **6**-13, **6**-51, **6**-52
  (*See also* Filter, linear-phase)
  input bandpass, subcarrier discriminator, **6**-41
  interpolation, **9**-8
  linear phase, **6**-13, **6**-51, **6**-52, **6**-57, **6**-58
  (*See also* Filter, Gaussian)
  loop, optimum, **9**-66
  phase-locked, **9**-64
  matched, PAM sync detector, **9**-17
  PFM, **9**-51 to **9**-54
  minimum phase, **9**-8
  output, optimum, phase-locked-demodulator, **9**-67
  Rauch, **6**-51
  side band rejection, **6**-57, **6**-58
  unmatched, PFM reception, **9**-51 to **9**-54
Filtering, digital, **12**-10, **12**-11
  premodulation, PAM, **5**-35

Finite differences, **A**-61 to **A**-69
Fire control, **15**-72, **15**-73
Firing angle, deboosting from orbit, **14**-32, **14**-33
Firing sector, missile range, **14**-27 to **14**-29
First order phase-locked loops, phase error variance, **9**-92
Flared-slot antenna, **4**-139
Flight path angle (*see* Burnout angle)
Flight readiness check, aircraft, **11**-57
Flight test, **12**-51 to **12**-56
  planning of, **12**-52
  program planning of, **12**-56
Flip-flops, **8**-18
Floating point for data reduction, **12**-40
"Florolube," **15**-108
Flotation, accelerometer suspension, **15**-97, **15**-98
  in gyro, **15**-102, **15**-108, **15**-116
Flow, transducer, gas, **3**-14
Flow chart, program, **12**-39
Flow-velocity transducer, **3**-21
Flowmeter, **3**-3, **3**-12
Fluctuation noise in PDM system, **7**-32
Fluid viscous-shear law, **15**-94
Flush mounted antenna, **4**-123, **4**-127, **4**-128, **4**-130
Flutter, **7**-39, **12**-20
  in data recording, **11**-34
  recorder, **6**-33, **6**-34
Flux, radiation, **10**-15
Flux-dose plot, **10**-14, **10**-15
Flux linkage, **3**-15
Flyby, planetary, **14**-28 to **14**-30
Flywheel gating, **7**-28, **7**-32
FM, **1**-8
  average signal power, **2**-38
  average unmodulated power, **2**-38
  capture effect, **6**-11
  carrier improvement threshold, **7**-5
  demodulator, phase-locked, **9**-67 to **9**-69
  feedback receiver, **9**-93, **9**-94
  multiplexing, **6**-13
  noise improvement, **6**-10, **6**-11
  noise spectrum, **6**-10
  receiver, phase-locked loop, **2**-54
  signal spectrum, **6**-10
  telemetry system, communication properties, **9**-2 to **9**-5
  theory, **6**-7 to **6**-13
  thresholding, **6**-12, **6**-13
  wideband improvement ratio, **6**-11
FMFB (frequency modulation feedback), **4**-92, **16**-17, **16**-18
FM-FM, adjacent channel interference, **6**-15
  AIA standard, **6**-5
  allowable percentage deviation, **6**-13
  automatic calibration, **6**-52, **6**-54
  channel linearity, **6**-3, **6**-14
  channel zero error, **6**-19
  component spurious output, **6**-14
  constant bandwidth system, **6**-2, **6**-5, **6**-17
  crosstalk, **6**-13, **6**-14
  deviation level adjustment, **6**-16
  distortion in RF link, **6**-16
  dynamic characteristics, **6**-3
  dynamic error sources, **6**-3
  equivalent information rate, **4**-32
  in industrial telemetry; **13**-35
  intermodulation products, **6**-15
  IRIG standard system, **6**-4
  multiplex peaking, **6**-16
  preemphasis schedule, **6**-16 to **6**-18

INDEX

FM-FM, proportional bandwidth system, **6**-2, **6**-4, **6**-17
  sources of error, **6**-14 to **6**-20
  static characteristics, **6**-3
  static error sources, **6**-3
  subcarrier frequency deviation, permissible, **6**-3 to **6**-6
  subcarrier harmonic distortion, **6**-14
  system functions, **6**-2, **6**-3
  tape-speed variation error, **6**-19
  and telemetry, IRIG definition of, **5**-27
  standards for, **5**-27 to **5**-31
  systems for, **1**-8, **4**-18, **4**-21
  error sources in, **9**-24
  typical performance levels, **6**-2
  typical system applications, **6**-3
  zero drift, **6**-3, **6**-14
Foam encapsulent, **10**-78
Foam potting, **6**-27
Fokker-Planck equation, **9**-79
Fokker-Planck method, phase-locked loop analysis, **9**-77 to **9**-82
Footprint, reentry, **14**-32
Force-balance accelerometer, **15**-95, **15**-96, **15**-98
Force transducer, **3**-17, **3**-22, **3**-30 to **3**-32
Forcer, accelerometer, **15**-95
Format, antenna pattern, **4**-131
  common-language, used with PCM, **8**-48
  computer instruction, **11**-36
  computer output, **12**-46
  control, **12**-23, **12**-24
  data, **12**-13
  data output, **12**-4
  data processor words, **11**-37, **11**-38
  instruction, data processor, **11**-36, **11**-38
    ELECOM 1000, **11**-38
    IBM 650, **11**-38
  instruction and command, **11**-47
  Mariner telemetry word and bit, **14**-102
  PACM, **9**-27
  PAM composite signal, **9**-13
  PCM commutated data, **8**-16
  PCM output, **8**-43, **8**-44, **8**-48
  PFM, **9**-51, **9**-52
  standardization, **12**-40
  telemetry modulation, **6**-27, **6**-28, **6**-31 to **6**-33
  translation, **12**-23
Fort Churchill, **3**-47
Forter-Seeley discriminator, **6**-31, **6**-32
Forward velocity, rocket, **14**-39, **14**-40
Foucault pendulum, **15**-120, **15**-121
Fourier analysis, of periodic functions, A-35 to A-37
  of sampled data, **12**-35
Fourier coefficient of periodic functions, A-37
Fourier integral, **2**-9
Fourier transform, **2**-9, A-38
  of pulses, **2**-47
Four-out-of-eight code, **13**-29
Frame, and PCM, **8**-30
  and PDM telemetry, **7**-3
Frame counter, PCM, **8**-17, **8**-46
Frame rate, and PACM, **9**-27
  and PDM, **5**-31
  and television, for remote handling, **15**-155
Frame structure, PCM, **5**-36, **5**-37
Frame synchronization, and PCM, **8**-38, **8**-39, **8**-42
  and PDM, **7**-28, **7**-29
Free gyro, **15**-120

Free-space path loss, **4**-92
  and isotropic links, **4**-97 to **4**-100
  in solar system, from earth, **14**-96
  (*See also* r-f free-space loss)
Freely mounted gyro, **15**-118
Frequency, carrier, radio-telemetry, **4**-94
  highest significant, **9**-6, **13**-26
  instantaneous, **6**-12
  natural, phase-locked loop, **6**-46
    imperfect integrator, **9**-77
  for space communication, **4**-94
  test range, **4**-94
Frequency adjustment, power system, **13**-45
Frequency allocation, **5**-4 to **5**-24
  communications satellite, **16**-4
  international authority, **5**-4
  radio, IRIG, **5**-3, **5**-8, **5**-27, **5**-47 to **5**-50
  ITU, **5**-6, **5**-7
Frequency analysis for data reduction, **12**-34, **12**-35
Frequency assignment, range, **5**-8
Frequency authority, range, **5**-8
Frequency band, allocated telemetry, **7**-27
  assigned, ITU defined, **5**-14
  nomenclature, ITU, **5**-15
Frequency change rate, natural frequency, and loop, **9**-75
  of reference, and phase-locked loop tracking, **9**-75
Frequency-code system for pulse code telemetering, **13**-36
Frequency converter, **4**-44 to **4**-48
Frequency coordination, r-f, **5**-4 to **5**-24
  working group for, IRIG, **5**-2
Frequency counter, analog-to-digital converter, **12**-19, **12**-20
Frequency deviation, PAM signals, **5**-33
Frequency division, **2**-37
Frequency-division multiplex, communications satellite, **16**-22
  telemetry, **5**-27 to **5**-31
Frequency domain, **2**-19
Frequency error, initial and phase-lock, second-order loop, **9**-75
  phase error plot normalized, second order phase-locked loop, **9**-72, **9**-76
Frequency locus, gain-phase plane, **15**-61
Frequency meters, **13**-36
  constant amplitude, **13**-37
Frequency modulation, by band-limited, white, Gaussian signal, **2**-45
  carrier spectrum, PCM, **8**-50 to **8**-52
  by Gaussian signal, **2**-45
  necessary bandwidth, ITU, **5**-17, **5**-18
  by sinusoid, **2**-45
  by square wave, **2**-45
  wideband, **16**-16 to **16**-18
Frequency modulation feedback in communications satellite, **16**-17, **16**-18
Frequency modulator, **6**-22, **6**-23
Frequency multiplier, **6**-25, **6**-26
  communications satellite repeater, **16**-20
Frequency-to-number analog-to-digital conversion, **11**-22 to **11**-24
Frequency range, transducer, **3**-6
Frequency requirements, IRIG, telemetry transmitters and receivers, **5**-48 to **5**-50
Frequency response, **2**-20
  flat amplitude filter, **6**-52
  FM-FM channel, **5**-27
  linear phase filter, **6**-52
  required by measurements, **4**-9
  second-order system, **15**-98, **15**-99

INDEX

Frequency separation, space transmitters and receivers, **5**-8
Frequency sharing, space and ground services, **5**-8, **5**-23
Frequency-shift-keying, in industrial telemetry, **13**-32
  PCM, **8**-49, **8**-50
Frequency-shift telegraphy, ITU defined, **5**-10
Frequency spectrum, continuous, **2**-9
  discrete, **2**-8
  PCM, **2**-49
  PDM-FM, **2**-49
Frequency standards, IRIG, **5**-48 to **5**-50
Frequency sweep, adiabatic, **2**-45
Frequency sweeping range measurement, **3**-44
Frequency telemetering system, **13**-35 to **13**-37
  power frequency type, **13**-36
Frequency telemetry, **13**-2
Frequency tolerance, **5**-13
  ITU, **5**-19 to **5**-21
Frequency translation, **1**-9, **6**-55 to **6**-59
  adjacent channel interference, **6**-59, **6**-60
  double sideband, **6**-59
  FM-FM system, **6**-13
  performance with recorder, **6**-60, **6**-61
  reference carrier frequency, **6**-55, **6**-58
  spectra, **6**-56
  switch type, **6**-56
  telemetry system, **6**-55 to **6**-62
Frequency uncertainty, receive-transmit, **4**-107
Friction, Coulomb, **15**-102
Frits, fusing, strain gage, **3**-36
FSK (frequency-shift keying), **8**-49
  in industrial telemetry, **13**-32
Fuel cell, **10**-100 to **10**-107, **14**-71
  characteristics, **10**-106
  comparative weight of, **10**-103
Full duplex links, **13**-15
Fully oriented satellite, **14**-34
Function generation by feedback, **15**-75, **15**-76
Functional determinant (Jacobian), A-15
Functional layout, PAM-FM ground station, **9**-14
Functions, of automatic checkout equipment, **11**-57, **11**-58
Fungus, test specification, **10**-6, **10**-9

Gage, electrolytic, **3**-12
Gage factor, **3**-34
Gain, absolute, **5**-14
  antenna, **4**-112, **4**-113
  communications satellite, **16**-14
  of digital filter, **12**-31
  ground antenna, **4**-102
  helical antenna, **4**-102, **4**-103
  isotropic, **5**-14
  ITU definition, **5**-14
  maximum, **4**-113
  non-oriented satellite reflector, **16**-8
  relative, **5**-14
  space telemetry, **14**-95 to **14**-97
  transistor, available power, **10**-29
    current, Ai, **10**-29
    and h and r parameters, **10**-25, **10**-28, **10**-29
    insertion power, **10**-28
    operating power, **10**-29
    with series feedback, **10**-32
    with shunt feedback, **10**-32

Gain, transistor, transducer power, **10**-28
  voltage, Av, **10**-29
  and z and y parameters, **10**-30
  vehicle antenna, **4**-97
Gain-frequency plot, transfer function, **15**-41, **15**-42
Gain margin, control system, **15**-40 to **15**-48
Gain-phase plot, **15**-61
  control system design, **15**-40, **15**-48
  control system with lags, **15**-85, **15**-86
  in sampled-data system, **15**-55
Gain stabilization, transistor, **10**-25, **10**-30 to **10**-32
Galactic coordinates, **14**-6
Gamma distribution, A-50, A-51
Gamma function, A-26
Gamma per square cm per sec, **14**-66
Gamma radiation, **10**-14
Gap effect, PDM recording, **7**-37
Gap scatter, recording, **5**-39
Gas content, planetary atmosphere, **10**-12
Gate, balanced, **4**-82, **4**-83
  diode, **4**-83
  electronic commutator, **4**-80, **4**-84
  exclusive OR, **8**-19
  FET, **4**-83, **4**-84
  high level, **4**-82 to **4**-84
  low level, **4**-82, **4**-83
  sequential PAM, **9**-10, **9**-11
  transistor, **4**-81, **4**-82
    direct coupled, **9**-11
    sampling, **9**-9 to **9**-11
Gated-switch frequency converter, **4**-45, **4**-46
Gateway in communications satellite system, **16**-21
Gauss interpolation method for data reduction, **12**-33
Gauss quadrature formulae, A-66, A-67
Gaussian distribution, A-46 to A-49
Gaussian noise, **13**-28
  effect of, on PCM, **8**-53
Gee, **15**-12
Geiger-Mueller counter, **3**-13
  counting rates, **3**-14
  quenching, **3**-14
Gemini program, **14**-11
General Electric Company, **3**-52, **13**-14, **13**-28
Generator allocation, **13**-44, **13**-45
Generator speed control, power system, **13**-45
Geocentric vertical, **15**-121, **15**-122
Geodetic instrumentation system, **12**-4
Geographic vertical, **15**-122
Geometric distribution, A-45
Geometric progression, A-24
Geophysical environment, effect of, on spacecraft design, **14**-60 to **14**-62
George, P. A., **9**-85
Germanium photo transducer, **3**-26, **3**-29
Gestalt in remote-handling, **15**-142
Ghosts in radio-telemetry, **4**-7
Giga, **5**-15
Gill's method, **12**-36
Gimbal, **15**-103
  in stable platform system, **15**-126, **15**-127
Gimbal lock, gyro, **15**-126
Glossary, telemetry terms, IRIG, **5**-57 to **5**-67
G-M (Geiger-Mueller) counter, **3**-13, **3**-14
Goddard Space Flight Center (NASA), **14**-75
Golay, **9**-34

INDEX

Grasping device, **15**-162
  space remote-handling, **15**-152
Grasping motion, manipulator, **15**-145
Gravitation, **14**-6
Gravitational constant, **1**-22
  earth, **14**-35
Gravitational parameter, **14**-17, **14**-18, **14**-21
Gravitational variation, with altitude, **14**-60
Gravity cell, **10**-84
Gravity gradient, **14**-34
Gravity gradient stabilization, satellite, **16**-8, **16**-14
Gray body, space temperature of, **10**-16
Gray code, **9**-59
  for discrete transmission, analog data, **9**-31, **9**-58
  for position analog-to-digital converter, **11**-21
Ground, floating, **8**-10
Ground plane, antenna pattern, **4**-115, **4**-116
Ground station, PCM, **8**-30, **8**-48
  PDM, **7**-27 to **7**-31
Grounding, electrical, **4**-36 to **4**-38
Group synchronization, **8**-31, **8**-36 to **8**-43
Grove, W. R., **10**-100
Gruen, W. M., **9**-59, **9**-77
Guidance, **15**-4 to **15**-19
  beam rider, **15**-8, **15**-9
  command, **15**-7, **15**-8
  dead reckoning, **15**-9, **15**-11
  defined, **15**-5
  homing, **15**-13, **15**-14
  inertial, **15**-11, **15**-12
  maneuver strategy, **15**-14 to **15**-19
  radio navigation, **15**-12, **15**-13
  system, **15**-7 to **15**-14
  terminology, **15**-5
  vehicular, **15**-4 to **15**-19
Gyro, **15**-99 to **15**-127
  angular momentum, **15**-100, **15**-101
  bearings, **15**-102, **15**-103
  captive, **15**-100
  classification, **15**-99, **15**-100
  compass, **15**-120, **15**-121
  construction, **15**-100 to **15**-103
  control, **15**-99
  directional, **15**-100, **15**-117, **15**-119 to **15**-121
    missile application, **15**-10
  drift rate, **15**-100 to **15**-102
  drives, **15**-101
  freely mounted, **15**-118
  gimbal lock, **15**-126
  HIG, **15**-116, **15**-117
  instrument, **15**-99
  integrating, **15**-100, **15**-115 to **15**-117
  as integrating accelerometer, **15**-123, **15**-124
  master, **15**-99
  precessional velocity, **15**-104
  rate, **15**-100, **15**-105 to **15**-115
    drift errors of, **15**-114, **15**-115
    missile application of, **15**-10
  restraint, spin-axis, **15**-100
  rotor balance, **15**-102, **15**-103
  single-degree of freedom, **15**-100, **15**-103 to **15**-118
  spin velocity, **15**-101
  stabilizing, **15**-99
  theory of operation, **15**-103 to **15**-105
  torquer, **15**-116 to **15**-118
  two-degree-of-freedom, **15**-118 to **15**-124
  vector relations in, **15**-103

Gyro, vertical, **15**-100, **15**-121 to **15**-123
Gyro-stabilized platform, **15**-124 to **15**-127
Gyrolube, **15**-108
Gyroscopic integrating accelerometer, **15**-123, **15**-124

h parameter, **10**-24 to **10**-30
  common base, **10**-25
    and r and other h parameters, **10**-26
  common-collector, and r and other h parameters, **10**-26
  common emitter, and r and other h parameters, **10**-27
  and r parameter relationships, **10**-26, **10**-27
  typical values of, and transistor, **10**-30
Half adder, **8**-19
  unipolar transistor, **10**-26
Half-cell reactions, **10**-79, **10**-80
Half duplex link, **13**-15
Half-split analog-to-digital conversion, **10**-65 to **10**-68
Hall coefficient, **3**-18
Hall effect, **3**-18
Hamming, **13**-28
Handling, remote, **15**-144
"Handyman" manipulator, **15**-146
Hard copy, data presentation, **11**-48, **11**-50
Hard-wire command link in remote handling, **15**-158, **15**-159
Harmonic analysis, **A**-35 to **A**-37, **A**-63, **A**-65
Hartley, **2**-22
Head, direct recording, **5**-39
Heading gyro, **15**-119, **15**-120
Hearing in remote handling, **15**-155, **15**-156
Heat balance of earth experiment, **14**-8
Heat balance equation in space, **14**-67
Heating rate, nose cone of entry vehicle, **14**-30, **14**-31
Helical antenna, **4**-129
Helical antenna gain, **4**-102, **4**-103
Hertz, defined, **5**-15
Heuristic program, computer, **11**-5
HF band, defined, **5**-15
HIG (hermetically sealed integrating gyro) gyro, **15**-116, **15**-117
  for local vertical, **15**-123
High-level gate, **4**-82, **4**-83
High-level signal conditioning, **8**-5
High-speed testing for quick-look, **12**-43
High temperature fuel cell, **10**-103
Hohmann transfer ellipse, **14**-18 to **14**-20
Hold circuit, **15**-50
Holes, antenna pattern, **4**-114
Homing guidance, **15**-13, **15**-14
Hoover, C. W., Jr., **9**-38
Horizontal polarization, **4**-113
  of source and image, **4**-115
Horn antenna, **4**-130
Hostile environment, **15**-138, **15**-139
  operational analysis of, **15**-142
Hostile-environment system, **15**-138 to **15**-142
  (*See also* Remote-handling system)
Hot cell, nuclear, **15**-140
Hot test, military specifications for, **10**-5, **10**-7
Hour angle, **14**-58
Hour circle, **14**-6, **14**-58
Housekeeping functions in observatory satellite, **14**-89
Hub communications network, **13**-15, **13**-16
Hub subnetwork, **13**-17
Huffman, D. A., **13**-28

INDEX

Human engineering of film reader, **12**-16
Human factors analysis in remote handling, **15**-164, **15**-166, **15**-167
Human tolerances, environmental, **14**-65, **14**-69
Humidity, test specifications for, **10**-5, **10**-9
Humidity transducer, **3**-11, **3**-12
Hunting, **7**-39
Hurwitz test of stability, **15**-32, **15**-33
Hybrid parameter, transistor, **10**-18, **10**-19, **10**-24, **10**-30
Hydraulic amplifier, **15**-133
Hydraulic instrumentation signal, **11**-9
Hydraulic prime mover, **15**-132
Hydraulic pump, **15**-135
Hydraulic transmissions, **15**-135
Hydrocarbon fuel cell, **10**-104, **10**-105
Hydrox fuel cell, **10**-103, **10**-104
Hyperbolic Doppler, **3**-48
Hyperbolic functions, **A**-22
Hyperbolic orbit, **14**-17, **14**-18, **14**-22
Hyperbolic partial differential equations, **12**-36
Hyperbolic radio navigation, **15**-13
Hyperbolic velocity, **14**-6
HYPERDOP (hyperbolic Doppler), **3**-48
Hypergeometric probability distribution, **A**-43
Hypergolic, **14**-6
Hyperthermal free molecular flow theory, **14**-36
Hysteresis, in analog-to-digital converter, **11**-15
  magnetic, **11**-28
  in magneto strictive transducer, **3**-23
Hysteresis curve, **3**-15
Hysteresis error, rate gyro, **15**-110
Hysteresis loop in magnetic recording, **11**-26

IBM, and data link noise test, **13**-28
  650, **11**-38
  704, **11**-37, **11**-45
  709, **11**-37
  1620, **11**-50
  7090, **11**-37, **11**-46
Ice crystals, upper atmospheric, **10**-12
Identity matrix, **A**-11
i-f (intermediate frequency) amplifier, filter, **6**-30, **6**-31
i-f (intermediate frequency) bandwidth, **6**-31
  and sampling rate, **5**-33
  telemetry receiver, IRIG recommendations, **5**-32, **5**-33
IFF (identification friend or foe), **11**-71
IGY (International Geophysical Year) satellites, **14**-73
Illumination in remote vision, **15**-153
Image formation display, **11**-53, **11**-54
Image rejection, **6**-30
  and predetection recording, **6**-33
Image tube, cathode ray, **11**-53
  shaped beam, **11**-54
Image type telemetry and observatory satellite, **14**-90
Immersion, test specification, **10**-6
Impact prediction, real-time, **12**-26
Impedance, forward transfer, transistor, **10**-19
  input, in sampling gate, **9**-11
  transistor, **10**-19, **10**-20
  output, transistor, **10**-19
  reverse transfer, transistor, **10**-19
  switch, **8**-11

Impedance, transistor, **10**-19
  and h and r parameters, **10**-25, **10**-28
  with series feedback, **10**-32
  with shunt feedback, **10**-32
  and z and y parameters, **10**-30
Improvement ratio, **2**-51, **2**-52
Improvement threshold, **2**-53
Impulse function, **A**-33 to **A**-35
Impulse noise, and PAM-FM performance, **9**-3
  in PDM system, **7**-31, **7**-32
Impulse response, **2**-20
  second-order system, **15**-98, **15**-99
Impulse sampler, **15**-48
Impulse system, telemetry, **13**-13
Impulse telemetry, **13**-2
Impulse torsinal for attitude control, **14**-38
Inclination, **14**-6, **14**-57
  communications satellite, **16**-26, **16**-27
Incremental cost control, power system, **13**-45
Incremental interval in numerical integration, **12**-36
Incremental resistance, transistor, **10**-21 to **10**-24
Indeterminate forms, values of, **A**-15, **A**-16
Index register (XR), ALU, **11**-39
Indexing, data processor, **11**-44
Indicator, telemetry receiver, **6**-32
Indirect addressing, data processor, **11**-44, **11**-45
Indium dry cell, **10**-83
Inductance, **3**-16
  mutual, **3**-16
Inductance-controlled oscillator, **10**-43 to **10**-46
Inductance transducer, **3**-19, **3**-20
Induction (electrostatic) transducer, **3**-14
Inductor, printed circuit, **10**-78
Inductosyn, **15**-108
Industrial control system, **13**-3 to **13**-7
Industrial telemetry, **4**-3, **13**-1 to **13**-46
  addressing in, **13**-16
  and links, central station communications, **13**-16
  in subnetwork, **13**-17
  and modulation system, **13**-2
Inertial guidance, **15**-11, **15**-12
Inertial reference system, **15**-124
Inertial space, **15**-104, **15**-124
Inertial stable platform, **15**-124
In-flight calibration, PDM, **5**-32
In-flight checkout, **11**-73
Information, **1**-4
  in industrial telemetry, **13**-24 to **13**-26
  of a message, **13**-24
  mutual, **13**-25
  selective, **13**-24
  semantic, **13**-24
  unit of, **2**-21
Information capacity, communications satellite channel, **16**-16
  as telemetry system criterion, **9**-19 to **9**-21
Information efficiency, **2**-55
Information feedback control, automatic checkout, **11**-70
Information flow, automatic checkout equipment, **11**-57, **11**-58
Information processing system, telemetry, **12**-50 to **12**-52
Information rate, **2**-23
  capacity of telemetry system types, **4**-30 to **4**-32

INDEX

Information rate, continuous output modulation system, **2**-55
  continuous source, **2**-32
  of digital telemetry signal, **9**-20
  of discrete vs. analog channel, **9**-56, **9**-57
  and interrupted analog signal, **9**-21
  and normalized error, telemetry, **9**-21
  PCM, **2**-55
  phase-locked discriminator, peak phase error limited, **9**-69
  for signal of given rms error, **9**-20
  uniform peak-limited signal in Gaussian noise, **9**-20
Information source, continuous, **2**-28
Information transfer rate, **4**-27, **4**-30
Initial frequency error and phase-lock, second order loop, **9**-75
Initial modulation, **1**-8
Initial value problem, numerical solution of, **A**-68, **A**-69
Injection point to lunar trajectory, **14**-50, **14**-51
Injection velocity, **14**-6, **14**-18 to **14**-21, **14**-22 to **14**-26
Input-data for data reduction, **12**-4
Input impedance, signal conditioner, **4**-39
Input-output control, data processor, **11**-45
Input-output register (I-O), **11**-39
Input-output unit and data processor, **11**-36, **11**-45 to **11**-48
Input range, transducers, **3**-4
Insertion loss, receiver, **6**-28
Inside-out electric motor, **15**-101
Instantaneous frequency, **5**-16, **6**-11
Instantaneous sampling, **4**-53
Instantaneous slope uncertainty, **11**-15
Instantaneous speed error, recorder, **5**-53
Instruction, computer format, **11**-36
  single-address, **11**-36, **11**-38
  data processor, **11**-38
Instruction counter (IC), ALU, **11**-39
  and computer, **11**-37
Instruction (I) cycle, **11**-39
Instruction register (IR), **11**-39
Instrument gyro, **15**-99
Instrumentation, definition of, **1**-2
  loop, **1**-7
  signals of, to data acquisition system, **11**-9
  systems of, as data reduction inputs, **12**-4
  optical, **12**-4
Instrumentation Society of America (*see* ISA)
Instruments as sensors in remote-handling system, **15**-157
Insulation resistance, commutator, **4**-72, **4**-73, **4**-86
Integral calculus, **A**-16 to **A**-18
Integrated circuits, **4**-84, **10**-73 to **10**-77
  phase-shift oscillator, **10**-74, **10**-75
  shift register, **10**-74 to **10**-76
Integrated communications system, space, **14**-95
Integrated micromodule, **10**-73
Integrated pairs, transistor, **4**-82
Integrated Range Mission, **5**-1
Integrated S-band System, **14**-95
Integrating gyro, **15**-100, **15**-115 to **15**-117
  and equation of measurement, **15**-116
  and errors, **15**-117
Integration, **A**-16 to **A**-18
  in data reduction, **12**-34

Integration, of data reduction and flight test activity, **12**-50, **12**-51, **12**-59
  by parts, **A**-16
  signal, coded-phase coherent system, **9**-42
Integrator, analog computer, **15**-66, **15**-67
  imperfect, in phase-locked loops, **9**-77
Intensity-penetration, X rays, auroral, **10**-17
Interconnection channel, supervisory control system, **13**-41
Inter-Department Radio Advisory Committee, **5**-4, **5**-8
Interface, product, pipeline, **13**-43
Interference, adjacent channel, FM carrier system, **9**-25, **9**-26
  amplitude modulated carrier, **9**-4, **9**-5
  fixed-pitch beat note, **9**-5
  harmful ITU definition of, **5**-14
  multipath, **6**-35
  radio, between services, **5**-8
  signal, in data acquisition, **11**-11
  in missile telemetry, **4**-6
  and wide bandwidth system, **16**-16
Interferometer, **3**-45, **3**-46
  crossed-baseline, **3**-46
  and film reading, **12**-16
Interlock, communications subsets, **13**-19
Intermediate data read-out, **12**-43 to **12**-45
Intermodulation, phase, **6**-10
Intermodulation products, errors caused by, **6**-15, **6**-16
Internal gimbal system, **15**-126, **15**-127
International frequency allocations, r-f, **5**-5 to **5**-7
International frequency register, **5**-8
International Radio Consultative Committee, **5**-4
International Telecommunications Union (*see* ITU)
Interplanetary distances, **14**-16
Interplanetary missions, trajectory analysis, of, **14**-17
Interpolation, **4**-58, **4**-59
  in commutation, **4**-89
  in data reduction, **12**-33
Interpolation errors, PDM, **7**-33 to **7**-35
Interpolation filter, **8**-57, **8**-58, **9**-8
Interpolation formulas, **A**-62 to **A**-64
Interpolation polynomial, **A**-62, **A**-63
Interpreted data presentation, **11**-48
Inter-Range Instrumentation Group, **5**-1 to **5**-4
(*See also* IRIG)
Interrogation technique, automatic check-out equipment, **11**-70 to **11**-72
Interrupt, program, data processor, **11**-43, **11**-44
Intersection, **15**-78
Interval sampling, **4**-53
Intrinsic conduction, **10**-96
Inverse matrix, **A**-10
Inverse-mean time, first order loop, **9**-81, **9**-82
Inverse trigonometric functions, **A**-22, **A**-23
I-O unit (*see* Input-output unit)
Ion-concentration gradient, **10**-80
Ion-exchange-membrane fuel cell, **10**-104 to **10**-107
Ion sheath, **10**-11
Ionization attenuation dropout, **12**-57
Ionization chamber, **3**-14
Ionized layer in atmosphere, **10**-11
Ionosphere, **14**-6

INDEX

IRAC (Inter-Department, Radio Advisory Committee), **5**-4, **5**-8
IRIG (Inter-Range Instrumentation Group), **4**-94, **4**-131
  document No. 106-60 of, **5**-25 to **5**-67
  and frequency allocations, **5**-8
  organization of, **5**-1, **5**-2
  responsibilities of, **5**-1
  standardizing procedure of, **5**-3
  standards of, **5**-25 to **5**-67
    FM/FM or FM/PM, **5**-27 to **5**-31
    magnetic tape recorder, **5**-37 to **5**-47
    PAM-FM-FM, **5**-29 to **5**-31
    PAM-FM telemetry, **5**-32 to **5**-36
    PCM-FM or PM, **5**-36, **5**-37
    PCM recording, **5**-42 to **5**-46
    PDM-FM, PDM-PM, **5**-31
    PDM-FM-FM, **5**-31, **5**-32
    PDM-FM system, **7**-9, **7**-11, **7**-22
    PDM recording, **5**-42
    radio frequency, **5**-47 to **5**-50
    receiver IF bandwidth, **5**-33
    recorder/reproducer, **5**-37 to **5**-47
    telemetry, **5**-25 to **5**-67
ISA (Instrumentation Society of America), **3**-42, **5**-3
  code of, for centralized pipeline control, **13**-41
  and transducer compendium, **3**-42
Isoelasticity in gyro, **15**-111, **15**-112
Isotropic antenna gain, **4**-97
Isotropic antenna pattern, defined, **4**-112
Isotropic gain, antenna, ITU definition, **5**-14
ITS (Integrated Trajectory System), **3**-51
ITU (International Telecommunications Union), **4**-94, **5**-5, **5**-8
  and Administrative Radio Conference, **5**-4
  and CCIR, **5**-4
  and frequency bands, **5**-15
  and frequency tolerances, **5**-19 to **5**-21
  and radio regulations, **5**-7, **5**-8
  and rules for frequency assignment, **5**-8
  and spurious emission tolerances, **5**-22

Jackson, J. L., **9**-88
Jacobian, **A**-14, **A**-15
Jacobs, I, **9**-34, **9**-35
JATO rocket, **12**-55
Jaffe, R., **9**-59, **9**-65
JB-2 pilotless aircraft, **4**-3
Jet Propulsion Laboratory, **14**-75
Joint probability distribution function, second, **2**-12
Joule effect, **3**-21
Jump feedback, counter, **8**-21
Jump program, data processor, **11**-43
Junction error, thermocouple, **3**-41
Jupiter, characteristics of, **14**-53, **14**-55 to **14**-57
  and missile stable platform, **15**-126

Karma strain-gage wire, **3**-35, **3**-37
Kelvin balance telemetering system, **13**-31
Kepler's laws, **14**-6, **14**-7, **14**-17
Kernel, Volterra functional expansion, **9**-82
Key addressing, **11**-25
Keyer, PDM, **7**-12, **10**-61 to **10**-64
Kinetic energy of orbital objects, **14**-21
Kiniplex, **13**-32
Kotel'nikov, **9**-30, **9**-32 to **9**-34, **9**-40

LACE, **14**-50
Ladder network for analog-to-digital converter, **11**-18
Lagrange's interpolation method, **12**-33, **A**-62
Lalande cell, **10**-84, **10**-85
Lamp array display, **11**-53
Languages, programming, **11**-4
LaPlace transforms, **15**-21, **15**-22, **A**-36, **A**-39, **A**-40
  bilateral, **2**-10
  and Z transform, **15**-50, **15**-51
LaPlace's distribution, **A**-50, **A**-51
Laser communications, **14**-71, **14**-72
Latitude, **14**-6
Launch angle, **14**-22 to **14**-26
  (*See also* Burnout angle)
Launch vehicle, performance of, **14**-41, **14**-42, **14**-46
Launch velocity, **14**-47
Launch weight, **14**-39, **14**-40
  vs. burnout velocity, **14**-46
Launch window, **14**-7, **14**-17, **14**-52
Lead-acid cell, **10**-84, **10**-86, **10**-89, **10**-90
Lead sulfide photo transducer, **3**-26
Learning by operator in remote handling, **15**-166
Leased line, **13**-13
Leased network, **13**-15
Least-squares fit, **12**-31
LeCarbone cell, **10**-84
Leclanche cell, **10**-82, **10**-85
Leeds and Northrup Company, automatic dispatch equipment of, **13**-44
Legendre codes and frame synchronization, **8**-39
Legendre polynomials, **1**-9
Lehan, F. W., **9**-66
Lenz's law, **3**-11, **3**-23
LES (lower end-scale value), **11**-15
Levels, signal power, r-f links, **4**-95 to **4**-97
LF band, defined, **5**-15
L'Hôpital's rule, **A**-15
Libration, **14**-7
Life, commutator, **4**-69
Life support, **14**-70
Lifetime, earth orbital, **14**-26
  of photo electrons and photoconductive transducer, **3**-25
Lift coefficient, **15**-7
Lift modulation to reduce entry deceleration, **14**-32
Lift-off time, planned, **14**-51
Lift-to-drag ratio, **14**-32
Lifting entry (reentry), **14**-32
Light intensity transducer, **3**-13
Light-year, **14**-7
Likelihood function, **A**-51
Likelihood ratio, **A**-57
Limit cycle, **15**-56, **15**-61
  and second-order phase-locked loop, **9**-75
Limiter, **6**-31, **6**-32
  and subcarrier discriminator, **6**-43
Limits, frequently used, **A**-16
Line of nodes, **14**-7, **14**-54
Line printer, **11**-49, **11**-50
Line replaceable unit, **11**-59
Line-of-sight course, **15**-14, **15**-15
Line-of-sight distances, aircraft, **4**-96
  satellite, **4**-97
Linear differential equations and LaPlace transform solution, **A**-41
Linear interpolation, **12**-33
Linear model, phase-locked loop analysis, **9**-61, **9**-62

INDEX

Linear polarization, antenna, 4-113
Linear system, 2-19
Linearity, 3-4
  in ramp and errors in analog-to-digital conversion, 8-24
  and transducer, 3-4
Linearization of non-linear system, 15-56
Link, open, 13-17
  r-f, coverage by vehicle, 4-131 to 4-135
Liquid level indicator, 3-3, 3-12, 3-20
Liquid propellant, 14-43
  and rocket vehicles, 14-41
Lissajous pattern, 6-38
Little Joe, 14-42
Live register, 11-37, 11-44
Local feedback and remote handling, 15-162
Local vertical, 15-121, 15-122
Lock-on, phase-locked loops, 9-70 to 9-77
  first-order, 9-70, 9-71
  imperfect integrator, 9-77
  and natural frequency, 9-77
  second-order, 9-71 to 9-77
  swept-frequency, 9-76
Locking rate and synchronization loop, 14-101
Locomotion, indoor, for remote handling, 15-150
  orbital, for remote handling, 15-152, 15-153
  outdoor, remote handling, 15-151
  subsystem, 15-144, 15-149 to 15-152
  undersea, 15-151, 15-152
Log-periodic antenna, 4-114, 4-126, 4-127
Logarithms, A-4
Logger, data, 13-5
Logging, automatic, 13-39
Logic, circuits, basic, 15-80
  counter, 8-18
  feedback, 8-18
  fixed, in process control, 13-4, 13-9
  sync, PAM demultiplexer, 9-16, 9-17
Logistic requirement for man in space, 14-70
Longitude, 14-7
  celestial, 14-5
Longitudinal interference, 11-11
Longitudinal recording, 11-25
Longitudinal redundancy check, 13-8, 13-9
Loop filter for bit synchronization, 8-34, 8-35
Loop noise bandwidth, 9-63
  threshold, 9-68
Loop range, 3-43
Loop-Vee® antenna, 4-122, 4-123
Loran, 15-12
Loss, in electrochemical cells, 10-80
  of lock, first order loop, 9-81, 9-82
  r-f, free space, 4-97 to 4-101
  transmitter, transmission line, r-f, 4-96, 4-97
Low-level control computer system, 13-5
Low-level gate, 4-82, 4-83
Low-level high speed multiplexing, 11-10
Low-level signal conditioning, 8-5
Low-level signal in data acquisition, 11-10
LRU (line-replaceable units), 11-59
Lunar manned mission, 14-11
Lunar mission trajectory analysis, 14-16, 14-17
Lunar orbit, 14-23
Lyman Alpha experiment, Explorer VII, 14-79
Lyman Alpha spectral line, 10-14

Mach number, 14-7, 14-49
Machine cycles, data processor, 11-39 to 11-42
MacLaurin's series, A-24
Macro-order, 13-8
MAD (Manipulation and Disassembly) nuclear facility, 15-140
Magnesium cell, 10-82, 10-85
Magnesyn, 15-108
Magnetic core storage in satellite, 14-77
Magnetic drum and PCM bit rate, 5-36
Magnetic field, of earth, 14-35
  parameters for, 3-15, 3-16
Magnetic recorder for PDM, 7-37 to 7-40
Magnetic recording, 11-25 to 11-28
Magnetic tape, 11-5
  in data reduction system, 12-21, 12-22
  frequency modulated, 12-22
Magnetic-tape recorder, performance specifications of, for Scout satellite, 14-86
  reproducer standards of, 5-37, 5-47
  reproducer tests of, 5-51 to 5-57
  for satellite, 14-76, 14-82
Magnetic-tape units, 11-28 to 11-30
Magnetic torque on satellite attitude, 14-35, 14-36
Magnetic transducer, 3-14
  and Hall effect, 3-18, 3-19
Magnetizing force, 3-15
Magneto-elastic transducer (see Magnetostrictive transducer)
Magnetometer, 3-21
Magnetomotive force, 3-16
Magnetoresistive transducer, 3-21
Magnetostrictive transducer, 3-21, 3-22
Maintenance space telemetering, 5-13
Manchester representation, 8-29
Maneuvers, in earth orbits, 14-18 to 14-20
  and strategies, 15-14 to 15-19
Manipulation, 15-144
  device for, design criteria for, 15-145
  remote, 15-138
Manipulator, master-slave, 15-140, 15-146
  servo-controlled, 15-146
  underwater, 15-141
Manipulator controls, 15-147 to 15-149
Manipulatory subsystem and remote handling, 15-144 to 15-149
Manned space mission, 14-10, 14-11
Manual data entry and data acquisition system, 11-9, 11-10
Mapping from orbit, 14-15, 14-16
Mariner Mars, 1964, 14-95, 14-97
  and telemetry parameters, 14-102
Mariner II, 14-95
Mariner program, 14-12
Markov process, 9-78
Mars, characteristics, 14-53, 14-55 to 14-57
  orbits of, 14-24
Marshall Space Flight Center (NASA), 14-75
M-ary PCM, 8-48
Maser amplifier and noise performance, 16-5
Mass expulsion for attitude control, 14-36 to 14-39
Mass ratio of rocket, 14-7, 14-39, 14-45, 14-47
Master gyros, 15-99
Master-slave manipulator, 15-140, 15-146
Master station, in supervisory control system, 13-41
Matched filter, coherent, 8-53 to 8-56
  (See also Correlation detector)

INDEX

Matched filter, for detection, **14**-98
  and PAM sync detector, **9**-17
  phase-locked loop as, **9**-65 to **9**-70
  and pulse compressing, **9**-37
  in telemetry, **9**-2
Mathematical formulae, **A**-1 to **A**-70
Mathematical tables, **A**-70 to **A**-94
Mathematics, **A**-1 to **A**-70
Matrices, **A**-9 to **A**-11
Matrix, decoding, **8**-17
  diode, **4**-80
  electronic commutator, **4**-79, **4**-80
  operation decoder, **11**-42
  operations, **A**-10
  resistance, **4**-80
  transistor parameter interrelations, **10**-20
Maximal length shift-register code, **8**-39
Maximum-length sequence generator, **8**-19, **8**-20
Maximum-likelihood estimate, **A**-52
Maximum-likelihood estimator, **9**-32
Maximum-likelihood in trajectory determination, **12**-36
Maxwell's equations, **4**-115
Mean, **A**-46
Mean absolute error, **15**-48
Mean radial error, **A**-49
Mean-squared error, **15**-48
Mean system error, **15**-48
Mean time-weighted absolute error, **15**-48
Mean time-weighted error, **15**-48
Measurand, **1**-6, **1**-7, **5**-62
Measurement, limitations of, by radio-telemetry systems, **4**-7, **4**-8
  remote, classification of, **13**-3
    incremental, **13**-3
    requirements of, **4**-9
  telemetered, **3**-2
Mechanical chopper amplifier, **4**-41, **4**-42
Mechanical commutation, **4**-62 to **4**-78
Mechanical conversion factors, **1**-12 to **1**-20
Mechanical systems and electrical circuit equivalents, **15**-63, **15**-64
Mechanical time constant and a-c servo motor, **15**-132
Mechanical tolerances of recorder, electronic test for, **5**-56
Mechanically activated batter, **10**-91, **10**-92
Memory, and data-acquisition system, **11**-7
  magnetic-core, **8**-48
  unit read only, **11**-7
Memory buffer register (MBR), ALU, **11**-39
Mercury cell, **10**-82, **10**-85
Mercury characteristics, **14**-53, **14**-55 to **14**-57
Mercury jet switch, **4**-74
Mercury program, **14**-11
Merging of scientific data and coordinates, **14**-87, **14**-88
Merging program, **12**-42
Meridian, celestial, **14**-5
Mescator projection, **11**-49
Mesh communications network, **13**-15, **13**-16
Mesh reflector, satellite, **16**-9
Message, addressing of, and industrial telemetry links, **13**-16
  amplitude of, peak, **2**-50
  antipodal, **9**-41
  coded, **9**-39, **9**-40
  command, remote control, **13**-8
  control, **13**-4
  and error control, **13**-8
  orthogonal, **9**-41
Message digit, **2**-37

Message exchange, data processor, **11**-36
Message function, **2**-44
Message spectrum and square law, **2**-43, **2**-44
Message structure and process communications, **13**-7
Message variance, **2**-46
Metadyne as constant-current source, **15**-130
Metal deposition and printed circuit, **10**-77
Metal oxide field effects transistor, **4**-84
Metallizing of film capacitor, **10**-72
Meteorites, **10**-17
Meteorological instrumentation system, **12**-4
Meter, **10**-17, **10**-17
  for data presentation, **11**-51
MF band, defined, **5**-15
Microlock, **3**-48
Microlock receiver, **14**-75, **14**-82
Micrometeorites, **10**-17
  detection of, by Vanguard III, **14**-76, **14**-78
  penetration capability of, **14**-61, **14**-62
Micromodule, **10**-72, **10**-73
Microoperations and data processing, **11**-42, **11**-43
Microphonic noise in transmitter, **10**-53
Microphonism, receiver, **6**-18
Microprogramming in data processing, **11**-43
Microsyn, **15**-108, **15**-116
Microsyn accelerometer transducer, **15**-98
Microsyn transducer, theory of operation of, **15**-108
Microwave equipment in pipeline telemetry, **13**-19
MIDAS (missile-intercept data-acquisition system), **3**-51
Midcourse corrections, **14**-22
  of trajectory, **14**-17
  of velocity, to planned mission, **14**-52
Middleton, D., **9**-40
MIDOT (multiple interferometer determination of trajectory), **3**-51
MIL-E-4970 specification, **10**-4, **10**-5
MIL-E-5272 specification, **10**-4, **10**-5
MIL-E-5400 specification, **10**-4, **10**-5
Military satellite communications, **16**-21
Military systems checkout, **11**-56, **11**-57
Milky Way, **14**-52
Miller integrator, **7**-30
Milne's method, initial value problem, **A**-69
MIL-STD-202 standard, **10**-4
MIL-STD-810 standard, **10**-5
MIL-T-5422 specification, **10**-4, **10**-5
Miniaturization, **1**-9, **6**-27, **10**-69, **10**-79
  and telemetry, **10**-3
Minimal polynomials, Boolean, **A**-12
Minimum energy trajectory, **14**-26, **14**-27
  and range deboosting from orbit, **14**-32, **14**-33
MINITRACK, **3**-51
  and constraints on satellite telemetry, **14**-74, **14**-77
Minors, **A**-7
Minuteman telemetry, **8**-3
MIRAN (missile ranging), **3**-49
Missile, antennas of, **4**-111 to **4**-134
  checkout of, **11**-56, **11**-57
  definition of, **15**-5
  trajectory measurement of, **3**-42
  (*See also* Electronic trajectory measurement)
Missile command computer, **15**-73

Missile ranging (*see* MIRAN)
Missing pulse protection, **5**-42
Mission constraints in space telemetry, **14**-95
Mission payload capability of existing rockets, **14**-46
MISTRAM (high-precision trajectory-measuring system), **3**-52
M.I.T. Instrumentation Laboratory, **15**-108, **15**-116
Mobility, electron, **10**-37
  hole, **10**-37
Modern industrial telemetry, **13**-17
  (*See also* Subsets)
Modular construction, **6**-27
Modular data processing, **12**-38 to **12**-40
Modularization, **10**-72, **10**-73
Modulated signals and frequency spectra, **2**-43 to **2**-48
Modulation, **1**-4, **2**-37 to **2**-42
  AM, defining equation for, **2**-38
  amplitude, **2**-38
  angle, **6**-7, **6**-8
  carrier, PCM, **8**-49
    IRIG recommendation, **5**-37
    pulse, IRIG standard, **5**-36
  in communications satellites, **16**-16 to **16**-22
    comparison of, for communications satellite, **16**-19
  designation of, ITU, **5**-14
  digital, **13**-26 to **13**-28
    and methods of data representation, **8**-29
    and synchronization requirement, **9**-39
  discrete, **9**-31
  equivalent discrete, **9**-31
  format of, **6**-27 to **6**-29, **6**-31 to **6**-33
  frequency (FM), defining equation for, **2**-38
    sidebands of, **6**-9
  multi-dimensional, **13**-35
  multimodal, **13**-35
  of one carrier by another, phase, **6**-10
  phase (PM), **2**-39
  pulse, **2**-39
  pulse amplitude, **2**-39
  pulse-code, **2**-41, **2**-48
    in industrial telemetry, **13**-26
  random access, **16**-22
  and relationship between phase and frequency, **6**-8, **6**-9
  and sideband relationships, **6**-7 to **6**-9
  and sinusoidal carrier, **2**-38, **2**-39
Modulation converter as communications satellite repeater, **16**-20
Modulation error and FM feedback receiver, **9**-94
Modulation following, and phased-locked loop, linear, **9**-65 to **9**-70
  and phase-plane analysis of loop, **9**-73 to **9**-76
Modulation index, FM, **6**-9
  permissible in phase-locked discriminators, **6**-46, **6**-47
  and quasi-linear receiver performance, **9**-89, **9**-91
  signal input, first order loop, **4**-86
  for threshold improvement by phase-locked discriminator, **6**-47
Modulation system, for deep space missions, **14**-97, **14**-98
  for industrial telemetry, **13**-21
  for information efficiency, reasons for low, **2**-55

Modulation system, for information handling capabilities, **2**-54, **2**-55
Modulation technique, choice of, for satellite telemetry, **14**-85
Modulator, balanced, **6**-58, **6**-61
  in control system, **15**-83
  FM, for PDM system, **7**-23
  PCM and bandwidth reducing, **4**-14, **4**-15
    and output spectrum, **8**-50
  reactance, **10**-54
  telemetry transmitter, **6**-22 to **6**-24
Modules and program for data reduction, **12**-38 to **12**-40
Modulo-2 adder, in shift-register coder, **9**-45
Modulo 2 summation, **8**-14
MOL (manned orbiting laboratory) program, **14**-11
Momentum, of body in orbital motion, **14**-20 to **14**-22
  storage of, **14**-36 to **14**-39
Monisoelasticity, **15**-111
Monitor, d-c voltage, **4**-49, **4**-50
  scope, PDM, **7**-31
  step function, **4**-50
Monopoles, antenna, **4**-118, **4**-119
Monopropellant for attitude control, **14**-38, **14**-43
Moon, noise temperature of, **4**-105
  orbital characteristics of, **14**-54
  physical characteristics of, **14**-53, **14**-54
MOPTARS (multiobject phase-tracking and ranging system), **3**-50
MOSFET (metal oxide field effects transistor), **4**-84
Motion, earth orbital, **14**-17 to **14**-21
  manipulating, **15**-145
  orbital, equations of, **14**-17, **14**-18
Motors, control, pneumatic, **15**-136
  electric, inside-out, **15**-101
  hydraulic, servo, **15**-133 to **15**-135
Moving boundaries in data processing, **12**-37
MTBF and mechanical commutators, **4**-75, **4**-77, **4**-86
Multichannel commutation, **4**-55, **4**-56
Multicoder, **7**-17 to **7**-21
  high level, **7**-17, **7**-18
  low level, **7**-18, **7**-19
Multidimensional probability distribution, **A**-46
Multimode checkout, **11**-61
Multipath, effects of, **4**-93
  in radiotelemetry, **4**-7
Multiple-access, **16**-29
  and communications satellites, **16**-3, **16**-19, **16**-21, **16**-22
  by frequency division, **16**-22
  by time division, **16**-22
Multiple-angle trigonometric formula, **A**-21, **A**-22
Multiple antennas around vehicle, **4**-118
Multiple-control computer system, **13**-5
Multiple phase transmission, **13**-32
Multiplex configuration design, FM, **6**-13
Multiplex modulation methods, comparisons of, **7**-7, **7**-8
Multiplex peaking, **6**-16
Multiplexed signals, and amplitude-modulation systems, **2**-48
  and frequency-modulation systems, **2**-48 to **2**-49
  frequency spectra of, **2**-48
Multiplexed wire command link in remote handling, **15**-159

Multiplexer, cascaded, **4**-23
  data acquisition, **11**-10
  data processor, **11**-40
  diode bridge gate, **10**-65
  high level, **8**-6
    PAM, **9**-11
  low level, PAM, **9**-11
  PAM, **4**-9, **9**-11
  PCM, **8**-11, to **8**-15
  relay, **8**-12
  r-f, **4**-20
  and switch configurations, **8**-14, **8**-15
  time, electronic, **10**-55 to **10**-61
  time division, **6**-2
Multiplexing, **1**-4, **2**-37, **2**-42, **2**-43
  composite systems of, **2**-43
  frequency, in PCM telemeters, **4**-17
  frequency-division, **2**-42, **2**-43, **6**-2
  in industrial telemetry, **13**-30, **13**-31
  PAM-FM, **9**-2, **9**-3
  in process control, **13**-7
  single sideband, **6**-61, **9**-3
  SSB (single sideband), **6**-61, **9**-3
  time division, **2**-43
Multiplexing unit, randomly addressable, **11**-34
Multiplication as example of data processing operation, **11**-40, **11**-41
Multiplier, analog, **15**-75
  diode-squarer, **15**-77
  quarter-square, **15**-77
Multiplier-quotient (M-Q) in ALU, **11**-39
Multiprocessors, **11**-45
Multistaging, **14**-44 to **14**-46
Multiunit system and process control, **13**-9
Multivibrator as voltage-controlled oscillator, **10**-46 to **10**-49
Multiwire command link in remote handling system, **15**-158
Mutual visibility, area of, **16**-27, **16**-28
  communications satellite, **16**-5
  duration, **16**-24
  effect on multiple access, **16**-21
  maximum, **16**-26
  probability of, **16**-24
Mylar, **11**-25
  capacitor, **10**-71
Mylar base tape, **12**-22

N-body problem, **14**-17
Narrow-band telemetry, OGO specification, **14**-92
NASA (National Aeronautics and Space Agency), **5**-2
National Telemetering Conference, **5**-3
Natural frequency, and loop in quasi-linear receiver, **9**-90
  and loop and rate of reference frequency change, **9**-75
  and phase-locked loop, second-order, **9**-64
  and rate gyro, **15**-106
  and transducer, **3**-6
Naval Ordnance Missile Test Facility, **5**-1
Naval Ordnance Test Station, **5**-1
Naval Research Laboratory, **3**-51, **14**-75
Navigation, definition of, **15**-5
NBS (National Bureau of Standards) strain gage, **3**-37, **3**-38
Near earth manned space program, **14**-11
Necessary bandwidth, ITU, computation, **5**-16 to **5**-19
Neptune, characteristics of, **14**-56 to **14**-58
Network, communications, **13**-15, **13**-16
  tree, and checkout equipment, **11**-63
Neutralization cell, **10**-85
Neutron flux, equivalent to rem per hour, **14**-66
Neutron radiation, **10**-14, **10**-15
Newton-Cotes quadrative formulae, **A**-66
Newton-Gregory interpolation formulae, **A**-63
Newton's interpolation formula, **A**-62
Newton's interpolation method for data reduction, **12**-33
Newton's laws of motion, **14**-7, **14**-39
Newton's viscous shear law, **15**-94
Neyman-Pearson criteria, statistical, **A**-57
Nichome V strain-gage wire, **3**-36
Nickel-cadmium battery, **14**-71
Nickel-cadmium cell, **10**-86, **10**-89 to **10**-91
Nickel-iron cell, **10**-87, **10**-89 to **10**-91
Nimbus satellite telemetry, **14**-91
Node, **14**-7
No-go indicator, **11**-60, **11**-64
Noise, acoustical, **10**-10
  from aliasing, **9**-72
  amplitude, effect of, on PDM signal, **7**-5
  bandwidth, equivalent, **2**-20
  receiver, **4**-93, **4**-107, **4**-108, **4**-110
  burst, **13**-28
  in communications satellite, **16**-3 to **16**-5
  in data reduction, **12**-29
  in digital transmission, **13**-28
  distortion, **9**-73
  edge, **4**-61
  fluctuation, **9**-23, **9**-24
    effect of, on PDM system, **7**-5
  in FM-FM system, affecting preemphasis schedule, **6**-16, **6**-17
  galactic, **4**-103, **4**-104
  galaxy, **16**-4
  Gaussian, white, **13**-28
  and Gaussian output statistics, **9**-53
  impulse, effect of, on PDM system, **7**-5
    in PCM, **8**-60
  intensity of, **9**-30
  on-time, **7**-13
  output and phase-locked discriminator for limited peak error, **9**-68
  and overmodulation drop-out, **9**-24
  in PDM system, **7**-31, **7**-32, **7**-36
    and telemetry, **7**-4, **7**-5
  and performance of phase-locked loop, **9**-62
  in photo emissive device, **3**-27
  potentiometer, **3**-33, **3**-34
  in pulse-coded system, **13**-28
  quantization, **9**-24
  Rayleigh, and output statistic, **9**-53
  receiver, **4**-103 to **4**-110
  single frequency, **13**-28
  single-impulse, **13**-28
  of spectral power, normalized in phase-locked loop, **9**-63
  spike, **13**-28
  switching, **4**-73, **4**-74, **4**-86
  temperature of, of receiver, **4**-94
  tests for, and data link, **13**-28
Noise factor, receiver, **6**-29
Noise figure, **2**-53
  of PDM receiver, **7**-23, **7**-25
  of receiver, **4**-94
Noise filter and PAM data detector, **9**-15
Noise immunity, and PAM ground station, **9**-16
  and PCM, **8**-4
Noise level, receiver, calculation, **4**-108, **4**-110
Noise power, density of, per cps, **8**-53

INDEX

Noise power, per unit BW, **9**-30
Noise spectral density, **9**-30
  in space communications system, **14**-96, **14**-97
Noise spectrum, FM, **6**-10
Noise temperature, equivalent, of first state, **16**-5
Noncoherent AM and FSK, efficiency of, for PCM, **9**-33
Noncoherent demodulation, PCM, **8**-53 to **8**-56
Noncontact recording, **11**-27
Nonlifting entry, planetary, **14**-30, **14**-31
Nonlinear computer, **15**-73 to **15**-76
Nonlinear equation, in data reduction, **12**-36
Nonlinear system, control of, **15**-55 to **15**-63
  with delay, **15**-86
  simulation of, **15**-67, **15**-70
Nonlinearity correction in data reduction, **12**-6
Nonmagnetic, definition of, **3**-15
Non-return-to-zero recording, IRIG standing, **5**-45
Non-return-to-zero representation, PCM, **8**-29
Nonsaturating recording, **11**-28
Nonsingular matrix, A-11
Nonstationarity, PCM signals, **8**-36
Normal distribution, A-46, A-49
Normal distribution area, A-71
Normal distribution curve, table of ordinates, A-72
Normal surface-reflection coefficient, **14**-36
Normalized autocorrelation function of random process, **2**-15
Nose cone telemetry, **12**-55
  and trajectory determination, **12**-54
NOT (Naval Ordnance Test) operation, **15**-80
Nova, **14**-38, **14**-41
Nozzle characteristics, rocket, **14**-44
NRL (Naval Research Laboratory) telemeter, **4**-22
NRZ (non-return-to-zero) data recording, **11**-31
  IRIG standard of, **5**-45
NRZ representation and PCM, **8**-29
NRZI (non-return-to-zero on I) data recording, **11**-32
NRZ(L) representation, **8**-29
  PCM, **8**-29
NRZ(M) representation, PCM, **8**-29
NRZ(S) representation, PCM, **8**-29
Nuclear cell, **10**-99 to **10**-101
  characteristics of, **10**-101
Nuclear facilities, **15**-190
Nuclear hostile environment, **15**-139 to **15**-141
Nuclear power for space, **14**-71
Nuclear propulsion, **14**-43, **14**-45, **14**-48
Null point, **14**-17
Nulling telemetering system, **13**-33, **13**-34
Numerical analysis, **12**-58
  calculations for, A-61 to A-69
  and harmonic analysis, A-63, A-65
  integrating, A-66, A-67
  tables for, A-70 to A-91
  and techniques for data processing, **12**-29
  for nonlinear system solutions, **15**-57
nvt product of (number of particles per unit volume and their velocity), **10**-15
Nyquist criterion and control system stability, **15**-33 to **15**-36
Nyquist diagram for control system with delay, **15**-85
Nyquist loci, **15**-35

Nyquist sampling theorem, **13**-26
Nyquist stability criterion and pulse transfer functions, **15**-54

Observatory satellite telemetry, **14**-89 to **14**-93
Obstacle sensing in remote handling, **15**-156
Occupancy of state, A-42
Octal arithmetic in data processor, **11**-42
Odd-parity code, **13**-29
Off current in source, **4**-89
  (*See also* Back current)
Office of Civil and Defense Mobilization, **5**-4
Offset, zero signal, **4**-89
Offset error, **4**-39
  limiter, **6**-31, **6**-32
Offset potential, **7**-19
Offset voltage, in gating, **4**-82, **4**-84
  and transistor switch, **8**-14
Ohm's law, **3**-31
Oil-damping, accelerometers, **15**-94
Oil viscosity, **15**-94
Oliver, B. M., **9**-5
Omni-directional antenna pattern, defined, **4**-112
On current in source, **4**-89
  (*See also* Back current)
On line processors in data presentation, **11**-49
On-off control, **15**-62
On-off pneumatic mode and attitude control system, **14**-37 to **14**-39
Open-loop system and process control, **13**-4, **13**-6, **13**-10
Open-sleeve stub antenna, **4**-120
Operand of computer instruction, **11**-36
Operation-decoder matrix of data processor control logic, **11**-42
Operation designator and computer instruction, **11**-36
Operational amplifier, **15**-64, **15**-65
Operational readiness check, **11**-56
Operational security and communication-based control system, **13**-39 to **13**-41
Operator-guide and pipeline control, **13**-39
Oppositions, Earth-Mars, **14**-16
Optical code disk system for analog-to-digital conversion, **11**-22
Optimal receiver, phase-locked, for band-limited, phase-encoded white Gaussian signal, **9**-98, **9**-90
Optimization, in FM carrier telemetry system, **9**-22 to **9**-25
  Wiener, **9**-88
  criterion for, **9**-65 to **9**-67
Optimum coding, **2**-32
Optimum filter, loop, phase-locked demodulator, **9**-66
  output, phase-locked demodulator, **9**-67
OR operation, **15**-78, **15**-79
Orbital assembly, **14**-45, **14**-47, **14**-48
Orbital Astronomical Observatory programmer, **14**-87, **14**-91
Orbital-belt system, **16**-10
  (*See also* West Ford)
Orbital configuration, least cost, **16**-23
  modified random, **16**-27
  polar, **16**-27
  random, **16**-26
Orbital elements, **14**-8, **14**-26
Orbital equations, **14**-17
Orbital laboratories, **14**-11
Orbital lifetime, earth, **14**-26, **14**-27
Orbital mapping, **14**-15
Orbital mechanics, **14**-20 to **14**-26, **16**-23
Orbital motion around earth, **14**-17 to **14**-21

Orbital transfer at planetary entry, **14**-30
Orbiting Geophysical Observatory telemetry system, **14**-90 to **14**-93
Orbits, constant energy of, **14**-21 to **14**-26
  constant momentum of, **14**-22 to **14**-26
Ordinary differential equations, numerical integration of, **12**-36
Organization of data reduction effort, **12**-50 to **12**-52
Ornstein, L. S., **9**-78
Orthogonal-axis reference system, **15**-4, **15**-5
Orthogonal coding in space communications, **14**-97
Orthogonal function, **1**-9
Orthogonal message, **9**-41
Orthogonal polynomial, **12**-31
Orthogonal QPPM, **9**-36
Orthogonal sine wave function, **9**-50
Orthogonal spatial coordinate system, **15**-124
Orthogonal system, discrete communications, **9**-34
  and PCM, **8**-4
Orthogonal television camera for remote handling, **15**-154
Orthonormal exponentials, **2**-5
Oscillator, Colpitts, **10**-44
  and control logic data processor, **11**-42
  crystal offset, **6**-23
  current-controlled, **10**-52
  digitally timed, **9**-51
  inductance-controlled, **10**-43 to **10**-46
  local, spectral purity, **6**-30
  noise frequency modulated, **9**-95
  parallel-T, **10**-60
  phase-shift, **4**-12, **4**-13
  resistance-controlled, **10**-42, **10**-43
  subcarrier, **6**-36 to **6**-40
  telemetry transmitter, **6**-24, **6**-25
  tetrode, **10**-55
  variable reactance, **10**-49, **10**-51
  voltage-controlled, **9**-60, **9**-61, **10**-46 to **10**-52
Oscillogram, data processing, **12**-16 to **12**-18
  reader, fixed-scale, **12**-16
  variable-scale, **12**-16
Oscillograph chart, **12**-4
Oscillograph recorder setup, **12**-27
Outage, and communications satellite, **16**-23, **16**-26
  duration of, in communications satellite, **16**-28, **16**-29
Output circuits and PAM ground station, **9**-17
Output range, transducers, **3**-4
Output report program generator, **12**-46, **12**-47
Output signal-to-noise ratio, **2**-54
  and PCM, **2**-54
Output voltage, piezoelectric, **3**-31, **3**-32
Overdamping, transducer, **3**-7
Overlay for calibration, **12**-17
Overshoot, second order system, **15**-37
Overvoltage, **10**-80
Oxidation-reduction reaction, **10**-79

Pacific Missile Range (PMR), **5**-1
  firing sector of, **14**-27, **14**-28
Packaging, **10**-69 to **10**-79
  telemetry transmitter, **6**-27
Packed counts, **12**-48
PACM, **9**-18 to **9**-29
  functional description of, **9**-28
  system design of, **9**-26 to **9**-28
Pade table for exponential functions, **15**-84

PAM (pulse-amplitude modulation), **1**-8, **2**-46, **9**-2 to **9**-18, **13**-21, **13**-22
  and average signal power, **2**-40
  and average unmodulated power, **2**-41
  and channel distortion, **4**-56 to **4**-58
  format of, **4**-58
  frame and pulse structure of, **5**-56
  and IF bandwidth, **5**-56
  mathematical representation of, **2**-39
  in PDM telemetry, **7**-3
  quantized, **13**-22
  (*See also* PAM-FM; PAM-FM-FM)
PAM signal, **4**-56 to **4**-58
  format of, IRIG specification for, **5**-30
PAM spectrum, **7**-34, **7**-35, **8**-57
PAM synchronization, **5**-56, **5**-57
PAM theory, **9**-5 to **9**-9
PAM-FM (pulse-amplitude modulation-frequency modulation), **9**-2 to **9**-18
  and airborne package, **9**-9 to **9**-13
  and airborne system, **9**-11 to **9**-13
  applicability of, areas of, **9**-2, **9**-4, **9**-5
  communication properties of, **9**-2 to **9**-5
  error sources in, **9**-24
  and ground station equipment, **9**-13 to **9**-18
  and ground station layout, **9**-14
  and impulse noise to signal ratio, **9**-3
  IRIG standards for, **5**-32 to **5**-36
  and transmitter deviation, **5**-56
PAM-FM-FM, **7**-8, **7**-9
  IRIG standards for, **5**-29 to **5**-31
Paper feed in plotter, **11**-52
Paper tape, for PCM format control, **8**-48
  verified in automatic checkout, **11**-65, **11**-66
Paper-tape processor, **11**-4
Paper-tape punch, **11**-49, **11**-50
Parabolic orbit, **14**-17, **14**-18, **14**-22
Parabolic partial differential equations, **12**-36
Parallax error, and meter, **11**-52
Parallax film reader, **12**-16
Parallax in remote vision, **15**-154
Parallel computer (*see* Parallel data processor)
Parallel data processor, **11**-37, **11**-38, **11**-45
Parallel recording, **5**-43, **11**-27
Parallel-tone system, **13**-36, **13**-37
Paramagnetism, **3**-16
Parameter, external telemetry, **9**-23
  internal telemetry, **9**-23
  monitored, pipelines, **13**-43
Parameter stability equations, transistor, **10**-38
Parametric amplifier and noise performance, **16**-5
PARDOP, **3**-48
Parity, **13**-29
Parity check, **2**-35, **13**-40
Parity checking, in magnetic tape records, **11**-30
  and PCM for quick-look, **8**-44, **8**-45
  and PCM systems, **8**-4, **8**-48
Parking orbit, **14**-19, **14**-46
Parks, R. J., **9**-66
Parsec, **14**-8
Parseval's theorem, **9**-63
Partial differential equations in data reduction, **12**-36
Particles, solid, space, **10**-17
Partitions, **A**-41
Pascal's distribution, **A**-46
Passive attitude control of satellite, **16**-9
Passive satellite, ITU definition of, **5**-13

INDEX

Patchboard in PCM, **8**-46, **8**-47
Patching program for data reduction, **12**-42
Path loss, free space, **4**-97 to **4**-101
   in solar system, from earth, **14**-96
Path-loss ratio in West Ford type system, **16**-12
Patterns, antenna, **4**-112
   contour plotting of, **4**-131, **4**-134
Payload fraction (*see* Payload ratio)
Payload performance, of aerospace airplane, **14**-49
   of launch vehicles, **14**-46, **14**-47
Payload ratio, **14**-45
   and orbital assembly, **14**-48
   and rockets, **14**-45, **14**-47
PCM (pulse-code modulation), **1**-8, **4**-14 to **4**-18, **8**-2 to **8**-5
   advantages of, **8**-3 to **8**-5
   and analog signal inputs, **8**-2
   applications of, in telemetry, **8**-3 to **8**-5
   and average signal power, **2**-41, **2**-42
   binary, **8**-49
   carrier modulation, IRIG standards for, **5**-37
   coherent, **8**-4
   and communications efficiency, **8**-4, **9**-32 to **9**-34
   and communications satellites, **16**-18, **16**-19
   and compatibility with computers, **8**-5
   and data reception, **8**-52 to **8**-56
   and data-recording, **11**-33
   detection methods of, efficiency of, **9**-33
   and error expectancy vs. signal-to-noise ratio, **10**-69
   in Explorer VI, **14**-78
   IRIG standards for, **5**-36, **5**-37
   and M-ary, **8**-48
   for observatory satellites, **14**-90, **14**-93
   and premodulation filtering, IRIG recommendation, **5**-51
   and receiver bandwidth vs. bit rate, **5**-51
   and recording, IRIG standards for, **5**-42 to **5**-46
   and satellite telemetry, **14**-85
   self-timed, **8**-31
   synchronization pattern of, IRIG recommendation for, **5**-51
   system elements of, **8**-2, **8**-3
   transmitted by PAM, **9**-13
   and variable word length, **8**-28
   and word and frame structure, IRIG, **5**-51
PCM-FM (pulse-code modulation–frequency modulation), error sources in, **9**-24
PCM-FSK (pulse-code modulation–frequency-shift keying), **9**-34
PCM-PM-PM (pulse-code modulation–phase modulation–phase modulation), **8**-2
PCM-PSK (phase-code modulation–phase-shift keying), **9**-34
PDM (pulse-duration modulation), **1**-8, **2**-40, **2**-47, **7**-1, **13**-21, **13**-22
   and allocated r-f bands, **7**-27
   and average power, **2**-41
   and average signal power, **2**-41
   and commutation circuits, **7**-15 to **7**-17
   crosstalk in, **7**-32, **7**-33
   errors in, from dynamic contact resistance, **7**-36
   on FM subcarrier, **7**-8
   ground-station equipment for, **7**-27
   IRIG format for, **7**-11

PDM (pulse-duration modulation), IRIG standards for, **7**-22
   mathematical representation of, **2**-40
   minimum receivable power for, **7**-25
   and samples per cycle, **7**-34
   and sampling and interpolation errors, **7**-33 to **7**-35
   signal format of, IRIG, **5**-31
   single edge, **2**-48
   symmetrical, **2**-47
   system specifications for, **7**-9 to **7**-11
   transmitter power required for, **7**-27
   transmitting equipment of, **7**-11
   and video-improvement threshold, **7**-4
   (*See also* PDM-FM; PDM-FM-FM)
PDM converter, **7**-12
PDM keyer, **10**-61 to **10**-64
PDM recording, **7**-36, **7**-41, **11**-33
   IRIG standards for, **5**-42
PDM spectra, **7**-34, **7**-35
PDM-FM (pulse-duration modulation–frequency modulation) **4**-21, **4**-22, **7**-5 to **7**-8
   advantages, **7**-5
   and airborne equipment, **7**-11
   bandwidth requirements for, **7**-6
   carrier specifications for, **7**-11
   and channel threshold signal-to-noise ratio, **7**-10
   comparison of, with other systems, **7**-7, **7**-8
   with PDM-PM, **7**-8
   error sources in, **9**-24
   in industrial telemetry, **13**-35
   and radio links, **7**-22 to **7**-27
   and significant parameter relationships, **7**-10
   and video bandwidth, **7**-7
   and wideband gain, **7**-7
   (*See also* PDM)
PDM-FM-FM (pulse-duration modulation–frequency modulation–frequency modulation), **7**-8, **7**-9
   IRIG standards, **5**-31, **5**-32
PDM-PM (pulse-duration modulation–phase modulation), comparison with PDM-FM, **7**-8
Peak envelop power, ITU definition of, **5**-14
Peak error criterion in phase-locked demodulator design, **9**-67 to **9**-70
Peak power in communications satellites, **16**-19
Peal, strain-gage adhesive, **3**-35
Pedestal, **4**-89
   insertion, **4**-67
Peltier effect, **10**-96
Pen plotter, **11**-52, **11**-53
Pen recorder, **12**-25, **12**-26
Pendulous accelerometer, **15**-96 to **15**-98
Per-channel testing in FM-FM systems, **6**-38, **6**-39
Performance commutator in telemetry, **4**-86
Performance criteria in telemetry systems, **9**-19 to **9**-22
Performance crossover in analog and digital telemetry, **9**-24
Performance of rocket vehicles, **14**-41, **14**-42, **14**-46
Periapsis, **14**-8
Perifocus, **14**-8, **14**-17, **14**-18
Perigee, **14**-8, **14**-26
Perihelion, **14**-8, **14**-57
Period, of lunar orbits, **14**-23
   of Mars orbits, **14**-24

INDEX

Period, of near-earth elliptical orbits, **14**-26
  orbital, **14**-17, **14**-19 to **14**-21
  of solar orbits, **14**-25
  of Venus orbits, **14**-24
Permeability, **3**-15
Permeance, **3**-16
Permutation, **A**-41
  frame-sync code, **8**-39, **8**-41
Permuting program for data reduction, **12**-42
PFM (pulse-frequency modulation), **1**-9, **9**-48 to **9**-54
  and correlation coefficients, for words, **9**-49
  for small satellites, **14**-88, **14**-89
  system of, in Explorer VIII, **14**-81, **14**-82
Phase characteristic, filter, **6**-31
Phase-coherent, definition of, **9**-41
Phase-coherent system, coded, **9**-39 to **9**-48
  uncoded, **9**-48
Phase comparator, **6**-45
Phase correction, digital links, **13**-12
Phase-delay range measurement, **3**-43, **3**-44
Phase-demodulation in coded phase-coherent system, **9**-41
Phase detection and PCM bit synchronization, **8**-33, **8**-34
Phase error, first-order loop, probability density, **9**-78, **9**-79
  in Fokker-Planck analysis of loop, **9**-78
  mean-square, phase-locked loop, **9**-63
  demodulator, **9**-66
  modulation, phase-locked-loop, **9**-63, **9**-64
  non-linear second-order phase-locked loop, **9**-75
  peak, phase-locked-loop, **9**-64
  demodulator, **9**-68
  probability density, first-order loop, **9**-78
  second-order loop, **9**-81
  transient phase-locked loops, **6**-47
  Volterra functional expansion, **9**-82 to **9**-84
Phase-error-frequency-error plot, **9**-72 to **9**-76
Phase-frequency plot, transfer functions of, **15**-41, **15**-42
Phase jitter, bit synchronizer, **8**-36
  phase-locked loop, **9**-63
  sub-carrier, in coded phase coherent telemetry, **14**-101
Phase linearity, SS-FM system, **6**-61
Phase lock (*see* Lock-on, phase-locked loops)
Phase-locked demodulator, **6**-31
  optimized, **9**-66
  optimum output signal filter, **9**-67
  peak error constrained, **9**-67
Phase-locked discriminator, and loop filter, **6**-44, **6**-45
  and loss of lock, **6**-45, **6**-46
  modulation limitations of, **6**-45 to **6**-47
Phase-locked loop, **9**-59 to **9**-97
  and analytic difficulty, sources of, **9**-61
  application of, **9**-59, **9**-60
  and bandwidth for locking to bit rate, **8**-35
  and Bode-diagram, **6**-45, **6**-46
  and comparison of non-linear methods, **9**-91, **9**-92
  and cutoff frequency, **6**-46
    (*See also* Natural frequency)
  and damping factor, **6**-46
  and demodulation, **9**-59, **9**-60
  demodulator, **4**-13

Phase-locked loop, and equivalent-gain analysis, **9**-77, **9**-78, **9**-86 to **9**-96
  and Fokker-Planck analysis, **9**-77 to **9**-82
  functional configuration of, **9**-60, **9**-61
  and gain, closed-loop, **9**-63
  and hold-in range, **8**-35
  and imperfect integrator (filter), **9**-77
  linear model (analysis) of, **9**-61 to **9**-70
  and lock-on of first-order loops, **9**-70, **9**-71
  and loop gain for synchronization, **8**-35
  and loop noise bandwidth, **9**-63
  and maximum phase error, **6**-47
  and modulation following, **9**-65 to **9**-70
  and natural frequency, **6**-46
  non-linear analysis of, with noiseless inputs, **9**-70 to **9**-77
  and noisy signals, **9**-77 to **9**-96
  and optimal receiver of band-limited phase-encoded white Gaussian signal, **9**-89
  and optimum signal tracking loop, **9**-65
  order of, **9**-61
  and PAM ground station sync, **9**-16
  and PCM bit synchronizer, **8**-31, **8**-32
  random frequency modulated, **9**-85, **9**-86
  and random signal performance, **9**-63, **9**-64
  second order, with band-limited phase-encoded white Gaussian signal, **9**-90
  and signal tracking, **9**-64, **9**-65
  systems of, **9**-2, **9**-59 to **9**-96
  use of, **9**-59, **9**-60
  for telemetry synchronization, **14**-101
  theory of operation of, **9**-60, **9**-61
  and tracking, **9**-59, **9**-60
  transient phase error in, for step frequency, **6**-47
  and Volterra functional analysis, **9**-77, **9**-82 to **9**-86
Phase-locked receiver, compared with FM feedback receiver, **9**-93, **9**-94
  for Explorer I data recovery, **14**-76
  PDM, **7**-25, **7**-26
Phase margin and control system, **15**-40 to **15**-48
Phase meters, **3**-50
Phase-modulated carrier, PCM spectrum, **8**-50, **8**-51
Phase-modulated PDM, **7**-8
Phase-modulated subsets, **13**-17
Phase modulation, **1**-8, **2**-45, **2**-46
  RZ type, for data recording, **11**-33
  (*See also* PM)
Phase modulator, **6**-22, **6**-23
  high-pass configuration, **6**-22, **6**-23
Phase-plane analysis and second-order phase-locked loop, **9**-72 to **9**-77
Phase-plane method, **15**-58
Phase-plane plot, and constant reference frequency, second-order loop, **9**-72, **9**-73
  and first-order phase-locked loop, **9**-70, **9**-71
  normalized, and second-order loop, **9**-72 to **9**-76
  and phase-locked loop, **9**-70 to **9**-77
  and reference frequency changing second-order loop, **9**-73 to **9**-76
Phase-rate velocity measurement, **3**-44, **3**-45
Phase-sensitive demodulator, **4**-48, **4**-49
Phase shift of digital filter, **12**-31
Phase-shift keying, **8**-49, **8**-50
Phase-shift oscillator, **4**-12, **4**-13
  integrated circuit, **10**-74 to **10**-76
  RC, **10**-43
  as VCO, **10**-49 to **10**-51**

30 INDEX

Phase-tracking and ranging system, multi-object, **3**-50, **3**-51
Phase variance and tracking phase-locked-loop, **9**-65
Phase velocity, surface wave, **4**-129
Phased-array, self-steering, **16**-15
Phasing, commutator poles, **4**-64
Phasor as electronic commutator, **13**-31, **13**-32
Phasor diagram, **6**-7
Philco Division of Ford Motor Company, **9**-18
Photoconductive material, **3**-24 to **3**-26
Photoconductive transducer, **3**-24 to **3**-26
Photodiode, **3**-25
Photoelectric nuclear battery, **10**-100
Photoelectric transducer, **3**-24 to **3**-29
Photoelectric yield, **3**-25
  (*See also* Quantum yield; Specific sensitivity)
Photoemissive tube, **3**-25
Photoengraving, printed circuit, **10**-77
Photography, documentary, **12**-4
  engineering, sequential, **12**-4
  metric, **12**-4
Photojunction, nuclear cell, **10**-99
Photomultiplier, **3**-25, **3**-28
Photoscanning, **14**-15
Phototransistor, **3**-25
Phototube, gas-filled, **3**-27
  high-vacuum, **3**-27
Photovoltaic cells, **3**-28, **10**-92
Photovoltaic transducer and temperature sensitivity, **3**-29
Physical constants, **1**-22
Physiological constraints in space flight, **14**-6
Pick off (*see* Transducer)
Pickup (*see* Transducer)
Pierce, J. R., **9**-5
Piezo strain coefficient, **3**-32
Piezo stress coefficient, **3**-32
Piezoelectric crystal, characteristics of, **3**-32
  transducer, **3**-30 to **3**-32
Piezoelectric transducer and accelerometer, **15**-93
Pilotless aircraft, **4**-2, **4**-3
Pipeline, **13**-39
  balance control of, **13**-43
  control of, **13**-42 to **13**-44
  centralized, **13**-42
  closed-loop, **13**-39
  operator-guide, **13**-39
  parameters of, monitored, **13**-43
  and telemetry and data presentation, **11**-49, **11**-54
Piston-type hydraulic motor, **15**-134
Planar arm in remote handling, **15**-146, **15**-149
Planar spiral antenna, **4**-123 to **4**-125
Planck's constant, **1**-22
Planetary capture of space vehicle, **14**-30
Planetary encounter, **14**-27 to **14**-30
Planetary exploration, **14**-10
  programs for, **14**-12
Planets, orbital characteristics of, **14**-57
  physical characteristics of, **14**-56
  solar distances of, **10**-16
  surface temperature of, **10**-16
Planning, astronautical, **14**-49 to **14**-52
  for experiments, **12**-12
Plasma, **14**-8
  sheath attenuation, **4**-99
Plateau voltage, ionization counters, **3**-13, **3**-14

Plotter, for data presentation, **11**-52, **11**-53
  high speed, **12**-23, **12**-24, **12**-45
Plotting, by computer printer, **12**-45, **12**-46
  data, **12**-7
Plug board data processor, **11**-4
Plumb bob, **15**-122
PM (phase modulation), **1**-8, **2**-45, **2**-46
  average power in, **2**-37
  defining equation for, **2**-39
  and instantaneous phase, **2**-39
  (*See also* Phase modulation)
PMR (Pacific Missile Range), **5**-1
p-n junction nuclear cell, **10**-99
PN (pseudo-noise) sequence, **14**-98 to **14**-101
  (*See also* Pseudo-noise sequence)
Pneumatic prime mover, **15**-135, **15**-136
Pocket plate and nickel-cadmium battery, **10**-86, **10**-89 to **10**-91
Poisson distribution, **A**-43, **A**-44
Poisson's ratio, **8**-34
Polar orbit of communications satellite, **16**-2
Polaris telemetry, **8**-3
Polarity reversal, power, **4**-38
Polarization, antenna, **4**-7, **4**-113 to **4**-117
  receiving, high-gain, **4**-20
  diversity in, **4**-114
  and link gain, **4**-103
  effects of, on electrochemical cell, **10**-82
  in electrolytic gage, **3**-12
  in PDM-FM, **7**-11
Polarization factor, **4**-134
Polarization loss, **4**-98 to **4**-100, **4**-114
Polarized relay in command link, **15**-160
Pole-zero test and control system stability, **15**-34
Poles, **15**-28, **15**-29, **15**-34, **15**-38
  bandpass filter, **6**-41, **6**-42
Polled system, communications to central, **13**-16
Polling, communications network, **13**-19
Polonium-210, **10**-99
Polynomial, and least-squares, **12**-31
  orthogonal, **12**-31
  smoothing by, **12**-31
Polyrod antenna, **4**-130
Population, statistical, **A**-52, **A**-53
Portel Telemetry System, **4**-11, **4**-14
Position, on celestial sphere, **14**-58
  command systems, in remote-handling control, **15**-147, **15**-166
  to number, analog-to-digital converter, **11**-20 to **11**-22
  telemetering systems, **13**-2, **13**-37 to **13**-39
Post multiplex conditioner, **8**-6
Potential, couple, electrochemical, **10**-79
  energy, orbital objects, **14**-21
Potentiometer, **3**-33
  with accelerometer, **15**-91
  inductance, **3**-19
Potentiometer effect, **4**-85
Pounds delivered to orbit by launch vehicle, **14**-41, **14**-42, **14**-46
Power, average signal, **2**-38 to **2**-42
  capabilities of, in Explorer satellites, **14**-74
  density of, average, **2**-16
  effective radiated, **5**-14
  flow of, electric, telemetering of, **13**-45
  gain in, transistor, G, **10**-28, **10**-29
  interchange control of, **13**-44
  levels of, in active communications satellite, **16**-14
  limitations of, PDM-FM, **7**-11
  limits of, ITU, earth stations in space bands, **5**-23

Power, mean, of transmitter, ITU definition of, **5**-14
　peak envelope, ITU definition of, **5**-14
　solar, **4**-25
　source weight, space mission, **10**-103
　sources of, **10**-79 to **10**-108
　　for space use, **14**-58, **14**-71
　　and telemetry system requirements, **4**-7
　　unmodulated, **2**-38 to **2**-42
Power flux density limits, from space stations, ITU, **5**-24
Power-frequency product, **6**-26, **6**-27
Power ground, **4**-36
Power relationships, transmitter, CCIR recommendations on, **5**-14
Power requirements vs. accuracy in FM telemetry systems, **9**-4
Power-series expansion, **A**-25
Power subsystems and remote-handling, **15**-144, **15**-167
Power supply, and signal conditioner, **4**-51, **4**-52
　and telemetry transmitter, **6**-27
Power-system control, **13**-44, **13**-45
Powers, mathematical, **A**-4
PPM (pulse-position modulation), **2**-46, **4**-22 to **4**-24, **13**-21, **13**-22
　average signal power of, **2**-41
　biorthogonal, **9**-36
　mathematical representation of, **2**-41
　multiposition, **9**-34 to **9**-39
　quantized, **9**-34 to **9**-39
PPM-AM telemetry system, **4**-22 to **4**-24
Preamplifier for PCM telemeter, **10**-64
Precession, **14**-8
　apparent, **15**-120
　of earth magnetic field, **14**-35
　of spin stabilized bodies, **14**-34
Precessional velocity, gyro, **15**-104, **15**-119
Precision, measurement, **11**-13
Predetection filtering, **6**-35
Predetection recording, **4**-25, **4**-26, **6**-28, **6**-32 to **6**-36, **11**-33
　in PAM systems, **9**-14
　and recorder noise and flutter, **6**-33
　set-up procedures for, **6**-36
Predetection video recording, **5**-47
　(See also Predetection recording)
Preemphasis, in FM-FM systems, **6**-15, **6**-16
　and frequency-translation system, **6**-61
　subcarrier, FM-FM system, **5**-28
Preemphasis schedules, and constant-bandwidth system, **6**-17
　and FM-FM systems, **6**-16 to **6**-18
　and IRIG system, **6**-16
Preemptive system, communications to control, **13**-16
Premodulation filtering, PAM, IRIG requirements for, **5**-35, **5**-57
　PCM, IRIG recommendations for, **5**-37, **5**-51
Premultiplex conditioner, **8**-6
Premultiplex filtering and PCM, **8**-6
Preprocessing in data-acquisition equipment, **11**-8
Pre-programmed command in remote handling, **15**-162
Preselector, receiver, **6**-28, **6**-29
Pressure, measurement requirements of, **4**-9
Pressure stress, aerodynamic, **14**-36
Pressure transducer, **3**-17, **3**-30 to **3**-32
Primary cell, **10**-82 to **10**-85
Prime frame, **8**-17

Prime mover, and control system, **15**-127, **15**-136
　electric, **15**-132 to **15**-135
　hydraulic, **15**-132 to **15**-135
　pneumatic, **15**-136
Printed circuit, **10**-77, **10**-78
Printer, in checkout equipment, **11**-64
　for data presentation, **11**-50, **11**-51
　high speed, **4**-11
Printer-plotters, **12**-23, **12**-24
Private-line facilities, **13**-15
Probability, conditional, and transmitted message, **9**-20
　cumulative, phase error, first-order loop, **9**-80
　dropout, phase-locked discriminator, **9**-70
　error, analog message, **9**-20
　　bit, **9**-30, **9**-31
　　in PCM, **8**-52, **8**-53, **9**-20, **9**-32
　　code word reception, **9**-41
　　in PCM system, **16**-18
　　in QPPM, **9**-35
　　word, in biorthogonally coded system, **9**-47
　　　bit by bit detection (uncoded), **9**-48
　　　in coded phase-coherent system, **9**-46 to **9**-48
　　　in orthogonally coded system, **9**-46
　　　sample, in sampled analog system, **9**-56
　　　in uncoded phase-coherent system, **9**-48
　message, **9**-20, **13**-24
　service, **16**-26, **16**-29
　transitional, **8**-52
Probability density, signal plus Gaussian noise, **9**-53
　signal plus Raleigh noise, **9**-53
　steady-state phase error, first-order loop, **9**-79
　　second-order loop, **9**-81
Probability density function, Gaussian, **2**-14
　normal (see Probability density function, Gaussian)
　Raleigh, **2**-14
　second order, **2**-12
　uniform, **2**-14
Probability distribution, **A**-43 to **A**-50
Probability distribution function, **2**-11
Probability theory, **2**-11
Probable error and reduced data, **12**-60
Probing functions, calibration, **11**-13
Process control, **1**-6, **13**-3, **13**-7
　communication-based, **13**-39 to **13**-45
Processing gain, **9**-38
　satellite, **16**-3
　wideband FM, **16**-17
Product tracking, pipeline, **13**-43
Program, computer, **11**-4, **11**-35, **11**-48
　diagnostic, **12**-21
　heuristic, **11**-5
　modules, for data reduction, **12**-38 to **12**-40
　preventive maintenance, **12**-21
　space, **14**-11, **14**-12
Programmed automatic sequenced checkout, **11**-61 to **11**-64
Programmed systems, remote-handling of, **15**-139
Programmer, for automatic checkout, **11**-61 to **11**-64
　for data acquisition system, **11**-7
　OGO specification for, **14**-93
　for PACM, **9**-28
　for PAM, **9**-11, **9**-13

Programmer, for PCM, **8**-3, **8**-15 to **8**-23
 for satellite telemetry, **14**-83, **14**-87
Programming, **11**-48, **12**-9
 for data reduction, **12**-14, **12**-37 to **12**-48
 languages for, **11**-4
 in remote handling, **15**-162
 (*See also* Program, computer)
Progressions, mathematical, **A**-24, **A**-25
Projection display system, **11**-53
Promethium-147, **10**-100
Propagation, **4**-92, **4**-97 to **4**-99
Propagation loss, atmospheric, **4**-99, **4**-100
Propagation-time range measurement, **3**-43, **3**-44
Propellant, characteristics of, **14**-43
Propellant consumption, aerospace airplane, **14**-49
Proportional counter, **3**-14
Proportional feedback in control system, **15**-46
Proportional navigation, **15**-17, **15**-18
Proprioception in remote handling, **15**-156
Propulsion, **14**-39 to **14**-49
Proton intensity in earth orbits, **14**-63
Proton precession magnetometer, **14**-76
Prototype sampled-data system, **15**-77, **15**-78
Pseudo-noise sequence in space telemetry, **14**-98 to **14**-100
Pseudo-random binary sequence, **14**-98 to **14**-101
Pseudo-random codes for PCM frame synchronization, **8**-39
Pseudo-random sequence, **8**-3
PSK (phase-shift keying), **8**-49, **8**-50, **14**-97, **14**-98
PSO (*see* Phase-shift oscillator)
Pull-in range, phase-locked loop, first-order, **9**-71
 imperfect integrator, **9**-77
Pull-in time, phase-locked loop, first-order, **9**-71
 second-order loop, constant input signal frequency, **9**-72
Pulse, Fourier transform, **2**-47
 Gaussian, **2**-47
 single polarity, **2**-46, **2**-48
Pulse-amplitude modulation (*see* PAM)
Pulse-carrier modulation, IRIG standard for, **5**-36
Pulse-code interrogation, automatic checkout, **11**-70
Pulse-code modulation (*see* PCM)
Pulse compression, **9**-37
Pulse-double polarity, **2**-46, **2**-48
Pulse-duration modulation (*see* PDM)
Pulse duration and PDM, IRIG specifications for, **5**-31, **5**-32
Pulse-frequency modulation (*see* PFM)
Pulse modulation, **13**-13
 necessary bandwidth, ITU, **5**-18
Pulse-position modulation (*see* PPM)
Pulse-repetition rate and PDM, **7**-3
Pulse rise time and PDM, IRIG standard for, **5**-31
Pulse-stretching and PDM systems, **7**-34
Pulse-time interrogation, automatic checkout, **11**-71, **11**-72
Pulse transfer function, **15**-51 to **15**-53
 and root-locus technique, **15**-54, **15**-55
 stability of, **15**-53
Pulse transforms, Fourier, **2**-47
Pulse waveforms, **2**-47
Pulse-width jitter, **7**-12
Pulse-width modulation (PWM), **1**-8, **7**-1

Pulse-width scatter, **7**-12
Pump, hydraulic, **15**-133
Punched card, **11**-51
 in automatic checkout, **11**-65
 in data conversion, **12**-6
Punched paper tape, **11**-50
 and control of automatic checkout, **11**-61, **11**-64, **11**-66
 used in data conversion, **12**-6
 for recording in Explorer III, **14**-79
Pursuit course, **15**-15
PWM (pulse-width modulation), **1**-8, **7**-1

q, samples per cps, **9**-18, **9**-21, **9**-22, **9**-26
 and error sources, **9**-21
QPPM (quantized pulse-position modulation), **9**-34 to **9**-39
 coherent, **9**-36
Quantitative commands in space system, **14**-102
Quantization, **2**-6
 channel, **9**-30
 error, **8**-26
 and communication efficiency, **9**-31, **9**-32
 telemetry signal, **13**-22
Quantized PAM, **13**-22
Quantized pulse-position modulation, **9**-34 to **9**-39
Quantum efficiency, light detector, **14**-72
Quantum noise, laser, **14**-72
Quantum yield, **3**-25
Quarter-square multiplier, **15**-25
Quartz-crystal, frequency stability of, **6**-24
Quartz-crystal resonator, **6**-23, **6**-24
Quartz-crystal transducer, **3**-30 to **3**-32
Quasi-linear phase-locked loop, optimum performance of, **9**-89
Quasi-linear receiver, **9**-87
Quasi-passive communications satellite, **16**-14
Queuing in communications satellite, **16**-21, **16**-22
Quick-look, **4**-10, **12**-5
 and data presentation, **11**-52
 for data processing check, **12**-43
 in PCM, **8**-44
 printed, **12**-45
Quick-look processing, **11**-8, **12**-26, **12**-27

r parameters, transistor, **10**-21
 and h parameter relationships, **10**-26, **10**-27
Rad, defined, **14**-65
Radar, **3**-53 to **3**-56
 beacon, **3**-54, **3**-55
 tracking, range performance, **3**-55
 ITU defined, **5**-11
Radar signal processing, **9**-37
Radar tracking, **3**-54
Radiation, Inc., multistylus printer, **4**-11
Radiation, cosmic, **10**-13
 damage by, to electronic components, **10**-14
 solar cell, **14**-64
 dosage of, annual in typical earth orbits, **14**-62
 effects of, on man, **14**-65
 and flux, **10**-15
 to dose conversion, **14**-66
 gamma, **10**-14
 neutron, **10**-14
 outer space, **10**-12, **10**-13

INDEX 33

Radiation, shielding, **10**-15, **10**-16
  solar-heat, **10**-16
  thresholds of, for electronic component, **14**-63
  transducers for, **3**-13
Radiation efficiency factor, **4**-112
Radiation pressure, **14**-34, **14**-35, **14**-64
Radiation resistance, **4**-94
Radio astronomy, **5**-12
  interference protection of, **5**-8
Radio command links in remote handling, **15**-159, **15**-160
Radio determination, **5**-11
Radio frequency, allocations for, **5**-5 to **5**-7
  IRIG standards for, **5**-47 to **5**-50
Radio isotope thermoelectric generator, **10**-98
Radio links, **4**-95 to **4**-110
  and telemetry, **4**-92 to **4**-111
Radio location, **5**-11
Radio navigation, **15**-13
  ITU defined, **5**-11
Radio noise map, **4**-104 to **4**-107
Radio regulations, by ITU, **5**-7, **5**-8
Radio service, **5**-8 to **5**-13
Radio station, ITU defined, **5**-10 to **5**-12
Radio system, ITU defined, **5**-10 to **5**-12
Radio telemetry, **4**-1 to **4**-135
Radio telemetry system design, **4**-5 to **4**-7
Radio telemetry systems, **4**-11 to **4**-24
Radio terms, ITU defined, **5**-10
Radiosonde, **1**-3, **5**-12
  and data extrapolation, **12**-33
Ragazzini, J. R., **13**-23
Railroads and data transmission, **13**-15
Ramjet engine, **14**-49, **14**-50
Ramp comparison and analog-to-digital converter, **11**-17, **11**-18
Ramp driving function, **15**-67, **15**-70
Ramp encoder, **8**-23
Ramp generator, **10**-63
RAMPART radar, **3**-53
Random access interrogation and automatic checkout, **11**-70
Random access modulation, **16**-22
Random noise in PDM systems, **7**-32
Random process, stationary, **2**-11 to **2**-45
Random sample, **A**-51
  statistics, **A**-52 to **A**-54
Random torque errors, gyro, **15**-114, **15**-115
Random variable, **2**-10
Random-walk process in oscillator, **9**-95
Range (*see* Distances; Transmission)
Range Commanders Conference, **5**-1, **5**-2
Range frequency assignments, **5**-8
Range to impact, trajectory, **14**-27
Range Management Agency, **5**-2
Range measurement, **11**-13
Range performance, radar beacon tracking, **3**-55
Range safety, **14**-27
Range telemetry, **4**-4, **4**-5
Range time code, format, **12**-40, **12**-41
  processing, PCM, **8**-48
Ranger 7, **14**-54
Ranging frequencies, **3**-44
Rate, cycle skipping, first-order loop, **9**-81
Rate command systems and remote-handling control, **15**-148, **15**-166
Rate-counting circuit, **4**-45
Rate gyro, **15**-100, **15**-105 to **15**-115
  acceleration insensitive errors, **15**-114
  compared to accelerometer, **15**-105
  construction, **15**-107
  damping factor, **15**-106, **15**-107, **15**-109

Rate gyro, equation of measurement, **15**-105
  error sources, **15**-109 to **15**-115
  missile application, **15**-10
  for missile attitude determination, **12**-55
  natural frequency, **15**-106
  random drift error, **15**-114, **15**-115
  steady-state response, **15**-105, **15**-107
  torque balance, **15**-115, **15**-117 to **15**-119
  transducer, output, **15**-108
  transfer function, **15**-106
  transient response, **15**-106, **15**-107
Ratio system for position telemetering, **13**-37 to **13**-39
Rationalizing fractions, **A**-4
Rauch filter, **6**-51
RAYDIST, **3**-50
Rayleigh distribution, **A**-49
RB (return-to-bias) representation, PCM, **8**-29
RBE (relative biological effectiveness), **14**-65
RC (resistance-controlled) oscillator, **10**-43
Reaction, electrochemical, **10**-80
Reaction engine, **14**-8
Reaction wheel, momentum storage, **14**-36
Reactor refueling, **15**-140, **15**-141
Read-only memory, **11**-7, **11**-11
Read out, non-destructive, of recorders, **11**-27
Readers, metric film, **12**-14
  one-axis, **12**-14
  oscillogram, **12**-14
  two-axis, **12**-14
Readiness checkout, **11**-56, **11**-57
Reading resolution, recorder, **11**-27
Real time, **1**-6
  command loop presentation, **14**-87
  communications system, **13**-19
  data, **4**-3, **4**-4
  display, **12**-25, **12**-26
  display, **12**-5
  PAM, **9**-15 to **9**-18
  PCM, **8**-49
  telemetry processing, **11**-47, **11**-48
Received power, **4**-92
Received signal power, from antenna contour pattern, **4**-134
  satellite, **16**-12
  terminal, active communications satellite, **16**-13
    distributed passive satellite, **16**-10 to **16**-12
    passive satellite system, **16**-5, **16**-6
  West Ford type system, **16**-9 to **16**-12
Receiver, correlation detection, **7**-25
  discriminator, **6**-31, **6**-32
  double conversion, **6**-28
  dynamic range, **6**-29
  FM, with feedback, **9**-93, **9**-94
    for PDM system, **7**-23, **7**-24
  FM-FM, **4**-19
  i-f bandwidth, **6**-31
  limiter, **6**-31, **6**-32
  linearity, **6**-29, **6**-30
  noise factor, **6**-29
  PDM phase-locked, **7**-25
  PDM-FM, **4**-22
  phase nonlinearity distortion, **6**-31
  PPM-AM, **4**-23
  pre-selector, **6**-29
  quasi-linear, with optimal transfer function, **9**-89, **9**-90
    second order loop filter, band-limited, phase encoded white Gaussian signal, **9**-90

Receiver, simple FM telemetry, **4**-12
  superheterodyne, **6**-28
  telemetry, **6**-27 to **6**-36
    capture ratio of, **6**-31
    functional description of, **6**-27, **6**-28
    typical performance of, **6**-36
    windows, **6**-29
Receiver bandwidths, telemetry, PAM-FM signals, **5**-32, **5**-33
Receiver IF bandwidth, and bit rate, PCM, **5**-36
  telemetry, IRIG standards, **5**-33
Receiver noise, **4**-103 to **4**-110
Receiver noise bandwidth, **4**-107, **4**-108, **4**-110
Receiver noise level, calculation of, **4**-108, **4**-110
Receiver sensitivity, telemetry, **4**-93
Receiver temperature, effective, **4**-93
Rechtin, E., **9**-59, **9**-65
Reconstitution, PCM systems, **16**-18, **16**-19
Record, in data processor, **11**-30, **11**-47
  punched card, **11**-51
Record gaps in digital computer tapes, **11**-31
Recorder, **11**-24 to **11**-34
  bias level, **6**-36
  instantaneous speed error, IRIG recommended test, **5**-54, **5**-55
  low-inertia, servo controlled, **6**-20, **6**-34
  magnetic, in down range telemetry, **4**-5
    dynamic range, **7**-41
    principle of operation, **11**-25 to **11**-28
    wideband, **9**-14
  magnetic-tape (*see* Recorder, magnetic)
  moving-coil galvanometer, **11**-53
  PDM, **7**-36 to **7**-41
  post detection, PCM, **8**-49
  pre-detection, PCM, **8**-44
  speed errors, **6**-19, **6**-20, **6**-33, **6**-34
    (*See also* Tape-speed errors)
  tolerances test, IRIG electronic method, **5**-56
  (*See also* Recording)
Recorder-amplifier, PDM, **7**-38
Recording, PCM, IRIG standards, **5**-42 to **5**-46
  PDM, IRIG standards, **5**-42
    without interpolation, **7**-33
  single carrier FM, **5**-46, **5**-47
Recording errors from tape tension, **7**-39
Recording head positioning in plotters, **11**-53
Recording media for data reduction input, **12**-4
Recording tapes, telemetry, record gaps in, **11**-31
Recruit, **14**-42
Rectangular distribution, A-50, A-51
Redox fuel cell, **10**-103, **10**-104
Redstone missile, stable platform, **15**-126
Redundancy, **2**-23
Redundancy check, longitudinal, **13**-8, **13**-9
Redundancy data, **4**-10
Redundancy elimination processing, **8**-28
Redundancy methods and security system, **13**-40
Redundant time, orbit to ground communications, **14**-71
Redundant transducers, **4**-8
Reed, I., **9**-44
Reed switch, **8**-12
Reentry, **10**-11, **14**-8
  planetary non-lifting, **14**-30

Reentry charts, **14**-30, **14**-31
Reentry dynamics, planetary, **14**-27 to **14**-30
Reference mount, **15**-124
Reference point, regulator, **4**-52
Reference system, body, **15**-4, **15**-5
  inertial, **15**-124
Reference temperature control, thermocouples, 8-41
Reflected binary code, **11**-21
Reflection, Doppler, **3**-48
Reflection cross section, passive communications satellite, **16**-7
Reflectors, mesh sphere, **16**-9
  non-oriented, **16**-7
  resonant dipole, **16**-9
  satellite, **16**-7 to **16**-9
  spherical shell, **16**-8
Reflex-arc, remote-handling, **15**-168
Refractive index, air, effect on propagation, **4**-92
Regeneration, data, **13**-11
Regenerative PCM (*see* Reconstitution, PCM systems)
Regenerative system, subnetwork, industrial telemetry, **13**-17
Regions, ITU, **5**-7, **5**-9
Register, ALU, **11**-39
  computer, **11**-37
Reiger, S., **16**-23
Relative biological effectiveness of radiation, **14**-65
Relative gain, antenna, ITU definition of, **5**-14
Relative stability, control system, **15**-36
Relay logic circuits, basic, **15**-80
Relay multiplexer, **8**-12
Relays, latching, for multiplexing, **8**-12
Reliability, **14**-9, **14**-13 to **14**-15
  commutators, mechanical, **4**-74 to **4**-77
Reliability proof method, commutators, **4**-77
Reluctance, **3**-16
Reluctance transducers in accelerometers, **15**-93
Rem, defined, **14**-63
Remnant magnetism in recording, **11**-26, **11**-27
Remote calibration in FM-FM systems, **6**-54
Remote control, **4**-3, **15**-1 to **15**-159
  components for, **15**-88 to **15**-136
  defined, **15**-2, **15**-3
  industrial, **13**-3 to **13**-7
  of instrumentation, **14**-86
  in space, **14**-95
Remote control arms, **15**-146, **15**-147
Remote-control systems, industrial, **13**-1 to **13**-46
  and remote handling, **15**-139
Remote feedback and remote handling, **15**-162
Remote handling, **15**-137 to **15**-169
  applications of, **15**-168, **15**-169
  arm design in, **15**-146, **15**-147
  bandwidth requirements for, **15**-155, **15**-161
  computer controlled, **15**-161 to **15**-164
  and control systems, **15**-147 to **15**-149, **15**-166
  design of, **15**-166
  data rate, command, **15**-159, **15**-160
  frame rate requirements for, **15**-155
  and human factors analysis, **15**-164, **15**-166, **15**-167

INDEX 35

Remote handling, and operational analysis, **15**-142
  research in, **15**-167, **15**-168
  in space orbit, **15**–151, **15**–152
  systems of, **15**-137, **15**-142
    requirements for, **15**-144
  visual requirements for, **15**-153, **15**-155
Remote-handling system, **15**-138 to **15**-142
  and direct vision, **15**-139
  equipment for, **15**-142 to **15**-144
  programmed, **15**-139
  remote controlled, **15**-139
  and subsystems, **15**-142 to **15**-144
Remote measurement, **4**-7
  (*See also* Telemetry)
Remote process control in multiunit system, **13**-9, **13**-10
Remote reset automatic control, **13**-91
Remote station, **13**-41
  on pipeline, **13**-44
Remote supervisory control, **13**-41
Remote vision for remote-handling, **15**-153 to **15**-155
Remotized vehicle, **15**-150, **15**-151
Remotizing, **15**-145, **15**-150
Rendezvous, orbital, **14**-9, **14**-20, **14**-21
Reorientation of spin stabilized bodies, **14**-34
Rep, **14**-65
Repair procedures and military system, **11**-57
Repeatability, **11**-15
  of film reader, **12**-15, **12**-16
Repeater, data-link, **13**-11
  delayed, **16**-2
  real-time, **16**-2
  regenerative, **13**-17
  satellite, **16**-20
  (*See also* Communications satellite)
Representation, signal, **2**-4
Reproduce electronics, **7**-38
Requirements, system, in radio telemetry, **4**-2 to **4**-7
  telemetry, of missile program, **9**-19
Reserve battery, **10**-91 to **10**-93
Reserve cell, **10**-91 to **10**-93
Reset in automatic checkout, **11**-65
Resistance, electrolytic, **10**-80, **10**-82
  leakage, in gates, **4**-81, **4**-86
  radiation, **4**-94
  saturation, in gates, **4**-81
Resistance-controlled oscillator circuit, **10**-42, **10**-43
Resistance matrix, **4**-80
Resistance telemetry system, **13**-33
Resistance thermometer, **3**-38
Resistance transducer, **3**-31
Resistive temperature gage, **3**-38
Resistivity, variable, gages, **3**-38
Resistor, printed circuit, **10**-78
Resolution, film reader, **12**-15
  mapping, **14**-15
  measurement of, **11**-13
  oscillogram readers, **12**-16
  transducer, **3**-5
Resolver, **15**-73
Resonant-circuit frequency converter, **4**-47
Resonant frequency, second-order transfer function, **15**-42
  transducer, **3**-6
Responder, pulse-time interrogation, **11**-72
Response, accelerometer, **15**-98, **15**-99
  angular displacement, rate gyro, **15**-106
  second-order system, **15**-98, **15**-99

Response time, photo tubes, **3**-27, **3**-28
  transducer, **3**-8
Restraint, gyro spin axis, **15**-100
  integrating gyro, **15**-115
Retrodirective array, **16**-15
  image-frequency, **16**-15
  pilot tone, **16**-16
Return-to-bias representation in PCM, **8**-29
Return-to-zero representation, in data recording, **11**-31
  and PCM, **8**-29
Reza, F. M., **13**-26
Reza's theorem, **13**-26
r-f (radio-frequency) amplifier, filter, **6**-29, **6**-30
r-f free-space loss, **4**-97 to **4**-101
r-f link design in PDM system, **7**-26, **7**-27
RFI filter and telemetry transmitter, **6**-27
Richman, D., **9**-59
Right ascension, **14**-9, **14**-58
Rigid body motion, **15**-4, **15**-5
Rigidity in gyros, **15**-111 to **15**-114, **15**-118, **15**-119
Ring, multivibrator, commutator, **10**-58, **10**-59
Ring counter, **8**-18, **8**-19
Ripple-through counter, **8**-17, **8**-18
Rise time, **3**-6, **3**-7, **6**-37, **6**-39
Rochelle, R. W., **9**-49
Rochelle crystal, characteristics of, **3**-32
Rocket equations, **14**-39 to **14**-44
Rocket performance design, payload vs. launch velocity, **14**-47
Rocket propellant, **14**-93
Rocket vehicle, **14**-38 to **14**-47
  design charts for, **14**-45 to **14**-47
  dimensions of, **14**-38
  for space projects, **14**-38, **14**-41, **14**-42
Rocket velocity equation, **14**-39, **14**-40
Roentgen, **10**-15, **14**-63, **14**-65, **14**-66
  defined, **14**-65
Roentgen equivalent per man, **14**-65
Roll-indicator, missile, **10**-43
ROMOTAR, **3**-49
Root-locus technique, control system, **15**-38 to **15**-40
  with delay, **15**-85
  pulse transfer functions, **15**-54, **15**-55
Roots, A-4
Rotary hydraulic prime mover, **15**-133, **15**-134
Rotating amplifier as constant-current source, **15**-130
Rotational system electrical analog, **15**-65
Rotor, gyro, **15**-100 to **15**-102
Rotor balance, gyro, **15**-102, **15**-103
Rotorace, **15**-101
Round-off errors, **12**-29
Routh-Hurwitz criteria for linear system stability, **15**-29 to **15**-33
Routh test of stability, **15**-29 to **15**-31
Ruben, S., **10**-83
RUM (remote underwater manipulator), **15**-141
Runge-Kutta method, **12**-36
  for initial value problem, A-68
RZ (*see* Return-to-zero representation)

Salt spray, test specifications for, **10**-6, **10**-8
Sample average, A-53, A-54
Sample variance, A-53, A-54
  of frame-sync code, PCM, **8**-39 to **8**-41
Sample-and-hold circuit, **8**-2
  and PCM encoder, **8**-26, **8**-27

Sampled-data, **12**-29 to **12**-37
  control system, **15**-76 to **15**-78
    prototype response, **15**-77, **15**-78
  in industrial telemetry, **13**-24
  systems of, **15**-48 to **15**-55
    with delay, **15**-86
  in teleoperating systems, **13**-23
  (*See also* Sampling)
Samples, and combinations, A-42
  discrete, **2**-5, **2**-6
Samples per cps (*see* Samples-per-cycle)
Samples-per-cycle(q), **4**-27, **9**-18, **9**-21, **9**-22
  and error moment criterion, **9**-21
  and PDM systems, **7**-34
Sampling, **4**-53 to **4**-55, **13**-23, **13**-24
  instantaneous, PDM, **7**-33
  minimum rate, **9**-6
  PCM, **8**-56, **8**-57
  plus and minus, **9**-7
  in pulse-amplitude modulation, **9**-6
  (*See also* Sampling theorem)
Sampling circuit, PAM multiplexer, **9**-9
Sampling errors, PDM, **7**-33 to **7**-35
Sampling function, ideal, **8**-57
  non-ideal, **8**-59
Sampling gates, PAM, **9**-9 to **9**-11
Sampling rate, for frequency analysis, **12**-34
  mechanical commutators, **4**-66, **4**-67
  and message bandwidth, **4**-27
  minimum, and signal bandwidth, **9**-6
  and PACM system, **9**-27
  and PAM, and IF bandwidth, **5**-32, **5**-33
    stability, IRIG standard for, **5**-34
  and PDM telemetry, **7**-3
    IRIG standards for, **5**-31, **5**-32
    stability, in commutators, **4**-86
    IRIG standard for, **5**-34
Sampling-and-smoothing system, PAM, **9**-6
Sampling switches, non-wiping, **4**-74
Sampling theorem, **2**-6, **13**-25, **13**-26, **15**-49, **15**-50
Sand and dust, test specification for, **10**-6, **10**-8
Sanders, R. W., **8**-4, **9**-31, **9**-69, **9**-70
Sandia Corporation, **5**-2
SARA (ship angle and range), **3**-53
Satellite, communications, **16**-1 to **16**-32
  Vanguard III, **9**-49
Satellite Communications Service, **5**-8
Satellite-relay communications, **16**-1 to **16**-32
Satellite system constraints on telemetry system, **14**-73, **14**-74
Satellite telemetry, systems used for, **14**-75 to **14**-83
Saturable reactor, **3**-24
Saturation, **15**-56, **15**-58
  simulation, **15**-70, **15**-72
Saturation recording, **11**-26
Saturn, **14**-38, **14**-41, **14**-46
  planetary characteristics of, **14**-53, **14**-55 to **14**-57
Sauereisen cement, strain gage, **3**-36
Save register (SAR), ALU, **11**-39, **11**-44
Sawtooth encoders and analog-to-digital converters, **12**-19
Scale change, data, **11**-7
Scale factor, in converter, **11**-15
  amplifier, **12**-5
  variable, over range, **11**-15
Scale height, **14**-9, **14**-30, **14**-31
Scatter, pulse-width, **7**-12
Scientific space mission, **14**-10
Scimitar antenna, **4**-126

Scintillation counter, **3**-13
SCO (*see* Subcarrier oscillator)
Score, **16**-2
Scout, **14**-38, **14**-42
Screen printing of printed circuit, **10**-77
Sculling error, **15**-127
Sealed storage battery, **10**-86
Sealing, hermetic, electrochemical cell, **10**-82
Search coil (*see* Magnetometer)
Search routines in remote handling, **15**-162
Second-order phase-locked-loop, **9**-64
  and comparative phase error variance, **9**-92
Second-order systems, **15**-37 to **15**-48, **15**-98, **15**-99
  response in, rate gyro, **15**-106
Secondary cell, **10**-84 to **10**-92
Secondary-emission, nuclear battery, **10**-100
SECOR (sequential ranging), **3**-49
Security, data, **13**-27, **13**-28
  operational, communications-based control system, **13**-39 to **13**-41
Seebeck effect, **10**-96
Select instruction and data processor, **11**-47
Selective symbol display, **11**-53
Selenium photo-transducer, **3**-24 to **3**-26, **3**-29
Self-discharge, **10**-80
Self-stabilization of transistor parameters, **10**-37 to **10**-42
Self-steering, phased-array, **16**-15
Self-timed PCM system, **8**-31
Semiactive homing system, **15**-13
Semi-automatic data reduction system, **12**-14 to **12**-18
  data rate, **12**-19
Semi-conductor, capacitor, **10**-70
  integrated circuit, **10**-74 to **10**-76
  material, in integrated circuit, **10**-73
  thermistor, **3**-39
Semi-conductor strain gage, **3**-38
Semi-conductor switch, **8**-13 to **8**-15
  and multiplexed impedances, **8**-15
Semi-conductor transducer, **3**-12
Sensing, extra-human, in remote handling, **15**-157
Sensitivity, control systems and effect of feedback, **15**-25 to **15**-27
  of phototubes, **3**-27, **3**-28
  receiver, **4**-93
  transducer, **3**-5
Sensitivity error, **6**-19
Sensors, **3**-2
  attitude of, **14**-34
  position of, **3**-42
  (*See also* Transducer)
Sensory subsystem in remote handling, **15**-144, **15**-152 to **15**-158
Sequence generator, maximum length, **8**-19
Sequencer, automatic checkout, **11**-59
  PAM demultiplexer, **9**-15
Sequences, erroneous, in counters, **8**-19
Sequencing, automatic, in checkout equipment, **11**-61
  programmed automatic, in checkout equipment, **11**-61 to **11**-64
  semi-automatic, in checkout equipment, **11**-60
Sequencing circuit, PAM multiplexer, **9**-9
Sequential ranging, **3**-49
SERF (Sandia Engineering Reactor Facilities) nuclear facility, **15**-140
Serial data processor, **11**-37
Serial network, bridged, **13**-19

INDEX

Serial-to-parallel converter, PCM, **8**-43 to **8**-45
Serial recording, **11**-27
Serial subnetwork, **13**-17
Series expansion, **A**-24, **A**-25
Series feedback for transistor amplifier stabilization, **10**-32
Servo-balance and analog-to-digital converter, **12**-19, **12**-20
Servo-controlled manipulator, **15**-146
Servo-mechanism, **13**-6
Servomotor, d-c, **15**-128, **15**-130, **15**-132
Servo-multiplier, **15**-75
Shaft-position encoder, **3**-11
Shannon, C. E., **8**-4, **9**-5, **9**-6, **9**-22, **9**-30, **13**-23, **13**-25, **16**-16
Shannon-Nyquist theorem, **15**-49
Shannon's limit, channel capacity, **9**-89 to **9**-91
Shannon's sampling theorem, **13**-26
Shaped-beam image tube, **11**-54
Shear, strain-gage adhesive, **3**-35
Shearing stress, aerodynamic, **14**-36
Sheppard's correction, **A**-54
SHF band, defined, **5**-15
Shield ground, **4**-37
Shielded-loop antenna, **4**-13
Shielding, electrical, **4**-37, **4**-38
 nuclear, **15**-140
 radiation, **10**-15
Shift-register, code generator, **9**-45
 integrated circuit, **10**-74 to **10**-76
Shift-register code recognizer, frame synchronization, **8**-41
Shift-register counter, **8**-18
Shingled solar cell, **10**-95
Shock, test specification for, **10**-6, **10**-7
Shockley, W., **10**-33
Short-circuit current-amplification factor, α, transistor, **10**-20, **10**-21
Shuler-tuning, **15**-116, **15**-123
Shunt feedback for transistor amplifier stabilization, **10**-25, **10**-30 to **10**-32
Side lobes, contribution of, to noise temperature, **4**-106
 surface-wave antenna, **4**-129
Sideband power in FM-FM system affecting multiplex design, **6**-18, **6**-19
Sideband splash, amplitude modulation, **9**-5
Sidereal time, **14**-9
Signal, analog, degradation, **11**-11
 telemetry information rate, **9**-20
 antipodal, **9**-48
 band limited, **9**-6, **13**-26
 data, characteristic of missile programs, **9**-19
 telemetry, characteristics, **9**-4
 digital, telemetry, information rate, **9**-20
 interrupted analog, information rate, **9**-20, **9**-21
 multisymbol, **9**-40
 output, phase-locked discriminator, peak phase error limited, **9**-68
 time-limited, **13**-26
 uniform peak limited, in Gaussian noise, **9**-20
Signal acquisition in down-range telemetry, **4**-5
Signal conditioner, circuits, **4**-41 to **4**-52
 PCM, **8**-5 to **8**-7
Signal conditioning, **4**-35 to **4**-52
 amplifiers, **4**-41 to **4**-44
 data acquisition system, **11**-9 to **11**-11
 design requirements, **8**-7

Signal conditioning, input and output data, typical, **4**-40
 for PCM synchronization, **8**-32, **8**-33
 in PDM systems, **7**-22
Signal energy per bit, **8**-53, **9**-30
Signal flow-graph, for feedback analysis, **15**-22 to **15**-24
 gyro, **15**-106, **15**-116
Signal format, PCM, IRIG, **5**-36, **5**-37
Signal generators for automatic checkout, **11**-58
Signal ground, **4**-37
Signal injection for automatic checkout, **11**-59
Signal-to-noise (*see* Signal-to-noise ratio)
Signal-to-noise constraints, in satellite data recovery system, **14**-84, **14**-85
 in satellite telemetry, **14**-74
Signal-to-noise improvement, PAM-FM, **9**-3
Signal-to-noise intensity in space communications system, **14**-96
Signal-to-noise intensity ratio, **9**-23
Signal-to-noise ratio, coded phase-coherent system, **9**-42
 coherent, loop, **9**-79, **9**-85, **9**-86
 in discrete vs. analog channel, **9**-57
 in discrete channel, sampled analog system, **9**-56
 first-order loop, **9**-79
 in loop bandwidth, **9**-79, **9**-85
 and optimal quasi-linear receiver, **9**-89, **9**-90
 and PCM, **8**-53, **8**-56
 communications satellites, **16**-18
 phase locked discriminator, peak phase error limited, **9**-68
 satellite, **16**-12
 second-order loop quasi-linear receiver, **9**-90, **9**-91
 threshold, **2**-51, **2**-52
 wideband FM, **16**-17
Signal power, received, quasi-linear receiver, **9**-87
Signal power level, r-f links, **4**-95 to **4**-97
Signal-processing, discriminator output, **6**-51
 radar, **9**-37
 receiver, **6**-28, **6**-32
Signal representation, sinusoidal, **2**-7
Signal simulator for automatic checkout, **11**-58
Signal spectrum, FM, as function of modulation index, **6**-10
Signal suppression, FM, **6**-13
Significant digits, **11**-14
Silicon boron, **3**-39
Silicon carbide, **3**-39
Silicon phototransducer, **3**-26, **3**-29
Silicon solar cell, **10**-92, **10**-94
Silicon stain gage, **3**-33
Silver chloride cell, **10**-83
Silver-oxide cell, **10**-97, **10**-89 to **10**-91
Silver-zinc battery, **10**-84, **14**-71
Simple sequencer, type of, checkout equipment, **11**-19
Simplex operation, **5**-10
Simpson's rule, **A**-66
 in data reduction, **12**-34
Simulation, block diagram, **15**-67
 control system, **15**-63
 with delay, **15**-86
 transfer functions, **15**-71
Simulator, **15**-63
Sin x/x function, **9**-6, **9**-7

INDEX

Sinc function, **A**-25
Sine integral, **A**-26
Sine integral table, **A**-75
Sine potentiometer, **3**-33
Single-address computer, **11**-36, **11**-38
Single carrier FM recording, IRIG standard for, **5**-46, **5**-47
Single channel detector in phase-coherent telemetry system, **14**-101
Single-degree-of-freedom gyro, **15**-100, **15**-103, **15**-118
Single-ended signal, **8**-5
Single mode checkout, **11**-60
Single sideband modulation, **1**-9, **6**-61
  in communications satellites, **16**-18 to **16**-20
  multiplexing, **9**-3
Sintered plate, nickel-cadmium battery, **10**-86, **10**-89, **10**-90
Skew in data recording, **11**-34
Sky background maps, radio noise, **4**-104 to **4**-107
Sky noise temperature from oxygen and water vapor, **4**-106, **4**-108
Slant range, maximum, in communications satellite, **16**-6
Sleeve-stub antenna, **4**-120
Slew search mode, high-speed, in paper tape readers, **11**-67
Slicer, PCM, **8**-31 to **8**-33
Slicing, noise errors, **7**-4
  for pulse regeneration, **7**-4
Slot antenna, **4**-127, **4**-128
Slow-neutron, radiation, **10**-14, **10**-15
Small satellite telemetry, **14**-88 to **14**-90
Smoothing, **4**-10
  data, **12**-9, **12**-31
  polynomial, **12**-31
Snap reactor, **14**-71
Solar cell, **10**-92, **10**-94, **10**-95, **14**-71
  and radiation damage, **14**-64
Solar constant, **14**-9
Solar energy, **10**-13
Solar flare data, **14**-61, **14**-64, **14**-66, **14**-68, **14**-79
Solar-heat radiation, **10**-16
Solar orbit, **14**-25
Solar power, **4**-25
Solar proton, **14**-66, **14**-68
Solar radiant power density, **10**-92
Solar radiation torque, **14**-34, **14**-35
Solar system, basic characteristics of, **14**-52, **14**-53
  free space path loss from earth, **14**-96
Solaris, underwater remote handling system, **15**-141
Solid propellant vehicle, **14**-42
Solid state display, **11**-49 to **11**-54
Solid state modulator in communications satellite, **16**-14
Solid tantalum capacitor, **10**-71
Sonar-sensing in remote-handling system, **15**-157
Sorting program, **12**-42
Space-communication frequencies, **4**-94
Space frequency allocations, **5**-5 to **5**-7
Space hostile environment, **15**-139
Space mission, communications parameters, **14**-97
  orbital assembly for, **14**-48
  telemetry requirements for, **14**-2, **14**-10 to **14**-12
Space mission planning, **14**-49 to **14**-52
Space power system, **14**-71
Space program, costs of, **14**-10, **14**-12, **14**-13
  scientific, **14**-12
Space services, special, ITU rules for, **5**-23
Space system, basic design information on, **14**-2 to **14**-72
  radio, ITU defined, **5**-12, **5**-13
Space systems checkout, **11**-57
Space Technology Laboratories, **14**-75
Space telemetering, ITU defined, **5**-13
Space telemetry, **5**-13, **14**-95 to **14**-104
Space temperatures, of black body, **10**-16
  of gray body, **10**-16
Space transmitters, receivers, frequency separation, **5**-8
Span, measurement, **11**-13
Sparks, **10**-53
Spatial stabilization, **15**-124
Special purpose telemetry, **14**-92
Specific impulse, **14**-9, **14**-39, **14**-40, **14**-43, **14**-44
  of attitude control propellant, **14**-38
Specific sensitivity, **3**-25
  (*See also* Quantum yield)
Specifications, environmental test, **10**-4 to **10**-7
  signal conditioning, typical, **4**-40
  for voltage-to-number analog-to-digital converters, **11**-19
Spectral bandwidth, effective for random processes, **2**-19
Spectral density, **2**-43, **2**-44
  AM, **2**-49
  FM, **2**-49
  modulated pulses, **2**-46
  modulated sinusoids, **2**-44
  of random processes, **2**-16, **2**-19
  (*See also* Spectrum)
Spectral sensitivity, **3**-25, **3**-27
Spectrum, PCM modulator output, **8**-50, **8**-51
  plus-and-minus sampled signal, **9**-7
  pulsed sine wave, **9**-50
  sampled signal (function), **9**-7
Spectrum analyser, analog, **12**-35
Spectrum analysis, **12**-34, **12**-35
Spectrum occupancy and telemetry systems, **9**-25, **9**-26
Spectrum sharing in communications satellite, **16**-30
Spectrum utilization in telemetry systems, **4**-32
Speed control, commutator, **4**-68
  signal, recorder, **5**-41
Speed error test, recorder, **5**-51 to **5**-56
  IRIG recommended, **5**-55
Speed errors, recorder, defined, **5**-52
Speed-level changer, generator, power systems, **13**-45
SPHEREDOP (Spherical Doppler), **3**-48
Spherical Doppler, **3**-48
Spike antenna, **4**-120
Spike noise, **13**-28
Spin stabilization, **14**-34
  of communications satellite, **16**-14
Spin velocity, **14**-34
  gyro, **15**-101
Split-field series servo motor, **15**-130
Split-gate phase detector, **8**-34
Split-phase representation PCM, **8**-29
Spray-printing of printed circuits, **10**-77, **10**-78
Spring, bow, accelerometer, **15**-91
  flat helical, accelerometer, **15**-91
  torsion, accelerometer, **15**-96
  rate gyro, **15**-105

INDEX 39

Spring constant in accelerometer, **15**-89
Spurious emissions, ITU definition of, **5**-14
   level tolerances, ITU, **5**-22
Spurious response testing, FM system, **6**-34, **6**-35
Spurious signal in missile, **4**-6
Spurious-signal radiation specifications, PDM-FM, **7**-11
Square loop recording material, **11**-26
Square root function generator, **15**-76
SS-FM system, **6**-55, **6**-61, **6**-62
   data channels response, **6**-62
   frequency reference, **6**-62
   performance, **6**-62
Stability, control system, **15**-27 to **15**-36
   pulse transfer functions, **15**-53, **15**-54
   system, **15**-27 to **15**-29
     s plane test of, **15**-28, **15**-29
Stabilization, **15**-5
   gravity gradient, **14**-35
   spin, **14**-34
   transistor amplifier, external, **10**-33
   transistor parameter, **10**-25, **10**-30 to **10**-42
Stabilizing gyros, **15**-99
Stable platforms, inertial, **15**-124, **15**-127
   single-degree of freedom gyros, **15**-124, **15**-125
   two-degree of freedom gyros, **15**-125
Stable point, phase-plane, phase-locked loops, **9**-70, **9**-72, **9**-75
Stable solutions in data processing, **12**-60, **12**-61
Stages, rocket, liquid propellant, **14**-41
   solid propellant, **14**-42
Stagnation point, heating rate of, **14**-31
Standard cell, electrochemical, **10**-34
Standard deviation, **2**-13
   of reduced data, **12**-60
   sample, **A**-53
Standardization, of data reduction process, **12**-27
   program for, **12**-40
Standardizing organizations, **5**-1 to **5**-4
Standards, IRIG, **5**-25 to **5**-67
   FM-FM or FM-PM, **5**-27 to **5**-31
   PAM-FM telemetry, **5**-32 to **5**-36
   PCM-FM or PM, **5**-36, **5**-37
   PDM-FM, **5**-31, **5**-32
   PDM-FM-FM, **5**-32
   PDM-PM, **5**-31, **5**-32
   receiver IF bandwidth, **5**-33
   recorder, reproducer, **5**-37 to **5**-47
   telemetry, **5**-1
   and telemetry system selection, **4**-33
   where mandatory, **5**-2, **5**-3
Stars, map of, **14**-59
Start-stop system and industrial telemetry links, **13**-17
State University of Iowa, **14**-15
States, occupancy of, **A**-42
Station check in communications-based control, **13**-41
Stationary satellite, **16**-2
   ITU defined, **5**-13
Statistic, **A**-52
Statistical analysis in data reduction, **12**-35
Statistical averaging, **4**-10
Statistical hypothesis, **A**-57
Statistical independence, **2**-12
Statistical parameters, **2**-10, **2**-13, **2**-14
   second-order, **2**-14
Statistical relative frequency, **A**-52
Statistical signals for control system test, **15**-48

Statistics, mathematical, **A**-50 to **A**-61
   output, matched filter, **9**-53
   unmatched filter, **9**-53
Steady-state errors in control-system, **15**-36
Steady-state response in control system, **15**-36, **15**-37
Steffensen's interpolation formula, **A**-64
Stenotyper printer, **11**-51
Step driving function, **15**-67, **15**-70
Step function, **A**-32 to **A**-34
   for calibration, **12**-16
Step function converter, **4**-50
Step function modulation, demodulated by phase-locked loops, **6**-45
Step response, second-order system, **15**-98, **15**-99
Stepper, commutator drive, **4**-69, **4**-70
Stereophonic sound in remote handling, **15**-156
Stereoscopic television for remote handling, **15**-154, **15**-155
Stiction, **15**-91, **15**-92, **15**-102
   errors, **7**-39
Stieltjes integrals, **A**-34
Stirling's expansions, **A**-27
Stirling's interpolation formula, **A**-64
Stoichiometric rocket fuel, **14**-9
Storage, cyclical, data processor, **11**-38
   devices for, **11**-24
   for large time-bandwidth processing, **9**-37, **9**-38
   paper tape checkout system orders, **11**-65
Storage battery, **10**-84 to **10**-92
Storage register (SR), ALU, **11**-39
Storage selection circuit (SSC), data processor, **11**-40
Storage tank, pneumatic, **15**-136
Store-and-forward satellite, **16**-2
Stored cold gas for attitude control, **14**-38
Stored-program computer, **11**-4
Stored-program decommutator, PCM, **8**-48
Stored-program plugboard, **11**-48
Stored-program processor, **11**-34
Strain, measurement requirements for, **4**-9
Strain-gage suspension, accelerometer, **15**-91
Strain-gage wire, characteristics of, **3**-35
   and low temperature coefficient, **3**-35
Strain gages, **3**-34 to **3**-38
   accelerometer, **15**-92
   alloy film, **3**-37
   bonded, **3**-35
   deposited metal, **3**-37
   foil type, **3**-35 to **3**-37
   high-temperature, **3**-37, **3**-38
   low-temperature, **3**-38
   National Bureau of Standards (*see* NBS strain gage)
   semiconductor, **3**-38
   temperature compensation, **3**-36, **3**-37
Strapped down system, inertial, **15**-124
Strip chart, **12**-14
Strip-line circuit, **6**-27
Strip printer, **11**-51
Stripping, **13**-41 to **13**-43
Strontium-90, **10**-99
Structure ratio, **14**-9, **14**-45
Stubs, antenna, **4**-118, **4**-119
Student distribution, table of probabilities, **A**-73
Student's ratio, **A**-54
Student's t distribution, **A**-56
Subcarrier, **6**-2, **6**-3
Subcarrier bands, IRIG FM-FM system, **5**-27, **5**-28

Subcarrier center frequency, selection of, **6**-17, **6**-18
Subcarrier discriminator, bandpass input filters, **6**-41
  carrier-frequency ratio, **6**-40
  comparative performance, **6**-48 to **6**-50
  double-pulse averaging, **6**-43
  output-signal processing, **6**-51
  performance requirements, **6**-40
  phase-locked, **6**-41, **6**-43 to **6**-47
  pulse-averaging, **6**-41, **6**-43
Subcarrier frequency, deviation, permissible, **6**-3, **6**-6
Subcarrier modulation, **1**-8
Subcarrier oscillator, **6**-2, **6**-36 to **6**-40
  charge-controlled, **6**-37
  dynamic test set-up, **6**-39
  harmonic distortion testing, **6**-39
  milliwatt controlled, **6**-37
  multivibrator type, **6**-37
  performance parameters, **6**-38, **6**-39
  reactive element type, **6**-37
  transducer-modulated, **6**-37
  voltage controlled, **6**-36, **6**-37
Subcommutation, **4**-89, **8**-22, **11**-11
  IRIG standard, **5**-34, **5**-35
  PACM, **9**-27
  in PAM demultiplexer, **9**-17
  in PCM, **8**-16
  PCM, IRIG recommendations, **5**-37
Subframe, **8**-17
Submultiplexing, PAM, **9**-12, **9**-13
Subnetwork, communication, serial bridged, **13**-19
  hub, **13**-19
  industrial telemetry links, **13**-17 to **13**-19
Suboperations, computer, **11**-42
Subroutine, in checkout equipment, **11**-67
  computer, **12**-9
  in process control, **13**-8
  program, **12**-38
    (*See also* Modules and program for data reduction)
Subsatellite point, **16**-6, **16**-23, **16**-24
Subscript notation in transistor parameters, **10**-24
Subsets, industrial telemetry links, **13**-17
  phase modulated, **13**-17
Successive approximation, analog-to-digital converter, **12**-19, **12**-20
  in numerical data processing, **12**-29
Successive comparison, analog-to-digital converter, **11**-18, **11**-19
Summer, analog computer, **15**-66, **15**-67
Summing amplifier for FM-FM telemetry, **10**-52, **10**-53
Sun, **14**-53, **14**-54
  noise temperature of, **4**-104, **4**-105
Suomi, V. E., **14**-81
Supercommutation, **4**-89
  IRIG standards, **5**-34, **5**-35
  PACM, **9**-27
  PAM, **9**-12
  PCM, IRIG recommendation, **5**-37
Superheterodyne receiver, double conversion, **6**-28
Superinsulation, **14**-9, **14**-68, **14**-70
Supermultiplexing, **8**-15
Superposition, and non-linear systems, **15**-56
  in phase-locked loops, **9**-61
Supervisory control system, **13**-39 to **13**-45
  automatic, **13**-41
  semi-automatic, **13**-41
  utility system, **4**-3

Supplementary data in data reduction operation, **12**-13
Surface-reflection coefficient, **14**-36
Surface temperature, planetary, **10**-16
Surface-wave antenna, **4**-129, **4**-130
Surveyor program, **14**-12
Suspension, accelerometer, **15**-91
  cantilever, **15**-91
  pivot, **15**-96, **15**-97
  single-point, **15**-97
  strain gage, **15**-91
  single-point, accelerometer, **15**-97
Swastika antenna, **4**-121, **4**-122
Swept frequency technique, phase-locked-loop, **9**-76
Swimming vehicles for remote undersea operations, **15**-151
Switch, commutation, **8**-11
  diode, **8**-12, **8**-13
  electromechanical, **8**-11, **8**-12
  electronic, **8**-12 to **8**-14
  high-contact-resistance, **4**-60
  low-contact-resistance, **4**-60
  mercury jet, **8**-2
  reed, **8**-12
  semiconductor, **8**-13, **8**-14
  stepper, **8**-12
  telemetering, **4**-60, **4**-61
  transistor, **8**-14
Switch impedances, **8**-11
Switching assignments for data reduction, **12**-25
Switching center, **13**-13
  automatic, **13**-14
  manual, **13**-14
Switching or distribution network, data processor, **11**-42, **11**-43
Switching functions in automatic checkout, **11**-62
Syllable, PCM, **8**-30
  (*See also* Word, PCM)
Symbol-display method, **11**-53
Symbol stamp in plotter, **11**-52, **11**-53
Symbols, code, **13**-27
  mathematical, glossary, **A**-1 to **A**-3
  orthogonal, **9**-41
  PFM, **9**-50
Symmetry limiter, **6**-31, **6**-32
Sync alarm, PAM ground station, **9**-17
Sync detector, PAM ground station, **9**-15, **9**-17
Sync generator, **8**-2
Sync pulse selector, **7**-28, **7**-29
Sync separator, PAM demultiplexer, **9**-15, **9**-16
Synchro system, **13**-37 to **13**-39
Synchronization, for automatic digitization, **12**-20
  bit, **8**-30 to **8**-36, **13**-31
  character, **13**-31
  in decommutation, **4**-87
  frame, **8**-30, **8**-38 to **8**-43
    PDM ground station, **7**-28, **7**-29
  group, **8**-31, **8**-36 to **8**-43
  in industrial telemetry, **13**-30, **13**-31
  mechanical, **7**-13, **7**-14
  PACM, **9**-28
  PAM, **4**-58, **5**-56
    IRIG specification, **5**-34
    PAM-FM-FM, **5**-30
  PCM, **5**-37, **5**-51
    integrated word and frame system, **8**-43
    OGO specification, **14**-93
  PDM, **7**-4

INDEX

Synchronization, requirements in digital system, **9**-39
  in space communications system, **14**-98
  subsets, serial, **13**-17
  word, **8**-30, **8**-37
Synchronization code, PCM, **8**-22, **8**-23
Synchronization errors, PDM, **7**-32
Synchronization pattern, PCM, IRIG recommendation for, **5**-51
Synchronization phase-locked loop for space telemetry, **14**-101
Synchronous data processor, **11**-38
Synchronous satellite, **14**-9
Synchros, **3**-24
Syncom, **16**-1
Synodic month, **14**-9
Synthetic-mica strain gage, **3**-36
System applications and radio telemetry, **4**-2 to **4**-5
System design, and radio telemetry, **4**-5 to **4**-11
  and satellite telemetry, **14**-84, **14**-85
System generated inputs and data acquisition system, **11**-10
System selection, telemetry, **4**-26 to **4**-35

t distribution, **A**-54, **A**-56
  table of probabilities, **A**-73
T equivalent circuits, transistor, **10**-21
T parameters, **10**-27
  (*See also* r parameters)
Table look-up, for arithmetic data processing, **11**-42
  in data conversion, **12**-6
Tachometer, **3**-21
Tactile sensing in remote handling, **15**-156
Tag addressing, **11**-25
Tangential surface-reflection coefficient, **14**-36
Tanker modules for orbital assembly, **14**-45, **14**-47
Tantalum capacitor, **10**-71, **10**-72
Tape, magnetic, direct recording standards, **5**-39
    duplicating, in data reduction, **12**-21
    PCM recording standards, **5**-43, **5**-44
  mark, **11**-30
  paper, for processor control, **11**-34
  programming, for automatic checkout, **11**-65 to **11**-67
  reader, automatic checkout, **11**-62, **11**-65
  speed compensation, **6**-52 to **6**-54
    IRIG standards for, **5**-41
Tape-speed errors, **6**-19
  PDM, **7**-38, **7**-39
Tape-speed ratio, **14**-86
Tape transports for magnetic recording, **11**-28 to **11**-30
Taylor's expansion, **A**-24
Teal, **10**-33
Teflon, capacitor, **10**-71
Telecommunication, ITU definition of, **5**-10
Telecontrol, **1**-4
Telegon, **15**-108
Telegraph for telemetry, **13**-14
  capacity, **13**-2
Telemetering (*see* Telemetry)
Telemetry, **1**-2
  adaptive, **8**-28
  applications of, **1**-5
  in automatic checkout, **11**-56
  bibliographies of, **1**-24, **1**-25
  and data presentation, **11**-48, **11**-49

Telemetry, and data processing, staff, functions of, **12**-51, **12**-52
  and data reduction, **12**-2 to **12**-62
  and equipment selection parameters, **4**-32 to **4**-34
  history of, **1**-1 to **1**-3
  ITU definition of, **5**-10
  milestones in, **1**-3
  parameters, Mariner 1964, **14**-102
  publications sources on, **1**-25
  receivers for, **6**-27 to **6**-36
  standards for, **5**-1
  and Standards Coordination Committee, **5**-3
  systems of, advanced radio, **9**-18 to **9**-60
    bandwidth, **2**-49 to **2**-51
    classification of, **1**-6, **13**-21
    for data reduction systems, **12**-4
    FM carrier, comparison, **9**-24
    error sources in, **9**-24
    optimization, **9**-22 to **9**-25
    spectrum occupancy, **9**-25, **9**-26
  terms on, IRIG glossary of, **5**-57 to **5**-67
  and tracking and command system, **14**-95, **14**-102, **14**-103
  and transmitters, mechanical and thermal design of, **6**-27
  wired, **1**-6
Telemetry frequencies, United States military, **4**-94
Telemetry tapes from satellite experiments, **14**-74
Telemetry Working Group, IRIG, **5**-2
Teleoperating system, **13**-4, **13**-9
  analysis of, **13**-23, **13**-24
  central, **13**-10
Telereader, **5**-54
Teletype, for Explorer VI data recovery, **14**-79
  and punched paper tape, **11**-50
  for telemetry, **13**-14
Television, closed-loop, for remote-handling, **15**-153 to **15**-155
Telstar, parameters of, **16**-13
  remote repair in, **11**-73
Temperature, and altitude, **14**-60
  correction, strain gage, **3**-36, **3**-37
  equivalent circuits, incremental, transistor, **10**-33 to **10**-35
    with biasing, **10**-35
  gage, platinum, **3**-38
  limits, electronic components, **10**-17
  measurement requirements of, **4**-9
  sensitivity, thermistor gages, **3**-39
  in space, black body, **10**-16
    equilibrium, **14**-67, **14**-68
  stabilization, transistor, **10**-33 to **10**-37
  transducer, **3**-12
Tera, **5**-15
Terminal velocity correction to planned mission, **14**-52
Terminology, astronautical, **14**-3 to **14**-10
  guidance and control, **15**-5
  telemetry, **5**-57 to **5**-67
Ternary coding of remote handling commands, **15**-160
Tertiary code, **13**-27
Test, environmental, **10**-7 to **10**-9
Test facilities, United States Government, **4**-35
Test points, in automatic checkout, **11**-58
Test range frequencies, **4**-94
Test ranges, missile and space, **4**-35
Test recommendations, IRIG, magnetic tape recorder, **5**-51 to **5**-56

INDEX

Test sequence modification, in checkout equipment, **11**-66
Test specifications, environmental, **10**-5 to **10**-7
Tests of significance, statistical, A-58 to A-61
Texture sensing in remote handling, **15**-156
Thallium sulfide phototransducer, **3**-26
Thermajunction nuclear cell, **10**-99
Thermal activated battery, **10**-91, **10**-92
Thermal conductivity, **14**-68
Thermionic cell, **10**-98
Thermistor, **3**-39
   temperature compensation, transistor, **10**-34, **10**-36
Thermocouple, **3**-40, **3**-41
   emf, various couples, **3**-40
Thermocouple strain gage, **3**-36
Thermoelectric generator, **10**-96 to **10**-98
Thermoelectricity, **10**-96
Thickness gage, variable reluctance, **3**-23
Thor, **14**-38, **14**-41, **14**-46
Thorad, **14**-38, **14**-41
Threat evaluation, telemetry use for, **4**-3
Three address instruction, data processor, **11**-38
Three axis controlled satellite, **14**-34
Three camera system, remote vision, **15**-155
Threshold, analog modulation, **9**-39
   carrier-improvement, **7**-5, **7**-6
   FM feedback receiver, **9**-93
      input RF carrier required for, **7**-25
   frequency, photo emissivity, **3**-25
   improvement, phase-locked discriminators, **6**-47
   PAM-FM, **9**-3
   PDM, bandwidth required for, **7**-6
   video improvement, **7**-5
   phase-lock demodulation, **9**-88
   power, **2**-52
   quasi-linear receiver, optimal, **9**-89
   second-order loop, **9**-90
   signal to noise power ratio, **2**-53, **2**-54
   sync loop, coded-phase coherent system, **14**-101
   wideband FM system, **16**-17
   (*See also* Thresholding)
Thresholding, phase-locked discriminator, **6**-44
   pulse averaging discriminator, **6**-43, **6**-44
   (*See also* Threshold)
Through-put maximization, **13**-4
Thrust, **14**-9
Thrust coefficient, **14**-42, **14**-44
Thrust cutoff transient, in attitude control, **14**-33
Thrust launch vehicle, typical, **14**-41, **14**-42
Thrust rocket, **14**-39 to **14**-41
Thrust starting transient, in attitude control, **14**-33
Thrust vector control, **14**-33, **14**-34
Tie-line facilities of power system, **13**-45
Tikhonov, V. I., **9**-77
Tilt in PDM system, **7**-23
Time, base, **12**-5
   error, **6**-19
   generator, PACM, **9**-18
   in data reduction, **12**-40 to **12**-42
   delay in, in communications satellite, **16**-2
      in control system, **15**-81 to **15**-86
      in FM-FM system, **9**-19
      and process control, **13**-4, **13**-6, **13**-7
      realization, **15**-78, **15**-86
      sources of, in control system, **15**-83

Time, delay in, in space remote control system, **14**-95
   variation, in filter, **6**-31
   pull-in (*see* Pull-in time)
   (*See also* Timing)
Time-bandwidth product, **9**-36 to **9**-39
   and QPPM, **9**-37
   and signal to noise ratio, **9**-37
   and storage devices, **9**-38
Time codes, **12**-21
   for computer tape, **12**-41, **12**-42
   PCM processing, **8**-48
   range, **12**-40, **12**-41
Time constant, servomotor, **15**-132
   transducer, **3**-8
Time division, **2**-37
   multiplex, communications satellite, **16**-22
   remote handling commands, **15**-160
   telemetry, IRIG standards, **5**-31 to **5**-37
Time domain, **2**-19
Time-error correction in power system, **13**-45
Time-and-frequency-to-number analog-to-digital conversion, **11**-22 to **11**-24
Time function, **2**-8
Time histories, in flight testing, **12**-53, **12**-54
   lag in (*see* Time, delay in)
   and scientific data presentation, **14**-88
Time multiplexing, in PCM telemeters, **4**-18
Time register and data processor, **11**-42
Time variable, in plotter, **11**-52
Timed tape reader for checkout sequencing, **11**-65
Timing, and automatic checkout equipment, **11**-64, **11**-65
   clock for, in checkout equipment, **11**-68
   in industrial telemetry, **13**-30, **13**-31
   and multicontact switch, **4**-61
   (*See also* Time)
Timing signal, PDM-FM, **7**-13, **7**-14
Tiros satellites, **14**-85 to **14**-87, **14**-90
   and data presentation, **11**-49
Titan, **14**-38, **14**-41, **14**-42, **14**-46
   satellite of Saturn, **14**-55
   stable platform, **15**-126
   and telemetry, **8**-2
Titanium-alloy dry cell, **10**-83
Tone-signal, **13**-32
"Torn" tape data-link system, **13**-14
Torque-balance rate gyro, **15**-115, **15**-117 to **15**-119
Torque-speed curve and electrical prime mover, **15**-129 to **15**-131
Torquer, **15**-132
   and accelerometer, **15**-98
   in gyro system, **15**-116 to **15**-118
Torques, affecting spacecraft attitude, **14**-34 to **14**-39
   in gyro, **15**-104, **15**-114
   and rate gyro, **15**-106
Torsion bar, and accelerometer, **15**-96
   and rate gyro, **15**-107
Tourmaline, crystal, **3**-32
TRAC(E), **3**-48
Trace crossover (*see* Crossover)
Track(s), recorder, direct recording standards, **5**-39
Tracking, in satellite telemetry system, **14**-76, **14**-77, **14**-79, **14**-91
   signal, **6**-28, **6**-29
      accelerating phase, **9**-65

INDEX

Tracking, signal, constant frequency, **9**-64, **9**-65
  optimum tracking loop, **9**-65
  by phase-locked-loop, **9**-64
    subcarrier, **9**-64
  space system, **14**-95
  systems for, as data reduction systems inputs, **12**-4
  velocity, by phase-locked-loop, **9**-64
Trajectory, analysis of, **14**-16, **14**-17
  change of, during flyby, **14**-28 to **14**-30
  constant energy, **14**-21 to **14**-26
  constant momentum, **14**-22 to **14**-26
  coordinate system, IRIG defined, **4**-131, **4**-132
  determination of, and data processing procedure, **12**-36, **12**-37
  lofted, **14**-26, **14**-27
  measurement of, **3**-42
    by interferometer, **3**-45, **3**-46
    by relative phase, **3**-45
  minimum energy, **14**-26, **14**-27
  phase-plane, phase-locked loop, **9**-70 to **9**-77
  planned, mission, **14**-49
  sub-orbital, **14**-26, **14**-27
Transcendental computer, **15**-73 to **15**-76
Transducer, **1**-7, **3**-1 to **3**-42, **11**-14
  accelerometer, **15**-91, **15**-94, **15**-98
  active, **3**-3
  basic types, **3**-3
  bridge type, frequent failure mode, **8**-7
  capacitance, **3**-9
  capacitance-to-voltage, **3**-11
  commercially available, **3**-42
  defined, **3**-2
  dielectric, **3**-11
  digital, **3**-11
    in checkout equipment, **11**-72
  displacement, **3**-3, **3**-10, **3**-12, **3**-13, **3**-17, **3**-21
  eddy current, **3**-11, **3**-12
  electronic, **3**-12
  gyro, **15**-108, **15**-116
  inductance, **3**-19, **3**-20
  passive, **3**-3
  performance characteristics of, **3**-4, **3**-8
  permeance, **3**-23, **3**-24
  piezoelectric, **6**-37
  principles of operation of, **3**-8 to **3**-42
  for radio telemetry, **4**-5
  selection of, **4**-8
  variable-reluctance, **3**-23, **3**-24, **15**-108
  and VCO's, **10**-46, **10**-47
  voltage, of, **4**-37
  in voltage telemetering system, **13**-33
Transfer, characteristics, transducers, **3**-4
  maneuver, **14**-9, **14**-18 to **14**-21
  program, data processor, **11**-43
  resistance, transistor, **4**-82
  tree, unipolar transistor, **10**-77
  velocity, **14**-19 to **14**-21
Transfer function, **2**-20, **15**-21
  and a-c servomotor, **15**-131, **15**-132
  and accelerometer, **15**-90
  closed loop, and loop noise bandwidth B_L, **9**-67, **9**-68
  and phase-locked-loop, **9**-63, **9**-64
  communication link, **13**-11 to **13**-13
  and d-c servomotor, **15**-128
  exponential, **15**-83 to **15**-86
    algebraic approximation, **15**-83, **15**-84
    MacLaurin series approximation, **15**-84
  and FM feedback receiver, **9**-93
  and hydraulic prime mover, **15**-133

Transfer function, and integrating gyro, **15**-115
  and loop filter, peak error design, **9**-67
    phase-locked-loop, **9**-64
  and optimal phase locked receiver, **9**-89
  and optimum loop filter, demodulating, **9**-66
  and optimum realizable linear filter, **9**-88
  output, **9**-67
  and phase-locked discriminator, peak error constrained, **9**-68
  and phase-locked-loop, **9**-61
  and process control, **13**-6
  rate gyro, **15**-106
  second-order, **15**-37, **15**-40
  simulated, **15**-71
  torque-balance rate gyro, **15**-118
  and transmission system, **13**-11
  and Wiener optimized loop filter, **4**-66
Transformations, data, **11**-7, **11**-8
  transistor equivalent circuit, **10**-22 to **10**-24
Transforms, Fourier, pulse waveforms, **2**-47
  Volterra kernel, **9**-85
Z, **13**-23
Transient measurement, stress and temperature, **3**-37, **3**-38
Transient response, **3**-7
  design, control system, **15**-37 to **15**-40
  (*See also* Frequency response)
Transistor, **10**-18 to **10**-42
  admittance definitions of, **10**-19
  and amplifier stabilization, **10**-25, **10**-30 to **10**-37
  and bias network, stabilizing, **10**-38, **10**-39
  common-base, **10**-20, **10**-21
  common-collector, **10**-21
  common emitter, **10**-20, **10**-21
    incremental temperature equivalent circuit, **10**-34
  complementary, for sampling gates, **9**-10, **9**-11
  connection types, **10**-20
  equivalent circuit, **10**-20 to **10**-24
    parameters, **10**-18 to **10**-20
  field effects in PDM, **7**-21
  and hybrid parameters, **10**-24 to **10**-30
  impedance definitions of, **10**-19
  and incremental resistances, **10**-21 to **10**-24
    temperature equations, **10**-34
  integrated, **4**-82
  inverted, **4**-82
  notation, parameter, **10**-24
  parameter, **10**-18 to **10**-20
    hybrid, **10**-19
    interrelations, **10**-20
    open-circuit impedance, **10**-19
    short-circuit admittance, **10**-19
  parameter stability equations, **10**-38
  parameter stabilization, **10**-25, **10**-30 to **10**-42
  small signal, **10**-37 to **10**-42
  printed circuit, **10**-78
  small signal analysis, **10**-19, **10**-24, **10**-33
  switch, **8**-14
    in PAM gates, **9**-10
  T equivalent circuit, **10**-21
  in telemetry, **4**-24
  temperature equivalent circuit, incremental, **10**-33 to **10**-35
  temperature stabilization, **10**-33 to **10**-37
    indirect methods of, **10**-40, **10**-41
    and input conductance, **10**-41

INDEX

Transistor, temperature stabilization, preferred procedure for, **10**-39, **10**-40
   and thermistor, **10**-40
   thermistor temperature compensated, **10**-36
   transfer resistance, **4**-82
   (*See also* Transistorization; Transistorized)
Transistorization, telemetry, **10**-3
Transistorized amplifier, **4**-43, **4**-44
Transistorized chopper, **10**-70
Transistorized PCM system, **10**-64, **10**-69
Transistorized phase-shift VCO, **10**-50, **10**-51
Transistorized reactance modulator, **10**-54
Transistorized telemetering circuits, **10**-42 to **10**-69
Transistorized vibrator, **10**-70
Transit time in photo conductive transducer, **3**-25
Translational system, electrical analog, **15**-65
Translator, communications satellite, **16**-20
   PDM, **7**-30
Transmission, of binary data over continuous channel, **9**-58, **9**-59
   discrete, of continuous data, **9**-31
   of sampled analog data, **9**-54 to **9**-57
   in industrial telemetry, **13**-32
   multiple phase, **13**-32
Transmission delay in stationary communications satellite, **16**-31
Transmission designation, ITU, **5**-14
Transmission line loss, receiver, **4**-103
   transmitter, **4**-96, **4**-97
Transmission link, industrial, **13**-11 to **13**-13
Transmitter, **6**-20 to **6**-27
   FM, **4**-12, **6**-20 to **6**-22
   specifications, for typical, **6**-21
   FM-FM, **4**-19
   functional components of, **6**-20, **6**-22 to **6**-27
   PAM-FM, **9**-11
   PDM-FM, **4**-21
   PM, **6**-20 to **6**-22
   PPM-AM, **4**-23
   r-f, **10**-53
   transistorized telemetry, **10**-53 to **10**-55
   vhf, **10**-54, **10**-55
Transponder, radar, **3**-55
Transportation cost, **14**-9
Transportation log, **15**-70, **15**-81 to **15**-86
Transverse interference, **11**-11
Transverse recording, **11**-25
Trapezoidal rule, in data reduction, **12**-34
   numerical integration, A-66
Trapping, program, data processor, **11**-43, **11**-44
Travelling wave tube, communications satellite, **16**-14
   noise performance, **16**-5
Tree, decoding, for checkout equipment, **11**-62 to **11**-64
Trial voltage error, feedback encoder, **8**-26
TRIDOP, **3**-47
Trigger, circuit, PDM keyer, **10**-62
   magnetic, **7**-13 to **7**-15
Trigonometric functions, A-18 to A-22
Trigonometric polynomials, **12**-31
Tri-level representation, PCM, **8**-29
Triple level protection, remote station, **13**-42
Triple modulation, **1**-9, **7**-8
True anomaly, **14**-9
True vertical, **15**-122

Truncation errors, **12**-29, **12**-36
Truth tables, **15**-78 to **15**-80
Tunnel diode amplifier, **16**-5
Turbojet engine, **14**-49, **14**-50
Turborocket engine, **14**-50
Turn-bank instrument, **15**-100
Turnstile antenna, **4**-121
TV tube display in telemetry data presentation, **11**-49, **11**-53, **11**-54
TVC (*see* Thrust vector control)
Two address instruction, data processor, **11**-38
Two-degree-of-freedom gyros, **15**-118 to **15**-124
   theory of operation of, **15**-118, **15**-119
Two-out-of-five code, in data processor, **11**-37
   security system, **13**-40
Two-phase induction servo motor, **15**-130 to **15**-132
TWT (*see* Travelling wave tube)
Typewriter, **11**-51

UDOP (UHF Doppler), **3**-47
UES (upper end-scale) value, **11**-15
UHF band, defined, **5**-15
Uhlenbeck, G. E., **9**-78
Uncoded binary system, **14**-97
Underdamping, transducer, **3**-7
Undersea hostile environment, **15**-139, **15**-141
Underwater manipulator, **15**-141
Unidirectional antenna pattern, **4**-112
Uniform distribution, A-50, A-51
Union, **15**-78
Union Switch and Signal Company, **13**-39
Unipolar transistor, **10**-76, **10**-77
Unit impulse, **2**-19
U.S. Air Force environmental specifications, **10**-4
U.S. Army Signal Research and Development Laboratory, **9**-18, **14**-75
U.S.-U.K. Scout telemetry, **14**-89, **14**-90
Units, **1**-10 to **1**-23
   atomic, **1**-21 to **1**-23
   Barn, **1**-21
   basic metric, **1**-10, **1**-11
   conversion factors, **1**-10 to **1**-23
   electrical, **1**-21
   Gauss, **1**-21
   Gilbert, **1**-21
   Henry, **1**-10, **1**-21
   Newton, **1**-10
   Oersted, **1**-20
Universal time code, **12**-41
Unmanned space mission, **14**-10 to **14**-12
Uplink (active communications satellite), **16**-12
Uranus, characteristics of, **14**-53, **14**-56 to **14**-58
Use factor of communications satellite, **16**-21, **16**-22
User requirements, telemetry, missile, **9**-19
Utility, data transmission, **13**-15
   system control, **13**-44, **13**-45

V-1, German, **4**-3
Vacuum deposition, **10**-78
Vacuum environment, **10**-12
Validity, data reduction results, **12**-59, **12**-60
Vanadium pentoxide cell, **10**-83
Van Allen, J., **14**-81

INDEX 45

Van Allen belts, **10**-12, **10**-13
Van Atta, L., **16**-14
Van der Pol, B., **6**-41
Vanguard modulation system, **14**-88, **14**-89
 (*See also* PFM)
Vanguard satellite, **1**-9, **14**-74, **14**-75
Vanguard II, telemetry system of, **14**-76
Vanguard III, data processing system of,
  14-77, **14**-78
 satellite, **9**-49
 telemetry and instrumentation system of,
  14-76
Van Meter, D., **9**-40
Van Trees, H. L., **9**-82, **9**-92
Varactor, diode modulator, in retrodirec-
  tive satellite, **16**-15
Varactor multiplier, **6**-25, **6**-26
Varactor phase-modulator, **6**-22, **6**-23
Variable-focal length camera for remote
  vision, **15**-155
Variable-reactance oscillator, current controlled, **10**-52
 as VCO, **10**-49, **10**-51
Variable-reluctance transducer, **3**-20, **3**-23,
  3-24, **15**-108
 push-pull, **3**-23
 ratio-output, **3**-23
Variance, **2**-13, A-46, A-47
 in discrete channel, **9**-56
 minimum, for white Gaussian noise, **9**-88
 noisy signals, **9**-91, **9**-92
 non-linear phase-locked loop, Develet's
  model, **9**-91
  equivalent gain model, **9**-91
  Van Trees' model, **9**-91
 phase error, first-order loop, by Fokker-
  Plank method, **9**-80
  by Volterra expansion method, **9**-84,
   9-85
 phase-locked loop noise, **9**-63
 of quasi-linear phase-locked receiver,
  4-87
 quasi-linear receiver, second-order, **9**-90,
  9-91
 for random-frequency modulation, first-
  order loop, by Volterra method,
  9-85, **9**-86
 second-order loop, by Volterra expansion, **9**-85
Varicap, **4**-13
VCO (voltage-controlled oscillator), **6**-36,
 6-37, **6**-43 to **6**-45, **8**-33, **8**-34, **9**-60,
 9-61, **10**-46 to **10**-52
 for bit synchronizer, **8**-33, **8**-34
 and frequency modulator, **6**-22, **6**-23
 multivibrator type, **10**-46 to **10**-49
 in phase-locked loop, **9**-60 to **9**-77, **9**-86,
  9-87, **9**-93 to **9**-96
 phase-shift type, **10**-49 to **10**-51
VCXO, **6**-22, **6**-23
 and frequency modulator, **6**-22, **6**-23
Vehicle, remotized, **15**-150, **15**-151
Vehicle coordinate system, IRIG defined,
 4-131, **4**-132
Vehicular antenna, **4**-111, **4**-134
 characteristics of specific types of, **4**-117
Velocimeter, **3**-48
Velocity, circular orbit, **14**-5, **14**-19, **14**-20,
 14-22 to **14**-26
 injection into orbit, **14**-6, **14**-18 to **14**-26
  of light, **1**-22
 total, to reach orbit, **14**-19 to **14**-21
 transducer, **3**-21
 transfer, orbital, **14**-19 to **14**-21
Velocity ratio of rockets, **14**-45

Velocity requirements, of lunar mission,
 14-51
 of Mars mission, **14**-51
 of rocket vs. weight delivered to mission,
  14-46
 of Venus mission, **14**-51
Venn diagram, **15**-78, **15**-79, A-12
Venus, **14**-53, **14**-55 to **14**-57
Venus orbits, altitude vs. object velocity
 and launch angle, **14**-24
Verified paper tape, **11**-66
Vernal equinox, **14**-9, **14**-57, **14**-58
Vertical, local, **15**-121, **15**-122
Vertical circle, **14**-10
Vertical gyro, **15**-100, **15**-121 to **15**-123
 and Shuler-tuned system, **15**-123
Vertical polarization, **4**-113
 of source and image, **4**-116
Vertical recording, **11**-25
VHF band, defined, **5**-15
Vibration, effect of, on transducer, **3**-8
 measurement requirements of, **4**-9
 test specification of, **10**-7, **10**-8
Vibrator, transistor, **10**-70
Video-improvement threshold, **7**-4
Video recording, **5**-47
Video threshold, **7**-7
 (*See also* Video-improvement threshold)
Villari effect, **3**-21
Violin string effect, **12**-22
 in recorders, **7**-39
Viscosity, oil, **15**-94
 rate gyros, **15**-109
Viscous friction in accelerometer, **15**-90
Viscous restraint, **15**-115
Viscous-shear law, Newton's, **15**-94
Visibility, area of mutual, communications
 satellites, **16**-23, **16**-24
 mutual (*see* Mutual visibility)
 probability of, communications satellite,
  16-24
Visual displays, large scale, telemetry data,
 11-49
Viterbi, A. J., **8**-59, **9**-39, **9**-60, **9**-67, **9**-77,
 9-81, **9**-91
VLF band, defined, **5**-15
Voltage, measurement requirements of, **4**-9
Voltage comparator, regenerative, **10**-62
Voltage controlled crystal oscillator (*see*
 VCXO)
Voltage controlled oscillator (*see* VCO)
Voltage gain, A_v, transistor, **10**-29
Voltage sensitivity coefficient, piezoelec-
 tric, **3**-31, **3**-32
Voltage standing wave ratio (*see* VSWR)
Voltage telemetering system, **13**-2, **13**-3,
 13-32 to **13**-35
Volterra functional analysis, phase-locked
 loops, **9**-82 to **9**-86
Volterra functional expansion, phase error,
 first-order loop, **9**-82 to **9**-84
Volterra kernel transforms, **9**-85
Voltmeter, true rms, **4**-50, **4**-51
Voyager, **14**-95
 program for, **14**-12
VSWR (voltage standing wave ratio),
 antenna, **4**-114
 receiver input, **6**-36

Wafer micromodule, **10**-73
Wagon wheel communications network,
 13-15, **13**-16
Walking vehicle, remote controlled, **15**-151
Water activated battery, **10**-91, **10**-92

INDEX

Water hammer, pipeline, **13**-44
Waveforms, commutator, **4**-62
  PDM, **7**-3
  pulse, Fourier transforms for, **2**-47
Wavelength and energy, relationships of, in atomic particles, **1**-23
Wavelength nomenclature, ITU, **5**-15
Weapons systems checkout, **11**-56, **11**-57
Weddle's rule, numerical integration, **A**-66
Weight, filter, **12**-31, **12**-32
  ratio, stable platform, **15**-126
  telemetry, **4**-7
Weighting factors, **2**-4 to **2**-8
West Ford, **16**-9 to **16**-12
  and communications system parameters, **16**-12
Wet primary cell, **10**-84
Whip antenna, **4**-119, **4**-120
Whistler-like signals from Vanguard III, **14**-78
White noise, **2**-18
  band-limited, **2**-18
White Sands, **3**-47
White Sands Missile Range, **5**-1
Wide-bandwidth system, in communications satellite, **16**-16
Wideband antenna, **4**-114, **4**-123 to **4**-129
Wideband frequency modulation, **16**-16 to **16**-18
  signal to noise ratio, **16**-17
Wideband gain, **2**-51, **7**-10
  PAM-FM, **9**-3
  PDM without interpolation, **7**-34
  PDM-FM, **7**-7, **7**-10
Wideband improvement ratio, FM systems, **6**-11
Wideband telemetry, OGO specification, **14**-92
Wiener-Hopf solutions, **9**-83
Wiener optimization, **9**-65 to **9**-67
Wiener optimum filter, **12**-32
Window, launch, **14**-7, **14**-17
Wire, strain-gage, **3**-35
Wire printer, **11**-51
Word, PCM, **8**-30
  PFM, **9**-50

Word detection, **14**-97, **14**-100 to **14**-102
Word-error probability, biorthogonally coded system, **9**-47, **9**-48
  orthogonal coded system, **9**-46, **9**-47
  PFM, **9**-53, **9**-54
Word structure, PCM, IRIG standard, **5**-36, **5**-37
Work function, **3**-25
Working groups, IRIG, **5**-2
Wow, **7**-39
  in data recording, **11**-34
Wow and flutter, **6**-52, **12**-22
  compensation, **4**-21
  effect, IRIG defined, **5**-53
  test, IRIG, **5**-55, **5**-56
Writing definition, recorder, **11**-27
WSMR (White Sands Missile Range), **5**-1

X rays in space auroras, **10**-17
X-Y plot, **4**-10
X-Y plotter, high speed, **12**-45
  output sample, **12**-48
X-Y readers, **12**-14
X^2 distribution, table of probabilities, **A**-74
Xerography, **10**-24
XY counter matrix, commutator, **4**-80, **4**-81

Young's modules, crystals, **3**-30, **3**-32
Yovits, M. C., **9**-88

Z transforms, **13**-23, **13**-24, **15**-50 to **15**-52
Zadeh, L. A., **13**-23
Zamboni pile, **10**-83
Zener diode, **10**-70
Zero, **15**-28, **15**-29, **15**-34, **15**-38
Zero-offset, **12**-5, **12**-6
Zero order hold, **15**-50
Zero uncertainty and analog-to-digital converter, **11**-15
Zodiac, **14**-10
Zone bits, **11**-37
Zoomar lens for remote vision, **15**-155